COMMON ELEMENTS

Name	Symbol	Approx. mass	Common ox. nos.	Name	Symbol	Approx. mass	Common ox. nos.
aluminum	Al	27.0	+3	magnesium	Mg	24.3	+2
antimony	Sb	121.8	−3, +3, +5	manganese	Mn	54.9	+2, +3, +4, +6, +7
arsenic	As	74.9	−3, +3, +5	mercury	Hg	200.6	+1, +2
barium	Ba	137.3	+2	nickel	Ni	58.7	+2, +3
bismuth	Bi	209.0	+3	nitrogen	N	14.0	−3, +3, +5
bromine	Br	79.9	−1, +1, +3, +5, +7	oxygen	O	16.0	−2
calcium	Ca	40.1	+2	phosphorus	P	31.0	+3, +5
carbon	C	12.0	+2, +4	platinum	Pt	195.1	+2, +4
chlorine	Cl	35.5	−1, +1, +3, +5, +7	potassium	K	39.1	+1
chromium	Cr	52.0	+2, +3, +6	silicon	Si	28.1	+2, +4
cobalt	Co	58.9	+2, +3	silver	Ag	107.9	+1
copper	Cu	63.5	+1, +2	sodium	Na	23.0	+1
fluorine	F	19.0	−1	strontium	Sr	87.6	+2
gold	Au	197.0	+1, +3	sulfur	S	32.1	−2, +4, +6
hydrogen	H	1.0	−1, +1	tin	Sn	118.7	+2, +4
iodine	I	126.9	−1, +1, +3, +5, +7	titanium	Ti	47.9	+2, +3, +4
iron	Fe	55.8	+2, +3	tungsten	W	183.8	+6
lead	Pb	207.2	+2, +4	zinc	Zn	65.4	+2

COMMON IONS

Name	Symbol	Name	Symbol
aluminum	Al^{3+}	lead(II)	Pb^{2+}
ammonium	NH_4^+	magnesium	Mg^{2+}
barium	Ba^{2+}	mercury(I)	Hg_2^{2+}
calcium	Ca^{2+}	mercury(II)	Hg^{2+}
chromium(III)	Cr^{3+}	nickel(II)	Ni^{2+}
cobalt(II)	Co^{2+}	potassium	K^+
copper(I)	Cu^+	silver	Ag^+
copper(II)	Cu^{2+}	sodium	Na^+
hydronium	H_3O^+	tin(II)	Sn^{2+}
iron(II)	Fe^{2+}	tin(IV)	Sn^{4+}
iron(III)	Fe^{3+}	zinc	Zn^{2+}
acetate	$C_2H_3O_2^-$	hydrogen sulfate	HSO_4^-
bromide	Br^-	hydroxide	OH^-
carbonate	CO_3^{2-}	hypochlorite	ClO^-
chlorate	ClO_3^-	iodide	I^-
chloride	Cl^-	nitrate	NO_3^-
chlorite	ClO_2^-	nitrite	NO_2^-
chromate	CrO_4^{2-}	oxide	O^{2-}
cyanide	CN^-	perchlorate	ClO_4^-
dichromate	$Cr_2O_7^{2-}$	permanganate	MnO_4^-
fluoride	F^-	peroxide	O_2^{2-}
hexacyanoferrate(II)	$Fe(CN)_6^{4-}$	phosphate	PO_4^{3-}
hexacyanoferrate(III)	$Fe(CN)_6^{3-}$	sulfate	SO_4^{2-}
hydride	H^-	sulfide	S^{2-}
hydrogen carbonate	HCO_3^-	sulfite	SO_3^{2-}

Authoritative, Comprehensive, and Relevant to Today's Students

MODERN CHEMISTRY

ANNOTATED TEACHER'S EDITION

Holt, Rinehart and Winston
Harcourt Brace Jovanovich

Austin • Orlando • San Diego • Chicago • Dallas • Toronto

Authors

Nicholas D. Tzimopoulos, Ph.D.
Director of Curriculum/Science Education for the Lexington Public Schools, Lexington, MA, and Adjunct Professor at PACE University in New York. Dr. Tzimopoulos was a former chemistry teacher and Chairman on the Science Department for the Public Schools of Tarrytowns, North Tarrytown, New York.

H. Clark Metcalfe
Formerly teacher of chemistry at Winchester-Thurston School, Pittsburgh, Pennsylvania, and Head of the Science Department, Wilkinsburg Senior High School, Wilkinsburg, Pennsylvania.

John E. Williams
Former teacher of chemistry and physics at Newport Harbor High School, Newport Beach, California, and Head of the Science Department, Broad Ripple High School, Indianapolis, Indiana.

Joseph F. Castka
Formerly Assistant Principal for the Supervision of Physical Science, Martin Van Buren High School, New York City, and Adjunct Professor of General Science and Chemistry, C.S. Post College, Long Island University, New York.

Cover Photograph: Light fuel oil using the technique of phase contrast photography.
Manfred Kage/Peter Arnold

Copyright © 1993 by Holt, Rinehart and Winston, Inc. All rights reserved. No part of this publication may be reproduced or transmitted in any form or by any means, electronic or mechanical, including photocopy, recording, or any information storage and retrieval system, without permission in writing from the publisher. Requests for permission to make copies of any part of the work should be mailed to: Permissions Department, Holt, Rinehart and Winston, Inc. 8th Floor, Orlando, Florida 32887

Material from earlier editions Copyright © 1990, 1986, 1982 by Holt, Rinehart and Winston, Inc. All rights reserved.

For permission to reprint copyrighted material, grateful acknowledgement is made to the following sources:

CRC Press, Inc.: From *CRC Handbook of Chemistry and Physics,* 69th Edition, page B-88; Editor-in-Chief: Robert C. Weast, Ph.D. Copyright © 1988 by CRC Press, Inc., Boca Raton, Florida.

Doubleday, a division of Bantam, Doubleday, Dell Publishing Group, Inc.: From *Radiation Alert, A Consumer's Guide to Radiation* by David I. Poch. Copyright © 1985 by Energy Probe Research Foundation.

Printed in the United States of America

ISBN 0-03-075961-7

1 2 3 4 5 6 7   036   96 95 94 93 92

Contributing Writers

Robert Becker
Chemistry Teacher
Kirkwood High School
Kirkwood, MO

Dava C. Coleman
Chemistry Teacher
Cedar Shoals High School
Athens, GA

Penney Sconzo
Chemistry Resource Teacher
The Westminster Schools
Atlanta, GA

Other Contributors

Henry Altmiller, Ph.D.
Professor of Chemistry
St. Edward's University
Austin, TX

Eva Lou Apel
Westlake High School
Austin, TX

Karen Arms, Ph.D.
Author and Consultant
Savannah, GA

Jane Bowser
Science Department Chairperson
East Davidson High School
Lexington, NC

Londa L. Borer, Ph.D.
Professor of Chemistry
California State University
Sacramento, CA

Sally Young Busboom, M.S.
Chemistry Teacher
Anderson High School
Austin, TX

Mary E. Castellion
Chemistry Editor and Author
Norwalk, CT

Earl B. Chappell, III, Ed.D.
Virginia Beach City Public Schools
Virginia Beach, VA

Camillo Cimis
Science and Editorial Consultant
Stockholm, NJ

Theron M. Cole, Jr., Ph.D.
Editor, NASA Tech Briefs
Brooklyn, NY

Cheryl Epperson
Chemistry Teacher
Flour Bluff High School
Corpus Christi, TX

Edmund J. Escudero
Chemistry Teacher
Summit Country Day School
Cincinnati, OH

George Gross
Chemistry Teacher
Union High School
Union, NJ

Marilyn Lawson
Chemistry Teacher and Science
    Department Chairperson
Gregory Portland High School
Portland, TX

Zoe A. Godby Lightfoot
Chemistry Teacher and Science
    Department Chairperson
Carbondale Community High School
Carbondale, IL

Charles W. McLaughlin, Ph.D.
Central High School
St. Joseph, MO

David J. Olney
Lexington High School
Lexington, MA

Ron Perkins, M.S.T.
Sr. Chemistry Teacher
Greenwich High School
Greenwich, CT

Janella Rachal, Ph.D.
Director, Bureau of Evaluation and
    Analytical Services
Louisiana Department of Education
Baton Rouge, LA

Ruth Rand
The Latin School of Chicago
Chicago, IL

Patricia J. Smith
Chemistry Teacher
Air Academy High School
USAF Academy, CO

Joseph F. Tarello, M.S.
Chemistry Teacher
Somerville High School
Somerville, MA

Claudia K. Viehland
Chemistry Teacher
Chaminade College Prep.
St. Louis, MO

Sally Ann Vonderbrink, Ph.D.
St. Xavier High School
Cincinnati, OH

Contents

NEW Features for this Edition T14
Problem Solving T18
Laboratory Program T22
Chapter Reviews T26
Teacher's Edition Format T30
Supplements to Modern Chemistry T32
Audiovisual Resources for
    Modern Chemistry T34
Developing Reading and Writing Skills T36
Cooperative Learning T38
Providing for Individual Differences T40
Reteaching and Enrichment Strategies T43
Master Materials List for Desktop Investigations,
    Microscale Investigations, Equipment Notes,
    and Materials for Demonstrations T46

Unit 1 Introduction to Chemistry and Matter 2

Chapter 1 **Matter, Energy, and Change**
Planning Chart 4A
Answers and Solutions 4F
Periodic Table Videodisc Barcodes 4G
1.1 What Is Chemistry? 5
1.2 Matter and Energy 10
1.3 Classification of Matter 19
1.4 The Chemical Elements 23
Chapter 1 Review 28

Chapter 2 **Measurements and Solving Problems**
Planning Chart 30A
Answers and Solutions 30D
2.1 Units of Measurement 31
2.2 Heat and Temperature 40
2.3 Using Scientific Measurements 46
2.4 Solving Quantitative Problems 55
Chapter 2 Review 60
Chemistry and Careers: Fire Chemist, Science Writer 64
Focus on Reasoning: College Entrance Exam Preparation 65
Intra-Science: The Portable Power of the Battery 66

Unit 2 Organization of Matter 68

Chapter 3 Atoms: The Building Blocks of Matter
Planning Chart 70A
Answers and Solutions 70D
3.1 The Atom: From Philosophical Idea to Scientific Theory 71
3.2 The Structure of the Atom 75
3.3 Weighing and Counting Atoms 82
Chapter 3 Review 93

Chapter 4 Arrangement of Electrons in Atoms
Planning Chart 96A
Answers and Solutions 96D
Periodic Table Videodisc Barcodes 96G
4.1 Refinements of the Atomic Model 97
4.2 Quantum Numbers and Atomic Orbitals 108
4.3 Electron Configurations 112
Chapter 4 Review 122

Chapter 5 The Periodic Law
Planning Chart 126A
Answers and Solutions 126D
Periodic Table Videodisc Barcodes 126G
5.1 History of the Periodic Table 127
5.2 Electron Configuration and the Periodic Table 132
5.3 Electron Configuration and Periodic Properties 143
Chapter 5 Review 157

Chapter 6 Chemical Bonding
Planning Chart 160A
Answers and Solutions 160E
6.1 Introduction to Chemical Bonding 161
6.2 Covalent Bonding and Molecular Compounds 164
6.3 Ionic Bonding and Ionic Compounds 175
6.4 Metallic Bonding 179
6.5 The Properties of Molecular Compounds 182
Chapter 6 Review 191
Chemistry and Careers: Laser Chemist, Chemical Technician 194
Focus on Reasoning: College Entrance Exam Preparation 195
Intra-Science: How to Replace Asbestos 196

Unit 3 Language of Chemistry 198

Chapter 7 **Chemical Formulas and Chemical Compounds**
Planning Chart 200A
Answers and Solutions 200E
Periodic Table Videodisc Barcodes 200H
7.1 Chemical Names and Formulas 201
7.2 Oxidation Numbers 212
7.3 Using Chemical Formulas 216
7.4 Determining Chemical Formulas 224
Chapter 7 Review 229

Chapter 8 **Chemical Equations and Reactions**
Planning Chart 232A
Answers and Solutions 232E
Periodic Table Videodisc Barcodes 232H
8.1 Chemical Equations 233
8.2 Types of Chemical Reactions 244
8.3 Activity Series of the Elements 251
Chapter 8 Review 256

Chapter 9 **Stoichiometry**
Planning Chart 260A
Answers and Solutions 260D
9.1 Introduction to Stoichiometry 261
9.2 Ideal Stoichiometric Calculations 265
9.3 Limiting Reactants and Percent Yield 272
Chapter 9 Review 279
Chemistry and Careers: Analytical Pharmaceutical Chemist, Chemical Plant Manager 282
Focus on Reasoning: College Entrance Exam Preparation 283
Intra-Science: The Amazing Polyhexamethyleneadipamide 284

Unit 4 Phases of Matter 286

Chapter 10 **Representative Gases**
Planning Chart   288A
Answers and Solutions   288D
Periodic Table Videodisc Barcodes   288F
10.1 Oxygen and Ozone   289
10.2 Hydrogen   294
10.3 Nitrogen and Ammonia   298
10.4 Carbon Dioxide and Carbon Monoxide   303
Chapter 10 Review   311

Chapter 11 **Physical Characteristics of Gases**
Planning Chart   314A
Answers and Solutions   314C
11.1 The Kinetic Theory of Matter   315
11.2 Qualitative Description of Gases   320
11.3 Quantitative Description of Gases   322
Chapter 11 Review   338

Chapter 12 **Molecular Composition of Gases**
Planning Chart   342A
Answers and Solutions   342C
12.1 Volume-Mass Relationships of Gases   343
12.2 The Ideal Gas Law   350
12.3 Stoichiometry of Gases   356
12.4 Effusion and Diffusion   360
Chapter 12 Review   366

Chapter 13 **Liquids and Solids**
Planning Chart   370A
Answers and Solutions   370C
Periodic Table Videodisc Barcodes   370G
13.1 Liquids   371
13.2 Solids   377
13.3 Changes of State   383
13.4 Water   390
Chapter 13 Review   395
Chemistry and Careers: Occupations in the Compressed-Gas Industries, Chemical Patent Attorney   398
Focus on Reasoning: College Entrance Exam Preparation   399
Intra-Science: Preserving the Ozone Layer   400

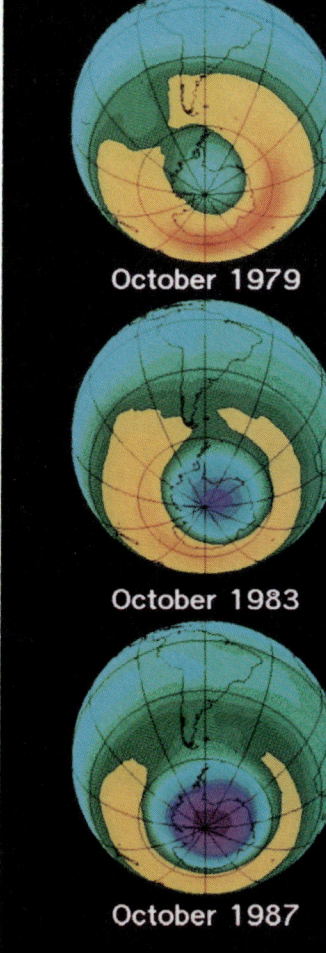

October 1979

October 1983

October 1987

T7

Unit 5 Solutions and Their Behavior 402

Chapter 14 **Solutions**
Planning Chart 404A
Answers and Solutions 404D
Periodic Table Videodisc Barcodes 404J
14.1 Types of Mixtures 405
14.2 The Solution Process 410
14.3 Concentration of Solutions 420
14.4 Colligative Properties of Solutions 426
Chapter 14 Review 434

Chapter 15 **Ions in Aqueous Solutions**
Planning Chart 438A
Answers and Solutions 438D
15.1 Ionic Compounds in Aqueous Solution 439
15.2 Molecular Electrolytes 447
15.3 Properties of Electrolyte Solutions 450
Chapter 15 Review 457

Chapter 16 **Acids and Bases**
Planning Chart 460A
Answers and Solutions 460E
Periodic Table Videodisc Barcodes 460H
16.1 Acids 461
16.2 Bases and Acid-Base Reactions 470
16.3 Relative Strengths of Acids and Bases 475
16.4 Oxides, Hydroxides, and Acids 479
16.5 Chemical Reactions of Acids, Bases, and Oxides 483
Chapter 16 Review 487

Chapter 17 **Acid-Base Titrations and pH**
Planning Chart 490A
Answers and Solutions 490D
17.1 Concentration Units for Acids and Bases 491
17.2 Aqueous Solutions and the Concept of pH 496
17.3 Acid-Base Titrations 505
Chapter 17 Review 517
Chemistry and Careers: Food Chemist, Aroma Chemist 520
Focus on Reasoning: College Entrance Exam Preparation 521
Intra-Science: Detergents and the Environment 522

Unit 6 Chemical Reactions 524

Chapter 18 **Reaction Energy and Reaction Kinetics**
Planning Chart 526A
Answers and Solutions 526E
Periodic Table Videodisc Barcodes 526H
18.1 Thermochemistry 527
18.2 Driving Force of Reactions 536
18.3 The Reaction Process 541
18.4 Reaction Rate 549
Chapter 18 Review 558

Chapter 19 ☘ **Chemical Equilibrium**
   Planning Chart   562A
   Answers and Solutions   562D
   19.1   The Nature of Chemical Equilibrium   563
   19.2   Shifting Equilibrium   569
   19.3   Equilibria of Acids, Bases, and Salts   576
   19.4   Solubility Equilibrium   583
   Chapter 19 Review   592

Chapter 20 ☘ **Oxidation-Reduction Reactions**
   Planning Chart   595A
   Answers and Solutions   595E
   Periodic Table Videodisc Barcodes   596H
   20.1   The Nature of Oxidation and Reduction   597
   20.2   Balancing Redox Equations   602
   20.3   Oxidizing and Reducing Agents   608
   20.4   Electrochemistry   611
   Chapter 20 Review   623
   Chemistry and Careers: Industrial Technician, Environmental Chemist   626
   Focus on Reasoning: College Entrance Exam Preparation   627
   Intra-Science: Chemiluminescence   628

Unit 7   *Carbon and Its Compounds* 630

Chapter 21 ☘ **Carbon and Hydrocarbons**
   Planning Chart   632A
   Answers and Solutions   632F
   21.1   Abundance and Importance of Carbon   633
   21.2   Organic Compounds   639
   21.3   Hydrocarbons   643
   21.4   Representative Hydrocarbons and Polymers   655
   Chapter 21 Review   661

Chapter 22 ☘ **Substituted Hydrocarbons**
   Planning Chart   664A
   Answers and Solutions   632F
   22.1   Alcohols   665
   22.2   Halocarbons   670
   22.3   Ethers   673
   22.4   Aldehydes and Ketones   674
   22.5   Carboxylic Acids   676
   22.6   Esters   679
   Chapter 22 Review   682

Chapter 23   **Biochemistry**
   Planning Chart   684A
   Answers and Solutions   684D
   23.1   The Chemistry of Life   685
   23.2   Proteins   686
   23.3   Carbohydrates   689
   23.4   Lipids   691
   23.5   Nucleic Acids   695
   23.6   Biochemical Pathways   700
   Chapter 23 Review   704
   Chemistry and Careers: Agricultural Chemist, Biochemist   706
   Focus on Reasoning: College Entrance Exam Preparation   707
   Intra-Science: Custom-Designed Plants   708

Unit 8 Descriptive Chemistry 710

Chapter 24 **The Metals of Groups 1 and 2**
Planning Chart   712A
Answers and Solutions   712D
Periodic Table Videodisc Barcodes   712E
24.1 The Alkali Metals   713
24.2 The Alkaline-Earth Metals   717
24.3 Sodium: One of the Active Metals   720
Chapter 24 Review   727

Chapter 25 **The Transition Metals**
Planning Chart   728A
Answers and Solutions   728D
Periodic Table Videodisc Barcodes   728E
25.1 The Transition Elements   729
25.2 Iron, Cobalt, and Nickel   736
25.3 Copper, Silver, and Gold   744
Chapter 25 Review   749

Chapter 26 **Aluminum and the Metalloids**
Planning Chart   750A
Answers and Solutions   750C
Periodic Table Videodisc Barcodes   750D
26.1 Introduction to Aluminum and the Metalloids   751
26.2 Aluminum   753
26.3 Representative Metalloids   758
Chapter 26 Review   763

Chapter 27 **Sulfur and Its Compounds**
Planning Chart   764A
Answers and Solutions   764C
Periodic Table Videodisc Barcodes   764E
27.1 Elemental Sulfur   765
27.2 Important Compounds of Sulfur   770
Chapter 27 Review   776

Chapter 28 **The Halogens**
Planning Chart   778A
Answers and Solutions   778C
Periodic Table Videodisc Barcodes   778E
28.1 The Halogen Family   779
28.2 Fluorine   781
28.3 Chlorine   782
28.4 Bromine   787
28.5 Iodine   790
Chapter 28 Review   792
Chemistry and Careers: Metallurgist and Materials Specialist, Chemical Supplier   794
Focus on Reasoning: College Entrance Exam Preparation   795
Intra-Science: The Miracle Fluorocarbon Resin   796

Unit 9 Nuclear Reactions 798

Chapter 29 ★ **Nuclear Chemistry**
Planning Chart 800A
Answers and Solutions 800C
Periodic Table Videodisc Barcodes 800D
29.1 The Composition and Structure of the Nucleus 801
29.2 The Phenomenon of Radioactivity 804
29.3 Applications of Radioactivity 816
29.4 Energy from the Nucleus 817
Chapter 29 Review 822

Microscale Investigations 824

Microscale Investigations Teacher's Guide 826-1

1. Conservation of Mass 830
2. Mixture Separation 832
3. Measuring the Specific Heats of Metals 834
4. Conductivity 837
5. Colored Precipitates 840
6. Pin-the-Formulas-on-the-Compounds 843
7. Generating and Collecting O_2 845
8. Generating and Collecting H_2 848
9. Testing Reaction Combinations of H_2 and O_2 851
10. "Wet" Dry Ice 854
11. Measuring the Triple Point Pressure of CO_2 855
12. Water Purification 858
13. Testing Water 860
14. Is It an Acid or a Base? 864
15. How Much Calcium Carbonate is in an Eggshell? 866
16. Clock Reactions 869
17. Measuring K_a for Acetic Acid 871
18. Acid-Catalyzed Iodination of Acetone 873

Appendix A Useful Tables 876
Appendix B Logarithms 886
Appendix C Study Skills for Chemistry 888

Glossary 892
Index 901
Credits 910

Focus on Reasoning Answers 910A

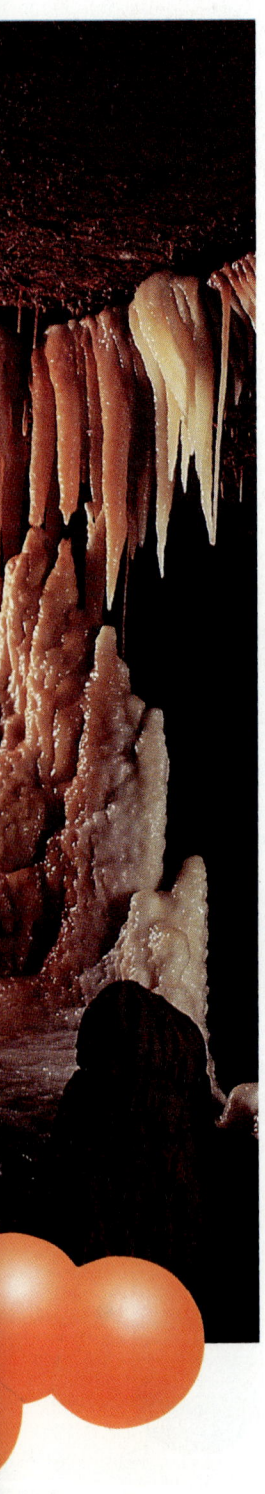

Desktop Investigations

Physical and Chemical Changes 18
Density of Pennies 38
Constructing a Model 74
The Wave Nature of Light: Interference 102
Designing Your Own Periodic Table 131
Models of Chemical Compounds 215
Balancing Equations Using Models 249
Limiting Reactants in a Cookie Recipe 273
Properties of Carbon Dioxide 307
Diffusion 361
Surface Tension of Liquids and Water Solutions 374
Solutions, Colloids, Suspensions 408
Acids and Bases in Your Kitchen 476
Testing for Acid Precipitation 509
Factors Influencing Reaction Rate 555
Redox Reactions 600
Enzyme Action on Starch 690
Paramagnetism of Nickel 730
Silver Halides 789

Features

Chemistry and Careers
Fire Chemist, Science Writer 64
Laser Chemist, Chemical Technician 194
Analytical Pharmaceutical Chemist, Chemical Plant Manager 282
Occupations in the Compressed-Gas Industries, Chemical Patent Attorney 398
Food Chemist, Aroma Chemist 520
Industrial Technician, Environmental Chemist 626
Agricultural Chemist, Biochemist 706
Metallurgist and Materials Scientist, Chemical Supplier 794

Chemistry and Health
Carbon Monoxide Poisoning 309
Electrolytes in Your Body 447
"Acid Stomach" and Antacids 481
Polyunsaturated Fats and Pour-on Margarines 668
Group 1 Ions in Medication and Diet 715
Medicines Made from Sulfur 769
Radium — Cause of and Cure for Cancer 806

Chemistry and the Environment
The Nitrogen Cycle   300
Acid Precipitation and Lakes   504
Chlorohydrocarbons   640
Strontium: Fireworks and Fallout   719
The Smell of Sulfur Compounds   772
Measuring Radiation Exposure   818

Chemistry and Its Applications
Scientific Journals: Getting the Word Out   264
Liquid Crystals   376
Swimming Pool Chemistry   570
Not-So-Blue Jeans   610
Octane Ratings   646
Can a Can Become a Plane?   754
PVC Pipes   784

Chemistry and Technology
"Looking" at Atoms   81
Hydrazine, a Liquid Rocket Fuel   254
Measuring and Reaching Low Temperatures   322
Separating Isotopes   363
Superconductors   382
Fresh Water from the Sea   428
Fuels and Energy Content   531
Toward Human Gene Transplants   699
Gallium Arsenide, Lasers, and Computer Chips   761

Chemistry in Detail
Interconversion of Matter and Energy   15
Millikan's Oil-Drop Experiment   79
Heisenberg's Uncertainty Principle   107
Exceptions to the Octet Rule   168
Resonance   174
Atomic Symbols and Molecular Formulas   221
Molecular Motion   316
Boiling Liquids   451
Discovering How to Discover Elements   791

T13

Modern Chemistry
The Most Informative High School Chemistry Text You Can Use

This new edition of *Modern Chemistry* continues to provide the most comprehensive coverage of chemical principles available for the first year high school course. Using *Modern Chemistry,* you can be sure that your students will get a thorough foundation in theoretical and descriptive chemistry with the added benefit of a strategic problem-solving strand. To keep pace with changing classroom needs, you'll find that this new edition of *Modern Chemistry* places greater emphasis on developing process and critical thinking skills.

Thorough Content Coverage in Just The Right Sequence

- Atomic structure is covered early to lay the appropriate foundation for teaching formula writing, nomenclature, chemical equations, and stoichiometry.

- Quantitative concepts, like stoichiometry, are developed using a four step method for analyzing and solving problems.

- No other high school text provides such thorough coverage of descriptive chemistry.

NEW for this edition!

Microscale Investigations supplement the Desktop Investigations
(see page T22)
18 new Investigations using microscale equipment and techniques

Improved Chapter Reviews
(see page T26)
more critical thinking, cross-discipline connections

College Entrance Exam Preparation
(see page T26)
students can sharpen their reasoning skills in preparation for college entrance exams

Improved Planning Charts in the Teacher's Edition
(see page T30)
management of the entire program is provided at-a-glance

Correlation to Videotapes and Videodisc
(see page T34)
World of Chemistry Videotapes
Shakhashiri Chemical Demonstrations Videotapes
The Periodic Table Videodisc with Barcodes in the Teacher's Edition

New Supplements

Teaching Transparencies
(see page T32)

Teacher's Resource Binder
(see page T32)

Modern Chemistry
Well-Organized Chapters Provide the Structure for Success

Each chapter is subdivided into sections that open with a set of clearly defined Objectives. Students know from the start exactly what they are expected to learn.

Chemistry vocabulary is highlighted with boldface type and clearly defined when introduced.

Authoritative, straightforward explanations are used to present key concepts.

Section Reviews measure mastery of the Section Objectives.

Tables are used extensively to visually organize data and provide concrete models for abstract ideas.

Cautions, in the margins, alert students to possible hazards associated with working with certain chemicals.

T17

Modern Chemistry
Problem-Solving Tool Kit: A Four-Step Model That Ensures Student Success

True mastery of the material is demonstrated when students can apply mathematical skills to solving chemical problems. Solving problems goes beyond the rote memorization of a worked-out problem solution. It requires the ability to *analyze* the problem, make a *plan* for reaching a solution, apply mathematical models if needed, and verify the reasonableness of an answer. *Modern Chemistry* provides the structural framework students need to develop logical reasoning skills using a four-step approach. The steps focus on developing strategies specific to each type of problem to be solved. Continuous exposure to good strategies enables students to attack any type of problem with success.

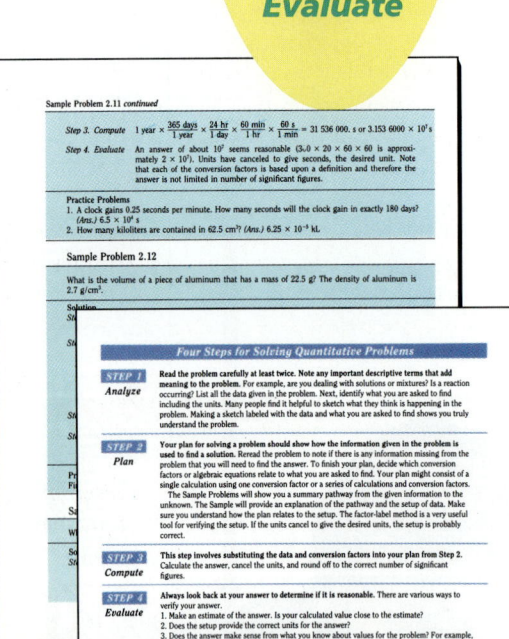

Four Phases to Mastery

1. **Sample Problems**
2. **Practice Problems**
3. **Section Review**
4. **Problems in the Chapter Review**

Most *Problems* sections provide two levels of difficulty. Those problems which can be easily solved using a Sample Problem from the chapter as a model have item numbers printed in black. Those problems with item numbers in red require students to apply their knowledge to new situations or combine several concepts and ideas to solve a multistep problem.

When making assignments it is suggested that you assign the second level problems once you are confident that students have mastered the first set. Assigning difficult problems before students are ready to tackle them only results in frustration and undermines students' confidence in themselves.

Ancillaries

Modern Chemistry Problem Solving Worksheets in the *Teacher's Resource Binder* are designed to carry through the problem solving philosophy covered in Chapter 2 of the text. Studies show that some students are unsuccessful with chemistry problems because they use *Compute* as the first phase of their solution method. These students usually have not analyzed the problem and therefore the equation or formula they apply has no meaning. To overcome this practice, the worksheets take students carefully through the four steps — analyze, plan, compute, and evaluate — so students see a solution unfold using a structured approach.

T19

Problem-Solving Tool Kit, *continued*

Using Calculators and Computers in the Chemistry Classroom

Calculators

Though a scientific calculator is certainly useful, the computations for most of the problems in this text are easily handled by an inexpensive four function calculator. At the beginning of the year impress upon students that a scientific calculator will not improve their grades.

Though there is no formal instruction on the use of calculators to solve problems in the Sample Problems throughout the text, it is assumed that students will be using calculators. It is also assumed that the teacher is evaluating a student's ability to execute a rational plan for solving a particular type of problem rather than the mathematical accuracy of a computation. Because so many types of calculators are available, it is a good idea to take time in the beginning of the year for students to become familiar with the functions of their calculators.

Programmable Calculators

If you have a programmable calculator, it is suggested that you make its use an integral part of your problem-solving instruction. Using a programmable calculator requires that students carefully plan the series of steps and operations required to solve a problem. This instructional tool forces students to focus on the process used to achieve an answer. It is suggested that you use the programmable calculator for large group instruction when you are reviewing the Sample Problems in the text. If enough programmable calculators are available for small cooperative groups (3-5 students), you can assign similar types of problems and compare the problem solving methods the individual groups employ. This teaching strategy allows students to see that there may be alternative pathways to solving a particular type of problem.

Computer Spreadsheets

Commercial spreadsheet programs can be used in the same way as a programmable calculator for students to solve various types of problems. For example, cooperative groups and computers can be used when covering Chapter 11 on the gas laws. Each group could be assigned the task of programming a spreadsheet to solve a particular type of gas law problem. Each group should review another group's work to check for flaws in the reasoning process used.

Spreadsheets are also widely used for performing lab calculations. Lab groups can be assigned the responsibility of setting up the program calculations for a particular experiment on a rotating basis. This strategy gets students more involved in the mechanics of the experiment and generally makes the results far more meaningful for the group that did the programming.

Computer-Assisted Instruction/Lab Simulations

There are a number of good tutorial programs available to help students with basic skills. CAI software is best used by an individual student at a computer. Lab simulations work best when groups are small (2 students) as with an actual lab experiment. Since programs become outdated quickly, it is suggested that you obtain current catalogs from the companies listed on page T35 before ordering any software. Most companies will allow you to preview materials on a 30-day basis with no obligation to purchase if the software does not meet your needs.

Modern Chemistry
The Most Realistic, Flexible Laboratory Program Anywhere

The *Modern Chemistry Laboratory Program* provides flexibility to meet a variety of scheduling needs and equipment limitations. The materials in the student text along with the *Laboratory Experiments* manual provide a total of 81 hands-on experiences. The lab program is realistic in providing experiments using microscale techniques which greatly reduce preparation time, reagent cost, and chemical waste. The program is flexible in that you can choose Investigations ranging from simple home experiments to microscale experiments and the traditional laboratory experiments that are a standard part of the curriculum.

Safety is emphasized and students should become familiar with the safety logos which are found on page 828. Additional safety information for teachers can be found on page 826-2 of this Teacher's Edition. It is suggested that you purchase the *Total Science Safety System* software for thorough coverage in this area. See page T35.

Modern Chemistry Laboratory Experiments

The 44 Investigations in this softbound manual provide additional hands-on opportunities to sharpen process skills. Seventeen of the Investigations can be performed using microscale procedures. The format of the labs parallels that of the *Microscale Investigations* in the student text.

Modern Chemistry Laboratory Experiments is available in both a consumable and nonconsumable format.

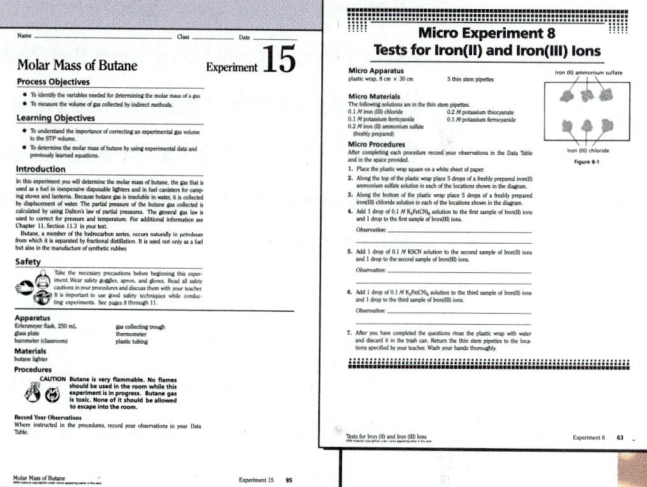

Desktop Investigations provide students with experiments that can be done at home or in a classroom. Each *Desktop Investigation* opens with a *Question* that is answered by collecting data in the *Procedure* section. The *Discussion* section directs students in forming generalizations and conclusions from their observations.

New **Microscale Investigations** provide thorough development of chemical principles using an inquiry philosophy and microscale quantities of reagents. The process skills thoroughly developed in the procedures are highlighted at the beginning of each Investigation. Mastery of the objectives for the Investigation is evaluated through the comprehensive set of questions at the end of each *Microscale Investigation*.

The *Extension* sections provide numerous opportunities for students to test additional variables or design their own Investigations. For additional information on the *Microscale Investigations*, see page 826-1.

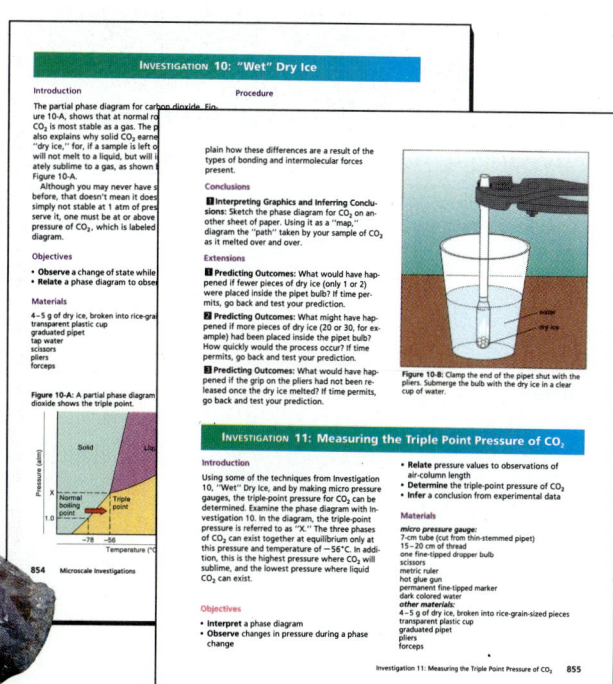

T23

Modern Chemistry
Thorough Development of Process Skills

The listing that follows describes the various process skills developed in the program. In addition to the Desktop and Microscale Investigations, many of these skills are highlighted in the Critical Thinking and Projects sections of the Chapter Reviews. The chart gives you an overview of process skill development for the in-text labs.

Observing is the process of using one or more of the five senses to perceive objects or events as they occur.

Measuring is the process of making quantitative observations. Measurements in chemistry have units.

Organizing is the process of placing data from observations and measurements in logical order. Data is more useful when it is organized to show trends for the purpose of making generalizations or forming conclusions.

Process Skills in Microscale Investigations

Process Skill	Inv #	1	2	3	4	5	6	7	8	9	10	11	12	13	14	15	16	17	18
Observing		•	•	•	•	•	•	•	•	•	•	•	•	•	•	•	•	•	•
Measuring		•		•	•		•			•		•				•	•	•	•
Organizing		•		•	•	•	•	•				•		•	•	•	•		•
Classifying			•										•		•				•
Hypothesizing				•	•	•	•	•	•	•		•		•				•	•
Predicting		•		•	•	•	•	•				•		•			•	•	•
Analyzing		•		•	•	•	•		•	•	•	•	•	•	•	•	•	•	•
Designing Experiments		•	•	•	•	•	•	•					•	•	•	•	•	•	•
Inferring		•		•	•	•	•	•	•	•	•	•	•	•	•	•	•	•	•
Modeling		•	•		•		•							•		•			•
Communicating		•		•		•	•	•			•					•			•

Process Skills in Desktop Investigations

Process Skill	Text Pg.#	18	38	74	102	131	215	249	273	307	361	374	408	476	509	555	600	690	730	789
Observing		•		•	•		•			•		•	•	•	•	•	•	•	•	•
Measuring		•	•					•	•	•	•		•		•	•	•	•		•
Organizing		•	•			•		•	•	•	•		•	•	•	•	•	•	•	•
Classifying		•					•	•					•		•	•	•	•	•	•
Hypothesizing			•	•					•			•		•			•	•	•	•
Predicting					•		•		•											
Analyzing			•	•				•	•		•	•	•	•	•	•	•	•	•	•
Designing Experiments																				
Inferring		•		•	•		•	•	•	•	•	•	•	•	•	•	•	•	•	•
Modeling				•	•	•	•													
Communicating																				

Classifying is the process of grouping or ordering items according to the similarities and differences among the items by using established classes or by developing new organizational patterns.

Hypothesizing is the process of constructing statements about phenomena that can be tested. Evidence can either support or refute a hypothesis but can never prove it true beyond all doubt.

Predicting is the process of stating in advance the result that will be obtained from testing a hypothesis. Together hypotheses and predictions often take the form of an if/then statement. (If the hypothesis is true, then expect to observe....) When the prediction turns out to be true, the hypothesis is supported; when the prediction turns out to be false, the hypothesis is refuted.

Analyzing is the process of determining the reliability of data and whether the data support or refute the hypothesis. Scientists analyze data in many ways, including interpreting graphs, using statistics, comparing data to those gained on other studies, determining the relationships between variables, and finding sources of error.

Analysis consistently occurs in the *Investigations,* in solving problems, and in answering the *Critical Thinking* items.

Designing Experiments is the process of thinking through a problem, developing a hypothesis, and designing an organized procedure to test that hypothesis.

Students are frequently asked to design experiments as *Extensions* in the *Microscale Investigations* and as *Projects* in the Chapter Review.

Inferring is the process of drawing conclusions about objects or evernts based on related facts or premises but not on direct observation. Inferences are based of data collected, reasoning, or past experience.

Students often make inferences in the *Investigations* and as part of the *Critical Thinking* items.

Modeling is the process of making visual, mathematical, or verbal constructs that help explain data. A model is a simplified representation or simulation of a process. Some models, such as a model airplane, are physical representations. Other models are symbolic, such as a computer generated animation showing drug molecules attaching to cell receptors.

Students are frequently asked to apply models in anwering items in the *Critical Thinking* sections.

Communicating is the process of sharing information with others in written or spoken form. Scientists often give papers and speeches to share the nature and results of their work. Communication of information provides the scientific community the opportunity to accept or reject one's conclusions via repeated testing to verify results.

Lab Reports and *Research and Writing* items in **Modern Chemistry** provide the opportunities for developing communication skills.

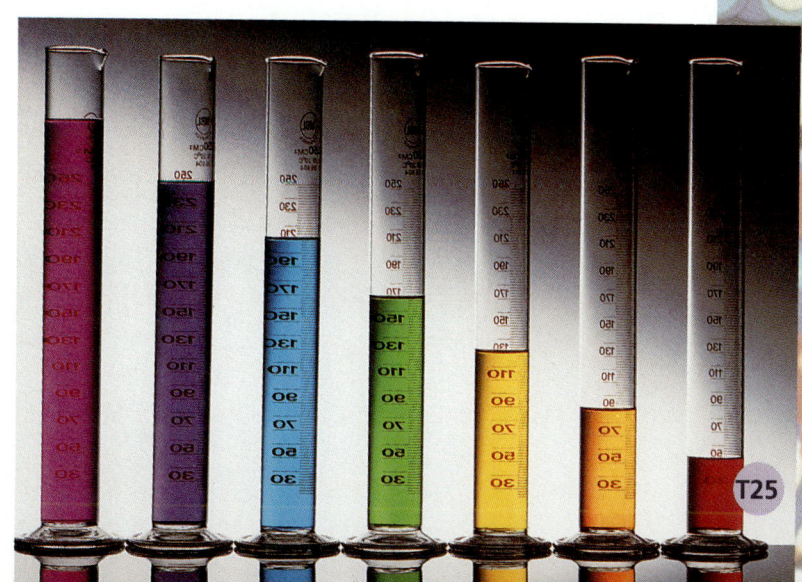

Modern Chemistry
Comprehensive Evaluation to Assure You That Students Are Learning

Objectives
The *Section Objectives* that open each lesson focus on the information that should be mastered for that section.

Section Reviews
The end-of-section review ties back to the *Section Objectives*.

Chapter Reviews
The end-of-chapter evaluation thoroughly assesses the section objectives.

Focus on Reasoning
These end-of-unit features focus on the assessment of reasoning skills to prepare students for the ACT College Entrance Exam Science Reasoning test. Students analyze experimental situations to form valid conclusions. Coverage of chapter content is not a requirement for completing the reasoning items.

Chapter Review Components
Vocabulary provides a listing of the boldface terms within the chapter. All terms are referenced to the page where they are defined.

Reviewing Concepts establishes whether the student can recall the factual knowledge presented in the chapter. Since the sections that follow require more complex thinking processes, students who cannot complete Reviewing Concepts should be put through a reteaching session.

Problems are categorized by topic. Within each section, you will find two levels of problems. The first level provides problems for a topic that are similar to the *Sample Problems* presented in the text and are considered drill in solving fundamental chemistry problems. Students should be made aware of this hint as it will be useful for them when they are completing problem assignments outside of class.

The second set of problems is highlighted with red item numerals. These problems require students to apply the problem-solving process they learned in the chapter to a new situation or to a problem requiring a multistep solution.

Critical Thinking items are labeled with the critical thinking skill required in formulating an answer. You will notice that many of the labels are similar to the process skills developed in the *Modern Chemistry Microscale Investigations*.

Cross Connections items highlight the integration of chemical knowledge with other subject areas. These items generally provide excellent opportunities for library research.

Projects consist of both laboratory and library research ideas to extend the material presented in the chapter to further challenge students.

Modern Chemistry

Features That Highlight the Relevance of Chemistry

The classroom can be an ideal arena in which to discuss conflicts between scientific research and the ethical concerns of society. Such discussions also encourage the development of critical-thinking skills.

Science, Technology, and Society

A popular way to introduce issues in the chemistry classroom is to have students collect newspaper and magazine articles about chemistry and discuss them in class. The problems discussed in some of the Intra-Science features and some of the suggestions in the Research and Writing section of the Chapter Reviews can be the basis of such study. Encourage students to carefully research all arguments of both sides of the issue before attempting to take a position. Formulating a position on an issue should be the result of a logical evaluation process. Discussions of the issue should center on presenting enough information so that students can evaluate several points of view. This approach will encourage an open attitude toward formulating opinions and arriving at decisions.

Chemistry and Careers pages show some of the professions that utilize a knowledge of chemistry. Students realize that what they are learning forms the knowledge base for some interesting professions.

Chemistry and Health features show the connection between chemical principles and the human body. These pages give you an opportunity to introduce and discuss some biochemical topics before Chapter 23.

Chemistry and the Environment features show how chemicals both positively and adversely affect the environment.

Chemistry and Its Applications features show how chemistry knowledge is the basis for the development of new products that are familiar to students.

Chemistry and Technology features focus on new research and its applications in science and industry.

Chemistry in Detail features provide an in-depth look at a topic from the chapter.

Crossing Disciplines

No field of study, including chemistry, exists in isolation from the others. Chemistry has been and continues to be the basis for discoveries in other disciplines. Most important biological research concerns studying processes at the molecular level.

The Intra-Science features provide an excellent springboard for classroom discussions about the interconnections among the sciences. In addition, the Cross Connections items in each Chapter Review provide fascinating topics for further study which show the link between chemistry and other fields of study, some of which are outside the field of science.

Intra-Science features show how chemists work with other scientists to answer important questions that cross the boundaries of chemistry.

Modern Chemistry
Efficient Instruction Starts with a Functional Teacher's Edition

Your *Annotated Teacher's Edition* for *Modern Chemistry* consists of four parts.

1. Teacher's Introductory material preceding Unit 1
2. Chapter Interleaves that precede each chapter
3. Teacher's Guide to the Microscale Investigations program which begins on page 826-1
4. Teacher's Annotations throughout the text overprinted on the student pages

Chapter Interleaves

Chapter Planning Charts are your road map to using the various components of the *Modern Chemistry* program in your teaching plans. Each text section is split into in-text materials and outside resources. *Other Resources* refers to supplements available with *Modern Chemistry* but not as part of the *Teacher's Resource Binder*. The text sections are coded by ability level. After selecting those sections you want to teach, you know immediately which materials to incorporate into your lesson plans.

- **Basic** - coverage of fundamental principles only
- **Average** - coverage of principles needed for adequate preparation for post-secondary chemistry studies
- **Honors** - comprehensive coverage of principles designed to challenge students with aptitudes for science and plans to major in science at the post-secondary level
- **Suggested time frames** help you in pacing the course to allow appropriate time for mastery of content and skills. The first time allotment is a suggestion based on the 180-day curriculum. The second time allotment is the number of days suggested for an extended curriculum.

Introductions clarify the teaching goals of the chapter and suggest ways to involve students.

Teaching Strategies for each section include many hints, motivational activities, analogies, and discussion guides. Numerous reteaching ideas can be found within the strategies section.

	SECTION/TEXT FEATURES		RESOURCES	

BASIC	5.1 HISTORY OF THE PERIODIC TABLE
	Demo 1
	Periodic Table Videodisc
	Desktop Investigation
	Section Review
	Chapter Review
	• Questions 1–7
		127
		126B
		126G
		131
		131
		157
			Teacher's Resource Binder
			  Section Review Worksheets
			Other Resources
			  Blackline Master 16
				15

BASIC	5.2 ELECTRON CONFIGURATION AND THE PERIODIC TABLE
	Demo 2
	Periodic Table Videodisc
	Section Review
	Chapter Review
	• Questions 8–19, Problems 1–6
		132
		126C
		126G
		121
		122
			Teacher's Resource Binder
			  Section Review Worksheets
			Other Resources
			  World of Chemistry Video 7
			  Transparency 10
				16

AVERAGE	5.3 ELECTRON CONFIGURATION AND THE PERIODIC PROPERTIES
	Demo 3
	Section Review
	Chapter Review
	• Questions 20–30, Problems 7–13
		143
		126C
		156
		157
			Teacher's Resource Binder
			  Section Review Worksheets
			Other Resources
			  Transparencies 11, 12
			  Blackline Masters 17, 18, 19, 20, 21, 22, 23
				17

	CHAPTER END ACTIVITIES Critical Thinking, Cross Connections, R… Writing, Projects		*Teacher's Resource Binder* Test Booklet	21, 23

A comprehensive collection of **Demonstrations** is included to give students the concrete, motivating experiences they need to understand abstract ideas. Additional demonstrations can be found on videotape. See page T34.

Background information and preparation notes for all **Desktop Investigations** are found within the chapter interleaves.

References and Resources provide you with a complete listing of books, periodicals, AV aids, and computer software that can be used to enrich the topics of the chapter.

Answers and Solutions give worked out solutions to all questions and problems in the Chapter Review.

Barcode Correlation to *The Periodic Table* Videodisc helps you instantly access images that expand the examples given in the text. For more information, see page T35.

Teacher's Guide to the Microscale Investigations

provides complete instructions for setting up the lab prior to running a particular Investigation. Use of equipment and new techniques are thoroughly explained. Special safety precautions and disposal instructions are provided where appropriate. All answers to questions for the students' lab reports are also included.

Teacher Annotations

supply insightful background information, interesting science facts, safety notes, useful teaching suggestions and answers to the Section Reviews.

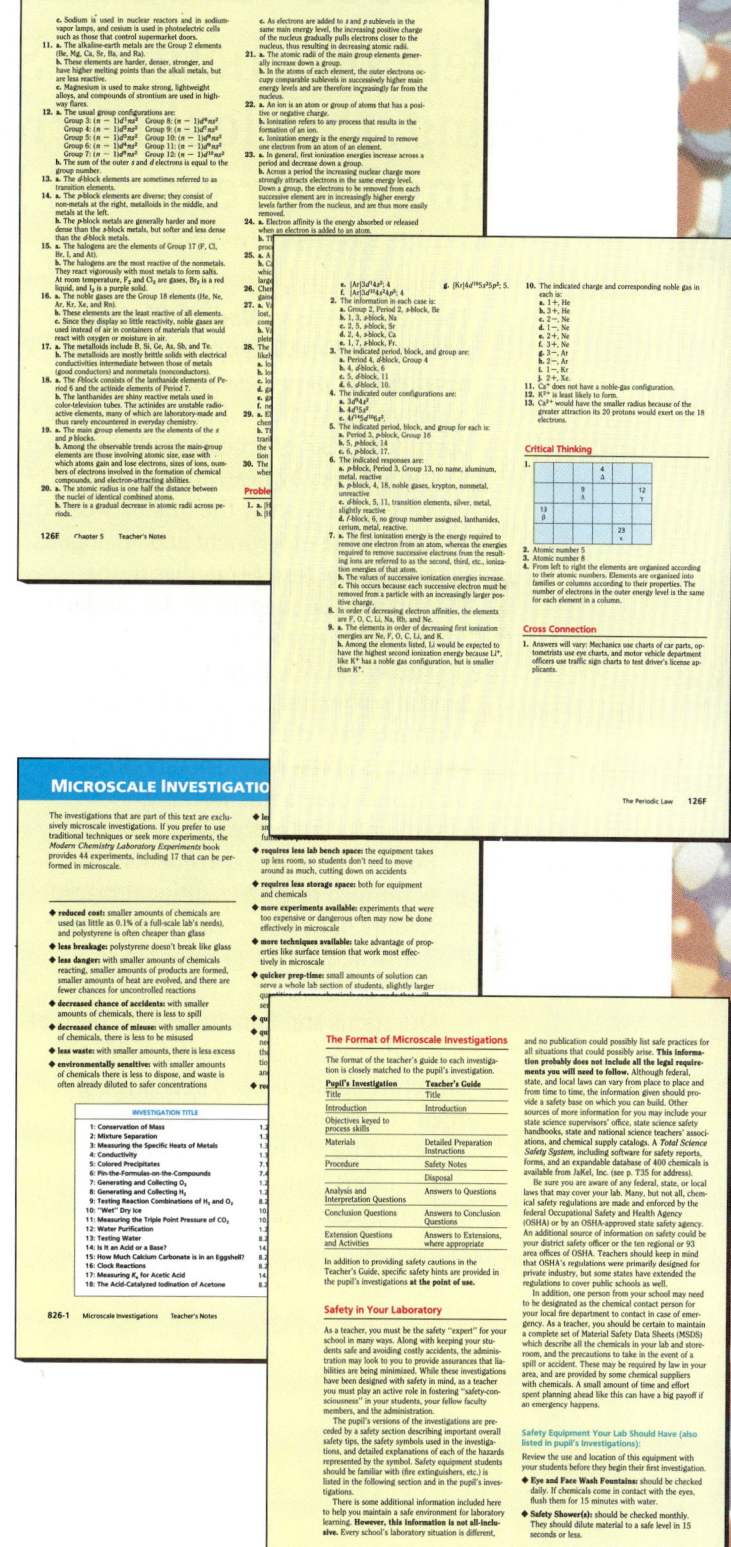

Modern Chemistry

Complete Instructional Support for Planning, Developing Process Skills, Assessing Progress, and Motivating Students

Laboratory Experiments

The experiments in ***Modern Chemistry Laboratory Experiments*** focus on the development of process skills through hands-on experiences. Each experiment opens with a set of process skill objectives that is followed by Learning Objectives which generally show the link between the content of the experiment and analogous content in the text. Strategies in the margin provide suggestions for students to practice the process skills.

The laboratory manual contains a special section on safety, including safety contracts that students sign to show their awareness of the responsibility they have for maintaining a safe working environment for everyone.

Seventeen of the experiments can be performed using a traditional macro procedure or as microscale experiments. Full instructions are given for each approach.

Questions and Conclusions sections at the end of each experiment provide the structure students need in analyzing their results and relating them to the experiment objectives. General Conclusions items relate the results of the experiment to students' daily experiences.

Tests

Modern Chemistry Blackline Master Tests provides two tests for each chapter and unit of the text. The test and alternate test for a chapter or unit covers the same material using different problems and questions. Two comprehensive final exams are also provided.

A variety of question types are used including multiple choice, fill-in, matching, problems and calculations, and essay. Tables, graphs, and diagrams are included when necessary to test material covered in this test book.

Test items are keyed to the Section Objectives in the ***Modern Chemistry*** text. The correlation to specific objectives can be found with the answer keys in the back of the text booklet.

Teaching Transparencies and Masters

Modern Chemistry Teaching Transparencies package is available for use with an overhead projector. The package includes 60 full color transparencies taken from the illustrations in ***Modern Chemistry***. Illustrations were selected for their instructional value in making abstract ideas more concrete through the use of illustrative models. Because chemistry is a quantitiative science and the ability to interpret data is a critical skill many of the quantitative tables from the text have been reproduced in blackline master form for you to photocopy or make into blackline transparencies.

The use of each transparency and blackline master is keyed in the *Planning Charts* in the Teacher's Edition and in the *Lesson Plan Book*.

Teacher's Resource Binder

Modern Chemistry Teacher's Resource Binder provides a wealth of supplemental materials, in copymaster form, to extend the text. Included in the binder, you will find the following.

- **Lesson Plan Guide** which contains detailed lesson plans for each section of the text using a lesson cycle format.
- **Problem Solving Worksheets** which provide practice using the four-step model in the student text. Each problem set contains one fully developed sample, one guided sample, and three problem variations for each concept.
- **Laboratory Experiments** manual
- **Section Reviews** workbook
- **Tests** (with Alternate Tests)
- **Answer Keys**

Computer Test Generator

The Computer Test Generator for Modern Chemistry contains 5000 test items that are correlated to the *Section Objectives* in the text. The software allows you to construct multiple formats of a chapter test that is customized to meet your teaching objectives.

The test generators are available for use on Apple,® Macintosh® and IBM PC® and compatible computers.

Section Review Worksheets

This workbook is designed to take students through the major concepts of each text section by answering a variety of questions. The questions on each 2-page worksheet measure mastery of the Section Objectives. The questions progress from simple recall of facts through interpretation and application level. These pages have the flexibility to be used as review guides in preparation for a test, as pretests, homework, and quizzes. The actual format resembles that of the Chapter Tests.

Audiovisual Materials
Bringing the Chemical World into the Classroom

World of Chemistry Videotapes

This Annenberg/CPB Project series provides a comprehensive overview of chemical principles for a first year chemistry course. The tapes include interviews with scientists, demonstrations, animations, and scenes from industrial settings. The process of science is emphasized in each tape as students experience first-hand how chemical principles become the basis for solving some of today's technological problems. In addition, students get a historical perspective on the importance of chemical knowledge in the development of useful products. The tapes are correlated to the topics in *Modern Chemistry* in the *Planning Charts* that begin each chapter in your Teacher's Edition. Each tape is approximately 30 minutes of playing time.

A *Study Guide* to accompany *World of Chemistry* videotapes is available from Holt, Rinehart and Winston. The *Study Guide* lesson plan for each tape gives you an outline of important points to look for while watching the tape and thorough background information on each of these topics. The lesson plan for each videotape closes with a series of hands-on activities and questions that reinforce the ideas on the videotape. (Annenberg/CPB Project; see T35)

Shakhashiri Chemical Demonstrations Videotapes

This series of three videotapes contains 49 demonstrations performed by Bassam Shakhashiri. These demonstrations highlight some of the particularly spectacular reactions in chemistry and as such provide a motivational setting to develop chemical principles through inquiry. The demonstrations are correlated to the topics in *Modern Chemistry* in the *Planning Charts* for each chapter in your Teacher's Edition.

An *Instructor's Guide* to accompany the *Shakhashiri Chemical Demonstrations Videotapes* is available from Holt, Rinehart and Winston. The Instructor's Guide provides a description of each demonstration, the chemical background needed to understand the demonstration, further references, and some questions to ask in conjunction with each demonstration. This guide makes the demonstration videos extremely effective teaching tools.

World of Chemistry Videotapes

1. Chemistry
2. Color
3. Measurement: The Foundation of Chemistry
4. Modeling the Unseen
5. States of Matter
6. The Atom
7. The Periodic Table
8. Chemical Bonds
9. Molecular Architecture
10. Signals from Within
11. The Mole
12. Water
13. The Driving Forces
14. Molecules in Action
15. The Busy Electron
16. The Proton in Chemistry
17. The Precious Envelope
18. The Chemistry of the Earth
19. Metals
20. On the Surface
21. Carbon
22. The Age of Polymers
23. Proteins: Structure and Function
24. The Genetic Code
25. Chemistry and the Environment
26. Futures

The Periodic Table Videodisc

The Periodic Table Videodisc is available from *Journal of Chemical Education Software*. (JCE: Software, see below) The videodisc contains images of most of the elements, their uses, and their reactions with air, water, acids, and bases. Students become far more aware of the properties of the elements when they can see these properties first hand. The ability to access images almost instantaneously in any order makes this visual chemical database an extremely valuable teaching tool. To assist you in using the videodisc in the classroom, barcodes are provided in the *Modern Chemistry Annotated Teacher's Edition* in the Chapter Interleaves for those chapters in which the videodisc is a useful resource. Images can be accessed by frame number as well as bar code. The images are grouped by student-text page numbers, so you know where the images would be most useful in illustrating the subject matter of the text.

Audiovisual and Software Suppliers

The Aluminum Association
Educational Services
818 Connecticut Avenue, NW
Washington, D.C. 20006

Annenberg/CPB Project
P.O. Box 2345
South Burlington, VT 05407
1-800-LEARNER

Barclay School Supplies
166 Livingston Street
Brooklyn, NY 11201

Cambridge Development Laboratory
42 4th Avenue
Waltham, MA 02154

Central Scientific Company
11222 Melrose Avenue
Franklin Park, IL 60131-1364

Chemistry According to ROF
Richard O. Fee
5634 South Kensington Avenue
LaGrange, IL 60525

Coronet Films and Video
108 Wilmot Road
Deerfield, IL 60015

Dow Corning Corporation
P.O. Box 1767, Dept. No. 01-58
Midland, MI 48640

Educational Images, Ltd.
P.O. Box 367
Lyons Falls, NY 13368

Films for the Humanities and Sciences
Box 2053
Princeton, NJ 08543

Fisher Scientific Company
Educational Materials Division
4901 West LeMoyne Street
Chicago, IL 60651

Flinn Scientific, Inc.
P.O. Box 219
Batavia, IL 60510-0219

Frey Scientific Company
905 Hickory Lane
Mansfield, OH 44905

Human Relations Media
175 Tomkins Avenue
Pleasantville, NY 10570-9973

Journal of Chemical Education Software (JCE: Software)
Project Seraphim/JCE
Department of Chemistry
Eastern Michigan University
Ypsilanti, MI 48197

Learning Arts
P.O. Box 179
Wichita, KS 67201

Modern Talking Picture Service
5000 Park Street North
St. Petersburg, FL 33709

National Science Teacher's Association
NSTA Special Publications
1742 Connecticut Avenue, NW
Washington, D.C. 20009

Prentice Hall Media
150 White Plains Road
Tarrytown, NY 10591

Queue, Inc.
(also available through Queue: COMPress, HRM, Knowledge Factory, Learning by Practice, and Programs for Learning)
562 Boston Avenue
Bridgeport, CT 06610

Random House Media
Department 119
P.O. Box 126
Stamford, CT 06904

Sargent Welch
7300 N. Linden Avenue
Skokie, IL 60077

Science Kit and Boreal Labs
777 East Park Drive
Tonawanda, NY 14150

Total Science Safety System
JaKel, Inc.
585 Southfork Drive
Waukee, IA 50263

TV Ontario/TVO Video
143 West Franklin St., Suite 206
Chapel Hill, NC 27516

University of Akron
Audiovisual Services
302 East Buchtel Avenue
Akron, OH 44325

Walch Science
J. Weston Walch, Publisher
321 Valley Street
P.O. Box 658
Portland, ME 04104-0568

Ward's Modern Learning Aids
Ward's Natural Science Establishment, Inc.
5100 West Henrietta Road
P.O. Box 92912
Rochester, NY 14692

Reading and Writing
Developing Skills in the Chemistry Classroom

Reading

Recent research has shown that the time spent in helping students prepare to read is critical to developing not only basic comprehension but also an understanding of more complex relationships such as cause and effect. Two strategies are particularly important in the prereading phase of instruction: building on prior knowledge and establishing a purpose for reading.

Building on Prior Knowledge

The amount of prior knowledge that students have about a topic directly influences their comprehension. The more students know about a topic, the easier it will be for them to grasp new information about it. Helping students identify the information they already have about a topic before reading assists students in relating new information to existing knowledge. Research strongly suggests that students tend to maintain misconceptions they may have about a topic. If the text information seems to conflict with their preconceptions, students may ignore or reject the new information. Students are able to discard or modify misconceptions, however, when the teacher provides direct instruction that clarifies and corrects their misconceptions before they read the text.

Establishing a Purpose for Reading

One way to establish a purpose is to make a study guide with questions that students can answer as they read. You can also help students learn to make their own study guides. First, teach students to preview a chapter by looking at all the chapter headings, illustrative material, and boldfaced terms and phrases before reading the text. This technique helps students gain a feel for the chapter and helps them build a basic structure for the new information. Then show them how to use the objectives, captions, headings, and vocabulary words to devise study-guide questions to answer later as they read the chapter. Following this method, students will read with a purpose in mind — a purpose of their own devising.

Reading Strategies

If students have been prepared for the reading assignment by pre-reading activities, they will likely retain more of what they read. But many students have no real strategies for reading text materials, other than simply rereading difficult sections of a chapter. They often have little awareness of such higher-level cognitive strategies as organizing, reflecting upon, and evaluating what they read. You can take three major steps to help your student learn and practice effective reading strategies.

1. Help students understand how science textbooks organize information. Usually each subsection of a chapter contains one main idea, stated first or last. Supporting details follow or precede the main idea in a recognizable pattern. Also show students that key phrases mark the specific types of relationship among the ideas being presented. For example, key words such as "like" or "different" indicate comparison and contrast. Being aware of cues such as these gives students a familiar conceptual framework into which they can place the new information presented in the text.

2. Point out the basic features of Modern Chemistry. The easier it is for students to find information, the more they will get out of the textbook. Very early in the course, review with students the parts of the textbook and the functions of each part. Point out the reference guides in the front and back of the textbook, such as the Table of Contents, which provides a content outline for the text. Students should also note *Appendix C: Study Skills for Chemistry,* the *Glossary,* and *Index.* Show students how to use these aids to locate desired information. Also encourage student to take advantage of such pedagogical aids as the Section Objectives which are designed to help students prioritize the information presented in the chapter. Point out that main ideas, definitions, and key equations often appear in boldface or italic type and that important concepts are illustrated in captioned charts, labeled drawings, or photographs.

3. Model effective reading and study practices for students. During class lectures or discussions about textbook material, point out to students the various levels of comprehension they should achieve in reading *Modern Chemistry:* literal understanding, interpretation, and application. At the literal level of understanding, students comprehend the information explicitly stated but may not realize its significance. Interpretation involves understanding what is implied (but not actually stated) and its importance. Using the information in a new context and applying the understanding gained from the reading is the highest level of understanding. When you ask questions during class, point out which level of understanding your question addresses. Student will eventually become so familiar with each level that they can devise similar questions on their own, a process that makes them independent learners.

Writing

Studies show that writing is an effective tool for improving reading. As students write, they are creating a text for others to read. The more they write, the more they are able to recognize compositional devices they encounter in the text. The *Cross Connections* and *Reseach* and *Writing* sections of the *Chapter Review* offer students many opportunities to communicate their understanding of chemistry in writing.

Many students strive for verbosity in their writing, perhaps because they have had teachers who emphasized length and vocabulary more than content. But complex sentence structure and five-syllable words make understanding difficult. Scientific writing, in particular, suffers because the subject matter and technical jargon are often unfamiliar to the reader (and sometimes to the writer as well!). The most important advice to give students about scientific writing is to strive for clarity and accuracy. These characteristics can often be achieved with simple vocabulary and short sentences. A useful approach might be to have students imagine they are writing for students three years younger than themselves.

You might also bring in examples of both effective and ineffective scientific writing. Show students example of clear logical arguments and explanations used in effective writing, and point out how ineffective scientific writing uses inexact wording, circular reasoning, and often jumps to conclusions.

Cooperative Learning
Modeling the Environment of the Scientist

Allowing students to share their learning with one another is a strategy that has been used since the days of the one-room schoolhouse. Currently, the concept has been refined and its merits clarified for today's classrooms. One of the benefits of cooperative learning is that student can experience science in action. Cooperative learning models the real scientific environment, where scientists often work in laboratory teams and collaborate on solving difficult problems. It makes the classroom a place for experimentation, where the free flow of ideas replicates that which scientists engage in and where tentative ideas can be bandied about. Working in cooperative groups also allows students to learn together about the excitement and frustration involved in scientific inquiry.

Benefits of Cooperative Learning

Cooperative learning teaches students to ask the kinds of questions that scientists ask. It builds confidence in students' abilities to ask these questions as independent, self-sufficient investigators. It makes the classroom a place for wondering and curiosity rather than one where the learned pass on tidy formulations to the unlearned. It removes chemistry from its pedestal as a discipline that only the brilliant can successfully employ. Cooperative learning shows that chemistry is a learning process and a way of thinking rather than a set of universal principles students must memorize.

Cooperative learning empowers and involves students. It raises their self-esteem because they are learning something on their own through cooperation rather than being handed prepackaged knowledge. It helps students become self-sufficient, self-directed, lifelong learners. In a cooperative learning environment, students are less dependent on you for knowledge.

In addition, working in teams gives students a sense of belonging and shared responsibility. It can increase the motivation of both low and high achievers. Low achievers finally feel that they can play a vital role in a process they may once have associated with failure. High achievers benefit from the self-esteem gained by helping to raise the level of group accomplishment.

Using Cooperative Groups in the Classroom

Establishing cooperative learning in the classroom requires you to relinquish some control, so students can become responsible for constructing knowledge. Working in groups to prove and investigate ideas, answer questions, and draw conclusions about observations allows students to discover and discuss concepts in their own language. When students learn through cooperation, the knowledge derived becomes their own.

The *Desktop Investigations, Microscale Investigations,* and the *Experiments* in the Laboratory Manual provide excellent opportunities for cooperative group work. Even some of the *Demonstrations* in this Teacher's Edition can be modified to become cooperative learning activities. Concept mapping, which is described in *Appendix C,* also makes an excellent cooperative activity. Finished concept maps can then be used as study guides for exams. When students work together to build maps, deficiencies in their understanding can be quickly identified.

In preparing for a cooperative learning activity in your chemistry class, consider the following information.

Group Size - Group size may vary in cooperative learning activities. If your class is attempting cooperative learning for the first time, you may find it easier to keep group size to two or three members. Small groups require fewer cooperative skills on the part of students. Assign groups so that each has a leader and is balanced according to academic abilities.

Group Goal - Students need to understand what is expected of them. Identify the group goal, whether it be to master specific objectives or to create a product such as a chart, report, or booklet. In addition, you should identify and explain the specific cooperative skills required for each activity.

Individual Accountability - Individual accountability means that each group member is responsible for knowing the assigned material. In addition, each group member is responsible for the learning of other group members.

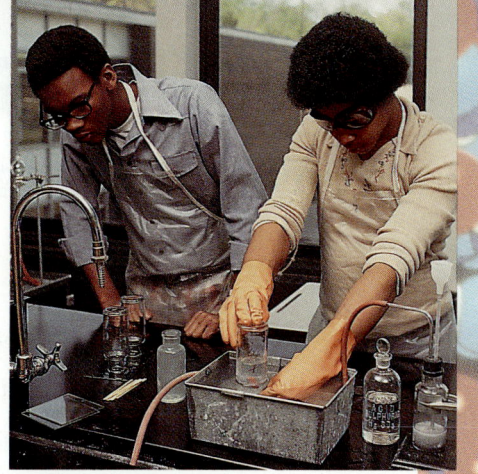

Establish your grading scheme so that if all group members achieve a preset criterion, then each member receives points. Or you might award both individual and group grades. This makes students responsible for one another as well as for themselves.

Positive Interdependence - A learning activity becomes cooperative only when group members realize that no one member can be successful unless all members of the group are successful. Therefore, students must work with and depend upon one another to achieve group goals. Positive interdependence may be achieved through the division of labor, resources, or roles among the members of the group. You may want to assign and rotate different roles for laboratory reports (for example, writer, experimenter, observer, recorder, statistician, editor). Or have each student in a group become an "expert" on some topic. Each member could then prepare and teach a lesson to the other group members or to the entire class.

Rewards for the group, such as the same grade for every member of the group, may be an additional incentive. This interdependence helps develop social and communication skills and helps students realize that if they work together, they might produce a better end-product than if they had worked alone. Students' ultimate success is a function of how well they work together.

Making Science Accessible
Providing for Individual Differences

Obviously, each individual in any given class is unique. This uniqueness translates into a student body that represents a variety of differences.

In most cases these differences provide the richness that make teaching interesting and in many respects unpredictable. Certain differences, however, require additional preparation in order to weave them successfully into your teaching. In many cases, a minimal amount of forethought will be enough to help you address these individual needs. In other cases, more preparation may be necessary. In all cases, the effort you put forth to adapt your teaching techniques to the needs of your student will provide great dividends in terms of student understanding, motivation, and success.

Teaching Students Who Have Physical Impairments

Professionals classify physical impairments into two broad categories. Orthopedic impairments, caused by disorders such a cerebral palsy and muscular dystrophy, primarily affect the skeletal and muscular systems. Health impairments refer to heart condition, diabetes, and other acute or chronic disorders that limit the student's strength and alertness.

If a student is confined to a wheelchair, check to make sure the classroom is as accessible as possible — that aisles are wide enough to accommodate the chair and that equipment is within reach. You may wish to enlist the aid of other students in the class to physically assist the disabled student as necessary. For student who cannot write for long periods of time, taped lectures, laptop computers to facilitate note-taking, and oral exams may help.

Continue to encourage hands-on activities, even though such activities may be more difficult for orthopedically impaired students. Adapt laboratory surfaces so that they are adjustable to various heights. Use a mobile demonstration table so that it can be moved to different areas of the room for maximum visibility.

Teaching Students Who Have Visual Impairments

Visually impaired students include students with limited vision and students who are totally blind. For those students who have some vision, the use of magnifiers and closed circuit television make text materials and exams readily accessible. In the classroom, seat the visually-impaired student near the front of the room, away from glare if possible, to maximize the view of both you and the chalkboard. Say what you write on the chalkboard. All visual materials should be explained in detail as they are presented.

Students who are completely blind should be allowed to become familiar with the classroom layout before the first class session begins. Promptly inform visually-impaired students of any changes in laboratory layout. Blind students may also use computers equipped with voice synthsizers, or an Optacon, an electronic device that converts printed material into raised letters. In addition, science measurement kits with Braille markings are available.

Teaching Students Who Have Hearing Impairments

Hearing-impaired students have either partial or total hearing loss. If you have a partially hearing-impaired student in your class, remember to always face the class while speaking. Minimize classroom noise, and arrange seating in a circle or semicircle so that hearing-impaired students can see others. This arrangement facilitates speechreading. Speak in simple, direct language and avoid digressions or sudden changes in the topic. During class discussions, periodically summarize what student are saying and repeat students' questions before answering them. Use visual media such as filmstrips, overhead projectors, and closed-captioned films if available. You might arrange a buddy system in which another student provides copies of notes and assignments.

Students who are completely deaf may require a sign language interpreter. If so, let the student and the interpreter determine the most convenient seating arrangement. When asking the student a question, be sure to look at the student, not at the interpreter. If the student also has a speech impairment, group assignments for oral reports or other alternatives should be considered.

Teaching Students Who Have Speech Impairments

Speech impairments include stuttering, voice disorders, and other disorders that interfere with a person's ability to communicate. Educators have had great success in mainstreaming speech-impaired students into regular classrooms.

Perhaps the most important quality to exhibit when teaching speech-impaired students is patience. For example, resist the temptation to finish sentences for a student who stutters. Also attend to such nonverbal cues as facial expression and body language. Making committee assignments for oral reports and using multisensory materials when possible will create a more comfortable learning environment for the speech-impaired student.

Teaching Students Who Have Learning Disabilities

Learning disabilities are believed to be caused by disorders of the central nervous system that obstruct a person's listening, reasoning, communication, or mathematical abilities. It is estimated that two percent of all adolescents have some type of learning disability, making this the most common type of handicap. In teaching learning disabled students, provide a supportive and structured environment in which rules and assignments are clearly stated. Use familiar words and short, simple sentences, repeating or rephrasing when necessary.

Depending on the severity of the disability, students may require extra time to complete exams or may need to tape-record lectures and answers to exam questions. For those who have difficulty organizing materials, you might provide chapter or lecture outlines for students to fill in. Using peer tutors can also be an effective strategy.

If available, computers can be an extremely useful tool for reviewing material and for exercising learning skills. With computer-assisted instruction, students receive simplified instructions, proceed in small manageable steps, and receive immediate reinforcement and feedback.

Providing for Individual Differences, *continued*

Teaching Students Who Have Behavior Disorders

Behavior disorders refer to emotional disturbances or behaviors that hinder a student's overall functioning. Students may exhibit any of a variety of behaviors, ranging from aggression to a complete lack of interaction with others.

Behavioral psychologists disagree on the best way to deal with students who have behavior disorders. In general, however, a fair and consistent, yet flexible approach is recommended. Make sure to clearly state rules and expectations. Reinforce desirable behavior or even approximations of such behavior, and ignore or mildly admonish undesirable behavior. Because learning disabilities often accompany behavior disorders, you might also wish to refer to the guidelines for learning disabilities.

Teaching Students Who Have Limited English Proficiency

Students who have limited English proficiency (LEP) or students who speak English as a second language (ESL) offer special challenges in your instruction for chemistry. It is vitally important from the outset that you evaluate the reading and speaking level of your students so that you may develop your instructional materials accordingly. Many school systems have teachers whose main responsibility is to instruct students with limited English proficiency. You may be able to use this teacher as a resource person. The following strategies will help you assist LEP and ESL students in your class.

- Provide students with a list of the important words before you discuss them in class. This will give students more time to do any necessary translations.
- Allow students to work with partners or in small groups. In this way, your students can help one another learn the material. Not only will your LEP or ESL students benefit from having personalized help, but the other students will benefit from the process of helping to teach the information. The other students will also become more aware of the difficulties of overcoming a language barrier, which will make them more considerate and understanding.
- Speak slowly, restate phrases, and repeat key words frequently.
- Use as many illustrations, photographs, and diagrams as possible.
- Use videotapes and films to help clarify concepts.
- Provide for oral or pictorial responses to test questions.

Teaching Students Who Are Gifted in Science

Gifted students are able to read the textbook rapidly and comprehend it easily, and they have the quickest grasp of the more abstract concepts of chemistry. How do you maintain the interest of gifted students without leaving the others behind? First, you may wish to recommend outside readings to these student on topics in which they have expressed an interest. Second, you may assign different questions and problems to gifted students. Third, gifted students may be given an opportunity to perform independent research that they can share with their classmates. They could contribute the results of their research during cooperative-learning activities or serve as leaders of these activities. And finally, the natural curiosity of gifted students may be encouraged by field trips, especially related to areas of particular interest.

Concept Mapping
A Valuable Reteaching Strategy

Concept Maps help students visualize the cognitive relationships among major topics and supporting ideas in the text.

Drawing concept maps forces students to examine relationships among concepts and as such is an excellent reteaching strategy. The use of concept mapping helps students move text information from their short-term to long-term memory. Concept mapping is an especially effective strategy for students who learn better through visual stimuli. In addition, developing concept maps in cooperative groups provides the peer tutoring that students in a reteaching phase need.

Constructing Concept Maps
Students are given some basic background in concept mapping techniques in *Appendix C: Study Skills for Chemistry*.

Evaluating Concept Maps
When working with students and evaluating maps, emphasize that there is no single correct concept map for an idea. Each map should be evaluated on the following.

1. Adequate branching — Students' first maps generally look similar to outlines, and as such, they tend to lack adequate branching. Provide incentives for students to turn in highly branched maps. They will work hard on developing this skill.

2. Accurate links — Is the link between concepts meaningful? Does it correctly represent the relationship? Can complete ideas be traced through several links? Are there sufficient cross links? Here again, provide incentives for student to turn in maps that show good cross linking.

3. Concept selection — Concepts should be succinctly represented. They should not be longer than three words. Concepts should not be repeated in a map. A repeating concept is best represented by cross links.

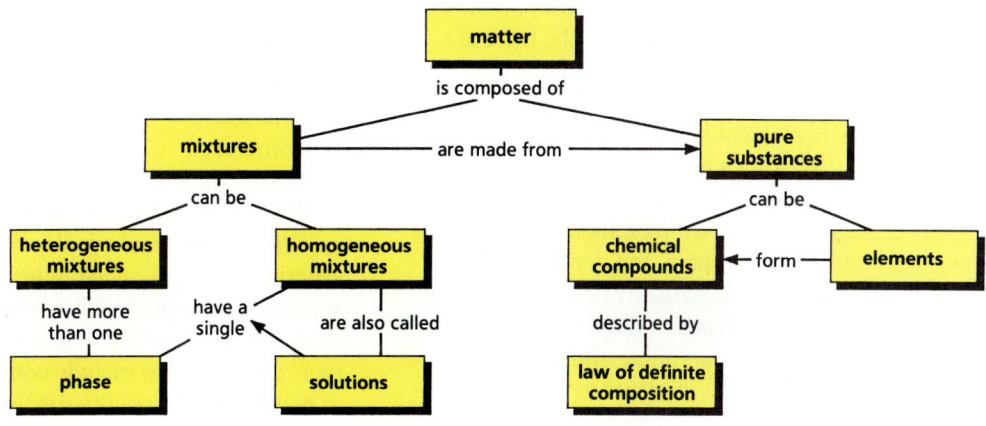

Extension and Enrichment
Opportunities for Inquiry and Exploration

Going "beyond the basics" is an essential part of teaching chemistry, for it gives students a chance to learn more on their own. Motivating activities, guest speakers, demonstrations, films, and field trips not only stimulate interest and curiosity but also prepare students for what they are about to read.

Library Research

The library research report provides an opportunity for students to work independently and to improve their research and writing skills. When possible, let students pick their own research topics. Students do better work on subjects in which they are genuinely interested. As a compromise between allowing students to select their own topics and assigning topics, students could select topics from a list you have compiled. Make sure that the titles are specific and that their degree of complexity is appropriate to the achievement level of your students.

Writing a research report will give students an opportunity to use a variety of up-to-date sources, including reference texts, newspapers, and popular and scientific magazines. Encourage students to use the *Reader's Guide to Periodical Literature,* microfilm, and database indexes. Allow students to write letters to specific sources for additional information. In some instances, students may be able to conduct interviews.

Impress upon students the importance of accurately citing sources used in their research. For each source, students should record bibliographic information, which can appear as either on-page or end-of-paper citations. Any direct quotations should be enclosed in quotation marks.

Once students have compiled their sources, they can construct a working outline of their report. You might want to approve the outlines before they begin writing the body of the report. Following approval, students can begin writing. You might want to check students' drafts at various intervals during the writing process. Students should not only summarize but also offer their own conclusions about their research findings in the reports.

Lab Research Reports

Lab research reports should include the following:
- title page
- introduction
- library research
- hypothesis
- a list of materials and of procedures
- observations and data collected
- conclusions
- footnotes
- bibliography

The report should clearly state the specific question and hypothesis that the experiment was designed to test. Data should be recorded in tables and graphs to show the results of the experiment. Lab reports should also include conclusions as to whether the results supported or refuted the original hypothesis.

Experiment Design and Development

The *Extensions* following the *Microscale Investigations* in *Modern Chemistry* encourage students to design and carry out experiments of their own. These experiments should consider questions that can be answered by observation. All student-designed experiments should be approved by the teacher before they are carried out. Review the proposed experiment for safety, practicality, and repeatability.

In each experiment, students should develop a hypothesis and then test that hypothesis using lab procedures they have learned. Students will have to decide on what their independent and dependent variables will be, how they will vary the independent variables, and how they will measure changes in the dependent variables. Controls for each experiment must be established.

Dramatic Presentations

You may want to give students the opportunity to express their knowledge of chemistry topics in creative ways. Dramatic presentations can be both illustrative and exciting. Skits can be based on historical characters or on personifications of chemical phenomena. You may want to experiment with having students take turns teaching small sections of the text or offering suggestions for demonstrations.

Field Trips and Guest Speakers

Taking the class on a field trip and inviting guest speakers to address current topics are both good ways to keep the class interesting. At the same time, you will reinforce the message that chemistry is at work in your community. Such visits could be valuable in showing students what careers are available in chemistry.

Master Materials List
What will you need?

The section that follows includes five materials lists for the course. The first four itemize the materials and amounts you will need for fifteen lab groups to perform the in-text Microscale Investigations and the Desktop Investigations. The fifth list details the materials needed for the Demonstrations mentioned in the "Teaching Strategies" of the Teacher's Edition interleaf pages.

The materials lists for the Desktop Investigations and Microscale Investigations include separate lists for Chemicals, Lab Equipment, Miscellaneous Materials, and a special list for two Desktop Investigations that are based on equipment and materials found in most household kitchens, and should not be performed in a laboratory setting.

Chemicals

This list shows the chemicals needed to for a class of thirty students working in pairs to perform all the investigations found in the textbook, both the Desktop Investigations found within many of the chapters and the Microscale Investigations bound in the back of the book. In the far right column, Microscale Investigations are identified by a number, and Desktop Investigations by their page number in the text.

Although some preparation instructions for solution preparations are given in this list, some substances may be best ordered as ready-made solutions. This may be a safer way to handle compounds of cobalt and nickel, whose dust is believed to be a carcinogen if inhaled. As solutions, these substances are much less dangerous. In addition, most of these investigations require a ready supply of deionized or distilled water, tap water, and paper towels.

Legend
Microscale Investigation numbers in parentheses (see "lactic acid," in the chemical list, for example) indicate materials needed only for an extension activity. If you do not assign this extension activity to your students, they will not need the material indicated to perform the investigation.

Chemical	Amount Needed	Microscale Inv. # or Desktop Pg.
acetic acid solution, 1.0 M	2.3 mL glacial acetic acid in 50 mL of solution	17
acetone solution, 4.0 M	36.9 mL acetone in 125 mL of solution	18
ammonia solution, dilute	1.5 mL concentrated (14.8 M) $NH_3(aq)$ in 75 mL of solution	p. 18
barium nitrate solution, 0.5 M	3.27 g $Ba(NO_3)_2$ in 25 mL of solution	13
calcium carbonate	100 g $CaCO_3$	14
calcium hydroxide (saturated solution)	5.0 g $Ca(OH)_2$ in 1.0 L of solution, filtered	p. 307
calcium sulfate	0.1 g $CaSO_4$	13
cobalt(II) chloride solution, 0.1 M	0.60 g $CoCl_2 \cdot 6H_2O$ in 25 mL of solution	5
copper foil	144 cm^2	p. 555

Chemical	Amount Needed	Microscale Inv. # or Desktop Pg.
copper(II) chloride	5.30 g $CuCl_2 \cdot 2H_2O$	6 & 5
copper(II) sulfate solution, 0.2 M	1.25 g $CuSO_4 \cdot 5H_2O$ in 25 mL of solution	5
dry ice, CO_2	130 g	10 & 11
glycerin	1.2 L	p. 374
graphite dust	50 g	4
hydrochloric acid	85 mL concentrated HCl	4, 8, 9, 14, 15, 17, & 18
hydrogen peroxide, 3%	900 mL 3% H_2O_2	7, 9, & p. 600
iodine solution, 0.0012 M	0.10 g I_2 in 125 mL of solution (saturated)	18
iodine, tincture of	4 mL	p. 18
iron filings	115 g	2 & 14
iron(III) chloride	185 g $FeCl_3 \cdot 6H_2O$	5, 6, 13, & 14
iron(III) chloride solution, 0.1 M	2.02 g $FeCl_3 \cdot 6H_2O$ in 75 mL of solution	p. 18
iron(III) sulfate solution, 0.1 M	1.40 g $Fe_2(SO_4)_3 \cdot 9H_2O$ in 25 mL of solution	5
lactic acid, 0.50 M	2.2 mL of 85% lab grade lactic acid in 50 mL of solution	(17)
limewater (see calcium hydroxide)		
magnesium ribbon	180 cm	p. 555
manganese dioxide	15 g	p. 600
methyl orange indicator solution	0.1 g methyl orange in 50 mL denatured alcohol and 50 mL of water	14
mineral oil	15 mL	12
nickel(II) chloride solution, 0.1 M	0.60 g $NiCl_2 \cdot 6H_2O$ in 25 mL of solution	5
nickel(II) nitrate	45 g $Ni(NO_3)_2$	p. 730
nickel(II) sulfate solution, 0.2 M	1.32 g $NiSO_4 \cdot 6H_2O$ in 25 mL of solution	5
pH paper sensitive to 12-14 range	1 roll	(5)
pH hydrion paper narrow range ($\pm$0.2-0.3) or pH meter	1 roll pH paper or 15 pH meters)	14 & p. 509
phenolphthalein solution	0.47 g phenolphthalein in 24 mL denatured alcohol and 24 mL water	6, 14, 15, & p. 18
plaster of paris (see calcium sulfate)		
potassium iodate	1.07 g KIO_3	16
potassium iodide	9.3 g KI	7 & 9
rubbing alcohol (70% isopropanol solution)	500 mL	p. 374
silver nitrate solution, 0.1 M	0.85 g $AgNO_3$ in 50 mL of solution (store in brown bottle to avoid decomposition)	13 & p. 789
sodium bicarbonate (baking soda)	1 kg $NaHCO_3$	1 & p. 307
sodium borate (borax)	30 g $Na_2B_4O_7$	
sodium bromide	21 g NaBr	p. 789
sodium chloride (table salt)	77 g NaCl	2, 4, p. 374, & p. 789
sodium hydroxide	24 g NaOH	5, 6, & 14
sodium iodide	30 g NaI	p. 789
sodium metabisulfate	0.05 g $Na_2S_2O_5$	16
sodium oxalate solution, 0.5 M	1.68 g $Na_2C_2O_4$ in 25 mL of solution	13
sodium sulfate	50 g Na_2SO_4	4
sodium thiocyanate solution, 0.5 M	1.01 g NaSCN in 25 mL of solution	13
starch, water soluble	55 g	16, p. 18, & p. 408
sugar (sucrose)	250 g $C_{12}H_{22}O_{11}$	4 & p. 408
sulfur (powder)	50 g	4
sulfuric acid, 1.0 M	1.3 mL of 1.0 M H_2SO_4	16
tin foil	50 g	4
tin(IV) oxide	100 g SnO_2	4
zinc (mossy)	60 g	8 & 9
zinc (sheet)	144 cm^2	p. 555

Laboratory Equipment

This list shows the lab and classroom equipment needed to for a class of thirty students working in pairs to perform all the investigations found in the textbook, both the Desktop Investigations found within many of the chapters and the Microscale Investigations bound in the back of the book. In the far right column, Microscale Investigations are identified by a number, and Desktop Investigations by their page number in the text.

Even if you are unable to purchase all of the microscale equipment, it is possible to achieve some of the benefits of small-scale investigations by improvising with traditional glassware. Instead of using 24-well microplates and 8-well microstrips, small test tubes may be used as reaction vessels. Traditional medicine droppers or pasteur pipets may be able to replace micropipets (beral pipets) in some situations.

In addition to the equipment described in this list, most of these investigations require a ready supply of deionized or distilled water, tap water, and paper towels. Unless indicated as described in the legend, all materials are re-usable.

Legend

An asterisk (*) indicates an amount that may be made smaller if resources dictate.

A dagger (†) indicates equipment that the teacher (or even students) can make out of simple materials. Instructions for these are included immediately following these materials lists, and at the point of use in the teacher's guide to the Microscale Investigations.

Entries in plain type indicate materials that can be re-used.

Entries in boldface indicate materials that will be used up to make specialized pieces of equipment. Once made, these specialized pieces of equipment can be re-used year after year.

Entries in boldface and italics indicate materials that will be consumed during the investigation.

Microscale Investigation numbers in parentheses (see "lactic acid," in the chemical list, for example) indicate materials needed only for an extension activity. If you do not assign this extension activity to your students, they will not need the material indicated to perform the investigation.

Laboratory Equipment	Amount Needed	Microscale Inv. # or Desktop Pg.
balances, triple-beam or electronic	15*	1,3, (5), 15, p. 18, & p. 74
beakers (150 mL)	45	15, p. 307, p. 730, & p. 789
beakers (250 mL)	60	p. 18, p. 374, & p. 690
beakers (400 mL)	105	3, 7, 8, 9, p. 307, & p. 408
beakers or glass containers (500 mL)	75	p. 509
boiling chips	for 15 hot water baths	3
bottles, droppers	15	4, 7, 8, & 9
bottles, micro solution (50 mL)	15	15
bulbs, collection (incremented)†	30	9
bulbs, dropper (fine-tipped)†	150	6, 11, (13), 15, & 16
bunsen burners	15	3, 7, 8, p. 555, p. 600, & p. 690
clock with a second hand	one (or several stopwatches)	16, 18, & p. 361
conductivity testers† (LED or tone-emitting)	15	4, (13), 14, & 17
dessicators	15	15
drying oven	1	15
evaporating dish	15	p. 374
eye droppers	15	p. 374 & p. 789
filter funnels	15	2, (5), & (13)
filter paper	**300 pieces**	2, (5), (13), 15, & p. 789
flashlight and screen (slide projector, etc.)	15	12, p. 102, & p. 408

Laboratory Equipment	Amount Needed	Microscale Inv. # or Desktop Pg.
flask, erlenmeyer (125 mL)	1	12
forceps or tweezers	15 pairs	2, 7, 8, 9, 10, 11, p. 374, & p. 600
freezer	1	(13)
graduated cylinders (10 mL)	15	9, 15, p. 374, & p. 555
graduated cylinders (100 mL)	15	1, 3, p. 18, p. 307, & p. 374
magnets	15	2 & p. 730
microplate lids, 24-well (usually available with microplates)	15	13
microplates, 24-well	15	14, 17, & 18
microstrips, 8-well, flat-bottom	15	2, 5, 6, & 16
mortars and pestles	15*	15 & p. 18
overhead projector	1	13
petri dishes with lids	15	2, 7, 8, 9, & 12
pH meters (or 1 roll of pH paper, narrow range)	15*	14 & p. 509
piezoelectric sparking elements†	4-5	9
pipets, graduated	**90**	1, 7, 8, 9, & 12
pipets, graduated	15	10, 11, & 12
pipets, jumbo	**30**	2 & 12
pipets, thin-stemmed	**45**	11 & 15
pipets, thin-stemmed	150	2, 4, 5, 12, 14, 17, & 18
refrigerator or ice bath	1	(9)
ring clamps	15	3, p. 690, & p. 789
ringstands	15	3, p. 690, & p. 789
rulers (metric, thin, transparent and plastic)	15	11, p. 74, p. 102, p. 215 & p. 509
scissors	15* pairs	2, 3, 10, 11, p. 102, & p. 730
spatulas	15	4
spot plates	15	4
stirring rods, glass (small)	15	4, p. 307, p. 374, p. 600, & p. 690
stopper pegs, to fill holes in size 0 or 00 one-holed stoppers	105	7 & 8
stopper, rubber, one-holed (fits 125 mL erlenmeyer flask)	1	12
stoppers, rubber, one-holed (size 0 or 00)	105	2, 7, & 8
strainers (small - fine mesh)	15*	12
test tube clamps	15	3 & p. 600
test tube racks	15	12
test tubes (13 × 100)	105	2, 7, 8, 12, & p. 690
test tubes (16 × 150)	90	p. 555, p. 600, & p. 789
test tubes (large—20 × 150)	75	3, p. 18, & p. 600
thermometers (alcohol-filled, not mercury)	15	3 & (13)
tongs	15 pairs	3, 15, & p. 555
vials with caps	**30**	7, 8, & 9
wash basins	4	2 & 12
wax pencils	15*	p. 690
weighing paper	**30**	15
wire (12 AWG)	180 cm*	12
wire gauze squares	15	3 & p. 690
wood splints	**175**	2, 7, 8, 12, p. 600, & p. 690

Miscellaneous Materials

This list shows the non-lab materials needed for a class of thirty students working in pairs to perform all the investigations found in the textbook, both the Desktop Investigations found within many of the chapters and the Microscale Investigations bound in the back of the book. In the far right column, Microscale Investigations are identified by a number, and Desktop Investigations by their page number in the text.

In addition, most of these investigations require a ready supply of deionized or distilled water, tap water, and paper towels. Unless indicated as described in the legend, the materials are re-usable.

> **Legend**
> **An asterisk (*)** indicates an amount that may be made smaller if resources dictate.
> **Entries in plain type** indicate materials that can be re-used.
> **Entries in boldface** indicate materials that will be used up to make specialized pieces of equipment. Once made, these specialized pieces of equipment can be re-used year after year.
> **Entries in boldface and italics** indicate materials that will be consumed during the investigation.
> **Microscale Investigation numbers in parentheses** (see "lactic acid," in the chemical list, for example) indicate materials needed only for an extension activity. If you do not assign this extension activity to your students, they will not need the material indicated to perform the investigation.

Material	Amount Needed	Microscale Inv. # or Desktop Pg.
aluminum foil	one roll	2, p. 102, & p. 730
ammonia, household cleaning	**250 mL**	p. 361
antacid tablet (effervescent)	**15**	p. 18
ash (from paper)	**15 g**	p. 555
aspirin tablets	**15 (325 mg each)**	p. 600
bleach, liquid (5.25% NaClO solution)	**175 mL**	p. 600
cellophane	one roll	2
candles	30	p. 18 & p. 307
cans, 2-lb. coffee	15	p. 74
cardboard or posterboard, white	**15 35 cm × 35 cm pieces**	p. 102
chalk, white	**30 3-cm pieces (10 sticks of chalk)**	(15) & p. 18
chalk, yellow	**2 sticks of chalk**	4
charcoal, activated (aquarium grade)	**50 g**	12
charcoal	2 or 3 chunks	4
clay, modeling (red, yellow, green and blue)	**7 lb. (total)**	p. 215 & p. 408
coffee grounds, sifted	**0.25 tsp**	12
containers, with lids (such as resealable plastic food containers)	30	p. 361
cooking oil	**2.0 L**	p. 374 & p. 408
corn starch	**135 g**	p. 690
cotton balls	**15**	2
detergent, liquid	**5 mL**	p. 374
duct tape	1 roll	p. 74
eggshells	**15**	15
food coloring	**32 mL**	11 & p. 408
garlic powder, sifted	**0.25 tsp**	12

T50

Material	Amount Needed	Microscale Inv. # or Desktop Pg.
gelatin, plain	*45 g*	p. 408
glass beads	1.15 kg	(3)
glue gun (hot)	15*	1 & 11
gumdrops, large (at least two colors)	**750**	p. 215
gumdrops, small (at least two other colors)	**750**	p. 215
index cards	**350**	p. 131
iron nails	15	(4)
jars or similar containers	**150**	4
manila folders	**15**	p. 108
markers, permanent (fine-tipped)	5*	9 & 11
masking tape	*1 roll*	p. 102
matches	*160*	p. 18, p. 307, & p. 555
milk of magnesia	*75 mL*	p. 18
miscellaneous objects that fit in a 2-lb. coffee can	15	p. 74
needle	15	1 & p. 374
"nickels," U.S. five-cent coins of various ages	75	p. 730
paper clips	**25**	2 & p. 555
pennies (made before 1982)	600	p. 18
pennies (made after 1982)	600	p. 18
perfume or cologne	250 mL	p. 361
plastic bags, resealable (zippered)	15	(1) & 2
plastic cups (transparent)	**20**	1, 10, & 11
plastic cups (foam)	**30**	3
plastic forks	10	2
plastic spoons	15*	2 & 12
pliers or clamps	15	10 & 11
poppy seeds	**15 g**	2
potatoes	*30*	p. 555
sand	115 g	2 & 12
sandpaper	15 small pieces	p. 555
shells (clam, oyster, snail etc.)	**15**	(15)
socks	**15**	**p. 74**
soda bottles, 2-L, plastic with caps	**15**	1
starch, laundry, spray	1 can	18
steel wool	*enough for 150 small tufts*	7, p. 555, p. 600, & p. 730
straws (plastic)	*25*	2 & p. 307
sugar cubes	*30*	p. 555
tape dispensers	15*	2
thread	**360 cm**	1 & 11
thumb tacks	15*	p. 102
tissue paper	*one roll*	2 & 12
toothpicks	*750*	4, 14, p. 215, & p. 248
vinegar (5% acetic acid solution)	*6.0 L*	1, p. 18, p. 307, p. 555, p. 600, & p. 690
water, bottled, spring	*15 mL*	13
water, local river or lake	*15 mL*	13
wax	75 g	(4)
wax paper	15 large pieces	p. 374
white paper	30 sheets	14 & 18
wire hooks (small)†	**15**	1

Kitchen Items

Two Desktop Investigations, p. 273 and p. 476, involve activities that are better suited to a kitchen environment, either at home, or perhaps in a school cafeteria kitchen. Items for 15 lab groups to perform these investigations are listed in the chart that follows.

> **Legend**
> **An asterisk (*)** indicates an amount that may be made smaller if resources dictate.
> **Entries in plain type** indicate materials that can be re-used.
> **Entries in boldface** indicate materials that will be used up to make specialized pieces of equipment. Once made, these specialized pieces of equipment can be re-used year after year.
> **Entries in boldface and italics** indicate materials that will be consumed during the investigation.

Item	Amount Needed	Microscale Inv. # or Desktop Pg.
beakers (small), jars or similar small containers	400	p. 476
baking powder	*15.0 tsp (5 tbsp)*	p. 476
baking soda	*20.0 tsp (¼ cup)*	p. 273 & p. 476
bleach	*75 mL (⅜ cup)*	p. 476
blender	*15**	p. 476
brown sugar	*7½ cups*	p. 273
cabbage (red)	*5 heads*	p. 476
chocolate chips	*20 cups*	p. 273
cookie sheets	15	p. 273
detergent (dishwasher)	*75 mL (⅜ cup)*	p. 476
detergent (laundry)	*75 mL (⅜ cup)*	p. 476
dishwashing liquid	*75 mL (⅜ cup)*	p. 476
eggs	*15*	p. 273
fabric softener	*75 mL (⅜ cup)*	p. 476
flour	*22½ cups*	p. 273
laundry stain remover	*75 mL (⅜ cup)*	p. 476
lemon juice	*75 mL (⅜ cup)*	p. 476
margarine	*20 sticks (10 cups)*	p. 273
mayonnaise	*75 mL (⅜ cup)*	p. 476
measuring spoons and cups	15 sets	p. 273 & p. 476
milk	*75 mL (⅜ cup)*	p. 476
mineral water	*75 mL (⅜ cup)*	p. 476
mixing bowls	15	p. 273
salt	*7½ tsp*	p. 273
soft drinks	*75 mL (⅜ cup)*	p. 476
spoons, wooden (for mixing)	15	p. 273
strainer	15	p. 476
sugar	*7½ cups*	p. 273
vanilla	*15 tsp (5 tbsp)*	p. 273
vinegar, cider	*75 mL (⅜ cup)*	p. 476
vinegar, white	*75 mL (⅜ cup)*	p. 476

Equipment Notes for Microscale Investigations

Hints for making certain pieces of equipment for the Microscale Investigations are given below. These procedures are also reprinted at the point of use in the Teacher's Guide to the Microscale Investigations. Some procedures describe ways to cheaply make items that are also available commercially, and others describe methods for creating specialized pieces of equipment that must be made prior to performing the Microscale Investigations. Once made, these items may be reused from year to year. Several of these specialized pieces of equipment may be handy to have around the lab for other uses besides the Microscale Investigations.

Although these instructions are not included in the Pupil's Edition of the Microscale Investigations, if there is plenty of time and the students are capable of it, you may want to consider having them make their own equipment according to these instructions.

Fine-tipped Dropper Bulbs (used in Investigations 6, 11, 13, 15, and 16)

Materials 150 thin-stemmed pipets
pliers, two sets

Preparation Although these bulbs are available commercially, they may be made quite easily out of thin-stem pipets. Stretch the pipets using the pliers and then cut them off. (See Figure T0-A) For very accurate drop sizes, place the dropper bulb in a tubing screw-clamp. Each turn of the screw will force out a reproducible drop.

Conservation of Mass Apparatus and Plungers (used in Investigation 1)

Materials 15 2-L plastic soda bottles with caps
15 hooks or length of wire
15 graduated pipets
120 cm thread cut into 15 8-cm lengths
needle

Preparation Hook-insert caps:
1) If you are making your own hooks, bend the wire into small hooks that will fit inside of a soda bottle's neck
2) Use hot-melt glue to attach the wire hook to inside center of the cap. (See Figure T1-A)

Figure T0-A

Figure T1-A

Preparation Plungers:
1) Cut across the top of the bulbs of the graduated pipets. Also cut off the tips of the pipets, as shown in Figure T1-A.
2) Thread the needle, heat it up, and make a hole near the end of the graduated pipet.
3) Bring the needle and thread through the hole.
4) Tie a knot in the thread so that the microplunger has a thread loop through the hold in the pipet stem.

Conductivity Tester (used in Investigations 4 and 17)

Materials
15 1-kΩ resistors ($\frac{1}{4}$ W)
15 LED's (light-emitting diodes)
15 9-V battery with battery clips
15 film canisters with lids
4.5 m of hook-up wire (22 GA, solid): cut into 15 10-cm lengths and 15 20-cm lengths
15 3.5-cm lengths of straw or tubing to fit over LED's
nail or ice-pick
electrician's tape
glue gun (optional)

Preparation
1) Strip 1.5 cm of insulation from the ends of the hook-up wires and from the battery clip leads. Cut the resistor leads down to half of their length.
2) Poke two holes near the edge of the canister cap about as far apart as the two LED leads are. Insert the leads through the holes, and spread them apart slightly.
3) Connect the circuit as shown in Figure T4-A. **Be sure the RED battery clip lead is connected to the LONGER LED lead.** The black battery clip lead should connect to the resistor, and the other end of the resistor should connect to the 10-cm length of wire. The shorter LED lead should be connected to the 20-cm wire. Bend the leads up and tape them against the inside of the lid, so that they are not touching each other and there are no exposed wires.
4) Wrap tape around the resistor and its exposed wires.
5) Poke two holes in the side of the film canister about 1 cm apart, near the bottom of the film canister. Insert the hook-up wires through these holes.
6) Connect the battery and place it in the canister, snapping the lid shut. Adjust the protruding hook-up wires so they extend out about 5 cm. (Optional: place drops of hot-melt glue around the LED and the hook-up wires so that the canister is water-proof).
7) Wrap electrician's tape around the straw and place it over the LED to provide a "dark tunnel" through which the LED may be viewed for testing weak electrolytes.

Figure T4-A

Gas Micro-Generators and Collection Bulbs (used in Investigations 7, 8, and 9)

Materials
30 vials with caps
30 graduated pipets

For O$_2$ generators:
500 mL 3% H$_2$O$_2$ (store-bought hydrogen peroxide is fine, provided it is "fresh")
4-8 dropper bottles containing 1.0 M KI (8.3 g in 50 mL of solution or 4.15 g in 25 mL of solution will make far more than your class will use, so keep it for next time. If dropper bottles are not available, the KI solution may be placed in cut-off thin-stem pipets. One bulb can hold from 80-120 drops, possibly enough for the entire class.)

For H₂ generators:
60 g mossy zinc
500 mL of 1.0 M HCl (20.6 mL conc. HCl in 250 mL of solution)

Preparation of gas generators

1) Cut the tapered tips off the graduated pipets, and they may be used as nozzles for the gas of gas generators generators. Another way to make nozzles is to take a thin-stemmed pipet, and, with a steady pull, stretch it to a much narrower diameter. With this method, each stem can provide two nozzles.
2) Punch or drill holes (or melt them with a hot nail) in the center of the vial cap and slip the tapered tips into the holes. If this segment is inserted until the flared-end section wedges into the hole, a relatively air-tight junction can be formed.
3) For the oxygen generator, the concentrations listed above for the H_2O_2 solution (3%) and the KI solution (1.0 M) are meant only as guidelines. You may wish to adjust them as needed to moderate the reaction rate. Having the solutions at (or slightly above) room temperature will also help speed the reaction.
4) For the hydrogen generator, break up the zinc into pieces. Put about 2 g zinc into 30 vials. (The vials should be about $\frac{1}{6}$ full.) **Prepare the metal shortly before the lab. Read all safety notes that come with your supply of zinc. It should not be pulverized or powdered, since the metal dust can be flammable. Remember that if zinc dust gets damp in a closed container, an explosion may occur. Keep your zinc dry until lab time.** A few drops of $CuSO_4$ solution will help catalyze the Zn/HCl reaction. In addition, the HCl solution concentrations may be altered to moderate the reaction rate.
5) Use petri dishes to catch the water displaced by the gases during the investigations.

Preparation of collection bulbs for Investigation 9:

1) Cut the bulbs from the pipets used to make nozzles. (See Figure T9-A.)
2) To calibrate the bulbs, fill one completely with water. (If you have trouble filling the bulb completely, or the students do later, you may want to consider using thin stem pipets to deliver the water into the bulbs instead of having the water squeezed up from the petri dish.)
3) Squeeze all the water out into a 10-mL graduated cylinder and record the volume.
4) Empty the graduated cylinder and refill the bulb with water. Into the graduated cylinder, squeeze out of the bulb $\frac{1}{6}$ of the volume of water you recorded earlier. Mark the water level on the surface of the bulb when held mouth-downward. (You may have to dry the side of the bulb first.)
5) Squeeze out a second sixth of the volume, and mark the level again. Repeat for the remainder of the water.
6) Once one bulb has been calibrated, rest a wood splint across it, with the splint's end even with the end of the bulb. Mark off the splint at the same places that the bulb is marked.
7) Use the splint as a template for the rest of the bulbs. Once calibrated, the bulbs can be re-used from class to class and year to year.

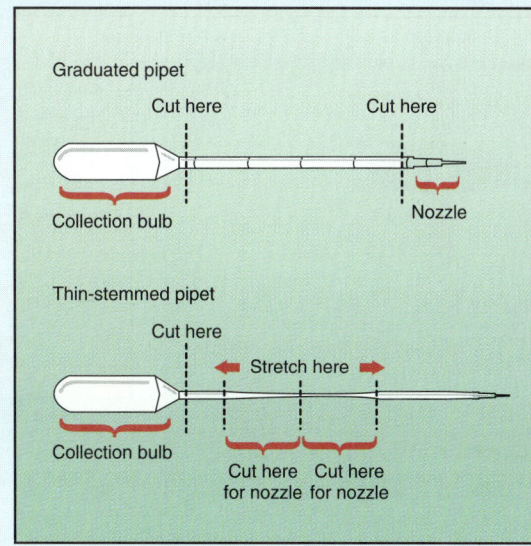

Figure T9-A

Piezoelectric Sparker Construction (used in Investigation 9):

Materials
4-5 piezoelectric/butane charcoal/fireplace lighters (such as Aim 'n Flame, etc.)
50 cm of 24 AWG speaker hook-up wire (solid not stranded: Radio Shack #278-1509)
4-5 film canister lids
electrician's tape
hot glue gun and glue
scissors
wire strippers
$\frac{1}{8}$" drill or 6-D common nail

Construct the sparker from a piezoelectric charcoal/fireplace lighter found in many stores for just three or four dollars. Once wired as described in the following instructions, it can be used in place of a Tesla coil for a wide variety of chemical demonstrations. If the wire and tape are removed, the sparker can be used as a bunsen burner lighter or for any other demonstration requiring a quick and safe light (Mg ribbons, exploding cans, etc.).

1) Using the scissors, split the end of a 10-cm length of speaker wire down the center for a distance of 3.5 cm. Strip the last 2.3 cm of insulation from both sides of the fork, as shown in Figure T9-B.
2) Drill a $\frac{1}{8}$" hole (or alternately, melt a hole with a hot 6D common nail) through the center of the film canister lid.
3) Slide the wire through, and secure the lid in place about 4 cm from the unforked end of the wire with hot glue or other water-proof adhesive. The film canister lid serves both as launch support pad and as a splash guard.
4) Slide one of the two stripped wire ends down into the small butane nozzle of the charcoal lighter, all the way down, so that the insulation is against the nozzle. (Rotate the wire as you insert it.)
5) Lay the other stripped end along the metal barrel of the lighter and secure it in place with 3-4 wrappings of electrician's tape.
6) To use, put the bulb of hydrogen and oxygen gas over the wire, aim and pull the trigger. Rather than arcing at the butane nozzle, the device will arc at the tip of the wire (the path of least resistance).

Separatory Bulbs (used in Investigation 12)

The separatory bulbs may be made by cutting off all but 4-5 mm of the stem from a graduated or micro-tip pipet, as shown in Figure T12-B.

Long-Stemmed Micro Funnels (used in Investigations 2 and 12)

The micro-funnels may be made by cutting the top section off the bulb of a jumbo-sized pipet, as shown in Figure T12-C.

Figure T9-B

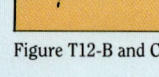

Figure T12-B and C

Demonstration Supplies

Entries appear in alphabetical order. Numbers in parentheses after an entry indicate the quantity needed for the demonstration. Numbers in brackets indicate the chapter where the demonstration is found.

1,1,2-trichloro-1,2,2,-trifluoroethane [21,22]
1-butanol (*n*-butyl alcohol) [22]
2-butanol (*sec*-butyl alcohol) [22]
2-methyl-2-propanol (*tert*-butyl alcohol) [22]
acetic acid [1]
acetic acid, 0.1 *M* [15]
acetic acid, glacial [15,22]
acetone [9,22]
alcohol, ethyl or isopropyl, 70% [14]
alligator clips [15]
aluminum [3,8,9]
aluminum foil [3,24]
aluminum foil, heavy strip [26]
aluminum nitrate, 0.1 *M* [8]
aluminum soda can [11]
aluminum sulfate, 0.1 *M* [16]
ammeter, 0-200 mA [15]
ammonia water (conc.) [11,12,15,16]
ammonia water (dil.) [1,16,21]
ammonia water, 0.1 *M* [15,17]
ammonia, household [16]
ammonium chloride [14,21]
ammonium thiocyanate [18]
antacid, liquid form [16]
apples (2) [20]
baby food jars (49) [8]
balance [1,9]
balance, (decigram) [3,14]
balloon, round [4]
balloons, 15" diameter (3) [9]
balloons, oblong (2) [6]
balloons, oblong—different colors (3) [4]
balloons, oblong—same color (4) [4]
balloons, small (2) [1,18]
balloons, thin oblong (2) [4]
barium chloride [24]
barium hydroxide [18]
barium oxide [24]
batteries 6-V (2) [15]
battery charger, 6-V/12-V DC [20]
battery, 9-V [25,28]
beakers, 50-mL (49) [1,7,8,20,21,22]
beaker, 100-mL [27]
beakers, 125-mL (2) [21]
beakers, 150-mL (4) [16,17]
beakers, 250-mL (10) [3,13,14,15,16,20,25]
beaker, 400-mL [1,22]
beakers, 600-mL [10,24]
beakers, 1000-mL (12) [1,2,11,12,14,17,18,23]
beral micropipets [8]
bottle caps (12) [9]
bottles (10) [9]
bromphenol blue [17]
bromthymol blue [8,17]
Bunsen burner [6,9,11,14,21,24-27]
burets (2) [9,17]
burette stand [9]
butyric acid [22]
cafeteria tray [6]
calcium, metal [5,24]
calcium carbonate [16,25]
calcium chloride [14,24]
calcium hydroxide [16,24]
calcium oxide [11,24]
candle [11,18]
cardboard [4]
celsius thermometers (4) [2,14]
ceramic cup, porous unglazed, 35 × 80mm [12,20]
ceramic pad [1]
cereal, iron-enriched [25]
charcoal, sample [3]
clock, small with battery [20]
club soda (2 bottles) [14]
club soda (4 12-oz bottles) [14]
cobalt, pellets [8]
cobalt nitrate [26]
cobalt sulfate [25]
cobalt(II) chloride [25,26]
cobalt(II) chloride, 0.1 *M* [8]
cobalt(II) nitrate 0.1 *M* [8]
coffee can, 3-lb. [6]
conductivity detector or meter [17]
connecting wires (5) [20]
construction paper, white/colored [25]
cooking oil [22]
copper [3,7,8]
copper shot [1]
copper strips (5) [20]
copper wire [9]
copper wire, No. 20, insulated [15]
copper wires (2) [20]
copper(I) oxide [3]
copper(II) chloride crystals [26]
copper(II) nitrate, 0.1 *M* [8, 20]
copper (II) oxide [3]
copper(II) sulfate pentahydrate [22,25]
copper(II) sulfate, 0.1 *M* [8,20,25]
cork [24]
cornstarch [21]
cotton [12]
cranberry juice [16]
crucible and crucible cover [24]
crucible tongs [25]
cyaclume light sticks (4) [1,18]
cyclohexane [21]
cyclohexene [21,22]
decanoic acid [22]
deflagrating spoon [6]
detergent solution [24]
diffraction grating [4]
distilled water [1,5,6,14-17,21-24]
distilled water bottle [18]
drain cleaner [16]
drinking straw [24]
dropping bottles [11]
dropping bottles, 125-mL (2) [16]
dry ice [17]
egg, boiled [4]
electrodes (2) [20]
electrodes, carbon [25,28]
ethanol, 95% [6,14,15,22]
evaporating dish [15,22]
eyedroppers (2) [6,16]
face shield [as needed]
felt, colored [3]

Demonstration Supplies, *continued*

filter paper [16,22,24]
filters, colored cellophane [25]
flashlight [25]
flask, Dewar [10]
flask, Erlenmeyer, 125-mL [14]
flask, Erlenmeyer, 250-mL (3) [9,10,17,18,22]
flask, Erlenmeyer, 500-mL (3) [1,9,12,17]
flask, Florence [9]
flux [20]
fructose solution, 1% [23]
funnel [10,16,24]
gas-collecting bottles [10]
gelatin (3 pkgs.) [23]
gelatin capsules (pharmaceutical) [25]
gelatin, fruit flavored [14,16]
glass, long-stem [10]
glass bend, 5×20 cm [16]
glass bends, 5×5 cm (3) [14]
glass bend, right-angle [10]
glass container, 2-L [27]
glass plate [14]
glass tubing [12]
gloves [as needed]
glucose [22]
glucose solution, 1% [23]
glycerine [15]
goggles [all]
graduated cylinder, 10-mL [6,16]
graduated cylinder, 25-mL [22]
graduated cylinder, 100-mL (4) [2,14,16,17]
graduated cylinder, 500-mL [1]
graduated cylinder, tall [26]
grapefruit juice [16]
graphite [3]
"grow beast" [1]
gumdrop kits composed of
15 large gumdrops—one color,
36 small gumdrops—different colors,
51 toothpicks,
the gumdrop-toothpick model from Demonstration 1 [21]
gumdrops, large—3 colors [6,15,21]
gumdrops, small—2 colors [6,15]
hard water [24]
heat lamp [9]
heat source [13]
heater, with temperature control [7]
helium [1]

Hoffman apparatus [8]
hot plate [1,9,22,25]
hydrochloric acid, (conc.) [9,11,12,15,16,25]
hydrochloric acid (dil.) [1,16]
hydrochloric acid, $0.1\ M$ [7,8,9,15,17]
hydrochloric acid $1\ M$ [16,20,24]
hydrochloric acid, $6\ M$ [7,10,17,25,27]
hydrogen [1,12,18]
hydrogen peroxide, 20% [3]
hydrogen peroxide, 3% [10]
ice [14]
ice cubes [7,9]
ice, 100-g blocks (2) [2]
iodine [7,9,14,21,22]
iron [3,8]
iron nail [16]
iron ring [21,24]
iron(II) sulfate [23,25,26]
iron(III) chloride [26]
iron(III) chloride, $0.1\ M$ [16,25]
iron(III) chloride, $0.5\ M$ [21]
iron(III) nitrate, $0.1\ M$ [8]
iso-amyl alcohol [22]
kerosene [21]
lab apron [all]
lab coat [as needed]
lamp holder, threaded [15]
lead [8]
lead chloride, $0.1\ M$ [8]
lead(II) nitrate, $0.1\ M$ [8]
lead(IV) oxide [9]†
leads with alligator clips [20,25,28]
lemon [20]
lemon juice [16]
lemon-lime soda [14,16]
Life Savers®, wintergreen [4]
lightbulb, 6.3-V/150-mA [15]
lime [16]
limewater [16]
lithium chloride [14]
lithium hydroxide (dil.) [1]
litmus paper, red/blue [16,24]
Lugol's solution [23]
lycopodium powder [6]
lye [16]
magnesium [24]
magnesium chloride [24]
magnesium oxide [24]
magnesium ribbon [5,6,24]
magnesium turnings [9]
magnet [10,25]

magnetic stirrer and magnet [24,25]
magnifying lens [4]
manganese [25]
manganese sulfate [25]
manganese(IV) oxide [3,25,26]
marker, felt-tip, permanent [14]
marshmallows, large white [6]
marshmallows, small colored [6]
masking tape [18]
matches [1,10,18]
medicine droppers [8]
mercury vapor lamp [4]
meter stick [1,18]
methyl alcohol [22]
methyl orange [17]
methyl red [17]
methylene blue [23]
microplate, 24-well [8]
milk [16]
milk container, plastic (1-gal.) [11]
milk of magnesia [14,16]
milk of magnesia tablets [24]
mineral oil [21]
modeling clay, 4 colors [15]
molecular model kit (optional) [21]
mustard [16]
n-amyl alcohol [22]
nails (2) [3,4]
nail, zinc-coated [20]
naphthalene [21]
nickel, shot [8]
nickel(II) nitrate [26]
nickel(II) nitrate, $0.1\ M$ [8]
nitric acid (conc.) [1]
nitric acid (dil.) [1,16]
nitric acid, $0.1\ M$ [7]
nitrogen, liquid form [10]
octyl alcohol [22]
oleic acid [6]
orange juice [16]
oven cleaner [16]
overhead projector [5,8,20,22,23,24,25]
oxalic acid, $0.300\ M$ [20]
oxygen [1,10,18]
pH meter [17]
paint can and lid (1 pt.) [18]
pan, 2-L [14]
paper [1,4]
paper cup, 5-oz [22]
paper towels [2]
pennies, minted after 1982 (100) [3,25]

pennies, minted before 1982 (100) [3,25]
petri dish, glass, 100-mm [8,9,20,23]
phenol red [17]
phenolphthalein [1,5,8,9,17,20,24,28]
phosphoric acid (dil.) [16]
phosphoric acid, 0.1 M [7]
pickle juice [16]
piezo sounder [20]
pineapple, fresh [23]
pipet, 1-mL [6]
plastic storage bags, sealable (1-gal.) [16]
platinum or nichrome wire [26]
pliers [4]
pneumatic trough (or plastic dishpan) [14]
popsicle stick, wooden or plastic [15]
poster board, black [23]
potassium chloride [24]
potassium hydroxide (dil.) [1]
potassium iodide, 0.1 M [20]
potassium nitrate [14,15,21]
potassium permanganate [10,21,22]
potassium permanganate 0.000 40 M [7]
potassium permanganate 0.005 M [7]
potassium permanganate 0.01 M [20]
potassium permanganate 0.05 M [7]
potassium sodium tartrate [22]
potassium thiocyanate, 0.01 M [25]
power supply, 6-V/12-V DC [8]
prism [4]
raisins [1]
reagent bottles (4) [16]
Red Hots® [6]
ring stand [12,14,21,24]
ring stand clamps (2) [12]
rock salt [14]
room deodorizer, spray type [13]
rubber delivery tubes (3) [10,14]
rubber hose [18]
rubber stopper, 1-hole (3) [12,14]
rubber stopper, 2-hole [10,12]
rubber stopper, solid [16,18]
rubber stoppers (8) [21]
rubber stoppers, #3 solid (4) [14]
rubber stoppers, #5 (3) [22]

ruler, clear, plastic, metric [3,6]
salad dressing, oil and vinegar [14]
salicyclic acid [22]
salt solution [1]
sandwich bags, plastic [3,13,22]
sand paper [5]
scissors [21]
sharp file [25]
silicon [3]
silver nitrate, 0.1 M [21,28]
slide projector [4,14,27]
soap solution [24]
soft drink bottles, 2-L plastic (2) [14,24]
soft drink bottles, 1-L (3) [14]
soda (clear soft drink) [1]
soda straws, long [17]
soda-straw balance [25]
sodium bisulfite, 0.1 M [7]
sodium carbonate [24]
sodium chloride [7,25,28]
sodium chloride, 0.1 M [8,15,21]
sodium hydroxide pellets [8,9,15,17,21,22,24]
sodium hydroxide, 0.1 M [15]
sodium hydroxide, 1 M [16]
sodium hydroxide, 2 M [16]
sodium hydroxide, 6 M [25]
sodium hydroxide, 8 M [7]
sodium nitrate [7,24]
sodium phosphate [7]
sodium silicate solution [22,26]
sodium sulfate [7,8]
sodium tetraborate [26]
sodium thiosulfate pentahydrate [14,17]
sodium thiosulfate, 0.1 M [9]
sodium-lead alloy, 90% Pb/10% Na [24]
solder [20]
soldering gun [20]
sour milk or buttermilk [16]
spray bottle [24]
starch liquid, from pasta [14]
starch solution [20]
starch solution, 1% [23]
steel wool [20,25]
stirring rods (7) [1,14,16,21,22,25]
stoppers, cork or rubber [12]
straight pins (2) [4]
string [18]
strontium chloride [24]
sucrose [14,15,21,27]
sucrose solution, 1% [23]
sulfur, [3,6]

sulfur, powdered [27]
sulfuric acid (conc.) [22,27]
sulfuric acid (dil.-3 M) [1,10,16,23]
sulfuric acid, 0.1 M [7,9]
superabsorbent diaper [1]
switch, push-button [15]
syringe, 50-mL or 100-mL [11]
talcum powder [6]
tape [4,6,18,21]
tea bag [14]
test tube holder [22]
test tube rack [22]
test tube, 10-mL [8]
test tube, large Pyrex® [10]
test tubes, (5) [9,17,23,24,25]
test tubes, large (8) [16,21,27]
test tubes, medium (7) [3,5,14,16,21,22]
thermometer [13,24]
thistle tube [10]
thread [25]
thumbtack [11]
thymolphthalein [17]
tongs [6,11,27]
toothpicks, round or heavy [6,15,21]
transparencies [5,21]
triangle [24]
trough or large pan [10]
U-tube [28]
universal indicator [7,17,24]
urea [14]
vinegar [16]
voltmeter, 0-5 V [20]
watch glasses (7) [7,9,16,22,25]
water bath [7]
Waterlock J-500 [1]
Waterlock J-550 [1]
wide bore glass tubing [12]
wide-mouth container [26]
wire gauze [21]
wire gauze with ceramic center [14]
wood (block) [18]
wooden splints [9,10,16,22]
yeast, dry [21]
yogurt [16]
zinc nitrate [26]
zinc nitrate, 0.1 M [8,20]
zinc strips (5) [20]
zinc, granular [25]
zinc, mossy [3,7,8,10]

Unit 1

Chapter 1
Matter, Energy, and Change 4

Chapter 2
Measurements and Solving Problems 30

Introduction to Chemistry and Matter

At the NASA Lewis Research Center, I am involved in developing computer codes that use the laws of thermodynamics to model combustion processes. These codes have become useful tools for solving a wide range of problems in chemistry and engineering.

Space exploration depends on the science of chemistry. Chemists, for example, helped produce the rocket fuels that provide 4.4 million pounds of thrust to the space shuttle at liftoff.

Scientists also apply chemical principles to meet our society's need for food, clothing, building materials, medicines, and many other essentials. This unit introduces you to chemistry, a science that is an integral part of our lives.

Bonnie J. McBride
Bonnie J. McBride
Physical Chemist
NASA Lewis Research Center

The Space Shuttle *Discovery* at liftoff

CHAPTER 1
Matter, Energy, and Change

PLANNING CHART

Approximate Time: 9/11 days

	SECTION/TEXT FEATURES		RESOURCES	

B A S I C	**1.1 WHAT IS CHEMISTRY?**
	Demos 1, 2, 3, 4, 5, 6
	Section Review
	Chapter Review
	• Questions 1–5
		5
		4A–C
		10
		28
			Teacher's Resource Binder
			   Section Review Worksheets
			Other Resources
			   Shakhashiri Video Demo 11
			   World of Chemistry Video 1
			   Blackline Master 2
				1

B A S I C	**1.2 MATTER AND ENERGY**
	Demo 7
	Periodic Table Videodisc
	Microscale Investigation 1
	Section Review
	Chapter Review
	• Questions 6–19, Problems 1–2
		10
		4D
		4G
		830
		19
		28
			Teacher's Resource Binder
			   Lab Experiments 1, 2
			   Section Review Worksheets
			Other Resources
			   Shakhashiri Video Demos 20, 23
			   Transparency 1
				15, 31
				2

B A S I C	**1.3 CLASSIFICATION OF MATTER**
	Microscale Investigation 2
	Section Review
	Chapter Review
	• Questions 20–21, Problems 3–4
		19
		832
		23
		28
			Teacher's Resource Binder
			   Section Review Worksheets
			Other Resources
			   Transparency 2
				3

B A S I C	**1.3 THE CHEMICAL ELEMENTS**
	Periodic Table Videodisc
	Section Review
	Chapter Review
	• Question 22
		23
		4G
		27
		28
			Teacher's Resource Binder
			   Section Review Worksheets
			Other Resources
			   World of Chemistry Video 19
				4

	CHAPTER END ACTIVITIES Critical Thinking, Cross Connections, Research and Writing, Projects		*Teacher's Resource Binder* Test Booklet *Computer Test Generator*	1, 3

Teaching Strategies

Introduction

Let students experience the "wonder" of chemistry by giving them a glimpse of a few of the reactions they will study during the course of the year. Encourage them to think of explanations, but leave detailed explanations until later. You are simply whetting their appetites.

Demonstration 1: The Ira Remsen Reaction
PURPOSE: To introduce a chemical change using a historical anecdote
MATERIALS: At lease one pre-1982 penny, 500-mL Erlenmeyer flask, concentrated nitric acid
PROCEDURE: While reading the text that follows, place the penny in the Erlenmeyer flask and add 20 mL of concentrated nitric acid. Place the flask on a table where all may view the reaction.

While reading a textbook of chemistry, I came upon the statement, "nitric acid acts upon copper." I was getting tired of reading such absurd stuff, and I was determined to see what this meant. Copper was more or less familiar to me, for copper cents were then in use. I had seen a bottle marked "nitric acid" on a table in the doctor's office where I was then "doing time." I did not know its peculiarities, but the spirit of adventure was upon me. Having nitric acid and copper, I had only to learn what the words "act upon" meant. The statement "nitric acid acts upon copper" would be something more than mere words. In the interest of knowledge, I was even willing to sacrifice one of the few copper cents then in my possession. I put one of them on the table, opened the bottle marked nitric acid, poured some of the liquid on the copper and prepared to make an observation. But what was this wonderful thing which I beheld? The cent was already changed, and it was no small change either! A green-blue liquid foamed and fumed over the cent and over the table. The air in the neighborhood of the performance became colored dark red. A great colored cloud arose. This was disagreeable and suffocating. How should I stop this? I tried to get rid of the objectionable mess by picking it up and throwing it out of the window. I learned another fact. Nitric acid not only acts upon copper, but it acts upon fingers. The pain led to another unpremeditated experiment. I drew my fingers across my trousers and another fact was discovered. Nitric acid acts upon trousers. Taking everything into consideration, that was the most impressive experiment and, relatively, probably the most costly experiment I have ever performed. It was a revelation to me. It resulted in a desire on my part to learn more about that remarkable kind of action. Plainly, the only way to learn about it was to see its results, to experiment, to work in a laboratory.

Have the students pay special attention to the discussion questions that follow the Desktop Investigation.

SAFETY CONSIDERATIONS: Point out to your students that ALL colored gases are poisonous. Use of the large flask will concentrate the gas and not allow it to diffuse through the room. This demonstration should be done under a hood or using an improvised funnel vent connected to an aspirator on a faucet. Handle the nitric acid carefully. Point out that it is a key laboratory acid, but it must be handled with extreme caution. Do not dispose of the solution. After adding the 200 mL of distilled water, point out that blue is a characteristic color of dissolved copper. Pour the solution into a clear glass bottle and put it on display. You will be able to refer to it many times in future lessons.

Demonstration 2: Cyalume Light Stick
PURPOSE: To demonstrate that some chemical reactions give off light and that the rate of the reaction depends on the temperature of the system
MATERIALS: cyalume light stick, 1000-mL beaker with ice water, a 1000-mL beaker with hot (boiling) water
PROCEDURE: Remove the light stick from its protective wrapper and ask the students if they have ever seen or used one. Explain the construction of the light stick—two chemicals one of which is kept in a sealed glass vial. The reaction begins by bending the light stick and breaking the vial allowing the two chemicals to mix. The reaction produces an intense light. The color depends on the type of light stick. Yellow-green, red, blue, and white light sticks, are available. Turn off the room lights and place the light stick in the beaker of cold water. Note the dimming of the light evolved. Remove the light stick and place it in the beaker with the boiling water. Note the increase in the intensity of the light. Point out that temperature is critical to the speed at which a reaction occurs. Low temperatures slow down reactions while high temperatures speed up reactions.

Demonstration 3: Superabsorbent Polymers
PURPOSE: To demonstrate a relatively new product with amazing properties and a variety of uses
MATERIALS: Distilled water, 400-mL beaker, Waterlock J-550 or Waterlock J-500, a "grow beast," and a superabsorbent diaper

PROCEDURE: Add 200 mL of distilled water to the 400-mL beaker. Sprinkle one half teaspoon of the polymer into the beaker. Swirl the contents; wait a few seconds, and turn the beaker over. The contents should have solidified and the gelled water remains in the beaker.

Students are amazed by this reaction. Ask the students if they have ever seen or used such a product. If no one volunteers, show them a diaper and a "grow beast." The product is the main ingredient in the new varieties of superabsorbent diapers. The toy industry uses the polymer to make the "grow beasts"—the plastic toys that expand to 200 times their original size when placed in water. Commercial airlines use the polymers to make fuel-filter elements. The polymer takes up any water that might have mixed with the fuel. The petroleum industry uses them as "superslurpers" to clean up oil spills.

After allowing the students to observe the beaker, sprinkle some salt (sodium chloride) on the gel. It immediately begins to liquefy. Swirling the beaker will show the change in properties. In place of the solid salt, 20 mL of a concentrated salt solution may be poured into the beaker. The salt interferes with the absorbing capacity of the polymer.

Ask students about the absorbing capacity of the polymer. Does it make a difference if tap water were used instead of distilled, or if hot water were used instead of cold? How well is urine absorbed in the diapers since urine contains dissolved salts? You might at this time place a "grow beast" in a wide-mouth gallon jar filled with distilled water and observe its growth during the coming week. The jar and "beast" may then be put on display.

SAFETY AND DISPOSAL: The gelled Waterlock polymer may be kept as a display. It may be safely washed down the drain after it has been poisoned with the salt.

Demonstration 4: Dancing Raisins
PURPOSE: To stimulate thought and to demonstrate density and buoyancy
MATERIALS: Cold, clear sodapop, raisins, a large beaker or 500-mL graduate cylinder
PROCEDURE: Pour the cold, clear soda into the container. Add a dozen raisins. Observe. The raisins begin to rise to the surface where they appear to "dance" and then sink, only to rise once again, repeating the process many times.

Have the students attempt to explain what they are observing. But, again, do not spend much time on the discussion. (Gas bubbles are trapped by the convoluted surface of a raisin, and thus lower the density of the whole. As the raisin reaches the surface, it begins to lose the gas bubbles, the density increases, and it sinks to the bottom again.)

Demonstration 5: Bursting Balloons
PURPOSE: To demonstrate the synthesis of water from hydrogen and oxygen gas while demonstrating that some reactions take place violently
MATERIALS: Balloons, a source of helium, hydrogen, and oxygen gas
PROCEDURE: Fill three balloons with (1) helium, (2) pure hydrogen, and (3) a mixture of hydrogen and oxygen. Tie a string to each balloon and allow them to float over the desk. Ask the students if they know what gas(es) might be in the balloons. Most are familiar with helium balloons. Few will have seen a

hydrogen balloon. Ask if any know the story of the German zeppelin the Hindenburg, destroyed by lightning in the late 30's. Dim the room lights and with a candle taped to the end of a meter stick, ignite the gas in each of the balloons.
CAUTION: Be sure and have the students cover their ears. You should use ear protectors yourself. The helium balloon will pop with very little noise. The hydrogen balloon will burst with a muffled explosion and a ball of flame. The hydrogen/oxygen balloon will explode with a very loud noise.
SAFETY AND DISPOSAL: Have the students cover their ears. Small balloons should be used and should not be overfilled. They will make a loud noise.

1.1 What is Chemistry?

The first day demonstrations should have whetted the appetites of the students. After distributing the textbooks during the second day of class, discuss the introduction to chemistry. Point out how chemistry is everywhere. Ask the students for examples where they have come in contact with chemistry in action. The text lists a number, but the students should be able to come up with their own. Define chemistry and relate the definition to the previous day's demonstrations.

To help students see the wide scope of chemistry in their daily lives, display several household chemicals—a variety of plastics, fibers, fertilizers, motor oil, car polish, medicines, and vitamins. Chemistry not only has provided us with many useful products, but manufacturing processes have left us a legacy of problems with waste disposal, ozone depletion, water contamination, etc. Point out that solutions to the problems facing society will be found by chemists themselves. Note the importance of chemistry as a career option.

Encourage students to bring in newspaper and magazine articles that bear on chemical products or issues. You may wish to make this an assignment.

Discuss the traditional branches of chemistry, but emphasize that these boundaries are not as clear today as they were in the past. As biology has moved to the molecular level, the border between biology and chemistry is disappearing. Chemistry is becoming ever more important.

Point out that scientific discoveries are made in two ways: *(1)* by accident and *(2)* through the use of scientific methods. Since there are so many fascinating stories surrounding accidental discoveries, discuss a few important discoveries made by accident, such as the discoveries of penicillin, saccharin, dynamite, quinine, petroleum jelly, and teflon. During your discussion, use Pasteur's comment that "Chance favors the prepared mind" to emphasize that unless a scientist recognizes the importance of a discovery, it will be meaningless.

Define Scientific Method. Ask the students to give you examples of the scientific method. Most will give the example of experimentation. Prior to discussing generalizations, you should cover order in the universe, the principle that allows broad generalizations to be made from a rather limited number of observations because of the belief that everything in the universe follows a given pattern. Call to the students' attention the fact that they have been generalizing for many years. For example, they learned early in life that every time they dropped something heavier than air, it fell to the floor. Ask what would happen if a beaker were dropped in the lab. All would agree that it would fall to the floor and break even though no one had ever seen that particular beaker before. This points out that after a limited number of trials, one can accurately predict future outcomes because of order in the universe. Remind students that scientists are always looking for patterns and keep pointing this out throughout the course.

In discussing the scientific method, stress the importance of observations in collecting data. Observations are simple statements of facts and do not imply anything that cannot be measured. These observations or information can be quantitative or qualitative. To distinguish between the two, show your students a candle and have them list observations. Describing the candle as a white, cylindrical solid makes use of qualitative information. A meter stick can provide quantitative information such as length and width of the candle. Qualitative information may become quantitative if it can be compared to other measurable information. For example, to say that the candle is made up of a white solid soft enough to be scratched by a fingernail adds a quantitative element to the observation. To say that the lit candle gives off light is a qualitative observation, but to add that it gives off enough light to be visible in a well-lit room makes the information quantitative.

In making observations, encourage the students to pretend that they are radio broadcasters. Their verbal descriptions must be so complete and graphic that a person hearing or reading the descriptions will be able to visualize the object.

Next it is important to help students distinguish between a theory and a law. Make sure that you make the distinction that a law tells what happens and a theory explains why it happens.

Demonstration 6: Effect of phenolphthalein on acids and bases
PURPOSE: To elicit a generalization from students
MATERIALS: Dilute solutions of several different acids and bases (concentration is not critical) LiOH, NaOH, KOH, NH_4OH, HCl, HNO_3, H_2SO_4, $HC_2H_3O_2$ in eight different beakers; phenolphthalein; stirring rod
PROCEDURE: Tell the students the names of each of the solutions. Add a few drops of phenolphthalein to all but the LiOH and HCl solutions. Explain that the first four are classed as bases and the final four as acids. Explain that acids and bases will be studied later. Ask the students to make a broad generalization concerning the observation they have just made. Have them predict what will happen if phenolphthalein is added to LiOH, a base, and to the HCl, an acid.

1.2 Matter and Energy

Define matter. Use various objects in the room as examples of matter. Solids and liquids should pose no problems. To demonstrate that gases are also a form of matter, determine the mass of an empty balloon. Blow up the balloon and determine the new mass. Although the difference in mass does not reflect the actual mass of the air in the balloon, the increase in mass should convince the students that gases are composed of matter.

Define inertia. Ask the students for examples of inertia from their own experiences. Action of cars on wet or icy roads is a good example. Others may have seen video of the astronauts pushing objects while in orbit. Use this last example to distinguish between mass and weight. Stress the fact that weight depends on the gravitational attraction. The mass, however, remains the same. Stress that chemists are concerned with determining the mass of substances. Point out that on a platform balance, mass is determined by comparing an unknown mass to a set of known masses. Next show the students a spring balance. Bathroom and/or kitchen scales are usually spring balances. Gravity pulling on the object being weighed determines the reading on the scale. If the same object could be weighed on the moon, the platform balance would give the same mass, but the spring balance would give a smaller reading because of the weaker gravitational field on the moon.

Ask the students to state the law of conservation of matter. It should have been covered in earlier courses. The law applies to "ordinary chemical and physical changes," not to nuclear reactions which are beyond the scope of the high school lab.

Define energy. Point out that scientists have some difficulty with this definition because it is not precise. Since all energy can be classified as kinetic or potential, use a transparency or the chalkboard and have the students classify the forms of energy listed in the chapter. The chart might look as follows:

Type of Energy	Kinetic	Potential
Chemical energy		X
Mechanical (including sound)	X	X
Electrical	X	X
Radiant (heat, light, etc.)	X	
Nuclear	X	X

As before, have the students state the law of conservation of energy. It too applies to ordinary laboratory situations and must be modified when discussing nuclear reactions.

Chemistry in Detail: Interconversion of Matter and Energy

This feature on the interconversion of matter and energy gives you the opportunity to tie the two conservation laws into one more general law. Break the law into two parts: (1) matter and energy are interchangeable and (2) the total amount of matter and energy in the universe is constant. Although Einstein formulated his theory in 1905, it was not confirmed in the lab until nuclear reactions were carried out. The power of a nuclear bomb arises from the conversion of a small amount of mass into a very large supply of energy.

List the four common states of matter. Some students may not know of the fourth, the plasma state. Point out that plasmas require temperatures too high (1×10^7 °C) to be reached in a high school lab. Have sample of solids and liquids as well as some empty containers on the desk. Show that the solids do not change shape when moved from container to container. The liquids do not change volume but they do conform to the shape of the container. To confirm that gases fill the container in which they are placed, spray some cologne or deodorizer into a beaker covered with a watch glass. Remove the cover and have students sitting near the desk indicate when the odor is noticeable. The cologne or deodorant will eventually fill the room.

Two distinctions must be made regarding the properties of matter. Students must be able to distinguish between physical and chemical properties; between intensive and extensive physical properties, and between physical and chemical changes.

Demonstration 7: Physical-chemical properties/changes

PURPOSE: To help students distinguish between physical and chemical properties and changes, and intensive physical and extensive physical properties

MATERIALS: Paper, matches, large watch glass or ceramic pad, balance, copper shot (any metal available in shot form), graduated cylinder, hot plate, 1-L beaker, water, ice

PROCEDURE: To distinguish between physical versus chemical properties and changes:

(1) Show students a piece of white paper. Have them describe it in terms of color, physical state, size, etc.

(2) Cut the paper in half. Ask if any of the physical properties have changed. Note that a physical change, cutting the paper, does not change the identity of the material.

(3) Holding one half of the paper with a pair of tongs over the ceramic pad or large watch glass, light a match and bring the match to the paper. Is burning a physical or chemical change? Is the ability of paper to burn a physical or chemical property? In this case, a new substance does form.

To distinguish between intensive physical and extensive physical properties:

(1) Prior to class measure approx. 25 ml of metallic shot. (2) Show the shot to the students and have the students describe it. (3) Place the shot on the balance, and determine its mass. (4) Pour the shot into a dry graduated cylinder, and determine its volume. (5) Divide the sample into two unequal portions. (6) Ask the students to list the properties that have changed. Intensive properties such as color and melting point have not changed. Mass and volume have changed. The extensive properties have changed; the intensive properties did not. (7) If time is available, you can determine the densities of a small and a large sample by (a) determining the mass of each and (b) determining the volume by pouring each into a graduated cylinder half-filled with water. Show the students how to determine the volume of the solid and show that the ratio of mass to volume, i.e. the density does not change. This may be used to introduce density before actually discussing it in the text, or, if you wish, you may delay this part of the demonstration until later and use it to reinforce the distinction between intensive and extensive properties.

Physical changes associated with a phase change.

(1) Place the 1-L beaker on a hot plate. Add about 200 mL of water and cover with a watch glass. (2) Place several ice cubes on the watch glass. (3) Turn on the hot plate and allow the water to boil and the steam to condense on the cold watch glass. (If this the first time your students are seeing a hot plate, explain its use and any safety precautions that must be taken.) (4) Point out that a phase change is a physical change, whereas burning is a chemical change. The physical change can be reversed by simply changing the temperature while the chemical change results in new compounds.

SAFETY CONSIDERATIONS: Wear your goggles. Be careful when burning the paper. Be sure that the large watch glass or ceramic plate are below the burning paper. The other materials may be reused.

Indications of chemical reaction are best presented in the context of the first Desktop Investigation. It also serves as the ideal time to introduce your students to the laboratory and to the required safety rules.

Matter, Energy, and Change

Desktop Investigation: Physical and Chemical Changes

Go over the procedure with your students. Stress the importance of recording all observations in the data table. After they have performed the exercise, elicit from them any observations that indicated a chemical change. Relate these to the ones listed in the text. Define endothermic and exothermic changes in the context of their observations.

1.3 Classification of Matter

Introduce this section using as many examples as you can. Have available on your desk a variety of samples of matter. For mixtures you might use a bottle of salad dressing, a piece of granite, a candy bar, and a copper-sulfate solution. For pure substances include water, sugar, salt, copper, and other element samples. Have the students describe each and begin to place them into the classification scheme for matter.

Although not discussed in this part of the text, some students will be aware of colloidal suspensions. Explain that if particles can be made small enough, the motion of the water molecules will keep them suspended. Milk is the best example.

Care must be taken in distinguishing the various types of mixtures and pure substances. You may wish to evaporate the liquid from two beakers. One beaker might contain distilled water (nothing remains), the second, a salt solution (the salt remains). If an electrolysis apparatus is available, electrolyze the water and show that it can be decomposed into simpler components. Stress that once one reaches the atomic or elemental level, one may no longer subdivide the matter in question with the chemical means available in the high school lab.

1.4 The Chemical Elements

As an introduction to the Periodic Table, if a recording of the song, "The Elements" from the album *An Evening Wasted* (Reprise Records) by Tom Lehrer is available, use it to introduce the section on the elements.

In this section you are introducing one of the most important tools a chemist has. Understanding how to use the Periodic Table is one of the skills your students must develop. Hand out Periodic Tables on which the students may write and take notes. First divide the table in two. Show them where the metals, nonmetals, and metalloids are found. Further subdivide the table into the main group or representative elements, the transition metals, and the rare earths. Distinguish between periods and families, and have them write down the names of the most common families, i.e. alkali metals, alkaline earths, halogens, and noble gases. Do not get bogged down with the group numbering system. Use the 1–18 adopted internationally.

List on the board those elements whose symbols are not related to their modern names. Encourage the students to begin learning them. The symbols of the elements are analogous to the letters of the alphabet. Just as we use letters to make up words, chemists use the symbols of the elements to write the formulas of compounds, and, finally, just as words are used to make up sentences, chemists use the formulas in writing chemical equations.

A Periodic Table with photographs of each element would be a great aid in distinguishing types of elements. List properties of metals and then contrast the properties of nonmetals. Show how metalloids fall in between.

Have available samples of the two typical elements discussed in the text. **CAUTION:** White or yellow phosphorus is very dangerous and may not be allowed in your school. Use only red phosphorus and show pictures of the other form. If phosphorus is not available, use roll sulfur as your nonmetal sample.

Stress with the students that this has been a first look at the Periodic Table and the wealth of information that it contains. As the year goes on, each chapter will unlock more secrets and make the Periodic Table ever more useful.

As a special exercise, you may wish to have your students pick and report on a favorite element. If your students are seated alphabetically, point out that they are seated according to a classification system. At some future date, you may wish to seat your students according to some other property, e.g. home telephone numbers, and have them try to discover the pattern. Anything that you do to impress on the students the importance of the Periodic Table will bear fruit in the long run.

References and Resources

Books and Periodicals

Alyea, Hubert and Frederic B. Dutton. *Tested Demonstrations in Chemistry*, Easton, PA: Journal of Chemical Education, 1965. (Facsimile edition available from Flinn Scientific).

Borgford, Christie L. and Lee R. Summerlin. *Chemical Activities: A Sourcebook for Science Teachers*, Washington, D.C.: The American Chemical Society, 1988.

Gensler, Walter J. "Physical Versus Chemical Change," *Journal of Chemical Education*, Vol. 47, No. 2, Feb. 1970, pp. 154–156.

Redlich, Otto. "Intensive and Extensive Properties," *Journal of Chemical Education*, Vol. 47, No. 2, Feb. 1970, pp. 154–156.

Shakashiri, Bassam Z. *Chemical Demonstrations*, Vol. 1, Madison: University of Wisconsin Press, 1983.

Shakhashiri, Bassam Z. *Chemical Demonstrations*, Vol. 2, Madison: University of Wisconsin Press, 1985.

Strong, Laurence E. "Differentiating Physical and Chemical Changes," *Journal of Chemical Education*, Vol. 47, No. 10, Oct. 1970, pp. 689–690.

Summerlin, Lee R., et al. "Introducing Chemistry," "Physical Changes," and "Reactions of Some Elements," *Chemical Demonstrations—A Sourcebook for Teachers*, Vol. 2, Washington, D.C.: American Chemical Society, 1987.

Talesnick, Irwin. *Idea Bank Collation—A Handbook for Science Teachers*, Kingston, Ont.: S17 Science Supplies and Services Company, Inc., 1984.

Webb, Michael J. "Physical and Chemical Change: What's the Difference?," *The Science Teacher*, Vol. 49, No. 3, March 1982, pp. 39–40.

Answers and Solutions

Reviewing Concepts

1. **a.** Science is a systematic and organized collection of facts.
   b. Scientific facts are used to provide explanations, to make predictions, and to manage events.
2. **a.** Chemistry is the study of the composition of materials and changes in the composition of materials.
   b. Chemistry is involved in the manufacture of synthetic fabrics, in the growing of fruits and vegetables, and in water purification.
3. The dictionary defines *chemical* as a substance produced by or used in a chemical process. According to this definition, everything is made of chemicals. As used in the advertising slogan cited, however, the word "chemical" refers to particular substances associated with undesirable circumstances. In fact, although some chemicals (in the dictionary sense) are harmful, many are not.
4. **a.** In science, a law is a generalization that describes behavior in nature.
   b. Laws in science are generally expressed by mathematical equations or concise statements.
5. **a.** Measuring is the collecting of numerical information about the object, material, or process under study.
   b. Experimenting is carrying out sequences of observations under controlled conditions.
   c. A hypothesis is a testable statement.
   d. A theory is a broad generalization that accounts for a class of known facts or phenomena.
6. **a.** A moving object resists a change in direction or speed, whereas an object at rest resists being moved.
   b. Inertia varies directly with the quantity of matter to be moved.
7. **a.** Matter is anything that has mass and occupies space.
   b. Sand, paper, air, and chalk are examples of matter.
8. Matter cannot be either created or destroyed in ordinary chemical or physical changes.
9. **a.** Energy is the ability to cause change.
   b. Chemistry deals mainly with the chemical energy stored in matter and either the effects on matter or the production from matter of thermal, radiant, or electrical energy.
10. Options **a** and **d** illustrate kinetic energy, whereas **b** and **c** illustrate potential energy.
11. Energy can be converted from one form to another, but it cannot be either created or destroyed in ordinary chemical or physical changes.
12. The particles of matter in a solid are closely packed and rigidly held in fixed positions, those in a liquid are closely packed but are not bound in fixed positions, and those in a gas are very far apart and have no ordered positions at all.
13. **a.** A physical property is one that can be observed or measured without altering the identity of a material.
    b. Color, melting point, and boiling point are examples of physical properties.
14. **a.** Extensive properties depend on the amount of matter present, intensive properties do not.
    b. Extensive: mass, length, volume; Intensive: density, ductility, color
15. A chemical property relates to the ability of a substance to undergo a change that alters the identity of a substance.
16. **a.** Physical changes do not result in a change in identity. Chemical changes involve conversions into different substances.
    b. Three examples of physical changes are cutting a wire, grinding a solid, and melting a solid. Examples of chemical changes include the burning of wood, the rusting of iron, the tarnishing of silver.
17. Examples **a**, **b**, **d**, and **e** are physical changes, whereas **c** is a chemical change.
18. Observable changes that generally indicate that a chemical reaction has occurred include the evolution of heat and light, the production of a gas, and the formation of a precipitate.
19. **a.** In exothermic processes, heat is released, whereas in endothermic processes, heat is absorbed.
    b. exothermic: hydrogen and oxygen $\rightarrow$ water and heat; endothermic: calcium carbonate + heat $\rightarrow$ calcium oxide + carbon dioxide
20. Any sample of matter can be classified as being either a pure substance or a mixture. Pure substances are either elements or compounds. An element is a substance that cannot be separated into other substances by any ordinary chemical change. A chemical compound is a pure substance that can be made to decompose chemically into two or more simpler substances. A mixture is a combination of two or more kinds of matter, each of which retains its own characteristic composition and properties.
21. Pure substances differ from mixtures in the following ways.
    • Every sample of a given pure substance has exactly the same characteristic properties; this is not the case in a given mixture.
    • Every sample of a given pure substance has exactly the same composition, whereas mixtures have variable compositions.
    • A pure substance cannot be separated into other substances without changing its identity and properties. In a mixture, the components do not lose their identity and can be separated by physical means.
22. **a.** An element cannot be separated into other substances by ordinary chemical changes, but a compound can undergo decomposition in this way.
    b. Aluminum, oxygen, and copper are elements; sugar and water are compounds.

Problems

1. Example **b** is an exothermic process, whereas **a**, **c**, and **d** are endothermic.

2. **a.** The kinetic energy of object A is twice that of B.
   b. Object A has a kinetic energy four times that of B.
   c. Object A has a kinetic energy eight times that of B.
3. Examples **a** and **c** illustrate heterogeneous matter, whereas **b** and **d** represent homogeneous matter.
4. **a.** Texture, color, and density could possibly be used.
   b. Color, density, and melting point could possibly be used.
   c. Color, density, and texture could possibly be used.
   d. Density and viscosity could possibly be used.

Critical Thinking

1. Answers will vary. If water quality was not governmentally controlled, each home owner would have to be responsible for the quality of their source of water. This would probably include testing the water monthly for lead and bacteria, as well as, adding helpful ions such as fluorine. Each home would also have a water purifier. Intestinal diseases would be more common.
2. The objects would have the same mass and weight as long as they were on the same planet and in the same location on that planet together.

Cross Connections

1. Answers will vary. The data table should include the material name, melting point, density, resistance to environmental elements, length of life, and cost.

Videodisc Selections
from *The Periodic Table Videodisc* (JCE Software)

step forward

step reverse

CHAPTER 1
pg 13 Physical states of elements

49875
solid - gold (s)

35179
liquid - bromine (s)

4537
gas - nitrogen (s)

pg 14 Physical properties determine uses of elements

36664
copper wire (s)

36337
lithium in batteries (s)

36585
titanium in pigments (s)

36657
nickel in batteries, coins,
    Alnico magnets, jewelry (s)

pg 14 Chemical properties of some elements

115
hydrogen - combustion with O2
    to form water (m)

17334
potassium - reaction with water (m)

page 16 Properties of reactions

7535
magnesium + oxygen to produce
    heat and light (m)

37311
strontium + oxygen to produce
    heat and light (m)

29501
copper + nitric acid to produce
    a gas (m)

31167
zinc + acids to produce hydrogen (m)

pg 17 Exothermic reactions

17334
potassium + water (m)

7535
magnesium + oxygen (m)

115
hydrogen + oxygen (m)

pg 24 Classifications of elements

metals

268
lithium (s)

4565
sodium (s)

16833
potassium (s)

7523
magnesium (s)

18433
calcium (s)

50720
lead (s)

21230
chromium (s)

25876
cobalt (s)

28541
copper (s)

39192
silver (s)

49852
tungsten (s)

49863
platinum (s)

Matter, Energy, and Change **4H**

49875
gold (s)

nonmetals

3822
carbon (s)

4537
nitrogen (s)

12307
phosphorus (s)

4543
oxygen (s)

13140
sulfur (s)

14228
chlorine (s)

35171
bromine (s)

45067
iodine (s)

metalloids

3816
boron (s)

11574
silicon (s)

32429
germanium (s)

33218
arsenic (s)

42664
antimony (s)

pg 26 Reactions of copper and uses of phosphorus

28546
copper + air (m)

29501
copper + acids (m)

36482
phosphorus in food (s)

36486
phosphates in detergents (s)

41   Chapter 1   Teacher's Notes

Teacher's Notes

Chapter 1
Matter, Energy, and Change

SECTION PREVIEW
1.1 What is Chemistry?
1.2 Matter and Energy
1.3 Classification of Matter
1.4 The Chemical Elements

The light, noise, and heat generated by fireworks are all caused by matter and energy involved in chemical changes.

1.1 What Is Chemistry?

> **SECTION OBJECTIVES**
> - Define chemistry.
> - Identify some applications of chemistry in everyday life.
> - List the major areas of chemistry.
> - Describe the scientific method.

Human beings are naturally curious. We wonder and ask questions about what we see and experience. The phenomena of nature provide a never-ending source of questions such as: What is fire made of? What happens to the air we breathe? Why do objects fall toward the earth? While these three questions and many more like them have been answered, countless questions remain: How was the universe formed? Can heart disease be cured? Is there a material that superconducts at room temperature?

Posing questions about nature and carefully gathering the facts to provide answers is the work of natural scientists. Over the centuries, natural scientists have contributed to a greater understanding of our world and our experiences. This knowledge has enabled us to make tremendous changes in the way we live. Today, for example, laser-disk players are widely used to reproduce music with stunning clarity. Positron-emission tomography is used in hospitals to harmlessly probe the human brain for signs of disease. By using genetic engineering, bacteria can be designed to produce life-saving insulin for diabetics. None of these were possible when most of today's high school seniors were born.

These changes bring many benefits, but they also provide challenges to all of us. One of these challenges is cost. Many new and sophisticated medical techniques are expensive. Can society afford to make them available for everyone? Another challenge is the risk that genetically engineered life forms may pose to humans. Attempting to understand words and phrases like *laser*, *superconductivity*, *positron-emission tomography*, *pollution*, and *genetic engineering* is another challenge.

To successfully evaluate and meet these challenges, members of today's society must have a greater understanding of natural science than ever before.

> Have students list other benefits and contributions of science, as well as other potential problems and hazards attributed to science.

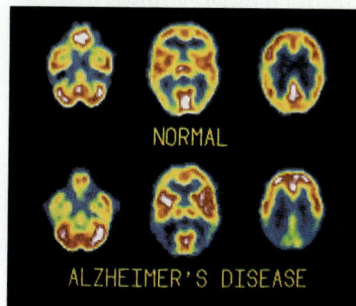

Figure 1-1 Color-enhanced images produced by *Positron Emission Tomography* show differences between normal brains and brains affected by Alzheimer's disease.

Chemistry is a Physical Science

The natural sciences fall into two general categories—the biological sciences and the physical sciences. The biological sciences are concerned primarily with living things, and the physical sciences are concerned primarily with nonliving things, including rocks, the stars, electricity, the weather, energy from the sun, and the composition of all materials. Chemistry is a physical science.

Chemistry *is the study of the composition, structure, and properties of substances and the changes they undergo.* This definition may suggest to you that chemistry has little to do with everyday life. This is not true. Your way of life would be radically different without the practical applications of chemistry. Imagine the contents of your closet if all clothing made of synthetic fabrics were removed. Chances are, over half your wardrobe would be gone. Imagine a supermarket offering only fruits and vegetables grown without manufactured fertilizers or pesticides. The quantities and varieties offered would be far fewer. Imagine drinking water from your tap that had not been purified. The unpurified water would probably make you sick. Try to imagine a world without gasoline or heating oil. It would be very different from the world we live in.

Chemistry in Modern Society

In generations past, the design and construction of houses, public buildings, and vehicles was limited by the available materials. Physicians had medicines that were discovered only by accident or by laborious trial and error. Today, new substances can be created to meet specific needs in architecture, aircraft and automobile design, medicine, and many other human endeavors.

The use of chemistry to manipulate the composition and behavior of substances is also important in meeting the challenges of our complex society. To feed an expanding world population and keep it healthy is an ever-present challenge. New ways must be found to generate and use energy efficiently. Dwindling supplies of natural resources must be replaced with other substances. Ways to recycle materials such as plastics (Figure 1-3) rather than allowing them to clutter and foul our environment are needed.

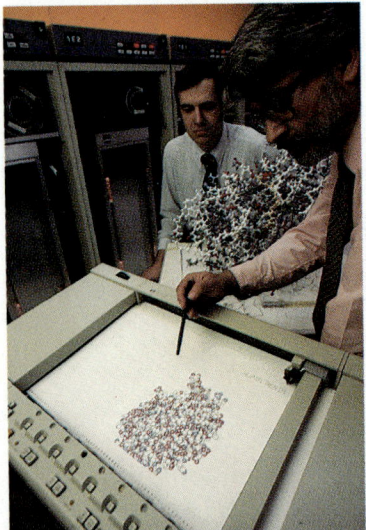

Figure 1-2 The scientists pictured above are studying the structure of renin, a chemical produced by the kidneys that causes blood pressure to rise. Their hope is to find a blood-pressure medicine that works by inhibiting the action of renin.

Figure 1-3 Plastic bottles (left) can be recycled and turned into raw material (right) for new plastic products.

Other examples are: new adhesives, pharmaceuticals, TV sets, insecticides, fertilizers, textiles, etc.

Encourage students to bring in newspaper and magazine articles that deal directly or indirectly with chemistry. Try to find those that place chemistry in a favorable light.

What is a Chemical?

Have you ever seen an advertisement that says, "All natural—no chemicals added"? Perhaps it was in reference to a fruit drink or a snack of some kind. What, exactly, is a chemical and what does the advertisement mean? One dictionary defines *chemical* as a substance produced by or used in a chemical process. Sugar is produced by plants from carbon dioxide and water— definitely a chemical process. We separate the sugar from the plants, purify it, and add it to foods. Is sugar a chemical? It certainly is to a chemist, and he or she can describe its exact composition. But what about water? It can be found everywhere. It can also be made in a laboratory by combining hydrogen and oxygen. Water is indeed a chemical. The point is that chemistry is the study of any and all materials. With sufficient investigation, the composition of every material can be described in terms of chemistry—everything is made of "chemicals."

Unfortunately, the word *chemical* has become associated with undesirable circumstances—pollution, cancer, poisonings, and accidents that result in spills of harmful substances. Some chemicals do, indeed, have unpleasant and dangerous properties, and contact with them must be avoided. Other chemicals cure diseases, maintain the food supply, and contribute to our comfort. One outcome of your study of chemistry should be an ability to evaluate uses of the word *chemical*. A substance that is a "chemical" is not, by definition, harmful. Nor can anything else be "free of chemicals."

Branches of Chemistry

Chemistry is a very broad subject. Most chemists would describe themselves as working in one of the following five major areas of the science:

1. *Organic chemistry*—the study of substances containing carbon and hydrogen, and their derivatives
2. *Inorganic chemistry*—the study of all substances not classified as organic chemicals, which includes the chemistry of all substances containing elements other than carbon
3. *Physical chemistry*—the study of the properties, transformations, and interrelationships of energy and matter
4. *Biochemistry*—the study of all substances and processes that occur in living things
5. *Analytical chemistry*—the identification of substances and the qualitative and quantitative determination of the composition of materials

Chemists also work in numerous specialized branches of chemistry. Nuclear chemists study the properties and transformations of matter that involve the atomic nucleus. Polymer chemists concentrate on understanding and making polymers—the raw materials for, among many other things, plastic objects. Some chemists become specialists in interdisciplinary fields, such as bioinorganic chemistry, which is the study of the role of inorganic materials in living things, or pharmaceutical chemistry, which is the chemistry of drugs, medicines, and vitamins.

Chemistry is applied to the large-scale production of useful materials by chemical engineers who are involved in the design, construction, and operation of equipment required in the chemical manufacturing processes.

Originally, "organic" chemicals were thought to be made only within living things. We now know this to be incorrect, but the name remains in use.

Current boundaries among the major areas of science are not as clear as they were in the past.

The atomic nucleus and nuclear chemistry are described in later chapters.

The Scientific Method

The general approach to investigations in chemistry is no different from that in any other natural science. Some important scientific discoveries come about by accident. Others are the result of brilliant new ideas. Most scientific knowledge, however, is the result of carefully planned investigations carried out by trained scientists, like the analytical chemist pictured in Figure 1-4.

The aim of specific investigations is usually to solve a problem or answer a question about a natural phenomenon. For instance, an analytical chemist may carry out an investigation to determine the exact composition of certain substances that have shown the ability to kill cancer cells in rats while leaving normal cells unharmed.

The ultimate goal of scientific investigations is to explain and predict natural phenomena. Scientists investigating cancer, for example, would like to explain how cancers develop and predict what substances will kill human cancer cells and leave normal cells unharmed.

The way scientists carry out their investigations is often referred to as the "scientific method." The **scientific method** *is defined as a logical approach to the solution of problems that lend themselves to investigations by observing,*

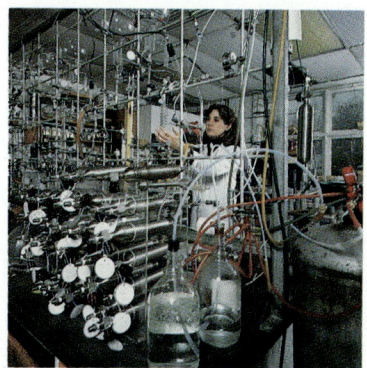

Figure 1-4 A chemist at Lamont-Doherty Geological Observatory, NY carries out a carefully planned experiment.

Matter, Energy, and Change

generalizing, theorizing, and testing. The processes involved in observing, generalizing, theorizing, and testing are explained below.

Observing Observing substances and events is essential to all natural sciences. Practical chemistry began with observations of materials found in nature, such as sand that could be converted to glass and rocks that yielded copper or iron when heated.

Collecting data goes hand in hand with observation. The data may be written descriptions of what is observed. The data may also be the result of *measuring* the object, material, or process under study. *Numerical information is referred to as* **quantitative information**. *Nonnumerical information is referred to as* **qualitative information**. Modern science relies on instruments of many kinds to extend the boundaries of what our senses can detect and measure. Figure 1-5 shows an image made by an electron microscope that allows details invisible to the naked eye to be "seen."

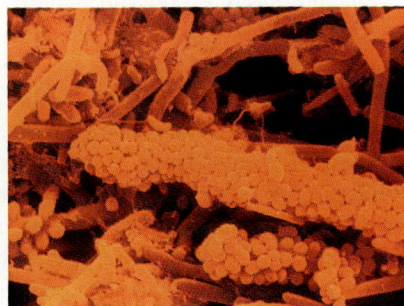

Figure 1-5 The scanning electron microscope (SEM) at the left can achieve magnifications of 100 million times the size of the specimen. A false color image *(right)* of bacteria on an unbrushed human tooth.

Scientists know that observing is most productive when conditions that affect the observations are brought under their control. Chemists do most of their observing in the controlled environment of the laboratory. *Experimenting* is carrying out sequences of observations under controlled conditions. Experimentation is the foundation upon which modern science is built.

Before new experiments are planned and carried out, most scientists turn to the scientific literature, which is extensive. They study the reported observations of others working in the same area of science. In this way, new experiments can be designed to extend knowledge and understanding. When experiments are completed, each scientist has the responsibility of *communicating* results and new information to others.

Generalizing Once observations have been made and data collected, the process of generalizing gets under way. *Organizing data* by making graphs, charts, or tables may be necessary. *Analyzing data* by using statistics, mathematical relationships, or other techniques is often necessary. Frequently, computers are used to organize and analyze data. If possible, general principles will be developed that help in *classifying* the data.

The goal is to find relationships among the accumulated data. Scientists hope that determining such relationships will enable formulation of a **hypothesis** *which is a testable statement.* A hypothesis is the basis for making predictions and carrying out further experiments. *A generalization that describes a wide variety of behaviors in nature is known as a* **law**. Laws may be expressed by mathematical equations or concise statements. Scientific laws describe, but do not explain, natural phenomena. If a scientific law is observed to be violated, then the law must be changed because it is not a valid description of nature. Two of the laws you will study in chemistry are *Boyle's law* and the *periodic law*.

Theorizing Once successful predictions can be made in an area of science, it becomes possible to move on to the task of determining *how* something occurs. To do so, scientists often engage in *constructing models*. Such models are usually not physical objects, like airplane models. Rather, they are visual, verbal, or mathematical models that show relationships among data or events. Frequently, a scientific model relates observed behavior to familiar and well-understood phenomena. Modern chemistry is based upon a model that states that all matter is composed of tiny particles.

In many cases, a model that is successfully applied may become part of a **theory**, *which is a broad generalization that explains a body of known facts or phenomena*. Theories can never be proven but they are considered successful if they can predict the results of many new experiments. Two major theories in chemistry that you will study in later chapters are the atomic theory and the kinetic–molecular theory.

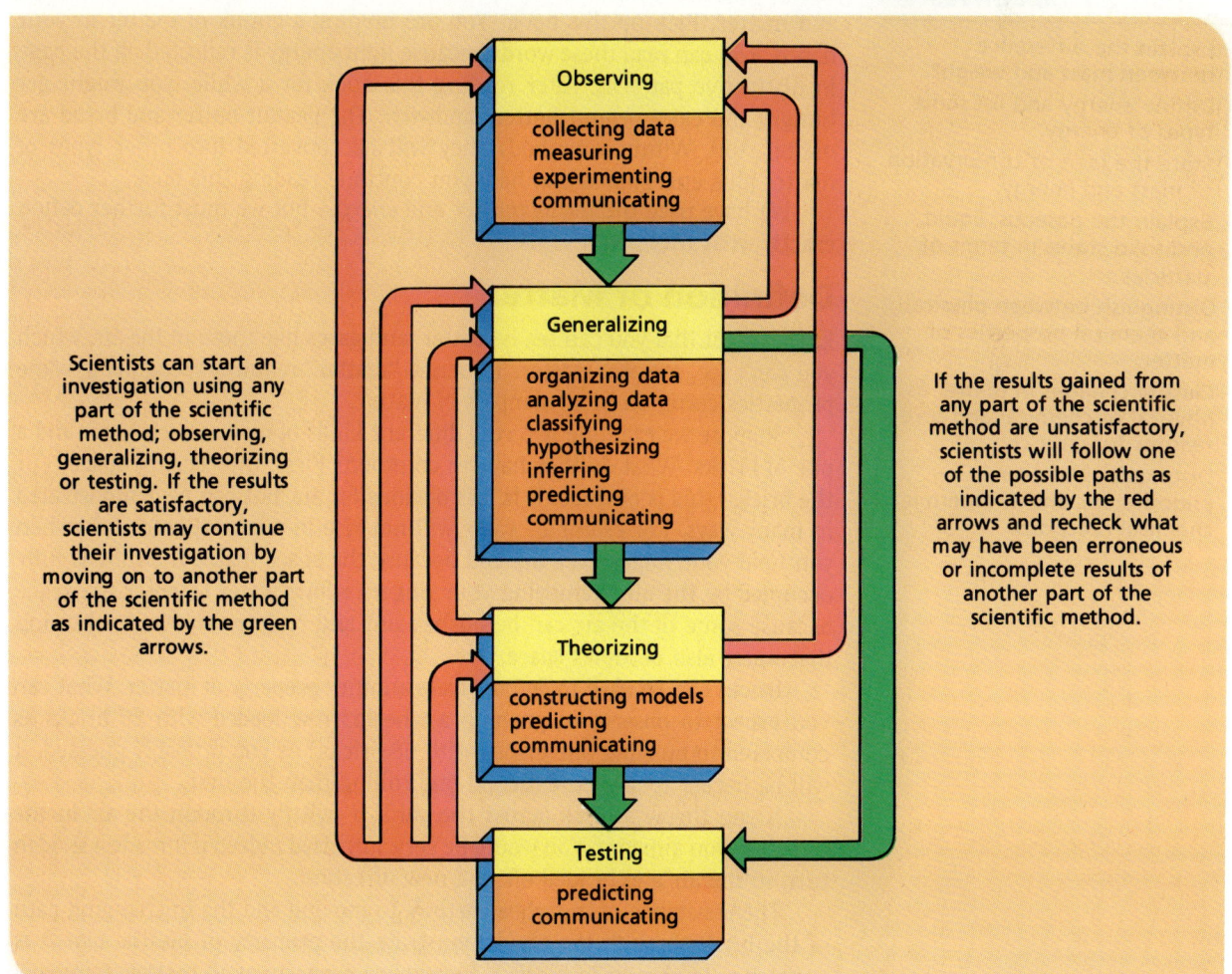

Figure 1-6 A flow chart of the scientific method

1. *(a)* Biological and physical *(b)* Physical
2. *(a)* A substance produced by or used in a chemical process. *(b)* Many are possible. Sugar (sucrose) and salt (sodium chloride)
3. Organic, inorganic, physical, biochemistry, and analytical
4. Observing, generalizing, theorizing, testing

Testing Through every stage of scientific investigation, testing takes place continuously. Scientists test data and the results of experiments by carrying out new experiments. Frequently, testing is based upon *predicting* the outcome of experiments based upon hypotheses, laws, or theories. No hypothesis, law, or theory is *ever* free from testing. All are constantly open to modification or abandonment based on new evidence. A summary of the scientific method and its processes is given in Figure 1-6.

Section Review

1. *(a)* What are the two categories of natural sciences? *(b)* Into which category does chemistry fall?
2. *(a)* What is a chemical? *(b)* Give two examples of chemicals.
3. List the five major areas of chemistry
4. What are four major activities of the scientific method?

1.2 Matter and Energy

SECTION OBJECTIVES

- Explain the difference between mass and weight.
- Define energy and list some types of energy.
- State the laws of conservation of mass and energy.
- Explain the gaseous, liquid, and solid states in terms of particles.
- Distinguish between physical and chemical properties of matter.
- Classify changes in matter as physical or chemical and give reasons for your choices.
- Distinguish between endothermic and exothermic chemical reactions.

Everything that you see or sense is either matter or the interaction of energy and matter. Pick up this book. You are holding a chunk of matter in your hands. You can read these words because light energy is reflected off the page in distinctive patterns. After reading this book for a while, you might get hungry and eat a peanut butter sandwich. The peanut butter and bread are matter, too. Within your body they will be converted into other types of matter plus energy that can help you continue reading this book.

We have used the terms *matter* and *energy,* but we must further define exactly what they mean.

Definition of Matter

Every object that you can see is matter, and gases like those in the air, which you can't see, are also matter. To define "matter" requires recognizing some properties common to all samples of matter.

Suppose we examine two very different kinds of matter such as air and a pile of bricks. What do they have in common? First, they occupy space. For the bricks, this is obvious. It is less obvious for air, but can be demonstrated in many ways. For example, water will not rise in a tube (Figure 1-7) when you hold your finger over the end because the space inside the tube is fully occupied by the air. Removing your fingers allows water to enter the space because some of the air can be pushed out and replaced by water, showing that water also occupies space.

Bricks can be used to demonstrate another property of matter. What can you expect to observe in pushing a wheelbarrow loaded with 20 bricks as compared to pushing one loaded with 60 bricks? As you probably guessed, it will be harder to start the second one rolling than the first.

If you throw a brick across the yard, it will fly through the air in the direction you aimed it. You would be very surprised indeed if it made a sudden turn in the air and headed off in a new direction.

The resistance of the wheelbarrows to moving and the unchanging path of the brick through the air demonstrate the property of **inertia**, *which is resistance to change in motion.* Inertia is a property of all matter. A moving

10 Chapter 1

object resists a change in direction or being stopped; an object at rest resists being moved.

Inertia varies directly with the quantity of matter to be moved. To explain the greater effort needed to move the larger load of bricks, however, it is doubtful one would say "Sixty bricks have more inertia than 20 bricks." Most likely, you would say "Sixty bricks weigh more than 20 bricks." Now we must find out what this means in the specific language of science.

Weight *is a measure of the earth's gravitational attraction for matter.* Hanging an object on a spring scale shows how the earth pulls that object down (Figure 1-8). The more matter in the object, the stronger the force exerted on it and the further it is pulled down. Comparing the weight of a vessel containing a vacuum and one containing a gas would demonstrate that gases, like bricks, can be weighed.

Scientists make a distinction between weight and mass that is usually ignored in everyday life. An astronaut weighs less on the moon than on earth. But on earth, on the moon, or in orbit, the amount of matter in the astronaut's body is the same. Here is where mass comes in. **Mass** *is a measure of the quantity of matter.*

The interchangeable use of the terms *mass* and *weight* seems unavoidable and ordinarily it causes no problems. If the masses of two objects are the same, they have equal weights while in the same location.

Now that you know something about the properties of matter, the usual definition should be more meaningful to you: **Matter** *is anything that has*

Figure 1-7 Water rises in the open tube *(top)*, but the trapped air inside the closed tube *(bottom)* does not allow the water to rise.

Figure 1-8 These spring scales show how strongly each object is attracted by the gravitational pull of the earth. Which object is attracted the most and which is attracted the least?

Matter, Energy, and Change

> The word "ordinary" is very important. Only in nuclear reactions does one encounter a change in mass. The "lost" mass is known as "the mass defect."

mass and occupies space. The **law of conservation of mass** is fundamental to chemistry. It states: *Matter cannot be either created or destroyed in ordinary chemical or physical changes.* Determining the mass of a sample of matter before and after a change allows this law to be tested.

Imagine using a spark to ignite a match in a glass jar that is so tightly sealed that no matter can enter or escape. As the match burns, flammable chemicals in the head of the match and the wood of the matchstick are transformed into water, carbon dioxide, and other substances. Some oxygen from the air in the jar is "used up" as it combines with materials from the match. If matter is being created or destroyed, the mass of the closed jar plus its contents must change. Measurements taken before and after the match has burned would show that the total mass remains the same, however.

The phrase "in ordinary chemical and physical changes" must be added to the law of conservation of mass because one circumstance is known in which matter is changed into energy, as described in the Chemistry in Detail feature on page 15. In your experiences in everyday life and in the chemistry lab, however, you can expect that matter will not be created or destroyed.

Definition of Energy

Energy is available from a variety of sources. Consider an automobile. It uses chemical energy stored in the battery and in gasoline as its primary sources of energy. The stored energy is converted to other forms of energy as it is needed. Some of the energy conversions in starting a car and driving away are the following:

Chemical energy in battery → *electrical energy* in starter and spark plugs.
Electrical energy in starting motor → *mechanical energy* to crank engine.
Electrical energy in spark plug ignites gasoline/air mixture in piston.
Chemical energy from gasoline/air mixture → mechanical energy as gasoline explodes and drives pistons down.
Mechanical energy in engine → mechanical energy in wheels.
Mechanical energy from engine → electrical energy in alternator.
Electrical energy from alternator → *sound energy* for horn, *light (radiant) energy* for headlights.

Look over Figure 1-9. Note the change that is occurring in each transformation. **Energy** *is the ability to cause change or the ability to do work.* Chemistry deals mainly with the chemical energy stored in matter and with either the effects of energy on matter or the production of thermal energy, radiant energy, or electrical energy from matter.

All energy is classified as either kinetic energy or potential energy. *The energy of an object in motion is* **kinetic energy**. A baseball in flight, a car rolling down the street, and a book falling off the edge of the desk all have *kinetic energy*. The kinetic energy a moving object has depends on the mass of the object and its velocity. *The energy that an object has because of its position or composition is classified as* **potential energy**. Potential energy can be thought of as stored energy, waiting to be released. The energy in gasoline is chemical potential energy. Water held behind a dam has gravitational potential energy. When water is released from the dam, its potential energy is converted to kinetic energy as it falls to a lower level. At the lower level it possesses less potential energy.

Figure 1-9 Numerous energy conversions take place when a car is started.

Law of Conservation of Energy

Accounting for all of the energy present before and after an event is not as easy as keeping track of mass by weighing. But scientists who have done the necessary "energy bookkeeping" are confident in the **law of conservation of energy**: *Energy can be converted from one form to another, but it cannot be created or destroyed in ordinary chemical or physical changes.*

States of Matter

A piece of wood placed on a table retains its shape and volume. To change either the shape or volume of the piece of wood would require applying considerable external force. *The* **solid state** *is the state of any matter that has a definite shape and volume.*

Suppose some water is poured onto the table. It flows out over the surface because it does not retain its shape. A liquid takes the shape of its container. *The* **liquid state** *is the state of any matter that has a definite volume and the ability to flow and take the shape of a container.*

When an automobile tire is inflated with air, the air assumes the shape and volume of the fully inflated tire. If a blowout occurs, the air escapes and immediately expands in volume. *The* **gaseous state** *is the state of any matter that has neither a definite shape nor a definite volume.*

When materials are heated to extremely high temperatures, like those believed to exist in the center of the sun, they become a plasma. *The* **plasma state** *is a gaseous system composed of positively-charged particles and negatively-charged electrons.*

The volume of an object is the amount of space it occupies

Matter, Energy, and Change

Figure 1-10 The distance between molecules in a substance and the rigidity with which they hold their positions determine whether the substance is a solid, a liquid, or a gas. The plasma state of matter involves substances at such high temperatures (10^8 K) that they are a mixture of charged particles, electrons, and atoms or molecules. Like gases, particles in a plasma are relatively far apart.

Water is one of many substances that can exist in all three common states of matter. A block of ice is a solid. When it melts, it becomes a liquid. When liquid water evaporates, it becomes a gas.

Figure 1-10 illustrates how chemists explain the differences between solids, liquids, gases, and plasmas according to a model that states that matter is composed of tiny particles. The particles in a solid are closely packed and rigidly held in fixed positions, giving a solid its definite volume and shape.

The particles in a liquid are also closely packed. Because they remain close together, liquid volume changes very little with pressure or temperature. The particles in a liquid are not bound in fixed positions, however. They can "slide" past and around each other, which allows a liquid to take the shape of its container and flow when poured.

The particles in a gas under ordinary conditions of temperature and pressure are widely separated. Their positions have no order at all and, because they are constantly in motion, they quickly expand into any space available. Plasmas are somewhat like gases, but are composed of electrically charged particles.

Properties and Changes of Matter

Most chemical investigations are related to the properties of matter. **Properties** *are characteristics that enable us to distinguish one kind of matter from another.*

Physical properties Melting point and boiling point are examples of physical properties. *A* **physical property** *can be observed or measured without altering the identity of a material.* Many physical properties can be described both qualitatively and quantitatively. Qualitatively, for example, a box can be described as "large." With the aid of a ruler, the size can be given a quantitative description by specifying the height, width, and length of the box. Measuring the size of the box causes no change in its identity. It is the same box after it is measured.

Physical properties are either intensive or extensive. **Extensive physical properties** *depend on the amount of matter present and include mass, length, and volume.* **Intensive physical properties** *do not depend on the amount of matter present and include melting point, boiling point, density, ductility, malleability, color, crystalline shape, and refractive index.*

Any change in a property of matter that does not result in a change in identity is called a **physical change**. Cutting wire, grinding a solid, or allowing a gas to expand are physical changes. Observations would show that the substances changed in these ways retain their own characteristic properties, which is evidence that their identities are unchanged.

Changes of state are *changes between the four states of matter.* Changes of state are physical changes. Melting and boiling are changes of state, as are the reverse processes, freezing and liquefaction of a gas. These changes do not alter the identity of the substance.

Chemical properties In contrast to a physical property, a **chemical property** *refers to the ability of a substance to undergo a change that alters its identity.* Chemical properties are observed when substances are converted to new substances with different characteristic properties. The tendency of wood

Chemistry in Detail

Interconversion of Matter and Energy

During the early 1900s, a revolution occurred in how scientists view matter and energy. Experiments on the speed of light that were then in progress provided results that could not be explained by the theories and laws of the time. Albert Einstein (1879–1955) is the best known of many scientists who rose to the challenge of finding new explanations for the observed behavior of matter and energy.

In 1905, Einstein suggested that matter and energy are related. His now-famous equation

$$E = mc^2$$

states this relationship. In the equation, E represents

energy, m represents the amount of matter (or mass), and c is the speed of light. Many experiments have established the validity of this equation.

The equation shows that matter can be converted to energy and energy to matter. The conversion factor, c^2, has a very large value, 9×10^{16} meters^2 per second^2, which means that a very small amount of matter can be converted to a very large amount of energy. The world has seen evidence for this conversion in nuclear weapons and nuclear power plants, such as shown in the photograph to the left.

The evidence for the interconversion of matter and energy is taken into account in the *law of conservation of mass and energy*, which states: Matter and energy are interchangeable, and the total mass and energy in the universe is constant.

---

to burn in air is a chemical property. We find out that a material has this property by observing the material burning and being converted into different substances. When a match burns completely, the materials in the match no longer exist as they did—the particles of which each was composed are transformed into particles of different substances. Rusting iron, leaves changing color in the fall, silver that is tarnishing, milk turning sour—all these commonly observed phenomena are evidence of chemical changes.

Any change in which one or more substances are converted into different substances with different characteristic properties is described as a **chemical change** *or a* **chemical reaction**. *The substance or substances that undergo a chemical reaction are the* **reactants**. *The new substance or substances produced by a chemical reaction are the* **products**. In the chemical reaction between sodium, an explosive metal, and chlorine, a poisonous gas, sodium chloride—known to everyone as "salt"—is formed. The vastly different characteristic properties of the reactants and product are apparent in Figure 1-11.

$$\text{reactants} \rightarrow \text{product}$$
$$\text{sodium} + \text{chlorine} \rightarrow \text{sodium chloride (table salt)}$$

Arrows like those above are used to indicate *chemical* or, sometimes, *physical* changes and are read as "yields" or "gives."

Take a wooden safety match and break it into two pieces. This is a physical change. Strike the match and allow the wood to burn. This is a chemical change.

Figure 1-11 Chlorine *(left)* and sodium *(center)* react chemically to form sodium chloride *(right)*. While chlorine is a poisonous gas, and sodium an explosive metal, the product, sodium chloride, is ordinary table salt.

Figure 1-12 Sodium reacts in chlorine gas to form sodium chloride.

Figure 1-13 Bubbles of carbon dioxide form when vinegar is poured on sodium bicarbonate (baking soda).

A solution in which water is the solvent is referred to as an *aqueous solution*.

Indications of chemical reaction To know for certain that a chemical reaction has taken place requires evidence that one or more substances have been changed in identity. Absolute proof of such a change can be provided only by chemical analysis of the products. Certain easily observed changes strongly suggest that a chemical reaction has occurred, however.

1. Evolution of heat and light. A change in matter that releases energy as both heat and light is good evidence that a chemical reaction has taken place. If you burn gas for cooking in your house, you see evidence of the reaction of natural gas with oxygen every day. Figure 1-12 shows the result of the reaction between the elements sodium and chlorine.

Heat evolution without accompanying light also indicates a possible chemical reaction. But heat alone is not as certain an indication of chemical change because some physical changes also release heat.

2. Production of a gas. The evolution of gas bubbles when two substances or solutions are mixed is a possible indication of a chemical reaction. When sodium bicarbonate (baking soda) is added to vinegar, for example, bubbles of carbon dioxide gas form immediately, as shown in Figure 1-13.

3. Formation of a precipitate. Many interesting and useful chemical reactions take place between substances that are dissolved in water. If a solid appears after two solutions have been mixed, a chemical reaction most likely has occurred. *A solid that separates from a solution is known as a* **precipitate**. Figure 1-14 shows the result of mixing a solution containing sodium chloride (table salt) with a solution containing silver nitrate.

Energy and changes in matter Chemical and physical changes are accompanied by energy changes. Energy may be either released or absorbed. The reaction of hydrogen with oxygen to form water produces a large amount of heat, which can be thought of as another reaction product:

$$\text{hydrogen} + \text{oxygen} \rightarrow \text{water} + \text{heat}$$

The change of state from liquid to solid water releases heat and can be represented as

$$\text{liquid water} \rightarrow \text{ice} + \text{heat}$$

A process that releases heat is **exothermic**. The reaction shown in Figure 1-15 is exothermic.

According to the law of conservation of energy, the energy released as heat in an exothermic process must come from somewhere. Prior to the process, the energy exists in the reactants. In the reaction of hydrogen with oxygen, chemical potential energy stored in the reactants was released as kinetic energy. In the freezing of water, the water particles give up some of their potential energy to the surroundings in the form of heat.

Figure 1-14 A precipitate of silver chloride forms when a solution of silver nitrate is poured into a solution of sodium chloride.

Figure 1-15 Sulfur (yellow) and zinc (grey) are mixed at left. When the mixture is heated, an exothermic reaction takes place which gives off heat and light (right).

One way to think of energy changes in chemical reactions is to picture a boulder at the top of a hill, as shown in Figure 1-16. The boulder has potential energy because of its position; this is converted to kinetic energy as the boulder rolls downhill. The reactants in an exothermic chemical reaction also "fall down a potential energy hill," converting potential energy into an equal amount of kinetic energy released as heat.

Figure 1-16 As a boulder rolls down a hill (left), potential energy is released as shown in the diagram (middle). When hydrogen and oxygen combine to form water, potential energy is released as shown in the diagram (right).

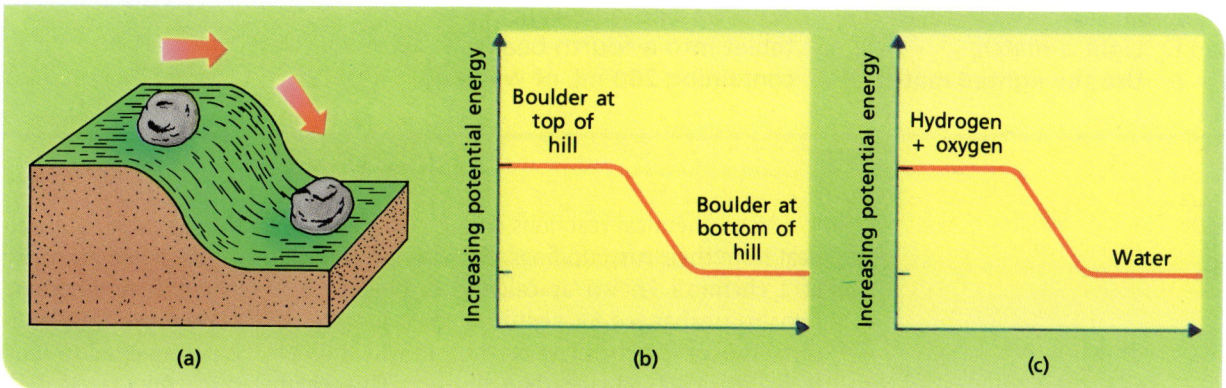

Matter, Energy, and Change

Desktop Investigation

Physical and Chemical Changes

Question
How can you distinguish a physical change from a chemical change?

Materials
matches
candle
chalk (calcium carbonate)
vinegar (dilute acetic acid)
soluble starch
water
tincture of iodine
Alka-Seltzer® tablet
phenolphthalein solution
milk of magnesia
  (suspension of magnesium hydroxide)
dilute ammonia solution
iron(III) chloride solution
four 250-mL beakers
mortar and pestle
two test tubes

Procedure
Record all your observations at the end of each step in a data table. **CAUTION:** Be careful when lighting the match; be sure to dispose of it properly. Remove all combustible materials from the work area. Wear safety glasses and an apron.
1. Light a match.
2. Use the lighted match to

ignite a candle sitting in a 250-mL beaker.

3. Powder a 3-cm piece of chalk with a mortar and pestle.

4. React the powdered chalk with 10 mL of vinegar in a second beaker.

5. Add 0.5 g of soluble starch to a third beaker containing 100 mL of water.

6. Add two drops of tincture of iodine to the starch solution.

7. Drop one Alka-Seltzer® tablet into a fourth beaker containing 200 mL of water.

8. Add three drops of phenolphthalein to a test tube containing 5 mL of milk of magnesia.

9. Add 5 mL of ammonia solution to a second test tube containing 5 mL of iron(III) chloride solution.

Discussion
1. List four different types of evidence you observed in this lab that suggest a chemical change occurred.

2. List three types of physical changes you observed in this lab.

Many chemical reactions do not take place unless they are able to absorb heat from their surroundings. For example, limestone is composed primarily of a chemical known as calcium carbonate. It is obvious that limestone remains unchanged for centuries when it is left undisturbed. Heating limestone causes the chemical reaction in which limestone is converted to lime (calcium oxide) plus carbon dioxide. This chemical reaction is used in industry for the production of *lime*, a chemical that is used in many processes such as water and waste treatment and paper production.

limestone   + heat ⟶   lime   + carbon dioxide
(calcium carbonate)   (calcium oxide)

Heat is also required to melt ice.

ice + heat → liquid water

A process that absorbs heat is **endothermic**. An endothermic reaction is equivalent to moving a boulder from the bottom to the top of a hill. It only happens with the input of energy.

Section Review

1. What is inertia?
2. Distinguish between *(a)* weight and mass *(b)* kinetic and potential energy *(c)* extensive and intensive physical properties of matter *(d)* physical and chemical properties of matter *(e)* physical and chemical changes *(f)* exothermic and endothermic processes.

1.3 Classification of Matter

Matter exists in countless forms. However, each sample of matter can be classified as one of two types: a pure substance or a mixture. Pure substances are either elements or compounds, both of which have uniform composition and properties throughout a given sample and from one sample to another. Some mixtures also have compositions and properties that are uniform throughout a given sample, but the composition and properties of a mixture can vary from one sample to another. The composition and properties of many mixtures are not uniform, that is, they vary from one part of the mixture to another. All matter can thus be classified according to the uniformity of composition and properties of a given sample.

Mixtures

Most samples of matter are mixtures. The air is a mixture of oxygen, nitrogen, carbon dioxide, other gases, and water. Brass is a solid mixture of copper and zinc. Gasoline is a mixture that may contain only a few or as many as 20 different substances. Grapefruit is a mixture of about 2000 different substances. Some other familiar examples of mixtures are granite, vegetable soup, milk, soft drinks, and sugar syrup.

A **mixture** *is a combination of two or more kinds of matter each of which retains its own composition and properties.* The different kinds of matter in a mixture can be separated by physical means. The properties of a mixture are a combination of the properties of the different kinds of matter it contains. Unlike a pure substance, the composition of a mixture must be specified because it can contain varying amounts of different kinds of matter.

Mixtures may be either heterogeneous or homogeneous. *In a* **heterogeneous mixture**, *the composition and properties are not uniform—they differ*

---

1. Resistance to change in motion
2. *(a)* Weight: measure of gravitational attraction for matter; mass: measure of the quantity of matter present *(b)* Kinetic: energy of motion; potential: stored energy *(c)* Extensive properties depend on the amount of substance present; intensive properties are independent of the amount of substance present. *(d)* Physical properties can be measured or observed without altering the identity of the material; chemical properties are related to chemical changes that alter the identity of the material. *(e)* Physical changes are changes in the properties of matter that do not result in a change in identity; chemical changes are changes in which one or more substances are converted into different substances with different characteristic properties. *(f)* Exothermic refers to a release of energy; endothermic refers to the absorption of energy.

SECTION OBJECTIVES

- Distinguish between homogeneous and heterogeneous matter.
- Distinguish between a pure substance and a mixture.
- List the types of homogeneous matter.
- State the law of definite composition and explain its meaning for analysis of chemical compounds.

Have samples of pure substances and mixtures (solutions, compounds, and elements) available on the desk.

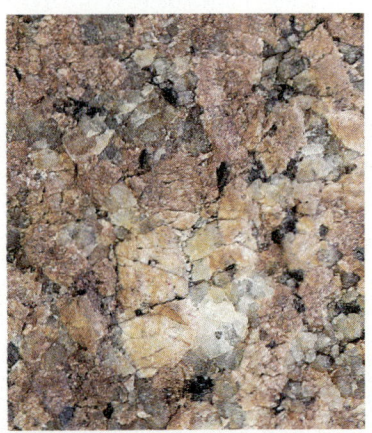

Figure 1-17 Granite is a mixture of different minerals such as quartz, mica, and feldspar.

from point to point in the mixture. Vegetable soup and granite are heterogeneous mixtures. Granite, shown in Figure 1-17, is a mixture of different minerals. Physical examination reveals that granite consists of nonuniformly distributed matter identified as the minerals quartz, mica, and feldspar. Crystals of each mineral have their own appearance, composition, and properties. Some materials must be examined quite closely to reveal that they are not uniform. Milk, for example, appears to be the same throughout, but under a microscope, suspended globules of fat can be seen.

Portions of matter that have both the same chemical properties and the same physical properties are called a **phase**. Phases might be different states of matter, such as ice and liquid water. Phases also might have the same state, such as the different crystalline phases in granite.

Other mixtures, such as the air we breathe or the beverages we drink, do not have distinct regions. These are homogeneous mixtures. *In a* **homogeneous mixture**, *the composition and properties are uniform throughout the mixture.* Such mixtures are the same from point to point. Consider the result of stirring some sugar into a glass of water. As the sugar dissolves, it spreads throughout the water and seems to disappear. Examination under a microscope would fail to reveal phases of different substances. Each teaspoonful of the mixture contains the same amount of dissolved sugar and will taste as sweet as the next teaspoonful.

Homogeneous mixtures are also called **solutions**. Solutions are found in the gaseous and solid states as well as in the liquid state. A sample of air is a gaseous solution; a sample of sea water is a liquid solution; a sample of brass is a solid solution.

The components of any mixture can be separated from each other by physical methods. That is, no chemical changes take place when mixtures are separated. The separated substances have the same characteristic properties and identities as before they were mixed together. Separating the components of a mixture may not always be as easy as it is for the mixture shown in Figure 1-18, but it is always possible. A sugar and water solution can be separated by carefully evaporating the water and collecting it. The crystalline sugar would be left behind, as shown in Figure 1-19. How many ways can you think of to separate a mixture of salt and pepper?

Pure Substances

Some kinds of matter display the same properties throughout. A teaspoonful of sugar taken from any spot in a large sugar bowl has the same properties, such as melting point and sweet taste, as a sample taken from any other spot in the bowl. The same properties would be found in any teaspoonful of pure sugar, no matter what its source. Chemical analysis shows that these kinds of matter have the same composition throughout a sample. Such kinds of matter are known as pure substances. A **pure substance** is a *homogeneous sample of matter that has the same composition and properties, whatever its source.* Examples include water, sugar, oxygen, silver, the graphite in a pencil, and the mercury in a thermometer. Each of these substances differs from one another in composition and can be identified by their physical and chemical properties. Pure substances are similar to homogeneous mixtures in that they both have uniform compositions throughout. But each pure substance always has the same composition in every sample, while the

Figure 1-18 The components of a mixture can be separated by using differences in their physical properties. Here a mixture is separated by filtration after mixing it in a liquid in which one component (but not the other) is soluble.

composition of a homogeneous mixture can vary from sample to sample. Pure substances differ from mixtures in three major ways.

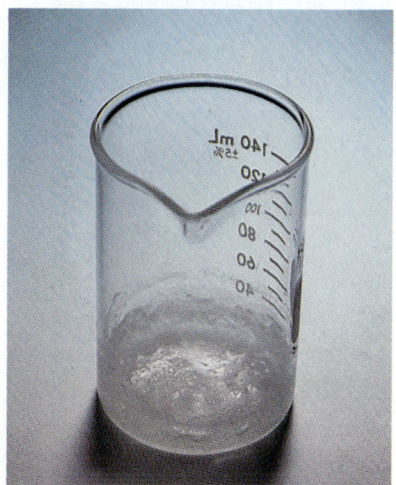

Figure 1-19 Sugar is added to water (left) to form a homogeneous mixture or solution (middle). When the water is evaporated the sugar is left behind (right).

1. Every sample of a given pure substance has exactly the same characteristic properties. Pure substances can be identified by determining their characteristic physical and chemical properties. The properties of a mixture of two substances vary with the amounts of each that are present in the mixture.

2. Every sample of a given pure substance has exactly the same composition. Unlike mixtures, which have variable compositions, pure substances have fixed compositions. Pure water is always composed of 11.2% hydrogen combined with 88.8% oxygen by mass.

3. A pure substance cannot be separated into other substances without changing its identity and properties. A pure substance cannot be broken down by any process that is based on a physical property. For example, when water is evaporated, filtered, melted, or frozen, it remains water. To separate water into the elements hydrogen and oxygen that compose it, a chemical change is necessary. As shown in Figure 1-20, hydrogen and oxygen—substances totally unlike water in their properties—can be produced when electrical current is passed through a sample of water.

Elements and chemical compounds Pure substances are either elements or compounds. Among the first questions a chemist asks in identifying an unknown material are, "Is this a mixture or a pure substance?" and "What is the composition of this material?" Experiments must be performed to determine whether the material can be divided into other substances by either physical or chemical means.

Early in the history of chemistry, it was discovered that certain substances could not be separated into other substances by *any* means. These substances became known as "elements." A few elements, including gold, silver, copper, and sulfur have been known since ancient times because they are found as pure substances in nature. Eventually, about ninety elements were found to exist naturally on our planet. In recent times, other elements have been made artificially.

Figure 1-20 Water can be decomposed into hydrogen and oxygen by electrolysis. In the apparatus shown above hydrogen forms at the negative electrode (left) and oxygen forms at the positive electrode (right).

Matter, Energy, and Change

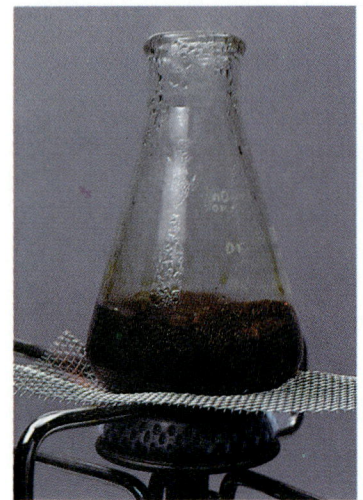

Figure 1-21 Sugar decomposes into carbon and water when heated. A small quantity of sugar heated in a flask melts and turns brown. Eventually only a black carbon residue and water droplets remain.

Figure 1-22 The classification of matter.

At the time this book is being written, a total of 109 elements are known. *An **element** is a pure substance that cannot be decomposed by ordinary chemical change.* An element is composed only of one type of matter. Pure gold, for example, is 100% gold. All matter found on earth is composed of the naturally occurring elements.

Most pure substances are not elements, but chemical compounds. Water is a chemical compound. In Figure 1-20 the process by which water can be broken down by a chemical reaction to the elements hydrogen and oxygen is shown. We say that, in water, hydrogen and oxygen are "chemically combined," indicating that they can be separated only by a chemical reaction. *A **chemical compound** is a pure substance that can be decomposed into two or more simpler substances by a chemical change.*

Sugar is also a chemical compound. If a small quantity of sugar is heated in a flask as shown is Figure 1-21, the sugar first melts and changes color. Eventually, only a black residue remains and a few drops of a clear, colorless liquid can be seen near the opening of the flask. The black substance is carbon, an element, and the colorless liquid is water. The sugar has been broken down chemically into two simpler substances, an element and a compound.

A landmark in the history of chemistry was the formulation of the **law of definite composition**: *A chemical compound contains the same elements in exactly the same proportions by mass regardless of the size of the sample or source of the compound.* Experiments show that pure sugar is composed of 42.1% carbon, 51.4% oxygen, and 6.5% hydrogen by mass. *Every pure chemical compound consists of two or more elements combined chemically in definite percentages.* The law of definite composition was formulated in

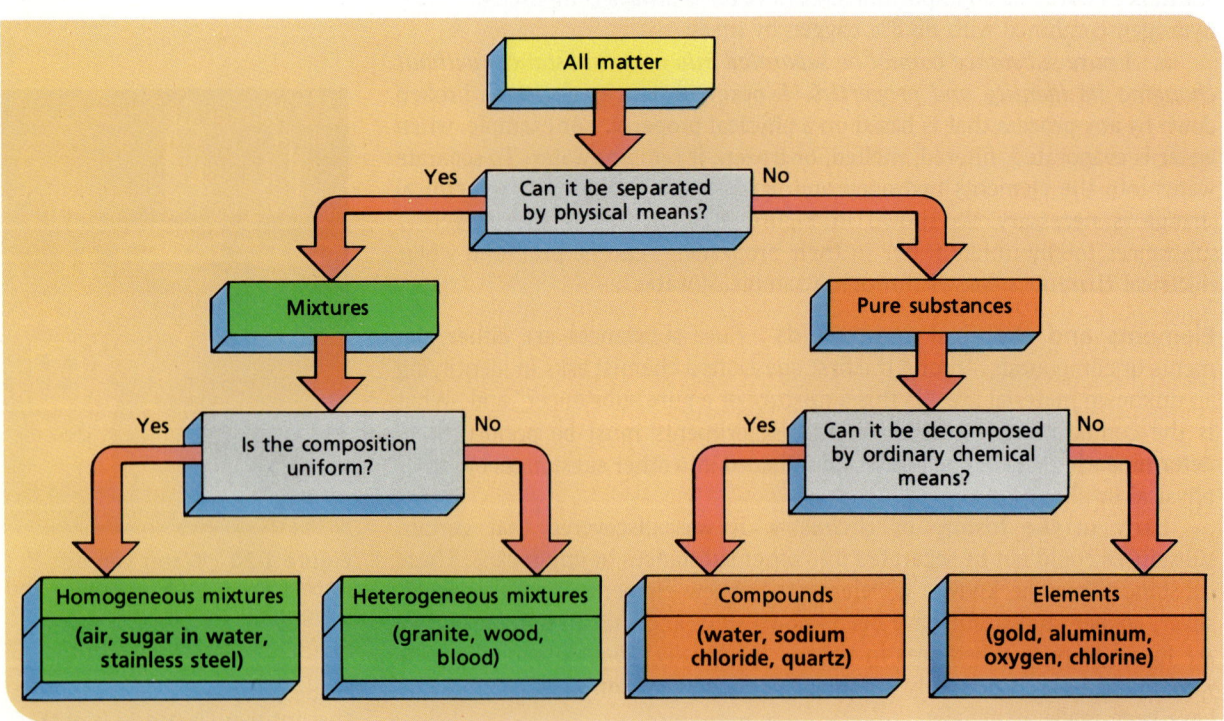

1799 by Joseph Louis Proust (1754–1826), who showed that copper carbonate ($CuCO_3$), whether prepared in the laboratory or obtained from natural sources, always contains the same elements—copper, carbon, and oxygen—in the same proportion by mass. Analytical chemists are able to identify the types and quantities of elements combined in any chemical compound. Figure 1-22 summarizes the classification of matter.

Section Review

1. Distinguish between: *(a)* heterogeneous and homogeneous matter *(b)* a pure substance and a mixture.
2. Identify two types of homogeneous matter.
3. State the law of definite composition.

1.4 The Chemical Elements

The chemical elements are the building blocks of matter. Anyone who studies or uses chemistry, therefore, must be familiar with their properties. Each element has its own characteristic properties. Early chemists made many attempts to organize the elements into groups with similar properties. Eventually, a table was devised in which all elements having similar properties could be grouped together in the same column.

Introduction to the Periodic Table

Figure 1-23 is a modern version of this table, known as the Periodic Table. Each block contains the name of an element and the symbol for its name. *The vertical columns of elements in the Periodic Table are referred to as* **groups** *or* **families** *and are numbered consecutively from 1 to 18.* The elements of Group 1, for example, are lithium (Li), sodium (Na), potassium (K), rubidium (Rb), cesium (Cs), and francium (Fr). Elements within a group have many similar properties.

The horizontal rows of elements in the Periodic Table are referred to as **periods** *and are numbered 1 to 7 from the top down.* Elements close to each other in the same period are more similar than those further apart. For instance, the properties of the elements of Group 15 vary significantly from those of Group 1 but are somewhat similar to the properties of Group 16 elements. The two rows of elements placed apart from the others that appear below the table belong to Periods 6 and 7. They are placed at the bottom simply to keep the table conveniently narrow.

Symbols for the Elements

The majority of elements have one- or two-letter symbols derived from their names, such as H for hydrogen, O for oxygen, or Si for silicon. Table 1-1 is a list of the names and symbols of some common elements.

In Period 7 of the table there are some three-letter symbols. The elements represented by these symbols are all synthetic and have been created for the first time in recent years. These symbols are temporary until the International Union of Pure and Applied Chemistry (IUPAC) gives them official names.

The Swedish chemist Jöns Jakob Berzelius (1779–1848) introduced a system in which elements were symbolized by letters. In earlier times, pictorial symbols were used for elements, or what were thought to be elements.

In 1799, the French chemist Joseph Louis Proust (1754–1826) first stated the law of constant proportions, or the law of definite composition.

1. *(a)* Homogeneous matter is uniform throughout; heterogeneous is not. *(b)* A mixture can be subdivided into two or more components by physical means; a pure substance cannot.
2. Solutions and pure substances.
3. A chemical compound contains the same elements in exactly the same proportions by mass regardless of the size of the sample or the source of the compound.

SECTION OBJECTIVES

- Name the common elements, given their symbols.
- Write the symbols of the common elements, given their names.
- Describe the arrangement of the periodic table.
- Discuss the differences between metals, nonmetals, and metalloids.

TABLE 1-1 ELEMENTS WITH SYMBOLS BASED ON OLDER NAMES

Modern Name	Symbol	Older Name
antimony	Sb	stibium
copper	Cu	cuprum
gold	Au	aurum
iron	Fe	ferrum
lead	Pb	plumbum
mercury	Hg	hydrargyrum
potassium	K	kalium
silver	Ag	argentum
sodium	Na	natrium
tin	Sn	stannum
tungsten	W	wolfram

Periodic Table of the Elements

[Periodic table figure showing elements classified as Metals, Nonmetals, Metalloids, and Noble Gases, organized by Groups 1-18 and Periods 1-7, with the Lanthanide Series (elements 58-71) and Actinide Series (elements 90-103) shown separately below.]

Figure 1-23
The Periodic Table.

Types of Elements

General similarities among the properties of different elements provide one way of classifying them. Three general classes of elements are the metals, the nonmetals, and the metalloids. See Figure 1-23. Notice that most of the elements are metals.

Metals You would be able to recognize many metals by characteristic properties with which you are already familiar. Most metals have a characteristic *luster*—that is, they shine like silver. Copper and gold have a metallic luster and they also have distinctive colors. Metals are good reflectors of heat and light. Most metals are *ductile*; they can be drawn out into fine wire. As illustrated in Figure 1-24, most are also *malleable*; they can be hammered or rolled into thin sheets. Many have a high *tensile strength*; they resist breaking when pulled. Different metals have the metallic properties of ductility, malleability, and high tensile strength to varying degrees.

The property that most clearly distinguishes metals from other types of elements is the ease with which they conduct heat and electricity. *A* **metal** *is an element that is a good conductor of heat and electricity.* All metals except for mercury are solids under ordinary conditions. Mercury is a liquid at room temperature and pressure.

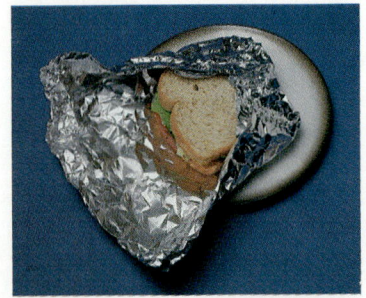

Figure 1-24 Because aluminum foil is malleable, it resists breaking when reshaped and can be used to wrap sandwiches.

24   Chapter 1

Nonmetals In contrast to metals, nonmetals are very poor conductors of heat and electricity. Under ordinary conditions, some nonmetals—such as oxygen, chlorine, and nitrogen—are gases. Bromine, however, is a dark red liquid. Those nonmetals that are solids, carbon and sulfur, for example, are neither malleable nor ductile but instead are brittle.

A **nonmetal** *is an element that is a poor conductor of heat and electricity.* As shown in Figure 1-23, there are far fewer nonmetals than metals.

Metalloids In the Periodic Table, the metals appear at the left and the nonmetals are on the right. Between them lies a small class of elements known as the "metalloids," which are somewhat like metals, but not entirely like them. In appearance, metalloids such as arsenic and antimony, shown in Figure 1-25, are lustrous metals. But in many properties, they are more like nonmetals. Under ordinary conditions, all are solids.

The ordinary conditions referred to are room temperature (25°C) and the pressure exerted by the atmosphere at sea level (1 atm). You will learn more about pressure and "standard" conditions in Chapter 11.

The metalloids are also referred to as *semimetals* or *semiconducting elements.* The property of semiconductivity is discussed in Section 26.3.

Figure 1-25 Elemental antimony (left) and arsenopyrite (right), a mineral which contains the element arsenic.

Silicon and germanium are much better conductors of electricity than the nonmetal sulfur, but much poorer conductors than the metals silver and copper. Silicon, germanium, and the other metalloids are *semiconductors.*

A **metalloid** *is an element that has some properties characteristic of metals and others characteristic of nonmetals, and is a semiconductor.* The microelectronics revolution, which has resulted in desktop computers, handheld calculators, digital watches, and hundreds of other items that are changing our lifestyles, is based upon semiconducting elements.

Noble Gases The elements in Group 18 are known as noble gases. They do not fit into other classes because of their nonreactive natures.

Gallium and cesium are two low melting metals. Under normal lab conditions on a warm day, both would be liquids. Cesium melts at 28.5°C and gallium at 29.78°C. Since body temperature is 37°C, samples of the two metals (sealed in glass vials) would melt if held in one's hand.

Two Typical Elements

The properties of the elements and their compounds are the basis for what is sometimes called "descriptive chemistry." Descriptive chemistry is involved more with observable facts than explanations. What does an element look like? How does it behave when combined with other substances? What are its physical properties? What are some of the ways in which its properties can be put to use? Answers to questions such as these are at the heart of the descriptive chemistry of the elements.

As an introduction to descriptive chemistry, we will look at two elements in the following paragraphs. They are chosen from two of the three general classes of elements shown in the Periodic Table in Figure 1-23. Copper is a well-known metal. Phosphorus is a nonmetal that has widespread applications.

Copper—a metal Copper is found naturally both as a free element and in minerals such as those shown in Figure 1-26, in which it is combined with sulfur (chalcocite and chalcopyrite) or with oxygen or a compound of oxygen and carbon (cuprite, malachite). Although it is needed in extremely small amounts, copper is an essential element in the human diet. In elemental form, it is not known to be toxic.

 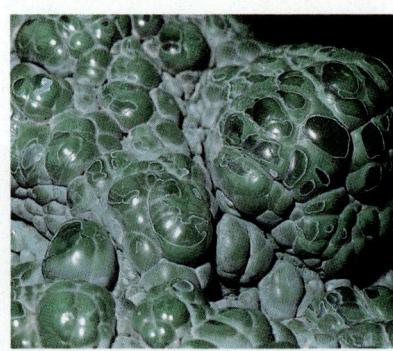

Figure 1-26 Copper is found naturally in a number of different forms. Two forms, free element (left) and combined with oxygen as *malachite* (right), are shown here.

Copper has a characteristic reddish color, a metallic luster, high boiling and melting points, and other properties considered typical for metals. It is readily formed into wire, sheets, and tubing. Copper allows electricity to pass through it with little loss of energy. This makes copper the metal of choice for electrical wiring, its principle commercial application. The resistance of copper to weathering and corrosion makes it desirable for water pipes, roofing materials, and other applications in plumbing and construction.

Copper remains unchanged in pure, dry air at room temperature. When heated, it reacts with the oxygen in the air. Copper also reacts easily with sulfur and the elements of Group 17 in the Periodic Table, which includes chlorine and bromine. Copper that is exposed to the weather for a long time reacts with oxygen, carbon dioxide, and sulfur compounds to produce a green coating on the copper surface. This coating is responsible for the green color of the Statue of Liberty and other copper monuments. Copper is mixed with other metals to form many useful alloys, including both brass and bronze.

The following nonmetals are solid at room temperature.

carbon
phosphorus
sulfur
iodine
astatine

Phosphorus—a nonmetal Phosphorus is one of the five nonmetals that are solid rather than gaseous at room temperature. Pure phosphorus is known in two common forms that are quite different from each other in their properties. Red phosphorus, a dark red powder with a melting point (mp) of 597°C, remains relatively unchanged in air, is not toxic, and can be handled with reasonable safety. White phosphorus is a waxy, low-melting solid with a mp of 44°C, ignites spontaneously in air, causes serious burns when it comes into contact with the skin, and is extremely toxic. White phosphorus is usually stored under water, as shown in Figure 1-27.

Phosphorus is much too reactive to be found free in nature. It is present in huge quantities in deposits known as phosphate rock, that contain chemical compounds composed of phosphorus, oxygen, and calcium. All living things contain phosphorus, and require it as a nutrient. One of the major uses of phosphorus compounds is in agricultural fertilizers. For this purpose, phosphates can be manufactured directly from phosphate rock. In living things, biochemical compounds containing phosphorus are essential in the storage and production of energy.

Phosphates, phosphoric acid, and other chemical compounds containing phosphorus have many applications. Many phosphorus compounds are not toxic and are added to food products, including soft drinks and baking powders. Cleaning agents such as laundry detergents, dishwasher detergents, and soaps contain phosphorus compounds. Phosphorus compounds also play important roles in water softening and water purification. The flammability of phosphorus and certain phosphorus compounds allows them to be used in matches and fireworks. Interestingly, some phosphorus compounds are also excellent flame retardants and are used in fabric for clothing.

Section Review

1. In reference to the Periodic Table, what are *(a)* groups? *(b)* periods?
2. Write the chemical symbols for: *(a)* nitrogen *(b)* calcium *(c)* zinc *(d)* oxygen *(e)* sulfur *(f)* radon.
3. Name the element represented by the symbol: *(a)* He *(b)* Al *(c)* Br *(d)* Ni *(e)* Ag *(f)* Sn.
4. What are the three general classes of elements?

Chapter Summary

- Chemistry is a physical science with five major branches: organic chemistry, inorganic chemistry, physical chemistry, biochemistry, and analytical chemistry.
- The scientific method involves observing, generalizing, theorizing, and testing.
- Mass and weight measure different properties. Weight is a measure of the earth's gravitational attraction for matter. Mass is a measure of the quantity of matter.
- The law of conservation of mass and energy states that the total mass and energy in the universe is constant.
- A physical property can be observed or measured without altering the identity of a material.
- A chemical property refers to the ability of a substance to undergo a change that alters the identity of the substance.
- A chemical reaction is likely to have taken place if one or more of the following changes occur: evolution of heat and light, production of a gas, formation of a precipitate.
- A quantity of matter may be classified as either a pure substance or a mixture.
- Mixtures are either homogenous or heterogenous.
- Pure substances are either compounds or elements.
- The Periodic Table arranges the elements according to their properties. Elements with similar properties are found in the same column.

Figure 1-27 White phosphorus actually has a yellowish color and is stored under water since it ignites spontaneously in air. Red phosphorus is in the dish.

1. *(a)* Vertical columns
   (b) Horizontal rows
2. *(a)* N *(b)* Ca *(c)* Zn
   (d) O *(e)* S *(f)* Rn
3. *(a)* Helium *(b)* Aluminum
   (c) Bromine *(d)* Nickel
   (e) Silver *(f)* Tin
4. Metals, nonmetals, and metalloids

In some areas, the use of phosphates in household detergents has been limited to protect natural bodies of water. Waste phosphates that find their way into streams and lakes can cause excessive growth of algae and aquatic plants. This can harm other living things in these bodies of water.

Chapter 1 Review

Matter, Energy, and Change

VOCABULARY

changes of state (14)
chemical change or chemical reaction (15)
chemical compound (22)
chemical property (14)
chemistry (5)
element (22)
endothermic (19)
energy (12)
exothermic (17)
extensive physical property (14)
gaseous state (13)
groups or families (23)
heterogeneous mixture (19)
homogeneous mixture (20)
hypothesis (8)
inertia (10)
intensive physical property (14)
kinetic energy (12)
law of conservation of energy (13)
law of conservation of mass (12)
law of definite composition (22)
law (8)
liquid state (13)
mass (11)
matter (11)
metal (24)
metalloid (25)
mixture (19)
nonmetal (25)
period (23)
phases (20)
physical change (14)
physical property (14)
plasma (13)
precipitate (16)
potential energy (12)
products (15)
properties (14)
pure substance (20)
qualitative information (8)
quantitative information (8)
reactants (15)
scientific method (7)
solid state (13)
solution (20)
theory (9)
weight (11)

REVIEWING CONCEPTS

1. a. What is science?
   b. How does science affect the way we live? (1.1)
2. a. What is chemistry?
   b. Give three examples of the many applications of chemistry in everyday life. (1.1)
3. Distinguish between the dictionary definition of the word *chemical* and the definition implied in such advertising slogans as, "All natural—no chemicals added." (1.1)
4. a. What is the meaning of the term *law* as used in science?
   b. How are laws in science generally expressed? (1.1)
5. Define:
   a. measuring   c. hypothesis
   b. experimenting d. theory (1.1)
6. a. Explain how inertia is demonstrated in both moving objects and objects at rest.
   b. What is the relationship between inertia and the quantity of matter in an object? (1.2)
7. a. What is matter?
   b. Among the following, which are examples of matter: sand, paper, heat, air, sound, light, and chalk?
8. State the law of conservation of mass. (1.2)
9. a. What is energy?
   b. Describe chemistry in terms of energy. (1.2)
10. Identify the following as examples of either kinetic or potential energy:
    a. a moving car
    b. a coiled spring
    c. food to be eaten
    d. water flowing downstream (1.2)
11. State the law of conservation of energy. (1.2)
12. Describe the model that explains the differences between solids, liquids, and gases. (1.2)
13. a. What is a physical property?
    b. List three examples of physical properties. (1.2)
14. a. Distinguish between intensive and extensive physical properties.
    b. Give three examples of each. (1.2)
15. What is a chemical property? (1.2)
16. a. Distinguish between physical and chemical changes.
    b. List three examples of each. (1.2)

17. Classify each of the following as a physical or chemical change:
    a. melting ice
    b. chopping wood
    c. burning paper
    d. dissolving salt in water
    e. boiling water (1.2)
18. List three observable changes that would generally indicate that a chemical reaction has occurred. (1.2)
19. a. Distinguish between exothermic and endothermic processes.
    b. Give an example to illustrate each. (1.2)
20. List and define the two basic categories into which all matter can be classified. (1.3)
21. List three ways in which pure substances differ from mixtures. (1.3)
22. a. What is the basic difference between an element and a compound?
    b. Identify each of the following as elements or compounds: aluminum, sugar, oxygen, copper, and water. (1.4)

PROBLEMS

▶ **Matter and Energy**

1. Identify each of the following as exothermic or endothermic processes:
   a. the melting of ice
   b. the burning of charcoal
   c. photosynthesis
   d. the vaporization of water
2. The formula for determining the kinetic energy of a moving object is: $KE = \frac{1}{2}mv^2$, where m = the mass of the object and v = its velocity. Based on this equation, compare the relative kinetic energies of objects A and B given the following information:
   a. Object A has a mass twice that of B, but both are moving with the same velocity.
   b. Object A has a velocity twice that of B, but both have the same mass.
   c. Both the mass and velocity of object A are twice that of object B.

▶ **Classifying Matter**

3. On the basis of general appearance, identify each of the following examples of matter as heterogeneous or homogeneous:
   a. notebook paper
   b. pencil lead
   c. fresh-squeezed lemonade
   d. silver
4. Given samples of the following pairs of substances, name two physical properties that could be used to distinguish between each of the paired materials:
   a. sand and salt
   b. copper and silver
   c. cork and lead
   d. water and motor oil

CRITICAL THINKING

1. **Inferring Conclusions** Water purification is one chemical process that most people take for granted. Describe possible implications to society if water purification was not governmentally controlled.
2. **Evaluating Ideas** On earth, if two objects have the same mass, they have the same weight as long as they are in the same location. Evaluate the validity of this statement if the objects were on another planet in the solar system.

CROSS CONNECTIONS

1. **Architecture/Engineering** You are elected to chair a committee to propose a building plan for a new concession stand to be built near the football stadium. Part of your job will be to evaluate various building materials such as wood, brick, and metal for their strength and durability. Prepare a table to compare the physical properties of the building materials your committee will consider suitable for the project.

RESEARCH AND WRITING

1. Prepare a report concerning of the role of semiconducting elements in the microelectronics revolution.

CHAPTER 2
Measurement and Solving Problems

PLANNING CHART

Approximate Time: 11/12 days

	SECTION/TEXT FEATURES		RESOURCES	
BASIC	**2.1 UNITS OF MEASUREMENT**	**31**	*Teacher's Resource Binder*	
	Desktop Investigation	38	Problem Solving Worksheets	
	Section Review	40	Section Review Worksheets	5
	Chapter Review	60	*Other Resources*	
	• Questions 1-14, Problems 1–4		Blackline Masters 3, 4, 5, 6	
AVERAGE	**2.2 HEAT AND TEMPERATURE**	**40**	*Teacher's Resource Binder*	
	Demo 1	30B	Problem Solving Worksheets	
	Microscale Investigation 3	834	Lab Experiments 5, 6	47, 51
	Section Review	45	Section Review Worksheets	6
	Chapter Review	60	*Other Resources*	
	• Questions 15–20, Problems 5–11		Blackline Master 7	
AVERAGE	**2.3 USING SCIENTIFIC MEASUREMENTS**	**46**	*Teacher's Resource Binder*	
	Section Review	55	Problem Solving Worksheets	
	Chapter Review	60	Section Review Worksheets	7
	• Questions 21–28, Problems 12–17		*Other Resources*	
			World of Chemistry Video 3	
			Blackline Masters 8, 9	
BASIC	**2.4 SOLVING QUANTITATIVE PROBLEMS**	**55**	*Teacher's Resource Binder*	
	Section Review	59	Problem Solving Worksheets	
	Chapter Review	60	Section Review Worksheets	8
	• Questions 29–31, Problems 18–23			
	CHAPTER END ACTIVITIES		*Teacher's Resource Binder*	
	Critical Thinking, Cross Connections, Research and Writing, Projects		Test Booklet	5, 7
			Computer Test Generator	

Teaching Strategies

2.1 Units of Measurement

Bring the metric system home to the students by displaying a variety of household products. Most household products now carry both English and metric equivalents. Have the students examine their pantries and medicine cabinets at home and bring in other examples.

Stress from the beginning that chemists do not work with pure numbers. To report 125.3 is meaningless unless one adds a unit of measurement, 125.3 grams, for example. Any number measured or derived in the chemistry course will consist of a number *and* a unit of measurement. Students must get used to always writing the units.

If time permits you may wish to give a historical background to the metric system and to the International System of Units. In any case, list and use each of the three fundamental SI units discussed. Have the students relate the size of each unit to something with which they are familiar. For example, a 2 meter-tall basketball player would probably play center on the school's basketball team.

Introduce the factor-label method by having students convert between units they are accustomed to using, e.g. yard to inches. Most can tell you that there are 120 in. in 10 ft., but they don't know how to set it up in the factor-label form. Do it for them. Move on to a conversion from days to seconds. Show that if the problem is set up correctly, the unit found in a numerator is cancelled by the unit in the following denominator. The unit remaining is the unit requested by the problem. For example, to convert 1 day to seconds, set up the following:

$$1 \text{ day} \times \frac{24 \text{ hours}}{1 \text{ day}} \times \frac{60 \text{ minutes}}{1 \text{ hour}} \times \frac{60 \text{ s}}{1 \text{ minute}} = 86\,400 \text{ s}$$

Not only are you demonstrating the factor-label method, you are also introducing the concept of exact conversion factors and building a case for the four-step quantitative problem-solving method that will be introduced in Section 2.4. Explain also the international system used for writing numbers. Commas are no longer used to separate groups of three digits. A space is used instead. Many students will resist having to write the units. They consider it a waste of time when problems are relatively simple. Require them to write a unit after every number.

To illustrate the importance of units, write the following on the chalkboard: $36 + 16 = 1$. When students protest say, "Oh, I left out the units, but that doesn't matter, does it?" Then add the units: 36 weeks + 16 weeks = 1 year.

Continue practicing conversions as you describe the derived SI units. Explain to the students that the text is using a set of modified SI units. Some of the SI units are too large to be of practical use in chemistry. For example, few chemists ever encounter volumes measured in cubic meters. The following calculation may bring this home. For water:

$$1 \text{ m}^3 = 1\,000\,000 \text{ cm}^3 = 1\,000\,000 \text{ g} = 1000 \text{ kg}$$
$$1000 \text{ kg} = 1 \text{ metric ton} = 2200 \text{ lbs.}$$

Have the students set up the proper conversion factors. Most students will need help in converting cubic meters to cubic centimeters. Explain that everything in the conversion factor must be cubed, both the number and the unit.

$$1 \text{ m}^3 = 100 \text{ cm}^3$$
$$(1 \text{ m})$$
$$= 100 \text{ cm} \times 100 \text{ cm} \times 100 \text{ cm} = 1\,000\,000 \text{ cm}^3$$
$$(1 \text{ m}) \quad (1 \text{ m}) \quad (1 \text{ m})$$

The same problem arises when discussing density. The SI unit of kg/m^3 is not commonly used. We shall use g/cm^3 for solids and liquids and g/L for gases.

Desktop Investigation: Density of Pennies

Proceed with this as a way of bringing home the concept of density. Pennies minted prior to 1982 were made of an alloy containing 95% copper and 5% zinc. Penny hoarding as the price of copper rose in the 1970's forced Congress to adopt a change in which pennies would be made of a barrel composed of an alloy of 99.2% zinc and 0.8% copper onto which was electroplated a minimum thickness of 0.000 2 inches of copper. The overall composition of these pennies is 97.6% zinc and 2.4% copper. The pre-1982 pennies have a mass of 3.1 grams while the post-1982 pennies only a mass of 2.5 grams. When doing the experiment, stay away from 1982 pennies. These vary in composition.

For students who have difficulty with density problems, introduce the following mnemonic aid.

From the triangle students can derive the following relationships:
$D = m/v$ $m = Dv$ $v = m/D$

To work a density problem, two of the three variables must be known. The third, the unknown, can then be calculated.

A density column of ordinary substances, such as corn syrup, glycerin, water, salad oil, alcohol, and solids such as rubber erasers, plastic, cork, etc. would be good to use. Another simple, yet effective density "interest center" is a cylinder (1000 mL) 3/4 filled with a saturated NaCl solution. Float an egg in the salt water, and then carefully fill with tap water. The egg will appear to be suspended and will remain in position for weeks.

2.2 Heat and Temperature

To help students distinguish the difference between heat and temperature, perform the following demonstration as an introduction to the topic.

Demonstration 1: Heat vs Temperature

PURPOSE: To help students distinguish between heat and temperature

MATERIALS: Four 1–L beakers half filled with water at approximately the following temperatures: beaker 1 at 0°C (ice water), beakers 2 and 3 at 20°C (approx. room temperature), and beaker 4 at 40°C; four Celsius thermometers, paper towels, two graduated cylinders, two 100-g blocks of ice.

PROCEDURE: Invite two or three volunteers to come to the front of the room. Do not tell them the temperature of the water in each beaker. Have each student place one hand in beaker 1 and their other hand in beaker 4. Next have them remove their hands from beakers 1 and 4 and place the hand that was in beaker 1 into beaker 2 and the hand that was in beaker 4 into beaker 3. On the basis of this data, ask them how they think the temperatures of beakers 2 and 3 compare with one another. Most will indicate that beaker 2 is warmer when in fact they are the same. Use the thermometers to show that the temperatures of beakers 2 and 3 were indeed the same.

Pour 10 mL of boiling water (100°C) onto one of the blocks of ice, pour 100 mL of boiling water on the other block. Though the water is the same temperature in both instances, the effect on the blocks of ice is much different.

In discussing the difference between heat and temperature, point out that temperature is an intensive physical property since it measures the kinetic energy of particles while heat is an extensive property because it is the sum total of the kinetic energy of the sample and depends on the size of the sample. Using the examples in the text, define temperature, heat, and heat flow, especially as the latter applies to the body. Heat flowing into the body is sensed as warm, while heat flowing away from the body is sensed as cool.

To explain the various temperature scales, draw three thermometers on the board or on a transparency. Label each as Fahrenheit, Celsius, and Kelvin. Draw two horizontal lines, one representing the freezing point of water and the other the boiling point of water. Show that there are 100 divisions between the freezing and boiling points on the Celsius and Kelvin scales and 180 divisions on the Fahrenheit scale. Then 100 Celsius degrees must equal 180 Fahrenheit degrees.

Although the students are most familiar with the calorie, the text will be using the SI unit of heat, the joule. Emphasize that the calorie is a very small unit and the joule an even smaller unit.

Remind the students that temperature is an intensive property while heat is extensive. Heat lost or gained by objects can be determined if one knows the change in temperature, the mass of the object, and a new intensive property known as the specific heat of the object. Metals which conduct heat very well have low specific heats while non-conductors, like the plastic handles on frying pans, have high specific heats.

2.3 Using Scientific Measurements

Use the dartboard analogy on page 46 to distinguish between precision and accuracy. Follow this discussion with a set of data representing masses of a beaker. Show how precision and accuracy apply to scientific measurements.

To help students comprehend significant figures, have them read measurements from meter sticks, graduated cylinders, and thermometers. In each case, they have to estimate between the two smallest marks on the instrument if they are to report the measurement to the limits of the measuring tool. Two students may read that final place differently.

The greatest problem students encounter in determining significant digits is deciding whether or not beginning or trailing zeros are significant. The following mnemonic may help:

If the number contains a decimal point, draw an arrow from the decimal point LEFT to RIGHT stopping at the first NON-ZERO digit. It and the remaining digits are significant. Add to the digits to the left of the decimal point. (To include a geographical twist: If a decimal point is *p*resent, draw an arrow from the *P*acific Ocean.)

If the number does not contain a decimal point, draw an arrow from RIGHT to LEFT stopping at the first NON-ZERO digit. It and the remaining digits are significant. (Continuing the geographical twist: If a decimal point is *a*bsent, draw an arrow from the *A*tlantic Ocean.)

Examples: 60.300 grams (5 significant figures)
86 000 miles/s (2 significant figures)

The rounding rule presented in the text will most likely be new to your students. They are accustomed to rounding number 5 or larger up to the next number. The rule that they will find strange applies only to rounding when the final digit is a 5. Rather than automatically rounding up, students must look to the digit preceding the 5, if it is an even digit drop the 5, if it is an odd digit round up. The following example may help clarify the issue.

If the numbers 10.5 and 11.5 were known to three significant digits, their average would be (10.5 + 11.5)/2 or 22.0/2 or 11.0. If, on the other hand, 10.5 and 11.5 were known only to two significant digits, should one round off the numbers and then take the average, the following would occur.

(a) Using the recommended system 10.5 would round to 10 and 11.5 would round to 12. The average of 10 and 12 is 11.

(b) Using the old system of always rounding up, 10.5 would round to 11 and 11.5 would round to 12. The average of 11 and 12 is 11.5 or, rounding, 12.

The method recommend by the text (rules 4 and 5) keeps large amounts of data from being skewed too high.

Go over the rules for adding and subtracting and multiplying and dividing numbers with significant figures. The students must be able to classify the problem they are working on as one (adding/subtracting) or the other (multiplying/dividing) before they can apply the correct rule. When adding and subtracting, the decimal points must line up. Perform the required operation and make sure that the answer is rounded off to the leftmost uncertain digit appearing in the original numbers. Multiplication and division is a bit easier. Simply determine the number of significant digits in each of the numbers being multiplied and/or divided. Make sure to round the answer to the number representing the fewest number of significant digits. When performing such calculations students will run across exact conversion factors, and they will try to round off numbers incorrectly. Point out the presence of exact conversion factors and that such factors do not limit the number of significant digits in the answer.

Students with a poor algebra background will usually encounter difficulties when working with exponential notation. You may wish to give a pre-test as a diagnostic tool. If students ask why they have to use scientific notation, write a number like Avogadro's number (6.02×10^{23}) on the board including all of the zeros. Write a very small number, one outside the range of most calculators, e.g. the mass of an electron (9.1×10^{-28} g). The only practical way to use such numbers is by expressing them in scientific notation.

2.4 Solving Quantitative Problems

The text proposes a four step method:
(1) analyze the problem
(2) plan the route to solve the problem
(3) compute the answer
(4) evaluate the answer

Stress that steps *(1)* and *(2)* are the key steps. They demand an understanding of the chemistry required to find the route to the solution. The final two steps are little more than arithmetic, yet they are important if one is to report the correct answer. Most students initially find the process too time consuming and too difficult. Work some examples for them and then allow them to work some of the practice problems while you observe. The more practice they get, the less difficult will later problem-solving chapters appear.

Conclude the unit by distinguishing between direct and inverse proportions. The definitions may appear strange to them. Two variables are directly proportional if dividing them gives a constant value, and two variable are inversely proportional if multiplying them gives a constant value.

References and Resources

Books and Periodicals

Asten, A.V. "Standards of Measurement," *Scientific American*, Vol. 218, No. 6, 1968.

Calder, Lord Ritchie. "Conversion to the Metric System," *Scientific American*, Vol. 233, No. 1, 1970.

Fields, Lawrence D. and Stephen J. Hawks. "Minimizing Significant Figure Fuzziness," *Journal of College Science Teaching*, Vol. XVI, No. 1, Sept./Oct. 1986, pp. 30–34.

Guymon, E. Park, et al. "Teaching Significant Figures Using a Learning Cycle," *Journal of Chemical Education*, Vol. 63, No. 9, Sept. 1986, pp. 786–787.

Norris, A.C. "SI Units in Physico-Chemical Calculations," *Journal of Chemical Education*, Vol. 48, No. 12, Dec. 1971, pp. 797–800.

Pinkerton, R.C. and C.E. Gleit. "The Significance of Significant Figures," *Journal of Chemical Education*, Vol. 44, No. 232, 1967.

Whitmer, Joan C. "Are Your Students Proportionality Literate?," *The Science Teacher*, Vol. 54, No. 8, Nov. 1987, pp. 37–39.

Computer Software

Groves, Paul, Science Skills—Disk 1 "Uncertainties and Measurements?," The Mole Company, Flinn.

Smith, S.G., R. Chabay, and E. Kean. Introduction to General Chemistry, Disk 9, The Metric System, COMPress.

Answers and Solutions

Reviewing Concepts

1. **a.** Scientists around the world use the International System of Units.
   b. The abbreviation "SI" is generally used.
   c. Many SI units were taken from the metric system.
2. **a.** 6500 **e.** 0.920 400
   b. 78 420 **f.** 127.800 00
   c. 207 891 **g.** 42 600.3500
   d. 0.3500
3.

Quantity	Unit	Symbol
length	meter	m
mass	kilogram	kg
time	second	s
temperature	kelvin	K
amount of substance	mole	mol
electric current	ampere	A
luminous intensity	candela	cd

4. Examples of approximately one-meter distances include the width of a doorway, the length of a baseball bat, and the distance that an adult can cover in one large step.
5. **a.** cm or dm **e.** m or dm
   b. m or dam **f.** km
   c. dm **g.** Gm or Mm
   d. mm **h.** μm
6. **a.** kg **b.** g **c.** mg **d.** g
7. The factor-label method is a problem-solving method based upon treating calculation units like algebraic factors.
8. A conversion factor is a ratio that is derived from the equality between two different units and that can be used to convert from one unit to the other.
9. **a.** A derived unit is a unit that can be obtained from combinations of fundamental units.
   b. Two examples of important derived units in chemistry are volume and density.
10. **a.** The derived SI unit for area is the square meter (m^2).
    b. The derived SI unit for volume is the cubic meter (m^3).
11. The SI unit for volume, the cubic meter, corresponds to a cube whose sides are each one meter (the SI length unit) long.
12. **a.** A liter is a unit of volume equal to 1000 cm^3.
    b. One cubic centimeter is equivalent to a volume of one milliliter.
13. **a.** mL or cm^3 **b.** L **c.** mL **d.** m^3
14. **a.** The density of a material is its mass divided by its volume, or its mass per unit volume.
    b. Density = $\dfrac{mass}{volume}$, or $D = \dfrac{m}{V}$
    c. The SI unit for density is kg/m^3; other metric units for density include g/cm^3 and g/mL.
15. Since many substances increase slightly in volume as their temperatures increase, densities determined at different temperatures will vary somewhat.
16. The greater the energy of the particles, the higher the temperature of the matter.
17. Heat always flows spontaneously from matter at a higher temperature to matter at a lower temperature.
18. The two fixed points on which the Celsius scale is based are the freezing point (0°C) and boiling point (100°C) of water.
19. This means that the banana would provide 130 kilocalories of heat if it were completely burned (metabolized) by the body.
20. **a.** Heat capacity is the quantity of heat needed to raise the temperature of a sample of matter by 1C°.
    b. Specific heat is the quantity of heat required to raise the temperature of 1 g of a substance by 1C°.
21. Uncertainty in measurements can result from limitations in accuracy or precision. Accuracy is the closeness of a measurement to a true or accepted value; precision is agreement among the numerical values of a set of measurements made in the same way.
22. The measurements in **a** and **c** are precise. Those in **b** are not.
23. The location of the estimated digit depends upon the smallest subdivision of the measurement scale being used.
24. **a.** 4 **b.** 3 **c.** 2 **d.** 4
    e. 6 **f.** 1 **g.** 3 **h.** 2
    i. 2 **j.** 2 **k.** 4 **l.** 2
    m. 7 **n.** 5
25. **a.** 2.7 g **b.** 47.37 mL **c.** 8.4 mL
    d. 4.16 L **e.** 20 km **f.** 0.049 mg
    g. 0.064 mL
26. **a.** The initial sum or difference should be rounded off so that the final digit is in the same place as the leftmost uncertain digit.
    b. The initial product or quotient should be rounded off to the same number of significant figures as in the measurement that has the fewest significant figures.
27. **a.** Exact conversion factors are those that have no uncertainty and do not limit the number of digits in a calculation.

b. Examples of exact conversion factors are $\frac{100 \text{ cm}}{1 \text{ m}}$, $\frac{1000 \text{ g}}{1 \text{ kg}}$, and $\frac{4.184 \text{ J}}{1 \text{ cal}}$.

28. Scientific notation is a shorthand technique in which numbers are written in the form $M \times 10^n$ where M is a number greater than or equal to 1, but less than 10, and n is an integer.

29. The stepwise method of problem solving consists of the following: Step 1: Analyze the problem; Step 2: Plan a route to solve the problem; Step 3: Compute the answer; Step 4: Evaluate the answer.

30. Examples are temperature, mass, and volume.

31. a. Two variables are directly proportional to each other if dividing one by the other gives a constant value. Two variables are inversely proportional to each other if their product has a constant value.

b. direct: $\frac{x}{y} = k$ or $x = ky$ inverse: $xy = k$ or $y = \frac{k}{x}$

c. Values of directly proportional quantities produce a linear graph; inversely proportional quantities produce a hyperbola.

Problems

1. a. First, express each in comparable units, such as meters:
1.2 m = 1.2 m
$750 \text{ cm} \times \frac{1 \text{ m}}{100 \text{ cm}} = 7.50 \text{ m}$
$0.005 \text{ km} \times \frac{1000 \text{ m}}{1 \text{ km}} = 5 \text{ m}$
$65 \text{ dm} \times \frac{1 \text{ m}}{10 \text{ dm}} = 6.5 \text{ m}$
$2000 \text{ mm} \times \frac{\text{m}}{1000 \text{ mm}} = 2 \text{ m}$

Thus, the values ranked from largest to smallest are: 750 cm, 65 dm, 0.005 km, 2000 mm, and 1.2 m

b. Convert each into comparable units, such as grams:
$450 \text{ mg} \times \frac{\text{g}}{1000 \text{ mg}} = 0.45 \text{ g}$
$3.8 \text{ cg} \times \frac{\text{g}}{100 \text{ cg}} = 0.038 \text{ g}$
$0.27 \text{ dg} \times \frac{\text{g}}{10 \text{ dg}} = 0.027 \text{ g}$
$0.50 \text{ g} = 0.50 \text{ g}$
$0.000 \, 47 \text{ kg} \times \frac{1000 \text{ g}}{\text{kg}} = 0.47 \text{ g}$

Thus, the values ranked from largest to smallest are: 0.50 g, 0.000 47 kg, 450 mg, 38 cg, and 0.27 dg.

c. Convert each into comparable units, such as liters:
2.5 L = 2.5 L
$22 \text{ cL} \times \frac{\text{L}}{100 \text{ cL}} = 0.22 \text{ L}$
$13 \text{ dL} \times \frac{\text{L}}{10 \text{ dL}} = 1.3 \text{ L}$
$870 \text{ mL} \times \frac{\text{L}}{1000 \text{ mL}} = 0.87 \text{ L}$
$175 \text{ cm}^3 \times \frac{\text{mL}}{\text{cm}^3} \times \frac{\text{L}}{1000 \text{ mL}} = 0.175 \text{ L}$

Thus, the values ranked from largest to smallest are: 2.5 L, 13 dL, 870 mL, 22 cL, and 175 cm³.

2. apples: $3 \times 125 \text{ g} \times \frac{\text{kg}}{1000 \text{ g}} = 0.375 \text{ kg}$

bananas: $5 \times 115 \text{ g} \times \frac{\text{kg}}{1000 \text{ g}} = 0.575 \text{ kg}$

pears: $2 \times 75 \text{ g} \times \frac{\text{kg}}{1000 \text{ g}} = 0.15 \text{ kg}$

peaches: $4 \times 65 \text{ g} \times \frac{\text{kg}}{1000 \text{ g}} = 0.26 \text{ kg}$

melon: $1750 \text{ g} \times \frac{\text{kg}}{1000 \text{ g}} = 1.75 \text{ kg}$

grapes: $5.5 \text{ hg} \times \frac{\text{kg}}{10 \text{ hg}} = 0.55 \text{ kg}$

sum = 3.66 kg

3. a. $30 \text{ L} \times \frac{1000 \text{ mL}}{\text{L}} = 30\,000 \text{ mL}$

b. $12 \text{ dL} \times \frac{100 \text{ mL}}{\text{dL}} = 1200 \text{ mL}$

c. $4.5 \text{ cL} \times \frac{10 \text{ mL}}{\text{cL}} = 45 \text{ mL}$

d. $6.4 \text{ cm}^3 \times \frac{\text{mL}}{\text{cm}^3} = 6.4 \text{ mL}$

4. a. $22.4 \text{ L} \times \frac{1000 \text{ mL}}{\text{L}} = 22\,400 \text{ mL}$

b. $22.4 \text{ L} \times \frac{1000 \text{ mL}}{\text{L}} \times \frac{\text{cm}^3}{\text{mL}} = 22\,400 \text{ cm}^3$

5. a. $d = \frac{m}{V}$ $d = \frac{240.00 \text{ g}}{60.0 \text{ cm}^3} = 4.00 \text{ g/cm}^3$

b. $d = \frac{m}{V}$
$m = d \times V$
$m = 11.3 \frac{\text{g}}{\text{cm}^3} \times 4.6 \text{ cm}^3 = 52 \text{ g}$

c. $d = \frac{m}{V}$ $V = \frac{m}{d}$
$V = \frac{14.50 \text{ g}}{13.6 \text{ g/ml}} = 1.07 \text{ mL}$

6. a. $T(K) = t(°C) + 273$
$T(K) = 0°C + 273 = 273 K$

b. $T(K) = 27°C + 273 = 300 K$

c. $T(K) = 72°C + 273 = 345 K$
d. $T(K) = -50°C + 273 = 223 K$

7. a. $t(°C) = T(K) - 273$
$t(°C) = 300 K - 273 = 27°C$
b. $t(°C) = 350 K - 273 = 77°C$
c. $t(°C) = 273 K - 273 = 0°C$
d. $t(°C) = 150 K - 273 = -123°C$

8. a. $25.0 \text{ cal} \times \dfrac{4.184 \text{ J}}{\text{cal}} = 105 \text{ J}$

b. $105 \text{ J} \times \dfrac{\text{kJ}}{1000 \text{ J}} = 0.105 \text{ kJ}$

9. a. $1.42 \text{ kJ} \times \dfrac{1000 \text{ J}}{\text{kJ}} = 1420 \text{ J}$

b. $1420 \text{ J} \times \dfrac{\text{cal}}{4.184 \text{ J}} = 339 \text{ cal}$

c. $339 \text{ cal} \times \dfrac{\text{kcal}}{1000 \text{ cal}} = 0.339 \text{ kcal}$

10. $c = \dfrac{q}{m \times \Delta t}$

$c = \dfrac{34 \text{ J}}{25 \text{ g} \times (33°C - 21°C)}$

$c = \dfrac{34 \text{ J}}{25 \text{ g} \times 12 C°} = 0.11 \dfrac{\text{J}}{\text{g} \cdot \text{C}°}$

11. $q = m \times c \times \Delta t$

$q = 55 \text{ g} \times \dfrac{0.39 \text{ J}}{\text{g} \cdot \text{C}°} \times (40°C - 24°C)$

$q = 55 \text{ g} \times \dfrac{0.39 \text{ J}}{\text{g} \cdot \text{C}°} \times 16 C° = 340 \text{ J}$

12. a. $4.07 \text{ g} + 1.863 \text{ g} = 5.933 \text{ g} = 5.93 \text{ g}$
b. $3127.55 \text{ cm} - 784.2 \text{ cm} = 2343.35 = 2343.4 \text{ cm}$
c. $0.067 \text{ mL} + 1.01 \text{ mL} + 2.5 \text{ mL} = 3.577 \text{ mL}$
$= 3.6 \text{ mL}$
d. $36.427 + 12.5 \text{ m} + 6.33 \text{ m} = 55.257 \text{ m}$
$= 55.3 \text{ m}$

13. a. $9.40 \text{ cm} \times 2.6 \text{ cm} = 24.44 \text{ cm}^2 = 24 \text{ cm}^2$
b. $8.08 \text{ dm} \times 5.3200 \text{ dm} = 42.9856 \text{ dm}^2 = 43.0 \text{ dm}^2$
c. $1.50 \text{ g}/2 \text{ cm}^3 = 0.75 \text{ g}/\text{cm}^3 = 0.8 \text{ g}/\text{cm}^3$
d. $0.084\,21 \text{ g}/0.640 \text{ mL} = 0.131\,578\,1 \text{ g/mL}$
$= 0.132 \text{ g/mL}$
e. $4.00 \text{ m} \times 0.020 \text{ m} \times 1.57 \text{ m} = 0.1256 \text{ m}^3$
$= 0.13 \text{ m}^3$

f. $\dfrac{21.50 \text{ g}}{4.06 \text{ cm} \times 1.8 \text{ cm} \times 0.905 \text{ cm}}$
$= 3.250\,8082 \text{ g/cm}^3 = 3.3 \text{ g/cm}^3$

14. a. $900 = 9 \times 10^2$
b. $750\,000 = 7.5 \times 10^5$
c. $93\,000\,000 = 9.3 \times 10^7$
d. $0.000\,082 = 8.2 \times 10^{-5}$
e. $0.000\,403 = 4.03 \times 10^{-4}$
f. $6500 \times 10^7 = 6.5 \times 10^{(7+3)} = 6.5 \times 10^{10}$
g. $0.000\,015 \times 10^8 = 1.5 \times 10^{(8-5)} = 1.5 \times 10^3$
h. $320\,000 \times 10^{-9} = 3.2 \times 10^{(-9+5)}$
$= 3.2 \times 10^{-4}$

i. $0.0097 \times 10^{-4} = 9.7 \times 10^{(-4-3)} = 9.7 \times 10^{-7}$

15. a. $5 \times 10^3 = 5000$
b. $7.8 \times 10^4 = 78\,000$
c. $2 \times 10^{-5} = 0.000\,02$
d. $3.81 \times 10^{-4} = 0.000\,381$
e. $4.07 \times 10^7 = 40\,700\,000$
f. $6.09 \times 10^{-9} = 0.000\,000\,006\,09$

16. a. 6.7×10^2
b. 9.5×10^{-5}
c. 2.34×10^4
d. 8.07×10^{-6}
e. 3.7×10^4

17. a. 6×10^{12}
b. $15.96 \times 10^7 = 1.596 \times 10^8 = 1.6 \times 10^8$
c. $12.504 \times 10^{-10} = 1.2504 \times 10^{-9}$
$= 1.3 \times 10^{-9}$
d. $18.06 \times 10^{12} = 1.806 \times 10^{13} = 1.8 \times 10^{13}$
e. $4.2 \times 10^2 = 4 \times 10^2$
f. $1.875 \times 10^8 = 2 \times 10^8$

18. *Step 1. Analyze*
Given: speed = 40.0 km/1.0 h
time = 5.5 h
Unknown: distance traveled
Step 2. Plan
speed, time → distance

$\text{speed}\left(\dfrac{\text{km}}{\text{h}}\right) \times \text{time (h)} = \text{distance (km)}$

Step 3. Compute

$\dfrac{40.0 \text{ km}}{1.0 \text{ h}} \times 5.5 \text{ h} = 220 \text{ km}$

Step 4. Evaluate
The answer is close to an estimated value of 200 km (40 × 5). Units have canceled to give kilometers, and the answer is correctly rounded to two significant figures.

19. *Step 1. Analyze*
Given: mass = 8.50 g
density = 0.24 g/cm³
Step 2. Plan
mass, density → volume

$d\left(\dfrac{\text{g}}{\text{cm}^3}\right) = \dfrac{m\,(\text{g})}{V\,(\text{cm}^3)} \quad V = \dfrac{m\,(\text{g})}{d\left(\dfrac{\text{g}}{\text{cm}^3}\right)}$

Step 3. Compute

$V = \dfrac{8.50 \text{ g}}{0.24 \text{ g/cm}^3} = 35 \text{ cm}^3$

Step 4. Evaluate
The value of the answer is close to an estimated value of 40 cm³ $\left(\dfrac{8}{0.2}\right)$. Units have canceled as desired and the answer is correctly rounded to two significant figures.

20. $V = l \times w \times h$
$V = 12.00 \text{ cm} \times 8.50 \text{ cm} \times 0.50 \text{ cm}$
$V = 51 \text{ cm}^3$

$$d = \frac{m}{V}$$

$$d = \frac{457.00 \text{ g}}{51 \text{ cm}} = 9.0 \text{ g/cm}^3$$

21. $d = \dfrac{m}{V}$

    $d = \dfrac{32.50 \text{ g}}{3.10 \text{ cm}^3}$

    $d = 10.5 \text{ g/cm}^3$

    Based on the information in Table A-14, the unknown could be silver.

22. First, determine the specific heat of the unknown:

    $$c = \frac{q}{m \times \Delta t}$$

    $c = \dfrac{20.4 \text{ J}}{6.67 \text{ g} \times 7.0\text{C°}}$ $c = 0.44 \text{ J/(g·C°)}$

    Based on the table of specific heats, the substance is most likely iron.

23. **a.** A direct proportion
    b. As Rankine temperature increases, Kelvin temperature increases at a constant rate such that;
    $T(K) = \frac{5}{9}T(°R)$.

Critical Thinking

1. $\dfrac{180°F}{100°C} = \dfrac{1.8°F}{1.0°C}$

   $t(°C) = \dfrac{1.0°C}{1.8}[t(°F) - 32°F]$

   $t(°C) = \dfrac{[t(°F) - 32°F]}{1.8}$

2. The water in the pool will be slow to increase its temperature under the summer sun and slow to decrease its temperature as cool fall temperatures lower the temperature of the air.

3. **a.** Cork   **b.** Gasoline   **c.** Oak Wood   **d.** Water
   e. Aluminum   **f.** Mercury   **g.** Gold

Cross Connections

1. Answers will vary.
   - Using a skinfold caliper, the length of fat on the underside of the arm is used as an indicator of fat percentage.
   - Using a water displacement technique, the volume of a body is determined. A separate mass determination is made. The density of the body is compared to the density of muscles and a percentage is determined.

2. Answers will vary. A comparison of the densities shows that the oil remains on top of the water. For aquatic life near the surface, oil can collect on skin or block sunlight needed for photosynthesis. Oil would also be able to wash up on shore due to the tides or waves. Since oil is flammable and sits on top of the water, fires are a constant threat.

Teacher's Notes

Measurements and Solving Problems **30H**

Chapter 2
Measurements and Solving Problems

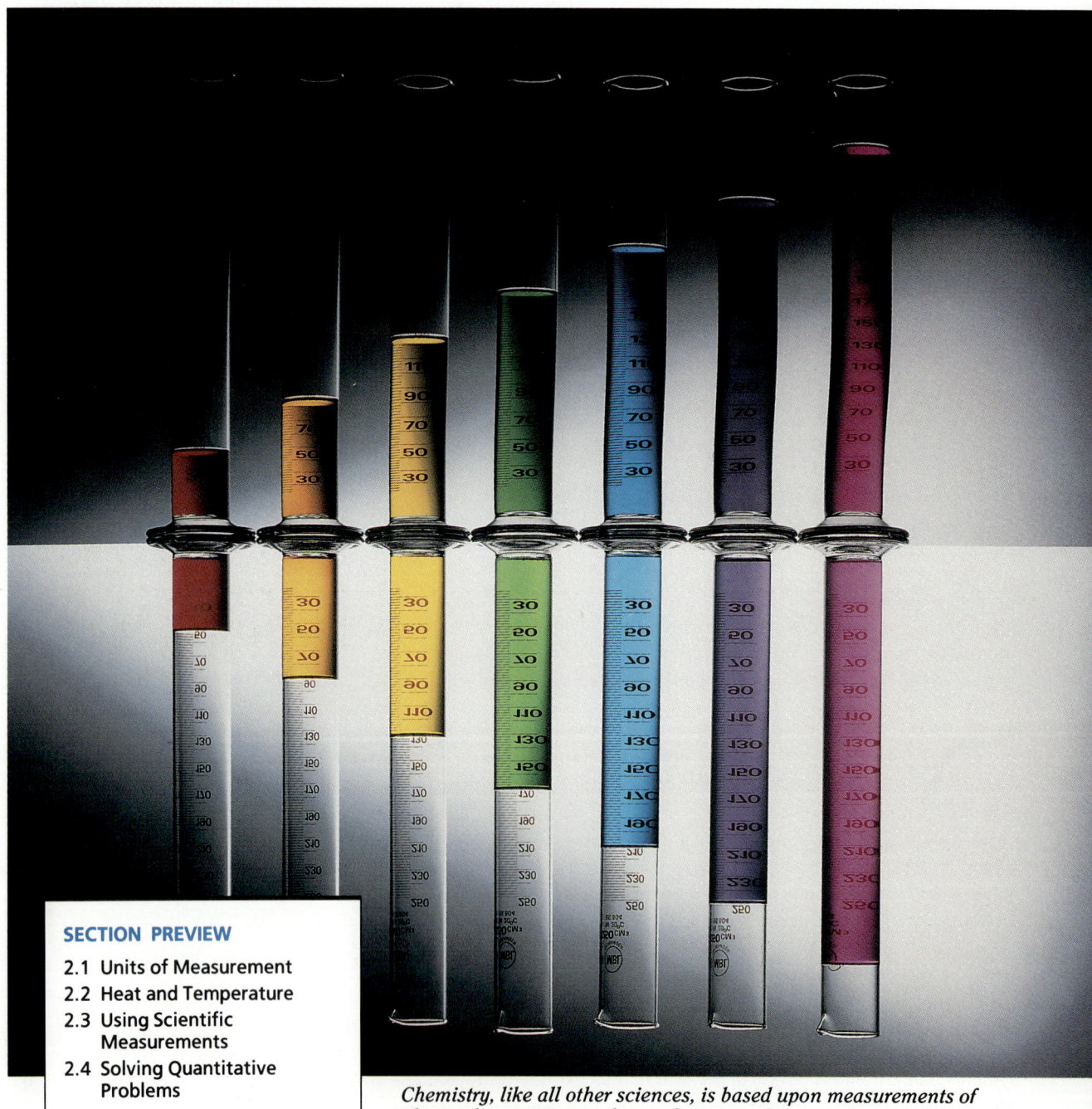

SECTION PREVIEW

2.1 Units of Measurement
2.2 Heat and Temperature
2.3 Using Scientific Measurements
2.4 Solving Quantitative Problems

Chemistry, like all other sciences, is based upon measurements of observed properties, such as volume or color.

2.1 Units of Measurement

You have probably had the experience of ordering a large pizza in a restaurant only to discover that what is called "large" on the menu is smaller than you expected. A less ambiguous way of ordering would have been to ask for a 16-inch pizza.

A scientific description should be unambiguous, that is, it should mean the same thing to people all over the world. This section introduces the system of measurement scientists have agreed to use in order to avoid confusion and facilitate communication.

The SI Measurement System

Many systems of measurement are used throughout the world. In the United States, most people are familiar with the English system that uses units of measurement such as inches, ounces, and pounds. Units of measurement are used to compare whatever is to be measured with something of a previously defined size. For instance, to measure the length of this book, you could compare its length with a ruler marked off in centimeters. However, before the ruler could be made, the length of the centimeter would have to be defined. A **unit of measurement** *is a physical quantity of a defined size.* Scientists from all countries have agreed to use the measurement system called *Le Système International d'Unités* (the International System of Units), usually abbreviated **SI.** This system, which consists of seven basic units and other units derived from them, was adopted in 1960 by the General Conference on Weights and Measures.

Science is now in a period of transition between the use of non-SI units and SI units. Therefore, some non-SI units are used in this book. The reasons for this are explained as the units are introduced in the text.

Various international and national organizations have the responsibility for defining the sizes of units and governing their use. Certain units are defined by choosing **standards of measurements,** *which are objects or natural phenomena of constant value, easy to preserve and reproduce, and practical in size.* In the United States, the National Institute of Standards and Technology in Washington, D.C. plays a major role in maintaining standards of measurement (Figure 2-1).

If it were not for the use of standards as a basis for comparison, references to sizes of objects might mean completely different things to different people. Suppose the unit *foot* was not based on a standard, universally accepted length, but instead depended on the size of the feet of the person making the measurement. A person with very large feet might measure the length of a field as 100 feet. A person with very small feet might measure the same length as 200 feet.

It is also important that standards remain unchanged over time. Standards are chosen because they do not change and their dimensions are exactly reproducible whenever measurements are made.

The result of every measurement is a number plus a unit. As you will see throughout this chapter, units are usually represented in writing by symbols that are abbreviations of the unit names. For example, "ten centimeters" is written 10 cm.

SECTION OBJECTIVES

- Name SI and other common units of length, mass, and time.
- Perform unit conversions using the factor-label method.
- Name SI and other common units of volume and density.
- Perform density calculations.

Figure 2-1 Prototype kilogram K-20. This cylinder of platinum-iridium alloy was sent to the United States in 1889 by the International Bureau of Weights and Measures in France and still serves as the U.S. standard of mass.

A number without a corresponding unit is of little use in chemistry.

Figure 2-2 A woman and a girl use their feet as measurement standards. How can their measurements be made more useful?

This marks a departure from previous editions.

To calibrate a measuring device means to adjust its scale reading to agree with the size of a standard.

In this book, we follow an international system for writing numbers. Commas are not used in numbers. Counting from the decimal point, numbers with five or more digits on either side of the decimal point have a space rather than a comma after each group of three digits. For example, 45 000 (*not* 45,000) is used to represent "forty-five thousand." In this example, note the space after three digits to the left of the implied decimal point. Take another example. The number "four hundred and fifty thousand" is expressed as 450 000 (*not* 450,000). For numbers less than 1, spaces follow every three digits counting to the right from the decimal point—for example, 0.435 98 and 0.045 237.

Fundamental SI Units

Seven SI units are fundamental units, or base units. A **fundamental unit** *is one that is defined by a physical standard of measurement.* The seven SI units are shown in Table 2-1. All other SI units can be derived from these fundamental units. A discussion of the fundamental units for length, mass, and time follows. Later in this chapter, the SI unit for temperature is discussed. Another fundamental unit, the mole, is discussed in Chapter 3.

TABLE 2-1	SI FUNDAMENTAL UNITS	
Quantity	Unit	Symbol
Length	meter	m
Mass	kilogram	kg
Time	second	s
Temperature	kelvin	K
Amount of substance	mole	mol
Electric current	ampere	A
Luminous intensity	candela	cd

SI is a decimal (base 10) system. Therefore, larger and smaller units for each type of measurement can be created by multiplying the fundamental unit by the appropriate power of 10. A prefix is then added to the name of the unit, indicating the power of 10 by which the unit has been multiplied. These prefixes can be used with any SI unit. Refer to Table 2-2 for their meanings. For example, a length of 1 centimeter (cm) is equal to 10^{-2} meter, or 0.01 meter.

TABLE 2-2 SI PREFIXES

Prefix	Symbol	Exponential Factor	Meaning	Example
tera	T	10^{12}	1 000 000 000 000	1 terameter (Tm) = 1×10^{12} m
giga	G	10^{9}	1 000 000 000	1 gigameter (Gm) = 1×10^{9} m
mega	M	10^{6}	1 000 000	1 megameter (Mm) = 1×10^{6} m
kilo	k	10^{3}	1000	1 kilometer (km) = 1000 m
hecto	h	10^{2}	100	1 hectometer (hm) = 100 m
deca	da	10^{1}	10	1 decameter (dam) = 10 m
deci	d	10^{-1}	1/10	1 decimeter (dm) = 0.1 m
centi	c	10^{-2}	1/100	1 centimeter (cm) = 0.01 m
milli	m	10^{-3}	1/1000	1 millimeter (mm) = 0.001 m
micro	μ	10^{-6}	1/1 000 000	1 micrometer (μm) = 1×10^{-6} m
nano	n	10^{-9}	1/1 000 000 000	1 nanometer (nm) = 1×10^{-9} m
pico	p	10^{-12}	1/1 000 000 000 000	1 picometer (pm) = 1×10^{-12} m

Length The SI standard unit for length is the **meter.** The international standard for the meter (m) is the distance light travels in a vacuum during a time interval of 1/299 792 458 of a second. Most of us use meter sticks or other measuring devices with lengths that have been set on the basis of this standard. To estimate a length in meters, you can remember that one meter is about the width of an average doorway.

When measuring lengths shorter than a meter, it is often convenient to use centimeters. For longer distances, the kilometer (km) is more useful. One kilometer is equal to 1000 m. The dimensions of this book would most conveniently be measured in centimeters, whereas meters would be more useful for measuring the dimensions of your classroom. For measuring distances between cities, as shown in Figure 2-3, kilometers are usually the most practical unit.

Mass The SI standard unit for mass is the **kilogram.** The international standard for the kilogram (kg) is a platinum-iridium cylinder kept at the International Bureau of Weights and Measures in the French town of Sèvres. Copies of this standard are used to calibrate precision balances all over the world. You may find it helpful to know that this book has a mass of about 2 kilograms.

The gram (g) is a more convenient unit than the kilogram for measuring small masses. One gram is equal to 10^{-3} kg (0.001 kg). Figure 2-4 shows the gram can be used to measure small objects, such as pencils and school rings. For smaller objects, the milligram (mg) is useful. One milligram is equivalent to 0.000 001 kg or 0.001 g.

Figure 2-3 The distance between two cities is conveniently expressed in kilometers, although in the United States most people still use the more common unit of miles.

Measurements and Solving Problems

Figure 2-4 The gram is a convenient unit of mass for relatively small items, such as a pencil or a school ring.

Time *The SI standard unit for time is the* **second.** The second (s) is commonly defined as $\frac{1}{60}$ of a minute. The definition of a minute depends on the definition of an hour, which in turn is based on the duration of a day. A day is the length of time it takes the earth to rotate exactly once on its axis. Because this time varies slightly, a more precise standard is needed for the SI unit. The SI standard for the second is the duration of 9 192 631 770 periods of a particular radiation emitted by cesium-133 atoms. This interval of time is always exactly the same. In everyday life, where such precise time measurements are unnecessary, the SI second can be considered equal to the common second.

Units of Measurement in Calculations

Factor-label method The units given for a measured quantity are a very important part of that measurement and must be included in all calculations. Two times a 6-kg mass is 12 kg, not 12. This calculation should be written

$$2 \times 6 \text{ kg} = 12 \text{ kg}$$

Units in calculations are treated exactly like algebraic factors. If you are calculating the area of a rectangle 2 m on one side and 6 m on the other, the calculation should be written

$$2 \text{ m} \times 6 \text{ m} = 12 \text{ m}^2$$

The proper use of units in calculations is an extremely valuable aid to solving problems. Understanding how to use units and using them in every calculation in chemistry (and other subjects) will greatly decrease mistakes.

The use of units in calculations is referred to as the factor-label method. *The* **factor-label method** *is a problem-solving method based upon treating units in calculations as if they are algebraic factors.* The factor-label method is used in sample problems throughout this book. In the next section, the use of factor-label method in unit conversion is illustrated. You should use this valuable problem-solving technique throughout your study of chemistry.

Unit conversion As an introduction to unit conversion, consider a very simple, everyday type of problem. Suppose you need to know how many quarters there are in 75 dollars (Figure 2-5). We are going to solve the problem by the factor-label method.

You know that four quarters equal one dollar. You can express this as a simple mathematical equation: 4 quarters = 1 dollar. Because these two quantities are equal, their ratios are both equal to 1:

$$\frac{4 \text{ quarters}}{1 \text{ dollar}} = 1 \qquad \frac{1 \text{ dollar}}{4 \text{ quarters}} = 1$$

Each of these expressions is a conversion factor. *A* **conversion factor** *is a ratio derived from the equality between two different units and can be used to convert from one unit to the other.* Because conversion factors like those above are equal to 1, they can be multiplied by other factors in equations without changing the validity of the equations. The answer will just be in different units. When using conversion factors, set the problem up as follows:

quantity sought = quantity given × conversion factors.

Students have a difficult time associating the units with algebraic factors. Use of letters other than *a, b, c; m, n;* or *x, y, z* (the letters usually encountered in math problems) seems to confuse them.

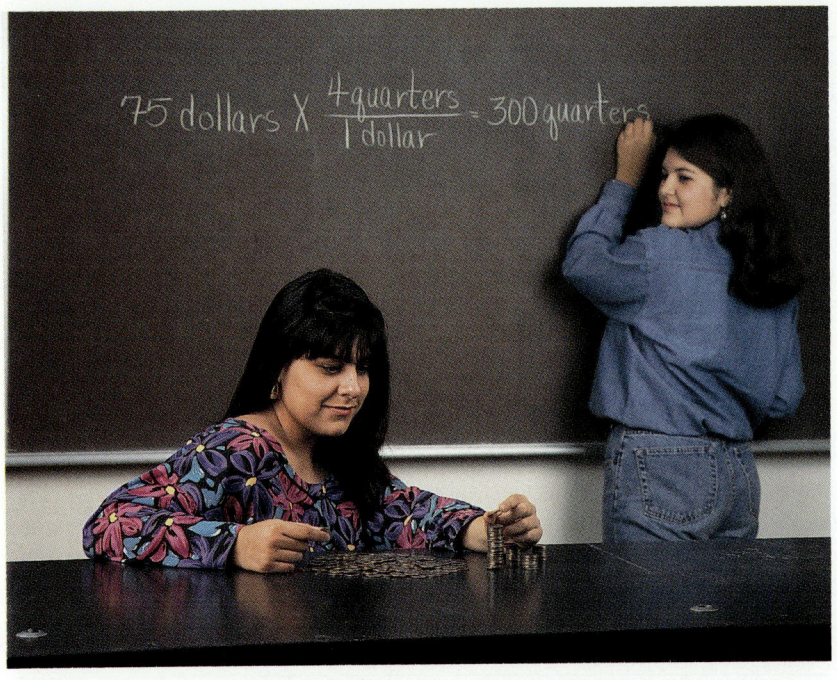

Figure 2-5 Each of these two students has a different approach to a quantitative problem. Which way would you choose to find the number of quarters in $75?

For example, to find out how many quarters there are in 75 dollars, we must carry out the unit conversion from dollars to quarters.

$$? \text{ quarters} = 75 \text{ dollars} \times \text{conversion factor}$$

The appropriate conversion factor is the one that allows cancellation of the unwanted unit and gives the answer in the desired unit. Set up the calculation with both conversion factors and cancel units as in algebra.

$$75 \text{ dollars} \times \frac{4 \text{ quarters}}{1 \text{ dollar}} = 300 \text{ quarters}$$

$$75 \text{ dollars} \times \frac{1 \text{ dollar}}{4 \text{ quarters}} = \frac{75 \text{ dollars}^2}{4 \text{ quarters}}$$

This shows that the first conversion factor is the correct one. The second gives the meaningless units of dollars²/quarters.

STUDY HINT

Note how the factor label method is used in Sample Problem 2-1.

Sample Problem 2.1

Express a mass of 92 cg in milligrams and grams.

Solution
Solving this problem requires conversion factors that relate centigrams to milligrams and grams. From Table 2-2 you get:

$$1 \text{ cg} = 10 \text{ mg} \quad \text{and} \quad 1 \text{ g} = 100 \text{ cg}$$

which give the conversion factors

$$\frac{1 \text{ cg}}{10 \text{ mg}} \quad \frac{10 \text{ mg}}{1 \text{ cg}} \quad \text{and} \quad \frac{1 \text{ g}}{100 \text{ cg}} \quad \frac{100 \text{ cg}}{1 \text{ g}}$$

Measurements and Solving Problems

Sample Problem 2.1 *continued*

Choosing in each case the conversion factor that leads to an answer in the desired units gives

$$92 \text{ cg} \times \frac{10 \text{ mg}}{1 \text{ cg}} = 920 \text{ mg} \quad \text{and} \quad 92 \text{ cg} \times \frac{1 \text{ g}}{100 \text{ cg}} = 0.92 \text{ g}$$

Practice Problems
1. Express a distance of 25 meters in centimeters and kilometers. *(Ans.)* 2500 cm, 0.025 km
2. Express a mass of 72.0 grams in milligrams, micrograms, and megagrams. *(Ans.)* 72 000 mg, 72 000 000 µg, 0.000 072 Mg
3. Express one week in terms of seconds. *(Ans.)* 604 800 s

Derived SI Units

The majority of SI units are derived units. *A **derived unit** is a unit that can be obtained from combinations of fundamental units.* Examples of some common SI derived units are given in Table 2-3. The units for pressure and energy are discussed in detail later in this book.

TABLE 2-3 DERIVED SI UNITS

Quantity	Unit	Symbol	Derivation	Units
Area	square meter	m^2	length × width	m × m
Volume	cubic meter	m^3	length × width × height	m × m × m
Density	kilogram per cubic meter	kg/m^3	$\frac{mass}{volume}$	$\frac{kg}{m^3}$
Speed (velocity)	meter per second	m/s	$\frac{length}{time}$	$\frac{m}{s}$
Force	newton	N	$\frac{mass \times length}{time \times time}$	$\frac{kg \times m}{s^2}$
Pressure	pascal	Pa	$\frac{force}{area}$	$\frac{N}{m^2} = \frac{kg}{m \times s^2}$
Energy	joule	J	force × length	$N \times m = \frac{kg \times m^2}{s^2}$

Examine the units given in the last column of Table 2-3. In each case, the quantity has units that are a combination of fundamental SI units. For example, speed, which is the distance in meters traveled in a given time in seconds, has the units of meters per second. The combinations of fundamental units for force, pressure, and energy include several different units. To simplify matters, therefore, each of these unit combinations is given a name of its own. For example, for the pascal (an SI pressure unit), 1 Pa = 1 kg/(m × s²).

Volume *The amount of space occupied by an object is called its **volume**.* The SI unit for volume is the cubic meter (m³). As shown in Table 2-3, this

1m³ of water has a mass of approx. 1000 kg or 2200 lbs. It represents a very large mass.

unit is derived from the fundamental SI unit of length, the meter, and corresponds to a cube whose sides are each 1 m long. Because this volume is often too large to be useful in normal laboratory measurements, the cubic centimeter (cm^3) is used more frequently in chemistry than the cubic meter.

For liquid and gas volume, an older metric unit, the liter, is often used, although it is not the official SI unit of volume. Imagine a cube with edges that are 10 cm long. The volume of this cube is 10 cm × 10 cm × 10 cm, or 1000 cm^3. *The* **liter (L)** *is a volume equal to 1000 cm^3.* One liter of water, or any other liquid, would fill the 1000-cm^3 cube. One milliliter (1 mL) is equivalent to 0.001 liter and is thus equivalent to a volume of 1 cm^3. These units—milliliter and cubic centimeter—are used interchangeably.

Most laboratory volumetric glassware, such as that shown in Figure 2-6, is calibrated and marked in liters or milliliters. These units are used throughout this book in references to liquid or gas volumes.

Figure 2-6 Most of the glassware used to measure volume in the chemistry laboratory is calibrated in liters or milliliters.

The liter is also equivalent to a cube with 1-decimeter sides. The volume of the cube, 1 dm^3, is equivalent to 1000 cm^3 or to 1 L.

Density What do we mean by "light" and "heavy"? Figure 2-7 shows what happens when a piece of lead and a piece of cork are placed into a beaker of water. We commonly say that lead sinks in water because "lead is heavy." Likewise, we say that cork floats on water because "cork is light." Figure 2-7 shows the same pieces of lead and cork on opposite pans of a balance. As you can see, the lead and the cork have the same mass. It would appear that the piece of cork is just as "heavy" as the piece of lead. However, the volume of the cork is much greater than that of the lead.

When we speak of lead as "heavy" and cork as "light," we are actually referring not to their masses but to their densities. *The* **density** *of a substance relates the mass of the substance to its volume, or its mass per unit volume.* Density is expressed mathematically as

$$\text{density} = \frac{\text{mass}}{\text{volume}} \quad \text{or} \quad d = \frac{m}{V}$$

where m is the mass of a material, V is its volume, and d is its density.

In SI units, density is expressed in kilograms per cubic meter (kg/m^3). More commonly, however, the density of a solid is expressed in grams per cubic centimeter (g/cm^3), the density of a liquid in grams per milliliter (g/mL), and the density of a gas is expressed in grams per liter (g/L). Because the volumes of solids and liquids change slightly with temperature and the volumes of gases change greatly with temperature, the densities also change.

Note that one liter is equal to one cubic decimeter (1 L = 1 dm^3).

The volume of most kinds of matter changes with temperature. To be accurate, devices for measuring volume must be calibrated at a given temperature. Room temperature (25°C) is usually used for this purpose.

Figure 2-7 Cork floats in water because it is less dense than water; lead sinks because it is denser than water (left). Comparison on a double-pan balance shows that the cork and lead have equal masses, although their volumes are different (right).

Measurements and Solving Problems

Desktop Investigation

Density of Pennies

Questions
Are the densities of pennies made before and after 1982 different? What does density tell you about the composition of pennies?

Materials
40 pennies made before 1982
40 pennies made after 1982
balance
100-mL graduated cylinder

Procedure
Record all your measurements in a data table.

1. Using a balance, determine the mass of 40 pennies minted before 1982. Repeat your measurement two more times. Average the results of the three trials to obtain the average mass of the 40 pennies.

2. Follow the same process as in Procedure 1 with the 40 pennies minted after 1982.

3. Pour 50 mL of water into a 100-mL graduated cylinder. Add the 40 pennies minted before 1982 to the graduated cylinder. Find the volume of the pennies. Repeat your measurement two more times, making sure to dry the pennies before each trial. Average the results of the three trials to obtain the average volume of the pennies.

4. Follow the same process as in Procedure 3 with the

pennies made after 1982.

5. Using the average volume and the average mass of the two sets of pennies, determine the average density of each set.

6. Review your data for any large differences between trials.

7. Compare the average experimental densities of the two sets of pennies with the densities of the metals listed in Appendix Table A-14, Properties of Common Elements.

Discussion

1. Why does the Procedure include three trials instead of one trial for each set of pennies?

2. (a) How would you recognize an inaccurate or incorrect set of data? (b) What might be some possible causes for such faulty data? (c) What would a scientist do with sets of data that appeared to be inaccurate or incorrect? Why?

3. On what experimental and accepted data do you base your conclusions about the composition of the two sets of pennies?

TABLE 2-4 DENSITIES OF SOME COMMON MATERIALS

Solids		Liquids		
Substance	Density at 20°C (g/cm^3)	Substance	Density (g/mL)	Temperature (°C)
Cork	0.24*	Gasoline	0.67*	20
Butter	0.86	Ether	0.736	0
Ice	0.92†	Alcohol, ethyl	0.791	20
Sugar	1.59	Kerosene	0.82	20
Bone	1.85*	Turpentine	0.87	20
Clay	2.20*	Water	1.00	4
Glass, common	2.60*	Sea Water	1.025	15
Granite	2.70*	Milk	1.031*	20
Diamond	3.26*	Chloroform	1.489	20
Lead	11.35	Mercury	13.6	20

* Average density   † Ice density measured at 0°C

Thus, when the density of a substance is given, the temperature at which it was measured must also be given. The densities of some common materials are listed in Table 2-4. Compare the densities of cork and lead. As shown by the experiments pictured in Figure 2-7, lead is denser than cork—in fact, about 47 times denser. The density of a material can be determined by measuring the mass and volume of a sample of any size. Although lead is a dense element, there are elements that are even denser. Osmium (Os), for example, is a metal that has a density of 22.57 g/cm^3, nearly twice that of lead. Osmium is the densest of all elements. Whatever type of material is involved, and whatever the size sample that is available in a density determination, the procedure is still essentially the same: simply measure the mass and volume of the sample, and divide the mass by the volume.

The change in gas volume with temperature is discussed in chapter 11.

Osmium, Os, is the densest element with a density of 22.57 g/cm^3.

Sample Problem 2.2

Find the density of a piece of aluminum with a volume of 4.0 cm^3 and a mass of 10.8 g. Is aluminum more or less dense than lead? Than cork? You can calculate the density of the sample of aluminum by substituting the given values for mass and volume in the equation for density:

$$\text{density} = \frac{\text{mass}}{\text{volume}} \quad \text{density} = \frac{10.8 \text{ g}}{4.0 \text{ cm}^3} = 2.7 \text{ g/cm}^3$$

The results of this calculation show that aluminum is denser than cork, (0.24 g/cm^3), but that it is less dense than lead, which has a density of 11.35 g/cm^3.

Practice Problems

1. What is the density of a block of marble with a mass of 594 g and a volume of 216 cm^3? *(Ans.)* 2.75 g/cm^3
2. A 40.0 cm^3 sample of quartz has a density of 2.65 g/cm^3. What is the mass of the quartz sample? *(Ans.)* 106 g
3. The density of a sample of cork is 0.24 g/cm^3. What is the volume of this sample if it has a mass of 36 g? *(Ans.)* 150 cm^3

1. Meter, kilogram, and second
2. *(a)* 2400 cm   *(b)* 0.024 km
3. *(a)* 5600 g   *(b)* 5 600 000 mg
4. 5.85 L
5. 2.75 g/cm³

Section Review

1. What are the fundamental SI units for length, mass, and time?
2. Express 24 meters in   *(a)* centimeters   *(b)* kilometers.
3. Express 5.6 kilograms in   *(a)* grams   *(b)* milligrams.
4. What volume of water, in liters, would be required to fill a tank 25 cm long, 18 cm wide, and 13 cm deep?
5. A piece of granite has a mass of 55 g and a volume of 20.0 cm^3. What is the density of this piece of granite?

2.2 Heat and Temperature

SECTION OBJECTIVES

- Define temperature and state the units in which it is measured.
- Convert temperatures between the Celsius scale and the Kelvin scale.
- Define heat and state its units.
- Perform specific-heat calculations.

The average person can tolerate temperatures up to 50°C. Touch can thus be used as a very rough indicator of temperature.

What do you mean when you say something feels hot or cold? Nerves in your skin send signals to your brain that seem to tell you whether something you touch is hot, cold, or warm. Whether something feels hot or cold does not indicate its temperature, however. You can prove this to yourself by a simple experiment. Put your left hand briefly into a container of cold water. At the same time, put your right hand into a container of very warm (*not* boiling) water. Then place both hands into a container of water that is at room temperature. Your left hand, which is colder than the room temperature water, will signal "warm." Your right hand, which is warmer than the room temperature water, will at the same time signal "cold."

The experiment just described demonstrates the difference between heat and temperature. The warm right hand signaled "cold" because energy was flowing away from it into water colder than your hand. The cold left hand signaled "warm" because energy was flowing into it from water that was warmer than the hand. Your hands detect the transfer of energy as heat. Yet, as indicators of temperature, the senses are clearly unreliable.

Definitions of Heat and Temperature

Temperature *is a measure of the average kinetic energy of the particles in a sample of matter.* Temperature indicates how hot or cold something is. The greater the kinetic energy of the particles in a sample of matter, the higher the temperature of the matter. To assign a numerical value to temperature, it is necessary to define a temperature scale.

Heat *or* **heat energy** *can be thought of as the sum total of the kinetic energies of the particles in a sample of matter.* Heat always flows spontaneously from matter at a higher temperature to matter at a lower temperature, as is shown in Figure 2-8. The temperature of the cold water in the right-hand beaker will increase as heat flows into it. Likewise, the temperature of the hot water in the left-hand beaker will decrease as heat flows away from it. When the temperature of the water in each beaker and the water in the pan are equal, heat will no longer flow in the system.

To understand the difference between temperature and heat, consider a bathtub full of warm water and a cup full of boiling water. The temperature of the water in the tub is lower than that of the water in the cup. Although the kinetic energy of each particle in the cup is greater than that of each particle in the tub, the total heat energy in the tub will be much greater than that in the cup, which has far fewer particles.

Figure 2-8 The direction of heat flow is determined by the temperature differences between the water in the two beakers and the basin. The transfer of heat out of the beaker on the left into the basin, and then from the basin into the beaker on the right will continue until the water in all three containers has reached the same temperature.

Measurement of temperature and temperature units Some properties of matter vary with temperature and therefore can be used to measure temperature. For example, most materials expand when heated and contract when cooled. The most familiar instruments for measuring temperature are thermometers containing mercury or alcohol (colored red). The measurement is based on the uniform expansion or contraction of the liquid with changing temperature.

If the thermometer comes into contact with a system at a higher temperature, heat is transferred to the thermometer. This heat transfer raises the temperature of the liquid and increases its volume, causing the liquid to rise in the thermometer. The liquid rises until the temperatures of the system and the thermometer are equal. Similarly, if the thermometer comes into contact with something at a lower temperature, the column of liquid falls as the liquid contracts. The liquid rises or falls next to a set of numbered marks, or calibrations, from which you can read the temperature on a specific temperature scale. The ability to measure temperature is thus based on heat transfer.

Several temperature scales have been developed. One of these, the Celsius scale, was developed by the Swedish astronomer Anders Celsius (1701–1744). He knew that water froze and boiled at certain temperatures, but at that time there were no agreed upon temperature values. He chose 0° to represent the freezing point of water and 100° as the boiling point. Between these two points on the Celsius scale of 0°C and 100°C are 100 equal parts, or degrees. *Each* **degree Celsius** *is a unit of temperature on the Celsius scale.* To represent a point of temperature one uses the expression °C, such as "The water temperature is 22°C." However, to explain an *interval* of Celsius degrees, the expression C° is used, as in "The water temperature has risen 4C°; it is now 26°C."

Another temperature scale was developed by William Thomson, Lord Kelvin (1824–1907), and bears his name. *The* **kelvin (K)** *is the unit of temperature on the Kelvin scale and is the fundamental SI unit for temperature.* The kelvin temperatures at which water freezes and boils are

"Normal" freezing and boiling points are those at 1 atm pressure.

Measurements and Solving Problems

Figure 2-9 A comparison of the Kelvin, Celsius, and Fahrenheit temperature scales.

273.15 K and 373.15 K, respectively. Figure 2-9 includes a comparison of the most common temperature scales. Note that on the Kelvin and Celsius scales, there are 100 units between the freezing and boiling points of water and the sizes of the degree divisions on both scales are equal. The following equation can be used to convert temperatures from the Celsius (t) to Kelvin (T) scales.

$$T(K) = t(°C) + 273.15$$

The equation shows that 273.15 must be added to any Celsius temperature to obtain the Kelvin temperature. For example, converting a room temperature of 25.00°C to the Kelvin scale gives

$$T(K) = 25.00°C + 273.15 = 298.15 K$$

Subtracting 273.15 allows conversion of a Kelvin temperature to a temperature on the Celsius scale. The equation is

$$t(°C) = T(K) - 273.15$$

For example, converting 315.00 K and 210.00 K to temperatures on the Celsius scale gives

$$t(°C) = 315.00 K - 273.15 = 41.85°C$$
$$t(°C) = 210.00 K - 273.15 = -63.15°C$$

In most calculations in this book, it is acceptable to use 273° rather than 273.15. This book uses both Celsius and Kelvin scales.

Another temperature scale with which you are probably more familiar is the Fahrenheit scale, which is also illustrated in Figure 2-9. In the United States, temperatures are frequently expressed in Fahrenheit degrees (°F). Water freezes at 32°F and boils at 212°F. As shown in Figure 2-9, 100 Celsius degrees are equal to 180 Fahrenheit degrees (the divisions between freezing and boiling). The Fahrenheit degree therefore, is equal to $\frac{100}{180}$ or $\frac{5}{9}$ the size of a Celsius degree. Figure 2-9 also shows that 0°C is equal to 32°F. Thus, the

Gabriel Daniel Fahrenheit (1686–1736), born in Danzig and settled in Holland, invented the mercury thermometer in 1714. For his zero point he chose the temperature produced by mixing equal amounts of snow and ammonium chloride. So that body temperature should be 96°, he chose 212° to be the boiling point of water. The absolute scale based on the Fahrenheit scale is known as the Rankine Scale.

relationship between Celsius and Fahrenheit temperatures is expressed by the equation

$$t(°C) = \frac{5}{9} \times [t(°F) - 32]$$

The conversion of, for example, the temperature on a hot summer's day of 95°F to a Celsius temperature gives

$$t(°C) = \frac{5}{9} \times (95°F - 32)$$
$$= \frac{5}{9} \times 63°F$$
$$= 35°C$$

Units of Heat

The SI unit of heat energy and of all forms of energy is the **joule** (pronounced "jool"). The joule (J) is not a fundamental unit. However, it is derived from the units for force and length, as you can see by looking at Table 2-3.

An older unit of heat, the calorie (cal), which is not an SI unit, was originally defined as the quantity of heat required to raise the temperature of 1 g of water from 14.5°C to 15.5°C. The calorie is now defined in terms of the SI unit: *one* **calorie** *is equal to 4.184 J.* The calorie and the joule are both rather small units for the quantities of heat typically dealt with in chemistry. Thus, the kilocalorie (kcal) and the kilojoule (kJ) are also commonly used as heat units.

The energy value of foods is reported in kilocalories, often (and somewhat confusingly) called Calories (with a capital C). For example, if a glass of whole milk contains 150 Calories, this means that the glass of milk will provide 150 kcal of energy if it is completely burned (metabolized) by the body. Table 2-5 lists the number of kilocalories in some common foods.

TABLE 2-5 AVERAGE ENERGY VALUES OF SOME COMMON FOODS

Food	Measure	Energy (kcal)	Energy (kJ)
apple	1 large	100	419
banana	1 medium	130	545
bread	1 slice	60	250
cottage cheese	100 cm³	93	390
cupcake	1 average	200	838
custard	100 mL	125	524
hamburger	113 g (1/4 lb) patty	420	1760
hot dog	1 average	138	578
honey	10 mL	67	280
ice cream	100 g	202	845
orange juice	200 mL	83	350
peach	1 medium	50	210
peanut butter	32 g (2 tbsp)	190	796
potato	1 medium	100	419
skim milk	200 mL	74	310
sugar (sucrose)	4 g (1 tsp)	16	67
whole milk	200 mL	133	557

Sample Problem 2.3

Convert 275 cal to joules and kilojoules.

Solution
Using the appropriate conversion factors based on 1 cal = 4.184 J and 1000 J = 1 kJ gives

$$275 \text{ cal} \times \frac{4.184 \text{ J}}{1 \text{ cal}} = 1150 \text{ J} \times \frac{1 \text{ kJ}}{1000 \text{ J}} = 1.15 \text{ kJ}$$

Practice Problems
1. Convert 55 cal to joules and kilojoules. *(Ans.)* 230 J, 0.23 kJ
2. Convert 669 J to calories. *(Ans.)* 160 cal

Heat Capacity and Specific Heat

How is heat measured? Consider a few liters of water. If the water is cooled, it loses heat to its surroundings. As a result, its temperature decreases. If the water is heated, its temperature increases. The quantity of heat lost or gained by water or any other substance can be found from the temperature change of that substance.

Three things determine the quantity of heat lost or gained during a temperature change—the nature of the matter changing temperature; the quantity, or mass, of matter changing temperature; and the size of the temperature change.

Heat capacity *is the amount of heat energy needed to raise the temperature of a given sample of matter by one Celsius degree (1C°)*. The amount of heat transferred depends upon the mass of the sample. To compare different materials, therefore, the heat capacities of equal masses, or *specific heats*, must be used. **Specific heat** *is the amount of heat energy required to raise the temperature of 1 g of a substance by one Celsius Degree (1C°)*. Values of specific heat can be given in units of joules per gram per Celsius degree [J/(g · C°)] or calories per gram per Celsius degree [cal/(g · C°)]. Table 2-6 gives the specific heats of some common substances. The specific heat of a substance can be determined by measuring the amount of heat energy gained or lost by a given mass that undergoes a measured temperature change.

TABLE 2-6 SPECIFIC HEATS OF SOME COMMON SUBSTANCES

Substance	Specific Heat J/(g · C°)	Substance	Specific Heat J/(g · C°)
Water (liquid)	4.184	Carbon, graphite	0.720
Water (ice)	2.092	Carbon, diamond	0.502
Water (steam)	2.013	Iron	0.444
Ethyl alcohol	2.452	Copper	0.385
Acetic acid	2.048	Silver	0.237
Sugar	1.250	Mercury	0.139
Aluminum	0.899	Gold	0.129
Salt (sodium chloride)	0.860	Lead	0.158

The equation is

$$\text{Specific heat} = \frac{\text{heat lost or gained (J or cal)}}{\text{mass (g)} \times \text{change in temperature (C°)}}$$

or, in abbreviated form,

$$c_p = \frac{q}{m \times \Delta t}$$

where c_p is the specific heat at a given pressure, q is the heat lost or gained, m is mass, and Δt represents the change in temperature

$$\Delta t = t_f - t_i = \text{final temperature} - \text{initial temperature}$$

Rearranging the equation for specific heat gives an equation that can be used to find the quantity of heat gained or lost with a change in temperature.

$$\text{Heat lost or gained (J)} = \text{mass (g)} \times c_p \text{ [J/(g·C°)]} \times \text{temperature change (C°)}$$

or

$$q = m \times c_p \times \Delta t$$

STUDY HINT

If the final temperature is lower than the initial temperature, Δt is negative. The quantity of heat lost by a system is also negative.

STUDY HINT

The specific heat of a substance also depends on whether the pressure or the volume is held constant. These differences are indicated by subscripts: c_p or c_v.

Sample Problem 2.4

A 4.0-g sample of glass was heated from 1°C to 41°C, a temperature increase of 40C°, and was found to have absorbed 32 J of heat.
(a) What is the specific heat of this type of glass?
(b) How much heat did the same glass sample gain when it was heated from 41°C to 71°C?

Solution
(a) The specific heat (c_p) of the glass is found from the equation above.

$$c_p = \frac{q \text{ (J)}}{m \text{ (g)} \times \Delta t \text{ (C°)}} = \frac{32 \text{ J}}{4.0 \text{ g} \times 40\text{C°}} = 0.20 \text{ J/(g·C°)}$$

(b) The rearranged specific-heat equation is used to find the heat gained when the glass was heated further.

$$q = m \times c_p \times \Delta t$$
$$= 4.0 \text{ g} \times 0.20 \text{ J/(g·C°)} \times (71°C - 41°C)$$
$$= 4.0 \text{ g} \times 0.20 \text{ J/(g·C°)} \times 30.\text{C°} = 24 \text{ J}$$

Practice Problem
1. Determine the specific heat of a material if a 12-g sample absorbed 48 J as it was heated from 20°C to 40°C. *(Ans.)* 0.20 J/(g·C°)

Section Review

1. Distinguish between temperature and heat, and state the SI units for each.
2. Give the names of the two temperature scales that are commonly used in chemistry.
3. Convert *(a)* 30°C to K *(b)* −20°C to K *(c)* 200K to °C.
4. Convert *(a)* 110 cal to J *(b)* 22 kJ to cal.
5. How much heat would be absorbed by 75 g of iron when heated from 22°C to 28°C?

1. Temperature is a measure of the average kinetic energy of the particles in a sample of matter. Heat is energy transferred between two systems at different temperatures. The SI unit for temperature is the kelvin (K); that for heat is the joule (J).
2. Celsius and Kelvin
3. (a) 303K (b) 253K (c) −73°C
4. (a) 460 J (b) 5300 cal
5. 200 J

Measurements and Solving Problems   45

2.3 Using Scientific Measurements

SECTION OBJECTIVES

- Distinguish between accuracy and precision.
- Determine the number of significant figures in measurements.
- Perform mathematical operations involving significant figures.
- Use scientific notation to write numbers and carry out arithmetical operations.

If a reported measurement is to be useful, it must include some indication of its uncertainty. The complete expression of a measured quantity must include the number value and the unit and also show how reliable the number is. This section discusses how measurements are reported and used so that the reliability of the numbers is understood by everyone.

Accuracy and Precision

The quality of a measurement depends on the measuring instrument and the skill of the person making the measurement. Uncertainty in measurements can result from limitations in accuracy or limitations in precision. The limitations may be those of the instrument or the experimenter, or both.

Accuracy *refers to the closeness of a measurement to the true or accepted value of the quantity measured.* An electronic balance that is often calibrated with a standard mass is likely to be more accurate than an old, mechanical balance that has been dropped many times.

Precision *refers to the agreement among the numerical values of a set of measurements of the same quantity made in the same way.* A chemist who frequently carries out a complex experiment is likely to have more precise results than someone just learning the experiment.

One way to visualize the difference between accuracy and precision is with some darts and a dartboard, as shown in Figure 2-10. The center of the bull's-eye represents the true or accepted value. The closer a dart is to the bull's-eye, the more accurate the throwing of the dart. The closer the darts in a group are to each other, the more precise that series of throws. A good darts player might produce a result like that on the left—both accurate and precise. On the other hand, the group of darts shown in the center was thrown with good precision, but poor accuracy. The player who threw the darts shown on the right had neither good accuracy nor good precision.

Suppose it is necessary to determine the density of a sample of chloroform ($CHCl_3$) at 20°C. One chemist obtained values of 1.495 g/mL, 1.476 g/mL, and 1.485 g/mL for the density of the sample. These measurements vary widely; the precision is poor. Another chemist repeated the measurements three times using the same equipment as the first chemist and found densities of 1.487

Figure 2-10 The darts at the center of the target show both accuracy and precision in throwing *(left)*. A tight cluster of darts indicates precision but not necessarily accuracy in throwing *(center)*. Some darts are thrown neither accurately nor precisely *(right)*.

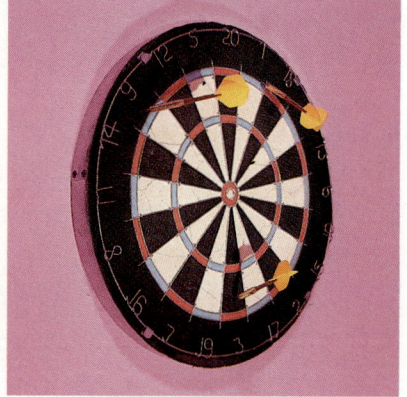

g/mL, 1.490 g/mL, and 1.488 g/mL. This group of measurements has good precision. Comparing these values with the accepted value of 1.489 g/mL for chloroform at 20°C shows that this group of measurements also has good accuracy. If a measurement can be compared to the correct value, its accuracy can be judged using percent error. **Percent error** *is calculated as follows:*

$$\text{Percent Error} = \frac{\text{Value}_{\text{accepted}} - \text{Value}_{\text{experimental}}}{\text{Value}_{\text{accepted}}} \times 100\%$$

Significant Figures

Suppose you were asked to measure the length of the nail shown in Figure 2-11 with the ruler drawn next to it. How long is the nail? It is longer than 3.6 cm, but how much longer? The tip of the nail appears to be less than halfway between 3.6 and 3.7, meaning that the nail is less than 3.65 cm long. Is it 3.61, 3.62, 3.63, or 3.64 cm long? The third digit in the length must be estimated. This estimated digit tells anyone reading the measurement that the nail is longer than 3.6 cm.

When scientists record and report the numerical values of their measurements, they want to indicate as much information as possible. They do not, however, want to give more information than is actually available and dependable. The compromise between these two goals is made by reporting only significant figures. *The* **significant figures** *in a measurement consist of all digits known with certainty plus one final digit, which is uncertain or estimated.*

In your study of chemistry, you will often make use of rulers, thermometers, graduated cylinders, or other instruments that have linear, graduated scales. For such measuring devices, the location of the estimated digit depends upon the smallest subdivision of the scale that appears on the device. An illustration of correctly reported measurements using thermometers with three different scales is shown in Figure 2-12. In each case, the final digit is an estimate of a digit between graduations on the scale. You can see that in correctly reported measurements, all digits retained are significant, but the last digit is uncertain.

Figure 2-11 Measuring the length of a nail.

It is unfortunate that the term *significant* was chosen. It leads students to believe that *nonsignificant* digits are not important, when, in fact, they indicate the size of the number. For example, 600 000 000.

Figure 2-12 Thermometer A has 10-degree calibrations, thermometer B has 1-degree calibrations, and thermometer C has 0.1-degree calibrations. Therefore the temperature can be read from thermometer A to two significant figures, from thermometer B to three significant figures, and from thermometer C, at temperatures higher than 10 degrees, to four significant figures. The underlined digit in each temperature is estimated.

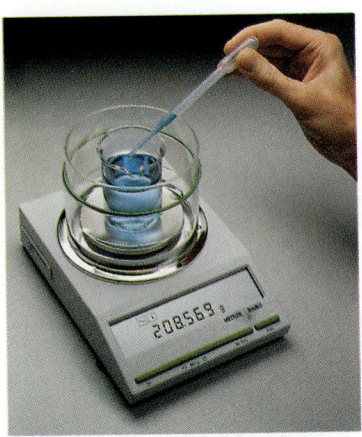

Figure 2-13 An electronic balance with a digital scale readable to 1 mg. Just place an item on the pan in such a balance and read its mass to 1 mg in an instant. The readout provides the final, estimated digit.

As a student of chemistry, you will use significant figures in three different ways:
1. You will make measurements and report them to others. Here, the number of significant figures is easy to determine. It will usually be the number of digits you are able to read with certainty from the measuring device, plus one digit that you estimate. Figure 2-13 shows a digital readout, which requires no estimate.
2. You will interpret measurements reported by others.
3. You will need to keep track of significant figures when measurements are used in calculations.

The following sections explain how to deal with the kinds of situations described in items 2 and 3.

Determining the number of significant figures Determining the number of significant figures in the number 4.643 is easy—just count the digits. This number has four significant figures. Some rules are needed to count significant figures in numbers containing zeros because the zeros can be significant or not, depending on their location. Table 2-7 gives the complete set of rules for counting significant figures, together with examples of the use of each rule.

TABLE 2-7 RULES FOR DETERMINING SIGNIFICANT FIGURES INVOLVING ZEROS

Rule	Examples of Measurements	Number of Significant Figures
1. All nonzero digits are significant.	438 g 26.42 m 1.7 cm 0.653 L	3 4 2 3
2. All zeros between two nonzero digits are significant.	506 dm 10 050 mL 900.43 kg	3 4 5
3. Zeros to the right of a nonzero digit, but to the left of an understood decimal point, are not significant. If such zeros are known to have been measured, however, they are significant and should be specified as such by inserting a decimal point to the right of the zero.	4830 km 60 g 4830. L 60. K	3 1 4 2
4. In numbers less than one and greater than negative one, zeros to the right of a decimal point that are to the left of the first non-zero digit are never significant. They are simply placeholders.	0.06 g 0.0047 L 0.005°C	1 2 1
5. In numbers less than one and greater than negative one, the zero to the left of the decimal is never significant. It is there to make sure the decimal point is not overlooked.	0.8 g	1
6. All zeros to the right of a decimal point and to the right of a nonzero digit are significant.	8.0 dm 16.40 g 35.000 L 1.60 sec	2 4 5 3

Sample Problem 2.5

How many significant figures are contained in each of the following measurements? Which significant-figure rule from Table 2-7 applies to each digit? *(a)* 37.4 g *(b)* 6070. dm *(c)* 0.009 03 km *(d)* 0.0540 cm

Solution

(a) With no zeros in the number, rule 1 applies to each digit.

    1 1 1   Rule
    37.4 g   three significant figures

(The numbers of the rules that apply to each digit are written above the digits.)

(b) The first zero is between two nonzero digits. The second zero is shown to be significant by the decimal point.

    1 2 1 3   Rules
    6070. dm   four significant figures

(c) The first three zeros are not significant because they precede the first nonzero digit. The next zero is significant because it is between nonzero digits.

    5 4 4 1 2 1   Rules
    0.009 03 km   three significant figures

(d) The first two zeros are not significant because they precede the first nonzero digit; the last zero has been added to show uncertainty and is significant

    5 4 1 1 6   Rules
    0.0540   three significant figures

Practice Problems
1. Determine the number of significant figures in each of the following:
   (a) 605.03 g *(Ans.)* 5   (c) 0.006 03 mL *(Ans.)* 3   (e) 450 m *(Ans.)* 2   (g) 300. dm *(Ans.)* 3.
   (b) 0.8030 L *(Ans.)* 4   (d) 20.00 cm *(Ans.)* 4   (f) 0.000 070 kg *(Ans.)* 2   (h) 350.0 K *(Ans.)* 4.
2. Write "two hundred meters" in numerical form with the indicated number of significant figures: (a) 1 *(Ans.)* 200 m   (b) 3 *(Ans.)* 200. m   (c) 4 *(Ans.)* 200.0 m

Rounding off The measurements to be used in a calculation often contain different numbers of significant figures. The answer must not have more significant figures than are justified by the measurements. Especially if you use an electronic calculator, the number of digits in your answer often suggests a greater degree of certainty than really exists. For example, the density of an object with a mass of 1.5 g and a volume of 4 mL can be calculated using the formula for density:

$$\text{density} = \frac{\text{mass}}{\text{volume}} = \frac{1.5 \text{ g}}{4 \text{ mL}} = 0.375 \text{ g/mL}$$

This answer, although arithmetically correct, contains three digits that, according to the rules, would be interpreted as significant. The initial measurements of 1.5 g and 4 mL contained, respectively, two and one significant figures. In division and multiplication, the answer must not have

Figure 2-14 Rounding off is especially important when using an electronic calculator. It would be misleading to report all digits in the calculator answer to the problem 96 divided by 7.

Students will find rules 4 and 5 for rounding off new and confusing. The rationale for the latter two rules is that always rounding up tends to skew a large volume of data to a higher-than-average value. Since there are 5 even and 5 odd digits, rounding down half the time and rounding up half the time gives a better average value.

more significant figures than the number in the calculation with the fewest significant figures. The answer of 0.375 given by the calculator must therefore be rounded off to only one significant figure, or 0.4 g/mL.

In order to reflect the true uncertainty of calculations with measurements, the number of significant figures in the answers should be determined according to specific rules discussed in the next section. First, however, it will be helpful to review the rules for rounding off numbers. These rules are given in Table 2-8. Note that the procedure for rounding off is governed by the digit to the immediate right of the last digit to be retained. Except in special cases, rounding off should be done when calculations are finished. Otherwise, a significant error may be introduced.

TABLE 2-8 RULES FOR ROUNDING OFF NUMBERS

If the digit immediately to the right of the last significant digit you want to retain is:	Then the last significant digit should:	Example (each rounded off to 3 significant figures):
1. Greater than 5	Be increased by 1	42.68 g → 42.7 g
2. Less than 5	Stay the same	17.32 m → 17.3 m
3. 5, followed by nonzero digit(s)	Be increased by 1	2.7851 cm → 2.79 cm
4. 5, not followed by nonzero digits(s), and preceded by an odd digit	Be increased by 1	4.635 kg → 4.64 kg (since 3 is odd)
5. 5, not followed by nonzero digits(s), and the preceding significant digit is even	Stay the same	78.65 mL → 78.6 mL (since 6 is even)

Sample Problem 2.6

Round off each of the following measurements to the indicated number of significant figures:
(a) 35.27 g to 3 significant figures
(b) 0.414 kL to 2 significant figures
(c) 87.257 dm to 3 significant figures
(d) 1.35K to 2 significant figures
(e) 6250 cm to 2 significant figures

Solution
(a) 35.27 g is rounded to 35.3 g because 7 is a number larger than 5 (Rule 1)
(b) 0.414 kL is rounded to 0.41 kL because the final 4 is a number less than 5 (Rule 2).
(c) 87.257 dm is rounded to 87.3 dm because the 5 is followed by a nonzero digit (Rule 3).
(d) 1.35 K is rounded to 1.4 K because the 5 is preceded by an odd number (Rule 4).
(e) 6250 cm is rounded to 6200 cm because the 5 is preceded by an even number (Rule 5). Note that zeros, though not significant, must be used here to retain the size of the number.

Practice Problems
1. Round off each of the following measurements to the indicated number of significant figures:
   (a) 6.42 g to 2 significant figures *(Ans.)* 6.4 g   (b) 7.535 mL to 3 significant figures *(Ans.)* 7.54 mL   (c) 4.681 cm to 2 significant figures *(Ans.)* 4.7 cm   (d) 56.45 kg to 3 significant figures *(Ans.)* 56.4 kg
2. Round off 279.55 m to the indicated number of significant figures:
   (a) 2 *(Ans.)* 280 m   (b) 3 *(Ans.)* 280. m   (c) 4 *(Ans.)* 279.6 m   (d) 1 *(Ans.)* 300 m.

Addition and subtraction involving significant figures   If 3 000 000 (1 significant figure) and 4 192 378 (7 significant figures) are added, the result is 7 192 378. This answer indicates 7 significant figures. All the zeros in the first addend, 3 000 000, are not significant and are therefore uncertain. *In addition or subtraction, the arithmetic result should be rounded off so that the final digit is in the same place as the leftmost uncertain digit.*

For example, in the following subtraction (the uncertain digit in each measurement is underlined)

$$213.6\underline{7}$$
$$\underline{9\underline{8}} \quad \text{leftmost uncertain digit}$$
$$11\underline{5}.6\underline{7}, \quad \text{rounded to } 11\underline{6}$$

the final digit in the answer must be in the ones place because in the numbers subtracted the leftmost uncertain digit is the 8 in 98.

Multiplication and division involving significant figures   *The arithmetic product or quotient should be rounded off to the same number of significant figures as in the measurement with the fewest significant figures.* Suppose you want to find the following product: 12.0 cm × 4.3 cm. The answer should have two significant figures, the same number as in 4.3:

$$12.0 \text{ cm} \times 4.3 \text{ cm} = 51.6 \text{ cm}^2, \text{ rounded to } 52 \text{ cm}^2$$

Sample Problem 2.7

Perform each of the following calculations and express the answers with the correct number of significant figures.   (a) 6.43 mL + 2.015 mL   (b) 2.50 g / 0.04 cm³.

Solution
(a)   6.4$\underline{3}$ mL

   + 2.01$\underline{5}$ mL

   8.4$\underline{45}$ mL,   rounded to 8.4$\underline{4}$ mL

(b) $\dfrac{2.50 \text{ g}}{0.04 \text{ cm}^3} = \underline{62.5}$ g/cm³, rounded to $\underline{60}$ g/cm³

Practice Problems
Perform the following calculations and express answers in the correct number of significant figures:
1. 6.821 m + 2.0 m + 0.45 m *(Ans.)* 9.3 m
2. 107.38 L − 65 L *(Ans.)* 42 L
3. 26.50 dm × 0.062 dm *(Ans.)* 1.6 dm²
4. 1.30 g / 0.02 cm³ *(Ans.)* 60 g/cm³

Exact conversion factors   Certain conversion factors do not limit the number of digits in the answer to a calculation. Consider the relationships

$$7 \text{ days} = 1 \text{ week} \qquad 100 \text{ cm} = 1 \text{ m}$$

There is no uncertainty in either of these statements because they are the result of definitions, not measurements. There are *exactly* 7 days in 1 week and *exactly* 100 cm in 1 m. *Exact conversion factors* have no uncertainty and do not limit the number of digits in a calculation. Exact numbers can be considered to have an unlimited number of significant figures. For example, in converting 7866 cm to meters

$$7866 \text{ cm} \times \dfrac{1 \text{ m}}{100 \text{ cm}} = 78.66 \text{ m}$$

STUDY HINT
Conversion factors based on units modified by SI prefixes are all exact conversion factors.

STUDY HINT
1 cal = 4.184 J is used as an exact conversion factor.

STUDY HINT
In algebra: $2a^5 + 6a^5 = 8a^5$.

the answer correctly has four significant digits because it is not limited by the 100-cm factor.

Suppose that, by counting, you find that there are 25 desks in a classroom. The conversion factor 25 desks/classroom would also be exact. Exact conversion factors result from definitions or counting.

Scientific Notation

In science, numbers can range from very small to very large, depending on what is being measured. A very large number that is extremely important in chemistry, called *Avogadro's number*, is

$$602\ 213\ 674\ 000\ 000\ 000\ 000\ 000$$

The mass of an electron, on the other hand, is a very small number:

$$0.000\ 000\ 000\ 000\ 000\ 000\ 000\ 000\ 000\ 9109 \text{ kg}$$

Because such numbers are tedious to read and write, scientists use a shorthand technique called scientific notation.

In **scientific notation,** *numbers are written in the form*

$$M \times 10^n$$

where M, the coefficient, is a number greater than or equal to 1 but less than 10 and n, the exponent, is an integer. For example, the speed of sound in water is about 1500 m/s. Written in scientific notation, this quantity is

$$1500 \text{ m/s} = (1.5 \times 1000 \text{ m/s}) = 1.5 \times 10^3 \text{ m/s}$$

An influenza virus is about 0.000 12 mm in diameter. In scientific notation, this quantity is

$$0.000\ 12 \text{ mm} = 1.2 \times \frac{1}{10\ 000} \text{ mm} = 1.2 \times 10^{-4} \text{ mm}$$

Avogardo's number and the mass of an electron are written in scientific notation as 6.022×10^{23} and 9.109×10^{-31} kg, respectively.

Figure 2-15 An illustration of how to express numbers in scientific notation. The number of places the decimal point must be moved gives the exponent for 10.

Scientific notation makes it easy to indicate the correct number of significant figures. In $M \times 10^n$, all digits, both zero and nonzero, expressed in M, are significant. For example, notice that the volume, 10 050 mL, in Table 2-7 is assumed to have four significant figures as indicated by the scientific notation: 1.005×10^4 mL. If this measurement was known to have five significant figures, it would be written 1.0050×10^4 mL. The distance from the earth to the sun is 149 740 000 km and has five significant figures. In scientific notation, it is 1.4974×10^8 km. Rounded off to three significant figures, the distance is 1.50×10^8 km.

Refer to Figure 2-15 as you read how to change a number from the long form to scientific notation:

1. Determine M by moving the decimal point in the original number to the left or to the right so that only one nonzero digit remains to the left of it.
2. Determine n by counting the number of places that the decimal point was moved. If moved to the left, n is positive; if moved to the right, n is negative.

To express 100 g as a two-significant-digit number, one would have to use scientific notation. As written, 100 g has only one significant figure. If one includes a decimal point, 100. g, then the number has 3 significant figures. To show two significant figures one would have to write it as 1.0×10^2 g.

Sample Problem 2.8

Write each of the following numbers in scientific notation. (a) 85 000 000 (b) 0.0009

Solution
(a) The decimal point in 85 000 000 must be moved 7 places to the left.

$$85\ 000\ 000 = 8.5 \times 10^7$$

(b) In 0.0009, the decimal point must be moved 4 places to the right, so the sign of the exponent must be negative.

$$0.0009 = 9 \times 10^{-4}$$

Practice Problems
Write each of the following numbers in scientific notation:

1. 74 000 *(Ans.)* 7.4×10^4
2. 0.000 005 *(Ans.)* 5×10^{-6}
3. 30 000 000 000 *(Ans.)* 3×10^{10}
4. 864 000 *(Ans.)* 8.64×10^5
5. 0.000 602 *(Ans.)* 6.02×10^{-4}
6. 4700×10^5 *(Ans.)* 4.7×10^8

To write out a number in the usual long form that is given in scientific notation, the above procedure is reversed.

Sample Problem 2.9

Write each of the following numbers in the usual long form. (a) 7×10^4 (b) 2.31×10^{-7}.

Solution
(a) The positive exponent shows that the decimal point must be moved to the right. Changing this to long form gives

$$7 \times 10^4 = 70\ 000$$

Sample Problem 2.9 *continued*

(b) The negative exponent shows that the decimal point must be moved to the left to give a number smaller than 1. Moving the decimal place to the left seven places gives

$$2.31 \times 10^{-7} = 0.000\,000\,231$$
$$\phantom{2.31 \times 10^{-7} = 0.000\,000\,2}765\,432\,1$$

Note that the zero before the decimal point is not included in counting the zeros that must be added.

Practice Problems
Express each of the following numbers in the usual, long form:
(1.) 5.3×10^4 *(Ans.)* 53 000   (2.) 7×10^{-5} *(Ans.)* 0.000 07   (3.) 4.21×10^{10} *(Ans.)* 42 100 000 000

STUDY HINT
In algebra:
$2a^7 \times 3a^{10} = 6a^{17}$.

Stress the difference in handling exponents when adding and subtracting and when multiplying and dividing.

STUDY HINT
In algebra: $\dfrac{8a^7}{2a^5} = 4a^2$

STUDY HINT
To restate numbers in scientific notation: if M is made larger, the exponent n becomes smaller; if M is made smaller, the exponent n becomes larger. The size of the change in n is equal to the number of places the decimal point is moved.

Mathematics involving scientific notation   Because many measurements in science are expressed in scientific notation, arithmetic operations involving such numbers are frequently necessary. The rules governing such operations are the same as the rules for exponents that you learned when you studied algebra.

In each case, the M part of the answer must be rounded to the correct number of significant figures. Then, if necessary, the decimal point can be moved and the exponent changed so that M is greater than or equal to 1 and less than 10.

1. Addition and subtraction of expressions in scientific notation can be performed only if the expressions have the same exponent. The coefficients themselves can then be added or subtracted, but the exponent stays the same. If the initial expressions do not have the same exponent, they must first be rewritten so that they do have the same exponent. Consider the following example:

$$\begin{array}{r} 6.3 \times 10^4 \\ +2.1 \times 10^5 \end{array} \quad \text{is written as} \quad \begin{array}{r} 0.63 \times 10^5 \\ +2.1 \times 10^5 \\ \hline 2.73 \times 10^5 \end{array}$$

$$\text{rounded to } 2.7 \times 10^5$$

2. In the multiplication of expressions involving scientific notation, the coefficients themselves are multiplied and the exponents are added. Thus $(4.5 \times 10^8) \times (3.17 \times 10^5) = 14.265 \times 10^{13}$, which is rounded to 14×10^{13}, and restated as 1.4×10^{14}. Note that, to restate the number, the decimal point was moved to the left one place and the exponent was increased by 1.

3. In the division of expressions involving scientific notation, the coefficients themselves are divided and the exponent in the denominator is subtracted from that in the numerator. As shown in the second of the following two examples, the exponent can change sign in division.

$$\frac{6.85 \times 10^7}{2.0 \times 10^3} = 3.425 \times 10^4, \text{ rounded to } 3.4 \times 10^4$$

$$\frac{9.9 \times 10^5}{3.0 \times 10^{10}} = 3.3 \times 10^{-5}$$

Sample Problem 2.10

Perform the indicated operations. Express all answers in scientific notation and with the correct number of significant figures. (a) $(3 \times 10^7) + (5 \times 10^6)$ (b) $(4.3 \times 10^8) \times (2.51 \times 10^{-14})$ (c) $(6.81 \times 10^{10}) \div (1.2 \times 10^{-4})$.

Solution
(a) Before performing the indicated addition, both expressions must be written in terms of the same exponent. For example, 5×10^6 can be rewritten as 0.5×10^7. Then $(3 \times 10^7) + (0.5 \times 10^7) = 3.5 \times 10^7 = 4 \times 10^7$.

(b) The coefficients must be multiplied and the exponents added:
$(4.3 \times 10^8) \times (2.51 \times 10^{-14}) = 10.793 \times 10^{-6}$, rounded to 11×10^{-6} and restated as 1.1×10^{-5}

(c) The coefficients must first be divided and the exponents then subtracted:
$(6.81 \times 10^{10}) \div (1.2 \times 10^{-4}) = 5.675 \times 10^{14}$, rounded to 5.7×10^{14}

Practice Problems
Perform the indicated operations and express all answers in scientific notation with the correct number of significant figures:
1. $(4.6 \times 10^5) + (3.2 \times 10^5)$ (Ans.) 7.8×10^5
2. $(5.7 \times 10^6) \times (2.8 \times 10^9)$ (Ans.) 1.6×10^{16}
3. $(7.7 \times 10^{-12}) \div (2.5 \times 10^5)$ (Ans.) 3.1×10^{-17}

Section Review

1. A student records volume readings of 8.42 mL, 8.41 mL, and 8.40 mL. What can you say about the precision of the student readings?
2. How many significant figures are indicated in each of the following measurements? (a) 49.35 g (b) 104.07 cm (c) 0.0082 kL (d) 20.00 m (e) 150 mL (f) 0.050 60 kg (g) 200K
3. Round off 785.6553 cm to the indicated number of significant figures: (a) 5 (b) 3 (c) 2 (d) 4 (e) 1.
4. Perform each of the following calculations and express all answers in the correct number of significant figures: (a) 430.62 m + 6.1 m + 10.5300 m (b) 750 cm × 24.32 cm (c) 804.00 g / 20 cm³.
5. Write each of the following numbers in scientific notation: (a) 45 000 (b) 0.000 0027 (c) 301×10^{21}.
6. Perform the following calculations. Express all answers with the correct number of significant figures and in scientific notation: (a) $(6.8 \times 10^5 \text{ g}) - (3.1 \times 10^4 \text{ g})$ (b) $(7.32 \times 10^4 \text{ dm}) \times (4.1 \times 10^{-10} \text{ dm})$ (c) $(9.05 \times 10^{-8} \text{ g}) \div (5.3 \times 10^{-3} \text{ mL})$.

Answers:
1. These measurements are precise.
2. (a) 4 (b) 5 (c) 2 (d) 4 (e) 2 (f) 4 (g) 1
3. (a) 785.66 cm (b) 786 cm (c) 790 cm (d) 785.7 cm (e) 800 cm
4. (a) 447.2 m (b) 18 000 cm² (c) 40 g/cm³
5. (a) 4.5×10^4 (b) 2.7×10^{-6} (c) 3.01×10^{23}
6. (a) 6.5×10^5 g (b) 3.0×10^{-5} dm² (c) 1.7×10^{-5} g/mL

2.4 Solving Quantitative Problems

Much of your work in chemistry will focus on solving quantitative problems. Many such problems will be word problems. Learning to solve a variety of problems with ease requires practice. The steps outlined in this section will help you analyze and solve even complicated word problems correctly.

SECTION OBJECTIVES
- Use the suggested four-step method to solve quantitative problems.
- Define and state equations for directly and indirectly proportional relationships.
- Plot graphs showing direct and indirect proportionality between two variables.

Four Steps for Solving Quantitative Problems

STEP 1
Analyze

Read the problem carefully at least twice. Note any important descriptive terms that add meaning to the problem. For example, are you dealing with solutions or mixtures? Is a reaction occurring? List all the data given in the problem. Next, identify what you are asked to find including the units. Many people find it helpful to sketch what they think is happening in the problem. Making a sketch labeled with the data and what you are asked to find shows you truly understand the problem.

STEP 2
Plan

Your plan for solving a problem should show how the information given in the problem is used to find a solution. Reread the problem to note if there is any information missing from the problem that you will need to find the answer. To finish your plan, decide which conversion factors or algebraic equations relate to what you are asked to find. Your plan might consist of a single calculation using one conversion factor or a series of calculations and conversion factors.

The Sample Problems will show you a summary pathway from the given information to the unknown. The Sample will provide an explanation of the pathway and the setup of data. Make sure you understand how the plan relates to the setup. The factor-label method is a very useful tool for verifying the setup. If the units cancel to give the desired units, the setup is probably correct.

STEP 3
Compute

This step involves substituting the data and conversion factors into your plan from Step 2. Calculate the answer, cancel the units, and round off to the correct number of significant figures.

STEP 4
Evaluate

Always look back at your answer to determine if it is reasonable. There are various ways to verify your answer.
1. Make an estimate of the answer. Is your calculated value close to the estimate?
2. Does the setup provide the correct units for the answer?
3. Does the answer make sense from what you know about values for the problem? For example, if you calculate the velocity of a car to be 5500 km/h, you know this answer is not reasonable. A car traveling at the maximum speed limit on a highway is only moving about 100 km/h.

The Sample Problems in this book provide strategies to solve routine problems in chemistry. You need to get in the habit of organizing your strategies so that your problem-solving ability improves.

Sample Problem 2.11

How many seconds are there in one year? Assume that there are exactly 365 days per year.

Solution

Step 1. Analyze   Given: 1 year
                  Unknown: seconds

Step 2. Plan   year ⟶ days ⟶ hours ⟶ minutes ⟶ seconds

Solving this problem requires using four conversion factors. Such a series of conversion factors can be combined in a single setup.

$$1 \text{ year} \times \frac{\text{days}}{\text{year}} \times \frac{\text{hours}}{\text{day}} \times \frac{\text{minutes}}{\text{hour}} \times \frac{\text{seconds}}{\text{minute}}$$

Sample Problem 2.11 continued

Step 3. Compute $1 \text{ year} \times \dfrac{365 \text{ days}}{1 \text{ year}} \times \dfrac{24 \text{ hr}}{1 \text{ day}} \times \dfrac{60 \text{ min}}{1 \text{ hr}} \times \dfrac{60 \text{ s}}{1 \text{ min}} = 31\,536\,000.\text{ s or } 3.153\,6000 \times 10^7 \text{s}$

Step 4. Evaluate An answer of about 10^7 seems reasonable ($300 \times 20 \times 60 \times 60$ is approximately 2×10^7). Units have canceled to give seconds, the desired unit. Note that each of the conversion factors is based upon a definition and therefore the answer is not limited in number of significant figures.

Practice Problems
1. A clock gains 0.25 seconds per minute. How many seconds will the clock gain in exactly 180 days? *(Ans.)* 6.5×10^4 s
2. How many kiloliters are contained in 62.5 cm³? *(Ans.)* 6.25×10^{-5} kL

Sample Problem 2.12

What is the volume of a piece of aluminum that has a mass of 22.5 g? The density of aluminum is 2.7 g/cm³.

Solution
Step 1. Analyze Given: mass = 22.5 g aluminum, density = 2.7 g/cm³
Unknown: volume of aluminum

Step 2. Plan mass, density $\longrightarrow$ volume

The mass and volume of a specific substance are related by the density of that substance, which is mass per unit volume. To find volume, the equation for density must be rearranged:

$$d \text{ (g/cm}^3) = \dfrac{m \text{ (g)}}{V \text{ (cm}^3)} \quad V \text{ (cm}^3) = \dfrac{m \text{ (g)}}{d \text{ (g/cm}^3)}$$

Step 3. Compute $V = \dfrac{22.5 \text{ g}}{2.7 \text{ g/cm}^3} = 8.3 \text{ cm}^3$

Step 4. Evaluate The value of the answer is reasonable compared to an estimated answer of 7 cm³ (21/3). Units have canceled as desired. The answer is correctly given with two significant figures, as in the factor 2.7 g.

Practice Problem
Find the mass of an object with a density of 1.8 g/cm³ and a volume of 4.5 cm³. *(Ans.)* 8.1 g

Sample Problem 2.13

What quantity of heat in calories is needed to increase the temperature of 35 g of copper by 7.2C°.

Solution
Step 1. Analyze Given: mass = 35 g copper
temperature change = 7.2C°
specific heat = 0.385 J/(g · C°) (from Table 2-6)
Unknown: heat gained or lost in calories

Sample Problem 2.13 *continued*

Step 2. Plan $m, \Delta t, c_p \longrightarrow q \text{ (joules)} \longrightarrow q \text{ (calories)}$

Solving this problem requires using specific heat. The equation for specific heat must be arranged and solved for heat. A conversion factor is also needed for converting joules to calories. The equation for specific heat can be multiplied by a unit conversion factor to convert joules into calories.

$$q = m \times c_p \times \Delta t \times \frac{\text{cal}}{\text{J}}$$

Step 3. Compute $35 \text{ g} \times \frac{0.384 \text{ J}}{\text{g} \times \text{C}°} \times 7.2°\text{C} \times \frac{1 \text{ cal}}{4.184 \text{ J}} = 23 \text{ cal}$

Step 4. Evaluate The answer is reasonable in comparison with an estimated value of 30 cal ($30 \times 8 \times 0.5 / 4$). Units have canceled to give the answers in calories, as desired. The answer is correctly limited to two significant figures by the factors 7.2 and 35.

Practice Problem
What quantity of heat, in kilocalories, is required to raise the temperature of 22.5 g of lead from 18°C to 30.°C? The specific heat of lead is 0.158 J/(g · C°). *(Ans.)* 1.0×10^{-2} kcal

Proportional Relationships

Solving quantitative problems often involves using proportional relationships between variables. *A* **variable** *is a quantity that can change in value.* For example, temperature is a variable.

Direct proportion *Two variables are* **directly proportional** *to each other if dividing one by the other gives a constant value.* In other words, directly proportional values have a constant ratio. For example, for a given substance, mass and volume are directly proportional to each other. The mass of a given substance divided by its volume is the density of that substance. The general equation for a direct proportion between two variables is

$$\frac{y}{x} = k$$

where x is one variable, y is the other variable, and the constant value of the ratio, known as the proportionality constant, is represented by k. A direct proportion can also be represented as

$$y = kx$$

In the equation for density, mass is equivalent to y, volume is equivalent to x, and k is equivalent to density:

$$\frac{\text{mass}}{\text{volume}} = \text{density or mass} = \text{density} \times \text{volume}$$

At 3.98°C, the density of water is equal to 1.00 g/mL. Figure 2-16 gives some values of the mass and volume of different samples of water at 3.98°C and a graph of these values. If the variables in a directly proportional relation are plotted, a straight line, or linear graph, is formed.

Figure 2-16 A graph of the mass of water at 3.98°C as a function of its volume. At this temperature, the density of water has the constant value of 1.00 g/mL. Mass and volume (at a constant temperature) are directly proportional variables. All directly proportional variables give straight-line graphs.

Inverse proportion *Two variable are* **inversely proportional** *to each other if their product has a constant value.* For example, the time it takes to travel a certain distance is inversely proportional to the speed at which you travel. If your speed increases, the time required to travel the fixed distance decreases.

The general equations for an inverse proportion between two variables are

$$xy = k \quad \text{or} \quad y = \frac{k}{x}$$

where x and y are the variables and k is the proportionality constant. The relationship between the speed and time it takes to travel a fixed distance is:

$$\text{speed} \times \text{time} = \text{distance} \quad \text{or} \quad \text{time} = \frac{\text{distance}}{\text{speed}}$$

Figure 2-17 gives some values of the time taken to travel a constant distance of 600 m at different speeds and a graph of these values. A graph of inversely proportional quantities is always of the shape shown. This shape is known as a hyperbola.

Figure 2-17 A graph of the time to travel a constant distance of 600 m as a function of the speed of travel. The time and speed of travel for a constant distance are inversely proportional variables.

Section Review

1. List and explain the four steps in problem solving.
2. Calculate the number of seconds in exactly one week.
3. What is the volume of a 12-g piece of gold? ($d = 19.3$ g/cm³)
4. An object travels at a speed of 7500 cm/s. How far will it travel in a day?
5. What type of relationship is illustrated between the variables A and B in each of the following: (a) As A increases, B decreases, and $A \cdot B = 35$, (b) As A increases, B increases, and $A/B = \frac{1}{2}$.

Chapter Summary

- All scientists have agreed to use the SI system of measurement.
- The SI system has seven fundamental units: meter, kilogram, second, kelvin, mole, ampere, and candela.
- The factor-label method is a valuable problem solving tool that treats units in calculations as if they are algebraic factors.
- Temperature is a measure of the average kinetic energy of the particles in a sample of matter.
- All measurements are subject to some uncertainty.
- Accuracy refers to the closeness of a measurement to the true or accepted value. Precision refers to the agreement among the numerical values of a set of measurements made in the same way.
- Significant figures convey the reliability or certainty of measurements.
- Scientific notation is a shorthand way of writing numbers with many decimal places and also a convenient way of indicating significant figures.
- Solving quantitative problems is simpler when done in a methodical way. This book uses a four-step method. The steps are: *(1)* Analyze *(2)* Plan *(3)* Compute and *(4)* Evaluate.
- Many quantities, or variables, that you will encounter in chemistry are related to each other by either a direct proportion or an inverse proportion.

1. *Step 1. Analyze* In this step, the problem is carefully read, and the given and unknown variables are identified.
   Step 2. Plan Appropriate conversion factors or algebraic equations are selected to relate the given to the unknown.
   Step 3. Compute All necessary values are substituted into the setup and the answer is calculated.
   Step 4. Evaluate The appropriateness of the answer is determined.
2. $6.048\ 00 \times 10^5$ s
3. $V = m/D = 0.62$ cm³
4. 6.5×10^8 cm
5. (a) Inverse proportion
   (b) Direct proportion

STUDY HINT

A variable with assigned values is referred to as an independent variable. A variable with determined values is referred to as a dependent variable. It is customary to plot the independent variable on the x axis and the dependent variable on the y axis.

Chapter 2 Review

Measurements and Solving Problems

VOCABULARY

accuracy (46)
calorie (43)
conversion factor (34)
degree Celsius (41)
density (37)
derived unit (36)
directly proportional (58)
factor-label method (34)
fundamental unit (32)
heat (40)
heat capacity (44)
inversely
 proportional (59)
joule (43)
kelvin (41)
kilogram (33)
liter (37)
meter (33)
precision (46)
second (34)
SI (31)
scientific notation (52)
significant figures (47)
specific heat (44)
standards of
 measurement (31)
temperature (40)
unit of
 measurement (31)
variable (58)
volume (36)

REVIEWING CONCEPTS

1. **a.** What system of measurement is used by scientists around the world?
   b. What abbreviation is used for the name of this system?
   c. From what system were many of its units derived? (2.1)

2. Write each of the following according to the international system for writing numbers:
   a. 6500
   b. 78420
   c. 207891
   d. 0.3500
   e. 0.920400
   f. 127.80000
   g. 42600.3500 (2.1)

3. List the seven SI fundamental units, identify the quantity measured by each, and write the corresponding symbol for each unit. (2.1)

4. Give three examples of a distance that is approximately one meter in length. (2.1)

5. Identify the SI unit(s) that would be most appropriate for measuring each of the following:
   a. the length of a new pencil
   b. the width of your classroom
   c. the length of your arm
   d. the diameter of pencil lead
   e. your height
   f. the distance from your home to the next town
   g. the distance from the earth to the sun
   h. the diameter of a bacterial cell (2.1)

6. Identify the SI unit(s) that would be most appropriate for expressing the mass of each of the following:
   a. a person
   b. a banana
   c. your daily vitamin C requirement
   d. a slice of cheese (2.1)

7. What is the factor-label method? (2.1)

8. What is a conversion factor? (2.1)

9. **a.** What is a derived unit?
   b. Give two examples of derived units that are important in chemistry. (2.1)

10. What is the derived SI unit and corresponding symbol for
    a. area   **b.** volume (2.1)

11. What is the relationship between the SI unit for volume and that for length? (2.1)

12. **a.** What is a liter?
    b. What is the relationship between cubic centimeters and milliliters? (2.1)

13. Identify the unit(s) that would be most appropriate for expressing the volume of each of the following:
    a. a glass of milk
    b. the amount of water required to fill a bathtub
    c. a child's dosage of cough syrup
    d. a truckload of sand (2.1)

14. **a.** What is meant by the density of a material?
    b. What formula can be used to find the density of an object?
    c. What are three SI or metric units that can be used to express density? (2.1)

15. Why do temperatures usually accompany the density values of many substances in charts or tables? (2.2)
16. How is temperature related to the energy of the particles in a sample of matter? (2.2)
17. What is the normal direction of heat flow? (2.2)
18. What are the two fixed points that form the basis of the Celsius temperature scale? (2.2)
19. Explain what is meant when we say that a banana contains 130 Calories. (2.2)
20. a. What is meant by the heat capacity of a material?
    b. What is specific heat? (2.2)
21. Identify and define the two limitations that contribute to uncertainty in measurements. (2.3)
22. Describe each of the following sets of measurements in terms of precision.
    a. 458.43 m, 458.41m, and 458.40 m
    b. 462.78 m, 451.33 m, and 475.89 m
    c. 468.49 m, 468.51 m, and 468.52 m (2.3)
23. What determines the location of the estimated digit in making a measurement? (2.3)
24. Determine the number of significant figures in each of the following measurements.
    a. 65.42 g
    b. 385 L
    c. 0.14 mL
    d. 709.2 m
    e. 5006.12 kg
    f. 400 dm
    g. 260. mm
    h. 0.47 cg
    i. 0.0068 km
    j. 7.0 cm^3
    k. 36.00 g
    l. 0.0070 kg
    m. 100.6040 L
    n. 340.00 cm (2.3)
25. Round off each of the following measurements to the indicated number of significant figures:
    a. 2.68 g to 2 significant figures
    b. 47.374 mL to 4 significant figures
    c. 8.35 mL to 2 significant figures
    d. 4.165 L to 3 significant figures
    e. 24 km to 1 significant figure
    f. 0.048 51 mg to 2 significant figures
    g. 0.063 50 mL to 2 significant figures (2.3)
26. What is the rule governing the number of significant figures in calculations involving
    a. addition and subtraction?
    b. multiplication and division? (2.3)
27. a. What is meant by exact conversion factors?
    b. Give three examples of such factors. (2.3)
28. What is scientific notation? (2.3)
29. List the four steps used in the suggested method of problem solving. (2.4)
30. Give three examples of variables. (2.4)
31. Distinguish between two variables that are directly proportional, as opposed to inversely proportional, in terms of
    a. the definition of each
    b. the general equation for each
    c. the type of graph that illustrates each (2.4)

PROBLEMS

▶ SI Measurement

1. Rank the measurements within each set from largest to smallest:
   a. 1.2 m, 750 cm, 0.005 km, 65 dm, and 2000 mm
   b. 450 mg, 3.8 cg, 0.27 dg, 0.50 g, and 0.000 47 kg
   c. 2.5 L, 22 cL, 13 dL, 870 mL, and 175 cm^3
2. Find the total mass, in kilograms, of the following grocery purchases: 3 apples (125 g each), 5 bananas (115 g each), 2 pears (75 g each), 4 peaches (65 g each), 1750 g of melon, 5.5 hectograms (hg) of grapes.
3. Express each of the following in milliliters:
   a. 30 L
   b. 12 dL
   c. 4.5 cL
   d. 6.4 cm^3
4. Express the volume of a 22.4 L container in
   a. milliliters
   b. cubic centimeters

▶ Density, Heat, and Temperature

5. Based on the given information determine the unknown quantity for the given samples.
   a. mass = 240.00 g; volume = 60.0 cm^3; density = ?
   b. density = 11.3 g/cm^3; volume = 4.6 cm^3; mass = ?

c. density = 13.6 g/mL; mass = 14.50 g; volume = ?

6. Convert the following temperatures from the Celsius to Kelvin scales:
   a. 0°C
   b. 27°C
   c. 72°C
   d. −50°C

7. Convert the following temperatures from the Kelvin to Celsius scales:
   a. 300K
   b. 350K
   c. 273K
   d. 150K

8. Convert 25.0 calories to
   a. J
   b. kJ

9. Convert 1.42 kJ to
   a. J
   b. cal
   c. kcal

10. Determine the specific heat of a substance if a 25 g sample absorbed 34 J as it was heated from 21°C to 33°C.

11. How much heat would be absorbed by 55 g of copper when heated from 24°C to 40°C. Refer to Table 2-6 on p. 44.

▶ **Significant Figures and Scientific Notation**

12. Perform each of the following calculations and express all answers in the correct number of significant figures:
    a. 4.07 + 1.863 g
    b. 3127.55 cm − 784.2 cm
    c. 0.067 mL + 1.01 mL + 2.5 mL
    d. 36.427 m + 12.5 m + 6.33 m

13. Perform each calculation and express all answers in the correct number of significant figures:
    a. 9.40 cm × 2.6 cm
    b. 8.08 dm × 5.3200 dm
    c. 1.50 g/2 cm^3
    d. 0.084 21 g/0.640 mL
    e. 4.00 m × 0.020 m × 1.57 m
    f. $\dfrac{21.50 \text{ g}}{4.06 \text{ cm} \times 1.8 \text{ cm} \times 0.905 \text{ cm}}$

14. Write each of the following numbers in scientific notation:
    a. 900
    b. 750 000
    c. 93 000 000
    d. 0.000 082
    e. 0.000 403
    f. 6500 × 10^7
    g. 0.000 015 × 10^8
    h. 320 000 × 10^{-9}
    i. 0.0097 × 10^{-4}

15. Write each of the following numbers in the usual, long form:
    a. 5 × 10^3
    b. 7.8 × 10^4
    c. 2 × 10^{-5}
    d. 3.81 × 10^{-4}
    e. 4.07 × 10^7
    f. 6.09 × 10^{-9}

16. Find each sum or difference. Express all answers in scientific notation and in the correct number of significant figures.
    a. 2.6 × 10^2 + 4.1 × 10^2
    b. 8.3 × 10^{-5} + 1.2 × 10^{-5}
    c. 7.43 × 10^4 − 5.09 × 10^4
    d. 9.30 × 10^{-6} − 1.23 × 10^{-6}
    e. 1.2 × 10^3 + 3.6 × 10^4

17. Find each product or quotient. Express all answers in scientific notation and in the correct number of significant figures.
    a. (3 × 10^5) (2 × 10^7)
    b. (4.2 × 10^4) (3.8 × 10^3)
    c. (5.21 × 10^{-6}) (2.4 × 10^{-4})
    d. (6.02 × 10^{23}) (3.0 × 10^{-11})
    e. (8.4 × 10^5)/(2 × 10^3)
    f. (7.5 × 10^6)/(4 × 10^{-2})

▶ **Proportional Relationships and Problem Solving**

18. A cyclist is able to maintain an average speed of 40.0 km per 1.0 hour. What distance could he travel in 5.5 hours? Use the suggested method of problem solving and label each step.

19. Use the suggested method of problem solving to determine the volume occupied by 8.50 g of cork. The density of cork is 0.24 g/cm^3.

20. A piece of copper 12.00 cm long, 8.50 cm wide, and 0.50 cm thick has a mass of 457.00 g. What is its density?

21. A 3.10 cm^3 sample of an unknown element is found to have a mass of 32.50 g. Based on the information in Appendix Table A-14,

which of the following elements could the unknown be: platinum, iron, gold, magnesium, silver, or copper?

22. If 20.4 J of heat are required to raise the temperature of 6.67 g of an unknown substance from 35.0°C to 42.0°C, determine which of the following would be the unknown by using the table of specific heats (Table 2-6): water, ice, aluminum, copper, iron, silver, or lead?

23. Construct a graph based on the following information with Rankine readings on the horizontal axis and the corresponding Kelvin readings on the vertical axis. The Rankine degree is the same as the Fahrenheit degree, but the Rankine scale "0" is set at absolute zero.
    a. What type of relationship is illustrated?
    b. Explain the meaning of this relationship.

Rankine scale (°R)	Kelvin scale (K)
491	273
540	300
581	323
671	373
360	200
311	173

CRITICAL THINKING

1. **Evaluating Methods** For converting a temperature from Fahrenheit to Celsius, the following formula is provided on page 43.

    $$t(°C) + 5/9 \times [t(°F) - 32]$$

    with $\frac{5}{9}$ specifying the difference in the degree intervals between the Fahrenheit and Celsius scales.

    Derive another formula for converting °F to °C, using 180°F/100°C as the ratio specifying the degree intervals between the °F and °C scales.

2. **Applying Ideas** In Table 2-6 on page 44, specific heats are provided for some common substances. You will notice that water has a very high specific heat of 4.184 J/g·C° Explain what that means for a 500 L swimming pool filled for the summer.

3. **Predicting Outcomes** Three liquids and four solids are placed in a 100 mL graduated cylinder. They arranged themselves from top to bottom according to their densities. Given the following density information, identify the liquids and the solids in the graduated cylinder.

Data Table

Substance	Density (g/mL)
Oak Wood	0.71
Gold	19.3
Water	1.00
Gasoline	0.67
Mercury	13.6
Aluminum	2.7
Cork	0.21

a. _____
b. _____
c. _____
d. _____
e. _____
f. _____
g. _____

CROSS CONNECTIONS

1. **Physical Fitness** A measure of physical fitness is the percentage of fat contained in your body. Research the methods that physical fitness experts use to determine the percentage of body fat in an athlete's body.

2. **Environment** The environmental devastation caused by an "oil spill" is not new to the public. Given that the density of petroleum-based oil is approximately 0.9 g/mL and the density of water is 1.0 g/mL, discuss the implications this has for aquatic life, possible oil fires, and clean-up operations.

Chemistry and Careers

Fire Chemist

If you're fascinated by fire and smoke, consider the employment opportunities at the National Institute of Standards and Technology (NIST).

One major area of study at NIST is the heat and light released by the many chemical reactions in a fire. NIST fire chemists study these reactions to understand the kinetics of flames and how fire spreads. Smoke research, another area of study at NIST, is useful in establishing national safety standards for smoke detection.

To work as a laboratory technician at NIST, you need a bachelor's degree in a science field. NIST fire chemists who become project leaders usually have master's or doctoral degrees. Whether you major in engineering or science, the more chemistry, physics, and mathematics you

learn, the better equipped you will be to undertake fire and smoke research.

Science Writer

The science writer must combine scientific knowledge with creativity to compose material for science magazines, newspapers, and books that the nonscientist can understand and enjoy.

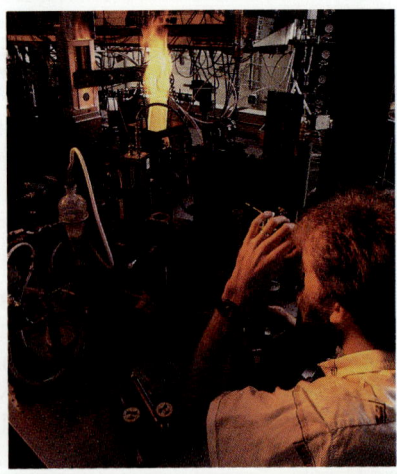

Many science writers work on a free lance basis, writing articles for newspapers and science magazines. Others work full time for book publishers, where they write and edit school textbooks or popular science books.

A significant portion of a science writer's time is spent doing research. Before writing about a subject, they read scientific journals and reports and consult with experts in the field. Once the research has been completed, science writers sometimes must prepare a number of drafts of the material until a satisfactory version is accepted by the publisher. Science writers must be able to produce interesting, concise prose; most have at least a bachelor's degree in English, journalism, or one of the sciences.

UNIT 1 • FOCUS ON REASONING

College Entrance Exam Preparation

▶ Experiment 1

A sample of gas A is placed in a sealed syringe. The sample has a constant mass and is kept at a constant temperature. The pressure on the gas is increased over a series of intervals and the resulting volume is recorded.

Interval	Pressure (atm)	Volume (cm³)
1	1.01	22.4
2	1.49	14.9
3	1.99	11.2
4	2.51	8.95
5	3.01	7.46
6	3.49	6.40
7	4.00	5.60

▶ Experiment 2

A 2.0×10^{-3} g sample of gas C is placed in a glass tube. The tube is sealed from the outside air using a movable plug. The glass tube is heated under constant pressure. The volume of gas C is measured at 10°C intervals.

Interval	Temperature (°C, K)	Volume (cm³)
1	10.1 283.1	1.45
2	19.9 292.9	1.50
3	30.1 303.1	1.55
4	40.1 310.1	1.59
5	49.9 319.9	1.64
6	60.0 330.0	1.69
7	70.0 340.0	1.74

1. Which of the following graphs would best represent the data from Experiment 1?

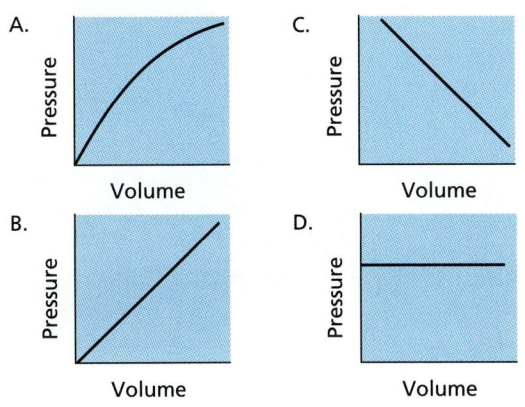

2. The purpose of Experiment 2 is to determine if there is a relationship between
   a. mass and volume.
   b. pressure and volume.
   c. temperature and volume.
   d. temperature, pressure, and mass.

3. Concerning the relationship between pressure and volume for Experiment 1, it appears that
   a. as the pressure on gas A increased, the volume of gas A increased.
   b. as the pressure on gas A decreased, the volume of gas A decreased.
   c. as the pressure on gas A increased, the volume of gas A decreased.
   d. for each incremental increase in pressure, there is a corresponding increase in volume.

4. Concerning the relationship between temperature and volume for Experiment 2, it appears that
   a. as the temperature of gas C was decreased, the volume of gas C increased.
   b. as the temperature of gas C was increased, the volume of gas C decreased.
   c. for each incremental increase in temperature there is a corresponding increase in volume.
   d. gas volume is not affected by gas temperature

5. On the basis of both experiments, what would you conclude would happen to the volume of a gas if the pressure were reduced and the temperature were increased?
   a. volume would be unaffected
   b. volume would be reduced
   c. volume would be increased
   d. cannot predict based on the results of the experiments

INTRA-SCIENCE

The Portable Power of the Battery

What does a World War II U-boat have in common with a pacemaker and a satellite? They all need portable power. You cannot connect a ship, a person, or a satellite to a power cord. You need batteries.

Batteries provide portable, stored energy that is used to start cars and to track Arctic wolves with special collars containing radios. Batteries run wristwatches and store solar energy in spacecraft. Without batteries, there would be no flashlights, no portable radios, and no pocket calculators.

A battery is a direct-current (dc) voltage source that converts chemical, solar, thermal, or nuclear energy into electrical energy. The first source of electric current was also the first battery. It was invented in 1800 by the Italian scientist Alessandro Volta.

In a chemical battery—the kind used in most ordinary electrical devices—chemical energy is transformed into electrical energy through an oxidation–reduction (redox) reaction. In redox reactions the atoms of one reactant give up one or more electrons (and thus are oxidized) to the other reactant (which thus is reduced). The two reactants do not come into direct contact with each other, so the reactant being oxidized cannot give electrons directly to the reactant being reduced. Instead, electrons flow out of the battery from the reactant being oxidized through an electrically conducting external circuit, and then back into the battery to the reactant being reduced. It is the electron current in the external circuit that powers the portable radio and lights the flashlight bulb.

Primary cells, such as flashlight batteries and the tiny silver oxide discs used in watches and calculators, can produce electricity only until one of the reactants is used up. They must then be discarded. Secondary cells, often called storage batteries, can be recharged by passing an electric current through the cell in the opposite direction to that followed by the current when the cell was discharging. This reverses the oxidation–reduction reaction, regenerating the original reactants of the cell.

The Connection to Biology

The crucial observation that led to Volta's invention of the battery was made in 1786 by the Italian biologist Luigi Galvani. After accidentally discovering that static electricity (electricity produced by friction) caused a contraction in the leg muscles of a newly killed frog, Galvani found that the same contraction occurred when frogs were hung on an iron support by copper hooks. Galvani thought that this contraction was caused by "animal electricity" in the muscles. Volta did not accept this view. He believed that the electricity that caused the muscular contraction had been produced by the contact of the two dissimilar metals: iron and copper.

Experiments stimulated by this correct idea led to his invention of the battery, which is sometimes called the "voltaic cell" in his honor. For his part, Galvani is honored by the term "galvanize."

Today, the battery has more than repaid its debt to biology. Battery-powered cameras, television camcorders, and tape recorders record animal behavior. Lions, tigers, and wolves have been fitted with collars containing battery-powered radio transmitters to better observe their behavior.

Biological cells act like miniature batteries. The double layer of lipids in a cell membrane separates aqueous solutions. Proteins embedded in the membrane conduct electrons from one solution to another. Tiny biological batteries that mimic the electrochemical action of cells are being tested. Sandwiched between two lipid layers, metals take the place of proteins. Biological batteries may one day be used in microcomputers that are made of organic molecules.

The Connection to Health Sciences

A wide variety of battery-powered devices are used to help people overcome medical problems. Wheelchairs, for example, are commonly powered by storage batteries. Tiny electric batteries that can provide power for up to ten years are used in pacemakers implanted in patients whose hearts do not beat properly. The pacemaker delivers electrical signals to the heart muscles so that they beat in the proper rhythm. Batteries power hearing aids that amplify sounds that are received by the inner ear. Recently, progress has been made toward the development of battery-powered devices that can electrically stimulate the auditory nerve of people who otherwise could not hear sounds, no matter how strongly they were amplified.

The Connection to Earth Science

Both storage batteries and solar-electric panels provide the power for the instruments in satellites that orbit the earth. Satellites monitor a wide variety of phenomena on the earth's surface, such as ocean currents, vegetation, weather patterns, and snow lines. Battery-powered strainmeters, tiltmeters, creepmeters, and other geophysical instruments monitor changes in the bedrock in earthquake zones to learn what conditions precede earthquakes. This may lead to an earthquake warning system. Storage batteries power underwater craft that carry oceanographers to the bottom of the ocean, where battery-powered light often reveals strange new species of animals or new geological features. Battery-powered in-

struments are carried into the upper atmosphere by balloons and rockets. They gather information that is used in weather forecasting and relay it to control stations by means of battery-powered transmitters.

Utility companies find batteries to be useful tools for "load leveling." Although many utility companies have a constant power source, such as hydroelectricity, the demand for power fluctuates. Energy can be stored by batteries during the night, when demand is low, and released during peak consumption. Utility companies also use batteries to store solar- and wind-generated electricity.

Unit 2

Chapter 3
Atoms: The Building Blocks of Matter   70

Chapter 4
Arrangement of Electrons in Atoms   96

Chapter 5
The Periodic Law   126

Chapter 6
Chemical Bonding   160

Organization of Matter

I first felt the excitement of uncovering new knowledge in a high school course called "Science Research." We formed a group to study ocean pollution off the Southern California coast where I lived. I found that research could sometimes be difficult and frustrating, but there is no experience as satisfying as discovering something that wasn't known before.

In this century, we have learned a great deal about the structure of atoms and molecules. The new knowledge has enabled scientists to produce new forms of matter such as plastics and other synthetic polymers.

I use chemistry to investigate the structures of some of the complicated molecules that are at the center of life. Our experiments make use of a simple inorganic chemical reaction to make "pictures" of DNA molecules. My work would not be possible without the basic concepts of matter you will learn in this unit.

*Thomas D. Tullius
Inorganic Chemist
The Johns Hopkins University*

Computer-generated image of a DNA double helix

CHAPTER 3
Atoms: The Building Blocks of Matter

PLANNING CHART

Approximate Time: 6/7 days

	SECTION/TEXT FEATURES		RESOURCES	
B A S I C	3.1 THE ATOM: FROM PHILOSOPHICAL IDEA TO SCIENTIFIC THEORY Demo 1 Desktop Investigation Section Review Chapter Review • Questions 1–3	71 70B 74 74 93	*Teacher's Resource Binder* Lab Experiment 4 Section Review Worksheets *Other Resources* *World of Chemistry* Video 4	43 9
B A S I C	3.2 THE STRUCTURE OF THE ATOM Demo 2 Section Review Chapter Review • Questions 4–13	75 70B 82 93	*Teacher's Resource Binder* Section Review Worksheets *Other Resources* Transparencies 3, 4	10
B A S I C	3.3 WEIGHING AND COUNTING ATOMS Demos 3, 4 Section Review Chapter Review • Questions 14–22, Problems 1–10	82 70C 92 93	*Teacher's Resource Binder* Problem Solving Worksheets Section Review Worksheets *Other Resources* *World of Chemistry* Video 11 Blackline Master 10	11
	CHAPTER END ACTIVITIES Critical Thinking, Cross Connections, Research and Writing, Projects		*Teacher's Resource Binder* Test Booklet *Computer Test Generator*	5, 7

Teaching Strategies

Introduction

Start this chapter by pointing out that atoms are invisible particles that no one has ever seen except for the very large atoms that have been photographed under a transmission-electron microscope. Even in this case, they only appear as tiny blobs of light. Yet this entire course in chemistry is based on their existence and their structure. With something this small, how can we begin to determine not only that they exist but also what they are made up of and how they are put together? To help your students understand how they can accept the existence of invisible things, ask the following questions. Does wind exist? Have you ever seen wind? Then how do you know it exists? If you go to a lake or pond, how might you know that deer, raccoons, or dogs have been there if you did not see them? Information that results from effects on or tracks left on other substances is known as indirect evidence. Much of the evidence that accounts for atoms, subatomic particles, and their relationship to one another is based on similar types of data. This chapter uses many examples of indirect evidence in discussing the atom and its structure.

A basic goal in teaching this chapter is to familiarize the students with the structure of an individual atom and show them how atoms of different elements and even certain atoms (isotopes) of the same element differ from one another. The approach used is a historical one, which makes it important to place a special emphasis on what the presently-accepted theory for each area is so students do not get the information included in original theories confused with that in modern theories. A second goal deals with indirect measurement, both in the explanation as to how historical developments were made and in counting out atoms using molar mass.

3.1 The Atom: From Philosophical Idea to Scientific Theory

After discussing Dalton's atomic theory from a historical viewpoint, go over the Modern atomic theory, which is stated as follows:
1. All matter is made up of very small particles called atoms.
2. Atoms of the same element have the same chemical properties, while atoms of different elements have different chemical properties.
3. While individual atoms of a given element may not all have the same mass (due to differences in nuclear structure), any

natural sample of the element will have a definite average mass that is characteristic of the element and different from that of any other element.
4. Compounds are formed when atoms of two or more elements unite, with each atom losing its characteristic properties as a result of this combination.
5. Atoms are not subdivided in physical or chemical reactions.

Help students compare and contrast Dalton's theory with the Modern theory. It is important for students to know what the presently accepted theory is. This also helps emphasize the changing nature of a theory as new data is discovered.

Demonstration 1: Compounds Exemplifying the Law of Multiple Proportions

PURPOSE: To show that compounds composed of the same elements, but in different proportions, are different compounds with different properties

MATERIALS: Copper(I) oxide, copper(II) oxide, 20% hydrogen peroxide*, manganese dioxide, water, colored felt, 2 250–mL beakers, 2 medium test tubes

PROCEDURE: Have students examine copper(I) oxide, Cu_2O, and copper(II) oxide, CuO, and note how they differ from one another in physical properties. (Copper(I) oxide is red and copper(II) oxide is black.)

Have students examine water, H_2O, and hydrogen peroxide, H_2O_2, and note how they are similar to one another in physical properties. (Both are colorless liquids.)

Show students that water and hydrogen peroxide differ greatly from one another in chemical reactions by doing the following:

(1) Put a piece of colored felt** in a beaker half filled with water, and put a second piece in a beaker half filled with 20% hydrogen peroxide. Allow it to stand overnight and observe results. (The hydrogen peroxide will bleach the felt, but the water will not affect it.)

(2) To 5 mL of hydrogen peroxide in a test tube, add manganese dioxide equal to the quantity of a rice grain. Repeat with 5 mL of water. (The H_2O_2 will bubble vigorously; the water does not.)

* 20% hydrogen peroxide is used to bleach hair and can be purchased at a drugstore.

** You can experiment with other pieces of material or yarn that might bleach equally well. It all depends upon the permanency of the dye. Organic dyes and organic substances such as hair work best.

Desktop Investigation: Constructing a Model

This Desktop could be done prior to the presentation of the materials in Section 3.2 so that students will have something to relate to as you discuss experiments regarding the discovery of electrons and the nucleus. Or it could be done after the completion of the discussion on the structure of the atom and then related to that discussion.

To make the sealed container, cut off the foot of a sock and attach the cut end of the sock to the top of an empty two-pound coffee can using duct tape (see Figure in text). Put one or more objects of various sizes, shapes, masses, and materials into the can and seal the cuff of the sock tightly with duct tape.

Try to use unfamiliar objects to reduce the student's tendency to attempt to guess the identity of the object rather than focusing on developing a model of the object. (A good source of such objects is old junk shops and antique stores. You can collect odd and unusual objects throughout the year, or ask members of your faculty to help you.) Number the cans, and make a key as to their contents. To save time, assign each student the task of making one "mystery can" according to the above suggestions. Have them turn it in to you with its contents identified prior to the lab.

You can provide the students with a data table or have them design their own to record their observations. Estimation of accuracy will improve considerably with practice, but use practice objects different from those in the sealed cans. Making mass/size estimations prior to the activity would probably work best. Remind the students of the importance of not looking into the can or removing the object(s) from the can by emphasizing that models are the visualizations of phenomenon that *cannot* be seen based on scientific data. When students are finished with their "mystery can," reseal the cuff end with tape so it will be ready for the next class.

A fourth discussion question you might use is:
4. Scientists often gather data or figure out answers by comparing their observations with known quantities. How did you improve your estimations of size or mass using a comparison method?

ANSWERS TO DISCUSSION QUESTIONS
1. In Part 1, all observations were made without contact with the object itself and, thus, were examples of indirect evidence (gathered through the sense of hearing). In Part 2, the observations were made by using a seldom-used sense: touch; but since contact with the object occurred, this would be an example of direct observations for qualitative data. However, because of the constraints in Part 2, all measurements still had to be estimations. Scientists gather and use both indirect evidence and estimations as data in their attempts to develop models.
2. Determination of mass, size, and number of objects resulted in quantitative data which was generally estimated. Determination of shape, material, and texture involved the gathering of qualitative data.
3. This will vary from student to student depending upon the nature of the object.
4. People can learn to use their fingers as crude devices for measuring the length of objects. As one works with objects of varying masses, one becomes more aware of how different masses "feel." The more one works with estimation, the more accurate the estimations become.

3.2 The Structure of the Atom

Demonstration 2: Indirect Measurement

PURPOSE: To help students understand how the size of small objects, like atoms and subatomic particles, can be determined by using indirect measurement

MATERIALS: *Modern Chemistry* textbook, clear plastic metric ruler

PROCEDURE: How could you determine the thickness of one page of your textbook using an ordinary ruler? After a student has suggested the solution of measuring the thickness of all the pages of the book with the ruler and dividing that amount by the total number of pages in the book, have all the students measure the book and make the necessary calculations. Then look at everyone's data, to determine if it all seems reasonable; and take an average of that which is determined to be acceptable. Those that look like serious errors can be re-checked. This is an excellent opportunity to learn the concept of indirect measurement and at the same time look at the degree that measurements can vary and how serious errors can be detected.

Answers should be in the area of 0.006–0.008 cm. You can use a micrometer to measure the thickness of a page directly, and compare the results with the indirect measurements.

When discussing Rutherford's experiment, caution students that what might at first appear to be a significant deviation from the norm in an experiment might not be error at all but very useful data; that is, in Rutherford's experiment, the few alpha particles that were deflected slightly and the few that were deflected back. A scientist has to be able to distinguish between error and meaningful deviations.

3.3 Weighing & Counting Atoms

Prior to your explanation of the individual atomic mass and average atomic mass, explain that *relative* means compared to something else. To aid in your discussion of individual atomic masses of isotopes and average atomic mass, you might refer to "Table of Isotopes" in Section B of the *CRC Handbook of Chemistry and Physics*. The individual atomic mass and percentage of natural abundance for every known nuclide is listed there. Students often ask, "How do you know what the atomic mass of a given nuclide is?" This a good time to let them look at this section in the handbook. They should find it interesting, and it is a chance to introduce them to the *CRC Handbook* as an excellent resource.

Students tend to get the following terms confused. A good way to help them is to put the following summary and table on the board or on a transparency.

Individual atomic mass—Exact mass in atomic mass units of one atom

Average atomic mass—Average mass in atomic mass units of all natural isotopes of an element

Mass number—Total number of protons and neutrons in a nucleus of an isotope OR the individual atomic mass rounded off to a whole number

Atomic number—number of protons in an atom of an element

Below is a table which helps explain each term:

Isotope	% in Nature	Atomic No.	Individual Atomic Mass	Mass No.	Average Atomic Mass
Ne-20	90.51%	10	19.992 44 u	20	
Ne-21	0.21%	10	20.993 84 u	21	20.179 u
Ne-22	9.22%	10	21.991 38 u	22	

A demonstration similar to the marble analogy discussed in the *Average Atomic Masses of Elements* section in the text follows.

Demonstration 3: Isotopes and Average Atomic Mass

PURPOSE: To show how different percentages of natural isotopes of a given element would affect the average atomic mass

MATERIALS: 100 pennies minted before 1982, 100 pennies minted after 1982, a decigram balance [Note: Pennies minted in 1982 are a mixture of both copper and zinc pennies; therefore, do not use them.]

PROCEDURE: Using the decigram balance, determine the mass of 100 pennies minted before 1982. To determine the mass of a single penny, divide the total mass by 100. (The mass of each penny is about 3.1 g. Answers of actual mass measurements should be in terms of four significant figures.) Is this direct or indirect measurement? (Indirect) Repeat the procedure with 100 pennies minted after 1982. (The individual mass is about 2.5 g.)

Have students predict (through mathematical calculations) the average mass of an "isotopic" mixture of 50% pre-1982 and 50% post-1982 pennies. Then do an actual weighing with 50 pennies minted before 1982 and 50 after 1982 and determine their average. (It should be about 2.8 g.) Repeat with a mixture of 25% before and 75% after (about 2.65 g.), 75% before and 25% after (about 2.95 g), or any other percentage you might wish to select.

Demonstration 4: Counting Out Atoms

PURPOSE: To determine the number of atoms in a few samples of different elements

MATERIALS: Aluminum, charcoal, copper, graphite, iron, silicon, sulfur, zinc, reclosable plastic bags, a penny minted before 1982, a nail, aluminum foil, decigram balance

PROCEDURE: Prior to class, weigh out 1 mol samples of aluminum (27.0 g.), carbon (12.0 g), copper (63.5 g), iron (55.8 g), silicon (28.1 g), sulfur (32.1 g), zinc (65.4 g), or any other element that you might have available. Place each sample in a reclosable plastic bag and label what element it contains. Have a few extra empty bags available for use in class.

Locate objects composed of the above elements such as a penny for Cu, a nail (Fe), a lump of charcoal (C), a piece of foil (Al), a chunk of S, a lump of Si, or a small piece of Zn.

When class begins, tell students that they are going to help you determine how many atoms there are in samples of various elements.

Start by holding up the bags of elements containing one mole of an element. (Do not tell them each bag contains one mole. That is for them to determine.) Ask the students: "How can I figure out how many atoms I have in these bags?" When a student suggests that you weigh them, you might ask some of them to determine the mass of the bag and its contents. Again ask "How can we determine the mass of the contents of this bag without removing it from the bag?" When someone suggests weighing several plastic bags to determine an average mass for the plastic bag, have other students do those mass determinations. Calculate the mass of each substance by subtracting the average mass of the plastic bag from the mass of the bag and its contents. The results will be very close to the molar mass of each element, and students should quickly conclude that each bag contains approximately 6.022×10^{23} atoms.

Ask your students: How many atoms are in this penny? Nail? Piece of aluminum foil? etc. Have them make weighings and calculate the number of atoms.

This demonstration can be easily changed into an inquiry lab in which you give each student a one mole bag of an element and a small piece of an element and ask them to devise a way to determine the number of atoms each contains. They can consult with one another as scientists often do; but do not tell them how to solve the problem—let them do that!

A poster on moles is available from the American Chemical Society, 1155 16th St. NW, Washington, D.C. 20036

References and Resources

Books and Periodicals

Bent, Henry A. "Should Atoms Be X-rated?" *Journal of Chemical Education*, Vol. 63, No. 10, Oct. 1986, pp. 878–9.

Herron, Dudley. "Rutherford and the Nuclear Atom," *Journal of Chemical Education*, Vol. 54, No. 8, Aug. 1977, p. 499.

Kolb, Doris. "But If Atoms Are So Tiny," *Journal of Chemical Education*, Vol. 54, No. 9, Sept. 1977, pp. 543–7.

Poskozim, Paul, et al. "Analogies for Avogadro's Number," *Journal of Chemical Education*, Vol. 63, No 2, Feb. 1986, pp. 125–6.

Rayner-Canham, Geoffrey, and Marelene Rayner-Canham. "The Shell Model of the Nucleus," *The Science Teacher*, Vol. 54, No. 1, Jan. 1987, pp. 18–21.

Records, Roger. "Developing Models: What is the Atom Really Like?" *Journal of Chemical Education*, Vol. 59, No. 4, April 1982, pp. 307–9.

Weinberg, Steven. *The First Three Minutes*, New York: Bantam Books, 1977.

Audiovisual Resources

The Atom and Atomic Mass—Nuclear Physics Series, filmstrip or slides, Educational Images, Ltd.

Experiments in Subatomic Particles, filmstrip, Educational Images, Ltd., 1988.

You in the Universe—The Atomic Universe, filmstrip or VHS and Beta cassettes, Educational Images, Ltd.

Computer Software

Atomic Structure, Apple 48K. Chemistry according to ROF. Project Seraphim, 1988. Apple II Disks, Vol. 2, Disk Nos. 1, 2, 4, 5 and Vol. 3, Disk Nos. 4, 5, 6.

SEI Atomic Structure, Apple 64K and IBM PC 128K. Queue, 1988.

Smith, S. G., R. Chabray, and E. Kean, *The Elements,* Apple 48K and IBM PC 128K; *Atomic Weights,* Apple 48K and IBM PC 128K. Queue, 1988.

Answers and Solutions

Reviewing Concepts

1. **a.** Since all chemical reactions simply involve the rearrangement of indivisible atoms, then matter, and its mass, can be neither created nor destroyed in such changes.
   b. If atoms of each element have their own characteristic masses, then compounds consisting of these elements should always have the same composition by mass.
   c. Since only whole atoms combine in chemical compounds, different compounds between the same two elements must result from the combination of different relative whole numbers of atoms.

2. Element AB would be expected to have a mass of 5 mass units.

3. **a.** The law of multiple proportions states that if two or more different compounds are composed of the same two elements, the masses of the second element combined with a certain mass of the first element can be expressed as ratios of small whole numbers.
   b. John Dalton is credited with the discovery of this law.

4. **a.** An atom is the smallest particle of an element that can exist either alone or in combination with other atoms.
   b. Atomic structure refers to the identity and arrangement of smaller particles within atoms.

5. All atoms consist of a very dense positively charged nucleus at the center, and a comparatively large region occupied by negatively charged particles surrounding the nucleus.

6. According to Thomson and Millikan, an electron is negatively charged, has a mass approximately 1/2000 that of a hydrogen atom, has the smallest possible negative electric charge, has a fixed charge-to-mass rate, and is present in atoms of all elements.

7. An electron is a negatively charged subatomic particle.

8. **a.** Rutherford is credited with the discovery of the atomic nucleus.
   b. The nucleus of an atom is its positively charged, dense, central portion, which contains nearly all of its mass but takes up only an insignificant fraction of its volume.
   c. According to Rutherford, an atom consists of a densely packed bundle of matter with a positive charge, called a nucleus, surrounded by mostly empty space. This nucleus contains nearly all of the mass of the atom but takes up only an insignificant fraction of its volume. He further proposed that the negatively charged electrons surround the positively charged nucleus like the rings around Saturn, but he could not explain what kept the electrons in motion around the nucleus.

9. Every atomic nucleus is made up of protons and neutrons. Protons are subatomic particles that have a positive charge of the same magnitude as the negative charge of an electron and are present in atomic nuclei. Neutrons are electrically neutral, subatomic particles found in atomic nuclei.

10. The identity of an atom is determined by the number of protons in its nucleus.

11.

Particle	Symbol	Mass	Actual Mass	Relative Charge
Electron	e^-	0	9.110×10^{-28} g	-1
Proton	p^+	1	1.673×10^{-24} g	$+1$
Neutron	n^0	1	1.675×10^{-24} g	0

12. a. Isotopes are atoms of the same element that have different masses.
  b. Isotopes of a particular element have the same number of protons.
  c. Isotopes of a particular element have different numbers of neutrons.

13.

Isotope	Name	No. of Protons	No. of Electrons	No. of Neutrons
H-1	protium	1	1	0
H-2	deuterium	1	1	1
H-3	tritium	1	1	2

14. a. The atomic number of an element is the number of protons in the nucleus of an atom of that element.
  b. The mass number of an isotope is the total number of protons and neutrons in its nucleus.
  c. The atomic number is at the lower left, whereas the mass number is at the upper left.
  d. The hyphen notation for this isotope is hydrogen-2.

15. A nuclide is a general term for any isotope of any element.

16. The hyphen notation for each isotope is:
  a. helium-4   **b.** oxygen-16   **c.** potassium-39

17. a. The standard is the carbon-12 atom.
  b. The relative atomic mass of carbon-12 has been arbitrarily assigned a value of exactly 12 atomic mass units.

18. a. One atomic mass unit is exactly 1/12 the mass of the carbon-12 atom, or $1.660\,540\,2 \times 10^{-24}$ g.
  b. The relative atomic mass of an atom is the mass of that atom expressed in atomic mass units, as compared to the atomic mass of C-12.

19. The indicated relative atomic masses would be:
  a. $4\,u$ **b.** $6\,u$ **c.** $144\,u$ **d.** $54\,u$

20. a. A mole is the amount of a substance that contains a number of particles equal to the number of atoms in exactly 12 grams of carbon-12.
  b. The symbol for the mole is *mol*.
  c. There are $6.022\,137 \times 10^{23}$ particles in one mole.
  d. The name given to the number of particles in one mole is Avogadro's number.

21. a. The molar mass of an element is the mass in grams of that element that is numerically equal to its mass on the relative atomic mass scale.
  b. The indicated molar masses are
  1. 12.011 g
  2. 20.179 g
  3. 55.847 g
  4. 238.029 g

Problems

1. a. 6.94 g Li   **b.** 27.0 g Al   **c.** 40.1 g Ca
  d. 55.8 g Fe   **e.** 12.01 g C   **f.** 107.9 g Ag

2. a. $3.00 \text{ mol Al} \times \dfrac{27.0 \text{ g Al}}{\text{mol Al}} = 81.0 \text{ g Al}$

  b. $4.25 \text{ mol Li} \times \dfrac{6.94 \text{ g Li}}{\text{mol Li}} = 29.5 \text{ g Li}$

  c. $1.38 \text{ mol N} \times \dfrac{14.0 \text{ g N}}{\text{mol N}} = 19.3 \text{ g N}$

  d. $8.075 \text{ mol Au} \times \dfrac{197.0 \text{ g Au}}{\text{mol Au}} = 1591 \text{ g Au}$

  e. $6.50 \text{ mol Cu} \times \dfrac{63.5 \text{ g Cu}}{\text{mol Cu}} = 413 \text{ g Cu}$

  f. $2.57 \times 10^8 \text{ mol S} \times \dfrac{32.1 \text{ g S}}{\text{mol S}} = 8.25 \times 10^9 \text{ g S}$

  g. $1.75 \times 10^{-6} \text{ mol Hg} \times \dfrac{201 \text{ g Hg}}{\text{mol Hg}} = 3.52 \times 10^{-4} \text{ g Hg}$

3. a. $40.1 \text{ g Ca} \times \dfrac{\text{mol Ca}}{40.1 \text{ g Ca}} = 1.00 \text{ mol Ca}$

  b. $11.5 \text{ g Na} \times \dfrac{\text{mol Na}}{23.0 \text{ g Na}} = 0.500 \text{ mol Na}$

  c. $5.87 \text{ g Ni} \times \dfrac{\text{mol Ni}}{58.7 \text{ g Ni}} = 0.100 \text{ mol Ni}$

  d. $150 \text{ g S} \times \dfrac{\text{mol S}}{32 \text{ g S}} = 4.7 \text{ mol S}$

  e. $2.65 \text{ g Fe} \times \dfrac{\text{mol Fe}}{55.8 \text{ g Fe}} = 0.0475 \text{ mol Fe}$

  f. $0.007\,50 \text{ g Ag} \times \dfrac{\text{mol Ag}}{108 \text{ g Ag}} = 6.94 \times 10^{-5} \text{ mol Ag}$

  g. $3.25 \times 10^5 \text{ g Pb} \times \dfrac{\text{mol Pb}}{207 \text{ g Pb}} = 1.57 \times 10^3 \text{ mol Pb}$

  h. $4.50 \times 10^{-12} \text{ g O} \times \dfrac{\text{mol O}}{16.0 \text{ g O}} = 2.81 \times 10^{-13} \text{ mol O}$

4. a. $6.022 \times 10^{23} \text{ atoms Ne} \times \dfrac{\text{mol Ne}}{6.022 \times 10^{23} \text{ atoms Ne}} = 1.000 \text{ mol Ne}$

b. 3.011×10^{23} atoms Mg $\times \dfrac{\text{mol Mg}}{6.022 \times 10^{23} \text{ atoms Mg}} = 0.5000$ mol Mg

c. 2.25×10^{25} atoms Zn $\times \dfrac{\text{mol Zn}}{6.02 \times 10^{23} \text{ atoms Zn}} = 3.74 \times 10^1$ mol Zn

d. 50.0 atoms Ba $\times \dfrac{\text{mol Ba}}{6.02 \times 10^{23} \text{ atoms Ba}} = 8.31 \times 10^{-23}$ mol Ba

5. Average atomic mass = $(0.003\ 37 \times 35.968\ u) + (0.000\ 63 \times 37.963\ u) + (0.996\ 00 \times 39.962\ u)$
Average atomic mass = $0.1212\ u + 0.0239\ u + 39.8022\ u$
Average atomic mass = $39.947\ u$

6. Expressing this relationship algebraically, and solving for x, the atomic mass of the other isotope,
$10.811\ u = (0.8020 \times 11.009\ u) + (0.1980 \times x)$
$10.811\ u = 8.829\ u + (0.1980 \times x)$
$1.982\ u = 0.1980 \times x$
$10.010\ u = x$

7. Expressing this relationship algebraically and solving for x,
$35.453\ u = (x \times 34.969\ u) + (1.000 - x)(36.966\ u)$
$35.453\ u = (34.969\ u \times x) + 36.966\ u - (36.966\ u \times x)$
$1.997\ u \times x = 1.513\ u$
$x = 0.757\ 64$
Thus, Cl-35 has a relative abundance of 75.764%, and Cl-37 has an abundance of 24.236%.

8. a. 1.50 mol Na $\times \dfrac{6.02 \times 10^{23} \text{ atoms Na}}{\text{mol Na}} = 9.03 \times 10^{23}$ atoms Na

b. 6.755 mol Pb $\times \dfrac{6.022 \times 10^{23} \text{ atoms Pb}}{\text{mol Pb}} = 4.068 \times 10^{24}$ atoms Pb

c. 0.250 mol Si $\times \dfrac{6.02 \times 10^{23} \text{ atoms Si}}{\text{mol Si}} = 1.50 \times 10^{23}$ atoms Si

9. a. 3.011×10^{23} atoms F $\times \dfrac{\text{mol F}}{6.022 \times 10^{23} \text{ atoms F}} \times \dfrac{19.00 \text{ g F}}{\text{mol F}} = 9.500$ g F

b. 1.50×10^{23} atoms Mg $\times \dfrac{\text{mol Mg}}{6.02 \times 10^{23} \text{ atoms Mg}} \times \dfrac{24.3 \text{ g Mg}}{\text{mol Mg}} = 6.05$ g Mg

c. 4.50×10^{12} atoms Cl $\times \dfrac{\text{mol Cl}}{6.02 \times 10^{23} \text{ atoms Cl}} \times \dfrac{35.5 \text{ g Cl}}{\text{mol Cl}} = 2.65 \times 10^{-10}$ g Cl

d. 8.42×10^{18} atoms Br $\times \dfrac{1 \text{ mol Br}}{6.02 \times 10^{23} \text{ atoms Br}} \times \dfrac{79.9 \text{ g Br}}{\text{mol Br}} = 1.12 \times 10^{-3}$ g Br

e. 25.0 atoms W $\times \dfrac{\text{mol W}}{6.02 \times 10^{23} \text{ atoms W}} \times \dfrac{184 \text{ g W}}{\text{mol W}} = 7.64 \times 10^{-21}$ g W

f. 1.00 atom Au $\times \dfrac{\text{mol Au}}{6.02 \times 10^{23} \text{ atoms Au}} \times \dfrac{197 \text{ g Au}}{\text{mol Au}} = 3.27\ 10^{-22}$ g Au

10. a. 5.40 g B $\times \dfrac{\text{mol B}}{10.8 \text{ g B}} \times \dfrac{6.02 \times 10^{23} \text{ atoms B}}{\text{mol B}} = 3.01 \times 10^{23}$ atoms B

b. 8.02 g S $\times \dfrac{\text{mol S}}{32.1 \text{ g S}} \times \dfrac{6.02 \times 10^{23} \text{ atoms S}}{\text{mol S}} = 1.50 \times 10^{23}$ atoms S

c. 1.50 g K $\times \dfrac{\text{mol K}}{39.1 \text{ g K}} \times \dfrac{6.02 \times 10^{23} \text{ atoms K}}{\text{mol K}} = 2.31 \times 10^{22}$ atoms K

d. 0.025 50 g Pt $\times \dfrac{\text{mol Pt}}{195.1 \text{ g Pt}} \times \dfrac{6.022 \times 10^{23} \text{ atoms Pt}}{\text{mol Pt}} = 7.871 \times 10^{19}$ atoms Pt

e. 1.00×10^{-10} g Au $\times \dfrac{\text{mol Au}}{197 \text{ g Au}} \times \dfrac{6.02 \times 10^{23} \text{ atoms Au}}{\text{mol Au}} = 3.06 \times 10^{11}$ atoms Au

Critical Thinking

1. Answers will vary. According to the law of definite composition, one atom of sulfur has a mass of 32.06 amu and one atom of oxygen has a mass of 15.99 amu, then the compound SO_2 contains 1 atom of sulfur and 2 atoms of oxygen and will always have a mass of 64 amu. According to the law of multiple proportions, sulfur and oxygen are not limited to combining 1 atom to 2 atoms. However, for any given mass of sulfur combined with oxygen, the masses of the two can be expressed as ratios of small whole numbers. If 32 grams of sulfur was used to produce SO_2 and SO_3, then 32 grams of oxygen would be required to produce SO_2 and 48 grams of oxygen would be required to produce SO_3.
2. Answers will vary. A possible model is shown below.

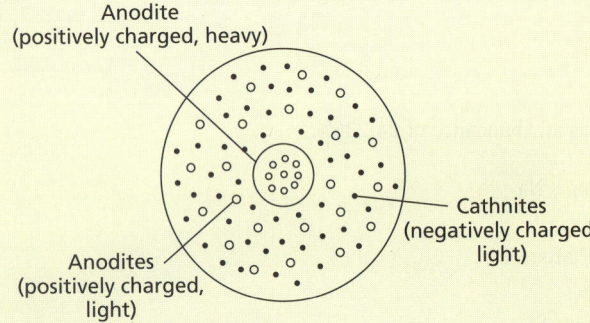

3. Relative atomic mass = 235.043 924.

   92 protons × 1.007 276 = 92.669 392 u
   92 electrons × 0.000 548 59 = 0.050 470 28 u
   143 neutrons × 1.008 665 = 144.239 095 u
   Total = 236.958 957 3 u

   236.958 957 3 − 235.043 924 = 1.915 033 3 u is lost

Cross Connections

1. Answers will vary: A radioactive isotope is substituted for the nonradioactive compound used by the organ being investigated. The radioactive compound can then be traced by the doctor using a detector sensitive to radioactivity. Common radioisotopes include chromium-51, iodine-131, gallium-67, selenium-75 and technetium-99m. Two new instruments that use radioisotopes are PET and SPECT.

Teacher's Notes

Atoms: The Building Blocks of Matter   **70H**

Chapter 3
Atoms: The Building Blocks of Matter

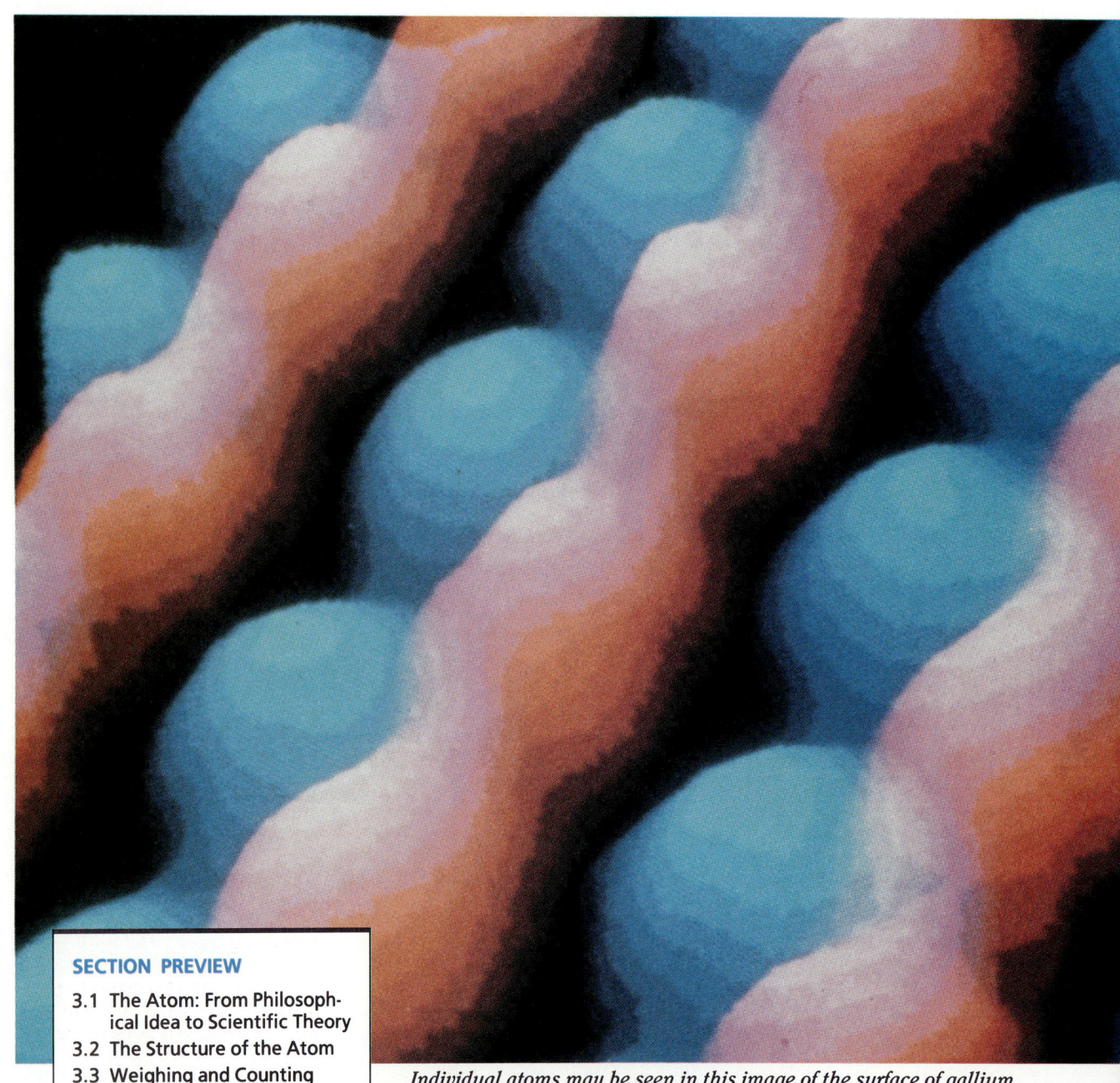

SECTION PREVIEW

3.1 The Atom: From Philosophical Idea to Scientific Theory
3.2 The Structure of the Atom
3.3 Weighing and Counting Atoms

Individual atoms may be seen in this image of the surface of gallium arsenide, produced by a scanning tunneling microscope. The blue spots are gallium atoms, and the red are arsenic atoms.

3.1 The Atom: From Philosophical Idea to Scientific Theory

SECTION OBJECTIVES

- Summarize the five essential points of Dalton's atomic theory.
- Explain the relationship between Dalton's atomic theory and the laws of conservation of mass and definite composition.
- Explain the law of multiple proportions.

The idea that matter is made up of simple, indivisible particles originated a long time ago. As early as 400 B.C., some Greek thinkers had the idea that matter could be divided into smaller and smaller particles until a basic particle of matter that could not be divided further was reached. Democritus (460–370 B.C.) called these particles "atoms." The word *atom* is from the Greek word meaning indivisible. However, Aristotle (384–322 B.C.), a Greek whose ideas had a large impact on Western civilization, did not believe in atoms, but rather that all matter was continuous. His view was accepted until the 18th century, when experimental evidence proved him wrong. Neither Democritus' nor Aristotle's theories were based on experimental evidence.

When you crush a lump of sugar, you can see that it is made up of many small particles of sugar. You may grind these particles into a very fine powder, but each tiny piece is still sugar. Now suppose you dissolve the sugar in water. The tiny particles disappear completely. Even if you look at the sugar-water solution through a microscope, you cannot see any sugar particles. But, if you taste the solution, you would know that the sugar is still there.

If you open a perfume bottle, you can smell the escaping perfume. Yet you cannot see gas particles in the air of the room, even if you use a powerful microscope. These observations and many others like them have led scientists to believe that the basic particles of matter must be very small.

Dalton's Atomic Theory

In the late 1700s, invention of the chemical balance gave chemists a tool for studying the composition of pure substances quantitatively. They discovered that a small number of elements are the components of many different substances. Investigations revealed how elements combine to form compounds and how compounds can be broken down into their constituent elements.

Observations made during the period resulted in the discovery of the law of conservation of mass and the law of definite composition, which were introduced in Chapter 1. An English schoolteacher, John Dalton (1766–1844), was the first to recognize that these laws and many experimental results could be explained by the existence of atoms.

In 1803, Dalton proposed an atomic theory that greatly expanded upon the concept of Democritus. Dalton was able to relate atoms to the measurable property of mass. Dalton's atomic theory can be summarized by the following statements:

1. All matter is composed of extremely small particles called atoms.
2. Atoms of a given element are identical in size, mass, and other properties; atoms of different elements differ in size, mass, and other properties.
3. Atoms cannot be subdivided, created, or destroyed.
4. Atoms of different elements can combine in simple, whole-number ratios to form chemical compounds.
5. In chemical reactions, atoms are combined, separated, or rearranged.

Figure 3-1 The smell from a perfume bottle will quickly travel throughout an area, even though you cannot see the perfume in the air. Observations such as this indicate that the basic units of matter are extremely small.

Reminder—The law of conservation of mass: Matter can be neither created nor destroyed in ordinary chemical or physical changes. The law of definite composition: Every chemical compound has a definite composition by mass.

John Dalton was interested in the composition and properties of gases in the atmosphere. He kept a daily record of the weather from 1787 until 1844.

In John Dalton's own words: "No new creation or destruction of matter is within the reach of chemical agency. We might as well attempt to introduce a new planet into the solar system, or to annihilate one already in existence, as to create or destroy a particle of hydrogen. All the changes we can produce consist in separating particles already in a state of cohesion or combination, and joining those that were previously at a distance."
—John Dalton: *A New System of Chemical Philosophy* (London, 1842).

STUDY HINT

Isotopes are defined and discussed in Section 3-3

Dalton's atomic theory was accepted and became the basis for further experiments because it successfully explained the laws of conservation of mass and definite composition, and other observations. Although exceptions to Dalton's atomic theory are now known, atomic theory has not been discarded. We now know that atoms are divisible and that a given element can have atoms with different masses, called *isotopes*. Atomic theory has been expanded to explain the new observations. The concepts that all matter is composed of atoms and that atoms of different elements differ in properties from atoms of other elements remain unchanged.

Atoms and the conservation of mass If atoms are indivisible and atoms of different elements can combine in chemical reactions then it *must* be that mass is conserved in chemical reactions. Let's apply these ideas to two imaginary elements, element A and element B. Suppose atoms of B weigh three times as much as atoms of A. If we assign a mass of 1 mass unit to atoms of A, then atoms of B will have 3 mass units.

In a chemical reaction, let's assume one atom of A and one atom of B combine to give a chemical compound, which we represent as AB:

$$A + B \rightarrow AB$$

According to atomic theory, neither atom can be subdivided in this reaction. Therefore, the total mass of the product can only be the sum of the masses of A and B

$$A + B \rightarrow AB$$
$$1 \text{ mass unit} + 3 \text{ mass units} = 4 \text{ mass units}$$

The reverse also would hold true in a reaction in which AB is separated into its constituent elements

$$AB \rightarrow A + B$$
$$4 \text{ mass units} = 1 \text{ mass unit} + 3 \text{ mass units}$$

Figure 3-2 The combination, separation, and rearrangement of atoms are depicted, showing the conservation of mass in chemical reactions.

If all chemical reactions involve the combination, separation, or rearrangement of indivisible atoms, then matter and its mass can be neither created nor destroyed in such changes. (See Figure 3-2)

Atoms and the definite composition of chemical compounds Figure 3-2 also illustrates how atomic theory accounts for the definite composition of chemical compounds by mass. If atoms of A and atoms of B always have their own characteristic masses, then compound AB must always have the same composition by mass. Whatever mass units are used and no matter how much of compound AB you have or where it came from, the mass ratio of A to B in the compound will be 1 to 3. Four grams of AB will contain 1 g of A combined with 3 g of B. One hundred grams of AB will contain 25 g of A and 75 g of B. In other words, compound AB will always have the composition of 25% by mass of substance A and 75% by mass of substance B.

Law of Multiple Proportions

Certain elements can combine to form two or more different chemical compounds. Carbon and oxygen provide a familiar example. They are known to form carbon monoxide (CO), which is the poisonous gas in automobile exhaust, and carbon dioxide (CO_2), which is given off by most living things and taken up by plants.

How does atomic theory explain the formation of different compounds by the same two elements? Quite simply, only whole atoms combine in chemical compounds between the same two elements. Therefore, different compounds formed by the same two elements must result from combinations of different relative numbers of atoms. The atoms of our elements A and B might combine not only one-to-one, as in AB, but in other ways. One atom of A and two atoms of B could form a compound written as AB_2. They might combine one-to-three in the compound AB_3.

The ratio of the masses of one element combined with a certain fixed mass of a second element in two compounds must be the same as the ratio of the numbers of atoms of the first element in the two compounds. Consider our two elements, A and B. Above, we noted that 1 g of A combines with 3 g of B to give 4 g of AB. If instead, these elements form AB_2, then 1 g of A would have to combine with twice as much, or 6 g of B, to give AB_2. In these two compounds the masses of B have the ratio of 1 to 2. This whole-number ratio is the same as the ratio of the numbers of atoms of B in the two compounds.

John Dalton made many similar observations and proposed the **law of multiple proportions** which states: *If two or more different compounds are composed of the same two elements, the masses of the second element combined with a certain mass of the first element can be expressed as ratios of small whole numbers.*

If carbon monoxide and carbon dioxide were each produced by the reaction of 12 g of carbon with oxygen, the resulting compounds would contain 16 g and 32 g of oxygen, respectively. The ratio of 16 g of oxygen to 32 g of oxygen = $\frac{1}{2}$. This indicates that, in carbon monoxide, there are half as many oxygen atoms for every carbon atom as there are in carbon dioxide. This result reflects the known compositions of carbon monoxide and carbon dioxide. Carbon monoxide contains one carbon atom and one oxygen atom. Carbon dioxide contains one carbon atom and two oxygen atoms.

Propose this problem. When magnesium burns in oxygen, the ratio is 3 mass units of magnesium to 2 mass units of oxygen to form 5 mass units of magnesium oxide. How many grams of oxygen are needed to react with 6 g of magnesium? (4 g) What would happen if there were 6 g of oxygen? (When the reaction is over, 2 g of oxygen would be left.)

For a given mass of carbon combined with oxygen to form carbon dioxide (CO_2: mass ratio, 1 C: 2.66 O) the mass of oxygen (O), in carbon dioxide is always twice that in carbon monoxide (CO: mass ratio, 1 C : 1.33 O). Every gram of carbon combines, therefore, with either 2.66 g or 1.33 g of oxygen.

Desktop Investigation

Constructing a Model

Question
How can you construct a model of an unknown object by (1) making inferences about an object that is in a closed container and (2) touching the object without seeing it?

Materials
sealed container
one or more objects that fit in the container
cm ruler
balance

Procedure
1. Your teacher will provide you with a can covered by a sock and sealed with tape. Record all your observations in a data table. Without unsealing the container, try to determine the number of objects, their mass, shape, size, plus the kind of material they are made of and their texture. To do this, you may carefully tilt and shake the can.
2. Remove the tape from the top of the sock. Do *not* look inside the container. Put one hand through the opening and make the same observations as in Procedure 1 by touching and holding the objects. To make more accurate estimations, practice estimating the size and mass of some known objects outside the can and then compare your estimates of the objects outside the can with actual measurements using a cm ruler and a balance.

Discussion
1. Scientists often use more than one method to gather data. How was this illustrated in the investigation?
2. Which of the observations you made were qualitative and which were quantitative?
3. Using the data you gathered, draw a model of the unknown object(s) and write a brief summary of your conclusions.

---

1. (a) Democritus is generally credited with the use of the term *atom* to describe the smallest particles of matter that can exist. (b) John Dalton proposed an atomic theory that related atoms to the measurable property of mass.
2. See bottom of page 71.
3. Laws of conservation of mass, definite composition, and multiple proportions
4. If two or more different compounds are composed of the same two elements, the masses of the second element combined with a certain mass of the first element can be expressed as ratios of small whole numbers.

The two compounds are represented by combining the atomic symbols as follows:

Compound Name	Compound Symbol
Carbon Monoxide	CO
Carbon Dioxide	CO_2

Section Review

1. Cite the major contributions of each of the following toward the present theory of the atom: (a) Democritus (b) John Dalton.
2. List the five essential points of Dalton's atomic theory.
3. What laws can be explained through the use of Dalton's theory?
4. State the law of multiple proportions.

3.2 The Structure of the Atom

You have thus far seen evidence that an **atom** is *the smallest particle of an element that can exist either alone or in combination with other atoms.* Experiments about the nature of matter continued into the 1800s as Dalton's atomic theory gained acceptance. The next developments on the pathway to modern atomic theory came with observations showing that atoms are divisible. Gradually, a model of atomic structure emerged. **Atomic structure** *refers to the identity and arrangement of smaller particles within atoms.*

Although atoms of various elements differ in the arrangements and numbers of particles that they contain, all atoms consist of two main regions. Within the atom, there is a very small region called the *nucleus* that is occupied by positively charged particles called *protons* and, in all but one case, neutral particles called *neutrons.* Surrounding the nucleus is a region occupied by negatively charged particles called *electrons.* The region occupied by electrons is very large compared to the size of the nucleus.

SECTION OBJECTIVES

- Summarize the observed properties of cathode rays that led to the discovery of the electron.
- Summarize the experiment conducted by Rutherford that led to the discovery of the nucleus.
- Describe the properties of protons, neutrons, and electrons.
- Define "atom" and "isotope."
- Describe the atomic structures of the isotopes of hydrogen.

It was not until 1939 that Lise Meitner figured out that an atom could indeed be subdivided. She and her associate Otto Hahn bombarded U-235 atoms with neutrons. Although Hahn suspected the uranium atoms were being split, it was Meitner who officially published the conclusion naming the process fission.

Figure 3-3 A simple cathode-ray tube.

Discovery of the Electron

The first evidence for the existence of smaller particles within atoms came in studies of the relationship between electricity and matter. Numerous experiments were performed in the late 1800s in which electrical current was passed through different gases at very low pressures. These experiments were carried out in glass tubes like those shown in Figure 3-3. An electrical current passes through the tube from the *cathode,* the electrode (plate) connected to the negative terminal of the battery, to the *anode,* the electrode (plate) connected to the positive electrode. Such tubes, called cathode-ray-tubes, were the forerunners of television tubes.

Cathode rays and electrons Gases at atmospheric pressure do not conduct electricity very well. At the very low pressures in cathode-ray-tubes,

In a television tube, a cathode ray constantly undergoes deflections that result from the varying magnetic field within the tube. The beam strikes a coated screen, where it creates a luminescent image.

Atoms: The Building Blocks of Matter

Figure 3-4 A paddle-wheel cathode-ray tube

The different colors result from the different energies that electrons have in different kinds of atoms. The electrons, however, are identical in all atoms.

current flows readily through a gas. A series of experiments established a number of facts about what happens in such tubes: *(1)* Different gases glow with different colors as current passes through the tube. *(2)* The part of the glass tube directly opposite the cathode glows. *(3)* An object placed between the cathode and the opposite end of the tube casts a shadow on the glass. *(4)* A paddle wheel between the electrodes rolls along on its rails from the cathode toward the anode (Figure 3-4).

The cathode ray is composed of particles (as shown by the paddle wheel experiment) that pass from the cathode to the anode. Further experiments gave additional information: *(5)* Cathode rays are deflected by a magnetic field in the same manner as a wire carrying electrical current, which was known to have a negative charge. *(6)* The rays are deflected away from a negative electrode. The observations of cathode-ray behavior led to the hypothesis that cathode rays consist of negatively charged particles. A brilliant series of experiments carried out in 1897 by the English physicist Sir Joseph John Thomson (1856–1940) strongly supported the hypothesis that cathode rays consist of negatively charged particles. Thomson used the apparatus shown in Figure 3-5 to measure the ratio of the charge of cathode-ray particles to their mass. He found that this ratio was always the same, regardless of the metal used to make the cathode or of the gas in the cathode-ray tube. Thomson concluded that all cathode rays were composed of identical negatively charged particles, which came to be called electrons. **Electrons** *are negatively charged subatomic particles.*

Charge and mass of the electron The ratio that Thomson obtained in his experiments was extremely high. This ratio indicated that the electron had a very small mass compared to its charge. In 1909, experiments conducted by the American physicist Robert A. Millikan (1868–1953) showed that the mass of the electron is in fact approximately one two-thousandth the mass of a hydrogen atom, which is the smallest atom known. This experiment is discussed further in the Chemistry in Detail feature beginning on page 79. More accurate experiments conducted since then indicate that the electron has a mass of 9.109×10^{-28} g, which is $\frac{1}{1837}$ the mass of the simplest type of hydrogen atom.

Millikan's experiments also confirmed that the electron has the smallest possible negative electric charge, which is what Thomson and others had hypothesized. All other negative charges are whole-number multiples of the charge of the electron. Thus, an electric charge can be twice or three times that of an electron but never one-half or three-halves of that charge. Because cathode rays always have the same properties, regardless of the element used to produce them, it was concluded that electrons are present in atoms of all elements.

The electrons in cathode rays could come only from within the atoms of the gas or the metal cathode. Cathode-ray experiments had provided, therefore, clear-cut evidence that atoms are divisible. Two other inferences could be made about atomic structure based on what was learned about electrons. These are: *(1)* Because atoms are electrically neutral, they must contain a positive charge to balance the negative electrons; and *(2)* Because electrons are so much smaller in mass than atoms, atoms must contain additional particles that account for most of their mass.

Chapter 4 is devoted to the arrangement of electrons in atoms. This arrangement plays a very important role in determining the properties of matter.

Figure 3-5 A CRT with a calibrated luminescent screen can measure the charge-to-mass ratio of a charged particle.

Discovery of the Atomic Nucleus

In 1908 and 1909, New Zealander Ernest Rutherford (1871–1937), working in England with his associates Hans Geiger (1882–1945) and Ernest Marsden (1889–1970), studied the bombardment of thin metal foils with fast-moving positively charged particles. The experiments were carried out using the setup shown in Figure 3-6. Almost all the fast-moving particles passed through the foil with only a slight deflection. This result is to be expected in bombarding the 10 000-atom-thick foil if the mass and positive charge of the atoms is uniformly distributed.

Geiger and Marsden also checked for the possibility of wide-angle deflections. Much to their surprise, they found that roughly one in eight thousand particles actually ricocheted back toward the source, as shown in Figure 3-6. As Rutherford later exclaimed, it was "as if you had fired a 15-inch [artillery] shell at a piece of tissue paper and it came back and hit you."

Rutherford thought about this experimental result until, after two years, he finally saw the solution. He reasoned that the fast-moving particles must be repelled by some powerful force within the atom. He reasoned further that

After Rutherford received the 1908 Nobel Prize in chemistry for his work with radioactivity, he discovered the proton. He was also the first person that was able to carry out the alchemist's dream of transmutation when he changed nitrogen atoms into oxygen atoms via nuclear reactions.

Figure 3-6 Rutherford and his team used an apparatus like this in experiments that led him to consider the existence of the atomic nucleus.

whatever caused this repulsion must occupy a very small amount of space, since only a very few particles shot into the foil ran into it. Rutherford concluded that this force must be caused by a very densely packed bundle of matter with a positive electric charge. As shown in Fig 3-7, this bundle of matter is the nucleus. *The **nucleus** is the positively charged, dense central portion of the atom that contains nearly all of its mass but takes up only an insignificant fraction of its volume.* If the nucleus had the diameter of a dime, the distance to the outer edge of the atom would be about one-half the length of a football field! Thus, the atom is mostly empty space.

Where, then, were the electrons? Rutherford suggested, although he had no supporting evidence, that the electrons surrounded the positively charged nucleus like planets around the sun. He could not explain, however, what kept the electrons in motion around the nucleus.

Composition of the Atomic Nucleus

Except for the nucleus of the simplest type of hydrogen atom (discussed in the next section) every atomic nucleus is made up of two kinds of particles,

Rutherford discovered the proton in 1919 as a particle produced in the bombardment of atoms of several elements.

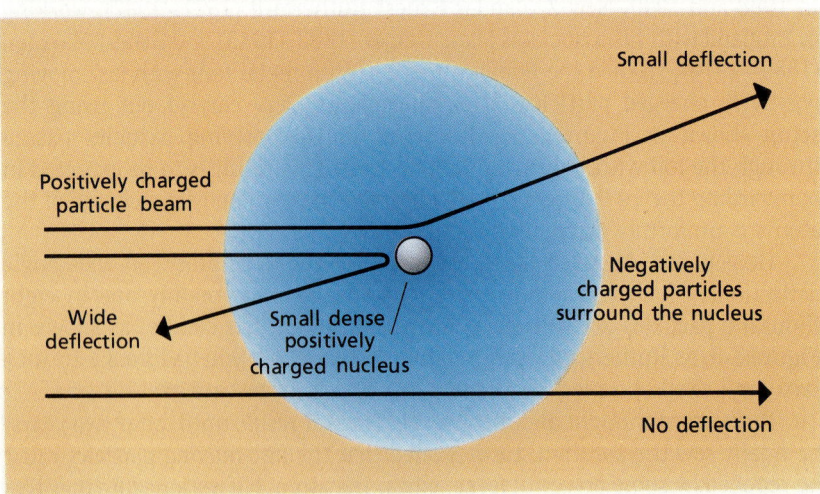

Figure 3-7 Rutherford and his team bombarded metal foil with a beam of positively charged particles. Something deflected them; Rutherford suggested that it was a dense, positively charged nucleus.

Chemistry in Detail

Millikan's Oil-Drop Experiment

The problem of determining the actual mass and charge of the electron was solved in 1909 by Robert A. Millikan of the University of Chicago using the apparatus shown in the figure to the right.

A mist of fine droplets of oil was allowed to fall slowly through the air in the chamber. Some of these droplets were negatively charged, having acquired extra electrons dislodged from gases in the air by X rays. A microscope outside the chamber allowed Millikan to observe the relatively few droplets that fell through the small central hole in the upper metal plate.

When there was no voltage between the two metal plates, the falling droplets were acted on only by the force of gravity and by friction with the air. Millikan carefully adjusted the voltage between the metal plates so that the downward gravitational force on a droplet was countered by an upward force between the charged droplet and the charged metal plates. At this

voltage, the drop is held motionless in the air. The upward force was proportional to the product of the charge of the droplet and the voltage between the two plates. The charge on each droplet could then be calculated from the voltage between the two plates. It is proportional to the charge required to balance the gravitational force on the droplet.

Millikan found that all the negative charges on the falling droplets are whole-number multiples of the same negative charge. He assumed that this was the smallest possible negative charge, and, therefore, the charge of a single electron. Using this value for the charge of the electron, he then used Thomson's charge-to-mass ratio for the electron to calculate the mass of the electron by using the following relationship:

$$\text{mass of electron} = \frac{\text{charge of electron}}{\text{charge-to-mass ratio of electron}}$$

protons and *neutrons*. Protons have a mass of 1.673×10^{-24} g, which is 1836 times larger than the mass of an electron and $\frac{1836}{1837}$ of the mass of the lightest and simplest type of hydrogen atom, which consists of a single-proton nucleus with a single electron moving about it. Most of the mass of such a hydrogen atom is due to the mass of the proton. **Protons** *are subatomic particles that have a positive charge equal in magnitude to the negative charge of an electron and are present in atomic nuclei.* The electrical neutrality of atoms is due to the presence of equal numbers of protons and electrons.

Neutrons were discovered in 1932 by the English scientist James C. Chadwick (1891–1974), who had worked with Rutherford.

Protons alone do not account for the masses of atoms of elements other than the simplest hydrogen atom. These atoms also contain **neutrons,** *which are electrically neutral, subatomic particles found in atomic nuclei.* The mass of a neutron is 1.675×10^{-24} g—about the same mass as that of a proton.

Physicists have now identified many other subatomic particles. However, particles other than electrons, protons, and neutrons play little role in the properties of matter that are of interest in chemistry. Table 3-1 summarizes the properties of electrons, protons, and neutrons.

TABLE 3-1 PROPERTIES OF SUBATOMIC PARTICLES

Particle	Symbols	Relative Electric Charge	Mass Number	Relative Mass*	Actual Mass
electron	e^-, $_{-1}^{0}e$	-1	0	$0.000\ 548\ 6\ u$	9.109×10^{-28} g
proton	p^+, $_{1}^{1}H$	$+1$	1	$1.007\ 276\ u$	1.673×10^{-24} g
neutron	n°, $_{0}^{1}n$	0	1	$1.008\ 66\ u$	1.675×10^{-24} g

* $1\ u = 1.660\ 5402 \times 10^{-24}$ g

STUDY HINT
Chemical changes do not involve the creation or change in identity of atoms.

Changes in which atoms are changed in identity are referred to as nuclear reactions or nuclear changes and are discussed in Chapter 29.

An *element* is a substance composed of atoms all of which contain the same number of protons.

Tritium, which is found in the atmosphere, might be formed when nitrogen is bombarded by neutrons from cosmic rays.

$_{7}^{14}N + _{0}^{1}n \rightarrow _{6}^{12}C + _{1}^{3}H$

It is used as a tracer, to make luminous paint, and in the production of hydrogen bombs. Both deuterium and tritium may be used as nuclear fuels for controlled nuclear fusion reactions in the future.

Particles that have the same electric charge generally repel one another. Nevertheless, over 100 protons can exist close together in a nucleus. When two protons are very close to each other, there is a strong attraction between them. Proton–proton attractive forces, neutron–neutron attractive forces, and proton–neutron attractive forces exist when these pairs of particles are very close together. *These short-range proton–neutron, proton–proton, and neutron–neutron forces hold the nuclear particles together and are referred to as* **nuclear forces.**

The nuclei of atoms of different elements differ in numbers of protons and therefore in positive charge. The nuclei of atoms of different elements also have different masses, although the difference in mass is sometimes very slight.

The Isotopes of Hydrogen

The number of protons in the nucleus determines the identity of an atom. All hydrogen atoms contain only one proton. An atom that contains more than one proton in its nucleus is an atom of an element other than hydrogen.

Like those of all naturally occurring elements, however, hydrogen atoms can contain varying numbers of neutrons. Three types of hydrogen atoms are known. The most common type of hydrogen is sometimes called *protium*. The nucleus of a protium atom consists of *one proton only*, and has one electron moving about it.

In addition to protium, which makes up 99.985% of naturally occurring hydrogen, there are two other known types of hydrogen. One is *deuterium,* which occurs in 0.015% of all hydrogen in nature. Each deuterium atom has a nucleus containing one proton and one neutron. The third form of hydrogen is known as *tritium,* which is radioactive. It exists in very small amounts in nature, but can be prepared artificially in a nuclear reaction. Each

Chemistry and Technology

"Looking" at Atoms

Scientists have long believed that atoms exist because the atomic theory successfully explains observable properties of matter. Now various modern techniques provide images of atoms on a computer screen or on film. When transmission electron microscopy (TEM) was first developed in the 1930s by Ernst Ruska, it could magnify objects 12 thousand ×. Since then, TEM has achieved magnifications of 2 million × and can be used to show the arrangement of relatively large atoms (left photo above). A thin specimen (about 1000 atoms thick) is illuminated from the rear by an electron beam. Electrons emerging through the front of the sample are focused by a magnetic lens onto camera film or a fluorescent screen.

Images of atoms on the surface of a material are provided by a technique in microscopy called scanning-tunneling microscopy (STM).

The STM can achieve magnifications of up to 100 million. It was developed in 1981 by Gerd Binnig and Heinrich Rohrer. A precise mechanism causes an extremely thin needle-like probe to move along the surface of the sample in parallel, closely spaced lines. A high-voltage electric current is maintained between the sample surface and the needle tip. This current is related to the distance between the tip and the atoms on the sample surface.

The same distance is always maintained between the tip and the surface atoms. In this way, the needle precisely follows the uneven contour of the surface. The up-and-down motion of the tip is processed by a computer, which generates an image of the sample surface like that shown in the STM image of silicon on the right. Ernst Ruska, Gerd Binnig, and Heinrich Rohrer shared the 1986 Nobel Prize in physics for their work on microscopy.

tritium atom contains one proton, two neutrons, and one electron, as shown schematically in Figure 3-8.

Protium, deuterium, and tritium are described as *isotopes* of hydrogen. Atoms of each have one proton, but they have zero, one, and two neutrons, respectively, and therefore have different total masses. **Isotopes** *are atoms of the same element that have different masses.* In all three types of hydrogen atoms, the charge of the single proton is balanced by that of one electron.

The Sizes of Atoms

The nuclei of atoms have radii of about 1×10^{-13} cm. They have a density of about 2×10^8 metric tons/cm^3. In a unit that is more convenient for the sizes of atoms, nuclei have radii of about 0.001 pm (picometers—1 pm = 10^{-12} m). By contrast, atomic radii range from about 40–270 pm. To get some idea of the extreme smallness of the picometer, consider the fact that 1 cm is the same fractional part of 10^3 km (about 600 miles) as 100 pm is of 1 cm.

The radius of an atom is the distance from the center of the nucleus to the outer edge of the region occupied by electrons, which are rapidly moving around the nucleus. It is convenient to think of the region occupied by the electrons as an electron cloud—a cloud of negative charge. In the next chapter, you will learn more about the nature of the electron cloud.

Protium

Deuterium

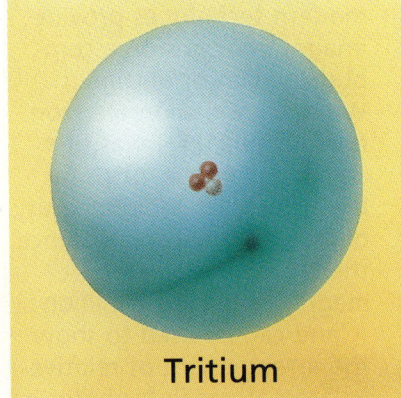
Tritium

Figure 3-8 Hydrogen and its isotopes have different nuclei.

Section Review

1. Define each of the following: *(a)* atom *(b)* electron *(c)* nucleus *(d)* protons *(e)* neutrons *(f)* isotopes.
2. Cite at least one contribution of each of the following toward the development of the present atomic theory: *(a)* Thomson *(b)* Millikan *(c)* Rutherford.
3. Compare and contrast the three types of subatomic particles in terms of location in the atom, mass, and relative charge.
4. How do the three isotopes of hydrogen compare in terms of the numbers of subatomic particles in each?

3.3 Weighing and Counting Atoms

Consider neon (Ne), the gas in many illuminated signs. Neon is one of the minor components of air, since dry air contains only about 0.002% neon. Yet about 5×10^{17} atoms of neon are present in each breath you inhale. Atoms are much too small to be weighed individually or counted in most experiments. How, then, can chemists quantitatively study elements and the chemical compounds made from them?

In this section, you will be introduced to some of the concepts used to link observed chemical behavior with tiny atomic and subatomic particles.

Answers to Section Review:

1. *(a)* The smallest particle of an element that can exist either alone or in combination with other atoms *(b)* A negatively charged subatomic particle *(c)* The positively charged, dense, central portion of an atom that contains nearly all of its mass but takes up only an insignificant fraction of its volume *(d)* Subatomic particles of the nucleus that have a positive charge of the same size as the negative charge of an electron *(e)* Electrically neutral, subatomic particles found in atomic nuclei *(f)* Atoms of the same element that have different masses
2. *(a)* Thomson's work provided strong support for the hypothesis that cathode rays are negatively charged particles *(b)* Millikan's work confirmed that the electron has the smallest possible negative electric charge *(c)* Rutherford discovered the nucleus and described its contents.

SECTION OBJECTIVES

- Define "atomic number" and "mass number" and describe how they apply to isotopes and nuclides.
- Determine the number of protons, neutrons, and electrons in a nuclide, given the identity of the nuclide.

Atomic Numbers and Mass Numbers

*The **atomic number** (Z) of an element is the number of protons in the nucleus of each atom of that element.* An element consists of atoms, all of which have the same number of protons in their nuclei. Atoms of different elements contain different numbers of protons.

In the Periodic Table (inside the back cover), the elements are placed from left to right in order of increasing atomic numbers. At the top of the table is hydrogen (H), which has atomic number 1. Atoms of the element hydrogen, whether they be protium, deuterium, or tritium atoms, have one proton in each nucleus. Next in order are helium (He) with two protons in each nucleus; lithium (Li) with three protons, beryllium (Be) with four protons; and so on.

The atomic number alone identifies an element. If you want to know which element has, for example, atomic number 47, consulting the Periodic Table would show that it is silver (Ag). All silver atoms contain 47 protons in their nuclei. Because atoms are neutral, we know from the atomic number that all silver atoms must also contain 47 electrons.

Atoms of different isotopes have different masses. Identifying an *isotope* requires knowing both the name or atomic number of the element and the mass of the isotope. *The **mass number** is the total number of protons and neutrons in the nucleus of an isotope.* The three isotopes of hydrogen described earlier have mass numbers 1, 2, and 3, as shown below.

	Atomic Number (Number of Protons)	Number of Neutrons	Mass Number
protium	1	0	0 + 1 = 1
deuterium	1	1	1 + 1 = 2
tritium	1	2	1 + 2 = 3

Two methods are used to designate isotopes in writing. In the first method, the mass number is added with a hyphen to the name of the element. Tritium, for example, is described as hydrogen-3. We will refer to this method as "hyphen notation." The uranium isotope used in fuel for nuclear power plants has a mass number of 235 and is known as uranium-235.

To determine the composition of, for example, a uranium-235 nucleus, one first finds from the Periodic Table the atomic number of uranium, which is 92. The number of neutrons is then found by subtracting the atomic number from the mass number:

mass number − atomic number = ?
235 (protons + neutrons) − 92 protons = 143 neutrons

A uranium-235 nucleus contains 92 protons and 143 neutrons.

In the second method of designating isotopes, the atomic number is added to the symbol for the element as a subscript on the left and the mass number is added as a superscript on the left. For hydrogen-3 and uranium-235, these nuclear symbols are

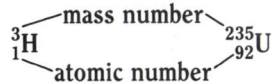

- Distinguish between *relative* atomic mass and *average* atomic mass.
- Calculate the average atomic mass of an element given the relative abundances of each isotope of the element.
- Define a mole in terms of Avogadro's number; define *molar mass*.
- Solve problems involving mass, mole, and the number of atoms of an element.

Section Review Answers
continued

3. Electrons are located in energy levels surrounding the nucleus of an atom, they have a mass of 9.110×10^{-28} g and a relative charge of 1−. Protons are located in the nucleus, have a mass of 1.673×10^{-24} g, and have a relative charge of 1+. Neutrons are also in the nucleus, have a mass of 1.675×10^{-24} g, and have no charge.
4. All contain one proton and one electron, but protium contains no neutrons, deuterium contains one neutron, and tritium contains two.

Table 3-2 gives the names, symbols, and composition of the isotopes of hydrogen and helium. **Nuclide** *is a general term for any isotope of any element*. For example, we could say that Table 3-2 lists the composition of five different nuclides.

TABLE 3-2 ISOTOPES OF HYDROGEN AND HELIUM

Isotope	Symbol	Number of Protons	Number of Electrons	Number of Neutrons
hydrogen-1 (protium)	$^{1}_{1}H$	1	1	0
hydrogen-2 (deuterium)	$^{2}_{1}H$	1	1	1
hydrogen-3 (tritium)	$^{3}_{1}H$	1	1	2
helium-3	$^{3}_{2}He$	2	2	1
helium-4	$^{4}_{2}He$	2	2	2

Sample Problem 3.1

How many protons, electrons, and neutrons are there in an atom of chlorine-37?

Solution

Step 1. Analyze   Given: name and mass number of chlorine-37
Unknown: numbers of protons, electrons, and neutrons

Step 2. Plan

atomic number ⟶ no. of protons and electrons
mass number ⟶ no. of neutrons

The atomic number of chlorine can be found in the Periodic Table, and this is equal to the number of protons and also the number of electrons. The number of neutrons must be found by subtracting the atomic number from the mass number.

mass number of chlorine-37 − atomic number of chlorine
= number of neutrons in chlorine-37

Step 3. Compute   mass number − atomic number = 37 (protons plus neutrons) − 17 protons
= 20 neutrons

An atom of chlorine-37 contains 17 electrons, 17 protons, and 20 neutrons.

Step 4. Evaluate   The number of protons equals the number of electrons and the sum of the protons and neutrons equals the given mass number as they should.

Practice Problems

1. How many protons, electrons, and neutrons are in an atom of bromine-80?   *(Ans.)* 35 protons, 35 electrons, and 45 neutrons
2. Write the nuclear symbol for carbon-13.   *(Ans.)* $^{13}_{6}C$
3. Write the hyphen notation for the element that contains 15 electrons and 15 neutrons.   *(Ans.)* phosphorus-30

Relative Atomic Masses

Masses of atoms measured in grams are very small. An atom of oxygen-16, for example, has a mass of 2.657×10^{-23} g. It is only in this century that scientists have been able to measure masses this small. Although we now know the actual masses of atoms, for most chemical calculations it is more convenient to use *relative* masses.

As you learned in Chapter 2, scientists use standards of measurement that are constant and the same everywhere. In order to set up a relative scale of *atomic mass*, one atom is arbitrarily chosen as the standard and assigned a relative mass value. The masses of all the other atoms are then expressed in relation to this defined standard. The carbon-12 atom has been chosen as the standard by the international organization of scientists that governs units of measurement. A single atom of the nuclide carbon-12 is arbitrarily assigned a mass of exactly 12 atomic mass units. *The **atomic mass unit** is symbolized by u.* **One atomic mass unit**, *1 u, is exactly $\frac{1}{12}$ the mass of a carbon-12 atom, or $1.660\,5402 \times 10^{-24}$ g. The atomic mass of a carbon-12 is exactly 12 u. The mass of an atom expressed in atomic mass units is called the **atomic mass** of the atom.*

The hydrogen-1 atom actually does have a relative atomic mass of *about* $\frac{1}{12}$ that of the carbon-12 atom, or about $1\,u$. The precise value of the relative atomic mass of a hydrogen-1 atom is $1.007\,825\,u$. An oxygen-16 atom has about $\frac{4}{3}$ the mass of a carbon-12 atom. Careful measurements show the relative atomic mass of oxygen-16 to be $15.994\,915\,u$. The mass of a magnesium-24 atom is found to be slightly less than twice that of a carbon-12 atom. Its atomic mass is $23.985\,042\,u$. The atomic mass of any nuclide is determined by comparison with the mass of the carbon-12 atom.

Some additional examples of the relative atomic masses of the naturally occurring isotopes of several elements are given in Table 3-3. Isotopes of an element may occur naturally, or they may be made in the laboratory (*artificial isotopes*). Although isotopes have different masses, they do not differ significantly in their chemical behavior.

The masses of subatomic particles can also be expressed on the atomic mass scale (see Table 3-1). The atomic mass of the electron is $0.000\,548\,59\,u$. That of the proton is $1.007\,276\,u$ and that of the neutron is $1.008\,665\,u$. Note that the proton and neutron masses are close to but not equal to 1. You have learned that the mass number is the total number of protons and neutrons in the nucleus of an atom. You can now see that the mass number and relative atomic mass of a given nuclide are quite close to each other. They are not identical because the proton and neutron masses deviate slightly from $1\,u$ and the atomic masses include electrons. Also, as explained in Chapter 29, a small amount of mass is changed to energy in the creation of a nucleus from protons and neutrons.

Frédéric and Irène Joliot-Curie received the 1935 Nobel Prize in chemistry for discovering a way to make the first artificial isotopes by bombarding stable atoms with neutrons or deuterons. As a result of their work, radioactive isotopes can be made for all elements.

Average Atomic Masses of Elements

As illustrated for the elements listed in Table 3-3, most elements occur naturally as mixtures of isotopes. The percentage of each isotope in the naturally occurring element is nearly always the same, no matter where the element is found.

STUDY HINT
The isotopes of hydrogen are unusual in having distinctive names. Most isotopes are identified only by their mass numbers.

Atoms: The Building Blocks of Matter

TABLE 3-3 ATOMIC MASSES AND ABUNDANCES OF SEVERAL NATURALLY OCCURRING ISOTOPES

Isotope	Mass Number	Percent Natural Abundance	Relative Atomic Mass (u)	Average Atomic Mass* of Elements (u)
hydrogen-1	1	99.985	1.007 825	1.007 94
hydrogen-2	2	0.015	2.0140	
carbon-12	12	98.90	12.0 (by definition)	12.011
carbon-13	13	1.10	13.003 355	
oxygen-16	16	99.762	15.994 915	
oxygen-17	17	0.038	16.999 131	15.9994
oxygen-18	18	0.200	17.999 160	
copper-63	63	69.17	62.939 598	63.546
copper-65	65	30.83	64.927 793	
cesium-133	133	100	132.905 429	132.905
uranium-235	235	0.720	235.043 924	238.024
uranium-238	238	99.280	238.050 784	

*Values, established by international agreement, may differ slightly from calculated averages.

A *nonweighted* average of the marble masses is (2 g + 3 g) ÷ 2 = 2.5 g.

Atomic masses for the elements, which are included in the Periodic Table are averages for these naturally occurring mixtures of isotopes. *Weighted averages* account for the percentages of each isotope of a given element that is present.

As a simple example of a weighted average, suppose you had a box containing two sizes of marbles. If 25% of the marbles weigh 2.00 g each and 75% of the marbles weigh 3.00 g each, how could the weighted average be found? You could count the marbles, calculate the total mass of the mixture, and divide by the total number of marbles. Suppose you had 100 marbles. The calculations would go as follows: 25 marbles × 2.00 g = 50 g and 75 marbles × 3.00 g = 225 g. Adding these masses gives 50 g + 225 g = 275 g. Dividing by 100 gives an average marble mass of 2.75 g. A simpler method is to multiply the mass of each marble by the decimal fraction representing its percentage in the mixture.

$$2.0 \text{ g} \times 0.25 + 3.0 \text{ g} \times 0.75 \text{ g} = 2.75 \text{ g}$$

Average atomic mass *is the weighted average of the atomic masses of the naturally occurring isotopes of an element.* As illustrated, the average atomic mass depends on both the mass and the relative abundance of each isotope. For example, naturally occurring copper consists of 69.17% copper-63, which has an atomic mass of 62.939 598 u, and copper-65, which occurs with a proportion of 30.83% and has an atomic mass of 64.927 793 u. The average atomic mass of copper can be calculated by multiplying the atomic mass of each isotope by its relative abundance (expressed in decimal form) and adding the results:

$$0.6917 \times 62.939\ 598\ u + 0.3083 \times 64.927\ 793\ u = 63.55\ u$$

The calculated average atomic mass of naturally occurring copper is 63.55 u.

STUDY HINT

You may sometimes see the term *atomic weight* used synonymously with *atomic mass*.

The average atomic masses of the elements are important quantities to chemists because they indicate relative mass relationships in chemical reactions. You can find the average atomic masses of the elements in the Periodic Table inside the back cover of this book. What is the average atomic mass of silicon (Si)? What is the average atomic mass of lead (Pb)?

The average atomic masses are included for the elements listed in Table 3-3. In common use, we do not bother to distinguish between relative and average atomic masses, for it is clear from the context which is intended. The **atomic mass** *of a nuclide is the relative atomic mass of atoms of that nuclide;* the atomic mass of an element is the average atomic mass of the naturally occurring mixture of isotopes of the element.

As illustrated in Table 3-3, most atomic masses are known to four or more significant figures. In calculations in this book (other than addition), atomic masses of the elements are rounded before the calculation is carried out to the full number of digits that are known.

The Mole, Avogadro's Number, and Molar Mass

The relative atomic mass scale makes it possible to know how many atoms of an element are present in a given sample of the element with a measurable mass. To understand the composition of chemical compounds and what happens in chemical reactions, however, relative numbers of atoms and other chemical species must be noted.

Imagine that you could place a single carbon atom on one side of a balance. How many helium atoms would be needed on the other side to balance the carbon atom? The atomic mass of the element carbon is 12.01 u (to two decimal places) and that of helium is 4.00 u. Three helium atoms would just about balance the carbon atom, as shown schematically in Figure 3-9. If a second carbon atom were added to the first, then six helium atoms would be needed to achieve balance. No matter what number of carbon atoms were placed on the balance, three times as many helium atoms would be needed. Also, whatever mass of carbon were to be balanced, an equal mass of helium would contain three times as many atoms.

Figure 3-9 The relative masses of carbon and helium atoms

Figure 3-10 Approximately one molar mass of each of three elements: carbon (as graphite, *top*), iron nails *(center)*, and copper wire *(bottom)*.

Tell students that you counted out 6.02×10^{23} atoms of iron and sulfur for them. Then, show them a plastic bag containing 55.8 g of iron and one with 32.1 g of sulfur.

Next, imagine that you have placed 12.01 g of carbon on the balance. But instead of equal masses of helium atoms, now you would like to have the number of helium atoms present equal to the number of carbon atoms present. Note that, in the balance to the far right in Figure 3-9, when the numbers of atoms of carbon and helium are equal, the masses in grams are not. The number of carbon atoms in the balance has a mass of 12.01 g, but the equal number of helium atoms has a mass of 4.00 g.

The mole Three very important concepts—the mole, Avogadro's number, and molar mass—provide the basis for relating masses in grams to numbers of atoms. The first, the mole, is the SI unit for amount of substance. *The* **mole** *is the amount of a substance that contains the same number of particles as the number of atoms in exactly 12 g of carbon-12.* The abbreviation for the mole is **mol**. The mole is a counting unit, just like one dozen. We don't usually order 12 or 24 ears of corn; we order one dozen or two dozen. In writing, a chemist might refer to 1 mol of carbon, or 2 mol of iron, or 2.567 mol of calcium. In this chapter you will see how the mole applies to chemical compounds.

Avogadro's number The number of particles in a mole has been experimentally determined in a number of ways. The best modern value is $6.022\ 137 \times 10^{23}$ particles. This number is of such importance to chemistry that it is named in honor of the Italian scientist Amedeo Avogadro (1776–1856), whose ideas were crucial in the early understanding of chemistry. **Avogadro's number**—$6.022\ 137 \times 10^{23}$—*is the number of particles in exactly one mole of a pure substance.* Exactly 12 g of carbon-12 contains an Avogadro's number of carbon-12 atoms. For most purposes, Avogadro's number is rounded to 6.022×10^{23}.

To get a sense of how large Avogadro's number is, imagine this situation: Everyone living on earth (5 billion people) is to help count the atoms in one mole of an element. If each person counted continuously at the rate of one atom per second, it would require about 4 million years for all the atoms to be counted.

Molar mass Can you suggest how many grams of helium contain an Avogadro's number of helium atoms? If you decided it is 4.00 g, you are correct. Similarly, 6.941 g of lithium (Li) or 200.59 g of mercury (Hg) would contain the Avogadro's number of atoms. These convenient quantities are molar masses of these elements. *The* **molar mass** *is the mass in grams of one mole of a pure substance.* The mass of one mole of atoms, in grams, is numerically equal to the atomic mass of the element in atomic mass units.

Molar masses of elements contain equal numbers of atoms. One mole of a substance is one molar mass of that substance. An alternate definition of the *mole* is the amount of a substance that contains an Avogadro's number of particles, or chemical units. Here the chemical unit is the atom. For elements such as those shown in Figure 3-10, one molar mass contains one mole of atoms of the naturally occurring isotopes. For individual nuclides, one molar mass contains one mole of atoms of the nuclide.

The molar mass is used like a conversion factor in chemical calculations. For a specific substance, a known number of grams can be converted to moles or a known number of moles can be converted to grams. In such calculations, molar mass has the units of grams per mole. For example, to find how many grams of helium is equal to 2 mol of helium, you would use the dimensional method introduced in Chapter 2. The molar mass of helium is taken as 4.0 g He/mol He (to the required number of significant figures).

$$2.0 \text{ mol He} \times \frac{4.0 \text{ g He}}{\text{mol He}} = 8.0 \text{ g He}$$

Figure 3-11 shows the conversions of molar mass, moles, and Avogadro's number.

Sample Problem 3.2

What is the mass in grams of 3.50 mol of the element copper (Cu)?

Solution

Step 1. Analyze Given: 3.50 mol Cu
Unknown: mass of Cu in grams

Step 2. Plan moles ⟶ mass

The molar mass of copper is the necessary factor for converting from moles of copper to grams of copper. It is the mass in grams numerically equal to the atomic mass for copper given in the Periodic Table.

$$\text{mol Cu} \times \frac{\text{grams Cu}}{\text{mol Cu}} = \text{grams Cu}$$

Step 3. Compute
$$3.50 \text{ mol Cu} \times \frac{63.5 \text{ g Cu}}{\text{mol Cu}} = 222 \text{ g Cu}$$

Step 4. Evaluate The number of moles of copper was given to three significant figures; the molar mass was therefore rounded to three significant figures. The size of the answer is about right, for it is somewhat more than 3.5 times 60.

Practice Problems
1. What is the mass in grams of 2.25 mol of the element iron (Fe)? *(Ans.)* 126 g Fe
2. What is the mass in grams of 0.375 mol of the element potassium (K)? *(Ans.)* 14.7 g K

Sample Problem 3.3

A chemist produced 11.9 g of aluminum (Al). How many moles of aluminum has been produced?

Solution

Step 1. Analyze Given: 11.9 g Al
Unknown: moles of Al

Step 2. Plan mass ⟶ moles

The molar mass of aluminum is the conversion factor for finding moles of aluminum from a known mass of aluminum in grams. As shown in Figure 3-11, going from mass in grams to number of moles is done by dividing by molar mass. In using the dimensional method, it is easiest to do this by inverting the factor to moles per gram and multiplying.

Atoms: The Building Blocks of Matter

Sample Problem 3.3 *continued*

$$\text{grams Al} \times \frac{\text{mol Al}}{\text{grams Al}} = \text{moles Al}$$

Step 3. Compute

$$11.9 \text{ g Al} \times \frac{1 \text{ mol Al}}{27.0 \text{ g Al}} = 0.441 \text{ mol Al}$$

Step 4. Evaluate — The molar mass and the answer are correctly rounded to three significant figures. The answer is reasonable because 11.9 g is somewhat less than half of 27.0 g.

Practice Problems
1. How many moles of calcium (Ca) are contained in 5.00 g of calcium? *(Ans.)* 0.125 mol Ca
2. How many moles of gold (Au) are contained in 3.60×10^{-10} g of gold? *(Ans.)* 1.83×10^{-12} mol Au

Figure 3-11 shows that Avogadro's number can be used to find the number of atoms of an element from number of moles or to find the number of moles from the number of atoms of an element. While these types of problems are less common in chemistry than converting between moles and grams, they are useful in demonstrating the meaning of Avogadro's number. Note that, in these calculations, Avogadro's number is expressed as units of atoms per mole.

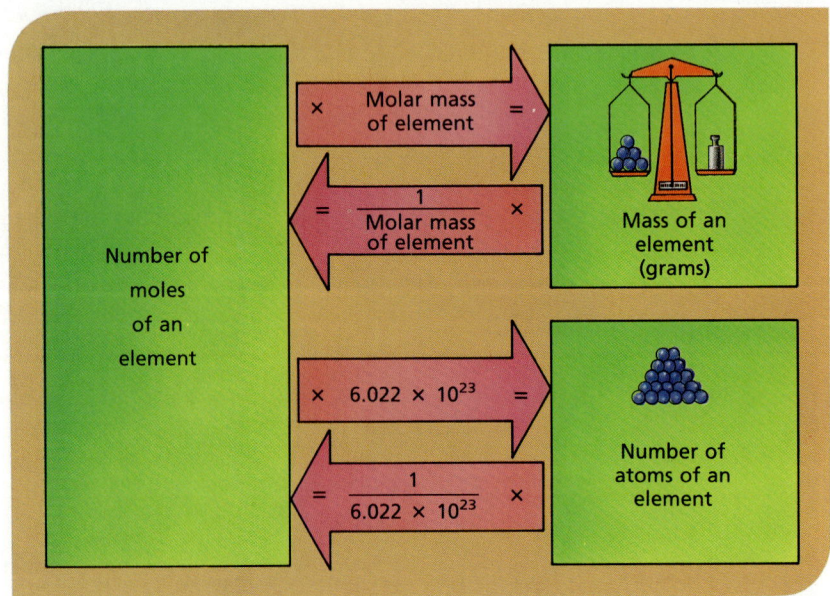

Figure 3-11 Relationships between mass, the number of moles, and the number of atoms of an element in a sample

Sample Problem 3.4

How many moles of silver (Ag) are in 3.01×10^{23} atoms?

Sample Problem 3.4 *continued*

Solution

Step 1. Analyze Given: 3.01×10^{23} atoms of silver
Unknown: moles of silver

Step 2. Plan atoms ⟶ moles
Number of atoms is converted to number of moles by dividing by Avogadro's number (see Figure 3.11). In setting up the problem, we will use the inverted Avogadro's number factor so that we can multiply and cancel units.

$$\text{Ag atoms} \times \frac{\text{mol Ag}}{\text{Avogadro's number of Ag atoms}}$$

Step 3. Compute

$$3.01 \times 10^{23} \text{ Ag atoms} \times \frac{1 \text{ mol Ag}}{6.02 \times 10^{23} \text{ Ag atoms}} = 0.500 \text{ mol Ag}$$

Step 4. Evaluate The answer is correct—units cancel correctly and the number of atoms is exactly one-half of Avogadro's number.

Practice Problems

1. How many moles of lead (Pb) are equivalent to 1.50×10^{12} atoms? *(Ans.)* 2.49×10^{-12} mol Pb
2. How many moles of tin (Sn) are equivalent to 2500 atoms? *(Ans.)* 4.2×10^{-21} mol Sn
3. How many atoms of aluminum (Al) are contained in 2.75 mol? *(Ans.)* 1.66×10^{24} atoms Al

Sample Problem 3.5

What is the mass in grams of 1.20×10^8 atoms of copper?

Solution

Step 1. Analyze Given: 1.20×10^8 atoms of copper
Unknown: mass in grams

Step 2. Plan atoms ⟶ moles ⟶ mass

As indicated in Figure 3.11, the given number of atoms must first be converted to moles by dividing by Avogadro's number. The unknown mass in grams can then be found by multiplying by the molar mass of copper.

$$\text{Cu atoms} \times \frac{\text{mol Cu}}{\text{Avogadro's no. of Cu atoms}} \times \frac{\text{grams Cu}}{\text{mol Cu}} = \text{grams Cu}$$

Step 3. Compute

$$1.20 \times 10^8 \text{ Cu atoms} \times \frac{1 \text{ mol Cu}}{6.02 \times 10^{23} \text{ Cu atoms}} \times \frac{63.5 \text{ g Cu}}{\text{mol Cu}} = 1.27 \times 10^{-14} \text{ g Cu}$$

Step 4. Evaluate Units cancel correctly to give the answer in grams. The size of the answer is reasonable—10^8 has been divided by 10^{23}.

Practice Problems

1. What is the mass in grams of 7.5×10^{15} atoms of nickel (Ni)? *(Ans.)* 7.4×10^{-7} g Ni
2. How many atoms of sulfur (S) are contained in 4.00 g? *(Ans.)* 7.50×10^{22} atoms S
3. What mass of gold (Au) contains the same number of atoms as 9.0 g of aluminum (Al)? *(Ans.)* 67 g Au

1. (a) The number of protons in the nuclei of atoms of that element (b) The total number of protons and neutrons in a nucleus. (c) The mass of an atom expressed in atomic mass units. (d) The weighted average of the masses of the naturally occurring isotopes of an element. (e) The amount of a substance that contains a number of particles equal to the number of atoms in exactly 12 grams of carbon-12 (f) The number of particles in exactly one mole of a pure substance (g) The mass in grams of an element that is numerically equal to its mass on the relative atomic mass scale

2. (a) 11, 11, 12 (b) 20, 20, 20 (c) 29, 29, 35 (d) 47, 47, 61

3. (a) $^{28}_{14}$Si; silicon-28 (b) $^{56}_{26}$Fe; iron-56 (c) $^{138}_{56}$Ba; barium-138

4. (a) 39.098 u (b) 39.098 g

5. (a) 28.0 g N (b) 17.8 g Cl

6. (a) 0.500 0 mol Mg (b) 0.249 mol F

7. (a) 1.50 × 10^{24} atoms Zn (b) 7.52 × 10^{22} atoms C

Section Review

1. Define each of the following: (a) atomic number (b) mass number (c) relative atomic mass (d) average atomic mass (e) mole (f) Avogadro's number (g) molar mass.

2. Determine the number of protons, electrons, and neutrons in each of the following isotopes: (a) sodium-23 (b) calcium-40 (c) $^{64}_{29}$Cu (d) $^{108}_{47}$Ag.

3. Write the nuclear symbol and hyphen notation for each of the following neutral isotopes: (a) mass number of 28 and atomic number of 14 (b) 26 protons and 30 neutrons (c) 56 electrons and 82 neutrons.

4. To three decimal places, what is the (a) relative atomic mass and (b) molar mass of the element potassium, K?

5. What is the mass in grams of: (a) 2.00 mol N (b) 3.01 × 10^{23} atoms Cl?

6. How many moles are represented by: (a) 12.15 g Mg (b) 1.50 × 10^{23} atoms F?

7. How many atoms are contained in: (a) 2.50 mol Zn (b) 1.50 g C?

Chapter Summary

- Scientists since the ancient Greeks have created theories of atoms. In the nineteenth century, John Dalton proposed a five-point theory of atoms that can still explain properties of many chemicals and compounds today.
- Matter and its mass can be neither created nor destroyed in chemical reactions.
- Cathode ray tubes supplied the first evidence of the existence of *electrons*, which are negatively charged subatomic particles having relatively little mass.
- Rutherford found evidence for the existence of the atomic nucleus—a positively charged, very dense core within the atom—by using a positively charged particle beam aimed at metal foil.
- Atomic nuclei are composed of protons, that have an electric charge of 1+, and, in all but one case, neutrons, that have no electric charge.
- Isotopes of an element differ by the number of neutrons in their nuclei.
- Atomic nuclei have radii of about 0.001 pm (pm = picometers; 1 pm = 10^{-12} m), while atoms have radii of about 40–270 pm.
- The atomic number of an element is the number of protons in the nuclei of the atoms of that element.
- The mass number is equal to the total number of protons and neutrons in the nucleus of an atom of that element.
- The relative atomic mass unit (u) is based on the carbon-12 atom and is a convenient unit for measuring the mass of atoms.
- The average atomic mass of an element is found by calculating the weighted average of the atomic masses of the naturally occurring isotopes of the element.
- Avogadro's number is equal to 6.022 137 × 10^{23} particles per mole of a pure substance. It is the number of atoms in 12 grams of carbon-12. A sample that contains a number of particles equal to Avogadro's number contains a mole of those particles.

CHAPTER 3 REVIEW

Atoms: The Building Blocks of Matter

VOCABULARY

atom (75)
atomic mass (85)
atomic mass unit (85)
atomic number (83)
atomic structure (75)
average atomic mass (86)
Avogadro's number (88)
electrons (76)
isotopes (81)
law of multiple
 proportions (73)

mass number (83)
molar mass (88)
mole (88)
neutrons (80)
nuclear forces (80)
nucleus (78)
nuclide (84)
protons (79)

REVIEWING CONCEPTS

1. Explain each of the following in terms of Dalton's atomic theory:
   a. the law of conservation of mass
   b. the law of definite composition
   c. the law of multiple proportions

2. According to the law of conservation of mass, if element A has a mass of 2 mass units and element B has a mass of 3 mass units, what mass would be expected for element AB? (3.1)

3. a. State the law of multiple proportions.
   b. Who is credited with this law? (3.1)

4. a. What is an atom?
   b. What is meant by the atomic structure of an atom? (3.2)

5. What two major regions make up all atoms? (3.2)

6. Cite at least four properties of electrons that evolved from the experiments of Thomson and Millikan.

7. What is an electron? (3.2)

8. a. Who is credited with the discovery of the atomic nucleus?
   b. What is the nucleus of an atom?
   c. Summarize Rutherford's description of an atom. (3.2)

9. Identify and define the two kinds of particles contained in the nucleus of an atom. (3.2)

10. What one factor determines the identity of a particular atom? (3.2)

11. Copy and complete the following table concerning the properties of subatomic particles:

Particle	Symbol	Mass Number	Actual Mass	Relative Charge
Electron				
Proton				
Neutron				

12. a. What are isotopes?
    b. How are the isotopes of a particular element alike?
    c. How are they different? (3.2)

13. Copy and complete the following table relative to the three isotopes of hydrogen (H-1, H-2, and H-3): (3.2)

Isotope	Name	Number of Protons	Number of Electrons	Number of Neutrons
H-1				
H-2				
H-3				

14. a. What is the meaning of the term *the atomic number of* an element?
    b. What is the mass number of an isotope?
    c. In the nuclear symbol for deuterium (^2_1H), identify the atomic number and the mass number.
    d. Write its hyphen notation. (3.3)

15. What is a nuclide? (3.3)

16. Use the Periodic Table and the information given below to write the hyphen notation for each isotope described:
    a. atomic number = 2, mass number = 4
    b. atomic number = 8, mass number = 16
    c. atomic number = 19, mass number = 39 (3.3)

Atoms: The Building Blocks of Matter   93

17. a. In the development of the relative scale for atomic masses, what atom is used as the standard?
    b. What is its relative mass assignment? (3.3)
18. a. What is an atomic mass unit (symbol u)?
    b. What is the relative atomic mass of an atom? (3.3)
19. What would the atomic mass of an atom be if its mass is approximately
    a. $\frac{1}{3}$ that of C-12?
    b. $\frac{1}{2}$ that of C-12?
    c. 12 × that of C-12?
    d. 4.5 times as much as C-12? (3.3)
20. a. What is the SI definition of the mole?
    b. What is the abbreviation for mole?
    c. How many particles are contained in a mole?
    d. What name is given to the number of particles in a mole? (3.3)
21. a. What is the molar mass of an element?
    b. To three decimal places, write the molar masses for: *(1)* carbon *(2)* neon *(3)* iron *(4)* uranium. (3.3)

PROBLEMS

▶ The Mole and Molar Mass

1. What is the mass, in grams, of each of the following?
   a. 1.00 mol Li
   b. 1.00 mol Al
   c. 1.00 molar mass Ca
   d. 1.00 molar mass Fe
   e. 6.022×10^{23} atoms C
   f. 6.022×10^{23} atoms Ag
2. Determine the mass, in grams, of each of the following:
   a. 3.00 mol Al
   b. 4.25 mol Li
   c. 1.38 mol N
   d. 8.075 mol Au
   e. 6.50 mol Cu
   f. 2.57×10^8 mol S
   g. 1.75×10^{-6} mol Hg
3. How many moles of atoms are represented by each of the following?
   a. 40.1 g Ca
   b. 11.5 g Na
   c. 5.87 g Ni
   d. 150 g S
   e. 2.65 g Fe
   f. 0.007 50 g Ag
   g. 3.25×10^5 g Pb
   h. 4.50×10^{-12} g O
4. How many moles of atoms are equivalent to each of the following?
   a. 6.022×10^{23} atoms Ne
   b. 3.011×10^{23} atoms Mg
   c. 2.25×10^{25} atoms Zn
   d. 50.0 atoms Ba

▶ Relative Atomic Mass

5. Three isotopes of argon occur in nature,— $^{36}_{18}Ar$, $^{38}_{18}Ar$, and $^{40}_{18}Ar$. If the relative atomic masses and abundances of each of these isotopes are as follows, calculate the average atomic mass of argon to three decimal places argon-36 (35.968 u; 0.337%), argon-38 (37.963 u; 0.063%), and argon-40 (39.962 u; 99.600%).
6. If naturally occurring boron is 80.20% B-11 (atomic mass = 11.009 u) and 19.80% of some other isotopic form of boron, what must the atomic mass of this second isotope be in order to account for the 10.811 u average atomic mass of boron? (Express to three decimal places.)
7. If the only two naturally occurring isotopes of chlorine, Cl-35 and Cl-37, have relative atomic masses of 34.969 u and 36.966 u respectively, what must be the approximate abundance of each in nature if the average atomic mass of chlorine is 35.453 u? (Express the final percentages to three decimal places.)

▶ Number of Atoms in a Sample

8. How many atoms are contained in each of the following?

    a. 1.50 mol Na
    b. 6.755 mole Pb
    c. 0.250 mol Si

9. What is the mass, in grams, of each of the following?
    a. 3.011×10^{23} atoms F
    b. 1.50×10^{23} atoms Mg
    c. 4.50×10^{12} atoms Cl
    d. 8.42×10^{18} atoms Br
    e. 25.0 atoms W
    f. 1.00 atoms Au

10. Determine the number of atoms contained in each of the following:
    a. 5.40 g B
    b. 8.02 g S
    c. 1.50 g K
    d. 0.025 50 g Pt
    e. 1.00×10^{-10} g Au

CRITICAL THINKING

1. **Organizing Ideas** Using a chemical compound as an example, describe the difference between the law of definite proportions and the law of multiple proportions.

2. **Constructing Models** As described on pages 75 through 80, the structure of the atom was determined through the observations made from painstaking experimental research. Suppose that a series of experiments were conducted and the following observations were made instead:
    a. When an electric current was passed through gas under pressure the area of the tube opposite the anode glows with a particular color. A paddlewheel placed in the tube rolls from the anode toward the cathode when the current is on. What charge do the particles passing through the gas possess?
    b. When the particles described in the first observation were fired at a very thin piece of metal foil, 50% of the particles passed through, 25% never emerged on the opposite side of the foil, and 25% were deflected at varying angles with no definite pattern. From these observations, name and describe the particles and draw a model of the atom.

3. **Inferring Relationships** How much mass is converted into energy during the creation of the nucleus of a $^{235}_{92}U$ atom?

CROSS CONNECTIONS

1. **Medicine** The study of atomic structure and the nucleus produced a new field of medicine, nuclear medicine. Describe the use of radioactive tracers to detect and treat diseases.

RESEARCH AND WRITING

1. Prepare a report on the series of experiments conducted by Sir James Chadwick that led to the discovery of the neutron.

2. What are the functions of cyclotrons, synchrotrons, and linear accelerators? What subatomic particles have been discovered through the use of these devices?

3. Write a report on the contributions of Amedeo Avogadro that led to the determination of the value of Avogadro's number.

4. Trace the development of the electron microscope and cite some of its many uses.

PROJECT

1. Observe a cathode ray tube in operation and write a description of your observations.

CHAPTER 4
Arrangement of Electrons in Atoms

PLANNING CHART
Approximate Time: 5/6 days

	SECTION/TEXT FEATURES		RESOURCES	
A V E R A G E	**4.1 REFINEMENTS OF THE ATOMIC MODEL** Demos 1, 2 *Periodic Table* Videodisc Desktop Investigation Section Review Chapter Review • Questions 1–14, Problems 1–8	97 96B 96G 102 108 122	*Teacher's Resource Binder* Lab Experiment 7 Section Review Worksheets **Other Resources** *Shakhashiri* Video Demos 5, 6, 8 Transparencies 5, 6	55 12
B A S I C	**4.2 QUANTUM NUMBERS** Demo 3 Section Review Chapter Review • Questions 15–22	108 96B 111 122	*Teacher's Resource Binder* Section Review Worksheets **Other Resources** *World of Chemistry* Videos 2, 6 Transparency 7 Blackline Master 11	13
B A S I C	**4.3 ELECTRON CONFIGURATIONS** Demo 4 Section Review Chapter Review • Questions 23–25, Problems 9–13	112 96C 121 122	*Teacher's Resource Binder* Section Review Worksheets **Other Resources** Transparencies 8, 9 Blackline Masters 12, 13, 14, 15	14
	CHAPTER END ACTIVITIES Critical Thinking, Cross Connections, Research and Writing, Projects		*Teacher's Resource Binder* Test Booklet *Computer Test Generator*	17, 19

Teaching Strategies

Introduction

Remind students that scientists are always looking for patterns in nature (order in the universe). This chapter is an excellent example of such a pursuit—looking for the patterns followed in the electron arrangement of neutral atoms in the ground state. Mention that atoms can become charged ions which are not neutral and not always in the ground state, but that this chapter deals mainly with the electron configurations of unexcited neutral atoms. Tell the students that this is an essential chapter; the rest of the course is based on understanding it.

This is a good time to explain that althrough experimentation is the most well-known scientific method used in research, two equally important methods are observation and mathematical calculation. Distinguish between mathematics as a tool for expressing relationships when working with the experimental or observation method and mathematics calculations as a scientific method to figure out answers to problems, such as Einstein used in the development of the theory of relativity and Schrödinger and Heisenberg in the development of quantum mechanics. Since quantum mechanics is used to explain much of the electron structure of the atom, a good portion of the information in this chapter was gained through the mathematical calculation method and then supported by experimental evidence.

This is a difficult chapter with many concepts that students may not be capable of fully grasping. However, to determine the electron structure of the different atoms, it is necessary to cover concepts such as Heisenberg's uncertainty principle, Hund's rule, Pauli's exclusion principle, and the wave-particle duality theory. They are also important because they add excitement, beauty, and intrigue to chemistry. Hopefully, they help students understand that everything in science is not concrete and it sometimes needs to be looked at in different ways and with different rules than they are accustomed to!

4.1 Refinements of the Atomic Model

Desktop Investigation: The Wave Nature of Light: Interference

ANSWERS TO DISCUSSION QUESTIONS
1. Yes, there are light and dark circular areas on the screen.

Answers to Discussion Questions

2. Actually, two phenomena are responsible for forming the pattern on the screen. They are (1) diffraction which is the spreading of a wave disturbance into a region behind an obstruction and (2) interference which is the mutual effect of two light beams that results in the loss of intensity in certain regions and the reinforcement of intensity in other regions resulting in the pattern of concentric circles of light and dark areas seen on the screen. Since diffraction and interference are properties shown by waves, one would conclude that light is acting like a wave in this experiment.

Light shows the wave property of interference when a beam of light is projected through a pinhole onto a screen because it produces a pattern of concentric circles (similar to that seen in a bull's eye) on the screen.

Demonstration 1: Exciting Atoms by Mechanical Means

PURPOSE: To show that atoms give off light when excited by mechanical means
MATERIALS: A pair of pliers, wintergreen Life Savers®
PROCEDURE: Crush a wintergreen mint suddenly between the jaws of a pair of pliers in a dark room. What did you see happen? (Flashes of blue-white light were produced.)

Why do you think this happened? (The electrons of atoms in the candy were excited as it was crushed. As the electrons of the excited atoms dropped back to their ground state, they gave off energy in the form of blue-white light.) What kind of energy was used to excite these atoms? (Mechanical energy)

An additional interesting note is that similar results occur if you watch someone chew wintergreen Life Savers® in the dark. An interesting experiment that students could design and carry out at home would be to test to see (1) if other brands of wintergreen-flavored mints produce light when they are crushed and (2) if spearmint- or peppermint-flavored candies produce light when they are crushed.

Demonstration 2: Continuous and Line Spectra

PURPOSE: To show students how to produce a continuous spectrum and a bright-line spectrum
MATERIALS: Magnifying lens, mercury vapor lamp, piece of cardboard with a 1 mm slit, prism or diffraction grating, screen, slide projector
PROCEDURE: To produce a continuous spectrum, pass a beam of light from the slide projector through a narrow slit in a piece of cardboard, and project the image of this slit on the screen. Then put a prism or diffraction grating between the cardboard and the screen, and adjust the angle of projection so the spectrum appears on the screen.

What kind of spectrum is this? (Continuous) Why wasn't this spectrum produced the first time? (The prism or diffraction grating was not there to spread out the different colors in the light.) How does the prism spread these colors out? (Light of different frequencies is bent different amounts as it passes through a prism or diffraction grating.)

To produce a line emission spectrum, substitute a mercury vapor lamp and a magnifying lens for the slide projector. **CAUTION:** Since the mercury vapor lamp produces ultraviolet rays, it should be shielded so that no one can look directly at the light source. What kind of spectrum is this? (A line emission spectrum—sometimes called a bright-line spectrum.) What element does this line spectrum identify? (If possible, show the students pictures of line spectra of several elements including mercury so they can match the spectrum on the screen with the mercury spectrum in the picture.) Many supply houses have spectroscopy kits which contain several inexpensive spectroscopes and liquids to test. These can be used for flame tests and for seeing the characteristic line spectra produced by atoms of different elements. Once you purchase the kits, you can make your own solutions to replace the ones in the kit and continue to use the spectroscopes, making the purchase of the kit a good investment.

Chemistry in Detail: Heisenberg's Uncertainty Principle

Douglas Rickard's article—"A Model of Uncertainty" in **References and Resources** (at the end of this chapter), is an excellent way to explain and demonstrate Heisenberg's Uncertainty Principle.

4.2 Quantum Numbers and Atomic Orbitals

Students often have difficulty understanding the three-dimensional nature of orbitals. Many of them were taught Bohr's orbit theory in lower grades and have a hard time giving up that idea for the more complex orbital theory. The following demonstration will help.

Demonstration 3: Shapes of Orbitals

PURPOSE: To help students visualize the 3-D shapes of the s, p, d, and f-orbitals and of the s- and p-sublevels
MATERIALS: 1 round balloon, 3 oblong balloons of different colors, 4 oblong balloons of the same color, 3 thin oblong balloons, 4 additional oblong balloons
PROCEDURE: Ask students to help by blowing up balloons. Hold up a round balloon. What orbital does this represent? (an s-orbital) (See Figure T4-1a.) Point out that this also represents an s-sublevel.

Twist an oblong balloon in the middle for a p-orbital. (See Figure T4-1b.) How do you recognize a p-orbital? (It has two lobes.) Do the same for two other oblong balloons of different colors. Twist the three together as in Figure T4-1c to make a p-sublevel. Then using the three different colored balloons, you can point out the p_x, the p_y, and the p_z, orbitals.

Twist two oblong balloons of the same color in the middle, and put them together for a d-orbital. (See Figure T4-1d.) What orbital does this represent? (a d-orbital) How many lobes does it have? (4) For the one odd d-orbital—d_{z^2}—curl a thin oblong balloon around the center of a twisted oblong balloon. (See Figure T4-1e.) Sometimes students want to know why the fifth d-orbital is different in shape from the rest. The answer is: This is the shape of the space that is left in the d-sublevel.

For the f-orbitals with eight lobes twist four oblong balloons of the same color together at their centers. (See Figure T4-1f.) For the three f-orbitals in Figure T4-1g, curl two thin oblong balloons around the center of a twisted oblong balloon.

Models of the orbitals can be purchased from many of the supply houses. However, balloons seem to be a better choice because they are cheap and they require student involvement. You could make up sets of balloons, and have students work in small groups and let each student group make all the orbitals.

(a) s-level + sublevel (b) p-orbital (c) p-sublevel
(d) d-orbital (e) d_{z^2}-orbital
(f) f-orbital (g) f-orbital

Figure T4-1

There is one problem with balloon models, or any orbital model, which is they make it appear that there is a definite point at which the orbital ends. Be sure to emphasize to students that each orbital does not have a definite point at which it ends but the probability of finding a given electron beyond a certain area becomes less and less likely as one moves further out.

Another problem is that after the shapes of the orbitals with accompanying explanations have been presented, students are often confused as to how they all fit together in the atom. Use explanations, chalkboard diagrams, layered transparencies (to show how different sublevels overlap one another) and analogies to help them see how these orbitals fit together to make up sublevels, energy levels and finally the electron cloud. The infant toy that consists of barrels within barrels can be used as an illustration.

An excellent analogy to help students understand (1) the 3-D nature of the electron cloud and (2) that energy levels, sublevels and orbitals are subdivisions of the electron cloud follows. Use an apartment building analogy to represent the electron cloud. Each floor represents an energy level with the first floor being closest to the ground, as the first energy level is closest to the nucleus. Each floor is subdivided into apartments, which represent sublevels; and each apartment is made up of rooms representing orbitals. Apartments have different numbers of rooms: one-room efficiency—s sublevel, three room—p sublevel, five room—d sublevel, and the luxurious seven room—the f sublevel. All rooms have one or two electrons "living" in them.

4.3 Electron Configurations

Students are often puzzled as to how one can be sure that the two electrons in a given orbital are able to spin in opposite directions. They ask the question: If the first electron is spinning in a clockwise direction and the entering one is also spinning in a clockwise direction, how can it go into the orbital. (Of course, we know electrons do not enter atoms one at a time.) The question is good in that it points out how two electrons in the same orbital always manage to spin in opposite directions as is stated in Pauli's exclusion principle. The following demonstration clears this dilemma up nicely.

Demonstration 4: Orbital Electron Pairs Spin in Opposite Directions

PURPOSE: To show how electrons in the same orbital are always able to spin in opposite directions

MATERIALS: 1 boiled peeled egg, 2 nails, 1 narrow strip of paper with arrows going in one direction, 2 small pieces of paper marked N and S, 2 straight pins, tape

PROCEDURE: Put one nail in each end of the boiled egg. (See Figure T4-2) Attach the piece of paper marked N to one nail and S to the other nail. This is to represent the north and south poles of the spinning electrons. (Spinning electrons produce magnetic fields and thus act like miniature magnets with north and south poles. This helps cut down on the repulsion of the negative particles.)

Attach the narrow strip of paper with arrows around the center of the egg using straight pins and spin it in the direction indicated by the arrows. Then show students that if the spinning electron simply flips over it is spinning in the opposite direction.

Figure T 4-2

Figuring out the electron configurations of atoms is very difficult for many students; so do several examples and have them do many practice problems. Use the three different methods given in the text for electron configurations. When students are figuring out electron configurations using the diagonal rule, they often want to write the sublevels in the order that they fill rather than keeping each set of sublevels together in their energy levels. Emphasize that, in this text, sublevels must be filled in the order designated by the diagonal rule, but sublevels in the same energy level must be kept together. Other texts do it in order of filling. In the end, the main concern is the number of electrons in the outermost energy level as is shown by the electron-dot notation.

References and Resources

Books and Periodicals

Bent, Henry. "Should Orbitals Be X-rated in Beginning Chemistry Courses?" *Journal of Chemical Education*, Vol. 61, No. 5, May 1984, pp. 421–3.

Burdman, Clark S. "The Significance of Arrows in Orbital Diagrams," *Journal of Chemical Education*, Vol. 63, No. 4, April 1986, p. 320.

Festa, Roger R. "Wolfgang Pauli (1900–1958): A Brief Anecdotal Biography," *Journal of Chemical Education*, Vol. 58, No. 3, March 1981, pp. 273–4.
Haendler, Bianca L. "Presenting the Bohr Atoms," *Journal of Chemical Education*, Vol. 59, No. 5, May 1982, pp. 372–6.
Lamb, William. "Idea Bank—An Alternative Method for Flame Tests," *The Science Teacher*, Vol. 51, No. 3, March 1984, p. 61.
Morwick, James. "Should Orbitals Be Taught in High School?" *Journal of Chemical Education*, Vol. 56, No. 4, April 1979, pp. 262–3.
Rickard, Douglas. "A Model of Uncertainty," *Journal of Chemical Education*, Vol. 63, No. 10, Oct. 1986, p. 833.
Wiger, George and Melvin Dutton. "A Lecture Demonstration Model of the Quantum Mechanical Atom," *Journal of Chemical Education*, Vol. 58, No. 10, Oct. 1981, pp. 801–2.
Zukav, Gary. *The Dancing Wu Li Masters,* New York: Bantam Books, 1979.

Audiovisual Resources

Chemistry 103–Dissecting the Atom, sound filmstrip, Learning Arts, 1988.
Electron Configuration, VHS cassette, Learning Arts, 1988.

Computer Software

Atomic Structure Series #3, Apple 48K. Queue, 1988. Several programs relating to this chapter.
Fee, Richard. *Electron Configurations*, Apple 48K. Chemistry According to ROF, 1988.
General Chemistry 1A—Topic 2: Atomic Theory and Structure, Apple 128K (4 disks). Queue, 1988.
Project Seraphim, 1988, Apple II, Disks-Vol. 1. Disk No. 6, Vol. 2, Disk No. 2

Answers and Solutions

Reviewing Concepts

1. **a.** Examples include X rays, ultraviolet light, visible light, infrared light, microwaves, and radio waves.
   b. Electromagnetic radiation moves through a vacuum at a constant speed of 3.0×10^{10} cm/s.
   c. In a vacuum, wavelength and frequency are inversely proportional; as wavelength decreases, frequency increases, and vice versa. The equation relating them is $c = \lambda v$, where c is the speed of light, λ is wavelength, and v is frequency.
2. **a.** A continuous spectrum is one in which all wavelengths within a given range are included.
   b. When sunlight passes through a prism, it is spread out into an array of colors according to wavelength.
3. **a.** The electromagnetic spectrum consists of all electromagnetic radiation arranged according to increasing wavelength.
   b. Shorter wavelengths are measured in nanometers, whereas longer wavelengths are measured in centimeters or meters.
   c. The SI unit for frequency is the hertz. One hertz is equal to one wave per second.
4. The frequency range of visible light is approximately 10^{15} Hz and its wavelength ranges from 400 to 700 nm.
5. The colors of light in the visible spectrum, in order of increasing frequency, are red, orange, yellow, green, blue, and violet.
6. The wave theory could not explain the emission of visible light by matter and the absorption of light by matter.
7. **a.** $E = hv$ where E is the energy of the photon, v is the frequency of the light emitted, and h is Planck's constant.
   b. Energy and frequency are directly proportional.
8. The theory of light that best explains each is:
   a. the wave theory
   b. the particle theory
   c. the particle theory.
9. **a.** Atoms can be made to emit electromagnetic radiation through heating, which causes their energies to increase. Then, when the atoms return to their original energy states, the added energy is given off in the form of electromagnetic radiation.
   b. This phenomenon is seen in the production of colored light in neon signs and in fireworks.
10. The ground state is the state of lowest energy of an atom, whereas an excited state is one in which an atom has a higher energy than it does in its ground state.
11. In a line emission spectrum, such as that emitted by excited hydrogen atoms, only photons of certain energies are emitted. These produce the individual lines observed in the hydrogen emission spectrum. In a continuous spectrum, such as that emitted by the sun, photons encompassing a broad range of energies are emitted, so that countless lines corresponding to these energies are produced. The result appears as a solid band of light.
12. According to Bohr, a line emission spectrum is produced when an electron drops from a higher-energy orbit to one with lower energy, emitting a photon whose energy is equal to the difference in energy between the orbits.
13. Bohr's model of the hydrogen atom did not explain the spectra of atoms that have more than one electron, nor did it fully explain the chemical behavior of atoms.
14. **a.** An orbital is a three-dimensional region about the nucleus in which there is a high probability that a particular electron is located.
    b. Orbitals can be pictured as clouds showing the region of probable electron locations. The sizes and shapes of electron clouds depend on the energies of the electrons that occupy them.
15. Whereas Bohr's model worked only for the hydrogen atom, Schrödinger's mathematical model applies to all atoms. The essential difference btween the two models involves the issue of certainty. Bohr described definite orbits occupied by electron particles, whereas Schrödinger treated electrons as waves having a certain probability of being found at various distances from the

nucleus, in what he called orbitals. The two models are similar from the standpoint of energy, in that Schrödinger's orbitals, like Bohr's orbits, associate the energy of an electron with its location relative to the nucleus, and the production of specific levels of electromagnetic radiation with electrons dropping from higher-energy to lower-energy orbitals. The most probable location for the electron of hydrogen, as described by Schrödinger, is at a distance from the nucleus exactly equal to that of Bohr's lowest-energy orbit.

16. a. Principal quantum numbers indicate the main energy levels surrounding a nucleus.
    b. The principal quantum number is symbolized by n.
    c. The term "shells" is used to refer to the main energy levels surrounding a nucleus.
    d. As the value of n increases, the distance of the main energy levels from the nucleus increases, and their energy increases.

17. a. The orbital quantum number indicates the shape of an orbital.
    b. Sublevels or subshells are the regions within main energy levels that are occupied by the differently shaped orbitals.
    c. The number of possible orbital shapes in each main energy level is equal to the value of the principal quantum number.
    d. The first four orbital shapes are designated by the symbols s, p, d, and f.

18. The responses are:
    a. 1 sublevel, an s orbital
    b. 2 sublevels, s and p orbitals
    c. 3 sublevels, s, p, and d orbitals
    d. 4 sublevels, s, p, d, and f orbitals
    e. 7 sublevels

19. a. The magnetic quantum number indicates the orientation of an orbital about the nucleus.
    b. The number of possible orientations for the s, p, d, and f orbitals are 1, 3, 5, and 7, respectively.
    c. Subscripts are used to indicate the orientation of the orbital in terms of the x, y, and z axes of a three-dimensional coordinate system centered on the nucleus. The p_x notation refers to a p orbital along the x axis, p_y indicates a p orbital along the y axis, and p_z indicates a p orbital along the z axis.

20. a. The number of orbitals within each main energy level is equal to the square of the principal quantum number, n.
    b. The third main energy level contains nine orbitals; the fifth contains 25.

21. a. The spin quantum number indicates which of the two possible orientations an electron has in an orbital.
    b. There are two possible values for this number, $+\frac{1}{2}$ and $-\frac{1}{2}$.
    c. The spin quantum number is significant because each orbital can hold no more than two electrons, which must have opposite spins. This limitation thus also determines the maximum number of electrons that can occupy each main energy level.

22. The number of electrons for each main energy level is:
    a. 2   b. 8   c. 18   d. 32
    e. 50   f. 72   g. 98   h. $2n^2$.

23. a. According to the Aufbau principle, an electron occupies the lowest-energy orbital that can receive it.
    b. In a many-electron atom, the lowest-energy orbital is filled first, and then electrons are added to the orbital with the next-lowest energy, and so on, until all of the electrons in the atom have been placed in orbitals.

24. a. Hund's rule states that orbitals of equal energy are each occupied by one electron before any orbital is occupied by a second electron.
    b. By placing as many single electrons as possible in separate orbitals in the same energy level, electron-electron repulsion is minimized and more-favorable arrangements result because of lower energy.

25. a. According to the Pauli exclusion principle, no two electrons in the same atom can have the same four quantum numbers.
    b. The two different values of the spin quantum number permit two electrons of opposite spins to occupy the same orbital.
    c. Two electrons in the same orbital would have the same principal, orbital, and magnetic quantum number, but would differ in spin quantum number.

26. a. Electron-dot notation shows only those electrons in the highest, or outermost main energy level.
    b. The highest occupied energy level is the electron-containing main energy level with the highest principal quantum number.
    c. Inner-shell electrons are electrons not in the highest occupied energy level.

27. The highest occupied energy level in each of the indicated atoms is the
    a. 1st   b. 2nd   c. 3rd   d. 4th   e. 5th.

28. The electron-dot symbols are:
    a. $\cdot \dot{Z} \cdot$ b. $: \dot{Z} :$ c. $: \ddot{Z} :$ d. $\cdot \dot{Z} \cdot$

29. a. Oxygen contains eight electrons.
    b. Its atomic number is 8.
    c. The highest occupied energy level is the second main energy level.
    d. Its orbital notation is

    $\underline{\uparrow\downarrow} \quad \underline{\uparrow\downarrow} \quad \underline{\uparrow\downarrow} \quad \underline{\uparrow} \quad \underline{\uparrow}$
    $1s 2s 2p 2p 2p$

    e. Its electron dot symbol is $:\dot{O}:$
    f. It contains two innershell electrons.
    g. They are located in the $1s$ orbital.

30. The indicated responses are
    a. 15   b. 15   c. third

    d. $\underline{\uparrow\downarrow} \quad \underline{\uparrow\downarrow} \quad \underline{\uparrow\downarrow} \quad \underline{\uparrow\downarrow} \quad \underline{\uparrow\downarrow} \quad \underline{\uparrow\downarrow} \quad \underline{\uparrow} \quad \underline{\uparrow} \quad \underline{\uparrow}$
    $1s 2s 2p 2p 2p 3s 3p 3p 3p$

    e. $\cdot \dot{P} :$ f. 10   g. in $1s$, $2s$, and $2p$ orbitals.

31. a. The [Ne]$3s^2$ configuration denotes an element consisting of the Ne electron configuration with the $3s^2$ electrons added, for a full configuration of $1s^2 2s^2 2p^6 3s^2$.
    b. The designated element is Mg.

32. The shorthand electron configurations are:
    a. [Ne]$3s^2 3p^5$ b. [Ar]$4s^2$.

96E Chapter 4 Teacher's Notes

33. a. The noble gases are the Group-18 elements helium, neon, argon, krypton, xenon, and radon.
b. A noble-gas configuration is an outer main energy level fully occupied by eight electrons.

34. The indicated electron configurations and electron-dot symbols are as follows:

a. $1s^2 2s^2 2p^6 3s^1$ [Ne]$3s^1$ Na·

b. $1s^2 2s^2 2p^6 3s^2 3p^1$ [Ne]$3s^2 3p^1$ ·Al·

c. $1s^2 2s^2 2p^6 3s^2 3p^3$ [Ne]$3s^2 3p^3$ ·P:

d. $1s^2 2s^2 2p^6 3s^2 3p^6$ [Ne]$3s^2 3p^6$:Ar:

e. $1s^2 2s^2 2p^6 3s^2 3p^6 3d^{10} 4s^2 4p^5$ [Ar]$3d^{10} 4s^2 4p^5$:Br:

f. $1s^2 2s^2 2p^6 3s^2 3p^6 3d^{10} 4s^2 4p^6 5s^2$ [Kr]$5s^2$ ·Sr·

35. The indicated elements are as follows:
a. B **b.** F **c.** Mg **d.** Si **e.** Cl **f.** K **g.** Fe.

Problems

1. $c = \lambda\nu$

$\lambda = \dfrac{c}{\nu}$

$\lambda = \dfrac{2.998 \times 10^8 \text{ m/s}}{7.500 \times 10^{12} \text{/s}} = \boxed{3.997 \times 10^{-5} \text{ m}}$

2. $c = \lambda\nu$

$\nu = \dfrac{c}{\lambda}$

$\nu = \dfrac{2.998 \times 10^{10} \text{ cm/s}}{4.257 \times 10^{-7} \text{ cm}} = \boxed{7.043 \times 10^{16} \text{/s or Hz}}$

3. $E = h\nu$
$= 6.63 \times 10^{-34} \text{ J·s} \times 3.55 \times 10^{17} \text{/s}$
$= \boxed{2.35 \times 10^{-16} \text{ J}}$

4. $E = h\nu$

$\nu = \dfrac{E}{h}$

$\nu = \dfrac{1.55 \times 10^{-24} \text{ J}}{6.63 \times 10^{-34} \text{ J·s}} = \boxed{2.34 \times 10^9 \text{/s or Hz}}$

5. Solving each equation for ν yields:

$\nu = \dfrac{c}{\lambda}$

$\nu = \dfrac{E}{h}$

Equating the two expressions,

$\dfrac{c}{\lambda} = \dfrac{E}{h}$

$E = \dfrac{hc}{\lambda}$

6. Since the speed of light (c) is 3.00×10^8 m/s,

$\dfrac{8.00 \times 10^7 \text{ km}}{3.00 \times 10^8 \text{ m/s}} \times \dfrac{10^3 \text{ m}}{\text{km}} = 2.67 \times 10^2 \text{ s}$

[Note that the frequency of the radiowave did not enter into the calculations.]

7. a. $E = h\nu$

$\nu = \dfrac{E}{h}$

$\nu = \dfrac{3.37 \times 10^{-19} \text{ J}}{6.63 \times 10^{-34} \text{ J·s}} = 5.08 \times 10^{14} \text{/s or Hz}$

b. $\lambda = \dfrac{c}{\nu} = \dfrac{3.00 \times 10^8 \text{ m/s}}{5.08 \times 10^{14} \text{/s}}$

$= 5.91 \times 10^{-7} \text{ m}$

8. Recalling that 1 Å = 10^{-8} cm,

$1.00 \times 10^{-2} \text{ Å} \times \dfrac{10^{-8} \text{ cm}}{\text{Å}} = 1.00 \times 10^{-10} \text{ cm}$

Substituting into $E = \dfrac{hc}{\lambda}$

$E = \dfrac{6.63 \times 10^{-34} \text{ J·s} \times 3.00 \times 10^{10} \text{ cm/s}}{1.00 \times 10^{-10} \text{ cm}}$

$= 1.99 \times 10^{-13} \text{ J}$

9. In many-electron atoms, the $4s$ orbital has a lower energy than does the $3d$ orbital. Thus, according to the Aufbau principal, the $4s$ orbital will fill before the $3d$ orbitals.

10. The order of filling is: $1s$, $2s$, $2p$, $3s$, $3p$, $4s$, $3d$, $4p$, $5s$, $4d$, $5p$, $6s$, $4f$, $5d$, $6p$, $7s$, $5f$, $6d$, $7p$.

11. The shorthand electron configurations are as follows:

a. [Ar]$3d^1 4s^2$ h. [Xe]$4f^{14} 5d^2 6s^2$
b. [Ar]$3d^6 4s^2$ i. [Xe]$4f^{14} 5d^{10} 6s^2$
c. [Ar]$3d^{10} 4s^2$ j. [Xe]$4f^{14} 5d^{10} 6s^2 6p^2$
d. [Ar]$3d^{10} 4s^2 4p^3$ k. [Xe]$4f^{14} 5d^{10} 6s^2 6p^5$
e. [Kr]$4d^{10} 5s^2 5p^2$ l. [Rn]$6d^1 7s^2$
f. [Kr]$4d^{10} 5s^2 5p^6$ m. [Rn]$5f^{14} 6d^1 7s^2$.
g. [Xe]$5d^1 6s^2$

12. The electron-dot symbols are:

a. ·Zn· **b.** ·Sn· **c.** :Xe: **d.** ·Pb· **e.** :At:

13. The unusual configurations of chromium and copper are due to the lower energy associated with d sublevels that are half-filled in chromium and completely filled in copper.

Critical Thinking

1. • As a wave, light would be emitted continuously and this light would be at a certain frequency. When viewed as a particle, light is emitted from a hot object in specific amounts called photons. Planck's equation shows us that energy and frequency are directly proportional. Thus, the different colors of visible light given off by hot materials is dependent on the amount of energy being absorbed by the material.
• Because light is a form of energy, it could knock lose electrons from metals. However the wave theory could not explain why only light with certain frequencies could cause electron emission. By understanding the particle properties of light, you can picture a metal sur-

face being struck by a tiny stream of photons, each containing a quantum of energy equal to $h\nu$. A certain amount of energy is needed by the electron to overcome the forces that bind it to the metal. Thus, an electron will not be emitted unless the photon is high frequency radiation.

2. Hydrogen produces an absorption spectrum that is a continuous visible light spectrum interrupted by black lines at 656.3 nm, 486.1 nm, 434.1 nm, 410.1 nm, and 364.6 nm.

3. Answers will vary: Students may infer that they have located a new element present in a star's outer layers. To verify its existence, students may suggest that the first method would be to rule out any mechanical malfunction in the spectrograph and as it was transmitted to earth from the Hubble. They may also suggest that it would be prudent to examine all future spectra to see if the unknown element appears again.

4. $c = \nu\lambda$

$$\nu = \frac{3.00 \times 10^8 \text{ m/s}}{284 \text{ nm}} \times \frac{10^9 \text{ nm}}{1 \text{ m}}$$

$$= 1.056 \times 10^{15}/\text{s}$$

$$E_{\text{photon}} = h\nu$$
$$= 6.63 \times 10^{-34} \text{ J} \cdot \text{s}(1.056 \times 10^{15}/\text{s})$$
$$= 7.00 \times 10^{-19} \text{ J}$$

The work function of chromium is 7.00×10^{-19} J.

Cross Connections

1. Answers will vary: Students should include on the design specifications the size, colors to be used, and the electrical connections. Neon glows a brilliant orange-red. Other colors can be achieved by mixing neon with argon or mercury vapor, and by using colored tubing.

Videodisc Selections from
The Periodic Table Videodisc (JCE Software)

step forward

step reverse

CHAPTER 4

pg 99 Photoelectric effect

36945
cesium in photocell   (s)

pg 100 Emission of light by excited atoms

37311
strontium + oxygen   (m)

18438
calcium + oxygen   (m)

7535
magnesium + oxygen   (m)

36444
magnesium in flashbulbs/flares   (s)

36438
sodium vapor lamp   (s)

37179
mercury vapor lamp   (s)

36426
neon in lighted sign   (s)

36318
He-Ne laser   (s)

36532
argon in lighted sign (s)

36759
krypton in laser (s)

36519
argon in laser (s)

Teacher's Notes

Chapter 4
Arrangement of Electrons in Atoms

SECTION PREVIEW
4.1 Refinements of the Atomic Model
4.2 Quantum Numbers and Atomic Orbitals
4.3 Electron Configurations

Electrons occupy specific regions in space around the nucleus of an atom, much as these colored shapes are arranged in this view of a kaleidoscope.

4.1 Refinements of the Atomic Model

In the preceding chapter, you followed the evolution of the atomic theory up to the Rutherford model, which consists of a nucleus surrounded by electrons. Although it was an improvement over previous models, Rutherford's model was incomplete. It did not explain how negatively charged electrons could fill the space surrounding the positively charged nucleus. Because oppositely charged particles attract each other, what prevented the electrons from being drawn into the nucleus?

In order to understand how a new atomic model was developed that did not have this and other shortcomings, you need to know something about the nature of light. At this point, you may wonder what light has to do with atoms. As you will see, light and electrons in atoms have some properties in common. Discoveries about the nature of light led directly to a revolutionary view of the nature of matter and of the atom.

Wave–Particle Nature of Light

At the time of Rutherford, electrons were pictured as particles, and light was pictured as waves. Many properties of electrons and light can be explained by these models. A revolution occurred in science in the early 1900s when it was discovered that electrons have certain wavelike properties and that light has certain particle-like properties. To fully explain all observations, it is necessary to describe electrons and light as having a dual wave–particle nature. If you find this difficult to accept and understand, you are no different than many scientists during the period when the idea was introduced. The wave–particle model has withstood many experimental tests and is now well-established. No doubt it will remain so until someone comes up with an observation of either light or electrons that the wave–particle model fails to explain.

Electromagnetic radiation Visible light is one kind of electromagnetic radiation. Some other kinds of electromagnetic radiation are X rays, ultraviolet and infrared light, microwaves, and radio waves. **Electromagnetic radiation** *is a form of energy that exhibits wavelike behavior as it travels through space.* You are familiar with waves moving through water. A duck sitting on the surface moves up and down as waves pass through the water beneath it. The waves pass by, though the duck remains in place.

Electromagnetic radiation moves through a vacuum at a constant speed of 3.0×10^{10} centimeters per second (cm/s). The same value may be used for the speed of light in air, which is largely a vacuum. According to present theories, this is the fastest speed possible in nature.

Electromagnetic radiation has the measurable wave properties of wavelength and frequency. One wavelength is illustrated in Figure 4-1 as the distance from the base of one wave to the base of the next. **Wavelength** (λ) *is the distance between corresponding points on adjacent waves.* For radiation of a specific energy, this distance is the same throughout the entire succession of waves.

SECTION OBJECTIVES

- Discuss the dual wave–particle nature of light.
- Explain the mathematical relationship among the velocity, wavelength, and frequency of electromagnetic radiation.
- Discuss the significance of the line emission spectrum of hydrogen to the model of atomic structure.
- Describe the Bohr model of the hydrogen atom.
- Distinguish between an orbit and an orbital.

Whether electrons and light behave as matter, by showing particle characteristics, or behave as energy, exhibiting wave-like properties, depends upon the type of experiment being carried out. For more detail, refer to *The Dancing Wu Li Masters* by Zukav.

Arrangement of Electrons in Atoms

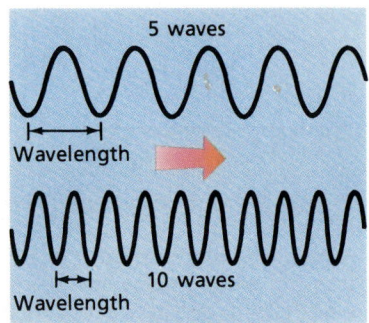

Figure 4-1 Wavelength and frequency are inversely related.

STUDY HINT

λ is lower case Greek lambda; v is lower case Greek nu.

STUDY HINT

Inverse proportion was discussed in Section 2.4.

Frequency (v) *is defined as the number of waves that pass a given point in a specific amount of time, usually one second.* The higher the frequency of water waves, the more often the duck sitting on the surface will bob up and down. Frequency and wavelength are mathematically related to each other. When waves are closer together, as shown at the bottom of Figure 4-1, more of them pass a given point in the same amount of time (because of light's constant speed).

The relationship between frequency and wavelength is given by the following equation,

$$c = \lambda v$$

where c is the speed of light, λ is the wavelength, and v is the frequency. Because c is the same for all electromagnetic radiation, the product λv is a constant, and λ is inversely proportional to v. In other words, as the wavelength of light decreases, the frequency increases (see Figure 4-1) and vice versa.

When sunlight passes through a prism (Figure 4-2), it is spread out according to its wavelength. The resulting array of colors is an example of a **continuous spectrum**, *a spectrum in which all wavelengths within a given range are included.* At either end of the visible spectrum are other types of

Figure 4-2 A prism separates white light into its component colors, creating a rainbow effect.

radiation that we cannot see. Visible light is only a small part of the total electromagnetic spectrum, which is shown in Figure 4-3. The **electromagnetic spectrum** *consists of all electromagnetic radiation, arranged according to increasing wavelength.* The upper scale in Figure 4-3 gives wavelength (λ) in units of length based on the meter. For visible light and radiation of shorter wavelengths, the unit is the nanometer. For radiation of longer wavelengths, it is the centimeter or the meter. The lower scale gives the frequency (v) in the SI unit for frequency, the *hertz*. One hertz is equal to 1 wave per second.

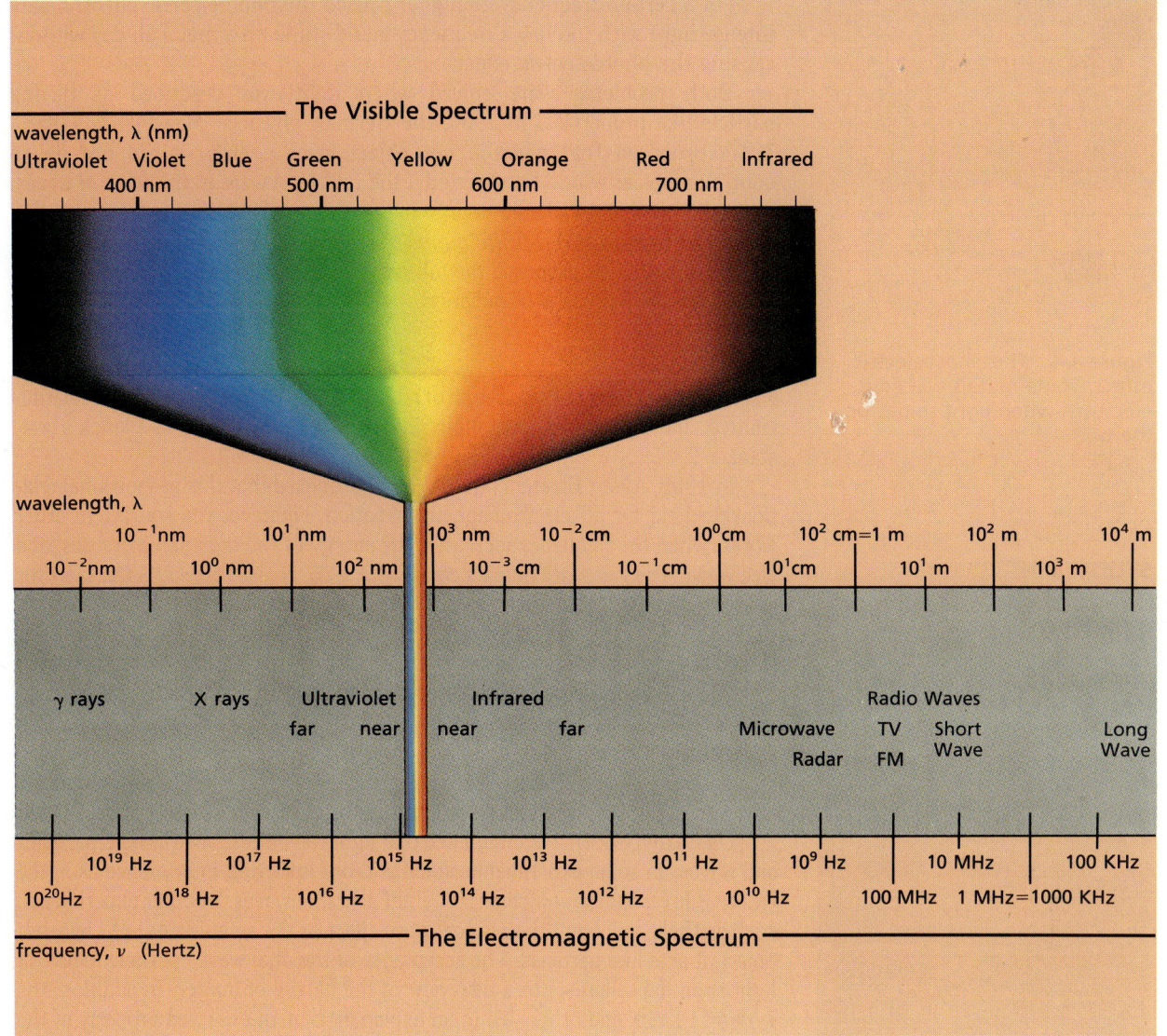

Figure 4-3 Visible light represents only a small portion of the electromagnetic spectrum.

Light as particles As the twentieth century began, scientists encountered two properties of light that could not be explained by wave theory. One involved the emission of visible light by matter, and the other the absorption of light by matter.

When an object made of a material that does not burn, such as iron, is heated, it gives off light. First, a red glow appears. As the temperature (and therefore the energy added) increases, the red glow changes to yellow and then to white. We sometimes speak of this condition as "white hot." The wave theory incorrectly predicted that extremely hot bodies would emit mainly (invisible) ultraviolet light.

A problem was also encountered in explaining the **photoelectric effect**, *which is the emission of electrons by certain metals when light shines on them.* It was agreed that light, being a form of energy, could knock loose an electron from a metal surface. But wave theory could not explain why only

To help students understand a *quantum,* use the following analogy: If a soda costs 45 cents and you put 40 cents in a soda machine, you will not get the soda when you push the button. Only if the required "packet" of money is put in, will you get a soda.

Arrangement of Electrons in Atoms

Figure 4-4 The photoelectric effect. Some metals will emit electrons when light shines on them.

STUDY HINT

One photon of radiation of frequency v has energy hv. Total energy delivered by this radiation would be 1 hv, 2 hv, 3 hv, 4 hv, ...

Figure 4-5 Hydrogen atoms are excited when high-voltage electric current is passed through a glass tube containing hydrogen gas. The lavender glow is characteristic of hydrogen gas.

light of a certain frequency or higher caused the photoelectric effect. A very intense light with too low a frequency could shine on a metal all day without causing the photoelectric effect.

Both problems were solved when light was described as having particle-like properties. In 1900, the German physicist Max Planck (1858–1947) proposed that when a hot object loses energy, it did not do so continuously, as would be expected if the radiation were in the form of waves. Instead, the object radiated energy in small, specific amounts that he called *quanta*. A **quantum** *is a finite quantity of energy that can be gained or lost by an atom.* Planck proposed the following relationship between a quantum of energy and the frequency of radiation:

$$E = hv$$

where E is the energy of a quantum of radiation, v is the frequency of the radiation emitted, and h is a fundamental physical constant now known as Planck's constant.

In 1905, Albert Einstein (1879–1955) proposed that this proportional relationship held for all electromagnetic radiation, whatever the source. In other words, when the frequency increases, the energy of the radiation increases, and vice versa. The value of Planck's constant, 6.626×10^{-34} J · s, is the same for all electromagnetic radiation. *An individual quantum of light is called a* **photon.** Photons are thought of as particles of radiation. Radiation is emitted and absorbed only in whole numbers of photons.

Einstein proposed that the absorption of photons, each with a specific energy, could explain the photoelectric effect. The photoelectric effect will not occur if the frequency, and therefore the energy of each photon, is too low to dislodge an electron. Even a few photons of light of the necessary frequency, however, deliver enough energy to produce the effect (Figure 4-4).

Gradually, evidence accumulated that light can be described as both particles and waves. In some experiments, such as those involving interference as in the Desktop Investigation on page 102, light acts like waves. In other experiments, such as those involving the photoelectric effect or the emission of light by a hot object, it acts like particles. The acceptance of the dual wave–particle nature of light soon led scientists to a new view of the nature of matter. In addition, the work of Planck and others led to an explanation of an observed property of the hydrogen atom—the hydrogen spectrum.

When atoms in the gaseous state are heated, their potential energy increases. Then, almost immediately, the atoms return to their original energy states, giving off the added energy in the form of electromagnetic radiation as they do so. The production of colored light in neon signs and in fireworks are familiar illustrations of this process. Visible light or other electromagnetic radiation is emitted when an atom passes from a state of higher potential energy to a state of lower potential energy. *The state of lowest energy of an atom is its* **ground state.** *A state in which it has a higher potential energy than it has in its ground state is an* **excited state.** Excited hydrogen atoms can be produced by passing a high-voltage electric current through hydrogen gas. When this happens, the gas glows with a characteristic lavender color, as shown in Figure 4-5. When a narrow beam of this lavender light is passed through a prism, the light is separated into four narrow lines, each of a different color. Figure 4-6 shows this hydrogen spectrum, which is called a line spectrum, or a line-emission spectrum. Each of the colored lines is produced by light of a different wavelength.

As previously discussed, light of a particular wavelength has a definite frequency ($\nu = c/\lambda$), and thus, it also has a definite energy ($E = h\nu$). Each line in the hydrogen spectrum must therefore be produced by emission of photons with certain energies. This would be the result if excited hydrogen atoms could only have specific potential energies. Whenever an excited hydrogen atom drops from such a specific excited state to its ground state or to a low-energy excited state, it emits a photon. The energy of this photon ($E = h\nu$) is equal to the difference in energy between the initial state and the final state, as illustrated in Figure 4-7.

During the late nineteenth and early twentieth centuries, additional series of lines were discovered in the ultraviolet and infrared regions of the hydrogen spectrum. The wavelengths of some of the lines in the hydrogen spectrum—known as the Lyman, Balmer, and Paschen series, after their discoverers—are shown in Figure 4-8. There appeared to be a mathematical relationship between the various energy states of a hydrogen atom. The problem facing scientists was to provide a model of the hydrogen atom that could account for these energy states and, therefore, for the hydrogen-atom spectrum.

Figure 4-6 A bright-line spectrum is produced from hydrogen atoms. (Colors shown are not the exact match of the spectrum.)

Figure 4-7 When an excited atom with energy E_2 returns to energy E_1, it releases a photon having energy $E_2 - E_1 = E = h\nu$.

Figure 4-8 Representative lines in the hydrogen spectrum. The small letter below each line indicates which of the energy-level transitions produces it.

Einstein, using Planck's work, suggested the particle-like behavior of light. He used this premise for his work on the photoelectric effect. For this work, not for his theory of relativity, he received the 1921 Nobel Prize in physics.

Desktop Investigation

The Wave Nature of Light: Interference

Question
Does light show the wave property of interference when a beam of light is projected through a pinhole onto a screen?

Materials
scissors
manila folders
thumb tack
masking tape
aluminum foil
white posterboard
  or cardboard
flashlight

Procedure
Record all your observations.
1. To make the pinhole screen, cut an 20-cm × 20-cm square from a manila folder. In the center of the square, cut a 2-cm square hole. Cut a 7-cm × 7-cm square of aluminum foil. Using a thumbtack, make a pinhole in the center of this foil square. Tape the aluminum foil over the 2-cm square hole, making sure the pinhole is centered as shown in the diagram above.
2. Use white posterboard to make a projection screen 35-cm × 35-cm.
3. In a dark room, center the light beam from a flashlight on the pinhole. Hold the flashlight about 1 cm from the pinhole. The pinhole screen should be about 50 cm from the projection screen, as shown in the diagram. Adjust the distance to form a sharp image on the projection screen.

Discussion
1. Did you observe interference patterns on the screen?
2. As a result of your observations, what do you conclude about the nature of light?

Bohr Model of the Hydrogen Atom

The problem of explaining the hydrogen-atom spectrum was solved in 1913 by the Danish physicist Niels Bohr (1885–1962). Bohr, who had worked with Rutherford, proposed a model of the hydrogen atom as follows: The single electron of the hydrogen atom can circle the nucleus only in allowed paths or *orbits*. When the electron is in each of these allowed orbits, the atom has a definite, fixed energy. Its lowest total energy (kinetic plus potential energy) occurs when it is in the orbit closest to the nucleus. This orbit is separated from the nucleus by a relatively large empty space, where the electron cannot exist. The orbit therefore provides the outer dimension of the hydrogen atom in its ground state. The total energy of the electron increases as it moves into orbits that are successively farther from the nucleus. The highest energy level corresponds to the state in which the electron is sufficiently far from the nucleus that it is no longer attracted by it.

The various electron orbits or atomic energy levels in Bohr's model can be compared to the rungs of a ladder. When you are standing anywhere on a ladder, your feet are on one rung or another. You cannot stand between rungs. The potential energy corresponds to standing on the first rung, the second rung, and so forth. Your potential energy cannot correspond to standing in mid-air between two rungs. The potential energy of your body with respect to the ground is thus subdivided into small, definite amounts—the energy is *quantized*. In the same way, an electron can be in one orbit or another, but not in-between.

How can this model of the hydrogen atom explain the observed spectral lines? While in an orbit, the electron neither gains nor loses energy, But it can move to a higher energy orbit by gaining an amount of energy exactly equal to the difference in energy between the higher energy orbit and the initial lower energy orbit. When a hydrogen atom is in an excited state, its electron is in a higher energy orbit.

Line emission spectra can then be explained as follows: *A* **line spectrum** *is produced when an electron drops from a higher-energy orbit to a lower energy orbit.* As the electron drops, a photon is emitted that has an energy ($E = h\nu$) equal to the difference in energy between the initial higher energy orbit and the final lower energy orbit. The absorption and emission of radiation according to the Bohr model of the hydrogen atom is illustrated in Figure 4-9.

By using the measured values for the mass and charge of the electron, together with Planck's constant, Bohr was able to calculate the energies that an electron would have in the allowed energy levels for the hydrogen atom. This enabled him to show mathematically how the various spectral series of hydrogen could be produced. Bohr's calculated values agreed exactly with the experimentally observed values for the lines in each series. For example, the Lyman spectral series was shown to be the result of electrons dropping from various higher-energy levels to the ground-state energy level. The origins of several of the series of lines in the hydrogen atom spectrum are shown in Figure 4-10.

The success of Bohr's model of the hydrogen atom in explaining observed spectral lines led many scientists to conclude that a similar model could be applied to all atoms. Scientists soon recognized, however, that Bohr's approach did not explain the spectra of atoms with more than one electron and that it did not fully explain the chemical behavior of atoms.

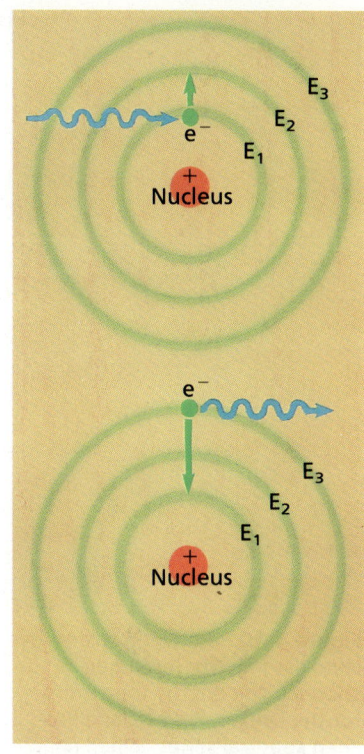

Figure 4-9 According to the Bohr model of the hydrogen atom, the electron can only be in orbits corresponding to the energies, E_1, E_2, E_3, and so forth.

Spectroscopy

Spectroscopy is a useful tool for studying the structure of atoms and the composition of matter in general. By studying the emission spectra of certain substances as described below, the identity of atoms in a substance can be determined.

Use of a spectroscope The spectroscope is an instrument that separates light into a spectrum that can be examined. One type of spectroscope consists of a glass prism and a collimator tube that focuses a narrow beam of light rays

Figure 4-10 An electron energy-level diagram for hydrogen showing the energy transitions for the Lyman, Balmer, and Paschen spectral series. When an electron falls from a higher to a lower energy level, a photon is emitted. The greater the difference in energy levels, the greater will be the energy of the emitted photon.

on the prism. It also has a small telescope for examining the light that passes through the prism. When white light passes through a triangular prism, a band of colors called a continuous spectrum appears. This effect is caused by the unequal bending of light of different wavelengths as it enters and emerges from the prism. A continuous spectrum is shown in the top band of Figure 4-11.

Origin of spectral lines Light rays of different frequencies are bent different amounts when passed through a prism. Light energy of the shortest wavelengths (highest frequencies) is bent most, forming the deep violet color seen at one end of the visible spectrum. Light energy of the longest wavelengths (lowest frequencies) is bent least, forming the deep red color characteristic of the other end of the visible spectrum. Between these two extremes there is a gradual blending from one color to the next. It is possible to recognize six elementary colors: red, orange, yellow, green, blue, and violet.

When substances are heated or energized sufficiently, electrons are raised to higher energy levels by heat energy. Whenever one of these electrons falls back into an available lower energy level, a photon of energy with a specific wavelength is released. Every substance has a characteristic set of possible wavelengths. Each of these photons contribute to a "line" in the spectrum of the element. For example, vaporized sodium atoms produce a spectrum consisting of two narrow yellow lines that are very close together (seen in the ordinary spectroscope as a single yellow line). Potassium atoms produce two red lines and a violet line. Lithium atoms yield intense red and yellow lines and weak blue and violet lines. The spectra produced by excited atoms of different elements are as distinct as fingerprints.

Bright lines in the visible portion of the spectrum account for the flame coloration produced by certain metals. The color seen is the combination of emitted light of different wavelengths.

Quantum Model of The Atom

To many scientists, the biggest shortcoming of Bohr's theory was that the very idea of the quantum appeared to contradict common sense. Why could the electron in the hydrogen atom orbit the nucleus only in a relatively small number of allowed paths? Why weren't there limitless orbits of slightly different energies? Scientists searched for some insight into the nature of matter that would explain why atomic energy states are quantized.

The key to a general atomic theory that explains the quantization of electron energies was provided in 1924 by the French physicist Louis de Broglie (1892–1987). Inspired by the dual, wave–particle nature of light, de Broglie proposed that electrons might also have a wave–particle nature. De Broglie pointed out that Bohr's proposed quantized electron orbits were consistent with known wave behavior. For example, waves confined in a space have only certain frequencies. Reflected waves of all other frequencies destroy themselves by destructive interference and disappear. If electrons are thought of as waves confined to the space around an atomic nucleus, this could explain why only certain frequencies, and therefore only certain energies, of electrons are possible. De Broglie's suggestion that electrons have wave properties was soon confirmed by experiments demonstrating the interfer-

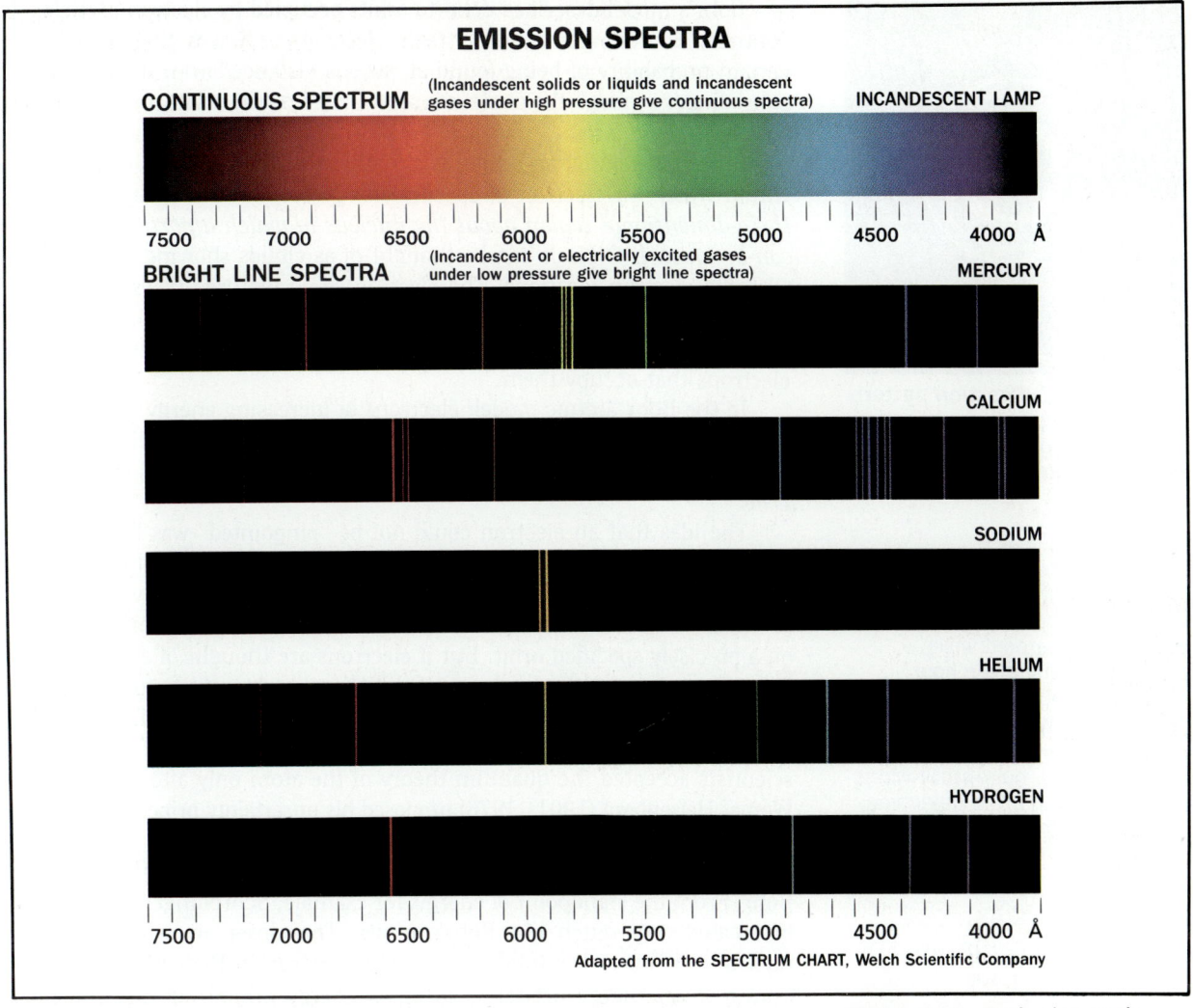

Figure 4-11 Each element's atoms will produce a unique emission spectrum when heated.

ence between electron waves. For instance, when light waves pass through a small opening, like a pinhole in the Desktop Investigation on page 102, the transmitted light waves are bent, or diffracted. Passing electrons through the narrow spaces between atoms in crystalline materials produced diffraction patterns like those previously known only for light waves. (Figure 4-12)

The Austrian physicist Erwin Schrödinger (1887–1961) was impressed by de Broglie's hypothesis that electrons have wave properties. In 1926, Schrödinger devised an equation that treated electrons moving around nuclei as waves. This now-famous equation, known as the Schrödinger wave equation, laid the foundation for modern quantum theory. **Quantum theory** *describes mathematically the wave properties of electrons and other very small particles.*

Schrödinger's wave equation is a mathematical model that applies to all atoms, unlike Bohr's model, which worked only for the hydrogen atom. The quantization of electron energies is a natural outcome of the Schrödinger equation. Only specific energies provide solutions to the equation.

Arrangement of Electrons in Atoms

Figure 4-12 Diffraction pattern produced by passing a beam of electrons through many small crystals of aluminum oriented at random.

To develop the idea of probable location, use the analogy of a teacher who constantly moves about the classroom while teaching. At any given hour during a school day, there is a high probability that the teacher will be found somewhere in that assigned room, but exactly where in the room is hard to determine. The same is true of electrons in their orbitals.

Bohr's model described definite orbits occupied by electron particles. In contrast, Schrödinger's model treats electrons as waves that have only a certain probability of being found at various distances from the nucleus. In Schrödinger's model, the nucleus is surrounded not by orbits, but by what are described as *orbitals*.

Strictly speaking, an orbital is a mathematical description of the region within which an electron of specific energy can be found. *An **orbital** is a three-dimensional region about the nucleus in which a particular electron can be located.* Orbitals can be thought of as clouds showing the region of probable electron locations. The electron-cloud picture allows us to relate orbitals to something familiar and to make drawings like those in Figure 4-13. The sizes and shapes of electron clouds depend on the energies of the electrons that occupy them.

In the Bohr atomic model, electrons of increasing energy occupy orbits further and further from the nucleus. Atomic orbitals also fall into similar quantized regions of increasing energy, which are referred to as *main energy levels*.

The idea that an electron could not be "pinpointed" was Schrödinger's revolutionary contribution to the model of the atom. The Bohr model had attempted to explain electron behavior as though electrons were physical particles. It made sense to think of the electron as being at a particular point on a precisely specified orbit. But if electrons are thought of as waves, then Schrödinger's electron-cloud or probability approach is easier to accept. After all, waves are not physical bodies that can be precisely located.

This view of matter contradicts our ordinary experience of reality. Many scientists accepted the quantum theory of the atom only after the physicist Werner Heisenberg (1901–1976) proposed his uncertainty principle, discussed in the Chemistry in Detail feature on the following page.

In many other ways, Schrödinger's atomic model is similar to that of Bohr. From the standpoint of energy, for example, Schrödinger's orbitals can be treated very much like Bohr's orbits. The closer an orbital is to the nucleus, the lower the total energy of an electron in that orbital. To jump from a lower-energy orbital to a higher-energy orbital, an electron must absorb a quantity of energy precisely equal to the energy difference between the two orbitals. When an electron drops from a higher-energy orbital to a lower-energy orbital, it gives off electromagnetic radiation. The energy of the emitted radiation is precisely equal to this energy difference. Finally, the most probable location for the single electron in a hydrogen atom is at a distance from the nucleus exactly equal to that of Bohr's lowest-energy orbit.

Figure 4-13 Two ways of showing electrons around atoms. In (a) the probability of finding the electron is proportional to the density of the cloud. In (b) the surface within which the electron can be found a certain percentage of the time, i.e., 90% is shown.

Chemistry in Detail

Heisenberg's Uncertainty Principle

The *uncertainty principle* was first stated in 1927 by the German physicist Werner Heisenberg (1901–1976). According to Heisenberg, it is not possible to measure precisely both the velocity and the position of an electron at the same time. (Velocity is the speed and direction of motion of an object.) The following baseball analogy may help you understand this principle, which is widely accepted by scientists today.

At the crack of the bat, an outfielder can predict almost where a fly ball is going to land, run to that spot, and be ready to catch it. The outfielder instinctively predicts the path that the ball will follow by observing its position after it is hit. The outfielder sees the ball by means of photons reflected off the ball and into his eyes. The photons do not affect the path of the ball because the ball is much more massive than the photons striking it.

Now suppose that a submicroscopic outfielder is trying to catch an electron. This imaginary outfielder also observes objects by means of reflected photons. But unlike a baseball, an electron is so small that its path is significantly changed each time it is struck by a photon. The outfielder would be able to tell only where the electron had been when it was struck by the last photon. The velocity, and thus, the path of the electron could not be predicted.

If the submicroscopic outfielder tried to avoid this problem by using lower

energy light so as not to disturb the path of the electron, a new difficulty would appear. Low energy light would have such a low frequency and long wavelength that it would not be possible for the outfielder to tell the exact direction from which the reflected light came. (You may have noticed that it is much easier to locate the source of a high frequency, like that of a flute, than that of a low-frequency sound like that of a bass drum.) The outfielder would therefore not know where the electron was when it reflected the light.

As illustrated in the baseball analogy, it is not possible to measure both the position and the path of an electron. Since the path of an electron depends partly on its energy, scientists cannot measure both the energy and the position of an electron. **Heisenberg's uncertainty principle** states that it is not possible to know both the velocity and the position of a particle at the same time.

Arrangement of Electrons in Atoms

1. It did not explain how negatively charged electrons could fill the space surrounding a positively charged nucleus.
2. $c = \lambda\nu$
3. (a) A form of energy that exhibits wavelike behavior as it travels through space (b) The distance between corresponding points on adjacent waves (c) The number of waves that pass a given point in a specified amount of time, usually one second (d) A finite quantity of energy that can be gained or lost by an atom (e) A quantum of light
4. Light has both wave and particle characteristics; in some experiments light acts like a wave, whereas in others it acts like a particle.
5. Refer to the section *Bohr Model of the Hydrogen Atom*, p. 102.

To help students remember the different orbital shapes: *s* orbitals are spherical, *p* orbitals—peanut shaped, *d* orbitals—double peanut shaped, and *f* orbitals—flower shaped. The letters actually stand for words describing bright-line spectra given off by atoms that have been excited to different levels. They are *s*harp, *p*rincipal, *d*iffuse, and *f*undamental.

Section Review

1. What was the major limitation of Rutherford's model of the atom?
2. Write and label the equation that relates the speed, wavelength, and frequency of electromagnetic radiation.
3. Define *(a)* electromagnetic radiation *(b)* wavelength *(c)* frequency *(d)* quantum *(e)* photon.
4. What is meant by the dual wave–particle nature of light?
5. Describe the Bohr model of the hydrogen atom.

4.2 Quantum Numbers and Atomic Orbitals

This section introduces the terminology, symbols, and rules used to associate individual electrons with specific atomic orbitals. The concepts introduced here will be used again and again in later chapters as you learn more about the characteristic properties of matter.

Quantum Numbers

More than one set of numbers is needed to describe the quantization of electron energies. **Quantum numbers** *are numbers that specify the properties of atomic orbitals and of their electrons.* The first three quantum numbers are derived from the Schrödinger equation. They indicate the region occupied by a given orbital in terms of *(1)* distance from the nucleus, *(2)* orbital shape, and *(3)* orbital position with respect to the three-dimensional *x*, *y*, and *z* axes. These three quantum numbers are whole numbers and are related to each other. The fourth quantum number has only two values and is needed to specify one of two possible orientations of an electron within an orbital. As you read the following descriptions of the quantum numbers, refer to the appropriate columns in Table 4-1.

Principal quantum number The **principal quantum number**, *symbolized by n, indicates the main energy levels surrounding a nucleus.* These energy levels, sometimes referred to as *shells*, are the quantum-theory equivalent of the Bohr orbits. Values of *n* are whole numbers only—1, 2, 3, and so on. The main energy level with $n = 1$ is the one closest to the nucleus. As *n* increases, the distance of the main energy levels from the nucleus increases and their energy increases. The known elements utilize main energy levels with $n = 1$ to 7 in their ground states.

Orbital quantum number The **orbital quantum number** *indicates the shape of an orbital.* Within each main energy level beyond the first, orbitals with different shapes occupy different regions. The regions are referred to as *sublevels*, or *subshells*.

As shown in the second column of Table 4-1, the number of subshells, or possible orbital shapes, in each main energy level is equal to the value of the principal quantum number. The first four orbital quantum numbers are designated in ascending order by the letters *s*, *p*, *d*, and *f* (the *s* sublevel

TABLE 4-1 QUANTUM NUMBER RELATIONSHIPS IN ATOMIC STRUCTURE

Principal Quantum Number: Main Energy Level (n)	Orbitals (Sublevels): n Orbital Shapes; n Sublevels	Number of Orbitals per Sublevel	Number of Orbitals per Main Energy Level (n^2)	Number of Electrons per Sublevel	Number of Electrons per Main Energy Level ($2n^2$)
1	s	1	1	2	2
2	s p	1 3	4	2 6	8
3	s p d	1 3 5	9	2 6 10	18
4	s p d f	1 3 5 7	16	2 6 10 14	32

is the lowest in energy; the p sublevel is higher in energy than the s sublevel; and so on.)

The s orbitals are spherical, the p orbitals have dumbbell shapes, and the d orbitals (with one exception) have four lobes. These are all shown in Figure 4-14. The f orbital shapes are too complex to treat here.

In the first energy level (n=1), only one sublevel is allowed. The first energy level therefore has only an s sublevel. In the second energy level (n=2), two different sublevels are possible. The second energy level thus has s and p orbitals. The third energy level (n=3) has three sublevels, and therefore has s, p, and d orbitals. The fourth energy level (n=4) has four possible sublevels and therefore has s, p, d, and f orbitals. In an nth main energy level, orbitals of n shapes are possible.

Each sublevel or type of orbital is designated by the principal quantum number followed by the letter of the sublevel. For example, the 1s sublevel is the s sublevel in the first main energy level, the 2s sublevel is the s orbital in the second main energy level. A 4d orbital is an orbital in the d sublevel of the fourth main energy level. How would you designate the p sublevel in the third main energy level (see margin)? How many other sublevels are in the same main energy level with this one (see margin)?

STUDY HINT

Principal quantum number	Types of orbitals
1	1s
2	2s, 2p
3	3s, 3p, 3d
4	4s, 4p, 4d, 4f

Figure 4-14 s, p and d orbitals have different shapes.

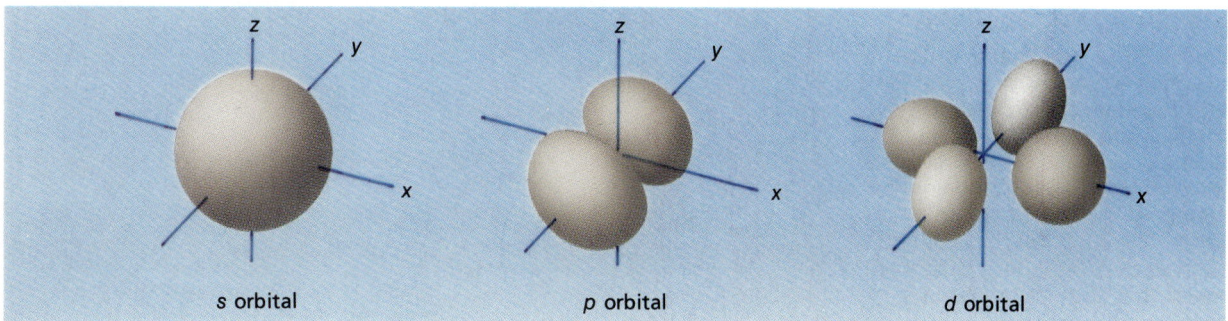

s orbital   p orbital   d orbital

Arrangement of Electrons in Atoms

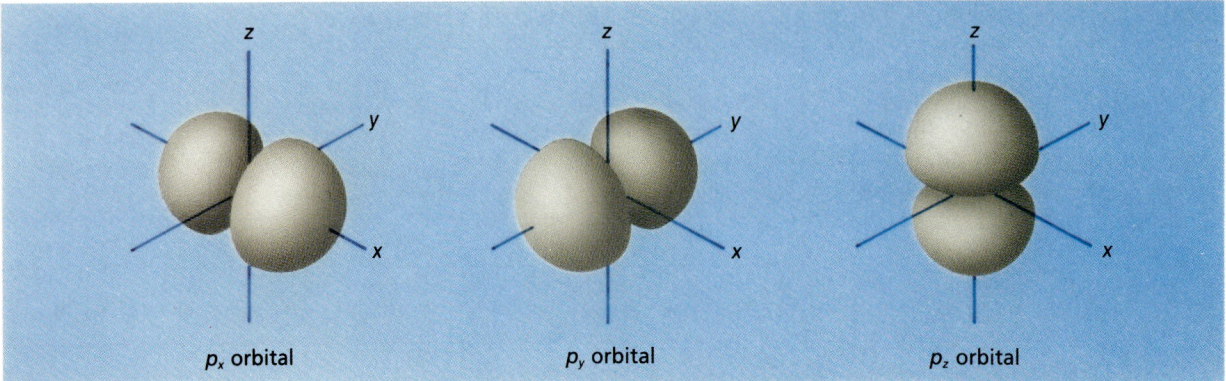

Figure 4-15 There are three orientations for *p* orbitals.

Magnetic quantum number *The **magnetic quantum number** indicates the orientation of an orbital about the nucleus.* There is only one possible orientation for an *s* orbital, which is a sphere centered on the nucleus, like the orbitals drawn in Figure 4-14. There is therefore only one *s* orbital in each *s* sublevel. The *p* orbitals can have three different orientations. The lobes extend along the *x*, *y*, and *z* axes of a three-dimensional (Cartesian) coordinate system, as shown in Figure 4-15. Each *p* sublevel therefore contains three different *p* orbitals. These usually are designated as p_x, p_y, and p_z orbitals, where the subscripts *x*, *y*, and *z* indicate the locations of the *p* orbitals.

There are five different *d* orbitals (Figure 4-16) in each *d* sublevel, as shown in Table 4-1. There are seven different *f* orbitals in each *f* sublevel.

The combined effect of the orbital and magnetic quantum numbers is that, with increasing main energy levels there are larger numbers of orbitals. Table 4-1 should convince you that the number of orbitals in each main

To help students understand how all parts of the electron cloud fit together and its 3-D nature, use the analogy of an apartment building to represent the electron cloud. Each floor is an energy level, each apartment is a sublevel, and each room is an orbital with one or two people (electrons) living in it.

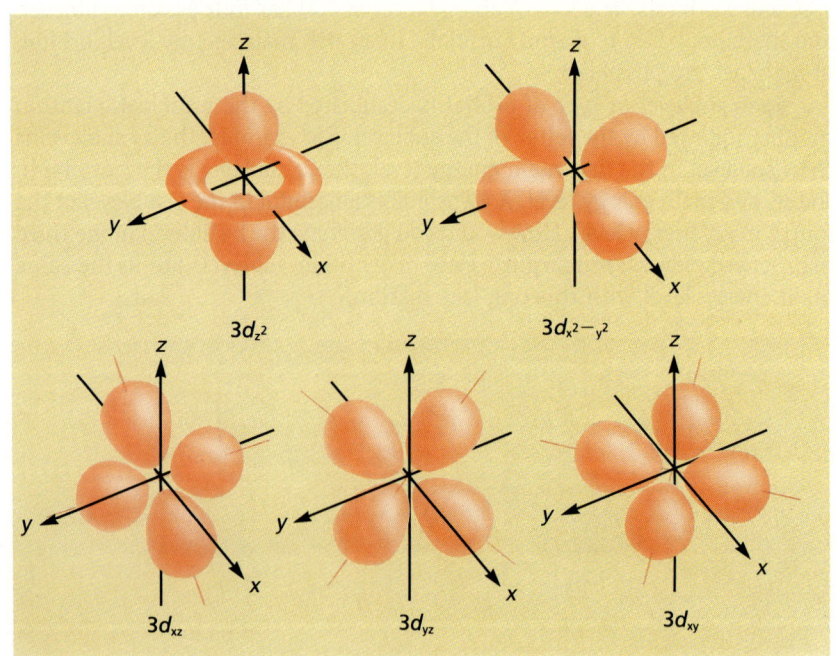

Figure 4-16 The spatial orientations and shapes of *d* orbitals. Each orbital can accommodate a maximum of two electrons. One electron in a d_{yz} orbital has an equal probability of being found in any one of the four lobes. The subscripts *xy*, *yz*, and *xz* in each orbital designation mean that the orbital lies in that plane.

energy level equals the square of the principal quantum number. What is the total number of sublevels and orbitals in the second energy level and of what types are they (see margin, p. 109)?

Spin quantum number   *The **spin quantum number**, has only two possible values—(+½, −½)—which indicate two possible states of an electron in an orbital.* Of all the quantum numbers, electron spin has the most familiar physical significance. You can picture an electron spinning on an imaginary axis in much the same way that the earth spins on its axis, thus generating its magnetic field. Each electron thus behaves as though it were a tiny bar magnet with a north and south pole.

Magnetism   All magnetic phenomena stem from the motions of electrons about the nuclei of atoms. **Diamagnetism** *is the property of a substance whereby it is weakly repelled by a magnetic field.* The magnetic field causes the electrons in any substance to move in such a way that they produce an opposing magnetic field that is repelled by the original magnetic field. *Paramagnetism* is a weak attraction between magnetic fields and substances whose atoms have a more or less uneven electron distribution around their nuclei. This attraction is slightly stronger than the repulsive diamagnetism, so that paramagnetic substances are weakly attracted by magnetic fields. Oxygen is among the common substances that are paramagnetic. *Ferromagnetism,* such as that shown by ordinary bar magnets, is a strong form of paramagnetism, resulting from all of the atoms in the ferromagnetic substance becoming oriented with their like magnetic poles in the same direction.

Electrons in Each Main Energy Level

The rules that govern the location of electrons in the ground state of atoms of every element are discussed in the next section. The first three quantum numbers completely indicate the energy level and shape of each orbital. The spin quantum number is significant, as you will see, because each orbital can hold no more than two electrons, which must have opposite spins. This limitation determines the maximum number of electrons that can occupy each main energy level, as shown in Table 4-1.

The first main energy level has only a single *s* orbital, and therefore, will hold only 2 electrons. The second main energy level can accommodate 2 electrons in its single *s* orbital and 2 electrons in each of its three *p* orbitals, for a total occupancy of 8 electrons. Can you explain how the third main energy level can hold 18 electrons (see margin)? Examine Table 4-1 to convince yourself that the maximum number of electrons in each main energy level is equal to $2n^2$.

Section Review

1. Define   *(a)* main energy levels   *(b)* quantum numbers.
2. *(a)* What general information about atomic orbitals is provided by the four sets of quantum numbers?   *(b)* List the four quantum numbers.
3. Identify and explain the meanings associated with the possible values for each of the four quantum numbers in a set.

Point out that in the first energy level, the *s*-orbital, the *s*-sublevel, and the first energy level are the same thing.

STUDY HINT
Filled third main energy level ($n = 3$) has 2 e⁻ in 3*s* orbital, 6 e⁻ in 3*p* orbitals, 10 e⁻ in 3*d* orbitals.
2 e⁻ + 6 e⁻ + 10 e⁻ = 18 e⁻. $2n^2 = 2 \times (3)^2 = 2 \times 9 = 18$

1. *(a)* The quantized regions of increasing energy into which atomic orbitals fall   *(b)* Sets of numbers that specify the properties of atomic orbitals and their electrons.
2. *(a)* The distance of an orbital from the nucleus, the shape of the orbital, its position with respect to the three-dimensional axes, and the state of the electrons within that orbital   *(b)* The principal, orbital, magnetic, and spin quantum numbers.
3. (1) Principal quantum number *n* (1, 2, 3, and so on): distance of the main energy level from the nucleus (2) Orbital quantum number *(s, p, d,* and *f)*: various orbital shapes (3) Magnetic quantum number (the subscripts *x, y,* and *z*): orbital orientation with respect to a three-dimensional coordinate system (4) Spin quantum number ($+\frac{1}{2}$ and $-\frac{1}{2}$): the two possible electron states within an orbital

4.3 Electron Configurations

SECTION OBJECTIVES

- State the Aufbau principle, Hund's rule, and the Pauli exclusion principle.
- Describe the arrangement of electrons around the atoms of any element using orbital notation, electron-configuration notation or electron-dot notation.
- Build up the electron configuration for atoms of any element, given the atomic number or identity of the element.
- Describe the noble-gas configuration and write it for any noble gas.

Electron configuration *is the arrangement of electrons in atoms.* Atoms of different elements have different numbers of electrons. Therefore, there is a distinctive electron configuration for atoms of each element. Electrons in atoms, like all systems in nature, tend to assume arrangements that have the lowest possible energies. A few simple rules, combined with the quantum number relationships discussed in Section 4.2, permit assignment of these stable, or ground-state, electron configurations.

In this section you will learn how to build up electron configurations for the ground state of any particular atom. The process begins with the hydrogen atom, with its single electron. Electrons are added one by one according to the rules. As you follow the changing electron configurations with increasing atomic number, you will begin to discover the relationship between electron configurations and the Periodic Table. Real atoms are not, however, built up by adding protons and electrons one at a time, and it is wise to remember this.

Rules Governing Electron Configurations

The first rule shows the order in which vacant orbitals will be occupied. According to the **Aufbau** ('building up' in German) **principle**, *an electron occupies the lowest-energy orbital that can receive it.* In hydrogen atoms, the electron is in the orbital with the lowest energy of all orbitals—the 1s orbital. The 2s and 2p orbitals are next highest in energy and will be filled in that order—2s, then 2p.

In multielectron atoms, with the third main energy level ($n = 3$), sublevels in different main energy levels begin to overlap as shown in Figure 4-17. The order of increasing energy for the sublevels is summarized on the lefthand axis. Note, for example, that the 4s sublevel is lower in energy than the 3d sublevel. In building up the electron configurations for atoms of different elements,

Figure 4-17 Relative energies of orbitals in isolated multielectron atoms. Each box represents an orbital capable of holding two electrons. The available orbitals with the lowest energies are filled first.

therefore, the 4s orbital fills before any electrons enter the 3d orbitals. Once the 3d orbitals are fully occupied by 10 electrons, which sublevel will be occupied next?

The second rule requires placing as many single electrons as possible in separate orbitals in the same sublevel. In this way, electron–electron repulsion is minimized and the electron arrangements have a lower energy. According to **Hund's rule,** *orbitals of equal energy are each occupied by one electron before any one orbital is occupied by a second electron, and all electrons in singly occupied orbitals must have the same spin.* Applying this rule shows, for example, that one electron will enter each of the three p orbitals in a main energy level before a second electron enters any of them. What is the maximum number of unpaired electrons in a d sublevel (see margin)? The third rule is the one we referred to in noting the importance of the spin quantum number. According to the **Pauli exclusion principle,** *no two electrons in the same atom can have the same set of four quantum numbers.* Different combinations of the first three quantum numbers apply to each different orbital. The two values of the spin quantum number permit two electrons of opposite spins to occupy the same orbital.

Representing Electron Configurations

Three methods, or notations, are used to indicate electron configurations. Two of these notations indicate the ground-state electron configuration. The third indicates only the outermost electron arrangement, which is the most important in the formation of chemical compounds. These three notations are illustrated for atoms of hydrogen (H) and helium (He).

The single electron in ground-state hydrogen is in the lowest-energy orbital, the 1s orbital. The electron can have either one of its two possible spin states. Helium has two electrons, which are paired in the lowest-energy orbital, the 1s orbital.

Orbital notation In orbital notation, an unoccupied orbital is represented by a line ____. An orbital containing one electron is represented as ↑___ (or ↓___). An orbital containing two electrons is represented as ↑↓___, showing the electrons paired and with opposite spins. The lines are labeled with the principal quantum number and subshell letter. The orbital notations for hydrogen and helium are, therefore,

$$\text{H} \quad \frac{\uparrow}{1s} \qquad \text{He} \quad \frac{\uparrow\downarrow}{1s}$$

Electron-configuration notation. Electron-configuration notation eliminates the lines and arrows of orbital notation. Instead, the number of electrons in a sublevel is shown by adding superscripts to the sublevel designation. The hydrogen configuration is represented by $1s^1$. The superscript ^{1} indicates that one electron is present in the orbital of the first energy level ($n = 1$). The helium configuration is represented by $1s^2$, the superscript ^{2} indicating that there are two electrons in the orbital of the first energy level.

Electron-dot notation Electron-dot notation shows only electrons in the highest, or outermost, main energy level. The **highest occupied energy level** *is the electron-containing main energy level with the highest principal*

STUDY HINT

Reading up the lefthand axis in Figure 4-17 shows that the 4p sublevel is next to be occupied after the 3d sublevel.

If given a choice between a private room (orbital) and a semi-private room, electrons prefer private rooms, provided they do not have to exert any more energy.

STUDY HINT

A d sublevel has 5 orbitals and can accommodate 5 unpaired electrons.

Two electrons occupying the same orbital are known as an electron pair, or an orbital pair.

STUDY HINT
Electron-dot notation was invented by the American chemist, Gilbert Newton Lewis (1875–1946).

STUDY HINT

Number of outer electrons	Electron-dot symbol
1	X·
2	·X·
3	·X· (with dot above)
4	·X· (with dots above and below)
5	·X:
6	:X:
7	:X: (with additional dot)
8	:X: (fully surrounded)

quantum number. For hydrogen and helium atoms, there is only one occupied energy level.

Dots representing the number of electrons in the highest occupied energy level are placed around the symbol for the element. The electron-dot symbols for hydrogen and helium are

$$H\cdot \qquad He:$$

Helium atoms have a fully occupied 1s sublevel, therefore, the next electron must enter the next highest vacant sublevel, which is the 2s sublevel (see Figure 4-17). For lithium, with atomic number 3 and three electrons, the electron dot symbol is

$$Li\cdot$$

The two electrons in the 1s sublevel are no longer in the outermost main energy level. They have become **inner-shell electrons**, *which are electrons not in the highest occupied energy level.* For atoms of all elements beyond helium, the atomic symbol in electron-dot notation represents the nucleus plus any inner-shell electrons.

Additional dots after the first can be added in several different arrangements. In this book, for elements of the second or higher periods, a second dot will be placed at the left, a third dot at the top, and a fourth dot at the bottom. This arrangement does not always correspond to the ground-state electron configuration. Instead, it provides a clue about how the atom will react with other atoms (see Chapter 6). The dot arrangements for five, six, seven, and eight electrons are also shown in the margin.

Sample Problem 4.1

The electron configuration of boron is $1s^2 2s^2 2p^1$. *(a)* How many electrons are present in a boron atom? What is the atomic number for boron? Write the orbital notation for boron. *(Ans.)* From the electron-configuration notation for boron, each atom contains $2 + 2 + 1 = 5$ electrons. Since the number of protons equals the number of electrons in a neutral atom, boron has 5 protons and, thus, has an atomic number of 5. To write out the orbital notation, we first draw the lines representing orbitals:

$$\underline{\quad} \quad \underline{\quad} \quad \underline{\quad} \quad \underline{\quad} \quad \underline{\quad}$$
$$1s \qquad 2s \qquad 2p_x \qquad 2p_y \qquad 2p_z$$

Next, we add arrows showing the electron locations. The first two electrons occupy $n = 1$ energy level and fill the 1s orbital:

$$\underline{\uparrow\downarrow} \quad \underline{\quad} \quad \underline{\quad} \quad \underline{\quad} \quad \underline{\quad}$$
$$1s \qquad 2s \qquad 2p_x \qquad 2p_y \qquad 2p_z$$

The next three electrons enter the $n = 2$ main-energy level. As shown by the electron-configuration notation, two electrons occupy the 2s orbital and one, a p orbital.

$$\underline{\uparrow\downarrow} \quad \underline{\uparrow\downarrow} \quad \underline{\uparrow} \quad \underline{\quad} \quad \underline{\quad}$$
$$1s \qquad 2s \qquad 2p_x \qquad 2p_y \qquad 2p_z$$

(b) Write the electron-dot symbol for boron. How many inner-shell electrons does boron have? In which orbital are they located? *(Ans.)* A boron atom has three electrons in its outermost occupied energy level, the $n = 2$ level. The electron-dot symbol for boron, is therefore,

$$\cdot \dot{B} \cdot$$

The two remaining electrons in a boron atom are inner-shell electrons in the 1s orbital.

Sample Problem 4.1 *continued*

> **Practice Problems**
> 1. The electron configuration of nitrogen is $1s^2 2s^2 2p^3$. *(a)* How many electrons are present in a nitrogen atom? What is the atomic number of nitrogen? Write the orbital notation for nitrogen. *(Ans.)* 7, 7, $\frac{\uparrow\downarrow}{1s} \ \frac{\uparrow\downarrow}{2s} \ \frac{\uparrow}{2p_x} \ \frac{\uparrow}{2p_y} \ \frac{\uparrow}{2p_z}$ *(b)* Write the electron-dot symbol for nitrogen. How many inner-shell electrons does nitrogen have? *(Ans.)* $\cdot\ddot{N}\!:$, 2
> 2. The electron configuration of fluorine is $1s^2 2s^2 2p^5$. *(a)* What is the atomic number of fluorine? What is its highest occupied energy level? How many of its *p* orbitals are filled? How many unpaired electrons does a fluorine atom contain? *(Ans.)* 9, 2nd, 2, 1 *(b)* Write the electron-dot symbol for fluorine. How many inner-shell electrons does fluorine have? *(Ans.)* $:\ddot{F}\!:$, 2

Elements of the Second and Third Periods

Second-period elements As shown above, the first main energy level is filled in the first-period elements, hydrogen and helium. The ground-state configurations in Table 4-2 illustrate how the Aufbau principle, Hund's rule, and the Pauli exclusion principle are applied in the second period. (A diagram that aids in applying the Aufbau principle is given in Figure 4-18.)

According to the Aufbau principle, the first added electron enters the *s* sublevel in the second main energy level. Thus, lithium (Li) has the $1s^2 2s^1$ configuration. The fourth electron in beryllium (Be) atoms must pair up in the 2*s* sublevel because this sublevel is of lower energy than the 2*p* sublevel.

With the 2*s* sublevel filled, the 2*p* sublevel, which has three vacant orbitals of equal energy, can be occupied. Hund's rule applies here, as clearly shown in the orbital notation. Each of the three *p* orbitals is occupied in turn by a single electron in boron (B), then carbon (C), then nitrogen (N) atoms.

The next electron must either pair up with another electron in one of the 2*p* orbitals or enter the third main energy level, where it could be alone in an orbital. The Aufbau principle shows that the 2*p* orbitals must be filled before higher-energy orbitals are occupied. The Pauli exclusion principle shows that the next electron can pair up in a 2*p* orbital by having the opposite spin from the first electron. Oxygen (O) atoms, therefore, have the configuration $1s^2 2s^2 2p^4$. This notation does not directly show electron pairing in the 2*p* sublevel. Based on the rules, however, we know that of four electrons in a *p* sublevel, two are paired and two are not.

The remaining two 2*p* orbitals are filled in fluorine (F) and neon (Ne) atoms, respectively. Atoms like those of neon, which have the *s* and *p* sublevels of their highest main energy level filled with eight electrons, are said to have an *octet* of electrons. Note in the Periodic Table inside the back cover than neon is the last element in the second period.

Third-period elements After the outer octet is filled in neon atoms, the filling sequence begins again. The next electron enters the *s* sublevel in the $n = 3$ main energy level. Thus, sodium (Na) atoms have the configuration $1s^2 2s^2 2p^6 3s^1$. Compare this sodium atom configuration with that of neon atoms in Table 4-2. The first 10 electrons in a sodium atom have the neon configuration, $1s^2 2s^2 2p^6$. To simplify writing the notation, the symbol for neon, enclosed in square brackets, is used to represent the complete neon

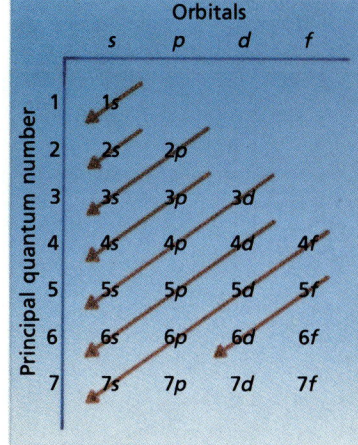

Figure 4-18 A useful diagram for remembering the order in which orbitals fill. Move from the head of each arrow over to the base of the next arrow.

TABLE 4-2 ELECTRON CONFIGURATIONS OF ATOMS OF SECOND-PERIOD ELEMENTS SHOWING THREE NOTATIONS

Element	Symbol	Orbital Notation 1s 2s 2p	Electron-Configuration Notation	Electron-Dot Notation
Lithium	Li	↑↓ ↑ __ __ __	$1s^2 2s^1$	Li·
Beryllium	Be	↑↓ ↑↓ __ __ __	$1s^2 2s^2$	·Be·
Boron	B	↑↓ ↑↓ ↑ __ __	$1s^2 2s^2 2p^1$	·Ḃ·
Carbon	C	↑↓ ↑↓ ↑ ↑ __	$1s^2 2s^2 2p^2$	·Ċ·
Nitrogen	N	↑↓ ↑↓ ↑ ↑ ↑	$1s^2 2s^2 2p^3$	·N̈·
Oxygen	O	↑↓ ↑↓ ↑↓ ↑ ↑	$1s^2 2s^2 2p^4$	:Ö·
Fluorine	F	↑↓ ↑↓ ↑↓ ↑↓ ↑	$1s^2 2s^2 2p^5$	:F̈:
Neon	Ne	↑↓ ↑↓ ↑↓ ↑↓ ↑↓	$1s^2 2s^2 2p^6$	:N̈ë:

configuration, $[Ne] = 1s^2 2s^2 2p^6$. This permits writing $[Ne]3s^1$ for the sodium configuration.

The last element in the third period is argon (Ar). Like neon, the element above it in Group 18, argon has an outermost octet of electrons, $[Ne]3s^2 3p^6$. In fact, each Group 18 element other than He has an electron octet in its highest energy level (see Periodic Table). *The Group 18 elements (helium, neon, argon, krypton, xenon, and radon) are called the* **noble gases**. The symbol for each is used, as just described for neon, to simplify the electron configuration notation for elements that follow it. *A* **noble gas configuration** *is an outer main energy level fully occupied, in most cases, by eight electrons.*

Sample Problem 4.2

(a) Write the electron-dot notation for silicon (Si). How many inner-shell electrons does a silicon atom contain?   *(Ans.)* The configuration of silicon is $[Ne]3s^2 3p^2$. (See Table 4-3.) A silicon atom has four electrons in the highest energy level, and therefore the electron dot symbol is

$$·\text{Si}·$$

All of the electrons represented by [Ne] are inner-shell electrons. The total number, as shown in Table 4-3, is 10.

(b) Identify the element with atoms that have five electrons in the *p* sublevel in their third main energy level. What is the total number of electrons in the third main energy level in these atoms? Write the electron-dot symbol for this element. Name the element in the second period with the same number of outermost electrons.   *(Ans.)* Table 4-3 shows that a chlorine atom, with configuration $[Ne]3s^2 3p^5$, has five electrons in its 3p sublevel. There are seven electrons in the third main energy level. Therefore, chlorine has the electron-dot symbol

$$:\ddot{\text{Cl}}:$$

Fluorine (F), as shown in Table 4-2, also has seven outermost electrons.

Sample Problem 4.2 *continued*

> **Practice Problems**
> 1. *(a)* Write the electron-dot notation for sulfur (S). How many inner-shell electrons does a sulfur atom contain? *(Ans.)* :$\ddot{\text{S}}$:, 10
>   *(b)* Identify the element with atoms that have three electrons in the *p* sublevel in their third main energy level. What is the total number of electrons in the third main energy level in these atoms? Write the electron-dot symbol for this element. Name the element in the second period with the same number of outermost electrons. *(Ans.)* P, 5, ·$\ddot{\text{P}}$:, N
> 2. *(a)* Given the shorthand electron configuration [Ne]$3s^23p^1$, which element does this represent? Write its complete electron configuration. How many inner-shell electrons do its atoms contain? Write its electron-dot symbol. *(Ans.)* Al, $1s^22s^22p^63s^23p^1$, 10, ·$\dot{\text{Al}}$·
>   *(b)* Write both the complete and shorthand electron configurations for argon. How many outer-shell electrons do atoms of argon contain? Write its electron-dot symbol. How many unpaired electrons are contained in its electron-configuration notation? *(Ans.)* $1s^22s^22p^63s^23p^6$, [Ne]$3s^23p^6$, 8, :$\ddot{\text{Ar}}$:, none

TABLE 4-3 ELECTRON CONFIGURATIONS OF ATOMS OF THIRD-PERIOD ELEMENTS

Name	Symbol	Atomic Number	Number of Electrons in Sublevels					Shorthand Notation
			1s	2s	2p	3s	3p	
Sodium	Na	11	2	2	6	1		[Ne]$3s^1$
Magnesium	Mg	12	2	2	6	2		[Ne]$3s^2$
Aluminum	Al	13	2	2	6	2	1	[Ne]$3s^23p^1$
Silicon	Si	14	2	2	6	2	2	[Ne]$3s^23p^2$
Phosphorus	P	15	2	2	6	2	3	[Ne]$3s^23p^3$
Sulfur	S	16	2	2	6	2	4	[Ne]$3s^23p^4$
Chlorine	Cl	17	2	2	6	2	5	[Ne]$3s^23p^5$
Argon	Ar	18	2	2	6	2	6	[Ne]$3s^23p^6$

Elements of the Fourth and Fifth Periods

The electron configurations of atoms in the fourth-period elements are shown in Table 4-4. The period begins by filling the 4s orbital, the empty orbital of lowest energy. Thus, the first element in the fourth period is potassium (K), which has the electron configuration [Ar]$4s^1$. The next element is calcium (Ca), which has the electron configuration [Ar]$4s^2$. The electron-dot symbols for potassium and calcium are

$$\text{K}\cdot \qquad \cdot\text{Ca}\cdot$$

With the 4s sublevel filled, the 4p and 3d sublevels have the next available vacant orbitals. Figure 4-17 shows that the 3d sublevel is lower in energy than the 4p sublevel, and therefore the five 3d orbitals are next to be filled in the ten elements from scandium (Z = 21) to zinc (Z = 30).

Scandium (Sc) has the electron configuration [Ar]$3d^14s^2$. Titanium (Ti), has the configuration [Ar]$3d^24s^2$, and vanadium (V), has the configuration [Ar]$3d^34s^2$. To this point, three electrons with the same spin have been added to three separate *d* orbitals, as expected from Hund's rule.

Arrangement of Electrons in Atoms

TABLE 4-4 ELECTRON CONFIGURATION OF ATOMS OF ELEMENTS IN THE FOURTH PERIOD

Name	Symbol	Atomic Number	Number of Electrons in Sublevels Above 2p					Electron Configuration Notation
			3s	3p	3d	4s	4p	
Potassium	K	19	2	6		1		[Ar]$4s^1$
Calcium	Ca	20	2	6		2		[Ar]$4s^2$
Scandium	Sc	21	2	6	1	2		[Ar]$3d^14s^2$
Titanium	Ti	22	2	6	2	2		[Ar]$3d^24s^2$
Vanadium	V	23	2	6	3	2		[Ar]$3d^34s^2$
Chromium	Cr	24	2	6	5	1		[Ar]$3d^54s^1$
Manganese	Mn	25	2	6	5	2		[Ar]$3d^54s^2$
Iron	Fe	26	2	6	6	2		[Ar]$3d^64s^2$
Cobalt	Co	27	2	6	7	2		[Ar]$3d^74s^2$
Nickel	Ni	28	2	6	8	2		[Ar]$3d^84s^2$
Copper	Cu	29	2	6	10	1		[Ar]$3d^{10}4s^1$
Zinc	Zn	30	2	6	10	2		[Ar]$3d^{10}4s^2$
Gallium	Ga	31	2	6	10	2	1	[Ar]$3d^{10}4s^24p^1$
Germanium	Ge	32	2	6	10	2	2	[Ar]$3d^{10}4s^24p^2$
Arsenic	As	33	2	6	10	2	3	[Ar]$3d^{10}4s^24p^3$
Selenium	Se	34	2	6	10	2	4	[Ar]$3d^{10}4s^24p^4$
Bromine	Br	35	2	6	10	2	5	[Ar]$3d^{10}4s^24p^5$
Krypton	Kr	36	2	6	10	2	6	[Ar]$3d^{10}4s^24p^6$

The 1s, 2s and 2p sublevels are filled in atoms of each element shown. The configuration for these filled sublevels is $1s^22s^22p^6$.

The Aufbau Principal does not work for atoms of all elements; some elements that deviate are Cr, Cu, Ag, Au, Pd, Pt, and Mo.

Somewhat surprisingly, chromium (Cr), has the electron configuration [Ar]$3d^54s^1$. Not only did the added electron go into the fourth 3d orbital, but an electron was also moved from the 4s orbital into the fifth 3d orbital, leaving the 4s orbital with a single electron. The electron configuration of chromium atoms is contrary to the Aufbau principle, but it is thought to be of lower energy than that resulting from the Aufbau principle.

A chromium atom with the [Ar]$3d^54s^1$ configuration has six orbitals that each contain only one electron. All six electrons have the same spin. This is very stable according to Hund's rule. Moreover, the 3d and 4s sublevels are exactly half-filled, and special stability is associated with half-filled (and also completely filled) sublevels. These two factors are apparently strong enough to decrease the total energy of this configuration.

Manganese (Mn) has the electron configuration [Ar]$3d^54s^2$. The added electron goes to the lower-energy 4s orbital, completely filling this orbital while leaving the 3d orbitals still half-filled. Beginning with the next element, electrons pair up in d orbitals, giving iron (Fe) the configuration [Ar]$3d^64s^2$; cobalt (Co), [Ar]$3d^74s^2$; and nickel (Ni), [Ar]$3d^84s^2$. Next is copper, with the electron configuration [Ar]$3d^{10}4s^1$. An electron moves from the 4s orbital to pair with the electron in the fifth 3d orbital. This gives the 3d sublevel a completely-filled, stable configuration and the 4s orbital a half-filled, stable configuration.

In zinc (Zn) atoms, the 4s sublevel is filled to give the electron configuration [Ar]$3d^{10}4s^2$. In atoms of the next six elements, electrons add one by one to the three 4p orbitals. According to Hund's rule, one electron is added to each of the three 4p orbitals before electrons are paired in any 4p orbital. The electron-dot notations for the atoms of the elements from zinc to krypton (Kr) are

·Zn· ·Ga· ·Ge· ·Äs: :S̈e: :B̈r: :K̈r:

In the 18 elements of the fifth period, sublevels fill in the same order as in elements of the fourth period. Successive electrons are added first to the 5s orbital, then to the 4d orbitals, and finally to the 5p orbitals, as shown in Table 4-5. There are occasional deviations from the predicted configurations here also. The deviations differ from those for fourth-period elements; in each case it is believed that the preferred configuration has a lower energy.

TABLE 4-5 ELECTRON CONFIGURATIONS OF ATOMS OF ELEMENTS IN THE FIFTH PERIOD

Name	Symbol	Atomic Number	Number of Electrons in Sublevels above 3d					Electron Configuration Notation
			4s	4p	4d	5s	5p	
Rubidium	Rb	37	2	6		1		$[Kr]5s^1$
Strontium	Sr	38	2	6		2		$[Kr]5s^2$
Yttrium	Y	39	2	6	1	2		$[Kr]4d^15s^2$
Zirconium	Zr	40	2	6	2	2		$[Kr]4d^25s^2$
Niobium	Nb	41	2	6	4	1		$[Kr]4d^45s^1$
Molybdenum	Mo	42	2	6	5	1		$[Kr]4d^55s^1$
Technetium	Tc	43	2	6	5	2		$[Kr]4d^55s^2$
Ruthenium	Ru	44	2	6	7	1		$[Kr]4d^75s^1$
Rhodium	Rh	45	2	6	8	1		$[Kr]4d^85s^1$
Palladium	Pd	46	2	6	10			$[Kr]4d^{10}$
Silver	Ag	47	2	6	10	1		$[Kr]4d^{10}5s^1$
Cadmium	Cd	48	2	6	10	2		$[Kr]4d^{10}5s^2$
Indium	In	49	2	6	10	2	1	$[Kr]4d^{10}5s^25p^1$
Tin	Sn	50	2	6	10	2	2	$[Kr]4d^{10}5s^25p^2$
Antimony	Sb	51	2	6	10	2	3	$[Kr]4d^{10}5s^25p^3$
Tellurium	Te	52	2	6	10	2	4	$[Kr]4d^{10}5s^25p^4$
Iodine	I	53	2	6	10	2	5	$[Kr]4d^{10}5s^25p^5$
Xenon	Xe	54	2	6	10	2	6	$[Kr]4d^{10}5s^25p^6$

All sublevels up to 3d are filled. The configuration for these filled sublevels is $1s^22s^22p^63s^23p^63d^{10}$.

Sample Problem 4.3

(a) Write both the complete and shorthand electron configurations for iron (Fe). *(Ans.)* The complete configuration of iron is $1s^22s^22p^63s^23p^63d^64s^2$, whereas its shorthand notation is $[Ar]3d^64s^2$.

(b) How many electron-containing orbitals are in an atom of iron? How many of these orbitals are filled? How many unpaired electrons are indicated in the electron-configuration notation of an iron atom? *(Ans.)* An iron atom contains 1 1s orbital, 1 2s orbital, 3 2p orbitals, 1 3s orbital, 3 3p orbitals, 5 3d orbitals, and 1 4s orbital for a total of 15. Eleven orbitals are filled. There are 4 unpaired electrons.

Practice Problems
1. *(a)* Write the complete and shorthand electron configuration for iodine (I). Write its electron-dot symbol. How many inner-shell electrons does an iodine atom contain? *(Ans.)* $1s^22s^22p^63s^23p^63d^{10}4s^24p^64d^{10}5s^25p^5$, $[Kr]4d^{10}5s^25p^5$, :Ï:, 46

Arrangement of Electrons in Atoms

Sample Problem 4.3 *continued*

(b) How many electron-containing orbitals are in an atom of iodine? How many of these orbitals are filled? How many unpaired electrons are indicated in the electron-configuration notation of an iodine atom?
(Ans.) 27, 26, 1

2. (a) Write the shorthand electron configuration for tin (Sn). Write its electron-dot symbol. How many unpaired electrons are indicated in the electron-configuration notation for tin?
(Ans.) [Kr]$4d^{10}\,5s^25p^2$, ·Ṡn·, 2

(b) How many electron-containing *d* orbitals are there in an atom of tin? Name the element in the fourth period whose atoms have the same number of highest energy level electrons as tin. Write the electron-dot symbol for that element.
(Ans.) 10, Germanium, ·Ġe·

Elements of the Sixth and Seventh Periods

The sixth period, consisting of 32 elements, is much longer than the others. To build up electron configurations for elements of this period, electrons are added first to the 6s orbital, in cesium (Cs) and barium (Ba) atoms. Then, an electron is added to a 5d orbital in lanthanum (La) atoms.

With the next element, cerium (Ce), the 4f orbitals begin to fill, giving cerium atoms the [Xe]$4f^15d^16s^2$ configuration. Electrons are then added to the 4f orbitals in the next 13 elements. The 5d orbitals are next filled and the period is completed by filling the 6p orbitals. Because the 4f and the 5d orbitals are very close in energy, numerous unexpected variations in configurations occur as these orbitals are filling. The electron configurations of the sixth-period elements can be found in the Periodic Table inside the back cover.

The seventh period is incomplete and consists largely of unstable, radioactive elements. These will be discussed in Chapter 29.

Sample Problem 4.4

(a) Write the complete electron configuration of a rubidium atom using the electron-configuration notation and the shorthand electron-configuration notation.
(Ans.) $1s^22s^22p^63s^23p^63d^{10}4s^24p^65s^1$, [Kr] $5s^1$

(b) Write the electron-dot symbol for rubidium and identify the elements in the second, third, and fourth periods that have the same number of highest energy level electrons. (Ans.) Rubidium has one electron in its highest energy level (the fifth). Its electron-dot symbol is, therefore Rb·. The elements with the same outermost configuration are, in the second period, lithium (Li); in the third period, sodium (Na); and in the fourth period, potassium (K).

Practice Problems

1. (a) Write both the complete and shorthand-electron configurations for a barium atom.
(Ans.) $1s^2\,2s^22p^63s^23p^63d^{10}4s^24p^64d^{10}5s^25p^66s^2$, [Xe]$6s^2$

(b) Write the electron-dot symbol for barium and identify the elements in the second, third, fourth, and fifth periods that have the same number of highest-energy level electrons.
(Ans.) ·Ba·, Be, Mg, Ca, and Sr

Sample Problem 4.4 *continued*

2. *(a)* Identify the element whose electron-dot symbol contains five dots in its fifth and highest-energy level. Write its shorthand electron-configuration notation. *(Ans.)* Sb, $[Kr]4d^{10}5s^25p^3$
(b) Name the fourth-period element whose atoms have seven dots in their electron-dot notation. Which element would have that same number of outermost electrons in its sixth main-energy level? *(Ans.)* Bromine, Astatine

Section Review

1. *(a)* What is electron configuration? *(b)* What three principles guide the development of the electron configuration of an atom?
2. What three methods are used to represent the arrangement of electrons in atoms?
3. What is an octet of electrons?
4. Write both the complete and shorthand electron configurations, as well as the orbital and electron-dot notation for *(a)* C *(b)* Ne *(c)* S.
5. Identify each of the indicated elements: *(a)* $1s^22s^22p^63s^23p^3$ *(b)* $[Ar]4s^1$ *(c)* contains four electrons in its third and outer main energy level *(d)* contains one paired and three single electrons in its electron-dot symbol in the fourth and outer main-energy level *(e)* the first element to contain a *d* electron.

Chapter Summary

- Quantum theory was developed early in this century to explain such observations as the photoelectric effect and the emission spectrum of hydrogen.
- Quantum theory states that electrons can exist in atoms only at specified energy levels called orbitals.
- When an electron moves from one orbital to an orbital with lower energy, a photon is emitted whose energy equals the energy difference between the two orbitals.
- An electron in an atom can move from one orbital to a higher energy orbital within the atom only by absorbing an amount of energy exactly equal to the difference between the two orbitals.
- The four quantum numbers that describe the properties of electrons in atomic orbitals are the principal quantum number, the orbital quantum number, the magnetic quantum number, and the spin quantum number.
- Electrons occupy atomic orbitals in the ground state of an atom according to the Aufbau principle, Hund's rule, and the Pauli exclusion principle.
- Electron configurations can be written using three different types of notations. They are orbital notation, electron-configuration notation, and electron-dot notation.
- Some electron configurations of larger atoms such as chromium do not strictly follow the Aufbau principle, but the entire ground-state configuration that results is thought to be of a lower energy than that predicted by the Aufbau principle.

1. *(a)* The arrangement of electrons in atoms *(b)* The Aufbau principle, Hund's rule, and the Pauli exclusion principle
2. Orbital notation, electron-configuration notation, and electron-dot notation
3. The filling of the *s* and *p* sublevels of an atom's highest main energy level with eight electrons
4. The indicated responses are as follows:
(a) $1s^22s^22p^2$, $[He]2s^22p^2$
$\underline{\uparrow\downarrow}\,\underline{\uparrow\downarrow}\,\underline{\uparrow}\,\underline{\uparrow}\,\underline{}$, $\cdot\dot{C}\cdot$
1s 2s 2p 2p 2p

(b) $1s^22s^22p^6$, $[He]2s^22p^6$
$\underline{\uparrow\downarrow}\,\underline{\uparrow\downarrow}\,\underline{\uparrow\downarrow}\,\underline{\uparrow\downarrow}\,\underline{\uparrow\downarrow}$, $:\!\ddot{Ne}\!:$
1s 2s 2p 2p 2p

(c) $1s^22s^22p^63s^23p^4$, $[Ne]3s^23p^4$
$\underline{\uparrow\downarrow}\,\underline{\uparrow\downarrow}\,\underline{\uparrow\downarrow}\,\underline{\uparrow\downarrow}\,\underline{\uparrow\downarrow}$
1s 2s 2p 2p 2p
$\underline{\uparrow\downarrow}\,\underline{\uparrow\downarrow}\,\underline{\uparrow}\,\underline{\uparrow}$, $:\!\ddot{S}\!:$
3s 3p 3p 3p

5. The indicated elements are: *(a)* P *(b)* K *(c)* Si *(d)* As *(e)* Sc

CHAPTER 4 REVIEW

Arrangement of Electrons in Atoms

VOCABULARY

Aufbau principle (112)
continuous spectrum (98)
diamagnetic (111)
electromagnetic radiation (97)
electromagnetic spectrum (98)
electron configuration (112)
excited state (100)
frequency (98)
ground state (100)
highest occupied energy level (113)
Hund's rule (113)
inner-shell electrons (114)
line emission spectrum (103)
magnetic quantum number (100)
noble gases (116)
noble gas configuration (116)
orbital (106)
orbital quantum number (108)
Pauli exclusion principle (113)
photoelectric effect (99)
photon (100)
principal quantum number (108)
quantum (100)
quantum numbers (108)
quantum theory (105)
spin quantum number (111)
wavelength (97)

REVIEWING CONCEPTS

1. a. List five examples of electromagnetic radiation.
   b. What is the speed of all forms of electromagnetic radiation in a vacuum? (4.1)
   c. Relate frequency and wavelength.

2. a. What is a continuous spectrum?
   b. Cite a familiar example of such a spectrum. (4.1)

3. a. What is the electromagnetic spectrum?
   b. What units are used to denote wavelength?
   c. Define the SI unit for frequency. (4.1)

4. What is the frequency and wavelength range of visible light? (4.1)

5. List the colors of light in the visible spectrum in order of increasing frequency. (4.1)

6. In the 20th century, what two properties of light were encountered that could not be explained by wave theory? (4.1)

7. a. Write and label the components of the equation developed by Planck.
   b. How are energy and frequency related? (4.1)

8. Which theory of light, the wave or particle, best explains:
   a. the interference of light
   b. the photoelectric effect
   c. the emission of energy by a hot object? (4.1)

9. a. How can atoms be made to emit electromagnetic radiation?
   b. Cite two illustrations of this phenomenon. (4.1)

10. Distinguish between the ground state and the excited state of an atom. (4.1)

11. Distinguish between a line emission spectrum and a continuous spectrum. (4.1)

12. According to the Bohr model of the hydrogen atom, how is a line emission spectrum produced? (4.1)

13. Cite the two major shortcomings of the Bohr model of the atom. (4.1)

14. a. What is an orbital?
    b. Describe an orbital in terms of an electron cloud. (4.1)

15. Cite the major similarities and differences between Schrödinger's model of the atom and that proposed by Bohr. (4.2)

16. a. What is the principal quantum number?
    b. How is it symbolized?
    c. What are shells?
    d. How does n relate to the energy and distance of a main energy level from the nucleus? (4.2)

17. a. What information is given by the orbital quantum number?
    b. What are sublevels or subshells?

122   Chapter 4

c. What is the relationship between the value of n and the number of orbitals in a given main energy level?
d. List, in order of increasing energy, the first four orbital quantum numbers. (4.2)

18. For each of the following values of n, indicate the numbers and types of sublevels possible for that main energy level:
    a. $n = 1$
    b. $n = 2$
    c. $n = 3$
    d. $n = 4$
    e. $n = 7$
    (number only). (4.2)

19. a. What information is given by the magnetic quantum number?
    b. How many orientations are possible per subshell for the s, p, d, and f orbitals?
    c. Explain and illustrate the notation for distinguishing among the different p orbitals in a subshell. (4.2)

20. a. What is the relationship between n and the total number of orbitals in a main energy level?
    b. How many total orbitals are contained in the 3rd main energy level? in the 5th? (4.2)

21. a. What information is given by the spin quantum number?
    b. What are the possible values for this quantum number?
    c. What is the overriding significance of the spin quantum number? (4.2)

22. How many electrons could be contained in each of the following main energy levels if n is:
    a. 1
    b. 2
    c. 3
    d. 4
    e. 5
    f. 6
    g. 7
    h. n? (4.2)

23. a. What is stated in the Aufbau principle?
    b. Explain the meaning of this principle in terms of an atom with many electrons. (4.3)

24. a. What is stated in Hund's rule?
    b. What is the basis for this rule? (4.3)

25. a. State the Pauli exclusion principle.
    b. What is the significance of the spin quantum number?
    c. How would two electrons in the same orbital compare in terms of the values of their respective four quantum numbers? (4.3)

26. a. What information is provided by electron-dot notation?
    b. What is meant by the highest-occupied energy level in an atom?
    c. What are inner-shell electrons? (4.3)

27. Determine the highest-occupied energy level in:
    a. He
    b. Be
    c. Al
    d. Ca
    e. Sn. (4.3)

28. Write the electron-dot symbol for the unidentified element Z if it contains the following number of highest-energy level electrons:
    a. 3 b. 6 c. 8 d. 5. (4.3)

29. Given the electron configuration for oxygen as $1s^2 2s^2 2p^4$, respond to each of the following:
    a. How many electrons are in each atom?
    b. What is the atomic number of this element?
    c. What is the highest occupied energy level?
    d. Write its orbital notation.
    e. Write its electron-dot symbol.
    f. How many inner-shell electrons does the atom contain?
    g. In which orbital(s) are these inner-shell electrons located? (4.3)

30. Repeat Question 29 for an atom of phosphorus with an electron configuration of $1s^2 2s^2 2p^6 3s^2 3p^3$. (4.3)

31. a. What information is given by the shorthand electron configuration [Ne]$3s^2$?
    b. What element does this represent? (4.3)

32. Write the shorthand electron configurations for
    a. Cl b. Ca. (4.3)

33. a. What are the noble gases?
    b. What is a noble-gas configuration? (4.3)

34. Write both the complete and shorthand electron configurations, as well as the electron-dot symbols, for each of the following:
    a. Na
    b. Al
    c. P
    d. Ar
    e. Br
    f. Sr (4.3)

Arrangement of Electrons in Atoms

35. Identify each atom on the basis of its electron configuration:
a. $1s^2 2s^2 2p^1$
b. $1s^2 2s^2 2p^5$
c. $[Ne]3s^2$
d. $[Ne]3s^2 3p^2$
e. $[Ne]3s^2 3p^5$
f. $[Ar]4s^1$
g. $[Ar]3d^6 4s^2$ (4.3)

PROBLEMS

▶ Photons and Electromagnetic Radiation

1. If the speed of light is 2.998×10^8 m/s, calculate the wavelength of the electromagnetic radiation whose frequency is 7.500×10^{12} Hz.

2. Determine the frequency of light with a wavelength of 4.257×10^{-7} cm.

3. Determine the energy, in joules, of a photon whose frequency is 3.55×10^{17} Hz.

4. What is the frequency of a radio wave with an energy of 1.55×10^{-24} J/photon?

5. Derive an equation expressing E in terms of h, c, and λ, using the two equations $E = h\nu$ and $c = \lambda\nu$.

6. How long would it take a radiowave with a frequency of 7.25×10^5 Hz to travel from Mars to earth if this distance between the two planets is approximately 8.00×10^7 km?

7. When sodium is heated, a yellow spectral line whose energy is 3.37×10^{-19} J/photon is produced.
   a. What is the frequency of this light?
   b. What is its wavelength?

8. Cobalt-60 ($^{60}_{27}$Co) is an artificial radioisotope that is produced in a nuclear reactor for use as a gamma-ray source in the treatment of certain types of cancer. If the wavelength of the gamma radiation from a Co-60 source is 1.00×10^{-2} Å, determine its energy.

▶ Orbitals and Electron Configuration

9. In multielectron atoms, why do $4s$ orbitals fill with electrons before $3d$ orbitals?

10. According to Figure 4-18, list the order in which orbitals fill, from the $1s$ to the $7p$ orbital.

11. Write the shorthand electron configurations for each of the following:
    a. Sc
    b. Fe
    c. Zn
    d. As
    e. Sn
    f. Xe
    g. La
    h. Hf
    i. Hg
    j. Pb
    k. At
    l. Ac
    m. Lr

12. Write the electron-dot symbols for the following elements:
    a. Zn
    b. Sn
    c. Xe
    d. Pb
    e. At

13. Explain the basis for the "deviant" electron configurations of chromium and copper.

CRITICAL THINKING

1. **Applying Models** Explain why the wave theory of light could not account for the following two phenomena:
   a. Visible light is given off when nonflammable substances are heated; first red, then yellow, finally white.
   b. When light of a certain frequency strikes certain metals they emit electrons.

2. **Inferring Relationships** In the emission spectrum of hydrogen shown on page 105, Figure 4-11, each colored line is produced by the emission of photons with specific energies. Substances also produce an absorption spectrum when continuous electromagnetic radiation passes through the substance. Certain wavelengths are absorbed. Using the diagram

Hydrogen Spectrum (Visible light)

Light Source · H₂ Gas · Slit · Prism · Continuous Line Spectrum

below, predict what the wavelengths of the absorption lines for hydrogen will be.

3. **Inferring Conclusions** As an astronomical physicist, you are engaged in the analysis of emission spectra obtained from Hubble Space Telescope transmissions. The spectra are from other solar systems. On one particular spectrum you notice a strange spectral pattern that you are unable to identify. Make an inference as to what is happening and suggest several methods for verifying your inference.

4. **Applying Models** The photoelectric effect describes the emission of electrons from certain metals when they are radiated by a light source. The minimum energy needed to remove an electron from the metal is called the work function and is a characteristic of the metal. For example, Chromium (Cr) metal will emit electrons when the wavelength of the radiation is 284 nm or less. Calculate the work function for chromium. (Hint: You will need to use the two equations that describe the relationships among wavelength, frequency, speed of light, and Planck's constant.)

CROSS CONNECTIONS

1. **Art** Neon signs cost less to operate than ordinary electric lights, and their light penetrates fog better. Recently, artists have begun to express their ideas in neon sculptures. In the southwest, neon lights shaped like cacti are very popular. Gather information on the construction of neon signs and design a neon sculpture for your bedroom. Include all of the information that an artist would need to sculpt your creation.

RESEARCH AND WRITING

1. Prepare a report concerning the photoelectric effect, and cite some of its practical uses. Explain the basic operation of each device or technique mentioned.

2. Research the evolution of solar cells, and cite some of their uses. What are the advantages and disadvantages of using solar cells as an energy source?

PROJECT

1. Prepare a two-column table listing those properties of light that can best be explained by the wave theory in one column, and those best explained by the particle theory in the second column. You may want to consult a physics textbook for reference.

CHAPTER 5
The Periodic Law

PLANNING CHART

Approximate Time: 5/7 days

	SECTION/TEXT FEATURES		RESOURCES	
BASIC	**5.1 HISTORY OF THE PERIODIC TABLE** Demo 1 *Periodic Table* Videodisc Desktop Investigation Section Review Chapter Review • Questions 1–7	127 126B 126G 131 131 157	**Teacher's Resource Binder** Section Review Worksheets **Other Resources** Blackline Master 16	15
BASIC	**5.2 ELECTRON CONFIGURATION AND THE PERIODIC TABLE** Demo 2 *Periodic Table* Videodisc Section Review Chapter Review • Questions 8–19, Problems 1–6	132 126C 126G 121 122	**Teacher's Resource Binder** Section Review Worksheets **Other Resources** World of Chemistry Video 7 Transparency 10	16
AVERAGE	**5.3 ELECTRON CONFIGURATION AND THE PERIODIC PROPERTIES** Demo 3 Section Review Chapter Review • Questions 20–30, Problems 7–13	143 126C 156 157	**Teacher's Resource Binder** Section Review Worksheets **Other Resources** Transparencies 11, 12 Blackline Masters 17, 18, 19, 20, 21, 22, 23	17
	CHAPTER END ACTIVITIES Critical Thinking, Cross Connections, Research and Writing, Projects		**Teacher's Resource Binder** Test Booklet **Computer Test Generator**	21, 23

Teaching Strategies

Introduction

Placing the emphasis on *organized knowledge*, point out that the Periodic Table is the most useful means of organizing or classifying the elements. Throughout the table, the elements tend to follow patterns in their similarities and differences. Thus, the Periodic Table is a valued source of information both for chemists and for scientists in other fields. Mention to the students that perhaps as they observed you teaching chemistry and answering their questions, they often saw you look at the Periodic Table. Perhaps they thought that you were somehow reading the answer from the table because you knew where to look. Then tell them this is partially true; with a limited amount of knowledge, you can obtain a great deal of information from the Periodic Table. Tell them that they too will be able to do this after completing this chapter.

Possibly you have already introduced students to the Periodic Table as a useful tool. If so, ask: In what ways can you already use the Periodic Table as a tool? Then move on to the other things they will be able to do. On a transparency, show the following outline of information that can be obtained from the Periodic Table.

A. Determine:
(1) the number of elements (only on up-to-date tables)
(2) the types of elements (metals, nonmetals, semiconductors, and noble gases)
(3) whether elements are natural or synthetic
(4) the state of the element at a given temperature, usually 20°C, generally indicated by the color of its chemical symbol
(5) the atomic number and atomic weight of an element
(6) the number of energy levels in atoms of different elements
(7) the number of electrons in the outer energy level
(8) the overall electron configuration of an atom

B. Predict:
(1) the physical properties of elements
(2) the chemical properties of elements
(3) the relative sizes of atoms
(4) the relative ionization energies and electron affinities of atoms

Tell the students that this list will continue to grow as more is learned throughout the year.

5.1 History of the Periodic Table

Remind the students that because scientists are always looking for patterns in nature (order in the universe), early chemists were looking for some way to organize the elements. Before the establishment of the Periodic Table, approximately 60 elements were recognized, but no one was able to discern the relationships between them.

Demonstration 1: How Mendeleev Discovered the Periodic Table

PURPOSE: To show the thinking Mendeleev used in figuring out the periodic table

MATERIALS: Transparencies, overhead projector

PROCEDURE: An excellent way to help students understand how Mendeleev made his discoveries is as follows: Prior to class make a transparency with elements known to Mendeleev in the block format shown in Table T5-1.

TABLE T5-1						
Li	Be	B	C	N	O	F
Na	Mg	Al	Si	P	S	Cl
K	Ca	not discovered		As	Se	Br

(Note that Ga, Ge, and the noble gases are left out because they had not been discovered yet; the transition elements in series 4 are omitted because they would confuse the issue. Point out to the students that this is a simplified presentation.) Fill in each block with the element's atomic weight (no atomic numbers, for they had not yet been discovered) and some information that helps show relationships between that element and other elements in its family. See examples in Table T5-2:

TABLE T5-2		
Li 7 Soft, Silvery, active metal 1 atom of Li reacts with 1 atom of Cl	**Be 9.4** Hard, silvery, active metal 1 atom of Be reacts with 2 atoms of Cl	**F 19** Pale yellow gas Very strong nonmetallic properties 1 atom of F reacts with 1 atom of K
Na 23 Soft, silvery, very active metal 1 atom of Na reacts with 1 atom of Cl	**Mg 24** Hard, silvery, active metal 1 atom of Mg reacts with 2 atoms of Cl	**Cl 35.5** Yellow-green gas Strong nonmetallic properties 1 atom of Cl reacts with 1 atom of K

After you have made the transparency, cut the blocks of the 20 elements into individual parts. You now have simulations of Mendeleev's cards. Now mix the blocks so they are in no particular order. In class, tell the students: "Mendeleev listed each of the 63 known elements and their properties on individual cards, which he pinned on his laboratory walls, and started searching for some type of order. Let's see if we can duplicate this process."

Scatter the blocks on the overhead. If there is a Periodic Table in the room, ask the students not to look at it during the exercise. "Let's arrange the elements according to their increasing atomic weights, just as Mendeleev did." While doing this, read the properties on the cards. The students will quickly catch the repetition in the similarity of characteristics.

Be sure to leave spaces for the missing elements just as Mendeleev did. Point out that Mendeleev not only left spaces for these elements but predicted their characteristics. To help them understand how he did this, ask: "What would you predict are the characteristics for Rb? For Sr?" (Point out their position in the Periodic Table.) Mendeleev stuck with his idea of the Periodic Table and not only predicted the existence of missing elements, but accurately predicted their properties.

A simulation that is even more fun, if you have a flair for dramatics, is to announce to the students the day before beginning this chapter that you are going to have a guest speaker the next day. Then, with the aid of the theater department, dress up as Mendeleev. It can be handled in the lecture or interview format. Mendeleev was quite a character, so he lends himself to such a presentation. The students will love it! The publications by Jaffe (1976), McBryde (1974), Trimble (1981), and van Spronsen (1981) shown in "References and Resources" will provide the background necessary for the presentation.

Desktop Investigation: Formulating Your Own Periodic Table

The elements were designated by letters, rather than by their names, so that students would not just look at the present Periodic Table to figure out the answers. To keep the exercise simple, only a limited amount of information is given about each element, and the transition elements in Period 3 are left out. The information given was probably of the type available to Mendeleev; this is why the average atomic masses are whole numbers. All elements listed were known to Mendeleev at the time of his work. You may also suggest to the students that they make up fictitious names and chemical symbols for their elements; they might use their last names, such as Smithium or Jacksonium. They could design their own cards, or you could make photocopies or dittoed cards with the format shown in Table T5-3.

TABLE T5-3
Name of element _____ Average atomic mass _____ Type _____
State _____ Density _____ Boiling point _____ Melting point _____
Other physical properties _____
Chemical formulas of compounds _____ _____ _____
Other useful information _____

As students are working on their cards, you might wish to show them actual samples of the elements, if it is safe to do so, and pictures of the dangerous ones.

After students have completed their Periodic Table and determined the actual names of the elements, they might look up additional information about the element in their textbook, encyclopedia, or the *CRC Handbook of Chemistry and Physics*. Table T5-4 presents a general format for the Periodic Table of the elements given in this excercise.

The Periodic Law **126B**

TABLE T5-4 A SIMULATION OF MENDELEEV'S PERIODIC TABLE						
Element G Lithium	Element D Beryllium	Element L Boron	Element Q Carbon	Element P Nitrogen	Element N Oxygen	Element K Fluorine
Element M Sodium	Element B Magnesium	Element E Aluminum	Element I Silicon	Element S Phosphorous	Element J Sulfur	Element C Chlorine
Element A Potassium	Element F Calcium	Element X Not found yet	Element Y Not found yet	Element O Arsenic	Element H Selenium	Element R Bromine

Answers to Discussion Questions

1. (a) Atomic numbers were not discovered until about 45 years later. (b) See Table T5-4.
2. Seven families and three periods
3. Element X will be a silver metal with an average atomic mass of about 69–70. When combined with chlorine, it will have the formula XCl_3; when combined with oxygen, it will have the formula X_2O_3. Element Y will be a silver metal with an average atomic mass of about 71–72. When combined with chlorine, it will have the formula YCl_4; when combined with oxygen, it will have the formula YO_2.

5.2 Electron Configuration and the Periodic Table

Demonstration 2: Reactivity of the Alkaline–Earth Metals

Purpose: To compare the reactions of water with magnesium and calcium
Materials: Calcium, magnesium ribbons, phenolphthalein, distilled water, 2 medium-sized test tubes, sand paper.
Caution: Wear goggles and lab coat at all times.
Procedure: Put 10 mL of water into each of the two test tubes. Add two drops of phenolphthalein to each tube. To one test tube, add a 5-cm piece of freshly sand-papered magnesium ribbon rolled into a coil. To a second test tube, add a piece of calcium about equal to $\frac{1}{2}$ of a pea. (The magnesium reacts very slowly, forming hydrogen gas bubbles and turning the water-phenolphthalein light pink. The calcium is more reactive, fizzes, and turns the water-phenolphthalein dark pink.)
The equations for these reactions are:

$$Mg + 2H_2O \rightarrow Mg(OH)_2 + H_2(g)$$
$$Ca + 2H_2O \rightarrow Ca(OH)_2 + H_2(g)$$

Ask students what they think would happen if a small piece of strontium were added to the water solution. (It would be even more reactive than Ca, but similar in that it turns the water solution dark pink and fizzes rapidly as it produces hydrogen.)
The alkali metals also react with water to form hydrogen and are even more reactive than the alkaline-earth metals. This reactivity often prohibits the use of the alkali metals in class demonstrations. In class discussion, you can compare the reactivities of the alkali metals with the alkaline-earth metals.
Students sometimes have difficulty understanding the definition of d-block elements. A transparency with orbital notation drawings similar to Table T5-5 will help. Discuss additional transition elements, such as Y, Zr, and Nb.
Point out that the electron configurations of these elements are all the same except for the d-sublevel in the next-to-outermost energy level.

TABLE T5-5 The d-BLOCK ELEMENTS OR TRANSITION ELEMENTS						
	1s 2s 2p		3s 3p		3d	4s
Mn	↑↓ ↑↓ ↑↓ ↑↓ ↑↓		↑↓ ↑↓ ↑↓ ↑↓		↑ ↑ ↑ ↑ ↑	↑↓
Fe	↑↓ ↑↓ ↑↓ ↑↓ ↑↓		↑↓ ↑↓ ↑↓ ↑↓		↑↓ ↑ ↑ ↑ ↑	↑↓
Co	↑↓ ↑↓ ↑↓ ↑↓ ↑↓		↑↓ ↑↓ ↑↓ ↑↓		↑↓ ↑↓ ↑ ↑ ↑	↑↓

The sources for three good puzzles that will help familiarize students with the Periodic Table are listed in "References and Resources": Nelson (1984), Nelson (1985), and Wolfgram (1987).

5.3 Electron Configuration and Periodic Properties

Demonstration 3: Ionization Energy

Purpose: To show that the farther away an electron in the outer energy level is from the nucleus, the smaller amount of energy needed to remove it
Materials: The teacher and two students
Procedure: One student will act as a nucleus, and the teacher plays the part of the electron in the outer energy level. The second student functions as the ionization energy. For an electron close to the nucleus (1st or 2nd energy level), grip the student's arm close to but below the elbow and have him or her do the same. Hold tight! Then have the second student try to dislodge you by pushing on your shoulder. (Fairly gently—of course!) Change the grip to wrist-to-wrist (3rd and 4th energy level), hand-shake hold for 5th energy level, and hold curled tips of fingers for 6th or 7th energy level. At the 6th or 7th energy level, very little pushing is needed to remove the "electron."
To help students understand + and − electron affinities, liken the concept to buying and selling. Those atoms (like Cl) that really need another electron will pay a lot for it. Those (like Na) that are really not interested in getting another e^-, but that have space in their orbital, will pay a small amount. Those (like Mg) that already have a small amount of stability in their filled s sublevel or with total stability (like Ne) would say: "What do you mean—do I want to buy an e^-? You'll have to *pay* me to take one!"

References and Resources

Books and Periodicals

Ciparick, Joseph. "Element X," *ChemMatters*, Vol. 5, 1987, pp. 8–9.

Fernelius, W. Conard. "Some Reflections on the Periodic Table and Its Use," *Journal of Chemical Education*, Vol. 63, 1986, pp. 263–66.

Guenther, William B. "An Upward View of the Periodic Table," *Journal of Chemical Education*, Vol. 64, 1987, pp. 9–10.

Hogg, John T. "Moseley at Oxford," *Journal of Chemical Education*, Vol. 52, 1975, pp. 325–26.

Jaffe, Bernard. "Mendeleev—Siberia Breeds a Prophet," *Crucibles: The Story of Chemistry*, Dover Publications, New York, 1976.

Krishnan, C.V. "The New Format for the Periodic Table of Elements: Concerns of a High School Chemistry Teacher," *Journal of Chemical Education*, Vol. 64, 1987, p. 558.

Mason, Joan. "Periodic Contractions Among the Elements or On Being the Right Size," *Journal of Chemical Education*, Vol. 65, 1988, pp. 17–20.

McBryde, J.W. "The Periodic Law," *CHEM13*, 1974, p. 2.

Nelson, Bronwyn. "The Elements and Their Organization," *ChemMatters*, Vol. 2, 1984, p. 16.

Nelson, Bronwyn. "Periodically Puzzling," *ChemMatters*, Vol. 3, 1985, p. 16.

Ratzlaff, Becky. "Love Always, Francium," *ChemMatters*, Vol. 3, 1985, p. 13.

Sanderson, R.T. "Principles of Electronegativity, Part 1. General Nature," *Journal of Chemical Education*, Vol. 65, 1988, pp. 112–18.

Saturnelli, Annette. "Setting the Periodic Table," *The Science Teacher*, Vol. 52, 1985, pp. 46–49.

Strong, Frederick. "Revised Atomic Form Periodic Table," *Journal of Chemical Education*, Vol. 62, 1985, p. 456.

Trimble, R. F. "Mendeleev's Discovery of the Periodic Law," *Journal of Chemical Education*, Vol. 58, 1981, p. 28.

van Spronsen, Jan W. "Mendeleev as a Speculator," *Journal of Chemical Education*, Vol. 58, 1981, pp. 790–91.

Wolfgram, Dale. "Family Resemblance," *ChemMatters*, Vol. 5, 1987.

Audiovisual Resources

Chemistry in Today's World Series—Metals and the Halogens (Slide sets + cassette) Educational Images, 1988.

Experiments in Chemistry Series—Chemistry of Columns I & II and Chemistry of Halogens (Filmstrips) Educational Images, 1988.

Periodic Table (VHS) Learning Arts, 1988. "The Periodic Table Videodisk: Reactions of the Elements," Project Seraphim/JCE Software.

Computer Software

Chemistry Courseware—Periodic Table, Apple II 64K, Barclay, 1988.

The Chemistry Help Series—Periodic Table, Apple II 48K, Barclay, 1988.

Chemistry Series—Periodic Table, Apple II 48K, TRS-80 III & IV, 32K, Barclay, 1988.

General Chemistry 1A—Topic 4: The Periodic Table, Apple II 128K, Queue, 1988.

Journal of Chemical Education: Software. "The One Computer Classroom," (Apple II).

"Discoverer: Exploring the Properties of the Chemical Elements," (MS-DOS/PC). Available from Project Seraphim

Periodic Table Seraphim Vol. 3, Disk 6, Project Seraphim, 1988.

Fee, Richard. *Periodic Table*, Apple 48K. Chemistry According to ROF, 1988.

Nagel, Edgar. *The Periodic Table: A Computer-Assisted Lecture Aid*, Apple II 48K, Queue, 1988.

Smith, Stanley. *The Elements*, Apple II 48K and IBM PC 128K, Queue, 1988.

Answers and Solutions

Reviewing Concepts

1. According to the periodic law, when the elements are arranged in order of increasing atomic number, elements with similar properties recur at regular intervals.
2. The Periodic Table is an arrangement of elements in order of their atomic numbers such that elements with similar properties fall in the same column.
3. **a.** The lanthanides are the 14 elements with atomic numbers from 58 (Ce) to 71 (Lu) in Period 6. The actinides are the 14 radioactive elements with atomic numbers from 90 (Th) to 103 (Lr) in Period 7.
   b. The rare earth elements are the 14 lanthanides.
4. Periodicity with respect to atomic number can be observed in any group of elements. The atomic number of each successive element in a group is 8, 8, 18, 18, and 32 higher than that of the preceding element, depending on the number of elements in the group.
5. The lack of chemical reactivity among the noble gases is a result of the fact that the highest-energy levels are fully occupied in noble-gas atoms.
6. **a** For Groups 1, 2, and 13–18, the outer electron configurations of all members of the group are identical.
   b. Many of the properties of elements are based on their outermost electron configurations. Inner-shell electrons have little influence on properties.
7. The length of each period in the table is determined by the number of sublevels that can be filled with electrons.
8. The period in which an element appears in the table corresponds to its highest occupied main energy level.
9. **a.** The letter of the block indicates the type of sublevel that is being filled in successive elements of that block.
   b. The *s*-block consists of the elements in Groups 1 and 2, the *d*-block encompasses Groups 3–12, the *p*-block is made up of Groups 13–18, and the *f*-block consists of the elements in Periods 6 and 7 that are located between Groups 3 and 4.
10. **a.** The alkali metals are the elements in Group 1 (Li, Na, K, Rb, Cs, and Fr).
    b. They are extremely reactive metals, react with increasing vigor with water (and thus must be stored under kerosene), are silvery in appearance, are soft enough to be cut with a knife, and impart bright colors to a flame.

c. Sodium is used in nuclear reactors and in sodium-vapor lamps, and cesium is used in photoelectric cells such as those that control supermarket doors.
11. a. The alkaline-earth metals are the Group 2 elements (Be, Mg, Ca, Sr, Ba, and Ra).
b. These elements are harder, denser, stronger, and have higher melting points than the alkali metals, but are less reactive.
c. Magnesium is used to make strong, lightweight alloys, and compounds of strontium are used in highway flares.
12. a. The usual group configurations are:
Group 3: $(n-1)d^1ns^2$ Group 8: $(n-1)d^6ns^2$
Group 4: $(n-1)d^2ns^2$ Group 9: $(n-1)d^7ns^2$
Group 5: $(n-1)d^3ns^2$ Group 10: $(n-1)d^8ns^2$
Group 6: $(n-1)d^4ns^2$ Group 11: $(n-1)d^9ns^2$
Group 7: $(n-1)d^5ns^2$ Group 12: $(n-1)d^{10}ns^2$
b. The sum of the outer s and d electrons is equal to the group number.
13. a. The d-block elements are sometimes referred to as transition elements.
14. a. The p-block elements are diverse; they consist of non-metals at the right, metalloids in the middle, and metals at the left.
b. The p-block metals are generally harder and more dense than the s-block metals, but softer and less dense than the d-block metals.
15. a. The halogens are the elements of Group 17 (F, Cl, Br, I, and At).
b. The halogens are the most reactive of the nonmetals. They react vigorously with most metals to form salts. At room temperature, F_2 and Cl_2 are gases, Br_2 is a red liquid, and I_2 is a purple solid.
16. a. The noble gases are the Group 18 elements (He, Ne, Ar, Kr, Xe, and Rn).
b. These elements are the least reactive of all elements.
c. Since they display so little reactivity, noble gases are used instead of air in containers of materials that would react with oxygen or moisture in air.
17. a. The metalloids include B, Si, Ge, As, Sb, and Te.
b. The metalloids are mostly brittle solids with electrical conductivities intermediate between those of metals (good conductors) and nonmetals (nonconductors).
18. a. The f-block consists of the lanthanide elements of Period 6 and the actinide elements of Period 7.
b. The lanthanides are shiny reactive metals used in color-television tubes. The actinides are unstable radioactive elements, many of which are laboratory-made and thus rarely encountered in everyday chemistry.
19. a. The main group elements are the elements of the s and p blocks.
b. Among the observable trends across the main-group elements are those involving atomic size, ease with which atoms gain and lose electrons, sizes of ions, numbers of electrons involved in the formation of chemical compounds, and electron-attracting abilities.
20. a. The atomic radius is one half the distance between the nuclei of identical combined atoms.
b. There is a gradual decrease in atomic radii across periods.
c. As electrons are added to s and p sublevels in the same main energy level, the increasing positive charge of the nucleus gradually pulls electrons closer to the nucleus, thus resulting in decreasing atomic radii.
21. a. The atomic radii of the main group elements generally increase down a group.
b. In the atoms of each element, the outer electrons occupy comparable sublevels in successively higher main energy levels and are therefore increasingly far from the nucleus.
22. a. An ion is an atom or group of atoms that has a positive or negative charge.
b. Ionization refers to any process that results in the formation of an ion.
c. Ionization energy is the energy required to remove one electron from an atom of an element.
23. a. In general, first ionization energies increase across a period and decrease down a group.
b. Across a period the increasing nuclear charge more strongly attracts electrons in the same energy level. Down a group, the electrons to be removed from each successive element are in increasingly higher energy levels farther from the nucleus, and are thus more easily removed.
24. a. Electron affinity is the energy absorbed or released when an electron is added to an atom.
b. The electron affinity is negative for an exothermic process, and positive for an endothermic process.
25. a. A cation is a positive ion, and an anion is a negative ion.
b. Cations are always smaller than the atoms from which they were formed, whereas anions are always larger than the neutral atoms.
26. Chemical compounds form because electrons are lost, gained, or shared between atoms of different elements.
27. a. Valence electrons are the electrons available to be lost, gained, or shared in the formation of chemical compounds.
b. Valence electrons are generally located in incompletely filled main energy levels.
28. The loss or gain and the number of electrons most likely to be involved in each case is:
a. lost, 1
b. lost, 2
c. lost, 3
d. gained, 2
e. gained, 1
f. neither lost nor gained, 0.
29. a. Electronegativity is the tendency of an atom in a chemical compound to attract shared electrons.
b. The most electronegative element, fluorine, is arbitrarily assigned an electronegativity value of 4.0, and the values for the other elements are calculated in relation to this value.
30. The most electronegative group is the halogens, whereas the least electronegative is the alkali metals.

Problems

1. a. [He]$2s^1$; 2   c. [Ne]$3s^2$; 3
   b. [He]$2s^22p^4$; 2   d. [Ne]$3s^23p^3$; 3

e. [Ar]$3d^{1}4s^{2}$; 4   g. [Kr]$4d^{10}5s^{2}5p^{2}$; 5.
   f. [Ar]$3d^{10}4s^{2}4p^{5}$; 4
2. The information in each case is:
   a. Group 2, Period 2, s-block, Be
   b. 1, 3, s-block, Na
   c. 2, 5, s-block, Sr
   d. 2, 4, s-block, Ca
   e. 1, 7, s-block, Fr.
3. The indicated period, block, and group are:
   a. Period 4, d-block, Group 4
   b. 4, d-block, 6
   c. 5, d-block, 11
   d. 6, d-block, 10.
4. The indicated outer configurations are:
   a. $3d^{5}4s^{2}$
   b. $4d^{1}5s^{2}$
   c. $4f^{14}5d^{10}6s^{2}$.
5. The indicated period, block, and group for each is:
   a. Period 3, p-block, Group 16
   b. 5, p-block, 14
   c. 6, p-block, 17.
6. The indicated responses are:
   a. p-block, Period 3, Group 13, no name, aluminum, metal, reactive
   b. p-block, 4, 18, noble gases, krypton, nonmetal, unreactive
   c. d-block, 5, 11, transition elements, silver, metal, slightly reactive
   d. f-block, 6, no group number assigned, lanthanides, cerium, metal, reactive.
7. a. The first ionization energy is the energy required to remove one electron from an atom, whereas the energies required to remove successive electrons from the resulting ions are referred to as the second, third, etc., ionization energies of that atom.
   b. The values of successive ionization energies increase.
   c. This occurs because each successive electron must be removed from a particle with an increasingly larger positive charge.
8. In order of decreasing electron affinities, the elements are F, O, C, Li, Na, Rb, and Ne.
9. a. The elements in order of decreasing first ionization energies are Ne, F, O, C, Li, and K.
   b. Among the elements listed, Li would be expected to have the highest second ionization energy because Li$^+$, like K$^+$ has a noble gas configuration, but is smaller than K$^+$.
10. The indicated charge and corresponding noble gas in each is:
    a. 1+, He
    b. 3+, He
    c. 2−, Ne
    d. 1−, Ne
    e. 2+, Ne
    f. 3+, Ne
    g. 3−, Ar
    h. 2−, Ar
    i. 1−, Kr
    j. 2+, Xe.
11. Ca$^+$ does not have a noble-gas configuration.
12. K^{2+} is least likely to form.
13. Ca^{2+} would have the smaller radius because of the greater attraction its 20 protons would exert on the 18 electrons.

Critical Thinking

1.

			4 Δ	
		9 Λ		12 γ
13 β				
			23 ε	

2. Atomic number 5
3. Atomic number 8
4. From left to right the elements are organized according to their atomic numbers. Elements are organized into families or columns according to their properties. The number of electrons in the outer energy level is the same for each element in a column.

Cross Connection

1. Answers will vary: Mechanics use charts of car parts, optometrists use eye charts, and motor vehicle department officers use traffic sign charts to test driver's license applicants.

Videodisc Selections from
The Periodic Table Videodisc (JCE Software)

step
forward

step
reverse

CHAPTER 5

pg 136 Alkali and alkaline earth metals

268
lithium (s)

1058
lithium + water (m)

274
lithium + air (m)

4565
sodium (s)

5104
sodium + water (m)

4578
sodium + air (m)

16826
potassium (s)

17334
potassium + water (m)

16838
potassium + air (m)

37287
rubidium (s)

46095
cesium (s)

36343
uses of beryllium (s)

7523
magnesium (s)

7535
magnesium + air (m)

36445
uses of magnesium (s)

18426
calcium (s)

19244
calcium + water (m)

18438
calcium + air (m)

36555
marble (s)

37306
strontium (s)

37311
strontium + air (m)

38127
strontium + water (m)

46113
barium (s)

46119-46843
barium + air (m)

46844
barium + water (m)

pg 137 Hydrogen and helium

109
hydrogen (s)

262
helium (s)

pg 138 Transition elements

20739
scandium (s)

20754
titanium (s)

21231
chromium (s)

21242
chromium + air (m)

22055-22751
chromium + acids (m)

22760
manganese (s)

22764
manganese + air (m)

23281-24071
manganese + acids (m)

24079
iron (s)

24084
iron + air (m)

The Periodic Law **126H**

24981
iron + acids (m)

25876
cobalt (s)

25881
cobalt + air (m)

26677-27398
cobalt + acids (m)

27399
nickel (s)

27411
nickel + air (m)

28046
nickel + acids (m)

28534
copper (s)

28546
copper + air (m)

29501
copper + acids (m)

30270
zinc (s)

30274
zinc + air (m)

31167
zinc + acids (m)

39162
molybdenum (s)

39192
silver (s)

39821
cadmium (s)

39826
cadmium + air (m)

40532
cadmium + acids (m)

49853
tungsten (s)

49863
platinum (s)

49875
gold (s)

49888
mercury (s)

49893
mercury + air (s)

49899
mercury + acids (m)

pg 140 p-block elements

3816
boron (s)

8255
aluminum (s)

36462
uses of aluminum (s)

32417
gallium (s)

41154
indium (s)

50708
thallium (s)

3822
carbon (s)

11574
silicon (s)

32429
germanium (s)

41160
tin (s)

50713
lead (s)

4537
nitrogen (s)

12307
phosphorus (s)

36482
uses of phosphorus (s)

33218
arsenic (s)

42664
antimony (s)

51683
bismuth (s)

4543
oxygen (s)

The Periodic Law

36410
uses of oxygen (s)

13133
sulfur (s)

36507
uses of sulfur (s)

33878
selenium (s)

43477
tellurium (s)

14228
chlorine (s)

35178
bromine (s)

45067
iodine (s)

Teacher's Notes

Chapter 5
The Periodic Law

SECTION PREVIEW

5.1 History of the Periodic Table

5.2 Electron Configuration and the Periodic Table

5.3 Electron Configuration and Periodic Properties

Periodic patterns repeat at specific intervals like the motifs in this artwork, Symmetry Drawing E 85. *(©1952 M. C. Escher/Cordon Art/ Baarn Holland)*

5.1 History of the Periodic Table

From our present viewpoint, it is hard to imagine the confusion among chemists during the middle of the nineteenth century. By 1860, more than 60 elements had been discovered. Chemists had to learn the properties of these elements as well as those of the many compounds that they formed—a difficult task. To make matters worse, there was no method for accurately determining atomic masses or the numbers of atoms of different elements combined in chemical compounds. Different chemists used different atomic masses for the same elements, which resulted in their determining different compositions for the same compounds. This made it difficult for one chemist to understand the reported results of another. The development of the Periodic Table and an understanding of atomic masses helped eliminate some of this confusion.

Mendeleev and Chemical Periodicity

In September 1860, a group of chemists assembled at the First International Congress of Chemists in Karlsruhe, Germany. The goal of the Congress was to review various scientific matters on which there was little agreement. One topic on the agenda was the measurement of atomic masses. Another was how to determine the composition of compounds by using atomic masses. Chemists hoped that communication on these matters, which had become almost impossible, could be improved.

At the Congress, the Italian chemist Stanislao Cannizzaro (1826–1910) presented a method for measuring atomic masses and interpreting the results of the measurements. Cannizzaro's method enabled scientists to agree on standard values for atomic masses. Scientists then began to search for generalizations. Could any relationships be found among atomic masses and other properties of the elements? Could predictions about such properties be made based on these atomic masses?

Dmitri Mendeleev (1834–1907), a Russian chemist, had heard about the new atomic masses discussed at the Karlsruhe Congress. Mendeleev was writing a chemistry textbook and decided to include the new values. In his book, Mendeleev was hoping to organize the elements according to their properties. He went about this much as you might organize information for a research paper. He placed the name of the known element on a card together with the atomic mass of the element and a list of its observed properties. He then arranged the cards according to various properties and looked for trends or patterns.

Mendeleev noticed that when the elements were arranged in order of increasing atomic mass, similar properties appeared at regular intervals. Such a repeating pattern is referred to as "periodic." The second hand of a watch, for example, passes over any given mark at periodic, 60-second intervals.

Mendeleev created a table in which elements with similar properties were grouped together—a periodic table of the elements. His first periodic table, shown in Figure 5-2, was published in 1869.

Note that Mendeleev placed iodine (I, atomic mass 127) after tellurium (Te, atomic mass 128), even though this meant not listing the elements in

SECTION OBJECTIVES

- Explain the role of Mendeleev in the development of the Periodic Table.
- Explain how the periodic law can be used to predict the physical and chemical properties of elements.
- Describe the modern Periodic Table.
- Describe how the elements belonging to a group of the Periodic Table are related by atomic number.

In the nineteenth century and for a long time afterward, the masses of atoms were referred to as "atomic weights."

Cannizzaro's method, which requires measuring the densities of gases, is discussed in Chapter 12.

Figure 5-1 The human heartbeat is an example of a repeating pattern. The periodicity of this pattern is especially evident on an electrocardiogram.

The Periodic Law 127

Figure 5-2 Mendeleev's first published periodic table. The periods are the vertical columns and the groups are the horizontal rows (the reverse of the modern Periodic Table). The numbers after the symbols of the elements are the atomic masses accepted by Mendeleev. The question marks at atomic mass 45, 68, and 70 were later identified as the elements scandium, gallium, and germanium.

Mendeleev, surprisingly, never won a Nobel Prize; the monumental importance of his work was not recognized in his lifetime. In 1906, the year before his death, he was nominated for the Nobel Prize in chemistry but lost by one vote to Henri Moissan (1852–1907), who discovered fluorine.

Moseley shot high-speed electrons at different metal targets in an evacuated tube, producing X rays of different frequencies. When he graphed the square roots of the frequencies against the numbers representing the elements' positions on the periodic chart, he obtained a straight line. He called the number representing each element's position "the atomic number." Later it was discovered that this number was the number of positive charges or protons found in a neutral atom of that element.

order of increasing atomic mass. In this way, tellurium could be placed in the same group as elements with similar properties such as oxygen (O), sulfur (S), and selenium (Se). Iodine could also, then, be placed in the same group as fluorine (F), chlorine (Cl), and bromine (Br), which have properties similar to those of iodine.

Mendeleev's procedure left several empty spaces in his periodic table. In 1871, Mendeleev boldly predicted the existence and properties of the elements that would fill three of these spaces. By 1886, all three had been discovered. Their properties were found to be strikingly similar to those predicted by Mendeleev, as shown in Table 5-1. The success of his predictions persuaded most chemists to accept Mendeleev's periodic table and gave him historical credit as the discoverer of periodicity. Two questions remained, however: *(1)* Why was it that most elements could be arranged in the order of increasing atomic mass, but a few could not? *(2)* What is the reason for chemical periodicity? Careful experimentation, observation, and analysis of data would eventually reveal the answers to these questions as you will discover in the remainder of this chapter.

Moseley and the Periodic Law

The first question was not answered until over forty years after Mendeleev's first periodic table had been published. In 1911, the English scientist Henry Gwyn-Jeffreys Moseley (1887–1915) discovered why a few elements could not be arranged according to their atomic masses. Moseley, who was working with Rutherford, performed a series of experiments on 38 different metals. By analyzing his data, he discovered a previously unrecognized pattern: The positive charge of the nucleus increases by one unit from one element to the next when the elements are arranged as they are in the Periodic Table.

To emphasize that the Modern Periodic Table arranges elements according to increasing atomic numbers, rather than masses, have students examine the atomic numbers on the Periodic Table to see if any are out of numerical sequence. Then have them do the same with atomic masses. They will find that Ar and K, Co and Ni, and Te and I are reversed in terms of atomic mass, but all elements are in order in terms of atomic numbers.

TABLE 5-1 PROPERTIES OF ELEMENTS PREDICTED BY MENDELEEV

	Mendeleev's Predictions	Modern Values
	Eka*-aluminum	Gallium (discovered 1875)
Atomic mass	68	69.7
Density (g/cm^3)	5.9	5.9
Melting point	Low	29.8°C
Oxide formula	E$_2$O$_3$	Ga$_2$O$_3$
Oxide density (g/cm^3)	5.5	5.9
	Eka-boron	Scandium (discovered 1879)
Atomic mass	44	44.96
Oxide formula	E$_2$O$_3$	Sc$_2$O$_3$
Oxide density (g/cm^3)	3.5	3.9
	Eka-silicon	Germanium (discovered 1886)
Atomic mass	72	72.6
Density (g/cm^3)	5.5	5.3
Oxide formula	EO$_2$	GeO$_2$
Oxide density (g/cm^3)	4.7	4.2

*Mendeleev used the prefix *eka* to designate certain then-unknown elements that he assumed would follow in order beyond a specified element. *Eka* is delivered from a Sanscrit word meaning "one" or "one beyond."

Moseley's work led to both the modern definition of atomic number and the recognition that atomic number, not atomic mass, is the basis for the organization of the Periodic Table.

Moseley's studies provided experimental justification for Mendeleev's ordering of the Periodic Table by properties rather than strictly by atomic mass. For example, the order Mendeleev chose for tellurium and iodine is justified by the atomic numbers of 52 for tellurium and 53 for iodine. A similar reversal of argon (Ar, atomic mass 39.95, atomic number 18) and potassium (K, atomic mass 39.10, atomic number 19) can be seen in the modern Periodic Table.

Today, Mendeleev's principle of chemical periodicity is correctly stated in what is known as the **periodic law:** *the physical and chemical properties of the elements are periodic functions of their atomic numbers.* In other words,

STUDY HINT

Atomic numbers were introduced in Section 3.3.

The Periodic Law **129**

> **STUDY HINT**
> Locate the noble gases in the Periodic Table inside the back cover.
>
> Helium was discovered in 1868 by Pierre Janssen (1824–1907) as he was studying a solar eclipse. However, helium was not accepted as an element until it was discovered that it also exists on earth.
>
> Ramsay received the 1904 Nobel Prize in Chemistry for his discovery of the noble gases.
>
> **STUDY HINT**
> Scandium and yttrium are sometimes considered rare earth elements since their properties are similar.

when the elements are arranged in order of increasing atomic number, elements with similar properties recur at regular intervals.

The Modern Periodic Table

The periodic table has undergone extensive change since Mendeleev's time. Chemists have discovered new elements and, in more recent years, synthesized them in the laboratory. Each of the more than forty new elements, however, can be placed in a group of other elements with similar properties. *The* **Periodic Table** *is an arrangement of the elements in order of their atomic numbers so that elements with similar properties fall in the same column.*

The noble gases Perhaps the most significant addition to the Periodic Table came with the discovery of the noble gases. Argon (Ar) was discovered in 1894 by the English physicists John William Strutt (Lord Rayleigh; 1842–1919) and William Ramsay (1852–1916). Although argon makes up about 1 percent of the atmosphere, it had escaped notice because of its total lack of chemical reactivity. Helium (He), another noble gas, had been detected earlier using the spectrum of sunlight. In 1894 Ramsay discovered that helium also exists on earth.

In order to fit argon and helium into the Periodic Table, Ramsay found it necessary to propose a new group. He placed this group between the groups now known as Group 17 (the fluorine family) and Group 1 (the Lithium family). In 1898 Ramsay discovered two more noble gases, krypton (Kr) and xenon (Xe). The final noble gas, radon (Rn), was discovered in 1900 by the German scientist Friedrich Ernst Dorn (1848–1916).

The next step in the development of the Periodic Table was completed in the early 1900s when the puzzling chemistry of the "rare earth" elements was finally understood. Because the rare earths are so similar in properties, their separation and identification required the efforts of many chemists during the next 100 years. Today, the 14 different rare-earth elements are known as the lanthanides. *The* **lanthanides, or rare-earth elements,** *are the 14 elements with atomic numbers from 58 (cerium, Ce) to 71 (lutetium, Lu).*

The actinide elements Another landmark in the development of the Periodic Table was the discovery of the actinide elements, all of which are radioactive. *The* **actinides** *are the 14 elements with atomic numbers from 90 (thorium, Th) to 103 (lawrencium, Lr).* The lanthanides and actinides belong in periods 6 and 7, respectively, of the Periodic Table between the elements of Groups 3 and 4.

Periodicity Periodicity with respect to atomic number can be observed in any group of elements. Consider, for example, the noble gases of Group 18. The first noble gas is helium, which has an atomic number of 2. The elements following helium in atomic number have completely different properties until the next noble gas, neon (Ne), which has an atomic number of 10, is reached. The remaining noble gases in order of increasing atomic number are argon (atomic number 18), krypton (atomic number 36), xenon (atomic number 54), and radon (atomic number 86). The difference in atomic number between each noble gas is shown in Figure 5-3. Below the noble gases in Figure 5-3 are the elements of Group 1. These are all silvery metals of low density that react strongly with water. As you can see, the differences in atomic number between the Group 1 metals follow exactly the same pattern

Figure 5-3 Comparison of the differences in atomic numbers for Group 18 and Group 1. The differences between successive elements is 8, 8, 18, 18, and 32. Groups 1, 2, and 13–18 follow a similar pattern.

Desktop Investigation

Designing Your Own Periodic Table

Question
Using information similar to that available to Mendeleev, can you design your own periodic table?

Materials
index cards

Procedure
1. Write the information available for each element on an index card. The following information is appropriate: a letter of the alphabet (A, B, C, etc.) to identify each element; average atomic mass; state; density; melting point; boiling point; and some other readily observable physical properties. Do not write the name of the element on the index card, but keep a separate list of the letters you have assigned to each element.

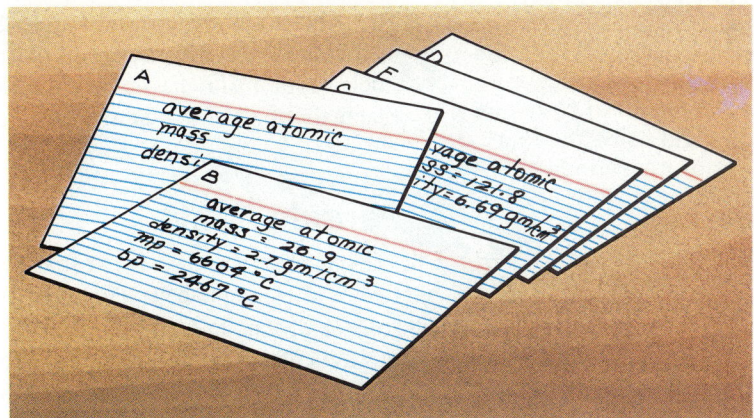

2. Organize the cards for the elements in a logical pattern as you think Mendeleev might have done.

Discussion
1. Keeping in mind that the information you have is similar to that available to Mendeleev in 1869, (a) why are atomic masses given instead of atomic numbers? (b) can you identify each element by name?
2. How many families or groups are in your periodic table? How many periods or series?
3. Predict the characteristics of any missing elements. Once you have finished, check your work using your separate list of elements and a Periodic Table.

---

as for the noble gases. Starting with the first member of Groups 13–17 a similar pattern is repeated: The atomic number of each successive element is 8, 18, 18, and 32 higher than that of the element above it.

In Section 5.2, you will see that the second mystery presented by Mendeleev's periodic table—the reason for periodicity—is explained by the arrangement of the electrons around the nucleus.

Section Review

1. Who is credited with each of the following: (a) development of a method that led to the determination of standard atomic masses (b) discovery of chemical periodicity (c) development of atomic numbers as the basis for organization of the Periodic Table?
2. State the periodic law.
3. Name three sets of elements added to the Periodic Table after Mendeleev's time.
4. How do the atomic numbers of the elements within Groups 1, 2, and 13–18 of the Periodic Table differ?

1. (a) Cannizzaro (b) Mendeleev (c) Moseley
2. The physical and chemical properties of the elements are periodic functions of their atomic numbers.
3. The noble gases, the lanthanide elements, and the actinide elements
4. The atomic numbers of successive elements differ by 8, 8, 18, 18, and 32, in that order.

The Periodic Law

5.2 Electron Configuration and the Periodic Table

SECTION OBJECTIVES

- Describe the relationship between electrons in sublevels and the length of each period.
- Locate and name the four blocks of the Periodic Table. Explain the reasons for these names.
- Discuss the relationship between group configurations and group numbers.
- Describe the location in the Periodic Table and the general properties of the alkali metals, the alkaline-earth metals, the halogens, and the noble gases.

The noble gases are known to undergo very few chemical reactions. This lack of chemical reactivity is a result of the fully occupied highest energy levels in noble gas atoms. Electron configurations can be used to explain the chemical reactivity of all of the elements.

Groups, Periods, and Blocks of the Periodic Table

In Chapter 4, you discovered that atoms of fluorine, chlorine, and all other Group 17 elements have identical outer electron configurations. The same is true for Groups 1, 2, and 13–18. In a few cases for Groups 3–12, the arrangements of electrons in the highest occupied energy levels are similar, but not identical for elements in the same group. The outermost electrons play the largest role in determining chemical properties. Inner-shell electrons have little influence on properties of an element because they do not come into contact with electrons of other atoms.

The length of each period in the Periodic Table is determined by the sublevel being filled with electrons as shown in Table 5-2. The first energy level holds only two electrons in its 1s sublevel. This accounts for hydrogen and helium in the first period. The second main-energy level holds two electrons in a 2s sublevel and six electrons in a 2p sublevel. These eight electrons account for the eight elements in the second period. In the eight elements of the third period, 3s and 3p sublevels are being filled. Filling 3d and 4d sublevels in addition to s and p sublevels adds 10 elements to the fourth and fifth periods, which therefore include a total of 18 elements each. Filling 4f sublevels in addition to s, p, and d sublevels adds 14 elements to the sixth period, for a total of 32 elements in the sixth period. As new artificial elements are created, the 23 known elements in period 7 could be extended to 32. Compare these numbers of elements in each period with the relationships shown in Figure 5-3.

The period of an element can be determined from its electron configuration. For example, arsenic (As) has the configuration $[Ar]3d^{10}$

Figure 5-4 Key to the information provided in the Periodic Table in Figure 5-6 and also in the back cover.

TABLE 5-2 RELATIONSHIP BETWEEN PERIOD LENGTH AND SUBLEVELS BEING FILLED IN THE PERIODIC TABLE

Period Number	Number of Elements in Period	Energy Sublevels in Order of Filling
1	2	1s
2	8	2s 2p
3	8	3s 3p
4	18	4s 3d 4p
5	18	5s 4d 5p
6	32	6s 4f 5d 6p
7	23 (to date)	7s 5f 6d etc.

$4s^2 4p^3$. The "4" in "$4p^3$" indicates that the highest occupied energy level is the fourth energy level. Arsenic is therefore in the fourth period in the Periodic Table.

Refer to the modern Periodic Table in Figure 5-6 and the electron configurations given there, as you read the rest of this chapter.

It is helpful to divide the Periodic Table into four blocks: the *s* block, the *p* block, the *d* block, and the *f* block, as shown in Figure 5-5. Refer to Figure 5-5 as you read about the elements of *s*, *p*, *d*, and *f* blocks in this section.

The name of each block is based on whether an *s*, *p*, *d*, or *f* sublevel is being filled in successive elements of that block.

The *s*-Block Elements: Groups 1 and 2

The elements of the *s* block are all active metals, the Group 1 metals being more active than those of Group 2. Some *s*-block elements are shown in Figure 5-7.

To help students become familiar with the arrangement of the elements on the Periodic Table, give them a blank table so they can fill in information such as chemical symbol, atomic number, and atomic mass of the elements in each group, block, or period being discussed.

STUDY HINT
The properties of the Group 1 metals are discussed in Section 5.2 and Chapter 24.

Figure 5-5 Sublevel blocks of the Periodic Table.

Periodic Table of the Elements

Figure 5-6 Periodic Table of the Elements

Metals
▢ Alkali Metals
▢ Alkaline-Earth Metals
▢ Transition Metals
▢ Other Metals

Metalloids
▢ Metalloids

Nonmetals
▢ Halogens
▢ Other Nonmetals

Noble Gases
▢ Noble Gases

Group 10	Group 11	Group 12	Group 13	Group 14	Group 15	Group 16	Group 17	Group 18
								2 **He** Helium 4.002 60 $1s^2$
			5 **B** Boron 10.81 $[He]2s^22p^1$	6 **C** Carbon 12.0111 $[He]2s^22p^2$	7 **N** Nitrogen 14.0067 $[He]2s^22p^3$	8 **O** Oxygen 15.9994 $[He]2s^22p^4$	9 **F** Fluorine 18.998 403 $[He]2s^22p^5$	10 **Ne** Neon 20.179 $[He]2s^22p^6$
			13 **Al** Aluminum 26.981 54 $[Ne]3s^23p^1$	14 **Si** Silicon 28.0855 $[Ne]3s^23p^2$	15 **P** Phosphorus 30.973 76 $[Ne]3s^23p^3$	16 **S** Sulfur 32.06 $[Ne]3s^23p^4$	17 **Cl** Chlorine 35.453 $[Ne]3s^23p^5$	18 **Ar** Argon 39.948 $[Ne]3s^23p^6$
28 **Ni** Nickel 58.69 $[Ar]3d^84s^2$	29 **Cu** Copper 63.546 $[Ar]3d^{10}4s^1$	30 **Zn** Zinc 65.39 $[Ar]3d^{10}4s^2$	31 **Ga** Gallium 69.72 $[Ar]3d^{10}4s^24p^1$	32 **Ge** Germanium 72.59 $[Ar]3d^{10}4s^24p^2$	33 **As** Arsenic 74.9216 $[Ar]3d^{10}4s^24p^3$	34 **Se** Selenium 78.96 $[Ar]3d^{10}4s^24p^4$	35 **Br** Bromine 79.904 $[Ar]3d^{10}4s^24p^5$	36 **Kr** Krypton 83.80 $[Ar]3d^{10}4s^24p^6$
46 **Pd** Palladium 106.42 $[Kr]4d^{10}5s^0$	47 **Ag** Silver 107.868 $[Kr]4d^{10}5s^1$	48 **Cd** Cadmium 112.41 $[Kr]4d^{10}5s^2$	49 **In** Indium 114.82 $[Kr]4d^{10}5s^25p^1$	50 **Sn** Tin 118.71 $[Kr]4d^{10}5s^25p^2$	51 **Sb** Antimony 121.75 $[Kr]4d^{10}5s^25p^3$	52 **Te** Tellurium 127.60 $[Kr]4d^{10}5s^25p^4$	53 **I** Iodine 126.905 $[Kr]4d^{10}5s^25p^5$	54 **Xe** Xenon 131.29 $[Kr]4d^{10}5s^25p^6$
78 **Pt** Platinum 195.08 $[Xe]4f^{14}5d^96s^1$	79 **Au** Gold 196.967 $[Xe]4f^{14}5d^{10}6s^1$	80 **Hg** Mercury 200.59 $[Xe]4f^{14}5d^{10}6s^2$	81 **Tl** Thallium 204.383 $[Xe]4f^{14}5d^{10}6s^26p^1$	82 **Pb** Lead 207.2 $[Xe]4f^{14}5d^{10}6s^26p^2$	83 **Bi** Bismuth 208.980 $[Xe]4f^{14}5d^{10}6s^26p^3$	84 **Po** Polonium (209) $[Xe]4f^{14}5d^{10}6s^26p^4$	85 **At** Astatine (210) $[Xe]4f^{14}5d^{10}6s^26p^5$	86 **Rn** Radon (222) $[Xe]4f^{14}5d^{10}6s^26p^6$

Mass numbers in parentheses are those of the most stable or most common isotope.

63 **Eu** Europium 151.96 $[Xe]4f^76s^2$	64 **Gd** Gadolinium 157.25 $[Xe]4f^75d^16s^2$	65 **Tb** Terbium 158.925 $[Xe]4f^96s^2$	66 **Dy** Dysprosium 162.50 $[Xe]4f^{10}6s^2$	67 **Ho** Holmium 164.930 $[Xe]4f^{11}6s^2$	68 **Er** Erbium 167.26 $[Xe]4f^{12}6s^2$	69 **Tm** Thulium 168.934 $[Xe]4f^{13}6s^2$	70 **Yb** Ytterbium 173.04 $[Xe]4f^{14}6s^2$	71 **Lu** Lutetium 174.967 $[Xe]4f^{14}5d^16s^2$
95 **Am** Americium (243) $[Rn]5f^77s^2$	96 **Cm** Curium (247) $[Rn]5f^76d^17s^2$	97 **Bk** Berkelium (247) $[Rn]5f^97s^2$	98 **Cf** Californium (251) $[Rn]5f^{10}7s^2$	99 **Es** Einsteinium (252) $[Rn]5f^{11}7s^2$	100 **Fm** Fermium (257) $[Rn]5f^{12}7s^2$	101 **Md** Mendelevium (258) $[Rn]5f^{13}7s^2$	102 **No** Nobelium (259) $[Rn]5f^{14}7s^2$	103 **Lr** Lawrencium (260) $[Rn]5f^{14}6d^17s^2$

The Periodic Law

Sodium

Calcium

Figure 5-7 Sodium and calcium are members of the *s*-block of the Periodic Table.

Hydrogen and helium are the simplest and most abundant elements in the universe. Hydrogen makes up 76% of the mass in the universe, and helium makes up 23%.

The highest occupied energy level in atoms of each Group 1 element contains a single *s* electron. For example, the configurations of lithium and sodium are [He]$2s^1$ and [Ne]$3s^1$. As you will see in Section 5-3, the ease with which the single electron is lost helps to make the Group 1 metals very reactive. Using *n* for the number of the highest occupied energy level, the outer or *group configuration* of the Group 1 and 2 elements is written ns^1 and ns^2 respectively.

The alkali metals *The elements of Group 1 of the Periodic Table (lithium, sodium, potassium, rubidium, cesium, and francium) are known as the* **alkali metals.** The alkali metals combine vigorously with many nonmetals. Because they are extremely reactive, alkali metals are not found in nature as free elements. All alkali metals react with water, with increasing vigor for each element down the group, to produce hydrogen gas and aqueous solutions of substances known as alkalis. Alkalis are common and useful chemicals that can, for example, dissolve fats to produce soaps. Because of their extreme reactivity with air or moisture, the alkali metals are usually stored under kerosene.

All of the alkali metals have a silvery appearance and are soft enough to cut with a knife. Lithium (Li), sodium (Na), and potassium (K) are less dense than water. Proceeding down the column, alkali metals melt at successively lower temperatures; except for lithium, the alkali metals have melting points lower than the boiling point of water.

Molten sodium is used as a heat-transfer fluid in nuclear reactors, and sodium vapor lamps are efficient light sources. The photoelectric effect occurs readily in cesium. Therefore, cesium (Cs) is used in photoelectric cells, such as those that will open a supermarket door as you approach the door. Compounds of the Group 1 elements sodium and potassium play essential roles in living things, where they are present in body fluids that fill and surround all cells.

The alkaline-earth metals *The elements of Group 2 of the Periodic Table (beryllium, magnesium, calcium, strontium, barium, and radium) are called* **the alkaline-earth metals.** Atoms of each alkaline-earth metal contain a pair of electrons in the outermost *s* sublevel. Thus, for Group 2, the group configuration is ns^2. The Group 2 metals are all harder, denser, stronger, and have higher melting points than the alkali metals. Although they are less reactive than the alkali metals, the alkaline-earth metals are also too reactive to be found in nature as free elements.

Compounds containing calcium are among the most common materials in the earth's crust. The minerals limestone and marble are forms of calcium carbonate, a compound of calcium, carbon (C), and oxygen (O). Compounds containing magnesium (Mg) are also abundant. Sea water contains large quantities of these compounds. Beryllium (Be) is a very rare element. Its source is the gemstone *beryl*. Crystals of colorless or brown beryl are used as the source of metallic beryllium. The light blue-green form of beryl, called aquamarine, and the deep green form, called emerald, are valued for their beauty. Radium (Ra), a naturally occurring radioactive element first discovered in 1898, is extremely rare and is a minor component of uranium ore.

Magnesium is widely used to make strong, lightweight alloys for aircraft, truck bodies, and ladders. The chief use of strontium (Sr) is in fireworks and flares for highway danger signals. Compounds of calcium are extremely important biologically. Teeth and bones consist of calcium compounds.

Hydrogen and helium The electron configurations of hydrogen and helium are $1s^1$ and $1s^2$, respectively. Neither hydrogen nor helium are metals. Both are colorless, odorless gases. Hydrogen, although it is sometimes placed with Group 1 at the left of the Periodic Table, is a unique element with properties that do not closely resemble those of any group.

Why is helium a member of Group 18 rather than Group 2? Like the Group 2 elements, helium has an ns^2 group configuration. (Helium is placed in Group 18, however, because of its nonreactive nature.) Here, also, electron configuration governs properties. As in the other noble gases, the highest occupied energy level in helium is filled, in this case by two rather than eight electrons. As you learned in Chapter 4, there is a special stability associated with a completely filled highest occupied energy level. By contrast, the Group 2 metals have no special stability; their highest occupied energy levels are not filled since they have an empty p sublevel.

STUDY HINT
The properties of the Group 1 and Group 2 metals are further discussed in Chapter 24.

Francium, the heaviest and most active alkali metal, is found in minute amounts in uranium minerals. Because it has a half-life of 22 minutes, there is probably less than one ounce of francium to be found at any one time. Francium was discovered in 1939 by Marguerite Perey (1909–1975), who was the first woman to be elected to the French Academy of Sciences.

Sample Problem 5.1

(a) Without looking at the Periodic Table, give the group, period, and block in which the element that has the following electron configuration is located: $[Xe]6s^2$.

(b) Without looking at the Periodic Table, write the electron configuration for the element in the third period in Group 1. Is this element likely to be a more or less active element than that described in part *(a)*?

Solution

(a) The element is in Group 2, as shown by the outer, or group, configuration of ns^2. It is in the sixth period, as shown by the highest principal quantum number of 6, in its configuration. The element is in the *s* block.

(b) A third-period element has filled $1s$, $2s$, and $2p$ sublevels (see Table 5-2). The third main-energy level, with $n = 3$, is its highest occupied energy level. An element in Group 1 has single electron in its highest *s* sublevel (group configuration ns^1). Therefore, this element has the configuration

$$1s^2 2s^2 2p^6 3s^1 \text{ or } [Ne]3s^1$$

Because it is in Group 1 (the alkali metals), this element is likely to be more active than that described in part *(a)*, which is an alkaline-earth metal from Group 2.

Practice Problems
1. Without looking at the Periodic Table give the group, period, and block in which the element with the following shorthand electron configuration is located: $[Kr]5s^1$ *(Ans.)* Group 1, period 5, and *s* block.
2. (a) Without looking at the Periodic Table, write both the group and complete electron configurations for the element in the fourth period in Group 2.   *(Ans.)* ns^2, $1s^2 2s^2 2p^6 3s^2 3p^6 4s^2$
   (b) Refer to the table to identify the indicated element in part *(a)* and then write its shorthand notation.   *(Ans.)* Ca, $[Ar]4s^2$
   (c) How does the reactivity of the element in *(a)* compare with that in Group 1 of the same period?   *(Ans.)* Because this element is in Group 2, it is likely to be less active than that in Group 1 of the same period.

The Periodic Law

Mercury

Tungsten

Vanadium

Figure 5-8 Mercury, tungsten and vanadium are all members of the *d*-block of the Periodic Table.

Because of their low reactivity, Pd, Pt, and Au are sometimes called the noble metals.

STUDY HINT

The properties of the *d*-block metals are covered further in Chapter 25.

STUDY HINT

Materials that are referred to as *reactive* quickly or easily undergo chemical changes. Materials of low reactivity undergo few chemical changes and do so only with difficulty (often with addition of energy).

The *d*-Block Elements: Groups 3–12

In the *d*-block elements the filling of the *ns* sublevels in Groups 1 and 2 is followed by the addition of electrons to the *d* sublevels of the next lower energy level. Some *d*-block elements are shown in Figure 5-8. Atoms of the Group 3 elements each have one electron in the *d*-block sublevel corresponding to the $(n-1)$ energy level just below *n*. This is in addition to the two *ns* electrons from Group 2. For Group 3, the group configuration is $(n-1)d^1ns^2$. Nine more electrons are added across each period. Atoms of the Group 12 elements hold 10 electrons in the *d* sublevel plus two electrons in the *ns* sublevel. The group configuration for Group 12 is $(n-1)d^{10}ns^2$.

In Groups 4–11, some deviations from orderly *d* sublevel filling occur, as you saw in Chapter 4. As a result, elements in these *d*-block groups, unlike those in *s*-block and *p*-block groups, do not necessarily have identical electron configurations. For example, in Group 10, nickel (Ni) has the electron configuration [Ar]$3d^84s^2$, palladium (Pd) has the configuration [Kr]$4d^{10}5s^0$, and platinum (Pt) has the configuration [Xe]$4f^{14}5d^96s^1$. Notice, however, that the sum of the outer *s* and *d* electrons is equal to the group number. This is true for all *d*-block elements within the same group.

The d-block elements are all metals with typical metallic properties and are referred to as **transition elements.** They are good conductors of electricity and have a high luster when freshly cut or polished. The metals of the *d*-block are harder, denser, and with the exception of mercury, have higher melting points that the *s*-block metals. Palladium, platinum, and gold are among the least reactive of all the elements.

The *d*-block metals are typically less reactive than the alkali metals and the alkaline-earth metals. Some, such as gold (Au), silver (Ag), and copper (Cu) in Group 11; palladium (Pd) and platinum (Pt) in Group 10; and iridium (Ir) in Group 9 are commonly found in nature as free elements. These elements are so nonreactive that they do not easily form compounds.

Perhaps the most important of the *d*-block metals is iron, which is mixed with carbon to make steel. Other elements can be included in the mixture to make different varieties of steel. Some elements used in steel making are vanadium (V), chromium (Cr), manganese (Mn), cobalt (Co), nickel (Ni), molybdenum (Mo), and tungsten (W).

Tungsten has the highest melting point of all the metals and is used for the filaments of incandescent light bulbs. Copper, one of the best electrical conductors, is widely used for electrical wiring. Silver, gold and platinum are well known for their use in jewelry. Silver compounds, which are highly sensitive to light, are essential to photographic film.

Sample Problem 5-2

An element has the electron configuration [Kr]$4d^5 5s^1$. Without looking at the Periodic Table, identify the period, block, and group in which this element is located. Then, consult the Periodic Table and identify this element and the others in its group (excluding the period).

Solution
The number of the highest occupied energy level is 5, showing that the element is in the fifth period. There are 5 electrons in the d sublevel, which is therefore incompletely filled since this sublevel can hold 10 electrons. Therefore, the element is in the d-block. For d-block elements, the sum of the electrons in the ns sublevel and the $(n-1)d$ sublevel is equal to the group number, giving $5 + 1 = 6$. This is the fifth period element in Group 6. The element is molybdenum and the others in Group 6 are chromium and tungsten.

Practice Problems
1. Without looking at the Periodic Table, identify the period, block, and group in which the element with the following electron configuration is located: [Ar]$3d^8 4s^2$. *(Ans.)* 4, d block, and 10
2. *(a)* Without looking at the Periodic Table write the outer electron configuration for the fifth period element in Group 12. *(Ans.)* $4d^{10} 5s^2$ *(b)* Refer to the table to identify that element and to write its shorthand electron configuration. *(Ans.)* Cd, [Kr]$4d^{10} 5s^2$

Fluorine   **Chlorine**

Bromine

Iodine

Figure 5-9 Fluorine, chlorine, bromine, and iodine are all members of the p-block of the Periodic Table.

The *p*-Block Elements: Groups 13–18

The p-block elements include Groups 13–18 (Figure 5-5). There are six p-block elements in each period except the first. Because electrons add to a p sublevel only after the s sublevel in the same energy level is filled, atoms of all p-block elements contain two electrons in the ns sublevel. For Group 13 elements, the added electron enters the np sublevel, giving a group configuration of $ns^2 np^1$. Atoms of Group 14 elements contain two electrons in the p sublevel, giving $ns^2 np^2$ for the group configuration. This pattern continues in Groups 15–18. In Group 18 the stable noble gas configuration of $ns^2 np^6$ is reached. Some p-block elements are shown in Figure 5-9. *The p-block elements together with the s-block elements are called the* **main-group elements.**

STUDY HINT

Note that the number of s plus p electrons in atoms of p-block elements is equal to the last digit in the group number.

TABLE 5-3 RELATIONSHIPS AMONG GROUP NUMBERS, BLOCKS, AND ELECTRON CONFIGURATIONS

Group Number	Group Configuration	Block	Comments
1, 2	$ns^{1,2}$	s	One or two electrons in ns sublevel
3–12	$(n-1)d^{0-10}ns^{0-2}$	d	Sum of electrons in ns and $(n-1)d$ levels equals group number
13–18	ns^2np^{1-6}	p	Sum of electrons in ns and np sublevels equals group number minus 10. Number of electrons in np sublevel equals group number minus 12.

STUDY HINT

The properties of the metalloids and aluminum, a p-block metal, are discussed in Chapter 26.

For many years, the noble gases were called "the inert gases" because it was thought they would not react to form compounds. However, in 1962 this idea was shown to be incorrect when $XePtF_6$ was formed. Since then, other compounds of Xe, Kr, and Ra have been produced.

STUDY HINT

Carbon and organic chemistry are discussed in Chapter 21 and 22, and the properties of sulfur are discussed in Chapter 27.

STUDY HINT

Chapter 28 is devoted to the chemistry of the halogens.

For atoms of all p block elements, the total number of electrons in the highest occupied level is equal to the group number minus 10. For example, bromine, in Group 17, has $17 - 10 = 7$ electrons ($4s^2 4p^5$) in its highest energy level. Because atoms of all p-block elements contain two electrons in the ns sublevel, bromine must have 5 electrons in the p sublevel. The electron configuration of bromine (Br) is therefore $[Ar]3d^{10}4s^24p^5$. The relationships among group numbers and electron configurations for all the groups are summarized in Table 5-3.

The properties of elements of the p block vary greatly. At its right-hand end, the p block includes all of the nonmetals except hydrogen and helium. All six of the metalloids (boron, silicon, germanium, arsenic, antimony, and tellurium) are also in the p block. To the left and bottom of the block, there are eight p-block metals. The location of the nonmetals, metalloids, and metals in the p block are shown by distinctive colors in Figure 5-6 and in the Periodic Table in the inside back cover of the book.

The millions of compounds that contain carbon and hydrogen form the basis of organic chemistry. The nonmetals carbon, nitrogen, oxygen, sulfur, and phosphorus are present in living things and in a great many industrial chemicals. Oxygen also has a central role in the inorganic world. It forms compounds with almost every element and is present in most rocks and minerals.

The elements of Group 17 (fluorine, chlorine, bromine, iodine, and astatine) are known as the **halogens**. The halogens are the most reactive of the nonmetals. They react vigorously with most metals to form the type of compounds known as "salts." The most common salt is ordinary table salt, which contains the halogen chlorine combined with the alkali metal, sodium. As you will see, the reactivity of the halogens is based on the presence of seven electrons, one electron short of the stable noble gas configuration, in their outer energy levels. Fluorine and chlorine are gases at room temperature, bromine is a red liquid, and iodine is a dark purple solid. The halogens are used in bleaches, water purification, photography, insecticides, plastics, and many other practical applications.

The Group-18 elements, the noble gases, are the least reactive of all the elements. They have many uses that take advantage of this lack of reactivity. For example, noble gases are used instead of air in containers with materials that would react with oxygen or moisture in the air.

The metalloids, or semiconducting elements, fall on both sides of a line separating nonmetals and metals in the *p* block. They are mostly brittle solids with some properties of metals and some of nonmetals. The metalloid elements have electrical conductivity intermediate between that of metals, which are good conductors, and nonmetals, which are nonconductors. The electrical properties of semiconducting materials made from metalloid elements make them essential components of computers.

The metals of the *p* block are generally harder and denser than the alkaline earth metals, ns^2 block metals, but softer and less dense than the *d*-block metals. With the exception of bismuth, these metals are sufficiently reactive to be found in nature only in the form of compounds. Once obtained as free metals, however, they are stable in the presence of air. Aluminum is the most important of the *p*-block metals because it forms strong, lightweight alloys with many uses in building construction and aircraft production.

Sample Problem 5.3

(a) Without looking at the Periodic Table, write the outer electron configuration for the element in the second period of Group 14.

(b) Name this element. Is it a metal, a nonmetal, or metalloid?

Solution

(a) Because the group number is higher than 12, this element must be in the *p* block. The total number of electrons in the highest occupied *s* and *p* sublevels is therefore $14 - 10 = 4$. With two electrons in the *s* sublevel, two electrons must also be present in the 2*p* sublevel, giving an outer electron configuration of $2s^2 2p^2$.

(b) This element is carbon (C), which is a nonmetal.

Practice Problems

1. *(a)* Without looking at the Periodic Table, write the outer electron configuration for the third period element in Group 17.   *(Ans.)* $3s^2 3p^5$

   (b) Name the element and identify it as a metal, nonmetal, or metalloid.   *(Ans.)* chlorine, nonmetal

2. *(a)* Without looking at the Periodic Table, identify the period, block, and group of an element with the electron configuration: $[Ar]3d^{10}4s^2 4p^3$ *(Ans.)* 4, *p*-block, Group 15.

   (b) Name this element and identify it as a metal, nonmetal, or metalloid.   *(Ans.)* arsenic, metalloid

The *f*-Block Elements: Lanthanides and Actinides

In the Periodic Table, the *f*-block elements are wedged between Groups 3 and 4 in Periods 6 and 7. Their position reflects the fact that, after lanthanum, the 4*f* sublevel is filled before electrons are added to the 5*d* sublevel. With seven 4*f* orbitals to be filled with two electrons each, there are a total of 14 *f*-block elements between lanthanum (La) and hafnium (Hf) in Period 6. The lanthanides, or "rare earths" (which are not actually rare), are all shiny reactive metals. With modern separation techniques, all are readily available and practical uses have been found for them. For example, many of the phosphors that glow in color television tubes are compounds of lanthanide elements. Figure 5-10 shows one mineral from which lanthanides are recovered.

Bastnasite

Figure 5-10 Thorium, an element in the *f*-block portion of the Periodic Table is contained in the mineral bastnasite shown above.

STUDY HINT
Radioactivity and the synthesis of new elements by nuclear reactions are discussed in Chapter 29.

There are also 14 *f*-block elements, the actinides, between actinium and element 104 (Unq) in Period 7. In these elements the 5*f* sublevel is being filled with 14 electrons. The actinides are all unstable and radioactive. The first four actinides (thorium, Th, through neptunium, Np) have been found naturally on earth. The remaining actinides are known only as laboratory-made elements. Because of their radioactivity, the actinides are not encountered in everyday chemistry.

Sample Problem 5.4

The electron configurations of atoms of four elements are written below. For each, name the block and group in the Periodic Table in which it is located; name the element (consult the Periodic Table); identify it as a metal, nonmetal, or metalloid; and describe it as likely to be of high reactivity or of low reactivity.
(a) $[Xe]4f^{14}5d^96s^1$ (b) $[Ne]3s^23p^5$ (c) $[Ne]3s^23p^6$ (d) $[Xe]4f^66s^2$

Solution
(a) The 4*f* sublevel is filled with 14 electrons. The 5*d* sublevel is partially filled with 9 electrons. Therefore, this is a *d*-block element. The element is platinum (Pt) which has a low reactivity.
(b) The incompletely filled *p* sublevel shows this to be a *p*-block element. With a total of 7 electrons in the *ns* and *np* sublevels, this is a Group 17 element, a halogen. The element is chlorine and is likely to be highly reactive.
(c) This element has the noble gas configuration and thus is in Group 18 in the *p* block. The element is argon (Ar), an unreactive nonmetal and a noble gas.
(d) The incomplete 4*f* sublevel shows that it is an *f*-block element and a lanthanide. Group numbers are not assigned to the *f* block. The element is samarium (Sm). The lanthanides are all reactive metals.

Practice Problems
For each of the following, identify the block, period, group, group name (where appropriate), element name, element type (metal, nonmetal, or metalloid), and relative reactivity (high or low):
1. $[He]2s^22p^5$ *(Ans.)* *p* block, 2, 17, halogens, fluorine, nonmetal, highly reactive
2. $[Ar]3d^{10}4s^1$ *(Ans.)* *d* block, 4, 11, transition elements, copper, metal, low reactivity
3. $[Kr]5s^1$ *(Ans.)* *s* block, 5, 1, alkali metals, rubidium, metal, highly reactive

1. They have identical outer electron configurations and thus have similar properties.
2. *s* block, *p* block, *d* block, and *f* block
3. (a) alkali metals (b) alkaline-earth metals (c) transition elements (d) halogens (e) noble gases
4. *s*-block: ns^1 and ns^2 configurations designate elements in Groups 1 and 2, respectively; *p*-block: the ns^2np^1 through ns^2np^6 configurations designate elements in Groups 13 through 18, respectively; *d*-block: $(n-1)d^1ns^2$ through $(n-1)d^{10}ns^2$ configurations denote elements in Groups 3 through 12, respectively.

Section Review

1. What is the relationship among the elements within a specific group (for the *s* and *p* block elements)?
2. Into what four blocks can the Periodic Table be divided to illustrate the relationships between electron configurations and Periodic Table placement?
3. What name is given to each of the following groups of elements on the Periodic Table: (a) Group 1 (b) Group 2 (c) Groups 3–12 (d) Group 17 (e) Group 18.
4. What is the relationship between group configuration and group number for elements in the *s*, *p*, and *d* blocks?
5. Without looking at the Periodic Table write the outer configuration for the fourth period element in Group 15.
6. Without looking at the Periodic Table identify the period, block, and group of the element: $[Ar]3d^74s^2$.

5. $3d^{10}4s^24p^3$
6. fourth period, *d* block, Group 9

5.3 Electron Configuration and Periodic Properties

Thus far, you have seen that the elements are grouped in the Periodic Table according to their physical and chemical properties and also, as shown in Section 5-2, their electron configurations. In this section, the relationship between the periodic variation in properties and electron configurations will be further explored.

Atomic Radii

Electrons occupy the relatively large region around the nucleus of an atom. The electron cloud determines the size of an atom. The size of an atom is not a definite quantity; however, the boundary is somewhat fuzzy and may vary. For these reasons and also because it is not possible to measure the size of an isolated atom, chemists compare the sizes by measuring the distance between atoms in an element.

The **atomic radius** *is one-half the distance between the nuclei of identical atoms joined in a molecule.* For example, the distance between the nuclei of two iron (Fe) atoms in a sample of solid iron can be measured, as shown in Figure 5-11. The radius of an iron atom is then taken as one-half this distance.

Period trends Figure 5-12 gives the atomic radii of the elements and Figure 5-13 presents this information graphically. There is a gradual decrease in atomic radii across the periods, for example, examine the decrease in atomic radii across the second period from lithium (Li) to neon (Ne).

The trend to smaller atoms across a period is caused by the increasing positive charge of the nucleus. As electrons add to *s* and *p* sublevels in the same main-energy level, they are gradually pulled closer to the more highly charged nucleus. This increased pull results in decreasing atomic radii. The attraction of the nucleus is offset somewhat by repulsion among the increased number of electrons in the same outer energy level, however. As a result, the decrease in radii between neighboring atoms in each period grows smaller, as shown by the graph in Figure 5-13.

Group trends The atomic radii of the main-group elements generally increase as one reads down a group. For example, examine the radii of the Group 1 elements in Figures 5-12 and 5-13. In the atoms of each element, the outer electrons occupy *s* sublevels in successively higher main-energy levels and are therefore farther from the nucleus.

Next, examine the radii of the Group 13 elements. Note that gallium (Ga) atoms have slightly smaller radii than aluminum (Al) atoms, although gallium follows aluminum in the group. Recall the atomic number differences given in Figure 5-3. Unlike aluminum, gallium is preceded in its period by the ten *d*-block elements. The expected increase in radius is overcome in gallium atoms by the higher nuclear charge. Gallium has a slightly smaller atomic radius than zinc (Zn), the element immediately to its left. This decrease is as predicted based on trends across the period.

SECTION OBJECTIVES

- Define atomic and ionic radii, ionization energy, electron affinity, and electronegativity.
- Compare the periodic variations of atomic radii, ionization energy, and electronegativity and state the reasons for these variations.
- Define valence electrons and state how many are present in atoms of each *s* and *p* block elements.
- Describe electron configurations of common ions of elements in the *s* and *p* blocks.
- Compare the atomic radii, ionization energies, and electronegativities of the *d*-block elements with those of the main-group elements.

Figure 5-11 Atomic radius is defined as one-half the distance between the nuclei of two identical atoms combined in an element or chemical compound. For iron atoms, the radius is 124 picometers (pm).

Periodic Table of Atomic Radii

Figure 5-12 Atomic radii decrease from left to right across a period and increase down a group.

Sample Problem 5.5

(a) Among the elements magnesium (Mg), chlorine (Cl), sodium (Na), and phosphorus (P), which has the largest atomic radius? Explain in terms of Periodic Table trends.

(b) Among the elements calcium (Ca), beryllium (Be), barium (Ba), and strontium (Sr), which has the largest atomic radius? Explain in terms of Periodic Table trends.

Solution

(a) These elements are all in the third period. Of the four, sodium has the lowest atomic number and is the first element in the period. Sodium has the largest atomic radius, as expected because atomic radii decrease across the period. (b) These elements are all in Group 2. Of the four, barium has the highest atomic number and is furthest down the group. From its position, we would expect it to have the largest atomic radius, and this is correct.

Practice Problems

1. Among the elements Li, O, C, F, and N, identify the one with the: *(a)* largest atomic radius *(Ans.)* Li *(b)* smallest atomic radius. *(Ans.)* F
2. Among the elements Br, At, F, I, and Cl, identify the one with the: *(a)* smallest atomic radius *(Ans.)* F *(b)* largest atomic radius. *(Ans.)* At

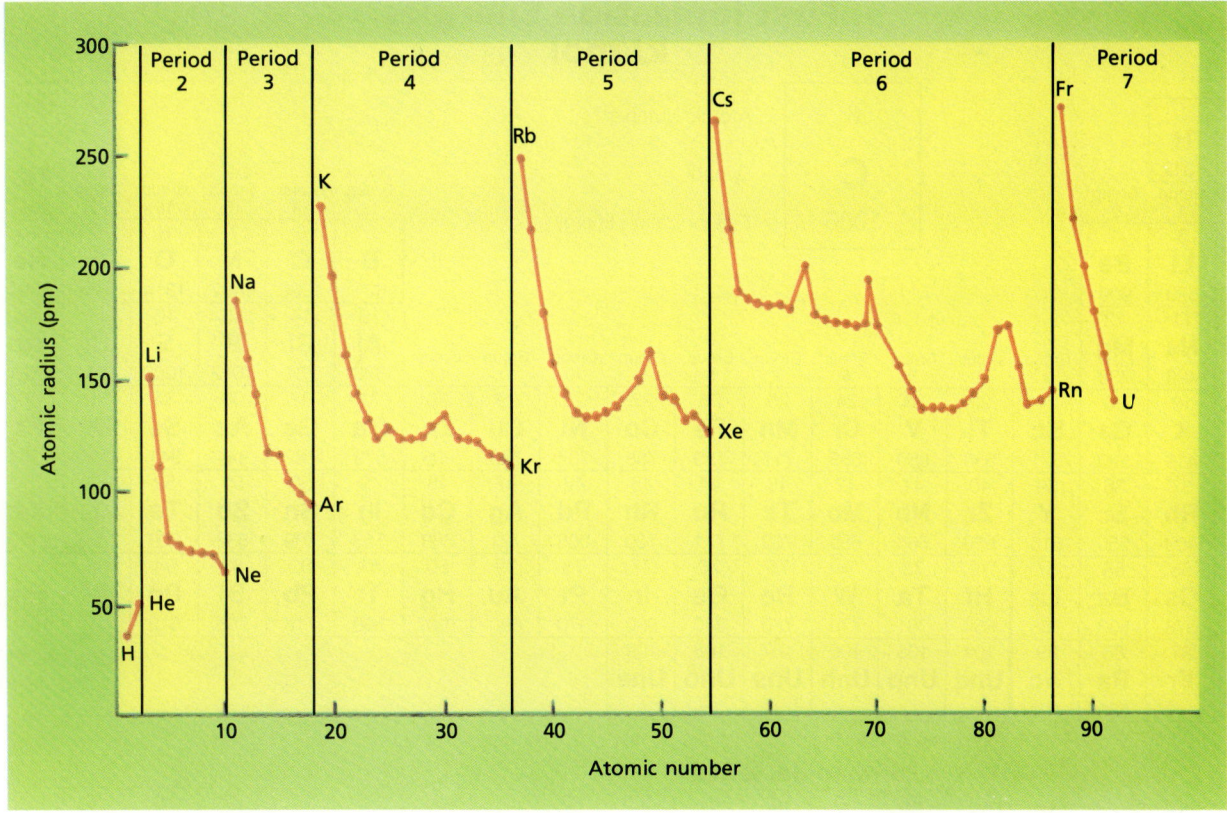

Figure 5-13 Plot of atomic radius versus atomic number

Ionization Energy

An electron can be removed from an atom when enough energy is supplied. Using A as a symbol for an atom of any element, the process can be represented as follows:

$$A + energy \rightarrow A^+ + e^-$$

The A^+ represents an ion of element A with a single positive charge, referred to as a 1+ charge. *An ion is an atom or group of atoms that has a positive or negative charge.* Sodium, for example, readily forms an Na^+ ion. *Any process that results in the formation of an ion is referred to as* **ionization.**

To compare the ease with which atoms of different elements give up electrons, chemists compare ionization energies. *The energy required to remove one electron from an atom of an element is the* **ionization energy** *(or first ionization energy).* Measurements of ionization energy are made on atoms in the gas phase to eliminate the influence of nearby atoms. Figure 5-14 gives first ionization energies in kilojoules per mole (kJ/mol) for the elements. Figure 5-15 presents this information graphically.

Period trends Examine the ionization energies in Figures 5-14 and 5-15 for the first and last elements in each period. You can see that the Group 1 metals have the lowest first ionization energies: that is, they lose electrons most easily. This ease of electron loss is a major reason for the high reactivity of the Group 1 (alkali) metals. The Group 18 elements, the noble gases, have

First Ionization Energies
kJ/mol

Group 1	Group 2											Group 13	Group 14	Group 15	Group 16	Group 17	Group 18
1 **H** 1312																	2 **He** 2372
3 **Li** 520	4 **Be** 900											5 **B** 801	6 **C** 1086	7 **N** 1402	8 **O** 1314	9 **F** 1681	10 **Ne** 2081
11 **Na** 496	12 **Mg** 738	Group 3	Group 4	Group 5	Group 6	Group 7	Group 8	Group 9	Group 10	Group 11	Group 12	13 **Al** 578	14 **Si** 787	15 **P** 1012	16 **S** 1000	17 **Cl** 1251	18 **Ar** 1521
19 **K** 419	20 **Ca** 590	21 **Sc** 631	22 **Ti** 658	23 **V** 650	24 **Cr** 653	25 **Mn** 717	26 **Fe** 759	27 **Co** 758	28 **Ni** 737	29 **Cu** 746	30 **Zn** 906	31 **Ga** 579	32 **Ge** 762	33 **As** 944	34 **Se** 941	35 **Br** 1140	36 **Kr** 1351
37 **Rb** 403	38 **Sr** 550	39 **Y** 616	40 **Zr** 660	41 **Nb** 664	42 **Mo** 685	43 **Tc** 702	44 **Ru** 711	45 **Rh** 720	46 **Pd** 805	47 **Ag** 731	48 **Cd** 868	49 **In** 558	50 **Sn** 709	51 **Sb** 832	52 **Te** 869	53 **I** 1008	54 **Xe** 1170
55 **Cs** 376	56 **Ba** 503	57 **La** 538	72 **Hf** 654	73 **Ta** 761	74 **W** 770	75 **Re** 760	76 **Os** 840	77 **Ir** 880	78 **Pt** 870	79 **Au** 890	80 **Hg** 1007	81 **Tl** 589	82 **Pb** 716	83 **Bi** 703	84 **Po** 812	85 **At** —	86 **Rn** 1038
87 **Fr** —	88 **Ra** 509	89 **Ac** 490	104 **Unq**	105 **Unp**	106 **Unh**	107 **Uns**	108 **Uno**	109 **Une**									

Atomic Number: 6
Symbol: C
First Ionization Energy: 1086

Lanthanide Series

58 **Ce** 528	59 **Pr** 523	60 **Nd** 530	61 **Pm** 536	62 **Sm** 543	63 **Eu** 547	64 **Gd** 592	65 **Tb** 564	66 **Dy** 572	67 **Ho** 581	68 **Er** 589	69 **Tm** 597	70 **Yb** 603	71 **Lu** 523
90 **Th** 590	91 **Pa** 570	92 **U** 590	93 **Np**	94 **Pu**	95 **Am**	96 **Cm**	97 **Bk**	98 **Cf**	99 **Es**	100 **Fm**	101 **Md**	102 **No**	103 **Lr**

Actinide Series

Figure 5-14 Periodic table of the elements showing first ionization energies in kilojoules per mole (kJ/mol)

the highest ionization energies and do not lose electrons easily. As you might guess, the low reactivity of the noble gases is partly based on this difficulty of removing electrons.

The trend across the periods for the main-group elements is toward a general increase in ionization energy. This increase is caused by the increasing nuclear charge, which more strongly attracts electrons in the same energy level. Thus, increasing nuclear charge is responsible for both *increasing* ionization energy and decreasing radii across the periods.

Note that, in general, nonmetals have higher ionization energies than metals. The elements of Group 1, the alkali metals, have the lowest ionization energies and the elements of Group 18, the noble gases, have the highest ionization energies.

Group trends Among the main-group elements, the ionization energy generally decreases down the groups as the size of the atoms increases. The electrons removed from atoms of each succeeding element are in higher energy levels. Therefore they are farther from the nucleus and more easily removed. Also, as atomic number increases going down a group, more and more electrons lie between the nucleus and the electrons in the highest

Figure 5-15 Plot of first ionization energy versus atomic number

occupied energy levels. The result is that the outer electrons are shielded from the positive charge of the nucleus. These influences combine to outweigh the attraction for electrons of the increasing nuclear charge.

Formation of positive ions With sufficient energy, electrons can be removed from ions as well as from neutral atoms. The energies for removal of additional electrons from an atom in a gas are referred to as the "second ionization energy," "third ionization energy," and so on.

Table 5-4 shows the first five ionization energies for the elements of Periods 1, 2, and 3. You can see that the second ionization energy is always higher than the first, the third is always higher than the second, and so on. These increases are to be expected. Each successive electron must be removed from an ion with a larger positive charge.

The first ionization energies in Table 5-4 show that removing a single electron from an atom of a Group 18 element is much harder than removing an electron from atoms of other elements in the same period. This special stability of the noble gas electron configuration is also shown by *ions* with the electron configurations of noble gases. In Table 5-4, look at the large increase between the first and second ionization energies of lithium (Li), between the second and third ionization energies of beryllium, between the third and fourth ionization energies of boron (B), and between the fourth and fifth ionization energies of carbon (C). In each case, the jump in ionization energy occurs when the noble gas configuration is reached. For example, the removal of one electron from a lithium atom ($[He]2s^1$) leaves the helium noble gas configuration. The removal of four electrons from a carbon atom ($[He]2s^22p^2$) also leaves the helium configuration. A bigger table would show that this trend continues across the entire period.

To help students visualize how ionization energies increase with each successive removal of an electron, use the analogy of adults (the protons) looking after energetic children (the electrons). The fewer children per adult, the more control the adults have over the children. With fewer children, more energy is required to take them away.

The Periodic Law **147**

TABLE 5-4 FIRST FIVE IONIZATION ENERGIES (IN KJ/MOL) FOR ELEMENTS OF PERIODS 1–3

	Period 1		Period 2							
	H	He	Li	Be	B	C	N	O	F	Ne
I	1312	2372	520	900	801	1086	1402	1314	1681	2081
II		5250	7298	1757	2427	2353	2856	3388	3374	3952
III			11 815	14 849	3660	4621	4578	5300	6050	6122
IV				21 007	25 026	6223	7475	7469	8408	9370
V					32 827	37 830	9445	10 990	11 023	12 178

	Period 3							
	Na	Mg	Al	Si	P	S	Cl	Ar
I	496	738	578	787	1012	1000	1251	1521
II	4562	1451	1817	1577	1903	2251	2297	2666
III	6912	7733	2745	3232	2912	3361	3822	3931
IV	9544	10 540	11 578	4356	4957	4564	5158	5771
V	13 353	13 628	14 831	16 091	6274	7013	6540	7238

Sample Problem 5.6

Consider two main-group elements A and B. Element A has a first ionization energy of 419 kJ/mol. Element B has a first ionization energy of 1000 kJ/mol. For each element, decide if it is more likely to be a metal or a nonmetal, and if it is more likely to be in the s block or p block. Which element is more likely to form a positive ion in a chemical compound?

Solution
Element A has quite a low ionization energy, meaning that A atoms lose electrons easily. Therefore, element A is most likely to be a metal in the s block, because ionization energies increase across the periods. Element B has quite a high ionization energy, meaning that B atoms lose electrons with difficulty. We would expect element B to be a nonmetal, meaning that it must be in the p block. Element A is more likely to form a positive ion, because it has a relatively low ionization energy and it is a metal.

Practice Problems
Consider the four hypothetical main-group elements R, S, T, and U with the indicated outer electron configurations: R—$3s^23p^5$ S—$3s^1$ T—$4d^{10}5s^25p^5$ U—$4d^{10}5s^25p^1$

1. Identify the block location of each hypothetical main-group element and whether each is a metal, nonmetal, or metalloid. *(Ans.)* R is a p-block nonmetal, S is an s-block metal, T is a p-block nonmetal, and U is a p-block metal.
2. Which of these elements are in the same *(a)* period? *(Ans.)* S and R, U and T *(b)* group? *(Ans.)* R and T
3. Which would you expect to have *(a)* the highest first ionization energy? *(Ans.)* R *(b)* the lowest first ionization energy? *(Ans.)* S
4. Which of these hypothetical main-group elements would you expect to have the highest second ionization energy? *(Ans.)* S
5. Which is most likely to form a 1+ ion in a chemical compound? *(Ans.)* S

Electron Affinity

Neutral atoms can acquire electrons. The energy involved when an atom gains an electron is referred to as the electron affinity. **Electron affinity** *is the energy change that occurs when an electron is acquired by a neutral atom.* Many atoms readily add electrons and release energy, which classifies this as an exothermic process:

$$A + e^- \rightarrow A^- + \text{energy}$$

By convention, the quantity of energy released in an exothermic process is represented by a negative number.

Some atoms, however, must be "forced" to gain an electron by the addition of energy:

$$A + e^- + \text{energy} \rightarrow A^-$$

The quantity of energy absorbed in such an endothermic process is represented by a positive number. An ion produced in this way is likely to be unstable and to lose the added electron spontaneously.

Figure 5-16 shows the electron affinity in kilojoules per mole for the main-group elements. Figure 5-17 presents these data in graphic form. Note that there are both positive and negative values.

Period trends For each period, halogen atoms (Group 17) gain electrons most easily, as shown by the large negative values for their electron affinities. The ease with which halogen atoms gain electrons is a major reason for the high reactivities of the halogens. Electron affinities become more negative across each period within the *p* block, as electrons add to the same *p* sublevel

Figure 5-16 Periodic table of electron affinities. Values in parentheses are calculated; values not in parentheses are experimental.

Values in parentheses are calculated; values not in parentheses are experimental.

The Periodic Law

Figure 5-17 Plot of electron affinity versus atomic number. The data are plotted from the experimental and estimated values shown in Figure 5-16.

of atoms with increasing nuclear charge. An exception to this trend occurs between Groups 14 and 15. Compare the electron affinities of carbon ([He]$2s^2 2p^2$) and nitrogen ([He]$2s^2 2p^3$). Adding an electron to a carbon atom gives a half-filled sublevel. This occurs more easily than adding a second electron to an already half-filled p sublevel of a nitrogen atom.

Atoms of Group 2 and Group 18 elements are the least likely to gain electrons as shown by the positive values for their electron affinities (Figure 5-17). Here, electrons must be forced into atoms with filled outer sublevels, ns^2 for Group 2 and ns^2np^6 for Group 18 (with the exception of helium, $1s^2$).

Group trends Trends for electron affinity values within groups are not as regular as ionization energies and they are more difficult to explain. In some main groups, however, especially from the third to the sixth periods, there is a tendency for electrons to add with greater difficulty as one proceeds down the group. You may wonder why the increased nuclear charge down the groups does not increase the attraction for added electrons. Recall that atomic radii also increase as one moves down the group and it is generally more difficult to add electrons to larger atoms than to smaller atoms.

Formation of negative ions For an isolated ion in the gas phase, it is always more difficult to add a second electron to an already negatively charged ion. Therefore, second electron affinities are all positive. However, ions with negative charges higher than 1 are present in chemical compounds. Electron affinities are measured on isolated atoms in the gas phase. Other factors contribute to the energy balance in the formation of chemical compounds in the liquid or solid states. Certain *p*-block nonmetals tend to form negative ions that have noble gas configurations. The halogens do so by adding one electron. For example, chlorine (Cl, [Ne]$3s^23p^5$) achieves the configuration of the next highest noble gas, argon (Ar), by adding an electron to give Cl$^-$ ([Ne]$3s^23p^6$). Atoms of Group 16 elements are present in many compounds as 2- ions. For example, oxygen ([He]$2s^22p^4$) achieves the configuration of the next highest noble gas, neon, by adding two electrons (O^{2-}[He]$2s^22p^6$). Nitrogen achieves a neon configuration by adding three electrons to give N^{3-}.

Ionic Radii

Figure 5-18 shows the radii of some of the most common ions of the elements—ions present in many chemical compounds.

A positive ion is known as a **cation.** The formation of a cation by the loss of one or more electrons always leads to a decrease in radius. As expected, with the highest energy level electrons removed, the size of the remaining electron cloud is smaller. Also, the remaining electrons are drawn closer to the nucleus by its unbalanced positive charge.

A negative ion is known as an **anion.** The formation of an anion by the addition of one or more electrons always leads to an increase in radius. In the common anions shown in Figure 5-18, enough electrons have been added to

Figure 5-18 The ionic radii of the ions most common in chemical compounds. Cations are smaller, and anions are larger, than the atoms from which they are formed.

Periodic Table of Ionic Radii

give anions with noble gas configurations. The total positive charge of the nucleus is unchanged. The electron cloud spreads out because there is a smaller attractive force on the electrons and a greater repulsion between them.

Periodic trends Within each period, the metals at the left form cations and the nonmetals at the right form anions. There is a regular decrease of cation radii with increasing atomic number. This decrease is to be expected as electrons are removed one by one from the highest energy level as the nuclear charge increases.

Beginning with Group 15, anions are more common than cations. There is also a regular decrease of radii because the nuclear charge is increasing and the number of added electrons is decreasing.

Group trends In proceeding down the groups, the outer electrons for both cations and anions are in higher energy levels, as they are for atoms. Therefore, there is a gradual increase of ionic radii down the groups as there is for atomic radii.

Valence Electrons

As you will learn in detail in Chapter 6, chemical compounds form because electrons are lost, gained, or shared between atoms of different elements. The electrons that interact in this manner are those in the highest energy levels. These are the electrons most subject to the influence of nearby atoms or ions. *The electrons available to be lost, gained, or shared in the formation of chemical compounds are referred to as* **valence electrons.** For example, the electron lost from the $3s$ sublevel of Na to form Na^+ is a valence electron. Valence electrons are often located in incompletely filled main-energy levels.

For main-group elements, electrons in the s and p sublevels of the highest occupied energy level are the valence electrons. The inner electrons are all in filled energy levels and are too closely held by the nucleus to be involved in compound formation.

The Group 1 and Group 2 elements have one and two valence electrons, respectively, as shown in Table 5-5. The elements of Groups 13–18 have a number of valence electrons equal to the group number minus 10. In some cases, both the s and p sublevel valence electrons of the p-block elements are involved in compound formation. In other cases, only the electrons from the p sublevel are involved.

STUDY HINT

Note that electronegativity is a property of atoms in compounds and thus differs from ionization energy and electron affinity, which are properties of isolated atoms.

Electronegativity

Valence electrons are responsible for holding atoms together in chemical compounds. Also, the properties of many chemical compounds are strongly influenced by the concentration of negative charge closer to one atom than to another. It is useful to have a measure of how strongly one atom attracts the electrons of another atom when they are combined in a compound.

TABLE 5-5 VALENCE ELECTRONS IN MAIN-GROUP ELEMENTS								
Group number:	1	2	13	14	15	16	17	18
Valence electrons:	1	2	3	4	5	6	7	8

Linus Pauling (1901–   ), one of the most widely known American chemists, developed the concept of electronegativity, a scale for assigning numerical values to the tendency of an atom to attract electrons. **Electronegativity** *is a measure of the ability of an atom in a chemical compound to attract electrons.* The most electronegative element, fluorine, is arbitrarily assigned an electronegativity value of four. Electronegativity values for the other elements are then calculated in relation to this value.

The electronegativities of the elements are shown in Figure 5-19 and plotted in graphical form in Figure 5-20. The graph shows that electronegativity values vary periodically.

Periodic trends Across each period, there is a gradual increase in electronegativity. Nonmetals are generally more electronegative than metals. The active alkali and alkaline earth metals at the left of each period are the least electronegative elements. Their atoms have very little attraction for electrons in chemical compounds. The halogens at the right are the most electronegative elements. Halogen atoms attract electrons in chemical compounds most strongly of all the elements.

STUDY HINT
In general, metals are less electronegative than nonmetals.

Linus Pauling received the 1954 Nobel Prize in Chemistry for his work in molecular structure and chemical bonding (which includes electronegativity). He received the 1962 Nobel Peace Prize for his efforts in support of the Nuclear Test Ban Treaty of 1963. Pauling, along with Marie Curie (1867–1934), is one of the few people to receive two Nobel Prizes.

Figure 5-19 Electronegativities of the elements according to the Pauling scale.

Periodic Table of the Electronegativities

Group 1	Group 2	Group 3	Group 4	Group 5	Group 6	Group 7	Group 8	Group 9	Group 10	Group 11	Group 12	Group 13	Group 14	Group 15	Group 16	Group 17	Group 18
1 H 2.1																	2 He
3 Li 1.0	4 Be 1.5											5 B 2.0	6 C 2.5	7 N 3.0	8 O 3.5	9 F 4.0	10 Ne
11 Na 0.9	12 Mg 1.2											13 Al 1.5	14 Si 1.8	15 P 2.1	16 S 2.5	17 Cl 3.0	18 Ar
19 K 0.8	20 Ca 1.0	21 Sc 1.3	22 Ti 1.5	23 V 1.6	24 Cr 1.6	25 Mn 1.5	26 Fe 1.8	27 Co 1.8	28 Ni 1.8	29 Cu 1.9	30 Zn 1.6	31 Ga 1.6	32 Ge 1.8	33 As 2.0	34 Se 2.4	35 Br 2.8	36 Kr 3.0
37 Rb 0.8	38 Sr 1.0	39 Y 1.2	40 Zr 1.4	41 Nb 1.6	42 Mo 1.8	43 Tc 1.9	44 Ru 2.2	45 Rh 2.2	46 Pd 2.2	47 Ag 1.9	48 Cd 1.7	49 In 1.7	50 Sn 1.8	51 Sb 1.9	52 Te 2.1	53 I 2.5	54 Xe 2.6
55 Cs 0.7	56 Ba 0.9	57 La 1.1	72 Hf 1.3	73 Ta 1.5	74 W 1.7	75 Re 1.9	76 Os 2.2	77 Ir 2.2	78 Pt 2.2	79 Au 2.4	80 Hg 1.9	81 Tl 1.8	82 Pb 1.8	83 Bi 1.9	84 Po 2.0	85 At 2.2	86 Rn 2.4
87 Fr 0.7	88 Ra 0.9	89 Ac 1.1	104 Unq	105 Unp	106 Unh	107 Uns	108 Uno	109 Une									

Lanthanide Series

58 Ce 1.1	59 Pr 1.1	60 Nd 1.1	61 Pm 1.1	62 Sm 1.1	63 Eu 1.1	64 Gd 1.1	65 Tb 1.1	66 Dy 1.1	67 Ho 1.1	68 Er 1.1	69 Tm 1.1	70 Yb 1.1	71 Lu 1.2
90 Th 1.3	91 Pa 1.5	92 U 1.7	93 Np 1.3	94 Pu 1.3	95 Am 1.3	96 Cm 1.3	97 Bk 1.3	98 Cf 1.3	99 Es 1.3	100 Fm 1.3	101 Md 1.3	102 No 1.3	103 Lr

Actinide Series

The Periodic Law

Figure 5-20 Plot of electronegativity versus atomic number for Periods 1–6

Group trends Within each main group, electronegativities either decrease down the group or have very similar values. This trend is in keeping with the sizes of the atoms. Larger, more easily ionized atoms attract electrons less strongly than smaller, less easily ionized atoms. Cesium and francium, which have the largest atomic radii and the lowest ionization energies, are the least electronegative elements. Fluorine, the most electronegative element, has a very small radius and a high ionization energy.

Sample Problem 5.7

Among the elements gallium (Ga), bromine (Br), and calcium (Ca), which has the highest electronegativity? Explain in terms of Periodic Table trends. Will the element of highest electronegativity have the largest or smallest atomic radius among the elements listed?

Solution
The elements are all in Period 4. Bromine has the highest atomic number and is furthest to the right in the period. It should have the highest electronegativity, for electronegativity increases across the periods. Bromine should have the smallest atomic radius because smaller atoms have higher electronegativity, and because radii tend to decrease across a period.

Practice Problems
Consider the five hypothetical main group elements V, W, X, Y, and Z with the indicated outer electron configurations:

$V = 2s^2 2p^5$ $W = 4d^{10} 5s^2 5p^5$ $X = 2s^2 2p^2$ $Y = 5d^{10} 6s^2 6p^5$ $Z = 2s^2 2p^4$

Sample Problem 5.7 *continued*

> 1. Identify the block location for each, and then determine which elements are in the same group and which are in the same period. *(Ans.)* All are *p*-block: V, X, and Z are in the same period and V, W, and Y are in the same group.
> 2. Which would you expect to *(a)* have the highest electron affinity? *(Ans.)* V *(b)* form with a 1− charge ion? *(Ans.)* V, W, Y *(c)* have the highest electronegativity *(Ans.)* V
> 3. Compare the ionic radius of the typical ion formed by the element W with the radius of its neutral atom. *(Ans.)* The ionic radius would be larger.
> 4. Which element(s) would be expected to contain seven valence electrons? *(Ans.)* V, W, and Y

Periodic Properties of the *d*- and *f*-Block Elements

Up to now, the fairly regular periodic trends of the main-group elements have been considered. Many properties of the *d*-block elements (which are all metals) vary less and with less regularity than those of the main-group elements. You can see that the curves in Figures 5-13 and 5-15 flatten where the *d*-block elements fall in the middle of Periods 4–6.

Atoms of the *d*-block elements contain from zero to two electrons in the highest occupied energy level. Electrons are added to the *d* sublevel of the next highest energy level because it is close in energy to the outermost level for these elements. Electrons in both the *ns* sublevel and the $(n-1)d$ sublevel are therefore available to interact with their surroundings. As a result, electrons in the incompletely filled *d* sublevels are responsible for many characteristic properties of the *d*-block elements.

Atomic radii   The atomic radii of the *d*-block elements generally decrease across the periods. However, this decrease is less than that for the main-group elements because the electrons added to the $(n-1)d$ sublevel shield the outer electrons from the nucleus. Also, note in Figure 5-13 that the radii dip to a low and then increase slightly across each of the four periods that contain *d*-block elements. As the number of electrons in the *d* sublevel increases, the radii increase because of the repulsion among the electrons.

In the sixth period, the *f*-block elements fall between lanthanum (Group 3) and hafnium (Group 4). Because of the increase in atomic number that occurs from lanthanum to hafnium, the atomic radius of hafnium is actually slightly less than that of zirconium (Zr), which is immediately above it. The radii of elements following hafnium in the sixth period vary with increasing atomic number in the usual manner.

Ionization energy   Ionization energies generally increase across the periods with the decreasing size of the atoms, as they do for the main-group elements. In contrast to the decrease down the main groups, however, the first ionization energies of the *d*-block elements increase as one proceeds down each group. The electrons available for ionization in the outer *s* sublevels are apparently less shielded from the increasing nuclear charge by electrons in the incomplete $(n-1)d$-sublevels. It is also harder to pull away the outer electrons in atoms of the Period-6 elements, which are about the same size as those of the Period-5 elements but with an increased nuclear charge.

Ion formation and ionic radii The order in which electrons are removed from all atoms is *exactly the reverse of the order given by the electron-configuration notation*. In other words, electrons in the highest occupied sublevel are always removed first. For the *d*-block elements this means that although newly added electrons are in the *d* sublevels, the first electrons to be removed are those in the highest occupied *s* sublevels. For example, iron (Fe), which has the electron configuration $[Ar]3d^6 4s^2$, first loses a $4s$ electron to form the Fe^+ ion ($[Ar]3d^6 4s^1$). The Fe^+ ion can then lose the second $4s$ electron to form the Fe^{2+} ion ($[Ar]3d^6$). The Fe^{2+} ion can then lose a $3d$ electron to form the Fe^{3+} ion ($[Ar]3d^5$). The iron ions most commonly found in chemical compounds are the Fe^{2+} and the Fe^{3+} ions.

The Group-3 elements form only ions with a 3+ charge in compounds. Copper forms 1+ and 2+ ions and silver forms 1+ ions. Most other *d*-block elements commonly form 2+ ions in compounds although some, such as iron, also form 3+ ions. As expected, the cations have smaller radii than the atoms. Comparing 2+ ions across the periods shows a decrease in size that parallels the decrease in atomic radii.

Electronegativity The *d*-block elements all have electronegativity values between 1.1 and 2.4, as shown in Figure 5-20. Only the active metals of Groups 1 and 2 have lower electronegativities. Across the periods there is a general trend for electronegativity values to increase as radii decrease and vice versa. In other words, small atoms tend to attract electrons more strongly than large atoms.

Section Review

1. State the general period and group trends among main-group elements with respect to each of the following: *(a)* atomic radii *(b)* first ionization energy *(c)* electron affinity *(d)* ionic radii *(e)* electronegativity.
2. Among the main-group elements, what is the relationship between group number and the number of valence electrons among group members?
3. *(a)* In general, how do the periodic properties of the *d*-block elements compare with those of the main-group elements? *(b)* Explain the comparisons made in *(a)*.

Chapter Summary

- Elements in the Periodic Table with similar chemical properties fall in the same column.
- The rows in the Periodic Table are referred to as *periods*.
- The columns in the Periodic Table are referred to as *groups*.
- As you move to the right in the Periodic Table, elements in the *s*, *p*, *d*, and *f* blocks respectively gain an electron in their *s*, *p*, *d*, or *f* sublevel.
- The Periodic Table can be used to display trends in the following properties of the elements—electron affinity, electronegativity, ionization energy, atomic radii, and ionic radii.

---

1. *(a)* generally decrease across periods and increase down groups *(b)* generally increase across periods and decrease down groups *(c)* generally increase across periods among Groups 13–17 and decrease down groups, but numerous exceptions are observed *(d)* generally decrease across periods and increase down groups *(e)* gradually increases across periods and either decreases or is very similar in value down groups
2. For Groups 1 and 2, the number of valence electrons is equal to the group number; for Groups 13–18, the number of valence electrons is equal to the group number minus 10.
3. *(a)* vary less, and with less regularity, than main-group elements *(b)* generally attributed to the presence of electrons in incompletely filled *d* sublevels in these atoms.

CHAPTER 5 REVIEW

The Periodic Law

VOCABULARY

actinides (130)
alkali metals (136)
alkaline earth metals (136)
anion (151)
atomic radius (143)
cation (151)
electron affinity (149)
electronegativity (153)
halogens (140)
ion (145)
ionic radius (151)
ionization (145)
ionization energy (145)
lanthanides (rare earth elements) (130)
main-group elements (139)
periodic law (129)
Periodic Table (130)
transition elements (138)
valence electrons (152)

REVIEWING CONCEPTS

1. Explain the meaning of the periodic law. (5.1)
2. What is the Periodic Table? (5.1)
3. **a.** Distinguish between the lanthanide and actinide elements by definition and table location.
   b. What are the rare earth elements? (5.1)
4. How is periodicity with respect to atomic number demonstrated within Periodic Table groups? (5.1)
5. Why are the noble gases relatively nonreactive? (5.1)
6. **a.** How do the electron configurations of the elements within a group compare (for main groups only)?
   b. What is the relationship between the electron configuration of an element and the properties of that element? (5.1)
7. What determines the length of each period in the table? (5.1)
8. What is the relationship between the electron configuration of an element and the period in which that element appears in the table? (5.2)
9. **a.** What information is provided by the specific block location of an element?
   b. Identify, by number, the groups located within each of the four block areas. (5.2)
10. **a.** Which elements are designated as the alkali metals?
    b. List four of their characteristic properties.
    c. Cite two of their common uses. (5.2)
11. **a.** Which elements are designated as the alkaline-earth metals?
    b. How do their characteristic properties compare with those of the alkali metals?
    c. Cite at least two common uses of alkaline-earth metals. (5.2)
12. **a.** Write the usual group configuration notation for each d-block group.
    b. How do these group numbers relate to the number of outer s and d electrons?
13. What name is sometimes used to refer to the entire set of d-block elements? (5.2)
14. **a.** What types of elements make up the p block?
    b. How do the properties of the p block metals compare with those in the s and d blocks? (5.2)
15. **a.** Which elements are designated as the halogens?
    b. List three of their characteristic properties. (5.2)
16. **a.** Which elements are designated as the noble gases?
    b. What is the most significant property of these elements?
    c. What use is generally made of that property? (5.2)
17. **a.** Which elements are metalloids?
    b. Describe their characteristic properties. (5.2)
18. **a.** Which elements make up the f-block in the Periodic Table?

The Periodic Law 157

b. Describe the characteristic properties and primary uses of these elements. (5.2)

19. a. What are the main-group elements?
    b. What trends can be observed across the various periods within the main-group elements? (5.2)

20. a. What is meant by atomic radius?
    b. What trend is observed among the atomic radii of main-group elements across a period?
    c. How can this trend be explained? (5.3)

21. a. What trend is observed among the atomic radii of main-group elements down a group?
    b. How can this trend be explained? (5.3)

22. Define each of the following:
    a. ion
    b. ionization
    c. ionization energy (5.3)

23. a. How do the first ionization energies of main-group elements vary across a period and down a group?
    b. Explain the basis for each trend. (5.3)

24. a. What is electron affinity?
    b. What signs are associated with electron affinity values and what are their significance? (5.3)

25. a. Distinguish between a cation and an anion.
    b. How do the sizes of each compare with those of their neutral atoms from which they were formed? (5.3)

26. What causes chemical compounds to form? (5.3)

27. a. What are valence electrons?
    b. Where are such electrons generally located? (5.3)

28. For each of the following groups, indicate whether electrons are more likely to be lost or gained in compound formation and the number of such electrons typically involved:
    a. Group 1   d. Group 16
    b. Group 2   e. Group 17
    c. Group 13   f. Group 18 (5.3)

29. a. What is electronegativity?
    b. How are electronegativity values assigned to the Periodic Table elements? (5.3)

30. Identify the most and least electronegative group or groups of elements on the Periodic Table. (5.3)

PROBLEMS

▶ **Electron Configuration and Periodic Properties**

1. Write the shorthand electron configuration and indicate the period in which each belongs?
   a. Li   c. Mg   e. Sc   g. Sn
   b. O   d. P   f. Br

2. Based on the information given below, give the group, period, block, and identity of each element described:
   a. $[He]2s^2$ d. $[Ar]4s^2$
   b. $[Ne]3s^1$ e. $[Rn]7s^1$
   c. $[Kr]5s^2$

3. Without looking at the Periodic Table identify the period, block, and group in which each of the following elements is located:
   a. $[Ar]3d^24s^2$ c. $[Kr]4d^{10}5s^1$
   b. $[Ar]3d^54s^1$ d. $[Xe]4f^{14}5d^96s^1$

4. Without looking at the Periodic Table write the expected outer electron configuration for each of the following elements:
   a. Period 4, Group 7   c. Period 6, Group 12
   b. Period 5, Group 3

5. Without looking at the Periodic Table identify the period, block, and group in which each of the following is located:
   a. $[Ne]3s^23p^4$
   b. $[Kr]4d^{10}5s^25p^2$
   c. $[Xe]4f^{14}5d^{10}6s^26p^5$

6. For each of the following identify the block, period, group, group name (where appropriate), element name, element type, and relative reactivity:
   a. $[Ne]3s^23p^1$ c. $[Kr]4d^{10}5s^1$
   b. $[Ar]3d^{10}4s^24p^6$ d. $[Xe]4f^{15}d^16s^2$

▶ **Ionization, Electron Affinity, and Electronegativity**

7. a. Distinguish among the first, second, and third ionization energies of an atom.

b. How do the values of successive ionization energies compare?
c. Why does this occur?

8. Without looking at the electron affinity table, arrange the elements listed in order of *decreasing* electron affinities: C, O, Li, Na, Rb, Ne, and F.

9. a. Without looking at the ionization energy table, arrange the elements listed in order of decreasing first ionization energies: Li, O, C, K, Ne, and F.
b. Which would you expect to have the highest second ionization energy? Why?

10. For each element listed below, determine the charge of the ion most likely to be formed and the identity of the noble gas whose electron configuration is thus achieved:
  a. Li   **f.** Al
  b. B   **g.** P
  c. O   **h.** S
  d. F   **i.** Br
  e. Mg   **j.** Ba

11. Which among the following does not have a noble gas configuration: Na^+, Rb^+, O^{2-}, Br^-, Ca^+, Al^{3+}, S^{2-}?

12. Which of the following ions is least likely to form: I^-, Sr^{2+}, Al^{3+}, Cl^-, O^{2-}, K^{2+}?

13. The two ions K^+ and Ca^{2+} each have 18 electrons surrounding the nucleus. Which would you expect to have the smaller radius? Why?

CRITICAL THINKING

As a member on the newly inhabited space station Alpha, you are given the task of organizing information on newly discovered elements as it comes from the laboratory. To date, 5 elements have been discovered and have been assigned names from the Greek alphabet.

1. Applying Models Using the data collected so far on the planetary elements analyzed create a periodic table based on the elements' properties.

Element Name	Atomic No.	Atomic Mass	Properties
Epsilon	23	47.333	nonmetal, very reactive, produces a salt when combined with a metal, gaseous state
Beta	13	27.011	metal, very reactive, soft solid, low melting point
Gamma	12	25.356	nonmetal, gaseous element, extremely unreactive
Delta	4	7.988	nonmetal, very abundant, forms compounds with most other elements
Lambda	9	16.174	metal, solid state, good conductor, high luster, hard and dense

2. Predicting Outcomes Using your newly created periodic table, predict the atomic number of an element with an atomic mass of 11.289 that has nonmetallic properties and is very reactive.

3. Predicting Outcomes Predict the atomic number of an element having an atomic mass of 15.023 that exhibits metallic properties, but is softer than lambda and harder than beta.

4. Analyzing Information Analyze your periodic table for periodic trends and describe the trends that you identify.

CROSS CONNECTIONS

1. Health Just as chemists use the periodic table as their source of basic information about the elements, dietitians use a nutritional chart as their source of information about nutrients in food groups. Construct a list of other occupations that use a particular chart as a source of basic information as chemists use the periodic table.

RESEARCH AND WRITING

1. Prepare a report tracing the evolution of the current Periodic Table. Cite the chemists involved and the major contribution of each.

2. Write a report describing the contributions of Glenn Seaborg toward the discovery of many of the actinide elements.

CHAPTER 6
Chemical Bonding

PLANNING CHART

Approximate Time: 8/10 days

	SECTION/TEXT FEATURES		RESOURCES	
BASIC	**6.1 INTRODUCTION TO CHEMICAL BONDING**	**161**	*Teacher's Resource Binder*	
	Demo 1	160A	Section Review Worksheets	18
	Section Review	164	*Other Resources*	
	Chapter Review	191	Transparency 13	
	• Questions 1–6, Problems 1–4			
BASIC	**6.2 COVALENT BONDING AND MOLECULAR COMPOUNDS**	**164**	*Teacher's Resource Binder*	
	Demos 2, 3	160B–C	Lab Experiments 23, 42	131, 213
	Section Review	174	Section Review Worksheets	19
	Chapter Review	191	*Other Resources*	
	• Questions 7–14, Problems 5–7		Transparency 14	
			Blackline Master 24	
BASIC	**6.3 IONIC BONDING AND IONIC COMPOUNDS**	**175**	*Teacher's Resource Binder*	
	Section Review	179	Section Review Worksheets	20
	Chapter Review	191	*Other Resources*	
	• Questions 15–18		World of Chemistry Video 8	
			Transparency 15	
AVERAGE	**6.4 METALLIC BONDING**	**179**	*Teacher's Resource Binder*	
	Demo 4	160C	Section Review Worksheets	21
	Microscale Investigation 4	837		
	Section Review	181		
	Chapter Review	191		
	• Questions 19–21			
AVERAGE	**6.5 THE PROPERTIES OF MOLECULAR COMPOUNDS**	**182**	*Teacher's Resource Binder*	
	Demo 5	160D	Section Review Worksheets	22
	Section Review	190	*Other Resources*	
	Chapter Review	191	World of Chemistry Video 9	
	• Questions 22–30, Problems 8–14		Transparencies 16, 17	
			Blackline Master 25	
	CHAPTER END ACTIVITIES		*Teacher's Resource Binder*	
	Critical Thinking, Cross Connections, Research and Writing, Projects		Test Booklet	25, 27
			Computer Test Generator	

Teaching Strategies

6.1 Introduction to Chemical Bonding

Students sometimes have difficulty understanding how a compound with a percent composition of 50% can be polar covalent and one with 51% can be ionic. Electronegativity, percent of ionic character, and the 0-4%, 5-50%, 51-99% cut-off points are convenient indicators to give an idea of the type of bonds we are dealing with. To help students understand the concept of chemical bonding, have them visualize the following situation: Take a given space and put black at one end and white at the other. Then, starting with the black, gradually and imperceptively fade it into dark grey, medium grey, lighter grey until it blends into white. Could we draw a line in the middle and say everything to the left is black and everything to the right is white—even though we know it is all different gradations of grey?

Demonstration 1: Bond Formation as an Exothermic Reaction

PURPOSE: To show the energy changes that occur in the formation of ionic bonding and covalent bonding
MATERIALS: Bunsen burner, coffee can—3 lb, deflagrating spoon, magnesium ribbon, sulfur, tongs

PROCEDURE: Part 1: (This is a dramatic twist to the burning of magnesium ribbon.) Goggles and a lab coat should be worn. Darken the room as much as possible. CAUTION: Remind students not to look directly at the magnesium ribbon when it ignites. Holding a 6-cm piece of Mg ribbon with tongs, ignite it in a burner flame and immediately put it in a coffee can while it continues to burn. Students will see the reflected light produced as the Mg burns. Describe the energy change that occurred here. (Energy in the form of light was definitely released.) What kind of bond was formed? (Tell students that the magnesium reacts with both the oxygen and the nitrogen in the air, thus two different compounds are formed in this reaction.) Have students figure out that both compounds have ionic bonds by using electronegativities.

$$2Mg + O_2 \rightarrow 2MgO \text{ and } 3Mg + N_2 \rightarrow Mg_3N_2$$

Part 2: Put a small piece of sulfur in a deflagrating spoon, and ignite it under a hood. Students will see that it burns with a blue flame. Describe the energy change that occurred here. (Energy in the form of light was given off.) What kind of bond was formed? (Polar covalent) Remind students that most, but not all, bond formation is exothermic.

6.2 Covalent Bonding and Molecular Compounds

In 1932, Langmuir received the Nobel Prize in chemistry for his work on monomolecular films, which are films that are assumed to have a thickness of one molecule. Using a method similar to his makes it possible to determine the approximate size of a molecule. This is also an excellent demonstration because the size is determined by indirect measurement. Before doing the demonstration, go over the formula for determining the volume of a cylinder, $V = \pi r^2 h$. Point out that if you knew the volume of a cylinder and its diameter, d, you could easily determine its height or thickness, $h = V/\pi r^2$. Do a couple of sample problems determining the thickness of a cylinder when its volume and diameter are known. Hold up a piece of filter paper and ask students if this is a cylinder. This helps develop the idea that cylinders can be very short.

Demonstration 2: Size of a Molecule
PURPOSE: To learn how the size of a molecule can be determined experimentally
MATERIALS: Cafeteria tray (approximately 30 cm × 40 cm), a 1-mL pipet (a 10-mL graduate and an eyedropper can be substituted for the pipet), a plastic ruler, distilled water, 95% ethyl alcohol, lycopodium powder or fine talcum powder, oleic acid
PROCEDURE: Prior to class: (1) Clean all equipment several times with detergent, and rinse well with water. *Clean, dry equipment is a must!* (2) Dissolve 5 mL of oleic acid in 95 mL of ethyl alcohol. Make a second solution by dissolving 5 mL of the first solution in 45 mL of ethyl alcohol. The second solution, which will be the one used in the demo, should contain 0.005 mL of oleic acid/mL of solution. To determine the volume of oleic acid/drop, use the pipet and count the number of drops/mL of solution 2. Volume of oleic acid/drop = 0.005 mL/number of drops per mL of solution. (Assuming 20 drops/mL, the volume of oleic acid/drop would be 0.0025 mL.) (3) Refrigerate sufficient distilled water one day prior to the demonstration. Add enough distilled water to cover the tray to a depth of 0.5 cm. Put a small amount of lycopodium or talcum powder on a piece of paper; gently blow the powder over the water. Examine the water surface carefully to make sure that there is only a thin dusting of powder spread across the water. Now add one drop of the second oleic acid solution at the center of the tray. Gently tap the tray so that the oleic acid film is spread out over the largest possible area, without being broken. The oleic acid layer should assume a roughly circular shape. If the equipment is dirty, an irregular, star-shaped figure may result. In this case, clean the tray and start over. After waiting several minutes for the alcohol to evaporate, make several measurements of the diameter of the oleic acid film in cm and take an average of these readings. Using r = average diameter/2, calculate the thickness of the film; this should turn out to be somewhere in the area of 10^{-7} cm. Assuming it is monomolecular, the thickness will be the diameter of an oleic acid molecule. Even if the demonstration does not turn out perfectly, the students will get the idea.

Demonstration 3: Bond Energies of the Hydrogen Halides
PURPOSE: To help students understand bond energies and, by using models, to show students how to use the Bond Length and Bond Energy Tables to explain why certain reactions take place
MATERIALS: Bond Length and Bond Energy Tables, small gumdrops—same color, large gumdrops in 3 colors, round or heavy toothpicks, tape
PROCEDURE: (1) As you propose the following problem to the students, write on the chalkboard the information about each hydrogen halide. (2) As you discuss each molecule, have students assemble a gumdrop-toothpick model of it. Each model of a hydrogen halide should consist of a small gumdrop representing hydrogen and a large gumdrop for the halide. Make each halogen a different color. Break 2 cm off of each end of four toothpicks and tape them together for the H-Cl bond; break 1 cm off each end of two toothpicks and tape for the H-Br bond; and use one full length toothpick for the H-I bond. (See Figure T6-1.) To show bonds breaking, let students break the toothpick bonds and determine which is easiest and hardest to break—indicating which needs the least and most bond energy. Also have students identify how the ease of breaking relates to the length of the toothpick.

Figure T6-1

Chemical Bonding **160B**

(3) A problem that shows differences in bond energies follows: When a heated glass rod is thrust into a test tube of HCl(g), the HCl shows no signs of decomposing. Continuous external heating of the HCl also gives the same results. When a heated glass rod is thrust into a test tube of HBr(g), there is no reaction; but continuous external heating results in the production of red-brown Br_2, indicating the breaking of the H-Br bonds. When a heated glass rod is put into HI, a violet-colored gas is produced, indicating the presence of I_2 vapor due to the breaking of the H-I bonds. Why do you think the bonds of the HI molecules were broken by the hot glass rod and those of the HCl and HBr were not? Hint: Look at Table 6-1 Bond Lengths and Bond Energies. What is the bond energy of the H-Cl bond? Of the H-Br bond? Of the H-I bond? Explain. To introduce the idea of unequal sharing (polar covalent bonds), personalize the atoms a little. You can explain that chlorine promised *faithfully* to share her electron with hydrogen, but she never intended to share *equally*. Chlorine is a pig! (it has a high electronegativity and thus strong pull for e⁻). She keeps her electron and hydrogen's a greater percentage of the time. Then she says, "Here, Hydrogen, I'll let you have my electron and yours for just a little while, but I want them right back." Since hydrogen is weaker (weak pull for e⁻ due to a lower electronegativity), there's nothing he can do. Remind students that what is really true is that the density of the electron cloud is greater near the chlorine and less near the hydrogen.

A spectacular demonstration of the formation of a covalent bond can be found in Shakhashiri, "Reactions of Zinc and Iodine," *Chemical Demonstrations*, Vol. 1, pp. 49–50.

6.3 Ionic Bonding and Ionic Compounds

Use the following drawing on a transparency or chalkboard to help review Lewis structures and as a visual reference for the following explanation of ionic bonding.

[Energy level diagram for $^{23}_{11}Na°$ and $^{35}_{17}Cl°$ showing 1s 2s 2p 3s and 3p 3s 2p 2s 1s orbitals, with 11p⁺ 12n° and 17p⁺ 18n°; second diagram showing 1st 2nd 3rd energy levels with 2e⁻) 8e⁻) 1e⁻) and (7e⁻ (8e⁻ (2e⁻ 18n°; Na• and :Cl:]

A successful means of introducing ionic bonding is with talking atoms. Referring to the above drawing, tell the students that if these atoms could talk, here's what they might say: (*Include a little dramatics.*)

Na: "Drat, if I could get rid of this one stupid electron in my 3rd energy level, the 2nd energy level would be my outermost energy level and I would be a happy atom—I mean chemically stable with a complete outer energy level of eight electrons! I'd sell it cheap!" (*ionization energy*)
Cl: "If I could just get one more electron, I would have a complete outer energy level. I'd be willing to pay lots for it!" (*electron affinity*)
 [Na is walking along and meets Cl.]
Na: "Hi, Chlorine, do you know anyone who would like an electron?"
Cl: "Do I ever! I'll take it. It's all I need to complete my outer energy level!"
Na: "Well, then, here it is, Chlorine. Now I'm positively charged, and I feel so stable!" (Using the energy level diagram, show why this is so. Count the number of electrons and compare it to the number of protons in the nucleus.)
Cl: "Thanks, Sodium. I'll take good care of it. Now I'm negatively charged, and do I ever feel stable."
Na: "Say, do you know what, Chlorine? Suddenly I feel very attracted to you."
Cl: "Sodium, I feel exactly the same way."
 [Na and Cl move together]
Both: "Ah, togetherness!"

You could make this into a student skit. However you do it, when your atoms start talking, your students start listening! To help the students fully visualize what happened in this process, add the following diagram to the first one.

[Diagram: Na⁺ with 1st 2nd levels 11p⁺ 12n° 2e⁻) 8e⁻); Cl⁻ with 3rd 2nd 1st levels (8e⁻ (8e⁻ (2e⁻ 17p⁺ 18n°; Na⁺ :Cl:⁻]

A dramatic demonstration of ionic bonding is found in Shakhashiri, "Reaction of Sodium and Chlorine," *Chemical Demonstrations*, Vol. 1, pp. 61–63.

To help students distinguish between ionic, polar covalent, and nonpolar covalent bonds, recreate the following chart:

Types of Bonds Compared to Candy Bar Exchange and Sharing	
Type of Bond	
Ionic	Generous George has a candy bar he really doesn't want. Grabby Gertie, who already has one candy bar but wants more, gives him a little kiss, and he readily hands it over to her.
Pure or Nonpolar	Considerate Carl has a chocolate bar, and Caring Carol has a caramel nut bar. Carl shares half of his chocolate bar with Carol, and Carol shares half of her caramel nut bar with Carl. EQUAL SHARING!
Polar	Tricky Tony has a chocolate bar, and Sweet Sue has a coconut bar. Tony shares ONLY $\frac{1}{4}$ of his chocolate bar with Sue and keeps $\frac{3}{4}$ for himself. Sue gives Tony $\frac{3}{4}$ of her coconut bar and keeps only $\frac{1}{4}$ for herself. UNEQUAL SHARING!

6.4 Metallic Bonding

Demonstration 4: Structure of Metals

PURPOSE: To use a model to show how one plane of ions in metallic solids slide over another one without encountering resistance or breaking of bonds

MATERIALS: Gumdrops, toothpicks
PROCEDURE: To make the model, construct each layer by connecting 16 gumdrops with toothpicks—producing a layer of 4 gumdrops × 4 gumdrops. Make several layers and stack them together but do not connect the layers. Then you can show the students how the layers slide over one another when a horizontal force is applied to them. Since no resistance or breaking of bonds occurs, the structure does not shatter.

6.5 The Properties of Molecular Compounds

As you discuss the different theories explaining the shapes of molecules, use models to enhance your discussion. The models suggested below not only show shape, but also show position of lone pair electrons.

Demonstration 5: The Shapes of Molecules
PURPOSE: To help understand molecular polarity by using models to show the shapes of molecules and position of lone pair electrons in them
MATERIALS: Small colored marshmallows, large white marshmallows, Red Hots,® toothpicks, 2 oblong balloons
PROCEDURE: *(1)* To make a water molecule, use a large marshmallow for the oxygen and attach two small ones for the hydrogens, as shown below. To show the position of the two lone pair electrons, moisten four Red Hots® and stick them to the end of the large marshmallow opposite the hydrogens. The presence of the lone electrons on models should help students better understand how that end can be negative.

(2) For ammonia use one large marshmallow for nitrogen, three small marshmallows for the hydrogens, and two Red Hots® for the lone electron pair.

(3) A very simple way to show the tetrahedral shape of methane is to blow up two oblong balloons, twist them in the center, twist them around one another in the center, and let them flip to their most stable position.

Desktop Investigation: [Note: Models of Chemical Compounds in Chapter 7 could be used here.]

References and Resources

Books and Periodicals

Halpern, March. "A Simple Inexpensive Model for Student Discovery of VSEPR," *Journal of Chemical Education*, Vol. 56, No. 8, Aug. 1979, p. 531.

Jense, William B. "Abegg, Lewis, Langmuir and the Octet Rule," *Journal of Chemical Education*, Vol. 61, No. 3, March 1984, pp. 191–200.

Kapelios, S. and A. Mavrides. "Electronegativity: A Mnemonic Rule," *Journal of Chemical Education*, Vol. 64, No. 11, Nov. 1987, p. 941.

Mickey, Charles. "Molecular Geometry," *Journal of Chemical Education*, Vol. 57, No. 3, March 1980, pp. 210–12.

Pauling, Linus. "C.N. Lewis and the Chemical Bond," *Journal of Chemical Education*, Vol. 61, No. 3, March 1984, pp. 201–3.

Sanderson, R.T. "A Needed Replacement for the Customary Description of Chemical Bonding," *Journal of Chemical Education*, Vol. 59, No. 5, May 1982, p. 376.

Sanderson, R.T. "Principles of Electronegativity, Part 1, General Nature 1," *Journal of Chemical Education*, Vol. 65, No. 2, Feb. 1988, pp. 112–8.

Audiovisual Resources

Chem Study—Chemical Bonding, and *Shapes and Polarities of Molecules* (16mm film or VHS-video) Central Scientific Co., 1988

Chemistry 104—Bonding of Atoms, sound filmstrip, and *Chemistry 109—Forces in Solids, Liquids, and Gases*, VHS cassette or sound filmstrip, Learning Arts, 1988.

Experiments in Chemistry Series—The Size of Molecules, filmstrip, Educational Images, Ltd., 1988.

Introduction to Chemistry—Ionic and Covalent Bonds, filmstrip, Fisher Scientific-EMD, 1988.

Solo-Learn—Polar Covalence, Introduction to Chemical Bonding, and *Bond Type and Properties of Matter*, sound filmstrips, Central Scientific Co., 1988.

Computer Software

Atomic Structure 3, Apple II 64K and TRS-80 48K, Queue, 1988.

General Chemistry 1A–Chemical Bonding and *Solids and Liquids*, 128K, Queue, 1988.

Chemistry Courseware–Chemical Bonding, Apple II 48K, Barclay, 1988.

Chemistry Series–Bonding in Molecules and *Bonding between Molecules*, Apple II 48K and TRS-80 32K, Barclay, 1988.

Chemistry Help Series–Atomic Structure and Bonding, Apple II 48K, Barclay.

Fee, Richard. *Bonding*, Apple 48K. Chemistry According to ROF, 1988.

SEI Physical Chemistry, Apple 48K and IBM PC, 128K, Queue, 1988.

VideoChem, Apple 64K and IBM PC, 128K, Queue, 1988.

Answers and Solutions

Reviewing Concepts

1. **a.** A chemical bond is a link between atoms resulting from the mutual attraction of their nuclei and electrons.
   b. Chemical bonds are classified by how valence electrons are distributed around the nuclei of the combined atoms.
2. Metals tend to lose electrons to form positive ions, whereas nonmetals tend to gain electrons to form negative ions.
3. The two major bonding types are ionic and covalent. An ionic bond is the chemical bond resulting from electrostatic attraction between positive and negative ions. A covalent bond is a chemical bond resulting from the sharing of electrons between bonded atoms.
4. In general, the greater the electronegativity difference between two atoms is, the more ionic is the bond between them.
5. **a.** Polar refers to bonds that have an uneven distribution of charge.
   b. A polar-covalent bond is a covalent bond in which there is an unequal attraction for the shared electrons. A nonpolar-covalent bond is a covalent bond in which the bonding electrons are shared equally by the bonded atoms.
6. **a.** In general, atoms will form a chemical bond if their potential energy is lowered by the change.
   b. Chemical-bond formation is an energy-releasing process.
7. **a.** Bond length is determined by the distance at which potential energy is at a minimum; it is the point at which balance is attained between attraction and repulsion in a stable covalent bond.
   b. Bond energies become larger as bond lengths become shorter.
8. The electrons in a covalent bond occupy overlapping orbitals; each electron is free to move around in either of the orbitals, but both generally spend more time in the space between the two nuclei.
9. An unshared pair, or lone pair, is a pair of electrons that is not involved in bonding, but instead belongs exclusively to one atom.
10. The indicated number of valence electrons in each is:
    a. 1   **b.** 7   **c.** 2   **d.** 6
    e. 3   **f.** 5   **g.** 4   **h.** 8
11. In a Lewis structure, the most electropositive atom is usually the central atom of the arrangement. When carbon is present in the compound, it is usually found as the central atom, even in the case of CH_3I, where the electronegativities of C and I are both 2.5.
12. A single-covalent bond is a covalent bond between two atoms produced by the sharing of one pair of electrons, such as in H_2O. A double-covalent bond is a covalent bond between two atoms produced by the sharing of two pairs of electrons, such as in C_2H_4. A triple-covalent bond is a covalent bond between two atoms produced by the sharing of three pairs of electrons, such as in N_2.
13. The need for a multiple bond is evident if there are not enough valence electrons to complete octets by adding unshared pairs.
14. **a.** A polyatomic ion is a charged group of covalently bonded atoms.
    b. NH_4^+ and ClO_3^- are examples.
    c. Polyatomic ions are normally found combined in chemical compounds with ions of opposite charge.
15. **a.** An ionic compound is composed of positive and negative ions combined so that the positive and negative charges are equal.
    b. Most ionic compounds occur as solids composed of crystals.
16. **a.** A formula unit is the simplest unit indicated by the formula of any compound.
    b. One formula unit of calcium fluoride (CaF_2) consists of one calcium ion plus two fluoride ions.
    c. One formula unit would consist of two potassium ions and one sulfide ion (K_2S).
17. **a.** Lattice energy is the energy released when one mole of an ionic crystalline compound is formed from gaseous ions.
    b. The greater the lattice energy is, the stronger is the ionic bond.
18. **a.** Ionic compounds generally have higher melting and boiling points than do molecular compounds, and they do not tend to vaporize as readily at room temperature as do molecular compounds.
    b. The differences in their properties are generally due to differences in how strongly the individual particles are held together in the compounds.
    c. Ionic compounds are hard and brittle crystalline solids. In the molten state or dissolved in water they conduct electricity. Many are soluble in water.
19. **a.** Metals are better conductors of heat. They are more easily deformed than many solid molecular or ionic compounds, and they are electrical conductors in the solid state.
    b. The unusual electrical conductivity of metals in the solid state is due to the presence of highly mobile electrons.
20. Metal atoms contain a smaller number of valence electrons than of vacant highest energy-level orbitals, they have low ionization energies, and they have low electronegativities.
21. **a.** A metallic bond is a chemical bond resulting from the attraction between positive ions and surrounding mobile electrons.
    b. The strength of a metallic bond varies with the nuclear charge and the number of valence electrons.
22. **a.** According to the VSEPR theory, molecules are classified according to how many electron pairs surround a central atom.
    b. Both would be expected to be linear.
23. According to the VSEPR theory, the following molecular geometries are indicated:

a. linear
b. triangular planar
c. tetrahedral
d. triangular bipyramidal
e. octahedral.

24. a. The position of unshared electron pairs does not determine the shape of the molecule, although the presence of such pairs may somewhat alter the bond angles.
b. In predicting molecular shapes, double bonds are treated in the same way as single bonds.

25. a. Hybrid orbitals are orbitals of equal energy produced by the combination of two or more orbitals in the same atom.
b. The number of hybrid orbitals produced is always equal to the number of orbitals that have combined.

26. a. Intermolecular forces are the forces of attraction between molecules.
b. These forces are generally weaker than those involved in ionic and metallic bonding.
c. The strongest intermolecular forces act between polar molecules.

27. a. Dipole-dipole forces are forces of attraction between polar molecules.
b. The polarity of a molecule depends upon both the polarity of the individual bonds and on how the bonds are oriented with respect to each other.

28. a. An induced dipole is one produced in a nonpolar molecule when its electrons are momentarily attracted by a polar molecule.
b. These forces account for the solubility of nonpolar molecular oxygen in water, a process necessary for aquatic life.

29. a. Hydrogen bonding is the intermolecular attraction between a hydrogen atom bonded to a strongly electronegative atom (F, O, or N) and an unshared pair of electrons on another strongly electronegative atom (F, O, or N).
b. It is strong because the great electronegativity difference between an atom of H and one of F, O, or N bonded to it makes the bond highly polar. The hydrogen atom thus has a positive charge approaching that of a proton, and this, coupled with the small size of the hydrogen atom, allows it to attract quite strongly and approach closely an unshared pair of electrons on an adjacent molecule.

30. a. London dispersion forces are intermolecular attractions resulting from the constant motion of electrons and the creation of instantaneous dipoles and induced dipoles.
b. London forces increase with increasing masses of atoms or molecules.

Problems

1.

Bond	Electronegativity difference	Bond type	More negative atom
a. H and I	2.5 − 2.1 = 0.4	Polar covalent	I
b. Ba and S	2.5 − 0.9 = 1.6	Polar covalent	S
c. K and Br	2.8 − 0.8 = 2.0	Ionic	Br
d. Zn and O	3.5 − 1.6 = 1.9	Ionic	O
e. At and I	2.5 − 2.2 = 0.3	Nonpolar covalent	I
f. S and I	2.5 − 2.5 = 0.0	Nonpolar covalent	Identical

2. The order would be K-Br, Zn-O, Ba-S, H-I, At-I, and S-I.

3. a. Cl $1s^2\,2s^2\,2p^6\,3s^2\,3p^5$ (with one unpaired 3p electron)
Cl $1s^2\,2s^2\,2p^6\,3s^2\,3p^5$

b. O $1s^2\,2s^2\,2p^4$ (two unpaired 2p electrons)
O $1s^2\,2s^2\,2p^4$

c. P $1s^2\,2s^2\,2p^6\,3s^2\,3p^3$ (three unpaired 3p electrons)
P $1s^2\,2s^2\,2p^6\,3s^2\,3p^3$

d. H $1s^1$

F $1s^2\,2s^2\,2p^5$

4. a. Na· + Na· + :S̈: → Na$^+$ + Na$^+$ + :S̈:$^{2-}$
b. ·Ca· + :Ö: → Ca^{2+} + :Ö:$^{2-}$
c. ·Al· + ·Al· + :S̈: + :S̈: + :S̈: →
Al^{3+} + Al^{3+} + :S̈:$^{2-}$ + :S̈:$^{2-}$ + :S̈:$^{2-}$

5. a. Li·   **b.** ·Ca·   **c.** :C̈l: **d.** :Ö:
e. ·C̈·   **f.** ·P̈:   **g.** ·Al·   **h.** :S̈:

6. a.
:F̈:
:F̈:C̈:F̈:   or   :F̈—C—F̈:
:F̈:
:F̈:

b. H:S̈e:H   or   H—S̈e—H

c.
:Ï:
:Ï:N̈:Ï: or   :Ï—N̈—Ï:

d.
:B̈r:
:B̈r:Si:B̈r: or   :B̈r—Si—B̈r:
:B̈r:
:B̈r:

e.
H
H:C̈:C̈l: or   H—C—C̈l:
H
H

7. a. Ö::Ö   or   Ö=Ö
b. :N::N: or   :N≡N:

c. $\ddot{\text{O}}::\text{Si}::\ddot{\text{O}}$ or $\ddot{\text{O}}=\text{Si}=\ddot{\text{O}}$

d. $:\text{C}::\text{O}:$ or $:\text{C}\equiv\text{O}:$

e. $\ddot{\text{O}}::\ddot{\text{S}}:\ddot{\text{O}}:$ or $\ddot{\text{O}}=\ddot{\text{S}}-\ddot{\text{O}}:$

8. According to the VSEPR theory, the following geometries are indicated:
   a. triangular pyramidal b. bent c. bent.

9. The carbon atom in CH_4 contains four valence electrons, two paired in the $2s$ orbital and two unpaired in $2p$ orbitals. Hybridization of the $2s$ and three $2p$ orbitals produces four new, identical orbitals of equal energy. The resulting four sp^3 orbitals are oriented tetrahedrally, 109.5° apart, and contain one electron. Each orbital then overlaps with the electron-containing orbital of a hydrogen atom to form four covalent bonds.

10. The direction of the dipole is toward:
    a. F b. Cl c. Br d. I.

11. The indicated bonds are:
    a. nonpolar   b. polar   c. polar
    d. nonpolar   e. polar   f. polar.

12. The molecules are:
    a. polar   b. nonpolar   c. nonpolar
    d. polar   e. nonpolar.

13. a. $:\ddot{\text{Cl}}:\ddot{\text{S}}:\ddot{\text{Cl}}:$ (AB_2E_2, bent)

    b. $:\ddot{\text{I}}:\ddot{\text{P}}:\ddot{\text{I}}:$ with $:\ddot{\text{I}}:$ above (AB_3E, triangular pyramidal)

    c. $:\ddot{\text{Cl}}:\ddot{\text{O}}:\ddot{\text{Cl}}:$ (AB_2E_2, bent)

    d. $\text{H}:\ddot{\text{N}}:\ddot{\text{Cl}}:$ with H below (AB_3E, triangular pyramidal)

    e. $:\ddot{\text{Cl}}:\text{Si}:\ddot{\text{Cl}}:$ with $:\ddot{\text{Cl}}:$ above and $:\ddot{\text{Br}}:$ below (AB_4, tetrahedral)

    f. $\ddot{\text{O}}::\ddot{\text{N}}:\ddot{\text{Cl}}:$ (AB_2E, bent)

14. a. $\left[:\ddot{\text{O}}:\text{N}:\ddot{\text{O}}:\right]^-$ with $:\ddot{\text{O}}:$ below (AB_3, triangular planar)

    b. $\left[\text{H}:\ddot{\text{N}}:\text{H}\right]^+$ with H above and H below (AB_4, tetrahedral)

    c. $\left[:\ddot{\text{O}}:\text{S}:\ddot{\text{O}}:\right]^{2-}$ with $:\ddot{\text{O}}:$ above and $:\ddot{\text{O}}:$ below (AB_4, tetrahedral)

    d. $\left[:\ddot{\text{O}}:\ddot{\text{Cl}}:\ddot{\text{O}}:\right]^-$ (AB_2E_2, bent)

Critical Thinking

1. The bond lengths grow shorter as the number of bonds increases from single to triple.

Teacher's Notes

Chemical Bonding 160H

Chapter 6
Chemical Bonding

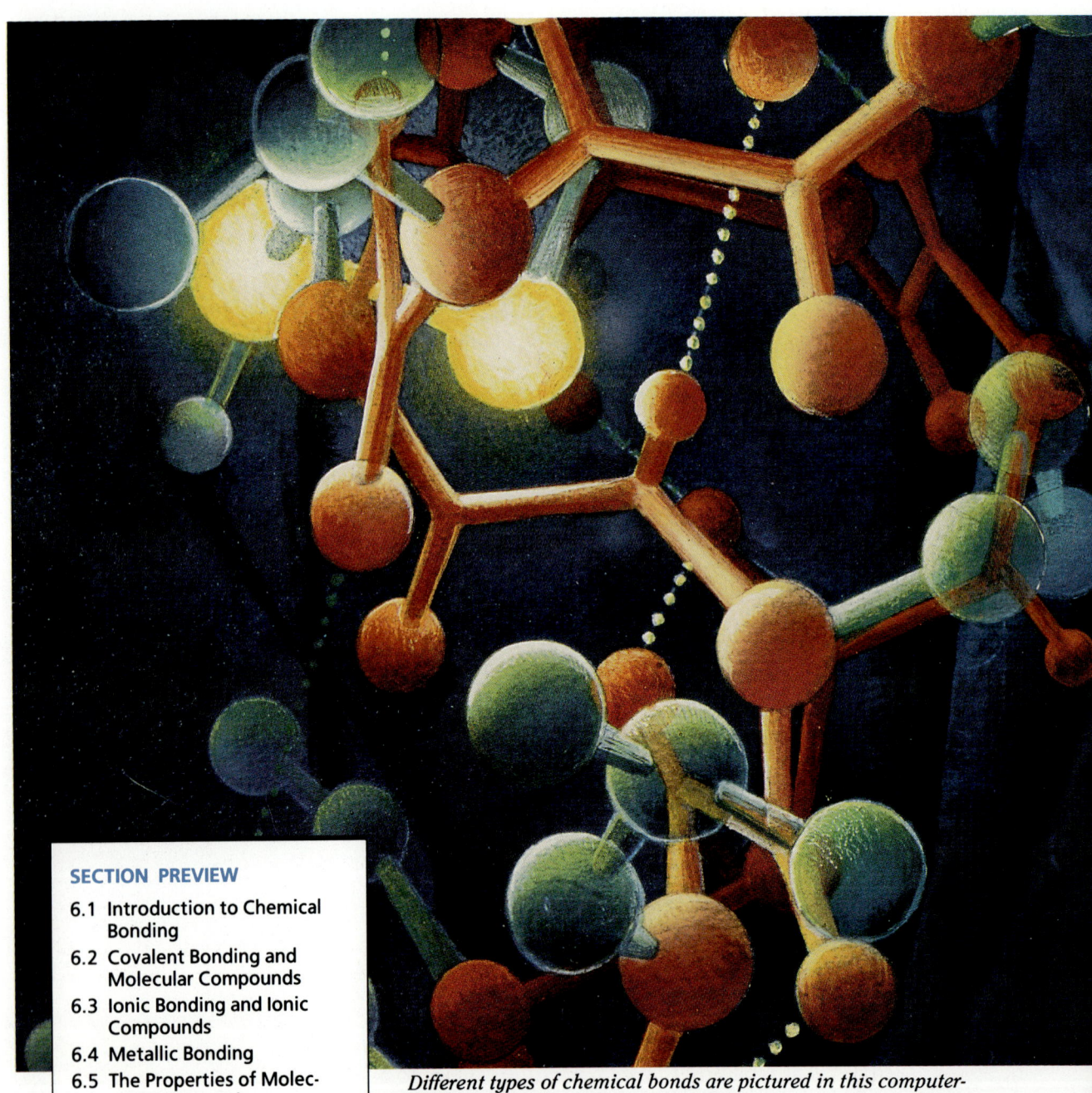

SECTION PREVIEW

6.1 Introduction to Chemical Bonding
6.2 Covalent Bonding and Molecular Compounds
6.3 Ionic Bonding and Ionic Compounds
6.4 Metallic Bonding
6.5 The Properties of Molecular Compounds

Different types of chemical bonds are pictured in this computer-generated image of the enzyme crambin, found in the seeds of a cabbage-like plant.

6.1 Introduction to Chemical Bonding

Isolated and electrically neutral atoms are rare in nature. Ordinarily, only noble-gas atoms exist independently. Atoms of other elements are usually combined with each other or with atoms of different elements.

Atoms are held together by electrostatic attraction between positively charged nuclei and negatively charged electrons. This attraction permits two atoms to be held together by a **chemical bond**, *which is a link between atoms that results from the mutual attraction of their nuclei for electrons.* Chemical bonds are classified by the way in which valence electrons are distributed around the nuclei of the combined atoms.

Types of Chemical Bonds

In Chapter 5 we described how atoms of main group elements can gain or lose electrons to form ions with noble-gas electron configurations. Metals tend to lose electrons to form positive ions, and nonmetals tend to gain electrons to form negative ions. Many chemical compounds are composed of ions; in these compounds the chemical bond is an ionic bond. An **ionic bond** *is the chemical bond resulting from electrostatic attraction between positive and negative ions.* In a *purely* ionic bond, one atom has completely given up one or more electrons, and another atom has gained them—as illustrated for two atoms that each have one unpaired electron at the top in Figure 6-1.

SECTION OBJECTIVES

- Define chemical bond.
- Describe ionic and covalent bonding.
- Classify bonds according to electronegativity differences.
- Explain why most chemical bonds are neither purely ionic nor purely covalent.

Electrostatics is the study of the properties and behavior of stationary electrical charges.

Figure 6-1 In ionic bonding, electrons are transferred from one atom to another, and positive and negative ions are formed. In covalent bonding, an electron pair is shared between two atoms.

In the second major type of chemical bonding, called covalent bonding, neither bonding atom completely loses or gains an electron or electrons. A **covalent bond** *is a chemical bond resulting from the sharing of electrons between two atoms.* A covalent bond in which two electrons are shared is represented by a pair of electron dots, as shown at the bottom right-hand corner in Figure 6-1. In a purely covalent bond, the shared electrons are "owned" equally by the two atoms.

Chemical bonds between unlike atoms are never completely ionic and rarely completely covalent. Bonds can be anywhere in the range between these two extremes, depending upon how strongly the bonded atoms attract electrons.

Chemical Bonding **161**

STUDY HINT

Electronegativity was introduced in Section 5.3, and electronegativity values are given in Figure 5-19.

Linus Pauling, who received the 1954 Nobel Prize in chemistry for his work on molecular structure, was the first chemist who felt that chemical bonds could show degrees of both a covalent and ionic nature. This led to the development of electronegativity values.

Figure 6-2 The variation of bond type with percent ionic character and the electronegativity difference between the bonded atoms.

The degree to which bonds are ionic or covalent can be estimated by comparing the electronegativities of the bonded atoms. The more two atoms differ in electronegativity, the more ionic the bond between them. In other words, the electrons spend more time close to the bonded atom that attracts them more strongly and hence cause that atom to partially resemble an anion and the other atom, a cation.

Figure 6-2 can be used to classify bonds according to electronegativity differences. The electronegativity of one bonded atom is subtracted from that of the other. For example, the electronegativity difference between a cesium (Cs) atom and a fluorine (F) atom is $4.0 - 0.7 = 3.3$. According to Figure 6-2, a cesium–fluorine bond is an ionic bond. In fact, it is one of the most highly ionic bonds known.

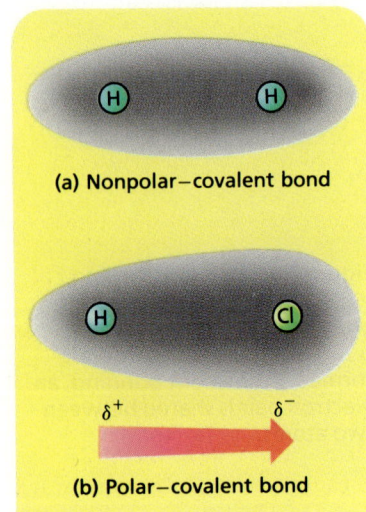

Figure 6-3 Comparison of the electron density in (a) a nonpolar, hydrogen–hydrogen bond and (b) a polar, hydrogen–chlorine bond. Because a chlorine atom is more electronegative than a hydrogen atom, the electron density in the hydrogen–chlorine bond is greater around the chlorine atom.

Bonds that have an ionic character of 50% or less are classified as covalent bonds. A bond between identical atoms is completely covalent. Hydrogen, for example, exists in nature not as isolated atoms, but as *pairs* of atoms held together by covalent bonds, H:H. The hydrogen–hydrogen bond has 0% ionic character. It is a **nonpolar-covalent bond**, *a covalent bond in which the bonding electrons are shared equally by the bonded atoms, with a resulting balanced distribution of electrical charge.* Bonds having 0%–5% ionic character, corresponding to electronegativity differences of roughly 0 to 0.3, are generally considered nonpolar-covalent bonds. For example, because the electronegativity difference between hydrogen (H) and boron (B) is 0.1, they form a bond that is essentially nonpolar.

In bonds with significantly different electronegativities, the electrons are more attracted to the more electronegative atom. Such bonds are **polar**, meaning that they have an uneven distribution of charge. Covalent bonds having 5%–50% ionic character are classified as polar. A **polar-covalent bond** *is a covalent bond in which the united atoms have an unequal attraction for the shared electrons.*

Nonpolar and polar-covalent bonds are compared in the sketches in Figure 6-3 of the electron density in hydrogen–hydrogen and hydrogen–chlorine bonds. Hydrogen and chlorine atoms combine to produce the compound known as hydrogen chloride (HCl). The electronegativity difference between chlorine and hydrogen atoms is $3.0 - 2.1 = 0.9$, indicating formation of a polar-covalent bond. The electrons in this bond spend more of their time near the more electronegative chlorine atom than near the hydrogen atom, as indicated in Figure 6-3b. Consequently, the chlorine end of the bond has a relative surplus of electrons and a partial negative charge, indicated by writing δ−. The hydrogen end of the bond then has an equal partial positive charge, δ+.

In addition to ionic and covalent bonding, there is a third major type of bonding—metallic bonding. In solid or liquid metals, metal atoms give up electrons, as in ionic compounds. The liberated electrons however, are free to move throughout the material, rather than being held in place in negative ions.

In general, atoms of nonmetals form covalent bonds with each other, atoms of metals form metallic bonds with each other, and atoms of metals form ionic bonds with atoms of nonmetals. There are many exceptions, however. One common and important exception is the formation of polar-covalent bonds between metals and nonmetals that do not differ greatly in electronegativity.

An alloy is a solid-solution or combination of metals. Alloys are discussed in Section 14-1.

Sample Problem 6.1

Use electronegativity differences and Figure 6.2 to classify bonds between sulfur and the following elements: hydrogen, cesium, and chlorine. Which atom in each bond will be more negative? The electronegativity of sulfur is 2.5. The more electronegative end of the atom in each bond will be the atom with the larger electronegativity.

Bond from sulfur to	Electronegativity difference	Bond type	More electro-negative atom
hydrogen	2.5 − 2.1 = 0.4	polar covalent	sulfur
cesium	2.5 − 0.7 = 1.8	ionic	sulfur
chlorine	3.0 − 2.5 = 0.5	polar covalent	chlorine

Practice Problems

Use electronegativity differences and Figure 6.2 to classify bonds between chlorine and the following elements: *(a)* calcium *(b)* aluminum *(c)* silver *(d)* bromine
Indicate the more negative atom in each bond

Bond from chlorine to	Electronegativity difference	Bond type	More electro-negative atom
(a) Ca	3.0 − 1.0 = 2.0	ionic	Cl
(b) Al	3.0 − 1.5 = 1.5	polar covalent	Cl
(c) Ag	3.0 − 1.9 = 1.1	polar covalent	Cl
(d) Br	3.0 − 2.8 = 0.2	nonpolar covalent	Cl

Why Chemical Bonding Occurs

The fascinating variety of materials in the world around us is possible because chemical bonds unite atoms of the elements in so many different combinations. Some elements are found in nature only in chemical compounds. Even the atoms of elements that can be found in nature in uncombined form—such as oxygen and nitrogen in the air, or gold and copper in the earth's crust—do not exist as independent atoms. They are bonded together.

The spontaneity of chemical reactions is discussed further in Chapter 18.

If their potential energy is lowered by the change, two atoms will form a chemical bond. Throughout nature, changes that decrease potential energy are favored. Books fall off desks, and the result is lower potential energy for the books. Most atoms have lower potential energy in bonds than as independent atoms.

Chemical-bond formation is often an energy-releasing process. Experiments also show that the reverse—breaking chemical bonds—is often an energy-absorbing process. Atoms separated by breaking a chemical bond have a higher total potential energy than when they were bonded. Whether or not a given chemical reaction occurs spontaneously is partly dependent on whether or not forming new bonds in the products produces enough energy to break bonds in the reactants.

Covalent, ionic, and metallic bonding all decrease the potential energy of the combined atoms. The next section discusses how this occurs in covalent bond formation.

1. Ionic, covalent, and metallic
2. The degree to which bonds are ionic or covalent can be estimated from their electronegativity difference
3. (a) ionic (b) polar covalent (c) nonpolar covalent
4. I—Br, Cu—S, and H—F

Eight elements—H_2, N_2, O_2, F_2, Cl_2, Br_2, I_2, and At_2—have diatomic molecules. An easy way to remember them is that seven of them form a right angle on the Periodic Table.

Section Review

1. What are the three major types of bonds?
2. What is the role of electronegativity in the determination of the ionic or covalent character of a bond?
3. What type of bond would be expected between (a) H and F 1.9 (b) Cu 0.6 and S (c) I and Br? 0.3
4. List the three bonding pairs referred to in the previous question in order of increasing ionic characters.

6.2 Covalent Bonding and Molecular Compounds

SECTION OBJECTIVES

- Explain the relationships among potential energy, distance between approaching atoms, bond length, and bond energy.
- State the octet rule.
- List the six basic steps used in writing Lewis structures.
- Explain how to determine Lewis structures for polyatomic ions or molecules containing multiple bonds.
- Write the Lewis structure for a molecule or polyatomic ion, given the identity of the atoms combined and other specific information.

Atoms held together by covalent bonds stay together as a unit and can be separated only by a chemical reaction in which the bonds are broken. A **molecule** *is the smallest unit quantity of matter which can exist by itself and retain all the properties of the original substance.*

A molecule may consist of a single atom, as in helium and argon, of two or more of the same atoms, as in oxygen and sulfur, or of two or more different atoms, as in water and sugar. Because the number of molecules that exist as single atoms is very small, we usually think of molecules as a group of bonded atoms. *A molecule containing two atoms is called a* **diatomic molecule.** The molecules in Figure 6-3 are examples of diatomic molecules.

Each water molecule consists of one oxygen atom joined by a separate covalent bond to each of two hydrogen atoms. *A chemical compound whose simplest formula units are molecules is known as a* **molecular compound.**

A **chemical formula** *represents the relative numbers of atoms of each kind in a chemical compound by using atomic symbols and numerical subscripts.* The chemical formula for the element hydrogen is H_2, the chemical formula for hydrogen chloride is HCl, and the chemical formula for water is H_2O. When they represent molecular compounds, these formulas are known as molecular formulas. *A* **molecular formula** *shows the types and numbers of atoms combined in a single molecule.*

Anywhere from two atoms to millions of them can be joined together by covalent bonds. They may be atoms of the same or different elements. A diamond crystal, for example, is a giant molecule containing millions of carbon atoms united by covalent bonds. Molecular compounds are everywhere around us. The chemicals of living things are predominantly molecular, and the processes of life primarily involve making and breaking covalent bonds. The molecular combinations of carbon, hydrogen, and oxygen alone account for more than 90% of all known compounds.

Molecules range in size from 62 to 23,000 picometers.

Formation of Covalent Bonds

What is a covalent bond? One way to answer this question is to consider a simple example, the formation of a hydrogen–hydrogen bond. Picture two hydrogen atoms separated by a large enough distance that they have no influence on each other. The potential energy at this point, shown below in Figure 6-4, is arbitrarily set at zero.

Figure 6-4 The relationship between potential energy and the distance between two hydrogen atoms. The separate atoms, at potential energy = 0, approach each other from right to left. A bond forms at the distance of minimum potential energy.

Each hydrogen atom contains a positively charged proton surrounded by a negatively charged electron in a spherical 1s orbital. As the atoms approach, there are several influences on the potential energy, as illustrated in Figure 6-5. By itself repulsion between the like charges of the two protons and the two electrons would cause an increase in potential energy as the atoms approach each other. By itself, attraction between the opposite charges of the proton of each atom and the electron of the other atom would cause a decrease in potential energy as the atoms approach each other.

When the two hydrogen atoms first "see" each other, mutual attraction between the electron of one hydrogen atom and the proton of the other dominates. As shown by the dip in the curve in Figure 6-4, the potential energy decreases. Eventually, a point is reached where the attractive and repulsive forces are balanced and potential energy is at a minimum. At this point, which is represented by the bottom of the valley in the curve, a stable hydrogen molecule forms. A closer approach of the atoms, shown at the left in Figure 6-4, would cause a sharp rise in potential energy as repulsion becomes increasingly greater than the attraction.

The valley in the curve in Figure 6-4 represents the balance between attraction and repulsion in a stable covalent bond. The two bonded atoms

Figure 6-5 Attraction and repulsion between electrons and nuclei of two atoms. If attraction outweighs repulsion, a bond forms at a potential energy minimum. If repulsion outweighs attraction, no bond forms.

Chemical Bonding **165**

vibrate back and forth a bit, but as long as their potential energy remains at this minimum, they are covalently bonded to each other. **Bond length** *is the average distance between two bonded atoms,* that is, the distance of minimum potential energy. The bond length of the hydrogen–hydrogen bond is 74 pm (1 picometer = 10^{-12} m).

The potential-energy change in bond formation is the difference between zero for the isolated atoms and the point of minimum potential energy. This amount of energy, shown in Figure 6-4 as the distance between the zero level and the bottom of the valley, is released in formation of a hydrogen–hydrogen bond. The same amount of energy must be added to separate the bonded atoms. **Bond energy** *is the energy required to break a chemical bond and form neutral atoms.* Bond energies are usually reported for breaking one mole of bonds in isolated molecules in the gaseous state and are given in kilojoules per mole. The addition of 436 kJ of energy is needed to break the hydrogen–hydrogen bonds in one mole of hydrogen molecules.

The general energy relationships described here for formation of a hydrogen–hydrogen bond apply to all covalent bonds. Bond lengths and bond energies vary with the types of atoms that have combined, however, as illustrated by the values in Table 6-1. The energy of a bond between two different types of atoms varies somewhat depending on what other bonds the atoms have formed. As you can see in the table, bond energies become larger as bond lengths become shorter.

The electrons in a covalent bond are pictured as occupying overlapping orbitals, represented for the hydrogen molecule at the top in Figure 6-6. Each electron is free to move around in either of the orbitals, but both electrons spend more time in the space between the two nuclei than elsewhere. Thus, in the region of orbital overlap, there is an increased electron density. The attraction of the nuclei to the high electron density in this region holds the two atoms together.

By sharing electrons, each atom has achieved the stable, noble-gas configuration of helium, $1s^2$. Orbital notation is used to illustrate this configuration as shown on the left. Noble gases are so-called because they do not react readily with other elements. The notation for each separate atom shows one unpaired electron. A box drawn around the two orbitals containing

Molecules in the gaseous state are not influenced by neighboring molecules because adjacent molecules are far apart.

TABLE 6-1	BOND LENGTHS AND BOND ENERGIES FOR SINGLE COVALENT BONDS				
Bond	Bond Length (pm)	Bond Energy (kJ/mol)	Bond	Bond Length (pm)	Bond Energy (kJ/mol)
H—H	74	436	C—C	154	347
F—F	128	159	C—N	147	293
Cl—Cl	198	243	C—O	143	351
Br—Br	228	192	C—H	110	414
I—I	266	151	C—Cl	176	330
H—F	92	569	C—Br	193	284
H—Cl	127	431	N—N	140	159
H—Br	142	368	N—H	98	389
H—I	161	297	O—H	94	464

Figure 6-6 Representations of formation of the hydrogen–hydrogen bond in molecular hydrogen (H₂) by a drawing of orbital overlap and by electron-dot symbols that show electron-pair sharing

these unpaired electrons represents the overlapping orbitals, which are both effectively filled by the shared pair of electrons.

Like hydrogen, the halogens all normally consist of diatomic molecules. The orbital notation for formation of a fluorine molecule from two fluorine atoms is given below. By sharing their unpaired electrons, each fluorine atom in the molecule has its highest energy level completely filled by eight electrons, giving the noble gas configuration of neon [He]$2s^2 2p^6$. The molecular formulas for the diatomic halogen molecules are fluorine—F_2; chlorine—Cl_2; bromine—Br_2; iodine—I_2; and astatine—At_2.

Fluorine atoms → Fluorine molecules

← Bonding electron pair

The orbital diagram below shows that in hydrogen chloride, (HCl), the hydrogen achieves a helium configuration by sharing an electron pair, and the chlorine atom achieves the argon configuration. Notice that chlorine, by sharing one electron with hydrogen, has 8 electrons in its outer energy level.

Hydrogen and chlorine atoms

Hydrogen chloride molecule

The Octet Rule

With the exception of helium atoms, each noble gas atom has eight electrons in its highest occupied energy level. In halogen molecules, each atom has achieved a noble-gas configuration by sharing an electron pair (see above). The same is true of hydrogen and of chlorine in hydrogen chloride shown above.

Chemical Bonding **167**

Chemistry in Detail

Exceptions to the Octet Rule

In many molecules, a central atom is surrounded by fewer than an octet of electrons.

Beryllium (Be) and boron (B) from Period 2 of the Periodic Table form such molecules. The Lewis structures for the beryllium fluoride (BeF$_2$) molecule and for the boron trifluoride (BF$_3$) molecule are given in Table 6-4 as examples of molecules with AB$_2$ and AB$_3$ geometry. (The reasons why these molecules have these shapes will be considered in Section 6.5.) In beryllium fluoride, the central beryllium atom is surrounded by only four valence electrons. Each of the beryllium atom's two valence electrons pairs with a valence electron of a fluorine atom to form a total of two covalent bonds. In boron trifluoride, the central boron atom is surrounded by only six valence electrons. Here, there are three valence electrons from boron and one each from the three covalently bonded fluorine atoms.

Beryllium and boron compounds do tend to form additional bonds to complete the octet around the central beryllium or boron atom. For example, when boron trifluoride reacts with ammonia (NH$_3$), the unshared pair of electrons from the nitrogen atom forms a covalent bond between the nitrogen atom and the boron atom as shown below.

Atoms of elements of the third and higher periods of the Periodic Table can be surrounded by *more* than four pairs of valence electrons when they combine with atoms of the highly electronegative elements fluorine, oxygen, and chlorine. In these cases of "expanded valence," bonding involves electrons in *d* orbitals as well as in *s* and *p* orbitals. (Such bonding *d* orbitals are not available in Periods 1 and 2.) The Lewis structures for several molecules in which the central atom has expanded valence are shown below.

The octet rule is also violated when the total number of valence electrons in a molecule is odd. For example, in nitrogen monoxide (NO), nitrogen has five and oxygen has six valence electrons, a total of eleven.

Gilbert Lewis (1875–1946), an American theoretical chemist, proposed the Octet Rule and electron-dot symbols (also known as Lewis structures) to show arrangement of valence electrons in atoms.

In many compounds, atoms of main group elements form bonds so that each atom has an octet of electrons—four pairs of electrons in the outermost *s* and *p* orbitals. Such bond formation is described by the **octet rule:** *Chemical compounds tend to form so that each atom, by gaining, losing, or sharing electrons, has an octet of electrons in its highest occupied energy level.* For example, nitrogen, which is in Group 15, has the outer configuration $2s^2 2p^3$. With five electrons in its highest energy level, a nitrogen atom needs three more electrons to complete an octet. Nitrogen forms three covalent bonds in many molecules. The three electrons contributed by other atoms fill the outer energy level.

There are, however, certain exceptions to the octet rule. These are discussed in the Chemistry in Detail feature above.

Lewis Structures

Electron-dot symbols are used to represent molecules, as illustrated in Figures 6-1 and 6-6. The symbol for a hydrogen *atom* is H. The hydrogen molecule is shown as H:H, where the pair of dots represents the shared electron pair of the covalent bond.

For a halogen molecule, say that of chlorine, Cl_2, the electron-dot symbols of two chlorine atoms are combined to give :C̈l:C̈l:. Here also the pair of dots in the center represents the covalent bond. Also, each chlorine atom is surrounded by three pairs of electrons that are not shared in bonds. An **unshared pair**, *also called a lone pair, is a pair of electrons that is not involved in bonding, but instead belongs exclusively to one atom.*

The pair of dots representing a covalent bond is often replaced by a long dash, for example H—H instead of H:H or :C̈l—C̈l: instead of :C̈l:C̈l:. These are all **Lewis structures**: *formulas in which atomic symbols represent nuclei and inner-shell electrons, dot-pairs or dashes between two atomic symbols represent electron pairs in covalent bonds, and dots adjacent to only one atomic symbol represent unshared electrons.*

It is common to indicate only the electrons that are shared, using dashes for bonds. A **structural formula** *indicates the kind, number, arrangement, and bonds of the atoms in a molecule.* For example, Cl–Cl and H–Cl are structural formulas. The structural formula of a larger molecule, ethane, shows that the molecule contains two carbon atoms and six hydrogen atoms, arranged so that the carbon atoms are bonded to each other and three hydrogen atoms are bonded to each carbon atom.

```
    H   H
    |   |
H — C — C — H
    |   |
    H   H
```

The Lewis structures for many molecules, especially those composed of main group elements, can be drawn if one knows the composition of the molecule and which atoms are bonded to each other. The following Sample Problem illustrates the basic steps for writing Lewis structures. The molecule described in this problem contains bonds with single shared electron pairs. A single covalent bond, or a **single bond** *is a covalent bond produced by the sharing of one pair of electrons between two atoms.*

Sample Problem 6.2

Draw the Lewis structure of iodomethane (CH_3I), in which the carbon atom is the central atom.

Step 1.   *Determine the type and number of atoms in the molecule*
              One carbon atom, one iodine atom, and three hydrogen atoms

Step 2.   *Draw the electron-dot symbols for each type of atom*
              Carbon is from Group 14 and has 4 valence electrons; iodine is from Group 17 and has 7 valence electrons; and hydrogen has 1 valence electron.

$$\cdot \ddot{C} \cdot \qquad : \ddot{I} : \qquad H \cdot$$

Chemical Bonding

Sample Problem 6.2 *continued*

Step 3. *Determine the total number of valence electrons in the atoms to be combined:*

$$C\ 1 \times 4\,e^- = 4\,e^-$$
$$I\ 1 \times 7\,e^- = 7\,e^-$$
$$H\ 3 \times 1\,e^- = \underline{3\,e^-}$$
$$14\,e^-$$

Step 4. *Arrange the atoms to form a skeleton structure for the molecule and connect the atoms by electron-pair bonds*
Usually, the most electropositive atom is central. In this case, carbon and iodine have the same electronegativity, but we know that carbon is the central atom (it usually is).

$$\begin{array}{c} H \\ .. \\ H:C:I \\ .. \\ H \end{array}$$

Step 5. *Add unshared pairs of electrons so that each nonmetal atom (except hydrogen) is surrounded by eight electrons*

$$\begin{array}{c} H \\ .. \ .. \\ H:C:I: \\ .. \ .. \\ H \end{array} \quad \text{or} \quad \begin{array}{c} H \\ | \\ H-C-I: \\ | \\ H \end{array}$$

Step 6. *Count the electrons in the structure to be sure that the number of valence electrons used equals the number available.*
There are 8 electrons in the four covalent bonds and 6 electrons in the three unshared pairs, giving the correct total of 14 valence electrons.

Practice Problems

1. Draw the Lewis structure of ammonia (NH_3), which is present in household ammonia cleanser. The molecule contains one nitrogen atom and three hydrogen atoms.

   (Ans.) $\begin{array}{c} .. \\ H:N:H \\ .. \\ H \end{array}$ or $\begin{array}{c} .. \\ H-N-H \\ | \\ H \end{array}$

2. Draw the Lewis structure of iodine monochloride, which contains one iodine atom and one chlorine atom.

   (Ans.) :Ï:Cl: or :Ï—Cl:

3. Draw the Lewis structure of silicon tetrafluoride, which contains one silicon atom and four fluorine atoms.

   (Ans.) $\begin{array}{c} :F: \\ :F:Si:F: \\ :F: \end{array}$ or $\begin{array}{c} :F: \\ | \\ :F-Si-F: \\ | \\ :F: \end{array}$

4. Draw the Lewis structure for hydrogen sulfide, which contains two hydrogen atoms and one sulfur atom.

   (Ans.) H:S:H or H—S—H

Multiple Covalent Bonds

Atoms of some elements, especially carbon, nitrogen, and oxygen, can share more than one pair of electrons. A double covalent bond, or simply a **double**

bond, *is a covalent bond between two atoms produced by the sharing of two pairs of electrons.* In ethane, for example, two electron pairs are simultaneously shared by two carbon atoms:

$$\begin{array}{c}H\\ \\ H\end{array}\!\!\!\!C::C\!\!\!\!\begin{array}{c}H\\ \\ H\end{array} \quad \text{or} \quad \begin{array}{c}H\\ \\ H\end{array}\!\!\!\!C\!=\!C\!\!\!\!\begin{array}{c}H\\ \\ H\end{array}$$

A double bond is shown either by two side-by-side pairs of dots or by two parallel dashes. All four electrons in a double bond "belong" to both atoms.

A nitrogen atom, which has five valence electrons, can acquire three electrons to complete an octet by sharing three pairs of electrons with another nitrogen atom.

$$\cdot\ddot{N}\!: \;+\; \cdot\ddot{N}\!: \;\rightarrow\; \begin{array}{c}:N\!::\!:N:\\ \text{or}\\ :N\!\equiv\!N:\end{array}$$

A triple covalent bond, or simply a **triple bond**, *is a covalent bond produced by the sharing of three pairs of electrons between two atoms.* Elemental nitrogen, like hydrogen and the halogens, normally exists as diatomic molecules (N_2). The triple bond in nitrogen molecules is nonpolar, as are the single bonds in hydrogen and halogen molecules. The orbital diagrams for the atom for nitrogen and the nitrogen molecule, given below, illustrate that

Nitrogen molecule

atoms of nonmetals in multiple covalent bonds also follow the octet rule. Each nitrogen atom in the molecule has eight valence electrons, six that it shares with the other nitrogen atom, plus its own unshared pair of electrons.

Carbon forms a number of compounds containing triple bonds. For example, a carbon–carbon triple bond is present in acetylene, familiar for its use in welding and cutting torches.

$$H:C\!::\!:C:H \quad \text{or} \quad H\!-\!C\!\equiv\!C\!-\!H$$

Double and triple bonds are referred to as **multiple bonds,** or multiple covalent bonds. Table 6-2 compares the bond lengths and bond energies for some single, double, and triple bonds. Double bonds in general have higher bond energies and are shorter than single bonds. Triple bonds are yet stronger and shorter.

In writing Lewis structures for molecules that contain carbon, nitrogen, or oxygen, one must remember that multiple bonds between pairs of these atoms are possible. The need for a multiple bond becomes obvious if there are not enough valence electrons to complete octets by adding unshared pairs. Step 6b in the following example illustrates how to deal with this situation.

STUDY HINT
Some common multiple bonds:
$C\!=\!C$, $C\!\equiv\!C$, $C\!\equiv\!N$, $C\!=\!O$, $N\!\equiv\!N$, $N\!=\!O$

Chemical Bonding

Formaldehyde is believed to cause cancer and other problems. Because of its bacterial-killing nature, it was used to preserve animals in biology labs. However, now other safer chemicals are sometimes used as preservatives. If formalin-preserved specimens are used, they should be washed in slowly running water for eight hours before being used.

TABLE 6-2 COMPARISON OF BOND LENGTHS AND BOND ENERGIES FOR SINGLE AND MULTIPLE COVALENT BOND

Bond	Bond Length (pm)	Bond Energy (kJ/mol)	Bond	Bond Length (pm)	Bond Energy (kJ/mol)
C—C	154	347	C—O	143	351
C=C	134	611	C=O	120	745
C≡C	121	837	C≡O	113	1075
C—N	147	293	N—N	140	159
C≡N	115	891	N=N	125	428
			N≡N	110	946

Sample Problem 6.3

Draw the Lewis structure for formaldehyde, which in water solution is used as an antiseptic known as formalin. The formaldehyde molecule has the chemical formula CH_2O.

Step 1. Determine the type and number of atoms.
The formula shows one carbon atom, two hydrogen atoms, and one oxygen atom.

Step 2. Draw the electron-dot symbols for each type of atom.
Carbon from Group 14 has four valence electrons; oxygen, which is in Group 16, has six valence electrons; and hydrogen has only one electron.

$$\cdot \overset{\cdot}{\underset{\cdot}{C}} \cdot \qquad H\cdot \qquad :\overset{\cdot}{\underset{\cdot\cdot}{O}}:$$

Step 3. Determine the total number of valence electrons in the atoms to be combined.

$$\begin{array}{rl} C\ 1 \times 4e^- = & 4e^- \\ O\ 1 \times 6e^- = & 6e^- \\ 2H\ 2 \times 1e^- = & \underline{2e^-} \\ & 12e^- \end{array}$$

Step 4. Arrange the atoms to form a skeleton structure for the molecule and connect the atoms by electron-pair bonds.
The carbon atom must be the central atom.

$$\begin{array}{c} H \\ \cdot\cdot \\ H:C:O \end{array}$$

Step 5. Add unshared pairs of electrons so that each nonmetal atom (except hydrogen) is surrounded by eight electrons.

$$\begin{array}{c} H \\ \cdot\cdot\ \cdot\cdot \\ H:C:O: \\ \cdot\cdot\ \cdot\cdot \end{array}$$

Step 6a. Count the electrons in the Lewis structure to be sure that the number of valence electrons used equals the number available.
The structure above has 6 electrons in covalent bonds and 8 electrons in four lone pairs, for a total of 14 electrons. The structure has two valence electrons too many.

Sample Problem 6.3 *continued*

Step 6b. If too many electrons have been used, subtract one or more lone pairs until the total number of valence electrons is correct, and then move one or more lone electron pairs to existing bonds (between C, N, or O) until the outer shells of all atoms are completely filled. Subtract the lone pair of electrons from the carbon atom. Then move one lone pair of electrons from the oxygen to the bond between carbon and oxygen forming a double bond.

$$H:\overset{H}{\underset{..}{C}}::\overset{..}{\underset{..}{O}} \quad \text{or} \quad H-\overset{H}{\underset{|}{C}}=\overset{..}{\underset{..}{O}}$$

There are 8 electrons in covalent bonds and 4 electrons in lone pairs, for a total of 12 valence electrons.

Practice Problems
1. Draw the Lewis structure for carbon dioxide (CO_2). *(Ans.)* $\overset{..}{\underset{..}{O}}=C=\overset{..}{\underset{..}{O}}$
2. Draw the Lewis structure for hydrogen cyanide, which contains one hydrogen atom, one carbon atom, and one nitrogen atom. *(Ans.)* H—C≡N:

Polyatomic Ions

Certain groups of atoms combine to form ions rather than neutral molecules. A **polyatomic ion** *is a charged group of covalently bonded atoms.* The covalent bonds in a polyatomic ion are no different from other covalent bonds. The charge of the ion results from an excess of electrons (negative charge) or a shortage of electrons (positive charge).

The most common positively charged polyatomic ion is the ammonium ion, which contains one nitrogen atom and four hydrogen atoms and has a single positive charge. Its formula is NH_4^+, sometimes written $[NH_4]^+$ to show that the group of atoms as a whole has a charge of 1+.

In the ammonium ion, there is a total positive charge of 11+ from the seven protons in the nitrogen atom (atomic number 7) plus the four protons in the four hydrogen atoms (atomic number 1).

Nitrogen ($1s^2 2s^2 2p^3$) has a total of seven electrons (five are valence electrons and two are inner-shell electrons). There are three electrons contributed from the hydrogen atoms, giving a total of ten electrons, or a negative charge of 10−. (One of the hydrogen atoms is simply a proton, and thus does not contribute an electron.) Polyatomic ions do not exist in isolation but are found combined in chemical compounds with ions of opposite charge. Electron-dot formulas for the ammonium ion and some common negative polyatomic ions—the nitrate, sulfate, and phosphate ions—are shown below. To find the Lewis structure for polyatomic ions, follow the usual six steps, but with the following exception. At the end of the third step, add the appropriate number of e^- to the total number of valence electrons if the ion is negatively charged. If the ion is positively charged, subtract the appropriate number of e^- from the total number of valence electrons.

ammonium ion   nitrate ion   sulfate ion   phosphate ion

Chemistry in Detail

Resonance

Some molecules cannot be represented adequately by a single Lewis structure. An example is the ozone molecule. Ozone is a form of oxygen found in the upper part of the atmosphere. There, it absorbs much of the incoming ultraviolet radiation from the sun, helping to protect people from skin cancer and sunburn. The ozone molecule consists of three atoms of oxygen. Experimental measurements show that ozone is a bent molecule. Following the octet rule, the ozone molecule can be represented by the two Lewis structures shown below. (The double-headed arrow is reserved for use with resonance hybrids.)

O=O—O: ↔ :O—O=O

Neither structure, however, can be correct by itself because each indicates that the ozone molecule has two different types of O—O bonds—one single and the other double—whereas experiments show that both bonds are identical.

Chemists once speculated that ozone had one of these structures for half the time and the other structure the other half. They thought that it was constantly "resonating" from one structure to the other. Today, scientists say that ozone has a *single* structure that is the *average* of these two. **Resonance** *refers to bonding in molecules that cannot be correctly represented by a single Lewis structure.*

The different Lewis structures that together represent such molecules are called *resonance structures* or *resonance hybrids*.

Another example of a molecule that should be represented by resonance structures is sulfur trioxide. In sulfur trioxide, three resonance structures all obey the octet rule as shown above.

As illustrated by ozone and sulfur trioxide, the atoms in resonance structures all keep the same positions. The only difference is in the distribution of electrons. Lone-pair electrons in one resonance structure become a bonding pair in another. Simultaneously, a bonding pair of electrons in the first resonance structure becomes a lone pair in the other.

---

1. *(a)* The distance between two bonded atoms, that is, the distance of minimum potential energy *(b)* The energy required to break a chemical bond and form neutral atoms
2. Chemical compounds tend to form so that each atom, by gaining, losing, or sharing electrons, has an octet of electrons in its highest occupied energy level.
3. Formulas in which atomic symbols represent nuclei and inner-shell electrons, dots, or dashes represent electron pairs in covalent bonds, and dots represent unshared e⁻.
4. *(a)* One pair *(b)* Two pairs *(c)* Three pairs

Section Review

1. Define *(a)* bond length *(b)* bond energy.
2. State the octet rule.
3. What are Lewis structures?
4. How many pairs of electrons are shared in *(a)* a single bond *(b)* a double bond *(c)* a triple bond?
5. Draw the Lewis structures for molecules consisting of: *(a)* one bromine atom and one iodine atom *(b)* one carbon atom, three hydrogen atoms, and one bromine atom *(c)* one hydrogen atom, one chlorine atom, and two carbon atoms *(d)* the sulfate ion (SO_4^{2-}).

5. *(a)* :Br:I: *(b)* H:C:Br: with H above and H below   *(c)* H:C::C:Cl: *(d)* $[:O:S:O:]^{2-}$ with :O: above and :O: below

174 Chapter 6

6.3 Ionic Bonding and Ionic Compounds

SECTION OBJECTIVES

- Compare the meaning of a chemical formula for a molecular compound and one for an ionic compound.
- Discuss the arrangements of ions in a crystal.
- Define lattice energy and explain its significance.
- List and compare the distinctive properties of ionic and molecular compounds.

Most of the rocks and minerals that make up the earth's crust consist of positive and negative ions held together by ionic bonding. Sodium chloride, which is common table salt, is found in nature as rock salt. It is a typical chemical compound composed of ions. The sodium ion (Na^+) has a charge of $1+$. The chloride ion (Cl^-) has a charge of $1-$. These ions combine in a one-to-one ratio—Na^+Cl^-—because this ratio allows each positive charge to be balanced by a negative charge. The chemical formula for sodium chloride is ordinarily written NaCl.

An **ionic compound** *is composed of positive and negative ions combined so that the positive and negative charges are equal.* Most ionic compounds are crystalline solids, such as the familiar crystals of table salt. Some examples of naturally occurring crystalline substances are shown in Figure 6-7. A crystal of any ionic compound is a three-dimensional network of positive and negative ions mutually attracted to one another. As a result, the formula for an ionic compound does not represent an independent unit that can be isolated and examined. It instead represents the simplest ratio, or smallest numbers, of the combined ions that will give electrical neutrality.

A **formula unit** *is the simplest collection of atoms from which a compound's formula can be established.* One formula unit of sodium chloride (NaCl) is one sodium ion plus a chloride ion. The number of ions in one formula unit depends on the charges of the ions combined. For example, the magnesium ion has a charge of $2+$ (Mg^{2+}). In the ionic compound magnesium chloride, two chloride ions, which each have a $1-$ charge, must be present for each magnesium ion present in order for neutrality to be maintained. The formula of magnesium chloride is $MgCl_2$. The chemical formula of an ionic compound shows the ratio of the ions present in a sample of the compound of any size. The symbols for a few of the more common ions are given in the margin. How many Al^{3+} and Br^- ions would be in one formula unit of the ionic compound formed by these two types of ions?

STUDY HINT

Remember that when describing an ionic compound, you should not refer to "molecules," but rather to formula units.

STUDY HINT

Some common ions:
Na^+, K^+
Mg^{2+}, Ca^{2+}, Ba^{2+}
O^{2-}, S^{2-}
F^-, Cl^-, Br^-, I^-

Figure 6-7 Some crystalline minerals, such as aluminum oxide, are ionic compounds.

Chemical Bonding

Formation of Ionic Bonds

Electron-dot structures can be used to demonstrate the changes that take place in the formation of ionic bonds. Ionic compounds do not ordinarily form by combination of isolated atoms, but for the moment consider a sodium atom and a chlorine atom approaching each other. The two atoms are neutral and have one and seven valence electrons, respectively.

$$\text{Na·} \qquad\qquad \text{:}\ddot{\text{Cl}}\text{:}$$
sodium atom chlorine atom

We have already seen that atoms of sodium and the other alkali metals readily lose one electron and atoms of chlorine and the other halogens readily gain one electron. Combination of the two atoms to give one formula unit of sodium chloride can be represented as

$$\text{Na·} + \text{:}\ddot{\text{Cl}}\text{:} \longrightarrow \text{Na}^+ + \text{:}\ddot{\text{Cl}}\text{:}^-$$
sodium atom chlorine atom sodium cation chlorine anion

An electron is transferred from the sodium atom to the chlorine atom. The result is that each atom is transformed into an ion with a noble-gas configuration. In the combination of calcium with fluorine, two fluorine atoms are needed to accept the two valence electrons given up by calcium, which is in Group 2.

$$\text{·Ca·} + \text{:}\ddot{\text{F}}\text{:} + \text{:}\ddot{\text{F}}\text{:} \longrightarrow \text{Ca}^{2+} + \text{:}\ddot{\text{F}}\text{:}^- + \text{:}\ddot{\text{F}}\text{:}^-$$
calcium atom fluorine atoms calcium cation fluorine anions

Electron transfer can also be represented by orbital notation, as illustrated below for sodium and chlorine atoms. The orbital notation shows that the valence electron lost by the sodium atom came from a 3s orbital and entered the 3p orbital in chlorine that contained a single electron.

> **Emphasize** that an ion in an ionic crystal is surrounded by ions of opposite charge; it is impossible to say that it belongs with any one of its neighbors more than another. A sodium chloride crystal model would help illustrate this.

STUDY HINT

Charges of ions are not usually included in formulas for ionic compounds.

Figure 6-8 Potential energy decreases as ions assume their positions in the orderly array of ions in a crystal.

Recall how hydrogen atoms form diatomic molecules with bond lengths that have a minimum potential energy. In an ionic compound the equivalent potential energy minimum is found in the orderly arrangement of ions in the crystal, as represented in Figure 6-8. According to the laws of electrostatics, like charges repel each other and opposite charges attract each other. The force of attraction or repulsion decreases as the charges move farther apart. In an ionic crystal, there is attraction between oppositely charged ions and between the nuclei and electrons of adjacent atoms. Also, there is repulsion between the valence-shell electrons of adjacent ions. The distance between ions in a crystal and their arrangement represents a balance among all these forces of attraction and repulsion.

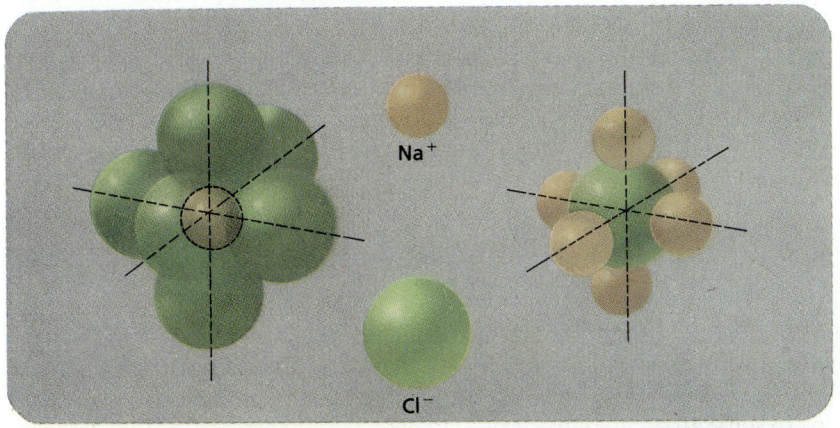

Figure 6-9 The packing of Na^+ and Cl^- ions in the NaCl crystal. In the model on the left, the six Cl^- ions are clustered about one Na^+ ion that cannot be seen here, but whose location is indicated by the dashed outline.

Within a sodium chloride crystal, as illustrated in Figure 6-9, each sodium cation is surrounded by six chlorine anions and each chlorine anion is also surrounded by six sodium cations. Attraction between the adjacent oppositely charged ions is much greater than repulsion by other ions of the same charge, which are further away. Figure 6-10 shows two views of the arrangement of ions in sodium chloride.

The three-dimensional arrangements of ions and the strength of attraction between them varies with the sizes and charges of the ions and the numbers of ions of different charges. For example, in calcium fluoride (CaF_2), which is known commonly as the mineral *fluorite* (Figure 6-11), there are two anions for each cation. Each calcium cation is surrounded by eight

Figure 6-10 Two representations of the sodium and chloride ions in crystalline sodium chloride. The ions *(left)* shown with their electron density regions just touching, as they do in the crystal. An expanded view *(right)* of the positions of the ions in the crystal.

Chemical Bonding 177

Figure 6-11 Crystals of the mineral fluorite, which is calcium fluoride (CaF_2).

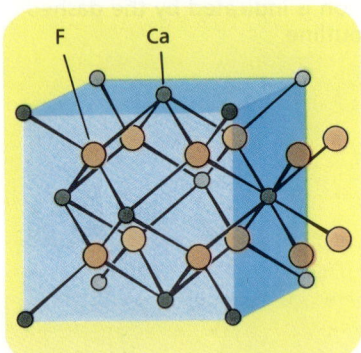

Figure 6-12 The arrangement of the ions in crystalline calcium fluoride (CaF_2).

Figure 6-13 Shattering a crystal. Attraction between positive and negative ions causes planes of ions in a crystalline ionic compound to resist motion. When struck with sufficient force, ions of the same charge approach each other, causing repulsion, and the crystal shatters along the planes.

fluoride anions, and each fluoride ion is surrounded by four calcium cations (see Figure 6-12).

To compare the strengths of ionic bonds, chemists compare the amounts of energy released when isolated ions in a gas come together to form a crystalline solid. **Lattice energy** *is the energy released when one mole of an ionic crystalline compound is formed from gaseous ions.* A few lattice energies are given in the margin on the opposite page. Calculations based on experimental data make it possible to find lattice energy values, which are useful for comparing bond strengths in ionic compounds. Actual ionic compound formation does not always take place by combination of individual ions, however. In a reaction between elements that form an ionic compound, the energy needed to form ions from atoms is supplied by the energy released when the ions form the lower structure of the crystal lattice, which has a relatively low potential energy.

A Comparison of Ionic and Molecular Compounds

The ions in an ionic compound are bound together by ionic bonding and held in place in a crystalline solid. The strong attraction between positive and negative charges is the force that holds ions together in ionic compounds. In a molecular compound, the covalent bonds are quite strong, but the forces of attraction between individual molecules are weaker than the forces of ionic bonding. The differences in the attraction between individual particles in molecular and ionic compounds give rise to different properties in the two types of compounds.

The melting point, boiling point, and hardness of a compound depend upon how strongly the molecules are attracted to each other. Many molecular compounds melt at relatively low temperatures because the forces of attraction between the molecules are not very strong.

Because the forces holding ions together are stronger than those holding molecules together, ionic compounds have generally higher melting and boiling points than do molecular compounds. Also, they do not vaporize as readily at room temperature as many molecular compounds do.

Ionic compounds are hard and brittle crystalline solids. The ions are so positioned that the slightest shifting of one row of ions relative to another causes an enormous buildup of repulsive forces, as shown in Figure 6-13.

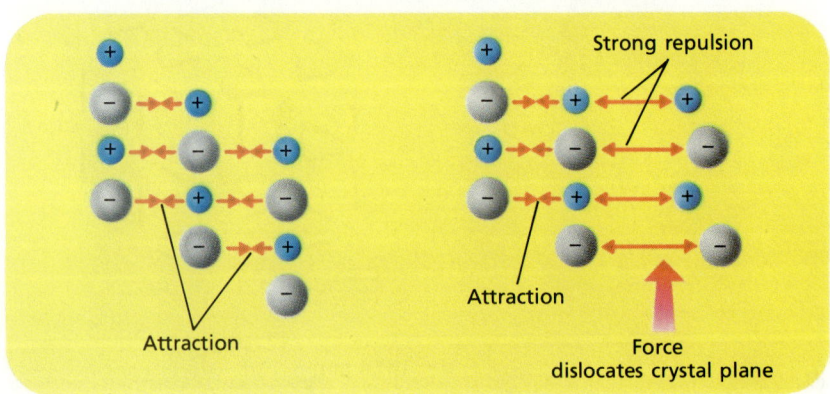

These forces make it difficult for one layer to move relative to another, causing ionic compounds to be hard. If one layer does move, however, the repulsive forces make the layers part completely, causing ionic compounds to be brittle.

In the molten state, or when dissolved in water, ionic compounds are electrical conductors because the ions are free to carry electrical current. In the solid state, the ions cannot move, and the compounds are not electrical conductors. Many ionic compounds are soluble in water. During dissolution, the ions separate from each other and become surrounded by water molecules. Many ionic compounds do not dissolve in water, however, because the attraction of the water molecules cannot overcome the attraction between the ions.

Section Review

1. Give two examples of ionic compounds.
2. Use electron-dot structures to demonstrate the formation of ionic bonds involving *(a)* Li and Cl *(b)* Ca and I.
3. Distinguish between ionic and molecular compounds in terms of the basic units that make up each.
4. When compared to compound B, compound A has lower melting and boiling points, and vaporizes more readily. If one is an ionic compound and the other is molecular, determine the identity of each.

6.4 Metallic Bonding

The properties of metals are quite different from those of either ionic or molecular compounds. These differences suggest that the bonding in metals is also quite different from ionic and covalent bonding. Metals are better conductors of heat and are more easily deformed than many solid molecular or ionic compounds. Most importantly, metals are electrical conductors in the solid state—much better conductors than even molten ionic compounds. This property suggests the presence in solid metals of highly mobile charged particles: the valence-level electrons. Such mobility is not possible in molecular compounds, in which electrons are localized in electron-pair bonds between neutral atoms. Nor is it possible in solid ionic compounds, in which electrons are bound to individual ions and the ions are held in place in crystals.

Formation of Metallic Bonds

Metal atoms contain a smaller number of valence electrons than vacant highest energy level orbitals. In *s*-block and *d*-block metals, there are one or two valence electrons in *s* sublevels, leaving three vacant *p* orbitals in the highest energy levels. In metals of the third and higher periods, there are also vacant *d* orbitals.

Metal atoms also have relatively low ionization energies and electronegativities in common. They easily give up electrons, but do not attract electrons in bonds as strongly as nonmetals. Two adjacent metal atoms consequently, would form at best only a very weak covalent bond.

Some lattice energy values, in kJ/mol:
LiF: -1032
NaF: -922
NaCl: -769
KCl: -718
CsCl: -660

1. Sodium chloride (NaCl) and magnesium chloride ($MgCl_2$)
2. *(a)* $Li^. + :\ddot{Cl}: \longrightarrow$
   $Li^+ + :\ddot{Cl}:^-$
   (b) $.Ca^. + :\ddot{I}: + :\ddot{I}: \longrightarrow$
   $Ca^{2+} + :\ddot{I}:^- + :\ddot{I}:^-$
3. Ionic compounds consist of positive and negative ions bound together by ionic bonding and held in place in a crystalline solid. Molecular compounds are made up of molecules that move around independently and are capable of independent existence.
4. Compound A is molecular and B is ionic.

SECTION OBJECTIVES

- Describe the electron-sea model of metallic bonding.
- Explain why metals are malleable and ductile, but ionic-crystalline compounds are not.
- Explain why metal surfaces are shiny.

Metal atoms do have room in their vacant highest energy level orbitals to share many electrons from surrounding atoms, however. In sodium, for example, the atoms are packed closely together. Each atom is surrounded by eight other sodium atoms. Eight electron pair bonds cannot form. Therefore, each sodium atom ($[Ne]3s^1$) is pictured as contributing its one valence electron to a region surrounding the atoms, leaving behind a sodium ion. The electrons are delocalized—they do not belong to any one ion, but are free to move about. All of these valence electrons are shared by all of the atoms. (See Figure 6-14).

Figure 6-14 Electron-sea model of bonding in metals. The bonding strength increases with the number of electrons contributed by each atom.

A **metallic bond** *is a chemical bond resulting from the attraction between positive ions and surrounding mobile electrons.* The electrons are often referred to as forming an *electron sea*.

The strength of metallic bonding varies with the nuclear charge and the number of valence electrons. To compare bonding strengths in metals, the amounts of heat needed for vaporization are compared. By vaporizing a metal, the bonded atoms in the normal (usually solid) state are converted to individual metal atoms in the gaseous state. The higher the heat of vaporization, therefore, the stronger the metallic bonding.

Some heats of vaporization for metals are given in Table 6-3. Compare, for example, sodium and magnesium. Sodium atoms in the metal give up one electron and magnesium atoms give up two electrons. The attraction between cations and the surrounding electrons should be stronger in magnesium than in sodium because the magnesium cations have twice the charge of the sodium cations and because there are twice as many electrons attracting the cations in magnesium as there are in sodium. The higher heat of vaporization that magnesium has compared to that of sodium shows this to be the case.

Nature of Metals

Whenever you think of metals, you probably think of substances that are lustrous or reflective and that are good conductors of heat and electricity. The electron-sea model of metallic bonding is successful in explaining these intensive properties of metals. The high electrical conductivity is provided by

TABLE 6-3	HEATS OF VAPORIZATION OF SOME METALS (KJ/MOL)		
Row of the Periodic Table		Element	
second	Li 134.7	Be 224	B 540
third	Na 89.6	Mg 132	Al 284
fourth	K 77.08	Ca 153	Sc 306
fifth	Rb 69	Sr 153	Y 390
sixth	Cs 66	Ba 149	La 400

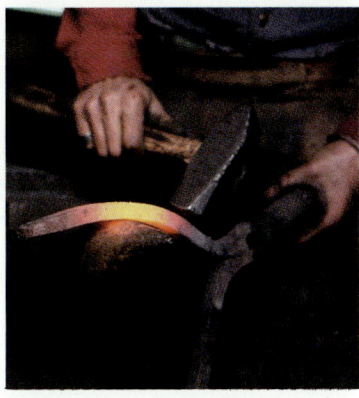

Figure 6-15 This blacksmith relies on the malleability of metals to beat iron into the shape of useful tools.

the motion of the free electrons through the solid. The high thermal conductivity of metals—the rapid transport of heat—also is accounted for by the mobile electrons. When a metal bar is heated at one end, the electrons at that end increase in kinetic energy and rapidly flow through the metal, transporting heat energy with them. In contrast, in molecular or ionic compounds heat can only be passed along stepwise from neighbor to neighbor.

Most metals are also characteristically easy to work and form into desired shapes. Two important characteristics of this type are called *malleability* and *ductility*. **Malleability** *is the state of being able to be shaped or extended by beating with a hammer, rolling, or otherwise exerting physical pressure that results in a change in contour* (see Figure 6-15). **Ductility** *is the state of being able to be drawn, pulled, or extruded through a small opening to produce a wire.* The malleability and ductility of metals are possible because metallic bonding is not directional, but uniform throughout the solid. One plane of ions in a metal can slide past another, without encountering any resistance or breaking any bonds. By contrast, shifting the layers of an ionic crystal causes the bonds to break and the crystal to shatter, as shown in Figure 6-13.

To reflect light and appear lustrous, a material must be able to absorb and re-emit light of many wavelengths. This occurs in metals because electrons in the electron sea have a wide range of possible energies. As a result, metals can absorb and re-emit light over a wide range of wavelengths giving them a characteristically shiny appearance.

Section Review

1. Describe the electron-sea model of metallic bonding.
2. What is the relationship between metallic-bonding strength and heat of vaporization?
3. Explain why metals are malleable and ductile, whereas ionic crystals are not.

STUDY HINT

For an introduction to properties of metals, see Section 1.4.

1. Metallic bonding is generally viewed as the result of the mutual sharing of many electrons by many atoms. Each atom contributes its valence electrons to a region surrounding the atoms, leaving behind positive ions. These electrons are then free to move about in what is often called an "electron sea," where they are shared by all of the atoms.
2. In general, the higher the heat of vaporization, the stronger the metallic bonding.
3. Metallic bonding is not directional but uniform throughout the solid. One plane of ions in a metal can slide past another without any resistance or breaking of bonds as would be the case if layers of an ionic crystal were shifted.

6.5 The Properties of Molecular Compounds

SECTION OBJECTIVES

- Explain the VSEPR theory.
- Predict the shapes of molecules or polyatomic ions using VSEPR.
- Explain how the shapes of molecules are accounted for by hybridization.
- Describe dipole–dipole forces, hydrogen bonding, an induced dipole, and London dispersion forces.
- Explain what determines molecular polarity.

The properties of molecules depend not only upon bonding, but also upon the arrangement of the atoms in space—molecular geometry. Understanding molecular geometry is useful, because many properties of molecular compounds are strongly influenced by it. Together with bond polarity, geometry determines molecular polarity. This polarity, in turn, strongly influences the forces that act between molecules in liquids and solids.

Two different, equally successful approaches to predicting molecular geometry are presented in this section. It is not unusual for two or more theories or models for the same phenomenon to stand the test of experiment. In such a case, scientists do not hesitate to switch back and forth, using whichever model best fits a given situation.

VSEPR Theory

Diatomic molecules like those of hydrogen (H_2), chlorine (Cl_2), and hydrogen chloride (HCl) must be linear. For more complicated molecules, prediction of geometry requires knowing the locations of electrons in bonds.

Where would you expect the bonding electrons to be in a molecule of beryllium fluoride (F—Be—F)? For each beryllium–flourine bond, the pair of electrons creates a region of high electron density next to the fluorine atom. Because electrons repel each other, the two pairs of bonding electrons stay as far away from each other as possible. The geometry that results is linear. All three atoms lie on a straight line with the two beryllium–fluorine bonds 180° apart as illustrated in Figure 6-16a.

The VSEPR theory for predicting molecular geometry is based upon the simple assumption that electrons in molecules repel each other. The letters VSEPR stand for "valence-shell, electron-pair repulsion," referring to the repulsion between pairs of bonding valence electrons in the highest occupied energy level (the "valence shell"). **VSEPR theory** states that *electrostatic repulsion between the valence-level electron pairs surrounding an atom causes these pairs to be oriented as far apart as possible.*

To use VSEPR, molecules are classified according to how many electron pairs surround a central atom. Beryllium fluoride is an AB_2 molecule, where A represents the central atom (Be) and B_2 represents the two atoms bonded to the central atom (2F).

What about an AB_3 molecule? Three A-B bonds stay farthest apart by pointing to the corners of an equilateral triangle, giving 120° angles between the bonds. Boron trifluoride (BF_3), shown in Figure 6-16b, is found by experiment to have the predicted triangular planar geometry.

The A atoms in AB_4 molecules follow the octet rule and have eight electrons in their highest energy levels. Experiment has shown that methane (CH_4) for example, is a symmetrical molecule with a carbon atom at the center of a regular tetrahedron. Each carbon–hydrogen bond points to one of the four corners of the tetrahedron and all four bond angles are 109.5°, as shown in Figure 6-16c. This is exactly the arrangement in which repulsion among the four bonding electron pairs in an AB_4 molecule is at a minimum.

Figure 6-16 Ball and stick models showing the shapes of AB_2, AB_3 and AB_4 molecules according to VSEPR.

TABLE 6-4 VSEPR AND MOLECULAR GEOMETRY

Molecular Shape	Atoms Bonded to Central Atom	Lone Pairs of Electrons	Type of Molecule	Formula Example	Lewis Structure
Linear	2	0	AB_2	BeF_2	F—Be—F
Bent	2	1	AB_2E	$SnCl_2$	Sn, :Cl: :Cl:
Triangular planar	3	0	AB_3	BF_3	:F: B :F: :F:
Tetrahedral	4	0	AB_4	CH_4	H—C—H, H, H
Triangular pyramidal	3	1	AB_3E	NH_3	N, H H H
Bent	2	2	AB_2E_2	H_2O	Ö, H H
Triangular bipyramidal	5	0	AB_5	PCl_5	:Cl: :Cl—P :Cl: :Cl:
Octahedral	6	0	AB_6	SF_6	:F: :F: :F: S :F: :F: :F:

The shapes of AB_2 to AB_6 molecules are summarized in Table 6-4. The Bs in such molecules need not be single atoms, nor need they be identical. The shape will still be based upon that given in the table. Different sizes of B groups can distort the bond angles, however, making some larger or smaller than those given in the table.

To use Table 6-4 to predict molecular geometry, it is necessary to know the Lewis structure for a molecule and classify it according to the number of bonds surrounding the central atom.

Chemical Bonding 183

Sample Problem 6.4

Use VSEPR theory to predict the molecular geometry of aluminum trichloride ($AlCl_3$).

First, the Lewis structure must be written for aluminum trichloride ($AlCl_3$) molecules. Aluminum is in Group 13 and has three valence electrons, [·Al·]; chlorine is in Group 17 and has seven valence electrons, [:Cl̈:]. The total number of available valence electrons is 24 e^- (3 e^- from aluminum and 21 e^- from chlorine). The following Lewis structure uses all 24 e^-.

$$:\ddot{\ddot{Cl}}:$$
$$:\ddot{\ddot{Cl}}:Al:\ddot{\ddot{Cl}}:$$

As explained in the Chemistry Notebook on page 168, this molecule is an exception to the octet rule. Aluminum trichloride is an AB_3 type of molecule. According to VSEPR theory, therefore, it should have triangular–planar geometry.

Practice Problems
1. Use VSEPR theory to predict the molecular geometry of each of the following: *(a)* HI *(Ans.)* linear *(b)* CBr_4 *(Ans.)* tetrahedral *(c)* BF_3 *(Ans.)* triangular-planar

Ammonia (NH_3) and water (H_2O) molecules illustrate two more common molecular shapes. The central atoms of these molecules (N and O) are both from *p*-block elements and follow the octet rule. The molecules, however, are not tetrahedral because unshared electron pairs are present.

The Lewis structure of ammonia in Table 6-4 shows that the molecule contains one unshared electron pair. The general formula for molecules like ammonia is AB_3E, where E represents the unshared electron pair. In an ammonia molecule, illustrated in Figure 6-17a, the three hydrogen atoms occupy three corners of a tetrahedron and the unshared pair occupies the fourth corner. The shape of the molecule depends only upon the positions of the four atoms, not on the position of the unshared pair. Consequently, the molecular geometry is that of a pyramid with a triangular base.

The water molecule, as shown by its Lewis structure in Table 6-4, has two unshared electron pairs. It is an AB_2E_2 molecule. Here, the oxygen atom is at the center of a tetrahedron, with two corners occupied by hydrogen atoms

Figure 6-17 The location of bonds and unshared electrons are shown for ammonia *(a)* and water *(b)*. Ball and stick models show the shapes of these molecules.

(a) Ammonia (NH_3)

(b) Water (H_2O)

and two by the unshared pairs (Figure 6-17b). The shape of the molecule is determined by the three atoms—it is referred to as a "bent" molecule. In Figure 6-17, note that the bond angles in ammonia and water are somewhat less than the 109.5° of a perfect tetrahedron. These angles are decreased because the unshared electron pairs repel electrons more strongly than do bonding electron pairs. Table 6-4 includes an example of an AB_2E type molecule, which results when a central atom forms two bonds and retains one unshared electron pair.

Lewis structures and Table 6-4 can be used together to predict the shapes of many molecules and polyatomic ions. Note that in predicting the shape of a molecule, double bonds are treated in the same way as single bonds.

STUDY HINT
Carbon dioxide, :O=C=O:

Chlorate ion, $\begin{bmatrix} :\ddot{O} \diagup \overset{\overset{\ddot{Cl}}{|}}{\underset{:\ddot{O}:}{}} \diagdown \ddot{O}: \end{bmatrix}^-$

Sample Problem 6.5

Use VSEPR theory to predict the shape of (a) carbon dioxide (CO_2) and (b) the chlorate ion (ClO_3^-).

Solution
(a) The Lewis structure of carbon dioxide shows two carbon–oxygen double bonds and no unshared electron pairs on the carbon atom. This is an AB_2-type molecule and is predicted to be linear.
(b) The chlorate ion has three oxygen atoms surrounding a central chlorine atom and, as shown by the Lewis structure, there is an unshared electron pair on the chlorine atom. The chlorate ion is classified as of the AB_3E type and predicted to have triangular–pyramidal geometry, with the three oxygen atoms at the base of the pyramid and the chlorine atom at the top.

Practice Problems
1. Use VSEPR theory to predict the molecular geometry of (a) hydrogen cyanide H—C≡N:
(b) formaldehyde (H_2CO) see Sample Problem 6.3. (Ans.) (a) linear, (b) triangular planar
2. Use VSEPR theory to predict the molecular geometries of the molecules whose Lewis structures are given below:
(a) :F̈—S̈—F̈: (Ans.) (a) bent   (b) :C̈l—P̈—C̈l: (Ans.) (b) triangular pyramidal
                                          |
                                       :C̈l:

Hybridization

VSEPR theory is an excellent guide to predicting and understanding the shapes of molecules. It does not reveal, however, any relationship between geometry and the orbitals occupied by bonding electrons. For this, a different model of bonding and molecular geometry is needed. One such model is based upon **hybridization**, *which is the mixing of two or more atomic orbitals of similar energies on the same atom to give new orbitals of equal energies.*

Methane (CH_4) provides a good example of how hybridization accounts for bonding and the shape of a molecule. The orbital notation for a carbon atom

C ↑↓   ↑↓   ↑ ↑ __
   1s   2s   2p

shows that it has four valence electrons, two in the 2s orbital and two in 2p orbitals. It is difficult to picture how these four electrons could form four

You might point out that hybrid orbitals are special types of *atomic* orbitals that have energy intermediate between s- and p- orbitals. An analogy to use might be mixing blue clay (s orbital) and yellow clay (p orbital) to get green clay (an sp, sp^2, or sp^3 orbital).

Figure 6-18 sp^3 Hybridization of atomic orbitals on a carbon atom. One s and three p orbitals combine to give four sp^3 hybrid orbitals. As is always the case in hybridization, the hybrid orbitals are at an energy level in between the levels of the orbitals that have combined.

equivalent, tetrahedrally arranged covalent bonds by orbital overlap with orbitals of four other atoms. Two of the electrons are already paired with each other. Also, the 2s orbital and the 2p orbitals have very different shapes as you learned in Section 4.2.

To account for the bonding in methane, the 2s and 2p orbitals on the carbon atom are assumed to combine and become rearranged, that is, to *hybridize*. The relative energies of the original and new orbitals on a carbon atom are represented in Figure 6-18. Note that the energy of the hybrid sp^3 orbital is different than that of the 2p or 2s orbital. Hybridization of the single 2s and the three 2p orbitals yields four new, identical orbitals.

Hybrid orbitals *are orbitals of equal energy produced by the combination of two or more orbitals on the same atom.* The number of hybrid orbitals produced is always equal to the number of orbitals that have combined. The four orbitals formed by combination of one s orbital and three p orbitals are designated sp^3 hybrid orbitals.

The superscript 3 indicates that three p orbitals were included in the hybridization (the superscript 1 on the s is not written, but understood).

The four sp^3 orbitals are oriented tetrahedrally, 109.5° apart, and account for the tetrahedral molecular geometry around any carbon atom with four single bonds. Each hybrid orbital is occupied by one electron from carbon and can overlap with an electron-containing atomic or hybrid orbital on another atom to form a covalent bond. Such bonding with sp^3 hybrid orbitals is illustrated in Figure 6-19a for methane.

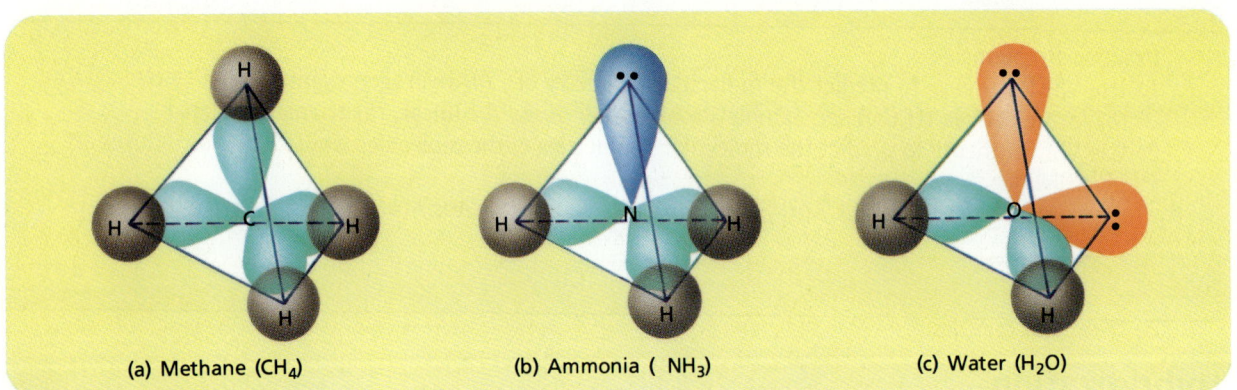

Figure 6-19 Bonds formed by overlap of sp^3 orbitals on carbon, nitrogen, and oxygen atoms with 1s orbitals on hydrogen atoms.

Hybridization also explains the bonding and geometry of many molecules formed by Group 15 and 16 elements. The sp^3 hybridization of a nitrogen atom ([He]$2s^2 2p^3$) yields four hybrid orbitals—one orbital that contains a pair of electrons and three orbitals that each contain an unpaired electron that can form a single bond, as in ammonia (Figure 6-19b). Similarly two of the four sp^3 hybrid orbitals on an oxygen atom ([He]$2s^2 2p^4$) are occupied by two electron pairs, and two are occupied by single electrons that can form single bonds, as in water (Figure 6-19c).

The linear and triangular planar geometry of molecules, such as those of beryllium fluoride and boron fluoride (illustrated in Table 6-4), is made possible by hybridization involving only one and two of the available empty p orbitals, respectively. The geometry of sp, sp^2, and sp^3 hybridization is summarized in Table 6-5.

TABLE 6-5 GEOMETRY OF HYBRID ORBITALS

Atomic Orbitals Hybridization	Type of Hybridization	Number of Hybrid Orbitals	Geometry
s, p	sp	2	180° linear
s, p, p	sp²	3	120° planar
s, p, p, p	sp³	4	109.5° tetrahedral

Since Johannes van der Waals studied the attractive forces that exist between molecules of gases and the liquids they formed upon condensation, certain intermolecular forces, dipole–dipole forces, and London dispersion forces are sometimes collectively referred to as van der Waals forces. In 1910, he received the Nobel Prize in physics for this work.

TABLE 6-6 BOILING POINTS AND BOND TYPES

Bond Type	Substance	bp (1 atm, °C)
noble gas	He	−269
	Ar	−186
nonpolar molecular	H_2	−253
	O_2	−183
	Cl_2	−34
	Br_2	59
	CH_4	−164
	CCl_4	77
	C_6H_6	80
polar molecular	PH_3	−88
	NH_3	−33
	H_2S	−61
	HF	19.5
	HCl	−85
	ICl	97
ionic	NaCl	1413
	MgF_2	2239
metallic	Cu	2567
	Fe	2750
	W	5660

Intermolecular Forces

The boiling point is the highest temperature, at a given pressure, at which a substance can exist as a liquid. At this temperature, the forces of attraction between particles are not strong enough to overcome the kinetic energy of the particles, which increases with increasing temperature. At the boiling point, adjacent particles in a liquid pull away from each other and enter the gas phase. The strengths of the attraction between particles of various substances, therefore, can be compared by comparing boiling points. The higher the boiling point, the stronger the interparticle forces.

The forces of attraction between molecules are known as **intermolecular forces**. Intermolecular forces vary in strength, but are generally weaker than the bonds that join atoms in molecules, ions in ionic compounds, or metal atoms in solid metals. In Table 6-6, compare the boiling points of the metals and ionic compounds with those of the other substances. You can see that they are much higher, indicating that ionic bonding and metallic bonding are much stronger than intermolecular forces.

Dipole–dipole forces and molecular polarity The strongest intermolecular forces act between polar molecules. *Equal but opposite charges separated by a short distance create a* **dipole**. The direction of a dipole is represented by an arrow pointing toward the negative pole with a tail crossed to indicate that it lies in the direction of the positive pole. For the hydrogen chloride molecule, the dipole is indicated as follows, with the negative end at the more electronegative chlorine atom:

Dipole–dipole forces *are forces of attraction between polar molecules.* The negative region in one polar molecule attracts the positive region in adjacent molecules, and so on throughout a liquid or solid. The effect of dipole–dipole forces is seen by comparing the boiling points of iodine chloride (I-Cl) and bromine (Br-Br). The boiling point of polar iodine chloride is 97°C. The boiling point of nonpolar bromine is 59°C. The dipole–dipole forces between molecules of ICl are illustrated schematically in Figure 6-20.

Figure 6-20 Dipole–dipole forces in iodine monochloride. An attractive force exists where the heads of two arrows meet.

The polarity of diatomic molecules like hydrogen chloride is determined by just one bond. For molecules containing more than two atoms, molecular polarity depends upon both the polarity of the individual bonds and the orientation of bonds.

The water molecule, for example, has two hydrogen–oxygen bonds. The more electronegative oxygen atom is the negative pole of both bonds. The polarities of these two bonds combine to make the water molecule highly polar, as indicated in Figure 6-21. The ammonia molecule is also polar, because the dipoles of the three nitrogen–hydrogen bonds extend in the same direction.

A polar molecule can *induce* a dipole in a nonpolar molecule by momentarily attracting its electrons. The result is an intermolecular force that is somewhat weaker than the dipole–dipole force. Such forces account, for example, for the solubility of nonpolar, molecular oxygen in water, a process necessary for aquatic life. The negative pole of a water molecule repels the outer electrons of an adjacent oxygen molecule. The oxygen molecule, then, has an induced positive pole on the side toward the water molecule and an induced negative pole on the opposite side. The result is an attraction to the water molecule.

In some molecules, the individual bond dipoles oppose one another, causing the resulting molecular polarity to be zero. Carbon dioxide and carbon tetrachloride are molecules of this type, as indicated by the illustrations in Figure 6-21.

Hydrogen bonding A particularly strong type of dipole–dipole force accounts for the unusually high boiling points of hydrogen-containing compounds, such as hydrogen fluoride (HF), water (H_2O) and ammonia (NH_3).

Hydrogen bonding *is an intermolecular attraction between a hydrogen atom and an unshared pair of electrons on a strongly electronegative atom (such as fluorine, oxygen, or nitrogen) in another molecule.* The large

Linus Pauling was the first chemist to recognize that the hydrogen bond was different in nature and strength from the covalent bond.

Water   Ammonia   Carbon tetrachloride (no molecular dipole)   Carbon dioxide (no molecular dipole)

electronegativity differences between hydrogen atoms and fluorine, oxygen, or nitrogen atoms makes the bonds connecting them highly polar. The hydrogen atom in such a bond has a positive charge approaching that of a proton. Moreover, the small size of the hydrogen atom allows an unshared pair of electrons on an adjacent molecule to approach very closely.

Hydrogen bonds are usually represented by dotted lines connecting the hydrogen-bonded hydrogen to the unshared electron pair of the electronegative atom to which it is attracted, as illustrated at right for water.

The effect of hydrogen bonding is shown by comparing the boiling points in Table 6-6 of phosphine (PH_3) and hydrogen-bonded ammonia (NH_3) or of hydrogen sulfide (H_2S) and that of hydrogen-bonded water (H_2O) at 100°C. Hydrogen bonding plays an important role in determining the properties of water, including its high surface tension.

London dispersion forces   Even noble-gas atoms and molecules that are nonpolar are attracted to each other when the temperature becomes low enough. The weak intermolecular forces solely responsible for the attraction between such atoms or molecules are known as London dispersion forces. The explanation of these forces was first provided in 1930 by Fritz London (1900–1954).

In any molecule, polar or nonpolar, the electrons are in continuous motion. The result is, that at a given instant, the electron distribution may be

Figure 6-21 Some examples of the effect of molecular shape on molecular polarity. In water and ammonia molecules, the bond polarities all extend in the same direction, causing the molecules to be polar. In the carbon–tetrachloride (CCl_4), and carbon–dioxide (CO_2) molecules, the bond polarities extend symmetrically in different directions and cancel one another.

Hydrogen bonding in water

STUDY HINT
The very important role of hydrogen bonding in the properties of water is discussed in Section 13.4.

Figure 6-22 The surface tension of water is the result of hydrogen bonding. The leaf at left is supported by the surface tension of water. At right, water droplets are spherical due to the surface tension of water. If water had zero surface tension, the droplets would not be uniform.

Figure 6-23 Helium can become temporarily polar under the right conditions. If an instantaneous, temporary dipole develops in a helium atom *(top)*, it can induce a dipole in a neighboring atom *(bottom)*.

1. The spatial arrangement of the atoms bonded together in a molecule
2. The VSEPR theory and/or the Theory of Hybridization
3. *(a)* Ö::Si::Ö (AB$_2$—linear)

   (b) :Cl:C:Cl: (AB$_4$—tetrahedral)
   (with :Cl: above and :Cl: below)

4. In sp^3 hybridization, one *s* and three *p* orbitals combine and become rearranged to form four new, identical orbitals of equal energies.
5. *(a)* Hydrogen bonding affects the properties of water molecules. *(b)* Such bonding causes water to have an unusually high boiling point and to display high surface tension.

uneven, creating a positive pole in one part of the molecule and a negative pole in another. This instantaneous and temporary dipole can induce a dipole in an adjacent molecule, as illustrated in Figure 6-23. The two molecules are then held together for an instant by the weak attraction between the temporary dipoles. **London dispersion forces** *are intermolecular attractions resulting from the constant motion of electrons and the creation of instantaneous dipoles and induced dipoles.*

London forces act between all atoms and molecules, but are the only intermolecular forces acting among noble-gas atoms and nonpolar molecules. In Table 6-6, it can be seen that noble gases and nonpolar–molecular compounds have the lowest boiling points. Since London forces are dependent upon the motion of electrons, they are stronger as the number of electrons in a molecule is greater. In other words, London forces increase with increasing masses of atoms or molecules. Comparison of the boiling points of the gases helium (He) and argon (Ar), hydrogen (H_2) and oxygen (O_2), and chlorine (Cl_2) and bromine (Br_2) shows the effect of increasing mass on increasing London dispersion forces.

Section Review

1. What is meant by molecular geometry?
2. What two theories can be used to predict molecular geometry?
3. Draw the Lewis structure and use the VSEPR theory to predict the molecular geometry of *(a)* SiO_2 *(b)* CI_4.
4. Explain what is meant by sp^3 hybridization.
5. *(a)* What type of intermolecular force plays a major role in accounting for various properties of water molecules? *(b)* Explain.

Chapter Summary

- Two or more atoms may form ionic, covalent, or polar-covalent bonds depending on the difference in electronegativity between the atoms.
- The octet rule states that chemical compounds tend to form so that each atom, by gaining, losing or sharing electrons has eight electrons in its highest occupied energy level.
- Double bonds occur when atoms share two pairs of electrons in a covalent bond. Triple bonds occur when atoms share three pairs of electrons in a covalent bond.
- Because of the strong attraction between positively and negatively charged ions, materials containing ionic bonds tend to be harder and more brittle and to have higher boiling points than materials containing covalently bonded atoms.
- The "electron sea" formed in metallic bonding gives metals their properties of high electrical and thermal conductivity, malleability, ductility, and luster.
- VSEPR theory uses the fact that electron pairs strongly repel each other and tend to be oriented as far apart as possible to predict shapes of molecules.
- Hybridization uses the fact that orbitals within an atom can mix to form orbitals of equal energy in order to predict the shapes of molecules.
- Intermolecular forces such as dipole–dipole forces, hydrogen bonding, and London dispersion forces can exist between certain types of molecules.

Chapter 6 Review

Chemical Bonding

VOCABULARY

bond energy (166)
bond length (166)
chemical bond (161)
chemical formula (164)
covalent bond (161)
diatomic molecule (164)
dipole (187)
dipole–dipole forces (188)
double bond (170)
ductility (181)
formula unit (175)
hybrid orbitals (186)
hybridization (185)
hydrogen bonding (188)
intermolecular forces (187)
ionic bond (161)
ionic compound (175)
lattice energy (178)
Lewis structure (169)
London dispersion forces (190)
malleability (181)
metallic bond (180)
molecular compound (164)
molecular formula (164)
molecule (164)
multiple bond (171)
nonpolar–covalent bond (162)
octet rule (168)
polar (162)
polar-covalent bond (162)
polyatomic ion (173)
resonance (174)
single bond (169)
structural formula (169)
triple bond (171)
unshared pair (169)
VSEPR theory (182)

REVIEWING CONCEPTS

1. **a.** What is a chemical bond?
   b. Upon what basis are the various chemical bonds classified? (6.1)
2. Among main group elements, what ion formation tendencies are generally observed? (6.1)
3. Identify and define the two major types of chemical bonding. (6.1)
4. What is the relationship between electronegativity and the ionic character of a chemical bond? (6.1)
5. **a.** What is the meaning of the term polar, as applied to chemical bonding?
   b. Distinguish between polar- and nonpolar-covalent bonds. (6.1)
6. **a.** In general, what determines whether atoms will form chemical bonds?
   b. What energy change is usually associated with bond formation? (6.1)
7. **a.** What determines bond length?
   b. How are bond energies and bond lengths related? (6.2)
8. Describe the general location of the electrons in a covalent bond. (6.2)
9. As applied to covalent bonding, what is meant by an unshared or lone pair of electrons? (6.2)
10. Determine the number of valence electrons contained in an atom of each:
    a. H   **c.** Mg   **e.** Al   **g.** C
    b. F   **d.** O   **f.** N   **h.** Ne (6.2)
11. When drawing Lewis structures, how do you determine the central atom of the arrangement? (6.2)
12. Distinguish among single-, double-, and triple-covalent bonds by defining each and providing an illustration of each type. (6.2)
13. In writing Lewis structures, how is the need for multiple bonds generally determined? (6.2)
14. **a.** What is a polyatomic ion?
    b. Give two examples of polyatomic ions.
    c. How do such ions normally occur in nature? (6.2)
15. **a.** What is an ionic compound?
    b. In what form do most ionic compounds occur? (6.3)
16. **a.** What is a formula unit?
    b. What are the components of one formula unit of CaF_2?
    c. How many K^+ and S^{2-} ions would be in one formula unit of the ionic compound formed by these ions? (6.3)

Chemical Bonding

17. **a.** What is lattice energy?
    b. In general, what is the relationship between lattice energy and the strength of an ionic bond? (6.3)

18. **a.** In general, how do ionic and molecular compounds compare in terms of melting points, boiling points, and ease of vaporization?
    b. What accounts for the observed differences in their properties?
    c. Cite three physical properties of ionic compounds. (6.3)

19. **a.** How do the properties of metals differ from those of both ionic and molecular compounds?
    b. What specific property of metals accounts for their unusual electrical conductivity? (6.4)

20. What properties of metals contribute to their tendency to form metallic bonds? (6.4)

21. **a.** What is a metallic bond?
    b. What determines the strength of a metallic bond? (6.4)

22. **a.** How is the VSEPR theory used to classify molecules?
    b. What molecular geometry would be expected for F_2 and HF? (6.5)

23. According to the VSEPR theory, what molecular geometries are associated with the following types of molecules:
    a. AB_2 **d.** AB_5
    b. AB_3 **e.** AB_6 (6.5)
    c. AB_4

24. Describe the role of each of the following in predicting molecular geometries:
    a. unshared electron pairs
    b. double bonds (6.5)

25. **a.** What are hybrid orbitals?
    b. What determines the number of hybrid orbitals produced by an atom? (6.5)

26. **a.** What are intermolecular forces?
    b. How do these forces compare in strength to those in ionic and metallic bonding?
    c. Where are the strongest intermolecular forces found? (6.5)

27. **a.** What are dipole–dipole forces?
    b. What determines the polarity of a molecule? (6.5)

28. **a.** What is meant by an induced dipole?
    b. What is the everyday importance of this type of intermolecular force? (6.5)

29. **a.** What is hydrogen bonding?
    b. What accounts for its extraordinary strength? (6.5)

30. **a.** What are London dispersion forces?
    b. How are London forces related to atomic and molecular masses?

PROBLEMS

▶ Chemical Bond Character

1. Determine the electronegativity difference, the bond type, and the more electronegative atom with respect to bonds formed between the following pairs of atoms:
   a. H and I   **d.** Zn and O
   b. Ba and S   **e.** At and I
   c. K and Br   **f.** S and I

2. List the bonding pairs described in Problem 1 in order of increasing covalent character.

3. Use orbital notation to illustrate the bonding in each of the following:
   a. a chlorine molecule
   b. an oxygen molecule
   c. a phosphorus molecule
   d. a molecule of hydrogen fluoride (consisting of one atom of each element)

4. Use electron-dot structures to demonstrate the formation of ionic bonds involving:
   a. Na and S   **b.** Ca and O   **c.** Al and S

▶ Lewis Structures

5. For each of the following, use electron dot notation to illustrate the number of valence electrons present in one atom of each element listed:
   a. Li   **e.** C
   b. Ca   **f.** P
   c. Cl   **g.** Al
   d. O   **h.** S

6. Draw Lewis structures for each of the following molecules:
   a. contains one C and four F atoms
   b. contains two H and one Se atom
   c. contains one N and three I atoms
   d. contains one Si and four Br atoms
   e. contains one C and three H, and one Cl atom.

7. Draw Lewis structures for each of the following molecules:
   a. contains two oxygen atoms
   b. contains two nitrogen atoms
   c. contains one silicon atom and two oxygen atoms
   d. contains one carbon atom and one oxygen atom
   e. contains one sulfur atom and two oxygen atoms

▶ **VSEPR Theory and Molecular Geometry**

8. According to the VSEPR theory, what molecular geometries are associated with the following types of molecules?
   a. AB_3E b. AB_2E_2 c. AB_2E

9. Use hybridization to explain the bonding in methane (CH_4).

10. For each of the following polar molecules, indicate the direction of the resulting dipole:
    a. H—F
    b. H—Cl
    c. H—Br
    d. H—I

11. Determine whether each of the following bonds would be polar or nonpolar:
    a. H—H
    b. H—O
    c. H—F
    d. Br—Br
    e. H—Cl
    f. H—N

12. On the basis of the individual bond polarity and orientation, determine whether each of the molecules would be polar or nonpolar:
    a. H_2O
    b. I_2
    c. CH_4
    d. NH_3
    e. CO_2

13. Draw the Lewis structure for each of the following molecules and then use the VSEPR theory to predict the molecular geometry of each:
    a. SCl_2
    b. PI_3
    c. Cl_2O
    d. NH_2Cl
    e. $SiCl_3Br$
    f. $ONCl$

14. Draw the Lewis structure for each of the following polyatomic ions and then use VSEPR theory to identify the ion type and subsequent geometry of each:
    a. NO_3^-
    b. NH_4^+
    c. SO_4^{2-}
    d. ClO_2^-

CRITICAL THINKING

1. **Inferring Relationships** In Table 6-1 on page 166, bond lengths are given for common single covalent bonds. However the length of a bond varies depending on the other bonds in the molecule. Predict and compare the lengths of the C-C bond in the following molecules. Explain your answer.

C_2H_6 C_2H_4 C_2H_2

RESEARCH AND WRITING

1. Prepare a report on the work of Linus Pauling.
   a. Discuss his work on the nature of the chemical bond.
   b. Linus Pauling has been an advocate for the use of vitamin C as remedy for colds. Evaluate Pauling's claims. Determine if there is any scientific evidence that indicates whether vitamin C is an effective cold remedy.

PROJECT

1. Identify ten common substances in and around your home, and indicate whether you would expect these substances to contain ionic, covalent, or metallic bonds.

Chemistry and Careers

Laser Chemist

Laser technology is a rapidly growing field for chemists. Laser applications range from scanning labels at the grocery store to welding detached retinas in eye surgery.

Every compact disk player has a tiny laser that senses information on the surface of the disk and is used to translate it into music. An exciting new application is in the computer industry, where gallium arsenide lasers can serve as extremely precise tools for creating the tiny circuits on computer chips.

Employment in laser chemistry looks promising. If you want to develop new applications for lasers, you need to understand how lasers operate, which requires a background in chemistry and physics. For some technical jobs in laser chemistry, an associate's degree is sufficient. To be a laser chemist, however, you must earn at least a bachelor's degree.

Chemical Technician

Chemical technicians are in high demand these days because of the growth of chemical industries. Before a new chemical product such as an improved pain killer or a solvent is ready for marketing, it must undergo a battery of tests that are usually conducted by technicians.

A chemical technician sets up laboratory equipment such as glassware, balances, burners, and computers according to a chemist's specifications. The technician then mixes chemical solutions, conducts experiments, and collects data. For example, a technician who works in a laboratory in which materials are tested for corrosion resistance collects and records acids, bases, salts, and other chemicals that are to be used to test the materials. Then, the technician runs tests and collects data on degrees of corrosion.

A chemical technician may specialize in one area, but most are trained to work in a wide range of fields. Strong interests in chemistry and math are helpful to people who wish to become chemical technicians. In most cases, a technician has at least two years of college or technical training. An interest in computers is useful because laboratory data are entered into computers for analysis.

UNIT 2 • FOCUS ON REASONING

College Entrance Exam Preparation

The atomic mass of an element represents a weighted average of the atomic masses of an element's individual isotopes. Isotopes of an element have the same numbers of electrons and protons, but different numbers of neutrons. Therefore, isotopes of the same element have different masses. An instrument called the mass spectrometer can be used to separate the isotopes and determine the percent abundance of each isotope in an element sample.

Element X is analyzed using a mass spectrometer. The isotopes of element X in a natural sample are converted to ions by being bombarded by electrons. A high-speed beam of these element-X ions splits into three parts when it passes through the magnetic field of the mass spectrometer. Light ions are deflected more than heavy ones. The three-part beam produces three peaks on the recording device signifying isotopes with masses 20, 21, and 22 respectively. The area represented by each peak is a measure of the relative abundance (in perentage) of the isotope.

Isotope	Atomic Mass	Abundance
X-20	20.00	90.92%
X-21	21.00	0.26%
X-22	22.00	8.82%

1. One conclusion that can be drawn from the data is that
   a. out of every 10 000 atoms of element X, 26 would be X-21.
   b. out of every 22 atoms of element X, 8.82 would be X-22.
   c. X-20 has the most neutrons of the three isotopes.
   d. X-22 is the most abundant isotope.

2. The average atomic mass of element X would be closest to the mass of which isotope?
   a. X-20
   b. X-21
   c. X-22
   d. no closer to one isotope than another.

3. The isotope deflected most by the magnetic field is
   a. X-20.
   b. X-21.
   c. X-22.
   d. No one isotope is deflected more than another.

4. Which of the following could be used to calculate the average atomic mass of element X? atomic mass of element X =

   a. $20.00 \times \frac{(x)}{100} + 21.00 \times \frac{(x)}{100} + 22 \times \frac{(x)}{100}$

   b. $20.00 \times \frac{(x-20)}{100} + 21.00 \times \frac{(x-21)}{100} + 22.00 \times \frac{(x-22)}{100}$

   c. $20.00 \times \frac{90.92}{100} + 21.00 \times \frac{0.26}{100} + 22.00 \times \frac{8.82}{100}$

   d. $\frac{20.00 + 21.00 + 22.00}{3} \times 100$

INTRA-SCIENCE

How to Replace Asbestos

When it was established beyond a doubt that asbestos was a health hazard, the call went out for a substitute. The "want ad" might have read something like this:

> Must be noncombustible and resistant to decay, vermin, and most acids. Must withstand continuous heat of up to 500°C and peak temperatures of up to 2000°C. Must have strong, yet flexible fibers that can be spun and woven. Must be cheap.

Asbestos is a general name for six types of rock-forming minerals. The most commonly used form is a fibrous serpentine called chrysolite, which consists of magnesium, silicon, hydrogen, and oxygen.

The Connection to Physics

The most important physical property of asbestos is its fibrous form, shown in the photograph above, which permits it to be spun into yarn, cloth, or tape. Fibers are removed from crushed rock by equipment that blasts and extracts, separating them from the rock matrix in which they are embedded. The resulting product is fireproof, a good thermal insulator, and has a tensile strength almost as great as some types of steel. Such physical properties led to the use of asbestos in automobile brake linings, fire barriers inside walls, insulator blankets around heaters and pipes, and roofing and floor-covering materials. There are more than 200 major industrial applications of asbestos. Replacing it with a safer substance has not been easy. However, dozens of different materials have been developed to meet specific needs.

The asbestos cement used in walls, ceilings, and fire doors is being replaced by cement composites containing glass, steel, and carbon, and by glass-reinforced gypsum and polyester. Pipes that previously might have been made of asbestos cement are now made of glass-reinforced cement, reinforced concrete, or metal.

Prefabricated asbestos insulating boards are being replaced by boards made of reinforced calcium silicate, reinforced portland cement, rock wool, and ceramic-fiber material. Vermiculite minerals and perlite, a volcanic glass, have proved adequate in insulated boards and in sprayed and loose-fill insulation.

Finding new materials to replace asbestos in brake and clutch linings has been a problem. Silicon nitride was used in the brake pads of the *Concorde* airliner prototypes, but was so expensive that in the production planes, reinforced carbon composites were used instead. Steel fibers are cheaper still, but pose an abrasion problem. So do glass fibers.

In the electrical industry, substitutes include oil-impregnated paper and cloth, epoxy resins, resin-impregnated mica and glass fabrics, and thermosetting materials such as polyamide. The choice between the substitutes depends on the heat resistance required.

In floor tiles and roofing materials, asbestos has been replaced by a variety of readily available substances, chiefly cellulose fibers.

The Connection to Health Sciences

The ancient Egyptians and Romans valued asbestos for its fireproofing ability, but it did not become commercially important as a building material until the 19th century. Russia and Canada became the world's leading producers. In the 1920s it was discovered that asbestos caused a severe lung disease, which was appropriately named asbestosis.

The disease asbestosis is caused by fibers of asbestos that are inhaled. These fibers range in length from 5 to 100 mm and may remain in the respiratory system—especially the lungs—indefinitely. The disease appears to be more

the result of fiber size and shape than of chemical properties. The irritating fibers cause an inflammatory reaction in the lungs, as seen in the X ray above, that results in scarring. If enough scarring occurs, the lungs may become less elastic and breathing more difficult. In the 1960s U.S. scientists found a link between asbestos and cancer. We now know that even a low level of exposure to asbestos is unsafe.

People working directly with asbestos are not the only ones endangered. Dust on clothing worn home from a worksite exposes the entire family to danger. Even the asbestos inside the walls of buildings presents a hazard. As scientists search for acceptable substitutes, the Environmental Protection Agency is working toward a total phaseout of all asbestos products.

One concern was that substitute fibers might be as hazardous as asbestos, but preliminary studies of likely replacements such as glass, vermiculite, and perlite have shown no health hazard. The biggest drawback in the use of these materials is increased cost.

197

Unit 3

Chapter 7
Chemical Formulas and Chemical Compounds 200

Chapter 8
Chemical Equations and Reactions 232

Chapter 9
Stoichiometry 260

Language of Chemistry

I always liked chemistry and physics classes in school, but wanted a career that would let me use these sciences in a practical way. As a chemical engineer, I can use science to solve real world problems. Now, I help Du Pont's researchers choose the best way to make quality products at the lowest cost.

Chemical engineers use the principles you will learn in this unit to operate chemical plants, making sure the right chemicals are flowing in the right directions, in the right amounts, and at the right temperatures to produce the desired chemical reactions. We use a special language of formulas and equations to communicate about these processes—the language of chemistry.

R. A. Cantini
Senior Consulting Engineer
E. I. du Pont de Nemours & Company

Petroleum refining plant near Texas City, Texas

CHAPTER 7
Chemical Formulas and Chemical Compounds

PLANNING CHART

Approximate Time: 12/15 days

	SECTION/TEXT FEATURES		RESOURCES	
BASIC	**7.1 CHEMICAL NAMES AND FORMULAS**	**201**	**Teacher's Resource Binder**	
	Demos 1, 2	200B–C	Lab Experiment 8	61
	Periodic Table Videodisc	200H	Section Review Worksheets	23
	Microscale Investigation 5	840	**Other Resources**	
	Section Review	212	Transparencies 18, 19, 20	
	Chapter Review	229	Blackline Masters 26, 27, 28, 29	
	• Questions 1–15, Problems 1–4			
BASIC	**7.2 OXIDATION NUMBERS**	**212**	**Teacher's Resource Binder**	
	Demo 3	200C	Section Review Worksheets	24
	Desktop Investigation	215	**Other Resources**	
	Section Review	216	Blackline Master 29	
	Chapter Review	229		
	• Questions 16–19, Problems 5–6			
BASIC	**7.3 USING CHEMICAL FORMULAS**	**216**	**Teacher's Resource Binder**	
	Section Review	223	Section Review Worksheets	25
	Chapter Review	229	**Other Resources**	
	• Questions 20–21, Problems 7–11		Blackline Master 30	
BASIC	**7.4 DETERMINING CHEMICAL FORMULAS**	**224**	**Teacher's Resource Binder**	
	Microscale Investigation 6	843	Lab Experiment 10	71
	Section Review	228	Section Review Worksheets	26
	Chapter Review	229		
	• Questions 22–23, Problems 12–17			
	CHAPTER END ACTIVITIES		**Teacher's Resource Binder**	
	Critical Thinking, Cross Connections, Research and Writing, Projects		Test Booklet	33, 35
			Computer Test Generator	

Teaching Strategies

7.1 Chemical Names and Formulas

An important basic concept in chemistry is presented in this chapter; a chemical compound is represented by a unique chemical formula. The formula consists of symbols for each element in the compound and subscripts which indicate the number of atoms of each element in the compound.

If the symbols are changed, the formula represents a different compound; if the subscripts are changed, the formula represents a different compound. Students often have difficulty recognizing the importance of maintaining the subscripts once a formula has been determined. You might ask your students to "engrave in stone" correct formulas, or to put boxes around them, so they know that once determined, the formulas cannot be changed. Elements are also represented by symbols and subscripts, but more about that later.

Begin this chapter by bringing to class many compounds and elements for your students to observe. A good collection might be H_2O, H_2O_2, KI, NaI, S_8, Na_2S, Na_2SO_4, Cu_2SO_4, $CuSO_4$, FeO, Fe_2O_3, CH_4, CH_3OH, K_2CrO_7, K_2CrO_4. Have students list their observations of each substance along with its formula. You might distinguish between H_2O and H_2O_2 by adding a few mL of each to a dilute solution of $KMnO_4$. This will not only vividly demonstrate the differences in the chemical properties of H_2O and H_2O_2 but will prepare students for the fact that a single element can exist in various oxidation states.

All of chemistry is encoded in the Periodic Table. It is the job of the chemistry student to learn to decode this information. If you can, give each one of your students a notebook copy of the Periodic Table. The Sargent Welch table is two-sided, and contains excellent information about ionization potential and electronegativity. Require your students to bring this table to class each day. If you cannot supply them with a supplementary table, there is one in the back of their textbook which can be photocopied.

Refer to the Periodic Table constantly in this unit. Point to the class table and indicate that column 1 elements always form 1+ ions, column 2 elements always form 2+ ions, column 3 elements always form 3+ ions and column 17 elements usually form 1− ions. Emphasize that positive ions have more protons than electrons and that negative ions have more electrons than protons. Students manage to construct creative but false theories to explain the charge on particles.

Do a particle count, making use of the Periodic Table to determine the number of protons in the nucleus of each element:

Na	yields	Na+
11 protons		11 protons
11 electrons		10 electrons

Cl_2	yields	$2Cl^-$
17 protons	(each Cl atom)	17 protons
17 electrons		18 electrons

Continually remind students that electrons are negatively charged and protons are positively charged. Atoms are neutral: ions are charged particles. The charge on the ion is determined by the balance of positive charges contributed by the protons and the negative charges contributed by the electrons. When an atom loses an electron, the resulting ion is positive because it has more protons than electrons. Students have difficulty with the concept that an atom loses a particle and becomes positive.

All students are familiar with common table salt. Use this compound to illustrate a few points about binary ionic compounds. Ions are charged particles, but the compounds they form are neutral. When writing the formula for the compound, the positive ion is written first and the negative ion, which is written second, has an -ide ending. Refer to the Periodic Table as you are presenting this. Touching the salt and not receiving an electric charge might convince students that the compound is neutral even though it is composed of positive and negative charges. Students will be convinced that the compound does form charged particles if you compare the results of a conductivity test in salt water with the results of a conductivity tester dipped in distilled water. If you also test a sugar solution for conductivity, this demonstration will set the stage for discussing the difference between ionic and molecular compounds.

If space-filled styrofoam models are available, this is an excellent time to use them. NaCl illustrates the ionic nature of the compound as well as showing the difference in size between the positive and negative ions.

If students understand that an ionic compound is composed of charged particles combined to form a neutral substance, they will not have too much trouble writing formulas. Go over the rules carefully. Give the class lots of sample problems. Refer to the Periodic Table as you do that. Diagram the formation of compounds on the chalkboard.

Demonstration 1: The synthesis of a binary iodide compound

PURPOSE: To demonstrate the formation of a compound from two elements
[Note: This reaction will be done quantitatively in Chapter 9. At this time, emphasize the qualitative evidence that a reaction has occurred.]
MATERIALS: 50-mL beaker, 0.5 g of metal—use mossy zinc or coiled copper wire, 0.5 g of iodine crystals, watch glass, ice cube, water bath or heater with temperature control
PROCEDURE: Place 0.5 g of metal in a beaker. Pass the beaker around the classroom. Ask each student to record the properties of the metal. Add iodine crystals to the beaker. Place a watch glass over the beaker and an ice cube on the watch glass. Heat gently over a water bath or on a heater set at a low temperature. Allow to remain for fifteen minutes.

Iodine will sublime resulting in a beautiful purple vapor.
CAUTION: Iodine vapors are toxic—but iodine is the least toxic of all halogens—the ice cube will contain the vapors so that students are not exposed to them. This demonstration should be done in a hood. Some of the vapor will react with the metal. Excess iodine will recrystallize on the bottom of the watch glass. Remove ice cube from watch glass. Discard meltage. Ask one student to carry the watch glass around the room so that students will be able to observe the iodine crystals that have formed. Remove metal with a tweezers and place in a petri dish. Pass around the room so that each student can observe the iodide compound that has formed on the surface of the metal.

A compound has formed from the combination of two elements. The properties of the compound are different than the properties of the elements. A chemical reaction has occurred. In the course of the reaction, electrons have been transferred from the metal to the iodine. Before the iodine received the electrons from the metal, the bond was broken that held the two iodine atoms together into one diatomic elemental molecule. Each iodine atom received one of the metallic electrons.

Write the reaction on the board.

$$Cu + I_2 \rightarrow CuI_2$$

Do a particle count.

Cu	$\rightarrow$	Cu^{2+}	+ 2 electrons
29 protons		29 protons	
29 electrons		27 electrons	

2 electrons	+	I_2	$\rightarrow$	$2I^-$
		53 protons		53 protons
each atom		53 electrons		54 electrons

$$Cu^{2+} + 2I^- \rightarrow CuI_2$$

Use models to simulate this reaction. The methods and materials for using common materials to make models are described in the Desktop Investigation.

Naming compounds can be tricky, but is actually easier than all of the rules seem to indicate. The rules are important and memorization should be encouraged. But, you can try to prevent many of your students' difficulties by showing them how

Chemical Formulas and Chemical Compounds

much they already know. They will probably be surprised how much they have learned about naming elements and compounds just through their brief introduction to chemistry in Chapters 1–6. You can encourage them to apply this knowledge by writing formulas on the board and asking for their names. Start with what they may already be familiar with. Also, utilize the tables in the chapter and other resources that you may have. This is an easy way to become familiar with most of the common monatomic and polyatomic ions. Remember that practice makes perfect. Help your students to become familiar with the common names.

A binary compound consists of two elements. To name *binary ionic compounds*, the following rules are used: *(1)* Name the metal first. An *-ide* is added to the stem of the nonmetal (ex. sodium chloride). *(2)* If the metal exhibits more than one oxidation state, a Roman numeral is used to indicate which oxidation state of the metal is present in the compound. The Roman numeral is placed in parentheses following the name of the metal. An *-ide* ending is added to the stem of the nonmetal as in *(1)* above (ex. FeO, iron(II) oxide; Fe_2O_3, iron(III) oxide). To name *binary molecular compounds*, prefixes are used to indicate the number of atoms of the nonmetal. The rules for naming binary ionic compounds are then applied (e. g. PCl_3, phosphorous trichloride; PCl_5, phosphorous pentachloride).

Chemistry is based on observations and problem solving. A lot of memorizing is usually not a good strategy, but learning the names, formulas, and charges of polyatomic ions is the exception. Students have to memorize this, and the sooner the better. One hint that you might give students regarding the charge on ions is that elements in odd numbered columns in the Periodic Table form oxyions that have an odd negative charge (1− or 3−) such as PO_4^{3-}, NO_3^-, NO_2^-, and IO_3^-, and that elements in even numbered columns in the Periodic Table form oxyions that have an even negative charge (2−) such as CO_3^{2-} and SO_4^{2-}.

If students have memorized the table of polyatomic ions, writing formulas for compounds containing polyatomic ions should not be more difficult than writing formulas for binary compounds. Emphasize that parentheses must be placed around the polyatomic ion if two or more exist in the compound. At this point, make sure that students know how to analyze a formula for the correct number of atoms.

CuI_2	1 copper	2 iodine	
NaCl	1 sodium	1 chlorine	
Na_2SO_4	2 sodium	1 sulfur	4 oxygen
$Al_2(SO_4)_3$	2 aluminum	3 sulfur	12 oxygen

Many demonstrations can be done effectively with small quantities of chemicals in 50-mL beakers on an overhead projector projected on a screen. The identity of the material can be included on a transparency prepared for use with 50-mL beakers. The transparency should include a circle for each beaker that you use. Place the beaker on the appropriate circle. You can prepare your master by placing your 50-mL beakers on a piece of paper and tracing around them.

Demonstration 2: Observations of Aqueous Solutions of Acids and Salts

PURPOSE: To observe some of the properties of aqueous solutions of acids and salts
MATERIALS: 8 50-mL beakers; 0.1-*M* solutions of sulfuric acid, nitric acid, phosphoric acid, hydrochloric acid; 4 small vials containing sodium sulfate, sodium nitrate, sodium phosphate, sodium chloride; universal indicator
PROCEDURE: Add 20 mL of each acid to a 50-mL beaker placed on an overhead projector with projection on screen. Add 2 drops of universal indicator to each acid solution. Record color. Observe and record physical properties of the salts. Add 20 mL of water to the four empty beakers. Add 0.1 g of each salt to a beaker and stir. Record observations. Add 2 drops of universal indicator to each salt solution. Record color. Compare the color of the indicator in the acid solution to the color of the indicator in the corresponding salt solution. Write the formulas of the acids and the salts.

7.2 Oxidation Numbers

Begin this topic by referring to the Periodic Table. If you are using the Sargent Welch Table, refer to the Pauling electronegativity values on page 2. At this point, it is a good idea to expose students to the difference between charge (the actual transfer of electrons) and oxidation state (the distribution of electrons). Do not expect them to thoroughly understand this concept yet. They will review it again in Chapter 20 (and throughout their experiences with chemical reactions). Do make sure they know all of the rules for assigning oxidation numbers.

The various oxidation states of manganese are indicated by a variety of identifying colors. The following demonstration can be done at this time so that students can observe changing oxidation states. Emphasize the formula, the oxidation state of the ion, the oxidation state of the manganese, and the color of the compound. Do not ask students to write equations for these reactions at this time. You can redo this demonstration and discuss the underlying chemical phenomena with Chapter 20.

Demonstration 3: Oxidation States of Manganese

PURPOSE: To observe and relate the colors and oxidation states of manganese
MATERIALS: 0.1 *M* $NaHSO_3$ (dissolve 1.04 g $NaHSO_3$ in 100 mL of H_2O), 0.05 *M* $KMnO_4$ (dissolve 0.79 g powdered $KMnO_4$ in 100 mL H_2O), 0.005 *M* $KMnO_4$ (mix 10 mL of solution 2 with 90 mL of H_2O), 0.000 40 *M* $KMnO_4$ (mix 20 mL of solution 3 with 230 mL of H_2O), 6 *M* HCl, 8 *M* NaOH.
PROCEDURE: Add 20 mL of 0.000 40 *M* $KMnO_4$ in each of five 50-mL beakers on an overhead projector. Add the HCl, NaOH, and $NaHSO_3$ by drops as needed to each beaker. Use stirring rods to mix solutions after addition of each reagent.

Drops HCL	Drops NaOH	Drops $NaHSO_3$	Color	Product	Oxidation State of Manganese
0	0	0	purple	MnO_4^{1-}	7
8	0	8	colorless	Mn^{2+}	2
0	0	4	orange	MnO_2	4
0	8	1	green	MnO_4^{2-}	6

It is a bit difficult to obtain the green MnO_4^{2-}. Too little $NaHSO_3$ will produce a blue coloration due to the incomplete reduction of $KMnO_4$. This is really due to absorption of light by two chromophores, MnO_4^{1-} and MnO_4^{2-} leaving blue as the only visible color. If you get this result, add as small a drop as possible and mix. If too much $NaHSO_3$ is added, manganese will be reduced to the orange MnO_2 stage. If this happens, try again. Getting the green Mn^{6+} is well worth the effort.

The attractive force for electrons is referred to as the electronegativity of an element. The greater the electronegativity, the stronger the affinity for electrons. Those elements on the right side of the Periodic Table are more electronegative than those on the left because their electron shells are more nearly complete. Those on the left side have relatively incomplete electron shells and are thus more inclined to donate them. Also, elements lower in the same column of the Periodic Table are larger and, therefore, less electronegative. This is because of the reduced force of attraction between the nucleus (positive charge) and the electrons of the outer energy level. This is due to the increased distance between the nucleus and the outer electron energy level, and because of the shielding effect of core electrons. (See Chapter 5, Section 5.3 regarding electronegativity.)

Have students determine the oxidation states of lots of atoms in various ionic states and compounds. Again, show them many different chemicals (e. g. Na_2CrO_4, $Na_2Cr_2O_7$). First determine the oxidation state of the ion. Base this on the fact that Na is always $+1$, and there are two sodium ions in the compounds. Determine the oxidation states of the chromate and dichromate ions. After determining the oxidation state of these ions, and assigning -2 to each oxygen atom, the oxidation state of chromium in each ion can be determined. The same strategy can be used to determine the oxidation states of the central atom in all other polyatomic ions.

Desktop Investigation: Models of Chemical Compounds

Making models of chemical compounds gives students an excellent opportunity to observe the structure of a variety of ionic and molecular substances. Use this investigation to introduce the shapes of molecules, the arrangement of atoms in molecules, the arrangements of ions in crystals, and the differences between ionic and molecular compounds. Students are so accustomed to seeing representations of molecules in two dimensions that it is very important for them to have as many opportunities as possible to see three dimensional representations of these substances. Emphasize that atoms, ionic lattices, and molecules exist in three dimensions, not two. Supplement this Desktop Investigation with a display of other three-dimensional materials. Use styrofoam models if you have them.

This is a good opportunity to use computer displays of molecules.

7.3 Using Chemical Formulas

Once students learn about the combining of ions to form compounds and the oxidation state of atoms, they are ready to start working with the essential quantitative elements used in understanding these compounds. These essential quantitative elements are: *formula mass*, *molar mass*, and *percent composition*. Formula mass and molar mass are straightforward concepts if correct compound formulas are used. They are essential concepts in order to begin work with chemical reactions. Students who have become familiar with formulas should have little difficulty with this section. Some frustrations may arise as a result of using atomic masses. (These are found in the Appendix, the Periodic Table inside the back cover, and approximate masses are found in the table on the inside front cover.) Emphasis should be given to the use, and the importance, of significant figures. It should also be emphasized that small errors to the hundredth or thousandth of a gram are relatively insignificant. Some students may be greatly relieved to know this. This is expecially important when your students begin to work problems, particularly if they have some anxiety already.

Students generally do not have any difficulties, other than formula writing, with these percent composition problems. Once in a while they will make a mathematical error in calculations. A quick and easy way to check this, assuming they have started with the right formula, is to check the resulting percentages to see if they add up to 100%.

Experiment 10, *Water Crystallization and Formula of a Hydrate*, gives students the opportunity to determine the number of moles of water that are bound to one mole of anhydrous salt.

7.4 Determining Chemical Formulas

Students sometimes have difficulty with this analysis of composition data. You might not want to present this as fundamental material. Three-element compounds and molecular formulas that differ from empirical formulas might be considered enrichment. The common mistakes that students make in determining the formula from data are *(1)* to forget to convert percent to grams, and *(2)* to forget to convert grams to moles. Remind students to keep enough significant figures when determining number of moles. This would be an appropriate unit for supplementary computer tutorials.

References and Resources

Books and Periodicals

Fernelius. "Numbers in Chemical Names," *Journal of Chemical Education*, Vol. 59, No. 11, 1982, p. 965.

Kolb, Doris. "Chemical Formulas, Part I: Development," *Journal of Chemical Education*, Vol. 55, No. 1, 1978, p. 44.

Kolb, Doris. "The Chemical Formula, Part II: Determination," *Journal of Chemical Education*, Vol. 55, No. 2, 1978, p. 109.

Meyers. "Electronegativity, Bond Energy, and Reactivity," *Journal of Chemical Education*, Vol. 56, No. 11, 1979, p. 711.

Pearson, Robert and George Arnold. "Manganese Color Reactions," *Journal of Chemical Education*, Vol. 65, 1988, p. 451.

Shakhashiri, Bassam Z. "Reactions of Fe and S," (p. 55) and "Reaction of Al and Br," (p. 68), and "Reaction of Na and Cl," (p. 61), *Chemical Demonstrations*, Vol. 1, Madison: University of Wisconsin Press, 1963. *The Mole Concept*, VHS cassette, TV Ontario.

Computer Software

Fundamental Skills for General Chemistry 3 (ELEMENT, IONS, NOMEN, BALEQ), Apple II, TRS-80, Programs for Learning.

Howbert, J. Jeffrey. *Molecular Animator*, Apple II, IBM, COMPress.

Inorganic Nomenclature II, Apple II, IBM, Intellectual Software.

Michel-Evleth, William H., *Language of Chemistry; The Personal Chemical Nomenclature Learning System*, Apple II, COMPress.

Project Seraphim, Disk #303, NSF Science and Engineering Education, Dept. of Chemistry, Eastern Michigan University, Ypsilanti, MI 48197.

Smith, S.G., Ruth Chabay, and Elizabeth Kean. *Percent Composition*, Apple II, IBM, COMPress

Topic 3: Formulas, Moles, Stoichiometry, Apple II, Knowledge Factory.

Weyh, John A., Joseph R. Crook, and Les N. Hauge, *Mole Concept*, Version 2.0, IBM, COMPress.

Answers and Solutions

Reviewing Concepts

1. The indicated formulas and systematic names are:
   a. $NaHCO_3$, sodium hydrogen carbonate
   b. $Mg(OH)_2$, magnesium hydroxide
   c. $MgSO_4 \cdot 7H_2O$, magnesium sulfate hepta hydrate
   d. $CaCO_3$, calcium carbonate
   e. NaOH, sodium hydroxide
   f. CH_3OH, methanol
2. A chemical formula can represent one molecule, one mole, or one molar mass of the indicated material.
3. a. Monatomic ions are ions formed from a single atom.
   b. Examples include Na^+, Mg^{2+}, and N^{3-}.
4. The ion most often formed in each case is:
   a. K^+ b. Ca^{2+} c. S^{2-}
   d. Cl^- e. Ba^{2+} f. Br^-
5. The indicated configurations are:
   a. Na = $[Ne]3s^1$, Na^+ = [Ne]
   b. Cl = $[Ne]3s^23p^5$, Cl^- = [Ar]
   c. Ca = $[Ar]4s^2$, Ca^{2+} = [Ar]
   d. O = $[He]2s^22p^4$, O^{2-} = [Ne]
6. The indicated formulas are:
   a. Na^+ b. Al^{3+} c. Cl^-
   d. N^{3-} e. Fe^{2+} f. Fe^{3+}
7. The indicated names are:
   a. potassium ion   d. chloride ion
   b. magnesium ion   e. oxide ion
   c. aluminum ion   f. calcium ion
8. The indicated formulas are:
   a. NaI   b. CaS   c. $ZnCl_2$ d. BaF_2
   e. Li_2O f. CrF_2 g. NiO   h. Fe_2O_3
9. The indicated compound names are:
   a. potassium chloride   d. sodium fluoride
   b. calcium bromide   e. silver iodide
   c. magnesium iodide   f. lithium oxide
10. The less electronegative element is given first in the names and formulas of binary molecular compounds.
11. The indicated names are:
    a. carbon dioxide
    b. carbon tetrachloride
    c. phosphorus pentachloride
    d. carbon disulfide
    e. sulfur dioxide
    f. phosphorus tribromide
    g. dinitrogen tetroxide
    h. carbon tetraiodide
    i. diarsenic pentoxide
12. The indicated formulas are:
    a. CBr_4 b. SiO_2 c. CO
    d. SO_3 e. P_2O_5 f. As_2S_3
13. Binary acids, such as HCl and HF, are solutions in water of binary molecular compounds containing hydrogen and, usually, one of the halogens. Oxyacids are acids, such as H_2SO_4 and HNO_3, containing hydrogen, oxygen, and a third element.
14. a. A salt is an ionic compound composed of a cation and the anion from an acid.
    b. Two common salts are KCl and $MgSO_4$.
15. The indicated acids are:
    a. hydrofluoric acid   d. sulfuric acid
    b. hydrobromic acid   e. phosphoric acid
    c. nitric acid
16. The indicated names are:
    a. iron(II)   b. iron(III)   c. lead(II)
    d. lead(IV)   e. tin(II)   f. tin(IV)
17. a. carbon(IV) bromide   d. sulfur(VI) oxide
    b. silicon(IV) oxide   e. phosphorus(V) oxide
    c. carbon(II) oxide   f. arsenic(III) sulfide
18. The indicated formulas are:
    a. PI_3 b. SCl_2 c. CS_2
19. a. Oxidation numbers are assigned numbers that show the general distribution of electrons among bonded atoms.
    b. Oxidation numbers are useful in naming compounds, in writing formulas, and in studying certain types of chemical reactions.
20. a. The formula mass of any compound or polyatomic ion is the sum of the atomic masses of all atoms represented in the formula.
    b. Formula mass is expressed in atomic mass units (u).
21. The molar mass of a compound is the mass, in grams, numerically equal to the formula mass of the compound.
22. Finding the simplest formulas from percent-composition requires a knowledge of composition by mass, composition in moles, and the smallest whole-number ratio of atoms.
23. The subscripts in the molecular formula are whole-number multiples of those in the simplest formula.

Problems

1. The indicated formulas and charges are:
   a. NH_4^+ b. $C_2H_3O_2^-$ c. OH^- d. CO_3^{2-}
   e. SO_4^{2-} f. PO_4^{3-} g. Cu^{2+} h. Sn^{2+}
   i. Fe^{3+} j. Cu^+ k. Hg_2^{2+} l. Hg^{2+}

2. The indicated names are:
   a. ammonium ion
   b. chlorate ion
   c. hydroxide ion
   d. sulfate ion
   e. nitrate ion
   f. carbonate ion
   g. phosphate ion
   h. acetate ion
   i. hydrogen carbonate ion
   j. chromate ion

3. The indicated formulas are:
   a. NaF   b. CaO   c. K_2S d. $MgCl_2$
   e. $AlBr_3$ f. Li_3N g. FeO

4. The indicated names are:
   a. sodium chloride
   b. magnesium oxide
   c. potassium fluoride
   d. calcium sulfide
   e. lithium bromide
   f. potassium hydroxide
   g. silver nitrate
   h. iron(II) hydroxide
   i. tin(II) nitrate
   j. cobalt(II) nitrate
   k. iron(III) phosphate
   l. mercury(I) sulfate
   m. mercury(II) phosphate

5. The assigned oxidation numbers are:
   a. $+1, -1$ f. $+1, -1$
   b. $+4, -2$ g. $+1, -1$
   c. $+3, -1$ h. $+5, -2$
   d. $+4, -2$ i. $+5, -2$
   e. $+2, -1$ j. $+1, +5, -2$

6. The assigned oxidation numbers are:
   a. $+5, -2$ e. $+5, -2$
   b. $-3, +1$ f. $+7, -2$
   c. $+7, -2$ g. $+2, -2$
   d. $+4, -2$ h. $+6, -2$

7. a. $6 \text{ C atoms} \times \dfrac{12.011 \, u}{\text{C atom}} = 72.0666 \, u$

   $12 \text{ H atoms} \times \dfrac{1.007\,94 \, u}{\text{H atom}} = 12.0953 \, u$

   $6 \text{ O atoms} \times \dfrac{15.9994 \, u}{\text{O atom}} = 95.9964 \, u$

   Formula mass of $C_6H_{12}O_6 = 180.1583 \, u$

   b. $1 \text{ Ca atom} \times \dfrac{40.08 \, u}{\text{Ca atom}} = 40.08 \, u$

   $4 \text{ C atoms} \times \dfrac{12.0111 \, u}{\text{C atom}} = 48.0444 \, u$

   $6 \text{ H atoms} \times \dfrac{1.007\,94 \, u}{\text{H atom}} = 6.047\,64 \, u$

   $4 \text{ O atoms} \times \dfrac{15.9994 \, u}{\text{O atom}} = 63.9976 \, u$

   Formula mass of $Ca(C_2H_3O_2)_2 = 158.17 \, u$

   c. $1 \text{ N atom} \times \dfrac{14.0067 \, u}{\text{N atom}} = 14.0067 \, u$

   $4 \text{ H atoms} \times \dfrac{1.007\,94 \, u}{\text{H atom}} = 4.031\,76 \, u$

   Formula mass of $NH_4^+ = 18.0385 \, u$

   d. $1 \text{ Cl atom} \times \dfrac{35.453 \, u}{\text{Cl atom}} = 35.453 \, u$

   $3 \text{ O atoms} \times \dfrac{15.9994 \, u}{\text{O atom}} = 47.9982 \, u$

   Formula mass of $ClO_3^- = 83.451 \, u$

8. The indicated numbers are:
   a. 1 mol K^+ and 1 mol NO_3^-; 1 mol K, 1 mol N, and 3 mol O
   b. 2 mol Na^+ and 1 mol SO_4^{2-}; 2 mol Na, 1 mol S, and 4 mol O
   c. 1 mol Ca^{2+} and 2 mol OH^-; 1 mol Ca, 2 mol O, and 2 mol H
   d. 2 mol NH_4^+ and 1 mol SO_3^{2-}; 2 mol N, 8 mol H, 1 mol S, and 3 mol O
   e. 3 mol Ca^{2+} and 2 mol PO_4^{3-}; 3 mol Ca, 2 mol P, and 8 mol O
   f. 2 mol Al^{3+} and 3 mol CrO_4^{2-}; 2 mol Al, 3 mol Cr, and 12 mol O.

9. a. $1 \text{ mol K} \times \dfrac{39.0983 \text{ g K}}{\text{mol K}} = 39.0983 \text{ g K}$

   $1 \text{ mol N} \times \dfrac{14.0067 \text{ g N}}{\text{mol N}} = 14.0067 \text{ g N}$

   $3 \text{ mol O} \times \dfrac{15.9994 \text{ g O}}{\text{mol O}} = 47.9982 \text{ g O}$

   Molar mass of $KNO_3 = \boxed{101.1032 \text{ g}}$

   b. $2 \text{ mol Na} \times \dfrac{22.989\,77 \text{ g Na}}{\text{mol Na}} = 45.979\,54 \text{ g Na}$

   $1 \text{ mol S} \times \dfrac{32.06 \text{ g S}}{\text{mol S}} = 32.06 \text{ g S}$

   $4 \text{ mol O} \times \dfrac{15.9994 \text{ g O}}{\text{mol O}} = 63.9976 \text{ g O}$

   Molar mass of $Na_2SO_4 = \boxed{142.04 \text{ g}}$

   c. $1 \text{ mol Ca} \times \dfrac{40.08 \text{ g Ca}}{\text{mol Ca}} = 40.08 \text{ g Ca}$

   $2 \text{ mol O} \times \dfrac{15.9994 \text{ g O}}{\text{mol O}} = 31.9988 \text{ g O}$

   $2 \text{ mol H} \times \dfrac{1.007\,94 \text{ g H}}{\text{mol H}} = 2.015\,88 \text{ g H}$

   Molar mass of $Ca(OH)_2 = \boxed{74.09 \text{ g}}$

   d. $2 \text{ mol N} \times \dfrac{14.0067 \text{ g N}}{\text{mol N}} = 28.0134 \text{ g N}$

   $8 \text{ mol H} \times \dfrac{1.007\,94 \text{ g H}}{\text{mol H}} = 8.063\,52 \text{ g H}$

   $1 \text{ mol S} \times \dfrac{32.06 \text{ g S}}{\text{mol S}} = 32.06 \text{ g S}$

   $3 \text{ mol O} \times \dfrac{15.9994 \text{ g O}}{\text{mol O}} = 47.9982 \text{ g O}$

   Molar mass of $(NH_4)_2SO_3 = \boxed{116.14 \text{ g}}$

   e. $3 \text{ mol Ca} \times \dfrac{40.08 \text{ g Ca}}{\text{mol Ca}} = 120.24 \text{ g Ca}$

$2 \text{ mol P} \times \dfrac{30.973\,76 \text{ g P}}{\text{mol P}} = 61.947\,52 \text{ g P}$

$8 \text{ mol O} \times \dfrac{15.9994 \text{ g O}}{\text{mol O}} = 127.9952 \text{ g O}$

Molar mass of $Ca_3(PO_4)_2$ = $\boxed{310.18 \text{ g}}$

f. $2 \text{ mol Al} \times \dfrac{26.981\,54 \text{ g Al}}{\text{mol Al}} = 53.963\,08 \text{ g Al}$

$3 \text{ mol Cr} \times \dfrac{51.996 \text{ g Cr}}{\text{mol Cr}} = 155.988 \text{ g Cr}$

$12 \text{ mol O} \times \dfrac{15.9994 \text{ g O}}{\text{mol O}} = 191.9928 \text{ g O}$

Molar mass of $Al_2(CrO_4)_3$ = $\boxed{401.944 \text{ g}}$

10. a. Molar mass of NaCl = 58.443 g

$1.000 \text{ mol NaCl} \times \dfrac{58.44 \text{ g NaCl}}{\text{mol NaCl}} = \boxed{58.44 \text{ g NaCl}}$

b. Molar mass of H_2O = 18.0153 g

$2.000 \text{ mol } H_2O \times \dfrac{18.02 \text{ g } H_2O}{\text{mol } H_2O} = \boxed{36.04 \text{ g } H_2O}$

c. Molar mass of $Ca(OH)_2$ = 74.09 g

$3.500 \text{ mol } Ca(OH)_2 \times \dfrac{74.09 \text{ g } Ca(OH)_2}{\text{mol } Ca(OH)_2}$
= $\boxed{259.3 \text{ g } Ca(OH)_2}$

d. Molar mass of $Ba(NO_3)_2$ = $\boxed{261.34 \text{ g}}$

$0.625 \text{ mol } Ba(NO_3)_2 \times \dfrac{261 \text{ g } Ba(NO_3)_2}{\text{mol } Ba(NO_3)_2}$
= $\boxed{163 \text{ g } Ba(NO_3)_2}$

11. a. Molar mass of H_2O = 18.0153 g

$4.50 \text{ g } H_2O \times \dfrac{\text{mol } H_2O}{18.0 \text{ g } H_2O} = \boxed{0.250 \text{ mol } H_2O}$

b. Molar mass of $Ba(OH)_2$ = 171.34 g

$471.6 \text{ g } Ba(OH)_2 \times \dfrac{\text{mol } Ba(OH)_2}{171.3 \text{ g } Ba(OH)_2}$
= $\boxed{2.753 \text{ mol } Ba(OH)_2}$

c. Molar mass of $Fe_3(PO_4)_2$ = 357.48 g

$129.68 \text{ g } Fe_3(PO_4)_2 \times \dfrac{\text{mol } Fe_3(PO_4)_2}{357.48 \text{ g } Fe_3(PO_4)_2}$
= $\boxed{0.362\,76 \text{ mol } Fe_3(PO_4)_2}$

12. a. Molar mass of NaCl = 58.443 g

$\% \text{ Na} = \dfrac{22.989\,77 \text{ g Na}}{58.443 \text{ g NaCl}} \times 100\% = \boxed{39.337\% \text{ Na}}$

$\% \text{ Cl} = \dfrac{35.453 \text{ g Cl}}{58.443 \text{ g NaCl}} \times 100\% = \boxed{60.663\% \text{ Cl}}$

b. Molar mass of $AgNO_3$ = 169.873 g

$\% \text{ Ag} = \dfrac{107.868 \text{ g Ag}}{169.873 \text{ g } AgNO_3} \times 100\%$
= $\boxed{63.4992\% \text{ Ag}}$

$\% \text{ N} = \dfrac{14.0067 \text{ g N}}{169.873 \text{ g } AgNO_3} \times 100\% = \boxed{8.245\,40\% \text{ N}}$

$\% \text{ O} = \dfrac{47.9982 \text{ g O}}{169.873 \text{ g } AgNO_3} \times 100\% = \boxed{28.2553\% \text{ O}}$

c. Molar mass of $Mg(OH)_2$ = 58.320 g

$\% \text{ Mg} = \dfrac{24.305 \text{ g Mg}}{58.320 \text{ g } Mg(OH)_2} \times 100\%$
= $\boxed{41.675\% \text{ Mg}}$

$\% \text{ O} = \dfrac{31.9988 \text{ g O}}{58.320 \text{ g } Mg(OH)_2} \times 100\% = \boxed{54.868\% \text{ O}}$

$\% \text{ H} = \dfrac{2.015\,88 \text{ g H}}{58.320 \text{ g } Mg(OH)_2} \times 100\% = \boxed{3.4566\% \text{ H}}$

13. Molar mass of $CuSO_4 \cdot 5H_2O$ = 249.68 g

Mass of $H_2O = \dfrac{5 \text{ mol } H_2O}{\text{mol } CuSO_4 \cdot 5H_2O} \times \dfrac{18.0153 \text{ g } H_2O}{\text{mol } H_2O}$

= 90.0765 g H_2O/mol $CuSO_4 \cdot 5H_2O$

$\% H_2O = \dfrac{90.0765 \text{ g } H_2O}{249.68 \text{ g } CuSO_4 \cdot 5H_2O} \times 100\%$

= 36.077% H_2O

14. a. $100.0 \text{ g NaCl} \times \dfrac{39.337 \text{ g Na}}{100 \text{ g NaCl}} = \boxed{39.34 \text{ g Na}}$

b. $140.0 \text{ g } Mg(OH)_2 \times \dfrac{54.868 \text{ g O}}{100 \text{ g } Mg(OH)_2} = \boxed{76.82 \text{ g O}}$

15. $63.50 \text{ g Ag} \times \dfrac{\text{mol Ag}}{107.9 \text{ g Ag}} = 0.5885 \text{ mol Ag}$

$8.25 \text{ g N} \times \dfrac{\text{mol N}}{14.0 \text{ g N}} = 0.589 \text{ mol N}$

$28.26 \text{ g O} \times \dfrac{\text{mol O}}{16.00 \text{ g O}} = 1.766 \text{ mol O}$

$\dfrac{0.5885 \text{ mol Ag}}{0.5885} : \dfrac{0.589 \text{ mol N}}{0.5885} : \dfrac{1.766 \text{ mol O}}{0.5885}$

1.000 mol Ag : 1.00 mol N : 3.001 mol O
The simplest formula is $AgNO_3$.

16. $52.20 \text{ g C} \times \dfrac{\text{mol C}}{12.01 \text{ g C}} = 4.346 \text{ mol C}$

$13.00 \text{ g H} \times \dfrac{\text{mol H}}{1.008 \text{ g H}} = 12.90 \text{ mol H}$

$34.80 \text{ g O} \times \dfrac{\text{mol O}}{16.00 \text{ g O}} = 2.175 \text{ mol O}$

$\dfrac{4.346 \text{ mol C}}{2.175} : \dfrac{12.90 \text{ mol H}}{2.175} : \dfrac{2.175 \text{ mol O}}{2.175}$

= 1.998 mol C : 5.931 mol H : 1.000 mol O
The simplest formula is C_2H_6O.

17. $0.365 \text{ g Na} \times \dfrac{\text{mol Na}}{23.0 \text{ g Na}} = 0.0159 \text{ mol Na}$

$0.221 \text{ g N} \times \dfrac{\text{mol N}}{14.0 \text{ g N}} = 0.0158 \text{ mol N}$

$0.752 \text{ g O} \times \dfrac{\text{mol O}}{16.0 \text{ g O}} = 0.0470 \text{ mol O}$

$$\frac{0.0159 \text{ mol Na}}{0.0158} : \frac{0.0158 \text{ mol N}}{0.0158} : \frac{0.0470 \text{ mol O}}{0.0158}$$
= 1.01 mol Na : 1.00 mol N : 2.97 mol O
The simplest formula is $NaNO_3$.

$$0.29 \text{ g O} \times \frac{1 \text{ mol O atoms}}{16.00 \text{ g O}} = 0.018 \text{ mol O atoms}$$

$$\frac{0.018 \text{ mol Ni atoms}}{0.018 \text{ mol O atoms}} = \frac{1 \text{ Ni}}{1 \text{ O}}$$

Empirical formula for the nickel oxide compound is NiO.

Critical Thinking

1. Because the oxide ion has a charge of −2, the sulfur has a charge of +6. The formula would be SO_3, also called sulfur trioxide. In the molecule there is one sulfur atom and 3 oxygen atoms. The oxidation state for sulfur is +6 and for oxygen, a −2.
2. Mass of Nickel = 1.05 g
   Mass of nickel oxide compound = 1.34 g
   Mass of oxygen = .29 g

   $$1.05 \text{ g Ni} \times \frac{1 \text{ mol Ni atoms}}{58.69 \text{ g Ni}} = 0.018 \text{ mol Ni atoms}$$

Cross Connections

Answers will vary:

1.

Chemical Nomenclature	Linnaeus System
Name is associated with elements that make up the compound	Names are based on physical and chemical characteristics of the organism
Name consists of two words	Names consists of two words, genus and species
Names are not capitalized	Genus name is capitalized, but not the species
Roman numerals are used sometimes	Numerals are never used

Videodisc Selections
from *The Periodic Table Videodisc* (JCE Software)

step forward

step reverse

pg 205 Formation of aluminum oxide

8267
aluminum + oxygen (m)

CHAPTER 7

pg 202 Formation of water

115
hydrogen + oxygen (m)

Teacher's Notes

Chapter 7
Chemical Formulas and Chemical Compounds

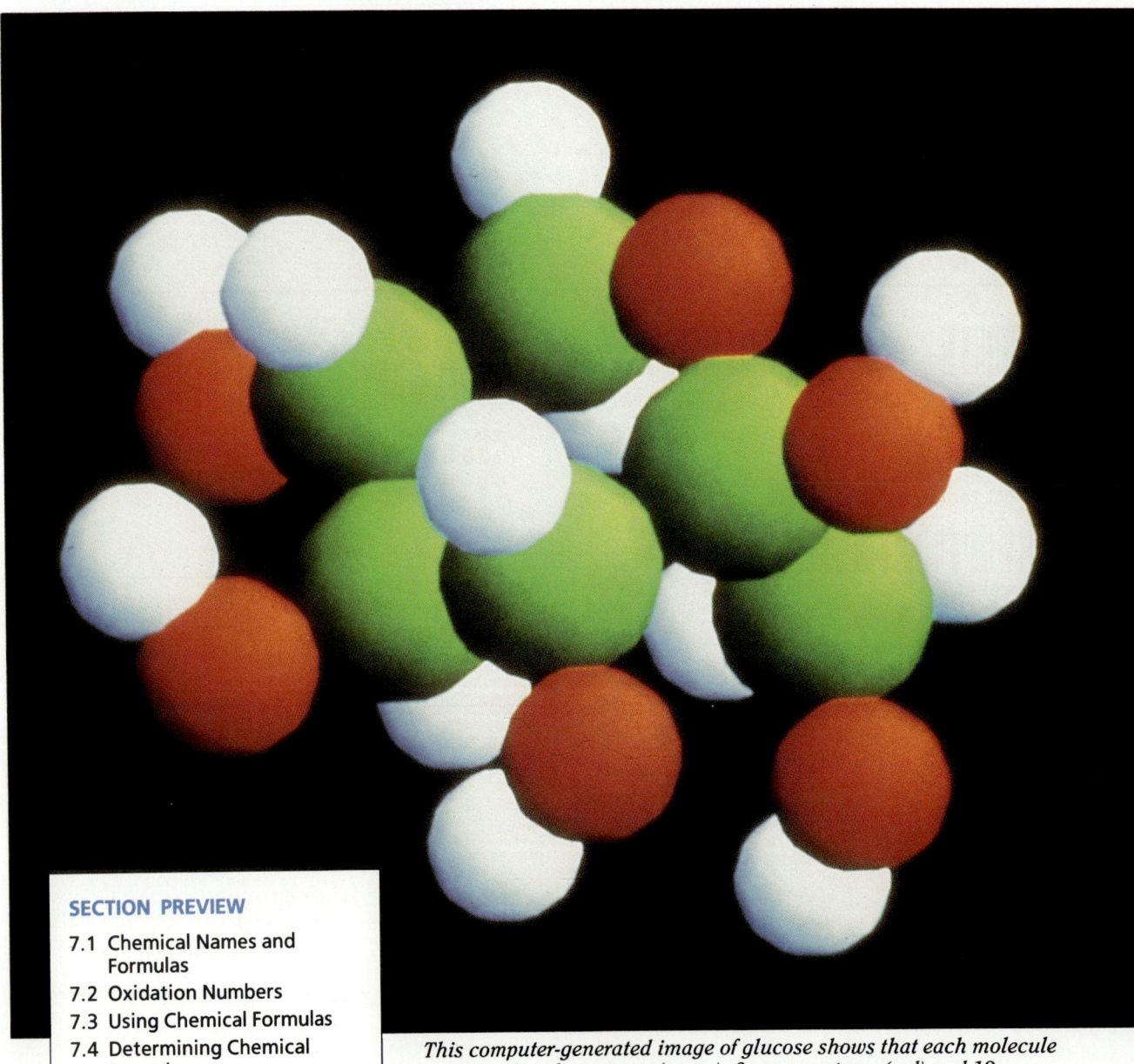

SECTION PREVIEW

7.1 Chemical Names and Formulas
7.2 Oxidation Numbers
7.3 Using Chemical Formulas
7.4 Determining Chemical Formulas

This computer-generated image of glucose shows that each molecule contains 6 carbon atoms (green), 6 oxygen atoms (red) and 12 hydrogen atoms (white).

7.1 Chemical Names and Formulas

With all of the compounds of all of the elements to be identified, systematic methods for writing formulas and naming compounds are necessary. In this section you will be introduced to the rules that apply to simple chemical compounds. Table 7-1 lists the common and systematic names for a number of well-known substances. Certain common names such as "milk of magnesia" or "lime" remain in everyday use. As you can see, common names usually give no information about chemical composition.

Significance of a Chemical Formula

Figure 7-1 uses the formulas of octane, a hydrocarbon that is a component of gasoline, and aluminum sulfate, a compound used in dyeing fabrics, to illustrate chemical formulas. Like all hydrocarbons, octane is a molecular compound. One octane molecule is composed of 8 carbon atoms and 18 hydrogen atoms.

Aluminum sulfate contains the sulfate ion and is clearly an ionic compound. Recall that ionic compounds consist of huge numbers of positive and negative ions held together by mutual attraction. The formula therefore represents one *formula unit*—not just a molecule. One formula unit of aluminum sulfate consists of 2 aluminum ions and 3 sulfate ions. Note in the figure how the parentheses are used. They surround the polyatomic ion as a unit. The subscript 3 applies to the entire unit within the parentheses. One formula unit of aluminum sulfate is therefore composed of 2 aluminum atoms, 3 sulfur atoms, and 12 oxygen atoms.

A correctly written chemical formula must represent the known facts about the analytically determined composition of a compound. Care must be

SECTION OBJECTIVES

1. Explain the significance of a chemical formula.
2. Determine the formula of an ionic compound between any two given ions.
3. Explain the two systems for distinguishing different ionic compounds of the same two elements.
4. Name an ionic compound, given its formula.
5. Using prefixes, name a binary molecular compound from its formula.
6. Write the formula of a binary molecular compound, given its name.
7. List the names and formulas of the common laboratory acids.

C_8H_{16}, *octene,* is a compound with properties different from octane.

TABLE 7-1 SOME FAMILIAR SUBSTANCES WITH THEIR SYSTEMATIC AND COMMON NAMES

Formula	Systematic Name	Common Name
Hg	mercury	quicksilver
S	sulfur	brimstone
$NaHCO_3$	sodium hydrogen carbonate	baking soda
CaO	calcium oxide	lime
$CaCO_3$	calcium carbonate	limestone, calcite, marble
$MgSO_4 \cdot 7H_2O$	magnesium sulfate hepta hydrate	epsom salts
N_2O	dinitrogen oxide	laughing gas
NaOH	sodium hydroxide	lye, caustic soda
HCl	hydrochloric acid	muriatic acid
$Mg(OH)_2$	magnesium hydroxide	milk of magnesia
NaCl	sodium chloride	table salt
$CaSO_4 \cdot \frac{1}{2}H_2O$	calcium sulfate hemihydrate	plaster of paris
CO_2	carbon dioxide	dry ice (solid CO_2)
CH_3OH	methanol	wood alcohol
CH_3CH_2OH	ethanol	grain alcohol

Chemical Formulas and Chemical Compounds

Figure 7-1 Interpretation of chemical formulas

taken that subscripts are correct. The two formulas below

$$H_2O \qquad H_2O_2$$
$$\text{water} \qquad \text{hydrogen peroxide}$$

represent different compounds with very different characteristic properties, as illustrated in Figure 7-2.

Chemical formulas are often used to represent amounts of substances. The formula H_2O can represent one molecule of water. As you learned in chapter 3, the mole is equal to 6.02×10^{23}. The formula H_2O can also represent one mole of water molecules. Using the definition of the mole, the mass in grams of one mole of water molecules or one mole of any compound can be calculated. The formula H_2O is also sometimes used to represent one molar mass of water molecules.

$$H_2O$$
one water *molecule*
one *mole* of water molecules
one *molar mass* of water molecules

Section 7.3 is devoted to the quantitative meaning of chemical formulas.

Figure 7-2 The same elements can make different compounds with different chemical properties. Water, (H_2O), *(left)*, is unreactive with green felt cloth, but hydrogen peroxide (H_2O_2) bleaches the green dye in the felt.

Monatomic Ions

Many main group elements form ions with noble-gas configurations by either gaining or losing electrons. The Periodic Table is therefore a guide to the ions formed by these elements. Group 1 metals lose one electron to give 1+ ions, such as Na^+. Group 2 metals lose two electrons to give 2+ ions such as Mg^{2+}. Of the Group 13 elements, aluminum is the only one commonly found as an ion. Aluminum forms the 3+ ion, Al^{3+}. *Ions formed from a single atom are known as* **monatomic ions.**

With Group 14, the loss of enough electrons to give a noble-gas configuration becomes difficult. The Group 14 metals tin and lead, therefore, form 2+ ions in many compounds. These 2+ ions result from loss of two of the four valence electrons. Tin and lead also both form some compounds in which all four valence electrons are involved in polar-covalent bonding.

The nonmetals of Groups 15, 16, and 17 form ions by gaining electrons rather than losing electrons. In ionic compounds, nitrogen forms the 3− anion, N^{3-}. The three added electrons plus the five outermost electrons in nitrogen atoms give a completed outermost octet. Similarly, oxygen and sulfur, from Group 16, form 2− ions, and the halogens, from Group 17, form 1− ions. These common ions of main-group elements are shown in Periodic Table form in Table 7-2. Elements from the *d*-block form ions with 2+, 3+ (or, in a few cases, 1+) charges.

The symbols and names of the common monatomic ions are organized according to their charges in Table 7-3. Positive ions are named by the element name followed by "ion" as in

K^+ potassium ion Mg^{2+} magnesium ion

Remind students of the following: Metallic ions contain more protons than electrons and are thus positively charged ions (cations):

Na^+	11 protons	10 electrons
Mg^{2+}	12 protons	10 electrons
Al^{3+}	13 protons	10 electrons

Remind students of the following: The negative ions formed by nonmetals contain more electrons than protons and are thus negatively charged ions (anions):

N^{3-}	7 protons	10 electrons
O^{2-}	8 protons	10 electrons
F^-	9 protons	10 electrons

TABLE 7-2 SOME COMMON IONS OF MAIN GROUP ELEMENTS*

Group 1	Group 2	Group 13	Group 14	Group 15	Group 16	Group 17	Group 18
Li^+				N^{3-}	O^{2-}	F^-	
Na^+	Mg^{2+}	Al^{3+}			S^{2-}	Cl^-	
K^+	Ca^{2+}					Br^-	
Rb^+	Sr^{2+}		Sn^{2+}			I^-	
Cs^+	Ba^{2+}		Pb^{2+}				

*All of the ions shown except Sn^{2+} and Pb^{2+} have noble-gas configurations.

STUDY HINT
The formula of an ionic compound shows the simplest whole-number ratio of ions.

Chemical Formulas and Chemical Compounds

TABLE 7-3 SOME COMMON MONATOMIC IONS

1+ Charge		2+ Charge		3+ Charge	
cesium ion	Cs^+	barium ion	Ba^{2+}	aluminum ion	Al^{3+}
copper(I) ion	Cu^+	cadmium ion	Cd^{2+}	chromium(III) ion	Cr^{3+}
lithium ion	Li^+	calcium ion	Ca^{2+}	iron(III) ion	Fe^{3+}
potassium ion	K^+	chromium(II) ion	Cr^{2+}		
rubidium ion	Rb^+	cobalt(II) ion	Co^{2+}		
silver ion	Ag^+	copper(II) ion	Cu^{2+}		
sodium ion	Na^+	iron(II) ion	Fe^{2+}		
		lead(II) ion	Pb^{2+}		
		magnesium ion	Mg^{2+}		
		manganese(II) ion	Mn^{2+}		
		*mercury(I) ion	Hg_2^{2+}		
		mercury(II) ion	Hg^{2+}		
		nickel(II) ion	Ni^{2+}		
		tin(II) ion	Sn^{2+}		
		zinc ion	Zn^{2+}		

1− Charge		2− Charge		3− Charge	
bromide ion	Br^-	oxide ion	O^{2-}	nitride ion	N^{3-}
chloride ion	Cl^-	sulfide ion	S^{2-}		
fluoride ion	F^-				
iodide ion	I^-				

*The mercury(I) ion exists as two Hg^+ ions joined together by a covalent bond and is written as Hg_2^{2+}.

Negative ions are named by dropping the ending of the element name and adding "ide" to it, as in

F fluor*ine* F^- fluor*ide* ion
N nitr*ogen* N^{3-} nitr*ide* ion

Binary Ionic Compounds

In ionic compounds, ions must be combined so that the total positive and negative charges are equal. The formulas for many ionic compounds can be written once the identities of the combined ions are known.

Compounds composed of two different elements are known as **binary compounds.** What is the formula of the binary compound formed by the ions of magnesium and bromine? Magnesium is in Group 2 and forms the Mg^{2+} ion. Bromine is in Group 17 and forms the Br^- ion. Two bromide ions are required to balance the 2+ charge of the magnesium ion. The formula must therefore show one magnesium ion and two bromide ions. It is customary to write the symbol for the positive ion first.

Ions combined: Mg^{2+}, Br^-, Br^- *Chemical formula:* $MgBr_2$

The subscript 1 is understood when no subscript is present and is not written. The subscript 2 shows that two Br^- ions are present in each formula unit. The formula shows that the simplest whole-number ratio of the ions in this compound is 1:2. Note that charges of ions are not usually included in chemical formulas of compounds.

As an aid to writing formulas of ionic compounds, the positive and negative charges can be "crossed over" to give subscripts. For example, the formula for the compound formed by the aluminum ion, Al^{3+}, and the oxide ion, O^{2-} is found as follows:

Step 1: Write the symbols for the ions side by side, with the positive ion (or most metallic element) first.

$$Al^{3+} \quad O^{2-}$$

Step 2: Cross over the charge values to give subscripts.
Use the value of each charge as the subscript for the other ion.

$$Al_2^{3+} \quad O_3^{2-}$$

Step 3: Check the subscripts and write the formula. Multiplying the charge by the subscript shows that the charge on two Al^{3+} ions ($2 \times 3+ = 6+$) equals the charge on three O^{2-} ions ($3 \times 2- = 6-$). The correct formula is therefore

$$Al_2O_3$$

Ionic compounds are named by combining the names of the positive and negative ions, as illustrated in Figure 7-3. The name of the positive ion is given first. Al_2O_3 is therefore named aluminum oxide.

The names and charges of the ions listed in Table 7-3 should be memorized as an aid to writing and interpreting chemical formulas.

No element commonly forms negative ions of two different charges.

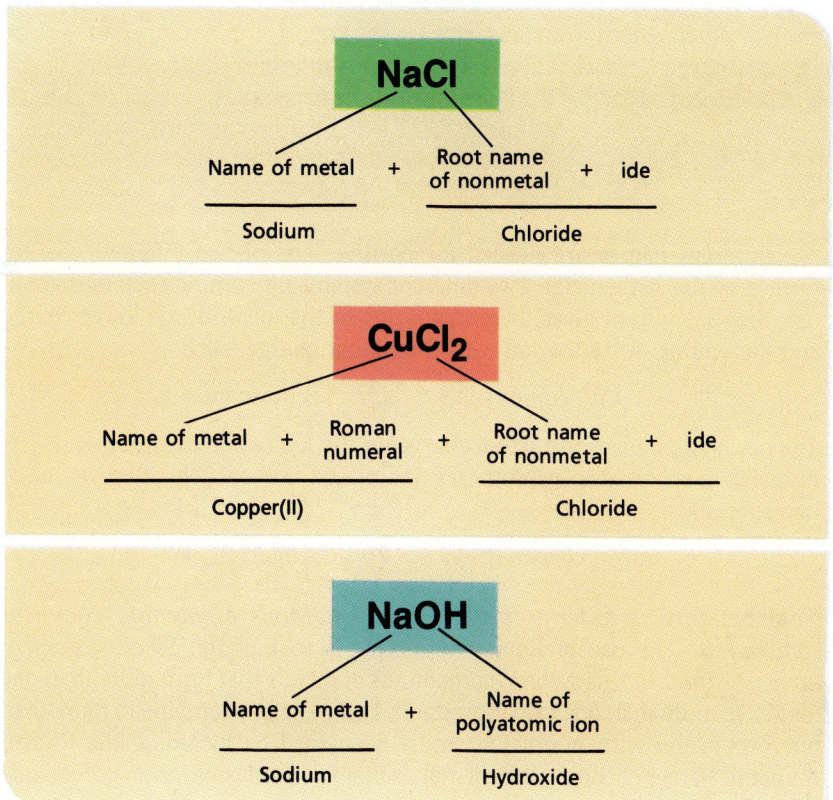

Figure 7-3 Naming ionic compounds

Sample Problem 7.1

Write the formulas for the binary ionic compounds formed between (a) zinc and iodine and (b) zinc and sulfur.

Solution

Step 1. *Write the symbols for the ions side by side, with the positive ion first.*
(a) Zn^{2+} I^- (b) Zn^{2+} S^{2-}

Step 2. *Cross over the charge values to give subscripts.*
(a) Zn_1^{2+} I_2^- (b) Zn_2^{2+} S_2^{2-}

Step 3: *Check the subscripts and write the formula.*
(a) The subscripts are correct because they give equal total charges of $1 \times 2+ = 2+$ and $2 \times 1- = 2-$. The subscript 1 is not written, so the formula is ZnI_2. Combining the cation and anion names gives this compound the name *zinc iodide*.
(b) The subscripts are correct because they give equal total charges of $2 \times 2+ = 4+$ and $2 \times 2- = 4-$. Remember that ionic compounds exist as formula units represented by the smallest whole-number ratio of ions in a compound. Therefore the correct formula is ZnS.

Practice Problems

1. Write formulas for the binary ionic compounds formed between each of the following:
   (a) potassium and iodine *(Ans.)* KI (b) magnesium and chlorine *(Ans.)* $MgCl_2$ (c) sodium and sulfur *(Ans.)* Na_2S (d) calcium and oxygen *(Ans.)* CaO (e) aluminum and sulfur *(Ans.)* Al_2S_3.
2. Name the binary ionic compounds indicated by each of the following chemical formulas:
   (a) BaO *(Ans.)* barium oxide (b) AgCl *(Ans.)* silver chloride (c) KBr *(Ans.)* potassium bromide (d) ZnO *(Ans.)* zinc oxide

Different names are needed for positive ions of two different charges formed by the same metal. Two different systems for naming such ions are in use. The old system gives the ending *-ous* to the ion with the lower charge and the ending *-ic* to the ion with the higher charge, such as

Cu^+ cup*rous* ion Cu^{2+} cup*ric* ion

The new system gives the actual charge on the ion as a Roman numeral. The numeral is enclosed in parentheses and placed immediately after the metal name:

Cu^+ copper(I) ion Cu^{2+} copper(II) ion

STUDY HINT
To use the Stock system for ionic compounds, you must know which metals form more than one cation.

Nomenclature is a term that refers to methods of naming chemical compounds. The use of Roman numerals is part of the *Stock system of nomenclature*. In this system, compounds of metals that form more than one ion (have more than one oxidation state), and therefore combine in more than one proportion with nonmetals, must be named by including the Roman numeral. Names of compounds of metals that form only one ion *do not* include a Roman numeral.

Figure 7-4 Compounds of different ions of the same metal may look different. Compare lead(II) oxide *(left)*, lead(I) oxide *(center)* and lead(III) oxide.

Sample Problem 7.2

Write the formula and give the name in the Stock system for the compound formed by the ions Cr^{3+} and F^-.

Solution

Step 1. Write the symbols for the ions side by side, with the positive ion first.
$$Cr^{3+} \quad F^-$$

Step 2. Cross over the charge values to give subscripts.
$$Cr_1^{3+} \quad F_3^-$$

Step 3. Check the subscripts and write the formula.
The total positive charge is $1 \times 3+ = 3+$. The total negative charge is $3 \times 1- = 3-$. The formula is therefore
$$CrF_3$$

To name this compound correctly, it is necessary to know that chromium forms more than one ion. Since that is the case, this compound is named *chromium(III) fluoride* according to the Stock system.

Practice Problems

1. Write the formula and give the name in the Stock system for the compounds formed between each of the following: *(a)* Cu^{2+} and I^- *(Ans.)* CuI_2, copper(II) iodide *(b)* Fe^{2+} and O^{2-} *(Ans.)* FeO, iron(II) oxide *(c)* Pb^{2+} and Cl^- *(Ans.)* $PbCl_2$, lead(II) chloride *(d)* Hg^{2+} and S^{2-} *(Ans.)* HgS, mercury(II) sulfide *(e)* Sn^{2+} and F^- *(Ans.)* SnF_2, tin(II) fluoride.
2. Name the following compounds: *(a)* CuO *(Ans.)* copper(II) oxide *(b)* CoF_3 *(Ans.)* cobalt(III) fluoride *(c)* SnI_4 *(Ans.)* tin(IV) iodide *(d)* FeS *(Ans.)* iron(II) sulfide *(e)* Hg_2Cl_2 *(Ans.)* mercury(I) chloride.

TABLE 7-4 SOME POLYATOMIC IONS

1+ Charge		1− Charge	
Ammonium ion	NH_4^+	acetate ion	$C_2H_3O_2^-$
		bromate ion	BrO_3^-
		chlorate ion	ClO_3^-
		chlorite ion	ClO_2^-
		cyanide ion	CN^-
		hydrogen carbonate ion (bicarbonate ion)	HCO_3^-
		hydrogen sulfate ion	HSO_4^-
		hydroxide ion	OH^-
		hypoiodite ion	IO^-
		iodate ion	IO_3^-
		nitrate ion	NO_3^-
		nitrite ion	NO_2^-
		perchlorate ion	ClO_4^-
		permanganate ion	MnO_4^-

2− Charge		3− Charge	
carbonate ion	CO_3^{2-}	phosphate ion	PO_4^{3-}
chromate ion	CrO_4^{2-}	phosphite ion	PO_3^{3-}
dichromate ion	$Cr_2O_7^{2-}$		
oxalate ion	$C_2O_4^{2-}$		
peroxide ion	O_2^{2-}		
sulfate ion	SO_4^{2-}		
sulfite ion	SO_3^{2-}		
thiosulfate ion	$S_2O_3^{2-}$		

Compounds Containing Polyatomic Ions

Some polyatomic ions are listed in Table 7-4. All but the ammonium ion are negatively charged. Note that most of these are **oxyanions**—*polyatomic ions that contain oxygen.*

In several cases, two different oxyanions are formed by the same two elements. Nitrogen and oxygen, for example, are combined in both NO_3^- and NO_2^-. The ion with the larger number of oxygen atoms is named with the *-ate* ending and the ion with the smaller number of oxygen atoms is named with the *-ite* ending.

$$NO_3^- \qquad NO_2^-$$
nitrate ion   nitrite ion

Compounds containing polyatomic ions are named in the same manner as binary ionic compounds. The cation name is given first and followed by the anion name. For example, the two compounds formed with silver by the nitrate and nitrite anions are named silver nitrate ($AgNO_3$) and silver nitrite ($AgNO_2$). When more than one polyatomic ion is present in a compound, the formula for the ion is surrounded by parentheses as illustrated in Figure 7-1 for aluminum sulfate, $Al_2(SO_4)_3$. This formula says that an aluminum sulfate formula unit has 2 aluminum ions and 3 sulfate ions.

STUDY HINT
The system for naming oxyanions is discussed in Chapter 16.

Hints for relating structure of polyatomic ions to its name: If the ion has an -ide ending, it is monatomic (example: S^{2-} is sulfide). If the ion has an -ate ending or an -ite ending, it is polyatomic and contains oxygen (example: SO_4^{2-} is sulfate, SO_3^{2-} is sulfite).

Sample Problem 7.3

Write the formula for tin(IV) sulfate.

Solution

Step 1. Write the symbols for the ions side by side, with the positive ion first.

$$Sn^{4+} \ (SO_4)^{2-}$$

Step 2. Cross over the charge values to give subscripts.

$$Sn_2^{4+} \ (SO_4)_4^{2-}$$

Step 3. Check the subscripts and write the formula.
The total positive charge is $2 \times 4+ = 8+$. The total negative charge is $4 \times 2- = 8-$. The charges are equal. Because ionic compounds are represented in the smallest possible whole-number ratio, the ratio 1:2 should be used instead of 2:4. The correct formula is therefore

$$Sn(SO_4)_2$$

Practice Problems
1. Write formulas for each of the following ionic compounds: *(a)* Sodium iodide *(Ans.)* NaI *(b)* calcium chloride *(Ans.)* $CaCl_2$ *(c)* potassium sulfide *(Ans.)* K_2S *(d)* lithium nitrate *(Ans.)* $LiNO_3$ *(e)* copper(II) sulfate *(Ans.)* $CuSO_4$ *(f)* sodium carbonate *(Ans.)* Na_2CO_3 *(i)* barium phosphate *(Ans.)* $Ba_3(PO_4)_2$.
2. Name each of the following ionic compounds using both systems where appropriate:
(a) Ag_2O *(Ans.)* silver oxide *(b)* $Ca(OH)_2$ *(Ans.)* calcium hydroxide *(c)* $Zn(NO_3)_2$ *(Ans.)* zinc nitrate *(d)* $KClO_3$ *(Ans.)* potassium chlorate *(e)* NH_4OH *(Ans.)* ammonium hydroxide *(f)* $FeCrO_4$ *(Ans.)* iron(II) chromate *(g)* $Pb_3(PO_4)_2$ *(Ans.)* lead(II) phosphate

Figure 7-5 Naming binary molecular compounds using prefixes

Binary Molecular Compounds

Unlike ionic compounds, molecular compounds contain discrete covalently-bonded molecular units. The old system for naming binary molecular compounds, which is based on prefixes, is illustrated in Figure 7-5. For CCl_4, the *tetra* in the name carbon *tetra*chloride shows that four chloride atoms are present. The two oxides of carbon (CO and CO_2) are

The alternate system for naming binary molecular compounds is explained in Section 7.2.

TABLE 7-5 NUMERICAL PREFIXES	
Number	Prefix
$\frac{1}{2}$	hemi-
1	mono-
2	di-
3	tri-
4	tetra-
5	penta-
6	hexa-
7	hepta-
8	octa-
9	nona-
10	deca-

The order of nonmetals in formulas is generally P, H, C, S, I, Br, N, Cl, O, F.

The -o or -a at the end of a prefix is often dropped before another vowel, e.g., monoxide, pentoxide.

named "carbon *mon*oxide" and "carbon *di*oxide." In these names the prefixes *mon-* and *di-* indicate one and two oxygen atoms, respectively.

The least electronegative element is given first in the names and formulas of binary molecular compounds. As a result, oxygen and the halogens are usually given second, as in carbon tetrachloride and the carbon oxides. The order of nonmetals in binary compound names and formulas is generally that listed in the margin.

The prefixes that specify number of atoms (or sometimes other groups) are listed in Table 7-5. These prefixes are used according to the following rules when naming binary molecular compounds:

1. A prefix is used with the name of the first element (the one that is less electronegative) only if more than one atom of that element is present.
2. The second element is named by combining *(a)* a prefix if more than one compound can be formed by the two elements, *(b)* the root of the name of the second element, and *(c)* the ending *-ide*.

With few exceptions the ending *-ide* shows that a compound contains only two elements.

The use of the rules is illustrated by the names of the six oxides of nitrogen, listed in Table 7-6. For example, note the application of rule 1 in the name *nitrogen monoxide* for NO. No prefix is needed with "nitrogen" because only one nitrogen atom is present. A prefix is needed with the single oxygen atom, as specified by rule 2, however, because nitrogen forms more than one oxide. Take a moment to review the prefixes in the other names in Table 7-6.

TABLE 7-6	BINARY COMPOUNDS OF NITROGEN AND OXYGEN		
Formula	Prefix-Based Name	Formula	Prefix-Based Name
N_2O	dinitrogen monoxide	NO_2	nitrogen dioxide
NO	nitrogen monoxide	N_2O_4	dinitrogen tetroxide
N_2O_3	dinitrogen trioxide	N_2O_5	dinitrogen pentoxide

Sample Problem 7.4

(a) Give the name for As_2O_5. *(b)* Write the formula for oxygen difluoride.

Solution

(a) The compound contains two arsenic atoms, so the first word in the name is "*di*arsenic." The five oxygen atoms are indicated by adding the prefix *pent-* to the word "oxide." The complete name is diarsenic pentoxide.

(b) The first symbol in the formula is that for oxygen. Oxygen is first in the name because it is less electronegative than fluorine. Since there is no prefix, there must be only one oxygen atom. The prefix *di-* in *difluoride* shows that there are two fluorine atoms in the molecule. The formula is OF_2.

Practice Problems

1. Name each of the following binary molecular compounds: *(a)* SO_3 *(Ans.)* sulfur trioxide *(b)* CF_4 *(Ans.)* carbon tetrafluoride *(c)* NO_2 *(Ans.)* nitrogen dioxide *(d)* ICl_3 *(Ans.)* iodine trichloride *(e)* P_2O_5 *(Ans.)* diphosphorus pentoxide.
2. Write formulas for each of the following: *(a)* carbon tetraiodide *(Ans.)* CI_4 *(b)* phosphorus trichloride *(Ans.)* PCl_3 *(c)* sulfur dioxide *(Ans.)* SO_2 *(d)* dinitrogen trioxide *(Ans.)* N_2O_3.

Figure 7-6 Common laboratory acids

1756 Henry Cavendish (1731–1810) discovered hydrogen by experimenting with the reaction of metals with acids.

Acids and Salts

Acids are a distinctive class of chemical compounds about which you will learn much more in later chapters. The acids found on almost every laboratory shelf (Figure 7-6) are either binary acids or oxyacids. Each is a molecular compound that contains one or more hydrogen atoms.

The common binary acids are solutions in water of binary molecular compounds containing hydrogen and one of the halogens—fluorine, chlorine, bromine, or iodine. The binary compound HCl, for example, is named hydrogen chloride. In water solution, HCl is known as *hydrochloric acid*.

Oxyacids *are acids containing hydrogen, oxygen, and a third element.* The common binary and oxyacids are listed in Table 7-7. Many of the common polyatomic ions are produced by the loss of the hydrogen ions from oxyacids. A few examples of the relationship between oxyacids and oxyanions are the following:

sulfuric acid	H_2SO_4	sulfate ion	SO_4^{2-}
nitric acid	HNO_3	nitrate ion	NO_3^{-}
phosphoric acid	H_3PO_4	phosphate ion	PO_4^{3-}

An ionic compound composed of a cation and the anion from an acid is often referred to as a **salt**. Table salt, sodium chloride (NaCl), contains the anion from hydrochloric acid. Calcium sulfate ($CaSO_4$) is a salt containing an anion from sulfuric acid.

CAUTION: Laboratory acids must be handled with care and according to instructions. They can burn the skin.

TABLE 7-7 COMMON BINARY ACIDS AND OXYACIDS

HF	Hydrofluoric acid	HNO_3	Nitric acid
HCl	Hydrochloric acid	HNO_2	Nitrous acid
HBr	Hydrobromic acid	H_3PO_4	Phosphoric acid
HI	Hydriodic acid	H_2CO_3	Carbonic acid
H_2SO_4	Sulfuric acid	$HC_2H_3O_2$	Acetic acid
H_2SO_3	Sulfurous acid		

Some salts contain anions in which one or more hydrogen atoms from the acid are retained. Such anions are named by adding the word *hydrogen* or the prefix *bi-* to the anion name. The best-known such anion is formed by loss of one hydrogen atom from carbonic acid, H_2CO_3.

$$HCO_3^-$$
hydrogen carbonate ion
bicarbonate ion

Section Review

1. It is used to represent a chemical compound and indicates the types and numbers of atoms or ions present in the formula unit of that compound.
2. (a) $AlBr_3$ (b) Na_2O (c) MgI_2 (d) PbO (e) SnI_2 (f) Fe_2S_3 (g) $Cu(NO_3)_2$ (h) $(NH_4)_2SO_4$
3. (a) sodium iodide (b) magnesium sulfide (c) calcium oxide (d) potassium sulfide (e) copper(I) bromide (f) iron(II) chloride
4. (a) BaS (b) NaOH (c) $Pb(NO_3)_2$ (d) $KMnO_4$ (e) $Fe_2(SO_4)_3$

1. What is the significance of a chemical formula?
2. Write formulas for the compounds formed between (a) aluminum and bromine (b) sodium and oxygen (c) magnesium and iodine (d) Pb^{2+} and O^{2-} (e) Sn^{2+} and I^- (f) Fe^{3+} and S^{2-} (g) Cu^{2+} and NO_3^- (h) NH_4^+ and SO_4^{2-}.
3. Name each of the following compounds using the Stock system: (a) NaI (b) MgS (c) CaO (d) K_2S (e) CuBr (f) $FeCl_2$.
4. Write formulas for each of the following compounds: (a) barium sulfide (b) sodium hydroxide (c) lead(II) nitrate (d) potassium permanganate (e) ferric sulfate.

7.2 Oxidation Numbers

SECTION OBJECTIVES

- List the rules for assigning oxidation numbers.
- Give the oxidation number for each element in the formula of a chemical compound.
- Name binary compounds using oxidation numbers and the Stock system.

Ionic charges show the distribution of electrons in ionic compounds. Rules have been developed for assigning numbers that similarly show electron distributions for atoms in any type of compound.

Oxidation numbers, also called **oxidation states**, *are assigned to the atoms in molecules, including molecular ions, to show the general distribution of electrons among the bonded atoms.* In assigning oxidation numbers, "ownership" of the bonding electrons is given to the more electronegative atom in each bond. Unlike ionic charges, oxidation numbers do not have an exact physical meaning. In some cases oxidation numbers are quite arbitrary. However, oxidation numbers are useful in naming compounds, in writing formulas, and, as will be discussed in Chapter 20, in studying certain types of chemical reactions.

Assigning Oxidation Numbers

A comparison among sodium fluoride (NaF), hydrogen fluoride (HF), and water (H_2O) illustrates the basis for oxidation numbers. In sodium fluoride, the sodium and fluorine have ionic charges of 1+ and 1−. In hydrogen fluoride (HF), the bond is polar, with a partial negative charge on the fluorine atom and a partial positive charge on the hydrogen atom. If HF were an ionic compound in which an electron was fully transferred to the fluorine atom, H would have a 1+ charge and F would have a 1− charge. In a water molecule, the oxygen atom is more electronegative than the hydrogen atoms. If H_2O were an ionic compound, the oxygen atom would have a charge of 2− and the hydrogen atoms would each have a charge of 1+.

If the following set of rules for assigning oxidation numbers are followed, the resulting oxidation numbers will appropriately indicate the distribution of electrons in the bonded atoms of the compound being discussed.
1. An uncombined element has an oxidation number of zero. Thus, the atoms of Na, O_2, P_4, and S_8 all have oxidation numbers of zero.
2. A monatomic ion has an oxidation number equal to the charge of the ion. Thus, the ions Na^+, Ca^{2+}, and Cl^- have oxidation numbers of $+1$, $+2$, and -1, respectively.
3. Fluorine has an oxidation number of -1 in all of its compounds, because it is the most electronegative element.
4. Oxygen has an oxidation number of -2 in almost all compounds (except in peroxides such as H_2O_2, where it is -1, in superoxides such as KO_2, where it is $-\frac{1}{2}$; and in compounds with halogens, such as OF_2, where it is $+2$).
5. Hydrogen has an oxidation number of $+1$ in all compounds with more electronegative elements, and an oxidation number of -1 in compounds with metals known as hydrides, such as NaH and CaH_2.
6. The more electronegative element in a binary compound is assigned the number equal to the charge it would have if it were an ion.
7. The algebraic sum of the oxidation numbers of all atoms in a neutral compound is zero.
8. The algebraic sum of the oxidation numbers of all atoms in a polyatomic ion is equal to the charge of the ion.

Rules 7 and 8 make it possible to assign oxidation numbers where they are not known. The total of the known and unknown oxidation numbers must satisfy rule 7 or 8, as illustrated in Sample Problem 7.5.

Elements with greater electronegativity tend to be on the right of the Periodic Table and attract electrons, while the elements with the least electronegativity are found on the left and are more likely to donate electrons.

These rules have to be memorized. Tell students that memorization is usually not a good strategy to use in chemistry, that problem solving is. However, occasionally, some information *has* to be memorized.

An *algebraic sum* is a sum of positive and negative numbers.

Sample Problem 7.5

Assign oxidation numbers to each atom in the following compounds or ions. (a) UF_6 (b) $GeCl_2$ (c) ClO_3^- (d) H_2SO_4

Solution
(a) In UF_6, fluorine has oxidation number -1 as it does in all compounds in which it is present. Therefore, uranium has oxidation number $+6$.

$$\overset{+6\,-1}{UF_6} \quad \text{Oxidation no. total} = (+6) + (6 \times -1) = 0$$

(b) In $GeCl_2$, chlorine is assigned the oxidation number it would have as an ion, -1, because it is the more electronegative of the two combined elements. Therefore, germanium in this compound has oxidation number $+2$.

$$\overset{+2\,-1}{GeCl_2} \quad \text{Oxidation no. total} = (+2) + (2 \times -1) = 0$$

(c) In ClO_3^-, oxygen has oxidation number -2. The total oxidation number due to the three oxygen atoms is therefore -6. For the chlorate ion to have a $1-$ charge, chlorine must be assigned an oxidation number of $+5$.

$$\overset{+5\,-2}{ClO_3^-} \quad \text{Oxidation no. total} = (+5) + (3 \times -2) = -1$$

Chemical Formulas and Chemical Compounds

Sample Problem 7.5 *continued*

(d) In H_2SO_4, according to the rules, hydrogen has oxidation number +1 and oxygen has oxidation number −2. With a total positive oxidation number of +2 for the two hydrogen atoms, and a total negative oxidation number of −8 for the four oxygen atoms, sulfur must have an oxidation number of +6.

$$\underset{H_2SO_4}{\overset{+1\ +6\ -2}{}} \quad \text{Oxidation no. total} = (2 \times +1) + (+6) + (4 \times -2) = 0$$

Practice Problems
Assign oxidation numbers to each atom in the following compounds or ions:
1. HCl *(Ans.)* +1, −1;
2. CF_4 *(Ans.)* +4, −1;
3. PCl_3 *(Ans.)* +3, −1;
4. SO_2 *(Ans.)* +4, −2;
5. HNO_3 *(Ans.)* +1, +5, −2;
6. SiO_2 *(Ans.)* +4, −2;
7. KH *(Ans.)* +1, −1;
8. P_2O_5 *(Ans.)* +5, −2;
9. $HClO_3$ *(Ans.)* +1, +5, −2.

UF_6 — uranium(VI) fluoride
$GeCl_2$ — germanium(II) chloride
ClO_3^- — chlorate ion

Using Oxidation Numbers for Formulas and Names

Table 7-8 lists the common oxidation numbers for the nonmetals. A more extensive list of oxidation numbers is given in Appendix Table A-14. These numbers can sometimes be used in the same manner as ionic charges to determine formulas. Suppose, for example, you want to know the formula of a binary compound between sulfur and oxygen. From the common +4 and +6 oxidation states of sulfur, you could expect that sulfur might form SO_2 or SO_3. Both are known compounds. As with formulas for ionic compounds, of course, a formula must represent facts. Oxidation numbers alone cannot be trusted to predict whether or not a compound exists.

In Section 7.1 we introduced the use of Roman numerals as ionic charges in the Stock system of nomenclature. This system is actually based upon oxidation numbers. The Roman numeral system of distinguishing between compounds of the same element applies whether or not the compound is ionic. The names for the first three compounds in Sample Problem 7.5 are listed in the margin. These names could be assigned without knowing whether the compounds are molecular or ionic.

The Stock system is an alternative to the prefix system for naming binary molecular compounds. In the two systems, the two oxides of sulfur would be named sulfur dioxide/sulfur(IV) oxide and sulfur trioxide/sulfur(VI) oxide.

TABLE 7-8 COMMON OXIDATION NUMBERS OF NONMETALS THAT HAVE VARIABLE OXIDATION STATES

Group 14	Carbon	+2, +4
Group 15	Nitrogen	−3, +3, +5
	Phosphorus	+3, +5
Group 16	Sulfur	−2, +4, +6
Group 17	Chlorine	−1, +1, +3, +5, +7
	Bromine	−1, +1, +3, +5, +7
	Iodine	−1, +1, +3, +5, +7

Desktop Investigation

Models of Chemical Compounds

Question
Using common materials, can you make models of molecular and ionic compounds?

Materials
red, blue, green, and yellow modeling clay
large and small gumdrops in different colors
toothpicks
centimeter ruler

Procedure
1. Using different colors of clay and toothpicks, make molecular models of the following substances: hydrogen chloride, water, hydrogen peroxide, ammonia, boron trifluoride, carbon dioxide, and sulfur dioxide. As you make the models, check Figure 5-12 in Chapter 5 to make sure the balls of clay represent each atom correctly according to scale. Refer to the diagram (right) to help you see the way that the atoms are arranged in the different molecules.

2a. Attach gum drops of two different sizes and colors to one another with toothpicks to make an ionic crystal of NaCl similar to the one in the diagram (above). Use the larger gumdrops for the chlorine ions and the smaller ones for the sodium ions.

2b. Using clay and toothpicks, make models of what you think a formula unit of $FeBr_2$, $FeBr_3$, Cu_2O and CuO might look like. Even though the sizes of Fe^{3+} and Cu^+ ions are not shown in Figure 5-18 in Chapter 5, you can estimate their sizes by comparing with the known sizes of the Fe^{2+} and Cu^{2+} ions.

Discussion
1. How do molecular compounds and ionic compounds differ from one another structurally?

2. Iron(II) bromide and iron(III) bromide are composed of the same elements; the same is true of copper(I) oxide and copper(II) oxide. How could these compounds be differentiated?

The international body that governs nomenclature has endorsed the Stock system, which is definitely more practical for more complicated compounds. The prefix-based names and the Stock system names are still used interchangeably for many simple compounds. A few more examples of names in both systems are given below.

1815 William Prout (1785–1850) pointed out that the atomic weights of the elements were all multiples of that of hydrogen, which was the lightest known element.

1. (a) +1, −1 (b) +4, −1
   (c) +1, −2 (d) +3, −1
   (e) +4, −2 (f) +1, −1
   (g) +1, +4, −2 (h) +3, −2
   (i) +6, −2, (j) +3, −2
   (k) +5, −2
2. (a) carbon(IV) iodide
   (b) sulfur(VI) oxide
   (c) arsenic(III) sulfide
   (d) nitrogen(VI) oxide

SECTION OBJECTIVES

- Calculate the formula mass or molar mass of any given compound.
- Use molar mass to convert between mass in grams and amount in moles of a chemical compound.
- Give the number of molecules, formula units, or ions in a given molar amount of a chemical compound.
- Calculate the percent composition of a given chemical compound.

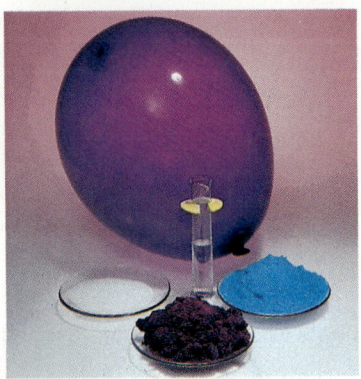

Figure 7-7 One mole of nitrogen (balloon), water (graduated cylinder) $CuSO_4 \cdot 5H_2O$ (blue substance) $CoCl_2 \cdot 6H_2O$ (reddish substance) and NaCl (white substance).

	Old System	Stock System
PCl_3	Phosphorus trichloride	phosphorus(III) chloride
PCl_5	Phosphorus pentachloride	phosphorus(V) chloride
N_2O	Dinitrogen monoxide	nitrogen(I) oxide
NO	Nitrogen monoxide	nitrogen(II) oxide
PbO_2	Lead dioxide	lead(IV) oxide
Mo_2O_3	dimolybdenum trioxide	molybdenum(III) oxide

Section Review

1. Assign oxidation numbers to each atom in the following compounds or ions: (a) HF; (b) CI_4 (c) H_2O (d) PI_3 (e) CS_2 (f) Na_2O_2 (g) H_2CO_3 (h) NO_2^- (i) SO_4^{2-} (j) ClO_2^- (k) IO_3^-
2. Name each of the following binary molecular compounds according to the Stock system: (a) CI_4 (b) SO_3 (c) As_2S_3 (d) NO_3.

7.3 Using Chemical Formulas

Chemical compounds are represented by combinations of chemical symbols called chemical formulas. A chemical formula indicates the elements present in a compound and the relative numbers of atoms of each element in the compound. In chemical formulas, the elements are given by their symbols and the relative numbers of atoms of each element by numeral subscripts. Where no subscript is written, it means that there is one atom of the element in the formula unit of that substance.

In this section, you will learn how to use the chemical formula to calculate the formula mass, the molar mass, and the percent composition by mass of a compound.

Formula Masses and Molar Masses

One hydrogen atom has (on average) a mass of 1.007 94 u on the relative atomic mass scale. One oxygen atom has (on average) a mass of 15.9994 u on the relative atomic mass scale. From the chemical formula, H_2O, we know that one water molecule is composed of two hydrogen atoms and one oxygen atom. The mass of a water molecule in atomic mass units is found therefore, by summing the masses of the three atoms combined in the molecule.

$$2 \text{ H atoms} \times \frac{1.007\,94\,u}{\text{H atom}} = 2.015\,88\,u$$

$$1 \text{ O atom} \times \frac{15.9994\,u}{\text{O atom}} = 15.9994\,u$$

$$\text{Formula mass of } H_2O = 18.0153\,u$$

The mass of a water molecule (on average) on the atomic mass scale is 18.0153 u.

The procedure illustrated above, applied to any chemical formula, gives the mass of the unit represented by the formula. The unit might be a molecule, a formula unit, or a polyatomic ion. *The **formula mass** of any compound or polyatomic ion is the sum of the average atomic masses of all the atoms represented in the formula.*

The mass of the water molecule is correctly referred to as a molecular mass. Often, the terms *molecular mass* or *molecular weight* are used with the same meaning as *formula mass*. You may see these terms in other books. Strictly speaking, however, the mass of one NaCl formula unit is not a molecular mass because NaCl is an ionic compound. Therefore, we prefer the more general term, and will use "formula mass" throughout this book.

Atomic masses are given to varying numbers of decimal places in the Periodic Table or tables like that in Appendix Table A-3. Therefore, the question arises of how to deal with significant figures in calculating formula masses. In the following problems, we have used atomic masses as given in Appendix Table A-3. The formula masses are rounded off according to the rules for addition of significant figures. If a formula mass is to be used in a further calculation, it is then limited to the number of digits needed in the calculation.

Sample Problem 7.6

Find the formula mass of potassium chlorate, $KClO_3$.

Solution

$$1 \text{ K atom} \times \frac{39.0983 \; u}{\text{K atom}} = 39.0983 \; u; \quad 1 \text{ Cl atom} \times \frac{35.453 \; u}{\text{Cl atom}} = 35.453 \; u;$$

$$3 \text{ O atoms} \times \frac{15.9994 \; u}{\text{O atom}} = 47.9982 \; u; \quad \text{Formula mass of } KClO_3 = 122.550 \; u$$

Practice Problems
Find the formula mass of each of the following:
1. H_2SO_4 *(Ans.)* 98.07 u 2. $Ca(NO_3)_2$ *(Ans.)* 164.09 u 3. PO_4^{3-} *(Ans.)* 94.9714 u

In Chapter 3 you learned that the number of atoms in one molar mass of a given element is equal to 6.022×10^{23}. For example, one molar mass of calcium (Ca) is 40.08 g and contains 6.022×10^{23} atoms. One molar mass of lead (Pb) is 207.2 g and also contains 6.022×10^{23} atoms.

The same relationship applies to compounds. The molar mass of a chemical compound is the mass in grams numerically equal to the formula mass of the compound. This molar mass is equivalent to the sum of the molar masses of all the elements present in the chemical formula. The formula mass of water was calculated just above as 18.0153 u. The molar mass of water is therefore 18.0153 g. One mole of H_2O molecules contains 2 mol of H atoms and 1 mol of O atoms. The 2 mol of H atoms have a mass of 2.015 88 g, and the 1 mol of O atoms has a mass of 15.9994 g. One mole of water has a mass of 18.0153 g and contains 6.022×10^{23} water molecules, i.e. Avogadro's number of molecules. This is the same number of molecules as there are atoms in one mole of calcium.

One mole of an ionic compound contains one mole of formula units. Potassium chlorate ($KClO_3$) was found in Sample Problem 7.6 to have a formula mass of 122.550 u. Taken in grams, this is the mass of one mole, one Avogadro's number, of $KClO_3$ formula units. But note that the total number of individual particles in one mole of potassium chlorate is twice

1811 Count Amedeo Avogadro (1776–1856) proposed that all gases contain the same number of particles per unit volume. Dalton and Berzelius rejected the proposal, and the idea was not in the mainstream of chemical thought until 1860. Ampère upheld Avogadro's hypothesis.

1860 Stanislao Cannizzaro (1826–1910) attended the First International Congress of Chemistry and convinced many attending of the validity of Avogadro's theory. He saw that if Avogadro was correct, molar masses of gases could be determined from their density when compared to the density of hydrogen. The formula for various gasses could then be inferred from the molar mass.

Avogadro's number. One mole of $KClO_3$ is composed of one mole of K^+ ions and one mole of ClO_3^- ions.

Several more examples of the meaning of "molar mass" for molecules and ionic compounds are given in Figure 7-7.

Sample Problem 7.7

What is the molar mass of barium nitrate, $Ba(NO_3)_2$?

Solution
One mole of barium nitrate contains 1 mol of Ba^{2+} ions and 2 mol of NO_3^- ions, giving a total of 3 mol of ions. One mole of barium nitrate contains 6 mol of oxygen atoms.

$$1 \text{ mol Ba} \times \frac{137.33 \text{ g Ba}}{\text{mol Ba}} = 137.33 \text{ g Ba}$$

$$2 \text{ mol N} \times \frac{14.0067 \text{ g N}}{\text{mol N}} = 28.0134 \text{ g N}$$

$$6 \text{ mol O} \times \frac{15.9994 \text{ g O}}{\text{mol O}} = 95.9964 \text{ g O}$$

Molar mass of $Ba(NO_3)_2$ = 261.34 g

Practice Problems
1. Determine the number of moles of ions and the number of moles of atoms present in one mole of each of the following: *(a)* $NaNO_3$ *(Ans.)* 1 mol Na^+ and 1 mol NO_3^-; 1 mol Na, 1 mol N, and 3 mol O *(b)* $Ba(OH)_2$ *(Ans.)* 1 mol Ba^{2+} and 2 mol OH^-; 1 mol Ba, 2 mol O, and 2 mol H.
2. Find the molar mass of each of the compounds listed in the preceding question. *(a)* *(Ans.)* 84.9947 g *(b)* *(Ans.)* 171.34 g

Use of molar mass was introduced in Section 3.3.

Molar Mass as a Conversion Factor

The molar mass of a compound is a conversion factor that relates mass in grams to number of moles. Conversions of this type for elements and compounds are summarized in Figure 7-9. Molar mass as a conversion factor, you will recall, has the units of grams per mole. To convert a known number of moles to mass in grams the calculation is set up as follows:

$$\text{moles} \times \frac{\text{grams}}{\text{mol}} = \text{mass in grams}$$

To convert a known mass in grams to number of moles, the mass must be divided by the molar mass. This is done by multiplying by the molar mass in its inverted form so that units are easily cancelled.

$$\text{mass in grams} \times \frac{1 \text{ mol}}{\text{grams}} = \text{moles}$$

Note that the molar masses used in solving Sample Problems 7.8 to 7.11 have been calculated using all known digits and then rounded to the number of significant figures required by the problem.

STUDY HINT
Remember to account for the effect of parentheses when finding formula masses.

Figure 7-9 Relationships between mass, moles, and numbers of molecules or atoms.

Sample Problem 7.8

What is the mass in grams of 2.50 mol of oxygen gas?

Solution

Step 1. Analyze   Given: 2.50 mol O_2
Unknown: mass O_2, in grams

Step 2. Plan   moles $O_2 \rightarrow$ mass O_2

The molar mass of O_2 is the conversion factor needed to convert moles to mass.

$$\text{moles } O_2 \times \frac{\text{grams } O_2}{\text{mol } O_2} = \text{grams } O_2$$

Step 3. Compute   First, the molar mass of O_2 must be found:

$$2 \text{ mol } O \times \frac{15.9994 \text{ g}}{\text{mol } O} = 31.9988 \text{ g} = \text{molar mass of } O_2$$

Then, the calculation as in Step 2 can be carried out.

$$2.50 \text{ mol } O_2 \times \frac{32.0 \text{ g } O_2}{\text{mol } O_2} = 80.0 \text{ g } O_2$$

Step 4. Evaluate   The answer is correctly given to three significant figures and is close to an estimated value of 75, (2.5 × 30).

Sample Problem 7.9

Ibuprofen, the active ingredient in many nonprescription pain relievers, has the molecular formula $C_{13}H_{18}O_2$. The molar mass of ibuprofen is 206.2860 g/mol. *(a)* If the tablets in a bottle contain a total of 33 g of ibuprofen, how many moles of ibuprofen are in the bottle? *(b)* How many molecules of ibuprofen are in the bottle? *(c)* What is the total mass in grams of carbon in this much ibuprofen?

Solution

Step 1. Analyze Given: 33 g of $C_{13}H_{18}O_2$,
molar mass 206.2860 g
Unknown: *(a)* moles $C_{13}H_{18}O_2$ *(b)* molecules $C_{13}H_{18}O_2$ *(c)* total mass of C

Step 2. Plan *(a)* mass → moles   *(b)* moles → molecules
(c) moles $C_{13}H_{18}O_2$ → moles C → mass C

Part *(a)* requires using the molar mass of $C_{13}H_{18}O_2$ as a conversion factor. Then, knowing the moles of $C_{13}H_{18}O_2$, Avogadro's number is the conversion factor needed in part *(b)* to find the number of molecules. For part *(c)*, the two conversion factors needed are the moles of carbon per mole of $C_{13}H_{18}O_2$ and the molar mass of carbon.

(a) grams $C_{13}H_{18}O_2 \times \dfrac{1 \text{ mol } C_{13}H_{18}O_2}{\text{grams } C_{13}H_{18}O_2} = $ mol $C_{13}H_{18}O_2$

(b) moles $C_{13}H_{18}O_2 \times \dfrac{6.022 \times 10^{23} \text{ molecules}}{\text{mol}} = $ molecules $C_{13}H_{18}O_2$

(c) moles $C_{13}H_{18}O_2 \times \dfrac{\text{moles C}}{\text{mol } C_{13}H_{18}O_2} \times \dfrac{\text{grams C}}{\text{mol C}} = $ grams C

Step 3. Compute *(a)* 33 g $C_{13}H_{18}O_2 \times \dfrac{1 \text{ mol } C_{13}H_{18}O_2}{210 \text{ g } C_{13}H_{18}O_2} = 0.16$ mol $C_{13}H_{18}O_2$

(b) 0.16 mol $C_{13}H_{18}O_2 \times \dfrac{6.0 \times 10^{23} \text{ molecules}}{\text{mol}} = 9.6 \times 10^{22}$ molecules $C_{13}H_{18}O_2$

(c) 0.16 mol $C_{13}H_{18}O_2 \times \dfrac{13 \text{ mol C}}{\text{mol } C_{13}H_{18}O_2} \times \dfrac{12 \text{ g C}}{\text{mol C}} = 25$ g C

The bottle contains 0.16 mol of ibuprofen, which is 9.6×10^{22} molecules of ibuprofen. The ibuprofen contains 25 g of carbon.

Step 4. Evaluate Checking each step shows that the arithmetic is correct, significant figures have been used correctly, and units have cancelled as desired.

Practice Problems

1. What is *(a)* the molar mass of copper(II) nitrate? *(Ans.)* 187.556 g *(b)* the mass in grams of 6.25 mol of copper(II) nitrate? *(Ans.)* 1180 g
2. How many moles are represented by each of the following: *(a)* 11.0 g CO_2 *(Ans.)* 0.250 mol CO_2 *(b)* 6.60 g $(NH_4)_2SO_4$ *(Ans.)* 0.0500 mol $(NH_4)_2SO_4$ *(c)* 4.5 kg $Ca(OH)_2$ *(Ans.)* 61 mol $Ca(OH)_2$.
3. How many molecules are contained in: *(a)* 25.0 g H_2SO_4 *(Ans.)* 1.53×10^{23} molecules H_2SO_4 *(b)* 125 g of sugar $(C_{12}H_{22}O_{11})$ *(Ans.)* 2.20×10^{23} molecules $C_{12}H_{22}O_{11}$.

Chemistry in Detail

Atomic Symbols and Molecular Formulas

It is common to shorten expressions we use repeatedly. The alchemists used symbols to represent the few elements they knew and the common compounds they manipulated in their attempts to turn lead into gold. Unfortunately, their concern for secrecy led each alchemist to use a separate set of symbols. Only a few alchemical symbols became generally accepted.

John Dalton (1766-1844) realized that atomic theory provided a natural way to represent compounds. He invented a set of elemental symbols, and then combined them to represent what he believed were the molecules of various compounds. Thus he represented a molecule of nitrogen(I) oxide (laughing gas) as a simple circle (oxygen) and two circles that each contained a vertical line (nitrogen).

Because it is difficult to create simple, unique figures for all of the elements, Dalton used letters inside circles for some symbols. Jöns Jakob Berzelius (1779-1848) realized that the circles were superfluous, and recommended the one- and two-letter symbols we now use. Since then, we have added subscripts, parentheses, dots for hydrates, and lines representing bonds. To handle complicated molecules, we even resort to three-dimensional perspective with drawings generated by computer.

The standard set of atomic symbols and molecular formulas transcends language barriers and thus can be used internationally.

	Alchemy	Dalton	Present
Oxygen	(Unknown)	○	O
Hydrogen	(Unknown)	⊙	H
Nitrogen	(Unknown)	⊘	N
Sulfur	☿	Ⓢ	S
Mercury	☿	✳	Hg
Iron	♂	Ⓘ	Fe
Water	≈	⊙○⊙	H_2O
Nitrogen(I) oxide	(Unknown)	⊘○⊘	N_2O

Chemical Formulas and Chemical Compounds

1799 Joseph Proust (1754–1826) experimentally demonstrated that the composition of a compound contained definite proportions by mass of its elements. This specific composition was not dependent on the manner in which the compound was formed.

Percent Composition

It is frequently useful to know the percent composition by mass of a chemical compound. The percent of iron in iron(III) oxide (Fe_2O_3) may be used to calculate the mass of iron in an ore. Or the percent of oxygen in potassium chlorate ($KClO_3$) may be useful if it is to be used as a source of oxygen.

The **percent composition** of a compound is the percent by mass of each element in the compound. The percent composition is the same, no matter what the size of the sample.

The molar mass of a compound is 100% of the mass represented by the formula. The mass percent of an element is therefore the mass of the element in one molar mass of the compound divided by the molar mass or formula mass of the compound and multiplied by 100%.

$$\frac{\text{mass of element}}{\text{molar mass of compound}} \times 100\% = \% \text{ of element}$$

Sample Problem 7.10

Find the percent composition of copper(I) sulfide (Cu_2S), which occurs as the copper ore chalcocite.

Solution

Step 1. Analyze Given: formula, Cu_2S
Unknown: percent composition of Cu_2S

Step 2. Plan formula → molar mass → % of each element

The molar mass of the compound must be found. Then the mass of each element present is used to calculate the percent of each element

Step 3. Compute

$$2 \text{ mol Cu} \times \frac{63.546 \text{ g Cu}}{1 \text{ mol Cu}} = 127.09 \text{ g Cu}$$

$$1 \text{ mol S} \times \frac{32.06 \text{ g S}}{1 \text{ mol S}} = 32.06 \text{ g S}$$

Molar mass of Cu_2S = 159.15 g

$$\frac{127.09 \text{ g Cu}}{159.15 \text{ g Cu}_2\text{S}} \times 100\% = 79.855\% \text{ Cu}$$

$$\frac{32.06 \text{ g S}}{159.15 \text{ g Cu}_2\text{S}} \times 100\% = 20.14\% \text{ S}$$

Step 4. Evaluate A good check is to see if the results add up to 100%. (Because of rounding off or experimental errors, the total may not always be exactly 100%.)

Sample Problem 7.11

As some compounds crystallize from a water solution, they trap water molecules. Sodium carbonate forms such a hydrate, in which 10 water molecules are present for every formula unit of sodium carbonate. Find the percent of water in sodium carbonate decahydrate, $Na_2CO_3 \cdot 10H_2O$, which has a molar mass of 286 g.

Sample Problem 7.11 *continued*

Solution

Step 1. Analyze Given: chemical formula, $Na_2CO_3 \cdot 10H_2O$
 molar mass of $Na_2CO_3 \cdot 10H_2O$
 Unknown: percent H_2O

Step 2. Plan formula → mass H_2O per mole $Na_2CO_3 \cdot 10H_2O$ → % water

The mass of water per mole of sodium carbonate decahydrate must be found using atomic masses and then divided by the mass of 1 mol of $Na_2CO_3 \cdot 10H_2O$.

Step 3. Compute First, we must find the mass of 10 mol of water:

$$20 \text{ mol H} \times \frac{1.007\,94 \text{ g H}}{\text{mol H}} = 20.1588 \text{ g H}$$

$$10 \text{ mol O} \times \frac{15.9994 \text{ g O}}{\text{mol O}} = 159.994 \text{ g O}$$

Mass of H_2O per mole $Na_2CO_3 \cdot 10H_2O$ = 180.153 g

$$\text{Percent mass of } H_2O \text{ per mol } Na_2CO_3 \cdot 10 H_2O = \frac{180 \text{ g } H_2O}{286 \text{ g } Na_2CO_3 \cdot 10H_2O} \times 100\%$$

$$= 62.9\% \text{ } H_2O$$

Step 4. Evaluate Checking shows the arithmetic to be correct and units to cancel as desired. The value of the answer is reasonably larger than an estimate of 50% (which would be the answer for 143 g of water).

Practice Problems
1. Find the percent composition of each of the following compounds.
   (a) $PbCl_2$ *(Ans.)* 74.51% Pb and 25.50% Cl
   (b) $Ba(NO_3)_2$ *(Ans.)* 52.548% Ba, 10.719% N, and 36.732% O
2. Find the percent of water in $ZnSO_4 \cdot 7H_2O$ *(Ans.)* 43.85% H_2O
3. Magnesium hydroxide is 54.87% oxygen by mass.
   (a) How many grams of oxygen would be contained in 175 g of the compound? *(Ans.)* 96.0 g O
   (b) How many moles of oxygen does this represent? *(Ans.)* 6.00 mol O

Section Review

1. Determine both the formula mass and molar mass of the compound $(NH_4)_2CO_3$.
2. How many moles of each ion and how many of each atom are present in 1 mol of $(NH_4)_2CO_3$?
3. What is the mass in grams of 3.25 mol $Fe_2(SO_4)_3$?
4. How many moles are present in a 250. g sample of hydrogen nitrate HNO_3?
5. How many molecules of aspirin ($C_9H_8O_4$) are contained in a 100.0 mg tablet of aspirin?
6. Find the percent composition of all of the elements found in ammonium carbonate, $(NH_4)_2CO_3$.

1. Formula mass = 96.086 2 u
   Molar mass = 96.086 2 g
2. One mole of $(NH_4)_2CO_3$ contains 2 mol NH_4^+ ion and 1 mol CO_3^{2-} ion. One mole of this compound consists of 2 mol N, 8 mol H, 1 mol C, and 3 mol O atoms.
3. 1.30×10^3 g $Fe_2(SO_4)_3$
4. 3.97 mol HNO_3
5. 3.342×10^{20} molec $C_9H_8O_4$
6. N = 29.1544%, H = 8.391 96%, C = 12.5003%, O = 49.9533%

7.4 Determining Chemical Formulas

SECTION OBJECTIVES

- Define simplest formula, and explain how the term applies to ionic and molecular compounds.
- Find a simplest formula from either percent or mass composition.
- Explain the relationship between the simplest formula and the molecular formula of a given compound.
- Find a molecular formula from a simplest formula.

When a new substance is prepared in the laboratory or is discovered in nature, its chemical formula is unknown. Qualitative analysis may first be done to determine what elements are present. Next, quantitative analysis may be performed to determine percent composition and the simplest formula. For a molecular compound, molar mass must then be found.

Calculation of Simplest Formulas

The first step in determining a molecular formula is to find the simplest, or empirical, formula of the compound. Composition data make this possible. *The* **simplest formula** *or* **empirical formula** *consists of the symbols for the elements combined, with subscripts showing the smallest whole-number ratio of the atoms.*

Only simplest formulas can be found from composition data. For example, diborane is a compound that is very useful to organic chemists working to synthesize new medicines and other materials. Diborane, which is 78% boron and 22% hydrogen, is known to have the molecular formula B_2H_6. However, the compound BH_3—a short-lived, unstable substance—has the same percent composition as B_2H_6.

For ionic compounds, which are not made of discrete molecules, the simplest formula is the accepted formula. Although the simplest (empirical) formula indicates the ratio of different atoms in a compound, it does not necessarily indicate the actual numbers of atoms present in each molecule. Many molecular compounds have molecular formulas that are whole-number multiples of the empirical formula, as in the case of diborane, which has the empirical formula BH_3 and the molecular formula B_2H_6.

Calculation of a simplest formula requires that composition data, if found as percent, be converted to composition in grams. An easy way to do this is to start with 100 g of the compound. For diborane, for example, the percent composition shows that 100 g contains 78 g of boron and 22 g of hydrogen. Next, the composition in grams is converted to composition in moles by using molar masses. For diborane,

$$78 \text{ g B} \times \frac{1 \text{ mol B}}{11 \text{ g B}} = 7.1 \text{ mol B} \quad 22 \text{ g H} \times \frac{1 \text{ mol H}}{1.0 \text{ g H}} = 22 \text{ mol H}$$

These values give a mole ratio of 7.1 mol B : 22 mol H. The formula could be written $B_{7.1}H_{22}$, but this is not a proper formula with smallest whole-number subscripts. The smallest ratio of atoms (equal to smallest ratio of moles) is found by dividing each molar amount by the smallest number:

$$\frac{7.1 \text{ mol B}}{7.1} : \frac{22 \text{ mol H}}{7.1}$$

$$1.0 \text{ mol B} : 3.1 \text{ mol H}$$

Sometimes the ratio consists of whole numbers, and sometimes, because of rounding or experimental error, the ratio consists of numbers close to whole numbers. The differences from whole numbers may be ignored and the nearest whole number taken. The calculation shows that the simplest formula of diborane is BH_3.

In summary, finding a simplest formula from percent composition requires finding the following:

> Composition by mass
> Composition in moles
> Smallest whole-number ratio of atoms

Sample Problem 7.12

Analysis shows a compound to contain 26.56% potassium, 35.41% chromium, and 38.03% oxygen. Find the simplest formula of this compound.

Solution

Step 1. Analyze   Given: percent composition: 26.56% K, 35.41% Cr, and 38.03% O.
Unknown: simplest formula

Step 2. Plan   % composition → composition in mass → composition in moles → smallest whole-number ratio of atoms

Step 3. Compute   Composition by mass: 26.56 g K, 35.41 g Cr, 38.03 g O

Composition in moles: $26.56 \text{ g K} \times \dfrac{1 \text{ mol K}}{39.10 \text{ g K}} = 0.6793 \text{ mol K}$

$35.41 \text{ g Cr} \times \dfrac{1 \text{ mol Cr}}{52.00 \text{ g Cr}} \times 0.6810 \text{ mol Cr}$

$38.03 \text{ g O} \times \dfrac{1 \text{ mol O}}{16.00 \text{ g O}} = 2.377 \text{ mol O}$

Smallest whole-number ratio of atoms:

$$\dfrac{0.6793 \text{ mol K}}{0.6793} : \dfrac{0.6810 \text{ mol Cr}}{0.6793} : \dfrac{2.377 \text{ mol O}}{0.6793}$$

$$1.000 \text{ mol K} : 1.003 \text{ mol Cr} : 3.499 \text{ mol O}$$

A formula of $KCrO_{3.5}$ would not be correct. When the ratio at this point contains numbers that are not close to whole numbers, the numbers are converted to fractions and the fractions cleared. Here, 3.499 is $3\frac{1}{2}$. Expressing the ratio in halves and clearing fractions gives

$$1 \text{ mol K} : 1 \text{ mol Cr} : 3.5 \text{ mol O}$$

$$1 \text{ mol K} : 1 \text{ mol Cr} : 3\tfrac{1}{2} \text{ mol O}$$

$$2 \text{ mol K} : 2 \text{ mol Cr} : 7 \text{ mol O}$$

The simplest formula of this compound is $K_2Cr_2O_7$.

Step 4. Evaluate   The best evaluation of a problem such as this is to review each step by checking the calculations and the unit cancellations. Each step in this problem appears to be correct.

Sample Problem 7.13

Analysis of a 10.150 g sample of a compound known to contain only phosphorus and oxygen yields 5.717 g of oxygen. What is the simplest formula of this compound?

Solution

Step 1. Analyze Given: sample mass = 10.150 g
oxygen mass = 5.717 g
Unknown: simplest formula

Step 2. Plan Composition by mass → composition in moles → smallest whole-number ratio of atoms
The mass of oxygen is given. The mass of phosphorus is found by subtracting the oxygen mass from the sample mass.

Step 3. Compute Sample mass − oxygen mass = 10.150 g − 5.717 g = 4.433 g P

Composition by mass: 4.433 g P, 5.717 g O
Composition in moles:

$$4.433 \text{ g P} \times \frac{1 \text{ mol P}}{30.97 \text{ g P}} = 0.1431 \text{ mol P}$$

$$5.717 \text{ g O} \times \frac{1 \text{ mol O}}{16.00 \text{ g O}} = 0.3573 \text{ mol O}$$

Smallest whole-number ratio of atoms:

$$\frac{0.1431 \text{ mol P}}{0.1431} : \frac{0.3573 \text{ mol O}}{0.1431}$$

1.000 mol P : 2.497 mol O
2 mol P : 5 mol O

Step 4. Evaluate Converting the ratio to fractions and multiplying by 2 gives the smallest whole-number ratio of 2 : 5. The simplest formula is P_2O_5. Checking each step shows correct use of significant figures and unit cancellation. The formula is reasonable because +5 is a common oxidation state of phosphorus.

Practice Problems

1. A compound is found to contain 63.52% iron and 36.48% sulfur. Find its simplest formula. *(Ans.)* FeS
2. Find the simplest formula of a compound found to contain 32.38% sodium 22.65% sulfur, and 44.99% oxygen. *(Ans.)* Na_2SO_4
3. Analysis of 20.0 grams of a compound containing only calcium and bromine indicates that 4.00 grams of calcium are present. What is the simplest formula of this compound? *(Ans.)* $CaBr_2$.
4. If 3.50 g of iron react with 1.50 g of oxygen, what is the simplest formula of the compound formed? *(Ans.)* Fe_2O_3

Calculation of Molecular Formulas

An experimentally determined simplest formula may or may not be the correct molecular formula. In introducing this section, we found the simplest formula of diborane to be BH_3. Any multiple of BH_3, such as B_2H_6, B_3H_9, B_4H_{12}, and so on represents the same ratio of boron to hydrogen atoms. How can the correct molecular formula be found?

It is not possible to decide which is the correct molecular formula unless the molecular mass of the substance has been determined. The relationship between the simplest formula and the molecular formula is

$$(\text{simplest formula})_x = \text{molecular formula}$$

where x is a whole-number multiple of the simplest formula. Therefore the formula masses have the same relationship

$$(\text{simplest-formula mass})_x = \text{molecular-formula mass}$$

For our diborane example, experiment would show the formula mass of diborane to be 27.67 u. The formula mass for the simplest formula, BH_3, is 13.83 u. Dividing the experimental molar mass by the simplest formula mass gives

$$x = \frac{27.67\ u}{13.83\ u} = 2.000$$

The value of x is 2, showing that the molecular formula is

$$(BH_3)_2 = B_2H_6$$

Sample Problem 7.14

In Sample Problem 7.13, the simplest formula of a compound of phosphorus and oxygen was found to be P_2O_5. Experiment shows that the formula mass of this compound is 283.889 u. What is the molecular formula of this compound? How would it be named in the prefix system and in the Stock system?

Solution

Step 1. Analyze Given: simplest formula
Unknown: molecular formula

Step 2. Plan
$$x = \frac{\text{molecular-formula mass}}{\text{simplest-formula mass}}$$
$$(\text{simplest formula})_x = \text{molecular formula}$$

Step 3. Compute The simplest formula mass is found by the usual method to be 141.945 u.

$$\frac{283.889\ u}{141.945\ u} = 1.999\ 99 \quad (P_2O_5)_2 = P_4O_{10}$$

The compound P_4O_{10} would be named tetraphosphorus decoxide or phosphorus(V) oxide.

Step 4. Evaluate Checking the arithmetic shows it to be correct.

Practice Problems
1. Determine the molecular formula of a compound having a simplest formula of CH and a formula mass of 78.110 u. *(Ans.)* C_6H_6
2. A sample of a compound with a formula mass of 34.00 u is found to consist of 0.44 g of hydrogen and 6.92 g of oxygen. Find its molecular formula. *(Ans.)* H_2O_2
3. A compound with a formula mass of 42.08 u is found to be 85.64% C and 14.36% H by mass. Find its molecular formula. *(Ans.)* C_3H_6

1. Na₂SO₃
2. Fe₂S₃
3. K₂CrO₄
4. N₂O₅
5. N₂O₄

Section Review

1. A compound is found to contain 36.48% sodium, 25.41% sulfur, and 38.02% oxygen. Find its simplest formula.
2. Find the simplest formula of a compound found to contain 53.70% iron and 46.30% sulfur.
3. Analysis of a compound indicates that it contains 1.04 g K, 0.70 g Cr, and 0.82 g O. Find its simplest formula.
4. If 4.04 g of N combine with 11.46 g O to produce a compound with a formula mass of 108.0 u, what is the molecular formula of this compound?
5. The formula mass of a compound is 92 u. Analysis of a sample of the compound indicates that it contains 0.606 g N and 1.390 g O. Find its molecular formula.

Chapter Summary

- Empirical or simplest chemical formulas indicate how many atoms of each type are combined in the simplest unit of a chemical compound.
- In ionic compounds, the charge of each ion in the compound may be used to determine the simplest chemical formula for the compound.
- Positive monatomic ions are named by the element name followed by the word *ion;* negative monatomic ions are named by dropping the ending of the element name and adding *-ide* to it.
- Compounds composed of two different elements are known as binary compounds and are named by combining the names of the positive and negative ions.
- The Stock system of nomenclature is used for compounds of metals that have more than one oxidation state.
- Compounds containing polyatomic ions are named in the same manner as binary ionic compounds.
- In inorganic molecular compounds, oxidation numbers of each element in the compound may be used to determine the compound's simplest chemical formula.
- Oxidation numbers (oxidation states) are used to indicate the general distribution of electrons among bonded atoms and are assigned according to a set of eight rules.
- By knowing the oxidation numbers, names can be assigned to compounds without knowing whether they are ionic or molecular.
- Formula mass, molar mass, and percent composition can be calculated from the chemical formula for a compound.
- The percent composition of a compound is the percent by mass of each element in the compound no matter what the size of the sample.
- Molar mass can be used as a conversion factor between number of moles and mass of a given compound or element.
- A simplest or empirical chemical formula shows the simplest whole number ratio of atoms in a given compound.
- Each molecule of a compound contains a whole-number multiple of the atoms in the simplest chemical formula. In many cases this whole-number multiple is equal to one.

CHAPTER 7 REVIEW

Chemical Formulas and Chemical Compounds

VOCABULARY

binary
  compound (204)
empirical
  formula (224)
formula mass (216)
monatomic ions (203)
nomenclature (206)
oxidation
  numbers (212)

oxidation states (212)
oxyacids (211)
oxyanions (208)
percent composition (222)
salt (211)
simplest formula (224)

REVIEWING CONCEPTS

1. Write the chemical formula and systematic name for each of the following:
   a. baking soda
   b. milk of magnesia
   c. epsom salts
   d. limestone
   e. lye
   f. wood alcohol (7.1)

2. What three different quantitative meanings can be given by a chemical formula? (7.1)

3. a. What are monatomic ions?
   b. Give three examples of monatomic ions. (7.1)

4. Using only the Periodic Table, write the formula of the ion most typically formed from each of the following elements:
   a. K
   b. Ca
   c. S
   d. Cl
   e. Ba
   f. Br (7.1)

5. Write the shorthand electron configurations for each of the following element–ion pairs:
   a. Na and Na$^+$
   b. Cl and Cl$^-$
   c. Ca and Ca^{2+}
   d. O and O^{2-} (7.1)

6. Write formulas and charges for each of the following:
   a. sodium ion
   b. aluminum ion
   c. chloride ion
   d. nitride ion
   e. iron(II) ion
   f. iron(III) ion (7.1)

7. Name each of the following monatomic ions:
   a. K$^+$
   b. Mg^{2+}
   c. Al^{3+}
   d. Cl$^-$
   e. O^{2-}
   f. Ca^{2+} (7.1)

8. Write formulas for the binary ionic compounds formed between:
   a. sodium and iodine
   b. calcium and sulfur
   c. zinc and chlorine
   d. barium and fluorine
   e. lithium and oxygen
   f. Cr^{2+} and F$^-$
   g. Ni^{2+} and O^{2-}
   h. Fe^{3+} and O^{2-} (7.1)

9. Name each of the following binary ionic compounds:
   a. KCl
   b. CaBr$_2$
   c. MgI$_2$
   d. NaF
   e. AgI
   f. Li$_2$O (7.1)

10. In naming and writing formulas for molecular compounds, what determines the order in which the component elements are written? (7.1)

11. Name the following binary molecular compounds:
    a. CO$_2$
    b. CCl$_4$
    c. PCl$_5$
    d. CS$_2$
    e. SO$_2$
    f. PBr$_3$
    g. N$_2$O$_4$
    h. CI$_4$
    i. As$_2$O$_5$ (7.1)

12. Write formulas for the following binary molecular compounds:
    a. carbon tetrabromide;
    b. silicon dioxide;
    c. carbon monoxide;
    d. sulfur trioxide
    e. diphosphorus pentoxide
    f. diarsenic trisulfide (7.1)

13. Distinguish between binary and oxyacids, and give two examples of each. (7.1)

14. a. What is a salt?
    b. Give two examples (7.1)

15. Name each of the following acids:
    a. HF
    b. HBr
    c. HNO$_3$
    d. H$_2$SO$_4$
    e. H$_3$PO$_4$ (7.1)

Chemical Formulas and Chemical Compounds

16. Name each of the following ions according to the Stock system:
    a. Fe^{2+}
    b. Fe^{3+}
    c. Pb^{2+}
    d. Pb^{4+}
    e. Sn^{2+}
    f. Sn^{4+} (7.2)

17. Name each of the binary molecular compounds in Question 12 using the Stock system. (7.2)

18. Write formulas for each of the following:
    a. phosphorus(III) iodide
    b. sulfur(II) chloride
    c. carbon(IV) sulfide (7.2)

19. a. What are oxidation numbers?
    b. What useful function do oxidation numbers serve? (7.2)

20. a. Define formula mass.
    b. In what units is formula mass expressed? (7.3)

21. What is meant by the molar mass of a compound? (7.3)

22. What three types of information are needed in order to find simplest formulas from percent composition data? (7.4)

23. What is the relationship between the simplest formula and the molecular formula of a compound? (7.4)

PROBLEMS

▶ **Nomenclature and Chemical Formulas**

1. Write the formula and charge for each of the following polyatomic ions:
    a. ammonium ion
    b. acetate ion
    c. hydroxide ion
    d. carbonate ion
    e. sulfate ion
    f. phosphate ion
    g. Cu(II) ion
    h. Sn(II) ion
    i. Fe(III) ion
    j. copper(I) ion
    k. mercury(I) ion
    l. mercury(II) ion

2. Name each of the following ions:
    a. NH_4^+
    b. ClO_3^-
    c. OH^-
    d. SO_4^{2-}
    e. NO_3^-
    f. CO_3^{2-}
    g. PO_4^{3-}
    h. $C_2H_3O_2^-$
    i. HCO_3^-
    j. CrO_4^{2-}

3. Write formulas for each of the following compounds:
    a. sodium fluoride
    b. calcium oxide
    c. potassium sulfide
    d. magnesium chloride
    e. aluminum bromide
    f. lithium nitride
    g. iron(II) oxide

▶ **Oxidation Number and Stock System**

4. Name each of the following ionic compounds using the Stock system:
    a. NaCl
    b. MgO
    c. KF
    d. CaS
    e. LiBr
    f. KOH
    g. $AgNO_3$
    h. $Fe(OH)_2$
    i. $Sn(NO_3)_2$
    j. $Co(NO_3)_2$
    k. $FePO_4$
    l. Hg_2SO_4
    m. $Hg_3(PO_4)_2$

5. Assign oxidation numbers to each atom in the following compounds:
    a. HI
    b. CO_2
    c. PBr_3
    d. GeS_2
    e. OF_2
    f. KH
    g. H_2O_2
    h. As_2O_5
    i. P_4O_{10}
    j. H_3PO_4

6. Assign oxidation numbers to each atom in the following ions:
    a. NO_3^-
    b. NH_4^+
    c. ClO_4^-
    d. CO_3^{2-}
    e. PO_4^{3-}
    f. MnO_4^-
    g. $S_2O_3^{2-}$
    h. $Cr_2O_7^{2-}$

▶ **Mole Relationships and Percent Composition**

7. Find the formula mass of each of the following:
    a. glucose, $C_6H_{12}O_6$
    b. calcium acetate, $Ca(C_2H_3O_2)_2$
    c. the ammonium ion, NH_4^+
    d. the chlorate ion, ClO_3^-

8. Determine the number of moles of ions and then the number of moles of each type of

atom present in one mole of each of the following:
a. KNO_3
b. Na_2SO_4
c. $Ca(OH)_2$
d. $(NH_4)_2SO_3$
e. $Ca_3(PO_4)_2$
f. $Al_2(CrO_4)_3$

9. Determine the molar mass of each compound listed in Problem 8.
10. What is the mass in grams of each of the following:
    a. 1.000 moles NaCl
    b. 2.000 moles H_2O
    c. 3.500 moles $Ca(OH)_2$
    d. 0.625 moles $Ba(NO_3)_2$
11. How many moles are represented by each of the following:
    a. 4.50 g H_2O
    b. 471.6 g $Ba(OH)_2$
    c. 129.68 g $Fe_3(PO_4)_2$
12. Find the percent composition of each compound:
    a. NaCl
    b. $AgNO_3$
    c. $Mg(OH)_2$
13. Find the percent of water in $CuSO_4 \cdot 5H_2O$.
14. Based on the percent composition information obtained in Problem 12, determine the number of grams of the specified element contained in each of the following:
    a. Na in 100.0 g NaCl
    b. O in 140.0 g $Mg(OH)_2$
15. Find the simplest formula of a compound containing 63.50% silver, 8.25% nitrogen, and 28.26% oxygen.
16. Find the simplest formula of a compound found to contain 52.20% carbon, 13.00% hydrogen, and 34.80% oxygen.
17. A 1.338 g sample of a compound contains 0.365 g of Na, 0.221 g of N, and 0.752 g of O. Determine its simplest formula.

CRITICAL THINKING

1. **Analyzing Information** The compound sulfur (VI) oxide is produced in the atmosphere through a reaction of sulfur dioxide and oxygen. Sulfur dioxide is a primary air pollutant. List all of the chemical information you can by analyzing the formula for sulfur (VI) oxide.

2. **Analyzing Data** In the laboratory a sample of pure nickel was placed in a clean, dried, and massed crucible. The crucible was heated so that the nickel would react with the oxygen in the air. After the reaction appeared complete, the crucible was allowed to cool and the mass determined. The crucible was heated, cooled, and its mass was determined again to be certain that the reaction was complete. The following data was collected:

    Mass of crucible = 30.02 g
    Mass of nickel and crucible = 31.01 g
    Mass of the nickel oxide compound and crucible = 31.36 g

    Perform the following calculations based on the data given above.

    Mass of nickel =
    Mass of nickel oxide compound =
    Mass of oxygen =

    Based upon your calculations, what is the empirical formula for the nickel oxide compound?

CROSS CONNECTIONS

1. **Biology** Biologists involved in the naming of newly discovered organisms use a system structured very much like chemists do in naming compounds. The system used by biologists is called the Linnaeus system after its creator, Carolus Linnaeus. Research this system in a biology textbook and then note similarities and differences between the Linnaeus system and the chemical nomenclature.

RESEARCH AND WRITING

1. Read the labels on various brands of multivitamin and multimineral tablets. Using both the common and systematic names, prepare a list of the vitamins contained in the tablets.

CHAPTER 8
Chemical Equations and Reactions

PLANNING CHART

Approximate Time: 12/15 days

SECTION/TEXT FEATURES		RESOURCES	
BASIC	**8.1 CHEMICAL EQUATIONS** 233 Demos 1, 2, 3 232A–C Section Review 244 Chapter Review 256 • Questions 1–8, Problems 1–7	*Teacher's Resource Binder* Section Review Worksheets *Other Resources* Transparency 21 Blackline Master 31	27
BASIC	**8.2 TYPES OF CHEMICAL REACTIONS** 244 Demos 4, 5 232C *Periodic Table* Videodisc 232H Desktop Investigation 249 Section Review 251 Chapter Review 256 • Questions 9–14, Problems 8–17	*Teacher's Resource Binder* Problem Solving Worksheets Section Review Worksheets	28
BASIC	**8.3 ACTIVITY SERIES OF THE ELEMENTS** 251 Demo 6 232D Section Review 255 Chapter Review 256 • Questions 15–21, Problems 18–21	*Teacher's Resource Binder* Lab Experiment 9 Problem Solving Worksheets Section Review Worksheets *Other Resources* Blackline Master 32	67 29
	CHAPTER END ACTIVITIES Critical Thinking, Cross Connections, Research and Writing, Projects	*Teacher's Resource Binder* Test Booklet *Computer Test Generator*	37, 39

Teaching Strategies

Introduction

An equation is a shorthand expression for describing a chemical change. Qualitative and quantitative information is contained in it. The equation not only describes what substances are consumed and what new substances are formed in a reaction, but also describes the relative amounts of reactant(s) and product(s).

It is important for students to understand just how powerful equations are. All of the chemical reactions that occur can be described by equations. To help students fully grasp the concept that equations represent actual phenomena that occur in the real world and not just on the pages of textbooks, it is important for them to have many experiences with chemical changes. Many of these can be done in the classroom as demonstrations, Desktop Investigations, or laboratory experiments.

8.1 Chemical Equations

Many chemical changes are part of everyday experience and students should become aware of these reactions: hydrocarbons are oxidized, iron rusts, and nonmetal acids dissolve in water, forming acid precipitation. Also there is an entire world of biochemical reactivity (see Chapter 23), some of which is neatly locked away in cells, but some of which is evident outside of cells: starch is hydrolyzed to sugar, CaO and CO_2 combine to form the $CaCO_3$ of eggshells, the proteins of our food are broken down to amino acids which are then used to form other proteins which are used in the body. Encourage students to determine from their own observations what substances are consumed and what substances are produced. You can facilitate this with the appropriate use of demonstrations.

Demonstration 1: A Single-Replacement Reaction

PURPOSE: To demonstrate that a metal (zinc) can replace an ion (Cu^{2+}) in solution and that the results of this reaction (elemental Cu) can be determined from easily observable experimental evidence

MATERIALS: 0.5 g of mossy zinc, 20 mL 0.1 *M* copper(II) sulfate solution, a 100-mm petri dish, an overhead projector

PROCEDURE: Have students examine the zinc ingot and record their observations. (At this point in the year, they will have to be reminded to record all observations carefully.) Pass the zinc around the room so that each student has the opportunity to observe its hardness, luster, approximate size. Ask students to observe and record the color of the copper sulfate solution.

Place a petri dish on an overhead projector. Pour approximately 25 mL of copper sulfate solution into the petri dish until the level of the solution is about 5 mm deep. Place the zinc in the solution. Observe the reaction that will begin immediately. (The zinc will begin to turn black. The color of the solution will begin to fade.)

Cover the petri dish and put aside until the following class. Observe the changes that have occurred. The solution is no longer blue due to reduction of the copper ion; the zinc will be obviously reduced in size; metallic copper will appear.

Encourage students to record all of their observations. Bolster their confidence in their ability to make observations and to interpret them. Although all of the students will see that the solution is no longer blue, they may be hesitant to record that information or to even consider that as a valid observation.

Tell students that the blue color of the solution is an indication of the presence of the copper ion. Alternatively, you can show students several stock solutions of copper(II) compounds and allow them to make that interpretation themselves.

Ask the class to explain the significance of the disappearance of the blue solution. Pass the covered petri dish around the room and ask students to determine what substance has been produced. Emphasize that they do have evidence to make that determination.

When the class has determined that the copper ions have "disappeared," that metallic copper has appeared, and that much of the zinc has "disappeared," write the following half-reactions on the chalkboard.

$$Cu^{2+} + electrons \rightarrow Cu$$
$$Zn \rightarrow Zn^{2+} + electrons$$

Although you do not want to formally introduce the concept of oxidation and reduction at this point, it is not too early to discuss:
1. Chemical reactions are based on a rearrangement of electrons.
2. In a chemical reaction, mass must be conserved.
3. In a chemical reaction, charge must be conserved.
4. In a chemical reaction, when electrons move from one substance to another, the number of electrons that leave one substance must be identical to the number of electrons that go to the other substance. This establishes the need to adjust coefficients in chemical equations.

When the following points have been established, write the complete equation on the board.

$$Zn + CuSO_4 \rightarrow ZnSO_4 + Cu$$

Additional demonstrations of single replacement reactions using this procedure can be very helpful. If a copper penny is placed in silver nitrate solution, the copper will replace the silver ion and the penny will become plated with silver. Since silver loses one electron and copper gains two, coefficients will have to be adjusted to balance the equation. A very attractive demonstration based on the Cu/Ag^+ reaction is to coil a piece of copper wire and place it upright in silver nitrate solution in a petri dish, covered with an inverted beaker. Glistening particles of the silver will adhere to the copper. This is a logical extension of the $Zn-Cu^{2+}$ demonstration.

A third demonstration could be the generation of hydrogen by the action of zinc on HCl or H_2SO_4. Since hydrogen is diatomic, this will be an opportune time to emphasize the importance of subscripts in representing the composition of elements as well as compounds.

Demonstration 2: The Formation of a Base from a Metal Oxide

PURPOSE: To demonstrate that evidence of a chemical reaction can be detected by indirect means
MATERIALS: Beaker, water, phenolphthalein, calcium oxide
PROCEDURE: Add a few drops of phenolphthalein to a 50-mL beaker that is half filled with water. Add approximately 0.005 g calcium oxide. [Hints: This simple demonstration can be done in a variety of ways. If you want to use small amounts of material, use a 50-mL beaker placed on an overhead projector. If you want to add a bit of drama to the demonstration, use a large beaker or cylinder placed on a magnetic stirrer. Add the calcium oxide as the stirrer is mixing the reactants at low speed.]

Demonstration 3: A Double-Replacement Reaction

PURPOSE: To demonstrate formation of a precipitate as evidence that chemical reaction has taken place
MATERIALS: 0.1 M $Pb(NO_3)_2$, 0.1 M NaCl, beaker, test tube, or petri dish
PROCEDURE: Mix together 5 mL of each solution. A precipitate will form immediately.

Double-replacement reactions resulting in the formation of a precipitate can be used extensively in this unit on balancing equations. Actually seeing the reaction take place is a stimulus to students to balance the equation. Continue to ask students how they know that a reaction has taken place.

To entice students to participate in the process, you might ask them to choose the color of the product in the reaction. By mixing a few mL of the appropriate stock solutions in a beaker or test tube, you can convince students that there is beauty and excitement in chemistry.

Color	Compound
White	PbI_2, $PbCl_2$, $BaCO_3$, $BaSO_4$
Blue	$Cu(OH)_2$
Green	$Ni(OH)_2$
Pink	$Co(OH)_2$, $MnCO_3$, $Mn(OH)_2$
Red	$Fe(SCN)_3$

Before proceeding to Section 8.2, students should be able to understand the following concepts:
1. A chemical equation represents a chemical reaction.
2. The reactants are written on the left side of the equation and the products are written on the right side of the equation.
3. A chemical equation must be balanced. When an equation is balanced, mass is conserved and charge is conserved.

$CoCl_2 + 2NaOH \rightarrow Co(OH)_2 + 2NaCl$

4. The formulas in a chemical reaction identify the reactants and the products. The subscripts as well as the symbols in the formula are unique to each compound. The subscripts in the formulas cannot be changed: coefficients have to be adjusted to balance the equation.
5. Atoms are rearranged during the course of a chemical reaction. The number of atoms in a single formula unit or molecule of a reactant is not necessarily the same as the number of atoms in a single molecule or formula unit of a product. (Because we emphasize that subscripts cannot be changed in order to balance an equation, some students have the misconception that the subscript must "stay with the symbol" and cannot be changed from reactant to product.)

$$Zn + 2HCl \rightarrow ZnCl_2 + H_2$$

This equation is a good example of that principle.
6. Even when a molecule of product contains the same number of atoms of a specific element as a molecule of reactant, the atoms are rearranged during the course of the reaction.

$$2H_2 + O_2 \rightarrow 2H_2O$$

Most students will recognize that, when water is formed from hydrogen and oxygen, the bonds between the oxygen are broken. But some will not realize that the bonds between the hydrogen are broken as well. Diagram this reaction on the board using structural formulas and use molecular models to show the structure of the hydrogen molecules, the oxygen molecule, and the water molecule.

8.2 Types of Chemical Reactions

Continue to do at least one simple demonstration each day as you proceed to teach students how to master the skill of equation balancing. Emphasize that the equation represents the chemical reaction. Students will develop a better understanding of this if they continue to have experience with many chemical reactions.

Introduce the concept that the equation contains quantitative information as well as qualitative information. The equation represents the relationship between moles and masses or reactants and products in a chemical reaction.

Translate the following equation into three sentences and have the class join you in chorus:

$$Pb(NO_3)_2 + 2NaCl \rightarrow PbCl_2 + 2NaNO_3$$

"One formula unit of lead nitrate reacts with two formula units of sodium chloride to produce one formula unit of lead chloride and two formula units of sodium nitrate."

"6.02×10^{23} formula units of lead nitrate react with 12.04×10^{23} formula units of sodium chloride to produce 6.02×10^{23} formula units of lead chloride and 12.04×10^{23} units of sodium nitrate."

"One mole of lead nitrate reacts with two moles of sodium chloride to produce one mole of lead chloride and two moles of sodium nitrate."

Continue to demonstrate as many of the reactions used in this chapter as possible. Most of your demonstrations will be quick, easy, and qualitative. Also take the time to do one or two quantitative demonstrations.

Demonstration 4: Ratios of Ions

PURPOSE: To determine the formula of a compound and to establish a balanced equation for the reaction

$$Co^{2+}(aq) + OH^-(aq) \rightarrow Co(OH)_2(s)$$

MATERIALS: 0.1 M $CoCl_2$, 0.1 M NaOH, phenolphthalein, 24-well microplate or 10–mL test tube, thin-stem micropipettes or medicine droppers. (The microplate and thin-stem pipettes are the basic pieces of equipment used for microchemistry techniques. Both types of equipment were developed by biologists and have been utilized by Lewis, Maloney, Russo, and others for student lab procedures that require less material and time than more traditional techniques. Microplates and thin-stem pipettes are available from Flinn and other scientific supply houses.)

PROCEDURE: *(1)* Add ten drops of cobalt chloride solution to a 10-mL test tube. (Alternatively, you can use a well in a 24-well microplate.) *(2)* Add one drop of phenolphthalein solution to the cobalt chloride solution. *(3)* Add NaOH solution, one drop at a time, carefully counting drops. The phenolphthalein indicator will turn to red when there is an excess of NaOH. Repeat steps 1–3 five times. If test tubes are used, they will have to be cleaned after each titration. If microplates are used, the five titrations can be done simultaneously in 5 different wells.

Determine the ratio of drops of sodium hydroxide to drops of cobalt chloride in each titration. Find the average ratio of the five titrations and round to the nearest whole number ratio.

The ratio of drops of NaOH to drops of $CoCl_2$ will indicate the relative number of OH ions to cobalt ions in the product. Write the balanced equation for this reaction based on the drop ratio determined above. Compare it to the correct equation for the reaction.

Demonstration 5: The Electrolysis of H_2O

PURPOSE: To show that the electrolytic decomposition of water produces two gases in the molar proportion of 2:1
MATERIALS: Hoffman apparatus, sodium sulfate, bromthymol blue, a 6 V/12 V d-c power supply
[Note: If platinum electrodes are available, use them in this demonstration in order to get the 2:1 gas volume ratio of hydrogen and oxygen that is predicted. Graphite electrodes can be used for this electrolysis but some adsorption of oxygen on the surface of the anode will interfere with quantitative results.]
PROCEDURE: In a large beaker on a magnetic stirrer prepare a dilute solution of sodium sulfate. Explain to the students that pure water is a very poor conductor of electricity. A small amount of sodium sulfate must be added to the water to conduct the current that will produce the decomposition reaction.

Add a generous amount of bromthymol blue. The water solution is expected to turn green. If your distilled water is not neutral, you may get a yellow or blue solution. Briefly explain the nature of the indicator and the reason for the color that you observe. Explain that you are using the indicator to develop a better understanding of the mechanism of the reaction.

Attach your power supply to your electrodes and begin your electrolysis. A 6 V/12 V battery charger is an excellent source of electrical energy and will give you two distinct reaction rates. Take this opportunity to tell your class that water does not decompose spontaneously but requires an input of energy for the reaction to procede, in this case that energy is electrical energy.

Since heating water results in the formation of a vapor, some students might confuse the physical change associated with adding thermal energy to water with the chemical change associated with adding electrical energy. You can demonstrate the difference experimentally by collecting the gases produced in electrolysis and the vapor produced by boiling and performing a glowing splint test.

Results: $2H_2O + \text{electrical energy} \rightarrow 2H_2 + O_2$

As soon as the current is applied, bubbles of hydrogen will form at the cathode and bubbles of oxygen will appear at the anode. The volume of hydrogen produced will be twice the volume of oxygen produced, as predicted.

The solution surrounding the anode will become yellow due to the presence of hydronium ions produced, and the solution surrounding the cathode will turn blue due to the presence of the hydroxide ions produced.

Anode: $6H_2O \rightarrow O_2 + 4H_3O^+(aq) + 4e^-$
Cathode: $4H_2O + 4e^- \rightarrow 2H_2 + 4OH^-(aq)$

(Do not show reactions with an unequal number of electrons; that is why you need $4H_2O$ at the cathode.)

Overall: $10H_2O(\ell) \rightarrow 2H_2(g) + O_2(g) + 4H_3O^+(aq) + 4OH^-$

This is not an appropriate time to explain the anode and the cathode reactions in detail. It is sufficient to review the simplified reaction, to state that some acid is produced at the anode and some base is produced at the cathode, and that the reaction will be studied again in more detail. Use this demonstration again when gas laws are introduced, and a third time when oxidation-reduction is formally presented. It will be helpful to students to have some familiarity with this reaction when new concepts are introduced. Chemistry is not a fragmented science: one concept builds on another. Using the same demonstration to present several concepts illustrates that continuity.

8.3 Activity Series of the Elements

Demonstration 6: Reactions in an Activity Series

PURPOSE: To determine the relative ease with which metal atoms lose electrons to form ions in aqueous solutions, to give students an opportunity to determine the products of a reaction from a knowledge of the reactants and from the observations of a reaction, and to introduce students to the qualitative concept of rate through observations of slow and fast reactions

MATERIALS: Metals: Al, Zn, Fe, Co, Ni, Pb, Cu; CAUTION: several of these metals are hazardous when powdered. Avoid making metal dust. Solutions: 0.1 M solutions of the following: $Al(NO_3)_3$, $Zn(NO_3)_2$, $Fe(NO_3)_3$, $Co(NO_3)_2$, $Ni(NO_3)_2$, $Pb(NO_3)_2$, $Cu(NO_3)_2$, and HCl; 49 small containers: 50–mL beakers, petri dishes, or baby food jars

[Many of the chemicals used in this investigation are available in hardware stores and supermarkets so that students can continue this investigation at home. Baby food jars can be used as reaction containers.]

PROCEDURE: Have students prepare a grid as indicated.

Activity Series

TIME	Al/Al^{3+}	Zn/Al^{3+}	Fe/Al^{3+}	Co/Al^{3+}	Ni/Al^{3+}	Pb/Al^{3+}	Cu/Al^{3+}
Three Min	N.R.	N.R.	N.R.	N.R.	N.R.	N.R.	N.R.
Thirty Min							
One Day							

TIME	Al/Zn^{2+}	Zn/Zn^{2+}	Fe/Zn^{2+}	Co/Zn^{2+}	Ni/Zn^{2+}	Pb/Zn^{2+}	Cu/Zn^{2+}
Three Min							
Thirty Min							
One Day							

TIME	Al/Fe^{3+}	Zn/Fe^{3+}	Fe/Fe^{3+}	Co/Fe^{3+}	Ni/Fe^{3+}	Pb/Fe^{3+}	Cu/Fe^{3+}
Three Min							
Thirty Min							
One Day							

TIME	Al/Co^{2+}	Zn/Co^{2+}	Fe/Co^{2+}	Co/Co^{2+}	Ni/Co^{2+}	Pb/Co^{2+}	Cu/Co^{2+}
Three Min							
Thirty Min							
One Day							

TIME	Al/Ni^{2+}	Zn/Ni^{2+}	Fe/Ni^{2+}	Co/Ni^{2+}	Ni/Ni^{2+}	Pb/Ni^{2+}	Cu/Ni^{2+}
Three Min							
Thirty Min							
One Day							

TIME	Al/H^+	Zn/H^+	Fe/H^+	Co/H^+	Ni/H^+	Pb/H^+	Cu/H^+
Three Min							
Thirty Min							
One Day							

TIME	Al/Cu^{2+}	Zn/Cu^{2+}	Fe/Cu^{2+}	Co/Cu^{2+}	Ni/Cu^{2+}	Pb/Cu^{2+}	Cu/Cu^{2+}
Three Min							
Thirty Min							
One Day							

Label containers according to the solutions indicated on the grid sheet. Pour 15 mL of each solution in the appropriately labeled container. If you are doing this as a demonstration, use 50-mL beakers or small petri dishes.

If students are doing this as an activity, test tubes, or baby food jars can be used. This procedure can also be followed using microchem techniques in 2 24-well microplates if 5 mL of solution is used in each well.

Add approximately 0.5 grams of the metal indicated to the appropriately labeled container. Record observations within the first 5 minutes of the reaction, thirty minutes later, and on the following day.

It is not necessary to do all of this investigation at one time. It is important for students to keep an accurate record of the results of the reaction even if you choose to do the demonstration over a period of several weeks.

Students can construct a fairly accurate activity series based on their observations of these reactions by doing the following:
1. By counting the number of reactions in each *column* of the record sheet, tabulate the number of ions that each metal replaces.
2. Based on the number of reactions for each metal, rate each of the metals from the greatest to the least number of ions it will replace. In case of a tie, consider the rate of the reactions.
3. By counting the number of reactions in each *row* of the record sheet, tabulate the number of metals that will replace each ion.
4. Based on the number of reactions for each ion, rate the ions from the highest to the lowest tendency to attract electrons. In case of a tie, consider the rate of the reactions.
5. Compare the two lists.
6. Of the metals tested, which one has the highest tendency to donate electrons to ions tested?
7. Of the metals tested, which one has the lowest tendency to donate electrons to ions tested?
8. Of the ions tested, which one has the highest tendency to attract electrons from the metals tested?
9. Of the ions tested, which one has the lowest tendency to attract electrons from the metals tested?
10. Based on an element's relatively high tendency to donate electrons to ions when that element is in its ionic state, construct an activity series of elements from greatest reducing power to lowest.

References and Resources

Books and Periodicals

Festa, Roger. "Joseph Priestley, Preeminent Amateur Chemist," *Journal of Chemical Education*, Vol. 64, No. 9, 1987, pp. 745–747.

Computer Software

Smith, Stanley G., Ruth Chabay, and Elizabeth Keen. *Chemical Formulas and Equations*, Apple II, IBM, COMPress.

Weyh, John A., Joseph R. Crook, and Les N. Hauge. *Chemical Reactions*, IBM, COMPress.

Answers and Solutions

Reviewing Concepts

1. A correctly written chemical equation must represent the known facts, contain the correct formulas of the reactants and products, and satisfy the law of conservation of matter.
2. **a.** A coefficient is a number that appears in front of a formula in a chemical equation.
   b. The number of atoms of each type in the formula is multiplied by the coefficient.
3. An example of a word equation is:
   methane + oxygen → carbon dioxide + water. A formula equation is $CH_4(g) + O_2(g) \rightarrow CO_2(g) + H_2O(g)$. A chemical equation is:
   $CH_4(g) + 2O_2(g) \rightarrow CO_2(g) + 2H_2O(g)$.
4. Both word and formula equations have only qualitative meaning; they do not express the quantities of reactants used or products formed.
5. **a.** An aqueous solution is a solution in water.
   b. A catalyst is a substance that accelerates a chemical reaction but can be recovered unchanged.
   c. A reversible reaction is a chemical reaction in which the products re-form the original reactants.
6. The indicated formulas are:

   a. KOH
   b. $Ca(NO_3)_2$
   c. Na_2CO_3
   d. CCl_4
   e. $MgBr_2$
   f. SO_2
   g. $(NH_4)_2SO_4$.

7. Four useful guidelines are:
   • balance the different types of atoms one at a time,
   • first balance the types of atoms that appear only once on each side of the equation,
   • balance polyatomic ions that appear on both sides of the equation as single units, and
   • balance H atoms and O atoms last.

8. The number of atoms indicated in each is:
   a. $N = 6$
   b. $H = 4, O = 2$
   c. $H = 4, N = 4, O = 12$
   d. $Ca = 2, O = 4, H = 4$
   e. $Ba = 3, Cl = 6, O = 18$
   f. $Fe = 5, N = 10, O = 30$
   g. $Mg = 12, P = 8, O = 32$
   h. $N = 4, H = 16, S = 2, O = 8$
   i. $Al = 12, C = 18, O = 54$.

9. • In synthesis, or composition, reactions, two or more substances combine to form a new substance. Equation: $A + X \rightarrow AX$.

- In decomposition reactions, a single compound undergoes a reaction that produces two or more simpler substances. Equation: AX → A + X.
- In single-replacement, or displacement, reactions, one element replaces another element in a compound. Equation: A + BX → AX + B or Y + BX → BY + X.
- In double-replacement, or ionic, reactions, the ions of two compounds exchange places in an aqueous solution to form two new compounds. Equation: AX + BY → AY + BX.
- In combustion reactions, a substance combines with oxygen, releasing a large amount of energy in the form of light and heat. Equation: $C_3H_8(g) + 5O_2(g) \rightarrow 3CO_2(g) + 4H_2O(g)$.

10. Most decomposition reactions take place only when energy in the form of heat or electricity is added.
11. Electrolysis is the decomposition of a substance by an electric current.
12. **a.** Most single-replacement reactions take place in aqueous solutions.
    b. Single replacement reactions usually involve less energy than do synthesis or decomposition reactions.
13. Hydrocarbons are compounds made up of carbon and hydrogen only.
14. The products are carbon dioxide and water.
15. **a.** The ability of an element to react
    b. The greater the activity of an element, the more readily it reacts.
16. **a.** An activity series is a set of elements listed according to the ease with which they undergo certain chemical reactions.
    b. The elements are usually listed in an order determined by a tendency to undergo single-replacement reactions. Any element in such a series can replace an element below it.
17. **a.** The activity series of metals is based on the ease with which metal atoms lose electrons to form ions in aqueous solution; the higher a metal is in the series, the more easily its atoms lose electrons, and thus the greater the reactivity of that metal.
    b. The greater the distance between two metals in the activity series is, the more likely it is that one will replace the other in a reaction.
18. - An element will replace from a compound in aqueous solution any of those elements below it in the activity series; the larger the interval between the elements in the activity series, the greater is the tendency for the replacement reaction to occur.
    - Any metal above magnesium will replace hydrogen from water.
    - Any metal above cobalt will replace hydrogen from steam.
    - Any metal above hydrogen will react with acids, replacing hydrogen.
19. **a.** The higher a metal is in the activity series, the more readily it will undergo synthesis with oxygen.
    b. Any metal above silver reacts with oxygen, forming oxides; those near the top react rapidly. Any metal below mercury forms oxides only indirectly (not by direct reaction with O_2).
20. **a.** The more active a metal, the more strongly it holds onto oxygen in an oxide, and, therefore, the more strongly the oxide resists decomposition into its elements upon heating.
    b. Oxides of metals below copper decompose with heat alone, oxides of metals below chromium yield free metals when heated with hydrogen, and oxides of metals above iron resist conversion to free metals when heated with hydrogen.
21. - Elements near the top of the activity series are never found free in nature.
    - Elements near the bottom of the series are often found free in nature.

Problems

1. **a.** $3.0 \text{ mol } O_2 \times \dfrac{4 \text{ mol } NH_3}{3 \text{ mol } O_2} = 4.0 \text{ mol } NH_3$

    $3.0 \text{ mol } O_2 \times \dfrac{2 \text{ mol } N_2}{3 \text{ mol } O_2} = 2.0 \text{ mol } N_2$

    $3.0 \text{ mol } O_2 \times \dfrac{6 \text{ mol } H_2O}{3 \text{ mol } O_2} = 6.0 \text{ mol } H_2O$

    b. $8.0 \text{ mol } NH_3 \times \dfrac{3 \text{ mol } O_2}{4 \text{ mol } NH_3} = 6.0 \text{ mol } O_2$

    $8.0 \text{ mol } NH_3 \times \dfrac{2 \text{ mol } N_2}{4 \text{ mol } NH_3} = 4.0 \text{ mol } N_2$

    $8.0 \text{ mol } NH_3 \times \dfrac{6 \text{ mol } H_2O}{4 \text{ mol } NH_3} = 12 \text{ mol } H_2O$

    c. $1.0 \text{ mol } N_2 \times \dfrac{4 \text{ mol } NH_3}{2 \text{ mol } N_2} = 2.0 \text{ mol } NH_3$

    $1.0 \text{ mol } N_2 \times \dfrac{3 \text{ mol } O_2}{2 \text{ mol } N_2} = 1.5 \text{ mol } O_2$

    $1.0 \text{ mol } N_2 \times \dfrac{6 \text{ mol } H_2O}{2 \text{ mol } N_2} = 3.0 \text{ mol } H_2O$

    d. $0.40 \text{ mol } H_2O \times \dfrac{4 \text{ mol } NH_3}{6 \text{ mol } H_2O} = 0.27 \text{ mol } NH_3$

    $0.40 \text{ mol } H_2O \times \dfrac{3 \text{ mol } O_2}{6 \text{ mol } H_2O} = 0.20 \text{ mol } O_2$

    $0.40 \text{ mol } H_2O \times \dfrac{2 \text{ mol } N_2}{6 \text{ mol } H_2O} = 0.13 \text{ mol } N_2$

2. **a.** $2KBr + BaI_2 \rightarrow 2KI + BaBr_2$
    b. $2HCl + Mg(OH)_2 \rightarrow MgCl_2 + 2H_2O$
    c. $2HNO_3 + Ca(OH)_2 \rightarrow Ca(NO_3)_2 + 2H_2O$

3. **a.** $H_2 + Cl_2 \rightarrow 2HCl$
    b. $2Al + Fe_2O_3 \rightarrow Al_2O_3 + 2Fe$
    c. $Pb(C_2H_3O_2)_2 + H_2S \rightarrow PbS + 2HC_2H_3O_2$

4. **a.** $Ca(OH)_2 + (NH_4)_2SO_4 \rightarrow CaSO_4 + 2NH_3 + 2H_2O$
    b. $2C_2H_6 + 7O_2 \rightarrow 4CO_2 + 6H_2O$
    c. $2Cu_2S + 3O_2 \rightarrow 2Cu_2O + 2SO_2$
    d. $2NaCl + 2H_2SO_4 + MnO_2 \rightarrow MnSO_4 + Na_2SO_4 + Cl_2 + 2H_2O$

5. **a.** LiO_2 is an incorrect formula. $4Li + O_2 \rightarrow 2Li_2O$
    b. H_2Cl_2 is an incorrect formula. $H_2 + Cl_2 \rightarrow 2HCl$

c. MgO_2 is an incorrect formula, and the equation as written is not balanced. $MgCO_3 \rightarrow MgO + CO_2$
d. I is an incorrect formula (diatomic I_2), and the equation is not balanced. $2NaI + Cl_2 \rightarrow 2NaCl + I_2$

6. a. $4Al + 3O_2 \rightarrow 2Al_2O_3$
   b. $P_4O_{10} + 6H_2O \rightarrow 4H_3PO_4$
   c. $Fe_2O_3 + 3CO \rightarrow 2Fe + 3CO_2$

7. The correct formulas are:
   a. Fe_2O_3 b. NH_3 c. P_2O_5 d. NO_2

8. a. $2Mg + O_2 \rightarrow 2MgO$
   b. $4Fe + 3O_2 \rightarrow 2Fe_2O_3$
   c. $2Li + Cl_2 \rightarrow 2LiCl$
   d. $Ca + I_2 \rightarrow CaI_2$

9. a. sodium + oxygen → sodium oxide
      $4Na + O_2 \rightarrow 2Na_2O$
   b. copper + oxygen → copper(I) oxide
      $4Cu + O_2 \rightarrow 2Cu_2O$
      copper + oxygen → copper(II) oxide
      $2Cu + O_2 \rightarrow 2CuO$
   c. potassium + chlorine → potassium chloride
      $2K + Cl_2 \rightarrow 2KCl$
   d. magnesium + fluorine → magnesium fluoride
      $Mg + F_2 \rightarrow MgF_2$

10. a. $2HgO \xrightarrow{\Delta} 2Hg + O_2$
    b. $2H_2O(\ell) \xrightarrow{electric\ current} 2H_2(g) + O_2(g)$
    c. $2Ag_2O \xrightarrow{\Delta} 4Ag + O_2$
    d. $MgCl_2 \xrightarrow{electric\ current} Mg + Cl_2$

11. a. $Zn + Pb(NO_3)_2 \rightarrow Pb + Zn(NO_3)_2$
    b. $2K + Ba(C_2H_3O_2)_2 \rightarrow Ba + 2KC_2H_3O_2$
    c. $2Al + 3NiSO_4 \rightarrow 3Ni + Al_2(SO_4)_3$
    d. $2Na + 2H_2O \rightarrow 2NaOH + H_2$

12. a. $AgNO_3(aq) + NaCl(aq) \rightarrow NaNO_3(aq) + AgCl(s)$
    b. $Mg(NO_3)_2(aq) + 2KOH(aq) \rightarrow 2KNO_3(aq) + Mg(OH)_2(s)$
    c. $(NH_4)_2S(aq) + ZnCl_2(aq) \rightarrow 2NH_4Cl(aq) + ZnS(s)$
    d. $3LiOH(aq) + Fe(NO_3)_3(aq) \rightarrow 3LiNO_3(aq) + Fe(OH)_3(s)$

13. a. $CH_4 + 2O_2 \rightarrow CO_2 + 2H_2O$
    b. $2C_3H_6 + 9O_2 \rightarrow 6CO_2 + 6H_2O$
    c. $C_5H_{12} + 8O_2 \rightarrow 5CO_2 + 6H_2O$

14. a. $H_2 + I_2 \rightarrow 2HI$; synthesis
    b. $2Li + 2HCl \rightarrow 2LiCl + H_2$; single replacement
    c. $Na_2CO_3 \rightarrow Na_2O + CO_2$; decomposition
    d. $2HgO \rightarrow 2Hg + O_2$; decomposition
    e. $Mg(OH)_2 \rightarrow MgO + H_2O$; decomposition
    f. $Cu + Cl_2 \rightarrow CuCl_2$; synthesis
    g. $Ca(ClO_3)_2 \rightarrow CaCl_2 + 3O_2$; decomposition
    h. $2Li + 2H_2O \rightarrow 2LiOH + H_2$; single replacement
    i. $PbCO_3 \rightarrow PbO + CO_2$; decomposition

15. a. $Mg(OH)_2 \rightarrow MgO + H_2O$
    b. $Pb(OH)_2 \rightarrow PbO + H_2O$
    c. $2LiClO_3 \rightarrow 2LiCl + 3O_2$
    d. $Ba(ClO_3)_2 \rightarrow BaCl_2 + 3O_2$
    e. $Ni(ClO_3)_2 \rightarrow NiCl_2 + 3O_2$

16. a. $C_3H_8 + 5O_2 \rightarrow 3CO_2 + 4H_2O$
    b. $2C_4H_{10} + 13O_2 \rightarrow 8CO_2 + 10H_2O$
    c. $C_5H_{12} + 8O_2 \rightarrow 5CO_2 + 6H_2O$

17. a. $Zn + S \rightarrow ZnS$; synthesis
    b. $Ca + 2NaNO_3 \rightarrow 2Na + Ca(NO_3)_2$; single replacement
    c. $2AgNO_3 + K_2S \rightarrow 2KNO_3 + Ag_2S$; double replacement
    d. $2NaI \rightarrow 2Na + I_2$; decomposition
    e. $2C_6H_6 + 15O_2 \rightarrow 12CO_2 + 6H_2O$; combustion
    f. $2C_{10}H_{22} + 31O_2 \rightarrow 20CO_2 + 22H_2O$; combustion

18. a. $2K(s) + CuCl_2(aq) \rightarrow 2KCl(aq) + Cu(s)$
    b. $Zn(s) + Pb(NO_3)_2(aq) \rightarrow Zn(NO_3)_2(aq) + Pb(s)$
    c. $Cl_2(g) + 2KI(aq) \rightarrow 2KCl(aq) + I_2(s)$
    d. no reaction
    e. $Ba(s) + 2H_2O(\ell) \rightarrow Ba(OH)_2(aq) + H_2(g)$

19. a. $2Ca(s) + O_2(g) \rightarrow 2CaO(s)$
    b. $2Ni(s) + O_2(g) \rightarrow 2NiO(s)$
    c. no reaction
    d. $4Al(s) + 3O_2(g) \rightarrow 2Al_2O_3(s)$

20. a. $2Ag_2O(s) \rightarrow 4Ag(s) + O_2(g)$
    b. no reaction
    c. no reaction
    d. $CuO(s) + H_2(g) \rightarrow Cu(s) + H_2O(\ell)$

21. The indicated element in each case is:
    a. K   b. Al   c. Cr   d. F_2
    e. Ag   f. Cl_2 g. Sr   h. F_2.

Critical Thinking

1. A chemical formula represents the atoms that are bonded together to form the compound. The subscripts indicate the number of atoms of a particular element in the compound. Compounds obtain their chemical and physical properties from the arrangement of and type of atoms that form them. Changing the subscript is adding or removing atoms that results in a change of chemical and physical properties, thus their identity.

2. The most reactive metals are located on the left side of the chart in Groups 1 and 2. These metals have 1 (Group 1) or 2 (Group 2) electrons in their outer energy level. The energy requirement to achieve a stable configuration is much less to lose the electrons rather than to gain electrons. The least reactive metals are in the lower right of the transition metals. The d sublevel of the 5th energy level is more than half full in these. They are much more stable than metals on the left side of the periodic table.

3. Q—most reactive
   A
   Z
   J—least reactive

4. Answers will vary: Generally experiments should be planned to observe reactions between each metal and each solution. One method of doing this is to obtain 3 samples of each metal and to prepare 3 test tubes of each solution. Each metal would then be dropped into a test tube of each solution except for the solution containing its own ion.

Cross Connections

1. $59.75

Videodisc Selections from
The Periodic Table Videodisc (JCE Software)

step forward

step reverse

CHAPTER 8

pg 244 Reactions of elements with oxygen

115
hydrogen + oxygen (m)

7535
magnesium + oxygen (m)

18438
calcium + oxygen (m)

37311
strontium + oxygen (m)

46119
barium + oxygen (m)

20759
titanium + oxygen (m)

22764
manganese + oxygen (m)

28546
copper + oxygen (m)

8267
aluminum + oxygen (m)

3835
carbon + oxygen (m)

12319
phosphorus + oxygen (m)

13145
sulfur + oxygen (m)

33890
selenium + oxygen (m)

Teacher's Notes

Chapter 8
Chemical Equations and Reactions

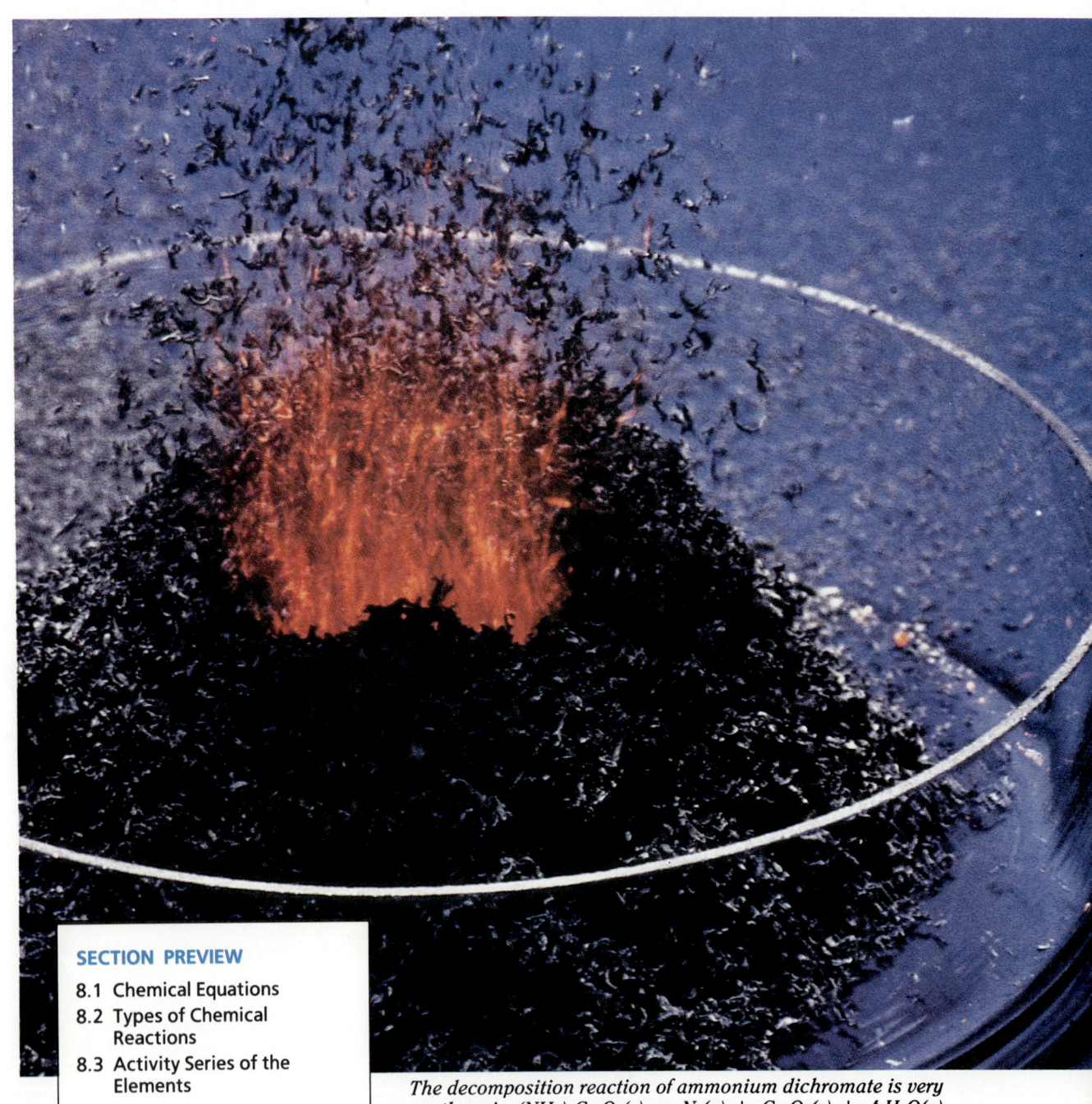

SECTION PREVIEW
- 8.1 Chemical Equations
- 8.2 Types of Chemical Reactions
- 8.3 Activity Series of the Elements

The decomposition reaction of ammonium dichromate is very exothermic. $(NH_4)_2Cr_2O_7(s) \rightarrow N_2(g) + Cr_2O_3(s) + 4\,H_2O(g)$

8.1 Chemical Equations

Chemical reactions are described by chemical equations. For example, the formulas of reactants and products can be written in the following form:

$$Fe_2O_3 + 2Al \rightarrow 2Fe + Al_2O_3$$

This equation shows that the *reactants* iron(III) oxide and aluminum yield the *products* iron and aluminum oxide. This strongly exothermic reaction is shown in the photograph on the preceding page. When fully understood, every chemical change can be summarized by a chemical equation.

A **chemical equation** *represents, with symbols and formulas, the reactants and products in a chemical reaction.*

Reading and Writing Chemical Equations

A chemical equation is of value only if it is correct in all details. The requirements outlined below must be met by every chemical equation.

Requirements for chemical equations Whenever you read or write a chemical equation, you must keep the following requirements in mind.

1. The equation must represent the known facts. All the reactants and products must be identified. This information is obtained by chemical analysis in the laboratory or from sources that give the results of experiments.

2. The equation must contain the correct formulas of the reactants and products. To read or write a chemical equation, you must apply what you learned in Chapter 7 about symbols and formulas. Knowledge of the common oxidation states of the elements and methods of writing formulas will enable you to supply formulas for reactants and products if they are not available. Recall that the elements listed in Table 8-1 are found primarily as *diatomic molecules*, such as H_2 and O_2. Each of these elements must be represented in an equation by its molecular formula. Other elements in the elemental state are usually represented by their atomic symbols without any subscripts. In the equation that appears at the beginning of this section, for example, iron (Fe) and aluminum (Al) were represented in this manner.

3. The law of conservation of mass must be satisfied. The law of conservation of mass was introduced in Chapter 1. Atoms are neither

SECTION OBJECTIVES

- List three requirements for a correctly written chemical equation.
- Translate chemical equations into sentences.
- Write a word equation and a formula equation, given a description of a chemical reaction.
- List three things you can determine about chemical reactants and products from a chemical equation.
- Balance a formula equation by inspection.

Reactants and products are identified by formulas. All compounds contain more than one element.

Some elements exist as polyatomic molecules. Their formulas must represent this (e.g., S_8, P_4, O_2, N_2, F_2, Cl_2, Br_2, I_2, and H_2).

TABLE 8-1 ELEMENTS THAT NORMALLY EXIST AS DIATOMIC MOLECULES

Element	Symbol	Molecular Formula	Physical State at Room Temperature
hydrogen	H	H_2	gas
nitrogen	N	N_2	gas
oxygen	O	O_2	gas
fluorine	F	F_2	gas
chlorine	Cl	Cl_2	gas
bromine	Br	Br_2	liquid
iodine	I	I_2	solid

Chemical Equations and Reactions

Coefficients that are simple fractions ($\frac{1}{2}$, $\frac{1}{3}$, . . .) are sometimes used for special purposes (e.g., in Chapter 18).

A subscript is a small whole number that represents how many atoms of each element are in molecules and formula units. Subscripts are assigned when the identity of the reactants and products are determined. Once assigned, they cannot be changed.

created nor destroyed in ordinary chemical reactions. The same number of atoms of each element must appear on each side of a correctly written chemical equation. To equalize numbers of atoms, coefficients are added where necessary. A **coefficient** *is a small whole number that appears in front of a formula in a chemical equation.* Two coefficients of 2 appear in the equation written at the beginning of this section. Placing a coefficient in front of a formula specifies the relative number of moles of the substance.

Word and formula equations The first step in writing a chemical equation is identifying the facts to be represented. It may be helpful to write a **word equation**, *an equation in which the reactants and products in a chemical reaction are represented by words.*

As an example, consider the reaction of methane with oxygen. Methane is the principal component of natural gas. Experiments show that when methane burns in air, it combines with oxygen to produce carbon dioxide and water. Thus, methane and oxygen are the reactants, and carbon dioxide and water are the products.

The word equation for the reaction of methane and oxygen is

$$\text{methane} + \text{oxygen} \rightarrow \text{carbon dioxide} + \text{water}$$

This verbal equation is a brief statement of the experimental facts. A word equation does not give the quantities of reactants used or products formed; thus, it does not tell the whole story. A word equation has only qualitative (descriptive) meaning. The arrow → is read as "react to yield," or "yield." The equation above is read, "Methane and oxygen react to yield carbon dioxide and water," or simply, "Methane and oxygen yield carbon dioxide and water."

The next step in writing a chemical equation is to replace the names of the reactants and products with symbols and formulas. Methane (CH_4) is a molecular compound composed of one carbon atom and four hydrogen atoms. Oxygen is one of the elements that exist as diatomic molecules; it is represented as O_2. The correct formulas for carbon dioxide and water are CO_2 and H_2O.

A **formula equation** *represents the reactants and products of a chemical reaction by their symbols or formulas.* The formula equation for the reaction of methane and oxygen is

$$CH_4(g) + O_2(g) \rightarrow CO_2(g) + H_2O(g) \text{ (not balanced)}$$

The *g*'s in parentheses after the four formulas indicate that the corresponding substances are in the gaseous state. Like a word equation, a formula equation gives no information about amounts of reactants or products. A formula equation is also a qualitative statement.

The formula equation meets two of the three requirements for a correct chemical equation by representing the facts and showing the correct symbols and formulas for the reactants and products. To complete the process of writing a correct equation, the law of conservation of matter must be taken into account. The relative amounts of reactants and products represented in the equation must be adjusted so that the number and type of atoms is the same on both sides of the equation. This process, called "balancing an equation," is carried out by inserting coefficients. Once it is balanced, a formula equation is a correctly written chemical equation.

Look again at the formula equation for the combustion of methane:

$$CH_4(g) + O_2(g) \rightarrow CO_2(g) + H_2O(g) \text{ (not balanced)}$$

Counting hydrogen atoms shows that there are four hydrogen atoms in the reactants but only two in the products. Two hydrogen atoms are therefore needed on the right side of the equation to balance it. They can be added by placing the coefficient 2 in front of the chemical formula H_2O. The number of atoms of each type is multiplied by the coefficient. Thus, $2H_2O$ represents *four* H atoms and *two* O atoms. If no coefficient is written, the coefficient is assumed to be *one*. By placing a 2 in front of the formula for water in the equation for the combustion of methane, we can balance the H atoms.

$$CH_4(g) + O_2(g) \rightarrow CO_2(g) + 2H_2O(g) \text{ (partially balanced)}$$

You may have considered changing the subscript in the formula of water from H_2O to H_4O to add two more hydrogen atoms to the right side of the equation, but this would have been a mistake. Subscripts in chemical formulas must not be changed while balancing an equation. When you change the subscripts of a chemical formula, you change the very *identity* of the compound. The compound H_4O is not a product in the combustion of methane (in fact, there is no such compound.) Changing subscripts of compounds is not the correct way to balance an equation. Only inserting coefficients is appropriate because, while doing so specifies the relative amounts of reactants and products, it does not change their identity.

We must next count the number of oxygen atoms. There are now four oxygen atoms on the right of the arrow in the partially balanced equation. Yet there are only two oxygen atoms on the left of the arrow in the reactants. By placing the coefficient 2 in front of the formula for oxygen on the left side of the equation, we can increase the number of oxygen atoms to four. The correct chemical equation for the combustion of methane—*a balanced formula equation*—can be summarized in Figure 8-1 as

$$CH_4(g) + 2O_2(g) \rightarrow CO_2(g) + 2H_2O(g)$$

How can you be certain that the coefficients are correct? You could confirm them by carefully carrying out the reaction in the laboratory.

Figure 8-1 The reaction of methane with oxygen to give carbon dioxide and water. The molecular models illustrate that, in a balanced formula equation (as required by the law of conservation of matter), the number of atoms of each element in the reactants equals the number of atoms of each element in the products.

TABLE 8-2 SYMBOLS USED IN CHEMICAL EQUATIONS

Symbol	Explanation
$\rightarrow$	"Yields" indicates result of reaction.
$\rightleftarrows$	Used in place of a single arrow for reversible reaction
(s)	A reactant or product in the solid state
$\downarrow$	Alternative to (s); used only for a precipitate (a solid) formed in a solution
(ℓ)	A reactant or product in the liquid state
(aq)	A reactant or product in an aqueous solution (dissolved in water)
(g)	A reactant or product in the gaseous state
$\uparrow$	Alternative to (g); used only for a gaseous product
$\xrightarrow{\Delta}$ or $\xrightarrow{heat}$	Reactants are heated
$\xrightarrow{2\ atm}$	Pressure at which reaction is carried out, in this case 2 atm
$\xrightarrow{pressure}$	Pressure exceeding normal atmospheric pressure
$\xrightarrow{0°C}$	Temperature at which reaction is carried out, in this case 0°C
$\xrightarrow{MnO_2}$	Formula of catalyst used to alter the rate of the reaction, in this case, manganese dioxide

> Sometimes a solid (s) is represented by the letter "c," which comes from the Latin expression for solid food: *cibus firmus.*

Experiments are unnecessary, however. If the values are consistent with the law of conservation of matter, the coefficients must be correct. Table 8-2 summarizes symbols that are used in chemical equations. Sometimes a gaseous product is indicated by an arrow pointing upwards, $\uparrow$, instead of (g), as shown above. A downward arrow, $\downarrow$, is sometimes used to show the formation of a precipitate during a reaction in aqueous solution.

The conditions under which a reaction takes place are often indicated by placing information above or below the reaction arrow. The word *heat*, symbolized by a capital Greek delta (Δ) indicates that the reactants must be heated. The specific temperature at which a reaction occurs may also be written over the arrow. For some reactions, it is important to specify the pressure at which the reaction occurs or that the pressure is above normal. Many reactions are speeded up and can take place at lower temperatures in the presence of a catalyst (a substance that changes the rate of a chemical reaction but can be recovered unchanged). To show that a catalyst must be present, the formula for the catalyst (or the word *catalyst*) is written over the reaction arrow.

In many reactions, once products have begun to form, they immediately begin to react to give back the reactants. In other words, the reverse reaction also occurs. Depending upon the specific reaction and the conditions, this may occur to a greater or lesser extent. A **reversible reaction** *is a chemical reaction in which the products re-form the original reactants.* The reversibility of a reaction such as that between iron and steam is indicated by writing two arrows pointing in opposite directions.

$$3Fe(s) + 4H_2O(g) \rightleftarrows Fe_3O_4(s) + 4H_2(g)$$

> Reversible reactions, which are common and important, are discussed in Chapter 19.

In the late 18th century, this reaction was commonly used to prepare hydrogen, by passing the steam through a hot gun barrel.

With a knowledge of all the symbols and formulas used, it is possible to translate a chemical equation into a sentence. For example, the equation

$$2HgO(s) \xrightarrow{\Delta} 2Hg(\ell) + O_2(g)$$

can be translated in this manner: When heated, solid mercury(II) oxide yields liquid mercury and gaseous oxygen. The equation

$$C_2H_4(g) + H_2(g) \xrightarrow[Pt]{pressure} C_2H_6(g)$$

is read as follows: Under pressure and in the presence of a platinum catalyst, gaseous ethene and hydrogen form gaseous ethane.

> The arrow in a chemical reaction is read as "gives," "yields," "produces," or "forms." The plus sign in a chemical reaction does not imply mathematical addition. The + sign on the reactant side of the equation can be read as "reacts with."

Throughout this chapter we will include the symbols for phases (i.e., s, ℓ, and g) in balanced formula equations. You should be able to interpret these symbols when they are used and supply them when the necessary information is available.

Sample Problem 8.1

Write word and formula equations for the chemical reaction occurring when solid sodium oxide is added to water at room temperature and forms sodium hydroxide (dissolved in the water). Include symbols for phases in the formula equation.

Solution

The word equation must show the reactants, sodium oxide and water, to the left of the arrow and the product, sodium hydroxide, to the right of the arrow.

$$\text{sodium oxide} + \text{water} \rightarrow \text{sodium hydroxide}$$

This word equation is converted to a formula equation by replacing each name with a formula. To do this requires knowing (or looking up) that sodium has the +1 oxidation state, that oxygen usually has the −2 oxidation state, and that a hydroxide ion has a −1 charge.

$$Na_2O + H_2O \rightarrow NaOH \text{ (not balanced)}$$

Adding phase symbols and coefficients gives

$$Na_2O(s) + H_2O(\ell) \rightarrow 2NaOH(aq)$$

Sample Problem 8.2

Translate the following chemical equation into a sentence:

$$PbCl_2(aq) + Na_2CrO_4(aq) \rightarrow PbCrO_4(s) + 2NaCl(aq)$$

Chemical Equations and Reactions

Sample Problem 8.2 *continued*

Solution
Each of the reactants on the left side of the chemical equation is an ionic compound and is named according to the rules for such compounds. Both reactants are in aqueous solution. One product is a solid (a precipitate) and the other remains in solution. The meaning of this equation is as follows: Aqueous solutions of lead(II) chloride and sodium chromate react to produce a precipitate of lead(II) chromate plus an aqueous solution of sodium chloride.

Practice Problems
1. Write word and formula equations for each of the following chemical reactions. Include symbols for phases when indicated.
   (a) Calcium reacts with sulfur to produce calcium sulfide. *(Ans.)* calcium + sulfur → calcium sulfide; $Ca + S \rightarrow CaS$
   (b) Hydrogen reacts with fluorine to produce hydrogen fluoride. *(Ans.)* hydrogen + fluorine → hydrogen fluoride; $H_2 + F_2 \rightarrow 2HF$
   (c) Aluminum metal (solid) reacts with aqueous zinc chloride to produce zinc metal (solid) and aqueous aluminum chloride. *(Ans.)* aluminum + zinc chloride → zinc + aluminum chloride; $2Al(s) + 3ZnCl_2(aq) \rightarrow 3Zn(s) + 2AlCl_3(aq)$
2. Translate the following chemical equations into sentences:
   (a) $CS_2(s) + 3O_2(g) \rightarrow CO_2(g) + 2SO_2(g)$ *(Ans.)* Solid carbon disulfide reacts with oxygen gas to produce carbon dioxide gas and sulfur dioxide gas.
   (b) $NaCl(aq) + AgNO_3(aq) \rightarrow NaNO_3(aq) + AgCl(s)$ *(Ans.)* Aqueous solutions of sodium chloride and silver nitrate react to produce aqueous sodium nitrate and solid silver chloride.

Significance of a Chemical Equation

Chemical equations are of great significance to quantitative work in chemistry. The arrow in a chemical equation (meaning "yields") is like an equals sign, and thus chemical equations are similar to algebraic equations in that both express equalities. Chemical equations can be interpreted quantitatively in three different ways frequently used in chemistry.

1. Amounts of reactants and products. A chemical equation as usually written shows the smallest numbers of atoms, molecules, or ions that will satisfy the law of conservation of matter in a given chemical reaction. Because of the manner in which moles and molar masses are defined, a chemical equation also provides information about moles and masses of reactants and products. Recall that the mole is equal to 6.02×10^{23} atoms, molecules, or formula units. The molar mass of a substance is the mass in grams of one mole of that substance.

The equation for the reaction of hydrogen with chlorine, for example, provides the following information about molecules, moles, and masses of the reactants and products.

$$H_2(g) + Cl_2(g) \rightarrow 2HCl(g)$$

1 molecule of hydrogen reacts with 1 molecule of chlorine to yield 2 molecules of hydrogen chloride

1 mol of hydrogen reacts with 1 mol of chlorine to yield 2 mol of hydrogen chloride

STUDY HINT
You may wish to review the introduction to the mole in Section 3.3.

2 g of hydrogen (1 × molar mass) reacts with 71 g
of chlorine (1 × molar mass) to yield
73 g of hydrogen chloride (2 × molar mass)

The relative number of moles of each substance is given by the coefficients. Going from molecules to moles is like multiplying the equation by Avogadro's number, as shown in Figure 8-2. The relative masses of reactants and products are then known because the molar mass is the mass in grams of one mole.

2. Ratios of reactants and products. The second important interpretation of chemical equations is possible because coefficients indicate relative, not absolute, amounts of reactants and products. From the equation in Figure 8-2, we know that 1 mol of hydrogen reacts with 1 mol of chlorine to produce 2 mol of hydrogen chloride, giving mole ratios of reactants and products of:

1 mol H_2 : 1 mol Cl_2 : 2 mol HCl

We can therefore see that, for example, 20 mol of hydrogen would react with 20 mol of chlorine to give 40 mol of hydrogen chloride. Similarly, the equation shows the mass ratios of reactants to products. Hydrogen combines with chlorine to form hydrogen chloride in mass ratios of roughly

2 g H_2 : 71 g Cl_2 : 73 g HCl

Twice as much hydrogen (4 g H_2) would therefore react with twice as much chlorine (142 g Cl_2) to give twice as much hydrogen chloride (146 g HCl), and so on.

3. Reverse reactions. The third important quantitative interpretation of chemical equations is possible because, as for algebraic equations, the equality can be interpreted in the reverse direction as well as in the direction written. The equation in Figure 8-2 shows a reaction in which hydrogen chloride is broken down to give back the elements from which it formed. For example, 2 mol of hydrogen chloride would yield 1 mol of hydrogen plus 1 mol of chlorine. All of the information given is valid for the reverse reaction.

We have discussed several important ways in which chemical equations provide information. There are also several important kinds of information *not* provided by chemical equations, however. They give no indication of whether or not a reaction will actually occur. Chemical equations can be written for reactions that do not necessarily take place. In Sections 8.2 and 8.3,

Mass ratios are derived by multiplying mole ratios by the molar mass of each substance.
1 mol H_2 : 1 mol Cl_2 : 2 mol HCl
(1 × 2.0) : (1 × 35) : (2 × 36)
2.0 g H_2 : 71 g Cl_2 : 73 g HCl

Figure 8-2 The reaction of hydrogen and chlorine to give hydrogen chloride, showing the various ways to interpret a chemical equation

Chemical Equations and Reactions

You will learn methods for balancing more complicated equations in Chapter 20.

Water can be made conductive by adding a small amount of salt that dissolves to give ions in solution.

some guidelines are given about types of simple reactions that can be expected to occur. Later chapters will provide further guidelines for other types of reactions. In all cases, experiment forms the basis for the guidelines used in predicting whether or not chemical reactions will occur.

Chemical equations also give no information about the speed with which reactions occur or the actual pathway atoms or ions take in moving from reactants to products. These aspects of chemical reactions are discussed in Chapter 18.

Balancing Chemical Equations

Most of the equations that you will now be expected to balance can be balanced by inspection. The following procedure demonstrates how to master this skill by using a step-by-step approach. The equation for the reaction shown in Figure 8-3 will be used.

Figure 8-3 The decomposition of water by an electrical current to give hydrogen and oxygen. Indicators (substances that change color when the reaction occurs) have been added to show that reactions are occurring at the electrodes.

When an electric current is passed through water that has been made slightly conductive, the water molecules breakdown to yield hydrogen and oxygen.

Step 1. Identify the names of reactants and products and write a word equation. The word equation for the reaction shown in Figure 8-3 is

water → hydrogen + oxygen

Step 2. Substitute correct formulas for the names of the reactants and products, and write a formula equation. The formula for water is H_2O (which is correct according to the oxidation number of +1 for hydrogen and −2 for oxygen). Both hydrogen and oxygen exist as diatomic molecules and their correct formulas are H_2 and O_2. The formula equation for the breakdown of water is

$$H_2O \rightarrow H_2 + O_2 \text{ (not balanced)}$$

Step 3. Balance the formula equation according to the law of conservation of mass. This is done by trial and error. Coefficients are changed and the numbers of atoms counted on both sides of the equation until the numbers of each type of atom are the same for products and reactants. The trial-and-error method is made easier by use of the following guidelines.

1800 William Nicholson (1753–1815), using a battery designed by Volta, decomposed water into its elements.

- Balance the different types of atoms one at a time.
- First, balance the types of atoms that appear only once on each side of the equation.
- Balance polyatomic ions that appear on both sides of the equation as single units.
- Balance H atoms and O atoms after atoms of all other elements have been balanced.

Checking the formula equation in our example shows that there are two oxygen atoms on the right and only one on the left. To balance oxygen atoms, the number of H$_2$O molecules must be increased. Placing the coefficient 2 before H$_2$O gives the necessary 2 oxygen atoms on the left.

$$2H_2O \rightarrow H_2 + O_2 \text{ (partially balanced)}$$

The coefficient 2 in front of H$_2$O, however, has upset the balance of hydrogen atoms. Placing a coefficient of 2 in front of hydrogen (H$_2$) on the right gives an equal number of hydrogen atoms (4) on both sides of the equation.

$$2H_2O(\ell) \rightarrow 2H_2(g) + O_2(g)$$

Step 4. Count atoms to be sure that the equation is balanced. By counting, verify that equal numbers of atoms of each type appear on both sides of the arrow.

$$2H_2O(\ell) \rightarrow 2H_2(g) + O_2(g)$$
$$(4H, 2O \rightarrow 4H, 2O)$$

Figure 8-4 The reaction of zinc with aqueous hydrochloric acid to give zinc chloride and hydrogen.

Occasionally at this point, the coefficients are not the smallest possible whole numbers. When this happens, the coefficients should be converted to the smallest possible whole numbers.

Sample Problem 8.3

The reaction of zinc with aqueous hydrochloric acid produces a solution of zinc chloride and hydrogen gas, as shown in Figure 8-4. Write a balanced chemical equation for this reaction.

Solution

Step 1. Analyze   Write the word equation.

$$\text{zinc + hydrochloric acid} \rightarrow \text{zinc chloride + hydrogen}$$

Step 2. Plan   Write the formula equation.

$$Zn(s) + HCl(aq) \rightarrow ZnCl_2(aq) + H_2(g) \text{ (not balanced)}$$

Step 3. Compute   Adjust the coefficients. Note that both chlorine and hydrogen appear in only one substance on each side of the equation, but in unequal numbers of atoms. To balance chlorine we must use a coefficient of 2 for HCl. Two molecules of hydrogen chloride, 2HCl, also gives the required two hydrogen atoms on the left. Finally, note that there is one zinc atom on each side in the formula equation, so no further coefficients are needed.

$$Zn(s) + 2HCl(aq) \rightarrow ZnCl_2(aq) + H_2(g)$$

Sample Problem 8.3 *continued*

Step 4. Evaluate Count atoms to check balance.

$$Zn(s) + 2HCl(aq) \rightarrow ZnCl_2(aq) + H_2(g)$$
$$1Zn, 2H, 2Cl \rightarrow 1Zn, 2H, 2Cl$$

The equation is balanced.

Practice Problems
Write word, formula, and balanced chemical equations for each of the following reactions:
1. Sodium reacts with sulfur to produce sodium sulfide. *(Ans.)* Word: sodium + sulfur → sodium sulfide; Formula: Na + S → Na_2S; Balanced: 2Na + S → Na_2S
2. The reaction between magnesium and hydrochloric acid produces magnesium chloride and hydrogen. *(Ans.)* Word: magnesium + hydrochloric acid → magnesium chloride + hydrogen; Formula: Mg + HCl → $MgCl_2$ + H_2; Balanced: Mg + 2HCl → $MgCl_2$ + H_2
3. Aqueous nitric acid reacts with aqueous magnesium hydroxide to produce aqueous magnesium nitrate and water. *(Ans.)* Word: nitric acid + magnesium hydroxide → magnesium nitrate + water; Formula: $HNO_3(aq)$ + $Mg(OH)_2(aq)$ → $Mg(NO_3)_2(aq)$ + $H_2O(\ell)$; Balanced: $2HNO_3(aq)$ + $Mg(OH)_2(aq)$ → $Mg(NO_3)_2(aq)$ + $2H_2O(\ell)$

Balancing chemical equations by inspection will become easier as you continue your study of chemistry and gain experience. Learn to avoid the most common mistakes of *(1)* neglecting to be sure that each formula is correct and *(2)* trying to balance an equation by changing subscripts. Keep in mind that once a formula is correctly written, *subscripts cannot be added, deleted, or changed.*

You should be able to dispense with writing the word equation and the separate steps as you become more experienced. But *do not* dispense with the final step of counting atoms to be sure the equation is balanced.

Sample Problem 8.4

Aluminum carbide (Al_4C_3) reacts with water to produce methane gas and aluminum hydroxide. Write a chemical equation for this reaction.

Solution
The reactants are aluminum carbide and water. The products are methane and aluminum hydroxide.

$$Al_4C_3(s) + H_2O(\ell) \rightarrow CH_4(g) + Al(OH)_3(s) \text{ (not balanced)}$$

There are 4 Al atoms on the left, which indicates the need for a coefficient of 4 for $Al(OH)_3$.

$$Al_4C_3(s) + H_2O(\ell) \rightarrow CH_4(g) + 4Al(OH)_3(s) \text{ (partially balanced)}$$

Leaving H and O to balance last, we next balance C atoms. With 3 C atoms on the left, a coefficient of 3 must be added for CH_4 on the right.

$$Al_4C_3(s) + H_2O(\ell) \rightarrow 3CH_4(g) + 4Al(OH)_3(s) \text{ (partially balanced)}$$

It is best to balance O atoms next, since each appears only once on each side of the equation. On the left, there is 1 O atom. On the right, there are 12 O atoms—3 in each of the 4 $Al(OH)_3$ formula units. A coefficient of 12 for H_2O balances O atoms.

$$Al_4C_3(s) + 12H_2O(\ell) \rightarrow 3CH_4(g) + 4Al(OH)_3(s)$$

Sample Problem 8.4 *continued*

This leaves the H atoms to be balanced. There are 24 H atoms on the left (2 in each of 12 water molecules). On the right there are 12 H atoms in methane (4 in each of 3 molecules) plus 12 in aluminum hydroxide (3 in each of 4 formula units) for a total of 24 H atoms on the right. Hydrogen atoms are balanced.

$$Al_4C_3(s) + 12H_2O(\ell) \rightarrow 3CH_4(g) + 4Al(OH)_3(s)$$
$$4Al, 3C, 24H, 12O \rightarrow 4Al, 3C, 24H, 12O$$

The equation is balanced.

Sample Problem 8.5

Aluminum sulfate and calcium hydroxide are used in a water purification process. When added to water, they dissolve and react to produce two insoluble products, aluminum hydroxide and calcium sulfate. These products settle out taking suspended solid impurities with them. Write a balanced chemical equation for the reaction.

Solution

Each of the reactants and products is an ionic compound. Knowing the charges of the ions involved permits writing the formula equation

$$Al_2(SO_4)_3 + Ca(OH)_2 \rightarrow Al(OH)_3 + CaSO_4 \text{ (not balanced)}$$

Calcium appears only once on each side of the equation and is therefore balanced. There are two aluminum atoms on the left and one on the right. Placing a coefficient of 2 in front of $Al(OH)_3$ gives equal number of aluminum atoms on both sides of the equation.

$$Al_2(SO_4)_3 + Ca(OH)_2 \rightarrow 2Al(OH)_3 + CaSO_4 \text{ (partially balanced)}$$

Next, checking SO_4^{2-} ions shows that there are three SO_4^{2-} ions on the left side of the equation and only one on the right side. Placing the coefficient 3 in from of $CaSO_4$ gives an equal number of SO_4^{2-} ions on each side.

$$Al_2(SO_4)_3 + Ca(OH)_2 \rightarrow 2Al(OH)_3 + 3CaSO_4 \text{ (partially balanced)}$$

There are now three calcium atoms on the right, however, so calcium must now be balanced. By placing the coefficient 3 in front of $Ca(OH)_2$, we have an equal number of calcium atoms on each side. This last step also gives an equal number (6) of OH^- ions on both sides of the equation.

$$Al_2(SO_4)_3(aq) + 3Ca(OH)_2(aq) \rightarrow 2Al(OH)_3(s) + 3CaSO_4(s)$$
$$2Al, 3SO_4^{2-}, 3Ca, 6OH^- \rightarrow 2Al, 3SO_4^{2-}, 3Ca, 6OH^-$$

The equation is balanced.

Practice Problems

Write chemical equations for each of the following reactions:
1. Sodium combines with chlorine to produce sodium chloride. *(Ans.)* $2Na + Cl_2 \rightarrow 2NaCl$
2. When metallic copper reacts with aqueous silver nitrate the products are aqueous copper(II) nitrate and metallic silver. *(Ans.)* $Cu(s) + 2AgNO_3(aq) \rightarrow Cu(NO_3)_2(aq) + 2Ag(s)$
3. The reaction between solid iron(III) oxide and carbon monoxide gas produces metallic iron and carbon dioxide gas. *(Ans.)* $Fe_2O_3(s) + 3CO(g) \rightarrow 2Fe(s) + 3CO_2(g)$

1. A word equation uses words to represent the reactants and products in a chemical reaction; a formula equation uses their symbols or formulas; a chemical equation is a balanced formula equation.
2. Sulfuric acid + sodium hydroxide → sodium sulfate + water
   $H_2SO_4(aq) + NaOH(aq) \rightarrow Na_2SO_4(aq) + H_2O(\ell)$
3. Aqueous solutions of nitric acid and magnesium hydroxide react to produce a solution of magnesium nitrate in water.
4. Hydrogen sulfide + oxygen → sulfur dioxide + water
   $H_2S(g) + O_2(g) \rightarrow SO_2(g) + H_2O(g)$
   $2H_2S(g) + 3O_2(g) \rightarrow 2SO_2(g) + 2H_2O(g)$

SECTION OBJECTIVES

- Define and give general equations for synthesis, decomposition, single-replacement, and double-replacement reactions.
- Classify a reaction as synthesis, decomposition, single-replacement, double-replacement, or combustion.
- List three types of synthesis reactions and six types of decomposition reactions.
- List four types of single-replacement reactions and three types of double-replacement reactions.
- Predict the products of simple reactions, given the reactants.

5. (a) $2NH_4Cl + Ca(OH)_2 \rightarrow CaCl_2 + 2NH_3 + 2H_2O$
   (b) $2C_6H_{14} + 19O_2 \rightarrow 12CO_2 + 14H_2O$

CAUTION
Never look directly at the flame of burning magnesium. This can cause eye damage.

Section Review

1. Distinguish among word, formula, and chemical equations.
2. Write word and formula equations for the reaction in which aqueous solutions of sulfuric acid and sodium hydroxide react to form aqueous sodium sulfate and water.
3. Translate the following chemical equation into a sentence:
   $$2HNO_3(aq) + Mg(OH)_2(aq) \rightarrow Mg(NO_3)_2(aq) + 2H_2O(\ell).$$
4. Write the word, formula, and chemical equations for the reaction between hydrogen sulfide gas and oxygen gas if the products are sulfur dioxide gas and water vapor.
5. Write the chemical equation for each of the following reactions:
   (a) ammonium chloride + calcium hydroxide → calcium chloride + ammonia + water (b) hexane (C_6H_{14}) + oxygen → carbon dioxide + water.

8.2 Types of Chemical Reactions

Thousands of chemical reactions occur in living systems, in industrial processes, and in chemical laboratories. Based upon observation and analysis, chemical equations can be written to describe these many and varied reactions. Often it is necessary to predict the products formed by the combination of reactants. At this stage in your study of chemistry, you may be given the reactants in certain simple types of reactions and asked to predict the products. You must either remember what the products might be or else be able to reason out the answer from general knowledge about types of reactions.

Memorizing the equations for thousands of chemical reactions would be a very difficult task. It is therefore more useful and realistic to organize reactions according to various similarities and regularities.

There are several different ways of classifying chemical reactions. No single scheme is entirely satisfactory. The classification scheme described in this section provides an introduction to simple chemical reactions and is helpful in predicting the products of many reactions. Five types of chemical reactions are introduced: synthesis, decomposition, single-replacement, double-replacement, and combustion. In later chapters you will be introduced to categories useful in classifying other types of chemical reactions.

Synthesis Reactions

In a **synthesis**, *or* **composition**, *reaction two or more substances combine to form a new compound.* This type of reaction is represented by the following general equation:

$$A + X \rightarrow AX$$

A and X can be elements or compounds. AX is a compound. The following examples illustrate several kinds of synthesis reactions.

Reactions of elements with oxygen A simple type of synthesis reaction is the combination of an element with oxygen to produce an oxide of the element. For example, when a thin strip of magnesium metal is placed in a

244 Chapter 8

Bunsen burner flame, it burns with bright white light. When the metal strip is completely burned, only a fine white powder of magnesium oxide is left. This chemical reaction, shown in Figure 8-5, is represented by the following equation:

$$2Mg(s) + O_2(g) \rightarrow 2MgO(s)$$

Iron reacts with oxygen to produce two different iron oxides.

$$2Fe(s) + O_2(g) \rightarrow 2FeO(s)$$
$$4Fe(s) + 3O_2(g) \rightarrow 2Fe_2O_3(s)$$

Reactions of two nonmetals The reaction of two nonmetals forms a covalent compound. For example, hydrogen reacts with oxygen to form water, and sulfur reacts with oxygen to form sulfur dioxide as shown by the equations

$$2H_2(g) + O_2(g) \rightarrow 2H_2O(g)$$
$$S(s) + O_2(g) \rightarrow SO_2(g)$$

Other nonmetals also react with oxygen. For example, when carbon is burned in air, carbon dioxide is produced.

$$C(s) + O_2(g) \rightarrow CO_2(g)$$

Reactions of metals with nonmetals other than oxygen The reaction of a metal with a nonmetal often produces an ionic compound (a salt). Sodium plus chlorine gives this type of synthesis reaction.

$$2Na(s) + Cl_2(g) \rightarrow 2NaCl(s)$$

The Group 17 elements, the halogens, undergo synthesis reactions with many different metals. Fluorine, in particular, is so reactive that it combines with almost all metals, for example, with cobalt, sodium, and uranium.

$$Co(s) + F_2(g) \rightarrow CoF_2(s)$$
$$2Na(s) + F_2(g) \rightarrow 2NaF(s)$$
$$U(s) + 3F_2(g) \rightarrow UF_6(g)$$

Sodium fluoride (NaF) is added to municipal water supplies to provide fluoride ions, which help to prevent tooth decay in the people who drink it. Natural uranium is converted to uranium(VI) fluoride (UF_6) as the first step in the production of uranium for use in nuclear power plants. Phosphorus and sulfur are also reactants in many synthesis reactions. For example, the reaction between sodium and molten sulfur is the basis for the sodium–sulfur battery, which has been investigated as a source of power for electric motors for automobiles.

$$2Na(s) + S(\ell) \rightarrow Na_2S(s) + energy$$

Synthesis reactions of oxides Oxides of active metals react with water to produce metal hydroxides. For example, calcium oxide reacts with water to form calcium hydroxide.

$$CaO(s) + H_2O(\ell) \rightarrow Ca(OH)_2(s)$$

Calcium oxide (CaO), known as lime or quicklime, is manufactured in bulk quantities. The addition of water to lime to produce calcium hydroxide,

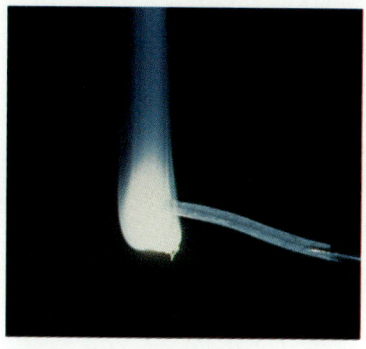

Figure 8-5 The synthesis reaction of magnesium and oxygen to give magnesium oxide. $2Mg(s) + O_2(g) \rightarrow 2MgO(s)$.

If the supply of oxygen is limited, carbon monoxide (CO) also forms when carbon is burned in air.

1789 Lavoisier (1743–1794) published *Elementary Treatise on Chemistry* in 1789. His extensive collection of laboratory glassware was carefully illustrated by Mme. Lavoisier. Lavoisier studied the combustion of many substances and was able to conclude that various elements such as mercury and phosphorous increased in mass when burned in air. If the experiment was performed over water in a closed container, one-fifth of the volume of the air was consumed.

Ca(OH)₂, known as slaked lime, is a crucial step in the conversion of lime (mixed with other materials) to mortar and cement.

Many nonmetal oxides undergo a synthesis reaction with water to produce a solution of an oxyacid. For example, sulfur trioxide (SO_3) combines with water to produce sulfuric acid (H_2SO_4)

$$SO_3(g) + H_2O(\ell) \rightarrow H_2SO_4(aq)$$

The statue shown in Figure 16-19 was damaged by acid precipitation, which is formed as rain falls through air polluted with SO_3.

Certain metal oxides and nonmetal oxides react with each other in synthesis reactions to form salts. For example, calcium sulfate is formed by the reaction of calcium oxide and sulfur trioxide.

$$CaO(s) + SO_3(g) \rightarrow CaSO_4(s)$$

Decomposition Reactions

*In a **decomposition reaction** a single compound undergoes a reaction that produces two or more simpler substances.* Decomposition reactions are the opposite of synthesis reactions and are represented by the following general equation:

$$AX \rightarrow A + X$$

AX is a compound, and A and X can be elements or compounds.

Most decomposition reactions take place only when energy in the form of heat or electricity is added. Examples of several types of decomposition reactions are given in the following examples.

Decomposition of binary compounds The simplest kind of decomposition reaction is the decomposition of a binary compound into its elements. When heated, oxides of the less active metals decompose in this manner. For example, oxygen was discovered by Joseph Priestley in 1774 when he heated mercury(II) oxide to produce mercury and oxygen. (Figure 8-6).

$$2HgO(s) \xrightarrow{\Delta} 2Hg(\ell) + O_2(g)$$

Electrolysis *The decomposition of a substance by an electric current is called **electrolysis**.* Passing an electric current through water will decompose it

Priestley also discovered NH_3, HCl, N_2O, NO_2, H_2S, SO_2, and CO. He recognized the O_2–CO_2 cycles in nature and was a codiscoverer of photosynthesis.

Figure 8-6 The decomposition of mercury(II) oxide to form mercury metal and oxygen.

$$2HgO(s) \xrightarrow{\Delta} 2Hg(\ell) + O_2(g)$$

into its constituent elements, hydrogen and oxygen, as was shown in Figure 8-3.

$$2H_2O(\ell) \xrightarrow{\text{electric current}} 2H_2(g) + O_2(g)$$

Decomposition of metal carbonates When a metal carbonate is heated, it breaks down to produce a metal oxide and carbon dioxide gas. For example, calcium carbonate decomposes to calcium oxide and carbon dioxide.

$$CaCO_3(s) \xrightarrow{\Delta} CaO(s) + CO_2(g)$$

Decomposition of metal hydroxides Most metal hydroxides decompose when heated to give metal oxides and water. For example, calcium hydroxide decomposes to form calcium oxide and water.

$$Ca(OH)_2(s) \xrightarrow{\Delta} CaO(s) + H_2O(g)$$

Sodium hydroxide and potassium hydroxide are exceptions to this generalization.

Decomposition of metal chlorates When a metal chlorate is heated it decomposes to a metal chloride and oxygen. For example, potassium chlorate, ($KClO_3$) decomposes to give potassium chloride and oxygen.

$$2KClO_3(s) \xrightarrow[MnO_2(s)]{\Delta} 2KCl(s) + 3O_2(g)$$

Notice the $MnO_2(s)$ written below the arrow. In this reaction, MnO_2 is a catalyst, which increases the rate of the reaction and also helps the decomposition of $KClO_3$ take place at a lower temperature than would be needed without the presence of MnO_2.

Decomposition of acids Certain acids decompose to nonmetallic oxides and water. Carbonic acid is unstable and decomposes readily at room temperature to form carbon dioxide and water.

$$H_2CO_3(aq) \rightarrow CO_2(g) + H_2O(\ell)$$

When heated, sulfuric acid decomposes into sulfur trioxide and water.

$$H_2SO_4(aq) \xrightarrow{\Delta} SO_3(g) + H_2O(\ell)$$

Single-Replacement Reactions

In a **single-replacement** *or* **displacement-reaction***, one element replaces a similar element in a compound.* Most single-replacement reactions take place in aqueous solution. The amount of energy involved in this type of reaction is usually smaller than the amount involved in synthesis or decomposition reactions. Single-replacement reactions can be represented by the following general equations:

$$A + BX \rightarrow AX + B$$

or

$$Y + BX \rightarrow BY + X$$

A, B, X, and Y are elements and AX, BX, and BY are compounds.

1807 Sir Humphrey Davy (1778–1829) isolated potassium and sodium by constructing a battery and using it as an energy source for the electrolysis of molten salts. Hydrogen was released when he dropped small pieces of the metals in water.

Figure 8-7 The replacement of hydrogen in aqueous hydrochloric acid by magnesium: $Mg(s) + 2HCl(aq) \rightarrow H_2(g) + MgCl_2(aq)$

Replacement of a metal in a compound by a more reactive metal
When metallic aluminum is placed in a solution of iron(II) nitrate, the aluminum atoms replace iron ions, forming metallic iron plus a solution of aluminum nitrate.

$$2Al(s) + 3Fe(NO_3)_2(aq) \rightarrow 3Fe(s) + 2Al(NO_3)_3(aq)$$

Replacement of hydrogen in water by a metal The most active metals (such as those in Group 1) react vigorously with water to produce metal hydroxides and hydrogen. For example, sodium reacts with water to form sodium hydroxide and hydrogen gas.

$$2Na(s) + 2H_2O(\ell) \rightarrow 2NaOH(aq) + H_2(g)$$

Less active metals, such as iron, react with steam to form a metal oxide and hydrogen gas.

$$3Fe(s) + 4H_2O(g) \rightarrow Fe_3O_4(s) + 4H_2(g)$$

Replacement of hydrogen in an acid by a metal The more active metals react with certain acidic solutions (such as hydrochloric acid and dilute sulfuric acid). The metal replaces the hydrogen in the acid, and the reaction products are a metal compound (a salt) and hydrogen gas. For example, when magnesium metal reacts with hydrochloric acid, as shown in Figure 8-7, the reaction products are hydrogen gas and aqueous magnesium chloride.

$$Mg(s) + 2HCl(aq) \rightarrow H_2(g) + MgCl_2(aq)$$

Replacement of halogens In another type of single-replacement reaction, one halogen replaces another halogen in a compound. Fluorine is the most active halogen, and it can replace any of the other halogens in their compounds. Each halogen is less active than the one above it in the Periodic Table. Each halogen can therefore replace those below it, but not those above it in the group (Group 17). Thus, while chlorine can replace bromine in potassium bromide, it cannot replace fluorine in potassium fluoride. The reaction of chlorine with potassium bromide produces bromine and potassium chloride whereas the combination of fluorine and sodium iodide produces sodium fluoride and solid iodine.

$$Cl_2(g) + 2KBr(aq) \rightarrow 2KCl(aq) + Br_2(\ell)$$
$$F_2(g) + 2NaI(aq) \rightarrow 2NaF(aq) + I_2(s)$$

Double-Replacement Reactions

In **double-replacement reactions** or **ionic reactions,** *the ions of two compounds exchange places in an aqueous solution to form two new compounds.* One of the compounds formed is usually a precipitate (a very slightly soluble compound), an insoluble gas that bubbles out of the solution, or a molecular compound, usually water. The other compound is often soluble and remains dissolved in solution. A double-replacement reaction is represented by the following general equation

$$AX + BY \rightarrow AY + BX$$

A, X, B, and Y in the reactants represent ions. AY and BX represent ionic or molecular compounds.

Desktop Investigation

Balancing Equations Using Models

Question
How can molecular models and formula-unit ionic models be used to demonstrate: (1) the rearrangement of atoms in chemical reactions; and (2) the law of conservation of mass as shown by balancing equations?

Materials
large and small gumdrops in at least 4 different colors
toothpicks

Procedure
1. Make models to represent the reactions shown. Use different colors of gumdrops to represent atoms of different elements.

2. Make sure you have the same number of each type atom on each side of the equation. Use toothpicks to connect the "atoms" of your models.

3. Reaction (a) and Reaction (b) in each group are unbalanced equations; balance them using your models. Determine the products for Reaction (c) in each group using models, and then complete and balance the equation.

4. Work Reactions (1a) and (5a) using the figures in this book to guide you in model making.

5. Classify each group of chemical reactions listed in (1) through (5).

(1) The reactions in this group are classified as ___.
  (a) $H_2 + Cl_2 \rightarrow HCl$
    (see Figure 8-2)
  (b) $Mg + O_2 \rightarrow MgO$
    (see Figure 8-5.)
  (c) $BaO + H_2O \rightarrow$

(2) The reactions in this group are classified as ___.
  (a) $H_2CO_3 \rightarrow CO_2 + H_2O$
  (b) $KClO_3 \rightarrow KCl + O_2$
  (c) $H_2O \xrightarrow{electricity}$
    (see Figure 8-3)

(3) The reactions in this group are classified as ___.
  (a) $Ca + H_2O \rightarrow Ca(OH)_2 + H_2$
  (b) $KI + Br_2 \rightarrow KBr + I_2$
  (c) $Zn + HCl \rightarrow$
    (see Figure 8-4)

(4) The reactions in this group are classified as ___.
  (a) $AgNO_3 + NaCl \rightarrow AgCl + NaNO_3$
  (b) $FeS + HCl \rightarrow FeCl_2 + H_2S$
  (c) $H_2SO_4 + KOH \rightarrow$

(5) The reactions in this group are classified as ___.
  (a) $CH_4 + O_2 \rightarrow CO_2 + H_2O$
    (see Figure 8-1)
  (b) $CO + O_2 \rightarrow CO_2$
  (c) $C_3H_8 + O_2 \rightarrow$

Discussion
1. What are the five categories used to classify most reactions? What is the advantage of using this classification method?

2. How does model making help illustrate the law of conservation of matter?

3. Some reactions can be placed in more than one category. Can you give an example of such a reaction in this investigation?

Figure 8-8 The double-replacement reaction between lead(II) nitrate and potassium iodide in aqueous solution: $Pb(NO_3)_2(aq) + 2KI(aq) \rightarrow PbI_2(s) + 2KNO_3(aq)$

Formation of a precipitate The formation of a precipitate occurs when the positive ions of one reactant combine with the negative ions of another reactant and form an insoluble or slightly soluble compound. For example, when a solution of potassium iodide is added to a solution of lead(II) nitrate, a yellow precipitate separates from the mixture, as shown in Figure 8-8.

$$Pb(NO_3)_2(aq) + 2KI(aq) \rightarrow PbI_2(s) + 2KNO_3(aq)$$

The precipitate of lead(II) iodide (PbI_2) forms as a result of the very strong attractive forces between the positive Pb^{2+} ions and the negative I^- ions. The other product is the water-soluble salt potassium nitrate (KNO_3). Potassium ions and nitrate ions do not take part in the reaction. They remain in solution as aqueous ions. The principles that help identify which ions form a precipitate and which ions remain in solution are developed in Chapter 15.

Formation of a gas Double-replacement reactions also occur when one of the products is an insoluble gas that bubbles out of the mixture. For example, iron(II) sulfide reacts with aqueous hydrogen chloride to form hydrogen sulfide gas and iron(II) chloride.

$$FeS(s) + 2HCl(aq) \rightarrow H_2S(g) + FeCl_2(aq)$$

Formation of water In some double-replacement reactions, a molecular compound such as water is one of the products. For example,

$$HCl(aq) + NaOH(aq) \rightarrow NaCl(aq) + H_2O(\ell)$$

Combustion Reactions

In a **combustion reaction***, a substance combines with oxygen releasing a large amount of energy in the form of light and heat.* The burning of natural gas, propane, and wood are examples of combustion reactions.

When hydrocarbons burn completely, they form carbon dioxide and water. **Hydrocarbons** *are compounds made up of carbon and hydrogen.* For example, the combustion of propane gas, shown in Figure 8-9, produces light and heat with the formation of carbon dioxide and water according to the following equation.

$$C_3H_8(g) + 5O_2(g) \rightarrow 3CO_2(g) + 4H_2O(g)$$

STUDY HINT

Reactions between acids and hydroxides are discussed in Chapters 16 and 17.

Figure 8-9 The combustion of propane gas can produce enough heat to boil water. $C_3H_8(g) + 5O_2(g) \rightarrow 3CO_2(g) + 4H_2O(g)$

Section Review

1. List the five types of chemical reactions.
2. Complete and balance each of the following reactions identified by type: *(a)* synthesis: $Li_2O + H_2O \rightarrow$ ___ *(b)* decomposition: $Mg(ClO_3)_2 \rightarrow$ ___ *(c)* single-replacement: $Na + H_2O \rightarrow$ ___ *(d)* double-replacement: $HNO_3 + Ca(OH)_2 \rightarrow$ ___ *(e)* combustion: $C_5H_{12} + O_2 \rightarrow$ ___ .
3. Classify each of the following reactions as synthesis, decomposition, single-replacement, double-replacement, or combustion: *(a)* $N_2(g) + 3H_2(g) \rightarrow 2NH_3(g)$ *(b)* $2Li(s) + 2H_2O(\ell) \rightarrow 2LiOH(aq) + H_2(g)$ *(c)* $2NaNO_3(s) \rightarrow 2NaNO_2(s) + O_2(g)$ *(d)* $2C_6H_{14}(\ell) + 19O_2(g) \rightarrow 12CO_2(g) + 14H_2O(\ell)$ *(e)* $NH_4Cl(s) \rightarrow NH_3(g) + HCl(g)$ *(f)* $BaO(s) + H_2O(\ell) \rightarrow Ba(OH)_2(aq)$ *(g)* $AgNO_3(aq) + NaCl(aq) \rightarrow AgCl(s) + NaNO_3(aq)$.
4. Complete and balance each of the following equations and identify each as to category: *(a)* $Br_2(\ell) + KI(aq) \rightarrow$; *(b)* $Zn(s) + HCl(aq) \rightarrow$; *(c)* $Ca(s) + Cl_2(g) \rightarrow$; *(d)* $NaClO_3(s) \xrightarrow{\Delta}$; *(e)* $C_8H_{18}(\ell) + O_2(g) \rightarrow$; *(f)* $CuCl_2(aq) + Na_2S(aq) \rightarrow$.

Answers:
1. Synthesis, decomposition, single replacement, double replacement, and combustion
2. *(a)* $Li_2O + H_2O \rightarrow 2LiOH$
   (b) $Mg(ClO_3)_2 \rightarrow MgCl_2 + 3O_2$
   (c) $2Na + 2H_2O \rightarrow 2NaOH + H_2$
   (d) $2HNO_3 + Ca(OH)_2 \rightarrow Ca(NO_3)_2 + 2H_2O$
   (e) $C_5H_{12} + 8O_2 \rightarrow 5CO_2 + 6H_2O$
3. *(a)* Synthesis
   (b) Single replacement
   (c) Decomposition
   (d) Combustion
   (e) Decomposition
   (f) Synthesis
   (g) Double replacement
4. *(a)* $Br_2(\ell) + 2KI(aq) \rightarrow 2KBr(aq) + I_2(s)$; single replacement *(b)* $Zn(s) + 2HCl(aq) \rightarrow ZnCl_2(aq) + H_2(g)$; single replacement *(c)* $Ca(s) + Cl_2(g) \rightarrow CaCl_2(s)$; synthesis *(d)* $2NaClO_3(s) \rightarrow 2NaCl(s) + 3O_2(g)$; decomposition *(e)* $2C_8H_{18}(\ell) + 25O_2(g) \rightarrow 16CO_2(g) + 18H_2O(g)$; combustion *(f)* $CuCl_2(aq) + Na_2S(aq) \rightarrow 2NaCl(aq) + CuS(s)$; double replacement

8.3 Activity Series of the Elements

The ability of an element to react is referred to as the *activity* of the element. The more readily an element reacts with other substances, the greater its activity. An **activity series** *is a list of elements organized according to the ease with which they undergo certain chemical reactions.* The elements are usually listed in an order determined by single-replacement reactions. The most active element, placed at the top in the series, can replace every element below it. An element further down can replace those elements that follow it, but not those above it.

Fluorine is the most active halogen. Activity decreases as one proceeds down this group (Group 17) in the Periodic Table. In discussing single-replacement reactions in Section 8.2, it was noted that each halogen will react to replace those below it. An activity series for the halogens, therefore, lists the elements in the same order as in Group 17. This is shown in Table 8.3 on the following page.

Being able to write a chemical equation does not necessarily mean that a reaction will actually take place. Activity series like those shown in Table 8-3 are useful aids in predicting whether or not certain chemical reactions will occur. For example, based on the activity series for metals in Table 8-3, one would predict that aluminum can replace zinc, and therefore that the following reaction does occur:

$$2Al(s) + 3ZnCl_2(aq) \rightarrow 3Zn(s) + 2AlCl_3(aq)$$

Cobalt, however, cannot replace sodium, and therefore the reaction below does *not* occur:

$$Co(s) + 2NaCl(aq) \rightarrow 2Na(s) + CoCl_2(aq)$$

SECTION OBJECTIVES

- Explain the significance of an activity series.
- List the generalizations based on the activity series that apply to single-replacement reactions.
- List the generalizations based on the activity series that apply to synthesis reactions.
- List the generalizations based on the activity series that apply to decomposition of oxides.
- Use the activity series to predict whether or not a given reaction will occur and what the products will be.

TABLE 8-3 ACTIVITY SERIES OF THE ELEMENTS	
Activity of Metals	
Li Rb K Ba Sr Ca Na	Can react with cold H_2O and acids, replacing hydrogen.
Mg Al Mn Zn Cr Fe Cd	Can react with steam and acids, replacing hydrogen.
Co Ni Sn Pb	Can react with acids, replacing hydrogen.
H_2 Sb Bi Cu Hg	React with oxygen, forming oxides.
Ag Pt Au	Fairly unreactive. Form oxides only indirectly.
Activity of Halogen Nonmetals	
F_2 Cl_2 Br_2 I_2	

Principles of the Activity Series

Like many other aids used to predict the products of chemical reactions, activity series are based upon experiment. Some metals (potassium, for example) react vigorously with water and acids, replacing hydrogen to form new compounds. Other metals such as iron or zinc (see Section 8.2) replace hydrogen in acids such as hydrochloric acid but react with water only when it is hot enough to become steam. Nickel will not react with steam but does react with acids, replacing hydrogen. Gold will not react with water (either as a liquid or as steam), or with any single acid to replace hydrogen. Such experimental observations led chemists to arrange metals in an activity series as shown in Table 8-3.

Metal atoms lose electrons to the positively charged ions of any metal below it in the activity series. The greater the distance between the two elements in the activity series, the more likely the displacement reaction will take place.

Single-replacement reactions of metals occur when the atoms of one metal give up electrons to the positive ions of another metal in solution. The positive ions gaining electrons become metal atoms, and the atoms giving up electrons become positive ions.

Consider the reaction between a strip of zinc and copper ions in solution, shown in Figure 8-10. A piece of zinc placed in an aqueous solution of copper(II) sulfate becomes coated with copper. These changes show that a reaction does occur, as would be predicted from the activity series.

$$Zn(s) + Cu^{2+}SO_4^{2-}(aq) \rightarrow Zn^{2+}SO_4^{2-}(aq) + Cu(s)$$

The position of zinc in the table indicates that the zinc atoms have a greater tendency to lose electrons than do copper atoms. Zinc atoms therefore transfer electrons to the copper ions and form Zn^{2+} ions.

$$Zn \rightarrow Zn^{2+} + 2e^-$$

The copper ions gain the electrons and form Cu metal atoms.

$$Cu^{2+} + 2e^- \rightarrow Cu$$

Notice the SO_4^{2-} ions do not participate in the reaction. During this reaction, the fading of the blue color caused by Cu^{2+} ions in aqueous solution and the coating of zinc by copper can both be observed. The formation of Zn^{2+} ions is also occurring, but it cannot be seen.

When metallic copper is placed in a solution of zinc sulfate ($ZnSO_4$), no reaction is observed. Here, too, the result can be predicted from the activity series. Copper is a less active metal than zinc and does not give up its electrons to Zn^{2+} ions in solution. Copper therefore does not replace zinc ions from a solution of zinc compound, but will replace the ions of a metal below it in the activity series, as illustrated in Figure 8-11 on the following page.

A piece of copper was placed in a solution of silver nitrate. The copper is more active than silver, and replaces silver ions in solution.

$$Cu(s) + 2Ag^+NO_3^-(aq) \rightarrow Cu^{2+}(aq) + 2NO_3^-(aq) + 2Ag(s)$$

In this reaction, there is also visual evidence of both the gain and loss of electrons as Cu atoms transfer electrons to Ag$^+$ ions, and the Ag metal crystallizes out of solution. The copper atoms give up electrons and become Cu^{2+} ions, as observed by the blue appearance of the solution.

Useful Generalizations Based on the Activity Series

We have seen that the activity series of metals helps to predict whether or not single replacement reactions of metals with metal ions and of metals with water and acids such as hydrochloric acid will occur. Generalizations 1–4 summarize how the activity series can be used for reactions of these types.

1. An element will replace from a compound in aqueous solution any of those elements below it in the activity series. The larger the interval between the elements in the activity series, the greater the tendency for the replacement reaction to occur.
2. Any metal above magnesium replaces hydrogen from water.
3. Any metal above cobalt replaces hydrogen from steam.
4. Any metal above hydrogen reacts with acids, replacing hydrogen. The synthesis reactions of metals with oxygen also occur more readily, the higher a metal is placed in the activity series. Generalizations 5 and 6 apply to such synthesis reactions.
5. Any metal above silver reacts with oxygen, forming oxides; those near the top react rapidly.
6. Any metal below mercury forms oxides only indirectly (i.e., not by reaction with O$_2$).

The more active a metal, the more strongly it holds onto oxygen in an oxide and therefore, the more strongly the oxide resists decomposition into its elements upon heating. Generalizations 7–9 summarize how the activity series can be used to predict the outcome of the decomposition of oxides or their conversion to free metals in single replacement reactions with hydrogen such as

$$H_2(g) + NiO(s) \rightarrow Ni(s) + H_2O(\ell)$$

7. Oxides of metals below copper decompose with heat alone.
8. Oxides of metals below chromium yield metals when heated with hydrogen.

Figure 8-10 The replacement of Cu^{2+} ions in aqueous solution by zinc, a metal that is above copper in the activity series: Zn(s) + CuSO$_4$(aq) → ZnSO$_4$(aq) + Cu(s)

Figure 8-11 The replacement of Ag$^+$ ions in aqueous solution by copper, a metal that is above silver in the activity series: Cu(s) + 2AgNO$_3$(aq) → Cu(NO$_3$)$_2$(aq) + 2Ag(s) Notice the appearance of the blue color in the right beaker due to the presence of aqueous copper(II) ions.

Chemistry and Technology

Hydrazine, a Liquid Rocket Fuel

The Raschig process for forming hydrazine (N₂H₄) has three steps:

Step 1
2NaOH(aq) + Cl₂(g) → NaClO(aq) + NaCl(aq) + H₂O(ℓ)

Step 2
NaClO(aq) + NH₃(aq) → NH₂Cl(aq) + NaOH(aq)

Step 3
NH₂Cl(aq) + NH₃(aq) + NaOH(aq) → N₂H₄(aq) + NaCl(aq) + H₂O(ℓ)

Notice that one of the products of each step shown above serves as a reactant in the next step and that hydrazine is not formed until the last step. (Additional reactions, known as "side reactions," also occur but are not shown above.)

Sodium hypochlorite (NaClO, the active ingredient in household bleach) is produced in the first reaction. Chloramine (NH₂Cl) is produced in the second. Both are crucial to the Raschig process, but are consumed immediately after their formation by the next step. This means that they don't have to be transported or stored, which is advantageous because chloramine is dangerous.

The net chemical equation for the Raschig process is the sum of the equations above:
2NaOH(aq) + Cl₂(g) + 2NH₃(aq) → N₂H₄ (aq) + 2NaCl(aq) + 2H₂O(ℓ)

The byproducts of the main reaction summarized in the equation above are salt and water. Other byproducts are ammonium chloride and nitrogen, produced by a side reaction between hydrazine and chloramine. This and other side reactions reduce the typical overall yield of hydrazine to 65 percent.

9. Oxides of metals above iron resist conversion to the free metal when heated with hydrogen.

   The most active metals are not likely to remain uncombined with other substances for very long. Some are so active that they must be isolated from the air when they are stored. Therefore, generalizations 10 and 11 can be made about the occurrence of free elements in nature.
10. Elements near the top of the series are never found free in nature.
11. Elements near the bottom of the series are often found free in nature.

Sample Problem 8.6

For each of the possible reactions listed below, describe the generalization based on the activity series that applies. Use the generalization to predict whether or not the reaction will take place and what the products will be. *(a)* Zn(s) + H₂O(ℓ) → ___. *(b)* Sn(s) + O₂(g) → ___. *(c)* Cd(s) + Pb(NO₃)₂(aq) → ___. *(d)* Cu(s) + HCl(aq) → ___. *(e)* NiO(s) + H₂(g) $\xrightarrow{\Delta}$ ___.

Sample Problem 8.6 *continued*

Solution
(a) This is a reaction between a metal and water—not steam, which would be shown as $H_2O(g)$. A metal more active than magnesium will replace hydrogen from water. Zinc is *below* magnesium according to the activity series. No reaction will occur.
(b) Any metal more active than silver will react with oxygen to form an oxide. Tin is above silver. Therefore, a reaction will occur and the product will be a tin oxide, either SnO or SnO_2.
(c) An element will replace from a compound in aqueous solution any element below it in the activity series. Cadmium is above lead and therefore a reaction will occur to produce lead (Pb) and cadmium nitrate, $Cd(NO_3)_2$.
(d) A metal more active than hydrogen will replace hydrogen from an acid. Copper is not above hydrogen in the series. Therefore, no reaction will occur.
(e) An oxide of a metal less active than chromium will form the free metal when heated with hydrogen. Therefore, a reaction will occur and the products will be nickel (Ni) and water (H_2O).

Practice Problems
1. Predict whether or not each of the following reactions will occur:
   (a) $Cr(s) + H_2O(\ell) \rightarrow$ _____ (Ans.) No
   (b) $Pt(s) + O_2(g) \rightarrow$ _____ (Ans.) No
   (c) $Pb(s) + HNO_3(aq) \rightarrow$ _____ (Ans.) Yes,
       $Pb(s) + 2HNO_3(aq) \rightarrow Pb(NO_3)_2(aq) + H_2(g)$.
   (d) $Mg(s) + H_2O(g) \rightarrow$ _____ (Ans.) Yes,
       $Mg(s) + 2H_2O(g) \rightarrow Mg(OH)_2(aq) + H_2(g)$.
   (e) $BaO(s) \xrightarrow{\Delta}$ _____ (Ans.) No
2. Identify the element that replaces hydrogen from cold water, but cannot replace calcium from its compounds. (Ans.) Na

Section Review

1. In what ways is the activity series useful in predicting chemical behavior?
2. Predict whether or not the following reactions will occur. (a) $Ni(s) + H_2O(\ell) \rightarrow$ _____ (b) $Br_2(\ell) + KI(aq) \rightarrow$ _____ (c) $Au(s) + HCl(aq) \rightarrow$ _____ (d) $Cd(s) + HCl(aq) \rightarrow$ _____ (e) $Mg(s) + Co(NO_3)_2(aq) \rightarrow$ _____ (f) $FeO(s) \rightarrow$ _____.
3. Write the balanced chemical equation for each of the reactions in Question 2 that can occur.

1. To predict whether or not certain chemical reactions will occur, as well as the products of those reactions that do occur.
2. (a) No (b) Yes (c) No (d) Yes (e) Yes (f) No
3. $Br_2(\ell) + 2KI(aq) \rightarrow 2KBr(aq) + I_2(s)$
   $Cd(s) + 2HCl(aq) \rightarrow CdCl_2(aq) + H_2(g)$
   $Mg(s) + Co(NO_3)_2(aq) \rightarrow Mg(NO_3)_2(aq) + Co(s)$

Chapter Summary

- A correct chemical equation will identify all known reactants and products. It contains the correct formulas for the reactants and products and is balanced.
- Synthesis reactions are represented by the general equation $A + X \rightarrow AX$.
- Decomposition reactions are represented by the general equation $AX \rightarrow A + X$.
- Single-replacement reactions are represented by the general equations $A + BX \rightarrow AX + B$, and $Y + BX \rightarrow BY + X$.
- Double-replacement reactions are represented by the general equation $AX + BY \rightarrow AY + BX$.
- Activity series list the elements in order of their activity and are useful in predicting whether or not a chemical reaction will occur.
- Chemists have determined the activity series by experiments.

CHAPTER 8 REVIEW

Chemical Equations and Reactions

VOCABULARY

activity series (251)
chemical equation (232)
coefficient (234)
combustion reaction (250)
decomposition reaction (246)
double-replacement reaction (ionic reaction) (248)
electrolysis (246)
formula equation (234)
hydrocarbons (250)
reversible reaction (237)
single-replacement reaction (displacement reaction) (247)
synthesis reaction (composition reaction) (244)
word equation (234)

REVIEWING CONCEPTS

1. List the three requirements for a correctly written chemical equation. (8.1)
2. a. What is meant by the term coefficient in relation to a chemical equation?
   b. How does the presence of a coefficient affect the number of atoms of each type in the formula that it precedes? (8.1)
3. Give one example each of a word equation, formula equation, and chemical equation. (8.1)
4. What limitations are associated with the use of both word and formula equations? (8.1)
5. Define each of the following:
   a. aqueous solution
   b. catalyst
   c. reversible reaction (8.1)
6. Write formulas for each of the following compounds:
   a. potassium hydroxide
   b. calcium nitrate
   c. sodium carbonate
   d. carbon tetrachloride
   e. magnesium bromide
   f. sulfur dioxide
   g. ammonium sulfate (8.1)
7. What four guidelines are useful in balancing an equation? (8.1)
8. How many atoms of each type are represented in each of the following?
   a. $3N_2$
   b. $2H_2O$
   c. $4HNO_3$
   d. $2Ca(OH)_2$
   e. $3Ba(ClO_3)_2$
   f. $5Fe(NO_3)_2$
   g. $4Mg_3(PO_4)_2$
   h. $2(NH_4)_2SO_4$
   i. $6Al_2(CO_3)_3$ (8.1)
9. Define and give general equations for the five types of chemical reactions. (8.2)
10. How are most decomposition reactions initiated? (8.2)
11. What is electrolysis? (8.2)
12. a. In what environment do most single replacement reactions occur?
    b. How do single-replacement reactions compare with synthesis and decomposition reactions in terms of the amount of energy generally involved? (8.2)
13. What are hydrocarbons? (8.2)
14. Complete the general pattern for the complete combustion of hydrocarbons by identifying the types of products usually formed:

    hydrocarbon + oxygen → (8.2)

15. a. What is meant by the "activity" of an element?
    b. What is the relationship between the activity of an element and the ease with which it reacts with other substances? (8.3)
16. a. What is an activity series of elements?
    b. What is the basis for the ordering of the elements in the activity series? (8.3)
17. a. What is the chemical principle upon which the activity series of metals is based?
    b. What is the significance of the distance between two metals in the activity series? (8.3)
18. Based on the activity series of metals, summarize the four generalizations that relate to predictions of whether single-replacement re-

actions of metals with metal ions, and of metals with water and acids, will occur. (8.3)

19. **a.** How does the position of a metal in the activity series affect its ability to react with oxygen through synthesis?
b. What two generalizations (as provided in the text discussion) apply to such synthesis reactions? (8.3)

20. **a.** How does the position of a metal in the activity series affect the tendency of its oxide to undergo decomposition?
b. What three activity series generalizations apply to the ability of metal oxides to decompose or be converted to free metals in single replacement reactions with hydrogen? (8.3)

21. Based on the activity series, what two generalizations can be made concerning the occurrence of free elements in nature? (8.3)

PROBLEMS

▶ Chemical Equations

1. Given the chemical equation below and the number of moles of the indicated reactant or product, determine the number of moles of all remaining reactants and products:
$4NH_3 + 3O_2 \rightarrow 2N_2 + 6H_2O$
a. 3.0 mol O_2 **c.** 1.0 mol N_2
b. 8.0 mol NH_3 **d.** 0.40 mol H_2O

2. Write the chemical equation that relates to each of the following word equations:
a. potassium bromide + barium iodide → potassium iodide + barium bromide
b. hydrochloric acid (HCl) + magnesium hydroxide → magnesium chloride + water
c. nitric acid (HNO_3) + calcium hydroxide → calcium nitrate + water.

3. Write balanced chemical equations for each of the following:
a. $H_2 + Cl_2 \rightarrow HCl$
b. $Al + Fe_2O_3 \rightarrow Al_2O_3 + Fe$
c. $Pb(C_2H_3O_2)_2 + H_2S \rightarrow PbS + HC_2H_3O_2$

4. Balance each of the following:
a. $Ca(OH)_2 + (NH_4)_2SO_4 \rightarrow CaSO_4 + NH_3 + H_2O$
b. $C_2H_6 + O_2 \rightarrow CO_2 + H_2O$
c. $Cu_2S + O_2 \rightarrow Cu_2O + SO_2$
d. $NaCl + H_2SO_4 + MnO_2 \rightarrow MnSO_4 + Na_2SO_4 + Cl_2 + H_2O$

5. The following equations are incorrect in some way. Identify the error and then correctly balance each:
a. $Li + O_2 \rightarrow LiO_2$
b. $H_2 + Cl_2 \rightarrow H_2Cl_2$
c. $MgCO_3 \rightarrow MgO_2 + CO_2$
d. $NaI + Cl_2 \rightarrow NaCl + I$

6. Write chemical equations for each of the following:
a. Aluminum reacts with oxygen to produce aluminum oxide.
b. Phosphoric acid is produced through the reaction between tetraphosphorus decoxide and water.
c. Iron(III) oxide combines with carbon monoxide to produce iron and carbon dioxide.

7. Each of the following equations is correctly balanced. Determine the subscripts (x and y) needed to complete the specified formulas:
a. $6HCl + Fe_xO_y \rightarrow 2FeCl_3 + 3H_2O$
b. $3H_2 + N_2 \rightarrow 2N_xH_y$
c. $4P + 5O_2 \rightarrow 2P_xO_y$
d. $N_2 + 2O_2 \rightarrow 2N_xO_y$

8. For each of the following synthesis reactions, identify the missing reactant(s) or product(s) and then balance the resulting equation:
a. $Mg + \underline{} \rightarrow MgO$
b. $\underline{} + O_2 \rightarrow Fe_2O_3$
c. $Li + Cl_2 \rightarrow \underline{}$
d. $Ca + \underline{} \rightarrow CaI_2$

▶ Types of Chemical Reactions

9. Complete the following synthesis reactions by writing both word and chemical equations for each:
a. sodium + oxygen → $\underline{}$
b. copper + oxygen → (two possibilities) $\underline{}$ and $\underline{}$
c. potassium + chlorine → $\underline{}$
d. magnesium + fluorine → $\underline{}$

10. Complete and balance the equation for each of the following decomposition reactions:
    a. $HgO \xrightarrow{\Delta}$
    b. $H_2O(\ell) \rightarrow$ (electrolysis)
    c. $2Ag_2O \xrightarrow{\Delta}$
    d. $MgCl_2 \rightarrow$ (electrolysis)

11. Complete and balance the equations for each of the following single replacement reactions:
    a. $Zn + Pb(NO_3)_2 \rightarrow$ ___
    b. $K + Ba(C_2H_3O_2)_2 \rightarrow$ ___
    c. $Al + NiSO_4 \rightarrow$ ___
    d. $Na + H_2O \rightarrow$ ___

12. Complete and balance the equations for the following double-replacement reactions, each of which occur, given normal conditions:
    a. $AgNO_3(aq) + NaCl(aq) \rightarrow$ ___
    b. $Mg(NO_3)_2(aq) + KOH(aq) \rightarrow$ ___
    c. $(NH_4)_2S(aq) + ZnCl_2(aq) \rightarrow$ ___
    d. $LiOH(aq) + Fe(NO_3)_3(aq) \rightarrow$ ___

13. Complete and balance the equation for each of the following combustion reactions:
    a. $CH_4 + O_2 \rightarrow$ ___
    b. $C_3H_6 + O_2 \rightarrow$ ___
    c. $C_5H_{12} + O_2 \rightarrow$ ___

14. Write and balance each of the following equations and then identify each as to type:
    a. hydrogen + iodine → hydrogen iodide
    b. lithium + hydrochloric acid → lithium chloride + hydrogen
    c. sodium carbonate → sodium oxide + carbon dioxide
    d. mercury(II) oxide → mercury + oxygen
    e. magnesium hydroxide → magnesium oxide + water
    f. copper + chlorine → copper(II) chloride
    g. calcium chlorate → calcium chloride + oxygen
    h. lithium + water → lithium hydroxide + hydrogen
    i. lead(II) carbonate → lead(II) oxide + carbon dioxide

15. Identify the compound that could theoretically undergo decomposition to produce the following products and then balance the final equation:
    a. magnesium oxide and water
    b. lead(II) oxide and water
    c. lithium chloride and oxygen
    d. barium chloride and oxygen
    e. nickel chloride and oxygen.

16. In each of the following combustion reactions, identify the missing reactant(s), product(s) or both, and then balance the resulting equation:
    a. $C_3H_8 +$ ___ $\rightarrow$ ___ $+ H_2O$
    b. $C_4H_{10} +$ ___ $\rightarrow$ ___ $+$ ___
    c. ___ $+ 8O_2 \rightarrow 5CO_2 + 6H_2O$

17. Complete and balance each of the following reactions observed to occur, and then identify each as to type:
    a. zinc + sulfur → ___
    b. calcium + sodium nitrate → ___
    c. silver nitrate + potassium sulfide → ___
    d. sodium iodide $\xrightarrow{\Delta}$ ___
    e. benzene (C_6H_6) + oxygen → ___
    f. decane ($C_{10}H_{22}$) + oxygen → ___

▶ **Activity Series**

18. Use the activity series to predict whether each of the following single-replacement reactions will occur. Write the chemical equations for those predicted to occur.
    a. $K(s) + CuCl_2(aq) \rightarrow$ ___
    b. $Zn(s) + Pb(NO_3)_2(aq) \rightarrow$ ___
    c. $Cl_2(g) + KI(aq) \rightarrow$ ___
    d. $Cu(s) + FeSO_4(aq) \rightarrow$ ___
    e. $Ba(s) + H_2O(\ell) \rightarrow$ ___

19. Use the activity series to predict whether each of the following synthesis reactions will occur and write the chemical equations for those predicted to occur:
    a. $Ca(s) + O_2(g) \rightarrow$ ___
    b. $Ni(s) + O_2(g) \rightarrow$ ___
    c. $Au(s) + O_2(g) \rightarrow$ ___
    d. $Al(s) + O_2(g) \rightarrow$ ___

20. Use the activity series to predict whether each of the following reactions will occur and write the chemical equations for those predicted to occur:
    a. $Ag_2O(s) \xrightarrow{\Delta}$ ___
    b. $PbO(s) \xrightarrow{\Delta}$ ___

   c. $NiO(s) \xrightarrow{\Delta}$ ___
   d. $CuO(s) + H_2(g) \xrightarrow{\Delta}$ ___

21. Based on the activity series of metals and halogens, which element within each pair is more likely to replace the other in a compound?
   a. K and Na   **e.** Au and Ag
   b. Al and Ni   **f.** Cl_2 and I_2
   c. Bi and Cr   **g.** Fe and Sr
   d. Cl_2 and F_2 **h.** I_2 and F_2

CRITICAL THINKING

1. **Organizing Information** When balancing an equation, changing the subscript of a chemical formula is not an option because by changing the subscript you change the chemical identity of the substance. Explain why changing the subscript changes the identity of the substance.

2. **Inferring Relationships** Activity series are prepared by comparing single displacement reactions between metals. Based upon observations, the metals can be ranked as to their ability to react. However, reactivity can be explained by the ease with which atoms of metals lose electrons. Using information from the activity series, identify the location on the periodic table of the most reactive metals and the least reactive metals. Based upon your knowledge of electron configurations and periodic trends, infer possible explanations for their reactivity and position on the periodic table.

3. **Analyzing Results** Formulate an activity series for the following hypothetical elements, A, J, Q, and Z using the reaction information provided.

   $A + ZX \rightarrow AX + Z$
   $JA + ZX \rightarrow$ no reaction
   $Q + AX \rightarrow QX + A$

4. **Designing an Experiment** Design a set of experiments that will enable you to create an activity series for the following metals and solutions:
Aluminum and aluminum chloride
Chromium and chromium(III) chloride
Iron and iron(II) chloride
Magnesium and magnesium chloride

CROSS CONNECTIONS

1. **Accounting** Just as chemists balance chemical equations to gain information about the nature and amounts of products formed, accountants do the same with people's finances. Calculate the state of Andrea's budget from the following account ledger information.

Check Register	From: 10/19/91 To: 10/27/91		
Date/Type	Num	Payee/Memo	Clr	Amount
10/19/91 Deposit		Beginning Balance	X	.00
10/19/91 Deposit		Wages from part-time job		150.00
10/19/91 Check		Purchased new outfit for date		−79.97
10/22/91 Check		Dinner and Movie with Friends		−13.86
10/25/91 Deposit		Allowance		20.00
10/27/91 Check		Gas for car		−13.00
10/27/91 Check		School supplies		−3.42

RESEARCH AND WRITING

1. Trace the evolution of the current practice of fluoridation of municipal water supplies. What advantages and disadvantages are associated with this practice?

PROJECT

1. Explain how a soda-acid fire extinguisher works and determine the chemical equation for the reaction. Check your house and adjacent structures for the location of fire extinguishers and ask your local fire department to verify the effectiveness of the extinguishers.

CHAPTER 9
Stoichiometry

PLANNING CHART
Approximate Time: 10/14 days

	SECTION/TEXT FEATURES		RESOURCES	
BASIC	9.1 **INTRODUCTION TO STOICHIOMETRY** Demos 1, 2 Section Review Chapter Review • Questions 1–4, Problem 1	261 260A–B 265 279	*Teacher's Resource Binder* Problem Solving Worksheets Section Review Worksheets	30
AVERAGE	9.2 **IDEAL STOICHIOMETRIC CALCULATIONS** Demos 3, 4 Section Review Chapter Review • Question 5, Problems 2–13	265 260B–C 271 279	*Teacher's Resource Binder* Lab Experiment 11 Problem Solving Worksheets Section Review Worksheets	75 31
AVERAGE	9.3 **LIMITING REACTANTS AND PERCENT YIELD** Demos 5, 6 Section Review Chapter Review • Questions 6–12, Problems 14–23	272 260C 278 279	*Teacher's Resource Binder* Problem Solving Worksheets Section Review Worksheets	32
	CHAPTER END ACTIVITIES Critical Thinking, Cross Connections, Research and Writing, Projects		*Teacher's Resource Binder* Test Booklet *Computer Test Generator*	41, 43

Teaching Strategies

Introduction

A two-prong approach can be very effective in this chapter. Here is an opportunity to spark your students' interest as the work gets more quantitative, or to excite them with the phenomena of chemistry. You can develop an enthusiasm for this subject and help them to master stoichiometry by adding several quick demonstrations every day. A majority of the reactions that are the basis of the problems in this chapter can be easily demonstrated to your class. (Some of your students might take the responsibility for this work.) The ideal stoichiometry calculations will be more meaningful to your students if they see these reactions, even if they are not done quantitatively.

The second prong in approaching the topic of stoichiometry is to emphasize a very systematic procedure for solving problems. Emphasize that the first step in solving these problems is to *solve a balanced equation*. The equation describes the chemistry. If you are actually doing the chemistry, students will tend to be more receptive to this idea. Make sure that students include all units in their problems and cancel carefully. If students have difficulty interpreting problems, have them underline interrogative words. Show them that they can always determine what information is being given and what is being asked for.

9.1 Introduction to Stoichiometry

Joseph Black (1728–1799) discovered that when $CaCO_3$ was heated, CaO and CO_2 were formed. Furthermore, if CaO was exposed to air, $CaCO_3$ was formed. Black, and his student Daniel Rutherford, studied this reversible reaction quantitatively. Their experiments, designed to measure the redistribution of mass during the course of a reaction, laid the groundwork for modern stoichiometry.

When students are first introduced to stoichiometry, they often think that quantitative information about a reaction must be given in order to determine a relative amount of product. Throughout this chapter, emphasize that if they know the amount of one substance in a reaction, they can determine the amounts of all the other substances through stoichiometric calculations.

Demonstration 1: The Production of Oxygen by Thermal Decomposition of an Oxide
PURPOSE: To observe chemical change resulting from decomposition of an oxide
MATERIALS: 0.3 g lead(IV) oxide, test tube, Bunsen burner, wooden splint

PROCEDURE: Place 0.3 g lead(IV) oxide in a test tube and heat over a Bunsen burner. CAUTION: This demonstration should only be done under a hood in good working condition. Light a wooden splint and blow out the flame. Quickly hold the splint inside the test tube.

The appearance of the powdered material will change as the oxide decomposes into lead and oxygen. The splint will glow inside the test tube indicating the presence of oxygen.

Demonstration 2: Single-Replacement Reaction

[This reaction can be demonstrated in a petri dish placed on an overhead projector and projected on a screen.]
PURPOSE: To show the reaction of a metal in acid
MATERIALS: 100-mm glass petri dish, 0.5 g aluminum, 15 mL 0.1 M H_2SO_4, (Zn and HCl could be used as an alternative.)
PROCEDURE: Pour 15 mL of H_2SO_4 into a petri dish placed on an overhead projector. Add a piece of aluminum.

The aluminum will replace the hydrogen in the acid. Bubbles of hydrogen will evolve as the reaction proceeds.

As an additional demonstration, add a few drops of universal indicator to the acid before you add the aluminum. The indicator will change color from red towards the more neutral color range as the hydrogen ion is consumed.

9.2 Ideal Stoichiometric Calculations ✳

This section focuses on quantitative relationships in chemical reactions. It is important for students to have an opportunity to become familiar with some of the equipment and procedures that are used in stoichiometry. The following demonstration provides that opportunity. Use quantitative methods and record all masses accurately.

Demonstration 3: Formation of Copper Iodide

PURPOSE: To determine the formula of the product formed from the reaction of copper (Cu) and iodine (I_2) and to write the equation for the reaction

$$Cu(s) + I_2(s) \rightarrow CuI_2$$

MATERIALS: Copper wire (1.5–2.0 g), iodine crystals (1.5–2.0 g), 0.1 M sodium thiosulfate solution, acetone, Erlenmeyer or Florence flask, hot plate, balance (0.01 g digital balance works well), watch glass, ice cubes, heat lamp (optional)
PROCEDURE: Reproduce the following data/calculation table on the chalkboard, or give as handouts to your students.

(1) Place a few crystals of iodine in the bottom of a flask. Place the flask on the hot plate and heat gently. A watch glass holding an ice cube should be placed on the mouth of the flask to contain the iodine vapors that will be generated. CAUTION: Iodine vapors are toxic. The ice cube on the watch glass will contain the vapors so that students are not exposed to them. This demonstration should be done in a hood.

(2) Clean and sand the surface of the copper wire. Make a coil of the copper wire leaving a convenient length of the wire to act as a handle when the wire is lowered into the flask.

(3) Lower the copper into the flask containing the iodine vapors and allow it to remain for one minute. *(4)* Remove the copper wire and immerse it in sodium thiosulfate solution. After one minute, remove the wire from the cleaning solution, and rinse it in distilled water. *(5)* Immerse the copper wire in acetone. Remove excess acetone, and repeat the acetone immersion two more times. *(6)* Air dry the prepared copper wire. *(7)* Determine the mass of the prepared copper wire and record the mass in line 1 of the data table.

(8) Immerse the copper wire in the flask containing the iodine vapors. *(9)* After two minutes, remove the reacted copper wire from the reaction flask. *(10)* Hold the copper wire in front of a heat lamp in order to remove excess free iodine. Determine the mass of the wire which is now a mixture of unreacted copper and newly formed copper iodide. Record the mass on line 2 of the data table.

(11) Immerse the copper wire in 0.1 M sodium thiosulfate and rinse in acetone as before. The copper iodide will be removed in this process. *(12)* Determine the mass of the dried copper wire and record on line 3 of the data table. This is the mass of the unreacted copper.

Repeat steps 8–12 several times if you can. If you do not have enough class time to repeat the procedure, you might give this work to a student to do as a special project.

Copper iodide will form on the surface of the copper after it has been immersed in iodine vapors. The mass of the newly formed copper iodide and the unreacted copper can be determined. All copper iodide is then removed by the sodium thiosulfate. The acetone will remove any excess water leaving pure, unreacted copper. The mass of the residual copper can finally be determined.

Data	
1. Mass of copper	
2. Mass of reacted copper and copper iodide	
3. Mass of remaining copper	
Calculations	
4. Mass of copper iodide (Data line 2 − data line 3)	
5. Mass of copper consumed (Data line 1 − data line 3)	
6. Mass of iodine in cpd (Data line 4 − data line 5)	
7. Moles of copper consumed to form copper iodide	
8. Moles of iodine in product	
9. Molar ratio copper/iodine	

If you have the opportunity to collect data from several trials, determine the mean values of the mass of iodine from all trials and the mean values of the mass of copper from all trials. Base the calculations for the molar ratio of copper to iodine on the mean mass values.

Except for a highly gifted honors class, do not emphasize the calculations at this time.

With this demonstration, as with all of the other demonstrations that accompany this chapter, it is important for you to keep sight of your goals. You are primarily trying to give your students an experimental basis for equation writing. They should be able to develop some intuitive concept of the nature of a chemical reaction. They should recognize the different kinds of evidence for chemical reactions: direct or indirect color change, formation of a precipitate, temperature change, evolution of a gas. They should also start to become familiar with the methods that chemists use to determine the identity of a product; that includes determining molar ratios. The quantitative procedures are not included at this point to teach stoichiometry, although they are good preparation for those manipulations, rather, they are included to give students some actual experiences with fundamental laboratory procedures.

Demonstration 4: An Acid–Base Titration
Purpose: To determine the amount of acid required to neutralize a given amount of base
Materials: two burettes, burette stand, 25-50 mL 0.1 M NaOH, 25-50 mL 0.1 M HCl, phenolphthalein, 250-mL Erlenmeyer flask
Procedure: HCl will be titrated against NaOH.
Place burettes in burette stand. Fill burette on left with acid and fill burette on right with base. Read both burettes. Add about 20 mL of acid to the Erlenmeyer flask. Read the burette accurately. Add a few drops of phenolphthalein to the acid in the flask. Add NaOH to the flask containing the HCl. Swirl the flask. Continue to add base until the presence of the red color of phenolpthalein indicates that the titration is complete. Read the NaOH burette accurately.

Compare the volume of acid added to the flask to the volume of base added to the flask. Tell students that this procedure allows investigators to determine the number of moles of acid in a solution if the number of moles of base in solution are known. Likewise, if the number of moles of acid in solution are known, the unknown number of moles of base in a solution can be determined.

[Note: The procedure is being done now to acquaint students with the methods of acid–base stoichiometry, not to actually determine the molarity of a base.]

9.3 Limiting Reactants and Percent Yield
Many students have difficulty solving limiting reactant problems. It might be wise to omit this section for your fundamental classes and to spend some additional time working on mass–mass problems with them.

The Desktop Investigation and the demonstration that follows give students examples of limiting reactants. Help students grasp the concept that a deficit in the amount of one reactant will limit the amount of product that can be produced whether that product is a batch of cookies, capped bottles, or a chemical compound. After students understand that when one reactant is used up, no more product can be formed, carefully explain the strategy described for solving limiting reactant problems.

Students often fail to recognize limiting reactant problems. However, if a limiting reactant problem is designated as such, they will be able to solve it successfully. Consider this when you prepare a test that covers this material.

Demonstration 5: How many bottles can a bottle capper cap?
Materials: One dozen bottle caps, ten bottles
Procedure: Invite a student to help you with this demonstration. Ask the students to cap as many bottles as possible. Of course you have a limited supply of bottles. Students never forget this demonstration, so refer to it while reviewing problems.

Dr. Larry Peck, of Texas A&M University, presented the following demonstration at CAST '86.

Demonstration 6: The Limiting Reactant
Purpose: To demonstrate the limiting effect of the amount of magnesium in a reaction
Materials: 3 500-mL flasks, 3 15″-diameter balloons, magnesium turnings, 12 M HCl, water
Procedure: CAUTION: Always add **acid** to **water.** Keep flames away from H_2-filled balloons. In flask 1, place 52.5 mL of H_2O and 7.5 mL HCl. In flask 2, place 45 mL H_2O and 15 mL HCl. In flask 3, place 37.5 mL H_2O and 22.5 mL HCl. Into each of three balloons, place 2.19 g Mg. Being careful not to spill any Mg into the acid, stretch balloons over the mouth of the flasks, allowing them to dangle over the side of the flasks. Simultaneously hold the balloons up and shake the Mg into the flasks and observe.

Flask 1 contains 0.09 mol Mg and 0.09 mol HCl, so half of the Mg will remain unreacted. Flask 2 contains 0.09 mol Mg and 0.18 mol HCl. All of the Mg reacts because it is a stoichiometric mixture. Balloon 2 will be larger than balloon 1.

Flask 3 contains 0.09 mol Mg and 0.27 mol HCl. In this case, the magnesium is the limiting reactant. Therefore, balloon 3 will be the same size as balloon 2.

References and Resources

Books and Periodicals
Bowman and Schull. "Mysterious Stoichiometry," *Journal of Chemical Education,* Vol. 52, No. 3, 1975, p. 186.
Gizara. "Bridging the Stoichiometric Gap," *Science Teacher,* Vol. 48, No. 4, 1981, p.36.
Maciel. "Chemistry that's Chinese, not Greek, to Students!" *Journal of Chemical Education,* Vol. 57, No. 10, 1980, p. 727.
Morrison. "Emphasis on Stoichiometry and Structure," *Journal of Chemical Education,* Vol. 55, No. 4, 1978, p. 255.
Russo, Tom. "Microchemistry for High School General Chemistry," Educational Concepts, 1986.
Vipond. "Stoichiometry and Chemical Equations", *Science Teacher,* Vol. 43, No. 5, 1978, p. 48.

Computer Software
Chemistry 102 #3: Mole: Its Meaning #4: Mole: Learning by Practice, 1986.
Gold, Marvin. "Principles of Stoichiometry: Part I—Principles of Stoichiometry (Sandwich Analogy), Part II—Stoichiometry in

Chemical Reactions (Ammonia Production)," COM4031A, AII, 48K, 2 disks, COMPress.
Project Seraphim, Disks #305 and #306, NSF Science and Engineering Education, Dept. of Chemistry, Eastern Michigan University, Ypsilanti, MI 48197.

Weyh, J.A., J.R. Crook, and L.N. Hauge. "Chemical Stoichiometry, Version 2.0," COM4010B, I, 256K, CGA, disk, COMPress.
Weyh, J.A., J.R. Crook, and L.N. Hauge. "Mole Concept, Version 2.0," COM 4012B, I, 256K, CGA, disk, COMPress.

Answers and Solutions

Reviewing Concepts

1. **a.** Composition stoichiometry involves the mass relationships of elements in chemical compounds, whereas reaction stoichiometry involves the mass relationships among reactants and products in a chemical reaction.
   b. An example of a composition stoichiometry calculation involves the mass-mole relationship. A typical reaction stoichiometry calculation involves the use of a chemical equation to determine the amounts of reactants needed to produce a given amount of product.
2. **a.** A mole ratio is a conversion factor that relates the number of moles of any two substances involved in a chemical reaction.
   b. This information is obtained directly from the chemical equation.
3. The coefficients in a chemical equation represent the relative numbers of moles of the substances involved in a reaction.
4. **a.** $2Ca + O_2 \rightarrow 2CaO$

   1. $\dfrac{2 \text{ mol Ca}}{1 \text{ mol } O_2}$ or $\dfrac{1 \text{ mol } O_2}{2 \text{ mol Ca}}$

   2. $\dfrac{2 \text{ mol Ca}}{2 \text{ mol CaO}}$ or $\dfrac{2 \text{ mol CaO}}{2 \text{ mol Ca}}$

   3. $\dfrac{1 \text{ mol } O_2}{2 \text{ mol CaO}}$ or $\dfrac{2 \text{ mol CaO}}{1 \text{ mol } O_2}$

   b. $Mg + 2HF \rightarrow MgF_2 + H_2$

   1. $\dfrac{1 \text{ mol Mg}}{2 \text{ mol HF}}$ or $\dfrac{2 \text{ mol HF}}{1 \text{ mol Mg}}$

   2. $\dfrac{1 \text{ mol Mg}}{1 \text{ mol MgF}_2}$ or $\dfrac{1 \text{ mol MgF}_2}{1 \text{ mol Mg}}$

   3. $\dfrac{1 \text{ mol Mg}}{1 \text{ mol H}_2}$ or $\dfrac{1 \text{ mol H}_2}{1 \text{ mol Mg}}$

   4. $\dfrac{2 \text{ mol HF}}{1 \text{ mol MgF}_2}$ or $\dfrac{1 \text{ mol mgF}_2}{2 \text{ mol HF}}$

   5. $\dfrac{2 \text{ mol HF}}{1 \text{ mol H}_2}$ or $\dfrac{1 \text{ mol H}_2}{2 \text{ mol HF}}$

   6. $\dfrac{1 \text{ mol MgF}_2}{1 \text{ mol H}_2}$ or $\dfrac{1 \text{ mol H}_2}{1 \text{ mol MgF}_2}$

5. **a.** Molar mass is the mass (in grams) of one mole of a substance.
   b. Molar mass is the conversion factor that relates the mass of a substance to the number of moles of that substance.
6. Ideal stoichiometry calculations do not take into account those factors that can affect the relative amounts of reactants needed or products actually produced in chemical reactions; they deal with the amounts of reactants or products under ideal conditions. Real stoichiometry calculations do take into account actual laboratory or industrial conditions.
7. **a.** The limiting reactant is the reactant that controls the amount of product formed in a chemical reaction, whereas the excess reactant is the substance that is not used up completely in a reaction.
   b. The limiting reactant is used up.
8. **a.** The theoretical yield is the maximum amount of product that can be produced from a given amount of reactant. The actual yield is the measured amount of a product obtained from a reaction.
   b. The actual yield is less than the theoretical yield.
9. The percent yield is the ratio of the actual yield to the theoretical yield multiplied by 100 percent.
10. Actual yields are often less than theoretical yields because: reactions sometimes do not go to completion, impure reactions produce less product than expected, competing side reactions may consume some of the reactants, and some are lost when products collected in impure form are purified.
11. The efficiency of a reaction is expressed in terms of percent yield $\dfrac{\text{actual yield}}{\text{theoretical yield}} \times 100\%$

Problems

1. Na_2CO_3: $\dfrac{105.99 \text{ g } Na_2CO_3}{1 \text{ mol } Na_2CO_3}$ or $\dfrac{1 \text{ mol } Na_2CO_3}{105.99 \text{ g } Na_2CO_3}$

   $Ca(OH)_2$: $\dfrac{74.09 \text{ g } Ca(OH)_2}{1 \text{ mol } Ca(OH)_2}$ or $\dfrac{1 \text{ mol } Ca(OH)_2}{74.09 \text{ g } Ca(OH)_2}$

   NaOH: $\dfrac{40.00 \text{ g NaOH}}{1 \text{ mol NaOH}}$ or $\dfrac{1 \text{ mol NaOH}}{40.00 \text{ g NaOH}}$

   $CaCO_3$: $\dfrac{100.09 \text{ g } CaCO_3}{1 \text{ mol } CaCO_3}$ or $\dfrac{1 \text{ mol } CaCO_3}{100.09 \text{ g } CaCO_3}$

2. **a.** $2H_2 + O_2 \rightarrow 2H_2O$

   $5.0 \text{ mol } H_2O \times \dfrac{2 \text{ mol } H_2}{2 \text{ mol } H_2O} = 5.0 \text{ mol } H_2$

   b. $5.0 \text{ mol } H_2O \times \dfrac{1 \text{ mol } O_2}{2 \text{ mol } H_2O} = 2.5 \text{ mol } O_2$

Stoichiometry **260D**

3. **a.** $2C_2H_6 + 7O_2 \rightarrow 4CO_2 + 6H_2O$

   $4.50 \text{ mol } C_2H_6 \times \dfrac{7 \text{ mol } O_2}{2 \text{ mol } C_2H_6} = 15.8 \text{ mol } O_2$

   b. $4.50 \text{ mol } C_2H_6 \times \dfrac{4 \text{ mol } CO_2}{2 \text{ mol } C_2H_6} = 9.00 \text{ mol } CO_2$

   $4.50 \text{ mol } C_2H_6 \times \dfrac{6 \text{ mol } H_2O}{2 \text{ mol } C_2H_6} = 13.5 \text{ mol } H_2O$

4. **a.** $H_2SO_4 + 2NaOH \rightarrow Na_2SO_4 + 2H_2O$

   b. $0.75 \text{ mol NaOH} \times \dfrac{1 \text{ mol } H_2SO_4}{2 \text{ mol NaOH}} \times \dfrac{98 \text{ g } H_2SO_4}{\text{mol } H_2SO_4} = 37 \text{ g } H_2SO_4$

   c. $0.75 \text{ mol NaOH} \times \dfrac{1 \text{ mol } Na_2SO_4}{2 \text{ mol NaOH}} \times \dfrac{140 \text{ g } Na_2SO_4}{\text{mol } Na_2SO_4} = 53 \text{ g } Na_2SO_4$

   $0.75 \text{ mol NaOH} \times \dfrac{2 \text{ mol } H_2O}{2 \text{ mol NaOH}} \times \dfrac{18 \text{ g } H_2O}{\text{mol } H_2O} = 14 \text{ g } H_2O$

5. $2Na + Cl_2 \rightarrow 2NaCl$

   $25.0 \text{ mol NaCl} \times \dfrac{2 \text{ mol Na}}{2 \text{ mol NaCl}} \times \dfrac{23.0 \text{ g Na}}{\text{mol Na}} = 575 \text{ g Na}$

   $25.0 \text{ mol NaCl} \times \dfrac{1 \text{ mol } Cl_2}{2 \text{ mol NaCl}} \times \dfrac{71.0 \text{ g } Cl_2}{\text{mol } Cl_2} = 888 \text{ g } Cl_2$

6. **a.** $Cu + 2AgNO_3 \rightarrow 2Ag + Cu(NO_3)_2$

   $2.25 \text{ g Ag} \times \dfrac{\text{mol Ag}}{108 \text{ g Ag}} \times \dfrac{1 \text{ mol } Cu(NO_3)_2}{2 \text{ mol Ag}} = 0.0104 \text{ mol } Cu(NO_3)_2$

   b. $2.25 \text{ g Ag} \times \dfrac{\text{mol Ag}}{108 \text{ g Ag}} \times \dfrac{1 \text{ mol Cu}}{2 \text{ mol Ag}} = 0.0104 \text{ mol Cu}$

   $2.25 \text{ g Ag} \times \dfrac{\text{mol Ag}}{108 \text{ g Ag}} \times \dfrac{2 \text{ mol } AgNO_3}{2 \text{ mol Ag}} = 0.0208 \text{ mol } AgNO_3$

7. **a.** $Fe_2O_3(s) + 3CO(g) \rightarrow 2Fe(s) + 3CO_2(g)$

   $4.00 \text{ kg } Fe_2O_3 \times \dfrac{10^3 \text{ g}}{\text{kg}} \times \dfrac{\text{mol } Fe_2O_3}{160. \text{ g } Fe_2O_3} \times \dfrac{3 \text{ mol CO}}{\text{mol } Fe_2O_3}$
   $= 75.0 \text{ mol CO}$

   b. $75.0 \text{ mol CO} \times \dfrac{2 \text{ mol Fe}}{3 \text{ mol CO}} = 50.0 \text{ mol Fe}$

   $75.0 \text{ mol CO} \times \dfrac{3 \text{ mol } CO_2}{3 \text{ mol CO}} = 75.0 \text{ mol } CO_2$

8. $CO(g) + 2H_2(g) \rightarrow CH_3OH(g)$

   $100.0 \text{ kg } CH_3OH \times \dfrac{10^3 \text{ g}}{\text{kg}} \times \dfrac{\text{mol } CH_3OH}{32.04 \text{ g } CH_3OH} \times \dfrac{1 \text{ mol CO}}{\text{mol } CH_3OH} \times \dfrac{28.01 \text{ g CO}}{\text{mol CO}} = 8.742 \times 10^4 \text{ g CO}$

   $100.0 \text{ kg } CH_3OH \times \dfrac{10^3 \text{ g}}{\text{kg}} \times \dfrac{\text{mol } CH_3OH}{32.04 \text{ g } CH_3OH} \times \dfrac{2 \text{ mol } H_2}{\text{mol } CH_3OH} \times \dfrac{2.016 \text{ g } H_2}{\text{mol } H_2} = 1.258 \times 10^4 \text{ g } H_2$

9. $2NO + O_2 \rightarrow 2NO_2$

   a. $384 \text{ g } O_2 \times \dfrac{1 \text{ mol } O_2}{32.0 \text{ g } O_2} \times \dfrac{2 \text{ mol } NO_2}{\text{mol } O_2} \times \dfrac{46.0 \text{ g } NO_2}{\text{mol } NO_2} = 1.10 \times 10^3 \text{ g } NO_2$

   b. $384 \text{ g } O_2 \times \dfrac{1 \text{ mol } O_2}{32.0 \text{ g } O_2} \times \dfrac{2 \text{ mol NO}}{\text{mol } O_2} \times \dfrac{30.0 \text{ g NO}}{\text{mol NO}} = 720. \text{ g NO}$

10. **a.** $2NaOH + CO_2 \rightarrow Na_2CO_3 + H_2O$

    $925.0 \text{ g } CO_2 \times \dfrac{\text{mol } CO_2}{44.01 \text{ g } CO_2} \times \dfrac{2 \text{ mol NaOH}}{\text{mol } CO_2}$
    $= 42.04 \text{ mol NaOH}$

    b. $42.04 \text{ mol NaOH} \times \dfrac{1 \text{ mol } Na_2CO_3}{2 \text{ mol NaOH}} = 21.02 \text{ mol } Na_2CO_3$

    $42.04 \text{ mol NaOH} \times \dfrac{1 \text{ mol } H_2O}{2 \text{ mol NaOH}} = 21.02 \text{ mol } H_2O$

    [Note: It is not necessary to use the results of part **a** to find the answers to part **b**. The original reactant mass value could again be used to calculate the moles of each product as well.]

11. $AgNO_3 + NaBr \rightarrow AgBr + NaNO_3$

    a. $4.50 \text{ mol } AgNO_3 \times \dfrac{1 \text{ mol NaBr}}{\text{mol } AgNO_3} \times \dfrac{103 \text{ g NaBr}}{\text{mol NaBr}} = 464 \text{ g NaBr}$

    b. $4.50 \text{ mol } AgNO_3 \times \dfrac{1 \text{ mol AgBr}}{\text{mol } AgNO_3} \times \dfrac{188 \text{ g AgBr}}{\text{mol AgBr}} = 846 \text{ g AgBr}$

12. $H_2SO_4 + 2NaHCO_3 \rightarrow 2CO_2 + Na_2SO_4 + 2H_2O$

a. $150.0 \text{ g } H_2SO_4 \times \dfrac{\text{mol } H_2SO_4}{98.07 \text{ g } H_2SO_4} \times \dfrac{2 \text{ mol } NaHCO_3}{\text{mol } H_2SO_4} = 3.059 \text{ mol } NaHCO_3$

b. $3.059 \text{ mol } NaHCO_3 \times \dfrac{2 \text{ mol } CO_2}{2 \text{ mol } NaHCO_3} = 3.059 \text{ mol } CO_2$

$3.059 \text{ mol } NaHCO_3 \times \dfrac{1 \text{ mol } Na_2SO_4}{2 \text{ mol } NaHCO_3} = 1.530 \text{ mol } Na_2SO_4$

$3.059 \text{ mol } NaHCO_3 \times \dfrac{2 \text{ mol } H_2O}{2 \text{ mol } NaHCO_3} = 3.059 \text{ mol } H_2O$

13. $2C_7H_6O_3(s) + C_4H_6O_3(\ell) \rightarrow 2C_9H_8O_4(s) + H_2O(\ell)$

a. $75.0 \text{ mol } C_7H_6O_3 \times \dfrac{2 \text{ mol } C_9H_8O_4}{2 \text{ mol } C_7H_6O_3} \times \dfrac{180. \text{ g } C_9H_8O_4}{\text{mol } C_9H_8O_4} \times \dfrac{\text{kg}}{10^3 \text{ g}} = 13.5 \text{ kg } C_9H_8O_4$

b. $75.0 \text{ mol } C_7H_6O_3 \times \dfrac{1 \text{ mol } C_4H_6O_3}{2 \text{ mol } C_7H_6O_3} \times \dfrac{102 \text{ g } C_4H_6O_3}{\text{mol } C_4H_6O_3} \times \dfrac{\text{kg}}{10^3 \text{ g}} = 3.82 \text{ kg } C_4H_6O_3$

c. $75.0 \text{ mol } C_7H_6O_3 \times \dfrac{1 \text{ mol } H_2O}{2 \text{ mol } C_7H_6O_3} \times \dfrac{18.0 \text{ g } H_2O}{\text{mol } H_2O} \times \dfrac{\text{mL } H_2O}{0.998 \text{ g } H_2O} \times \dfrac{L}{10^3 \text{ mL}} = 0.676 \text{ L } H_2O$

14. a. $2.0 \text{ mol HCl} \times \dfrac{1 \text{ mol NaOH}}{\text{mol HCl}} = 2.0 \text{ mol NaOH}$ $2.5 \text{ mol NaOH} \times \dfrac{1 \text{ mol HCl}}{\text{mol NaOH}} = 2.5 \text{ mol HCl}$

Because only 2.0 mol of the 2.5 mol of NaOH are needed to react with the 2.0 mol HCl, the HCl must be the limiting reactant.

b. $2.5 \text{ mol Zn} \times \dfrac{2 \text{ mol HCl}}{1 \text{ mol Zn}} = 5.0 \text{ mol HCl}$ **c.** $4.0 \text{ mol Fe(OH)}_3 \times \dfrac{3 \text{ mol } H_2SO_4}{2 \text{ mol Fe(OH)}_3} = 6.0 \text{ mol } H_2SO_4$

$6.0 \text{ mol HCl} \times \dfrac{1 \text{ mol Zn}}{2 \text{ mol HCl}} = 3.0 \text{ mol Zn}$ $6.5 \text{ mol } H_2SO_4 \times \dfrac{2 \text{ mol Fe(OH)}_3}{3 \text{ mol } H_2SO_4} = 4.3 \text{ mol Fe(OH)}_3$

Since only 2.5 mol Zn are available, the Zn is the limiting reactant.   More than the required 6.0 mol of H_2SO_4 is available, so $Fe(OH)_3$ is the limiting reactant.

15. a. 2.5 mol NaOH available − 2.0 mol NaOH needed = 0.5 mol excess NaOH
b. 6.0 mol HCl available − 5.0 mol HCl needed = 1.0 mol excess HCl
c. 6.5 mol H_2SO_4 available − 6.0 mol H_2SO_4 needed = 0.5 mol excess H_2SO_4

16. a. $2.0 \text{ mol HCl} \times \dfrac{1 \text{ mol NaCL}}{1 \text{ mol HCl}} = 2.0 \text{ mol NaCl}$ **c.** $4.0 \text{ mol Fe(OH)}_3 \times \dfrac{1 \text{ mol } Fe_2(SO_4)_3}{2 \text{ mol Fe(OH)}_3}$

$2.0 \text{ mol HCl} \times \dfrac{1 \text{ mol } H_2O}{1 \text{ mol HCl}} = 2.0 \text{ mol } H_2O$ $= 2.0 \text{ mol } Fe_2(SO_4)_3$

b. $2.5 \text{ mol Zn} \times \dfrac{1 \text{ mol } ZnCl_2}{1 \text{ mol Zn}} = 2.5 \text{ mol } ZnCl_2$ $4.0 \text{ mol F (OH)}_3 \times \dfrac{6 \text{ mol } H_2O}{2 \text{ mol Fe(OH)}_3} = 12 \text{ mol } H_2O$

$2.5 \text{ mol Zn} \times \dfrac{1 \text{ mol } H_2}{1 \text{ mol Zn}} = 2.5 \text{ mol } H_2$

17. a. $Cu + 2AgNO_3 \rightarrow Cu(NO_3)_2 + 2Ag$ **c.** $2.50 \text{ mol Cu} \times \dfrac{1 \text{ mol } Cu(NO_3)_2}{1 \text{ mol Cu}}$

$2.50 \text{ mol Cu} \times \dfrac{2 \text{ mol } AgNO_3}{1 \text{ mol Cu}} = 5.00 \text{ mol } AgNO_3$ $= 2.50 \text{ mol } Cu(NO_3)_2$

$5.50 \text{ mol } AgNO_3 \times \dfrac{1 \text{ mol Cu}}{2 \text{ mol } AgNO_3} = 2.75 \text{ mol Cu}$ $2.50 \text{ mol Cu} \times \dfrac{2 \text{ mol Ag}}{1 \text{ mol Cu}} = 5.00 \text{ mol Ag}$

The Cu is thus the limiting reactant.   **d.** $2.50 \text{ mol } Cu(NO_3)_2 \times \dfrac{188 \text{ g } Cu(NO_3)_2}{\text{mol } Cu(NO_3)_2}$

b. 5.50 mol $AgNO_3$ available − 5.00 mol $AgNO_3$ needed = 0.50 mol excess $AgNO_3$ $= 470 \text{ g } Cu(NO_3)_2$

$5.00 \text{ mol Ag} \times \dfrac{108 \text{ g Ag}}{\text{mol Ag}} = 540. \text{ g Ag}$

18. a. $3H_2SO_4 + 2Al(OH)_3 \rightarrow Al_2(SO_4)_3 + 6H_2O$

$30.0 \text{ g } H_2SO_4 \times \dfrac{\text{mol } H_2SO_4}{98.1 \text{ g } H_2SO_4} = 0.306 \text{ mol } H_2SO_4$

$25.0 \text{ g } Al(OH)_3 \times \dfrac{\text{mol } Al(OH)_3}{78.0 \text{ g } Al(OH)_3} = 0.321 \text{ mol } Al(OH)_3$

$0.306 \text{ mol } H_2SO_4 \times \dfrac{2 \text{ mol } Al(OH)_3}{3 \text{ mol } H_2SO_4} = 0.204 \text{ mol } Al(OH)_3$

The limiting reactant is thus H_2SO_4.

b. $[0.321 \text{ mol } Al(OH)_3 - 0.204 \text{ mol } Al(OH)_3] \times \dfrac{78.0 \text{ g } Al(OH)_3}{\text{mol } Al(OH)_3} = 9.13 \text{ g excess } Al(OH)_3$

c. $0.306 \text{ mol } H_2SO_4 \times \dfrac{1 \text{ mol } Al_2(SO_4)_3}{3 \text{ mol } H_2SO_4} \times \dfrac{342 \text{ g } Al_2(SO_4)_3}{\text{mol } Al_2(SO_4)_3} = 34.9 \text{ g } Al_2(SO_4)_3$

$0.306 \text{ mol } H_2SO_4 \times \dfrac{6 \text{ mol } H_2O}{3 \text{ mol } H_2SO_4} \times \dfrac{18.0 \text{ g } H_2O}{\text{mol } H_2O} = 11.0 \text{ g } H_2O$

19. a. $2N_2H_4 + (CH_3)_2N_2H_2 + 3N_2O_4 \rightarrow 6N_2 + 2CO_2 + 8H_2O$

$1200. \text{ kg } N_2H_4 \times \dfrac{10^3 \text{ g}}{\text{kg}} \times \dfrac{1 \text{ mol } N_2H_4}{32.05 \text{ g } N_2H_4} = 3.744 \times 10^4 \text{ mol } N_2H_4$

$1000. \text{ kg } (CH_3)_2N_2H_2 \times \dfrac{10^3 \text{ g}}{\text{kg}} \times \dfrac{1 \text{ mol } (CH_3)_2N_2H_2}{60.10 \text{ g } (CH_3)_2N_2H_2} = 1.664 \times 10^4 \text{ mol } (CH_3)_2N_2H_2$

$4500. \text{ kg } N_2O_4 \times \dfrac{10^3 \text{ g}}{\text{kg}} \times \dfrac{1 \text{ mol } N_2O_4}{92.01 \text{ g } N_2O_4} = 4.891 \times 10^4 \text{ mol } N_2O_4$

$3.744 \times 10^4 \text{ mol } N_2H_4 \times \dfrac{1 \text{ mol } (CH_3)_2N_2H_2}{2 \text{ mol } N_2H_4} = 1.872 \times 10^4 \text{ mol } (CH_3)_2N_2H_2$

The compound $(CH_3)_2N_2H_2$ *seems* to limit this reaction thus far.

$1.664 \times 10^4 \text{ mol } (CH_3)_2N_2H_2 \times \dfrac{3 \text{ mol } N_2O_4}{1 \text{ mol } (CH_3)_2N_2H_2} = 4.992 \times 10^4 \text{ mol } N_2O_4$

The N_2O_4 is thus the actual limiting reactant. As such, it is the fuel component first used up.

b. The amount of water thus produced is:

$4.891 \times 10^4 \text{ mol } N_2O_4 \times \dfrac{8 \text{ mol } H_2O}{3 \text{ mol } N_2O_4} \times \dfrac{18.02 \text{ g } H_2O}{\text{mol } H_2O} \times \dfrac{\text{kg}}{10^3 \text{ g}} = 2.350 \times 10^3 \text{ kg } H_2O$

20. a. % yield $= \dfrac{\text{actual yield}}{\text{theoretical yield}} \times 100\% = \dfrac{15.0 \text{ g}}{20.0 \text{ g}} \times 100\% = 75.0\%$ yield

b. % yield $= \dfrac{22.5 \text{ g}}{24.0 \text{ g}} \times 100\% = 93.8\%$ yield

c. % yield $= \dfrac{4.75 \text{ g}}{5.00 \text{ g}} \times 100\% = 95.0\%$ yield

21. a. % yield $= \dfrac{\text{actual yield}}{\text{theoretical yield}} \times 100\%$

actual yield $= \dfrac{\text{theoretical yield} \times \text{\% yield}}{100\%}$

actual yield $= \dfrac{(12.0 \text{ g})(90.0\%)}{100\%} = 10.8 \text{ g}$

b. actual yield $= \dfrac{(8.50 \text{ g})(70.0\%)}{100\%} = 5.95 \text{ g}$

c. actual yield $= \dfrac{(3.45 \text{ g})(48.0\%)}{100\%} = 1.66 \text{ g}$

22. $5.00 \text{ kg } NH_3 \times \dfrac{10^3 \text{ g}}{\text{kg}} \times \dfrac{1 \text{ mol } NH_3}{17.0 \text{ g } NH_3} \times \dfrac{4 \text{ mol } NO}{4 \text{ mol } NH_3} \times 80.0\% = 235 \text{ mol } NO$

$235 \text{ mol } NO \times \dfrac{2 \text{ mol } NO_2}{2 \text{ mol } NO} \times 80.0\% = 188 \text{ mol } NO_2$

$$188 \text{ mol NO}_2 \times \frac{2 \text{ mol HNO}_3}{3 \text{ mol NO}_2} \times \frac{63.0 \text{ g HNO}_3}{\text{mol NHO}_3} \times 80.0\% = 6.32 \times 10^3 \text{ g HNO}_3$$

Critical Thinking

1. The chemical equation does provide information related to quantitative amounts of reactants and products involved in the reaction, but it cannot describe detailed reaction information. Thus, the chemical equation can provide the theoretical yield, which would be obtained under ideal conditions. The actual yield is always less than the amount predicted by the chemical equation. The actual yield and the theoretical yield can be used to calculate the percent yield.
2. Answers will vary:
   Use only a clean, dry crucible
   Make mass determinations carefully
   Add an excess of oxygen
   Take at least 2 final mass determinations with additional heating between the two weights to be certain that the reaction has gone to completion
   Remove any byproducts that may form from the reaction that will interfere with the final product

Cross Connections

1. Answers will vary.

Teacher's Notes

Chapter 9
Stoichiometry

SECTION PREVIEW

9.1 Introduction to Stoichiometry

9.2 Ideal Stoichiometric Calculations

9.3 Limiting Reactants and Percent Yield

Chemists can plan the outcomes of chemical reactions by strictly controlling the amounts of reactants that they use.

9.1 Introduction to Stoichiometry

Much of our knowledge of chemistry is based on the careful quantitative analyses of substances involved in chemical reactions. *The branch of chemistry that deals with the mass relationships of elements in compounds and the mass relationships between reactants and products in chemical reactions is known as* **stoichiometry**. The word *stoichiometry* comes from the Greek words *stoichion*, which means element, and *metron*, which means measure.

Applications of Stoichiometry

You are already familiar with one aspect of stoichiometry: **Composition stoichiometry** *involves the mass relationships of elements in chemical compounds.* The calculations involving particle–mass relations in Chapter 3 are examples of composition stoichiometry.

Another aspect of stoichiometry, called reaction stoichiometry, is concerned with chemical reactions. **Reaction stoichiometry** *involves the mass relationships among reactants and products in a chemical reaction.* It is based on chemical equations and the law of conservation of matter. A chemical equation can provide much information. For example, chemical equations indicate the amounts of reactants needed to produce a given amount of product, or the amount of one reactant needed to react with a given amount of another reactant. All reaction-stoichiometry calculations require knowing the chemical equation for the reaction under study.

SECTION OBJECTIVES

1. Define *stoichiometry* and distinguish between composition and reaction stoichiometry.
2. Define *mole ratio* and describe its role in stoichiometry calculations.
3. Give the mole ratio for any two substances in a chemical equation.
4. Name the four types of reaction-stoichiometry calculations.

Emphasize that if students know the amount of one substance in a reaction, they can determine the amounts of all of the other substances. They might be given one reactant and be asked to determine the amount of a second reactant needed to complete the reaction, or they might be given the amount of a product and be asked to determine the amount of reactant that is needed to produce it.

Figure 9-1 Copper sulfate pentahydrate *(left)*, when heated, decomposes to copper sulfate *(right)* and water vapor.
$CuSO_4 \cdot 5H_2O(s) \xrightarrow{\Delta} CuSO_4(s) + 5H_2O(g)$

Reaction-Stoichiometry Problems

All of the reaction-stoichiometry problems in this chapter can be classified as one of four types: mole–mole, mole–mass, mass–mole, or mass–mass problems. In each of these types, the given information is the amount of one or more substances involved in a chemical reaction. The amounts are usually expressed in moles or grams. Along with the chemical equation for the reaction, this information permits calculation of the amounts of the other

Stoichiometry **261**

substances involved in the reaction. As you will see in later chapters, you might be given the amount of reactants or products in units other than moles or grams. All stoichiometry problems are solved by converting the given units into moles or grams and then applying the methods described here.

Routes for solving reaction-stoichiometry problems Figure 9-2 shows the pathways or routes by which the four types of problems may be solved. The use of the conversion factors needed to convert between the blocks, namely the molar mass and the mole ratio, are discussed later in this section. Molar masses are needed for mass–mole or mole–mass conversions. Mole ratios are needed for mole–mole conversions between two substances in a chemical reaction.

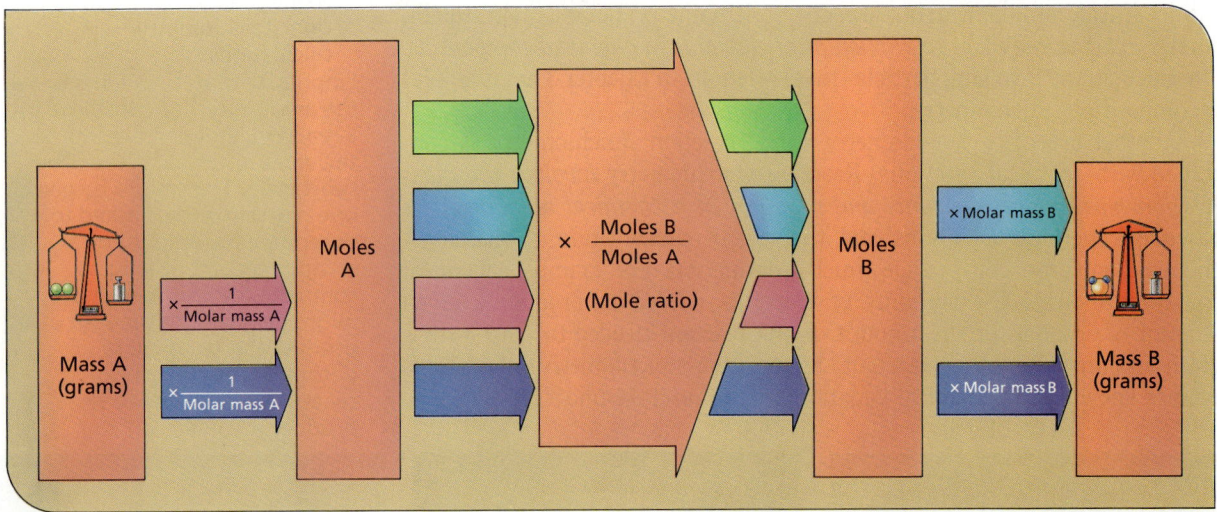

Figure 9-2 Solving the four types of reaction-stoichiometry problems

1. In *mole–mole problems*, you are given the number of moles of one substance and asked to calculate the number of moles of another substance in the chemical reaction. According to Figure 9-2, the general route is

    moles of A $\rightarrow$ moles of B

2. In *mole–mass problems*, you are given the number of moles of one substance and asked to calculate the mass of another substance in the chemical reaction. According to Figure 9-2, the general route is

    moles of A $\rightarrow$ moles of B $\rightarrow$ mass of B

Remind students to refer to the equation to get the information needed to relate moles of B to moles of A.

3. In *mass–mole problems*, you are given the mass of one substance and asked to calculate the number of moles of another substance in the chemical reaction. According to Figure 9-2, the general route is

    mass of A $\rightarrow$ moles of A $\rightarrow$ moles of B

4. In *mass–mass problems*, you are given the mass of one substance and asked to calculate the mass of another substance in the chemical reaction. According to Figure 9-2, the general route is

    mass of A $\rightarrow$ moles of A $\rightarrow$ moles of B $\rightarrow$ mass of B

Mole ratio As illustrated in Figure 9-2, each of the four types of reaction-stoichiometry problems requires the use of a mole ratio. A **mole ratio** is a *conversion factor that relates the number of moles of any two substances involved in a chemical reaction.* This information is obtained directly from the balanced chemical equation.

Consider, for example, the chemical equation for the electrolysis of aluminum oxide to produce aluminum and oxygen:

$$2Al_2O_3(\ell) \rightarrow 4Al(s) + 3O_2(g)$$

Recall from Chapter 8 that the coefficients in a chemical equation satisfy the law of conservation of matter and represent the relative numbers of moles of reactants and products. Therefore, 2 mol of aluminum oxide decompose to produce 4 mol of aluminum metal and 3 mol of oxygen gas. These relationships can be expressed in the following mole ratios:

$$\frac{2 \text{ mol } Al_2O_3}{4 \text{ mol } Al} \quad \text{or} \quad \frac{4 \text{ mol } Al}{2 \text{ mol } Al_2O_3}$$

$$\frac{2 \text{ mol } Al_2O_3}{3 \text{ mol } O_2} \quad \text{or} \quad \frac{3 \text{ mol } O_2}{2 \text{ mol } Al_2O_3}$$

$$\frac{4 \text{ mol } Al}{3 \text{ mol } O_2} \quad \text{or} \quad \frac{3 \text{ mol } O_2}{4 \text{ mol } Al}$$

For the decomposition of aluminum oxide, the appropriate mole ratio would be used as a conversion factor to convert a given number of moles of one substance to the corresponding number of moles of another substance. To determine the number of moles of aluminum that can be produced from 13.0 mol of aluminum oxide, the mole ratio needed is that of Al to Al_2O_3.

$$\text{mol } Al_2O_3 \times \frac{\text{mol } Al}{\text{mol } Al_2O_3} = \text{mol } Al$$

$$13.0 \text{ mol } Al_2O_3 \times \frac{4 \text{ mol } Al}{2 \text{ mol } Al_2O_3} = 26.0 \text{ mol } Al$$

Moles are the "currency" of chemistry. Moles of one substance can be "cashed in" for moles of another substance. If some of your students have lived in or visited other countries, refer to the exchange of currency when you use this analogy.

Figure 9-3 Industrial production of aluminum via the electrolysis of aluminum oxide

Stoichiometry

Chemistry and its Applications

Scientific Journals: Getting the Word Out

To stay up-to-date on any topic, from racing cars to antique furniture, people read magazines that cover their fields of interest. This is also true for scientists, whose specialty periodicals are called journals. Very recent scientific findings are not contained in books, but in journal articles.

Since there are thousands of journals, a scientist often uses a special index to find articles about a subject. The most widely used index in chemistry is *Chemical Abstracts*, a weekly publication that provides short summaries of articles recently published in all the major disciplines of chemistry. Computers can also be used to search the enormous number of cross-referenced entries in *Chemical Abstracts*.

Scientists write journal articles to gain recognition from their peers in the scientific community and establish that they were the first to find a particular result. Scientists read articles written by other scientists to keep

abreast of the latest developments in areas of research. For instance, a scientist in the United States working on heat-resistant polymers might read a recent article written by a German scientist who is working in the same field. The United States scientist might gain fresh ideas from the German scientist's work.

Also, the questions and comments that often follow the publication of a journal article are helpful to the author and the readers. Finally, a scientist's results must be verified by others before they are accepted, and this can happen only if others find out about the work and try to duplicate the results.

Journals contain articles that are reviewed for scholarship and scientific accuracy. Newspapers also contain articles about science. Although they normally do not have review boards, they can present some information about current scientific developments.

Because mole ratios are exact, they do not limit the number of significant figures in a calculation. The number of significant figures in the answer is therefore determined only by the number of significant figures of any measured quantities in a particular problem.

Molar mass Recall from Chapter 7 that the molar mass is the mass, in grams, of one mole of a substance. The molar mass is the conversion factor that relates the mass of a substance to the number of moles of that substance, as illustrated in Figure 9-2. To solve mass–mole problems, mole–mass

264 Chapter 9

problems, and mass–mass problems, the appropriate molar masses must therefore be found.

Returning to the previous example, the decomposition of aluminum oxide, what are the molar masses of the substances involved in this reaction? Written as conversion factors, these molar masses are

$$\frac{102 \text{ g Al}_2\text{O}_3}{\text{mol Al}_2\text{O}_3} \quad \text{or} \quad \frac{1 \text{ mol Al}_2\text{O}_3}{102 \text{ g Al}_2\text{O}_3}$$

$$\frac{27.0 \text{ g Al}}{\text{mol Al}} \quad \text{or} \quad \frac{1 \text{ mol Al}}{27.0 \text{ g Al}} \quad\quad \frac{32.0 \text{ g O}_2}{\text{mol O}_2} \quad \text{or} \quad \frac{1 \text{ mol O}_2}{32.0 \text{ g O}_2}$$

For example, to find the number of grams of aluminum equivalent to 12.0 mol of aluminum, the calculation would be as follows:

$$\text{moles Al} \times \frac{\text{grams Al}}{\text{mol Al}} = \text{grams Al}$$

$$12.0 \text{ mol Al} \times \frac{27.0 \text{ g Al}}{\text{mol Al}} = 324 \text{ g Al}$$

Section Review

1. What is stoichiometry?
2. List the four basic types of reaction-stoichiometry problems.
3. What is the role of mole ratio in reaction stoichiometry problems?
4. For each of the following chemical equations, write all possible mole ratios: (a) $2\text{HgO}(s) \rightarrow 2\text{Hg}(\ell) + \text{O}_2(g)$ (b) $4\text{NH}_3(g) + 6\text{NO}(g) \rightarrow 5\text{N}_2(g) + 6\text{H}_2\text{O}(\ell)$ (c) $2\text{Al}(s) + 3\text{H}_2\text{SO}_4(aq) \rightarrow \text{Al}_2(\text{SO}_4)_3(aq) + 3\text{H}_2(g)$.

1. The branch of chemistry that deals with the mass relationships of elements in compounds, and the mass relationships among reactants and products in chemical reactions
2. Mole-mole, mole-mass, mass-mole, and mass-mass problems
3. Mole ratio is used as a conversion factor to convert a given number of moles of one substance to the corresponding number of moles of another substance.
4. (a) Refer to p. 263 for illustration of mole ratios.

9.2 Ideal Stoichiometric Calculations

In the previous section, four types of reaction-stoichiometry calculations were introduced—mole–mole, mole–mass, mass–mole, and mass–mass calculations. The chemical equation plays a very important part in these calculations, since the mole ratio is obtained directly from it.

Chemical equations help us understand and make predictions about chemical reactions without having to actually work the reactions in the laboratory. Chemical equations do have certain limitations, however. They do not reveal the conditions under which reactions actually occur. Furthermore, as you learned in Chapter 8, chemical equations can be written to describe chemical changes that do not occur. Finally, chemical equations do not tell us how fast reactions occur or the pathways of the reactions.

Factors not revealed by chemical equations can also affect the relative amounts of reactants needed or products actually produced in chemical reactions. These limitations are not taken into account in this section. The reaction-stoichiometry calculations illustrated here are therefore theoretical. They tell us the amounts of reactants needed and the products that are yielded for a given chemical reaction under ideal conditions. By *ideal conditions*, we refer to the complete conversion of all reactants into products. Rarely are

SECTION OBJECTIVES

- Calculate the amount in moles of a reactant or product, given the amount in moles of a different reactant or product.
- Calculate the mass of a reactant or product, given the amount in moles of a different reactant or product.
- Calculate the amount in moles of a reactant or product, given the mass of a different reactant or product.
- Calculate the mass of a reactant or product, given the mass of a different reactant or product.

Stoichiometry 265

ideal conditions met in the laboratory or in industry. Theoretical stoichiometry calculations serve a very important function nonetheless. For example, it is useful to know the maximum amount of product possible before a reaction is carried out in the laboratory. Some stoichiometry calculations take into account actual laboratory and industrial conditions to make a more accurate determination of the quantities involved in a reaction.

Solving stoichiometry problems requires practice. Please do not think of the reaction stoichiometry problems presented here as four new types of problems to solve. See them instead as extensions of the composition stoichiometry problems you solved in Chapters 3 and 7. Practice working the sample problems in the rest of this section. You will discover that a logical, systematic approach to solving such problems can be quite successful.

Mole–Mole Calculations

In mole–mole stoichiometry problems, you are asked to calculate the number of moles of one substance that will react with or be produced from another substance. According to Figure 9-2, the route for a mole–mole problem is

$$\text{moles of substance A} \rightarrow \text{moles of substance B}$$

This route requires only one conversion factor—the mole ratio of B to A. To solve this type of problem, simply multiply the *known* quantity (the number of moles of A) by the appropriate conversion factor, which in this case is the mole ratio B/A.

$$\text{given information} \times \text{conversion factor} = \text{unknown information}$$

$$\text{moles A} \times \text{mole ratio of } \frac{\text{B}}{\text{A}} = \text{moles B}$$

Figure 9-4 Lithium hydroxide is used in spacecraft to chemically remove carbon dioxide from the cabin. Here, astronauts Daniel Brandenstein *(left)* and Guion Buford *(right)* check the stowage area for LiOH canisters on the space shuttle *Challenger* in 1983.

Sample Problem 9.1

In a spacecraft, the carbon dioxide exhaled by astronauts can be removed by its reaction with lithium hydroxide (LiOH) according to the following chemical equation

$$CO_2(g) + 2LiOH(s) \rightarrow Li_2CO_3(s) + H_2O(\ell)$$

How many moles of lithium hydroxide are required to react with 20 mol of CO_2, the average amount exhaled by a person each day?

Solution

Step 1. Analyze   Given: chemical equation amount of CO_2 = 20 mol
Unknown: amount of LiOH in moles

Step 2. Plan   moles of CO_2 → moles of LiOH

Only one conversion factor is needed—the mole ratio of LiOH to CO_2. This factor can be chosen from the given chemical equation

$$\text{moles of } CO_2 \times \frac{\text{moles LiOH}}{\text{moles } CO_2} = \text{moles LiOH}$$

Step 3. Compute   Substitute the values in the equation in Step 2 and compute the answer.

$$20 \text{ mol } CO_2 \times \frac{2 \text{ mol LiOH}}{1 \text{ mol } CO_2} = 40 \text{ mol LiOH}$$

Step 4. Evaluate   The answer is correctly given to one significant figure as in the factor 20 mol CO_2. Cancellation of the units leaves *mol LiOH*, exactly as the problem asked for.

Practice Problems

1. Ammonia (NH_3) is widely used as a fertilizer and in many household cleaners. How many moles of ammonia are produced when 6 mol of hydrogen gas react with an excess of nitrogen gas? *(Ans.)* 4 mol NH_3
2. The decomposition of potassium chlorate ($KClO_3$) is used as a source of oxygen in the laboratory. How many moles of potassium chlorate are needed to produce 15 mol of oxygen? *(Ans.)* 10. mol $KClO_3$

Mole–Mass Calculations

In mole–mass stoichiometry problems, you are asked to calculate the mass (usually in grams) of a substance that will react with or be produced from a given number of moles of a second substance.

Referring back to Figure 9-2, the route for mole–mass problems is

$$\text{moles A} \rightarrow \text{moles B} \rightarrow \text{mass B}$$

This route requires two conversion factors—the mole ratio of substance B to substance A and the molar mass of substance B. To solve this kind of problem, you simply multiply the known quantity, the number of moles of A, by the appropriate conversion factors.

$$\text{moles A} \times \text{mole ratio of } \frac{B}{A} \times \text{molar mass of B} = \text{mass of B}$$

STUDY HINT

At this point it would be useful to review balancing chemical equations, in Section 8.1.

Stoichiometry

Sample Problem 9.2

In photosynthesis, plants use energy from the sun to combine carbon dioxide and water, forming glucose ($C_6H_{12}O_6$) and oxygen. What mass, in grams, of glucose is produced when 3.00 mol of water react with carbon dioxide?

Solution

Step 1. Analyze Given: names of reactants and products
amount of H_2O = 3.00 mol
Unknown: mass of $C_6H_{12}O_6$ produced in grams

Step 2. Plan moles of H_2O → moles $C_6H_{12}O_6$ → mass $C_6H_{12}O_6$

To solve this problem requires writing the chemical equation and finding two conversion factors—the mole ratio of $C_6H_{12}O_6$ to H_2O and the molar mass of $C_6H_{12}O_6$.

$$\text{moles } H_2O \times \frac{\text{moles } C_6H_{12}O_6}{\text{moles } H_2O} \times \frac{\text{grams } C_6H_{12}O_6}{\text{mole } C_6H_{12}O_6} = \text{grams } C_6H_{12}O_6$$

Step 3. Compute Write the chemical equation for the reaction that occurs during the photosynthesis process.

$$6CO_2(g) + 6H_2O(\ell) \rightarrow C_6H_{12}O_6(s) + 6O_2(g)$$

$$3.00 \text{ mol } H_2O \times \frac{1 \text{ mol } C_6H_{12}O_6}{6 \text{ mol } H_2O} \times \frac{180 \text{ g } C_6H_{12}O_6}{1 \text{ mol } C_6H_{12}O_6} = 90.0 \text{ g } C_6H_{12}O_6$$

Step 4. Evaluate The answer is correctly rounded off to three significant figures, as in 3.00 mol H_2O. Cancellation of the units in the problem as set up leaves "g $C_6H_{12}O_6$" as the units for the answer. This is correct since the problem asked how many grams of glucose are produced. The answer is reasonable, for it is one-half of 180, and 3/6 = 1/2.

Sample Problem 9.3

What mass, in grams, of carbon dioxide is needed to react with 3.00 mol of H_2O in the photosynthesis reaction described in Sample Problem 9.2?

Solution

Step 1. Analyze Given: chemical equation
amount of H_2O = 3.00 moles
Unknown: amount of CO_2 needed in grams

Step 2. Plan moles of H_2O → moles CO_2 → mass CO_2

Two conversion factors are needed—the mole ratio of CO_2 to H_2O and the molar mass of CO_2.

$$\text{moles } H_2O \times \frac{\text{moles } CO_2}{\text{moles } H_2O} \times \frac{\text{grams } CO_2}{\text{mole } CO_2} = \text{grams } CO_2$$

Sample Problem 9.3 *continued*

Step 3. Compute   The chemical equation from Sample Problem 9.2 is $6CO_2(g) + 6H_2O(l) \rightarrow C_6H_{12}O_6(s) + 6O_2(g)$

$$3.00 \text{ mol } H_2O \times \frac{6 \text{ mol } CO_2}{6 \text{ mol } H_2O} \times \frac{44.0 \text{ g } CO_2}{\text{mol } CO_2} = 132 \text{ g } CO_2$$

Step 4. Evaluate   The answer is correctly given to three significant figures, as in 3.00 mol H_2O. Cancellation of the units in the problem as set up leaves "g CO_2." This is correct since the problem asked for the amount needed, in grams, of CO_2. The answer is close to an estimate of 120, (3 × 40).

Practice Problems
1. When magnesium burns in air, it combines with oxygen to form magnesium oxide according to the following equation: $2Mg(s) + O_2(g) \rightarrow 2MgO(s)$. What mass in grams of magnesium oxide is produced from 2.00 mol of magnesium?   *(Ans.)* 80.6 g MgO
2. What mass in grams of oxygen combines with 2.00 mol of magnesium in this same reaction? *(Ans.)* 32.0 g O_2

In Sample Problems 9.2 and 9.3, the conversion sequence was from moles of one substance to mass of a second substance. Next, some problems that require conversion from mass into moles are explained.

Mass–Mole Calculations

In mass–mole stoichiometry problems, you are asked to calculate the amount, in moles, of one substance that will react with or be produced from a given mass of another substance. Referring back to Figure 9-2, the route for a mass–mole problem is

$$\text{mass A} \rightarrow \text{moles A} \rightarrow \text{moles B}$$

This route also requires two conversion factors: the molar mass of substance A and the mole ratio of substance B to substance A. To solve this type of problem, simply multiply or divide the known quantity, the mass of A, by the appropriate conversion factors as follows.

$$\frac{\text{mass A}}{\text{molar mass of A}} \times \text{mole ratio of } \frac{B}{A} = \text{moles B}$$

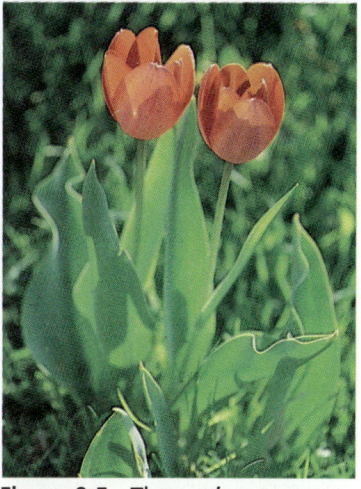

Figure 9-5 The sun's energy drives a reaction in green plants that turns carbon dioxide and water into glucose and oxygen.

Sample Problem 9.4

The industrial solvent carbon disulfide (CS_2) is produced through the following reaction between coke (C) and sulfur dioxide (SO_2): $C(s) + SO_2 \rightarrow CS_2(l) + CO(g)$ (unbalanced). If 8.00 g of SO_2 reacts, (a) how many moles of CS_2 are formed and   (b) how many moles of CO are formed?

Solution
Step 1. Analyze   Given: chemical equation
mass of SO_2 = 8.00 g
Unknown: *(a)* amount of CS_2 produced, in moles
(b) amount of CO produced, in moles

Stoichiometry

Sample Problem 9.4 continued

Step 2. Plan
(a) mass SO_2 → moles SO_2 → moles CS_2
(b) mass SO_2 → moles SO_2 → moles CO

Three conversion factors are needed, the molar mass of SO_2, the mole ratio of CS_2 to SO_2, and the mole ratio of CO to SO_2. To convert from mass of SO_2 to moles of SO_2, the SO_2 molar mass is used in its inverted form.

(a) $\text{grams } SO_2 \times \dfrac{1 \text{ mol } SO_2}{\text{grams } SO_2} \times \dfrac{\text{moles } CS_2}{\text{moles } SO_2} = \text{moles } CS_2$

(b) $\text{grams } SO_2 \times \dfrac{1 \text{ mol } SO_2}{\text{grams } SO_2} \times \dfrac{\text{moles CO}}{\text{moles } SO_2} = \text{moles CO}$

Step 3. Compute
The balanced chemical equation is:

$$5C(s) + 2SO_2(g) \rightarrow CS_2(l) + 4CO(g)$$

(a) $8.00 \text{ g } SO_2 \times \dfrac{1 \text{ mol } SO_2}{64.1 \text{ g } SO_2} \times \dfrac{1 \text{ mol } CS_2}{2 \text{ mol } SO_2} = 0.0624 \text{ mol } CS_2$

(b) $8.00 \text{ g } SO_2 \times \dfrac{1 \text{ mol } SO_2}{64.1 \text{ g } SO_2} \times \dfrac{4 \text{ mol CO}}{2 \text{ mol } SO_2} = 0.250 \text{ mol CO}$

Step 4. Evaluate
The answers are correctly given to three significant figures. Cancellation of units in the two problems set up leaves *mol CS_2* and *mol CO*, respectively. This is correct, since the problem asked for the number of moles of CS_2 and CO.

Practice Problems

Oxygen was discovered by Joseph Priestley in 1774 when he decomposed mercury(II) oxide to its constituent elements by heating it.

1. How many moles of mercury(II) oxide (HgO) are needed to produce 125 g of oxygen (O_2)? *(Ans.)* 7.81 mol HgO
2. How many moles of mercury are produced? *(Ans.)* 7.81 mol Hg

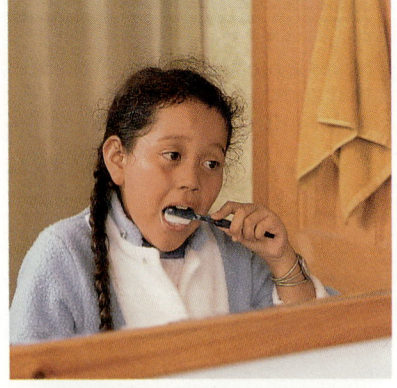

Figure 9-6 Stannous fluoride (SnF_2) is used in some toothpastes to help prevent cavities. (See Sample Problem 9.5.)

Mass–Mass Calculations

In mass–mass stoichiometry problems, you are asked to calculate the number of grams of one substance that is required to react with or be produced from a given number of grams of a second substance involved in the chemical reaction. For this reason, mass–mass calculations involve a more real-life situation than do mole–mole, mole–mass, and mass–mole calculations. You can never measure moles directly. You are generally required to calculate the number of moles of a substance from its mass, which you can measure in the lab. Mass–mass problems can be viewed as the combination of individual mole–mole, mass–mole, or mole–mass problems. Examine Figure 9-2 to prove this to yourself. The route for solving mass–mass problems is

mass A → moles A → moles B → mass B

Three conversion factors are needed to solve mass–mass problems: the molar mass of substance A, the mole ratio B/A, and the molar mass of B.

Sample Problem 9.5

Tin(II) fluoride (SnF_2), or stannous fluoride, is used in some home dental treatment products. It is made by the reaction of tin with hydrogen fluoride according to the equation

$$Sn(s) + 2HF(g) \rightarrow SnF_2(s) + H_2(g)$$

How many grams of SnF_2 are produced from the reaction of 30.00 g of HF with Sn?

Solution

Step 1. Analyze Given: chemical equation, amount of HF = 30.00 g
Unknown: amount of SnF_2 produced in grams

Step 2. Plan mass HF → moles HF → moles SnF_2 → mass SnF_2

Conversion factors needed: molar masses of HF and SnF_2 and SnF_2: HF mole ratio.

$$\text{grams HF} \times \frac{1 \text{ mol HF}}{\text{grams HF}} \times \frac{\text{moles SnF}_2}{\text{moles HF}} \times \frac{\text{grams SnF}_2}{\text{mol SnF}_2} = \text{grams SnF}_2$$

Step 3. Compute

$$30.00 \text{ g HF} \times \frac{1 \text{ mol HF}}{20.01 \text{ g HF}} \times \frac{1 \text{ mol SnF}_2}{2 \text{ mol HF}} \times \frac{156.7 \text{ g SnF}_2}{1 \text{ mol SnF}_2} = 117.5 \text{ g SnF}_2$$

Step 4. Evaluate The answer is correctly rounded off to four significant figures. Cancellation of the units leaves $g \: SnF_2$. The answer is close to an estimated value of 120.

Practice Problems

Laughing gas (nitrous oxide, N_2O) is sometimes used as an anesthetic in dental work. It is produced when ammonium nitrate is decomposed according to the reaction $NH_4NO_3(s) \rightarrow N_2O(g) + 2H_2O(\ell)$.
1. How many grams of NH_4NO_3 are required to produce 33.0 g of N_2O? *(Ans.)* 60.0 g NH_4NO_3
2. How many grams of water are produced in this reaction? *(Ans.)* 27.0 g H_2O

Section Review

1. Balance the equation below, and then, based on the number of moles of each reactant or product given, determine the corresponding number of moles of all remaining reactants and products involved: $NH_3 + O_2 \rightarrow N_2 + H_2O$. (a) 4 mol NH_3 (b) 4 mol N_2 (c) 4.5 mol O_2
2. Hydrogen gas can be produced through the following unbalanced reaction: $Mg(s) + HCl(aq) \rightarrow MgCl_2(aq) + H_2(g)$. If 2.50 moles of magnesium react completely, (a) what mass of HCl is consumed? (b) what mass of each product is produced?
3. Acetylene gas (C_2H_2) is produced as a result of the following reaction: $CaC_2(s) + 2H_2O(\ell) \rightarrow C_2H_2(g) + Ca(OH)_2(aq)$. (a) If 32.0 grams of CaC_2 are consumed in this reaction, how many moles of H_2O are needed? (b) How many moles of each product would be formed?
4. When sodium chloride reacts with silver nitrate, silver chloride precipitates. What mass of AgCl is produced from 75.0 g of $AgNO_3$?
5. Acetylene gas (C_2H_2), used in welding, produces an extremely hot flame when it burns in pure oxygen according to the reaction: $2C_2H_2(g) + 5O_2(g) \rightarrow 4CO_2(g) + 2H_2O(g)$. How many grams of each product are produced when 2.50×10^4 g of C_2H_2 burns completely?

1. $4NH_3 + 3O_2 \rightarrow 2N_2 + 6H_2O$ (a) 3 mol O_2, 2 mol N_2, 6 mol H_2O (b) 8 mol NH_3, 6 mol O_2, 12 mol H_2O (c) 6.0 mol NH_3, 3.0 mol N_2, 9.0 mol H_2O
2. $Mg(s) + 2HCl(aq) \rightarrow MgCl_2(aq) + H_2(g)$ (a) 182 g HCl (b) 238 g $MgCl_2$, 5.05 g H_2
3. (a) 0.998 mol H_2O (b) 0.499 mol C_2H_2, 0.499 mol $Ca(OH)_2$
4. 63.1 g AgCl
5. 8.47×10^4 g CO_2, 1.73×10^4 g H_2O

9.3 Limiting Reactants and Percent Yield

SECTION OBJECTIVES

1. Define *limiting reactant*.
2. Describe the method for determining which of two reactants is a limiting reactant.
3. Calculate the amount in moles or mass of a product, given the amounts in moles or masses of two reactants, one of which is in excess.
4. Define *theoretical yield*, *actual yield*, and *percent yield*.
5. Calculate percent yield, given actual yield and quantity of a reactant.

In this section, you will be introduced to two types of stoichiometry calculations that deal with actual reactions common in the laboratory. In the laboratory, reactants are combined in amounts different from the precise amounts required for complete reaction. The quantity of product separated after a reaction is finished is often less than the maximum quantity possible from the given amounts of reactants. In other words, real stoichiometric calculations are rarely as straightforward as ideal stoichiometric calculations.

Limiting Reactant

All the problems that you have seen thus far involved calculations in which you sought to find the amount of a substance that will react with or be produced from a given amount of a second substance. However, in the laboratory a reaction is rarely carried out with exactly required amounts of each of the reactants. In most cases one or more reactants is present in excess; that is, in more than the exact amounts required to react with the given amount of the other reactant according to the chemical equation.

When all of one reactant is used up, no more product can be formed, even if there is more of other reactants available. The substance that is completely used up first in a reaction is called the limiting reactant. *The* **limiting reactant** *is the reactant that limits the amount of the other reactants that can combine, and the amount of products formed in a chemical reaction. The substance that is not used up completely in a reaction is sometimes called the* **excess reactant.**

The concept of the limiting reactant is analogous to the relationship between the number of people that want to take a certain airplane flight and the number of seats available in the airplane. If there are 400 people that want to travel on the flight and only 350 seats are available, then only 350 people can go on the flight. The number of seats on the airplane limits the number of people that can travel. There are 50 people in excess.

You can apply the same reasoning to chemical reactions. Consider the reaction between carbon and oxygen to form carbon dioxide.

$$C(s) + O_2(g) \rightarrow CO_2(g)$$

According to the equation, one mole of carbon reacts with one mole of oxygen to form one mole of carbon dioxide. Suppose you could mix 5 mol of C with 10 mol of O_2 and allow the reaction to take place. There is twice as much oxygen as needed to react with the carbon. In this situation carbon is the limiting reactant and it limits how much CO_2 is formed. Oxygen here is the excess reactant, and 5 mol of O_2 will be unchanged at the end of the reaction. In real laboratory reactions, the reactant in excess is usually one that is easily obtainable and/or least expensive.

Given the amounts of each of two reactants, A and B, how can you decide which is the limiting reactant? You must discover which of the reactants requires more of the other than is available. The following method is recommended. Choose one of the two reactants; let's say you choose reactant

When you are told that one reactant is in excess, you do not have to be concerned with its quantity. The reaction is limited by the limiting reactant.

Figure 9-7 In a game of musical chairs, the chairs can be considered the limiting reactant.

Desktop Investigation

Limiting Reactants in a Cookie Recipe

Question
What are the limiting reactants in a cookie recipe? The limiting reactant sets a limit on the amount of product that can be produced.

Materials
0.5 cup sugar
0.5 cup brown sugar
1⅓ stick (0.67 cup) margarine
1 egg
0.5 tsp salt
1 tsp vanilla
0.5 tsp baking soda
1.5 cup flour
1.33 cup chocolate chips
mixing bowl
wooden spoon
measuring spoons and cups
cookie sheet

Procedure
1. Mix sugars and margarine together until smooth.
2. Add egg, salt, and vanilla and mix well.
3. Stir in baking soda, flour, and chocolate chips. Chill for best results.
4. Make dough into balls 3 cm in diameter; place on ungreased cookie sheet.
5. Bake at 350°F for about 10 minutes, until pale brown. Yields 24 chocolate-chip cookies.

Discussion
Suppose you have only the following materials: a dozen eggs, 4 fluid ounces of vanilla, a pound of salt, a pound of baking soda, 3 cups of chocolate chips, 5 pounds of flour, a pound of margarine, 5 pounds of sugar, and 2 pounds of brown sugar.

1. For each ingredient, calculate how many batches could be prepared if all of that ingredient were consumed. (For example, based on the 12 eggs available, at one egg per batch, 12 batches are possible.)
2. To determine the limiting reactants, identify the ingredient(s) that will produce the fewest batches.
3. How many batches can be produced from these materials?

A. Calculate the amount in moles of the other reactant, B, that is required by A. (This step is similar to solving a mole–mole or mass–mole problem.) Then compare the calculated amount with the amount of B actually available. If more is required than is available, B is the limiting reactant. If less is required, the reactant that you started with (reactant A) is the limiting reactant.

Finding which of two reactants is the limiting reactant is not difficult, but requires a bit of practice. The general method is illustrated in Sample Problems 9.6 and 9.7.

Sample Problem 9.6

Silicon dioxide (quartz) is usually quite stable but reacts with hydrogen fluoride according to the following equation.

$$SiO_2(s) + 4HF(g) \rightarrow SiF_4(g) + 2H_2O(\ell)$$

If 2.0 mol of HF is combined with 4.5 mol of SiO_2, which is the limiting reactant?

Solution

Step 1. Analyze
Given: chemical equation
amount of HF = 2.0 mol
amount of SiO_2 = 4.5 mol
Unknown: limiting reactant

Step 2. Plan
moles of HF → moles of SiO_2 or moles of SiO_2 → moles of HF

The given amount of either reactant is used to calculate the required amount of the other reactant. The calculated amount is then compared with the amount actually available, and the limiting reactant can be identified. We will choose to calculate the moles of SiO_2 required by the given amount of HF.

$$\text{moles HF} \times \frac{\text{moles } SiO_2}{\text{moles HF}} = \text{moles } SiO_2$$

Step 3. Compute

$$2.0 \text{ mol HF} \times \frac{1 \text{ mol } SiO_2}{4 \text{ mol HF}} = 0.50 \text{ mol } SiO_2$$

Under ideal conditions, the given amount of HF (2.0 mol) will require 0.50 mol of SiO_2 for complete reaction. Because the amount of SiO_2 available (4.5 mol) is more than the amount required (0.50 mol), the limiting reactant is HF.

Step 4. Evaluate
The calculated amount of SiO_2 is correctly given to two significant figures. Since each mole of SiO_2 requires 4 mol of HF, it is reasonable that HF is the limiting reactant, because the molar amount of HF available is less than half that of SiO_2.

Practice Problem

Some rocket engines use a mixture of hydrazine (N_2H_4) and hydrogen peroxide (H_2O_2) as the propellant system. The reaction is given by the equation

$$N_2H_4(\ell) + 2H_2O_2(\ell) \rightarrow N_2(g) + 4H_2O(g)$$

(a) Which is the limiting reactant in this reaction when 0.750 mol of N_2H_4 reacts with 0.500 mol of H_2O_2? *(Ans.)* H_2O_2 (b) How much of the excess reactant, in moles, remains unchanged? *(Ans.)* 0.500 mol N_2H_4 (c) How much of each product, in moles, is formed? *(Ans.)* 0.250 mol N_2, 1.00 mol H_2O

Sample Problem 9.7

The coating you see on a corroded iron object that has been left in moist soil is black iron oxide (Fe_3O_4). This substance can also be made in the laboratory by the reaction between iron and steam, according to the following equation:

$$3Fe(s) + 4H_2O(g) \rightarrow Fe_3O_4(s) + 4H_2(g)$$

Sample Problem 9.7 *continued*

(a) When 36.0 g of H_2O reacts with 167 g of Fe, which is the limiting reactant? How much of the excess reactant is used when the reaction is completed?

(b) What mass in grams of black iron oxide is produced?

Solution

Step 1. Analyze

Given: chemical equation
mass of H_2O = 36.0 g
mass of Fe = 167 g
Unknown: limiting reactant
mass of Fe_3O_4, in grams

Step 2. Plan

mass $H_2O \to$ moles H_2O; mass Fe $\to$ moles Fe

To determine which is limiting reactant

moles $H_2O \to$ moles Fe
moles limiting reactant $\to$ moles $Fe_3O_4 \to$ mass Fe_3O_4

(a) It is necessary to convert both given masses to moles. Then, choose one to calculate the needed amount of the other and determine which is the limiting reactant. We have chosen H_2O. The mole ratio Fe: H_2O is from the balanced equation

$$\text{grams } H_2O \times \frac{1 \text{ mole } H_2O}{\text{grams } H_2O} = \text{moles } H_2O$$

$$\text{grams Fe} \times \frac{1 \text{ mole Fe}}{\text{grams Fe}} = \text{moles Fe}$$

$$\text{moles } H_2O \times \frac{\text{moles Fe}}{\text{moles } H_2O} = \text{moles Fe}$$

(b) To find the maximum amount of Fe_3O_4 that can be produced, the given number of moles of the limiting reactant must be used in what is simply a mole–mass stoichiometry problem

$$\text{moles limiting reactant} \times \frac{\text{moles } Fe_3O_4}{\text{moles limiting reactant}} \times \frac{\text{grams } Fe_3O_4}{1 \text{ mol } Fe_3O_4} = \text{grams } Fe_3O_4$$

Step 3. Compute

$$36.0 \text{ g } H_2O \times \frac{1 \text{ mol } H_2O}{18.0 \text{ g } H_2O} = 2.00 \text{ mol } H_2O$$

$$167 \text{ g Fe} \times \frac{1 \text{ mol Fe}}{55.8 \text{ g Fe}} = 2.99 \text{ mol Fe}$$

$$2.00 \text{ mol } H_2O \times \frac{3 \text{ mol Fe}}{4 \text{ mol } H_2O} = 1.50 \text{ mol Fe}$$

(a) The required 1.50 mol of Fe is less than the 2.99 mol of Fe available, so H_2O is the limiting reactant. 2.99 − 1.50 = 1.49 mol Fe not used.

(b) $$2.00 \text{ mol } H_2O \times \frac{1 \text{ mol } Fe_3O_4}{4 \text{ mol } H_2O} \times \frac{232 \text{ g } Fe_3O_4}{\text{mol } Fe_3O_4} = 116 \text{ g } Fe_3O_4$$

Step 4. Evaluate

Three significant digits have been correctly carried through each calculation. At each step, units have cancelled correctly. Both questions asked have been

Stoichiometry

Sample Problem 9.7 continued

answered by identifying the limiting reactant and calculating the mass of product Fe_3O_4. The mass of Fe_3O_4 found in the last step is close to an estimated answer of 115, which is one-half of 230.

Practice Problems
1. Zinc and sulfur react to form zinc sulfide according to the following: $8Zn(s) + S_8(s) \rightarrow 8ZnS(s)$.
   (a) If 2.00 moles of Zn are heated with 1.00 mole of S_8, identify the limiting reactant. *(Ans.)* Zn
   (b) How many moles of excess reactant remain? *(Ans.)* 0.750 mol S_8 remains
   (c) How many moles of the product are formed? *(Ans.)* 2.00 mol ZnS
2. Carbon reacts with steam (H_2O) under certain conditions to produce hydrogen and carbon monoxide. (a) If 2.40 mol of carbon react with 3.10 moles of steam, identify the limiting reactant. *(Ans.)* C (b) How many moles of each product are formed? *(Ans.)* 2.40 mol H_2 and 2.40 mol CO (c) What mass of each product is formed? *(Ans.)* 4.85 g H_2 and 67.2 g CO
3. Aluminum reacts with hydrochloric acid according to the following unbalanced equation: $Al + HCl \rightarrow AlCl_3 + H_2$. (a) If 18 g of Al are combined with 75 g of HCl, which reactant is the limiting reactant? *(Ans.)* Al (b) What mass of each product is formed? *(Ans.)* 87 g $AlCl_3$ and 2.0 g H_2

Percent Yield

The amounts of products calculated in the stoichiometry problems in this chapter so far represent theoretical yields. The **theoretical yield** *is the maximum amount of product that can be produced from a given amount of reactant.* In most chemical reactions, the amount of product obtained is less than the theoretical yield. *The measured amount of a product obtained from a reaction is called the* **actual yield** *of that product.*

Chemists, like the one shown in Figure 9-8, strive to vary the conditions of a reaction so that the largest possible actual yield is obtained. There are many reasons why the actual yield may be less than the theoretical yield. Some of the reactant might take part in competing side reactions that reduce the amount of the desired product. Also once a product is formed it is usually

Figure 9-8 One goal of the chemist is to get the maximum possible yield from a reaction

collected in impure form. Often, during the purification process some of the product is lost.

Chemists are usually interested in the efficiency of a reaction. The efficiency is expressed by comparing the actual and theoretical yields. *The* **percent yield** *is the ratio of the actual yield to the theoretical yield, multiplied by 100 percent.*

$$\text{percent yield} = \frac{\text{actual yield}}{\text{theoretical yield}} \times 100\%$$

Sample Problem 9.8

Chlorobenzene (C_6H_5Cl) is used in the production of many important chemicals such as aspirin, dyes, and disinfectants. One industrial method of preparing chlorobenzene, used in the chemical industry, is the reaction between benzene (C_6H_6) and chlorine, represented by the equation

$$C_6H_6(\ell) + Cl_2(g) \rightarrow C_6H_5Cl(s) + HCl(g)$$

When 36.8 g of C_6H_6 react with an excess of Cl_2, the actual yield of C_6H_5Cl is 38.8 g. What is the percent yield of C_6H_5Cl?

Solution

Step 1. Analyze Given: chemical equation
mass of C_6H_6 = 36.8 g
mass of Cl_2 = excess
actual yield of C_6H_5Cl = 38.8 g
Unknown: percent yield of C_6H_5Cl

Step 2. Plan mass C_6H_6 → moles C_6H_6 → moles C_6H_5Cl → mass C_6H_5Cl → percent yield

First it is necessary to do a mass–mass calculation from which can be found the theoretical yield of C_6H_5Cl.

$$\text{grams } C_6H_6 \times \frac{\text{mole } C_6H_6}{\text{grams } C_6H_6} \times \frac{\text{mole } C_6H_5Cl}{\text{mole } C_6H_6} \times \frac{\text{grams } C_6H_5Cl}{\text{mole } C_6H_5Cl}$$
$$= \text{grams } C_6H_5Cl \text{ (theoretical yield)}$$

Then the percent yield can be found.

$$\text{percent yield } C_6H_5Cl = \frac{\text{actual yield}}{\text{theoretical yield}} \times 100\%$$

Step 3. Compute
$$36.8 \text{ g } C_6H_6 \times \frac{1 \text{ mol } C_6H_6}{78.1 \text{ g } C_6H_6} \times \frac{1 \text{ mol } C_6H_5Cl}{1 \text{ mol } C_6H_6} \times \frac{113 \text{ g } C_6H_5Cl}{\text{mol } C_6H_5Cl}$$
$$= 53.2 \text{ g } C_6H_5Cl \text{ (theoretical yield)}$$

$$\text{percent yield} = \frac{38.8 \text{ g}}{53.2 \text{ g}} \times 100\% = 72.9\%$$

Step 4. Evaluate The answer is correctly rounded off to three significant figures, as in 36.8 g C_6H_6. The units have cancelled correctly. The theoretical yield is close to an estimated value of 50 g, (one-half 100 g). The percent yield is close to an estimated value of 70%, (70/100 × 100%).

Sample Problem 9.8 continued

Practice Problems

1. Methanol can be produced through the reaction of CO and H_2 in the presence of a catalyst:

$$CO(g) + 2H_2(g) \xrightarrow{catalyst} CH_3OH(\ell)$$

If 75.0 g of CO reacts to produce 68.4 g CH_3OH, what is the percent yield of CH_3OH? *(Ans.)* 79.8%

2. Aluminum reacts with excess copper(II) sulfate according to the reaction given below. If 1.85 g of Al react and the percent yield of Cu is 56.6%, what mass of Cu is produced?
$Al(s) + CuSO_4(aq) \rightarrow Al_2(SO_4)_3(aq) + Cu(s)$ (unbalanced) *(Ans.)* 3.70 g

1. (a) O_2 is the limiting reactant.
(b) 0.667 mol CS_2 remains
(c) 0.333 mol CO_2, 0.667 mol SO_2
2. (a) H_2O is the limiting reactant. (b) 0.333 mol Mg remains (c) 19.4 g $Mg(OH)_2$, 0.674 g H_2
3. (a) The CuO is the limiting reactant (b) 15.9 g Cu
4. 93.8% yield

Section Review

1. Carbon disulfide burns in oxygen to yield carbon dioxide and sulfur dioxide according to the following chemical equation: $CS_2(\ell) + 3O_2(g) \rightarrow CO_2(g) + 2SO_2(g)$. (a) If 1.00 mol of CS_2 is combined with 1.00 mol of O_2, identify the limiting reactant. (b) How many moles of excess reactant remain? (c) How many moles of each product are formed?
2. Metallic magnesium reacts with water to produce magnesium hydroxide and hydrogen gas. (a) If 16.2 g of Mg are combined with 12.0 g of H_2O, what is the limiting reactant? (b) How many moles of the excess reactant are left? (c) How many grams of each product are formed?
3. (a) What is the limiting reactant when 19.9 g of CuO react with 2.02 g of H_2 according to the following: $CuO(s) + H_2(g) \rightarrow Cu(s) + H_2O(g)$? (b) How many grams of Cu are produced?
4. Quicklime, CaO, can be prepared by roasting limestone, $CaCO_3$, according to the following reaction: $CaCO_3(s) \xrightarrow{\Delta} CaO(s) + CO_2(g)$. When 2.00×10^3 g of $CaCO_3$ are heated, the actual yield of CaO is 1.05×10^3 g. What is the percent yield?

Chapter Summary

- Reaction stoichiometry involves the mass relationships between reactants and products in a chemical reaction.
- There are four types of reaction stoichiometry that may need to be determined in a chemical reaction. They are mole–mole, mole–mass, mass–mole or mass–mass. Each of the four relationships can be found using one of four problem-solving routes.
- A *mole ratio* is the conversion factor that relates the number of moles of any two substances in a chemical reaction. It is one of the quantities that must be known in order to determine any one of the possible relationships being considered.
- In actual reactions the reactants are combined in proportions different from the precise proportions required for complete reaction. Given certain quantities of reactants, the quantity of product is always less than the maximum possible.
- The limiting reactant controls the amount of product formed.
- Percent yield = actual yield ÷ theoretical yield × 100%.

CHAPTER 9 REVIEW

Stoichiometry

VOCABULARY

actual yield (276)
composition stoichiometry (261)
excess reactant (272)
limiting reactant (272)
mole ratio (263)
percent yield (276)
reaction stoichiometry (261)
stoichiometry (261)
theoretical yield (276)

REVIEWING CONCEPTS

1. **a.** Distinguish between composition stoichiometry and reaction stoichiometry.
   b. Give one example of each type of calculation. (9.1)
2. **a.** Explain the concept of "mole ratio" as used in reaction stoichiometry problems.
   b. What is the source of this value? (9.1)
3. What information is given by the coefficients in a chemical equation? (9.1)
4. For each of the following chemical equations, write all possible mole ratios:
   a. $2Ca + O_2 \rightarrow 2CaO$
   b. $Mg + 2HF \rightarrow MgF_2 + H_2$ (9.1)
5. **a.** What is molar mass?
   b. What is its role in reaction stoichiometry? (9.2)
6. Distinguish between ideal and real stoichiometry calculations. (9.3)
7. **a.** Distinguish between the limiting reactant and the excess reactant in a chemical reaction.
   b. Which is used up? (9.3)
8. **a.** Distinguish between the theoretical and actual yields in stoichiometry calculations.
   b. How do the values of the theoretical and actual yields generally compare? (9.3)
9. What is meant by the percent yield of a reaction? (9.3)
10. Why are actual yields generally less than those calculated theoretically? (9.3)
11. How do chemists express the efficiency of a reaction? (9.3)

PROBLEMS

▶ **General Stoichiometry**

1. Given the chemical equation $Na_2CO_3 + Ca(OH)_2 \rightarrow 2NaOH + CaCO_3$, determine (to two decimal places), the molar masses of all substances involved, and then write these as conversion factors.
2. Hydrogen and oxygen react under a specific set of conditions to produce water according to the following: $2H_2(g) + O_2(g) \rightarrow 2H_2O(g)$.
   a. How many moles of hydrogen would be required in order to produce 5.0 mol of water?
   b. How many moles of oxygen would be required?
3. **a.** If 4.50 mol of ethane (C_2H_6) undergo combustion according to the unbalanced reaction $C_2H_6 + O_2 \rightarrow CO_2 + H_2O$, how many moles of oxygen are required?
   b. How many moles of each product are formed?
4. Sulfuric acid reacts with sodium hydroxide according to the following: $H_2SO_4 + NaOH \rightarrow Na_2SO_4 + H_2O$.
   a. Balance the equation for this reaction.
   b. What mass of H_2SO_4 would be required to react with 0.75 mol of NaOH?
   c. What mass of each product is formed by this reaction?
5. Sodium chloride is produced from its elements through a synthesis reaction. What mass of each reactant would be required to produce 25.0 mol of sodium chloride?
6. Copper reacts with silver nitrate through single replacement.
   a. If 2.25 g of silver are produced from the

Stoichiometry **279**

reaction, how many moles of copper(II) nitrate are also produced?
b. How many moles of each reactant are required in this reaction?

7. Iron is generally produced from iron ore through the following reaction in a blast furnace: $Fe_2O_3(s) + CO(g) \rightarrow Fe(s) + CO_2(g)$.
a. If 4.00 kg of Fe_2O_3 are available to react, how many moles of CO are needed?
b. How many moles of each product are formed?

8. Methanol (CH_3OH) is an important industrial compound that is produced from the following reaction: $CO(g) + H_2(g) \rightarrow CH_3OH(g)$. What mass of each reactant would be needed to produce 100.0 kg of methanol?

9. Nitrogen combines with oxygen in the atmosphere during lightning flashes to form nitrogen monoxide (NO) which then reacts further with O_2 to produce nitrogen dioxide, NO_2.
a. What mass of NO_2 is formed when NO reacts with 384 g of O_2?
b. How many grams of NO are required to react with this amount of O_2?

10. As early as 1938, the use of NaOH was suggested as a means of removing CO_2 from the cabin atmosphere of a lunar-bound spacecraft according to the following reaction: $NaOH + CO_2 \rightarrow Na_2CO_3 + H_2O$.
a. If the average human body discharges 925.0 g of CO_2 per day, how many moles of NaOH would be needed each day?
b. How many moles of each product would be formed?

11. The double-replacement reaction between silver nitrate and sodium bromide produces silver bromide, a component of photographic film.
a. If 4.50 mol of silver nitrate react, what mass of sodium bromide is required?
b. What mass of silver bromide is formed?

12. In a soda-acid fire extinguisher, concentrated sulfuric acid reacts with sodium hydrogen carbonate to produce carbon dioxide, sodium sulfate and water.
a. How many moles of sodium hydrogen carbonate would be needed to react with 150.0 g of sulfuric acid?
b. How many moles of each product would be formed?

13. Aspirin ($C_9H_8O_4$) is produced through the following reaction of salicylic acid ($C_7H_6O_3$) and acetic anhydride ($C_4H_6O_3$): $C_7H_6O_3(s) + C_4H_6O_3(\ell) \rightarrow C_9H_8O_4(s) + H_2O(\ell)$.
a. What mass of aspirin (in kg) could be produced from 75.0 mol of salicylic acid?
b. What mass of acetic anhydride (in kg) would be required?
c. At 20°C, how many liters of water would be formed?

▶ **Limiting Reactant**

14. Given the reactant amounts specified in each chemical equation, determine the limiting reactant in each case:
a. HCl + NaOH → NaCl + H_2O
   2.0 mol   2.5 mol
b. Zn + 2HCl → $ZnCl_2$ + H_2
   2.5 mol   6.0 mol
c. $2Fe(OH)_3$ + $3H_2SO_4$ → $Fe_2(SO_4)_3$ + $6H_2O$
   4.0 mol   6.5 mol

15. For each reaction specified in Problem 14 above, determine the number of moles of excess reactant that remains.

16. For each reaction specified in Problem 14 above, calculate the number of moles of each product formed.

17. a. If 2.50 mol of copper and 5.50 mol of silver nitrate are available to react by single replacement, identify the limiting reactant.
b. Determine the number of moles of excess reactant remaining.
c. Determine the number of moles of each product formed.
d. Determine the mass of each product formed.

18. Sulfuric acid reacts with aluminum hydroxide by double replacement.
a. If 30.0 g of sulfuric acid react with 25.0 g of aluminum hydroxide, identify the limiting reactant.

b. Determine the mass of excess reactant remaining.
   c. Determine the mass of each product formed.

19. The energy used to power one of the Apollo lunar missions was supplied by the following overall reaction: $2N_2H_4 + (CH_3)_2N_2H_2 + 3N_2O_4 \rightarrow 6N_2 + 2CO_2 + 8H_2O$. For that phase of the mission when the lunar module ascended from the surface of the moon, a total of 1200. kg of N_2H_4 were available to react with 1000. kg of $(CH_3)_2N_2H_2$ and 4500. kg of N_2O_4.
   a. For this portion of the flight, which of the allocated components was used up first?
   b. How much water, in kilograms, was left in the lunar atmosphere through this reaction?

▶ **Percent Yield**

20. From theoretical and actual yields of the various chemical reactions given below, calculate the percent yield for each:
   a. theoretical yield = 20.0 g, actual yield = 15.0 g
   b. theoretical yield = 24.0 g, actual yield = 22.5 g
   c. theoretical yield = 5.00 g, actual yield = 4.75 g

21. From the theoretical and percentage yields given below, determine the actual yields:
   a. theoretical yield = 12.0 g, percent yield = 90.0%
   b. theoretical yield = 8.50 g, percent yield = 70.0%
   c. theoretical yield = 3.45 g, percent yield = 48.0%

22. The Ostwald Process for producing nitric acid from ammonia consists of the following steps:

$4NH_3(g) + 5O_2(g) \rightarrow 4NO(g) + 6H_2O(g)$
$2NO(g) + O_2(g) \rightarrow 2NO_2(g)$
$3NO_2(g) + H_2O(g) \rightarrow 2HNO_3(aq) + NO(g)$

If the yield in each step is 80.0%, how many grams of nitric acid can be produced from 5.00 kg of ammonia?

CRITICAL THINKING

1. **Relating Ideas** The chemical equation is a good source of information concerning a reaction. Explain the relationship that exists between the actual yield of a reaction product and the chemical equation of the product.

2. **Analyzing Results** Very seldom are chemists able to achieve a 100% yield of a product from a chemical reaction. However, the yield of a reaction is important usually because of the expense involved in producing less product. For example, when magnesium metal is heated very hot in a crucible, the product magnesium oxide, MgO, is formed. Based upon your analysis of the reaction, describe some of the actions you would take to increase your percent yield. The reaction is as follows:

$$2\,Mg(s) + O_2(g) \rightarrow 2MgO(s)$$

CROSS CONNECTIONS

1. **Art** Just as reactants combine in certain proportions to form a product, colors can be combined to create other colors. Artists do this all the time to find just the right color for their paintings. Using poster paint, determine the proportions of primary colors used to create the following colors. Remember your proportions should be such that anyone could mix the color perfectly. (Hint: Don't forget to record the amount of the primary color and water used when you mix them.)

RESEARCH AND WRITING

1. Prepare a report describing the work of the alchemists. What were their major contributions to modern chemistry? What techniques and procedures were employed by these early chemists?

Chemistry and Careers

Analytical Pharmaceutical Chemist

Analytical pharmaceutical chemists ensure that the compounds used as medicines (called pharmaceuticals) are safe. This work starts with a study of the composition of the raw materials used. The chemist must look for a variety of impurities, such as halogenated hydrocarbons, arsenic, lead, and mercury.

Sensitive electronic equipment is used to detect trace impurities in medications. In the past, it was difficult to detect 1 g of impurity in a kg of material, but chemists can now detect a few parts per billion.

In recent years, drug products have been labeled with expiration dates that reflect how long a pharmaceutical takes to lose 10% of its potency. Although 90% of a pharmaceutical may remain active, the decomposition products could be toxic. Outdated pharmaceuticals must be discarded. A chemist conducts tests to determine how fast a product decomposes.

Analytical pharmaceutical chemists work for private companies, government agencies, or independent testing laboratories. Technical training through a junior college program is a minimum requirement for more advanced jobs, but most labs require a bachelor's degree in chemistry for such positions. Because working with electronic instrumentation is now part of the job, knowledge of computer programming and electronics is desirable.

Chemical Plant Manager

A plant manager at a chemical manufacturing company carries considerable responsibility. He or she oversees the ordering of materials, makes production schedules, and takes responsibility for plant safety and staff training.

Since the plant manager is ultimately responsible for everything that occurs at the plant, knowledge of chemistry is vital. The manager must be able to answer questions and make decisions quickly. For example, during a chemical process, what side reactions might occur? How might the waste materials be recycled into useful materials? What could go wrong during production? If a chemical spill should occur, how might it affect the environment?

The manager of a chemical plant must have a bachelor's degree in chemistry. Also, chemical manufacturing companies usually require their plant managers to have advanced training in business administration and personnel management.

UNIT 3 • FOCUS ON REASONING

College Entrance Exam Preparation

If two substances react chemically, their behavior is described by the law of conservation of mass. Mass is conserved in all chemical reactions. If substance A combines completely with substance B to form AB, the mass of AB will be the sum of the masses of A and B. Using this information chemists can determine combining ratios of atoms in chemical reactions and then write balanced equations for reactions.

A chemist investigating the reactive properties of substance X and Y runs a series of reactions. Substances X and Y combine to form substance Z. Masses of the reactants X and Y and the product Z are carefully measured. The amount of Z formed is generally not the sum of the masses of X and Y. The chemist assumes that for each reaction either X or Y is in excess.

▶ **Experiment 1**   **Experiment 2**

Substance	Mass (in grams)					
	X	Y	Z	X	Y	Z
	10.00	2.00	3.29	2.00	10.00	2.54
	10.00	4.00	6.59	4.00	10.00	10.17
	10.00	6.00	9.89	6.00	10.00	15.26
	10.00	8.00	13.16	8.00	10.00	16.50
	10.00	10.00	16.47	10.00	10.00	16.50
	10.00	12.00	19.77	12.00	10.00	16.49
	10.00	14.00	23.07	14.00	10.00	16.50
	10.00	16.00	25.20	16.00	10.00	16.51
	10.00	18.00	25.19	18.00	10.00	16.51
	10.00	20.00	25.21	20.00	10.00	16.50

1. Concerning the relationship between the yield of substance Z in Experiment 1, it appears that
   a. as the mass of Y increased, the yield of Z increased.
   b. the yield of Z increases as the mass of Y increases until it approaches an optimum level.
   c. the mass of X has no effect on the yield of Z.
   d. the yield of Z depends on the mass of X.

2. If Experiment 1 were run again using 10.0 grams of X and 15.0 grams of Y, the expected yield of Z would be

   a. greater that 25.19 grams
   b. between 23.07 grams and 25.19 grams
   c. 16.47 grams
   d. less than 16.47 grams

3. Which of the following graphs would represent the results of the mass of X vs. Z for Experiment 2?

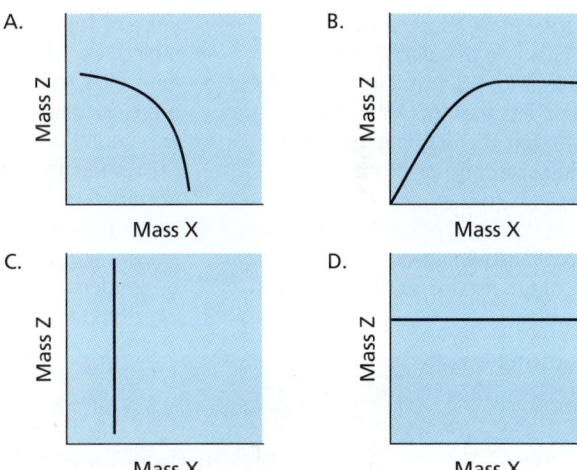

4. At what point in Experiment 1 does the mass of X limit the yield of Z?
   a. at all points
   b. 10.0 grams of X, 10.0 grams of Y, 16.47 grams of Z
   c. 10.0 grams of X, 14.0 grams of Y, 23.07 grams of Z
   d. 10.0 grams of X, 16.0 grams of Y, 25.19 grams of Z

5. At what point in Experiment 2 does the mass of Y limit the yield of Z?
   a. at all points
   b. 6.0 grams of X, 10.0 grams of Y, 15.26 grams of Z
   c. 8.0 grams of X, 10.0 grams of Y, 16.50 grams of Z
   d. 16.0 grams of X, 10.0 grams of Y, 16.51 grams of Z

INTRA-SCIENCE

The Amazing Polyhexamethyleneadipamide

The impossible-to-pronounce word in the title above is the proper name of one of the most important discoveries of all time. A skier schussing down a mountain depends on it; so do fishermen casting their lines and drivers driving on the interstate. We now know this discovery as "nylon."

When nylon was in the early stages of development in the 1930s, it was called Fiber 66 or Rayon no. 66, but it soon became apparent that this fiber was so different that its name should be distinct from that of rayon. Suggestions for names included "neosheen," "pontella," and "lustrol." For two years, various committees wrestled with the name problem.

Meanwhile, lab research indicated that Fiber 66 would make women's hosiery that did not run ("no-run" stockings). To make pronunciation easier, "no-run" was changed to nylon, and the new fiber finally had a name.

Nylon is a product of polymerization, a chemical process in which relatively small molecules unite to form very large, chainlike molecules. Natural organic polymers include starch, cellulose, and proteins. Diamonds and graphite are natural inorganic polymers.

Nylon is one type of synthetic polyamide fiber made of four elements—carbon, hydrogen, nitrogen, and oxygen. It is produced by condensation polymerization, in which a diamine such as hexameth-

ylene diamine, $H_2N(CH_2)_6NH_2$, reacts with a dibasic acid, $HOOC(CH_2)_4COOH$. As the two molecules join, a molecule of water is formed. The reaction continues, linking additional molecules until a very long chain of polyamides is produced. Different types of nylon are produced by varying the acid and amine building blocks.

The Connection to Other Sciences

During World War II, nylon began replacing twisted silk thread for sutures in surgery, and finely woven nylon fabric served as filters for blood plasma. Nylon's strength and resistance to moisture, mildew, and impact made it ideal for use in parachutes and protective vests.

Today's hospitals have hundreds of nylon products. A plastic wheel, now standard on many wheelchairs, is not only lighter, but can take greater vertical or lateral impact than a wire-spoked wheel. Surgical-staple extractors of nylon resin are made to match the surgeon's hand. The resin provides the stiffness needed for control and the elongation required for the springlike functioning of the tension arm.

The Connection to Physics

The physical properties of a polymer depend on both the characteristics of its building blocks and its total length. Long chains overlap more and have greater cohesive force than shorter chains. Nylons are strong, durable, elastic, have a low coefficient of friction, and retain these properties over a temperature range of $-100°–110°C$.

Nylon was synthesized first as a fiber, then as a plastic. The first molded nylon products, introduced in 1939, were textile machinery gears. This was a perfect opportunity to demonstrate nylon's potential as an engineering thermoplastic—the gears did not require lubrication. Today, car makers use nylon in many forms. Automotive timing gears, for example, are coated with nylon to prolong their life. Because of its high resistance to hydrocarbons, nylon is used for emission-control canisters and is found in electrical connectors, wire jackets, and gears for windshield wipers and speedometers. Glass-reinforced nylons are employed in engine fans, radiator headers, brake-fluid reservoirs, and valve covers.

The Connection to Other Sciences

Nylon's light weight, combined with its strength and durability, make it ideal for people who work outdoors. Packs, clothing, tarpaulins, and tents incorporate nylon fabric. Nylon lines are used by foresters who climb trees and by rock-climbing geologists.

Many field scientists use equipment made of nylon. Biologists use radio collars with nylon bands to track animals in their natural habitats. Ornithologists use mist nets made of nylon to capture birds. The mesh of the nets is very fine and blends in with the background. Flying birds become entangled in the nets and can be removed for weighing and banding, and for parasite studies.

Because it is hygroscopic (it has a propensity to attract water), nylon is often found in humidistats as the mechanism for moisture control. Because nylon resists fire, the electrical industry uses it for plugs, coil forms, wiring devices, terminal blocks, wire jacketing, and antenna-mounting devices.

Unit 4

Chapter 10
Representative Gases 288

Chapter 11
Physical Characteristics of Gases 314

Chapter 12
Molecular Composition of Gases 342

Chapter 13
Liquids and Solids 370

PHASES OF MATTER

I majored in chemistry in college, but didn't know exactly what sort of chemistry I wanted to spend my life working on. A professor at my college made some laboratory measurements of a chemical reaction that is important to the atmosphere of Jupiter. I was fascinated to hear that chemistry doesn't just take place in test tubes, but in the atmosphere of an entire planet. From then on I was hooked on atmospheric chemistry.

I study the earth's atmosphere and try to understand the influence that human beings may have upon it. I'm especially interested in the depletion of the ozone layer. The polar stratospheric clouds you see at right are important in the chemistry of stratospheric ozone. My work involves experimental projects to gather data on the amount of ozone and other chemicals in the atmosphere, and theoretical studies in which computer models are used to simulate and interpret the data.

Understanding the chemistry of the gases, liquids, and solids of our atmosphere will help us to find ways of protecting it from damaging pollutants.

Susan Solomon
Atmospheric Chemist
National Oceanic and
    Atmospheric Administration

Polar stratospheric clouds over Kiruna, Sweden

CHAPTER 10
Representative Gases

PLANNING CHART

Approximate Time: 3/4 days

	SECTION/TEXT FEATURES		RESOURCES	
B A S I C	**10.1 OXYGEN AND OZONE** Demos 1, 2 *Periodic Table* Videodisc Microscale Investigation 7 Section Review Chapter Review • Questions 1–7, Problem 1	**289** 288A–B 288F 845 294 311	*Teacher's Resource Binder* Section Review Worksheets **Other Resources** *Shakhashiri* Video Demos 20, 33, 38 *World of Chemistry* Video 17	33
H O N O R S	**10.2 HYDROGEN** Demo 3 *Periodic Table* Videodisc Microscale Investigation 8 Section Review Chapter Review • Questions 8–14, Problems 2, 3, 5	**294** 288B 288F 848 297 311	*Teacher's Resource Binder* Section Review Worksheets **Other Resources** *Shakhashiri* Video Demos 9, 21	34
H O N O R S	**10.3 NITROGEN AND AMMONIA** *Periodic Table* Videodisc Section Review Chapter Review • Questions 15–23	**298** 288F 302 311	*Teacher's Resource Binder* Section Review Worksheets **Other Resources** *Shakhashiri* Video Demos 3, 4, 6, 26	35
H O N O R S	**10.4 CARBON DIOXIDE AND CARBON MONOXIDE** Desktop Investigation Section Review Chapter Review • Questions 24–33, Problems 4, 6	**303** 307 310 311	*Teacher's Resource Binder* Section Review Worksheets **Other Resources** *Shakhashiri* Video Demos 2, 37	36
	CHAPTER END ACTIVITIES Critical Thinking, Cross Connections, Research and Writing, Projects		*Teacher's Resource Binder* Test Booklet **Computer Test Generator**	49, 51

Teaching Strategies

10.1 Oxygen and Ozone

Few students are aware of the paramagnetic nature of the oxygen molecule. The traditional double bond between two oxygen atoms needed to complete the octet rule does not take into account the two unpaired electrons. If the materials are available, the following demonstration is dramatic and clearly shows the paramagnetic properties of oxygen.

Demonstration 1: Paramagnetism in Oxygen
PURPOSE: To show that oxygen molecules are paramagnetic and thus must contain unpaired electrons

MATERIALS: A very strong magnet (the World War II surplus radar magnets are ideal, but hard to find), a dewar flask, liquid nitrogen, large pyrex test tube, a source of oxygen gas

PROCEDURE: *(1)* Attach a piece of 6-mm glass tubing to a length of rubber tubing. *(2)* Connect the other end of the rubber tubing to the regulator on the oxygen tank. *(3)* Fill the dewar flask with liquid nitrogen. *(4)* Place the large test tube in the dewar flask. *(5)* Insert the glass tubing in the test tube. Open the valve on the oxygen tank and maintain a very slow flow of oxygen. It will take about 5 minutes to collect about 50 mL of liquid oxygen. *(6)* Pour some liquid nitrogen between the poles of the magnet. Note that the nitrogen is diamagnetic and flows right through the poles. *(7)* Remove the test tube containing the liquid oxygen from the dewar flask and note that it has a blue color caused by the unpaired electrons. (8) Slowly pour the liquid oxygen between the cooled poles of the magnet. The oxygen will remain suspended between the poles, demonstrating the presence of unpaired electrons and the resulting paramagnetism.

Demonstration 2: Preparation of Oxygen
PROCEDURE: CAUTION: Wear goggles, gloves, and a lab apron. (1) Dissolve 3.0 g $KMnO_4$ in 60 mL of 3 M sulfuric acid. (2) Pour 100 mL 3% hydrogen peroxide into the 600-mL beaker. (3) Pour 5 mL of the $KMnO_4$ solution into the beaker. Note the formation of bubbles and evolution of a gas. (4) Light a wooden splint and blow out the flame, leaving a few glowing embers. (5) Add another 5-mL portion of $KMnO_4$ to the beaker and immediately insert the glowing splint. The splint should burst into flames. The procedure may be repeated.

Note that 3% hydrogen peroxide was used. Thirty percent hydrogen peroxide must not be used as it is a very strong oxidizing agent.

Note that 3% hydrogen peroxide was used. Thirty percent hydrogen peroxide must not be used as it is a very strong oxidizing agent.

10.2 Hydrogen

In discussing the occurrence, preparation, and properties of hydrogen, again take into account the students' previous experiences. Hydrogen makes up 76% of the universe and helium 23%. Point out that the sun utilizes a fusion reaction in which hydrogen is converted to helium to generate the vast amounts of energy on which life on earth depends. We have harnessed the power of the sun in the thermonuclear hydrogen bomb and are working on harnessing the same energy in fusion reactors. Although hydrogen is placed in Group 1 in the Periodic Table, its properties are different from other Group 1 elements.

Demonstration 3: Preparation and Properties of Hydrogen
PURPOSE: To demonstrate the preparation and properties of hydrogen gas
MATERIALS: 30 g of mossy zinc, 6 M hydrochloric acid, 250-mL Erlenmeyer flask and 2-hole rubber stopper, long stem glass funnel or thistle tube, right angle glass bend, rubber tubing, gas-collecting bottles, large pan, matches
PROCEDURE: CAUTION: Be very careful with open flames and H_2 gas. Insert funnel and glass bend through the stopper and connect rubber tubing to bend and collecting bottles. Place zinc in the flask and cover with H_2O. Secure stopper to flask with tip of funnel below water line. Add the HCl in 10-mL increments. Collect the hydrogen in the gas-collecting bottles by water displacement.

Collect several 200-mL bottles of hydrogen. The bottles should be inverted and the mouth of the bottle placed on a wet glass plate on the table top. Ask the students to explain the necessity of the inverted bottles. Care must be taken that only hydrogen is present in the bottles. If pure hydrogen is present, the gas will burn quietly at the mouth of the bottle. If a mixture of hydrogen and oxygen is present, the mixture explodes with a loud pop. To demonstrate the difference between burning and supporting combustion, light a small candle. Bring the candle to the mouth of an inverted bottle filled with the hydrogen. The hydrogen burns at the mouth of the bottle with an almost invisible blue flame. Slowly insert the candle into the bottle. The hydrogen extinguishes the flame. Slowly withdraw the candle. The flame at the mouth reignites the candle. Some condensation should have formed on the inside of the bottle. Write the equation for the combustion of hydrogen. Water is a product of the reaction. A very important point should be made at this time. Oxygen gas does not burn, but it does support combustion. Hydrogen gas burns, but it does not support combustion. (Both the fuel and a source of oxygen are needed to burn something.)

Place the mouth of a second bottle of hydrogen over the mouth of an empty (filled with air) bottle. Invert the two. The hydrogen in the bottom bottle will now diffuse into the upper bottle. Separate the two bottles. Invert the lower bottle and place it on the table. Light a candle and bring the flame to the mouth of the inverted bottle that was originally filled with air. A loud pop occurs. Hydrogen diffused into the bottle and mixed with oxygen present in the air. This procedure may be repeated with the remaining bottle. Note the difference between the sound and the rate of the reaction when pure hydrogen burns and when a mixture of hydrogen and oxygen burns. Discuss the Hindenburg disaster.

After completing the demonstrations, have the students classify them as representing physical or chemical properties of hydrogen. Density and diffusion are physical, while burning and not supporting combustion are chemical.

10.3 Nitrogen and Ammonia

Nitrogen accounts for about 80% of the atmosphere. The triple bond in the nitrogen molecules is strong enough for nitrogen gas to be considered the "poor man's inert or noble gas." This section is best introduced by telling the story about Fritz Haber and Germany's desperate situation in World War I. Explosives at that time used nitrates and nitric acid as major raw materials. Chilean nitrates were an integral part of Germany's war effort. The Allied naval blockade could have deprived Germany of a key raw material and hastened the end of the war. Haber, however, had developed the process, now named after him, for producing ammonia from the nitrogen of the air. The ammonia was easily converted to the nitrates needed for the production of explosives. The very air we breathe became the source of German gunpowder. Many social issues may be discussed at this time. The importance of nitrogen in fertilizers, in feeding underdeveloped nations, in explosives, nitrogen fixation, and in air and water pollution. The students may be given an assignment to find articles that touch on these points.

A number of methods for preparing oxygen-free nitrogen are available. Placing moistened steel wool on a bottle of air inverted in a pan of water may be the simplest. Liquid nitrogen may be obtained from a college, hospital, doctor's office, or welding supply store. Dewar flasks for transporting the nitrogen may be leased from chemical suppliers. Should liquid nitrogen be available, a number of demonstrations would serve as an introduction to the gas laws. Students are fascinated by the liquid nitrogen.

When introducing ammonia, be sure to distinguish between ammonia and the ammonium ion. Writing the formulas on the board and displaying molecular models of each will help.

10.4 Carbon Dioxide and Carbon Monoxide

Introduce the study of carbon dioxide by having a piece of dry ice on the desk. You may place it in a large graduated cylinder filled with water to which an acid–base indicator has been added. One of the universal indicators works well.

The "greenhouse effect" and its importance afford you another opportunity to begin discussing carbon dioxide and its properties. Encourage students to bring in articles that discuss the greenhouse effect.

Bubbling carbon dioxide or blowing one's breath into a beaker with limewater is a traditional demonstration for showing the presence of carbon dioxide. You may extend the demonstration by continuing to bubble carbon dioxide into the limewater. After a short period of time, the precipitate dissolves and the limewater is once again clear. Most students will not have seen this latter demonstration. As discussed in the text, the excess carbon dioxide reacts with the calcium carbonate precipitate and forms the soluble calcium–hydrogen carbonate. You may go on and explain how this system is important in cave formation and within the blood.

Write equations and discuss each of the processes for producing carbon dioxide. Most of the processes should have been encountered in earlier science courses.

Carbon monoxide may be introduced while discussing the toxic effects of carbon dioxide. Both are poisonous, but react differently. In discussing the structure of carbon monoxide, you may wish to mention that the structure is isoelectronic with that of nitrogen gas.

You may end the section by pointing out that, despite all of its dangerous features, carbon monoxide serves as a key resource in producing more complex organic compounds. Its role may become more important as petroleum supplies dwindle.

Desktop Investigation: Properties of Carbon Dioxide

If you wish your students to carry this out at home, the following modifications can be made: A 4-ounce juice glass can be substituted for a 150-mL beaker, a pint jar for a 400-mL beaker, and a soda straw for the piece of glass tubing. 1/4 cup of baking soda can be substituted for 50 g of baking soda, and 1/4 cup of vinegar for 50 mL of vinegar. (These are not exact equivalents, but handy amounts to substitute for good results.)

Limewater is made by adding 5.0 g of calcium hydroxide to 1.0 L of distilled water. Shake and allow the undissolved solids to settle. Decant or filter the saturated solution to another bottle.

ANSWERS TO DISCUSSION QUESTIONS

1. Carbon dioxide can be poured from one beaker to another because it is denser than air and thus can displace the air.

2. One experiment that might be proposed, but does not work when attempted, is to bubble exhaled air through limewater for 15 minutes. There is not enough carbon dioxide to redissolve the calcium carbonate precipitate. However, this is a good hypothetical solution to the problem. One that does work is bubbling exhaled air through two different containers of 50 mL of limewater for one minute. Allow the containers to stand until the precipitate settles out. Carefully decant the supernatant liquid and add 100 mL of distilled water to one container and 100 mL of club soda to the second container. (Club soda is a supersaturated solution of carbon dioxide after the bottle is opened.) Stir each container 2 to 3 minutes and observe that the precipitated calcium carbonate dissolves in the club soda, but not in the distilled water.

3. (a) $NaHCO_3(aq) + HC_2H_3O_2(aq) \rightarrow NaC_2H_3O_2(aq) + H_2O(\ell) + CO_2(g)$
(b) $Ca^{2+}(aq) + 2OH^-(aq) + CO_2(g) \rightarrow CaCO_3(s) + H_2O(\ell)$

References and Resources

Books and Periodicals

Baugh, Mark. "What's So Bad About Oxygen?," *ChemMatters*, Vol. 4, No. 1, Feb. 1986, pp. 10–11.

Borgford, Christie L. and Lee R. Summerlin. "Gases," *Chemical Activities: A Sourcebook for Science Teachers*, Washington, D.C.: The American Chemical Society, 1988.

Davenport, Derek A. "The Back Burner—Joseph Priestley and the All-American Lunch," *ChemMatters*, Vol. 1, No. 1, Feb. 1983, pp. 14–15.

Dinga, Gustav P. "Hydrogen: The Ultimate Fuel and Energy Carrier," *Journal of Chemical Education*, Vol. 65, No. 8, Aug. 1986, pp. 688–691.

Linner, Marilyn. "Hydrogen and Helium," *ChemMatters*, Vol. 3, No. 3, Oct. 1985, pp. 4–7.

Martins, George F. "Percent Oxygen in Air," *Journal of Chemical Education*, Vol. 64, No. 9, Sept. 1987, pp. 809–810.

Ponnamperuma, Cyril. "The Atmosphere of Planet Earth," *ChemMatters*, Vol. 1, No. 1, Feb. 1983, pp. 10–12.

Shakhashiri, Bassam Z. *Chemical Demonstrations*, Vol. 2, Madison: University of Wisconsin Press, 1985.

Summerlin, Lee R., et al. "Gases," *Chemical Demonstrations— A Sourcebook for Teachers*, Vol. 1, Second Edition, Washington, D.C.: American Chemical Society, 1988.

Summerlin, Lee R., et al. "Gases," *Chemical Demonstrations— A Sourcebook for Teachers*, Vol. 2, Washington, D.C.: American Chemical Society, 1987.

Audiovisual Resources

"Out of the Air: 1," part of the series *Chemistry in Action*, VHS cassette, color, 20 min., Films for the Humanities, Inc.

"Out of the Air: 2," part of the series *Chemistry in Action*, VHS cassette, color, 20 min., Films for the Humanities, Inc.

Computer Software

Bauder, Donald. *Haber Tech: A Simulation of the Industrial Synthesis of Ammonia*. Apple, Queue. (Principles of chemical equilibrium, reaction kinetics, formulas, equations, and gas laws)

Edens, R. and K. Shaw. *HABER: Ammonia Synthesis*, Apple, CONDUIT. (Production of ammonia by the Haber process)

Answers and Solutions

Reviewing Concepts

1. **a.** Oxygen exists as both ordinary oxygen gas (O_2) and ozone (O_3).
   b. Oxygen gas is necessary for the basic life process of aerobic respiration, in which it combines with molecules from food to produce the energy needed for life. Ozone, found mostly in the upper atmosphere, absorbs most of the ultraviolet portion of the light radiated from the sun to the earth.
2. The singly bonded oxygen atoms contain two unpaired electrons whose spins do not cancel. This accounts for the attraction of oxygen by a magnet.
3. Chlorofluorocarbons (CFCs), used in aerosols and refrigerants, may be damaging the ozone layer.
4. Oxygen composes 49.5% of the total mass of the earth's crust, waters, and atmosphere; it makes up 20% of the earth's atmosphere by volume; 47% of the earth's crust is oxygen by mass.
5.
6. **a.** An oxide is a compound containing the simple oxide anion, which has an oxidation number of -2. A peroxide is a compound containing the O_2^{2-} anion, in which oxygen has the oxidation number -1. A superoxide is a binary compound containing the O_2^- anion, in which oxygen has the uncommon oxidation number $-\frac{1}{2}$.
   b. The bond character in an oxide depends on the difference in electronegativities between oxygen and the element with which it is combined; the greater the difference in electronegativities is, the more ionic, or less covalent, is the bond.
7. 1. Decomposition of hydrogen peroxide:
      $2H_2O_2(aq) \xrightarrow{MnO_2} 2H_2O(\ell) + 2O_2(g)$
   2. Oxidation of water by sodium peroxide:
      $2Na_2O_2(s) + 2H_2O(\ell) \rightarrow 4NaOH(aq) + O_2(g)$
   3. Thermal decomposition of potassium chlorate:
      $2KClO_3(s) \xrightarrow[MnO_2]{\Delta} 2KCl(s) + 3O_2(g)$
   4. Electrolysis of water:
      $2H_2O(\ell) \rightarrow 2H_2(g) + O_2(g)$
   5. Liquefaction of air: This method is the most common industrial method for preparing oxygen. Ordinary air is liquefied by compression and cooling, and then nitrogen is separated from the oxygen by fractional distillation.
8. **a.** The most abundant element in the universe is hydrogen.
   b. On earth, free elemental hydrogen exists as a gas consisting of diatomic H_2 molecules. In combined form, hydrogen exists mainly as water.
   c. It exists in the sun and stars as elemental hydrogen in the plasma state.
9. Hydrogen represents an estimated 90% of all the atoms in the universe. The earth's crust contains 0.88% hydrogen by mass. Water contains 11% hydrogen by mass.
10. The major commercial source of hydrogen is the decomposition of hydrocarbons.
11. Unlike the other Group 1 elements, hydrogen exists in the form of diatomic H_2 molecules, and as such is a poor conductor of heat and electricity. Unlike other Group 1 elements, it is a gas and is colorless, odorless, and tasteless.
12. Free hydrogen atoms are extremely reactive, and thus combine readily with each other to form hydrogen molecules or with other elements to form various compounds.
13. The nearly 100 other elements were produced from hydrogen nuclei by nuclear fusion in the cores of the stars.
14. Hydrogen can be prepared:
    1. From acids by replacement:
       $Zn(s) + H_2SO_4(aq) \rightarrow ZnSO_4(aq) + H_2(g)$
    2. From water by replacement:
       $2Na(s) + 2H_2O(\ell) \rightarrow 2NaOH(aq) + H_2(g)$
    3. From the electrolysis of water:
       $2H_2O(\ell) \xrightarrow{elec.} 2H_2(g) + O_2(g)$
    4. From the decomposition of hydrocarbons:
       $CH_4(g) + H_2O(g) \xrightarrow{Ni} CO(g) + 3H_2(g)$
15. Nitrogen is present in the DNA molecules that control heredity and it is a major component of all amino acids.
16. **a.** The earth's atmosphere is approximately 80% molecular nitrogen, N_2, by volume.
    b. Elemental nitrogen is a colorless, odorless, and tasteless gas that is slightly less dense than air. At temperatures lower than $-195.8°C$ at 1 atm, it condenses to a colorless liquid; it freezes to a white solid at $-209.9°C$.
17. **a.** Because of the high energy required to break the nitrogen–nitrogen triple bond, it is hard to separate the nitrogen atoms of the N_2 molecule so that they can enter into new combinations.
    b. Since it is relatively nonreactive, many food items are packaged in the presence of nitrogen rather than air to prevent the interaction of these products with the oxygen in air, since this interaction often leads to loss of flavor. Due to its nontoxic and nonpolluting characteristics, nitrogen is used as a propellant in aerosol cans.
18. **a.** Nitrogen fixation is the process through which atmospheric nitrogen gas (N_2) is converted into the nitrogen compounds that can be used by plants.
    b. Since most natural nitrogen fixation is carried out by bacteria on the roots of legumes, such plants can restore nitrogen to soil that has become depleted of it by other crops.
19. **a.** The Lewis structure is:

    H
    H : N : H

    b. The ammonia molecule is polar covalent, and, as a result of hydrogen bonding, ammonia has higher boiling and freezing points than expected. Ammonia is a colorless gas, has a strong odor, is less dense than air, and is extremely soluble in water.

20. Ammonia is used to revive fainting victims; as household ammonia, it is a useful grease-cutting agent; it is important as a base and as an intermediate in various organic reactions; it is also used in the production of artificial fertilizers and of explosives.

21. **a.** Ammonia is continually produced naturally by the fixation of atmospheric nitrogen and by the decay of organic matter.
    b. Industrially, ammonia is produced by decomposition of ammonium compounds:
    1. $Ca(OH)_2(s) + 2NH_4Cl(s) \xrightarrow{\Delta} CaCl_2(s) + 2NH_3(g) + 2H_2O(g)$
    2. The Haber process:
    $N_2(g) + 3H_2(g) \rightleftarrows 2NH_3(g)$

22. **a.** Nitrogen is the major component of the earth's atmosphere.
    b. The most important compound of this element is ammonia.
    c. Ammonia is used as a fertilizer, as a refrigerant, and as an intermediate in the production of such compounds as nitric acid, nylon, and antimalarial drugs.

23. Ammonia, which may have been abundant in the early atmosphere of the earth, is now converted to other substances in living organisms as fast as it is produced. Also because it is so low in mass, ammonia can escape from the earth's gravitational field to a significant extent.

24. The polar carbon monoxide molecule has four resonance structures, each of which contribute different percentages to the molecule.
    a. $^-$:C::O:$^+$ 50%
    b. $^+$:C:Ö:$^-$ 10%
    c. :C::O: 20%
    d. :C::O: 20%

25. The "greenhouse effect" is the trapping of reflected energy by the carbon dioxide in the atmosphere. Increased levels of carbon dioxide in the atmosphere (through the increased burning of carbon-containing fuels, or the destruction of forests that normally consume carbon dioxide) could cause an increase in global temperature that could melt the polar ice caps and flood sea-level coastal cities. Other areas could become dry and unproductive.

26. Dry ice, or solid carbon dioxide, has a very low temperature ($-78.5°C$) and at atmospheric pressure, passes directly from the solid to the gaseous state without melting.

27. In a liquid carbon dioxide fire extinguisher, a stream of carbon dioxide snow cools the burning material and also shuts off the supply of oxygen needed for combustion.

28. Carbon dioxide molecules are linear and thus nonpolar. The carbon–oxygen bonds are shorter than ordinary carbon–oxygen double bonds. The molecule has the following resonance hybrid structure.

$$^-:Ö:C::O:^+$$
$$:Ö::C::O: \leftarrow \rightarrow :O::C::Ö:$$
$$^+:O::C:Ö:^-$$

29. Carbon monoxide is produced naturally by decaying plants, by live algae, by the oxidation of atmospheric methane, and by volcanic action.

30. **a.** The average concentration is 0.5 ppm.
    b. This is 0.01 of the level considered toxic.
    c. In the United States, as much as 90% of the human-produced carbon monoxide comes from the incomplete combustion of carbon-containing fuels in automobile engines.

31. Carbon monoxide is a colorless, odorless, tasteless gas; it is slightly less dense than air, and only slightly soluble in water.

32. **a.** Carbon dioxide does not support combustion of most substances.
    b. It decomposes into carbon and oxygen.
    $2Mg(s) + CO_2(g) \rightarrow 2MgO(s) + C(s)$
    c. When dissolved in water, a small amount of the dissolved carbon dioxide unites with water to form carbonic acid.
    $H_2O(\ell) + CO_2(aq) \rightleftarrows H_2CO_3(aq)$
    d. It reacts to form a carbonate, which may or may not precipitate out depending upon the identity of the positive ion.
    $Ca^{+2}(aq) + 2OH^-(aq) + CO_2(g) \rightarrow CaCO_3(s) + H_2O(\ell)$

33. **a.** The carbon monoxide acts as a reducing agent.
    $Fe_2O_3(s) + 3CO(g) \rightarrow 2Fe(s) + 3CO_2(g)$
    b. When burned, carbon monoxide is oxidized to carbon dioxide.
    $2CO(g) + O_2(g) \rightarrow 2CO_2(g)$
    c. The reduction of carbon monoxide produces methanol.
    $CO(g) + 2H_2(g) \rightarrow CH_3OH(\ell)$

Problems

1. **a.** $2Mg(s) + O_2(g) \rightarrow 2MgO(s)$
   b. $Cs(s) + O_2(g) \rightarrow CsO_2(s)$
   c. $H_2(g) + O_2(g) \rightarrow H_2O_2(\ell)$

2. **a.** $Zn(s) + 2HF(aq) \rightarrow ZnF_2(aq) + H_2(g)$
   b. $Mg(s) + H_2SO_4(aq) \rightarrow MgSO_4(aq) + H_2(g)$
   c. $2Fe(s) + 3H_2SO_4(aq) \rightarrow Fe_2(SO_4)_3(aq) + 3H_2(g)$
   d. $Mg(s) + 2HCl(aq) \rightarrow MgCl_2(aq) + H_2(g)$
   e. $2Rb(s) + 2HOH(\ell) \rightarrow 2RbOH(aq) + H_2(g)$
   f. $Mg(s) + 2HOH(\ell) \rightarrow Mg(OH)_2(aq) + H_2(g)$

3. **a.** $2H_2(g) + O_2(g) \rightarrow 2H_2O(g); H = +1, O = -2$
   b. $H_2(g) + Cl_2(g) \rightarrow 2HCl(g); H = +1, Cl = -1$
   c. $2Na(s) + H_2(g) \rightarrow 2NaH(s); Na = +1; H = -1$
   d. $Ba(s) + H_2(g) \rightarrow BaH_2(s); Ba = +2, H = -1$

4. **a.** $MgCO_3(s) \rightarrow MgO(s) + CO_2(g)$
   b. 1. $MgCO_3(s) + 2HCl(aq) \rightarrow MgCl_2(aq) + H_2CO_3(aq)$
       2. $H_2CO_3(aq) \rightarrow H_2O(\ell) + CO_2(g)$

5. **a.** $Zn(s) + H_2SO_4(aq) \rightarrow ZnSO_4(aq) + H_2(g)$

   $13.1 \text{ g Zn} \times \dfrac{\text{mol Zn}}{65.4 \text{ g Zn}} \times \dfrac{1 \text{ mol } H_2}{\text{mol Zn}} \times \dfrac{2.02 \text{ g } H_2}{\text{mol } H_2}$
   $= 0.405 \text{ g } H_2$

   b. $2Na(s) + 2H_2O(\ell) \rightarrow 2NaOH(aq) + H_2(g)$

   $50.0 \text{ g } H_2O \times \dfrac{\text{mol } H_2O}{18.0 \text{ g } H_2O} \times \dfrac{1 \text{ mol } H_2}{2 \text{ mol } H_2O}$

$$\times \frac{2.02 \text{ g H}_2}{\text{mol H}_2}$$
$$= 2.81 \text{ g H}_2$$

6. a. $2\text{NaHCO}_3(aq) + \text{H}_2\text{SO}_4(aq) \rightarrow \text{Na}_2\text{SO}_4(aq) + 2\text{H}_2\text{O}(\ell) + 2\text{CO}_2(g)$

$$1.500 \text{ kg NaHCO}_3 \times \frac{10^3 \text{ g}}{\text{kg}} \times \frac{\text{mol NaHCO}_3}{84.01 \text{ g NaHCO}_3}$$

$$\frac{1 \text{ mol H}_2\text{SO}_4}{2 \text{ mol NaHCO}_3} \times \frac{98.07 \text{ g H}_2\text{SO}_4}{\text{mol H}_2\text{SO}_4} = 875.5 \text{ g H}_2\text{SO}_4$$

b. $875.5 \text{ g H}_2\text{SO}_4 \times \dfrac{\text{mol H}_2\text{SO}_4}{98.07 \text{ g H}_2\text{SO}_4} \times \dfrac{2 \text{ mol CO}_2}{\text{mol H}_2\text{SO}_4}$

$$\times \frac{44.0099 \text{ g CO}_2}{\text{mol CO}_2} = 785.8 \text{ g CO}_2$$

[Note: It is not necessary to use the results of part **a** to find the answers to part **b**. The original reactant mass value could again be used to calculate the mass of each product as well.]

Critical Thinking

1. The superoxide contains the O_2^- anion with an oxidation state of $-\frac{1}{2}$. All Group 2 metals carry a 2+ oxidation state. Combinations of the 2+ metal ion with the $-$ oxygen ion will yield BaO (oxide) or BaO_2 (peroxide). It is not possible to have a combination that will yield a $-\frac{1}{2}$ oxidation state on the oxygen.

2. In a photochemical smog, nitric oxide (NO) from automobile exhaust is oxidized to NO_2 either in air or in the automobile engines. Then, NO_2 is decomposed into NO and free oxygen in sunlight. The free oxygen undergoes a reaction with O_2 to form O_3. The O_3 then decomposes NO_2 into NO and O_2. The higher amount of O_3 present in photochemical smog is due to the presence of NO and NO_2 from automobile exhaust. NO serves a function of increasing the decomposition of O_3.

3. Answers will vary.

Cross Connections

1. Answers will vary. The converter is essentially a reaction container where certain catalysts have been placed to convert nitrogen oxides to molecular nitrogen and hydrocarbons to carbon dioxide and water.

Videodisc Selections
from *The Periodic Table Videodisc* (JCE Software)

step reverse

step forward

CHAPTER 10

pg 289 Oxygen

4543
oxygen (s)

pg 290 Formation of oxides

115
hydrogen + oxygen (m)

18438
calcium + oxygen (m)

37311
strontium + oxygen (m)

46119
barium + oxygen (m)

20759
titanium + oxygen (m)

22764
manganese + oxygen (m)

7535
magnesium + oxygen (m)

28546
copper + oxygen (m)

8267
aluminum + oxygen (m)

3835
carbon + oxygen (m)

12319
phosphorus + oxygen (m)

13145
sulfur + oxygen (m)

33890
selenium + oxygen (m)

pg 294 Hydrogen

109
hydrogen (s)

pg 296 Reactions of hydrogen

115
hydrogen + oxygen (m)

pg 297 Preparation of hydrogen

31167
zinc + acids (m)

5104
sodium + water (m)

17334
potassium + water (m)

19244
calcium + water (m)

10310
aluminum + sodium hydroxide (m)

pg 298 Nitrogen

4537
nitrogen (s)

Teacher's Notes

Chapter 10
Representative Gases

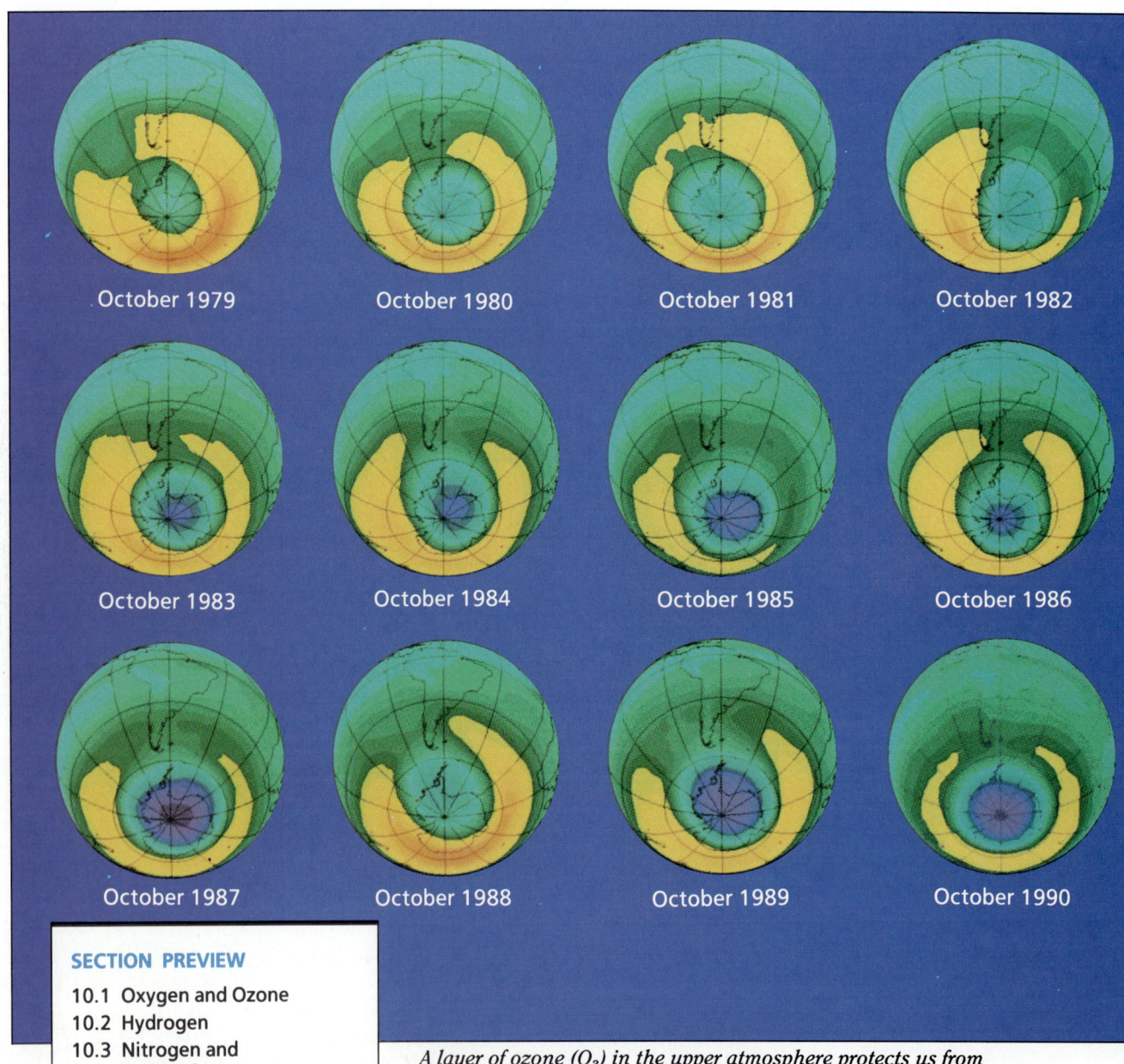

SECTION PREVIEW
- 10.1 Oxygen and Ozone
- 10.2 Hydrogen
- 10.3 Nitrogen and Ammonia
- 10.4 Carbon Dioxide and Carbon Monoxide

A layer of ozone (O_3) in the upper atmosphere protects us from harmful solar radiation, but scientists have discovered a hole appearing in the layer over Antarctica. This plot of ozone distribution from NASA's Nimbus-7 *satellite shows the fluctuation of the ozone hole over several years.*

10.1 Oxygen and Ozone

Two types of gaseous oxygen molecules are present in the earth's atmosphere. One of these molecules is composed of two oxygen atoms and is called, simply, oxygen. The other type of molecule is composed of three oxygen atoms and is called ozone. Oxygen and ozone gas are allotropes. *An **allotrope** is one of the two or more forms of an element that exist in the same physical state. The existence of two or more different molecular forms of an element, such as O_2 and O_3, is called* **allotropy.** Molecular allotropes differ in the arrangement and number of atoms in each molecule and therefore have different physical and chemical properties. To avoid confusion, the term "oxygen" will be used to refer only to ordinary molecular oxygen (O_2), not ozone (O_3).

Oxygen gas is required for the basic life process of aerobic respiration. Respiration occurs in the cells of most living things. In aerobic respiration, oxygen combines with food molecules to produce the energy needed for life. The presence of oxygen in the air is also necessary for the combustion of the various fuels—petroleum, coal, natural gas, and wood—that we use to run our automobiles, generate our electricity, and heat our homes.

In the upper atmosphere, ozone plays an equally crucial role. Ozone absorbs most of the ultaviolet portion of the light radiated from the sun to the earth. Ultraviolet light that reaches the earth's surface causes harmful chemical reactions in plant and animal tissues. Many scientists are concerned about the possible destruction of the ozone layer in the upper atmosphere. This destruction may be caused by chlorofluorocarbons, or CFCs, chemical compounds that escape into the atmosphere during their use in aerosols, refrigerators, and air conditioners.

Occurrence of Oxygen

Oxygen is the most abundant element in the earth's crust, waters, and atmosphere. It composes 49.5% of the total mass of these components. Free or uncombined oxygen makes up about 20% of the earth's atmosphere by volume. Free oxygen consists of covalently bonded O_2 molecules. Additional free oxygen is dissolved in the earth's waters, where it is essential for the respiration of aquatic life.

Most of the earth's oxygen exists as combined oxygen, oxygen that has united with other elements to form compounds. For example, water contains about 89% oxygen by mass. The earth's crust is about 47% oxygen by mass. Oxygen is chemically combined with silicon, aluminum, iron, calcium, magnesium, titanium, and other elements to form most of the earth's rocks and minerals.

Structure of the Oxygen Molecule

According to the guidelines introduced in Section 4-3, the Lewis structure for oxygen is

$$:\ddot{O}::\ddot{O}:$$

This representation of the oxygen molecule fails to account for certain experimentally determined properties of O_2, however. For example, when

SECTION OBJECTIVES

- Discuss the importance of oxygen and ozone in the atmosphere.
- Compare the structure of ordinary oxygen with that of ozone.
- Describe the physical and chemical properties of oxygen and ozone.
- Discuss the industrial and laboratory preparations of oxygen and ozone.

STUDY HINT

CFCs and the ozone layer are discussed in the Intra-Science article, "Preserving the Ozone Layer" on pages 400–401.

Figure 10-1 Many aerosol sprays, such as deodorants and hair sprays, no longer use chlorofluorocarbons.

Representative Gases **289**

STUDY HINT
Resonance is discussed in a Chemistry Notebook in Section 6.2.

liquid oxygen is allowed to flow past a magnet, it is attracted to the magnetic field. As discussed in Section 4.2, a substance that is attracted to a magnetic field is said to be paramagnetic. To be paramagnetic, oxygen would need two unpaired electrons, the spins of which do not cancel. The following resonance hybrid structures of the O₂ molecule better explain oxygen's paramagnetic properties and other observed properties of oxygen:

$$:\ddot{\text{O}}\cdot\cdot\ddot{\text{O}}: \longleftrightarrow :\ddot{\text{O}}::\ddot{\text{O}}: \longleftrightarrow :\dot{\text{O}}:::\dot{\text{O}}:$$

Physical Properties of Oxygen

At normal temperature and pressure, pure oxygen is a colorless, odorless, tasteless gas that is slightly denser than air. Like all other gases, oxygen can be liquefied. At temperatures lower than −118.6°C and at pressures greater than 49.8 atm (5.05 MPa), it becomes a pale blue liquid. At normal atmospheric pressure, liquid oxygen boils at −183.0°C. When cooled below −218.4°C, liquid oxygen freezes to become a pale-blue crystalline solid. Physical constants of oxygen are listed in Table 10-1.

TABLE 10-1 PHYSICAL PROPERTIES OF OXYGEN (O_2) AND OZONE (O_3)		
	Oxygen (O_2)	Ozone (O_3)
Molar mass	31.9988 g	47.9982 g
Boiling point	−183.0°C	−111.9°C
Melting point	−218.4°C	−192.7°C
Density (0°C, 1 atm)	1.429 g/L	2.144 g/L
Bond length (O—O)	0.1208 nm	0.1278 nm
Bond angle	180°	117°

Chemical Properties of Oxygen

Oxygen is one of the most active elements. With the exception of the lighter noble gases, oxygen forms compounds with all other elements. For example, when exposed to air, Group 1 metals combine directly with oxygen to form three types of oxygen compounds called oxides, peroxides, and superoxides. *An **oxide** is a compound containing the simple oxide anion (O^{2-}), which has an oxidation number of −2.* The bond character in an oxide depends on the difference in electronegatives between oxygen and the element with which it is combined. The greater the difference in electronegativities, the more ionic, or less covalent, the bond.

Oxygen is the second most electronegative element. Because Group 1 and Group 2 metals have low electronegativities, oxides of Group 1 and Group 2 metals are highly ionic. Among the Group 1 metals, only lithium combines directly with oxygen to form an oxide.

$$4Li(s) + O_2(g) \rightarrow 2Li_2O(s)$$

Sodium, which is more reactive than lithium and the Group 2 metals, burns in oxygen to form sodium peroxide. *A **peroxide** is a compound containing the O_2^{2-} anion, in which oxygen has the oxidation number −1.*

Important dates in the history of oxygen:
15th century Leonardo da Vinci (Italian artist, 1452–1519) writes that air has several constituents, one of which supports combustion.
1773–1774 Karl Wilhelm Scheele (Swedish chemist, 1742–1786) and Joseph Priestley (English chemist, 1733–1804) independently discover oxygen.
1775–1777 Antoine-Laurent Lavoisier (French chemist, 1743–1794) first recognized oxygen as an element. He developed the modern theory of combustion and demolished the phlogiston theory. Lavoisier coined the name "oxygen" from the Greek *oxys*, "sharp or sour" and *geinomai*, "I produce."

$$2Na(s) + O_2(g) \rightarrow Na_2O_2(s)$$

The still more reactive Group 1 metals—potassium, rubidium, and cesium—burn in oxygen to form superoxides. A **superoxide** is a binary compound containing the O_2^- anion, in which oxygen has the uncommon oxidation number $-1/2$.

$$K(s) + O_2(g) \rightarrow KO_2(s)$$

The Group 2 metals react spontaneously with pure oxygen to form highly ionic oxides. For example, barium reacts with oxygen to form barium oxide as follows:

$$2Ba(s) + O_2(g) \rightarrow 2BaO(s)$$

Figure 10-2 Iron burns brilliantly in pure oxygen to form the iron oxide Fe_3O_4.

Although most of the remaining metals react relatively slowly with oxygen at normal temperatures, many react quickly when heated. Because these metals have electronegativities that are close to the electronegativity of oxygen, the bonding in the resulting oxides is generally polar covalent. In the open air, moist iron reacts with oxygen to form hydrated iron(III) oxide, commonly called rust, in the following reactions.

$$2Fe(s) + O_2(g) + 2H_2O(\ell) \rightarrow 2Fe(OH)_2(s)$$

$$4Fe(OH)_2(s) + O_2(g) + 2H_2O(\ell) \rightarrow 4Fe(OH)_3(s)$$

$$2Fe(OH)_3(s) \longrightarrow Fe_2O_3 \cdot H_2O(s) + 2H_2O(\ell)$$

As shown in Figure 10-2, however, red-hot steel wool burns brilliantly when plunged into pure oxygen. This reaction produces another iron oxide (Fe_3O_4), called *black magnetite*.

$$3Fe(s) + 2O_2(g) \rightarrow Fe_3O_4(s)$$

Black magnetite is a mixed oxide of Fe_2O_3 and FeO and is referred to as an iron(II, III) oxide.

Because they have electronegativities only slightly lower than the electronegativity of oxygen, nonmetals that burn in pure oxygen or in air produce covalently bonded oxides. The combustion of hydrogen and oxygen to produce water, and the combustion of sulfur in air to produce sulfur dioxide, are examples of this type of reaction.

$$2H_2(g) + O_2(g) \rightarrow 2H_2O(\ell)$$

$$S(s) + O_2(g) \rightarrow SO_2(g)$$

Sulfur dioxide is also produced in the combustion of sulfur-containing fuels, such as high-sulfur coal. When SO_2 gas is released into the atmosphere, it is further oxidized and combines with water in the air to form sulfuric acid. The presence of sulfuric acid in the air is a major cause of acid precipitation. Figure 10-3 shows one harmful effect of acid precipitation on the environment.

Uses and Preparation of Oxygen

Pure gaseous oxygen has a variety of important applications, some of which are illustrated in Figure 10-4. Oxygen can be used to remove impurities from molten steel. The oxygen gas unites with carbon and other impurities. The resulting compounds escape as gases or solids that float to the surface of the molten steel in the form of removable slag.

Figure 10-3 The effects of acid precipitation caused by the combustion of sulfur-containing fuels have caused the death of the trees in this forest.

(a) (b) (c) (d)

Figure 10-4 *(a)* Impurities are burned off molten steel in a basic-oxygen furnace by a stream of oxygen gas. *(b)* Portable oxygen equipment is used in the emergency treatment of people with breathing difficulties. *(c)* Oxygen in the air kills harmful anaerobic bacteria in a sewage-treatment plant. *(d)* An oxygen–acetylene torch is used to weld steel.

CAUTION: If you demonstrate this reaction, be sure to use a 3% solution of H_2O_2. A 30% solution is too reactive to be used safely.

Metals are commonly welded together by the use of an oxygen–acetylene torch. A burning mixture of oxygen gas and acetylene gas can have a temperature as high as 3500°C, which is hot enough to melt most common metals.

The combustion of liquid oxygen and liquid hydrogen provides the energy to lift huge spacecraft into orbit. Oxygen is also used to treat people with breathing difficulties or heart problems and to purify water by killing harmful anaerobic bacteria and by destroying odor-causing substances. In the discussion that follows, some of the most common laboratory and industrial preparations of oxygen are described.

Decomposition of hydrogen peroxide Oxygen can be prepared safely and conveniently in the laboratory by decomposing hydrogen peroxide in the presence of a catalyst, powdered manganese dioxide. The laboratory apparatus for this reaction is shown in Figure 10-5. The reaction that takes place can be written as follows:

$$2H_2O_2(aq) \xrightarrow{MnO_2} 2H_2O(\ell) + O_2(g)$$

Figure 10-5 Oxygen can be prepared in the laboratory by the decomposition of hydrogen peroxide in the presence of a catalyst (manganese dioxide). The oxygen is collected by water displacement. This same apparatus can be used to prepare oxygen by allowing water to drop onto sodium peroxide.

Oxidation of water by sodium peroxide When water is allowed to drop onto sodium peroxide in a generator similar to that shown in Figure 10-5, oxygen is liberated as follows:

$$2Na_2O_2(s) + 2H_2O(\ell) \rightarrow 4NaOH(aq) + O_2(g)$$

Thermal decomposition of potassium chlorate When potassium chlorate is mixed with the catalyst manganese dioxide and heated, it decomposes into oxygen and potassium chloride.

$$2KClO_3(s) \xrightarrow[MnO_2]{\Delta} 2KCl(s) + 3O_2(g)$$

Electrolysis of water Figure 10-6 shows a standard laboratory apparatus used for the electrolysis of water. Electrolysis of water is the decomposition of water into hydrogen gas and oxygen gas by the passage of a direct current of electricity through the water. As the current flows, oxygen gas collects at the positive terminal; at the same time hydrogen gas collects at the negative terminal.

$$2H_2O(\ell) \xrightarrow{\text{electrical energy}} 2H_2(g) + O_2(g)$$

Industrially, this process yields oxygen of very high purity.

Liquefaction of air The most common industrial preparation of oxygen involves the liquefaction of ordinary air by compression and cooling. Nitrogen is then separated from oxygen as follows. When liquid air at −200°C is allowed to warm up, nitrogen, which boils at −195.8°C, boils off first. Oxygen, which boils at −183.0°C, and a few trace gases remain. This process is called fractional distillation. The oxygen is separated from the trace gases by further fractional distillation and condensation.

Formation and Properties of Ozone

You may already have observed the formation of ozone and not realized it. For example, have you ever smelled an irritating or pungent odor after a lightning storm or near electrical machinery such as a photocopier? If so, what you smelled was probably ozone.

The structure and properties of ozone Like ordinary molecular oxygen (O_2), ozone (O_3) is a gas at room temperature. Oxygen is colorless and odorless, however, and ozone is a poisonous blue gas with a pungent odor. Physical constants of ozone are listed in Table 10-1.

Figure 10-7 compares the shapes of an oxygen molecule and an ozone molecule. The ozone molecule is triatomic and bent. The O—O—O bond angle is about 116.5°. The O—O bonds in ozone have a bond length of 128 pm and an energy of 301 kJ/mol. These bonds are shorter and stronger than an O—O single bond, but longer and weaker than an O=O double bond. The ozone molecule is therefore represented by two resonance hybrid structures.

Formation of ozone Ozone is formed from oxygen in a two-step process. First, energy is added to molecular oxygen to produce atomic oxygen. Next,

Figure 10-6 Water is decomposed by electrolysis into its constituent elements, oxygen and hydrogen.

STUDY HINT
Electrolysis was introduced in Section 8.2.

CAUTION: Sodium peroxide must be handled carefully. It is a very strong oxidizing agent and a high fire and explosion risk. It should not come in contact with any organic materials, powdered metals, alcohols, acids, or water. It is toxic, irritating, and corrosive to body tissues.

Oxygen molecule

Ozone molecule

Figure 10-7 An oxygen molecule consists of two atoms of oxygen. An ozone molecule consists of three atoms of oxygen.

SECTION OBJECTIVES

- Discuss the uses and abundance of hydrogen.
- Describe the physical and chemical properties of hydrogen.
- Discuss the industrial and laboratory preparations of hydrogen.

1. (a) See definition on page 289. (b) Both gases at room temperature
2. Absorbs most of the ultraviolet light radiated from the sun which can harm plant and animal tissue
3. Removes impurities from molten steel, welds metals together, used in spacecraft engines, used to treat people with breathing difficulties or heart problems, and purifies water supplies
4. Ozone is a poisonous blue gas with a pungent odor; oxygen is a colorless and odorless gas and is necessary for life.
5. See "Formation of Ozone," above.

each free oxygen atom combines with an oxygen molecule. The overall equation for the formation of ozone is

$$3O_2(g) + \text{energy} \rightarrow 2O_3(g)$$

The energy required in the first step may be provided by ultraviolet radiation, as in the case in the formation of the ozone layer in the upper atmosphere. The energy can also be provided by electrical discharges, which explains why the smell of ozone can be noticed after lightning storms or near electrical machinery. A typical laboratory ozone generator produces ozone by passing a high-voltage electric current through air or oxygen.

Section Review

1. (a) What is an allotrope? (b) Why are oxygen and ozone considered to be allotropes?
2. Why is ozone in the upper atmosphere important to life on earth?
3. Describe five important technological applications of oxygen.
4. Compare the physical properties of oxygen and ozone.
5. Describe the two steps in the formation of ozone and write the overall equation for this process.

10.2 Hydrogen

Hydrogen is the first element in the Periodic Table and the most abundant element in the universe. It is estimated that 90% of all the atoms in the universe are hydrogen atoms, which constitute 70% of the mass of the universe. The earth's surface layer, the crust, however, contains only 0.88% hydrogen by mass. On the earth's surface, free elemental hydrogen exists in negligible quantities as a gas consisting of diatomic hydrogen molecules H_2. The main engines of the space shuttle shown in Figure 10-8 burn liquid hydrogen as fuel. Pure hydrogen, however, is not used as a fuel in ordinary situations, such as heating a home or running a car, because of its tendency to explode. Some scientists speculate that gaseous hydrogen could replace natural gas as a fuel for homes and industries. If better techniques to control its explosiveness are developed, hydrogen gas could be an efficient nonpolluting source of energy for a "hydrogen economy" of the future.

Occurrence of Hydrogen

Some scientists hypothesize that hydrogen nuclei were the first atomic nuclei formed from subatomic particles as the universe cooled down from the "big bang" that may have marked its origin. Today, the universe still consists mainly of hydrogen, although nearly 100 other elements have been produced by nuclear reactions in the cores of stars. Our sun and the other stars are largely elemental hydrogen in the plasma state.

Very little free elemental hydrogen exists on earth. Some traces of elemental hydrogen are released into the air by volcanoes, but the density of H_2 gas is so low that it rises into the upper atmosphere and escapes into space. The huge amount of water on the earth's surface is 11% hydrogen by mass. In addition, hydrogen is a constituent of many minerals and all living things. Petroleum, natural gas, wood, and soft coal are rich in hydrogen. The combination of hydrogen with oxygen provides much of the heat produced by the combustion of these fuels.

Physical Properties of Hydrogen

Unlike the Group 1 elements, hydrogen ordinarily exists in the form of H_2 molecules. As a typical molecular substance, molecular hydrogen is a poor conductor of heat and electricity.

Under ordinary conditions, hydrogen exists as a colorless, odorless, and tasteless gas. Hydrogen gas—the least dense of all gases—has a density only 7% that of air. (At 0°C and 1 atm, the density of hydrogen is 0.08987 g/L and the density of air is 1.293 g/L.) Because of its low solubility in water, hydrogen can be collected in the laboratory by water displacement, as shown in Figure 10-9. When cooled below −252.9°C, hydrogen gas becomes a liquid. Liquid hydrogen is the least dense of all liquids, having a density only 7% that of liquid water. When cooled below −259.1°C, liquid hydrogen becomes a solid. Solid hydrogen is the least dense solid, being only 8% as dense as liquid water. Some physical properties of hydrogen are listed in Table 10-2.

Figure 10-8 Liquid hydrogen is the fuel used in the space shuttle's main engines. Liquid oxygen is used as the oxidizing agent for the very rapid fuel combustion needed during lift-off.

Figure 10-9 Laboratory setup for the preparation of hydrogen. Any metal more active than hydrogen can be used to replace the hydrogen from any one of many acids. The metal is put in the wide-mouth generator bottle, and the acid is added through the funnel tube. Hydrogen is collected by water displacement. This preparation must be performed under a hood.

TABLE 10-2 PHYSICAL PROPERTIES OF HYDROGEN (H_2)	
Molar mass	2.0158 g
Boiling point	−252.9°C
Melting point	−259.1°C
Density (0°C, 1 atm)	0.0899 g/L
Bond length (H—H)	0.074 611 nm

Representative Gases

Chemical Properties of Hydrogen

There are more compounds of hydrogen than of any other element. Hydrogen reacts with both metals and nonmetals.

Reaction with nonmetals Hydrogen's electronegativity of 2.1 is close to that of most other nonmetals. As a result, hydrogen reacts with nonmetals to form molecular compounds with single covalent bonds. The bonds range in polarity from the nonpolar H—P bond to the highly polar H—F bond.

Although free hydrogen atoms are extremely reactive, ordinary molecular hydrogen is not very reactive at room temperature. A mixture of hydrogen and oxygen must be heated to 800°C or ignited by an electric spark to make the two gases combine. They then combine explosively to form water according to the following equation:

$$2H_2(g) + O_2(g) \rightarrow 2H_2O(g)$$

Hydrogen and chlorine do not combine when they are mixed in the dark. They react explosively in sunlight to form hydrogen chloride, however. Hydrogen chloride can also be prepared by burning a stream of hydrogen gas in chlorine. The equation for the reaction of hydrogen with chlorine is

$$H_2(g) + Cl_2(g) \rightarrow 2HCl(g)$$

Reaction with metals Hydrogen reacts with the least electronegative metals to form ionic hydrides. *An **ionic hydride** is a compound consisting of hydrogen and one other element that is less electronegative.* Each hydrogen atom receives an electron from the less electronegative element to form the hydride ion, H^-. The hydrides of the Group 1 and Group 2 metals are prepared by heating the metals in the presence of hydrogen. The preparations of the white, crystalline, ionic hydrides of sodium and barium are represented by the following equations:

$$2Na(s) + H_2(g) \rightarrow 2NaH(s)$$

$$Ba(s) + H_2(g) \rightarrow BaH_2(s)$$

*A **covalent hydride** is a compound consisting of hydrogen and one other element that is more electronegative.* For example, hydrogen burns in the presence of bromine to form hydrogen bromide.

$$H_2(g) + Br_2(\ell) \rightarrow 2HBr(g)$$

Uses and Preparation of Hydrogen

Among the important industrial uses of hydrogen is its combination with liquid vegetable oils to produce hydrogenated solid or semisolid fats, such as margarine. Many dietary experts advise against the use of hydrogenated oils, which some research has linked to heart disease.

Hydrogen gas is sometimes used as a reducing (deoxidizing) agent. Its use as a fuel—particularly in fuel cells that produce electricity—is increasing. Hydrogen gas is also used to remove sulfur impurities from petroleum and coal. The hydrogen combines with sulfur to produce hydrogen sulfide gas (H_2S) which can be removed easily. The remainder of this section discusses the common laboratory and commercial preparations of hydrogen.

CAUTION: The mixing of hydrogen and chlorine in a glass tube is a popular but dangerous demonstration because of the possibility of the glass tube shattering. The glass tube should be covered with plastic netting as a safety measure.

Important dates in the history of hydrogen:
1671 Robert Boyle (English physicist and chemist, 1627–1691), along with other 17th century scientists, observed that a flammable gas was produced by the action of dilute sulfuric acid on iron.
1766 Henry Cavendish (English chemist and physicist, 1731–1810) observed that hydrogen gas was lighter than air. He reacted iron, zinc, and tin with several acids, establishing hydrogen's properties.
1781 Cavendish disproved that water is an element by forming it from the combustion of hydrogen in oxygen.
1783 Lavoisier coined the name "hydrogen" from two Greek words, *hydro* and *genes*, meaning "water former."
1878 Sir Joseph Norman Lockyer (English astronomer, 1836–1920) detected hydrogen spectroscopically in the sun's corona.

From acids by replacement The usual laboratory method of preparing hydrogen involves the replacement of hydrogen from acids by metals above it in the activity series. For example, iron, zinc, or magnesium react with hydrochloric acid or sulfuric acid to produce hydrogen. The equations for the chemical reactions of zinc with sulfuric acid (H_2SO_4) and hydrochloric acid (HCl) are

$$Zn(s) + H_2SO_4(aq) \rightarrow ZnSO_4(aq) + H_2(g)$$

$$Zn(s) + 2HCl(aq) \rightarrow ZnCl_2(aq) + H_2(g)$$

In each of these two reactions, an atom of zinc replaces two atoms of hydrogen, which unite to form gaseous molecular hydrogen, H_2.

From water by replacement Group 1 and Group 2 metals react with water, displacing hydrogen and forming a base containing the hydroxide ions. For example, sodium reacts vigorously with water. Each sodium atom replaces one of the hydrogen atoms in a molecule of water, as indicated by the following equation:

$$2Na(s) + 2H_2O(\ell) \rightarrow 2NaOH(aq) + H_2(g)$$

Potassium is more active than sodium and displaces hydrogen from water so vigorously that the heat of the reaction ignites the hydrogen, as shown in Figure 10-10. The reaction is as follows:

$$2K(s) + 2H_2O(\ell) \rightarrow 2KOH(aq) + H_2(g)$$

The Group 2 metal magnesium displaces hydrogen slowly from boiling water, whereas the more reactive Group 2 metal calcium displaces hydrogen from cold water. In each case, each atom of metal displaces two atoms of hydrogen—one from each of two molecules of water—as shown in the following equation:

$$Ca(s) + 2H_2O(\ell) \rightarrow Ca(OH)_2(aq) + H_2(g)$$

Electrolysis of water In the electrolysis of water, water is decomposed to hydrogen and oxygen. Electrolysis is discussed in Section 10.1.

Decomposition of hydrocarbons Hydrocarbons are commonly derived from petroleum or natural gas. They are the major commercial source of hydrogen. If a hydrocarbon such as methane (CH_4) reacts with steam in the presence of a nickel catalyst at a temperature of about 850°C, a mixture of hydrogen and carbon monoxide is produced.

$$CH_4(g) + H_2O(g) \xrightarrow{Ni} CO(g) + 3H_2(g)$$

When the mixture of gaseous products is cooled and compressed, the carbon monoxide liquefies and can be separated from the hydrogen gas. (The reverse of this reaction can be used to synthesize methane.)

Section Review

1. What are some important industrial uses of hydrogen?
2. Discuss the physical properties of hydrogen.
3. How does hydrogen react with *(a)* nonmetals and *(b)* metals? Write one balanced equation for each type of reaction.
4. What is a hydride?

Figure 10-10 Potassium, like other Group 1 metals, reacts vigorously with water. The light for this photograph was produced by dropping a small amount of potassium into a beaker of water.

1. Hydrogen combines with vegetable oils to produce solid or semisolid fats. Hydrogen gas is used as a reducing agent, as a fuel, and to remove impurities from petroleum and coal.
2. Under ordinary conditions, hydrogen exists as a colorless, odorless, and tasteless gas. Hydrogen is the least dense of all gases and has low solubility in water. Molecular hydrogen is a poor conductor of heat and electricity.
3. *(a)* Forms molecular compounds with single covalent bonds
$2H_2(g) + O_2(g) \rightarrow 2H_2O(g)$
$2H_2(g) + Cl_2(g) \rightarrow 2HCl(g)$
(b) Forms ionic hydrides
$2Na(s) + H_2(g) \rightarrow 2NaH(s)$
$Ba(s) + H_2(g) \rightarrow BaH_2(s)$
4. A compound consisting of hydrogen and one other element that is less electronegative

10.3 Nitrogen and Ammonia

SECTION OBJECTIVES

- Discuss the importance and abundance of nitrogen in the atmosphere.
- Compare the structure of molecular nitrogen with that of ammonia.
- Describe the physical and chemical properties of nitrogen and ammonia.
- Discuss the preparation of nitrogen and ammonia.
- Describe the nitrogen cycle.

Nitrogen in the form of gaseous molecular nitrogen (N_2), is the major component of the earth's atmosphere. Although nitrogen in this form is nonreactive, it is important to life on earth. Nitrogen is found in all living things on earth. It is present in the DNA molecules that control heredity. It is also a major component of amino acids, which are the building blocks of proteins. Proteins are present in living tissue and make up the enzymes that control most chemical reactions in living things. The nitrogen cycle is described in the Chemistry and the Environment feature on page 300.

The most important nitrogen compound is ammonia (NH_3) which is the chief intermediate between atmospheric nitrogen and the millions of nitrogen-containing compounds. Ammonia is used as a fertilizer, a refrigerant, and as an intermediate in the production of nitric acid, nylon, antimalarial drugs, and other nitrogen-containing compounds. Other uses of nitrogen are illustrated in Figure 10-11.

Occurrence of Nitrogen

The earth's atmosphere consists of approximately 80% molecular nitrogen (N_2) by volume. Combined nitrogen in the form of ammonia, nitrates, nitrites, and nitric acid is widely distributed on earth in living things and in natural deposits of sodium nitrate and potassium nitrate.

TABLE 10-3 PHYSICAL PROPERTIES OF NITROGEN (N_2)	
Molar mass	28.0134 g
Boiling point	−195.8°C
Melting point	−209.9°C
Density (0°C, 1 atm)	1.251 g/L
Bond length (N—N)	0.109 75 nm

Physical Properties of Nitrogen

Under normal conditions, elemental nitrogen (N_2) is a colorless, odorless, and tasteless gas that is slightly less dense than air. At temperatures lower than −195.8°C at 1 atm, nitrogen gas condenses to a colorless liquid. Liquid nitrogen freezes to a white solid at −209.9°C. Physical properties of nitrogen are listed in Table 10-3.

Structure and Chemical Properties of Nitrogen

Elemental nitrogen normally exists in the form of nonpolar, triple-bonded N_2 molecules. Because of the high energy required to break the nitrogen-nitrogen triple bond, it is hard to separate the nitrogen atoms of the N_2 molecule so that they can enter into new combinations. This explains why elemental nitrogen is relatively nonreactive. The formation of many nitrogen compounds requires the net addition of energy, which is then stored in the compounds. This stored energy has a strong tendency to be released. As a result, these compounds are often unstable and sometimes explosive.

Figure 10-11 Fruit protected by a "blanket" of nitrogen (*top*). Liquid nitrogen is used to keep blood frozen, for later use in surgery (*bottom*).

Oxides of nitrogen are also formed during the high-temperature, high-pressure combustion in automobile engines. These oxides can combine with water in the atmosphere to produce acids, that sometimes fall to earth as acid precipitation. *The energy from ultraviolet light from the sun causes nitrogen oxides to react with unburned hydrocarbons from automobile exhaust or with oxygen in the air to produce ozone and other irritating chemicals that, together, are referred to as* **photochemical smog.** The catalytic converters required on all new cars help reduce this type of air pollution by deoxidizing nitrogen oxides to nitrogen and by completely oxidizing hydrocarbons to carbon dioxide and water.

At high temperatures, nitrogen combines directly with such metals as magnesium, titanium, and aluminum to produce nitrides. Magnesium will burn in nitrogen gas to form Mg_3N_2, magnesium nitride. In nitrides, nitrogen exists in the form of the nitride ion (N^{3-}).

Uses and Preparation of Nitrogen

The major uses of nitrogen are a direct result of its nonreactivity. Many foods are packaged in the presence of pure nitrogen rather than air. This is done to prevent the interaction of these products with the oxygen in air because in some foods, the presence of oxygen results in loss of flavor. Nitrogen is now being used as the propellant in aerosol cans because it is both nontoxic and nonpolluting.

Nitrogen is convenient to use for many of these applications because of its ready availability and ease of production. The industrial production of nitrogen involves the liquefaction of air, as described in Section 10.1.

Nitrogen can be prepared in the laboratory by heating certain ammonia-containing compounds. For example, heating a water solution of ammonium nitrite (NH_4NO_2) produces nitrogen and water according to the following equation:

$$NH_4^+(aq) + NO_2^-(aq) \rightarrow N_2(g) + 2H_2O(\ell)$$

Occurrence of Ammonia

Ammonia is abundant in the atmospheres of the outer planets Jupiter, Saturn, and Neptune. In fact, clouds of solid ammonia surround these planets. The top layer of Jupiter's colorful and violent atmosphere as photographed by the *Voyager* spacecraft (Figure 10-12), is made up of ammonia crystals. On Earth, ammonia is continually being produced naturally by the fixation of atmospheric nitrogen and by the decay of organic matter, as described in the Chemistry Notebook "The Nitrogen Cycle" on page 300. As fast as it is produced, however, it is converted into other substances in living organisms.

Structure and Properties of Ammonia

The Lewis structure for ammonia is

$$H:\overset{..}{\underset{H}{N}}:H$$

As discussed in Chapter 6, the ammonia molecule has the shape of a pyramid. The three hydrogen atoms form the base, the nitrogen atom is in the center,

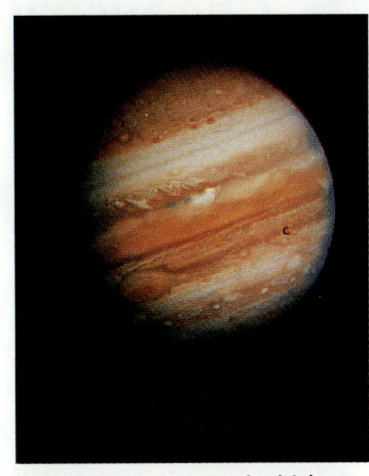

Figure 10-12 Jupiter's thick atmosphere is composed of swirling clouds of ammonia crystals.

The extremely high energy of the triple bond in N_2, 946 kJ/mol, gives it an inert gas nature. Lithium is the only element to combine with N_2 at room temperature.

Representative Gases

Chemistry and the Environment

The Nitrogen Cycle

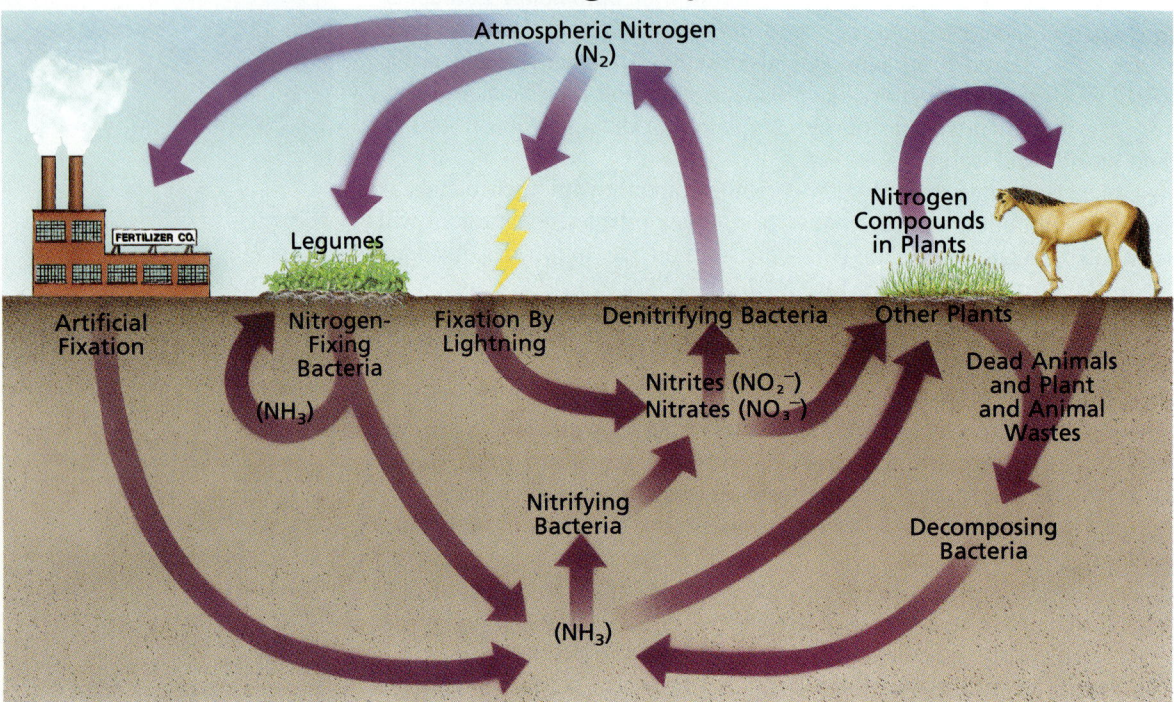

The process by which nitrogen continuously moves back and forth between living things and the atmosphere is called the *nitrogen cycle*. Through **nitrogen fixation,** *atmospheric nitrogen gas (N_2) is converted into the nitrogen compounds that can be used by plants.*

Most natural nitrogen fixation is carried out by bacteria on the roots of legumes. A legume is a plant belonging to the pea, bean, clover, or alfalfa family. In this case, the nitrogen is fixed in the form of ammonia (NH_3). Ammonia can be incorporated into plants either directly or after having been converted to nitrate ions (NO_3^-) and/or nitrite (NO_2^-) by nitrifying bacteria. Legumes used in crop rotation restore nitrogen to soil that has been depleted of it by other crops. In Asia, the fertility of rice paddies is maintained by free-living, nitrogen-fixing blue-green algae.

A great deal of natural nitrogen fixation also occurs by means of lightning, which causes atmospheric nitrogen and oxygen to react to form nitrate ions. These nitrates are dissolved in rainwater and carried down to the soil.

Animals consume nitrates in plants. Nitrogen compounds in dead plants and animals are then decomposed by bacteria into ammonia compounds.

Nitrogen is fixed artificially on a large scale by the direct combination of nitrogen and hydrogen to form ammonia in the Haber process discussed on page 302. Much of the resulting ammonia is then industrially converted to nitrates. Both ammonia and inorganic nitrates are widely used as artificial fertilizers.

Nitrogen is returned to the atmosphere through the action of denitrifying bacteria, which convert nitrates in the soil to gaseous nitrogen and thus complete the nitrogen cycle.

and the electron pair is at the apex. The ammonia molecule is therefore polar-covalent. In liquid and solid ammonia, hydrogen bonds link the hydrogen atoms of ammonia molecules to the nitrogen atoms of neighboring ammonia molecules. Because of this hydrogen bonding, ammonia has higher boiling and freezing points than would otherwise be expected for a molecule of its formula mass.

> **STUDY HINT**
> Formula mass is discussed in Chapter 7.

Under normal conditions, ammonia is a colorless gas with such a characteristically strong odor that it has traditionally been used to revive fainting victims. However, in high concentration it is very harmful. Ammonia is less dense than air and is extremely soluble in water, with which it forms strong hydrogen bonds. You are probably familiar with household ammonia, which is a useful grease-cutting agent. Some physical properties of ammonia are listed in Table 10-4.

TABLE 10-4 PHYSICAL PROPERTIES OF AMMONIA (NH_3)

Molar mass	17.0304 g
Boiling point	−33.35°C
Melting point	−77.7°C
Density (0°C, 1 atm)	0.771 g/L
Bond length (N—H)	0.100 8 nm
Bond angle (H—N—H)	107.3°

Chemically, ammonia is important as a base and as an intermediate in important natural, laboratory, and industrial organic reactions. At high temperatures, ammonia decomposes into nitrogen (N_2) and hydrogen (H_2). In the presence of a hot platinum-wire catalyst, ammonia also burns in air or oxygen to produce nitrogen and water, as is shown by the following equation:

$$4NH_3(g) + 3O_2(g) \xrightarrow[Pt]{\Delta} 2N_2(g) + 6H_2O(\ell)$$

Uses and Preparation of Ammonia

Ammonia is important in the production of artificial fertilizers. Until about a hundred years ago, farmers commonly added naturally occurring nitrogen compounds to the soil to increase crop production. For example, during the nineteenth century, sodium nitrate from Chile was imported by the United States and European countries for use as fertilizer. Then, in 1913, the German chemist Fritz Haber (1868–1934) invented a process by which nitrogen could be extracted from the air to make ammonia that could be used as an artificial fertilizer.

In 1985, more moles of ammonia were produced than of any other compound. In 1986, nitrogen gas replaced ammonia as the compound most produced. Sulfuric acid continues to be the first in terms of mass.

Decomposition of ammonium compounds In the most common laboratory preparation of ammonia, a mixture of calcium hydroxide and ammonium chloride is heated in a test tube fitted with an L-shaped delivery tube, as shown in Figure 10-13. Ammonia is so soluble in water that, unlike most other gases, it cannot be collected by water displacement. It is collected by the displacement of air in an inverted test tube, instead. The equation for this reaction is

$$Ca(OH)_2(s) + 2NH_4Cl(s) \xrightarrow{\Delta} CaCl_2(s) + 2NH_3(g) + 2H_2O(g)$$

Representative Gases

Figure 10-13 When a mixture of ammonium chloride and calcium hydroxide is heated in a test tube, ammonia is given off. The ammonia is collected by the downward displacement of air because it is less dense than air. This acid–base indicator paper turns blue in the presence of ammonia, a base. This reaction should be performed under a hood.

Figure 10-14 A flow chart of the Haber process. Nitrogen and hydrogen are compressed and then enter a catalyst chamber. Ammonia produced here is condensed into a liquid in the cooler. The uncombined nitrogen and hydrogen are recirculated through the compressor and catalyst chamber.

Industrial preparation of ammonia Ammonia is formed as an industrial by-product in the production of coal gas by heating bituminous coal in the absence of air. Most ammonia is produced industrially by the Haber process, however. *The* **Haber process** *is the catalytic synthesis of ammonia from nitrogen gas and hydrogen gas.*

$$N_2(g) + 3H_2(g) \xrightleftharpoons{\text{catalyst}} 2NH_3(g)$$

This reaction is reversible and is carried out at temperatures of 400°–500°C and at pressures as high as 1000 atm (101 MPa) using a catalyst that is a mixture of porous iron and oxides of other metals. Under these conditions, about 40 to 60 percent of the reacting gases are changed to ammonia in each pass through the catalyst chamber, as shown in the chart in Figure 10-14. The ammonia is separated from the unreacted nitrogen and hydrogen by dissolving it in water or by cooling and liquefying it. The unreacted nitrogen and hydrogen are recompressed and again passed over the catalyst. Eventually, all the nitrogen and hydrogen are converted to ammonia in this way. More than 10 billion kilograms of ammonia are produced each year in the United States.

Section Review

1. Explain the importance of nitrogen to life on earth.
2. Compare the structures of nitrogen (N_2) and ammonia (NH_3).
3. Compare the physical properties of nitrogen and ammonia.
4. What is photochemical smog?
5. How is nitrogen prepared industrially?
6. Explain what is meant by the nitrogen cycle.

10.4 Carbon Dioxide and Carbon Monoxide

Depending on the reaction conditions, oxygen and carbon can combine to form either carbon dioxide or carbon monoxide. These two compounds have extremely different properties. Carbon monoxide is a highly toxic compound produced both in nature and as a by-product of many human activities. Carbon dioxide, however, plays a key role in supporting life on the earth. It is one of two ingredients used by plants in photosynthesis. **Photosynthesis** *is the process by which plants use energy from the sun in the presence of chlorophyll (a catalyst) to combine carbon dioxide with water to produce carbohydrates.* The overall equation for this reaction is

$$3CO_2(g) + 3H_2O(\ell) \xrightarrow[\text{chlorophyll}]{\text{energy (sunlight)}} C_3H_6O_3(s) + 3O_2(g)$$

In effect, photosynthesis converts the radiant energy from sunlight into stored chemical energy in a carbohydrate molecule. The carbohydrate produced by photosynthesis is converted by living things into a wide variety of other carbon-containing substances and ultimately provides energy for all life processes on earth.

Occurrence of Carbon Dioxide

Carbon dioxide makes up only about 0.03 percent of the earth's atmosphere by volume. This percentage is gradually increasing, however. The increased amounts of carbon dioxide come from the burning of carbon-containing fuels, such as coal. The destruction of forests throughout the world also causes an increase in atmospheric carbon dioxide. These forests remove carbon dioxide from the atmosphere through photosynthesis. Some scientists fear that the increased levels of carbon dioxide may be responsible for a gradual increase in the average temperature of the atmosphere. The earth continuously absorbs and reflects radiant energy from the sun. A certain temperature balance is maintained. Carbon dioxide gas in the atmosphere tends to trap some of the reflected energy. This is known as the "greenhouse effect." Adding carbon dioxide to the atmosphere is like adding a layer of insulation to a house. The house gets warmer.

Even a small increase in global temperature could be harmful. The polar ice caps might melt and raise the level of the oceans, thus flooding coastal cities. Also, areas that are now valuable farmland might become dry and unproductive because of changes in global weather patterns caused by a rise in global temperature.

Structure of Carbon Dioxide

Carbon dioxide molecules are linear, with the two oxygen atoms bonded on opposite sides of the carbon atom, as shown in Figure 10-15. Because of this arrangement, the identical polarities of the individual carbon–oxygen bonds are opposed and the molecule is nonpolar. Scientists have determined experimentally that the carbon–oxygen bonds in carbon dioxide are shorter

C_3O_2, carbon suboxide, is also known. It is a gas (bp 6°C) prepared by the dehydration of malonic acid with P_2O_5 at 300°C.

SECTION OBJECTIVES

- Discuss the importance and relative abundance of carbon dioxide and carbon monoxide in the atmosphere.
- Compare the structure of carbon dioxide with that of carbon monoxide.
- Describe the physical and chemical properties of carbon dioxide and carbon monoxide.

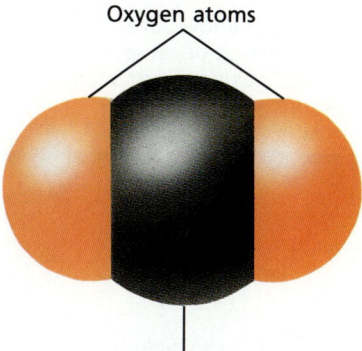

Figure 10-15 A carbon dioxide molecule is linear and consists of one carbon atom and two oxygen atoms.

1. In DNA molecules; major component of amino acids
2. N_2: nonpolar, triple-bonded molecules. NH_3: covalent molecule in the shape of a pyramid with the three hydrogen atoms forming the base, the nitrogen atom at the center, and an electron pair at the apex
3. N_2: colorless, odorless, tasteless, slightly less dense than air. NH_3: colorless, strong odor, less dense than air, extremely soluble in water
4. Ozone and other irritating chemicals produced by the reaction, catalyzed by ultraviolet light, of nitrogen oxides and unburned hydrocarbons from automobile exhaust
5. Liquefaction of air
6. Nitrogen continuously moves back and forth between living things and the atmosphere.

Figure 10-16 Carbon dioxide can be poured down this stair of candles because of its high density and slow rate of diffusion. The candles are extinguished because carbon dioxide gas does not support combustion.

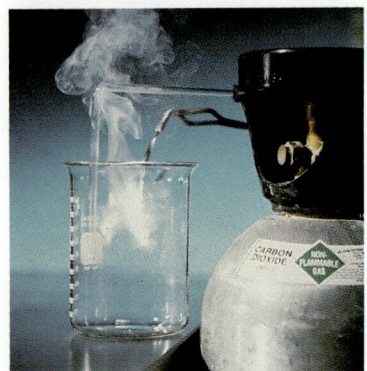

Figure 10-17 Although carbon dioxide does not ordinarily support combustion, a burning magnesium ribbon decomposes CO_2 into carbon and oxygen. The oxygen supports the continued combustion of the magnesium.

than an ordinary carbon–oxygen double bond. Thus, carbon dioxide is often represented by the following four resonance hybrid structures:

$$:\overset{-}{\ddot{O}}:C::O:^{+}$$
$$:\ddot{O}::C::\ddot{O}: \longleftrightarrow :O::C::\ddot{O}:$$
$$^{+}:O::C:\ddot{O}:^{-}$$

Each of these structures contributes equally to the actual structure. Such a resonance hybrid has the carbon–oxygen bond distance and energy that are actually observed in carbon dioxide molecules. Either of the Lewis structures shown on the left or right, above, is sufficiently accurate a representation of the carbon dioxide molecule for most purposes.

Physical Properties of Carbon Dioxide

Under ordinary conditions, carbon dioxide is a colorless gas with a faintly irritating odor and a slightly sour taste. The sour taste, characteristic of acids, actually results from the production of carbonic acid when carbon dioxide dissolves in the water in saliva. Carbon dioxide is about 1.5 times as dense as air, which means it can be poured downward into a vessel of air and displace the air. This property of carbon dioxide is illustrated in Figure 10-16, in which the candles are being extinguished by the CO_2 gas pouring into the container. Because of its density, carbon dioxide sometimes collects in mines and can cause miners to suffocate.

At temperatures lower than 31°C and at pressures higher than 72.9 atm (7.39 MPa), gaseous carbon dioxide condenses to a liquid. When exposed to normal atmospheric pressure, liquid carbon dioxide evaporates rapidly. Enough heat is lost through evaporation to cause the remaining carbon dioxide to solidify. At normal atmospheric pressure, solid carbon dioxide (dry ice) passes directly to the gaseous state at −78.5°C. Some physical constants of carbon dioxide are listed in Table 10-5.

TABLE 10-5 PHYSICAL PROPERTIES OF CARBON DIOXIDE (CO_2)	
Molar mass	44.010 g
Boiling point	−78.5°C (sublimes)
Melting point	−56.6°C (at 5.2 atm)
Density (0°C, 1 atm)	1.977 g/L
Bond length (C—O)	0.120 7 nm
Bond angle (OCO)	180°

Chemical Properties of Carbon Dioxide

Carbon dioxide is a stable gas. It does not burn and does not support combustion. A burning magnesium ribbon, as shown in Figure 10-17, is hot enough to decompose carbon dioxide into carbon and oxygen. The oxygen produced supports the continued combustion of the magnesium. The overall equation for these two reactions is:

$$2Mg(s) + CO_2(g) \rightarrow 2MgO(s) + C(s)$$

When carbon dioxide dissolves in water, a small amount of the dissolved carbon dioxide unites with water to form carbonic acid according to the following equation:

$$H_2O(\ell) + CO_2(g) \leftrightarrows H_2CO_3(aq)$$

When carbon dioxide is passed into a water solution of a hydroxide, it reacts to form a carbonate

$$CO_2(g) + 2OH^- \rightarrow CO_3^{2-} + H_2O$$

If the positive ion of the hydroxide forms an insoluble carbonate, it is precipitated. As shown in the Desktop Investigation, page 307, the test for the presence of carbon dioxide involves the reaction of carbon dioxide with calcium hydroxide, $Ca(OH)_2$, to form a white precipitate of calcium carbonate, $CaCO_3$. The following ionic equation shows this reaction:

$$Ca^{2+}(aq) + 2OH^-(aq) + CO_2(g) \rightarrow CaCO_3(s) + H_2O(\ell)$$

If excess carbon dioxide gas is bubbled through the resulting mixture of water and precipitated calcium carbonate, the precipitate disappears. The excess CO_2 gas reacts with the precipitate and water to form soluble calcium hydrogen carbonate, $Ca(HCO_3)_2$, according to the following ionic equation:

$$CaCO_3(s) + H_2O(\ell) + CO_2(g) \rightarrow Ca^{2+}(aq) + 2HCO_3^-(aq)$$

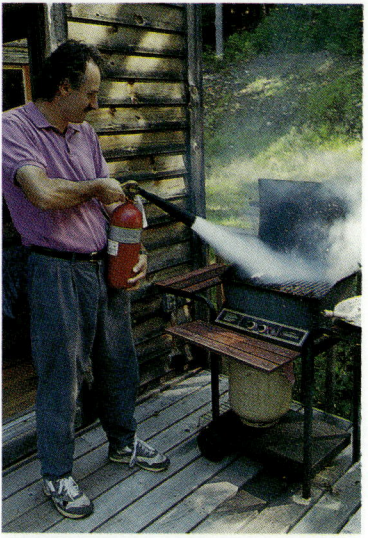

Figure 10-18 A liquid carbon dioxide fire extinguisher is effective in putting out oil and electrical fires. It cools the burning material and shuts off the supply of oxygen needed for combustion.

Uses and Preparation of Carbon Dioxide

Dry ice, which is used to keep ice cream and other foods frozen during shipping, is solid carbon dioxide. Dry ice is convenient to use for this purpose because it has a very low temperature ($-78.5°C$) and because at atmospheric pressure it does not melt to form a liquid, as does ice. Instead, it changes directly from a solid to a gas without going through the liquid state.

Liquid carbon dioxide fire extinguishers are widely used, especially for putting out oil or electrical fires. Figure 10-18 shows how a stream of liquid carbon dioxide cools the burning material and also shuts off the supply of oxygen needed for burning.

Leavening agents, such as yeast and baking powder, produce bubbles of carbon dioxide gas. These bubbles cause dough to rise and produce a lighter, less dense bread. Soft drinks are carbonated by forcing carbon dioxide into them under pressure. When this pressure is released, as when a soft-drink bottle is opened, the carbon dioxide gas bubbles out of solution.

A common laboratory preparation of carbon dioxide is discussed in the Desktop Investigation on page 307. In numerous reactions, carbon dioxide is a byproduct and thus does not have a specific commercial preparation. The following reactions are examples of commercial sources of carbon dioxide.

Burning carbon-containing substances Carbon dioxide is one of the products of the complete combustion in oxygen or air of any material that contains carbon in the presence of oxygen or air.

$$C(\text{combined}) + O_2(g) \rightarrow CO_2(g)$$

Representative Gases

Reaction of steam and natural gas Carbon dioxide is formed as an industrial by-product when natural gas, which is mostly methane (CH_4), reacts with steam in the presence of metallic oxide catalysts at temperatures between 500°C and 1000°C.

$$CH_4(g) + 2H_2O(g) \xrightarrow[\text{catalyst}]{\Delta} 4H_2(g) + CO_2(g)$$

The primary purpose of this reaction is to prepare hydrogen for making synthetic ammonia. The carbon dioxide is separated from the hydrogen by dissolving the carbon dioxide in cold water under high pressure.

Fermentation of sugar Carbon dioxide is a by-product in the production of alcohols from sugars through the process of fermentation. This reaction is catalyzed by enzymes produced by certain types of yeast. The overall equation for the fermentation reaction of glucose ($C_6H_{12}O_6$) to produce ethyl alcohol (C_2H_5OH) is

$$C_6H_{12}O_6(aq) \rightarrow 2C_2H_5OH(aq) + 2CO_2(g)$$

Heating carbonates When calcium carbonate (in limestone, marble, or seashells, for example) is heated strongly, calcium oxide and carbon dioxide are produced. The calcium oxide produced, known as quicklime or lime, is used for making plaster and mortar. The carbon dioxide produced is a by-product. The equation is

$$CaCO_3(s) \xrightarrow{\Delta} CaO(s) + CO_2(g)$$

Reaction of acid and carbonates One of the simplest ways to produce carbon dioxide is the reaction between vinegar and baking soda, as in the Desktop Investigation on page 307. Another common laboratory preparation of carbon dioxide involves the addition of dilute hydrochloric acid to marble chips. In the first stage of the reaction, the marble and the hydrochloric acid undergo a double displacement reaction:

$$CaCO_3(s) + 2HCl(aq) \rightarrow CaCl_2(aq) + H_2CO_3(aq)$$

In the second stage of the reaction, the unstable carbonic acid decomposes to form water and carbon dioxide.

$$H_2CO_3(aq) \rightarrow H_2O(\ell) + CO_2(g)$$

Geologists make use of the reaction between carbonates and acids to test for the presence of limestone and marble, both of which are mostly $CaCO_3$.

Respiration and decay Carbon dioxide is produced naturally by the processes of respiration and decay. Respiration is the reaction by which foods, usually carbohydrates, are oxidized in living things to release energy. Decay is the breakdown of the tissue of dead plants and animals into simpler substances.

Occurrence of Carbon Monoxide

Carbon monoxide is a poisonous gas produced naturally by decaying plants, some live algae, the oxidation of atmospheric methane, and volcanoes. As a result, carbon monoxide is a trace constituent of even the least polluted air. The average concentration of carbon monoxide in the atmosphere is about

CO is classified as toxic, while CO_2 is nontoxic but suffocating.

Desktop Investigation

Properties of Carbon Dioxide

Question
How is carbon dioxide made, what are some of its chemical and physical properties, and what is the common test for its presence?

Materials
candle
baking soda (sodium hydrogen carbonate)
vinegar
limewater
matches
balance
100-mL graduated cylinder
three 150-mL beakers
400-mL beaker
stirring rod
plastic straw

Procedure
Record all of your results in a data table. Remove all combustible materials from the work area. Wear safety glasses and an apron.
1. Put the candle into the 150-mL beaker and light it. Put 50 g baking soda into a 400-mL beaker. Slowly add 50 mL vinegar to this second beaker and stir. After foaming slows, carefully pour only the gas formed into the beaker that contains the lighted candle. What gas do you conclude is formed? Why?
2. Repeat the first part of Step 1, but this time do not pour the gas into a beaker containing a candle. Instead, after the reaction has slowed, pour the gas into 50 mL limewater (a saturated solution of calcium hydroxide) in a 150-mL beaker, and stir for one minute. What is the result? Bubble some exhaled air through a plastic straw into another beaker containing 50 mL limewater for one minute. What is the result? What substance does this reaction indicate is present in exhaled breath?

Discussion
1. Why can carbon dioxide gas be poured from one beaker to another?
2. After the reaction between limewater and carbon dioxide is complete, if excess carbon dioxide is bubbled through the mixture of the precipitate and water, the precipitate will redissolve. Can you design an experiment to verify this?
3. Write equations to show (a) the formation of carbon dioxide and (b) the result of bubbling carbon dioxide through limewater.

Representative Gases **307**

0.5 parts per million (ppm), which is only about one-hundredth the toxic concentration of carbon monoxide.

Toxic—even lethal—concentrations of carbon monoxide are commonly produced by the incomplete combustion of carbon-containing fuels. In the United States, as much as 90 percent of the human-produced carbon monoxide comes from automobile engines. In a crowded city on a calm day, the street-level concentration of carbon monoxide can reach unhealthy levels. Fortunately, still unidentified natural processes remove carbon monoxide from the atmosphere and prevent its average concentration from approaching a dangerous level.

Figure 10-19 The large number of cars that are concentrated in a city are responsible for most of the high concentration of carbon monoxide found there.

Structure of Carbon Monoxide

The structure of carbon monoxide can be represented by the following resonance structures :

$$^+:C:\ddot{O}:^-$$
$$^-:C::O:^+ \longleftrightarrow \quad \longrightarrow :C::\ddot{O}:$$
$$:C::\ddot{O}:$$

Unlike carbon dioxide, these four structures do not contribute equally. The hybrid is estimated to be 10 percent $^+:C:\ddot{O}:^-$, 20 percent each $:C::\ddot{O}:$ and $:C::\ddot{O}:$, and 50 percent $^-:C:::O:^+$. As indicated by this Lewis structure and by the fact that carbon and oxygen differ considerably in electronegativity, the carbon monoxide molecule is polar. The carbon atom is slightly negative and the oxygen atom is slightly positive.

CO is isoelectronic with N_2. Note the similarities in properties.

	CO	N_2
mp(°C)	−199	−210
bp(°C)	−191.5	−196

Properties and Uses of Carbon Monoxide

Unfortunately, carbon monoxide is difficult to detect. It is a colorless, odorless, and tasteless gas. It is slightly less dense than air and only slightly soluble in water. Some physical properties of carbon monoxide are shown in Table 10-6.

Chemistry and Health

Carbon Monoxide Poisoning

Every year, thousands of people are killed by carbon monoxide, which is often produced in toxic concentrations by the incomplete combustion of carbon-containing substances. The major sources of carbon monoxide are automobile engines and faulty home heaters, both of which produce potentially deadly quantities of carbon monoxide. Carbon monoxide is also present in cigarette smoke. In addition, perhaps the majority of deaths resulting from fires are caused by carbon monoxide poisoning. Whatever the source of carbon monoxide, it can have toxic effects, including headaches, nausea, and impairment of vision, hearing, and mental functions whenever its concentration in the surrounding air reaches

50 ppm. Its effects can be lethal when its concentration remains at 1600 ppm for two hours or at 1000 ppm for four hours. To prevent these harmful effects, carbon monoxide should never be released into enclosed or inadequately ventilated spaces.

Carbon monoxide produces its toxic, deadly effect by combining with some of the blood's hemoglobin. In fact, carbon monoxide binds more readily with hemoglobin than does oxygen. Once combined with carbon monoxide, hemoglobin cannot carry out its usual role of transporting oxygen to the body's cells. This loss of oxygen transport can cause asphyxiation, and ultimately coma or death, for a normally healthy person.

Carbon monoxide acts as a reducing agent in the production of iron, copper, and other metals from their oxides. An example of this is indicated in the following equation describing the reduction of iron(III) oxide:

$$Fe_2O_3(s) + 3CO(g) \rightarrow 2Fe(s) + 3CO_2(g)$$

Carbon monoxide is an important component of artificial industrial fuel gases. When these gases are burned, the carbon monoxide is oxidized to carbon dioxide.

$$2CO(g) + O_2(g) \rightarrow 2CO_2(g)$$

The industrial organic solvent methanol (CH_3OH), which can be used as a paint remover, is prepared by the reduction (hydrogenation) of carbon monoxide.

$$CO(g) + 2H_2(g) \rightarrow CH_3OH(\ell)$$

Carbon monoxide is also used in the production of many other useful organic compounds.

TABLE 10-6
PHYSICAL PROPERTIES OF CARBON MONOXIDE (CO)

Molar mass	28.010 g
Boiling point	−191.5°C
Melting point	−199°C
Density (0°C, 1 atm)	1.250 g/L
Bond length (C—O)	0.133 nm

1. *(a)* 0.03% *(b)* About 0.5 parts per million (ppm)
2. The process by which plants use energy from the sun to combine carbon dioxide with water to produce glucose
$6CO_2(g) + 12H_2O(\ell) \rightarrow C_6H_{12}O_6(s) + 6O_2(g) + 6H_2O(\ell)$
3. Carbon dioxide molecules are linear, with two oxygen atoms bonded on opposite sides of the carbon atom. As a result, the molecule is nonpolar. In carbon monoxide, the carbon and oxygen is united by a triple bond. The molecule is polar, with the carbon atom being slightly negative and the oxygen atom slightly positive.
4. Under ordinary conditions, carbon dioxide is a colorless gas with a faintly irritating odor and a slightly sour taste. It is about 1.5 times as dense as air. Carbon monoxide is a colorless, odorless, and tasteless gas that is slightly less dense than air and only slightly soluble in water.
5. Combines in the lungs with hemoglobin. Hemoglobin then cannot transport oxygen to the body's cells.

Section Review

1. *(a)* What is the percentage of carbon dioxide in the earth's atmosphere? *(b)* What is the concentration of carbon monoxide in the earth's atmosphere?
2. What is photosynthesis? Write the equation for this process.
3. Compare the structure of carbon dioxide (CO_2) and carbon monoxide (CO).
4. Compare the physical properties of carbon dioxide and carbon monoxide.
5. How does carbon monoxide interfere with oxygen transport in the body?

Chapter Summary

- Oxygen gas (O_2) and ozone (O_3) are allotropes of the element oxygen.
- An oxide is a compound containing oxygen where oxygen has an oxidation number of -2.
- Oxygen is one of the most active elements and is essential for life.
- Ozone is a poisonous blue gas. A high-altitude layer of ozone in the atmosphere absorbs harmful UV radiation.
- Hydrogen is the most abundant element in the universe.
- Hydrogen reacts with nonmetals forming single covalent bonds. It also reacts with metals to form ionic hydrides.
- Molecular nitrogen (N_2) is the major component of the earth's atmosphere. Nitrogen-containing compounds are found in all living things on earth.
- Ammonia is a colorless gas under normal conditions and is an important ingredient in the production of artificial fertilizers.
- Carbon monoxide is a highly toxic compound. It is used in the manufacture of metals and in the production of some industrial chemicals.
- Carbon dioxide is a colorless gas under normal conditions. It is one of two ingredients used by green plants to produce glucose.
- Tests for the presence of various gases are based on the properties of each gas. Some of these tests are listed in Table 10-7.

TABLE 10-7	TESTS FOR SOME COMMON GASES
H_2	Burns with a pale blue flame to produce water.
CO	Burns with a blue flame to produce CO_2.
CO_2	Forms a white precipitate when bubbled through limewater.
N_2	Does not burn or support combustion. A piece of magnesium ribbon ignited in air will continue to burn in N_2 to produce Mg_3N_2; when water is added, NH_3 is formed, which can be identified by its odor.
O_2	A glowing splint will ignite in O_2. All O_2 will be absorbed when it stands over an aqueous solution of Na_2S.

Chapter 10 Review

Representative Gases

VOCABULARY

allotrope (289)
allotropy (289)
covalent hydride (296)
Haber process (302)
ionic hydride (296)
nitrogen fixation (300)
oxide (290)
peroxide (290)
photochemical smog (299)
photosynthesis (303)
superoxide (291)

REVIEWING CONCEPTS

1. **a.** What two forms of oxygen are present in the atmosphere?
   b. How does each affect our everyday lives? (10.1)
2. Explain the attraction of the oxygen molecule by a magnet. (10.1)
3. What types of materials may be damaging the ozone layer in the upper atmosphere? (10.1)
4. Cite data reflecting the abundance of oxygen in the earth's crust, waters, and atmosphere. (10.1)
5. Draw the two resonance forms of ozone. (10.1)
6. **a.** Distinguish among oxides, peroxides, and superoxides.
   b. What determines the bond character in an oxide? (10.1)
7. Identify and write balanced chemical equations (where appropriate) that illustrate five methods used to prepare oxygen. (10.1)
8. **a.** What is the most abundant element in the universe?
   b. In what form does this element normally exist on Earth?
   c. In what form does this element exist in our sun and in other stars? (10.2)
9. Cite data reflecting the abundance of the element from Question 8a in the universe and on Earth. (10.2)
10. What is the major commercial source of hydrogen? (10.2)
11. Though it appears in Group 1 on the Periodic Table, hydrogen differs from the other Group 1 elements in a number of ways. Give at least three differences. (10.2)
12. Why is most hydrogen found on the earth in combined form, rather than as separate hydrogen atoms? (10.2)
13. According to the "big bang" theory, what is the role of hydrogen in the origin of the other elements in the universe? (10.2)
14. Identify and write sample balanced equations to illustrate four methods used to prepare hydrogen. (10.2)
15. Give two examples of the role of nitrogen in the functioning of the human body. (10.3.)
16. **a.** What is the percent of nitrogen in the earth's atmosphere?
    b. List five physical properties of elemental nitrogen. (10.3)
17. **a.** Why is elemental nitrogen relatively nonreactive?
    b. How has this property been put to use? (10.3)
18. **a.** What is nitrogen fixation?
    b. Why are legumes often used in crop rotation schemes? (10.3)
19. **a.** Draw the Lewis structure for ammonia.
    b. List at least four properties of ammonia. (10.3)
20. Cite at least three important uses of ammonia. (10.3)
21. **a.** How is ammonia produced in nature?
    b. How is it produced industrially? Identify and write balanced equations to illustrate two identical methods used to prepare ammonia. (10.3)
22. **a.** Which element is the major component of the earth's atmosphere?

b. What is the most important compound in which this element appears?
c. List three uses of this compound. (10.3)

23. Explain why ammonia, which is abundant in the atmospheres of Jupiter, Saturn, and Neptune, does not accumulate on Earth. (10.3)

24. Draw the resonance structures of carbon monoxide. What percentage does each contribute to the structure of the CO molecule? (10.4)

25. Explain the "greenhouse effect" and its possible effect on the earth's climate. (10.4)

26. What properties of dry ice make it particularly useful in shipping frozen foods? (10.4)

27. How does a liquid carbon dioxide fire extinguisher put out a fire? (10.4)

28. Describe and draw the three resonance hybrid structures of carbon dioxide. (10.4)

29. What four natural processes result in the production of carbon monoxide? (10.4)

30. **a.** What is the average concentration of carbon monoxide in the atmosphere?
b. How does this compare to the level that is considered toxic?
c. In the United States, what is the major source of human-produced carbon monoxide? (10.4)

31. List five physical properties of carbon monoxide. (10.4)

32. Describe and write balanced equations (where appropriate) to illustrate the chemical properties of carbon dioxide in each of the following reactions:
a. ability to support combustion of most substances,
b. response to burning magnesium,
c. reaction with water,
d. passage through a water solution of a hydroxide. (10.4)

33. Describe and write balanced equations (where appropriate) to illustrate the chemical properties of carbon monoxide in each of the following reactions:
a. reaction with metal oxides

b. combustion (reaction with oxygen)
c. reduction (reaction with hydrogen). (10.4)

PROBLEMS

▶ **Reactions of Representative Gases**

1. Write the balanced chemical equations for the reactions between each of the following:
a. magnesium and oxygen
b. cesium and oxygen (to produce a superoxide)
c. hydrogen and oxygen (to produce a peroxide).

2. Write the balanced chemical equations for the production of hydrogen from each of the following chemical combinations:
a. zinc and hydrofluoric acid
b. magnesium and sulfuric acid
c. iron and sulfuric acid to form iron (III) sulfate
d. magnesium and hydrochloric acid
e. rubidium and water
f. magnesium and water.

3. Write the balanced equations for the reactions between hydrogen and each of the following and determine the oxidation numbers of the atoms in each product
a. oxygen   **c.** sodium
b. chlorine   **d.** barium

4. Write the balanced chemical equation for the production of carbon dioxide from each of the following chemical processes:
a. heating of magnesium carbonate
b. reaction of magnesium carbonate and hydrochloric acid.

▶ **Stoichiometry**

5. Determine the mass of hydrogen that would be produced through each of the following:
a. the replacement reaction between sulfuric acid and 13.1 g zinc
b. from 50.0 g water by single replacement with sodium.

6. **a.** How many grams of sulfuric acid are required to react with 1.500 kg sodium hydro-

gen carbonate in a soda-acid fire extinguisher given the following equation for the reaction?

$$2NaHCO_3(aq) + H_2SO_4(aq) \rightarrow Na_2SO_4(aq) + 2H_2O(\ell) + 2CO_2(g)$$

b. What mass of carbon dioxide is produced in the above reaction?

CRITICAL THINKING

1. **Analyzing Conclusions** Group 2 metals form only oxides and peroxides when reacted with oxygen. Can you provide evidence why Group 2 metals do not form superoxides?

2. **Analyzing Information** In the earth's atmosphere, ozone is produced through electrical discharges, diffusion from the stratosphere, and photochemical smog. Its concentration ranges from 0 to 0.01 ppm in nonpolluted air. Average concentration in photochemical smog is 0.5 ppm. Hypothesize one possible explanation for this difference.

3. **Evaluating Viewpoints** A thinning in the ozone layer over the South Pole has been documented by photographs taken by orbiting satellites equipped with the Total Ozone Mapping Spectrometer. This thinning of the ozone layer occurs annually during the spring (October in the Southern Hemisphere) when the layer decreases by almost 50% of its total in August. The release of chlorofluoromethanes into the atmosphere has been determined to be the cause of the thinning by some groups. Other groups deny that there is a problem. Research the two positions and evaluate the validity of their data.

CROSS CONNECTIONS

1. **Environmental Science** Emission controls were first set by the Environmental Protection Agency in 1970. Limits were set for hydrocarbon, nitrogen oxides, and carbon monoxide emissions. The limits were to be gradually imposed between 1970 and 1975. Initial actions, such as modifying the air-fuel ratio, adjusting timing on spark plugs, placing carbon canisters in the engine to absorb gasoline vapors, and designing exhaust gas recirculation systems were not enough to meet the new standards. The catalytic converter was developed to meet the standards. Research the design of a catalytic converter and describe how it operates.

2. **Health** Prepare a report comparing the physical and chemical properties of hydrogenated fats versus nonhydrogenated fats. Discuss the dietary aspects of each.

3. **History** Research the impact of the development of the Haber process on World War I.

RESEARCH AND WRITING

1. Prepare a report concerning the development, operation, and current uses of fuel cells.

2. Write a report on the international efforts to ban the use of chlorofluorocarbons worldwide.

3. Animals are continually converting oxygen and glucose into carbon dioxide and water during respiration. Will the supply of oxygen ever run out? To answer this, research the role of carbon dioxide and oxygen in the animal and plant life on earth. Compare your findings to what you have learned about the nitrogen cycle in this chapter.

4. Prepare a report on the potential impact of the destruction of the ozone layer on plant and animal life.

PROJECT

1. Examine the labels of various food items in your kitchen that contain fats or oils. Indicate which contain hydrogenated, as opposed to nonhydrogenated, fats. Compare your list with those of your classmates and discuss the implications of your findings.

CHAPTER 11
Physical Characteristics of Gases

PLANNING CHART
Approximate Time: 5/6 days

	SECTION/TEXT FEATURES		RESOURCES	
B A S I C	**11.1 THE KINETIC THEORY OF MATTER** Demo 1 Section Review Chapter Review • Questions 1–7	315 314B 319 338	*Teacher's Resource Binder* Section Review Worksheets *Other Resources* *World of Chemistry* Video 10 Transparency 22	37
B A S I C	**11.2 QUALITATIVE DESCRIPTION OF GASES** Demo 2 Section Review Chapter Review • Questions 8–11	320 314B 321 338	*Teacher's Resource Binder* Section Review Worksheets *Other Resources* *World of Chemistry* Video 5 Transparencies 23, 24	38
B A S I C	**11.3 QUANTITATIVE DESCRIPTION OF GASES** Demos 3, 4 Microscale Investigation 11 Section Review Chapter Review • Questions 12–16, Problems 1–29	322 314B 855 337 338	*Teacher's Resource Binder* Problem Solving Worksheets Section Review Worksheets *Other Resources* *Shakhashiri* Video Demos 28, 29, 30 Transparencies 25, 26 Blackline Master 33	39
	CHAPTER END ACTIVITIES Critical Thinking, Cross Connections, Research and Writing, Projects		*Teacher's Resource Binder* Test Booklet *Computer Test Generator*	53, 55

Teaching Strategies

Introduction

Chapter 10 introduced the students to the specific properties of a number of gases. Chapter 11 begins the study of properties common to all gases as well as the underlying assumptions that give rise to the properties. Central to the discussion is the kinetic theory of gases. From the kinetic theory, qualitative relationships among the four properties of all gases are explored prior to formulating the quantitative relationships that exist and which are collectively known as the gas laws.

11.1 The Kinetic Theory of Matter

Before beginning the discussion of the kinetic theory, open a bottle of a strong cologne or perfume, or perhaps release a brief spray of room deodorizer. Have the students raise their hands when they are able to detect the odor. As a pattern develops, have the students try and explain what must be occurring at the molecular level. In short, the invisible particles of gas must be moving through the air.

Be sure that the students understand the concept of an ideal gas, the molecules of which are considered points (i.e., occupy no volume), and experience no forces of interaction. The concept of an ideal gas simplifies the model used. Stress that all gases are real, but that they conform to the model of an ideal under certain conditions (high temperature and low pressure).

List the five assumptions of the kinetic theory. Emphasize the meaning of an elastic collision—no friction or energy loss in the collision. Stress that the kinetic energy is directly related to the Kelvin temperature.

Absolute zero is defined as the temperature at which all translational motion stops. Doubling the Kelvin temperature doubles the kinetic energy. Later we shall see that the speed at which gas molecules move is also related to the Kelvin temperature. Qualitatively, at high temperatures gas molecules move faster than at lower temperatures.

Key to the study of gases is the extremely large number of particles that are in any gas sample. Such large samples allow one to make statistically accurate calculations. Many statements discuss the "average kinetic energy;" the statistical average is implied.

Demonstration 1: Kinetic Theory and the Nature of Gases

PURPOSE: To help the students visualize certain characteristic physical properties of gases

MATERIALS: A 50-mL or 100-mL syringe, a candle, one-gallon empty plastic milk container, dropping bottles of conc. HCl and conc. NH_3, a 1-L beaker filled with water

PROCEDURE: *(a) Expansion* Half fill the syringe with air and seal the tip with a serum cap. Pull on the plunger. The air fills the larger volume. *(b) Low Density* Partially fill a balloon with air. Place it in a beaker of water. The balloon floats. *(c) Fluidity* Light a candle and place it on the desk. Stand back about 5 feet and point the open end of a plastic one gallon milk container at the flame. Strike the bottom of the jug. If properly aimed, a pulse of air flows from the container and extinguishes the flame. *(d) Compressibility* Push in on the syringe plunger. The gas compresses and fills a smaller volume. *(e) Diffusion* Place a drop of concentrated ammonia near a drop of concentrated hydrochloric acid on a watch glass. The formation of a white smoke, NH_4Cl, indicates where the diffusing gases combine to form a solid.

SAFETY AND DISPOSAL: Handle the concentrated ammonia and hydrochloric acid carefully. After the reaction, rinse the watch glass with water. The other materials should be saved for further demonstrations.

Ask the students if they know of any substance which, when cooled sufficiently, will not condense and eventually freeze. Van der Waals showed that real gas particles do take up space and do attract each other and the container walls. Conditions which bring molecules close together (very high pressures) and which cause molecules to move slowly enough for the attractive forces to become substantial (low temperatures) lead to nonideal behavior.

Chemistry and Technology: Measuring and Reaching Low Temperatures

The technology feature answers the question sometimes posed by students who realize that a mercury or alcohol thermometer cannot be used to measure very low temperatures because the liquid in the thermometer would freeze.

11.2 Qualitative Description of Gases

Write the four properties of a gas on the chalkboard or overhead projector. Point out that relationships between any two can be studied and described if the remaining two properties are held constant.

Point out that most of the relationships can be predicted from day-to-day experiences. A formula need not be memorized. Use of common sense is the best approach.

Demonstration 2: Gas relationships

PURPOSE: To show the qualitative properties of gases
MATERIALS: a syringe, half-filled with air
PROCEDURE: Demonstrate the particular relationships among the qualitative properties of gases by using the table below. It lists the combinations of variable properties and constant properties and an action, using the half-filled syringe.

Variables	Constants	Action
PV	nT	Squeeze on the plunger of the syringe. *Volume decrease as pressure increases.*
TV	nP	Put the syringe in a beaker of hot water. The plunger moves outward. *Volume increases as the temperature increases.*
PT	nV	If the syringe is heated and the volume expands, the only way to return to the original volume is to exert a larger pressure. *Pressure increases as temperature increases.*
Pn	TV	Add gas to the syringe. Pressure must be added to bring the syringe back to the original volume. *Pressure increases as the number of particles increase.*
Vn	PT	Add gas to the syringe. The volume increases. If the syringe is held vertically, the pressure of the atmosphere and gravity pulling on the syringe plunger remain constant. *Volume increases as the number of particles increases.*

11.3 Quantitative Description of Gases

Based on the qualitative relationships discussed above, students should notice that only two types of relationships were observed: *(1)* as one property was increased, the other property also increased—a direct relationship, and *(2)* as one property was increased, the other property decreased—an inverse relationship. The quantitative formulation of the gas laws is little more than expression the relationships as either direct or inverse.

Demonstration 3: Area and Pressure

PURPOSE: To show the effect of area on the pressure exerted by an object (gravitational force)
MATERIALS: *Modern Chemistry* textbook, thumbtack
PROCEDURE: Place the thumbtack point up on the palm of your hand. Balance the textbook on the point. Now turn the thumbtack point down against the palm of your hand. Ask the students to predict what might happen if you were to place the textbook on the thumbtack. Most will agree that the thumbtack would be driven into your hand. **CAUTION:** Do NOT actually place the book on the thumbtack. On an overhead projector or chalkboard, show that reducing the size of the volume (the denominator) will increase the size of the fraction (the pressure). The numerator (the gravitational force exerted by the book) remains constant.

Demonstration 4: Atmospheric Pressure and Gases

PURPOSE: To show the difference between atmospheric pressure and the pressure exerted by a gas
MATERIALS: an empty aluminum soda can, beaker, tongs, water, Bunsen burner
PROCEDURE: Hold the can with a pair of beaker tongs, add 10 mL of water to the can, and heat the can over a Bunsen-burner

flame. After about one minute of vigorous boiling, turn the can over and place the *mouth* of the can in a beaker of cold water. The can immediately implodes. Perform the demonstration a second time, but this time have your students explain what might be happen during each step of the process. (During the boiling process the air originally in the can is replaced with steam. Placing the mouth of the can in the water causes the steam to condense. A vacuum results. The few vapor molecules left in the can are not able to balance the atmospheric pressure and the walls of the can collapse.)

References and Resources

Books and Periodicals

"Calculating Chemistry—Gases," *ChemMatters*, Vol. 1, No. 1, Feb. 1983, p. 7.

Dolan, Arthur W. "What's the Big Deal About an Empty Bottle? Some Simple Demonstrations with Air," *The Science Teacher*, Vol. 54, No. 7, Oct. 1987, pp. 50–51.

Garde, Ira Batra. "An Easy Approach for Reading Manometers to Determine Gas Pressure: The Analogy of the Child's Seesaw," *Journal of Chemical Education*, Vol. 63, No. 9, Sept. 1986, pp. 796–797.

Rose, Diane. "Charles' Law: Students Develop Their Own Procedure," *Journal of Chemical Education*, Vol. 64, No. 8, Aug. 1987, pp. 712–713.

Shakhashiri, Bassam Z. "Physical Behavior of Gases," *Chemical Demonstrations*, Vol. 2, Madison: University of Wisconsin Press, 1985.

Computer Software

Moving Molecules, HRM Software. (Changes in pressure and temperature affect the behavior of molecules in gases, liquids, and solids)

Smith, S.G., R. Chabay, and E. Kean. Introduction to General Chemistry, Disk 7, *Gas Laws*, COMPress. (Ideal Gases, Atmospheric Pressure, Boyle's law, Charles' law, and Pressure—Temperature)

Answers and Solutions

Reviewing Concepts

1. It is based on the idea that particles of matter are always in motion and that this motion has consequences.
2. An ideal gas is an imaginary gas that conforms perfectly to all of the assumptions of the kinetic theory.
3. • Gases consist of large numbers of tiny particles that are far apart in comparison to their size.
   • The particles of a gas are in constant motion, moving rapidly in straight lines in all directions.
   • The collisions between particles of the gas and between particles and container walls are elastic collisions.
   • There are no forces of attraction or repulsion between the particles of a gas.
   • The average kinetic energy of the particles of a gas is directly proportional to the Kelvin temperature of the gas.
4. The molecules of gases are much farther apart.
5. An elastic collision is one in which there is no net kinetic energy loss.
6. **a.** $KE = \frac{1}{2}mv^2$, where m is the mass of the particle, and v is its velocity.
   b. The average velocities and kinetic energies of gas molecules increase with an increase in temperature and decrease with a decrease in temperature.
7. **a.** Diffusion is the spontaneous mixing of the particles of two substances due to their random motion.
   b. The rate of diffusion of one gas through another is largely dependent upon the speeds of the intermingling gas particles.
   c. In general, the more massive the gas particle is, the slower is its speed and the more slowly it diffuses through another gas.
   d. Effusion is the process by which gas particles under pressure pass through a very small opening from one container to another.
8. **a.** A gas in a closed container exerts pressure because its moving molecules hit the walls of the container.
   b. At constant temperature, the pressure exerted by a fixed quantity of gas increases as the volume decreases.
9. **a.** As temperature increases, the volume occupied by a fixed quantity of a gas must also increase in order to maintain constant pressure. Conversely, as the temperature of a gas decreases, its volume must also decrease in order to maintain that constant pressure.
   b. This effect is observed because, at increasingly higher temperatures, the gas molecules move faster and collide with the walls of the container more frequently and with more force. Thus, the only way to maintain constant pressure is to increase volume, thereby reducing the number and frequency of collisions per unit of wall area and offsetting the greater force of collisions at the higher temperature.
10. **a.** At constant volume, as temperature increases, so does the pressure exerted by a given quantity of a gas.
    b. As temperature increases, the average kinetic energy and velocity of the gas molecules increase, so that they experience more frequent and more energetic collisions with the walls of their container. Since the volume occupied by the gas is fixed, the pressure must increase.
11. **a.** At constant pressure and temperature, the volume of a gas varies directly with the number of molecules.
    b. As the number of molecules increases, the number of collisions per unit of wall area would increase if volume were held constant. The volume of the container must therefore increase if the collision rate, and thus the pressure, are held constant. If the number of molecules

of gas decreases under the same conditions, the volume must also decrease.

12. **a.** Atmospheric pressure is the pressure exerted by the atmosphere, or blanket of air surrounding the earth.
b. The pressure of the atmosphere is due to the weight of the gases composing the atmosphere. The force of gravity holds the atmosphere close to the earth.
c. Atmospheric pressure at sea level is equal to 1.01325×10^5 N/m², or 10.1325 N/cm² which is about the weight of a one-kilogram mass per square centimeter of surface, or 9.8 N/cm².

13. **a.** According to Torricelli, the height of the mercury in the tube rises or falls until the pressure exerted by its weight is equal to that exerted by the atmosphere. At sea level, the force per unit area exerted by the atmosphere is approximately equal to the force exerted by a column of mercury 760 mm in height.
b. The comparable height maintained by a column of water would be about 34 feet.
c. The reason for the difference in heights is that mercury is 14 times as heavy as water, so it exerts 14 times more pressure than does a comparable column of water.

14. **a.** Three commonly used units of pressure are millimeters of mercury (mm Hg), torrs (1 torr = 1 mm Hg), and pascals.
b. One atmosphere of pressure is equal to 1.01325 bar of pressure.
c. A pascal is defined as the pressure exerted by a force of one newton (N) acting on an area of one square meter; 1 Pa = 1 N/m².
d. One standard atmosphere of pressure is defined as 1.01325×10^5 Pa, or 101.325 kPa.

15. **a.** The Celsius equivalent of absolute zero is $-273.15°C$.
b. This temperature is the lowest temperature possible.
c. The average kinetic energy of the molecules of a gas is directly proportional to the Kelvin temperature of the gas.

16. **a.** The partial pressure of a gas is the pressure that it exerts in a mixture of gases.
b. The pressure that each gas exerts in the mixture is independent of the pressures of the other gases present.

Problems

1. **a.** 14×760 mm = 11 000 mm
b. $\dfrac{1}{1.40}$ or 0.714×760 mm = 543 mm

2. **a.** 1.25 atm $\times \dfrac{760 \text{ mm Hg}}{\text{atm}} = 950.$ mm Hg
b. 2.48×10^{-3} atm $\times \dfrac{760 \text{ mm Hg}}{\text{atm}} = 1.88$ mm Hg
c. 4.75×10^4 atm $\times \dfrac{760 \text{ mm Hg}}{\text{atm}} = 3.61 \times 10^7$ mm Hg
d. 7.60×10^6 atm $\times \dfrac{760 \text{ mm Hg}}{\text{atm}} = 5.78 \times 10^9$ mm Hg

3. **a.** 125 mm Hg $\times \dfrac{\text{torr}}{1 \text{ mm Hg}} = 125$ torr
b. 3.20 atm $\times \dfrac{1.01325 \times 10^5 \text{ Pa}}{\text{atm}} = 3.24 \times 10^5$ Pa
c. 5.38 kPa $\times \dfrac{\text{atm}}{101.325 \text{ kPa}} \times \dfrac{760 \text{ mm Hg}}{\text{atm}}$
    = 40.4 mm Hg

4. **a.** $0.°C + 273 = 273$K
b. $27°C + 273 = 300.$K
c. $-50.°C + 273 = 223$K
d. $-273°C + 273 = 0.$K

5. **a.** 273K $- 273 = 0.°C$
b. $350.$K $- 273 = 77°C$
c. $100.$K $- 273 = -173°C$
d. $20.$K $- 273 = -253°C$
e. $0.$K $- 273 = -273°C$

6. **a.** $P_1V_1 = P_2V_2$
$V_2 = \dfrac{P_1V_1}{P_2} = \dfrac{350. \text{ mm Hg} \times 200. \text{ mL}}{700. \text{ mm Hg}} = 100.$ mL
b. $P_1V_1 = P_2V_2$
$V_1 = \dfrac{P_2V_2}{P_1} = \dfrac{0.48 \text{ atm} \times 435 \text{ mL}}{0.75 \text{ atm}} = 280$ mL
c. $P_1V_1 = P_2V_2$
$P_1 = \dfrac{P_2V_2}{V_1} = \dfrac{(180 \text{ torr})(1.8 \times 10^3 \text{ L})}{2.4 \times 10^5 \text{ L}} = 1.4$ torr

7. $P_1V_1 = P_2V_2$
$V_2 = \dfrac{P_1V_1}{P_2} = \dfrac{325 \text{ mm Hg} \times 240. \text{ mL}}{550 \text{ mm Hg}} = 140$ mL

8. $P_1V_1 = P_2V_2$
$P_2 = \dfrac{P_1V_1}{V_2} = \dfrac{22.5 \text{ kPa} \times 155 \text{ cm}^3}{90.0 \text{ cm}^3} = 38.8$ kPa

9. **a.** $P_1V_1 = P_2V_2$
$V_2 = \dfrac{P_1V_1}{P_2} = \dfrac{P_1 \times 450.0 \text{ mL}}{2P_1} = 225.0$ mL
b. $P_1V_1 = P_2V_2$
$V_2 = \dfrac{P_1V_1}{P_2} = \dfrac{P_1 \times 450.0 \text{ mL}}{0.25P_1} = 1800.$ mL

10. $P_1V_1 = P_2V_2$
$V_2 = \dfrac{P_1V_1}{P_2} = \dfrac{(575 \text{ mm Hg})(1.00 \times 10^6 \text{ mL})}{(1.25 \text{ atm})(760 \text{ mm Hg/atm})}$
    $= 6.05 \times 10^5$ mL

11. **a.** $T_1 = 27°C + 273 = 300.$K; $T_2 = 77°C + 273 = 350.$K
$\dfrac{V_1}{T_1} = \dfrac{V_2}{T_2}$
$V_2 = \dfrac{V_1 T_2}{T_1} = \dfrac{80.0 \text{ mL} \times 350.\text{K}}{300.\text{K}} = 93.3$ mL

b. $T_2 = 127°C + 273 = 400.K$

$$\frac{V_1}{T_1} = \frac{V_2}{T_2}$$

$$T_1 = \frac{V_1 T_2}{V_2} = \frac{125 \text{ L} \times 400.\text{K}}{85.0 \text{ L}} = 588\text{K}$$

c. $T_1 = -33°C + 273 = 240.\text{K};$
$T_2 = 160.°C + 273 = 433\text{K}$

$$\frac{V_1}{T_1} = \frac{V_2}{T_2}$$

$$V_1 = \frac{V_2 T_1}{T_2} = \frac{54.0 \text{ mL} \times 240.\text{K}}{433\text{K}} = 29.9 \text{ mL}$$

12. $T_1 = 67°C + 273 = 340.\text{K}$

$$\frac{V_1}{T_1} = \frac{V_2}{T_2}$$

$$T_2 = \frac{V_2 T_1}{V_1} = \frac{50.0 \text{ mL} \times 340.\text{K}}{140.0 \text{ mL}} = 121\text{K}$$

13. $T_1 = 0.°C + 273 = 273\text{K};$
$T_2 = 130.°C + 273 = 403\text{K}$

$$\frac{V_1}{T_1} = \frac{V_2}{T_2}$$

$$V_2 = \frac{V_1 T_2}{T_1} = \frac{275 \text{ mL} \times 403\text{K}}{273\text{K}} = 406 \text{ mL}$$

14. $T_1 = 47°C + 273 = 320.\text{K};$
$T_2 = 77°C + 273 = 350.\text{K}$

$$\frac{P_1}{T_1} = \frac{P_2}{T_2}$$

$$P_2 = \frac{P_1 T_2}{T_1} = \frac{250. \text{ mm Hg} \times 350.\text{K}}{320.\text{K}} = 273 \text{ mm Hg}$$

15. $T_1 = 27°C + 273 = 300.\text{K}$

$$\frac{P_1}{T_1} = \frac{P_2}{T_2}$$

$$T_2 = \frac{P_2 T_1}{P_1} = \frac{1.125 \text{ atm} \times 300.\text{K}}{0.625 \text{ atm}} = 540.\text{K}$$

16. $T_1 = -73°C + 273 = 200.\text{K}$

$$\frac{P_1}{T_1} = \frac{P_2}{T_2}$$

$$T_2 = \frac{T_1 P_2}{P_1} = \frac{200.\text{K} \times 2P_1}{P_1} = 400.\text{K or } 127°C$$

17. $T_1 = 47°C + 273 = 320.\text{K};$
$T_2 = 107°C + 273 = 380.\text{K}$

$$\frac{P_1 V_1}{T_1} = \frac{P_2 V_2}{T_2}$$

$$V_2 = \frac{P_1 V_1 T_2}{P_2 T_1} =$$

$$\frac{780 \text{ mm Hg} \times 2.20 \text{ L} \times 380.\text{K}}{600. \text{ mm Hg} \times 320.\text{K}} = 3.4 \text{ L}$$

18. $T_1 = 35°C + 273 = 308\text{K};$
$T_2 = 57°C + 273 = 330.\text{K}$

$$\frac{P_1 V_1}{T_1} = \frac{P_2 V_2}{T_2}$$

$$P_2 = \frac{P_1 V_1 T_2}{V_2 T_1} =$$

$$\frac{550 \text{ mm Hg} \times 350 \text{ mL} \times 330.\text{K}}{425 \text{ mL} \times 308\text{K}} = 490 \text{ mm Hg}$$

19. $T_1 = -23°C + 273 = 250.\text{K}$

$$\frac{P_1 V_1}{T_1} = \frac{P_2 V_2}{T_2}$$

$$T_2 = \frac{P_2 V_2 T_1}{P_1 V_1} =$$

$$\frac{210. \text{ kPa} \times 1.30 \text{ L} \times 250.\text{K}}{150. \text{ kPa} \times 1.75 \text{ L}} = 260.\text{K}$$

20. $T_1 = 40.°C + 273 = 313\text{K};$
$T_2 = 60.°C + 273 = 333\text{K}$

$$\frac{P_1 V_1}{T_1} = \frac{P_2 V_2}{T_2}$$

$$P_1 = \frac{P_2 V_2 T_1}{V_1 T_2} =$$

$$\frac{1.40 \text{ atm} \times 1250 \text{ mL} \times 313\text{K}}{820. \text{ mL} \times 333\text{K}} = 2.01 \text{ atm}$$

21. $T_1 = 17°C + 273 = 290.\text{K}$

$$\frac{P_1 V_1}{T_1} = \frac{P_2 V_2}{T_2}$$

$$T_2 = \frac{P_2 V_2 T_1}{P_1 V_1} = \frac{(8.10 \times 10^4 \text{ Pa})(720. \text{ cm}^3)(290.\text{K})}{(7.75 \times 10^4 \text{ Pa})(850. \text{ cm}^3)}$$

$$T_2 = 275\text{K, or } -16°C$$

22. $T_1 = 22°C + 273 = 295\text{K};$
$T_2 = -52°C + 273 = 221\text{K}$

$$\frac{P_1 V_1}{T_1} = \frac{P_2 V_2}{T_2}$$

$$V_2 = \frac{P_1 V_1 T_2}{P_2 T_1} =$$

$$\frac{740. \text{ mm Hg} \times 250. \text{ L} \times 221\text{K}}{0.750 \text{ atm} \times \frac{760 \text{ mm Hg}}{\text{atm}} \times 295\text{K}} = 243 \text{ L}$$

23. $T_1 = 295\text{K}$

$$\frac{P_1 V_1}{T_1} = \frac{P_2 V_2}{T_2}$$

$$T_2 = \frac{P_2 V_2 T_1}{P_1 V_1} = \frac{0.475 \text{ atm} \times 400.\text{L} \times 295\text{K}}{740.\text{mm Hg} \times \frac{1 \text{ atm}}{760 \text{ mm Hg}} \times 250.\text{L}}$$

$$T_2 = 230.\text{K or } -43°C$$

24. $T_1 = 20.°C + 273 = 293\text{K};$
$T_2 = 0.°C + 273 = 273\text{K}$

$$\frac{P_1 V_1}{T_1} = \frac{P_2 V_2}{T_2}$$

$$V_2 = \frac{P_1V_1T_2}{P_2T_1} =$$

$$\frac{(9.95 \times 10^4 \text{ Pa})(505 \text{ cm}^3)(273\text{K})}{1 \text{ atm} \times \frac{1.013\,25 \times 10^5 \text{ Pa}}{\text{atm}} \times 293\text{K}} = 462 \text{ cm}^3$$

$$\frac{462 \text{ cm}^3}{\text{breath}} \times \frac{15 \text{ breaths}}{\text{min}} \times \frac{60 \text{ min}}{\text{h}} \times \frac{24 \text{ h}}{\text{day}} \times \frac{\text{m}^3}{10^6 \text{ cm}^3}$$
$$= 9.98 \text{ m}^3$$

25. $P_T = P_{CO_2} + P_{N_2} + P_{O_2}$
    $P_{O_2} = P_T - P_{CO_2} - P_{N_2} =$
    $760.000 \text{ mm Hg} - 0.285 \text{ mm Hg} - 593.525 \text{ mm Hg}$
    $P_{O_2} = 166.190 \text{ mm Hg}$

26. $P_T = P_{O_2} + P_{H_2O}$
    $P_{O_2} = P_T - P_{H_2O} =$
    $730.0 \text{ mm Hg} - 17.5 \text{ mm Hg} = 712.5 \text{ mm Hg}$

27. $P_T = P_{gas} + P_{H_2O}$
    $P_{gas} = P_T - P_{H_2O} =$
    $742.0 \text{ mm Hg} - 42.2 \text{ mm Hg} = 699.8 \text{ mm Hg}$

28. $P_{gas} = P_{total} - P_{H_2O} =$
    $752.0 \text{ mm Hg} - 12.8 \text{ mm Hg} = 739.2 \text{ mm Hg}$
    Since temperature is constant, $P_1V_1 = P_2V_2$.
    Thus,
    $$V_2 = \frac{P_1V_1}{P_2} = \frac{739.2 \text{ mm Hg} \times 175 \text{ mL}}{770.0 \text{ mm Hg}} = 168 \text{ mL}$$

29. $P_{Ar} = 780.0 \text{ mm Hg} - 23.8 \text{ mm Hg} = 756.2 \text{ mm Hg}$
    $T_1 = 25 \text{ °C} + 273 = 298\text{K};$
    $T_2 = 0.\text{°C} + 273 = 273\text{K}$
    $$\frac{P_1V_1}{T_1} = \frac{P_2V_2}{T_2}$$
    $$V_2 = \frac{P_1V_1T_2}{P_2T_1} =$$
    $$\frac{756.2 \text{ mm Hg} \times 120. \text{ mL} \times 273\text{K}}{760 \text{ mm Hg} \times 298\text{K}} = 109 \text{ mL}$$

Critical Thinking

1. **a.** At low temperature, a larger fraction of the molecules have a low velocity, whereas, at the higher temperature, a larger fraction of the molecules have a higher velocity. The peak of the second curve is lower because the velocities are more widely distributed.
   b. As the temperature is increased further, the resulting distribution should shift farther to the right and peak at an even lower level than was observed in the two previous curves. Thus, a larger fraction of the molecules would have higher velocities.
2. The average kinetic energy values can remain the same because, at the same temperature, the higher the molecular mass of a gas is, the more slowly its molecules move.
3. If all gases behaved ideally, the individual gas particles would never exert the attractive forces on one another that are needed to form liquids or solids.
4. This apparent inconsistency can be explained by the fact that both the pressure due to the column of mercury in the tube and that exerted by the atmosphere on the surface of mercury in the pot can be thought of as being felt at the imaginary surface in the tube that is level with the mercury in the pot outside the tube. Since both forces are acting on the same surface, the size of that surface is not a factor in determining the eventual pressures.
5. $P_{total} = P_{Hg} + P_{atm}$
   $60 \text{ mm Hg} + 760 \text{ mm Hg} = 820 \text{ mm Hg}$
6. **a.** The temperature for B is greater than A
   b. The average kinetic energy is greater for B.
   c. Intermolecular attractive forces are the same.
   d. Not enough information to tell.
   e. Since a higher temperature implies a higher pressure, the pressure for B is greater than A, for a given volume.

Cross Connections

1. Most tire gauges measure pressure in pounds per square inch.

Teacher's Notes

Chapter 11
Physical Characteristics of Gases

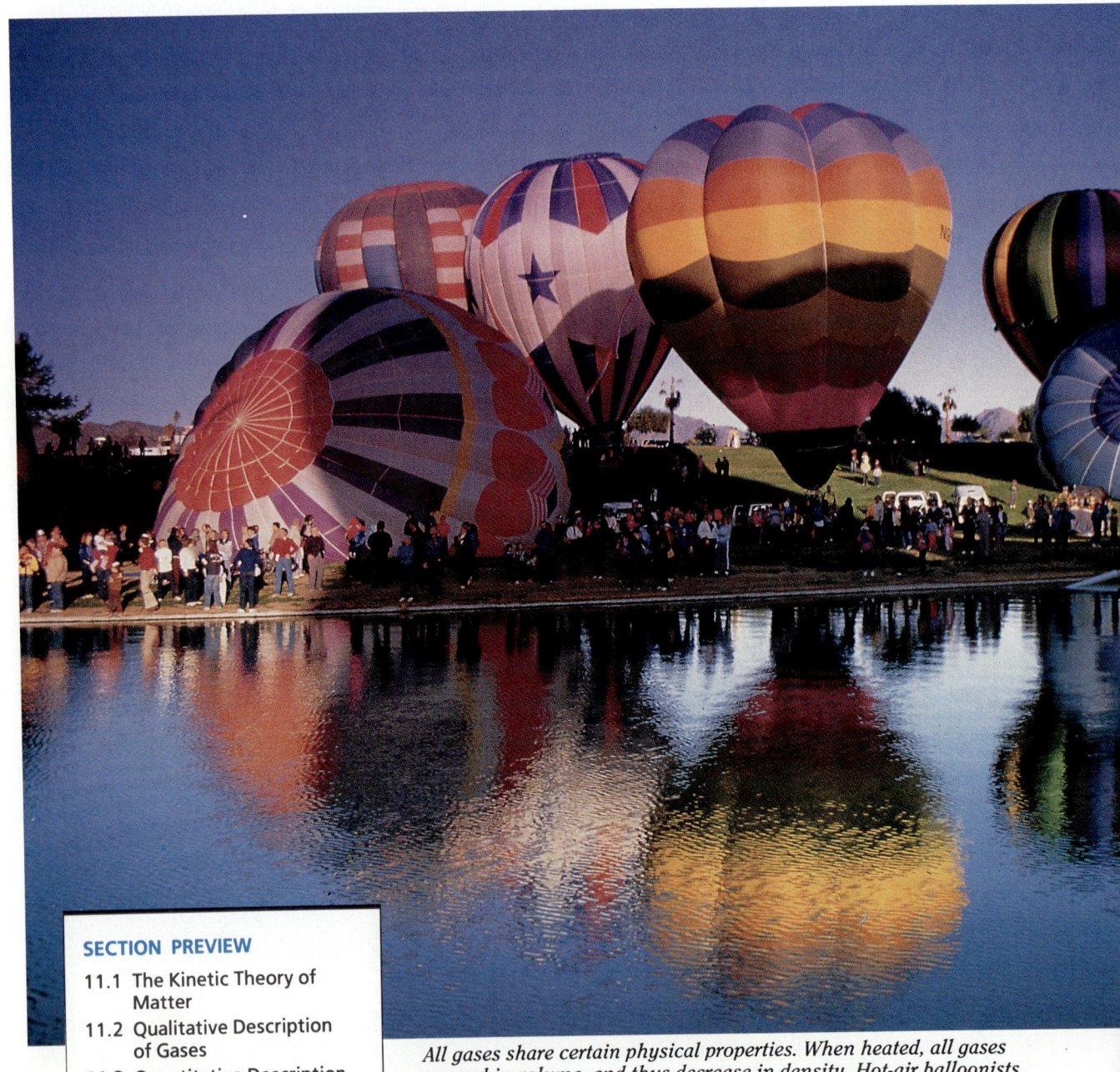

SECTION PREVIEW
11.1 The Kinetic Theory of Matter
11.2 Qualitative Description of Gases
11.3 Quantitative Description of Gases

All gases share certain physical properties. When heated, all gases expand in volume, and thus decrease in density. Hot-air balloonists take advantage of this property of gases when they use burners to inflate balloons.

11.1 The Kinetic Theory of Matter

In Chapter 1, matter is described briefly as existing on earth in the form of solids, liquids, and gases. Although it is impossible to observe directly the behavior of individual particles, scientists have been able to study large groups of these particles as they are combined in solids, liquids, or gases.

In the second half of the nineteenth century, scientists developed the kinetic theory of matter to explain the behavior of the atoms and molecules that make up matter. The word *kinetic* comes from the Greek word *kinetikos*, which means "moving." The kinetic theory of matter can be used to explain the properties of gases, liquids, and solids in terms of the energy possessed by the particles of matter and the forces that act between the particles of matter. *The* **kinetic theory** *is based on the idea that particles of matter are always in motion and that this motion has consequences.* In this section, you will study the kinetic theory as it applies to gases.

SECTION OBJECTIVES

1. State the kinetic theory of matter and describe how it explains certain properties of matter.
2. List the five assumptions of the kinetic theory of gases. Define the terms *ideal gas* and *real gas*.
3. Describe each of the characteristic properties of gases: expansion, low density, fluidity, compressibility, and diffusion.
4. Describe the conditions under which a real gas deviates from ideal behavior.

The Kinetic-Molecular Theory of Gases

The kinetic-molecular theory of gases provides a model of an ideal gas that helps us to understand the behavior of gas molecules and the physical properties of gases. *An* **ideal gas** *is an imaginary gas that conforms perfectly to all of the assumptions of the kinetic theory.*

The kinetic theory of gases is based upon the following five assumptions:

1. *Gases consist of large numbers of tiny particles.* These particles, usually molecules or atoms, typically occupy a volume about 1000 times larger than that occupied by the same number of particles in the liquid or solid state. Thus, molecules of gases are much farther apart than those of liquids or solids. Most of the volume occupied by a gas is empty space. This accounts for the lower density of gases compared to liquids and solids, and the fact that gases are easily compressible.

2. *The particles of a gas are in constant motion, moving rapidly in straight lines in all directions, and thus possess kinetic energy.* The kinetic energy of the particles overcomes the attractive forces between them except near the temperature at which the gas condenses and becomes a liquid. Gas particles travel in random directions at high speeds, (See Figure 11-1).

3. *The collisions between particles of a gas and between particles and container walls are elastic collisions. An* **elastic collision** *is one in which there is no net loss of kinetic energy.* Kinetic energy is transferred between two particles during collisions, but the total kinetic energy of the two particles remains the same, at constant temperature and volume.

4. *There are no forces of attraction or repulsion between the particles of a gas.* You can think of ideal gas molecules as behaving like small billiard balls. They move very fast, and when they collide they do not stick together, but immediately bounce apart.

5. *The average kinetic energy of the particles of a gas is directly proportional to the Kelvin temperature of the gas.* The kinetic energy of a particle (or any other moving object) is given by the equation

$$KE = \frac{1}{2} mv^2$$

Figure 11-1 Particles of gases are in constant motion, moving rapidly in all directions

Ideal gases do not exist in nature. Molecules of ideal gases are point sources—they occupy no volume, and there are no forces of interaction between molecules. Some gases approach ideality under condition of high temperature and low pressure.

Theories and models are discussed in Chapter 1.

Physical Characteristics of Gases **315**

Chemistry in Detail

Molecular Motion

You know that temperature is a measure of the average kinetic energy, or energy of motion, of molecules. There are actually three different kinds of molecular motion.

The first kind of motion—and the one you probably tend to think of when the word motion is mentioned—is called "translation." Translation of a molecule is a change in position of the molecule as a whole. The motion can be in any direction, but all parts of the molecule move in the same direction and there is no change in the distances between the atoms within the molecule. (See (a) in the diagram at right.)

In the second type of motion, the molecule as a whole does not change its position, but simply turns "in place." This type of motion is called rotation. (See (b) in the diagram.) Note that different parts of the rotating molecule move in different directions, but that the distances between the atoms within it do not change.

In the third type of motion, the distances between atoms within the molecule may change as bonds stretch or contract. This motion is called vibration. (See (c) in the diagram.) Vibration in more complex molecules can include "bond bending," or changes in the angles between atoms.

(a) Translation (b) Rotation (c) Vibration
Bond stretching Bond bending

All these forms of motion involve kinetic energy. Temperature is actually a direct measure of kinetic energy of translation. Increased temperature can provide energy for rotation or vibration, but is not a measure of the actual kinetic energies of these motions.

As the temperature of a gaseous sample decreases, the translational motion of its molecules decreases. Liquification and solidification occur on further cooling. Even in the solid state, translational motion continues as long as the temperature remains above zero Kelvin. Although the molecules cannot move freely through space in the solid state, they oscillate back and forth, and thus have translational energy. (See (a) in the diagram.) At absolute zero, however, all translational motion ceases. At this point, kinetic energy of translation equals zero and the lowest possible temperature has been reached.

STUDY HINT

Recall that velocity is simply directional speed. It is equal to the displacement (directed distance) divided by time.

where m is the mass of the particle and v is its velocity. Because all the particles of a specific gas have the same mass, their kinetic energies are dependent only on their velocities. The average velocities and kinetic energies of gas molecules will increase with an increase in temperature and decrease with a decrease in temperature.

Figure 11-2 shows the distribution of molecular velocities at two different temperatures. You can see from Figure 11-2 that at the higher temperature, a larger fraction of the molecules has a higher velocity.

Samples of all gases at the *same* temperature have the same average kinetic energy value. Lighter gas molecules, however, such as hydrogen molecules, have higher average velocities than do heavier gas molecules, such as oxygen molecules at the same temperature.

The Kinetic Theory and the Nature of Gases

The kinetic theory applies only to ideal gases. Although ideal gases do not actually exist, the behavior of many gases is close to ideal in the absence of very high pressures or very low temperatures. In the following sections, you will see how characteristic physical properties are explained by the assumptions of the kinetic theory.

Expansion Gases do not have a definite shape or a definite volume. They fill the entire volume of any container in which they are enclosed and assume its shape. A gas transferred from a 1-L vessel to a 2-L vessel will quickly expand to fill the entire 2-L volume. According to the kinetic theory, gas molecules are independent particles moving rapidly in all directions (Assumption 2) without significant attraction or repulsion between them (Assumption 4).

Fluidity Because the attractive forces between gas particles are negligible (Assumption 4), *gas particles glide easily past one another.* This ability to flow causes gases to show mechanical behavior similar to that of liquids. Because *liquids and gases flow, they are referred to collectively as* **fluids.**

Low density The density of a substance in the gaseous state is about 1/1000 the density of the same substance in the liquid or solid state because the particles are so much farther apart in the gaseous state (Assumption 1). For example, oxygen gas has a density of .001 429 g/mL at 0°C and 1 atmosphere pressure. As a liquid at −183°C, oxygen has a density of 1.149 g/mL. Figure 11-3 shows the relative volumes of a given amount of oxygen as a gas, a liquid, and a solid.

Figure 11-2 Molecular speed distribution in a gas measured at different temperatures. The ranges of molecular velocity and kinetic energy in a warm or a cold gas are about the same. At the higher temperature, a greater percentage of the molecules have a higher velocity and a higher kinetic energy of motion.

STUDY HINT

Recall that density equals mass divided by volume.

Figure 11-3 Relative volumes of equal amounts of oxygen as a gas, a liquid, and a solid at 1 atmosphere of pressure.

A flat or empty automobile tire is actually full of air at atmospheric pressure.

Compressibility During the compression of a gas, the gas particles, which are initially very far apart (Assumption 1), are crowded closer together. Under sufficient compression, the volume of a given sample of a gas can be decreased thousands of times. The steel cylinders containing nitrogen, oxygen, or other gases under pressure that are widely used in industry illustrate this point. Such cylinders have an internal volume of about 55 L.

When they are returned "empty" at ordinary pressures, they contain about 55 L of gas (the volume of the cylinder), although when they were delivered "full" they may have had 100 times as many molecules of gas compressed within the same cylinder.

Diffusion Gases spread out and mix with one another without stirring and in the absence of circulating currents. If the stopper is removed from a container of ammonia, the presence of this gas, which irritates the eyes, nose, and throat, soon becomes evident. Eventually, the ammonia mixes uniformly with the air in the room, as the random and continuous motion of the ammonia molecules (Assumption 2) carries them throughout the available space. *The spontaneous mixing of the particles of two substances because of their random motion is referred to as* **diffusion.**

Suppose a sample of nitrogen dioxide gas is released into a tube that is already filled with air. As shown in Figure 11-4, brown nitrogen dioxide (NO_2) gas diffuses out of the flask and into the attached tube in a matter of minutes. Because of collisions, diffusion into an enclosed space is slowed, but never prevented, by the presence of other gases in the space.

Figure 11-4 Diffusion of nitrogen dioxide gas

Figure 11-5 A bottle filled with hydrogen gas is placed above a bottle filled with air. After several minutes, the molecules of hydrogen have diffused rapidly into the lower jar.

The rate of diffusion of one gas through another depends on three properties of the intermingling gas particles: their speeds, their diameters, and the attractive forces between them. Hydrogen gas diffuses rapidly into other gases at the same temperature because its molecules are smaller and move about with greater speed than the larger, heavier molecules of the other gases, as shown in Figure 11-5.

Whereas diffusion is a process by which particles of a gas sample spread out spontaneously and mix with other gases, **effusion** *is a process by which gas particles under pressure pass through a very small opening from one container to another.* The rates of effusion of different gases are directly proportional to their molecular velocities.

Deviations of Real Gases from Ideal Behavior

In 1873, Johannes van der Waals (1837–1923) proposed that real gases deviate from the behavior expected of an ideal gas because *(1)* particles of real gases occupy space, and *(2)* particles of real gases exert attractive forces on each other. *A* **real gas** *is a gas that does not behave completely according to the assumptions of the kinetic theory.*

Most real gases behave like ideal gases when their molecules are sufficiently far apart and have sufficiently high kinetic energy. The kinetic theory does not hold true under all conditions of temperature and pressure, however. Experiments show that gas molecules at very high pressures or very low temperatures, or both, do not remain far apart. Under such conditions, all gases will deviate considerably from ideal gas behavior. This is illustrated in Figures 11-6 and 11-7.

The kinetic theory is more likely to hold true for gases composed of particles that have little attraction for each other. The noble gases, such as helium (He) and neon (Ne), show essentially ideal gas behavior over a wide range of temperatures and pressures. These gases consist of molecules that are monatomic and thus, nonpolar. Gases such as nitrogen (N_2) and hydrogen (H_2) have nonpolar, diatomic molecules and these are also found to show gas behavior that is close to ideal under many conditions. The more polar its molecules are, the more a gas deviates from ideal gas properties. Substances such as ammonia (NH_3) and water (H_2O) in the gaseous state deviate from ideal behavior to a large degree because they are composed of highly polar molecules.

Figure 11-6 Under high pressure, gas molecules move closer together.

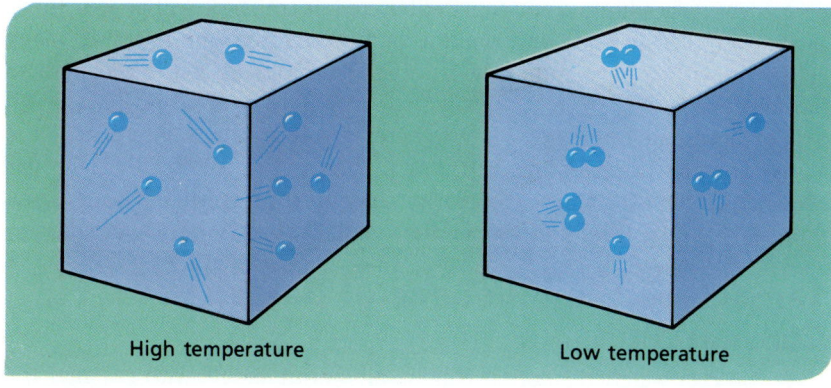

High temperature   Low temperature

Figure 11-7 Particles of gases at high temperatures (left) are far apart. At low temperatures (right), gas particles tend to stick together because the forces of attraction between them exert a greater influence than does their energy of motion.

Section Review

1. Explain each of the following characteristic properties of gases according to the kinetic theory: *(a)* expansion *(b)* low density *(c)* fluidity *(d)* compressibility *(e)* diffusion.
2. Describe the conditions under which a real gas is most likely to behave ideally.
3. State the two factors that van der Waals proposed to explain why real gases deviate from ideal behavior.
4. Which of the following gases would you expect to deviate the most from ideal behavior? *(a)* He *(b)* O_2 *(c)* H_2 *(d)* H_2O *(e)* N_2 *(f)* HCl *(g)* NH_3.

1. Refer to *The Kinetic Theory and the Nature of Gases*, p.317, for discussion on expansion, fluidity, low density, compressibility, and diffusion.
2. Under conditions of high temperature and low pressure
3. *(1)* Gas particles occupy space. *(2)* Gas particles exert attractive forces on each other.
4. *(d)* H_2O

11.2 Qualitative Description of Gases

SECTION OBJECTIVES

- Explain why the measurable quantities volume, pressure, temperature, and number of molecules of gas are needed to describe properly the state, or condition, of a gas.
- Name the variables that one must hold constant in order to study the (1) pressure–volume, (2) temperature–volume (3) pressure–temperature, (4) pressure–number of molecules, and (5) volume–number of molecules relationships of gases.
- Explain the relationship between the variables in each of the five pairs listed above in terms of the kinetic theory.

Suppose you have a one-liter bottle of air. How much air do you have? The expression *a liter of air* means little unless the temperature and pressure at which the air is measured are known. A liter of air can be compressed to a few mL of volume; it can also be allowed to expand to fill the volume of an auditorium.

To fully describe the state, or condition, of a gas, you need to use four measurable quantities: *(1)* volume, *(2)* pressure, *(3)* temperature, and *(4)* quantity or number of molecules. Mathematically simple relationships connect these four quantities. If you know the value of any three, you can calculate the value of the fourth. Such a calculated value will be fairly accurate for most gases over a very wide range of conditions.

In the next section, the mathematical relationships between volume, pressure, temperature, and number of molecules of gas will be examined, and examples of how to calculate an unknown value from known information will be given.

Pressure and Volume at Constant Temperature

The pressure exerted by a gas in a container is caused by moving molecules hitting the walls of the container. What happens when a gas in a one-liter container is placed into a one-fourth liter container? There will be four times as many molecules per unit volume and, therefore, four times as many collisions with a given unit of wall area. The pressure will thus increase fourfold.

As represented in Figure 11-8, the pressure exerted by a fixed quantity of gas at constant temperature becomes higher as the volume that the gas occupies becomes smaller. Similarly, the pressure exerted by the same number of gas molecules is lower if the volume available to the gas is larger.

Temperature and Volume of Gases at Constant Pressure

Have you ever watched a hot-air balloon billow out and fill with gas after the flame at its base is lighted? Balloonists such as those in Figure 11-17 are making use of a physical property of gases: if pressure is constant, gases can expand when heated and contract when cooled.

When the temperature increases, the volume occupied by a fixed quantity of gas must increase if the pressure is to remain constant. At the higher temperature, the gas molecules move faster and collide with the walls of the container more frequently and with more force. As illustrated in Figure 11-9, if the volume is increased, the molecules travel farther before reaching the walls. The number of collisions and their frequency per unit of wall area decreases. This smaller frequency of collisions offsets the greater force of collisions at the higher temperature. The pressure thus remains constant.

On the other hand, when the temperature decreases, the volume occupied by a fixed amount of gas must decrease if the pressure is to remain

P vs *V* at constant *n* and *T*
Where *n* = moles—the SI unit for quantity of substance

T vs *V* at constant *n* and *P*

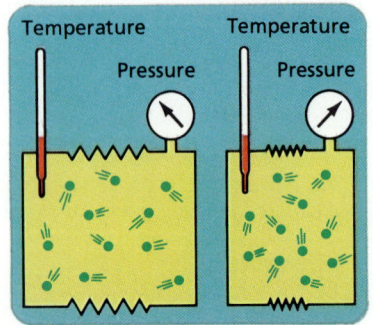

Figure 11-8 At a constant temperature, the pressure that a gas exerts increases as the volume of the gas decreases.

constant. At the lower temperature, the gas molecules move more slowly and they collide with the walls with less force. If the volume is decreased, however, the frequency of collisions per wall area increases, offsetting the force of the collisions. The pressure can therefore remain constant.

Pressure and Temperature

Recall that the temperature of a gas is an indication of the average kinetic energy of its molecules. The higher the temperature of a gas, the more kinetic energy its molecules have, and the more rapidly they move about. We have already seen that the pressure exerted by a gas is due to the collisions of the gas molecules with its container walls. As the temperature increases, gas molecules have more kinetic energy and undergo more frequent and more energetic collisions with the walls with more force per collision. As shown in Figure 11-10, if the volume of the same number of molecules remains constant with an increase in temperature, the pressure must increase. It follows that if the temperature is lowered, the pressure, exerted by the same number of molecules at constant volume, is lowered. This is true because the molecules move more slowly and exert less force per collision at lower temperature.

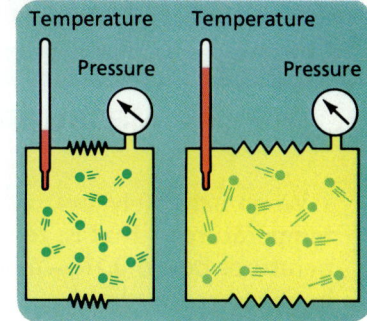

Figure 11-9 At a constant pressure, the volume of a gas increases as the temperature increases.

Relationship Between Pressure and Moles

At a given volume and temperature, the number of collisions of gas molecules with the container walls per unit time depends only on the number of molecules of the gas. The more molecules, the more collisions there are per unit time and therefore the higher the pressure. We make use of this property whenever we inflate bicycle or automobile tires. A cylinder that contains twice as many gas molecules as another cylinder has twice the pressure, assuming equal volume and temperature.

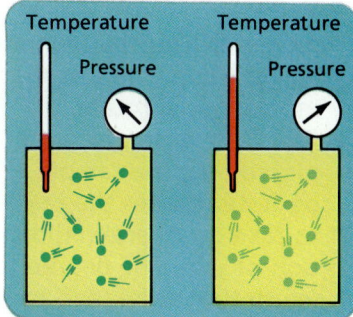

Figure 11-10 At a constant volume, the pressure a gas exerts increases as the temperature of the gas increases.

Relationship Between Volume and Moles

The volume of a gas depends on its temperature, its pressure, and the number of molecules present. If the temperature and pressure are constant, the volume of a gas varies directly with the number of molecules. As the number of molecules increases, the number of collisions per unit area of the container walls would increase if volume were held constant. The collision rate (and therefore the pressure) is kept constant by an increase in volume. If the number of molecules decreases, and the pressure and temperature are held constant, the volume must decrease.

Section Review

1. Name the four measurable quantities by which the state, or condition, of a gas can be described.
2. Describe qualitatively the relationships between each of the following: *(a)* pressure and volume at constant temperature and number of molecules, *(b)* temperature and volume at constant pressure and number of molecules, *(c)* pressure and temperature at constant volume and number of molecules, and *(d)* pressure and number of molecules at constant volume and temperature.

1. Pressure, volume, temperature, and amount of gas
2. *(a)* Inversely proportional
   (b) Directly proportional
   (c) Directly proportional
   (d) Directly proportional

Chemistry and Technology

Measuring and Reaching Low Temperatures

When extremely low temperatures—that is, temperatures below 0.1K—must be measured, ordinary alcohol- or mercury-filled thermometers are useless because the liquid in them freezes before reaching these temperatures. One device for measuring these extremely low temperatures is a nuclear resonance thermometer. This thermometer makes use of the fact that atomic nuclei rotate, or "spin". A strong magnet placed near the sample to be tested causes the nuclei to take on a certain direction of spin. The degree to which the nuclei tend to respond to this influence depends upon how low the kinetic energy of the sample is, and thus is a measure of temperature.

The magnetic spin properties of nuclei are useful not only in measuring very low temperatures, but also in reaching them. In a process called adiabatic demagnetization, the sample is first cooled in a liquid-helium bath in the presence of a powerful magnet, which causes the spinning nuclei in the sample to line up or become magnetized in certain ways. Then, the sample is removed from the bath and insulated so that no heat can flow into it. The sample begins to demagnetize itself, as the nuclei take on their usual spins. The nuclei require energy for this demagnetization and they get it by converting the kinetic energy of their atoms to other forms. Thus, as the kinetic energy falls, so does the temperature. Temperatures as low as 0.000 016K can be reached through this process.

Reaching very low temperatures is important in studying atomic structure and detecting subatomic particles.

11.3 Quantitative Description of Gases

SECTION OBJECTIVES

- Define pressure, explain how it is measured, and state the standard conditions of temperature and pressure.
- State Boyle's law and use it to calculate volume–pressure changes at a fixed temperature.
- Discuss the significance of absolute-zero temperature. Use the Kelvin scale in calculations.

Scientists have been studying physical properties of gases for hundreds of years. In 1662, Robert Boyle (1627–1691) discovered that the pressure and volume of a gas are related to each other mathematically. The observations of Boyle and others led to development of **gas laws**—*simple mathematical relationships between the volume, temperature, pressure, and quantity of a gas.* These gas laws were established long before the ideal-gas model and kinetic-molecular theory were proposed.

Pressure

Before examining the gas laws, we must first understand *pressure*—how it is defined and how it is measured.

Pressure and Force If we put some air into a rubber balloon, the balloon increases in size because of the pressure created by collisions between the air molecules inside the balloon and the inside walls of the balloon. **Pressure** *is defined as the force per unit area on a surface:*

$$\text{Pressure} = \frac{\text{force}}{\text{area}}$$

Consider a ballet dancer who has a mass of 51 kg (Figure 11-11). A mass of 51 kg exerts a force of 500 N on the earth's surface. *The SI unit for force is the* **newton,** *abbreviated* N. *It is the force that will increase the speed of a 1-kilogram mass by 1 meter per second for every second that it is applied.* No matter how the dancer stands, she exerts a force of 500 N against the floor. The pressure she exerts against the floor depends on the area of contact, as shown by the relationship above. When she rests her weight on the soles of both feet, the area of contact with the floor is about 325 cm^2. The pressure, or force per unit area, when she stands in this manner is 500 N/325 cm^2, or about 1.5 N per cm^2. When she stands on her toes, or "on point," as shown in Figure 11-15, the total area of contact of these two toes with the floor is 13 cm^2. Thus, when she balances in this way, the pressure exerted is equal to 500 N/13 cm^2—roughly 38.5 N/cm^2. And when she stands on one toe, the pressure she exerts is twice that, or 77 N/cm^2. It is clear that the same force applied to a smaller area results in a greater pressure.

Gas molecules exert pressure on any surface with which they collide. The pressure exerted by a gas depends on the number of molecules present, the temperature, and the volume in which the gas is confined.

The pressure of gases must not be confused with atmospheric pressure —that is, the pressure exerted by the atmosphere. The atmosphere—the blanket of air surrounding the earth—is composed of about 78% nitrogen, 21% oxygen, and 1% other gases, such as carbon dioxide and argon. The pressure of the atmosphere is due to the weight of the gases composing the atmosphere.

Atmospheric pressure at sea level is about equal to the weight of a 1-kilogram mass per square centimeter of surface, or 9.8 N per square centimeter. All objects on earth are subjected to the pressure of the atmosphere.

- State Charles' law, and use it to calculate volume–temperature changes at a fixed pressure.
- State Gay-Lussac's law, and use it to calculate pressure–temperature changes at a fixed volume.
- Use the combined gas law to calculate volume–temperature–pressure changes.
- State Dalton's law of partial pressures, and use it to calculate partial pressures and total pressures.

Force is equal to the mass of the object acted upon times the acceleration caused by the force:
force = mass × acceleration.

1 newton (N) = 1 kg · m/s^2

STUDY HINT

One newton is equal to the force that must be exerted to accelerate a 1 kg mass at the rate of 1 m/s².

Figure 11-11 The pressure exerted upon the floor by the ballet dancer is dependent upon the area of contact.

Physical Characteristics of Gases

Figure 11-12 A gas can collapses when the air inside it is removed by a vacuum pump.

To understand the concept of gas pressure and its magnitude, consider the gasoline can shown in Figure 11-12. The "empty" can contains about 5 g of air. The atmosphere exerts a pressure of 9.8 N per square centimeter on the outside of the can. If the can measures 15 cm × 10 cm × 28 cm, it has a total area of 1700 cm^2, and the total force is greater than 1.0 metric tons of weight. To demonstrate the size of the force exerted by the 5 g of air inside the can, we remove the air from the can by using a vacuum pump, as shown in Figure 11-12. In the absence of a balancing outward force exerted by air inside, the force due to atmospheric pressure immediately crushes the can.

Measuring pressure *A device used to measure atmospheric pressure is called a* **barometer.** The word barometer comes from the Greek *baros,* which means "weight," and *metron,* which means "to measure." The first type of barometer, illustrated in Figure 11-13, was introduced by Evangelista Torricelli (1608–1647) in 1643. Torricelli wondered why water pumps could raise water to a height of only about 34 feet. He thought that the height must be associated with the accepted idea that air had weight. Thus, he reasoned

STUDY HINT

There are two types of manometers. In one type, the mercury column is pushed upward into a tube that contains a vacuum. In another type, the mercury column is pushed upward into a tube that is open to the atmosphere. In the latter type, atmospheric pressure must be taken into account when calculating the pressure exerted by a gas.

Figure 11-13 Torricelli discovered that the pressure of the atmosphere supported a column of mercury 760 mm above the surface of the mercury in the pot.

that liquid mercury, which is about 14 times as dense as water, could be raised only 1/14 as high as water. To test this idea, Torricelli sealed a long glass tube at one end and filled it with mercury. Holding the open end with his thumb, he inverted the tube into a pot of mercury without allowing any air to enter the tube. When he removed his thumb, the mercury in the tube dropped to a height of about 30 in., or 760 mm, above the surface of the mercury in the pot. He repeated the experiment with tubes of different diameters and lengths longer than 760 mm. In every case, the mercury dropped to a height of about 760 mm.

The space above the mercury in such a tube contains a little mercury vapor, but is nearly a vacuum. The column of mercury in the tube retains its height because the atmosphere exerts a pressure on the surface of the mercury outside the tube. The pressure is transmitted through the mercury to an imaginary surface in the tube that is level with the mercury outside the tube. The mercury in the tube pushes downward because of its weight, which is due to the gravitational force. The mercury in the tube thus falls until the pressure exerted by its weight is equal to the pressure exerted by the atmosphere. The height of the mercury in the tube depends on the pressure—the force per unit area exerted by the atmosphere on the mercury in the pot. The pressure of the atmosphere is proportional to the height of the mercury column supported in the barometer tube. It is this relationship that allows the height of the mercury in a barometer tube to be used to measure atmospheric pressure.

From experiments like those of Torricelli, it is known that at sea level at 0°C, the average pressure of the atmosphere is equal to the pressure that could support a 760-mm column of mercury. At any given place on the earth, the specific atmospheric pressure depends on the elevation and the weather conditions at the time. If the atmospheric pressure is greater than the average at sea level, the height of the mercury column in a barometer will be greater than 760 mm. If the atmospheric pressure is less, the height of the mercury column will be less than 760 mm high.

Figure 11-14 In the manometer shown here, the pressure exerted by the gas in the container causes the column of mercury to rise upward. The difference in height of mercury in the two arms of the U-tube can be used to find the gas pressure.

Units of pressure A number of different units are used to measure pressure. Some pressure units are based on force per unit area, some on mass per unit area, and others on the height of a liquid column. Because atmospheric pressure is traditionally measured by a mercury barometer, scientists often express pressure in terms of the height of a column of mercury. *Thus, a common unit of pressure is* **millimeters of mercury**, *symbolized* **mm Hg**. The average atmospheric pressure at sea level and 0°C is 760 mm Hg. *To honor Torricelli for his invention of the barometer, a pressure unit named after him, the* **torr**, *has been introduced. By definition,* 1 torr = 1 mm Hg.

High pressures are often measured in units of atmospheres. *One* **atmosphere of pressure (atm)** *is defined as being equal to the average sea level pressure at 0°C. Thus,* 1 atm = 760 mm Hg or 760 torr.

In the SI system, pressure is expressed by a derived unit named the pascal (Pa) for Blaise Pascal (1623–1662), a French scientist who studied pressure in fluid systems. *One* **pascal** *is defined as the pressure exerted by a force of one newton (1 N) acting on an area of one square meter.*

$$1 \text{ Pa} = 1 \text{ N/m}^2$$

In many cases it is convenient to express pressure in kilopascals (kPa): 1 kPa = 1000 Pa. The standard atmosphere (1 atm) is equal to $1.013\ 25 \times 10^5$ Pa, or 101.325 kPa.

The pressure units used in this book are summarized in Table 11-1. Other common units for expressing pressure include *pounds per square inch* (psi) and *bars* (1 bar = 1×10^5 Pa).

TABLE 11-1   UNITS OF PRESSURE

Unit	Symbol	Definition/Relationship
pascal	Pa	SI pressure unit 1 Pa = 1 kg/(m × s²) = 1 N/m²
millimeter of mercury	mm Hg	Pressure that supports a 1-mm mercury column in a barometer
torr	torr	1 torr = 1 mm Hg
atmosphere	atm	Average atmospheric pressure at sea level and 0°C 1 atm = 760 mm Hg = 760 torr = 1.013 25 × 10⁵ Pa = 101.325 kPa

Sample Problem 11.1

The atmospheric pressure in Denver, Colorado, on the average is 0.830 atm. Express this pressure in *(a)* millimeters of mercury (mm Hg) and *(b)* kilopascals (kPa).

Solution
Step 1. Analyze   Given: P = 0.830 atm $\dfrac{760 \text{ mm Hg}}{1 \text{ atm}}$ (definition)

$\dfrac{101.325 \text{ kPa}}{1 \text{ atm}}$ (definition)

Unknown: P in mm Hg; P in kPa

Sample Problem 11.1 *continued*

Step 2. Plan	(a) atm → mm Hg	
	(b) atm → kPa	
	(a) mm Hg = atm × $\dfrac{\text{mm Hg}}{\text{atm}}$	
	(b) kPa = atm × $\dfrac{\text{kPa}}{\text{atm}}$	
Step 3. Compute	(a) 0.830 atm × $\dfrac{760 \text{ mm Hg}}{1 \text{ atm}}$ = 631 mm Hg	
	(b) 0.830 atm × $\dfrac{101.325 \text{ kPa}}{1 \text{ atm}}$ = 84.1 kPa	
Step 4. Evaluate	Units have canceled to give the desired units, and answers are properly expressed to the correct number of significant digits. The known pressure is roughly 80% of atmospheric pressure. The results are therefore reasonable because each is roughly 80% of 1 atm pressure as expressed in the new units.	

Practice Problem
Convert a pressure of 1.75 atm to *(a)* kPa   *(b)* torr.   *(Ans.)* 177 kPa, 1330 torr

Standard temperature and pressure   The volume of a gas depends upon temperature and pressure. Therefore, in order to compare volumes of gases, it is necessary to know the temperature and pressure at which the volumes are measured. *To aid in such comparisons, scientists have agreed upon standard conditions of exactly 1 atm pressure and 0°C. These conditions of* **standard temperature and pressure** *are commonly abbreviated* **STP**.

Boyle's law: Pressure–Volume Relationship

The English chemist and physicist Robert Boyle was the first scientist to make careful measurements that showed the quantitative relationship between the pressures and volumes of gases. He discovered that doubling the pressure on a sample of gas at constant temperature reduces its volume one-half. Tripling the gas pressure reduces its volume to one-third of the original, as illustrated in Figure 11-15. On the other hand, reducing the pressure on a gas by one-half allows the volume of the gas to double.

Data that are representative of pressures and volumes measured for a constant mass of gas at constant temperature are shown in Table 11-2. Notice that the product of the pressure and volume is a constant, k, for each pair of corresponding values. A constant product is found whenever there is an inverse relationship between two variables. Plotting the values of such inversely proportional variables gives a curve like that in Figure 11-16 for volume versus pressure. This volume–pressure relationship is referred to as **Boyle's law:** *the volume of a fixed mass of gas varies inversely with the pressure at constant temperature.*

Mathematically, Boyle's law is expressed as

$$V = k \frac{1}{P} \quad \text{or} \quad PV = k$$

STUDY HINT
In an inverse relationship, one variable increases in value as the other decreases in value. The graph of an inverse relationship is generally a hyperbola.

Physical Characteristics of Gases

Figure 11-15 At constant temperature, the volume of a gas will be reduced in proportion to increases in pressure.

where k is constant, its value depending only on the quantity of gas and the temperature. If the pressure of a given gas sample at constant temperature changes, the volume will change, but the quantity *pressure times volume* will remain equal to the same value of k. Using P_1 and V_1 to indicate initial conditions and P_2 and V_2 to indicate new conditions,

$$P_1V_1 = k \qquad P_2V_2 = k$$

Combining these two relationships gives the following very useful statement of Boyle's law:

$$P_1V_1 = P_2V_2$$

If three of the four values P_1, V_1, P_2, and V_2 are known, this equation can be used to calculate the fourth. For example, if 1.0 L of gas at 1.0 atm ($V_1 = 1.0$ L, $P_1 = 1.0$ atm) is allowed to expand at constant temperature to 5.0 L ($V_2 = 5.0$ L), the new pressure (P_2) is

$$P_2 = \frac{P_1V_1}{V_2}$$
$$= \frac{1.0 \text{ atm} \times 1.0 \text{ L}}{5.0 \text{ L}} = 0.20 \text{ atm}$$

The volume has increased five times and the pressure has decreased to one-fifth the original pressure.

Figure 11-16 The variation of volume with pressure at constant temperature.

TABLE 11-2 PRESSURE–VOLUME DATA (AT CONSTANT MASS AND TEMPERATURE)

Volume (mL)	Pressure (atm)	Pressure × Volume, k (mL × atm)
1200	0.5	600
600	1.0	600
300	2.0	600
200	3.0	600
150	4.0	600
120	5.0	600
100	6.0	600

Sample Problem 11.2

A sample of oxygen gas collected in the laboratory occupies a volume of 150 mL when its pressure is 720 mm Hg. What volume will the gas occupy at a pressure of 750 mm Hg if the temperature remains constant?

Solution

Step 1. Analyze Given: $V_1 = 150$ mL
$P_1 = 720$ mm Hg $P_2 = 750$ mm Hg
Unknown: V_2

Step 2. Plan $P_1, V_1, P_2 \rightarrow V_2$

Rearrange the equation for Boyle's law ($P_1V_1 = P_2V_2$) to obtain V_2.

$$V_2 = \frac{P_1V_1}{P_2}$$

Step 3. Compute Substitute values for P_1, V_1, and P_2 to obtain the new volume, V_2.

$$V_2 = \frac{P_1V_1}{P_2} = \frac{(720 \text{ mm Hg})(150 \text{ mL})}{750 \text{ mm Hg}} = 140 \text{ mL}$$

Step 4. Evaluate When the pressure is increased slightly at constant temperature, the volume decreases slightly, as expected. Units cancel to give milliliters, as desired.

Practice Problem

A balloon filled with helium gas has a volume of 500 mL at a pressure of 1 atm. After the balloon is released, it reaches an altitude of 6.5 km, where the pressure is 0.5 atm. Assuming that the temperature has remained the same, what volume does the gas occupy at this height? *(Ans.)* 1000 mL

Charles' Law: Temperature–Volume Relationship

The quantitative relationship between temperature and volume was first demonstrated by the French scientist Jacques Charles (1746–1823) in 1787. Charles' experiments revealed that all gases expand and contract to the same extent when heated through the same temperature interval. Furthermore, Charles found that the change in volume amounts to $\frac{1}{273}$ of the original volume for each Celsius degree, when the pressure remains the same and the initial temperature is 0°C. Raising the temperature to 1°C causes the gas volume to increase by $\frac{1}{273}$ of the volume it had at 0°C. A 10C°-temperature increase causes the volume to expand by $\frac{10}{273}$ of the original volume at 0°C. If the temperature is increased by 273C°, the volume increases by $\frac{273}{273}$ of the original—that is, the volume doubles.

The same regularity of volume change is observed in cooling a gas at constant pressure. At 0°C, a 1C° decrease in temperature decreases the original volume by $\frac{1}{273}$. At this rate of volume decrease, a gas cooled from 0°C to −273°C would have zero volume. Real gases cannot be cooled to −273°C. Before they reach this temperature, intermolecular forces exceed the kinetic energy of the molecules and the gases condense to liquids or solids.

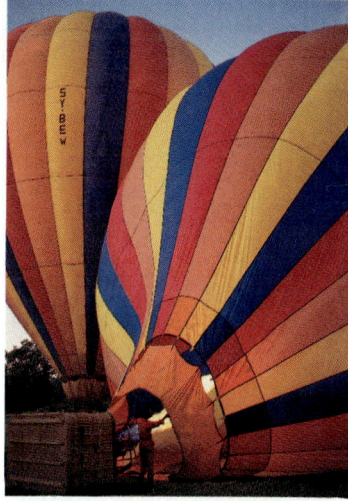

Figure 11-17 These hot air balloonists are taking advantage of the fact that gases expand upon heating.

Physical Characteristics of Gases **329**

If doubling the temperature doubles the volume of a gas, to what temperature must a gas be raised to double its volume if the original sample is at 0°C? This is a nonsense question. Two times zero is zero. Only the kelvin temperature scale may be used.

TABLE 11-3 TEMPERATURE–VOLUME DATA (AT CONSTANT MASS AND PRESSURE)

Temperature (°C)	Volume (mL)
273	1092
100	746
10	566
1	548
0	546
−1	544
−73	400
−173	200
−223	100
−273	0

Figure 11-18 The variation in volume with Kelvin temperature at constant pressure.

STUDY HINT

In a direct relationship, the value of one variable decreases (or increases) as the value of the other variable decreases (or increases) proportionally. The graph of a direct relationship is linear.

The data in Table 11-3 illustrate the temperature–volume relationship at constant pressure for a gas sample that has a volume of 546 mL at 0°C. Recall that if the gas is warmed 1C°, it expands $\frac{1}{273}$ its original volume. Each 1C°-temperature change thus causes a volume change of 546 mL/273, or 2 mL in the gas sample mentioned above. Raising the temperature to 100°C increases the volume by $\frac{100}{273}$ of 546 mL, or by 200 mL.

Note that in Table 11-3, the volume does not increase in direct proportion to the Celsius temperature. For example, increasing the temperature tenfold does not cause a proportional increase in the volume.

You learned in Section 2.2 that the Kelvin temperature scale starts at a temperature that corresponds to −273°C, or, more precisely, −273.15°C. Lord Kelvin realized the significance of this temperature, identifying it as the lowest temperature possible. *The temperature −273.15°C is referred to as* **absolute zero**, *and is given a value of zero on the Kelvin scale.* This gives the relationship you learned earlier, K = 273.15 + °C. In many calculations, 273.15 is rounded off to 273K.

Because the Kelvin scale has its zero point at the lowest possible temperature, the average kinetic energy of the molecules of a gas is directly proportional to the Kelvin temperature of the gas. As a result, the volume and Kelvin temperature are directly proportional to each other, as shown in Table 11-4. For example, quadrupling the Kelvin temperature from 50K to 200K causes the volume to quadruple. Plotting the volume and Kelvin temperature data of Table 11-4 gives a straight line, as shown in Figure 11-18, which is characteristic of plots of directly proportional variables.

Today, the relationship between the Kelvin temperature of a gas and its volume is known as **Charles' law:** *The volume of a fixed mass of gas varies directly with the Kelvin temperature at constant pressure.* Mathematically, Charles' law may be expressed as

$$V = kT \text{ or } \frac{V}{T} = k$$

where k is a constant, the value of which depends only on the quantity of gas and the pressure, and T is the Kelvin temperature. For the ratio V/T for any set of volume–temperature values (where mass is known), k is always the

**TABLE 11-4 VOLUME–TEMPERATURE DATA
(AT CONSTANT MASS AND PRESSURE)**

Volume (mL)	Celsius Temperature (°C)	Kelvin Temperature (K)	V/T or k (mL/K)
1092	273	546	2
746	100	373	2
566	10	283	2
548	1	274	2
546	0	273	2
544	−1	272	2
400	−73	200	2
100	−223	50	2
0	−273	0	—

same. The relationship thus obtained is a useful expression of Charles' law that can be applied to most volume–temperature problems involving gases.

$$\frac{V_1}{T_1} = \frac{V_2}{T_2}$$

where V_1 and T_1 represent one set of conditions, and V_2 and T_2 represent a different set of conditions. When three of the four values V_1, T_1, V_2 and T_2 are known, this equation can be used to calculate the fourth.

Sample Problem 11.3

A sample of neon gas occupies a volume of 752 mL at 25°C. What volume will the gas occupy at 50°C if the pressure remains constant?

Solution

Step 1. Analyze   Given: $V_1 = 752$ mL
$T_1 = 25°C + 273 = 298$K
$T_2 = 50°C + 273 = 323$K
Unknown: V_2

Step 2. Plan $V_1, T_1, T_2 \rightarrow V_2$

Because the gas remains at constant pressure, an increase in temperature will cause an increase in volume. Therefore, the gas obeys the Charles' law. It is *very important* to remember that temperature must be expressed in degrees Kelvin in all problems involving Charles' law. To obtain V_2, rearrange the equation for Charles' law.

$$V_2 = \frac{V_1 T_2}{T_1}$$

Step 3. Compute   Substitute values for V_1, T_1, and T_2 to obtain the new volume, V_2.

$$V_2 = \frac{V_1 T_2}{T_1} = \frac{752 \text{ mL} \times 323 \text{K}}{298 \text{K}} = 815 \text{ mL}$$

Physical Characteristics of Gases

Sample Problem 11.3 continued

Step 4. Evaluate   As expected, the volume of the gas increases as the temperature increases. Units cancel to yield milliliters, as desired. The answer contains the appropriate number of significant digits, and is reasonably close to an estimated value of 813:

$$\frac{750 \times 325}{300}.$$

Practice Problem
1. A helium-filled balloon has a volume of 2.75 L at 20.°C. The volume of the balloon decreases to 2.46 L after it is placed outside on a cold day. What is the outside temperature?
   (Ans.) 262K, or −11°C

Figure 11-19   The variation in pressure with Kelvin temperature at constant volume

Gay-Lussac's Law: Pressure–Temperature Relationship

What would you predict regarding the quantitative relationship between pressure and temperature at constant volume? You have seen that pressure is the result of collisions of molecules with container walls and that the average kinetic energy of molecules is directly proportional to Kelvin temperature. Thus, for a fixed quantity of gas at constant volume, the pressure should be directly proportional to the Kelvin temperature. For every degree of temperature increase or decrease, the pressure of a confined gas increases or decreases respectively $\frac{1}{273}$ of the pressure of the gas at 0°C. Joseph Gay-Lussac (1778–1850) is given credit for recognizing this in 1802.

The data plotted in Figure 11-19 illustrate **Gay-Lussac's law:** *The pressure of a fixed mass of gas varies directly with the Kelvin temperature at constant volume.* Mathematically, Gay-Lussac's law is expressed as

$$P = kT \text{ or } \frac{P}{T} = k$$

where k is a constant that depends on the volume of the sample and the quantity of the gas, and T is the temperature in Kelvin. For a given mass of gas at constant volume, the ratio $\frac{P}{T}$ is the same for any set of pressure-temperature values. Unknown values can be found using this relationship.

$$\frac{P_1}{T_1} = \frac{P_2}{T_2}$$

when values are known for three of the four quantities involved.

Sample Problem 11.4

The gaseous contents in an aerosol can are under a pressure of 3.00 atm at 25°C. Directions on the can caution the user to keep the can in a place where the temperature does not exceed 52°C. What would the pressure of the gas in the aerosol can be at 52°C?

Sample Problem 11.4 *continued*

Solution		
Step 1. Analyze	Given:	$P_1 = 3.00$ atm $T_1 = 25°C + 273 = 298K$ $T_2 = 52°C + 273 = 325K$
	Unknown:	P_2
Step 2. Plan	\multicolumn{2}{l}{Because the gaseous contents remain at the constant volume of the can, an increase in temperature will cause an increase in pressure. The gaseous contents, therefore, obey Gay-Lussac's law.}

$$P_1, T_1, T_2 \rightarrow P_2$$

Rearrange Gay-Lussac's law to obtain P_2:

$$P_2 = \frac{P_1 T_2}{T_1}$$

Step 3. Compute   Substitute values for P_1, T_2, and T_1 to obtain the new pressure, P_2.

$$P_2 = \frac{(3.00 \text{ atm})(325K)}{298K} = 3.27 \text{ atm}$$

Step 4. Evaluate   As expected, a temperature increase at constant volume causes the pressure of the contents in the can to increase. Units cancel correctly. The answer contains the proper number of significant digits and is reasonably close to an estimated value of 3.25, calculated as $\frac{3 \times 325}{300}$.

Practice Problem
Before a trip from New York to Boston, the pressure in an automobile tire is 1.8 atm at 20°C. At the end of the trip, the pressure gauge reads 1.9 atm. What is the new Celsius temperature of the air inside the tire? *(Ans.)* 36°C.

The Combined Gas Law

A gas sample often undergoes simultaneous changes in temperature, pressure, and volume. When this happens, three variables must be dealt with at once. Boyle's law, Charles' law and Gay-Lussac's law can be combined into a single expression that describes a relationship that is useful in such situations. *This relationship, known as the* **combined gas law,** *expresses the relationship between pressure, volume, and temperature of a gas when the amount of gas is fixed.* In mathematical form, the combined gas law can be expressed

$$\frac{PV}{T} = k$$

where k is constant and depends on the amount of gas. The combined gas law can also be written as

$$\frac{P_1 V_1}{T_1} = \frac{P_2 V_2}{T_2}$$

As in the individual gas laws, the subscripts above indicate two different sets of conditions, and T represents the temperature in Kelvin. From this expression, any one value can be calculated if the other five are known. Note that each of the individual gas laws (Charles' law, Boyle's law, or Gay-Lussac's law) can be obtained from the combined gas law when the appropriate variable is kept constant. When temperature is constant ($T_1 = T_2$), Boyle's law is obtained; when pressure is constant ($P_1 = P_2$), Charles' law is obtained; and when volume is constant ($V_1 = V_2$), Gay-Lussac's law is obtained.

Sample Problem 11.5

A helium-filled balloon has a volume of 50.0 L at 25°C and 820. mm Hg. What volume will it occupy at 650. mm Hg and 10.°C?

Solution

Step 1. Analyze Given: $V_1 = 50.0$ L $T_2 = 10°C + 273 = 283$K
$T_1 = 25°C + 273 = 298$K
$P_1 = 820.$ mm Hg $P_2 = 650.$ mm Hg
Unknown: V_2

Step 2. Plan

$$V_1, T_1, T_2, P_1, P_2 \rightarrow V_2$$

Because the gas is subjected to changes in both temperature and pressure, we must apply the combined gas law. Rearrange the combined gas law,

$$\frac{P_1 V_1}{T_1} = \frac{P_2 V_2}{T_2}, \text{ to give } V_2 = \frac{P_1 V_1 T_2}{P_2 T_1}$$

Step 3. Compute Substitute the known values into the equation to obtain a value for V_2:

$$V_2 = \frac{820. \text{ mm Hg} \times 50.0 \text{ L} \times 283\text{K}}{650. \text{ mm Hg} \times 298\text{K}} = 59.9 \text{ L}$$

Step 4. Evaluate As expected, the net result of the two changes gives an increase in the volume, from 50.0 L to 59.9 L. The result is expected because the pressure decreases much more than the temperature decreases. Units cancel appropriately. The answer is correctly expressed to three significant digits and is reasonably close to an estimated value of 66.6:

$$\frac{800 \times 50 \times 300}{600 \times 300}$$

Practice Problems

1. The volume of a gas is 27.5 mL at 22.0°C and 740. mm Hg pressure. What will be its volume at 15.0°C and 755 mm Hg pressure? *(Ans.)* 26.3 mL
2. A 700-mL gas sample at STP is compressed to a volume of 200 mL and the temperature is increased to 30°C. What is the new pressure of the gas? *(Ans.)* 2.95×10^3 mm Hg

Dalton's Law of Partial Pressures

John Dalton, the English chemist who proposed the atomic theory, also made an important contribution to the study of mixtures of gases. He found that, in the absence of a chemical reaction, the pressure of a mixture of gases is equal to the sum of the individual pressures exerted by each gas alone. For example, suppose we have a 1.0-L container filled with some oxygen gas that exerts a pressure of 1.0 atm at 0°C. In another 1.0-L container, we have some nitrogen gas that exerts a pressure of 1.0 atm at 0°C. When these gas samples are combined in the same 1.0-L container, the total pressure of the mixture is found to be 2.0 atm at 0°C. The pressure that each gas exerts in the mixture is independent of the other gas or gases present. *The pressure of each gas in a mixture is called the **partial pressure** of that gas.* The relationship between the total pressure of a mixture of gases and the individual pressures of the component gases is now known as **Dalton's law of partial pressures:** *The total pressure of a mixture of gases is equal to the sum of the partial pressures of the component gases.* The law holds true regardless of the number of different gases that have been combined. Dalton's law may be expressed as

$$P_T = P_1 + P_2 + P_3 + \ldots$$

where P_T is the total pressure of the mixture and $P_1, P_2, P_3, \ldots$ are the partial pressures of component gases 1, 2, 3, ...

Dalton's law is easily understood in terms of the kinetic-molecular theory. The rapidly moving particles of each gas in a mixture have equal opportunities to collide with the container walls. Therefore, each exerts a pressure independent of that of the other gases present. The total pressure is the result of the total number of collisions per unit of wall area in a given time. (Note that because gas molecules move independently, the combined gas law, as well as Dalton's law, can be applied to gas mixtures.)

Gases collected by water displacement Gases generated in the laboratory are often collected over water, as shown in Figure 11-20. Dalton's law of partial pressures has a practical application in calculating the volume of gases collected in this manner. A gas collected by water displacement is not pure, but is always mixed with water vapor. This is because water molecules at the liquid surface always evaporate and mingle with the gas molecules. Water vapor, like other gases, exerts a pressure, known as *water vapor pressure*.

Figure 11-20 Gases are often collected in the laboratory by water displacement.

To determine the total pressure of the gas and water vapor inside the collection bottle, the bottle is raised until the water levels inside and outside the bottle are the same. At this point, the total pressure inside the bottle is the same as the atmospheric pressure. According to Dalton's law of partial pressures,

$$P_T = P_{atm} = P_{gas} + P_{H_2O}$$

To determine the partial pressure of the dry gas collected, one must subtract the vapor pressure of the water at the given temperature from the total pressure. The atmospheric pressure (P_{atm}) at the time of the experiment is read from a barometer in the laboratory. Water vapor pressure varies with temperature. The value of P_{H_2O} at the temperature of the experiment is read from a standard reference table like that given in Table A-7 of this book.

Gases generated in the laboratory can also be collected over mercury. Since mercury does not evaporate markedly at room temperature, it is not necessary to subtract its vapor pressure from the total pressure to determine the pressure of the gas that is being collected.

Sample Problem 11.6

Oxygen from the decomposition of potassium chlorate ($KClO_3$) was collected by water displacement. The barometric pressure and the temperature during the experiment were 731.0 mm Hg and 20.0°C, respectively. What was the partial pressure of the oxygen collected?

Solution

Step 1. Analyze   Given: $P_T = P_{atm} = 731.0$ mm Hg
$P_{H_2O} = 17.5$ mm Hg (vapor pressure of water at 20.0°C, from Table A-7)
$P_T = P_{atm} = P_{O_2} + P_{H_2O}$
Unknown: P_{O_2}

Step 2. Plan
$$P_{atm}, P_{H_2O} \rightarrow P_{O_2}$$

The partial pressure of oxygen collected over water must be found by subtracting the partial pressure of water vapor from the atmospheric pressure, according to Dalton's law of partial pressures.

$$P_{O_2} = P_{atm} - P_{H_2O}$$

Step 3. Compute   Substituting values for P_{atm} and P_{H_2O} gives P_{O_2}

$$= 731.0 \text{ mm Hg} - 17.5 \text{ mm Hg} = 713.5 \text{ mm Hg}$$

Step 4. Evaluate   As expected, the oxygen partial pressure is less than atmospheric pressure, but much larger than the partial pressure of water vapor at this temperature. The answer has the appropriate number of places and is reasonably close to an estimated value of 713, (730–17).

Practice Problem

When hydrogen gas is collected over water at 20.0°C, the levels of water inside and outside the gas-collection bottle are the same. The partial pressures of hydrogen was determined to be 742.5 mm Hg. What was the barometric pressure at the time the gas was collected? *(Ans.)* 760.0 mm Hg

Section Review

1. Define pressure and explain how it is measured.
2. Convert the following pressures to pressures in standard atmospheres: *(a)* 151.98 kPa *(b)* 456 mm Hg *(c)* 912 torr.
3. What are the standard conditions for gas measurements?
4. State Boyle's law, Charles' law, and the combined gas law in mathematical terms.
5. A sample of helium gas has a volume of 200.0 mL at 730 mm Hg pressure. What pressure, in atm, is needed to reduce the volume at constant temperature to 50.0 mL?
6. A certain quantity of gas has a volume of 0.750 L at 298K. At what temperature, in degrees Celsius, would this quantity of gas be reduced to 0.500 L at constant pressure?
7. An aerosol can contains gases under a pressure of 4.50 atm at 20.0°C. If the can is left on a hot sandy beach, the pressure of the gases increases to 4.80 atm. What is the Celsius temperature on the beach?
8. Discuss the significance of the absolute-zero temperature.
9. The volume of a sample of methane gas measures 350. mL at 27.0°C and 810. mm Hg. What is the volume in liters at −3.0°C and 650. mm Hg pressure?
10. A certain mass of oxygen was collected over water when potassium chlorate was decomposed by heating. The volume of the oxygen sample collected was 720 mL at 25.0°C and a barometric pressure of 755 mm Hg. What would the volume of the oxygen be at STP?

1. Force per unit area; measured using either a barometer (atmospheric pressure) or a manometer (pressure in a closed container)
2. *(a)* 1.4999 atm *(b)* 0.600 atm *(c)* 1.20 atm
3. (1) 1 atm pressure (2) 273K
4. *(1)* $P_1V_1 = P_2V_2$ *(2)* $V_1/T_1 = V_2/T_2$ *(3)* $P_1V_1/T_1 = P_2V_2/T_2$
5. 3.8 atm
6. −74°C
7. 40.°C
8. Absolute-zero is the temperature at which average molecular kinetic energy is at a minimum.
9. 0.393 L
10. 655 mL O_2

Chapter Summary

- The kinetic theory of matter provides an explanation of the properties of gases, liquids, and solids in terms of the forces between the particles of matter and the energy these particles possess.
- The kinetic–molecular theory of gases describes a model of an ideal gas. Although ideal gases do not actually exist, the behavior of most gases is close to ideal except at very high pressures and low temperatures.
- Gases consist of large numbers of tiny, fast-moving particles that are far apart in comparison to their size. Their average kinetic energy is in direct proportion to the Kelvin temperature of the gas.
- Gases exhibit expansion, fluidity, low density, compressibility, and diffusion.
- Gases are described by four measurable quantities: pressure, volume, temperature, and number of molecules. Relationships among the first three of these four quantities are: Boyle's law—$PV = k$; Charles' law—$V = kT$; Gay-Lussac's law—$P = kT$; and the combined gas law—$PV/T = k$.
- A gas exerts pressure on the walls of its container. In a mixture of gases, the total pressure is equal to the sum of the partial pressures of each gas.
- Conditions of standard temperature and pressure (abbreviated STP) allow us to accurately compare volumes of different gases.
- A barometer measures the pressure of the atmosphere. The pressure of a gas in a closed container can be measured by a manometer.

Chapter 11 Review

Physical Characteristics of Gases

VOCABULARY

absolute zero (330)
atmosphere of pressure (atm) (326)
barometer (324)
Boyle's law (327)
Charles' law (330)
combined gas law (333)
Dalton's law of partial pressure (335)
diffusion (318)
effusion (318)
elastic collision (315)
fluids (315)
gas laws (322)
Gay-Lussac's law (332)
ideal gas (315)
kinetic theory (315)
millimeters of mercury (mm Hg) (326)
newton (323)
partial pressure (335)
pascal (326)
pressure (323)
real gas (319)
standard temperature and pressure (STP) (326)
torr (326)

REVIEWING CONCEPTS

1. Upon what two factors is the kinetic theory based? (11.1)
2. What is an ideal gas? (11.1)
3. State the five basic assumptions of the kinetic theory. (11.1)
4. How do gases compare with liquids and solids in terms of the distance between the molecules? (11.1)
5. What is an elastic collision? (11.1)
6. a. Write and label the equation that relates the average kinetic energy of the particles in a gas to the velocity of those particles.
   b. What is the relationship among the temperature, velocity, and kinetic energy of gas molecules? (11.1)
7. a. What is diffusion?
   b. What factors affect the rate of diffusion of one gas through another?
   c. What is the relationship between the mass of a gas particle and the rate at which it diffuses through another gas?
   d. What is effusion? (11.1)
8. a. Why does a gas in a closed container exert pressure?
   b. At constant temperature, what is the relationship between the pressure exerted by a fixed quantity of gas and the volume occupied by that gas? (11.2)
9. a. At constant pressure, what is the relationship between temperature and the volume occupied by a given quantity of gas?
   b. Explain the reason for this effect. (11.2)
10. a. At constant volume, what is the relationship between temperature and the pressure exerted by a given quantity of gas?
    b. Explain why this effect is observed. (11.2)
11. a. At constant pressure and temperature, how is the volume occupied by a gas in a closed container related to the number of molecules of gas present?
    b. Why is this observed? (11.2)
12. a. What is atmospheric pressure?
    b. Why does the atmosphere exert pressure on the earth?
    c. In terms of force per unit area, what is the value of atmospheric pressure at sea level? (11.3)
13. a. According to Torricelli's experiments, why does a column of mercury in a tube inverted into a pot of mercury retain a height of 760 mm at sea level?
    b. What comparable height would be maintained by a column of water inverted into a pot of water?
    c. What accounts for the difference in the heights of the mercury and water columns? (11.3)
14. a. Identify three sets of units typically used to express pressure.
    b. Convert one atmosphere to bars.
    c. What is a pascal?

d. What is the SI equivalent of one standard atmosphere of pressure? (11.3)

15. a. What Celsius temperature represents absolute zero?
b. What is the significance of this temperature?
c. What is the relationship between Kelvin temperature and the average kinetic energy of the molecules of a gas? (11.3)

16. a. Explain what is meant by the partial pressure of each gas within a mixture of gases.
b. How do the partial pressures of each of the gases in a mixture affect each other?

PROBLEMS

▶ Pressure and Temperature Conversions

1. If the atmosphere can support a column of mercury 760 mm in height at sea level, what comparable heights of each of the following (in mm) could be supported, given the relative density values cited?
a. water, whose density is approximately $\frac{1}{14}$ that of Hg
b. a substance with a density 1.40 times that of Hg

2. Convert each of the following into pressure readings expressed in millimeters of mercury:
a. 1.25 atm
b. 2.48×10^{-3} atm
c. 4.75×10^4 atm
d. 7.60×10^6 atm

3. Convert each of the following into the unit specified:
a. 125 mm Hg into torr
b. 3.20 atm into Pa
c. 5.38 kPa into mm Hg.

4. Convert each of the following Celsius temperatures to Kelvin temperatures:
a. 0.°C
b. 27°C
c. −50.°C
d. −273°C

5. Convert each of the following Kelvin temperatures to Celsius temperatures:
a. 273K
b. 350.K
c. 100.K
d. 20.K
e. 0.K

▶ Boyle's Law

6. Use Boyle's law to solve for the missing value in each of the following:
a. $P_1 = 350.$ mm, $V_1 = 200.$ mL, $P_2 = 700.$ mm, $V_2 = ?$
b. $P_1 = 0.75$ atm, $V_2 = 435$ mL, $P_2 = 0.48$ atm, $V_1 = ?$
c. $V_1 = 2.4 \times 10^5$ L, $P_2 = 180$ torr, $V_2 = 1.8 \times 10^3$ L, $P_1 = ?$

7. If the pressure exerted on a 240. mL sample of hydrogen gas at constant temperature is increased from 325 mm to 550 mm, what will be the final volume of the sample?

8. A flask containing 155 cm³ of hydrogen was collected under a pressure of 22.5 kPa. What pressure would have been required in order for the volume of the gas to have been 90.0 cm³ at constant temperature?

9. A gas has a volume of 450.0 mL at standard pressure. If the temperature is held constant, what volume would the gas occupy if the pressure were
a. doubled
b. reduced to one-fourth of its original value.

10. If a sample of oxygen that occupies 1.00×10^6 mL at 575 mm is subjected to a pressure of 1.25 atm, what will be the final volume of the sample if the temperature is held constant?

▶ Charles' Law

11. Use Charles' law to solve for the missing value in each of the following:
a. $V_1 = 80.0$ mL $t_1 = 27°C$, $t_2 = 77°C$, $V_2 = ?$
b. $V_1 = 125$ L, $V_2 = 85.0$ L, $t_2 = 127°C$, $T_1 = ?$
c. $t_1 = -33°C$, $V_2 = 54.0$ mL, $t_2 = 160.°C$, $V_1 = ?$

12. A sample of air has a volume of 140.0 mL at 67°C. To what temperature must the gas be lowered to reduce its volume to 50.0 mL at constant pressure?

13. At standard temperature a gas has a volume of 275 mL. If the temperature is increased to

130.°C, but the pressure is held constant, what is its new volume?

▶ Gay-Lussac's Law

14. A sample of hydrogen at 47°C exerts a pressure of 250. mm. If the gas is heated to 77°C at constant volume, what will its new pressure be?
15. To what temperature must a sample of nitrogen at 27°C and 0.625 atm be heated so that its pressure becomes 1.125 atm at constant volume?
16. If the pressure on a gas at −73°C is doubled but its volume is held constant, what will its final temperature be in degrees Celsius?

▶ Combined Gas Law

17. A sample of gas at 47°C and 780 mm pressure occupies of volume of 2.20 L. What volume would this gas occupy at 107°C and 600. mm pressure?
18. A 350-mL air sample collected at 35°C has a pressure of 550 mm. What pressure will the air exert if it is allowed to expand to 425 mL at 57°C?
19. A gas measures 1.75 L at −23°C and 150. kPa. At what temperature would the gas occupy 1.30 L at 210. kPa?
20. A sample of oxygen at 40.°C occupies 820. mL. If this sample occupies 1250 mL at 60.°C and 1.40 atm, what was its original pressure?
21. A gas at 7.75×10^4 Pa and 17°C occupies a volume of 850. cm³. At what temperature, in degrees Celsius, would the gas occupy 720. cm³ at 8.10×10^4 Pa?
22. A meteorological balloon contains 250. L of He at 22°C and 740. mm Hg. If the volume of the balloon can vary according to external conditions, what volume would it occupy at an altitude at which the temperature is −52°C and the pressure is 0.750 atm?
23. The balloon in the previous problem will burst if its volume reaches 400. L. Given the initial conditions specified in that problem, at what temperature, in degrees Celsius, will the balloon burst, if its pressure at that point is 0.475 atm?
24. The normal respiratory rate for a human being is 15 breaths per minute. If the average volume of air for each breath is 505 cm³ at 20.°C and 9.95×10^4 Pa, what volume of air, in cubic meters and at STP, does an individual breathe in one day?

▶ Dalton's Law of Partial Pressures

25. Three of the primary components of air are carbon dioxide, nitrogen, and oxygen. In a sample containing a mixture of these gases at one atmosphere of pressure, the partial pressures of carbon dioxide and nitrogen are given as P_{CO_2} = 0.285 mm and P_{N_2} = 593.525 mm. What is the partial pressure of oxygen?
26. Determine the partial pressure of oxygen collected by water displacement if the water temperature is 20.0°C and the total pressure of the gases in the collection bottle is 730.0 mm Hg.
27. A sample of gas is collected over water at a temperature of 35.0°C when the barometric pressure reading is 742.0 mm Hg. What is the partial pressure of the dry gas?
28. A sample of oxygen is collected in a 175-mL container over water at 15°C. If the barometer reads 752.0 mm Hg, what volume would the dry gas occupy at 770.0 mm and 15°C?
29. If 120. mL of argon is collected over water at 25°C and 780.0 mm, compute the volume of the dry argon at STP.

CRITICAL THINKING

1. **Applying Models**
   a. Explain the distribution of molecular velocities in a gas at two different temperatures, as illustrated in Figure 11-2.

b. If the temperature of the gas is increased further, what would you anticipate the resulting kinetic energy curve would look like?

2. **Relating Ideas** Explain how different gases in a mixture can have the same average kinetic energy value, even though the masses of their individual particles differ.

3. **Inferring Conclusions** If all gases behaved as ideal gases under all conditions of temperature and pressure, there would be no solid or liquid forms of these substances. Explain.

4. **Relating Ideas** Pressure is defined as force per unit area. Yet in his experiments with the barometer, Torricelli found that the diameter of the tub (and, correspondingly, the surface area of contact between the mercury in the tube and the mercury in the tub exposed to the atmosphere) did not affect the height of mercury that was supported by the atmosphere at sea level. Explain this seemingly inconsistent observation in view of the relationship between pressure and surface areas.

5. **Interpreting Graphics** Examine Boyle's J-tube apparatus. If $h = 60$ mm, what is the pressure exerted by the gas labeled V_{gas}?

6. Velocity distribution curves are shown in the graph for the same gas under two different conditions, A and B. Compare the behavior of the gas under Conditions A and B in relation to:

a. temperature.
b. average kinetic energy.
c. intermolecular attractive forces.
d. gas volume.
e. gas pressure.

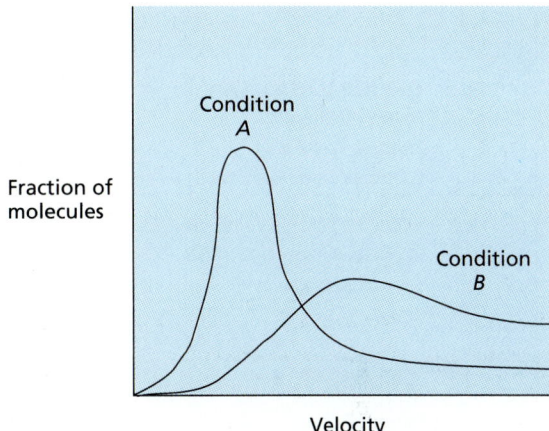

CROSS CONNECTIONS

1. **Consumer Skills** Air pressure of the tires on a car should be checked regularly for safety reasons and to prevent uneven wear on the tires. Find out what the tire gauge measures.

RESEARCH AND WRITING

1. Prepare a report on the evolution of the modern submarine. Include a discussion of the technology that enables the submarine to withstand the tremendous pressure of the deep ocean. Also report on the equipment used to ensure a sufficient supply of oxygen for submarine crew members.

PROJECT

1. Design and conduct a meteorological study to examine the interrelationships among barometric pressure, temperature, humidity, and other pertinent weather variables. Prepare a report explaining your results.

CHAPTER 12
Molecular Composition of Gases

PLANNING CHART

Approximate Time: 9/11 days

	SECTION/TEXT FEATURES		RESOURCES	
A V E R A G E	**12.1 VOLUME-MASS RELATIONSHIPS OF GASES** Section Review Chapter Review • Questions 1–4, Problems 1–13	343 349 366	*Teacher's Resource Binder* Lab Experiments 13, 14, 15 Section Review Worksheets Transparency 27	83, 91, 95 40
A V E R A G E	**12.2 THE IDEAL GAS LAW** Section Review Chapter Review • Questions 5–6, Problems 14–20	350 355 366	*Teacher's Resource Binder* Problem Solving Worksheets Section Review Worksheets	41
A V E R A G E	**12.3 STOICHIOMETRY OF GASES** Microscale Investigation 9 Section Review Chapter Review • Question 7, Problems 21–37	356 851 360 366	*Teacher's Resource Binder* Problem Solving Worksheets Section Review Worksheets *Other Resources* *World of Chemistry* Video 11	42
A V E R A G E	**12.4 EFFUSION AND DIFFUSION** Demos 1, 2 Desktop Investigation Section Review Chapter Review • Question 8, Problems 38–41	360 342B–C 361 365 366	*Teacher's Resource Binder* Problem Solving Worksheets Section Review Worksheets *Other Resources* *Shakhashiri* Video Demo 32	43
	CHAPTER END ACTIVITIES Critical Thinking, Cross Connections, Research and Writing, Projects		*Teacher's Resource Binder* Test Booklet *Computer Test Generator*	57, 59

Teaching Strategies

12.1 Volume — Mass Relationships of Gases

The introduction presents Dalton's and Gay-Lussac's work with combining volumes of gases and Avogadro's interpretation of the data. It is very important that the students understand the importance of Avogadro's insight. He was able to predict molecular behavior from macroscopic properties and to predict the composition of molecules without being able to see or prove their existence.

Use the chalkboard to lead the students through the reasoning process Avogadro might have used.
Data: Hydrogen gas and oxygen gas react to form water vapor
$$2 \text{ volumes } + 1 \text{ volume} \rightarrow 2 \text{ volumes}$$
If equal volumes at the same temperature and pressure contain the same number of particles, then
$$2 \text{ molecules } + 1 \text{ molecule} \rightarrow 2 \text{ molecules}$$
At this time, the structures (atomic composition) for the molecules of hydrogen, oxygen, and water vapor had not been determined.

If one assumes that each molecule of *(1)* oxygen contains a total of **X** atoms *(2)* hydrogen contains a total of **Y** atoms, and *(3)* water vapor contains a total of **Z** atoms, then from the above molecular equation, one can show that

$$2\mathbf{X} \text{ atoms of hydrogen} + \mathbf{Y} \text{ atoms of oxygen} \rightarrow 2\mathbf{Z} \text{ atoms of water vapor}$$

Now, applying some basic arithmetic, regardless of what **X** and **Z** are, 2**X** and 2**Z** must be even numbers (two times an odd or an even number gives an even number).

$$\underset{\text{even}}{2\mathbf{X}} + \mathbf{Y} \rightarrow \underset{\text{even}}{2\mathbf{Z}}$$

Y must also be an even number because only an even number can be added to an even number to give an even number. The simplest value of **Y** would then be two. The oxygen molecule must consist of at least two oxygen atoms. Once one has determined that oxygen is diatomic, one can combine oxygen with other gas molecules and begin to catalog those molecules which must also be diatomic.

Since a mole of any gas can be expressed in terms of its molar mass, one can specify the volume of a mole of gas. Thus all gases should occupy the same molar volume at the same temperature and pressure. Students object to this conclusion because large molecules should take up more space than smaller molecules. Remind the students that the distance separating gaseous molecules is very large in comparison to the size of individual molecules. Doubling the size of a molecule still leaves a very large distance between it and the next molecule.

The molar volume of a gas at standard temperature and pressure is 22.4 L. To bring this home to the students, display a 22.4 liter box. If time allows, challenge the students to construct and bring in their own container that measures 22.4 L in volume.

Avogadro's principle is next related to the density of gases. Introduce the topic by having the students reason as follows:

Density is defined as mass/volume, but under conditions of constant temperature and pressure, volume can be related to number of gaseous molecules, thus density is related to mass/number of molecules. If the number of molecules is to be taken to be one mole, then the mass becomes the molar mass, and the original volume related to the molar volume. Thus the density of any gas can be expressed as the ratio of its molar mass to its molar volume (22.4 liters if at STP).

12.2 The Ideal Gas Law

Review Chapter 11, especially the qualitative relationships between the four quantities needed to describe a gas sample—volume, temperature, pressure, and number of particles. Write down the expression of the general gas law developed earlier.

$$\frac{P_1V_1}{n_1T_1} = \frac{P_2V_2}{n_1T_1}$$

Review how each of the gas laws can be derived from the above general expression if two of the terms are kept constant. Now if the second set of conditions were assumed to be one mole of a gas at STP, the relationship becomes

$$\frac{P_1V_1}{n_1T_1} = \frac{(1\ \text{atm})(22.4\ \text{L})}{(1\ \text{mol})(273\text{K})} = 0.0821\ \frac{\text{L}\cdot\text{atm}}{\text{K}\cdot\text{mol}}$$

If the constant calculated is represented by the letter R, then

$$\frac{P_1V_1}{n_1T_1} = R \quad \text{or} \quad P_1V_1 = Rn_1T_1$$

This may now be generalized to any set of conditions as
$$PV = nRT$$
Should the students ever forget the value of R, this procedure will help them to calculate it on their own.

The remaining section presents samples of the many problems associated with the ideal gas law. Impress upon the students that the problems studied in Chapter 11 consisted of a gas sample and two sets of conditions. One could predict what might happen to the gas sample if some kind of stress were placed on the gas. The ideal gas law now allows them to calculate one of the variables knowing the other three. In actuality, the above derivation points out that the ideal gas law can be seen as an application of the general gas law. The second set of conditions is always one mole of gas at STP.

The ideal gas law $PV = nRT$ can be rearranged: $\frac{n}{V} = \frac{P}{RT}$

But n/V can be expressed as mass/V if one multiplies n, the number of moles, by the molar mass of the gas. Thus the equation may be rewritten as
$$d = MP/RT$$

Another series of problems are once again presented. These problems will give the students the practice in problem solving that will help them in future chapters.

12.3 Stoichiometry of Gases

Calculations studied in Chapter 9 may now be extended to include not only masses and moles of reactants and products, but volumes of each when gases are part of the reaction. A complete flow chart of the process may be presented to the students.

If both A and B are gases, Avogadro's principle allows us to directly relate volume and moles. If the coefficients of the equation were 1A and 2B then one mole of A produces two moles of B or 1 L A produces 2 L B.

Students are a bit wary of this simplification at first, and probably will continue to follow the conversion method learned in Chapter 9. Work each type of problem with them and help them to categorize the problem before beginning to work out a solution.

12.4 Effusion and Diffusion

The concluding section on Graham's law combines the derivation, problem solving, a Desktop Investigation, and a Technology feature on the technique of separating isotopes using different diffusion rates.

The text presents the derivation beginning with the definition of kinetic energy. The math is a bit more difficult because students are faced with an inverse square (or square root) relationship. The key qualitative concept is that light gases move more rapidly than heavier slower gases. Average energies are the same.

The analogy of a very slow-moving train and a bullet fired from a rifle might help. The slow-moving train contains a large kinetic energy because of the large mass. The bullet, on the other hand, has a very small mass, but moves very rapidly. Since the velocity term is squared, the kinetic energy of the bullet is also very large. You may wish to bring up other examples from athletics. In football, a smaller lineman will have to be faster than a heavier player if he hopes to compensate for his small mass.

The difference between effusion and diffusion is also very important. Effusion requires a small aperture through which gas will pass. Diffusion involves the mixing of two gases in a container.

Demonstration 1: Rate of Diffusion of Gases

PURPOSE: To compare the rate of diffusion of two gases
MATERIALS: conc. HCl and conc. NH$_3$(aq), 0.05 to 1-m length of wide bore glass tubing (2 cm), cork or rubber stoppers to fit the glass tubing, cotton, two rings stands and clamps
PROCEDURE: Suspend the piece of glass tubing horizontally on the demonstration table. Take small pieces of cotton and staple them into the ends of the rubber stoppers. (As an alternative, use a cork borer to bore out a hole in the small end of the cork stoppers. The cotton may be placed in the hollow.) Add a few drops of HCl to the cotton on one stopper, and a few drops of NH$_3$(aq) to the other. Simultaneously insert the two stoppers into the glass tube. Depending on the length of the glass tube, a white ring will form where the ammonia and hydrogen chloride vapors meet. The white ring is solid ammonium chloride.

Conditions in the demonstration are such that Graham's law

will not apply. The ammonia gas will move a greater distance than the hydrogen chloride. You are showing that lighter molecules travel more rapidly. You might wish to calculate the ratio of the square roots of the molar masses of HCl and NH_3 and to predict where the gases should have met should Graham's law have been followed completely.

SAFETY AND DISPOSAL: Observe usual lab precautions when handling the conc. acid and base. Use dropping bottles if possible. Acid and base remaining on the cotton may be rinsed down the sink. The glass may be cleaned by pouring water through it.

If a ready source of hydrogen gas is available, the following is a very dramatic demonstration and may be used to replace the preceding one.

Demonstration 2: Comparative Effusion Rates

PURPOSE: To show that lighter gases effuse more rapidly
MATERIALS: porous unglazed ceramic cup (the kind used in electrochemical cells), one-hole rubber stopper to fit the cup, a 500-mL Erlenmeyer flask, two-hole rubber stopper to fit the flask, glass tubing, 1-L beaker, a source of hydrogen gas
PROCEDURE: Cut a piece of glass tubing 10 cm longer than the flask's length. Bend the tubing so that the part extending outside the flask has a 120° angle. Insert this tube into the two-hole rubber stopper. Insert a straight piece of tubing into both the one-hole stopper and the two-hole stopper's other hole, so that each end protrudes less than 1 cm through the stoppers. Add water to the Erlenmeyer flask, leaving several cm of air in the neck. Insert the two-hole stopper into the flask, so that the bent tube is well under the water's surface. Insert the stopper on the other end of the straight tube into the porous cup.

Collect hydrogen in the inverted 1-L beaker. Carefully place the hydrogen-filled inverted beaker over the porous cup. Water immediately flows out of the flask through the bent tubing. Carefully remove the beaker. Bubbles of air enter the flask through the tubing that produced the jet of water. Once more, place the beaker over the porous cup, and water will flow from the bent tubing, but not as vigorously as before.

Have students explain the demonstration as a competition between molecules of air (nitrogen and oxygen) and hydrogen, using Graham's law. Hydrogen molecules are so small they rapidly move through the porous cup into the flask where there is no hydrogen. These hydrogen molecules increase the pressure in the flask, forcing water out the bent tube. When the beaker is removed, hydrogen molecules move from the porous cup to the room, leaving a vacuum in the flask. Atmospheric pressure pushes air back into the flask.

SAFETY AND DISPOSAL: Be careful when handling hydrogen gas. Once constructed, the demonstration apparatus may be stored for later use.

References and Resources

Books and Periodicals

Borgford, Christie L. and Lee R. Summerlin. "Gases," *Chemical Activities: A Sourcebook for Science Teachers*, Washington, D.C.: The American Chemical Society, 1988.

Scott, Arthur F. "The Invention of the Balloon and the Birth of Modern Chemistry," *Scientific American*, Vol. 250, No. 1, Jan. 1984, pp. 126–137.

Audiovisual Resources

Campbell, J. Arthur. *Gas Pressure and Molecular Collisions*, A Chem Study Film, B&W, 21 min., VHS cassette, Central Scientific Co. (Shows the effects of varying the number of molecules per unit of volume and of varying the temperature)

Pimentel, George C. *Gases and How They Combine*, A Chem Study Film, Color, 22 min., VHS cassette, Central Scientific Co. (Simple integer volume ratios lead to Avogadro's hypothesis)

Answers and Solutions

Reviewing Concepts

1. **a.** The law applies only to gas volumes measured at the same temperature and pressure.
   b. At the same temperature and pressure, the volume of any given gas varies directly with the number of molecules.
2. **a.** Gas volume is directly proportional to the number of moles of gas at constant temperature and pressure.
   b. $V = kn$, where n is the number of moles of gas and k is a constant.
3. At STP; 22.4 L of all gases contain the same number of molecules (6.022×10^{23}), but they have different masses that correspond to their respective molar masses.
4. In order to compare the densities of gases, temperature and pressure must be specified because the volume of a specific number of gas molecules varies with temperature and pressure.
5. **a.** $PV = nRT$
   b. The ideal gas law relates pressure, volume, temperature, and number of moles for a given sample of gas under any combination of conditions.
6. **a.** The ideal gas law applies to any situation in which a single gas sample undergoes no change in conditions and values are known for three of the four variables, P, V, T, and n.
   b. Care must be taken to match the units of the known quantities and the units of R.
7. **a.** The molar and volume ratios are the same and are both given by the coefficients in the chemical equation.
   b. The volume ratios can be used only if all volumes are

measured at the same temperature and pressure.
8. a. Diffusion is involved in any movement or mixing of gases. Effusion occurs when molecules of a gas in a container that has a small opening pass through the opening.
b. The rates are governed by the velocities of the molecules, which, in turn, are dependent upon the square roots of the masses of the molecules.

Problems

1. a. $\dfrac{1.08 \times 10^{23} \text{ molec}}{5.00 \text{ L } O_2} \times \dfrac{1 \text{ L } O_2}{\text{L } H_2} \times 5.00 \text{ L } H_2$
$= 1.08 \times 10^{23}$ molec

b. $\dfrac{1.08 \times 10^{23} \text{ molec}}{5.00 \text{ L } O_2} \times \dfrac{1 \text{ L } O_2}{\text{L } CO_2} \times 5.00 \text{ L } CO_2$
$= 1.08 \times 10^{23}$ molec

c. $\dfrac{1.08 \times 10^{23} \text{ molec}}{5.00 \text{ L } O_2} \times \dfrac{1 \text{ L } O_2}{\text{L } NH_3} \times 10.00 \text{ L } NH_3$
$= 2.16 \times 10^{23}$ molec

2. a. $1.00 \text{ mol } O_2 \times \dfrac{6.02 \times 10^{23} \text{ molec } O_2}{\text{mol } O_2}$
$= 6.02 \times 10^{23}$ molec O_2

b. $2.50 \text{ mol He} \times \dfrac{6.02 \times 10^{23} \text{ molec He}}{\text{mol He}}$
$= 1.50 \times 10^{24}$ molec He

c. $0.0650 \text{ mol } NH_3 \times \dfrac{6.02 \times 10^{23} \text{ molec } NH_3}{\text{mol } NH_3}$
$= 3.91 \times 10^{22}$ molec NH_3

d. $\dfrac{11.5 \text{ g } NO_2}{\text{mol } NO_2} \times \dfrac{\text{mol } NO_2}{46.0 \text{ g } NO_2}$
$\times \dfrac{6.02 \times 10^{23} \text{ molec } NO_2}{\text{mol } NO_2}$
$= 1.50 \times 10^{23}$ molec NO_2

3. a. $2.25 \text{ mol } Cl_2 \times \dfrac{70.9 \text{ g } Cl_2}{\text{mol } Cl_2} = 160. \text{ g } Cl_2$

b. $3.01 \times 10^{23} \text{ molec } H_2S \times \dfrac{\text{mol } H_2S}{6.02 \times 10^{23} \text{ molec } H_2S}$
$\times \dfrac{34.1 \text{ g } H_2S}{\text{mol } H_2S} = 17.0 \text{ g } H_2S$

c. $25.0 \text{ molec } SO_2 \times \dfrac{\text{mol } SO_2}{6.02 \times 10^{23} \text{ molec } SO_2}$
$\times \dfrac{64.1 \text{ g } SO_2}{\text{mol } SO_2} = 2.66 \times 10^{-21} \text{ g } SO_2$

4. a. $1.00 \text{ mol } O_2 \times \dfrac{22.4 \text{ L } O_2}{\text{mol } O_2} = 22.4 \text{ L } O_2$

b. $3.50 \text{ mol } F_2 \times \dfrac{22.4 \text{ L } F_2}{\text{mol } F_2} = 78.4 \text{ L } F_2$

c. $0.0400 \text{ mol } CO_2 \times \dfrac{22.4 \text{ L } CO_2}{\text{mol } CO_2} = 0.896 \text{ L } CO_2$

d. $1.20 \times 10^{-6} \text{ mol He} \times \dfrac{22.4 \text{ L He}}{\text{mol He}}$
$= 2.69 \times 10^{-5}$ L He

5. a. $22.4 \text{ L } N_2 \times \dfrac{\text{mol } N_2}{22.4 \text{ L } N_2} = 1.00 \text{ mol } N_2$

b. $5.60 \text{ L } Cl_2 \times \dfrac{\text{mol } Cl_2}{22.4 \text{ L } Cl_2} = 0.250 \text{ mol } Cl_2$

c. $0.125 \text{ L Ne} \times \dfrac{\text{mol Ne}}{22.4 \text{ L Ne}} = 5.58 \times 10^{-3} \text{ mol Ne}$

d. $70.0 \text{ mL } NH_3 \times \dfrac{\text{L } NH_3}{10^3 \text{ mL } NH_3} \times \dfrac{\text{mol } NH_3}{22.4 \text{ L } NH_3}$
$= 3.12 \times 10^{-3}$ mol NH_3

6. a. $11.2 \text{ L } H_2 \times \dfrac{\text{mol } H_2}{22.4 \text{ L } H_2} \times \dfrac{2.02 \text{ g } H_2}{\text{mol } H_2} = 1.01 \text{ g} H_2$

b. $2.80 \text{ L } CO_2 \times \dfrac{\text{mol } CO_2}{22.4 \text{ L } CO_2} \times \dfrac{44.0 \text{ g } CO_2}{\text{mol } CO_2}$
$= 5.50$ g CO_2

c. $15.0 \text{ mL } SO_2 \times \dfrac{\text{L } SO_2}{10^3 \text{ mL } SO_2} \times \dfrac{\text{mol } SO_2}{22.4 \text{ L } SO_2}$
$\times \dfrac{64.1 \text{ g } SO_2}{\text{mol } SO_2} = 0.0429 \text{ g } SO_2$

d. $3.40 \text{ cm}^3 F_2 \times \dfrac{\text{mL } F_2}{\text{cm}^3 F_2} \times \dfrac{\text{L } F_2}{10^3 \text{ mL } F_2} \times \dfrac{\text{mol } F_2}{22.4 \text{ L } F_2}$
$\times \dfrac{38.0 \text{ g } F_2}{\text{mol } F_2} = 5.77 \times 10^{-3} \text{ g } F_2$

7. a. $8.00 \text{ g } O_2 \times \dfrac{\text{mol } O_2}{32.0 \text{ g } O_2} \times \dfrac{22.4 \text{ L } O_2}{\text{mol } O_2} = 5.60 \text{ L } O_2$

b. $3.50 \text{ g CO} \times \dfrac{\text{mol CO}}{28.0 \text{ g CO}} \times \dfrac{22.4 \text{ L CO}}{\text{mol CO}} = 2.80 \text{ L CO}$

c. $0.0170 \text{ g } H_2S \times \dfrac{\text{mol } H_2S}{34.1 \text{ g } H_2S} \times \dfrac{22.4 \text{ L } H_2S}{\text{mol } H_2S}$
$= 0.0112$ L H_2S

d. $2.25 \times 10^5 \text{ kg } NH_3 \times \dfrac{10^3 \text{ g } NH_3}{\text{kg } NH_3} \times \dfrac{\text{mol } NH_3}{17.0 \text{ g } NH_3}$
$\times \dfrac{22.4 \text{ L } NH_3}{\text{mol } NH_3} = 2.96 \times 10^8 \text{ L } NH_3$

8. a. $\dfrac{28.0 \text{ g}}{\text{mol}} \times \dfrac{\text{mol}}{22.4 \text{ L}} = 1.25 \text{ g/L}$

b. $\dfrac{70.9 \text{ g}}{\text{mol}} \times \dfrac{\text{mol}}{22.4 \text{ L}} = 3.17 \text{ g/L}$

c. $\dfrac{28.0 \text{ g}}{\text{mol}} \times \dfrac{\text{mol}}{22.4 \text{ L}} = 1.25 \text{ g/L}$

d. $\dfrac{64.1 \text{ g}}{\text{mol}} \times \dfrac{\text{mol}}{22.4 \text{ L}} = 2.86 \text{ g/L}$

9. a. $\dfrac{2.00 \text{ g}}{1.50 \text{ L}} \times \dfrac{22.4 \text{ L}}{\text{mol}} = 29.9 \text{ g/mol}$

b. $\dfrac{3.25 \text{ g}}{2.30 \text{ L}} \times \dfrac{22.4 \text{ L}}{\text{mol}} = 31.7 \text{ g/mol}$

c. $\dfrac{0.620 \text{ g}}{250. \text{ mL}} \times \dfrac{10^3 \text{ mL}}{\text{L}} \times \dfrac{22.4 \text{ L}}{\text{mol}} = 55.6 \text{ g/mol}$

10. a. $\dfrac{P_1V_1}{T_1} = \dfrac{P_2V_2}{T_2}$

$V_2 = \dfrac{P_1V_1T_2}{P_2T_1} =$

$\dfrac{740. \text{ mm Hg} \times 250. \text{ mL} \times 273\text{K}}{760 \text{ mm Hg} \times 300.\text{K}} = 222 \text{ mL}$

b. $\dfrac{0.205 \text{ g}}{222 \text{ mL}} \times \dfrac{10^3 \text{ mL}}{\text{L}} \times \dfrac{22.4 \text{ L}}{\text{mol}} = 20.7 \text{ g/mol}$

11. a. First, find volume at STP:
$T_1 = 50.°\text{C} + 273 = 323\text{K}$
$T_2 = 0.°\text{C} + 273 = 273\text{K}$

$\dfrac{P_1V_1}{T_1} = \dfrac{P_2V_2}{T_2}$

$V_2 = \dfrac{P_1V_1T_2}{P_2T_1} =$

$\dfrac{840. \text{ mm Hg} \times 6.51 \text{ m}^3 \times 273\text{K}}{760 \text{ mm Hg} \times 323\text{K}} = 6.08 \text{ m}^3$

Next, find molar mass:

$\dfrac{12.5 \text{ kg}}{6.08 \text{ m}^3} \times \dfrac{10^3 \text{ g}}{\text{kg}} \times \dfrac{\text{m}^3}{10^6 \text{ cm}^3} \times \dfrac{\text{cm}^3}{\text{mL}} \times \dfrac{10^3 \text{ mL}}{\text{L}}$

$\times \dfrac{22.4 \text{ L}}{\text{mol}} = 46.1 \text{ g/mol}$

b. NO_2, which has a molar mass of 46.0 g/mol, is most likely to be the unknown.

12. First, find P_{gas}:
$P_{\text{gas}} = P_{\text{total}} - P_{H_2O} =$
$650.0 \text{ mm Hg} - 17.5 \text{ mm Hg} = 632.5 \text{ mm Hg}$
Then, find volume at STP:
$T_1 = 20.°\text{C} + 273 = 293\text{K}$
$T_2 = 0.°\text{C} + 273 = 273\text{K}$

$\dfrac{P_1V_1}{T_1} = \dfrac{P_2V_2}{T_2}$

$V_2 = \dfrac{P_1V_1T_2}{P_2T_1} = \dfrac{632.5 \text{ mm Hg} \times 850. \text{ mL} \times 273\text{K}}{760 \text{ mm Hg} \times 293\text{K}}$

$= 659 \text{ mL}$

Finally, find molar mass:

$\dfrac{0.925 \text{ g}}{659 \text{ mL}} \times \dfrac{10^3 \text{ mL}}{\text{L}} \times \dfrac{22.4 \text{ L}}{\text{mol}} = 31.4 \text{ g/mol}$

13. First, find P_{gas}:
$P_{\text{gas}} = 575.0 \text{ mm Hg} - 42.2 \text{ mm Hg}$
$= 532.8 \text{ mm Hg}$
Next, find volume at STP:
$T_1 = 35°\text{C} + 273 = 308\text{K}$
$T_2 = 0.°\text{C} + 273 = 273\text{K}$

$V_2 = \dfrac{P_1V_1T_2}{P_2T_1}$

$= \dfrac{532.8 \text{ mm Hg} \times 500. \text{ cm}^3 \times 273\text{K}}{760 \text{ mm Hg} \times 308\text{K}}$

$= 311 \text{ cm}^3$

Finally, find molar mass:

$\dfrac{0.750 \text{ g}}{311 \text{ cm}^3} \times \dfrac{\text{cm}^3}{\text{mL}} \times \dfrac{10^3 \text{ mL}}{\text{L}} \times \dfrac{22.4 \text{ L}}{\text{mol}}$

$= 54.0 \text{ g/mol}$

14. a. $PV = nRT$

$P = \dfrac{nRT}{V}$

$= \dfrac{1.35 \text{ mol} \times 0.0821 \dfrac{\text{L} \cdot \text{atm}}{\text{mol} \cdot \text{K}} \times 320.\text{K}}{2.50 \text{ L}}$

$= 14.2 \text{ atm}$

b. $PV = nRT$

$P = \dfrac{nRT}{V}$

$= \dfrac{0.86 \text{ mol} \times 0.0821 \dfrac{\text{L} \cdot \text{atm}}{\text{mol} \cdot \text{K}} \times 300.\text{K}}{4.75 \text{ L}}$

$= 4.5 \text{ atm}$

c. $T = 57°\text{C} + 273 = 330.\text{K}$

$P = \dfrac{nRT}{V} = \dfrac{2.15 \text{ mol} \times 0.0821 \dfrac{\text{L} \cdot \text{atm}}{\text{mol} \cdot \text{K}} \times 330.\text{K}}{7.50. \text{ mL} \times \dfrac{\text{L}}{10^3 \text{ mL}}}$

$= 77.7 \text{ atm}$

15. a. $PV = nRT$

$V = \dfrac{nRT}{P} = \dfrac{2.00 \text{ mol} \times 0.0821 \dfrac{\text{L} \cdot \text{atm}}{\text{mol} \cdot \text{K}} \times 300.\text{K}}{1.25 \text{ atm}}$

$= 39.4 \text{ L}$

b. $T = 37°\text{C} + 273 = 310.\text{K}$

$P = 550. \text{ mm Hg} \times \dfrac{\text{atm}}{760 \text{ mm Hg}} = 0.724 \text{ atm}$

$V = \dfrac{nRT}{P}$

$= \dfrac{0.425 \text{ mol} \times 0.0821 \dfrac{\text{L} \cdot \text{atm}}{\text{mol} \cdot \text{K}} \times 310.\text{K}}{0.724 \text{ atm}}$

$= 14.9 \text{ L}$

c. $T = 57°\text{C} + 273 = 330.\text{K}$

$P = 675 \text{ mm Hg} \times \dfrac{\text{atm}}{760 \text{ mm Hg}} = 0.888 \text{ atm}$

$n = 4.00 \text{ g } O_2 \times \dfrac{\text{mol } O_2}{32.0 \text{ g } O_2} = 0.125 \text{ mol } O_2$

$$V = \frac{nRT}{P}$$

$$= \frac{0.125 \text{ mol } O_2 \times 0.0821 \frac{L \cdot atm}{mol \cdot K} \times 330.K}{0.888 \text{ atm}}$$

$$= 3.81 \text{ L}$$

16. a. $PV = nRT$

$$n = \frac{PV}{RT} = \frac{1.06 \text{ atm} \times 1.25 \text{ L}}{0.0821 \frac{L \cdot atm}{mol \cdot K} \times 250.K}$$

$$= 0.0646 \text{ mol}$$

b. $T = 27°C + 273 = 300.K$

$$n = \frac{PV}{RT} = \frac{0.925 \text{ atm} \times 0.80 \text{ L}}{0.0821 \frac{L \cdot atm}{mol \cdot K} \times 300.K}$$

$$= 0.030 \text{ mol}$$

c. $T = -50°C + 273 = 223K$

$$P = 700. \text{ mm Hg} \times \frac{atm}{760 \text{ mm Hg}} = 0.921 \text{ atm}$$

$$n = \frac{PV}{RT} = \frac{0.921 \text{ atm} \times 750. \text{ mL} \times \frac{L}{10^3 \text{ mL}}}{0.0821 \frac{L \cdot atm}{mol \cdot K} \times 223K}$$

$$= 0.0377 \text{ mol}$$

17. a. $PV = nRT$

$$n = \frac{PV}{RT}$$

$$= \frac{1.75 \text{ atm} \times 5.60 \text{ L}}{0.0821 \frac{L \cdot atm}{mol \cdot K} \times 250.K} = 0.477 \text{ mol}$$

$$0.477 \text{ mol } O_2 \times \frac{32.0 \text{ g } O_2}{\text{mol } O_2} = 15.3 \text{ g } O_2$$

b. $P = 700. \text{ mm Hg} \times \frac{atm}{760 \text{ mm Hg}} = 0.921 \text{ atm}$

$T = 27°C + 273 = 300.K$

$$n = \frac{PV}{RT}$$

$$= \frac{0.921 \text{ atm} \times 3.50 \text{ L}}{0.0821 \frac{L \cdot atm}{mol \cdot K} \times 300.K} = 0.131 \text{ mol}$$

$$0.131 \text{ mol } NH_3 \times \frac{17.0 \text{ g } NH_3}{\text{mol } NH_3} = 2.23 \text{ g } NH_3$$

c. $P = 625 \text{ mm Hg} \times \frac{atm}{760 \text{ mm Hg}} = 0.822 \text{ atm}$

$T = -53°C + 273 = 220.K$

$$n = \frac{PV}{RT} = \frac{0.822 \text{ atm} \times 125 \text{ mL} \times \frac{L}{10^3 \text{mL}}}{0.0821 \frac{L \cdot atm}{Mol \cdot K} \times 220.K}$$

$$= 5.69 \times 10^{-3} \text{ mol}$$

$$5.69 \times 10^{-3} \text{ mol } SO_2 \times \frac{64.1 \text{ g } SO_2}{\text{mol } SO_2}$$

$$= 0.365 \text{ g } SO_2$$

18. a. $n = \frac{PV}{RT}$

$$= \frac{1.14 \text{ atm} \times 1.12 \text{ L}}{0.0821 \frac{L \cdot atm}{mol \cdot K} \times 280.K} = 0.0555 \text{ mol}$$

$$M = \frac{0.650 \text{ g}}{0.0555 \text{ mol}} = 11.7 \text{ g/mol}$$

b. $T = 37°C + 273 = 310.K$

$$n = \frac{PV}{RT}$$

$$= \frac{0.840 \text{ atm} \times 2.35 \text{ L}}{0.0821 \frac{L \cdot atm}{mol \cdot K} \times 310.K} = 0.0776 \text{ mol}$$

$$M = \frac{1.05 \text{ g}}{0.0776 \text{ mol}} = 13.5 \text{ g/mol}$$

c. $T = -23°C + 273 = 250.K$

$$P = 785 \text{ mm Hg} \times \frac{atm}{760 \text{ mm Hg}} = 1.03 \text{ atm}$$

$$n = \frac{PV}{RT} = \frac{1.03 \text{ atm} \times 750. \text{ mL} \times \frac{L}{10^3 \text{ mL}}}{0.0821 \frac{L \cdot atm}{mol \cdot K} \times 250.K}$$

$$= 0.0376 \text{ mol}$$

$$M = \frac{0.432 \text{ g}}{0.0376 \text{ mol}} = 11.5 \text{ g/mol}$$

19. $T = -18°C + 273 = 255K$

$$P = 1650 \text{ mm Hg} \times \frac{1 \text{ atm}}{760 \text{ mm Hg}} = 2.17 \text{ atm}$$

$$M = \frac{dRT}{P} = \frac{\frac{3.20 \text{ g}}{L} \times 0.0821 \frac{L \cdot atm}{mol \cdot K} \times 255K}{2.17 \text{ atm}}$$

$$= 30.9 \text{ g/mol}$$

20. $d = \frac{1.40 \text{ g}}{cm^3} \times \frac{cm^3}{mL} \times \frac{10^3 \text{ mL}}{L} = 1.40 \times 10^3 \text{ g/L}$

$$M = \frac{dRT}{P}$$

$$T = \frac{MP}{dR} = \frac{2.00\frac{g}{mol} \times 1.30 \times 10^9 \text{ atm}}{1.40 \times 10^3 \frac{g}{L} \times 0.0821 \frac{L \cdot atm}{mol \cdot K}}$$

$= 2.26 \times 10^7 \text{K}$

$t = 2.26 \times 10^7 \text{K} - 273 = 2.26 \times 10^7 \text{°C}$

21. $2CO(g) + O_2(g) \rightarrow 2CO_2(g)$

   a. $1.0 \text{ L CO} \times \frac{1 \text{ L } O_2}{2 \text{ L CO}} = 0.50 \text{ L } O_2$

   b. $1.0 \text{ L CO} \times \frac{2 \text{ L } CO_2}{2 \text{ L CO}} = 1.0 \text{ L } CO_2$

22. $2C_2H_2(g) + 5O_2(g) \rightarrow 4CO_2(g) + 2H_2O(g)$

   a. $75.0 \text{ L } CO_2 \times \frac{2 \text{ L } C_2H_2}{4 \text{ L } CO_2} = 37.5 \text{ L } C_2H_2$

   b. $75.0 \text{ L } CO_2 \times \frac{2 \text{ L } H_2O}{4 \text{ L } CO_2} = 37.5 \text{ L } H_2O$

   c. $75.0 \text{ L } CO_2 \times \frac{5 \text{ L } O_2}{4 \text{ L } CO_2} = 93.8 \text{ L } O_2$

23. $CS_2(\ell) + 3O_2(g) \rightarrow CO_2(g) + 2SO_2(g)$

   $450. \text{ mL } O_2 \times \frac{1 \text{ mL } CO_2}{3 \text{ mL } O_2} = 150. \text{ mL } CO_2$

   $450. \text{ ml } O_2 \times \frac{2 \text{ mL } SO_2}{3 \text{ mL } O_2} = 300. \text{ mL } SO_2$

24. $2Mg(s) + O_2(g) \rightarrow 2MgO(s)$

   a. $22.4 \text{ L } O_2 \times \frac{mol \text{ } O_2}{22.4 \text{ L } O_2} = 1.00 \text{ mol } O_2$

   $1.00 \text{ mol } O_2 \times \frac{2 \text{ mol MgO}}{1 \text{ mol } O_2} = 2.00 \text{ mol MgO}$

   b. $11.2 \text{ L } O_2 \times \frac{mol \text{ } O_2}{22.4 \text{ L } O_2} = 0.500 \text{ mol } O_2$

   $0.500 \text{ mol } O_2 \times \frac{2 \text{ mol MgO}}{1 \text{ mol } O_2} = 1.00 \text{ mol MgO}$

   c. $1.40 \text{ L } O_2 \times \frac{mol \text{ } O_2}{22.4 \text{ L } O_2} = 0.0625 \text{ mol } O_2$

   $0.0625 \text{ mol } O_2 \times \frac{2 \text{ mol MgO}}{1 \text{ mol } O_2} = 0.125 \text{ mol Mg}$

   $0.125 \text{ mol MgO} \times \frac{40.3 \text{ g MgO}}{\text{mol MgO}} = 5.04 \text{ g MgO}$

25. $CuO(s) + H_2(g) \rightarrow Cu(s) + H_2O(g)$

   a. $5.60 \text{ L } H_2 \times \frac{mol}{22.4 \text{ L}} = 0.250 \text{ mol } H_2$

   b. $0.250 \text{ mol } H_2 \times \frac{1 \text{ mol Cu}}{1 \text{ mol } H_2} = 0.250 \text{ mol Cu}$

   c. $0.250 \text{ mol Cu} \times \frac{63.5 \text{ g Cu}}{\text{mol Cu}} = 15.9 \text{ g Cu}$

26. $2KI(aq) + Cl_2(g) \rightarrow 2KCl(aq) + I_2(g)$

   a. $8.50 \text{ L } I_2 \times \frac{\text{mol } I_2}{22.4 \text{ L } I_2} = 0.379 \text{ mol } I_2$

   b. $0.379 \text{ mol } I_2 \times \frac{2 \text{ mol KI}}{1 \text{ mol } I_2} = 0.758 \text{ mol KI}$

   c. $0.758 \text{ mol KI} \times \frac{166 \text{ g KI}}{\text{mol KI}} = 126 \text{ g KI}$

27. $2Fe(OH)_3(s) \rightarrow Fe_2O_3(s) + 3H_2O(g)$

   a. $0.75 \text{ L } H_2O \times \frac{mol}{22.4 \text{ L}} = 0.034 \text{ mol } H_2O$

   $0.034 \text{ mol } H_2O \times \frac{2 \text{ mol Fe(OH)}_3}{3 \text{ mol } H_2O}$

   $= 0.023 \text{ mol Fe(OH)}_3$

   $0.023 \text{ mol Fe(OH)}_3 \times \frac{107 \text{ g Fe(OH)}_3}{\text{mol Fe(OH)}_3}$

   $= 2.4 \text{ g Fe(OH)}_3$

   b. $0.034 \text{ mol } H_2O \times \frac{1 \text{ mol } Fe_2O_3}{3 \text{ mol } H_2O} = 0.011 \text{ mol } Fe_2O_3$

   $0.011 \text{ mol } Fe_2O_3 \times \frac{160. \text{ g } Fe_2O_3}{\text{mol } Fe_2O_3} = 1.8 \text{ g } Fe_2O_3$

28. $Fe(s) + H_2SO_4(aq) \rightarrow FeSO_4(aq) + H_2(g)$

   $650. \text{ mL } H_2 \times \frac{L}{10^3 \text{ mL}} \times \frac{mol}{22.4 \text{ L}} = 0.0290 \text{ mol } H_2$

   $0.0290 \text{ mol } H_2 \times \frac{1 \text{ mol } FeSO_4}{1 \text{ mol } H_2} = 0.0290 \text{ mol } FeSO_4$

   $0.0290 \text{ mol } FeSO_4 \times \frac{152 \text{ g } FeSO_4}{\text{mol } FeSO_4} = 4.41 \text{ g } FeSO_4$

29. $2Al(s) + 6HCl(aq) \rightarrow 2AlCl_3(aq) + 3H_2(g)$

   a. $13.5 \text{ g Al} \times \frac{\text{mol Al}}{27.0 \text{ g Al}} = 0.500 \text{ mol Al}$

   b. $0.500 \text{ mol Al} \times \frac{3 \text{ mol } H_2}{2 \text{ mol Al}} = 0.750 \text{ mol } H_2$

   c. $0.750 \text{ mol } H_2 \times \frac{22.4 \text{ L}}{\text{mol}} = 16.8 \text{ L } H_2$

30. $2C_8H_{18}(g) + 25O_2(g) \rightarrow 16CO_2(g) + 18H_2O(g)$

   a. $25.0 \text{ L } C_8H_{18} \times \frac{25 \text{ L } O_2}{2 \text{ L } C_8H_{18}} = 312 \text{ L } O_2$

   $312 \text{ L } O_2 \times \frac{100 \text{ L air}}{20.9 \text{ L } O_2} = 1.49 \times 10^3 \text{ L air}$

   b. $25.0 \text{ L } C_8H_{18} \times \frac{16 \text{ L } CO_2}{2 \text{ L } C_8H_{18}} = 200. \text{ L } CO_2$

   $25.0 \text{ L } C_8H_{18} \times \frac{18 \text{ L } H_2O}{2 \text{ L } C_8H_{18}} = 225 \text{ L } H_2O$

31. $CO(g) + 2H_2(g) \rightarrow CH_3OH(g)$

   a. $450. \text{ mL CO} \times \frac{2 \text{ mL } H_2}{1 \text{ mL CO}} = 900. \text{ mL } H_2 \text{ are needed}$

$825 \text{ mL H}_2 \times \dfrac{1 \text{ mL CO}}{2 \text{ mL H}_2} = 412 \text{ mL CO}$ are needed

Thus the CO is present in exccess.
b. The excess amount is $450. \text{ mL} - 412 \text{ mL}$
$= 38.0 \text{ mL CO}$

c. $825 \text{ mL H}_2 \times \dfrac{1 \text{ ml CH}_3\text{OH}}{2 \text{ mL H}_2} = 412 \text{ mL CH}_3\text{OH}$

32. $CH_4(g) + 2O_2(g) \rightarrow CO_2(g) + 2H_2O(g)$

$P = 730. \text{ mm Hg} \times \dfrac{1 \text{ atm}}{760 \text{ mm Hg}} = 0.961 \text{ atm}$

$T = 20°C + 273 = 293K$

$n = \dfrac{PV}{RT} = \dfrac{0.961 \text{ atm} \times 45.0 \text{ L}}{0.0821 \dfrac{\text{L} \cdot \text{atm}}{\text{mol} \cdot \text{K}} \times 293K} = 1.80 \text{ mol}$

$1.80 \text{ mol CH}_4 \times \dfrac{1 \text{ mol CO}_2}{1 \text{ mol CH}_4} = 1.80 \text{ mol CO}_2$

$1.80 \text{ mol CO}_2 \times \dfrac{44.0 \text{ g CO}_2}{\text{mol CO}_2} = 79.2 \text{ g CO}_2$

$1.80 \text{ mol CH}_4 \times \dfrac{2 \text{ mol H}_2\text{O}}{1 \text{ mol CH}_4} = 3.60 \text{ mol H}_2\text{O}$

$3.60 \text{ mol H}_2\text{O} \times \dfrac{18.0 \text{ g H}_2\text{O}}{\text{mol H}_2\text{O}} = 64.8 \text{ g H}_2\text{O}$

33. $2Al(s) + 6NaOH(aq) \rightarrow 3H_2(g) + 2Na_3AlO_3(aq)$

$P_{H_2} = 740.0 \text{ mm Hg} - 10.9 \text{ mm Hg} = 729.1 \text{ mm Hg}$

$P = 729.1 \text{ mm Hg} \times \dfrac{1 \text{ atm}}{760 \text{ mm Hg}} = 0.959 \text{ atm}$

$T = 12.5°C + 273 = 286K$

$n = \dfrac{PV}{RT} = \dfrac{0.959 \text{ atm} \times 7.50 \text{ L}}{0.0821 \dfrac{\text{L} \cdot \text{atm}}{\text{mol} \cdot \text{K}} \times 286K} = 0.306 \text{ mol}$

$0.306 \text{ mol H}_2 \times \dfrac{2 \text{ mol Al}}{3 \text{ mol H}_2} = 0.204 \text{ mol Al}$

$0.204 \text{ mol Al} \times \dfrac{27.0 \text{ g Al}}{\text{mol Al}} = 5.51 \text{ g Al}$

34. $N_2(g) + 3H_2(g) \rightarrow 2NH_3(g)$

$T = 550.°C + 273 = 823K$

$10.0 \text{ kg N}_2 \times \dfrac{10^3 \text{ g N}_2}{\text{kg N}_2} \times \dfrac{\text{mol N}_2}{28.0 \text{ g N}_2} = 357 \text{ mol N}_2$

$357 \text{ mol N}_2 \times \dfrac{2 \text{ mol NH}_3}{1 \text{ mol N}_2} = 714 \text{ mol NH}_3$

$PV = nRT$

$V = \dfrac{nRT}{P} = \dfrac{714 \text{ mol} \times 0.0821 \dfrac{\text{L} \cdot \text{atm}}{\text{mol} \cdot \text{K}} \times 823K}{250. \text{ atm}}$

$= 193 \text{ L NH}_3$

35. $4C_3H_5(NO_3)_3(\ell) \rightarrow 12CO_2(g) + 6N_2(g) + O_2(g) + 10H_2O(g)$

$500. \text{ g C}_3\text{H}_5(\text{NO}_3)_3 \times \dfrac{\text{mol C}_3\text{H}_5(\text{NO}_3)_3}{227 \text{ g C}_3\text{H}_5(\text{NO}_3)_3}$

$= 2.20 \text{ mol C}_3\text{H}_5(\text{NO}_3)_3$

$2.20 \text{ mol C}_3\text{H}_5(\text{NO}_3)_3 \times \dfrac{12 \text{ mol CO}_2}{4 \text{ mol C}_3\text{H}_5(\text{NO}_3)_3}$

$= 6.60 \text{ mol CO}_2$

$2.20 \text{ mol C}_3\text{H}_5(\text{NO}_3)_3 \times \dfrac{6 \text{ mol N}_2}{4 \text{ mol C}_3\text{H}_5(\text{NO}_3)_3}$

$= 3.30 \text{ mol N}_2$

$2.20 \text{ mol C}_3\text{H}_5(\text{NO}_3)_3 \times \dfrac{1 \text{ mol O}_2}{4 \text{ mol C}_3\text{H}_5(\text{NO}_3)_3}$

$= 0.550 \text{ mol O}_2$

$2.20 \text{ mol C}_3\text{H}_5(\text{NO}_3)_3 \times \dfrac{10 \text{ mol H}_2\text{O}}{4 \text{ mol C}_3\text{H}_5(\text{NO}_3)_3}$

$= 5.50 \text{ mol H}_2\text{O}$

Total number of moles of gases produced $= 6.60$ mol $+ 3.30$ mol $+ 0.550$ mol $+ 5.50$ mol $= 16.0$ mol of gases

$16.0 \text{ mol gas} \times \dfrac{22.4 \text{ L}}{\text{mol}} = 358 \text{ L of gases}$

36. $SO_2(g) + 2H_2S(g) \rightarrow 2H_2O(\ell) + 3S(s)$

$4.50 \times 10^5 \text{ kg S} \times \dfrac{10^3 \text{ g}}{\text{kg}} \times \dfrac{\text{mol S}}{32.1 \text{ g S}}$

$= 1.40 \times 10^7 \text{ mol S}$

$1.40 \times 10^7 \text{ mol S} \times \dfrac{1 \text{ mol SO}_2}{3 \text{ mol S}}$

$= 4.67 \times 10^6 \text{ mol SO}_2$

$1.40 \times 10^7 \text{ mol S} \times \dfrac{2 \text{ mol H}_2\text{S}}{3 \text{ mol S}} = 9.33 \times 10^6 \text{ mol H}_2\text{S}$

$P = 730. \text{ mm Hg} \times \dfrac{1 \text{ atm}}{760 \text{ mm Hg}} = 0.961 \text{ atm}$

$T = 22°C + 273 = 295K$

$PV = nRT$

$V_{SO_2} = \dfrac{nRT}{P}$

$= \dfrac{4.67 \times 10^6 \text{ mol} \times 0.0821 \dfrac{\text{L} \cdot \text{atm}}{\text{mol} \cdot \text{K}} \times 295K}{0.961 \text{ atm}}$

$= 1.18 \times 10^8 \text{ L SO}_2$

$V_{H_2S} = \dfrac{nRT}{P}$

$= \dfrac{9.33 \times 10^6 \text{ mol} \times 0.0821 \dfrac{\text{L} \cdot \text{atm}}{\text{mol} \cdot \text{K}} \times 295K}{0.961 \text{ atm}}$

$= 2.35 \times 10^8 \text{ L H}_2\text{S}$

37. $CaC_2(s) + 2H_2O(\ell) \rightarrow C_2H_2(g) + Ca(OH)_2(aq)$

$$3.25 \text{ g } CaC_2 \times \frac{\text{mol } CaC_2}{64.1 \text{ g } CaC_2} = 0.0507 \text{ mol } CaC_2$$

$$0.0507 \text{ mol } CaC_2 \times \frac{1 \text{ mol } C_2H_2}{1 \text{ mol } CaC_2}$$

$$= 0.0507 \text{ mol } C_2H_2$$

$P_{C_2H_2} = 740.0$ mm Hg $- 14.5$ mm Hg $= 725.5$ mm Hg

$$725.5 \text{ mm Hg} \times \frac{1 \text{ atm}}{760 \text{ mm Hg}} = 0.955 \text{ atm}$$

$T = 17°C + 273 = 290.K$
$PV = nRT$

$$V = \frac{nRT}{P}$$

$$= \frac{0.0507 \text{ mol} \times 0.0821 \frac{\text{L} \cdot \text{atm}}{\text{mol} \cdot \text{K}} \times 290.\text{K} \times \frac{10^3 \text{ mL}}{\text{L}}}{0.955 \text{ atm}}$$

$$= 1.26 \times 10^3 \text{ mL}$$

38. a. $\dfrac{\text{rate of effusion}_A}{\text{rate of effusion}_B} = \dfrac{\sqrt{M_B}}{\sqrt{M_A}}$

$$\frac{\text{rate}_{H_2}}{\text{rate}_{N_2}} = \frac{\sqrt{28.0 \text{ g/mol}}}{\sqrt{2.02 \text{ g/mol}}} = 3.72$$

b. $\dfrac{\text{rate}_{F_2}}{\text{rate}_{Cl_2}} = \dfrac{\sqrt{71.0 \text{ g/mol}}}{\sqrt{38.0 \text{ g/mol}}} = 1.37$

39. $\dfrac{\text{velocity}_{H_2}}{\text{velocity}_{Ne}} = \dfrac{\sqrt{M_{Ne}}}{\sqrt{M_{H_2}}} = \dfrac{\sqrt{20.179 \text{ g/mol}}}{\sqrt{2.0159 \text{ g/mol}}} = 3.16$

40. $\dfrac{v_{Cl_2}}{v_{SO_2}} = \dfrac{\sqrt{M_{SO_2}}}{\sqrt{M_{Cl_2}}}$

$$v_{SO_2} = \frac{v_{Cl_2}\sqrt{M_{Cl_2}}}{\sqrt{M_{SO_2}}}$$

$$= \frac{0.0380 \text{ m/s}\sqrt{70.9 \text{ g/mol}}}{\sqrt{64.1 \text{ g/mol}}} = 0.0400 \text{ m/s}$$

41. $\dfrac{\text{effusion}_{He}}{\text{effusion}_X} = \dfrac{\sqrt{M_X}}{\sqrt{M_{He}}}$

$$\sqrt{M_X} = \frac{v_{He}\sqrt{M_{He}}}{v_X}$$

Substituting 6.50 for $\dfrac{v_{He}}{v_X}$:

$\sqrt{M_X} = 6.50\sqrt{M_{He}}$
$M_X = (6.50\sqrt{M_{He}})^2 = 42.2 \times M_{He}$
$= 42.2 \times 4.00 \text{ g/mol} = 169 \text{ g/mol}$

Critical Thinking

1. From Boyle's law, $V \propto \dfrac{1}{P}$; from Charles' law $V \propto T$; and from Avogadro's principle, $V \propto n$. Since a quantity that is proportional to each of two or more other quantities is also proportional to their product:

$$V \propto \frac{1}{P} \times T \times n; V = \frac{kTn}{P}.$$

Replacing k with R, and rearranging, $PV = nRT$ (the ideal gas law).

2. a. The combined gas law is used when a gas changes from one set of P-V-T conditions into another.
  b. The ideal gas law is used to compute one of the four missing quantities (P, V, T, or n) when the others are given under the same set of conditions.

3. $\dfrac{\text{rate of effusion}_A}{\text{rate of effusion}_B} = \dfrac{\sqrt{M_B}}{\sqrt{M_A}} = \dfrac{\sqrt{\text{density}_B}}{\sqrt{\text{density}_A}}$

4. Avogadro's principle states that equal volumes of gases at the same temperature and pressure contain equal numbers of molecules. Gay-Lussac noted that volume proportions in gas reactions were simple definite proportions and the relative volumes could be of any proportionate size. The relationship between these two ideas is volume and the number of molecules. For any proportional change in the volumes of gas reactants at a constant temperature and pressure, the number of molecules available to react would also be reduced or increased by the same proportion that the reacting volumes were reduced or in-

5. One possible response could be to run the reaction for water at the same temperature and pressure using six different volumes of reactants that are in the ratio of 2 to 1, hydrogen to oxygen. The volume of water vapor formed in each case should approximate the volume of hydrogen as the reactant if the relationship holds true.

6. a. No, if the gases were at the same temperature and pressure there would be equal numbers of particles in each container.
  b. gas C
  c. gas C
  d. they are equal

7. graph A

Teacher's Notes

Chapter 12
Molecular Composition of Gases

SECTION PREVIEW

12.1 Volume–Mass Relationships of Gases
12.2 The Ideal Gas Law
12.3 Stoichiometry of Gases
12.4 Effusion and Diffusion

Carbon dioxide gas is one of the products formed when this antacid tablet reacts with water.

12.1 Volume–Mass Relationships of Gases

The volumes taken up by gases under certain conditions can often reveal information about other properties of the gases. In this section, you will examine the relationships that exist among volumes of gases that react with each other. You will also learn about the ways in which volume, mass, density, and molar mass are interrelated.

The first important step toward the recognition of these relationships was taken when early chemists discovered the law of definite composition. In the early 1800s, Dalton developed the atomic theory by applying the law of definite composition to his investigation of the masses of combining substances. At the same time, the French chemist and early balloonist Joseph Louis Gay-Lussac was experimenting with gaseous substances (Figure 12-1). He was particularly interested in measuring and comparing the volumes of gases that react with each other in various chemical reactions.

Gay-Lussac discovered that gases combine in small, whole-number ratios by volume. This simple observation opened the door to solving several problems that had stalled progress in chemistry. It was recognized that some gases, such as oxygen, are composed not of isolated *atoms*, but of diatomic *molecules*. Gay-Lussac's work made possible the comparison of densities and relative molecular masses of gases, a discovery that led to the first correct determination of the relative scales of atomic mass.

Measuring and Comparing the Volumes of Reacting Gases

One of Gay-Lussac's earliest investigations of the volume relationship of gases involved a chemical reaction between hydrogen and oxygen. He observed that, when measured at the same temperature and pressure, 2 L of hydrogen react with 1 L of oxygen to form 2 L of water vapor.

$$\text{hydrogen gas} + \text{oxygen gas} \rightarrow \text{water vapor}$$
$$2\text{ L} \qquad\qquad 1\text{ L} \qquad\qquad 2\text{ L}$$

This equation can be written in a more general form to show the simple and definite 2:1:2 relationship between the volumes of the reactants and product.

$$\text{hydrogen gas} + \text{oxygen gas} \rightarrow \text{water vapor}$$
$$2\text{ volumes} \quad 1\text{ volume} \quad 2\text{ volumes}$$

A relationship between volumes like that above applies to any units of volume—2 mL and 1 mL and 2 mL; or 2 m^3, 1 m^3, and 2 m^3. The relative volumes can be of any proportionate size—for example, 4 mL, 2 mL, and 4 mL; or 600 L, 300 L, and 600 L.

In another reaction with gases, Gay-Lussac noticed similar simple and definite proportions by volume.

$$\text{hydrogen gas} + \text{chlorine gas} \rightarrow \text{hydrogen chloride gas}$$
$$1\text{ L} \qquad\qquad 1\text{ L} \qquad\qquad 2\text{ L}$$
$$1\text{ volume} \quad\; 1\text{ volume} \quad\; 2\text{ volumes}$$

SECTION OBJECTIVES

1. State the law of combining volumes.
2. State Avogadro's principle, and explain its significance.
3. Define standard molar volume of a gas, and use it as a conversion factor to calculate gas masses and volumes.
4. Calculate the molar mass and density of a gas by using standard molar volume.

The law of definite composition states that each compound has a definite composition by mass. This law is described in Section 1.3.

Figure 12-1 A balloon of a given volume would have about the same number of moles of molecules whether it was filled with air, hydrogen, or oxygen, provided the gases were at the same pressure and temperature.

Molecular Composition of Gases

John Dalton (1766–1844), English chemist; his love of meteorology led to his study of air and properties of gases which naturally led to his atomic theory.

Dalton's atomic theory is discussed in Section 3.1.

Gay-Lussac summarized the results of his experiments in 1808 in a statement known today as **Gay-Lussac's law of combining volumes of gases,** which states that *at a constant temperature and pressure, the volumes of gaseous reactants and products can be expressed as ratios of small whole numbers.* This law applies only to gas volumes measured at the same temperature and pressure.

Avogadro's Principle

Recall an important point of Dalton's atomic theory: atoms are indivisible. Dalton also believed that one atom of one element always combines with one atom of another element to form a single particle of the product. He thus thought that gaseous elements exist in the form of isolated single atoms. To explain the volume relationship observed by Gay-Lussac in terms of Dalton's theory would require that atoms be subdivided.

In 1811, Avogadro found a way to explain Gay-Lussac's simple ratios of combining gas volumes without violating Dalton's concepts of indivisible atoms. He rejected Dalton's idea that reactant elements are in monatomic form when they combine to form products and reasoned that molecules could contain more than one atom. Avogadro's explanation, known today as **Avogadro's principle,** states *that equal volumes of gases at the same temperature and pressure contain equal numbers of molecules.* It follows that at the same temperature and pressure, the volume of any given gas varies directly with the number of molecules.

Consider the reaction of hydrogen and chlorine to produce hydrogen chloride, illustrated in Figure 12-2.

hydrogen gas + chlorine gas → hydrogen chloride gas
1 volume   1 volume   →   2 volumes

According to Avogadro's principle, equal volumes of hydrogen and chlorine contain the same number of molecules.

Accepting Dalton's idea that matter is composed of indivisible atoms, but rejecting his belief in universally monatomic elements, Avogadro concluded

Figure 12-2 Because hydrogen molecules combine with chlorine molecules in a 1-to-1 ratio to produce two volumes of hydrogen chloride, the molecules of hydrogen and chlorine must be diatomic.

that the hydrogen and chlorine components must each consist of two or more atoms joined together. The simplest assumption—that hydrogen and chlorine molecules are composed of two atoms each—leads to the following equation for the reaction of hydrogen with chlorine:

$$\begin{array}{ccccc} H_2(g) & + & Cl_2(g) & \rightarrow & 2HCl(g) \\ \text{1 molecule} & & \text{1 molecule} & & \text{2 molecules} \\ \text{1 volume} & & \text{1 volume} & & \text{2 volumes} \end{array}$$

The simplest hypothetical formula for hydrogen chloride (HCl) indicates that the molecule contains one hydrogen atom and one chlorine atom. The simplest formulas for hydrogen and chlorine must be H_2 and Cl_2, respectively. Determination of the molecular masses of these gases has shown that these formulas are indeed correct.

Avogadro's reasoning applied equally well to the combining volumes and molecular composition of hydrogen and oxygen in the reaction to produce water vapor. The simplest hypothetical formula for oxygen must indicate two oxygen atoms and the simplest possible molecule of water must indicate two hydrogen atoms and one oxygen atom per molecule. Further experiments eventually showed that all of the active gaseous elements except the noble gases normally exist as diatomic molecules.

Using the number of moles to represent numbers of molecules, Avogadro's principle indicates that gas volume is directly proportional to the number of moles of gas (at constant temperature and pressure). The equation for this relationship is

$$V = kn$$

where n is the number of moles of gas and k is a constant. As illustrated below, the coefficients in a chemical reaction involving gases indicate the numbers of molecules, the numbers of moles, and the relative volumes of reactants and products.

Review direct and inverse proportions.

$$\begin{array}{ccccc} 2H_2(g) & + & O_2(g) & \rightarrow & 2\ H_2O(g) \\ \text{2 molecules} & & \text{1 molecule} & & \text{2 molecules} \\ \text{2 moles} & & \text{1 mole} & & \text{2 moles} \\ \text{2 volumes} & & \text{1 volume} & & \text{2 volumes} \end{array}$$

Molar Volume of Gases

Recall that one mole contains Avogadro's number of molecules (6.022×10^{23}). One mole of oxygen (O_2) contains 6.022×10^{23} diatomic oxygen molecules and has a mass of 31.9988 g. One mole of hydrogen (H_2) contains the same number of hydrogen molecules but has a mass of 2.015 88 g. Helium (He) is a monatomic gas, yet one mole of helium gas contains Avogadro's number of helium atoms and has a mass of 4.002 60 g.

Avogadro's number is introduced in Section 3.3.

Because one-mole quantities of all molecular substances contain the Avogadro number of molecules, one mole of any gas contains the same number of molecules as one mole of any other gas. According to Avogadro's principle, each gas will occupy the same volume at the same temperature and pressure (Figure 12-3). Experiments show that one mole of any gas has the same volume at STP as one mole of any other gas. *This volume, known as the* **standard molar volume of a gas,** *is the volume occupied by one mole of*

Molecular Composition of Gases   **345**

Figure 12-3 One-mole quantities of different gases occupy 22.4 L at STP, yet have different masses.

Molar volumes at other temperatures

Temp (°C)	Molar volume (L/mol)
25	24.5
100	30.6

a gas at STP, and has been found to be 22.414 10 L. As illustrated in Figure 12-3, 22.414 10 L of every gas contains the same number of molecules, but the masses of these volumes are different for different gases (in calculations we will use 22.4 L as the molar volume). The mass of each is equal to the molar mass of the gas—the mass of 1 mol of molecules.

It is often more convenient to measure the volume of a gas than it is to measure its mass. If you remember that molar volume applies only at STP, you can use 1 mol/22.4 L as a conversion factor to find the number of moles, and therefore the mass, of a given volume of a known gas. The molar volume of a gas can be used to find the volume, at STP, of a known number of moles or a known mass of a gas.

Sample Problem 12.1

A chemical reaction is expected to produce 0.0680 mol of oxygen gas. What volume in liters will be occupied by this gas sample at STP?

Solution

Step 1. Analyze Given: 0.0680 mol of O_2
Unknown: volume of O_2 in liters (at STP)

Step 2. Plan moles of $O_2 \rightarrow$ volume of O_2 (at STP)

The standard molar volume can be used to find the volume of a known molar amount of a gas at STP.

$$\text{moles} \times \frac{22.4 \text{ L}}{\text{mole}} = O_2 \text{ volume in L}$$

Step 3. Compute $0.0680 \text{ mol} \times \dfrac{22.4 \text{ L}}{\text{mol}} = 1.52 \text{ L}$

Step 4. Evaluate The answer is close to an estimated value of 1.4, computed as 0.07×20. Units have canceled to yield liters, as desired. The result is correctly carried to three significant figures.

Sample Problem 12.2

A chemical reaction produced 98.0 mL of sulfur dioxide gas (SO_2) at STP. What was the mass (in grams) of the gas produced?

Solution

Step 1. Analyze Given: $V = 98.0$ mL of SO_2 (at STP)
Unknown: mass of SO_2 in grams

Step 2. Plan SO_2 volume in mL (at STP) → SO_2 volume in L (at STP) → SO_2 moles → SO_2 mass

$$mL \times \frac{1 \text{ L}}{1000 \text{ mL}} \times \frac{1 \text{ mol } SO_2}{22.4 \text{ L}} \times \frac{g \, SO_2}{1 \text{ mol } SO_2}$$
$$= g \, SO_2$$

Step 3. Compute
$$98.0 \text{ mL} \times \frac{1 \text{ L}}{1000 \text{ mL}} \times \frac{1 \text{ mol } SO_2}{22.4 \text{ L}} \times \frac{64.1 \text{ g } SO_2}{1 \text{ mol } SO_2}$$
$$= 0.280 \text{ g } SO_2$$

Step 4. Evaluate The result is correctly limited to three significant figures. Units cancel correctly to give the answer in grams. The known volume is roughly $\frac{1}{200}$ of the molar volume (22 400 mL/200 = 112 mL). The answer is reasonable: the mass should be roughly $\frac{1}{200}$ of the molar mass (64 g/200 = 0.32 g).

Practice Problem
What is the mass of 1.33×10^4 mL of oxygen gas (O_2) at STP? *(Ans.)* 19.0 g

Gas Density

The densities of different gases represent the masses of equal numbers of their respective molecules measured at STP. From the gas laws, you know that the volume of a specific number of gas molecules varies with temperature and pressure. For example, at higher temperatures, a gas tends to occupy more volume and become less dense. If you are to compare the densities of different gases, you must be given the density values for the same temperature and pressure. The density of a gas is usually given in grams per liter at STP. For example, hydrogen has a density at STP of 0.0899 g/L.

As is true for solids and liquids, the density of a gas can be found by dividing the mass of the gas sample by its volume. Unlike solid and liquid densities, however, a gas density at STP can also be found from the molar mass alone. The molar volume is the necessary conversion factor. The relationships in terms of the units are as follows:

$$\text{Density (at STP)} = \frac{\text{molar mass}}{\text{molar volume}} = \frac{\text{g/mol}}{22.4 \text{ L/mol}}$$

$$= \frac{g}{\text{mol}} \times \frac{1 \text{ mol}}{22.4 \text{ L}}$$

Volume–pressure and volume–temperature relationships are described by Boyle's law and Charles' law, respectively.

Figure 12-4 A balloon filled with helium rises because helium is less dense than air. Why does the balloon filled with carbon dioxide sink?

Sample Problem 12.3

What is the density of CO_2 gas in grams per liter at STP?

Solution

Step 1. Analyze Given: chemical formula, CO_2
Unknown: density of CO_2 at STP

Step 2. Plan chemical formula → molar mass → density (at STP)

$$\frac{g}{1 \text{ mol}} \times \frac{1 \text{ mol}}{22.4 \text{ L}} = \frac{g}{L}$$

Step 3. Compute

$$\frac{44.0 \text{ g}}{1 \text{ mol}} \times \frac{1 \text{ mol}}{22.4 \text{ L}} = 1.96 \text{ g/L}$$

Step 4. Evaluate Units cancel correctly and the answer is properly limited to three significant figures. The magnitude of the answer is reasonable for a gas density and is close to an estimated value of 2.0, (44 ÷ 22).

Practice Problem
What is the density of SO_2 gas in grams per liter at STP? *(Ans.)* 2.86 g/L

Finding Molar Mass from Volume at STP

The combined gas law is discussed in Section 11.3

The molar mass of a gas or any substance that vaporizes easily without decomposing can be determined by use of the molar volume. The density of the gas at STP can be determined by finding either the volume occupied by a given mass of the gas or the mass that occupies a known volume. Then the relationship given above for density, molar mass, and molar volume is used to solve for the molar mass:

$$\text{density (at STP)} = \frac{\text{molar mass}}{\text{molar volume}}$$

$$\text{molar mass} = \text{density (at STP)} \times \text{molar volume}$$

$$= \frac{g}{L} \times \frac{22.4 \text{ L}}{\text{mol}}$$

Sample Problem 12.4

What is the molar mass of a gas that has a density of 1.28 g/L at STP?

Solution

Step 1. Analyze Given: density (at STP) = 1.28 g/L
Unknown: molar mass

Step 2. Plan Density (at STP) → molar mass.
Density at STP, and molar volume can be used to find the molar mass.

Step 3. Compute Result: $\dfrac{1.28 \text{ g}}{1 \text{ L}} \times \dfrac{22.4 \text{ L}}{1 \text{ mol}} = 28.7 \text{ g/mol}$

Sample Problem 12.4 *continued*

Step 4. Evaluate Units cancel correctly. The answer is given to three significant figures and is close to an estimated value of 22, or 1 × 22.

Sample Problem 12.5

A 0.519-g gas sample is found to have a volume of 200. mL at STP. What is the molar mass of this gas? Is this gas propane (molar mass = 44.1 g/mol) or butane (molar mass = 58.1 g/mol)?

Solution

Step 1. Analyze Given: mass and volume of gas at STP, 0.519 g and 200. mL
Unknown: molar mass

Step 2. Plan mass/volume (at STP) → density (at STP) → molar mass

Because the volume is known at STP, the molar volume can be used to find the molar mass. The factor for converting milliliters to liters must be included.

$$\frac{g}{mL} \times \frac{1000 \text{ mL}}{L} \times \frac{22.4 \text{ L}}{mol} = \frac{g}{mol}$$

Step 3. Compute
$$\frac{0.519 \text{ g}}{200. \text{ mL}} \times \frac{1000 \text{ mL}}{L} \times \frac{22.4}{mol} = 58.1 \text{ g/mol}$$

Step 4. Evaluate Units cancel correctly and the answer is properly given to three significant figures. The answer has the correct value for one of the two possible gases and is close to an estimated value of 55, ($\frac{0.5}{200} \times 1000 \times 22$). The gas is butane.

Sample Problem 12.6

A 1.25 g sample of the gaseous product of a chemical reaction was found to have a volume of 350. mL at 20.0°C and 750. mm Hg. What is the molar mass of this gas?

Solution

Step 1. Analyze Given: mass of gas = 1.25 g
pressure = 750. mm Hg; volume = 350. mL; temperature = 20.0°C
Unknown: molar mass

Step 2. Plan temperature (°C) → temperature (K); pressure, volume, temperature
→ volume (at STP) → density (at STP) → molar mass

$$t(°C) + 273 = T(K)$$

The measured volume is not at STP. Before the molar mass can be found from the density and standard molar volume, therefore, the volume at STP must be found by solving the combined gas law for V_2.

$$V_2(\text{mL}) = \frac{P_1 V_1 T_2}{P_2 T_1}$$

$$\frac{g}{(\text{mL})} \times \frac{1000 \text{ mL}}{L} \times \frac{22.4 \text{ L}}{mol} = \frac{g}{mol}$$

Sample Problem 12.6 *continued*

Step 3. Compute

$$20°C + 273 = 293K$$

$$V_2 = \frac{750. \text{ mm Hg} \times 350. \text{ mL} \times 273K}{760 \text{ mm Hg} \times 293K}$$

$$= 322 \text{ mL at STP}$$

$$= \frac{1.25 \text{ g}}{322 \text{ mL}} \times \frac{1000 \text{ mL}}{1 \text{ L}} \times \frac{22.4 \text{ L}}{1 \text{ mol}} = 87.0 \text{ g/mol}$$

Step 4. Evaluate Units cancel correctly to give molar mass in grams per mole. The answer is properly limited to three significant digits and is close to an estimated value of 73, or $\frac{1}{300} \times 1000 \times 22$.

Practice Problems
1. What is the molar mass of a gas whose density at STP is 2.08 g/L? *(Ans.)* 46.6 g/mol
2. What is the molar mass of a gas if a 1.39-g sample of the gas has a volume of 375 mL at 22°C and 755 mm Hg? *(Ans.)* 90.2 g/mol

1. At constant temperature and pressure, the volumes of gaseous reactants and products can be expressed as ratios of small whole numbers.
2. Equal volumes of gases at the same temperature and pressure contain equal numbers of molecules. This led to the discovery of diatomic molecules and allowed balanced equations with gaseous reactants and products to be interpreted in terms of molecules, moles, and volumes.

SECTION OBJECTIVES

- State the ideal gas law.
- Derive the ideal gas constant and discuss its units.
- Using the ideal gas law, calculate one of the quantities—pressure, volume, temperature, amount of gas—when the other three are known.
- Using the ideal gas law, calculate the molar mass or density of a gas.
- Reduce the ideal gas law to Boyle's law, Charles' law, and Avogadro's principle. Describe the conditions under which each applies.

Section Review

1. Explain Gay-Lussac's law of combining volumes.
2. State Avogadro's principle and explain its significance.
3. Define molar volume.
4. How many moles of oxygen gas are there in 135 L of oxygen at STP?
5. What volume (in mL) at STP will be occupied by 0.0035 mol of methane (CH_4)?
6. What is the density of nitrogen gas in grams per liter at STP?
7. What is the molar mass of a gas if 0.445 g of the gas occupies a volume of 2.80 L at 25°C and 740 mm Hg pressure?

12.2 The Ideal Gas Law

You learned in Section 11.2 that the values of four quantities—pressure, volume, temperature, and the number of moles of a gas—are needed to completely describe a gas sample. A single equation relating pressure, volume, temperature, and number of moles for a given sample of gas under any combination of conditions can be derived mathematically. *The* **ideal gas law** *is a relationship between pressure, volume, temperature, and the number of moles of a gas.*

Derivation of the Ideal Gas Law

Combining Boyle's law, Charles' law, and Avogadro's principle into a single expression gives a general equation which relates the four measurable quantities: pressure, volume, temperature, and moles of gas present. A general equation describing this expression can be obtained by considering the individual gas laws. This general equation can be used to calculate unknown information about gas samples.

3. The volume occupied by one mole of a gas at STP, or 22.4 L
4. 6.03 mol
5. 78 mL
6. 1.25 g/L
7. 3.99 g/mol

Boyle's law: At constant temperature, the volume of a given mass of gas is inversely proportional to the pressure:

$$V \propto \frac{1}{P}$$

Charles' law: At constant pressure the volume of a given mass of gas is directly proportional to the Kelvin temperature:

$$V \propto T$$

Avogadro's principle: At constant pressure and temperature, the volume of a given mass of gas is directly proportional to the number of moles:

$$V \propto n$$

Because a quantity that is proportional to each of three separate quantities is also proportional to their product, combining the three gas laws above gives:

$$V \propto \frac{1}{P} \times T \times n$$

Mathematically, we can change a proportionality to an equality by introducing a constant, k. Thus we have:

$$V = k \times \frac{1}{P} \times T \times n$$

The constant, k, has the same value for all gases whose behavior approaches that of an ideal gas. If we substitute R for k and rearrange the variables, we obtain the equation for the ideal gas law:

$$V = \frac{nRT}{P} \text{ or } PV = nRT$$

The ideal gas law equation states that the volume of a gas varies directly with the number of moles or molecules of a gas and its Kelvin temperature and varies inversely with its pressure. Under ordinary conditions, most gases exhibit behavior that is close to ideal and the equation can be applied with reasonable accuracy.

A close look will reveal that the ideal gas law reduces to Boyle's law, Charles' law, Gay-Lussac's law, or Avogadro's principle when the appropriate variables are held constant. For example, if n and T are constant, the product nRT is constant because R is also constant. In this case, the ideal gas law reduces to $PV = k$, which is Boyle's law.

The Ideal Gas Constant

In the ideal gas equation, *the constant R is known as the* **ideal gas constant.** Its value depends upon the units chosen for pressure, volume and temperature. Measured values of P, V, T, and n for a gas at conditions that are close to ideal can be used to calculate R. Recall from Section 12.1 that the volume of one mole of an ideal gas at STP (1 atm and 273.15K) is 22.414 10 L. Substituting these values and solving the ideal gas law equation for R gives

$$R = \frac{PV}{nT} = \frac{1 \text{ atm} \times 22.414\ 10 \text{ L}}{1 \text{ mol} \times 273.15\text{K}} = 0.082\ 0574\ \frac{\text{L} \cdot \text{atm}}{\text{mol} \cdot \text{K}}$$

This calculated value of R, usually rounded to 0.0821 L·atm/(mol·K), can be used in ideal gas law calculations only when the volume is given in liters, the

Figure 12-5 The ideal gas law demonstrates the interrelationship between the pressure, volume, number of molecules, and temperature of an ideal gas.

Intermediate steps in the derivation

$V = k_1/P \times k_2T \times k_3n = k_1k_2k_3 \times 1/P \times T \times n$

if $R = k_1k_2k_3$, then
$V = R \times 1/P \times T \times n$, or $PV = nRT$

Other values for R are:

R = 8.31 joules/(mol·K) = 62.4 liter mm Hg/(mol·K) = 1.99 calories/(mol·K) = 8.31 volt coulombs/(mol·K)

Molecular Composition of Gases **351**

TABLE 12-1 NUMERICAL VALUES OF THE GAS CONSTANT, R

Units of R	Numerical value of R	Units of P	Units of V	Units of T	Units of n
$\dfrac{L \cdot mm\,Hg}{mol \cdot K}$	62.4	mm Hg	L	K	mol
$\dfrac{L \cdot atm}{mol \cdot K}$	0.082 0574	atm	L	K	mol
$\dfrac{J}{mol \cdot K}$	8.314[a]	atm	L	K	mol
$\dfrac{L \cdot kPa}{mol \cdot K}$	8.314[b]	kPa	L	K	mol

$$R = \dfrac{PV}{nT}$$

(a) 1 L·atm = 101.325 J
(b) SI units

pressure is given in atmospheres, and the temperature is given in kelvins. The value of R depends on units chosen for n, P, V, and T. See Table 12-1.

Using the Ideal Gas Law

Finding P, V, T, or n from the ideal gas law The ideal gas law applies to situations in which a single gas sample undergoes no change in conditions and values are known for three of the four variables—P, V, T, or n. CAUTION: Always be sure to match the units of the known quantities and the units of R. We shall be using $R = 0.0821$ L·atm/(mol·K). If necessary, volumes must be converted to liters, pressures to atmospheres, temperature to kelvins, and masses to numbers of moles before using the ideal gas law.

Sample Problem 12.7

What is the pressure in atmospheres exerted by a 0.500-mol sample of nitrogen in a 10.0-L container at 298K?

Solution

Step 1. Analyze Given: $V = 10.0$ L; $n = 0.500$ mol; $T = 298$K
Unknown: P

Step 2. Plan $n, V, T \rightarrow P$

Because the gas sample undergoes no changes in conditions, the ideal gas law can be rearranged and used to find the pressure:

$$P = \dfrac{nRT}{V}$$

Step 3. Compute $P = \dfrac{0.500 \text{ mol} \times 0.0821 \text{ L·atm/(mol·K)} \times 298\text{K}}{10.0 \text{ L}} = 1.22 \text{ atm}$

Sample Problem 12.7 *continued*

Step 4. Evaluate All units cancel correctly to give the result in atmospheres. The answer is properly limited to three significant figures and is close to an estimated value of 1.5 (computed as $\frac{0.5 \times 0.1 \times 300}{10}$).

Sample Problem 12.8

What is the volume in liters occupied by 0.250 mol of oxygen at 20.0°C and 740. mm Hg pressure?

Solution

Step 1. Analyze Given: $P = 740.$ mm Hg; $n = 0.250$ mol; $t(°C) = 20.0°C$
Unknown: V

Step 2. Plan $P, n, T \rightarrow V$

The ideal gas law may be rearranged to solve for V for this oxygen sample that undergoes no change in conditions.

$$V = \frac{nRT}{P}$$

To use our chosen value of R [0.0821 L·atm/(mol·K)] in the equation, the units of P and $t(°C)$ must first be converted into atmospheres and kelvins.

$$P = \text{mm Hg} \times \frac{1 \text{ atm}}{760 \text{ mm Hg}}; \quad T(K) = t(°C) + 273$$

Step 3. Compute
$$P = 740. \text{ mm Hg} \times \frac{1 \text{ atm}}{760 \text{ mm Hg}} = 0.974 \text{ atm}$$

$$T(K) = 20.0°C + 273 = 293K$$

$$V = \frac{0.250 \text{ mol} \times 0.0821 \text{ L·atm/(mol·K)} \times 293K}{0.974 \text{ atm}} = 6.17 \text{ L}$$

Step 4. Evaluate Units cancel to give liters, as desired. The answer is correctly limited to three significant digits and is close to an estimated value of 6 (calculated as $0.2 \times 0.1 \times 300$).

Sample Problem 12.9

What mass of chlorine (Cl_2), in grams, is contained in a 10.0-L tank at 27°C and 3.50 atm of pressure?

Solution

Step 1. Analyze Given: identity of gas, Cl_2
$P = 3.50$ atm; $V = 10.0$ L; $t(°C) = 27°C$
Unknown: mass of Cl_2

Sample Problem 12.9 *continued*

Step 2. Plan

$P, V, T \rightarrow$ moles of Cl_2 (n) $\rightarrow$ mass of Cl_2 in grams

The ideal gas law may be rearranged to solve for n after temperature is converted to kelvins. The number of moles is then converted to grams.

$$n = \frac{PV}{RT}; m \text{ (in g)} = n \times \frac{g}{mol}$$

$$T(K) = 27.0°C + 273 = 300.K$$

Step 3. Compute

$$n = \frac{3.50 \text{ atm} \times 10.0 \text{ L}}{0.0821 \text{ L} \cdot \text{atm/(mol} \cdot \text{K)} \times 300K} = 1.42 \text{ mol}$$

$$\text{mass of } Cl_2 = 1.42 \text{ mol} \times \frac{70.9 \text{ g } Cl_2}{\text{mol}} = 101 \text{ g } Cl_2$$

Step 4. Evaluate In each calculation, units cancel to leave those that are required. The result of each calculation is correctly given to three significant figures. The answers are close to the estimated values.

Practice Problems
1. What pressure, in atmospheres, is exerted by 0.325 mol of hydrogen gas (H_2) in a 4.08-L container at 35°C? *(Ans.)* 2.01 atm
2. What is the mass, in grams, of oxygen (O_2) in a 12.5-L container at 45°C and 7.22 atm? *(Ans.)* 111 g

Density is explained in Section 2.1.

Finding molar mass or density from the ideal gas law You can also use the ideal gas law to calculate the molar mass or density of a gas sample. Suppose that the pressure, volume, and temperature are known for a gas sample of a given mass. The number of moles (n) in the sample can be calculated using the ideal gas law. Then the molar mass (grams per mole) can be calculated by dividing the known mass by the number of moles.

An equation showing the relationship between and among density, pressure, temperature, and molar mass can be derived from the ideal gas law. The number of moles (n) is equal to mass (m) divided by molar mass (M). Substituting m/M for n into $PV = nRT$ gives

$$PV = \frac{mRT}{M} \text{ or } M = \frac{mRT}{PV}$$

Density (d) is mass (m) per unit volume (V). Writing this definition in the form of an equation gives $d = \frac{m}{V}$. You can see that $\frac{m}{V}$ appears in the equation on the right just above. Introducing density into that equation gives

$$M = \frac{mRT}{PV} = \frac{dRT}{P}$$

Solving for density gives

$$d = \frac{MP}{RT}$$

You can see that the density of a gas varies directly with molar mass and pressure and inversely with Kelvin temperature.

Sample Problem 12.10

At 28°C and 740. mm Hg pressure, 1.00 L of an unidentified gas has a mass of 5.16 g. What is the molar mass of this gas?

Solution

Step 1. Analyze Given: $P = 740.$ mm Hg; $V = 1.00$ L; $t(°C) = 28°C$; $m = 5.16$ g
Unknown: molar mass

Step 2. Plan $P, V, T \rightarrow n \rightarrow M$

Pressure and temperature must be converted to atmospheres and kelvins, respectively. The number of moles equivalent to 5.16 g of the unknown gas is found from the rearranged ideal gas law:

$$n = \frac{PV}{RT}$$

The molar mass (mass of one mole of gas) is then found by dividing the known mass by the number of moles calculated.

$$M = \frac{m \text{ (in g)}}{n \text{ (number of moles)}}$$

Step 3. Compute

$$P = 740. \text{ mm Hg} \times \frac{1 \text{ atm}}{760 \text{ mm Hg}} = 0.974 \text{ atm}$$

$$T(K) = 28°C + 273 = 301 K$$

$$n = \frac{0.974 \text{ atm} \times 1.00 \text{ L}}{0.0821 \text{ L} \cdot \text{atm}/(\text{mol} \cdot \text{K}) \times 301 K} = 0.0394 \text{ mol}$$

$$M = \frac{5.16 \text{ g}}{0.0394 \text{ mol}} = 131 \text{ g/mol}$$

Step 4. Evaluate Units cancel as needed. The answer is correctly given to three significant digits and is close to an estimated value of 125 (calculated as $5 \div 0.04$).

Practice Problems

1. What is the molar mass of a gas if 0.427 g of the gas occupies a volume of 125 mL at 20.0°C and 745 mm Hg pressure? *(Ans.)* 83.8 g/mol
2. What is the density of a sample of ammonia gas (NH_3) if the pressure is 705 mm Hg and the temperature is 63.0°C? *(Ans.)* 0.572 g/L

Section Review

1. What is the volume in liters of 2.00 g of CS_2 vapor at 276 mm Hg pressure and 70.0°C?
2. Why must the units P, V, T, and n be appropriate to those of the ideal gas constant in solving problems? What are the units for R if P is in pascals, T is in Kelvins, V is in liters, and n is in moles?
3. What is the molar mass of 1.25-g sample of gas that occupies a volume of 1.00 L at a pressure 730. mm Hg and a temperature of 27.0°C?

1. 1.9 L
2. If the units of P, V, T and n are different from those in which R is given, the units will not cancel in the program. The units are L·pascals/K·mole.
3. 32.1 g/mole

12.3 Stoichiometry of Gases

SECTION OBJECTIVES

- Explain how Gay-Lussac's law and Avogadro's principle apply to the volumes of gases in chemical reactions
- Use a chemical equation to specify volume ratios for gaseous reactants and/or products
- Use volume ratios, standard molar volume, and the gas laws where appropriate to calculate volumes, masses, or molar amounts of reactants or products in reactions involving gases

In Chapter 9, you learned the relationships between moles and masses of reactants and products in chemical reactions. You saw that the numerical coefficients in a balanced chemical equation reveal the relative numbers of moles of any two substances involved in a chemical reaction.

Frequently, one or more reactants or products are gases. When the information known about gases in chemical reactions includes volumes, pressures, and temperatures rather than masses and molar amounts, the combined gas law, the standard molar volume, or the ideal gas law can be used to solve stoichiometric problems involving chemical reactions.

The discoveries of Gay-Lussac and Avogadro described in Section 12.1 can be applied in the stoichiometry of reactions involving gases. For reactants or products that are gases, the coefficients in chemical equations indicate volumes and volume ratios in addition to molar amounts and mole ratios. For example, consider the reaction of carbon monoxide with oxygen to give carbon dioxide:

$$2CO(g) + O_2(g) \rightarrow 2CO_2(g)$$

2 molecules	1 molecule	2 molecules
2 mol	1 mol	2 mol
2 volumes	1 volume	2 volumes

The concept of stoichiometry is introduced in Section 9.1.

The possible volume ratios can be expressed as:

$$(1) \; \frac{2 \text{ vol CO}}{1 \text{ vol } O_2} \text{ or } \frac{1 \text{ vol } O_2}{2 \text{ vol CO}} \quad (2) \; \frac{2 \text{ vol CO}}{2 \text{ vol } CO_2} \text{ or } \frac{2 \text{ vol } CO_2}{2 \text{ vol CO}} \quad (3) \; \frac{1 \text{ vol } O_2}{2 \text{ vol } CO_2} \text{ or } \frac{2 \text{ vol } CO_2}{1 \text{ vol } O_2}$$

Volumes can, of course, be compared in this way only if each is measured at the same temperature and pressure.

Volume–Volume Calculations

Volume–volume problems are simple to solve. When the volume of a gas involved in a chemical reaction is known and the volume of another reactant or product must be found, volume ratios like those given above are used exactly like mole ratios. The route for solving the problem is

$$\text{gas volume A} \rightarrow \text{gas volume B}$$

The problem setup is similar to that of the mole-mole calculations given in Section 9.2.

Sample Problem 12.11

The complete combustion of propane (C_3H_8), a gas that is in tanks and used for cooking and heating in some areas (see Figure 12-6), occurs according to the following equation:

$$C_3H_8(g) + 5O_2(g) \rightarrow 3CO_2(g) + 4H_2O(g)$$

Assuming all volume measurements are made at the same temperature and pressure, (*a*) What will be the volume in liters of oxygen required for the complete combustion of 0.350 L of propane; and (*b*) What will be the volume of carbon dioxide produced in the reaction?

Sample Problem 12.11 *continued*

Solution

Step 1. Analyze Given: balanced chemical equation
volume of propane = 0.350 L
Unknowns: *(a)* volume of $O_2(g)$ *(b)* volume of $CO_2(g)$

Step 2. Plan *(a)* vol $C_3H_8 \rightarrow$ vol O_2 *(b)* vol $C_3H_8 \rightarrow$ vol CO_2

Because all volumes are to be compared at the same temperature and pressure, volume ratios can be used like mole ratios to find the unknowns.

Step 3. Compute *(a)* $0.350 \text{ L } C_3H_8 \times \dfrac{5 \text{ L } O_2}{1 \text{ L } C_3H_8} = 1.75 \text{ L } O_2$

(b) $0.350 \text{ L } C_3H_8 \times \dfrac{3 \text{ L } CO_2}{1 \text{ L } C_3H_8} = 1.05 \text{ L } CO_2$

Step 4. Evaluate Each result is correctly given to three significant figures. The answers are reasonably close to estimated values of 2 (calculated as 0.4×5), and 1.2 (calculated as 0.4×3), respectively.

Practice Problem
Assuming all volume measurements are made at the same temperature and pressure, what volume of hydrogen gas is needed to react completely with 4.55 L of oxygen gas to produce water vapor? *(Ans.)* 9.10 L

Volume–Mass and Mass–Volume Calculations

In many cases, stoichiometric calculations will involve both masses and gas volumes. Sometimes the volume of a reactant or product is given and the mass of a second gaseous substance is unknown. On the other hand, a mass amount may be known and a volume may be the unknown. The calculations require routes such as

gas volume A $\rightarrow$ moles A $\rightarrow$ moles B $\rightarrow$ mass B

or

mass A $\rightarrow$ moles A $\rightarrow$ moles B $\rightarrow$ gas volume B.

To find the unknown information in cases such as these, one must know the conditions under which both the known and unknown gas volumes have been measured. At STP, the conversion factors 1 mol/22.4 L and 22.4 L/mol can be used to convert volume to moles and moles to volume, respectively.

The ideal gas law is very useful for calculating values at nonstandard conditions. For example, the number of moles of gas together with the known pressure and temperature conditions can be substituted into the ideal gas law and the gas volume calculated directly. If the volume of a gas under nonstandard conditions of temperature and pressure is known, the ideal gas law can be used to calculate the number of moles of the gas. Sample Problem 12.12 is a volume–mass calculation that illustrates use of molar volume as a conversion factor. Sample Problem 12.13 is a mass–volume calculation in

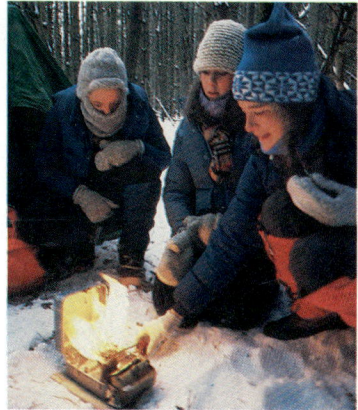

Figure 12-6 Teenage campers in Fairlee, Vermont using a portable propane grill. Propane is often used at trailer parks and at campsites for cooking, water heating, space heating, and refrigeration.

which the ideal gas law is used to find the unknown volume of a gas at nonstandard conditions.

Sample Problem 12.12

Calcium carbonate ($CaCO_3$), which is also known as limestone, is calcined to produce calcium oxide (lime), an industrial chemical with a wide variety of uses ranging from glass manufacture to sewage treatment. The balanced equation for the calcination is

$$CaCO_3(s) \rightarrow CaO(s) + CO_2(g)$$

How many grams of calcium carbonate must be decomposed to produce 5.00 L of carbon dioxide at STP?

Solution

Step 1. Analyze Given: Balanced chemical equation
desired volume of $CO_2(g)$ produced = 5.00 L (STP)
Unknown: mass of $CaCO_3(s)$ in grams

Step 2. Plan vol $CO_2 \rightarrow$ moles $CO_2 \rightarrow$ moles $CaCO_3 \rightarrow$ mass $CaCO_3$

The known volume is given at STP. Standard molar volume provides the necessary factor to convert the known volume to moles for use with mole ratios from the balanced equation. (Note that volume ratios do not apply here because calcium carbonate is a solid.)

$$\text{L } CO_2 \times \frac{1 \text{ mol } CO_2}{22.4 \text{ L } CO_2} \times \frac{1 \text{ mol } CaCO_3}{1 \text{ mol } CO_2} \times \frac{\text{g } CaCO_3}{1 \text{ mol } CaCO_3}$$

Step 3. Compute $5.00 \text{ L } CO_2 \times \dfrac{1 \text{ mol } CO_2}{22.4 \text{ L } CO_2} \times \dfrac{1 \text{ mol } CaCO_3}{1 \text{ mol } CO_2} \times \dfrac{100. \text{ g } CaCO_3}{\text{mol } CaCO_3} = 22.3 \text{ g } CaCO_3$

Step 4. Evaluate Units all cancel correctly. The answer is properly given to three significant figures and is close to an estimated value of 25 (computed as $5 \times \frac{1}{20} \times 100$).

Sample Problem 12.13

Tungsten, a metal that is used in light bulb filaments (Fig. 12.7), is produced industrially by the following reaction of tungsten oxide with hydrogen:

$$WO_3(s) + 3H_2(g) \rightarrow W(s) + 3H_2O(\ell)$$

How many liters of hydrogen at 35°C and 745 mm Hg are needed to react completely with 875 g of tungsten oxide?

Solution

Step 1. Analyze Given: balanced chemical equation
reactant mass = 875 g $WO_3(s)$
conditions for gaseous reactant: P = 745 mm Hg and $t(°C)$ = 35°C
Unknown: volume of H_2 at known non-standard conditions

Sample Problem 12.13 *continued*

Step 2. Plan mass WO_3 → moles WO_3 → moles H_2 → volume H_2

Moles of H_2 are found by converting mass of WO_3 to moles and then using the mole ratio. Because the conditions for the unknown volume are not STP, the ideal gas law is used to find the volume from the calculated number of moles of H_2. Pressure and temperature must therefore be converted first to atmospheres and kelvins in the usual manner.

Step 3. Compute

$$875 \text{ g } WO_3 \times \frac{1 \text{ mol } WO_3}{232 \text{ g } WO_3} \times \frac{3 \text{ mol } H_2}{1 \text{ mol } WO_3} = 11.3 \text{ mol } H_2$$

$$P = 745 \text{ mm Hg} \times \frac{1 \text{ atm}}{760 \text{ mm Hg}} = 0.980 \text{ atm}$$

$$T(K) = 35°C + 273 = 308 K$$

$$V = \frac{nRT}{P} = \frac{11.3 \text{ mol} \times 0.0821 \text{ L·atm/(mol·K)} \times 308 K}{0.980 \text{ atm}}$$

$$= 292 \text{ L}$$

Step 4. Evaluate Unit cancellations are correct in each calculation, as is the use of three significant figures for each result. The answer is reasonably close to an estimated value of 330 (computed as $\frac{11 \times 0.1 \times 300}{1}$).

Practice Problems

1. What mass of sulfur must be used to produce 12.61 L of sulfur dioxide at STP according to the equation

$$S(s) + O_2(g) \rightarrow SO_2(g) \quad (Ans.) \text{ 18.0 g}$$

2. How many liters of carbon monoxide at 27°C and 188 mm Hg can be produced from the burning of 65.5 g of carbon according to the equation

$$2C(s) + O_2(g) \rightarrow 2CO(g) \quad (Ans.) \text{ 544 L}$$

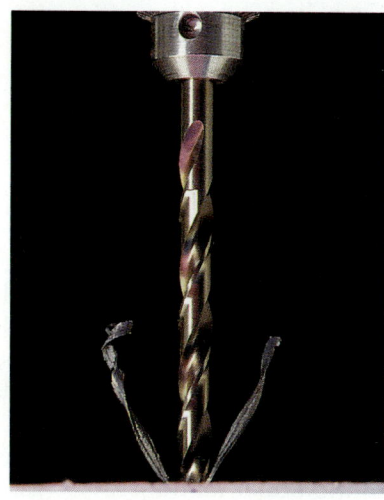

Figure 12-7 The exceptional strength of tungsten at very high temperatures makes it well-suited for use in light-bulb filaments and other industrial applications, such as high-speed drills. The chemical symbol for tungsten, **W**, comes from wolfram, the German name for this element.

1. 100. L NH₃
2. 2.24 L H₂
3. 0.821 g Na
4. V = 9.25 L O₂

Section Review

1. How many liters of ammonia gas can be formed from the reaction of 150. L of hydrogen gas, assuming that there is complete reaction of hydrogen with excess nitrogen gas, and that all measurements are made at the same temperature and pressure?
2. How many liters of H_2 at STP can be produced by the reaction of 4.60 g of Na and excess water?
3. How many grams of Na are needed to react with H_2O to liberate 400. mL of H_2 at STP?
4. What volume of oxygen in liters can be collected at 750. mm Hg pressure and 25.0°C when 30.6 g of $KClO_3$ decomposes by heating?

12.4 Effusion and Diffusion

The constant motion of gas molecules causes them to spread out to fill any container in which they are placed. The gradual mixing of two gases due to their spontaneous, random motion is known as *diffusion*. When a gas is confined in a container that has a very small opening, its molecules will randomly encounter the opening and pass through it in a process called *effusion* (Figure 12-8). In this section, you will learn how a chemical law describing diffusion and effusion can be used to discover the molar mass of an unknown gas.

SECTION OBJECTIVES

- State Graham's law of effusion or diffusion.
- Determine the relative rates of effusion of two gases of known molar masses.

Figure 12-8 Molecules of two gases diffuse into each other after the barrier has been lifted in the box at the left. Gas molecules exit the box on the right by means of effusion.

Kinetic energy and temperature are discussed in Section 11.1.

See Section 11.1 for a discussion of molecular motion.

Graham's Law of Effusion or Diffusion

The rates of effusion and diffusion are governed by the velocities of gas molecules. The velocity of a gas varies inversely with its mass. Lighter molecules move faster than heavier molecules at the same temperature. An experiment demonstrating velocities of the different gases is shown in Figure 12-9.

Recall that the average kinetic energy of the molecules in any gas is determined only by the temperature and is given by $\frac{1}{2}mv^2$. For two different gases, A and B, at the same temperature

Desktop Investigation

Diffusion

Question
How rapidly do gases of different molecular mass diffuse?

Materials
household ammonia
perfume or cologne
two small, wide-mouthed plastic containers with lids (such as plastic food-storage containers)
clock or watch with second hand

Procedure
Record all of your results in a data table.

1. Outdoors or in a room separate from the one in which you will carry out the rest of the investigation, pour a small amount (roughly a tablespoon) of ammonia water into a plastic container. Place the lid onto the container. Pour roughly the same amount of perfume or cologne into a second container. Cover them.

2. Take the two samples you have prepared into a large, draft-free large room. Place the two about 12 to 15 feet apart and at the same height. Remove the lids at as close to the same time as you can, and immediately position yourself midway between the open containers as shown above.

3. Note whether you smell the ammonia or the perfume first. Record how long this takes. Also, record how long it takes the vapor of the other substance to reach you. Air the room after you have finished.

Discussion
1. What do the times that the two vapors took to reach you show about their diffusion rates?

2. The molar mass of ammonia is about one-tenth the molar mass of most of the fragrant substances in the perfume or cologne. What does this tell you about the qualitative relationship between molar mass and diffusion rate?

$$\frac{1}{2} M_A v_A^2 = \frac{1}{2} M_B v_B^2$$

where M_A and M_B represent the molar masses of gases A and B and v_A and v_B represent their molecular velocities. Multiplying the equation by 2 gives

$$M_A v_A^2 = M_B v_B^2$$

To compare the velocities of the two gases, the equation is first rearranged to give the velocities as a ratio:

Figure 12-9 The rate of diffusion of hydrogen is greater than that of an average air molecule.

$$\frac{v_A^2}{v_B^2} = \frac{M_B}{M_A}$$

The square root of each side of the equation is then taken to give

$$\frac{v_A}{v_B} = \frac{\sqrt{M_B}}{\sqrt{M_A}}$$

This equation shows that the molecular velocities of two different gases are inversely proportional to the square roots of their molar masses. Because the rates of effusion are directly proportional to molecular velocities, we can express the equation as

$$\frac{\text{rate of effusion of A}}{\text{rate of effusion of B}} = \frac{\sqrt{M_B}}{\sqrt{M_A}}$$

The Scottish chemist Thomas Graham (1805–1869) performed experiments in which he studied the effusion and diffusion of gases. The equation derived above is a mathematical statement of Graham's discoveries and describes the rates of both effusion and diffusion. **Graham's law of effusion (or diffusion)** *states that the rates of effusion (or diffusion) of gases at the same temperature and pressure are inversely proportional to the square roots of their molar masses.*

Applications of Graham's Law

Graham's experiments dealt with the densities of gases. Because the density of a gas varies directly with its molar mass, it should be clear to you that the square roots of the molar masses in the equation above can be replaced by the square roots of the gas densities. For the rates of effusion of two gases, A and B, the mathematical relationships are

$$\frac{\text{rate of effusion}_A}{\text{rate of effusion}_B} = \frac{\sqrt{M_B}}{\sqrt{M_A}} = \frac{\sqrt{\text{density}_B}}{\sqrt{\text{density}_A}}$$

The mathematical expression for Graham's law can be used in a number of variations, as shown in Table 12-2. In every case, the ratio of some property

---

TABLE 12-2 GRAHAM'S LAW EQUATION

$$\frac{\text{rate of effusion of A}}{\text{rate of effusion of B}} = \frac{\sqrt{M_B}}{\sqrt{M_A}} \qquad \frac{\text{distance traveled by A}}{\text{distance traveled by B}} = \frac{\sqrt{M_B}}{\sqrt{M_A}}$$

$$\frac{\text{molecular velocity of A}}{\text{molecular velocity of B}} = \frac{\sqrt{M_B}}{\sqrt{M_A}} \qquad \frac{\text{rate of effusion of A}}{\text{rate of effusion of B}} = \frac{\sqrt{\text{density}_B}}{\sqrt{\text{density}_A}}$$

$$\frac{\text{effusion time of A}}{\text{effusion time of B}} = \frac{\sqrt{M_A}}{\sqrt{M_B}}$$

Chemistry and Technology

Separating Isotopes

As you know from Graham's law, the rate of effusion or diffusion of a gas varies inversely with the square root of the molar mass. The English physicist Francis William Aston (1877–1945) was the first to use this fact to separate isotopes, or atoms of the same atomic number but different atomic mass. Aston placed naturally occurring neon gas, a mixture of mostly neon-20 and neon-22, into porous clay pipes. Atoms of the lower-mass isotope, neon-20, tended to effuse more readily and rapidly through the pores than did atoms of the higher-mass neon-22. The effused gas, which was richer in neon-20, was separated from the gas that remained in the tube. By repeating the process, increasingly pure samples of neon-20 and neon-22 were produced.

Later, this method was improved upon by use of a "cascade" arrangement. In this method, shown in the figure above, the gas is passed into a series of separate porous containers. The portions of the gas that have diffused to one end are passed on to containers farther along and undiffused portions are returned to containers farther back. The process is repeated over and over, and isotopes are progressively separated and concentrated.

This method is used to collect the isotope uranium-235, which is useful in nuclear-energy production. The U-235 makes up only about 0.7% of all uranium, the rest being mostly U-238. The uranium that is to undergo separation occurs in the form of a sample of uranium hexafluoride gas (UF_6), which contains both isotopes. The $^{235}UF_6$, being less massive than the $^{238}UF_6$, effuses more readily. After more than a thousand repetitions of the basic diffusion step, the lighter isotope can be separated out with a high degree of purity.

The diffusion process separates isotopes of many different elements. In most cases, isotopes of different radioactivity are separated for use in nuclear energy, health sciences, geology, and analytical chemistry.

Molecular Composition of Gases 363

(for example, rate of effusion, molecular velocity, and so on) is equal to the square root of a ratio of the molar masses or densities.

The experiment illustrated in Figure 12-10 demonstrates the relationship between distance traveled and molar mass. When ammonia (NH_3) molecules and hydrogen chloride (HCl) molecules diffuse toward each other from opposite ends of a glass tube as shown, a white ring will form at the point where the two gases meet and a chemical reaction takes place. The white ring is solid ammonium chloride (NH_4Cl) and it forms at the end of the tube closest to the HCl. The lighter NH_3 molecules (molar mass 17.03 g) diffuse faster than the heavier HCl molecules (molar mass 36.46 g), as shown by the longer distance traveled by the NH_3 molecules before the two gases meet.

Figure 12-10 Plugs of absorbant cotton moistened with ammonia (NH_3) and hydrogen chloride (HCl) have been placed at opposite ends of the glass tube several minutes earlier. Why does the white ring of solid ammonium chloride form at the end closest to the HCl?

The principle expressed by Graham's law and its variations has a number of applications. For example, differences in the rates of effusion of gases are used to separate light isotopes from heavy isotopes (as discussed in Technology, above). Graham's law also provides a method for determining molar masses. The rate of effusion of a gas of known molar mass can be compared to the effusion rate of an unknown gas at the same temperature and pressure. The unknown molar mass can then be calculated using Graham's law.

Sample Problem 12.14

Compare the rates of effusion of hydrogen and oxygen at the same temperature and pressure.

Solution

Step 1. Analyze   Given: identities of two gases, H_2 and O_2
Unknown: relative rates of effusion

Step 2. Plan   molar mass ratio → rate of effusion ratio

The ratio of the rates of effusion of two gases at the same temperature and pressure can be found from Graham's law.

$$\frac{\text{rate of effusion}_A}{\text{rate of effusion}_B} = \frac{\sqrt{M_B}}{\sqrt{M_A}}$$

Sample Problem 12.14 *continued*

Step 3. Compute
$$\frac{\text{rate of effusion of H}_2}{\text{rate of effusion of O}_2} = \frac{\sqrt{M_{O_2}}}{\sqrt{M_{H_2}}} = \frac{\sqrt{32}}{\sqrt{2}} = \sqrt{\frac{32}{2}} = \sqrt{\frac{16}{1}} = 4$$

Hydrogen diffuses 4 times faster than oxygen.

Step 4. Evaluate The result is correctly limited to one significant figure and is equivalent to an estimated value of 4 (calculate $\sqrt{32}/\sqrt{2}$).

Practice Problem
A sample of hydrogen effuses through a porous container 8.94 times faster than an unknown gas. What is the molar mass of the unknown gas? *(Ans.)* 161 g/mol

Section Review

1. Compare diffusion and effusion.
2. State Graham's law of effusion, and explain its meaning.
3. Calculate the molar mass of a gas that effuses at 1.59 times the effusion rate of carbon dioxide.

1. Effusion is the process by which gases confined in a container pass through a very small opening in the container. Diffusion is the process by which two gases mix together in a container.
2. Graham's law of effusion (or diffusion) states that the rates of effusion (or diffusion) of gases at the same temperature and pressure are inversely proportional to the square roots of their molar masses.
3. $M = 17.4$ g/mol

Chapter Summary

- Gay-Lussac's law of combining volumes of gases states that the volumes of reacting gases and their products can be expressed as ratios of small whole numbers when they are measured at the same temperature and pressure.
- Avogadro's principle states that equal volumes of gases at the same temperature and pressure contain equal numbers of molecules. At constant temperature and pressure, the volume of a gas is thus directly proportional to the number of moles of the gas.
- Gay-Lussac's law and Avogadro's principle can be used to show that molecules of the active elemental gases are diatomic.
- The volume occupied by one mole of any gas at STP is called the standard molar volume. The standard molar volume of all gases is experimentally determined to be 22.414 10 L at standard temperature and pressure.
- Charles' law, Boyle's law, and Avogadro's principle can be combined to create the ideal gas law, a useful equation that can be used to describe a gas sample under any combination of conditions.
- The ideal gas law is stated mathematically as $PV = nRT$. It can be used to calculate the pressure, volume, temperature, or number of moles of a gas sample when three of these four variables are known.
- The combined gas law, the standard molar volume, and the ideal gas law can be used to solve stoichiometry calculations involving gases. These calculations may be those in which the known and unknown quantities are both volumes and those in which the known and unknown quantities are volume and mass or molar amount.
- Graham's law of diffusion (or effusion) states that the rates of diffusion (or effusion) of gases at the same temperature and pressure are inversely proportional to their molar masses. Graham's law reflects the fact that less massive molecules diffuse or effuse faster than do more massive ones.

CHAPTER 12 REVIEW

Molecular Composition of Gases

VOCABULARY

Avogadro's principle (344)
Gay-Lussac's law of combining volumes of gases (344)
Graham's law of effusion (or diffusion) (362)
ideal gas constant (351)
ideal gas law (350)
standard molar volume of a gas (345)

REVIEWING CONCEPTS

1. a. What restrictions are associated with the use of Gay-Lussac's law of combining volumes? (12.1)
   b. At the same temperature and pressure, what is the relationship between the volume of a gas and the number of molecules present? (12.1)
2. According to Avogadro,
   a. what is the relationship between gas volume and the number of moles of gas at constant temperature and pressure?
   b. What is the mathematical expression denoting this relationship? (12.1)
3. What is the relationship between the number of molecules and the mass of 22.4 L of different gases at STP? (12.1)
4. When stating gas density values, why is it necessary that the corresponding temperature and pressure be specified? (12.1)
5. a. Write the equation for the ideal gas law.
   b. What relationship is expressed in the ideal gas law? (12.2)
6. a. In what situation does the ideal gas law apply?
   b. What precautions involving units must be taken in using this law? (12.2)
7. a. In a balanced chemical equation, what is the relationship between the molar ratios and the volume ratios of gaseous reactants and products?
   b. What restriction applies to the use of the volume ratios in solving stoichiometry problems? (12.3)
8. a. Distinguish between diffusion and effusion.
   b. At a given temperature what factor determines the rates at which different molecules undergo these processes? (12.4)

PROBLEMS

▶ Molar Volume and Gas Density

1. If a 5.00-L sample of O_2 at a given temperature and pressure contains 1.08×10^{23} molecules of O_2, how many molecules would be contained in each of the following at the same temperature and pressure?
   a. 5.00 L H_2 b. 5.00 L CO_2 c. 10.00 L NH_3
2. How many molecules are contained in each of the following?
   a. 1.00 mol O_2
   b. 2.50 mol He
   c. 0.0650 mol NH_3
   d. 11.5 g NO_2
3. Find the mass of each of the following:
   a. 2.25 mol Cl_2
   b. 3.01×10^{23} molecules H_2S
   c. 25.0 molecules SO_2
4. What is the volume, in liters, of each of the following at STP?
   a. 1.00 mol O_2
   b. 3.50 mol F_2
   c. 0.0400 mol CO_2
   d. 1.20×10^{-6} mol He
5. How many moles are contained in each of the following at STP?
   a. 22.4 L N_2
   b. 5.60 L Cl_2
   c. 0.125 L Ne
   d. 70.0 mL NH_3
6. Find the mass, in grams, of each of the following at STP:
   a. 11.2 L H_2
   b. 2.80 L CO_2
   c. 15.0 mL SO_2
   d. 3.40 cm^3 F_2
7. Find the volume, in liters, of each of the following at STP:
   a. 8.00 g O_2
   b. 3.50 g CO
   c. 0.0170 g H_2S
   d. 2.25×10^{-5} kg NH_3

8. Find the density of each of the following gases in grams per liter at STP.
   a. N_2 b. Cl_2 c. CO   d. SO_2

9. Using the masses and corresponding volumes of each of the gases listed that were collected at STP, find their molar masses.
   a. 2.00 g and 1.50 L
   b. 3.25 g and 2.30 L
   c. 0.620 g and 250. mL

10. A sample of a gas collected at 300. K and 740. mm Hg pressure has a mass of 0.205 g and occupies a volume of 250. mL.
    a. What volume would this gas occupy at STP?
    b. What is its molar mass?

11. A 6.51-m^3 container is filled with an unknown gas at 50.°C and 840. mm Hg pressure. The gas has a mass of 12.5 kg.
    a. Find its molar mass using the combined gas law.
    b. Among the gases listed below, which is likeliest to be the unknown gas — H_2, N_2, O_2, NH_3, SO_2, NO_2, CO_2, or H_2S?

12. Find the molar mass of a gas using the combined gas law if 850. mL of the gas collected over water at 650.0 mm Hg and 20.°C has a mass of 0.925 g.

13. A 500.-cm^3 flask contains 0.750 g of a gas collected over water at 35°C and 575.0 mm Hg. Use the combined gas law to find the molar mass of this gas.

▶ **Ideal Gas Law**

14. Calculate the pressure, in atmospheres, exerted by each of the following:
    a. 2.50 L of HF containing 1.35 mol at 320.K
    b. 4.75 L of NO_2 containing 0.86 mol at 300.K
    c. 750. mL of CO_2 containing 2.15 mol at 57°C

15. Calculate the volume, in liters, occupied by each of the following:
    a. 2.00 mol of H_2 at 300.K and 1.25 atm
    b. 0.425 mol of NH_3 at 37°C and 550. mm Hg
    c. 4.00 g of O_2 at 57°C and 675 mm Hg

16. Determine the number of moles of gas contained in each of the following.
    a. 1.25 L at 250.K and 1.06 atm
    b. 0.80 L at 27°C and 0.925 atm
    c. 750. mL at −50.°C and 700. mm Hg

17. Find the mass of each of the following.
    a. 5.60 L of O_2 at 1.75 atm and 250.K
    b. 3.50 L of NH_3 at 700. mm Hg and 27°C
    c. 125 mL of SO_2 at 625 mm Hg and −53°C

18. Find the molar mass of each gas measured at the specified conditions:
    a. 0.650 g occupies 1.12 L at 280.K and 1.14 atm
    b. 1.05 g occupies 2.35 L at 37°C and 0.840 atm
    c. 0.432 g occupies 750. mL at −23°C and 785 mm Hg

19. If the density of an unknown gas is 3.20 g/L at −18°C and 1650 mm Hg, what is the molar mass of this gas?

20. One method for estimating the temperature of the center of the sun is based on the assumption that the center consists of gases that have an average molar mass of 2.00 g/mol. If the density of the center is 1.40 g/cm^3 at a pressure of 1.30×10^9 atm, calculate its temperature in degrees Celsius.

▶ **Gas Stoichiometry**

21. Carbon monoxide reacts with oxygen to produce carbon dioxide. If 1.0 L of carbon monoxide reacts with oxygen,
    a. how many liters of oxygen are required?
    b. how many liters of carbon dioxide are produced?

22. Acetylene gas (C_2H_2) undergoes combustion to produce carbon dioxide and water vapor.
    a. How many liters of C_2H_2 are required to produce 75.0 L of CO_2?
    b. What volume of H_2O is produced?
    c. What volume of O_2 is required?

23. If liquid carbon disulfide reacts with 450. mL of oxygen to produce the gases carbon dioxide and sulfur dioxide, what volume of each product is produced?

24. Balance the chemical equation
    $Mg(s) + O_2(g) \rightarrow MgO(s)$ and then, based on the quantity of reactant or product given, de-

Molecular Composition of Gases

termine the corresponding quantities of the specified reactants or products, assuming that the system is at STP:
a. 22.4 L O_2 = ___ mol O_2 → ___ mol MgO
b. 11.2 L O_2 = ___ mol O_2 → ___ mol MgO
c. 1.40 L O_2 = ___ mol O_2 → ___ mol MgO

25. Assume that 5.60 L of $H_2(g)$ at STP react with CuO(s) according to the following reaction:
$CuO(s) + H_2(g) \rightarrow Cu(s) + H_2O(g)$
a. How many moles of $H_2(g)$ react?
b. How many moles of Cu(s) are produced?
c. How many grams of Cu(s) are produced?

26. Assume that 8.50 L of $I_2(g)$ are produced according to the following reaction that takes place at STP:
$2KI(aq) + Cl_2(g) \rightarrow 2KCl(aq) + I_2(g)$
a. How many moles of $I_2(g)$ are produced?
b. How many moles of KI(aq) were used?
c. How many grams of KI(aq) were used?

27. Solid iron(III) hydroxide decomposes to produce iron(III) oxide and water vapor. If 0.75 L of water vapor are produced at STP,
a. how many grams of iron(III) hydroxide were used?
b. how many grams of iron(III) oxide are produced?

28. If 650. mL of hydrogen gas is produced through a replacement reaction involving solid iron and sulfuric acid (H_2SO_4) at STP, how many grams of iron(II) sulfate are also produced?

29. Assume that 13.5 g of Al(s) react with HCl(aq) according to the following balanced equation at STP:
$2Al(s) + 6HCl(aq) \rightarrow 2AlCl_3(aq) + 3H_2(g)$
a. How many moles of Al react?
b. How many moles of H_2 are produced?
c. How many liters of H_2 at STP are produced?

30. If air is 20.9% oxygen by volume,
a. how many liters of air are needed to complete the combustion of 25.0 L of octane vapor (C_8H_{18})?
b. what volume of each product is produced?

31. Methanol (CH_3OH) is made by causing carbon monoxide and hydrogen gases to react at high temperature and pressure. If 450. mL of CO and 825 mL of H_2 are mixed,
a. which reactant is present in excess?
b. how much of that reactant remains?
c. what volume of CH_3OH is produced?

32. If 45.0 L of natural gas, which is essentially methane (CH_4), undergoes complete combustion at 730 mm Hg and 20°C, how many grams of each product are formed?

33. How many grams of aluminum metal must react with aqueous sodium hydroxide to produce 7.50 L of hydrogen collected over water at 12.5°C and 740.0 mm Hg if the other product is aqueous Na_3AlO_3?

34. A modified Haber process for making ammonia is conducted at 550.°C and 250. atm. If 10.0 kg of nitrogen are used, what volume of ammonia is produced?

35. When nitroglycerin, $C_3H_5(NO_3)_3$, explodes, the products are carbon dioxide, nitrogen, oxygen, and water vapor. If 500. g of nitroglycerin explodes at STP, what is the total volume, at STP, for all gases produced?

36. The principal source of sulfur is in the form of deposits of free sulfur occurring mainly in volcanically active regions. The sulfur was initially formed by the reaction between the two volcanic vapors SO_2 and H_2S to form $H_2O(\ell)$ and S(s). What volume of each gas, at 730. mm Hg and 22°C, was needed to form a sulfur deposit of 4.50×10^5 kg on the slopes of a volcano in Hawaii?

37. A 3.25-g sample of solid calcium carbide (CaC_2) reacts with water to produce acetylene gas (C_2H_2) and aqueous calcium hydroxide. If the acetylene was collected over water at 17°C and 740.0 mm Hg, how many milliliters of acetylene were produced?

▶ **Effusion and Diffusion**

38. Quantitatively compare the rates of effusion of the following pairs of gases at the same temperature and pressure:
a. hydrogen and nitrogen
b. fluorine and chlorine.

39. What is the ratio of the velocities of hydro-

gen molecules and neon atoms at the same temperature and pressure?

40. At a certain temperature and pressure, chlorine molecules have a velocity of 0.0380 m/s. What is the velocity of sulfur dioxide molecules under the same conditions?

41. A sample of helium effuses through a porous container 6.50 times faster than does unknown gas X. What is the molar mass of the unknown gas?

CRITICAL THINKING

1. **Relating Ideas** Explain the basis upon which Boyle's law, Charles' law, and Avogadro's principle can be combined to give the ideal gas law.
2. **Evaluating Methods** Explain the specific kind of problem situation in which each of the following would apply:
   a. the combined gas law
   b. the ideal gas law.
3. **Relating Ideas** Write the composite expression relating the rates of effusion, molar masses, and densities of two different gases, A and B.
4. **Evaluating Ideas** Pages 343 and 344 of the text state that Gay-Lussac's law of combining volumes holds true for relative volumes at any proportionate size. Explain how this proportionality exists using Avogadro's Principle.
5. **Designing Experiments** Design an experiment to prove that the proportionality described in item 4 exists.
6. **Interpreting Concepts** The diagrams that follow represent equal volumes of four different gases.

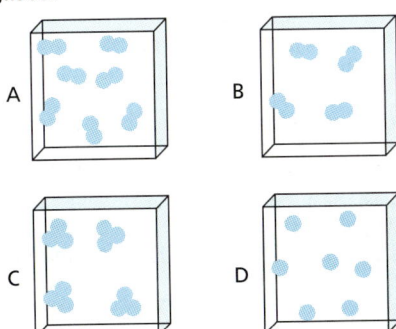

Use the diagrams to answer the following:
a. Are these gases at the same temperature and pressure? How do you know?
b. If the molecular mass of gas B is 38 g/mol and that of gas C is 46 g/mol, which gas is more dense?
c. To make the densities of gases B and C equal, which gas could expand in volume?
d. If the densities of gas A and C are equal, what is the relationship between their molecular masses?

7. **Interpreting Concepts** Which of the following graphs represents a constant density for a gas under conditions of increasing temperature?

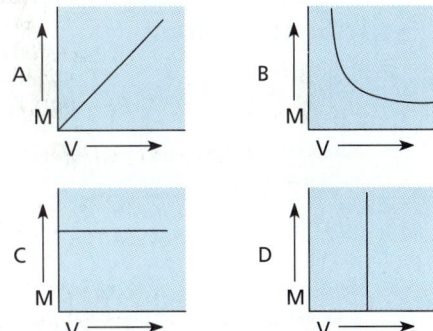

CROSS CONNECTIONS

1. **Sports** Research the causes of the bends. What precautions do SCUBA divers take to prevent the bends?

RESEARCH AND WRITING

1. Explain the processes involved in the liquefaction of gases. What substances that are gases under normal room conditions are typically used in the liquid form? Why?
2. Research the relationship between explosives and the establishment of Nobel Prizes. Prepare a report that describes your findings.

PROJECT

1. Over a typical day, record every instance in which you encounter the diffusion or effusion of gases (example: smelling perfume).

CHAPTER 13
Liquids and Solids

PLANNING CHART
Approximate Time: 5/6 days

	SECTION/TEXT FEATURES		RESOURCES	
B A S I C	**13.1 LIQUIDS** Desktop Investigation Section Review Chapter Review • Questions 1–3	371 374 376 395	***Teacher's Resource Binder*** Section Review Worksheets ***Other Resources*** *Shakhashiri* Video Demos 22, 35, 36	44
B A S I C	**13.2 SOLIDS** *Periodic Table* Videodisc Section Review Chapter Review • Questions 4–7	377 370G 382 395	***Teacher's Resource Binder*** Problem Solving Worksheets Section Review Worksheets ***Other Resources*** *Shakhashiri* Video Demo 10 *World of Chemistry* Video 5	45
B A S I C	**13.3 CHANGES OF STATE** Demos 1, 2 Microscale Investigation 10 Section Review Chapter Review • Questions 8–15, Problems 1–5	383 370B-C 854 390 395	***Teacher's Resource Binder*** Problem Solving Worksheets Lab Experiment 16 Section Review Worksheets ***Other Resources*** Transparencies 28, 29	99 46
B A S I C	**13.4 WATER** Section Review Chapter Review • Questions 16–17, Problems 6–8	390 393 395	***Teacher's Resource Binder*** Section Review Worksheets ***Other Resources*** *Shakhashiri* Video Demo 39 *World of Chemistry* Video 12	47
	CHAPTER END ACTIVITIES Critical Thinking, Cross Connections, Research and Writing, Projects		***Teacher's Resource Binder*** Test Booklet ***Computer Test Generator***	61, 63

Teaching Strategies

13.1 Liquids

Begin by reviewing the kinetic theory of gases. Summarize those aspects that change for liquids and solids. These latter two phases have molecules so close together that they are incompressible. Kinetic energy is restricted, and intermolecular forces become significant. These are the changes that give liquids and solids their properties.

You can illustrate the incompressibility of liquids by drawing 25 mL of water into a syringe and squeezing the plunger. (Make sure there is no air present, and do not squeeze too hard.) Explain that automobile brakes function because of the incompressibility of the brake fluid. When air is trapped in the line, the brake pedal goes to the floor and no pressure is applied to the brakes. One has to pump the brakes, reducing the volume of air in the line, before the brakes begin to hold.

To show evaporation and boiling (and to provide some data for a later discussion of phase changes) half fill a 250-mL beaker with warm water. Place a cold watch glass on the beaker. Some of the water vapor condenses. Remind students that evaporation is an endothermic process.

Draw some of the warm water into the 25-mL syringe. Make sure no air is present. Pull back on the syringe and the water boils. Do not go into detail at this time. Repeat the demonstration and have the students note that reducing the pressure by pulling back on the syringe causes the water to boil. You may have them touch the syringe to note that the water is not at 100°C, and yet it boiled.

Remind the students that all of these macroscopic properties are the result of interactions of molecules, which in turn can be traced back to the number and arrangement of valence electrons.

13.2 Solids

As you introduce solids, note that intermolecular forces are so strong that only vibrational kinetic energy is evident. Quickly move through each of the properties discussed in the text and contrast them with the same properties in the liquid state. Some students may have a difficult time distinguishing between crystalline and amorphous solids and supercooled liquids and solids. Have examples of each on the desk. Copper sulfate and alum crystals, as well as minerals such as galena, may be used as examples of crystalline solids. A plastic bottle is a good example of an amorphous solid.

Crystal models and styrofoam spheres are excellent illustrations of crystalline solids and the concept of unit cells. Instead of asking students to memorize specific crystal structures, stress the idea that the forces of attraction determine the type of crystal. The unit cell concept may be demonstrated with eight styrofoam balls in a simple cubic packing. With a sharp knife, cut off the parts that belong to a single cell.

Have examples of each crystal type available on the desk. If mineral samples of quartz, mica, and asbestos are available, you can explain the importance of three-dimensional, planar, and linear bonding in each. A candle may be used to illustrate the covalent molecular crystal.

Use models of nonpolar molecules (methane, hydrogen, iodine) and polar molecules (HCl, NH_3, H_2O) to explain the difference in properties. This is a good time to review intermolecular forces of attraction and remind students of the importance of hydrogen bonding.

13.3 Changes of State

To emphasize the equilibrium concept, use the analogy of a sports team. If basketball is used, one notes that during the course of a game there are always 10 players on the court and another fixed number of players on the bench. As the game progresses, players enter the court from the bench while at the same time an equal number of players leave the court and return to the bench. During the course of the game, the actual players on the court may not be the same ones who began the game, but the number on the court has not varied.

Football, soccer, and baseball also work. If students are familiar with ice hockey, players substitute on the fly with no break in the action. The sports team analogy emphasizes the idea that a system in equilibrium must be a closed system, i.e. the total number of particles must be constant.

Equilibrium as it applies to changes of state should be presented with the students giving much of the information while the teacher catalogs and makes inferences. An uncovered beaker half-filled with water serves as a good starting point. Students will observe that with time the level of the water drops. It evaporates. Particles of water leave the liquid phase and enter the gaseous phase. This can be diagramed on an overhead projector or chalkboard. If a cover is placed on the beaker, the particles that enter the gaseous phase remain in the beaker, colliding with the walls, and eventually return to the liquid phase. At some point, the number of particles leaving the surface and entering the gaseous state equals the number striking the surface and returning to the liquid state. Equilibrium has been established, but the process continues on the molecular level with no apparent external macroscopic change. Pose this question: "What would happen if the water were heated?" Most students would agree that more particles would evaporate. Relate this to the kinetic theory. Temperature is directly related to the kinetic energy of the particles. If the volume of the container were reduced, as in a syringe, students would agree that with less space some of the particles in the gaseous state would condense. Diagram this condition on the board.

Liquid and solid deodorizers may be used to reinforce the concept of vapor pressure. Kinetic theory explains what is happening on a molecular level. Even the solid **deodorizer** must have molecules escaping from the surface, since an odor can be detected. Use the graphs in the text to show the relationship between the observed physical phenomena and the graphical interpretation of the phenomena.

The following demonstration may help to bring all of these ideas together.

Demonstration 1: Phase Change

PURPOSE: To allow the students to examine a phase change from liquid to vapor

MATERIALS: Zip-lock sandwich bag, can of dry spray room deodorizer

PROCEDURE: Open the zip-lock bag. Place the nozzle of the spray can inside the bag. Closing the mouth of the bag around the can, hold down the nozzle for approximately three to five seconds and spray the deodorizer into the bag. The bag gets very cold. Seal the bag. You should have captured several milliliters of a very cold liquid in the bag.

Holding the corner of the bag that contains the liquid causes the liquid to boil vigorously. Several bags may be prepared and passed around the room.

As the liquid boils, the volume of the zip-lock bag increases. Pressure may build until the bag begins to leak. Your students should be able to observe the large difference between the volume of a particular mass of liquid and the volume of the same mass of vapor.

Condensation of water vapor from the air may be demonstrated by allowing the liquid in the bag to flow from one corner to the opposite corner. One should be able to observe a trail of ice on the outside of the bag that follows the path taken by the liquid butane.

Any dry spray deodorizer will work. With the removal of halogenated hydrocarbons (freons) as propellants because of their adverse effect on the ozone layer, manufacturers of aerosol sprays had to find another propellant. Most manufacturers are now using butane. Many of the labels do not list butane as an ingredient, but they do indicate that the product is flammable. Butane boils at 0.5°C. At room pressure, butane is a gas. It can be liquefied at room temperature by exerting a greater than atmospheric pressure on the gas. Once liquefied it will remain a liquid until the pressure is lowered: in this case, by pressing the nozzle and allowing some of the gas to escape. Since evaporation is an endothermic process, the container and the target toward which the spray is directed get very cold. When sprayed into the zip-lock bag, the temperature drops below −5°C, and the butane liquefies at room pressure.

The bag and liquid also may be used to qualitatively show the effect of the external pressure on the boiling point. As the bag fills with gas, the liquid does not boil as readily. Opening the bag and allowing the gas to escape reduces the pressure and the boiling point of the liquid. You may wish to make the demonstration quantitative by collecting some of the liquid in a syringe. You may more easily demonstrate the effect of pressure on the boiling point.

The demonstration is a simple one, but it fascinates most students. Be sure to warn them of the danger of spraying the contents near a flame or near electrical equipment. Directing the spray toward exposed skin may cause the equivalent of frostbite.

Dispose of the deodorizer according to the instructions on the can. The zip-locks bags may simply be opened and the fragrance allowed to escape into the room.

After working with the bags or syringes, the concept of molar heat of vaporization may be introduced. Refer once again to the difference between intensive and extensive properties. By agreeing on one mole of material, one obtains an intensive property that can be tabulated and made available to other chemists.

Experiment 5 in *Modern Chemistry Laboratory Experiments* discusses freezing and melting in terms of entropy. The students will not cover this concept until later in the course, but you may introduce the topic, using the following very simple demonstration.

Demonstration 2: Phase Changes in Ice and Water

PURPOSE: To introduce the concepts of heat of fusion and heat of vaporization and to observe the temperature behavior of a system undergoing phase changes

MATERIALS: A 250-mL beaker half-filled with an ice and water mixture, a thermometer, and a heat source

PROCEDURE: Stir the ice-water mixture with the thermometer, record the temperature and begin warming the beaker. The temperature remains constant even though heat is flowing into the beaker. Have the students explain what is occurring based on the kinetic theory. (The energy is being used to overcome the hydrogen bonds and other attractive forces present in the ice molecules.)

Once the ice has melted, the temperature begins to rise. Again have the students explain the temperature increase. (Now water molecules are beginning to show an increase in kinetic energy—vibration, rotation, translation—manifested by the temperature increase.)

At the boiling point, the thermometer again seems to malfunction because it no longer records a temperature increase even though heat flows into the system. Once again have the students describe what might be happening on the molecular level. (The energy this time is used to overcome the attractive forces between liquid molecules and to give them enough energy to enter the gaseous state.)

The students need to be aware that heat is being transferred in both sets of phase changes although the temperature does not change. These energy changes on the molecular level are what the text calls heat of fusion and heat of vaporization.

If a thermistor connected to a computer is available, the same demonstration may be carried out with the change in temperature shown on the computer monitor. Students are able to see in real time that the temperature no longer changes at each of the phase changes.

Observe the usual safety precautions when warming a liquid.

The concept of phase diagram, triple point, and critical temperature and pressure for water may now be introduced. The difference between the boiling point and normal boiling point needs to be clarified. The syringe half-filled with warm water may be used once again to show that water will boil under reduced pressure. Glass "hand boilers" or "love meters," which may be found in novelty stores and catalogs, are also good illustrations of this phenomenon. The normal boiling point is the temperature at which a liquid boils when the pressure is exactly one atmosphere.

13.4 Water

The closing section on water gives you an opportunity to review the concepts presented in this chapter, as well as concepts involving electronic structures, bonding, electronegativity, polarity of bonds, and molecular shapes.

Use molecular models to explain some of the unique properties of water: less dense in solid state than in liquid, a very high heat of vaporization, and a very large temperature range in which it is a liquid.

Encourage students to bring in newspaper and magazine articles that focus on the importance of water in our daily lives.

Answers and Solutions

Reviewing Concepts

1. A fluid is a substance that can flow and therefore conform to the outline of its container.
2. Surface tension is a force that tends to pull adjacent parts of a liquid's surface together, making the surface less penetrable by solid bodies.
3. Water that falls to the earth in the form of rain and snow originally evaporated from the earth's oceans, lakes, and rivers. Human perspiration cools our bodies when evaporated into the surrounding air.
4. • Solids can maintain a definite shape without a container. Crystalline solids are geometrically regular, whereas amorphous solids are not.
  • Solids have definite volumes that change only slightly with changes in temperature or pressure; this property is due to the close packing of the solid particles.
  • Solids are nonfluid because they are held in relatively fixed positions.
  • Whereas crystalline solids have definite melting points, some amorphous solids do not. Instead, as they are heated, amorphous solids such as glass and plastic gradually soften to become thick, sticky liquids. Solids melt at temperatures at which the kinetic energies of their particles can overcome the interactive forces between them.
  • Solids have high densities, due to the close packing of their particles.

- Solids are incompressible, due to the already close packing of their particles.
- Solids have a slow rate of diffusion, due to the low mobility of their particles.

5. A supercooled liquid is a substance that retains certain liquid properties even at temperatures at which it appears to be solid. Glass, since it softens to become a thick, sticky liquid upon being heated and because of its ability to flow, is sometimes classified as a supercooled liquid.

6. A crystal lattice is the total three-dimensional array of points that describes the arrangement of the particles of a crystal. A unit cell is the smallest portion of the crystal lattice that reveals the three-dimensional pattern of the entire lattice.
   a. Rubber, glass, plastics, synthetic fibers, and amorphous forms of metal alloys
   b. Answers will vary.

7. **a.** The ionic crystal lattice consists of positive and negative ions arranged in a characteristic regular pattern. In covalent network crystals, the lattice sites contain single atoms, each of which is covalently bonded to its nearest neighboring atoms. The metallic crystal lattice consists of positive metal cations surrounded by valence electrons that are donated by the metal atoms and belong to the crystal as a whole. The attraction between the metal cations and the electrons binds the crystal together. The crystal lattice of covalent molecular crystals consists of covalently bonded molecules held together by van der Waals forces. If these molecules are nonpolar, the bonding is because of weak dispersion forces; however, if these molecules are polar, they are held together by both dispersion forces and stronger dipole-dipole forces.
   b. Ionic crystals are hard and brittle, have high melting points, and are good insulators. Covalent network crystals are hard and brittle, have high melting points, and are nonconductors. Metallic crystals have high electric conductivity, but their melting points vary greatly. Covalent molecular crystals have low melting points, are relatively soft and easily vaporized, and are good insulators.

8. The indicated equilibrium vapor pressures are:
   a. 50 mm Hg   **d.** 375 mm Hg
   b. 375 mm Hg   **e.** 90 mm Hg
   c. 460 mm Hg

9. **a.** Sublimation is the change of state directly from a solid to a gas.
   b. Dry ice and iodine are examples.

10. The normal freezing point of a substance is the temperature at which the solid and liquid are in equilibrium at 1 atm pressure.

11. With increasing temperature, the rate of evaporation of a liquid increases, thus increasing the number of particles in the vapor phase. This disturbs the liquid-vapor equilibrium and causes an increase in the rate of condensation. Once the equilibrium is reestablished, however, a greater number of particles are in the vapor phase than there were before, thus increasing the vapor pressure.

12. The stronger the attractive forces among the liquid particles are, the smaller the percentage of liquid particles is that can escape to the vapor phase at any given temperature. Since the equilibrium vapor pressure of the liquid is dependent on the relative number of particles in the vapor phase, the pressure above the liquid that has strong attractive forces will be less than that above the liquid in which the comparable forces are weaker.

13. **a.** As atmospheric pressure increases, so does the boiling point of a liquid, because the liquid must be heated to a higher temperature before its equilibrium vapor pressure is equal to atmospheric pressure.
    b. Both the liquid and its vapor have the same constant temperature during boiling.
    c. The energy from the constant supply of heat is used to overcome the attractive forces between the liquid molecules during the liquid-to-gas change, rather than to increase the kinetic energy of the liquid or vapor molecules, and is stored in the vapor as potential energy.

14. The greater the attraction between the liquid particles is, the greater is the energy that is required to overcome that attraction, and the higher is the molar heat of vaporization.

15. As the degree of attraction between the solid particles increases, more heat must be absorbed by the melting solid in order to overcome the attractive forces and separate the solid particles, resulting in an increase in the molar heat of fusion.

16. A water molecule consists of two atoms of hydrogen and one atom of oxygen united by polar covalent bonds. The angle between the two hydrogen-oxygen bonds is about 105°.

17. At room temperature, pure water is a transparent, odorless, tasteless, and almost colorless liquid. It freezes and melts at 0°C at 1 atm; its molar heat of fusion is 6.008 kJ/mol; water expands 11% in volume as it freezes; it has its maximum density at 3.98°C; at a pressure of 1 atm, it boils at 100°C; its molar heat of vaporization is 40.79 kJ/mol.

Problems

1. **a.** $5.00 \text{ mol } H_2O \times \dfrac{40.79 \text{ kJ}}{\text{mol } H_2O} = 204 \text{ kJ}$

   b. $45.0 \text{ g } H_2O \times \dfrac{\text{mol } H_2O}{18.0 \text{ g } H_2O} \times \dfrac{40.79 \text{ kJ}}{\text{mol } H_2O} = 102 \text{ kJ}$

   c. $8.45 \times 10^{50} \text{ molec } H_2O$
   $\times \dfrac{\text{mol } H_2O}{6.02 \times 10^{23} \text{ molec } H_2O} \times$
   $\dfrac{40.79 \text{ kJ}}{\text{mol } H_2O} = 5.73 \times 10^{28} \text{ kJ}$

2. **a.** $12.75 \text{ mol } H_2O \times \dfrac{6.008 \text{ kJ}}{\text{mol } H_2O} = 76.60 \text{ kJ}$

   b. $6.48 \times 10^{20} \text{ kg } H_2O \times \dfrac{10^3 \text{ g } H_2O}{\text{kg } H_2O} \times \dfrac{\text{mol } H_2O}{18.0 \text{ g } H_2O} \times$

$\dfrac{6.008 \text{ kJ}}{\text{mol H}_2\text{O}} = 2.16 \times 10^{23}$ kJ

3. $\dfrac{36.5 \text{ kJ}}{0.433 \text{ mol}} = 84.3$ kJ/mol

4. **a.** $71.8 \text{ g} \times \dfrac{1 \text{ mol}}{259.0 \text{ g}} = 0.277$ mol

   b. $\dfrac{4.31 \text{ kJ}}{0.277 \text{ mol}} = 15.6$ kJ/mol

5. **a.** $83.2 \text{ kJ} \times \dfrac{1 \text{ mol}}{3.811 \text{ kJ}} = 21.8$ mol

   b. $\dfrac{5519 \text{ g}}{21.8 \text{ mol}} = 253$ g/mol

6. Ice:

   $5.00 \text{ cm}^3 \times \dfrac{0.917 \text{ g}}{\text{cm}^3} \times \dfrac{6.02 \times 10^{23} \text{ molecules}}{18.0 \text{ g}}$

   $= 1.53 \times 10^{23}$ molecules

   Liquid water:

   $5.00 \text{ cm}^3 \times \dfrac{1.00 \text{ g}}{\text{cm}^3} \times \dfrac{6.02 \times 10^{23} \text{ molecules}}{18.0 \text{ g}}$

   $= 1.67 \times 10^{23}$ molecules

   The liquid water sample thus contains $1.67 \times 10^{23} - 1.53 \times 10^{23}$, or 1.4×10^{22}, more molecules than does the sample of ice.

7. From the problem above, the liquid water contains 1.67×10^{23} molecules.

   Water vapor:

   $5.00 \text{ cm}^3 \times \dfrac{1 \text{ L}}{1000 \text{ cm}^3} \times \dfrac{1 \text{ mol}}{22.4 \text{ L}}$

   $= 2.23 \times 10^{-4}$ mol
   2.23×10^{-4} mol $\times 6.02 \times 10^{23}$ molecules/mol
   $= 1.34 \times 10^{20}$ molecules

   The liquid water thus contains $1.67 \times 10^{23} - 1.34 \times 10^{20}$, or $1.67 \times 10^{23} - 0.00134 \times 10^{23}$, which equals (allowing for significant figures) 1.67×10^{23} mole molecules. Thus, the $H_2O(g)$ contains a negligible number of molecules when compared with the $H_2O(\ell)$.

8. Liquid water:

   $100. \text{ cm}^3 \times \dfrac{1.00 \text{ g}}{\text{cm}^3} \times \dfrac{1 \text{ mol}}{18.0 \text{ g}} \times 6.008$ kJ/mol

   $= 33.4$ kJ
   (released when the liquid water is frozen)

   Steam:

   $33.4 \text{ kJ} \times \dfrac{1 \text{ mol}}{40.58 \text{ kJ}} = 0.823$ mol (required to release

   33.4 kJ when condensed)
   $T = 100°C + 273 = 373$ K
   $PV = nRT$

   $V = \dfrac{nRT}{P} = \dfrac{0.823 \text{ mol} \times 0.0821 \dfrac{\text{L} \cdot \text{atm}}{\text{mol} \cdot \text{K}} \times 373 \text{ K}}{1.00 \text{ atm}}$

   $= 25.2$ L

   $0.823 \text{ mol} \times \dfrac{18.0 \text{ g}}{\text{mol}} = 14.8$ g

   A considerably larger volume of steam is thus required to release the same amount of heat. However, the mass of this larger volume of steam is considerably less than that of the liquid water.

Critical Thinking

1. A steel needle can rest on a water surface because the surface tension of the water makes the surface less penetrable by solid bodies. A boat floats in water because it displaces a volume of water whose weight is equal to that of the boat.
2. Our bodies are cooled by the evaporation of water-based perspiration that accumulates on our skin. On a humid day, water collects on the skin faster than it can evaporate, because of the high rate of condensation of water from the atmosphere. On a dry day, however, the rate of condensation is lower than the rate of evaporation, and the water does not collect. Thus, one is more comfortable on a hot, dry day than on a hot, humid day.
3. Diamond is a three-dimensional covalent network solid in which the three-dimensional nature of the covalent bonding accounts for the extreme hardness. Graphite, on the other hand, is a planar network solid in which the covalent bonding occurs in only two dimensions. The molecular sheets of graphite are held together only by van der Waals forces. These intersheet bonds are relatively weak and allow the adjacent sheets to slide past each other readily, accounting for the softness of graphite.
4. The indicated responses are:
   a. higher   **c.** higher
   b. higher   **d.** lower
5. The indicated responses are:
   a. increased   **c.** increased
   b. increased   **d.** decreased
6. The energy that is continually being removed during freezing is actually a loss of potential energy resulting from the closer approach of particles that attract each other and not of kinetic energy that would be reflected in the temperature of the system.
7. According to Le Châtelier's principle, the addition of heat to the ice-water system pushes the equilibrium to the right (toward liquid water), whereas the removal of heat shifts it to the left. As long as both ice and liquid water are present, the addition or removal of heat is simply reflected in the phase change that involves potential energy, and not in the kinetic energy of the overall system.
8. **a.** More solid would sublime.
   b. More vapor would solidify.
   c. More vapor would solidify.
   d. More solid would vaporize.
9. Decrease the pressure.
10. Methane is a nonpolar molecule and does not undergo hydrogen bonding. Water, on the other hand, is polar

and undergoes extensive hydrogen bonding. Higher temperatures are required in order to overcome the resultant attractive forces and convert water from a liquid into a vapor.

11. Yes. A container of water can freeze and boil at the same time if it is attached to a vacuum pump and the pressure over the water is greatly reduced. At its triple point, the water will boil and its temperature will fall to its freezing point.

12. At 6 atm of pressure and $-100°C$, CO_2 is a solid and remains a solid until a temperature of $-50°C$ is reached. At $-50°C$, CO_2 solid melts to form CO_2 liquid. At a temperature of $-40°C$, CO_2 liquid boils.

Videodisc Selections from
The Periodic Table Videodisc (JCE Software)

step forward

step reverse

CHAPTER 13

pg 380 Ionic crystals

36543
potassium chlorate (matches) and KCl (s)

pg 380 Covalent network crystals

36378
carbon (s)

36474
opal (contains Al3+, Si4+, and O2- in lattice) (s)

11568
silicon (s)

pg 381 Metallic crystals

36462
aluminum (s)

36579
titanium (s)

36867
silver (s)

pg 381 Covalent molecular crystals

3817
boron (s)

13134
sulfur (s)

45067
iodine (s)

pg 381 Amorphous solids

36360
glass (s)

36513
plastic (s)

36711
gallium semiconductors (s)

Liquids and Solids **370F**

Teacher's Notes

Chapter 13
Liquids and Solids

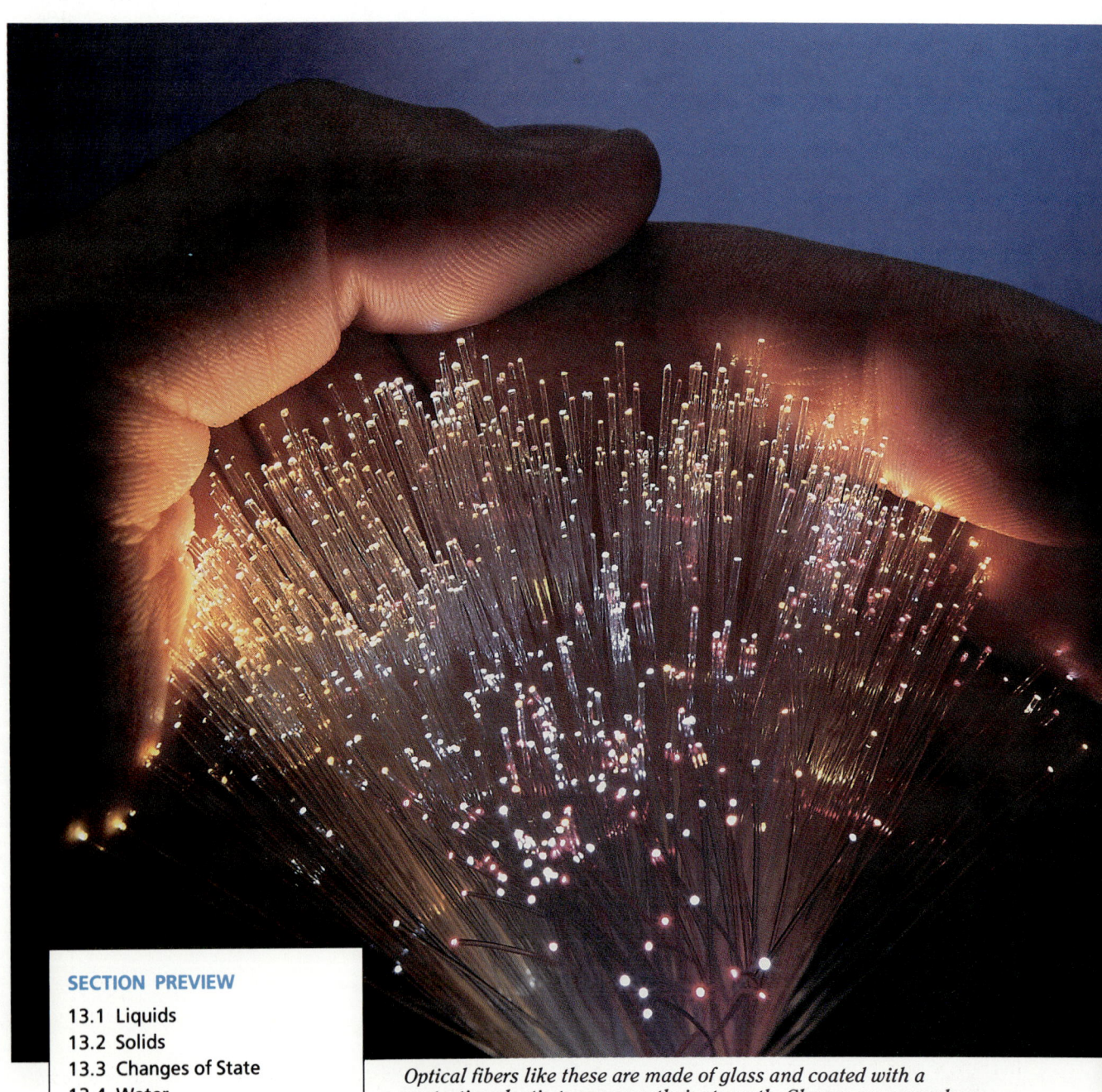

SECTION PREVIEW
13.1 Liquids
13.2 Solids
13.3 Changes of State
13.4 Water

Optical fibers like these are made of glass and coated with a protective plastic to preserve their strength. Glasses are amorphous solids that have some properties of solids and some properties of liquids.

13.1 Liquids

The water in the breakers crashing on an ocean beach and the molten lava rushing down the sides of a volcano are both examples of matter in the liquid state. When you think of the earth's oceans, lakes, and rivers, the fluids in the bodies of all living things and the many liquids you use every day, it is hard to believe an important fact about liquids—they are probably the *least* common state of matter in the universe. Liquids are less common than solids, gases, or plasmas because the liquid state of any substance can exist only within a relatively narrow range of temperatures and pressures.

In this section, you will examine the properties of the liquid state and compare them with those of the solid state and the gaseous state. These properties will be discussed in terms of a model of liquids at the submicroscopic level.

Kinetic-Theory Description of the Liquid State

How do we describe a liquid? One way to describe a liquid is to note that a liquid is a form of matter that has no definite shape, but instead takes the shape of its container. If you were to observe various types of liquids, you could include other properties in your definition of a liquid.

In order to accurately describe liquids and understand their properties, it is helpful to recall the kinetic molecular theory and the particle model of matter. To do this, think of liquids in terms of the motion, arrangement, and attractive forces of particles. Figure 13-1 is a diagram of liquid particles and the analogous model of gases discussed in Chapter 11.

Like gases and solids, liquids can be seen as consisting of particles—ions, atoms, or molecules. What makes a substance a liquid is not the nature of its particles, but the arrangement and behavior of those particles.

Like gas particles, liquid particles are in constant motion. Liquid particles are closer together and lower in energy than gas particles, however, and therefore the attractive forces between them are more effective than those between gas particles. This attraction between liquid particles is caused by the intermolecular forces discussed in Chapter 6: dipole–dipole forces, London dispersion forces, and hydrogen bonding. Remember that these dispersion forces are not very effective in gases because their particles are widely separated. Because the particles of a liquid are more closely packed than those of a gas, the movement of particles in a liquid is limited.

Liquids are more ordered, or less disordered, than gases. This is because of the effective intermolecular forces and the resulting decrease in mobility of the liquid particles. The kinetic energy of liquid particles is high enough so that they are not bound together as are solid particles, however. Particles of liquids can change position with other liquid particles, although they tend to do so more slowly and less often than do gas particles.

Properties of Liquids and the Particle Model

In many ways, the liquid state is intermediate between the solid state and the gaseous state. At normal atmospheric pressure (760 mm Hg or 101.325 kPa), most solids pass through the liquid state before turning to gas as their

SECTION OBJECTIVES

- Describe the motion of liquid particles according to the kinetic theory.
- Discuss the properties of liquids in terms of the particle model.

STUDY HINT

Temperature and pressure are discussed in Chapters 1, 11, and 12.

The earth receives just enough energy from the sun to maintain most of the water on earth in the liquid state.

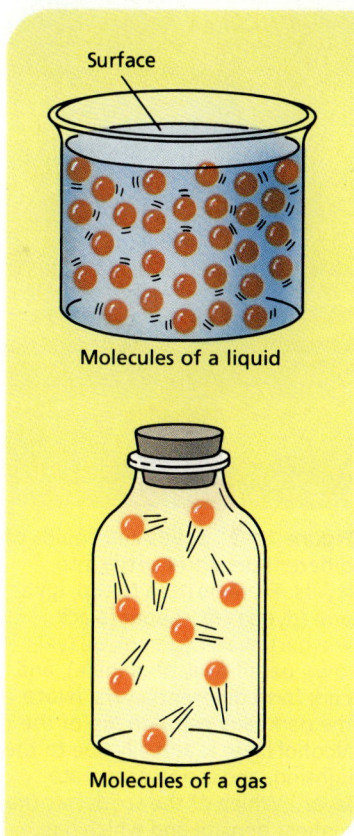

Figure 13-1 This diagram compares the particle models of liquids (top) and gases (bottom).

temperature is increased. Most gases also pass through the liquid state before solidifying as their temperature is decreased. Some properties of liquids closely resemble those of solids, whereas others closely resemble those of gases.

Now you will take a closer look at some properties of liquids. These properties will be discussed in terms of the particle model of liquids.

Definite volume Unlike gases, which expand indefinitely unless enclosed by a container, liquids have a definite volume. This volume changes only slightly with changes in temperature or pressure.

Molar volumes of liquids
$10^1 - 10^2$ mL
Molar volumes of gases
$10^4 - 10^5$ mL
Since $d \div m/V$ then
$d_{liq} = 1000 \times d_{gas}$

Fluidity Like gases, but unlike solids, liquids can flow. According to the particle model of liquids, the particles are not bound together in fixed positions, but can move about constantly. This particle mobility explains why liquids and gases are referred to as fluids. *A* **fluid** *is a substance that can flow and therefore conform to the outline of its container.* Although most liquids naturally flow downhill because of gravity, flow can be in other directions as well. For example, liquid helium at temperatures near absolute zero has the unusual property of being able to flow uphill. Because they are fluids, liquids must be enclosed in containers. Unlike a gas, which presses against all the walls of its container, however, a liquid can have one free surface. Its other surfaces must be supported by the container walls.

Relatively high density At normal atmospheric pressure, most substances are thousands of times denser in their liquid state than in their gaseous state. This density is a result of the closer packing of liquid particles, as described by the particle model. Most substances are slightly (about 10%) less dense in their liquid state than in their solid state. As will be discussed in Section 13.4, water is one of the few substances that decreases in density when it solidifies.

At the same temperature and pressure, different liquids can differ in density by factors of almost 20. A dramatic demonstration of these differences in density is shown in Figure 13-2. The densities of the five liquids shown differ so greatly that the liquids settle out in layers.

Figure 13-2 Solids and liquids of different densities settle into layers. From bottom to top, glycerin (dyed red), rubber (black washer), water (green), oil (yellow), paraffin, alcohol (blue), and cork form different layers. (Note: the dark blue color on top of the alcohol layer is actually due to the lighting, and is not a separate layer.) Which of the solids has the highest density and which has the lowest? Which of the liquids have the highest and lowest densities?

Relative incompressibility When liquid water at 20°C is compressed by a pressure of 1000 atm (roughly 100 MPa), its volume decreases by only 4%. This behavior of liquid water is typical of all liquids and is similar to the behavior of solids. In contrast, the volume of a gas under a pressure of 1000 atm would be only 1/1000 of its volume at normal atmospheric pressure. Liquids are much less compressible than gases because liquid particles are more closely packed together than gas particles. Liquids are used in the braking systems of motor vehicles, as shown in Figure 13-3, because of their relative incompressibility and because they have the ability to transmit pressure equally in all directions.

Dissolving ability *When one substance is added to another so that, after mixing, only one physical state is observed, the substance has been* **dissolved.** For example, after you add a teaspoon of solid sugar crystals to a glass of liquid water and stir thoroughly, you no longer see solid sugar crystals. The mixture of sugar and water appears to be all liquid. Liquids are the most common dissolving agents and among these, liquid water is the most

Figure 13-3 This diagram shows how liquids are used in the braking systems of automobiles.

generally useful. The dissolving ability of liquids is explained by the motion of their particles and the attractive forces they may exert on the particles of the substance being dissolved.

Ability to diffuse As described in Chapter 11, gases diffuse and mix with the particles of other gases. Liquids also diffuse, as shown in Figure 13.4. Any liquid will gradually diffuse throughout any other liquid in which it can be dissolved. As in gases, diffusion in liquids occurs because of the constant, random motion of particles. Yet diffusion is much slower in liquids than in gases. This is because each forward-moving particle of closely packed liquid soon strikes another particle, changing its path. Also, the attractive forces between the particles of a liquid impede their movement. As the temperature of a liquid is increased, diffusion occurs more rapidly because the average kinetic energy and, therefore, the average speed of the particles is increased.

STUDY HINT
The dissolving process is discussed in Chapter 14.

Figure 13-4 Like gases, the two liquids in this beaker diffuse. The liquid food coloring and water gradually diffuse to form a uniform mixture.

Liquids and Solids

Desktop Investigation

Surface Tension of Liquids and Water Solutions

Question
Do different liquids have different degrees of surface tension; what effect do dissolved substances have on the surface tension of water?

Materials
water
glycerin
cooking oil
rubbing alcohol (70% isopropyl alcohol)
table salt (sodium chloride)
liquid detergent
250-mL beaker
10- and 100-mL graduated cylinders
evaporating dish
tweezers
needle
eye dropper
stirring rod
wax paper

Procedure
Remember to wear safety glasses and an apron. Record all of your results in a data table.

1. Add 70 mL of water to an evaporating dish. Using tweezers, carefully place a needle on the surface of the water, as shown in the illustration. What happens? Empty the dish and dry it and the needle and tweezers. Repeat this procedure using *(a)* glycerin and *(b)* cooking oil instead of water.

2. Repeat Procedure 1, but replace water in the dish with the following solutions. Make sure that you clean and dry the dish, needle, and tweezers between each procedure. *(a)* 60 mL of water + plus 10 mL of alcohol *(b)* 50 mL of water + 20 mL alcohol *(c)* 70 mL water + a few grains of

table salt *(d)* 70 mL water + 1 g table salt *(e)* 70 mL water + 1 drop of liquid detergent.

3. In your data table, include the answers to these questions: *(a)* Does the needle rest on the surface of the liquid or does it sink? *(b)* How long does it remain on the surface? *(c)* Describe the appearance of the liquid surface with the needle resting on it.

4. Place three drops of each liquid or solution used in Procedures 1 and 2 together on a large piece of wax paper. Describe the shape of each drop. Is it rounded or flat? Be sure to use a different area of the wax paper for each different liquid or solution.

Discussion

1. Which liquids have the strongest surface tension? The weakest? How do you know?

2. How do dissolved substances affect the surface tension of water?

3. Did you find any relationship between the ability of a needle to rest on the surface of a liquid or solution and the shapes of the drops of the different liquids or solutions?

Surface tension Surface tension, *a property common to all liquids, is a force that tends to pull adjacent parts of a liquid's surface together, thus making the surface less penetrable by solid bodies.* For example, the surface tension of water can keep a steel needle on the water's surface, as illustrated in the Desktop Investigation above. Surface tension is a result of the attractive forces between particles of a liquid as described by the particle model of matter discussed in Section 11.1. It gives a taut appearance to a liquid's free surface and acts to minimize its area. As a result of surface tension, liquid droplets tend to take on a spherical shape because a sphere has the smallest possible surface area for a given volume. An example of this phenomenon is shown in Figure 13-5.

Capillary action, the tendency of a liquid to rise up a narrow tube, is closely related to surface tension. Capillary action allows liquids to rise up a piece of slender glass tubing or a straw. It is the result of the attractive forces between particles of a liquid as well as forces of attraction between the liquid particles and the surface of the glass. Capillary action is responsible for the meniscus, the concave liquid surface that you see in a straw, test tube, or drinking glass.

Tendency to evaporate and to boil *The process by which a liquid or solid changes to a gas is* **vaporization.** *Evaporation is a form of vaporization.* **Evaporation** *is the process by which particles escape from the surface of a nonboiling liquid and enter the gaseous state.*

A small amount of liquid bromine has just been added to the bottle shown in Figure 13-6. Within a few minutes, the air above the liquid bromine is reddish. Some bromine molecules have escaped from the surface of the liquid and gone into the gaseous state, becoming bromine vapor, which mixes with the air. A similar phenomenon occurs when you apply cologne to your wrist. Within seconds, you become aware of the cologne's odor. Cologne scent molecules evaporate from your skin and diffuse through the air to your nose.

Evaporation can be explained by the kinetic energy of the particles in a liquid. At any instant, the separate particles of a liquid have different kinetic energies. Some particles have higher-than-average energies and move faster. Other particles have lower-than-average energies and move slower. Surface particles with higher-than-average energies may overcome the intermolecular forces that bind them to the liquid and "escape" into the gaseous state.

Evaporation is a crucial process in nature. All of the water that falls on the earth in the form of rain and snow originally evaporated from the earth's oceans, lakes, and rivers. The evaporation of perspiration plays an important role in keeping us cool. Perspiration, which is mostly water, cools us by absorbing body heat when it evaporates.

Boiling is the change of a liquid to bubbles of vapor that appear throughout the liquid when the equilibrium vapor pressure of the liquid equals the atmospheric pressure. The temperature at which a liquid boils is called the *boiling point* of that liquid; boiling is related to evaporation. For most liquids, the boiling point is higher than room temperature. Both evaporation and boiling are discussed in detail in Section 13.3.

Tendency to solidify When a liquid is cooled, the average energy of its particles may become low enough that attractive forces hold the particles together in an orderly arrangement. The substance then becomes a solid. The physical change of a liquid to a solid is called *freezing*. Perhaps the best-known example of freezing is the change of liquid water to solid water—or ice—at 0°C. This property of water is discussed more fully in Section 13.4. Another familiar example of the tendency of liquids to solidify is the freezing of liquid candle wax at room temperature. All liquids freeze, although not necessarily at temperatures we encounter in our everyday lives. Ethyl alcohol, for example, freezes at −112°C.

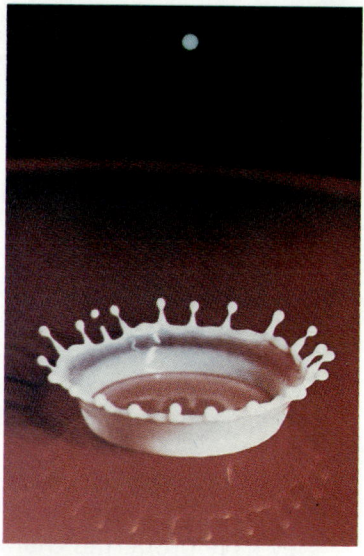

Figure 13-5 This photo shows that, like all liquids, milk forms spherically shaped drops as a result of surface tension.

Figure 13-6 Liquid bromine will gradually diffuse into the air above its surface.

Liquids and Solids

Chemistry and its Applications

Liquid Crystals

The unique properties of liquid crystals have intrigued scientists since the discovery of these crystals nearly 100 years ago. Today, liquid-crystal displays (LCDs) are widely used in pocket calculators, digital watches, and clocks.

Liquid crystals derive their name from the fact that they have properties characteristic of both liquids and solids. In the liquid-crystal state, substances can transmit and reflect light in the same way that crystals do, while at the same time taking on the form of a liquid.

There are three types of liquid crystals: nematic, smectic, and cholesteric. Each is defined according to its molecular arrangement. In nematic liquid crystals, the cigar-shaped molecules are parallel to each other.

The attractive forces between molecules in nematic crystals are strong enough for the molecules to move together as a group. Because their molecules change directions in response to an electric field, nematic liquid crystals are most often used in LCDs. The numbers and letters in the LCD are made up of small glass tubes filled with liquid crystals. As an electric charge passes through the glass, the crystals rotate in the direction of the resulting electric field. When the electric field is not present, the display is lighted. When the field is present, light is blocked in certain areas resulting in a numeric display.

In smectic liquid crystals, the molecules are arranged in planes or layers. The molecules of cholesteric liquid crystals are also arranged in layers. Because they are sensitive to temperature changes, both smectic and cholesteric liquid crystals are used in thermometers.

1. See the discussions of kinetic theory and the particle model at the beginning of this section.
2. See the discussion of properties of liquids.
3. Refer to the coverage of the particle model earlier in this section.
4. See "Tendency to evaporate and to boil."

Section Review

1. Describe the liquid state according to the kinetic theory and particle model.
2. List the properties of liquids.
3. How does the particle model explain the following properties of liquids: *(a)* their relatively high density   *(b)* their dissolving ability   *(c)* their ability to evaporate?
4. Compare vaporization, evaporation, and boiling.

13.2 Solids

"Solid as a rock" is a common expression that intuitively describes a solid—something that is hard or unyielding, and that has a definite shape and volume. Many things besides rocks are solids; in fact, solids are far more common than liquids.

In this section, you will examine the properties of the solid state and compare them with those of the liquid and gaseous states. As with liquids and gases, the properties of solids are explained in terms of the kinetic theory.

SECTION OBJECTIVES

- Describe the motion of solid particles according to the kinetic theory.
- Discuss the properties of solids in terms of the particle model.
- Distinguish between the two types of solids

Kinetic Theory Description of the Solid State

Like gases and liquids, solids consist of particles. The particles in a solid may be ions, atoms, or molecules. Figure 13-7 compares the structure of solids, liquids, and gases. The average kinetic energy of the particles in solids is lower than that of the particles in gases and liquids. Because they have less kinetic energy than gases and liquids, solid particles have less motion than liquid particles. They also have much less motion than gas particles. As shown in Figure 13-7, the particles of a solid do not undergo overall changes in position, but rather move back and forth about fixed positions. Solid particles are more closely packed than liquid or gas particles, and therefore intermolecular forces between particles are even more important in solids. Dipole-dipole attractions, ion-dipole forces, London dispersion forces, and hydrogen bonding are all much stronger in solids than in liquids or gases. These forces help to hold solid particles in their positions. Because solid particles are held in relatively fixed positions, solids are more ordered than liquids and much more ordered than gases. The importance of order and disorder in physical and chemical changes is discussed in Chapter 18.

Properties of Solids and the Particle Model

The properties of solids can be explained in terms of the particle model of solids—that is, in terms of the arrangement and motion of solid particles and the attractive forces between them. There are two types of solids: crystalline solids and amorphous solids. *Most solids are* **crystalline solids**—*they consist of crystals. A* **crystal** *is a substance in which the particles are arranged in an orderly, geometric, repeating pattern.* Examples of crystalline solids are shown in Figure 13-8. Other solids, including glass and plastics, are non-crystalline and are called *amorphous solids. An* **amorphous solid** *is one in which the particles are arranged randomly.* These two types of solids are discussed further in this section.

Definite shape Unlike liquids and gases, all solids can maintain a definite shape without a container. In addition, crystalline solids are geometrically regular. Even the fragments of a shattered crystalline solid have distinct geometric shapes. On the other hand, while amorphous solids do maintain a definite shape (at least temporarily), they do not have the distinct geometric shapes of crystalline solids. Instead, they can have nearly any shape. For example, glass can be molded into any shape, and, if it is shattered, its fragments can have a wide variety of irregular shapes.

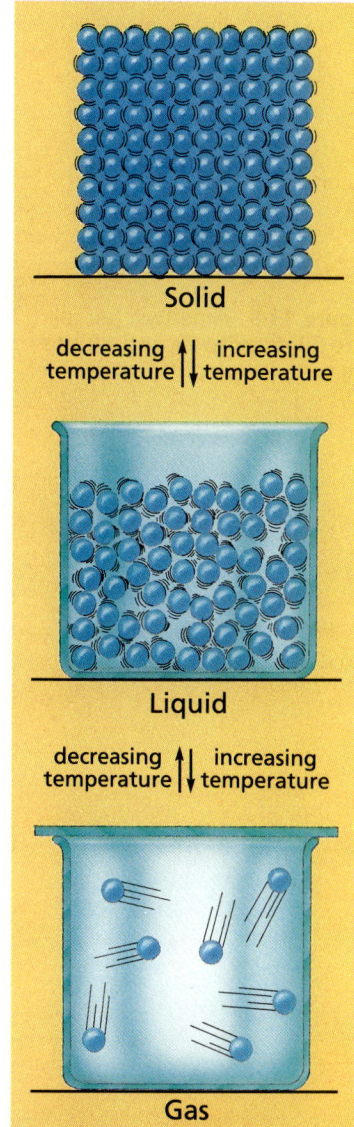

Figure 13-7 This diagram shows particles of a solid, a liquid, and a gas.

Liquids and Solids

Figure 13-8 Amethyst (a), calcite (b), fluorite (c), and azurite (d) are examples of crystals.

Definite volume Like liquids but unlike gases, all solids have a definite volume. The volume of a solid changes only slightly with a change in temperature or pressure. Solids have definite volume because their particles are packed closely together. There is very little empty space into which the particles can be compressed, and the particles do not have a tendency to move apart. Unlike liquids and gases, all the surfaces of a solid can be free.

Nonfluidity Unlike liquids and gases, crystalline solids generally do not flow. This is because their particles, as described by the particle model, are held in relatively fixed positions. Amorphous solids also tend to retain their shapes. Some amorphous solids do flow, although usually at a very slow rate. A distinctive group of amorphous solids is the glasses, which are made by cooling certain molten materials in such a manner that they do not crystallize but remain in an amorphous state.

Definite melting point At a given pressure, a pure crystalline solid melts abruptly at a particular temperature. *The temperature at which a solid becomes a liquid is its* **melting point.** At this temperature, the kinetic energies of the particles within the solid overcome the intermolecular forces between them, allowing the particles to break out of their positions in the crystal. In contrast, amorphous solids like glass and plastics have no definite melting point. Instead they gradually soften as they are heated, to become thick, sticky liquids. Because of this, and also because of their ability to flow, amorphous solids such as glass are sometimes classified as **supercooled liquids;** *these are substances that retain certain liquid properties even at temperatures at which they appear to be solid.* These properties exist because the particles in amorphous solids are not arranged in an orderly, uniform crystal structure, but instead are arranged randomly, like the particles in a liquid. Unlike the particles in a true liquid, however, the particles in amorphous solids are not constantly changing their positions. An amorphous solid can be thought of as being much like a liquid that has been cooled enough so that it no longer flows.

High density In general, most substances are at their densest in the solid state, which is typically slightly denser than the liquid state. Solids are denser than liquids or gases because the particles of a solid are more closely packed than those of a liquid or a gas. Solid hydrogen is the least dense solid; it has a density of about 1/320 that of the densest element, osmium (Os).

Incompressibility Solids are even less compressible than liquids. For all practical purposes, solids can in fact be considered incompressible. Wood and cork are solids that *seem* compressible, but they are not. They contain pores that are filled with air. When subjected to intense pressure, the pores are compressed, not the wood or cork.

Slow rate of diffusion If a zinc (Zn) plate and a copper (Cu) plate are clamped together for several months, a small amount of diffusion between the particles of zinc and copper will occur. This observation shows that diffusion does occur in solids. The rate of diffusion is millions of times slower in solids than in liquids, however.

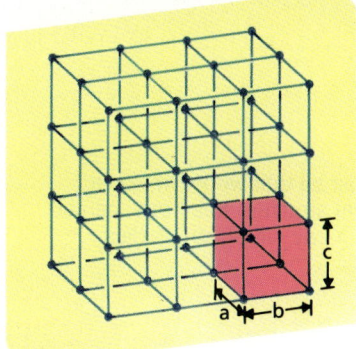

Figure 13-9 The colored cube at the lower right of this crystal lattice is a unit cell.

Crystalline Solids

Earlier, you learned the definition of the two types of solids: crystalline and amorphous. Now you will examine types of crystalline solids more closely. Later in this section, you will read more about amorphous solids.

Classification of crystals by arrangement and shape Crystalline solids exist either as single crystals or as groups of crystals fused together. Figure 13-8 shows four different types of crystals. Scientists can determine the arrangement of the particles of a crystal by mathematical analysis of the crystal's diffraction pattern; a unique pattern that results when X rays or other forms of electromagnetic radiation pass around the edges of particles.

The total three-dimensional array of points that describes the arrangement of the particles of a crystal is called a **crystal lattice.** Such a lattice is shown in Figure 13-9. *The smallest portion of a crystal lattice that reveals the three-dimensional pattern of the entire lattice is called a* **unit cell.** Each crystal lattice is constructed of many unit cells packed solidly together. As a result of this construction, a crystal has the same symmetry as its unit cell. Unit cells can have any one of seven types of symmetry, a characteristic that enables scientists to classify crystals by shape. This classification by shape is a part of the science of crystallography. Any crystal can be described as an example of one of the seven crystalline systems shown in Figure 13-10.

Binding forces in crystals Crystal-lattice structures can also be described in terms of the types of particles making up the crystals and the types of chemical bonding between them. According to this method of classifying crystal lattices, there are four types of crystals, which are listed in Table 13-1. You may wish to refer to this table as you read the following discussion.

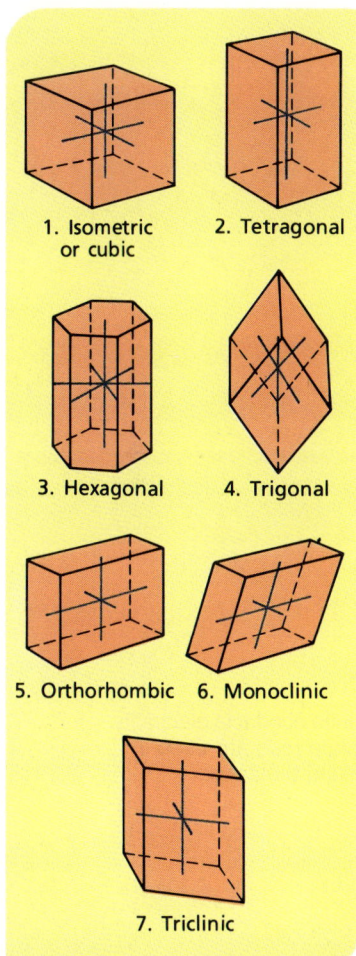

Figure 13-10 This schematic diagram shows the seven basic crystal systems.

Liquids and Solids **379**

Almost all that we know about crystal structure has been learned from X-ray diffraction studies.

TABLE 13-1 MELTING POINTS AND BOILING POINTS OF REPRESENTATIVE CRYSTALLINE SOLIDS

Type of Substance	Substance	Melting Point (°C)	Boiling Point (1 atm, °C)
ionic	NaCl	801	1413
	MgF_2	1266	2239
covalent network	$(SiO_2)_x$	1610	2230
metallic	Hg	−39	357
	Cu	1083	2567
	Fe	1535	2750
	W	3410	5660
nonpolar covalent molecular	H_2	−259	−253
	O_2	−218	−183
	CH_4	−182	−164
	CCl_4	−23	77
	C_6H_6	6	80
polar covalent molecular	NH_3	−78	−33
	H_2O	0	100

1. Ionic crystals. As discussed in Chapter 6, the ionic crystal lattice consists of positive and negative ions arranged in a characteristic regular pattern. The ions can be monatomic or polyatomic. Generally, ionic crystals are formed when Group 1 or Group 2 metals combine with Group 16 or Group 17 nonmetals or the nonmetallic polyatomic ions. The strong binding forces between the positive and negative ions in the crystal lattice cause ionic crystals to be hard and brittle, to have high melting points, and to be good insulators.

2. Covalent network crystals. In covalent network crystals, the lattice sites contain single atoms, each of which is covalently bonded to its nearest neighboring atoms. The covalent bonding extends throughout a network that includes a very large number of atoms. Three-dimensional covalent network solids, including diamond (C_x), quartz $(SiO_2)_x$, shown in Figure 13-11, silicon carbide $(SiC)_x$, and many oxides of transition metals, are essentially giant molecules. The subscript x in these formulas indicates simply that the chain of components extends indefinitely. These network solids are very hard and brittle. They have rather high melting points, and are nonconductors or semiconductors. Planar network solids, such as graphite (C_x), shown in Figure 13-11, consist of molecular sheets. Each sheet can be thought of as a single molecule consisting of atoms held together by many covalent bonds. The adjacent separate sheets of molecules, in turn, are held together by weak intermolecular forces between atoms in the different sheets. Graphite is commonly used as a lubricant because the weak intersheet bonds of this planar network solid allow adjacent sheets to slide past each other readily. For this same reason, graphite is used as pencil "lead." As you write, a thin layer of graphite is deposited onto the paper.

Graphite conducts electricity. Graphite rods are used in flashlight batteries.

3. Metallic crystals. As discussed in Chapter 6, the metallic crystal lattice consists of positive ions of the metal surrounded by a cloud of valence electrons, that are donated by the metal atoms and belong to the crystal as a whole. The binding force is the attraction between the positive ions and the electrons. At any instant, each electron and the positive metallic ions immediately surrounding it are attracting each other electrostatically. The freedom of the valence electrons to migrate throughout the crystal lattice explains the high electric conductivity of metals. As you can see from Table 13-1, the melting points of different metallic crystals vary greatly.

4. *Covalent molecular crystals.* The crystal lattice of a covalent molecular substance consists of covalently bonded molecules that are held to one another by intermolecular forces. If the molecules are nonpolar—for example, hydrogen (H_2), iodine (I_2), methane (CH_4), and benzene (C_6H_6)—then the bonding between molecules, resulting from London dispersion forces, is weak. Crystals of the noble gases are held together by the same type of weak forces. In this case, because the noble gases are monatomic, these forces act between individual atoms. If the molecules in a covalent molecular crystal are polar—for example, water (H_2O) and ammonia (NH_3)—they are held together by both dispersion forces and the somewhat stronger dipole-dipole forces. The forces that hold either polar or nonpolar molecules together in the crystal lattice are much weaker than the covalent chemical bonds that bind the atoms within the molecules. Covalent molecular crystals thus have low melting points, are easily vaporized, are relatively soft, and are good insulators. Ice crystals, the most familiar molecular crystals, are discussed in Section 13.4.

Figure 13-11 Covalent network crystals include three-dimensional covalent network solids, such as quartz *(left),* and planar network solids, such as graphite *(right).*

Amorphous Solids

Noncrystalline solids, such as rubber, glass, plastics, and synthetic fibers, are called *amorphous solids*. The word "amorphous" comes from the Greek for "without shape (*a* [without] + *morphe* [form])." Unlike crystals, amorphous solids do not have a regular, natural shape, but instead take on whatever shape is imposed on them. The particles making up amorphous solids are not arranged in a uniform crystal lattice, but are arranged randomly, somewhat like the particles in a liquid.

There are hundreds of types of plastic and glass, which have thousands of important applications. Glass is used for everything from fiberglass automo-

Chemistry and Technology

Superconductors

In early 1987, the scientific community was excited by the discovery of a new class of superconducting materials that offer no resistance to an electric current when cooled to 90K. This temperature can be reached conveniently using inexpensive liquid nitrogen (boiling point, 77K). Previously discovered superconductors functioned at temperatures that could be reached only by using liquid helium (boiling point, 4.2K), which is expensive and inconvenient to handle.

Many of the new superconductors are in the family of materials known as ceramic oxides. They have the empirical formula $YBa_2Cu_3O_x$ and are related to the perovskite class of minerals. In a perfect crystal of this class, x (the number of oxygen atoms) = 9. The superconductors are imperfect crystals, however, in which the domain of x extends from approximately 6.5 to 7.2.

The new superconductors may someday be used to transmit electricity hundreds of miles without loss of energy in the form of heat. They could also be made into powerful electromagnets that could suspend railroad trains in mid-air above their tracks. These magnetically levitated trains could travel without friction, reaching very high speeds. Powerful electromagnets could also produce the strong magnetic fields required to confine the hot plasmas that are needed to generate energy by controlled nuclear fusion.

1. Solids consist of particles (ions, atoms, or molecules). Solid particles have lower kinetic energy and are less mobile than liquid or gas particles. The particles of a solid do not change position but move back and forth about fixed positions. Solid particles are more closely packed than liquid or gas particles.
2. (a) Crystalline solids melt abruptly at a definite temperature called the melting point because the kinetic energy of the solid particles overcomes the strong

bile bodies to optical fibers that transmit telephone conversations by means of light waves. Amorphous forms of metal alloys can be prepared by very rapid cooling of thin films of melted metal. The cooling occurs so rapidly that the metal atoms do not have time to arrange themselves in a crystalline pattern. The resulting "metallic glasses" are strong, flexible, and much more resistant to corrosion than crystalline alloys of the same composition.

Section Review

1. Describe the solid state according to kinetic theory and particle model.
2. How does the particle model explain (a) the definite melting point (b) the definite volume (c) the high density of solids?
3. What is the difference between amorphous and crystalline solids?
4. What are the four types of crystals?

13.3 Changes of State

Matter on earth can exist in any of three states—gas, liquid, or solid—and can change from one state to another. Table 13-2 lists the possible changes of state. In this section, you will examine these changes of state and the factors that govern them.

SECTION OBJECTIVES

- Explain the relationship between equilibrium and changes of state.
- Predict changes in equilibrium using Le Châtelier's Principle.
- Explain what is meant by equilibrium vapor pressure.
- Describe the processes of boiling, freezing, melting, and sublimation.
- Interpret phase diagrams.

TABLE 13-2 POSSIBLE CHANGES OF STATE

Change of State	Name	Example
solid → liquid	melting	ice → water
solid → gas	sublimation	dry ice → CO_2 gas
liquid → solid	freezing	water → ice
liquid → gas	vaporization	liquid bromine → bromine vapor
gas → liquid	condensation	water vapor → water

Equilibrium

In chemistry, **equilibrium** *is a dynamic condition in which two opposing physical or chemical changes occur at equal rates in a given closed (constant mass) system.* As an analogy, think of a busy department store at holiday time. When the doors are first opened in the morning, customers begin to enter the store. As time passes, some customers finish their shopping and leave the store, while new customers enter. At this point, the store is at a kind of equilibrium. The number of customers inside the store remains constant as long as new customers enter the store at the same rate that other customers leave. As closing time approaches, people leave the store at a faster rate than other people enter; equilibrium no longer exists.

Equilibrium is a very important chemical concept. Here you will learn about it in relation to changes of state. In Chapter 19, you will study equilibrium in terms of chemical reactions.

Equilibrium and changes of state Consider the evaporation of a liquid in a closed container in which, initially, there is a vacuum over the liquid, as shown in Figure 13-12. Assume that the contents of the container maintain the same temperature as the surroundings, which is normal room temperature (approximately 22°C). The liquid and the container are the **system,** that is, *the sample of matter being studied.* Initially, there is a single phase inside this system—the liquid phase. *A* **phase** *is any part of a system that has uniform composition and properties.*

A molecule at the upper surface of the liquid can overcome the attraction of neighboring molecules and evaporate if its energy is high enough. Molecules that have entered the vapor phase behave as typical gas molecules. Some of these vapor molecules move down toward the liquid surface and condense, however, or reenter the liquid phase. **Condensation** *is the process by which a gas changes to a liquid.*

As long as the temperature and surface area of the liquid remain constant, the rate at which its molecules enter the vapor phase remains constant. The rate at which molecules pass from the vapor phase to the liquid

A chemical equilibrium is a closed system, but one in which the total number of particles is a constant. The store represents an open system because new customers are entering; the total number of customers is not a constant, but an ever increasing number.

Section Review Answers
continued

interactive forces holding them together and the particles can break out of their positions in the crystal. (b) The solid particles are packed closely together. (c) The particles of a solid are more closely packed than those of a liquid or a gas.
3. The particles of amorphous solids are not arranged in a uniform crystal lattice, but are arranged randomly, somewhat like the particles in a liquid.
4. Ionic crystals, covalent network crystals, metallic crystals, covalent molecular crystals

Liquids and Solids

Figure 13-12 A liquid-vapor equilibrium developing in a closed system. *(a)* At first there is only liquid present but molecules begin to evaporate. *(b)* Evaporation continues at the same rate. There is a relatively low concentration of vapor. Vapor molecules are condensing to liquid, however. The rate of condensation is lower than the rate of evaporation. *(c)* The concentration of vapor has increased to the point at which the rates of evaporation and condensation have become equal. Equilibrium has been reached. The amounts of liquid and of vapor now remain constant.

phase also depends upon the concentration of molecules in the vapor phase, however. **Concentration** *refers to the number of particles per unit volume.* Initially, the concentration of molecules and therefore, the rate of condensation are zero, as shown in Figure 13-12a. As time passes and evaporation continues, the concentration of molecules in the vapor phase increases. This increase in concentration results in an increase in the rate at which molecules reenter the liquid phase, as shown in Figure 13-12b.

Eventually, the rate at which molecules condense equals the rate at which molecules evaporate, as shown in Figure 13-12c. In other words, equilibrium has been reached.

Whenever a liquid changes to a vapor, as it does when it evaporates, it absorbs heat energy from its surroundings. Evaporation can therefore be represented as

$$\text{liquid} + \text{heat energy} \rightarrow \text{vapor}$$

Whenever a vapor condenses, it gives off heat energy to its surroundings. Condensation can therefore be represented as

$$\text{vapor} \rightarrow \text{liquid} + \text{heat energy}$$

The liquid-vapor equilibrium can be represented as

$$\text{liquid} + \text{heat energy} \rightleftarrows \text{vapor}$$

The "double-yields" sign in the equation above represents a reaction at equilibrium. This means that the reaction can be read in either direction. The forward reaction is represented when the equation is read from left to right: liquid + heat energy → vapor. The reverse reaction is represented when the equation is read from right to left: vapor → liquid + heat energy.

Le Châtelier's principle In 1888, the French chemist Henri Louis Le Châtelier (1850–1936) developed an important principle that enables scientists to predict how a change in the conditions of a system at equilibrium will affect the equilibrium. **Le Châtelier's principle** can be stated as follows: *When a physical or chemical system at equilibrium is disturbed by application of a stress (such as a change in concentration, pressure, or temperature), it attains a new equilibrium position that minimizes the stress.*

Le Châtelier's principle can be used to predict how the liquid-vapor equilibrium in the system just discussed changes when the temperature of the system is increased from 20°C to 50°C. The equilibrium of the system can be represented as

$$\text{liquid} + \text{heat energy} \rightleftarrows \text{vapor}$$

According to Le Châtelier's principle, the system will adjust itself to counteract an increase in temperature. In this case, the forward reaction is endothermic—it absorbs energy. This reaction will be favored in order to counteract the rise in temperature. The forward reaction temporarily proceeds at a faster rate than the reverse reaction, until finally a new equilibrium is reached. At this new, higher-temperature equilibrium, the concentration of vapor is higher than it was at the lower-temperature equilibrium. That is, the equilibrium system has shifted to the right, favoring the products. At the new equilibrium point, the reverse reaction, condensation, occurs at a faster rate than it did at the old equilibrium point so that the rates of the reverse and forward reaction are equal.

Suppose, on the other hand, that the temperature of the system in equilibrium at 20°C is lowered to 5°C. According to Le Châtelier's principle, the system will adjust itself to counteract the decrease in temperature. Here the reverse reaction will be favored because it is exothermic and releases energy. Equilibrium is shifted to the left and reestablished at 5°C with a vapor concentration less than that at 20°C.

Suppose the mass in the system and the temperature of this equilibrium system remain the same, but the volume of the system suddenly increases. What happens to the equilibrium? Initially, the concentration of molecules in the vapor phase decreases because the same number of vapor-phase molecules now occupy a larger volume. Because the concentration of vapor molecules decreases, fewer vapor molecules strike the liquid surface and return to the liquid phase at any given time. The rate of condensation therefore decreases. The rate of evaporation, which remains the same as it was before, is now higher than the rate of condensation. As a result, an additional net evaporation of liquid counteracts the imposed decrease in vapor concentration by increasing the number of vapor molecules. When equilibrium is reestablished, the concentration of molecules in the vapor phase is the same as in the system of smaller volume. Because there was a net movement of molecules from the liquid phase to the vapor phase, the number of molecules in the liquid phase has decreased. The equilibrium has therefore shifted to the right.

The system will never be able to reestablish the original equilibrium conditions. A system partially counteracts the stress before reestablishing equilibrium.

Equilibrium Vapor Pressure of a Liquid

The vapor molecules in equilibrium with a liquid in a closed system exert a pressure proportional to the vapor concentration. *The pressure exerted by the molecules of a vapor which are in equilibrium with its corresponding liquid at a given temperature is called the* **equilibrium vapor pressure** *of the liquid.* Solids can also have an equilibrium vapor pressure, since some of the solid can change directly to vapor. Figure 13-13 shows the apparatus used to measure the equilibrium vapor pressure of a solid or liquid.

When the equilibrium vapor pressure of water is graphed, as in Figure 13-14, the result is a vapor pressure-vs.-temperature curve for liquid water.

Liquids and Solids

Figure 13-13 This apparatus can be used to measure the equilibrium vapor pressure of a liquid or a solid.

Liquid water can exist in equilibrium with water vapor only up to a temperature of 374.1°C. As discussed later in this chapter, neither liquid water nor solid water can exist at temperatures above 374.1°C. The curve in Figure 13-14 shows that the equilibrium vapor pressure of water increases as temperature increases, although not by direct proportion. This finding is consistent with Le Châtelier's principle since the concentration of water vapor, and therefore the equilibrium vapor pressure of water, increases in response to temperature increases. At any given temperature, water vapor exerts a specific vapor pressure.

The increase in equilibrium vapor pressure with increasing temperature can be explained in terms of the particle models for the liquid and gaseous states. Increasing the temperature of a liquid increases its average kinetic energy, which in turn increases the number of its molecules that have enough energy to escape from the liquid phase into the vapor phase. This increased evaporation rate increases the concentration of molecules in the vapor phase, thus disturbing the liquid-vapor equilibrium. But the increased concentration of molecules in the vapor phase then increases the rate at which molecules reenter the liquid phase. Soon equilibrium is reestablished, but at a higher equilibrium vapor pressure.

The strength of the attractive forces is independent of temperature. Higher temperatures with resultant higher kinetic energies make these forces less effective.

Every liquid has a specific equilibrium vapor pressure at a given temperature. This is because all liquids have characteristic forces of attraction between their particles. The stronger these attractive forces are, the smaller is the percentage of liquid particles that can escape to the vapor phase at any given temperature, and therefore, the lower is the equilibrium vapor pressure. Volatile liquids, which evaporate readily, have relatively weak forces between their particles. Ether is a typical volatile liquid. Nonvolatile liquids, which evaporate slowly, have relatively strong attractive forces between particles. Molten ionic compounds are examples of nonvolatile liquids.

Figure 13-14 This graph shows the vapor pressure curve for water.

Be careful not to confuse the equilibrium vapor pressure of a liquid with the pressure of a vapor that is not in equilibrium with its corresponding liquid. The equilibrium vapor pressure of a liquid depends only on temperature. The pressure of a vapor that is not in equilibrium with the corresponding liquid obeys the gas laws, like any other gas, and is inversely proportional to its volume at constant temperature, according to Boyle's law.

Boiling

Equilibrium vapor pressures can be used to explain the common phenomenon of boiling. **Boiling** *is the conversion of a liquid to a vapor, within the liquid as well as at its surface when the equilibrium vapor pressure of the liquid is equal to the atmospheric pressure.*

A liquid does not boil if its equilibrium vapor pressure is lower than the atmospheric pressure. As the temperature of the liquid is increased, the equilibrium vapor pressure also increases. Finally, a temperature is reached at which the equilibrium vapor pressure is equal to the atmospheric pressure. This temperature is known as the boiling point. *The* **boiling point** *of a liquid is the temperature at which the equilibrium vapor pressure of the liquid is equal to the atmospheric pressure.* The lower the atmospheric pressure is, the lower is the boiling point; the higher the atmospheric pressure is, the higher is the boiling point. This is the reason why foods take longer to cook at high elevations where atmospheric pressures are lower than at sea level. This principle is the basis for the operation of the pressure cooker shown in Figure 13-15.

The pressure cooker increases the boiling temperature of the water inside by allowing a certain amount of steam pressure to build up over the surface of the boiling water. In contrast, a device called a vacuum evaporator brings about boiling at lower-than-normal temperatures. Vacuum evaporators are used to remove water from milk and sugar solutions. Under reduced pressure, the water boils away at a temperature low enough to avoid scorching the milk or sugar. This process is used to prepare evaporated milk and sweetened condensed milk.

At normal atmospheric pressure (760 mm Hg, or 101.3 kPa [0.1 MPa]), the boiling point of water is exactly 100°C. This temperature is known as the normal boiling point of water. It takes longer to hard-boil an egg in Denver, Colorado, than it does in Boston, Massachusetts, however, because water boils at a lower temperature in Denver. Food cooks faster in a pressure cooker because water boils at a higher temperature. The normal boiling points of water and some other liquids are shown in the graph in Figure 13-16. Note that the normal boiling point of each liquid is the temperature at which the liquid's equilibrium vapor pressure equals 760 mm Hg.

If you were to measure carefully the temperature of a boiling liquid and its vapor, you would find that both are at the same constant temperature. As you know from experience, heat must be continuously added in order to keep a liquid boiling. A pot of boiling water stops boiling almost immediately after it is removed from the stove. Since the temperature and, therefore, the average kinetic energy of the particles in a system at the boiling point remain constant despite the continuous addition of heat, where does the added heat energy go? It is used to overcome the attractive forces between liquid molecules during the liquid-to-gas change, and is stored in the vapor as potential energy. *The amount of heat energy required to vaporize one mole of liquid at its boiling point is known as the liquid's* **molar heat of vaporization.** The magnitude of the molar heat of vaporization is a measure of the attraction between the particles of the liquid. The stronger this attraction is, the more energy is required to overcome it, and, therefore, the higher is the molar heat of vaporization. Each liquid has a characteristic molar heat of vaporization. Compared to other liquids, water has an unusually

The strength of the attractive forces is independent of temperature. Higher temperatures with resultant higher kinetic energies make these forces less effective.

Figure 13-15 A pressure cooker makes use of elevated pressure to raise boiling point and shorten cooking time.

Figure 13-16 The normal boiling point of each of these liquids is the temperature at which the liquid's equilibrium vapor pressure equals 760 mm Hg (shown here as a dotted line).

high molar heat of vaporization. This property, which you will learn more about in the next section, makes water a very effective cooling agent. When water evaporates from your skin, the escaping molecules carry a great deal of heat away with them.

Freezing and Melting

The physical change of a liquid to a solid is called **freezing**. Freezing involves a loss of energy, in the form of heat, by the liquid:

$$\text{liquid} \longrightarrow \text{solid} + \text{heat energy}$$

In the case of a pure substance, this change occurs at constant temperature. *The normal* **freezing point** *is the temperature at which the solid and liquid are in equlibrium at 1 atm (101.3 kPa) pressure.* This observation indicates that the particles of the liquid and the solid have the same average kinetic energy. Therefore, the energy loss during freezing is a loss of potential energy that was present in the liquid. At the same time as energy decreases, there is a significant increase in order, since the solid state of a substance is much more ordered than the liquid state, even at the same temperature.

The reverse physical change, melting, also occurs at constant temperature. **Melting** *is the physical change of a solid to a liquid.* The **melting point** *is the temperature at which the transition from the solid phase to the liquid phase occurs.* During melting, the melting solid continuously absorbs heat:

$$\text{solid} + \text{heat energy} \longrightarrow \text{liquid}$$

The amount of heat energy required to melt one mole of solid at its melting point is its **molar heat of fusion.** The heat absorbed by the melting solid increases the potential energy of the solid as its particles are pulled apart in opposition to the attractive forces tending to hold them together. At the same time, there is a significant decrease in order.

For pure, crystalline solids, the melting point and the freezing point are the same. At equilibrium at the melting or freezing point, melting and freezing proceed at equal rates according to the following general equilibrium equation:

$$\text{solid} + \text{heat energy} \rightleftarrows \text{liquid}$$

At normal atmospheric pressure, the temperature of a system containing ice and liquid water will remain at 0°C as long as both ice and water are present, no matter what the surrounding temperature. As predicted by Le Châtelier's principle, adding heat to such a system shifts the equilibrium to the right, increasing the proportion of liquid water and decreasing that of ice. Only after all the ice has melted will the addition of heat increase the temperature of the system.

Under sufficiently low temperature and pressure conditions, a liquid cannot exist. Under such conditions, a solid substance exists in equilibrium with its vapor instead of its liquid:

$$\text{solid} + \text{heat energy} \rightleftarrows \text{vapor}$$

The change of state from a solid directly to a gas is known as **sublimation.** Among the common substances that sublime at ordinary temperature are dry ice and iodine. Ordinary ice sublimes slowly at temperatures lower than its melting point (0°C). This is how a thin layer of snow can eventually disappear, despite the fact that the temperature remains below 0°C. Frost-free refrigerators make use of sublimation by periodically slightly warming their freezer compartments, and then using a blower to remove molecules of water as they sublime.

Phase Diagrams

A **phase diagram** *is a graph of temperature versus pressure that indicates the conditions under which gaseous, liquid, and solid phases of a particular substance exist.* A phase diagram also reveals how the states of a system change with changing temperature or pressure.

Figure 13-17 shows the phase diagram for water over a range of temperatures and pressure conditions. Note the three curves, AB, AD, and AC. Each point on curve AC indicates the simultaneous temperature and pressure conditions under which liquid water and water vapor can exist together at equilibrium. Similarly, curve AD indicates the temperature and pressure conditions under which ice and liquid water are in equilibrium. Curve AB indicates the temperature and pressure conditions under which ice and water vapor coexist at equilibrium. Point A is called the triple point of water. *The* **triple point** *of a substance indicates the temperature and pressure conditions at which the solid, liquid, and vapor of the substance can coexist at equilibrium.* Point C is called the critical point of water. The **critical point** *of a substance indicates the* **critical temperature** *(t_c), the temperature above which the substance cannot exist in the liquid state.* The critical temperature of water is 374.1°C. Above this temperature, water cannot be liquified, even if thousands of atmospheres of pressure are applied. *The critical point also indicates the* **critical pressure** *(P_c), the lowest pressure required for the substance to exist as a liquid at the critical temperature.*

The phase diagram in Figure 13-17 also indicates the normal boiling point and the normal freezing point of a substance. As shown by the slope of line AD, ice melts at a higher temperature with decreasing pressure. As the

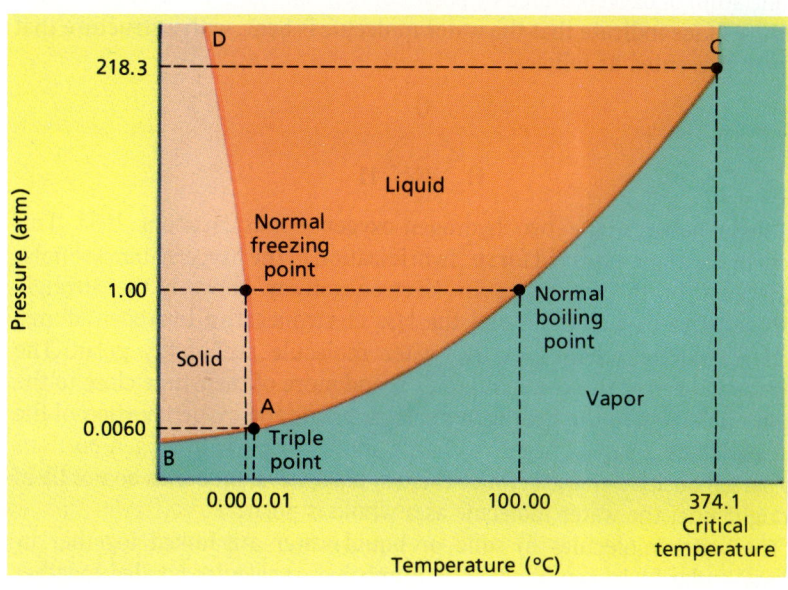

Figure 13-17 The phase diagram for water. Water vapor cannot be converted to a liquid beyond the critical temperature (374.1°C), no matter how high the pressure exerted on the vapor. The critical pressure (218.3 atm) is the lowest pressure required for the substance to exist as a liquid at the critical temperature.

1. A dynamic condition in which two opposing physical or chemical changes occur at equal rates in the same closed system.
2. If any of the factors determining an equilibrium are changed, the system will adjust itself in a way that tends to minimize that change.
3. With an increase, the equilibrium reaction shifts to the right. This forward (left-to-right) reaction absorbs energy and thus counteracts the rise in temperature. With a decrease, the reaction shifts to the left. The reverse reaction releases energy and thus counteracts the decrease in temperature.

SECTION OBJECTIVES

- Describe the structure of a water molecule.
- Discuss the physical properties of water and explain how these are determined by the structure of water.

4. The pressure exerted by a vapor in equilibrium with its corresponding liquid at a given temperature
5. The temperature at which the equilibrium vapor pressure is equal to the atmospheric pressure
6. (a) The temperature and pressure conditions under which ice, water vapor, and liquid water all exist together at equilibrium (b) The temperature above which water cannot exist in the liquid state, as well as the lowest pressure required for water to exist as a liquid at this temperature

STUDY HINT

Orbitals, hybridization, and chemical bonding are treated in Chapters 4 and 6.

Δe.n. = e.n.(oxygen) − e.n.(hydrogen)
Δe.n. = 3.5 − 2.1 = 1.4

temperature drops below the triple point, the temperature of sublimation decreases with decreasing pressure. The freeze-drying of foods is accomplished by vaporizing the ice and lowering the pressure after freezing.

Section Review

1. What is equilibrium?
2. State Le Châtelier's principle.
3. What happens when a liquid-vapor system at equilibrium experiences an increase in temperature? What happens when it experiences a decrease in temperature?
4. What is the equilibrium vapor pressure of a liquid?
5. What is the boiling point of a liquid?
6. In the phase diagram for water, what is (a) the triple point and (b) the critical point?

13.4 Water

Water is a familiar substance in all three physical states: solid (ice), liquid (liquid water), and gas (water vapor). On earth, water is by far the most abundant liquid. Oceans, rivers, and lakes cover about 75 percent of the earth's surface and significant quantities of water are frozen in glaciers. Water is an essential component of all organisms; 70 percent to 90 percent of the mass of living things is water. The chemical reactions of life processes take place in water, and water is frequently also a reactant or product in such reactions. In order to understand better the importance of water, you will take a closer look first at its structure and then at its properties.

Structure of Water

As discussed in Chapter 6, water molecules consist of two atoms of hydrogen and one atom of oxygen united by polar covalent bonds. Studies of the crystal structure of ice indicate that the water molecule is bent, with a structure that can be represented as follows:

$$\begin{array}{c} O \\ \diagup \diagdown \\ H 105° H \end{array}$$

The angle between the two hydrogen-oxygen bonds is about 105°. This is close to the angle expected for sp^3 hybridization of the oxygen-atom orbitals.

Because of the difference in electronegativity between the strongly electronegative oxygen atom and the less electronegative hydrogen atoms, the H—O covalent bonds in the water molecule are highly polar. The electrons in each of the covalent bonds spend more of their time close to the oxygen nucleus than the hydrogen nucleus, giving the oxygen portion of the water molecule a partial negative charge and leaving the hydrogen portions with partial positive charges. Since the two polar-covalent bonds do not lie in a straight line, the water molecule as a whole is polar.

The water molecules in solid or liquid water are linked together by intermolecular hydrogen bonds. The number of molecules bonded together decreases with increasing temperature because increases in kinetic energy makes such bond formation difficult. Nevertheless, there are usually from four to eight molecules per group in liquid water, as shown in Figure 13-18. If it were not for the formation of these molecular groups, water would be a gas at room temperature. In contrast, nonpolar molecules such as methane, CH_4, which are similar in size and mass to water molecules, do not undergo hydrogen bonding and are gases at room temperature.

Ice consists of H_2O molecules arranged in the hexagonal latticework shown in Figure 13-18. The spaces between molecules in this pattern are filled with empty space and account for the relatively low density of ice. As ice is heated, the increased energy of the atoms and molecules causes them to move and vibrate more vigorously. When the melting point of ice is reached, the energy of the atoms and molecules is so great that the rigid open-lattice structure of the ice crystals breaks down. The ice turns into liquid water.

The hydrogen bonds between molecules of liquid water at 0°C are longer and more flexible than those between molecules of ice at the same temperature. Because of the flexibility of hydrogen bonds and because the rigid lattice structure of ice has broken down, groups of water molecules tend to crowd together and liquid water is denser than ice.

Figure 13-18 The structure of liquid water (left) and ice (right)

As water is warmed from 0°C, the breakage of some of the hydrogen bonds enables water molecules to crowd closer together. At the same time, the increased energy of the water molecules (owing to the rise of temperature) allows them to overcome molecular attractions more effectively and gradually spread apart. Water molecules are as tightly packed as possible at 3.98°C; hence, we say that water has its maximum density at 3.98°C. At temperatures above 3.98°C, the increasing energy of the water molecules causes them to move farther apart as temperatures continue to rise. Thus, the density of water steadily decreases at temperatures above 3.98°C.

As temperature approaches the boiling point, groups of water molecules must absorb enough energy to break up into single molecules before the water boils. Because of hydrogen bonding between water molecules, a large amount of heat is needed, and water's boiling point is relatively high (100°C).

Physical Properties of Water

At room temperature, pure water is a transparent, odorless, tasteless, and almost colorless liquid. Any observable odor or taste is caused by impurities such as dissolved minerals, liquids, or gases.

As shown by its phase diagram in Figure 13-17, at a pressure of 1 atm (101.3 kPa), water freezes and melts at 0°C. The molar heat of fusion of ice, 6.008 kJ/mol, is relatively large compared to the molar heats of fusion of other solids. Water has the unusual property of expanding 11% in volume as it freezes because its molecules form a rigid latticework, as described earlier. As a result, ice at 0°C has a density of about 0.917 g/cm^3, compared with a density of 0.999 84 g/cm^3 for liquid water at 0°C.

This is why ice floats in liquid water. The low density of ice is particularly important in the case of large bodies of water. If ice did not float, the water in lakes and ponds of temperate climates would freeze solid during the winter, killing nearly all the living things in them.

When liquid water at 0°C is warmed, its density increases until its temperature reaches 3.98°C. At this temperature, water molecules are as tightly packed as they can possibly be. Thus, water is densest at 3.98°C. This behavior is also unusual compared with that of most other substances. Above 3.98°C, water behaves more typically by gradually expanding as its temperature increases. This is because of molecular motion due to an increase in kinetic energy that accompanies the rise in temperature. At 3.98°C, one milliliter of water has a mass of 0.999 973 g.

Under a pressure of 1 atm (101.3 kPa), water boils at 100°C. At this temperature, water's molar heat of vaporization is 40.79 kJ/mol. Both the boiling point and the molar heat of vaporization of water are quite high compared with those of nonpolar substances with comparable molecular masses, such as methane (CH$_4$). The high molar heat of vaporization makes water useful for household heating systems. Many homes are heated by steam (vaporized water) which condenses in radiators, releasing great quantities of heat, as shown in Figure 13-19.

Sample Problem 13.1

How much heat energy is absorbed when 47.0 g of ice melts at its given melting point (i.e., at STP)? How much heat energy is absorbed when this same mass of liquid water boils?

Solution

Step 1. Analyze Given: mass = 47.0 g H$_2$O(s)
mass = 47.0 g H$_2$O(ℓ)
Unknown: heat energy

Step 2. Plan

$$\text{mass} \rightarrow \text{mol} \rightarrow \text{heat energy}$$

$$\frac{g}{g/mol} = \text{mol}$$

mol × molar heat of fusion or vaporization = heat energy
mol × 6.008 kJ/mol = kJ (on melting)
mol × 40.79 kJ/mol = kJ (on vaporizing or boiling)

Sample Problem 13.1 *continued*

Step 3. Compute
$$\frac{47.0 \text{ g}}{18.0 \text{ g/mol}} = 2.61 \text{ mol}$$

2.61 mol × 6.008 kJ/mol = 15.7 kJ (on melting)

2.61 mol × 40.79 kJ/mol = 106 kJ (on vaporizing or boiling)

Step 4. Evaluate Units have canceled appropriately. The answers have the proper number of significant digits and are reasonably close to estimated values of 18, (3 × 6) and 120, (3 × 40), respectively.

Practice Problems
1. What quantity of heat energy is released when 506 g of liquid water freezes? *(Ans.)* 169 kJ
2. What mass of steam is required to release 4.97×10^5 kJ of heat energy on condensation? *(Ans.)* 2.19×10^5 g

1. Because of the difference in electronegativity between the strongly electronegative oxygen atom and the less electronegative hydrogen atoms, the H—O covalent bonds in the water molecule are highly polar. Since the two polar–covalent bonds do not lie in a straight line, the water molecule is polar.
2. The structure allows for the formation of molecular groups which in turn give water its high heat of fusion, high boiling point, and high heat of vaporization.
3. They are linked together by intermolecular hydrogen bonds in both liquid water and ice. In the liquid, there are usually four to eight molecules per group, while in the solid all the molecules are arranged in a hexagonal latticework.

Figure 13-19 A steam heating system is an efficent way to heat a home because of the great amount of heat contained in steam.

4. Because, at 0°C, it is less dense than liquid water. If ice did not float, the water in lakes and ponds would freeze solid, killing nearly all the living things in them.

Section Review

1. Why is a water molecule polar?
2. How is the structure of water responsible for some of water's unique characteristics?
3. Describe the arrangement of molecules in liquid water and in ice.
4. Why does ice float? Why is this phenomenon important?

Chapter Summary

- The particles of a liquid are closer together and lower in energy than those of a gas. They are more ordered than those of a gas and less ordered than those of a solid.
- Liquids have a definite volume, a fairly high density, and are relatively incompressible. Like gases, liquids can flow and are thus considered to be fluids. Liquids can dissolve other substances.
- Liquids have the ability to diffuse. Unlike gases or solids, they exhibit surface tension and can evaporate or boil. A liquid is said to freeze when it becomes a solid.
- The particles of a solid have less kinetic energy than those of a liquid or gas. They move back and forth less than particles of liquids or gases and are more densely packed.
- Solids have a definite shape and may be crystalline or amorphous. They have a definite volume and are nonfluid. High density, incompressibility, extremely slow rates of diffusion, and definite melting points are additional characteristics of solids.
- A crystal lattice is the total three-dimensional array of points that describes the arrangement of the particles of a crystal. A crystal can be classified into one of seven crystalline systems on the basis of the three-dimensional shape of its unit cells. A crystal can also be described as one of four types on the basis of the kinds of particles it contains and the types of chemical bonding between its particles.
- Amorphous solids do not have a regular shape, but instead take on whatever shape is imposed on them.
- A liquid in a closed system will gradually reach a liquid–vapor equilibrium as the rate at which molecules condense equals the rate at which they evaporate.
- When two opposing physical or chemical changes occur at equal rates in the same closed system, such a system is said to be at equilibrium. Le Châtelier's principle states that if any of the factors determining an equilibrium are changed, the system will adjust itself in a way that tends to minimize that change.
- The pressure exerted by a vapor in equilibrium with its corresponding liquid at a given temperature is its equilibrium vapor pressure. A liquid begins to boil when the equilibrium vapor pressure of the liquid is equal to the atmospheric pressure. The amount of heat energy required to vaporize one mole of liquid at its boiling point is known as the liquid's molar heat of vaporization.
- A liquid is said to freeze when it changes to a solid. Freezing involves a loss of energy, in the form of heat, by the liquid. Melting is the physical change of a solid to a liquid. The amount of heat required to melt one mole of solid at its melting point is its molar heat of fusion.
- Water is a polar-covalent compound. The water molecule is bent in shape and has a partial negative charge near its oxygen atom and a partial positive charge near each of its hydrogen atoms. This unique structure is responsible for water's relatively high molar heat of fusion, boiling point and molar heat of vaporization. It is also the reason why water expands upon freezing and is able to float.

CHAPTER 13 REVIEW

Liquids and Solids

VOCABULARY

amorphous solid (377)
boiling (387)
boiling point (387)
concentration (384)
condensation (383)
critical point (389)
critical pressure (389)
critical temperature (389)
crystal (377)
crystal lattice (379)
crystalline solid (377)
dissolved (372)
equilibrium (383)
equilibrium vapor pressure (385)
evaporation (375)
fluid (372)
freezing (388)
freezing point (388)
Le Châtelier's principle (384)
melting (388)
melting point (378)
molar heat of fusion (388)
molar heat of vaporization (387)
phase (383)
phase diagram (389)
sublimation (388)
supercooled liquid (378)
surface tension (374)
system (383)
triple point (389)
unit cell (379)
vaporization (375)

REVIEWING CONCEPTS

1. What is a fluid? (13.1)
2. What is surface tension?
3. Give two reasons why evaporation is such a crucial process in nature. (13.1)
4. List seven properties of solids and explain each in terms of the particle model of solids. (13.2)
5. Why is glass sometimes classified as a supercooled liquid? (13.2)
6. Distinguish between a crystal lattice and a unit cell.
   a. List four common examples of amorphous solids.
   b. Cite some of the uses of glass. (13.2)
7. a. List and describe the four types of crystals in terms of the nature of and type of bonding between their component particles.
   b. What physical properties are associated with each type of crystal? (13.2)
8. Using Figure 13-16, determine the approximate equilibrium vapor pressure of each of the following at the specified temperature:
   a. water at 40°C
   b. water at 80°C
   c. ether at 20°C
   d. alcohol at 60°C
   e. glycerol at 140°C (13.3)
9. a. What is sublimation?
   b. Give two examples of common substances that sublime at ordinary temperatures. (13.3)
10. What is meant by the normal freezing point of a substance? (13.3)
11. Explain why the vapor pressure of a liquid increases with increasing temperature. (13.3)
12. Explain the relationship between the attractive forces between the particles in a liquid and the equilibrium vapor pressure of that liquid. (13.3)
13. a. Explain the relationship between atmospheric pressure and the actual boiling point of a liquid.
    b. During continual boiling at this pressure, what is the relationship between the temperature of the liquid and that of its vapor?
    c. How can this phenomenon be explained? (13.3)
14. Explain the relationship between the magnitude of the molar heat of vaporization of a liquid and the degree of attraction between the particles of that liquid. (13.3)
15. Explain the relationship between the molar heat of fusion of a solid and the degree of attraction between its particles. (13.3)
16. Describe the basic structure of a water molecule. (13.4)
17. List at least eight physical properties of water. (13.4)

PROBLEMS

▶ **Molar Heat**

1. The standard molar heat of vaporization for water is 40.79 kJ/mol. How much energy would be required to vaporize each of the following?
   a. 5.00 mol H_2O
   b. 45.0 g H_2O
   c. 8.45×10^{50} molecules H_2O

2. The molar heat of fusion for water is 6.008 kJ/mol. How much energy would be required to melt each of the following?
   a. 12.75 mol ice
   b. 6.48×10^{20} kg ice

3. Calculate the molar heat of vaporization of a substance given that 0.433 mol of the substance absorbs 36.5 kJ of energy when it is vaporized.

4. Given that a substance has a molar mass of 259.0 g/mol and 71.8 g of the substance absorbs 4.307 kJ when it melts,
   a. calculate the number of moles in the 71.8-g sample.
   b. calculate the molar heat of fusion.

5. a. Calculate the number of moles in a liquid sample of a substance that has a molar heat of fusion of 3.811 kJ/mol, given that the sample releases 83.2 kJ when it freezes.
   b. Calculate the molar mass of this substance if the mass of the sample is 5519 g.

▶ **Water**

6. Which contains more molecules of H_2O — 5.00 cm³ of ice at 0°C or 5.00 cm³ of liquid water at 0°C? How many more?

7. Which contains more molecules of H_2O — 5.00 cm³ of liquid water at 0°C or 5.00 cm³ of water vapor at STP? How many more?

8. What volume and mass of steam at 100°C and 1.00 atm would, when condensed, release the same amount of heat as 100. cm³ of liquid water when frozen? What do you note, qualitatively, about the relative volumes and masses of steam and liquid water required to release the same amount of heat?

CRITICAL THINKING

1. **Analyzing Results** Distinguish between the phenomena involved when a steel needle rests on a water surface and when a boat floats in water.

2. **Applying Ideas** Explain the role of the humidity of the air in determining our relative comfort on a hot day.

3. **Applying Models** Both diamond and graphite consist only of carbon atoms, yet these materials have very different properties. Explain the basic reasons for these observed differences.

4. **Predicting Outcomes** The following liquid-vapor system is at equilibrium at a given temperature.

   $$\text{liquid} + \text{heat energy} \rightleftharpoons \text{vapor}$$

   If the temperature is increased and equilibrium is established at the higher temperature, how does the final value of each of the following compare with its initial value? (In each case, answer either: higher, lower or the same.)
   a. the rate of evaporation
   b. the rate of condensation
   c. the final concentration of vapor molecules
   d. the final number of liquid molecules

5. **Applying Models** Given a liquid-vapor equilibrium system, decide whether the temperature of that system should be increased or decreased in order to make each of the following changes in the process of establishing equilibrium at the new temperature:
   a. an increased final rate of evaporation
   b. an increased final concentration of vapor
   c. an increased final rate of condensation
   d. an increased final number of liquid molecules

6. **Interpreting Concepts** During the freezing of a substance, energy is continually being removed from that substance, yet the temperature of the liquid-solid system remains constant. Explain this phenomenon.

7. **Applying Models** At normal atmospheric pressure, the temperature of an ice-water

system will remain at 0°C as long as both ice and liquid water are present, no matter what the surrounding temperature. Explain this in terms of Le Châtelier's principle.

8. **Interpreting Graphics** Given a sample of water at any point on curve AB in Figure 13-17, what effect would each of the following changes have on that sample?
   a. increasing the temperature at constant pressure
   b. increasing the pressure at constant temperature
   c. decreasing the temperature at constant pressure;
   d. decreasing the pressure at constant temperature.

9. **Predicting Outcomes** Given a sample of water at any point on curve AD in Figure 13-17, how could more of the liquid water in that sample be converted into a solid without changing the temperature?

10. **Interpreting Concepts** Explain why methane (CH_4), which is similar in size and mass to water, is a gas at room temperature, whereas water is a liquid at that temperature.

11. **Designing Experiments** Is it possible for a container of water to simultaneously freeze and boil? Describe an experimental set-up that could demonstrate this.

12. **Interpreting Graphics** Using the phase diagram for CO_2, describe all the phase changes that would occur when CO_2 is heated from $-100°C$ to $-10°C$ at a constant pressure of 6 atm.

CROSS CONNECTIONS

1. **Art** Ceramics are formed from silicates found in the earth. Research the use of ceramics as containers and art forms in the history of ancient Greece, Egypt, and Rome.

2. **Biology** How is capillary action important to plants (hint: think of very large plants, such as trees).

3. **Home Economics** Prepare a report about the adjustments that must be made when cooking and baking at high elevations. Collect instructions for high elevation adjustments from the packages of prepared mixes at the grocery store. Explain why changes must be made in recipes that will be prepared at high elevations. Check your library for cookbooks containing information about food preparation at high elevations.

RESEARCH AND WRITING

1. Write a brief report about the glass-making process. If possible, observe a glassblower at work and record the step-by-step procedures involved.

2. Consult reference materials at the library and prepare a report on the process of freeze-drying, describing its history and its applications.

PROJECTS

1. Compile separate lists of crystalline and amorphous solids found in your home. Compare your lists with those of your classmates.

2. Design an experiment to grow crystals of various safe, common household materials. Record the conditions under which each type of crystal is best grown.

Chemistry and Careers

Occupations in the Compressed-Gas Industries

The major method of producing pure gases is by compressing and cooling air until it liquefies. This liquid is then fractionally distilled to extract the desired gases. The process requires special equipment but generally does not require chemical expertise from the equipment operators.

The liquefied gases are extremely cold. People who work with compressed gases must understand the changes of pressure and volume that occur as the gases warm and the changes that occur in materials that come in contact with cold, liquified gases. Pipes can shatter, since many metals get brittle when they are cold. Joints will leak if they have not been sealed with materials that can withstand extreme temperature changes.

The manufacture and handling of nonatmospheric gases require chemical knowledge as well as common sense. The chemical reaction used to produce the gas must be carefully monitored to avoid potentially dangerous side reactions. For example, chlorine gas reacts with nearly any metallic surface it touches. Leaks of carbon monoxide can lead to fatalities because, in its early stages, CO poisoning is easily confused with the onset of a bad cold. Plans must be developed for coping with spills and leaking pipes.

Opportunities abound in the compressed-gas industries for people with a background or interest in chemistry. A college degree is usually not necessary, but some technical training is essential.

Chemical Patent Attorney

Students who have an interest in both the law and chemistry may find that chemical patent law is an attractive career. Chemical patent attorneys are becoming increasingly important to industry.

In addition to doing the painstaking work needed to establish a valid patent, chemical patent attorneys help decide whether a product is sufficiently different to justify filing a patent application. Deciding if an idea is new enough to patent or whether it infringes on existing patents is often difficult. Even if a chemical compound is already known, it may be possible to patent a new method of making it.

To become a chemical patent attorney, you must study chemistry as part of your undergraduate college program and then obtain a law degree.

UNIT 4 • FOCUS ON REASONING

College Entrance Exam Preparation

X-ray diffraction studies of metal crystals show various forms of unit cells. The unit cell is the smallest structural unit that repeated through 3-dimensions gives the structure of the crystal. Three simple unit cells found in metals are shown below.

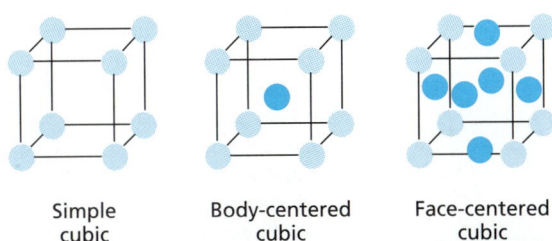

Simple cubic   Body-centered cubic   Face-centered cubic

The dimensions of each unit cell can be described mathematically using the following.

$$s = \text{length of the edge}$$
$$r = \text{atomic radius}$$

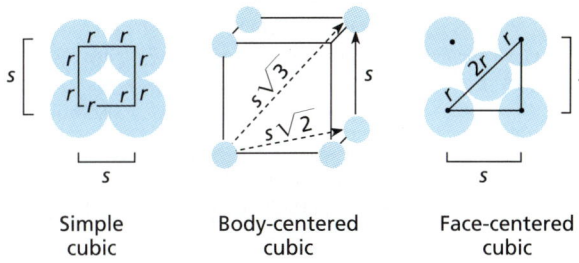

Simple cubic   Body-centered cubic   Face-centered cubic

The following reasoning is used to assign a number of atoms to each type of unit cell.

- An atom at the center of the face of a cube is shared by another cube that touches that face. Only half of that atom can be assigned to a given cell.

- An atom at the corner of a cube forms part of eight different cubes that touch at a particular point. Therefore, only one-eighth of a corner atom belongs to a particular cell.

1. Which of the following relationships applies to the simple cubic cell?
   a. $2r = s$ c. $r = s$
   b. $8r = 2s$ d. $2s = r$

2. Which of the following relationships applies to the face-centered cubic cell?
   a. $2s = 4r$ c. $4r = s\sqrt{2}$
   b. $2r = s$ d. $3r = s$

3. Based on the information given, what would you conclude to be the number of atoms assigned to the simple cubic cell?
   a. 4 corner atoms × 1/4 = 1 atom per cube
   b. 8 corner atoms × 1/8 = 1 atom per cube
   c. 8 corner atoms × 1/4 = 2 atoms per cube
   d. 8 corner atoms × 1/32 = 1/4 atom per cube

4. What would you conclude to be the number of atoms assigned to the face-centered cubic cell?
   a. 4 corner atoms × 1/4 = 1 atom per cube
   b. 8 corner atoms × 1/8 = 1 atom per cube
   c. 8 corner atoms × 1/8 + 6 face atoms × 1/2 = 4 atoms per cube
   d. 8 corner atoms × 1/8 + 1 center atom = 2 atoms per cube

5. If a metal crystal showing a body-centered cubic arrangement consists of one million unit cells, how many atoms would be found in the crystal?
   a. 1 million   c. 4 million
   b. 2 million   d. 8 million

INTRA-SCIENCE

Preserving the Ozone Layer

In 1988 a giant U.S. chemical company announced that it would voluntarily phase out production of all chlorofluorocarbons (CFCs). The reason: to protect the earth's ozone shield.

It was a significant announcement. Although it had been suspected for some time that CFCs were damaging the ozone layer, and although an international accord in 1987 had called for a freeze on production levels, E.I. du Pont de Nemours & Company, a chemical concern, was more than merely complying with the agreement's terms. The chemical company was ready to reduce, rather than merely freeze, its level of CFC production.

Why is ozone so important? No substance is more crucial to life than ordinary molecular oxygen (O_2), which makes up about 20 percent of the volume of the lower atmosphere, where it is crucial to respiration. Ozone (O_3) is a different molecular form of the element oxygen, and is equally important to life. A layer of ozone in the upper atmosphere guards living things on earth against deadly solar radiation. CFCs, such as those used as propellants in aerosol cans, are believed to be threatening this protective layer.

Ironically, ozone is a pollutant at the earth's surface—a major ingredient of photochemical smog. Scientists are trying to minimize the amount of ozone in the lower atmosphere and at the same time trying to preserve the ozone layer in the upper atmosphere.

Ozone is a bluish gas that is produced by the process in which O_2 is broken up into atomic oxygen, followed by the combination of the single atoms of oxygen (O) with O_2 to produce O_3. On the earth's surface, the production of ozone commonly involves an electrical spark that passes through air or pure oxygen. The pungent smell that can be noticed near a spark-producing electrical apparatus or during a lightning storm is that of ozone.

The Connection to Biology and Health Sciences

Because of its strong oxidizing character, ozone is toxic to animals, including humans. Los Angeles smog can contain up to 0.5 parts per million (ppm) of ozone, a concentration sufficiently high to irritate eyes and respiratory tracts. This ozone is formed by the action of sunlight on oxygen in the presence of impurities in the air. At the altitudes flown by intercontinental aircraft, cabin levels of ozone can reach 1.2 ppm unless reduced by carbon filters and ozone-decomposing equipment.

Although ozone itself is toxic to life, its presence in the upper atmosphere protects life because ozone absorbs ultraviolet radiation at a wavelength range of 240–320 nanometers (a nanometer is one billionth of a meter— 10^{-9} m).

400

Ultraviolet radiation can penetrate 60 feet or more into the ocean and kill one-celled organisms that are at the base of the ocean food chain. The loss of significant amounts of these organisms could result in a collapse of the ocean ecosystem.

Ultraviolet radiation also kills the surface cells of plants. A 15-percent drop in atmospheric ozone could result in $3 billion worth of crop damage in the United States.

Sunburn, premature aging of the skin, and skin cancer are consequences of increased ultraviolet radiation. The Environmental Protection Agency estimates that a 1- percent increase in ultraviolet radiation could cause the incidence of skin cancer to increase by 2 percent.

The Connection to Physics

The production of ozone in the upper atmosphere and at the earth's surface depends upon the absorption of ultraviolet light. In the first step of the production of ozone in the upper atmosphere, ultraviolet light, which has a wavelength lower than 240 nm, is absorbed by O_2. This absorbed light is changed into an equivalent amount of energy, which causes the oxygen molecule to break apart into two oxygen atoms. Ozone is then produced by the reaction of O with O_2.

Once ozone molecules have been formed, they absorb ultraviolet light in the 240–320 nm range. The energy released from this absorbed light causes each ozone molecule to decompose to produce a molecule O_2 and an atom of O. These two species then recombine to form a new molecule of ozone. In this way, the ozone layer is maintained to perform its shielding role.

On the earth's surface, nitrogen dioxide (NO_2), which is produced in automobile engines by the combination of atmospheric oxygen with atmospheric nitrogen, absorbs ultraviolet light with a wavelength lower than 395 nm. The released energy causes the NO_2 molecule to decompose to NO and O. The resulting atomic oxygen can combine with O_2 to form ozone or can combine with unburned hydrocarbons from automobile exhaust to produce a variety of irritating compounds.

The Connection to Earth Science

The British scientist Joseph Farman began monitoring the levels of ozone over Antarctica in 1957 using instruments positioned at Halley Bay. In 1985 he shocked the scientific community with his report that there had been a decline in these ozone levels every year since 1978. His results were confirmed by measurements made from NASA's *Nimbus-7* satellite. Scientists immediately sought an explanation for this "ozone hole."

Today there is evidence that this hole may be caused by human-made chemicals such as CFCs. Because they are nontoxic and chemically inert, these compounds have been widely used as refrigerants and as propellants in aerosol cans. When these substances are released to the atmosphere, however, their inertness ensures that they are not readily destroyed. As a consequence, they eventually find their way into the upper atmosphere.

Scientists hypothesize that, once in the upper atmosphere, CFCs are decomposed by ultraviolet light to produce chlorine, which destroys ozone. Recent experiments in which a NASA aircraft collected air samples high over Antarctica revealed a chlorine level 500 times greater than normal.

Unit 5

Chapter 14
Solutions 404

Chapter 15
Ions in Aqueous Solutions 438

Chapter 16
Acids and Bases 460

Chapter 17
Acid–Base Titration and pH 490

SOLUTIONS AND THEIR BEHAVIOR

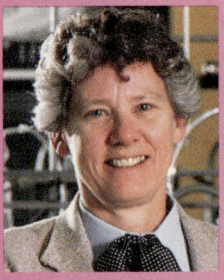

My high school chemistry teacher, John St. Sure, hooked me on chemistry with a class dominated by his twin loves: pranks, and technical excellence. His perennial stock of anecdotes about his former students who had made good in science or engineering delivered a powerful message about the impotence of mere enthusiasm unsupported by hard work.

My present work in analytical chemistry is devoted to developing better techniques for measuring trace amounts of substances in mixtures known as solutions. The aim is to have better tools to measure and understand our environment.

Solutions are an important part of our environment. For example, the oxygen and carbon dioxide dissolved in oceans, lakes, and rivers allow aquatic life to flourish. Aquatic plants liberate tremendous amounts of oxygen as they break down carbon dioxide during respiration. This helps maintain a life-sustaining balance between oxygen and carbon dioxide in our atmosphere.

Janet Osteryoung
Janet G. Osteryoung
Analytical Chemist
State University of New York at Buffalo

Coral reef in the Red Sea

CHAPTER 14
Solutions

PLANNING CHART

Approximate Time: 9/11 days

	SECTION/TEXT FEATURES		RESOURCES	
BASIC	**14.1 TYPES OF MIXTURES** Demos 1,2 *Periodic Table* Videodisc Desktop Investigation Microscale Investigation 12 Section Review Chapter Review • Questions 1–2	405 404A–B 404J 408 858 410 434	*Teacher's Resource Binder* Section Review Worksheets *Other Resources* *Shakhashiri* Video Demo 40	48
BASIC	**14.2 THE SOLUTION PROCESS** Demos 3, 4, 5, 6 Section Review Chapter Review • Questions 3–13, Problems 1–2	410 404B–C 419 434	*Teacher's Resource Binder* Lab Experiment 18 Section Review Worksheets *Other Resources* *Shakhashiri* Video Demos 3, 4, 6, 10, 41 Transparency 30 Blackline Master 34, 35, 36	107 49
BASIC	**14.3 CONCENTRATION OF SOLUTIONS** Demo 7 Section Review Chapter Review • Questions 14–15, Problems 3–22	420 404C 425 434	*Teacher's Resource Binder* Problem Solving Worksheets Section Review Worksheets	50
HONORS	**14.4 COLLIGATIVE PROPERTIES OF SOLUTIONS** Demo 8 Section Review Chapter Review • Question 16, Problems 23–35	426 404D 433 434	*Teacher's Resource Binder* Problem Solving Worksheets Section Review Worksheets *Other Resources* *Shakhashiri* Video Demo 42	51
	CHAPTER END ACTIVITIES Critical Thinking, Cross Connections, Research and Writing, Projects		*Teacher's Resource Binder* Test Booklet *Computer Test Generator*	69, 71

Teaching Strategies

Introduction

Introduce this chapter by telling students they are going to learn why salt is added to ice in an ice-cream freezer and to icy roads in wintertime, why the temperature goes up on a candy thermometer as you cook the candy, how to remove stains from clothing, why soda fizzes when the cap is removed from the bottle, why warm soda goes flat quickly, the types of foods that are colloids, and other exciting things.

14.1 Types of Mixtures

The following demonstration involves making two common solutions with seemingly illogical but interesting results!

Demonstration 1: 1 + 1 Does Not Always = 2
PURPOSE: To determine the volume of a solution made up of equal volumes of its components
MATERIALS: 2 250-mL beakers, Celsius thermometer, distilled water, 95% ethanol, 4 100-mL graduates, stirring rod, sucrose
PROCEDURE: Fill four 100-mL graduates with 50 mL of ethyl alcohol, 50 mL of distilled water, 50 mL of sucrose, and 50 mL

of hot water (about 70°C). Ask your students what they think the volume of each of the following solutions will be when *(1)* you pour the 50 mL of ethyl alcohol into the graduate containing 50 mL of water, and *(2)* you mix the 50 mL of sucrose and 50 mL of hot water in a beaker? Carry out the two experiments, and determine the volume of the resulting mixtures. Be sure the students are aware of the original volumes of the components of the mixtures. (The alcohol and water mixture will be about 92 mL and the sucrose and water solution about 75 mL.) Ask why they think the volumes are smaller? (Possible explanation: The smaller water molecules can slip into the spaces between the bigger alcohol and sucrose molecules. Also, spaces existed between the sucrose crystals.)

You could try similar experiments with other substances that are highly soluble in one another.

Demonstration 2: Common Solutions, Colloids, and Suspensions

PURPOSE: To determine whether certain common substances found around the home are solutions, colloids, or suspensions
MATERIALS: 2 bottles of club soda in clear glass bottles, lemon-lime soda in a clear glass bottle, 70% ethyl or isopropyl alcohol, fruit gelatin, iodine crystals, milk of magnesia, oil and vinegar salad dressing, starch liquid remaining after macaroni or noodles are cooked, tap water, slide projector, 6 medium test tubes
PROCEDURE: Fill each of the six test tubes three-fourths full of club soda, 70% alcohol, milk of magnesia, starch liquid, fruit gelatin (prepare ahead and refrigerate), tap water. Put the milk of magnesia in several hours prior to the demonstration as it takes awhile for it to settle out.

Remove the labels from the unopened soda bottles and the oil and vinegar dressing. Examine the contents of the test tubes plus the unopened club soda, lemon-line soda, and the oil and vinegar dressing, and classify them as *(1)* suspensions or *(2)* solutions or colloids. (The milk of magnesia, the oil and vinegar salad dressing, and the club soda in the test tube are all suspensions because they settle upon standing. Actually, the gas in the soda rises. The starch liquid is cloudy but does not settle out; thus, it is a colloid.)

Check the remaining clear mixtures by shining the beam of a projector on them in a darkened room to see if they show the Tyndall effect. (The fruit gelatin should show the Tyndall effect; hopefully, the tap water will not. The remaining substances are solutions.)

Add 2 or 3 crystals of iodine to the 70% alcohol solution, and check it for the Tyndall effect. (negative result—solution)

Desktop Investigation Solutions, Colloids, Suspensions

If students wish to do this lab at home, the following substitutions can be made for equipment and measuring devices: pint jars for beakers, small juice glasses for test tubes, 1 cup for 250 mL, 1 tablespoon—15 mL, 1 teaspoon—5 mL, $1\frac{1}{2}$ tsp—12 g sucrose, 1 tsp—3 g starch, $1\frac{1}{2}$ tsp—5 g clay, $\frac{1}{4}$ tsp—2 g borax, 1 tsp—3 g gelatin. If juice glasses are used for test tubes, they should be filled about $\frac{2}{3}$ full.

Have students make a data table to record the results of their experiment in. The data table (with answers) might look similar to the one below.

DATA TABLE

Water Mixture of	Appearance	Separation occurs	Path of light visible	Conclusion
sucrose	clear	no	no	solution
starch	slightly cloudy	no	yes	colloid
clay	two layers	yes	—	suspension
food coloring	clear	no	no	solution
sodium borate	clear	no	no	solution
oil	two layers	yes	—	suspension
gelatin	clear-cloudy	no	yes	colloid

ANSWERS TO DISCUSSION QUESTIONS

1. The water mixtures of sucrose, food coloring, and sodium borate are solutions; those of starch and gelatin are colloids; and those of clay and oil are suspensions.

2. Solutions are clear, do not settle out or separate on standing, and do not show a visible light beam path. Colloids are sometimes clear or cloudy, do not settle out or separate on standing, but the path of a light beam is always clearly visible in them. Suspensions settle out or separate on standing.

14.2 The Solution Process

Demonstration 3: Solution Equilibrium

PURPOSE: To explain solution equilibrium using a common saturated solution
MATERIALS: Glass, stirring rod, ice, sucrose, tea bag, water
PROCEDURE: Have a glass of iced tea with too much sugar added to it; thus, the excess is on the bottom of the glass. Think back to our discussion on physical equilibrium in Chapter 13. Is any of the sugar on the bottom dissolving after you have stirred the tea thoroughly? What happens every time another sugar molecule dissolves? Why is this another example of physical equilibrium? What is the relationship between a saturated solution and solution equilibrium? What is wrong with the definition "A saturated solution is one that cannot dissolve any more solute"? (The definition makes a saturated solution sound like a static situation rather the continual movement between particles as they take turns dissolving and recrystallizing.)

Demonstration 4: Solubility of Gases in Liquids

PURPOSE: To show the effects of temperature and pressure on the solubility of gases in liquids
MATERIALS: 4 12-oz bottles of club soda, 3 5×5 cm glass bends, glass plate, 3 1-hole stoppers to fit soda bottles, 3 rubber delivery tubes, 3 1-liter soda bottles, 2 1-liter beakers, pneumatic trough (or a plastic dishpan or similar container), ice, Bunsen burner, ring stand and ring, wire gauze with ceramic center
PROCEDURE: Part 1: Chill one bottle of soda several hours before the experiment. Ten minutes before class, start to heat 350 mL of water in a 1-L beaker, and fill the trough and the three 1-L soda bottles with water.

Solutions **404B**

When you are ready to start, cover the mouth of each bottle with a glass plate, and carefully invert it in the trough. Have two students help hold the bottles. Assemble all the apparatus as shown in Figure T14-1 so that the stopper assembly can be put into the soda bottles as soon as they are opened and gas collection started immediately. Fill the second beaker with ice.

Open the one chilled bottle and the two room-temperature bottles of soda, and put the stopper assemblies in immediately. Place the chilled bottle in a beaker with ice and one of the room temperature bottles in the beaker containing the hot (not boiling) water. Leave the other one at room temperature. At the end of fifteen minutes, have students note the differences of the water levels in each gas-collecting bottle. How do you explain the results? (The solubility of gases in liquids decreases as the temperature increases. Therefore, you would expect soda in the hot water to produce the most carbon dioxide and the one in the ice, the least.)

Figure T14-1

Part 2: Have students observe an unopened bottle of soda. Open the bottle, and have students note what occurs. (The soda in the unopened bottle looks like water. After opening, little bubbles of gas immediately appear.) How do you account for the differences? (Opening soda releases the pressure and thus decreased the solubility of the gas in the liquid.)

Demonstration 5: Supersaturated Solutions

PURPOSE: To show the properties of a supersaturated solution
MATERIALS: Sodium thiosulfate pentahydrate, 4 medium test tubes, 4 #3 solid rubber stoppers, 250-mL beaker, ring stand and ring, wire gauze with ceramic center. Bunsen burner
PROCEDURE: Using a water bath, heat 4 or 5 test tubes half filled with sodium thiosulfate pentahydrate ($Na_2S_2O_3 \cdot 5H_2O$) until the crystals are completely melted forming a colorless liquid. Allow the test tubes and their contents to cool to room temperature in an upright position, and then stopper to prevent contamination. To reestablish equilibrium, drop a small crystal of $Na_2S_2O_3 \cdot 5H_2O$ into the supersaturated solution, and crystals start to form immediately. Let students *lightly* touch the sides of the tube to observe the exothermic nature of this recrystallization. Stopper the tube and save for next year. Extra samples should be made because sometimes samples recrystallize on cooling. The extras assure that you will have a successful sample.

For a dramatic variation of this demonstration, refer to Shakhashiri's "Crystallization from Supersaturated Solutions of Sodium Acetate," *Chemical Demonstrations*, Vol. 1, p. 27.

Demonstration 6: Heats of Solution

PURPOSE: To show the effects of solutes having positive and negative heats of solution on the temperature of the solutions they form with water
MATERIALS: Ammonium chloride, calcium chloride, lithium chloride, potassium nitrate, 4 medium test tubes, stirring rod, distilled water
PROCEDURE: Put 10 mL of water into each of four test tubes. Have students note the temperature of the water both by touching the outside of the test tubes and by a thermometer reading. Add 5 g of potassium nitrate to the first tube, 5 g of lithium chloride to the second, 5 g of ammonium chloride to the third, and 5 g of calcium chloride to the last. Have students note the changes in temperature and relate these changes to the heats of solution on Table 14-5 in the text. (KNO_3 and NH_4Cl both caused a decrease in temperature as would be expected since they both have a $+\Delta H_s$, while LiCl, having a $-\Delta H_s$, and $CaCl_2$ both caused an increase in temperature.) On the basis of the data gathered from the demonstration, ask whether calcium chloride, which is not listed on the table, would have a positive or negative ΔH_s? (Since $CaCl_2$ gave off heat, it would have a $-\Delta H_s$.)

A practical application of heat of solution follows: Hot and cold packs, which are sometimes used for treating injuries when hot water or ice is not available, are perfect examples of heat of solution put to a practical use. They consist of an outer plastic bag containing a dry chemical plus an inner pouch containing water and a dye. The inner pouch must be firmly squeezed to release the water so that it can mix with the dry chemical and produce heat or cold. In the cold packs the dry chemical is ammonium nitrate and in the hot ones, calcium chloride. If you wish to use them in a demonstration, they are available from medical supply outlets or sports stores. Your athletic department might even have them! **CAUTION:** If you use them in a demonstration, the hot packs produce enough heat to cause burns, and the cold ones should not have prolonged contact with the body, so handle with care.

14.3 Concentration of Solutions

A little consumer economics which applies percent by mass solutions to everyday life follows: If you have a choice of buying two different brands of bleach costing different amounts, you should check the labels, and if the % of NaClO is the same, buy the cheaper brand. (Throughout chemistry, emphasize reading labels. As students learn more chemistry, they are capable of becoming better consumers.)

The following demonstration is a detailed version of the one suggested in the TE annotation on page 423.

Demonstration 7: Molar and Molal Solutions

PURPOSE: To show how molar solutions and molal solutions differ from one another
MATERIALS: Decigram balance, permanent felt tip marker, 100-mL graduated cylinder, 2 2-L plastic soda bottles, sucrose, water
PROCEDURE: Prior to class, weigh out two 1-mol samples (342 g) of sucrose. Fill one soda bottle with 1000 mL of water. Mark the level of the water with the marking pen, and transfer the water to the second bottle. Again, mark the level of the water, discard the water.

In class, have the students calculate the mass of one mole of sucrose. Tell them the preweighed sample of sucrose has a mass of 342 g. Ask how many moles of sucrose are present? (1 mol) Put the 1 mol of sucrose in one bottle, and add enough warm water (not boiling) to reach the 1-L mark on the bottle. Ask what the volume of the solution is? (1 L) What would you call this solution? (a 1-molar solution)

Weigh the second empty plastic bottle on the decigram balance. Then fill it with water to the 1-L mark and weigh it again. Have students calculate the mass of the water. What is the mass of the water? (1000 g or 1 kg)

Remind students that the reason 1 L of water has a mass of 1 kg is because water has a density of 1 g/mL at 4°C. (The warm water would vary slightly from this if precise measuring devices were used. However, the devices used here are not very precise.) Further, point out that 1 kg of most liquids does NOT occupy a volume of 1 L. For example, 1 kg of ethanol (density 0.789 g/mL) occupies a volume of 1.27 L while 1 kg of glycerol (density 1.26 g/mL) occupies a volume of 0.794 L. Now add the 1 mol of sucrose to the 1 kg of water and shake until dissolved. Have students estimate the volume of the solution. (About 1.3 L) Ask, what would you call this solution? (a 1-molal solution) What is the difference between a 1 M and a 1 m solution? (A molar solution always has a volume of 1 L. The volume of a molal solution depends upon the density of the solvent and the amount of solute used; thus, its final volume varies greatly.)

14.4 Colligative Properties of Solutions

For an activity that will help your students understand why salt is added to ice in an ice-cream freezer and at the same time make you popular with them (and maybe even other people in the school) do "Team Ice Cream" in *Chemical Activities, Teacher Edition*, by Borgford and Summerlin, American Chemical Society, Washington, D.C., 1988, pp. 299–301.

Demonstration 14.8: Determination of the Molecular Weight of Urea

PURPOSE: To determine the molar mass of urea (H_2NCONH_2)
MATERIALS: 125-mL Erlenmeyer flask, 2-L pan, Celsius thermometer, ice, urea, rock salt, distilled water
PROCEDURE: Dissolve 10 g of urea in 50 mL of water in an Erlenmeyer flask. Submerge the flask in a pan of chipped-ice and salt mixture (1 L chipped ice and 100 mL rock salt—add more salt if a lower temperature is needed). Stir the mixture continuously with the thermometer until freezing is almost complete. At that point, note the temperature. (The temperature should be about −6.2°C.) Have students calculate the molar mass of urea. (Should be about 60 g/mol)

References and Resources

Books and Periodicals

Borgford, Christie and Lee Summerlin. "Mayonnaise: An Edible Emulsion," *Chemical Activities, Teacher Edition*, Washington, D.C.: American Chemical Society, 1988, pp. 292–3.

Davenport, Derek. "When Push Comes to Shove: Disturbing the Equilibrium" *ChemMatters*, Vol. 3, No. 1, Feb. 1985, pp. 15–6.

"Gas Laws and Scuba Diving," *ChemMatters*, Vol. 1, No. 1, Feb. 1983, pp. 4–6.

Goldsmith, Robert H. "A Simple Tyndall Effect Experiment," *Journal of Chemical Education*, Vol. 65, No. 7, July 1988, p. 623.

Grosser, Arthur E. "Emulsified Sauces," and "Ice Cream and Sherbert," *The Cookbook Decoder*, New York: Warner Books, 1981, pp. 166–174.

Nordstrom, Brian H. "The Effect of Polarity on Solubility," *Journal of Chemical Education*, Vol. 61, No. 11, Nov. 1984, p. 109.

Russel, Joan M. "Simple Models for Teaching Equilibrium and LeChâtelier's Principle," *Journal of Chemical Education*, Vol. 65, No. 10, Oct. 1988, pp. 871-2.

Shakhashiri, Bassam Z. "Chemical Cold Pack," "Chemical Hot Pack," and "Heat of Solution of LiCl," *Chemical Demonstrations*, Vol. 1, Madison Wisconsin: University of Wisconsin Press, 1983.

Talesnick, Irwin, ed. "Solubility in Different Solvents," Idea 404, *Idea Bank Collation*, Kingston, Ontario: Science Supplies and Services Co. Ltd., 1984.

Audiovisual Resources

Chemistry 110—Looking at the Solution and *Chemistry 111—Molarity of Solutions*, sound filmstrip or VHS cassette, Learning Arts, 1988.

Chemistry: Part 4, "Solutions," VHS cassette or film, Coronet Films.

Computer Software

Chemistry Help Series-Solutions, Apple II 48K, Barclay School Supplies, 1988.

Chemistry Courseware–Solutions, Apple II 64K and SEI Solutions, Apple 64K and IBM PC 128K, Queue, 1988.

Fee, Richard. *Liquids, Solids, Solutions*, Apple 48K, Chemistry According to ROF, 1988.

Project Seraphim, Volume 5, Disk 2, Apple II, 1988.

Smith, S.G., R. Chabray, and E. Kean. *Solutions*, Apple 48K and IBM PC 128K, Queue, 1988.

Answers and Solutions

Reviewing Concepts

1. a. The Tyndall effect is the scattering of light by colloidal particles dispersed in a transparent medium.
b. The visibility of a headlight beam on a foggy night represents an illustration of this effect.
c. This phenomenon can be used to distinguish between a solution and a colloid.

2. Shine a beam of light through the mixture. If it is a solution, no scattering of light will be observed. If it is a colloid, the light will be scattered. A suspension will settle out.
3. a. Solution equilibrium is the state in which the opposing processes of the dissolving and crystallizing of a solute occur at equal rates.
   b. The factors that determine the equilibrium point are the nature of the solute, the nature of the solvent, and the temperature. In the case of a gaseous solution, pressure is also a factor.
4. a. A saturated solution is one that contains the maximum possible amount of dissolved solute at solution equilibrium under the existing conditions.
   b. A residual quantity of undissolved solute remains in contact with the solution.
   c. An unsaturated solution is one that contains less solute than does a saturated solution under the existing conditions.
5. a. The solubility of a substance is the amount of that substance that is dissolved at solution equilibrium in a specific amount of solvent at a specified temperature.
   b. Temperature and pressure (for gases) must be specified.
6. There is no relationship between rate of dissolving and solubility.
7. a. The rule of thumb for predicting solubility is "like dissolves like."
   b. In general, the rule means that polar substances dissolve in polar solvents, and nonpolar substances dissolve in nonpolar solvents.
8. a. The solubility of a gas in a liquid is directly proportional to its pressure above the liquid.
   b. This is a statement of Henry's law.
   c. The amount of gas that will dissolve increases.
9. a. As the temperature increases, the solubility of a gas usually decreases.
   b. In general, as temperature increases, the solubility of a solid in a liquid increases.
10. The indicated solubility levels are:
    a. 80 g/100 g H_2O
    b. 110 g/100 g H_2O
    c. 36 g/100 g H_2O
11. The indicated temperature readings are:
    a. 15°C   b. 50°C   c. 32°C
12. a. When the dissolving process is exothermic, the total heat content of the solution is lower than that of its separate components; when the process is endothermic, the heat content of the solution is higher.
    b. When the process is exothermic, the solution warms; when it is endothermic, the solution cools.
    c. When the solution formation process is exothermic, the heat of solution is negative; when it is endothermic, the heat of solution is positive.
    d. When the process is exothermic, increased temperatures usually decrease solubility; when it is endothermic, increased temperatures generally increase solubility. Most solids have increased solubility at increased temperatures.
13. a. $AgNO_3(s) + H_2O(\ell) + 22.8$ kJ/mol $\rightarrow$ solution
    b. The dissolving process is endothermic; crystallization is exothermic.
    c. The temperature drops.
    d. The two rates are equal.
    e. The rate of dissolving will initially increase at a faster pace than the rate of crystallization because the dissolving process is endothermic and will serve to reduce the stress placed on the system.
    f. The amount will increase if additional solute is available.
    g. The rate of crystallizing will increase to relieve the stress, and the solubility of the solute will then decrease.
14. a. A dilute solution contains a relatively small amount of solute in a solvent, whereas a concentrated solution contains a relatively large amount of solute in a solvent.
    b. These terms are not directly related to solubility; a saturated solution of a substance that is not very soluble could, in fact, be very dilute.
15. a. Percent by mass would be used for practical purposes when only the mass of the solute, and not its molar mass, is needed.
    b. Molarity would be used when it is important to know the molar mass of solute in a given volume of solution.
    c. Molality is preferred when it is important to know the relative numbers of solute and solvent particles.
16. a. The presence of the nonvolatile solute lowers the vapor pressure of the solvent in which it is dissolved.
    b. The freezing point of the resulting solution is lower than that of the pure solvent.
    c. The boiling point of the solution is higher than that of the pure solvent.

Problems

1.

   a. Solubility increases with increasing temperature.
   b. At 30°C, solubility is approximately 250 g/100 g H_2O; at 55°C, solubility is approximately 380 g/100 g H_2O; at 75°C, solubility is approximately 540 g/100 g H_2O.

c. This solubility would be observed at approximately 35°C.
  d. The solution would be unsaturated; the solution would be saturated.

2. Solubility at 60°C = 110 g; at 10°C = 20 g. The amount of precipitate is therefore 110 g − 20 g = 90 g.

3. a. % by mass = $\dfrac{10.0 \text{ g sugar}}{10.0 \text{ g sugar} + 40.0 \text{ g water}} \times 100\%$
= 20.0% sugar

  b. % by mass = $\dfrac{0.63 \text{ g AgNO}_3}{0.63 \text{ g AgNO}_3 + 12.00 \text{ g H}_2\text{O}} \times 100\% = 5.0\%$ AgNO$_3$

4. a. % by mass = $\dfrac{\text{solute mass}}{\text{solute mass} + \text{solvent mass}} \times 100\%$

Let x = solute mass (mass of NaCl)

$10.0\% = \dfrac{x}{x + 50.0 \text{ g}} \times 100\%$

$\dfrac{10.0\%}{100\%} = \dfrac{x}{x + 50.0 \text{ g}}$

$0.100 = \dfrac{x}{x + 50.0 \text{ g}}$

$0.100(x + 50.0 \text{ g}) = x$
$0.100x + 5.00 \text{ g} = x$
$5.00 \text{ g} = 0.900x$
$5.56 \text{ g} = x$

∴ solute mass = 5.56 g NaCl

  b. Let x = mass of KOH

$2.4\% = \dfrac{x}{x + 8.5 \text{ g}} \times 100\%$

$\dfrac{2.4\%}{100\%} = \dfrac{x}{x + 8.5 \text{ g}}$

$0.024 = \dfrac{x}{x + 8.5 \text{ g}}$

$0.024(x + 8.5 \text{ g}) = x$
$0.024 x + 0.204 \text{ g} = x$
$0.204 \text{ g} = 0.976x$
$0.21 \text{ g} = x$

∴ mass of KOH = 0.21 g KOH

5. a. $\dfrac{3.50 \text{ mol H}_2\text{SO}_4}{\text{L}} \times 1.00 \text{ L} \times \dfrac{98.1 \text{ g H}_2\text{SO}_4}{\text{mol H}_2\text{SO}_4} = 343$ g H$_2$SO$_4$

  b. $\dfrac{1.75 \text{ mol Ba(NO}_3)_2}{\text{L}} \times 2.50 \text{ L} \times \dfrac{261 \text{ g Ba(NO}_3)_2}{\text{mol Ba(NO}_3)_2} = 1140$ g Ba(NO$_3$)$_2$

  c. $\dfrac{0.250 \text{ mol KOH}}{\text{L}} \times 500. \text{ mL} \times \dfrac{\text{L}}{10^3 \text{ mL}} \times \dfrac{56.1 \text{ g KOH}}{\text{mol KOH}} = 7.01$ g KOH

6. a. $\dfrac{20.0 \text{ g NaOH}}{2.00 \text{ L}} \times \dfrac{\text{mol NaOH}}{40.0 \text{ g NaOH}} = \dfrac{0.250 \text{ mol NaOH}}{\text{L}} = 0.250\ M$ NaOH

  b. $\dfrac{14.0 \text{ g NH}_4\text{Br}}{150. \text{ mL}} \times \dfrac{\text{mol NH}_4\text{Br}}{97.9 \text{ g NH}_4\text{Br}} \times \dfrac{10^3 \text{ mL}}{\text{L}} = \dfrac{0.953 \text{ mol NH}_4\text{Br}}{\text{L}} = 0.953\ M$ NH$_4$Br

  c. $\dfrac{32.7 \text{ g H}_3\text{PO}_4}{500. \text{ mL}} \times \dfrac{\text{mol H}_3\text{PO}_4}{98.0 \text{ g H}_3\text{PO}_4} \times \dfrac{10^3 \text{ mL}}{\text{L}} = \dfrac{0.667 \text{ mol H}_3\text{PO}_4}{\text{L}} = 0.667\ M$ H$_3$PO$_4$

7. a. $\dfrac{2.20 \text{ mol NaOH}}{\text{L}} \times 65.0 \text{ mL} \times \dfrac{\text{L}}{10^3 \text{ mL}} = 0.143$ mol NaOH

  b. $0.143 \text{ mol NaOH} \times \dfrac{40.0 \text{ g NaOH}}{\text{mol NaOH}} = 5.72$ g NaOH

8. $\dfrac{26.42 \text{ g (NH}_4)_2\text{SO}_4}{50.00 \text{ mL}} \times \dfrac{\text{mol (NH}_4)_2\text{SO}_4}{132.1 \text{ g (NH}_4)_2\text{SO}_4} \times \dfrac{10^3 \text{ mL}}{\text{L}} = \dfrac{4.000 \text{ mol (NH}_4)_2\text{SO}_4}{\text{L}} = 4.000\ M$ (NH$_4$)$_2$SO$_4$

9. $\dfrac{12.0 \text{ mol HCl}}{\text{L}} \times \dfrac{100. \text{ mL}}{2.00 \text{ L}} \times \dfrac{\text{L}}{10^3 \text{ mL}} = \dfrac{0.600 \text{ mol HCl}}{\text{L}} = 0.600\ M$ HCl

10. $\dfrac{\text{L } 16.0\ M\text{HNO}_3}{16.0\ \text{mol HNO}_3} \times 750.\ \text{mL dilute solution} \times \dfrac{\text{L dilute solution}}{10^3\ \text{mL dilute solution}} \times \dfrac{0.500\ \text{mol HNO}_3}{\text{L dilute solution}} \times \dfrac{10^3\ \text{mL } 16.0\ M\text{HNO}_3}{\text{L } 16.0\ M\text{HNO}_3}$
    $= 23.4\ \text{mL } 16.0\ M\text{HNO}_3$

11. $\dfrac{\text{L}}{0.54\ \text{mol AgNO}_3} \times \dfrac{10^3\ \text{mL}}{\text{L}} \times \dfrac{1\ \text{mol AgNO}_3}{170\ \text{g AgNO}_3} \times 0.34\ \text{g AgNO}_3 = 3.7\ \text{mL}$

12. $2\text{H}_3\text{PO}_4 + 3\text{Ca(OH)}_2 \rightarrow \text{Ca}_3(\text{PO}_4)_2 + 6\text{H}_2\text{O}$

    $\dfrac{6.00\ \text{mol H}_3\text{PO}_4}{\text{L}} \times 750.\ \text{mL} \times \dfrac{\text{L}}{10^3\ \text{mL}} = 4.50\ \text{mol H}_3\text{PO}_4$

    $4.50\ \text{mol H}_3\text{PO}_4 \times \dfrac{1\ \text{mol Ca}_3(\text{PO}_4)_2}{2\ \text{mol H}_3\text{PO}_4} \times \dfrac{310.\ \text{g Ca}_3(\text{PO}_4)_2}{\text{mol Ca}_3(\text{PO}_4)_2} = 698\ \text{g Ca}_3(\text{PO}_4)_2$

    $4.50\ \text{mol H}_3\text{PO}_4 \times \dfrac{6\ \text{mol H}_2\text{O}}{2\ \text{mol H}_3\text{PO}_4} \times \dfrac{18.0\ \text{g H}_2\text{O}}{\text{mol H}_2\text{O}} = 243\ \text{g H}_2\text{O}$

13. $3\text{H}_2\text{SO}_4 + 2\text{Al(OH)}_3 \rightarrow \text{Al}_2(\text{SO}_4)_3 + 6\text{H}_2\text{O}$

    $\dfrac{2.50\ \text{mol Al(OH)}_3}{\text{L}} \times 250.\ \text{mL} \times \dfrac{\text{L}}{10^3\ \text{mL}} = 0.625\ \text{mol Al(OH)}_3$

    $0.625\ \text{mol Al(OH)}_3 \times \dfrac{3\ \text{mol H}_2\text{SO}_4}{2\ \text{mol Al(OH)}_3} \times \dfrac{\text{L}}{18.0\ \text{mol H}_2\text{SO}_4} \times \dfrac{10^3\ \text{mL}}{\text{L}} = 52.1\ \text{mL (of } 18.0\ M\text{H}_2\text{SO}_4)$

14. Since 1 mol AgNO$_3$ yields 1 mol Ag, mol AgNO$_3$ = mol Ag. Thus,

    $\dfrac{0.250\ \text{g Ag}}{75.0\ \text{mL}} \times \dfrac{1\ \text{mol Ag}}{108\ \text{g Ag}} \times \dfrac{10^3\ \text{mL}}{1\ \text{L}} = 0.0309\ M(\text{AgNO}_3\ \text{solution})$

15. **a.** $\dfrac{4.50\ \text{mol H}_2\text{SO}_4}{\text{kg H}_2\text{O}} \times 1.00\ \text{kg H}_2\text{O} \times \dfrac{98.1\ \text{g H}_2\text{SO}_4}{\text{mol H}_2\text{SO}_4} = 441\ \text{g H}_2\text{SO}_4$

    b. $\dfrac{1.00\ \text{mol HNO}_3}{\text{kg H}_2\text{O}} \times 2.00\ \text{kg H}_2\text{O} \times \dfrac{63.0\ \text{g HNO}_3}{\text{mol HNO}_3} = 126\ \text{g HNO}_3$

    c. $\dfrac{3.50\ \text{mol MgCl}_2}{\text{kg H}_2\text{O}} \times 0.450\ \text{kg H}_2\text{O} \times \dfrac{95.2\ \text{g MgCl}_2}{\text{mol MgCl}_2} = 150.\ \text{g MgCl}_2$

16. **a.** $\dfrac{294.3\ \text{g H}_2\text{SO}_4}{1.000\ \text{kg H}_2\text{O}} \times \dfrac{\text{mol H}_2\text{SO}_4}{98.07\ \text{g H}_2\text{SO}_4} = \dfrac{3.001\ \text{mol H}_2\text{SO}_4}{\text{kg H}_2\text{O}} = 3.001\ m\ \text{H}_2\text{SO}_4$

    b. $\dfrac{63.0\ \text{g HNO}_3}{0.250\ \text{kg H}_2\text{O}} \times \dfrac{\text{mol HNO}_3}{63.0\ \text{g HNO}_3} = \dfrac{4.00\ \text{mol HNO}_3}{\text{kg H}_2\text{O}} = 4.00\ m\ \text{HNO}_3$

    c. $\dfrac{10.0\ \text{g NaOH}}{300.\ \text{g H}_2\text{O}} \times \dfrac{\text{mol NaOH}}{40.0\ \text{g NaOH}} \times \dfrac{10^3\ \text{g H}_2\text{O}}{\text{kg H}_2\text{O}} = \dfrac{0.833\ \text{mol NaOH}}{\text{kg H}_2\text{O}} = 0.833\ m\ \text{NaOH}$

17. $\dfrac{17.1\ \text{g C}_{12}\text{H}_{22}\text{O}_{11}}{275\ \text{g H}_2\text{O}} \times \dfrac{\text{mol C}_{12}\text{H}_{22}\text{O}_{11}}{342\ \text{g C}_{12}\text{H}_{22}\text{O}_{11}} \times \dfrac{10^3\ \text{g H}_2\text{O}}{\text{kg H}_2\text{O}} = \dfrac{0.182\ \text{mol C}_{12}\text{H}_{22}\text{O}_{11}}{\text{kg H}_2\text{O}} = 0.182\ m\ \text{C}_{12}\text{H}_{22}\text{O}_{11}$

18. $\dfrac{\text{kg H}_2\text{O}}{0.500\ \text{mol Ca(NO}_3)_2} \times \dfrac{\text{mol Ca(NO}_3)_2}{164\ \text{g Ca(NO}_3)_2} \times 75.5\ \text{g Ca(NO}_3)_2 = 0.921\ \text{kg H}_2\text{O}$

19. $\dfrac{1.25\ \text{mol C}_6\text{H}_{12}\text{O}_6}{\text{kg H}_2\text{O}} \times \dfrac{180.\ \text{g C}_6\text{H}_{12}\text{O}_6}{\text{mol C}_6\text{H}_{12}\text{O}_6} \times 750.\ \text{g H}_2\text{O} \times \dfrac{\text{kg H}_2\text{O}}{10^3\ \text{g H}_2\text{O}} = 169\ \text{g C}_6\text{H}_{12}\text{O}_6$

20. $\dfrac{1.75\ \text{mol C}_2\text{H}_5\text{OH}}{1\ \text{kg H}_2\text{O}} \times 250.\ \text{g H}_2\text{O} \times \dfrac{\text{kg H}_2\text{O}}{10^3\ \text{g H}_2\text{O}} \times \dfrac{46.1\ \text{g C}_2\text{H}_5\text{OH}}{\text{mol C}_2\text{H}_5\text{OH}} = 20.2\ \text{g C}_2\text{H}_5\text{OH}$

21. $\dfrac{\text{kg H}_2\text{O}}{0.450\ \text{mol NaCl}} \times \dfrac{1\ \text{L H}_2\text{O}}{1\ \text{kg H}_2\text{O}} \times \dfrac{1\ \text{mol NaCl}}{58.4\ \text{g NaCl}} \times 65.0\ \text{g NaCl} = 2.47\ \text{L H}_2\text{O}$

22. $\dfrac{3.00\ \text{mol HNO}_3}{\text{kg H}_2\text{O}} \times \dfrac{2.25\ \text{kg H}_2\text{O}}{5.75\ \text{kg H}_2\text{O}} = 1.17\ m\ \text{HNO}_3$

23. **a.** $\Delta t_f = K_f m = \dfrac{-1.86\text{C}°\cdot\text{kg H}_2\text{O}}{\text{mol C}_6\text{H}_{12}\text{O}_6} \times \dfrac{1.50\ \text{mol C}_6\text{H}_{12}\text{O}_6}{\text{kg H}_2\text{O}} = -2.79\text{C}°$

b. $\Delta t_f = K_f m = \dfrac{-1.86\text{C}° \cdot \text{kg H}_2\text{O}}{\text{mol C}_{12}\text{H}_{22}\text{O}_{11}} \times \dfrac{\text{mol C}_{12}\text{H}_{22}\text{O}_{11}}{342 \text{ g C}_{12}\text{H}_{22}\text{O}_{11}} \times \dfrac{171 \text{ g C}_{12}\text{H}_{22}\text{O}_{11}}{1.00 \text{ kg H}_2\text{O}} = -0.930\text{C}°$

c. $\Delta t_f = K_f m = \dfrac{-1.86\text{C}° \cdot \text{kg H}_2\text{O}}{\text{mol C}_{12}\text{H}_{22}\text{O}_{11}} \times \dfrac{\text{mol C}_{12}\text{H}_{22}\text{O}_{11}}{342 \text{ g C}_{12}\text{H}_{22}\text{O}_{11}} \times \dfrac{77.0 \text{ g C}_{12}\text{H}_{22}\text{O}_{11}}{400. \text{ g H}_2\text{O}} \times \dfrac{10^3 \text{ g H}_2\text{O}}{\text{kg H}_2\text{O}} = -1.05\text{C}°$

24. a. $\Delta t_f = K_f m.$

Therefore, $m = \dfrac{\Delta t_f}{K_f} = \dfrac{-0.930\text{C}° \cdot \text{mol sugar}}{-1.86\text{C}° \cdot \text{kg H}_2\text{O}} = 0.500 \; m$ sugar

b. $m = \dfrac{-3.72\text{C}° \cdot \text{mol sugar}}{-1.86\text{C}° \cdot \text{kg H}_2\text{O}} = 2.00 \; m$ sugar

c. $m = \dfrac{-8.37\text{C}° \cdot \text{mol sugar}}{-1.86\text{C}° \cdot \text{kg H}_2\text{O}} = 4.50 \; m$ sugar

25. a. $\Delta t_f = K_f m = \dfrac{-1.86\text{C}° \cdot \text{kg H}_2\text{O}}{\text{mol C}_6\text{H}_{12}\text{O}_6} \times \dfrac{\text{mol C}_6\text{H}_{12}\text{O}_6}{180. \text{ g C}_6\text{H}_{12}\text{O}_6} \times \dfrac{20.0 \text{ g C}_6\text{H}_{12}\text{O}_6}{250. \text{ g H}_2\text{O}} \times \dfrac{10^3 \text{ g H}_2\text{O}}{\text{kg H}_2\text{O}} = -0.827\text{C}°$

26. a. $\dfrac{0.51\text{C}° \cdot \text{kg H}_2\text{O}}{\text{mol C}_6\text{H}_{12}\text{O}_6} \times \dfrac{2.5 \text{ mol C}_6\text{H}_{12}\text{O}_6}{\text{kg H}_2\text{O}} = 1.3\text{C}°$

b. $\dfrac{0.51\text{C}° \cdot \text{kg H}_2\text{O}}{\text{mol C}_6\text{H}_{12}\text{O}_6} \times \dfrac{\text{mol C}_6\text{H}_{12}\text{O}_6}{180 \text{ g C}_6\text{H}_{12}\text{O}_6} \times \dfrac{3.20 \text{ g C}_6\text{H}_{12}\text{O}_6}{1.00 \text{ kg H}_2\text{O}} = 0.0091\text{C}°$

c. $\dfrac{0.51\text{C}° \cdot \text{kg H}_2\text{O}}{\text{mol C}_{12}\text{H}_{22}\text{O}_{11}} \times \dfrac{\text{mol C}_{12}\text{H}_{22}\text{O}_{11}}{340 \text{ g C}_{12}\text{H}_{22}\text{O}_{11}} \times \dfrac{20.0 \text{ g C}_{12}\text{H}_{22}\text{O}_{11}}{500. \text{ g H}_2\text{O}} \times \dfrac{10^3 \text{ g H}_2\text{O}}{\text{kg H}_2\text{O}} = 0.060\text{C}°$

27. a. $\Delta t_b = t_b\text{solution} - t_b\text{solvent} = 100.25°\text{C} - 100.00°\text{C} = 0.25\text{C}°$
$\Delta t_b = K_b m$

$m = \dfrac{\Delta t_b}{K_b} = \dfrac{0.25\text{C}°}{0.51\text{C}° \cdot \text{kg H}_2\text{O/mol solute}} = 0.49 \; m$ solute

b. $\Delta t_b = 101.53°\text{C} - 100.00°\text{C} = 1.53\text{C}°$
$\Delta t_b = K_b m$

$m = \dfrac{\Delta t_b}{K_b} = \dfrac{1.53\text{C}°}{0.51\text{C}° \cdot \text{kg H}_2\text{O/mol solute}}$
$= 3.0 \; m$ solute

c. $\Delta t_b = 102.805°\text{C} - 100.000°\text{C} = 2.805\text{C}°$
$\Delta t_b = K_b m$

$m = \dfrac{\Delta t_b}{K_b} = \dfrac{2.805\text{C}°}{0.51\text{C}° \cdot \text{kg H}_2\text{O/mol solute}}$
$= 5.5 \; m$ solute

28. $\Delta t_f = K_f m$

$m = \dfrac{\Delta t_f}{K_f} = \dfrac{-1.5\text{C}° \cdot \text{mol solute}}{-7.27\text{C}° \cdot \text{kg phenol}} = 0.21 \; m;$ $\dfrac{\text{kg phenol}}{0.21 \text{ mol solute}} \times \dfrac{5.00 \text{ g solute}}{150. \text{ g phenol}} \times \dfrac{10^3 \text{ g}}{\text{kg}} = 160 \text{ g solute/mol solute}$

29. $\Delta t_b = t_b\text{solution} - t_b\text{solvent} = 185.5°\text{C} - 184.4°\text{C} = 1.1\text{C}°$
$\Delta t_b = K_b m$

$m = \dfrac{\Delta t_b}{K_b} = \dfrac{1.1\text{C}° \cdot \text{mol solute}}{3.52\text{C}° \cdot \text{kg aniline}} = 0.31 \; m;$ $\dfrac{\text{kg aniline}}{0.31 \text{ mol solute}} \times \dfrac{7.55 \text{ g solute}}{125 \text{ g aniline}} \times \dfrac{10^3 \text{ g}}{\text{kg}} = 190 \text{ g solute/mol solute}$

30. $\Delta t_f = K_f m$

$m = \dfrac{\Delta t_f}{K_f}$

$m = \dfrac{-20°\text{C}}{-1.86\text{C}° \cdot \text{kg H}_2\text{O/mol C}_2\text{H}_4(\text{OH})_2}$

$m = 10.8 \text{ mol C}_2\text{H}_4(\text{OH})_2/\text{kg H}_2\text{O}$
$m = 10.8 \; m \; \text{C}_2\text{H}_4(\text{OH})_2$

$\dfrac{10.8 \text{ mol C}_2\text{H}_4(\text{OH})_2}{\text{kg H}_2\text{O}} \times 500. \text{ g H}_2\text{O} \times \dfrac{\text{kg}}{10^3 \text{ g}} \times \dfrac{62.1 \text{ g C}_2\text{H}_4(\text{OH})_2}{\text{mol C}_2\text{H}_4(\text{OH})_2} = 335 \text{ g C}_2\text{H}_4(\text{OH})_2$

31. $\Delta t_f = K_f m$

$$K_f = \frac{\Delta t_f}{m} = \frac{-1.90\text{C}°}{m}$$

$$K_f = \frac{-1.90\text{C}°}{\dfrac{7.24 \text{ g } C_2Cl_4H_2}{115 \text{ g benzene}} \times \dfrac{\text{mol } C_2H_2Cl_4}{168 \text{ g } C_2H_2Cl_4} \times \dfrac{10^3 \text{ g benzene}}{\text{kg benzene}}} = -5.07\text{C}°/m$$

32. $\Delta t_f = \dfrac{-40.00\text{C}° \cdot \text{kg camphor}}{\text{mol solute}} \times \dfrac{1.500 \text{ g solute}}{35.00 \text{ g camphor}} \times \dfrac{\text{mol solute}}{125.0 \text{ g solute}} \times \dfrac{10^3 \text{ g camphor}}{\text{kg camphor}} = -13.71\text{C}°$

New freezing point:
$\Delta t_f = t_f \text{solution} - t_f \text{solvent}$
$t_f \text{solution} = \Delta t_f + t_f \text{solvent} = -13.71\text{C}° + 178.40°\text{C} = 164.69°\text{C}$

33. $\Delta t_f = K_f m$

$$m = \frac{\Delta t_f}{K_f} = \frac{-6.51\text{C}°}{-1.86\text{C}° \cdot \text{kg H}_2\text{O/mol solute}} = 3.50 \, m \text{ solute}$$

$\Delta t_b = K_b m$
$= 0.51\text{C}°/m \times 3.50 \, m = 1.8\text{C}°$

Since $\Delta t_b = t_b \text{solution} - t_b \text{solvent}$, then
$t_b \text{solution} = \Delta t_b + t_b \text{solvent} = 1.8\text{C}° + 100.00°\text{C} = 101.8°\text{C}$

34. $\Delta t_f = K_f m$

$$m = \frac{\Delta t_f}{K_f} = \frac{-0.568\text{C}°}{-1.79\text{C}° \cdot \text{kg ether/mol solute}} = 0.317 \text{ mol solute/kg ether}$$

$$\frac{\text{kg ether}}{0.317 \text{ mol solute}} \times \frac{21.2 \text{ g solute}}{740. \text{ g ether}} \times \frac{10^3 \text{ g ether}}{\text{kg ether}} = 90.4 \text{ g solute/mol solute}$$

35. a. $\Delta t_f = K_f m$

$$m = \frac{\Delta t_f}{K_f} = \frac{-0.372\text{C}° \cdot \text{mol solute}}{-1.86\text{C}° \cdot \text{kg H}_2\text{O}} = 0.200 \, m \text{ solute}$$

$$\frac{\text{kg H}_2\text{O}}{0.200 \text{ mol solute}} \times \frac{9.00 \text{ g solute}}{250. \text{ g H}_2\text{O}} \times \frac{10^3 \text{ g H}_2\text{O}}{\text{kg H}_2\text{O}} = 180. \text{ g solute/mol solute}$$

b. The molar masses of CH_3OH, C_6H_6, C_2H_5OH, $C_6H_{12}O_6$, and $C_{12}H_{22}O_{11}$, are 32.0, 78.1, 46.1, 180., and 342 g/mol respectively. Since $C_6H_{12}O_6$ is the only substance with a molar mass that is close to the experimentally determined value, $C_6H_{12}O_6$ most likely represents the solute.

Critical Thinking

1. Data Table 1 — Sample 1 = Solution
Sample 2 = Suspension or Colloid
Sample 3 = Colloid
Sample 4 = Colloid
Data Table 2 — Filtrate 1 = Solution
Filtrate 2 = Colloid
Filtrate 3 = Colloid
Filtrate 4 = Suspension
Final Test: Examine with a High Power Microscope.

Videodisc Selections from
The Periodic Table Videodisc (JCE Software)

step
forward

step
reverse

pg 407 Alloys

36640
iron alloys (s)

36658
Alnico magnets (s)

37174
mercury dental amalgams (s)

CHAPTER 14

pg 405 Solutions

36480
phosphates found in milk,
   sodas, juices (s)

36507
sulfuric acid + water (s)

36513
hydrochloric acid + water (s)
bleach

Teacher's Notes

Chapter 14
Solutions

SECTION PREVIEW

14.1 Types of Mixtures
14.2 The Solution Process
14.3 Concentration of Solutions
14.4 Colligative Properties of Solutions

In nature, pure liquids are very rare. Nearly all liquids, such as these brightly-colored fluids, exist as solutions, homogeneous mixtures of two or more substances.

14.1 Types of Mixtures

It is unlikely that you will encounter many substances of high purity. Almost all the objects or materials that you see, touch, eat, drink, or use are made of mixtures.

It is easy to determine that some materials are mixtures. For example, soil is a mixture of various substances, including small rocks and decomposed animal and plant matter. You can see this by picking up some soil in your hand and taking a close look. Milk, on the other hand, does not appear to be a mixture, but in fact it is. Milk is comprised principally of fats, proteins, milk sugar, and water. Looking at milk under a microscope, you can see round droplets of fat measuring from 1 to 10 micrometers in diameter. Also, irregularly shaped casein (protein) particles about 0.2 micrometers wide can be seen. Both milk and soil are examples of heterogeneous mixtures since their composition is not uniform throughout. Soil, milk, and other heterogeneous mixtures are shown in Figure 14-1.

SECTION OBJECTIVES

- Distinguish between heterogeneous and homogeneous mixtures.
- Distinguish between electrolytes and nonelectrolytes.
- Compare the properties of suspensions, colloids, and solutions.

Other everyday examples of heterogeneous mixtures are tossed salad, buttered popcorn, vegetable soup, and chocolate-almond candy bars.

Figure 14-1 Four different mixtures *(left to right)*: a handful of soil, milk magnified 500 times, nuts and bolts, whole wheat bread

Sugar and water form a mixture in which the sugar molecules are interspersed among the water molecules. To distinguish the components of the mixture you would have to inspect the mixture molecule by molecule.

Above the molecular scale, the mixture of sugar and water is uniform throughout and is considered to be a homogeneous mixture.

Mixtures differ according to the size of the particles in the mixture. Based on this, mixtures are separated into suspensions, colloids, and solutions. Solutions contain the smallest particles and suspensions contain the largest. Colloids contain intermediate size particles.

Solutions

Suppose a lump of sugar is dropped into a glass of water. You know from experience that the sugar will dissolve. Sugar is described as "soluble in water." By **soluble** we mean *capable of being dissolved.*

What happens as the sugar dissolves? The lump gradually disappears as sugar molecules leave the surface of their crystals and intermingle with water molecules. Eventually all of the sugar molecules become uniformly distributed among the water molecules, as indicated by the equally sweet taste of any part of the mixture. Examination of a sugar–water mixture, even under a microscope, would not reveal any particles of sugar. Such a mixture is called a "solution." A **solution** *is a homogeneous mixture of two or more substances in a single phase.* In a solution, atoms, molecules, or ions are thoroughly

Solutions are homogeneous mixtures. Suspensions and colloids are heterogeneous mixtures.

Have you ever looked at a movie projector beam in a dark room and seen the suspended dust particles in the beam of light?

mixed, with the result that the mixture has the same composition and properties throughout.

The nature of solutions In the simplest type of solution, such as a sugar–water solution, the molecules of one substance are distributed around the molecules of another substance. *The dissolving medium in a solution is called the* **solvent,** *and the substance dissolved in a solution is called the* **solute.** In the sugar–water solution, water is the solvent and sugar is the solute.

Substances that dissolve in water are divided into two types according to whether they yield molecules or ions in solution. When an ionic compound dissolves, the positive and negative ions separate from each other and become surrounded by water molecules. These solute ions are free to move, making it possible for an electric current to pass through the solution. *A substance that dissolves in water to give a solution that conducts electric current is called an* **electrolyte.** Sodium chloride (NaCl), is an electrolyte, as is any soluble ionic compound. Certain highly polar molecular compounds, such as hydrogen chloride (HCl), are also electrolytes because the molecules split into ions (H_3O^+ and Cl^-) when dissolved in water.

By contrast, a solution containing neutral solute molecules does not conduct electric current because no mobile charged particles are available. *A substance that dissolves in water to give a solution that does not conduct an electric current is called a* **nonelectrolyte.** Sugar is a nonelectrolyte. An experiment that demonstrates the difference between electrolytes and nonelectrolytes is shown in Figure 14-2.

Figure 14-2 Apparatus for testing the conductivity of solutions. If the bulb is lit, the solution conducts electricity. A solution of copper(II) sulfate *(right)* conducts electricity, while a sugar solution *(left)* does not.

Types of solutions Solutions are formed in all three states—the gaseous, liquid, and solid states. The nine possible combinations of gases, liquids, and solids in solutions are summarized in Table 14-1. In each example, one component is designated as the solvent and one as the solute. Occasionally, however, these terms have little meaning. For example, in a 50%–50% solution of ethyl alcohol (C_2H_5OH), and water (H_2O), it would be difficult to say which is the solvent and which is the solute.

TABLE 14-1 TYPES OF SOLUTIONS

Solute	Solvent	Example
gas	gas	air
gas	liquid	soda water
gas	solid	hydrogen in platinum
liquid	gas	water vapor in air
liquid	liquid	alcohol in water
liquid	solid	mercury in copper
solid	gas	sulfur vapor in air
solid	liquid	sugar in water
solid	solid	copper in nickel

Figure 14-3 A brass bell. Brass is an alloy of copper and zinc.

In a solution, the dissolved particles are so small that they cannot be seen. They remain mixed with the solvent indefinitely, so long as the existing conditions remain unchanged. If a liquid solution is poured through a filter paper, both the solute and the solvent will pass through the paper. The solute particle dimensions are those of atoms, molecules, and ions—from about 0.01 to 1 nm. It is useful to break solutions into three main types: gaseous, liquid, and solid solutions.

1. Gaseous solutions. Molecules in a gas are far apart and in constant motion. When two or more gases are mixed, therefore, the molecules quickly become uniformly intermingled. Although we do not commonly describe them as such, all mixtures of gases are solutions.

The molecules of any liquid or solid that evaporate enter the gas phase and mix with the molecules there. In a sense, air is a solvent for all such evaporated substances.

2. Liquid solutions. Soft drinks are solutions of a gas, carbon dioxide, in water in which solid flavoring materials are also dissolved. Automobile radiators contain solutions of ethylene glycol, a liquid, in water. Vinegar is acetic acid dissolved in water. Water-soluble solids are present in solutions that we use for washing dishes and clothing.

3. Solid solutions. Many metal objects are actually solid solutions of metals, called alloys. **Alloys** *are solid solutions in which the atoms of two or more metals are uniformly mixed.* By properly choosing the proportions of each metal in the alloy, many desirable properties can be obtained. For example, alloys can have higher strength, greater resistance to corrosion and higher melting points than pure elemental metals (Figure 14-3).

Gas–solid solutions form when small atoms of elements like hydrogen, boron, or nitrogen can intermingle with metal atoms. Hydrogen dissolves especially well in transition metals such as platinum and palladium.

In solutions of substances that are mutually soluble in each other, such as isopropyl alcohol and water or nitrogen and oxygen, the substance that is present in the larger amount is called the solvent.

Water is considered to be the universal solvent. Can you think why?

Some examples of common alloys are sterling silver—92% silver and 8% copper, jewelry gold—gold and copper (percentages vary depending upon the number of karats), white gold—gold containing nickel and possibly tin, zinc, or copper, and the amalgam for filling teeth—silver and mercury.

Solutions

Desktop Investigation

Solutions, Colloids, Suspensions

Question
How can you distinguish between solutions, colloids, and suspensions?

Materials
water
sucrose (table sugar)
soluble starch
clay
red food coloring
sodium borate (borax)
cooking oil
plain gelatin
7 beakers, 400-mL
flashlight

Procedure
Record your observations in a data table. Wear safety glasses and an apron.
1. Prepare seven mixtures, each containing 250 mL of water and one of the following substances: 12 g of sucrose, 3 g of starch, 5 g of clay, 2 mL of food coloring, 2 g of sodium borate, 50 mL of oil, and 3 g of gelatin. To make the gelatin mixture, first soften the gelatin with 65 mL of cold water, and then add 185 mL of boiling water.

2. Observe the seven mixtures and their characteristics. What is the appearance of each mixture after stirring? Which mixtures separate and which remain mixed after stirring?

3. Transfer 10 mL of each mixture that does not separate after stirring to individual test tubes and shine a flashlight on the mixture in a dark room. Is the path of the light beam clearly visible in the mixture? (See the diagram.)

Discussion
1. Using the information on your data table, decide whether each mixture is a solution, a colloid, or a suspension.
2. How can solutions, colloids, and suspensions be distinguished?

Suspensions

At the opposite extreme from the solute particles in solutions are particles so large that they can be seen and will settle out unless the mixture is constantly stirred or agitated. Think of a jar of muddy water. If left undisturbed, particles of soil collect on the bottom of the jar. Such a mixture is a **suspension**—*a heterogeneous mixture of a solvent-like substance with particles that slowly settle out.* The soil particles are huge and much more dense when compared to the water molecules. Their interaction with the water molecules is not strong enough to prevent them from falling to the bottom of the container. Particles over 1000 nm in diameter—1000 or more times larger than atoms, molecules, or ions—form suspensions. The particles in suspension can be separated from the heterogeneous mixture by passing the mixture through a filter.

Colloids

Particles that are intermediate in size between those in solutions and suspensions form mixtures known as colloidal dispersions, or simply **colloids.** Particles between 1 nm and 1000 nm in diameter form colloids. After large particles settle out of muddy water, the water is often still cloudy because colloidal particles remain dispersed in the water. Pouring the cloudy mixture through a filter will not remove the colloidal particles, and the mixture will remain cloudy. The particles in a colloidal dispersion are small enough to be kept in suspension by the constant movement of the surrounding molecules.

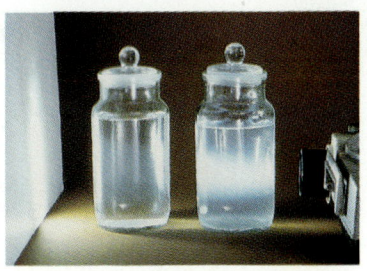

Figure 14-4 The jar at the left contains a mixture of water and sodium chloride. The jar at right contains a mixture of gelatin in water. Which mixture is a colloidal suspension and which is a solution?

Colloids form in all phases. The colloidal particles make up the *dispersed phase,* and the solventlike phase is called the *dispersing medium.* Examples of the various types of colloids are given in Table 14-2. Note that some familiar terms, such as *emulsion* or *foam,* refer to specific types of colloids. For example, mayonnaise is an emulsion of oil droplets in water; the egg yolk in it acts as an emulsifying agent, which helps to keep the oil droplets dispersed.

TABLE 14-2 COLLOIDAL SUSPENSIONS

Class of Colloid	Phases
sol	solid dispersed in liquid
gel	solid network extending throughout liquid
liquid emulsion	liquid dispersed in liquid
foam	gas dispersed in liquid
aerosol	
smoke	solid dispersed in gas
fog	liquid dispersed in gas
smog	solid and liquid dispersed in gas
solid emulsion	liquid dispersed in solid

Many colloids appear homogeneous because the individual particles cannot be seen with the naked eye. The particles are, however, large enough to scatter light. You have probably noticed how a headlight beam becomes visible on a foggy night. This effect, known as the Tyndall effect, occurs only when light is scattered by colloidal particles dispersed in a transparent medium. The Tyndall effect can be used to distinguish between a solution and a colloid, as demonstrated in Figure 14-4.

The distinctive properties of solutions, colloids, and suspensions are summarized in Table 14-3. The individual particles of a colloid can be detected under a microscope if a bright light is cast on the specimen at a right angle. The particles, which appear as tiny specs of light are seen to move rapidly in a random motion. This motion is due to bombardment by rapidly moving molecules and is called Brownian motion after its discoverer, Robert Brown (1773–1858). Brown did not know what caused this motion and suggested that the particles were alive.

Thomas Graham, an English chemist, started the study of colloids in 1861. Since the first colloids were gelatin or glue, he used the Greek word *kolla,* meaning glue, as the root for their name.

TABLE 14-3 PROPERTIES OF SOLUTIONS, COLLOIDS, AND SUSPENSIONS

Solutions	Colloids	Suspensions
Homogeneous	Heterogeneous	Heterogeneous
0.01–1 nm solute particles: atoms, ions, molecules	1–1000 nm dispersed particles: particles or large molecules	Over 1000 nm suspended particles: large particles or aggregates
Does not separate on standing	Does not separate on standing	Particles settle out
Cannot be separated by filtration	Cannot be separated by filtration	Can be separated by filtration
No light scattering	Scatters light (Tyndall effect)	May scatter light, but not transparent

1. (a) Suspension
   (b) Solution (c) Colloid
2. (a) Heterogeneous
   (b) Homogeneous
3. (a) Conducting (b) Non-conducting (c) Electrolytes
4. A suspension. Unless constantly agitated, it will separate. Let the two mixtures stand until the sand settles out of the water. The sugar remains mixed. Note that light is not scattered by the transparent liquid phase.

Section Review

1. Classify the following as either suspensions, colloids, or solutions. Explain. *(a)* sand mixed with water *(b)* brass doorknob *(c)* atmospheric cloud.
2. Classify the following as either a heterogeneous or homogeneous mixture. Explain. *(a)* orange juice *(b)* tap water
3. What type of solution does *(a)* an ionic solute form in water? *(b)* a non-ionic solute generally form in water? *(c)* What are substances called whose water solutions conduct electricity?
4. In what type of mixture is it easiest to separate the component substances? Why? State one way to prove that a mixture of sugar and water is a solution and a mixture of sand and water is not.

14.2 The Solution Process

Factors Affecting the Rate of Dissolving

Suppose it is a hot day, you are preparing a glass of iced tea, and you are in a hurry. The only available sugar is in large cubes, and the tea solution has already cooled to room temperature. How can you speed up dissolving the sugar in the tea? Which kitchen equipment shown in Figure 14-5 will be of help?

Increasing the surface area of the solute We noted earlier that sugar dissolves as sugar molecules leave their crystal surface and mingle with water molecules. The same is true for any solid solute in a liquid solvent: molecules or ions of the solute are attracted away from the surface and into solution.

If dissolution occurs at the surface, clearly it can be speeded up if the surface area is increased. Is crushing the sugar cubes one of your suggestions? In general, the more finely divided a substance is, the greater the surface area per unit mass and the more quickly it will dissolve.

Agitating the solution Dissolved solute tends to build up in the solvent close to the solid. Stirring or shaking helps to disperse the solute particles

SECTION OBJECTIVES

- List and explain three factors that influence the rate of dissolving of a solid in a liquid.
- Explain solution equilibrium and distinguish among saturated, unsaturated, and supersaturated solutions.
- Explain the meaning of "like dissolves like" in terms of polar and nonpolar substances.
- List the three interactions that contribute to the heat of solution, and explain what causes dissolving to be exothermic or endothermic.
- Compare the effects of temperature and pressure on solubility.

Figure 14-5 Using the kitchen utensils shown, how would you increase the rate of dissolving of sugar in iced tea?

and brings fresh solvent into contact with the surface. Thus, the effect of stirring is similar to that of crushing a solid—contact between the solvent and the solute surface is increased.

Heating the solvent Most likely you have noticed that sugar and many other materials dissolve more quickly in warm water than in cold water. As the temperature of the solvent increases, solvent molecules move around faster; their average kinetic energy increases. Therefore, collisions at the higher temperatures between the solvent molecules and the solute are more frequent and of higher energy than at lower temperatures. This causes solute molecules to leave the surface at a faster rate.

Figure 14-6 illustrates these three methods of increasing the dissolving rate.

Solubility

If you add spoonful after spoonful of sugar to tea—even if it is powdered sugar added to hot tea with vigorous stirring—a point will come at which no more sugar will dissolve. For every solute–solvent combination at a given temperature, there is a limit to the amount of solute that can be dissolved.

When solid sugar is first dropped into the water, sugar molecules begin to depart from the solid surface and move about at random in the liquid solvent. Eventually, enough sugar molecules are in solution that some collide with the solid surface and crystallize (return to the solid crystal). If the amount of solid sugar present is less than the limit that can dissolve, sugar molecules will leave the solid surface faster than they reenter it. All of the sugar molecules present will eventually be in solution. If more sugar is present than can dissolve, the rate at which solute molecules go into solution

A way to help students understand the dynamic nature of a saturated solution: If you *could* tie red tags on all the molecules in pile of undissolved sugar in a saturated sugar solution, after a while there would be sugar molecules with red tags in solution and sugar molecules without tags in the undissolved sugar due to the fact that they are continually dissolving and recrystallizing. Can we tie red tags on molecules? In a way—we can tag molecules by using radioactive atoms in them.

Figure 14-6 Three methods for increasing the rate at which a solute dissolves

Figure 14-7 A saturated solution contains the equilibrium concentration of solute under existing conditions, as shown in this bottle.

and the rate at which they return to the solid surface will become equal and a state of dynamic equilibrium is established.

$$\text{solute} + \text{solvent} \rightleftarrows \text{solution}$$

Solution equilibrium *is the physical state in which the opposing processes of dissolving and crystallizing of a solute occur at equal rates.* The point at which equilibrium is reached for any solute–solvent combination is difficult to predict precisely and depends upon the nature of the solute, the nature of the solvent, and the temperature.

A solution that contains the maximum amount of dissolved solute is described as a **saturated solution**. (How can you tell that the solution pictured in Figure 14-7 is saturated?) At 20°C, 204 g of sugar is the maximum amount that will dissolve at equilibrium in 100 g of water. If solid sugar is added to the solution just described, it falls to the bottom and does not appear to dissolve because an equilibrium has been established between molecules leaving and entering the solid phase. If more water is added to the saturated solution, then more sugar will dissolve in it. *A solution that contains less solute than a saturated solution under the existing conditions is an* **unsaturated solution**.

When a hot, saturated solution of sugar is allowed to cool, the excess solute usually, but not always, separates as the solution cools. If this does not occur, a **supersaturated solution**—*a solution that contains more dissolved solute than a saturated solution under the same conditions*—is produced. A supersaturated solution is not likely to remain supersaturated. Crystals will eventually form in it until equilibrium is reestablished at the lower temperature. A supersaturated solution can be made by preparing a saturated solution of sodium thiosulfate ($Na_2S_2O_3$) or sodium acetate ($NaC_2H_3O_2$) in hot water, filtering the hot solution, and allowing the filter solution to stand undisturbed as it cools. Dropping a small crystal of the solute into the supersaturated solution ("seeding") or disturbing the solution will cause a rapid formation of crystals by the excess solute.

The **solubility** *of a substance is the amount of that substance required to form a saturated solution with a specific amount of solvent at a specified temperature.* We say that the solubility of sugar is 204 g per 100 g of water at 20°C. The temperature must be specified because solubility varies with temperature. For gases, the pressure must also be specified. Solubilities must be determined experimentally; and they vary widely, as illustrated for three compounds in Figure 14-8. Solubility values are recorded in chemical handbooks and are usually given as grams of solute per 100 g of solvent or per 100 mL of solvent at a given temperature.

Figure 14-8 A comparison of the masses of three common solutes that can be dissolved in 100 g of water at 0°C. Convert these quantities to moles of solute per 100 g of water and compare.

In the preceding paragraphs, we listed factors that speed up the rate of dissolving. The speed at which a solid dissolves is *unrelated* to solubility. The maximum amount of solute that enters into solution and reaches equilibrium is always the same under the same conditions.

Factors Affecting Solubility

The extent to which a given solute dissolves in a solvent depends on the identity of the solute and solvent and also on the existing conditions of pressure and temperature.

Types of solvents and solutes Sodium chloride is highly soluble in water, but gasoline is not. On the other hand, gasoline mixes readily with benzene (C_6H_6), but sodium chloride does not. What is the basis for such differences in solubility?

"Like dissolves like" is a roughly useful rule for predicting whether or not one substance dissolves in another. The necessary "alikeness" depends on the type of bonding, the polarity or nonpolarity of molecules, and the intermolecular forces that are found in the two mixed substances.

Water molecules are strongly attracted to each other because they interact by hydrogen bonding. According to the "like dissolves like" rule, water should be a good solvent for any substance composed of polar

Gases lose energy when changing to the liquid phase.

You might wish to review the sections on molecular polarity and intermolecular forces in Chapter 6.

molecules. The molecules are more likely to be drawn away from each other by polar water molecules than by a nonpolar solvent. The positively and negatively charged regions in polar solute molecules are attracted to the oppositely charged regions of the water molecules as diagrammed in Figure 14-9. Molecules that can take part in hydrogen bonding are especially soluble. The large solubility of sugar is aided by the presence in each sugar molecule of O—H bonds that can form hydrogen bonds with water molecules.

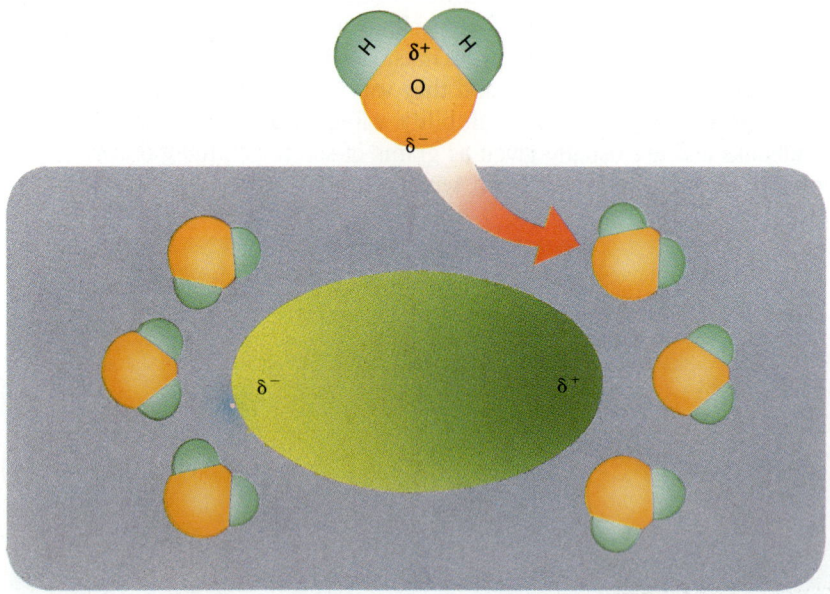

Figure 14-9 Polar molecules are hydrated as they dissolve in water.

See Dalton's law of partial pressures: Section 11.3

At the opposite extreme are liquids composed of molecules that are nonpolar and do not form hydrogen bonds. Carbon tetrachloride (CCl_4) molecules are symmetrical and, despite the polar C—Cl bonds, are nonpolar. A carbon tetrachloride molecule can exert only relatively weak, London forces on other molecules. Would we expect water or sodium chloride to dissolve in carbon tetrachloride? No, we would not. There is too little similarity in bonding, polarity, and intermolecular forces.

Substances that are not soluble in each other are **immiscible.** Carbon tetrachloride and water are examples of immiscible substances as are oil and water. When you shake a bottle of salad dressing, oil droplets become dispersed in the water. As soon as you stop shaking, the strong attraction between the water molecules squeezes out the oil droplets, and separate layers form.

Substances that are mostly nonpolar, however—such as fats, oils and greases—are quite soluble in carbon tetrachloride. Carbon tetrachloride was once widely used as a household cleaning agent. It is now considered too toxic for such use.

Gasoline contains mainly nonpolar hydrocarbons and is also an excellent solvent for fats, oils, and greases. The major intermolecular forces acting between the nonpolar molecules are the relatively weak, London forces.

The "like dissolves like" rule predicts that ethanol (ethyl alcohol, C_2H_5OH) should be soluble in water, and the prediction is correct. Ethanol molecules are polar because of the single electronegative oxygen atom, and they can form hydrogen bonds with water molecules between the oxygen atom of water and the hydrogen of ethanol's hydroxyl group. The intermolecular forces are so similar that unlimited quantities of these two liquids mix together freely. *Two substances that are mutually soluble in all proportions are said to be* **miscible.** Nonpolar substances with very similar intermolecular forces, such as gasoline and benzene, are also completely miscible. The nonpolar molecules experience no strong forces of attraction or repulsion and therefore mingle freely with little change in energy.

Ethanol is intermediate in polarity between water and carbon tetrachloride (Figure 14-10) and is not as good a solvent for polar or ionic substances as water. Sodium chloride is only slightly soluble in ethanol. On the other hand, ethanol is a better solvent than water for less polar substances because the molecule has a region containing only nonpolar carbon–hydrogen bonds.

$$\text{Nonpolar region} \quad H-\underset{\underset{H}{|}}{\overset{\overset{H}{|}}{C}}-\underset{\underset{H}{|}}{\overset{\overset{H}{|}}{C}}-O-H \quad \text{polar region}$$

Diethyl ether ($C_2H_5OC_2H_5$), for example, has one electronegative atom but no possibility for hydrogen bonding. It is completely miscible with ethanol but only partially miscible with water.

Variations on this are to be expected, however. The small, nonpolar carbon tetrachloride molecules are able to dissolve in the somewhat polar ethanol, for example. The ions in some ionic compounds are so strongly attracted to each other that the compounds are only very slightly soluble in water.

Pressure Changes in pressure have very little effect on the solubilities of liquids or solids in liquid solvents. Gas solubilities in liquids, however, always increase with increasing pressure.

When a gas is in contact with the surface of a liquid, gas molecules can enter the liquid. As the amount of dissolved gas increases, some molecules begin to escape and reenter the gas phase. An equilibrium is eventually established between the rates at which gas molecules enter and leave the liquid phase. As long as this equilibrium is undisturbed, the solubility of the gas in the liquid is unchanged.

$$\text{gas} + \text{solvent} \rightleftarrows \text{solution}$$

Increasing the pressure of the solute gas above the solution puts stress on the equilibrium. Molecules collide with the liquid surface more often. The increase in pressure is partially offset by an increase in the rate of gas molecules entering the solution. In turn, the increase in the amount of dissolved gas causes an increase in the rate at which molecules escape from the liquid surface and become vapor. Eventually, equilibrium is restored at a higher gas solubility. As expected from Le Châtelier's principle, an increase in gas pressure has caused the equilibrium to shift so that fewer molecules are in the gas phase.

The solubility of a gas in a liquid is directly proportional to the partial pressure of that gas on the surface of the liquid. This is a statement of **Henry's**

Spot removal techniques depend upon the "like dissolves like" rule. To remove ballpoint-pen stains from clothing, pre-spot with isopropyl alcohol. The nonpolar hydrocarbon part dissolves the nonpolar ink and the OH part dissolves in water (and detergent), thus removing the ink.

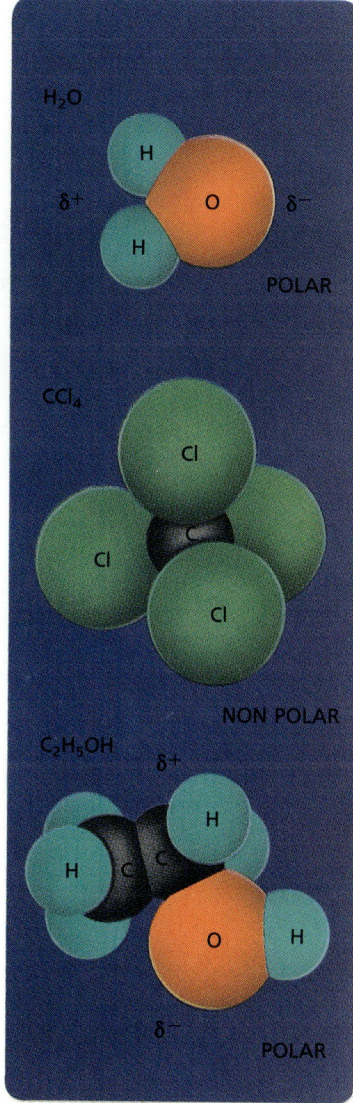

Figure 14-10 Water *(top)* is polar. Carbon tetrachloride *(middle)* is nonpolar. Ethanol *(bottom)* is intermediate in polarity between water and carbon tetrachloride.

In 1826, Thomas Graham discovered that very soluble gases, such as NH_3 and CO_2, do not obey Henry's law.

A question students often ask is: Why does tapping a can of soda which has been shaken before opening keep it from bubbling over? Perhaps because tapping the can causes the undissolved, trapped bubbles in the soda to rise to the surface allowing the gas at the surface to expand and escape upon opening. (If trapped bubbles from shaking expand in the liquid upon opening, the liquid will bubble over.)

Figure 14-11 A carbonated soft drink effervesces when the bottle is opened and the pressure is reduced.

law, named after the English chemist William Henry (1775–1836). Henry's law applies to gas-liquid solutions at constant temperature.

Recall that when a mixture of gases is confined in a space of constant volume at a constant temperature, each gas exerts the same pressure it would if it alone occupied the space. Assuming that the gases do not react in any way when in solution, each gas dissolves to the extent it would if no other gases were present.

The increased solubility of gases with increased pressure is put to use in making carbonated beverages. At the bottling plant, carbon dioxide gas (CO_2) is forced into solution in flavored water at a pressure of 5–10 atm. The gas-in-liquid solution is then sealed in bottles or cans. When the cap is removed, the pressure is reduced to 1 atm, and some of the carbon dioxide escapes as gas bubbles. *The rapid escape of a gas from a liquid in which it is dissolved is known as* **effervescence** (Figure 14-11).

Temperature You have already learned that temperature affects the rates at which solutes dissolve. Temperature can also have a large effect on the amounts of solute that dissolve in a given amount of solvent.

First, let's consider gas solubility. Increasing the temperature usually decreases gas solubility. When you first heat water and before boiling begins, you can often see small bubbles. These bubbles are formed by the gases that were in solution in the colder water and are being driven out. As the temperature increases, the average kinetic energy of the molecules in solution increases. A greater number of solute molecules are able to escape from the attraction of solvent molecules and return to the gas phase. Therefore, at higher temperatures equilibrium is reached with fewer gas molecules in solution. (Figure 14-12)

The effect of temperature on the solubility of solids in liquids is more difficult to predict than the effect on gas solubility. Most often, increasing the temperature increases the solubility of solids. However, an equivalent temperature increase may result in a large increase in solubility in one case and a slight increase in another. In Table 14-4 and Figure 14-13, compare the effect of temperature on the solubilities of potassium nitrate (KNO_3) and

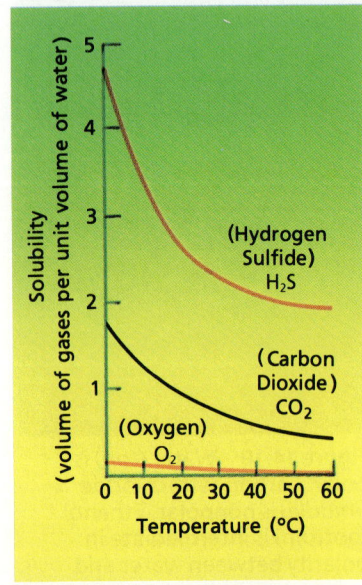

Figure 14-12 The solubility of gases in liquid decreases with increasing temperature.

TABLE 14-4 SOLUBILITY OF SOLUTES AS A FUNCTION OF TEMPERATURE
Grams of solute per 100 grams of H₂O

Substance	Temperature, °C					
	0°	20°	40°	60°	80°	100°
AgNO₃	122	216	311	440	585	733
Ba(OH)₂	1.67	3.89	8.22	20.9	101	—
C₁₂H₂₂O₁₁	179	204	238	287	362	487
Ca(OH)₂	0.189	0.173	0.141	0.121	0.094	0.07
Ce₂(SO₄)₃	20.8	10.1	—	3.87	—	—
KCl	28.0	34.2	40.1	45.8	51.3	56.3
KI	128	144	162	176	192	206
KNO₃	13.9	31.6	61.3	106	167	245
Li₂CO₃	1.54	1.33	1.17	1.01	0.85	0.72
NaCl	35.7	35.9	36.4	37.1	38.0	39.2
NaNO₃	73	87.6	102	122	148	180
Yb₂(SO₄)₃	44.2	37.5 at 10°	17.2 at 30°	10.4	6.4	4.7
CO₂(gas at SP)	0.335	0.169	0.0973	0.058	—	—
O₂(gas at SP)	0.00694	0.00537	0.00308	0.00227	0.00138	0.00

sodium chloride (NaCl). About 14 g of potassium nitrate will dissolve in 100 g of water at 0°C. The solubility of potassium nitrate increases by more than 150 g KNO₃/100 g H₂O when the temperature is raised to 80°C. Under similar circumstances, the solubility of sodium chloride increases by only about 2 g NaCl/100 g H₂O. In some cases, solubility of a solid *decreases* with an increase in temperature. For example, between 0°C and 60°C the solubility of cerium sulfate, Ce₂(SO₄)₃, decreases by about 17 g. Here, too, the size of the change varies.

Heats of Solution

The formation of a solution is accompanied by an energy change. If you dissolve some potassium iodide, KI, in water, you will find that the outside of the container becomes cold to your touch. But if you dissolve some lithium chloride (LiCl), in the same way, the outside of the container will be hot to your touch. The formation of a solid–liquid solution can apparently either absorb heat (KI in water) or release heat (LiCl in water).

The amount of heat energy absorbed or released when a solute dissolves in a specific amount of solvent is the **heat of solution.** It is given in kilojoules per mole (kJ/mol) of solute dissolved in a specified amount of solvent. Table 14-5 lists some heats of solution. You can see that there are both positive values (a change that absorbs heat) and negative values (a change that releases heat).

Endothermic—positive heat of solution:

solute + solvent + heat → solution

Exothermic—negative heat of solution:

solute + solvent → solution + heat

During solution formation, solvent and solute particles experience changes in the forces attracting them to other particles. There are three

Figure 14-13 Solubility curves

Figure 14-14 Three steps in solution formation

The colligative properties of solutions of electrolytes are discussed in Chapter 15.

different interactions that contribute to the heat of solution. Before dissolving begins, solvent molecules are held together by intermolecular forces (solvent–solvent attraction). In the solute, molecules are held together by intermolecular forces (solute–solute attraction). During solution formation, these forces are disrupted. Energy is required to separate solute molecules and solvent molecules from their neighbors.

In the solution, particles acquire new neighbors and solute–solvent attraction takes over. *A solute molecule surrounded by solvent molecules* (as in Figure 14-9) *is said to be* **solvated.** Intermolecular attraction brings the solvent and solute molecules together, and energy is released in the process.

Solution formation can, thus be pictured as the result of the three interactions summarized in Figure 14-14.

1. Solute–solute attraction is broken up, requiring energy.
2. Solvent–solvent attraction is broken up, requiring energy.
3. Solute–solvent attraction is formed, releasing energy.

The heat of solution is positive (heat is absorbed) when the sum of *(1)* and *(2)* is greater than *(3)*. The heat of solution is negative (heat is released) when the sum of *(1)* and *(2)* is less than *(3)*. Here is the basis for the effect of increasing temperatures on the solubilities of solids: In a saturated solution,

Figure 14-15 Two beakers of water at the same temperature *(left).* A salt added to the left beaker raises its temperature *(right).* Why?

there is a dynamic equilibrium between undissolved and dissolved solute. Le Châtelier's principle applies to this situation. Adding energy by raising the temperature of the solution causes the rate of dissolving to change in the direction that consumes energy. If dissolving is endothermic, more solute dissolves:

$$\text{solute} + \text{solvent} + \text{heat} \rightleftharpoons \text{solution} + \text{heat}$$

We saw that heating usually decreases the solubility of a gas, so dissolving of gases must usually be exothermic. The heat of solution values in Table 14-5 show this to be correct.

TABLE 14-5 HEATS OF SOLUTION (kJ/mol solute in 200 moles H_2O [(s) = solid, (ℓ) = liquid, (g) = gas at STP]

Substance	Heat of Solution	Substance	Heat of Solution
$AgNO_3$(s)	+22.77	KOH(s)	−54.59
CO_2(g)	−19.91	LiCl(s)	−35.0
$CuSO_4$(s)	−67.81	Li_2CO_3(s)	−12.8
$CuSO_4 \cdot 5H_2O$(s)	+11.51	$MgSO_4 \cdot 7H_2O$(s)	+15.9
$HC_2H_3O_2$(ℓ)	−1.59	NaCl(s)	+4.27
HCl(g)	−74.26	$NaNO_3$(s)	+21.1
HI(g)	−29.39	NaOH(s)	−41.6
H_2SO_4(ℓ)	−74.30	$Na_2SO_4 \cdot 10H_2O$(s)	+78.53
KCl(s)	+17.58	NH_3(g)	−34.7
$KClO_3$(s)	+42.03	NH_4Cl(s)	+16.2
KI(s)	+21.4	NH_4NO_3(s)	+25.5
KNO_3(s)	+35.7		

In the gaseous state, molecules are so far apart that there are virtually no intermolecular forces of attraction between them. Therefore, the solute–solute interaction has little effect on the heat of solution of a gas. Energy is released when a gas dissolves in a liquid because attraction between solute gas and solvent molecules outweighs the energy needed to separate solvent molecules.

This section has shown that intermolecular forces can explain factors affecting the rate of dissolving, variation in the solubility of a substance in a particular solvent and the energy changes that occur during the solution process.

Section Review

1. Why can you expect a packet of sugar to dissolve faster in hot tea than iced tea?
2. Explain how you would prepare a saturated solution of sugar in water. How would you then make it a supersaturated solution?
3. Explain why ethanol will dissolve in water and carbon tetrachloride will not.
4. When a solute molecule is solvated, is heat released or absorbed?
5. A warm and a cold bottle of soda pop are opened. Which will effervesce more?

1. The molecules in hot tea move around faster than in ice tea. Therefore, there are more collisions between the tea and sugar molecules in the surface of the sugar grain. Sugar molecules leave the surface of the grains faster in hot tea than in cold tea.
2. Continually add sugar to hot water until undissolved grains remain at the bottom of the container. This is a saturated solution of sugar and water. Set the saturated solution aside until it cools to room temperature. Once it cools, it will be a supersaturated solution of sugar and water.
3. Ethanol and water are each polar molecules. The negatively charged region of the ethanol molecules are attracted to the positively charged region of the water molecules and vice versa. Carbon tetrachloride is nonpolar.
4. Intermolecular attraction brings the solvent and solute molecules together. Energy is released in this process.
5. The warm bottle, because the dissolved CO_2 gas is less soluble in warm liquid than in cold liquid.

Solutions

14.3 Concentration of Solutions

SECTION OBJECTIVES

- Define concentration using molarity, molality, and percent by mass.
- Given the concentration of a solution, find the amount of solute in a given amount of solution.
- Given the concentration of a solution, find the amount of solution that contains a given amount of solute.

The **concentration** *of a solution is a measurement of the amount of solute in a given amount of solvent or solution.* Some medications are solutions of drugs—a one-teaspoonful dose at the correct concentration might cure the condition; the same dose in the wrong concentration might kill the patient.

In this section, we introduce three different ways of expressing the concentrations of solutions: molarity, molality, and percent by mass.

Sometimes solutions are referred to as "dilute" or "concentrated." These are relative terms. "Dilute" means that there is a relatively small amount of solute in a solvent. "Concentrated," on the other hand, means that there is a relatively large amount of solute in a solvent. Note that these terms are unrelated to the degree to which a solution is saturated. A saturated solution of a substance that is not very soluble might be very dilute.

Percent by Mass

Percent by mass *of a solute in solution is the number of grams of solute dissolved in 100 g of solution.* For example, the percent by mass of a solution made from 10 g of sodium hydroxide (NaOH) dissolved in 90 g of water is found as follows:

$$\text{Percent by mass} = \frac{\text{mass of solute}}{\text{mass of solute} + \text{mass of solvent}} \times 100\%$$

$$= \frac{10 \text{ g NaOH}}{10 \text{ g NaOH} + 90 \text{ g H}_2\text{O}} \times 100\%$$

$$= 10\% \text{ NaOH (by mass)}$$

Note that the mass of the solution is the sum of the masses of the solute and solvent. Percent concentration is common for solutions used for practical purposes—household or industrial cleaning, killing pests in the garden, medical applications. Percent-by-mass concentration is based only on the mass of a solute and is unrelated to its chemical formula or molar mass.

> Explain that *dilute* and *concentrated* are very indefinite terms by using the example of strong and weak coffee. When people talk about strong coffee, they really mean concentrated coffee; weak would be dilute. What is concentrated coffee to one person may be dilute to another and vice-versa. In science, solutions have to be able to be duplicated exactly to be useful.

> Liquid bleach is 5.25% NaClO, vinegar is 5–6% $HC_2H_3O_2$, and medicinal hydrogen peroxide is 3% H_2O_2.

Sample Problem 14.1

A solution of sodium chloride is prepared by dissolving 5 g of salt in 550 g of water. What is the concentration of this solution given as percent by mass?

Solution

Step 1. Analyze
Given: mass of solute and identity = 5 g NaCl
mass of solvent and identity = 550 g H_2O
Unknown: percent by mass concentration of solution

Step 2. Plan
solvent mass, solute mass → solution mass → percent by mass of solute

The mass of solute must be divided by the mass of the solution and multiplied by 100%.

$$\frac{\text{mass of solute}}{\text{mass of solute} + \text{mass of solvent}} \times 100\% = \% \text{ solute (by mass)}$$

Sample Problem 14.1 continued

Step 3. Compute

$$\frac{5 \text{ g NaCl}}{5 \text{ g NaCl} + 550 \text{ g H}_2\text{O}} \times 100\% = 0.9\% \text{ NaCl}$$

Step 4. Evaluate The answer is rounded to one significant digit because it is limited by the given mass of the solute. If 5 g of solute were dissolved in 500 g of solvent, a 1% solution would be produced. It is reasonable, therefore, that 5 g of solute in somewhat more solvent should give a solution of somewhat smaller concentration.

Practice Problems
1. What is the percent by mass of a solution prepared by dissolving 4.0 g of acetic acid in 35 g of water? *(Ans.)* 10.% acetic acid
2. Determine the percent by mass of a solution prepared by dissolving 32.5 g of glucose in 155 g of water. *(Ans.)* 17.3% glucose

Molarity

Molarity *is the number of moles of solute in one liter of solution.* To find the molarity of a solution you must know the molar mass of the solute.

For example, a "one-molar" solution of sodium hydroxide (NaOH) contains one mole of NaOH in every liter of solution. The symbol for molarity is M and the concentration of the preceding solution is written $1\ M$ NaOH.

One mole of sodium hydroxide (NaOH) has a mass of 40.0 g. This quantity of sodium hydroxide dissolved in enough water to make exactly 1.00 L of solution gives a $1\ M$ solution. If 20.0 g of NaOH, which is 0.500 mol, is dissolved in 1.00 L of solution, a $0.500\ M$ NaOH solution is produced:

$$\text{molarity} = \frac{\text{number of moles of solute}}{\text{number of liters of solution}}$$

$$= \frac{0.500 \text{ mol NaOH}}{1.00 \text{ L}}$$

$$= 0.500\ M \text{ NaOH}$$

If twice the molar mass of NaOH (80.0 g) is dissolved in enough water to make 1 L of solution, a $2\ M$ solution is produced. The molarity of any solution can be calculated by dividing the number of moles of solute by the number of liters of solution.

Note that a one molar solution is *not* made by adding 1 mol of solute to 1 L of *solvent*. In such a case, the final total volume would be slightly different from 1 L. Instead, 1 mol of solute is first dissolved in less than 1 L of solvent. Then, the resulting solution is carefully diluted with more solvent to bring the *total volume* to 1 L, as illustrated in Figure 14-16.

Another method for describing the concentration of a solution, called *molality*, will be discussed following sample problems on molarity. Molarity is useful when the quantity of solute participating in a chemical reaction taking place in solution is of interest. Any required molar quantity of a solute can be selected by measuring out the appropriate volume of a solution of known molarity.

Solutions

Figure 14-16 Two methods for the preparation of a 1 molar solution. 1 mol of solute can be added to a 1-liter volumetric flask or 0.5 mol can be added to a 500-mL volumetric flask *(top left)*; enough water is then added to dissolve the solute in each flask and the solutions are agitated *(top right)*. Enough water is added to bring the volume up to the 1-liter mark on one flask and the 500-mL mark on the other *(bottom)*. Each solution contains the same proportion of solute to solution.
$$\frac{1 \text{ mol solute}}{1.0 \text{ L solution}} = \frac{0.5 \text{ mol}}{0.5 \text{ L}} = 1 \text{ mol/L}$$

Sample Problem 14.2

What is the molarity of 3.50 L of solution that contains 90.0 g of sodium chloride (NaCl)?

Solution

Step 1. Analyze Given: solute mass and solute formula = 90.0 g NaCl
               solution volume = 3.50 L of solution
Unknown: molarity of the solution

Step 2. Plan   You only need one conversion (mass to moles of solute) to arrive at your answer.

mass of solute → number of moles of solute → molarity

$$\text{mass solute (g)} \times \frac{1}{\text{molar mass of solute (g/mol)}} \times \frac{1}{\text{volume of solution (L)}}$$

$$= \text{molarity of solution}$$

Step 3. Compute $90.0 \text{ g NaCl} \times \frac{1 \text{ mol NaCl}}{58.4 \text{ g NaCl}} \times \frac{1}{3.50 \text{ L of solution}} = \frac{0.440 \text{ mol NaCl}}{1 \text{ L of solution}} = 0.440 \, M \text{ NaCl}$

Step 4. Evaluate   Because each factor involved is limited to three significant digits, the answer should have three significant digits, which it does. The units cancel correctly to give the desired moles of solute per liter of solution, which is the unit of molarity. Because the numerical calculation looks roughly like 90/200 (slightly less than 0.5), the final answer seems reasonable.

Sample Problem 14.3

How many moles of HCl are present in 0.8 L of a 0.5-M HCl solution?

Solution

Step 1. Analyze   Given: volume of solution = 0.8 L
molarity of solution = 0.5 M HCl
Unknown: moles of HCl in given volume

Step 2. Plan   The molarity indicates the moles of solute that is in one liter of solution. Given the volume of the solution, the number of moles of solute can then be found.

molarity, volume → moles solute

molarity (moles of solute/liter of solution) × volume (liters of solution)
= moles of solute

Step 3. Compute $\dfrac{0.5 \text{ mol HCl}}{1 \text{ L of solution}} \times 0.8 \text{ L of solution} = 0.4 \text{ mol HCl}$

Step 4. Evaluate   The answer is correctly given to one significant digit. The units cancel correctly to give the desired label, moles. There should be less than 0.5 moles HCl, since less that 1 L of solution was used.

Practice Problems

1. What is the molarity of a solution composed of 5.85 g of potassium iodide (KI) dissolved in enough water to make 0.125 L of solution?
   (Ans.) 0.282 M KI
2. How many moles of H_2SO_4 are present in 0.500 L of a 0.150 M H_2SO_4 solution?
   (Ans.) 0.0750 mol

Molality

Molality *is the concentration of a solution expressed in moles of solute per kilogram of solvent.* A solution that contains 1 mol of solute, ammonia (NH_3), for example, dissolved in exactly 1 kg of solvent, is a "one molal" solution. The symbol for molality is m, and the concentration of this solution is written as 1 m NH_3.

One mole of ammonia has a molar mass of 17.0 g, and 17.0 g of NH_3 dissolved in 1 kg of water gives one-molal ammonia solution. If 8.50 g of ammonia, which is one-half mole of ammonia, is dissolved in exactly 1 kg of water, a 0.500 m NH_3 solution is prepared.

$$\text{molality} = \dfrac{\text{no. of moles solute}}{\text{mass of solvent (kg)}} = \dfrac{0.500 \text{ mol } NH_3}{1 \text{ kg } H_2O} = 0.500 \, m \, NH_3$$

If 34.0 g of ammonia, which is 2 mol, is dissolved in 1 kg of water, a 2 m solution of NH_3 is produced. The molality of any solution can be found by dividing the number of moles of solute by the mass in kilograms of the solvent in which it is dissolved. Note that if grams of solvent are known, the quantity of solvent must be converted to kilograms by multiplying by the factor 1 kg/10^3 g.

> To help students understand the difference between molarity and molality, make a 1 M solution of sucrose by adding enough water to 342 g of sucrose to make a liter of solution. Then add 342 grams of sucrose to 1 kg of water, and determine the volume of this 1 m solution. It should form approximately 1.3 L of solution.

Later in this chapter, you will see that concentrations are given in molality when studying properties of solutions related to vapor pressure and temperature changes. Molality and percent composition by mass are useful since they do not change their value with changes in temperature. Molarity can vary with temperature because solution volume varies with temperature.

For a dilute solution, where water is the solvent, molarity and molality have similar values. To see why this is true, imagine a certain number of moles of a solute, A, dissolved in a given amount of water. The molarity and molality of this solution are, respectively;

$$M = \frac{\text{moles of A}}{\text{volume of solution}} \qquad m = \frac{\text{moles of A}}{\text{mass of solvent}}$$

When there are relatively few particles of solute compared to solvent in a solution, the volume and mass of the solution are essentially the volume and mass of the solvent. Also, water has a density of 1 kg/L. The mass of a sample of water, in kilograms, is numerically equal to its volume in liters. Therefore, for a very dilute aqueous solution, the volume of solution is approximately equal to the mass of the solution. Since the numerators of the above expression for m and M are equal and their denominators are approximately equal, the expressions are approximately equal.

Sample Problem 14.4

A solution contains 17.1 g of sucrose (table sugar, $C_{12}H_{22}O_{11}$) dissolved in 125 g of water. Find the molal concentration of this solution.

Solution

Step 1. Analyze   Given: solute mass and formula = 17.1 g $C_{12}H_{22}O_{11}$
solvent mass and identity = 125 g water
Unknown: molal concentration

Step 2. Plan   To find molality you must solve for moles of solute per kilogram of solvent. The given mass of solute must be converted to moles and the mass in grams of solvent must be converted to kilograms.

mass solute, mass in grams of solvent → molar mass of solute, mass in kilograms of solvent → molality

$$\frac{\text{mass solute (g solute)}}{\text{mass solvent (g solvent)}} \times \frac{1}{\text{molar mass of solute}} \left(\frac{\text{mol solute}}{\text{g solute}}\right) \times \frac{10^3 \text{g solvent}}{\text{kg solvent}}$$

$$= \text{molality} \left(\frac{\text{mol solute}}{\text{kg solvent}}\right)$$

Step 3. Compute

$$\frac{17.1 \text{ g } C_{12}H_{22}O_{11}}{125 \text{ g water}} \times \frac{1 \text{ mol } C_{12}H_{22}O_{11}}{342 \text{ g } C_{12}H_{22}O_{11}} \times \frac{10^3 \text{ g water}}{1 \text{ kg } H_2O} =$$

$$\frac{0.400 \text{ mol } C_{12}H_{22}O_{11}}{\text{kg } H_2O} = 0.400 \ m \ C_{12}H_{22}O_{11}$$

Step 4. Evaluate   The answer is correctly given to three significant digits. A numerical estimate of $(2 \times 10^4) \div (4 \times 10^4)$ indicates an answer of approximately 0.5. The label *mol solute/kg solvent* is correct for molality, the desired measurement.

Sample Problem 14.5

A solution of iodine (I_2) in carbon tetrachloride (CCl_4) is used when iodine is needed for certain chemical tests. How much iodine must be added to prepare a 0.480 m solution of iodine in CCl_4 if 100.0 g of CCl_4 are used?

Solution

Step 1. Analyze Given: molality of solution and solute formula = 0.480 m I_2
mass of solvent and identity = 100.0 g CCl_4
Unknown: mass of solute

Step 2. Plan The molality gives you the moles of solute, which can be converted to the mass of solute when you determine the molar mass.

molality → molar mass of solute → mass of solute

$$\text{molality}\left(\frac{\text{mol solute}}{\text{kg solvent}}\right) \times \text{molar mass of solute}\left(\frac{\text{g solute}}{\text{mol solute}}\right)$$

× mass of solvent (kg solvent) = mass of solute (g solute)

Step 3. Compute

$$\frac{0.480 \text{ mol } I_2}{\text{kg } CCl_4} \times \frac{254 \text{ g } I_2}{\text{mol } I_2} \times 100.0 \text{ g } CCl_4 \times \frac{1 \text{ kg}}{10^3 \text{ g}} = 12.2 \text{ g } I_2$$

Step 4. Evaluate The answer has three significant digits, as do the factors in the calculation. The units cancel, leaving the desired label, g I_2. A numerical estimate of 125/10, or 12.5, would support the computation.

Practice Problems

1. What is the molality of a solution composed of 2.55 g of acetone, $(CH_3)_2CO$, dissolved in 200. g of water? *(Ans.)* 0.219 m acetone
2. What quantity, in grams, of methanol (CH_3OH) is required to prepare a 0.244 m solution in 400. g of water? *(Ans.)* 3.12 g CH_3OH
3. How many grams of $AgNO_3$ are needed to prepare a 0.125 m solution in 250. mL of water? *(Ans.)* 5.31 g $AgNO_3$
4. What is the molality of a solution containing 18.2 g HCl and 250. g of water? *(Ans.)* 1.99 m

Section Review

1. What concentration units give the number of moles of solute in a given volume of solution?
2. A gas–gas solution with a fixed amount of solute and solvent by mass is heated and increases volume. Which units of concentration will not change? Why?
3. Five grams of sugar, $C_{12}H_{22}O_{11}$, are dissolved in one liter of water. What is the concentration of this solution in percent by mass, molality, and molarity?
4. Two solutions of sugar, $C_{12}H_{22}O_{11}$, and water are in different bottles. One is a one mol*ar* solution and the other is a one mol*al* solution. In percent by mass, which solution is more concentrated?

1. Molarity
2. Molality and percent by mass. Unlike molarity, molality and percent by mass do not depend on volume.
3. By mass: 0.5%, molality: 0.01 m, molarity: 0.01 M (providing the mixture forms 1 L of solution).
4. A 1 M solution would have 342 g of sugar plus just enough water to make 1 L of solution; the 1 m solution would have 342 g of sugar plus 1 kg (1 L at 4°C) of water. Since both solutions contain the same amount of sugar, the *1 M* solution, having less water, would be more concentrated.

14.4 Colligative Properties of Solutions

SECTION OBJECTIVES

- List three colligative properties and describe how each is caused.
- Write the expression for freezing-point depression and boiling-point elevation. Define and give the units for each term in the expressions.
- Given appropriate information, calculate freezing-point depression, boiling-point elevation, or solution molality.
- Describe an experimental method for determining molar mass using a colligative property.
- Calculate molar mass from freezing-point depression or boiling-point elevation data.

Imagine that it is a very cold day in the middle of winter, like that pictured in Figure 14-17. A steady flow of cars and trucks moves down the highway. The operation of these vehicles is dependent on the steady flow of water in their cooling systems. How can this water remain fluid? What is the purpose of the salt that has been sprinkled on the sidewalk and road? A scientific understanding of the behavior of solutions helps us to continue our daily activities despite the winter cold. Using the properties of solutions, we can counteract nature by lowering the freezing point of water.

Figure 14-17 The cooling systems of most cars depend on liquid water. How can these cars operate at temperatures below the freezing point of water?

Simplest formulas and molecular formulas were discussed in Chapters 6 and 7.

Vapor-Pressure Lowering

To understand how the freezing point of water is lowered, we must go back to our study of vapor pressure—the pressure caused by molecules that have escaped from the liquid phase to the gaseous phase. Vapor pressure is a measure of the tendency of molecules to escape from a liquid or a solid. Experiments show that the vapor pressure of a solvent containing a nonvolatile solute is lower than the vapor pressure of the pure solvent at any temperatures. By **nonvolatile,** *we mean that a substance that has little tendency to become a gas under the existing conditions.*

Figure 14-18 illustrates the effect a nonvolatile solute has on the pressure of a solvent. Compare the top curve for the pure solvent with the middle curve for a dilute solution of a nonvolatile solute. Vapor pressure has been lowered, and the solution remains liquid over a larger temperature range. At the left, you can see that the freezing point of the solution (the point where the two curves intersect) is lower than that of the pure water. At the right, you can see that the boiling point of the solution is higher than that of pure water.

Have you noticed we have said nothing about the identity of the solute? That is because it does not matter. Each solute particle at the liquid surface has the same effect, whether it is an ion or a molecule and whatever its composition. *A property that depends on the number of solute particles but*

426 Chapter 14

Figure 14-18 Vapor pressure of water as a function of temperature plotted for pure water and water solutions of molalities *X* and *2X*. Freezing-point depressions and boiling-point elevations are proportional to the molal concentrations of the solutions.

is independent of their nature is called a **colligative property.** In Figure 14-18, you can see that adding a nonvolatile solute has *lowered* the vapor pressure of the solvent, *lowered* the freezing point of the solution, and *raised* the boiling point of the solution—these are all colligative properties.

In studying colligative properties, solution concentrations are given in molality, which is a useful method for expressing concentration because it does not change its value with temperature changes. Our discussion of colligative properties in this chapter is limited to dilute solutions of molecular solutes—that is, nonelectrolytes. Dilute solutions of equal molality of any molecular solute in the same solvent have equal vapor-pressure lowering and other colligative properties. As represented by the bottom curve in Figure 14-18, if the molality is doubled, the vapor-pressure lowering, freezing-point depression, and boiling-point elevation are all doubled. Colligative properties are directly proportional to the molal concentration of a molecular solute.

Figure 14-19 A solution of purple dye and water *(left)* is frozen *(right)*. The water (solvent) freezes but the dye (solute) does not.

Solutions 427

Chemistry and Technology

Fresh Water From the Sea

Approximately 97% of the earth's water is contained in its oceans. Human beings have tried for centuries to develop an efficient, inexpensive method to desalinate seawater so that it can be used for drinking and cooking.

One of the oldest techniques for removing salt from water is distillation. Even today, distillation provides about three-fourths of the world's desalinated water. In saltwater distillation, the water is heated and vaporized, and the vapor is then condensed into freshwater. Though effective, distillation is an expensive way to desalinate water.

Newer methods use semipermeable membranes to separate salt from water. One of these is reverse osmosis, in which pressure is used to force salty water against a membrane. The salt is left behind as the water passes through the membrane. Another method that uses membranes is electrodialysis. (See the diagram.) In electrodialysis, an electric current is sent through the salt water, drawing the positive and negative ions of salt through separate membranes. Pure water remains between the two membranes.

Membrane processes are less effective than distillation methods of desalination and are usually used to desalinate brackish inland water, rather than seawater. The Yuma Desalting Plant on the United States-Mexico border, for example, uses reverse osmosis to purify water from the Colorado River for irrigation.

Freezing-Point Depression

The ability of motor vehicles to start and run on a cold winter day depends upon the addition of ethylene glycol or other antifreeze substances to the cooling water. All antifreezes act by lowering the vapor pressure of the solution and, as a result, they lower the freezing point.

When a solution containing nonvolatile solute freezes, only pure solvent solidifies. At the freezing point there is an equilibrium between a solid and a liquid. This means that the pure solid solvent's vapor pressure is the same as that of the liquid solution. The curve at the left in Figure 14-18 (see previous page) shows solid–vapor equilibrium. You can see that the solution must be at a lower temperature before the vapor pressures of the solid and of the solution become equal.

The freezing point of a 1 molal solution of any molecular solute in water is found by experiment to be lowered by 1.86C°. That is, when one mole is dissolved in 1 kg of water, the freezing point of the solution is −1.86°C instead of 0.00°C. For any concentration of a molecular solute in water, this decrease can be determined by using the value 1.86C°/molal. This value is called a **molal freezing-point constant (K_f),**—*the freezing-point depression of the solvent in a 1 molal solution of a nonvolatile, molecular solute.* Each solvent has its own characteristic molal freezing-point depression constant.

The values of K_f for some common solvents are given in Table 14-6. These values are most accurate for dilute solutions at 1 atmosphere of pressure. Some variations are introduced in the value of K_f (and K_b) at other pressures and with more concentrated solutions.

Reminder: °C is a temperature point; C° is a temperature interval.

TABLE 14-6 MOLAL FREEZING-POINT AND BOILING-POINT CONSTANTS

Solvent	Normal f.p. (°C)	Molal f.p. constant, K_f (C°/molal)	Normal b.p. (°C)	Molal b.p. constant, K_b (C°/molal)
acetic acid	16.6	−3.90	118.5	3.07
acetone	−94.8	—	56.00	1.71
aniline	−6.1	−5.87	184.4	3.52
benzene	5.48	−5.12	80.15	2.53
carbon disulfide	−111.5	−3.80	46.3	2.34
carbon tetrachloride	−22.96	—	76.50	5.03
ethanol	−114.5	—	78.26	1.22
ether	−116.3	−1.79	34.42	2.02
naphthalene	80.2	−6.9	218.0	5.65
phenol	40.9	−7.27	181.8	3.56
water	0.00	−1.86	100.0	0.51

The use of candy thermometers is based on the boiling-point elevation of water by sugar. As the water boils away in making candy, the solution becomes more concentrated and its boiling point and temperature increases. The thermometer indicates when you should stop cooking your candy.

Because molality is equal to moles of solute per kg of solvent, the units for K_f, for example, for water can be expressed as follows:

$$K_f = \frac{-1.86 C°}{\text{mol solute/kg H}_2\text{O}} = \frac{-1.86 C° \cdot \text{kg H}_2\text{O}}{\text{mol solute}}$$

The **freezing-point depression,** Δt_f, is the difference between the freezing points of the pure solvent and a nonelectrolyte solution in that solvent, and is directly proportional to the molal concentration of the solution: $\Delta t_f \propto m$. If the molal concentration is doubled, the freezing-point depression is doubled. The proportionality constant for this relationship is K_f, giving

$$\Delta t_f = K_f m$$

With K_f in the units shown above and m expressed in moles solute/kg solvent, Δt_f is found in Celsius degrees. Sample Problems 14-6 and 14-7 illustrate the use of this expression.

Sample Problem 14.6

What is the freezing-point depression of water in a solution of 17.1 g of sucrose, ($C_{12}H_{22}O_{11}$), and 200. g of water?

Sample Problem 14.6 *continued*

Solution

Step 1. Analyze Given: solute mass and chemical formula = 17.1 g $C_{12}H_{22}O_{11}$
solvent mass and identity = 200. g water
Unknown: freezing point depression

Step 2. Plan The molal-freezing point constant for water can be found in Table 14-6. Using the formula for freezing-point depression, $\Delta t_f = K_f m$, the following route can be used.

solute mass, solvent mass → molality of solution, molar freezing-point constant → freezing-point depression

$$\Delta t_f = K_f m$$

$$K_f \left(\frac{\text{in } C° \cdot \text{kg solvent}}{\text{mol solute}} \right) \times \frac{\text{mass of solute (g)}}{\text{mass of solvent (g)}} \times \frac{(\text{mol of solute})}{\text{molar mass solute (g)}} \times \frac{10^3 \text{ g}}{\text{kg}}$$
$$= \Delta t_f \, (C°)$$

Step 3. Compute $\dfrac{-1.86 C° \cdot \text{kg water}}{\text{mol } C_{12}H_{22}O_{11}} \times \dfrac{17.1 \text{ g } C_{12}H_{22}O_{11}}{200. \text{ g water}} \times \dfrac{1 \text{ mol } C_{12}H_{22}O_{11}}{342 \text{ g } C_{12}H_{22}O_{11}} \times \dfrac{10^3 \text{ g}}{\text{kg}}$
$= -0.465 C°$

Step 4. Evaluate As shown, the units cancel correctly to give the answer in C°. The answer is correctly given to three significant digits because there is no factor in the calculation with fewer significant digits.

Sample Problem 14.7

A water solution containing an unknown quantity of a molecular solute is found to have a freezing point of $-0.23°C$. What is the molal concentration of the solution?

Solution

Step 1. Analyze Given: solvent identity = water
freezing point of solution = $-0.23°C$
Unknown: molality of the solution

Step 2. Plan Look up the value of K_f, the molal-freezing-point constant for water. Recall the equation for freezing-point depression.

$$\Delta t_f = K_f m \quad \text{Solve for molality, } m.$$

$$m = \frac{\Delta t_f}{K_f}$$

Then the route is: freezing point, temperature change → molality

Step 3. Compute $-0.23°C \div \dfrac{(-1.86 C°)}{1 \, m} = 0.12 \, m$

Step 4. Evaluate As shown by the unit cancellation, the answer gives the molality, as desired. The answer is properly limited to two significant digits.

Sample Problem 14.7 *continued*

> **Practice Problems**
> 1. A solution consists of 10.3 g of glucose ($C_6H_{12}O_6$), dissolved in 250. g of water. What is the freezing-point depression of the solvent?   *(Ans.)* $-0.426C°$
> 2. In a laboratory experiment, the freezing point of a water solution of glucose is found to be $-0.325°C$. What is the molal concentration of this solution?   *(Ans.)* $0.175\ m$

Boiling-Point Elevation

The boiling point of a liquid is the temperature at which the vapor pressure of the liquid is equal to the prevailing atmospheric pressure. Therefore, a change in either the atmospheric pressure or the vapor pressure of the liquid will cause a corresponding change in the boiling point.

The boiling-point elevation of a 1 molal solution of any molecular solute in water has been found by experiment to be $0.51C°$. That is, when one mole of a nonelectrolyte, such as sugar, is dissolved in 1 kg of water, the boiling point of the solution is found to be $100.51°C$. For different solvents, the boiling-point elevation of 1 molal solutions have different values. *The* **molal boiling-point constant (K_b)** *is the boiling-point elevation of the solvent in a 1 molal solution of a nonvolatile, molecular solute.* For water, the molal boiling-point constant is

$$K_b = \frac{0.51C°}{\text{mol solute/kg } H_2O} = 0.51C° \cdot \text{kg } H_2O/\text{mol solute}$$

Figure 14-20 Many recipes call for adding salt to water. Why?

Some other values for K_b are included in Table 14-6. Like the freezing-point constants, these values are the most accurate for dilute solutions.

The **boiling-point elevation**, Δt_b, *is the difference between the boiling points of the pure solvent and a nonelectrolyte solution in that solvent, and is directly proportional to the molal concentration of the solution.* The direct proportionality between boiling-point elevation and molality is given by

$$\Delta t_b = K_b m$$

When K_b is expressed in $C°/(\text{mol solute/kg solvent})$ and m is expressed in moles of solute/kg of solvent, Δt_b is the boiling-point elevation in $C°$. This expression can be used in problem solving to find Δt_b if m is known and to find m if Δt_b is known in the same ways illustrated for freezing-point depression.

Determination of Molar Mass of a Solute

Because colligative properties depend only on the number of solute particles and not on their identity, boiling-point elevation and freezing-point depression can be used to determine the molar mass of soluble, nonvolatile nonelectrolytes.

An experiment is performed in which a known mass of a molecular solute is dissolved in a known mass of a solvent. It must be a solvent with which the solute does not react chemically. The freezing-point depression or boiling-point elevation of the solvent is then measured. This experimental value is used to calculate the molality of the solution. Then, the calculated molality

Solutions

and the known solute mass are used to find the molar mass of the solute. The molality, mass of solute, and mass of solvent are related to the molar mass of the solute as follows:

$$\text{molality} = \frac{\text{mass of solute (g)}}{\text{molar mass of solute} \times \text{mass of solvent (kg)}}$$

or

$$\text{molar mass of solute} = \frac{\text{mass of solute (g)}}{\text{molality} \times \text{mass of solvent (kg)}}$$

The equations that express the relationship between the molality and the change in boiling and freezing points can be expanded to include the factor of molar mass.

$$\Delta t_b = K_b m \quad \text{or} \quad m = \frac{\Delta t_b}{K_b} \quad \text{and} \quad \Delta t_f = K_f m \quad \text{or} \quad m = \frac{\Delta t_f}{K_f}$$

Thus,

$$\frac{\text{mass of solute (g)}}{\text{molar mass of solute} \times \text{mass of solvent (kg)}} = \frac{\Delta t_b}{K_b}$$

and

$$\frac{\text{mass of solute (g)}}{\text{molar mass of solute} \times \text{mass of solvent (kg)}} = \frac{\Delta t_f}{K_f}$$

Rearranging gives expressions for the molar mass of solute

$$\text{molar mass of solute} = \frac{\text{mass of solute (g)} \times K_b}{\Delta t_b \times \text{mass of solvent (kg)}}$$

and

$$\text{molar mass of solute} = \frac{\text{mass of solute (g)} \times K_f}{\Delta t_f \times \text{mass of solvent (kg)}}$$

This method is illustrated in Sample Problem 14.8.

Sample Problem 14.8

In order to determine the molar mass of sulfur, 1.8 g of sulfur (S), is dissolved in 100. g of naphthalene ($C_{10}H_8$): $K_f = -6.9 C°/\text{molal}$; Table 14-6. The freezing-point depression of the naphthalene, was thereby determined. The value found for Δt_f was $-0.48 C°$. What is the molar mass of sulfur? Suggest a molecular formula for the element sulfur.

Solution

Step 1. Analyze   Given: mass and identity of solute = 1.8 g S
mass and identity of solvent = 100. g $C_{10}H_8$
value of $\Delta t_f = -0.48 C°$
value of $K_f = 6.9 C°/\text{molal}$
Unknown: molar mass (and molecular formula) of S

Step 2. Plan   mass of solute, mass of solvent, freezing-point depression of solvent, $K_f \rightarrow$
molar mass of sulfur, possible molecular formula

$$\text{molar mass of sulfur (g/mol)} = \frac{\text{mass of solute (g S)} \times K_f \text{ (C°·kg } C_{10}H_8/\text{mol S)}}{\Delta t_f \text{ (C°)} \times \text{mass of solvent (kg } C_{10}H_8)}$$

Sample Problem 14.8 *continued*

Step 3. Compute

$$\frac{1.8 \text{ g S} \times (-6.9\text{C}°\cdot\text{kg C}_{10}\text{H}_8/\text{mol S})}{-0.48\text{C}° \times 100.\text{ g C}_{10}\text{H}_8} \times \frac{1000 \text{ g}}{1 \text{ kg}} = 260 \text{ g S/mol S}$$

Knowing that the atomic mass of sulfur is 32 g/mol, this value for the molar mass indicates that the molecular formula for sulfur is

$$S_8 \ (260 \text{ g/mol} \div 32 \text{ g/mol} = 8).$$

Step 4. Evaluate   In each expression in Step 3, the units cancel appropriately to yield the desired label, g/mol. The calculated answer is limited to two significant digits. Numerically, a predicted answer would be 280, which is close to the final answer.

Practice Problems
1. When 1.56 g of an unknown, nonvolatile solute is dissolved in 200. g of water, the freezing-point depression of the solvent is $-0.453\text{C}°$. Determine the molar mass of the solute. *(Ans.)* 32.0 g/mol
2. If 1.84 g of a molecular solute is dissolved in 150 g of acetic acid, the boiling point of the solvent is elevated $0.60\text{C}°$. What is the molar mass of the solute? *(Ans.)* 63 g/mol

Section Review

1. What colligative properties are responsible for the following phenomena and how are they caused? *(a)* Antifreeze keeping water flowing in a car's cooling system when the air temperature is below freezing. *(b)* Ice melting on sidewalks after salt has been spread on them.
2. Two moles of a nonelectrolyte solute are added to 1 kg of an unknown solvent. The solution is observed to freeze at $7.9\text{C}°$ below its normal freezing point. What is molal freezing-point constant of the unknown solvent? Can you guess the identity of the solvent?
3. A solution of methanol and water contains 1.56 g of methanol dissolved in 200 g of water. Describe how you would determine the molar mass of methanol given the above materials.

1. *(a)* Antifreeze lowers the freezing point of water by lowering its vapor pressure, thus keeping it in the liquid state. *(b)* Salt, such as NaCl, causes ice to melt because the Na^+ and Cl^- ions lower the freezing point by lowering its vapor pressure. The more solute particles available, the more the freezing point is lowered.
2. $7.9\text{C}°/2 = 4.0\text{C}°$. Since acetic acid has a molal freezing-point constant of $-3.90\text{C}°/m$, it is probably acetic acid.
3. Determine the freezing point of the solution and then the number of moles of solute using $\Delta t_f/K_f = m$. Then
   molar mass of solute = $\frac{\text{mass of solute (g)}}{m \times \text{mass of solvent (kg)}}$

Chapter Summary

- Most samples of matter we encounter are mixtures.
- A mixture is classified as either a solution, a suspension, or a colloid depending on the size of the particles in the mixture.
- A solute will dissolve at a rate that depends on the surface area of the solute, how vigorously the solution is mixed, and the temperature of the solvent.
- The solubility of a substance indicates how much of that substance will dissolve in a specific amount of solvent under certain conditions.
- The solubility of a substance depends on the temperature.
- The solubilities of gases in liquids increases with increased pressure.
- The overall energy change in solution formation is called the heat of solution.
- The concentration of a solution can be stated in the three different ways: percent by mass, molarity, and molality.
- Colligative properties of solutions depend only on the number of solute particles present.

Chapter 14 Review

Solutions

VOCABULARY

alloy (407)
boiling-point elevation (431)
colligative property (427)
colloid (409)
concentration (420)
effervescence (416)
electrolyte (406)
freezing-point depression (429)
heat of solution (417)
Henry's law (415)
immiscible (414)
miscible (415)
molal boiling-point constant (431)
molal freezing-point constant (429)
molality (423)
molarity (421)

percent by mass (420)
nonelectrolyte (406)
nonvolatile substance (426)
saturated solution (412)
solubility (413)
soluble (405)
solute (406)
solution (405)
solution equilibrium (412)
solvated (418)
solvent (406)
supersaturated solution (412)
suspension (408)
unsaturated solution (412)

REVIEWING CONCEPTS

1. **a.** What is the Tyndall effect?
   b. Cite a common example of this effect.
   c. Identify one application of this effect (14.1)

2. Given an unknown mixture consisting of two or more substances, explain one technique that could be used to determine whether that mixture is a true solution, a colloid, or a suspension. (14.1)

3. **a.** What is solution equilibrium?
   b. What factors determine the point at which a given solute-solvent combination reaches equilibrium? (14.2)

4. **a.** What is a saturated solution?
   b. What visible evidence indicates that a solution is at its saturation point?
   c. What is an unsaturated solution? (14.2)

5. **a.** What is meant by the solubility of a substance?
   b. What condition(s) must be specified in citing solubility levels? (14.2)

6. What is the relationship between the rate at which a substance dissolves and the solubility of that substance? (14.2)

7. **a.** What rule of thumb is useful for predicting whether or not one substance dissolves in another?
   b. In general, what does the rule mean in terms of polar and nonpolar solutes and solvents? (14.2)

8. **a.** How does pressure affect the solubility of a gas in a liquid?
   b. What law is a statement of this relationship?
   c. If the pressure of a gas above a liquid is increased, what happens to the amount of the gas that will dissolve in the liquid, if all other conditions remain constant? (14.2)

9. What is the effect of temperature on the solubility of:
   a. a gas in a liquid?
   b. a solid in a liquid? (14.2)

10. Based on Figure 14-13, determine the solubility of each of the following in grams of solute per 100 grams H_2O.
    a. $NaNO_3$ at 10°C
    b. KNO_3 at 60°C
    c. NaCl at 50°C (14.2)

11. Based on Figure 14-13, at what temperature would each of the following solubility levels be observed?
    a. 140 g KI/100 g H_2O
    b. 115 g $NaNO_3$/100 g H_2O
    c. 50 g KNO_3/100 g H_2O (14.2)

12. Compare endothermic and exothermic dissolving processes in terms of:
    a. the heat content of each solution relative to the combined heat content of the initial components of each
    b. the change in solution temperature that occurs during the dissolving process
    c. the sign of the heat of solution

434 Chapter 14

d. the effect of increased temperature on the solubility of most solids dissolved in a liquid. (14.2)

13. The heat of solution for $AgNO_3$ is $+22.8$ kJ/mol.
   a. Write the equation that represents the dissolving of $AgNO_3$ in water.
   b. What energy change is associated with the dissolving process; the crystallizing process?
   c. As $AgNO_3$ dissolves, what change occurs in the temperature of the solution?
   d. When the system is at equilibrium, how do the rates of dissolving and crystallizing compare?
   e. If the solution is then heated, how will rates of dissolving and crystallizing be affected? Why?
   f. How will the increased temperature affect the amount of solute that can be dissolved?
   g. If the solution is allowed to reach equilibrium and then cooled, how will the system be affected? (14.2)

14. a. Distinguish between dilute and concentrated as related to solution strength.
   b. How do these terms relate to solubility? (14.3)

15. Under what circumstances would solution concentrations be expressed in terms of:
   a. percent by mass   **c.** molality (14.3)
   b. molarity

16. How does the presence of a nonvolatile solute affect each of the following properties of the solvent into which the solute is dissolved?
   a. vapor pressure
   b. freezing point
   c. boiling point (14.4)

PROBLEMS

▶ Solubility

1. Plot a solubility graph for $AgNO_3$ from the following data, with grams of solute (by increments of 50) per 100 grams of H_2O on the vertical axis, and with temperature in °C on the horizontal axis.

Grams solute per 100 g H_2O	Temperature, °C
122 g	0°
216 g	20°
311 g	40°
440 g	60°
585 g	80°
733 g	100°

   a. How does the solubility of $AgNO_3$ vary with the temperature of the water?
   b. Estimate the solubility of $AgNO_3$ at 30°C, 55°C, and 75°C.
   c. At what temperature would the solubility of $AgNO_3$ be 275 g per 100 g of H_2O?
   d. If 100 g of $AgNO_3$ were added to 100 g of H_2O at 10°C, would the resulting solution be saturated or unsaturated? What would occur if 325 g of $AgNO_3$ were added to 100 g of H_2O at 35°C?

2. If a saturated solution of KNO_3 in 100 g of H_2O at 60°C is cooled to 10°C, approximately how many grams of the solute would precipitate out of the solution? (Use Figure 14-13).

▶ Percent by Mass

3. Find the percent by mass of each of the following solutions:
   a. 10.0 g of sugar in 40.0 g of water
   b. 0.63 g of $AgNO_3$ in 12.00 g of water

4. Find the mass of each solute present in each of the following solutions.
   a. 50.0 g of solvent in a 10.0% by mass NaCl solution
   b. 8.5 g of solvent in a 2.4% by mass KOH solution

▶ Molarity

5. Determine the number of grams of solute needed to make solutions of the following volumes and concentrations:
   a. 1.00 L of a 3.50 M solution of H_2SO_4
   b. 2.50 L of a 1.75 M solution of $Ba(NO_3)_2$
   c. 500. mL of a 0.250 M solution of KOH

6. Determine the molarity of each of the following solutions.
   a. 20.0 g NaOH in enough H_2O to make 2.00 L of solution
   b. 14.0 g NH_4Br in enough H_2O to make 150. mL of solution
   c. 32.7 g H_3PO_4 in enough H_2O to make 500. mL of solution

7. a. How many moles of NaOH are contained in 65.0 mL of a 2.20 M solution of NaOH in H_2O?
   b. How many grams of NaOH does this represent?

8. What is the molarity of a solution made by dissolving 26.42 g of $(NH_4)_2SO_4$ in enough H_2O to make 50.00 mL of solution?

9. If 100. mL of a 12.0 M HCl solution is diluted to 2.00 L, what is the molarity of the final solution?

10. How many milliliters of 16.0 M HNO_3 would be required to prepare 750. mL of a 0.500 M solution?

11. How many milliliters of 0.54 M $AgNO_3$ would contain 0.34 g of pure $AgNO_3$?

12. What mass of each product results if 750. mL of 6.00 M H_3PO_4 react according to the following?

$$H_3PO_4 + Ca(OH)_2 \rightarrow Ca_3(PO_4)_2 + H_2O$$

13. How many milliliters of 18.0 M H_2SO_4 would be required to react with 250. mL of 2.50 M $Al(OH)_3$ if the products are aluminum sulfate and water?

14. If 75.0 mL of a $AgNO_3$ solution reacts with enough Cu to produce 0.250 g of Ag by single replacement, what was the molarity of the initial $AgNO_3$ solution if $Cu(NO_3)_2$ is the other product?

▶ **Molality**

15. Determine the number of grams of solute needed to make each of the following solutions:
    a. a 4.50 m solution of H_2SO_4 in 1.00 kg H_2O
    b. a 1.00 m solution of HNO_3 in 2.00 kg H_2O
    c. a 3.50 m solution of $MgCl_2$ in 0.450 kg H_2O

16. Determine the molality of each of the following solutions:
    a. 294.3 g H_2SO_4 in 1.000 kg H_2O
    b. 63.0 g HNO_3 in 0.250 kg H_2O
    c. 10.0 g NaOH in 300. g H_2O

17. A solution is prepared by dissolving 17.1 g of sucrose ($C_{12}H_{22}O_{11}$), in 275 g of H_2O. What is the molality of that solution?

18. How many kilograms of H_2O must be added to 75.5 g of $Ca(NO_3)_2$ to form a 0.500 m solution?

19. How many grams of glucose ($C_6H_{12}O_6$), must be added to 750. g of H_2O to make a 1.25 m solution?

20. A solution of ethyl alcohol (C_2H_5OH) in water is 1.75 m. How many grams of C_2H_5OH are contained per 250. g of water?

21. In how many liters of water should 65.0 g of NaCl be dissolved to make a 0.450 m solution?

22. If a 3.00 m solution of HNO_3 containing 2.25 kg H_2O is diluted so that the resulting solution contains 5.75 kg H_2O, what is the molality of that final solution?

▶ **Colligative Properties**

23. Determine the freezing-point depression of H_2O in each of the following solutions.
    a. a 1.50 m solution of $C_6H_{12}O_6$ in H_2O
    b. 171 g of $C_{12}H_{22}O_{11}$ (sucrose) in 1.00 kg H_2O
    c. 77.0 g of $C_{12}H_{22}O_{11}$ in 400. g H_2O

24. Determine the molality of each solution of sugar in water given the following freezing-point depressions.
    a. $-0.930C°$ b. $-3.72C°$ c. $-8.37C°$

25. A solution contains 20.0 g of $C_6H_{12}O_6$ in 250. g of water.
    a. What is the freezing-point depression of the solvent?
    b. What is the freezing point of the solvent?

26. Determine the boiling-point elevation of H_2O in each of the following solutions.
    a. a 2.5 m solution of $C_6H_{12}O_6$ in H_2O
    b. 3.20 g $C_6H_{12}O_6$ in 1.00 kg H_2O
    c. 20.0 g $C_{12}H_{22}O_{11}$ in 500. g H_2O

27. Determine the molality of each water solution given the following boiling points:
    a. 100.25°C   b. 101.53°C   c. 102.805°C
28. What is the molar mass of a molecular solute when 5.00 g of the solute dissolved in 150. g of phenol lowers the freezing point 1.5C°?
29. Determine the molar mass of an unknown nonelectrolytic solute when 7.55 g of the solute dissolved in 125 g of aniline raises the boiling point to 185.5°C.
30. How many grams of antifreeze, $C_2H_4(OH)_2$, would be required per 500. g of water to prevent the water from freezing down to -20.0°C?
31. Pure benzene (C_6H_6), freezes at 5.45°C. A solution containing 7.24 g of $C_2Cl_4H_2$ in 115 g of benzene (specific gravity = 0.879) freezes at 3.55°C. Based on these data, what is the molal freezing-point constant for benzene?
32. The freezing point of pure camphor is 178.40°C; its molal freezing-point constant is -40.00C°/molal. If 1.500 g of a solute having a molar mass of 125.0 g were dissolved in 35.00 g of camphor, what would be the resulting freezing point of the camphor?
33. Find the boiling point of an aqueous solution that freezes at -6.51°C.
34. The molal freezing-point constant for ether is -1.79C°/molal. When 21.2 g of a solute is dissolved in 740. g of ether, the freezing point of the ether is lowered by 0.568C°. What is the molar mass of that solute?
35. A solution containing 9.00 g of a solute dissolved in 250. g of water freezes at -0.372°C.
    a. What is the molar mass of the solute?
    b. Among the following, which formula most likely represents the solute: CH_3OH, C_6H_6, C_2H_5OH, $C_6H_{12}O_6$, or $C_{12}H_{22}O_{11}$?

CRITICAL THINKING

1. **Predicting Outcomes** You have been investigating the nature of suspensions, colloids, and solutions and have collected the following observational data on four unknown samples. Infer whether the samples are solutions, suspensions, or colloids from the data.

Data Table 1—Samples

Sample	Color	Clarity (clear or cloudy)	Settle Out	Tyndall Effect
1	Green	Clear	No	No
2	Blue	Cloudy	Yes	Yes
3	Colorless	Cloudy	No	Yes
4	White	Cloudy	No	Yes

Data Table 2—Filtrate of Samples

Sample	Color	Clarity (clear or cloudy)	Filter Paper	Tyndall Effect
1	Green	Clear	None	No
2	Blue	Cloudy	Gray Solid	Yes
3	Colorless	Cloudy	None	Yes
4	White	Clear	White Solid	No

Based upon your inferences in Data Table 1, you decide to conduct one more test of the particles. You filter the samples and then re-examine the filtrate. You obtain the data found in Data Table 2. Infer the classifications based upon the data in Table 2. What would you use as your final test to verify your conclusions?

RESEARCH AND WRITING

1. Find out how much salt a large northern city, such as New York or Chicago, uses on its streets in a typical winter. What environmental problems result from this use of salt? What substitutes are being used for salt to melt ice and snow?

PROJECT

1. If your laboratory is suitably equipped, obtain a water-soluble unknown from your instructor and attempt to determine its molar mass by using the freezing-point depression approach.

CHAPTER 15
Ions in Aqueous Solutions

PLANNING CHART

Approximate Time: 4/6 days

	SECTION/TEXT FEATURES		RESOURCES	
BASIC	**15.1 IONIC COMPOUNDS IN AQUEOUS SOLUTIONS** Demos 1, 2 Microscale Investigation 13 Section Review Chapter Review • Questions 1–8, Problems 1–7	439 438B 860 446 457	*Teacher's Resource Binder* Lab Experiment 19 Section Review Worksheets *Other Resources* *Shakhashiri* Video Demos 11, 12, 13, 14 Transparency 31	111 52
BASIC	**15.2 MOLECULAR ELECTROLYTES** Demos 3, 4 Section Review Chapter Review • Questions 9–12	447 438C 450 457	*Teacher's Resource Binder* Section Review Worksheets *Other Resources* Transparency 32	53
AVERAGE	**15.3 PROPERTIES OF ELECTROLYTE SOLUTIONS** Demo 5 Section Review Chapter Review • Questions 13–19, Problems 8–17	450 438D 456 457	*Teacher's Resource Binder* Section Review Worksheets *Other Resources* Transparencies 33, 34	54
	CHAPTER END ACTIVITIES Critical Thinking, Cross Connections, Research and Writing, Projects		*Teacher's Resource Binder* Test Booklet *Computer Test Generator*	73, 75

Teaching Strategies

Introduction

Remind students that Chapter 14 dealt mainly with nonelectrolytes and that this chapter deals almost exclusively with electrolytes. It is a difficult and highly theoretical chapter. Go slowly, making sure students understand one concept before proceeding to the next.

The overall objectives for this chapter are *(1)* to study the Theory of Ionization from a historical and a modern-day viewpoint *(2)* to determine the differences between ionic and polar–covalent electrolytes and learn how to differentiate between the two in the laboratory *(3)* to learn how to write equations for dissociation, ionization, and reactions involving ions, and *(4)* to discuss the effects of electrolytes on electric conductivity as well as on freezing and boiling points of solutions.

15.1 Ionic Compounds in Aqueous Solution

Students often confuse the terms *conductor, nonconductor, electrolyte,* and *nonelectrolyte.* Point out that conductors and nonconductors can be pure substances (elements and compounds) or mixtures, but electrolytes and nonelectrolytes are the solutes of water solutions. Thus, either a solution of NaCl and water would be a conductor, but the NaCl would be an electrolyte, or a solution of sugar and water would be a nonconductor, but the sugar would be the nonelectrolyte. These distinctions are important to understanding concepts throughout the chapter. To help students visualize these differences, put Table T15-1 (shown below) on a transparency. This transparency will be very helpful throughout the chapter, and students will be able to understand it better as future sections are discussed.

CONDUCTORS			
Pure Substances		**Mixtures**	
Elements	Compounds	Alloys	Electrolytic Solutions
All metals Cu, Ag, Fe	All ionic compounds in liquid state NaBr(ℓ), KNO$_3$(ℓ)	Stainless steel, Sterling silver	Water solutions of NaBr, HCl, NH$_3$

NONCONDUCTORS		
Pure Substances		Mixtures
Elements	Compounds	Nonelectrolytic Solutions
All non-metals S_8, I_2, P_4	All covalent compounds in liquid state $HBr(\ell)$, $Al_2Cl_6(\ell)$ All solid compounds sucrose(s), NaBr(s), AlBr$_3$(s), KNO$_3$(s)	Water solutions of sucrose, isopropyl alcohol, ethyl alcohol, glycerin

To help lay a foundation for this chapter, do the following demonstration as an interest-catching introduction.

Demonstration 1: Conductivity of Pure Substances and Solutions

PURPOSE: To help students differentiate between conductors, nonconductors, electrolytes, and nonelectrolytes

MATERIALS: Acetic acid, glacial; ammonia water, concentrated; ethanol (95% denatured will be an acceptable substitute for absolute ethanol); hydrochloric acid, concentrated; glycerin; sodium chloride; sodium hydroxide; sucrose; water, distilled; ammeter, 0-200-milliampere; 2 batteries, 6-volt; 10 beakers, 250-mL; bulb, 6.3-V/150-mA; clips, alligator; copper wire, No. 20, insulated and No. 12, bare; lamp holder, threaded; popsicle stick, wooden or plastic; switch, push button

PROCEDURE: *(a)* Assemble a set-up like the one shown in Figure T15-1. It is very simple and inexpensive to make. All parts can be obtained from your local electronics store. All connections are made with No. 20 insulated Cu wire.

Figure T15-1

(b) Make up 0.1 *m* solutions of each of the following substances by dissolving the indicated amounts in 200 mL of distilled water: sucrose—7 g; glycerin—2.3 mL; sodium chloride—1.2 g; ammonia—1.4 mL of concentrated NH$_4$OH; acetic acid—1.2 mL glacial acetic acid; hydrochloric acid—1.7 mL concentrated HCl; and NaOH—0.8 g.

(c) Measure out 200 mL of the following pure substances – distilled water, glycerin, and ethanol.

(d) First test the pure substances and then the solutions, using the following steps: *(1)* immerse the electrodes of the conductivity apparatus in 200 mL of each material being tested; *(2)* close the switch; *(3)* check to see if the bulb lights up and if

so, whether it is bright or dim; *(4)* take a reading from the ammeter; *(5)* open the switch; *(6)* remove the material being tested; *(7)* rinse the electrodes with distilled water and dry with a paper towel after each immersion. *Note:* The bulb does not light up for some poor conductors, but the degree of conductivity can be noted on the ammeter. After each test ask: Is this a conductor or a nonconductor? Would it be an electrolyte or a nonelectrolyte? (Answers will vary. The pure substances can be only conductors or nonconductors. After students identify the solutions as conductors or nonconductors, have them identify the solutes as electrolytes or nonelectrolytes.)

This section offers an excellent opportunity to emphasize some characteristics of outstanding scientists and at the same time show that scientists, too, are human beings. We often view scientists as older, experienced individuals, but Arrhenius was just a young man working on his doctorate in chemistry. He was obviously a creative thinker—a man far ahead of his time who believed in himself. He received only discouragement from his chemistry professor, Cleve, and his other professors at the university. When he presented his revolutionary theory to Cleve, the response was, "You have a new theory? That is very interesting. Goodbye." When he persisted in using the theory of ionization for his dissertation, he was severely punished with a fourth-class rating—the lowest given. This could have ended his career in chemistry. However, he wrote to several scientists in Europe seeking help.

Finally, Ostwald, a German chemist, came to Arrhenius's rescue by actually making the long trip to Stockholm to give his support. Twenty years later, Arrhenius was properly rewarded with the Nobel Prize for his brilliant work. Students love this story! Ask your students how they would evaluate Cleve's reaction in terms of scientific thinking. (According to one version, Cleve explained his reason for failing to listen to Arrhenius: He had heard so many theories of ionization – all of which were soon proven wrong or inadequate – that he thought this was just another unacceptable student theory.) To add a little dramatics, and at the same time to help students understand part three of the theory, hold up a beaker containing a solution of Na$_2$SO$_4$ and stress the fact that it contains billions and billions of electrically charged particles (+ and − ions). Ask if anyone is brave enough to stick a finger into the beaker to find out which, if any, of the charged particles is in excess? Are there more positive ions than negative ions? Since there are twice as many positive ions as negative ones in this solution, why don't you get shocked? Can you give an example of a solution that has twice as many negative ions as positive ones and an example of a solution with an equal number of positive and negative ions? (Mg(NO$_3$)$_2$ and SrCO$_3$) In each case, what is true about the number of positive and negative charges?

Emphasize that dissociation occurs only when ionic solutes are dissolved in water. Ask how someone can determine whether a compound had ionic bonds? (If it is a binary compound, use the electronegativity difference and percent of ionic character rules. If it is a compound containing a polyatomic ion, it will probably be ionic, unless it is an acid. In the laboratory, see if the compound conducts an electric current when fused.)

Demonstration 2: How Dissociation Takes Place

PURPOSE: To show how water molecules cluster around an ion and pull it into the solution by separating it from the other ions

Ions in Aqueous Solutions

MATERIALS: Large gumdrops—two colors, small gumdrops—two colors, toothpicks

PROCEDURE: To make a model of a sodium chloride crystal, attach large gumdrops (for chlorine ions) to small ones (for sodium ions) with toothpicks using Figure T15-2a as a guide. Have a few of the gumdrops on one corner loosely attached for easy removal. Using Figure T15-2b as a guide, make water molecules using small gumdrops for oxygen and 1/4 of a small gumdrop for hydrogen. Using the Na^+Cl^- model and the models of water molecules, show how water molecules cluster around a sodium ion; how they separate the Na^+ from the other ions, and pull it into the solution. Repeat with a Cl^-. The water molecules can be attached to the ions by using small pieces of toothpicks. While demonstrating with the models, ask how a water molecule would approach an Na^+ (with its negative end —O). What about Cl^- (with its positive end—H)?

Figure T15-2

Students may have difficulty writing ionic equations showing dissociation. (They don't have as many problems until ionization is introduced; then they forget what to do.) Giving them a general equation, having them write several sample equations, and asking questions helps alleviate these problems. General equation: $P^+N^-(s) \rightarrow P^+(aq) + N^-(aq)$. Why are + and − charges written in P^+N^-? (To show that the compound is composed of ions) How do you know the compound is a solid, (s)? (Because of the strong forces of attraction between particles of ionic compounds, they all exist as solids at room temperature.) Why do you write (aq) next to the ions on the product side? (To show that they are hydrated) Now give students several sample problems, for example: Write an ionic equation to show the dissociation of calcium nitrate.

15.2 Molecular Electrolytes

Demonstration 3: How Ionization Takes Place

PURPOSE: To show how water molecules cluster around a polar molecule and pull it apart into ions during the solution process

MATERIALS: Modeling clay, four colors

PROCEDURE: Using modeling clay, make a model of hydrogen chloride and water. (You could use gumdrops also.) Use Figure T15-3 as a guide. After you have clustered the water molecules around the hydrogen chloride molecule, have the water molecules pull it apart into ions. Have students explain how this was different from Demonstration 2.

Demonstration 4: The Difference between Dissociation and Ionization

PURPOSE: To show how dissociation and ionization can be differentiated in the laboratory

MATERIALS: Acetic acid, glacial; potassium nitrate; water, distilled; 3 beakers, 250-mL; porcelain evaporating dish; conducting apparatus used for Demonstration 1

PROCEDURE: Part I Dissociation. To show that ionic compounds are conductors in the liquid state, and electrolytes when dissolved in water, use the simulated fused KNO_3 (described below),

Figure T15-3

and a 0.1 M solution of KNO_3 (made by dissolving 2.0 g in 200 mL of water) using the conductivity apparatus from Demonstration 1.

Because of the possible danger involved in fusing the low-melting compound, KNO_3, to show it as a conductor in the liquid state, use a simulation. Dissolve a few crystals of KNO_3 in 100 mL of distilled water in an evaporating dish (should give about 75-milliamp reading on ammeter). Tell students that this is what the test for fused KNO_3 would look like.

Why does the fused (melted) KNO_3 conduct an electric current? (The heat energy allows the potassium ions and the nitrate ions to separate from one another and move around; thus, they are able to conduct electricity.) Help students write equations to show what happened when KNO_3 was melted and when it was dissolved in water.

Fused: $KNO_3(s) \rightarrow K^+(\ell) + NO_3^-(\ell)$
Water Solution: $KNO_3(s) \rightarrow K^+(aq) + NO_3^-(aq)$

Part II Ionization. To show that certain covalent compounds that are nonconductors in the liquid state become electrolytes when placed in water, test 200 mL of glacial acetic acid and 200 mL of 0.1 M acetic acid (1.2 mL of glacial acetic acid/200 mL of distilled water) with the conductivity apparatus.

CAUTION: Glacial acid must be handled with care. If you do not have glacial acetic acid, a good alternative to simulate the demonstration is using 200 mL of distilled water in place of the glacial acetic acid and 200 mL of white vinegar in place of the 0.1 M solution.

Why won't glacial acid, which is 99.5% pure acetic acid, conduct an electric current? (In its pure state, its particles are molecules; ions must be present in the liquid state for a substance to conduct an electric current.) Why does the solution of acetic acid conduct an electric current? (During ionization, the water molecules pull the acetic acid molecules apart into hydronium ions and acetate ions.)

To help students write ionic equations for ionization reactions, use the general formula: $PN(s, l, g) + H_2O \rightarrow P \cdot H_2O^+ + N^-(aq)$.

Why are no charges written in PN? (PN is a molecule; no ions exist. Students often ask: (1) How do I know what state PN is in? (Use these rules: hydrogen halides—HF, HCl, HBr, and HI—are gases at room temperature. Most covalent salts such as the aluminum halides are solids.) (2) Why is water included in the reactive side for ionization reactions and not dissociation reactions? (Probably because it is needed to form hydronium ions when acids ionize. To be consistent, therefore, it is included in all ionization reaction equations.)

After dissociation and ionization have been completely covered, it is very important to help students differentiate between the two processes by asking questions such as: What type of particles does a solute that dissociates have before it is dissolved in water? (Ions) How could you prove in the lab that ions exist? (Fuse the solute, and test it for electrical conductivity.) What do the water molecules do to these + and − ions when the solute is dissolved? (The water molecules separate the ions from one another. The ions then form individual hydrated ions.) What type of particles does a solute that ionizes have before it is dissolved in water? (Polar molecules) What do the water molecules do to these polar molecules? (Pull them apart into ions.) What kind(s) of solute particles would you expect to find in a solution in which ionization has taken place? (+ and − ions and molecules.) A thought question: A substance is tested in the lab and found not to conduct an electric current in its solid state, but it conducts one in its liquid state. When this substance is dissolved in water, will the solution conduct an electric current? Explain.

15.3 Properties of Electrolyte Solutions

Demonstration 5: Strong and Weak Electrolytes

PURPOSE: To show how to differentiate between strong and weak electrolytes on the basis of electric conductivity

MATERIALS: 0.1 M solutions of acetic acid, ammonia water; hydrochloric acid; sodium chloride; sodium hydroxide; 5 beakers, 250-mL; conductivity apparatus from Demonstration 1

PROCEDURE: Using the conductivity apparatus, retest the following 0.1 M solutions from Section 15.1: NaCl, HCl, NaOH, $HC_2H_3O_2$, and NH_4OH. This time have students pay close attention to the readings from the ammeter. Which of these solutions are strong electrolytes? weak electrolytes? (Strong—NaCl, HCl, and NaOH; weak—$HC_2H_3O_2$ and NH_4OH.) In the weak electrolyte $HC_2H_3O_2$, what particles would you expect to find in solution? (H_3O^+, $C_2H_3O_2^-$, and $HC_2H_3O_2$ molecules) Which of these would be in excess? (The molecules) In the strong electrolyte HCl, what particles would be in excess? (The H_3O^+ and the Cl^-) To emphasize that compounds with covalent bonds can be strong or weak electrolytes: What do the electrolytes HCl and $HC_2H_3O_2$ have in common? How are they different from one another? (They both ionize and form H_3O^+. HCl ionizes to a large extent and thus is a strong electrolyte; $HC_2H_3O_2$ ionizes to a small extent and thus is a weak electrolyte.)

To help students understand that water solutions of electrolytes never completely ionize or dissociate and that the more dilute the solution, the greater the degree of ionization or dissociation, use a hypothetical problem such as: At what temperature would a 1 m solution of H_2SO_4 freeze assuming ionization to be 60%? Write an equation to determine the number of particles formed from each molecule. It works best if you start with an equation showing the ionization of H_2SO_4 and then rewrite it using 100 H_2SO_4 molecules. $H_2SO_4 + 2H_2O \rightarrow 2H_3O^+ + SO_4^{2-}$. Remind students that 60% ionization means that 60 out of every hundred molecules ionize, therefore:

$100 H_2SO_4 + 120 H_2O \rightarrow 120 H_3O^+ + 60 SO_4^{2-} + 40 H_2SO_4$.

100 molecules → 120 ions (+ ions) + 60 (− ions) + 40 molecules or 100 molecules → 220 particles or 1 molecule → 2.2 particles. Therefore, the freezing point of the solution will be lowered 2.2 times as much as a nonelectrolyte of the same molality or the freezing point = 0°C − 2.2 × −1.86C° = −4.09°C.

After discussing the Debye-Hückel theory, reemphasize that even in solutions of ionic solutes, there are two kinds of particles: (1) free (+ or −) ions and (2) groups of ions or ion clusters. Each ion cluster acts like a single particle in the way that it affects freezing-point depressions and boiling-point elevations.

References and Resources

Books and Periodicals

Jaffe, Bernard. "Svante Arrhenius," *Crucibles: The Story of Chemistry*, New York: Dover Publications, 1976, pp. 140-154.

Kauffman, George B. "Svante August Arrhenius, Swedish Pioneer in Physical Chemistry," *Journal of Chemical Education*, Vol. 65, No. 5, May 1988, pp. 437–8.

Kauffman, George B. and John Baxter. "Hydrated Cations in the General Chemistry Course," *Journal of Chemical Education*, Vol. 58, No. 4, April 1981, pp. 349–53.

Monroe, Manus and Karl Abrams. "A Perspective on Solubility Rules," *Journal of Chemical Education*, Vol. 61, No. 10, Oct. 1984, p. 885.

Parrott, Mary E. "Electrography: A Metal Detective Story," *The Science Teacher*, Vol. 50, No. 5, May 1983, pp. 52–53.

Talesnick, Irwin, "Solubility," *Idea Bank Collation,* Idea 248, Kingston, Ontario: Science Supplies and Services Co. Ltd., 1984.

Tanis, David. "Underground Sculpture," *ChemMatters*, Vol. 2, No. 1, Feb. 1984, pp. 10–11.

Audiovisual Resources

Computer Software General Chemistry 1A—Solutions, 128K, Queue, 1988.

Experiments in Chemistry—Ionization, filmstrip, Educational Images, Ltd.

Introduction to Chemistry—Ionization and Dissociation in Solution, filmstrip, Fisher-Scientific-EMD, 1988.

Answers and Solutions

Reviewing Concepts

1. The theory of ionization assumes that:
   - electrolytes in solution exist in the form of ions
   - an ion is an atom or a group of atoms that carries an electric charge
   - in the aqueous solution of an electrolyte, the total charge of the positive ions equals the total charge of the negative ions.

2. **a.** Hydration is the solution process in which water is the solvent.
   b. The number of water molecules surrounding a hydrated ion depends mainly upon the size and charge of the ion.

3. **a.** Heats of solution are determined by the magnitude of the solute-solute, solvent-solvent, and solute-solvent interactions.
   b. The heat of solution of an ionic compound is the overall result of the energy changes occurring during the separation of ions from the crystal, the separation of water molecules from each other, and the hydration of the separated ions. Energy is required for the first two processes, but is released in the third.

4. **a.** Heat of hydration is the energy released when ions become surrounded by water molecules.
   b. More energy is usually released in the hydration of smaller ions than of larger ones, and in the hydration of ions with higher charges.

5. **a.** Dissociation is the separation of ions that occurs when an ionic compound dissolves.
   b. The 1 M solution of KCl contains one mole each of K^+ and Cl^- ions; the BaF_2 solution contains one mole of Ba^{2+} and two moles of F^- ions; the $Mg(NO_3)_2$ solution contains one mole of Mg^{2+} ions and two moles of NO_3^- ions.

6.
   a. soluble   **f.** soluble
   b. soluble   **g.** soluble
   c. insoluble **h.** insoluble
   d. insoluble **i.** insoluble
   e. insoluble **j.** soluble

7. K^+(aq) and NO_3^-(aq) are the spectator ions.

8. A net ionic equation includes only those compounds and ions that undergo a chemical change in a reaction, and does not include spectator ions.

9. When molecules with polar-covalent bonds are dissolved in water, the dipoles of the water molecules and the oppositely charged regions of the solute molecules are attracted to each other such that one or more bonds in the solute molecules are broken. The more electronegative atom retains the bonding electron pair and forms a negative ion; the less electronegative atom forms a positive ion.

10. **a.** Ionization refers to the formation of ions from solute molecules by the action of the solvent; it essentially means the creation of ions where there were none.
    b. When an ionic compound dissolves through dissociation, the ions that were already present and associated with each other in the crystal become dissociated from each other. On the other hand, when a molecular compound dissolves in a polar solute, ions are formed where none existed in the undissolved compound.

11. **a.** A strong electrolyte is a solute that is present in aqueous solution entirely as hydrated ions; to whatever extent such substances dissolve in water, they yield only ions. A weak electrolyte is an ionizing solute that yields a relatively low concentration of ions in aqueous solution.
    b. Hydrogen chloride and aluminum chloride are examples of strong electrolytes, whereas hydrogen fluoride and acetic acid are weak electrolytes.

12. In general, the extent to which a solute ionizes in solution depends on the strength of the bonds within the molecules of the solute. If the strength of the bonds within the solute molecule is less than the attractive forces of the water dipoles, then the covalent bond breaks and the molecule is separated into ions.

13. **a.** In order for an aqueous solution to be conductive, its solute must be capable of transporting an electric charge. Whereas the ions of electrolytes are charge carriers, the neutral molecules of nonelectrolytes are not, and such substances are thus nonconductive.
    b. The degree to which an electrolyte solution conducts is related to the solute ion concentration; the higher the concentration of ions in an aqueous solution is, the more easily a solution conducts electric current.

14. Strong and weak molecular electrolytes differ in the degree of ionization, whereas dilute and concentrated solutions differ in the amount of solute dissolved in a given quantity of a solvent.

15. **a, b, d,** and **f** would conduct.

16. **a.** Nonvolatile electrolytes in aqueous solutions lower the freezing point and raise the boiling point sometimes two, three, or more times as much as nonvolatile nonelectrolytes in solutions of the same molality.
    b. Both the freezing-point depression and the boiling-point elevation of a solvent by a solute depend on the relative numbers of dissolved particles of solute present, and not on their identity. When electrolytes dissolve, they produce two, three, or more moles of solute particles for each mole of compound dissolved. Such substances would thus be expected to have two, three, or more times the effect on colligative properties in solutions of the same molality.

17. As the concentration of the aqueous nonvolatile solution decreases, the molal freezing-point depression and boiling-point elevation approach a whole number multiple of K_f and K_b, respectively. However, the result is never quite what it is predicted to be.

18. The more strongly the ions attract each other, the smaller is the apparent degree of ionization based on colligative-property measurements.

19. A concentrated solution of a weak electrolyte contains a relatively large quantity of solute dissolved in a given amount of solvent, but that solute is only slightly ion-

ized. On the other hand, a dilute solution of a strong electrolyte contains a relatively small quantity of solute dissolved in a given amount of solvent, but that solute is ionized to a relatively large degree.

Problems

1. a. $KI(s) \rightarrow K^+(aq) + I^-(aq)$;
   b. $NaNO_3(s) \rightarrow Na^+(aq) + NO_3^-(aq)$;
   c. $MgCl_2(s) \rightarrow Mg^{2+}(aq) + 2Cl^-(aq)$;
   d. $Na_2SO_4(s) \rightarrow 2Na^+(aq) + SO_4^{2-}(aq)$
2. a. 1 mol K^+, 1 mol I^-; 2 mol ions
   b. 1 mol Na^+, 1 mol NO_3^-; 2 mol ions
   c. 1 mol Mg^{2+}, 2 mol Cl^-; 3 mol ions
   d. 2 mol Na^+, 1 mol SO_4^{2-}; 3 mol ions
3. a. $Sr(NO_3)_2(s) \rightarrow Sr^{2+}(aq) + 2NO_3^-(aq)$
       0.50 mol   0.50 mol   1.0 mol
       1.5 mol solute ions
   b. $Na_3PO_4(s) \rightarrow 3Na^+(aq) + PO_4^{3-}(aq)$
       0.50 mol   1.5 mol   0.50 mol
       2.00 mol solute ions
   c. $K_2S(s) \rightarrow 2K^+(aq) + S^{2-}(aq)$
       0.275 mol   0.550 mol   0.275 mol
       0.825 mol solute ions
   d. $Al_2(SO_4)_3(aq) \rightarrow 2Al^{3+}(aq) + 3SO_4^{2-}(aq)$
       0.15 mol   0.30 mol   0.45 mol
       0.75 mol solute ions
4. a. $Co_2S_3(s) \rightleftarrows 2Co^{3+}(aq) + 3S^{2-}(aq)$
   b. $PbS(s) \rightleftarrows Pb^{2+}(aq) + S^{2-}(aq)$
   c. $PbCrO_4(s) \rightleftarrows Pb^{2+}(aq) + CrO_4^{2-}(aq)$
5. a. 1. $AgNO_3(aq) + NaI(aq) \rightarrow AgI(s) + NaNO_3(aq)$
   (balanced chemical equation)
   2. $Ag^+(aq) + NO_3^-(aq) + Na^+(aq) + I^-(aq) \rightarrow$
   $AgI(s) + Na^+(aq) + NO_3^-(aq)$
   (overall ionic equation)
   3. Spectator ions: Na^+ and NO_3^-
   Possible precipitate: AgI
   4. $Ag^+(aq) + I^-(aq) \rightarrow AgI(s)$
   (net ionic equation)
   b. 1. $HgCl_2(aq) + K_2S(aq) \rightarrow HgS(s) + 2KCl(aq)$
   2. $Hg^{2+}(aq) + 2Cl^-(aq) + 2K^+(aq) + S^{2-}(aq) \rightarrow HgS(s) + 2K^+(aq) + 2Cl^-(aq)$
   3. Spectator ions: Cl^- and K^+ Possible precipitate: HgS
   4. $Hg^{2+}(aq) + S^{2-}(aq) \rightarrow HgS(s)$
   c. 1. $2Al(NO_3)_3(aq) + 3Ba(OH)_2(aq) \rightarrow 2Al(OH)_3(s) + 3Ba(NO_3)_2(aq)$
   2. $2Al^{3+}(aq) + 6NO_3^-(aq) + 3Ba^{2+}(aq) + 6OH^-(aq) \rightarrow 2Al(OH)_3(s) + 3Ba^{2+}(aq) + 6NO_3^-(aq)$
   3. Spectator ions: NO_3^- and Ba^{2+} Possible precipitate: $Al(OH)_3$
   4. $Al^{3+}(aq) + 3OH^-(aq) \rightarrow Al(OH)_3(s)$
   d. 1. $3CuCl_2(aq) + 2(NH_4)_3PO_4(aq) \rightarrow Cu_3(PO_4)_2(s) + 6NH_4Cl(aq)$
   2. $3Cu^{2+}(aq) + 6Cl^-(aq) + 6NH_4^+(aq) + 2PO_4^{3-}(aq) \rightarrow Cu_3(PO_4)_2(s) + 6NH_4^+(aq) + 6Cl^-(aq)$
   3. Spectator ions: NH_4^+ and Cl^- Possible precipitate: $Cu_3(PO_4)_2$
   4. $3Cu^{2+}(aq) + 2PO_4^{3-}(aq) \rightarrow Cu_3(PO_4)_2(s)$
6. 1. $Mg_3(PO_4)_2(aq) + 3Pb(NO_3)_2(aq) \rightarrow Pb_3(PO_4)_2(s) + 3Mg(NO_3)_2(aq)$
   2. $3Mg^{2+}(aq) + 2PO_4^{3-}(aq) + 3Pb^{2+}(aq) + 6NO_3^-(aq) \rightarrow Pb_3(PO_4)_2(s) + 3Mg^{2+}(aq) + 6NO_3^-(aq)$
   3. $3Pb^{2+}(aq) + 2PO_4^{3-}(aq) \rightarrow Pb_3(PO_4)_2(s)$

$$13.0 \text{ g } Mg_3(PO_4)_2 \times \frac{\text{mol } Mg_3(PO_4)_2}{263 \text{ g } Mg_3(PO_4)_2} = 0.0494 \text{ mol } Mg_3(PO_4)_2$$

$$0.0494 \text{ mol } Mg_3(PO_4)_2 \times \frac{1 \text{ mol } Pb_3(PO_4)_2}{1 \text{ mol } Mg_3(PO_4)_2} = 0.0494 \text{ mol } Pb_3(PO_4)_2$$

$$0.0494 \text{ mol } Pb_3(PO_4)_2 \times \frac{812 \text{ g } Pb_3(PO_4)_2}{\text{mol } Pb_3(PO_4)_2} = 40.1 \text{ g } Pb_3(PO_4)_2$$

7. a. $3NH_4OH(aq) + Al(NO_3)_3(aq) \rightarrow 3NH_4NO_3(aq) + Al(OH)_3(s)$
   b. $3NH_4^+(aq) + 3OH^-(aq) + Al^{3+}(aq) + 3NO_3^-(aq) \rightarrow 3NH_4^+(aq) + 3NO_3^-(aq) + Al(OH)_3(s)$
   c. Spectator ions: NH_4^+ and NO_3^-
   Precipitate: $Al(OH)_3$
   d. $Al^{3+}(aq) + 3OH^-(aq) \rightarrow Al(OH)_3(s)$
   e. $3NH_4OH(aq) + Al(NO_3)_3(aq) \rightarrow 3NH_4NO_3(aq) + Al(OH)_3(s)$
       3 mol   1 mol   3 mol   1 mol

This is a limiting-reactant problem. The next step is to determine the limiting reactant.

$$\frac{0.250 \text{ mol NH}_4\text{OH}}{\text{L solution}} \times 150. \text{ mL solution} \times \frac{\text{L solution}}{10^3 \text{ mL solution}} = 0.0375 \text{ mol NH}_4\text{OH}$$

$$\frac{0.400 \text{ mol Al(NO}_3)_3}{\text{L solution}} \times 125 \text{ mL solution} \times \frac{\text{L solution}}{10^3 \text{ mL solution}} = 0.0500 \text{ mol Al(NO}_3)_3$$

$$0.0375 \text{ mol NH}_4\text{OH} \times \frac{1 \text{ mol Al(NO}_3)_3}{3 \text{ mol NH}_4\text{OH}} = 0.0125 \text{ mol Al(NO}_3)_3$$

$$0.0125 \text{ mol Al(OH)}_3 \times \frac{78.0 \text{ g Al(OH)}_3}{1 \text{ mol Al(OH)}_3} = 0.975 \text{ g Al(OH)}_3$$

f. Since NH_4OH is the limiting reactant,
0.0500 mol $Al(NO_3)_3$ available − 0.0125 mol $Al(NO_3)_3$ required = 0.0375 mol $Al(NO_3)_3$ remains in excess.

8. $\Delta t_f = K_f m$

a. $\dfrac{-1.86 \text{C}° \cdot \text{kg solvent}}{\text{mol particles}} \times \dfrac{1.00 \text{ mol KI}}{\text{kg solvent}} \times \dfrac{2 \text{ mol particles}}{1 \text{ mol KI}} = -3.72\text{C}°$

b. $\dfrac{-1.86 \text{C}° \cdot \text{kg solvent}}{\text{mol particles}} \times \dfrac{1.00 \text{ mol CaCl}_2}{\text{kg solvent}} \times \dfrac{3 \text{ mol particles}}{1 \text{ mol CaCl}_2} = -5.58\text{C}°$

c. $\dfrac{-1.86 \text{C}° \cdot \text{kg solvent}}{\text{mol particles}} \times \dfrac{1 \text{ mol Ba(NO}_3)_2}{\text{kg solvent}} \times \dfrac{3 \text{ mol particles}}{1 \text{ mol Ba(NO}_3)_2} = -5.58\text{C}°$

d. $\dfrac{-1.86 \text{C}° \cdot \text{kg solvent}}{\text{mol particles}} \times \dfrac{1 \text{ mol C}_{12}\text{H}_{22}\text{O}_{11}}{\text{kg solvent}} \times \dfrac{1 \text{ mol particles}}{1 \text{ mol C}_{12}\text{H}_{22}\text{O}_{11}} = -1.86\text{C}°$

9. $\Delta t_f = K_f m$

$$\frac{-1.86 \text{C}° \cdot \text{kg solvent}}{\text{mol ions}} \times \frac{0.015 \text{ mol AlCl}_3}{\text{kg solvent}} \times \frac{4 \text{ mol ions}}{1 \text{ mol AlCl}_3} \rightarrow -0.11\text{C}°$$

or $\dfrac{-1.86 \text{C}°}{m \text{ ions}} \times 0.015 \, m \text{ AlCl}_3 \times \dfrac{4 \, m \text{ ions}}{1 \, m \text{ AlCl}_3} \rightarrow -0.11\text{C}°$

10. $\Delta t_f = K_f m$

$$\frac{-1.86 \text{C}° \cdot \text{kg H}_2\text{O}}{\text{mol ions}} \times \frac{\text{mol NaCl}}{58.4 \text{ g NaCl}} \times \frac{85.0 \text{ g NaCl}}{450. \text{ g H}_2\text{O}} \times \frac{2 \text{ mol ions}}{1 \text{ mol NaCl}} \times \frac{10^3 \text{ g H}_2\text{O}}{\text{kg H}_2\text{O}} = -12.0\text{C}°$$

$\Delta t_f = t_f\text{solution} - t_f\text{solvent}$
$t_f\text{solution} = \Delta t_f + t_f\text{solvent} = -12.0\text{C}° + 0.0°\text{C} = -12.0°\text{C}$

11. $\dfrac{0.51 \text{C}° \cdot \text{kg H}_2\text{O}}{\text{mol ions}} \times \dfrac{\text{mol BaCl}_2}{210 \text{ g BaCl}_2} \times \dfrac{25.0 \text{ g BaCl}_2}{0.150 \text{ kg H}_2\text{O}} \times \dfrac{3 \text{ mol ions}}{1 \text{ mol BaCl}_2} = 1.2\text{C}°$

$\Delta t_b = t_b\text{solution} - t_b\text{solvent}$
$t_b\text{solution} = \Delta t_b + t_b\text{solvent} = 1.2\text{C}° + 100.0°\text{C} = 101.2°\text{C}$

12. $\Delta t_b = K_b m$

$m = \dfrac{\Delta t_b}{K_b} = \dfrac{0.65 \text{C}° \cdot \text{mol ions}}{0.51 \text{C}° \cdot \text{kg H}_2\text{O}} \times \dfrac{1 \text{ mol KI}}{2 \text{ mol ions}} = 0.64 \, m \text{ KI}$

or $\dfrac{m \text{ ions}}{0.51\text{C}°} \times \dfrac{1 \, m \text{ KI}}{2 \, m \text{ ions}} \times 0.65°\text{C} = 0.64 \, m \text{ KI}$

13. $\Delta t_f = t_f\text{solution} - t_f\text{solvent} = -2.65°\text{C} - 0.00°\text{C} = -2.65\text{C}°$
$\Delta t_f = K_f m$

$m = \dfrac{\Delta t_f}{K_f} = \dfrac{-2.65 \text{C}° \cdot \text{mol ions}}{-1.86 \text{C}° \cdot \text{kg H}_2\text{O}} \times \dfrac{\text{mol Ba(NO}_3)_2}{3 \text{ mol ions}} = 0.475 \, m \text{ Ba(NO}_3)_2$

14. $\Delta t_f = K_f m$

$$\frac{-1.86 \text{C}° \cdot \text{kg H}_2\text{O}}{\text{mol ions}} \times \frac{0.250 \text{ mol NaCl}}{1.00 \text{ kg H}_2\text{O}} \times \frac{2 \text{ mol ions}}{1 \text{ mol NaCl}} = -0.930\text{C}°$$

$\Delta t_f = t_f\text{solution} - t_f\text{solvent}$
$t_f\text{solution} = \Delta t_f + t_f\text{solvent}$
$= -0.930\text{C}° + 0.000°\text{C} = -0.930°\text{C}$

15. The effect on freezing point is dependent upon the number and concentration of dissolved particles:
    a. 0.01 mol NaI → 0.02 mol ions
    b. 0.01 mol $CaCl_2$ → 0.03 mol ions
    c. 0.01 mol K_3PO_4 → 0.04 mol ions
    d. 0.01 mol $C_6H_{12}O_6$ → 0.01 mol molecules

    Since an increasing concentration in particles causes a corresponding increase in the change of the freezing point, then the order is: **d, a, b, c**.

16. $\Delta t_f = K_f m$

    $$m = \frac{\Delta t_f}{K_f} = \frac{-3.72\text{C}°\cdot\text{mol ions}}{-1.86\text{C}°\cdot\text{kg H}_2\text{O}} \times \frac{1 \text{ mol MgSO}_3}{2.00 \text{ mol ions}} = 1.00 \ m \text{ MgSO}_3$$

    Since 1.00 mol $MgSO_3 \cdot ?H_2O$ = 212 g/mol and 1 mol $MgSO_3$ = 104 g/mol, then there is 108 g of H_2O in the hydrate in the above solution.

    $$108 \text{ g H}_2\text{O} \times \frac{\text{mol H}_2\text{O}}{18.0 \text{ g H}_2\text{O}} = 6 \text{ mol H}_2\text{O}$$

    Therefore, the formula is $MgSO_3 \cdot 6H_2O$.

17. $\Delta t_f = K_f m$

    $$\frac{-1.86\text{C}°}{m \text{ ions}} \times 1.00 \ m \text{ MgI}_2 \times \frac{3 \ m \text{ ions}}{1 \ m \text{ MgI}_2} = -5.58\text{C}°$$

    According to the Debye-Hückel theory, there are probably many clusters of hydrated ions acting as a single ion, thereby effectively reducing the concentration of the individual ions. It is also possible that an MgI^+ species may exist due to strong solute-solute attraction. Also, there may not be enough water to hydrate all of the ions.

Critical Thinking

1. a. $\Delta t_f = t_f\text{solution} - t_f\text{solvent} = -1.15°\text{C} - 0.25°\text{C} = -1.40\text{C}°$
    $\Delta t_f = K_f m$

    $$m = \frac{-1.40\text{C}°\cdot\text{mol ions}}{-1.86\text{C}°\cdot\text{kg H}_2\text{O}} \times \frac{1 \text{ mol KNO}_3}{2 \text{ mol ions}} = 0.376 \ m \text{ KNO}_3$$

    b. $\frac{0.415 \text{ g KNO}_3}{10.00 \text{ mL H}_2\text{O}} \times \frac{1 \text{ mol KNO}_3}{101 \text{ g KNO}_3} \times \frac{10^3 \text{ mL H}_2\text{O}}{\text{kg H}_2\text{O}} = 0.411 \ m \text{ KNO}_3$

    $$\text{\%Difference} = \frac{|\text{observed value} - \text{actual value}|}{\text{actual value}}$$

    $$= \frac{|0.376 \ m - 0.411 \ m|}{0.411 \ m} \times 100\% = 8.52\%$$

2. Ion clustering does occur, resulting in fewer particles than expected. The clustering effect increases as the concentration increases.

Chapter 15
Ions in Aqueous Solutions

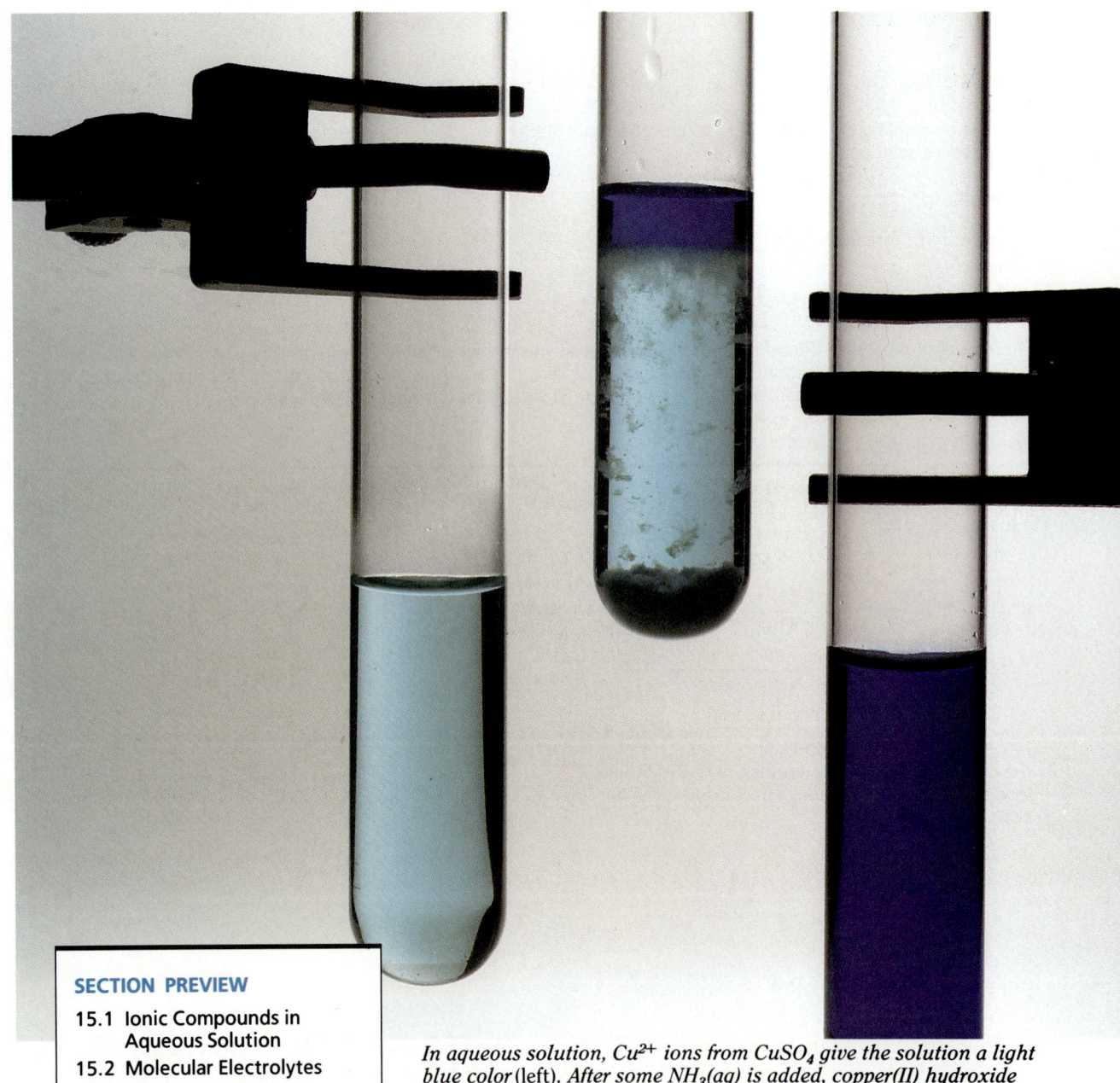

SECTION PREVIEW
- 15.1 Ionic Compounds in Aqueous Solution
- 15.2 Molecular Electrolytes
- 15.3 Properties of Electrolyte Solutions

In aqueous solution, Cu^{2+} ions from $CuSO_4$ give the solution a light blue color (left). After some $NH_3(aq)$ is added, copper(II) hydroxide precipitates (center). As more ammonia water is added, the precipitate redissolves, forming a dark blue color typical of the $Cu(NH_3)_4^{2+}$ complex ion.

15.1 Ionic Compounds in Aqueous Solution

Ionic compounds are soluble only in solvents that have polar molecules, such as water. Some of the properties of solutions containing ionic compounds are different from solutions containing only molecular compounds. In order to understand the unique properties of ionic solutions, we will look briefly at the theories and observations of Michael Faraday (1791–1867) and Svante Arrhenius (1859–1927). Their work forms the basis of the modern theory of ions in **aqueous solution**, *a solution in which water is the solvent*.

Theory of Ionization

Michael Faraday was a self-taught English physicist who did many experiments with electricity and matter in the early nineteenth century. Little was then known of the structure of the atom, but Faraday became convinced that atoms are "associated with electrical powers, to which they owe their most striking qualities." Faraday coined the words *electrolyte* for substances that conduct electricity when dissolved in water and *nonelectrolyte* for substances that do not conduct electricity in aqueous solution. Faraday conducted experiments with electric currents in electrolytic solutions. The elemental products formed at the electrodes led him to suggest that with current, some compounds separate into electrically charged atoms, which he called ions. He mistakenly believed the ions formed at the electrodes during electrolysis.

The colligative properties of electrolytes, however, did not fit well with Faraday's theory. In 1887, the Swedish chemist Svante Arrhenius introduced the theory of ionization, beginning with an assumption different from Faraday's. Arrhenius assumed that the passage of electricity in an aqueous solution was not necessary for ion formation. His description of many properties of electrolytes showed this assumption was correct. He concluded that ions were produced by "ionization" of molecules in aqueous solution.

Arrhenius considered the ions to be electrically charged. The solution as a whole contained equal numbers of positive and negative charges. He also considered ionization to be complete only in very dilute solutions. In more concentrated solutions, Arrhenius believed ions to be in equilibrium with solute molecules that were not ionized.

As chemists gained a better understanding of crystals and water molecules, some of the original concepts developed by Arrhenius were either modified or replaced. Most importantly, the explanation of the behavior of electrolytes in concentrated solutions was brought into agreement with what had been learned about ionic compounds. It is a great tribute to Arrhenius that his original theory served for so long. Present knowledge of the crystalline structure of ionic compounds was not available to him when, at the age of 24, he wrote his thesis on ionization.

From earlier chapters in this book, you have already learned more about the structure of the atom and chemical bonds than Arrhenius knew when he proposed his theory of ionization. You have seen that *before* they dissolve, solid ionic compounds contain equal amounts of positive and negative charge. You know that all ionic compounds have characteristic lattice structures that

SECTION OBJECTIVES

- Describe the solution process for an ionic compound.
- Define *hydration* and write equations representing ion hydration.
- Explain what contributes to the heat of solution of an ionic compound and what determines whether the solution process is exothermic or endothermic.
- Write equations for dissolving soluble ionic compounds in water.
- Write equations for solubility equilibria of very slightly soluble ionic compounds.
- Predict whether a precipitate will form when solutions of soluble ionic compounds are combined.
- Write net ionic equations for precipitation reactions.

STUDY HINT

The relationship between electrical current and chemical reactions is explored in Chapter 20.

The colligative properties of electrolyte solutions are discussed in Section 15.3.

Arrhenius developed the Theory of Ionization at the age of 24 as his doctoral dissertation. His professors at the University of Uppsala thought it was a radical idea and granted him a doctor of philosophy degree with the lowest possible rating. Twenty years later he was granted the Nobel Prize in chemistry for his creative work.

Ions in Aqueous Solutions

The electrical neutrality of an ionic solution is an example of the law of conservation of charge.

STUDY HINT
A review of Section 6.3 will be helpful at this point.

The general term for solute–solvent interaction is solvation (Section 14.2).

are determined by the relative sizes and charges of their ions. In solution, these ions simply separate from each other. Certain polar molecular compounds are also electrolytes because they are converted into ions as they dissolve. In this chapter we examine the solution process for ionic compounds and molecular electrolytes. We also explore some of the properties of solutions of such compounds.

Dissolving Ionic Compounds

The polarity of water molecules plays an important role in the formation of solutions of ionic compounds in water. The charged ends of water molecules attract the ions in the ionic compounds and bring them into solution. The energy released when an ion is attracted to a water molecule also affects the heat of solution of the ionic compound in water.

The solution process for ionic compounds Suppose we were to drop a few crystals of sodium chloride into a beaker of water. At the crystal surfaces, water molecules come into contact with Na^+ and Cl^- ions. The positive ends of the water molecules are attracted to Cl^- ions; the negative ends are attracted to Na^+ ions. The attraction between water molecules and the ions is strong enough to draw the ions away from the crystal surface and into solution, as illustrated in Figure 15-1. *This solution process with water as the solvent is referred to as* **hydration.** The ions are said to be *hydrated.* The attraction between ions and water molecules is strong enough that each ion in solution remains surrounded by water molecules. As hydrated ions diffuse into the solution, other ions are exposed and attracted away from the crystal surface. In this manner the entire crystal gradually dissolves, and hydrated ions become uniformly distributed in the solution.

The number of water molecules surrounding each hydrated ion depends mainly upon the size and charge of the ion. Some ions in water solution are known to be associated with a specific number of water molecules. For others, the degree of hydration varies. A dynamic equilibrium is established between hydrated ions and water molecules. For Na^+, for example, this equilibrium can be represented as

Figure 15-1 Table salt, sodium chloride, dissolves as water hydrates its sodium and chloride ions.

$$Na^+ + nH_2O(\ell) \rightleftharpoons [Na(H_2O)_n]^+$$

where n is the number of water molecules.

If the water is evaporated from a solution of sodium chloride, the solution process is reversed. The Na^+ and Cl^- ions begin to reunite and form a crystal when enough water molecules have evaporated for the solution to become saturated. Unlike sodium chloride, some ionic substances form crystals that incorporate water molecules. Certain anions and/or cations hold onto water molecules as they return to the crystal surface. Such crystalline compounds, known as *hydrates,* retain specific numbers of water molecules and are represented by formulas such as $CuSO_4 \cdot 5H_2O$. When a crystalline hydrate is dissolved in water, the water of hydration returns to the solvent. The solution is no different, therefore, from that of a nonhydrated crystalline compound—it contains hydrated ions and water (Figure 15-2).

STUDY HINT
The similar process for dissolving polar molecules is illustrated in Figure 14-9.

To help students understand that hydrates and their anhydrous forms are different from one another, show them samples of crystalline hydrates and their powdery anhydrous forms; ask them to note the differences.

Figure 15-2 1 mol (250 g) of $CuSO_4 \cdot 5H_2O$ (blue crystals) is added to a quantity of water to produce a 1 M solution. 1 mol (160 g) of $CuSO_4$ (white crystals) is also added to a greater quantity of water to make a 1 M solution. Both solutions that result are similar in all respects. Each molecule of $CuSO_4 \cdot 5H_2O$ contributes 5 molecules of water to the solution. Therefore, less water is needed to make a 1 M aqueous solution using $CuSO_4 \cdot 5H_2O$.

Heat of solution for ionic compounds The same three interactions (solute–solute, solvent–solvent, and solute–solvent) contribute to the heats of solution of both ionic compounds and molecular compounds. The heat of solution of an ionic compound is the overall result of the energy changes occurring during the separation of ions from the crystal, the separation of water molecules from each other, and the hydration of the separated ions. Energy is required to separate ions away from the crystal lattice. Energy is also required to separate polar water molecules.

By contrast, hydration is a process in which opposite charges are drawn together. Therefore, energy is always released in hydration. *The energy released when ions become surrounded by water molecules is called the* **heat of hydration.** More energy is usually released in the hydration of smaller ions than large ones. For example, compare the radii and heat of hydration of the lithium, sodium, and chloride ions:

Ion	Radius	Heat of Hydration
Li^+	68 pm	-523 kJ/mol
Na^+	95 pm	-418 kJ/mol
Cl^-	181 pm	-361 kJ/mol

Also, more energy is released in the hydration of ions with higher charges, as illustrated by comparing the radius and heat of hydration of the magnesium ion with the values above for Li^+ and Na^+:

Mg^{2+} radius: 65 pm   heat of hydration: -1949 kJ/mol

The Mg^{2+} ion is about the same size as the Li^+ ion, but has a heat of hydration almost four times larger.

Note: Although the levels of solution in the two volumetric flasks in the photograph on the right above are not at the same height, the volume inside each flask is equal to 1 L. Volumetric flasks are individually calibrated and, therefore, volume indications are not in the same location on the stem from flask to flask.

Ions in Aqueous Solutions

STUDY HINT

The interactions that contribute to the heat of solution are shown in Figure 14-14.

Whether dissolving an ionic compound *releases* heat (exothermic process) or *absorbs* heat (endothermic process) depends upon the relative amounts of energy needed *(1)* to separate the ions in the crystal from each other and *(2)* to hydrate the ions. The energy contribution from separating water molecules is always the same.

As illustrated in Figure 15-3, if more energy is released during ion hydration than is needed to separate ions and water molecules, the solution process is exothermic. If less energy is released during ion hydration than is needed to separate ions and water molecules, the solution process is endothermic.

Figure 15-3 During the dissolution of an ionic compound, if more energy is released in hydrating ions than is used to pull them apart, (a) the process produces heat or is exothermic. When the reverse is true, (b) heat is absorbed, and the solution process is endothermic.

Exothermic solution process: negative heat of solution
Endothermic solution process: positive heat of solution

A simple research project that students could run would be to see how the dissolving of a hydrate and its anhydrous form affects the temperature of the resulting solution. Will both forms lower it or raise it? If so, how much? Or if one raises it, will the other lower it?

Emphasize that in dissociation, the ions already exist; all the water molecules have to do is to separate them.

The heats of solution of several ionic compounds are included in Table 14-5. Heat is released in the hydration of lithium chloride, but is absorbed in the hydration of sodium chloride. The difference results from the larger heat of hydration of the small Li^+ ion. Ammonium nitrate (NH_4NO_3), which is composed of polyatomic ions of larger radii than either sodium or chloride ions, has a much larger positive heat of solution (25.5 kJ/mol) than that of sodium chloride (4.27 kJ/mol). Dissolving a compound like ammonium nitrate absorbs so much heat from the solvent that the solvent temperature drops noticeably. The temperature change that occurs as ammonium nitrate is dissolved has been put to use in cold packs for the treatment of athletic injuries.

Dissociation *The separation of ions that occurs when an ionic compound dissolves is called* **dissociation.** The equations representing the dissociation of sodium chloride and calcium chloride in water are:

$$NaCl(s) \xrightarrow{H_2O} Na^+(aq) + Cl^-(aq)$$
$$CaCl_2(s) \xrightarrow{H_2O} Ca^{2+}(aq) + 2Cl^-(aq)$$

As usual, (s) indicates a solid species and (aq) indicates a species in an aqueous solution. These equations, like all others, must be balanced.

Whenever ions in solution are studied, one must be aware of the number of ions produced per formula unit. One formula unit of calcium chloride gives three ions in solution, compared with two ions in solution for each formula unit of sodium chloride. A 1 M solution of sodium chloride contains 1 mol of Na^+ ions and 1 mol of Cl^- ions.

$$NaCl(s) \xrightarrow{H_2O} Na^+(aq) + Cl^-(aq)$$
$$\text{1 mol} \qquad \text{1 mol} \quad \text{1 mol}$$

A 1 M solution of calcium chloride contains 1 mol of Ca^{2+} ions and 2 mol of Cl^- ions—a total of 3 mol of ions.

$$CaCl_2(s) \xrightarrow{H_2O} Ca^{2+}(aq) + 2Cl^-(aq)$$
$$\text{1 mol} \qquad \text{1 mol} \quad \text{2 mol}$$

Sample Problem 15.1

Write the equation for dissolving aluminum sulfate, $Al_2(SO_4)_3$, in water. How many moles of aluminum ions and sulfate ions are produced by dissolving 1 mol of aluminum sulfate? What is the total number of moles of ions produced by dissolving 1 mol of aluminum sulfate?

Solution

Step 1. Analyze Given: solute amount and formula = 1 mol $Al_2(SO_4)_3$
solvent identity = water
Unknown: *(a)* equation for dissociation of solute in solvent
(b) moles of aluminum ions and sulfate ions
(c) total number of moles of solute ions produced

Step 2. Plan Solute amount and formula → balanced dissociation equation → mole relationship → total moles of solute ions

Step 3. Compute *(a)* $Al_2(SO_4)_3(s) \rightarrow 2Al^{3+}(aq) + 3SO_4^{2-}(aq)$
(b) 1 mol   2 mol   3 mol
(c) 2 mol Al^{3+} + 3 mol SO_4^{2-} = 5 mol of solute ions

Step 4. Evaluate The ion species (Al^{3+} and SO_4^{2-}) are correctly identified. Since one formula unit of $Al_2(SO_4)_3$ produces 5 ions, then 1 mol of $Al_2(SO_4)_3$ must also produce 5 mol of ions.

Practice Problems

1. Write the equation for dissolving ammonium chloride in water. How many moles of ammonium ion are produced when one mole of ammonium chloride is dissolved?
(*Ans.*) $NH_4Cl(s) \rightarrow NH_4^+(aq) + Cl^-(aq)$, 1 mol NH_4^+
2. Write the equation for dissolving sodium sulfide in water. What is the total number of moles of ions formed by dissolving one mole of sodium sulfide?   (*Ans.*) $Na_2S(s) \rightarrow 2Na^+(aq) + S^{2-}(aq)$, 3 mol of ions
3. Write the equation for dissolving barium nitrate, $Ba(NO_3)_2$. How many moles of barium and nitrate ions are produced by dissolving 0.5 mol of barium nitrate?
(*Ans.*) $Ba(NO_3)_2(s) \rightarrow Ba^{2+}(aq) + 2NO_3^-(aq)$, 0.5 mol Ba^{2+} and 1 mol NO_3^-

Solubility Equilibria

Ionic compounds have different degrees of solubility. No ionic compound is completely insoluble, however. Even the least soluble ionic compounds dissociate to some extent when placed in a polar solvent such as water.

Very slightly soluble ionic compounds When a very slightly soluble ionic compound is placed in water, an equilibrium is established between the solid compound and its ions in solution, as shown in Figure 15-4. The result is the formation of a saturated solution. Some concentration of positive and

Predicting the concentrations of ions in solution in equilibrium with slightly soluble ionic compounds is explained in Section 19.4.

negative ions will be present in solution whenever water is in contact with such a compound. The equilibrium situation is represented by using the equilibrium arrow in dissociation equations like those previously written for completely soluble compounds. The solubility equilibria for three ionic compounds of low solubility are:

$$AgCl(s) \rightleftarrows Ag^+(aq) + Cl^-(aq)$$
$$Fe(OH)_3(s) \rightleftarrows Fe^{3+}(aq) + 3OH^-(aq)$$
$$Ag_2S(s) \rightleftarrows 2Ag^+(aq) + S^{2-}(aq)$$

Precipitation reactions Both potassium chloride (KCl) and silver nitrate (AgNO$_3$) are quite soluble in water. Their solubilities at 20°C are 4.5 mol/L for KCl and 13 mol/L for AgNO$_3$. Their dissociation equations are:

$$KCl(s) \rightarrow K^+(aq) + Cl^-(aq)$$
$$AgNO_3(s) \rightarrow Ag^+(aq) + NO_3^-(aq)$$

What is likely to happen when solutions of potassium chloride and silver nitrate are combined? If a precipitate forms, a double-replacement reaction will occur. The two possible precipitates are potassium nitrate (KNO$_3$) and silver chloride (AgCl). To decide whether a precipitate can form, we must know the solubilities of these two compounds. Unlike KCl and AgNO$_3$, silver chloride has a low solubility, 1×10^{-5} mol/L at 20°C. To discover whether potassium nitrate (or any other compound with which we are unfamiliar) is a possible precipitate, we must consult a table of solubilities.

As listed in the margin, compounds are referred to as "soluble" if more than 1 g dissolves in 100 g of water, "slightly soluble" if 0.1–1 g dissolves in 100 g of water, and "insoluble" if the solubility is less than 0.1 g in 100 g of water. Consulting Table 15-1, we notice that KNO$_3$ is soluble in water.

We can therefore predict that, when solutions of potassium chloride and silver nitrate are combined, potassium nitrate will not precipitate and silver chloride will. When silver and chloride ions in solution are brought together in large enough concentrations, their attraction to each other is greater than their attraction to surrounding water molecules. In such a situation, precipitation occurs, as illustrated in Figure 15-5. Crystals of AgCl form in the solution and settle to the bottom of the container.

The balanced formula equation for the double-replacement reaction between potassium chloride and silver nitrate in solution is used to show that silver chloride precipitates and potassium nitrate remains in solution.

$$KCl(aq) + AgNO_3(aq) \rightarrow KNO_3(aq) + AgCl(s)$$

Net ionic equations Double-replacement reactions and other reactions of ions in aqueous solution are usually represented by what are known as "net ionic equations." To write a net ionic equation, one first writes an equation in which all soluble compounds are shown as dissociated ions in solution. For the precipitation of silver chloride described above, the ionic equation is

$$K^+(aq) + Cl^-(aq) + Ag^+(aq) + NO_3^-(aq) \rightarrow$$
$$AgCl(s) + K^+(aq) + NO_3^-(aq)$$

Note that the complete formula for silver chloride is written because it is so slightly dissociated that it is considered insoluble. Potassium ion (K$^+$) and nitrate ion (NO$_3^-$) appear on both sides of this equation, showing that they

Figure 15-4 When a very slightly soluble ionic compound is placed in water, ions entering and leaving solution will reach equilibrium. At that point, the solution is called *saturated*.

Point out to students that some saturated solutions can be exceedingly dilute!

Rough rules of solubility
soluble: >1 g/100 g H$_2$O
slightly soluble:
0.1–1 g/100 g H$_2$O
insoluble: <0.1 g/100 g H$_2$O

STUDY HINT
Examples of double-replacement reactions are given in Section 8.2.

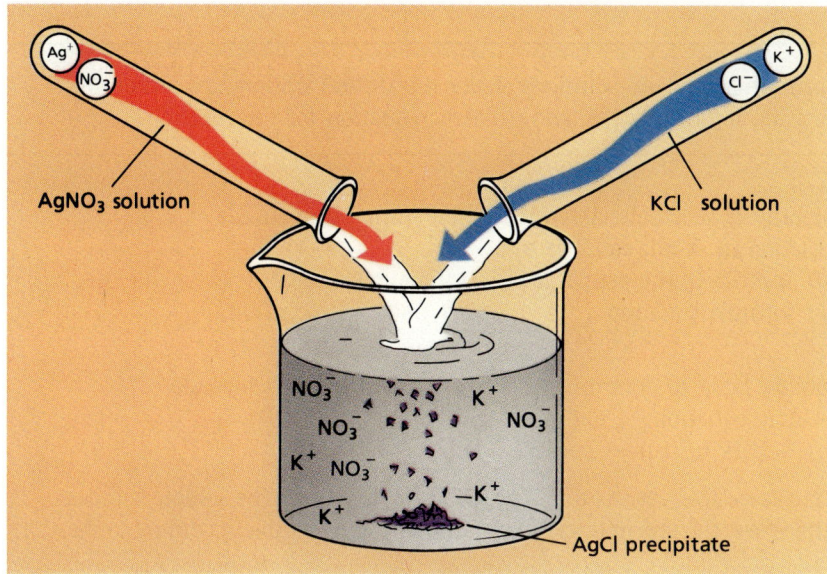

Figure 15-5 Solutions of silver nitrate and potassium chloride react to produce the insoluble precipitate, silver chloride.

Only dissociated substances are written as ions in net ionic equations. Gases, pure liquids, or solids, if present, are represented by complete formulas.

TABLE 15-1
SOLUBILITY OF IONIC COMPOUNDS

1. Common sodium, potassium, and ammonium compounds are soluble in water.
2. Common nitrates, acetates, and chlorates are soluble.
3. Common chlorides are soluble except those of silver, mercury(I), and lead. [Lead(II) chloride is soluble in hot water.]
4. Common sulfates are soluble except those of calcium, barium, strontium, and lead.
5. Common carbonates, phosphates, and silicates are insoluble except those of sodium, potassium, and ammonium.
6. Common sulfides are insoluble except those of calcium, barium, strontium, magnesium, sodium, potassium, and ammonium.

have not undergone any chemical change and are still present in their original form. *Ions that do not take part in a chemical reaction and are found in solution both before and after the reaction are referred to as* **spectator ions.**

A **net ionic equation** *includes only those compounds and ions that undergo a chemical change in a reaction in an aqueous solution and does not include spectator ions.* To convert an ionic equation into a net ionic equation, the spectator ions are removed from both sides of the equation. Eliminating the K^+ and NO_3^- ions from the ionic equation above gives the following net ionic equation.

$$Ag^+(aq) + Cl^-(aq) \rightarrow AgCl(s)$$

This net ionic equation applies to any reaction in which a precipitate of silver chloride forms when the ions are combined in solution. Additional examples of precipitation reactions and net ionic equations are given in Figure 15-6.

Figure 15-6 Precipitation reactions. Silver nitrate solution added to potassium iodide solution yields a dark precipitate of silver iodide *(left)*.
$Ag^+(aq) + I^-(aq) \rightarrow AgI(s)$
Lead nitrate solution added to sodium chromate solution yields a yellow precipitate of lead chromate *(right)*.
$Pb^{2+}(aq) + CrO_4^{2-}(aq) \rightarrow PbCrO_4(s)$

Ions in Aqueous Solutions

Sample Problem 15.2

Identify the precipitate that will form when aqueous solutions of zinc nitrate and ammonium sulfide are combined. Write the formula equation, ionic equation, and net ionic equation for the reaction.

Solution

Step 1. Analyze Given: identity of reactants: zinc nitrate and ammonium sulfide
reaction medium: aqueous solution
Unknown: (a) identity of the precipitate (c) ionic equation
(b) formula equation (d) net ionic equation

Step 2. Plan consult solubility table to determine identity of precipitate → formula equation → ionic equation → net ionic equation
$Zn(NO_3)_2(aq) + (NH_4)_2S(aq) \rightarrow ZnS(?) + 2NH_4NO_3(?)$

Step 3. Compute (a) Consulting Table 15-1 reveals that zinc sulfide is not one of the soluble sulfides and is, therefore, the precipitate. Ammonium nitrate is soluble according to the table.
(b) $Zn(NO_3)_2(aq) + (NH_4)_2S(aq) \rightarrow ZnS(s) + 2NH_4NO_3(aq)$
(c) $Zn^{2+}(aq) + 2NO_3^-(aq) + 2NH_4^+(aq) + S^{2-}(aq) \rightarrow$
$ZnS(s) + 2NH_4^+(aq) + 2NO_3^-(aq)$
(d) According to the complete ionic equation above, the ammonium and nitrate ions do not change during the reaction and are, therefore, spectator ions. The net ionic equation is: $Zn^{2+}(aq) + S^{2-}(aq) \rightarrow ZnS(s)$.

Step 4. Evaluate The formulas for the reactants have been correctly written. Assuming a double-replacement reaction, the products have been correctly identified. On rechecking the table, zinc sulfide is the insoluble product. This is reasonable because most nitrates are very soluble and most sulfides are not.

Practice Problems

1. Will a precipitate form if solutions of potassium sulfate and barium nitrate are combined? If so, write the net ionic equation for the reaction. *(Ans.)* Yes; $Ba^{2+}(aq) + SO_4^{2-}(aq) \rightarrow BaSO_4(s)$
2. Will a precipitate form if solutions of barium chloride and sodium sulfate are combined? If so, identify the spectator ions and write the net ionic equation. *(Ans.)* Yes; Na^+ and Cl^-; $Ba^{2+}(aq) + SO_4^{2-}(aq) \rightarrow BaSO_4(s)$
3. Write the net ionic equation for the precipitation of nickel(II) sulfide. *(Ans.)* $Ni^{2+}(aq) + S^{2-}(aq) \rightarrow NiS(s)$

---

1. Attraction between the positive ends of H₂O molecules and Cl⁻ ions and between the negative ends and the Na⁺ ions pulls the ions away from the crystal and into solution.
2. An exothermic process since energy is released as the ions are surrounded by water molecules.
3. $Sr(NO_3)_2(s) \rightarrow Sr^{2+}(aq) + 2NO_3^-(aq)$
Strontium ions: 0.5 mol; nitrate ions: 1 mol

Section Review

1. How does the polarity of water molecules affect the solvation process of the ionic compound sodium chloride (NaCl) in water?
2. Is hydration an endothermic or an exothermic process?
3. Write the equation for dissolving $Sr(NO_3)_2$ in water. How many moles of strontium ions and nitrate ions are produced by dissolving 0.5 mol of strontium nitrate?
4. Will a precipitate form if solutions of magnesium acetate and strontium chloride are combined? If so, write the net ionic equation for the reaction.

4. No, both potential products—$MgCl_2$ and $Sr(C_2H_3O_2)_2$—are soluble.

> ## *Chemistry and Health*

Electrolytes in Your Body

The human body is mostly an electrolytic solution with water as the solvent and potassium ions (K^+), sodium ions (Na^+), calcium ions (Ca^{2+}), and organic ions as the electrolytes. The concentration of these ions serves an important role in the functioning of your nervous system.

Positive- and negative-ion concentrations in solution in the body are controlled by "ion pumps" and cell membranes that at one moment allow ions to pass through and at other moments bar passage to those same ions.

For instance, when a nerve cell is stimulated, Na^+ ions are allowed inside the nerve cell membrane, causing a higher concentration of positively charged ions inside the cell than outside the cell. When a nerve returns to the

resting state, Na^+ ions are pumped outside causing a higher concentration of Na^+ ions outside the cell than inside. Unlike the stimulated cell membranes, the resting cell membrane will not allow Na^+ ions to pass through.

Stimulation of one end of the nerve cell causes a change in voltage across the cell membrane. This is because the concentration of positive and negative ions has changed. The voltage change disturbs the adjacent part of the nerve cell, allowing Na^+ ions to diffuse through the membrane of the adjacent part of the cell. This happens repeatedly in adjacent sections of the cell and the impulse travels along the nerve cell.

Communication between nerve cells uses a different mechanism, but once the next cell is stimulated, Na^+ ions are allowed inside the cell and the process is restarted.

15.2 Molecular Electrolytes

Molecular compounds, as well as ionic compounds, can form electrolytic solutions. Usually such compounds are composed of polar covalent molecules. Unlike ionic compounds in solution, however, only a few of the particles of a molecular electrolyte in solution are charged.

The solution process for molecular electrolytes When a polar molecule is dissolved in water, the water dipoles exert attractive forces on the oppositely charged regions of the solute molecule. If they exceed the bonding force, the covalent bond breaks and the molecule is separated into simpler (charged) fragments. The charged fragments disperse into the solvent as hydrated ions.

SECTION OBJECTIVES

- Describe the solution process for a molecular electrolyte and contrast dissociation and ionization.
- Draw the structure of the hydronium ion and discuss why it is used to represent the hydrogen ion in solution.
- Distinguish between strong and weak electrolytes and give examples of each.

Ions in Aqueous Solutions

STUDY HINT

A review of the discussion of covalent bonds in Section 6.2 and the discussion of intermolecular forces in Section 6.5 will be helpful at this point.

To explain ionization, use a tug-of-war analogy. The water molecules actually *tear* the polar solute molecules apart into ions by pulling on them on opposite sides until they break apart. You might even develop this into a little skit.

STUDY HINT

Acids, an important class of chemical compounds, are discussed in Chapters 16 and 17.

The bonding electron pair stays with the more electronegative atom, forming a negative ion. The less electronegative atom is left with a positive charge, forming a positive ion.

$$A\text{—}B \rightarrow A^+ + :B^-$$

The formation of ions from solute molecules by the action of the solvent is called **ionization.** The more general meaning of this term is the creation of ions where there were none. Note that the terms *dissociation* and *ionization* refer to different processes. When an ionic compound dissolves, the ions that were already present and associated with each other in the crystal become dissociated from each other. When a molecular compound dissolves in a polar solvent, ions are formed where none existed in the undissolved compound.

Hydrogen chloride (HCl) is a molecular compound that ionizes in aqueous solution. It is one of a series of compounds of hydrogen with the members of Group 17 (known as halogens), which include fluorine, chlorine, bromine, and iodine. The hydrogen halides are all molecular compounds with polar, single covalent bonds. All are gases and very soluble in water.

Experiments show that pure liquefied hydrogen chloride (bp–84.9°C) does not conduct an electric current. A solution of hydrogen chloride in a nonpolar solvent such as benzene (C_6H_6) is also nonconductive. A solution of hydrogen chloride in water, by contrast, does conduct an electric current. The attraction between the polar H—Cl bond and the water-molecule dipoles is strong enough to break the bond. Hydrogen and chloride ions in solution are formed by the ionization of the hydrogen chloride molecule (Figure 15-7).

$$HCl \rightarrow H^+(aq) + Cl^-(aq)$$

Like all ions in aqueous solution, the ions formed by a molecular solute are hydrated. The heat released during the hydration of the ions provides the energy needed to break the covalent bonds.

Figure 15-7 When a hydrogen chloride molecule ionizes in water, its hydrogen ion bonds covalently to a water molecule. A hydronium ion and a chloride ion are formed.

Hydrogen chloride in aqueous solution is known as hydrochloric acid. The hydrogen chloride molecules are 100% ionized—the solution contains only ions. The properties of the solution are distinctly different from those of hydrogen chloride gas or liquid. Many of these properties are characteristic of all acids that produce hydrogen ions in aqueous solution.

The hydronium ion The hydrogen ion (H^+) is a bare hydrogen nucleus—only a proton. As such it is much smaller than any other ion. The H^+ ion attracts other molecules or ions so strongly that it does not normally have any independent existence. The ionization of hydrogen chloride in water is better described as a chemical reaction in which a proton is transferred directly from HCl to a water molecule, where it becomes covalently bonded to oxygen. The process is represented in Figure 15-7. *The H_3O^+ ion is known as the* **hydronium ion.** The hydronium ion has three equivalent covalent bonds between oxygen and hydrogen atoms and an overall charge of 1+.

The equation for the reaction between hydrogen chloride and water to form hydrochloric acid is often written, as follows, to show the formation of the hydronium ion rather than H^+:

$$HCl(g) + H_2O(\ell) \rightarrow H_3O^+(aq) + Cl^-(aq)$$

The hydration of the H^+ ion to form the hydronium ion is a very favorable reaction. The energy released makes a large contribution to the energy needed to ionize a molecular solute. Many compounds classified as molecular electrolytes contain hydrogen and form H_3O^+ in an aqueous solution.

Strong and weak electrolytes Hydrogen chloride (HCl), hydrogen bromide (HBr), and hydrogen iodide (HI) are 100% ionized in dilute aqueous solution. The product solutions are known as hydrochloric, hydrobromic, and hydriodic acids, respectively. *Any substance that is present in dilute aqueous solution entirely as hydrated ions as referred to as a* **strong electrolyte.** Hydrogen chloride, many other acids, and ionic compounds are strong electrolytes. The distinguishing feature of strong electrolytes is that, to whatever extent they dissolve in water, they yield only ions. An ionic salt may not dissolve to an appreciable extent, but that amount that does dissolve exists solely as hydrated ions in solution. For example, silver chloride (AgCl) is almost insoluble, with a solubility of only 0.0001 g/L. However, it is considered a strong electrolyte because the small amount that does dissolve exists in the form of ions.

Certain metal halides that are molecular compounds are also strong electrolytes. Aluminum chloride has the formula Al_2Cl_6 in the liquid and vapor phases and is a very poor conductor as a liquid. The Al^{3+} ion, however, has a strong tendency to become hydrated, and the hydration energy contributes to breaking the aluminum–chlorine bonds. An aqueous solution of aluminum chloride is a good conductor, showing that the compound is a strong electrolyte that ionizes as follows:

$$Al_2Cl_6(\ell) + 12H_2O(\ell) \rightarrow 2[Al(H_2O)_6]^{3+}(aq) + 6Cl^-(aq)$$

Hydrogen fluoride (HF) dissolves in water to give an acid solution known as hydrofluoric acid. For hydrogen fluoride, the process of dissolving and the properties of the solution are somewhat different than for the other hydrogen halides. Fluorine atoms are the smallest halogen atoms. The hydrogen–fluorine bond is not only highly polar, but also much stronger than the bonds between hydrogen and the other halogens. When hydrogen fluoride dissolves, some molecules ionize. But the reverse reaction—the transfer of H^+ ions back to F^- ions to form hydrogen fluoride molecules—also takes place. The result is an equilibrium in which the concentration of dissolved hydrogen fluoride molecules remains high and the concentration of H_3O^+ and F^- ions remains low.

$$HF(aq) + H_2O(\ell) \rightleftharpoons H_3O^+(aq) + F^-(aq)$$

In an aqueous solution, the majority of the HF molecules are present as dissolved HF molecules.

Students understand the H_3O^+ formula for a hydronium ion better if you first write it as $H^+ \cdot H_2O$.

Actually, chemists have found evidence that the hydronium ion exists in a hydrated form in most strong acids. A possible formula for the hydronium ion is $H_9O_4^+$, or $H_3O^+ \cdot 3H_2O$.

Ions in Aqueous Solutions

Most organic acids (all of which have carbon in their formulas), such as acetic acid, are weak electrolytes.

A **weak electrolyte** *is a solute that yields a relatively low concentration of ions in aqueous solution.* Hydrogen fluoride is a weak electrolyte. Acetic acid is another. In a 0.1 M solution of acetic acid ($HC_2H_3O_2$), only 1% of the acetic acid molecules ionize.

$$HC_2H_3O_2(aq) + H_2O(\ell) \rightleftarrows H_3O^+(aq) + C_2H_3O_2^-(aq)$$

Ammonia (NH_3) is a molecular compound that is a gas at ordinary temperatures and is extremely soluble in water. The aqueous solution of ammonia conducts electric current poorly, showing that the concentrations of ions is low. Ammonia is a weak electrolyte that reacts with water to give a low concentration of ammonium (NH_4^+) and hydroxide (OH^-) ions.

$$NH_3(g) + H_2O(\ell) \rightleftarrows NH_4^+(aq) + OH^-(aq)$$

Ammonia solutions in water were once called ammonium hydroxide solutions because of the presence of the ammonium and hydroxide ions. The name most likely persists because it is etched into many old reagent bottles. Actually, it is inappropriate because the compound NH_4OH has never been found to exist. It is probably better to call its solution ammonia water—$NH_3 \cdot H_2O$.

In general, the extent to which a solute ionizes in solution depends on the strength of the bonds within the molecules of the solute and the strength of attraction to solvent molecules. If the strength of the bonds within the solute molecule is less than the attractive forces of the water dipoles then the covalent bond breaks and the molecule is separated into ions.

Section Review

1. Explain why HCl is a strong electrolyte and HF is a weak electrolyte.
2. What property of the atoms in a polar–covalent compound determines whether they will gain or lose an electron when they are ionized by a polar solvent.
3. How is a hydronium ion formed from water molecules and what is its net charge?

1. Because the H—F bond is much stronger than the H—Cl bond, the attraction of water molecules is not sufficient to prevent recombination of H^+ and F^-
2. Electronegativity: atoms with higher electronegativity will attract electrons and form negative ions; those with lower electronegativity will form positive ions.
3. The hydronium ion is formed when a hydrogen ion (H^+), or a bare proton, is attracted to the negative (or oxygen) end of the molecule: its net charge is 1+.

15.3 Properties of Electrolyte Solutions

SECTION OBJECTIVES

- Explain why electrolytes in solution conduct electricity. Write an equation that describes the ionization process in pure water.
- Explain the expected and experimentally observed colligative properties of electrolyte solutions.
- Calculate the expected freezing-point depression of an electrolytic solution.

For an aqueous solution to be conductive, its solute must be capable of transporting an electric charge. The ions of an electrolyte are charge carriers, and the magnitude of the current that an electrolyte solution conducts is related to the solute ion concentration.

Conductivity of Solutions

Figure 15-8 shows an apparatus for testing the conductivity of solutions. The electrodes are conductors attached to a power supply that make electric contact with the test solution. For a current to pass through the light bulb filament, the test solution must provide a conducting path between the two electrodes. A

> ## Chemistry in Detail

Boiling Liquids

When a solid is dissolved in a liquid, the vapor pressure of the liquid is lowered because the surface area from which molecules can evaporate is reduced. Raoult's law gives the vapor pressure:

$$P = xP°$$

where P, the solvent's vapor pressure above a solution, is equal to the product of x, the solvent's mole fraction in the solution and $P°$, the pure solvent's vapor pressure.

When both solute and solvent are volatile, each acts as solute to the other and reduces the vapor pressure. If the inter-molecular forces between substances A and B are about the same as the A–A and B–B forces, the total vapor pressure can be predicted by using Raoult's law twice:

$$P = x_A P_A° + x_B P_B°$$

The boiling point of a liquid mixture can be estimated by finding the temperature at which this equation yields a value of P equal to the atmospheric pressure.

When the A–B forces are weaker than the A–A or B–B forces, both molecules escape from a solution more easily than from a pure liquid. Mixtures of such liquids have a higher vapor pressure than predicted by the equation, and therefore have a lower boiling point than expected.

Hydrogen bonds between water and ethanol cause their mixtures to vaporize less readily than expected and to have higher boiling points. A mixture that is 95% ethanol and 5% water cannot be separated by simple distillation; such a mixture is called an *azeotrope*.

nonconducting solution is like an open switch between the electrodes. Thus, there can be no current in the circuit.

The light bulb glows brightly if a liquid that is a good conductor is tested. For a moderately conducting liquid, however, the light bulb is not as brightly lit. If a liquid is a poor conductor, the light bulb does not glow at all.

Whether the ions are from a strong electrolyte or a weak one, the higher the concentration of ions in an aqueous solution, the more easily a solution will conduct electric current. Therefore, to compare the conductivities of strong and weak electrolytes, the conductivities of solutions of equal concentration must be compared. In Figure 15-8, the conductivities of pure water, acetic acid (a weak electrolyte), and hydrochloric acid (a strong electrolyte) are contrasted. Hydrochloric acid is a good conductor because it dissociates completely in solution, and causes the light bulb to glow brightly. Acetic acid is a moderate conductor because it is not completely ionized in solution, and the light bulb is not as brightly lit. Water is a poor conductor, so the light bulb is not lit.

Reference to a "strong electrolyte" or a "weak electrolyte" must not be confused with reference to a solution as "concentrated" or "dilute." Strong

Figure 15-8 The brightness of the lamp filament indicates comparative conductivities of solutions tested. Distilled water *(left)* is a nonconductor. Dilute acetic acid, a solution of a weak electrolyte *(center)*, is a moderate conductor. Dilute hydrochloric acid, a solution of a strong electrolyte *(right)*, is a good conductor.

In everyday usage, the terms "strong" and "weak" are synonymous with concentrated and dilute, i.e., strong or weak coffee. Be sure to emphasize that the terms "strong" and "weak" are used in a very different way in chemistry; they refer to the ability of solutes to produce ions, not the concentration of the solution.

The hydronium ion is a hydrated proton.

and weak molecular electrolytes differ in the degree of ionization. Concentrated and dilute solutions differ in the amount of solute dissolved in a given quantity of a solvent. Hydrochloric acid is always a strong electrolyte. This is true even in a solution that is 0.000 01 M—very dilute. By contrast, acetic acid is always a weak electrolyte, even in a 5 M solution.

The ionization of water If pure water is placed in the apparatus shown in Figure 15-8, the light will not shine at all. On the basis of this experiment, water would be judged a nonconductor and a nonelectrolyte.

You might be wondering why we are constantly cautioned about the danger of dropping hair dryers or other electric appliances into the bathtub. The reason is that tap water commonly contains a high enough concentration of dissolved ions to make it a significantly better conductor than pure water. More sensitive tests for conductivity than those using light bulbs show that even very pure water has a slight, but measurable, conductivity. At ordinary temperatures, approximately two water molecules out of one billion are ionized. The ionization occurs, as illustrated in Figure 15-9, when one water molecule gives up a proton (a hydrogen ion) to another water molecule to form a *hydronium ion*. This process is:

$$H_2O(\ell) + H_2O(\ell) \rightleftarrows H_3O^+(aq) + OH^-(aq)$$

Figure 15-9 The ionization of pure water, a very weak electrolyte. Out of one billion water molecules, only two will ionize.

Water is a *very* weak electrolyte and it is such a poor conductor that, except in special circumstances, it can be considered a nonconductor. In the next chapter, however, you will see that the ionization of water, although slight, plays an important role in the chemistry of acids and bases.

Colligative Properties of Electrolyte Solutions

Colligative properties depend on the relative numbers of dissolved particles and not on their identity. We have already discussed how nonvolatile molecular solutes affect the solvent by lowering the vapor pressure, depressing the freezing point, and elevating the boiling point. Early investigators were puzzled by experiments in which certain substances depressed the freezing point or elevated the boiling point more than expected. Knowing more about dissociation and ionization than Michael Faraday did, you can predict how these properties change when the solute is an electrolyte.

Freezing-point depression in electrolyte solutions The molal freezing-point constant (K_f) for water of -1.86C°/m is derived from experiments with nonelectrolytes in water solution. Electrolytes have a greater influence on the freezing points of solvents than do nonelectrolytes. A 0.1 m solution of sodium chloride lowers the freezing point nearly twice as much as a 0.1 m solution of sugar. A 0.1 m solution of potassium sulfate (K_2SO_4) or calcium chloride ($CaCl_2$) lowers the freezing point nearly three times as much as a 0.1 m solution of sugar. The effect on boiling points is similar. Nonvolatile electrolytes in aqueous solutions raise the boiling point nearly two, three, or more times as much as nonelectrolytes in solutions of the same molality.

Consider that five sugar molecules dissolve to produce five particles in solution. Five hydrogen chloride molecules, however, dissolve to produce ten particles in solution. This is the basis for the results of experiments on the colligative properties of electrolytes: electrolytes produce two or three or more moles of solute particles for each mole of compound dissolved. For the same molal concentrations of sugar and hydrogen chloride, we would expect the effect on colligative properties to be twice as large for hydrochloric acid as for sugar.

What about barium nitrate, $Ba(NO_3)_2$? Each formula unit of barium nitrate yields three ions in solution.

$$Ba(NO_3)_2(s) \rightarrow Ba^{2+}(aq) + 2NO_3^-(aq)$$

We would expect a $Ba(NO_3)_2$ solution of a given molality to lower the freezing point of its solvent three times as much as a nonelectrolyte solution of the same molality.

> **STUDY HINT**
> A review of the discussion of colligative properties in Section 14.4 will be helpful at this point.

Sample Problem 15.3

What is the expected change in the freezing point of water in a solution of 62.5 g of barium nitrate, $Ba(NO_3)_2$, in 1.00 kg of water?

Solution

Step 1. Analyze Given: solute mass and formula = 62.5 g $Ba(NO_3)_2$
solvent mass and identity = 1.00 kg water
$\Delta t_f m = K_f m$
Unknown: expected freezing-point depression

Step 2. Plan mass of solvent, mass of solute $\rightarrow$ molality $\rightarrow$ molality in terms of number of ions in solution $\rightarrow$ expected freezing-point depression

Sample Problem 15.3 *continued*

$$\frac{\text{mass of solute (g)}}{\text{mass of solvent (kg)}} \times \frac{1 \text{ mol solute}}{\text{molar mass of solute (g)}} = \text{molality of solution} \left(\frac{\text{mol}}{\text{kg}}\right)$$

$$\text{molality of solution} \left(\frac{\text{mol}}{\text{kg}}\right) \times \text{molality conversion} \left(\frac{\text{mol ions}}{\text{mol}}\right) \times K_f \left(\frac{\text{C°} \cdot \text{kg H}_2\text{O}}{\text{mol ions}}\right)$$

$$= \text{expected freezing-point depression}$$

This problem is similar to Sample Problem 14.6 except that we are using an ionic solute rather than a nonionizing molecular solute. The number of particles in solution will therefore equal the number of ions of the solute.

Step 3. Compute

$$\frac{62.5 \text{ g Ba(NO}_3)_2}{\text{kg H}_2\text{O}} \times \frac{1 \text{ mol Ba(NO}_3)_2}{261 \text{ g Ba(NO}_3)_2} = 0.239 \frac{\text{mol Ba(NO}_3)_2}{\text{kg H}_2\text{O}}$$

$$\frac{0.239 \text{ mol Ba(NO}_3)_2}{\text{kg H}_2\text{O}} \times \frac{3 \text{ mol ions}}{\text{mol Ba(NO}_3)_2} \times \frac{-1.86\text{C°} \cdot \text{kg H}_2\text{O}}{\text{mol ions}} = -1.33\text{C°}$$

Step 4. Evaluate The units cancel properly to give the desired answer in Celsius degrees. The answer is correctly given to three significant digits. Noticing that the mass of the solute is approximately one-half of its molar mass, and would give 1.5 mol of ions in the 1 kg of solvent, the estimated answer of $1.5 \times -1.86\text{C°} = -2.8\text{C°}$ supports our computation.

Practice Problems
1. What is the expected freezing-point depression for a solution that contains 2.0 mol of sodium chloride dissolved in 1.0 kg of water? *(Ans.)* -7.4C°
2. What is the expected freezing-point depression of water for a solution that contains 150 g of sodium chloride dissolved in 1.0 kg of water? *(Ans.)* -9.6C°

Table 15-2 presents data on the freezing points of solutions of electrolytes that yield two or three ions per formula unit. The actual freezing-point lowering was measured for each solute at the five concentrations shown. The molal freezing-point depressions listed were calculated by dividing the observed freezing-point depression by the molality of the solution.

For an ionic compound that gives two ions per formula unit, we would predict the freezing-point lowering to be $2 \times -1.86\text{C°} = -3.72\text{C°}$ for a one molal solution. Examine the data in the table for potassium chloride (KCl). As the concentration of the solution decreases, the molal freezing-point depression comes closer and closer to the expected value of -3.72C°. But it does not quite reach it even in a $0.001\ m$ solution.

For each solute in Table 15-2, the situation is the same. The more dilute the solution, the closer the molal freezing-point depression comes to a whole-number multiple of that for a nonelectrolytic solute.

Colligative-property measurements can be used to find the percent ionization of weak electrolytes.

Apparent degree of ionization For many years, colligative property measurements like those in Table 15-2 prevented chemists from deciding whether a compound was completely ionized in a water solution or was a weak electrolyte. For example, based on the freezing-point depressions in Table 15-2, KCl is apparently 93% ionized and $MgSO_4$ is apparently 60% ionized in $0.1\ m$ solutions in water.

TABLE 15-2 MOLAL FREEZING-POINT DEPRESSIONS FOR AQUEOUS SOLUTIONS OF IONIC SOLUTES

	Concentration (in mol solute/kg H$_2$O)				
Solute	0.1	0.05	0.01	0.005	0.001
	Values are observed freezing-point depressions in C° for concentrations indicated divided in each case by the molality.				
$2K_f = 2 \times -1.86\text{C}° = -3.72\text{C}°$					
AgNO$_3$	−3.32	−3.42	−3.60	—	—
KCl	−3.45	−3.50	−3.61	−3.65	−3.66
KNO$_3$	−3.31	−3.43	−3.59	−3.64	—
LiCl	−3.52	−3.55	−3.60	−3.61	—
MgSO$_4$	−2.25	−2.42	−2.85	−3.02	−3.38
$3K_f = 3 \times -1.86\text{C}° = -5.58\text{C}°$					
BaCl$_2$	−4.70	−4.80	−5.03	−5.12	−5.30
Ba(NO$_3$)$_2$	−4.25	—	−5.01	—	−5.39
CaCl$_2$	−4.83	−4.89	−5.11	—	—
Cd(NO$_3$)$_2$	−5.08	—	−5.20	−5.28	—
CoCl$_2$	−4.88	−4.92	−5.11	−5.21	—

Today it is recognized that attractive forces exist between ions in aqueous solution. The more strongly the ions attract each other, the smaller the apparent degree of ionization based on colligative-property measurements. Ions in more concentrated solutions attract each other more strongly because they are closer together. The attraction between hydrated ions in solution is small compared to those in the crystalline solid. Forces of attraction do interfere with the movements of the aqueous ions, even in quite dilute solutions, however. Only in very dilute solutions is the average distance between the ions large enough and the attraction between ions small enough

for the solute ions to move about freely. This is illustrated in Figure 15-10 and in Table 15-3, which lists the observed freezing-point depressions of increasingly dilute solutions of sodium chloride, the calculated freezing-point depressions per mole of sodium chloride, and the apparent degree of ionization at each concentration.

Figure 15-10 In a concentrated solution *(right)*, ions are closer together and attract each other more strongly. As a result, the compound in a concentrated solution appears to be less ionized.

Peter Debye, a Dutch-born American physical chemist, is most well-known for his work with electrolytes, ionization, and dissociation. He explained the role of water molecules in the process of dissolving an ionic solid. In 1936, he was awarded the Nobel Prize in chemistry for his research on dipole moments and molecular structure.

TABLE 15-3 INFLUENCE OF CONCENTRATION ON FREEZING POINT OF AQUEOUS SOLUTIONS OF NaCl

Concentration of NaCl (m)	Freezing-point depression (C°)	Freezing-point depression/molal NaCl (C°/m)	Apparent degree of ionization
0.100	−0.346	−3.46	93%
0.0100	−0.0361	−3.61	97%
0.00100	−0.00366	−3.66	98%
0.000100	−0.000372	−3.72	100%

In 1923, Peter Debye and E. Hückel introduced a theory to account for the attraction between dissociated ions of ionic solids in dilute aqueous solutions. According to this theory, each solute ion is surrounded, on average, by more ions of opposite charge than of like charge. This clustering effect hinders the movements of solute ions. A cluster of hydrated ions may act as a single unit rather than as individual ions. Thus, the effective concentration of each ion, or its "activity," is less than expected on the basis of the number of ions known to be present. A solution dilute enough that the ions have the expected activity is referred to as an "ideal solution."

Ions of higher charge attract other ions very strongly. They therefore have lower activity and lower effective concentrations than ions with smaller charge. Note in Table 15-2 that the molal freezing-point depressions for magnesium sulfate ($MgSO_4$), with its 2+ and 2− charged ions, are lower than those for the compounds with 1+ and 1− charged ions.

In concentrated solutions, ion activity may be further decreased by a shortage of solvent molecules. In such a case, clusters of unhydrated ions may act as single solute units.

Section Review

1. How would the apparatus in Figure 15-8 perform if ammonia were used as a test solution? If hydrobromic acid were tested?
2. Compare the conductivity of tap water and distilled water.
3. Why is rock salt used during the making of homemade ice cream?

1. With dilute ammonia, the light would be dim; with dilute hydrobromic acid, it would be bright.
2. Distilled water is such a poor conductor it is considered to be a nonconductor; tap water, being a very dilute solution of various mineral electrolytes, is a fairly good conductor.
3. Rock salt is added to the ice in the freezer to lower its freezing point; the lower temperature causes the ice cream to freeze.

Chapter Summary

- The heat of solution of an ionic compound is a combination of the energy needed to: *(1)* separate the solute ions, *(2)* separate the solvent molecules, and *(3)* hydrate the solute ions.
- When two different ionic solutions are mixed, a precipitate may form if any one of the possible compounds formed by the solute ions is insoluble.
- A molecular compound with polar–covalent bonds may ionize in a water solution if the attraction of the polar water molecules is strong enough to overcome the strength of the polar–covalent bonds of the solute molecules.
- Solutes that produce ions in solution are called electrolytes. Electrolytes have a greater effect on the freezing and boiling points of solvents than do nonelectrolytes.

Chapter 15 Review

Ions in Aqueous Solutions

VOCABULARY

aqueous solution (439)
dissociation (442)
heat of hydration (441)
hydration (440)
hydronium ion (449)
ionization (448)
net ionic equation (445)
spectator ions (445)
strong electrolyte (449)
weak electrolyte (450)

REVIEWING CONCEPTS

1. List the three basic assumptions of the modern theory of ionization. (15.1)
2. a. Define the term *hydration*.
   b. What determines the number of water molecules surrounding each hydrated ion? (15.1)
3. a. What interactions contribute to the heats of solution of ionic and molecular compounds?
   b. Name the three energy changes and the direction of each, in the determination of the heat of solution of an ionic compound. (15.1)
4. a. Define heat of hydration.
   b. What are the effects of the size and charge of an ion on its heat of hydration? (15.1)
5. a. Define dissociation.
   b. How many moles of ions are contained in 1 L of a 1 M solution of KCl? of BaF_2? of $Mg(NO_3)_2$? (15.1)
6. Use Table 15-1 to predict the relative solubility of each of the following compounds (indicate whether each is soluble or insoluble).
   a. KCl
   b. $NaNO_3$
   c. AgCl
   d. $BaSO_4$
   e. $Ca_3(PO_4)_2$
   f. $Pb(ClO_3)_2$
   g. $(NH_4)_2S$
   h. $PbCl_2$ (in cold water)
   i. FeS
   j. $Al_2(SO_4)_3$ (15.1)
7. Identify the spectator ions in the reaction between KCl and $AgNO_3$ in an aqueous solution. (15.1)
8. What is a net ionic equation? (15.1)
9. Describe the solution process for molecular electrolytes. (15.2)
10. a. What is ionization?
    b. Distinguish between ionization and dissociation. (15.2)
11. a. Define and distinguish between strong and weak electrolytes.
    b. Give two examples of each type. (15.2)
12. What is the relationship between the molecular bond strength of a solute and the extent to which the solute ionizes in solution. (15.2)
13. a. What determines whether an aqueous solution will be conductive?
    b. What determines the degree to which an electrolyte solution conducts? (15.3)
14. Distinguish between the two sets of terms *strong* and *weak,* as opposed to *dilute* and *concentrated,* when used to describe electrolyte solutions. (15.3)
15. If aqueous solutions of each of the following were tested with the apparatus shown in Figure 15-8, which would you expect to conduct electricity?
    a. NaCl
    b. KCl
    c. $C_{12}H_{22}O_{11}$
    d. $Mg(NO_3)_2$
    e. CCl_4
    f. $Al_2(SO_4)_3$ (15.3)
16. a. Compare the effects of nonvolatile electrolytes with the effects of nonvolatile nonelectrolytes on the freezing and boiling points of solvents into which they are dissolved.
    b. Why are such differences observed? (15.3)
17. What is the relationship between the concentration of an aqueous nonvolatile electrolyte solution and the observed molal freezing-point depression and boiling-point elevation? (15.3)
18. What is the relationship between the attractive forces between ions in aqueous solutions and the apparent degree of ionization of those ions? (15.3)
19. Explain the difference between a concentrated solution of a weak electrolyte and a dilute solution of a strong electrolyte. (15.3)

Ions in Aqueous Solutions **457**

PROBLEMS

▶ Solution Process

1. Write the equation for the dissolving of each of the following ionic compounds in water:
   a. KI
   b. $NaNO_3$
   c. $MgCl_2$
   d. Na_2SO_4

2. For the compounds listed in the previous problem, determine the number of moles of each ion produced, as well as the total number of moles produced when 1 mol of each compound dissolves in water.

3. Write the equation for dissolving each of the following in water and then indicate the total number of moles of solute ions formed:
   a. 0.50 mol strontium nitrate
   b. 0.50 mol sodium phosphate
   c. 0.275 mol potassium sulfide
   d. 0.15 mol aluminum sulfate.

4. For each of the following slightly soluble compounds, write the equations for the solubility equilibria:
   a. $Co_2S_3(s)$ b. $PbS(s)$ c. $PbCrO_4(s)$

▶ Precipitation Reactions

5. Use Table 15-1 to write the balanced chemical equation, to write the overall ionic equation, to identify the spectator ions and possible precipitates, and to write the net ionic equation for each of the following reactions:
   a. silver nitrate(aq) + sodium iodide(aq) →
   b. mercury(II) chloride(aq) + potassium sulfide(aq) →
   c. aluminum nitrate(aq) + barium hydroxide(aq) →
   d. copper(II) chloride(aq) + ammonium phosphate(aq) →

6. Magnesium phosphate and lead(II) nitrate react in aqueous solutions by double replacement. If 13.0 g of magnesium phosphate reacts, what is the maximum amount of precipitate that could be found? Write the balanced chemical equation, the overall ionic equation, and the net ionic equation for this reaction.

7. Given that 150. mL of 0.250 M NH_4OH is mixed with 125 mL of 0.400 M $Al(NO_3)_3$, complete each of the following exercises:
   a. Write the balanced chemical equation for the reaction.
   b. Write the ionic equation for the reaction.
   c. Identify the spectator ions and the insoluble product.
   d. Write the net ionic equation.
   e. Find the maximum number of grams of precipitate that could be produced through this reaction.
   f. Determine the number of moles of excess reactant that remains.

▶ Colligative Properties

8. Given 1.00 m aqueous solutions of each of the following substances, what is the expected change in the freezing point of the solvent?
   a. KI
   b. $CaCl_2$
   c. $Ba(NO_3)_2$
   d. $C_{12}H_{22}O_{11}$

9. What is the anticipated change in the freezing point of an aqueous solution that is 0.015 m $AlCl_3$?

10. What is the freezing point of a solution containing 85.0 g of NaCl dissolved in 450. g of water?

11. Determine the boiling point of a solution made by dissolving 25.0 g of barium chloride in 0.150 kg of water.

12. The change in the boiling point of an aqueous solution of potassium iodide is 0.65C°. Determine the apparent molal concentration of potassium iodide.

13. The freezing point of an aqueous solution of barium nitrate is −2.65°C. Determine the apparent molal concentration of barium nitrate.

14. Calculate the predicted freezing point of 1.00 kg of H_2O to which 0.250 mol of NaCl has been added.

15. Given 1.0 kg of 0.01 m aqueous solutions of

each of the following, arrange these solutions in order beginning with the one that is expected to cause the least change in the freezing point of the solution:
a. NaI   c. K_3PO_4
b. $CaCl_2$ d. $C_6H_{12}O_6$ (glucose)

16. A solution made by dissolving hydrated $MgSO_3$ crystals in 1.00 kg of water lowered the freezing point of the water by 3.72C°. Determine the chemical formula of this solute if its molar mass is 212 grams.

17. Experimental data for a 1.00 m MgI_2 aqueous solution indicate an actual change in the freezing point of water of −4.78C°. Determine the predicted change in the freezing point of water. Suggest reasons for any discrepancies that may exist between the experimental and the predicted values.

CRITICAL THINKING

1. **Applying Models**
   a. You are conducting a freezing point determination in the laboratory using an aqueous solution of KNO_3. The observed freezing point of the solution is −1.15°C. Using a pure water sample, you recorded the freezing point of the pure solvent on the same thermometer as 0.25°C. Determine the molal concentration of KNO_3. Assume that there is no loss in ion activity according to the Debye-Hückel theory.
   b. You are not satisfied with the result in part *a* of this question because you suspect that you cannot ignore the effect of ion interaction. You take a 10.00-mL sample of the solution. After carefully evaporating the water from the solution, you obtain a mass of 0.415 g KNO_3. Determine the actual molal concentration of KNO_3 and the percentage difference between the predicted and the actual concentrations of KNO_3.

2. **Analyzing Information** The observed freezing-point depression for electrolyte solutions is sometimes less than the calculated value. Why does this occur and is the error greater for concentrated or dilute solutions?

CROSS CONNECTIONS

1. **Biology** Osmotic pressure is another colligative property. The osmotic pressure of a dilute solution can be calculated as follows.

   $\pi = MRT$
   $\pi =$ osmotic pressure
   $M =$ concentration in moles per liter
   $R =$ ideal gas constant
   $T =$ absolute temperature of the solution

   How does the osmotic pressure equation compare to the ideal gas law?

2. **Biology** Research the role of electrolytes and electrolytic solutions in your body. Find out how electrolytes work in the functioning of nerves and muscles. What are some of the health problems that can arise from an imbalance of electrolytes in body fluids?

RESEARCH AND WRITING

1. Use the work of Michael Faraday and Svante Arrhenius as examples to explain the scientific method.

PROJECTS

1. Using a low-voltage dry cell, assemble an apparatus similar to the one shown in Figure 15-8. Secure several unknown aqueous solutions of equal molality from your instructor and use the apparatus to distinguish the electrolytes from the nonelectrolytes. Among those identified as electrolytes, rank their relative strengths as conductors, from good conductors down to poor conductors.

2. Using equal volumes of the unknown solutions from the preceding activity, explain how you could use the freezing-point depression concept to distinguish the electrolytes from the nonelectrolytes. Explain further how you could determine the number of ions contained per molecule among the solutes identified as electrolytes. Design and conduct an experiment to test your theories.

CHAPTER 16
Acids and Bases

PLANNING CHART

Approximate Time: 6/9 days

	SECTION/TEXT FEATURES		RESOURCES	
BASIC	**16.1 ACIDS**	**461**	*Teacher's Resource Binder*	
	Demo 1	460A	Section Review Worksheets	55
	Periodic Table Videodisc	460H		
	Section Review	469		
	Chapter Review	487		
	• Questions 1–9, Problem 1			
BASIC	**16.2 BASES AND ACID-BASE REACTIONS**	**470**	*Teacher's Resource Binder*	
	Demos 2, 3	460B	Section Review Worksheets	56
	Periodic Table Videodisc	460H		
	Microscale Investigation 14	864		
	Section Review	475		
	Chapter Review	487		
	• Question 10, Problems 2–5			
BASIC	**16.3 RELATIVE STRENGTH OF ACIDS AND BASES**	**475**	*Teacher's Resource Binder*	
	Demo 4	460C	Section Review Worksheets	57
	Desktop Investigation	476	**Other Resources**	
	Section Review	479	*World of Chemistry* Video 16	
	Chapter Review	487	Blackline Master 37	
	• Questions 11–17			
BASIC	**16.4 OXIDES, HYDROXIDES, AND ACIDS**	**479**	*Teacher's Resource Binder*	
	Demo 5	460C	Lab Experiment 43	217
	Periodic Table Videodisc	460H	Section Review Worksheets	58
	Section Review	483	**Other Resources**	
	Chapter Review	487	Blackline Master 38	
	• Questions 18–20			
BASIC	**16.5 CHEMICAL REACTIONS OF ACIDS, BASES, AND OXIDES**	**483**	*Teacher's Resource Binder*	
	Demo 6	460D	Section Review Worksheets	59
	Periodic Table Videodisc	460H		
	Section Review	486		
	Chapter Review	487		
	• Question 21, Problems 6–16			
	CHAPTER END ACTIVITIES		*Teacher's Resource Binder*	
	Critical Thinking, Cross Connections, Research and Writing, Projects		Test Booklet	77, 79
			Computer Test Generator	

Demonstration 1: Identifying Acids

PURPOSE: To show how acids can be identified

MATERIALS: Red and blue litmus paper, dilute hydrochloric acid, dilute nitric acid, dilute phosphoric acid, dilute sulfuric acid, vinegar, mustard, pickle juice, orange juice, grapefruit juice, lemon juice, cranberry juice, sour milk or buttermilk, yogurt, lemon-lime soda, fruit-flavored gelatin in liquid form, milk, distilled water, 4 250-mL beakers, 4 stirring rods, 4 reagent bottles.

PROCEDURE: CAUTION: Wear goggles, gloves, and a lab coat during the preparation and performance of this demonstration. Remember, *always* add concentrated acid to water, slowly with stirring. Work with concentrated HCl under the fume hood.

The concentration of the dilute hydrochloric, nitric, phosphoric, and sulfuric acids can vary. If you do not have them prepared, you can make each dilute acid by dissolving 5 mL of the desired concentrated acid in 100 mL H_2O in a 250-mL beaker and storing it in a glass-stoppered reagent bottle.

Inorganic Acids: Tell the students the first group of acids are examples of inorganic acids. Pour a small amount of dilute hydrochloric acid in a 50-mL beaker. Write the name and the formula of the acid. Then test it with red and blue litmus paper. Repeat for nitric, sulfuric, and phosphoric acids. Ask your students how they could determine if a substance is an inorganic acid by using litmus paper. (If blue litmus turns red it is an acid.) Tell students that the color change from blue to red indicates that the substance is an acid. Red litmus will stay red in neutral substances like water. Blue litmus will remain blue in neutral substances. Again ask, How do you think you can recognize an inorganic acid from its formula? (One or more hydrogens are the first element in the formula.)

Organic Acids: Using red and blue litmus paper, test the following substances to determine if they contain acids: vinegar, mustard, pickle juice, orange juice, grapefruit juice, lemon juice, cranberry juice, sour milk or buttermilk, yogurt, lemon-lime soda, fruit-flavored gelatin in liquid form, and milk. After each substance is tested, write its name, the name of the acid it contains, and its formula. Acetic acid, CH_3COOH, is in vinegar, mustard, and pickle juice.

Citric acid is in the juice of oranges, lemons, grapefruits, gelatins, and most sodas. Lactic acid is in sour milk, buttermilk, and yogurt. Benzoic acid is in cranberry juice.

$$\begin{array}{ccc} \text{COOH} & & \\ | & & \\ CH_2-C-CH_2 & CH_3-CH-COOH & C_6H_5-COOH \\ | \ | \ | & | & \\ \text{COOH OH COOH} & \text{OH} & \\ \text{Citric acid} & \text{Lactic acid} & \text{Benzoic acid} \end{array}$$

All the substances will turn blue litmus paper red except milk. Continue with the following questions. "Do organic acids affect litmus in the same way as inorganic acids?" (Yes, all acids turn blue litmus red.) "Since neither blue nor red litmus change color in milk, what does this mean?" (Milk is a neutral substance.) "If you needed sour milk, how could you make it quickly?" (Add an acid such as vinegar or lemon juice.) "What would happen if milk was used in place of water in making fruit gelatin?" (The citric acid in the gelatin would curdle the milk just like any other acid.) "Have you tasted all the foods we tested today? What taste do all the foods containing acids have in common?" (All foods containing acids have a sour taste.) Tell students that a sour taste is characteristic of all acids, not just organic acids.

CAUTION: Do not allow students to taste the foods, and remind them the taste test is not acceptable in the lab.

Ask the students to compare the list of inorganic acids and their formulas and the organic acids and their formulas and ask: "How can you tell the difference between an inorganic acid and an organic acid by looking at their formulas?" (Organic acids have carbon in their formulas and have one or more −COOH groups.) Organic acids will be discussed in more detail in Chapter 22.

16.2 Bases and Acid–Base Reactions

Introduce bases in a manner similar to acids by using the following demonstration.

Demonstration 2: Identifying Bases in Common Substances

PURPOSE: To show how bases can be identified using litmus paper and how they can be identified by their formulas

MATERIALS: Household ammonia, drain cleaner, red and blue litmus paper, lime, lye, liquid antacid medication, milk of magnesia, oven cleaner, distilled water, 3 150-mL beakers

PROCEDURE: CAUTION: Wear goggles, gloves, and a lab coat. Many of these substances are very caustic. Have students wear goggles and aprons, and make sure all containers are tightly closed.

Test each of the substances with red and blue litmus paper. If it is in solid form, dissolve a small amount in distilled water in a beaker before testing. After testing, have the students read the label to try to determine which base it contains. Write the name of the product, the base it contains, and the formula of the base. Students should observe that all the substances will turn red litmus paper blue and conclude that bases turn red litmus blue. They should also note that bases contain one or more hydroxide ions in their formulas. Ammonia water, however, is an exception; the equation $NH_3(g) + H_2O \rightarrow NH_4^+(aq) + OH^-(aq)$ shows the solution contains hydroxide ions.

Sometimes a substance tests positive for a base when no base is present. This is because it contains a salt with basic properties. A good example of this is automatic dishwashing detergent, which contains Na_2CO_3.

Tell students they will learn why this happens in Chapter 17.

Demonstration 3: Neutralization Reactions

PURPOSE: To show that a reaction between an acid and a base forms a salt and water

MATERIALS: 1 M sodium hydroxide solution, 1 M hydrochloric acid, calcium hydroxide, red and blue litmus paper, distilled water, 4 150-mL beakers, 2 eyedroppers, stirring rod, 2 watch glasses

PROCEDURE: CAUTION: Wear goggles, gloves, and lab coat. Work with concentrated HCl under a hood.

Make the 1 M solution of sodium hydroxide by dissolving 4.0 g of NaOH in enough distilled water to make 100 mL of solution and the 1 M solution of HCl by dissolving 8.3 mL of concentrated HCl in enough distilled water to make 100 mL of solution. Test the HCl and NaOH solutions with red and blue litmus paper. (The HCl has no affect on the red but turns the blue litmus red, and the NaOH has no affect on the blue but turns the red litmus blue.)

To a 150-mL beaker, add 10 mL of HCl solution and 9.5 mL of NaOH solution. Stir thoroughly and test the mixture with red and blue litmus paper. (The red stays red, and the blue turns red.) Continue to add NaOH, one drop at a time, stirring after each addition and checking with red and blue litmus paper until the blue stays blue and the red stays red. Then ask, "Why doesn't the litmus paper change color?" (Because the solution is neutral—there are no hydronium or hydroxide ions present; thus they must have reacted to form water.) Put 2 mL of the neutral solution in a watch glass, and allow the water to evaporate until the next day. Have students observe the white residue. "What is the white substance on the evaporating dish?" (NaCl) "Can you write an equation for the reaction?"

$$NaOH + HCl \rightarrow H_2O + NaCl$$

The following reaction is a common neutralization reaction

used to sweeten acid soil by adding calcium hydroxide (commonly referred to as lime or slaked lime) to the soil. Add 1 g of solid calcium hydroxide to a beaker containing 5 mL of distilled water. Test the mixture with litmus paper. (It will turn red litmus paper blue.) Slowly add dilute hydrochloric acid to the mixture until almost all the $Ca(OH)_2$ has reacted. (The white solid $Ca(OH)_2$ appears to dissolve.) When there is still a tiny amount of unreacted $Ca(OH)_2$, test the solution with litmus. (Blue should stay blue, and red should turn blue.) At that point, start adding the HCl drop by drop until all the $Ca(OH)_2$ is gone. Test again. Red litmus should stay red and blue should stay blue. Add 2 mL of the neutral solution to an evaporating dish and allow it to evaporate until the next day. A white residue of calcium chloride should remain on the evaporating dish. Have students write an equation to show what happened.

$$Ca(OH)_2 + 2HCl \rightarrow CaCl_2 + 2H_2O$$

Interesting note: The word *alkali* comes from two words that mean *plant ashes*. Wood ashes contain potash (K_2CO_3) which was one of the first alkalies known in ancient times. Two others were soda (Na_2CO_3), made by evaporating alkaline water; and lime (CaO), made by roasting seashells. Caustic soda (NaOH) and caustic potash (KOH) were made by reacting soda or potash with lime.

People who have acid soil and a wood-burning stove could use their wood ashes to neutralize their soil. Potassium carbonate is another example of a salt with basic properties.

A good opportunity to review the difference between dissociation and ionization is in the discussion of what determines the strength of bases as compared to what determines the strength of acids. Point out that all acids before they are placed in water have polar molecules; therefore, they ionize in water, and the degree to which they ionize determines their strength. However, most bases (unlike acids) are ionic compounds that dissociate in water, which is why solubility in water is what determines their strength. (NH_3 is the only exception.)

Desktop Investigation: Acids and Bases in Your Kitchen

Preparing cabbage juice to use as an indicator is as simple as dropping wedges of red cabbage into a blender with a cup of water and processing it. Pouring off the water yields about one cup of indicator. In lieu of cabbage juice, the old standby is litmus paper. Suggest to your students that they experiment with other deeply colored juices. Elderberry juice, for example, is deep red when acidic and green when basic.

A discussion of this investigation in class could include the observation that most people enjoy sour, salty, or sweet tastes, but not bitter tastes. Most bases taste bitter, so it's unlikely that any of their foods will be strongly basic.

ANSWERS TO DISCUSSION QUESTIONS:
1. Most cleaning products are basic. There are a few glass cleaning products that contain acetic acid or HCl.
2. Most foods are neutral. Some foods are slightly acidic, such as carbonated soda, lemon juice, vinegar, mayonnaise, and pickle juice.
3. Most strongly basic products such as dishwasher detergent, oven cleaner, and drain cleaner carry consumer warnings.

16.3 Relative Strengths of Acids and Bases

In the discussion of Brønsted acids and bases, it is pointed out that unlike traditional acids and bases, Brønsted acid-base reactions do not need water to take place. The following demonstration shows this very well.

Demonstration 4: Reaction of a Brønsted Acid and Base

PURPOSE: To show a Brønsted acid-base reaction that takes place in the absence of water
MATERIALS: Concentrated ammonia water, concentrated hydrochloric acid, plastic storage bag with a snap-and-seal top (1 gallon capacity), 2 250-mL beakers, 10-mL graduate
PROCEDURE: CAUTION: Wear goggles, gloves, and a lab coat.

Using a fume hood, add 2 mL of concentrated hydrochloric acid to a 250-mL beaker. To a second beaker, add 2 mL of concentrated ammonia water. Place both beakers in a plastic storage bag and seal the top. Within a short time, the two volatile liquids will produce gases, and the area where the gases meet and react will appear white because of the formation of white solid ammonium chloride. Ask students to write the equation for the reaction, and identify the acid, the base, the conjugate acid, the conjugate base, and the conjugate acid-base pairs.

$$\underbrace{HCl(g) + NH_3(g)}_{\text{acid} \quad \text{base}} \rightarrow \underbrace{NH_4^+}_{\text{conjugate acid}} + \underbrace{Cl^-}_{\text{conjugate base}}$$

(conjugate acid–base pairs indicated: HCl/Cl⁻ and NH₃/NH₄⁺)

16.4 Oxides, Hydroxides, and Acids

It is sometimes difficult for students to understand the definitions of amphoteric oxides and hydroxides. The following demonstration makes it very clear.

Demonstration 5: Which Hydroxide Is Amphoteric?

PURPOSE: To learn how to recognize amphoteric hydroxides
MATERIALS: 0.1 M aluminum sulfate solution, dilute ammonia water, dilute hydrochloric acid, 0.1 M iron(III) chloride solution, 2 M sodium hydroxide solution, distilled water, 2 125-mL dropping bottles, 100-mL graduate, 10-mL graduate, 3 reagent bottles, stirring rod, 4 medium test tubes
PROCEDURE: CAUTION: Wear goggles, gloves, and a lab coat.

(1) Make the solutions listed in materials by dissolving: 6.7 g $Al_2(SO_4)_3 \cdot 18H_2O$/100 mL water in a dropping bottle; 20 mL concentrated HCl/80 mL water in a reagent bottle; 20 mL concentrated $NH_3 \cdot H_2O$/80 mL water in a reagent bottle; 2.7 g $FeCl_3 \cdot 6H_2O$/100 mL water in a dropping bottle; and 8 g NaOH/100 mL water in a reagent bottle.

(2) Add 6 mL of aluminum sulfate solution to each of two test tubes. To each test tube add 3 mL of dilute ammonia water.

What is the result? (A pinkish-white precipitate forms.) Write an equation for the reaction:

$$Al_2(SO_4)_3(aq) + 6\,NH_3 \cdot H_2O(aq) \rightarrow$$
$$2Al(OH)_3(s) + 3(NH_4)_2SO_4(aq)$$

Allow the precipitate to settle and decant the clear liquid off the top. To the precipitate in one test, add dilute HCl while stirring until all the Al(OH)$_3$ has reacted. The Al(OH)$_3$ will appear to dissolve. Repeat with the second tube using the NaOH solution instead of the HCl. What happened in each case? (The Al(OH)$_3$ disappeared.) Did it dissolve or react chemically? (It reacted chemically.) Since the Al(OH)$_3$ reacted with both the HCl and the NaOH, what kind of hydroxide is it? (An amphoteric hydroxide.) Write chemical equations to show what happened.

$$Al(OH)_3 + 3HCl \rightarrow AlCl_3 + 3H_2O$$
$$Al(OH)_3 + NaOH \rightarrow NaAlO_2 + 2H_2O$$

(3) Repeat Step 2, substituting FeCl$_3$ solution for the Al$_2$(SO$_4$)$_3$ solution. In this case a brown precipitate forms. The equation for the reaction is:

$$FeCl_3 + 3NH_3 \cdot H_2O \rightarrow Fe(OH)_3(s) + 3NH_4Cl$$

The addition of the HCl to the Fe(OH)$_3$ caused it to disappear and form a yellow liquid, but the NaOH had no effect. Thus, Fe(OH)$_3$ is not an amphoteric hydroxide. The equations for the reactions are:

$$Fe(OH)_3 + 3HCl \rightarrow FeCl_3 + 3H_2O$$
$$Fe(OH)_3 + NaOH \rightarrow \text{no reaction}$$

16.5 Chemical Reactions of Acids, Bases, and Oxides

Many buildings and sculptures are made of marble, which is a form of CaCO$_3$. Some fancy trims, outdoor furniture, and most automobiles contained iron or an iron alloy. Acid precipitation is composed of sulfuric and/or nitric acids. The following demonstration shows what happens when acid precipitation comes in contact with marble and iron.

Demonstration 6: Some Effects of Acid Precipitation

PURPOSE: To show the effects of acid on a carbonate and a metal and to relate these reactions to acid precipitation
MATERIALS: Marble–calcium carbonate, limewater, iron nail, dilute sulfuric acid, wooden splint, 2 beakers, reagent bottle, dropping bottle, filter paper, funnel, 3 large test tubes, rubber stopper to fit one test tube, glass bend (5 cm × 20 cm)
PROCEDURE: Make dilute sulfuric acid by slowly adding 15 mL of concentrated sulfuric acid to 85 mL of water in a beaker. When cool, store in a reagent bottle. Make limewater by adding 0.5 g of Ca(OH)$_2$ to 100 mL of water in a beaker. Stir thoroughly and filter into a 125-mL dropping bottle.

Put a nail into a test tube, and cover with sulfuric acid. Let the reaction continue until the next day. Bubbles of gas will form in the acid. By the next day the nail is partly "eaten" into by the acid. The equation for the reaction is:

$$Fe + H_2SO_4 \rightarrow FeSO_4 + H_2(g)$$

Put 3 or 4 chips of marble into a test tube. To a second test tube, add 20 mL of limewater. Assemble as shown in Figure T16–1. When the assembly is ready, take the rubber stopper out of the test tube with the chips, add dilute sulfuric acid, and restopper. Bubbles of carbon dioxide will start to form.

Figure T16-1

As the CO$_2$ bubbles through the limewater, a white precipitate forms, which is a positive test for CO$_2$. The equations for the reactions are:

Marble and acid: $CaCO_3 + H_2SO_4 \rightarrow CaSO_4(s) + H_2O + CO_2(g)$

Limewater and CO$_2$: $Ca(OH)_2 + CO_2 \rightarrow CaCO_3(s) + H_2O$.

How do the acid reactions compare with similar ones involving acid precipitation? Since acid precipitation is not as concentrated as the acid used in the lab, the reactions involving acid precipitation are slower. But over a period of years, serious damage occurs to objects made of marble or iron.

References and Resources

Books and Periodicals

Borgford, Christie L., and Lee R. Summerlin. "Flower Pigments as Acid–Base Indicators," pp. 92–94; and "Distillation of Vinegar," pp. 266–68. *Chemical Activities, Teacher Edition*, Washington, D.C.: American Chemical Society, 1988.

Bybee, Roger. "Acid Rain: What's the Forecast?," *The Science Teacher*, Vol. 51, 1984, pp. 36–40, 45–47.

Bybee, Roger, et al. "The Acid Rain Debate," *The Science Teacher*, Vol. 51, 1984, pp. 50–55.

Charola, A. Elena. "Acid Rain Effects on Stone Monuments," *Journal of Chemical Education*, Vol. 64, 1987, pp. 436–37.

D'Orso, Jennifer, and Diane Burnett. "Mood Lipstick," *ChemMatters*: Vol. 3, 1985, p. 12.

Jensen, William B. "Acids and Bases: Ancient Concepts in Modern Science," *ChemMatters*, Vol. 1, 1983, pp. 14–15.

Kauffman, George B."The Brønsted–Lowry Acid–Base Concept," *Journal of Chemical Education*, Vol. 65, 1988, pp. 28–31.

Keever, Diane. "Acids and Bases," *Science Teacher*, Vol. 53, 1986, p. 173.

Kolb, Doris. "Chemical Principles Revisited: Acids and Bases," *Journal of Chemical Education*, Vol. 55, 1976, pp. 459–62.

MCA Staff. "An Acid Can Be Basic," *Journal of Chemical Education*, Vol. 56 , 1979, pp. 529–31.

Naylor, Robert W., and David Blackman. "A Physical Model to Demonstrate Acid–Base Conjugate Pairs," *Journal of Chemical Education*, Vol. 58, 1981, p. 1017.

Shakhashiri, Bassam Z. "Dehydration of Sugar by Sulfuric Acid,"

Chemical Demonstrations, Vol. 1, Madison: University of Wisconsin Press, 1983, pp. 77–78.

Shakhashiri, Bassam Z. "Antacids," *ChemMatters,* Vol. 1, 1983, p. 6

Summerlin, Lee R., Christie L. Borgford, and Julie E. Ealy. "White Wine or Grape Juice," pp. 169–70; and "Disappearing Ink," p. 176, *Chemical Demonstrations, Vol. 2,* 2nd ed., Washington, D.C.: American Chemical Society, 1988.

Talesnick, Irwin. "Addition of Acid to Water," Idea No. 89; "Excess Stomach Acid Treatment," Idea No. 52. *Idea Bank Collation: A Handbook for Science Teachers,* Kingston, Ontario, Canada: A17 Science Supplies and Services Co. Ltd., 1984.

Tanis, David. "Underground Sculpture," *ChemMatters,* Vol. 2, 1984, pp. 10–11.

Audiovisual Resources

Acid Rain, VHS, and *Special Topics in Chemistry, Volume I- Solutions, Acids, and Bases,* VHS, Random House Media, Stamford, CT, 1989.

All About Acids and Bases, VHS or sound filmstrip; *Acid Rain,* sound filmstrip; and *Acid Rain: History and Dangers,* VHS, Learning Arts, Wichita, KS, 1988–1989.

Newton, David. *Acids and Bases,* Visual Masters, Walch Science, Portland, Maine, 1988.

Computer Software

Chemistry Series (J&S Software) Acid Base Theories, Apple II, 48K; and *The Chemistry Help Series—Acids & Bases,* Apple II, 48K, Barclay School Supplies, Brooklyn, NY, 1988.

Experiments in Chemistry, Apple II, 48K and IBM PC, 128K; *Chemistry Simulations I—The Manufacture of Sulfuric Acid,* Apple II, 48K, Queue, Bridgeport, Conn. 1988.

Fee, Richard O. *Acids, Bases, Salts 1 and 2,* Apple II, 48K, Chemistry According to ROF, LaGrange, IL, 1988.

Answers and Solutions

Reviewing Concepts

1. **a.** Strong acids are strong electrolytes that ionize 100% in dilute aqueous solutions, whereas weak acids are weak electrolytes.
   b. HCl and HNO_3 are examples of strong acids, and HF and H_3PO_4 are examples of weak acids.

2. Monoprotic acids, such as HCl, donate only one proton per molecule; diprotic acids, such as H_2SO_4, donate two protons per molecule; triprotic acids, such as H_3PO_4, donate three protons per molecule.

3. **a.** Organic acids are organic compounds that contain an acid group, the carboxyl group (-COOH); mineral acids are acids made from minerals.
   b. Acetic acid is an example of an organic acid, and sulfuric acid is an example of a mineral acid.

4. **a.** Binary acids are acids that are compounds of hydrogen and a second element; oxyacids are acids that are compounds of hydrogen, oxygen, and a third element. Binary acids are named using the prefix *hydro-* followed by the root of the name of the second element, and then the suffix *-ic*. In naming oxyacids, a prefix and a suffix are added to the root name of the third element. Generally, one acid in the series defined as the *-ic* acid and named accordingly, and then the acid with one more oxygen per molecule than the *-ic* acid is named by adding the prefix *per-*. The one containing one less oxygen per molecule than the *-ic* acid is given the suffix *-ous,* and then the acid containing two fewer oxygen atoms per molecule than the *-ic* acid has the prefix *hypo-* in addition to the suffix *-ous.*
   b. Examples of binary acids are HF, HCl, and HBr. Typical oxyacids include $HClO_3$, HNO_3, and H_2SO_4.

5. The names of the acids indicated are:
   a. hydrochloric acid
   b. hydriodic acid
   c. nitric acid
   d. sulfuric acid
   e. chloric acid
   f. hydrofluoric acid
   g. nitrous acid
   h. perchloric acid
   i. phosphoric acid
   j. hypochlorous acid
   k. carbonic acid.

6. The indicated formulas are:
   a. HF
   b. HNO_3
   c. HI
   d. $HClO_3$
   e. HBr
   f. H_3PO_4
   g. HClO
   h. HNO_2
   i. H_2SO_4
   j. H_3PO_3

7. Common acids produced in the United States are sulfuric acid (H_2SO_4), phosphoric acid (H_3PO_4), nitric acid (NHO_3), hydrochloric acid (HCl), and acetic acid ($HC_2H_3O_2$). Sulfuric acid is the number-one industrial chemical in terms of volume production per year. It is a dense oily liquid with a high boiling point used in the production of fertilizers, in petroleum refining, in metallurgy, and in automobile batteries.

   Phosphoric acid is the second-largest volume industrial acid, and, in pure form, it is a low-melting solid. It is used in making fertilizers and animal feed, and as a flavoring agent in beverages. Nitric acid is the third-largest volume industrial acid. In pure form it is a volatile, unstable liquid. It is important in the manufacture of fertilizers, explosives, rubber, plastics, dyes, and drugs. Hydrochloric acid is an important laboratory acid. In concentrated form it is 37.2 percent hydrogen chloride (by mass) in water. It is used in the pickling of iron to remove surface impurities, as a cleaning agent, in food processing, and in the activation of oil wells. It is also produced in the stomach as an aid to digestion. Acetic acid, an organic acid, is a clear colorless liquid. It is produced by the fermentation of malt, barley, and fruit juices. It is found in household vinegar and is used industrially to make plastics and as a solvent.

8. The strength of an acid depends on the degree of ionization in aqueous solution, not on the amount of hydrogen in the molecule. Hydrochloric acid, which ionizes completely, is considered to be a strong acid, while phosphoric acid, which ionizes only slightly, is considered weak.

9. Pure hydrogen chloride gas consists of covalently bonded molecules, with no ions present. When hydrogen chloride is dissolved in a nonpolar solvent, still no hydrogen ions are produced. However, in a dilute aqueous solution, the hydrogen chloride becomes highly ionized because of the hydrating action of the water dipole. This results in the production of hydronium ion, H_3O^+, which accounts for the acidic properties of the resulting solution.

10. **a.** The strength of a base depends on its solubility and on the concentration of OH^- ions provided in solution. It is unrelated to the number of hydroxide ions per mole of the undissolved compound.
    b. Sodium hydroxide is strongly basic, calcium hydroxide is moderately basic, and aqueous ammonia is weakly basic.

11. **a.** A conjugate base is the particle that remains after an acid has given up a proton.
    Example: $HF(aq) + H_2O(\ell) \rightarrow H_3O^+(aq) + F^-(aq)$
                acid   conjugate base
    The F^- is the conjugate base of HF.
    b. A conjugate acid is the particle formed by the addition of a proton to a base.
    Example: $HF(aq) + H_2O(\ell) \rightarrow H_3O^+(aq) + F^-(aq)$
                         base   conjugate acid
    The H_3O^+ is the conjugate acid of H_2O.

12. **a.** The stronger an acid, the weaker its conjugate base.
    b. The stronger a base, the weaker its conjugate acid.

13. **a.** Proton-transfer reactions favor the production of the weaker acid and the weaker base.
    b. The extent of a proton-transfer reaction depends upon the relative strengths of the acids and the bases involved. For such a reaction to approach completion, the reactants must be much stronger as an acid and a base than the products.

14. **a.** Amphoteric substances are species that can react as either an acid or a base.
    b. Hydrogen sulfate ion. In the first ionization of H_2SO_4, the HSO_4^- is the conjugate base of H_2SO_4 since it can accept a proton from H_3O^+ in the reverse reaction. However, in a dilute H_2SO_4 solution, the HSO_4^- ionizes further to form the SO_4^{2-} ion, so it acts like an acid in this situation.

15. **a.** $HC_2H_3O_2$ (acid) and $C_2H_3O_2^-$ (conjugate base); H_2O (base) and H_3O^+ (conjugate acid)
    b. H_2O (acid) and OH^- (conjugate base); HCO_3^- (base) and H_2CO_3 (conjugate acid)
    c. HCN (acid) and CN^- (conjugate base); SO_4^{2-} (base) and HSO_4^- (conjugate acid)

16. The indicated responses are:
    a. HNO_3 **b.** H_2S **c.** HS^- **d.** NO_3^-.

17. By definition, strong acids and bases are substances that are highly ionized in solution. To remain highly ionized, the corresponding conjugate base and acid, respectively, are apparently too weak to compete successfully with the initial acid and base; otherwise, the initial reactants would not be classified as strong. Thus, the strength of an acid or base is rated in terms of the extent to which the corresponding conjugate base or acid remains ionized in solution.

18. **a.** Basic anhydrides are oxides that react with water to form alkaline solutions, whereas acid anhydrides are oxides that react with water to form an acid.
    b. Acid anhydrides are oxides of nonmetals, whereas basic anhydrides are oxides of active metals.

19. **a.** In general, the active metals form basic oxides, and the nonmetals form acidic oxides. The elements that lie between these typically form amphoteric oxides.
    b. The bonding in amphoteric oxides is intermediate between ionic and covalent bonding, and, as a result, these oxides have behavior intermediate between that of acidic and basic oxides. A typical amphoteric oxide, aluminum oxide, is only slightly soluble in water and dissolves in both acidic and alkaline solutions.

20. The nature of the compound is dependent upon the type of bonding to the oxygen atom in the OH group, the size and electronegativity of the atom bonded to the OH group, and the number of other oxygen atoms bonded to the atom connected to the OH group. In general, when the bond to the OH group is ionic and involves metal atoms with large radii and low electronegativities, the resulting compound will be basic. For the compound to be acidic, the bond must be highly polar such as that formed by small, highly electronegative nonmetals. Additionally, the greater the number of other oxygen atoms bonded to the atom connected to the OH group, the more acidic the compound.

21. **a.** Nonoxidizing acids provide hydronium ions that react with metals above hydrogen in the activity series to produce salts and hydrogen gas. On the other hand, in the reactions of oxidizing acids with metals throughout the entire activity series, the entire acid molecule is likely to be involved.
    b. Hydrochloric acid and dilute sulfuric acid are nonoxidizing acids, and hot concentrated sulfuric acid and nitric acid are typical oxidizing acids.

Problems

1. **a.** 1. $H_2SO_4(aq) + H_2O(\ell) \rightarrow H_3O^+(aq) + HSO_4^-(aq)$
      2. $HSO_4^-(aq) + H_2O(\ell) \rightleftarrows H_3O^+(aq) + SO_4^{2-}(aq)$
   b. The degree of ionization in the first stage is much greater than that in the second.

2. **a.** HCl(aq) + KOH(aq) → KCl(aq) + H$_2$O(ℓ)
   b. H$_3$O$^+$(aq) + Cl$^-$(aq) + K$^+$(aq) + OH$^-$(aq) → K$^+$(aq) + Cl$^-$(aq) + 2H$_2$O(ℓ)
   c. H$_3$O$^+$(aq) + OH$^-$(aq) → 2H$_2$O(ℓ)
3. **a.** H$_3$PO$_4$(aq) + 3NaOH(aq) → Na$_3$P$_4$(aq) + 3H$_2$O(ℓ)
   b. 3H$_3$O$^+$(aq) + PO$_4^{3-}$(aq) + 3Na$^+$(aq) + 3OH$^-$(aq) → 3Na$^+$(aq) + PO$_4^{3-}$(aq) + 6H$_2$O(ℓ)
   c. H$_3$O$^+$(aq) + OH$^-$(aq) → 2H$_2$O(ℓ)
   (was 3)   (was 3)   (was 6)
4. **a.** HCl(aq) + NaOH(aq) → NaCl(aq) + H$_2$O(ℓ)
   Ionic = H$_3$O$^+$(aq) + Cl$^-$(aq) + Na$^+$(aq) + OH$^-$(aq) → Na$^+$(aq) + Cl$^-$(aq) + 2H$_2$O(ℓ)
   Net Ionic = H$_3$O$^+$(aq) + OH$^-$(aq) → 2H$_2$O(ℓ)
   b. HNO$_3$(aq) + KOH(aq) → KNO$_3$(aq) + H$_2$O(ℓ)
   Ionic = H$_3$O$^+$(aq) + NO$_3^-$(aq) + K$^+$(aq) + OH$^-$(aq) → K$^+$(aq) + NO$_3^-$(aq) + 2H$_2$O(ℓ)
   Net Ionic = H$_3$O$^+$(aq) + OH$^-$(aq) → 2H$_2$O(ℓ)
   c. Ca(OH)$_2$(aq) + H$_2$SO$_4$(aq) → CaSO$_4$(aq) + 2H$_2$O(ℓ)
   Ionic = Ca^{2+}(aq) + 2OH$^-$(aq) + 2H$_3$O$^+$(aq) + SO$_4^{2-}$ → Ca^{2+}(aq) + SO$_4^{2-}$(aq) + 4H$_2$O(ℓ)
   Net Ionic = H$_3$O$^+$(aq) + OH$^-$(aq) → 2H$_2$O(ℓ)
   d. 3Mg(OH)$_2$(aq) + 2H$_3$PO$_4$(aq) → Mg$_3$(PO$_4$)$_2$(aq) + 6H$_2$O(ℓ)
   Ionic = 3Mg^{2+}(aq) + 6OH$^-$(aq) + 6H$_3$O$^+$(aq) + 2PO$_4^{3-}$(aq) → 3Mg^{2+}(aq) + 2PO$_4^{3-}$(aq) + 12H$_2$O(ℓ)
   Net Ionic = H$_3$O$^+$(aq) + OH$^-$(aq) → 2H$_2$O(ℓ)
5. 2H$_3$PO$_4$(aq) + 3Mg(OH)$_2$(aq) → Mg$_3$(PO$_4$)$_2$(aq) + 6H$_2$O(ℓ)
   Ionic = 6H$_3$O$^+$(aq) + 2PO$_4^{3-}$(aq) + 3Mg^{2+}(aq) + 6OH$^-$(aq) → 3Mg^{2+}(aq) + 2PO$_4^{3-}$(aq) + 12H$_2$O(ℓ)
   Net Ionic = H$_3$O$^+$(aq) + OH$^-$(aq) → 2H$_2$O(ℓ)
6. **a.** Since MgO reacts with water according to the following equation, it is a basic anhydride by definition.
   MgO(s) + H$_2$O(ℓ) → Mg(OH)$_2$(s)
7. **a.** Zn(s) + 2HCl(aq) → ZnCl$_2$(aq) + H$_2$(g)
   Zn(s) + 2H$_3$O$^+$(aq) → Zn^{2+}(aq) + H$_2$(g) + 2H$_2$O(ℓ)
   b. 2Al(s) + 3H$_2$SO$_4$(aq) → Al$_2$(SO$_4$)$_3$(aq) + 3H$_2$(g)
   2Al(s) + 6H$_3$O$^+$(aq) → 2Al^{3+}(aq) + 3H$_2$(g) + 6H$_2$O(ℓ)
8. **a.** MgO(s) + H$_2$SO$_4$(aq) → MgSO$_4$(aq) + H$_2$O(ℓ)
   b. CaO(s) + 2HCl(aq) → CaCl$_2$(aq) + H$_2$O(ℓ)
   c. Al$_2$O$_3$(s) + 6HNO$_3$(aq) → 2Al(NO$_3$)$_3$(aq) + 3H$_2$O(ℓ)
   d. 3ZnO(s) + 2H$_3$PO$_4$(aq) → Zn$_3$(PO$_4$)$_2$(aq) + 3H$_2$O(ℓ)
9. **a.** BaCO$_3$(s) + 2HCl(aq) → BaCl$_2$(aq) + H$_2$O(ℓ) + CO$_2$(g)
   b. MgCO$_3$(s) + 2HNO$_3$(aq) → Mg(NO$_3$)$_2$(aq) + H$_2$O(ℓ) + CO$_2$(g)
   c. Na$_2$CO$_3$(s) + H$_2$SO$_4$(aq) → Na$_2$SO$_4$(aq) + H$_2$O(ℓ) + CO$_2$(g)
   d. 3CaCO$_3$(s) + 2H$_3$PO$_4$(aq) → Ca$_3$(PO$_4$)$_2$(aq) + 3H$_2$O(ℓ) + 3CO$_2$(g)
10. **a.** CO$_2$(g) + 2KOH(aq) → K$_2$CO$_3$(aq) + H$_2$O(ℓ)
    b. CO$_2$(g) + KOH(aq) → KHCO$_3$(aq)
11. **a.** CaO(s) + CO$_2$(g) → CaCO$_3$(s)
    b. 6BaO(s) + P$_4$O$_{10}$(s) → 2Ba$_3$(PO$_4$)$_2$(s)
    c. SrO(s) + SO$_3$(g) → SrSO$_4$(s)
12. **a.** CO$_2$(g) + H$_2$O(ℓ) → H$_2$CO$_3$(aq)
    b. SO$_2$(g) + H$_2$O(ℓ) → H$_2$SO$_3$(aq)
    c. CaO(s) + H$_2$O(ℓ) → Ca(OH)$_2$(s)
    d. SO$_3$(g) + H$_2$O(ℓ) → H$_2$SO$_4$(aq)
13. Ca(s) + 2HCl(aq) → CaCl$_2$(aq) + H$_2$(g)
    Ca(s) + 2H$_3$O$^+$(aq) → Ca^{2+}(aq) + H$_2$(g) + 2H$_2$O(ℓ)
14. SO$_2$ + $\frac{1}{2}$O$_2$ → SO$_3$

    $$3.50 \times 10^8 \text{ kg SO}_2 \times \frac{10^3 \text{g}}{\text{kg}} \times \frac{\text{mol SO}_2}{64.1 \text{ g SO}_2} = 5.46 \times 10^9 \text{ mol SO}_2$$

    $$5.46 \times 10^9 \text{ mol SO}_2 \times \frac{1 \text{ mol SO}_3}{1 \text{ mol SO}_2} = 5.46 \times 10^9 \text{ mol SO}_3$$

    SO$_3$ + H$_2$O → H$_2$SO$_4$

    $$5.46 \times 10^9 \text{ mol SO}_3 \times \frac{1 \text{ mol H}_2\text{SO}_4}{1 \text{ mol SO}_3} = 5.46 \times 10^9 \text{ mol H}_2\text{SO}_4$$

    $$5.46 \times 10^9 \text{ mol H}_2\text{SO}_4 \times \frac{98.1 \text{ g H}_2\text{SO}_4}{\text{mol H}_2\text{SO}_4} \times \frac{\text{kg}}{10^3 \text{ g}} = 5.36 \times 10^8 \text{ kg H}_2\text{SO}_4$$

15. a. $Zn(s) + H_2SO_4(aq) \rightarrow ZnSO_4(aq) + H_2(g)$

$$\frac{6.00 \text{ mol } H_2SO_4}{L} \times \frac{L}{10^3 \text{ mL}} \times 100. \text{ mL} = 0.600 \text{ mol } H_2SO_4$$

$$0.600 \text{ mol } H_2SO_4 \times \frac{1 \text{ mol } ZnSO_4}{1 \text{ mol } H_2SO_4} = 0.600 \text{ mol } ZnSO_4$$

$$0.600 \text{ mol } ZnSO_4 \times \frac{161.4 \text{ g } ZnSO_4}{\text{mol } ZnSO_4} = 96.8 \text{ g } ZnSO_4$$

b. $0.600 \text{ mol } H_2SO_4 \times \frac{1 \text{ mol } H_2}{1 \text{ mol } H_2SO_4} = 0.600 \text{ mol } H_2$

$$0.600 \text{ mol } H_2 \times \frac{22.4 \text{ L } H_2}{\text{mol } H_2} = 13.4 \text{ L } H_2$$

16. $CaCO_3(s) + 2HCl(aq) \rightarrow CO_2(g) + CaCl_2(aq) + H_2O(\ell)$

a. $1500. \text{ mL } CO_2 \times \frac{L}{10^3 \text{ mL}} \times \frac{\text{mol } CO_2}{22.41 \text{ L } CO_2} = 0.066\ 93 \text{ mol } CO_2$

$$0.066\ 93 \text{ mol } CO_2 \times \frac{1 \text{ mol } CaCO_3}{1 \text{ mol } CO_2} = 0.066\ 93 \text{ mol } CaCO_3$$

$$0.066\ 93 \text{ mol } CaCO_3 \times \frac{100.1 \text{ g } CaCO_3}{\text{mol } CaCO_3} = 6.700 \text{ g } CaCO_3$$

b. $0.066\ 93 \text{ mol } CaCO_3 \times \frac{2 \text{ mol } HCl}{1 \text{ mol } CaCO_3} = 0.1339 \text{ mol } HCl$

$$0.1339 \text{ mol } HCl \times \frac{L}{2.00 \text{ mol } HCl} = 0.066\ 95 \text{ L or } 66.95 \text{ mL}$$

Critical Thinking

1. Answers will vary: Students should provide examples of acids not containing oxygen such as HCl or HBr. They should also discuss the different ways in which an acid can be defined that can either narrow or broaden the definition.
2. Answers will vary.

Cross Connections

1. Sulfuric acid is involved in many technical and manufacturing processes. Typically, the more highly industrialized a particular country, the greater the level of sulfuric acid consumed by that country in the course of its everyday industrial activities. Thus, sulfuric acid production can actually serve as a measure of the degree of industrialization and economic activity of that country.

Videodisc Selections
from *The Periodic Table Videodisc* (JCE Software)

step forward

CHAPTER 16

pg 461 Acids

36360
boric acid (s)

step reverse

36480
solutions of phosphoric acid, citric acid, lactic acid (s)

36507
sulfuric acid (s)

36513
hydrochloric acid (s)

pg 461 Acid reactions with metals

1978
lithium + acids (m)

6036
sodium + acids (m)

17611
potassium + acids (m)

7857
magnesium + acids (m)

20119
calcium + acids (m)

38650
strontium + acids (m)

47293
barium + acids (m)

31167
zinc + acids (m)

9040
aluminum + acids (m)

22055
chromium + acids (m)

23281
manganese + acids (m)

24981
iron + acids (m)

26677
cobalt + acids (m)

28046
nickel + acids (m)

pg 470 Bases

36390
ammonia (s)

36444
magnesium hydroxide (s)

pg 480 Amphoterism

9040
aluminum + acids (m)

10310
aluminum + bases (m)

pg 484 Reactions of acids, bases, oxides

31167
zinc + acids (m)

460l   Chapter 16   Teacher's Notes

Teacher's Notes

Chapter 16
Acids and Bases

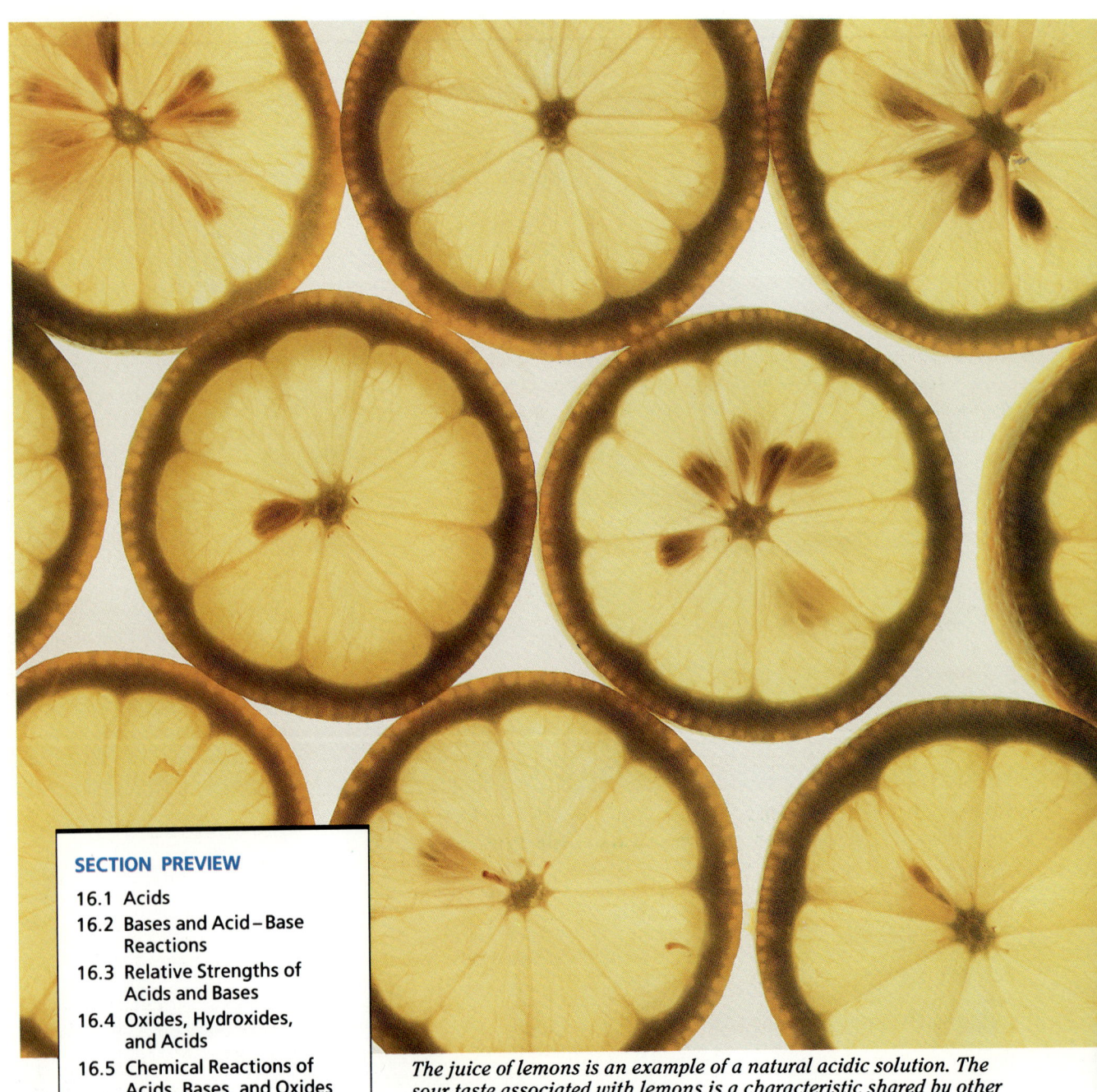

SECTION PREVIEW

16.1 Acids
16.2 Bases and Acid–Base Reactions
16.3 Relative Strengths of Acids and Bases
16.4 Oxides, Hydroxides, and Acids
16.5 Chemical Reactions of Acids, Bases, and Oxides

The juice of lemons is an example of a natural acidic solution. The sour taste associated with lemons is a characteristic shared by other acids.

16.1 Acids

How many foods can you think of that are sour? Chances are that almost all the foods on your list (such as those in Figure 16-1) owe their sour taste to an acid. Lemons, oranges, and grapefruit contain *citric acid*. Apples contain *malic acid*. Vinegar, which can be produced by the fermentation of hard cider or grape juice, contains *acetic acid*. Almost all fruits contain some kind of acid. Sour milk contains *lactic acid*. *Tartaric acid* is present in wine, and *phosphoric acid* imparts a tart flavor to many carbonated beverages.

General Properties of Aqueous Acids

Acids were first recognized as a distinct class of compounds by the sourness and other common properties of their aqueous solutions. These properties are as follows:

   1. Acids have a sour taste. You are familiar with the taste of acids in the foods mentioned above. These foods contain weak acids in aqueous solution. Solid acids, such as citric acid that is often found in candy, and acetylsalicylic acid, which is aspirin, taste sour as they dissolve in saliva. Many laboratory acids, especially in concentrated solutions, are very corrosive (they destroy skin and clothing) and are powerful poisons.

   2. Acids contain hydrogen, and some react with active metals to liberate hydrogen gas (H_2). Metals above hydrogen in the activity series of metals (see Section 8.3) undergo single-replacement reactions with certain acids to produce hydrogen gas.

   3. Acids change the color of dyes known as "acid-base indicators." For example, litmus dye, which is extracted from lichens, is an acid-base indicator. If a strip of paper permeated with blue litmus is dipped into an acid solution, it turns red, as demonstrated in Figure 16-2.

   4. Acids react with bases to produce salts and water. When chemically equivalent amounts of acids and bases react, the three properties described above are "neutralized"—they disappear. The reaction products are water and an ionic compound called a salt.

   5. Acids are electrolytes. Aqueous solutions of acids conduct electric current. (See Figure 16-3.)

Definitions of Acids

Traditional acids   Arrhenius included a description of acids in his theory of ionization. He suggested that the characteristic properties of acids

SECTION OBJECTIVES

- List five general properties of aqueous acids.
- Define and give an example of a traditional acid, a Brønsted acid, and a Lewis acid.
- Explain the difference between a strong acid and a weak acid, and give an example of each.
- Explain and use the systems for naming common binary acids and oxyacids.
- Write the electron-dot structures for common oxyacids.
- Name five acids commonly found in industry and/or the laboratory, and describe their nature (strong or weak, the number of ionizable hydrogen atoms, whether inorganic or organic, etc.).

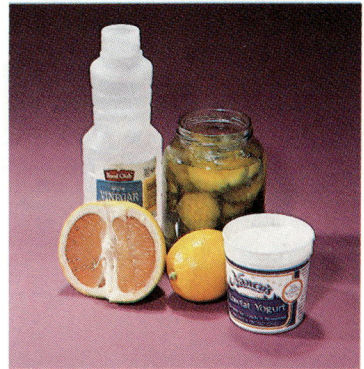

Figure 16-1 Common acidic foods. If a food tastes sour, it probably is acidic.

Figure 16-2 Acids can change the color of many materials. Here, lemon juice, an acid, changes the color *(left to right)* of tea and the acid-base indicator litmus paper.

Acids and Bases   **461**

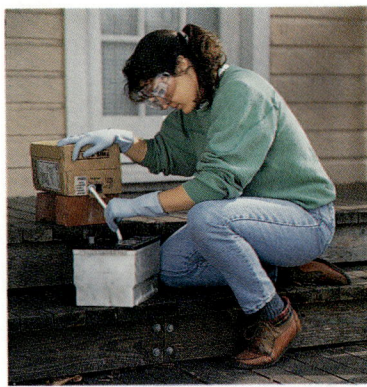

Figure 16-3 An automobile battery contains acid.

STUDY HINT
The solvent–solute attraction is so strong between water and some acids that the acid molecules "break" into ions.

The typical hydrogen ion, (H⁺) is an atomic nucleus containing only one proton. Other isotopes of hydrogen do contain one or two neutrons in their nuclei.

Often two or more scientists who are working independently come up with the same idea at the same time. This was the case when Brønsted and T. M. Lowry developed their acid–base concept. It might be interesting for students to try to find out why Brønsted is often given more credit for the idea than Lowry.

described above are caused by the presence of the hydrogen ion (H⁺) in aqueous solution.

The observations of Arrhenius are the basis for the **traditional definition of an acid:** *a chemical compound that contains hydrogen and ionizes in aqueous solution to form hydrogen ions*. These acids are molecular compounds with ionizable hydrogen atoms. Their water solutions are known as *aqueous acids*.

As explained in our earlier study of the hydration of ions (Section 15.2), the hydrogen ion in aqueous solution is best represented as H_3O^+, the hydronium ion. Using the hydronium ion, the formation of, for example, a nitric acid solution is shown by the following equation:

$$HNO_3(\ell) + H_2O(\ell) \rightarrow H_3O^+(aq) + NO_3^-(aq)$$
nitric acid

Brønsted acids In 1923, Danish chemist J. N. Brønsted and English chemist T. M. Lowry independently introduced extensions of the traditional acid definition. We will refer to this as the Brønsted definition of an acid.

A **Brønsted acid** *is a molecule or ion that is a proton donor.* In the equations above and below you can see that HNO_3 and HCl donate protons to H_2O to form H_3O^+. In aqueous solution, all traditional acids donate protons to water and are Brønsted acids also. In addition, other species that donate protons are included in the category of Brønsted acids.

Hydrogen chloride dissolved in ammonia transfers protons to the solvent much as it does in water.

$$HCl(g) + H_2O(\ell) \rightarrow H_3O^+(aq) + Cl^-(aq)$$
$$HCl(g) + NH_3(\ell) \rightarrow NH_4^+(aq) + Cl^-(aq)$$

A proton is transferred from the hydrogen chloride molecule to the ammonia molecule to form the ammonium ion, just as a proton is transferred to the water molecule to form the hydronium ion. The similarity of these reactions is very clear when electron-dot formulas are used.

$$H\!:\!\ddot{\underset{..}{Cl}}\!: \;+\; H\!:\!\underset{\underset{H}{|}}{\ddot{O}}\!: \;\longrightarrow\; \left[H\!:\!\underset{\underset{H}{|}}{\ddot{O}}\!:\!H\right]^+ \;+\; :\!\ddot{\underset{..}{Cl}}\!:^-$$

$$H\!:\!\ddot{\underset{..}{Cl}}\!: \;+\; H\!:\!\underset{\underset{H}{|}}{\ddot{N}}\!:\!H \;\longrightarrow\; \left[H\!:\!\underset{\underset{H}{|}}{\ddot{N}}\!:\!H\right]^+ \;+\; :\!\ddot{\underset{..}{Cl}}\!:^-$$

In both cases, hydrogen chloride is a Brønsted acid, even though no hydronium ions form in the second reaction.

Water is also a Brønsted acid. In the reaction with ammonia, for example,

$$H_2O(\ell) + NH_3(g) \rightleftharpoons NH_4^+(aq) + OH^-(aq)$$

the water molecule donates a proton to the ammonia molecule.

In the Brønsted scheme, the hydronium ion is also an acid and not just an ion produced by an acid. As will be illustrated shortly, the hydronium ion is the actual proton donor in reactions of aqueous acid solutions.

Lewis acids A third acid classification based on bonding and structure includes substances that do not contain hydrogen ions. This definition of acids was introduced in 1923 by G. N. Lewis, the same American chemist whose name was given to electron-dot structures. Lewis wanted to avoid a definition of acids that depended upon the presence of any particular element. His definition emphasizes the role of electron pairs in chemical reactions. *A **Lewis acid** is an atom, ion, or molecule that is an electron-pair acceptor.*

The Lewis definition is the broadest of the three acid definitions we have given. It applies to any substance that can combine with another by accepting an electron pair to form a covalent bond. The bare proton (hydrogen ion) is a Lewis acid in all reactions in which it forms a covalent bond, such as this one:

$$H^+ + H{:}\overset{..}{\underset{H}{N}}{:}H \longrightarrow \left[H{:}\overset{..}{\underset{H}{N}}{:}H \right]^+$$

$$H^+ + {:}NH_3 \longrightarrow NH_4^+$$

Lewis acids do not need to involve hydrogen ions. A compound in which the central atom has three valence electrons and forms three covalent bonds reacts as a Lewis acid by forming an additional covalent bond to complete an electron octet. Boron trifluoride, for example, is an excellent Lewis acid. It forms a fourth covalent bond with many molecules and ions in the manner shown below for the reaction with fluoride ion:

$$:\!\!\overset{..}{\underset{..}{F}}\!\!:\overset{..}{\underset{..}{B}} \; + \; :\!\overset{..}{\underset{..}{F}}\!:^- \longrightarrow \left[:\!\!\overset{..}{\underset{..}{F}}\!\!:\overset{..}{\underset{..}{B}}:\overset{..}{\underset{..}{F}}\!\!: \right]^-$$

$$BF_3(aq) + F^-(aq) \longrightarrow BF_4^-(aq)$$

The Lewis definition of acids applies to species in all phases. For example, boron trifluoride is a Lewis acid in the gas-phase combination with ammonia:

$$BF_3(g) + NH_3(g) \longrightarrow F_3BNH_3(s)$$

Types of aqueous acids A general reference to an "acid" usually is a reference to either a substance that produces hydronium ions in water or a solution containing hydronium ions. All of the pure acids are molecular electrolytes. Their molecules are sufficiently polar so that the water molecules attract away one or more of their hydrogen ions to leave behind anions, negatively charged ions.

An acid that is a strong electrolyte and ionizes 100% in aqueous solution is a **strong acid**. Hydrogen chloride is a strong acid. The common strong acids are listed in Table 16-1. Each of the strong acids—perchloric acid ($HClO_4$), hydrochloric acid (HCl), and nitric acid (HNO_3)—is assumed to ionize completely to give up one proton (hydrogen ion). *An acid that can donate one proton (hydrogen ion) per molecule is known as a* **monoprotic acid**.

The proton, or hydrogen ion, is always a Lewis acid because, having no electrons, it can "accept" an electron pair.

Figure 16-4 Most acids are corrosive substances.

Acids and Bases

TABLE 16-1 COMMON AQUEOUS ACIDS

Strong Acids		Weak Acids	
$HClO_4$	$\rightarrow H^+ + ClO_4^-$	HSO_4^-	$\rightleftarrows H^+ + SO_4^{2-}$
HCl	$\rightarrow H^+ + Cl^-$	H_3PO_4	$\rightleftarrows H^+ + H_2PO_4^-$
HNO_3	$\rightarrow H^+ + NO_3^-$	HF	$\rightleftarrows H^+ + F^-$
H_2SO_4	$\rightarrow H^+ + HSO_4^-$	$HC_2H_3O_2$	$\rightleftarrows H^+ + C_2H_3O_2^-$
		H_2CO_3	$\rightleftarrows H^+ + HCO_3^-$
		H_2S	$\rightleftarrows H^+ + HS^-$
		HCN	$\rightleftarrows H^+ + CN^-$
		HCO_3^-	$\rightleftarrows H^+ + CO_3^{2-}$

Sulfuric acid (H_2SO_4) is a **polyprotic acid**—*an acid that can donate more than one proton per molecule.* Sulfuric acid ionizes in two stages. In its first ionization, sulfuric acid is a strong acid and is completely converted to hydrogen sulfate ions (HSO_4^-).

$$H_2SO_4(aq) + H_2O(\ell) \rightarrow H_3O^+(aq) + HSO_4^-(aq)$$

The hydrogen sulfate ion is a weak electrolyte, however, and establishes the following equilibrium in solution:

$$HSO_4^-(aq) + H_2O(\ell) \rightleftarrows H_3O^+(aq) + SO_4^{2-}(aq)$$

The extent to which this second ionization occurs depends on the amount of dilution: the more dilute the solution, the more hydrogen sulfate ions will convert to sulfate ions. All stages of ionization of a polyprotic acid occur in the same solution. Sulfuric acid solutions may therefore contain H_3O^+, HSO_4^-, and SO_4^{2-} ions. The ionization of HSO_4^- ions may be complete only in very dilute solutions, giving the following overall reaction:

$$H_2SO_4(aq) + 2H_2O(\ell) \rightarrow 2H_3O^+(aq) + SO_4^{2-}(aq)$$

This equation is the sum of the two stages of ionization of sulfuric acid. In a complete neutralization reaction, sulfuric acid *is an acid that can donate two protons per molecule and is therefore known as a* **diprotic acid**.

Many acids, such as HSO_4^-, are weak electrolytes. In Section 15.2 we showed that hydrogen fluoride is also a weak electrolyte. In aqueous solution, both the ionization of HF and the reverse reaction occur simultaneously:

$$HF(aq) + H_2O(\ell) \rightleftarrows H_3O^+(aq) + F^-(aq)$$

Acids that are weak electrolytes are known as **weak acids**. The aqueous solution of a weak acid contains hydronium ions, anions, and dissolved acid molecules. The common weak acids are listed in Table 16-1. There is a wide variation in the degree that equivalent amounts of different weak acids convert to ions.

Phosphoric acid is a **triprotic acid**—*an acid able to donate three protons per molecule.*

$$H_3PO_4(aq) + H_2O(\ell) \rightleftarrows H_3O^+(aq) + H_2PO_4^-(aq)$$
$$H_2PO_4^-(aq) + H_2O(\ell) \rightleftarrows H_3O^+(aq) + HPO_4^{2-}(aq)$$
$$HPO_4^{2-}(aq) + H_2O(\ell) \rightleftarrows H_3O^+(aq) + PO_4^{3-}(aq)$$

As a general rule, the less concentrated the aqueous acid solution is, the greater the percentage of ionization. This percentage of ionization in very dilute solutions may range from an almost negligible amount to 100 percent.

The more dilute the solution, the more highly ionized are weak electrolytes.

Notice that phosphoric acid is a weak acid in each step of its ionization. The strength of an acid depends on the degree of ionization in aqueous solution, not on the amount of hydrogen in the molecule. A solution of phosphoric acid may contain H_3O^+, H_3PO_4, $H_2PO_4^-$, HPO_4^{2-}, and PO_4^{3-}. As with most polyprotic acids, the concentration of ions formed in the first ionization is the greatest, with lesser concentrations of the respective ions during each succeeding ionization.

Acetic acid is a weak acid that is commonly represented as $HC_2H_3O_2$. Its molecule contains four hydrogen atoms, but only one is ionizable. Acetic acid ionizes slightly in water to give hydronium ions and acetate ions ($C_2H_3O_2^-$).

$$HC_2H_3O_2(aq) + H_2O(\ell) \rightleftharpoons H_3O^+(aq) + C_2H_3O_2^-(aq)$$

Acetic acid is an **organic acid** *an organic compound that contains an acid group, the carboxyl group (—COOH).* An alternative formula for acetic acid, CH_3COOH, shows that it is the hydrogen atom in the carboxyl group that is "acidic" and forms the hydronium ion, as shown below:

$$CH_3COOH(aq) + H_2O(\ell) \rightleftharpoons H_3O^+(aq) + CH_3COO^-(aq)$$

Except for acetic acid, all of the common acids listed in Table 16-1 are inorganic acids. For centuries, four of these substances have been known as acids. *Because they were made from minerals, sulfuric acid, nitric acid, hydrochloric acid, and phosphoric acid are referred to as* **mineral acids.** Each plays an important role in the chemical industry.

Names and Structures of the Common Acids

Some acids contain only two elements and are called *binary acids;* others, such as oxyacids, contain three elements, and are called *ternary acids.*

Binary acids *A* **binary acid** *is an acid composed of hydrogen bonded to one of the more electronegative elements.* The common inorganic acids are either binary acids or oxyacids. The hydrogen halides—HF, HCl, HBr, HI—are all binary acids. The strength of a binary acid depends upon the polarity of the bond between hydrogen and the second element, and the ease with which the bond can be broken.

The name of a binary acid begins with the prefix *hydro-*. The root of the name of the second element follows this prefix. The name then ends with the suffix *-ic*. This scheme is illustrated by the examples given in Table 16-2. Each of the pure compounds listed in Table 16-2 is a gas. The water solutions of these compounds are known by the names given in the righthand column.

STUDY HINT

Organic compounds are carbon compounds, mostly derivatives of hydrocarbons. Chapters 21–23 are devoted to the chemistry of organic compounds.

The structure of the carboxyl group is

$$-C\begin{matrix}\diagup\!\!\!\!O\\\diagdown O-H\end{matrix}$$

Inorganic compounds include all compounds that do not contain carbon.

TABLE 16-2	NAMES OF BINARY ACIDS	
Formula	Name of Pure Substance	Name of Acid
HF	hydrogen fluoride	hydrofluoric acid
HCl	hydrogen chloride	hydrochloric acid
HBr	hydrogen bromide	hydrobromic acid
HI	hydrogen iodide	hydriodic acid
H_2S	hydrogen sulfide	hydrosulfuric acid

Point out to the students that knowing one piece of information sometimes helps you to know several others. For example, ask these questions: If HClO₃ is chloric acid, what would you call HBrO₃ and HIO₃? (Answer: bromic acid and iodic acid) Why? (Answer: Since Cl, Br, and I are halogens, we would expect them to be named in similar fashion.)

Oxyacids Sulfuric acid (H_2SO_4) is an example of an **oxyacid**—*an acid that is a compound of hydrogen, oxygen, and a third element.* An oxyacid is named by adding a prefix and a suffix to the root name of the third element.

The formulas and names of the series of oxyacids of chlorine illustrate the general method for naming such acids.

HClO	*hypo*-chlor-*ous* acid	-ic acid minus two O atoms
HClO₂	chlor-*ous* acid	-ic acid minus one O atoms
HClO₃	chlor-*ic* acid	
HClO₄	*per*-chlor-*ic acid*	-ic acid plus one O atom

The names are determined by defining one acid as the *-ic* acid, which is named by adding *-ic* to the root for the third element. For the chlorine oxyacids, this is chloric acid ($HClO_3$). An acid with one more oxygen atom per molecule than the *-ic* acid is named by adding the prefix *per-*, giving per-chlor-ic acid ($HClO_4$). The prefix *per-* is a contraction of *hyper-*, which means *above*.

The chlorine acid containing one less oxygen atom per molecule than the *-ic* acid is given the suffix *-ous*, which becomes chlor-ous acid, $HClO_2$. The chlorine acid that contains two fewer oxygen atoms per molecule than the *-ic* acid, has the prefix *hypo-* in addition to the suffix *-ous*, which becomes hypo-chlor-ous acid, HClO. The prefix *hypo-* means *below*.

To use the schemes for naming oxyacids, it is necessary only to know the formula and the name of the acid named with *-ic* in any series. You should memorize the following formulas and names for these members of the most common series of oxyacids: $HClO_3$ (chloric acid), H_2SO_4 (sulfuric acid), HNO_3 (nitric acid), H_3PO_4 (phosphoric acid), $HBrO_3$ (bromic acid). The names of some common oxyacids are also given in Table 16-3. The names of the polyatomic ions formed by the loss of all ionizable hydrogen atoms are given in the righthand column. The anion names have the same roots and prefixes

TABLE 16-3 NAMES OF OXYACIDS AND OXYANIONS

Formula	Name of acid	Name of anion
$HC_2H_3O_2$	acetic acid	acetate
H_2CO_3	carbonic acid	carbonate
HIO_3	iodic acid	iodate
HClO	hypochlorous acid	hypochlorite
$HClO_2$	chlorous acid	chlorite
$HClO_3$	chloric acid	chlorate
$HClO_4$	perchloric acid	perchlorate
HNO_2	nitrous acid	nitrite
HNO_3	nitric acid	nitrate
H_3PO_3	phosphorous acid	phosphite
H_3PO_4	phosphoric acid	phosphate
H_2SO_3	sulfurous acid	sulfite
H_2SO_4	sulfuric acid	sulfate

as the acid names. In the anion names, the ending *-ic* is changed to *-ate* and the ending *-ous* is changed to *-ite*.

All oxyacids are molecular electrolytes that contain one or more oxygen–hydrogen bonds. *The covalently bonded —OH group is referred to as a* **hydroxyl group.** These —OH groups are not to be confused with **hydroxide ions** (OH⁻) *which are found in the ionic compounds known as hydroxides*, such as sodium hydroxide (NaOH).

Figure 16-5 shows the electron-dot formulas of the four oxyacids of chlorine. Notice that each oxygen atom is bonded to the chlorine atom and that each hydrogen atom is bonded to an oxygen atom. Aqueous solutions of these molecules are acids because O—H bonds are broken as the hydrogen is attracted away by water molecules.

The ionizable hydrogen atoms in all polyprotic oxyacids are present in the hydroxyl groups only. For example, examine the structures of sulfuric acid and phosphoric acid in Figure 16-6. Four oxygen atoms surround the central sulfur and phosphorus atoms. The ionizable hydrogen atoms are bonded to oxygen atoms. The major exception to this structural pattern is phosphorous acid (H_3PO_3), which is a diprotic acid because one hydrogen atom is bonded directly to the phosphorus atom and is not ionizable.

Some Common Acids

Sulfuric acid, phosphoric acid, and nitric acid are among the ten chemicals manufactured in the largest volume each year in the United States. The major application of each is in the manufacture of agricultural products, principally fertilizers. Sulfuric acid is by far the number-one industrial chemical. It is a versatile chemical with many applications, and over 30 million metric tons are made each year in the United States. The quantity of sulfuric acid produced by a country can be taken as a measure of the size of its economy. Phosphoric acid is usually the second-largest volume industrial acid, and nitric acid the third.

Hydrochloric acid and acetic acid are manufactured in smaller quantities. Their different properties and costs render these two acids useful in ways that differ from those of the "big three" acids. Sulfuric acid, nitric acid, hydrochloric acid, and acetic acid are, however, all important laboratory chemicals.

The properties of each of these common acids as they are supplied in concentrated solutions are summarized in Table 16-4.

Figure 16-5 Electron-dot formulas of the four oxyacids of chlorine.

TABLE 16-4	PROPERTIES OF COMMERCIAL CONCENTRATED ACID SOLUTIONS		
Acid name	Percent by mass (%)	Molarity (mol/L)	Density (g/mL)
Sulfuric acid	98.0%	18	1.84
Phosphoric acid	85.5%	14.8	1.70
Nitric acid	70.4%	15.9	1.42
Hydrochloric acid	37.2%	12.1	1.19
Acetic acid	99.8%	17.4	1.05

Figure 16-6 Electron-dot formulas of phosphoric acid *(top)* and sulfuric acid *(bottom)*.

The evolution of heat may be so great when sulfuric acid is diluted in water that the solution boils. It is often necessary to cool the solution during the dilution process to avoid spattering and "boiling over."

CAUTION
To avoid heat evolution that causes spattering, never add water to sulfuric acid. In general, always add the acid to water.

Figure 16-7 Never add water to an acid. To dilute an acid safely, place a clean, dry glass rod into a container of water and decant the acid along the rod as shown.

Sulfuric acid In addition to its use in fertilizer manufacture, sulfuric acid is used in large quantities in petroleum refining and metallurgy. Sulfuric acid is also essential to a vast number of industrial processes, including the production of metals, paper, paint, dyes, detergents, and many chemical raw materials. It is the "battery acid" of automobile batteries.

Pure sulfuric acid is a dense, oily liquid that has a high boiling point. Concentrated sulfuric acid contains 98% sulfuric acid by mass combined with water. When concentrated sulfuric acid is diluted with water, a large amount of heat is released during the formation of hydrates, such as $H_2SO_4 \cdot H_2O$ and $H_2SO_4 \cdot 2H_2O$. The safe technique for diluting sulfuric acid is shown in Figure 16-7.

Because of its affinity for water, concentrated sulfuric acid is an effective dehydration (water-removing) agent. It can be used to remove water from gases with which it does not react. Sugar and other organic compounds that contain a 2-to-1 ratio of hydrogen to oxygen are also dehydrated by sulfuric acid. The reaction of sulfuric acid with sugar is quite dramatic, as shown in Figure 16-8. Skin contains organic compounds that are attacked by concentrated sulfuric acid, and it can cause serious burns.

Figure 16-8 Sugar and concentrated sulfuric acid, when mixed, react dramatically.

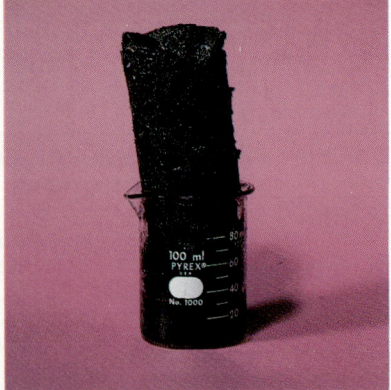

Phosphoric acid Pure phosphoric acid is a low-melting solid (mp 42°C). Commercially, phosphoric acid is available as "syrupy phosphoric acid," a viscous liquid that is 85 percent phosphoric acid by mass in water. Dilute phosphoric acid has a pleasant but sour taste and is not toxic. It is used as a flavoring agent in beverages and as a cleaning agent for dairy equipment.

Like nitrogen, phosphorus is an essential element for plants and animals. The bulk of all phosphoric acid produced each year is used directly in impure form for the manufacture of fertilizers and animal feed. Phosphoric acid is also important in the manufacture of detergents, ceramics, and phosphorus-containing chemicals.

Nitric acid Pure nitric acid is a volatile, unstable liquid rarely used in industry or in laboratories. Concentrated nitric acid contains about 70 percent by mass of the acid dissolved in water and is more stable. While pure nitric acid solutions are colorless, upon standing they gradually become yellow because of slight decomposition to form brown nitrogen dioxide gas.

$$4HNO_3(aq) \rightarrow 4NO_2(g) + O_2(g) + 2H_2O(\ell)$$

This acid has a suffocating odor and can cause serious burns to the skin. Small amounts of spilled nitric acid stain the skin yellow.

Nitric acid is used in making explosives (many of which are nitrogen-containing compounds), rubber, chemicals, plastics, dyes, and drugs.

Hydrochloric acid Dilute and concentrated hydrochloric acids are common laboratory acids. Common dilute hydrochloric acid is a 1:4 dilution of concentrated acid. Hydrochloric acid is also produced in the stomach, where secretions containing about 0.4 percent hydrochloric acid aid in digestion.

Industrially, hydrochloric acid is important for *pickling* of iron and steel, (the immersion of metals in acid solutions to remove surface impurities). It is also used in industry as a general cleaning agent, in food processing, in the activation of oil wells, in recovering magnesium from sea water, and in the production of other chemicals.

Acetic acid Pure acetic acid is a clear, colorless liquid with a pungent odor at room temperature. It forms crystals in a cold room, since its melting point is 17°C. Acetic acid is available as *glacial* acetic acid, which is 99.8 percent acetic acid by mass. Dilute acetic acid is by mass a 37 percent solution of acetic acid in water. The fermentation of malt, barley, and fruit juices produces acetic acid in vinegars that may retain the flavor of the starting material. Household vinegar contains 4 to 8 percent acetic acid.

Acetic acid is important industrially as a reactant in the synthesis of many chemicals used in making plastics and as a solvent. It is a raw material in the production of food supplements (e.g. lysine, an essential amino acid) and is used as a fungicide.

Section Review

1. What are five general properties of aqueous acids?
2. Which definition of the three acid definitions is the broadest? Why?
3. *(a)* Why are strong acids also strong electrolytes?   *(b)* Will a strong electrolyte also be a strong acid?
4. Name the following acids: *(a)* HBrO   *(b)* HBrO$_3$.

As a reminder, when you are in the laboratory and spill chemicals on your skin, you should rinse the affected area with water immediately and thoroughly. In the case of an acid, if a "tingling" sensation remains, this is your signal to continue rinsing the affected area.

Many proteins (those containing aromatic amino acids) turn yellow in the presence of nitric acid. This explains why small amounts of nitric acid stain skin yellow.

Technical grade, concentrated hydrochloric acid is sold in hardware stores under the name muriatic acid.

1. Sour taste; contain hydrogen, which is given off when certain acids react with metals; change litmus from blue to red; react with bases to produce salts and water; are electrolytes.
2. The Lewis definition, because it applies to any substance that can combine with another by accepting an electron pair to form a covalent bond
3. *(a)* Strong acids ionize 100% in aqueous solutions and thus are strong electrolytes.
   (b) Only if it is an acid to begin with
4. *(a)* Hypobromous acid
   (b) Bromic acid

Acids and Bases

16.2 Bases and Acid–Base Reactions

SECTION OBJECTIVES

- List five general properties of aqueous bases.
- Define and give an example of a traditional base, a Brønsted base, and a Lewis base.
- Define and recognize traditional, Brønsted, and Lewis acid-base reactions.
- Explain the difference between a strong base and a weak base, and give an example of each.

General Properties of Aqueous Bases

Several substances that have long been known as bases are commonly found in homes (Figure 16-9). Bases are excellent cleaning agents because they react with fats and oils and convert them to water-soluble substances. Household ammonia is an ammonia–water solution that is very useful for all types of general cleaning. Sodium hydroxide (NaOH), known by the common name *lye*, is present in some commercial products used to unclog sink drains and to clean ovens. Milk of magnesia is a suspension in water of magnesium hydroxide, $Mg(OH)_2$, which is not very water-soluble. It is used as an antacid (a neutralizer of excess hydrochloric acid in the stomach), a laxative, and as an antidote for ingestion of strong acids. Aluminum hydroxide, $Al(OH)_3$ is an antacid in medications such as Maalox® and Di-Gel®.

All of these common household products are bases of the type identified by Arrhenius in his study of ionization. The distinctive properties associated with such bases are as follows:

1. Bases have a bitter taste. As you may have noticed if you have ever gotten soap into your mouth, solutions of bases have a bitter taste. The stronger basic cleaning agents are caustic—they attack the skin and can cause severe burns. Each year, numerous tragic accidents occur when small children handle drain cleaners and similar substances that have been carelessly stored within easy reach.

2. Dilute aqueous solutions of bases feel slippery to the skin. You encounter this property of aqueous bases whenever you wash with soap.

3. Bases, like acids, change the color of dyes known as acid-base indicators. Aqueous bases cause litmus dye to change color from red to blue. This is the opposite of the color change caused by aqueous acids. Many dyes are acid-base indicators because they display different colors in acid and base solutions (See Figure 16-10). Phenolphthalein (feen-ol-thal-een) is colorless in the presence of acids and red (pink) in the presence of bases. Methyl orange is red in acid solutions and yellow in base solutions.

4. Bases react with acids to produce salts and water. When chemically equivalent quantities of acids and bases react, the previous three properties of bases are "neutralized"—they disappear. The reaction products are salts and water.

5. Bases are electrolytes. Aqueous solutions of bases conduct electricity.

Figure 16-9 Common household materials that contain bases.

Figure 16-10 A base will change the color an an acid–base indicator. Here, the base ammonium hydroxide (ammonia water) changes the color of an indicator from yellow to purple.

Definitions of Bases and Acid–Base Reactions

For each of the three definitions of acids given in Section 16.1, there is a related definition of a base and an acid–base reaction.

Traditional bases The general properties of bases listed above are based on Arrhenius' definitions of acids and bases. *The **traditional definition of a base** is a substance that contains hydroxide ions (OH^-) and dissociates to give hydroxide ions in aqueous solution.*

Traditional bases are all ionic compounds containing metal cations (positively-charged ions) and the hydroxide anion (OH^-). Sodium hydroxide (NaOH) and potassium hydroxide (KOH) are common laboratory bases. They are water soluble and dissociate as shown by the following equations:

$$NaOH(s) \rightarrow Na^+(aq) + OH^-(aq)$$
$$KOH(s) \rightarrow K^+(aq) + OH^-(aq)$$

The aqueous solution of a traditional base contains only metal cations and hydroxide ions. All such bases are known as strong bases because they are strong electrolytes and their ions are completely dissociated in solution.

A solution that contains OH^- from a soluble base is referred to as **alkaline**. A highly concentrated solution of a strong base is strongly alkaline. Large amounts of heat are evolved when strong bases are dissolved, and the process must be carried out with care.

The word *alkaline* comes from *alkali*, an old name for substances with the general properties of bases. In Chapter 17 it is explained that aqueous alkaline solutions contain a higher concentration of OH^- than H_3O^+.

Neutralization reactions A traditional acid-base reaction is the reaction in aqueous solution between an acid that produces H_3O^+ and a hydroxide-producing base. Suppose a solution containing 1 mol of sodium hydroxide is added to a solution containing 1 mol of hydrochloric acid as in Figure 16-11. The formula equation is:

$$HCl(aq) + NaOH(aq) \rightarrow NaCl(aq) + H_2O(\ell)$$

Hydrochloric acid, sodium hydroxide, and sodium chloride are all strong electrolytes. The ionic and net ionic equations are therefore:

$$H_3O^+(aq) + Cl^-(aq) + Na^+(aq) + OH^-(aq) \rightarrow$$
$$Na^+(aq) + Cl^-(aq) + 2H_2O(\ell)$$
$$H_3O^+(aq) + OH^-(aq) \rightarrow 2H_2O(\ell)$$

The Na^+ and Cl^- are spectator ions. The only participants in the reaction are the hydronium ion and the hydroxide ion. Now we can begin to understand why the properties associated with acids and bases disappear when equivalent amounts react with each other. The H_3O^+ and OH^- ions are fully converted to water. A **neutralization** *is the reaction of hydronium ions and hydroxide ions to form water molecules.*

The net ionic equation for any reaction between a traditional acid and a traditional base shows only H_3O^+, OH^-, and H_2O. Consider another

STUDY HINT
The formula equations show that neutralization reactions are double-replacement reactions, which are introduced in Chapter 8.

Figure 16-11 When hydrochloric acid reacts with lye (sodium hydroxide), they produce a solution of table salt (sodium chloride) and water.

Acids and Bases

Figure 16-12 The color of a salt depends on the elements that compose it. Pictured here are four different salts, each a different color. The salts are, *clockwise from bottom,* copper nitrate, ammonium chromate, nickel sulfate, and potassium ferricyanide.

Set out several examples of salts, such as NaCl (white), $FeCl_3 \cdot 6H_2O$ (yellow), and $Cu(NO_3)_2 \cdot 3H_2O$ (blue). Remind students that some salts are very poisonous.

example—the reaction of a solution containing one mole of sulfuric acid with a solution containing two moles of potassium hydroxide.

$$H_2SO_4(aq) + 2KOH(aq) \rightarrow K_2SO_4(aq) + 2H_2O(\ell)$$
$$2H_3O^+(aq) + SO_4^{2-}(aq) + 2K^+(aq) + 2OH^-(aq) \rightarrow$$
$$2K^+(aq) + SO_4^{2-}(aq) + 4H_2O(\ell)$$
$$H_3O^+(aq) + OH^-(aq) \rightarrow 2H_2O(\ell)$$

The potassium and sulfate ions remain in solution. If the solvent water is evaporated, these ions are no longer separated from each other by water molecules. The cations and anions come together in a crystal, and the solid salt, potassium sulfate (K_2SO_4), precipitates as the solvent evaporates. *A* **salt** *is an ionic compound composed of a metal cation of an aqueous base and the anion from an aqueous acid*. Salts vary in solubility, and a slightly soluble salt would precipitate during a neutralization reaction (see Figure 16-12).

Brønsted bases; Brønsted acid-base reactions Earlier, we used the reaction of hydrogen chloride with aqueous ammonia to illustrate the definition of a Brønsted acid. In this reaction, ammonia is a **Brønsted base**—*a molecule or ion that is a proton acceptor*:

$$HCl(g) + NH_3(aq) \rightarrow NH_4^+(aq) + Cl^-(aq)$$
Brønsted   Brønsted
acid   base

The formation of ammonium chloride as a fine solid powder can be seen in the air when HCl and NH_3 mingle as gases (Figure 16-13).

$$HCl(g) + NH_3(g) \rightarrow NH_4Cl(s)$$

Gaseous ammonia is also a Brønsted base in its reaction with water:

$$H_2O(\ell) + NH_3(g) \rightarrow NH_4^+(aq) + OH^-(aq)$$
Brønsted   Brønsted
acid   base

Figure 16-13 HCl and NH_3 react to form a white cloud of ammonium chloride, NH_4Cl.

In this reaction, water is a proton donor and, therefore, is a Brønsted acid. The reactions of HCl and H_2O with NH_3 are representative Brønsted acid-base reactions. In a **Brønsted acid–base reaction**, *protons are transferred from one reactant (the acid) to another (the base).*

The traditional hydroxide bases such as NaOH are not, strictly speaking, Brønsted bases because, as compounds, they are not proton acceptors. The OH^- ion produced in solution is in each case the Brønsted base.

Lewis bases and Lewis acid–base reactions   A **Lewis base** *is an atom, ion, or molecule that is electron-pair donor.* An anion is a Lewis base in any reaction in which it forms a covalent bond by donating an electron pair, such as the fluoride anion in a typical Lewis acid–base reaction:

$$BF_3(aq) + :\ddot{F}:^-(aq) \rightarrow BF_4^-(aq)$$

A **Lewis acid–base reaction** *is the formation of one or more covalent bonds between an electron-pair donor and an electron-pair acceptor.*

Ammonia is a Lewis base in all reactions in which it forms a covalent bond; for example,

$$H^+(aq) + :NH_3(aq) \rightarrow H-NH_3^+(aq), \text{ or } NH_4^+(aq)$$
$$Ag^+(aq) + 2:NH_3(aq) \rightarrow [H_3N-Ag-NH_3]^+(aq) \text{ or } [Ag(NH_3)_2]^+(aq)$$

A comparison of the three acid–base definitions is given in Table 16-5.

TABLE 16-5 ACID-BASE SYSTEMS

Traditional	
Acid	H^+ or H_3O^+ producer
Base	OH^- producer

Brønsted	
Acid	proton (H^+) donor
Base	proton (H^+) acceptor

Lewis	
Acid	electron-pair acceptor
Base	electron-pair donor

Types of Bases

The bases present in aqueous alkaline solutions may be ionic hydroxides or compounds like ammonia that are Brønsted bases because they produce hydroxide ions by accepting protons from water molecules.

The hydroxides of the active metals in Groups 1 and 2 in the Periodic Table are all considered strong bases because they are strong electrolytes. Those that are water soluble can produce strongly alkaline solutions. The alkalinity of aqueous solutions of metal hydroxides depends upon their solubility and the concentration of OH^- ions produced in solution. It is unrelated to the number of hydroxide ions per mole of the undissolved compound.

Like sodium and potassium hydroxides, calcium hydroxide, $Ca(OH)_2$, is an ionic compound and a strong base that dissolves by the dissociation of its ions. Calcium hydroxide is only slightly soluble in water, however. Calcium hydroxide therefore produces a lower concentration of hydroxide ions in aqueous solution than might be expected from its chemical formula.

$$Ca(OH)_2(s) \rightleftarrows Ca^{2+}(aq) + 2OH^-(aq)$$

The solubilities of a number of metal hydroxides, all traditional bases, are given in Table 16-6. Most are not very soluble, including many hydroxides of the *d*-block metals, as shown in Figure 16-14 for iron(III) hydroxide.

We have seen that ammonia is a weak electrolyte.

$$NH_3(aq) + H_2O(\ell) \rightleftarrows NH_4^+(aq) + OH^-(aq)$$

The concentration of OH^- ions in an ammonia solution is relatively low, and ammonia is a *weak base*. The strong, water-soluble bases (sodium hydroxide, potassium hydroxide) and ammonia are the most common laboratory bases.

TABLE 16-6 SOLUBILITY OF METAL HYDROXIDES

	Hydroxide	Solubility g/100 g H$_2$O at 20°C
soluble (>1 g/100 g H$_2$O)	KOH	112
	NaOH	109
	LiOH	12.4
	Ba(OH)$_2$	3.89
slightly soluble (0.1–1 g/100 g H$_2$O)	Ca(OH)$_2$	0.173
insoluble (<0.1 g/100 g H$_2$O)	Pb(OH)$_2$	0.016
	Mg(OH)$_2$	0.0009
	Sn(OH)$_2$	0.0002
	Zn(OH)$_2$	negligible
	Cu(OH)$_2$	negligible
	Al(OH)$_3$	negligible
	Cr(OH)$_3$	negligible
	Fe(OH)$_3$	negligible

Many organic compounds that contain nitrogen atoms are also weak bases. For example, aniline, a component of many dyes, is a weak base:

$$C_6H_5NH_2(aq) + H_2O(\ell) \rightleftarrows C_6H_5NH_3^+(aq) + OH^-(aq)$$

Anions that have a mild attraction for protons are also weak Brønsted bases. Such anions can produce alkaline solutions in the absence of other substances with which they might ordinarily react. For example, the anion of acetic acid, the acetate ion, is a weak base:

$$C_2H_3O_2^-(aq) + H_2O(\ell) \rightleftarrows HC_2H_3O_2(aq) + OH^-(aq)$$

Figure 16-14 The hydroxides of d-block metals are practically insoluble in water as shown in this reaction.

Section Review

1. What are five general properties of aqueous bases? Name some common substances that fit one or more of these properties.
2. Given the following reactions: NaOH(s) → Na$^+$(aq) + OH$^-$(aq), HF + H$_2$O → F$^-$ + H$_3$O$^+$, H$^+$ + NH$_3$ → NH$_4^+$, which reaction involves a traditional base, a Brønsted base, and a Lewis base? Explain.

1. Bitter taste; feel slippery to the skin; change litmus from pink to blue, phenolphtalein from colorless to red, and methyl orange from orange to yellow; react with acids to produce salts and water; electrolytes
2. The NaOH is the traditional base because it yields OH$^-$ ions in a water solution. H$_2$O is the Brønsted base because it is the proton acceptor, and NH$_3$ is the Lewis acid because it is the electron-pair donor.

16.3 Relative Strengths of Acids and Bases

In the discussion of aqueous acids in Section 16.1, strong acids were defined as acids that ionize completely in water. Weak acids are weak electrolytes that ionize only partially. Relative strengths of acids and bases can be compared when their reactions are analyzed in terms of the Brønsted system.

SECTION OBJECTIVES

- Define conjugate acid, conjugate base, and conjugate acid-base pair.
- Write the formula for the conjugate acid of a base and for the conjugate base of an acid.
- Explain why the conjugate base of a strong acid is a weak base and why the conjugate acid of a strong base is a weak acid.
- Explain why proton-transfer reactions favor the production of the weaker acid and the weaker base.
- Define an amphoteric substance, and give an example.

Brønsted Acid–Base Pairs

The Brønsted definitions of acids and bases provide a basis for the study of proton-transfer reactions. Suppose that a Brønsted acid gives up a proton. The remainder of the acid particle is, in theory, capable of accepting back that proton. This remaining molecule or ion can act as a base and is known as a conjugate base.

A **conjugate base** *is the species that remains after a Brønsted acid has given up a proton.* The fluoride ion is the conjugate base of hydrogen fluoride.

$$\text{HF(aq)} + \text{H}_2\text{O}(\ell) \rightleftharpoons \text{H}_3\text{O}^+(\text{aq}) + \text{F}^-(\text{aq})$$
acid   conjugate base

The acetate ion is the conjugate base of acetic acid.

$$\text{HC}_2\text{H}_3\text{O}_2(\text{aq}) + \text{H}_2\text{O}(\ell) \rightleftharpoons \text{H}_3\text{O}^+(\text{aq}) + \text{C}_2\text{H}_3\text{O}_2^-(\text{aq})$$
acid   conjugate base

In these reactions, the water molecule is a Brønsted base that accepts a proton to form H$_3$O$^+$, which is an acid. A **conjugate acid** *is the species formed when a Brønsted base gains a proton.* Thus, the hydronium ion is the conjugate acid of water.

$$\text{HF(aq)} + \text{H}_2\text{O}(\ell) \rightleftharpoons \text{H}_3\text{O}^+(\text{aq}) + \text{F}^-(\text{aq})$$
       base   conjugate acid

In general, Brønsted acid-base reactions are equilibrium systems involving two interconvertible acid-base pairs, known as conjugate acid-base pairs.

$$\text{HF(aq)} + \text{H}_2\text{O}(\ell) \rightleftharpoons \text{H}_3\text{O}^+(\text{aq}) + \text{F}^-(\text{aq})$$
acid$_1$ base$_2$ acid$_2$ base$_1$

The subscripts designate the two conjugate acid–base pairs, *(1)* HF and F$^-$ and *(2)* H$_3$O$^+$ and H$_2$O. In every conjugate acid–base pair, the acid has one more proton than its conjugate base.

Point out to students that they can recognize conjugate acid-base pairs because one member of the pair has either one more or one less H$^+$ in its formula than has the other member. Remind them that H$^+$ is a proton.

Acids and Bases **475**

Desktop Investigation

Acids and Bases in Your Kitchen

Question
Which substances in the kitchen are acids and which are bases?

Materials
fresh red cabbage
tap water
foods and beverages, such as mayonnaise, baking powder, baking soda, white vinegar, cider vinegar, lemon juice, soft drinks, mineral water, and milk
cleaning products, such as dishwashing liquid, dishwasher detergent, laundry detergent, laundry stain remover, fabric softener, and bleach
blender
strainer
small containers
teaspoon
small beakers

Procedure
Remember to wear safety glasses and an apron. Record all your results in a data table.

1. To make an inexpensive and colorful indicator for acid/base detection, make red-cabbage juice. To do this, drop wedges of red cabbage into a blender with a cup of water. Process. Remove the cabbage from the water by straining. Save the water.

2. Assemble foods, beverages, and cleaning products to be tested.

3. If the substance being tested is a liquid, pour about 5 mL into a small container. If it is a solid, place a teaspoonful into a small beaker, moisten it with about 5 mL of water, wait a few minutes for some of the sample to dissolve, and then pour only the liquid into a small container.

4. Add a drop or two of the red-cabbage juice to the solution being tested and note the color. The solution will turn red if it is a strong acid and green if it is a strong base.

Questions
1. Are cleaning products acids, bases, or neutral?

2. What are acid/base characteristics of foods and beverages?

3. Did you find consumer warning labels on basic or on acidic products?

Unlike hydrofluoric acid, hydrochloric acid is completely ionized. The HCl molecules give up protons very readily and HCl is therefore a strong acid. It follows that the Cl^- ion has little tendency to attract and retain a proton. Consequently, the Cl^- ion is a very weak base.

$$HCl(g) + H_2O(\ell) \rightarrow H_3O^+(aq) + Cl^-(aq)$$
$$\text{acid} \quad \text{base} \quad \text{acid} \quad \text{weak base}$$

This observation leads to an important conclusion that follows naturally from the Brønsted definitions of acids and bases: *the stronger an acid, the weaker its conjugate base; the stronger a base, the weaker its conjugate acid.*

Relative Strengths of Acids and Bases in Chemical Reactions

Let's compare perchloric acid ($HClO_4$) and acetic acid ($HC_2H_3O_2$). Perchloric acid is a strong acid and is highly ionized. Even in concentrated solutions, the reaction to the right is virtually complete:

$$HClO_4(aq) + H_2O(\ell) \rightarrow H_3O^+(aq) + ClO_4^-(aq)$$

stronger acid stronger base weaker acid weaker base

The ClO_4^- ion is the conjugate base of $HClO_4$. It is too weak a base to compete successfully with H_2O as a base in acquiring a proton. The H_3O^+ ion is the conjugate acid of H_2O. It is too weak an acid to compete successfully in donating a proton with the $HClO_4$. There is little tendency, therefore, for the reaction to proceed to the left to recombine the $HClO_4$ and H_2O molecules.

In a comparable solution of acetic acid,

$$HC_2H_3O_2(aq) + H_2O(\ell) \leftarrow H_3O^+(aq) + C_2H_3O_2^-(aq)$$

weaker acid weaker base stronger acid stronger base

the H_3O^+ ion concentration is much lower: acetic acid is a weak acid. The $HC_2H_3O_2$ molecule does not compete very successfully with H_3O^+ ions in donating protons to a base. The H_2O molecules do not compete successfully with $C_2H_3O_2^-$ ions in accepting protons. The H_3O^+ ion is the stronger acid, and the $C_2H_3O_2^-$ ion is the stronger base. Thus, the reaction is more favorable from *right to left* as written above.

Note that in each reaction, the favored direction is toward the weaker acid and weaker base. This obervation leads to a second important conclusion based upon the Brønsted definitions: *proton-transfer reactions favor the production of the weaker acid and the weaker base.*

When a Brønsted acid and a Brønsted base are brought together in aqueous solution, a proton-transfer reaction can occur. The extent of the reaction depends upon the relative strengths of the acids and bases involved. For a proton-transfer reaction to approach completion, the reactants must be much stronger as an acid and a base than the products.

By comparing many different acids and bases, a table of relative acid and base strengths like Table 16-7 can be assembled. Note that the strongest acid listed ($HClO_4$) has the weakest conjugate base, ClO_4^-. The strongest base, the hydride ion (H^-), has the weakest conjugate acid (H_2). A violent proton-transfer reaction could result from bringing together the strongest acid and the strongest base in certain proportions. Such a reaction would be highly exothermic and dangerous. In fact, the reaction between hydride ions and water, a much weaker acid than perchloric acid, is quite vigorous, as shown in Figure 16-15.

To help students determine which direction will be favored in a reaction, they may follow three steps:
1) Identify all compounds in the reaction as either acids or bases.
2) Using Table 16-7, determine which of the two acids is the stronger and which of the two bases is the stronger.
3) The side that is favored is the one having the weaker acid and the weaker base. The weaker acid and the weaker base will always be on the same side.

TABLE 16-7 RELATIVE STRENGTHS OF ACIDS AND BASES

	Conjugate Acid	Formula	Conjugate Base	Formula	
↑ Increasing Acid Strength / ↓ Decreasing Acid Strength	perchloric	$HClO_4$	perchlorate ion	ClO_4^-	↑ Decreasing Base Strength / ↓ Increasing Base Strength
	hydrogen iodide	HI	iodide ion	I^-	
	hydrogen bromide	HBr	bromide ion	Br^-	
	hydrogen chloride	HCl	chloride ion	Cl^-	
	nitric	HNO_3	nitrate ion	NO_3^-	
	sulfuric	H_2SO_4	hydrogen sulfate ion	HSO_4^-	
	hydronium ion	H_3O^+	water	H_2O	
	hydrogen sulfate ion	HSO_4^-	sulfate ion	SO_4^{2-}	
	phosphoric	H_3PO_4	dihydrogen phosphate ion	$H_2PO_4^-$	
	hydrogen fluoride	HF	fluoride ion	F^-	
	acetic	$HC_2H_3O_2$	acetate ion	$C_2H_3O_2^-$	
	carbonic	H_2CO_3	hydrogen carbonate ion	HCO_3^-	
	hydrogen sulfide	H_2S	hydrosulfide ion	HS^-	
	ammonium ion	NH_4^+	ammonia	NH_3	
	hydrogen carbonate ion	HCO_3^-	carbonate ion	CO_3^{2-}	
	water	H_2O	hydroxide ion	OH^-	
	ammonia	NH_3	amide ion	NH_2^-	
	hydrogen	H_2	hydride ion	H^-	

Note in the table that water and hydrogen sulfate ion (HSO_4^-) can be either acids or bases. For example, in the first ionization of sulfuric acid,

$$H_2SO_4(aq) + H_2O(\ell) \rightarrow H_3O^+(aq) + HSO_4^-(aq)$$
$$\text{acid} \qquad \text{base} \qquad \text{acid} \qquad \text{base}$$

the HSO_4^- ion is a base—the conjugate base of H_2SO_4. It could accept a proton from H_3O^+ in the reverse reaction; however, HSO_4^- is such a weak base (Table 16-7) that this reaction occurs only very slightly. In a dilute

Figure 16-15 Calcium hydride reacts violently with water, releasing hydrogen gas.

sulfuric acid solution such as the one described below, the HSO_4^- ion ionizes further, yielding the sulfate ion and the hydronium ion shown in the following reaction.

$$HSO_4^-(aq) + H_2O(\ell) \rightleftarrows H_3O^+(aq) + SO_4^{2-}(aq)$$
$$\text{acid} \qquad \text{base} \qquad \text{acid} \qquad \text{base}$$

Here the HSO_4^- ion is a Brønsted acid. *Any species that can react as either an acid or a base is described as* **amphoteric**.

Section Review

1. The following reaction takes place:

$$H_2CO_3 + H_2O \rightleftarrows HCO_3^- + H_3O^+$$

 (a) Label the conjugate acid–base pairs. *(b)* Label reactants and products shown in the reaction above as *proton donor or proton acceptor* and either acidic or basic.
2. Label reactants and products in the reaction described in the previous question as either stronger or weaker acids or bases. *(b)* Determine which direction, forward or reverse, is favored in the reaction shown in the previous question.

1. (a) One conjugate acid-base pair is H_2CO_3 and HCO_3^- and the other is H_3O^+ and H_2O.
   (b) Proton donors (acidic)
   $$H_2CO_3 + H_2O \rightleftarrows HCO_3^- + H_3O^+$$
   Proton acceptors (basic)
2. (a) $H_2CO_3 + H_2O \rightleftarrows HCO_3^- + H_3O^+$
   acid   base   base   acid
   H_3O^+ is the stronger acid, HCO_3^- is the stronger base
   (b) The reverse reaction is favored.

16.4 Oxides, Hydroxides, and Acids

An oxide is a binary compound of oxygen and one other element. Oxides are known for almost all of the elements. The bonding in these compounds extends over the entire range from ionic to covalent. The relationships among oxides, hydroxides, and acids are discussed in this section.

Basic and Acidic Oxides

Oxides of the active metals are ionic compounds that contain oxide ions, O^{2-}. The oxide ion from any soluble oxide reacts immediately with water to form hydroxide ions.

$$O^{2-}(aq) + H_2O(\ell) \rightarrow 2OH^-(aq)$$

The active metals of Periodic Table Groups 1 and 2 form ionic oxides that react vigorously with water. A large amount of heat is released, and the reaction product is a metal hydroxide. For example,

$$Na_2O(s) + H_2O(\ell) \rightarrow 2NaOH(aq)$$

To the extent that the hydroxide formed is water soluble, the result is the production of an alkaline solution.

An oxide that reacts with water to form an alkaline solution is referred to as a basic oxide, or a **basic anhydride.** The term *anhydride* means "without water." Each basic oxide can be pictured as being formed by the removal of water from a hydroxide base, which is the reverse of the reactions of the oxides with water shown above. Oxides of the less active metals, such as

SECTION OBJECTIVES

- Define *acid anhydride* and *basic anhydride,* and give an example of each.
- Write the equations for the reactions of acid anhydrides and basic anhydrides with water.
- Discuss the general location in the Periodic Table of elements that form acidic, basic, and amphoteric oxides.
- Name some properties that determine whether an OH group will be basic or acidic.

magnesium, can be prepared by thermal decomposition to drive water off from their hydroxides.

$$Mg(OH)_2(s) \xrightarrow{\Delta} MgO(s) + H_2O(g)$$

Hydroxides of the very active metals of Group 1 are so stable that they do not decompose in this manner.

Nonmetals to the right in the Periodic Table form molecular oxides. For example, sulfur forms two oxides that are gases:

$$SO_2 \qquad SO_3$$
sulfur dioxide sulfur trioxide

In reactions typical of nonmetal oxides, each of the sulfur oxides reacts with water to form an oxyacid.

$$SO_2(g) + H_2O(\ell) \rightarrow H_2SO_3(aq)$$
sulfurous acid
$$SO_3(g) + H_2O(\ell) \rightarrow H_2SO_4(aq)$$
sulfuric acid

An oxide that reacts with water to form an acid is known as an acidic oxide, or an **acid anhydride**. As with the basic anhydrides, each acid anhydride can be pictured as being formed by the removal of water from the appropriate oxyacid, which is the reverse of the reactions of the oxides with water shown above. Sulfuric acid, for example, decomposes in this manner when heated.

$$H_2SO_4(aq) \xrightarrow{\Delta} H_2O(\ell) + SO_3(g)$$

> *When industries burn coal that contains sulfur, sulfur dioxide is often released into the atmosphere. Some of the sulfur dioxide is further converted into sulfur trioxide in the atmosphere. These two sulfur oxides then combine with rain and other forms of precipitation to form sulfurous acid and sulfuric acid, which contribute to acid precipitation.*

Amphoteric Oxides and Hydroxides

Table 16-8 lists in Periodic-Table form some common oxides of main group elements. You can see that the active metal oxides are basic and that the nonmetal oxides are acidic. Between these lies a group of oxides, the amphoteric oxides, with intermediate properties. The bonding in amphoteric oxides is intermediate between ionic and covalent bonding. As a result, oxides of this type have behavior intermediate between that of acidic oxides and basic oxides.

Aluminum oxide (Al_2O_3) is a typical amphoteric oxide. It is only slightly soluble in water and dissolves in both acidic and alkaline solutions. With hydrochloric acid, aluminum oxide forms a salt and water.

$$Al_2O_3(s) + 6HCl(aq) \rightarrow 2AlCl_3(aq) + 3H_2O(\ell)$$

> **Acidic oxides react with bases. Basic oxides react with acids.**

With aqueous sodium hydroxide, aluminum oxide forms a soluble ionic compound that contains aluminate ions, AlO_2^-. We will use the AlO_2^- formula rather than the more precise hydrated aluminate formula, $Al(OH)_4^-$.

$$Al_2O_3(s) + 2NaOH(aq) \rightarrow 2NaAlO_2(aq) + H_2O(\ell)$$

Next, consider aluminum hydroxide. In solution, the Al^{3+} ion is surrounded by six water molecules, $[Al(H_2O)_6]^{3+}$. As aqueous base is added to an Al^{3+} solution, the strongly basic OH^- ions attract protons from three of the water molecules of the hydrated aluminum ion. Aluminum hydroxide

Chemistry and Health

"Acid Stomach" and Antacids

Digestion in the stomach involves the gastric juices, hydrochloric acid and enzymes, that split food molecules. The stomach itself is not digested because it is lined with a protective mucus coating. The esophagus, however, is not protected by mucus. When the contents of the stomach splash up into it, you feel the sensation commonly called "heartburn." Sustained anxiety can cause hypersecretion of gastric juices, sometimes leading to ulcers.

To relieve heartburn, people often take antacids, which contain bases. Some antacids produce carbon dioxide as they react with stomach acid:

$$NaHCO_3 + HCl \rightarrow$$
$$NaCl + H_2O + CO_2$$
$$MgCO_3 + 2HCl \rightarrow$$
$$MgCl_2 + H_2O + CO_2$$

The more effective antacids contain hydroxides. Although insoluble, they neutralize acid on contact:

$$Mg(OH)_2 + 2HCl \rightarrow$$
$$MgCl_2 + 2H_2O$$
$$Al(OH)_3 + 3HCl \rightarrow AlCl_3 + 3H_2O$$

They have minor side effects that are often desirable:

magnesium hydroxide is a laxative, and aluminum hydroxide helps alleviate diarrhea. Over-the-counter heartburn remedies are often mixtures of hydroxides, to balance the side effects.

---

separates as a white jelly-like precipitate that includes the remaining water of hydration.

$$[Al(H_2O)_6]^{3+}(aq) + 3OH^-(aq) \rightarrow [Al(H_2O)_3(OH)_3](s) + 3H_2O(\ell)$$

If the addition of base continues, the precipitate dissolves with the formation of hydrated aluminate ions.

$$[Al(H_2O)_3(OH)_3](s) + OH^-(aq) \rightarrow [Al(H_2O)_2(OH)_4]^-(aq) + H_2O(\ell)$$

Writing this equation without the water, as is customary,

$$Al(OH)_3(s) + OH^-(aq) \rightarrow [Al(OH)_4]^-(aq)$$

shows clearly that aluminum hydroxide is soluble in aqueous base. In this reaction, aluminum hydroxide acts as an acid in the presence of a strong base.

Aluminum hydroxide is also soluble in aqueous acid. Written with and without the water of hydration, the equations are:

$$Al(H_2O)_3(OH)_3(s) + 3H_3O^+(aq) \rightarrow Al(H_2O)_6^{3+}(aq) + 3H_2O(\ell)$$
$$Al(OH)_3(s) + 3H_3O^+(aq) \rightarrow Al^{3+}(aq) + 6H_2O(\ell)$$

The three hydroxide ions in the aluminum hydroxide accept three protons

Water of hydration is usually omitted in net ionic equations.

from hydronium ions to form three water molecules. In this reaction, hydrated aluminum hydroxide reacts as a base in the presence of a strong acid.

Compare the equations for the reactions of Al(OH)$_3$ with acid and base with those above for the reactions of aluminum oxide with acid and base. These equations show that aluminum hydroxide, like aluminum oxide, is an amphoteric compound. In the presence of a strong acid, it acts as a base; in the presence of a strong base, it acts as an acid. Each of the elements that form an amphoteric oxide (Table 16-8) also forms an amphoteric hydroxide.

TABLE 16-8 PERIODICITY OF ACIDIC AND BASIC OXIDES OF MAIN GROUP ELEMENTS

Group Number						
1	2	13	14	15	16	17
Li$_2$O basic	BeO amphoteric	B$_2$O$_3$ acidic	CO$_2$ acidic	N$_2$O$_5$ acidic		
Na$_2$O basic	MgO basic	Al$_2$O$_3$ amphoteric	SiO$_2$ acidic	P$_4$O$_{10}$ acidic	SO$_3$ acidic	Cl$_2$O acidic
K$_2$O basic	CaO basic	Ga$_2$O$_3$ amphoteric	GeO$_2$ amphoteric	As$_4$O$_6$ amphoteric	SeO$_3$ acidic	
Rb$_2$O basic	SrO basic	In$_2$O$_3$ basic	SnO$_2$ amphoteric	Sb$_4$O$_6$ amphoteric	TeO$_3$ acidic	I$_2$O$_5$ acidic
Cs$_2$O basic	BaO basic	Tl$_2$O$_3$ basic	PbO$_2$ amphoteric	Bi$_2$O$_3$ basic		

Hydroxides, Acids, and Periodic Trends

Compounds that contain OH groups may be acidic, amphoteric, or basic. When the bond to the oxygen atom in the OH group is ionic, as in the oxides and hydroxides of the active metals, the compound is basic. The metal atoms that form basic hydroxides have large radii, low ionization energies, and low electronegativity. Electrons are easily attracted away from these metal atoms to form cations.

What determines whether a molecular compound containing OH groups will be acidic or amphoteric? For the compound to be acidic, it must be possible for a water molecule to attract away a hydrogen atom from an OH group. This occurs more easily when the O—H bond is very polar.

$$\overset{\delta-}{O}-\overset{\delta+}{H} \qquad \overset{\delta-}{:\!O}\!\!\diagup^{\overset{\delta+}{H}}_{H}$$

Any feature of a molecule that increases the polarity of the O—H bond increases the acidity of a molecular compound. The small, more highly electronegative atoms of nonmetals at the upper right in the Periodic Table form compounds with acidic hydroxyl groups, like the acids shown in Figures 16-5 and 16-6. These nonmetals do not form amphoteric hydroxides. Atoms of intermediate electronegativity and size tend to form amphoteric oxides and hydroxides. Notice the location of the amphoteric oxides in Table 16-8.

The acidic or basic behavior of a compound is also influenced by the number of other oxygen atoms bonded to the atom connected to the OH group. An increase in the number of such oxygen atoms is likely to cause the compound to be more acidic. The electronegative oxygen atoms draw electron density away from the O—H bond and make it more polar. In other words, the higher the oxidation number of the central atom, the more acidic the compound. It may not always be apparent from the formula or structure of a compound whether it has acidic, amphoteric, or basic properties. For example, chromium forms three different compounds containing OH groups:

basic	amphoteric	acidic
$Cr(OH)_2$	$Cr(OH)_3$	H_2CrO_4
chromium(II) hydroxide	chromium(III) hydroxide	chromic acid

Consider also the compounds shown in Figure 16-16. Acetic acid is a weak acid. Notice that in acetic acid, a second oxygen atom is bonded to the same carbon atom connected to the OH group. Ethanol is neither an acid nor a base in aqueous solution.

Figure 16-16 Acetic acid *(top)* and ethanol *(bottom)*

Section Review

1. Using the following equation
$$CaO + H_2O \rightarrow Ca^{2+} + 2OH^-$$
determine whether CaO is a basic or acidic anhydride.
2. Using the following equation
$$CO_2 + 2H_2O \rightarrow HCO_3^- + H_3O^+$$
determine whether CO_2 is a basic or acidic anhydride.
3. Explain why Group 1 metals tend to form basic anhydrides.
4. Explain how the presence of oxygen atoms in a compound may make the compound more acidic.

16.5 Chemical Reactions of Acids, Bases, and Oxides

Compounds with acidic properties and compounds with basic properties often react with each other. The acid-base reactions of traditional, Brønsted, and Lewis acids and bases were illustrated in Section 16.2.

In this section we list some other common reactions of traditional acids and bases as well as oxides. Even though several of these reactions were included in Chapter 8, it is worthwhile to review them here and recognize the similarities among them. One major similarity is that, in each type of reaction listed, one of the products is a salt.

1. Acids react with many metals. Solutions of acids, such as hydrochloric acid and dilute sulfuric acid, react with metals above hydrogen in the metal activity series. (See Table 8-3.) Single-replacement reactions occur

1. A basic anhydride, because OH^- ions are produced
2. An acidic anhydride, because H_3O^+ is produced
3. The active metals of Group 1 form ionic oxides that react immediately with water to form metallic hydroxides. An oxide that reacts with water to form a solution containing hydroxide ions is referred to as a basic anhydride.
4. An increase in the number of oxygen atoms in a compound causes it to be more acidic because the electronegative oxygen atoms pull the electron density away from the O—H bond and make it more polar; thus, the higher the oxidation number of the central atom, the more acidic the compound.

SECTION OBJECTIVES

- Describe four types of reactions involving aqueous acids.
- Describe two types of reactions involving aqueous bases.
- Describe three common types of reactions involving oxides.
- Predict the products of reactions involving aqueous acids, aqueous bases, and oxides.
- List eight types of reactions in which salts can be produced.

Figure 16-17 When concentrated nitric acid oxidizes metallic copper, it produces the poisonous reddish-brown gas nitrogen dioxide and a green solution of copper nitrate *(left)*. Zinc replaces hydrogen in sulfuric acid *(right)*, as can be seen by the formation of hydrogen gas bubbles.

with salts and hydrogen gas being produced. The single-replacement reaction between zinc and dilute sulfuric acid is typical.

$$Zn(s) + H_2SO_4(aq) \rightarrow ZnSO_4(aq) + H_2(g)$$

Hydrochloric acid and dilute sulfuric acid are nonoxidizing acids. Their reactions are those expected of the hydronium ion acting alone. The distinction between nonoxidizing and oxidizing acids is necessary because oxidizing acids react differently. Rather than just providing hydronium ions that react, the entire acid molecule is likely to be involved in the reactions of oxidizing acids. For example, when hot concentrated sulfuric acid (an oxidizing acid) reacts with zinc, some of the sulfuric acid is converted to hydrogen sulfide (H_2S). The other products are zinc sulfate and water.

Notice that hot, concentrated sulfuric acid is an oxidizing acid and that dilute sulfuric acid is *not* an oxidizing acid.

$$4Zn(s) + 5H_2SO_4(conc.) \rightarrow 4ZnSO_4(aq) + H_2S(g) + 4H_2O(\ell)$$

The most common oxidizing acids are hot concentrated sulfuric acid, nitric acid, and the chlorine-containing oxyacids. Nitric acid produces several different products depending upon the acid concentration and the reactants. In both of the following reactions, nitric acid is an oxidizing acid. With dilute nitric acid, the nitrogen-containing product is nitrogen monoxide (NO).

$$3Cu(s) + 8HNO_3(aq) \rightarrow 3Cu(NO_3)_2(aq) + 2NO(g) + 4H_2O(\ell)$$

With concentrated nitric acid, the nitrogen-containing product is nitrogen dioxide (NO_2). (See Figure 16-17.)

$$Cu(s) + 4HNO_3(conc.) \rightarrow Cu(NO_3)_2(aq) + 2NO_2(g) + 2H_2O(\ell)$$

Figure 16-18 When a piece of chalk, made primarily of calcium carbonate, is placed into a container of hydrochloric acid, it produces the salt calcium chloride, water, and carbon dioxide which bubbles out of solution. The yellow color results from a dye in the chalk.

Solutions of oxidizing acids react with metals both above and below hydrogen in the activity series.

2. Acids react with metal oxides. Metal oxides are either basic or amphoteric. In the reaction between an acid and a metal oxide, the metal oxide acts as a basic substance. The products are a salt and water—the same as those in a traditional acid-base reaction. For example, copper oxide reacts with dilute sulfuric acid to produce copper sulfate and water.

$$CuO(s) + H_2SO_4(dil.) \rightarrow CuSO_4(aq) + H_2O(\ell)$$

You may have recognized that this reaction follows the pattern of a double-replacement reaction (Chapter 8).

3. Acids react with carbonates. The products of the reaction of an acid with a carbonate are a salt, water, and carbon dioxide, which bubbles out of the solution as in Figure 16-18. With calcium carbonate the reaction is:

$$CaCO_3(s) + 2HCl(aq) \rightarrow CaCl_2(aq) + CO_2(g) + H_2O(\ell)$$

This reaction provides a simple test for carbonates. A sea shell or other material placed in an acid solution gives the fizzy production of CO_2 if a carbonate is present. The reaction of acids with carbonates plays a role in the deterioration of marble buildings and statues, as shown in Figure 16-19.

4. Hydroxides react with nonmetal oxides. You have seen that nonmetal oxides are acid anhydrides. The reaction of a hydroxide base with a nonmetal oxide is another acid-base reaction. The product is a salt or a salt and water, depending upon the relative quantities of reactants and their identity. For example, two moles of sodium hydroxide and one mole of carbon dioxide form sodium carbonate and water.

$$CO_2(g) + 2NaOH(aq) \rightarrow Na_2CO_3(aq) + H_2O(\ell)$$

One mole of sodium hydroxide and one mole of carbon dioxide produce sodium hydrogen carbonate.

$$CO_2(g) + NaOH(aq) \rightarrow NaHCO_3(aq)$$

Similar reactions between carbon dioxide and calcium hydroxide (Figure 16-20) are as follows:

$$CO_2(g) + Ca(OH)_2(aq) \rightarrow CaCO_3(s) + H_2O(\ell)$$
$$2CO_2(g) + Ca(OH)_2(aq) \rightarrow Ca(HCO_3)_2(aq)$$

5. Metal oxides react with nonmetal oxides. An oxygen-containing salt can be produced by the reaction between a basic metal oxide and an acidic nonmetal oxide. Water is not involved. Instead, the dry oxides are mixed and heated. Metal carbonates, phosphates, and sulfates are typical of salts that can be produced in this manner.

$$MgO(s) + CO_2(g) \rightarrow MgCO_3(s)$$
$$6CaO(s) + P_4O_{10}(s) \rightarrow 2Ca_3(PO_4)_2(s)$$
$$CaO(s) + SO_3(g) \rightarrow CaSO_4(s)$$

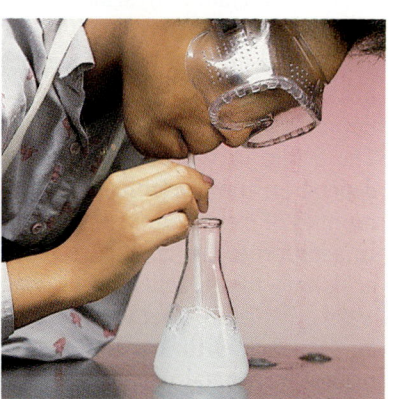

Figure 16-19 Acid precipitation has eroded the features of this marble statue.

Figure 16-20 Limewater *(left)* is a clear aqueous solution of calcium hydroxide, $Ca(OH)_2(aq)$. When carbon dioxide is added *(right)*, $CaCO_3(s)$ forms, making the water cloudy.

Acids and Bases

1. (a) Metal + acid → salt + hydrogen
   (b) Metallic oxide + acid → salt + water
   (c) (Metallic) carbonate + acid → salt + water + carbon dioxide
   (d) Nonmetal oxide + hydroxide → salt + water
   (e) Metallic oxide + nonmetallic oxide → salt
2. (a) $3Ca(s) + 8HNO_3(aq) \rightarrow 3Ca(NO_3)_2(aq) + 2NO(g) + 4H_2O(\ell)$
   (b) $2CO_2(g) + Ba(OH)_2(aq) \rightarrow Ba(HCO_3)_2(aq)$
   (c) $MgCO_3(s) + 2HCl(aq) \rightarrow MgCl_2(aq) + CO_2(g) + H_2O(\ell)$
   (d) $CaO(s) + H_2SO_4(dil.) \rightarrow CaSO_4(aq) + H_2O(\ell)$
   (e) $BaO(s) + SO_3(g) \rightarrow BaSO_4(s)$
   (f) $CO_2(g) + Sr(OH)_2(aq) \rightarrow SrCO_3(s) + H_2O(\ell)$

Section Review

1. Classify the following reactions as one of the five types of reactions discussed in this section. Identify the components of each reaction. (a) $Zn(s) + H_2SO_4(conc.) \rightarrow ZnSO_4(aq) + H_2(g)$ (b) $CuO(s) + H_2SO_4(dil) \rightarrow CuSO_4(aq) + H_2O(\ell)$ (c) $Na_2CO_3(s) + 2HCl(aq) \rightarrow 2NaCl(aq) + H_2O(\ell) + CO_2(g)$ (d) $CO_2(g) + Mg(OH)_2(aq) \rightarrow MgCO_3(s) + H_2O(\ell)$ (e) $CaO(s) + SO_3(g) \rightarrow CaSO_4(s)$.
2. Complete and balance each of the following: (a) $Ca(s) + HNO_3(aq) \rightarrow$ (b) $2CO_2(g) + Ba(OH)_2(aq) \rightarrow$ (c) $MgCO_3(s) + 2HCl(aq) \rightarrow$ (d) $CaO(s) + H_2SO_4(dil) \rightarrow$ (e) $BaO(s) + SO_3(g) \rightarrow$ (f) $CO_2(g) + Sr(OH)_2(aq) \rightarrow$.

Chapter Summary

- In general, acids have a sour taste, contain hydrogen, and react with active metals. They change the color of acid–base indicators, react with bases to produce salts and water, and are electrolytes.
- By definition, a traditional acid contains hydrogen and ionizes in aqueous solution to form hydrogen ions, a Brønsted acid is a proton donor, and a Lewis acid is an electron-pair acceptor.
- Acids are described as monoprotic, diprotic, or triprotic in accordance with whether they donate one, two, or three protons per molecule, respectively, in aqueous solutions.
- In general, bases have a bitter taste, feel slippery to the skin in dilute aqueous solutions, change the color of dyes known as acid–base indicators, react with acids to produce salts and water, and are electrolytes.
- By definition, a traditional base contains hydroxide (OH^-) ions and dissociates to give hydroxide ions in aqueous solution. A Brønsted base is a proton acceptor and a Lewis base is an electron-pair donor.
- The strength of an acid or a base is determined by the extent to which it ionizes or dissociates in aqueous solutions.
- In every Brønsted acid–base reaction, there are two acid-base pairs known as conjugate acid–base pairs.
- A strong acid has a weak conjugate base; a strong base has a weak conjugate acid.
- Proton-transfer reactions favor the production of weaker acids and bases.
- Basic and acidic anhydrides react with water to give alkaline and acidic solutions, respectively.
- Metals that are highly reactive form metal oxides that are basic. Nonmetal oxides are acidic. Oxides of elements between the active metals and nonmetals in the Periodic Table are amphoteric.
- The acidic or basic behavior of a molecule containing OH groups may depend on the electronegativity of other atoms in the molecule, and the number of oxygen atoms bonded to the atom connected to the OH group.
- In addition to the acid–base reactions of traditional, Brønsted, and Lewis acids and bases, acids, bases, and oxides can react in the following five ways: (1) acids react with metals (2) acids react with metal oxides (3) acids react with carbonates (4) hydroxides react with nonmetal oxides (5) metal oxides react with nonmetal oxides.

Chapter 16 Review

Acids and Bases

VOCABULARY

acid, Brønsted (462)
acid, Lewis (463)
acid, traditional (462)
acid anhydride (480)
acid–base reaction, Brønsted (473)
acid–base reaction, Lewis (473)
alkaline (471)
amphoteric (479)
base, Brønsted (472)
base, Lewis (473)
base, traditional (470)
basic anhydride (479)
binary acid (465)
conjugate acid (475)
conjugate base (475)
diprotic acid (464)
hydroxide ion (467)
hydroxyl group (467)
mineral acids (465)
monoprotic acid (463)
neutralization (471)
organic acid (465)
oxyacid (466)
polyprotic acid (464)
salt (472)
strong acid (463)
triprotic acid (464)
weak acid (464)

REVIEWING CONCEPTS

1. **a.** What distinguishes strong acids from weak acids?
   b. Give two examples of each. (16.1)

2. Distinguish among and give one example of a monoprotic, a diprotic, and a triprotic acid. (16.1)

3. **a.** Distinguish between organic and mineral acids.
   b. Give one example of each. (16.1)

4. **a.** Distinguish between binary acids and oxyacids in terms of their component elements and their naming systems.
   b. Give three examples of each. (16.1)

5. Name each of the following acids:
   a. HCl
   b. HI
   c. HNO_3
   d. H_2SO_4
   e. $HClO_3$
   f. HF
   g. HNO_2
   h. $HClO_4$
   i. H_3PO_4
   j. HClO
   k. H_2CO_3 (16.1)

6. Write formulas for each of the following acids:
   a. hydrofluoric acid
   b. nitric acid
   c. hydriodic acid
   d. chloric acid
   e. hydrobromic acid
   f. phosphoric acid
   g. hypochlorous acid
   h. nitrous acid
   i. sulfuric acid
   j. phosphorous acid (16.1)

7. Identify and describe the characteristic properties of five common acids produced in the United States and cite some of the typical uses of each. (16.1)

8. Explain why H_3PO_4, which contains three hydrogen atoms per molecule, is a weak acid, whereas, HCl, which contains only one hydrogen atom per molecule, is a strong acid. (16.1)

9. Explain why both pure HCl gas and HCl dissolved in a nonpolar solvent exhibit no acidic properties in the traditional sense, whereas HCl(aq) does exhibit such properties. (16.1)

10. **a.** What determines the strength of a base?
    b. Give one example each of solutions that are strongly, moderately, and weakly basic. (16.2)

11. Define and give an equation to illustrate:
    a. a conjugate base
    b. a conjugate acid (16.3)

12. What is the relationship between
    a. the strength of an acid and that of its conjugate base
    b. the strength of a base and that of its conjugate acid? (16.3)

13. **a.** What trend can be observed concerning the favored direction of proton-transfer reactions?
    b. What determines the extent to which a proton-transfer reaction occurs? (16.3)

14. **a.** What is meant by the term *amphoteric*?
    b. Give an example of a substance or ion with amphoteric characteristics. (16.3)

Acids and Bases   487

15. For each reaction listed, identify the proton donor (acid), the proton acceptor (base), and label each conjugate acid-base pair:
    a. $HC_2H_3O_2(aq) + H_2O(\ell) \rightleftharpoons H_3O^+(aq) + C_2H_3O_2^-(aq)$
    b. $HCO_3^-(aq) + H_2O(\ell) \rightleftharpoons H_2CO_3(aq) + OH^-(aq)$
    c. $HCN(aq) + SO_4^{2-}(aq) \rightarrow HSO_4^-(aq) + CN^-(aq)$. (16.3)

16. Based on the information given in Table 16-7, determine the following relative to HF, H_2S, HNO_3, and $HC_2H_3O_2$:
    a. the strongest acid
    b. the weakest acid
    c. the strongest conjugate base among those produced by the acids listed
    d. the weakest conjugate base among the four produced. (16.3)

17. Explain why the conjugate base of a strong acid is a weak base and the conjugate acid of a strong base is a weak acid. (16.3)

18. a. Distinguish between basic and acid anhydrides.
    b. In terms of their chemical makeup, how do they differ? (16.4)

19. a. In general, which Periodic-Table elements form acidic, basic, and amphoteric oxides?
    b. Describe some of the properties of a typical amphoteric oxide. (16.4)

20. What properties generally indicate whether compounds containing OH groups will be acidic, amphoteric, or basic? (16.4)

21. a. Distinguish between oxidizing and nonoxidizing acids in terms of their reactions with metals.
    b. Identify two acids of each type. (16.5)

PROBLEMS

▶ **Acid–Base Theory**

1. a. Write the two-step reaction that describes the ionization of sulfuric acid in a dilute aqueous solution.
   b. How do the degrees of ionization in the two steps compare?

2. Dilute HCl(aq) and KOH(aq) are mixed in chemically equivalent quantities. Write
   a. the chemical equation for the neutralization reaction,
   b. the ionic equation, and
   c. the net ionic equation.

3. Repeat Problem 2 with H_3PO_4(aq) and NaOH(aq).

4. Complete the following neutralization reactions. Balance each and then write the complete ionic and net ionic equation for each:
   a. $HCl(aq) + NaOH(aq) \rightarrow$
   b. $HNO_3(aq) + KOH(aq) \rightarrow$
   c. $Ca(OH)_2(aq) + H_2SO_4(aq)$ (very dilute) $\rightarrow$
   d. $Mg(OH)_2(aq) + H_3PO_4(aq)$ (very dilute) $\rightarrow$

5. Write the chemical equation, the ionic equation, and the net ionic equation for the neutralization reaction involving aqueous solutions of H_3PO_4 and $Mg(OH)_2$.

▶ **Reactions of Acids, Bases, and Oxides**

6. Write an equation that illustrates that MgO is a basic anhydride.

7. Write the chemical equation and net ionic equation for each of the following reactions involving nonoxidizing acids:
   a. $Zn(s) + HCl(aq) \rightarrow$
   b. $Al(s) + H_2SO_4(aq) \rightarrow$

8. Write the balanced chemical equation for each of the following reactions between the aqueous acid solution and metal oxide indicated.
   a. $MgO(s) + H_2SO_4(aq) \rightarrow$
   b. $CaO(s) + HCl(aq) \rightarrow$
   c. $Al_2O_3(s) + HNO_3(aq) \rightarrow$
   d. $ZnO(s) + H_3PO_4(aq) \rightarrow$

9. Write the balanced chemical equation for each of the following reactions between an acid and a carbonate:
   a. $BaCO_3(s) + HCl(aq) \rightarrow$
   b. $MgCO_3(s) + HNO_3(aq) \rightarrow$

c. $Na_2CO_3(s) + H_2SO_4(aq) \rightarrow$
d. $CaCO_3(s) + H_3PO_4(aq) \rightarrow$

10. Write the balanced chemical equation for the reaction involving the following hydroxide and nonmetal oxide:
    a. $CO_2(g) + 2KOH(aq) \rightarrow$
    b. $CO_2(g) + KOH(aq) \rightarrow$

11. Write the balanced chemical equation for the reactions involving the following metal oxides and nonmetal oxides:
    a. $CaO(s) + CO_2(g) \rightarrow$
    b. $BaO(s) + P_4O_{10}(s) \rightarrow$
    c. $SrO(s) + SO_3(g) \rightarrow$

12. Write the equation that illustrates each of the following:
    a. that carbon dioxide is the acid anhydride of carbonic acid
    b. that sulfur dioxide is the acid anhydride of sulfurous acid
    c. that calcium oxide is the basic anhydride of calcium hydroxide
    d. that sulfur trioxide is the acid anhydride of sulfuric acid.

13. Write the chemical and net ionic equations for the reaction between $Ca(s)$ and $HCl(aq)$.

▶ **Stoichiometry**

14. Acid precipitation is the term generally used to describe precipitation that is more acidic than normal. Among its many potential causes may be the formation of sulfuric and nitric acids from the various sulfur and nitrogen oxides produced in volcanic eruptions, forest fires, and thunderstorms. In a typical volcanic eruption, for example, 3.50×10^8 kg of SO_2 may be produced. If this amount of SO_2 were converted to H_2SO_4 according to the two-step process given below, how much H_2SO_4 (in kg) would be produced from such an eruption?

$$SO_2 + \tfrac{1}{2}O_2 \rightarrow SO_3$$
$$SO_3 + H_2O \rightarrow H_2SO_4$$

15. Zinc reacts with 100. mL of 6.00 M aqueous sulfuric acid through single replacement.

    a. How much zinc sulfate (in grams) is produced?
    b. How many liters of hydrogen gas would be released at STP?

16. A seashell (largely calcium carbonate) placed in a solution of HCl resulted in the production of 1500. mL of dry CO_2 gas at STP. The other products were $CaCl_2$ and H_2O.
    a. Based on this information, how many grams of $CaCO_3$ were consumed in the reaction?
    b. What volume of 2.00 M HCl solution was used in this reaction?

CRITICAL THINKING

1. **Analyzing Conclusions** Lavoisier experimented with substances called oxides, such as CO_2 and SO_2. He observed that they formed acidic solutions. His observations led him to infer that for a substance to exhibit acid behavior, it must contain oxygen. However, we know today that is not true. Provide evidence to refute Lavoisier's conclusion.

2. **Designing Experiments** Design an experiment that compares three brands of antacids to one another in terms of the speed of reaction and the amount of acid neutralized.

CROSS CONNECTIONS

1. **Economics** Explain how sulfuric acid production serves as a measure of a country's economy.

PROJECT

1. Antacids are designed to neutralize excess hydrochloric acid secreted by the stomach during digestion. Carbonates, bicarbonates, and hydroxides serve as the active ingredients in bringing about the neutralization reactions in the most widely used antacids. Examine the labels of several common antacids and identify the active ingredient(s).

Acids and Bases

CHAPTER 17
Acid-Base Titration and pH

PLANNING CHART

Approximate Time: 7/9 days

	SECTION/TEXT FEATURES		RESOURCES	
B A S I C	**17.1 CONCENTRATION UNITS FOR ACIDS AND BASES** Section Review Chapter Review • Questions 1–3, Problems 1–7	**491** 496 517	*Teacher's Resource Binder* Section Review Worksheets *Other Resources* Transparency 35	60
B A S I C	**17.2 AQUEOUS SOLUTIONS AND THE CONCEPT OF pH** Demo 1 Section Review Chapter Review • Questions 4–6, Problems 8–16	**496** 490B 505 517	*Teacher's Resource Binder* Problem Solving Worksheets Lab Experiment 20 Section Review Worksheets *Other Resources* *Shakhashiri* Video Demos 1, 3, 4, 6 Blackline Master 39, 40, 41	 115 61
B A S I C	**17.3 ACID-BASE TITRATIONS** Demos 2, 3, 4, 5 Desktop Investigation Microscale Investigation 15 Section Review Chapter Review • Questions 7–11, Problems 17–25	**505** 490C 509 866 516 517	*Teacher's Resource Binder* Problem Solving Worksheets Lab Experiments 21, 22 Section Review Worksheets *Other Resources* Transparency 36	 119, 125 62
	CHAPTER END ACTIVITIES Critical Thinking, Cross Connections, Research and Writing, Projects		*Teacher's Resource Binder* Test Booklet *Computer Test Generator*	81, 83

Teaching Strategies

Introduction

The concept of normality is often used in acid-base titrations, and also in other titration situations, such as for oxidation-reduction reactions. However, all of the calculations for titrations can be carried out using the molarity of the solutions instead of normality. Many teachers will want to skip Section 17.1, *Concentration Units for Acids and Bases,* which teaches the concepts of normality and equivalents, and begin with Section 17.2, *Aqueous Solutions and the Concept of pH.* In Section 17.3, *Acid-Base Titrations,* the last part of the section, which deals with normality and titrations, can also be omitted.

17.1 Concentration Units for Acids and Bases

Before students can understand the concept of equivalents, they must have a thorough understanding of the mole concept. Review the concepts of moles and molarity of solutions. Discuss the preparation of solutions such as 1 M NaCl, and 1 M sucrose, and make up solutions of these substances.

Discuss equivalents with the idea that equivalents are amounts of substances that completely react with each other, with no excess of either reactant. Develop the definitions of one equivalent of an acid and one equivalent of a base in proton transfer reactions. Write net ionic equations for the reactions of various strong acids and strong bases: $H_3O^+(aq) + OH^-(aq) \rightarrow 2H_2O(\ell)$, and point out that in these reactions it is only the hydronium and hydroxide ions that are taking part in the reaction, so it is logical to have a concentration unit which deals with just these substances. Have students calculate the equivalent mass of various acids and bases such as HCl, H_2SO_4, HNO_3, H_3PO_4, $Ba(OH)_2$, and KOH.

Define the term normality. Discuss normality of solutions in terms of the H_3O^+ or OH^- molarity, then show how this is the same as equivalents per liter. Use volumetric flasks of various sizes, and ask students how to calculate the mass of acids and bases needed to make up solutions of various normalities. Be sure that students can use the equation: N = equivalents/V (liters), and that they can derive the relationship: equivalents = $N \times V$ and can use it properly.

Illustrate dilution operations for preparing weaker concentrations from $1\,N$ solutions of various acids or bases. Develop the dilution relationship: $N_1V_1 = N_2V_2$. Illustrate that milliliters may be used instead of liters in this equation if both V_1 and V_2 are in milliliters. Have students read the data (percent HCl and density) on a label from a bottle of concentrated HCl. Have them calculate the mass of HCl in each milliliter of the concentrated acid, the molarity and the normality. Compare molarity and normality for various acid solutions. Work through the sample problems with the students.

17.2 Aqueous Solutions and the Concept of pH

Write the equation for self-ionization of water. Demonstrate this reaction using molecular models. Use both red and blue litmus paper in water to show that it is neutral, even though it does contain some H_3O^+ ions and OH^- ions. Using the chalkboard or overhead projector, go through the calculations in the text. Illustrate the use of square brackets to represent concentration in moles per liter: $[H_3O^+] = [OH^-] = 1.0 \times 10^{-7}$ mol/L. Give examples of H_3O^+ concentration and OH^- concentrations for acidic and basic solutions. Show how the ion product for water is calculated. Discuss the effect of the increase of temperature on the amount of ionization of water and thus on the ion product.

Illustrate the use of the ion product of water to calculate H_3O^+ and OH^- concentrations in solutions where the concentrations of strong acid or strong base are given. Show how the exponent for the hydronium ion concentration can be used to indicate an acid (smaller negative exponent than -7) or base (larger negative exponent than -7).

Write these values out in decimal form so that the students can relate the negative exponent to the size of the value. Define pH and show how much easier it is to describe hydronium ion concentration by pH rather than by a decimal value. Use the equation $pH = -\log[H_3O^+]$ to calculate the pH of pure water. Discuss the magnitude of the value of pH for acidic and basic solutions. Stress the idea that a pH change of 2 pH units means a change in hydronium and hydroxide ion concentrations of 10^2, or 100 fold. A pH change of 5 means a hydronium ion concentration change of 10^5, or 100 000 fold.

Many students will be intimidated by the use of a logarithmic function and feel that understanding pH is beyond their capabilities. Use the expression $[H_3O^+] = 10^{-pH}$ to help them reach an understanding of the relation between the size of the negative exponent, the value of pH, and the concentration that it expresses. Work with hydronium ion concentrations that are integral powers of 10. More able students should be able to handle the calculation of pH when H_3O^+ concentration is not an integral power of 10, but they will need practice and help in figuring out how to use their calculators. Go through some of the calculations with them as they push each key on their calculators.

Calculate the pH of solutions of strong acids and bases from the molar concentration. Determine both hydronium ion and hydroxide ion concentrations in these. Some students might find the concept of pOH useful: $pOH = -\log[OH^-]$. The relation between pH and pOH is: $pH + pOH = 14$.

Ask students to bring samples of various substances and check their pH using pH indicator paper. Alternatively, give them some pH paper, and ask them to report on the pH of various substances they find at home. Calculate the hydronium and hydroxide ion concentrations based on the pH values, and discuss the acidity or basicity of the products.

Chemistry and the Environment: Acid Precipitation and Lakes

Ask students to find current articles which deal with the topic of acid precipitation. Discuss the source of the acids that cause acid precipitation. Ask students to calculate the hydronium ion concentration of a lake when the pH is 5.0 or 4.5

Demonstration 1: Conductivity of Water

PURPOSE: To show the small amount of conductivity in distilled water
MATERIALS: Conductivity detector using a neon bulb or a 25-W bulb, or a sensitive conductivity meter (from the physics department), distilled water, $6\,M$ HCl
PROCEDURE: Wear goggles. Be careful to avoid shocks if you use a light-bulb conductivity detector. Demonstrate that distilled water has a measurable, but very small conductivity (sufficient to light the neon bulb, but not the 25-W bulb). Add a few drops of HCl solution to illustrate the change when a strong acid ionizes in water. Relate the low conductivity to the self-ionization of water.

17.3 Acid–Base Titrations

Show as many indicators as are available to the students, and demonstrate their color changes. Discuss the transition interval. Illustrate the four types of acid—base combinations and discuss the choice of indicator for each situation. Explain the mechanism by which indicators operate in terms of the weak acid dissociating in water. Demonstrate the use of pH indicator paper, and compare the color changes with universal indicator. Mix several indicators together to create a universal indicator: for example, combine methyl red, methyl orange, bromphenol blue and phenolphthalein. Extract indicators from natural products such as red cabbage, tea, blueberries, red roses, or from any flower petal and test for their color changes and transition intervals. Use a pH meter to determine the pH of a solution and compare the pH with that obtained using pH indicator paper or other indicators.

Discuss titration as a common technique used by analytical chemists to find the quantity of a substance present in a sample. Demonstrate the technique of titration using burets and common acids and bases. Discuss the necessity for a standard substance in the titration procedure, and define a standard solution. Mention the characteristics of substances which make them acceptable as standards for an acid or base, and show how a solution can be "standardized" using such a substance.

The calculations for titrations can be carried out using concentrations expressed as normality or molarity. If molarity is selected as the concentration unit for the acid and base, show how titration calculations are made. Point out the four steps necessary: find the moles of substance used in the titration, find mole ratio of acid and base from a balanced equation, find the moles of the unknown substance, and calculate the molarity of the unknown solution. Work through the sample problems in the text with the students. Steps 1, 3 and 4 can be combined in one sequence of unit conversion factors, using molarity as the conversion between moles and volume of a substance.

The choice of normality as a concentration unit simplifies the titration calculation, but the concept of normality is more difficult to understand than that of molarity. Demonstrate the use of the equation of calculation using normality: $V_a N_a = V_b N_b$. Work through the computations shown in the text.

There are a number of substances that students might like to analyze themselves by means of acid-base titrations. Some suggestions are to determine the amount of acid in orange juice, to analyze a vitamin C tablet to find the amount of ascorbic acid present, or to determine the amount of ammonia in a household cleaning product.

Testing a commercial product helps students appreciate that chemistry has relevance in their lives. Ask them to bring various brands of antacid tablets to analyze. Try to choose tablets that are white so that the indicator color is easy to see. Students can rank the tablets in relation to the amount of acid they neutralize by adding a few mL of cabbage juice indicator to each crushed tablet, which will give a distinct green color, and then dropping 5% white vinegar (0.837 M acetic acid) onto each tablet. After calibrating a dropper for drops per 10 mL, count and record the number of drops required to change the color of the indicator to red, its acid color. A fairly close measure of the moles of acid neutralized by the tablets can be obtained. Check the labels to see what substances are used in antacid tablets, and write the neutralization reactions. Many tablets will contain an acid-base buffer to keep the pH fairly constant.

Desktop Investigation: Testing for Acid Precipitation

The topic of acid precipitation is one that most students have heard of, but many will be surprised to find that they can determine if the rain in their area is acidic by means of a simple technique. Ask students to collect rain from the area where they live and at their school and compare the pH of rainwater from different sites. Make a chart of the pH of the rainwater in the local area.

ANSWERS TO DISCUSSION QUESTIONS:
1. 7.0 ± 0.2
2. 5.7. Dissolved carbon dioxide makes it slightly acidic.
3. Any precipitation having a pH below 5.6
4. Displacement of water by the ruler will affect measurement.
5. Determination must be done after experiment is carried out.
6. If pH is >5.6, no acid precipitation; <5.6, acid precipitation.

Demonstration 2: Behavior of Indicators
PURPOSE: To show the color changes that indicators undergo at various pH values
MATERIALS: solutions of several indicators: methyl orange, methyl red, bromphenol blue, phenolphthalein, thymolphthalein, universal indicator, or other indicators made from flowers, red cabbage, blueberries, etc.; HCl solution, 0.1 M; NaOH solution, 0.1 M; graduated cylinder, 100-mL; beakers and test tubes; distilled water
PROCEDURE: Wear goggles. Pour 100 mL of 0.1 M HCl into a beaker and label it pH 1. Combine 10 mL of 0.1 M HCl and 90 mL of distilled water, and put into a beaker marked pH 2. Take 10 mL of this solution (0.01 M HCl), mix with 90 mL of distilled water to form a solution with pH 3. Continue the sequential dilution, forming solutions with pH values of 4, 5, and 6. Use distilled water for pH 7. Dilute the 0.1 M NaOH solution in a similar fashion to prepare solutions with pH values from 13 to 8. Pour small amounts of these standard pH solutions into a series of test tubes marked with the proper pH, and add the same amount of indicator to each. Note the pH where the color change occurs, and the transition interval for the indicators.

Demonstration 3: Indicator Changes with Carbon Dioxide
PURPOSE: To provide a dramatic demonstration of indicator color changes
MATERIALS: 12 large beakers, 6 different indicators such as phenolphthalein, bromphenol blue, universal indicator, methyl orange, etc. dry ice, distilled water, 0.1 M NH$_3$(aq)
PROCEDURE: Wear goggles. Fill each beaker about three-fourths full of distilled water. Add phenolphthalein to two beakers and add enough NH$_3$(aq) to cause a definite color change to pink. Add the same amount of NH$_3$(aq) to all the other beakers. Set the beakers out in pairs of two, and add a different indicator to each. Then place a chunk of dry ice in one beaker of each pair. The dry ice will sublime, forming clouds of vapor, and the beaker will appear to be boiling. The slow production of acid will gradually cause the indicators to change color. The second beaker in each set is a control to show the color of the indicator in basic solution. The reaction is:

$$CO_2(g) + H_2O(\ell) \rightleftarrows H_2CO_3(aq) \rightleftarrows H_3O^+(aq) + HCO_3^-(aq)$$

Demonstration 4: A Breathtaking Color Change
PURPOSE: To show how carbon dioxide from one's breath can cause an indicator color change
MATERIALS: flasks, 250-mL or 500-mL; long soda straws, 6 M NH$_3$(aq), phenol red indicator solution or bromthymol blue indicator solution
PROCEDURE: Fill the flasks about two-thirds full of distilled water. Add one drop of 6 M NH$_3$(aq) and sufficient indicator to give an obvious color. Give student volunteers goggles to wear, and ask them to "blow the color away." Be sure that they blow into the flask, and do not drink the liquid. The carbon dioxide from their breath will form an acidic solution that will cause a color change in the indicator.

Demonstration 5: pH Change of an Acid-Base Titration
PURPOSE: To show how pH changes during an acid-base titration
MATERIALS: pH meter, buret, beaker, 0.1 M HCl, 0.1 M NaOH
PROCEDURE: Wear goggles. Carefully measure 25.0 mL of 0.1 M HCl into the beaker. Connect the pH meter, and record the pH of the solution. Continue to record the pH as 0.1 M NaOH is added and graph the resulting titration curve. Point out the very large change in pH that results when one drop of titrant is added at the equivalence point.

References and Resources

Borgford, Christie L. and Lee R. Summerlin, *Chemical Activities*. American Chemical Society, 1988.

Forster, M. "Plant Pigments as Acid–Base Indicators," *Journal of Chemical Education*, Vol. 55, No. 2, 1978, p. 107.

Kolb, D. "The pH Concept," *Journal of Chemical Education*, Vol. 56, No. 1, 1979, p. 49.

Mebane, R. and T. Rybolt, "Edible Acid–Base Indicators," *Journal of Chemical Education*, Vol. 62, No. 4, 1985, p. 285.

Schultz, C.W. and S.I. Spannuth. "Comparison of Strong and Weak Titration Curves," *Journal of Chemical Education*, Vol. 56, No. 3, 1979, p. 194.

Audiovisual Resources

"Acid–Base Indicators," CHEM Study 16mm film, Ward's Modern Learning Aids. Shows how a universal indicator is made and determines equilibrium constant for four indicators.

"Indicators and pH," University of Akron, Audio Visual Services.

Ward's Solo-Learn: 78-0700, "Introduction to Chemical Equivalents", Ward's Modern Learning Aids. A programmed film strip to teach chemical equivalents.

WCC 804A "Using Simple Logarithms, Logs and Antilogs," WCC 804B "Using Simple Logarithms, Multiplication and Division," WCC 813A "Titration Techniques, Part I," WCC 897 "Dilution and Titration Problems," WCC 815 "pH Measurements," WCC 866 "Titration Curves and Indicators," WCC 883 "pH, pOH and Buffers," filmstrips/cassettes, Prentice Hall Media.

Answers and Solutions

Reviewing Concepts

1. **a.** The mass of one equivalent of an acid is equal to the molar mass of that acid divided by the number of moles of H_3O^+ ions that can be supplied by one mole of the acid.
   b. One equivalent of a monoprotic acid, such as HCl, is equal to one mole of that acid. One equivalent of a diprotic acid, such as H_2SO_4, is equal to one-half mole of the acid. One equivalent of a triprotic acid, such as H_3PO_4, is equal to one-third mole of the acid.

2. **a.** The mass of one equivalent of a base is equal to the molar mass of that base divided by the number of OH^- ions supplied by each mole of that base.
   b. One mole of KOH supplies one equivalent of OH^- ions, one mole of $Ba(OH)_2$ supplies two equivalents of OH^- ions, and one mole of $Al(OH)_3$ supplies three equivalents of OH^- ions.

3. **a.** One equivalent of solute per liter of solution
   b. Three equivalents of solute per liter of solution
   c. 0.5 equivalent of solute per liter of solution

4. **a.** The negative of the common logarithm of the hydronium ion concentration
   b. $pH = -\log[H_3O^+]$
   c. The power to which ten must be raised to give the number. (Example: the log of 10^7 is 7.)

5. **a.** Neutral **b.** Basic **c.** Neutral **d.** Acidic
   e. Neutral **f.** Acidic **g.** Basic

6. **a.** 2.8 **b.** 3.1 **c.** 4.2
   d. 6.5 **e.** 3.1 **f.** 5.8
   g. 2.3 **h.** 10.5 **i.** 2.0

7. **a.** The end point of a titration is the equivalence point, or the point where equal amounts of H_3O^+ and OH^- ions are present.
   b. At the end point of the titration, a very rapid change in pH occurs.
   c. If the indicator selected changes color within the pH range in which the end point of the titration is reached, then it can provide visible evidence that neutralization has occurred.

8. **a.** The transition interval of an indicator is the pH range over which the indicator color change occurs.
   b. The choice of the indicator is based on the suitability of its transition interval for a given acid-base reaction.

9. • Strong-acid/strong-base: pH is about 7, and litmus or bromthymol blue would be suitable indicators.
   • Strong-acid/weak-base: pH is below 7, and methyl orange is a suitable indicator.
   • Weak-acid/strong-base: pH is above 7, and phenolphthalein is a suitable indicator.
   • Weak-acid/weak-base: pH may be either above or below 7, depending upon the reactant that is stronger. No indicator is satisfactory.

10. Through the use of a pH meter.

11. The pH is approximately 8.

Problems

1. **a.** $1 \text{ equiv HBr} = \dfrac{80.9 \text{ g HBr}}{\text{mol HBr}} \times \dfrac{1 \text{ mol HBr}}{1 \text{ mol H}_3\text{O}^+} = 80.9 \text{ g HBr/mol H}_3\text{O}^+$

   b. $1 \text{ equiv H}_2\text{S} = \dfrac{34.1 \text{ H}_2\text{S}}{\text{mol H}_2\text{S}} \times \dfrac{1 \text{ mol H}_2\text{S}}{2 \text{ mol H}_3\text{O}^+} = 17.0 \text{ g H}_2\text{S/mol H}_3\text{O}^+$

   c. $1 \text{ equiv H}_3\text{PO}_4 = \dfrac{98.0 \text{ g H}_3\text{PO}_4}{\text{mol H}_3\text{PO}_4} \times \dfrac{1 \text{ mol H}_3\text{PO}_4}{3 \text{ mol H}_3\text{O}^+} = 32.7 \text{ g H}_3\text{PO}_4/\text{mol H}_3\text{O}^+$

2. a. $1 \text{ equiv LiOH} = \dfrac{23.9 \text{ g LiOH}}{\text{mol LiOH}} \times \dfrac{1 \text{ mol LiOH}}{1 \text{ mol OH}^-} = 23.9 \text{ g LiOH/mol OH}^-$

b. $1 \text{ equiv Sr(OH)}_2 = \dfrac{122 \text{ g Sr(OH)}_2}{\text{mol Sr(OH)}_2} \times \dfrac{1 \text{ mol Sr(OH)}_2}{2 \text{ mol OH}^-} = 61.0 \text{ g Sr(OH)}_2/\text{mol OH}^-$

c. $1 \text{ equiv Al(OH)}_3 = \dfrac{78.0 \text{ g Al(OH)}_3}{1 \text{ mol Al(OH)}_3} \times \dfrac{1 \text{ mol Al(OH)}_3}{3 \text{ mol OH}^-} = 26.0 \text{ g Al(OH)}_3/\text{mol OH}^-$

3. a. $\dfrac{1.0 \text{ equiv HNO}_3}{\text{L}} \times 1.0 \text{ L} = 1.0 \text{ equiv HNO}_3$

b. $\dfrac{2.5 \text{ equiv NaOH}}{\text{L}} \times 1.0 \text{ L} = 2.5 \text{ equiv NaOH}$

c. $\dfrac{1.0 \text{ equiv Mg(OH)}_2}{\text{L}} \times 2.0 \text{ L} = 2.0 \text{ equiv Mg(OH)}_2$

d. $\dfrac{3.0 \text{ equiv HCl}}{\text{L}} \times 1.5 \text{ L} = 4.5 \text{ equiv HCl}$

e. $\dfrac{1.20 \text{ equiv K}_2\text{CO}_3}{\text{L}} \times 250. \text{ mL} \times \dfrac{\text{L}}{10^3 \text{ mL}}$
$= 0.300 \text{ equiv K}_2\text{CO}_3$

4. $\dfrac{1.20 \text{ equiv H}_2\text{SO}_4}{\text{L}} \times 500. \text{ mL} \times \dfrac{\text{L}}{10^3 \text{ mL}} = 0.600 \text{ equiv H}_2\text{SO}_4; \quad \dfrac{\text{L}}{1.50 \text{ equiv H}_2\text{SO}_4} \times 0.600 \text{ equiv H}_2\text{SO}_4$
$\times \dfrac{10^3 \text{ mL}}{\text{L}} = 400. \text{ mL}$

5. $\dfrac{0.200 \text{ mol H}_3\text{PO}_4}{\text{L}} \times 2.50 \text{ L} \times \dfrac{3 \text{ equiv H}_3\text{PO}_4}{1 \text{ mol H}_3\text{PO}_4} = 1.50 \text{ equiv H}_3\text{PO}_4; \quad \dfrac{\text{L}}{0.750 \text{ equiv H}_3\text{PO}_4} \times 1.50 \text{ equiv H}_3\text{PO}_4 = 2.00 \text{ L}$

6. a. $1.00 \text{ L} \times \dfrac{2.00 \text{ equiv HNO}_3}{\text{L}} \times \dfrac{1 \text{ mol HNO}_3}{1 \text{ equiv HNO}_3} \times \dfrac{63.0 \text{ g HNO}_3}{\text{mol HNO}_3} \times \dfrac{\text{mL conc. acid}}{1.42 \text{ g conc. acid}} \times \dfrac{100. \text{ g conc. acid}}{69.5 \text{ g HNO}_3}$
$= 128 \text{ mL conc. acid}$

b. $150. \text{ mL} \times \dfrac{\text{L}}{10^3 \text{ mL}} \times \dfrac{0.750 \text{ equiv HNO}_3}{\text{L}} \times \dfrac{1 \text{ mol HNO}_3}{1 \text{ equiv HNO}_3} \times \dfrac{63.0 \text{ g HNO}_3}{\text{mol HNO}_3} \times \dfrac{\text{mL conc. acid}}{1.42 \text{ g conc.acid}} \times \dfrac{100. \text{ g conc. acid}}{69.5 \text{ g HNO}_3}$
$= 7.18 \text{ mL conc. acid}$

c. $250. \text{ mL} \times \dfrac{\text{L}}{10^3 \text{ mL}} \times \dfrac{0.500 \text{ equiv HNO}_3}{\text{L}} \times \dfrac{1 \text{ mol HNO}_3}{1 \text{ equiv HNO}_3} \times \dfrac{63.0 \text{ g HNO}_3}{\text{mol HNO}_3} \times \dfrac{\text{mL conc. acid}}{1.42 \text{ g conc. acid}}$
$\times \dfrac{100. \text{ g conc. acid}}{69.5 \text{ g HNO}_3} = 7.98 \text{ mL conc. acid}$

7. a. $\dfrac{36.5 \text{ g HCl}}{\text{mol HCl}} \times \dfrac{1 \text{ mol HCl}}{1 \text{ equiv}} \times \dfrac{100 \text{ g conc. acid}}{37.5 \text{ g HCl}} \times \dfrac{\text{mL conc. acid}}{1.19 \text{ g conc. acid}} = 81.8 \text{ mL conc. acid/equiv}$

b. $\dfrac{81.8 \text{ mL conc. acid}}{\text{equiv}} \times \dfrac{1.00 \text{ equiv}}{\text{L dil soln}} \times 2.00 \text{ L dil soln} = 164 \text{ mL conc acid}$

c. $\dfrac{81.8 \text{ mL conc acid}}{\text{equiv}} \times \dfrac{0.750 \text{ equiv}}{\text{L dil soln}} \times 3.00 \text{ L dil soln} = 184 \text{ mL conc acid}$

8. $\text{pH} = -\log[\text{H}_3\text{O}^+]$
  a. $\text{pH} = -\log(1 \times 10^{-7}) = -(-7) = 7.0$ **c.** $\text{pH} = -\log(1 \times 10^{-12}) = -(-12) = 12.0$
  b. $\text{pH} = -\log(1 \times 10^{-3}) = -(-3) = 3.0$ **d.** $\text{pH} = -\log(1 \times 10^{-5}) = 5.0$

9. $\text{pH} = -\log[\text{H}_3\text{O}^+]$
  a. $\text{pH} = -\log(0.01) = -\log(1 \times 10^{-2}) = -(-2) = 2.0$
  b. $\text{pH} = -\log(0.001) = -\log(1 \times 10^{-3}) = -(-3) = 3.0$
  c. $\text{pH} = -\log(0.00001) = -\log(1 \times 10^{-5}) = -(-5) = 5.0$
  d. $\text{pH} = -\log(0.0001) = -\log(1 \times 10^{-4}) = -(-4) = 4.0$

10. a. Since pH $= -\log[H_3O^+]$, first find $[H_3O^+]$.
$[H_3O^+][OH^-] = 1 \times 10^{-14}$

$[H_3O^+] = \dfrac{1 \times 10^{-14}}{[OH^-]} = \dfrac{1 \times 10^{-14}}{1 \times 10^{-6}} = 1 \times 10^{-8}$

Next, find pH:
pH $= -\log[H_3O^+] = -\log(1 \times 10^{-8}) = -(-8)$
$= 8.0$

b. $[H_3O^+] = \dfrac{1 \times 10^{-14}}{[OH^-]} = \dfrac{1 \times 10^{-14}}{1 \times 10^{-9}} = 1 \times 10^{-5}$

pH $= -\log[H_3O^+] = -\log(1 \times 10^{-5}) = -(-5)$
$= 5.0$

c. $[H_3O^+] = \dfrac{1 \times 10^{-14}}{[OH^-]} = \dfrac{1 \times 10^{-14}}{1 \times 10^{-2}} = 1 \times 10^{-12}$

pH $= -\log[H_3O^+] = -\log(1 \times 10^{-12}) = -(-12)$
$= 12.0$

d. $[H_3O^+] = \dfrac{1 \times 10^{-14}}{[OH^-]} = \dfrac{1 \times 10^{-14}}{1 \times 10^{-7}} = 1 \times 10^{-7}$

pH $= -\log[H_3O^+] = -\log(1 \times 10^{-7}) = -(-7)$
$= 7.0$

11. $[H_3O^+][OH^-] = 1 \times 10^{-14}$

$[H_3O^+] = \dfrac{1 \times 10^{-14}}{[OH^-]}$

Now continue with parts **a–c** and then solve for pH.

a. $[H_3O^+] = \dfrac{1 \times 10^{-14}}{0.01} = \dfrac{1 \times 10^{-14}}{1 \times 10^{-2}} = 1 \times 10^{-12}$

pH $= -\log[H_3O^+] = -\log(1 \times 10^{-12}) = -(-12)$
$= 12.0$

b. $[H_3O^+] = \dfrac{1 \times 10^{-14}}{0.001} \times \dfrac{1 \times 10^{-14}}{1 \times 10^{-3}} = 1 \times 10^{-11}$

pH $= -\log[H_3O^+] = -\log(1 \times 10^{-11}) = -(-11)$
$= 11.0$

c. $[H_3O^+] = \dfrac{1 \times 10^{-14}}{0.0001} \times \dfrac{1 \times 10^{-14}}{1 \times 10^{-4}} = 1 \times 10^{-10}$

pH $= -\log[H_3O^+] = -\log(1 \times 10^{-10}) = -(-10)$
$= 10.0$

12. a. pH $= -\log[H_3O^+] = -\log(2.0 \times 10^{-5})$
$= -(\log 2.0 + \log 10^{-5}) = -(0.30 + (-5))$
$= -(0.30 - 5) = -(-4.70) = 4.70$

b. pH $= -\log[H_3O^+] = -\log(4.7 \times 10^{-7})$
$= -(\log 4.7 + \log 10^{-7}) = -(0.67 + (-7))$
$= -(0.67 - 7) = -(-6.33) = 6.33$

c. pH $= -\log[H_3O^+] = -\log(3.8 \times 10^{-3}) =$
$-(\log 3.8 + \log 10^{-3}) = -(0.58 + (-3)) =$
$-(0.58 - 3) = -(-2.42) = 2.42$

13. pH $= -\log[H_3O^+]$
$\log[H_3O^+] = -\text{pH}$
$[H_3O^+] = \text{antilog}(-\text{pH})$
a. $[H_3O^+] = \text{antilog}(-3) = 10^{-3}$
b. $[H_3O^+] = \text{antilog}(-7) = 10^{-7}$
c. $[H_3O^+] = \text{antilog}(-11) = 10^{-11}$
d. $[H_3O^+] = \text{antilog}(-5) = 10^{-5}$

14. pH $= -\log[H_3O^+]$
$\log[H_3O^+] = -\text{pH}$
$[H_3O^+] = \text{antilog}(-\text{pH})$
When $[H_3O^+]$ is known, you can find $[OH^-]$ by substituting into the equation: $[H_3O^+][OH^-] = 10^{-14}$, which becomes

$[OH^-] = \dfrac{10^{-14}}{[H_3O^+]}$

a. $[H_3O^+] = \text{antilog}(-\text{pH}) = \text{antilog}(-7) = 10^{-7}$

$[OH^-] = \dfrac{10^{-14}}{10^{-7}} = 10^{-7}$

b. $[H_3O^+] = \text{antilog}(-\text{pH}) = \text{antilog}(-11) = 10^{-11}$

$[OH^-] = \dfrac{10^{-14}}{10^{-11}} = 10^{-3}$

c. $[H_3O^+] = \text{antilog}(-\text{pH}) = \text{antilog}(-4) = 10^{-4}$

$[OH^-] = \dfrac{10^{-14}}{10^{-4}} = 10^{-10}$

d. $[H_3O^+] = \text{antilog}(-\text{pH}) = \text{antilog}(-6) = 10^{-6}$

$[OH^-] = \dfrac{10^{-14}}{10^{-6}} = 10^{-8}$

15. pH $= -\log[H_3O^+]$ $\log[H_3O^+] = -\text{pH}$ $[H_3O^+] = \text{antilog}(-\text{pH})$
a. $[H_3O^+] = \text{antilog}(-\text{pH}) = \text{antilog}(-4.23) = \text{antilog}(0.77-5) = 5.9 \times 10^{-5}$
b. $[H_3O^+] = \text{antilog}(-\text{pH}) = \text{antilog}(-7.65) = \text{antilog}(0.35-8) = 2.2 \times 10^{-8}$
c. $[H_3O^+] = \text{antilog}(-\text{pH}) = \text{antilog}(-9.48) = \text{antilog}(0.52-10) = 3.3 \times 10^{-10}$

16. a. $[H_3O^+] = \text{antilog}(-\text{pH}) = \text{antilog}(-2.70) = \text{antilog}(0.30-3) = 2.0 \times 10^{-3}$
b. $[H_3O^+][OH^-] = 1 \times 10^{-14}$

$[OH^-] = \dfrac{1.0 \times 10^{-14}}{[H_3O^+]} = \dfrac{1.0 \times 10^{-14}}{2.0 \times 10^{-3}} = 5.0 \times 10^{-12}$

c. Since HNO_3 is a monoprotic acid giving 1 mol of H_3O^+ ions per mol of HNO_3 when completely ionized in solution, $2.0 \times 10^{-3}\,M\,H_3O^+$ was, therefore, provided by a $2.0 \times 10^{-3}\,M$ solution. Thus, the HNO_3 solution is $2.0 \times 10^{-3}\,M$.

$$\frac{2.0 \times 10^{-3}\,\text{mol HNO}_3}{L} \times 5.50\,L \times \frac{63\,\text{g HNO}_3}{\text{mol HNO}_3} = 0.69\,\text{g HNO}_3$$

d. $0.69\,\text{g HNO}_3 \times \dfrac{100.\,\text{g conc. acid}}{69.5\,\text{g HNO}_3} \times \dfrac{\text{mL conc. acid}}{1.42\,\text{g conc. acid}} = 0.70\,\text{mL conc. acid}$

17. a. $\begin{array}{cccc}HCl & + & NaOH \rightarrow NaCl & + & H_2O \\ 1\,\text{mol} & & 1\,\text{mol}\quad\;\;1\,\text{mol} & & 1\,\text{mol}\end{array}$

$1.0\,\text{mol HCl} \times \dfrac{1\,\text{mol NaOH}}{1\,\text{mol HCl}} = 1.0\,\text{mol NaOH}$

b. $\begin{array}{cccc}HNO_3 & + & KOH \rightarrow KNO_3 & + & H_2O \\ 1\,\text{mol} & & 1\,\text{mol}\quad\;\;1\,\text{mol} & & 1\,\text{mol}\end{array}$

$0.75\,\text{mol KOH} \times \dfrac{1\,\text{mol HNO}_3}{1\,\text{mol KOH}} = 0.75\,\text{mol HNO}_3$

c. $\begin{array}{cccc}2HF & + & Ba(OH)_2 \rightarrow BaF_2 & + & 2H_2O \\ 2\,\text{mol} & & 1\,\text{mol}\qquad 1\,\text{mol} & & 2\,\text{mol}\end{array}$

$0.20\,\text{mol HF} \times \dfrac{1\,\text{mol Ba(OH)}_2}{2\,\text{mol HF}} = 0.10\,\text{mol Ba(OH)}_2$

d. $\begin{array}{cccc}3H_2SO_4 & + & 2Al(OH)_3 \rightarrow Al_2(SO_4)_3 & + & 6H_2O \\ 3\,\text{mol} & & 2\,\text{mol}\qquad 1\,\text{mol} & & 6\,\text{mol}\end{array}$

$0.60\,\text{mol Al(OH)}_3 \times \dfrac{3\,\text{mol H}_2SO_4}{2\,\text{mol Al(OH)}_3} = 0.90\,\text{mol H}_2SO_4$

18. $\begin{array}{cccc}H_2SO_4 & + & 2KOH \rightarrow K_2SO_4 & + & 2H_2O \\ 1\,\text{mol} & & 2\,\text{mol}\qquad 1\,\text{mol} & & 2\,\text{mol}\end{array}$

$\dfrac{0.0250\,\text{mol H}_2SO_4}{L} \times 15.0\,\text{mL} \times \dfrac{L}{10^3\,\text{mL}} = 0.000\,375\,\text{mol H}_2SO_4\,\text{used}$

$0.000\,375\,\text{mol H}_2SO_4 \times \dfrac{2\,\text{mol KOH}}{1\,\text{mol H}_2SO_4} = 0.000\,750\,\text{mol KOH reacted}$

$\dfrac{0.000\,750\,\text{mol KOH}}{10.0\,\text{mL}} \times \dfrac{10^3}{L} = 0.075\,\text{mol KOH/L} = 0.0750\,M\,\text{KOH}$

19. $\begin{array}{cccc}2HNO_3 & + & Ba(OH)_2 \rightarrow Ba(NO_3)_2 & + & 2H_2O \\ 2\,\text{mol} & & 1\,\text{mol}\qquad 1\,\text{mol} & & 2\,\text{mol}\end{array}$

$\dfrac{0.0175\,\text{mol Ba(OH)}_2}{L} \times 12.5\,\text{mL} \times \dfrac{L}{10^3\,\text{mL}} = 0.000\,219\,\text{mol Ba(OH)}_2\,\text{used}$

$0.000\,219\,\text{mol Ba(OH)}_2 \times \dfrac{2\,\text{mol HNO}_2}{1\,\text{mol Ba(OH)}_2} = 0.000\,438\,\text{mol HNO}_3\,\text{reacted}$

$\dfrac{0.000\,438\,\text{mol HNO}_3}{14.5\,\text{mL}} \times \dfrac{10^3\,\text{mL}}{L} = 0.0302\,\text{mol HNO}_3/L = 0.0302\,M\,HNO_3$

20. $\begin{array}{cccc}2H_3PO_4 & + & 3Ba(OH)_2 \rightarrow Ba_3(PO_4)_2 & + & 6H_2O \\ 2\,\text{mol} & & 3\,\text{mol}\qquad 1\,\text{mol} & & 6\,\text{mol}\end{array}$

$\dfrac{0.0230\,\text{mol H}_3PO_4}{L} \times 13.4\,\text{mL} \times \dfrac{L}{10^3\,\text{mL}} = 0.000\,308\,\text{mol H}_3PO_4\,\text{used}$

$0.000\,308\,\text{mol H}_3PO_4 \times \dfrac{3\,\text{mol Ba(OH)}_2}{2\,\text{mol H}_3PO_4} = 0.000\,462\,\text{mol Ba(OH)}_2\,\text{reacted}$

$\dfrac{0.000\,462\,\text{mol Ba(OH)}_2}{22.5\,\text{mL}} \times \dfrac{10^3\,\text{mL}}{L} = 0.0205\,\text{mol Ba(OH)}_2/L = 0.0205\,M\,Ba(OH)_2$

21. $V_a N_a = V_b N_b$

$V_b = \dfrac{V_a N_a}{N_b} = \dfrac{(20.0\,\text{mL})(0.75\,N)}{0.50\,N} = 30.\,\text{mL}$

22. $V_a N_a = V_b N_b$

$N_b = \dfrac{V_a N_a}{V_b} = \dfrac{(35\,\text{mL})(0.10\,N)}{21.8\,\text{mL}} = 0.16\,N$

23. $\dfrac{0.200\,\text{mol H}_2SO_4}{L} \times \dfrac{2\,\text{equiv H}_2SO_4}{1\,\text{mol H}_2SO_4} = 0.400\,N\,H_2SO_4 \qquad N_b = \dfrac{V_a N_a}{V_b} = \dfrac{12.5\,\text{mL} \times 0.400\,N}{27.5\,\text{mL}} = 0.182\,N$

24. $\dfrac{0.250 \text{ mol H}_3\text{PO}_4}{\text{L}} \times \dfrac{3 \text{ equiv H}_3\text{PO}_4}{1 \text{ mol H}_3\text{PO}_4} = 0.750 \, N \text{ H}_3\text{PO}_4$

$\dfrac{0.280 \text{ mol Ba(OH)}_2}{\text{L}} \times \dfrac{2 \text{ equiv Ba(OH)}_2}{1 \text{ mol Ba(OH)}_2}$
$= 0.560 \, N \text{ Ba(OH)}_2$

$V_a = \dfrac{V_b N_b}{N_a} = \dfrac{40.0 \text{ mL} \times 0.560 \, N}{0.750 \, N} = 29.9 \text{ mL}$

25. $\dfrac{0.350 \text{ mol H}_2\text{SO}_4}{\text{L}} \times \dfrac{2 \text{ equiv H}_2\text{SO}_4}{1 \text{ mol H}_2\text{SO}_4} = 0.700 \, N \text{ H}_2\text{SO}_4$

$N_b = \dfrac{V_a N_a}{V_b} = \dfrac{18.5 \text{ mL} \times 0.700 \, N}{12.5 \text{ mL}} = 1.04 \, N$

$\dfrac{1.04 \text{ equiv LiOH}}{\text{L}} \times 12.5 \text{ mL} \times \dfrac{\text{L}}{10^3 \text{ mL}} \times \dfrac{1 \text{ mol LiOH}}{1 \text{ equiv LiOH}} \times \dfrac{23.9 \text{ LiOH}}{\text{mol LiOH}} = 0.311 \text{ g LiOH}$

Critical Thinking

1. The two sudden changes of pH about 25 mL and 50 mL indicates a diprotic acid. The first change indicates the depletion of the first H+ that dissociates, and the second change to the second H+. The pH ends near 13 which corresponds to a 0.1 N NaOH solution.

Cross Connections

1. Healthy hair is flexible. Its flexibility depends upon the ability of the protein chains in the cuticle layer to shift. Scalp oils, the natural acid coating, and the structure of the hair help produce healthy hair. A shampoo with a basic pH can neutralize the acidic protective coating of the hair. A basic shampoo is also effective in eliminating scalp oils.

2. Answers will vary. The organic means used by most facilities are microorganisms and bacteria. Most of these organisms cannot survive in a pH of less than 5.0. Unless the waste water is placed in a holding tank until it can be neutralized, the microorganisms may not survive. At the same time, possible sources are investigated to determine what caused the decrease in pH.

Teacher's Notes

Chapter 17
Acid–Base Titration and pH

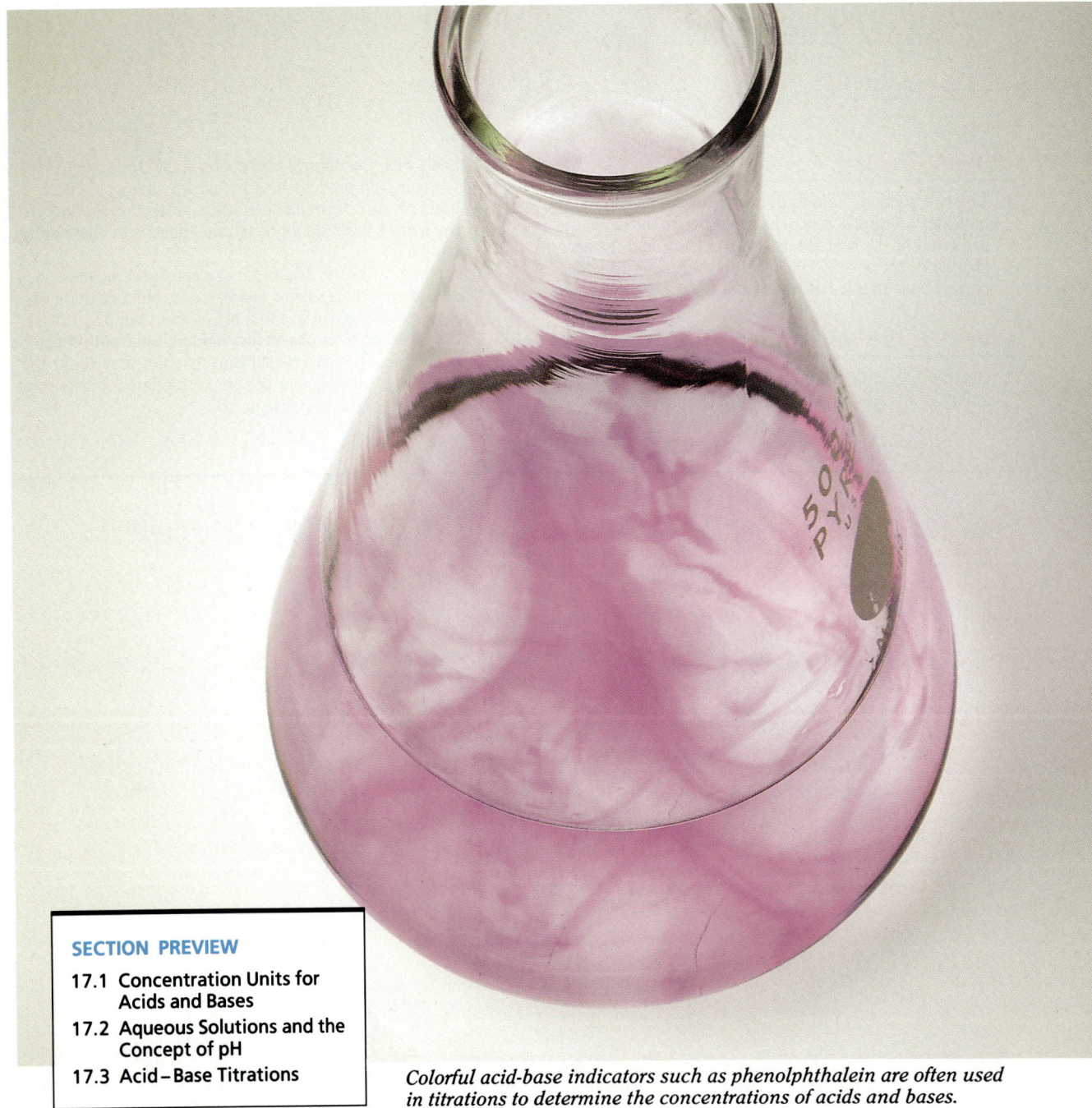

SECTION PREVIEW

17.1 Concentration Units for Acids and Bases
17.2 Aqueous Solutions and the Concept of pH
17.3 Acid–Base Titrations

Colorful acid-base indicators such as phenolphthalein are often used in titrations to determine the concentrations of acids and bases.

17.1 Concentration Units for Acids and Bases

In Chapter 14, you learned three ways to express concentration: percent by mass, molarity, and molality. Recall that molarity *(M)* is the number of moles of solute per liter of solution and that molality *(m)* is the number of moles of solute per kilogram of solvent. In this section you will learn about a method of expressing concentration that is useful in dealing with acids and bases.

Chemical Equivalents

Solution concentrations can be expressed in a way that allows for simple measurement of solute quantities that are chemically equivalent; that is, quantities of substances that have the same combining capacity in chemical reactions. *These quantities of solutes that have equivalent combining capacity are called* **equivalents** *(equiv).*

Consider the following equations. They show that one mole of HCl and one-half mole of H_2SO_4 are chemically equivalent in neutralization reactions with basic KOH. That is, they both react completely with the same quantity (in this case, one mole) of KOH.

$$HCl + KOH \rightarrow KCl + H_2O$$
$$1 \text{ mol} \quad 1 \text{ mol} \quad 1 \text{ mol} \quad 1 \text{ mol}$$

$$\tfrac{1}{2}H_2SO_4 + KOH \rightarrow \tfrac{1}{2}K_2SO_4 + H_2O$$
$$\tfrac{1}{2} \text{ mol} \quad 1 \text{ mol} \quad \tfrac{1}{2} \text{ mol} \quad 1 \text{ mol}$$

Since the mass of one mole of HCl is 36.5 g and the mass of one-half mole of H_2SO_4 is 49.0 g, then 36.5 g of HCl and 49.0 g of H_2SO_4 are chemically

SECTION OBJECTIVES

- Define *equivalent* and *equivalent mass*.
- Find the number of equivalents and the equivalent mass for a given acid or base in an acid–base reaction.
- State the relationship between normality and molarity.
- Calculate solution normality from solution concentration for a given acid or base.

STUDY HINT
It is helpful to review molarity, Section 14.3.

Another way to explain "chemically equivalent" is to state that equivalent quantities of two substances totally react with each other, and no excess of either reactant is left over.

HCl

Ionization of 1 hydrogen chloride molecule (HCl) produces 1 hydrogen ion (H^+) and 1 chloride ion (Cl^-). Ionization of 1 mole of hydrogen chloride molecules (HCl) produces 1 mole of hydrogen ions (H^+) and 1 mole of chloride ions (Cl^-).

H_2SO_4 → 2H^+ + SO_4^{2-}

Ionization of 1 hydrogen sulfate molecule produces 2 hydrogen ions (H^+) and 1 sulfate ion (SO_4^{2-}). Ionization of 1 mole of hydrogen sulfate molecules (H_2SO_4) produces 2 moles of hydrogen ions (H^+) and 1 mole of sulfate ions (SO_4^{2-}).

Figure 17-1 A mole of hydrogen sulfate produces twice as many hydrogen ions as a mole of hydrogen chloride.

Acid-Base Titration and pH

equivalent to each other in these neutralization reactions. Both of them are also chemically equivalent to 56.1 g (the mass of 1 mol) of KOH in these reactions.

In *proton-transfer reactions, the* **equivalent mass of an acid** *is the mass of the acid in grams that will provide one mole of protons in a reaction.* A method for determining the equivalent mass of an acid in neutralization reactions is shown in Sample Problem 17.1. Note that in each case the mass of *one equivalent* is numerically equal to the mass of one mole of the acid divided by the number of moles of protons (expressed in terms of H^+ or H_3O^+) that one mole of the acid can provide.

Remember, proton transfer refers to hydrogen ion (H^+) transfer.

Sample Problem 17.1

Determine the mass of one equivalent of HCl and H_2SO_4 in neutralization reactions.

Solution

Step 1. Analyze   Given: identity of substances: HCl and H_2SO_4
Unknown: *(a)* one equivalent mass of HCl; *(b)* one equivalent mass of H_2SO_4

Step 2. Plan   To find one equivalent mass of H_2SO_4 and HCl, the molar mass of each acid must be divided by the number of protons provided per mole of each acid.

molar mass, protons per mole of acid → equivalent mass

$$\text{molar mass} \left(\frac{g}{mol}\right) \times \frac{\text{mole substance (mol)}}{\text{mole protons (mol } H^+)} = \text{equivalent mass} \left(\frac{g}{mol\, H^+}\right)$$

Step 3. Compute   (a) $\dfrac{36.5 \text{ g HCl}}{\text{mol HCl}} \times \dfrac{1 \text{ mol HCl}}{1 \text{ mol } H^+} = 36.5 \text{ g HCl/mol } H^+$

(b) $\dfrac{98.1 \text{ g } H_2SO_4}{\text{mol } H_2SO_4} \times \dfrac{1 \text{ mol } H_2SO_4}{2 \text{ mol } H^+} = 49.0 \text{ g } H_2SO_4/\text{mol } H^+$

Step 4. Evaluate   Since one mole of HCl yields one mole of protons, it is reasonable that the molar mass of HCl is equal to one equivalent mass of the acid. Since one mole of H_2SO_4 yields two moles of protons, it is reasonable that the molar mass of H_2SO_4 is equal to twice the equivalent mass of the acid. One equivalent mass should therefore be equal to only half of one molar mass of H_2SO_4.

Practice Problems
1. Determine the mass of 1.00 equivalent of HNO_3. *(Ans.)* 63.0 g HNO_3/mol H^+
2. Determine the mass of 1.00 equivalent of H_2SO_3. *(Ans.)* 41.0 g H_2SO_3/mol H^+

Acetic acid, $HC_2H_3O_2$, contains 4 hydrogen atoms in the molecule. However, only one of these is capable of breaking away from the rest of the molecule in water solution. Therefore, $HC_2H_3O_2$ is monoprotic.

The three common acids, HCl, HNO_3, and $HC_2H_3O_2$ are monoprotic. One mole of each can supply one mole of H_3O^+ ions. One equivalent of a monoprotic acid therefore is the same as one mole of the acid. One mole of a diprotic acid, such as H_2SO_4, can supply two moles of H_3O^+ ions. For complete neutralization, one equivalent of a diprotic acid is the same as one-half mole of the acid. Phosphoric acid (H_3PO_4) is an example of a triprotic acid. It can furnish three moles of H_3O^+ ions per mole of acid. When completely neutralized, one equivalent of a triprotic acid is the same as one-third mole of the acid.

A similar relationship exists between chemical equivalents and moles of bases. One **equivalent mass of a base** *is the mass of the base in grams that will react with one mole of protons or will provide one mole of hydroxide (OH^-) ions in a reaction.* Consider again the type of neutralization reaction presented above. Suppose now that KOH is replaced with $Ca(OH)_2$ in the reaction with HCl. You can see, from the following equations, that one-half mole of $Ca(OH)_2$ is equivalent to one mole of KOH.

$$HCl + KOH \rightarrow KCl + H_2O$$
$$1 \text{ mol} \quad 1 \text{ mol} \quad 1 \text{ mol} \quad 1 \text{ mol}$$

$$HCl + \tfrac{1}{2} Ca(OH)_2 \rightarrow \tfrac{1}{2} CaCl_2 + H_2O$$
$$1 \text{ mol} \quad \tfrac{1}{2} \text{ mol} \quad \tfrac{1}{2} \text{ mol} \quad 1 \text{ mol}$$

Said another way, one mole of $Ca(OH)_2$ is equal to two equivalents of KOH. The equivalent masses of KOH and $Ca(OH)_2$ are found from the molar masses as illustrated for acids in Sample Problem 17.1.

$$\frac{56.1 \text{ g KOH}}{\text{mol KOH}} \times \frac{1 \text{ mol KOH}}{1 \text{ mol OH}^-} = 56.1 \text{ g KOH/mol OH}^-$$

$$\frac{74.1 \text{ g Ca(OH)}_2}{\text{mol Ca(OH)}_2} \times \frac{1 \text{ mol Ca(OH)}_2}{2 \text{ mol OH}^-} = 37.0 \text{ g Ca(OH)}_2/\text{mol OH}^-$$

In many chemical reactions, a diprotic or triprotic acid is not completely neutralized. In such a case, the number of moles of protons supplied per mole of acid is determined by the reaction that the acid undergoes. In fact, in all cases the number of equivalents of a given mass of reactant depends on the specific reaction involved. For example, suppose a solution containing one mole of H_2SO_4 is added to a solution containing one mole of NaOH. The product is the "acid salt" sodium hydrogen sulfate, which can be recovered by evaporating the water solvent. The equation is as follows:

$$H_2SO_4(aq) + NaOH(aq) \rightarrow NaHSO_4(aq) + H_2O(\ell)$$
$$1 \text{ mol} \quad\quad 1 \text{ mol} \quad\quad 1 \text{ mol} \quad\quad 1 \text{ mol}$$
$$98.1 \text{ g} \quad\quad 40.0 \text{ g}$$

Observe that the neutralization of H_2SO_4 is not complete. One mole of H_2SO_4 supplies only one mole of its two moles of protons (H^+) to the base. An acid salt containing HSO_4^- ions is the product. In this reaction, one equivalent of H_2SO_4 is the same as one molar mass, i.e. 98.1 g, of the acid.

Now, suppose that a solution containing one mole of phosphoric acid, H_3PO_4, is added to another containing one mole of NaOH. The reaction is shown below:

$$H_3PO_4(aq) + NaOH(aq) \rightarrow NaH_2PO_4(aq) + H_2O(\ell)$$
$$1 \text{ mol} \quad\quad 1 \text{ mol} \quad\quad 1 \text{ mol} \quad\quad 1 \text{ mol}$$
$$98.0 \text{ g} \quad\quad 40.0 \text{ g}$$

The product is the acid salt sodium dihydrogen phosphate, which can be recovered by evaporation. Since each H_3PO_4 unit in this reaction gives up only one of its three protons, one mole of H_3PO_4 is needed to supply one mole of protons to the base to form the acid salt product. In this reaction also, one equivalent of H_3PO_4 (98.0 g) is the same as one molar mass of this acid.

A specific reaction must be considered before an equivalent can be defined.

STUDY HINT
Remember, the mass of one equivalent depends on the reaction the substance is used in, not just the formula of the substance.

Figure 17-2 When phosphoric acid and sodium hydroxide react, any of three salts may be produced, depending on the relative number of moles of the reactants.

Normality

The concentration of solutions can be expressed by stating the quantity of the solute in equivalents rather than in moles. This measure of concentration is called normality (N). *The **normality** of a solution is the number of equivalents of solute per liter of solution.* A one-normal ($1\,N$) solution contains one equivalent of solute per liter of solution. A $0.25\,N$ solution contains 0.25 equivalent of solute per liter of solution. Three ways of expressing solution concentration are summarized in Table 17-1.

The normality of an acid or base solution is commonly expressed using the number of H_3O^+ or OH^- ions available for a complete neutralization. One mole of monoprotic hydrogen chloride has a mass of 36.5 g. As a reactant, this quantity of HCl can furnish one mole of protons. Thus, one mole of HCl in one liter of aqueous solution provides one equivalent of protons as H^+, or H_3O^+, ions. The concentration of this solution is $1\,N$.

TABLE 17-1 METHODS OF EXPRESSING CONCENTRATION OF SOLUTIONS

Name	Symbol	Solute Unit	Solvent Unit	Dimensions
molality	m	mole	kilogram solvent	$\dfrac{\text{mol solute}}{\text{kg solvent}}$
molarity	M	mole	liter solution	$\dfrac{\text{mol solute}}{\text{L solution}}$
normality	N	equivalent	liter solution	$\dfrac{\text{equiv solute}}{\text{L solution}}$

Sample Problem 17.2

Determine the normality of solutions that contain (a) 1 mol H_2SO_4 dissolved in 1 L of solution, (b) 0.01 mol of $Ca(OH)_2$ dissolved in 1 L of solution.

Solution
Step 1. Analyze Given: identity and amount of solutes, volume of solutions
Unknown: *(a)* Normality of H_2SO_4 solution;
(b) Normality of $Ca(OH)_2$ solution

Sample Problem 17.2 *continued*

Step 2. Plan Equivalents per mol, volume → normality

H_2SO_4 gives two equivalents of H^+ per mole and $Ca(OH)_2$ gives two equivalents of OH^- per mole

$$\frac{\text{equivalents}}{\text{mol}} \times \frac{\text{moles}}{\text{volume of solution}} = \frac{\text{equivalents}}{L} = \text{normality}$$

Step 3. Compute (a) $\dfrac{2 \text{ equivalents}}{1 \text{ mol } H_2SO_4} \times \dfrac{1 \text{ mol } H_2SO_4}{1 \text{ L}} = \dfrac{2 \text{ equivalents}}{1 \text{ L}} = 2\,N\;H_2SO_4$

(b) $\dfrac{2 \text{ equivalents}}{1 \text{ mol } Ca(OH)_2} \times \dfrac{0.01 \text{ mol } Ca(OH)_2}{1 \text{ L}} = \dfrac{0.02 \text{ equivalents}}{1 \text{ L}} = 0.02\,N\;Ca(OH)_2$

Step 4. Evaluate Since one mole of H_2SO_4 and one mole of $Ca(OH)_2$ yield 2 mol of H^+ and OH^- ions, respectively, solutions containing 1 mol/L and 0.01 mol/L must be $2\,N$ and $0.02\,N$.

Practice Problems
1. Determine the normality of a 0.5-L solution containing 2.5 equivalents of H_2SO_3. *(Ans.)* $5\,N$
2. Determine the normality of a 4-L solution that contains 0.06 mol of $Al(OH)_3$ *(Ans.)* $0.045\,N$

The Relationship of Normality to Molarity

An advantage in expressing solution concentration in terms of normality rather than molarity is that equal volumes of different solutions of the same normality are chemically equivalent; that is, they contain equal numbers of equivalents. In general, if one mole of a solute is also one equivalent, the molarity and normality of the solution are the same. A $1\,M$ HCl solution is thus a $1\,N$ solution. If one mole of solute is two equivalents, however, a $1\,M$ solution of that solute is $2\,N$. Therefore, a $0.01\,M\;H_2SO_4$ solution is $0.02\,N$. Similarly, a $0.01\,M\;H_3PO_4$ solution is $0.03\,N$ in a complete neutralization. Solutions of equal normality are chemically equivalent, volume for volume.

Normality is numerically related to molarity by the simple relationship

$$N = nM$$

where n is the number of equivalents per mole.

Sample Problem 17.3

What is the molarity of a $0.06\,N\;Sr(OH)_2$ solution?

Solution
Step 1. Analyze Given: normality of $Sr(OH)_2$ solution
Unknown: molarity of $Sr(OH)_2$ solution

Step 2. Plan Normality → equivalents per mole → molarity

$Sr(OH)_2$ has 2 equivalents of OH^- per mole

$$M = \frac{N}{n} = \frac{\text{equivalents}}{L} \times \frac{\text{mole}}{\text{equivalents}}$$

Acid-Base Titration and pH

Sample Problem 17.3 continued

Step 3. Compute $\dfrac{0.06 \text{ equivalents Sr(OH)}_2}{1 \text{ L}} \times \dfrac{1 \text{ mol Sr(OH)}_2}{2 \text{ equivalents Sr(OH)}_2} = \dfrac{0.03 \text{ mol Sr(OH)}_2}{1 \text{ L}}$

$= 0.03 \; M \; \text{Sr(OH)}_2$

Step 4. Evaluate The molarity of a substance that yields 2 equivalents per mole should be one-half its normality.

Practice Problems
1. Determine the normality of a 0.3 M H_3PO_4 solution. *(Ans.)* 0.9 N
2. Compute the molarity of a 0.45 N NaOH solution. *(Ans.)* 0.45 M

---

1. 40.0 g NaOH/mol OH^-
   85.7 g $Ba(OH)_2$/mol OH^-
2. Equal volumes of different solutions of the same normality contain equal numbers of equivalents. Solutions of equal normality are chemically equivalent, volume for volume.
3. $N = nM$, where n is the number of equivalents per mole.
4. Equal volumes of the same normality contain the same number of equivalents.

Section Review
1. Find the equivalent mass of NaOH and $Ba(OH)_2$ in acid–base reactions.
2. What is the relationship between equivalents and normality?
3. Explain the difference between molarity and normality.
4. Why would normality be a more useful measure of solution concentration when dealing with reactions between two different solutions?

17.2 Aqueous Solutions and the Concept of pH

You are probably used to thinking of water as an inactive medium that can be used to dissolve solutes and permit reaction between them. Actually, however, water has an important chemical property that makes it far from inactive. In the future, you will need to take this property into consideration when you deal with aqueous solutions—especially those containing acids or bases.

SECTION OBJECTIVES
- Describe the self-ionization of water.
- Define pH and give the pH of a neutral solution.
- Explain and use the pH scale.
- Given $[H_3O^+]$ or $[OH^-]$, find pH.
- Given pH, find $[H_3O^+]$ or $[OH^-]$.

Self-Ionization of Water

*Water, however pure, is an extremely weak electrolyte that undergoes **self-ionization**. In **self-ionization**,* two water molecules interact to produce a hydronium ion and a hydroxide ion by proton transfer. The conductivity of pure water results from this slight ionization of water molecules.

This fact can be demonstrated by testing water that has been highly purified by several different techniques. Electric conductivity measurements

Figure 17-3 Water is weakly ionized by self-ionization.

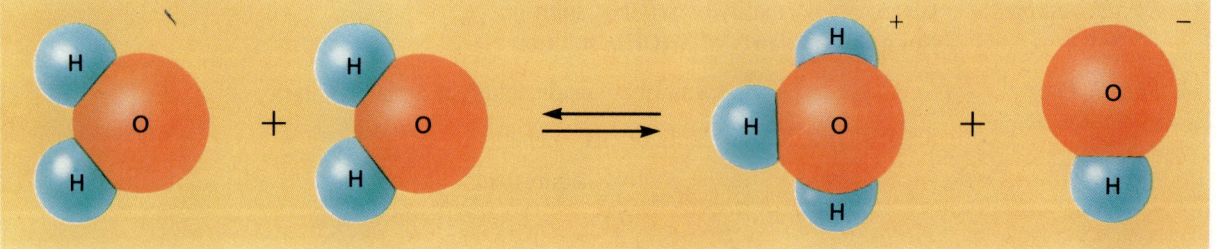

of such pure water show that it is indeed very slightly ionized to form H_3O^+ and OH^- ions. Concentrations of these ions in pure water are only one mole of each per 10^7 liters of water at 25°C.

You will find it is useful to express ion concentration as moles per liter rather than as moles per 10^7 liters.

$$\frac{1 \text{ mol } H_3O^+}{10^7 \text{ L } H_2O} = 10^{-7} \text{ mol } H_3O^+/\text{L } H_2O$$

The concentration of H_3O^+ ions (and OH^- ions) in water at 25°C is thus 10^{-7} M. In percentage, water is 2×10^{-7}% ionized.

Chemists use a standard notation to represent concentration in terms of moles per liter. The symbol or formula of the particular ion or molecule is enclosed in brackets, [ ]. For example, [H_3O^+] means "hydronium ion concentration in moles per liter," or "molar hydronium ion concentration." For the ionic concentrations in water at 25°C, [H_3O^+] is 10^{-7} M and [OH^-] is 10^{-7} M. Because the hydronium and hydroxide ion concentrations are the same, water is neutral: it is neither acidic nor basic. This neutrality prevails in any solution in which [H_3O^+] = [OH^-]. Thus, the word *neutral*, when applied to an aqueous solution, does not generally mean that there are no H_3O^+ or OH^- ions present. It means only that the concentrations of these ions are equal.

If the H_3O^+ ion concentration exceeds 10^{-7} mol/L, an aqueous solution is acidic. For example, a solution containing 10^{-5} mol H_3O^+ ion per liter is acidic ($10^{-5} > 10^{-7}$). If the OH^- ion concentration exceeds 10^{-7} mol per liter, the solution is basic. Thus, a solution containing 10^{-4} mol OH^- ions per liter is basic ($10^{-4} > 10^{-7}$).

In an aqueous solution, the following self-ionization equilibrium is taking place.

$$H_2O(\ell) + H_2O(\ell) \rightleftharpoons H_3O^+ + OH^-$$

The product of [H_3O^+] and [OH^-] remains constant in water and dilute aqueous solutions as long as the temperature does not change. According to LeChâtelier's principle, an increase in the concentration of either H_3O^+ or OH^- in an aqueous solution causes a decrease in the concentration of the other ion. Since an increase in either H_3O^+ or OH^- results in an exact proportional decrease in the other ion, it is also true that the *product* of [H_3O^+] and [OH^-] remains constant in water and dilute aqueous solutions. This constant will remain the same for a given temperature; however, a change in temperature results in a different constant. In water and dilute aqueous solutions at 25°C, the following relationship is valid:

$$K_w = [H_3O^+][OH^-] = 10^{-7} M \times 10^{-7} M = (10^{-7} M)^2 = 10^{-14} M^2$$

The ionization of water increases slightly with temperature increases. The ion product [H_3O^+][OH^-] which is equal to 1.0×10^{-14} M^2 is commonly used as a constant within the ordinary range of room temperatures.

The pH Scale

The range of solution concentrations encountered by chemists is very wide. It varies from above 10 M to below 10^{-15} M. Concentrations of 1 M or only

> In pure water, the concentrations of H_3O^+ and OH^- must be the same because one ion of each is formed when one water molecule breaks apart.

> If students are confused by the fact that $10^{-5} > 10^{-7}$, write the numbers out in decimal form: 0.000 01 > 0.000 0001.

> The product [H_3O^+][OH^-] is often referred to as the ionization constant of water (K_w).

somewhat less than 1 M are most commonly used, however. As stated above, the product of [H_3O^+] and [OH^-] is a constant at a given temperature. Therefore, if the concentration of either ionic species is known, the concentration of the other species can be determined using this information. For example, the OH^- ion concentration of a 0.01 M NaOH solution is 0.01, or 1×10^{-2} mole per liter. The H_3O^+ ion concentration of this solution is calculated as follows.

$$[H_3O^+][OH^-] = 10^{-14}\, M^2$$

$$[H_3O^+] = \frac{1 \times 10^{-14}\, M^2}{[OH^-]} = \frac{1 \times 10^{-14}\, M^2}{1 \times 10^{-2}\, M} = 1 \times 10^{-12}\, M$$

Similarly, the H_3O^+ and OH^- concentrations in a 0.0001 M HNO_3 solution, which is also 100% ionized, are found as follows.

Sample Problem 17.4

A 0.0001 M solution of HNO_3 has been prepared for a laboratory experiment. *(a)* Calculate the [H_3O^+] of this solution. *(b)* Calculate the [OH^-].

Solution

Step 1. Analyze   Given: Concentration and identity of solution = 0.0001 M HNO_3
Unknown: *(a)* [H_3O^+]   *(b)* [OH^-]

Step 2. Plan   concentration HNO_3 → concentration H_3O^+ → concentration OH^-

Since HNO_3 is a strong acid, which is essentially 100% ionized in dilute solutions, then the concentration of the hydronium ions is equal to the concentration of the acid: one molecule of acid produces one hydronium ion. Because the ion product of [H_3O^+][OH^-] is a constant, the [OH^-] can easily be determined by using the value for [H_3O^+].

(a) $\quad HNO_3 \;+\; H_2O \;\rightarrow\; H_3O^+ \;+\; NO_3^-$
$\quad\quad$ 1 mol $\quad$ 1 mol $\quad\quad$ 1 mol $\quad$ 1 mol $\quad$ (assuming 100% ionization)

$$\text{conc. of } HNO_3 \left(\frac{\text{mol } HNO_3}{\text{L solution}}\right) \times \frac{\text{mol } H_3O^+}{\text{mol } HNO_3} = \text{conc. of } H_3O^+ \left(\frac{\text{mol } H_3O^+}{\text{L solution}}\right)$$

(b) $[H_3O^+][OH^-] = 10^{-14}\, M^2$

$$[OH^-] = \frac{10^{-14}\, M^2}{[H_3O^+]}$$

Step 3. Compute   (a) $[H_3O^+] = \dfrac{0.0001 \text{ mol } HNO_3}{\text{L solution}} \times \dfrac{1 \text{ mol } H_3O^+}{1 \text{ mol } HNO_3} = 0.0001\, M\, H_3O^+ = 1 \times 10^{-4}\, M\, H_3O^+$

(b) $[OH^-] = \dfrac{1 \times 10^{-14}\, M^2}{[H_3O^+]} = \dfrac{1 \times 10^{-14}\, M^2}{1 \times 10^{-4}\, M} = 1 \times 10^{-10}\, M\, OH^-$

Step 4. Evaluate   Since one mole of HNO_3 produces one mole of hydronium ions, then a 0.0001 M HNO_3 solution will have a 0.0001 M concentration of H_3O^+. The answers are correctly expressed to one significant digit.

Sample Problem 17.4 *continued*

> **Practice Problems**
> 1. Determine the hydronium and hydroxide ion concentrations in a solution that is 10^{-4} *M* HCl.
>   *(Ans.)* $[H_3O^+] = 10^{-4} M$; $[OH^-] = 10^{-10} M$
> 2. Determine the hydronium and hydroxide ion concentrations in a solution which is 0.0010 *M* HNO$_3$.
>   *(Ans.)* $[H_3O^+] = 1.0 \times 10^{-3} M$; $[OH^-] = 1.0 \times 10^{-11} M$

Expressing acidity or basicity in terms of the concentration of H$_3$O$^+$ or OH$^-$ can become cumbersome. This is especially true in dilute solutions, whether decimal or scientific notation is used. Because it is more convenient, chemists use a quantity called pH to indicate the hydronium ion concentration of a solution. The letters *pH* stand for the French words *pouvoir hydrogène,* meaning "hydrogen power."

Numerically, *the* **pH** *of a solution is defined as the negative of the common (base = 10) logarithm of the hydronium ion concentration,* $[H_3O^+]$. The pH is expressed by the following equation:

$$pH = -\log [H_3O^+]$$

Figure 17-4 The pH of shampoo is adjusted during manufacturing to be slightly acidic or neutral; therefore, it is gentler to the hair.

As you may recall from your mathematics studies, the common logarithm of a number is the power to which 10 must be raised to give the number. For example, the logarithm of 10^7 is +7.

Pure water at 25°C contains a 10^{-7} *M* concentration of H$_3$O$^+$. The pH of water is therefore determined as follows:

$$pH = -\log [H_3O^+] = -\log (10^{-7}) = -(-7) = 7$$

Suppose the H$_3$O$^+$ ion concentration in a solution is greater than that in pure water, as is the case in acidic solutions. Then the concentration of H$_3$O$^+$ is higher than 10^{-7} *M*. Suppose, for example, that its value is 10^{-6} *M*. The pH of this acidic solution would then be

$$pH = -\log [H_3O^+] = -\log (10^{-6}) = -(-6) = 6$$

As you can see, the pH is a number smaller than 7. This is the case for all acidic solutions.

If you suppose, on the other hand, that the H$_3$O$^+$ ion concentration is smaller than that in pure water, the pH would then be a number larger than 7. This is the case for all basic solutions. To summarize, *if pH is less than 7, the solution is acidic. If pH is greater than 7, the solution is basic.*

Figure 17-5 Guess the pH of these common household products. Check your answers using Table 17-2.

The possible range of pH values of aqueous solutions generally falls between 0 and 14. The pH system is particularly useful in describing the acidity and basicity of solutions that are not too far from neutral. Near-neutral solutions include many food substances and many fluids, such as blood. The pH values of some common substances are given in Table 17-2.

Chemical solutions with pH values below 0 and above 14 can be prepared. For example, the pH of 6 *M* H$_2$SO$_4$ is between 0 and -1. The pH of 3-*M* KOH solution is approximately 14.5. Only common pH values in the 0–14 range will be considered in this course, however.

There are two basic types of pH problems that will concern you: (1) the calculation of pH when the $[H_3O^+]$ of a solution is known, and (2) the calculation of $[H_3O^+]$ when the pH of a solution is known.

Acid-Base Titration and pH

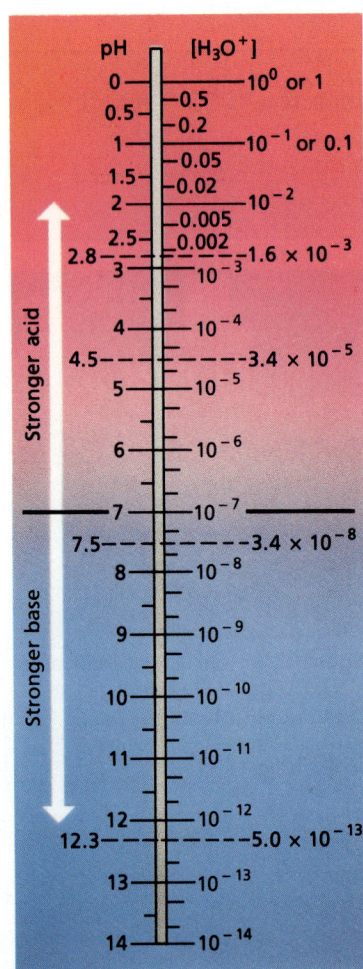

Figure 17-6 A comparison of scales showing pH and hydronium ion concentration

TABLE 17-2 APPROXIMATE pH OF SOME COMMON MATERIALS (AT 25 °C)

Material	pH	Material	pH
1.0 N HCl	0.1	bread	5.5
1.0 N H_2SO_4	0.3	rainwater	5.7
0.1 N HCl	1.1	potatoes	5.8
0.1 N H_2SO_4	1.2	milk	6.5
gastric juice	2.0	saliva	6.5–6.9
0.01 N H_2SO_4	2.1	pure water	7.0
lemons	2.3	eggs	7.8
vinegar	2.8	0.1 N $NaHCO_3$	8.4
0.1 N $HC_2H_3O_2$	2.9	seawater	8.5
soft drinks	3.0	milk of magnesia	10.5
apples	3.1	0.1 N NH_3	11.1
grapefruit	3.1	0.1 N Na_2CO_3	11.6
oranges	3.5	0.1 N NaOH	13.0
cherries	3.6	1.0 N NaOH	14.0
tomatoes	4.2	1.0 N KOH	14.0
bananas	4.6		

Calculations Involving pH

In the simplest of pH problems, the $[H_3O^+]$ of the solution is an integral power of 10, such as 1 M or 0.01 M. You can solve these problems simply by inspection. For example, the pH of a solution in which $[H_3O^+]$ is 10^{-5} M is 5. This is because the logarithm of 10^{-5} is equal to the exponent of 10, or -5 in this case. The pH, or negative of the logarithm, is thus $-(-5) = 5$.

$$pH = -\log [H_3O^+] = -\log (10^{-5}) = -(-5) = 5$$

In general, when $[H_3O^+]$ is an integral power of 10, the pH of a solution is the exponent of the hydronium ion concentration with the sign changed. Sample Problem 17.5 gives an example involving a base.

Sample Problem 17.5

What is the pH of a 0.001 M NaOH solution?

Solution

Step 1. Analyze Given: Identity and concentration of solution = 0.001 M NaOH
Unknown: pH of solution

Step 2. Plan concentration of base → concentration of OH^- → concentration of H_3O^+ → pH

Since NaOH is completely dissociated when it is dissolved in water, a 0.001 M NaOH solution produces an $[OH^-]$ equal to 0.001 M. The ion product of $[H_3O^+]$ and $[OH^-]$ is a constant, 1×10^{-14} M^2. By substitution, the $[H_3O^+]$ can be determined. The pH is then determined as in the above problem.

Sample Problem 17.5 continued

Step 3. Compute

$$[H_3O^+][OH^-] = 1 \times 10^{-14}\ M^2$$

$$[H_3O^+] = \frac{1 \times 10^{-14}\ M^2}{[OH^-]} = \frac{1 \times 10^{-14}\ M^2}{1 \times 10^{-3}\ M} = 1 \times 10^{-11}\ M$$

$$pH = -\log[H_3O^+] = -\log(1 \times 10^{-11}) = -(-11) = 11.0$$

Step 4. Evaluate The answer correctly indicates that NaOH forms a solution with pH > 7, which is basic.

Practice Problems
1. Determine the pH of a $10^{-3}\ M$ HCl solution. *(Ans.)* pH = 3
2. Determine the pH of a $10^{-5}\ M$ HNO$_3$ solution. *(Ans.)* pH = 5
3. Determine the pH of a $10^{-4}\ M$ NaOH solution. *(Ans.)* pH = 10
4. Determine the pH of a $10^{-2}\ M$ KOH solution. *(Ans.)* pH = 12

The preceding problems involve hydronium ion concentrations that are integral powers of 10. They are easily solved by inspection. Many problems, however, involve hydrogen ion concentrations that are not integral powers of 10. Solving such problems requires some basic knowledge of logarithms and exponents. Most scientific calculators have a "log" key. You may need to enter the number whose log is to be found and then press the log key, or the procedure may be in the reverse order. (Consult the instructions for your particular calculator. In the absence of a calculator, log tables must be used.) Note that the logarithm given by your calculator may have far more than the necessary number of digits.

The relationship between the pH and [H$_3$O$^+$] is shown on the scale of Figure 17-6. This scale can be used to estimate the pH from a known [H$_3$O$^+$] value. For example, suppose that you are given that the [H$_3$O$^+$] of a solution is $3.4 \times 10^{-5}\ M$ and you are asked to find the pH. Observe that 3.4×10^{-5} lies between 10^{-4} and 10^{-5} on the [H$_3$O$^+$] scale of Figure 17-6. The pH of the solution thus must be between 4 and 5. The value can be estimated to a second digit as 4.5 on the pH scale of Figure 17-6. You must carry out calculations as to obtain a more precise pH value. However, this reliable estimate of pH helps prevent you from making the sorts of errors that commonly occur in such calculations. Sample Problem 17.6 continues the actual calculation of the pH value for this example.

STUDY HINT
Logarithms are reviewed in Appendix B.

Point out to the students that pH is an exponential value. This means that if the pH changes by 3 units, the concentration of H$_3$O$^+$ changes by 10^3 or 1000 units.

Stress the fact that as the concentration of H$_3$O$^+$ increases, the pH decreases.

Sample Problem 17.6

What is the pH of a solution if [H$_3$O$^+$] is $3.4 \times 10^{-5}\ M$?

Solution

Step 1. Analyze   Given: Concentration of H$_3$O$^+$ = $3.4 \times 10^{-5}\ M$
Unknown: pH of solution

Step 2. Plan   concentration of H$_3$O$^+$ → pH

The only difference between this problem and previous pH problems is that you must determine the logarithm of 3.4 by using your calculator.

Sample Problem 17.6 continued

Step 3. Compute $pH = -\log [H_3O^+] = -\log (3.4 \times 10^{-5}) = -(\log 3.4 + \log 10^{-5}) = -(0.53 + (-5)) = -(0.53 - 5) = -(-4.47) = 4.47$

Step 4. Evaluate Since the pH of a 10^{-5} M H_3O^+ solution is 5, it is reasonable that a solution with a greater concentration of hydronium ions would be more acidic and have a pH less than 5.

Practice Problems
1. What is the pH of a solution if $[H_3O^+]$ is 6.7×10^{-4} mol/L? *(Ans.)* pH = 3.17
2. What is the pH of a solution with a hydronium ion concentration of 2.5×10^{-2} M? *(Ans.)* pH = 1.60

You have now learned to calculate the pH of a solution when given its hydronium ion concentration. Now suppose that you are given the pH of a solution instead. How can you determine its hydronium ion concentration? The equation for the pH in terms of the $[H_3O^+]$ is:

$$pH = -\log [H_3O^+]$$

Remember that the base of common logarithms is 10. The antilog of the common logarithm of a number is the power to which 10 must be raised to give that number. For the pH, this would be

$$\log [H_3O^+] = -pH$$

$$[H_3O^+] = \text{antilog} (-pH)$$

$$[H_3O^+] = 10^{-pH}$$

For an aqueous solution that has a pH of 2, for example, the $[H_3O^+]$ is equal to 10^{-2} M. As you can see, the exponent of 10 that gives the $[H_3O^+]$ is simply the negative of the pH. When the pH is 0, the $[H_3O^+]$ is 1 M, since 10^0 is 1. Sample Problem 17.7 involves another example of a pH value that is a positive integer. Sample Problem 17.8, which follows, involves a pH that is not an integral number. The procedure for finding an antilog varies with the calculator. There may be an "antilog" key, or a "10^x" key, or several keystrokes may be required.

Sample Problem 17.7

Determine the hydronium ion concentration of an aqueous solution that has a pH of 4.

Solution
Step 1. Analyze Given: pH = 4
Unknown: $[H_3O^+]$

Step 2. Plan pH → hydronium ion concentration

This problem requires that you rearrange the pH equation and solve for the $[H_3O^+]$. You must use antilogs.

Sample Problem 17.7 *continued*

Step 3. Compute

$$pH = -\log [H_3O^+]$$
$$\log [H_3O^+] = -pH$$
$$[H_3O^+] = \text{antilog } (-pH) = \text{antilog } (-4) = 10^{-4} \, M$$

Step 4. Evaluate It follows that if the pH of $10^{-4} \, M \, H_3O^+$ is 4, then a solution with a pH of 4 has a $[H_3O^+]$ equal to $10^{-4} \, M$. This problem is reasonably simple to double check by taking your answer and determining the pH.

Sample Problem 17.8

The pH of a solution is measured and determined to be 7.52. *(a)* What is the hydronium ion concentration? *(b)* What is the hydroxide ion concentration? *(c)* Is the solution acidic or basic?

Solution

Step 1. Analyze Given: pH of the solution = 7.52
Unknown: *(a)* $[H_3O^+]$ *(b)* $[OH^-]$ *(c)* Is solution acidic or basic?

Step 2. Plan

$$pH \rightarrow [H_3O^+] \rightarrow [OH^-]$$

This problem is very similar to previous pH problems. You will need to substitute values into the $pH = -\log [H_3O^+]$ equation and work the antilogs properly. Once the $[H_3O^+]$ is determined, the ion-product constant $[H_3O^+][OH^-]$ equals $1.0 \times 10^{-14} \, M^2$ may be used. The pH of the solution tells you about the acidity or basicity of the solution.

Step 3. Compute

(a) $pH = -\log [H_3O^+]$
$\log [H_3O^+] = -pH$
$[H_3O^+] = \text{antilog } (-pH) = \text{antilog } (-7.52) = 10^{-7.52} = 3.0 \times 10^{-8} \, M$

(b) $[H_3O^+][OH^-] = 1.0 \times 10^{-14} \, M^2$

$$[OH^-] = \frac{1.0 \times 10^{-14} \, M^2}{[H_3O^+]} = \frac{1.0 \times 10^{-14} \, M^2}{3.0 \times 10^{-8} \, M} = 3.3 \times 10^{-7} \, M$$

(c) A pH of 7.52 is slightly greater than a pH of 7. This means that the solution is slightly basic.

Step 4. Evaluate Since the solution is slightly basic, a hydroxide ion concentration slightly larger than $10^{-7} \, M$ is predicted. A hydronium ion concentration slightly less than $10^{-7} \, M$ is also predicted.

Practice Problems

1. The pH of a solution is determined to be 5. What is the hydronium ion concentration of this solution? *(Ans.)* $[H_3O^+] = 10^{-5} \, M$
2. The pH of a solution is determined to be 12. What is the hydronium ion concentration of this solution? *(Ans.)* $[H_3O^+] = 10^{-12} \, M$
3. The pH of an aqueous solution is measured as 1.50. *(a)* Calculate the $[H_3O^+]$. *(Ans.)* $[H_3O^+] = 3.2 \times 10^{-2} \, M$ *(b)* Determine the $[OH^-]$. *(Ans.)* $[OH^-] = 3.2 \times 10^{-13} \, M$
4. The pH of an aqueous solution is 3.67. Determine $[H_3O^+]$. *(Ans.)* $[H_3O^+] = 2.1 \times 10^{-4} \, M$

Chemistry and the Environment

Acid Precipitation and Lakes

Acid precipitation is caused by the combustion of fossil fuels, which produces oxides of nitrogen and sulfur that then dissolve in water to form nitric and sulfuric acid. Much of the northeastern United States and Canada is affected by acid precipitation.

A lake can withstand acid precipitation for a few years if the local soil is rich in alkaline materials. However,

when the substructure is granite, as is typical in New England, there is little protective buffering action, and the lake is susceptible to

permanent damage. Some adult fish begin to die if the pH drops below 5.0, and no fish can reproduce when it is below 4.5. A lake can have over 50 varieties of plankton at a pH of 7.0, but only 12 species can survive at a pH of 5.0, and none can survive below 4.5. Once the pH of a lake falls below 5.0, it is defined as acidic. Hundreds of lakes in the northeast are classified as acidic.

TABLE 17-3 RELATIONSHIP OF $[H_3O^+]$ TO $[OH^-]$ AND pH (AT 25 °C)

Solution	$[H_3O^+]$	$[OH^-]$	$[H_3O^+][OH^-]$	pH
0.02 M KOH	5.0×10^{-13}	2.0×10^{-2}	1.0×10^{-14}	12.30
0.01 M KOH	1.0×10^{-12}	1.0×10^{-2}	1.0×10^{-14}	12.00
pure H_2O	1.0×10^{-7}	1.0×10^{-7}	1.0×10^{-14}	7.00
0.001 M HCl	1.0×10^{-3}	1.0×10^{-11}	1.0×10^{-14}	3.00
0.1 M $HC_2H_3O_2$	1.3×10^{-3}	7.7×10^{-12}	1.0×10^{-14}	2.89

Table 17-3 shows the relationship between the hydronium ion and hydroxide ion concentrations, the product of these concentrations, and the pH for several solutions of typical molarities. Since KOH is a soluble ionic compound, it is completely dissociated when forming an aqueous solution. The molarity of each KOH solution indicates directly the $[OH^-]$. Since the product of $[H_3O^+]$ and $[OH^-]$ is the constant $10^{-14} M^2$ at 25°C, the $[H_3O^+]$ can be calculated. If the $[H_3O^+]$ is known, the pH can be determined as $-\log [H_3O^+]$.

Hydrochloric acid may be considered to be completely ionized in any aqueous solution that has a concentration below 1 M. The $[H_3O^+]$ concentration is therefore equal to the molar concentrations of the solution. For example, a 0.001 M HCl solution has a $[H_3O^+]$ equal to 0.001 M.

The weakly ionized acetic acid ($HC_2H_3O_2$) solution presents a different problem. Information about the concentrations of acetic acid molecules, hydronium ions, and acetate ions in the equilibrium mixture may be lacking. You can determine the hydronium ion concentration, however, by measuring the pH of the solution experimentally by using a pH meter. You can then calculate the $[H_3O^+]$ as the antilog $(-pH)$. (See Sample Problems 17.7 and 17.8)

Section Review

1. What is the concentration of hydronium and hydroxide ions in pure water at 25°C?
2. Why does the pH scale range from 0 to 14 in aqueous solutions?
3. Why does a pH of 7 represent a neutral solution?
4. Identify each of the following solutions as acidic or basic.
    (a) $[H_3O^+] = 1 \times 10^{-3} \frac{mol}{L}$ (b) $[OH^-] = 1 \times 10^{-4} \frac{mol}{L}$
    (c) pH = 5 (d) pH = 8.

1. In pure water at 25°C, $[H_3O^+] = 1 \times 10^{-7}\ M$; $[OH^-] = 1 \times 10^{-7}\ M$.
2. For water solutions, $[H_3O^+]$ generally falls between $1 \times 10^0\ M$ and $1 \times 10^{-14}\ M$. Since pH = $-\log[H_3O^+]$, a scale of 0 to 14 is convenient.
3. In neutral solutions, $[H_3O^+] = [OH^-]$. Since $[H_3O^+][OH^-] = 1 \times 10^{-14}\ M^2$, $[H_3O^+] = 1 \times 10^{-7}\ M$ and pH = 7.
4. (a) Acidic (b) Basic (c) Acidic (d) Basic

17.3 Acid–Base Titrations

Everything you have learned so far can be applied to the analysis of acidic and basic solutions and the reactions between them. In order for you to apply your knowledge in this way, however, you will need to find out about certain "tools." These "tools" will enable you to get general information on acid–base concentrations virtually at a glance.

SECTION OBJECTIVES

- Describe how an acid–base indicator functions.
- Explain the procedures involved in carrying out an acid–base titration.
- Calculate the molarity of a solution from titration data.
- Calculate the normality of a solution from titration data.

Indicators

The chemical "tools" just referred to are the substances called indicators. You have already read about the common acid–base indicator, litmus. As you may recall, litmus is red in acidic solutions and blue in basic solutions.

Indicators *are weak acid or base dyes whose colors are sensitive to pH, or hydronium ion concentration.* Chemists have a wide choice of indicators for use in the study of acids and bases and their reactions. In a given situation they are able to choose one that has different colors at certain pH values and that changes colors over the correct pH range for a particular reaction. *The pH range over which an indicator changes color is called its* **transition interval**. Table 17-4 gives the color changes and transition intervals for a number of common acid–base indicators. As you will see, the choice of

TABLE 17-4 INDICATOR COLORS

Indicator	Color			Transition Interval (pH)
	Acid	Transition	Base	
methyl violet	yellow	aqua	blue	0.0– 1.6
methyl yellow	red	orange	yellow	2.9– 4.0
bromphenol blue	yellow	green	blue	3.0– 4.6
methyl orange	red	orange	yellow	3.2– 4.4
methyl red	red	buff	yellow	4.8– 6.0
litmus	red	pink	blue	5.5– 8.0
bromthymol blue	yellow	green	blue	6.0– 7.6
phenol red	yellow	orange	red	6.6– 8.0
phenolphthalein	colorless	pink	red	8.2–10.6
thymolphthalein	colorless	pale blue	blue	9.4–10.6
alizarin yellow	yellow	orange	red	10.0–12.0

Figure 17-7 Indicators in acid–base titrations. An indicator's color change is most dramatic over a relatively narrow pH range, as marked by the black bars in the photograph.

indicator is based on the suitability of its transition interval for a given acid–base reaction.

Indicators are generally divided into three types, according to the pH at which they change color:

1. Those of the first type change color at about pH 7. Such indicators include litmus. The color-change interval for litmus, however, is inconveniently broad—pH 5.5–8.0. This makes for considerable imprecision when acid–base reactions are being studied. The indicator bromthymol blue performs more satisfactorily in this respect since its transition interval is relatively narrow—pH 6.0–7.6.

Bromthymol blue is yellow in acids, blue in bases, and green in solutions that are almost neutral. Indicators of this general type that undergo transition at about pH 7 are especially useful in the study of neutralization reactions between strong acids and strong bases. The neutralization of strong acids with strong bases produces a salt solution with a pH of approximately 7. The use of this type of indicator makes it easy to observe the point at which such neutralization occurs; that is, the point at which equal numbers of equivalents of acids and bases have combined to form products.

2. Indicators of the second type change color at pH lower than 7. Methyl orange is an example of this type. Such indicators are especially useful in studying neutralization reactions between strong acids and weak bases. These reactions produce salt solutions whose pH is lower than 7.

3. The third type of indicator changes color at pH higher than 7. Phenolphthalein is an example. Such indicators are especially useful when studying neutralization reactions between weak acids and strong bases. These reactions produce salt solutions whose pH is greater than 7.

You may be wondering what sort of indicator is useful when studying reactions between weak acids and weak bases. The surprising answer is "none at all!" The pH of the solution that results from complete neutralization reaction between weak acids and weak bases may be almost any value, depending on the relative strengths of the reactants. Without reasonably complete information, the transition interval of an indicator helps very little in determining whether reactions between such acids and bases are complete.

So-called universal indicators can be made by mixing together several different indicator solutions. The resulting solution can turn a number of colors, depending on the pH of the solution it is being used to test. Strips of paper can be soaked in universal indicator solution to produce what is called pH paper. This paper can take on almost any of the colors of the rainbow and thus, can be a fairly accurate way of distinguishing solutions having different pH values. Figure 17-8 shows the pH scale and the colors for this indicator paper.

The mechanism by which indicators operate is interesting. The indicators themselves are either weak acids or weak bases. In solution, the equilibrium of an acid-base indicator that is a weak acid can be represented by the following equation:

$$H\mathit{In} \rightleftarrows H^+ + \mathit{In}^-$$

The colors that an indicator displays result from the fact that HIn and In^- are different colors. For example, in the case of litmus, HIn is red and In^- is blue.

Figure 17-8 When pH paper touches an acidic or basic solution, it changes color. The pH of the solution may be estimated by comparison against the scale on the dispenser.

Students might be surprised to find that their distilled water is slightly acidic. This may be due to CO_2 from the air dissolving into the distilled water; $CO_2(g) + 2H_2O(\ell) \rightarrow HCO_3^-(aq) + H_3O^+$.

Acid-Base Titration and pH

If you study the equilibrium equation above in terms of Le Châtelier's principle and the Brønsted theory of acids, you will understand how indicators are able to operate the way they do. In sufficiently acidic solutions, the equilibrium is forced to the left by the presence of the many protons (H^+ ions) that the acid produces. Any In^- ions that are present are forced to act as Brønsted bases and thus accept these protons from the acid. The indicator is then present in largely nonionized form, HIn. It then has its characteristic acid-indicating color, which for litmus is red.

In sufficiently basic solutions, the indicator's equilibrium is forced to the right by the presence of the many hydroxide ions (OH^-) that the base produces. These OH^- ions combine with the H^+ ions that are produced by the indicator and force the indicator molecules to ionize further to try to offset the loss of H^+ ions. The indicator is thus present largely in the form of its anion, In^-. The solution now displays the characteristic base-indicating color of the indicator anion, which for litmus is blue.

Indicators such as those in methyl orange that change color at pH lower than 7 are simply stronger acids than the other types of indicators. They tend to ionize more completely than the others. The anions (In^-) that they produce are weaker Brønsted bases and have less tendency to accept protons from any acid being tested. These indicators therefore do not shift to their nonionized (HIn) form unless the concentration of H^+ is fairly high. Their color transition occurs at rather low pH. What can you say about the acid strength of indicators, such as phenolphthalein, that undergo transition in the higher pH range? If you said that they are weaker acids, you are correct.

Indicators can be used to compare the pH of different solutions. An indicator added to different solutions may show the same transition color, for example. If so, the solutions are considered to have the same pH. This is the basis for the common colorimetric determination of pH. A measured volume of a suitable indicator is added to each solution whose pH is to be determined. The color is then compared with that of the same indicator in solutions of known pH. By careful color comparison, the pH of a solution can be estimated with a precision of about 0.1 pH unit.

The pH of solutions can be found in ways other than by the use of indicators. Instrumental methods are generally used by chemists to make

Figure 17-9 An acid pushes the equilibrium of litmus to the left; the nonionized form predominates, and the litmus turns red. A base pushes the equilibrium to the right; the ionized form predominates, and the litmus turns blue.

Desktop Investigation

Testing for Acid Precipitation

Question
Do you have acid precipitation in your area?

Materials
rainwater
distilled water
500-mL glass containers
thin, transparent plastic metric ruler (± 0.1 cm)
hydrion-pH test paper: narrow-range, ± 0.2–0.3, or pH meter.

Procedure
Record all your results in a data table.

1. Each time it rains, set out five clean glass containers to collect the rainwater. If the rain continues longer than 24 hours, put out new containers at the end of each 24-hour period until the rain stops. (The same procedure can be used with snow if the snow is allowed to melt before measurements are taken. You may need to use larger containers if a heavy snowfall is expected.)

2. After the rain stops or at the end of each 24-hour period, measure with a thin plastic ruler the depth of the water collected to the nearest 0.1 cm, and test the water with the hydrion paper to determine its pH to the nearest 0.2–0.3.

3. Record the following information: (a) the date and time the collection was started; (b) the date and

time the collection was ended; (c) the location where the collection was made (town and state); (d) the amount of rainfall in cm; and (e) the pH of the rainwater.

4. Find the average pH of each collection that you have made for each rainfall, and record it in the data table.

5. Make collections on at least five different days. The more collections you make, the more reliable your data becomes.

6. For comparison, determine the pH of pure water by testing five samples of distilled water with hydrion paper. Record your results in a separate data table and then calculate an average pH for distilled water.

Discussion

1. What is the pH of distilled water?

2. What is the pH of normal rainwater? How do you explain this?

3. How is acid precipitation defined?

4. What are the drawbacks of using a ruler to measure the height of collected water?

5. Does the amount of rainfall or the time of day the sample is taken have an effect on its pH? Explain any variability among samples.

6. What conclusion can you draw from this investigation? Explain how your data support your conclusion.

Figure 17-10 pH colorimetry. A sample's pH is determined by adding indicator and comparing its color to those of standard solutions containing indicator.

rapid pH determinations. A pH meter (Figure 17-11) provides a convenient way of measuring the pH of a solution. The pH meter measures the voltage difference between a special electrode and a reference electrode, which are placed in the solution. The voltage changes as the hydronium ion concentration in the solution changes. In an acid–base neutralization reaction, a large change in voltage occurs at the point when equivalent quantities of the two reactants are present in the solution.

The word titration comes from the French word *titre* which means "a standard."

The Principle of Titration

The reaction that occurs between an acid and a base is neutralization. Recall that, in a neutralization reaction, the basic OH^- ion acquires a proton from the H_3O^+ ion to form a molecule of water.

$$H_3O^+ + OH^- \rightarrow 2H_2O$$

One mole of hydronium ions (19.0 g) and one mole of hydroxide ions (17.0 g) are chemically equivalent. They combine in a one-to-one ratio. Neutralization occurs when hydronium ions and hydroxide ions are supplied by reactants in equal numbers. The product for $[H_3O^+][OH^-]$ of $10^{-14}\ M^2$ is a constant for the aqueous acid and base solutions and for the solution that results when acids and bases mix and react. If 0.1 mol of gaseous HCl is dissolved in a liter of water for which $[H_3O^+]$ is $10^{-7}\ M$, the H_3O^+ ion concentration would rise to 0.1, or $10^{-1}\ M$. Since the product $[H_3O^+][OH^-]$ remains at $10^{-14}\ M^2$ and since $[H_3O^+]$ rises to $10^{-1}\ M$, the $[OH^-]$ must decrease from $10^{-7}\ M$ to $10^{-13}\ M$. The hydroxide ions are removed from the solution by combining with H_3O^+ ions to form water molecules. Almost 10^{-7} mol/L of H_3O^+ ions is also removed in this way. This quantity is only a small

Figure 17-11 A pH meter

portion (0.0001%) of the 0.1 mol of hydronium ions that are present in each liter of solution.

Now suppose that 0.1 mol (4.0 g) of solid NaOH is added to the liter of 0.1 M HCl solution. The NaOH dissolves and supplies 0.1 mol of hydroxide ions to the solution. Before the reaction proceeds significantly, both [H_3O^+] and [OH^-] are high. Their ion product is momentarily much greater than the constant value of $10^{-14}\ M^2$ for this dilute aqueous solution.

The ion-removal reaction (that is, neutralization) then proceeds quickly. Hydronium and hydroxide ions, which are present in equal numbers, combine until the product [H_3O^+] [OH^-] returns to the constant value of $10^{-14}\ M^2$. This neutralization process is one in which chemically equivalent quantities of hydronium ions and hydroxide ions have combined. Because the salt (NaCl) produced is the product of the neutralization of a strong acid and a strong base, the resulting solution is neither acidic nor basic, but it is neutral.

These preceding examples should help you understand the nature of the chemical reactions that occur between acids and bases as a solution of one of them is progressively added to a solution of the other. This progressive addition of an acid to a base (or a base to an acid) can be done in the laboratory to compare the concentrations of acid and base solutions. In such a case, it is called titration. **Titration** *is the controlled addition and measurement of the amount of a solution of known concentration that is required to react completely with a measured amount of a solution of unknown concentration.*

Titration provides a sensitive means of determining the relative volumes of acidic and basic solutions that are chemically equivalent. If the concentration of one solution is known, the concentration of the other solution can be calculated after determining the volume of one that is required to neutralize a given volume of the other. *A solution that contains a precisely known concentration of a solute is known as a* **standard solution.** Titration is an important laboratory procedure and is often used in analytical chemistry.

Let's consider an example of a titration. Suppose that successive additions of an aqueous base are made to a measured volume of an aqueous acid. Eventually, the acid is neutralized. With the continued addition of base, the solution becomes distinctly basic. The pH has now changed from a low to a high numerical value. If you were to monitor the change in pH, you would find that it occurs slowly at first, rapidly through the neutral point, and slowly again as the solution becomes basic. Typical pH curves for strong-acid/strong-base and weak-acid/strong-base titrations are shown in Figure 17-13.

Figure 17-12 The ion-removal reaction (neutralization). The water molecules produced are only slightly ionized.

The product [H_3O^+][OH^-] is often referred to as the ionization constant of water (K_w).

Figure 17-13 Acid-base titration curves: *(a)* strong-acid/strong-base; *(b)* weak-acid/strong-base.

In a neutralization reaction, the point at which there are equivalent quantities of hydronium and hydroxide ions is called the **equivalence point.** *The point in a titration where an indicator changes color is called the* **end point** *of the indicator.* Indicators are chosen that have color changes over ranges that include the pH of the equivalence point. Other methods such as pH meters can be used. At the equivalence point, they will show a large voltage change.

To determine the concentration of an "unknown" solution, one must titrate it using a "known" solution. To be certain of actually knowing the concentration of a "known" or "standard" solution, however, that solution must be compared to another standard solution whose concentration is well established. The standard solution is first prepared, and its volume is adjusted to the desired concentration. This concentration information is then refined by titrating the solution with a carefully measured quantity of a solution of a highly purified compound. *A highly purified compound, when used in solution to check the concentration of the known solution in a titration is known as a* **primary standard.** The actual concentration of the standard solution is established by this standardizing procedure.

Molarity and Titration

Burets, or graduated glass columns having good precision—like those shown in Figure 17-14—are commonly used to measure solution volumes during titration. Suppose that an aqueous solution of an unknown concentration of the base NaOH is added dropwise to 10.0 mL of a 0.01 M aqueous solution of hydrochloric acid that contains a few drops of a suitable indicator such as bromthymol blue. The addition of drops continues until the end point is reached. (The color of the solution changes from yellow to green.) The readout of the buret that contains the base indicates that 20.0 mL of the NaOH solution is used. How can these titration data be used to determine the molarity of the basic solution?

The equation for the neutralization reaction is

$$HCl(aq) + NaOH(aq) \rightarrow NaCl(aq) + H_2O(\ell)$$
$$\text{1 mol} \quad \text{1 mol} \quad \text{1 mol} \quad \text{1 mol}$$

The volume and molarity of the HCl solution are known. From these data, the quantity (in moles) of the HCl used for titrating can be determined.

$$\frac{0.01 \text{ mol HCl}}{L} \times 10.0 \text{ mL} \times \frac{1 \text{ L}}{1000 \text{ mL}} = 0.0001 \text{ mol HCl (used)}$$

The balanced chemical equation shows that one mole of NaOH is used for one mole of HCl. In other words, NaOH and HCl show chemical equivalence, mole for mole, in the reaction. The quantity of NaOH used in the titration, therefore, is also 0.0001 mole. From the titration, this quantity of base is furnished by 20.0 mL of the NaOH solution. The molarity of the NaOH solution can now be obtained as follows:

$$\frac{0.0001 \text{ mol NaOH}}{20.0 \text{ mL}} \times \frac{1000 \text{ mL}}{1 \text{ L}} = 0.005 \text{ mol NaOH/L or } 0.005 \, M \text{ NaOH}$$

Suppose the titration is repeated with the same "unknown" NaOH solution but this time with 10.0 mL of a 0.01 M solution of sulfuric acid (the diprotic acid H_2SO_4) as the standard acid solution. When titrated to the end

Figure 17-14 A titration stand

Strictly speaking, the "equivalence point" is the place where equivalent quantities of acid and base are present. The "end point" occurs when an indicator or pH meter signals the equivalence point. Ideally, they occur at the same time.

point, the base buret readout indicates that 40.0 mL of the NaOH solution is used. The equation for this reaction is

$$H_2SO_4(aq) + 2NaOH(aq) \rightarrow Na_2SO_4(aq) + 2H_2O(\ell)$$
$$\text{1 mol} \quad\quad\quad \text{2 mol} \quad\quad\quad \text{1 mol} \quad\quad \text{2 mol}$$

The amount of H_2SO_4 in the standard solution used in this titration was.

$$\frac{0.01 \text{ mol } H_2SO_4}{L} \times 10.0 \text{ mL} \times \frac{L}{1000 \text{ mL}} = 0.0001 \text{ mol } H_2SO_4 \text{ (used)}$$

The reaction equation shows that 2 mol of NaOH is required for each 1 mol of H_2SO_4. Therefore, the 40.0 mL of NaOH used to neutralize 0.0001 mol of H_2SO_4 contained 0.0002 mol of NaOH, as shown by the following calculation:

$$0.0001 \text{ mol } H_2SO_4 \times \frac{2 \text{ mol NaOH}}{1 \text{ mol } H_2SO_4} = 0.0002 \text{ mol NaOH}$$

The molarity of the NaOH solution is obtained as follows:

$$\frac{0.0002 \text{ mol NaOH}}{40.0 \text{ mL}} \times \frac{1000 \text{ mL}}{L} = 0.005 \text{ mol NaOH/L or } 0.005 \text{ M NaOH}$$

To summarize, the molarity of an aqueous base (or acid) of unknown concentration can be determined by titrating against an aqueous acid (or base) of known concentration. The following four steps are involved:

1. Determine the moles of acid (or base) from the standard solution used during the titration.
2. From a balanced chemical equation, determine the ratio of moles of acid (or base) to base (or acid).
3. Determine the moles of solute of the unknown solution used during the titration.
4. Determine the molarity of the unknown solution. Sample Problem 17.9 illustrates the above steps.

When the calculations for a titration are done by this method, molarity is used and not normality. Steps 2 and 3, a chemical equation and mole ratio conversion, are necessary to make the proper computation.

Sample Problem 17.9

In a titration, 27.4 mL of a standard solution of $Ba(OH)_2$ is added to a 20.0-mL sample of an HCl solution. The concentration of the standard solution is 0.0154 M. What is the molarity of the acid solution?

Solution

Step 1: Analyze   Given: volume, concentration, and identity of standard solution = 27.4 mL of 0.0154 M $Ba(OH)_2$
Unknown: molarity of acid solution.

Step 2: Plan   Use the above four steps as follows.
1. Volume of solution used → moles of base used

$$\frac{\text{mol Ba(OH)}_2}{L} \times \text{mL} \times \frac{1 \text{ L}}{1000 \text{ mL}} = \text{mol Ba(OH)}_2$$

2. Balanced chemical equation → ratio of moles of acid to moles of base
3. mole ratio, moles of base used → moles of acid used from unknown solution

Acid-Base Titration and pH

Sample Problem 17.9 *continued*

4. Volume of unknown solution, moles of solute in unknown solution → molarity of unknown solution

$$\frac{\text{moles of solute in unknown solution}}{\text{volume of unknown solution in mL}} \times \frac{1000 \text{ mL}}{1 \text{ L}}$$

= molarity of unknown solution

Step 3: Compute

1. $27.4 \text{ mL} \times \frac{0.0154 \text{ mol Ba(OH)}_2}{\text{L}} \times \frac{1 \text{ L}}{1000 \text{ mL}} = 4.22 \times 10^{-4} \text{ mol Ba(OH)}_2$

2. $\text{Ba(OH)}_2 + 2\text{HCl} \rightarrow \text{BaCl}_2 + 2\text{H}_2\text{O}$ 1 mol Ba(OH)$_2$: 2 mol HCl

3. $\frac{2 \text{ mol HCl}}{1 \text{ mol Ba(OH)}_2} \times 4.22 \times 10^{-4} \text{ mol Ba(OH)}_2 = 8.44 \times 10^{-4} \text{ mol HCl}$

4. $\frac{8.44 \times 10^{-4} \text{ mol HCl}}{20.0 \text{ mL}} \times \frac{1000 \text{ mL}}{1 \text{ L}} = 0.0422 \, M \text{ HCl}$

Step 4: Evaluate

The answer is correctly stated to three significant digits. A $0.0154 \, M$ Ba(OH)$_2$ solution is $0.0308 \, N$. We would expect the HCl solution to be greater than $0.0308 \, N$ (or, in the case of HCl, greater than $0.0308 \, M$) by a factor of roughly 1.4 since about 1.4 times more base solution was used than acid solution. An answer of approximately $0.043 \, M$ HCl (0.031×1.4) is therefore predicted.

Practice Problems

1. A 15.5-mL sample of $0.215 \, M$ KOH solution required 21.2 mL of aqueous acetic acid solution in a titration experiment. Calculate the molarity of the acetic acid solution. *(Ans.)* $0.157 \, M$ HC$_2$H$_3$O$_2$
2. By titration, 17.6 mL of aqueous H$_2$SO$_4$ just neutralized 27.4 mL of $0.0165 \, M$ LiOH solution. What was the molarity of the aqueous acid solution? *(Ans.)* $0.0128 \, M$ H$_2$SO$_4$

Normality and Titrations

When calculating with normality instead of molarity, it is not necessary to write the chemical equation and make the mole ratio conversion. These are "built into" the value for the normality.

Chemists often prefer to express acid and base concentrations in terms of normality rather than molarity. The advantage in doing so is that concentrations are expressed directly in terms of equivalents of solute per liter of solution. Solutions of the same normality are chemically equivalent, milliliter for milliliter. The normality of a given solution is its molarity times a whole-number factor. As we have seen earlier in this chapter, this factor depends on the substance and on the reaction in which the substance is involved.

A very simple relationship exists between volumes and normalities of the solutions used in titration. Take as an example a titration that requires 50.0 mL of a $0.100 \, N$ solution of NaOH to reach an end point with 10.0 mL of vinegar, a water solution of acetic acid (HC$_2$H$_3$O$_2$). The number of chemical equivalents of the acetic acid used is the product of the volume of the acid solution used, V_a, and the normality of the acid, N_a.

There exists an inverse relationship between volume and concentration. If the concentration increases, the volume needed to contain the same amount of substance decreases.

$$V_a \times N_a = \text{equiv}_a \quad (\text{L} \times \text{equiv/L} = \text{equiv})$$

Similarly, the number of chemical equivalents of NaOH used is the product of the volume of the base solution used, V_b, and the normality of the base, N_b.

$$V_b \times N_b = \text{equiv}_b \quad (L \times \text{equiv/L} = \text{equiv})$$

At the end point in the titration, equiv_a equals equiv_b. Therefore,

$$V_a N_a = V_b N_b$$

Sample Problem 17.10

A titration requires 50.0 mL of a 0.100 N NaOH solution to reach an end point with 10.0 mL of acetic acid solution. Determine the normal concentration of the acetic acid (vinegar) solution.

Solution

Step 1. Analyze Given: volume, concentration, and identity of standard solution = 50.0 mL of 0.100 N NaOH
volume and identity of acid solutions = 10.0 mL acetic acid
Unknown: normality of acetic acid solution

Step 2. Plan concentration of base, volume of base, volume of acid → concentration of acid

The equation $V_a N_a = V_b N_b$ will be used, substituting appropriately. This means that you should rearrange the equation algebraically to solve for the unknown quantity.

Step 3. Compute
$$V_a N_a = V_b N_b$$

$$N_a = \frac{V_b N_b}{V_a} = \frac{50.0 \text{ mL} \times 0.100 \, N}{10.0 \text{ mL}} = 0.500 \, N$$

Step 4. Evaluate The answer is correctly given to three significant digits. Since the volume of base required is five times the volume of acid, then the acid is five times more concentrated than the base, or 0.500 N as predicted.

Practice Problems

1. Determine the normal concentration of an unknown acid solution when 17.3 mL of this solution was used in titrating 31.2 mL of a 0.125 N NaOH solution. *(Ans.)* 0.225 N
2. Determine the normality of a NaOH solution when 45.1 mL of a 0.100 N HCl solution was used in titrating 31.0 mL of the base. *(Ans.)* 0.145 N

The acidity of vinegar results from presence of acetic acid. A 1.0 N acetic acid solution contains 1.0 equivalent of acid (H_3O^+) per liter of solution. In this case, 1.0 equivalent equals 1.0 mol (60 g) of acetic acid per liter of solution. The 0.500 N acetic acid solution described in the sample problem above must contain 30 g of $HC_2H_3O_2$ per liter. A liter of vinegar has a mass of approximately 1000 g. The sample of vinegar thus used contains 3.0% acetic acid by mass.

NaOH is a strong base, and $HC_2H_3O_3$ is a weak acid. The pH curve for this titration is shown in Figure 17-13. The end point occurs at a higher pH because the sodium acetate solution that is formed in the titration is slightly basic. Phenolphthalein would be a suitable indicator to use in this titration because it changes at a pH higher than 7. If a weak base and a strong acid are titrated, the end point would occur below a pH of 7 because of the acidic nature of the salt formed. Methyl orange would be a suitable indicator to use in this titration.

> The pH at the equivalence point is the pH of the salt solution that forms from the acid and base. Notice that when a weak acid or base is "neutralized" by a strong base or acid, the solution is not a neutral pH 7, but is slightly basic or acidic.

1. (a) Methyl red, methyl orange, or bromphenol blue. (b) Phenol red, phenolphthalein, or thymolphthalein.
2. Standard solutions of HCl, H_2SO_4, or $HC_2H_3O_2$.
3. $20.0 \text{ mL} \times \dfrac{0.0100 \text{ mol HCl}}{L} \times \dfrac{1 \text{ L}}{1000 \text{ mL}} =$
   2.00×10^{-4} mol HCl

   $HCl + NaOH \rightarrow H_2O + NaCl$
   2.00×10^{-4} mol HCl $\times$ $\dfrac{1 \text{ mol NaOH}}{1 \text{ mol HCl}} =$
   2.00×10^{-4} mol NaOH
   $\dfrac{2.00 \times 10^{-4} \text{ mol NaOH}}{30. \text{ mL}} \times \dfrac{1000 \text{ mL}}{1 L} = 6.7 \times 10^{-3}$ M NaOH
4. $V_a N_a = V_b N_b$
   $N_a = \dfrac{V_b N_b}{V_a} =$
   $\dfrac{20.0 \text{ mL} \times 0.10 \text{ N}}{12 \text{ mL}} =$
   0.17 N HCl

Section Review

1. Name an appropriate indicator for titrating a (a) strong acid and a weak base. (b) strong base and a weak acid.
2. You have a solution of an unknown concentration containing NaOH. Name three possible standard solutions that could be used to titrate the NaOH solution.
3. If 20.0 mL of 0.0100 M aqueous HCl is required to neutralize 30. mL of an aqueous solution of NaOH, determine the molarity of the NaOH solution.
4. If 20.0 mL of 0.10 N NaOH is required to neutralize 12 mL of aqueous HCl solution, what is the normality of the HCl solution?

Chapter Summary

- The quantities of solutes that have equivalent combining capacity in a chemical reaction are called equivalents.
- In a neutralization reaction, one equivalent of an acid provides one mole of H_3O^+ ions and reacts completely with one equivalent of a base, which provides one mole of OH^- ions.
- The equivalent mass of a substance is the mass (in grams) that provides one equivalent in a given chemical reaction.
- A solution with a normality of 1 contains one equivalent mass of solute per liter of solution.
- Normality is numerically related to molarity by the simple relationship $N = nM$, where n is the number of equivalents per mole.
- Pure water undergoes self-ionization to give 10^{-7} M H_3O^+ ions and 10^{-7} M OH^- ions per liter.
- $pH = -\log[H_3O^+]$
- Acids have a pH of less than 7, bases have a pH of greater than 7, and neutral solutions have a pH of 7.
- In weakly ionized acid solutions, $[H_3O^+]$ does not equal molar acid concentration, pH must be measured experimentally, and concentration calculated using antilog($-$pH).
- Indicators are divided into three types: (1) those that change color at a pH of 7 and are used to study the neutralization reactions between strong acids and strong bases (2) those that change color at a pH lower than 7 and are used to study the neutralization reactions between strong acids and weak bases (3) those that change color at a pH greater than 7 and are used to study the neutralization reactions between strong bases and weak acids.
- Titration uses a solution of known concentration (referred to as a *standard solution*), to determine the concentration of an "unknown" solution. The solutes in both the "standard" and the "unknown" solution must be known in order to calculate the molar concentration of the unknown solution from the titration data.
- When the molarity and volume of standard solution used in a titration are known, then the molarity of an unknown solution can be found, provided its volume has been measured.
- When a titration is complete, the volume of standard acid solution used (V_a) multiplied by its normality (N_a) is equal to the volume of unknown base solution (V_b) multiplied by its normality (N_b): $V_a N_a = V_b N_b$.

CHAPTER 17 REVIEW

Acid–Base Titration and pH

VOCABULARY

end point (512)
equivalence point (512)
equivalents (491)
equivalent mass (492)
indicator (505)
normality (494)
pH (499)
primary standard (512)
self-ionization (water) (496)
standard solution (511)
titration (511)
transition interval (505)

REVIEWING CONCEPTS

1. **a.** What is the relationship between the protic nature of an acid and the mass of one equivalent of that acid?
   b. Illustrate with one example each of a monoprotic, diprotic, and triprotic acid. (17.1)

2. **a.** What is the relationship between the number of OH⁻ ions in one mole of a base and the mass of one equivalent of that base?
   b. Illustrate with one example each of bases that contain one, two, and three moles of OH⁻ ions per mole. (17.1)

3. Express in words what is meant by a:
   a. 1 N solution
   b. 3 N solution
   c. 0.5 N solution (17.1)

4. **a.** Define what is meant by the pH of a solution.
   b. Write the equation for determining pH.
   c. Explain and illustrate what is meant by the common logarithm of a number. (17.2)

5. Identify each of the following solutions as acidic, basic, or neutral:
   a. $[H_3O^+] = 1 \times 10^{-7}$
   b. $[H_3O^+] = 1 \times 10^{-10}$
   c. $[OH^-] = 1 \times 10^{-7}$
   d. $[OH^-] = 1 \times 10^{-11}$
   e. $[H_3O^+] = [OH^-]$
   f. pH = 3
   g. pH = 13 (17.2)

6. What is the approximate pH of each of the following common substances?
   a. vinegar
   b. apples
   c. tomatoes
   d. milk
   e. grapefruit
   f. potatoes
   g. lemons
   h. milk of magnesia
   i. gastric juice (17.2)

7. **a.** What is meant by the end point of a titration?
   b. What can be observed about the pH of a solution at the end point of a titration?
   c. What is the role of an indicator in the titration process? (17.3)

8. **a.** What is meant by the transition interval of an indicator?
   b What is the basis upon which an indicator is selected for a particular titration experiment?

9. For each of the four possible types of acid–base titration combinations, indicate the approximate pH at the end point and one or more suitable indicators for detecting that end point. (17.3)

10. Other than through the use of indicators, how can the end point of a titration experiment or the pH of a solution be determined? (17.3)

11. An unknown solution is colorless when tested with phenolphthalein, but turns pink in the presence of the indicator phenol red. Based on this information, what is the approximate pH of this solution? (17.3)

PROBLEMS

▶ Equivalents and Normality

1. Determine the mass of one equivalent (in g/mol H_3O^+) to three significant digits for each of the following acids:
   a. HBr   **b.** H_2S **c.** H_3PO_4

2. Determine the mass of one equivalent (in g/mol OH⁻) to three significant digits for each one of the following bases:
   a. LiOH   **b.** $Sr(OH)_2$ **c.** $Al(OH)_3$

Acid–Base Titration and pH   **517**

3. Determine the number of equivalents needed to make each of the following solutions:
   a. 1.0 L of a 1.0 N solution of HNO_3
   b. 1.0 L of a 2.5 N solution of NaOH
   c. 2.0 L of a 1.0 N solution of $Mg(OH)_2$
   d. 1.5 L of a 3.0 N solution of HCl
   e. 250. mL of a 1.20 N solution of K_2CO_3

4. How many milliliters of 1.50 N H_2SO_4 solution would be required to prepare 500. mL of 1.20 N H_2SO_4 solution?

5. What volume of 0.750 N H_3PO_4 would be required to prepare 2.50 L of 0.200 M H_3PO_4 solution?

6. A given supply of concentrated nitric acid is 69.5% HNO_3 by mass and has a density of 1.42 g/mL. Based on this information, determine the number of milliliters of concentrated nitric acid that would be needed to prepare:
   a. 1.00 L of 2.00 N HNO_3 solution
   b. 150. mL of 0.750 N HNO_3 solution
   c. 250. mL of 0.500 N HNO_3 solution

7. Concentrated hydrochloric acid generally contains 37.5% HCl by mass and has a density of 1.19 g/mL. Based on this information, determine the number of milliliters of concentrated hydrochloric acid that would:
   a. contain 1.00 equivalent of HCl
   b. be required to prepare 2.00 L of 1.00 N HCl solution
   c. be needed to prepare 3.00 L of 0.750 N HCl solution.

▶ **pH Calculations**

8. Given the following $[H_3O^+]$ concentrations, determine the pH of each solution:
   a. 1×10^{-7} c. 1×10^{-12}
   b. 1×10^{-3} d. 1×10^{-5}

9. Determine the pH of each of the following solutions:
   a. 0.01 M HCl   c. 0.000 01 M HI
   b. 0.001 M HNO_3 d. 0.0001 M HBr

10. Given the following $[OH^-]$ concentrations, determine the pH of each solution:
    a. 1×10^{-6} c. 1×10^{-2}
    b. 1×10^{-9} d. 1×10^{-7}

11. Determine the pH of each solution:
    a. 0.01 M NaOH   b. 0.001 M KOH
    c. 0.0001 M LiOH

12. Determine the pH of each solution when given the following $[H_3O^+]$:
    a. 2.0×10^{-5}
    b. 4.7×10^{-7}
    c. 3.8×10^{-3}

13. Given the following pH values, determine the $[H_3O^+]$ for each solution:
    a. 3   c. 11
    b. 7   d. 5

14. Given the following pH values, determine the $[OH^-]$ for each solution:
    a. 7   c. 4
    b. 11   d. 6

15. Determine $[H_3O^+]$ for each of the solutions with the following pH values:
    a. 4.23   b. 7.65   c. 9.48

16. A nitric acid solution is found to have a pH of 2.70. Determine each of the following:
    a. the $[H_3O^+]$
    b. the $[OH^-]$
    c. the mass of anhydrous HNO_3 required to prepare 5.50 L of this solution
    d. the milliliters of concentrated acid needed to prepare the solution in c. (Concentrated nitric acid is 69.5% HNO_3 by mass and has a density of 1.42 g/mL.)

▶ **Titrations**

17. For each of the following acid–base titration combinations, determine the number of moles of the first substance listed that would be required to react with the given quantity of the second:
    a. NaOH with 1.0 mol HCl
    b. HNO_3 with 0.75 mol KOH
    c. $Ba(OH)_2$ with 0.20 mol HF
    d. H_2SO_4 with 0.60 mol $Al(OH)_3$

18. If 15.0 mL of 0.0250 M aqueous H_2SO_4 is required to neutralize 10.0 mL of an aqueous

solution of KOH, what is the molarity of the KOH solution?

19. In a titration experiment, a 12.5-mL sample of 0.0175 M Ba(OH)$_2$ just neutralized 14.5 mL of HNO$_3$ solution. Calculate the molarity of the HNO$_3$ solution.

20. If 13.4 mL of 0.0230 M H$_3$PO$_4$ is required to neutralize 22.5 mL of Ba(OH)$_2$ in a titration experiment, determine the molarity of the Ba(OH)$_2$ solution.

21. What volume of 0.50 N KOH would be needed to neutralize 20.0 mL of 0.75 N HNO$_3$ solution?

22. If 35 mL of 0.10 N acid solution is neutralized by 21.8 mL of a basic solution, what is the normality of the base?

23. If 12.5 mL of 0.200 M H$_2$SO$_4$ is required to neutralize 27.5 mL of KOH solution, what is the normality of the KOH solution?

24. How many milliliters of 0.250 M H$_3$PO$_4$ solution would be required to neutralize 40.0 mL of 0.280 M Ba(OH)$_2$?

25. If 18.5 mL of 0.350 M H$_2$SO$_4$ neutralizes 12.5 mL of aqueous LiOH solution, how many grams of LiOH were used to make the LiOH solution?

CRITICAL THINKING

1. **Interpreting Graphics** The following titration curve resulted from the titration of an unknown acid with 0.10 NaOH. Analyze the curve and make inferences related to the type of acidic solution titrated.

CROSS CONNECTIONS

1. **Health** Figure 17-4 shows several brands of shampoo and tells us that manufacturers of shampoos adjust the pH to be neutral or slightly acidic. Why would a basic pH be harmful to hair structure?

2. **Environment** Water quality depends upon the ability of waste water treatment facilities to reprocess waste water entering the treatment facility. Most waste water treatment facilities use living organisms to break down wastes. Due to acid rain, the acidic nature of wastewater can pose problems for a waste treatment plant. Make an inference as to how pH affects wastewater treatment.

PROJECTS

1. Conduct a titration experiment with a variety of brands of orange juice, and determine the percentage of citric acid in each.

2. Design and conduct an experiment to isolate and extract possible acid–base indicators from such potential sources as red cabbage, berries, and flower petals. Test the action of each indicator that you are able to isolate with known acidic, basic, and neutral solutions.

3. Design and conduct an experiment to study the pH of rain in various locations for an extended period of time. Coordinate your data-collection activities with your classmates so that the pH readings will be made in a variety of locations. Try to determine whether any patterns emerge due to location, season, time of day, degree of industrialization in the area, etc.

4. Design and conduct an experiment to determine the pH of a variety of brands of shampoo. Do any patterns emerge in terms of types of shampoos or contents of each, relative to the results?

Chemistry and Careers

Food Chemist

Chemists play an essential role in the preparation and packaging of foods so that they come to you without loss of color, flavor, or texture. Chemists from the U.S. Department of Agriculture and the U.S. Food and Drug Administration monitor ingredients and packaging processes to ensure that the food you eat is safe. Other chemists synthesize new flavor and aroma molecules, adding them to improve the appeal of the food.

Food chemists work in a variety of areas within the food industries. Nutrition chemists design diets for hospitals, and flavor chemists synthesize such essences as "watermelon flavor" for bubble gum. A recent project is the development of calorie-free substitutes for sugar and for fats such as shortening.

What sort of background do you need to be a food chemist? Chemistry and biochemistry are useful basic courses, but, depending on what you wish to focus on, you will need extra courses in cooking, nutrition, and physiology. With a high school diploma and some training as a lab technician, you may qualify as a technician in a food chemistry lab. Food chemists must have a college degree in chemistry.

Aroma Chemist

Aroma chemistry permeates many industries, just as its products permeate the air. The job of an aroma chemist is unusual because it combines science and aesthetics. These are two interests that are sometimes erroneously considered to be incompatible with each other. The aroma chemist, also known as a perfumer, mixes chemicals to develop pleasant fragrances.

Aroma chemists are hired by cosmetic or consumer-products companies to develop new fragrances that have certain characteristics. It usually takes about two years to arrive at a final combination of essential oils—derived from flowers, herbs, and occasionally animals—and base ingredients such as soaps, oils, waxes, or powders. As a consultant, a perfumer may specialize in one product line or develop fragrances as diverse as those found in cologne, window cleaner, and kitty litter.

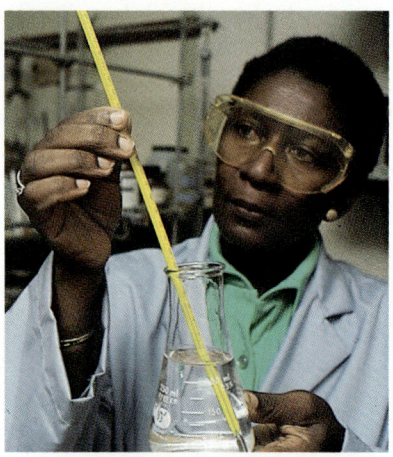

An analytical aroma chemist uses chromatography to separate the different aroma-producing compounds in a natural essence. The chemical structure of each compound is then determined by using various types of spectroscopy, including nuclear magnetic resonance. When the structure of each compound is known, it can be synthesized in the aroma laboratory and combined with other compounds in an aromatic mixture.

Most aroma chemists have a bachelor's degree in chemistry. A master's or doctoral degree in biochemistry and an interest in psychology are needed for aroma chemists who wish to understand how people perceive smells.

UNIT 5 • FOCUS ON REASONING

College Entrance Exam Preparation

A chemist isolates six different metals from ore found on the moon. To investigate some of the chemical properties of each metal, the metals are dissolved in various solvents and reacted with reagents containing the ions shown in the data table.

The reactivity tests were run at constant temperature. The formation of a precipitate shows that a product of the reaction is insoluble. The data are then used in formulating a set of properties for each metal.

1. Which reagent(s) give insoluble products when reacted with metal A?
   a. Cl^-.
   b. $C_2O_4^{2-}$.
   c. CO_3^{2-}.
   d. SCN^-.

2. If an unknown metal ion is reacted with CO_3^{2-} and forms a yellow-brown precipitate, it could be concluded that the unknown metal is
   a. metal C.
   b. metal D.
   c. metal E.
   d. metal F.

3. The metal that produces a soluble product when reacted with $Fe(CN)_6^{4-}$ is most likely
   a. metal C.
   b. metal D.
   c. metal E.
   d. metal F.

4. Which two metal ion samples appear to have the same chemical properties?
   a. metals A and B
   b. metals C and D
   c. metals E and F
   d. metals B and D

5. Which reagent produces insoluble products with all metal ion samples tested?
   a. Cl^-
   b. $C_2O_4^{2-}$
   c. CO_3^{2-}
   d. SCN^-

6. Chloride ion is added to a test tube with an unknown metal ion sample. A white precipitate forms. Which reagent would **not** be a good choice in determining whether the unknown was metal E or metal F?
   a. $C_2O_4^{2-}$
   b. CO_3^{2-}
   c. SCN^-
   d. $Fe(CN)_6^{4-}$

	Reagent Ion				
Metal Ion	Cl^-	$C_2O_4^{2-}$	CO_3^{2-}	SCN^-	$Fe(CN)_6^{4-}$
metal A	no reaction	no reaction	green ppt	no reaction	aqua ppt
metal B	no reaction	aqua ppt	blue ppt	no reaction	red-orange ppt
metal C	no reaction	white ppt	white ppt	no reaction	solution gels
metal D	no reaction	white ppt	white ppt	no reaction	solution gels
metal E	white ppt	white ppt	yellow-brown ppt	gray ppt	blue-green ppt
metal F	white ppt	white ppt	white ppt	black ppt	no reaction

INTRA-SCIENCE

Detergents and the Environment

Lake Tahoe is growing old before its time. Surrounded by the Sierra Nevada mountains of California, Lake Tahoe *was* like most other geologically young lakes—sparkling clear and cold, with few plants or plankton. In the past few decades, however, runoff containing organic wastes and phosphate detergents has nourished unwanted plants to the point that they have choked off other plant and animal life. Now, each spring the shores are covered by slimy green algae.

In the mid 1950s, soaps made of plant and animal fatty acids were replaced by synthetic detergents in American washing machines. Detergents give a cleaner, brighter wash than soaps, especially in "hard water," which is high in dissolved minerals.

Soaps reacted with dissolved calcium (Ca) and magnesium (Mg) ions in hard water to form insoluble salts, which leave a white scum on clothing. Detergents, on the other hand, remove dirt and grease and enter the wash water without leaving a residue.

Unfortunately the detergent-containing wastewater that entered the water systems through municipal sewage greatly altered the ecology and accelerated the aging of lakes and estuaries. Detergents have now been reformulated to avoid ecological damage, but much work remains to restore lakes to their original state and to eliminate other sources of water pollution.

The Connection to Physics

The cleansing action of detergents results from their ability not only to break up the oily film that binds dirt particles but also to weaken the cohesion between water molecules so that oil and dirt can enter the wash water. Detergent molecules contain a long, narrow hydrocarbon component that is attached to a group that has a negative charge. Because the hydrocarbon portion of detergent molecules adheres well to oil and because the charged portion of detergent molecules adheres well to water, oil is pulled off the clothing and into the wash water. Once in the water, each oil droplet is surrounded by a ball of detergent molecules. The hydrocarbon portions of the detergent molecules point inward, where they adhere to the oil molecules, which are also hydrocarbons. The negatively charged portion of the detergent molecules point outward into the water, which is a good solvent for charged molecules.

Once the oil has been taken up by the wash water, the dirt particles can also enter the water. This is facilitated by the action of the detergent in reducing the cohesive force between adjacent water molecules. By reducing the surface tension of water, detergents make it easier for the particles of dirt to push apart water molecules at the water-dirt interface and enter into the wash water, where they remain suspended.

The Connection to Biology and Health Sciences

Unlike soap, many of the first synthetic detergents were not biodegradable. The molecules of these synthetic detergents contained branched chains that could not be broken down into simple inorganic substances through the action of soil bacteria. Consequently, these detergents tended to collect in groundwater and contaminate it.

In contrast, detergents manufactured today are more similar to soaps made from natural fats. In the molecules of both, the hydrocarbon portions have skeletons in which carbon atoms are joined in a straight chain. Since soil bacteria are adapted to "eating" straight-chain hydrocarbons, they can prevent detergent molecules from accumulating.

Sodium phosphate pollution is the most severe environmental problem resulting from synthetic detergents, and it remains unresolved. Sodium phosphates are alkaline salts that form complexes with dissolved metal ions in the wash water so that these ions do not interfere with the process by which oil is emulsified.

When the phosphate concentration of a lake increases, so does the growth of algae and other plants. Plants and animals that were flourishing are forced out by other species. Lakes that were once clear become covered with a scum of blue-green algae. Decaying plants consume much of the lake's dissolved oxygen, and some species of fish die because there is not enough oxygen to sustain them.

Connection to Earth Science

Lakes are created by glaciers or rivers. Lakes age as sediment is deposited on the bottom and vegetation encroaches from the shore. The maturing of a lake (called eutrophication) normally takes thousands of years. The addition of phosphates greatly accelerates eutrophication by increasing the biomass and the rate of sedimentation. Because of pollution, the eutrophication of Lake Tahoe is occurring in only a few decades rather than thousands of years.

Many lakes, such as Lake Onondaga in New York, were redeemed when efforts were made to keep phosphate-laden wastes from flowing into them. Today, most detergents are free of phosphates. Phosphates have been replaced by safer but less effective substances, such as sodium carbonate.

Unit 6

Chapter 18
Reaction Energy
and Reaction Kinetics 526

Chapter 19
Chemical Equilibrium 562

Chapter 20
Oxidation–Reduction Reactions 596

CHEMICAL REACTIONS

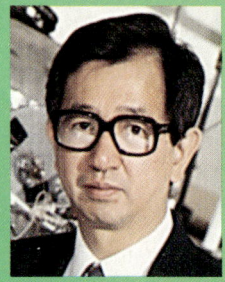

Growing up in Taiwan after World War II, I was exposed to the excitement of science transforming human society. But my decision to be a research chemist can perhaps be traced to the time when I had a chance to read the biography of Madame Curie in junior-high school. Her dedication to scientific research, her excitement about new discoveries, and her idealistic attitude convinced me that the life of a research chemist could be very beautiful.

My main research interest is understanding the exact details of how chemical reactions take place at the molecular level. Molecular collisions are responsible for chemical change. Since it is not possible to see the collisional events of microscopic atoms and molecules with the naked eye, we devise various complicated experimental apparatuses to try to "visualize" the reactive scattering processes of atoms and molecules. With tools such as the molecular beam apparatus and laser, scientists can study chemical reactions at the molecular and atomic levels in great detail.

Yuan T. Lee
Physical Chemist
University of California, Berkeley

Laser study of flame chemistry

CHAPTER 18
Reaction Energy and Reaction Kinetics

PLANNING CHART
Approximate Time: 5/7 days

	SECTION/TEXT FEATURES		RESOURCES	
A V E R A G E	**18.1 THERMOCHEMISTRY** Demo 1 Section Review Chapter Review • Questions 1–4, Problems 1–7	527 526B 536 558	*Teacher's Resource Binder* Lab Experiments 25, 26 Problem Solving Worksheets Section Review Worksheets *Other Resources* *Shakhashiri* Video Demo 7 Transparency 37 Blackline Masters 42, 43, 44	139, 143 63
A V E R A G E	**18.2 DRIVING FORCE OF REACTIONS** Demo 2 *Periodic Table* Videodisc Section Review Chapter Review • Question 5–8, Problems 8–10	536 526C 526H 540 558	*Teacher's Resource Binder* Problem Solving Worksheets Section Review Worksheets *Other Resources* *World of Chemistry* Video 13	 64
H O N O R S	**18.3 THE REACTION PROCESS** Section Review Chapter Review • Questions 9–13, Problems 11–12	541 549 558	*Teacher's Resource Binder* Section Review Worksheets *Other Resources* *Shakhashiri* Video Demo 46, 48 Transparencies 38, 39, 40, 41 Blackline Master 45	65
H O N O R S	**18.4 REACTION RATE** Demos 3, 4 Desktop Investigation Microscale Investigation 16 Section Review Chapter Review • Questions 14–15, Problems 13–14	549 526D 555 869 556 557	*Teacher's Resource Binder* Lab Experiment 27 Section Review Worksheets *Other Resources* *Shakhashiri* Video Demos 24, 44, 45, 46 *World of Chemistry* Videos 14, 20 Transparency 42	149 66
	CHAPTER END ACTIVITIES Critical Thinking, Cross Connections, Research and Writing, Projects		*Teacher's Resource Binder* Test Booklet *Computer Test Generator*	89, 91

Teaching Strategies

Introduction

Three questions confront a chemist studying a chemical reaction. *(1)* Will the reaction proceed as written? *(2)* How long will the reaction take to reach completion? *(3)* How much product will be obtained? Each question is answered by different but related fields of chemistry. Thermodynamics and the calculation of the free energy of the reaction answers the first question. Chemical kinetics or the study of rates of reactions answers the second, and, finally, the study of chemical equilibrium and the law of mass action answers the third. Many texts treat these as separate topics.

18.1 Thermochemistry

The term *heat content* is a confusing one. The term describes the energy stored within molecules. It represents the energy stored within the chemical bonds and, in a sense, measures the stability of the molecules. Explosives are carefully synthesized to contain large amounts of stored energy. This energy is released in the explosion or detonation of the material.

The physical state of the substance described is very important and must be written as part of the chemical reaction. Ask the question "What would cause a more severe burn, 10 g of liquid water at 100°C or 10 g of steam (water vapor) at 100°C?"

(The gaseous water-vapor molecules contain more kinetic energy than the molecules of the liquid. The steam would cause a more severe burn.)

As equations with the heat of reaction are presented, keep stressing that the law of conservation of energy is in effect. The total energy of the reactant side of the equation must equal the total energy of the product side. This total energy includes the heat that is either absorbed (when energy is a reactant) or released (when energy is a product). If the energy term is removed from the balanced equation, the total stored energies are no longer equal. If the heat term were removed from the reactant side, the products would have more energy. If the heat term were removed from the product side, the reactants would have more stored energy. Explain to the students that this stored energy is what is being described as heat content, ΔH. ΔH for a reaction is defined as the difference between the total heat content of the products minus the total heat content of the reactants. If the energy term is removed from the reactant side, ΔH must be positive, because the products have stored within them the energy written as a separate term on the reactant side. ΔH is a negative term for an exothermic reaction. The students must realize that there is no such thing as negative energy.

This is generally a very confusing issue and if it can be clarified early in the chapter, the students will have a less difficult time with changes in entropy and free energy. The sign of ΔH simply indicates heat flow. When positive, heat is flowing into the system and when negative, it is flowing away from the system.

Demonstration 1: Reaction between Hydrogen and Oxygen

PURPOSE: To illustrate thermochemistry by the combination of hydrogen with oxygen

MATERIALS: 2 small balloons filled with approximately 2:1 ratios of hydrogen and oxygen, candle, meter stick, matches, string, tape

PROCEDURE: Fill the balloons appropriately with the two gases and attach them to either end of the teachers desk with string. Discuss the concepts presented in this first part of the chapter in terms of the molecules in each of the balloons. Students are generally disappointed that nothing seems to happen even though the text indicates that hydrogen and oxygen form an explosive mixture and give off a very large amount of energy when they combine to form water. **CAUTION:** The following safety precautions must be followed: ears must be protected either by placing a piece of tissue in each or by cupping one's hands over the ears; teachers (and administrators) in neighboring classrooms need to be warned about the upcoming explosion; and, if any students are seriously affected by loud noises, they may be excused from the room.

Turn off the room lights and, using a candle taped to the end of a meter stick, touch one of the balloons with the flame. A very loud explosion and a brilliant yellow flash result. Have the students try to explain what they have seen. Many of the concepts that will be used later in the chapter may be drawn from them. Hydrogen and oxygen molecules were colliding in the balloon but not reacting (insufficient activation energy). Energy was needed to initiate the reaction. Although a small amount of heat was used to initiate the reaction, a large amount was given off—the reaction was definitely exothermic. Under room conditions, the reaction would be very slow, not even noticeable. After initiation, the reaction proceeded at an explosive pace.

At this point, introduce the concept of molar heat of formation, an intensive property that can be tabulated. Compare this term with the heat of reaction, an extensive property, presented earlier in the chapter.

Since all of these terms are defined by initial and final states, explain the rationale of assigning the heats of formation of elements in their standard state a value of zero. This does not mean that elements have no energy. It is a relative term that describes the difference in energy between the elements and the molecule produced. If ΔH is positive, the molecule contains more energy than the elements used to produce it. On the other hand, if ΔH is negative, the elements have more energy stored within them than does the molecule produced. Where does the extra energy go? It must be released as heat, sound, light, etc. The reaction must be exothermic.

With this background most students will begin to predict stability based on heats of reactions. If a synthesis process is very endothermic, the product is probably unstable. Manufacturers of explosives and fireworks go to great lengths to store a lot of energy within the chemical bonds of the molecules that make up the fireworks. A newspaper article describing the explosion of a fireworks manufacturing firm brings the concept home.

When calculating the heats of combustion, encourage the students not to take short cuts as they begin to learn the process. Show the students how they must take the tabulated data, change the sign depending on where the molecule occurs in the equation describing the reaction, and how they must multiply the tabulated value by the coefficient from the equation. Much drill at this time will make the presentation of Hess' law less difficult.

When presenting Hess' law, work out the initial examples in detail. First be sure that the students realize how the formulas in the equations must be shifted and then how the energy terms associated with those formulas are added algebraically. Stress that Hess's law is an extremely powerful tool. Energies for reactions that would be next to impossible to carry out in a lab, can be calculated from known heats of formation.

Working with the remaining hydrogen balloon and molecular models of hydrogen gas and oxygen gas, ask the students to explain what must occur before hydrogen and oxygen can react to form water. Explain how splitting apart an oxygen molecule requires energy and therefore is an endothermic process and how uniting the individual hydrogen and oxygen atoms as water molecules is an exothermic process.

18.2 Driving Force of Reactions

At this time, you might "accidentally" drop a book onto the floor. Ask the students why the book fell to the floor. All students will acknowledge the presence of gravity. Stress that in a gravitational field such as that found on Earth, all objects tend to move toward the center of the earth unless something, such as the floor, stands in the way. Use this analogy to explain that it is a fundamental tendency in nature to move to the most energetically stable position. In the case of a chemical reaction, considering only energy, the most stable state must be the state that contains the least amount of stored energy. Explain how the products of exothermic reactions meet this criteria and how many exothermic reactions are indeed spontaneous.

Pose, but don't answer, the question "Why don't the hydrogen and oxygen molecules in the balloon spontaneously react to form water?" The reaction is certainly exothermic and thus the water molecules must contain less energy than the original hydrogen and oxygen molecules. After posing the question, perform the following demonstration.

Demonstration 2: A Spontaneous Endothermic Reaction

PURPOSE: To introduce the concept of entropy as a driving force for chemical reactions
MATERIALS: Approximately 20 g of barium hydroxide ($Ba(OH)_2 \cdot 8H_2O$) and 10 g of ammonium thiocyanate (NH_4SCN) (Carefully grind the crystals of $Ba(OH)_2 \cdot 8H_2O$ into a fine powder with a mortar and pestle), 250-mL Erlenmeyer flask, solid rubber stopper for the flask, block of wood, distilled water bottle
PROCEDURE: Explain to the students that you will be carrying out a reaction involving two solids. Ask the students to describe what they observe. Show the vials containing the two solids to the students. Appear to clean the block of wood by pouring some distilled water onto the block of wood. The puddle should be about half the diameter of the base of flask.

Pour each of the solids into a flask. Stopper the flask and begin to swirl the contents. The solids appear to get wet and begin to form a liquid. As this happens, place the flask on the puddle of water on the block and hold up the block and flask for all to see. Continue to swirl the block and flask. After a minute or so, the flask adheres to the block of wood and a film of ice forms on the outside part of the flask in contact with the liquid. Feign surprise and try to remove the flask from the block of wood. (Don't try too hard. The ice layer binds the flask to wood to such an extent that the flask may break if too much pressure is applied.)

You may pass the block of wood around the room, allowing the students to feel the intense cold. **CAUTION:** Do not touch the base of the flask. Frostbite might occur if the mixture is placed directly on the skin. For this reason this reaction is not suitable for chemical ice packs.

Open the flask and waft the fumes towards the nose. The odor of ammonia will be detected. Thus two solids reacted to form a solution and a gas. The solution is formed from the water of hydration. The process absorbed so much energy that water in contact with the mixture froze. Explain how the solution and gaseous state contain particles in more random states than the original crystalline solids. Finally, explain that a second tendency in nature is the tendency toward the more random of two states. Such a movement always involves absorption of energy. This is a case of an entropy-driven reaction.

$$Ba(OH)_2 \cdot 8H_2O + 2NH_4SCN \rightarrow$$
$$Ba^{2+} + 2SCN^{-1} + 2NH_3 + 10H_2O$$

Summarize the information gleaned from this demonstration and from the examples of entropy driven reactions in the text by writing the phases of matter on the chalkboard or overhead projector.

lower	gas	more
energy	liquid	random
state	solid	state

Nature tends to move spontaneously to the lower energy state (solid) and to the more random state (gas). The two tendencies oppose each other. Whichever one wins out determines the course of the reaction. With this background, present the chemist's need for another term or function that takes both tendencies into consideration. Gibbs Free Energy, ΔG, is the function.

Point out that exothermic reactions move to lower energy states and that a negative ΔH is associated with exothermic reactions. A positive sign is required for the change in entropy because one moves from a less random to a more random state. Since the temperature of a material determines its physical state, an argument can be made for including temperature with the entropy term.

Define ΔG as the difference between the change in enthalpy (heat content) and the product of the Kelvin temperature and the change in entropy (randomness) of a system. For a reaction to proceed spontaneously as written (from reactant to product *or* left to right), the sign of ΔG must be negative.

When working the problems in the text, point out that tabulated values of ΔH and ΔS have different units. For our beginning study, ΔH is considered independent of temperature and is expressed in kJ/mol. ΔS is dependent on temperature and expressed in J/(mol·K). Students frequently forget to convert kJ to J or vice versa.

18.3 The Reaction Process

Return to the hydrogen–oxygen balloon to discuss the reaction process. Have the students once again try to describe what must be taking place on a molecular level. (Molecules are colliding with each other.) Bonds are not being broken because no reaction is evident at room conditions. Once the reaction is initiated, it goes to completion. Here is a reaction driven by the large decrease in enthalpy. All the heat evident in the reaction must have come from the original hydrogen and oxygen molecules. The water molecules must be at a lower enthalpy state. Entropy does not appear to be too important because one has gases as reactants and as a product.

At this point the concepts of reaction mechanism, collision theory, activation energy, and activated complex may be introduced using the hydrogen–oxygen reaction as well as the examples listed in the text. Draw from the students the following requirements: oxygen and hydrogen molecules must split apart so that the atoms can rearrange to form products; the molecules or atoms must come in contact with each other if they are to react; energy is required to split the molecules; as the molecules split and rearrange to form the product molecules, they must pass through some point in which they can return to the original molecules or move on to the product molecules.

18.4 Reaction Rate

Students will agree that the collision theory makes sense. In discussing the rate of reactions and factors that influence the rates, use a traffic analogy to bring the concepts home.

Most students will agree that more accidents (collisions) occur during rush hour when more cars are on the road (greater concentration). At the same time, the collisions during rush hour may not be as severe as collisions at other times because most traffic moves more slowly during rush hour (lower temperature and velocity of molecules). In a collision of a large car with a compact car, the smaller car is damaged more severely (nature of reactants). The analogy can continue with ice on the road acting as a catalyst and how the cars hit (orientation) will, in part, determine how much damage occurs.

A number of demonstrations are available to bring these concepts home to students. Grain-elevator explosions and some coal-mine explosions are due to the very large surface area of the grain residues and coal dust. If lycopodium powder is available, a little may be sprinkled into a Bunsen-burner flame. The resulting flash simulates what happens in a grain-elevator explosion.

Demonstration 3: Concentration vs. Reaction Rate

PURPOSE: To show the effect of concentration on the rate of a reaction

MATERIALS: A clean, one-pint paint can and lid, masking tape, short length of rubber hose, matches

PROCEDURE: Punch a hole with a can opener near the bottom of the can. Punch a smaller hole, approximately 1 cm diameter, in the lid. (The size of the upper hole determines the rate of the reaction.) Place the lid on the can and fill the can from the top with methane gas from a gas jet. Attach the short hose to the gas valve and allow the methane to flow until the odor is noticeable. Ask students why the can is filled from the top and not from the bottom (methane is less dense than air). After the can is filled, place a piece of tape over the two holes. The can may then be transported to a safe place, preferably one away from the students and not below any light fixtures.

Explain that you are going to light the methane in the can and the students are to try and explain what occurs. The lit methane burns with a bright yellow flame. It does not burn inside the can because no oxygen is present in the can. The flame remains on the outside of the can as long as the mixture in the can does not have enough oxygen to support combustion. Ask the students why two holes are needed. (Most will reply to replace the burning methane.) Depending on the size of the hole in the lid, reaction continues for several minutes. You may have to turn off the room lights as the flame gets smaller and burns with an almost invisible blue flame. As the flame disappears into the can, an explosion occurs, blowing off the lid. If the lid is pressed on tightly enough, the force of the explosion can send it to the ceiling. **CAUTION:** It is imperative that you and the students are not near the can at the end of the demonstration.

You should make the can and practice ahead of time. Enlarge the size of the hole in the lid so that the reaction time is between three and four minutes.

Demonstration 4: Temperature vs. Rate of Reaction

PURPOSE: To show that temperature determines the rate of a reaction

MATERIALS: Three cyalume light sticks, three 1000-mL beakers: one filled with boiling water, one with water at room temperature, and one with ice water

PROCEDURE: Initiate the cyalume reaction by bending the light sticks, breaking the inner glass ampule, and mixing the reagents. All three light sticks should glow with the same intensity.

Place all three light sticks in the water at room temperature. Note that all give off the same amount of light. Remove one from the beaker and place it in the beaker with the hot water. Remove a second and place it in the beaker with ice water. The brightness of the glow is indicative of the rate of the reaction. Take the brightly glowing light stick from the hot water and place it in the ice water, and do the reverse for the one that was in the ice water. Whichever stick is in the hot water will glow the brightest.

Explain that increasing the temperature also shortens the time the light stick will glow. Most yellow-green light sticks are rated for 12 hours at room temperature. The demonstration is a very dramatic one.

The demonstration books listed in the reference section include a number of other demonstrations that would be appropriate. The Desktop Investigation, *Factors Influencing Reaction Rate*, will allow students to study the factors influencing reaction rates on their own.

Now that the students have a qualitative grasp of the factors that influence reaction rates, explain that chemists need to quantitatively describe the rates of reactions. Rates are always positive numbers. Since reactions ordinarily take place in a series of steps, the slowest step must be the one that determines the overall rate. You may explain this in terms of a dishwashing analogy. If six students decide to wash glassware and two collect the dirty glassware, one washes and rinses and the others dry the glassware, the rate-determining step is the dishwashing step. The only way to increase the rate is to take one of the dryers and have him/her also wash glassware.

The rate-determining step must be determined experimentally and cannot be deduced from the balanced chemical equation. For a reaction to occur, individual steps will involve two-body collisions. Students will agree that even on a busy highway, seldom will three or four cars collide simultaneously. Most multiple car collisions occur in a chain reaction form with no more than two cars colliding at any one time.

Using the examples in the text, write out rate laws for the students.

References and Resources

Books and Periodicals

Borgford, Christie L. and Lee R. Summerlin. "Chemical Energy and Rates of Reaction" *Chemical Activities: A Sourcebook for Science Teachers*, The American Chemical Society, 1988.

Olney, David J. "Some Analogies for Teaching Rates/Equilibrium," *Journal of Chemical Education*, Vol. 65, No. 8, Aug. 1988, pp. 696–697.

Shakhashiri, Bassam Z. "Thermochemistry," *Chemical Demon-*

Spencer, J.N. and E.S. Holmboe. "Entropy and Unavailable Energy," *Journal of Chemical Education*, Vol. 60, No. 12, Dec. 1983, pp. 1018–1021.

Summerlin, Lee R., et al. "Kinetics," *Chemical Demonstrations—A Sourcebook for Teachers*, Vol. 1, Second Edition, Washington, D.C.: American Chemical Society, 1988.

Summerlin, Lee R., et al. "Kinetics and Equilibrium," *Chemical Demonstrations—A Sourcebook for Teachers*, Vol. 2, Washington, D.C.: American Chemical Society, 1987.

strations, Vol. 2, University of Wisconsin Press, 1985.

Treptow, Richard S. "LeChâtelier's Principle Applied to the Temperature Dependence of Solubility," *Journal of Chemical Education*, Vol. 61, No. 6, June 1984, pp. 499–502.

Audiovisual Resources

Eyring, Henry. *Introduction to Reaction Kinetics*, A Chem Study Film, Color, 13 min., 16mm film or VHS cassette, Central Scientific Co.

Powell, Richard E. *Catalysis*, A Chem Study Film, Color, 17 min., 16-mm film or VHS cassette, Central Scientific Co.

Answers and Solutions

Reviewing Concepts

1. **a.** Higher **b.** Lower
2. **a.** The heat of reaction is the quantity of heat released or absorbed during a chemical reaction and after the reactants return to their original temperature. The heat of formation is the heat released or absorbed when one mole of a compound is formed by combination of its elements. The heat of combustion is the heat released by the complete combustion of 1 mol of a substance.
   b. The heat of combustion is defined in terms of 1 mol of reactant. Heat of formation, on the other hand, is defined in terms of 1 mol of product.
3. **a.** ΔH = sum of ΔH°_f of products − sum of ΔH°_f of reactants
   b. ΔH°_f of compound X = sum of ΔH°_f of combustion products of X − ΔH_c of X.
4. The change in heat content of a reaction system is related to
   1. the change in the number of bonds breaking and forming, and
   2. the strengths of these bonds as the reactants form products.
5. Changes of phase in the direction of the solid phase are accompanied by decreases in entropy.
6. A rise in temperature generally favors an increase in entropy.
7. A negative ΔH and a positive ΔS always result in a negative ΔG that is indicative of a spontaneous reaction.
8. The temperature of a system is the dominant factor in determining the relative importance of the tendencies toward lower energy and higher entropy. At low temperatures, $T\Delta S$ is generally small in comparison to ΔH, so the reaction proceeds as the ΔH dictates. However, at high temperatures, when ΔS is positive, the $T\Delta S$ factor may be large enough to exceed the ΔH value and thus the entropy factor dictates the spontaneity of the reaction.
9. **a.** The collision theory is the set of assumptions regarding collisions and reactions.
   b. The collision must be energetic enough to supply the necessary activation energy and the colliding molecules must be oriented in a way that favors their efficient interaction.
10. **a.** The reactant particles must undergo an effective collision that raises the internal energies of the reactants to their minimum level for reaction.
    b. The activated complex occurs at the maximum-energy position along the reaction pathway.
11. The activation energy required for the endothermic change is greater than that required for the exothermic change by the amount of the heat of reaction of the system.
12. Since most reactions actually take place in a sequence of one-step processes, it is unlikely that the burning of ethane takes place in this single step.
13. $NO_2 + O \rightarrow NO + O_2$
14. The rate-determining step is the step that proceeds at the lowest rate, and thus determines the overall reaction rate.
15. $R = k[A]^n[B]^m$ . . . where R is the reaction rate, k is the rate constant, [A] and [B] . . . represent the molar concentrations of reactants, and n and m represent powers to which the concentrations are raised.

Problems

1. **a.** $\Delta H = -393.51$ kJ; exothermic
   b. $\Delta H = -890.31$ kJ; exothermic
   c. $\Delta H = +176$ kJ; endothermic
   d. $\Delta H = -44.02$ kJ; exothermic
   e. $\Delta H = -197.8$ kJ; exothermic
2. **a.** $H_2(g) + \frac{1}{2}O_2(g) \rightarrow H_2O(\ell) + 285.83$ kJ; exothermic
   b. $2Mg(s) + O_2(g) \rightarrow 2MgO(s) + 1200.0$ kJ; exothermic
   c. $I_2(s) + 62.4$ kJ $\rightarrow I_2(g)$; endothermic
   d. $3CO(g) + Fe_2O_3(s) \rightarrow 2Fe(s) + 3CO_2(g) + 24.7$ kJ; exothermic
   e. $2NO_2(g) + 114.2$ kJ $\rightarrow 2NO(g) + O_2(g)$; endothermic
   f. $C_2H_4(g) + 3O_2(g) \rightarrow 2CO_2(g) + 2H_2O(\ell) + 1411.0$ kJ; exothermic

3. a. $Ca(s) + Cl_2(g) \rightarrow CaCl_2(g) + 795.0$ kJ ($\Delta H_{reverse} = +795.0$ kJ)
  b. $2C(s) + H_2(g) + 226.73$ kJ $\rightarrow C_2H_2(g)$ ($\Delta H_{reverse} = -226.73$ kJ)
  c. $S(s) + O_2(g) \rightarrow SO_2(g) + 296.83$ kJ ($\Delta H_{reverse} = +296.83$ kJ)

4. a. Reaction: $CaCO_3(s) \rightarrow CaO(s) + CO_2(g)$
  $CaCO_3(s) \rightarrow Ca(s) + C(s) + \frac{3}{2}O_2(g); \Delta H = +1206.9$ kJ
  $Ca(s) + \frac{1}{2}O_2(g) \rightarrow CaO(s); \Delta H = -635.5$ kJ
  $C(s) + O_2(g) \rightarrow CO_2(g); \Delta H = -393.51$ kJ
  $CaCO_3(s) \rightarrow CaO(s) + CO_2(g); \Delta H = +177.9$ kJ
  $\Delta H = [\Delta H_f^o(CaO) + \Delta H_f^o(CO_2)] - [\Delta H_f^o(CaCO_3)]$
  $\Delta H = [(-635.5$ kJ$) + (-393.51$ kJ$)] - [-1206.9$ kJ$]$
  $\Delta H = +177.9$ kJ

  b. Reaction: $Ca(OH)_2(s) \rightarrow CaO(s) + H_2O(g)$
  $Ca(OH)_2(s) \rightarrow Ca(s) + O_2(g) + H_2(g); \Delta H = +986.6$ kJ
  $Ca(s) + \frac{1}{2}O_2(g) \rightarrow CaO(s); \Delta H = -635.5$ kJ
  $H_2(g) + \frac{1}{2}O_2(g) \rightarrow H_2O(g); \Delta H = -241.82$ kJ
  $Ca(OH)_2(s) \rightarrow CaO(s) + H_2O(g); \Delta H = +109.3$ kJ
  $\Delta H = [\Delta H_f^o(CaO) + \Delta H_f^o(H_2O)] - [\Delta H_f^o(Ca(OH)_2)]$
  $\Delta H = [(-635.5$ kJ$) + (-241.82$ kJ$)] - [-986.6$ kJ$]$
  $\Delta H = +109.3$ kJ

  c. Reaction: $Fe_2O_3(s) + 3CO(g) \rightarrow 2Fe(s) + 3CO_2(g)$
  $Fe_2O_3(s) \rightarrow 2Fe(s) + \frac{3}{2}O_2(g); \Delta H = +824.2$ kJ
  $3CO(g) \rightarrow 3C(s) + \frac{3}{2}O_2(g); 3[\Delta H = +110.53$ kJ$]$
  $3C(s) + 3O_2(g) \rightarrow 3CO_2(g); 3[\Delta H = -393.51$ kJ$]$
  $Fe_2O_3(s) + 3CO(g) \rightarrow 2Fe(s) + 3CO_2(g); \Delta H = -24.7$ kJ
  $\Delta H = [\Delta H_f^o(Fe) + 3\Delta H_f^o(CO_2)] - [\Delta H_f^o(Fe_2O_3) + 3\Delta H_f^o(CO)]$
  $\Delta H = [0 + 3(-393.51$ kJ$)] - [(-824.2$ kJ$) + 3(-110.53$ kJ$)]$
  $\Delta H = -24.7$ kJ

5. a. Reaction: $C_2H_6(g) + \frac{7}{2}O_2(g) \rightarrow 2CO_2(g) + 3H_2O(\ell)$
  $C_2H_6(g) \rightarrow 2C(s) + 3H_2(g); \Delta H = +84.68$ kJ
  $2C(s) + 2O_2(g) \rightarrow 2CO_2(g); 2[\Delta H = -393.51$ kJ$]$
  $3H_2(g) + \frac{3}{2}O_2(g) \rightarrow 3H_2O(\ell); 3[\Delta H = -285.83$ kJ$]$
  $C_2H_6(g) + \frac{7}{2}O_2(g) \rightarrow 2CO_2(g) + 3H_2O(\ell); \Delta H = -1559.83$ kJ
  $\Delta H = [2\Delta H_f^o(CO_2) + 3\Delta H_f^o(H_2O)] - [\Delta H_f^o(C_2H_6) + \Delta H_f^o(O_2)]$
  $\Delta H = [2(-393.51$ kJ$) + 3(-285.83$ kJ$)] - [-84.68$ kJ $+ 0]$
  $\Delta H = -1559.83$ kJ

  b. Reaction: $C_6H_6(\ell) + \frac{15}{2}O_2(g) \rightarrow 6CO_2(g) + 3H_2O(\ell)$
  $C_6H_6(\ell) \rightarrow 6C(s) + 3H_2(g); \Delta H = -48.50$ kJ
  $6C(s) + 6O_2(g) \rightarrow 6CO_2(g); 6[\Delta H = -393.51$ kJ$]$
  $3H_2(g) + \frac{3}{2}O_2(g) \rightarrow 3H_2O(\ell); 3[\Delta H = -285.83$ kJ$]$
  $C_6H_6(\ell) + \frac{15}{2}O_2(g) \rightarrow 6CO_2(g) + 3H_2O(\ell); \Delta H = -3267.05$ kJ
  $\Delta H = [6\Delta H_f^o(CO_2) + 3\Delta H_f^o(H_2O)] - [\Delta H_f^o(C_6H_6) + \Delta H_f^o(O_2)]$
  $\Delta H = [6(-393.51$ kJ$) + 3(-285.83$ kJ$)] - [+48.50$ kJ $+ 0]$
  $\Delta H = -3267.05$ kJ

6. $\Delta H_c(C_2H_5OH) = [2\Delta H_f^o(CO_2) + 3\Delta H_f^o(H_2O)] - [\Delta H_f^o(C_2H_5OH)]$
  $\Delta H_c(C_2H_5OH) = [2(-393.51$ kJ/mol$) + 3(-285.83$ kJ/mol$)] - [-277.0$ kJ/mol$]$
  $\Delta H_c(C_2H_5OH) = -1367.5$ kJ/mol

7. $S(s) + O_2(g) \rightarrow SO_2(g); \Delta H = -296.83$ kJ/mol
  $SO_2(g) + \frac{1}{2}O_2(\ell) \rightarrow SO_3(g); \Delta H = -98.91$ kJ/mol
  $SO_3(g) + H_2O(\ell) \rightarrow H_2SO_4(\ell); \Delta H = -132.4$ kJ/mol
  $H_2(g) + \frac{1}{2}O_2(g) \rightarrow H_2O(\ell); \Delta H = -285.83$ kJ/mol
  $S(s) + 2O_2(g) + H_2(g) \rightarrow H_2SO_4(\ell); \Delta H = -814.0$ kJ/mol

8. a. $\Delta G = \Delta H - T\Delta S$.
  $\Delta G = (+125$ kJ/mol$) - (293K)(0.0350$ kJ/(mol·K)$)$
  $\Delta G = +125$ kJ/mol $- 10.3$ kJ/mol
  $\Delta G = 115$ kJ/mol; not spontaneous

  b. $T = 127°C + 273 = 400.K$
  $\Delta G = \Delta H - T\Delta S$
  $\Delta G = (-85.2$ kJ/mol$) - (400.K)(0.125$ kJ/(mol·K)$)$
  $\Delta G = -85.2$ kJ/mol $- 50.0$ kJ/mol
  $\Delta G = -135$ kJ/mol; spontaneous

  c. $T = 500.°C + 273 = 773K$
  $\Delta G = \Delta H - T\Delta S$
  $\Delta G = (-275$ kJ/mol$) - (773K)(0.450$ kJ/(mol·K)$)$
  $\Delta G = -275$ kJ/mol $- 348$ kJ/mol
  $\Delta G = -623$ kJ/mol; spontaneous

9. $T = 25°C + 273 = 298K$
  $\Delta G = \Delta H - T\Delta S$

$\Delta G = (-393.51 \text{ kJ/mol}) - (298\text{K})(0.003\ 00 \text{ kJ/}(\text{mol} \cdot \text{K}))$
$\Delta G = -394.40 \text{ kJ/mol}$; spontaneous

10. $T = 27°\text{C} + 273 = 300.\text{K}$
$\Delta G = \Delta H - T\Delta S$
$\Delta G = (-74.8 \text{ kJ/mol}) - (300.\text{K})(-0.0809 \text{ kJ/}(\text{mol} \cdot \text{K}))$
$\Delta G = -50.6 \text{ kJ/mol}$; spontaneous

11. **a.** $\Delta H_{\text{forward}} = +80 \text{ kJ}$
$\Delta H_{\text{reverse}} = -80 \text{ kJ}$
$E_a = 100 \text{ kJ}$
$E_a' = 20 \text{ kJ}$

b. $\Delta H_{\text{forward}} = -40 \text{ kJ}$
$\Delta H_{\text{reverse}} = +40 \text{ kJ}$
$E_a = 20 \text{ kJ}$
$E_a' = 60 \text{ kJ}$

c. $\Delta H_{\text{forward}} = +10 \text{ kJ}$
$\Delta H_{\text{reverse}} = -10 \text{ kJ}$
$E_a = 70 \text{ kJ}$
$E_a' = 60 \text{ kJ}$

12. **a.** $\Delta H_{\text{reverse}} = -30 \text{ kJ}$
$E_a = 50 \text{ kJ}$

b. $\Delta H_{\text{reverse}} = +30 \text{ kJ}$
$E_a' = 50 \text{ kJ}$

c. $\Delta H_{\text{reverse}} = -10 \text{ kJ}$
$E_a = 50 \text{ kJ}$

d. $\Delta H_{\text{reverse}} = -95 \text{ kJ}$
$E_a = 115 \text{ kJ}$

e. $\Delta H_{\text{forward}} = -40 \text{ kJ}$
$E_a' = 70 \text{ kJ}$

13. $R = k[A]^2[B]$. Doubling A increases the rate to 4 times its original rate. Doubling B increases the rate to twice its original rate.

14. **a.** $R = k[A][B]^2$

b. $k = \dfrac{R}{[A][B]^2} = \dfrac{2.0 \times 10^{-4} \text{ mol/(L} \cdot \text{min)}}{(0.20 \text{ mol/L})(0.20 \text{ mol/L})^2} = 2.5 \times 10^{-2} \dfrac{L^2}{\text{mol}^2 \cdot \text{min}}$

c. $R = k[A][B]^2 = 2.5 \times 10^{-2} \dfrac{L^2}{\text{mol}^2 \cdot \text{min}} \times 0.30 \text{ mol/L} \times (0.30 \text{ mol/L})^2 = 6.8 \times 10^{-4} \dfrac{\text{mol}}{L \cdot \text{min}}$

Critical Thinking

1. Fuels are usually chosen for their energy content, the energy released per unit consumed, not the temperature of the flame. Cost is the prime criterion in most comparisons; transportation, ease of use, and cleanliness are also considerations.

2. As the formation of a single activated complex by the simultaneous collision of five molecules is highly improbable, the reaction mechanism must involve some sequence of simple steps.

3. **a.** This is the pathway by which the potential energy of the reactants is raised to the minimum level needed for effective collisions.
   b. The maximum-energy region of the minimum-energy pathway for reaction is reached when the activated complex is formed.

4. In such areas, the powdered nature of the dry combustible sawdust or grain allows for large surface areas of contact with oxygen so that the presence of even a spark could initiate a very rapid reaction that could result in an explosion.

5. To determine the reaction rate, measure the pressure of the reaction system. Since two moles of gas react to form one mole of gas product, the pressure will decrease as the reaction proceeds.

6. **a.** The rate is increased.
   b. The rate is doubled.
   c. The rate is cut in half.
   d. The rate is increased.

Videodisc Selections
from *The Periodic Table Videodisc* (JCE Software)

step forward

step reverse

CHAPTER 18

pg 537 Examples of spontaneous reactions

3835
carbon + oxygen (m)

7535
magnesium + oxygen (m)

24084
iron + oxygen (m)

32434
germanium + oxygen (m)

17334
potassium + water (m)

38127
strontium + water (m)

6036
sodium + acids (m)

26677
cobalt + acids (m)

Teacher's Notes

Chapter 18
Reaction Energy and Reaction Kinetics

SECTION PREVIEW

18.1 Thermochemistry
18.2 Driving Force of Reactions
18.3 The Reaction Process
18.4 Reaction Rate

A forest fire is a dramatic example of a chemical reaction that results in the release of much energy in the form of heat and light.

18.1 Thermochemistry

The role of energy in chemical reactions is of primary interest. Virtually every chemical reaction is accompanied by the absorption or the release of energy. To compare the quantities of energy associated with chemical reactions, the reactions are compared under clearly defined conditions. When this is done, a characteristic absorption or release of energy can be assigned to each.

During a chemical reaction, energy is used to break chemical bonds and energy is released when new bonds form. The overall energy change depends mainly on the relative strengths and numbers of bonds broken and formed.

Most often chemical reactions absorb or release energy in the form of heat. You learned in Chapter 13 that heat is also absorbed or released in physical changes, such as melting ice or condensing a vapor. **Thermochemistry** *is the study of the changes in heat energy that accompany chemical reactions and physical changes.*

Heat of Reaction

The heat absorbed or released in a chemical or physical change is measured in a device known as a **calorimeter.** For example, if a mixture of hydrogen and oxygen is ignited, water is formed and heat energy is released explosively. Known quantities of reactants are sealed in the reaction chamber, which is immersed in a known quantity of water in an insulated vessel. Therefore, the heat given off (or absorbed) during the reaction is determined from the temperature change of the known mass of surrounding water. For example, if the reaction is exothermic, the surrounding water will become warmer.

The quantity of heat given up during the formation of water from H_2 and O_2 is proportional to the quantity of water formed. No heat is supplied externally, except to ignite the mixture. The heat that is released comes from the reactants as they change into products. The heat content of the product (water) must be less than the heat content of the reactants before ignition. The equation for this reaction is written as

$$2H_2(g) + O_2(g) \rightarrow 2H_2O(g)$$

which expresses the fact that when 2 mol of hydrogen gas at room temperature are burned, 1 mol of oxygen gas is used, and 2 mol of water vapor are formed.

Writing the heat released as a reaction product gives the equation

$$2H_2(g) + O_2(g) \rightarrow 2H_2O(g) + 483.6 \text{ kJ}$$

This is a **thermochemical equation,** *an equation that includes the quantity of heat released or absorbed during the reaction as written.* The quantity of heat for this or any reaction depends on the amounts of reactants and products. Forming twice as much water vapor would release 2×483.6 kJ. Forming one-half as much water would release $1/2 \times 483.6$ kJ. The thermochemical equations for these reactions are

$$4H_2(g) + 2O_2(g) \rightarrow 4H_2O(g) + 967.2 \text{ kJ}$$
$$H_2(g) + \tfrac{1}{2}O_2(g) \rightarrow H_2O(g) + 241.8 \text{ kJ}$$

Fractional coefficients are sometimes needed in thermochemical equations.

SECTION OBJECTIVES

- Explain heat of reaction, heat of formation, and enthalpy.
- Solve problems involving heats of reaction, heats of formation, and heats of combustion.
- Explain the concept of bond energy and its relationship to heat of reaction.

STUDY HINT

One calorie of heat energy is absorbed by each gram of liquid water as its temperature increases by 1C°. The joule (J), rather than the calorie, is the SI unit of heat energy. One calorie equals exactly 4.184 J. Thus, a 1C° increase in the temperature of 1 g of water is brought about by the absorption of 4.184 J. One kilojoule (kJ) is equal to 1000 J.

Figure 18-1 Schematic diagram of an ignition calorimeter used for measuring nutritional calories

Figure 18-2 Change in heat content during a chemical reaction.

Δ always signifies the difference between a final state and the initial state. If energy is involved, a positive value indicates absorption of energy, and a negative value indicates a release of energy.

The **heat of reaction** *is the quantity of heat released or absorbed during a chemical reaction.* One way to think of heat of reaction is as the difference between the stored heat energy, or heat content, of the reactants and the products. As pictured in Figure 18-2a, in an exothermic reaction, the heat evolved is the difference between the higher heat content of the reactants and the lower heat content of the products. The situation is reversed when products have a higher heat content than reactants, as is the case in an endothermic reaction. The reverse reaction absorbs the same amount of heat that is released by the forward reaction. This is to be expected, for the difference between the heat content of reactants and products is unchanged. For the endothermic decomposition of water vapor, the thermochemical equation is

$$2H_2O(g) + 483.6 \text{ kJ} \rightarrow 2H_2(g) + O_2(g)$$

The states of reactants and products must always be included in thermochemical equations because they influence the overall amount of heat exchanged. The heat for the decomposition of water, for example, would be larger if we started with ice.

Chemical reactions are usually carried out in open vessels. Volume may change, but the pressure is equal to atmospheric pressure and does not change. The heat of a reaction under these conditions is represented by ΔH. The H is the symbol for a quantity called **enthalpy,** *the heat content of a system at constant pressure.* However, it is not very useful to talk just about heat content or enthalpy because we have no way to measure the enthalpy of a system. Only *changes* in enthalpy can be measured. The Greek letter Δ (a capital "delta") stands for "change in." Therefore, ΔH is read as "enthalpy change". An **enthalpy change** *is the amount of heat absorbed or lost by a system during a process at constant pressure.* The enthalpy change is always the difference between the enthalpies of the products and the reactants:

$$\Delta H = H_{products} - H_{reactants}$$

Thermochemical equations are usually written by including ΔH. In an exothermic reaction, ΔH is always given a minus sign.

$$2H_2(g) + O_2(g) \rightarrow 2H_2O(g) \qquad \Delta H = -483.6 \text{ kJ}$$

For an endothermic reaction, ΔH is always given a positive value.

$$2H_2O(g) \rightarrow 2H_2(g) + O_2(g) \qquad \Delta H = +483.6 \text{ kJ}$$

The sign convention is arbitrary, but it is logical. The heat of reaction is taken as negative when the heat content of a system decreases (see Figure 18-2a), and as positive when the heat content of a system increases (see Figure 18-2b).

Heat of Formation

Doing an experiment with a calorimeter every time a heat of reaction is needed is not very practical. Instead, the results of many hundreds of such measurements are recorded in chemistry reference books. By consulting these data and, if necessary, combining them as illustrated later in this section, chemists are able to calculate heats of reaction for many common reactions.

The formation of water from hydrogen and oxygen is a composition reaction—the formation of a compound from its elements. Thermochemical data are often recorded as the heats of such composition reactions. The **molar heat of formation** *is the heat released or absorbed when one mole of a compound is formed by combination of its elements.*

To make comparisons meaningful, heats of formation are given for the standard states of reactants and products—these are the states found at atmospheric pressure and, usually, room temperature (298 K). Thus, the standard state of water is liquid, not gas or solid. The standard state of iron is solid, not molten. To signify that a value represents measurements on substances in their standard states, a ° sign is added to the symbol, giving $\Delta H°$ for the standard heat of a reaction. Adding an $_f$ as in $\Delta H°_f$ further indicates a standard heat of formation.

Some standard heats of formation are given in Appendix Table A-13. Each is the heat of reaction for the synthesis reaction for *one mole* of the compound listed. The thermochemical equation to accompany each heat of reaction shows the formation of one mole of the compound from the elements in their standard states. For liquid water, sodium chloride, and sulfur dioxide, for example, the equations are

$$H_2(g) + \tfrac{1}{2}O_2(g) \rightarrow H_2O(\ell) \qquad \Delta H°_f = -285.8 \text{ kJ}$$
$$Na(s) + \tfrac{1}{2}Cl_2(g) \rightarrow NaCl(s) \qquad \Delta H°_f = -411.15 \text{ kJ}$$
$$S(s) + O_2(g) \rightarrow SO_2(g) \qquad \Delta H°_f = -296.83 \text{ kJ}$$

Elements in their standard states are *defined* as having $\Delta H°_f = 0$. A negative $\Delta H°_f$ indicates a substance that is more stable than the free elements.

Stability and Heat of Formation

You can see in Appendix Table A-13 that the majority of the heats of formation are negative.

A large amount of energy is released when a compound with a high negative heat of formation is formed. The same amount of energy is required to cause such a compound to decompose into its separate elements. This energy must be supplied by an external source. Such compounds are very stable. Once they start, the reactions forming them usually proceed spontaneously and vigorously. For example, the $\Delta H°_f$ of carbon dioxide is -393.51 kJ per mole of gas produced.

Compounds with relatively high values of heats of formation (that is, values that are positive or only slightly negative) are relatively unstable. For example, hydrogen sulfide (H_2S) has a heat of formation of -20.63 kJ/mol. It is not very stable and decomposes when heated. Hydrogen iodide (HI) is a colorless gas that decomposes somewhat when stored at room temperature. It has a relatively high heat of formation of $+26.48$ kJ/mol. As it decomposes, violet iodine vapor becomes visible throughout the container of the gas.

A compound with a high positive heat of formation is very unstable and may react or decompose explosively. For example, ethyne (C_2H_2) reacts explosively with oxygen. Nitrogen triiodide and mercury fulminate decompose explosively. The formation reactions of such compounds store a great deal of energy within them. Mercury fulminate ($HgC_2N_2O_2$) has a heat of formation of $+270$ kJ/mol. Its instability makes it useful as a detonator for explosives.

All "molar" quantities, e.g. molar heat of formation, are intensive properties.

Figure 18-3 Hydrogen iodide gas *(top)* decomposes into hydrogen and iodine *(bottom)*. Iodine vapor is violet.

Figure 18-4 The combustion of wood releases all its stored energy. Ashes have no stored energy.

Heat of Combustion

Whether used in a furnace, automobile, or rocket, fuels must be energy-rich. The products of their combustion reaction with oxygen are energy-poor substances. In these combustion reactions, the energy yield may be very high. The products of the chemical action may be of little importance compared to the quantity of heat energy given off.

The combustion of 1 mol of graphite, which is pure carbon (C), yields 393.51 kJ of heat energy.

$$C(s) + O_2(g) \rightarrow CO_2(g) \qquad \Delta H_c = -393.51 \text{ kJ}$$

TABLE 18-1	HEAT OF COMBUSTION	
Substance	Formula	ΔH_c
hydrogen (g)	H_2	-285.9
carbon (graphite) (s)	C	-393.51
carbon monoxide (g)	CO	-283.0
methane (g)	CH_4	-890.36
ethane (g)	C_2H_6	-1559.8
propane (g)	C_3H_8	-2219.9
butane (g)	C_4H_{10}	-2878.5
pentane (g)	C_5H_{12}	-3536.1
hexane (ℓ)	C_6H_{14}	-4163.1
heptane (ℓ)	C_7H_{16}	-4811.2
octane (ℓ)	C_8H_{18}	-5450.5
ethene (ethylene) (g)	C_2H_4	-1411.0
propene (propylene) (g)	C_3H_6	-2051.0
ethyne (acetylene) (g)	C_2H_2	-1299.6
benzene (ℓ)	C_6H_6	-3267.5
toluene (ℓ)	C_7H_8	-3908.7

ΔH_c = heat of combustion of the given substance. All values of ΔH_c are expressed as kJ/mol of substance oxidized to H_2O and/or $CO_2(g)$ at constant pressure and 25°C. (s) = solid, (ℓ) = liquid, (g) = gas.

Chemistry and Technology

Fuels and Energy Content

Acetylene burns with an unusually hot flame. Why isn't it always used for fuel? Fuels are usually chosen for their energy content, the energy released per unit mass consumed, not the temperature of the flame. However, cost is the prime criterion; consumers want the most joules per dollar.

When burned, coal releases 32.76 kJ/g. Coal is inexpensive for industrial use, but inconvenient for use in individual houses, since the cost for home delivery has to be included.

Natural gas, which is mostly methane, is easy to transport or pipe into homes.

Methane releases 55.50 kJ/g when it burns, more than coal. Methane is easier to use than coal, but it is more expensive in many industrial situations.

So why isn't acetylene always used as fuel? It releases 49.91 kJ/g when burned, less than natural gas. It is also expensive, since pressurized oxygen is needed to burn it efficiently. Moreover, acetylene is more dangerous than methane and it must be manufactured since it is not a component of natural gas or oil.

The heat released by the complete combustion of 1 mol of a substance is called the **heat of combustion** *of the substance.* Heat of combustion is defined in terms of *1 mol of reactant*. Heat of formation, on the other hand, is defined in terms of *1 mol of product*. The general heat-of-reaction notation—ΔH—applies to heats of combustion as well. The notation ΔH_c is used to refer specifically to heat of combustion, however, as shown in Table 18-1. A more extensive list is in Appendix Table A-2.

Carbon dioxide and water are the products of the complete combustion of organic compounds containing carbon, hydrogen, and oxygen. Knowing this, the thermochemical equation for the combustion of any compound listed in the Tables can be written by balancing the equation for the reaction of one mole of the compound. For propane, for example,

$$C_3H_8(g) + 5O_2(g) \rightarrow 3CO_2(g) + 4H_2O(\ell) \qquad \Delta H_c = -2219.9 \text{ kJ}$$

Calculating Heats of Reaction

Thermochemical equations can be rearranged, terms can be cancelled, and the equations can be added to give heats for reactions not included in the data tables. The basis for this type of calculation is known as **Hess's law:** *The overall enthalpy change in a reaction is equal to the sum of enthalpy changes for the individual steps in the process.* The energy difference between

Figure 18-5 Heat of reaction diagram for carbon dioxide and carbon monoxide

reactants and products, which is illustrated in Figure 18-5, is independent of the route taken to get from one to the other. In fact, measured heats of reaction can be combined to calculate heats of reaction that are difficult or impossible to actually measure.

To demonstrate how to apply Hess's law, we will work our way through the calculation of several heats of reaction. To start with a simple one, suppose you need to know the heat of reaction for the decomposition of ice to give hydrogen and oxygen. The available data are the ΔH_f° of liquid water as recorded in the table, and the heat of fusion, or melting, of ice, which is 6.0 kJ/mol.

$$H_2(g) + \tfrac{1}{2}O_2(g) \rightarrow H_2O(\ell) \qquad \Delta H_f^\circ = -285.8 \text{ kJ}$$
$$H_2O(s) \rightarrow H_2O(\ell) \qquad \Delta H = 6.0 \text{ kJ}$$

The reaction for which ΔH is desired is

$$H_2O(s) \rightarrow H_2(g) + \tfrac{1}{2}O_2(g) \qquad \Delta H = ?$$

The general principles for combining thermochemical equations are *(1)* to reverse the direction and/or *(2)* multiply the coefficients of the known equations so that when added together they give the desired thermochemical equation. In this case, all that is needed is to reverse the formation equation for water, remembering to change the sign of ΔH_f° from negative to positive, and add the equations:

$$H_2O(\ell) \rightarrow H_2(g) + \tfrac{1}{2}O_2(g) \qquad \Delta H = 285.8 \text{ kJ}$$
$$H_2O(s) \rightarrow H_2O(\ell) \qquad \Delta H = 6.0 \text{ kJ}$$
$$\overline{H_2O(s) \rightarrow H_2(g) + \tfrac{1}{2}O_2(g) \qquad \Delta H = 291.8 \text{ kJ}}$$

The decomposition of ice to hydrogen and oxygen requires 291.8 kJ per mole of ice.

As a somewhat more complex example, consider the heat of a reaction used to prepare ultra-pure silicon for the electronics industry:

$$2Mg(s) + SiCl_4(\ell) \rightarrow Si(s) + 2MgCl_2(s) \qquad \Delta H = ?$$

The available information is the heats of formation of the reactant ($SiCl_4$) and product ($MgCl_2$):

$$Si(s) + 2Cl_2(g) \rightarrow SiCl_4(\ell) \quad \Delta H_f^\circ = -687 \text{ kJ}$$
$$Mg(s) + Cl_2(g) \rightarrow MgCl_2(s) \quad \Delta H_f^\circ = -641 \text{ kJ}$$

These equations can be combined to give the desired ΔH for the formation of silicon. The first equation must be reversed because $SiCl_4$ is a reactant, not a product, in the new equation. The second equation need not be reversed, because $MgCl_2$ is a product. However, this equation must be multiplied by 2 because $2MgCl_2$ formula units are needed in the new equation. Using Hess' law, we can determine the enthalpy change, ΔH, for the overall reaction as shown below:

$$SiCl_4(\ell) \rightarrow Si(s) + 2Cl_2(g) \qquad \Delta H = 687 \text{ kJ}$$
$$2Mg(s) + 2Cl_2(g) \rightarrow 2MgCl_2(s) \qquad \Delta H = 2 \times -641 \text{ kJ} = -1282 \text{ kJ}$$
$$\overline{2Mg(s) + SiCl_4(\ell) \rightarrow Si(s) + 2MgCl_2(s) \quad \Delta H = -595 \text{ kJ}}$$

After doing many calculations of the type just illustrated, you would recognize that there is a general relationship between the heat of a reaction and the heats of formation of the reactants and products.

$$\Delta H = \text{Sum of } \Delta H_f^\circ \text{ of products} - \text{Sum of } \Delta H_f^\circ \text{ of reactants}$$

To find the necessary sums, the ΔH_f° values for each reactant and product must be multiplied by the coefficients of each in the desired equation. For the reaction of Mg with $SiCl_4$, applying this equation gives

$$\Delta H = [0 + (2 \times -641 \text{ kJ})] - [0 + (-687 \text{ kJ})] = \text{kJ}$$

Note the use of the assigned value of zero for the heats of formation of elements in their standard states.

Sample Problem 18.1

Calculate the heat of reaction for the combustion of nitrogen monoxide gas (NO) to form nitrogen dioxide gas (NO_2). Use the heat-of-formation data in Appendix Table A-13. Solve by combining the known thermochemical equations. Verify the result by using the general equation for finding heats of reaction from heats of formation.

Solution

Step 1. Analyze Given: $\frac{1}{2}N_2(g) + \frac{1}{2}O_2(g) \rightarrow NO(g)$ $\Delta H_f^\circ = 90.25$ kJ/mol
$\frac{1}{2}N_2(g) + O_2(g) \rightarrow NO_2(g)$ $\Delta H_f^\circ = 33.18$ kJ/mol

Unknown: ΔH for $NO(g) + \frac{1}{2}O_2(g) \rightarrow NO_2(g)$

Step 2. Plan ΔH_f° (NO), ΔH_f° (NO_2) $\rightarrow \Delta H_c$ (NO). The ΔH requested can be found by adding the ΔHs of the component reactions as specified in Hess's law such that $NO(g) + \frac{1}{2}O_2(g) \rightarrow NO_2(g)$ is the net result. Thus, reversing the first reaction and the sign of its ΔH yields $NO(g) \rightarrow \frac{1}{2}N_2(g) + \frac{1}{2}O_2(g)$ $\Delta H = -90.25$ kJ. Retaining the second, $\frac{1}{2}N_2(g) + O_2(g) \rightarrow NO_2(g)$ $\Delta H = +33.18$ kJ

Step 3. Compute

$NO(g) \rightarrow \frac{1}{2}N_2(g) + \frac{1}{2}O_2(g)$ $\Delta H = -90.25$ kJ

$\frac{1}{2}N_2(g) + O_2(g) \rightarrow NO_2(g)$ $\Delta H = +33.18$ kJ

$NO(g) + \frac{1}{2}O_2(g) \rightarrow NO_2(g)$ $\Delta H = -57.07$ kJ

(Note the cancellation of the $\frac{1}{2}N_2(g)$ and the partial cancellation of the $O_2(g)$.)

Step 4. Evaluate The unnecessary reactants and products cancel to give the desired equation.
Verification: $\Delta H = \Delta H_f(NO_2) - [\Delta H_f(NO) + 0] = 33.18$ kJ/mol $- 90.25$ kJ/mol $= -57.07$ kJ/mol

Practice Problem
1. Calculate the heat of reaction for the combustion of methane gas (CH_4) to form $CO_2(g) + H_2O(\ell)$. Write all equations involved. *(Ans.)* $\Delta H = -890.36$ kJ.

By the creative combination of available data, many types of thermochemical calculations are possible. Let's consider one in which a heat of formation and a heat of combustion are combined. When carbon is burned in a limited supply of oxygen, carbon monoxide is produced. In this reaction, carbon is probably first oxidized to carbon dioxide: $C(s) + O_2(g) \rightarrow CO_2(g)$. Then the carbon dioxide is reduced with carbon to give the carbon monoxide: $C(s) + CO_2(g) \rightarrow 2CO(g)$. Because these two reactions occur simultaneously, it is not possible to measure directly the heat of formation of $CO(g)$ from $C(s)$ and $O_2(g)$.

$$C(s) + \tfrac{1}{2}O_2(g) \rightarrow CO(g) \quad \Delta H_f^\circ = ?$$

However, we do know the heat of formation of carbon dioxide and the heat of combustion of carbon monoxide.

$$C(s) + O_2(g) \rightarrow CO_2(g) \quad \Delta H_f^\circ = -393.5 \text{ kJ}$$
$$CO(g) + \tfrac{1}{2}O_2(g) \rightarrow CO_2(g) \quad \Delta H_c = -283.0 \text{ kJ}$$

Reversing the second equation and adding gives the desired heat of formation of carbon monoxide.

$$C(s) + O_2(g) \rightarrow CO_2(g) \quad \Delta H_f^\circ = -393.5 \text{ kJ}$$
$$CO_2(g) \rightarrow CO(g) + \tfrac{1}{2}O_2(g) \quad \Delta H = 283.0 \text{ kJ}$$
$$\overline{C(s) + \tfrac{1}{2}O_2(g) \rightarrow CO(g) \quad \Delta H_f = -110.5 \text{ kJ}}$$

Figure 18-6 A possible mechanism for the water-gas reaction

The combined use of heats of formation and heats of combustion to find unknown heats of formation can also be generalized. The following equation applies: heat of formation of compound X = sum of heats of formation of products of combustion of compound X − heat of combustion of compound X.

Sample Problem 18.2

Calculate the heat of formation of pentane (C_5H_{12}) using the information on heats of formation in Appendix Table A-13 and on heats of combustion in Table 18-1. Solve by combining the known thermochemical equations. Verify the result by using the general equation for finding heats of formation from heats of combustion.

Solution

Step 1. Analyze

Given:
$$C(s) + O_2(g) \rightarrow CO_2(g) \qquad \Delta H_f^\circ = -393.51 \text{ kJ/mol}$$
$$H_2(g) + \tfrac{1}{2}O_2(g) \rightarrow H_2O(\ell) \qquad \Delta H_f^\circ = -285.83 \text{ kJ/mol}$$
$$C_5H_{12}(g) + 8O_2(g) \rightarrow 5CO_2(g) + 6H_2O(\ell) \quad \Delta H_c = -3536.1 \text{ kJ/mol}$$

Unknown: $\Delta H_f(C_5H_{12})$ for $5C(s) + 6H_2(g) \rightarrow C_5H_{12}(g)$

Step 2. Plan

$$\Delta H_f^\circ(CO_2), \Delta H_f^\circ(H_2O), \Delta H_c(C_5H_{12}) \rightarrow \Delta H_f(C_5H_{12})$$

To combine the equations given above according to Hess's law, the equation for combustion of C_5H_{12} must be reversed, the equation for formation of CO_2 must be multiplied by 5 to give 5C as reactant, and the equation for formation of H_2O must be multiplied by 6 to give $6H_2$ as reactant.

Step 3. Compute

$$5C(s) + 5O_2(g) \rightarrow 5CO_2(g) \qquad 5\Delta H_f^\circ = -1967.6 \text{ kJ}$$
$$6H_2(g) + 3O_2(g) \rightarrow 6H_2O(g) \qquad 6\Delta H_f^\circ = -1715.0 \text{ kJ}$$
$$5CO_2(g) + 6H_2O(\ell) \rightarrow C_5H_{12}(g) + 8O_2(g) \quad \Delta H = 3536.1 \text{ kJ}$$

$$5C(s) + 6H_2(g) \rightarrow C_5H_{12}(g) \qquad \Delta H_f = -146.5 \text{ kJ}$$

Step 4. Evaluate

The unnecessary reactants and products cancel to give the correct equation. The answer is reasonably close to an estimated value of -145, (-1970 kJ/mol -1715 kJ/mol + 3540 kJ/mol).

Verification:
$$\Delta H_f(C_5H_{12}) = 5\Delta H_f(CO_2) + 6\Delta H_f(H_2O) - \Delta H_c(C_5H_{12})$$
$$= 5(-393.51 \text{ kJ/mol}) + 6(-285.83 \text{ kJ/mol}) - (-3536.1 \text{ kJ/mol})$$
$$= -146.5 \text{ kJ/mol}$$

Practice Problems

1. Calculate the heat of formation of butane, C_4H_{10}, using the information in Appendix Table A-13 and Table 18-1. Write out the solution according to Hess's law. *(Ans.)* -124.7 kJ/mol
2. Calculate the heat of combustion of 1 mol of nitrogen, N_2, to form NO_2 using Appendix Table A-13. (Hint: Rearrange the general equation for ΔH_f of compound X to solve for ΔH_c.) *(Ans.)* $+66.36$ kJ/mol

Bond Energy and Reaction Heat

The change in heat content of a reaction system is related to *(1)* the change in the number of bonds breaking and forming, and *(2)* the strengths of these bonds as the reactants form products. The reaction for the formation of water gas shown in Figure 18-6 can be used to illustrate these relationships.

STUDY HINT
Thermodynamics is the study of heat and other forms of energy, and of heat-energy transfer.

STUDY HINT

A mixture of CO and H_2 is called water gas.

1. Heat content of a system at constant pressure
2. (a) Heat released or absorbed during a chemical reaction (b) Heat of reaction is the difference between the enthalpy, or heat content, of products and reactants
3. The more negative the heat of formation, the more stable the compound
4. Equations and heats of reaction are rearranged and added algebraically to give the desired reaction.

SECTION OBJECTIVES

- Explain the relationship between enthalpy change and the tendency of a reaction to occur.
- Explain the relationship between entropy change and the tendency of a reaction to occur.
- Discuss the concept of free energy and explain how the value of this quantity is calculated and interpreted.

The two oxygen-to-hydrogen bonds of each steam molecule must be broken, as must the carbon-to-carbon bonds of the graphite. Carbon-to-oxygen and hydrogen-to-hydrogen bonds must be formed. Recall that energy is absorbed when bonds break and is released when bonds form.

A hypothetical series of steps for the water-gas reaction is illustrated in Figure 18-6. Observe that a total heat input of 1632 kJ is required to break the bonds of 1 mol of graphite and 1 mol of steam. The formation of bonds in 1 mol of CO and 1 mol of H_2 releases 1498 kJ of heat energy. The net effect is that 134 kJ of heat must be supplied to the system from an external source. This quantity agrees quite closely with the experimental value of 131.3 kJ of heat input per mole of carbon used.

Section Review

1. What is meant by "enthalpy?"
2. *(a)* What is meant by "heat of reaction?" *(b)* How is heat of reaction related to the enthalpy, or heat content, of reactants?
3. Describe the relationship between a compound's stability and its heat of formation.
4. How can thermochemical equations be combined to calculate ΔH values according to Hess's law?

18.2 Driving Force of Reactions

Whether a reaction occurs and how it occurs are questions that have always concerned chemists. The answers to these questions involve a study of reactions conducted within the framework of the laws of thermodynamics. This is the realm of physical chemistry and physics. In this section, the concepts that chemists label collectively as the "driving force" of chemical reactions will be examined.

Enthalpy and Reaction Tendency

It is not surprising to see a ball roll down an incline unaided; it is just what would be expected. Potential energy is given up in this process, and the ball attains a more stable state at a lower energy level. It would be surprising, however, if the ball rolled up the incline by and of itself. From such experiences with nature, a basic rule for natural processes becomes apparent: There is a tendency for processes to occur that lead to a lower energy state. This tendency in nature is toward greater stability (resistance to change) in a system.

The great majority of chemical reactions in nature are exothermic. As the reactions proceed, energy is liberated and the products have less energy than the original reactants. With the above rule in mind, one would expect exothermic reactions to occur spontaneously; that is, that they would have the ability to proceed without the assistance of an external agent.

It should follow that endothermic reactions, in which energy is absorbed, would not occur spontaneously but would proceed only with the assistance of an external agent. Certainly energy must be expended on the ball to roll it up the incline. At the top of the incline the ball's potential energy is high and its stability is low.

Figure 18-7 Coal gasification plants, such as this one in Daggett, California, produce CO and H_2, called synthesis gas (syngas), by oxidation of the hydrocarbons found in various types of coal, petroleum coke, heavy fuel oil, or natural gas.

Some endothermic reactions *do* take place spontaneously, however. The production of water gas involves such a reaction. Steam is passed over white-hot coke (impure carbon), and the reaction proceeds spontaneously. It is not driven by any activity outside the reacting system. However, the process is endothermic, showing that the product gases, CO and H_2, collectively have a higher heat content than do the reactants, steam and carbon. The enthalpy change is positive, and therefore unfavorable, yet the reaction proceeds spontaneously.

$$H_2O(g) + C(s) \rightarrow CO(g) + H_2(g) \qquad \Delta H = +131.3 \text{ kJ}$$

A word about what chemists mean by *spontaneous* is in order. A spontaneous chemical reaction takes place without any continuing outside influence, such as heating, cooling, or stirring. A spontaneous change will keep going, once it gets started. As used in chemistry, *spontaneous* has nothing to do with speed. Some "spontaneous" chemical reactions are actually quite slow.

Entropy and Reaction Tendency

If a process is not driven by decreasing energy, what does drive it? To answer this question, consider a physical process that proceeds by and of itself with no energy change. Figure 18-8 shows two identical flasks connected by a valve. One flask is filled with ideal gas A and the other with ideal gas B. The entire system is at room temperature.

When the valve is opened, the two gases mix spontaneously until they are distributed evenly throughout the two containers. The gases will remain in this state indefinitely, with no tendency to separate. The temperature does not rise but remains constant throughout the mixing process. Thus, no exothermic process can have taken place. The total heat content, cannot have changed to a lower level. Clearly, the self-mixing process, since it is spontaneous, must be caused by a driving force other than the enthalpy-change tendency.

Another endothermic, but spontaneous, process is melting. An ice cube melts spontaneously at room temperature as heat is transferred from the

STUDY HINT
Natural processes tend toward maximum disorder.

Figure 18-8 The mixing of gases may occur without an energy change.

STUDY HINT
The bonding of water is discussed in Section 13.4.

Gibbs Free Energy was named for Josiah Willard Gibbs, (1839–1903). He was a professor of mathematical physics at Yale University from 1871–1903. He laid the foundations of many areas of thermodynamics.

warm air to the cool ice and to the ice water formed. The well-ordered structure of the ice crystal (the solid phase) is lost, and the less orderly liquid phase of higher energy content is formed.

What is observed here is a tendency for the ice to move into the less orderly liquid phase. This is a tendency for processes to occur that lead to a less orderly or a more disordered state. This tendency in nature is toward greater disorder in a system. A disordered system is one that lacks a regular arrangement of its parts. *The property that describes the disorder of a system is called* **entropy** (*S*). The more disordered or more random a system, the higher is its entropy. Liquid water has higher entropy than ice. A mixture of gases has higher entropy than a pure gas.

Thus, processes in nature are driven in two ways: toward lowest energy and toward highest entropy. When these two oppose each other, the dominant factor determines the direction of the spontaneous change. In the steam-plus-hot-carbon reaction, the temperature is high enough that the entropy factor overcomes the unfavorable energy-change factor. The spontaneous endothermic reaction occurs.

A system that can go from one state to another without an enthalpy change does so by becoming more disordered. Entropy increases, for example, when a gas expands, a solid or liquid dissolves, or the number of particles in a system increases.

Free Energy

The overall drive to spontaneous change, as you have seen, depends on two factors—the tendency toward lowest enthalpy and the tendency toward highest entropy. A function of the state of a reaction system has been defined to relate the enthalpy and entropy factors at a given temperature. *This combined enthalpy–entropy function is called the* **free energy** *of the system.* It is the function that simultaneously assesses both the enthalpy-change and entropy-change tendencies. In this sense, a process proceeds spontaneously in the direction that lowers the free energy of a system.

The net driving force of a chemical or physical change is called the free-energy change (ΔG) of the system. When the enthalpy change and the entropy change oppose each other, the direction of the net driving force is determined by the factor having the greater influence.

At a constant pressure and temperature, the **free-energy change**, ΔG, *of a system is defined as the difference between the change in enthalpy, ΔH, and the product of the Kelvin temperature and the entropy change, which is defined as $T\Delta S$.*

$$\Delta G = \Delta H - T\Delta S$$

In the equation above, ΔG is the change in free energy of the system, ΔH is the change in the enthalpy (heat content), T is the temperature (expressed as kelvins), and ΔS is the change in entropy. The product $T\Delta S$ and the quantities ΔG and ΔH all have the same dimensions, usually measured as kilojoules per mole. The dimensions of ΔS for use in this equation are usually kJ/(mol·K).

A chemical reaction proceeds spontaneously if it is accompanied by a decrease in free energy, that is, if the free energy of the products is lower than that of the reactants. In such a case, the free energy change in the system

(ΔG) is negative. In practical terms, free energy is the quantity of energy available to do useful work.

In exothermic reactions, ΔH has a negative value; in endothermic reactions, its value is positive. From the expression for ΔG, observe that the more negative ΔH is, the more negative ΔG is likely to be. Reaction systems that change from a high enthalpy state to a low one tend to proceed spontaneously. The more positive ΔS is, the more negative ΔG is likely to be. Thus, systems that change from well-ordered states to highly disordered states also tend to proceed spontaneously. Both of these tendencies in a given reaction system are assessed simultaneously in terms of the free-energy change—ΔG.

In some processes, ΔH is negative and ΔS is positive. Here, regardless of the relative magnitudes of ΔH and ΔS, the process will proceed spontaneously because ΔG is negative. Processes in which ΔH and ΔS have the same signs are more common.

Note that the sign of ΔG can be either positive or negative depending on the temperature T. The temperature of the system is the dominant factor that determines the relative importance of the tendencies toward lower energy and higher entropy.

Consider the water-gas reaction.

$$H_2O(g) + C(s) \rightarrow CO(g) + H_2(g) \qquad \Delta H = +131.3 \text{ kJ}$$

Here, a negative free-energy change requires that $T\Delta S$ be positive and greater than 131.3 kJ. This can be so if T is large, or if ΔS is large and positive, or both. Consider first the magnitude and sign of ΔS. Entropy increases (that is, ΔS is positive) when water gas is produced. In order to understand why, recall that the solid phase is well-ordered and the gaseous phase is random. One of the reactants, carbon, is a solid, and the other, steam, is a gas. In this case, however, both products are gases. The change from the orderly solid phase to the random gaseous phase involves an increase in disorder and thus an increase in entropy.

In general, a change from a solid to a gas will proceed with an increase in entropy. If all reactants and products are gases, an increase in the number of product particles increases entropy.

At low temperatures, whether ΔS is positive or negative, the product $T\Delta S$ may be small compared to ΔH. In such cases, the $T\Delta S$ term is nearly negligible, and the reaction proceeds as the enthalpy change predicts.

For the water-gas reaction, measurements show that ΔS is $+0.134$ kJ/(mol·K) at 25°C (298K). Thus

$$T\Delta S = 298\text{K} \times 0.134 \, \frac{\text{kJ}}{\text{mol}\cdot\text{K}} = 39.9 \text{ kJ/mol}$$

and

$$\Delta G = \Delta H - T\Delta S = +131.3 \text{ kJ/mol} - 39.9 \text{ kJ/mol}$$
$$= +91.4 \text{ kJ/mol}$$

Since ΔG has a positive value, the reaction that produces water gas is not spontaneous at the relatively low temperature of 25°C.

Increases in temperature tend to favor increases in entropy. When ΔS is positive, a high temperature gives $T\Delta S$ a large positive value. Certainly, at a high enough temperature $T\Delta S$ will be larger than ΔH, and ΔG will have a negative value.

The water-gas reaction occurs spontaneously at the temperature of white-hot carbon, approximately 900°C. Strictly speaking, the values of ΔH and ΔS vary slightly with different temperature. However, they are assumed to remain about the same at the reaction temperature of 900°C. Therefore the values can be used to calculate an approximate value of ΔG for this temperature. Retaining the value of ΔH as +131.3 kJ/mol, $T\Delta S$ is determined at 1173K (900°C) to be

$$\left(1173\text{K} \times 0.134 \frac{\text{kJ}}{\text{mol} \cdot \text{K}}\right) = 157 \text{ kJ/mol}$$

$$\Delta G = \Delta H - T\Delta S = +131.3 \frac{\text{kJ}}{\text{mol}} - \left(\frac{157 \text{ kJ}}{\text{mol}}\right)$$

$$= -26 \text{ kJ/mol}$$

The free-energy change for the water-gas reaction is negative at 900°C, and that is why the reaction is spontaneous at this temperature.

1. Large negative value of ΔH
2. The property that describes the disorder of a system. An increase in entropy is a driving force toward spontaneous reaction.
3. Crystalline solid to a solution of the solid in water. Change of phase from solid to liquid to gas. Decomposition of large molecules into smaller ones.

Sample Problem 18.3

For the reaction $NH_4Cl(s) \rightarrow NH_3(g) + HCl(g)$, at 25°C, $\Delta H = 176$ kJ/mol and $\Delta S = 0.285$ kJ/(mol·K). Calculate ΔG and decide if this reaction will be spontaneous in the forward direction at 25°C.

Solution
Step 1. Analyze   Given: $\Delta H = 176$ kJ/mol at 25°C
$\Delta S = 0.285$ kJ/(mol·K) at 25°C
Unknown: ΔG at 25°C

Step 2. Plan
$$\Delta S, \Delta H, T \rightarrow \Delta G$$

The value of ΔG can be calculated according to the equation

$$\Delta G = \Delta H - T\Delta S$$

Step 3. Compute
$\Delta G = 176$ kJ/mol $- [298\text{K} \times 0.285$ kJ/(mol·K)$]$
$= 176$ kJ/mol $- 84.9$ kJ/mol
$= 91$ kJ/mol

Step 4. Evaluate   The answer is reasonably close to an estimated value of 110, $[200 - (300 \times 0.3)]$. The positive value of ΔG shows that this reaction is not spontaneous at 25°C.

Practice Problems
1. In the reaction above, why does the entropy increase?   *(Ans.)* Gas formed from solid
2. What should be the effect of increasing temperature on the value of ΔG for this reaction?   *(Ans.)* At a large enough value of T, ΔG changes from positive to negative.

4. $\Delta G = \Delta H - T\Delta S$. Calculated from the difference of the change in enthalpy and the products of Kelvin temperature and the change in entropy of a system
5. A reaction for which $\Delta G < 0$ is spontaneous.

Section Review
1. What kind of enthalpy change tends to produce a spontaneous reaction?
2. What is entropy and how does it relate to spontaneity of reactions?
3. List several changes that result in entropy increase.
4. Define *free energy* and explain how it is calculated.
5. Explain the relationship between free energy and spontaneity of reactions.

18.3 The Reaction Process

Up to now, this chapter has dealt with aspects of chemical reactions that are independent of the actual routes by which reactions occur. For example, enthalpy change (ΔH) and entropy change (ΔS) depend only on the initial and final states of a reaction, irrespective of any intermediate processes that may have taken place. This section explores the energy pathway that a reaction follows and the changes that take place on the molecular level when substances interact.

SECTION OBJECTIVES
- Explain the concept of reaction mechanisms.
- Use the collision theory to interpret chemical reactions.
- Define *activated complex*.
- Relate activation energy and heat of reaction.

Reaction Mechanisms

Chemical reactions involve breaking existing chemical bonds and forming new ones. The relationships and arrangements of atoms in the products are different from those in the reactants. Colorless hydrogen gas consists of pairs of hydrogen atoms bonded together as diatomic molecules, H_2. Violet-colored iodine vapor is also diatomic, consisting of pairs of iodine atoms bonded together as I_2 molecules. A chemical reaction between these two gases at elevated temperatures produces hydrogen iodide (HI) a colorless gas. Hydrogen iodide molecules, in turn, tend to decompose and re-form hydrogen and iodine molecules. The reaction equations are

$$H_2(g) + I_2(g) \rightarrow 2HI(g)$$

and

$$2HI(g) \rightarrow H_2(g) + I_2(g)$$

Such equations indicate only what molecular species disappear as a result of the reactions and what species are produced. They do not show the pathway by which either reaction proceeds. That is, they do not show *the step-by-step sequence of reactions by which the overall chemical change occurs, called the* **reaction mechanism.**

A chemical system can usually be examined before a reaction occurs or after the reaction is over. Such an examination reveals nothing about the mechanism by which the action proceeded, however. For most chemical reactions, only the reactants that disappear and final products that appear are known. In other words, only the net chemical change is directly observable.

Sometimes, however, chemists are able to devise experiments that reveal the sequence of steps in a reaction mechanism. Each reaction step is usually a simple process. The equation for each step represents the *actual* atoms, ions, or molecules that participate in that step. Complicated chemical reactions take place in a sequence of simple steps. Even a reaction that appears from its balanced equation to be a simple process may actually occur in a sequence of steps.

The reaction between hydrogen gas and bromine vapor to produce hydrogen bromide gas is an example of a **homogeneous reaction,** *a reaction whose reactants and products exist in a single phase* —in this case, the gas phase. This reaction system is also an example of a homogeneous chemical system. In such a system, all reactants and products in all intermediate steps are in the same phase. The overall chemical equation for this reaction is

$$H_2(g) + Br_2(g) \rightleftharpoons 2HBr(g)$$

It might appear that one hydrogen molecule reacts with one bromine molecule to form the product in a simple, one-step-process. However, chemists have found that the reaction follows a more complex pathway. The initial forward reaction is the splitting of bromine molecules into bromine atoms.

$$(1) \quad Br_2 \rightleftarrows 2Br$$

The bromine atoms, in turn, initiate a sequence of two steps that can repeat themselves.

$$(2) \quad Br + H_2 \rightleftarrows HBr + H$$
$$(3) \quad H + Br_2 \rightleftarrows HBr + Br$$

Observe that the sum of steps 2 and 3 gives the net reaction as shown in the overall equation above.

The equation

$$H_2(g) + I_2(g) \rightleftarrows 2HI(g)$$

represents another homogeneous chemical system. Figure 18-9 illustrates the initial reactants and final products identified by this equation. For many years chemists thought this reaction was a simple one-step process. They assumed it involved interaction of two molecules, H_2 and I_2, in the forward direction and two HI molecules in the reverse reaction.

Initial State

Final State

Figure 18-9 The initial and final states of a mixture of hydrogen gas and iodine vapor, which react to form hydrogen iodide gas.

Experiments eventually showed, however, that a direct reaction between H_2 and I_2 does not take place. Chemists proposed alternative mechanisms for the reaction. The steps in each reaction mechanism had to add up to give the overall equation. Note that two of the species—I and H_2I—in the mechanism steps do not appear in the net equation. *Such species that appear* in some steps, but not in the net equation are known as **intermediates.** The first possible mechanism has a two-step pathway:

$$(1) \quad I_2 \rightleftarrows 2I$$
$$(2) \quad 2I + H_2 \rightleftarrows 2HI$$
$$\overline{\quad I_2 + H_2 \rightleftarrows 2HI \quad}$$

The second possible mechanism has a three-step pathway:

$$(1) \quad I_2 \rightleftarrows 2I$$
$$(2) \quad \cancel{I} + H_2 \rightleftarrows \cancel{H_2I}$$
$$(3) \quad \cancel{H_2I} + \cancel{I} \rightleftarrows 2HI$$
$$\overline{\quad I_2 + H_2 \rightarrow 2HI \quad}$$

In studying reaction mechanisms, chemists know that it is easier to prove what does *not* happen than to pin down what happens. No matter how simple a balanced equation, the reaction pathway may be complicated and difficult to determine. Some chemists believe that simple, one-step reaction mechanisms are unlikely, and that nearly all reactions are complicated.

Figure 18-10 Energy distribution among gas molecules at two different temperatures

Collision Theory

In order for reactions to occur between substances, their particles (molecules, atoms, or ions) must collide. Further, these collisions must result in interactions. *The set of assumptions regarding collisions and reactions is known as* **collision theory.** Chemists use this theory to interpret many facts they observe about chemical reactions.

From the kinetic theory you know that the molecules of gases are continuously in random motion. The molecules have kinetic energies ranging from very low to very high values. The energies of the greatest portion are near the average value for all the molecules in the system. When the temperature of a gas is raised, the average kinetic energy of the molecules is increased as is shown in Figure 18-10.

Consider what might happen on a molecular scale in one step of a homogeneous reaction system. We will analyze a proposed first step in the decomposition of hydrogen iodide.

$$HI + HI \rightleftarrows H_2 + 2I$$

According to the collision theory, the two HI molecules must collide in order to react. Further, they must collide while favorably oriented and with enough energy to disrupt the bonds of the molecules. If they do so, a reshuffling of bonds leads to the formation of the products: one H_2 molecule and two I atoms.

If collision is too gentle, the old bonds are not disrupted and new ones are not formed. The two molecules simply rebound from each other unchanged. This effect is illustrated in Figure 18-11(a).

Similarly, a collision in which the reactant molecules are poorly oriented has little effect. The angles and distances between certain atoms in the molecules are not those necessary for new bonds to form. The colliding molecules rebound without changing. A poorly oriented collision is shown in Figure 18-11(b).

Rebounding occurs because of the electron–electron repulsion.

(a) Collision too gentle   (b) Collision in poor orientation   (c) Effective collision

Figure 18-11 Possible collision patterns for HI molecules.

Thus, collision theory provides two reasons why a collision between reactant molecules may fail to produce a new chemical species: *(1)* the collision is not energetic enough to supply the required activation energy, and *(2)* the colliding molecules are not oriented in a way that enables them to react with each other. We will now look at what is necessary for an effective collision that produces a new chemical species.

Activation Energy

When a collision is suitably oriented and violent enough, an interpenetration of the electron clouds of the appropriate atoms in the colliding molecules occurs. Then, old bonds can be broken and new bonds can form. See Figure 18-11(c). Only collisions that occur with sufficient energy can be effective.

Consider the reaction for the formation of water from the diatomic gases oxygen and hydrogen. The heat of formation is quite high: $\Delta H = -285.8$ kJ/mol at 25°C. The free energy change is also large: $\Delta G = -241.0$ kJ/mol. Why, then, do hydrogen and oxygen not combine spontaneously and immediately to form water when they are mixed at room temperature?

Hydrogen and oxygen gases exist as diatomic molecules. By some reaction mechanism, the bonds of these molecular species must be broken in order for new bonds between oxygen and hydrogen atoms to be formed. Bond breaking is an endothermic process, and bond forming is exothermic. Even though the net process is exothermic, it appears that an initial energy kick is needed to start the reaction.

It might help to think of the reactants as lying in an energy trough. Before they can react to form water, they must be lifted from this trough even though the energy content of the product is lower than that of the reactants. Once the reaction is started, the energy released is enough to sustain the reaction by activating other molecules. Thus, the reaction rate keeps increasing. It is finally limited only by the time required for reactant particles to acquire the energy and make contact. Energy from a flame or a spark discharge, or the energy associated with high temperatures or radiations may start reactants along the pathway of reaction. This activation-energy concept is illustrated in Figure 18-12.

The reverse reaction is the decomposition of water molecules. The product water lies at an energy level that is lower than the one from which the reactants were lifted. The water molecules must be lifted from this deeper energy trough before they can decompose to form oxygen and hydrogen. The activation energy needed to start this endothermic reaction is greater than that required for the original exothermic change. The difference equals the amount of heat of reaction, ΔH, released in the original reaction. See Figure 18-13.

In the energy diagram shown in Figure 18-14, the difference in height of the two level portions of the diagram is compared to the energy of reaction, ΔH. Energies E_a and E_a' represent the activation energies of the forward and reverse reactions, respectively. These quantities are comparable to the heights of the pass above the high and low valleys.

The Activated Complex

The possession of high kinetic energy does not alone make molecules unstable. When molecules collide, some of this energy is converted into internal potential energy within the colliding molecules. If enough energy is converted, the molecules may be activated.

When particles collide with energy at least equal to the activation energy for the species involved, their interpenetration can disrupt existing bonds. New bonds can then form. In the brief interval of bond disruption and bond formation, the collision complex is said to be in a transition state. Some sort of partial bonding exists in this transitional structure. *A transitional structure that results from an effective collision and that persists while old bonds are breaking and new bonds are forming is called an* **activated complex**.

An activated complex is formed when an effective collision raises the internal energies of the reactants to their minimum level for reaction. (See Figure 18-14.) Both forward and reverse reactions go through the same activated complex. A bond that is broken in the activated complex for the forward reaction must be re-formed in the activated complex for the reverse reaction. Observe that the activated complex occurs at the maximum-energy position along the reaction pathway. In this sense, the activated complex defines the activation energy for the system. **Activation energy** *is thus the energy required to transform the reactants into the activated complex.*

In its brief existence, the activated complex has partial bonding that is characteristic of both reactant and product. In this state, it may respond to either of two possibilities: 1. It may re-form the original bonds and separate into the reactant particles or, 2. It may form new bonds and separate into product particles. Usually, the formation of products is just as likely as the formation of reactants. Do not confuse the activated complex with the relatively stable intermediate products of different steps of a reaction

Figure 18-12 Pathway of an exothermic reaction.

Figure 18-13 Pathway of an endothermic reaction.

Figure 18-14 Activation energies for the forward reaction E_a and the reverse reaction E_a' and the change in internal energy ΔH in a reversible reaction.

Figure 18-15 Possible activated-complex configuration, which could form either 2HI or $H_2 + 2I$

mechanism. The activated complex, unlike intermediate products, is a very-short-lived molecular complex in which bonds are in the process of being broken and formed.

A possible configuration of the activated complex in the hydrogen iodide reaction is shown in Figure 18-15. The broken lines represent some sort of partial bonding. This transitional structure may produce HI molecules or an H_2 molecule and two I atoms.

Figure 18-16 shows the energy profile for the hydrogen iodide (HI) decomposition, which occurs at elevated temperatures of 400–500°C. Of the 183-kJ activation energy, 171 kJ is available for excitation of the hydrogen (H_2) and iodide (I_2) molecules. Of this amount, 149 kJ is used in producing iodine atoms.

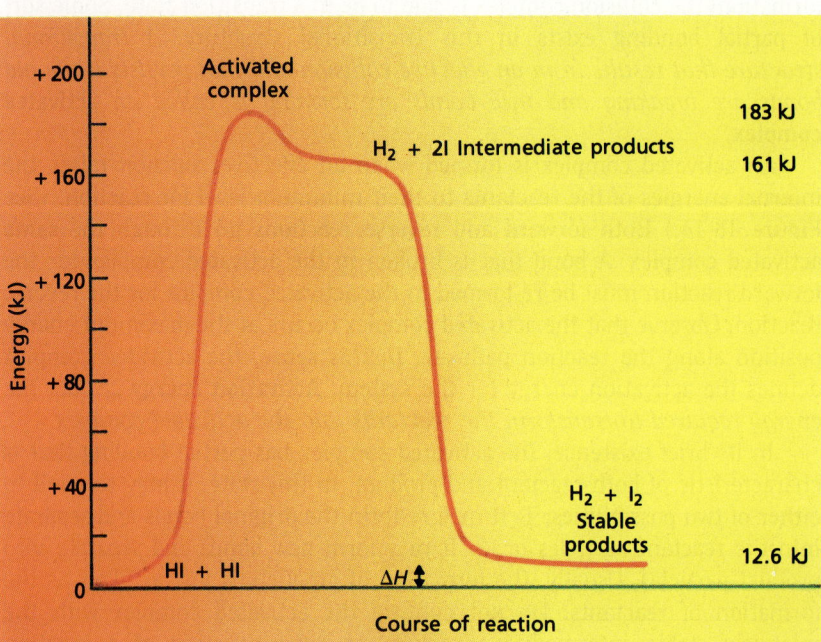

Figure 18-16 An energy profile. Results of kinetic studies of the decomposition of hydrogen iodide at high temperatures are graphed

Sample Problem 18.4

Copy the energy diagram below and label the reactants, products, ΔH, E_a, and E_a'. Determine the value of $\Delta H_{forward}$, $\Delta H_{reverse}$, E_a, and E_a'.

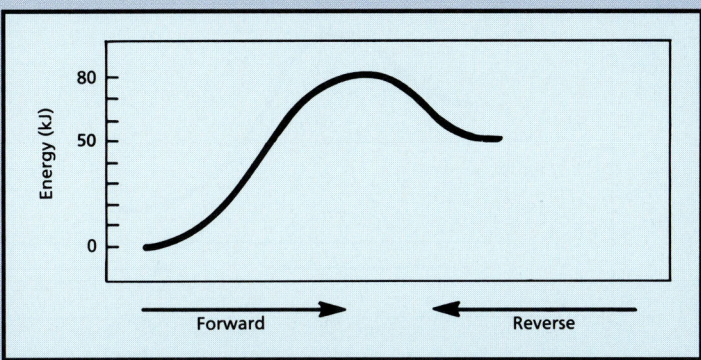

Solution

The reactant energy level is always at the left-hand end of such a curve and the product energy level is always at the right-hand end. The heat of reaction, ΔH, is the difference between these two energy levels. The activation energy, which differs in the forward and reverse directions, is the difference between the reactant or product energy level, and the peak in the curve—the minimum energy needed to achieve effective reaction.

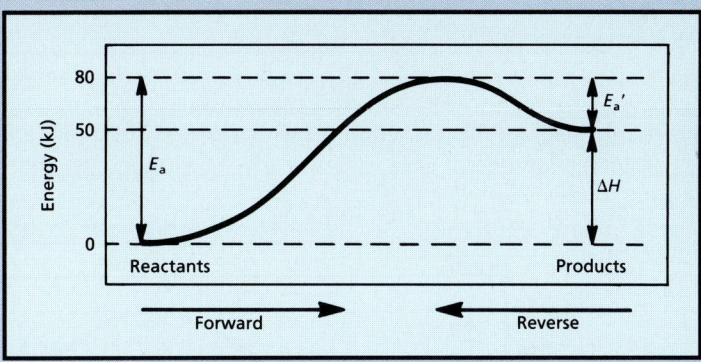

$$\Delta H_{forward} = 50 \text{ kJ} - 0 \text{ kJ} = +50 \text{ kJ}$$
$$\Delta H_{reverse} = 0 \text{ kJ} - 50 \text{ kJ} = -50 \text{ kJ}$$
$$E_a = 80 \text{ kJ} - 0 \text{ kJ} = 80 \text{ kJ}$$
$$E_a' = 80 \text{ kJ} - 50 \text{ kJ} = 30 \text{ kJ}$$

Sample Problem 18.4 *continued*

Practice Problems
1. Repeat the sample problem for the following energy diagram:

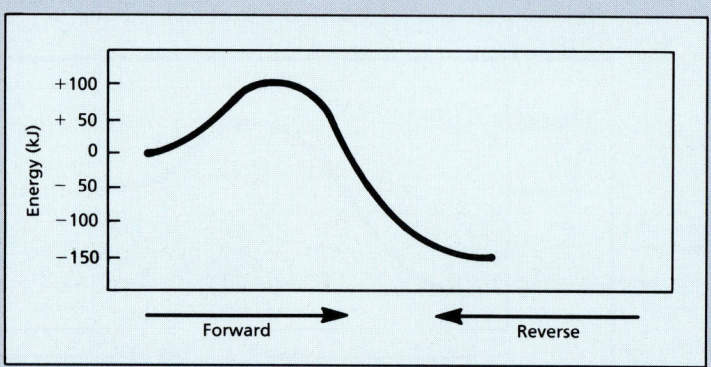

(Ans.)

$\Delta H_{\text{forward}}$ = -150 kJ
$\Delta H_{\text{reverse}}$ = $+150$ kJ
E_a = 100 kJ
E_a' = 250 kJ

2. Draw and label the energy diagram for a reaction for which $\Delta H = 30$ kJ/mol and $E_a = 40$ kJ/mol. Place the reactants at energy level zero. Give values of $\Delta H_{\text{forward}}$, $\Delta H_{\text{reverse}}$, and E_a'.

(Ans.)

$\Delta H_{\text{forward}}$ = $+30$ kJ
$\Delta H_{\text{reverse}}$ = -30 kJ
E_a' = 10 kJ

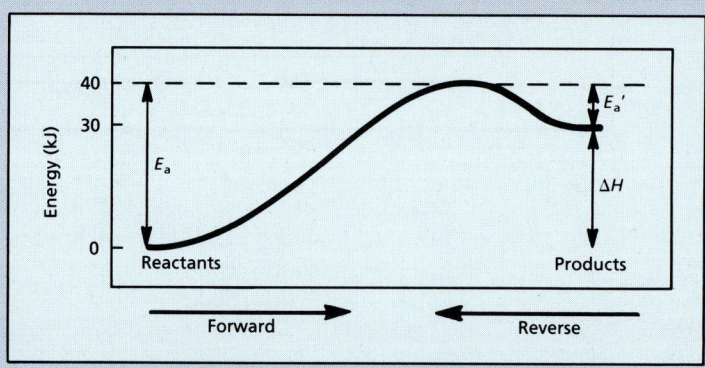

Section Review

1. What is meant by "reaction mechanism"?
2. What factors determine whether a molecular collision produces a reaction?
3. What is an activated complex?
4. What is activation energy?
5. How is activation energy related to the energy of reaction?

1. Sequence of steps by which a reaction occurs
2. Energy of the collision and orientation of the molecules during collision
3. Transitional structure that occurs as reactants form products
4. Energy required to transform reactants into the activated complex
5. The difference in activation energies of the forward and reverse reactions is the energy of reaction.

18.4 Reaction Rate

Chemists are interested in learning how rapidly reactants change into products in chemical reactions. *The change in concentration of reactants per unit time as a reaction proceeds is called the* **reaction rate.** *The branch of chemistry that is concerned with reaction rates and reaction mechanisms is called* **chemical kinetics.** As you will see as you read this concluding section of the chapter, the study of rates deals with the factors that affect rate and with the mathematical expressions that reveal the specific dependencies of rate upon concentration.

Rate-Influencing Factors

The rates of chemical reactions vary widely. Some reactions are over in an instant, and others take months or years to complete. The rate of reaction is measured by the amounts of reactants converted to products in a unit of time. Two conditions are necessary for reactions (other than simple decompositions) to occur at all. First, particles must come into contact. Second, this contact must result in appropriate interaction. Thus, the rate of a reaction depends on the collision frequency of the reactants and on the collision efficiency. An efficient collision is one with enough energy for activation and in which the reactant molecules are favorably oriented.

Changing conditions may affect either the collision frequency or the collision efficiency. Any such change affects the reaction rate. Five important factors influence the rate of chemical reaction.

Nature of reactants Substances vary greatly in their tendencies to react. For example, hydrogen combines vigorously with chlorine under certain conditions. Under the same conditions, it may react only feebly with nitrogen. Sodium and oxygen combine much more rapidly than iron and oxygen under similar conditions. Platinum and oxygen do not combine directly. Atoms, ions, and molecules are the particles of substances that react. Bonds are broken and other bonds are formed in reactions. The rate of reaction depends on the particular reactants and bonds involved.

Surface area Solid zinc reacts with aqueous hydrochloric acid and hydrogen gas according to the following equation.

$$Zn(s) + 2HCl(aq) \rightarrow ZnCl_2(aq) + H_2(g)$$

This reaction occurs at the surface of the zinc cube. A cube of zinc measuring 1 cm on each edge presents only 6 cm^2 of contact area to the solvent. The same amount of zinc in the form of a fine powder might provide a contact

SECTION OBJECTIVES

- Define *kinetics* and explain the two conditions necessary for chemical reactions to occur.
- Discuss the five factors that influence reaction rate.
- State what a catalyst is, and explain its effect on reaction rate.
- Explain and write rate laws for chemical reactions.

Rates are always positive terms. They are expressed as $\Delta(conc.)/\Delta t$. A change in concentration of product molecules is always positive since products are being formed. If rate is measured in terms of a change in concentration of a reactant molecule, a negative sign must precede the term because the concentration of a reactant decreases with time and $\Delta(conc.)$ is negative.

area on the order of 10^4 times the original area. Consequently, the reaction rate of the powdered solid is much higher.

A lump of coal burns slowly when kindled in air. The rate of burning can be increased by breaking the lump into smaller pieces, exposing new surfaces. If the piece of coal is powdered and then ignited while suspended in air, it burns explosively. Nickel in large pieces shows no noticeable oxidation in air, but finely powdered nickel reacts vigorously and spectacularly.

The reactions just mentioned are examples of heterogeneous reactions. **Heterogeneous reactions** *involve reactants in two different phases*. Such reactions can occur only when the two phases are in contact. Thus, the surface area of a solid (or liquid) reactant is an important rate consideration. Gases and dissolved particles do not have surfaces in the sense just described. In heterogeneous reactions, the reaction rate is proportional to the area of contact of the reaction substances.

Some chemical reactions between gases actually take place at the solid walls of the container. Others may occur at the surface of a solid or liquid catalyst. If the products are also gases, such a reaction system presents a problem of language. It is a heterogeneous reaction because it takes place at a surface between two phases, gas and solid. On the other hand, it is a homogeneous reaction because all the reactants and all the products are in one phase.

Concentration   Suppose a small lump of charcoal is heated in air until combustion begins. If it is then lowered into a bottle of pure oxygen, the reaction proceeds at a much higher rate. A substance that oxidizes in air reacts more vigorously in pure oxygen, as shown in Figure 18-17. The partial pressure of oxygen in air is approximately one-fifth of the total pressure. Pure oxygen at the same pressure as the air has five times the concentration of oxygen molecules.

This charcoal oxidation is a heterogeneous reaction system in which one reactant is a gas. Not only does the reaction rate depend on the amount of exposed charcoal surface, but it also depends on the concentration of the gas.

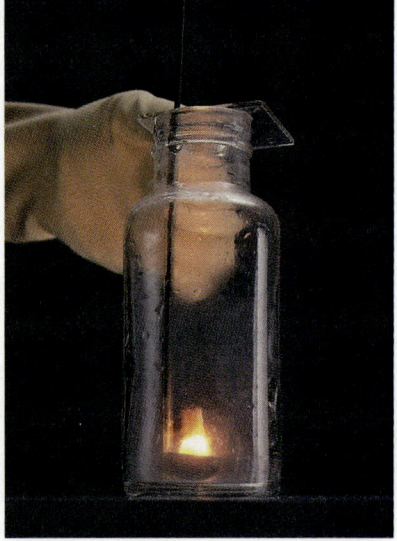

Figure 18-17 Carbon burns faster in oxygen *(right)* than in air because air is only about one-fifth oxygen.

Homogeneous reactions may involve reactants in gaseous or liquid solutions. The concentration of gases changes with pressure according to Boyle's law. In liquid solutions, the concentration of reactants changes if either the quantity of solute or the quantity of solvent is changed. Solids and liquids are practically noncompressible. Thus, it is not possible to change the concentration of pure solids and pure liquids to any measurable extent.

In homogeneous reaction systems, reaction rates depend on the concentration of the reactants. From collision theory, a rate increase might be expected if the concentration of one or more of the reactants is increased. Lowering the concentration should have the opposite effect. The specific effect of concentration changes on reaction rate, however, must be determined experimentally.

Increasing the concentration of substance A in reaction with substance B could increase the reaction rate or have no effect on it. The effect depends on the particular reaction. One cannot tell from the balanced equation for the net reaction how the reaction rate is affected by a change in concentration of reactants. Chemists account for these differences in behavior in terms of the reaction mechanisms.

> This is a key point: a balanced equation does not determine the reaction pathway.

Complex chemical reactions may take place in a series of simple steps. Instead of a single activated complex, there may be several activated complexes in sequence along the reaction pathway. Of these steps, the one that proceeds at the lowest rate determines the overall reaction rate. *This lowest-rate step is called the* **rate-determining step** *for the chemical reaction.*

> Emphasize that *lowest-rate* means *slowest* step.

Temperature The average kinetic energy of the particles of a substance is proportional to the temperature of the substance. Collision theory explains why a rise in temperature increases the rate of chemical reaction. According to this theory, a decrease in temperature lowers the reaction rate for both exothermic and endothermic reactions.

Near room temperature, the rates of many reactions roughly double or triple with a 10°C rise in temperature. This rule of thumb should be used with caution, however. The actual rate increase with a given rise in temperature must be determined experimentally.

Large increases in reaction rate are caused partly by the increase in collision frequency of reaction particles. However, for a chemical reaction to occur, the particles must also collide with enough energy to cause them to react. At higher temperatures, more particles possess enough energy to form the activated complex when collisions occur. In other words, more particles have the necessary activation energy. Thus, a rise in temperature produces an increase in collision energy and collision frequency.

Presence of catalysts Some chemical reactions proceed quite slowly. Frequently, their reaction rates can be increased dramatically by the presence of a catalyst. *A* **catalyst** *is a substance that changes the rate of a chemical reaction without itself being permanently consumed. The action of a catalyst is called* **catalysis.** Catalysts do not appear among the final products of reactions they accelerate. They may participate in one step along a reaction pathway and be regenerated in a later step. In large-scale and cost-sensitive reaction systems, catalysts are recovered and reused.

> The collision energy is a more important factor than the collision frequency.

A catalyst that is in the same phase as all the reactants and products in a reaction system is called a **homogeneous catalyst.** *When its phase is* different from that of the reactants, it is called a **heterogeneous catalyst**. Metals are often used as heterogeneous catalysts. The catalysis of many reactions is promoted by adsorption of reactants on the metal surfaces, which has the effect of increasing the concentration (and therefore the amount of contact) of the reactants.

Figure 18-18 An automobile's catalytic converter makes some chemicals in exhaust less toxic.

STUDY HINT

Enzymes are examples of biological catalysts. Each enzyme accelerates a specific chemical process, enabling the process to be completed typically within only a few hours. The same process uncatalyzed might require weeks for completion. See Section 23.2.

A tunnel through a mountain is analogous to a catalyst in a chemical reaction. The tunnel provides a new, easier path across the mountain. The starting and ending points are still the same.

Such catalytic actions are hindered by the presence of inhibitors. *An* **inhibitor** *is a substance that slows down a process*. Inhibitors, once referred to as negative catalysts, can interfere with the surface chemistry of heterogeneous catalysts and thus inhibit and eventually destroy the catalytic action. An example is the "poisoning" of metallic-oxide catalysts in the catalytic converters of automobile exhaust systems by leaded gasoline.

A catalyst provides an alternative pathway or reaction mechanism in which the potential-energy barrier between reactants and products is changed. The catalyst may be effective in forming an alternative activated complex that requires a lower activation energy, as suggested in the energy profiles of Figure 18-19.

Rate Laws for Reactions

The influence of reactant concentrations on reaction rate in a given chemical system has been mentioned. Measuring this influence is a challenging part of kinetics. Chemists determine the relationship between the rate of a reaction and the concentration of one reactant by first keeping the concentrations of other reactants and the temperature of the system constant. Then the reaction rate is measured for various concentrations of the reactant in question. A series of such experiments reveals how the concentration of each reactant affects the reaction rate.

The following homogeneous reaction is used as an illustration. The reaction is carried out in a vessel of constant volume and at an elevated constant temperature.

$$2H_2(g) + 2NO(g) \rightarrow N_2(g) + 2H_2O(g)$$

Here, four moles of reactant gases produce three moles of product gases. Thus, the pressure of the system diminishes as the reaction proceeds. The rate of the reaction can therefore be determined by measuring the change of pressure in the vessel with time.

Suppose a series of experiments is conducted using the same initial concentration of nitrogen monoxide but different initial concentrations of

Figure 18-19 If a reaction is catalyzed, the change in potential energy may differ from that of the uncatalyzed reaction.

hydrogen. The initial reaction rate is found to vary directly with the hydrogen concentration. That is, doubling the concentration of H_2 doubles the rate, and tripling the concentration of H_2 triples the rate. Therefore,

$$R \propto [H_2]$$

Here R is the reaction rate and $[H_2]$ is the concentration of hydrogen in moles per liter. The $\propto$ is a symbol that is read "is proportionate to."

Now suppose the same initial concentration of hydrogen is used but the initial concentration of nitrogen monoxide is varied. The initial reaction rate is found to increase fourfold when the NO concentration is doubled and ninefold when the concentration of NO is tripled. Thus, the reaction rate varies directly with the square of the nitrogen monoxide concentration.

$$R \propto [NO]^2$$

Since R is proportional to $[H_2]$ and to $[NO]^2$, it is proportional to their product.

$$R \propto [H_2][NO]^2$$

By introduction of an appropriate proportionality constant k, the expression becomes an equality.

$$R = k[H_2][NO]^2$$

*An equation that relates reaction rate and concentrations of reactants is called the **rate law** for the reaction.* It is applicable for a specific reaction at a given temperature. A rise in temperature causes an increase in the rate of nearly all reactions. The value of k usually increases as the temperature increases.

Collision theory indicates that the frequency of collisions between particles increases as the concentration of these particles is raised (Figure

Rate laws MUST be determined experimentally.

Reaction Energy and Reaction Kinetics

18-20). The reaction rate for any step in a reaction pathway is directly proportional to the frequency of collisions between the particles involved. It is also directly proportional to the collision efficiency.

Consider a single step in a reaction pathway. Suppose one molecule of gas A collides with one molecule of gas B to form two molecules of substance C. The equation for the step is

$$A + B \rightarrow 2C$$

One particle of each reactant is involved in each collision. Thus, doubling the concentration of either reactant will double the collision frequency. It will also double the reaction rate *for this step*. Therefore, the rate for the step is directly proportional to the concentration of A and of B. The rate law for the step becomes

$$R = k[A][B]$$

Now suppose the reaction is reversible. In the reverse step, two molecules of C must collide to form one molecule of A and one of B.

$$2C \rightarrow A + B$$

Thus the reaction rate for this reverse step is directly proportional to $[C] \times [C]$. The rate law for the step is

$$R = k[C]^2$$

A simple relationship may or may not exist between the chemical equation for a reaction and the rate law for that reaction. Observe that the power to which the molar concentration of each reactant or product is raised in the rate laws above corresponds to the coefficient for the reactant or product in the balanced equation. Such a relationship does not always hold, however. It holds if the reaction follows a simple one-step path, that is, if the reaction occurs at the molecular level exactly as written. A reaction may proceed in a series of steps, however, and in such reactions the rate law may be simply that for the slowest (rate-determining) step. The rate law may also be more complex, depending on the mechanism. Thus, the rate law for a reaction must be determined experimentally. It cannot be written based only on the balanced equation for the net reaction.

Again consider the reaction between hydrogen and nitrogen monoxide. The balanced equation for the net reaction is

$$2H_2(g) + 2NO(g) \rightarrow N_2(g) + 2H_2O(g)$$

Suppose this reaction occurs in a single step involving a collision between two H_2 molecules and two NO molecules. Doubling the concentration of either reactant should then quadruple the collision frequency and the rate. The rate law should then be proportional to $[H_2]^2$ as well as $[NO]^2$. It would take the form

$$R = k[H_2]^2[NO]^2$$

This equation is not in agreement with experimental results, and the assumed single-step reaction mechanism cannot be correct.

The actual experimentally determined rate law for the net reaction is

$$R = k[H_2][NO]^2$$

Figure 18-20 Under constant conditions, the collision frequency increases with the concentration of each reactant.

Desktop Investigation

Factors Influencing Reaction Rate

Question
How do the type of reactants, surface area of reactants, concentration of reactants, temperature of reactants, and catalysts affect the rates of chemical reactions?

Materials
Bunsen burner
paper ash
copper foil
graduate, 10-mL
magnesium ribbon
matches
paper clip
2 potatoes
sandpaper
steel wool
2 sugar cubes
white vinegar
sheet of zinc
6 test tubes, 16 x 150 mm
tongs

Procedure
Remove all combustible material from the work area. Wear safety glasses and an apron. Record all your results in a data table.

1. Add 10 mL of vinegar to each of three test tubes. To one test tube, add a 3-cm piece of magnesium ribbon; to a second, a 3-cm zinc strip; and to a third, a 3-cm copper strip. (All metals should be the same width.) If necessary, sandpaper the metals until shiny. (See drawing below.)

2. Using tongs, hold a paper clip in the hottest part of the burner flame for 30 seconds. Repeat with a ball of

steel wool 2 cm in diameter. (See drawing at upper right.)

3. To one test tube, add 10 mL of vinegar; to a second, 5 mL of vinegar plus 5 mL of water; to a third, 2.5 mL of vinegar plus 7.5 mL of water. To each of the three test tubes add a 3-cm piece of magnesium ribbon.

4. Select two potatoes that are approximately equal in shape and mass. Bake one at 300°F until it is thoroughly cooked. Bake the second at 350°F until it is thoroughly cooked.

5. Using tongs, hold a sugar cube and try to ignite it with a match. Then try to ignite it in a burner flame. Rub paper ash on a second cube and try to ignite it with a match.

Discussion
1. What are the rate-influencing factors in Procedures 1, 2, 3, 4, and 5?

2. What were the results from each of the five parts of the experiment? How do you interpret these results?

The reaction pathway must therefore, consist of more than one step. The slowest step, thought to be the collision of one H_2 molecule with two NO molecules, is rate-determining. The rate law found experimentally for the net reaction is the rate expression of this rate-determining step.

The general form for the rate law of reactions is

$$R = k[A]^n[B]^m \ldots$$

Here R is the reaction rate, k is the rate constant, and [A] and [B]... represent

1. Reaction rates and reaction mechanisms
2. (1) Nature of reactants
   (2) Surface area
   (3) Concentration
   (4) Temperature
   (5) Presence of a catalyst
3. A substance that changes the rate of a chemical reaction without itself being permanently consumed. A catalyst provides an alternative pathway or reaction mechanism in which the potential-energy barrier between reactants and products is changed. It lowers the activation energy.
4. An equation that relates the reaction rate to the concentrations of reactants. The rate law can be based on the balanced equation only if the reaction takes place in a single step. If the reaction requires more than one step, the rate law is based on the concentrations of the reactants involved in the slowest step.

the molar concentrations of reactants. The n and m are the respective powers to which the concentrations are raised, *based on experimental data*. Again it should be emphasized that one cannot assume that the coefficients in the balanced equation for a net reaction are the exponents in the rate law for the reaction.

The dependence of reaction rate on concentration of reactants was first recognized as a general principle by two Norwegian chemists, Cato Maximilian Guldberg (1836–1902) and Peter Waage (1833–1900), in 1867. They stated their discovery of this dependence as the **law of mass action:** *the rate of a chemical reaction is directly proportional to the product of the concentrations of reacting substances, each raised to the appropriate power.* This principle is interpreted in terms of the rate law for a chemical system in the modern science of chemical reaction kinetics.

Sample Problem 18.5

A reaction involving reactants X and Y was found to occur by a one-step mechanism: $X + 2Y \rightarrow XY_2$. Write the rate law for this reaction and then determine the effect of each of the following on the reaction rate: *(a)* doubling the concentration of X *(b)* doubling the concentration of Y *(c)* using one-third the concentration of Y.

Solution
Because the equation represents a single-step mechanism, the rate law can be written from the equation (otherwise it could not be). The rate will vary directly with the concentration of X, which has no coefficient in the equation, and will vary directly with the square of the concentration of Y, which has the coefficient of 2: $R = k[X][Y]^2$.
(a) Doubling the concentration of X will double the rate ($R = k[2X][Y]^2$).
(b) Doubling the concentration of Y will increase the rate fourfold ($R = k[X][2Y]^2$).
(c) Using one-third the concentration of Y will reduce the rate to one-ninth of its original value ($R = k[X][\frac{1}{3}Y]^2$).

Practice Problem
The rate of reaction involving L, M, and N is found to double if the concentration of L is doubled, to increase eightfold if the concentration of M is doubled, and to double if the concentration of N is doubled. Write the rate law for this reaction. *(Ans.)* $R = k[L][M]^3[N]$

Section Review

1. What is studied in the branch of chemistry that is known as chemical kinetics?
2. List the five important factors that influence the rate of chemical reactions.
3. What is a catalyst? Explain the effect of a catalyst on the rate of chemical reactions. How does a catalyst influence the activation energy required by a particular reaction?
4. What is meant by a rate law for a chemical reaction? Explain the conditions under which a rate law can be written from a chemical equation. When can a rate law not be written from a single step?

Chapter Summary

- Thermochemistry is the study of the changes in heat energy that accompany chemical reactions and physical changes.
- A thermochemical equation is an equation that includes the quantity of heat released or absorbed during the reaction as written.
- The heat of reaction is the quantity of heat released or absorbed during a chemical reaction.
- Enthalpy is the heat content of a system at constant pressure.
- An enthalpy change is the amount of heat absorbed or lost by a system in a process carried out at constant pressure.
- The heat of formation is negative for exothermic reactions and positive for endothermic reactions.
- Compounds with highly negative heats of formation tend to be stable; compounds with highly positive or only slightly negative heats of formation tend to be unstable.
- The standard molar heat of formation is the heat absorbed or released in formation of one mole of a compound from the elements in their standard states at 25°C.
- The heat released in a combustion reaction is called the heat of combustion.
- Heats of reaction can be found by using heats of formation of reactants and products.
- The heat of formation of a compound may be found by adding together the heats of formation of the products of combustion of the compound, and then subtracting the heat of combustion of the compound.
- The change in the heat content, or enthalpy, of a system is related to the number of bonds breaking, the number of bonds forming, and the strength of the bonds.
- Entropy is a measure of the disorder of a system.
- Free-energy change combines the effects of entropy and enthalpy changes and temperatures of a system, and is a measure of the overall drive to spontaneous change.
- A reaction will occur spontaneously if it causes a favorable change (decrease) in free energy. It will not occur if it causes an increase in free energy.
- The step-by-step process by which an overall chemical reaction occurs is called the reaction mechanism.
- In order for chemical reactions to occur, the particles of the reactants must collide.
- Activation energy is needed to loosen bonds so molecules can become reactive.
- An activated complex is formed when an effective collision between molecules of reactants raises their internal energies to the minimum level for a reaction to occur.
- The rate of reaction is influenced by the following factors: nature of reactants, surface area, temperature, concentration of reactants, and the presence of catalysts and inhibitors.
- The rates at which chemical reactions occur can sometimes be experimentally measured and expressed in terms of mathematical equations called rate laws.

Chapter 18 Review

Reaction Energy and Reaction Kinetics

VOCABULARY

activated complex (545)
activation energy (545)
calorimeter (527)
catalysis (551)
catalyst (551)
chemical kinetics (549)
collision theory (543)
enthalpy (528)
enthalpy change (528)
entropy (538)
free energy (538)
free-energy change (538)
heat of combustion (531)
heat of reaction (528)
Hess's law (531)
heterogeneous catalyst (552)
heterogeneous reaction (550)
homogeneous catalyst (551)
homogeneous reaction (541)
inhibitor (552)
intermediate (543)
law of mass action (556)
molar heat of formation (529)
rate-determining step (551)
rate law (553)
reaction mechanism (541)
reaction rate (549)
thermochemical equation (527)
thermochemistry (527)

REVIEWING CONCEPTS

1. How does the heat content of the products of a reaction system compare with the heat content of the reactants when the reaction is
   a. endothermic
   b. exothermic (18.1)

2. a. Distinguish between heats of reaction, formation, and combustion.
   b. Upon what basis are heats of formation and combustion defined? (18.1)

3. Write the equations that can be used to calculate
   a. heat of reaction from heats of formation
   b. heat of formation from heats of combustion (18.1)

4. What factors affect the value of ΔH in a reaction system? (18.1)

5. What kind of entropy change is observed in changes of phase in the direction of the solid? (18.2)

6. How does an increase in temperature affect the entropy of a system? (18.2)

7. According to the ΔG equation, what combination of ΔH and ΔS values always results in a spontaneous reaction? (18.2)

8. Explain the relationship between temperature and the spontaneity of reactions. (18.2)

9. a. What is the collision theory?
   b. According to this theory, what two conditions must be met in order for a collision between reactant molecules to be effective in producing new chemical species? (18.3)

10. a. What condition must be met in order for an activated complex to result from the collision of reactant particles?
    b. Where, in terms of energy, does the complex occur along a typical reaction pathway? (18.3)

11. In a reversible reaction, how does the activation energy required for the exothermic change compare with the activation energy required for the endothermic change? (18.3)

12. Would you expect the following reaction to represent the mechanism by which ethane (C_2H_6) burns? Why or why not? (18.3)
    $$C_2H_6(g) + \tfrac{7}{2}O_2(g) \rightarrow 2CO_2(g) + 3H_2O(g)$$

13. The decomposition of nitrogen dioxide $2NO_2 \rightarrow 2NO + O_2$ occurs in a two-step sequence at elevated temperatures. The first step is $NO_2 \rightarrow NO + O$. Predict a possible second step that, when combined with the first step, gives the complete reaction. (18.3)

14. What is meant by the rate-determining step for a chemical reaction? (18.4)

15. Write the general equation that expresses the rates of many chemical reactions, and label the various factors. (18.4)

PROBLEMS

▶ Reaction Heat

1. For each reaction listed below, determine the ΔH and type of reaction (endothermic or exothermic):
   a. $C(s) + O_2(g) \rightarrow CO_2(g) + 393.51$ kJ
   b. $CH_4(g) + 2O_2(g) \rightarrow CO_2(g) + 2H_2O(\ell) + 890.31$ kJ
   c. $CaCO_3(s) + 176$ kJ $\rightarrow CaO(s) + CO_2(g)$
   d. $H_2O(g) \rightarrow H_2O(\ell) + 44.02$ kJ
   e. $2SO_2(g) + O_2(g) \rightarrow 2SO_3(g) + 197.8$ kJ

2. Rewrite each reaction below with the ΔH value in its appropriate position and identify the reaction as endothermic or exothermic:
   a. $H_2(g) + \frac{1}{2}O_2(g) \rightarrow H_2O(\ell)$; $\Delta H = -285.83$ kJ
   b. $2Mg(s) + O_2(g) \rightarrow 2MgO(s)$; $\Delta H = -1200$ kJ
   c. $I_2(s) \rightarrow I_2(g)$; $\Delta H = +62.4$ kJ
   d. $3CO(g) + Fe_2O_3(s) \rightarrow 2Fe(s) + 3CO_2(g)$; $\Delta H = -24.7$ kJ
   e. $2NO_2(g) \rightarrow 2NO(g) + O_2(g)$; $\Delta H = +114.2$ kJ
   f. $C_2H_4(g) + 3O_2(g) \rightarrow 2CO_2(g) + 2H_2O(\ell)$; $\Delta H = -1411.0$ kJ

3. Use Appendix Table A-13 to write the reaction illustrating the formation of each of the following compounds from its elements. Write the ΔH as part of each equation and indicate the ΔH for the reverse reaction:
   a. $CaCl_2(s)$
   b. $C_2H_2(g)$ (ethyne or acetylene)
   c. $SO_2(g)$

4. Use heat of formation data given in Appendix Table A-13 to calculate the heat of reaction for each of the following. Solve each by combining the known thermochemical equations. Verify each result by using the general equation for finding heats of reaction from heats of formation.
   a. $CaCO_3(s) \rightarrow CaO(s) + CO_2(g)$
   b. $Ca(OH)_2(s) \rightarrow CaO(s) + H_2O(g)$
   c. $Fe_2O_3(s) + 3CO(g) \rightarrow 2Fe(s) + 3CO_2(g)$

5. Calculate the heats of reaction for combustion reactions in which ethane and benzene are the respective reactants and $CO_2(g)$ and $H_2O(\ell)$ are the products in each. Solve each by combining the known thermochemical equations using ΔH_f° values in Appendix Table A-13. Verify each result by using the general equation for finding heats of reaction from heats of formation.
   a. $C_2H_6(g) + O_2(g) \rightarrow$
   b. $C_6H_6(\ell) + O_2(g) \rightarrow$

6. The heat of formation of ethanol, C_2H_5OH, is -277.0 kJ/mol at 25°C. Calculate the heat of combustion of one mole of ethanol, assuming that the products are $CO_2(g)$ and $H_2O(\ell)$.

7. Calculate the heat of formation of $H_2SO_4(\ell)$ from the ΔH values for the combustion of sulfur to $SO_2(g)$ ($\Delta H = -296.83$ kJ/mol), the oxidation of SO_2 to $SO_3(g)$ ($\Delta H = -98.91$ kJ/mol), and the solution of SO_3 in $H_2O(\ell)$ (heat of solution $= -132.4$ kJ/mol) to give $H_2SO_4(\ell)$ at 25°C. Solve by combining the known thermochemical equations.

▶ Entropy and Free Energy

8. Based on the following values, compute ΔG values for each reaction, and predict whether the reaction will occur spontaneously.
   a. $\Delta H = +125 \frac{\text{kJ}}{\text{mol}}$, $T = 293$K, $\Delta S = 0.0350 \frac{\text{kJ}}{(\text{mol} \cdot \text{K})}$
   b. $\Delta H = -85.2 \frac{\text{kJ}}{\text{mol}}$, $t = 127°$C, $\Delta S = 0.125 \frac{\text{kJ}}{(\text{mol} \cdot \text{K})}$
   c. $\Delta H = -275 \frac{\text{kJ}}{\text{mol}}$, $t = 500°$C, $\Delta S = 0.450 \frac{\text{kJ}}{(\text{mol} \cdot \text{K})}$

9. The ΔS for the reaction shown, at 25°C, is 0.003 00 kJ/(mol · K). Calculate the ΔG for

this reaction and determine whether it will occur spontaneously at 25°C.

$$C(s) + O_2(g) \rightarrow CO_2(g) + 393.51 \text{ kJ}$$

10. When graphite reacts with hydrogen at 27°C, ΔH is -74.8 kJ and ΔS is -0.0809 kJ/(mol · K). Will this reaction occur spontaneously?

▶ **Reaction Diagrams**

11. For each of the energy diagrams provided below, label the reactants, products, ΔH, E_a, and E_a'. Also determine the values of ΔH for the forward and reverse reactions, and of E_a, and E_a'.

(a)

(b)

(c)
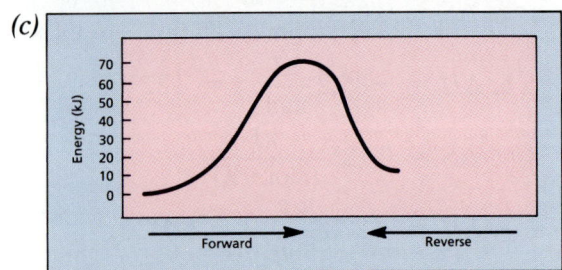

12. Draw and label energy diagrams depicting the following reactions, and determine all remaining values. Place the reactants at energy level zero.

a. $\Delta H_{\text{forward}} = +30$ kJ $E_a' = 20$ kJ
b. $\Delta H_{\text{forward}} = -30$ kJ $E_a = 20$ kJ
c. $\Delta H_{\text{forward}} = 10$ kJ $E_a' = 40$ kJ
d. $\Delta H_{\text{forward}} = 95$ kJ $E_a' = 20$ kJ
e. $\Delta H_{\text{reverse}} = 40$ kJ $E_a = 30$ kJ

▶ **Rate Law**

13. A reaction involving reactants A and B is found to occur in the one-step mechanism: $2A + B \rightarrow A_2B$. Write the rate law for this reaction and predict the effect of doubling the concentration of either reactant on the overall reaction rate.

14. A chemical reaction is expressed by the balanced chemical equation $A + 2B \rightarrow C$. Using the data that follows,
    a. determine the rate law for the reaction.
    b. calculate the value of the specific rate constant.
    c. if the initial concentrations of both A and B are 0.30 M, at what initial rate is C formed? Three reaction rate experiments yield the following data:

Experiment number	Initial [A]	Initial [B]	Initial rate of formation of C
1	0.20 M	0.20 M	2.0×10^{-4} M/min
2	0.20 M	0.40 M	8.0×10^{-4} M/min
3	0.40 M	0.40 M	1.6×10^{-3} M/min

CRITICAL THINKING

1. **Interpreting Concepts** Explain the relative importance, in fuel selection, of the level of energy contained in each unit mass of the reactants and products, as opposed to their specific ΔH_c values per mole or their flame temperature. What other factors should also be considered?

2. **Predicting Outcomes** The balanced equation for a rapid homogeneous reaction between two gases is as follows: $4A + B \rightarrow 2C + 2D$. Since the simultaneous collision of four molecules of one reactant with one molecule of the other reactant is extremely improbable, what would you predict about the nature of

the reaction mechanism for this reaction system?

3. **Evaluating Ideas**
   a. How can you justify calling the reaction pathway that is shown in Figure 18-14 the minimum-energy pathway for reaction?
   b. What significance is associated with the maximum-energy region of this minimum-energy pathway?

4. **Applying Models** Explain why there is a danger of explosion in such places as saw mills and grain elevators where large amounts of dry, powdered combustible materials are present.

5. **Evaluating Methods** What property would you measure in order to determine the reaction rate for the following reaction? Justify your choice.

$$2NO_2(g) \rightarrow N_2O_4(g)$$

6. **Interpreting Concepts** In a thought experiment, suppose two moles of hydrogen gas and one mole of iodine vapor are passed simultaneously into a one liter flask. The rate law for the forward reaction is $R = k[I_2][H_2]$. What is the effect on the rate of the forward reaction if
   a. the temperature is increased
   b. one mole of iodine vapor is added
   c. one mole of hydrogen is removed
   d. the volume of the flask is reduced (assume this is possible)

CROSS CONNECTIONS

1. **Health** Keep track of your food caloric intake for one day. Convert the total number of food Calories to joules (1 food Cal = 1000 cal = 4.184 kJ). Compare your totals with those of your classmates and members of your family. How does your caloric intake compare with that recommended by nutritionists?

RESEARCH AND WRITING

1. Carry out library research to obtain information on alternative measurement units used to express values of heat and energy. Also, find out how the quantities relate to SI metric units. Include information specifically on English units, such as the British Thermal Unit (BTU), and on typical BTU ratings of household appliances. Calculate how these ratings would be expressed in joules instead.

PROJECT

1. Design a simple calorimeter investigation to determine the molar heat of fusion of water. Use of the following materials: a large plastic foam cup with cover, a thermometer, a balance, water at room temperature, and an ice cube. Allow your teacher to review your design. Then carry out the investigation and write up a laboratory report, including your calculations and a comparison of your quantitative results with known values. Attempt to account for any disagreements between the experimental and actual values.

CHAPTER 19
Chemical Equilibrium

PLANNING CHART Approximate Time: 5/7 days

	SECTION/TEXT FEATURES		RESOURCES	
B A S I C	19.1 **THE NATURE OF CHEMICAL EQUILIBRIUM** Section Review Chapter Review • Questions 1–3, Problems 1–5	563 569 592	*Teacher's Resource Binder* Lab Experiment 29 Problem Solving Worksheets Section Review Worksheets *Other Resources* Transparency 43	159 67
A V E R A G E	19.2 **SHIFTING EQUILIBRIUM** Section Review Chapter Review • Questions 4–17	569 576 592	*Teacher's Resource Binder* Lab Experiment 28 Problem Solving Worksheets Section Review Worksheets *Other Resources* *Shakhashiri* Video Demo 8 *World of Chemistry* Video 14 Transparency 44 Blackline Master 46	153 68
A V E R A G E	19.3 **EQUILIBRIUM OF ACIDS, BASES, AND SALTS** Microscale Investigation 17 Section Review Chapter Review • Questions 18–22, Problems 6–7	576 871 583 592	*Teacher's Resource Binder* Problem Solving Worksheets Section Review Worksheets	 69
H O N O R S	19.4 **SOLUBILITY EQUILIBRIUM** Section Review Chapter Review • Questions 23–28, Problems 8–22	583 591 592	*Teacher's Resource Binder* Lab Experiment 30 Problem Solving Worksheets Section Review Worksheets	165 70
	CHAPTER END ACTIVITIES Critical Thinking, Cross Connections, Research and Writing, Projects		*Teacher's Resource Binder* Test Booklet *Computer Test Generator*	93, 95

Teaching Strategies

Introduction

The chapter on chemical equilibrium is one that students may find to be very difficult. Part of the difficulty is due to the mathematical basis of the problems, but the initial presentation of the concept may also add to the problem. The students have already been introduced to the key element of the chapter—the statement of the law of mass action. If the students are brought to realize that all of the different types of equilibrium problems they will encounter are different applications of the same law of mass action, they will probably fare better.

Begin with a restatement of the law of mass action for a general equation:

$$nA + mB \ldots \rightleftarrows xC + yD + \ldots$$

$$K = \frac{[C]^x[D]^y \ldots}{[A]^n[B]^m \ldots}$$

Since a system at equilibrium is being discussed, the temperature must be a constant and the system must be closed.

Under these conditions, pure solids and liquids that occur as reactants and products do not appear in the law of mass action expression because their "concentrations" are really expressions of density which is at constant temperature.

[ ] = mol/liter = mol/volume = mass/volume = density

Liquid water is a special case because one liter of water has a mass of approximately 1000 g or 55.5 mol/L. In most cases where water is involved in the reaction, so little water is actually used that the value 55.5 mol/L remains constant.

Only variables are retained in the law of mass action expression. Constants are incorporated into the constant, K.

If substances A, B, C, and D are gases, K_c would describe an equilibrium in which concentration is expressed in mol/L, while K_p would use partial pressures.

$$3H_2(g) + N_2(g) \rightleftarrows 2NH_3(g)$$

$$K_c = \frac{[NH_3(g)]^2}{[H_2(g)]^3[N_2(g)]}$$

$$K_p = \frac{P[NH_3(g)]^2}{P[H_2(g)]^3 P[N_2(g)]}$$

If A is an acid, then K becomes K_a. The same applies if A is base, K becomes K_b.

$$HC_2H_3O_2(aq) \rightleftarrows H^+(aq) + C_2H_3O_2^-(aq)$$

$$K_a = \frac{[H^+(aq)][C_2H_3O_2^-(aq)]}{[HC_2H_3O_2(aq)]}$$

or $HC_2H_3O_2(aq) + H_2O(\ell) \rightleftarrows H_3O^+(aq) + C_2H_3O_2^-(aq)$

$$K_a = \frac{[H_3O^+(aq)][C_2H_3O_2^-(aq)]}{[HC_2H_3O_2(aq)]}$$

In the case of a weak base,

$$NH_3(aq) + H_2O(\ell) \rightleftarrows NH_4^+(aq) + OH^-(aq)$$

$$K = \frac{[NH_4^+(aq)][OH^-(aq)]}{[NH_3(aq)][H_2O(\ell)]}$$

$$K[H_2O(\ell)] = \frac{[NH_4^+(aq)][OH^-(aq)]}{[NH_3(aq)]}$$

$$K_b = \frac{[NH_4^+(aq)][OH^-(aq)]}{[NH_3(aq)]}$$

If A represents water, $H_2O(\ell)$, then K becomes K_w.

$$H_2O(\ell) \rightleftarrows H^+(aq) + OH^-(aq)$$

$$K = \frac{[H^+(aq)][OH^-(aq)]}{[H_2O(\ell)]}$$

$$K[H_2O(\ell)] = [H^+(aq)][OH^-(aq)]$$

$$K_w = [H^+(aq)][OH^+(aq)]$$

or $H_2O(\ell) + H_2O(\ell) \rightleftarrows H_3O^+(aq) + OH^-(aq)$

$$K = \frac{[H^+(aq)][OH^-(aq)]}{[H_2O(\ell)]^2}$$

$$K[H_2O(\ell)]^2 = [H_3O^+(aq)][OH^-(aq)]$$

$$K_w = [H_3O^+(aq)][OH^-(aq)]$$

If A represents a slightly soluble ionic solid, K becomes K_{sp}.

$$AgCl(s) \rightleftarrows Ag^+(aq) + Cl^-(aq)$$

$$K = \frac{[Ag^+(aq)][Cl^-(aq)]}{[AgCl(s)]}$$

$$K[AgCl(s)] = [Ag^+(aq)][Cl^-(aq)]$$

$$K_{sp} = [Ag^+(aq)][Cl^-(aq)]$$

By initially outlining the chapter and how the same definition will be applied, you will make the students' task of trying to understand the concept less difficult.

Students who go on to take more advanced courses in chemistry will approach the same topic but base it on activities and activity coefficients. These should not be discussed at the introductory level. In the more advanced courses, students will learn of unit activities rather than constant densities.

As many demonstrations as are possible should be incorporated into the lectures and classroom discussions. Problem solving techniques are also very important. Students must work as many practice, review, and end-of-chapter problems as are needed to make them feel comfortable with them.

The demonstration manuals mentioned throughout this text unit have many demonstrations that are suitable and which may be used to supplement the ones described below.

19.1 The Nature of Chemical Equilibrium

Review physical equilibrium, dynamic equilibrium, equilibrium vapor pressure, Le Châtelier's principle, and solution equilibrium. Also review the role of free-energy change and the spontaneity of a reaction for the water–gas reaction. You may wish to use the data in that section for ΔS and ΔH and have the students calculate the temperature at which a closed system would reach equilibrium (about 980K).

Discuss Lavoisier's mercury(II) oxide bell-jar experiment and have students write the equation for this experiment and for equilibrium. Define chemical equilibrium and discuss the balance of the driving force of enthalpy change, ΔH, and the driving force of entropy change, ΔS.

Refer to Chapter 13 and discuss equilibrium vapor pressure (and saturated solutions) to show that for a given closed system, the particular equilibrium state depends upon temperature and, for systems involving gases, also depends on the pressure.

Discuss the ionization of dilute acetic acid as an example of a reaction reaching equilibrium far to the left since acetic acid is a weak acid. Discuss the dissociation of NaCl or the ionization of HCl as reactions that go far to the right. Discuss the chemist's problem of producing desirable products in terms of ammonia synthesis. Have students restate the five factors that determine the rate of chemical reactions.

Refer to the introduction in Chapter 13 and further develop the concept of the equilibrium concept. Much practice on the part of the students is needed. Follow the development in the text. Begin with the hypothetical equations and move to the real equations given. Have the students write out equilibrium constant expressions until they are comfortable with the practice. Point out that if concentrations are known, the equilibrium constant may be calculated, and that if the equilibrium constant is known, equilibrium concentrations can be calculated from the initial concentrations and the stoichiometry of the reaction.

The Technology feature on *Swimming Pool Chemistry* gives you the opportunity to discuss another equilibrium system.

19.2 Shifting Equilibrium

Raise the problem of how chemists may displace an equilibrium in order to increase the desired production. Review and restate Le Châtelier's principle. The text discusses the effect of changes in concentration, of changes in pressure where gases are involved, and the effect of temperature. Stress with the students that the only change that will affect the numerical value of the equilibrium constant is a change in temperature. An increase in temperature will favor the endothermic reaction. In all other cases, a new set of equilibrium concentrations or pressure will be established, but the numerical value of the equilibrium constant will be the same as before any changes occurred. Students will have a difficult time visualizing this. They should be led carefully through a description of the changes that occur on the molecular level and then have these changes related to the equilibrium expression that describes them. They should note the numerator (concentration of products) and the denominator (concentration of reactants). Catalysts deserve special attention. They affect the rate of a reaction, not the net amount of product. Catalysts have no effect on the numerical value of the equilibrium constant.

Explain to the students that there are signs that will help them predict whether a given reaction will run to completion. Two signs are readily apparent: (1) the formation of a gas and (2) the formation of a precipitate. A third is more difficult to see: (3) the formation of a slightly ionized product. Each of these should be demonstrated.

The section concludes with a discussion of the common ion effect. Treat it as just another case of the application of Le Châtelier's principle. Most students will be able to predict what should happen.

19.3 Equilibria of Acids, Bases, and Salts

A more detailed look at one specific type of equilibrium is now discussed. Refer back to your initial lecture. The one common aspect of the discussion is the dissociation of a slightly ionized, hydrogen-containing, water soluble compound—a weak acid. Because all equations have this in common, K_{eq} is now referred to as K_a.

As you proceed with your explanation, have your students note the difference in relative concentrations of reactants and products when K_a is very small. Very little product is present in the system. This will help you later when you broach the subject of buffers—systems in which both reactant and product (acid anion) concentrations are of the same magnitude. Buffers may be introduced as another case of the common ion effect.

Water is a very important case. Impress upon the students that it is so important that the equilibrium constant describing the dissociation of water has its own symbol, K_w. Refer back to your introductory lecture and point out, once again, that the water system is just another case of the application of the law of mass action.

The remaining sections discuss the hydrolysis of salts, anion hydrolysis, and cation hydrolysis. One method of helping students recall and organize the information is to have them reconstruct the acid and base from which the salt might have been formed. Using NaCl as an example, the cation, Na^+, is always derived from the base, and the anion, Cl^-, from the acid. To predict the base and acid, add the proper number of OH^-'s to the cation and the proper number of H^+'s to the anion. In the case of NaCl, the base must have been NaOH and the acid HCl. Since the salt formed from the neutralization of a strong acid and a strong base is neutral, NaCl is a neutral salt. Each of your predictions should be tested with an acid–base indicator. Use the following table as a mnemonic.

SA = strong acid   SB = strong base
wA = weak acid   wB = weak base

SB + SA → neutral salt (NaCl, or Na_2SO_4)
wB + SA → acid salt (NH_4Cl)
SB + wA → basic salt (NaF)
wB + wA → determined by the stronger of the two.

(One needs to know the K_a and K_b.)

For the mnemonic to be of any use, students must be aware of the common strong acids and strong bases.

The anion hydrolysis constant should again be discussed in terms of just another application of the law of mass action. When introducing cation hydrolysis, note the special importance of ions with a charge of 2+ and 3+. Ions of such solutions will always test acidic.

19.4 Solubility Equilibrium

Impress upon the students that no substance is completely insoluble. Certain ranges of solubility have been established. It is very important that after you derive the solubility product expression (again as a special case of the more general law of mass action), you point out to the students the general class of problems most frequently associated with solubility products: calculating the K_{sp} from solubility data, calculating solubilities given the K_{sp} for a reaction, and determining whether or not a precipitate will form, given the K_{sp} and initial concentration of the reactants.

References and Resources

Books and Periodicals

Kauffman, George B. "The Brønsted-Lowry Acid-Base Concept," *Journal of Chemical Education*, Vol. 65, No. 1, Jan. 1988, pp. 28–29.

Russell, Joan M. "Simple Models for Teaching Equilibrium and Le Châtelier's Principle," *Journal of Chemical Education*, Vol. 65, No. 10, Oct. 1988, pp. 871–872.

Audiovisual Resources

Campbell, J. Arthur. *Acid-Base Indicators*, A Chem Study Film, Color, 19 min., VHS cassette, Central Scientific Co. (Behavior of acid-base indicators according to the proton donor-acceptor theory)

Pimentel, George C. *Equilibrium*, A Chem Study Film, Color, 24 min., 16mm film or VHS cassette, Central Scientific Co. (The dynamic nature of equilibrium is examined)

Computer Software

"Chemical Equilibrium Series," Apple, Cambridge Development Laboratory.

Answers and Solutions

Reviewing Concepts

1. At the start, A and B are at their maximum concentrations, and no C and D exist. As A and B collide and react to form C and D, the concentrations of A and B decrease, while those of C and D increase. As the amounts of C and D increase, the rate at which they combine to re-form A and B increases, while the rate at which A and B combine decreases. Eventually, the two rates become equal and equilibrium is established. At that point, the concentrations of A, B, C, and D remain constant as long as conditions remain the same.

2. **a.** $K = \dfrac{[C]^x[D]^y \ldots}{[A]^n[B]^m \ldots}$

   b. The value of K provides information about the extent to which the reactants are converted into the products of a given reaction at the specified temperature.

3. **a.** Neither is favored.
   b. The reverse reaction is favored.
   c. The forward reaction is favored.

4. **a.** Forward **b.** Reverse **c.** Forward **d.** Reverse
5. **a.** Higher **b.** Higher **c.** Lower **d.** Lower
6. **a.** Forward **b.** Reverse

7. The concentrations of pure substances in the solid and liquid phases can be removed from the equilibrium constant expression because the equilibrium concentrations of such substances are not changed by adding or removing quantities of these materials. Since concentration is density-dependent, the densities of substances in these phases is constant regardless of the total amounts present.

8. A catalyst increases the rates of both the forward and reverse reactions at equilibrium, but does so to an equal extent such that the relative equilibrium amounts, and thus the value of K, are not affected.

9. **a.** Forward **b.** Forward **c.** Neither
   d. Forward **e.** Reverse **f.** Neither
   g. Neither **h.** Reverse **i.** Neither

10. **a.** HCl would increase; K would remain the same
    b. HCl would decrease; K would remain the same
    c. HCl would not change; K would remain the same
    d. HCl would increase; K would increase
    e. HCl would decrease; K would remain the same
    f. HCl would not change; K would remain the same
    g. HCl would not change; K would remain the same
    h. HCl would decrease; K would decrease
    i. HCl would not change; K would remain the same.

11. To the equilibrium system continually add more H_2 and F_2 while removing the HF as it is formed.

12. Once equilibrium is established, such changes have an equal effect on the numerator and denominator of the equilibrium constant expression so that the numerical ratio is not affected.

13. High pressure would shift the equilibrium to the right since the forward reaction converts three molecules into two, thereby effectively reducing the volume and relieving the stress imposed by the increased pressure.

14. High pressure would shift the equilibrium toward the increased production of methanol.

15. The yield of methanol would be increased by lowering the temperature of the system since the reaction to the right is exothermic. A rise in temperature would act as a stress on the system, and the equilibrium would shift in the direction of the endothermic reaction to relieve the stress. This would reduce the concentration of methanol.

16. **a.** High reactant concentrations, high pressure, low temperature
    b. High reactant concentrations, pressure not relevant, low temperature
    c. High reactant concentrations, pressure not relevant, high temperature
    d. High reactant concentrations, high pressure, low temperature
    e. High reactant concentrations, low pressure, high temperature

17. **a.** The rate of the reaction increases decidedly.
    b. Air is one-fifth oxygen; thus, pure oxygen at the same pressure as the air will provide 5 times the number of collisions between the reactants.

18. **a.** $K_w = [H_3O^+][OH^-]$ **b.** $K_w = 10^{-14}$

19. • Salts of strong acids and strong bases — Neither the cations of a strong base nor the anions of a strong acid hydrolyze appreciably in aqueous solutions. Such salts yield neutral aqueous solutions with a pH of 7. NaCl is an example.
    • Salts of weak acids and strong bases — The aqueous solutions of such salts are basic. The anions of the salt (the negative ions from the weak acid) undergo hydrolysis, while the cations of the salt (the positive ions from the strong base) do not. $NaC_2H_3O_2$ is an example.
    • Salts of strong acids and weak bases — The aqueous solutions of such salts are acidic. The cations of the salt (the positive ions from the weak base) undergo hydrolysis, while the anions do not. NH_4Cl is an example.
    • Salts of weak acids and weak bases — In aqueous solutions of such salts, both ions of the dissolved salt are hydrolyzed extensively and the solution can be acidic, basic, or neutral, depending on the salt involved. $NH_4C_2H_3O_2$ is an example.

20. Acetic acid is a weak electrolyte and is slightly ionized in water solution. Equilibrium is quickly established between acetic acid molecules and hydronium and acetate ions in the solution: $HC_2H_3O_2 + H_2O \rightleftarrows H_3O^+ + C_2H_3O_2^-$. Sodium acetate is an ionic salt and is therefore completely dissociated in water solution. When added to the acetic acid solution, the concentration of the acetate ion is increased, driving the equilibrium to the left. This removes hydronium ions from solution and increases the concentration of nonionized acetic acid molecules. Thus the solution is less acidic and has a higher pH value.

21. At equilibrium, the concentration of molecular acetic

acid is high and the concentrations of H_3O^+ and $C_2H_3O_2^-$ ions are low. Equilibrium is established early, and the low concentrations of ions in the equilibrium system show that acetic acid is a weak electrolyte.

22. a. $NH_3(aq) + H_2O(\ell) \rightleftarrows NH_4^+(aq) + OH^-(aq)$

$$K_b = \frac{[NH_4^+][OH^-]}{[NH_3]}$$

b. The value of K_b is small, showing that there are comparatively few NH_4^+ and OH^- ions at equilibrium.

23. Although a saturated solution contains the maximum concentration of solute possible at a given temperature in equilibrium with an undissolved substance, the actual concentration of dissolved solute may be high or low, depending on the solubility of that solute. Thus, saturated does not necessarily mean concentrated.

24. The extent to which reacting ions are removed from solution depends on the solubility of the compound formed and the degree of ionization, if the compound is soluble.

25. • A gaseous product is formed:
$H_3O^+(aq) + HCO_3^-(aq) \rightarrow 2H_2O(\ell) + CO_2(g)$.
• A precipitate is formed:
$Ag^+(aq) + Cl^-(aq) \rightarrow AgCl(s)$.
• A slightly ionized product is formed:
$H_3O^+(aq) + OH^-(aq) \rightarrow 2H_2O(\ell)$.

26. In general, a substance is said to be soluble if the solubility is greater than 1 g per 100 g of water, and insoluble if its solubility is less than 0.1 g per 100 g of water. Substances whose solubilities fall between these limits are termed slightly soluble.

27. Salts that are sparingly soluble in water, and generally described as insoluble, are typically involved.

28. If the ion product is less than the value of the given K_{sp}, the solution is unsaturated. However, if the ion product exceeds the K_{sp}, the solution is saturated and precipitation occurs until the ion concentrations decrease to equilibrium values.

Problems

1. a. $K = \dfrac{[C]}{[A][B]} = \dfrac{4.0}{(2.0)(3.0)} = 0.67$

   b. $K = \dfrac{[F][G]^3}{[D][E]^2}$ c. $K = \dfrac{[NH_3]^2}{[N_2][H_2]^3}$

   $K = \dfrac{(1.8)(1.2)^3}{(1.5)(2.0)^2}$ $K = \dfrac{(0.62)^2}{(0.45)(0.14)^3}$

   $K = 0.52$ $K = 310$

2. $K = \dfrac{[H_2O]^2[Cl_2]^2}{[HCl]^4[O_2]} = \dfrac{(5.8 \times 10^{-2})^2(5.8 \times 10^{-2})^2}{(1.2 \times 10^{-3})^4(3.8 \times 10^{-4})}$

   $K = 1.4 \times 10^{10}$

3. $K = \dfrac{[NH_3]^2}{[N_2][H_2]^3}$

   $[N_2] = [NH_3]^2 / K[H_2]^3$

   $[N_2] = \dfrac{(1.23 \times 10^{-4})^2}{(6.59 \times 10^{-3})(2.75 \times 10^{-3})^3}$

   $[N_2] = 110$

4. a. $K = \dfrac{[Zn^{2+}]}{[Ag^+]^2}$ d. $K = [Cd^{2+}][S^{2-}]$

   b. $K = [Pb^{2+}][I^-]^2$ e. $K = \dfrac{1}{[O_2]^3}$

   c. $K = \dfrac{[HCN][OH^-]}{[CN^-]}$

5. Since $K_{forward} = 40.0$, $K_{reverse} = \dfrac{1}{40.0} = 0.0250$

6. $[H_3O^+] = $ antilog$(-3.14) = 7.24 \times 10^{-4}$
   $[H_3O^+] = [X^-] = 7.24 \times 10^{-4}$
   $[HX] = 0.100 - 0.000\,0724 = 0.100$
   $K_a = \dfrac{[H_3O^+][X^-]}{[HX]} = \dfrac{(7.24 \times 10^{-4})^2}{0.100} = 5.24 \times 10^{-6}$

7. $K_h = \dfrac{K_w}{K_a} = \dfrac{1.00 \times 10^{-14}}{2.69 \times 10^{-9}} = 3.72 \times 10^{-6}$

8. a. $\dfrac{36.0 \text{ g NaCl}}{100.\text{ g}} \times \dfrac{10^3 \text{ g}}{L} \times \dfrac{\text{mol NaCl}}{58.4 \text{ g NaCl}}$

   $= 6.16$ mol NaCl/L

   b. $\dfrac{4.25 \times 10^{-3} \text{ g PbSO}_4}{100.\text{ g}} \times \dfrac{10^3 \text{ g}}{L} \times \dfrac{\text{mol PbSO}_4}{303 \text{ g PbSO}_4}$

   $= 1.40 \times 10^{-4}$ mol $PbSO_4$/L

   c. $\dfrac{1.84 \times 10^{-11} \text{ g Cu}_2\text{S}}{100.\text{ g}} \times \dfrac{10^3 \text{ g}}{L} \times \dfrac{\text{mol Cu}_2\text{S}}{159 \text{ g Cu}_2\text{S}}$

   $= 1.16 \times 10^{-12}$ mol Cu_2S/L

9. $[GH] = [G^+] = [H^-] = 0.0427$
   $K_{sp} = [G^+][H^-] = (0.0427)^2 = 1.82 \times 10^{-3}$

10. $[E^{2+}] = [J^{2-}] = 8.45 \times 10^{-6}$
    $K_{sp} = [E^{2+}][J^{2-}] = (8.45 \times 10^{-6})^2 = 7.14 \times 10^{-11}$

11. a. $\dfrac{1.2 \times 10^{-3} \text{ g CaCO}_3}{100.\text{ g}} \times \dfrac{10^3 \text{ g}}{L} \times \dfrac{\text{mol CaCO}_3}{1.0 \times 10^2 \text{ g CaCO}_3}$

    $= 1.2 \times 10^{-4}$ mol $CaCO_3$/L
    $CaCO_3(s) \rightleftarrows Ca^{2+}(aq) + CO_3^{2-}(aq)$
    $K_{sp} = [Ca^{2+}][CO_3^{2-}] = (1.2 \times 10^{-4})(1.2 \times 10^{-4})$
    $= 1.4 \times 10^{-8}$

    b. $\dfrac{2.4 \times 10^{-4} \text{ g BaSO}_4}{100.\text{ g}} \times \dfrac{10^3 \text{ g}}{L} \times \dfrac{\text{mol BaSO}_4}{230 \text{ g BaSO}_4}$

    $= 1.0 \times 10^{-5}$ mol $BaSO_4$/L
    $BaSO_4(s) \rightleftarrows Ba^{2+}(aq) + SO_4^{2-}(aq)$
    $K_{sp} = [Ba^{2+}][SO_4^{2-}] = (1.0 \times 10^{-5})(1.0 \times 10^{-5})$
    $= 1.0 \times 10^{-10}$

    c. $\dfrac{0.173 \text{ g Ca(OH)}_2}{100.\text{ g}} \times \dfrac{10^3 \text{ g}}{L} \times \dfrac{\text{mol Ca(OH)}_2}{74.1 \text{ g Ca(OH)}_2}$

    $= 2.33 \times 10^{-2}$ mol $Ca(OH)_2$/L
    $Ca(OH)_2(s) \rightleftarrows Ca^{2+}(aq) + 2OH^-(aq)$
    $K_{sp} = [Ca^{2+}][OH^-]^2$
    $K_{sp} = (2.33 \times 10^{-2})(2(2.33 \times 10^{-2}))^2$
    $= 5.06 \times 10^{-5}$

12. $K_{sp} = [M^{2+}][N^{2-}] = 8.1 \times 10^{-6}$
    Solubility $= [M^{2+}] = [N^{2-}] = \sqrt{K_{sp}} = \sqrt{8.1 \times 10^{-6}} = 2.8 \times 10^{-3}$ mol/L

13. **a.** $AgBr(s) \rightleftarrows Ag^+(aq) + Br^-(aq)$, $K_{sp} = [Ag^+][Br^-]$
    Since $[Ag^+] = [Br^-]$, $K_{sp} = [Ag^+]^2$
    $[Ag^+] = \sqrt{K_{sp}} = \sqrt{5.0 \times 10^{-13}} = 7.1 \times 10^{-7}$
    Solubility AgBr = 7.1×10^{-7} mol/L

    b. $CoS(s) \rightleftarrows Co^{2+}(aq) + S^{2-}(aq)$
    $K_{sp} = [Co^{2+}][S^{2-}]$
    Since $[Co^{2+}] = [S^{2-}]$, $K_{sp} = [Co^{2+}]^2$
    $[Co^{2+}] = \sqrt{K_{sp}} = \sqrt{3.0 \times 10^{-26}} = 1.7 \times 10^{-13}$
    Solubility CoS = 1.7×10^{-13} mol/L

14. **a.** $PbSO_4(s) \rightleftarrows Pb^{2+}(aq) + SO_4^{2-}(aq)$
    b. $K_{sp} = [Pb^{2+}][SO_4^{2-}]$
    c. Since $0.500\ M\ Pb(NO_3)_2$ is $0.0500\ M\ Pb^{2+}$,

    $$\frac{0.0500\ \text{mol}\ Pb^{2+}}{L} \times 25.0\ \text{mL} \times \frac{L}{10^3\ \text{mL}} = 1.25 \times 10^{-3}\ \text{mol}\ Pb^{2+}.$$

    Since $0.0400\ M\ Na_2SO_4$ is $0.0400\ M\ SO_4^{2-}$, $\dfrac{0.0400\ \text{mol}\ SO_4^{2-}}{L} \times 25.0\ \text{mL} \times \dfrac{L}{10^3\ \text{mL}} = 1.00 \times 10^{-3}\ \text{mol}\ SO_4^{2-}$

    For Pb^{2+}, $\dfrac{1.25 \times 10^{-3}\ \text{mol}\ Pb^{2+}}{50.0\ \text{mL}} \times \dfrac{10^3\ \text{mL}}{L} = 2.50 \times 10^{-2}\ \text{mol}\ Pb^{2+}/L$

    For SO_4^{2-}, $\dfrac{1.00 \times 10^{-3}\ \text{mol}\ SO_4^{2-}}{50.0\ \text{mL}} \times \dfrac{10^3\ \text{mL}}{L} = 2.00 \times 10^{-2}\ \text{mol}\ SO_4^{2-}/L$

    d. $[Pb^{2+}][SO_4^{2-}] = (2.50 \times 10^{-2})(2.00 \times 10^{-2}) = 5.00 \times 10^{-4}$
    K_{sp} for $PbSO_4$ at 25°C = 1.8×10^{-8}
    Since the ion product of 5.00×10^{-4} exceeds the given K_{sp}, a precipitate of $PbSO_4$ will form.

15. Since the ion product, 1.9×10^{-5}, exceeds the K_{sp} for $CaCO_3$, the precipitate will form.

16. $X_2Y(s) \rightleftarrows 2X^+(aq) + Y^{2-}(aq)$
    $K_{sp} = [X^+]^2[Y^{2-}]$
    $[X^+] = 2 \times 5.54 \times 10^{-7} = 1.11 \times 10^{-6}$
    $[Y^{2-}] = 5.54 \times 10^{-7}$
    $K_{sp} = (1.11 \times 10^{-6})^2(5.54 \times 10^{-7})$
    $= 6.83 \times 10^{-19}$

17. $T_3U_2(s) \rightleftarrows 3T^{2+}(aq) + 2U^{3-}(aq)$
    $K_{sp} = [T^{2+}]^3[U^{3-}]^2$
    $[T^{2+}] = 3(3.77 \times 10^{-20}) = 1.13 \times 10^{-19}$
    $[U^{3-}] = 2(3.77 \times 10^{-20}) = 7.54 \times 10^{-20}$
    $K_{sp} = (1.13 \times 10^{-19})^3(7.54 \times 10^{-20})^2 = (1.44 \times 10^{-57})(5.69 \times 10^{-39}) = 8.19 \times 10^{-96}$

18. $AgI(s) \rightleftarrows Ag^+(aq) + I^-(aq)$
    $K_{sp} = [Ag^+][I^-]$
    $8.3 \times 10^{-17} = (2.7 \times 10^{-10})[I^-]$

    $[I^-] = \dfrac{8.3 \times 10^{-17}}{2.7 \times 10^{-10}} = 3.1 \times 10^{-7}$

19. $Ca(OH)_2(s) \rightleftarrows Ca^{2+}(aq) + 2OH^-(aq)$
    $K_{sp} = [Ca^{2+}][OH^-]^2$

    For Ca^{2+}, $0.35\ L \times \dfrac{4.4 \times 10^{-3}\ \text{mol}\ Ca^{2+}}{L} = 1.5 \times 10^{-3}\ \text{mol}\ Ca^{2+}$

    $\dfrac{1.5 \times 10^{-3}\ \text{mol}\ Ca^{2+}}{(0.35\ L + 0.17\ L)} = 2.9 \times 10^{-3}\ \text{mol}\ Ca^{2+}/L$

    For OH^-, $0.17\ L \times \dfrac{3.9 \times 10^{-4}\ \text{mol}\ OH^-}{L} = 6.6 \times 10^{-5}\ \text{mol}\ OH^-$

    $\dfrac{6.6 \times 10^{-5}\ \text{mol}\ OH^-}{(0.35\ L + 0.17\ L)} = 1.3 \times 10^{-4}\ \text{mol}\ OH^-/L$

    Trial product: $[Ca^{2+}][OH^-]^2 = (2.9 \times 10^{-3})(1.3 \times 10^{-4})^2 = 4.9 \times 10^{-11}$
    Since the trial product is less than the K_{sp} of 5.5×10^{-6}, no precipitate is formed.

20. $CaSO_4(s) \rightleftarrows Ca^{2+}(aq) + SO_4^{2-}(aq)$
    $K_{sp} = [Ca^{2+}][SO_4^{2-}]$

    For Ca^{2+}: $\dfrac{0.0040\ \text{mol}\ Ca^{2+}}{L} \times 70.0\ \text{mL} \times \dfrac{L}{10^3\ \text{mL}} = 2.8 \times 10^{-4}\ \text{mol}\ Ca^{2+}$

$$\frac{2.8 \times 10^{-4} \text{ mol Ca}^{2+}}{100.0 \text{ mL}} \times \frac{10^3 \text{ mL}}{\text{L}} = 2.8 \times 10^{-3} \text{ mol Ca}^{2+}/\text{L}$$

For SO_4^{2-}: $\dfrac{0.010 \text{ mol SO}_4^{2-}}{\text{L}} \times 30.0 \text{ mL} \times \dfrac{\text{L}}{10^3 \text{ mL}} = 3.0 \times 10^{-4} \text{ mol SO}_4^{2-}$

$$\frac{3.0 \times 10^{-4} \text{ mol SO}_4^{2-}}{100.0 \text{ mL}} \times \frac{10^3 \text{ mL}}{\text{L}} = 3.0 \times 10^{-3} \text{ mol SO}_4^{2-}/\text{L}$$

Trial product: $[Ca^{2+}][SO_4^{2-}] = (2.8 \times 10^{-3})(3.0 \times 10^{-3}) = 8.4 \times 10^{-6}$
Since the trial product is less than the K_{sp} of 9.1×10^{-6}, no precipitate is formed.

21. $AgCl(s) \rightleftharpoons Ag^+(aq) + Cl^-(aq)$
    $K_{sp} = [Ag^+][Cl^-]$

    For Ag^+: $1.70 \text{ g AgNO}_3 \times \dfrac{108 \text{ g Ag}^+}{170. \text{ g AgNO}_3} = 1.08 \text{ g Ag}^+$

    $1.08 \text{ g Ag}^+ \times \dfrac{\text{mol Ag}^+}{108 \text{ g Ag}^+} = 0.0100 \text{ mol Ag}^+$

    $\dfrac{0.0100 \text{ mol Ag}^+}{200. \text{ mL}} \times \dfrac{10^3 \text{ mL}}{\text{L}} = 0.0500 \text{ mol Ag}^+/\text{L}$

    For: Cl^-: $14.5 \text{ g NaCl} \times \dfrac{35.5 \text{ g Cl}^-}{58.4 \text{ g NaCl}} = 8.81 \text{ g Cl}^-$

    $8.81 \text{ g Cl}^- \times \dfrac{1 \text{ mol Cl}^-}{35.5 \text{ g Cl}} = 0.248 \text{ mol Cl}^-$

    $\dfrac{0.248 \text{ mol Cl}^-}{200. \text{ mL}} \times \dfrac{10^3 \text{ mL}}{\text{L}} = 1.24 \text{ mol Cl}^-/\text{L}$

    Trial product: $[Ag^+][Cl^-] = (0.0500)(1.24) = 6.20 \times 10^{-2}$
    Since the trial product exceeds the K_{sp} of 1.8×10^{-10}, a precipitate does form.

22. $Fe(OH)_3(s) \rightleftharpoons Fe^{3+}(aq) + 3OH^-(aq)$
    $K_{sp} = [Fe^{3+}][OH^-]^3$

    For Fe^{3+}: $2.50 \text{ g Fe(NO}_3)_3 \times \dfrac{55.8 \text{ g Fe}^{3+}}{242 \text{ g Fe(NO}_3)_3} = 0.576 \text{ g Fe}^{3+}$

    $0.576 \text{ g Fe}^{3+} \times \dfrac{\text{mol Fe}^{3+}}{55.8 \text{ g Fe}^{3+}} = 1.03 \times 10^{-2} \text{ mol Fe}^{3+}$

    $\dfrac{1.03 \times 10^{-2} \text{ mol Fe}^{3+}}{100. \text{ mL}} \times \dfrac{10^3 \text{ mL}}{\text{L}} = 1.03 \times 10^{-1} \text{ mol Fe}^{3+}/\text{L}$

    For OH^-: $1.00 \times 10^{-20} \, M$ NaOH is $1.00 \times 10^{-20} \, M \, OH^-$
    Trial product: $[Fe^{3+}][OH^-]^3 = (1.03 \times 10^{-1})(1.00 \times 10^{-20})^3 = 1.03 \times 10^{-61}$
    Since the trial product is less than the K_{sp} of 1.5×10^{-36}, no precipitate is formed.

Critical Thinking

1. The concentrations of N_2 and O_2 will increase and the concentration of NO will decrease.
2. The solubilities are:
   $Pb(OH)_2$: 1.1×10^{-5} mol/L
   $Zn(OH)_2$: 3.6×10^{-6} mol/L
   MnS: 4.8×10^{-7} mol/L
   $PbCrO_4$: 1.4×10^{-8} mol/L

Teacher's Notes

Chapter 19
Chemical Equilibrium

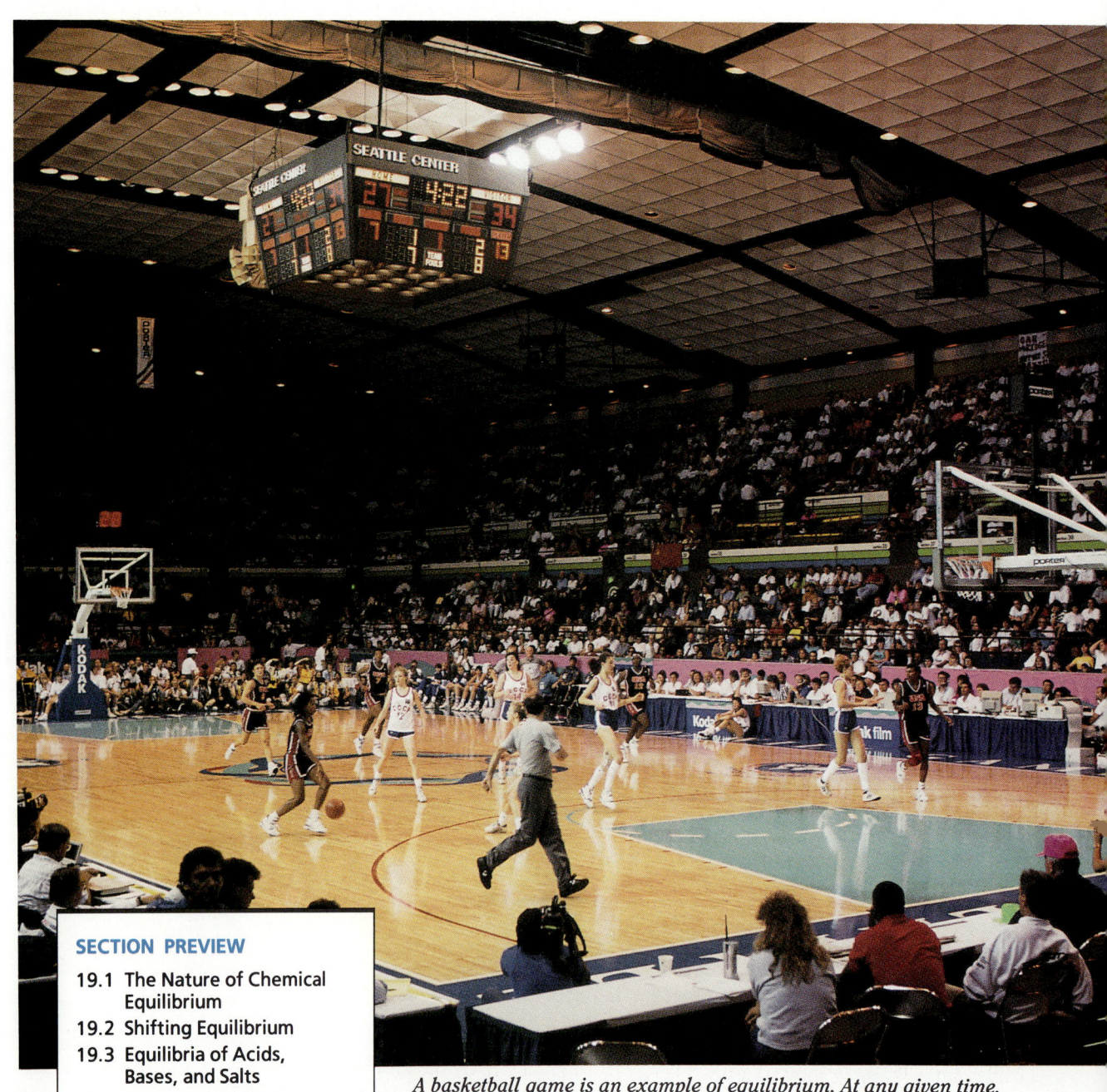

SECTION PREVIEW
19.1 The Nature of Chemical Equilibrium
19.2 Shifting Equilibrium
19.3 Equilibria of Acids, Bases, and Salts
19.4 Solubility Equilibrium

A basketball game is an example of equilibrium. At any given time, five players from each team are on the court, even though substitutions occur.

19.1 The Nature of Chemical Equilibrium

In earlier chapters, you learned about the equilibrium that can exist between certain physical processes. In the first section of this chapter, you will investigate the nature of the equilibrium that can exist between chemical processes. Then you will examine the mathematical relationships that govern equilibrium.

Reversible Reactions

The products formed in a chemical reaction may react to re-form the original reactants. *A chemical reaction in which the products can react to re-form the reactants is called a* **reversible reaction**. An example of such a reaction is shown in Figure 19-1. Essentially, all chemical reactions are considered to be reversible under suitable conditions. Theoretically, every reaction pathway can be traversed in two directions, forward and reverse. Some reverse reactions occur or are driven less easily than others, however. For example, a temperature near 3000°C is required for water vapor to decompose into measurable quantities of elemental hydrogen and oxygen.

SECTION OBJECTIVES

- Define *chemical equilibrium*.
- Explain the nature of the equilibrium constant.
- Write chemical equilibrium expressions and carry out calculations involving them.

Figure 19-1 A reversible reaction. Water vapor (steam) passed over the hot iron is reduced to hydrogen, and the iron is oxidized to iron(II) and iron(III) oxide. Hydrogen passed over the hot iron oxide is oxidized to water vapor, and the iron oxide is reduced to iron.

$4H_2O(g) + 3Fe(s) \rightarrow Fe_3O_4(s) + 4H_2(g)$ $4H_2(g) + Fe_3O_4(s) \rightarrow 3Fe(s) + 4H_2O(g)$

Mercury(II) oxide decomposes when heated strongly.

$$2HgO(s) \xrightarrow{\Delta} 2Hg(\ell) + O_2(g)$$

Mercury and oxygen combine to form mercury(II) oxide when heated gently.

$$2Hg(\ell) + O_2(g) \xrightarrow{\Delta} 2HgO(s)$$

Suppose mercury(II) oxide is heated in a closed container from which neither the mercury nor the oxygen can escape. Once decomposition has begun, the mercury and oxygen released can recombine to form mercury(II) oxide again. Thus, both reactions can proceed at the same time. Under these conditions, the rate of the composition reaction will eventually equal that of the decomposition reaction. Mercury and oxygen will combine to form

Chemical Equilibrium

mercury(II) oxide at the same rate as mercury(II) oxide decomposes into mercury and oxygen. The amounts of mercury(II) oxide, mercury, and oxygen can be expected to remain constant as long as these conditions persist. At this point, a state of equilibrium has been reached between the two chemical reactions. Both reactions continue, but there is no net change in the composition of the system. The equation for the reaction at equilibrium is written as follows to show the overall reversible reaction:

$$2HgO(s) \rightleftharpoons 2Hg(\ell) + O_2(g)$$

Chemical equilibrium *is a state of balance in which the rates of opposing chemical reactions are exactly equal.*

Equilibrium, A Dynamic State

> **STUDY HINT**
> Physical processes, such as evaporation and condensation, that can give rise to a state of equilibrium are discussed in Chapter 13.

In systems that are in equilibrium, opposing processes occur at the same time and at the same rate. For example, as you read in Chapter 13, the evaporation of a liquid in a closed vessel and the condensation of its saturated vapor can proceed at equal rates. The resulting equilibrium vapor pressure is a characteristic of the liquid at the prevailing temperature.

If an excess of sugar is placed in water, sugar molecules go into solution. Some of these molecules in turn separate from solution and rejoin the crystals. At saturation, molecules of sugar are crystallizing at the same rate that crystal molecules are dissolving.

The preceding are examples of physical equilibria. The opposing physical processes occur at exactly the same rate. Such an equilibrium, whether chemical or physical, is a dynamic state in which two opposing processes proceed simultaneously at the same rate.

Compounds such as sodium chloride are composed of ions. Their aqueous solutions consist of hydrated solute ions dispersed throughout the water solvent. When an excess of sodium chloride is placed in water, a saturated solution eventually results. A solution equilibrium is established in which the rate of association of ions reforming the crystals equals the rate of dissociation of ions from the crystals. The equilibrium for this reversible dissociation process is shown in the ionic equation

$$Na^+Cl^-(s) \rightleftharpoons Na^+(aq) + Cl^-(aq)$$

> In a chemical equation, formulas written to the left of the arrow are assumed to be the reactants, and the reaction proceeds from left to right.

Many chemical reactions are reversible under ordinary conditions of temperature and concentration. They may reach a state of equilibrium unless at least one of the substances involved escapes or is removed. In some cases, however, the forward reaction is nearly completed before the reverse reaction rate becomes high enough to equal it and establish equilibrium. Here, the products of the forward reaction (→) are favored, meaning that at equilibrium, there is a higher concentration or amount of products than of reactants. The favored reaction that produces this situation is referred to as a "reaction to the right" because the convention for writing chemical reactions is that *left-to-right* is forward and *right-to-left* is reverse. An example of such a system is

$$HBr + H_2O \rightleftharpoons H_3O^+ + Br^-$$

Note the inequality of the two arrow lengths.

In still other cases, the forward reaction is barely under way when the rate of the reverse reaction becomes equal to that of the forward reaction, and equilibrium is established. In these cases, the products of the reverse reaction (←) are favored and the original reactants are formed. That is, at equilibrium, there is a higher concentration of reactants than of products. The favored reaction that produces this situation is referred to as the "reaction to the left." An example of such a system is

$$H_2CO_3(aq) + H_2O \leftrightharpoons H_3O^+ + HCO_3^-$$

In still other cases, both forward and reverse reactions occur to nearly the same extent before chemical equilibrium is established. Neither reaction is favored and considerable concentrations of both reactants and products are present at equilibrium. An example is

$$H_3PO_4(aq) + H_2O \leftrightharpoons H_3O^+ + H_2PO_4^-$$

Chemical reactions are ordinarily used to convert available reactants into more-desirable products. Chemists try to produce as much of these products as possible from the reactants used. Chemical equilibrium may seriously limit the possibilities of a seemingly useful reaction. In dealing with equilibrium systems, it is important to remember the conditions that influence reaction rates: *(1)* the nature of reactants; *(2)* the surface area; *(3)* the concentration of reactants; *(4)* the temperature; and *(5)* the presence of a catalyst. Any factor that disturbs the rate of the forward or the reverse reaction upsets the equilibrium. In Section 19-2, you will see how a reaction at equilibrium responds to such stresses. But first, it is important to understand the quantitative nature of chemical equilibrium.

The only one we cannot change in the lab is *(1)* the nature of the reactants.

STUDY HINT
The factors that influence reaction rate are discussed in Section 18.4.

The Equilibrium Constant

Many chemical reactions are reversible and reach a state of equilibrium before the reactants are completely changed into products. Once at equilibrium, the concentrations of products and reactants remain constant.

Suppose two substances, A and B, react to form products C and D. In turn C and D react to produce A and B. Under appropriate conditions, equilibrium occurs for this reversible reaction. The equation for this hypothetical equilibrium reaction is

$$A + B \rightleftharpoons C + D$$

Initially, the concentrations of C and D are zero, and those of A and B are maximum. With time, as shown in Figure 19-2, the rate of the forward reaction decreases as A and B are used up. Meanwhile, the rate of the reverse reaction increases as C and D are formed. As these two reaction rates become equal, equilibrium is established. The individual concentrations of A, B, C, and D undergo no further change if conditions remain the same.

At equilibrium, the ratio of the mathematical product [C] × [D] to the mathematical product [A] × [B] for this reaction has a definite value at a given temperature. It is known as the equilibrium constant of the reaction and is designated by the letter K. Thus,

$$K = \frac{[C] \times [D]}{[A] \times [B]}$$

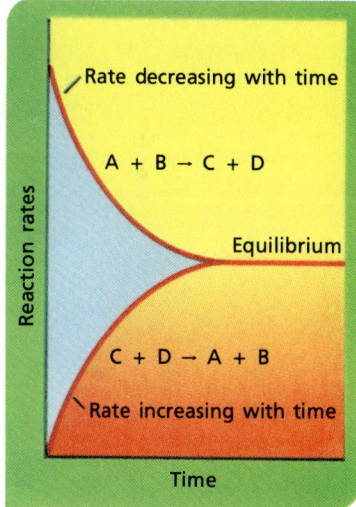

Figure 19-2 Reaction rates for the hypothetical equilibrium reaction system A + B ⇌ C + D. The rate of the forward reaction is represented by curve A + B. Curve C + D represents the rate of the reverse reaction. At equilibrium, the two rates are equal.

Chemical Equilibrium **565**

Observe that the concentrations of substances on the right side of the chemical equation are given in the numerator. These substances are the products of the forward reaction. The concentration of substances on the left side of the chemical equation are found in the denominator. These substances are the reactants of the forward reaction. Concentrations of reactants and products are given in moles per liter. The constant K is independent of the initial concentrations. It is, however, dependent on the temperature of the system.

The value of K for a given equilibrium reaction at a given temperature is important to chemistry. It shows the extent to which the reactants are converted into the products of the reaction. If K is equal to 1, the products of the concentrations in the numerator and denominator have the same value, and therefore, at equilibrium, these are roughly equal amounts of reactants and products. If the value of K is very small, the forward reaction occurs only very slightly before equilibrium is established, and the reactants are favored. A large value of K indicates an equilibrium in which the original reactants are largely converted to products. The numerical value of K for a particular equilibrium system is obtained experimentally. The chemist must analyze the equilibrium mixture and determine the concentrations of all substances.

A value of $K = 10$ is considered very large.

Suppose the balanced equation for an equilibrium reaction has the general form

$$nA + mB + \ldots \rightleftarrows xC + yD + \ldots$$

The equilibrium constant relationship at equilibrium (for this generalized reaction system) becomes

$$K = \frac{[C]^x[D]^y\ldots}{[A]^n[B]^m\ldots}$$

In general, then, *the **equilibrium constant**, K, is the ratio of the mathematical product of the concentrations of substances formed at equilibrium to the mathematical product of the concentrations of reacting substances, each concentration being raised to the power that is the coefficient of that substance in the chemical equation. The equation for K is sometimes referred to as the **chemical equilibrium expression or law.***

To illustrate, suppose a reaction system at equilibrium is shown by the equation

$$3A + B \rightleftarrows 2C + 3D$$

The equilibrium constant K is given by the chemical equilibrium expression

$$K = \frac{[C]^2[D]^3}{[A]^3[B]}$$

Note that the exponents in the expression correspond to the coefficients in the chemical equation.

Now consider an actual example. The reaction between H_2 and I_2 vapor in a closed flask at an elevated temperature is easy to follow by simply observing the rate at which the violet color of the iodine vapor diminishes (Figure 19-3). Suppose this reaction runs to completion with respect to iodine. If so, the color must disappear entirely, since the product, hydrogen iodide, is a colorless gas.

Figure 19-3 The production of hydrogen iodide gas from gaseous hydrogen and iodine. The violet color of iodine gas *(left)* becomes fainter *(center)* as the reaction consumes the iodine. The violet will not disappear but will reach a constant intensity when the reaction reaches equilibrium *(right)*.

The color does not disappear entirely, however, because the reaction is reversible. Hydrogen iodide decomposes to re-form hydrogen and iodine. The rate of this reverse reaction increases as the concentration of hydrogen iodide builds up. The concentrations of hydrogen and iodine decrease, meanwhile, as they are used up. The rate of the forward reaction decreases accordingly.

As the rates of the opposing reactions become equal, an equilibrium is established. A constant intensity of the color violet indicates that equilibrium exists between hydrogen, iodine, and hydrogen iodide. The net chemical equation for the reaction system at equilibrium is

$$H_2(g) + I_2(g) \rightleftarrows 2HI(g)$$

Thus, the chemical equilibrium expression is

$$K = \frac{[HI]^2}{[H_2][I_2]}$$

Chemists have carefully measured the concentrations of H_2, I_2, and HI in equilibrium mixtures at various temperatures. In some experiments, the flasks were filled with hydrogen iodide at known pressure. The flasks were held at fixed temperatures until equilibrium was established. In other experiments, hydrogen and iodine were the original substances.

Experimental data, together with the calculated values for K, are listed in Table 19-1. Experiments 1 and 2 began with hydrogen iodide. Experiments 3 and 4 began with hydrogen and iodine. Note the close agreement obtained for the numerical values of the equilibrium constant in each case.

TABLE 19-1 TYPICAL EQUILIBRIUM CONCENTRATION OF H_2, I_2, AND HI IN MOL/L AT 425°C

Experiment	$[H_2]$	$[I_2]$	$[HI]$	$K = \dfrac{[HI]^2}{[H_2][I_2]}$
(1)	0.4953×10^{-3}	0.4953×10^{-3}	3.655×10^{-3}	54.46
(2)	1.141×10^{-3}	1.141×10^{-3}	8.410×10^{-3}	54.33
(3)	3.560×10^{-3}	1.250×10^{-3}	15.59×10^{-3}	54.62
(4)	2.252×10^{-3}	2.336×10^{-3}	16.85×10^{-3}	53.97

The equilibrium constant K for this equilibrium reaction system at 425°C has the average value of 54.34. This value for K should hold for any system of H_2, I_2, and HI at equilibrium at this temperature. If the calculation for K yields a different result, there must be a reason. Either the H_2, I_2, and HI system has not reached equilibrium or the system is not at 425°C.

The balanced chemical equation for an equilibrium system yields the expression for the equilibrium constant. The data in Table 19-1 show that the

Chemical Equilibrium

validity of this expression is confirmed when the actual values of the equilibrium concentrations of reactants and product are determined experimentally. The values of K are calculated from these concentrations. No information concerning the kinetics of the reacting systems is required.

Once the value of the equilibrium constant is known, the equilibrium constant expression can be used to calculate concentrations of reactants or products at equilibrium. Suppose a system at equilibrium at 425°C is found to contain 1.5 mol/L each of H_2 and I_2. To find the concentration of HI in this system the chemical equilibrium expression can be rearranged

Note that the numerical value of K depends on the coefficients used to balance the chemical equation. Doubling the coefficients squares the numerical value of K.

$$K = \frac{[HI]^2}{[H_2][I_2]}$$

$$[HI] = \sqrt{K[H_2][I_2]}$$

and then solved, using the known K value.

$$[HI] = \sqrt{1.5 \times 1.5 \times 54.34}$$
$$= 11 \text{ mol/L}$$

Sample Problem 19.1

An equilibrium mixture of H_2, I_2, and HI gases at 425°C is determined to consist of 4.5647×10^{-3} mol/L of H_2, 7.378×10^{-4} mol/L of I_2 and 1.3544×10^{-2} mol/L of HI. What is the equilibrium constant for the system at this temperature?

Solution

Step 1. Analyze Given: $[H_2] = 4.5647 \times 10^{-3}$ mol/L
$[I_2] = 7.378 \times 10^{-4}$ mol/L
$[HI] = 1.3544 \times 10^{-2}$ mol/L

Unknown: K

Step 2. Plan $[H_2], [I_2], [HI] \rightarrow K$

The balanced chemical equation is $H_2(g) + I_2(g) \rightleftarrows 2HI(g)$

The chemical equilibrium expression is $K = \dfrac{[HI]^2}{[H_2][I_2]}$

Step 3. Compute $K = \dfrac{(1.3544 \times 10^{-2})^2}{(4.5647 \times 10^{-3})(7.378 \times 10^{-4})} = 54.47$

Step 4. Evaluate The answer has the correct number of significant digits and is close to an estimated value of 56, $\dfrac{(14)^2}{(5)(0.7)}$.

Practice Problems

1. An equilibrium mixture at 425°C is determined to consist of 1.83×10^{-3} mol/L of H_2, 3.13×10^{-3} mol/L of I_2, and 1.77×10^{-2} mol/L of HI. Calculate the equilibrium constant, K. *(Ans.)* 54.7
2. An equilibrium mixture at 425°C consists of 4.79×10^{-4} mol/L of H_2, 4.79×10^{-4} mol/L of I_2, and 3.53×10^{-3} mol/L of HI. Determine the equilibrium constant. *(Ans.)* 54.3

Section Review

1. What is meant by "chemical equilibrium"?
2. What is an equilibrium constant?
3. How does the value of an equilibrium constant relate to the relative quantities of reactants and products at equilibrium?
4. What is meant by a chemical equilibrium expression?
5. Write the chemical equilibrium expression for the reaction $4HCl(g) + O_2(s) \rightleftarrows 2Cl_2(g) + 2H_2O(g)$
6. At equilibrium at 2500K, [HCl] = 6.25 mol/L and [H$_2$] = [Cl$_2$] = 0.45 mol/L for the reaction $H_2(g) + Cl_2(g) \rightleftarrows 2HCl(g)$. Find the value of K.

1. State of balance in which the rates of opposing chemical reactions are exactly equal
2. Ratio of the mathematical product of the concentrations of substances formed at equilibrium to the mathematical product of the concentration of reacting substances each raised to the power of its coefficient in the balanced chemical equation
3. The larger the value of K, the larger the amount of products.
4. It is the equation for K.

5. $K = \dfrac{[Cl_2]^2[H_2O]^2}{[HCl]^4[O_2]}$
6. $K = \dfrac{[HCl]^2}{[H_2][Cl_2]} = \dfrac{(6.25)^2}{(0.45)(0.45)} = 190$

19.2 Shifting Equilibrium

In systems that have attained chemical equilibrium, opposing reactions are proceeding at equal rates. Any change that alters the rates of these reactions disturbs the original equilibrium. The system then seeks a new equilibrium state. By shifting an equilibrium in the desired direction, chemists can often increase production of important industrial chemicals. In this section, you will read about such shifts in equilibrium.

Le Châtelier's principle (Section 13.3) provides a means of predicting the influence of disturbing factors on equilibrium systems. As you may recall, Le Châtelier's principle states: *If a system at equilibrium is subjected to a stress, the equilibrium is shifted in the direction that relieves the stress.* This principle holds for all kinds of dynamic equilibria, physical as well as chemical. In applying Le Châtelier's principle to chemical equilibrium, three important stresses will be considered.

1. Change in concentration. From collision theory, an increase in the concentration of a reactant causes an increase in collision frequency. A reaction resulting from these collisions should then proceed at a faster rate. Consider the hypothetical reaction

$$A + B \rightleftarrows C + D$$

An increase in the concentration of A will shift the equilibrium to the right. Both A and B will be used up faster, and more of C and D will be formed. The equilibrium will be reestablished with a lower concentration of B. The equilibrium has shifted in such direction as to reduce the stress caused by the increase in concentration of A.

Similarly, an increase in the concentration of B drives the reaction to the right. An increase in either C or D shifts the equilibrium to the left. A decrease in the concentration of C or D has the same effect as an increase in the concentration of A or B, that is, it will shift the equilibrium to the right.

Changes in concentration have no effect on the value of the equilibrium constant. This is because such changes have an equal effect on the numerator and the denominator of the chemical equilibrium expression. Thus, all concentrations give the same numerical ratio for the equilibrium constant when equilibrium is reestablished.

2. Change in pressure. A change in pressure can affect only equilibrium systems in which gases are involved. According to Le Châtelier's principle, if

SECTION OBJECTIVES

- Discuss the factors that disturb equilibrium.
- Discuss conditions under which reactions go to completion.
- Describe the common-ion effect.

By shifting an equilibrium one means that the balance is shifted in favor of either the forward or the reverse reaction.

Chemical Equilibrium

▶ Chemistry and its Applications

Swimming Pool Chemistry

Killing bacteria in swimming pools is the job of the highly effective disinfectant, hypochlorous acid (HOCl). Tiny HOCl molecules are deadly to bacteria because they can easily penetrate cell walls and oxidize proteins. Too much HOCl, however, encourages algae growth and irritates eyes.

HOCl can be produced by bubbling corrosive chlorine gas through water, but this requires special equipment and is impractical for home pools. Solid sodium hypochlorite (NaClO, the active ingredient in household bleach) is a safer way to generate HOCl. The reaction is

$$NaClO(s) + H_2O(\ell) \rightleftharpoons HOCl(aq) + NaOH(aq)$$

Commercial pool products often contain calcium hypochlorite, Ca(ClO)$_2$, which reacts similarly:

$$Ca(ClO)_2 + 2H_2O(\ell) \rightleftharpoons 2HOCl(aq) + Ca(OH)_2$$

These equations show that adding sodium or calcium hypochlorite to a pool initially produces a basic solution, due to the sodium or calcium hydroxide. The HOCl also affects the pH, since it can dissociate to make the solution acidic:

$$HOCl(aq) \rightleftharpoons H^+(aq) + ClO^-(aq)$$

As the pool is used, sunlight causes the ClO$^-$ ion to be lost in other reactions, and the HOCl is consumed in the destruction of bacteria. The pH changes are therefore unpredictable. The pH in a pool should be checked daily and adjusted by adding HCl to lower it or NaHSO$_4$ to raise it. The ideal pH range is 7.2–7.8.

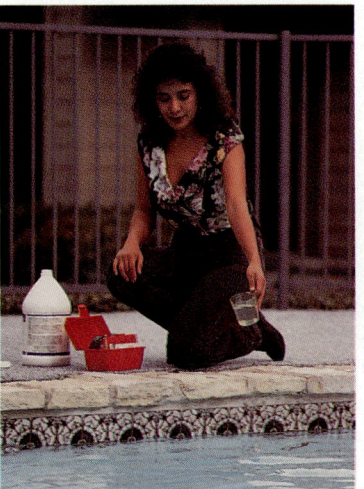

If moles of gaseous reactants and products are equal, pressure has no effect on equilibrium concentrations.

the pressure on an equilibrium system is increased, the reaction is driven in the direction that relieves the pressure.

The Haber process for catalytic synthesis of ammonia from its elements illustrates the influence of pressure on an equilibrium system.

$$N_2(g) + 3H_2(g) \rightleftharpoons 2NH_3(g)$$

The equation indicates that 4 molecules of the reactant gases form 2 molecules of ammonia gas. Suppose the equilibrium mixture is subjected to an increase in pressure. This pressure can be relieved by the reaction that produces fewer gas molecules and, therefore, a smaller volume. Thus the stress is *lessened* by the formation of ammonia. Equilibrium is shifted toward the right, favoring the production of ammonia. High pressure is desirable in this industrial process. Figure 19-4 illustrates the effect of pressure on this equilibrium system.

An increase in the partial pressure of H$_2$ or N$_2$ in the reaction vessel would shift the equilibrium to the right as this increase in pressure effects an increase in concentration of the resultant gas. Similarly, a decrease in the partial pressure of NH$_3$ gas in the reaction vessel would also shift the

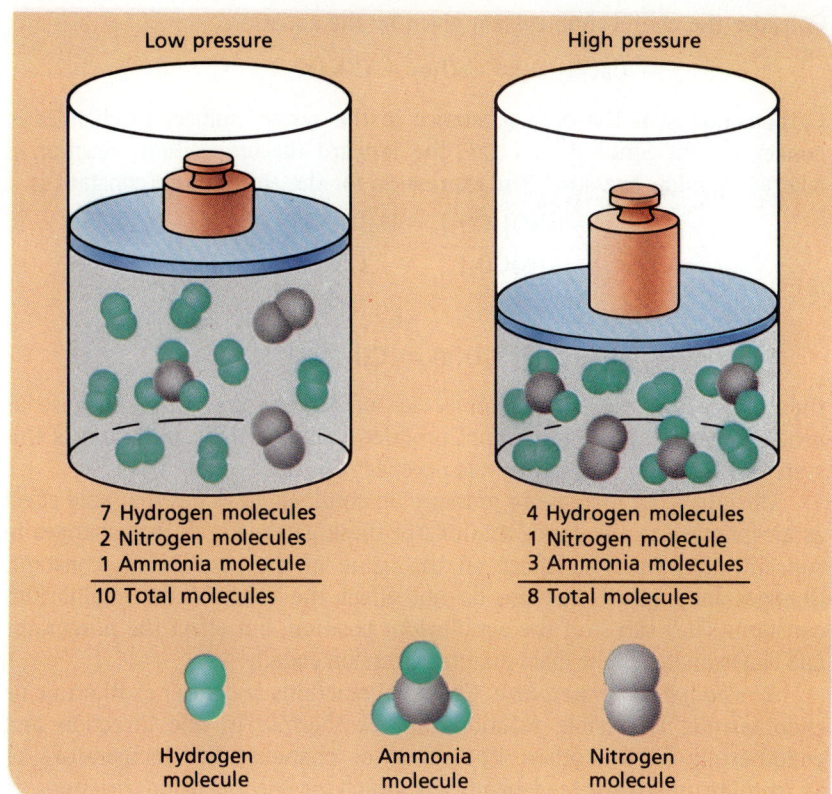

Figure 19-4 Increased pressure results in a higher yield of ammonia because the equilibrium shifts in the direction that produces fewer molecules.

equilibrium to the right. A change in the partial pressure (and thus the concentration) of at least one of the three reactant gases is necessary for the equilibrium position to be affected.

The introduction of an inert gas, such as helium into the reaction vessel for the synthesis of ammonia, would increase the total pressure in the vessel. However, both the partial pressures and the concentrations of the reactant gases present would not be altered by this kind of pressure change. Therefore, increasing pressure by adding a gas that is not a reactant cannot affect the equilibrium position of the reaction system.

In the Haber process, the ammonia produced is continuously removed by condensation to liquid. This condensation removes most of the product from the gas phase in which the reaction occurs. The change in concentration also tends to shift the equilibrium to the right. The Haber process is illustrated in Figure 19-5.

Many chemical processes involve heterogeneous reactions, reactions in which the reactants and products are in different phases. The *concentrations in equilibrium systems of pure substances in solid and liquid phases are not changed by adding or removing quantities of the substance.* This is because concentration is density-dependent, and the density of these phases is constant, regardless of the total amounts present. *A pure substance in a condensed phase, solid or liquid, can thus be removed from the expression for the equilibrium constant.* This is done by substituting the number *1* for its concentration, which is assumed to remain unchanged in the equilibrium system.

An inert gas does NOT alter the equilibrium.

Figure 19-5 The Haber process *(top)* for the production of ammonia and a bottle of ammonia *(bottom)*, a household cleaning agent. Application of Le Châtelier's principle to the equilibrium system, $N_2 + 3H_2 \rightarrow 2NH_3 + 92kJ$, suggests that high pressure and low temperature favors the yield of ammonia. Why, then, is a moderately high temperature used in this process?

Consider the equilibrium system given by the equation

$$CaCO_3(s) \rightleftarrows CaO(s) + CO_2(g)$$

Carbon dioxide is the only substance in the system subject to changes in concentration. Since it is a gas, the forward (decomposition) reaction is favored by a low pressure. The expression for the equilibrium constant is

$$K = \frac{[CaO][CO_2]}{[CaCO_3]} = \frac{[1][CO_2]}{[1]} = [CO_2]$$

In the reaction

$$CO(g) + H_2O(g) \rightleftarrows CO_2(g) + H_2(g)$$

there are equal numbers of molecules of gaseous reactants and gaseous products. Pressure change cannot produce a shift in the equilibrium of this system. Pressure change thus has no effect on the reaction.

Obviously, an increase in pressure on confined gases has the same effect as an increase in the concentrations of these gases. Recall that changes in concentration have no effect on the value of the equilibrium constant. Changes in pressure likewise do not affect the value of the equilibrium constant. They can shift the equilibrium position, but affect the numerator and denominator of the mass-action expression equally.

3. Change in temperature. Chemical reactions are either exothermic or endothermic. Reversible reactions are exothermic in one direction and endothermic in the other. The effect of changing the temperature of an equilibrium mixture depends on which of the opposing reactions is endothermic.

According to Le Châtelier's principle, addition of heat shifts the equilibrium so that heat is absorbed. This favors the endothermic reaction. The removal of heat favors the exothermic reaction. A rise in temperature increases the rate of any reaction. In an equilibrium system, however, the rates of the opposing reactions are raised unequally. Thus, *the value of the equilibrium constant for a given system is affected by the temperature.*

The synthesis of ammonia is exothermic.

$$N_2(g) + 3H_2(g) \rightleftarrows 2NH_3(g) + 92 \text{ kJ}$$

A high temperature is not desirable because it favors the decomposition of ammonia, the endothermic reaction. At low temperatures, however, the forward reaction is too slow to be commercially useful. The temperature used represents a compromise between kinetic and equilibrium requirements. It is high enough that equilibrium is established rapidly but low enough that the equilibrium concentration of ammonia is significant. Moderate temperature (about 500°C) and very high pressure (700–1000 atmospheres) produce a satisfactory yield of ammonia.

The reactions of the system are also accelerated by a suitable catalyst. Note, however, that catalysts have no effect on relative equilibrium amounts, but only on the rates at which equilibrium is reached. This is because catalysts increase the rates of forward and reverse reactions in a system to an equal extent. They therefore do not affect K.

Data on several equilibrium systems are listed in Table 19-2. The table also gives numerical values of the equilibrium constants at various temperatures. A very small K value means that the equilibrium mixture consists mainly of the substances on the left of the equation. If the value of K is large, the equilibrium mixture consists mainly of the substances on the right of the equation.

An increase in temperature always favors the endothermic reaction.

TABLE 19-2 EQUILIBRIUM CONSTANTS

Equilibrium System	Temperature (°C)	Value of K
$N_2(g) + 3H_2(g) \rightleftharpoons 2NH_3(g)$	350	2.66×10^{-2}
$N_2(g) + 3H_2(g) \rightleftharpoons 2NH_3(g)$	450	6.59×10^{-3}
$N_2(g) + 3H_2(g) \rightleftharpoons 2NH_3(g)$	727	2.37×10^{-3}
$H_2(g) + I_2(g) \rightleftharpoons 2HI(g)$	350	66.9
$H_2(g) + I_2(g) \rightleftharpoons 2HI(g)$	425	54.4
$H_2(g) + I_2(g) \rightleftharpoons 2HI(g)$	1123	45.9
$2H_2(g) + S_2(g) \rightleftharpoons 2H_2S(g)$	477	9.39×10^{-5}
$H_2(g) + CO_2(g) \rightleftharpoons H_2O(g) + CO(g)$	1727	4.40
$2H_2O(g) \rightleftharpoons 2H_2(g) + O_2(g)$	1727	5.31×10^{-10}
$2CO(g) + O_2 \rightleftharpoons 2CO_2(g)$	727	2.24×10^{22}
$C(s) + CO_2(g) \rightleftharpoons 2CO(g)$	1123	14.1
$Cu(s) + 2Ag^+(aq) \rightleftharpoons Cu^{2+}(aq) + 2Ag(s)$	25	2×10^{15}
$I_2(s) \rightleftharpoons 2I(g)$	727	3.76×10^{-5}
$2O_3(g) \rightleftharpoons 3O_2(g)$	1727	2.54×10^{12}
$N_2(g) \rightleftharpoons 2N(g)$	1000	1.31×10^{-31}

Reactions That Run to Completion

Many reactions are easily reversible under suitable conditions. A state of equilibrium may be established unless one or more of the products escapes or is removed. An equilibrium reaction may be driven in the preferred direction by application of Le Châtelier's principle.

Some reactants appear to be converted irreversibly into products if reactions proceed, on their own, in the forward direction. For example, potassium chlorate decomposition is essentially irreversible: no one has found a method of recombining potassium chloride and oxygen directly once the potassium chlorate decomposes. Sugar is decomposed into carbon and water by the application of heat. No single-step method of recombining these products is yet known.

Many compounds are formed by the chemical interaction of ions in solutions. Consider the mixing of solutions of soluble ionic compounds. Two pairings of ions are possible. In some cases, these pairings do not occur. If dilute solutions of sodium chloride and potassium bromide are mixed, for example, no reaction occurs. The resulting solution merely contains a mixture of Na^+, K^+, Cl^-, and Br^- ions. Association of ions occurs, in this case, only if enough water is evaporated to cause crystals to separate from solution. The result is then a mixture of NaCl, KCl, NaBr, and KBr.

With some combinations of ions, however, reactions do occur. Such reactions appear to run to completion in the sense that the ions are almost completely removed from solution. The extent to which reacting ions are

Chemical Equilibrium

removed from solution depends on: *(1)* the solubility of the compound formed and *(2)* the degree of ionization, if the compound is soluble. Thus, a product that escapes as a gas or is precipitated as a solid or is only slightly ionized effectively removes from solution the bulk of the reacting ions that make it up. Consider some specific examples of situations in which such ionic reactions run to completion.

1. Formation of a gas. Unstable substances formed as products of ionic reactions decompose spontaneously. An example is carbonic acid, the acid in carbonated water such as club soda, which yields a gas as a decomposition product.

$$H_2CO_3(aq) \rightarrow H_2O(\ell) + CO_2(g)$$

Carbonic acid is produced in the reaction between sodium hydrogen carbonate and hydrochloric acid, as shown by the equation

$$NaHCO_3(aq) + HCl(aq) \rightarrow NaCl(aq) + H_2CO_3(aq)$$

The aqueous solution of HCl contains H_3O^+ and Cl^- ions. A careful examination of the reaction suggests that the HCO_3^- ion, a Brønsted base, acquires a proton from the H_3O^+ ion, a Brønsted acid; Na^+ and Cl^- ions are merely spectator ions. The following equations may be more appropriate for this chemical reaction.

$$H_3O^+ + HCO_3^- \rightarrow H_2O(\ell) + H_2CO_3(aq)$$
$$H_2CO_3(aq) \rightarrow H_2O(\ell) + CO_2(g)$$

The net ionic equation is, thus, simply

$$H_3O^+ + HCO_3^- \rightarrow 2H_2O(\ell) + CO_2(g)$$

This reaction runs practically to completion because one of the products (CO_2) escapes as a gas.

The reaction between iron(II) sulfide and hydrochloric acid also goes virtually to completion because a gaseous product escapes. The reaction equation is

$$FeS(s) + 2HCl(aq) \rightarrow FeCl_2(aq) + H_2S(g)$$

The net ionic equation is

$$FeS(s) + 2H_3O^+(aq) \rightarrow Fe^{2+}(aq) + H_2S(g) + 2H_2O(\ell)$$

The hydrogen sulfide formed is only moderately soluble and is largely given off as a gas. The iron(II) chloride shown in the first equation could be obtained on evaporation of the water.

2. Formation of a precipitate. When solutions of sodium chloride and silver nitrate are mixed, a white precipitate of silver chloride immediately forms.

$$Na^+ + Cl^- + Ag^+ + NO_3^- \rightarrow Na^+ + NO_3^- + Ag^+Cl^-(s)$$

If chemically equivalent amounts of the two solutes are used, only Na^+ ions and NO_3^- ions remain in solution in appreciable quantities. All but a meager portion of the Ag^+ ions and Cl^- ions combine and separate from the solution as a precipitate of AgCl. This is because silver chloride is only very sparingly soluble in water. It separates by precipitation from what turns out to be a

Figure 19-6 When a sodium chloride solution is poured into a solution of silver nitrate, silver chloride, a white precipitate, is formed.

saturated solution of its particles. The reaction thus effectively runs to completion because an "insoluble" product is formed.

The only reaction is between the silver ions and chloride ions. Omitting the spectator ions, Na^+ and NO_3^-, the net ionic equation is simply

$$Ag^+(aq) + Cl^-(aq) \rightarrow Ag^+Cl^-(s)$$

Crystalline sodium nitrate could be recovered by evaporation of the water solvent.

3. Formation of a slightly ionized product. Neutralization reactions between H_3O^+ ions from aqueous acids and OH^- ions from aqueous bases result in the formation of water molecules, which are only slightly ionized. A reaction between HCl and NaOH illustrates this process. Aqueous HCl supplies H_3O^+ ions and Cl^- ions to the solution, and aqueous NaOH supplies Na^+ ions and OH^- ions.

$$H_3O^+ + Cl^- + Na^+ + OH^- \rightarrow Na^+ + Cl^- + 2H_2O$$

Neglecting the spectator ions, the net ionic equation is simply

$$H_3O^+(aq) + OH^-(aq) \rightarrow 2H_2O(\ell)$$

Because it is only slightly ionized, the water exists almost entirely as covalently bonded molecules. Thus, insofar as they are initially present in equal amounts, hydronium ions and hydroxide ions are almost entirely removed from the solution. Thus, the reaction effectively runs to completion because the product is only slightly ionized.

Common-Ion Effect

Suppose hydrogen chloride gas is bubbled into a saturated solution of sodium chloride. As the hydrogen chloride dissolves in sufficient quantity, sodium chloride separates as a precipitate. Since chloride ions but not sodium ions are common to both solutes, the concentration of chloride ions is initially increased, and that of sodium ions is not. In accordance with Le Châtelier's principle, sodium chloride crystals form, relieving the stress of added chloride. The concentration of sodium ions in the solution is lowered along with the total concentration of chloride ions; the overall amount of sodium chloride in solution is likewise lowered. Thus, increasing the concentration of chloride ions has the effect of decreasing the concentration of sodium ions. *This phenomenon, in which the addition of an ion common to two solutes brings about precipitation or reduced ionization is called the* **common-ion effect.** The initial equilibrium in this example is

$$Na^+Cl^-(s) \rightleftarrows Na^+(aq) + Cl^-(aq)$$

Additions of hydrogen chloride disturb this equilibrium and drive the reaction to the left. When the reaction is forced to the left, sodium chloride precipitates. This reduces the concentration of both the sodium and chloride ions in solution.

The common-ion effect is also observed when one ion species of a weak electrolyte is added in excess to a solution. Acetic acid ($HC_2H_3O_2$) is such an electrolyte. A 0.1-M $HC_2H_3O_2$ solution is only about 1.4% ionized. The ionic equilibrium is shown by the equation

$$HC_2H_3O_2(aq) + H_2O \rightleftarrows H_3O^+ + C_2H_3O_2^-$$

The common-ion effect is an application of LeChâtelier's principle.

Figure 19-7 One product of the reaction between hydrocholoric acid and sodium hydroxide is water.

STUDY HINT

The concept of pH, which is equal to $-\log$ of $[H_3O^+]$, is treated in Section 17.2.

1. A change in concentration, pressure, or temperature
2. Formation of a gas, a precipitate, or a slightly ionized product
3. The addition of an ion common to two solutes that brings about precipitation or reduced ionization

Small additions of sodium acetate (an ionic salt that is completely dissociated in water solution) to a solution containing acetic acid greatly increase the acetate-ion concentration. The equilibrium shifts in the direction that uses up acetate ions in accordance with Le Châtelier's principle. More molecules of acetic acid are formed and the concentration of hydronium ions is reduced. In general, the addition of a salt with an ion common to the solution of a weak electrolyte reduces the ionization of the electrolyte. A 0.1 M $HC_2H_3O_2$ solution that contains no sodium acetate has a pH of 2.9. A solution with 0.1 M concentrations of both acetic acid and sodium acetate has a pH of 4.6, indicating lower $[H^+]$ and thus lowered acetic acid ionization.

Section Review

1. Name three ways in which chemical equilibrium can be disturbed.
2. Describe three situations in which ionic reactions go to completion.
3. Describe the common-ion effect.

19.3 Equilibria of Acids, Bases, and Salts

SECTION OBJECTIVES

- Explain the concept of acid-ionization constants, and write acid-ionization equilibrium expressions.
- Define and state the value of the ion-product constant for water.
- Explain buffering.
- Define the hydrolysis constant.
- Compare cation and anion hydrolysis.

In Chapters 16 and 17, you read about the important classes of compounds called acids, bases, and salts. At the end of the preceding section of this chapter, you studied the effect of common ions on acid ionization. In this section you will look at the equilibrium-related characteristics of acidic and basic water solutions and at the behavior of salts dissolved in water.

Ionization Constant of a Weak Acid

About 1.4% of the solute molecules in a 0.1 M acetic acid solution are ionized at room temperature. The remaining 98.6% of the $HC_2H_3O_2$ molecules remain nonionized. Thus, the solution contains three species of particles in equilibrium—$HC_2H_3O_2$ molecules, H_3O^+ ions, and $C_2H_3O_2^-$ ions.

At equilibrium, the rate of the forward reaction

$$HC_2H_3O_2 + H_2O \rightarrow H_3O^+ + C_2H_3O_2^-$$

is equal to the rate of the reverse reaction

$$H_3O^+ + C_2H_3O_2^- \rightarrow H_2O + HC_2H_3O_2$$

The equilibrium equation is

$$HC_2H_3O_2 + H_2O \rightleftharpoons H_3O^+ + C_2H_3O_2^-$$

The equilibrium constant for this system expresses the equilibrium ratio of ions to molecules. From the equilibrium equation for the ionization of acetic acid,

$$K = \frac{[H_3O^+][C_2H_3O_2^-]}{[HC_2H_3O_2][H_2O]}$$

At the 0.1 M concentration, water molecules greatly exceed the number of acetic acid molecules. Without introducing a measurable error, one can assume that the mole concentration of H_2O molecules remains constant in such a solution. Thus the product $K[H_2O]$ is constant (since both K and $[H_2O]$ are constant).

$$K[H_2O] = \frac{[H_3O^+][C_2H_3O_2^-]}{[HC_2H_3O_2]}$$

By setting $K[H_2O] = K_a$

$$K_a = \frac{[H_3O^+][C_2H_3O_2^-]}{[HC_2H_3O_2]}$$

The term K_a is called the **acid-ionization constant.**

The equilibrium equation for the typical weak acid HB is

$$HB(aq) + H_2O(\ell) \rightleftarrows H_3O^+(aq) + B^-(aq)$$

From this equation, the expression for K_a can be written in the general form

$$K_a = \frac{[H_3O^+][B^-]}{[HB]}$$

How can the numerical value of the ionization constant K_a for acetic acid at a specific temperature be determined? First, the equilibrium concentrations of H_3O^+ ions, $C_2H_3O_2^-$ ions, and $HC_2H_3O_2$ molecules must be known. The ionization of a molecule of $HC_2H_3O_2$ in water yields one H_3O^+ ion and one $C_2H_3O_2^-$ ion. These concentrations can therefore be found experimentally by measuring the pH of the solution.

Suppose that precise measurements in an experiment show the pH of a 0.1000 M solution of acetic acid to be 2.876 at 25°C. The numerical value of K_a for $HC_2H_3O_2$ at 25°C can be determined as follows:

$$[H_3O^+] = [C_2H_3O_2^-] = 10^{-2.876} \text{ mol/L}$$

$$[H_3O^+] = [C_2H_3O_2] = \text{antilog}(-2.876) = \text{antilog}(0.124 - 3)$$
$$= 1.33 \times 10^{-3}$$

$$[HC_2H_3O_2]_{final} = 0.1000 \text{ mol/L} - 0.001\,33 \text{ mol/L}$$
$$= [HC_2H_3O_2]_{initial} - [H_3O^+] = 0.0987 \text{ mol/L}$$

$$K_a = \frac{[H_3O^+][C_2H_3O_2^-]}{[HC_2H_3O_2]}$$

$$K_a = \frac{(1.33 \times 10^{-3})^2}{9.87 \times 10^{-2}} = 1.79 \times 10^{-5}$$

An increase in temperature causes the equilibrium to shift according to Le Châtelier's principle. Thus K_a has a new value for each temperature. At constant temperature, an increase in the concentration of $C_2H_3O_2^-$ ions, through the addition of $NaC_2H_3O_2$, also disturbs the equilibrium. This disturbance causes a decrease in $[H_3O^+]$ and an increase in $[HC_2H_3O_2]$. Eventually, the equilibrium is reestablished with the same value of K_a. There is a higher concentration of nonionized acetic acid molecules and a lower concentration of H_3O^+ ions, however. Changes in the hydronium-ion concentration and in pH go together. In this example, the reduction in [H$_3$O$^+$] means an increase in the pH of the solution. Ionization data and constants for some dilute acetic acid solutions at 25°C are given in Table 19-3 (shown below).

TABLE 19-3 IONIZATION CONSTANT OF ACETIC ACID

Molarity	% Ionized	[H$_3$O$^+$]	[HC$_2$H$_3$O$_2$]	K_a
0.1000	1.33	0.00133	0.09867	1.79×10^{-5}
0.0500	1.89	0.000945	0.04906	1.82×10^{-5}
0.0100	4.17	0.000417	0.009583	1.81×10^{-5}
0.0050	5.86	0.000293	0.004707	1.82×10^{-5}
0.0010	12.60	0.000126	0.000874	1.82×10^{-5}

A solution containing both a weak acid and a salt of the acid can react with either an acid or a base. The pH of the solution remains nearly constant even when small amounts of acids or bases are added. Suppose an acid is added to a solution of the weak acid, acetic acid, and its salt, sodium acetate (NaC$_2$H$_3$O$_2$). Acetate ions react with the hydronium ions to form nonionized acetic acid molecules.

$$\text{C}_2\text{H}_3\text{O}_2^-(aq) + \text{H}_3\text{O}^+(aq) \rightarrow \text{HC}_2\text{H}_3\text{O}_2(aq) + \text{H}_2\text{O}(\ell)$$

The hydronium-ion concentration and the pH of the solution remain practically unchanged.

Suppose a small amount of a base is added to the original solution. The OH$^-$ ions of the base react with and remove hydronium ions, to form nonionized water molecules. Acetic acid molecules ionize and restore the equilibrium concentration of hydronium ions.

$$\text{HC}_2\text{H}_3\text{O}_2(aq) + \text{H}_2\text{O}(\ell) \rightarrow \text{H}_3\text{O}^+(aq) + \text{C}_2\text{H}_3\text{O}_2^-(aq)$$

The pH of the solution again remains practically unchanged.

A solution of a weak base containing a salt of the base behaves in a similar manner. The hydroxide-ion concentration (and the pH) of the solution remain essentially constant with small additions of acids or bases. Suppose a base is added to an aqueous solution of ammonia that also contains ammonium chloride. Ammonium ions remove the added hydroxide ions to form nonionized water molecules.

$$\text{NH}_4^+(aq) + \text{OH}^-(aq) \rightarrow \text{NH}_3(aq) + \text{H}_2\text{O}(\ell)$$

If a small amount of an acid is added to the solution instead, hydroxide ions from the solution remove the added hydronium ions to form nonionized water molecules. Ammonia molecules in the solution then ionize and restore the equilibrium concentration of hydronium ions and the pH of the solution.

$$\text{NH}_3(aq) + \text{H}_2\text{O}(\ell) \rightarrow \text{NH}_4^+(aq) + \text{OH}^-(aq)$$

The common ion salt in each of these solutions acts as a "buffer" against significant changes in pH in the solution. *Such solutions that can resist changes in pH are referred to as* **buffered solutions.**

Buffer action has many important applications in chemistry and physiology (see Figure 19-8). Many medicines are buffered to prevent large and potentially damaging changes in pH. Human blood is naturally buffered to maintain a pH of about 7.3. This is essential because large changes in pH would lead to serious disturbances of normal body functions or even death.

Figure 19-8 When acetic acid is buffered with sodium acetate *(left)*, the addition of a small amount of hydrochloric acid *(right)* does not affect the solution's pH as shown by the unchanging color of the indicator methyl violet.

Ionization Constant of Water

Pure water is a very poor conductor of electricity because it is only very slightly ionized according to the equation

$$H_2O(\ell) + H_2O(\ell) \rightleftharpoons H_3O^+(aq) + OH^-(aq)$$

The degree of ionization is slight. Equilibrium is quickly established with a very low concentration of H_3O^+ and OH^- ions.

Conductivity experiments with pure water at 25°C show that the concentrations of H_3O^+ and OH^- ions are both 10^{-7} moles per liter (see Figure 19-10). The expression for the equilibrium constant is

$$K = \frac{[H_3O^+][OH^-]}{[H_2O]^2}$$

A liter of water at 25°C contains

$$\frac{997 \text{ g } H_2O}{18.0 \text{ g/mol}} = 55.4 \text{ mol } H_2O$$

This concentration of water molecules, 55.4 mol/L, remains practically the same in all dilute solutions.

Figure 19-9 Many medicines and personal products are buffered to protect the body from possibly harmful pH changes.

Figure 19-10 Water is a poor conductor as shown by the high reading on the ohmmeter *(left)*, because it only ionizes to a very slight degree. Addition of HCl creates a solution with higher conductivity as indicated by the low reading on the ohmmeter *(right)*. An ohmmeter measures resistance which is the inverse of conductance.

Chemical Equilibrium

Thus, both $[H_2O]^2$ and K in the above equilibrium expression are constants. Their product is the ion-product constant for water. As you learned in Chapter 17, this ion product is the product of the molar concentrations of the H_3O^+ and OH^- ions. The ion-product constant is symbolized by K_w. Thus,

$$K_w = [H_3O^+][OH^-]$$

At 25°C,

$$K_w = 10^{-7} \times 10^{-7} = 10^{-14}$$

The product, K_w, of the molar concentrations of H_3O^+ and OH^- ions has this constant value not only in pure water but in all water solutions at 25°C. An acid solution with a pH of 4 has a $[H_3O^+]$ of 10^{-4} mol/L and thus, has a $[OH^-]$ of 10^{-10} mol/L (since $10^{-4} \times 10^{-10} = 10^{-14}$). A basic solution with a pH of 8 has an $[H_3O^+]$ concentration of 10^{-8} mol/L and thus has a $[OH^-]$ concentration of 10^{-6} mol/L.

Hydrolysis of Salts

When a salt is dissolved in water, the solution might be expected to be neutral. The aqueous solutions of some salts, such as NaCl and KNO_3, are indeed neutral; such solutions have a pH of 7.

Some other salts form aqueous solutions that are either acidic or basic, however. As salts, they are completely dissociated in aqueous solution; but, unlike salts whose aqueous solutions are neutral, their ions react chemically with the water solvent. The ions of a salt that do react with the solvent are said to hydrolyze in water solution. **Hydrolysis** *is a reaction between water and ions of a dissolved salt.*

Salts can be placed in four general categories, depending on their hydrolysis properties.

1. Salts of strong acids and strong bases. Recall that a salt is one product of an aqueous acid-base neutralization reaction. The salt consists of the positive ions (cations) of the base and negative ions (anions) of the acid. For example,

$$HCl(aq) + NaOH(aq) \rightarrow Na^+Cl^-(aq) + H_2O(\ell)$$

HCl(aq) is a strong acid, and NaOH(aq) is a strong base. Equivalent quantities of these two reactants yield a neutral aqueous solution of Na^+ and Cl^- ions; the pH of the solution is 7. Neither the Na^+ cation (the positive ion of the strong base) nor the Cl^- anion (the negative ion of the strong acid) undergoes hydrolysis in water solutions.

Similarly, KNO_3 is the salt of the strong acid HNO_3 and the strong base KOH. Measurements show that the pH of an aqueous KNO_3 solution is always very close to 7. Neither the cation of a strong base nor the anions of a strong acid hydrolyze appreciably in aqueous solutions.

2. Salts of weak acids and strong bases. The aqueous solutions of such salts are basic. Anions of the dissolved salt are hydrolyzed in the water solvent and the pH of the solution is raised, indicating that the hydroxide-ion concentration has increased. Observe that the anions of the salt that undergo hydrolysis are the negative ions forming the weak acid. The cations of the salt (the positive ions from the strong base) do not hydrolyze appreciably. An example of such a salt is sodium acetate ($NaC_2H_3O_2$).

STUDY HINT
The concept of strong and weak acids and bases is dealt with in Section 16.3. Recall that strong acids unlike weak ones, tend to ionize nearly completely aqueous solutions.

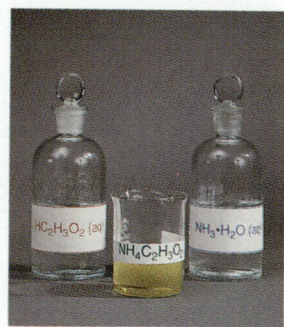

3. *Salts of strong acids and weak bases.* The aqueous solutions of such salts are acidic. Cations of the dissolved salt are hydrolyzed in the water solvent, and the pH of the solution is lowered, indicating that the hydronium-ion concentration has increased. Observe that, in this case, the cations of the salt that undergo hydrolysis are the positive ions from the weak base. The anions of the salt (the negative ions form the strong acid) do not hydrolyze appreciably. An example of such a salt is ammonium chloride (NH_4Cl).

4. *Salts of weak acids and weak bases.* In the aqueous solutions of such salts, both ions of the dissolved salt are hydrolyzed extensively and the solution can be either neutral, acidic, or basic, depending on the salt dissolved. If both ions of such a salt in aqueous solution are hydrolyzed equally, the solution remains neutral. Ammonium acetate ($NH_4C_2H_3O_2$) is such a salt. (See Figure 19-11.) In cases in which both the acid and the base are very weak indeed, the salt may undergo essentially compete decomposition to the products of hydrolysis. For example, when aluminum sulfide is placed in water, both a precipitate and a gas are formed. The reaction is

$$Al_2S_3(s) + 6H_2O(\ell) \rightarrow 2Al(OH)_3(s) + 3H_2S(g)$$

Both products are sparingly soluble in water and are removed from solution.

Figure 19-11 The acidity or basicity of solutions of various salts are tested using universal indicator solution, *from left to right:* $NaCl$, $NaC_2H_3O_2$, NH_4Cl, and $NH_4C_2H_3O_2$.

Hydrolysis often has important effects on the properties of solutions. Sodium carbonate (Na_2CO_3), known as washing soda, is widely used as a cleaning agent because of the basic properties of its water solution. (See Figure 19-12.) Sodium hydrogen carbonate, known as baking soda, forms a mildly alkaline solution in water and has many practical uses. Through the study of hydrolysis, you can now better understand a fact you were given in Chapter 17: the end point of a neutralization reaction can occur at a pH other than 7.

Now it is time to examine hydrolysis in more detail. First, you will learn more about anion hydrolysis, and then you will study cation hydrolysis.

Anion hydrolysis The cations of the salt of a weak acid and a strong base do not hydrolyze in water. Only anion hydrolysis occurs in an aqueous solution of such a salt. In the Brønsted sense, the anion of the salt is the conjugate base (proton acceptor) of the weak acid. These ions are basic enough to remove protons from some water molecules (proton donors) to form OH^- ions. An equilibrium is established in which the net effect of the anion hydrolysis is an increase in the hydroxide-ion concentrations, $[OH^-]$, of the solution. Using the arbitrary symbol B^- for the anion, this hydrolysis reaction is represented by the general anion hydrolysis equation that follows:

$$B^-(aq) + H_2O(\ell) \rightleftharpoons HB(aq) + OH^-(aq)$$

Figure 19-12 A polarized light photomicrograph of sodium carbonate (washing soda) crystals (9.5X).

Chemical Equilibrium   581

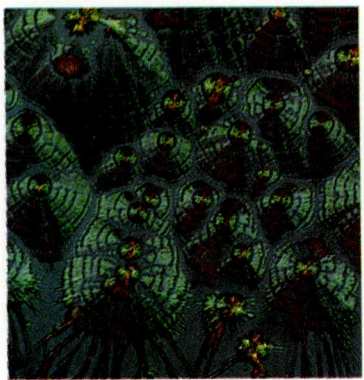

Figure 19-13 A polarized light photomicrograph of sodium hydrogen carbonate (baking soda) crystals (130X).

In the reaction to the right, the anion B^- acquires a proton from the water molecule to form the weak acid HB and hydroxide ion OH^-.

The equilibrium constant for an ion hydrolysis reaction is sometimes called **hydrolysis constant,** K_h. From this general anion hydrolysis equation, the expression for K_h is

$$K_h = \frac{[HB][OH^-]}{[B^-]}$$

You can see that a K_h has the same form as the general chemical-equilibrium expression, but without $[H_2O]$, which is essentially constant in dilute solutions. The hydrolysis constant K_h is also expressed by the ratio of the ion-product constant K_w for water to the ionization constant K_a for the weak acid.

$$K_h = \frac{K_w}{K_a}$$

The validity of this expression can be demonstrated in the following way. Since

$$K_w = [H_3O^+][OH^-] \quad \text{and} \quad K_a = \frac{[H_3O^+][B^-]}{[HB]}$$

Then,

$$\frac{K_w}{K_a} = \frac{[H_3O^+][OH^-]}{\frac{[H_3O^+][B^-]}{[HB]}} = \frac{[HB][OH^-]}{[B^-]} = K_h$$

Suppose sodium carbonate is dissolved in water and the solution is tested with litmus paper. The solution turns red litmus blue. The solution contains a higher concentration of OH^- ions than does pure water, and is basic.

The sodium ions (Na^+) in sodium carbonate do not undergo hydrolysis in aqueous solution, but the carbonate ions (CO_3^{2-}) react as a Brønsted base. A CO_3^{2-} anion acquires a proton from a water molecule to form the slightly ionized hydrogen carbonate ion, HCO_3^-, and the OH^- ion.

$$CO_3^{2-}(aq) + H_2O(\ell) \rightleftharpoons HCO_3^-(aq) + OH^-(aq)$$

The OH^- ion concentration increases until equilibrium is established. Consequently, the H_3O^+ ion concentration decreases since the product $[H_3O^+][OH^-]$ remains equal to the ionization constant K_w of water at the temperature of the solution. Thus the pH is *higher* than 7, and the solution is basic. In general, then, salts formed from weak acids and strong bases hydrolyze in water to form basic solutions.

Cation hydrolysis The anions of the salt of a strong acid and a weak base do not hydrolyze in water. Only cation hydrolysis occurs in an aqueous solution of such a salt.

An aqueous solution of ammonium chloride, NH_4Cl turns blue litmus paper red. This color change indicates that hydrolysis has occurred and the solution contains an excess of H_3O^+ ions. Chloride ions, the anions of the salt, show no noticeable tendency to hydrolyze in aqueous solution. Ammonium ions, the cations of the salt, donate protons to water molecules according to the equation

$$NH_4^+(aq) + H_2O(\ell) \rightleftharpoons H_3O^+(aq) + NH_3(aq)$$

Equilibrium is established with an increased H_3O^+ ion concentration. The pH is lower than 7, and the solution is acidic.

The metallic cations of some salts are hydrated in aqueous solution. These hydrated cations may donate protons to water molecules. Because of this cation hydrolysis, the solution becomes acidic. For example, aluminum chloride forms the hydrated cations

$$Al(H_2O)_6^{3+}$$

Copper(II) sulfate in aqueous solution forms the light-blue hydrated cations

$$Cu(H_2O)_4^{2+}$$

These ions react with water to form hydronium ions as follows:

$$Al(H_2O)_6^{3+} + H_2O(\ell) \leftrightarrows Al(H_2O)_5OH^{2+}(aq) + H_3O^+(aq)$$
$$Cu(H_2O)_4^{2+} + H_2O(\ell) \leftrightarrows Cu(H_2O)_3OH^+(aq) + H_3O^+(aq)$$

Cations such as $Cu(H_2O)_3OH^+$ ions may experience a secondary hydrolysis to a slight extent

$$Cu(H_2O)_3OH^+(aq) + H_2O(\ell) \leftrightarrows Cu(H_2O)_2(OH)_2(s) + H_3O^+(aq)$$

The resulting hydrated copper(II) hydroxide is not very soluble. The scummy appearance of reagent bottles in which copper(II) salt solutions are stored for a long time is caused by this slight secondary hydrolysis of $Cu(H_2O)_3OH^+$ cations. (See Figure 19-14.) Expressions for the hydrolysis constant K_h for cation acids are similar to those of anion bases. In general, salts formed from strong acids and weak bases hydrolyze in water to form acidic solutions.

Figure 19-14 When stored for a long time, a solution of copper(II) sulfate develops a scummy appearance. The $Cu(H_2O)_3OH^+$ cations hydrolyze, forming insoluble hydrated copper(II) hydroxide.

1. The ionization constant for a weak acid, K_a
2. $K_a = \dfrac{[H_3O^+][B^-]}{[HB]}$
3. Solutions that can resist changes in pH
4. The product of the molar concentrations of H_3O^+ and OH^-, $K_w = 1.0 \times 10^{-14}$
5. The reaction between water and ions of a dissolved salt. Cation hydrolysis produces an acid solution; anion hydrolysis produces a basic solution.

Section Review

1. What is meant by an acid-ionization constant?
2. How is an acid-ionization equilibrium expression written?
3. What is meant by the term buffered solution?
4. What is meant by the ion-product constant for water? What is the value of this constant?
5. What is hydrolysis? Compare cation and anion hydrolysis.

19.4 Solubility Equilibrium

Chemical-equilibrium considerations can be applied to solubilities of various substances. As you may recall from Chapter 14, solubility refers to the maximum amounts of a substance that will dissolve per unit volume of solution under various conditions. Solubility is generally expressed in units of moles of solute per liter of solution. In this section solubility is expressed in terms of equilibrium expressions, which are then used to predict whether precipitation occurs when solutions of various substances are combined.

Solubility Product

A saturated solution contains the maximum amount of solute possible at a given temperature in equilibrium with an undissolved excess of the

SECTION OBJECTIVES

- Explain what is meant by solubility-product constants, and calculate their values.
- Calculate solubilities, using solubility-product constants.
- Carry out calculations to predict whether precipitates will form.

substance. A saturated solution is not necessarily a concentrated solution. The concentration may be high or low, depending on the solubility of the solute.

A general rule is often used to express solubilities qualitatively. By this rule, a substance is said to be *soluble* if the solubility is *greater than* 1 g per 100 g of water. It is said to be *insoluble* if the solubility is *less than* 0.1 g per 100 g of water. Substances whose solubilities fall between these limits are described as *slightly soluble*.

Substances usually referred to as *insoluble* are, in fact, very sparingly soluble. An extremely small quantity of such a solute saturates the solution. Equilibrium is established with the undissolved excess remaining in contact with the solution. Equilibria between sparingly soluble solids and their saturated solutions are especially important in analytical chemistry.

You have observed that silver chloride precipitates when Ag^+ and Cl^- ions are placed in the same solution. Silver chloride is so sparingly soluble in water that it is described as insoluble. Its solution reaches saturation at a very low concentration of its ions. All Ag^+ and Cl^- ions in excess of this concentration eventually separate as an AgCl precipitate.

The equilibrium principles developed in this chapter apply to all saturated solutions of sparingly soluble salts. Consider the equilibrium system in a saturated solution of silver chloride containing an excess of the solid salt. The equilibrium equation is

$$AgCl(s) \rightleftarrows Ag^+(aq) + Cl^-(aq)$$

The equilibrium constant is expressed as

$$K = \frac{[Ag^+][Cl^-]}{[AgCl]}$$

Because the concentration of a pure substance in the solid or liquid phase remains constant, adding more solid AgCl to this equilibrium system does not change the concentration of the undissolved AgCl present (see Figure 19-15). Thus [AgCl] in the equation is set equal to 1, giving what is called the solubility-product constant K_{sp}. By combination of the two constants, this equation becomes

$$K[AgCl] = [Ag^+][Cl^-]$$

$$K_{sp} = [Ag^+][Cl^-]$$

This equation is the solubility-equilibrium expression for the reaction. It expresses the fact that the solubility-product constant K_{sp} of AgCl is the product of the molar concentrations of its ions in a saturated solution.

Calcium fluoride is another sparingly soluble salt. The equilibrium in a saturated CaF_2 solution is given by the following equation:

$$CaF_2(s) \rightleftarrows Ca^{2+}(aq) + 2F^-(aq)$$

The solubility-product constant is given by

$$K_{sp} = [Ca^{2+}][F^-]^2$$

Notice that this constant is the product of the molar concentration of Ca^{2+} ions and the molar concentration of F^- ions squared as required by the general chemical equilibrium expression.

> Nothing is truly "insoluble." Amounts dissolving may only be too small to measure.

Figure 19-15 A heterogeneous equilibrium in a saturated solution is not disturbed by adding more of the solid phase.

Similar equations apply to any sparingly soluble salt having the general formula M_aX_b. The equilibrium system in a saturated solution is shown by

$$M_aX_b(s) \leftrightarrows aM^{b+} + bX^{a-}$$

The solubility-product constant is expressed by

$$K_{sp} = [M^{b+}]^a[X^{a-}]^b$$

*The **solubility-product constant** of a substance is the product of the molar concentrations of its ions in a saturated solution, each raised to the power that is the coefficient of that ion in the chemical equation.*

From solubility data listed in Appendix Table A-12, 1.94×10^{-4} g of AgCl saturates 100 g of water at 20°C. One mole of AgCl has a mass of 143.3 g. The saturation concentration (solubility) of AgCl can therefore be expressed in moles per liter of water, which is roughly equal to moles per liter of solution:

$$\frac{1.94 \times 10^{-4} \text{ g AgCl}}{100. \text{ g H}_2\text{O}} \times \frac{1 \text{ g H}_2\text{O}}{\text{mL H}_2\text{O}} \times \frac{10^3 \text{ mL}}{\text{L}} \times \frac{1 \text{ mol AgCl}}{143.3 \text{ g AgCl}}$$
$$= 1.35 \times 10^{-5} \text{ mol/L}$$

The equilibrium equation is

$$\text{AgCl(aq)} \rightleftarrows \text{Ag}^+(\text{aq}) + \text{Cl}^-(\text{aq})$$

Silver chloride dissociates in solution, contributing equal numbers of aqueous Ag^+ and Cl^- ions. The ion concentrations in the saturated solution, are therefore 1.35×10^{-5} mol/L

$$[\text{Ag}^+] = 1.35 \times 10^{-5}$$

and

$$[\text{Cl}^-] = 1.35 \times 10^{-5}$$

$$K_{sp} = [\text{Ag}^+][\text{Cl}^-]$$
$$K_{sp} = (1.35 \times 10^{-5})(1.35 \times 10^{-5})$$
$$K_{sp} = (1.35 \times 10^{-5})^2$$
$$K_{sp} = 1.82 \times 10^{-10}$$

Chemical Equilibrium

This result is the solubility-product constant of AgCl at 20°C.

From Appendix Table A-12, the solubility of CaF_2 is 1.7×10^{-3} g/100 g of water at 25°C. Expressed in moles per liter as before, this concentration becomes 2.2×10^{-4} mol/L.

The equilibrium equation for a saturated solution of CaF_2 is

$$CaF_2(s) \rightleftarrows Ca^{2+}(aq) + 2F^-(aq)$$

CaF_2 dissociates in solution to yield twice as many F^- ions as Ca^{2+} ions. The ion concentrations in the saturated solution are

$$[Ca^{2+}] = 2.2 \times 10^{-4}$$
$$[F^-] = 2(2.2 \times 10^{-4}) = 4.4 \times 10^{-4}$$

Note carefully that at equilibrium at 25°C, $[Ca^2]$ equals the solubility of 2.2×10^{-4} mol/L, but $[F^-]$ equals twice the solubility, or 4.4×10^{-4} mol/L. The number of moles of positive and negative ions per mole of compound must always be accounted for in using K_{sp} and solubilities. Therefore

$$K_{sp} = [Ca^{2+}][F^-]^2$$
$$K_{sp} = (2.2 \times 10^{-4})(4.4 \times 10^{-4})^2$$
$$K_{sp} = (2.2 \times 10^{-4})(1.9 \times 10^{-7})$$
$$K_{sp} = 4.2 \times 10^{-11}$$

Thus, the solubility-product constant of CaF_2 is 4.2×10^{-11} at 25°C.

If the ion product $[Ca^{2+}][F^-]^2$ is less than the value of K_{sp} at a particular temperature, the solution is unsaturated. If the ion product is greater than the value for K_{sp}, CaF_2 precipitates. This precipitation reduces the concentrations of Ca^{2+} and F^- ions until equilibrium is established.

It is difficult to measure very small concentrations of a solute with precision. For this reason, solubility data from different sources may result in slightly different values of K_{sp} for a substance. Thus, calculations of K_{sp} ordinarily should be limited to two significant figures. Representative values of K_{sp} at 25°C for some sparingly soluble compounds are listed in Table 19-4.

Sample Problem 19.2

Calculate the solubility-product constant (K_{sp}) for copper(I) chloride (CuCl) given that the solubility of this compound at 25°C is 1.08×10^{-2} g/100. g H_2O.

Solution

Step 1. Analyze   Given: solubility of CuCl = 1.08×10^{-2} g CuCl/100. g H_2O
Unknown: K_{sp}

Step 2. Plan   solubility in g/100. g H_2O → solubility in mol/L
solubility in mol/L, solubility equilibrium
expression → K_{sp}

$$\frac{g\ CuCl}{100.\ g\ H_2O} \times \frac{1\ g\ H_2O}{1\ mL\ H_2O} \times \frac{10^3\ mL}{1\ L} \times \frac{1\ mol\ CuCl}{g\ CuCl} = \text{solubility in mol/L}$$

$$CuCl(s) \rightleftarrows Cu^+(aq) + Cl^-(aq)$$
$$[Cu^+] = [Cl^-] = \text{solubility in mol/L}$$
$$K_{sp} = [Cu^+][Cl^-]$$

Sample Problem 19.2 continued

Step 3. Compute Since the molar mass of CuCl is 99.0 g/mol,

$$\text{solubility} = \frac{1.08 \times 10^{-2}\,\text{g CuCl}}{100.\,\text{g H}_2\text{O}} \times \frac{10^3\,\text{g}}{\text{L}} \times \frac{\text{mol CuCl}}{99.0\,\text{g CuCl}} = 1.09 \times 10^{-3}\,\text{mol CuCl/L}$$

$$[\text{Cu}^+] = [\text{Cl}^-] = 1.09 \times 10^{-3}\,\text{mol/L}$$

$$K_{sp} = (1.09 \times 10^{-3})(1.09 \times 10^{-3}) = 1.19 \times 10^{-6}$$

Step 4. Evaluate The answer contains the proper number of significant digits, is approximately equal to an estimated value of 1.2×10^{-6}, $(1.1 \times 10^{-3})(1.1 \times 10^{-3})$, and is also close to the actual K_{sp} value given in Table 19-4.

Practice Problem
Calculate the solubility-product constant, K_{sp}, of lead(II) chloride ($PbCl_2$), which has a solubility of 1.0 g/100. g H₂O at 25°C. *(Ans.)* 1.9×10^{-4}

Calculating Solubilities

Solubility-product constants are computed from very careful measurements of solubilities and other solution properties. Once known, the solubility product is very helpful in determining the solubility of a sparingly soluble salt.

Suppose you wish to know how much barium carbonate ($BaCO_3$) can be dissolved in one liter of water solution at 25°C. From Table 19-4, K_{sp} for $BaCO_3$ has the numerical value 1.2×10^{-8}. The solubility equation is written as follows:

$$BaCO_3(s) \rightleftharpoons Ba^{2+}(aq) + CO_3^{2-}(aq)$$

Knowing the value for K_{sp} gives

$$K_{sp} = [Ba^{2+}][CO_3^{2-}] = 1.2 \times 10^{-8}$$

Therefore, $BaCO_3$ dissolves until the product of the molar concentrations of Ba^{2+} and CO_3^{2-} equals 1.2×10^{-8}.

The solubility-equilibrium equation shows that Ba^{2+} ions and CO_3^{2-} ions enter the solution in equal numbers as the salt dissolves. Thus,

$$[Ba^{2+}] = [CO_3^{2-}] = \text{solubility of BaCO}_3$$
$$K_{sp} = 1.2 \times 10^{-8} = (\text{solubility of BaCO}_3)^2$$
$$\text{solubility of BaCO}_3 = \sqrt{1.2 \times 10^{-8}}$$
$$\text{solubility of BaCO}_3 = 1.1 \times 10^{-4}\,\text{mol/L}$$

The solubility of $BaCO_3$ is 1.1×10^{-4} mol/L. Thus, the solution concentration is 1.1×10^{-4} M for Ba^{2+} ions and for CO_3^{2-} ions.

Sample Problem 19.3

Calculate the solubility of silver acetate ($AgC_2H_3O_2$) in mol/L, given the K_{sp} value for this compound listed in Table 19-4.

TABLE 19-4 SOLUBILITY-PRODUCT CONSTANTS K_{sp} AT 25°C

Salt	Ion Product	K_{sp}	Salt	Ion Product	K_{sp}
$AgC_2H_3O_2$	$[Ag^+][C_2H_3O_2^-]$	2.5×10^{-3}	CuCl	$[Cu^+][Cl^-]$	1.2×10^{-6}
AgBr	$[Ag^+][Br^-]$	5.0×10^{-13}	CuS	$[Cu^{2+}][S^{2-}]$	6.3×10^{-36}
Ag_2CO_3	$[Ag^+]^2[CO_3^{2-}]$	8.1×10^{-12}	FeS	$[Fe^{2+}][S^{2-}]$	6.3×10^{-18}
AgCl	$[Ag^+][Cl^-]$	1.8×10^{-10}	Fe_2S_3	$[Fe^{3+}]^2[S^{2-}]^3$	1.4×10^{-85}
AgI	$[Ag^+][I^-]$	8.3×10^{-17}	$Fe(OH)_3$	$[Fe^{3+}][OH^-]^3$	1.5×10^{-36}
Ag_2S	$[Ag^+]^2[S^{2-}]$	1.6×10^{-49}	HgS	$[Hg^{2+}][S^{2-}]$	1.6×10^{-52}
$Al(OH)_3$	$[Al^{3+}][OH^-]^3$	1.3×10^{-33}	$MgCO_3$	$[Mg^{2+}][CO_3^{2-}]$	3.5×10^{-8}
$BaCO_3$	$[Ba^{2+}][CO_3^{2-}]$	1.2×10^{-8}	$Mg(OH)_2$	$[Mg^{2+}][OH^-]^2$	1.5×10^{-11}
$BaSO_4$	$[Ba^{2+}][SO_4^{2-}]$	1.1×10^{-10}	MnS	$[Mn^{2+}][S^{2-}]$	2.5×10^{-13}
CdS	$[Cd^{2+}][S^{2-}]$	8.0×10^{-27}	$PbCl_2$	$[Pb^{2+}][Cl^-]^2$	1.9×10^{-4}
$CaCO_3$	$[Ca^{2+}][CO_3^{2-}]$	1.4×10^{-8}	$PbCrO_4$	$[Pb^{2+}][CrO_4^{2-}]$	2.8×10^{-13}
CaF_2	$[Ca^{2+}][F^-]^2$	4.2×10^{-11}	$PbSO_4$	$[Pb^{2+}][SO_4^{2-}]$	1.8×10^{-8}
$Ca(OH)_2$	$[Ca^{2+}][SO_4^{2-}]$	5.5×10^{-6}	PbS	$[Pb^{2+}][S^{2-}]$	8.0×10^{-28}
$CaSO_4$	$[Ca^{2+}][SO_4^{2-}]$	9.1×10^{-6}	SnS	$[Sn^{2+}][S^{2-}]$	1.2×10^{-25}
CoS	$[Co^{2+}][S^{2-}]$	3.0×10^{-26}	$SrSO_4$	$[Sr^{2+}][SO_4^{2-}]$	3.2×10^{-7}
Co_2S_3	$[Co^{3+}]^2[S^{2-}]^3$	2.6×10^{-124}	ZnS	$[Zn^{2+}][S^{2-}]$	1.6×10^{-24}

Sample Problem 19.3 continued

Solution

Step 1. Analyze   Given: $K_{sp} = 2.5 \times 10^{-3}$
Unknown: solubility of $AgC_2H_3O_2$

Step 2. Plan
$$K_{sp} \rightarrow [Ag^+] = \text{solubility of } AgC_2H_3O_2$$
$$K_{sp} = [Ag^+][C_2H_3O_2^-]$$
Since $[Ag^+] = [C_2H_3O_2^-]$
$$K_{sp} = [Ag^+]^2$$
$$[Ag^+] = \sqrt{K_{sp}}$$
solubility of $AgC_2H_3O_2 = [Ag^+]$

Step 3. Compute   Solubility of $AgC_2H_3O_2 = [Ag^+] = \sqrt{2.5 \times 10^{-3}} = 5.0 \times 10^{-2}$ mol/L

Step 4. Evaluate   The answer has the proper number of significant digits and is equal to an estimated value of 5.0×10^{-2} ($\sqrt{25 \times 10^{-4}}$).

Practice Problem
Calculate the solubility of cadmium sulfide (CdS) in moles/liter, given the K_{sp} value listed in Table 19-4. (Ans.) 8.9×10^{-14} mol/L

Precipitation Calculations

In the example used earlier in this section of calculating solubility, $BaCO_3$ served as the source of both Ba^{2+} and CO_3^{2-} ions. Since each mole of $BaCO_3$ yields one mole of Ba^{2+} and one mole of CO_3^{2-}, the concentrations of the two ions were equal. However, the equilibrium condition does not require that the two ion concentrations be equal. Equilibrium will still be established so that the ion product $[Ba^{2+}][CO_3^{2-}]$ not exceed the value of K_{sp} for the system.

Suppose unequal quantities of $BaCl_2$ and $CaCO_3$ are added to water. If the ion product $[Ba^{2+}][CO_3^{2-}]$ exceeds the K_{sp} of $BaCO_3$, a precipitate of $BaCO_3$ forms. Precipitation continues until the ion concentrations decrease to the point where $[Ba^{2+}][CO_3^{2-}]$ equals the K_{sp}.

Substances differ greatly in their tendencies to form precipitates when mixed in moderate concentrations. The photos in Figure 19-16 show the behavior of some negative ions in the presence of certain metallic ions. Note that some of the combinations have produced precipitates and some have not. The solubility product can be used to predict whether a precipitate forms when two solutions are mixed.

Figure 19-16
The behavior of some negative ions in the presence of certain metallic ions. The tendency for substances in mixed solutions to form precipitates is predictable.

Sample Problem 19.4

Will a precipitate form if 20.0 mL of 0.010 M $BaCl_2$ is mixed with 20.0 mL of 0.0050 M Na_2SO_4?

Solution

Step 1. Analyze Given: concentration of $BaCl_2$ = 0.010 M
volume of $BaCl_2$ = 20.0 mL
concentration of Na_2SO_4 = 0.0050 M
volume of Na_2SO_4 = 20.0 mL
Unknown: whether a precipitate forms

Step 2. Plan Concentration, volume → ion product → precipitate? The two possible new pairings of ions are NaCl and $BaSO_4$. Of these, $BaSO_4$ is a sparingly soluble salt. It will precipitate if the ion product $[Ba^{2+}][SO_4^{2-}]$ in the combined solutions exceeds K_{sp} for $BaSO_4$. From the table of solubility products (Table 19-4), the K_{sp} is found to be 1.1×10^{-10}. The solubility equilibrium equation is

$$BaSO_4(s) \rightleftharpoons Ba^{2+}(aq) + SO_4^{2-}(aq)$$

The solubility equilibrium expression is

$$K_{sp} = [Ba^{2+}][SO_4^{2-}] = 1.1 \times 10^{-10}$$

First $[Ba^{2+}]$ and $[SO_4^{2-}]$ in the new solution must be found. Then the ion product is calculated and compared with the K_{sp}.

Chemical Equilibrium

Sample Problem 19.4 continued

Step 3. Compute   Mole quantities of Ba^{2+} and SO_4^{2-} ions:

$$0.020 \text{ L} \times \frac{0.010 \text{ mol } Ba^{2+}}{\text{L}} = 0.000\,20 \text{ mol } Ba^{2+}$$

$$0.020 \text{ L} \times \frac{0.0050 \text{ mol } SO_4^{2-}}{\text{L}} = 0.000\,10 \text{ mol } SO_4^{2-}$$

Total volume of solution containing Ba^{2+} and SO_4^{2-} ions:

$$0.020 \text{ L} + 0.020 \text{ L} = 0.040 \text{ L}$$

Ba^{2+} and SO_4^{2+} ion concentrations:

$$\frac{0.000\,20 \text{ mol } Ba^{2+}}{0.040 \text{ L}} = 5.0 \times 10^{-3} \text{ mol } Ba^{2+}/\text{L}$$

$$\frac{0.000\,10 \text{ mol } SO_4^{2-}}{0.040 \text{ L}} = 2.5 \times 10^{-3} \text{ mol } SO_4^{2-}/\text{L}$$

Trial value of ion product:

$$[Ba^{2+}][SO_4^{2-}] = (5.0 \times 10^{-3})(2.5 \times 10^{-3})$$
$$= 1.2 \times 10^{-5}$$

The ion product is much greater than K_{sp} ($K_{sp} = 1.1 \times 10^{-10}$), so precipitation occurs.

Step 4. Evaluate   The answer contains the appropriate number of significant digits and is close to an estimated value of 1×10^{-5}, $(5 \times 10^{-3})(2 \times 10^{-3})$; since $10^{-5} > 10^{-10}$, precipitation should occur.

Practice Problems

1. Does a precipitate form when 100. mL of 0.0025 M $AgNO_3$ and 150. mL of 0.0020 M NaBr solutions are mixed?   *(Ans.)* AgBr ion product = 1.2×10^{-6}, K_{sp}(AgBr) = 5.0×10^{-13}; AgBr precipitates.

2. Does a precipitate form when 20. mL of 0.038 M $Pb(NO_3)_2$ and 30. mL of 0.018 M KCl solutions are mixed?   *(Ans.)* $PbCl_2$ ion product = 1.8×10^{-6}, K_{sp} ($PbCl_2$) = 1.9×10^{-4}; $PbCl_2$ does *not* precipitate.

The solubility-product principle can be very useful when applied to solutions of sparingly soluble substances. It *cannot* be applied very successfully to solutions of moderately soluble or very soluble substances. Also, many solubility-product constants are known only roughly because of difficulties involved in solubility measurements. Sometimes it is necessary to consider two equilibria simultaneously. For example, if either ion hydrolyzes, the salt will be more soluble than predicted by using only the solubility product constant. The solubility product is also sensitive to changes in solution temperature to the extent that the solubility of the dissolved substance is affected by such changes. All of these factors limit the conditions under which the principle can be applied.

Section Review

1. What is a solubility-product constant? How are such constants determined?
2. How are solubility-product constants used to calculate solubilities?
3. What is an ion product?
4. How are calculations to predict possible precipitation carried out?
5. What is the value of K_{sp} for Ag_2SO_4 if 5.40 g is soluble in 1.00 L of water?
6. Determine whether a precipitate will form if 20.0 mL of 1.00×10^{-7} M $AgNO_3$ is mixed with 20.0 mL of 2.00×10^{-9} M NaCl at 25°C.

1. The product of the molar concentrations of the ions in a saturated solution, each raised to the appropriate power. They are determined from very careful measurements of solubilities and other solution properties.
2. If the K_{sp} and equation are known, solubilities may be calculated from the K_{sp} expression. (See Sample problem 19.3.)
3. The product of the concentration of ions present in a solution each raised to the appropriate power
4. If the ion product exceeds the numerical value of K_{sp}, a precipitate will form.
5. 2.07×10^{-5}
6. Since the trial product (5.00×10^{-17}) is less than the K_{sp} of 1.8×10^{-10}, no precipitate is formed.

Chapter Summary

- A reaction system in which the forward and reverse reactions occur simultaneously and at the same rate is said to be in *equilibrium*. Both reactions continue, but there is no net change in the composition of the system.
- At equilibrium, the ratio of the product of the mole concentrations of substances formed to the product of the mole concentrations of reactants, each raised to the appropriate power, has a definite numerical value, K, which is the equilibrium constant at a given temperature. For values of K greater than 1, the products of the forward reaction are favored. For values of K less than 1, the reactants of the reverse reaction are favored.
- Any change that alters the rate of either the forward or reverse reaction disturbs the equilibrium of the system. According to Le Châtelier's principle, the equilibrium is shifted in the direction that relieves the stress.
- Catalysts increase the rates of forward and reverse reactions equally, and they do not shift an equilibrium or change the value of K.
- The common-ion effect is recognized when a solute containing ions like those of a reactant in an equilibrium system is added to the system. Le Châtelier's principle explains the response of the system to the stress.
- The equilibrium expression for the ionization constant of a weak acid HB is
$$K_a = \frac{[H_3O^+][B^-]}{[HB]}$$
- Salts formed from strong bases and weak acids produce aqueous solutions that are basic because of *anion hydrolysis*.
- Salts formed from strong acids and weak bases produce aqueous solutions that are acidic because of *cation hydrolysis*.
- Salts formed from strong acids and strong bases do not hydrolyze in water, and their solutions are neutral.
- Salts formed from weak acids and weak bases may produce neutral, acidic, or basic solutions, depending on the relative amounts of cation and anion hydrolysis. They may also hydrolyze completely in water solution.
- Ions of salts that are very sparingly soluble form saturated aqueous solutions at extremely low concentrations. The solubility-equilibrium expression for such salts yields a useful constant (K_{sp})—the solubility-product constant—which equals the product of the mole concentrations of solute ions in the saturated solution raised to appropriate exponential powers.

CHAPTER 19 REVIEW

Chemical Equilibrium

VOCABULARY

acid-ionization constant (577)
buffered solution (578)
chemical equilibrium (564)
chemical-equilibrium expression (566)
common-ion effect (575)
equilibrium constant (566)
hydrolysis (580)
hydrolysis constant (582)
reversible reaction (563)
solubility-product constant (585)

REVIEWING CONCEPTS

1. Explain how the concentrations of A, B, C, and D change from the time when A and B are first combined to the point at which equilibrium is established given A + B $\rightleftarrows$ C + D. (19.1)

2. a. Write the general expression for an equilibrium constant based on the equation: $nA + mB + ... \rightleftarrows xC + yD + ...$
   b. What information is provided by the value of K for a given equilibrium system at a specified temperature? (19.1)

3. In general, which reaction is favored (forward, reverse, or neither) if the value of K at a specified temperature is
   a. equal to 1
   b. very small
   c. very large (19.1)

4. Predict whether each of the following concentration changes would favor the forward or reverse reaction:
   $$2HI(g) \rightleftarrows H_2(g) + I_2(g)$$
   a. addition of HI
   b. addition of H_2
   c. removal of I_2
   d. removal of HI (19.2)

5. What will be the effect of each of the following changes on the new equilibrium concentration of H_2O?
   $$4HCl(g) + O_2(g) \rightleftarrows 2H_2O(g) + 2Cl_2(g)$$
   a. addition of HCl
   b. removal of Cl_2
   c. removal of O_2
   d. addition of Cl_2 (19.2)

6. Predict whether each of the following pressure changes would favor the forward or reverse reaction:
   $$2NO(g) + O_2(g) \rightleftarrows 2NO_2(g)$$
   a. increased pressure
   b. decreased pressure (19.2)

7. In heterogeneous reaction systems, what type of substances can be removed from the equilibrium constant expression? Why? (19.2)

8. Explain the effect of a catalyst on an equilibrium system. (19.2)

9. Predict the effect of each of the following on the indicated equilibrium system in terms of which reaction will be favored (forward, reverse, or neither):
   $$H_2(g) + Cl_2(g) \rightleftarrows 2HCl(g) + 184 \text{ kJ}$$
   a. addition of Cl_2
   b. removal of HCl
   c. increased pressure
   d. decreased temperature
   e. removal of H_2
   f. decreased pressure
   g. addition of a catalyst
   h. increased temperature
   i. decreased system volume (19.2)

10. How would parts **a** through **i** of Question 9 affect the new equilibrium concentration of HCl and the value of K at the new equilibrium? (19.2)

11. What concentration stresses can be placed on the following equilibrium system so that the maximum level of HF can be produced? (19.2)

$$H_2(g) + F_2(g) \rightleftarrows 2HF(g)$$

12. Changes in the concentrations of the reactants and products at equilibrium have no impact on the value of the equilibrium constant. Explain. (19.2)

13. What relative pressure (high or low) would result in the production of the maximum level of CO_2 according to the following? Explain. (19.2)

$$2CO(g) + O_2(g) \rightleftarrows 2CO_2(g)$$

14. What relative pressure (high or low) would increase the yield of methanol in the following equilibrium system? (19.2)

$$CO(g) + 2H_2(g) \rightleftarrows CH_3OH(g) \text{ (methanol)}$$

15. What relative temperature (high or low) would increase the yield of methanol in Question 14 if the former reaction has a $\Delta H = -101$ kJ. Explain. (19.2)

16. What relative conditions (reactant concentrations, pressure, and temperature) would favor a high equilibrium concentration of the underlined substance in each of the following equilibrium systems?
    a. $2CO(g) + O_2(g) \rightleftarrows \underline{2CO_2(g)} + 167$ kJ
    b. $Cu^{2+}(aq) + 4NH_3(aq) \rightleftarrows \underline{Cu(NH_3)_4^{2+}(aq)} + 42$ kJ
    c. $2HI(g) + 12.6 \text{ kJ} \rightleftarrows H_2(g) + \underline{I_2(g)}$
    d. $4HCl(g) + O_2(g) \rightleftarrows 2H_2O + \underline{2Cl_2(g)} + 113$ kJ
    e. $H_2O(\ell) + 42 \text{ kJ} \rightleftarrows \underline{H_2O(g)}$ (19.2)

17. A combustion reaction proceeding in air under standard pressure is transferred to an atmosphere of pure oxygen under the same pressure.
    a. What effect would you observe?
    b. How can you account for this effect? (19.2)

18. a. Write the ion-product constant expression for water.
    b. What is the value of this constant? (19.3)

19. List and distinguish among the four general categories of salts, based on their hydrolysis properties, and give an example of each. (19.3)

20. The pH of a solution containing both acetic acid and sodium acetate is higher than that of a solution containing the same concentration of acetic acid alone. Explain. (19.3)

21. The ionization constant K_a for acetic acid is 1.8×10^{-5} at 25°C. Explain the significance of this value. (19.3)

22. a. From the development of K_a described in Section 19.3, show how you would express an ionization constant K_b for the weak base NH_3.
    b. In this case, $K_b = 1.8 \times 10^{-5}$. What is the significance of this numerical value to equilibrium? (19.3)

23. A saturated solution is not necessarily a concentrated solution. Explain. (19.4)

24. What two factors determine the extent to which reacting ions are removed from solution? (19.4)

25. Identify the three conditions under which ionic reactions can run to completion and write an equation for each.

26. What rule of thumb is used to distinguish among soluble, insoluble, and slightly soluble substances? (19.4)

27. What is the major solubility characteristic of those types of substances typically involved in solubility equilibrium systems? (19.4)

28. What is the relationship between K_{sp} and the product of the ion concentrations in terms of determining whether a solution of those ions is saturated?

PROBLEMS

▶ **Equilibrium Constant**

1. Determine the value of the equilibrium constant for each reaction given, if the equilib-

rium concentrations are found to be those specified:
a. $A + B \rightleftarrows C$; $[A] = 2.0$; $[B] = 3.0$; $[C] = 4.0$
b. $D + 2E \rightleftarrows F + 3G$; $[D] = 1.5$; $[E] = 2.0$; $[F] = 1.8$; $[G] = 1.2$
c. $N_2(g) + 3H_2(g) \rightleftarrows 2NH_3(g)$; $[N_2] = 0.45$; $[H_2] = 0.14$; $[NH_3] = 0.62$

2. An equilibrium mixture at a specific temperature is found to consist of 1.2×10^{-3} mol/L HCl, 3.8×10^{-4} mol/L O_2, 5.8×10^{-2} mol/L H_2O, and 5.8×10^{-2} mol/L Cl_2 according to the following: $4HCl(g) + O_2(g) \rightleftarrows 2H_2O(g) + 2Cl_2(g)$. Determine the value of the equilibrium constant for this system.

3. At 450°C the value of the equilibrium constant for the following system is 6.59×10^{-3}. If $[NH_3] = 1.23 \times 10^{-4}$ and $[H_2] = 2.75 \times 10^{-3}$ at equilibrium, determine the concentration of N_2 at that point.

$$N_2(g) + 3H_2(g) \rightleftarrows 2NH_3(g)$$

4. Write equilibrium-constant expressions for each of the following heterogeneous systems:
a. $Zn(s) + 2Ag^+(aq) \rightleftarrows Zn^{2+}(aq) + 2Ag(s)$
b. $PbI_2(s) \rightleftarrows Pb^{2+}(aq) + 2I^-(aq)$
c. $CN^-(aq) + H_2O(\ell) \rightleftarrows HCN(aq) + OH^-(aq)$
d. $CdS(s) \rightleftarrows Cd^{2+}(aq) + S^{2-}(aq)$
e. $4Fe(s) + 3O_2(g) \rightleftarrows 2Fe_2O_3(s)$

5. The value of the equilibrium constant for the reaction below is 40.0 at a specified temperature. What would be the value of that constant for the reverse reaction under the same conditions? $H_2(g) + I_2(g) \rightleftarrows 2HI(g)$

▶ **Ionization Constants**

6. The pH of a $0.100M$ solution of an acid HX is 3.14. Calculate the value of K_a.

7. A weak acid has a K_a value of 2.69×10^{-9}. Calculate the value of the hydrolysis constant K_h. (Assume that $K_w = 1.00 \times 10^{-14}$.)

▶ **Solubility Product Constant**

8. Express the solubility information given for each substance below in moles per liter:
a. NaCl = 36.0 g/100. g H_2O at 20°C
b. $PbSO_4$ = 4.25×10^{-3} g/100. g H_2O at 25°C
c. Cu_2S = 1.84×10^{-11} g/100. g H_2O at 25°C

9. The ionic substance GH dissociates to form G^+ and H^- ions. The solubility of GH is 0.0427 mol/L. What is the value of the solubility-product constant K_{sp}?

10. The ionic substance EJ dissociates to form E^{2+} and J^{2-} ions. The solubility of EJ is 8.45×10^{-6} mol/L. What is the value of the solubility-product constant?

11. Calculate the solubility-product constant K_{sp} for each of the following, based on the solubility information provided:
a. $CaCO_3$ = 1.2×10^{-3} g/100. g H_2O at 20°C
b. $BaSO_4$ = 2.4×10^{-4} g/100. g H_2O at 20°C
c. $Ca(OH)_2$ = 0.173 g/100. g H_2O at 20°C

12. Calculate the solubility of a substance MN that ionizes to form M^{2+} and N^{2-} ions, given that $K_{sp} = 8.1 \times 10^{-6}$.

13. Use the K_{sp} values given in Table 19-4 to evaluate the solubility of each of the following in moles per liter:
a. AgBr   b. CoS

14. Complete each of the following relative to the reaction that occurs when 25.0 mL of 0.0500 M $Pb(NO_3)_2$ is combined with 25.0 mL of 0.0400 M Na_2SO_4 if equilibrium is reached at 25°C:
a. Write the solubility-equilibrium equation at 25°C.
b. Write the solubility-equilibrium expression for the net reaction.
c. Determine the number of moles and then the molar concentration of each in the solubility-equilibrium expression.
d. Determine whether a precipitate will form.

15. Follow the same sequence of steps outlined

in Problem 14 to determine whether a precipitate of $CaCO_3$ will form if 0.10 L of 0.025 M $Ca(NO_3)_2$ is mixed with 0.10 L of 0.0031 M Na_2CO_3 at 25°C. (See Table 19-4 for K_{sp} values.)

16. The ionic substance X_2Y ionizes to form X^+ and Y^{2-} ions. The solubility of X_2Y is 5.54×10^{-7} mol/L. What is the value of the solubility-product constant?

17. The ionic substance T_3U_2 ionizes to form T^{2+} and U^{3-} ions. The solubility of T_3U_2 is 3.77×10^{-20} mol/L. What is the value of the solubility product constant?

18. A saturated solution of AgI at 25°C contains 2.7×10^{-10} mol/L Ag^+. Find its I^- concentration.

19. Calculate whether a precipitate will form if 0.35 L of 0.0044 M $Ca(NO_3)_2$ and 0.17 L of 0.000 39 M NaOH are mixed at 25°C. (See Table 19-4 for K_{sp} values.)

20. Determine whether a precipitate will form if 70.0 mL of a 0.0040 M $Ca(NO_3)_2$ solution is mixed with 30.0 mL of a 0.010 M Na_2SO_4 solution at 25°C.

21. Determine whether a precipitate will form if 1.70 g of solid $AgNO_3$ and 14.5 g of solid NaCl are dissolved in 200. mL of water to form a solution at 25°C.

22. If 2.50 g of solid $Fe(NO_3)_3$ are added to 100. mL of a 1.0×10^{-20} M NaOH solution, will a precipitate form?

CRITICAL THINKING

1. **Predicting Outcomes** When gasoline burns in an automobile engine, nitric oxide is formed from oxygen and nitrogen. Nitric oxide is a major air pollutant. High temperatures are needed for the reaction, like those found in a combustion engine. The reaction is

$$N_2(g) + O_2(g) \leftrightarrow 2\ NO(g)$$

K for the reaction is 0.01 at 2000°C. If 4.0 moles of N_2, 0.1 moles of O_2, and 0.08 moles of NO are placed in a 1.0 L vessel, predict which reaction will be favored.

2. **Organizing Data** Examine the following K_{sp} values for each substance.

    $PbCrO_4$ 2.0×10^{-16}
    $Pb(OH)_2$ 1.2×10^{-15}
    $Zn(OH)_2$ 4.5×10^{-17}
    MnS 2.3×10^{-13}

Sequence the substances from greatest to least molar solubility.

CROSS CONNECTIONS

1. **Health** Research on the prevention of tooth decay has taken on a new dimension, thanks to equilibria and solubility. Tooth enamel is composed of the mineral hydroxyapatite, $Ca_5(PO_4)_3OH$, which has a K_{sp} of 6.8×10^{-37}. The molar solubility of hydroxypapatite is 2.7×10^{-5} mol/L. When hydroxyapatite is reacted with fluoride, the OH^- is replaced with the F^- ion on the mineral, forming fluorapatite, $Ca_5(PO_4)_3F$. The K_{sp} of fluorapatite is 1×10^{-60}. Calculate the solubility of fluorapatite in water. Given your calculations can you support the use of fluoridation of drinking water?

RESEARCH AND WRITING

1. Carry out library research on the formation of stalagmites and stalactites in caves. Investigate the equilibria processes involved in the formation of stalagmites and stalactites.

2. **a.** Investigate the life and work of the French scientist Henri Le Châtelier. How did he apply the scientific method to arrive at the principle which is named after him? What other important contributions did Le Châtelier make to science and technology?

   b. Carry out library research on Le Châtelier's principle. Explain how the principle can be determined using the laws of thermodynamics which you studied in Chapter 18.

CHAPTER 20
Oxidation-Reduction Reactions

PLANNING CHART

Approximate Time: 9/11 days

	SECTION/TEXT FEATURES		RESOURCES	
B A S I C	**20.1 THE NATURE OF OXIDATION AND REDUCTION** Demo 1 Desktop Investigation Section Review Chapter Review • Questions 1–3, Problems 1–2	597 596B 600 602 623	*Teacher's Resource Binder* Lab Experiment 31 Section Review Worksheets *Other Resources* Blackline Master 47	169 71
A V E R A G E	**20.2 BALANCING REDOX EQUATIONS** Section Review Chapter Review • Problems 3–9	602 607 623	*Teacher's Resource Binder* Section Review Worksheets	72
A V E R A G E	**20.3 OXIDIZING AND REDUCING AGENTS** Demo 2 Section Review Chapter Review • Questions 4–6	608 596B 611 623	*Teacher's Resource Binder* Lab Experiments 32, 33 Section Review Worksheets *Other Resources* *Shakhashiri* Video Demo 34 Blackline Master 48	173, 179 73
H O N O R S	**20.4 ELECTROCHEMISTRY** Demos 3, 4, 5, 6 *Periodic Table* Videodisc Section Review Chapter Review • Questions 7–12, Problems 10–15	611 596C–D 596H 622 623	*Teacher's Resource Binder* Problem Solving Worksheets Section Review Worksheets *Other Resources* *World of Chemistry* Video 15, 25 Transparencies 45, 46, 47, 48 Blackline Master 49	74
	CHAPTER END ACTIVITIES Critical Thinking, Cross Connections, Research and Writing, Projects		*Teacher's Resource Binder* Test Booklet *Computer Test Generator*	97, 99

Teaching Strategies

20.1 The Nature of Oxidation and Reduction

Here is an opportunity to emphasize that many chemical reactions involve a redistribution of electrons. In ionic species, we assume that an electron has been removed or added to an atom resulting in a protonic charge (+) or an electronic charge (−). In a molecular species, oxidation numbers are assigned according to the polarity of covalent bonds. For example, in water (H-O-H) the electron density of the O–H bond is greater towards the oxygen. Therefore, the oxygen is assigned an oxidation state of negative one, and the hydrogen is assigned an oxidation state of +1. The periodate ion, IO_4^-, has a charge of −1 because it has gained an electron from the environment and the total number of electrons exceeds the total number of protons by one (85 protons, 86 electrons). In this polyatomic ion, the electron density of the I–O bond is greater towards the oxygen. Therefore, each oxygen is assigned an oxidation state of −2, and the central iodine is assigned an oxidation state of +7. If IO_4^- is reduced to IO_3^-, the iodine will be assigned an oxidation state of +5.

$KMnO_4$ is an excellent oxidizing agent that goes through a series of color changes as it is reduced. The variety of colors produced in this reduction illustrates that a single element can have many oxidation states. Repeat Demonstration 3 on manganese described in the Chapter 7 Teaching Strategies. Incorporate this into your lesson explaining the rules that govern the assignment of oxidation numbers.

Demonstration 1: The Oxidation of I^-(aq) by MnO_4^-(aq)

PURPOSE: To observe evidence of a spontaneous redox reaction between a polyatomic ion and a monoatomic ion in aqueous solution

MATERIALS: 100-mm petri dish, 7 mL 0.1 M KI, 7 mL 0.01 M $KMnO_4$, 1 M HCl, starch (optional)

PROCEDURE: Add approximately 7 mL 0.1 M KI to a petri dish on an overhead projector. Add 0.01 M $KMnO_4$ to the same petri dish and mix the two solutions with a stirring rod. Add a few drops of 1.0 M HCl to the solutions. Stir.

The reaction will not proceed until the acid is added. Within a few minutes of adding HCl, the purple of the $KMnO_4$ will begin to fade. The appearance of an orange-brown precipitate indicating the presence of MnO_2 will develop along with the red-brown solution indicative of dissolved I_2. Spraying with starch or adding a few grains of soluble powdered starch will give positive evidence for the presence of iodine.

This simple demonstration illustrates many of the topics covered in this chapter. Students will observe that redox reactions are not limited to monoatomic ions and metallic elements. Additionally, they will see elements in a variety of oxidation states and they will observe that a reduction is accompanied by an oxidation. In this chapter, they will learn that redox reactions accompanied by the loss or gain of oxygen by a polyatomic ion will require the presence of acid or base. This point is illustrated by the above demonstration since the addition of acid (or base) is a requirement for the spontaneity of this reaction.

Equation:

$$2MnO_4^-(aq) + 6I^-(aq) + 8H^+(aq) \rightarrow 2MnO_2 + 3I_2 + 4H_2O$$

Many of the reactions demonstrated before in Chapters 7, 8, and 9 can be repeated now with emphasis on the transfer of electrons from reducing agent to oxidizing agent. Repeat some of the Activity Series reactions of Chapter 8. This time, refer to the orbital origin and destination of the electrons being transferred.

$Ni(s) \rightarrow Ni^{2+}(aq) + 2e^-$ (from s orbital at 4th energy level)

$2e^-$ (to s orbital at 4th energy level) $+ Cu^{2+}(aq) \rightarrow Cu(s)$

The electrolysis of water and the oxidation of alcohols can be used effectively to illustrate redox phenomena.

Desktop Investigation: Redox Reactions

This investigation gives students an opportunity to observe the oxidation and reduction of some familiar household items. Make sure to schedule this investigation so that students can follow the reactions over a twenty-four hour period. It is a good time to talk about rust and the dangers of allowing steel bridges to go unpainted. Experiment 32, *Factors that Influence Corrosion of Iron*, also illustrates this material. All solutions may be washed down the drain once they have been diluted with a ten-fold excess of water, provided your school is connected to a sanitary sewer system and treatment plant.

20.2 Balancing Redox Equations

Balancing equations is one of the primary tools used in chemistry. Accounting for electron shifts in half-reactions can simplify balancing procedures. Because students have only been exposed to simple reactions, they might resist learning redox methods for balancing. Give your students a reaction that appears to be really difficult initially, then help them to balance it with the methods carefully outlined in this section. Make sure to point out that mass and charge must be conserved; that the number of electrons lost by the reducing agent must equal the number of electrons gained by the oxidizing agent. Most of the reactions in this chapter include hydrogen or hydroxyl ions from water when needed. If they are not specified, hydrogen or hydroxyl ions from water can always be added if they are required to conserve the number of oxygen atoms or the charge in a reaction.

20.3 Oxidizing and Reducing Agents

Actually doing an electrochemical titration or diagramming it on the chalkboard is a good way to introduce the concept of equivalent mass. Make sure to point out that the equivalent mass of an element depends on the reaction in which it is participating. If Fe^0 is oxidized to Fe^{3+}, its equivalent mass is 55.8/3 g. If Fe is reduced from Fe^{3+} to Fe^{2+}, its equivalent mass is 55.8/1 g.

Equivalent Mass Addition

The reduction of a permanganate ion with oxalic acid can be demonstrated qualitatively or quantitatively. In this reaction, the permanganate ion is reduced to manganese dioxide. The manganese is reduced from +7 to +4; therefore, the equivalent weight of potassium permanganate is calculated by dividing the molar mass (158) of potassium permanganate by 3.

$$3H_2C_2O_4 + 2MnO_4^- \rightarrow$$
$$6CO_2 + 2OH^- + 2H_2O + 2MnO_2$$

Emphasize that the equivalent weight of a substance is dependent on the number of electrons lost or gained in a specific reaction. If potassium permanganate had been reduced to MnO_4^{2-}, the manganese is reduced from +7 to +6; therefore, the equivalent mass of potassium permanganate in this reaction would be 158.

Demonstration 2: The Reaction of Permanganate Ion with Oxalic Acid

PURPOSE: To show the reduction of the permanganate ion

MATERIALS: Overhead projector and screen, 50-mL beaker, 0.300 M oxalic acid, 0.01 M potassium permanganate, recently prepared (within the previous two weeks)

PROCEDURE: Add 10 mL of each solution to a 50-mL beaker on an overhead projector.

The characteristic purple color of the permanganate ion will fade within a few minutes as the ion is reduced. You can use this reaction qualitatively to introduce equivalent mass. This reaction is also the basis of an interesting rate study described in HRM's "Experiments in Colorimetry," which could be assigned to a student who would like to do some independent work.

Potentiometric titrations are usually not used as class laboratory investigations in a first year course, but could be used effectively as a demonstration or as an independent study. The $Fe^{2+}/Cr_2O_7^{2-}$ and the Fe^{2+}/Ce^{4+} systems can be studied with the help of the HRM software package "Experiments in Chemistry," Experiment 15.

The reactions involved can serve as illustrations for equivalent mass calculations.

$$6Fe^{2+} + Cr_2O_7^{2-} + 14H^+ \rightarrow$$
$$6Fe^{3+} + 2Cr^{3+} + 7H_2O$$
$$Fe^{2+}(aq) + Ce^{4+}(aq) \rightarrow$$
$$Fe^{3+} + Ce^{3+}$$

In both reactions, iron is the reducing agent. One mole of iron donates one mole of electrons to the oxidizing agent; therefore, the equivalent mass of iron is the molar mass of iron (55.8) divided by 1.

In the chromate reaction, each mole of chromium atoms receives 3 electrons, therefore, the equivalent mass of chromium in this reaction is the molar mass of chromium (52.0) divided by 3.

In the cerium reaction, since each mole of cerium receives one mole of electrons, the equivalent mass of cerium is the molar mass of cerium (140) divided by 1.

20.4 Electrochemistry

While chemistry students can acquire the ability to handle the algebraic manipulations concerning electrochemical cells, they often cannot relate that algebraic ability to understanding the fundamental chemical principles or the applications of electrochemistry. The best way to make sure that students can relate algebraic manipulations to chemical phenomena is to continue to present demonstrations and hands-on experiences in tandem with presentation of theoretical material.

The microchemical techniques that have been developed by Maloney, Lewis, Russo, and others enable you to introduce experiments that require very small amounts of material and reduced laboratory time. Microcomputer-based laboratory approaches developed by Tinker and by others allow students to convert the computer into a sophisticated scientific instrument capable of tracking redox reactions.

Electrical energy is released from the spontaneous flow of electrons from a substance that has less affinity for electrons to a substance that has a greater affinity. The greater the difference in attraction, the greater the energy released. This chemical process is the basis of the electrochemical cell. If electrons are going to flow from a substance that has greater affinity for electrons to a substance that has less affinity for electrons, the reaction will not be spontaneous and energy will have to be added to the chemical system. This process is the basis of the electrolytic cell.

An understanding of the fundamental process of electron flow is necessary for an understanding of electrochemistry. The flow of electrons from one substance to another can be demonstrated with single-replacement reactions in which one metal replaces a metallic ion in solution with the concomitant reduction of the ion to its neutral state. Select a few of the reactions used in the Chapter 8 Activity Series demonstration to illustrate the chemical phenomona underlying potential differences and electric current.

Students will be in a better position to understand electrochemical and electrolytic cells if they have an unambiguous understanding of a cell reaction. Once that firm foundation is laid, they can apply their understanding of electrochemistry to batteries, photovoltaic cells, electroplating and semiconductors.

After you have demonstrated the spontaneous reduction and oxidation that occurs in a simple replacement reaction, explain that the two half reactions, oxidation and reduction, have to be physically separated for the energy released to be harnessed to do work.

Ronald Ulrich has adapted an idea (from an article by Earl Zwicker) in developing two unique demonstrations based on the principles of potential difference, current, and redox reactions that will engage your students' interest and enthusiasm while utilizing electrochemical energy to do work.

Demonstration 3: Bio-Clock

PURPOSE: To show that electrochemical reactions can take place in nature

MATERIALS: small battery clock (the type that sticks on the dash of your car), 3 short connecting wires, soldering gun, solder, flux, 2 copper wires or flattened 1/8" copper strips, 2 zinc strips (1/4" × 5 cm) (or a zinc-coated nail), 2 apples

PROCEDURE: The clock can be separated by inserting a screwdriver into the groove and twisting. Notice where the + part of the battery is connected. Use a small Phillips screwdriver to remove the retaining metal piece. Solder the copper wires to the appropriate poles + and −. Solder one of the wires to one of the copper electrodes, the other to one of the zinc electrodes. Solder one end of the remaining wire to a copper strip, the other end to a zinc strip. Insert the two electrodes coming from the clock into two different apples. Insert the remaining wire (zinc and copper) next to the electrodes in the apples approximately 1/4" apart (the copper next to the zinc and the zinc next to the copper). Set the clock according to the directions that came with it. The clock is surprisingly accurate.

Demonstration 4: Screaming Lemon

PURPOSE: To show the electrochemical properties of a lemon
MATERIALS: Piezo sounder, 1.5-24 V d-c, 20 mA consumption; 1 zinc strip, 1/4" × 5 cm; 1 copper strip, 1/4" × 5 cm; two connecting wires, soldering iron, solder, flux, lemon
PROCEDURE: Solder the + and − ends of the Piezo Sounder to separate connecting wires. Solder the + wire to the copper strip. Solder the − wire to the zinc strip. Place the two strips approximately 1/4" apart into the lemon.

In addition to presenting these demonstrations, build a Daniell Cell, which is based on the spontaneous Zn/Cu^{2+} reaction that the class has seen before.

Demonstration 5: The Daniell Electrochemical Cell

PURPOSE: To build an electrochemical cell that is based on the spontaneous Zn/Cu^{2+} reaction and to measure the potential difference between the zinc and copper half-cells
MATERIALS: copper and zinc strips, approximately 4 × 150 mm; steel wool; 250-mL beaker; porous cup, 35 mm × 80 mm; 0.1 M solutions of copper nitrate (or copper sulfate) and zinc nitrate; voltmeter, 0-5 V −/+ 0.01 V; leads with alligator clips
PROCEDURE: Clean the metal strips with steel wool. Fill the porous cup 3/4 full with 0.1 M copper ion solution. Place the porous cup in the beaker and fill the beaker 3/4 full with 0.1 M zinc nitrate. Place the copper strip in the copper solution

and the zinc strip in the zinc solution. Connect the leads from the voltmeter to the two electrodes. If no reading is recorded on the voltmeter after adjusting the sensitivity, reverse the leads. Record the voltage.

A transparent voltmeter placed on an overhead projector will allow students to observe the measurable potential difference between the zinc and copper half-cells. Since students know from previous experience that the zinc is oxidized and the copper is reduced, they are in a good position to understand the terms *anode* (the electrode where oxidation takes place) and *cathode* (the electrode where reduction takes place). Once students are introduced to the Daniell cell, they can understand the chemistry fundamental to all batteries.

Encourage students to translate their understanding of electrochemical cells to electrolytic cells. In an electrochemical cell, electrons flow spontaneously, free energy is released, and that energy can be measured and utilized. In an electrolytic cell, energy is required to force electrons "uphill"; that is, to force electrons from a source that has a higher affinity for electrons to an electrode that has a lower affinity.

Carefully define the locations at which important events occur. The electrolysis of potassium iodide can attract students' attention and will illustrate the electrode reactions.

Demonstration 6: The Electrolysis of Potassium Iodide

PURPOSE: To illustrate electrode reactions and to define the locations at which electrochemical events occur

MATERIALS: 100-mm petri dish, overhead projector, screen, 0.1 M KI, 6 V/12 V battery charger, two electrodes, phenolphthalein, starch solution

EQUATIONS:

Anode: $2I^- \rightarrow I_2 + 2e^-$

$I^- + I_2 \rightarrow I_3^-$ (brown color)

Cathode: $2e^- + 2H_2O \rightarrow H_2(g) + 2OH^-$

PROCEDURE: Place the petri dish on the overhead projector. Add 20 mL 0.1 M KI solution. Add a few drops of phenolpthalein, and a few drops of starch solution to the KI solution. Attach electrodes to insulated leads of battery charger. Turn battery charger to 12 V (6 V if you want a slower reaction). Hold electrodes in KI solution.

I_2 can be detected at the anode by adding starch and detecting the characteristic blue-black color that develops. H_2 bubbles can be detected at the cathode. OH^- can be detected in solution near the cathode by adding a few drops of phenolpthalein.

Electrons are being removed from the iodide ions at the anode where they are oxidized to I_2. The electrons are being pumped from the anode, causing it to be positively charged. These electrons are being forced through an external circuit to the cathode where the hydrogen ions of the water successfully compete for the electrons with the sodium ions. Reduction of the hydrogen ions takes place at the cathode, which is the negative electrode because electrons have been pumped to it by the external energy source.

In both the electrolytic cell and the electrochemical cell, a closed circuit and electroneutrality must be maintained during the electrochemical reaction.

In both cases, the requirements for continuous conductivity in the internal circuit and electroneutrality in the cells are fulfilled by migrations of ions. In an electrolytic cell, when a salt solution, such as potassium iodide, is electrolyzed, there is a migration of the cations K^+ and H^+ and of the anions I^- and OH^-. In an electrochemical cell, a pathway must be provided for electrons to flow from the anode to the cathode. This pathway is provided by the external circuit, usually a copper wire. A pathway must also be provided for ions to flow between the electrolyte solutions which surround the electrodes. This is the internal circuit. The pathway can be provided by building one half-cell in a porous cup and placing that within the other half-cell. Alternatively, the pathway can be provided by an electrolyte bridge connecting two half-cells that are physically separated. The traditional bridge consists of a glass tube packed with glass wool and an ammonium nitrate solution. Hot dogs or pickles also make unique conducting devices.

References and Resources

Books and Periodicals

Baca and Lewis. "Electrochemistry in a Nutshell," *Journal of Chemical Education*, Vol. 55, No. 12, 1978, p. 804.

Campbell, J.A. "Kinetics—Early and Often," *Journal of Chemical Education*, Vol. 40, No. 11, 1963, pp. 578–583.

Ensman, E., T.R. Hacker, and R.A.D. Wentworth. "Vegetable Voltage and Fruit 'Juice': An Electrochemical Demonstration," *Journal of Chemical Education*, Vol. 65, No. 8, 1988, p. 727.

Gilbert, "Electrical Energy From Cells — A Corridor Demonstration," *Journal of Chemical Education*, Vol. 57, No. 3, 1980, p. 216.

Heldeman, Stephen, "The Electrolysis of Water: An Improved Demonstration Procedure," *Journal of Chemical Education*, Vol. 63, No. 9, 1986, p. 809.

Kam, T.T. "On the Nernst Equations of Mercury-Mercuric Oxide Half-Cell Electrodes," *Journal of Chemical Education*, Vol. 60, No, 2, 1983, p. 133.

Koch, Klaus R. "Oxidation by Mn_2O_7: An Impressive Demonstration of the Powerful Oxidizing Property of Dimanganeseheptoxide," *Journal of Chemical Education*, Vol. 59, No. 11, 1982, pp. 973–974.

Lipeles, Enid S. "The Chemical Contributions of Amadeo Avogadro," *Journal of Chemical Education*, Vol. 60, No. 2, 1983, pp. 127–128.

MacBeath, Marie E. and Andrew L. Richardson, "Tomato Juice Rainbow: A Colorful and Instructive Demonstration," *Journal of Chemical Education*, Vol. 63, No. 12, 1986, pp. 1092–1094.

Maloney, Barb and Robert Lewis. "Electrolysis of Potassium Iodide," unpublished, taken from Manjkow, Joseph and Dana Levine, "Electrodeposition of Nickel on Copper," *Journal of Chemical Education*, Vol. 63, No. 9, 1986, p. 809.

Microscaled Chemistry Workshop, ISTA, 1987.

Ophardt, Charles E. "Redox Demonstrations and Descriptive Chemistry Part 2: Halogens," *Journal of Chemical Education*, Vol. 64, No. 9, 1987, p. 807.

Parker, "50–Minute Experiment, An Appropriate Determination of the Avogadro Constant," *Journal of Chemical Education*, Vol. 57, No. 10, 1980, p. 735.

Pearson, Robert S. "Manganese Color Reactions," *Journal of Chemical Education*, Vol. 65, No. 5, 1988, pp. 451-452.

Russo, Tom, *Microchemistry for High School General Chemistry*, Educational Corp.,1985, (esp. Module 9: Electrochemistry).

"State of the Art Symposium: Electrochemistry," *Journal of Chemical Education*, Vol. 60, No. 4, 1983.

Stevens, George H., "Teaching Suggestions for Oxidation-Reduction," excerpted from *Journal of Chemical Education*, Feb. 1979.

Stroud, Linda M. and Denise Creech, "Effects of NO_2 and SO_2 on Plant Pigments," presented at the Southeast Regional Meeting of NSTA, Dec. 1988.

Ulrich, Ronald, "Bio Clock" and "Screamin' Lemon," adapted from Earl Zwicke (ed.), *The Physics Teacher*, Jan. 1988.

Weissman, "Batteries: The Workhorses of Chemical Energy," *CHEM*, Vol. 45, Oct. 1972.

Williams, L. Pearce, "Andre-Marie Ampère," *Scientific American*, Vol. 260, No. 1, 1989, pp. 90–97.

Computer Software

"Experiments in Colorimetry," Apple II, HRM.

Tinker, Robert F. and Diana Malone, "Experiment 12: Basic EMF Measurements – Fundamentals of Electrochemistry," Apple II, IBM, COMPress.

"Topic 7: REDOX-Acids and Bases," Apple II, Knowledge Factory.

"Topic 7: Electrochemistry Part I: Oxidation/Reduction, Part II: The Nernst Equation, The Relationship between delta E, delta G, and K, and the Activity Series," Apple II, Knowledge Factory.

Weyh, John A., Joseph R. Crook, and Les N. Hauge, "Oxidation-Reduction Reactions," IBM, COMPress.

Answers and Solutions

Reviewing Concepts

1. **a.** Oxidation processes are reactions in which the atoms or ions of an element attain a more positive (or less negative) oxidation state. Reactions in which the atoms or ions of an element attain a more negative (or less positive) oxidation state are described as reduction processes.
   b. Oxidation: $Na \rightarrow Na^+ + e^-$
   Reduction: $Cl_2 + 2e^- \rightarrow 2Cl^-$

2. a, b, c, f, g, i

3.

Equation	Oxidized	Reduced
a	Na	Cl_2
b	C	O_2
c	O^{2-}	H^+
f	O^{2-}	Cl^{5+}
g	H_2	Cl_2
i	Zn	Cu^{2+}

4. **a.** Lithium is the most active reducing agent of all the common elements.
   b. Due to their large atomic radii, the Group-1 metals have weak attraction for their valence electrons and form positive ions readily, thus making those lost electrons available to be gained by other substances that then undergo reduction.
   c. Fluorine is the most active oxidizing agent among the common elements.

5. **a.** Strongest = Ca; weakest = Cl^-
   b. Strongest = Al; weakest = Br^-
   c. Strongest = Na; weakest = F^-
   d. Strongest = NO_3^-; weakest = K^+
   e. Strongest = Cl_2; weakest = Zn^{2+}
   f. Strongest = F_2; weakest = Li^+

6. **a.** Yes **b.** Yes **c.** No
   d. Yes **e.** No

7. **a.** In an electrochemical cell, the redox reaction is spontaneous and the system of electrodes and electrolytes is used to transmit the electrical energy produced by the reaction. In an electrolytic cell the redox reaction is not spontaneous and thus current is used to drive the reaction through the electrodes and electrolytes used.
   b. In an electrochemical cell, the anode is the negative electrode, while the cathode is the positive electrode. In an electrolytic cell, the anode is the positive electrode, while the cathode is the negative electrode.

8. The specific products are determined by the relative ease with which the different particles present can be oxidized or reduced.

9. **a.** Electroplating is an electrolytic process in which a metal is deposited on a surface.
   b. The object to be plated serves as the cathode, while the plating metal is the anode.

10. **a.** The potential difference between the two electrodes is a measure of the energy required to move a certain electric charge between the electrodes.
    b. This difference can be measured by a voltmeter connected across the two electrodes, and is given in volts (V).

11. Whereas the potential difference across a voltaic cell is easily measured, there is no direct way to measure an individual electrode potential. By assigning an arbitrary potential to the standard reference electrode, a specific potential can then be determined for the other electrode of the complete cell made by combining the two half-cells involved.

12. **a.** A given half-cell potential is an indication of the tendency of the substance to undergo reduction.
    b. The larger the value of the potential for a given half-reaction, the greater is the tendency for it to undergo reduction and the less likely it is to undergo oxidation. The smaller the potential, the less the tendency for the substance to undergo reduction and the greater is its tendency to undergo oxidation instead.

Problems

1. a. Oxidation: $K \rightarrow K^+ + e^-$
  b. Reduction: $S + 2e^- \rightarrow S^{2-}$
  c. Oxidation: $Mg \rightarrow Mg^{2+} + 2e^-$
  d. Oxidation: $2F^- \rightarrow F_2 + 2e^-$
  e. Oxidation: $H_2 \rightarrow 2H^+ + 2e^-$
  f. Reduction: $O_2 + 4e^- \rightarrow 2O^{2-}$
  g. Reduction: $Fe^{3+} + e^- \rightarrow Fe^{2+}$
  h. Oxidation: $Mn^{2+} \rightarrow Mn^{7+} + 5e^-$

2. a. $\overset{0}{H}_2$ **b.** $\overset{+1\ -2}{H_2O}$ **c.** $\overset{0}{Al}$ **d.** $\overset{+2\ -2}{MgO}$
  e. $\overset{+3\ -2}{Al_2S_3}$ **f.** $\overset{+1+5-2}{HNO_3}$ **g.** $\overset{+1+6-2}{H_2SO_4}$ **h.** $\overset{+2-2+1}{Ca(OH)_2}$
  i. $\overset{+2+5-2}{Fe(NO_3)_2}$ **j.** $\overset{0}{O_2}$

3. a. $Fe + HCl \rightarrow FeCl_2 + H_2$
  Fe is oxidized; H^+ is reduced

  b. $\overset{0}{Fe} + \overset{+1-1}{HCl} \rightarrow \overset{+2-1}{FeCl_2} + \overset{0}{H_2}$

  c. Oxidation: $\overset{0}{Fe} \rightarrow \overset{+2}{Fe^{2+}} + 2e^-$
  Reduction: $2\overset{+1}{H^+} + 2e^- \rightarrow \overset{0}{H_2}$

  d. $Fe + 2HCl \rightarrow FeCl_2 + H_2$

4. a. $\overset{0}{Li} + \overset{+1\ -2}{H_2O} \rightarrow \overset{+1-2+1}{LiOH} + \overset{0}{H_2}$

  $2[\overset{0}{Li} \rightarrow \overset{+1}{Li^+} + e^-]$
  $2\overset{+1}{H^+} + 2e^- \rightarrow \overset{0}{H_2}$
  $2Li + 2H_2O \rightarrow 2LiOH + H_2$

  b. $\overset{0}{Al} + \overset{+1-1}{HCl} \rightarrow \overset{+3-1}{AlCl_3} + \overset{0}{H_2}$
  $2[\overset{0}{Al} \rightarrow \overset{+3}{Al^{3+}} + 3e^-]$
  $3[2\overset{+1}{H^+} + 2e^- \rightarrow \overset{0}{H_2}]$
  $2Al + 6HCl \rightarrow 2AlCl_3 + 3H_2$

  c. $\overset{+1-2}{H_2S} + \overset{+1+5-2}{HNO_3} \rightarrow \overset{+1+6-2}{H_2SO_4} + \overset{+4-2}{NO_2} + \overset{+1-2}{H_2O}$
  $8[\overset{+5}{N} + e^- \rightarrow \overset{+4}{N}]$
  $\overset{-2}{S} \rightarrow \overset{+6}{S} + 8e^-$
  $H_2S + 8HNO_3 \rightarrow H_2SO_4 + 8NO_2 + 4H_2O$

  d. $\overset{0}{Cu} + \overset{+1+5-2}{HNO_3} \rightarrow \overset{+2\ +5-2}{Cu(NO_3)_2} + \overset{+2-2}{NO} + \overset{+1-2}{H_2O}$
  $3[\overset{0}{Cu} \rightarrow \overset{+2}{Cu^{2+}} + 2e^-]$
  $2[\overset{+5}{N} + 3e^- \rightarrow \overset{+2}{N}]$
  $3Cu + 8HNO_3 \rightarrow 3Cu(NO_3)_2 + 2NO + 4H_2O$

  e. $\overset{+1+5-2}{KClO_3} \rightarrow \overset{+1-1}{KCl} + \overset{0}{O_2}$

$2[\overset{+5}{Cl} + 6e^- \rightarrow \overset{-1}{Cl^-}]$
$3[2\overset{-2}{O} \rightarrow \overset{0}{O_2} + 4e^-]$
$2KClO_3 \rightarrow 2KCl + 3O_2$

5. $\overset{0}{Zn} + \overset{+1+5-2}{HNO_3} \rightarrow \overset{+2\ +5-2}{Zn(NO_3)_2} + \overset{-3+1+5-2}{NH_3NO_3} + \overset{+1-2}{H_2O}$

$\overset{+5}{N} + 8e^- \rightarrow \overset{-3}{N}$
$4[\overset{0}{Zn} \rightarrow \overset{+2}{Zn^{2+}} + 2e^-]$
$4Zn + 10HNO_3 \rightarrow 4Zn(NO_3)_2 + NH_4NO_3 + 3H_2O$

6. $\overset{0}{Zn} + \overset{+1\ +6-2}{Na_2CrO_4} + \overset{+1-2+1}{NaOH} \rightarrow \overset{+1\ +2-2}{Na_2ZnO_2} + \overset{+1\ +3-2}{NaCrO_2}$
$+ \overset{+1-2}{H_2O}$

$2[\overset{+6}{Cr} + 3e^- \rightarrow \overset{+3}{Cr}]$
$3[\overset{0}{Zn} \rightarrow \overset{+2}{Zn^{2+}} + 2e^-]$
$3Zn + 2Na_2CrO_4 + 4NaOH \rightarrow$
$3Na_2ZnO_2 + 2NaCrO_2 + 2H_2O$

7. a. $K + H_2O \rightarrow KOH + H_2$

  b. $\overset{0}{K} + \overset{+1-2}{H_2O} \rightarrow \overset{+1}{K^+} + \overset{-2+1}{OH^-} + \overset{0}{H_2}$

  c. $2\overset{+1-2}{H_2O} + 2e^- \rightarrow \overset{0}{H_2} + 2OH^-$

  d. $\overset{0}{K} \rightarrow \overset{+1}{K^+} + e^-$

  e. $2H_2O + 2e^- \rightarrow H_2 + 2OH^-$
  $2K \rightarrow 2\overset{+1}{K^+} + 2e^-$

  f. $2K + 2H_2O \rightarrow 2KOH + H_2$

8. a. $HI + HNO_2 \rightarrow NO + I_2 + H_2O$
  $\overset{+1}{H^+} + \overset{-1}{I^-} + \overset{+1}{H^+} + \overset{+3-2}{NO_2^-} \rightarrow \overset{+2-2}{NO} + \overset{0}{I_2} + \overset{+1-2}{H_2O}$
  $2\overset{-1}{I^-} \rightarrow \overset{0}{I_2} + 2e^-$
  $2[\overset{+3-2}{NO_2^-} + e^- + 2H^+ \rightarrow \overset{+2-2}{NO} + H_2O]$
  $2HI + 2HNO_2 \rightarrow 2NO + I_2 + 2H_2O$

  b. $FeCl_3 + H_2S \rightarrow FeCl_2 + HCl + S$
  $\overset{+3}{Fe^{3+}} + 3\overset{-1}{Cl^-} + \overset{+1-2}{H_2S} \rightarrow \overset{+2}{Fe^{2+}} + 2\overset{-1}{Cl^-} + \overset{+1}{H^+}$
  $+ \overset{-1}{Cl^-} + \overset{0}{S}$
  $2[\overset{+3}{Fe^{3+}} + e^- \rightarrow \overset{+2}{Fe^{2+}}]$
  $\overset{+1-2}{H_2S} \rightarrow \overset{0}{S} + 2e^- + 2H^+$
  $2FeCl_3 + H_2S \rightarrow 2FeCl_2 + 2HCl + S$

  c. $SbCl_5 + KI \rightarrow KCl + I_2 + SbCl_3$
  $\overset{+5-1}{SbCl_5} + \overset{+1}{K^+} + \overset{-1}{I^-} \rightarrow \overset{+1}{K^+} + \overset{-1}{Cl^-} + \overset{0}{I_2} + \overset{+3-1}{SbCl_3}$

$\overset{+5-1}{SbCl_5} + 2e^- \rightarrow \overset{+3-1}{SbCl_3} + 2\overset{-1}{Cl^-}$

$2\overset{-1}{I^-} \rightarrow \overset{0}{I_2} + 2e^-$

$SbCl_5 + 2KI \rightarrow 2KCl + I_2 + SbCl_3$

d. $Ca(OH)_2 + NaOH + ClO_2 + C \rightarrow$
$NaClO_2 + CaCO_3 + H_2O$

$\overset{+2}{Ca^{2+}} + 2\overset{-2+1}{OH^-} + \overset{+1}{Na^+} + \overset{-2+1}{OH^-} + \overset{+4-2}{ClO_2} + \overset{0}{C} \rightarrow$

$\overset{+1}{Na^+} + \overset{+3-2}{ClO_2^-} + \overset{+2}{Ca^{2+}} + \overset{+4-2}{CO_3^{2-}} + \overset{+1-2}{H_2O}$

$\overset{0}{C} + 6\overset{-2+1}{OH^-} \rightarrow \overset{+4-2}{CO_3^{2-}} + 4e^- + 3\overset{+1-2}{H_2O}$

$4[\overset{+4-2}{ClO_2} + e^- \rightarrow \overset{+3-2}{ClO_2^-}]$

$Ca(OH)_2 + 4NaOH + 4ClO_2 + C \rightarrow 4NaClO_2 + CaCO_3 + 3H_2O$

9. a. $\overset{+1+6-2}{H_2SO_4} + \overset{0}{Zn} \rightarrow \overset{+2+6-2}{ZnSO_4} + \overset{+1-2}{H_2S} + \overset{+1-2}{H_2O}$

$\overset{+6}{S} + 8e^- \rightarrow \overset{-2}{S}$

$4[\overset{0}{Zn} \rightarrow \overset{+2}{Zn} + 2e^-]$

$4Zn + 5H_2SO_4 \rightarrow 4ZnSO_4 + H_2S + 4H_2O$

b. $10\overset{+1}{H^+}(aq) + \overset{+6-2}{SO_4^{2-}} + 8e^- \rightarrow \overset{+1-2}{H_2S} + 4\overset{+1-2}{H_2O}$

$4[\overset{0}{Zn} \rightarrow \overset{+2}{Zn^{2+}} + 2e^-]$

$4Zn + 10H^+(aq) + SO_4^{2-} \rightarrow 4Zn^{2+} + H_2S + 4H_2O$

or $4Zn + 5H_2SO_4 \rightarrow 4ZnSO_4 + H_2S + 4H_2O$

10. a. $Cu \rightarrow Cu^{2+} + 2e^-$ $E° = -0.34$ V
$2[Ag^+ + e^- \rightarrow Ag]$ $E° = +0.80$ V
$Cu + 2Ag^+ \rightarrow Cu^{2+} + 2Ag$ $E° = +0.46$ V

b. $Cd \rightarrow Cd^{2+} + 2e^-$ $E° = +0.40$ V
$Co^{2+} + 2e^- \rightarrow Co$ $E° = -0.28$ V
$Cd + Co^{2+} \rightarrow Cd^{2+} + Co$ $E° = +0.12$ V

c. $Ni^{2+} + 2e^- \rightarrow Ni$ $E° = -0.23$ V
$2[Na \rightarrow Na^+ + e^-]$ $E° = +2.71$ V
$2Na + Ni^{2+} \rightarrow 2Na^+ + Ni$ $E° = +2.48$ V

d. $Br_2 + 2e^- \rightarrow 2Br^-$ $E° = +1.06$ V
$2I^- \rightarrow I_2 + 2e^-$ $E° = -0.54$ V
$Br_2 + 2I^- \rightarrow 2Br^- + I_2$ $E° = +0.52$ V

11. a. $Mg \rightarrow Mg^{2+} + 2e^-$ $E° = +2.38$ V
$Sn^{2+} + 2e^- \rightarrow Sn$ $E° = -0.14$ V
$Mg + Sn^{2+} \rightarrow Mg^{2+} + Sn$ $E° = +2.24$ V
The reaction does occur spontaneously.

b. $3[K \rightarrow K^+ + e^-]$ $E° = +2.92$ V
$Al^{3+} + 3e^- \rightarrow Al$ $E° = -1.71$ V
$3K + Al^{3+} \rightarrow 3K^+ + Al$ $E° = +1.21$ V
The reaction does occur spontaneously.

c. $2[Li^+ + e^- \rightarrow Li]$ $E° = -3.04$ V
$Zn \rightarrow Zn^{2+} + 2e^-$ $E° = +0.76$ V
$2Li^+ + Zn \rightarrow 2Li + Zn^{2+}$ $E° = -2.28$ V
The reaction does not occur spontaneously.

d. $Cu \rightarrow Cu^{2+} + 2e^-$ $E° = -0.34$ V
$Cl_2 + 2e^- \rightarrow 2Cl^-$ $E° = +1.36$ V
$Cu + Cl_2 \rightarrow Cu^{2+} + 2Cl^-$ $E° = +1.02$ V
The reaction does occur spontaneously.

12.

	Actual Values	Hypothetical Values
$Br_2 + 2e^- \rightleftarrows 2Br^-$	$+1.06$ V	$(1.06 V - 0.54 V) = 0.52$ V
$I_2 + 2e^- \rightleftarrows 2I^-$	$+0.54$ V	0 V
$Al^{3+} + 3e^- \rightleftarrows Al$	-1.71 V	$(-1.71 V - 0.54 V) = -2.25$ V

a. $+0.52$ V
b. -2.25 V
c. original = $+1.06$ V $- 0.54$ V $= +0.52$ V
new = $+0.52$ V $- 0 = +0.52$ V
No change would be expected and computations verify that none would occur.

13. $Ni \rightarrow Ni^{2+} + 2e^-$ $E° = +0.23$ V
$2[Ag^+ + e^- \rightarrow Ag]$ $E° = +0.80$ V
$Ni + 2Ag^+ \rightarrow Ni^{2+} + 2Ag$ $E° = +1.03$ V
The Ni would gradually enter the solution as Ni^{2+} ions.

14. a. $2[Al \rightarrow Al^{3+} + 3e^-]$ $E° = +1.71$ V
$3[Zn^{2+} + 2e^- \rightarrow Zn]$ $E° = -0.76$ V
$2Al + 3Zn^{2+} \rightarrow 2Al^{3+} + 3Zn$ $E° = +0.95$ V
The Al spoon would begin to disintegrate as its atoms entered the solution as Al^{3+} ions.

b. $3[Zn \rightarrow Zn^{2+} + 2e^-]$ $E° = +0.76$ V
$2[Al^{3+} + 3e^- \rightarrow Al]$ $E° = -1.71$ V
$3Zn + 2Al^{3+} \rightarrow 3Zn^{2+} + 2Al$ $E° = -0.95$ V
The Zn strip could be used since the negative $E°$ value indicates that no reaction would occur.

15. $2[Al \rightarrow Al^{3+} + 3e^-]$ $E° = +1.71$ V
$3[Sn^{2+} + 2e^- \rightarrow Sn]$ $E° = -0.14$ V
$2Al + 3Sn^{2+} \rightarrow 2Al^{3+} + 3Sn$ $E° = +1.57$ V
The Al container could not be used to store the $Sn(NO_3)_2$ solution because the two substances would react. Al would enter the solution as Al^{3+} ions, causing the container to disintegrate.

Critical Thinking

1. In rechargeable cells, oxidation-reduction reactions characteristic of electrochemical cells are used as a source of electrical energy through the conversion of chemical energy into electrical energy during the discharge cycle. When such cells are being charged, electric energy is

converted to chemical energy by the oxidation-reduction reactions characteristic of electrolytic cells.
2. As the lead/acid battery discharges, the sulfuric acid electrolyte is consumed. The following equation illustrates this process.
$Pb(s) + PbO_2(s) + 4H^+(aq) + 2SO_4^{2-}(aq) \rightarrow 2PbSO_4(s) + 2H_2O(\ell)$
Sulfuric acid has a high density. As it is used up the density of the electrolyte in the cell decreases. An instrument called a hydrometer is used to measure the density of the fluid.
3. The Cr^{3+} is the species that is oxidized.

Cross Connections

1. Answers will vary.

Videodisc Selections
from *The Periodic Table Videodisc* (JCE Software)

step forward

step reverse

36867
silver in batteries (s)

CHAPTER 20

pg 612 Batteries

pg 615 Electroplated materials

36336
lithium-containing batteries (s)

36886
bolts plated with cadmium (s)

36622
Leclanche cell (s)

37156
gold-plated circuit board (s)

36681
zinc in batteries (s)

Teacher's Notes

Chapter 20
Oxidation–Reduction Reactions

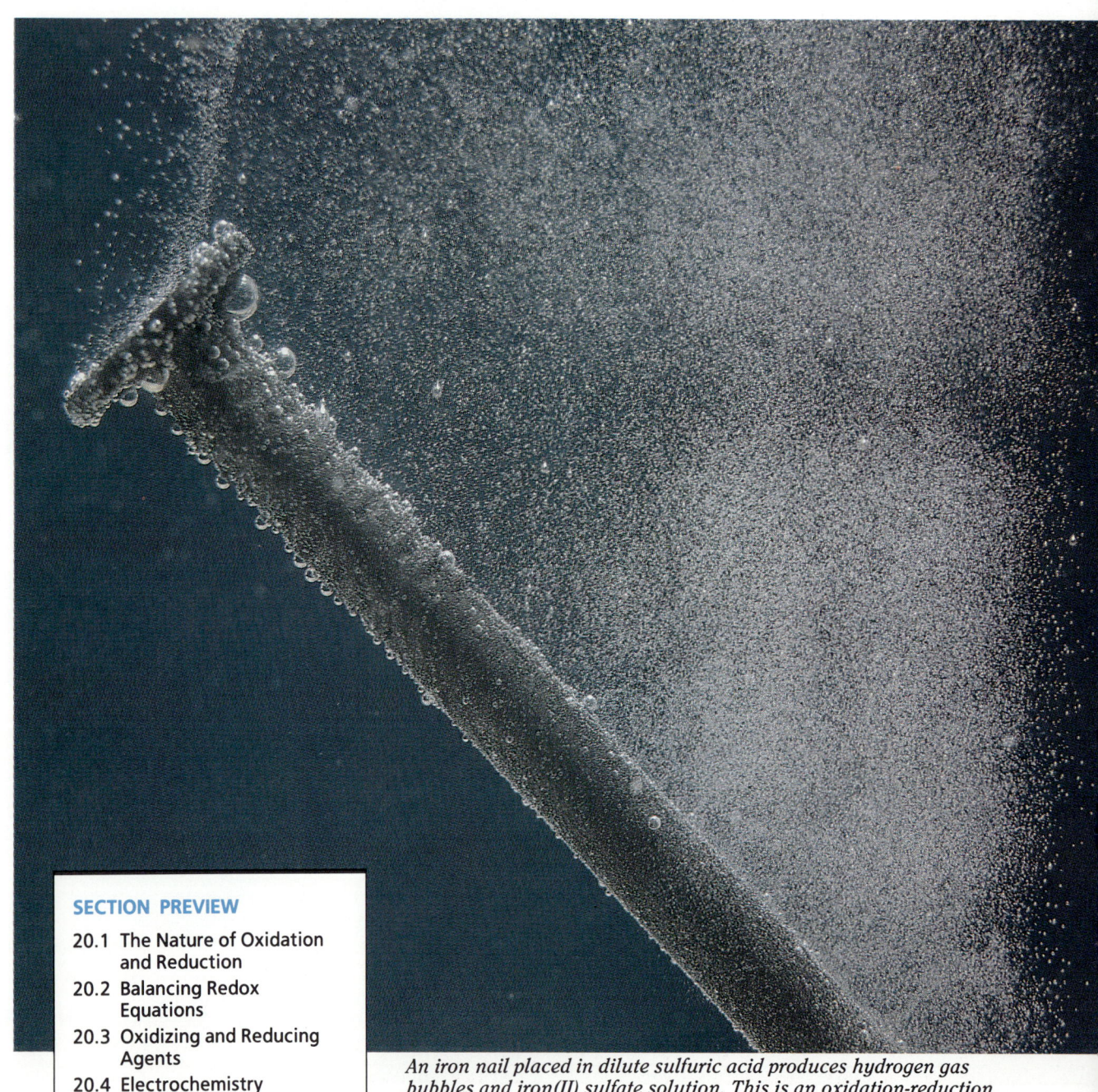

SECTION PREVIEW

20.1 The Nature of Oxidation and Reduction
20.2 Balancing Redox Equations
20.3 Oxidizing and Reducing Agents
20.4 Electrochemistry

An iron nail placed in dilute sulfuric acid produces hydrogen gas bubbles and iron(II) sulfate solution. This is an oxidation-reduction reaction.

20.1 The Nature of Oxidation and Reduction

In Chapter 7, you first learned about oxidation numbers, or numbers that are assigned to atoms and ions as a way of keeping track of electrons. Table 20-1 summarizes the set of rules by which oxidation numbers are assigned. In this section, you will learn to apply your knowledge of these numbers to an analysis of chemical reactions.

SECTION OBJECTIVES
- Define oxidation and reduction.
- Assign oxidation numbers to reactant and product species.
- Explain what is meant by oxidation–reduction reaction (redox reaction).

Oxidation

Reactions in which the atoms or ions of an element attain a more positive (or less negative) oxidation state were described in Section 7.2 as **oxidation** *processes. A species that undergoes a change in oxidation state in a positive mathematical direction is said to be* **oxidized**.

The combustion of metallic sodium in an atmosphere of chlorine gas illustrates the oxidation process (Figure 20-1). The product is sodium chloride, and the sodium–chlorine bond is ionic. The equation for this reaction is

$$2Na(s) + Cl_2(g) \rightarrow 2NaCl(s)$$

In this reaction, each sodium atom loses an electron and becomes a sodium ion.

$$Na \rightarrow Na^+ + e^-$$

The oxidation state of sodium has changed from the 0 state of the atom (Rule 1, Table 20-1) to the more positive +1 state of the ion (Rule 2). The sodium atom is said to be *oxidized* to the sodium ion. This change in oxidation state is indicated by the oxidation numbers placed above the symbol of the atom and the ion. For each sodium atom,

$$\overset{0}{Na} \rightarrow \overset{+1}{Na^+} + e^-$$

Figure 20-1 Sodium and chlorine react violently.

A second example of an oxidation process occurs when hydrogen burns in chlorine to form molecular hydrogen chloride gas. The hydrogen–chlorine bond is covalent. The equation for this reaction is

$$H_2(g) + Cl_2(g) \rightarrow 2HCl(g)$$

By Rule 1, the oxidation state of each hydrogen atom in the hydrogen molecule is 0. By Rule 5, the oxidation state of the combined hydrogen in the HCl molecule is +1. Because the hydrogen atom has changed from the 0 to the more positive +1 oxidation state, this change is an oxidation process.

$$\overset{0}{H_2} + Cl_2 \rightarrow 2\overset{+1}{HCl}$$

Reduction

Reactions in which the atoms or ions of an element attain a more negative (or less positive) oxidation state are recognized as **reduction** *processes. A species that undergoes a change in oxidation state in a negative sense is said to be* **reduced**. Consider the behavior of chlorine in the above reaction

> **TABLE 20-1 RULES FOR ASSIGNING OXIDATION NUMBERS**
>
> 1. An uncombined element has oxidation number zero.
> 2. A monatomic ion has an oxidation number equal to its charge.
> 3. Fluorine has oxidation number −1 in all compounds.
> 4. Oxygen has oxidation number −2 in almost all compounds (e.g., except in peroxides such as H_2O_2, where it is −1).
> 5. Hydrogen has oxidation number +1 in all compounds except those with metals, in which it has oxidation number −1 (e.g., in sodium hydride, NaH).
> 6. The more electronegative element in a binary compound is assigned the number equal to the charge it would have if it were an ion.
> 7. The algebraic sum of the oxidation numbers of all atoms in a neutral compound is zero.
> 8. The algebraic sum of the oxidation numbers of all atoms in a polyatomic ion is equal to the charge of the ion.

The reduced chlorine atom has 17 protons and 18 electrons.

with sodium and hydrogen. With sodium, each chlorine atom acquires an electron and becomes a chloride ion. The oxidation state of chlorine changes from the 0 state of the chlorine atom (Rule 1) to the more negative −1 state of the chloride ion (Rule 2). The chlorine atom is reduced to the chloride ion. For each chlorine atom,

$$\overset{0}{Cl} + e^- \rightarrow \overset{-1}{Cl^-}$$

In the reaction of chlorine with hydrogen, the pair of electrons shared by the hydrogen and chlorine atoms in the hydrogen chloride molecule is not shared equally. Rather, the pair of electrons is more strongly attracted to the chlorine atom because of its higher electronegativity. By Rules 5 and 6, chlorine is assigned an oxidation number of −1. Thus, the chlorine atom is changed from the 0 to the −1 oxidation state, a reduction process.

$$H_2 + \overset{0}{Cl_2} \rightarrow 2\overset{-1}{HCl}$$

Assignment of oxidation numbers to the atoms of covalent molecular species in this somewhat arbitrary way is useful in studying reactions. The numbers allow determination of the partial loss or gain of electrons in compounds that are not ionic. For the hydrogen chloride molecule, the oxidation number of the hydrogen atom is +1, and that of the chlorine atom is −1.

$$\overset{0}{H_2} + \overset{0}{Cl_2} \rightarrow 2\overset{+1\ -1}{HCl}$$

Decreases in charge or in oxidation number can be seen in terms of complete or partial gains of electrons. Reduction can thus be understood as a process in which electrons are gained.

Oxidation and Reduction as a Process

If oxidation occurs during a chemical reaction, then reduction must occur simultaneously. Furthermore, the amount of oxidation must match the amount of reduction. In other words, the total increase in oxidation number of the species being oxidized must equal the total decrease in oxidation number of

The number of electrons lost from the particles that have been oxidized must be gained by the particles that are reduced.

the species being reduced. *Any chemical process in which elements undergo a change in oxidation number is an* **oxidation–reduction reaction.** The name is often shortened to **"redox"** reaction.

Observe that as metallic sodium burns in chlorine gas, two sodium atoms are oxidized to Na^+ ions as one diatomic chlorine molecule is reduced to two Cl^- ions. Elemental sodium and elemental chlorine are chemically equivalent atom for atom and ion for ion. That is, for every electron lost by a sodium atom, one electron is gained by a chlorine atom.

$$\overset{0}{2Na} \rightarrow \overset{+1}{2Na^+} + 2e^- \quad \text{(oxidation)}$$
$$\overset{0}{Cl_2} + 2e^- \rightarrow \overset{-1}{2Cl^-} \quad \text{(reduction)}$$
$$\overset{0}{2Na} + \overset{0}{Cl_2} \rightarrow \overset{+1}{2Na^+} + \overset{-1}{2Cl^-} \quad \text{(redox reaction)}$$

Suppose a scheme is devised to recover metallic aluminum from an aluminum salt. In this scheme, sodium atoms are oxidized to Na^+ ions, and Al^{3+} ions are reduced to aluminum atoms. For these oxidation and reduction processes to be chemically equivalent, three Na^+ ions must be formed for each Al^{3+} ion reduced.

Figure 20-2 The combustion of antimony in chlorine is an oxidation–reduction reaction.

$$\overset{0}{3Na} \rightarrow \overset{+1}{3Na^+} + 3e^- \quad \text{(oxidation)}$$
$$\overset{+3}{Al^{3+}} + 3e^- \rightarrow \overset{0}{Al} \quad \text{(reduction)}$$
$$\overset{0}{3Na} + \overset{+3}{Al^{3+}} \rightarrow \overset{+1}{3Na^+} + \overset{0}{Al} \quad \text{(redox reaction)}$$

Observe that in the combined equation for each example, both electric charge and atoms are conserved, and that the oxidation and reduction processes are indeed equivalent.

Many of the reactions studied in elementary chemistry involve oxidation–reduction processes. Of the three examples considered so far in this chapter, two are composition reactions. (Figure 20-2 shows another composition redox reaction.) The third example is a replacement reaction. Not all composition reactions are oxidation–reduction reactions, however. For example, sulfur dioxide gas (SO_2) dissolves in water to form an acidic solution containing a low concentration of sulf*urous* acid (H_2SO_3).

$$\overset{+4-2}{SO_2} + \overset{+1-2}{H_2O} \rightarrow \overset{+1+4-2}{H_2SO_3}$$

Observe that the oxidation states of all elemental species remain unchanged in this composition reaction. It is therefore not a redox reaction.

When a solution of sodium chloride is added to a solution of silver nitrate, an ion-exchange reaction occurs, and a white precipitate of silver chloride separates.

$$\overset{+1}{Na^+} + \overset{-1}{Cl^-} + \overset{+1}{Ag^+} + \overset{+5-2}{NO_3^-} \rightarrow \overset{+1}{Na^+} + \overset{+5-2}{NO_3^-} + \overset{+1\,-1}{Ag^+Cl^-}$$

or, more simply, by omitting spectator ions,

$$\overset{+1}{Ag^+}(aq) + \overset{-1}{Cl^-}(aq) \rightarrow \overset{+1\,-1}{Ag^+Cl^-}(s)$$

Figure 20-3 As $KMnO_4$ solution is added to an acidic solution of $FeSO_4$, Fe^{2+} ions are oxidized to Fe^{3+} ions, and the red MnO_4^- ions are changed to colorless Mn^{2+} ions. When all Fe^{2+} ions are oxidized, MnO_4^- ions are no longer changed to colorless Mn^{2+} ions. Thus the first faint appearance of the MnO_4^- color indicates the end point of the titration.

Oxidation-Reduction Reactions

Desktop Investigation

Redox Reactions

Question
How do you recognize oxidation–reduction reactions in the laboratory?

Materials
aspirin, 325 mg tablet
liquid bleach, 5.25% NaClO solution
bunsen burner
forceps
H_2O_2 solution, 6 or 3%
manganese dioxide
stirring rod
steel wool
2 test tubes, 16 × 150 mm
5 test tubes, 25 × 150 mm
test tube clamp
white vinegar, 6% $HC_2H_3O_2$
water, distilled
wooden splints

Procedure
CAUTION: Goggles and apron should be worn during this investigation. Record all your results in a data table.
1. *(a)* Put 10 mL of hydrogen peroxide in a 16 × 150 mm test tube and add a small amount of manganese dioxide (equal to about half a pea). What is the result?
(b) After the reaction continues for 1 minute, test for oxygen by inserting a glowing wooden splint into the test tube. What is the result? (See diagram.)
2. Using forceps, heat a small roll of steel wool (about 3 cm in length) in the burner flame. What is the result?
3. *(a)* Repeat procedure 1*(a)*. *(b)* Heat a small roll of steel wool in the burner flame and immediately insert it into the test tube. What is the result?
4. Using 5 large test tubes, add *(a)* 10 mL of distilled water to the first; *(b)* 10 mL of bleach to the second;

(c) 10 mL of white vinegar to the third, *(d)* 9.5 mL of distilled water, 5 drops of bleach, and 5 drops of vinegar to the fourth (stir); *(e)* 1 aspirin tablet and 10 mL of distilled water to the fifth (stir). Lastly, add a 2.5 cm ball of steel wool to each test tube. Observe the reactions over a 24-hour period, noting the speed of the reactions and any color changes.
5. Your teacher will tell you how to dispose of your chemicals.

Discussion
1. Write general equations showing what happened in each of the reactions.
2. Assign oxidation numbers to the elements in the general equations for 1*(a)*, 2, 4*(b)*, and 4*(c)*, and write electronic equations to show which substance was oxidized and which was reduced.
3. How do you identify the oxidizing reaction and the reducing reaction? Give two examples.
4. Write a good conclusion for this experiment.

Peroxides are the only compounds in which oxygen has an oxidation state of -1 instead of -2. (peroxide ion: O_2^{2-}, hydrogen peroxide is H_2O_2: H—O—O—H.)

The oxidation state of each monatomic ion remains unchanged. This reaction is not an oxidation–reduction reaction.

Redox reactions sometimes involve polyatomic ions. An example is the redox reactions of permanganate ion (MnO_4^-), in which Mn has an oxidation number +7 (Rules 4 and 8). Under certain conditions, this MnO_4^- ion is changed to the *manganese(II) ion*, Mn^{2+}, as shown in Figure 20-3. Under other conditions, the MnO_4^- ion is changed to the nearly colorless manganate ion, MnO_4^{2-}, in which Mn has an oxidation number of +6 (Rules 4 and 8).

To correctly balance the equation for a redox reaction requires knowing all changes in oxidation state that occur in the reaction. The substance that is oxidized and the substance that is reduced must be identified. In the following electronic equations, the substances on the left are oxidized.

$$\overset{0}{Na} \rightarrow \overset{+1}{Na^+} + e^-$$
$$\overset{0}{Fe} \rightarrow \overset{+2}{Fe^{2+}} + 2e^-$$
$$\overset{+2}{Fe^{2+}} \rightarrow \overset{+3}{Fe^{3+}} + e^-$$
$$\overset{-1}{2Cl^-} \rightarrow \overset{0}{Cl_2} + 2e^-$$

The difference between oxidation numbers indicates the change in oxidation state. The chloride ion is assigned the oxidation number −1, a negative oxidation state. Oxidation results in the change of the oxidation numbers in a positive mathematical direction. The change from −1 to 0 is a change in a positive mathematical direction, as is the change from 0 to +1, from 0 to +2, or from +2 to +3.

In the electronic equations that follow, the substances on the left are reduced.

$$\overset{+1}{Na^+} + e^- \rightarrow \overset{0}{Na}$$
$$\overset{0}{Cl_2} + 2e^- \rightarrow \overset{-1}{2Cl^-}$$
$$\overset{+3}{Fe^{3+}} + e^- \rightarrow \overset{+2}{Fe^{2+}}$$
$$\overset{+2}{Cu^{2+}} + 2e^- \rightarrow \overset{0}{Cu}$$
$$\overset{0}{Br_2} + 2e^- \rightarrow \overset{-1}{2Br^-}$$

In the third equation, the iron(III) ion, oxidation number +3, is reduced to the iron(II) state, with oxidation number +2. The reduction results in the change of the oxidation number in a negative direction. The change from +3 to +2 is a change in a negative direction, as is the change from +1 to 0, 0 to −1, +2 to 0, and from 0 to −1.

Reactants and products in redox reactions are not limited to monatomic ions and uncombined elements. Molecular compounds or polyatomic ions containing elements that have more than one nonzero oxidation state can also be oxidized and reduced. For example, when sulfur dioxide (SO_2) is converted to sulfur trioxide (SO_3),

$$\cdots + \overset{+4}{SO_2} \rightarrow \overset{+6}{SO_3} + \cdots$$

sulfur is oxidized. (Usually we refer to the oxidation or reduction of the entire molecule or ion, e.g., "sulfur dioxide is oxidized to sulfur trioxide.") In a reaction in which permanganate ion (MnO_4^-) ion is converted to manganese dioxide (MnO_2),

$$\cdots + \overset{+7}{MnO_4^-} \rightarrow \overset{+4}{MnO_2} + \cdots$$

manganese is reduced.

Figure 20-4 Oxidation–reduction reactions. Top photo, zinc wrapped around iron nail, iron reduced; copper wrapped around iron nail, iron oxidized. Bottom photo, when a strip of zinc is immersed in an aqueous solution of copper(II) sulfate, what substance is oxidized? What is reduced?

Species with oxidation states of +7 always are good oxidizing agents.

1. (1) A reaction in which the atoms or ions of an element attain a more positive oxidation state (2) Na → Na$^+$ + e$^-$
2. (1) A reaction in which the atoms or ions of an element attain a more negative oxidation state (2) Cl + e$^-$ → Cl$^-$
3. See Table 20-1.
4. Any chemical process in which elements undergo a change in oxidation number

Section Review

1. What is oxidation? Give an example of this process.
2. What is reduction? Give an example.
3. How are oxidation numbers assigned?
4. What is meant by "redox reaction"?
5. Which of the following are oxidation–reduction reactions? Among those that are, identify the substance oxidized and that reduced.
   (a) $SO_3(g) + H_2O(\ell) \rightarrow H_2SO_4(aq)$ (b) $2KNO_3(s) \rightarrow 2KNO_2(s) + O_2(g)$
   (c) $H_2(g) + CuO(s) \rightarrow Cu(s) + H_2O(\ell)$
6. For each redox reaction identified in the previous question, write the electronic equations for the substances oxidized and reduced in each.

20.2 Balancing Redox Equations

SECTION OBJECTIVES

- Explain what must be conserved in redox equations.
- Balance redox equations by using the oxidation-number method.
- Balance redox equations by using the ion-electron method.

5. (a) Not oxidation-reduction
   (b) Yes, $\overset{-2}{O}$ is oxidized to $\overset{0}{O_2}$; $\overset{+5}{N}$ is reduced to $\overset{+3}{N}$
   (c) Yes, $\overset{0}{H_2}$; is oxidized to $\overset{+1}{H}$; $\overset{+2}{Cu}$ is reduced to $\overset{0}{Cu}$
6. (b) $\overset{-2}{2O} \rightarrow O_2 + 4e^-$; $\overset{+5}{N} + 2e^- \rightarrow \overset{+3}{N}$
   (c) $\overset{0}{H_2} \rightarrow 2\overset{+1}{H} + 2e^-$; $\overset{+2}{Cu} + 2e^- \rightarrow \overset{0}{Cu}$

A principal use of oxidation numbers is in balancing equations for oxidation–reduction reactions. These reactions are recognized by the fact that changes in oxidation states occur in reactants that undergo oxidation or reduction. In the equation-balancing process, accommodation is made for the conservation of charge as well as for the conservation of atoms. Equations for simple redox reactions can be balanced by inspection, which you learned to do in Chapter 8. Most redox equations, however, require more systematic methods.

The use of a pair of electronic equations to achieve electron balance in a redox equation has been demonstrated in two examples in Section 20.1. Two of these methods are illustrated here. The first method is simply an extension of the general balancing procedure you already know how to use. The second scheme is commonly used for ionic equations.

Oxidation-Number Method

In the oxidation-number method the electron shift is balanced for the particles oxidized and the particles reduced by balancing the pair of electronic equations that represents this shift. Then the coefficients of all reactants and products are adjusted to achieve a balanced equation.

The *oxidation-number method* (or, as it is sometimes called, the *electron-transfer method*) of balancing a redox equation includes the steps listed below. Observe that *Steps 2, 3,* and *4* represent additions to the steps you are accustomed to using in equation-balancing. *Step 1: Write the skeleton equation for the reaction.* To do this, you must know the reactants and products and must represent each by the correct formula. *Step 2: Assign oxidation numbers to all atoms in all species and determine what is oxidized and what is reduced. Step 3: Write the electronic equation for the oxidation process and the electronic equation for the reduction process. Step 4: Adjust the coefficients in both electronic equations so that the number of electrons lost equals the number gained. Step 5: Place these coefficients into the skeleton equation. Step 6: Supply the proper coefficients for the rest of the equation to satisfy the conservation of atoms.*

These steps are illustrated by application first to a very simple oxidation–reduction equation that is easily balanced by inspection. Gaseous hydrogen sulfide burns in air to form sulfur dioxide and water. These facts are used to write the skeleton equation.

Step 1:
$$H_2S + O_2 \rightarrow SO_2 + H_2O$$

Step 2: Next, oxidation numbers are assigned. Changes in oxidation numbers indicate that sulfur is oxidized from the -2 state to the $+4$ state; oxygen is reduced from the 0 state to the -2 state. The oxidation number of hydrogen remains the same; it therefore plays no part in the primary action of oxidation-reduction.

$$\overset{+1-2}{H_2S} + \overset{0}{O_2} \rightarrow \overset{+4-2}{SO_2} + \overset{+1-2}{H_2O}$$

Step 3: The change in oxidation state of sulfur requires the loss of 6 electrons: $(-2) - (+4) = -6$. The change in oxidation state of oxygen requires the gain of 2 electrons: $(0) + (-2) = -2$. The electronic equations for these two reactions are as follows.

$$\overset{-2}{S} \rightarrow \overset{+4}{S} + 6e^- \quad \text{(oxidation)}$$
$$\overset{0}{O} + 2e^- \rightarrow \overset{-2}{O} \quad \text{(reduction)}$$

Free oxygen is diatomic. Thus 4 electrons must be gained during the reduction of a molecule of free oxygen.

$$\overset{0}{O_2} + 4e^- \rightarrow 2\overset{-2}{O}$$

Step 4: The coefficients of the two electronic equations are now adjusted so that the number of electrons lost in the oxidation of sulfur is equal to the number gained in the reduction of oxygen. The smallest number of electrons common to both equations is 12. To show the gain and loss of 12 electrons in the two equations, the oxidation equation is multiplied by 2, and the reduction equation is multiplied by 3.

$$\overset{-2}{2S} \rightarrow \overset{+4}{2S} + 12e^-$$
$$\overset{0}{3O_2} + 12e^- \rightarrow 6\overset{-2}{O}$$

Step 5: Hence, the coefficients of H_2S and SO_2 are both 2, and the coefficient of O_2 is 3. Observe that the $6\overset{-2}{O}$ is divided between the two products SO_2 and H_2O. The 6 O atoms are accounted for by placing the coefficient 2 in front of each product formula including coefficients in the skeleton equation gives

$$2H_2S + 3O_2 \rightarrow 2SO_2 + 2H_2O$$

Step 6: The coefficients of the equation are adjusted in the usual way to satisfy the law of conservation of atoms. In this case, no further adjustments are needed; the equation is already balanced for both atoms and charges.

$$2H_2S + 3O_2 \rightarrow 2SO_2 + 2H_2O$$

As a second example, a redox equation that is slightly more difficult to balance will be used. This reaction occurs between manganese dioxide and hydrochloric acid (Figure 20-5). Water, manganese(II) chloride, and chlorine gas are formed.

Figure 20-5 The oxidation–reduction reaction between hydrochloric acid and manganese dioxide produces water, manganese(II) chloride and chlorine gas.

Oxidation-Reduction Reactions

Steps 1 and 2: The skeleton equation is written and oxidation numbers are assigned to the elements in the reaction.

$$\overset{+4\ -2}{MnO_2} + \overset{+1\ -1}{HCl} \rightarrow \overset{+1\ -2}{H_2O} + \overset{+2\ -1}{MnCl_2} + \overset{0}{Cl_2}$$

They show that Mn is reduced to $\overset{+4}{Mn}$ to $\overset{+2}{Mn^{2+}}$, and some of the $\overset{-1}{Cl^-}$ is oxidized to $\overset{0}{Cl}$. Hydrogen and oxygen do not take part in the oxidation–reduction reaction.

Step 3: The electronic equations are

$$\overset{-1}{2Cl^-} \rightarrow \overset{0}{Cl_2} + 2e^- \quad \text{(oxidation)}$$

$$\overset{+4}{Mn} + 2e^- \rightarrow \overset{+2}{Mn^{2+}} \quad \text{(reduction)}$$

Steps 4 and 5: The numbers of electrons lost and gained are the same. The coefficients are transferred to the skeleton equation, which becomes

$$MnO_2 + 2HCl \rightarrow H_2O + MnCl_2 + Cl_2$$

Step 6: The complete equation can now be balanced by inspection. Two additional molecules of HCl are needed to supply the two Cl^- ions of the $MnCl_2$. This balancing requires 2 molecules of water. These water molecules also account for the 2 oxygen atoms of the MnO_2. The final equation reads

$$MnO_2 + 4HCl \rightarrow 2H_2O + MnCl_2 + Cl_2$$

The equations for both of these examples could have been balanced without the use of electronic equations to balance electron shifts. The sample problem below involves a more complicated redox reaction, one in which use of electronic equations is necessary.

Sample Problem 20.1

The oxidation–reduction reaction between hydrochloric acid and potassium permanganate yields water, potassium chloride, manganese(II) chloride, and chlorine gas. Write the balanced equation.

Solution
We will balance this equation by using the stepwise, oxidation number method.

Step 1. $\quad HCl(aq) + KMnO_4(aq) \rightarrow H_2O(\ell) + KCl(aq) + MnCl_2(aq) + Cl_2(g)$

Step 2. $\quad \overset{+1\ -1}{HCl} + \overset{+1\ +7\ -2}{KMnO_4} \rightarrow \overset{+1\ -2}{H_2O} + \overset{+1\ -1}{KCl} + \overset{+2\ -1}{MnCl_2} + \overset{0}{Cl_2}$

Step 3. Chlorine is oxidized from -1 to 0 and manganese is reduced from $+7$ to $+2$. Note that in the electronic equations we include only the symbol for the element oxidized or reduced, a coefficent showing how many atoms of an element are oxidized or reduced, and the number of electrons needed to bring about the oxidation number change.

$$\overset{-1}{2Cl^-} \rightarrow \overset{0}{Cl_2} + 2e^- \quad \text{(oxidation)}$$

$$\overset{+7}{Mn} + 5e^- \rightarrow \overset{+2}{Mn^{2+}} \quad \text{(reduction)}$$

Sample Problem 20.1 *continued*

Step 4. To equalize the number of electrons, the first equation must be multiplied by 5 and the second by 2, to give 10e⁻ in both equations:

$$\overset{-1}{10Cl^-} \rightarrow \overset{0}{5Cl_2} + 10e^-$$

$$\overset{+7}{2Mn} + 10e^- \rightarrow \overset{+2}{2Mn^{2+}}$$

Step 5. $10HCl + 2KMnO_4 \rightarrow H_2O + KCl + 2MnCl_2 + 5Cl_2$

Step 6. By inspection, $2KMnO_4$ produces $2KCl$ and $8H_2O$. Now $2KCl$ and $2MnCl_2$ call for 6 additional molecules of HCl. The balanced equation becomes

$$16HCl(aq) + 2KMnO_4(aq) \rightarrow 8H_2O(\ell) + 2KCl(aq) + 2MnCl_2(aq) + 5Cl_2(g)$$

Inspection reveals that the equation is indeed balanced for atoms and for overall oxidation number. Balancing a redox equation should always be completed by an evaluation step. Check to be sure that the number of atoms of each element is the same on both sides of the equation and that the total oxidation number on each side equals zero (for a formula equation).

Practice Problem

1. A pure solution of nitric acid (HNO_3) is colorless but on standing becomes yellow because of slight decomposition that forms brown nitrogen dioxide, oxygen, and water. Using the oxidation-number method, balance the redox equation for this decomposition reaction.

(Ans.) $\overset{+1+5-2}{HNO_3(aq)} \rightarrow \overset{+4-2}{NO_2(g)} + \overset{0}{O_2(g)} + \overset{+1-2}{H_2O(\ell)}$

$\overset{+5}{4N} + 4e^- \rightarrow \overset{+4}{4N} \qquad \overset{-2}{2O} \rightarrow \overset{0}{O_2} + 4e^-$

$4HNO_3(aq) \rightarrow 4NO_2(g) + O_2(g) + 2H_2O(\ell)$

Ion-Electron Method

In this alternative scheme (also called the *half-reaction method*), the participating species are determined, as before, by assignment of oxidation numbers. Separate oxidation and reduction equations are balanced for both atoms and charge (adjusted so that equal numbers of electrons are gained and lost) and added together to give a balanced net ionic equation.

To compare the two balancing schemes, in Sample Problem 20.2, the ion-electron method is now used to balance the same equation balanced in Sample Problem 20.1 by the oxidation-number method. Sample Problem 20.2 shows the steps for the ion-electron method.

Sample Problem 20.2

Write a balanced equation for the reaction in Sample Problem 20.1 using the ion-electron method.

Solution
Step 1: Write the skeleton equation.

$$HCl(aq) + KMnO_4(aq) \rightarrow H_2O(\ell) + KCl(aq) + MnCl_2(aq) + Cl_2(g)$$

Sample Problem 20.2 *continued*

Step 2: Write the ionic equation.

$$H^+ + Cl^- + K^+ + MnO_4^- \rightarrow H_2O + K^+ + Cl^- + Mn^{2+} + 2Cl^- + Cl_2$$

Step 3: Assign oxidation numbers and retain only those species that include an element that changes oxidation state.

$$\overset{-1}{Cl^-} + \overset{+7}{MnO_4^-} \rightarrow \overset{+2}{Mn^{2+}} + \overset{0}{Cl_2}$$

Step 4: Write the equation for the reduction; balance it for both atoms and charge. Note the difference in this step from the oxidation number method. Entire formulas for the species oxidized or reduced are included (not just the atoms).

$$\overset{+7}{MnO_4^-} \rightarrow \overset{+2}{Mn^{2+}}$$

Observe from the initial equation that oxygen forms water. Account for the 4 oxygen atoms on the left by adding $4H_2O$ on the right.

$$MnO_4^- \rightarrow Mn^{2+} + 4H_2O$$

This addition now requires that $8H^+(aq)$ be added on the left, so that H atoms are balanced.

$$\underset{(-1)}{MnO_4^-} + \underset{(+8)}{8H^+(aq)} \rightarrow \underset{(+2)}{Mn^{2+}} + \underset{(0)}{4H_2O}$$

The equation is now balanced for atoms but not for charge; $5e^-$ must be added on the left side to balance the charge.

$$\underset{(-1)}{MnO_4^-} + \underset{(+8)}{8H^+(aq)} + \underset{(-5)}{5e^-} \rightarrow \underset{(+2)}{Mn^{2+}} + \underset{(0)}{4H_2O}$$

The reduction equation is now balanced for both atoms and charge.

Step 5: Write the equation for the oxidation and balance for atoms and charge.

$$\underset{(-2)}{\overset{-1}{2Cl^-}} \rightarrow \underset{(0)}{\overset{0}{Cl_2}}$$

To balance charge, $2e^-$ must be added on the right.

$$\underset{(-2)}{2Cl^-} \rightarrow \underset{(0)}{Cl_2} + \underset{(-2)}{2e^-}$$

Step 6: Multiply the oxidation and reduction equations by coefficients so that electrons gained equals electrons lost and add the equations. The reduction equation must be multiplied by 2 and the oxidation equation by 5 so that the number of electrons gained in reduction is equal to that lost in oxidation ($2e^- \times 5e^- = 5e^- \times 2e^- = 10e^-$)

$$2[MnO_4^- + 8H^+(aq) + 5e^- \rightarrow Mn^{2+} + 4H_2O]$$
$$5[2Cl^- \rightarrow Cl_2 + 2e^-]$$
$$\overline{2MnO_4^- + 16H^+(aq) + 10Cl^- + 10e^- \rightarrow 2Mn^{2+} + 8H_2O + 5Cl_2 + 10e^-}$$

The two $10e^-$ terms cancel, and the ion equation is balanced.

$$2MnO_4^- + 16H^+(aq) + 10Cl^- \rightarrow 2Mn^{2+} + 8H_2O + 5Cl_2$$

Sample Problem 20.2 continued

Step 7: Add whatever species are needed to balance the formula equation. You must include one K^+ ion for each MnO_4^- ion, and six more Cl^- ions must be added so there is one for each $H^+(aq)$ ion. The formula equation may then be expressed as follows:

$$2KMnO_4(aq) + 16HCl(aq) \rightarrow 2MnCl_2(aq) + 2KCl(aq) + 8H_2O(l) + 5Cl_2(g)$$

Checking shows that atoms and oxidation numbers are balanced.

Practice Problems

1. Copper reacts with hot, concentrated sulfuric acid to form copper(II) sulfate, sulfur dioxide, and water. Balance the equation using *(a)* the oxidation-number method and *(b)* the ion-electron method.

   (Ans.) *(a)* $\overset{+6}{S} + 2e^- \rightarrow \overset{+4}{S}$

   $\overset{0}{Cu} \rightarrow \overset{+2}{Cu^{2+}} + 2e^-$

   $Cu + 2H_2SO_4 \rightarrow CuSO_4 + SO_2 + 2H_2O$

   (b) $4H^+(aq) + \overset{+6}{SO_4^{2-}} + 2e^- \rightarrow \overset{+4}{SO_2} + 2H_2O$

   $\overset{0}{Cu} \rightarrow \overset{+2}{Cu^{2+}} + 2e^-$

   $\overline{Cu + 2H_2SO_4 \rightarrow CuSO_4 + SO_2 + 2H_2O}$

2. The products of a reaction between nitric acid and potassium iodide are potassium nitrate, iodine, nitrogen monoxide, and water. Balance the equation *(a)* using the oxidation-number method and *(b)* the ion-electron method.

   (Ans.) *(a)* $2\overset{+5}{N} + 6e^- \rightarrow 2\overset{+2}{N}$

   $6\overset{-1}{I^-} \rightarrow 3\overset{0}{I_2} + 6e^-$

   $8HNO_3 + 6KI \rightarrow 6KNO_3 + 3I_2 + 2NO + 4H_2O$

   (b) $8H^+(aq) + 2\overset{+5}{NO_3^-} + 6e^- \rightarrow 2\overset{+2}{NO} + 4H_2O$

   $6\overset{-1}{I^-} \rightarrow 3\overset{0}{I_2} + 6e^-$

   $\overline{8H^+(aq) + 2NO_3^- + 6I^- \rightarrow 3I_2 + 2NO + 4H_2O}$

   or

   $8HNO_3 + 6KI \rightarrow 6KNO_3 + 3I_2 + 2NO + 4H_2O$

Section Review

1. What two quantities must be conserved in redox equations?
2. What steps are involved in the oxidation-number method of equation-balancing?
3. What steps are involved in the ion-electron method of equation-balancing?
4. Use the oxidation number method to balance the following oxidation reduction reaction: $I_2O_5 + CO \rightarrow I_2 + CO_2$
5. Use the ion–electron method to balance the following oxidation–reduction reaction: $Na_2SnO_2 + Bi(OH)_3 \rightarrow Bi + Na_2SnO_3 + H_2O$

1. Charge and number of atoms
2. Refer to Section 20.2 for steps 1 through 6.
3. Refer to Section 20.2 for steps 1 through 7.
4. $I_2O_5 + 5CO \rightarrow I_2 + 5CO_2$
5. $3Na_2SnO_2 + 2Bi(OH)_3 \rightarrow 2Bi + 3Na_2SnO_3 + 3H_2O$

20.3 Oxidizing and Reducing Agents

SECTION OBJECTIVES

- Relate chemical activity to oxidizing and reducing strength.
- Explain the concept of auto-oxidation.

An **oxidizing agent** *attains a more negative oxidation state during an oxidation–reduction reaction.* A **reducing agent** *attains a more positive oxidation state.*

It follows that the substance oxidized is also the reducing agent, and the substance reduced is the oxidizing agent. In other words, in every redox reaction, there is one reducing agent and one oxidizing agent. Study the oxidation–reduction terms presented in Table 20-2. Refer to them in this section.

The relatively large atoms of the sodium family of metals make up Group 1 of the Periodic Table. These atoms have weak attraction for their valence electrons and form positive ions readily. They are very active reducing agents. According to electrochemical measurements, the lithium atom is the most active reducing agent of all the common elements. The lithium ion, on the other hand, is the weakest oxidizing agent of the common ions. The electronegativity scale suggests that Group-1 metals starting with lithium should become progressively more active reducing agents. Except for lithium, this is true. A possible basis for the unusual activity of lithium is discussed in Chapter 24.

Atoms of the Halogen family (Group 17 of the Periodic Table) have a strong attraction for electrons. They form negative ions readily and are very active oxidizing agents. The fluorine atom is the most highly electronegative atom. It is also the most active oxidizing agent among the elements. Because of its strong attraction for its own electrons, the fluoride ion is the weakest reducing agent.

The protons in the fluorine nucleus exert a strong attractive (coulombic) force on the electrons.

TABLE 20-2 OXIDATION–REDUCTION TERMINOLOGY

Term	Change in Oxidation Number	Change in Electron Population
oxidation	in a positive direction	loss of electrons
reduction	in a negative direction	gain of electrons
oxidizing agent	in a negative direction	gains electrons
reducing agent	in a positive direction	loses electrons
substance oxidized	in a positive direction	loses electrons
substance reduced	in a negative direction	gains electrons

The following species make excellent oxidizing agents because the central atoms in these polyatomic ions have high oxidation states: IO_4^-, ClO_4^-, MnO_4^-, NO_3^-, CrO_4^{2-}, and $Cr_2O_7^-$. The electron density of the covalent bonds that exists between the oxygen and the central atom is closer to the oxygen.

In Table 20-3 many familiar substances are arranged according to their activity as oxidizing and reducing agents. The left column of each pair shows the relative abilities of some metals to displace other metals from their compounds. Such displacements, which you first read about in Chapter 8, are oxidation–reduction processes. Zinc, for example, appears above copper. Thus, zinc is the more active reducing agent and displaces copper ions from solutions of copper compounds (Figure 20-6).

TABLE 20-3 RELATIVE STRENGTH OF OXIDIZING AND REDUCING AGENTS

Reducing Agents	Oxidizing Agents	Reducing Agents	Oxidizing Agents
Li Strong	Li^+ Weak	Cu	Cu^{2+}
K	K^+	I^-	I_2
Ca	Ca^{2+}	MnO_4^{2-}	MnO_4^-
Na	Na^+	Fe^{2+}	Fe^{3+}
Mg	Mg^{2+}	Hg	Hg_2^{2+}
Al	Al^{3+}	Ag	Ag^+
Zn	Zn^{2+}	NO_2^-	NO_3^-
Cr	Cr^{3+}	Br^-	Br_2
Fe	Fe^{2+}	Mn^{2+}	MnO_2
Ni	Ni^{2+}	SO_2	H_2SO_4 (conc)
Sn	Sn^{2+}	Cr^{3+}	$Cr_2O_7^{2-}$
Pb	Pb^{2+}	Cl^-	Cl_2
H_2	H_3O^+	Mn^{2+}	MnO_4^-
H_2S	S	F^- Weak	F_2 Strong

$$Zn(s) + Cu^{2+}(aq) \rightarrow Zn^{2+}(aq) + Cu(s)$$
$$\overset{0}{Zn} \rightarrow \overset{+2}{Zn^{2+}} + 2e^- \quad \text{(oxidation)}$$
$$\overset{+2}{Cu^{2+}} + 2e^- \rightarrow \overset{0}{Cu} \quad \text{(reduction)}$$

The copper(II) ion, on the other hand, is a more active oxidizing agent than the zinc ion.

Nonmetals and some important ions are included in the series in Table 20-3. Any reducing agent is oxidized by the oxidizing agents below it. Observe that F_2 displaces Cl^-, Br^-, and I^- ions from their solutions. Cl_2 displaces Br^- and I^- ions, Br_2 displaces I^- ions. The equation for the displacement of Br^- by Cl_2 is

$$Cl_2 + 2Br^-(aq) \rightarrow 2Cl^-(aq) + Br_2$$
$$\overset{-1}{2Br^-} \rightarrow \overset{0}{Br_2} + 2e^- \quad \text{(oxidation)}$$
$$\overset{0}{Cl_2} + 2e^- \rightarrow \overset{-1}{2Cl^-} \quad \text{(reduction)}$$

Permanganate ions (MnO_4^-) and dichromate ions ($Cr_2O_7^{2-}$) are important oxidizing agents. They are used mainly in the form of their potassium salts. In neutral or mildly basic solutions, $\overset{+7}{Mn}$ in permanganate ion is reduced to $\overset{+4}{Mn}$ in MnO_2. If a solution is strongly basic, manganate ion containing $\overset{+6}{Mn}$ is formed. In acid solutions, the $\overset{+7}{Mn}$ in permanganate ion is reduced to manganese(II) ion, Mn^{2+}, and $\overset{+6}{Cr}$ in dichromate ion is reduced to chromium(III) ion, Cr^{3+}.

Peroxide ions (O_2^{2-}) have a relatively unstable covalent bond between the two oxygen atoms. The electron-dot formula is written as

$$\left[:\overset{..}{\underset{..}{O}}:\overset{..}{\underset{..}{O}}: \right]^{2-}$$

Figure 20-6 Zinc displaces copper ions from a copper sulfate solution. Metallic copper precipitates. This reaction is the basis for the electrochemical cell.

Chemistry and its Applications

Not-So-Blue Jeans

If you want to fade your new blue jeans, rough them up a little with fine sandpaper before you wash them. This works because the blue jeans are colored by a special procedure that ensures their fading.

Most fabrics are colored by a thorough dye soaking, and the color becomes permanent. The cotton yarn that will be made into denim, however, is passed quickly through a solution of indigo (in a pale yellow soluble form, $C_{16}H_{12}N_2O_2$) so that only its surface gets wet. The core of the yarn never gets dyed, even with several passes through the indigo bath. Gradually the indigo reacts with oxygen in the air and changes to a navy blue compound that is insoluble in water as shown in the following reaction.

$$2C_{16}H_{12}N_2O_2 + O_2 \rightarrow 2C_{16}H_{10}N_2O_2 + 2H_2O$$

Jeans fade partly because some indigo washes out, but mostly because the thread surface gets worn off, exposing the white core. Rubbing your jeans with sandpaper just speeds up the process.

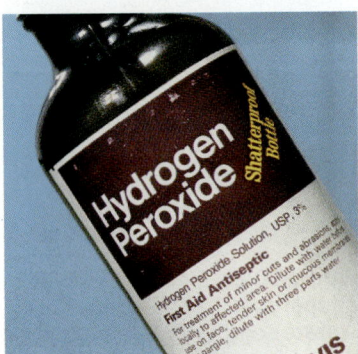

Figure 20-7 The antiseptic hydrogen peroxide

This structure represents an intermediate state of oxidation between free oxygen and oxides. The oxidation number of oxygen in the peroxide structure is -1.

Hydrogen peroxide (H_2O_2) is often used as an antiseptic (Figure 20-7). It decomposes by the oxidation *and* reduction of its oxygen. The products are water and molecular oxygen.

$$\overset{-1}{H_2O_2} + \overset{-1}{H_2O_2} \rightarrow 2\overset{-2}{H_2O} + \overset{0}{O_2}(g)$$

In this decomposition, half of the oxygen in the peroxide sample is *reduced* (from -1 to -2 oxidation state) to the oxide, forming water, and half is *oxidized* (from -1 to 0) to gaseous oxygen. *Such a process, in which a substance acts as both an oxidizing and a reducing agent and is self-oxidizing and self-reducing is called* **auto-oxidation**. Impurities in a water solution of hydrogen peroxide may catalyze this process.

Because of its oxidizing properties, hydrogen peroxide is used to bleach natural and artificial fibers and to purify gas and liquid waste materials produced by industrial plants. It is also used in uranium mining and metal extraction processes and as a starting material for preparing other compounds in which oxygen has an oxidation number of -1.

Section Review

1. Describe the chemical activity of the alkali metals and of the halogens on the basis of oxidizing and reducing strength.
2. What is meant by "auto-oxidation?"
3. Would Ca be oxidized by Cu^{2+}? Explain.
4. Would Cl_2 be reduced by I^-? Explain.

20.4 Electrochemistry

Oxidation–reduction reactions, like all reactions, involve energy changes. Since these reactions involve electron transfer, the net *release* or net *absorption* of energy can occur in the form of electrical energy rather than heat energy. As you will learn in this section, this property allows for a great many practical applications of redox reactions and also makes possible quantitative prediction and comparison of the oxidizing and reducing abilities of different substances. *The branch of chemistry that deals with these electricity-related applications of oxidation–reduction reactions is called* **electrochemistry**.

Electrochemical Cells

Oxidation–reduction reactions involve a transfer of electrons from the substance oxidized to the substance reduced. Sometimes, such reactions occur spontaneously and produce undesirable effects, such as the rusting of iron (Figure 20-8). Rust causes a tremendous amount of damage to metal structures every year. In light of the high cost of repairs and dwindling supplies of metal ores, research into rust prevention is a growing enterprise.

However, some spontaneous redox reactions can be used as sources of electric energy. If the reactants are in contact, the energy released during the electron transfer is in the form of heat. If the reactants are separated solutions and are connected externally by a wire conductor, the transfer of electrons takes place through the wire and the energy is released in the form of electrical energy. Solutions that conduct electricity, as you learned in Chapter 15, are classified as electrolytes. *A conductor used to establish electric contact with a nonmetallic part of a circuit, such as electrolytes, is known as an* **electrode**. *A system of electrodes and electrolytes in which a spontaneous redox reaction is used as a source of electrical energy is known as an* **electrochemical cell**. The transfer of electrons through a metallic conductor constitutes an electric current. Although, under these conditions, a part of the energy released during the electron transfer does appear as heat, the remainder is in fact available as electrical energy.

The dry cell (Figure 20-9) is a common source of electrical energy in the laboratory, as well as in countless devices from flashlights to portable radios. Small dry cells are familiar as so-called "flashlight batteries". A zinc cup serves as the negative electrode. A carbon rod serves as the positive electrode. The carbon electrode is surrounded by a mixture of manganese dioxide and powdered carbon. The electrolyte is a moist paste of ammonium chloride containing some zinc chloride. These components of the dry cell are illustrated in the schematic diagram of Figure 20-10.

1. The alkali metals are active reducing/weak oxidizing agents while the halogens are active oxidizing/weak reducing agents.
2. A process by which a substance acts as both oxidizing and reducing agent and is self-oxidizing and self-reducing
3. Yes, because Ca is a stronger reducing agent than Cu.

SECTION OBJECTIVES

- Describe the nature of electrochemical cells.
- Describe the nature of electrolytic cells.
- Discuss the electrolysis of water.
- Discuss the electrolysis of aqueous salt solutions.
- Explain the process of electroplating.
- Discuss the oxidation of rechargeable cells.
- Explain, use, and carry out calculations involving standard electrode potentials.

4. Yes, because Cl_2 is a stronger oxidizing agent than I_2.

Figure 20-8 Rust, the reddish corrosion product formed by electrochemical interaction between iron and atmospheric oxygen.

Figure 20-9 Many common batteries are zinc dry cells

When the external circuit is closed, zinc atoms are oxidized at the negative (zinc) electrode.

$$\overset{0}{\text{Zn}} \rightarrow \overset{+2}{\text{Zn}^{2+}} + 2e^-$$

Electrons flow through the external circuit to the positive (carbon) electrode. If manganese dioxide were not present, hydrogen gas would be formed at the carbon electrode by the reduction reaction shown in the following equation.

$$2\overset{+1}{\text{NH}_4^+} + 2e^- \rightarrow 2\text{NH}_3 + \overset{0}{\text{H}_2(g)}$$

Such a reaction that occurs at one electrode in an electrochemical cell is known as a **half-reaction**. However, manganese oxidizes the hydrogen gas to form water. Thus, manganese, rather than hydrogen, is reduced at the positive (carbon) electrode.

$$2\overset{+4}{\text{MnO}_2} + 2\text{NH}_4^+ + 2e^- \rightarrow \overset{+3}{\text{Mn}_2\text{O}_3} + 2\text{NH}_3 + \text{H}_2\text{O}$$

The ammonia that is formed reacts with Zn^{2+} ions to form complex $Zn(NH_3)_4^{2+}$ ions.

The electrode at which oxidation occurs is called the **anode**, *and the electrode at which reduction occurs is called the* **cathode.** In this context, because oxidation occurs at the negative electrode of an electrochemical cell (spontaneous reactions), the negative electrode of such a cell is an anode. Because reduction occurs at the positive electrode, the positive electrode of such a cell is a cathode. The zinc electrode (oxidation) of Figure 20-10 is the anode, and the carbon electrode (reduction) is the cathode of the zinc-carbon dry cell.

Electrolytic Cells

Some oxidation–reduction reactions do not occur spontaneously. Such reactions can be driven by means of electrical energy, however. *The process in which an electric current is used to drive an oxidation–reduction reaction is called* **electrolysis**. *An* **electrolytic cell** *is a system of electrodes and electrolytes in which a current is used to drive a nonspontaneous redox reaction.* A battery or other direct-current source is connected across the cell. One electrode, which is connected to the negative terminal of the battery, acquires a negative charge. The other electrode, connected to the positive terminal of the battery, acquires a positive charge.

Ion migration in the cell is responsible for the transfer of electric charge between electrodes. Positively charged ions migrate, or move, toward the negative electrode. Negatively charged ions migrate toward the positive electrode.

In an electrolytic cell, as in an electrochemical cell, reduction occurs at the cathode. In an electrolyte cell, however, this reduction occurs at the negative electrode. So, in this type of cell, the cathode is negatively charged. Electrons are removed from the cathode by the electrolyte in this process. Oxidation occurs at the anode, which in this case is the positive electrode, the electrode that acquires electrons from the electrolyte in the process. The chemical reactions at the electrodes complete the electric circuit between an external source of electric current and the cell. The closed-loop path

Figure 20-10 In a zinc dry cell, zinc is oxidized to Zn^{2+} at the negative electrode and manganese(IV) is reduced to manganese(III) at the positive electrode.

1800 Count Alessandro Volta (1745–1827) produced the first electric battery by stacking silver and zinc disks in a "pile," separating them by cardboard pads soaked in salt water, and attaching a metal wire at each end.

for electric current allows energy to be transferred from the external source to the electrolytic cell. This energy drives the electrode reactions in the electrolytic cell.

Electrolysis of water In the decomposition of water by electrolysis, energy is transferred from the energy source to the decomposition products. The reaction is endothermic, and the amount of energy required is 285.9 kJ/mol of water decomposed. The overall reaction is

$$2H_2O(\ell) + 571.8 \text{ kJ} \rightarrow 2H_2(g) + O_2(g)$$

Hydrogen gas is given up at the cathode (the reduction site), and oxygen gas is given up at the anode (the oxidation site).

A suitable electrolysis cell is shown in Figure 20-11. It consists of two inert platinum electrodes immersed in water. A very small amount of an electrolyte (such as H_2SO_4), is added to provide adequate conductivity. The electrodes are connected to a battery that supplies the electrical energy to drive the decomposition reaction forward.

The electric current in the external circuit consists of a flow of electrons. The electrode of the cell connected to the negative electrode of the battery acquires an excess of electrons and becomes the cathode of the electrolytic cell. The other electrode of the cell, which is connected to the positive electrode of the battery, loses electrons to the battery and becomes the anode of the electrolytic cell. The battery can be thought of as an electron pump supplying electrons to the cathode of the electrolytic cell and recovering electrons from the anode of the cell at equal rates.

Reduction occurs at the cathode, where, it is believed, water molecules acquire electrons directly. Hydrogen gas is the product of this electrochemical reaction.

cathode reaction *(reduction)*: $2H_2\overset{+1}{O}(\ell) + 2e^- \rightarrow 2OH^-(aq) + \overset{0}{H_2}(g)$

The OH^- ion concentration in the solution around the cathode rises.

Oxidation occurs at the anode; oxygen gas is the product. There are SO_4^{2-} ions, OH^- ions, and water molecules in the region about the anode. The OH^- ion concentration is quite low, and this ion is not likely to appear in the anode reaction. The SO_4^{2-} ions, also at low concentration, are more difficult to oxidize than are water molecules. For these reasons, chemists believe that water molecules are oxidized by giving up electrons directly to the anode.

anode reaction *(oxidation)*: $6H_2\overset{-2}{O}(\ell) \rightarrow 4H_3O^+(aq) + \overset{0}{O_2}(g) + 4e^-$

The H_3O^+ ion concentration in the solution around the anode rises.

The overall reaction is the sum of the net cathode and anode reactions. The equation for the cathode reaction is doubled because equal numbers of electrons must be involved in the oxidation and reduction processes.

cathode: $4H_2O(\ell) + 4e^- \rightarrow 4OH^-(aq) + 2H_2(g)$
anode: $6H_2O(\ell) \rightarrow 4H_3O^+(aq) + O_2(g) + 4e^-$
---
cell: $10H_2O(\ell) \rightarrow 2H_2(g) + O_2(g) + 4H_3O^+(aq) + 4OH^-(aq)$

Figure 20-11 The electrolysis of water. In electrolytic cells, reduction occurs at the negative electrode, and oxidation occurs at the positive electrode.

1832 Michael Faraday (1791–1867) described the following laws:
1. The mass of substance liberated at an electrode during electrolysis is proportional to the quantity of electricity driven through the solution.
2. The mass liberated by a given quantity of electricity is proportional to the atomic weight of the element liberated and inversely proportional to the valence of the element liberated.

Figure 20-12 Industrial chlorine production units

The solution around the cathode becomes basic because of the production of OH⁻ ions. The solution around the anode becomes acidic because of the production of H_3O^+ ions. Because of the ordinary mixing tendency in the solution, these ions can be expected to diffuse together eventually and form water. If this neutralization is complete, the net electrolysis reaction is

cell neutralization

cell: $10H_2O(\ell) \rightarrow 2H_2(g) + O_2(g) + 4H_3O^+(aq) + 4OH^-(aq)$

neutralization: $4H_3O^+(aq) + 4OH^-(aq) \rightarrow 8H_2O(\ell)$

net: $2H_2O(\ell) \rightarrow 2H_2(g) + O_2(g)$

Reduction at the cathode lowers the oxidation state of hydrogen from +1 to 0; the oxidation state of oxygen at the cathode remains −2. At the anode, oxidation raises the oxidations state of oxygen from −2 to 0; the oxidation state of hydrogen at the anode remains +1.

Electrolysis of aqueous salt solutions The electrode products from the electrolysis of aqueous salt solutions are determined by the relative ease with which the different particles present can be oxidized or reduced. In the case of aqueous NaCl, for example, Na^+ ions are more difficult to reduce at the cathode than H_2O molecules or H_3O^+ ions. Since a solution of NaCl is neutral, the H_3O^+ ion concentration remains very low (10^{-7} mol/L). Molecules of H_2O are therefore the particles that are reduced at the cathode.

Cathode reaction (reduction):

$$2\overset{+1}{H_2}O(\ell) + 2e^- \rightarrow 2OH^-(aq) + \overset{0}{H_2}(g)$$

Thus, Na^+ ions remain in solution, and hydrogen gas is released.

In general, metals that are easily oxidized form ions that are difficult to reduce. Thus, hydrogen gas, not sodium metal, is the cathode product in the preceding reaction. On the other hand, metals such as copper, silver, and gold are difficult to oxidize, and they form ions that are easily reduced. Aqueous solutions of their salts give up the metal at the cathode.

The choice for anode reaction in the aqueous NaCl electrolysis lies between Cl^- ions and H_2O molecules. The Cl^- ions are more easily oxidized, so Cl_2 gas is produced at the anode.

Anode reaction (oxidation):

$$2\overset{-1}{Cl^-}(aq) \rightarrow \overset{0}{Cl_2}(g) + 2e^-$$

Adding the cathode and anode equations gives the net reaction for the cell.

Net: $2H_2O(\ell) + 2Cl^-(aq) \rightarrow 2OH^-(aq) + H_2(g) + Cl_2(g)$

The electrolytic solution gradually changes from aqueous NaCl to aqueous NaOH as the electrolysis continues, as long as the Cl_2 gas is continuously removed from the cell (See Figure 20-12).

Aqueous Br^- ions and I^- ions are oxidized electrolytically in the same way as Cl^- ions. Their solutions give the free halogen at the anode. On the other hand, aqueous solutions of negative ions that do not participate in the oxidation reaction (NO_3^- ions, for example) give O_2 gas at the anode.

Electroplating *Inactive metals form ions that are more easily reduced than hydrogen. This fact makes possible an electrolytic process called* **electroplating**, *in which a metal is deposited on a surface.*

An electroplating cell contains a solution of a salt of the plating metal. It has an object to be plated (the cathode) and a piece of the plating metal (the anode). A silverplating cell, illustrated in Figure 20-13, contains a solution of a soluble silver salt and a silver anode. The cathode is the object to be plated. The silver anode is connected to the positive electrode of a battery or to some other source of direct current. The object to be plated is connected to the negative electrode. Silver ions are reduced at the cathode of the cell when electrons flow through the circuit.

$$\overset{+1}{Ag^+} + e^- \rightarrow \overset{0}{Ag}$$

Silver atoms are oxidized at the anode.

$$\overset{0}{Ag} \rightarrow \overset{+1}{Ag^+} + e^-$$

Silver ions are removed from the solution at the cathode and deposited as metallic silver. Meanwhile, metallic silver is removed from the anode as ions. This action maintains the Ag^+ ion concentration of the solution. Thus, in effect, silver is transferred from the anode to the cathode of the cell.

Rechargeable Cells

A rechargeable cell is a cell that, in effect, combines the oxidation–reduction chemistry of both the electrochemical cell and the electrolytic cell, which are compared in Figure 20-14. When the rechargeable cell is used as a source of

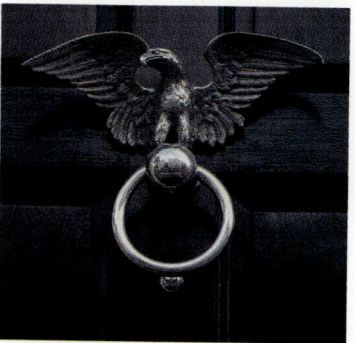

Figure 20-13 An electrolytic cell (top) used for silver plating. This door knocker is silver-plated.

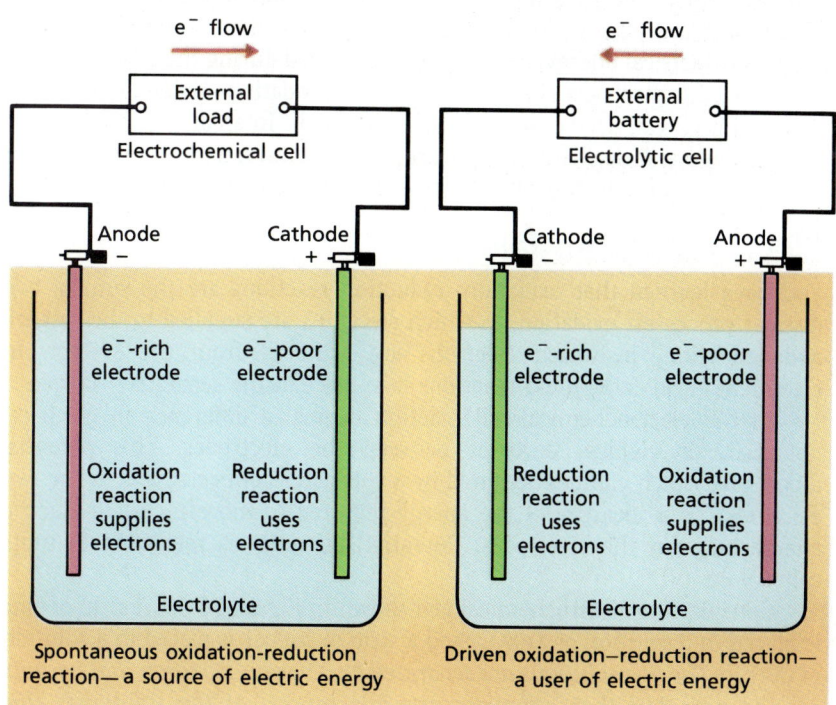

Figure 20-14 A comparison of electrochemical and electrolytic cells

Oxidation-Reduction Reactions

Figure 20-15 The rechargeable cells of a car battery produce electricity from reactions between lead(IV) oxide, lead, and sulfuric acid.

electrical energy (during the so-called discharge cycle), chemical energy is converted to electrical energy by the oxidation–reduction reactions characteristic of an electrochemical cell. When the cell is being recharged (charge cycle), electric energy is converted to chemical energy by the oxidation–reduction reactions of an electrolytic cell.

The standard 12-volt automobile battery consists of six lead(IV) oxide–lead–sulfuric acid rechargeable cells that are connected in series. The battery is charged by the conversion of electrical energy from an external source to stored chemical energy which occurs by an oxidation–reduction reaction in which each cell acts as an electrolytic cell. While the battery is used as a source of electrical energy, chemical energy stored during the charging cycle is converted to electrical energy by the reverse oxidation–reduction reaction, in which each cell acts as an electrochemical cell. In an automobile engine, these charge and discharge modes are regulated automatically to maintain the state of charge in the battery while the engine is in operation.

Electrode Potentials

You have learned that oxidation–reduction reactions are the sum of two distinct processes: oxidation, in which electrons are supplied to the system, and reduction, in which electrons are acquired from the system. In electrochemical cells, these reactions take place at the separate electrodes.

As the electrochemical-cell reaction begins, a difference in electrical potential, or voltage, develops between the electrodes. This potential difference, which can be measured by a voltmeter connected across the two electrodes, is a measure of the energy required to move a certain electric charge between the electrodes. Potential difference is measured in units called volts (V).

Consider the electrochemical cell shown in Figure 20-16. A strip of zinc is placed in a solution of $ZnSO_4$, and a strip of copper is placed in a solution of $CuSO_4$. *The two solutions are separated by a porous partition that permits ions to pass but that otherwise prevents mixing of the solutions. This*

The battery voltage is produced by a spontaneous electrochemical reaction. If the system is closed, that is, if reactions are reversible by providing electrons, the battery can be recharged.

Ions pass from one electrolytic half-cell to another so that overall electroneutrality is maintained in solution throughout the process.

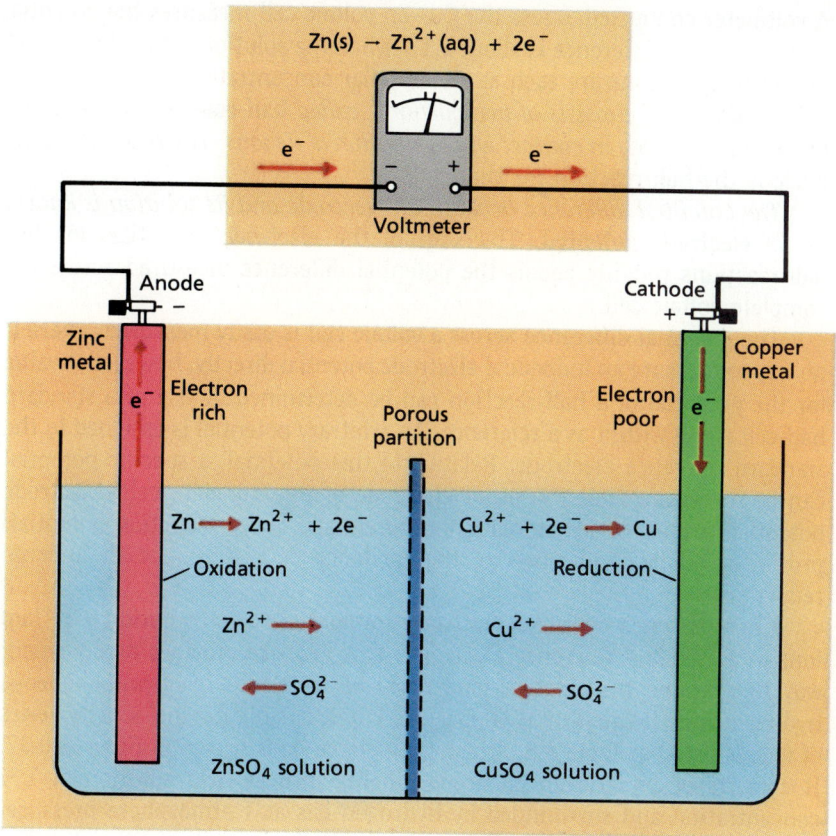

Figure 20-16 In a zinc-copper cell, zinc is oxidized and copper(II) ions are reduced.

arrangement, which is capable of generating an electric current in an external circuit connected between the electrodes, is called a **voltaic cell**.

In the two electrode reactions, the zinc electrode acquires a negative charge relative to the copper. The copper electrode becomes positively charged relative to the zinc. This fact reveals that zinc atoms have a stronger tendency to release electrons and enter the solution as ions than do copper atoms. Zinc is said to be more active, or more easily oxidized, than copper.

In the cell shown in Figure 20-16, then, the reaction at the surface of the zinc electrode is an oxidation.

$$\overset{0}{\text{Zn}}(s) \rightarrow \overset{+2}{\text{Zn}^{2+}}(aq) + 2e^-$$

The reaction at the surface of the copper electrode is a reduction.

$$\overset{+2}{\text{Cu}^{2+}}(aq) + 2e^- \rightarrow \overset{0}{\text{Cu}}(s)$$

As Zn^{2+} ions form, electrons accumulate on the zinc electrode, giving it a negative charge. As Cu atoms form, electrons are removed from the copper electrode, leaving it with a positive charge. Electrons also flow through the external circuit from the zinc electrode to the copper electrode. Here they replace the electrons removed, enabling Cu^{2+} ions to undergo reduction to Cu atoms. Thus, in effect, electrons are transferred from Zn atoms through the external circuit to Cu^{2+} ions. The overall reaction can be written as

$$Zn(s) + Cu^{2+}(aq) \rightarrow Zn^{2+}(aq) + Cu(s)$$

Oxidation-Reduction Reactions

Figure 20-17 A hydrogen electrode, the standard reference electrode for measuring electrode potentials. The electrode surface in contact with the solution is actually a layer of hydrogen adsorbed onto the surface of the platinum black.

A voltmeter connected across the Cu–Zn voltaic cell measures the potential difference. This difference is about 1.1 V when the solution concentrations of Zn^{2+} and Cu^{2+} ions are each at a one molar concentration (1 M).

A voltaic cell consists of two portions, called **half-cells,** *each consisting of a metal electrode in contact with a solution of its ions.* The reaction taking place is the half-reaction for that cell.

The potential difference between an electrode and its solution is known as its **electrode potential**. The sum of the electrode potentials for two half-reactions roughly equals the potential difference measured across the complete voltaic cell.

The potential difference across a voltaic cell is easily measured. There is no way to measure an individual electrode potential directly, however. A value for the potential of a half-reaction can be determined by use of a standard half-cell along with it as a reference. An arbitrary potential is assigned to the standard reference electrode. Relative to this potential, a specific potential can be determined for the other electrode of the complete cell. Electrode potentials are expressed as reduction potentials. These reduction potentials provide a reliable indication of the tendency of a substance to undergo reduction.

Chemists use a hydrogen electrode immersed in a 1 M solution of $H^+(aq)$ ions as a standard reference electrode. This practice provides a convenient way to examine the relative tendencies of metals to react with aqueous hydrogen ions [$H^+(aq)$ or $H_3O^+(aq)$]. It is responsible for the activity series of metals listed in Table 8-3. A hydrogen electrode is shown in Figure 20-17. It consists of a platinum electrode dipping into an acid solution of 1 M concentration and surrounded by hydrogen gas at 1 atmosphere pressure. This standard hydrogen electrode is assigned a potential of zero volts. The half-cell reaction is

$$\overset{0}{H_2}(g) \rightleftarrows 2\overset{+1}{H^+}(aq) + 2e^-$$

Since the potential of the hydrogen electrode is arbitrarily set at 0 V, the potential difference across the complete cell is attributed entirely to the electrode of the other half-cell. *A half-cell potential measured relative to a potential of zero for the standard hydrogen electrode is called a* **standard electrode potential**, *symbolized by* $E°$.

Suppose a complete cell consists of a zinc half-cell and a standard hydrogen half-cell, as in Figure 20-18. The potential difference across the cell measures the standard electrode potential (symbol: $E°$) of the zinc electrode relative to the hydrogen electrode (the zero reference electrode). It is found to be -0.76 V. The value is relative because the standard electrode potential of any electrode is given a negative value if this electrode has a negative charge relative to the standard hydrogen electrode. Electrons flow through the external circuit from the zinc to the hydrogen electrode. There, $H^+(aq)$ ions are reduced to H_2 gas.

This reaction indicates that the tendency for Zn^{2+} ions to be reduced to Zn atoms is 0.76 volt less than the tendency for $H^+(aq)$ ions to be reduced to H_2. The half-reaction (as a reduction) is

$$\overset{+2}{Zn^{2+}} + 2e^- \rightarrow \overset{0}{Zn} \qquad E° = -0.76 \text{ V}$$

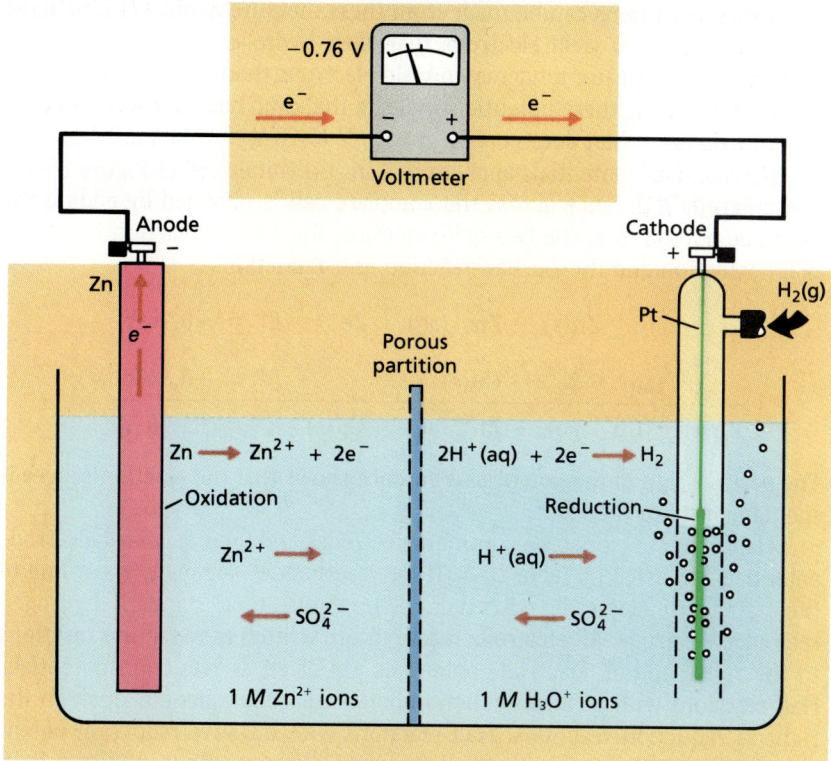

Figure 20-18 The electrode potential of a zinc half-cell is measured by coupling it with a standard hydrogen electrode.

This half-reaction has less tendency to occur than the half-reaction

$$\overset{+1}{2H^+}(aq) + 2e^- \rightarrow \overset{0}{H_2}(g)$$

by 0.76 V. This statement also means that the half-reaction (as an oxidation)

$$\overset{0}{Zn}(s) \rightarrow \overset{+2}{Zn^{2+}}(aq) + e^- \quad E° = +0.76 \text{ V}$$

has a greater tendency to occur by 0.76 volt than

$$\overset{0}{H_2}(g) \rightarrow \overset{+1}{2H^+}(aq) + 2e^-$$

Observe that the sign of the electrode potential is reversed when the Zn half-cell reaction is reversed and written as an oxidation.

A copper half-cell coupled with the standard hydrogen electrode gives a potential difference measurement of +0.34 V. This measurement indicates that Cu^{2+}(aq) ions are more readily reduced than H^+(aq) ions. *The value is positive because the standard electrode potential ($E°$) of an electrode is given a positive value if this electrode has a positive charge relative to the standard hydrogen electrode.* The half-reaction for copper (reduction) is

$$\overset{+2}{Cu^{2+}}(aq) + 2e^- \rightarrow \overset{0}{Cu}(s) \quad E° = +0.34 \text{ V}$$

This reaction has a greater tendency to occur than

$$\overset{+1}{2H^+}(aq) + 2e^- \rightarrow \overset{0}{H_2}(g)$$

by 0.34 V.

If Cu^{2+}, Zn^{2+}, and H^+ were competing for electrons, Cu^{2+} would get them because it exerts a stronger attractive force on the electrons.

Two observations can be made from these measurements: *(1)* Zinc has a greater tendency to yield electrons than does hydrogen by 0.76 V; and *(2)* Hydrogen has a greater tendency to yield electrons than does copper by 0.34 V. Taken together, these potentials indicate that zinc has a greater tendency toward oxidation than does copper by 1.10 V (0.76 V + 0.34 V).

How do these potentials apply to the Zn–Cu voltaic cell of Figure 20-16? The potential difference across the complete cell is obtained by adding the electrode potentials of the two half-reactions; the Zn half-reaction is written as an oxidation and the Cu half-reaction as a reduction.

$$\overset{0}{Zn}(s) \rightarrow \overset{+2}{Zn^{2+}}(aq) + 2e^- \quad E° = +0.76 \text{ V}$$
$$\overset{+2}{Cu^{2+}}(aq) + 2e^- \rightarrow \overset{0}{Cu}(s) \quad E° = +0.34 \text{ V}$$
$$\overline{Zn(s) + Cu^{2+}(aq) \rightarrow Zn^{2+}(aq) + Cu(s) \quad E° = +1.10 \text{ V}}$$

The positive sign of the potential difference shows that the reaction proceeds spontaneously to the right.

Half-reactions for some common electrodes and their standard electrode potentials are listed in Table 20-4. These reactions are arranged according to their standard electrode potentials ($E°$) relative to a standard hydrogen reference electrode. All electrode reactions are written as reduction reactions to the right, and all electrode potentials are given as reduction potentials. Half-reactions with positive reduction potentials occur spontaneously to the right as reduction reactions. Half-reactions with negative reduction potentials occur spontaneously to the left as oxidation reactions. When a half-reaction is written as an oxidation reaction, the sign of the electrode potential is changed.

TABLE 20-4 STANDARD ELECTRODE POTENTIALS (AS REDUCTION POTENTIALS)

	Standard Electrode Potential ($E°$) (in volts)		Standard Electrode Potential ($E°$) (in volts)
$F_2 + 2e^- \rightleftarrows 2F^-$	+2.87	$Fe^{3+} + 3e^- \rightleftarrows Fe$	−0.04
$MnO_4^- + 8H^+(aq) + 5e^- \rightleftarrows Mn^{2+} + 4H_2O$	+1.49	$Pb^{2+} + 2e^- \rightleftarrows Pb$	−0.13
$Au^{3+} + 3e^- \rightleftarrows Au$	+1.42	$Sn^{2+} + 2e^- \rightleftarrows Sn$	−0.14
$Cl_2 + 2e^- \rightleftarrows 2Cl^-$	+1.36	$Ni^{2+} + 2e^- \rightleftarrows Ni$	−0.23
$Cr_2O_7^{2-} + 14H^+(aq) + 6e^- \rightleftarrows 2Cr^{3+} + 7H_2O$	+1.33	$Co^{2+} + 2e^- \rightleftarrows Co$	−0.28
$MnO_2 + 4H^+(aq) + 2e^- \rightleftarrows Mn^{2+} + 2H_2O$	+1.21	$Cd^{2+} + 2e^- \rightleftarrows Cd$	−0.40
$Br_2 + 2e^- \rightleftarrows 2Br^-$	+1.06	$Fe^{2+} + 2e^- \rightleftarrows Fe$	−0.41
$Hg^{2+} + 2e^- \rightleftarrows Hg$	+0.85	$S + 2e^- \rightleftarrows S^{2-}$	−0.51
$Ag^+ + e^- \rightleftarrows Ag$	+0.80	$Cr^{3+} + 3e^- \rightleftarrows Cr$	−0.74
$Hg_2^{2+} + 2e^- \rightleftarrows 2Hg$	+0.80	$Zn^{2+} + 2e^- \rightleftarrows Zn$	−0.76
$Fe^{3+} + e^- \rightleftarrows Fe^{2+}$	+0.77	$Al^{3+} + 3e^- \rightleftarrows Al$	−1.71
$MnO_4^- + e^- \rightleftarrows MnO_4^{2-}$	+0.56	$Mg^{2+} + 2e^- \rightleftarrows Mg$	−2.38
$I_2 + 2e^- \rightleftarrows 2I^-$	+0.54	$Na^+ + e^- \rightleftarrows Na$	−2.71
$Cu^{2+} + 2e^- \rightleftarrows Cu$	+0.34	$Ca^{2+} + 2e^- \rightleftarrows Ca$	−2.76
$Cu^{2+} + e^- \rightleftarrows Cu^+$	+0.16	$Ba^{2+} + 2e^- \rightleftarrows Ba$	−2.90
$S + 2H^+(aq) + 2e^- \rightleftarrows H_2S(aq)$	+0.14	$K^+ + e^- \rightleftarrows K$	−2.92
$2H^+(aq) + 2e^- \rightleftarrows H_2$	0.00	$Li^+ + e^- \rightleftarrows Li$	−3.04

The magnitude of the electrode potential measures the tendency of the reduction half-reaction to occur as the equation is written in Table 20-4. The half-reaction at the bottom of the column has the least tendency toward reduction (adding electrons). Stated in another way, it has the greatest tendency to occur as an oxidation (yielding electrons). The half-reaction at the top of the column has the greatest tendency to occur as a reduction. Thus it has the least tendency to occur as an oxidation.

The higher a half-reaction's position is in the column, the greater is the tendency for its reduction reaction to occur. The lower a half-reaction is in the column, the greater is the tendency for the oxidation reaction to occur. For example, potassium (K) has a large negative electrode potential and a strong tendency to form K^+ ions. Thus, potassium is a strong reducing agent. Fluorine has a large positive electrode potential and a strong tendency to form F^- ions. Fluorine, then, is a strong oxidizing agent. If you compare the listings in Table 20-3 with those in Table 20-4, you will see that the former is a shortened and inverted form of the latter (inverted since activity series of elements are traditionally listed in descending order of reducing strength of metals).

Standard electrode potentials can be used to predict the direction in which an electrochemical reaction proceeds spontaneously. Sample Problem 20.3 illustrates the method that is followed. *Note that when a half-reaction is multiplied by a constant, the $E°$ value is not multiplied, but remains the same.* When a half-reaction is reversed, however, the sign of $E°$ is changed.

Figure 20-19 This Zn-Cu cell from a laboratory is being hooked up. Compare this picture to the diagram in Figure 20-16, and see if you can identify all of the parts.

Sample Problem 20.3

Two half-cells, one containing Fe^{3+} and Fe and the other containing Ag^+ and Ag are connected to form a voltaic cell. Use Table 20-4 to determine the direction of spontaneous reaction and the value of the electrode potential ($E°$).

Solution

Step 1. Analyze   Given: Fe^{3+}–Fe half-cell: $E° = -0.04$ V
                              Ag^+–Ag half cell: $E° = +0.80$ V
                    Unknown: reaction direction, $E°$

Step 2. Plan   half-cells → half-reaction → overall reaction → $E°$

Step 3. Compute   From the $E°$ values of the half-reaction, the combination most likely to result in a spontaneous reaction is for the Ag^+ – Ag half-cell to undergo reduction ($E° = +0.80$ V) and for the Fe–Fe^{3+} half-cell to undergo oxidation ($E° = +0.04$) since this is the combination that will result in a positive $E°$ value.

$$Fe(s) \rightarrow Fe^{3+}(aq) + 3e^- \qquad E° = +0.04 \text{ V}$$
$$3[Ag^+(aq) + e^- \rightarrow Ag(s)] \qquad E° = +0.80 \text{ V}$$
$$\overline{Fe(s) + 3Ag^+(aq) \rightarrow Fe^{3+}(aq) + 3Ag(s) \qquad E° = +0.84 \text{ V}}$$

The relative $E°$ value confirms that the overall reaction does proceed spontaneously as written since a positive value results.

Step 4. Evaluate   The answer contains the proper number of digits and its sign corresponds to that of the $E°$ of a spontaneous electrochemical reaction.

Sample Problem 20.3 *continued*

Practice Problems
Given the following half-cells, determine the overall electrochemical reaction that proceeds spontaneously and the value of $E°$.
1. $Ni^{2+}-Ni$, $Fe^{3+}-Fe^{2+}$ *(Ans.)* $Ni + 2Fe^{3+} \rightarrow Ni^{2+} + 2Fe^{2+}$, $E° = +1.00$ V
2. $Hg^{2+}-Hg$, $Cr^{3+}-Cr$ *(Ans.)* $3Hg^{2+} + 2Cr \rightarrow 3Hg + 2Cr^{3+}$, $E° = +1.59$ V

1. A system of electrodes and electrolytes in which a spontaneous redox reaction is used as a source of electrical energy
2. A system of electrodes and electrolytes in which a current is used to drive a nonspontaneous redox reaction
3. Refer to "Electrolysis of water" for discussion.
4. Refer to "Electrolysis of aqueous salt solutions" for discussion.
5. Refer to "Electroplating" for explanation.
6. A cell that combines the redox chemistry of both electrochemical and electrolytic cells
7. The potential difference between an electrode and its solution; its magnitude measures the tendency of reduction half-reactions to occur
8. $K + Na^+ \rightarrow K^+ + Na$
   $E° = +0.21$ V

Section Review

1. What is an electrochemical cell?
2. What is an electrolytic cell?
3. Discuss briefly the electrolysis of water.
4. Discuss briefly the electrolysis of an aqueous salt solution.
5. Explain the process of electroplating.
6. What is a rechargeable cell?
7. What is electrode potential, and how is it used to calculate information about an electrochemical reaction?
8. Given the Na^+-Na and K^+-K half cells, determine the overall electrochemical reaction that proceeds spontaneously and the $E°$ value.

Chapter Summary

- In the oxidation-number method for balancing equations, oxidation numbers are assigned to all atoms taking part in the reaction. Electronic equations representing the shift in electrons are then written and balanced. Coefficients from the balanced electronic equations are then used to balance the redox equation.
- In the ion-electron method for balancing equations, separate oxidation and reduction equations are balanced for atoms and charge. A balanced net ionic equation is then obtained.
- In redox reactions, the substance that is *reduced* acts as an *oxidizing* agent because it *acquires* electrons from the substance oxidized.
- The substance that is *oxidized* in redox reactions is a *reducing* agent because it *supplies* the electrons to the substance reduced.
- Some oxidation–reduction reactions occur spontaneously and may be used as sources of electrical energy when arranged in electrochemical cells. Other oxidation–reduction reactions that are not spontaneous can be driven by an external source of electric energy in a process called electrolysis.
- The oxidation or reduction reaction between an electrode and its electrolyte in an electrochemical cell is called a half-reaction.
- The potential difference between the electrode and its solution is called the electrode potential. The sum of the electrode potentials of the two half-reactions of a voltaic cell is roughly equal to the potential difference across the cell.
- Standard electrode potentials, which are measured relative to a standard hydrogen half-cell, indicate the relative strengths of substances as oxidizing and reducing agents.

Chapter 20 Review

Oxidation–Reduction Reactions

VOCABULARY

anode (612)
auto-oxidation (610)
cathode (612)
electrochemical cell (611)
electrochemistry (611)
electrode (611)
electrode potential (618)
electrolysis (612)
electrolytic cell (612)
electroplating (615)
half-cell (618)
half-reaction (612)
oxidation (597)
oxidized (597)
oxidizing agent (608)
oxidation-reduction reaction (599)
redox reaction (599)
reduced (597)
reducing agent (608)
reduction (597)
standard electrode potential (618)
voltaic cell (617)

REVIEWING CONCEPTS

1. **a.** Distinguish between the processes of oxidation and reduction.
   b. Write an equation to illustrate each. (20.1)

2. Which of the following are redox reactions?
   a. $2Na + Cl_2 \rightarrow 2NaCl$
   b. $C + O_2 \rightarrow CO_2$
   c. $2H_2O \rightarrow 2H_2 + O_2$
   d. $NaCl + AgNO_3 \rightarrow AgCl + NaNO_3$
   e. $NH_3 + HCl \rightarrow NH_4^+ + Cl^-$
   f. $2KClO_3 \rightarrow 2KCl + 3O_2$
   g. $H_2 + Cl_2 \rightarrow 2HCl$
   h. $H_2SO_4 + 2KOH \rightarrow K_2SO_4 + 2H_2O$
   i. $Zn + CuSO_4 \rightarrow ZnSO_4 + Cu$ (20.1)

3. For each oxidation-reduction reaction in Question 2, identify: the substance oxidized; and the substance reduced. (20.1)

4. **a.** Identify the most active reducing agent among all common elements.
   b. Why are all of the elements in its Periodic Table group very active reducing agents?
   c. Identify the most active oxidizing agent among the common elements. (20.3)

5. Based on Table 20-3, identify the strongest and weakest reducing agents among the substances listed within each of the following groupings:
   a. Ca, Ag, Sn, Cl^-
   b. Fe, Hg, Al, Br^-
   c. F^-, Pb, Mn^{2+}, Na
   d. Cr^{3+}, Cu^{2+}, NO_3^-, K^+
   e. Cl_2, S, Zn^{2+}, Ag^+
   f. Li^+, F_2, Ni^{2+}, Fe^{3+} (20.3)

6. Use Table 20-3 to respond to each of the following:
   a. Would Al be oxidized by Ni^{2+}?
   b. Would Cu be oxidized by Ag^+?
   c. Would Pb be oxidized by Na^+?
   d. Would F_2 be reduced by Cl^-?
   e. Would Br_2 be reduced by Cl^-? (20.3)

7. Distinguish between an electrochemical cell and an electrolytic cell in terms of
   a. the nature of the reaction involved
   b. the signs of the two electrodes. (20.4)

8. What determines the specific products that result from the electrolysis of an aqueous salt solution? (20.4)

9. **a.** What is electroplating?
   b. Distinguish between the nature of the anode and cathode in such a process. (20.4)

10. **a.** Explain what is meant by the potential difference between the two electrodes in an electrochemical cell.
    b. How, and in what units, is this potential difference measured? (20.4)

11. The standard hydrogen electrode is assigned an electrode potential of 0.00 V. Explain why this voltage is assigned. (20.4)

12. **a.** What information is provided by the electrode potential of a given half-cell?
    b. What does the relative value of the potential of a given half-reaction indicate about its oxidation-reduction tendency? (20.4)

PROBLEMS

▶ Redox Equations

1. Each of the following atom/ion pairs undergoes the oxidation number change indicated below. For each pair, determine whether oxidation or reduction has occurred, and then write the electronic equation indicating the corresponding number of electrons lost or gained:
   a. $K \rightarrow K^+$
   b. $S \rightarrow S^{2-}$
   c. $Mg \rightarrow Mg^{2+}$
   d. $F^- \rightarrow F_2$
   e. $H_2 \rightarrow H^+$
   f. $O_2 \rightarrow O^{2-}$
   g. $Fe^{3+} \rightarrow Fe^{2+}$
   h. $Mn^{2+} \rightarrow Mn^{7+}$

2. Determine the oxidation number of each atom indicated in the following:
   a. H_2
   b. H_2O
   c. Al
   d. MgO
   e. Al_2S_3
   f. HNO_3
   g. H_2SO_4
   h. $Ca(OH)_2$
   i. $Fe(NO_3)_2$
   j. O_2

3. Balance the oxidation-reduction equation below by using the oxidation-number method in response to each requested step:

   $$Fe + HCl \rightarrow FeCl_2 + H_2$$

   a. Write the skeleton equation for the reaction.
   b. Assign oxidation numbers to all atoms and determine what is oxidized and what is reduced.
   c. Write separate electronic equations for the oxidation and reduction processes, and balance each of them so that the number of electrons lost is equal to the number gained.
   d. Place the coefficients from the two electronic equations into the skeleton equation and then balance the overall equation so that atoms are conserved.

4. Use the oxidation-number method as outlined in Problem 3 to balance each of the following redox equations:
   a. $Li + H_2O \rightarrow LiOH + H_2$
   b. $Al + HCl \rightarrow AlCl_3 + H_2$
   c. $H_2S + HNO_3 \rightarrow H_2SO_4 + NO_2 + H_2O$
   d. $Cu + HNO_3 \rightarrow Cu(NO_3)_2 + NO + H_2O$
   e. $KClO_3 \rightarrow KCl + O_2$

5. Zinc reacts with nitric acid to form zinc nitrate, ammonium nitrate, and H_2O. Balance the redox equation by the oxidation-number method.

6. Zinc reacts with sodium chromate in a sodium hydroxide solution to form Na_2ZnO_2, $NaCrO_2$, and H_2O. Balance the redox equation using the oxidation-number method.

7. Balance the oxidation-reduction equation below by using the ion-electron method in response to each requested step:
   $K + H_2O \rightarrow KOH + H_2$.
   a. Write the skeleton equation for the reaction.
   b. Write the net ionic equation and assign oxidation numbers to all atoms to determine what is oxidized and what is reduced.
   c. Write the equation for the reduction and balance it for both atoms and charge.
   d. Write the equation for the oxidation and balance it for both atoms and charge.
   e. Adjust the oxidation and reduction equations by multiplying the coefficients as needed so that electrons lost equals electrons gained and add the two resulting equations.
   f. Add species as necessary to balance the overall formula equation.

8. Use the ion-electron method in Problem 7 to balance each of the reactions below.
   a. $HI + HNO_2 \rightarrow NO + I_2 + H_2O$
   b. $FeCl_3 + H_2S \rightarrow FeCl_2 + HCl + S$
   c. $SbCl_5 + KI \rightarrow KCl + I_2 + SbCl_3$
   d. $Ca(OH)_2 + NaOH + ClO_2 + C \rightarrow NaClO_2 + CaCO_3 + H_2O$

9. Hot, concentrated sulfuric acid reacts with zinc to form zinc sulfate, hydrogen sulfide, and water. Balance the equation, using
   a. the oxidation-number method;
   b. the ion-electron method.

▶ Electrochemical and Electrolytic Cells

10. Given the following half-cells, determine the overall electrochemical reaction that proceeds spontaneously and the value of $E°$:
    a. $Cu^{2+} - Cu, Ag^+ - Ag$
    b. $Cd^{2+} - Cd, Co^{2+} - Co$

c. $Na^+ \rightarrow Na$, $Ni^{2+} \rightarrow Ni$
d. $I_2 \rightarrow I^-$, $Br_2 \rightarrow Br^-$

11. Predict whether each of the following reactions will occur spontaneously as written by determining the $E°$ value for potential reaction. Write and balance the overall equation for each reaction that does occur.
    a. $Mg + Sn^{2+} \rightarrow$
    b. $K + Al^{3+} \rightarrow$
    c. $Li^+ + Zn \rightarrow$
    d. $Cu + Cl_2 \rightarrow$

12. Suppose chemists had chosen to make the $I_2 + 2e^- \rightleftarrows 2I^-$ half-cell the standard electrode and assigned it a potential of zero volts.
    a. What would be the $E°$ value for the $Br_2 + 2e^- \rightleftarrows 2Br^-$ half-cell?
    b. What would be the $E°$ value for the $Al^{3+} + 3e^- \rightleftarrows Al$ half-cell?
    c. How much change would be observed in the $E°$ value for the reaction involving $Br_2 + I^-$ using the I_2 half-cell as the standard?

13. If a strip of Ni is dipped into a solution of $AgNO_3$, what would be expected to occur? Explain, using $E°$ values and equations.

14. a. What would happen if an aluminum spoon were used to stir a solution of $Zn(NO_3)_2$?
    b. Could a strip of Zn be used to stir a solution of $Al(NO_3)_3$? Explain, using $E°$ values.

15. Can a solution of $Sn(NO_3)_2$ be stored in an aluminum container? Explain, using $E°$ values.

CRITICAL THINKING

1. **Applying Models** Explain how the oxidation-reduction chemistry of both the electrochemical cell and the electrolytic cell are combined in the chemistry of rechargeable cells.

2. **Applying Ideas** In lead/acid batteries, such as your car battery, it is possible to determine the charge of the battery by measuring the density of the battery fluid. Explain how this is possible.

3. **Interpreting Graphics** A galvanic cell is pictured to the right. The temperature reading is 25°C. Identify the species that is oxidized if current is allowed to flow.

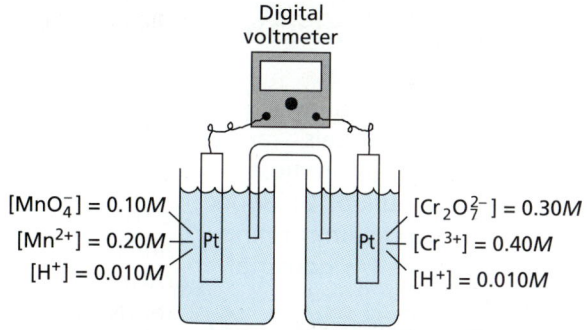

CROSS CONNECTIONS

1. **Consumer Skills** In our portable society, batteries have become a necessary power supply. As consumers, we want to purchase batteries that wil provide as much use time as possible. Advertisements tell us that some batteries last longer than others, but do they really? Design an investigation to answer the question, is there difference in usage time among the major brands of AA batteries? Add into your design a cost effectiveness component.

RESEARCH AND WRITING

1. Go to the library and find out what you can on the electroplating industry in the United States. What are the top three metals used for plating and how many metric tons of each are used in the United States each year for electroplating?

PROJECT

1. Take an inventory of the type of batteries used in your house. Find out the voltage supplied by each battery and what type of electrochemical reaction each uses. See if you can provide a reason why each type of electrochemical reaction is used in each case.

Chemistry and Careers

Industrial Technician

Thousands of chemical products—dyes, cleaning agents, drugs, and fuels—are produced every day. In most industries, the bulk of the work that keeps production running smoothly is done by industrial technicians.

Industrial technicians run the equipment that is used to make chemical products. A technician receives shipments of raw materials from other manufacturers, measures and mixes the materials, and runs production equipment. In electronic and pharmaceutical industries, in which the product must be kept sterile or dust-free, the technician controls remote or robotic equipment. Other technicians monitor temperature and pressure gauges during chemical processing. Mechanical skills help a technician spot and solve problems such as jamming or slippage, which may cause equipment to malfunction.

Most industrial technicians have a high school diploma or an associate's degree certifying their competence in a specific technical area. The ability to work well with others is important for industrial technicians who wish to become supervisors.

Environmental Chemist

An environmental chemist determines the identity and level of pollutants in various parts of our ecosystem.

To determine the composition of soil and water samples, environmental chemists use spectroscopy, titration, and other tools of analytical chemistry. Chemists who are interested in air pollution study the ozone layer, health hazards presented by leaded gasoline, and acid precipitation.

Analysis of waste is another major area of environmental work. Hazardous waste is often disposed of improperly. Sometimes it is dumped into open pits, where rain leaches soluble materials into nearby streams. Environmental chemists are needed to analyze problems and to propose treatments to prevent further pollution.

Many job opportunities exist for chemists interested in environmental issues: in the field gathering samples, in small businesses devoted to testing materials, in industries concerned about their toxic wastes, and at government research agencies. Most environmental chemists have at least a bachelor's degree in chemistry, though a high school diploma is sufficient for some technical jobs.

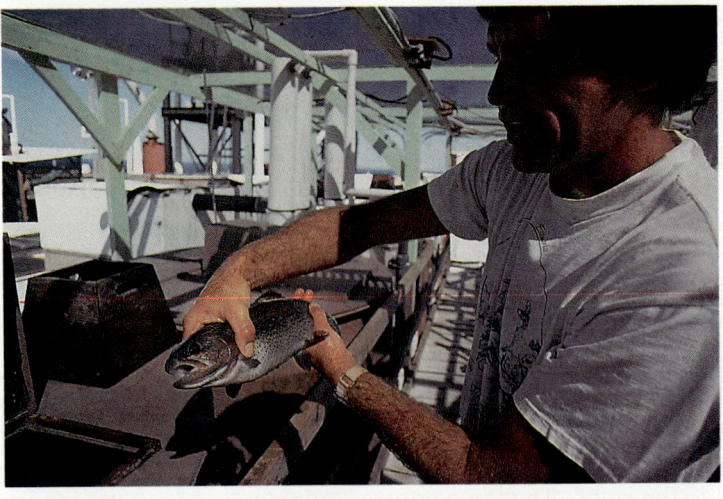

UNIT 6 • FOCUS ON REASONING

College Entrance Exam Preparation

All chemical reactions either absorb or release energy. The amount of energy absorbed or released depends upon the relative potential energies of the reactants and products. In addition, for a chemical reaction to occur, there is a minimum amount of energy that must be absorbed to get the reaction started. The minimum energy is called activation energy. The function of activation energy in the course of a reaction can be shown in the diagrams below.

Reaction 1

1. On the basis of these diagrams, one can conclude that the function of a catalyst is to
   a. increase the potential energy of the products.
   b. increase the potential energy of the reactants.
   c. shorten the reaction.
   d. reduce the activation energy.

2. Which of the following factors represents a reaction that releases energy?
   a. the products have a higher potential energy than the reactants
   b. the products have a lower potential energy than the reactants
   c. the products are formed over a very short period of time
   d. a catalyst is present to speed up the formation of product

3. To find the actual potential energy difference for reaction 2, you would subtract
   a. the highest potential energy point from the lowest potential energy point.
   b. potential energy of the product from the lowest potential energy point on the diagram.
   c. potential energy of the reactant from the potential energy of the product.
   d. potential energy of the reactant from a higher potential energy point on the diagram.

627

INTRA-SCIENCE

Chemiluminescence

As small children, we learn (sometimes painfully) that most sources of light—flames and incandescent lights, for example—are dangerously hot. When we get older, we are fascinated to discover that some substances are luminescent—that is, they emit light when they are at harmless temperatures. The most surprising type of luminescence is chemiluminescence, which is the reaction of two or more substances in complete darkness to produce light without becoming hot.

If hydrogen peroxide is added to a solution of the organic compound luminol, a chemical reaction produces a beautiful greenish blue luminescence. This continues until one or both of the reagents have been chemically consumed. Most chemiluminescent reactions are the result of oxidations in which an organic compound—such as luminol—is oxidized by oxygen, hydrogen peroxide, or ozone.

One familiar example of this is the operation of a Bunsen burner. The blue color of the innermost cone of a Bunsen burner flame is the result of a chemiluminescent reaction between oxygen from the air and natural gas. The innermost part of the flame is relatively cool. The heat of a Bunsen burner flame is produced when the products of this chemiluminescent reaction are further oxidized above the upper tip of the blue cone.

The Connection to Physics

Photons of light—particles of radiant energy—are produced when an atom or molecule changes from an excited state to a lower-energy state. This change involves an electron moving from an atomic or molecular orbital that is relatively far from the corresponding nucleus (where it has a relatively high energy) to an orbital closer to the nucleus (where it has a lower energy). The energy of the emitted photon is equal to the energy difference between the higher-energy and lower-energy orbital. The energy of the emitted photon determines whether the corresponding light is colored and, if so, what its color is.

Where the types of light production differ is in the method used to produce excited atoms or molecules. In ordinary incandescent lights and in flames, this excitation is caused by heat. In luminescence, however, atoms or molecules are excited without becoming hot. Sonoluminescence is a type of luminescence in which electrons of atoms and molecules are excited by high-frequency sound waves. In fluorescence and phosphorescence, atoms or molecules are excited by absorbing light. In fluorescence, the excited species re-emit light almost immediately (within 1/100 000 000 of a second). In phosphorescence, this re-emission occurs relatively slowly. Consequently, a phosphorescent object may glow for many minutes in the dark. In chemiluminescence, chemical reactions produce molecules that are in excited states to begin with. Light is emitted as these molecules drop to their lowest energy state.

The Connection to Biology

Long before the general phenomenon of chemiluminescence was recognized and understood, people had been fascinated by the light-producing ability of fireflies and glowworms (the larvae of fireflies), and by foxfire (the glow produced by certain fungi). This emission of light by living organisms is called bioluminescence.

During the last thirty years, it has been recognized that bioluminescence is a type of chemiluminescence. In each case of bioluminescence, a particular biochemical substance called luciferin is oxidized, usually by oxygen, in a reaction catalyzed by an enzyme called luciferase. Each bioluminescing species has its own type of luciferin and luciferase.

Many of the bioluminescing species are marine bacteria and fungi. The flashlight fish, found in the Pacific and Indian oceans, uses luminous bacteria, which it harbors in its light organ for a variety of purposes. It communicates with other flashlight fish by blinking—opening and closing a lid over the light organ. Feeding only on dark nights, the flashlight fish uses this light to find and capture its prey. It can also confuse and evade its predators by flashing the light on and off while zigzagging away.

The small crustacean *Cypridina*, found in Japanese waters, makes luciferin and luciferase in separate glands. When it is attacked, it squirts them into water, where they react, producing a ball of light. *Cypridina* can use this technique to escape predators, which will pursue the ball of light while the crustacean escapes.

During mating season, the firefly uses its light to attract the opposite sex. Each species of firefly has its own particular light signal so that it will not attract a firefly of a different species.

The Connection to Health Sciences

The bioluminescent reaction that occurs in the firefly requires adenosine triphosphate (ATP) in addition to luciferin and luciferase. Because even extremely dim light can be detected and measured, a mixture of luciferin and luciferase provides a sensitive method for detecting ATP, a crucial energy-carrying molecule present in most living cells. This bioluminescent reaction is used in cancer research to detect the quantity of ATP in cells under different conditions.

Unit 7

Chapter 21
Carbon and Hydrocarbons   632

Chapter 22
Substituted Hydrocarbons   664

Chapter 23
Biochemistry   684

Carbon and Its Compounds

I vividly remember washing and breaking many dishes during my childhood. If allowed to continue, we would have been eating off paper napkins. In time, however, my parents and I gained a great appreciation for the unbreakable nature of plastic cups and dinnerware.

While in high school, I was introduced to organic polymer chemistry and the notion that most plastics are the result of adding together a number of small molecules. This concept is comparable to building a long train from individual cars. Presently, I am synthesizing new, highly flexible, impact-resistant polymers that will be used in applications ranging from transparent airplane windows to high-gloss automotive paint.

Much of the effort in developing new polymers is toward mimicking the properties of natural substances. The subtle beauty of silk, for instance, has been valued in clothing for almost 5000 years. Silk fibers are made from the cocoons of silkworm larvae. Today, fabrics containing polymers such as polyesters can be produced with many of the prized qualitities of silk.

Van I. W. Stuart
Organic/Polymer Chemist
General Electric Corporation

Silkworm spinning cocoon

CHAPTER 21
Carbon and Hydrocarbons

PLANNING CHART

Approximate Time: 7/9 days

	SECTION/TEXT FEATURES		RESOURCES	
B A S I C	**21.1 ABUNDANCE AND IMPORTANCE OF CARBON** Demo 1 Section Review Chapter Review • Questions 1–10, Problem 1	633 632A 638 661	*Teacher's Resource Binder* Lab Experiment 24 Section Review Worksheets *Other Resources* Transparency 49	135 75
B A S I C	**21.2 ORGANIC COMPOUNDS** Demos 2, 3 Section Review Chapter Review • Questions 11–13, Problems 20–21	639 632B–C 643 661	*Teacher's Resource Binder* Section Review Worksheets *Other Resources* *World of Chemistry* Videos 21, 26 Transparency 50	76
B A S I C	**21.3 HYDROCARBONS** Demo 4 Section Review Chapter Review • Questions 14–24, Problems 2–19	643 632D 655 661	*Teacher's Resource Binder* Lab Experiment 40 Section Review Worksheets *Other Resources* Transparencies 51, 52 Blackline Master 50	205 77
B A S I C	**21.4 REPRESENTATIVE HYDROCARBONS AND POLYMERS** Section Review Chapter Review • Questions 25–28, Problems 23–24	655 600 661	*Teacher's Resource Binder* Section Review Worksheets *Other Resources* *Shakhashiri* Video Demos 15, 16, 17, 18, 19 *World of Chemistry* Videos 2, 9, 22 Blackline Master 51	78
	CHAPTER END ACTIVITIES Critical Thinking, Cross Connections, Research and Writing, Projects		*Teacher's Resource Binder* Test Booklet *Computer Test Generator*	105, 107

Teaching Strategies

Introduction

Start this chapter by pointing out that almost everything that will be discussed will be something that touches the students' lives in some way. Some of the substances you will discuss are: methane (the main component of natural gas); propane (bottled gas, which many of them use to heat their homes); gasoline (used to drive their cars); and kerosene (used to drive big diesel trucks, which deliver many of the commodities they buy in grocery stores, department stores, and at service stations). Many of the plastics and rubber they use are made from some component of crude oil or petroleum.

At the same time, much of the chapter will seem very foreign and different from the chemistry that has been discussed previously. Even though structural formulas have already been covered, the idea that every compound has its own individual structural formula and that many compounds have the same chemical formulas is difficult for students to grasp. Emphasize that similarity in structure enables many compounds to be grouped together and that similarity causes them to react in a similar way. In this chapter and the following chapter, students will see how over 4 000 000 compounds can be placed in a relatively small number of groups on the basis of similarity in structure. After they understand the basis for naming, they will begin to see the pattern for naming all classes of organic compounds.

21.1 Abundance and Importance of Carbon

Demonstration 1: Bonding of Carbon Atoms
PURPOSE: To demonstrate the bond angles of carbon atoms and

how they are able to link with one another to form chains, rings, plates, and networks

MATERIALS: Gumdrops, scissors, tape, toothpicks, transparencies, enlarged photocopied drawings of Figure T21-1

Figure T21-1

PROCEDURE: Prior to class, reproduce the drawing in Figure T21-1 on a sheet of paper $8\frac{1}{2} \times 11''$. You will be able to get four per page. Make as many photocopies as needed, and make transparencies using the photocopies of the drawings.

Have the students work in groups of two. First make a transparent model of a tetrahedron by cutting out the large equilateral triangle and bending up the sides of the tetrahedron. Tape two of the sides together, and leave one side open to insert the toothpick-gumdrop model of the carbon atom with its four covalent bonds. (See Figure T21-2.) Students may have to reinsert toothpicks a few times to get them at the correct angles so they go to the corners of the tetrahedron. Now the students have a model from which to work when making models through out this chapter.

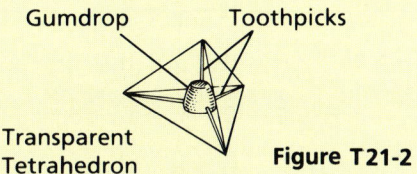

Figure T21-2

Using the first model as a guide, make additional carbon atoms with their bonds and join them in chains and rings, or even plates and networks, to demonstrate how they join with one another. (See Figure T21-3.) Save these models for further use in this chapter.

Figure T21-3

This demonstration lays the groundwork for understanding structural formulas of organic compounds.

As you discuss diamonds, you might use a diamond phonograph needle to scratch a piece of glass to illustrate its hardness. With certain cuts of fake diamonds it is sometimes difficult even for jewelers to determine if they are fake or not. The ultimate test is to discover if it can be scratched by another diamond. If it can be scratched, it is not a real diamond; if it cannot, it is real!

As you discuss graphite, exhibit a sample of powdered graphite and different substances or objects containing graphite, such as graphite lubricants and greases, graphite electrodes, pencil lead, and tennis rackets or fishing rods made of graphite fibers. You can make gumdrop models of graphite plates and use them to show how these plates slip over one another with ease. This explains why graphite feels greasy and is a good lubricant.

When you discuss coal, exhibit different types of coal. Along with the discussion of destructive distillation, show the different products of destructive distillation with their original form. For example, a piece of wood with a charcoal briquet, coal with coke, and a bone with boneblack. Remind students that they will be making charcoal in the lab. After they have completed the lab, you might ask them to design a similar lab to make coke.

21.2 Organic Compounds

It is very important that all discussions in this section and future sections dealing with the structure of organic molecules be accompanied by three-dimensional models that can be translated into two-dimensional structural formulas. Many types of commercial model kits are available. In addition to the ball-and-stick model kits, which represent all atoms as being the same size, newer, more sophisticated ones are now being produced with balls of varying sizes. The ball-and-stick models make it easier to see the arrangement of the atoms in the molecules. Since in reality the atoms are much closer together, the compact models more accurately represent the atoms in the molecule.

Most molecular model kits are fairly expensive. If your school can afford them, it is a good idea to have a ball-and-stick set and a compact set for demonstration purposes.

Let the students work along with you, building their own models using different colored gumdrops for different atoms, toothpicks for single-covalent bonds, and pipecleaners for double and triple bonds.

There are several advantages to using gumdrop models: *(1)* They are cheap enough that the students can build their own models. *(2)* You can quickly see if the students understand the three-dimensional nature of the discussion. *(3)* Gumdrops come in at least two sizes, so larger ones can be used for large atoms and smaller ones for small atoms. *(4)* Students can make ball-and-stick gumdrop models first and then push the gumdrops together into the compact form, which is the best representative of the molecule. Not even the commercial models can do this!

Demonstration 2: Building Models and Writing Structural Formulas of C_5H_{12} Isomers

PURPOSE: To help students develop a better understanding of isomers and structural formulas by building three-dimensional models of the three isomers having the formula C_5H_{12} and writing their structural formulas using the models

Carbon and Hydrocarbons **632B**

MATERIALS: Kits composed of 15 large gumdrops of one color, 36 small gumdrops of a different color, 51 toothpicks, the gumdrop-toothpick model of a carbon atom with its four hybrid covalent bonds from Demonstration 1, a commercial molecular model kit (optional)

PROCEDURE: If possible, have students work along with you using their own kits. Using the model from Demonstration 1 as a guide for bond angles, first build a continuous chain of five carbon atoms, letting the large gumdrops represent the carbon atoms. (See **Figure T21-4**.) Next attach small gumdrops, representing hydrogens, to all toothpicks with unattached ends. (See **Figure T21-5**.) Then write the structural formula for the compound.

Figure T21-4

Figure T21-5

Point out that the zig-zag nature of the three-dimensional model is represented as a straight-chain in the structural formula. Thus, in some ways the structural formula is not a good representation of the model or the actual molecule. To make the ball-and-stick model into a compact model, push the gumdrops together. (See Figure T21-6.)

Figure T21-6

Next make a model of the C_5H_{12} isomer having the structural formula:

$$\begin{array}{c} \text{H} \quad \text{H} \quad \text{H} \quad \text{H} \\ | \quad | \quad | \quad | \\ \text{H}-\text{C}-\text{C}-\text{C}-\text{C}-\text{H} \\ | \quad | \quad | \quad | \\ \text{H} \quad | \quad \text{H} \quad \text{H} \\ \text{H}-\text{C}-\text{H} \\ | \\ \text{H} \end{array}$$

By using the model and turning it in different directions, you can help students see that the following three formulas are the same as the first formula:

$$\begin{array}{c} \text{H} \quad \text{H} \quad \text{H} \quad \text{H} \\ | \quad | \quad | \quad | \\ \text{H}-\text{C}-\text{C}-\text{C}-\text{C}-\text{H} \\ | \quad | \quad | \quad | \\ \text{H} \quad \text{H} \quad | \quad \text{H} \\ \text{H}-\text{C}-\text{H} \\ | \\ \text{H} \end{array} \qquad \begin{array}{c} \text{H} \\ | \\ \text{H}-\text{C}-\text{H} \\ | \\ \text{H} \quad | \quad \text{H} \quad \text{H} \\ | \quad | \quad | \quad | \\ \text{H}-\text{C}-\text{C}-\text{C}-\text{C}-\text{H} \\ | \quad | \quad | \quad | \\ \text{H} \quad \text{H} \quad \text{H} \quad \text{H} \end{array}$$

and

$$\begin{array}{c} \text{H} \\ | \\ \text{H}-\text{C}-\text{H} \\ | \\ \text{H} \quad \text{H} \quad | \quad \text{H} \\ | \quad | \quad | \quad | \\ \text{H}-\text{C}-\text{C}-\text{C}-\text{C}-\text{H} \\ | \quad | \quad | \quad | \\ \text{H} \quad \text{H} \quad \text{H} \quad \text{H} \end{array}$$

Last, make a model of the isomer with the structural formula,

$$\begin{array}{c} \text{H} \\ | \\ \text{H} \quad \text{H}-\text{C}-\text{H} \quad \text{H} \\ | \quad | \quad | \\ \text{H}-\text{C}-------\text{C}-------\text{C}-\text{H} \\ | \quad | \quad | \\ \text{H} \quad | \quad \text{H} \\ \text{H}-\text{C}-\text{H} \\ | \\ \text{H} \end{array}$$

Demonstration 3: Differences Between Organic and Inorganic Compounds

PURPOSE: To show how organic compounds differ from inorganic compounds on the basis of (1) solubility in polar and nonpolar solvents, and (2) rate of reaction

MATERIALS: Ammonium chloride, dilute ammonia water, cornstarch, 0.5-M iron(III) chloride solution, kerosene, mineral oil, naphthalene, potassium nitrate, saliva, 0.1-M silver nitrate solution, 0.1-M sodium chloride solution, sucrose, distilled water, dry yeast, 2 125-mL beakers, burner, ringstand and iron ring, stirring rod, 8 rubber stoppers to fit large test tubes, 4 medium test tubes (15 × 150 mm), 8 large test tubes (25 × 150 mm), wire gauze. CAUTION: Do not allow $AgNO_3$ to come into contact with your skin or clothing. Wear gloves, goggles, and an apron.

PROCEDURE: *(1)* Add 20 mL of water to each of four large tubes. Check for solubility of each of the following by adding the indicated amount to the water: 0.5 g of NH_4Cl, 0.5 g KNO_3, 0.5 g of naphthalene, and 5 mL of kerosene. Stopper and shake each test tube. Which substances dissolved in the water? (NH_4Cl and KNO_3.) Why did this happen? (NH_4Cl and KNO_3 are inorganic compounds; inorganic compounds generally dissolve in water. Naphthalene and kerosene are organic compounds that do not dissolve in water.) *(2)* Repeat part *(1)* but substitute mineral oil for the water. Results? (The NH_4Cl and KNO_3 did not dissolve, but the naphthalene and the kerosene did dissolve.) Why? (Naphthalene and kerosene are organic compounds and mineral oil is an organic liquid; organic substances dissolve in other organic substances.) Like dissolves like: nonpolar solvents dissolve nonpolar substances; polar solvents dissolve polar and ionic substances. *(3)* Mix 5 mL of $FeCl_3$ solution with 5 mL of ammonia water in a medium test tube. Result? (A brown precipitate forms immediately.) *(4)* Mix 5 mL of NaCl solution with 1 mL of silver nitrate solution in a medium test tube. Result? (A white precipitate forms immediately.) *(5)* Add 5 g of dry yeast and 1 g of sucrose to 40 mL of water in a beaker and stir. Result? (Nothing immediately, but by the next day gas bubbles are forming, indicating a reaction is taking place.) *(6)* Before class, dissolve 2 g of cornstarch in 25 mL of cold water in a beaker and gently heat while stirring until it thickens. Cool. During class, add about 1 mL of saliva to the cool, thick starch mixture. Result? (Nothing happens immediately, but in 5-10 minutes it changes to a thin, watery mixture.) How do you explain these results? (The organic reactions in parts *(5)* and *(6)* proceed at much slower rates than the inorganic reactions in parts *(3)* and *(4)*.)

21.3 Hydrocarbons

To introduce this section, display several models of alkanes, alkenes, and alkynes. Remember, if you use gumdrop models, sections of pipecleaners work very nicely for the flexibility needed for double and triple bonds. By using models, students will be able to quickly see the differences and similarities between the hydrocarbon classes.

Demonstration 4: Tests for Alkanes and Alkenes
PURPOSE: To show how to differentiate between alkanes and alkenes on the basis of the reactivity of the single bond in alkanes and double bonds in alkenes
MATERIALS: 1,1,2-trichloro-1,2,2-trifluoroethane, cyclohexane, cyclohexene, iodine, potassium permanganate, sodium hydroxide, 4 50-mL beakers, 2 stirring rods
PROCEDURE: CAUTION: Wear goggles, gloves, and a lab coat or lab apron during the preparation and performance of this lab. NaOH and any solution containing NaOH is extremely corrosive to the skin. A great deal of heat is given off when NaOH is added to water. Cyclohexane and cyclohexene are extremely flammable. Store them under a hood until ready, and return them immediately when done. Be sure the room is well-ventilated.

Make the basic potassium permanganate solution by: *(1)* adding 4 g of NaOH to enough water to make 100 mL of solution; and *(2)* right before doing the lab, adding 0.16 g of $KMnO_4$ to the NaOH solution. (Make sure the solution is cool before adding the $KMnO_4$.)

Put 10 mL of cyclohexane and 10 mL of cyclohexene in two separate 50-mL beakers. Place each beaker near a white background so they can be seen. Add 3 mL of the basic $KMnO_4$ solution to the cyclohexane and 3 mL to the cyclohexene. Stir both solutions thoroughly. If no double bonds are present, the solution will remain purple; if double or triple bonds are present, it will turn brown. (Note: Depending on the purity of the alkene, the double-bonded compounds may turn green instead of brown. Either color change indicates the presence of an unsaturated double- or triple-bond.) Results: The cyclohexane and basic $KMnO_4$ solution will remain purple, but in the presence of the cyclohexene the color will change to brown (or possibly green). See unbalanced equations below:

cyclohexane + MnO_4^- → No reaction

$$R-CH=CH-R + MnO_4^- \rightarrow \underset{\underset{OH\ OH}{|\ \ \ |}}{R-CH-CH-R} + \underset{\text{brown}}{MnO_2} \text{ (or } \underset{\text{green}}{MnO_4^{2-}})$$

cyclohexene + MnO_4^- → cyclohexane-1,2-diol

What do the final colors indicate? (Cyclohexane has no double or triple bonds and cyclohexene has at least one double or triple bond. Of course, the name tells us it has a double bond.)

Using the fume hood, make the iodine solution by adding 3 g of iodine to 100 mL of 1,1,2-trichloro-1,2,2-trifluoroethane. Put 10 mL of cyclohexane and 10 mL of cyclohexene in two separate 50-mL beakers. Place each beaker on an overhead projector, and project on a screen. Add 3 mL of the iodine solution to the cyclohexane and 3 mL to the cyclohexene. Stir both solutions thoroughly. If no double bonds are present, the solution will not change color; if double bonds are present, it will turn colorless.

What do the final colors indicate? (Cyclohexane has no double or triple bonds, and cyclohexene does.)

The equations for the reactions follow:

$$\text{C}_6\text{H}_{12} + \text{I}_2 \longrightarrow \text{No reaction}$$

$$\text{C}_6\text{H}_{10} + \text{I}_2 \longrightarrow \text{C}_6\text{H}_{10}\text{I}_2$$

21.4 Representative Hydrocarbons and Polymers

The societal issue of plastic pollution is receiving a great deal of attention. The problems with plastic pollution are increasing yearly. Environmentalists point out some of the serious problems already plaguing our society:

(1) Plastics make up 30% of the volume of solid wastes in city garbage. Plastics, along with paper, are the fastest growing types of garbage. *(2)* All types of garbage are on the increase and places to put them on the decrease. *(3)* Most plastics are not biodegradable and will be in the environment for years. *(4)* Plastic pollution in the ocean is killing fish, turtles, birds, and mammals.

Students could research the problem and do short reports and position papers on it. Some solutions might be: *(1)* developing more biodegradable plastics, *(2)* recycling plastic by encouraging people to return plastic containers with bottle deposits and developing more ways to reuse plastic, *(3)* reducing plastic consumption.

You might wish to set a day aside to present some information on the problem to the class. This is definitely a problem that students will have to deal with in their lifetimes. Three good sources are: Seymour (1989), Thayer (1989), and Weisskopf (1988).

References and Resources

Books and Periodicals

Alper, Joseph. "Polymers," *ChemMatters*, Vol. 4, 1986, pp. 4–7.

Carraher, Charles E. Jr., and Raymond B. Seymour. "Polymer Structure—Organic Aspects (Definitions)," *Journal of Chemical Education*, Vol. 65, 1988, pp. 314–18.

Jaffe, Bernard. "Woehler: Urea without a Kidney," *Crucibles: The Story of Chemistry,* New York: Dover Publications, 1976, pp. 108–24.

Kauffman, George B. "Wallace Hume Carothers and Nylon, The First Completely Synthetic Fiber." *Journal of Chemical Education,* Vol. 65, 1988, pp 803–7.

Marsella, Gail. "Silly Putty," *ChemMatters,* Vol. 4, 1986, pp. 15–17.

Polymer Chemistry: A Teaching Package for Pre-College Teachers, Washington, D.C.: National Science Teachers Association, 1986.

Scheinberg, Stephen P. "Tyvek," *ChemMatters*, Vol. 4, 1986, pp. 8–10.

Seymour, Raymond B. "Alkanes: Abundant, Pervasive, Important, and Essential," *Journal of Chemical Education,* Vol. 66, 1989, pp. 59–63.

Shakhashiri, Bassam Z. "Combustion of Methane, Polystyrene," *Chemical Demonstrations: A Handbook for Teachers of Chemistry*, Vol. 1, Madison: The University of Wisconsin Press, 1983, pp. 241–42.

Shakhashiri, Bassam Z. "Determination of the Molecular Mass of the Gas from a Butane Lighter," *Chemical Demonstrations: A Handbook for Teachers of Chemistry*, Vol. 2, Madison: The University of Wisconsin Press, 1985, pp. 48–50.

Summerlin, Lee R., Christie L. Borgford, and Julie B. Ealy. "The Nonpolar Disk Game," p. 36; "Producing Methane Gas," p. 93; and "Alkanes versus Alkenes: Reaction of the Double Bond," pp. 94–95. *Chemical Demonstrations: A Sourcebook for Teachers*, Vol. 2, Washington, D.C.: American Chemical Society, 1988.

Summerlin, Lee R., and James L. Ealy, Jr. "Production of a Gas: Acetylene," p. 16; and "Synthetic Rubber," pp. 176–77, *Chemical Demonstrations: A Sourcebook for Teachers*, Vol. 1, Washington, D.C.: American Chemical Society, 1988.

Thayer, Ann M. "Solid Waste Concerns Spur Plastic Recycling Efforts," *Chemical & Engineering News*, Vol. 67, 1989, pp. 7–15.

Vernon, William. "Chocolate Chip Petroleum," *The Science Teacher*, Vol. 55, 1988, pp. 105–6.

Weisskopf, Michael. "Plastic Reaps a Grim Harvest in the Oceans of the World," *Smithsonian*, Vol. 18, 1988, pp. 58–67.

Wittcoff, Harold. "Nonleaded Gasoline: Its Impact on the Chemical Industry," *Journal of Chemical Education*, Vol. 1987, pp. 773–76.

Wood, Clair G. "Dissolving Plastic," *ChemMatters*, Vol. 5, 1987, pp. 12–14.

Young, Jay. "The Interrupted Party," *ChemMatters*, Vol. 2, 1984, pp. 4–5.

Audiovisual Resources

Chemistry 120—Organic Chemistry (sound filmstrips) Learning Arts, Wichita, KS, 1988.

Chemistry in Today's World Series—Carbon (Slides and cassette) Educational Images, Elmira, NY, 1988.

Chemistry Transparencies—Crystal structure of a diamond and graphite, Models of methane, ethane and propane, plus nine others, Frey Scientific, Mansfield, OH, 1988–89.

Chemistry Study Cards—Organic Chemistry: Organic Nomenclature and Organic Reactions, Frey Scientific, Mansfield, OH, 1988–9.

Organic Chemistry I and II. TV Ontario, Chapel Hill, NC, 27516.

Computer Software

Flash, Patrick, and Victor Bendall. *Organic Reaction Chemistry*, Apple II, 64K, Queue, Bridgeport, CN, 1988.

Groves, Paul. *An Introduction to Polymerization*, Apple II, National Science Teachers Association, Washington, D.C., 1985.

Smith, Stanley. *Alkanes and Alkenes and Arenes*, Apple II 48K and IBM PC 256K, Queue, Bridgeport, CN, 1988.

Chemistry Series (J & S Software) Organic Chemistry, Apple II, 48K, Barclay, Brooklyn, NY, 1988.

Organic Nomenclature 1, Apple II, 64K and TRS-80 48K, Queue, Bridgeport, CN, 1988.

SEI Organic Chemistry, Apple II, 64K and IBM PC, 128K, Queue, Bridgeport, CN, 1988.

Isomers, Disks AP301 & IB301; *Octane*, Disks AP803 & IB803; *Refinery*, Disk AP806; *Polymerization*, Disk AP705; and *Organic Nomenclature*, Disk AP705, Project Seraphim, Ypsilanti, MI, 1988. AP = Apple II; IB = IBM PC.

Answers and Solutions

Reviewing Concepts

1. Toward the four vertices of a regular tetrahedron, with the center of the atom at the center of the tetrahedron
2. Carbon atoms form covalent bonds readily with the atoms of other elements having similar electronegativities. Carbon atoms also link together in chains, rings, plates, and networks.
3. **a.** Allotropy is the existence of an element in two or more forms in the same physical phase.
   b. Diamond and/or graphite and/or fullerene
4. Its hardness
5. Heat is conducted through diamond by transfer of energy of vibration from atom to atom. This process is efficient because carbon atoms have a small mass and are strongly bonded to one another. The diamond structure has no mobile electrons; thus diamond is a nonconductor of electricity.
6. The carbon atom layers in graphite slide over one another easily, accounting for the softness. The atoms within the layers are strongly bonded and are difficult to pull apart in the direction of the layer. Hence, carbon fibers are strong.
7. **a.**

    b. Evidence indicates that all the bonds in graphite are the same. None of the structures has an independent existence. The actual structure is a resonance hybrid of the written structures. The carbon-carbon bonds are intermediate in character between single and double bonds.
8. The flat scalelike plates slide over one another easily. Graphite is not soluble, subject to corrosion, or affected by heat.
9. **a.** By long-time decomposition of buried vegetation at high temperature and pressure in the absence of air
   b. Peat, lignite, subbituminous coal, bituminous coal, and anthracite
   c. The length of decomposition time and increasingly higher temperatures and pressures
10. Because of its very large internal surface area
11. **a.** Compounds having the same molecular formula but different structures
    b.
12. The exact number and types of atoms in a molecule and how the atoms are bonded to each other
13. Most organic compounds do not dissolve in water; many inorganic compounds do so. Organic compounds are decomposed by heat more easily than most inorganic compounds. Organic reactions generally proceed at much slower rates than inorganic reactions. Organic reactions are greatly affected by reaction conditions. Organic compounds exist as molecules consisting of atoms joined by covalent bonds. Many inorganic compounds have ionic bonds.
14. **a.** C_nH_{2n+2} **b.** C_nH_{2n} **c.** C_nH_{2n-2}
15. **a.** C_7H_{16} **b.** C_7H_{14} **c.** C_7H_{12}
16. **a.** A series in which adjacent members differ by a constant unit
    b. Methane, ethane, propane, butane
    c. By using the Greek or Latin prefixes that show the number of carbon atoms in the molecule and the suffix *-ane*
17. $NaC_2H_3O_2(s) + NaOH(s) \rightarrow CH_4(g) + Na_2CO_3(s)$
18. **a.** $CH_4 + 2O_2 \rightarrow CO_2 + 2H_2O$
    b. $C_2H_4 + 3O_2 \rightarrow 2CO_2 + 2H_2O$
    c. $2C_2H_2 + 5O_2 \rightarrow 4CO_2 + 2H_2O$
    d. $2C_4H_6 + 11O_2 \rightarrow 8CO_2 + 6H_2O$

19. $CH_4 + Cl_2 \rightarrow CH_3Cl + HCl$; $CH_3Cl + Cl_2 \rightarrow CH_2Cl_2 + HCl$; $CH_2Cl_2 + Cl_2 \rightarrow CHCl_3 + HCl$; $CHCl_3 + Cl_2 \rightarrow CCl_4 + HCl$
20. **a.** Ethene **b.** Thermal cracking of petroleum
21.

$$\underset{H}{\overset{H}{>}}C=C-\underset{H}{\overset{H}{C}}-\underset{H}{\overset{H}{C}}-H + H_2 \xrightarrow{Ni} H-\underset{H}{\overset{H}{C}}-\underset{H}{\overset{H}{C}}-\underset{H}{\overset{H}{C}}-\underset{H}{\overset{H}{C}}-H$$

22. **a.** The small molecule from which a polymer is formed
 b. A molecule formed by combination of two identical molecules
 c. A very large molecule consisting of many repeating structural units
 d. The process of joining together a very large number of monomer molecules; the formation of a polymer
23. It is used in the synthesis of complex organic compounds and in the oxyacetylene torch.
24. **a.** They refer to the bonding in hydrocarbon molecules. Saturated molecules have only single bonds between carbon atoms, and unsaturated molecules have at least one double or triple bond.
 b. They refer to solutions. A saturated solution is one in which the concentration of solute is the maximum possible under existing conditions. An unsaturated solution is one in which the concentration of solute is less than the maximum possible under existing conditions.
25. • Mining and transportation to the surface, where it is heated in large ovens to drive out the oil
 • Underground heating and collection of the oil
26. The heating of rubber and other materials, always including sulfur, to a definite temperature for a definite time to improve its properties
27. It is little affected by oils and greases.
28. **a.** It stands for Styrene Butadiene Rubber, an all-purpose synthetic rubber.
 b. It can replace natural rubber for most purposes and is widely used for automobile tire treads.

Problems

1. $1.00 \text{ c} \times \dfrac{200 \text{ mg}}{1 \text{ c}} \times \dfrac{1 \text{ g}}{1000 \text{ mg}} \times \dfrac{1 \text{ cm}^3}{3.51 \text{ g}}$
 $= 0.0570 \text{ cm}^3$

2. **a.** $100.0 \text{ g } C_4H_6 \times \dfrac{1 \text{ mol } C_4H_6}{54.09 \text{ g } C_4H_6} = 1.849 \text{ mol } C_4H_6$
 b. $1.849 \text{ mol } C_4H_6 \times \dfrac{6.022 \times 10^{23} \text{ molecules}}{\text{mol}}$
 $= 1.113 \times 10^{24}$ molecules C_4H_6

13. $(12.000\ 00\ u + 13.003\ 35\ u + 14.003\ 24\ u)/3$
 $= 13.002\ 20\ u.$

3. Molar mass $C_{12}H_{24} = 168.324$ g
 $C = \dfrac{144.133 \text{ g C}}{168.324 \text{ g } C_{12}H_{24}} \times 100\% = 85.6283\%$ C
 $H = \dfrac{24.1906 \text{ g H}}{168.324 \text{ g } C_{12}H_{24}} \times 100\% = 14.3714\%$ H

4. $NaOH + NaC_2H_3O_2 \rightarrow CH_4 + Na_2CO_3$
 $136 \text{ g soda lime} \times \dfrac{15\% \text{ NaOH}}{100\% \text{ soda lime}} = 20.4 \text{ g NaOH}$
 $20.4 \text{ g NaOH} \times \dfrac{1 \text{ mol NaOH}}{40.0 \text{ g NaOH}} \times \dfrac{\text{mol } CH_4}{\text{mol NaOH}}$
 $\times \dfrac{22.4 \text{ L}}{\text{mol}} = 11.4 \text{ L } CH_4$

5. **a.** C_nH_{2n+2}
 $C_{22}H_{2(22)+2} = C_{22}H_{46}$ 46 H atoms
 b. C_nH_{2n}
 $C_{22}H_{2(22)} = C_{22}H_{44}$ 44 H atoms
 c. C_nH_{2n-2}
 $C_{22}H_{2(22)-2} = C_{22}H_{42}$ 42 H atoms

6. Molar mass of $C_5H_{12} = 72.2$ g
 $5.00 \text{ mol} \times 72.2 \text{ g/mol} = 361$ g

7. $C_3H_8 + 5O_2 \rightarrow 3CO_2 + 4H_2O$
 1 mol   3 mol
 $15.0 \text{ L } C_3H_8 \times \dfrac{3 \text{ mol } CO_2}{1 \text{ mol } C_3H_8} = 45.0 \text{ L } CO_2$

8. $C_6H_6 + 6Cl_2 \rightarrow C_6Cl_6 + 6HCl$
 $29.8 \text{ g } C_6Cl_6 \times \dfrac{1 \text{ mol } C_6Cl_6}{285 \text{ g } C_6Cl_6} \times \dfrac{6 \text{ mol } Cl_2}{1 \text{ mol } C_6Cl_6}$
 $= 0.627 \text{ mol } Cl_2$

9. $C_4H_4 + HCl \rightarrow C_4H_5Cl$
 $17.7 \text{ kg } C_4H_5Cl \times \dfrac{1 \text{ mol } C_4H_5Cl}{88.5 \text{ g } C_4H_5Cl} \times \dfrac{10^3 \text{ g}}{\text{kg}}$
 $\times \dfrac{1 \text{ mol HCl}}{1 \text{ mol } C_4H_5Cl} \times \dfrac{36.5 \text{ g HCl}}{\text{mol HCl}} \times \dfrac{1 \text{ kg}}{10^3 \text{ g}} = 7.30$ kg

10. $12.0 \text{ gal} \times \dfrac{4 \text{ qt}}{\text{gal}} \times \dfrac{946 \text{ mL}}{\text{qt}} \times \dfrac{0.692 \text{ g}}{\text{mL}} \times \dfrac{1 \text{ kg}}{1000 \text{ g}}$
 $= 31.4$ kg

11. Molar mass $C_6H_5NO_2 = 123.1$ g
 $\dfrac{14.01 \text{ g N}}{123.1 \text{ g } C_6H_5NO_2} \times 100\% \ C_6H_5NO_2 = 11.38\%$ N

12. $CaC_2 + 2H_2O \rightarrow C_2H_2 + Ca(OH)_2$
 $2.5 \text{ L } C_2H_2 \times \dfrac{1 \text{ mol}}{22.4 \text{ L}} \times \dfrac{1 \text{ mol } CaC_2}{1 \text{ mol } C_2H_2} \times \dfrac{64.1 \text{ g } CaC_2}{\text{mol } CaC_2}$
 $= 7.2 \text{ g } CaC_2$

14. $CH_4 + 414 \text{ kJ} \rightarrow CH_3 + H$ $\Delta H = 414$ kJ
 $Br_2 + 192 \text{ kJ} \rightarrow 2Br_2$ $\Delta H = 192$ kJ
 $CH_3 + Br \rightarrow CH_3Br + 284 \text{ kJ}$ $\Delta H = -284$ kJ
 $H + Br \rightarrow HBr + 368 \text{ kJ}$ $\Delta H = -368$ kJ
 $CH_4 + Br_2 \rightarrow CH_3Br + HBr$ $\Delta H = -46$ kJ

15. $\dfrac{93.75 \text{ g C}}{12.0 \text{ g/mol}} = 7.81 \text{ mol C}$ $\dfrac{6.25 \text{ g H}}{1.0 \text{ g/mol}} = 6.25 \text{ mol H}$ $\dfrac{7.81}{6.25} : \dfrac{6.25}{6.25} = 1.25 : 1.00 = 5 : 4$

Simplest formula = C_5H_4 Molar mass = $\dfrac{6.14 \text{ g solute} \times -5.12\text{C}°}{-2.45\text{C}° \times 100. \text{ g } C_6H_6} \times \dfrac{10^3 \text{ g}}{\text{kg}} = 128 \text{ g}$

$(C_5H_4)_x = 128 \text{ g}$
$x = 128 \div 64 = 2.0$
Molecular formula = $C_{10}H_8$

16. $81.7 \text{ g C} \times \dfrac{1 \text{ mol C}}{12.0 \text{ g C}} = 6.81 \text{ mol C}$ $18.3 \text{ g H} \times \dfrac{1 \text{ mol H}}{1.00 \text{ g H}} = 18.3 \text{ mol H}$ $\dfrac{6.81 \text{ mol C}}{6.81} : \dfrac{18.3 \text{ mol H}}{6.81}$

1 mol C : 2.69 mol H   Simplest formula = C_3H_8
3 mol C : 8 mol H
$C_3H_8 + O_2 \rightarrow CO_2$ (not balanced) $C_3H_8 + 5O_2 \rightarrow 3CO_2 + 4H_2O$ (balanced according to conservation of atoms)
Since the balanced equation shows 5 moles of O_2 for each mole of C_3H_8 and the volume relations are also known to be 5 volumes of O_2 for each volume of C_3H_8, C_3H_8 is shown to be the molecular formula.

17. Since the products of cracking are nearly equally divided between C_3H_6 and C_2H_4, half of the reactant would be used to produce hydrogen gas. By definition, if number of moles are equivalent, volumes of gases will be equal. Therefore, 12.5 kL of propane would produce 12.5 kL of H_2 gas.

$12.5 \text{ kL } C_3H_8 \times \dfrac{10^3 \text{ L}}{\text{kL}} \times \dfrac{1 \text{ mol}}{22.4 \text{ L}} = 558 \text{ mol } C_3H_8$ $558 \text{ mol } H_2 \times \dfrac{22.4 \text{ L}}{\text{mol } H_2} \times \dfrac{\text{kL}}{10^3 \text{ L}} = 12.5 \text{ kL}$

18. $60.0 \text{ g C} \times \dfrac{1 \text{ mol C}}{12.0 \text{ g C}} = 5.00 \text{ mol C}$ Formula mass of $C_3H_8O = 60.\ u$
   Formula mass of $(C_3H_8O)_x = 60.\ u$
$26.7 \text{ g O} \times \dfrac{1 \text{ mol O}}{16.0 \text{ g O}} = 1.67 \text{ mol O}$ $x = 60./60. = 1.0$
   Molecular formula = $(C_3H_8O)_1$ or C_3H_8O
$13.3 \text{ g H} \times \dfrac{1 \text{ mol H}}{1.01 \text{ g H}} = 13.2 \text{ mol H}$

$\dfrac{5.00 \text{ mol C}}{1.67} : \dfrac{1.67 \text{ mol O}}{1.67} : \dfrac{13.2 \text{ mol H}}{1.67}$

2.99 mol C : 1 mol; O : 7.90 mol H
Simplest formula = C_3H_8O

```
H   H   H              H   H   H              H       H   H
|   |   |              |   |   |              |       |   |
H—C—C—C—O—H        H—C—C—C—H          H—C—O—C—C—H
|   |   |              |   |   |              |       |   |
H   H   H              H   O   H              H       H   H
                           |
                           H
```

19. $V_{STP} = V_1 \times \dfrac{P_1}{P_{STP}} \times \dfrac{T_{STP}}{T_1} = 115 \text{ mL} \times \dfrac{738 \text{ mm} - 19.8 \text{ mm}}{760 \text{ mm}} \times \dfrac{273 K}{295 K} = 101 \text{ mL}$

(19.8 mm is the vapor pressure of water at 22°C from Appendix Table A-7.)

20.
```
    H   H   H   H   H   H
    |   |   |   |   |   |
H—C—C—C—C—C—C—H
    |   |   |   |   |   |
    H   H   H   H   H   H

    H   H   H   H   H
    |   |   |   |   |
H—C—C—C—C—C—H
    |   |   |   |
    H   |   H   H
      H—C—H
        |
        H

    H   H   H   H   H
    |   |   |   |   |
H—C—C—C—C—C—H
    |   |   |   |
    H   H   |   H   H
          H—C—H
            |
          H—C—H
            |
            H
```

```
        H
        |
      H—C—H
        |           H   H
        |           |   |
    H—C—C—C—H
        |           |   |
        H           H   H
        |
      H—C—H
        |
        H
```

```
        H
        |
      H—C—H
  H   H   |   H
  |   |       |
H—C—C—C—H
  |   |       |
  H   H   |   H
        |
      H—C—H
        |
        H
```

21.
```
        H   H   H   H   Cl
        |   |   |   |   |
    H—C—C—C—C—C—Cl     1,1,1-trichloropentane
        |   |   |   |   |
        H   H   H   H   Cl

        Cl  Cl  Cl  H   H
        |   |   |   |   |
    H—C—C—C—C—C—H      1,2,3-trichloropentane
        |   |   |   |   |
        H   H   H   H   H
```

22. (structures shown)

H−C(Cl)H−CH2−CH(Cl)−CH2−C(Cl)H−H 1,3,5-trichloropentane

Others are also possible.

23. $2H-C\equiv C-H \xrightarrow{Cu_2Cl_2, NH_4Cl}$ H2C=CH−C≡C−H

24. a. Yes **b.** Yes. The structural formula would be:

(structure of H2C=C(H)−C(H)=CH2 shown)

Critical Thinking

1. Ionic bonds are usually stronger than covalent bonds, and more energy (higher temperature) is needed to break them.

2. Almost all carbon compounds contain hydrogen. In addition, hydrogen is contained in acids, hydroxides, and ammonium compounds and in many other inorganic compounds.

3. From the general formula for the alkanes, C_nH_{2n+2}, the part of the formula mass that is hydrogen = $1.0(2n + 2)$. The total formula mass = $12.0(n) + 1.0(2n + 2)$. The fractional part of the total formula mass that is hydrogen = $\dfrac{1.0(2n+2)}{12(n) + 1.0(2n+2)} = \dfrac{2n+2}{14n+2} = \dfrac{n+1}{7n+1}$.

By inspection, as n assumes increasing integral values, the denominator will become larger faster than the numerator, and the value of the fraction will decrease. The percentage of hydrogen therefore decreases.

Teacher's Notes

Chapter 21
Carbon and Hydrocarbons

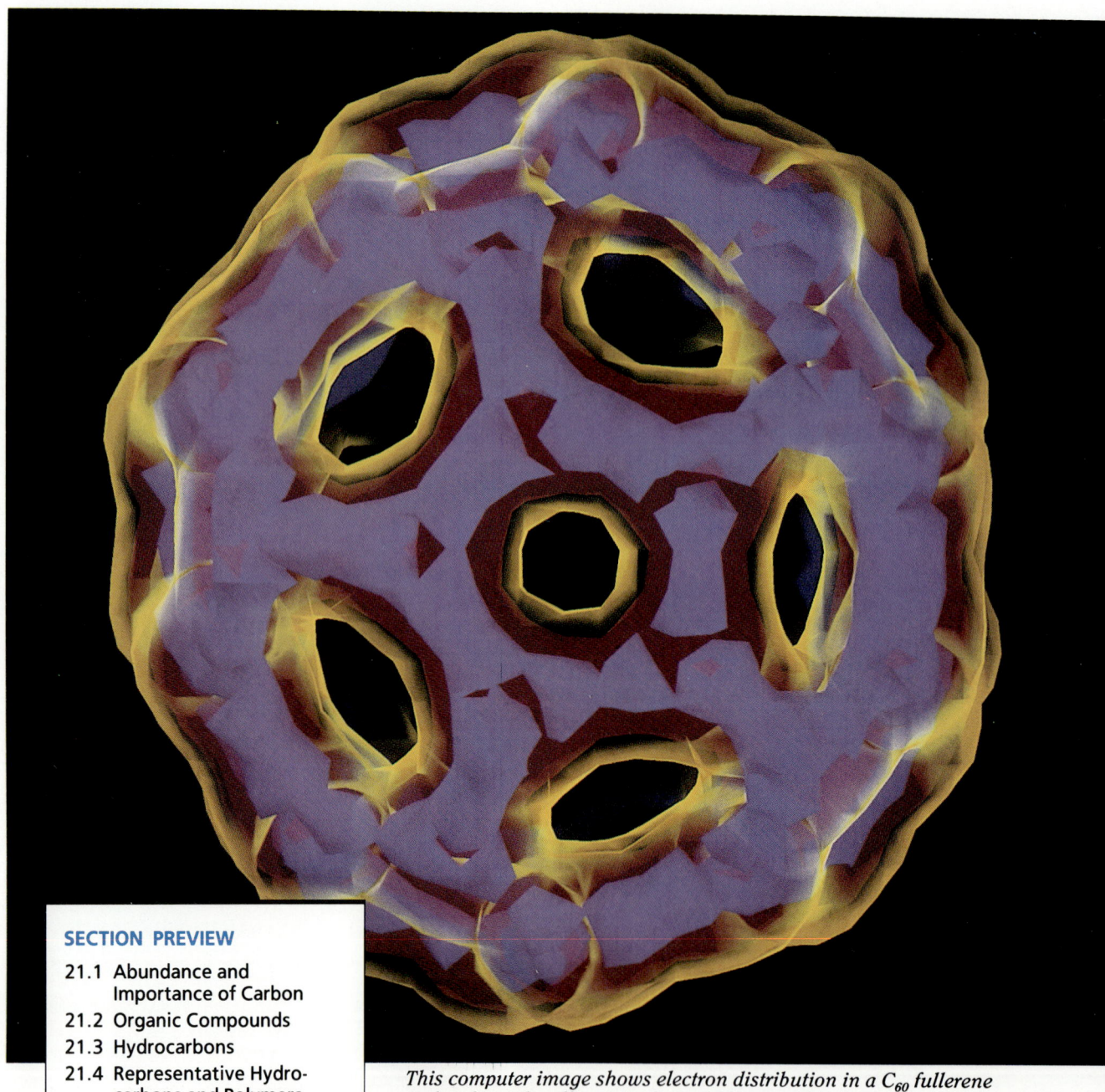

SECTION PREVIEW
- 21.1 Abundance and Importance of Carbon
- 21.2 Organic Compounds
- 21.3 Hydrocarbons
- 21.4 Representative Hydrocarbons and Polymers

This computer image shows electron distribution in a C_{60} fullerene molecule, with yellow, purple, and blue denoting increasing electron density.

21.1 Abundance and Importance of Carbon

Carbon is found in nature both as an element and in combined form. Although carbon ranks about seventeenth in abundance by mass among the elements in the earth's crust, it is exceedingly important because of its universal presence in living matter. Carbon is present in body tissues and in the foods you eat. It is found in common fuels such as coal, petroleum, natural gas, synthetic gas, and wood.

Structure and Bonding of Carbon

Carbon, the first member of Group 14, exhibits mostly nonmetallic properties. In its ground state, a carbon atom has an electronic configuration of $1s^2\,2s^2\,2p^2$. The two $1s$ electrons are tightly bound to the nucleus. The two $2s$ electrons and the two $2p$ electrons are the valence electrons. Carbon atoms show a very strong tendency to share electrons and form covalent bonds. This electron sharing usually has the effect of producing a stable highest-energy-level octet around a carbon atom.

The possession of four valence electrons makes it possible for a carbon atom to form four covalent bonds. These bonds are directed in space toward the four corners of a regular tetrahedron. As shown in Figure 21-1, the nucleus of the carbon atom is at the center of the tetrahedron. During a chemical reaction, all four of the carbon valence electrons are usually involved in bonding through sp, sp^2, or sp^3 hybridization. As shown in Figure 21-2, the ground-state electron configuration can be changed by the promotion of one of the $2s$ electrons to the unoccupied orbital of the $2p$ energy level. This change allows hybridization to produce four sp^3 orbitals that are equivalent. Each of these orbitals can form a single covalent bond. Carbon also forms double and triple covalent bonds.

Because the property of forming covalent bonds is so strong in carbon atoms, they join readily with other elements with similar electronegativities, such as hydrogen, oxygen, nitrogen, sulfur, phosphorus, and the halogens. They also link together with other carbon atoms in chains, rings, plates, and networks.

SECTION OBJECTIVES

- List the sources of carbon.
- Recognize the importance of carbon in chemistry.
- Discuss the structure and bonding of carbon atoms.
- Identify the allotropic forms of carbon.
- Name several amorphous forms of carbon.
- Describe the production, properties, and uses of various forms of carbon.

Figure 21-1 A carbon atom can form four covalent bonds and may be shown as a tetrahedron.

Figure 21-2 Carbon can form four sp^3 hybrid orbitals.

The role of carbon in living organisms is discussed in Chapter 23.

Hybridization is explained in Section 6.5.

The ability of carbon atoms to form chains by bonding to one another is known as *catenation*.

Allotropic Forms of Carbon

Carbon occurs in several solid allotropic forms, such as diamond and graphite. These two forms have dramatically different properties. In the form of **diamond,** *carbon is a colorless, crystalline solid with an extremely high density of 3.514 g/cm³.* **Graphite** *is a soft, black, crystalline form of carbon that is a fair conductor of electricity.* In addition, scientists are experimenting with a recently-discovered set of carbon allotropes, a class of spherically networked carbon atoms, which they have named **fullerenes.**

> Allotropes are two or more forms of the same element that exist in the same physical state and that have different chemical and physical properties.
>
> Borazon is the second-hardest substance known. It is formed when boron nitride, BN, is exposed to very high pressures, creating a tetrahedral structure similar to diamond.
>
> Diamond and graphite are also discussed in Section 13.2.

Figure 21-3 An uncut diamond *(left)* lacks the sparkle and brilliance of a cut diamond *(right).*

Diamond Over 95% of the world's natural diamonds are mined in Africa. Diamonds form when certain types of molten rock cool under conditions of extremely high temperature and pressure. When mined, diamonds do not have the shape or sparkle of gemstones. It is the art of cutting and polishing that gives diamonds their brilliant appearance, as shown in Figure 21-3. Because of the high refractive index of diamond, light passing into the diamond crystal is almost totally reflected, and is dispersed into various colors.

Synthetic diamonds are chemically identical to natural diamonds, but are produced in the laboratory. They are prepared by subjecting graphite and a metal that acts as solvent and catalyst to extremely high pressure (55,000 atm) and high temperature (2000°C) for several hours.

Diamond is the hardest material known. It is the densest form of carbon—about 3.5 times as dense as water. Both its hardness and density are due to its structure. Figure 21-4 shows that carbon atoms in diamond are covalently bonded in a strong, compact fashion. Note that each carbon atom is tetrahedrally oriented to its four nearest neighbors. The distances between the carbon nuclei are 154 pm. This type of structure gives the crystal its great strength in all three dimensions.

The rigidity of its structure gives diamond its hardness. The compactness, resulting from the small distances between nuclei, gives diamond its high density. The covalent network structure of diamond accounts for its extremely high melting point (greater than 3550°C). Since all the valence electrons are used in forming covalent bonds, none can migrate. This explains why diamond is a nonconductor of electricity. Because of its extreme hardness, the major industrial uses of diamond are in cutting, drilling, and grinding.

The unique structure of diamond also explains its extremely efficient ability to conduct heat. A perfect single diamond crystal conducts heat greater than five times as readily as silver or copper, which are the best metallic

Figure 21-4 In diamond, each carbon atom is bonded to four tetrahedrally oriented carbon atoms.

conductors. In diamond, heat is conducted by the transfer of energy of vibration from one carbon atom to the next. In a perfect single diamond crystal, this process is very efficient because the carbon atoms have a small mass. The forces binding the atoms together are strong and can easily transfer vibratory motion among the atoms.

Graphite Graphite is nearly as remarkable for its softness as diamond is for its hardness. It is easily crumbled and feels greasy, characteristics readily explained by its structure. The carbon atoms in graphite are arranged in layers of thin hexagonal plates, as shown in Figure 21-5. The distance between the centers of adjacent carbon atoms within a layer is 142 pm. This distance is less than the distance between adjacent carbon atoms in diamond. The distance between the centers of atoms in adjacent layers is 335 pm, however.

Figure 21-5 shows the bonding within a layer of graphite. Each carbon atom in a layer is bonded to only three other carbon atoms in that layer. This bonding consists of single and double covalent bonds between carbon atoms. When presented in this fashion, three different equivalent patterns appear. In each of these, some carbon-carbon bonds are single and others are double. There is, however, no experimental evidence that the bonds in a layer of graphite are of these two distinct types. On the contrary, the evidence indicates that the bonds are all the same. The layers of graphite have a resonance structure in which the carbon-carbon bonds are intermediate in character between single and double bonds. Each layer in graphite is a strongly bonded covalent network structure. As with diamond, this structure gives graphite a high melting point (3570°C). The strong bonds between the carbon atoms within a layer make graphite difficult to pull apart in the direction of the layer. As a result, carbon fibers, in which the carbon is in the form of graphite, are very strong.

The layers of carbon atoms in graphite are too far apart for the formation of covalent bonds between them. They are held together by weak London dispersion forces. The weak attraction between layers accounts for the softness of graphite and its greasy feel as one layer slides over another.

The average distance between carbon atoms in graphite is greater than the average distance in diamond; thus, graphite has a lower density. The mobile electrons in each carbon-atom layer make graphite a fairly good conductor of electricity, even though it is a nonmetal. Like diamond, graphite does not dissolve in any ordinary solvent. Similarly, it forms carbon dioxide when burned in oxygen.

Uses of graphite Graphite is a good lubricant, because it is not soluble, resists corrosion, and is not affected by heat. It is sometimes mixed with petroleum jelly or motor oil to form graphite lubricants, and can be used for lubricating machine parts that operate at temperatures that are too high for the usual petroleum lubricants. Graphite leaves a gray streak or mark when it is drawn across a sheet of paper. In making "lead" pencils (the name *lead* is merely poetic), graphite is powdered, mixed with clay, and then formed into sticks. The hardness of a pencil depends on the relative amount of clay that is used.

The most important use of synthetic graphite is in electrodes for electric-arc furnaces for making steel. Synthetic graphite electrodes are also

Figure 21-5 The arrangement of carbon atoms in graphite allows adjacent layers to slide past each other. Each carbon atom is bonded to three other carbon atoms.

used in the electrolysis of salt water for making chlorine and sodium hydroxide. Graphite does not react with acids, bases, or organic and inorganic solvents. These properties make it useful for a variety of processes and equipment in the food, chemical, and petroleum industries. Graphite is also used in nuclear reactors, as described in Chapter 29.

If certain synthetic fibers are combined with plastic resins and heated under pressure, they become carbon fibers. Carbon fibers are less dense than steel, but are stronger and stiffer. They are used in aircraft floor decking, wings, and wing flaps and in weather and communication satellites. In sporting goods, carbon fibers are used to make golf club shafts, tennis rackets, fishing rods (see Figure 21-6), and bicycle frames. Carbon fibers have been implanted in severely torn ligaments and tendons to help the natural process of reconstruction.

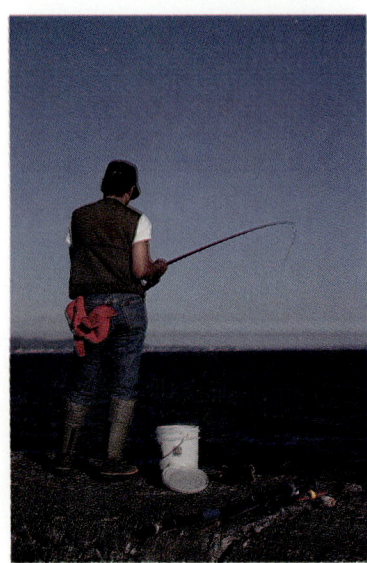

Figure 21-6 Carbon fibers give this graphite fishing rod its strength.

Amorphous Forms of Carbon

When substances that contain combined carbon are decomposed by heat, they leave **amorphous carbon**, *black, carbon-containing residues that have no crystalline shape.* Examples of amorphous carbon are coal, coke, charcoal, petroleum coke, and activated carbon. Studies of these substances' structures, made by x-ray diffraction, reveal that the various forms of so-called amorphous carbon contain regions in which the carbon atoms have an orderly arrangement. In charcoal, for example, the carbon atoms are arranged somewhat as they are in a layer of graphite.

Coal *Coal is a natural dark-brown or black solid that consists mostly of amorphous carbon with various organic and some inorganic compounds.* It is a solid, rocklike material that burns readily. Coal contains at least 50 percent carbon and varying amounts of moisture, plus volatile (easily evaporated) materials and noncombustible mineral matter (ash). It is our most plentiful fossil fuel.

Geologists believe that coal was formed about 300 million years ago during the Carboniferous Period, a time often referred to as "The Coal Age." Much of the North American land mass was then covered with shallow swamps filled with dense forests of ferns, horsetails, and primitive seed plants (see Figure 21-7). The moist, warm climate of the time supported luxurious plant growth. Tree ferns, giant club mosses, and other vegetation growing in these swamps supplied the material for the coal deposits we use today.

Peat *is a form of partially carbonized plant material that can be used as fuel.* Peat bogs are found extensively in Pennsylvania, Michigan, Wisconsin, and other northern states. The bogs in which it is found contain mosses, sedges, and other forms of vegetation that undergo partial decomposition in swampy land in the almost complete absence of oxygen. Peat burns with a smoky flame, and its heat content is rather low because the vegetable material within it becomes only partially converted to carbon and hydrocarbons. Peat contains a high percentage of moisture. In Ireland and Sweden, peat is used for domestic fuel. In the United States, peat is used mainly as a soil conditioner, not as fuel.

In some areas, upheavals of parts of the earth's crust buried extensive deposits of slowly decaying plant material during the Carboniferous Period. Once buried, this material was subjected to increased temperatures and

Figure 21-7 Lush vegetation during the Coal Age was the source of the coal, natural gas, and petroleum we use today.

pressures over millions of years. An early stage of carbon-containing material that formed under these conditions is **lignite**, *a low grade form of coal sometimes called "brown coal" because of its brownish-black color.* Lignite also burns with a smoky flame, and although its heat content is higher than that of peat, it is still significantly below that of the other coals. The moisture content of lignite is high.

The next stage of coal formation is subbituminous coal, a form of coal that is found extensively in Wyoming and other Western states. **Subbituminous coal** *is a form of coal that has a higher percentage of carbon and a lower percentage of moisture than lignite.* Subbituminous coal also has a higher heat content than lignite. It can be used for industrial heating, as in electric generating stations.

Bituminous coal *is another form of coal that appears to have been subjected to greater heat and pressure than subbituminous coal.* The percentage of carbon is high; the moisture content is low. Bituminous coal has a high heat content. It is widely distributed throughout the Appalachian region, the Midwest, and the Rocky Mountain states. It is used for home and industrial heating and for making coke and coal by-products.

Anthracite *is believed to have been subjected to greater temperatures and pressures than other coals.* It contains the least amount of volatile matter of all coals and has the highest percentage of carbon. Most anthracite is found in Pennsylvania. Anthracite coal has a higher heat content than any other coal.

Deep coal mining is a hazardous occupation, with one out of every 1000 underground miners killed each year and 30% of long-time coal miners developing "black lung" disease.

Coal gas is composed mainly of the three combustible gases—methane, hydrogen, and carbon monoxide. Coal tar can be separated by distillation into benzene, toluene, xylenes, naphthalene, phenol, and creosols.

Coke When bituminous coal is heated in a test tube, a flammable gas escapes that burns readily if mixed with air and ignited. A tarlike liquid also condenses on the upper walls of the test tube. If the heating is continued until all the volatile material is driven off, coke is left as a residue. **Coke** *is the solid residue left after volatile materials are removed from bituminous coal.*

Commercially, coke is prepared by destructive distillation of bituminous coal in by-product coke ovens, as shown in Figure 21-8. **Destructive distillation** *is a process by which a complex carbon-containing material, such as coal, is heated in a closed container in the absence of air or oxygen.* The volatile products are separated into coal gas, ammonia (NH_3), and coal tar. Coal gas can be used as a fuel. Ammonia is used in making fertilizers.

Figure 21-8 Coke is produced by destructive distillation of bituminous coal in by-product coke ovens.

Carbon and Hydrocarbons

Figure 21-9 Activated carbon is used in some home water-purification systems, such as this water faucet filter unit.

Adsorption is the adhesion of particles of gases, liquids, or solids to the surfaces of a solid.

1. Graphite, diamond, charcoals, soot, living matter, natural gas, coal, wood, petroleum, metal carbonates
2. It is present in all living matter.
3. The ground-state electron configuration can be changed by the promotion of one of the 2s electrons to the unoccupied orbital of the 2p energy level. This change produces 4 sp^3 orbitals that are equivalent.

Coal tar can be separated by distillation into materials used to make drugs, dyes, and explosives. The black pitch that remains after distillation of coal tar is used to surface roads.

About 60 million tons of coke are produced each year in the United States. Coke is a gray, porous solid that is harder and denser than charcoal. Coke burns with little flame, has a high heat content, and is also valuable as a fuel.

It is an excellent reducing agent and is widely used in obtaining iron, tin, copper, and zinc from oxide ores containing these metals. Coke readily reduces the metals in these oxides to yield relatively pure metals.

Charcoal Destructive distillation of wood yields several gases that can be burned. Wood also produces methanol (wood alcohol—CH_3OH)), acetic acid (CH_3COOH), and other volatile products. *The residue that remains after the destructive distillation of wood is* **charcoal**. Charcoal is prepared commercially by heating wood in retorts, closed containers in which substances are distilled or decomposed by heat.

Charcoal is a porous, black, brittle solid. It is odorless and tasteless. It is denser than water, but it often adsorbs enough gas to make it float on water. The ability of charcoal to adsorb a large quantity of gas is perhaps its most remarkable physical property. One cubic centimeter of freshly prepared willow charcoal adsorbs about 90 mL of ammonia gas. Charcoal has been used to adsorb toxic waste products from the blood of persons suffering from liver failure, from drug overdose, and from accidental swallowing of poisonous carbon–hydrogen compounds. It is also used in aquarium filters and to remove ozone from the cabin atmosphere of jet airplanes.

Petroleum coke *Petroleum coke is a form of amorphous carbon that is produced by the destructive distillation of an oily residue that is a remnant of petroleum refining.* Rods of petroleum coke are converted to synthetic graphite for use as positive electrodes in dry cells. They are also used in the production of aluminum by electrolysis.

Activated carbon *Activated carbon is a form of amorphous carbon that is prepared in a way that gives it a very large internal surface area.* This large surface area makes activated carbon exceptionally useful for the adsorption of liquids and gases. Activated carbon can be made from a variety of carbon-containing materials. Such a material is first destructively distilled and the resultant carbon is then treated with steam or carbon dioxide at about 100°C. These two processes produce a very porous form of carbon, which contains a very large internal surface area. This form of activated carbon can remove odors from the air circulated by large air conditioning systems. It is also used in municipal and industrial water treatment and in home water faucet units (see Figure 21-9) to adsorb harmful materials as well as objectionable odors and tastes.

Section Review

1. List several sources of carbon both in free and combined form.
2. What makes carbon an important element in the study of chemistry?
3. Explain how a carbon atom forms four covalent bonds when its valence configuration in the ground-state is $2s^2 2p^2$.

21.2 Organic Compounds

Organic compounds have one thing in common—they all contain carbon atoms. With few exceptions, *carbon-containing compounds are the subject of the study of* **organic chemistry**. Ionic compounds such as sodium carbonate (Na_2CO_3), and the oxides of carbon (CO, CO_2) are considered inorganic compounds.

The number of possible carbon compounds is virtually unlimited. Over 4 million natural and synthetic organic compounds are known, and about one hundred thousand new ones are isolated or synthesized each year. The fact that carbon is the backbone of more compounds than all of the other elements combined suggests that it has some unique characteristics. There are two major reasons for the occurrence of so many organic compounds.

1. Carbon atoms have the ability to link up with each other to form a virtually limitless number of chain, branched-chain, and ring structures of countless shapes and dimensions. This makes possible molecules in which thousands of carbon atoms are bonded to another. Not only are carbon atoms linked by single covalent bonds, but they are sometimes linked by double or triple covalent bonds. Also, carbon atoms form bonds with a large number of other elements.

2. Carbon atoms represented by a specific molecular formula may be arranged in several different ways, which creates compounds with different properties and structures. For example, the molecular formula C_2H_6O represents the molecular formula for both ethyl alcohol and dimethyl ether. Such compounds have the same molecular formula but different structures. *Compounds that have the same molecular formula but different structures are called* **isomers**. As the number of carbon atoms increases in a molecular formula, the number of possible isomers increases rapidly. For example, there are 18 isomers with the molecular formula C_8H_{18}, 35 with the molecular formula C_9H_{20}, and 75 with the molecular formula $C_{10}H_{22}$. Calculations show that with the molecular formula of just 40 carbon atoms and 82 hydrogen atoms ($C_{40}H_{82}$), there are theoretically 69 491 178 805 831 isomers. To distinguish one isomer from another, we must use structural formulas, the topic of our next discussion.

Although the number of possible organic compounds is virtually unlimited, the study of organic compounds can be organized into two major divisions: hydrocarbons and substituted hydrocarbons, the subject of Chapters 21 and 22.

Hydrocarbons *are the simplest organic compounds, composed only of carbon and hydrogen.* They are grouped into several families with altogether different characteristics—namely alkanes, alkenes, alkynes, and aromatic hydrocarbons. These families are based mainly on the type of bonding between carbon atoms. **Substituted hydrocarbons** *are those hydrocarbons that are classified according to the functional groups they contain. A* **functional group** *is an atom or group of atoms bonded to a hydrocarbon and is responsible for the specific properties of the resulting organic compound.* In some cases, the functional groups are substituted for H atoms in hydrocarbons, and in other cases for C atoms, thus the name substituted hydrocarbons. Because organic compounds with the same functional group

SECTION OBJECTIVES

- Give the reason for the occurrence of so many organic compounds.
- Describe the two major divisions of organic compounds.
- Recognize the need for structural formulas.
- Compare and contrast organic and inorganic compounds.

The structure for ethyl alcohol is

$$\begin{array}{ccc} H & H \\ | & | \\ H-C-C-OH \\ | & | \\ H & H \end{array}$$

The structure for dimethyl ether is

$$\begin{array}{ccc} H & H \\ | & | \\ H-C-O-C-H \\ | & | \\ H & H \end{array}$$

Give a few examples of functional groups, such as the hydroxyl group in ethyl alcohol and isopropyl alcohol, or the formyl group in acetaldehyde and propionaldehyde, or the carboxyl group in acetic acid and butyric acid. This will help lay the groundwork for the next chapter.

Chemistry and the Environment

Chlorohydrocarbons

Chlorohydrocarbons *are hydrocarbons in which chlorine atoms are substituted for one or more of the hydrogen atoms.* A wide variety of chlorohydrocarbon compounds were used extensively in industry and agriculture until it was found that use of many of these compounds may cause cancer, damage the kidney and liver, and result in ailments of the circulatory and respiratory systems.

The members of one group of chlorohydrocarbons, the chlorinated forms of methane, were widely used until their health hazards were recognized. Carbon tetrachloride (tetrachloromethane) was used in dry cleaning because it dissolves fats, oils, and greases. Chloroform (trichloromethane) is still used as a solvent in many paint strippers. Methyl chloride (chloromethane) was stored under pressure as a liquid and then sprayed onto skin as a local anaesthetic.

Many other chlorohydrocarbons are known to be dangerous to living organisms. DDT (dichlorodiphenyltrichloroethane) is a highly effective pesticide, but is now banned in many parts of the world. DDT becomes increasingly concentrated as it moves up the food chain and accumulates in animal tissues. It poisons fish and many other living organisms and interferes with bird reproduction. PCBs (polychlorinated bipenyls) were used in the United States as coolants and insulating fluids for electrical equipment. A herbicide called Agent Orange was used as a defoliant in Vietnam by the United States during its participation in the Vietnam War. A byproduct of the manufacture of PCBs and Agent Orange, the dioxins are a group of chlorinated aromatic hydrocarbons that are known to cause a wide variety of health problems. The town of Times Beach, Missouri, shown in the photograph below, was abandoned in 1983 because of dioxin contamination of the soil.

or groups behave similarly, we are able to classify them in terms of their functional groups.

Structural Formulas And Bonding

The molecular formula of a compound shows the exact number of atoms of each type of element present in a molecule. While the formula for sulfuric acid (H_2SO_4) gives enough information for most purposes in inorganic chemistry, a molecular formula such as that for sucrose ($C_{12}H_{22}O_{11}$) is not satisfactory in organic chemistry because there are over 100 different compounds that have the same molecular formula as sucrose.

For this reason, organic chemists use structural formulas to represent organic compounds. *A **structural formula** indicates the exact number and types of atoms present in a molecule and also shows how the atoms are bonded* to each other. *The various isomers with a specific molecular formula are referred to as* **structural isomers**.

Structural isomers may exist for hydrocarbons with four or more carbon atoms. For example, careful study of the molecular formula C_4H_{10} reveals the existence of two compounds that exhibit different properties. When we attempt to write structural formulas for C_4H_{10} we notice that atoms can be arranged to give two different structures.

```
    H   H   H   H              H       H       H
    |   |   |   |              |       |       |
H — C — C — C — C — H      H — C ——— C ——— C — H
    |   |   |   |              |       |       |
    H   H   H   H              H   H — C — H   H
                                       |
                                       H

        n-butane                     isobutane
```

These structural formulas represent molecules of two different compounds because the atoms are linked together in different ways, as shown in Figure 21-10. These two compounds are isomers of one another. The dash (—) in the structural formulas represents a pair of shared electrons forming a covalent bond. The *n* in the name *n*-butane stands for normal, which indicates a straight-chain structural formula—the four carbon atoms are in a continuous chain. The chain may be bent or twisted, but it is continuous. The formula of isobutane shows a continuous chain of three carbon atoms, with the fourth carbon atom attached to the second carbon of the chain.

Figure 21-10 *n*-butane *(left)* and isobutane *(right)* are isomers of butane.

Determination of an organic structural formula There are two different isomers of organic compounds having a molar mass of 46 g. In each, carbon represents 52.2% of the molar mass, hydrogen, 13%, and oxygen, 34.8%. One compound is a colorless liquid that boils at 78°C. The other is a colorless gas that condenses to liquid at −25°C under one atmosphere of pressure (760 mm Hg). Each has its own distinctive odor. How can their structural formulas be determined?

From the percent composition, you can calculate the empirical formula, which is C_2H_6O. Since this empirical formula has a formula mass of 46 *u*, it must also be the molecular formula of each compound. From what you have already learned about bonding, there are only two ways in which two

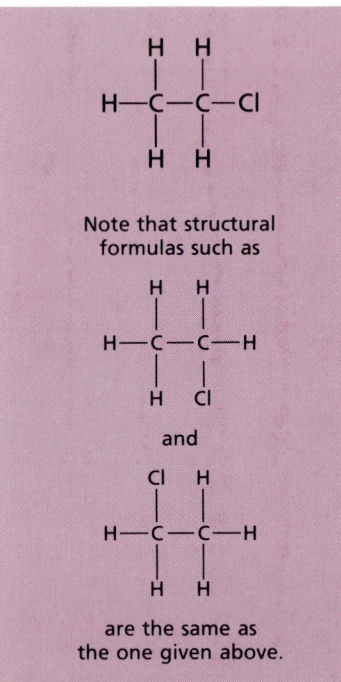

Figure 21-11 The structural formula for ethyl chloride (C₂H₅Cl) is written as shown here.

CAUTION
Many organic compounds are flammable and poisonous. Quite a few cause or are suspected of causing cancer. Some organic reactions are rapid and highly exothermic. A student should not perform any experiments with organic compounds without detailed laboratory directions, and then only with adequate ventilation and other safety precautions and under the supervision of an experienced instructor.

carbon atoms, six hydrogen atoms, and a single oxygen atom can combine, as shown below.

$$\begin{array}{cc} \text{Structure A} & \text{Structure B} \end{array}$$

Now the problem is to match these structures to the two compounds. If each compound is tested for reaction with metallic sodium, only the liquid compound reacts. In this reaction, hydrogen gas is given off. The amount of hydrogen given off is equal to one-sixth of the hydrogen that the compound contains. This evidence indicates that in the molecules of the liquid, one of the six hydrogen atoms is bonded in a way that is different from the others. Structure A is indicated.

Next, you discover that the liquid reacts with phosphorus trichloride and gives a product with the molecular formula C_2H_5Cl. In this reaction, chlorine has replaced both a hydrogen atom and an oxygen atom. Only one structural formula for C_2H_5Cl can be written, as shown in Figure 21-11. You may assume that the chlorine atom occupies the same position as the oxygen and hydrogen atoms that it replaced. Structure A is again indicated. Additional evidence can be found to indicate that the liquid does indeed have Structure A. This liquid substance is ethanol, or ethyl alcohol—C_2H_5OH. The gaseous substance has the other structural formula and is dimethyl ether—CH_3OCH_3.

Methods similar to those just described can be used to determine the structural formulas of other simple organic compounds. Complicated molecules are generally broken down into simpler molecules. From the structures of these simpler molecules, the structure of the complex molecule can be determined. Sometimes, simple molecules of known structure are combined to produce a complex molecule. A comparison of chemical and physical properties of compounds of unknown structure with those of known structure is sometimes helpful.

Differences Between Organic and Inorganic Compounds

The basic laws of chemistry are the same for organic and inorganic compounds. The behavior of organic compounds is somewhat different from that of inorganic compounds, however. Some of the most important differences are:

1. Most organic compounds do not dissolve in water. Many inorganic compounds will dissolve more or less readily in water. Organic compounds generally dissolve in other organic liquids, such as alcohol, chloroform, ether, or carbon tetrachloride.

2. Organic compounds are decomposed by heat more easily than most inorganic compounds. The decomposition (charring) of sugar when it undergoes moderate heating is familiar. Such charring upon heating is often a test for organic substances. By contrast, many inorganic compounds, such

TABLE 21-1 COMPARISON OF AN ORGANIC COMPOUND AND AN INORGANIC COMPOUND

Compound	Benzene (C_6H_6)	Table salt (NaCl)
formula	C_6H_6	NaCl
melting point	5.5°C	801°C
boiling point	80.1°C	1413°C
bonding	covalent	ionic
solubility in H_2O	insoluble	soluble
solubility in gasoline	soluble	insoluble

as common salt (sodium chloride), are simply vaporized without decomposition when exposed to high temperatures.

3. Organic reactions generally proceed at much slower rates than inorganic reactions. Such organic reactions often require hours or even days for completion. (Some organic reactions, such as those in living cells, may take place with great speed, however.) Most inorganic reactions occur almost as quickly as solutions of the reactants are brought together.

4. Organic reactions are greatly affected by reaction conditions. Many inorganic reactions follow well-known patterns. This makes it possible for you to learn about many inorganic reactions by studying the general types of reactions explained in Chapter 8. There are some general types of organic reactions, too. But changing the temperature or pressure or the nature of a catalyst can alter the identity of the products formed. The same organic reactants can form different products, depending on reaction conditions.

5. Organic compounds exist as molecules consisting of atoms joined by covalent bonds. Many inorganic compounds have ionic bonds.

6. Organic compounds may exist as isomers. Inorganic compounds rarely exhibit isomerism.

Section Review

1. Give two major characteristics of carbon atoms that make possible the occurrence of millions of organic compounds.
2. What is meant by the term *isomer*?
3. Define each of the following terms: *(a)* hydrocarbons *(b)* structural formula.
4. How many structural isomers have the molecular formula C_5H_{12}?
5. List several ways in which organic compounds are different from inorganic compounds.

1. Ability to link up with each other to form a virtually limitless (theoretically infinite) number of chain, branched-chain, and ring structures of countless shapes and dimensions. May be arranged in several different ways, which gives compounds with different properties and structures
2. Compounds that have the same molecular formula, but different structures
3. *(a)* Compounds composed only of carbon and hydrogen *(b)* Indicates the exact number and types of atoms present in a molecule and how the atoms are bonded to each other
4. Three
5. Most organic compounds do not dissolve in water, while many inorganic compounds do. Organic compounds are decomposed by heat more easily than most inorganic compounds. Organic reactions are slow, while inorganic reactions are fast. Organic reactions are greatly affected by reaction conditions, while inorganic reactions follow well-known patterns. Organic compounds consist of atoms joined by covalent bonds, while many inorganic compounds have ionic bonds. Organic compounds may exist as isomers, while inorganic compounds do not exhibit isomerism.

21.3 Hydrocarbons

Because they are the basic structures from which other organic compounds are derived, the hydrocarbons are studied first. Hydrocarbons are grouped into several different series of compounds. These groupings are based mainly on the type of bonding between carbon atoms.

1. *The* **alkanes** *(al-kaynes) are straight-chain or branched-chain hydrocarbons whose carbon atoms are connected by single covalent bonds:*

SECTION OBJECTIVES

- Classify hydrocarbons into several different groups and give examples of each.
- Draw structural formulas for hydrocarbon molecules if given their molecular formulas.

Carbon and Hydrocarbons

- Give systematic names of hydrocarbons if given their structural formulas.
- Discuss sources of alkanes and describe preparation methods of alkenes and ethyne.
- Describe reactions of alkanes, alkenes, alkynes, and benzene.
- Explain why alkanes and aromatic hydrocarbons react by substitution and alkenes and alkynes react by addition.
- Name and draw the structural formulas of aromatic hydrocarbons.

$$H-\underset{\underset{H}{|}}{\overset{\overset{H}{|}}{C}}-\underset{\underset{H}{|}}{\overset{\overset{H}{|}}{C}}-H$$

2. The **alkenes** (al-keens) are straight- or branched-chain hydrocarbons in which two carbon atoms in each molecule are connected by a double covalent bond:

$$\underset{H}{\overset{H}{\diagdown}}C=C\underset{H}{\overset{H}{\diagup}}$$

3. The **alkynes** (al-kynes) are straight- or branched-chain hydrocarbons in which two carbon atoms in each molecule are connected by a triple covalent bond:

$$H-C\equiv C-H$$

5. The **aromatic hydrocarbons** *are hydrocarbons with six-membered carbon rings that exhibit resonance.* These structures are sometimes represented by alternate single and double covalent bonds in six-membered carbon rings.

Alkanes

This series of organic compounds is sometimes called *the paraffin series* because **paraffin wax** *is a mixture of hydrocarbons of this series.* The word *paraffin* means "little attraction." Compared to the hydrocarbons of other series, the alkanes have low chemical reactivity. This stability results from their single covalent bonds. Because they have only single covalent bonds in each molecule, the alkanes are known as saturated hydrocarbons. **Saturated hydrocarbons** *are those in which each carbon atom in the molecule forms four single covalent bonds with other atoms.*

Table 21-2 lists a few members of the alkane series. The names of the first four members of this series follow no system. Beginning with pentane, the first part of each name is a Greek or Latin numerical prefix. This prefix indicates the number of carbon atoms. The name of each member ends in *-ane*, the same as the name of the series.

If you examine the formulas for successive alkanes, you will see a clear pattern. Each member of the series differs from the preceding one by the —CH_2 group.

TABLE 21-2 n-ALKANES

Alkane	Formula	Melting Point (°C)	Phase at Room Temperature	Number of Possible Isomers
Methane	CH_4	−183	gas	1
Ethane	C_2H_6	−172	gas	1
Propane	C_3H_8	−187	gas	1
Butane	C_4H_{10}	−135	gas	2
Pentane	C_5H_{12}	−130	liquid	3
Hexane	C_6H_{14}	−94	liquid	5
Heptane	C_7H_{16}	−91	liquid	9
Octane	C_8H_{18}	−57	liquid	18
Nonane	C_9H_{20}	−54	liquid	35
Decane	$C_{10}H_{22}$	−30	liquid	75

Figure 21-12 Methane *(top)*, ethane *(center)*, and propane *(bottom)* are the first three alkanes.

Compounds that differ in this fashion belong to a homologous series. A **homologous series** *is one in which adjacent members differ by a constant unit.* It is not necessary to remember the formulas for all members of a homologous series. Instead, a general formula, such as C_nH_{2n+2} for the alkanes, can be used to determine the formulas. Suppose a member of this series has 30 carbon atoms in its molecules. To find the number of hydrogen atoms, multiply 30 by 2, then add 2. The formula is $C_{30}H_{62}$.

Structures and systematic names of alkanes Names of the millions of organic compounds are derived from a systematic method developed by the International Union of Pure and Applied Chemistry (IUPAC). Systematic names can be derived from structural formulas. At the same time, structural formulas can be drawn from systematic names of organic compounds.

Each of the first three alkanes can have only one molecular structure, as shown by the formulas below. See the models in Figure 21-12.

Butane (C_4H_{10}), has two structural isomers

n-butane   2-methylpropane

Carbon and Hydrocarbons

Chemistry and its Applications

Octane Ratings

Have you ever wondered what the octane rating that you see on a gasoline pump means? The octane rating of a fuel is a measure of its burning efficiency and anti-knock properties. An automobile engine works properly when its fuel burns smoothly. If a fuel ignites spontaneously before the piston is in the proper position, power is lost and a knocking sound is heard.

Heptane is a straight-chain hydrocarbon that burns with a lot of knocking, so it is given an arbitrary octane rating of 0. Isooctane (also called 2,2,4-trimethylpentane) is a branched hydrocarbon that burns smoothly without knocking and is given a rating of 100. A gasoline mixture with an octane rating of 85 causes the same amount of knocking as a mixture of 85% isooctane and 15% heptane.

Until recently, small amounts of tetraethyl lead were added to fuels. This chemical inhibits knocking and thus increases the octane rating of a fuel. Unfortunately, tetraethyl lead directly contributes to air pollution and also ruins the catalytic converters that are installed on new cars to meet the reduced emissions requirements of the Clean Air Amendments of 1970. Other branched or ring hydrocarbons are now added to unleaded fuels to increase their octane ratings. This causes gasoline to be more expensive but the high performance coupled with a reduction in air pollution makes the extra cost worthwhile.

The basic part of the systematic name of an organic compound is the name of the longest carbon chain in the molecule. The straight-chain molecule is n-butane. Recall that n stands for normal—a structural formula in a continuous chain. The branched-chain molecule is named 2-methylpropane.

The longest continuous carbon chain in the second molecule shown above is three carbon atoms long, as in propane. One hydrogen atom attached to the second carbon atom in propane is replaced by the —CH_3 group. This is a substitution group called the methyl group and is indicated by the following structure:

$$\begin{array}{c} H \\ | \\ H-C-H \\ | \end{array}$$

The **methyl group** *is methane with one of the hydrogen atoms removed.* It is an example of an alkyl group. *An* **alkyl group** *is simply the neutral species formed when one hydrogen atom is removed from the parent alkane molecule.*

Table 21-3 lists a few alkyl groups. The methyl and other alkyl groups are neutral—they have the same number of protons and electrons. Carbon atoms in the main chain of the molecule are numbered. This numbering begins at the end of the molecule giving the carbon atoms with substitution groups the smallest numbers. Thus in the name 2-methylpropane, the 2 refers to the number of the carbon atom on which there is a substitution. Methyl is the substitution group. Propane is the parent hydrocarbon. The branched-chain butane is also commonly known as isobutane. The prefix *iso-* is needed to represent the alkyl group

$$CH_3-\underset{CH_3}{\overset{H}{C}}-$$

TABLE 21-3 ALKYL GROUPS	
Alkyl Group	Formula
methyl	CH_3
ethyl	C_2H_5
propyl	C_3H_7
butyl	C_4H_9
pentyl (amyl)	C_5H_{11}
hexyl	C_6H_{13}
heptyl	C_7H_{15}
octyl	C_8H_{17}
nonyl	C_9H_{19}
decyl	$C_{10}H_{21}$

This alkyl group, which sometimes is written as $-CH(CH_3)_2$, is derived from propane when one hydrogen atom is removed from the second carbon atom. The name that follows the prefix *iso-* indicates the number of carbon atoms.

There are three possible pentanes—C_5H_{12}. They are named: *n*-pentane, 2-methylbutane (not 3-methylbutane), and 2,2-dimethylpropane. In the name 2,2-dimethylpropane, the 2,2 refers to the position of both substitutions. The prefix *dimethyl* shows that the two substitutions are both methyl groups. Propane is the parent hydrocarbon. Which of the pentanes also has the name *isopentane*?

Just as the —CH_3 group derived from methane is the *methyl* group, the —C_2H_5 group that is derived from ethane is the *ethyl* group, and —C_3H_7 derived from propane is the *n*-propyl group. The *n*-butyl group (—C_4H_9), is derived from butane. The formula—C_5H_{11} is usually called the *amyl* group rather than the *pentyl* group. Other groups are given names following the general rule of dropping the *-ane* suffix and adding *-yl*. The alkyl groups of the first ten alkanes are listed in Table 21-3. The symbol **R—** is frequently used to represent an alkyl group in a formula.

Any straight-chained alkane that has a hydrogen replaced by a methyl group on the second carbon has the "iso-" prefix. For example: isopentane, isohexane, and isoheptane.

Sources and uses of alkanes Alkanes are generally found in petroleum and natural gas. It is fairly easy to separate the smaller members of the alkane series individually from petroleum and natural gas by fractional distillation, a technique that will be described in Section 21.4. The alkanes with higher boiling points are usually separated into mixtures with similar boiling points.

Methane is a colorless, nearly odorless gas that forms about 90% of natural gas. Pure methane can be separated from the other components of natural gas. Chemists sometimes prepare small amounts of methane in the laboratory by heating soda lime, which contains sodium hydroxide, with sodium acetate.

$$NaC_2H_3O_2(s) + NaOH(s) \rightarrow CH_4(g) + Na_2CO_3(s)$$

Ethane (C_2H_6) is a colorless gas that occurs in natural gas and is a product of petroleum refining. It has a higher melting point and boiling point

Coal can be converted into methane and other combustible gases using modern coal gasification processes that involve passing steam and oxygen through heated coal. The CO and H that are produced are then converted to methane in the presence of a catalyst. The result is a synthetic "natural" gas (SNG) with a composition and heat value similar to natural gas.

Carbon and Hydrocarbons

than methane. These properties are related to ethane's higher molar mass.

Reactions of alkanes **Combustion:** Because the alkanes make up a large proportion of our gaseous and liquid fuels, their most important reaction is combustion. Methane burns with a bluish flame.

$$CH_4 + 2O_2 \rightarrow CO_2 + 2H_2O$$

Ethane and other alkanes also burn in air to form carbon dioxide and water vapor.

$$2C_2H_6 + 7O_2 \rightarrow 4CO_2 + 6H_2O$$

Substitution: The alkanes react with halogens such as chlorine or bromine. In reactions such as these, one or more atoms of a halogen are substituted for one or more atoms of hydrogen. Therefore, the products are called substitution products. In the reaction shown below in Figure 21-13 an atom of bromine is substituted for one atom of hydrogen.

$$\begin{array}{c} H \\ | \\ H-C-H \\ | \\ H \end{array} + Br_2 \longrightarrow \begin{array}{c} H \\ | \\ H-C-Br \\ | \\ H \end{array} + HBr$$

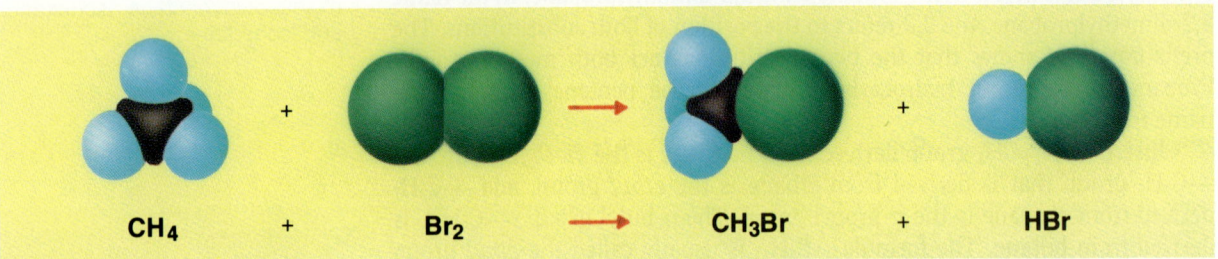

Figure 21-13 An atom of bromine is substituted for an atom of hydrogen during this substitution reaction.

CH₄ + Br₂ → CH₃Br + HBr

By supplying additional molecules of the halogen, a halogen atom can be substituted for each of the hydrogen atoms. Carbon tetrachloride (CCl_4), dichlorodifluoromethane (CCl_2F_2), and Teflon® $[CF_2CF_2]_n$, are examples of halogen-substituted alkanes.

Preparation of hydrogen: Propane reacts with steam in the presence of a nickel catalyst at a temperature of about 850°C.

$$C_3H_8 + 6H_2O \rightarrow 3CO_2 + 10H_2$$

To separate the carbon dioxide from the hydrogen, the carbon dioxide under pressure is easily dissolved in water, thus allowing the hydrogen to be collected.

Cycloalkanes The alkanes you have seen so far have been composed of open-ended chains of carbon atoms. Carbon and hydrogen atoms can also be arranged in a ring or cyclic structure. The rings may contain three, four, five, or a larger number of carbon atoms. Aromatic hydrocarbons, those that contain six-membered carbon rings, form an important class of compounds as you will see at the end of this section.

The physical and chemical properties of cyclic hydrocarbons are quite similar to those of their related open-chain hydrocarbons. The names of cycloalkanes are derived by adding the prefix *cyclo-* to the name of the open-chain molecule that has the same number of carbon atoms.

Alkenes

The alkenes, sometimes called the olefin series, are distinguished by a double covalent bond between two carbon atoms. Thus, the simplest alkene must have two carbon atoms. Its structural formula is

$$\begin{array}{c} H \\ \end{array} C = C \begin{array}{c} H \\ \end{array}$$
$$H H$$

and its name is ethene.

The name of an alkene comes from the name of the alkane with the same number of carbon atoms. Simply substitute the suffix *-ene* for the suffix *-ane*. Since ethane is the alkane with two carbon atoms, the alkene with two carbon atoms is named ethene. (This substance is also commonly called *ethylene*.) The general formula for the alkenes is C_nH_{2n}.

Ethene is the hydrocarbon commercially produced in greatest quantity in the United States. It is also the organic chemical industry's most important starting material.

> Every year the amount of ethene used in the U.S. increases. In 1987, 35 billion pounds was not sufficient to meet industrial demand. Seventy-five percent of the ethene produced is used to make plastics, 10 percent to make antifreeze, and 5 percent to make synthetic fibers.

Preparation of alkenes **Cracking alkanes:** The commercial method of producing alkenes is by cracking petroleum. **Cracking** *is a process in which complex organic molecules are broken up into simpler molecules.* This process involves the action of heat and sometimes involves a catalyst.

Cracking that uses heat alone is known as thermal cracking. During cracking, alkanes decompose in several ways and produce a variety of alkenes. A simple example is the thermal cracking of propane, which proceeds in either of two ways in nearly equal proportions.

$$C_3H_8 \xrightarrow{460\,°C} C_3H_6 + H_2$$

$$C_3H_8 \xrightarrow{460\,°C} C_2H_4 + CH_4$$

Alkenes (especially ethene), smaller alkanes, and hydrogen are typical products of the thermal cracking of alkanes. They can be separated, purified, and used as starting materials for making other organic compounds.

The cracking process that involves heat and requires a catalyst is catalytic cracking. The high-boiling fractions from petroleum distillation are catalytically cracked to produce smaller, lower-boiling hydrocarbons useful in gasoline. The catalysts used most frequently are oxides of silicon and aluminum. The cracking reactions produce smaller alkanes and alkenes with highly branched structures. Aromatic hydrocarbons are also formed.

Dehydration of alcohols: Ethene can be prepared in the laboratory by dehydrating (that is, removing water from) ethyl alcohol. Hot concentrated sulfuric acid is used as the dehydrating agent.

$$C_2H_5OH \xrightarrow[H_2SO_4]{170°C} C_2H_4 + H_2O$$

Reactions of alkenes **Addition:** *An organic compound that has one or more double or triple covalent bonds in each molecule is said to be* **unsaturated.** It is chemically possible to add other atoms directly to such molecules to form molecules of a new compound in what are known as addition reactions. For example, hydrogen atoms can be added to an alkene in the presence of a finely divided nickel catalyst. This reaction produces the corresponding alkane. Converting the carbon–carbon double bond to a single bond provides two new bond positions for the hydrogen atoms.

$$\begin{array}{c}H\\H\end{array}\!\!C\!=\!C\!\!\begin{array}{c}H\\H\end{array} + H_2 \xrightarrow{\text{Ni}} H\!-\!\underset{\underset{H}{|}}{\overset{\overset{H}{|}}{C}}\!-\!\underset{\underset{H}{|}}{\overset{\overset{H}{|}}{C}}\!-\!H$$

Halogen atoms can be added to alkene molecules. For example, two bromine atoms added to ethene form 1,2-dibromoethane as shown in Figure 21-14.

Figure 21-14 An addition reaction occurs when bromine atoms are added to ethene.

$$\begin{array}{c}H\\H\end{array}\!\!C\!=\!C\!\!\begin{array}{c}H\\H\end{array} + Br_2 \longrightarrow H\!-\!\underset{\underset{H}{|}}{\overset{\overset{Br}{|}}{C}}\!-\!\underset{\underset{H}{|}}{\overset{\overset{Br}{|}}{C}}\!-\!H$$

When the double bond between the carbon atoms is converted to a single bond, there is one bond position available for each bromine atom.

The name of the product is *1,2-dibromoethane*. The basic part of the name, ethane, is that of the related alkane with two carbon atoms. *Dibromo-* refers to the two bromine atoms that have been substituted for hydrogen atoms in ethane. The 1,2- means that one bromine atom is bonded to the first carbon atom and the other is bonded to the second carbon atom. An isomer, 1,1-dibromoethane, has the formula

$$Br\!-\!\underset{\underset{H}{|}}{\overset{\overset{Br}{|}}{C}}\!-\!\underset{\underset{H}{|}}{\overset{\overset{H}{|}}{C}}\!-\!H$$

A molecule of a hydrogen halide, such as hydrogen bromide, can be added to an alkene molecule. What is the name of the resultant product?

$$\begin{array}{c}H\\H\end{array}\!\!C\!=\!C\!\!\begin{array}{c}H\\H\end{array} + HBr \longrightarrow H\!-\!\underset{\underset{H}{|}}{\overset{\overset{H}{|}}{C}}\!-\!\underset{\underset{H}{|}}{\overset{\overset{Br}{|}}{C}}\!-\!H$$

Polymerization: A **polymer** *is a very large molecule composed of repeating structural units.* For example, molecules of ethene join together, or polymerize, at about 15°C and 20 atmospheres pressure in the presence of a catalyst. The resulting large molecules have molar masses of about 30 000 g. This polymerized material is called *polyethylene. The small molecules that join together to make a polymer are known as* **monomers.** Ethene is the monomer for polyethylene. During **polymerization,** *which is the reaction of monomers to form polyers,* the double bond in ethene becomes a single bond. Thus, the repeating unit in polyethylene is

$$-\underset{\underset{H}{|}}{\overset{\overset{H}{|}}{C}}-\underset{\underset{H}{|}}{\overset{\overset{H}{|}}{C}}-$$

and polyethylene is represented by $H(CH_2CH_2)_x H$, where x is a large number. Many monomers join together to make the polymer, which means "many units." Polyethylene is used in film for packaging, in electric insulation, in a variety of containers, and in tubing.

Alkylation: In **alkylation**, *gaseous alkanes and alkenes are combined.* For example, 2-methylpropane and 2-methylpropene combine in the presence of a sulfuric acid or anhydrous hydrogen fluoride catalyst. The result is a highly branched octane, 2,2,4-trimethylpentane, an important component of gasoline.

Combustion: The alkenes burn in oxygen. For example,

$$C_2H_4 + 3O_2 \rightarrow 2CO_2 + 2H_2O$$

Butadiene: an important alkadiene The **alkadienes** *(al-kah-dy-eens) are straight- or branched-chain hydrocarbons that have two double covalent bonds between carbon atoms in each molecule.* An example of an alkadiene is the substance butadiene shown here

$$\underset{H}{\overset{H}{\diagdown}}C=\underset{\underset{}{|}}{\overset{\overset{H}{|}}{C}}-\underset{\underset{}{|}}{\overset{\overset{H}{|}}{C}}=C\underset{H}{\overset{H}{\diagup}}$$

The *-ene* suffix indicates a double bond. The *-diene* suffix indicates two double bonds. The names of the alkadienes are derived in much the same way as those of the other hydrocarbon series. Butadiene must, therefore, have four carbon atoms and contain two double bonds in each molecule.

Actually, this structure is 1,3-butadiene, since the double bonds follow the first and third carbon atoms; the isomer 1,2-butadiene is so uncommon that 1,3-butadiene is usually called simply butadiene.

Butadiene is prepared by cracking petroleum fractions containing butane. It is used in the manufacture of SBR rubber, the most common type of synthetic rubber.

Alkynes

The alkynes are distinguished by a triple covalent bond between two carbon atoms. This series is sometimes called the acetylene series because the simplest alkyne has the common name acetylene. It has two carbon atoms, with the structural formula

$$H-C\equiv C-H$$

The names of the alkynes are derived from the names of the alkanes that have the same number of carbon atoms. The suffix *-yne* is substituted for *-ane*. Hence the systematic name for acetylene, the simplest alkyne, is ethyne. The general formula for the alkynes is

$$C_nH_{2n-2}$$

Ethyne, C_2H_2, is a minor component of the atmospheres of Jupiter and Saturn. Ethyne molecules have also been detected in interstellar space.

Preparation of ethyne **From calcium carbide:** Ethyne, a colorless gas, can be prepared by the action of water on calcium carbide (CaC_2). Calcium carbide is made from limestone ($CaCO_3$) in a series of operations. First, the limestone is heated in a kiln (oven) and CaO is produced.

$$CaCO_3(s) \xrightarrow{\Delta} CaO(s) + CO_2(g)$$

The calcium oxide is then heated with coke at 2000°C in an electric resistance furnace.

$$CaO(s) + 3C(s) \xrightarrow{\Delta} CaC_2(s) + CO(g)$$

Calcium carbide is an ionic compound with the electron-dot structure

$$Ca^{2+}$$
$$^-\!:\!C\!::\!C\!:^-$$

When it reacts with water, hydrogen replaces the calcium in the calcium carbide structure. The bonding of hydrogen and carbon in ethyne is covalent.

$$CaC_2(s) + 2H_2O(\ell) \rightarrow C_2H_2(g) + Ca(OH)_2(aq)$$

The preparation of ethyne from calcium carbide is both a commercial and laboratory method. An increasingly large percentage of the ethyne used commercially is obtained by the cracking of hydrocarbons at very high temperatures.

By partial oxidation of methane: This is an alternate commercial preparation of ethyne. Methane can be partially oxidized under controlled conditions to yield ethyne, carbon monoxide, and hydrogen.

$$6CH_4 + O_2 \xrightarrow{1500°C} 2C_2H_2 + 2CO + 10H_2$$

The source of methane is petroleum refining. The carbon monoxide and some of the hydrogen can be used to produce methanol. The balance of the hydrogen can be used as fuel to maintain the required reaction temperature.

Reactions of ethyne **Combustion:** Ethyne burns in air with a very smoky flame. Carbon, carbon dioxide, and water vapor are the products of combustion. With special burners, the combustion to carbon dioxide and water vapor is complete.

$$2C_2H_2 + 5O_2 \rightarrow 4CO_2 + 2H_2O$$

The oxyacetylene welding torch burns ethyne in the presence of oxygen and reaches a temperature of over 3200°C.

Figure 21-15 Calcium carbide and water, shown in the beaker, react together to produce ethyne. This reaction was used in miner's lamps like the one above to provide fuel for light.

Halogen addition: Ethyne is more unsaturated than ethene because of the triple bond. It is chemically possible to add two molecules of bromine to an ethyne molecule to form 1,1,2,2-tetrabromomethane.

$$H-C\equiv C-H + 2Br_2 \longrightarrow H-\underset{\underset{Br}{|}}{\overset{\overset{Br}{|}}{C}}-\underset{\underset{Br}{|}}{\overset{\overset{Br}{|}}{C}}-H$$

Dimerization: A principal use of ethyne is in the synthesis of more complex organic compounds. For example, two molecules of ethyne may combine and form the dimer, vinylacetylene. *A dimer is a molecule formed by the combination of two simple molecules.* This dimerization is brought about by passing ethyne through a water solution of copper(I) chloride and ammonium chloride (NH_4Cl). The solution acts as a catalyst.

$$2H-C\equiv C-H \xrightarrow[NH_4Cl]{Cu_2Cl_2} \overset{H}{\underset{H}{\diagdown}}C=\overset{H}{\overset{|}{C}}-C\equiv C-H$$

vinylacetylene

The —$CH=CH_2$ group is the **vinyl group.** Vinylacetylene is the basic raw material for producing Neoprene, a synthetic rubber.

Benzene and Aromatic Hydrocarbons

The aromatic hydrocarbons are generally obtained from coal tar and petroleum. The aromatic hydrocarbons are a group of hydrocarbons characterized by the presence of a benzene ring or related structure.

Benzene (C_6H_6) may be represented by the following resonance formula in which the two structures shown contribute equally.

The bonds in benzene are neither single bonds nor double bonds. Instead, each bond is a resonance hybrid bond. All of the carbon–carbon bonds in the molecule are the same. As a result, benzene and other aromatic hydrocarbons are not as unsaturated as the alkenes.

Because of the resonance structure of benzene, the benzene ring is sometimes written as

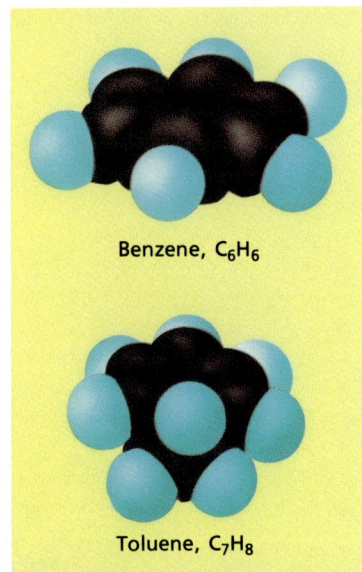

Figure 21-16 Aromatic hydrocarbons such as these are characterized by the presence of six-membered carbon rings.

CAUTION
Benzene is poisonous. Its vapors are harmful to breathe and are very flammable. It is now recommended that benzene *not* be used in school laboratories.

Prolonged exposure to benzene may destroy the bone marrow's ability to make blood cells and may also cause cancer.

The two main sources of benzene are petroleum and coal. Benzene is synthesized from petroleum alkanes and from the distillation of coal tar.

The resonance nature of its structure is understood. The —C_6H_5 group derived from benzene is the *phenyl group*.

Benzene is produced commercially from petroleum. It is a flammable liquid that is used as a solvent. Benzene is used in manufacturing many chemical products, including dyes, drugs, and explosives. Benzene has a strong, yet fairly pleasant, aromatic odor. It is less dense than water and only very slightly soluble in water.

Reactions of benzene Halogenation: In a process known as **halogenation,** *an organic compound reacts with a halogen to give a halogen-substituted compound.* For example, benzene reacts with bromine in the presence of iron to form the substitution product bromobenzene, or phenyl bromide.

$$\text{C}_6\text{H}_6 + \text{Br}_2 \xrightarrow{\text{Fe}} \text{C}_6\text{H}_5\text{Br} + \text{HBr}$$

Further treatment causes the successive substitution of other bromine atoms for hydrogen atoms. With complete substitution, hexabromobenzene is produced.

Nitration: In a process known as **nitration,** *an* —NO_2 *group is added to an organic compound when it is treated with nitric acid plus another acid.* For example, nitrobenzene is produced by treating benzene with concentrated nitric and sulfuric acids.

$$\text{C}_6\text{H}_6 + \text{HNO}_3 \xrightarrow{\text{H}_2\text{SO}_4} \text{C}_6\text{H}_5\text{NO}_2 + \text{H}_2\text{O}$$

Friedel-Crafts reaction: In the **Friedel-Crafts reaction**, *an alkyl group can be introduced into the benzene ring by using an alkyl halide in the presence of anhydrous aluminum chloride.*

$$\text{C}_6\text{H}_6 + \text{RCl} \xrightarrow{\text{AlCl}_3} \text{C}_6\text{H}_5\text{R} + \text{HCl}$$

Other aromatic hydrocarbons Toluene, another aromatic hydrocarbon, is shown in Figure 21-16. Most aromatic compounds contain one or more benzene rings. Toluene, or methylbenzene, is obtained from petroleum.

The xylenes or dimethylbenzenes, $C_6H_4(CH_3)_2$, are a mixture of three liquid isomers. The xylenes are used as starting materials for the production of polyester fibers, films, and bottles.

Ethylbenzene is produced by the Friedel-Crafts reaction of benzene and ethene in the presence of hydrogen chloride.

$$\text{C}_6\text{H}_6 + \text{C}_2\text{H}_4 \xrightarrow[\text{HCl}]{\text{AlCl}_3} \text{C}_6\text{H}_5\text{C}_2\text{H}_5$$

Ethylbenzene is treated with a catalyst of mixed metallic oxides to eliminate hydrogen and produce styrene. Styrene is used along with butadiene in making SBR synthetic rubber.

$$\text{C}_6\text{H}_5\text{C}_2\text{H}_5 \xrightarrow[\text{catalyst}]{650°C} \text{C}_6\text{H}_5\text{CH}{=}\text{CH}_2 + \text{H}_2$$

Styrene may be polymerized to polystyrene, a tough, transparent plastic. Polystyrene molecules have molar masses of about 500 000 g. Styrofoam®, a porous form of polystyrene, is used as a packaging and insulating material and for making throwaway beverage cups.

Section Review

1. List the basic features that characterize each of the following:
   (a) alkanes (b) alkenes (c) alkynes.
2. Draw all of the structural formulas that can represent C_4H_{10}.
3. Give the systematic name for each of the compounds whose formulas appear in Question 2.
4. Describe preparation methods for alkanes, alkenes, and ethyne.
5. Consider alkanes, alkenes, alkynes, and aromatic hydrocarbons. Which ones react by addition and which ones react by substitution? Explain why.
6. Give the structural formula of (a) butadiene (b) nitrobenzene (c) ethyl benzene

1. (a) Saturated hydrocarbons in straight or branched chains (b) Straight- or branched-chain hydrocarbons that contain a double C—C bond (C═C) (c) Straight- or branched-chain hydrocarbons that contain a triple C—C bond (C≡C)
2. See p. 645.
3. n-butane; 2-methylpropane
4. Alkanes are produced by fractional distillation from petroleum and natural gas. Alkenes are produced by cracking petroleum or dehydration of alcohols. Ethyne is produced by partial oxidation of methane or by action of water on calcium carbide.
5. Alkenes and alkynes react by addition because they contain multiple bonds, double and triple bonds respectively. Alkanes and aromatic hydrocarbons react by substitution.
6. (a) See p. 651. (b) See p. 654. (c) See p. 655.

21.4 Representative Hydrocarbons and Polymers

The remarkable development of organic chemistry during the last two centuries has given rise to many industrial applications that have raised the quality of life. The identification and modification of many natural products has played a key role in the development of new chemical products.

Natural Gas and Petroleum

Natural gas is a mixture of hydrocarbon gases and vapors found in porous formations in the earth's crust. It is mostly methane (CH_4) and frequently is

SECTION OBJECTIVES

- Recognize that natural gas and petroleum are complex mixtures of hydrocarbons
- Describe the formation of natural gas and petroleum.
- Identify products of the fractional distillation of petroleum.
- Identify several energy sources as substitutes for petroleum.
- Recognize rubber as an important hydrocarbon and describe the purpose of vulcanization.
- Describe the structure and uses of neoprene and SBR.

found with crude oil. Crude oil is **petroleum**, *a complex mixture of different hydrocarbons that varies greatly in composition.* The hydrocarbon molecules in petroleum contain from one to more than 50 carbon atoms.

Like coal, natural gas and petroleum were formed from the decay of plants and other organisms living in swampy areas millions of years ago. These plant remains became trapped beneath subsequent layers of new plant growth and slowly decomposed in the absence of oxygen. Later changes in the earth's surface subjected these carbon-rich plant remains to intense heat and pressure, resulting in natural gas and petroleum.

The processing of natural gas Up to about 97 percent of natural gas is methane (CH_4). Mixed with it are other hydrocarbons whose molecules contain between two and seven carbon atoms. These different hydrocarbons have different boiling points, and thus can be separated using a method called fractional distillation, described below for petroleum.

Hydrocarbons with three or four carbon atoms per molecule are sometimes separated and used for fuel as "bottled gas." Those hydrocarbons with five to seven carbon atoms per molecule are liquids at ordinary temperature. Their vapors can be condensed and used as solvents or in gasoline.

The hydrocarbons in petroleum are alkanes, cycloalkanes, and aromatics.

The methane in natural gas is still produced today by the same process that produced it millions of years ago. Methane production occurs in swamps, mud, sewage, and reclaimed garbage dumps.

The refining of petroleum Petroleum is refined by using fractional distillation to separate crude oil into different portions, each with unique properties suitable for certain uses. Distillation is evaporation followed by condensation of the vapors in a separate vessel. In **fractional distillation** *components of a mixture with different boiling points are collected separately.* No attempt is made to separate the petroleum into individual hydrocarbons as is done for the lower alkanes in natural gas. Instead, portions

TABLE 21-4 SUMMARY OF FRACTIONAL DISTILLATION OF PETROLEUM

Portion	No. of C atoms per molecule	Boiling point range (°C)	Uses
gas	C_1 to C_5	−164–30	fuel; hydrogen, gasoline by alkylation
petroleum ether	C_5 to C_7	20–100	solvent; dry cleaning
gasoline	C_5 to C_{12}	30–200	motor fuel
kerosene	C_{12} to C_{16}	175–275	fuel
fuel oil diesel oil	C_{15} to C_{18}	250–400	furnace fuel; diesel engine fuel; cracking
lubricating oils greases petroleum jelly	C_{16}	350	lubrication
paraffin wax	C_{20}	melts 52–57	candles; waterproofing; home canning
pitch tar		residue	road construction
petroleum coke		residue	electrodes

Figure 21-17 A pipe still and fractionating tower are shown here in cross-section.

that distill between certain temperature ranges are collected in separate receivers. Table 21-4 summarizes the characteristics of the different portions obtained during the fractional distillation of petroleum.

Petroleum refining is carried out in a pipe still and a fractionating tower, as shown in Figure 21-17. The crude oil is heated to about 370°C in the pipe still. At this temperature, nearly all the components of the crude oil are vaporized. The hot vapors are then discharged into the fractionating tower at a point near its base. Here, the portions with the highest condensation temperatures condense and are drawn off to collecting vessels. Portions with lower condensation temperatures continue to rise in the tower. As they rise, they are gradually cooled. In this way, the various portions reach their condensation temperatures at different levels. As they condense, the liquids collect in shallow troughs that line the inside of the tower.

Pipes carry away the overflow of condensed liquids from the troughs. The fraction that will become gasoline, together with the more volatile portions of the petroleum, passes as a gas from the top of the tower and is then liquefied in separate condensers. The liquids and remaining gaseous fractions will be subjected to other processes depending on what products are required. Figure 21-18 shows a petroleum refinery.

Petroleum Substitutes

Approximately 73 percent of the energy used in the United States comes from natural gas and petroleum. Within a few decades the known supplies of these important raw materials may be exhausted so it is essential that supplies of

Figure 21-18 The fractionating towers of this refinery are used to separate petroleum into its different components.

Carbon and Hydrocarbons **657**

natural gas and petroleum be conserved and that alternative sources of energy be developed.

Coal may possibly serve as a source for petroleum substitutes. There are two well-known methods used to convert coal to fuel gases and petroleum-like liquids. The Fischer-Tropsch process starts with a mixture of CO and H_2 formed by combining steam and hot coke. The resultant gases then react at suitable pressures and temperatures in the presence of a metallic catalyst to form a variety of straight-chain hydrocarbons. The amount of unsaturated products varies with the catalyst used. The Bergius process involves reacting finely powdered coal suspended in oil with hydrogen at very high pressure and high temperature, usually in the presence of a metallic catalyst. The complex, many-ring compounds in coal are broken down and liquid hydrocarbons are formed.

Oil shale, found in some western states, is another source of a petroleumlike liquid. In one process, the oil shale is mined and brought to the surface. There it is heated in large ovens to drive out the oil, which is then refined. Another method involves the underground heating and collection of the oil. More research is needed to develop economical production methods and resolve possible environmental problems.

Rubber

Rubber is an elastic hydrocarbon material obtained from rubber trees. Each tree daily yields about 30 grams of a milky fluid called *latex* that can be collected as shown in Figure 21-19. Latex contains about 35 percent rubber in colloidal suspension. When formic acid is added to latex, the rubber separates out as a curdlike mass. After being washed and dried, it is shipped to market as large sheets of crude rubber.

The simplest formula for rubber is $(C_5H_8)_x$. The "x" indicates a very large number. The C_5H_8 unit (below) is a monomer of rubber, which is a polymer made up of many C_5H_8 units. The units are joined in a zigzag chain and this accounts for rubber's elasticity.

rubber structural unit

Figure 21-19 A rubber tree being tapped to collect latex.

Compounding of rubber For commercial processing, raw rubber is mixed with a number of other materials in large batches. The ingredients in these batches vary according to the products to be made. Sulfur, however, is always one of the ingredients.

After mixing, the product is shaped either in a mold or from thin sheets. The whole mass is then vulcanized. **Vulcanization** *is a treatment process in which rubber is heated with sulfur*. During vulcanization, the sulfur atoms form bonds between adjacent rubber molecules. This makes rubber more elastic, less sticky, and longer wearing.

Neoprene—a synthetic rubber and its uses In 1910, scientists first produced hydrocarbon synthetic rubber by polymerizing isoprene (C_5H_8). This synthetic product was too expensive and was inferior to natural rubber from plantations in Southeast Asia. In 1931, a successful synthetic rubber called neoprene appeared on the market. **Neoprene** *is a synthetic rubber that is resistant to heat and flames, as well as to oil and chemicals.*

Hydrogen chloride can be added to vinylacetylene to yield chloroprene:

$$\underset{\text{vinylacetylene}}{\text{CH}_2=\text{CH}-\text{C}\equiv\text{C}-\text{H}} + \text{HCl} \rightarrow \underset{\text{chloroprene}}{\text{CH}_2=\text{CH}-\text{CCl}=\text{CH}_2}$$

Figure 21-20 Neoprene suits protect surfers from cold water.

The catalytic polymerization of chloroprene yields the unit shown below.

Oils and greases cause natural rubber to swell and rot; they have little effect on neoprene. Neoprene can hence be used in gasoline delivery hoses and in other objects (as shown in Figure 21-20) that must be flexible while resisting the action of hydrocarbons or temperature extremes.

$$-\text{CH}_2-\text{CCl}=\text{CH}-\text{CH}_2-$$

neoprene structural unit

SBR—a synthetic rubber and its uses SBR, Styrene Butadiene Rubber, is a good all-purpose synthetic rubber that can replace natural rubber for most purposes. Radial automobile tires are made from a mixture of about half natural rubber and half synthetic rubber. SBR is used often as the synthetic-rubber component in automobile tire treads because it resists wear better than other synthetic rubbers. It is made by churning butadiene and styrene together in soapy water. The churning is carried out at 5°C, using a catalyst. This causes the chemicals to polymerize and form SBR. The addition of an acid causes the rubber to separate in curdlike masses, which are washed and dried. A possible structural unit is shown in the structural formula below:

SBR structural unit

Figure 21-21 A worker inspects a sheet of rubber during the manufacturing process.

The remaining SBR, about 25 percent of total consumption, is used in a large number of products including hoses, seals, footwear, coated fabrics, and wire and cable insulation. A large quantity of SBR is sold as latex to be applied as carpet backing.

1. Methane
2. Probably formed by the decay of plants and animals living millions of years ago. Because of the changes in the earth's surface, these plants and animal residues were trapped in rock formations where they slowly decomposed under intense heat and pressure in the absence of atmospheric oxygen.
3. Petroleum ether, gasoline, kerosene, fuel oil, diesel oil, lubricating oils, petroleum jelly, paraffin wax, pitch, tar, petroleum coke
4. During vulcanization, which is heating of rubber with sulfur, the rubber becomes more elastic and loses its sticky qualities. Neoprene is resistant to heat, cold, and flames, as well as to oil and chemicals.

Section Review

1. What is the predominant constituent of natural gas?
2. How were natural gas and petroleum formed?
3. Name some of the refined products of the fractional distillation of petroleum.
4. Describe vulcanization and compare the characteristics of vulcanized rubber to those of neoprene.

Chapter Summary

- Carbon is an important constituent of many substances found in the natural world. A carbon atom contains four valence electrons and forms four covalent bonds with other atoms. Carbon atoms bond to other carbon atoms and to atoms of other elements with similar electronegativities.
- Diamond and graphite are two allotropic forms of carbon with dramatically different crystalline structures. Amorphous forms of carbon, such as coal, coke, and charcoal, lack a definite crystalline shape.
- Chemists refer to carbon-containing compounds as organic compounds. An almost infinite variety of organic compounds exists because of the carbon atom's ability to join with other carbon atoms and with other molecules. Carbon atoms can link up with hundreds or even thousands of other carbon atoms to form chains, branched chains, or ring structures.
- Carbon atoms represented by a specific molecular formula may be arranged in several different ways, creating compounds with different properties and structures. Compounds that have the same molecular formula but different structures are called isomers.
- Organic compounds that contain only carbon and hydrogen are called hydrocarbons. Hydrocarbons that contain functional groups in addition to atoms of carbon and hydrogen are called substituted hydrocarbons and are discussed in Chapter 22.
- Hydrocarbons are classified into four different groups based on differences in their carbon–carbon bonds. These four groups are the alkanes, alkenes, alkynes, and the aromatic hydrocarbons. Each group exhibits characteristic chemical properties.
- Alkanes are straight-chain or branched-chain hydrocarbons whose carbon atoms are connected by single covalent bonds. Alkenes are straight- or branched-chain hydrocarbons in which two carbon atoms in each molecule are connected by double covalent bond. Alkynes are straight- or branched-chain hydrocarbons in which two carbon atoms in each molecule are connect by a triple covalent bond. The aromatic hydrocarbons contain one or more benzene rings or related structures.
- Many hydrocarbons play important roles in energy and industry. Natural gas, petroleum, and coal are examples of hydrocarbons that are important sources of energy. Petroleum is refined by using fractional distillation to separate crude oil into different portions. Rubber, a natural hydrocarbon obtained from trees, can be chemically compounded to increase its strength. Synthetic forms of rubber such as neoprene and SBR are frequently used in place of natural rubber because they are more resistant to chemicals and extreme temperatures.

CHAPTER 21 REVIEW

Carbon and Hydrocarbons

VOCABULARY

activated carbon (638)
alkadienes (651)
alkanes (643)
alkenes (644)
alkyl group (647)
alkylation (651)
alkynes (644)
amorphous carbon (636)
anthracite (637)
aromatic hydrocarbons (644)
bituminous coal (637)
charcoal (638)
coal (636)
coke (637)
cracking (649)
destructive distillation (637)
diamond (634)
dimer (653)
fractional distillation (639)
Friedel-Crafts reaction (654)
fullerenes (634)
functional group (639)
graphite (634)
halogenation (654)
homologous series (645)
hydrocarbons (639)
isomer (639)
lignite (637)
methyl group (647)
monomer (651)
neoprene (659)
nitration (654)
organic chemistry (639)
paraffin wax (644)
peat (636)
petroleum (656)
petroleum coke (638)
polymer (651)
polymerization (651)
saturated hydrocarbon (644)
structural formula (640)
structural isomers (641)
subbituminous coal (637)
substituted hydrocarbons (639)
unsaturated (650)
vinyl group (653)
vulcanization (658)

REVIEWING CONCEPTS

1. How are the four covalent bonds of a carbon atom oriented in sp^3 hybridization? (21.1)
2. There are several times as many carbon compounds as noncarbon compounds. What property of carbon atoms makes this possible? (21.1)
3. **a.** Define allotropy.
   b. Name two allotropic forms of carbon. (21.1)
4. What property of diamond determines most of its industrial uses? (21.1)
5. Diamond is an excellent conductor of heat but a nonconductor of electricity. Why? (21.1)
6. Graphite is soft, yet carbon fibers in which the carbon is in the form of graphite are very strong. Why? (21.1)
7. **a.** Draw structural formulas for the three types of graphite layers.
   b. How is the actual structure of graphite an example of resonance? (21.1)
8. List several properties of graphite that make it useful as a lubricant. (21.1)
9. **a.** Explain how geologists believe coal was formed.
   b. In what sequence were the various types of coal probably formed?
   c. What conditions determine the type of coal formed? (21.1)
10. Why is activated carbon a useful adsorbent? (21.1)
11. **a.** What are isomers?
    b. Illustrate using structural formulas for the simplest alkene having isomers. (21.2)
12. What information is provided by a properly written structural formula? (21.2)
13. List five important differences between organic and inorganic compounds. (21.2)
14. What is the general formula for the members of the
    a. alkane series
    b. alkene series
    c. alkyne series? (21.3)
15. A certain hydrocarbon molecule contains seven carbon atoms. Give its formula if it is
    a. an alkane   **b.** an alkene
    c. an alkyne (21.3)
16. **a.** What is a homologous series?
    b. Name the first four members of the alkane series.
    c. How are the other members of the alkane series named? (21.3)

Carbon and Hydrocarbons **661**

17. Write the balanced equation for the laboratory preparation of methane. (21.3)
18. Write a balanced equation for the complete combustion of
    a. methane
    b. ethene
    c. ethyne
    d. butadiene (21.3)
19. Write equations for the step-by-step substitution of each of the hydrogen atoms in methane by chlorine. (21.3)
20. a. Which hydrocarbon is commercially produced in greatest quantity in the United States?
    b. What method is used for commercial production of this hydrocarbon? (21.3)
21. Butene reacts with hydrogen in the presence of a nickel catalyst. Write a structural-formula equation for the reaction. (21.3)
22. What is
    a. a monomer
    b. a dimer
    c. a polymer
    d. polymerization (21.3)
23. Give two uses for ethyne. (21.3)
24. a. What do the terms saturated and unsaturated mean when applied to hydrocarbons?
    b. What other meaning do these terms have in chemistry? (21.3)
25. Describe two methods of obtaining oil from oil shale. (21.4)
26. Define vulcanization. (21.4)
27. What advantage does neoprene have over rubber? (21.4)
28. a. What does SBR stand for?
    b. For what purposes is it used? (21.4)

PROBLEMS

▶ **Calculations with Carbon Compounds**

1. The jewelers' mass unit for diamond is the carat. By definition 1 carat equals exactly 200 mg. What is the volume of a 1.00-carat diamond? The density of diamond is 3.51 g/cm^3.

2. For 100.0 g of butadiene, calculate
    a. the number of moles
    b. the number of molecules
3. An alkene has the molecular formula $C_{12}H_{24}$. Determine its percent composition.
4. If a sample of soda lime contains 15% NaOH, how much methane, in mL at STP, can be produced by the reaction of 135 g of soda lime with sodium acetate ($NaC_2H_3O_2$)?
5. If an organic molecule contained 22 carbon atoms, how many hydrogen atoms would be present in the molecule if it were
    a. an alkane b. an alkene c. an alkyne
6. What is the mass, in grams, of 5.00 mol of pentane?
7. What volume of carbon dioxide is produced by the complete combustion of 15.0 L of propane? The volumes of carbon dioxide and of propane are measured under the same experimental conditions.
8. How many moles of chlorine are required for the production of 29.8 g of the completely substituted aromatic hydrocarbon hexachlorobenzene?
9. How many kilograms of hydrochloric acid are required for the production of 17.7 kg of chloroprene?
10. What is the mass in kilograms of 12.0 gallons of gasoline? Assume the gasoline is isooctane, which has a density of 0.692 g/mL.
11. Calculate the percentage of nitrogen in nitrobenzene.
12. How many grams of calcium carbide are required for the production of 2.5 L of ethyne at STP?
13. On a mythical planet in a far-off galaxy, the element carbon consists of equal numbers of atoms of C-12, atomic mass exactly 12 u; C-13, atomic mass, 13.003 35 u; and C-14, atomic mass, 14.003 24 u. What is the average atomic mass of carbon on this make-believe planet?
14. Calculate the energy change of the substitution reaction between 1 mol each of methane

and bromine molecules. Use the bond energies given in Table 6-1.

15. A compound consists of 93.75% carbon and 6.25% hydrogen. When 6.14 g are dissolved in 100. g of benzene, it lowers the freezing point of benzene 2.45C°. Determine the molecular formula of this compound. See Table 14-6 for necessary data.

16. A gaseous compound contains 81.7% carbon and 18.3% hydrogen. One volume of this gas reacts with five volumes of oxygen and yields three volumes of carbon dioxide. Knowing that oxygen molecules are diatomic, determine the molecular formula of this compound.

17. What volume of hydrogen will be given off as a byproduct of the cracking process if 25 kL of propane undergo cracking to form equal amounts of the alkenes propene and ethene.

18. A compound consists of 60.0% carbon, 26.7% oxygen, and 13.3% hydrogen. Its formula mass is 60. u. What are the possible structures for molecules of this chemical compound?

19. A volume of ethene (115 mL) is collected by water displacement at 22°C and 738 mm Hg pressure. What is the volume of the dry ethene at STP?

▶ Structural Formulas and Isomerism

20. Draw structural formulas for the five isomers of hexane.

21. Draw structural formulas for any three isomers of trichloropentane, and name the isomers whose formulas you draw.

22. The organic compounds 2-methylpropane and 2-methylpropene react to form 2,2,4-trimethylpentane. Write the structural formula equation for the reaction.

23. Ethyne molecules dimerize to form vinylacetylene. Write the structural formula equation for this reaction.

24. **a.** Is it geometrically possible for the four hydrogen atoms attached to the end carbon atoms in the 1,3-butadiene molecule to lie in the same plane?
b. If carbon-hydrogen bonds on adjacent singly bonded carbon atoms tend to repel each other, would it be likely for all six hydrogen atoms to lie in the same plane?

CRITICAL THINKING

1. **Inferring Conclusions** Why are organic compounds with covalent bonds usually less stable when heated than inorganic compounds with ionic bonds?

2. **Inferring Relationships** The element that appears in the greatest number of compounds is hydrogen. The element forming the second greatest number of compounds is carbon. Why are there more hydrogen compounds than carbon compounds?

3. **Relating Ideas** As the number of carbon atoms in an alkane molecule increases, does the percentage of hydrogen increase, decrease, or remain the same?

CROSS CONNECTIONS

1. **Health** Do library research and write a paper on the health hazards of some of the aromatic hydrocarbons.

2. **Architecture** Research the history and possible uses of fullerenes. Find out how the name for this class of compounds is related to R. Buckminster Fuller, the architect and inventor.

RESEARCH AND WRITING

1. *Chemical and Engineering News* publishes a list once a year of the fifty chemicals produced in largest quantity. Find out how many of the chemicals on this year's list are hydrocarbons and report their identity to the class.

2. Consult reference materials at the library and read about products made from hydrocarbons. Keep a list of the number of petroleum-related products you use in a single day.

CHAPTER 22
Substituted Hydrocarbons

PLANNING CHART

Approximate Time: 4/5 days

	SECTION/TEXT FEATURES		RESOURCES	
A V E R A G E	22.1 **ALCOHOLS** Demos 1, 2 Section Review Chapter Review • Questions 2–6, Problems 3–4, 17–18, 20, 21	665 664A–B 669 682	*Teacher's Resource Binder* Section Review Worksheets	79
H O N O R S	22.2 **HALOCARBONS** Demo 3 Section Review Chapter Review • Questions 7–9, Problems 2, 5–6, 14	670 664C 672 682	*Teacher's Resource Binder* Section Review Worksheets	80
H O N O R S	22.3 **ETHERS** Section Review Chapter Review • Problems 7–8, 15	673 673 682	*Teacher's Resource Binder* Section Review Worksheets	81
H O N O R S	22.4 **ALDEHYDES AND KETONES** Demo 4 Microscale Investigation 18 Section Review Chapter Review • Questions 10–11	674 664D 873 676 682	*Teacher's Resource Binder* Section Review Worksheets	82
H O N O R S	22.5 **CARBOXYLIC ACIDS** Demo 5 Section Review Chapter Review • Question 12, Problems 9–10, 16, 19	676 664D 678 682	*Teacher's Resource Binder* Lab Experiment 39 Section Review Worksheets	201 83
H O N O R S	22.6 **ESTERS** Demo 6 Section Review Chapter Review • Questions 1, 13–14, Problems 1, 11–13	679 664D 681 682	*Teacher's Resource Binder* Section Review Worksheets **Other Resources** *World of Chemistry* Video 20	84
	CHAPTER END ACTIVITIES Critical Thinking, Cross Connections, Research and Writing, Projects		*Teacher's Resource Binder* Test Booklet **Computer Test Generator**	109, 111

Teaching Strategies

22.1 Alcohols

Introduce this section by exhibiting several substances containing different alcohols, such as windshield cleaning fluid (methanol or methyl alcohol), liquid medicines (ethanol or ethyl alcohol), rubbing alcohol (2-propanol or isopropyl alcohol), antifreeze (1,2-ethanediol or ethylene glycol), and hand lotion (1,2,3-propanetriol or glycerol). Have students read the labels to determine what alcohols are present. Write the structural formula for each compound.

Demonstration 1: Oxidation of Alcohols
Purpose: To show the oxidation of alcohols and how this reaction helps to differentiate between primary or secondary alcohols and tertiary alcohols.
MATERIALS: 1-butanol (*n*-butyl alcohol), 2-butanol (sec-butyl

alcohol), 2-methyl-2-propanol (*tert*-butyl alcohol), potassium permanganate, distilled water, 3 50-mL beakers
PROCEDURE: CAUTION: Wear goggles, gloves, and a lab coat or lab apron during the preparation and performance of this demonstration. Make a solution of potassium permanganate by adding 0.16 g to 100 mL of water. Alcohols are flammable; make sure all flames are extinguished, and the room is well-ventilated.

Put 10 mL of each of the following alcohols into three separate beakers: 1-butanol, 2-butanol, and 2-methyl-2-propanol. Place each beaker near a white background so they can be seen. Add 3 mL of $KMnO_4$ to each alcohol and gently swirl the two liquids. (The 1-butanol and 2-butanol will turn brown because oxidation takes place. The 2-methyl-2-propanol will remain purple. Tertiary alcohols cannot be oxidized except under extreme conditions.)

The reactions may take several minutes before they turn brown. They may slowly turn red and finally brown. This is a good time to remind students that organic reactions take place more slowly than inorganic reactions.

Ask the students why the solutions of 1- and 2-butanol changed color. (They were oxidized by the $KMnO_4$, which was reduced to brown MnO_2.) What do they think the products of these reactions were? (1-butanol produced an aldehyde and 2-butanol, a ketone.) Ask the students to write equations for these reactions and the names of the organic compounds that were formed.

$$2\,\text{—C—C—C—C—OH} + O \xrightarrow{(KMnO_4)} 2\,\text{—C—C—C—C=O} + H_2O$$

butanal or butyraldehyde

Atomic oxygen is supplied by the $KMnO_4$.

$$2\,\text{—C—C—C—C—} + O \xrightarrow{(KMnO_4)} 2\,\text{—C—C—C—C—} + H_2O$$
$$\quad\quad\quad\quad\; OH \quad\quad\quad\quad\quad\quad\quad\quad\quad\quad O$$

methyl ethyl ketone or 2-butanone

Ask students how tertiary alcohols may be distinguished from primary or secondary alcohols. (Tertiary alcohols, unlike primary and secondary alcohols, are not oxidized by $KMNO_4$ in the presence of a base.) The color change of the $KMnO_4$ from purple to brown indicates that oxidation has occurred.

Demonstration 2: How Is a Super Ball Made?

PURPOSE: To make a polymer and to show that silicon also forms polymers.
MATERIALS: Sodium silicate (water glass) solution, 95% ethyl alcohol, 25–mL graduated cylinder, 5–ounce paper cup, plastic sandwich bag, wooden splint
PROCEDURE: CAUTION: Wear safety goggles, a lab coat or lab apron, and disposable plastic gloves.

Put 24 mL of sodium silicate solution in the paper cup. While stirring the sodium silicate solution with a wooden splint, add 6 mL of ethanol. Continue to stir until a solid substance is formed. Put the polymer in the palm of one hand. Gently press it with the palms of both your hands until a ball that does not crumble is formed. This takes a little time and patience. Occasionally moisten the ball by letting a small amount of water from a faucet run over it. When the ball no longer crumbles, bounce it! Store the ball in a plastic bag. If it starts to crumble, it can be reformed.

The formula for the final product is:

$$\begin{array}{ccccccc}
C_2H_5 & & C_2H_5 & & C_2H_5 & & C_2H_5 \\
| & & | & & | & & | \\
Si & -O- & Si & -O- & Si & -O- & Si \\
| & & | & & | & & | \\
C_2H_5 & & C_2H_2 & & C_2H_5 & & C_2H_5
\end{array}$$

Water is also formed. The liquid given off as the polymer is pressed is excess alcohol and a little water.

If students make their own super balls, you might have them devise additional experiments to see: (1) if the size of the ball has any effect on the height of its bounce; (2) the effect of different concentrations of ethyl alcohol; or (3) what the difference is between using isopropyl alcohol and ethyl alcohol.

22.2 Halocarbons

To introduce halocarbons, make a table on a transparency similar to the following:

NAME(S)	USES AND COMMENTS	STRUCTURAL FORMULA
Trichloromethane (Chloroform)	Previously used as an industrial solvent and anesthetic; use now limited because considered toxic and carcinogenic	Cl—C(H)(Cl)—Cl
Difluorodichloro-methane (Freon-12) 1,2-dichloro-1,1,2,2-tetrafluoroethane (Freon-114)	Coolant in refrigerators and air conditioners. Not toxic but is damaging earth's ozone layer	CF_2Cl_2 ; $ClF_2C\text{—}CF_2Cl$
1,4-dichlorobenzene	Insecticide, moth balls	Cl—C₆H₄—Cl
Polychlorinated biphenyls (PCBs)	Multitude of uses: hydraulic fluids, making paint and plastics, insulating fluid in transformers; may be most widespread chemical pollutant	(Cl-substituted biphenyl)
Polybrominated biphenyls (PBBs)	Formerly, flame retardant for clothing	(Br-substituted biphenyl)
Triiodomethane Iodoform	Antiseptic powder	I—CH—I (CHI₃)
Vinyl chloride	Monomer for poly-vinyl chloride plastic	$H_2C=CHCl$
tetrafluoroethane	Monomer for Teflon polymer	$F_2C=CF_2$
Dichlorodiphenyl-trichloroethane (DDT)	Insecticide; harmful to many animals; very long-lasting in the environment; illegal in U.S.	Cl—C₆H₄—CH(CCl₃)—C₆H₄—Cl
Hexachlorocyclo-hexane "Benzenehexa-chloride (BHC)"	Insecticide; long-lasting in the environment; causes environmental pollution	$C_6H_6Cl_6$

Substituted Hydrocarbons

Ask students what these compounds have in common, and what their general formula is. All have a carbon chain, frequently with hydrogens, and one or more halogens located on the chain. Their general formula is RX.

As is apparent from the chart and from the discussion of Freon in this chapter, many of the halocarbons are environmental pollutants that cause health problems that will affect life in the future. This would be a good place to assign reports (for extra credit or as a class project) on some of the problems caused by halocarbons.

Demonstration 3: Making a Halocarbon
PURPOSE: To make a halocarbon by adding a halogen to an alkene
MATERIALS: 1,1,2-trichloro-1,2,2-trifluoroethane, cyclohexene, iodine, 50-mL beaker, overhead projector
PROCEDURE: CAUTION: Wear goggles, gloves, and a lab coat or lab apron during the preparation and performance of this lab. Cyclohexene is extremely flammable. Store under a fume hood until ready to use, and return to hood as soon as the demonstration is completed.

Using a fume hood, add 3 g of iodine to 100 mL of 1,1,2-trichloro-1,2,2-trifluoroethane. Put 10 mL of cyclohexene into a 50-mL beaker. Put the beaker on the overhead projector and project onto a screen. Add 3 mL of the iodine solution to the cyclohexene and gently swirl. (The color disappears, indicating the addition of the iodine to the double bonds of the cyclohexene and the formation of a halocycloalkane.) Ask students to write an equation for the reaction and name the halocarbon.

22.3 Ethers
Remind students that there are many different types of ethers. Put the structural formulas and names for several different ethers on the board and ask the students to derive their general formula, ROR.

Emphasize that ethers are used mostly as solvents, though many people believe their main use is as anesthetics.

Remind students that ethers (like alcohols) burn easily because part of the oxygen needed for combustion is in the compound itself. Those that are volatile (the ethers with the smaller number of carbons in their chains) explode readily as they evaporate and mix with the air. An example equation for the combustion of methyl ethyl ether is:

$$2CH_3OC_2H_5 + 9O_2 \rightarrow 6CO_2(g) + 8H_2O(g)$$

22.4 Aldehydes and Ketones
Introduce this section by showing the students the general formulas for aldehydes and ketones and some simple examples of each. Display some common flavorings, spices, or odors whose flavor or odor is result of an aldehyde or ketone.

Demonstration 4: Fehling's Test for Aldehydes
PURPOSE: To show that aldehydes give a positive test with Fehling's solution
MATERIALS: Acetone, copper(II) sulfate pentahydrate, ethanol, glucose, potassium sodium tartrate (Rochelle salts), sodium hydroxide, distilled water, 400-mL beaker, 3 250-mL Erlenmeyer flasks, hot plate, 3 #5 rubber stoppers, 3 medium test tubes
PROCEDURE: CAUTION: Wear goggles, gloves, and a lab coat during the preparation and performance of this lab. Acetone and ethanol are extremely flammable; be sure there are no flames in the room.

(1) Dissolve 7 g of copper(II) sulfate pentahydrate in 100 mL of distilled water in an Erlenmeyer flask and insert stopper. In a second flask, dissolve 35 g of potassium sodium tartrate and 10 g of sodium hydroxide in 100 mL of distilled water and stopper. To a third flask, dissolve 5 g of glucose in 100 mL of distilled water. *(2)* Fill the beaker half full of water, and heat to boiling on a hot plate. To each of three test tubes, add 3 mL of copper sulfate solution and 3 mL of basic potassium sodium tartrate solution. To one test tube, add 1 mL of the glucose solution; to the second, 1 mL of acetone; and to the third, 1 mL of ethanol. Put all three test tubes in the boiling water bath for 5 minutes. *(3)* The formation of a brick-red precipitate is a positive test for an aldehyde. Glucose will give a positive test; the other two will not. Glucose contains an aldehyde group. The equation for the reaction is:

22.5 Carboxylic Acids

Introduce carboxylic acids by writing the names and formulas of several common substances that contain carboxylic acids, such as vinegar (acetic acid), CH_3COOH; yogurt or buttermilk (lactic acid),

$$CH_3CH-COOH;$$
$$\quad\quad\; |$$
$$\quad\quad OH$$

oranges, lemons, grapefruits (citric acid),

$$\quad\quad\quad\quad OH$$
$$\quad\quad\quad\quad\; |$$
$$HOOCCH_2-C-CH_2COOH;$$
$$\quad\quad\quad\quad\; |$$
$$\quad\quad\quad\; COOH$$

rhubarb and sauerkraut (oxalic acid), $HOOC-COOH$;

green apples (malic acid), $HOOC-CH_2-CH-COOH$;
$$\quad\quad\quad\quad\quad\quad\quad\quad\quad\quad\quad\; |$$
$$\quad\quad\quad\quad\quad\quad\quad\quad\quad\quad\; OH$$

and cranberries (benzoic acid), ⌬—COOH.

Ask students to locate the functional group that they have in common and write the general formula for carboxylic acids, RCOOH.

The following demonstration shows a reaction of an organic acid and makes a good lead-in for the next unit.

Demonstration 5: More Good Flavor-Making Esters

Purpose: To show how common flavorings are made
Materials: Glacial acetic acid, butyric acid, decanoic acid, salicylic acid, concentrated sulfuric acid, n-amyl alcohol, isoamyl alcohol, ethyl alcohol, methyl alcohol, octyl alcohol, 7 medium size test tubes, test tube rack, test tube holder, hot plate, 400-mL beaker, 7 stirring rods, filter paper, 7 watch glasses
Procedure: CAUTION: Wear a face shield, gloves, and a lab coat during the preparation and performance of this lab. Glacial acetic acid and concentrated sulfuric acid are corrosive. Butyric acid is an irritant and has a horrible smell that remains with you for hours if spilled on clothing or skin. Extinguish all flames, and be sure the room is well-ventilated.

Fill the beaker half full of water, and start to heat it to boiling on the hot plate. Number the test tubes. Use the chart to determine which acid and alcohol to put into each test tube. Add 2 mL of each alcohol and 2 mL of each liquid acid to each test tube as indicated on the chart. If the acid is a solid, add 1 g. Add 1 mL of concentrated sulfuric acid to each test tube. To mix the contents of the test tube, tap the bottom of the tube with your finger. Place the test tubes in the boiling water bath for a few minutes.

Cut a piece of filter paper into seven strips and place on seven watch glasses. Dip a clean stirring rod into each test tube and wet the filter paper by touching the stirring rod to the paper. Have the students try to identify the odor of the ester.

Tube	Acid	Alcohol	Odor of Ester	Name of Ester
1	butyric	n-amyl	apricot	n-amyl butyrate
2	butyric	ethyl	pineapple	ethyl butyrate
3	acetic	iso-amyl	banana	iso-amyl acetate
4	acetic	ethyl	apple	ethyl acetate
5	acetic	octyl	orange	octyl acetate
6	salicylic	methyl	wintergreen	methyl salicylate
7	decanoic	ethyl	grape	ethyl decanoate

22.6 Esters

Demonstration 6: Testing for Double Bonds in Oils

Purpose: To test for double bonds in cooking oils
Materials: 1,1,2-trichloro-1,2,2-trifluoroethane, cooking oil, iodine solution from Demonstration 3, 50-mL beaker, overhead projector
Procedure: CAUTION: Wear goggles, gloves, and a lab coat during the preparation and performance of this lab.

Put 10 mL of cooking oil into a 50-mL beaker. Put the beaker on an overhead projector and project on a screen. Add 3 mL of the iodine solution to the oil and stir thoroughly. (The color disappears, indicating the addition of the iodine to the double bonds of the oil. This shows that cooking oil has double bonds and thus is an unsaturated fat. The equation that indicates what happens with one oil follows:

$$\begin{array}{c}
H\quad\quad O \\
| \quad\quad\; \| \\
H-C-O-C-(CH_2)_7CH=CH(CH_2)_7CH_3 \\
|\quad\quad\; O \\
|\quad\quad\; \| \\
H-C-O-C-(CH_2)_7CH=CH(CH_2)_5CH_3 \quad + 3I_2 \rightarrow \\
|\quad\quad\; O \\
|\quad\quad\; \| \\
H-C-O-C-(CH_2)_7CH=CH(CH_2)_7CH_3 \\
|\\
H
\end{array}$$

$$\begin{array}{c}
O\quad\quad\quad\quad\quad H\; H \\
\| \quad\quad\quad\quad\quad\; |\;\; | \\
-C-O-C-(CH_2)_7-C-C-(CH_2)_7CH_3 \\
\quad\quad\quad\quad\quad\quad\quad\; |\;\; | \\
\quad\quad\quad O \quad\quad\quad\; H\; H \\
\quad\quad\quad \| \quad\quad\quad\quad\;\; |\;\; | \\
-C-O-C-(CH_2)_7-C-C-(CH_2)_5CH_3 \\
\quad\quad\quad\quad\quad\quad\quad\; |\;\; | \\
\quad\quad\quad\quad\quad\quad\quad\; H\; H \\
\quad\quad\quad\quad\quad\quad\quad\;\; |\;\; | \\
-C-O-C-(CH_2)_7-C-C-(CH_2)_5CH_3 \\
\quad\; \| \quad\quad\quad\quad\quad\; |\;\; | \\
\quad\; O \quad\quad\quad\quad\quad\; I\;\; I
\end{array}$$

Substituted Hydrocarbons

References and Resources

Books and Periodicals

Asimov, Isaac. *The World of Carbon,* New York: Collier Books, 1962.

Borgford, Christie L., and Lee R. Summerlin. "'Slime': A Shimmery, Clear Fluid Polymer," pp. 87–88; "Making a Super Ball," pp. 89–91; "Preparing a Detergent," pp. 152–54; and "Artificial Flavorings and Fragrances," pp. 237–39. *Chemical Activities,* Teacher's Edition, Washington, D.C.: American Chemical Society, 1988.

Breedlove, C. H. "Vanilla," *ChemMatters,* Vol. 6, 1988, pp. 8–9.

Casassa, E. Z. "The Gelation of Polyvinyl Alcohol with Borax," *Journal of Chemical Education,* Vol. 57, 1986, pp. 57–60.

Doheny, Anthony, and G. Marc Loudon. "The Effect of Free Radical Stability on the Rate of Bromination of Hydrocarbons," *Journal of Chemical Education,* Vol. 57, 1980, pp. 507–8.

Friedstein, Harriet G. "Basic Concepts of Culinary Chemistry," *Journal of Chemical Education,* Vol. 60, 1983, pp. 1037–38.

Gough, Michael. "Dioxin: From Plant Hormones to Agent Orange," *ChemMatters,* Vol. 6, 1988, pp. 9–12.

Gough, Michael. "Dioxin: Past War. Future Risk?" *ChemMatters,* Vol. 6, 1988, pp. 15–19.

Nagel, Mirian C. "Peroxides Can Be Treacherous," *Journal of Chemical Education,* Vol. 61, 1984, pp. 250–51.

Schultz, Emeric. "Pop-and-Sniff Experimentation," *Journal of Chemical Education,* Vol. 64, 1987, pp. 797–98.

Wood, Clair G. "Dissolving Plastic," *ChemMatters,* Vol. 5, 1987, pp. 12–14.

Zurer, Pamela S. "Studies on Ozone Destruction Expand Beyond Antarctic," *Chemical and Engineering News,* Vol. 66, 1988, pp. 16–25.

Audiovisual Resources

Chemistry 120—Organic Chemistry (Sound Filmstrips) Learning Arts, Wichita, KS, 1988.

Chemistry Study Cards—Organic Chemistry: Organic Nomenclature and Organic Reactions, Frey Scientific, Mansfield, OH, 1988–89.

The Ozone Layer: How Important Is It? (VHS) Random House Media, Stamford, CT, 1988.

Special Topics in Chemistry, Vol. II Organic Chemistry and Biochemistry, (VHS) Random House Media, Stamford, CT, 1988.

Computer Software

Flash, Patrick and Victor Bendall. *Organic Reaction Chemistry,* Apple II, 64K, Queue, Bridgeport, CT, 1988.

Chemistry Series (J & S Software) Organic Chemistry, Apple II, 48K, Barclay, Brooklyn, NY, 1988.

General Chemistry 1B, Topic 9: Organic Chemistry, Apple II, 128K, Queue, Bridgeport, CT, 1988.

SEI Organic Chemistry, Apple II, 64K and IBM PC 128K, Queue, Bridgeport, CT, 1988.

Smith, Stanley. *Substitution Reactions; Alcohols, Aldehydes and Ketones;* and *Carboxylic Acids,* Apple II 48K and IBM PC 256K, Queue, Bridgeport, CT, 1988.

Answers and Solutions

Reviewing Concepts

1. **a.** RX
   b. ROH
   c. ROR′
   d. RCHO
   e. RCOR′
   f. RCOOH
   g. RCOOR
   h. RCOOCH$_2$R′COOCHR″COOCH$_2$

2. As the formula mass increases, increasing masses are required per unit mass of water to produce the same freezing-point depression. The order of increasing formula mass is methanol, ethanol, and ethylene glycol. However, in this same order, the boiling points increase. Although a smaller mass of methanol is required for a freezing-point depression, it evaporates most readily. A larger mass of ethylene glycol is required, but it does not evaporate appreciably.

3. Alcohols do not yield hydroxide ions in water; alcohols are molecular and hydroxides are ionic compounds.

4. It destroys the cells of the optic nerve, causing blindness, and eventually death.

5. Denatured alcohol is a mixture composed principally of ethanol to which poisonous and nauseating materials have been added to make it unfit for consumption.

6. It is hygroscopic.

7. **a.** It is an excellent nonpolar solvent.
   b. It is used for dry-cleaning fabrics, removing grease from metals, extracting oils from seeds, and as an industrial solvent.
   c. Good ventilation must be provided.
   d. The vapors are poisonous.

8. Halocarbons react with water solutions of strong hydroxides to yield alcohols and the halide ion. Halocarbons react with benzenes to form alkyl benzenes and hydrogen chloride. Dichlorodifluoromethane is prepared from carbon tetrachloride and hydrofluoric acid.

9. Freons in the stratosphere are believed to react with and deplete the ozone that protects the earth's surface from excess ultraviolet radiation from the sun.

10. For making adhesives for plywood and particle board, and as resins for plastics

11. **a.** It is a ketone that is a colorless, volatile liquid and is widely used as a solvent.
    b. Dimethyl ketone, 2-propanone.

12. **a.** The carbon–oxygen bonds in formic acid are a single and a double bond because they have different lengths.

b. The carbon–oxygen bonds in sodium formate are alike because they have the same length.
c. The carbon–oxygen bonds in sodium formate are resonance hybrid bonds intermediate between single and double covalent bonds.
13. Dehydration of alcohols to ethers or alkenes; esterification
14. **a.** They are both esters of glycerol and long-carbon-chain acids.
    b. Oils are liquids at room temperature; fats are solids.
    c. Oils usually contain more unsaturated hydrocarbon chains than fats. The long-carbon-chain acids with double bonds produce esters having lower melting points.

Problems

1.
a. $Cl-\underset{\underset{Cl}{|}}{\overset{\overset{H}{|}}{C}}-Cl$ **b.** $H-\underset{\underset{H}{|}}{\overset{\overset{H}{|}}{C}}-\underset{\underset{H}{|}}{\overset{\overset{H}{|}}{C}}-\underset{\underset{H}{|}}{\overset{\overset{H}{|}}{C}}-H$

c. $H-\underset{\underset{H}{|}}{\overset{\overset{H}{|}}{C}}-\underset{\underset{H}{|}}{\overset{\overset{H}{|}}{C}}-O-\underset{\underset{H}{|}}{\overset{\overset{H}{|}}{C}}-\underset{\underset{H}{|}}{\overset{\overset{H}{|}}{C}}-H$

d. $H-\underset{\underset{H}{|}}{\overset{\overset{H}{|}}{C}}-\overset{\overset{O}{\|}}{C}-H$ **e.** $H-\underset{\underset{H}{|}}{\overset{\overset{H}{|}}{C}}-\overset{\overset{O}{\|}}{C}-\underset{\underset{H}{|}}{\overset{\overset{H}{|}}{C}}-\underset{\underset{H}{|}}{\overset{\overset{H}{|}}{C}}-H$

f. $H-\underset{\underset{H}{|}}{\overset{\overset{H}{|}}{C}}-\overset{\overset{O}{\|}}{C}-O-H$ **g.** $H-\overset{\overset{O}{\|}}{C}-O-\underset{\underset{H}{|}}{\overset{\overset{H}{|}}{C}}-\underset{\underset{H}{|}}{\overset{\overset{H}{|}}{C}}-H$

2. $-\underset{\underset{H}{|}}{\overset{\overset{H}{|}}{C}}-\underset{\underset{Cl}{|}}{\overset{\overset{H}{|}}{C}}-\underset{\underset{H}{|}}{\overset{\overset{H}{|}}{C}}-\underset{\underset{Cl}{|}}{\overset{\overset{H}{|}}{C}}-\underset{\underset{H}{|}}{\overset{\overset{H}{|}}{C}}-\underset{\underset{Cl}{|}}{\overset{\overset{H}{|}}{C}}-$

15. **a.** $54.5 \text{ C} \times \dfrac{1 \text{ mol C}}{12.0 \text{ g C}} = 4.54 \text{ mol C}$

$9.1 \text{ g H} \times \dfrac{1 \text{ mol H}}{1.0 \text{ g H}} = 9.1 \text{ mol H}$

$36.4 \text{ g O} \times \dfrac{1 \text{ mol O}}{16.0 \text{ g O}} = 2.28 \text{ mol O}$

$\dfrac{4.54 \text{ mol C}}{2.28} : \dfrac{9.1 \text{ mol H}}{2.28} : \dfrac{2.28 \text{ mol O}}{2.28}$

1.99 mol C : 3.99 mol H : 1 mol O
Simplest formula = C_2H_4O

b. $\Delta t_b = K_b m$

$m = \dfrac{78.87°C - 78.26°C}{1.22 C°\cdot\text{kg ethanol/mol solute}} = 0.50 \text{ mol solute/kg ethanol}$

molar mass solute = $\dfrac{11.0 \text{ g solute}}{(0.50 \text{ mol solute/kg ethanol})(1 \text{ kg}/10^3 \text{ g})(250 \text{ g ethanol})} = 88 \text{ g}$

3. $4\frac{1}{2}$
4. **a.** $C_2H_6 + Br_2 \rightarrow C_2H_5Br + HBr$
   b. $C_2H_4 + HBr \rightarrow C_2H_5Br$
   c. $C_2H_5OH + HBr \rightarrow C_2H_5Br + H_2O$
5. $CH_4 + 4Cl_2 \rightarrow CCl_4 + 4HCl$

   $CCl_4 + 2HF \xrightarrow{\text{catalyst}} CCl_2F_2 + 2HCl$
6. $C_3H_7OH + HI \rightarrow C_3H_7I + H_2O$
7. $2C_3H_7OH \xrightarrow{H_2SO_4} C_3H_7OC_3H_7 + H_2O$
8. **a.** $CH_3CHOHCH_2CH_3 + O$ (from oxidizing agent) $\rightarrow CH_3COCH_2CH_3 + H_2O$
   b. Methyl ethyl ketone or 2-butanone
9. **a.** $CH_3OH + CO \xrightarrow{\text{catalyst}} CH_3COOH$
   b. $C_2H_5OH + O_2 \rightarrow CH_3COOH + H_2O$ (catalyzed by bacterial enzymes). Equations for the stepwise oxidation of ethanol to acetaldehyde and then to acetic acid could also be given.
10. **a.** $H_2C_2O_4 + H_2O \rightleftarrows H_3O^+ + HC_2O_4^-$
    $HC_2O_4^- + H_2O \rightleftarrows H_3O^+ + C_2O_4^{2-}$
    b. Six moles
11. $C_4H_9OH + CH_3COOH \xrightarrow{H_2SO_4} CH_3COOC_4H_9 + H_2O$
12. $(C_{17}H_{35}COO)_3C_3H_5 + 3NaOH \rightarrow 3C_{17}H_{35}COONa + C_3H_5(OH)_3$
13. Molar mass of $C_{12}H_{25}OSO_2ONa = 288.38$ g

$\%\text{C} = \dfrac{144.13 \text{ g C}}{288.38 \text{ g}} \times 100\% = 49.98\% \text{ C}$

$\%\text{H} = \dfrac{25.199 \text{ g H}}{288.38 \text{ g}} \times 100\% = 8.74\% \text{ H}$

$\%\text{O} = \dfrac{63.998 \text{ g O}}{288.38 \text{ g}} \times 100\% = 22.19\% \text{ O}$

$\%\text{S} = \dfrac{32.06 \text{ g S}}{288.38 \text{ g}} \times 100\% = 11.12\% \text{ S}$

$\%\text{Na} = \dfrac{22.990 \text{ g Na}}{288.38 \text{ g}} \times 100\% = 7.97\% \text{ Na}$

14. $1 \text{ C atom} \times \dfrac{12.0111 \text{ u}}{\text{C atom}} = 12.0111 \text{ u}$

$3 \text{ Cl atoms} \times \dfrac{35.453 \text{ u}}{\text{Cl atom}} = 106.359 \text{ u}$

$1 \text{ F atom} \times \dfrac{18.998\,403 \text{ u}}{\text{F atom}} = 18.998\,403 \text{ u}$

Formula mass $CCl_3F = 137.369$ u

c. molar mass of $C_2H_4O = 44$ g
molar mass of $(C_2H_4O)_x = 88$ g
$x = 88 \text{ g}/44 \text{ g} = 2$
molecular formula = $C_4H_8O_2$
The compound could be ethyl acetate, $CH_3COOC_2H_5$.

16. $\text{pH} = -\log[\text{H}_3\text{O}^+] = -\log(9.4 \times 10^{-4}) = -(\log 9.4 + \log 10^{-4}) = -(0.97 - 4) = 3.03$

17. Molar mass $\text{C}_2\text{H}_5\text{OH} = 46.1$ g

$$\frac{0.750 \text{ mol } \text{C}_2\text{H}_5\text{OH}}{\text{L } \text{H}_2\text{O}} \times \frac{46.1 \text{ g } \text{C}_2\text{H}_5\text{OH}}{\text{mol } \text{C}_2\text{H}_5\text{OH}} \times 500. \text{ mL} \times \frac{1 \text{ L}}{1000 \text{ mL}} \times \frac{1 \text{ mL}}{0.789 \text{ g}} = 21.9 \text{ mL}$$

18. **a.** Molar mass $\text{C}_2\text{H}_4(\text{OH})_2 = 62.1$ g. Using one liter of ethylene glycol (1432 g) and one liter of water (1000 g), the mass ratio is

$$\frac{1432 \text{ g } \text{C}_2\text{H}_4(\text{OH})_2}{1000 \text{ g } \text{H}_2\text{O}} = \frac{23.1 \text{ mol } \text{C}_2\text{H}_4(\text{OH})_2}{1 \text{ kg } \text{H}_2\text{O}} = 23.1 \, m$$

$0.0°\text{C} - (23.1 \, m \times 1.86 \text{C°}/m) = -43.0°\text{C}$, the theoretical freezing point.

b. Freezing-point-depression calculations are based on a relationship that holds experimentally only for dilute solutions.

19. Molar mass $\text{C}_3\text{H}_5(\text{OH})_3 = 92.1$ g

$$\frac{0.400 \text{ mol } \text{C}_3\text{H}_5(\text{OH})_3}{\text{kg } \text{H}_2\text{O}} \times \frac{92.1 \text{ g } \text{C}_3\text{H}_5(\text{OH})_3}{\text{mol } \text{C}_3\text{H}_5(\text{OH})_3} \times 0.250 \text{ kg } \text{H}_2\text{O} = 9.21 \text{ g } \text{C}_3\text{H}_5(\text{OH})_3$$

20. $0.000°\text{C} - (0.400 \, m \times 1.86 \text{C°/molal}) = -0.744°\text{C}$
$100.00°\text{C} + (0.400 \, m \times 0.51 \text{C°/molal}) = 100.20°\text{C}$

Critical Thinking

1. It has a much higher boiling temperature than pure water.

Teacher's Notes

Substituted Hydrocarbons

Chapter 22
Substituted Hydrocarbons

SECTION PREVIEW

22.1 Alcohols
22.2 Halocarbons
22.3 Ethers
22.4 Aldehydes and Ketones
22.5 Carboxylic Acids
22.6 Esters

The plastic coating on these compact disks shown in the factory is one of the many products made from substituted hydrocarbons.

22.1 Alcohols

Alcohols *are substituted hydrocarbons that contain one or more hydroxyl units (—OH).* They are derived from saturated or unsaturated hydrocarbons by replacing a hydrogen atom with a hydroxyl group. The boiling points of alcohols are generally higher than those of their corresponding hydrocarbons because their molecules join together through the formation of hydrogen bonds between the hydrogen atom of the hydroxyl group of one molecule and the oxygen atom of the hydroxyl group of another molecule. Alcohols are good solvents and are useful in the preparation of other organic compounds.

Common names are used frequently for lower members of the alcohol series. A common name is derived by adding the word *alcohol* to the name of the alkyl group attached to the hydroxyl group. The IUPAC system is used to name the more complex alcohols. The systematic name of alcohols is derived by dropping the final *-e* of the alkane corresponding to the longest continuous chain of carbon atoms and adding *-ol*. The parent chain is numbered from the end nearer the hydroxyl group. The position of the hydroxyl group is often indicated by a number in front of the longest parent chain. Two common alcohols and their common and systematic names are

$$\overset{3\ \ \ 2\ \ \ 1}{CH_3CH_2CH_2OH} \qquad \overset{3\ \ \ 2\ \ \ 1}{CH_3\underset{OH}{CH}CH_3}$$

n-propyl alcohol   isopropyl alcohol
1-propanol   2-propanol

The chemical behavior of alcohols depends on the nature of the carbon atom to which the hydroxyl group is attached. Alcohols are classified as primary, secondary, or tertiary depending on the number of carbons bonded to the carbon atom bearing the —OH group. If one carbon is bonded to this carbon atom, the alcohol is primary; if two carbons are bonded, it is secondary; if three carbons are bonded, it is tertiary. An example of each type of alcohol is shown in Figure 22-1.

Preparation of Alcohols

1. Hydration of alkenes. Alcohols are formed when alkenes react with water in the presence of hydronium ions that are furnished by the catalytic action of H_2SO_4 or H_3PO_4. The reaction of ethene that is used commercially to produce an alcohol is

$$\underset{H}{\overset{H}{\diagdown}}C=C\underset{H}{\overset{H}{\diagup}} + H—O—H \xrightarrow[H_3PO_4]{400°C,\ 70\ atm} H—\underset{H}{\overset{H}{C}}—\underset{H}{\overset{H}{C}}—OH$$

This reaction produces C_2H_5OH, an alcohol known as *ethanol* or ethyl alcohol. Ethanol is commercially produced in operations such as the one shown in Figure 22-2.

SECTION OBJECTIVES

- Identify alcohols as organic compounds containing one or more hydroxyl functional groups.
- Describe some of the properties of alcohols.
- Describe methods of preparing alcohols, and write equations for some of these reactions.
- Give examples of some special alcohols and their properties, preparation, and uses.

n-butanol or 1-butanol

2-butanol

2-methyl isopropanol or 2 methyl-2-propanol

Figure 22-1 Alcohols may be classified as primary *(top)*, secondary *(middle)*, or tertiary *(bottom)*.

2. *Hydrolysis of halocarbons.* Halocarbons, which are covered in Section 22.2, react with water solutions of strong hydroxides to form the corresponding alcohols.

$$CH_3CH_2Cl + OH^- \rightarrow CH_3CH_2OH + Cl^-$$
ethyl chloride → ethanol

The halide ion (in this case, Cl^-) is displaced by the hydroxide ion.

Reactions of Alcohols

1. With sodium. Sodium reacts vigorously with ethanol, releasing hydrogen. A second product of the reaction is sodium ethoxide, C_2H_5ONa. This compound is recovered as a white solid after the excess ethanol is evaporated. This reaction is similar to the reaction of sodium with water.

Figure 22-2 This milling complex in Iowa produces ethanol from corn for use in fuel.

The reaction between sodium and ethanol is slower than the reaction between sodium and water. The reaction rate of alcohols decreases as the number of carbons in the alcohol molecule increases.

$$2\ H-\underset{\underset{H}{|}}{\overset{\overset{H}{|}}{C}}-\underset{\underset{H}{|}}{\overset{\overset{H}{|}}{C}}-OH + 2\ Na \longrightarrow 2\ H-\underset{\underset{H}{|}}{\overset{\overset{H}{|}}{C}}-\underset{\underset{H}{|}}{\overset{\overset{H}{|}}{C}}-ONa + H_2$$

2. With HX. Alcohols react with concentrated water solutions of hydrogen halides, particularly hydrobromic acid, HBr, and hydriodic acid, HI. These reactions produce halocarbons. In the example shown below, ethanol reacts with hydrobromic acid to produce ethyl bromide.

$$H-\underset{\underset{H}{|}}{\overset{\overset{H}{|}}{C}}-\underset{\underset{H}{|}}{\overset{\overset{H}{|}}{C}}-OH + HBr \longrightarrow H-\underset{\underset{H}{|}}{\overset{\overset{H}{|}}{C}}-\underset{\underset{H}{|}}{\overset{\overset{H}{|}}{C}}-Br + H_2O$$

3. Dehydration. Sulfuric acid is used as a dehydrating agent to remove the water produced by the reaction. Depending on the reaction conditions, ethanol dehydrated by hot, concentrated sulfuric acid yields either diethyl ether ($C_2H_5OC_2H_5$), which is covered in Section 22.3, or ethene (C_2H_4).

$$2H-\underset{\underset{H}{|}}{\overset{\overset{H}{|}}{C}}-\underset{\underset{H}{|}}{\overset{\overset{H}{|}}{C}}-OH \xrightarrow[\text{catalyst}]{H_2SO_4} H-\underset{\underset{H}{|}}{\overset{\overset{H}{|}}{C}}-\underset{\underset{H}{|}}{\overset{\overset{H}{|}}{C}}-O-\underset{\underset{H}{|}}{\overset{\overset{H}{|}}{C}}-\underset{\underset{H}{|}}{\overset{\overset{H}{|}}{C}}-H + H_2O$$

$$H-\underset{\underset{H}{|}}{\overset{\overset{H}{|}}{C}}-\underset{\underset{H}{|}}{\overset{\overset{H}{|}}{C}}-OH \xrightarrow[\text{catalyst}]{H_2SO_4} \underset{H}{\overset{H}{}}C=C\underset{H}{\overset{H}{}} + H_2O$$

The covalently bonded hydroxyl group (—OH) found in alcohols must not be confused with the ionically bonded hydroxide ion (OH^-). The hydroxyl group is a neutral species. Unlike hydroxide ions, the hydroxyl group does not exist as a stable ion in water solution.

4. Oxidation. Alcohols that have the hydroxyl group attached to the end carbon can be oxidized by hot copper(II) oxide. The product of such an oxidation is an aldehyde, RCHO.

$$R-\underset{\underset{H}{|}}{\overset{\overset{H}{|}}{C}}-OH + CuO \longrightarrow R-\underset{\underset{H}{|}}{C}=O + H_2O + Cu$$

666 Chapter 22

Low-molar-mass alcohols such as methanol are flammable and burn readily in air:

$$2CH_3OH + 3O_2 \rightarrow 2CO_2 + 4H_2O$$

5. Sulfation of long-chain alchohols. Under certain conditions, *sulfur may be added to alcohols,* a process that is called **sulfation**. 1-dodecanol ($C_{12}H_{25}OH$) is commonly called lauryl alcohol. It is obtained from coconut oil by hydrogenation and partial decomposition. Lauryl alcohol may be *sulfated* by treatment with sulfuric acid and then neutralized with sodium hydroxide. This process yields sodium lauryl sulfate ($C_{12}H_{25}OSO_2ONa$), a very effective cleaning agent and an ingredient in many shampoos.

$$C_{12}H_{25}OH + H_2SO_4 \rightarrow C_{12}H_{25}OSO_2OH + H_2O$$

$$C_{12}H_{25}OSO_2OH + NaOH \rightarrow C_{12}H_{25}OSO_2ONa + H_2O$$

Properties, Preparation, and Uses of Alcohols

1. Methanol. Methanol (CH_3OH) is a colorless liquid with a rather pleasant odor. It has a low density and boils at 64°C. *It is very poisonous, even when used externally.* If taken internally in small quantities, it causes blindness by destroying the cells of the optic nerve. Larger quantities cause death. Methanol is a good fuel and burns with a hot, smokeless flame.

All alcohols burn, even the solid ones. However, the shorter chain alcohols are much more volatile, and therefore their vapors ignite more readily.

All alcohols with one hydroxyl group are poisonous. Even breathing methanol for long periods of time can cause blindness and death. Ethanol is the least toxic of all the alcohols, but in large doses, even it can cause death.

Figure 22-3 Ethanol is produced commercially in large fermentation tanks *(left).* Both methanol and ethanol can be added to unleaded gasoline to form gasohol, a fuel that is used chiefly in automobile and truck engines *(right).*

Methanol is produced by the catalytic hydrogenation of carbon monoxide under pressure:

$$CO + 2H_2 \rightarrow CH_3OH$$

Methanol is used as a solvent and is added to ethanol to make it unfit for drinking. It serves as a starting material for preparing other organic compounds such as formaldehyde and acetic acid.

2. Ethanol. Ethanol (C_2H_5OH) is a colorless liquid that has a characteristic odor and a sharp, biting taste. It boils at 78°C, freezes at −117°C, and burns with a nearly colorless flame. Ethanol is a good solvent for many organic compounds that are insoluble in water.

Methanol is often called wood alcohol; this is because it is made as a byproduct of the destructive distillation of wood.

Chemistry and Health

Polyunsaturated Fats and Pour-on Margarines

We hear a lot these days about switching from saturated to polyunsaturated fats as part of a proper diet. Fats and oils are glycerides: glycerol molecules (see facing page) with one or more of the —OH groups replaced by a long-chain carboxylic acid (see Section 22.5). If these acids have only single bonds between the carbon atoms in the chain, they are considered saturated because they contain the maximum possible number of hydrogen atoms. If they have one carbon-carbon double bond, they are considered unsaturated. Polyunsaturated fats have carbon chains with more than one double bond. Polyunsaturated fats have low melting points because their chains bend at their double bonds and thus do not pack tightly together. Those that are liquid at room temperature may be called oils.

Polyunsaturated fats react readily at the double-bond sites and so are digested by our bodies more easily than saturated fats. Saturated fats may tend to be deposited in our arteries, leading to increased risk of high blood pressure, heart attacks, and strokes. Animal products usually contain saturated fats, such as lard and butter, that are solid at room temperature. Plant products generally contain unsaturated oils, such as corn oil and olive oil, that are liquid at room temperature. Changing from animal fats to vegetable oils may reduce health risks.

Margarine was first developed as a cheap replacement for butter, but new awareness of the dangers of saturated fats has caused increasing numbers of consumers to use margarines high in polyunsaturated oils, such as corn oil instead of butter. These polyunsaturated margarines and spreads are made by blending two or more fats or oils to produce a product of the desired consistency; some are even available in liquid forms. The bread and butter of yesterday may become tomorrow's bread and oil.

Ethanol is the alcohol in alcoholic beverages. It is a type of drug known as a depressant and slows down the activity of the central nervous system. Excessive use of alcohol over prolonged periods can damage the liver and cause a wide variety of other physical and psychological problems. In the United States, the production and distribution of ethanol is controlled by the federal government, and ethanol intended for use in alcoholic beverages is taxed. Ethanol that is manufactured for other uses is denatured, or made unfit to drink. **Denatured alcohol** *is a mixture composed primarily of ethanol and other poisonous and nauseating materials that have been added to make it unfit for consumption.*

Large quantities of ethanol are produced by hydrating ethene or by fermentation. The latter method occurs when yeast is added to a dilute solution of sugar at room temperature, or slightly warmer. In nature, ripe fruit often serves as the sugar source for fermentation. The yeast cells begin to grow and divide and, in the absence of oxygen, carry out fermentation. Yeast cells produce sucrase and zymase—two enzymes that act as catalysts in changing the sucrose into ethyl alcohol and carbon dioxide.

$$C_{12}H_{22}O_{11} + H_2O \rightarrow 4C_2H_5OH + 4CO_2$$

Although both processes are used in producing industrial alcohol, the hydration of ethene is less expensive than fermentation.

Another ethanol-preparation process under development involves catalytically reacting methanol with carbon monoxide and hydrogen. By this method, ethanol can be prepared starting with coal rather than petroleum.

3. Ethylene glycol. The compound ethylene glycol (1,2-ethanediol), $C_2H_4(OH)_2$, has the structural formula

```
    H  H
    |  |
H — C — C — H
    |  |
    OH OH
```

Figure 22-4 Ethylene glycol makes an excellent automotive antifreeze.

Ethylene glycol is used extensively as a "permanent" antifreeze in automobile radiators. Its boiling point (198°C) is so much higher than that of water that it does not readily evaporate or boil away. Ethylene glycol is poisonous; care should be taken to ensure that any spilled ethylene glycol is not left standing, for it is lethal to pets.

4. Glycerol. Glycerol (1,2,3-propanetriol), or glycerin, $C_3H_5(OH)_3$, has the structural formula

```
    H
    |
H — C — OH
    |
H — C — OH
    |
H — C — OH
    |
    H
```

Figure 22-5 Glycerol is an ingredient of these clear soaps.

It is a colorless, odorless, slow-flowing liquid with a sweet taste. It has a low vapor pressure and is hygroscopic (it uptakes water readily). It is used in making synthetic resins for paints and in cigarettes to keep the tobacco moist. It is also used in the manufacture of cellophane, nitroglycerin, and some soaps. Glycerol is an ingredient of cosmetics and drugs and many foods and beverages.

Section Review

1. Write the structural formulas for all alcohols having the molecular formula C_4H_9OH. Classify each as primary, secondary, or tertiary, and give the systematic name for each.
2. Why do alcohols have higher boiling points than the corresponding hydrocarbons?
3. How are alcohols prepared?
4. Name a specific alcohol and describe its properties, preparation, and uses.

1. See examples of structural formulas and rules for systematic naming of alcohols given in this section.
2. The alcohol molecules are associated through the formation of hydrogen bonds.
3. Hydration of alkenes or hydrolysis of halocarbons.
4. Accept answers as given in the subsection "Properties, Preparation, and Uses of Alcohols."

22.2 Halocarbons

SECTION OBJECTIVES

- Draw structural formulas of halocarbons and give their names.
- Describe methods of preparation of halocarbons.
- Write equations for the reactions of halocarbons.
- Give examples of some special halocarbons, their properties, preparation, and uses.

Halogens are discussed in Chapter 28.

Halocarbons or **alkyl halides** *are alkanes in which one or more halogen atoms—fluorine, chlorine, bromine, or iodine—are substituted for one or more hydrogen atoms in a hydrocarbon.* Because R— is often used to represent an alkyl group and —X any halogen, a halocarbon or alkyl halide may be represented by RX. The common name of a halocarbon is derived by adding the name of the halide—fluoride, chloride, bromide, or iodide—to the name of the alkyl group attached to the halide. The systematic name indicates the halogen by the prefixes *fluoro-, chloro-, bromo-,* and *iodo-*. The name consists of the name of the parent alkane preceded by the prefix of the halogen with a number to indicate its position. Two halides and their names are as follows:

$$\overset{3}{C}H_3\overset{2}{C}H_2\overset{1}{C}H_2Br$$

n-propyl bromide, or
1-bromopropane

$$\overset{1}{C}H_3\overset{2}{C}H\overset{3}{C}H_2\overset{4}{C}H\overset{5}{C}H_2\overset{6}{C}H_3$$
$$\quad\quad | \quad\quad\quad\; |$$
$$\quad\;\; CH_3 \quad\;\; Cl$$

4-chloro-2-methylhexane

Halocarbons are rare in nature, but many have been synthesized in the laboratory. They are easily converted to many other classes of organic compounds. Many of the halocarbons are excellent solvents. Other halocarbon compounds have been used as anesthetics, insecticides, aerosol propellants, fire extinguishers, and refrigerants.

Preparation of Halocarbons

1. Direct halogenation. Recall from your reading in Section 21.3 that halogens react with alkanes to form substitution products. For example, under suitable conditions halogen atoms can be substituted for each of the four hydrogen atoms in methane. This reaction occurs in four stages:

$$CH_4 + X_2 \rightarrow CH_3X + HX$$
$$CH_3X + X_2 \rightarrow CH_2X_2 + HX$$
$$CH_2X_2 + X_2 \rightarrow CHX_3 + HX$$
$$CHX_3 + X_2 \rightarrow CX_4 + HX$$

2. From alkenes and alkynes. Recall from Section 21.3 that alkenes and alkynes react with halogens or hydrogen halides to form halocarbons.

3. From alcohols. You learned earlier in this chapter that reactions of alcohols with hydrogen halides yield the corresponding halocarbons.

$$ROH + HX \rightarrow RX + H_2O$$

Reactions of Halocarbons

Halocarbons react with many molecules and ions. In some of these reactions, the halogen atom is replaced by another atom or group of atoms. For example, halocarbons react with water solutions of strong hydroxides to yield alcohols and the halide ion. The hydroxyl group from the hydroxide is substituted for the halogen atom in the halocarbon.

$$RX + OH^- \rightarrow ROH + X^-$$

Figure 22-6 Insecticides may contain chlorinated hydrocarbons. Here, insecticides are being used to control mosquitoes in Aswan, Egypt.

The Friedel-Crafts reaction [Section 21.3] is a halocarbon substitution reaction. Here, the phenyl group (C_6H_5—) is substituted for the halogen atom in the halocarbon.

$$RCl + C_6H_6 \xrightarrow{AlCl_3} C_6H_5R + HCl$$

Properties, Preparation, and Uses of Several Halocarbons

The chemical commonly known as carbon tetrachloride (CCl_4) is *tetrachloromethane*—a colorless, volatile, nonflammable liquid, and an excellent nonpolar solvent.

Carbon tetrachloride is prepared commercially by the direct chlorination of methane:

$$CH_4 + 4Cl_2 \rightarrow CCl_4 + 4HCl$$

Carbon tetrachloride is sometimes used for dry-cleaning fabrics, removing grease from metals, as a spot remover for clothes, and extracting oils from seeds. Its vapors are poisonous. Its most important use is in the preparation of a family of polyhalogenated derivaties of methane and ethane known as the Freon® halocarbons.

Dichlorodifluoromethane (CCl_2F_2) is the most important member of the Freon halocarbons. It is an odorless, nontoxic, nonflammable, easily liquefied gas that is prepared from carbon tetrachloride and hydrofluoric acid with antimony compounds as catalysts.

$$CCl_4 + 2HF \xrightarrow{catalyst} CCl_2F_2 + 2HCl$$

Because it has a low boiling point and is nontoxic and nonflammable, dichlorodifluoromethane is used as a refrigerant in mechanical refrigerators and air conditioners (Figure 22-7) and as a propellant in aerosol products. It has also been used as a bubble-making agent in manufacturing plastic foams.

Once released into the air, Freon halocarbons rise into the stratosphere, where ultraviolet radiation from the sun decomposes them and releases free halogen atoms. The halogen atoms catalyze the destruction of the ozone in the ozone layer, which protects us from cancer-causing ultraviolet radiation. Consequently, the production of Freon halocarbons has been limited, and their use as propellants has been discontinued in the United States.

Tetrafluoroethene (C_2F_4) can be polymerized to form a product that has a structure in which the following structual unit occurs again and again, forming long chains:

$$-\underset{\underset{F}{|}}{\overset{\overset{F}{|}}{C}}-\underset{\underset{F}{|}}{\overset{\overset{F}{|}}{C}}-$$

The tradename for this material is Teflon®. It is a very inactive, flexible substance that is stable up to about 325°C and has a very low coefficient of friction. It is

Carbon tetrachloride is thought to be a cancer-causing agent; it is toxic if taken internally, absorbed by the skin, or inhaled.

CAUTION:
There must be good ventilation when carbon tetrachloride is used. Its use in high school laboratories should be avoided.

Figure 22-7 An air conditioner uses recirculating Freon gas. The gas exits the compressor under high pressure and condenses to a liquid, giving up heat to the outdoors. It then passes through an expansion valve and enters the evaporator, where it becomes a gas and gets extremely cold. A fan pulls warm room air across the cold evaporator coils and blows cool air back into the room.

useful where heat-resistant, nonlubricated moving parts are needed and in many items as a "nonsticking" surface.

Ethylene dichloride ($ClCH_2CH_2Cl$), also called 1,2-dichloroethane, is made by chlorinating ethene, as described in Section 21.3. This liquid is used as a solvent and for making adhesives. Most of it, however, is converted to vinyl chloride:

$$Cl-\underset{\underset{H}{|}}{\overset{\overset{H}{|}}{C}}-\underset{\underset{H}{|}}{\overset{\overset{H}{|}}{C}}-Cl \xrightarrow{heat} \underset{H}{\overset{H}{}}C=C\underset{Cl}{\overset{H}{}} + HCl$$

Vinyl chloride is then polymerized to form polyvinyl chloride, which has the structural unit

$$-\underset{\underset{H}{|}}{\overset{\overset{H}{|}}{C}}-\underset{\underset{Cl}{|}}{\overset{\overset{H}{|}}{C}}-$$

Polyvinyl chloride (PVC) has a molar mass of about 1 500 000 g. It is used for plastic pipe, plastic siding for buildings, film, wire, and insulation.

Section Review

1. Write structural formulas for all halocarbons having the molecular formula C_4H_9Cl, and give the systematic name for each.
2. Write an equation for the preparation of each of the following halocarbons: *(a)* 2-chloropropane *(b)* dichloromethane.
3. Write an equation showing the preparation of ethyl alcohol from the reaction of a halocarbon with the water solution of a strong base.

22.3 Ethers

Ethers *are a group of compounds in which two hydrocarbon groups are bonded to the same atom of oxygen.* They may be represented by the general formula ROR′, in which R and R′ indicate two alkyl groups. These alkyl groups may or may not be the same.

Common names are usually used when ethers are named. They are usually named by giving the name of each alkyl group first, then the word "ether." Some common ethers and their names are shown below.

$CH_3OCH_2CH_3$ $CH_3CH_2OCH_2CH_3$
methyl ethyl ether   diethyl ether or ethyl ether

If both alkyl groups are the same, the prefix *di-* is usually used but the use of the prefix is not necessary.

SECTION OBJECTIVES
- Identify ethers. Draw structural formulas for ethers, and give their names.
- Describe the preparation of ethers.
- Describe the reactions, properties, and uses of ethers.

Preparation of Ethers

1. Dehydration of alcohols. Ethers can be prepared by dehydration of alcohols as described in Section 22.1. For example, diethyl ether can be prepared by heating ethanol and sulfuric acid to 140°C.

$$2CH_3CH_2OH \xrightarrow{H_2SO_4} CH_3CH_2OCH_2CH_3 + H_2O$$

2. Williamson synthesis. In this method, the reactants are the sodium salt of an alcohol and a halocarbon.

$$RO^-Na^+ + R'X \rightarrow ROR' + Na^+X^-$$

This method is used to prepare mixed ethers. In a mixed ether, the R and R′ groups are different.

Reactions, Properties, and Uses of Ethers

Ethers are relatively unreactive. However, when ethers are treated with hot concentrated inorganic acids (HI, HBr), they are cleaved:

$$CH_3OCH_2CH_3 + HI \rightarrow CH_3I + CH_3CH_2OH$$
$$CH_3CH_2OH + HI \rightarrow CH_3CH_2I + H_2O$$

The properties of ethers generally resemble those of the corresponding hydrocarbons. Like alcohols, ethers are extremely flammable. Ethers are only slightly soluble in water and are mainly used as solvents for many organic compounds. Diethyl ether slows down the operation of the central nervous system, hence its former widespread use as an anesthetic. One of the methyl butyl ethers, $(CH_3)_3COCH_3$, is used in gasoline to increase the octane rating.

Section Review

1. Draw structural formulas for each of the following ethers:
   (a) dimethyl ether   (b) ethyl *n*-propyl ether   (c) ethyl isopropyl ether.
2. Name each of the following ethers:
   (a) $CH_3CH_2CH_2OCH_2CH_3$ (b) $CH_3CH_2OCH_2CH_2CH_3$
3. Write an equation showing how dimethyl ether can be prepared from the dehydration of methanol.

1. (a) CH_3OCH_3
   (b) $CH_3CH_2OCH_2CH_2CH_3$
   (c) $CH_3CH_2OCHCH_3$
         |
         CH_3

2. (a) *n*-propyl ether or di-*n*-propyl ether
   (b) ethyl *n*-butyl ether

3. $2CH_3OH \xrightarrow{H_2SO_4} CH_3OCH_3 + H_2O$

22.4 Aldehydes and Ketones

SECTION OBJECTIVES

- Identify substances as aldehydes and ketones.
- Describe the preparation, uses, and reactions of aldehydes.
- Describe the preparation and uses of some ketones.

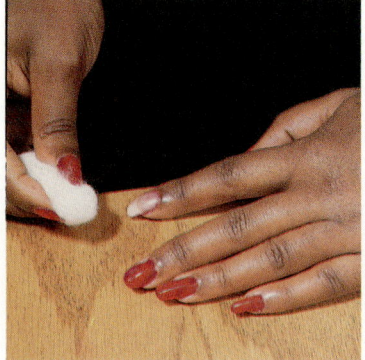

Figure 22-8 Acetone, an aldehyde, is the active ingredient in nail polish remover.

Aldehydes and ketones are substituted hydrocarbons characterized by the presence of the **carbonyl group,** *a group with the formula*

$$\underset{}{\overset{}{>}}C=O$$

If one of the bonds of the carbonyl group is attached to a hydrogen atom, the compound is an *aldehyde. In* **aldehydes,** *the carbon atom in the carbonyl group is attached to at least one hydrogen atom. The functional group of an aldehyde,*

$$\underset{H}{\overset{}{>}}C=O$$

is known as the **formyl group.** The other bond of the carbonyl carbon may bond to either a hydrogen or an alkyl group. The general formula for an aldehyde is

$$\underset{H}{\overset{R}{>}}C=O$$

When both bonds of the carbonyl group are attached to carbon atoms, the compounds are **ketones**. The general formula for ketones is

$$\underset{R'}{\overset{R}{>}}C=O$$

The alkyl groups R and R′ may be the same or different. Because both aldehydes and ketones contain the carbonyl group, their properties are usually similar.

Simple aldehydes are referred to by common names. The systematic name is derived by substituting *-al* for the final *-e* in the parent hydrocarbon chain name. Some aldehydes and their common and systematic names are:

$$H-C\overset{O}{\underset{H}{\nwarrow}} \qquad CH_3C\overset{O}{\underset{H}{\nwarrow}} \qquad CH_3CH_2CH_2C\overset{O}{\underset{H}{\nwarrow}}$$

formaldehyde   acetaldehyde   *n-* butyraldehyde
or   or   or
methanal   ethanal   butanal

The simpler ketones are usually referred to by common names. The names are derived by naming the alkyl groups bonded to the carboxyl group and adding the word *ketone.* Systematic names are given usually for very complex ketones. The suffix *-one* is substituted for the final *-e* in the name of the parent hydrocarbon chain. Three ketones and their names are:

$$\text{CH}_3-\overset{\overset{\text{O}}{\|}}{\text{C}}-\text{CH}_3 \qquad \text{CH}_3-\overset{\overset{\text{O}}{\|}}{\text{C}}-\text{CH}_2\text{CH}_3 \qquad \text{CH}_3\text{CH}_2-\overset{\overset{\text{O}}{\|}}{\text{C}}-\text{CH}_2\text{CH}_3$$

acetone or dimethyl ketone or propanone

methyl ethyl ketone or 2-butanone

diethyl ketone or 3-pentanone

Many natural substances are aldehydes and ketones. Some of these include flavors such as cinnamon, vanilla, almond, jasmine, clove oil, and the scents of many perfumes. Some of the sex hormones and some vitamins are aldehydes and ketones.

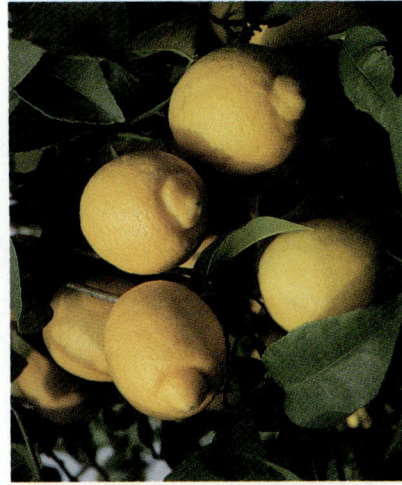

Figure 22-9 Vanilla beans *(left)* and lemons *(right)* each owe their characteristic flavors to a specific aldehyde.

Preparation and Uses of Aldehydes

The general method of preparing aldehydes, the oxidation of alcohols, was described in Section 22.1. A common process of this type involves passing methanol vapor and a regulated amount of air over heated copper. Formaldehyde (HCHO) is produced.

The molecular structures of some of the sex hormones are discussed in Section 23.4.

$$2\text{Cu} + \text{O}_2 \rightarrow 2\text{CuO}$$
$$\text{CH}_3\text{OH} + \text{CuO} \rightarrow \text{HCHO} + \text{H}_2\text{O} + \text{Cu}$$

The commercial preparation of formaldehyde requires the catalytic action of silver or an iron-molybdenum oxide. At room temperature, formaldehyde is a gas with a pungent, distinctive odor. Dissolved in water, it makes an excellent disinfectant. It is also used to preserve biological or medical specimens. The largest uses of formaldehyde are in making certain types of adhesives for plywood and particleboard and producing resins for plastics.

Preparation And Uses Of Ketones

Ketones can be prepared from alcohols that do not have the hydroxyl group attached to an end-carbon atom. For example, acetone (CH_3COCH_3) is prepared by the mild oxidation of 2-propanol ($\text{CH}_3\text{CHOHCH}_3$) with potassium dichromate ($\text{K}_2\text{Cr}_2\text{O}_7$) in water solution as the oxidizing agent.

Figure 22-10 The pungent odor of formaldehyde is familiar to those who have dissected biological specimens.

Substituted Hydrocarbons

$$H-\underset{\underset{H}{|}}{\overset{\overset{H}{|}}{C}}-\underset{\underset{H}{|}}{\overset{\overset{OH}{|}}{C}}-\underset{\underset{H}{|}}{\overset{\overset{H}{|}}{C}}-H + O \text{ (from oxidizing agent)} \longrightarrow H-\underset{\underset{H}{|}}{\overset{\overset{H}{|}}{C}}-\overset{\overset{O}{\|}}{C}-\underset{\underset{H}{|}}{\overset{\overset{H}{|}}{C}}-H + H_2O$$

Acetone is a colorless, volatile liquid that is widely used as a solvent.

Section Review

1. Identify each of the following compounds as an aldehyde or a ketone, and give the name for each:

   (a) $CH_3CH_2CH_2C\overset{\nearrow O}{\searrow H}$

   (b) $CH_3\underset{\underset{CH_3}{|}}{CH}\overset{\overset{O}{\|}}{C}-CH_2CH_3$

2. Write an equation that represents a method of preparing formaldehyde.
3. Describe the uses of formaldehyde and acetone.

Answers:
1. *(a)* aldehyde, butanal
   (b) ketone, ethyl isopropyl ketone/2-methyl-3-pentanone
2. Accept equation given in this section.
3. Accept uses given in this section.

22.5 Carboxylic Acids

SECTION OBJECTIVES

- Draw structural formulas of carboxylic acids and give their names.
- Describe the properties of carboxylic acids. Describe methods of preparation and reactions of carboxylic acids.
- Name and describe the specific acids discussed in this chapter, and describe their preparation, properties, and uses.

Carboxylic acids contain the **carboxyl group.**

$$-\overset{\overset{O}{\|}}{C}-OH$$

Carboxylic (organic) acids *are hydrocarbons in which two of the hydrogen atoms on an end carbon atom have been replaced by a doubly bonded oxygen, and the third hydrogen on the end carbon atom has been replaced by a hydroxyl group,* —*OH. The general formula for a carboxylic acid can be written as*

$$R-\overset{\overset{O}{\|}}{C}-OH$$

The R could be an H atom, designating the simplest carboxylic acid

$$H-\overset{\overset{O}{\|}}{C}-OH$$

The carboxyl group can also be written —COOH.

known as formic acid, or it could be an alkyl group. Carboxylic acids are highly polar, and their molecules form stable hydrogen bonds. As a result, organic acids have higher boiling points than alcohols of corresponding molar masses.

Because organic acids were known before a systematic nomenclature was developed, many of them are known by common names derived from a Greek or Latin name indicating the sources from which they were originally obtained. For example, formic acid was originally obtained from the distillation of ants and hence derives its name from the Latin word for ant, *formica*. The systematic system for naming organic acids uses the suffix *-oic* in place of the final *-e* of the alkane name and adds the word *acid*. Hence, acetic acid (CH_3COOH) is given the systematic name ethanoic acid. Butyric acid ($CH_3CH_2CH_2COOH$) is assigned the systematic name, butanoic acid.

Preparation of Carboxylic Acids

Carboxylic acids are most commonly prepared by the oxidation of primary alcohols, or by the oxidation of aldehydes.

$$RCH_2OH \xrightarrow{\text{oxidation}} R-\overset{\overset{O}{\|}}{C}-OH \quad R-\overset{\overset{O}{\|}}{C}-H \xrightarrow{\text{oxidation}} R-\overset{\overset{O}{\|}}{C}-OH$$

When oxidized, primary alcohols and aldehydes each produce acids with the same number of carbon atoms per molecule.

Commercially, carboxylic acids can be prepared from the reaction of CO and the alcohol possessing one fewer carbon atoms than the desired acid. The reaction is carried out under pressure and in the presence of dicobalt octacarbonyl, acting as a catalyst.

$$R-OH + CO \xrightarrow{Co_2(CO)_8} R-\overset{\overset{O}{\|}}{C}-OH$$

$$CH_3CH_2OH + CO \xrightarrow{Co_2(CO)_8} CH_3CH_2COOH$$

Reactions of Carboxylic Acids

1. Ionization. This reaction involves the hydrogen atom bonded to an oxygen atom in the carboxyl group. It is this hydrogen atom that ionizes in water solution, giving carboxylic acids their acid properties. The hydrogen atoms bonded to carbon atoms in these acids *never* ionize in water solution.

$$HCOOH + H_2O \rightleftarrows H_3O^+ + HCOO^-$$
$$CH_3COOH + H_2O \rightleftarrows H_3O^+ + CH_3COO^-$$

These equilibria yield low H_3O^+ ion concentrations. Therefore, carboxylic acids are generally weak acids.

2. Neutralization. An organic acid may be neutralized by a hydroxide. A salt is formed in a reaction similar to that undergone by inorganic acids.

$$CH_3COOH + NaOH \rightarrow CH_3COONa + H_2O$$

3. Esterification. In this reaction, an acid reacts with an alcohol to produce a substance known as an ester. We will study this reaction in Section 22.6.

Figure 22-11 Nylon, shown here being wound onto a stirring rod, is produced from the polymerization of adipic acid (a carboxylic acid) and hexamethylenediamine.

Preparation, Properties, and Uses of Carboxylic Acids

1. Acetic acid. Acetic acid, CH_3COOH, can be produced by the catalytic oxidation of acetaldehyde. Very pure acetic acid can be prepared by reacting methanol with carbon monoxide at 180°C and 30–40 atm pressure in the presence of a catalyst.

Substituted Hydrocarbons

Figure 22-12 Acetic acid is used to convert some of the hydroxyl (–OH) groups of cellulose into ester groups, resulting in cellulose acetate fibers, shown here in cross section.

$$CH_3OH + CO \xrightarrow{\text{catalyst}} CH_3COOH$$

Concentrated acetic acid is a colorless liquid that is a good solvent for some organic chemicals. It is used for making cellulose acetate, a basic material of many fibers and films.

Cider vinegar is made from apple cider that has fermented to "hard cider." The ethanol in hard cider is slowly oxidized in the presence of the oxygen in the air. This oxidation produces acetic acid. The reaction is catalyzed by enzymes produced by certain species of bacteria.

$$C_2H_5OH + O_2 \rightarrow CH_3COOH + H_2O$$

Pure acetic acid has a penetrating odor and produces burns if in contact with the skin. Acetic acid is the most commercially important organic acid. It is used extensively in the dyeing of textiles, and in the production of cellulose acetate, a major component of synthetic fibers. It is also an excellent solvent for organic compounds.

2. Formic acid. Formic acid, HCOOH, is prepared from sodium hydroxide solution and carbon monoxide under pressure. This reaction yields sodium formate (HCOONa).

$$NaOH + CO \rightarrow HCOONa$$

If sodium formate is carefully heated with sulfuric acid, formic acid distills off:

$$HCOONa + H_2SO_4 \rightarrow HCOOH + NaHSO_4$$

Formic acid is found in nature in stinging nettle plants and in certain ants. It is used as an acidifying agent in the textile industry.

3. Oxalic acid. Oxalic acid, (COOH)$_2$, is the first member of the dicarboxylic acid series. **Dicarboxylic acids** *are acids that have two carboxyl groups, and are said to be dibasic.* Oxalic acid is prepared by heating sodium formate in the presence of a strong base, and then acidifying sodium oxalate, the product of the reaction shown below.

$$2H-\overset{\overset{O}{\|}}{C}-O^-Na^+ \xrightarrow[\Delta]{NaOH} Na^+O^- -\overset{\overset{O}{\|}}{C}-\overset{\overset{O}{\|}}{C}-O^-Na^+ + H_2$$

Oxalic acid occurs in plants such as rhubarb and sorrel (wildflowers in the genus *Oxalis*). It is also found in urine and guano. Oxalic acid is used commercially for dyeing, calico printing, and in laundries for the removal of stains and ink spots.

Section Review

1. Write the structural formula for each of the following compounds:
   (a) pentanoic acid *(b)* butyric acid *(c)* isobutyric acid
   (d) 2-methyl butanoic acid.
2. Name each of the following compounds:

   (a) $CH_3CH_2CH_2CH_2CH_2CH_2C\overset{\displaystyle{\nearrow O}}{\searrow OH}$

   (b) $CH_3CHCH_2C\overset{\displaystyle{\nearrow O}}{\searrow OH}$ with CH_3 branch

---

1. (a)
   $$CH_3CH_2CH_2CH_2\overset{\overset{O}{\|}}{C}-OH$$

   (b)
   $$CH_3CH_2CH_2\overset{\overset{O}{\|}}{C}-OH$$

   (c)
   $$CH_3\underset{\underset{CH_3}{|}}{CH}\overset{\overset{O}{\|}}{C}-OH$$

Guano is manure produced by sea birds or bats. It is often used as fertilizer.

   (d)
   $$CH_3CH_2\underset{\underset{CH_3}{|}}{CH}\overset{\overset{O}{\|}}{C}-OH$$

2. (a) heptanoic acid
   (b) 3-methylbutanoic acid
   (c) 2-chlorobutanoic acid
   (d) propanoic acid

(c) CH₃CH₂CHCOOH
     |
     Cl

(d) CH₃CH₂COOH

3. Why do carboxylic acids have higher boiling points than alcohol with corresponding molecular masses?
4. Show the ionization reaction of acetic acid.

3. Carboxylic acids are highly polar and their molecules form stable hydrogen bonds. As a result, they have higher boiling points than alcohols of corresponding molecular masses.
4. $CH_3COOH + H_2O \leftrightarrow CH_3COO^- + H_3O^+$

22.6 Esters

Esters are similar to carboxylic acids, but have an alkyl group in place of the hydrogen atom of the carboxyl group.

An **ester** *may be represented by the general formula* $RC\overset{O}{\overset{\|}{-}}OR'$ *in which R′ is always an alkyl group and R can be either a hydrogen atom or an alkyl group*. Although the molecules of an ester are polar, like those of acids, they do not form hydrogen bonds with one another. As a result, esters have much lower boiling points than acids with corresponding molar masses. Ester molecules are able to form hydrogen bonds with water molecules and are somewhat soluble in water. Esters are excellent solvents for organic compounds. Many esters occur naturally and are responsible for the distinctive odors and flavors of fruits and flowers. Esters are used in perfumes and artificial flavoring.

Esters are easy to name. The name of the alkyl group (designated R′ in the general formula above) is given first and then is followed by the name of the acid that is represented by the carboxylic group before the replacement of the hydrogen atom by the alkyl group. The second part of the ester name is obtained by replacing the *-ic* of the acid name by *-ate*. The structural formulas and names of some common esters are as follows:

$$H-\overset{O}{\overset{\|}{C}}-OCH_2CH_3 \qquad CH_3\overset{O}{\overset{\|}{C}}-OCH_2CH_3 \qquad CH_3CH_2CH_2\overset{O}{\overset{\|}{C}}-OCH_2CH_3$$

  ethyl methanoate   ethyl ethanoate   ethyl-*n*-butanoate
         or   or
  ethyl formate   ethyl acetate

SECTION OBJECTIVES

- Draw structural formulas of esters and give their names.
- Describe properties of esters.
- Describe esterification
- Describe reactions of esters and define saponification.

Figure 22-13 The different flavors and aromas of pineapple, banana, orange, and apple are a result of different esters.

Preparation of Esters

An ester is produced when an organic acid reacts with an alcohol in a reaction called **esterification.** The reaction usually takes place in the presence of an inorganic acid, which acts as a catalyst, and can be represented by the general equation:

$$R-\overset{O}{\overset{\|}{C}}-OH + R'OH \underset{}{\overset{H_2SO_4}{\rightleftarrows}} R-\overset{O}{\overset{\|}{C}}-OR' + H_2O$$

Substituted Hydrocarbons **679**

As an example, ethyl acetate is the ester formed when acetic acid and ethanol react as shown below.

$$\underset{}{CH_3\overset{O}{\overset{\|}{C}}-OH} + HOCH_2CH_3 \xrightarrow{H_2SO_4} CH_3\overset{O}{\overset{\|}{C}}-OCH_2CH_3 + H_2O$$

Reactions of Esters

Esters are neutral compounds and do not exhibit acidic or basic properties. They undergo reactions in which the —OR′ group is replaced by another group. Esters can be decomposed by hydrolysis into the alcohol and acid from which they were derived. This hydrolysis occurs with dilute acid or metallic hydroxide solutions.

1. Acidic hydrolysis. When the hydrolysis of an ester is catalyzed by an acid such as HCl, esters yield a carboxylic acid and an alcohol.

$$CH_3\overset{O}{\overset{\|}{C}}-OCH_2CH_3 + H_2O \underset{\text{catalyst}}{\overset{H+}{\rightleftharpoons}} \underset{\text{acetic acid}}{CH_3\overset{O}{\overset{\|}{C}}-OH} + \underset{\text{ethanol}}{CH_3CH_2OH}$$

Figure 22-14 Polyesters are polymers containing many ester linkages. When a carboxylic acid with two carboxyl groups reacts with an alcohol with two hydroxyl groups, esterification may occur at both ends of each molecule, resulting in polyester.

This reaction is the reverse of esterification. In reality, the reverse reaction occurs to an appreciable extent, and the forward reaction does not go to completion.

2. Alkaline hydrolysis. When the hydrolysis of an ester is carried out in basic solution, the ester is converted to its alcohol and a salt of its acid.

$$R-\overset{O}{\overset{\|}{C}}-OR' + NaOH \longrightarrow R-\overset{O}{\overset{\|}{C}}-ONa + R'OH$$

Fats *and* **oils** *from plants or animals are esters of glycerol and long carbon chain acids.* The carbon chains of the acids usually contain from 12 to 20 carbon atoms. The structure of a fat or oil can be represented as

$$\begin{array}{c} RCOOCH_2 \\ | \\ R'COOCH \\ | \\ R''COOCH_2 \end{array}$$

R, R′, and R″ are saturated or unsaturated long-chain-hydrocarbon groups.

The only difference between a fat and an oil is the physical state of each at room temperature. Oils are liquids at room temperature; fats are solids. Oils usually contain more unsaturated hydrocarbon chains than fats. This is because long-carbon-chain acids with double bonds produce esters that have lower melting points and thus are liquid at room temperature. When natural fats are treated with sodium hydroxide, they are converted to glycerol and the sodium salt of the fatty acids. *This type of reaction, the hydrolysis of fats or oils using a solution of a strong hydroxide, is called* **saponification.** The term *saponification* was originally used to describe the alkaline hydrolysis of fats or oils in the manufacture of soaps.

The alkaline hydrolysis of fatty acid esters is represented by the general equation

RCOOCH$_2$
|
R'COOCH + 3NaOH → RCOONa + R'COONa + R"COONa + C$_3$H$_5$(OH)$_3$
|
R"COOCH$_2$

Soaps are generally made by hydrolyzing fats and oils with water heated to about 250°C in a closed container at high pressure. The long-chain carboxylic acids thus produced are neutralized with sodium hydroxide. This neutralization yields a mixture of the sodium salts that we recognize as soap.

Section Review

1. Give the structural formulas of the following compounds. *(a)* methyl acetate *(b)* ethyl methanoate *(c)* isopropyl butanoate *(d)* n-propyl acetate.
2. Give an example of an esterification reaction and write the equation for it.
3. Explain why esters have lower boiling points than acids with corresponding molecular masses.
4. Define *saponification* and write an equation to represent the reaction.

Chapter Summary

- Substituted hydrocarbons are hydrocarbons in which an atom or a group of atoms are substituted for one or more of the hydrogen atoms.
- Alcohols are alkanes in which one or more hydroxyl groups have been substituted for an equal number of hydrogen atoms. The general formula is ROH.
- Alcohols are prepared by the hydration of alkenes and from halocarbons by substitution. They react with sodium and with hydrogen halides and also undergo dehydration, oxidation, and sulfation, reactions.
- A halocarbon is an alkane in which a halogen atom is substituted for a hydrogen atom. The general formula for a halocarbon is RX.
- Halocarbons are prepared by the reaction of halogens with alkanes, alkenes, or alkynes and by the reaction of hydrogen halides with alkenes, alkynes, and alcohols. Many reactions of halocarbons are substitution reactions.
- Ethers are organic oxides with the general formula ROR'. They can be prepared by dehydrating alcohols.
- Aldehydes have the general formula RCHO. They are prepared by the controlled oxidation of alcohols having a hydroxyl group on a carbon at the end of the molecule.
- Ketones have the general formula RCOR'. They are prepared by oxidizing alcohols that do not have the hydroxyl group attached to the carbon at the end of the molecule.
- The general formula for carboxylic acids is RCOOH. They can be prepared by the oxidation of alcohols and aldehydes.
- Esters give fruits their characteristic flavors and odors. Their general formula is RCOOR'. Saponification is the hydrolysis of a fat using a solution of strong hydroxide and yields a mixture of sodium salts that makes up soap.

1. *(a)*
$$CH_3\overset{\overset{O}{\|}}{C}\!\!-\!\!OCH_3$$

(b)
$$HC\overset{\overset{O}{\|}}{}\!\!-\!\!OCH_2CH_3$$

(c)
$$CH_3CH_2CH_2\overset{\overset{O}{\|}}{C}\!\!-\!\!\underset{\underset{CH_3}{|}}{O}CHCH_3$$

(d)
$$CH_3\overset{\overset{O}{\|}}{C}\!\!-\!\!OCH_2CH_2CH_3$$

2. Accept all reactions between an alcohol and an organic acid that give the appropriate ester according to the general equation

$$R-\overset{\overset{O}{\|}}{\underset{\underset{OH}{}}{C}} + R'OH \xrightarrow{H_2SO_4}$$

$$R-\overset{\overset{O}{\|}}{\underset{\underset{OR'}{}}{C}} + H_2O$$

3. Ester molecules do not form hydrogen bonding with one another, unlike organic acids. As a result, esters have lower boiling points than the acids with corresponding molecular masses.

4. Hydrolysis of fats and oils using a solution of a strong hydroxide. Accept any reaction that is represented by the general equation

$$R-\overset{\overset{O}{\|}}{\underset{\underset{OR'}{}}{C}} + Na^+OH^- \xrightarrow{heat}$$

$$R-\overset{\overset{O}{\|}}{\underset{\underset{ONa}{}}{C}} + R'OH$$

Chapter 22 Review

Substituted Hydrocarbons

VOCABULARY

alcohol (665)
aldehyde (674)
alkyl halide (670)
carbonyl group (674)
carboxyl group (676)
carboxylic acid (676)
denatured alcohol (668)
dicarboxylic acid (678)
ester (679)
esterification (679)
ether (673)
fat (680)
formyl group (674)
halocarbon (670)
ketone (674)
oil (680)
saponification (680)
sulfation (667)

REVIEWING CONCEPTS

1. Write the general formula for a(n)
   a. halocarbon
   b. alcohol
   c. ether
   d. aldehyde
   e. ketone
   f. carboxylic acid
   g. ester
   h. fat (22.1–22.6)
2. On the basis of formula mass and boiling point, compare the advantages of methanol, ethanol, and ethylene glycol as antifreezes. (22.1)
3. Alcohols and metallic hydroxides both contain —OH groups. Why are their chemical properties so different? (22.1)
4. How does methanol affect the body? (22.1)
5. What is denatured alcohol? (22.1)
6. Why is glycerol an ingredient in many moisturizing skin lotions? (22.1)
7. a. What is an important physical property of carbon tetrachloride?
   b. What uses of carbon tetrachloride depend on this property?
   c. What precautions must be taken when using carbon tetrachloride?
   d. Why? (22.2)
8. Describe the types of halocarbon substitution reactions given in this chapter. (22.2)
9. How do Freon halocarbons affect the stratosphere? (22.2)
10. Give two uses for formaldehyde. (22.4)
11. a. Describe the characteristics of, and cite a major use for acetone.
    b. Give two other names that could be used for acetone. (22.4)
12. a. How do the two carbon–oxygen bonds in formic acid compare?
    b. How do the two carbon–oxygen bonds in sodium formate compare?
    c. Account for the difference. (22.5)
13. List the types of reactions given in this chapter in which sulfuric acid is used as a dehydrating agent. (22.6)
14. a. How are fats and oils alike?
    b. How do they differ?
    c. Why do they differ? (22.6)

PROBLEMS

▶ **Structural Formulas of Substituted Hydrocarbons**

1. Draw structural formulas for
   a. trichloromethane (chloroform)
   b. 1,3-dihydroxy-propane
   c. diethyl ether
   d. acetaldehyde
   e. methylethylketone
   f. acetic acid
   g. ethyl formate
2. Draw a structural formula showing the portion of a polymer formed from three molecules of vinyl chloride.

▶ **Reactions of Substituted Hydrocarbons**

3. How many molecules of oxygen are required for the complete combustion of one molecule of 2-propanol, $CH_3CHOHCH_3$? This compound is commonly called isopropyl alcohol.
4. Write an equation for the preparation of ethyl bromide starting with
   a. ethane   b. ethene   c. ethanol

5. Starting with methane, chlorine, and hydrogen fluoride, write equations for the preparation of dichlorodifluoromethane.

6. Write an equation for the preparation of propyl iodide from propyl alcohol.

7. Write an equation for the preparation of propyl ether from propyl alcohol.

8. a. Write an equation for the mild oxidation of 2-butanol, $CH_3CHOHCH_2CH_3$.
   b. Name the product formed.

9. Write a formula equation showing the preparation of acetic acid from
   a. methanol   b. ethanol

10. a. Write equations for the step-by-step complete ionization of oxalic acid.
    b. How many moles of sodium hydroxide are required for the complete neutralization of 3 mol of oxalic acid?

11. The ester n-butyl acetate is to be prepared from n-butyl alcohol, C_4H_9OH, and acetic acid. Write the equation for the preparation.

12. Write a balanced formula equation for the saponification with NaOH of a fat having the formula $(C_{17}H_{35}COO)_3C_3H_5$.

▶ **Calculations with Substituted Hydrocarbons**

13. What is the percentage composition (to two decimal places) of $C_{12}H_{25}OSO_2ONa$, sodium lauryl sulfate?

14. Calculate the formula mass of trichlorofluoromethane.

15. A compound is found to contain 54.5% carbon, 9.1% hydrogen, and 36.4% oxygen.
    a. Determine the simplest formula.
    b. The compound is soluble in ethanol. If 11.0 g of the compound dissolved in 250 g of ethanol boils at 78.87°C, what is the molecular mass of the compound?
    c. What is the molecular formula?

16. The hydronium ion concentration in 0.05 M acetic acid is 9.4×10^{-4} mol/L. What is the pH of the solution?

17. What volume of ethanol must be diluted with water to prepare 500. mL of 0.750 M solution? The density of ethanol is given as 0.789 g/mL.

18. The density of ethylene glycol is 1.432 g/mL.
    a. Calculate the theoretical freezing point of the water in a 50% (by volume) solution of ethylene glycol.
    b. The actual freezing point of such a solution is about −37°C. Account for any difference from your calculation.

19. Calculate the number of grams of glycerol that must be dissolved in 0.250 kg of water in order to prepare a 0.400 m solution.

20. Calculate the freezing point and boiling point of the water in Problem 20.

CRITICAL THINKING

1. **Applying Ideas** Why is a 50%–50% mixture of ethylene glycol and water better than water alone as a radiator fluid for extremely hot summer driving conditions?

CROSS CONNECTIONS

1. **History** Consult reference materials at the library and gather information about the different methods that various human cultures have used to synthesize ethanol.

2. **Health** Learn about the effects of dioxin on human health. Describe the link between dioxin and "Agent Orange," the herbicide used by the United States during our participation in the Vietnam War.

RESEARCH AND WRITING

1. Find out the meaning of "PCBs," the structural formula of a PCB, and the properties that made PCBs useful in electrical equipment such as transformers. Then find out why the use of PCBs in transformers is now banned.

CHAPTER 23
Biochemistry

PLANNING CHART

Approximate Time: 3/4 days

	SECTION/TEXT FEATURES		RESOURCES	
A V E R A G E	**23.1 THE CHEMISTRY OF LIFE** Section Review Chapter Review • Questions 1–3	685 686 704	*Teacher's Resource Binder* Section Review Worksheets	85
A V E R A G E	**23.2 PROTEINS** Demo 1 Section Review Chapter Review • Questions 4–7, Problem 1	686 684B 689 704	*Teacher's Resource Binder* Lab Experiment 41 Section Review Worksheets *Other Resources* *World of Chemistry* Video 23 Transparency 53 Blackline Master 52	209 86
A V E R A G E	**23.3 CARBOHYDRATES** Demo 2 Desktop Investigation Section Review Chapter Review • Questions 8–12	689 684B 690 691 704	*Teacher's Resource Binder* Section Review Worksheets *Other Resources* *Shakhashiri* Video Demos 27, 43 Transparencies 54, 55	87
A V E R A G E	**23.4 LIPIDS** Section Review Chapter Review • Questions 13–15, Problem 3	691 695 704	*Teacher's Resource Binder* Section Review Worksheets *Other Resources* Transparency 56 Blackline Master 53	88
A V E R A G E	**23.5 NUCLEIC ACIDS** Section Review Chapter Review • Questions 16–18, Problem 2	695 700 704	*Teacher's Resource Binder* Section Review Worksheets *Other Resources* *World of Chemistry* Video 24 Blackline Masters 54, 55, 56	89
H O N O R S	**23.6 BIOCHEMICAL PATHWAYS** Section Review	700 703	*Teacher's Resource Binder* Section Review Worksheets *Other Resources* Transparency 57 Blackline Master 57	90
	CHAPTER END ACTIVITIES Critical Thinking, Cross Connections, Research and Writing, Projects		*Teacher's Resource Binder* Test Booklet *Computer Test Generator*	113, 115

23.1 The Chemistry of Life

The evolution of life from small components is a fascinating story. You may wish to assign library research projects on this topic. Students might not be aware of the fact that the earth's atmosphere has changed from a reducing atmosphere to an oxidizing atmosphere. The precursors of proteins were smaller and less complex molecules that contained carbon and nitrogen.

23.2 Proteins

Proteins are found almost everywhere in the body and have a broad range of functions—hormonal, enzymatic, and structural (see Table 23-1 in the student text.). A good way to understand proteins is to know their composition and structure. Proteins have primary, secondary, tertiary, and quaternary levels of structure. Their functionality is determined by their structure, which can be altered by changes in the pH, temperature, and salinity of the protein's environment.

The basic building blocks of proteins are amino acids. There are at least 150 naturally occurring amino acids. Only 20 of these, however, are used to build genetically coded proteins. Although the relative number of specific amino acids as well as their sequence in the protein chain differs from protein to protein, proteins in all living systems are composed of the same 20 amino acids.

Demonstration 1: Bromelain: A Proteolytic Enzyme

PURPOSE: To demonstrate the proteolytic activity of an enzyme and to investigate the factors that will destroy that activity
MATERIALS: 3 packages of gelatin, fresh pineapple, four 1000-mL beakers
PROCEDURE: *(1)* Cut the pineapple in small pieces—approximately 1 cm × 1 cm size. *(2)* Divide the class into three groups. One group (the control group) will not add anything to the gelatin. One group will add fresh pineapple to the gelatin. One group will add boiled pineapple to the gelatin. *(3)* Prepare the gelatin according to the instructions on the package.

The gelatin prepared without pineapple will gel normally. The gelatin with boiled pineapple will also gel. The gelatin prepared with fresh pineapple, however, will not gel.

Bromelain is a mucoproteolytic agent found in fresh pineapple. It takes its name from the Bromeliad family of plants, of which the pineapple is a member. Gelatin is a protein that absorbs hot water and forms a gel as the water cools. Bromelain can break down gelatin, as well as many other proteins. Boiling the pineapple before addition to the gelatin denatures the proteolytic enzyme, preventing it from interfering with the properties of the gelatin.

Canned pineapple may be substituted for boiled pineapple in this demonstration because the canning process subjects the pineapple to heat and pressure.

23.3 Carbohydrates

Students may have heard the terms sucrose, glucose, and fructose, but may be confused about their role in the diet. Overhead transparencies are useful in distinguishing the disaccharide sucrose from the monosaccharides glucose and fructose.

Demonstration 2: Lugol's Test for Starch

PURPOSE: To detect the presence of starch using a simple chemical test
MATERIALS: five test tubes, Lugol's solution, 1% glucose solution, 1% fructose solution, 1% sucrose solution, 1% starch solution, distilled water
PROCEDURE: *(1)* Label five test tubes as follows: glucose, fructose, sucrose, starch, and water. *(2)* Place 3 mL of each solution in the corresponding test tube. *(3)* Add 3 mL of Lugol's solution to each test tube.

Lugol's solution contains I_2, which stains starch dark blue. The contents of the test tube containing the starch will turn dark blue immediately after addition of the Lugol's solution.

When the red-brown Lugol's solution is added to the other test tubes, no color change will be detected.

You may want to ask students to bring foods to class to test for the presence of starch. A positive test will be found in foods such as potatoes, crackers, bread, and cake. Some students may be surprised to find that sodas and candies do not contain starch.

Desktop Investigation: Enzyme Action on Starch

Nine grams of cornstarch is equal to one tablespoon. It is very important that students use a fresh, unused splint with each tube to avoid contamination.

DATA TABLE		
STARCH + SALIVA AND WATER	**STARCH + BOILED SALIVA AND WATER**	**STARCH + SALIVA AND VINEGAR**
Starch changes from semisolid to liquid	No change	No change

ANSWERS TO DISCUSSION QUESTIONS
1. The amylase changed the starch into a simpler form: the sugar maltose, which is soluble in water.

$$\text{starch} \xrightarrow{\text{amylase}} \text{maltose}$$

2. The heat and vinegar inactivated the amylase so that it had no effect on the starch. High temperatures (55°C or higher) and extremes in pH cause proteins to undergo total denaturation. Since enzymes (including amylase) are proteins, the heat and pH changes denatured the amylase. The reaction cannot be catalyzed and thus does not take place.
3. Since 1 mL of vinegar was added to one test tube containing saliva, 1 mL of water must be added to the other two so they will have equal volumes of liquid. The investigation is thus a controlled one.
4. The enzyme amylase, acting as a digestive catalyst, causes the starch to be broken down into smaller maltose units. High temperatures (100°C) and extremes in acid pH (which is caused by vinegar) denatures the enzyme and prevents the reaction from taking place.

23.4. Lipids

The term "polyunsaturated" is now a common household word, though most students do not know that the term refers to the long hydrocarbon chain in fatty acids. If the chain contains one or more double bonds, the fatty acid is not saturated with hydrogen and is referred to as unsaturated. Lipids of animal origin tend to be saturated and are solids at room temperature. Lipids derived from plants and fish are usually unsaturated and are liquids at room temperature. In general, people whose diets contain unsaturated fats suffer less from deposits on their arteries than people whose diets contain large amounts of saturated fats.

Students may find the subject of neurotransmitters particularly interesting. If there is sufficient time, you could discuss in detail a single neurotransmitter, such as serotonin.

Serotonin regulates several body mechanisms, including mood, alertness, and carbohydrate intake. In turn, the amount of serotonin produced is regulated by the proportion of carbohydrate in the diet. A high proportion of carbohydrate stimulates the rate at which tryptophan, an amino acid precursor, is converted to serotonin.

This type of biofeedback mechanism is common in regulating biochemical pathways in the body. These mechanisms work for most of us most of the time; sometimes they don't. Carbohydrate Craving Obesity is a syndrome that results from a severe aberration in the carbohydrate/serotonin feedback system. People with this syndrome suffer from both mood and appetite disorders. They tend to have bouts of depression in late afternoon or evening and eat large amounts of carbohydrate-rich snacks to compensate.

23.5 Nucleic Acids

Reading about the history of the discovery of DNA may help your students to understand how scientists synthesize information of many types and from many sources.

The discovery of the structure of DNA in 1953 is an example of a situation in which the function of a molecule could not be fully understood until the structure of the molecule was known. The discovery of the structure of DNA was not an isolated event. Following Friedrich Miescher's initial discovery of DNA, it took over one hundred years to elucidate its structure. The well-known double-helix model developed by Watson and Crick required the supporting evidence of the x-ray crystallography of Rosalind Franklin.

Students can read about the latest developments in genetic engineering in the popular press, even on the pages of the *Wall Street Journal*. Never before has biotechnology so quickly found its way into the marketplace. Many ethical questions have been raised: What role should animals have in genetic research? At what point is it appropriate to use a drug on humans? Should we use fetal tissue for experiments in cloning? Could we create a new microorganism that could endanger the human race?

These questions should be discussed in the classroom. Encourage students to bring relevant press releases to class. Think of asking a colleague in the humanities to be a cofacilitator in a discussion of these issues.

23.6 Biochemical Pathways

Emphasize to your students that energy is required for the growth and maintenance of cells. In a healthy individual, energy is extracted from food molecules when the molecules are broken down in the cell to smaller fragments. The energy released from the breakdown of food molecules is stored in the nucleotide adenosine triphosphate.

The energy in macromolecules has to come from somewhere. The ultimate source of energy is the sun. Green plants, algae, and some bacteria can capture the energy of the sun and convert light into chemical energy. The energy is stored in macromolecules produced by these photosynthesizing organisms and is utilized by consumers, organisms not capable of utilizing light directly for an energy source.

Demonstration 3: Oxidation of Methylene Blue

PURPOSE: To demonstrate the conversion of light energy into chemical energy through the reduction of an organic dye
MATERIALS: petri dish, 100-mm; overhead projector; black poster board, 150 mm × 150 mm; 10 g $FeSO_4$; 3 mL 3 M H_2SO_4; 500 mL H_2O; methylene blue
PROCEDURE: *(1)* Prepare an aqueous solution of methylene blue by adding 25 mg of methylene blue to 5 mL of water. This solution may be stored for several weeks. *(2)* Prepare an acidified solution of ferrous sulfate by adding 10 g of ferrous sulfate to 500 mL of water. Add 3 mL of 3 M sulfuric acid. This solution must be prepared on the day of the demonstration. Ferrous sulfate oxidizes easily to ferric sulfate, which would destroy the reducing properties of the iron solution. *(3)* Place a petri dish on an overhead projector. The projector light should be off. Position a piece of black poster board so that one-half of the petri dish is shielded from the light. *(4)* Pour 50 mL of the ferrous sulfate solution into the petri dish. *(5)* Add 4 to 6 drops of methylene blue to the ferrous sulfate. The solution should have a deep blue color. If it does not, add more methylene blue, one drop at a time, until the solution is deep blue. *(6)* Turn on the projector light. The solution in the half of the projector that is exposed to light will become colorless as the methylene blue is photochemically reduced. *(7)* Remove the black poster board. The solution in the protected half of the petri dish will still be deep blue. Color will begin to disappear as the remaining half of the solution absorbs light.

Remind students that light is the ultimate source of energy in living systems. It is absorbed by producers that store the energy from the sun in a variety of macromolecules that are passed along the food chain to consumers. Although students know about this process from their work in biology, they sometimes continue to think of heat as the source of all energy-absorbing reactions. The photochemical reduction of methylene blue clearly demonstrates that light can supply the necessary energy for a chemical reaction.

An analysis of the chemical structure of the oxidized and reduced form of methylene blue illustrates the relationship between form and function in organic molecules. This relationship has many analogies in living systems.

References and Resources

Books and Periodicals

Butler, P.J.G., and A. Klug. "The Assembly of a Virus," *Scientific American*, Vol. 239, 1978, pp. 62–69.

Dickerson, Richard E., and Irving Geis. *The Structure and Action of Proteins*, Benjamin Cummings Publishing Co., 1969.

Ferguson, Lloyd N. "Bio-Organic Mechanisms II: Chemoreception," *Journal of Chemical Eduction*, Vol. 58, 1981, pp. 456.

Freifelder, David. *Molecular Biology and Biochemistry: Problems and Applications*, W.H. Freeman and Co., 1978.

Goldsmith, Robert H. "A Tale of Two Sweeteners," *Journal of Chemical Education*, Vol. 64, 1987, p. 954.

The Hastings Center, 360 Broadway, Hastings-on-Hudson, New York 10706. Publications available to teachers at a nominal cost.

MacBeath, Marie E., and Andrew L. Richardson. "Tomato Juice Rainbow, A Colorful and Instructive Demonstration," *Journal of Chemical Education*, Vol. 63, 1986, pp. 1092–1095.

Ophardt, Charles E., and Paul F. Krause. "Blood Buffer Demonstration," *Journal of Chemical Education*, Vol. 60, 1983, p. 493.

Perutz, M.F. "The Hemoglobin Molecule," *Scientific American*, Vol. 211, 1964. pp. 64–76.

Piel, Jonathan, ed. "What Science Knows About AIDS: A Single Topic Issue," *Scientific American*, Vol. 259, 1988.

Portugal, Franklin H. and Jack S. Cohen. *A Century of DNA*, The MIT Press, 1977.

Sayre, A. "Rosalind Franklin and DNA," W.W. Norton, 1975.

Stent, G.S., and R. Calendar. *Molecular Genetics: An Introductory Narrative*, W.H. Freeman and Co., 1971.

Stroud, Robert M. "A Family of Protein-Cutting Proteins," *Scientific American*, Vol. 231, 1974, pp. 74–88.

Watson, James D. "The Double Helix," Atheneum, 1968.

Wurtmann, Richard and Judith Wurtmann. "Carbohydrates and Depression," *Scientific American*, Vol. 260, 1989, p. 68.

Computer Software

"Blood Sugar," ALB4605A, AII, 48K, Disk, available from Queue.

"Chemicals of Life II: Water, Carbohydrates, and Lipids," IBM02PC, I, Disk, available through Queue.

"Chemicals of Life III: Protein and Nucleic Acids," IBM03PC, I, Disk, available from Queue.

Currie, James and G. Scott Owen. "Molec," COM4084A, AII, 48K, Disk, available from Queue.

"Enzyme Investigations," Biology Explorations, HRM Software, HRM543A, available from Queue.

"Gene Machine," Genes and Heredity, HRM Software, HRM537A, AII, Disk, HRM537C, HRM537T, available from Queue.

Kleinsmith, Lewis J. "Protein Synthesis/Condon," COM4206A, Apple 128K, Disk; COM4206B, IBM, 256K, Disk; COM4206J, PC-Jr., 256K, Disk; available from Queue.

"Metabolic Pathways," ALB4614A, AII, 48K, Disk, available from Queue.

Zimmerman, S. Scott, "Stimulation of Hemoglobin Function," COM4200A (AII, 48K, Disk), available from Queue.

Answers and Solutions

Reviewing Concepts

1. N_2, CO_2, H_2O

2. C, H, N, O

3. Nearly all biochemical reactions take place in solution in water. Water makes up most of the mass of every living organism.

4. R_1—CHCOOH + R_2—CHCOOH →
   | |
   NH_2 NH_2

   R_1—CHC(=O)—N(H)—CHCOOH + H_2O
   | |
   NH_2 R_2

5. Enzyme action is specific because an enzyme binds its substrate in a "lock and key" fashion and each enzyme can bind only one or a few different substrate molecules.

6. Enzymes: catalyze chemical reactions
   Structural proteins: form part of the structure of a body or cell
   Hormones: carry messages that regulate bodily functions
   Blood transport proteins: carry things in the blood

7. pH, temperature

8. Energy (food) supply, food storage, structural support of a body or cell

9. [Open-chain and cyclic structures of glucose shown, with CH_2OH cyclic form]

10. Cellulose, chitin

11. Glycogen is the carbohydrate that stores energy in the liver of animals.

12. Cellulose is the structural carbohydrate that strengthens plant cell walls.

13. Fatty acids (oleic acid); waxes (beeswax, covering on a leaf); fats and oils (margarine, butter, olive oil, peanut oil, corn oil); steroids (estrogen, cortisone, testosterone)

14. The main use of waxes in living organisms is the waterproofing of the exterior of the organism.

15. Cholesterol: part of the structure of all biological membranes
    Estrogen: female sex hormone
    Androgens: male sex hormones
16. Sugar (ribose or deoxyribose), phosphate group, ring shaped nitrogen base
17. Nucleotides form hydrogen bonds between pairs of bases so the nucleotide sequence in one strand corresponds to the sequence in its partner. A nucleotide containing an adenine base pairs to one containing thymine, and a nucleotide containing a cytosine base pairs to one containing guanine.
18. Adenosine triphosphate (ATP)

Problems

1. The enzyme in beaker B was probably denatured by the heat.
2. TTTAGCCGT
3. 110 g of fat

Critical Thinking

1. Life depends on huge molecules held together by carbon–carbon bonds. Life based on silicon would be impossible because silicon–silicon macromolecules would fall apart in the presence of oxygen. Oxygen is vital to life. It is produced by plants during photosynthesis, which is indirectly responsible for the production of all the food on earth. Also, many metabolic reactions would be impossible if oxygen became stably bound to macromolecules, as it would if they were based on silicon.
2. When an egg is cooked, the protein in it is denatured, meaning that it loses its three-dimensional structure, although the chain of amino acids may remain the same.
3. Glucose. It is readily digested by the body.

Teacher's Notes

Chapter 23
Biochemistry

SECTION PREVIEW
23.1 The Chemistry of Life
23.2 Proteins
23.3 Carbohydrates
23.4 Lipids
23.5 Nucleic Acids
23.6 Biochemical Pathways

In this scanning-tunneling microscope image, magnified 3 300 000 times, each peak corresponds to a ridge in DNA's double-stranded helix.

23.1 The Chemistry of Life

The chemistries of plants, animals, bacteria, and fungi are impressive in two ways. First, these living things are more than half water. Second, what is not water is mainly composed of organic molecules that, as you have seen, all contain carbon. The biggest molecules on earth are organic. One kind of organic molecule is DNA. When it is straightened out, a molecule of DNA from a fruit fly is nearly 2 cm long—ten times as long as the fly itself!

We live surrounded by organic molecules; we touch them when we touch our noses or a clump of grass or a spider. All these things are alive, and they are kept alive by reactions between organic molecules. **Biochemistry** *is the study of the chemical reactions that go on in living things.* As people learn more about biochemistry, they are learning how to produce some of the molecules found in living things. We can now make substances such as hormones, vaccines, and even genes, in the laboratory—with the assistance of living organisms.

SECTION OBJECTIVES

- Understand how the chemistry of living things is unique.
- List the four elements that are most common in living organisms. List four main classes of biological macromolecules.
- Describe the formation of a macromolecule.

Inorganic Molecules and Water

The first living things are thought to have evolved more than three billion years ago from the chemical compounds that were common on earth at that time, such as water, ammonia, and carbon dioxide. These substances contain the elements hydrogen, oxygen, carbon, and nitrogen. Living things today are made up almost entirely of these elements. In addition, they contain small amounts of 18 other elements.

Water's main role in life is to act as a solvent. The chemical reactions in our bodies take place in a dilute solution that is much like sea water (Figure 23-1). The solution contains the ions of salts, such as potassium phosphate (K^+ and PO_4^{3-}), magnesium bicarbonate (Mg^{2+} and HCO_3^{2-}), and sodium chloride (Na^+ and Cl^-) dissolved in water. All substances that travel through the body, such as oxygen, hormones, and food, travel dissolved in a salt solution, and most biochemical reactions take place in this solution. When we digest food, store fat, or move our muscles, dozens of reactions take place between dissolved compounds. Water makes up most of the mass of every living organism. Chemists have spent many years discovering how to imitate the body's solution of salts in water. As a result of this research, many body parts can now be kept alive outside the body, bathed in salt solution. These body parts may be kidneys or hearts waiting to be transplanted, cells to be used for research, or eggs to be fertilized outside the body.

The chemistry of carbon is introduced in Section 21.1.

Although oxygen was present in CO_2, the atmosphere did not contain free oxygen molecules. A reducing atmosphere existed until organisms evolved that were capable of carrying on photosynthesis.

The chemical nature of the fluid inside our bodies is discussed in the Chemistry Notebook in Section 15.1.

Figure 23-1 Without water, there can be no life. These cells exist in a solution of salts in water.

Macromolecules

Large organic molecules are called **macromolecules** *(the prefix* macro *means large),* and they do many jobs in the body. Some of them are structural molecules that provide support in the bodies of plants and animals. The muscles that hold our bodies up and the wood that holds up a tree, are made of macromolecules. Other macromolecules store and release energy. Still others carry information that the body needs.

Most of the macromolecules essential to life belong to four main classes, three of which we know as types of food:

Biochemistry **685**

Proteins are chains of many amino acids that are linked together. The three-dimensional structure of proteins is very important to their function.

Lipids are the primary components of cell walls. They form bilayers with the hydrocarbon tails on the inside and the phosphate heads on the outside.

1. *Proteins*. Hair, tendons, ligaments, and silk are made of protein. Other proteins act as hormones, carry things around the body, and fight infections. Enzymes are proteins that control the body's chemical reactions.
2. *Carbohydrates*. Sugars, starches, and cellulose are carbohydrates.
3. *Lipids*. Fats, oils, waxes, and steroids are lipids, nonpolar substances, which do not dissolve in water.
4. *Nucleic acid*. The nucleic acids are DNA and RNA. In most organisms, DNA is used to store hereditary information, and RNA helps to assemble proteins.

Functional groups Organic molecules come in a vast variety of types and sizes. To make sense of this diversity, biochemists examine the functional groups that organic molecules contain. As you learned in Chapter 21, functional groups are clusters of atoms that behave in particular ways. The functional groups of a molecule determine the molecule's chemical properties. For instance, any compound containing a carboxyl group (—COOH) is acidic because the carboxyl group releases hydrogen ions in solution:

The carboxyl group is discussed in Section 22.5.

$$-COOH \rightarrow -COO^- + H^+$$

Polarity is discussed in Section 6.5.

The behavior of water as a solvent is discussed in Section 13.4 and Section 14.2.

As another example, C—H bonds are nonpolar. Therefore, a part of a molecule containing only carbon and hydrogen is also nonpolar. Such parts are not soluble in water.

Other functional groups contain polar bonds, particularly O—H, N—H, and C=O. Functional groups containing polar bonds are soluble in water. Macromolecules are so large that part of the molecule may be insoluble in water while the rest is soluble. Functional groups determine a molecule's solubility and determine what kinds of chemical reactions it will take part in.

1. The study of the chemical reactions that go on in living things
2. Structural, storing and releasing energy, carrying information
3. carboxyl group (—COOH), hydroxyl group (—OH), carbonyl group (C=O)

Section Review

1. What is biochemistry?
2. List three functions that different macromolecules perform in the body.
3. List three different functional groups found in organic molecules.

23.2 Proteins

SECTION OBJECTIVES

- Name the types of monomers from which proteins are made and describe how they are joined to form a peptide chain.
- List three functions of proteins in the body.
- Explain some functions of coenzymes and minerals in living things.
- Explain the effect of temperature and pH on reactions catalyzed by enzymes.

Proteins contain the elements carbon, oxygen, hydrogen, nitrogen, and usually some sulfur. They perform many different functions in the body (Table 23-1). **Proteins** *are macromolecules formed by condensation reactions between monomers called amino acids.*

Amino Acids

Every **amino acid** *has a carboxyl group (—COOH) and an amino group (—NH₂), both attached to the same carbon atom, which is also attached to a hydrogen (—H).* The reason one amino acid differs from another is that the fourth bond of this carbon is attached to a functional group (called an R group) that differs from one amino acid to another (Figure 23-2). The proteins of all organisms contain the same twenty common amino acids.

686 Chapter 23

TABLE 23-1	SOME FUNCTIONS OF PROTEINS	
Type	Example	Function
Enzymes	Amylase	Converts starch to glucose
	DNA polymerase I	Repairs DNA molecules
Structural proteins	Keratin	In hair, nails, hoofs
	Collagen	In tendons, cartilage
Hormones	Insulin	Regulate glucose metabolism
	Vasopressin	Slows water loss by kidneys
Transport proteins	Hemoglobin	Carries O_2 in blood

Insulin was one of the first proteins for which the protein structure was determined. Its amino acid sequence was discovered in the early 1950s.

Protein Structure

When a protein forms, the first step is a condensation reaction joining two amino acids by a **peptide bond**, *a type of bond that joins amino acids in proteins.* A peptide bond is formed by removing the OH from the carboxyl (—COOH) group of one amino acid and an H from the amino (—NH_2) group of another (Figure 23-2). Further peptide bonds join more amino acids to the chain. The finished peptide chain usually contains 100–300 amino acids. Each protein has its own unique sequence of amino acids. A complex organism has at least several thousand different proteins, each with a special structure and function.

Proteins are made up of long, unbranched chains of amino acids, but the final protein is quite a complex structure. This is because peptide chains fold up to form various structures. The most common of these structures is a coil formed by part of the chain and known as an alpha helix.

Each peptide chain may also link up with other chains to form the final protein. For instance, *insulin*, a hormone that helps the body to regulate the level of sugar in the blood, is made up of two linked chains. The chains are held together by S—S bonds between sulfur atoms in two cysteine amino acids (Figure 23-3). Insulin is one of the smaller proteins, containing only 51 amino acids. In contrast, hemoglobin, which carries oxygen in the blood, is a large protein consisting of four long chains with complicated three-dimensional structures. Proteins can be made to lose their shape by gentle heating or various chemical treatments. When they are returned to more normal surroundings, they fold or coil up again and join together, re-forming their original structure.

Changing even one amino acid can change a protein's structure and what it does. For example, the difference between normal hemoglobin and the hemoglobin that causes sickle cell anemia is just two amino acids.

Figure 23-2 The general structure of amino acids consists of an amino and a carboxyl group, both attached to the same carbon atom, which is also bonded to a hydrogen atom (top). This carbon's fourth bond is attached to an R group, which may be nonpolar, polar, or ionic. A peptide bond (bottom) forms by a condensation reaction between two amino acids. A peptide bond is one between the carboxyl carbon of one amino acid and the amino nitrogen of another.

(a) Structure of an amino acid

(b) Formation of a peptide bond between amino acids

Biochemistry

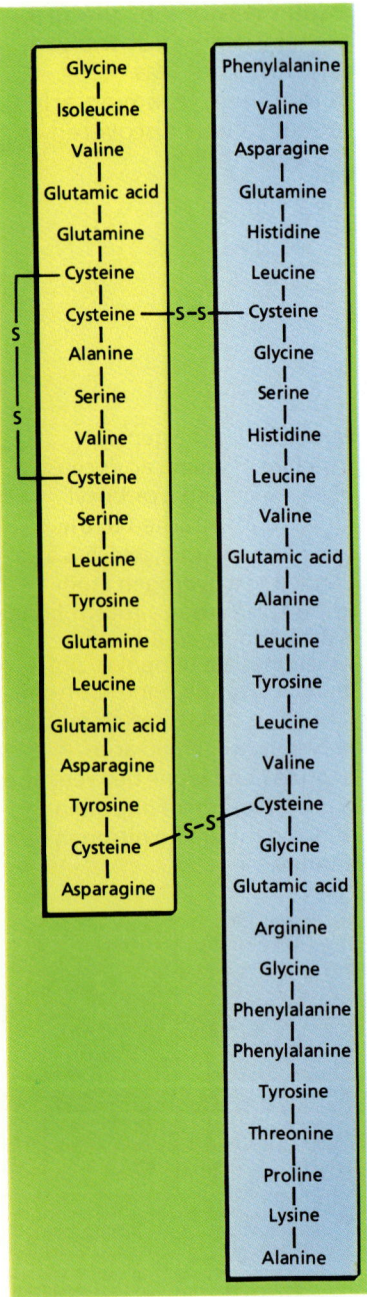

Figure 23-3 The amino acid sequence of the hormone insulin. This small protein is made up of two short polypeptide chains joined by "sulfur bridges," attachments between sulfur-containing cysteine amino acids.

Enzymes

Nearly all the chemical reactions in living things are controlled by **enzymes**, *protein molecules that act as catalysts*. A catalyst is a substance that speeds up a chemical reaction. An enzyme is rather like a robot on an assembly line, performing the same task over and over on *a particular set of parts called* **substrates**. *The molecules produced by an enzyme reaction are called the reaction's* **products**. Most enzymes have names ending with *-ase* and are named according to their specific substrates and the kinds of reactions they catalyze.

How does an enzyme speed up a reaction? Each enzyme combines with its substrates. An enzyme can bind only one or a few types of molecules in a lock-and-key fashion and hold them in the correct position for the reaction to occur. Parts of the enzyme pull on the substrates' bonds and loosen them. This lowers the energy of activation needed for the reaction to occur. As a result, the reaction occurs readily.

Vitamins, minerals, and coenzymes Some enzymes cannot bind to their substrates without the help of additional molecules. These may be *minerals* such as ions of calcium or iron, or *helper molecules called* **coenzymes** *that play accessory roles in enzyme catalyzed reactions*. Many vitamins are coenzymes or parts of coenzymes.

Vitamins are organic molecules that we cannot manufacture and hence need to eat in small amounts. You can see why we need vitamins and minerals in our diet—to permit our enzymes to work. You can also see why we need only small amounts of them. Minerals and coenzymes are not destroyed in biochemical reactions. Like enzymes, coenzymes and minerals can be used over and over again.

Factors that affect enzyme activity Various factors, notably temperature and pH, affect the rates of reactions catalyzed by enzymes. Many enzymes are affected by the pH of the surrounding solution. Most proteins work best in a solution of approximately neutral pH. (Most of our body cells have a pH of 7.4.) Some enzymes work in acidic or basic environments, however. Pepsin, the collective term for the digestive enzymes found in the human stomach, works best at the very acidic pH of about 1.5. Cells lining the stomach secrete hydrochloric acid (HCl) to produce this acid environment. When food travels down the digestive tract, it carries these enzymes out of the stomach into the intestine. Here the stomach enzymes stop working because the intestine contains sodium bicarbonate that raises the pH to about 8. The digestive enzymes that work in the intestine are formed by the pancreas and work best at pH 8.

Most chemical reactions, including enzyme reactions, speed up as it gets hotter. On the other hand, *high temperatures (above about 60°C) destroy, or* **denature** *proteins by breaking up their three-dimensional structure*. When we cook an egg or a piece of meat, we denature the proteins in the food. Proteins in the clear part of an egg turn white when denatured. Heating preserves food by denaturing the enzymes of organisms that cause decay. Milk is pasteurized by heating it to denature enzymes that would turn it sour. Refrigeration and freezing preserve food by cooling, which slows down the reactions of enzymes that cause decay.

Section Review

1. Name the monomers of proteins and draw a peptide bond.
2. What are the main elements found in proteins?
3. Why do temperature changes affect the activity of proteins?

1. Amino acids. See Figure 23-2 for an example of a peptide bond.
2. Carbon, oxygen, hydrogen, nitrogen, usually some sulfur
3. Can alter the three-dimensional structure of a protein

23.3 Carbohydrates

Carbohydrates *are sugars, starches, and related compounds.* Some carbohydrates supply energy. Others are structural molecules, building materials that shape the body. Carbohydrates tend to be polar, and thus will associate with water.

SECTION OBJECTIVES

- Name four carbohydrates.
- Describe the monomers from which carbohydrates are made.
- List three functions of carbohydrates in living organisms.
- Describe how carbohydrates are synthesized.

Simple Sugars

The monomers of carbohydrates are **monosaccharides**, *or simple sugars, such as fructose and glucose.* A monosaccharide contains carbon, hydrogen, and oxygen in about a 1:2:1 ratio (CH_2O).

Monosaccharides provide the body with ready energy. Sugary foods are quick-energy foods because sugars are easily converted by enzymes into molecules that the body can use for energy. Carbohydrates are transported in our blood mainly in the form of the six-carbon monosaccharide glucose, which the cells of the body take up and use for energy as they need it.

Carbohydrates are a good source of energy. If we eat more of them than our bodies can immediately use for energy, carbohydrates are then converted into fats to be stored in the body for later use.

Disaccharides and Polysaccharides

Two monosaccharides may be joined together to form one **disaccharide** (Figure 23-4). Sucrose (table sugar) and lactose (milk sugar) are disaccharides. A disaccharide can be hydrolyzed into the monosaccharides that formed it. *By a series of condensation reactions, many monosaccharides can be joined to form a polymer called a* **polysaccharide** *(commonly known as a* **complex carbohydrate**).

Three important polysaccharides made of glucose monomers are glycogen, starch, and cellulose. *Animals store energy in* **glycogen**. The liver and muscles remove glucose from the blood and condense it into glycogen, which can later be hydrolyzed back into glucose and used to supply energy as needed.

Starch *is the polysaccharide that is used to store energy in plants.* Starch consists of two kinds of glucose polymers. A plant hydrolyzes its starch into

Figure 23-4 Formation of a disaccharide. A condensation reaction between the monosaccharides glucose and fructose produces the disaccharide sucrose (table sugar) and water. This reaction is reversible by hydrolysis.

Biochemistry **689**

Desktop Investigation

Enzyme Action on Starch

Questions
What effect does the enzyme *amylase* have on the digestion of starch? What effect do heat and acids have on enzyme action?

Materials
cornstarch
distilled water
saliva
vinegar
beaker, 250-mL
stirring rod
6 test tubes (14 × 100 mm)
wax pencil
Bunsen burner
ringstand with ring
wire gauze-ceramic center
3 wooden splints

Procedure
Record all your results in a data table. Remember to wear safety glasses and an apron.

1. Add 9 g of cornstarch to a beaker containing 125 mL of cold distilled water. While stirring continuously, heat the mixture until it has boiled about two minutes.

2. Fill three test tubes with 5 mL each of the cornstarch mixture and allow them to cool. Label each tube near the rim with a wax pencil.

3. Spit equal amounts of saliva into each of 3 other test tubes. To two of these test tubes, add 1 mL of distilled water. Add 1 mL of vinegar to the third test tube. Label each tube appropriately.

4. Boil one of the test tubes containing saliva and water in a water bath for five minutes, as shown in the diagram.

5. Add the heated water and saliva to one of the test tubes that contains cornstarch. Add the unheated water and saliva to the second test tube that contains cornstarch. Add the vinegar and saliva to the third test tube that contains cornstarch. Label each tube appropriately. Stir each of these three test tubes with a new, unused wooden splint.

Discussion

1. What effect did amylase in the saliva have on the starch?

2. What effect did the heat and vinegar have on the amylase action? Explain.

3. Why was it necessary to add 1 mL of water to two of the test tubes containing saliva?

4. Write a good conclusion for this investigation.

It is the repetition and chemical stability of these polysaccharides that gives them their physical stability and, thus, their ability to form hard shells, bones, and fibers.

glucose. The plant uses some of this glucose for energy, and some as a building material to produce more cells.

The main building material made from glucose in plants is the structural polysaccharide **cellulose**, *which is probably the most common organic compound on earth* (Figure 23-5). Hydroxyl groups on the glucose monomers link different cellulose chains together, forming cellulose fibers. (The fibers of cotton consist almost entirely of cellulose.)

Humans cannot digest cellulose. Cellulose in our food passes through our digestive systems unchanged. It is important in our diet because it provides bulk (often called *fiber* or *roughage*), which stimulates the intestines and keeps things moving along.

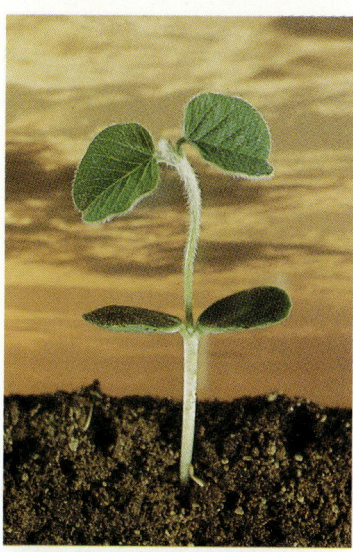

Figure 23-5 In cellulose, glucose monomers are linked into long, unbranched chains (left). Cellulose fibers are deposited in layers to form rigid plant cell walls that help to provide structural support for the entire plant (right).

Another structural polysaccharide is **chitin**, *which is composed of glucose with a nitrogen-containing group attached.* Chitin is an important component of the "armor plating" of crabs, spiders, insects, and their relatives.

Section Review

1. What is the difference between a disaccharide and a polysaccharide?
2. How are two monosaccharides joined to form a disaccharide?
3. Name four complex carbohydrates.
4. What is cellulose?

1. A disaccharide is formed from two monosaccharides joined together. A polysaccharide consists of many monosaccharides joined together.
2. By means of a condensation reaction
3. Glycogen, starch, cellulose, chitin
4. A large polysaccharide that consists of many glucose monomers joined together. It provides structural support for plants and is probably the most common organic compound on earth.

23.4 Lipids

Lipids *are a varied group of organic compounds that share one property: they are not very soluble in water.* This is because lipids contain a high proportion of C—H bonds. They dissolve in nonpolar organic solvents, such as ether, chloroform, and benzene, but not in water, which is polar. Most lipids are too small to be called macromolecules and they are not made of monomers, as other biological molecules are.

Lipids contain carbon, hydrogen, and oxygen, but the proportion of oxygen in lipids is much lower than in carbohydrates. Some lipids contain other elements, such as phosphorus and nitrogen.

Because they are insoluble in water, lipids are the main compounds in biological membranes. Membranes divide living things into many compartments, such as cells, each containing a particular salt solution and set of reactions.

Lipids are used as energy stores because a given weight of lipid can store much more energy than the same weight of carbohydrate or protein (Table 23-2). This is undoubtedly why lipids are used as food reserves in the body. Without lipid reserves, for instance, the migrations of many birds would be impossible. A small bird may almost double its body weight in the fall as it lays down fat reserves for the long flight. To carry the same energy reserve in the carbohydrate glycogen, it would have to get so fat that it could not fly.

SECTION OBJECTIVES

- Define the term *lipid*.
- List five different types of lipids.
- List four different roles that various lipids perform in the body.
- List the elements present in lipids.

Biochemistry **691**

Figure 23-6 The carboxyl group in the fatty acid shown above is at the bottom. The rest of the molecule is a nonpolar, hydrocarbon chain, insoluble in water. This is saturated because the carbon atoms of the hydrocarbon chain contain as many hydrogens as they can hold.

TABLE 23-2 ENERGY YIELD OF TYPES OF FOOD

Class	Composition	Energy Yield
Carbohydrates	CH_2O	3.8 kilocalories/gram
Lipids	CHO	9.3 kilocalories/gram
Proteins	CHON(S)	3.1 kilocalories/gram

Fatty Acids

Fatty acids are the simplest lipids. A **fatty acid** *consists of an unbranched chain of carbon and hydrogen atoms, with a carboxyl group at one end.* If all carbon atoms of the hydrocarbon chain are "filled" with as many hydrogens as they can possibly hold, the fatty acid is said to be a **saturated** fatty acid. A fatty acid with one or more double bonds in its hydrocarbon chain is said to be *unsaturated* because it could hold more hydrogens if a double bond were broken and two hydrogen atoms were attached to each carbon instead of one.

The bonds in a carboxyl group are polar, and so the carboxyl end of a fatty acid attracts water molecules. The carbon-hydrogen bonds of a lipid's hydrocarbon chain are nonpolar, however. The polar end will dissolve in water, and the other end will dissolve in nonpolar organic compounds. This behavior means that fatty acids will form membranes when they are dropped into water. It also gives soaps and detergents their cleaning power. Palmitic acid (16 carbons) and oleic acid (18 carbons) are common fatty acids.

Fatty acids are usually combined with other molecules to form substances such as **glycolipids** *(carbohydrate + lipid) or* **lipoproteins** *(lipid + protein).* They are also parts of more complex lipids.

Waxes

A **wax** *is made of esters of long fatty acids and long-chain monohydroxy alcohols.* These long chains make waxes solid at room temperature, with high melting points. Waxes form a waterproof coating on the outer surfaces of many organisms, for example on skin, hair, feathers, and leaves (Figure 23-7). Polishes and candles are made from waxes. Lanolin, another wax, is used in cosmetics.

Fats and Oils

Fats and oils are molecules formed by the esterification of three fatty acids with a molecule of the alcohol glycerol, as described in Section 22.6. Fats are solid at room temperature, whereas oils are liquid. **Saturated fats** *contain only saturated fatty acid groups.* **Unsaturated fats** *contain at least one fatty acid group with double bonds.* **Polyunsaturated fats** *are fats and oils with double bonds in their fatty acid chains.* There is some evidence that eating large amounts of saturated fat may increase the risk of heart attacks, so some cooking oils and margarine are advertised as "polyunsaturated." These are made of oils extracted from plants because oils generally contain more unsaturated fatty acids than do fats, and oils are more common in plants than in animals. (Examples are olive, corn, and peanut oils.)

Americans would probably be healthier if we ate less fat and oil. Part of the reason for this is that many people are overweight and a given weight of lipid provides more calories (energy) than the same weight of carbohydrate or protein. In addition, if the body contains too much of certain types of lipids, it may be more susceptible to some diseases. Fats perform many vital functions in the body, and some vitamins are soluble in fat but not in water. We need some fat in our diet.

Phospholipids

Phospholipids are the most important structural lipids because they are the chief lipids in biological membranes. **Phospholipids** *are lipids in which one (or two) of the fatty acids is replaced by a phosphate group.*

Steroids

Steroids differ from other lipids in structure. *A* **steroid** *is a lipid with a "skeleton" structure consisting of four carbon rings.* The most common steroid is cholesterol, shown in Figure 23-8, which has received some bad publicity because deposits of cholesterol in the body may cause gallstones or heart disease. **Cholesterol** *is a steroid that makes up part of every animal cell membrane and of the brain and nerves,* however. It is also the raw material from which steroid hormones are produced. We could not live without it. A buildup of cholesterol and other materials within the arteries makes it difficult for the arteries to expand and contract properly and blood moves through them with difficulty. Diets high in cholesterol increase the risk of heart disease.

Hormones *are chemical messengers that elicit specific responses from target cells.* Steroid hormones include cortisone, as well as sex hormones from the ovaries and testes—estrogens and androgens. These hormones control reproduction. They cause the differences between male and female bodies and affect sexual behavior. Synthetic derivatives of androgens are sometimes taken by athletes and weightlifters to promote muscle growth. Recent research, however, has shown that these steroid drugs can have serious side effects, including shrinking of the testes, breast development, and decreased sperm production in men; facial and body hair and menstrual irregularity in women; and high blood pressure in both sexes.

Figure 23-7 Flakes of wax form a waterproof covering on the surface of a leaf and prevent the plant from losing water into the surrounding dry air.

Hormones are basically regulatory in nature because they control certain biochemical activities in the body. An example is insulin, which is secreted by the pancreas in response to an increase in blood sugar levels. Insulin facilitates the transport of glucose into cells. Diabetes results from a deficiency of insulin.

Figure 23-8 Cholesterol (left) is an important component of animal cell membranes. The basic skeleton of carbon rings (yellow) is the same in all steroids, but other steroids have other groups attached to this skeleton. The accumulation of cholesterol on artery walls (right) may eventually block the passage of blood.

Biochemistry

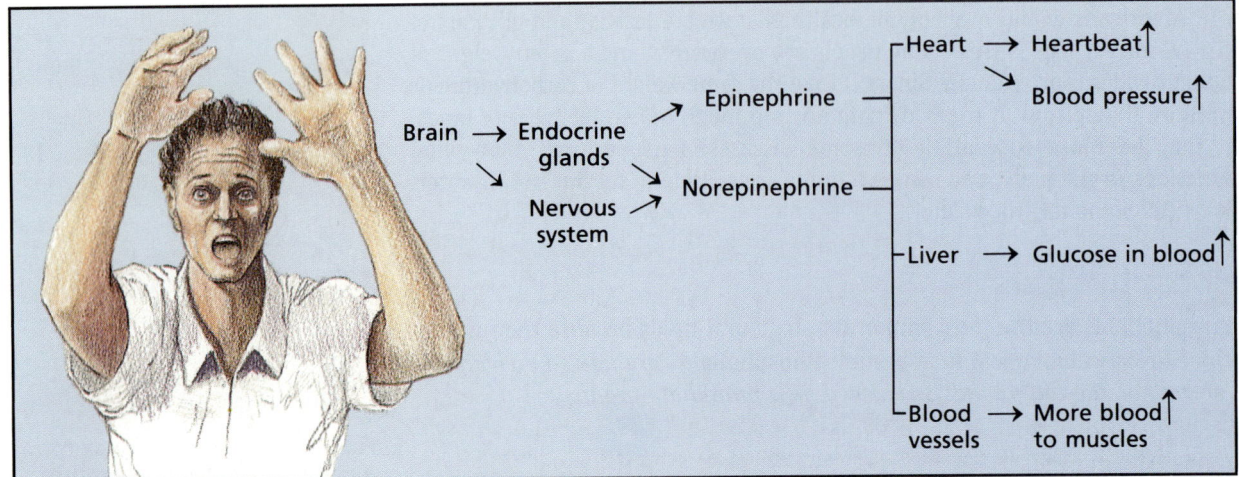

Figure 23-9 Fight or flight. When the brain registers excitement, glands and neurons produce the hormones epinephrine and norepinephrine, which travel around the body and get it ready for action.

Not all hormones are steroids. Many are proteins or amines. *(An **amine** is a base derived from ammonia in which one or more of the three hydrogens is replaced by an alkyl or aryl group.)* For instance, epinephrine (also called adrenalin) is an amine. **Epinephrine** *is the hormone enacted when you experience excitement. It speeds up the heart, increases the blood supply to the muscles, and generally gets the body ready for "fight or flight"* (Figure 23-9).

All the hormones mentioned so far are secreted by endocrine glands, special glands such as the thyroid, pancreas, and testes, which secrete the hormones they produce into the bloodstream. Until recently, biologists gave the name "hormone" only to substances produced by endocrine glands. But it is now becoming obvious that the body is full of all kinds of chemical

Figure 23-10 A synapse. The first neuron and part of a second neuron are shown. When the message in the first neuron reaches the synapse, neurotransmitter is released from the first neuron and travels across the synapse to stimulate the second neuron.

Histamine — Epinephrine — Acetylcholine

messengers carrying information from one place to another. These include local chemical messengers (short range chemical messengers not produced by endocrine glands) and neurotransmitters. *Neuro* means "nervous," and *the cells of the nervous system are called* **neurons.** When a message travels along a neuron, it travels as an electrical signal (which is why messages travel round the nervous system so rapidly). When a message reaches the end of a neuron, it is faced with a **synapse**—*a space between one neuron and the next* (Figure 23-10). *A chemical messenger called a* **neurotransmitter** *transmits the message from one neuron to the next across the synapse.* Several examples of neurotransmitters are shown in Figure 23-11.

Figure 23-11 The chemical structure of three neurotransmitters that are amines.

1. A type of organic compound insoluble in water, contains a high proportion of C—H bonds
2. Fats are solid at room temperature, oils are liquids.
3. Saturated means that the carbons hold as many hydrogens as possible. Unsaturated means that there are one or more double bonds in the hydrocarbon chain.
4. Because they are generally insoluble in water.
5. No. Many are proteins or amines.
6. A chemical that carries a message from one neuron to the next.

Section Review

1. What is a lipid?
2. What is the difference between a fat and an oil?
3. Describe what is meant by the terms *saturated* and *unsaturated*?
4. Why are all biological membranes made up largely of lipids?
5. Are all hormones lipids?
6. What is a neurotransmitter?

23.5 Nucleic Acids

Nucleic acids *are macromolecules that transmit genetic information.* Nucleic acids include the largest molecules made by organisms. There are two kinds. **Deoxyribonucleic acid (DNA)** *is the material that contains the genetic information that all organisms pass on to their offspring during reproduction.* This information includes instructions for making proteins and also for making the other nucleic acid, *ribonucleic acid* (RNA). **Ribonucleic acid (RNA)** *assists in protein synthesis by helping to coordinate the process of protein assembly.*

SECTION OBJECTIVES

- Describe the monomers of nucleic acids.
- Describe the structure of DNA.
- Describe the genetic code.
- Understand how DNA controls which proteins are to be synthesized.

Nucleotides

Nucleotides *are the monomers of nucleic acids.* A nucleotide has three parts: one or more phosphate groups, a sugar containing five carbon atoms, and a ring-shaped nitrogen base (Figure 23-12). RNA nucleotides contain the simple sugar ribose. DNA nucleotides contain deoxyribose (ribose stripped of one oxygen atom). Cells contain nucleotides with one, two, or three phosphate groups attached.

Besides being the monomers of nucleic acids, several nucleotides play other roles. For example, **adenosine triphosphate (ATP)** *is the nucleotide that supplies the energy for many metabolic reactions.*

Biochemistry   **695**

Figure 23-12 DNA is made up of nucleotides. One nucleotide unit, containing the base adenine, is shown at the left. In other DNA nucleotides, adenine is replaced by one of three other bases: thymine, cytosine, or guanine.

When enzymes join nucleotides to form nucleic acids, the phosphate group of one nucleotide is linked to the sugar of the next. This forms a long backbone of sugars and phosphates, with the bases sticking out at one side (Figure 23-13). This is what many RNA and some DNA molecules look like. But some nucleic acids coil up or join together to form more complicated structures held together by hydrogen bonds.

The bases in nucleotides attract each other in pairs, a phenomenon known as **base-pairing**. DNA, for instance, is made up of four different nucleotides—those containing the bases adenine (A), thymine (T), guanine (G), and cytosine (C). Adenine forms hydrogen bonds with thymine. Similarly, cytosine and guanine bond together. This *base-pairing* holds DNA nucleotides, and the DNA strands they lie in, together in pairs (Figure 23-14). *Paired strands of DNA then coil up to form a structure called a* **double helix.**

The Genetic Code

Base-pairing between nucleotides is one of the characteristics of DNA that permits it to store information. We can think of DNA's information as rather like a computer program: it tells the body what to do when certain things happen. Information in DNA, like computer information, is stored in code. But DNA is much more impressive than any computer. It stores more information in a smaller space than the most advanced gallium arsenide chip, it tells its owners how to reproduce themselves, and it even shows some of them how to learn to build computers!

DNA does all this by coding the information needed to build proteins. The genetic code is simple: a string of three DNA nucleotides codes for an amino acid. For instance, the nucleotide string TTC stands for the amino acid lysine. To read the DNA code, you find the string of nucleotides *TAC*, which means "start here," and read the nucleotides after the C in threes. For instance:

<u>TAC   TTC   CAA   CCT   TCG</u>
Means: START lysine valine glycine serine

This gives you the code for a chain of amino acids—in other words, the code for a protein. DNA contains coded information that tells the body what proteins to make. Some of the proteins are blood proteins or structural proteins. But the most important proteins are enzymes because these determine which chemical reactions go on in the body. The reason you are

Figure 23-13 A strand of DNA is is made up of nucleotides.

Figure 23-14 Two strands of nucleotides make up each DNA molecule. Here, a pair of strands, each four nucleotides long, is shown. The strands of both sugar-phosphate "backbones" lie on the outside of the molecule, with the nitrogen-containing bases meeting in the middle. Each base is held to a base on the opposite strand by hydrogen bonds (dashed lines). Adenine on one strand always pairs with thymine on the other, and cytosine always pairs with guanine.

chemically different from other human beings (unless you have an identical twin) is that you contain a unique collection of enzymes.

A **gene** *is a small segment of a DNA chain where the genetic instructions for making a specific protein are encoded.* Different living organisms contain different genes.

DNA Replication

DNA codes for protein structure, but it cannot actually make proteins. This is where base-pairing comes in. Remember that an A nucleotide on one DNA strand always bonds to a T nucleotide on the opposite strand, and a C bonds to a G. One strand of a DNA double helix has all the information needed to build a second strand because it consists of a string of nucleotides, each bearing one of the four bases A, T, C, or G. For instance, if the single strand is:

T—A—C—T—T—C—C—A—A—C—C—T—T—C—G

the missing second strand must be:

A—T—G—A—A—G—G—T—T—G—G—A—A—G—C

It is easy to synthesize the second strand because a free A nucleotide, floating around in the cell, will form hydrogen bonds with a T nucleotide in the single DNA strand. So the correct nucleotides line up opposite their base pairs, and the enzymes that build DNA merely have to join them together (Figure 23-15).

This is how DNA replicates (duplicates) itself. The two strands of the molecule separate. Each single strand then becomes the base upon which a new paired strand is built. DNA replication is vital to growth. Each of us starts with one set of genes (DNA) in the single fertilized egg cell from which we develop. But your body contains billions of cells. Each cell needs a copy of

The human body contains about 25 billion kilometers of the DNA double helix.

Figure 23-15 Replication of DNA. After the double-stranded DNA "unzips" along the hydrogen bonds between the strands, producing single strands, a nucleotide containing cytosine forms hydrogen bonds with a complementary nucleotide, containing guanine, on the original single strand. Then the enzyme DNA polymerase attaches the incoming cytosine nucleotide by its phosphate group to the sugar at the free end of the new, growing strand.

that DNA so that it has the information to make you you. Every time a new cell forms, it gets a complete copy of all your genes, made by duplicating the DNA.

There is another time when DNA must be duplicated—reproduction. When we pass on genes to our children, we do it by sending a copy of our genes to each sperm or egg we produce (Figure 23-16). All organisms, even the simplest ones, must transfer genetic information to their offspring. Plants pass on their genetic information in their seeds and pollen grains. Bacteria do it by dividing into two and leaving a copy of their DNA in each new bacterium.

The extraordinary process of DNA synthesis continues throughout an organism's life span. Remarkably, mistakes almost never occur. *An error in the sequence of DNA in a gene is known as* **mutation**. *Agents that can cause mutations, such as X rays or certain chemicals, are known as* **mutagens**. Some forms of cancer are caused by mutations.

RNA Synthesis

In addition to making copies of DNA, base-pairing permits DNA to make short RNA molecules containing the same information as parts of the DNA molecule. In this case, the RNA molecules do not stay bound to the DNA strands where they were produced. They leave the DNA and go off into the rest of the cell, carrying genetic messages to the locations where their information is required for protein synthesis.

Protein Synthesis

Proteins are made by structures inside the cell called **ribosomes**, *which are collections of RNA and protein molecules that act as the site of protein synthesis*. Ribosomes find out which proteins they are to make from RNA molecules called messenger RNAs. *A* **messenger RNA** *carries the coded information for a particular protein from DNA to a ribosome*. The ribosome decodes the RNA and makes the correct protein.

Figure 23-16 A human egg is shown here surrounded by sperm cells.

Chemistry and Technology

Toward Human Gene Transplants

Recent research indicates that we will one day be able to cure many inherited human diseases by gene transplants. For instance, beta thalassemia is a hereditary disease in which the blood cells produce defective hemoglobin. Victims suffer from severe anemia (lack of hemoglobin) and usually die in infancy. In one experiment, a human gene for hemoglobin was injected into fertilized eggs of mice with thalassemia. The eggs developed into adult mice with normal hemoglobin and passed the transplanted gene on to their offspring. In another experiment, scientists transplanted a rat growth hormone into mice, producing mice that are genetically identical to their normal counterparts except for their size (see photo).

The likelihood that we will one day be able to transplant genes from one human to another opens up many possibilities—both good and bad. It would mean that we could produce people with any genetic makeup we chose. We are not, and may never be, morally ready to cope with genetic control over the human population. Scientists believe that organizations such as churches and medical societies should start to consider the moral and ethical implications of genetic engineering now, before the techniques are ready to be used.

Genetic Engineering

The biochemistry of genes, known as **molecular genetics**, *is one of the most exciting and fast-moving branches of science.* Although we still have much to learn about genes, molecular genetics has already given rise to a new technology. We can isolate a desired gene and grow millions of copies of it. We can analyze these copies to find out the nucleotide sequence of the gene. We can find out the sequence of amino acids in the corresponding protein, and we can compare the structures of different genes. In several cases, we have even transferred functioning genes into the cells of plants and animals. Someday, we may be able to correct genetic defects with gene transplants. Current research in genetic diseases is seeking to find the precise locations of the one or more genes responsible for a disease.

A sequence of DNA can be spliced into bacterial chromosomes that are replicated every time the bacteria divide. Millions of copies of that gene can be made and harvested.

Insulin is an example of a hormone that has been synthesized using genetic engineering.

We can also make DNA to order. Many laboratories now have "gene machines" that can be programmed to produce short strands of DNA in any desired sequence. This tailor-made DNA is a useful tool for the study of DNA. It can also be used in protein synthesis experiments. By changing the genetic code so as to eliminate particular amino acids from a protein, biochemists can determine how the amino acids affect the function of the protein as a whole.

These feats are all part of **genetic engineering,** *the deliberate manipulation of genetic material.* Its applications, today or in the future, include making safer vaccines by engineering a weaker version of the disease-causing agent; producing chemicals synthesized by bacteria; producing enzymes for industry; cleaning up toxic wastes from industry, oil spills, and pesticide accidents; replacing defective genes; and creating pest-resistant plants and animals.

Mapping the human genome The availability of techniques for analyzing DNA makes it possible to think about mapping the entire human genome—analyzing human DNA so that we would know the relative positions of all human genes. In 1986, the United States committed itself to this formidable task. To print the nucleotide sequence for the entire human genome would require the equivalent of 200 Manhattan telephone books. It would take a large number of researchers years of time to assemble all this information. The effort would also require a lot of expensive equipment, both high-technology machinery to analyze the DNA, and powerful computers to handle the resulting data.

Many scientists question the value of committing so much money and talent to such a large project. Would we be doing it simply because it is now technically feasible? These people feel we would be better off working as in the past: focusing on specific genetic defects, pinpointing the responsible genes, and studying these genes and their neighbors to discover ways to alleviate the suffering of people born with those genetic defects.

Section Review

1. What is a gene?
2. Chemically, how does one organism differ from another?
3. How is genetic information coded in DNA?
4. What is the function of messenger RNA?

1. A length of DNA that codes for a particular protein
2. Each organism contains DNA that is chemically unique, and a unique collection of enzymes.
3. The information in DNA is used to specify the sequence of amino acids in a protein.
4. It helps to coordinate the process of protein synthesis by carrying the coded information for a particular protein to a ribosome.

SECTION OBJECTIVES

- Define metabolism.
- State how one reaction is linked to the next in a biochemical pathway.
- Describe the importance of ATP in metabolic pathways.

Biochemical thermodynamics is the study of the energy reactions in biochemical processes.

23.6 Biochemical Pathways

Biochemistry is much more than the study of macromolecules and their reactions. It is the study of the chemical reactions that allow organisms to do what they do. For example, what chemical reactions permit living organisms to respond to stimuli—by exploring, running away, or rolling into a ball?

The general answer to these questions is that living things use energy to power a vast number of chemical reactions. These reactions permit all living organisms to act in ways that nonliving things cannot. Every living cell is a busy biochemical factory. Here monomers are converted from one form to another, new macromolecules are built and old ones dismantled, and the energy required to run all this is extracted from food molecules.

Figure 23-17 Adenosine triphosphate (ATP) is made up of the base adenine, the sugar ribose, and three phosphate groups. Considerable energy is released when the bond attaching the last phosphate group (red squiggle) is hydrolyzed, converting the ATP into adenosine disphosphate, ADP.

All of an organism's biochemical reactions are collectively called its **metabolism** *and its reactions are organized into* **metabolic pathways**, *chemically-linked steps in which a series of enzymes converts substrate molecules, step by step, into a final product.* For example, in animals one pathway converts the six-carbon sugar glucose into the five-carbon sugar ribose, which is needed to make nucleotides. This conversion uses four different enzymes. Other metabolic pathways require more or fewer enzymes.

Glycolysis

As an example of a metabolic pathway, let us consider **glycolysis**, *one of a group of metabolic pathways by which animals turn their food into an energy supply.* The energy supply produced by most energy-producing pathways is the nucleotide, adenosine triphosphate (ATP) (Figure 23-17). When the bond holding the last of ATP's three phosphate groups is hydrolyzed, considerable energy is released:

$$\underset{\text{adenosine triphosphate}}{\text{ATP}} + \underset{\text{water}}{H_2O} \rightarrow \underset{\text{adenosine diphosphate}}{\text{ADP}} + \underset{\text{phosphate}}{H_3PO_4} + \text{energy}$$

The energy released by this reaction can be used to power reactions that cannot occur without an input of energy, such as the synthesis of macromolecules.

Cells can produce ATP by breaking down many different food molecules. One of the food molecules most commonly used to produce ATP is glucose, which is the substrate for glycolysis. Glycolysis is summarized in Figure 23-18. It consists of nine different reactions, each catalyzed by a different enzyme. Glycolysis adds to the amount of energy in the body by producing four molecules of ATP for each molecule of glucose that is broken down. However, note that glycolysis also uses up two existing molecules of ATP. This is because some of the reactions in the glycolysis series do not proceed spontaneously but are energetically "uphill." They need an input of energy from ATP to make them happen. One of the energy-using reactions is the first one, catalyzed by the enzyme hexokinase:

$$\text{Glucose} + \text{ATP} \xrightarrow{\text{hexokinase}} \text{glucose-6-phosphate} + \text{ADP}$$

Ask students to predict what would happen if a mistake occurred in the genetic code for hexokinase.

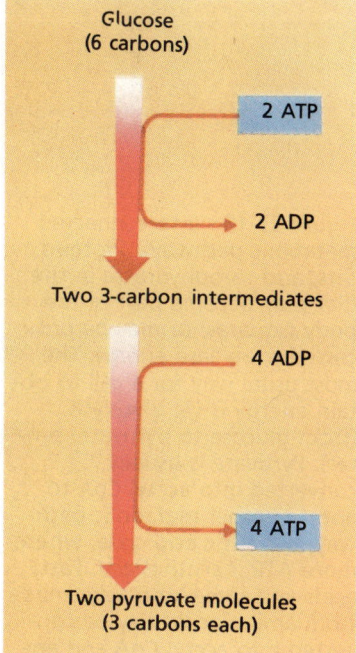

Figure 23-18 A summary of the nine reactions that make up glycolysis. Energy is needed to convert glucose into two 3-carbon molecules. The energy is supplied by a reaction that converts ATP to ADP. In the second half of glycolysis, two molecules of pyruvate are formed. These reactions release energy that is used to attach phosphate groups to 4 molecules of ADP, converting them into ATP.

Biochemistry

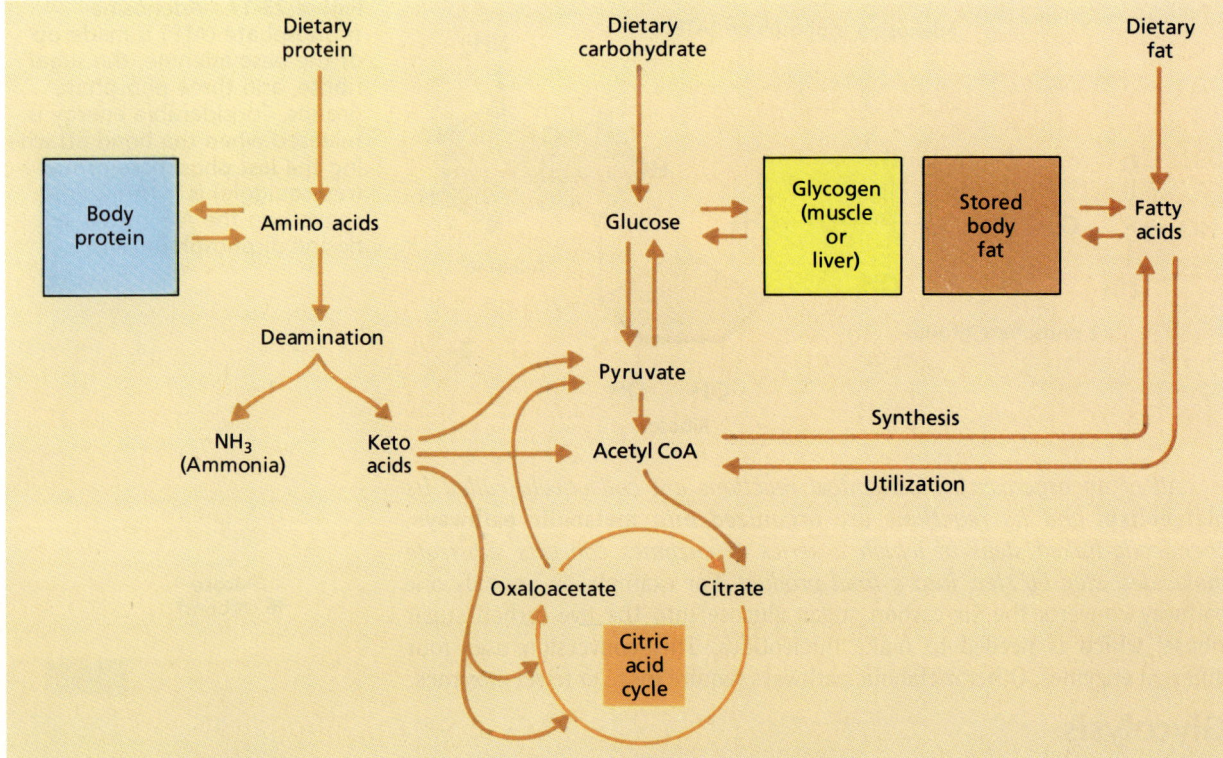

Figure 23-19 Interconnected metabolic pathways. Proteins, fats, and carbohydrates in the diet may become part of the body (squares) or may be processed to release energy. The most usual way for a cell to obtain energy is via glycolysis (from glucose to pyruvate, center). Pyruvate is usually converted into acetyl CoA to enter another metabolic pathway, the citric acid cycle, where more ATP is synthesized. Fatty acids and some keto acids made from amino acids may be converted into acetyl CoA and enter this pathway. Other keto acids enter as pyruvate or as molecules in the citric acid cycle. Both proteins and carbohydrates can also follow pathways that result in accumulation of body fat.

The pyruvate molecules that are the products of glycolysis are themselves the substrates for other metabolic pathways. Some of these generate still more ATP. Others use the pyruvate as substrates for building macromolecules.

Glucose is not the only food molecule that we use to produce energy. Energy comes from fat, proteins, and other carbohydrates. Figure 23-19 gives you an idea of the complex ways in which some metabolic pathways connect. For example, it shows that if we eat fat, the fat is hydrolyzed to fatty acids. These form the substrate for a metabolic pathway that produces a substance called acetyl CoA. Acetyl CoA, in turn, is a substrate for the pathway called the citric acid cycle. One product of the citric acid cycle is oxaloacetate that can be broken down into pyruvate. Pyruvate can be used as a monomer to build the storage polysaccharide glycogen. So this series of pathways, involving many different enzymes, has converted fat in our diet into a carbohydrate stored in the liver.

Organizing Metabolism

The various metabolic pathways interconnect in a complex pattern. Furthermore, each substrate occurs in very low concentrations. These tiny amounts of dozens of substrates must all be pushed into the right pathways when the body needs them, with no shortages or surpluses. How can these biochemical traffic patterns be controlled?

To bring order into the chaos, different metabolic pathways often occur in different areas. For instance, all the enzymes for one metabolic pathway may be arranged together in a group so that one enzyme's product is it neighbor's substrate. As the first enzyme releases its product, the second one binds the molecule before it can get away.

Membranes provide a second way to organize metabolic pathways. Many membranes separate compartments containing particular kinds of molecules and pathways.

Metabolic pathways can also be controlled by negative feedback control of a process by its product. When the pathway works too fast, too much product builds up. Some of the product molecules attach to molecules of the pathway's first enzyme and switch them off (just as hot air switches off the thermostat on a heater). This stops the entire pathway. Later, as the product is used up by other pathways, product molecules unbind from the start-up enzymes, and the enzymes go back to work.

Section Review

1. Define metabolism.
2. How does glycolysis contribute to a cell's energy store?

Chapter Summary

- Living organisms are made up of water, inorganic ions, and organic molecules. Organic macromolecules are polymers, each made up of many monomers.
- Large organic molecules are called macromolecules. Most of them fall into four main groups: carbohydrates, lipids, nucleic acids, and proteins.
- Proteins are made of monomers called amino acids. Different proteins have different sequences of amino acids. When linked together, these form polypeptides that have unique shapes.
- Enzymes are protein catalysts. Each enzyme catalyzes reactions between specific substrates. Enzymes enable organisms to carry out chemical reactions quickly. The activity of enzymes is affected by pH and temperature.
- Carbohydrates and lipids are composed mainly of carbon, hydrogen, and oxygen. Carbohydrates include monosaccharides, disaccharides, and polysaccharides. Some carbohydrates and lipids are important energy-storage compounds that may be broken down to release energy.
- Carbohydrates tend to be polar and associate with water, but lipids are nonpolar and so do not dissolve in water. Lipids include fatty acids, waxes, fats, oils, phospholipids, and steroids.
- Nucleic acids, such as DNA and RNA, and proteins play vital roles in directing an organism's growth, activity, and reproduction. DNA stores genetic information. RNA helps translate this information into proteins. Enzymes and structural proteins are examples of important proteins.
- Organisms extract useful energy from the energy stored in the chemical bonds of food molecules. The energy so released is used to make the cell's supply of ATP. ATP, in turn, donates the energy to various energy-requiring processes, such as metabolic reactions or production of new polymers.
- Glycolysis is an example of a metabolic pathway. During glycolysis, glucose is broken down to two molecules of pyruvate. Many other metabolic pathways feed into glycolysis, enabling cells to use many organic compounds other than glucose as food sources to generate usable energy in the form of ATP.

1. A term used to collectively describe all of an organism's biochemical reactions
2. It begins the process of glucose breakdown by converting glucose into pyruvate, which can then serve as a substrate for other metabolic pathways.

CHAPTER 23 REVIEW

Biochemistry

VOCABULARY

adenosine triphosphate (ATP) (695)
amine (694)
amino acid (686)
base-pairing (696)
biochemistry (685)
carbohydrate (689)
cellulose (690)
chitin (691)
cholesterol (693)
coenzyme (688)
complex carbohydrate (689)
denaturation (688)
deoxyribonucleic acid (DNA) (695)
disaccharide (689)
double helix (696)
enzyme (688)
epinephrine (694)
fatty acid (692)
gene (697)
genetic engineering (700)
glycogen (689)
glycolipid (692)
glycolysis (701)
hormone (693)
lipid (691)
lipoprotein (692)
macromolecule (685)
messenger RNA (698)
metabolic pathways (701)
metabolism (701)
molecular genetics (699)
monosaccharide (689)
mutagen (698)
mutation (698)
neuron (695)
neurotransmitter (695)
nucleic acid (695)
nucleotide (695)
peptide bond (687)
phospholipid (693)
polysaccharide (689)
polyunsaturated fat (692)
product (688)
protein (686)
ribonucleic acid (RNA) (695)
ribosome (698)
saturated fat (692)
saturated fatty acid (692)
starch (689)
steroid (693)
substrate (688)
synapse (695)
unsaturated fat (692)
wax (692)

REVIEWING CONCEPTS

1. What are some of the inorganic compounds from which living things are thought to have evolved over three billion years ago? (23.1)
2. What four elements make up the bulk of living organisms? (23.1)
3. How is water important to life? (23.1)
4. Write a reaction showing the formation of a peptide bond between two amino acids. (23.2)
5. Why is the action or an enzyme specific, meaning that each enzyme catalyzes reactions involving only one or a few substrates? (23.2)
6. List three functions of different proteins. (23.2)
7. Name two factors that can alter the activity of an enzyme. (23.2)
8. List three functions of carbohydrates. (23.3)
9. Draw the molecular structure of glucose. (23.3)
10. Name two structural carbohydrates. (23.3)
11. What is the function of glycogen? (23.3)
12. What is the function of cellulose? (23.3)
13. List the four classes of lipids and give an example of each. (23.4)
14. What is the function of waxes in living things? (23.4)
15. Name two lipids with different functions and state their uses in the body. (23.4)
16. List the three components that make up a nucleotide. (23.5)
17. How does the order of nucleotides in one strand of DNA reflect the order of nucleotides in the other strand? (23.5)
18. Name a nucleotide that is not part of a nucleic acid. (23.5)

PROBLEMS

1. You have a solution of an enzyme. You put half of it into each of two beakers containing identical substrate at equal concentrations. After waiting a while, you test both solutions and find that the substrate in beaker A has been changed but that the substrate in beaker B has not been acted on by the enzyme.

Beaker B has been sitting on a hot plate with the switch turned to "high." Why did the enzyme in beaker B probably not work?
2. A fragment of DNA has the nucleotide sequence AAATCGGCA. If another strand of DNA paired with this fragment, what would be its nucleotide sequence?
3. If you consume 1000. kilocalories of carbohydrate, how many grams of fat would you have to eat to supply the same amount of energy?

CRITICAL THINKING

1. **Interpreting Ideas** Science fiction tales sometimes feature life forms that are based on silicon rather than carbon. Silicon is much more abundant on earth than carbon. Like carbon, its atoms can bond to four other atoms. Bonds between two silicon atoms are unstable in the presence of O_2, but bonds between silicon and oxygen atoms are extremely stable and difficult to break. Is this the reason life is based on carbon not on silicon? Why or why not?
2. **Inferring Relationships** Explain what happens to the protein in an egg when the egg is cooked.
3. **Applying Ideas** A person who is too sick to eat may be given liquid nourishment intravenously. What carbohydrate is used for this purpose? Why?

CROSS CONNECTIONS

1. **Home Economics** When you cut an apple or banana, phenol oxidase enzymes in the injured areas quickly begin a "wound reaction," which results in the cut surfaces turning brown. Many cooks sprinkle lemon juice on sliced fruit to prevent it from discoloring in this way. Why does this work?
2. **Biology** Consult recent magazines and scientific journals in your library to learn the most current reports on the project to map the human genome. Share your findings with your classmates.

RESEARCH AND WRITING

1. Find an article in a newspaper or magazine about recent biochemical research. Use this chapter and a biochemistry text from the library to make notes for a talk in which you explain this research to your classmates.
2. Compile a list of chemicals that are considered to be mutagens. What mutagenic substances might you encounter in your daily life in your town or city?
3. Prepare a report on the dangers posed by the use of anabolic steroids to improve athletic performance.

PROJECTS

1. Keep a written record of all the foods you consume in a single day. Classify the foods as proteins, carbohydrates, or lipids.
2. After your class has read more about genetic engineering, have a debate on this new area of scientific research.
3. Make a list of some of the products in the grocery store that are fats or oils. Look up information in the library about each item. Which are saturated fats? Which are unsaturated? Group the products according to whether they are considered healthy or unhealthy.

Biochemistry

Chemistry and Careers

Agricultural Chemist

From seed to harvest, plants need the care and protection that chemists help provide. It's the agricultural chemist who develops new techniques to improve the yield and quality of crops. Seeds are often treated with pesticides to protect them until germination occurs. Herbicides keep down the growth of weeds that tend to crowd seedlings. And fertilizers provide missing nutrients.

If working outdoors appeals to you, you may consider

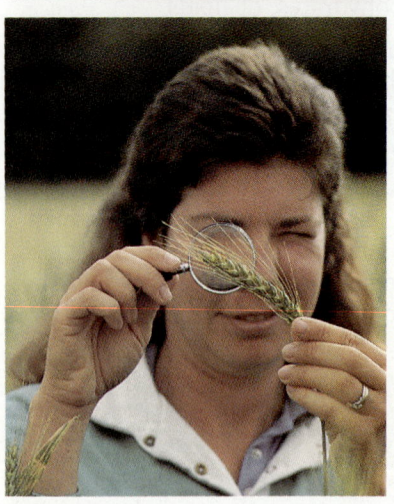

employment with the U.S. Department of Agriculture, doing field tests and consulting with farmers about their crops, animals, and water supplies. A high school diploma is required to work as a caretaker of plants and animals. Job advancement requires an associate's degree from a junior college, including as many chemistry and biology courses as possible. For a future in biotechnology, you need a bachelor's degree in chemistry, botany, ecology, molecular biology, genetics, or microbiology.

Biochemist

One of the most fascinating fields of science is biochemistry. A biochemist, such as the one pictured at upper right, works with substances that are present in living organisms. He or she might explore the pathways of an organism's metabolism or discover how closely related organisms share chemical similarities.

Perhaps the most active area of biochemistry is genetic engineering. Biochemists working in this area assist in "redesigning" DNA, a molecule in living cells that stores hereditary information and allows an organism to pass character traits from one generation to the next. Biochemists use chemical techniques to identify specific

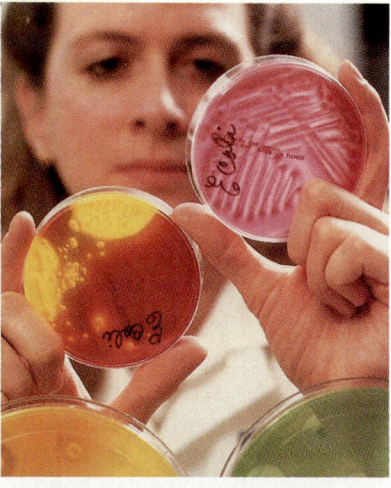

regions of DNA. These regions can then be isolated and added to the DNA within cells of a different organism, changing the organism's genetic characteristics. The resultant cells contain hybrid DNA and can be grown in the laboratory. In this way, cells can be given the instructions to manufacture a product that they would not produce under natural conditions. Biochemists can then use these techniques to produce valuable products, such as human insulin.

A few examples of the many research areas for biochemists are endocrinology, the study of hormones; oncology, the study of cancer; and cell biology, which includes study of the chemical reactions within cells.

A biochemist must have a master's or doctoral degree in chemistry or biology.

UNIT 7 • FOCUS ON REASONING

College Entrance Exam Preparation

Organic compounds in foods can be identified in the laboratory using substances called indicators. The color change of an indicator is usually a positive test for the presence of a certain type of organic compound.

To test for carbohydrates (the various classes of sugars and starches) two indicators are used. Benedict's solution will change from blue to orange in the presence of sugar. Iodine solution will change from yellow-brown to blue-black in the presence of starch.

Sudan III is an indicator that will turn from pink to red in the presence of a lipid. Biuret reagent can be used to test for proteins. Biuret reagent will change color from yellow to blue-violet in the presence of protein.

The following tests were conducted to determine the presence of various organic compounds. Five test tubes were labeled as follows: honey, egg white, potato, butter, and an unknown substance. Five milliliters of each food sample was added to each test tube.

▶ Carbohydrate Test

Ten drops of Benedict's solution were added to each test tube. The test tubes were placed in a hot water bath and heated for three minutes. The results of any change were recorded. The test tubes were washed and refilled with 5 mL of the samples on the labels. Five drops of iodine solution were added to each test tube and changes were recorded.

▶ Lipid Test

The test tubes were washed and refilled again with 5 mL of the food samples on the labels. Five drops of Sudan III were added to each test tube. The results of any change were recorded.

▶ Protein Test

The test tubes were washed and refilled to perform the last test. Five drops of Biuret reagent were added to each test tube and any color change observed was recorded.

1. If Biuret reagent is added to a food sample and a positive result occurs, the organic compound in the food sample is probably
   a. a sugar.   c. a lipid.
   b. a starch.   d. a protein.

2. Based on the data given, what classes of organic compound are present in the unknown?
   a. lipids and starch   c. carbohydrates
   b. sugar and starch   d. lipids and protein

3. What was the purpose of the laboratory experiment?
   a. to determine if food samples taste good
   b. to determine the indicator's colors change when subjected to laboratory testing
   c. to identify the unknown
   d. to identify organic compounds in foods using indicators

4. If you were to test other food samples for the presence of carbohydrates, what indicators would you use?
   a. Benedict's solution and Biuret reagent
   b. Sudan III and iodine solution
   c. Benedict's solution and iodine solution
   d. Sudan III and Biuret reagent

5. If a sample resulted in a positive test using iodine solution and a positive test using Biuret reagent, what can be concluded?
   a. the sample contained lipids
   b. the sample contained proteins and starch
   c. the sample contained carbohydrates
   d. the sample contained lipids and proteins

Substance	Sudan III color	Iodine color	Benedict's color	Biuret color
Honey			Orange	
Egg White				Blue-violet
Butter	Red			
Potato		Blue-black		
Unknown	Red			Blue-violet

INTRA-SCIENCE

Custom-Designed Plants

Custom-designed plants that are able to flourish in salty soil, resist drought, or survive deadly diseases are gradually becoming a reality. Tobacco plants that are resistant to herbicides are being developed. Strawberry plants sprayed with genetically altered bacteria are able to resist frost damage.

Ever since agriculture began, about 10 thousand years ago, farmers have tried to create improved breeds of plants. Parent plants with desirable characteristics were bred to each other to produce superior offspring. It was not until the mid 1800s, however, that a systematic study of inheritance patterns was made. Gregor Mendel, an Austrian monk, clarified the basis of heredity with his pea-plant experiments. Mendel's greatest contribution was the discovery that hereditary characteristics are determined by units (now called genes) rather than by a blending of parental characteristics.

A second major step in genetics was Watson and Crick's description of the DNA (deoxyribonucleic acid) molecule. Their work in the 1950s began a revolution in genetics, leading to the creation of artificial genes and new forms of life.

The molecular basis of heredity is the DNA molecule. DNA is made of long chains of nucleotides. A nucleotide consists of a molecule of the sugar deoxyribose, a phosphate group, and one of four bases—adenine, thymine, cytosine, or guanine. The sequence of bases acts as a code that regulates the sequence of amino acids in a protein. DNA controls the characeristics of an organism by controlling the synthesis of specific proteins.

Every plant has thousands of genes (a gene is a section of DNA). By changing the genes or inserting new ones, plants with new characteristics—custom-designed plants—can be created.

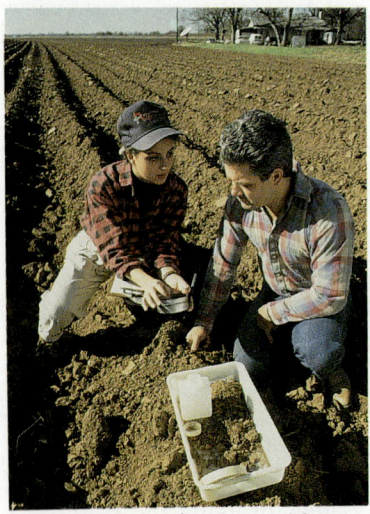

The Connection to Physics

The chemical nature of genes already present in an organism can be altered with radiation. Ultraviolet light, alpha-particles, beta-particles, high-energy electrons,

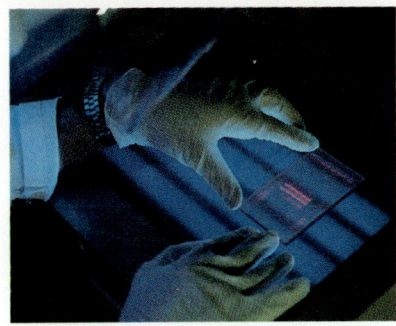

X rays, and other types of high-energy radiation can dislodge one or more electrons from the DNA of a gene. The ionized molecule is more reactive than normal and may combine with other substances. A nucleotide may be added or deleted, and this changes the sequence of bases. As a result, the sequence of amino acids in the protein produced by the gene will be different from the protein normally produced. Such a change in the amino acid sequence could improve the organism in some way. Conversely, the change could render the protein nonfunctional and kill the organism.

To insert foreign genes into organisms, genetic engineers use electricity to create tiny openings in the membranes of plant cells through which the genes can be inserted. This process, electroporation, has been used to transfer new genes into corn cells. The biggest stumbling block has been the difficulty of regenerating whole plants from genetically altered cells. If this problem can be solved, electroporation will enable scientists to transfer genes between unrelated species. More nutritious corn, or wheat that can survive floods, may be produced.

The Connection to Biology

Viruses and bacteria are used to transport new genes into the cells of plants. Bacteria of the genus *Agrobacterium* are most commonly used. These bacteria normally cause diseases in plants by inserting a plasmid, a ring of bacterial DNA, into plant cells. The plasmid

codes for the synthesis of a protein that causes plant cells to grow abnormally. Molecular biologists inserted a desirable gene into the bacterial plasmid. When the plant cell is infected by the bacterium, it also receives a copy of the desirable gene.

This process has been used to give tobacco plants greater resistance to herbicides that are used to kill weeds. The herbicide glyphosate kills plants by preventing the synthesis of an enzyme, EPSP synthetase, which is needed for making three essential amino acids. A mutant gene that produces a slightly different EPSP synthetase enzyme was inserted into cells of tobacco plants, using *Salmonella* bacteria. When fields of genetically altered tobacco plants are sprayed by glyphosate, the weeds are killed. The tobacco plants survive because they are still able to produce the essential amino acids.

Plant pathologists isolated the

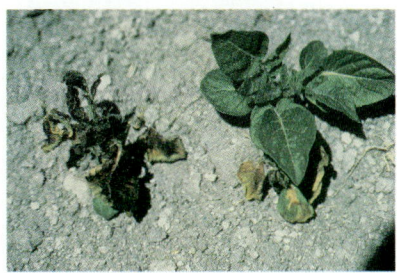

gene of a bacterium that causes ice crystals to form and destroy plant cells. They altered the gene that codes for the synthesis of the protein and inserted it into the bacteria. The new bacteria crowd out the old frost-forming type of bacteria, leaving the plants undamaged.

Scientists also learned how to form artificial genes by joining nucleotides together in long chains. The ability to make artificial genes, coupled with techniques to introduce genetic material into plant cells, opened the possibility of creating new life forms.

Unit 8

Chapter 24
The Metals of Groups 1 and 2   712

Chapter 25
The Transition Metals   728

Chapter 26
Aluminum and the Metalloids   750

Chapter 27
Sulfur and Its Compounds   764

Chapter 28
The Halogens   778

Descriptive Chemistry

Growing up in Butte, Montana, one of the major historical copper-producing areas in North America, one cannot help being interested in the mining and processing of copper ores. It wasn't until college, however, that I fully appreciated the chemical complexity of many of the modern processes. I am now a process metallurgist in a plant that uses a new method to treat previously untreatable gold ores.

In this unit you will learn about some of the more useful chemical elements, how they are recovered from nature, and how the science of chemistry is used to turn them into valuable products.

William J. Janhunen, Jr.
Process Metallurgist
Homestake Mining Company / McLaughlin Mine

Copper mine near Ruth, Nevada

CHAPTER 24
The Metals of Groups 1 and 2

PLANNING CHART

Approximate Time: 4/5 days

SECTION/TEXT FEATURES		RESOURCES	
24.1 THE ALKALI METALS	**713**	*Teacher's Resource Binder*	
Demos 1, 2	712A–B	Section Review Worksheets	91
Periodic Table Videodisc	712E	*Other Resources*	
Section Review	716	Blackline Master 58	
Chapter Review	727		
• Questions 1–5, Problems 1–4			
24.2 THE ALKALINE EARTH METALS	**717**	*Teacher's Resource Binder*	
Demos 3, 4, 5, 6, 7, 8, 9, 10	712B–C	Lab Experiment 34	181
Section Review	720	Section Review Worksheets	92
		Other Resources	
		Blackline Master 58	
24.3 SODIUM: ONE OF THE ACTIVE METALS	**720**	*Teacher's Resource Binder*	
Demos 11, 12	712C–D	Section Review Worksheets	93
Periodic Table Videodisc	712E	*Other Resources*	
Section Review	726	Transparency 58	
Chapter Review	727		
• Questions 6–12			
CHAPTER END ACTIVITIES		*Teacher's Resource Binder*	
Critical Thinking, Cross Connections, Research and Writing, Projects		Test Booklet	121, 123
		Computer Test Generator	

Teaching Strategies

24.1 The Alkali Metals

Students often confuse the properties of the elemental substances with the properties of the compounds formed by the elements. Help them keep in mind this distinction when discussing their properties. The elements are extremely reactive, and the compounds are much less reactive. Display samples of sodium and potassium stored under kerosene and show samples of some of the common compounds of the elements. Show samples of sodium chloride, sodium hydrogen carbonate, sodium hydroxide, potassium chloride, potassium iodide, and potassium carbonate. This will help students appreciate the difference in reactivity between the elements and the compounds.

To help students understand the concept of the unit cell, exhibit a model made from styrofoam balls or gumdrops and toothpicks of a body-centered cubic lattice. Discuss the mobility of free electrons and related properties.

Contrast the low melting points, low densities, and softness of alkali metals with the properties of the more familiar metals iron, nickel, copper, silver, and gold. Contrast the soft metallic lattice with a rigid salt crystal, showing that bending a metallic crystal forces rows of atoms to move past other identical rows of atoms, while bending a salt crystal would force positive and negative ions to move past one another.

Explain why free alkali metals are such vigorous reducing agents. Remind students that a good reducing agent is itself readily oxidized, losing electrons to the substance that is reduced. Show the relationship between the ease of oxidation and the ionization energy. Explain the unusual behavior of lithium in terms of small ion size, the consequent high charge density, and the hydration energy.

Demonstration 1: Flame Tests

PURPOSE: To show the colors produced when alkali metal compounds are excited in a flame

MATERIALS: Device made from 2-liter plastic soft drink bottle, cork, burner, and spray bottle; assemble as shown in Figure T24-1.

PROCEDURE: Spray solutions of alkali metal compounds into hole in bottle. Observe color in flame. Use solutions of NaCl, KCl, and Li_2CO_3. Compare the colors when NaCl, Na_2SO_4, KCl, and K_2SO_4 are used to verify that the color is produced by the alkali metal ion and not the negative ion. With this device the color is sufficiently long-lasting that students can observe the line spectra using an inexpensive hand spectroscope.

Figure T24-1

Demonstration 2: Reaction of Sodium with Water.

PURPOSE: To show the reactivity of sodium with water
MATERIALS: Beaker, sodium-lead alloy (90% lead, 10% sodium, obtainable from J. T. Baker Chemical Co., Phillipsburg, NJ), water, phenolphthalein, overhead projector
PROCEDURE: Wear safety goggles. Drop small pieces of sodium-lead alloy into a beaker of water containing some phenolphthalein on the overhead projector. This alloy is a convenient and safe source for reacting small amounts of sodium. Hydrogen gas is evolved, and the pink color of the solution shows that sodium hydroxide solution has also formed. Explain to students that the lead is a "moderator."

Chemistry and Health: Group 1 Ions in Medication and Diet

Discuss with the students the effects of salt concentration in cells as a means of regulating water retention of cells by the process of osmosis. Show containers of "No-Salt Salt" for seasoning food, and have the students read the labels to determine the contents. Most students will be aware of the fact that perspiration removes much sodium chloride from the body. Discuss the need to replace sodium chloride after strenuous exertion.

24.2 The Alkaline-Earth Metals

Compare and contrast the properties of the elements of Group 2 with those of Group 1. Students should recognize that this group of metals is also a very reactive family, though not as reactive as the Group 1 metals. Point out that the alkaline-earth metals are, like the alkali metals, never found free in nature but are always in a combined form. Display samples of the metals magnesium and calcium. Discuss the fact that these do not have to be stored under kerosene. Explain the difference in reactivity in terms of the energy required for these atoms to lose two electrons while the alkali metals need only lose one electron to form compounds.

Contrast the greater densities, hardness, and higher melting and boiling points of the alkaline-earth metals with the properties of the alkali metals. Explain the differences in terms of metallic crystal makeup. The alkaline-earth metal 2+ ions are more strongly attracted to the electron "gas" that permeates the metal than are the alkali metal 1+ ions. Compare the sizes of the atoms of Groups 1 and 2.

Display some of the common compounds of the alkaline-earth metals. You could show samples of $CaCO_3$, $Ca(OH)_2$, CaO, MgO, $Mg(OH)_2$, and $Ba(OH)_2$. Write the electron configurations of the elements, and discuss the reason for the large increase in ionization energy when the elements lose their third electrons. Use the concept of ionization energy to explain the increase in reactivity as one moves down the column in the Periodic Table.

Demonstration 3: Reactivity of Magnesium and Calcium with Water

PURPOSE: To compare the reactivity of magnesium and calcium with water
MATERIALS: Calcium metal, magnesium metal, two beakers, water, phenolphthalein
PROCEDURE: Wear goggles. Place a small piece of magnesium metal in a beaker that contains water and a few drops of phenolphthalein solution. Put a piece of calcium metal in another beaker with water and phenolphthalein solution. Observe that the calcium reacts fairly rapidly with evolution of hydrogen gas and formation of slightly soluble $Ca(OH)_2$, as shown by the phenolphthalein turning pink. The magnesium does not react. Relate the difference in reactivity to their positions in the Periodic Table.

Demonstration 4: Combination of Magnesium with Nitrogen

PURPOSE: To show that magnesium metal can react with nitrogen.
MATERIALS: Magnesium ribbon, crucible and crucible cover, red litmus paper, beaker, ring stand, triangle, burner
PROCEDURE: Wear goggles. Make a coil of a 5-cm piece of magnesium ribbon and place it in a crucible. Cover it loosely with the crucible cover. Support it on a ring stand and triangle, and heat it until the magnesium reacts. Allow it to cool. Examine the product and note the presence of white MgO and traces of dark Mg_3N_2. Add water, and the Mg_3N_2 will react to form NH_3 gas, which can be detected if a piece of moist red litmus is held over the crucible. Add some MgO to water in a beaker. $Mg(OH)_2$ will form, which can also be detected by the litmus paper.

Demonstration 5: Formation of Hydroxides from Alkaline-Earth Oxides

PURPOSE: To show that oxides of Ca, Mg, and Ba form hydroxides with water
MATERIALS: CaO, MgO, BaO, water, litmus paper, test tubes
PROCEDURE: Add the oxides to water, test with litmus paper to show that the solutions are basic. Write the equations for the reactions. Observe that the hydroxides are not very soluble.

The Metals of Groups 1 and 2

Demonstration 6: Flame Tests with Alkaline-Earth Compounds
PURPOSE: To show the colors produced when alkaline-earth metal compounds are excited in a flame
MATERIALS: Device made from 2-liter plastic soft drink bottle, cork, burner, and spray bottle (See Demonstration 1); solutions of alkaline-earth compounds such as magnesium chloride, calcium chloride, barium chloride, and strontium chloride
PROCEDURE: Spray solutions of alkaline-earth compounds into the hole in the bottle and observe the characteristic colors in the flame.

Demonstration 7: Reaction of Milk of Magnesia with Acid
PURPOSE: To show the gradual neutralization of acid by $Mg(OH)_2$
MATERIALS: Milk of Magnesia tablet, 600-mL beaker, distilled water, $1.0\,M$ HCl, universal indicator, magnetic stirrer
PROCEDURE: Place 300 mL of distilled water in the beaker on a magnetic stirrer. Add 6 drops of $1.0\,M$ HCl and 24 drops of the universal indicator. Drop in one tablet of Milk of Magnesia. Observe the color changes. The pH of the solution slowly changes from pH 4 to pH 10 as the tablet gradually neutralizes the acid. The indicator changes color from red to orange to yellow to green to blue and then violet.

Demonstration 8: Limewater Test for Carbon Dioxide
PURPOSE: To show how carbon dioxide precipitates calcium carbonate in limewater solution
MATERIALS: Saturated solution of calcium hydroxide (limewater), drinking straw
PROCEDURE: Have a student blow his or her breath into a solution of limewater. (Caution them not to drink the solution.) Observe the formation of the precipitate of $CaCO_3$. Write the equation for the reaction.

Demonstration 9: Hard Water
PURPOSE: To show the effects of hard water on soap and detergent solutions
MATERIALS: Distilled water, tap water, hard water containing a little calcium chloride and magnesium chloride, soap solution, detergent solution, test tubes
PROCEDURE: Place equal amounts of distilled water, tap water, and hard water in separate test tubes. Add soap solution, a few drops at a time, to each test tube and shake until a stable lather occurs. Observe that: distilled water lathers readily and appears clear; tap water forms a cloudy solution and needs more soap to form a lather; and hard water forms the most scum and requires the greatest amount of soap for a lather. Repeat with detergent instead of soap solution. Carry out Demonstration 10 to show how the hard water can be treated.

Demonstration 10: Softening Hard Water
PURPOSE: To show ways of softening hard water
MATERIALS: Hard water solutions from Demonstration 9, Na_2CO_3, soap solution
PROCEDURE: Add Na_2CO_3 to the hard water solution used in Demonstration 9 to precipitate $CaCO_3$ and $MgCO_3$. Repeat the test with soap solution to see how much is needed to form a stable lather. Compare these results with those of the unsoftened water in Demonstration 9.

Chemistry and Health: Strontium: Fireworks and Fallout
Students are always fascinated by fireworks. If you have done flame-test demonstrations, you can ask them to discuss different substances that will give various colors to fireworks.

Point out that strontium is very similar to calcium in its chemistry. Growing children absorb more calcium than adults and are thus more susceptible to having strontium-90 deposited in bones. Remind them that blood is produced in bone marrow, and discuss diseases of the blood and blood-forming tissues, such as leukemia.

24.3 Sodium: One of the Active Metals
Students may need some help in understanding the equations given for the various reactions. Explain the oxidation-reduction reaction that occurs in the Downs cell. Write the formulas for the ions oxide, O^{2-}, peroxide, O_2^{2-}, and superoxide, O_2^{-}. Discuss the uses of sodium. Explain production of tetraethyl lead from ethyl chloride and a sodium-lead alloy. Display and operate a sodium lamp. (Your physics department may have one.) Discuss the use of sodium as a heat-transfer agent in atomic reactors.

Display rock salt, table salt, and halite crystals. Examine NaCl crystals with a hand lens or microscope and compare them with a model of NaCl face-centered structure. Show that when rock salt crystals break, they fracture so that the faces are flat and the angles between the faces are 90°.

Exhibit forms of NaOH—flake, stick, and pellet—along with containers of commercial lye. Have students read the labels on cans of lye, which provide valuable information about the preparation of solutions, the uses of NaOH, and antidotes. Exhibit a can of any product used to clear clogged drains that contains NaOH in excess of 50 percent.

Demonstration 11: Dissolution of NaOH and Reaction of NaOH and Aluminum
PURPOSE: To show the exothermic nature of the dissolution of sodium hydroxide and the reaction that occurs between sodium hydroxide solution and aluminum
MATERIALS: Beaker, thermometer, sodium hydroxide pellets, aluminum foil
PROCEDURE: Wear safety goggles. Prepare approximately $3\,M$ NaOH by adding 12 g of NaOH pellets to 92 mL of water. Measure the temperature change as the dissolution progresses. (Remember that NaOH is caustic. Wash off any spills with copious amounts of water.) Add a few pieces of aluminum foil to the warm solution to show how the aluminum dissolves with the formation of hydrogen gas. Write the equation for the reaction. Discuss the use of aluminum in drain cleaners containing NaOH. Explain why oven cleaners can be used on porcelain or stainless steel ovens and utensils, but not on aluminum.

Display a chalkboard replica or transparency of the flow diagram of the Solvay process, and write out the appropriate equations. Exhibit commercial varieties of baking soda and washing soda.

Demonstration 12: Preparation of KNO_3

PURPOSE: To show the preparation of potassium nitrate and sodium chloride by crystallization from solution.
MATERIALS: $NaNO_3$, KCl, filter paper, funnel
PROCEDURE: CAUTION: Wear gloves, goggles, and lab apron. To 25 mL of water add 12 g $NaNO_3$ and 10 g KCl. Heat to complete solution and evaporate to one-half volume. Cool. Filter out the crystals (mostly sodium chloride). Examine with a hand lens or microscope, and compare with NaCl crystals. Evaporate filtrate further until a second crop of crystals forms (mostly KNO_3). Filter and examine these with a lens or microscope to see their crystal form, comparing with KNO_3.

References and Resources

Books and Periodicals

Alyea, Hubert N. "Burning Magnesium in Steam." *Tested Demonstrations in Chemistry*, 6th ed., 1965, p. 142. Reprint from the *Journal of Chemical Education*. Published by the Division of Chemical Education of the American Chemical Society.

Alyea, Hubert N. "Removal of Hardness and Deionization of Water by the Ion Exchange Method." *Tested Demonstrations in Chemistry*, 6th ed., 1965, p. 219. Reprint from the *Journal of Chemical Education*. Published by the Division of Chemical Education of the American Chemical Society.

Dandy, Alan J. "Chemistry of Cement." *CHEM*, Vol. 51, 1978, p. 13.

Gouge, Edward M. "A Flame Test Demonstration Device," *Journal of Chemical Education*, Vol. 65, 1988, p. 544.

Hall, Christopher. "On the History of Portland Cement," *Journal of Chemical Education*, Vol. 53, 1976, p. 222.

Hall, Christopher. "Chemistry of Fossilization," *Journal of Chemical Education*, Vol. 53, 1976, p. 270.

Summerlin, Lee R., Christie L. Borgford, and Julie B. Ealy. "Making Sodium Chloride from Sodium and Chlorine," *Chemical Demonstrations, A Sourcebook for Teachers*, Vol. 2, American Chemical Society, 1987, p. 56.

Summerlin, Lee R., Christie L. Borgford, and Julie B. Ealy. "Burning Magnesium in Carbon Dioxide." *Chemical Demonstrations: A Sourcebook for Teachers*, Vol. 2, American Chemical Society, 1987, p. 58.

Summerlin, Lee R., Christie L. Borgford, and Julie B. Ealy. "Reduction of Sand with Magnesium." *Chemical Demonstrations: A Sourcebook for Teachers*, Vol. 2, Americal Chemical Society, 1987, p. 193.

Audiovisual Resources

"Chemical Families," CHEM Study 16mm film, Ward's Modern Learning Aids. Shows the elements potassium, lithium, and sodium reacting with a number of nonmetals

"Chemicals from NaCl: 1" video, Films for the Humanities and Sciences, FFH1110F. Shows production of sodium hydroxide and chlorine by electrolysis of salt

"Chemicals from NaCl: 2" video, Films for the Humanities and Sciences, FFH1111F. Shows production of sodium metal from salt

"The Electrolysis of Brine": 16 mm film, PPG Industries, Inc., Free-loan film available from Modern Talking Picture Service. Shows production of chlorine and sodium hydroxide and their importance

"Salt—The Essence of Life": 16mm film, Salt Institute, Free-loan film available from Modern Talking Picture Service. Shows history, production, and uses of salt

Computer Software

"The Periodic Table Videodisc: Reactions of the Elements": CAV-type videodisc, Journal of Chemical Education Software, 1989. Shows each element in its most common forms; reactions with air, water, acids, bases; and some common applications

Answers and Solutions

Questions

1. The atoms of Group 1 elements have electron configurations of the preceding noble gases plus one additional electron in the next higher energy level. The ions have the electron configurations of the preceding noble gases.
2. Sodium is prepared by electrolysis of its fused chloride. Potassium is now prepared by the reaction of metallic sodium and potassium chloride.
3. Uses: in making tetraethyl lead for antiknock gasoline; as a heat-transfer agent; in making dyes; in making sodium-vapor lamps
4. Caustic substances, such as sodium hydroxide, are capable of changing some types of animal and vegetable matter into soluble materials by chemical means. Caustic substances have destructive effects on skin, hair, and wool. "Corrosive" is a more general term for substances that eat away by chemical action, such as the action of an acid on a metal.
5. *(a)* Common salt, limestone, and coal *(b)* Coal tar, sodium carbonate, sodium hydrogen carbonate, and calcium chloride
6. Molasses contains acids that react with baking soda to form carbon dioxide gas.
7. Sodium chloride is found in sea water, salt wells, and rock salt; Chile and Peru have sodium nitrate deposits; dry lake beds contain sulfates and borates; Sodium carbonate comes

b. Coal tar, sodium carbonate, sodium hydrogen carbonate, and calcium chloride
10. Molasses contains acids that react with baking soda to form carbon dioxide gas.
11. **a.** Under kerosene
    b. Because it reacts vigorously with water
12. $CaCO_3(s) \rightarrow CaO(s) + CO_2(s)$
    $CaO(s) + H_2O(\ell) \rightarrow Ca(OH)_2(s)$
    $NH_4^+(aq) + OH^-(aq) \rightarrow NH_3(g) + H_2O(\ell)$

Problems

1. $\quad$ 1.50 kg $\quad$ 1.50 kg $\quad$ X
   $NaNO_3 + KCl \rightarrow KNO_3 + NaCl$
   $\quad$ 1 mol $\quad\quad$ 1 mol $\quad\quad$ 1 mol
   1 mol $NaNO_3$ = 85.0 g $\quad$ 1 mol KCl = 74.6 g $\quad$ 1 mol KNO_3 = 101.1 g
   Because the reactants combine in a 1 : 1 mole ratio, KCl is in excess. The mass of KNO_3 produced will depend on the mass of $NaNO_3$ that is used.

   $$X = \frac{1.50 \text{ kg } NaNO_3}{85.0 \text{ g/mol}} \times \frac{1 \text{ mol } KNO_3}{\text{mol } NaNO_3} \times \frac{101.1 \text{ g}}{\text{mol}} = 1.78 \text{ kg } KNO_3$$

2. $Na_2CO_3 \cdot 10H_2O$ = 286 g/mol

   $$\% \ Na_2CO_3 = \frac{106 \text{ g/mol}}{286 \text{ g/mol}} \times 100\% = 37.1\%$$

   $$\$0.50/\text{kg} \times 0.371 = \frac{\$0.19}{\text{kg } Na_2CO_3}$$

3. $X = \dfrac{1.00 \times 10^3 \text{ kg } Na_2CO_3 \times 117 \text{ g NaCl}}{106 \text{ g } Na_2CO_3} = 1.10 \times 10^3$ kg NaCl

4. $\dfrac{1.1 \text{ g}}{500. \text{ mL}} \times \dfrac{\text{mol}}{40.0 \text{ g}} \times \dfrac{10^3 \text{ mL}}{\text{L}} = 0.055$ mol/L = 0.055 M

   $$[H_3O^+] = \frac{1.0 \times 10^{-14} \text{ mol}^2/\text{L}^2}{5.5 \times 10^{-2} \text{ mol/L}} = 1.8 \times 10^{-13} \text{ mol/L}$$

   pH = $-\log[H_3O^+] = -\log(1.8 \times 10^{-13}) = 12.74$

Critical Thinking

1. Common-ion effect: the presence of $NaC_2H_3O_2$ greatly increases the $[C_2H_3O_2^-]$ and shifts the $HC_2H_3O_2 + H_2O \rightleftarrows H_3O^+ + C_2H_3O_2^-$ equilibrium to the left. The $[H_3O^+]$ decreases and the pH rises.

Videodisc Selections
from *The Periodic Table Videodisc* (JCE Software)

step forward

step reverse

1058
lithium + water (m)

CHAPTER 24

pg 713 Alkali metals

274
lithium + air (m)

268
lithium (s)

1978-2827
lithium + acids (m)

2828
lithium + bases (m)

36326
uses of lithium (s)

36336
uses of lithium (s)

16826
potassium (s)

17334
potassium + water (m)

16838
potassium + air (m)

17611
potassium + acids (m)

18066
potassium + bases (m)

36544
uses of potassium (s)

37287
rubidium (s)

36779
uses of rubidium (photocells) (s)

46095
cesium (s)

36941
uses of cesium (atomic clock) (s)

pg 717 Alkaline earth metals

36343
uses of beryllium (s)

7523
magnesium (s)

7535-7856
magnesium + air (m)

7857-8254
magnesium + acids (m)

36445
uses of magnesium (s)

18426
calcium (s)

19244
calcium + water (m)

18438-19243
calcium + air (m)

The Metals of Groups 1 and 2

20119
calcium + acids (m)

36555
marble (s)

36550
Plaster of Paris, cement, bones (s)

37306
strontium (s)

37311
strontium + air (m)

38127
strontium + water (m)

38650
strontium + acids (m)

36784
fireworks (s)

46113
barium (s)

46119
barium + air (m)

46844
barium + water (m)

47293
barium + acids (m)

36952
uses of barium (s)

37203
radiation therapy (s)

pg 720 Sodium

4565
sodium (s)

5104
sodium + water (m)

4570
sodium + air (m)

6036
sodium + acids (m)

7040
sodium + bases (m)

36438
uses of sodium (s)

Teacher's Notes

Chapter 24
The Metals of Groups 1 and 2

SECTION PREVIEW
- 24.1 The Alkali Metals
- 24.2 The Alkaline-Earth Metals
- 24.3 Sodium: One of the Active Metals

When potassium metal is dropped into water, potassium hydroxide and hydrogen gas are formed in a violent reaction so exothermic that the hydrogen ignites. (Note: the pink color in the beaker is due to phenolphthalein indicating the presence of the hydroxide product.)

24.1 The Alkali Metals

The group of elements at the left of the Periodic Table includes lithium, sodium, potassium, rubidium, cesium, and francium. These elements are the *alkali metals,* sometimes called the Sodium family of elements. Because of their great chemical reactivity, none exist in nature in the elemental state. They are found in natural compounds as monatomic ions with a 1+ charge. Sodium and potassium are plentiful, lithium is fairly rare, and rubidium and cesium are rarer still. Francium does not exist as a stable element. Only trace quantities have been produced in certain nuclear reactions.

Structure and Properties

The Group-1 elements possess metallic characteristics to a high degree. Each has a silvery luster, is a good conductor of electricity and heat, and is ductile and malleable. These metals are relatively soft and can be cut with a knife. The properties of the Group-1 metals are related to their characteristic metallic crystal lattice structures, which are described in Chapter 13.

The crystal lattice consists of metallic ions with a 1+ charge. The lattice is built around a body-centered cubic unit cell in which each metallic ion is surrounded by eight nearest neighbors at the corners of the cube. Models of the body-centered cubic unit cell are shown in Figure 24-1. Valence electrons are not localized to one atom but are free to permeate the lattice structure. These "free" electrons belong to the metallic crystal as a whole. The mobility of the free electrons in Group-1 metals accounts for their high thermal and electric conductivity and silvery luster as well.

The softness, ductility, and malleability of the alkali metals are explained by the binding force in the metallic crystal lattice. Their low melting points and densities, together with their softness, distinguish these elements from the more familiar common metals. Some of the properties of the alkali metals are shown in Table 24-1.

The atoms of each element in Group 1 have a single electron in their highest energy level. Lithium atoms have the electron configuration $1s^2 2s^1$. All other Group-1 elements have next-to-highest energy levels containing eight electrons. An ion formed by removing the single-valence electron has the stable electron configuration of the preceding noble gas. For example, the sodium ion has the electron configuration $1s^2 2s^2 2p^6$. This matches the configuration of the neon atom. The electron configuration of the potassium ion, $1s^2 2s^2 2p^6 3s^2 3p^6$, matches the argon atom.

It is also possible to add an electron to a Group-1 atom to produce an anion having the electron pair in the *s* orbital of the highest energy level. An example is the Na^- anion which has the electron configuration $1s^2 2s^2 2p^6 3s^2$. Here the oxidation number of sodium is -1. Similar anions of potassium, rubidium, and cesium have also been produced.

Chemical Behavior

The very active free metals of the Sodium family are obtained by reducing the 1+ ions in their natural compounds. The metals are vigorous reducing agents. They have a weak attraction for their valence electrons.

Sodium makes up about 2.8% by mass of the earth's crust, potassium 2.6%, rubidium 0.009%, lithium 0.002%, and cesium 0.0002%.

SECTION OBJECTIVES

- Identify the alkali-metal elements.
- List some of the properties of the alkali-metal elements.
- Relate the properties of alkali metals to their characteristic metallic crystal lattice structures.
- Explain certain observed physical properties of the Group-1 elements in terms of their atomic properties.
- Explain why none of the alkali metals exist free in nature.

An alkali is a substance that forms a base and can neutralize an acid.

Figure 24-1 Body-centered cubic unit cell.

The Group-1 elements form hydroxides whose water solutions are extremely alkaline. Thus the Group-1 elements are called "alkali metals."

STUDY HINT
Crystalline structure is covered in Section 13.2

TABLE 24-1 PROPERTIES OF THE GROUP 1 METALS

	Li	Na	K	Rb	Cs	Fr
Atomic number	3	11	19	37	55	87
Atomic mass	6.941	22.989 77	39.0983	85.4678	132.905	223*
Electron Configuration	[He]$2s^1$	[Ne]$3s^1$	[Ar]$4s^1$	[Kr]$5s^1$	[Xe]$6s^1$	[Rn]$7s$
Melting point (°C)	180.5	97.8	63.65	38.89	28.5	27
Boiling point (°C)	1347	882.9	774	686	670	677
Density (g/cm³)	0.534	0.971	0.862	1.53	1.87	___
Ionization energy (kJ/mol)	520	496	419	403	376	___
Electronegativity	1.0	0.9	0.8	0.8	0.7	0.7
Atomic radius (pm)	152	186	227	248	265	270
Ionic radius (pm)	68	95	133	148	169	180
Oxidation number	+1	+1	+1	+1	+1	___
Crystal structure**	bcc	bcc	bcc	bcc	bcc	___
Hardness (Mohs scale)	0.6	0.4	0.5	0.3	0.2	___
Flame color	bright red	yellow	violet	purple	reddish violet	___

*atomic mass of most stable isotope   **body-centered cubic

STUDY HINT

You may wish to review ionization energies in Section 5.3

All alkali metals have relatively low first ionization energies, separated by a relatively large energy gap from the second ionization energy, as shown in Table 24-2. This suggests the presence of one easy-to-remove electron and accounts for the 1+ charge of ions formed by the Group-1 elements. Ionization energy decreases as the atom size increases going down the group. As shown in Figure 24-2, this decreasing energy with increasing size shows that it is easier to remove electrons from the highest energy levels of the heavier atoms. If you examine the net energy of reactions in which electrons are given up to other elements, however, you find that the lithium atom is the strongest reducing agent. Although it may be harder to remove an electron from the lithium atom, more energy is given back by the subsequent interaction of the lithium ion and its surroundings than by the larger alkali-metal ions with their surroundings.

Handling and storing the alkali metals is difficult because of their chemical behavior. They are usually stored in kerosene or some other liquid hydrocarbon because they react vigorously with water.

TABLE 24-2 FIRST AND SECOND IONIZATION ENERGIES AND THE SIZE OF ATOMS OF SELECTED GROUP-1 METALS

Element	Size of Atoms (pm)	First Ionization Energy (kJ/mol)	(eV)	Second Ionization Energy (kJ/mol)	(eV)
Li	152	520	5.34	7297	75.7
Na	186	496	5.14	4562	47.4
K	227	419	4.34	3071	31.9
Rb	248	403	4.18	2653	27.6
Cs	265	376	3.89	2423	25.1

Chemistry and Health

Group 1 Ions in Medication and Diet

The elements of Group 1 are important to the diet and in medicine because they form ionic compounds that usually dissolve well in water. If such ions are incorporated into large molecules, those molecules can then be dissolved and carried easily through the bloodstream. One example is sodium pentothal, an anesthetic popularly known as "truth serum."

Sodium and potassium ions are critical nutrients because they facilitate the transmission of nerve impulses and control the amount of water retained by cells.

Physicians often advise people with heart problems or high blood pressure to go on low-sodium diets to reduce the volume of water in their bodies. Diuretic drugs help to release excess water, but cause the loss of potas-

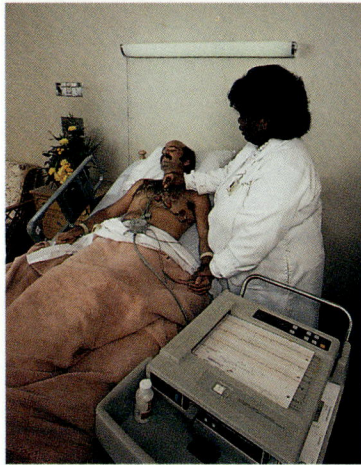

sium ions. Potassium-rich foods or supplements are sometimes needed to replace the loss.

The distribution of positively charged potassium ions and sodium ions and negatively charged organic ions across nerve cell membranes is an essential factor in the operation of the nervous system. This uneven distribution of ions creates a voltage across these membranes. When a nerve cell is stimulated, the ion concentration changes. This leads to a change in voltage across the nerve cell membrane. The resulting voltage change initiates the transmission of a nerve impulse.

In medical use, lithium carbonate is an ionic compound that is often prescribed to reduce nervous tremors, a symptom of Parkinson's disease. Frequently, it is referred to only as "lithium." Lithium carbonate is also the most effective drug for treating manic depressive disorder, a mental illness. The way it works is not well understood, but it is believed that lithium ions affect the way nerve cells communicate.

A small piece of potassium dropped into water, will react explosively, releasing H_2 to form a strongly basic hydroxide solution. The heat of reaction ignites the hydrogen that is produced.

$$2K(s) + 2H_2O(\ell) \rightarrow 2K^+(aq) + 2OH^-(aq) + H_2(g)$$

All of the ordinary compounds of the alkali metals are ionic, including their hydrides. Only lithium forms an oxide directly with oxygen. Sodium forms the peroxide instead, and the higher metals tend to form superoxides of the form $M^+O_2^-$. The ordinary oxides can be prepared indirectly, however. These oxides are basic anhydrides. They form hydroxides in water.

Nearly all the compounds of the alkali metals are quite soluble in water. The alkali-metal ions are colorless, have little tendency to hydrolyze in water solution, or to form polyatomic ions.

Alkali-metal peroxides and superoxides react with water to form oxygen gas and alkali-metal hydroxide solutions. This property is put to use in self-con-

A basic anhydride is a substance that will react with water to form a base.

Caution
The reaction of potassium with water is highly exothermic and can lead to an explosion. Do not attempt to use this reaction as an experiment or a demonstration.

The Metals of Groups 1 and 2   715

Figure 24-2 First ionization energies (shown in kJ/mol) decrease with increasing atomic radii for the atoms of Group-1 elements.

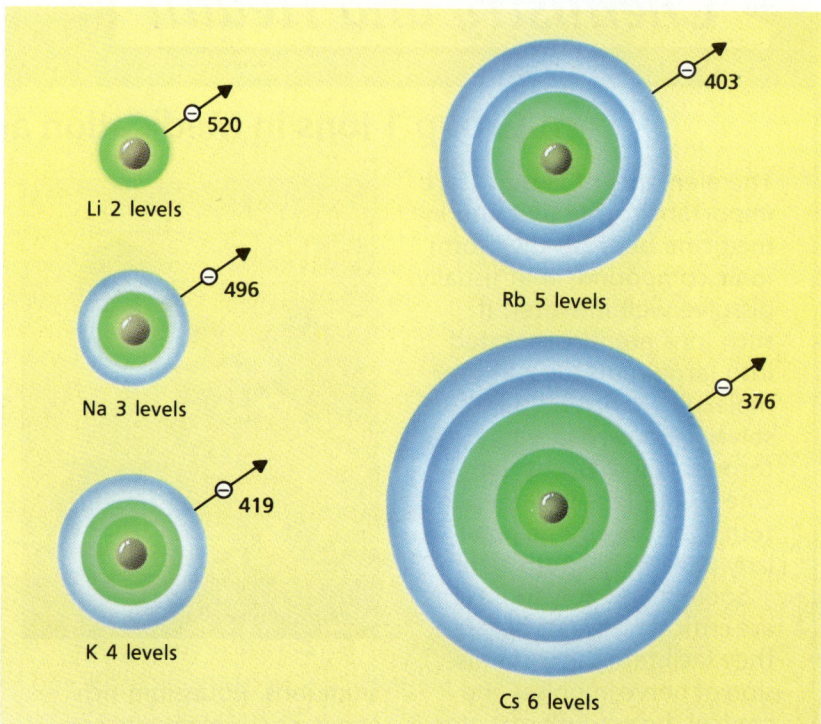

1. Lithium, sodium, potassium, rubidium, cesium, and francium
2. Silvery luster, good conductor of electricity and heat, ductile and malleable
3. Valance electrons form an electron "gas" that permeates the lattice structure. The mobility of the electrons accounts for the high thermal and electric conductivity, as well as the silvery luster. The softness, ductility, and malleability are explained by the binding force in the metallic crystal.
4. Ionization energy decreases as the atom size increases, making it easier to remove electrons from the highest energy levels of the heavier atoms. The outer energy-level electron is attracted less by the nuclear positive charge as the size of the atom gets larger.
5. They have a weak attraction for their valence electrons and therefore are easily oxidized to 1+ ions.

tained breathing appartus where breath is passed through a filter compartment containing KO_2. The moisture from the breath reacts with KO_2 to form oxygen gas. Carbon dioxide dissolves in the resulting hydroxide solution.

Compounds of the more important alkali metals are easily identified by flame tests. Flame tests are carried out by dipping a clean platinum wire into a solution containing a salt of the metal to be tested, then placing the wire in an oxidizing flame. Their compounds impart characteristic colors to a Bunsen flame. Sodium compounds color the flame yellow. Lithium compounds give a flame a red (carmine) color.

Potassium imparts a fleeting violet color to a Bunsen flame. The color comes from potassium atoms. The presence of sodium, however, masks the violet color of potassium in the flame. Potassium is detected in a mixture of sodium and potassium compounds by observing the colored flame through cobalt-blue glass. This glass filters out the yellow of the sodium flame, allowing the violet potassium flame to show clearly.

Section Review

1. Name the alkali metals.
2. Describe several properties of the alkali metals.
3. How are the properties of the Group 1 elements related to their characteristic metallic crystalline lattice structures?
4. Explain why it is easier to remove electrons from the heavier alkali atoms than it is from lithium.
5. Explain why none of the alkali metals exist in an uncombined state in nature.

24.2 The Alkaline-Earth Metals

The elements of Group 2 of the Periodic Table are beryllium, magnesium, calcium, strontium, barium, and radium and are known as the *alkaline-earth metals.*

Early chemists used the term "earth" for nonmetallic substances, such as the Group-2 metal oxides, that were practically insoluble in water and unchanged by strong heating. The Group-2 metal oxides give an alkaline reaction with water. Thus, these metal oxides have historically been called *alkaline earths,* and the metals *alkaline-earth metals.* Like the alkali metals, they are never found as free elements in nature. The metals must be recovered from their natural compounds. Many of their compounds are insoluble or slightly soluble and are found in the earth's crust. Their carbonates, phosphates, silicates, and sulfates are the most important deposits.

Beryllium and magnesium are commercially important light metals. In their chemical behavior, they resemble the corresponding alkali metals, lithium and sodium. Radium is important because it is radioactive. Radioactivity and other properties of radium are discussed in Chapter 29. The remaining three elements of Group 2—calcium, strontium, and barium—have similar properties. They are considered to be the typical alkaline-earth metals.

SECTION OBJECTIVES

- Identify the alkaline-earth metals and some of their sources.
- List some of the properties of the alkaline-earth metals.
- Name some alkaline-earth metal compounds and describe their properties.

Remind students that Group-1 metal oxides are all water soluble, forming bases in solution. They are also unchanged by strong heating.

One interesting use of beryllium is that it can be alloyed with copper to form a hard metal that is used for electrical contacts because it doesn't form sparks.

Structure and Properties

Each alkaline-earth element has two valence electrons beyond the stable configuration of the preceding noble gas. All form doubly charged ions of the M^{2+} type. Ions of the alkaline-earth metals thus have stable noble-gas structures. Their chemistry, like that of the alkali metals of Group 1, is generally uncomplicated.

The attraction between the metal ions and the electron "gas" of the alkaline-earth metal crystals is stronger than in the alkali metals. The alkaline-earth metals are therefore denser and harder than the corresponding alkali family metals, and they have higher melting and boiling points.

The atoms and ions of the alkaline-earth metals are smaller than those of the corresponding alkali metals because of their higher nuclear charge. For example, the magnesium ion (Mg^{2+}) has the same electron configuration as the sodium ion, Na^+. This configuration is $1s^2 2s^2 2p^6$. The nuclear charge of Mg^{2+} is 12+, and Na^+ has a nuclear charge of 11+. The higher nuclear charge of the Mg^{2+} ion attracts electrons more strongly and results in a smaller ionic radius. Some properties of Group-2 metals are listed in Table 24-3.

In Section 24.1 you learned that ionization energies of Group 1 metals decrease down the group as atomic size increases. This same relationship is seen in Group 2. The smaller atoms hold their outer electrons more securely than do the larger atoms.

Recall that ionization energy measures the tendency of an isolated atom to hold a valence electron. The first ionization energy relates to the removal of one electron from the neutral atom. The second ionization energy is that required to remove an electron from the ion with a 1+ charge (the atom that

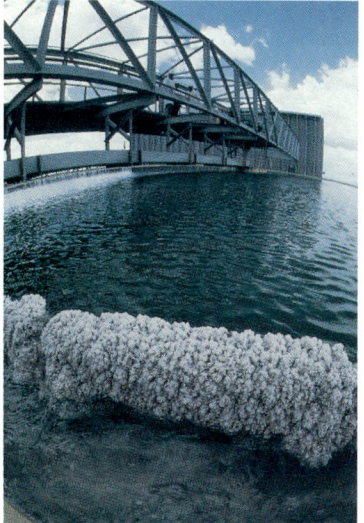

Figure 24-3 Beryllium processed in Utah. This metal produces alloys that are extremely elastic.

The alkaline-earth metals form 2+ ions that will have a greater attraction for the electron "gas" than the alkali metal 1+ ions.

The Metals of Groups 1 and 2 717

TABLE 24-3 PROPERTIES OF THE GROUP 2 METALS

	Be	Mg	Ca	Sr	Ba	Ra
Atomic number	4	12	20	38	56	88
Atomic mass	9.012 18	24.305	40.08	87.62	137.33	226.025
Electron configuration	[He]$2s^2$	[Ne]$3s^2$	[Ar]$4s^2$	[Kr]$5s^2$	[Xe]$6s^2$	[Rn]$7s^2$
Melting point (°C)	1280	649	839	769	725	700
Boiling point (°C)	2970	1090	1484	1384	1640	1140
Density (g/cm^3)	1.85	1.74	1.55	2.54	3.51	5 (?)
Atomic radius (pm)	111	160	197	215	217	220
Ionic radius (pm)	31	65	99	113	135	143
First ionization energy*	900	738	590	550	503	509
Second ionization energy*	1757	1451	1145	1058	958	979
Electronegativity	1.5	1.2	1.0	1.0	0.9	0.9
Oxidation number	+2	+2	+2	+2	+2	+2
Crystal structure**	hexagonal	hexagonal	fcc	fcc	fcc	—
Hardness (Mohs scale)	4	2.0	1.5	1.8	1.5	—
Flame Color	—	—	brick red	crimson	green	—

*(kJ/mol)   **face-centered cubic

has already had one valence electron removed). This second ionization energy is always higher than the first because the particle from which the electron is removed now has a positive charge.

If a third electron were to be removed from a Group-2 atom, it would come from the stable inner electron configuration (the noble-gas structure) of the 2+ ion. The predicted energy requirement would be much higher. In fact, the energies required for a third level of ionization of alkaline-earth elements are very high. They exceed the energies usually available in chemical reactions. The only forms observed are thus the 2+ ions of these elements. The first, second, and third ionization energies of the Group-2 elements are listed in Table 24-4.

Properties of Alkaline-Earth Metal Compounds

The alkaline-earth metals form hydrides, oxides or peroxides, and halides similar to those of the alkali metals. Almost all hydrides are ionic and contain H$^-$ ions. An exception is BeH$_2$. Binary compounds of beryllium have fairly

TABLE 24-4 IONIZATION ENERGIES OF THE GROUP-2 ATOMS

	First Ionization Energy (kJ/mol)	Second Ionization Energy (kJ/mol)	Third Ionization Energy (kJ/mol)
Be	900	1757	14849
Mg	738	1451	7733
Ca	590	1145	4912
Sr	550	1058	4205 (?)
Ba	503	958	3556 (?)
Ra	509	979	—

Chemistry and Health

Strontium: Fireworks and Fallout

Few people today have even heard of the element strontium (Sr), even though they have probably seen it in use many times. The bright red color in fireworks displays is due to the presence of strontium sulfate in the fireworks' fuel. Luminous paints sometimes contain strontium sulfide, which is prepared by heating strontium hydroxide with sulfur.

In the 1960s concern about the effects of a certain strontium isotope was a major political issue.

The isotope was released into the environment during atmospheric testing of nuclear weapons.

Natural strontium is a mixture of four stable isotopes. An unstable isotope, strontium-90, is produced by nuclear explosions.

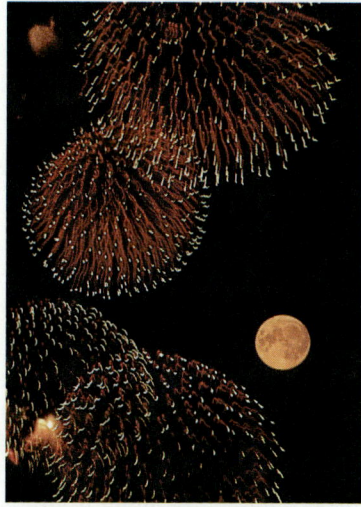

$^{235}_{92}U + n° \rightarrow\ ^{90}_{38}Sr + ^{143}_{54}Xe + 3n°$

Of the various components of nuclear fallout, the strontium-90 isotope presents the greatest hazard to people. When it enters the human body, strontium-90 is deposited in bones and can cause serious illnesses.

Concern about the possibility of strontium-90 finding its way into children's milk contributed to the Nuclear Test Ban Treaty of 1963, which prohibited atmospheric testing of nuclear weapons. Now, nuclear weapons tests are conducted deep underground.

---

strong covalent bonds and react with water to release hydrogen gas and form basic hydroxide solutions. Calcium hydride is often used as a laboratory source of hydrogen.

The oxides of the alkaline-earth metals have very high melting points. CaO and MgO are used as heat-resistant (refractory) materials. Beryllium oxide is amphoteric; otherwise, the oxides are basic. Recall, from Section 16.3, that any species that can react as either an acid or a base is *amphoteric*.

With the exception of beryllium oxide, the oxides are essentially ionic, but with more covalent character than the alkali-metal oxides. Strontium and barium form peroxides with oxygen, probably because of the large size of their ions.

The hydroxides are formed by adding water to the oxides. Except for $Be(OH)_2$, which is amphoteric, the hydroxides dissociate in water solution to yield OH^- ions. Hydroxides above barium are only slightly soluble in water, the solubility increasing with the metallic ion's size. Solutions of these hydroxides have low concentrations of OH^- ions.

The reactions are:
$BeH_2(s) + 2H_2O \rightarrow 2H_2(g) + Be(OH)_2$
$CaH_2(s) + 2H_2O \rightarrow 2H_2(g) + Ca(OH)_2$

Beryllium and its compounds are extremely toxic.

$Be(OH)_2$ can dissolve in both acid and base:
$Be(OH)_2(s) + 2H_3O^+(aq) \rightarrow Be^{2+}(aq) + 4H_2O$
$Be(OH)_2(s) + 2OH^-(aq) \rightarrow [Be(OH)_4]^{2-}(aq)$

The Metals of Groups 1 and 2

1. Beryllium, magnesium, calcium, strontium, barium, radium
2. In the earth's crust as carbonates, phosphates, silicates, and sulfates
3. Form ions with an +2 oxidation state. Denser and harder than the corresponding alkali metals and have higher melting and boiling points. Each atom bears a nuclear charge one greater than that of an atom of the corresponding alkali metal; thus, the two outer-level electrons in the alkaline-earth metals are held more firmly than the single outer-level electron of the alkali metals. The ionization energies decrease down the group as the atomic sizes increase.
4. Form hydrides, oxides or peroxides, and halides similar to those of the alkali metals. The oxides are essentially ionic but with more covalent character than the alkali-metal oxides.
5. Forms oxides that are amphoteric while the oxides of the other Group-2 metals are basic. Beryllium oxides are covalent while the others are mostly ionic in character. Unlike other Group-2 metal hydroxides beryllium hydroxide is not soluble and is amphoteric.

Figure 24-4 Elemental sodium is produced by the electrolysis of melted sodium chloride in a Downs cell. Chlorine is a by-product.

Section Review

1. Name the alkaline-earth metals.
2. Where are the alkaline-earth metals found in nature?
3. What are some of the properties of the alkaline-earth metals?
4. How does the reactivity of the alkaline-earth metals compare to that of the alkali metals?
5. How do the properties of beryllium compare to those of the other Group-2 metals?

24.3 Sodium: One of The Active Metals

Metallic sodium is never found free in nature. Compounds containing the Na^+ ion and sodium complexes are found in soil, natural waters, and in plants and animals, however. Because of the solubility of sodium compounds, it is almost impossible to find a sodium-free material. Vast quantities of sodium chloride are present in sea water and rock salt deposits. There are important deposits of sodium nitrate in Chile and Peru. The carbonates, sulfates, and borates of sodium are found in dry lake beds.

Preparation of Sodium

Sir Humphry Davy (1778–1829) prepared metallic sodium in 1807 by the electrolysis of fused sodium hydroxide. Today *sodium is prepared by the electrolysis of fused sodium chloride with an apparatus called the* **Downs cell** (Figure 24-4).

The process used in the Downs-cell apparatus requires the mixing of sodium chloride with calcium chloride. This results in the lowering of the melting point of the sodium chloride from 801°C to 580°C and in the simultaneous production of liquid sodium and chlorine gas. The liquid metallic sodium is collected under oil and is kept separate from the chlorine gas by an iron-gauze diaphragm.

The sodium is recovered by reducing Na^+ ions. This occurs at the negative electrode of the Downs cell. Each sodium ion acquires an electron from this electrode to form a neutral sodium atom.

$$Na^+ + e^- \rightarrow Na(\ell)$$

Chloride ions are oxidized at the positive electrode. Each chloride ion loses an electron to this electrode to form a neutral chlorine atom.

$$Cl^- \rightarrow Cl + e^-$$

But two chlorine atoms form the diatomic molecule of elemental chlorine gas. The net reaction at the positive electrode is

$$2Cl^- \rightarrow Cl_2(g) + 2e^-$$

In the overall cell reaction, electron-transfer balance is maintained between the electrodes. Two Na^+ ions are reduced for each Cl_2 molecule formed. The cell reaction can be written

$$\begin{aligned}\text{negative electrode: } & 2Na^+ + 2e^- \rightarrow 2Na(\ell) \\ \text{positive electrode: } & \underline{ 2Cl^- \rightarrow Cl_2(g) + 2e^-} \\ \text{cell: } & 2Na^+ + 2Cl^- \rightarrow 2Na(\ell) + Cl_2(g)\end{aligned}$$

Properties and Uses of Sodium

Sodium is a silvery-white, lustrous metal that tarnishes rapidly when exposed to air. It is very soft and has a lower density than water and a low melting point. A pellet of sodium dropped into water melts from the heat of the vigorous exothermic reaction that occurs. This reaction yields hydrogen gas and a strongly basic solution.

$$2Na(s) + 2H_2O(\ell) \rightarrow 2Na^+(aq) + 2OH^-(aq) + H_2(g)$$

When exposed to air, sodium unites with oxygen to form sodium peroxide (Na_2O_2). If sodium is supplied in excess, some sodium oxide (Na_2O) can be produced along with the bulk product, Na_2O_2. This production of Na_2O is a result of the strong reducing character of sodium atoms. Sodium oxide can also be formed by heating NaOH with sodium.

$$2NaOH + 2Na \xrightarrow{\Delta} 2Na_2O + H_2(g)$$

The superoxide of sodium, NaO_2, can be prepared indirectly. Superoxides have the O_2^- ions with one unpaired electron.

Sodium reacts with all aqueous acids. It burns in an atmosphere of chlorine gas, and directly forms sodium chloride.

As shown in Figure 24-5, a flame test of sodium compounds reveals a strong yellow color characteristic of vaporized sodium atoms. This yellow flame commonly identifies sodium. For comparison, Figure 24-5 also shows flame test for salts of other Group-1 elements.

Caution
Never add an alkali metal to water, an explosion can result.

Sodium is a good reducing agent because it loses its outer electron readily. It is easily oxidized to the 1+ ion, causing other substances to be reduced.

Figure 24-5 Flame tests of salts of *(left to right)* the elements lithium, potassium, sodium, rubidium.

Much of the sodium produced in the United States has been used in making tetraethyl lead, an antiknock additive for gasoline. This use is decreasing, however, as the production of leaded gasoline is being phased out to reduce the amount of lead in the atmosphere. Sodium is also used as a heat-transfer agent, in making dyes and organic compounds, and in sodium-vapor lamps.

Important Compounds of Sodium

Among the most important compounds of sodium are sodium chloride, sodium hydroxide, and sodium carbonate. Representative sodium compounds and their uses are shown in Table 24-5.

Sodium chloride   *Sodium chloride is found in sea water, in salt wells, and in deposits of* **rock salt,** *known as* **halite,** *which is mined in many places in the world.*

Pure sodium chloride is not deliquescent (tending to undergo gradual dissolution by the absorption of moisture from the air). Magnesium chloride is very deliquescent, however, and is usually present in sodium salt as an

Some nuclear power plants use liquid sodium as the agent to carry the heat generated in the reactor core to a secondary coil where the heat is used to turn water to steam to drive a turbine. Sodium's good thermal conductivity, large liquid temperature range, and high heat capacity make it ideal for this use.

Sodium chloride makes up 2.8% by mass of the ocean.

Figure 24-6 Native crystal formations of sodium chloride, known as halite, in Death Valley, California.

TABLE 24-5 REPRESENTATIVE SODIUM COMPOUNDS

Chemical Name	Common Name	Formula	Color	Uses
sodium tetraborate, decahydrate	borax	$Na_2B_4O_7 \cdot 10H_2O$	white	as a water softener; in making glass; as a flux
sodium carbonate, decahydrate	washing soda	$Na_2CO_3 \cdot 10H_2O$	white	as a water softener; in making glass
sodium hydrogen carbonate	baking soda	$NaHCO_3$	white	as a leavening agent in baking
sodium cyanide	prussiate of soda	$NaCN$	white	to destroy vermin; to extract gold from ores; in silver and gold plating; in case-hardening steel
sodium hydride	(none)	NaH	white	in cleaning scale and rust from steel forgings and castings
sodium nitrate	Chile saltpeter	$NaNO_3$	white or colorless	as a fertilizer
sodium peroxide	(none)	Na_2O_2	yellowish white	as an oxidizing and bleaching agent; as a source of oxygen
sodium phosphate, decahydrate	TSP	$Na_3PO_4 \cdot 10H_2O$	white	as a cleaning agent; as a water softener
sodium sulfate, decahydrate	Glauber's salt	$Na_2SO_4 \cdot 10H_2O$	white or colorless	in making glass; as a cathartic in medicine
sodium thiosulfate, pentahydrate	hypo	$Na_2S_2O_3 \cdot 5H_2O$	white or colorless	as a fixer in photography; as an antichlor
sodium sulfide	(none)	Na_2S	colorless	in the preparation of sulfur dyes; for dyeing cotton; to remove hair from hides

impurity. The presence of magnesium chloride explains why table salt becomes wet and sticky in damp weather.

Sodium chloride is essential in the diets of humans and animals and is present in certain body fluids. Perspiration contains considerable amounts of sodium chloride.

Sodium chloride is the cheapest compound of sodium and is used as a starting material in the production of most other sodium compounds. Some sodium compounds are most easily prepared directly from metallic sodium, however.

Sodium hydroxide Most commercial sodium hydroxide is produced by electrolysis of an aqueous sodium chloride solution. This electrolysis of aqueous NaCl is different from that of fused NaCl in the Downs cell. Chlorine gas is produced at the positive electrode, but hydrogen gas (instead of metallic

People sometimes put rice in salt shakers to absorb moisture in humid weather.

Commercial NaOH is called lye or caustic soda.

sodium) is produced at the negative electrode. The solution, meanwhile, becomes aqueous NaOH. Evidence indicates that water molecules acquire electrons from the negative electrode and are reduced to H_2 gas and OH^- ions.

The cell reaction for the electrolysis of an aqueous NaCl solution is

$$\begin{aligned}\text{electrode }(-) \quad & 2H_2O + 2e^- \rightarrow H_2(g) + 2OH^-(aq) \\ \text{electrode }(+) \quad & \underline{\; 2Cl^- \rightarrow Cl_2(g) + 2e^-} \\ \text{cell: } & 2H_2O + 2Cl^- \rightarrow H_2(g) + Cl_2(g) + 2OH^-(aq)\end{aligned}$$

As the Cl^- ion concentration diminishes, the OH^- ion concentration increases. The Na^+ ion concentration remains unchanged during electrolysis. Thus the solution is converted from aqueous NaCl to aqueous NaOH.

Sodium hydroxide, commonly known as **lye**, *converts some types of animal and vegetable matter into soluble materials by chemical action. It is very* **caustic**, *meaning that it has destructive effects on skin, hair, and wool.*

Sodium hydroxide is a white crystalline solid. It is marketed in the form of flakes, pellets, and sticks. It is very deliquescent and dissolves in the water that it removes from the air. It reacts with carbon dioxide from the air, producing sodium carbonate. Its water solution is strongly basic.

Sodium hydroxide reacts with fats, forming soap and glycerol. One of its important uses, therefore, is in making soap. It is also used in the production of rayon, cellulose film, and paper pulp, and in petroleum refining.

Some drain cleaners are made of sodium hydroxide and small pieces of aluminum. The alkali helps to dissolve grease, and the aluminum reacts with the NaOH solution to form a gas, causing agitation to help unblock the clogged drain.

Early pioneers obtained the alkali needed to make soap by saving wood ashes in a barrel open to rain. The rain water dissolved the alkali, mostly potassium hydroxide and potassium carbonate, and these would concentrate at the bottom of the barrel. The residue from the bottom would be cooked with animal fats to make soap.

Sodium carbonate and the Solvay process Almost all of the sodium carbonate produced in the United States comes from the mining of natural sodium carbonate deposits located in Wyoming or from the evaporation of brine from Searles Lake, California.

Sodium carbonate produced in the United States and much of that produced in the rest of the world is manufactured by the Solvay process. *The* **Solvay process** *uses brine (salty water) and limestone to produce sodium carbonate.* This process, a classic example of efficiency in chemical production, was developed in the 1860's by Ernest Solvay (1838–1922), a Belgian industrial chemist. We will describe the Solvay process here because it illustrates many chemical principles.

A flow diagram of the Solvay process is shown in Figure 24-7. The raw materials for the Solvay process are common salt, limestone, and coal. The salt is pumped as brine from salt wells. The thermal decomposition of limestone yields the carbon dioxide and calcium oxide needed in the process.

$$CaCO_3(s) \rightarrow CaO(s) + CO_2(g) \qquad \text{(Equation 1)}$$

Coal is converted into coke, gas, coal tar, and ammonia by destructive distillation. The coke and gas are used as fuel in the plant. The ammonia is also used in the process, and the coal tar is sold as a useful by-product.

To begin the process, a cold saturated solution of sodium chloride is further saturated with ammonia and carbon dioxide. The following reactions occur:

$$CO_2(g) + H_2O \rightarrow H_2CO_3$$

$$H_2CO_3 + NH_3(g) \rightarrow NH_4^+(aq) + HCO_3^-(aq)$$

$$\text{Net: } CO_2(g) + NH_3(g) + H_2O \rightarrow NH_4^+(aq) + HCO_3^-(aq)$$

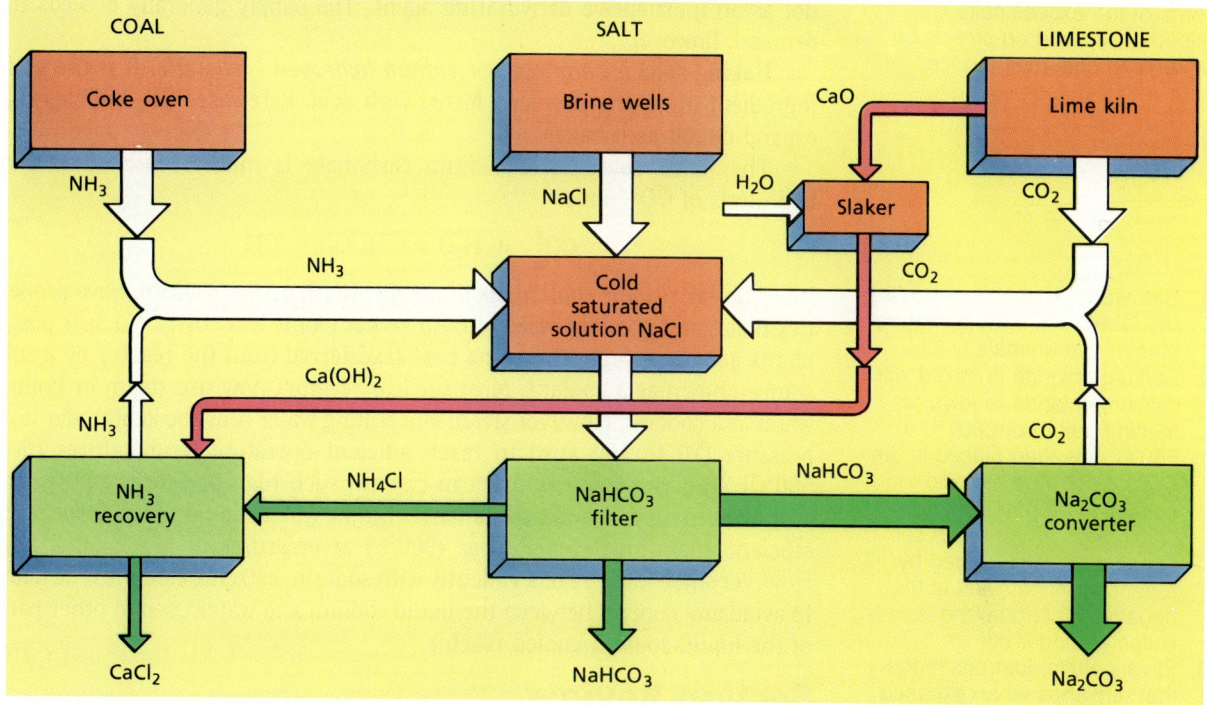

Figure 24-7 A flow diagram of the Solvay process.

The HCO_3^- ions form in a solution that has a high concentration of Na^+ ions. Because sodium hydrogen carbonate is only slightly soluble in this cold solution, it precipitates.

$$Na^+(aq) + HCO_3^-(aq) \rightarrow NaHCO_3(s)$$

The solution that remains contains NH_4^+ ions and Cl^- ions. The precipitated sodium hydrogen carbonate is filtered and dried. It is either sold as baking soda or converted into sodium carbonate by thermal decomposition.

$$2NaHCO_3(s) \rightarrow Na_2CO_3(s) + H_2O(g) + CO_2(g)$$

The dried sodium carbonate is an important industrial chemical called **soda ash**, which is used to manufacture a wide variety of materials including glass, chemicals, paper, and detergents.

The ammonia used in the process is more valuable than the sodium carbonate or sodium hydrogen carbonate. Hence, it must be recovered and used over again if the process is to be profitable.

The calcium oxide from Equation 1 is slaked, or crumbled, by adding water. Calcium hydroxide is formed in this reaction.

$$CaO(s) + H_2O(\ell) \rightarrow Ca(OH)_2(s)$$

The calcium hydroxide is added to the solution containing NH_4^+ ions and Cl^- ions to release the ammonia from the ammonium ion.

$$NH_4^+(aq) + OH^-(aq) \rightarrow NH_3(g) + H_2O(\ell)$$

The solution now contains mainly Ca^{2+} ions and Cl^- ions. Neither of these ions is recycled into the process. As a by-product, calcium chloride has some

Sodium carbonate is also called washing soda.

The Metals of Groups 1 and 2

Some of the excess calcium chloride is spread on streets on icy days to help melt the ice.

1. Sea water, rock salt deposits, dry lakes, soil, natural waters, plants, and animals
2. Sodium chloride is mixed with calcium chloride to lower its melting point from 801°C to 580°C. It is then placed in the Downs cell, and metallic sodium is prepared by the electrolysis of the fused salt. Liquid sodium is formed by reduction of Na$^+$ ions at the negative electrode and is collected under oil.
3. Silvery-white, lustrous metal that tarnishes when exposed to air; very soft, lower density than water, low melting point; unites with oxygen to form sodium peroxide when exposed to air; reacts vigorously with water to produce hydrogen gas and a strongly basic solution. Most sodium is used in making tetraethyl lead; also used in making dyes and organic compounds
4. (a) Found in sea water, salt wells, and in deposits of rock salt; can be prepared directly from metallic sodium and chlorine. Used as a starting material in the production of most other sodium compounds (b) Produced by electrolysis of an aqueous sodium chloride solution. Used in the manufacture of soaps, rayon, cellulose film, and paper pulp, and in petroleum refining (c) Mined from natural sodium carbonate deposits; often prepared by the Solvay process, most important use is the production of glass.

use as an inexpensive dehydrating agent. The supply generally exceeds the demand, however.

Baking soda *is composed of sodium hydrogen carbonate.* It is the main ingredient of baking powders. Mixed with acid, it releases CO_2 bubbles that expand dough as it bakes.

The water solution of sodium carbonate is mildly basic because of hydrolysis of CO_3^{2-} ions.

$$CO_3^{2-} + H_2O \rightleftharpoons HCO_3^- + OH^-$$

Nuclear reactor coolant alkali metals, in particular sodium, have properties that could possibly make nuclear power plants less costly. Nuclear power plants generate electricity using heat transferred from the reactor by a substance known as a coolant. Most nuclear reactors now use steam or boiling water as a coolant. However steam and boiling water must be kept under high pressure (70 to 150 atm) to reach efficient operating temperatures (over 500°C). Building reactors that can contain such high pressures adds to the cost of construction and maintenance. Unlike water, liquid sodium can reach efficient operating temperatures (900°C) at pressures of less than 2 atm. However since water reacts violently with sodium, extreme care must be taken to avoid any contact between the liquid sodium and water used in other parts of the liquid-sodium cooled reactor.

Section Review

1. Name several sources of sodium.
2. Describe the preparation of sodium by the electrolysis of sodium chloride in a Downs cell.
3. What are some of the properties and uses of the alkali metal sodium?
4. Give the preparation and uses of each of the following compounds: *(a)* sodium chloride   *(b)* sodium hydroxide   *(c)* sodium carbonate.

Chapter Summary

- Group-1 elements do not exist as free elements in nature; they are found only in natural compounds as monatomic positive ions.
- The elements of Group 1 in the Periodic Table are among the most chemically active metals.
- The elements of Group 1 are referred to collectively as the Sodium family and as the alkali metals.
- Group-1 elements are characterized by a single highest-energy-level electron added to the preceding noble-gas structure. These elements have relatively low ionization energies.
- Group-1 elements are soft, silvery-white metals that are good conductors of heat and electricity.
- The elements of Group 2 are the metals beryllium, magnesium, calcium, strontium, barium, and radium. They are known as the alkaline-earth metals.
- The alkaline-earth metals are characterized by two valence electrons beyond the preceding noble-gas structure. They are metals denser and harder than the alkali metals of Group 1 and have higher melting points.

Chapter 24 Review

The Metals of Groups 1 and 2

VOCABULARY

alkali metals (713)
alkaline-earth metals (717)
baking soda (726)
caustic (724)
Downs cell (720)
halite (722)
lye (724)
rock salt (722)
soda ash (725)
Solvay process (724)

REVIEWING CONCEPTS

1. Describe the flame tests for sodium and potassium. (24.1)
2. Describe:
   a. the metallic crystal lattice of the alkali metals
   b. the unit cell upon which this lattice is built. (24.1)
3. Why are sodium compounds more frequently used than other alkali compounds? (24.1)
4. Suppose you had a tremendous quantity of acid that had to be neutralized and that NaOH, KOH, and LiOH were all available at the same price per kilogram. Which of these three would you use? Why? (24.1)
5. Explain why potassium has a lower density than sodium, although it consists of heavier atoms. (24.1)
6. In the Solvay process, the reaction between sodium chloride and ammonium hydrogen carbonate runs to completion. Why? (24.3)
7. Why does a solution of sodium carbonate in water turn red litmus paper blue? (24.3)
8. Why does table salt become sticky in damp weather, although pure sodium chloride is not deliquescent? (24.3)
9. Identify:
   a. the raw materials used in the Solvay process
   b. the products and byproducts of the process. (24.3)
10. What leavening action occurs when molasses and baking soda are used in baking cookies? (24.3)
11. a. How is metallic sodium stored in the laboratory stockroom?
    b. Why must it be stored in this manner? (24.3)
12. Write three equations for the recovery of ammonia in the Solvay process. (24.3)

PROBLEMS

1. Given 1.50 kg of sodium nitrate and 1.50 kg of potassium chloride, how many grams of potassium nitrate can be produced by reacting these two substances?
2. If sodium carbonate decahydrate sells for 50 cents per kilogram, what is the anhydrous sodium carbonate worth per kilogram?
3. How many kilograms of sodium chloride are required to produce 1.00 metric ton of anhydrous sodium carbonate?
4. A solution is prepared by dissolving 1.1 g NaOH in water and diluting it to 500. mL. What is the pH of the solution?

CRITICAL THINKING

1. **Relating Ideas** A 0.1 M solution of $HC_2H_3O_2$ is found to have a pH of 2.9. A solution of 0.1 M $NaC_2H_3O_2$ is added to the acetic acid solution and the pH rises. Explain.

CROSS CONNECTIONS

1. **History** Research and write a report on the history and importance of fireworks in the Chinese culture.

RESEARCH AND WRITING

1. Investigate the advantages of hydrometallurgy over pyrometallurgy and make a short presentation to your class.

CHAPTER 25
The Transition Metals

PLANNING CHART

Approximate Time: 4/5 days

	SECTION/TEXT FEATURES		RESOURCES	
A V E R A G E	**25.1 THE TRANSITION ELEMENTS** Demos 1, 2, 3 Desktop Investigation *Periodic Table* Videodisc Section Review Chapter Review • Questions 1–4	729 728A–B 730 728E 736 749	*Teacher's Resource Binder* Lab Experiment 35 Section Review Worksheets	185 94
H O N O R S	**25.2 IRON, COBALT, AND NICKEL** Demos 4, 5 *Periodic Table* Videodisc Section Review Chapter Review • Question 5, Problems 1, 3–5	736 728B–C 728E 743 749	*Teacher's Resource Binder* Section Review Worksheets *Other Resources* Shakhashiri Video Demo 12 *World of Chemistry* Video 19 Blackline Master 59	95
H O N O R S	**25.3 COPPER, SILVER, AND GOLD** Demos 6, 7, 8 *Periodic Table* Videodisc Section Review Chapter Review • Questions 6–8, Problem 2	744 728C–D 728E 748 749	*Teacher's Resource Binder* Section Review Worksheets *Other Resources* Shakhashiri Video Demos 9, 13, 31 *World of Chemistry* Video 18 Blackline Master 59	96
	CHAPTER END ACTIVITIES Critical Thinking, Cross Connections, Research and Writing, Projects		*Teacher's Resource Binder* Test Booklet *Computer Test Generator*	125, 127

Teaching Strategies

25.1 The Transition Elements

It is helpful to students to develop a chart of the electron configurations of all the elements in Period 4. Point out that the transition metals are adding electrons to the next-to-highest energy level, close to the outermost level in energy. Mention the names of the transition elements and ask students if they know of substances formed from the various elements.

Compare the properties of the transition elements with those of the alkali metals and alkaline-earth metals. Emphasize the metallic properties of the transition elements: their shiny appearance, malleability, ductility, and good conduction of electricity. Show that all the metals have one or two electrons in their highest energy levels. Note the reduced reactivity of the transition elements. Explain their greater hardness and brittleness and higher melting and boiling points in terms of the covalent bonding resulting from the sharing of *d* electrons.

Map the melting points of the transition elements and show that they follow a trend—all have high melting points, increasing and then decreasing as electrons begin to half fill and then totally fill *d* orbitals. Show that these properties increase as one moves down the Periodic Table and that tungsten has the highest melting point of all. Point out the very high densities of osmium, iridium, and platinum. Have students calculate the mass of one cubic foot of osmium (634 kg or 1400 lb). Note the unusual behavior of mercury, the only metal that is a liquid at room temperature.

Define paramagnetism. Draw electron filling diagrams for iron, cobalt, and nickel to show the unpaired electrons in *d* orbitals.

Stress the horizontal similarities found in the transition metal groups.

Make a long-form version of the Periodic Table by cutting out the lanthanide and actinide series and pasting it in between Groups 2 and 3. Use this to discuss the electron configurations of the lanthanide and actinide series.

Discuss various oxidation states of the transition metals. Manganese can have oxidation states of +2, +3, +4, +6, and +7. Write out the electron configurations for these. Note that elements beyond manganese seldom have higher oxidation states.

Display samples of transition metal compounds that show a range of colors. Include compounds with various oxidation states of the same element, such as $MnSO_4$, MnO_2, and $KMnO_4$, and point out that the color is related to the oxidation state.

Carry out some demonstrations involving complex ion formation. Write the equations for the reactions and analyze the formulas to show the metal ion and the complexing ions. Determine the charge on the complex ion from the charges on the metal ion and the complexing groups.

Demonstration 1: Paramagnetism of Transition Metal Compounds

PURPOSE: To demonstrate that many transition metal compounds are paramagnetic

MATERIALS: Pharmaceutical gelatin capsules, thread, strong magnet, soda-straw balance or any other sensitive balance, transition metal compounds such as $MnSO_4$, $FeSO_4$, $CoSO_4$, MnO_2, Mn, and nontransition compounds such as NaCl, $CaCO_3$

PROCEDURE: Fill gelatin capsules with the compounds to be tested and suspend them from the balance. Bring the magnet under them and observe that the paramagnetic substances are pulled downward into the magnet, while the diamagnetic substances are not. Compare the effect of the magnet on substances with various numbers of unpaired electrons. For example, compare $MnSO_4$, which has five unpaired electrons, with MnO_2, which has three.

Demonstration 2: Colored Objects Absorb Light

PURPOSE: To demonstrate that colored objects reflect the color that is observed and absorb their complementary colors

MATERIALS: Flashlight, colored filters (colored cellophane, probably available from physics lab), construction paper in white and in various colors

PROCEDURE: In a darkened room, shine white light on white paper. Observe that white is reflected. Put a blue filter over a flashlight and shine on white paper. Now blue is reflected. Shine blue filter light on blue paper and again blue is reflected. Shine blue filtered light on yellow (the complement of blue) paper, and the paper appears black, showing that yellow absorbs the blue light. Reverse the colors and try with other combinations of complementary colors.

Demonstration 3: Dehydration of Copper(II) Sulfate Pentahydrate Crystals

PURPOSE: To show the color change of copper(II) sulfate pentahydrate crystals when dehydrated and then hydrated

MATERIALS: Fine crystals of copper(II) sulfate pentahydrate, 250-mL beaker, watch glass, hot plate

PROCEDURE: Place one to two grams of copper(II) sulfate pentahydrate crystals in the beaker. Cover with watch glass. Place on hot plate on medium setting. Observe water droplets on watch glass as crystals lose their water of hydration. Remove watch glass and continue to heat until the colorless anhydrous crystals are formed. Allow to cool. Add a few drops of water and observe the reappearance of the blue color as the crystals absorb water. Also note that heat is evolved as the hydration occurs.

Desktop Investigation: Paramagnetism of Nickel

The colorful, single displacement (replacement) reaction described in this Desktop Investigation is quite fascinating, especially when students observe that the "dust particles" formed respond to a magnet.

Many students believe that iron is the only magnetic substance. Three common metals, however, are magnetic: iron, cobalt, and nickel. Nickel-copper alloys were used in coins for many years because of the low cost of nickel and the durability of copper. However, the United States has decreased the amount of nickel in coins enough so that the "nickel" no longer has enough of that metal in it to respond to a magnetic field. Remember that nickel compounds are toxic, especially when inhaled as dust. The nickel metal formed may be disposed of in a hazardous waste landfill. Keep excess nickel nitrate solution for use next year, DO NOT POUR DOWN DRAIN.

ANSWERS TO DISCUSSION QUESTIONS
1. Nothing happened with nickel(II) nitrate or aluminum near the magnet.
2. The nickel dust responded to the magnetic field.
3. A single displacement (or replacement) reaction took place. The aluminum displaced the nickel from the compound.
4. No, the United States has decreased the amount of nickel in five-cent coins. Some older coins are attracted by magnets.

25.2 Iron, Cobalt, and Nickel

Point out the location of the metals iron, cobalt, and nickel in the Periodic Table and call attention to the horizontal similarities. Write out their electron configurations and compare the properties of the elements. Discuss the fact that iron can form a 3+ ion readily, leaving it with a half-filled $3d$ electron sublevel. Show the electron configurations of the 2+ ions of iron, cobalt, and nickel.

Show samples of the three metals and some of their alloys. Discuss ferromagnetism and the domain explanation. Test the metals and their alloys for attraction to a magnet.

Discuss the abundance of iron. List the various iron ores and their formulas. Give the various equations for the reactions in the blast furnace, pointing out the oxidation and reduction reactions that occur.

Emphasize steel production as a process in which impurities in blast furnace iron are removed by oxidation. The sources of oxygen are the elemental gas or the oxygen in the iron ore. Give the equations for the oxidation process and for slag formation.

Exhibit the oxides of iron and discuss their uses. Show a sample of $FeSO_4 \cdot 7H_2O$. Write the equation for its oxidation and explain how the addition of sulfuric acid and pieces of iron keep the Fe^{2+} in the reduced state.

Show a sample of $FeCl_3 \cdot 6H_2O$ and discuss its uses. Explain the equations for the hydrolysis of Fe^{3+} ions.

Review the complex ions $Fe(CN)_6^{4-}$ and $Fe(CN)_6^{3-}$, pointing out the difference in oxidation states of the iron. Show solutions of $K_4Fe(CN)_6$ and of $K_3Fe(CN)_6$. Explain the tests for the presence of the Fe^{2+} and the Fe^{3+} ions.

Demonstration 4: Rusting of Iron

PURPOSE: To discover the source of oxygen in rusting of iron
MATERIALS: Test tube, beaker, steel wool, water
PROCEDURE: Put a loose wad of steel wool into the bottom of a test tube. Moisten the steel wool with water. Place the test tube

The Transition Metals

upside down in a beaker containing some water. Note that the test tube will have the steel wool at the top and will be filled almost entirely with air that is trapped by the water in the beaker. After a day or two, check to see that rust has formed on the steel wool. Did the oxygen needed to form the rust come from the water or from the air? The raised level of water inside the test tube indicates that the air provided the oxygen to form the rust.

Demonstration 5: Finding Iron in Cereal
PURPOSE: To show the presence of iron in iron-enriched cereals
MATERIALS: Iron-enriched cereal, beaker, magnetic stirrer and stirring magnet, water
PROCEDURE: Add water to some iron-fortified cereal and stir until the cereal becomes soggy. Place a magnetic stirring bar in the cereal, and stir on a magnetic stirrer for 15-30 minutes. Remove the stirring bar and note the slivers of metallic iron covering the magnet. Discuss the reaction of metallic iron with the hydrochloric acid in the stomach to produce Fe^{2+} ions.

25.3 Copper, Silver, and Gold

Discuss the fact that the metals of the Copper family are not easily oxidized and are among the few metals that occur in the free state. Exhibit samples of the metals and of their ores.

Discuss the steps in copper metallurgy and write the equations for the reactions that occur.

Point out that Cu^{2+} compounds are the most important copper compounds. Show samples of Cu^{2+} compounds such as $CuSO_4 \cdot 5H_2O$, CuO, and $CuCO_3$. Discuss the hydrolysis of Cu^{2+} in water, and test the pH of water solutions of Cu^{2+}. Ask students to list the uses of copper. Explain the tests to show the presence of the Cu^{2+} ion.

Demonstration 6: Copper Plating
PURPOSE: To demonstrate plating of copper from solution
MATERIALS: Solution of $CuSO_4$, carbon electrodes, small nine-volt battery, beaker, wires and clips
PROCEDURE: Place the electrodes in the beaker with the $CuSO_4$ solution. Connect them to the terminals of the battery. Observe the formation of metallic copper on the cathode. Write the equation for the reduction of Cu^{2+} ions. Try using other substances for electrodes.

Demonstration 7: Floating Pennies
PURPOSE: To show the relative reactivity of copper and zinc
MATERIALS: Pennies minted after 1982, hydrochloric acid ($6M$), sharp file, beaker
PROCEDURE: Wear goggles. Use the file to make a sharp scratch in the edge of a penny on two opposite sides. The scratch must cut through the copper coating and expose the zinc layer underneath. Place several of these pennies in the hydrochloric acid and allow to react overnight. The copper will not react with the acid, but the zinc will, forming hydrogen gas and zinc chloride. The bubbles of hydrogen gas will collect inside the pennies and cause them to float. This is a demonstration that is a real attention getter!

Demonstration 8: Changing Pennies to Silver and Gold.
PURPOSE: To show the formation of a coating of zinc on copper, and the formation of a bronze alloy of copper and zinc, making pennies appear to be "silver" and "gold"
MATERIALS: Pennies minted before 1982, $6M$ NaOH solution, granular zinc, evaporating dish, hot plate, steel wool, crucible tongs, burner.
PROCEDURE: Wear goggles. Place about 5 g of granular zinc in an evaporating dish. Add enough $6M$ NaOH solution to cover the zinc and fill the dish about one-third full. Heat on a hot plate until near boiling. Clean a copper penny with steel wool. Hold the penny with crucible tongs by opposite edges in the dish for 3-4 minutes, until it appears "silver" in color as a zinc coating forms. Remove it, rinse with water, and blot dry. Prepare several of these "silver" pennies. Hold one of the "silver" pennies by opposite edges with crucible tongs in the flame of a burner. A gold color forms almost immediately as the zinc coating alloys with the copper. Remove the coin after 3-5 seconds, and wash and dry it. These pennies make great awards for students!

Dispose of Zn-NaOH solution by neutralizing it with acid and pouring it down the drain with large amounts of water. Spontaneous fires have occurred when a Zn-OH mixture was discarded in the trash with paper.

References and Resources

Books and Periodicals

Bonatti, Enrico. "Sea-floor Ore," *Scientific American*, Vol. 240, 1979, p. 95.

Bonatti, Enrico. "Chemistry of Iron, Raw Materials, and Steel," American Iron and Steel Institute, 150 West 42nd St., New York, NY 10017. Copies are provided free to schools.

Maddin, Robert, James D. Muhly, and Tamara S. Wheeler. "How the Iron Age Began," *Scientific American*, Vol. 237, 1977, p. 122.

Shakhashiri, Bassam Z., Glen E. Dureen, and Lloyd G. Williams. "Paramagnetism and Color of Liquid Oxygen; A Lecture Demonstration," *Journal of Chemical Education*, Vol. 57, 1980, p. 363.

Summerlin, Lee R., and James L. Ealy Jr. "The Silver Mirror Reaction," *Chemical Demonstrations, A Sourcebook for Teachers*, Vol. 1, American Chemical Society, 1985 p. 91.

Summerlin, Lee R., and James L. Ealy Jr. "Oxidation States of Manganese: Mn^{7+}, Mn^{6+}, Mn^{4+}, and Mn^{2+}," *Chemical Demonstrations, A Sourcebook for Teachers*, Vol. 1, American Chemical Society, 1985, p. 99.

Summerlin, Lee R., and James L. Ealy Jr. "Oxidation States of Vanadium: Reduction of V^{5+} to V^{2+}," *Chemical Demonstrations, A Sourcebook for Teachers*, Vol. 1, American Chemical Society, 1985, p. 108.

Summerlin, Lee R., Christie L. Borgford, and James L. Ealy, Jr. "Transition Metals and Complex Ions," *Chemical Demonstra-*

tions, *A Sourcebook for Teachers*, Vol. 2, American Chemical Society, 1987, pp. 69–86.

Viswanadham, Puligandla. "An Inexpensive Guoy Balance for Magnetic Susceptibility Determination," *Journal of Chemical Education*, Vol. 55, 1978, p. 54.

Audiovisual Resources

"Steel from Inland": 16-mm film or video, #13878, free-loan film from Modern Talking Picture Service. Shows how iron ore, coal, and limestone become steel

"Mining and the Environment": 16-mm film or videocassette, #16269, Free-Loan Film from Modern Talking Picture Service, 5000 Park Street North, St. Petersburg, FL 33709. Shows the operation of a copper and silver mine in Montana

"Vanadium—A Transition Element": CHEM Study 16-mm film, Ward's Modern Learning Aids. Shows different oxidation states of vanadium, electronic structure, and complex ion formation

Computer Software

"The Periodic Table Videodisc: Reactions of the Elements": CAV type videodisc, Journal of Chemical Education Software, 1989. Shows action sequences and still shots of each element in its most common forms; reactions with air, water, acids, and bases; some common applications

Answers and Solutions

Reviewing Concepts

1. The metallic character of transition elements is attributed to the small number of electrons in the outermost energy sublevel of their atoms.
2. Yellow, the complement of blue light
3. A polyatomic ion is a charged structure made up of two or more atoms covalently bonded. A complex ion is a charged structure composed of a central metal ion combined with a specific number of polar molecules or ions.
4. **a.** 0 **b.** +4 **c.** +2 **d.** +7 **e.** +6
5. **a.** It is reduced. **b.** It is purified.
6. All are found as the native metal. They are soft, dense, and inactive metals. They all have an oxidation state of +1 in some compounds.
7. In moist air, copper forms a basic copper carbonate coating. Sulfur dioxide in the air may also combine with copper, forming a green basic copper sulfate.
8. Enough gold and silver are recovered to pay for the process.

Problems

1. Formula mass of limonite = 106.9 g/mol

   $\dfrac{55.85 \text{ g Fe}}{106.9 \text{ g limonite}} \times 100\% = 52.25\%$ Fe

2. Each Ag atom forms 1 formula unit of $AgNO_3$.

   $100.\text{g Ag} \times \dfrac{1 \text{ mol Ag}}{108 \text{ g Ag}} \times \dfrac{1 \text{ mol AgNO}_3}{1 \text{ mol Ag}}$

   $\times \dfrac{170. \text{ g AgNO}_3}{\text{mol AgNO}_3} = 157 \text{ g AgNO}_3$

3. **a.** $\dfrac{112 \text{ g Fe}}{160. \text{ g Fe}_2\text{O}_3} \times 87.0\% \text{ Fe}_2\text{O}_3 = 60.9\%$ Fe

   b. $1.0 \times 10^6 \text{ t} \times 0.04 = 4.0 \times 10^4$ t

   c. Each mole of $CaCO_3$ yields 1 mol of CaO and each mole of CaO reacts with 1 mol of SiO_2.
   $(1.0 \times 10^6 \text{ t})(0.08) = 8 \times 10^4 \text{ t SiO}_2$

   $8.0 \times 10^4 \text{ t SiO}_2 \times \dfrac{10^6 \text{ g}}{\text{t}} \times \dfrac{1 \text{ mol SiO}_2}{60. \text{ g SiO}_2}$

   $\times \dfrac{1 \text{ mol CaCO}_3}{1 \text{ mol SiO}_2} \times \dfrac{100. \text{ g CaCO}_3}{\text{mol CaCO}_3} \times \dfrac{1 \text{ t}}{10^6 \text{ g}}$

   $= 1.3 \times 10^5 \text{ t CaCO}_3$

Critical Thinking

1. Iron(II) ions gradually oxidize to the +3 state, and the blue color characteristic of Fe ions present in two different oxidation states in the $KFeFe(CN)_6 \cdot H_2O$ product develops.
2. $K_4Fe(CN)_6$ forms the blue precipitate $KFeFe(CN)_6 \cdot H_2O$ with iron(III) ions. $K_3Fe(CN)_6$ forms the same blue precipitate with iron(II) ions.

Videodisc Selections
from *The Periodic Table Videodisc* (JCE Software)

step forward

step reverse

CHAPTER 25

pg 729 Transition elements and reactions

20739
scandium (s)

20754
titanium (s)

20759
titanium + air (m)

36592
uses of vanadium (s)

21231
chromium (s)

21242
chromium + air (m)

22055
chromium + acids (m)

36598
uses of chromium (s)

22760
manganese (s)

22764
manganese + air (m)

23281
manganese + acids (m)

36604
uses of manganese (s)

36609
uses of manganese (s)

36616
uses of manganese (s)

36622
uses of manganese (s)

30270
zinc (s)

30274
zinc + air (m)

31167
zinc + acids (m)

728E   Chapter 25   Teacher's Notes

36682
uses of zinc (s)

36700
uses of zinc (s)

39162
molybdenum (s)

36832
uses of molybdenum (s)

36838
uses of molybdenum (s)

39821
cadmium (s)

39826
cadmium + air (m)

40532
cadmium + acids (m)

36886
uses of cadmium (s)

49853
tungsten (s)

37090
uses of tungsten (s)

49863
platinum (s)

37132
uses of platinum (s)

37144
uses of platinum (s)

49888
mercury (s)

49893
mercury + air (s)

49899
mercury + acids (m)

37174
uses of mercury (s)

37180
uses of mercury (s)

pg 736 Iron, cobalt, and nickel

24079
iron (s)

The Transition Metals **728F**

24084
iron + air (m)

24981
iron + acids (m)

36634
uses of iron (s)

36640
uses of iron (s)

36628
uses of iron (s)

25876
cobalt (s)

25881
cobalt + air (m)

26677
cobalt + acids (m)

36646
uses of cobalt (s)

27399
nickel (s)

27411
nickel + air (m)

28046
nickel + acids (m)

36652
uses of nickel (s)

36658
uses of nickel (s)

pg 744 Copper, silver, and gold

28534
copper (s)

28546
copper + air (m)

29501
copper + acids (m)

36664
uses of copper (s)

39192
silver (s)

36868
uses of silver (s)

49875
gold (s)

37149
uses of gold (s)

37162
uses of gold (s)

37156 (s)
uses of gold

Teacher's Notes

Chapter 25
The Transition Metals

SECTION PREVIEW
25.1 The Transition Elements
25.2 Iron, Cobalt, and Nickel
25.3 Copper, Silver, and Gold

Transition metals such as gold, silver, copper, nickel, and zinc are used by many nations in minting coins.

25.1 The Transition Elements

Beginning with Period 4, ten subgroups of elements intervene between Group 2 and Group 13 and are called the **transition elements.** As the atomic number of the transition elements increases, there is an increase in the number of electrons in the next-to-highest energy level. This inner building of electronic structure among the transition elements can be interpreted as an interruption in the regular increase of highest-energy-level electrons from Group 1 to Group 18 across a period.

The transition elements (Figure 25-1), include gold, silver, and platinum, some of the most important elements in valuable jewelry. This category of elements also includes the iron used as the structural backbone of our buildings and bridges and the copper in millions of miles of wire bringing electricity to every part of the country.

SECTION OBJECTIVES
- Identify the transition elements on the Periodic Table.
- Describe the general properties of the transition elements.
- Describe ways in which transition elements differ from the alkali and alkaline-earth metals.
- Name the most common oxidation states of the first transition series.

Transition Elements

Group 2	Group 3	Group 4	Group 5	Group 6	Group 7	Group 8	Group 9	Group 10	Group 11	Group 12	Group 13
12 Mg 24.305											13 Al 26.98154
20 Ca 40.08	21 Sc 44.956	22 Ti 47.88	23 V 50.9415	24 Cr 51.996	25 Mn 54.938	26 Fe 55.847	27 Co 58.9332	28 Ni 58.69	29 Cu 63.546	30 Zn 65.39	31 Ga 69.72
38 Sr 87.62	39 Y 88.906	40 Zr 91.22	41 Nb 92.906	42 Mo 95.94	43 Tc (98)	44 Ru 101.07	45 Rh 102.906	46 Pd 106.42	47 Ag 107.868	48 Cd 112.41	49 In 114.82
56 Ba 137.33	57 La 138.906	72 Hf 178.49	73 Ta 180.948	74 W 183.85	75 Re 186.207	76 Os 190.2	77 Ir 192.22	78 Pt 195.08	79 Au 196.967	80 Hg 200.59	81 Tl 204.383
88 Ra 226.025	89 Ac 227.028	104 Unq (261)	105 Unp (262)	106 Unh (263)	107 Uns (262)	108 Uno (265)	109 Une (266?)				

Remember that the periods of the Periodic Table are horizontal rows; groups are vertical columns.

Figure 25-1 Transition metals are characterized by the buildup of *d*–sublevel electrons in the next-to-highest energy level.

General Properties of the Transition Elements

As you might expect, the properties of the elements remain similar going down each main group. Going horizontally across Periods 1–3 from Group 1 to Group 17, the properties change steadily from metallic to nonmetallic. In the transition region, however, horizontal similarities may be as great as (and in some cases greater than) similarities going down a subgroup.

All transition elements are metals. The metallic character is related to the presence of no more than one or two electrons in the highest energy levels of their atoms. These metals are generally harder and more brittle than the Group 1 and Group 2 metals. Their hardness is partially due to the covalent bonding resulting from the sharing of *d* electrons. Their melting points are higher than those of the Group 1 and Group 2 metals. Mercury (atomic number 80) is a familiar exception. It is the only metal that is liquid at room temperature. The increase in the melting points of the elements up through vanadium and chromium may be attributed to additional covalent bond formation resulting from the larger number of unpaired electrons. The transition metals are good conductors of heat and electricity. Such conduction is characteristic of a metallic crystal lattice permeated by an electron gas.

Tungsten, with a melting point of 3410°C, has the highest melting point of all the metals. The densest metal is osmium, which has a density of 22.4 g/cm³.

The only other element that is a liquid at room temperature is bromine.

Desktop Investigation

Paramagnetism of Nickel

Question
Can you demonstrate the magnetic character of a fourth-period transition element?

Materials
nickel(II) nitrate
aluminum foil
distilled water
five-cent coins of various ages
magnet
100-mL beaker
scissors
steel wool

Procedure
Record your results in a data table.
1. Test the solid nickel(II) nitrate and the aluminum foil for any response to a magnetic field.
2. Dissolve 2–3 g of nickel(II) nitrate in 10 mL of distilled water.
3. Cut a piece of aluminum foil about 4 cm × 4 cm; gently buff it with steel wool to expose a fresh surface.
4. Set the foil in the nickel(II) solution overnight.
5. The solid formed is nickel metal. Test its response to a magnetic field by moving the beaker over a magnet.
6. Test five-cent coins for their response to a magnet.
7. Your teacher will tell you how to dispose of your chemicals.

Discussion
1. What happened when the solid nickel(II) nitrate and aluminum were placed near the magnet?
2. What happened when the newly formed nickel was brought near the magnet?
3. What general type of chemical reaction took place?
4. Is there very much nickel in a five-cent coin? Are there differences between coins of different ages?

The chemical properties of transition elements are varied. Electrons of the next-to-highest d sublevel as well as those of the highest energy level may become involved in compound formation. Most transition metals exhibit variable oxidation states in forming compounds, and most of their compounds have color. They have a pronounced tendency to form complex ions. Many of their compounds are attracted into a magnetic field (a property called paramagnetism).

Transition metal similarities The number of electrons in the highest energy level of the transition elements remains fairly constant. It never exceeds two electrons as the number of electrons in the d sublevel increases across a period. Similarities often appear in a horizontal sequence of transition elements as well as vertically through a group. In some cases, horizontal similarity may exceed vertical similarity. The first five groups (Groups 3—7) are those headed by Sc, Ti, V, Cr, and Mn. Generally, similarities among these elements are recognized within each group. The next three groups (Groups 8—10) are headed by Fe, Co, and Ni. Here the similarities across each period are greater than those down each group. Iron, cobalt, and nickel resemble each other more than do iron, ruthenium, and osmium, the members of Group 8. The final two groups are headed by Cu and Zn. Here again, the greatest similarity is within each group.

Transition Metal Groups

The electron configuration of each transition element is shown in the Periodic Table inside the back cover. Compare the electron configuration of the Group 2 metals with those of the corresponding metals of the first transition group, the scandium (Sc) group. What differences do you observe? Now compare the electron configurations of elements across the first transition series.

There are minor irregularities in the $3d$-electron buildup at Cr (atomic number 24) and at Cu (atomic number 29). These apparent irregularities and those in the other periods of the transition elements are attributed to the extra stability associated with half-filled and completely filled sublevels.

The transition elements occupy the long periods in the Periodic Table. The choice of the first and last elements in the series depends somewhat on the definition used. Based on d- and s-electron configurations, the transition elements consist of the ten groups between Group 2 and Group 13. The elements of the first transition series are the most important, and they are the most abundant. Refer also to Appendix Table A-4 to see the electron arrangements of the transition elements.

Figure 25-2 These nodules, found on the floor of the Pacific Ocean, contain nickel, copper, cobalt and manganese. They might prove to be an important source of strategic transition metals.

Variable Oxidation States

Variable oxidation states are common among the transition metals. Some form different compounds in which they exhibit several oxidation states. The energies of the outermost d and s electrons do not differ greatly. The energies to remove these electrons are relatively low.

The d sublevel has five available orbitals that can hold ten electrons. The s sublevel has one orbital, and it can hold two electrons. Electrons occupy the d orbitals singly as long as unoccupied orbitals of similar energy exist. The s electrons and one or more d electrons can be used in bonding.

In metals of the first transition series, several $4s$ and $3d$ electrons may be transferred to or shared with other substances. Thus, several oxidation states become possible. Remember that the maximum oxidation state is limited by the total number of $4s$ and $3d$ electrons present.

Table 25-1 gives the metals' common oxidation states and $3d$- and $4s$-electron configurations. It uses orbital notations introduced in Section 4.3. In these notations __ represents an unoccupied orbital. The symbol ↑ represents an orbital occupied by one electron, and ↑↓ represents an orbital with an electron pair.

Observe that the maximum oxidation state increases to +7 for manganese and then decreases abruptly beyond manganese. The difficulty in forming the higher oxidation states increases toward the end of the row

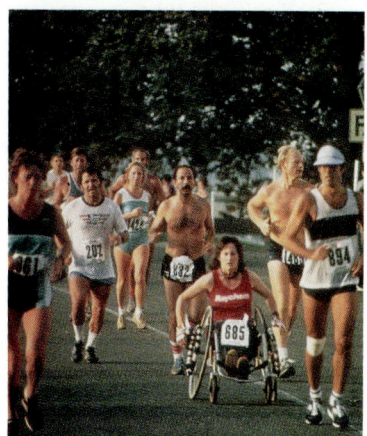

Figure 25-3 Chromium, magnesium, and nickel are ingredients of stainless steel, used to make durable rugged items such as wheelchairs.

Figure 25-4 "Setting out" the Cathedral of St. John the Divine. Zinc templates are used to trace the shape of each stone onto rough limestone blocks.

because of the general increase in ionization energy with atomic number. The higher oxidation states generally involve covalent bonding.

Manganese atoms lose two $4s$ electrons to become Mn^{2+} ions. Higher oxidation states involve $3d$ electrons. In the permanganate ion, MnO_4^-, manganese is in the $+7$ oxidation state, covalently bonded with the oxygen atoms.

When the electron configuration includes both $3d$- and $4s$-valence electrons, the $3d$ electrons are often lower in energy than the $4s$ electrons. Thus the first electron removed in an ionizing reaction is the one most loosely held, a $4s$ electron. This fact is illustrated by the step-by-step ionization of titanium. The ground-state configurations of the valence electrons are:

$$\begin{array}{ll} Ti & 3d^2 4s^2 \\ Ti^+ & 3d^2 4s^1 \\ Ti^{2+} & 3d^2 4s^0 \\ Ti^{3+} & 3d^1 \\ Ti^{4+} & 3d^0 \end{array}$$

Figure 25-5 Rhodochrosite is a rose-red mineral that consists primarily of manganese carbonate ($MnCO_3$).

TABLE 25-1 OXIDATION STATES OF FIRST-ROW TRANSITION ELEMENTS

Element	Electron Configuration $1s^2 2s^2 2p^6 3s^2 3p^6$	3d	4s	Common Oxidation States
scandium	Each transition	↑ _ _ _ _	↑↓	+3
titanium	element has all	↑ ↑ _ _ _	↑↓	+2 +3 +4
vanadium	these sublevels	↑ ↑ ↑ _ _	↑↓	+2 +3 +4 +5
chromium	filled in an	↑ ↑ ↑ ↑ ↑	↑	+2 +3 +6
manganese	argon	↑ ↑ ↑ ↑ ↑	↑↓	+2 +3 +4 +6 +7
iron	structure	↑↓ ↑ ↑ ↑ ↑	↑↓	+2 +3
cobalt		↑↓ ↑↓ ↑ ↑ ↑	↑↓	+2 +3
nickel		↑↓ ↑↓ ↑↓ ↑ ↑	↑↓	+2 +3
copper		↑↓ ↑↓ ↑↓ ↑↓ ↑↓	↑	+1 +2
zinc		↑↓ ↑↓ ↑↓ ↑↓ ↑↓	↑↓	+2

The Formation of Colored Compounds

A striking property of many compounds of transition metals is their color. Most colored inorganic compounds involve elements of the transition region of the Periodic Table, as shown in Figure 25-6. The color of the compounds and their solutions may vary depending on what other ions or groups are associated with the transition element. In general, these colored compounds are thought to have some electrons that are easily excited by selectively absorbing part of the energy of white light.

The color of transition metal compounds is associated with partially filled *d* orbitals. Note that Zn^{2+} compounds, which have completely filled *d* orbitals, are colorless. Compounds of Sc^{3+}, in which there are no *d* electrons, are also colorless.

Figure 25-6 Most compounds of transition metals are colored. The color depends on the metal, its oxidation state, and the anion with which it is combined.

As shown in Figure 4-3, the continuous spectrum of white light is composed of red, orange, yellow, green, blue, indigo, and violet colors of light. If light energy of a discrete band of wavelengths corresponding to a color region is absorbed by a substance, the remaining light reflected to the eye is the complement of the color absorbed. The energies of the wavelengths of white light that are absorbed are the energies required to raise *d* electrons to higher energy states.

Copper(II) sulfate pentahydrate crystals are blue when viewed in white light. Their aqueous solutions are also blue. They are blue because the energy of wavelengths corresponding to yellow light is absorbed. The light reflected (or transmitted) is blue light, the complement of yellow light. When these crystals are heated, the water of hydration is given up, and the anhydrous salt is neither blue nor crystalline. It is a white powder. If water is added to the anhydrous powder, it turns blue. Because many sulfate compounds are colorless, we can conclude that it must be the hydrated copper(II) ion, $Cu(H_2O)_4^{2+}$, that is blue.

The Formation of Complex Ions

The NO_3^- ion, SO_4^{2-} ion, and the NH_4^+ ion are examples of polyatomic ions. They are charged particles made up of more than a single atom. These familiar ions are covalent structures. Their net ionic charge is the algebraic sum of the oxidation numbers of all the atoms present. Because of their small size and their stability, they behave just like single-atom ions, and go through many reactions unchanged.

Polyatomic ions are described in Section 6.2.

The Transition Metals

Figure 25-7 Crystals of hydrated sulfates of nickel (green), iron (tan), cobalt (pink), and copper (blue).

Complex ions can be considered to be examples of Lewis acid-base reactions. They occur when a coordinate covalent bond forms as complexing agents donate electron pairs to a central metal ion.

Figure 25-8 *(below)* Solution of $CuSO_4$ in concentrated HCl; $[CuCl_4]^{2-}$(aq) ions give this solution a light green color. If H_2O is added to this solution, H_2O molecules replace Cl^- ions in the $[CuCl_4]^{2-}$ complex and the solution turns dark green. Anhydrous $CuSO_4$ is dissolved in water; $[Cu(H_2O)_4]^{2+}$(aq) ions give this solution a light blue color. Addition of NH_3(aq) causes the displacement of H_2O molecules by NH_3 molecules to form $[Cu(NH_3)_4]^{2+}$(aq) ions, giving the solution a deep blue color.

The name **complex ion** *is ordinarily restricted to an ionic species composed of a central metal ion combined with a specific number of polar molecules or ions.* Complex ions vary greatly in stability, but none is as stable as the common polyatomic ions mentioned above.

The most common complex ions are formed by ions of the transition metals combined with chemical species such as chloride ions (Cl^-), ammonia molecules (NH_3), water molecules (H_2O), and cyanide ions (CN^-). The transition metal cation is always the central ion in the complex. *The number of units covalently bonded to, or coordinated with, the central ion is the* **coordination number** *for that complex ion.* For example, the Fe^{2+} ion in the $Fe(CN)_6^{4-}$ complex has a coordination number of 6.

The hydrated Cu^{2+} ion, $Cu(H_2O)_4^{2+}$, discussed earlier in this section, is a complex ion. The Cu^{2+} ion is coordinated with 4 molecules of water of crystallization. It is this ionic species that crystallizes from an aqueous sulfate solution as the compound $CuSO_4 \cdot 5H_2O$.

When an excess of concentrated ammonia-water solution is added to an aqueous solution of Cu^{2+} ions (a $CuSO_4$ solution, for example), the light-blue color of the solution changes to a deep-blue color. The soluble $Cu(NH_3)_4^{2+}$ complex has been formed.

$$Cu(H_2O)_4^{2+} + 4NH_3(aq) \rightarrow Cu(NH_3)_4^{2+} + 4H_2O$$
$$\text{light blue} \qquad\qquad\qquad \text{deep blue}$$

   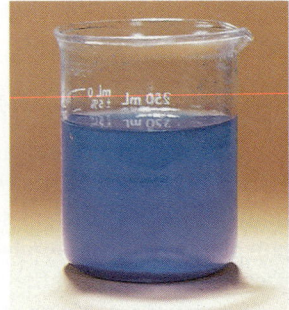

TABLE 25-2 SOME COMMON COMPLEX IONS

Coordinating Atom	Coordinating Group	Complex Ions	Color
N	H:N:H with H (ammonia)	$Ag(NH_3)_2^+$	—
		$Ni(NH_3)_4^{2+}$	blue
		$Co(NH_3)_6^{3+}$	blue
C	:C::N:$^-$	$Fe(CN)_6^{4-}$	yellow
		$Fe(CN)_6^{3-}$	red
S	:S:S:O:$^{2-}$ with :O: above and below	$Ag(S_2O_3)_2^{3-}$	—
N	:S::C::N:$^-$	$FeSCN^{2+}$	red

Zinc and cadmium form similar complex ions with ammonia. Table 25-2 gives the formulas and colors of some typical complex ions formed by transition metals.

If the ammonia-water solution is added to a solution of silver nitrate, the complex $Ag(NH_3)_2^+$ ion is formed.

$$Ag^+ + 2NH_3 \longrightarrow Ag(NH_3)_2^+$$

$$Ag^+ + 2:\!\!\overset{H}{\underset{H}{N}}\!\!:\!H \longrightarrow \left[H:\!\!\overset{H}{\underset{H}{N}}\!\!:\!Ag:\!\!\overset{H}{\underset{H}{N}}\!\!:\!H \right]^+$$

Brackets are used in this expression simply to enclose the electron-dot formula and indicate that the 1+ charge applies to the ion as a whole. They do not indicate a concentration expression.

Because complex ions are unstable to some extent, the reaction is reversible. The more stable the complex ion, the more quickly equilibrium is established between the reactions involving the formation and dissociation of the complex ions. At equilibrium the equation describing the decomposition of $Ag(NH_3)_2^+$ is

$$Ag(NH_3)_2^+ \rightleftarrows Ag^+ + 2NH_3$$

The equilibrium constant has the usual form.

$$K = \frac{[Ag^+][NH_3]^2}{[Ag(NH_3)_2^+]} = 7 \times 10^{-8}$$

In cases of complex ion equilibria, the equilibrium constant is called an **instability constant.** The larger the value of K, the more unstable is the complex ion.

Point out to students that these equations are written to show the complex ion breaking apart. Thus, the equilibrium constant is called the "instability" constant. For the reaction written in the reverse direction, the equilibrium constant is called the "formation" constant.

When metallic ions in solution form complexes, they are, in effect, removed from the solution as simple ions. Thus the formation of complex ions increases the solubility of metallic ions. The solubility of a precipitate may also be increased if cations of the substance form complex ions. For example, the formation of chloride complexes may increase solubility where, by common ion effect, a decrease would be expected.

Consider the equilibrium of sparingly soluble AgCl.

$$AgCl(s) \rightleftarrows Ag^+(aq) + Cl^-(aq)$$

Equilibrium constants are discussed in Section 19.1.

The common ion effect is discussed in Section 19.2.

The Transition Metals

Remind students that a large negative exponent means a very small number.

A small addition of Cl⁻ ions does shift the equilibrium to the left and lower the solubility as expected. If a large concentration of Cl⁻ ions is added, however, the solid AgCl dissolves. This apparent contradiction is explained by the formation of $AgCl_2^-$ complex ions.

The stability of complex ions is indicated by the small value of the equilibrium constants, K, as shown in Table 25-3. The smaller the value of K, the more stable is the complex ion and the more effectively it holds the central metallic ion as part of the soluble complex.

1. Between Group 2 and Group 13, beginning in Period 4.
2. Metals; all but mercury are hard and have high melting points; good conductors of heat and electricity; exhibit similarities horizontally and vertically down a group. Most exhibit variable oxidation states in forming compounds; have a pronounced tendency to form complex ions; form colored compounds.
3. Harder and have higher melting points than the alkali metals and alkaline-earth metals. Unlike the alkali metals and alkaline-earth metals, they exhibit variable oxidation states, are paramagnetic, and have colored compounds.
4. +1, +2, +3, +4, +5, +6, +7

TABLE 25-3 STABILITY OF COMPLEX IONS

Complex Ion	Equilibrium Reaction	K
$Ag(NH_3)_2^+$	$Ag(NH_3)_2^+ \rightleftharpoons Ag^+ + 2NH_3$	7×10^{-8}
$Co(NH_3)_6^{2+}$	$Co(NH_3)_6^{2+} \rightleftharpoons Co^{2+} + 6NH_3$	1×10^{-5}
$Co(NH_3)_6^{3+}$	$Co(NH_3)_6^{3+} \rightleftharpoons Co^{3+} + 6NH_3$	2×10^{-34}
$Cu(NH_3)_4^{2+}$	$Cu(NH_3)_4^{2+} \rightleftharpoons Cu^{2+} + 4NH_3$	4×10^{-13}
$Ag(CN)_2^-$	$Ag(CN)_2^- \rightleftharpoons Ag^+ + 2CN^-$	2×10^{-19}
$Au(CN)_2^-$	$Au(CN)_2^- \rightleftharpoons Au^+ + 2CN^-$	5×10^{-39}
$Cu(CN)_3^{2-}$	$Cu(CN)_3^{2-} \rightleftharpoons Cu^+ + 3CN^-$	1×10^{-35}
$Fe(CN)_6^{4-}$	$Fe(CN)_6^{4-} \rightleftharpoons Fe^{2+} + 6CN^-$	1×10^{-35}
$Fe(CN)_6^{3-}$	$Fe(CN)_6^{3-} \rightleftharpoons Fe^{3+} + 6CN^-$	1×10^{-42}
$FeSCN^{2+}$	$FeSCN^{2+} \rightleftharpoons Fe^{3+} + SCN^-$	8×10^{-3}
$Ag(S_2O_3)_2^{3-}$	$Ag(S_2O_3)_2^{3-} \rightleftharpoons Ag^+ + 2S_2O_3^{2-}$	6×10^{-14}

Section Review

1. Where are the transition elements found in the periodic table?
2. What are some of the properties of the transition elements?
3. How do the transition elements differ from alkali metals and alkaline-earth metals?
4. What are the common oxidation states of the elements in the first transition series?

25.2 Iron, Cobalt, and Nickel

SECTION OBJECTIVES
- Identify the members of the Iron Family, and describe their general characteristics.
- Describe the natural occurrence, refining methods, and compounds of iron.
- Outline the metallurgy of iron and steel.
- Describe the reactions and uses of Fe^{2+} and Fe^{3+} ions.

Iron, cobalt, and nickel, are elements of the first transition series. Each is the first member of a group of transition metals (Groups 8–10) that bears its name, as indicated in Figure 25-1. The similarities among iron, cobalt, and nickel are greater than those within each of the groups they head. Thus iron, cobalt, and nickel are referred to as the Iron family. The remaining six members of these three groups have properties similar to platinum and are considered to be members of the Platinum family. They are called **noble metals** *because they show little chemical activity*. They are rare and expensive.

General Characteristics

Iron is the most important of these elements. Alloys of iron, cobalt, and nickel are important structural metals. Some properties of the Iron family are listed in Table 25-4.

TABLE 25-4 THE IRON FAMILY

Element	Atomic Number	Atomic Mass	Electron Configuration	Oxidation Numbers	Melting Point (°C)	Boiling Point (°C)	Density (g/cm³)
iron	26	55.847	2,8,14,2	+2, +3	1535	2750	7.87
cobalt	27	58.9332	2,8,15,2	+2, +3	1495	2870	8.9
nickel	28	58.69	2,8,16,2	+2, +3	1453	2732	8.90

The Iron family is located in the middle of the transition elements. The next-to-highest energy level in these elements is incomplete: the $3d$ sublevels of iron, cobalt, and nickel contain only 6, 7, and 8 electrons, respectively, instead of the full number, 10. Each can exhibit the $+2$ and $+3$ oxidation states. The Fe^{2+} ion is easily oxidized to the Fe^{3+} ion by air and other oxidizing agents, but the $+3$ oxidation state occurs only rarely in cobalt and nickel. The electron configurations of iron, cobalt, and nickel sublevels are shown in Table 25-5. Notice that, in this table, paired and unpaired electrons of the $3d$ and $4s$ sublevels are represented by paired and unpaired dots. Compare this table with the electron configurations shown in Table 25-1.

TABLE 25-5 ELECTRON CONFIGURATIONS OF THE IRON FAMILY

Sublevel	1s	2s	2p	3s	3p	3d	4s
Maximum number of electrons	2	2	6	2	6	10	2
iron	2	2	6	2	6	:....	:
cobalt	2	2	6	2	6	::...	:
nickel	2	2	6	2	6	:::..	:

The two $4s$ electrons are removed easily, as is usually the case with metals. This removal forms the Fe^{2+}, Co^{2+}, or Ni^{2+} ions. In the case of iron, one $3d$ electron is also easily removed because the five remaining $3d$ electrons make a half-filled sublevel. When two $4s$ and one $3d$ electrons are removed, the Fe^{3+} ion is formed. It becomes increasingly difficult to remove a $3d$ electron from cobalt and nickel. This increasing difficulty is partly explained by the higher nuclear charges. Another factor is that neither $3d$ sublevel would be left at the filled or half-filled stage. Neither Co^{3+} nor Ni^{3+} ions are common. Cobalt atoms in the $+3$ oxidation state occur in complexes, however.

All three metals of the Iron family have *a strong magnetic property*. This property is commonly *known as* **ferromagnetism** because of the unusual extent to which it is possessed by iron. Cobalt is strongly magnetic, whereas nickel is the least magnetic of the three. The ferromagnetic nature of these metals is thought to be related to the similar spin orientations of their unpaired $3d$ electrons. Table 25-5 shows that atoms of iron have 4 unpaired $3d$ electrons, atoms of cobalt have 3 unpaired $3d$ electrons, and atoms of nickel have 2 unpaired $3d$ electrons.

The Transition Metals

Each spinning electron acts like a tiny magnet. Electron pairs are formed by two electrons spinning in opposite directions. The electron magnetisms of such a pair of electrons neutralize each other. In Iron family metals, *groups of atoms may be aligned and form small magnetized regions called* **domains**. Ordinarily, magnetic domains within the metallic crystals point in random directions. In this way, they cancel one another so that the net magnetism is zero. A piece of iron becomes magnetized when it is acted upon by an outside force that, for both paired and unpaired electrons, aligns the domains in the same direction.

Occurrence of Iron

Iron is the fourth element in abundance by mass in the earth's crust. Nearly 5 percent of the earth's crust is iron. It is the second most abundant metal, following aluminum. The known magnetic nature of the earth itself and the fact that meteors are known to contain iron suggests that the earth's core may consist mainly of iron.

Unfortunately, much of the iron in the crust of the earth cannot be removed profitably. Although iron is found in many minerals, only those from which iron can be profitably recovered by practical methods are considered to be iron ores.

Great deposits of *hematite* (Fe_2O_3) once existed in the Lake Superior region of the United States. They were the major sources of iron in this country for many years. These rich ore deposits have been depleted, and the iron industry now depends on medium- and low-grade ores, the most abundant of which is *taconite*.

Taconite has an iron content of roughly 25–50% in a chemically complex mixture. The main iron minerals that are present are hematite (Fe_2O_3) and magnetite (Fe_3O_4). Modern blast furnaces that are designed to reduce iron ore to iron require ores containing well above 60% iron. Thus, the raw ores must be concentrated by removing much of the waste materials. This is done at the mines before the ores are transported to the blast furnaces or smelters.

Taconite ores are first crushed and pulverized and then concentrated by a variety of complex methods. The concentrated ores are hardened into pellets for shipment to the smelters. Ore concentrates containing 90% iron are produced by the chemical removal of oxygen during the pelletizing treatment, as shown in Figure 25-9.

Iron is the most widely used of all metals. More iron is consumed than 14 times all other metals together.

Figure 25-9 Enrichment of low-grade iron ores. Almost all iron ore mined in the United States is concentrated before it is shipped to a blast furnace.

The blast furnace *Iron ore is converted to iron in a giant structure called a* **blast furnace**. The charge placed in the blast furnace consists of iron ore concentrate, coke, and a flux, which causes mineral impurities in the ore to melt more readily. The proper proportions for the charge are calculated by analyzing the raw materials. Usually the flux is limestone, because silica or sand is the most common impurity in the iron ore. Some iron ores contain limestone as an impurity; in such cases, the flux added is sand. A blast of hot air, sometimes enriched with oxygen, is forced into the base of the furnace, as shown in Figure 25-10.

Iron oxide is reduced in the blast furnace to iron. *The earthy impurities that remain are called* **slag**. *The iron recovered from a blast furnace is called* **pig iron**.

What happens inside the blast furnace when iron is produced? The actual chemical changes that occur are complex and are not entirely understood. First, the coke is ignited by the blast of hot air, and some of it burns to form carbon dioxide.

$$C(s) + O_2(g) \rightarrow CO_2(g)$$

When the oxygen is introduced, the carbon dioxide comes in contact with other pieces of hot coke and is reduced to carbon monoxide gas.

$$CO_2(g) + C(s) \rightarrow 2CO(g)$$

The carbon monoxide thus formed is actually the reducing agent that reduces the iron oxide to metallic iron.

$$Fe_2O_3(s) + 3CO(g) \rightarrow 2Fe(\ell) + 3CO_2(g)$$

This reduction probably occurs in steps as the temperature increases toward the bottom of the furnace. Some possible steps are

$$Fe_2O_3 \rightarrow Fe_3O_4 \rightarrow FeO \rightarrow Fe$$

The white-hot liquid iron collects in the bottom of the furnace and is removed every four or five hours. It may be cast in molds as pig iron or converted directly to steel.

In the central region of the furnace, the limestone decomposes to calcium oxide and carbon dioxide.

$$CaCO_3(s) \rightarrow CaO(s) + CO_2(g)$$

The calcium oxide combines with silica to form a calcium silicate slag.

$$CaO(s) + SiO_2(s) \rightarrow CaSiO_3(\ell)$$

This glassy slag melts readily and collects at the bottom of the furnace. Since it has a much lower density than liquid iron, it floats on top of the iron, thus preventing the reoxidation of the iron. The melted slag is tapped off every few hours and usually is thrown away.

Steel production The relatively high carbon content of iron recovered from the blast furnace makes it very hard and brittle. Phosphorus and sulfur are two other impurities. The phosphorus makes pig iron brittle at low temperatures. The sulfur makes it brittle at high temperatures.

The conversion of iron to steel is essentially a purification process in which impurities are removed by oxidation. This purification process is

Figure 25-10 A sectional view of a blast furnace.

Note that the oxidation number of iron in this sequence goes from Fe^{3+} in Fe_2O_3, to a combination of Fe^{3+} and Fe^{2+} in Fe_3O_4, to Fe^{2+} in FeO, and finally to Fe^0 in elemental Fe.

The Transition Metals

Figure 25-11 Steelmaking by the basic oxygen process. The furnace receives a charge of scrap steel *(upper left)* and a charge of molten iron from a blast furnace *(upper right)*. Oxygen is blown into the furnace and flux is added *(lower right)*. Molten steel is released from a tap hole *(lower left)* and transferred to a ladle. Alloying substances are then added according to precise specifications.

Figure 25-12 These marble-sized pellets are produced by direct reduction of iron ore.

carried out in a furnace at a high temperature. Several steps of this process are shown in Figure 25-11. Scrap steel that is being recycled is sometimes added, and iron ore may be added to supply oxygen for oxidizing the impurities. Oxygen gas is blown in for very rapid oxidation. Limestone or lime is included to form a slag with nongaseous oxides. Near the end of the process, selected alloying substances are added. By using different kinds and quantities of these added substances, the steel is given different desired properties.

The impurities in the pig iron are oxidized in the following way:

$$3C + Fe_2O_3 \rightarrow 3CO(g) + 2Fe$$
$$3Mn + Fe_2O_3 \rightarrow 3MnO + 2Fe$$
$$12P + 10Fe_2O_3 \rightarrow 3P_4O_{10} + 20Fe$$
$$3Si + 2Fe_2O_3 \rightarrow 3SiO_2 + 4Fe$$
$$3S + 2Fe_2O_3 \rightarrow 3SO_2(g) + 4Fe$$

Carbon and sulfur compounds escape as gases. The limestone flux decomposes as in the blast furnace.

$$CaCO_3 \rightarrow CaO + CO_2(g)$$

Calcium oxide and the oxides of the other impurities react to form slag.

$$P_4O_{10} + 6CaO \rightarrow 2Ca_3(PO_4)_2$$
$$SiO_2 + CaO \rightarrow CaSiO_3$$
$$MnO + SiO_2 \rightarrow MnSiO_3$$

Pure Iron

Pure iron is a metal that is seldom encountered. It is silver-white, soft, ductile, tough, and does not tarnish readily. It melts at 1535°C. Commercial iron contains carbon and other impurities that alter its properties. Cast iron melts at approximately 1150°C. All forms of iron corrode, or rust, in moist air, so iron is not a self-protective metal. The rust that forms is brittle and scales off, exposing the metal underneath to further corrosion. Iron does not rust in dry air or in water that is free of dissolved oxygen.

Iron is only moderately active chemically. Even so, it corrodes more extensively than more active metals such as zinc and aluminum. These metals form oxide coatings that adhere to the metal surface and protect it from further corrosion. The rusting of iron is a complicated (and not fully understood) electrochemical process that involves water, air, and carbon dioxide.

Rusting begins when water containing CO_2 in solution comes into contact with iron. Such a solution is acidic and reacts with iron to form Fe^{2+} ions. The iron(II) ions are oxidized to the iron(III) state as hydrated Fe_2O_3 (rust) by oxygen. Iron is made rust-resistant in several ways: *(1)* by alloying it with chromium, copper, or nickel; *(2)* by treating it to form a coating of Fe_3O_4, which adheres to and protects the surface; *(3)* by painting it, which protects the surface as long as the paint adheres and is not chipped or scratched; *(4)* by dip-coating it with a self-protective metal (for example galvanizing with zinc); *(5)* by plating it with nickel or chromium; or *(6)* by coating it with a material like porcelain.

Although dilute acids generally react readily with iron, concentrated nitric acid does not react with it, and concentrated sulfuric acid has little effect on iron. If fact, dipping iron into concentrated nitric acid makes the iron passive, or inactive, with other chemicals. Strong hydroxides do not react with iron.

Three oxides of iron Of the three oxides of iron, iron(II) oxide, FeO, is of little importance. It oxidizes rapidly when exposed to air to form iron(III) oxide, Fe_2O_3. This oxide is the important ore of iron. It is used as a cheap pigment for red paint and for grinding and polishing glass lenses and mirrors. Magnetic iron oxide known as magnetite, (Fe_3O_4), is an important ore. It is composed of Fe_2O_3 and FeO. Thus, it may be considered to be iron(II, III) oxide, and its formula can be written $FeO \cdot Fe_2O_3$.

Reactions of the Fe^{2+} ion Hydrated iron(II) sulfate, $FeSO_4 \cdot 7H_2O$, is the most useful compound of iron in the +2 oxidation state. It is used as a reducing agent and in medicine for iron tonics. Iron(II) sulfate can be prepared by the action of dilute sulfuric acid on iron. The crystalline hydrate loses water of hydration when exposed to air and turns brown because of oxidation. Iron(II) sulfate in solution is gradually oxidized to the iron(III) state by dissolved oxygen. The brown precipitate of basic iron(III) sulfate that

Figure 25-13 A worker inspects pipes at an iron oxide plant in West Virginia.

Galvanized iron is familiar as the metal used to make garbage cans and buckets.

The passivity of iron is easily destroyed by a scratch or by shock.

forms is evidence of this change:

$$4FeSO_4 + O_2 + 2H_2O \rightarrow 4Fe(OH)SO_4(s)$$

The Fe^{2+} ions can be kept in the reduced state by making the solution acidic with sulfuric acid and adding pieces of iron. Hydrated iron(II) ammonium sulfate, $Fe(NH_4)_2(SO_4)_2 \cdot 6H_2O$, is a better source of Fe^{2+} ions in the laboratory because it is stable in contact with air. Iron(II) salts are readily oxidized to iron(III) salts by the corresponding acid and an oxidizing agent. In the case of the nitrate, nitric acid meets both requirements.

$$3Fe(NO_3)_2 + 4HNO_3 \rightarrow 3Fe(NO_3)_3 + NO(g) + 2H_2O$$

Reactions of the Fe^{3+} ion Hydrated iron(III) chloride, $FeCl_3 \cdot 6H_2O$, is the most useful compound of iron in the +3 oxidation state. It is used as a mordant for dyeing cloth and as an oxidizing agent. The yellow crystalline hydrate is deliquescent. The hydrated Fe^{3+} ion, $Fe(H_2O)_6^{3+}$, is pale violet in color. This color usually is not seen, however, because of hydrolysis, which yields hydroxide complexes that are yellow-brown in color. The hydrolysis of Fe^{3+} ions in water solutions of its salts gives solutions that are acidic.

$$Fe^{3+} + 2H_2O \rightleftharpoons FeOH^{2+} + H_3O^+$$
$$FeOH^{2+} + 2H_2O \rightleftharpoons Fe(OH)_2^+ + H_3O^+$$
$$Fe(OH)_2^+ + 2H_2O \rightleftharpoons Fe(OH)_3 + H_3O^+$$

The hydrolysis is extensive when it occurs in boiling water. A blood-red colloidal suspension of iron(III) hydroxide is formed. This colloidal suspension can be produced by adding a few drops of $FeCl_3$ solution to a flask of boiling water.

Iron(III) ions are removed from solution by adding a solution containing hydroxide ions. A red-brown jelly-like precipitate of iron(III) hydroxide is formed.

$$Fe^{3+} + 3OH^- \rightarrow Fe(OH)_3(s)$$

If water is evaporated, red Fe_2O_3 remains.

Tests for iron ions Potassium hexacyanoferrate(II), $K_4Fe(CN)_6$ (also called potassium ferrocyanide), is a light yellow, crystalline salt. It contains the complex hexacyanoferrate(II) ion (ferrocyanide ion), $Fe(CN)_6^{4-}$. The iron is in the +2 oxidation state. The hexacyanoferrate(II) ion forms when an excess of cyanide ions is added to a solution of an iron(II) salt.

$$6CN^- + Fe^{2+} \rightarrow Fe(CN)_6^{4-}$$

Suppose KCN is used as the source of CN^- ions and $FeCl_2$ as the source of Fe^{2+} ions. The equation is

$$6KCN + FeCl_2 \rightarrow K_4Fe(CN)_6 + 2KCl$$

The iron of the $Fe(CN)_6^{4-}$ ion can be oxidized by chlorine to the +3 state. This reaction forms the hexacyanoferrate(III) ion (ferricyanide ion) $Fe(CN)_6^{3-}$.

$$2Fe(CN)_6^{4-} + Cl_2 \rightarrow 2Fe(CN)_6^{3-} + 2Cl^-$$

As the iron is oxidized from the +2 oxidation state to the +3 state, the chlorine is reduced from the 0 state to the −1 state.

Solutions containing cyanide ions are deadly poisons. Safety precautions should be followed when handling these solutions.

Figure 25-14 Potassium hexacyanoferrate(II) *(left)* and potassium hexacyanoferrate(III) *(right)* are salts containing iron.

The (II) following hexacyanoferrate tells you it is the Fe^{2+} ion in the complex. The (III) tells you it is the Fe^{3+} ion.

Using the potassium salt as the source of the $Fe(CN)_6^{4-}$ ions, the empirical equation is

$$2K_4Fe(CN)_6 + Cl_2 \rightarrow 2K_3Fe(CN)_6 + 2KCl$$

$K_3Fe(CN)_6$, potassium hexacyanoferrate(III) (known also as potassium ferricyanide), is a dark red crystalline salt.

Intense colors are observed in most compounds that have an element present in two different oxidation states. When iron(II) ions and hexacyanoferrate(III) ions are mixed and when iron(III) ions and hexacyanoferrate(II) ions are mixed, the same intense blue substance is formed.

$$Fe^{2+} + K^+ + Fe(CN)_6^{3-} + H_2O \rightarrow K\overset{+2}{Fe}\overset{+3}{Fe}(CN)_6 \cdot H_2O(s)$$

and

$$Fe^{3+} + K^+ + Fe(CN)_6^{4-} + H_2O \rightarrow K\overset{+3}{Fe}\overset{+2}{Fe}(CN)_6 \cdot H_2O(s)$$

In both reactions, the precipitate contains iron in the +2 and +3 oxidation states, and the same intense blue color is seen.

Iron(II) ions, Fe^{2+}, and hexacyanoferrate(II) ions, $Fe(CN)_6^{4-}$, form a white precipitate, $K_2FeFe(CN)_6$. Note that both Fe ions are in the same +2 oxidation state. This precipitate remains white if the oxidation of Fe^{2+} ions is prevented. Of course, on exposure to air, it begins to turn blue. Iron(III) ions (Fe^{3+}) and hexacyanoferrate(III) ions, $Fe(CN)_6^{3-}$, give a brown solution. Here, both Fe ions are in the same +3 oxidation state. The reactions given in the two preceding equations provide ways to detect the presence of iron in each of its two oxidation states.

1. Test for the Fe^{2+} ion. If a few drops of $K_3Fe(CN)_6$ solution is added to a solution of iron(II) sulfate, the characteristic intense blue precipitate $KFeFe(CN)_6 \cdot H_2O$ forms. Two-thirds of the potassium ions and all of the sulfate ions are merely spectator ions. The net reaction is

$$Fe^{2+} + K^+ + Fe(CN)_6^{3-} + H_2O \rightarrow KFeFe(CN)_6 \cdot H_2O(s)$$

2. Test for the Fe^{3+} ion. When a few drops of $K_4Fe(CN)_6$ solution are added to a solution of iron(III) chloride, the characteristic blue precipitate $KFeFe(CN)_6 \cdot H_2O$, is formed. The net reaction is

$$Fe^{3+} + K^+ + Fe(CN)_6^{4-} + H_2O \rightarrow KFeFe(CN)_6 \cdot H_2O(s)$$

Potassium thiocyanate, KSCN, provides another good test for the Fe^{3+} ion. It is often used to confirm the $K_4Fe(CN)_6$ test. A blood-red solution results from the formation of the complex $FeSCN^{2+}$ ion.

Section Review

1. What elements make up the Iron family?
2. Describe some of the properties of the Iron family.
3. Where does iron occur and what are some of its compounds?
4. How are limestone impurities removed from an iron ore?
5. What is the primary purpose of the reactions that occur during the conversion of iron to steel?
6. Write chemical equations to illustrate the reactions of Fe^{2+} and Fe^{3+}.
7. What precipitate serves to identify the presence of Fe^{2+} ion?

Blueprints: The basis for the blueprinting process is a photochemical reaction. Blueprint paper is first treated with a solution of iron(III) ammonium citrate and potassium hexacyanoferrate(III) and allowed to dry in the dark. A drawing on tracing paper is laid over the blueprint paper and exposed to light. Where light strikes the paper, Fe^{3+} ions are reduced to Fe^{2+} ions. The paper is dipped into water and the Fe^{2+} ions react with the hexacyanoferrate(III) ions to form the blue color in the paper. Where drawing lines covered the paper, no reduction of Fe^{3+} ions occurs and the paper rinses out white. Thus, a blueprint has white lines on blue paper.

1. Iron, cobalt, and nickel
2. Metals with low reactivity; can all show the +2 and +3 oxidation states; have high melting and boiling points; all form complex ions; all have strong magnetic properties. Ferromagnetism is the result of the similar spin orientations of their 3*d* electrons.
3. In the earth's crust, in meteors. Accept any compounds mentioned in Section 25.2.
4. By adding sand as the flux
5. Purification process
6. See equations in Section 25.2.
7. Blue precipitate of $KFeFe(CN)_6$ when $K_3Fe(CN)_6$ is added to a solution.

25.3 Copper, Silver, and Gold

SECTION OBJECTIVES

- Identify the elements of the Copper family, and describe their general characteristics.
- Describe the natural occurrence of and refining methods for copper and name compounds of copper and their uses.
- Describe the properties and uses of metallic copper.

Group 11, sometimes known as the Copper family, consists of copper, silver, and gold. All three metals appear below hydrogen in the electrochemical series. They are not easily oxidized and often occur in nature in the free, or native, state. Because of their pleasing appearance, durability, and scarcity, these metals have been highly valued since the time of their discovery. All have been used in ornamental objects and coins throughout history.

General Characteristics

The atoms of copper, silver, and gold have a single electron in their highest energy levels. Thus they often form compounds in which they exhibit the +1 oxidation state, and thereby resemble the Group 1 metals of the Sodium family.

Each metal of the Copper family has 18 electrons in the next-to-highest energy level. The d electrons in this next-to-highest energy level have energies that differ only slightly from the energy of the outer s electron. Thus, one or two of these d electrons can be removed with relative ease. For this reason, copper and gold often form compounds in which they exhibit the +2 and +3 oxidation states, respectively. In silver, the +2 oxidation state is reached only under extreme oxidizing conditions.

Copper, silver, and gold are very dense, ductile, and malleable. They are classed as heavy metals along with other transition metals in the central region of the Periodic Table. Some important properties of each metal are shown in Table 25-6.

Gold can be made into a very thin, flexible film that is used to coat statues or picture frames.

The major use of silver is in photographic film. A thin, light-sensitive coating on photographic film contains silver bromide. When it is exposed to light, a few of the silver ions are reduced to metallic silver. These ions sensitize the tiny silver bromide crystals in which they occur. Developing the film causes additional silver ions in the sensitized crystals to be reduced. An image is formed of dark areas of silver and light areas of unchanged AgBr in the unexposed areas. The unreacted AgBr is then removed so the developed film will no longer be light sensitive.

TABLE 25-6 THE COPPER FAMILY

Element	Atomic number	Atomic mass	Electron configuration	Oxidation numbers	Melting point (°C)	Boiling point (°C)	Density (g/cm³)
copper	29	63.546	2,8,18,1	+1, +2	1083.4	2567	8.96
silver	47	107.868	2,8,18,18,1	+1	961.9	2212	10.5
gold	79	196.967	2,8,18,18,32,1	+1, +3	1064.4	2807	19.3

Copper

Copper, alloyed with tin in the form of bronze, has been in use for over 5000 years. Native copper deposits lie deep underground and are difficult to mine. Most copper is thus obtained from its ores by either treating ores with extremely high temperatures or subjecting ores to aqueous solution at relatively low temperatures.

Sulfide ores of copper yield most of the supply of this metal. Chalcocite (Cu_2S), chalcopyrite ($CuFeS_2$), and bornite (Cu_3FeS_3), are the major sulfide ores. Malachite, $Cu_2(OH)_2CO_3$, and azurite, $Cu_3(OH)_2(CO_3)_2$, are basic carbonates of copper. Malachite is a rich green, and azurite is a deep blue. Besides serving as ores of copper, fine specimens of these minerals are sometimes polished for use as ornaments or in making jewelry.

Copper and its recovery The carbonate ores of copper are washed with a dilute solution of sulfuric acid, forming a solution of copper(II) sulfate. The metallic copper is then recovered by electrolysis. High-grade carbonate ores are heated in air to convert them to copper(II) oxide. The oxide is then reduced with coke, yielding metallic copper as the final product in the process.

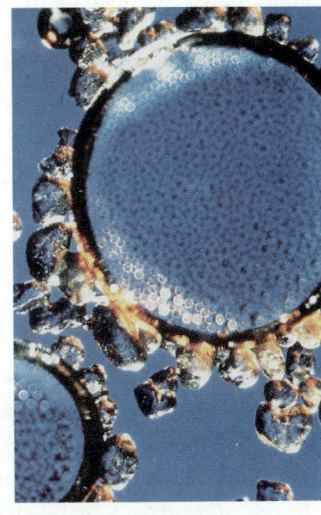

Figure 25-15 The oil-flotation process *(left)* for concentrating copper ore. Air bubbles carry ore particles to the surface. A photomicrograph *(right)* of a flotation bubble from the oil-flotation process

The sulfide ores are usually low-grade and require concentration before they can be refined profitably. The concentration is accomplished by oil-flotation which is illustrated in Figure 25-15. Earthy impurities associated with the ore are wetted by water, and the ore is wetted by oil. Air is blown into the mixture to form a froth. The oil-wetted ore floats to the surface in the froth. This treatment changes the concentration of the ore from about 2% copper to as high as 30% copper. Metallic copper is recovered from the concentrated ore by a process shown diagrammatically in Figure 25-16.

Figure 25-16 A flow diagram of the copper refining process

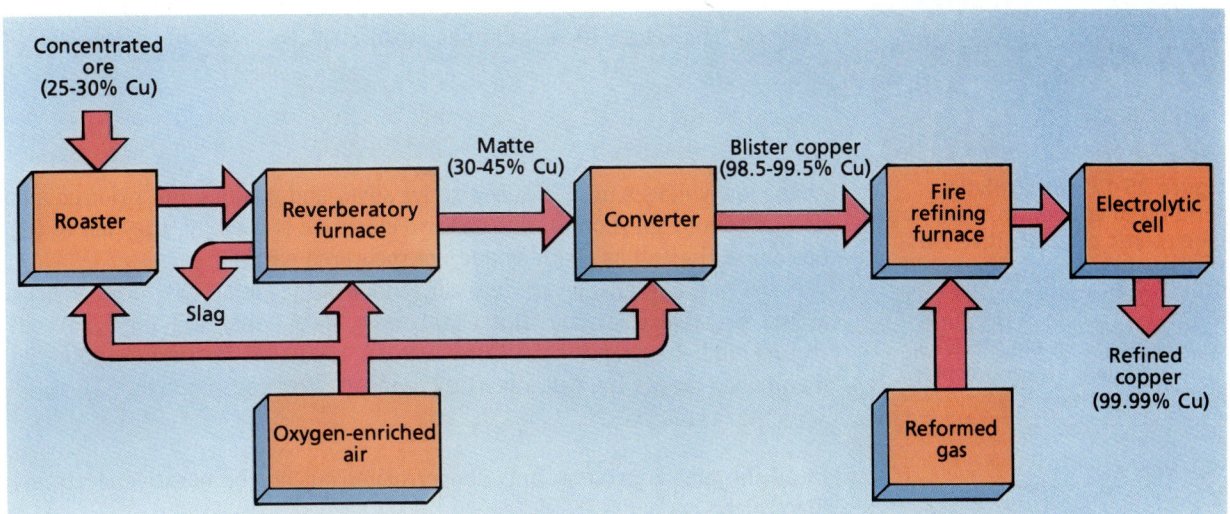

The Transition Metals 745

The concentrated ore is partially roasted to form a mixture of Cu_2S, FeS, FeO, and SiO_2. This mixture is known as calcine. The roasting process uses oxygen-enriched air and yields high-quality sulfur dioxide. This gas is converted to sulfuric acid, as described in Section 27.2. Calcine is fused with limestone in a furnace. Part of the iron is removed as a silicate slag. *The rest of the iron, together with the copper, forms a mixture of sulfides known as* **copper matte.** Copper matte is processed in a reverberatory furnace, and the end product contains 40% copper.

The melted matte is further refined in a converter that is supplied with oxygen-enriched air as the oxidizing agent. Sulfur from the sulfides as well as impurities of arsenic and antimony are removed as volatile oxides. Most of the iron is removed as slag. Some of the copper(I) sulfide is converted into copper(I) oxide. The copper(I) oxide then reacts with more copper(I) sulfide, forming metallic copper and sulfur dioxide. The following equations show the chemical reactions involved in this process.

$$2Cu_2S + 3O_2 \rightarrow 2Cu_2O + 2SO_2(g)$$
$$2Cu_2O + Cu_2S \rightarrow 6Cu + SO_2(g)$$

The molten copper is cast as blister copper of 98.5—99.5% purity. As the copper cools, dissolved gases escape and form blisters, hence the name. Impurities remaining are iron, silver, gold, and sometimes zinc. In the fourth step of the copper-refining process, blister copper is further purified in a fire-refining furnace or by electrolytic refining.

Electrolytic refining of copper Unrefined copper contains fairly large amounts of silver and gold. Thus the cost of its refining is offset by the recovery of these precious metals. Copper is largely used for making electric conductors, and very small amounts of impurities greatly increase the electric resistance.

In electrolytic refining, shown in Figure 25-17, sheets of pure copper are used as the negative electrodes in electrolytic cells. Large plates of impure copper are used as the positive electrodes. The electrolyte is a solution of copper(II) sulfate in sulfuric acid. A direct current at low voltage is used to operate the cell. During the electrolysis, copper and the other metals in the positive electrode that are above copper in the electrochemical series are oxidized. They enter the solution as ions.

$$Cu \rightarrow Cu^{2+} + 2e^-$$
$$Fe \rightarrow Fe^{2+} + 2e^-$$
$$Zn \rightarrow Zn^{2+} + 2e^-$$

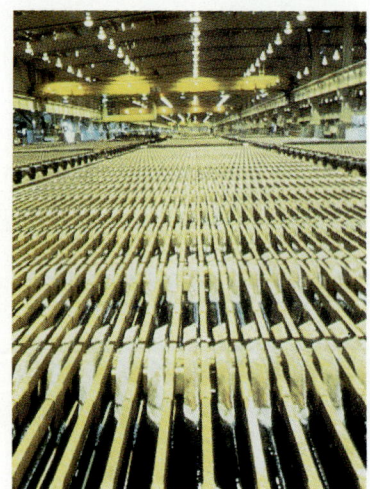

Figure 25-17 Electrolytic refining of copper. In these large electrolytic cells, positive electrodes of impure copper yield Cu^{2+} ions in solution. The Cu^{2+} ions are plated out on pure copper negative electrodes. Electrolytically refined copper is 99.99% pure.

At the low voltages used, the less active silver and gold are not oxidized and do not go into solution. As the positive electrode is used up, they fall to the bottom of the cell as a sludge and are recovered easily.

You may expect the various positive ions of the electrolyte to be reduced at the negative electrode. But H_3O^+ ions, Fe^{2+} ions, and Zn^{2+} ions all require higher voltages than Cu^{2+} ions for reduction. At the low potential maintained across the cell, only Cu^{2+} ions are reduced.

$$Cu^{2+} + 2e^- \rightarrow Cu$$

Of all the metals present, only copper plates out on the negative electrode. Electrolytic copper is 99.99% pure.

Properties and uses of copper and its compounds Copper is a soft, ductile, malleable, red metal with a density of 8.96 g/cm^3. It is second to silver in electric conductivity.

Copper compounds, in which copper exhibits the +1 oxidation state, are relatively unimportant as compared with those in which copper exhibits the +2 oxidation state. The best known copper(I) compounds are complexes or slightly soluble solids. These include CuI, CuCN, CuCl, and CuBr. Heated in air, copper forms a black coating of copper(II) oxide, CuO. Metallic copper and most copper compounds color a bunsen flame green.

Copper forms copper(II) salts that dissociate in water to give blue solutions. The color is characteristic of the hydrated copper(II) ion, $Cu(H_2O)_4^{2+}$. Adding an excess of ammonia to solutions containing this ion produces the deeper-blue complex ion, $Cu(NH_3)_4^{2+}$. Aqueous solutions of copper(II) salts are weakly acidic because of the mild hydrolysis of the Cu^{2+} ion.

$$Cu(H_2O)_4^{2+} + H_2O \rightarrow Cu(H_2O)_3OH^+ + H_3O^+$$

An excess of sulfur vapor forms a blue-black coating of copper(I) sulfide on hot copper. In moist air, copper tarnishes and forms a protective coating. This coating is a green basic carbonate, $Cu_2(OH)_2CO_3$. Sulfur dioxide in the air may also combine with copper. If so, a green basic sulfate, $Cu_4(OH)_6SO_4$, is produced. The green color seen on copper roofs is caused by the formation of these compounds.

Copper(II) compounds are much more common than copper(I) compounds. Copper(II) oxide is used to change alternating current to direct current. It is also used as an oxidizing agent in chemical laboratories.

Hydrated copper(II) sulfate, $CuSO_4 \cdot 5H_2O$, called blue vitriol, is an important copper compound. It is used to kill algae in reservoirs and to make certain pesticides. It is also used in electroplating and in preparing other copper compounds.

Because copper stands below hydrogen in the activity series, it does not replace hydrogen from acids. Thus it is not acted on by nonoxidizing acids such as hydrochloric acid and dilute sulfuric acid except very slowly when oxygen is present. The oxidizing acids, such as nitric and hot concentrated sulfuric, react vigorously with copper. Such reactions produce the corresponding copper(II) salts.

Because of its high thermal and electrical conductivity, and its ductility, malleability, and resistance to corrosion, copper has many uses. Copper is used in the manufacture of electrical wire, various instruments, transmission lines, and electrical machinery. It is also used in plumbing, roofing, screens, and gutters. A large amount of copper is used to make various alloys, including brass and bronze. The alloys are stronger, less expensive than copper, and easy to work with.

Tests for the Cu^{2+} ion A dilute solution of a copper(II) salt changes to a very deep blue color when an excess of ammonia is added. This color change is caused by the formation of complex $Cu(NH_3)_4^{2+}$ ions.

$$Cu^{2+} + 4NH_3 \rightarrow Cu(NH_3)_4^{2+}$$

The addition of $K_4Fe(CN)_6$ to a solution containing Cu^{2+} ions produces a red precipitate of $Cu_2Fe(CN)_6$. This precipitate is copper(II) hexacyanoferrate(II),

Gold is the third-best electrical conductor.

This black coating of copper(II) oxide is familiar to anyone who has cooked in a pot that has a copper-coated bottom.

The green coating is called a patina.

Brass is an alloy of copper and zinc. Bronze is an alloy of copper and tin.

1. Copper, silver, gold
2. Accept any properties shown in Table 25-8.
3. Found in sulfide ores, namely, chalcocite, Cu_2S; chalcopyrite, $CuFeS_2$; bornite, Cu_5FeS_4; malachite, $Cu_2(OH)_2CO_3$; and azurite, $Cu_3(OH)_2(CO_3)_2$. The recovery process depends on the ore. Accept any answers dealing with the recovery and refining processes discussed under the headings "Copper and its recovery" and "Electrolytic refining of copper."
4. CuO is used to change alternating current to direct current and as an oxidizing agent in chemical laboratories. $CuSO_4 \cdot H_2O$ is used to kill algae in reservoirs, to make certain pesticides, in electroplating, and in preparing other copper compounds.
5. High thermal conductivity, electrical conductivity, ductility, malleability, and resistance to corrosion. Used in the manufacture of electrical wire, instruments, machinery, plumbing, roofing, screens and gutters, and various alloys.

also known as copper(II) ferrocyanide. In a borax bead test, glasslike beads are formed when powdered borax is heated in a burner. Metallic contamination of the borax produces identifiable colors. If copper is present in a bead formed in an oxidizing flame, a clear blue color appears on cooling. The hot bead is green. A bead formed in a reducing flame is colorless when hot and an opaque red when cool.

Section Review

1. What elements constitute the Copper family?
2. Describe some properties that the elements of the Copper family exhibit.
3. Where does copper occur and how is it refined?
4. Name some important compounds of copper and describe their uses.
5. What properties make copper a very useful element? What are some of its uses?

Chapter Summary

- The transition metals consist of ten groups interposed between Group 2 and Group 13 in the Periodic Table. They are in this position because they represent an interruption of the regular increase in the number of highest-energy-level electrons from Group 1 to Group 18 in the long periods of the table.
- Chemical properties of transition metals are varied. Electrons of the next-to-highest d sublevel and the highest energy level may become involved in the formation of compounds.
- Most of the transition metals show different oxidation states in their compounds. They have strong tendencies to form complex ions. Many of their compounds are paramagnetic and are colored.
- With minor, apparent irregularities, the transition metals have two highest energy level electrons. In the first row of transition elements, chromium and copper have one $4s$ electron. The availability of $3d$ electrons tends to increase the number of oxidation states up to manganese in the middle of the row, after which the number of common oxidation states for each element decreases.
- The color of the compounds of transition elements depends to some extent on the ions or polyatomic groups associated with the element. The colors result from the ease with which d electrons can be excited by absorption of discrete quantities of light energy.
- Transition metals form many complex ions, with the transition element cation serving as the central ion in the complex. Certain ionic or molecular species are able to coordinate with the central ion to form the complex ion.
- The Iron family consists of iron, cobalt, and nickel. These metals are ferromagnetic.
- Iron forms compounds in two oxidation state, +2 and +3. Iron(II) ions are easily oxidized to the iron(III) state. Iron(II) compounds are sometimes used as reducing agents. Iron(III) compounds are used as oxidizing agents.
- The Copper family, or Group 11, consists of copper, silver, and gold. All are found as native metals in nature. Most copper used in commerce and industry is recovered from its sulfide ores by reduction and electrolytic refining.

Chapter 25 Review

The Transition Metals

VOCABULARY

blast furnace (739)
complex ion (734)
coordination number (734)
copper matte (746)
domain (738)
ferromagnetism (737)
instability constant (735)
noble metals (736)
pig iron (739)
slag (739)
transition elements (729)

REVIEWING CONCEPTS

1. Describe the structural similarity that determines the metallic character of the transition elements. (25.1)

2. Copper(II) sulfate crystals are blue. To what color does the light energy absorbed by the crystals correspond? (25.1)

3. Distinguish between a polyatomic ion and a complex ion. (25.1)

4. What is the oxidation number of manganese in
   a. Mn
   b. MnO_2
   c. $MnSO_4 \cdot 4H_2O$
   d. $KMnO_4$
   e. $BaMnO_4$ (25.1)

5. Chemically, what happens to iron
   a. in a blast furnace?
   b. during steel making? (25.2)

6. List five points of similarity among copper, silver, and gold. (25.3)

7. Why does the copper trim on roofs frequently acquire a green surface? (25.3)

8. Why does it usually pay to refine copper by electrolysis? (25.3)

PROBLEMS

1. What is the percentage of iron in a sample of limonite, $Fe(OH)_3$, to a figure of two decimal places?

2. How many grams of silver nitrate can be obtained by adding 100. g of pure silver to an excess of nitric acid?

3. a. A sample of hematite ore contains Fe_2O_3 87.0%, silica 8.0%, moisture 4.0%, other impurities 1.0%. What is the percentage of iron in the ore?
   b. What will be the loss in mass when 1.0×10^6 metric tons of the ore are heated to 200°C?
   c. How much limestone will be needed to combine with the silica in 1.0×10^6 metric tons of the ore?

CRITICAL THINKING

1. **Analyzing Results** When potassium hexacyanoferrate(II) is added to a solution of iron(II) sulfate, a white precipitate forms that gradually turns blue. Explain the color change.

2. **Designing Experiments** How can you detect an iron(II) and an iron(III) compound if both are present in the same solution?

CROSS CONNECTIONS

1. **Biology** Because of its central role as an oxygen carrier in living things, hemoglobin has been studied extensively. Investigate how the iron in the heme group binds oxygen and report your findings to the class.

RESEARCH AND WRITING

1. Investigate and report on the uses of ferromagnetic oxides in magnetic recording tape.

2. Do library research and investigate how silver halides are used in the photographic process. Summarize your findings in a report.

CHAPTER 26
Aluminum and the Metalloids

PLANNING CHART

Approximate Time: 3/4 days

	SECTION/TEXT FEATURES		RESOURCES	
H O N O R S	**26.1 INTRODUCTION TO ALUMINUM AND METALLOIDS** *Periodic Table* Videodisc Section Review Chapter Review • Question 1	**751** 750D 753 763	*Teacher's Resource Binder* Section Review Worksheets *Other Resources* Blackline Master 60	97
H O N O R S	**26.2 ALUMINUM** Demo 1 *Periodic Table* Videodisc Section Review Chapter Review • Questions 2–8, Problems 1–3	**753** 750B 750D 758 763	*Teacher's Resource Binder* Lab Experiment 36 Section Review Worksheets *Other Resources* *Shakhashiri* Video Demo 25 *World of Chemistry* Video 19	189 98
H O N O R S	**26.3 REPRESENTATIVE METALLOIDS** Demo 2 *Periodic Table* Videodisc Section Review Chapter Review • Questions 9–12, Problems 4–5	**758** 750B 750D 761 763	*Teacher's Resource Binder* Section Review Worksheets *Other Resources* *World of Chemistry* Video 18	99
	CHAPTER END ACTIVITIES Critical Thinking, Cross Connections, Research and Writing, Projects		*Teacher's Resource Binder* Test Booklet *Computer Test Generator*	129, 131

Teaching Strategies

26.1 Introduction to Aluminum and Metalloids

When discussing the properties and reactions of these elements, ask students to keep in mind which properties are typically metallic and which are nonmetallic. Physical properties of metals include shininess, malleability and ductility, and good conduction of heat and electricity. Chemical properties include low ionization energies, and formation of positive ions and ionic compounds.

Have the students identify the metalloids in the Periodic Table. Make comparisons of the physical properties of these elements: their melting points, boiling points, appearances, densities, oxidation states, and ionization energies. Determine which element has the smallest temperature range for the liquid phase (boron) and how this relates to the nonmetallic behavior of boron. Discuss other generalizations in this section, such as variations from metallic to nonmetallic in a period from left to right, and the increase in metallic properties going down a group. Use these generalizations to account for the diagonal similarities. Discuss the characteristics of the hydroxide, oxide, and hydride of aluminum.

26.2 Aluminum

Identify aluminum as a metal by its physical properties. Exhibit aluminum foil, wire, and other aluminum objects. Ask students to name items that contain aluminum and discuss the abundance of aluminum.

You may wish students to report on the history and discovery of aluminum and the coincidence of the Hall-Heroult discovery. Exhibit a set of charts and materials such as bauxite, refined alumina, clay, cryolite, and carbon electrodes. Review the chemical equations for oxidation and reduction that take place in the electrolytic cell used in the production of aluminum.

Discuss the invisible coating of aluminum oxide on aluminum metal and the advantage of this type of "built-in" corrosion protection. Explain the reaction for the hydrolysis of aluminum ions in soution. Test the pH of a water solution of an aluminum salt. Discuss the fact that aluminum will readily react with acids such as hydrochloric acid, forming aluminum ions and hydrogen gas, but will not react with oxidizing acids such as nitric acid because of the aluminum oxide coating. Show that the amphoteric aluminum will also react with bases to form the aluminate ion and hydrogen gas.

Discuss the thermite reaction and the enormous amount of heat released by this process. Explain how this reaction can be used to form molten iron that can be used for welding.

Point out the usefulness of substances containing aluminum oxide: the abrasives corundum and emery, the gems ruby and sapphire, and the refractory substances such as crucibles.

Ask students to write formulas for various compounds that can be classed as alums. The octahedral alum crystals are easy to grow from saturated solutions, and some students may like to see how large a crystal they can produce from a seed crystal of potassium aluminum sulfate dodecahydrate suspended by a thread in a saturated alum solution.

Chemistry and its Applications: Can a Can Become a Plane?

Most students know that aluminum can be recycled, but they will be surprised to find that the recycled metal cannot be used for all purposes. However, recycled aluminum can be purified and made into new products at a fraction of the cost of obtaining aluminum from its ore. Discuss the ecological advantages of recycling and types of substances that can be recycled.

Demonstration 1: Aluminum Displacement of Copper

PURPOSE: To demonstrate the relative reactivities of aluminum and copper

MATERIALS: Copper(II) chloride crystals, tall graduated cylinder, water, strip of heavy aluminum foil

PROCEDURE: Dissolve 4-5 teaspoons of the copper(II) chloride in about 500 mL of water and put it in a graduated cylinder. Twist the foil into an interesting shape and drop it into the solution. Observe the changes that occur. (The foil becomes coated with a dark substance—metallic copper. The green-blue color of the solution fades. The aluminum metal dissolves.) The reaction is exothermic. This demonstration provides a number of observations that students can make.

$$2Al(s) + 3Cu^{2+}(aq) \rightarrow 2Al^{3+}(aq) + 3Cu(s)$$

26.3 Representative Metalloids

Discuss the properties of silicon, noting that they are typical of a metalloid. Compare the properties of carbon bonded in diamond to those of silicon, which has a diamond-type structure. Point out that the larger silicon atoms cannot form the double bonds needed for a graphite-like structure. In silicon semiconductor chips, each silicon atom is tetrahedrally bonded to four other silicon atoms in a covalent network.

Emphasize the fact that silicon forms silicon-oxygen bonds more readily than silicon-silicon bonds, and that the silicates that make up so much of the earth's crust contain these silicon-oxygen bonds. Have students look up some of the types of silicates that might be found locally. Ask them to report on the manufacture of glass.

Discuss the meaning of the term *polymer*. Compare polymeric compounds of silicones to those of hydrocarbons. Note that the silicones are very heat resistant and are used as lubricants and oils.

Compare the properties of arsenic and antimony and their compounds. Discuss the uses of their compounds.

Chemistry and Technology: Gallium Arsenide, Lasers, and Computer Chips

The quantity of information etched into a computer chip will astound most students. They will easily appreciate the complexity of the processes that must be employed to produce the chips. Suggest that students look up other methods that have been used to produce these circuits and that they find out what information is etched into the chips.

Demonstration 2: A Silicate Garden

PURPOSE: To demonstrate the growth of insoluble silicates

MATERIALS: Wide-mouth container (a small fish bowl is a good choice), sodium silicate solution (also called water glass), various colored small crystals such as iron(III) chloride (brown), nickel(II) nitrate (dark green), copper(II) chloride (bright green), cobalt(II) chloride (dark blue), and zinc nitrate (white)

PROCEDURE: Dilute 1 part sodium silicate to 4 parts water. Fill the container with the solution. Drop in 3-5 small crystals (matchhead size).

Observe that the crystals seem to grow like stalagmites within a few seconds. As the salts are placed in the solution, a semipermeable membrane composed of insoluble silicate is formed around the crystal. Because the concentration is greater inside the membrane, water enters the membrane by osmosis. This causes the membrane sack to break. It ruptures upward where the pressure is the least. A new membrane forms and the crystal grows. This is a real attention-getting demonstration.

References and Resources

Books and Periodicals

Garrett, Alfred B. "The Flash of Genius-Discovery of the Transistor," *Journal of Chemical Education,* Vol. 40, 1963, p. 302.

Kolb, Doris, and Kenneth E. Kolb, "Chemistry of Glass," *Journal of Chemical Education,* Vol. 56, 1979, p. 604.

Sarquis, Mickey, "Arsenic, Old Myths," *Journal of Chemical Education,* Vol. 56, 1979.

Shakhashiri, Bassam Z. "Thermite Reaction," *Chemical Demonstrations,* Vol. 1, University of Wisconsin Press, 1983, p. 85.

Summerlin, Lee R., Christie L. Borgford, and Julie B. Ealy, "Reduction of Sand with Magnesium," *Chemical Demonstrations, A Sourcebook for Teachers,* Vol. 2, American Chemical Society, 1987, p. 193.

Summerlin, Lee R., Christie L. Borgford, and Julie B. Ealy, "Recycling Aluminum," *Chemical Demonstrations, A Sourcebook for Teachers,* Vol. 2, American Chemical Society, 1987, p. 54.

Summerlin, Lee R., and James L. Ealy, Jr., "Silicate Garden," *Chemical Demonstrations, A Sourcebook for Teachers,* Vol. 1, American Chemical Society, 1985, p.30.

Audiovisual Resources

"Aluminum": video FFH1108F, Films for the Humanities and Sciences. Describes the extraction of aluminum from bauxite and the production of various alloys.

The Aluminum Association, Educational Services. Free or inexpensive materials available, including wall charts, "The Aluminum Film Catalog," listing motion pictures and filmstrips, and resource guides for teachers.

Organosilicon Chemistry Course (slides, cassette, chart, etc.) from Dow Corning Corporation.

Computer Software

"The Periodic Table Videodisc: Reactions of the Elements": CAV type videodisc, Journal of Chemical Education Software, 1989. Shows action sequences and still shots of each element in its most common forms, and reacting with air, water, acids, bases, and some common applications.

Answers and Solutions

Reviewing Concepts

1. Elemental aluminum is metallic, having familiar properties of metals. However, such chemical properties as the amphoterism of aluminum hydroxide, $Al(OH)_3$, the tendency to form the negative aluminate ion, $Al(OH)_4^-$, and formation of covalent bonds in some compounds are metalloidal.
2. Many clays, rocks, and minerals; principal ore is bauxite.
3. Density 2.70 g per cm^3; ductile; malleable; not as strong as copper, brass, or steel; good conductor of electricity; can be welded, spun, or cast; difficult to solder.
4. **a.** Oxides of aluminum **b.** Al_2O_3 **c.** As abrasives
5. Aluminum oxide has a very high melting point. It can be dissociated by being dissolved in fused cryolite.
6. To produce a carbon-free metal. Also because aluminum is a more powerful reducing agent.
7. **a.** $2Al(s) + 6H_3O^+ + 6H_2O \rightarrow 2Al(H_2O)_6^{3+} + 3H_2(g)$
   b. $2Al(s) + 2OH^- + 6H_2O \rightarrow 2Al(OH)_4^- + 3H_2(g)$
   c. The protective oxide coating isolates the aluminum from the water.
8. The aluminum cannot be produced economically from the clay.
9. Second among the elements of the earth's crust.
10. Arsenic dust and vapor and most compounds of arsenic are very toxic.
11. **a.** A compound containing a chain of alternate silicon and oxygen atoms, with organic groups attached to the silicon atoms
    b. As a lubricant, as a varnish for the windings of electric motors, as a water repellent, and in automobile and furniture polishes.
12. Because of the hydrocarbon groups attached to the silicon atoms.

Problems

1. $2Al + Fe_2O_3 \rightarrow Al_2O_3 + 2Fe$

   $10.0 \text{ kg Fe} \times \dfrac{10^3 \text{ g}}{\text{kg}} \times \dfrac{1 \text{ mol Fe}}{55.8 \text{ g Fe}} \times \dfrac{2 \text{ mol Al}}{2 \text{ mol Fe}}$
   $\times \dfrac{27.0 \text{ g Al}}{\text{mol Al}} \times \dfrac{1 \text{ kg}}{10^3 \text{ g}} = 4.84 \text{ kg Al}$

   $10.0 \text{ kg Fe} \times \dfrac{10^3 \text{ g}}{\text{kg}} \times \dfrac{1 \text{ mol Fe}}{55.8 \text{ g Fe}} \times \dfrac{1 \text{ mol Fe}_2O_3}{2 \text{ mol Fe}}$
   $\times \dfrac{160. \text{ g Fe}_2O_3}{\text{mol Fe}_2O_3} \times \dfrac{1 \text{ kg}}{10^3 \text{ g}} = 14.3 \text{ kg Fe}_2O_3$

2. $NaAl(SO_4)_2 \cdot 12H_2O$, molar mass = 458 g/mol
   $27.0 \text{ g Al} \div 458 \text{ g} \times 100\% = 5.90\% \text{ Al}$

3. $2Al + 2NaOH + 6H_2O \rightarrow 2NaAl(OH)_4 + 3H_2$

   $50.0 \text{ g Al} \times \dfrac{1 \text{ mol Al}}{27.0 \text{ g Al}} = 1.85 \text{ mol Al}$

   $100. \text{ g NaOH} \times \dfrac{1 \text{ mol NaOH}}{40.0 \text{ g NaOH}} = 2.5 \text{ mol NaOH}$

   By inspection: NaOH is in excess.

   $1.85 \text{ mol Al} \times \dfrac{3 \text{ mol H}_2}{2 \text{ mol Al}} \times \dfrac{22.4 \text{ L}}{\text{mol}} = 62.2 \text{ L H}_2$

4. $Ca_2B_6O_{11} \cdot 5H_2O$, molar mass = 411 g
   6B, molar mass = 64.9 g
   $64.9 \text{ g B} \div 411 \text{ g} \times 100\% = 15.8\% \text{ B}$

5. $96.2 \text{ g As} \times \dfrac{1 \text{ mol As}}{74.9 \text{ g As}} = 1.28 \text{ mol As}$

   $3.85 \text{ g H} \times \dfrac{1 \text{ mol H}}{1.01 \text{ g H}} = 3.81 \text{ mol H}$

   $\dfrac{1.28 \text{ mol As}}{1.28} : \dfrac{3.81 \text{ mol H}}{1.28}$

   1.00 mol As : 2.98 mol H
   Simplest formula, AsH_3

   $\dfrac{3.48 \text{ g}}{1 \text{ L}} \times \dfrac{22.4 \text{ L}}{\text{mol}} = 78.0 \text{ g/mol}$

   molar mass AsH_3 = 77.9454
   The molecular formula is AsH_3.

Cross Connections

1. Availability of cheap electricity, usually generated by water power.

Videodisc Selections
from *The Periodic Table Videodisc* (JCE Software)

step forward

step reverse

CHAPTER 26

pg 751 Aluminum and metalloids

8255
aluminum (s)

3816
boron (s)

11574
silicon (s)

32429
germanium (s)

32434
germanium + air (m)

36718
uses of germanium (s)

33218
arsenic (s)

42664
antimony (s)

pg 753 Reactions and compounds of aluminum

8267
aluminum + air (m)

9040
aluminum + acids (m)

10310
aluminum + bases (m)

36462
uses of aluminum (s)

pg 758 Silicon

11574
silicon (s)

36468
uses of silicon (s)

36476
opal (s)

pg 760 Arsenic and antimony

36723
uses of arsenic (s)

36915
uses of antimony (s)

Chapter 26
Aluminum and the Metalloids

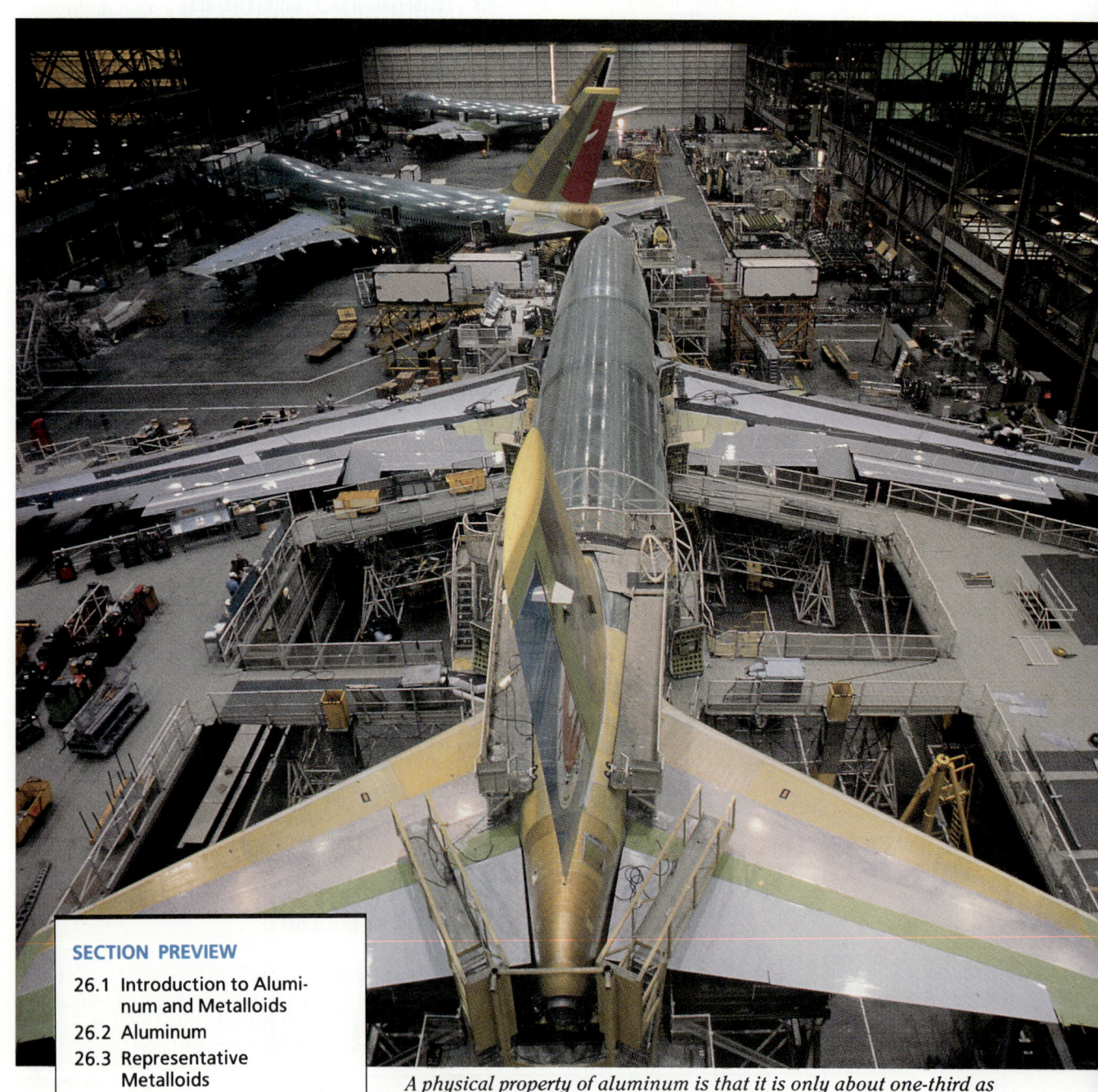

SECTION PREVIEW

26.1 Introduction to Aluminum and Metalloids

26.2 Aluminum

26.3 Representative Metalloids

A physical property of aluminum is that it is only about one-third as dense as steel, making it an ideal material for constructing aircraft.

26.1 Introduction to Aluminum and Metalloids

Some elements are neither distinctly metallic nor distinctly nonmetallic. Their properties are intermediate between those of metals and nonmetals. As a group, these elements are called *metalloids* or *semimetals*. In the Periodic Table, they occupy a diagonal region from the upper center toward the lower right as shown in Figure 26-1.

The metalloids are the elements boron, silicon, germanium, arsenic, antimony, and tellurium. Although aluminum and selenium are not included in the metalloids, they are included in this chapter because of their unique positions in the Periodic Table relative to the metalloids.

SECTION OBJECTIVES

- Name the elements known as metalloids and locate their position in the Periodic Table.
- Explain the ways in which aluminum can be related to the metalloids.
- Describe the general nature of the metalloids, and list specific properties of the metalloid elements.

The Nature of Aluminum

Elemental aluminum is distinctly metallic and possesses many of the familiar properties of metals. But aluminum forms the negative aluminate ion, $Al(OH)_4^-$. Its hydroxide, $Al(OH)_3$, is amphoteric. The oxide of aluminum is ionic, yet AlH_3 and $AlCl_3$ contain covalent bonds. Aluminum is so resistant to oxidation that it can be used as cookingware on hot stove burners. Yet it is easily oxidized in sodium hydroxide solution to yield hydrogen and sodium aluminate. In these respects, aluminum is similar to the metalloids.

Amphoteric hydroxides dissolve in both acids and bases:
$Al(OH)_3(s) + 3H^+(aq)$
$\rightarrow Al^{3+}(aq) + 3H_2O$
$Al(OH)_3(s) + OH^-(aq) \rightarrow$
$Al(OH)_4^-(aq)$

The Nature of Metalloids

Silicon, arsenic, and antimony are typical metalloids and are discussed in Section 26.3. Table 26-1 lists some properties of these elements together with those of aluminum and the other metalloids.

Figure 26-1 The metalloids are the elements *boron, silicon, germanium, arsenic, antimony, selenium,* and *tellurium*. Aluminum is a metal but it has some unique properties that relate it to the metalloids.

Aluminum and the Metalloids

TABLE 26-1 PROPERTIES OF ALUMINUM AND METALLOIDS

Element	Atomic Number	Electron Configuration	Oxidation States	Melting Point (°C)	Boiling Point (°C)	Density (g/cm³)	Atomic Radius (pm)	First Ionization Energy (kJ/mol)
boron	5	2,3	+3	2079	2550	2.34	79	801
aluminum	13	2,8,3	+3	660.4	2467	2.70	143	578
silicon	14	2,8,4	+2, +4	1410	2355	2.33	117	787
germanium	32	2,8,18,4	+2, +4	937.4	2830	5.32	122	762
arsenic	33	2,8,18,5	+3, +5, −3	sublimes		5.73	121	944
selenium	34	2,8,18,6	+4, +6, −2	217	685	4.5	116	941
antimony	51	2,8,18,18,5	+3, +5, −3	630.7	1750	6.69	141	832
tellurium	52	2,8,18,18,6	+2, +4, +6, −2	450	990	6.24	130	869

The amphoterism of aluminum hydroxide is discussed in detail in Section 16.4

The metalloids in general have atomic radii and ionization energies intermediate between those of the metals and the nonmetals. The electronegativities of these elements are also intermediate, most being approximately 2. The metalloids form covalent bonds in many compounds. Their most characteristic property is electrical conductivity that is low at ordinary temperatures, but increases with rising temperature. (The conductivity of metals decreases with rising temperature.)

Germanium is a moderately rare element. With the development of the transistor, germanium became important as a semiconductor material. It is chemically similar to silicon, the element above it in Group 14. Germanium is more metallic than arsenic just to its right in period four. The major oxidation state of germanium is +4.

Selenium and tellurium, members of Group 16, are somewhat rarer elements than germanium. Both have also become valuable because of their use in electronic devices. Selenium has several allotropes, including red, gray, and black forms. Black selenium, the commercially available form, is brittle but has a metallic luster. Selenium is a photoconductor—its ability to conduct electricity increases when light shines on it. This property is utilized in photoelectric cells and also in creating images in photocopying machines.

A semiconductor is a material that conducts electricity better than an insulator but not as well as a conductor. Semiconductors are used extensively in modern computer and other important electronic devices.

Selenium has positive oxidation states of +4, and +6. Like sulfur, just above in the group, it forms a dioxide, SeO_2, and an oxyacid, H_2SeO_4. Selenium also forms dihalides, such as Se_2Cl_2, and selenides, such as Na_2Se, which contain the selenide ion, Se^{2-}. Selenium and its compounds are used in metallurgy, glass and ceramics, pigments, and medicines. As expected, the metal-like characteristics of Group-16 elements increase down the group. Tellurium is more metallic than selenium above it and less metallic than polonium below it. This gradation in properties down the group is illustrated by the change in odor of the hydrogen compounds. Hydrogen oxide (water) is odorless. Hydrogen sulfide has the offensive odor of rotten eggs. The odor of hydrogen selenide is even more offensive, and that of hydrogen telluride is the foulest.

Although it can be amorphous, tellurium is more commonly a brittle, silvery, metal-like crystalline substance. It is classed as a semiconductor, and its chemistry is typically metalloidal. It appears with oxygen in both tellurite

(TeO_3^{2-}) and tellurate (TeO_4^{2-}) ions. In these ions, tellurium shows the +4 and +6 oxidation states, respectively. Tellurium combines covalently with oxygen and the halogens, to give compounds in which the +2, +4, and +6 oxidation states are observed. It forms tellurides (−2 oxidation state) with such elements as gold, hydrogen, and lead. In fact, tellurium is the only element combined with gold in nature.

Section Review

1. Which elements are known as metalloids and where they are located in the Periodic Table?
2. What general characteristics make aluminum similar to the metalloids?
3. Describe some of the important properties of the metalloids.
4. Describe some of the properties of selenium and tellurium.

1. Boron, silicon, germanium, arsenic, antimony, and tellurium are metalloids. They occupy a diagonal region from the upper center to the lower right in the Periodic Table.
2. Although elemental aluminum is distinctly metallic, it forms some compounds that exhibit ionic character and others in which it is covalently bonded. Furthermore, its compound, $Al(OH)_3$, is amphoteric.
3. Accept properties listed in Table 26-1 and properties given in the subsection "The Nature of Metalloids."
4. Accept properties of selenium and tellurium from Table 26-1 and those given in the subsection "The Nature of Metalloids."

26.2 Aluminum

Aluminum, atomic number 13, is the second member of Group 13. This group is headed by boron and includes gallium, indium, and thallium. All are typically metallic except boron, which is classed as a metalloid. The chemistry of boron differs from that of aluminum and the other Group 13 elements mainly because of the small size of the boron atom. Its chemistry resembles that of silicon and germanium more than it does that of aluminum. Much of the chemistry of aluminum is similar to that of the corresponding Group 2 metals, magnesium and beryllium.

Aluminum is a low-density metal; it is about one-third as dense as steel. The pure metal is used in chemical processes, in electronics, for forming jewelry, and as foil wrapping. Aluminum alloys are used in structural and industrial applications.

Aluminum is the most abundant metal in the earth's crust. Although it is found in many clays, rocks, and minerals, the only source for economical commercial production is **bauxite,** *a hydrated aluminum oxide ore.* As a source of aluminum, bauxite containing 40–60% aluminum oxide is required for present aluminum-recovery technology.

Major foreign sources for the recovery of bauxite are Australia, Jamaica, Surinam, and Guyana. Bauxite is also mined in Georgia, Alabama, Tennessee, and Arkansas.

SECTION OBJECTIVES

- Cite the occurrence, recovery, and uses of pure aluminum.
- Describe the properties of aluminum.
- Describe the *thermite reaction* and discuss its uses.
- Give the properties and uses of some important aluminum compounds.

Recovery of Aluminum

In what is known as the **Hall process** *aluminum is extracted by electrolyzing anhydrous aluminum oxide (refined bauxite) dissolved in molten* **cryolite,** *Na_3AlF_6.* The process requires a temperature slightly below 1000°C.

In the electrolytic cell, an iron box lined with graphite serves as the negative electrode, graphite rods serve as the positive electrode, and molten cryolite–aluminum oxide is the electrolyte. The cryolite is melted in the cell, and the aluminum oxide dissolves as it is added to the molten cryolite. The operating temperature of the cell is above the melting point of aluminum (660°C). The aluminum metal becomes liquid and sinks to the bottom of the cell where it is easily drawn off.

This process for extracting aluminum from its ores was independently discovered in 1886 by two 23-year-old men, Charles Hall in the United States and Paul Heroult in France. Before this time, pure aluminum was very rare and expensive. In the 1860s, Napoleon III of France used a fork made of aluminum which was considered more precious than the gold and silver tableware his guests used.

Aluminum and the Metalloids

Chemistry and its Applications

Can A Can Become A Plane?

Aluminum recycling has become increasingly common. What happens to the recycled aluminum cans? Can they end up as part of an airplane? Probably not, because the metal recovered from aluminum cans is not pure aluminum.

Until recently, titanium was found in recycled aluminum because the paint on aluminum cans contained titanium oxide. The metal ions accumulated in the recycled aluminum, building up to a high enough level to affect the quality of the aluminum being produced. Now cans are painted with organic dyes. Even so, recycled aluminum cans still contain small amounts of silicon (from glass), iron, and copper. These elements can stick together in particles as large as 10–50 micrometers. Such colloid particles can cause the recycled aluminum to crack when being rolled into sheets. Recycled aluminum cans cannot be used, therefore, for critical purposes such as aircraft parts.

The electrode reactions are complex and are not understood completely. Aluminum is reduced at the negative electrode. The graphite rods that serve as a positive electrode are gradually converted to carbon dioxide. This fact suggests that oxygen is formed at the positive electrode by oxidation of the O^{2-} ion. The following equations for the reaction mechanism summarize the oxidation–reduction processes in the cell.

negative electrode: $4Al^{3+} + 12e^- \rightarrow 4Al$
positive electrode: $6O^{2-} \rightarrow 3O_2 + 12e^-$
$3C + 3O_2 \rightarrow 3CO_2(g)$

Another important source of aluminum is scrap metal, which provides a valuable source of aluminum at a much lower cost than that of the electrolysis of aluminum oxides.

Properties of Aluminum

Aluminum has a density of 2.70 g/cm³ and a melting point of 660°C. It is ductile and malleable but is not as strong as brass, copper, or steel. Only silver, copper, and gold are better conductors of electricity. Aluminum can be welded, cast, or spun and can be soldered only by using a special solder.

Aluminum is a very active metal. The surface is always covered with a thin layer of aluminum oxide, which is not affected by air or moisture; aluminum is a self-protective metal that resists corrosion. At high temperatures, the metal combines with oxygen, releasing intense light and heat. Photoflash lamps contain aluminum foil or fine wire in an atmosphere of oxygen.

Aluminum is a very good reducing agent but is not as active as the Group 1 and Group 2 metals.

$$Al \rightarrow Al^{3+} + 3e^-$$

The Al^{3+} ion is quite small and carries a large positive charge. The ion hydrates vigorously in water solution and is usually written as the hydrated ion, $Al(H_2O)_6^{3+}$. Water solutions of aluminum salts are generally acidic because of the hydrolysis of $Al(H_2O)_6^{3+}$ ions.

$$Al(H_2O)_6^{3+} + H_2O \rightarrow Al(H_2O)_5OH^{2+} + H_3O^+$$

Water molecules are amphoteric but are very weak proton donors or acceptors. Water molecules that hydrate the Al^{3+} ion give up protons more readily, however. This increased activity results from the repulsion effect of the highly positive Al^{3+} ion. In the hydrolysis shown above, a water molecule succeeds in removing one proton from the $Al(H_2O)_6^{3+}$ ion.

Hydrochloric acid reacts with aluminum to form aluminum chloride and release hydrogen. The net ionic equation is

$$2Al(s) + 6H_3O^+ + 6H_2O \rightarrow 2Al(H_2O)_6^{3+} + 3H_2(g)$$

Nitric acid does not react readily with aluminum because of aluminum's protective oxide layer.

In basic solutions, aluminum forms aluminate ions, $Al(OH)_4^-$, and releases hydrogen.

$$2Al(s) + 2OH^- + 6H_2O \rightarrow 2Al(OH)_4^- + 3H_2(g)$$

If the base is sodium hydroxide, sodium aluminate, $NaAl(OH)_4$, is the soluble product, and hydrogen is given up as a gas.

$$2Al(s) + 2NaOH + 6H_2O \rightarrow 2NaAl(OH)_4 + 3H_2(g)$$

The oxide surface coating protects aluminum under normal atmospheric conditions and must be removed before the metal underneath can react with hydronium ions or with hydroxide ions. Hydroxide ions dissolve this oxide layer more readily than do hydronium ions. This explains why aluminum reacts more readily with hydroxides than with acids.

The Thermite Reaction

When a mixture of powdered aluminum and an oxidizing agent such as iron(III) oxide is ignited, the aluminum reduces the oxide to the free metal. This reduction is rapid and violent and yields a tremendous amount of heat. The sudden release of this heat energy produces temperatures from 3000 to 3500°C, enough to melt the iron. *Such a reaction between aluminum and the oxide of a less active metal is called the* **thermite reaction.** *The mixture of powdered aluminum and iron(III) oxide is called* **thermite.**

The Al_2O_3 coating on the aluminum is transparent and invisible.

Ductile: Capable of being drawn into wire. Malleable: Capable of being rolled into sheets

Solder is any metal alloy used when melted for joining or patching metal parts or surfaces.

The Al^{3+} ion is so small and highly charged that it has a very great charge density. This means that it will strongly attract electrons. The water molecules bonded to the aluminum ion have their electrons pulled so strongly toward the aluminum that the bonds between oxygen and hydrogen are weakened. Hydrogen ions readily break away.

Because HNO_3 is an oxidizing acid, it enhances the protective Al_2O_3 layer on the aluminum surface. Nitric acid can even be stored in aluminum containers. Nonoxidizing acids such as HCl will react with aluminum.

The formation of aluminum oxide is strongly exothermic. The reaction releases 1669 kJ of heat per mole of aluminum oxide formed. The heat of formation of iron(III) oxide is 820 kJ per mole. In the thermite reaction, the amount of heat released per mole of aluminum oxide formed equals the difference between these values. The net thermite reaction is considered to be the sum of these two separate reactions.

$$2Al + \tfrac{3}{2}O_2 \rightarrow Al_2O_3 \qquad \Delta H = -1669 \text{ kJ}$$
$$Fe_2O_3 \rightarrow 2Fe + \tfrac{3}{2}O_2 \qquad \Delta H = +820 \text{ kJ}$$
$$\overline{2Al + Fe_2O_3 \rightarrow 2Fe + Al_2O_3 \qquad \Delta H = -849 \text{ kJ}}$$

It is not practical to use aluminum to produce cheaper metals by the reduction of their oxides. The thermite reaction is often used to produce small quantities of carbon-free metal, however. A more important use of this reaction is to reduce metallic oxides that are not readily reduced with carbon. Chromium, manganese, titanium, tungsten, and molybdenum can be recovered from their oxides by the thermite reaction. All these metals are used in making alloy steels.

The very high temperature produced by the thermite reaction makes it useful in welding. Large steel parts, such as propeller shafts and rudder posts on ships or the crankshafts of heavy machinery, are repaired by thermite welding, as shown in Figure 26-2.

Some Compounds of Aluminum

Because of its amphoteric properties, aluminum and its oxide react both with acids and bases to form compounds with many useful properties. In this section, we limit our discussion to only a few aluminum compounds and their uses.

Aluminum oxide Bauxite, the chief ore of aluminum, contains the hydrated oxide $Al_2O_3 \cdot 2H_2O$. **Corundum** *and* **emery** *are minerals containing natural oxides of this metal that are useful as abrasives.* Emery is the abrasive of emery boards, emery paper, emery cloth, and emery grinding wheels.

The thermite reaction provides a convenient way to weld railroad tracks because no heavy tanks of gas or electrical equipment are needed.

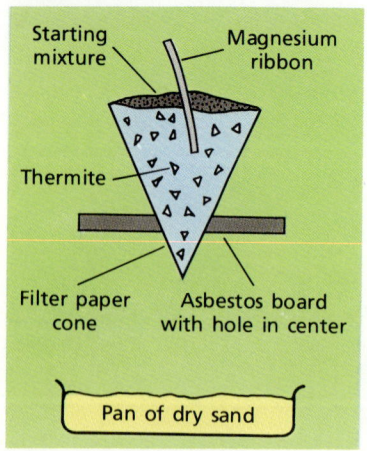

Figure 26-2 The very high temperature produced by the thermite reaction makes it useful in welding large steel parts, as in shipbuilding.

Anhydrous aluminum oxide, usually a white powder called **alumina**, *is a very inert substance having an extremely high melting point of 2027°C.* It is used in crucibles for high-temperature reactions.

Pure crystalline alumina, Al_2O_3, is clear and colorless. The **ruby** *and* **sapphire** *gemstones,* shown in Figure 26-3, *are composed of aluminum oxide contaminated with traces of certain impurities.* The presence of a trace quantity of chromium imparts the brilliant ruby-red color to a gem-quality crystal of Al_2O_3, yielding one of the most valuable gemstones found in nature whereas traces of titanium yield the blue sapphire.

Synthetic rubies and sapphires are produced by fusing aluminum oxide with the required trace contaminants in the flame of an oxyhydrogen blowtorch. As synthetic gemstones, they are used in jewelry. Because of the hardness of aluminum oxide, these synthetic crystals are also used as bearings (jewels) in watches and other precision instruments and as dies for drawing wires.

Scientists are finding new ways to strengthen and stiffen structural materials. One important development involves **fiber composites**, *monocrystalline strands or fibers of one substance embedded in and held in place by some other material.* An aluminum rod containing very hard, strong sapphire "whiskers" is quite unlike ordinary soft, ductile aluminum. Such a structure is six times stronger than aluminum and twice as stiff. These fiber composites enable engineers to greatly improve the strength-to-weight ratio of structural materials.

Alundum is a material made by melting and resolidifying Al_2O_3. It is used for making grinding wheels and other abrasives. It is also found in crucibles, funnels, tubing, and other pieces of laboratory equipment.

The alums *Alums* are common compounds that usually contain aluminum and have the general formula $M^+M^{3+}(SO_4)_2 \cdot 12H_2O$. The M^+ can be any one of several monopositive ions but is usually K^+, Rb^+, or NH_4^+. The M^{3+} can be any one of a number of tripositive ions such as Al^{3+}, Cr^{3+}, Co^{3+}, Fe^{3+}, or Mn^{3+}. Potassium aluminum sulfate, $KAl(SO_4)_2 \cdot 12H_2O$, is the most common alum. It is easy to see from this formula why alums are sometimes called double sulfates.

All alums, also called aluminum salts, have the same crystal structure. This crystal-structure requirement explains why some monopositive and tripositive ions do not form alums. The alum crystals are hydrated structures

Figure 26-3 Both synthetic and natural sapphires (left) and rubies (right) are composed of aluminum oxide, Al_2O_3. The blue color of sapphire is due to a trace of titanium. The red color of a ruby is due to a trace of chromium.

1. In many clays, rocks, and minerals. Recovered by electrolyzing anhydrous aluminum oxide dissolved in molten cryolite at a temperature slightly below 1000°C
2. Used in chemical processes, in electronics, for forming jewelry, and as foil wrapping. Aluminum alloys are used in structural and industrial applications.
3. Density of 2.70 g/cm³; melting point of 660°C; ductile and malleable; good conductor of electricity; very active metal that forms a protective layer of aluminum oxide on its surface;

SECTION OBJECTIVES

- Describe the occurrence, properties, and uses of silicon and its synthetic compounds.
- Discuss the occurrence and properties of arsenic and its compounds. Describe ways in which arsenic can be used.
- Describe the occurrence, properties, and uses of antimony and its compounds.

combines with oxygen at high temperatures; good reducing agent. Accept any other properties listed in the subsection "Properties of Aluminum."

4. A reaction between a mixture of powdered aluminum and the oxide of a less active metal that is strongly exothermic and produces very high temperatures. Often used to produce small quantities of carbon-free metals and to reduce oxides not readily produced by carbon. The high temperatures make it useful in welding heavy machinery.
5. Accept all answers from the subsection "Some Compounds of Aluminum."

in which six water molecules are coordinated to each metallic ion. This coordination number of six for both the monopositive and tripositive ions may restrict the kinds of metallic ions that can form alums.

Alums are used for water purification, as mordants, and in paper manufacture. $NaAl(SO_4)_2 \cdot 12H_2O$ is used as the acid compound in one type of baking powder.

Section Review

1. Where is aluminum found in nature, and how is it recovered?
2. What are some of the uses of aluminum and aluminum alloys?
3. List the physical and chemical properties of aluminum.
4. Describe the thermite reaction and list some important uses.
5. Name some aluminum compounds and their uses.

26.3 Representative Metalloids

Recall that the metalloids are elements that exhibit both metallic and nonmetallic properties to some extent. *Those elements and compounds that are characterized by a slight electric conductivity that increases with temperature are known as* **semiconductors**. This property is the reason for the use of metalloids in the electronics industry. Semiconductors permit the manufacture of tiny electronic devices such as transistors. A single computer memory chip, like those shown in Figure 26-4, may contain over a million transistors.

In this section, you will study the properties and uses of some of the metalloids.

Silicon

Silicon is as important in the mineral world as is carbon in living systems. Silicon dioxide and silicates make up about 87 percent of the material of the earth's crust. Silicon does not occur free in nature. It is found in many minerals in the form of SiO_2, as shown in Figure 26-5.

Properties of silicon Silicon atoms have four valence electrons. Silicon crystallizes with a tetrahedral-bond arrangement similar to that of carbon atoms in diamond. Atoms of silicon also have small atomic radii and tightly held electrons. Thus their ionization energy and electronegativity are fairly high. Silicon is a metalloid. It forms bonds with other elements that are essentially covalent. The electric conductivity of silicon is similar to that of the other semiconductors. Unlike carbon, silicon forms only single bonds. It forms silicon–oxygen bonds more readily than silicon–silicon or silicon–hydrogen bonds. However, much of its chemistry is similar to that of carbon. Silicon has much the same role in mineral chemistry as carbon has in organic chemistry.

Boron and silicon atoms have roughly similar small radii. This similarity allows them to be substituted for one another in glass, even though boron is a member of Group 13 and silicon of Group 14.

Because the relative size of atoms frequently defines properties, elements in diagonal positions in the Periodic Table (such as boron-silicon-arsenic)

Figure 26-4 Chunks of raw polysilicon are processed into silicon ingots, which are sliced and polished to form wafers. The silicon wafers can then be used to make semiconductor integrated circuits.

often have similar properties. In general, atom size decreases to the right from Groups 1 to 18. It increases down each group. This balancing effect results in atoms of similar size along diagonals in the Periodic Table.

Elemental silicon, of about 98 percent purity, is used in the production of alloys of iron, spring steel, copper, and bronze. It is also used in the manufacture of *silicones,* polmeric organosilicon compounds composed of silicon, carbon, oxygen, and hydrogen. Ultrapure silicon is used in microcomputers, transistors, and solar cells.

Extremely pure silicon that contains a trace of arsenic is used to make **n**-type semiconductors. Arsenic has one more valence electron than silicon, and this extra electron is a negative charge carrier that can diffuse through the crystal. A **p**-type semiconductor contains a trace of boron. With one less valence electron, a positive electron "hole" is formed, which can also diffuse through the crystal.

Silicones Silicon resembles carbon in the ability of its atoms to form chains. Unlike carbon chains, which are formed by carbon–carbon bonds, silicon chains are usually formed by silicon–oxygen bonds. In *silicones,* the silicon chain is bound together with oxygen atoms. Hydrocarbon groups are attached to the silicon atoms. Thus the silicones are part organic and part inorganic. By using different hydrocarbon groups, a variety of silicones can be produced. One silicone chain has the structure

$$\begin{bmatrix} & H & & H \\ & | & & | \\ H-&C-H & H-&C-H \\ & | & & | \\ -&Si-&O-&Si- \\ & | & & | \\ H-&C-H & H-&C-H \\ & | & & | \\ & H & & H \end{bmatrix}_x$$

The silicones are not greatly affected by heat. They have very good electric insulating properties and are water repellent. Water does not penetrate cloth treated with a silicone. Some silicones have the character of oils or greases and can be used as lubricants. Silicone varnishes are used to coat wires for the windings of electric motors. This insulation permits the electric motor to operate at high temperatures without developing short circuits. Silicones are used in automobile and furniture polishes.

Figure 26-5 Silicon, the second most abundant element. Its abundance is due to the wide distribution in nature of silicon dioxide (quartz rocks) and many different silicates.

Aluminum and the Metalloids

Figure 26-6 Amethyst is a form of silicon dioxide (SiO_2) crystals.

Arsenic

Arsenic occurs in nature principally in the form of sulfide minerals such as orpiment, As_2S_3, realgar, As_2S_2, and arsenopyrite, FeAsS. It is also found in the mineral arsenolite, As_2O_6, and in metal arsenides such as $FeAs_2$ and Co_2As.

Arsenic is allotropic and exists in two distinct crystalline forms, known as yellow arsenic and metallic arsenic. In the yellow form, arsenic is present as As_4 tetrahedral molecules. In the metallic form, it is present as sheets of atoms. Yellow arsenic is soft, has a density of 2.03 g/cm^3, and is metastable—it passes rapidly to the metallic form by heating or by the action of light. Metallic arsenic is a brittle, gray solid with density of 5.73 g/cm^3. Although it is a good conductor of heat, it is a very poor conductor of electricity.

Chemically, arsenic may act as a metal and form oxides and chlorides. It may also act as a nonmetal and form acids. When heated in moist air, arsenic sublimes and forms As_2O_3. Metallic arsenic has few commercial uses except as a hardening agent in alloys. A trace of arsenic is used to improve the hardness and sphericity of lead shots. Compounds of arsenic are used mostly for their toxicity as weed killers, sheep and cattle dips, and insecticides.

Antimony

Antimony is found in nature principally as the black sulfide ore, stibnite, Sb_2S_3. Both the metal and the ore have been used since antiquity for making vases and ornaments and in cosmetics. Antimony, like arsenic, exhibits the property of allotropy. It exists in the yellow and metallic forms. Yellow antimony is a soft solid with a density of 5.3 g/cm^3. This form of antimony is very unstable, and it exists only at very low temperatures. Metallic antimony is a dense, brittle, silver-white metalloid with a bright metallic luster and a density of 6.7 g/cm^3. It is less active than arsenic. When strongly heated in air, antimony forms a white oxide, Sb_2O_3.

An alloy of lead and antimony is used for the plates in storage batteries. This alloy is stronger and more resistant to acids than lead alone. Tartar emetic, or potassium antimony tartrate, $KSbOC_4H_4O_6$, is used as a mordant in the dyeing of cotton goods.

Metalloids in computers Silicon is a primary ingredient in the integrated circuits of computers. When precise quantities of the metalloids arsenic, antimony, or boron are added to high purity silicon, the electrical conductivity

The symbol for antimony is based on its Latin name, *stibium*.

CAUTION
The soluble compounds of antimony are almost as toxic as those of arsenic.

Figure 26-7 *Solar cells* (below) which are high purity silicon crystals treated with certain metalloids convert light to electricity. A diagram of a typical solar cell is shown at right.

Chemistry and Technology

Gallium Arsenide, Lasers, and Computer Chips

Gallium arsenide (GaAs) is a semiconductor similar to silicon, except that its crystals can be made to "lase"—to emit light of a single wavelength. By including some aluminum arsenide, a GaAs laser can be tuned to other wavelengths, as applications demand. The GaAs laser acts like a heat gun, focusing infrared light into an extremely small region for nanosecond intervals. This technique is extraordinarily useful for the manufacture of computer chips.

The procedure used now to make computer chips is to cover a wafer of silicon with a stencil that shows the wiring layout needed. The wafer is etched to expose a clean surface, and then a thin layer of metal is deposited into the etched lines. The stencil is lifted and the microscopic metal circuit is sealed with a polymer coating.

Gallium arsenide lasers will soon be used to "draw" a circuit on a silicon wafer that is surrounded by a gaseous mixture of tungsten hexafluoride and hydrogen. The energy of the laser will cause these gases to react to form tungsten metal, but only in the microscopic region where the laser is focused. Essentially, this makes the laser act like a pencil, drawing metal circuit lines wherever it traces. In the microchip industry, such a laser pencil will allow much greater flexibility in the design of microcircuits.

of the silicon sample can be changed dramatically, depending on the voltage that is applied. This voltage-sensitive conductivity was the basis for the operation of vacuum tubes which you may have seen in old radio or television sets. Circuits made of silicon have many of the same electronic properties of the old vacuum tubes but can operate using a tiny fraction of the power and space required by the vacuum tubes. This is how electronic engineers pack so much computing power into today's computers.

Section Review

1. Identify some properties of silicon and discuss its uses.
2. What are silicones and how are they used?
3. In what ways are antimony and arsenic the same? In what ways are they different?
4. Name some of the uses of arsenic and antimony.

Figure 26-8 Gallium arsenide (GaAs), as well as silicon, is an important ingredient in advanced electronic devices. The computer chip shown here is composed largely of GaAs. A single computer memory chip may contain over a million transistors.

Chapter Summary

- Metalloids have properties intermediate between those of metals and nonmetals. They occupy a diagonal region of the periodic table from upper center toward lower right. The metalloids include boron, silicon, germanium, arsenic, antimony, and tellurium. Selenium has some metalloid properties.
- Although certain of its properties tend to place aluminum among the metalloids, it is considered a metal. It forms negative aluminate ions and some covalent halides.
- Aluminum is distinctly metallic in many of its properties. It always has an oxide surface coating that protects it from corrosion by air and moisture.
- Aluminum is recovered by electrolysis of the aluminum oxide dissolved in molten cryolite. It is used as a low-density structural metal when alloyed with selected elements. Powdered aluminum mixed with iron(III) oxide and subsequently ignited reduces the oxide to the free metal in a rapid, violent combustion reaction known as the thermite reaction.
- Some natural or synthetic oxides of aluminum are very hard and are used as abrasives.
- Alums are double sulfates with the general formula $M^+M^{3+}(SO_4)_2 \cdot 12H_2O$.
- Silicon forms strong bonds with oxygen. Silicon–oxygen lattice chains are similar to carbon–carbon chains. Much of the chemistry of the element silicon is similar to that of carbon.
- Silicon chains bound together with oxygen atoms and having hydrocarbon groups attached to the silicon atoms are called silicones. Some silicones have properties similar to organic oils and greases.
- Arsenic and antimony are used in alloys. Arsenic is a hardening agent for lead. Antimony and lead are used in "antifriction" alloys and battery-plate alloys.
- Silicon mixed with precise amounts of the metalloids such as antimony, arsenic, or boron has electronic properties that can be used to construct integrated circuits.

CHAPTER 26 REVIEW

Aluminum and the Metalloids

VOCABULARY

alumina (757)
bauxite (753)
corundum (756)
cryolite (753)
emery (756)
fiber composite (757)
Hall process (753)
ruby (757)
sapphire (757)
semiconductor (758)
thermite (755)
thermite reaction (755)

REVIEWING CONCEPTS

1. Why is aluminum, a familiar structural metal, included in the study of metalloids? (26.1)
2. In what materials does aluminum occur in nature? (26.2)
3. What are the important physical properties of aluminum? (26.2)
4. a. What is the chemical composition of corundum and emery?
   b. Write the formula for these substances.
   c. What is their important use? (26.2)
5. Why must aluminum oxide be dissolved in molten cryolite before it can be decomposed by electricity? (26.2)
6. Why are certain metallic oxides reduced with aluminum rather than with carbon? (26.2)
7. What reaction occurs when aluminum is placed in
   a. hydrochloric acid solution
   b. sodium hydroxide solution?
   c. Why is there no reaction in neutral water? (26.2)
8. Why is bauxite imported from the West Indies and South America when almost any clay bank in the United States contains aluminum? (26.2)
9. How does silicon rank in abundance among the elements? (26.3)
10. Why must extreme care be used in handling arsenic and its compounds? (26.3)
11. a. What is a silicone?
    b. What are some of the important uses for silicones? (26.3)
12. Why would you expect silicones to be water repellent? (26.3)

PROBLEMS

1. How much aluminum and how much iron(III) oxide must be used in a thermite mixture to produce 10.0 kg of iron for a welding job?
2. What is the percentage of aluminum in sodium alum that crystallizes with 12 molecules of water of hydration?
3. How many liters of hydrogen can be prepared by the reaction of 50.0 g of aluminum metal and 100. g of sodium hydroxide in solution?
4. Calculate the percentage of boron in colemanite, $Ca_2B_6O_{11} \cdot 5H_2O$.
5. A compound contains 96.2% arsenic and 3.85% hydrogen. Its vapor is found to have a density of 3.48 g/L at STP. What is the molecular formula of the compound?

CROSS CONNECTIONS

1. **Geography** What geographic conditions affect the location of plants for the production of aluminum from purified bauxite?

RESEARCH AND WRITING

1. Perform library research and write a report on the current state of optical fibers. How are optical fibers used? Where do you encounter optical fibers in everyday life?
2. Describe the role of silicon in the formation of minerals. Conduct library research to discover why minerals such as quartz are found in a variety of colors.

CHAPTER 27
Sulfur and Its Compounds

PLANNING CHART
Approximate Time: 2/3 days

	SECTION/TEXT FEATURES		RESOURCES	
H O N O R S	**27.1 ELEMENTAL SULFUR** Demos 1, 2 *Periodic Table* Videodisc Section Review Chapter Review • Questions 1–3, Problem 1	765 764A–B 764E 769 776	*Teacher's Resource Binder* Section Review Worksheets **Other Resources** Transparency 59	100
H O N O R S	**27.2 IMPORTANT COMPOUNDS OF SULFUR** Demo 3 Section Review Chapter Review • Questions 4–12, Problems 2–15	770 764B 775 776	*Teacher's Resource Binder* Section Review Worksheets	101
	CHAPTER END ACTIVITIES Critical Thinking, Cross Connections, Research and Writing, Projects		*Teacher's Resource Binder* Test Booklet **Computer Test Generator**	133, 135

Teaching Strategies

Introduction
This chapter deals with the physical and chemical properties of sulfur and the chemistry of two important sulfur compounds: sulfur dioxide and sulfuric acid. The pollution of the atmosphere by sulfur dioxide and the formation of acid precipitation is a major environmental concern and will be of interest to students. Since more sulfuric acid is used each year than any other single chemical substance, its preparation and some of its uses should be discussed.

27.1 Elemental Sulfur
Exhibit samples of elemental sulfur in various forms such as flowers of sulfur and roll sulfur. Show minerals of sulfur such as gypsum and iron pyrite. If iron pyrite is available, point out the golden crystals that led to the name "fool's gold." Note the oxidation state of the sulfur in pyrite, gypsum, and other available minerals.

Discuss the Frasch process of obtaining sulfur. Explain how water can be heated to a temperature of 170°C when it is under great pressure. Refer to Figure 27-1 to explain the process for pumping the molten sulfur from underground.

Have students note the color and lack of odor of sulfur. Drop some sulfur in water to show the relative density of sulfur and its lack of solubility in water. Relate the solubility to the nonpolar nature of the sulfur molecule.

Make a ball-and-stick model of the S_8 molecule to show the ring structure. Discuss rhombic sulfur and monoclinic sulfur as allotropes containing the S_8 structure though stacked in different ways.

Discuss the behavior of sulfur when heated. Show how *lambda*-sulfur consists of S_8 molecules in the liquid state. Break open the ball-and-stick model to show what happens upon further heating, and explain the viscosity of *mu*-sulfur in terms of the tangling of the open chains of sulfur. Explain how further heating shortens the chains so that the viscosity decreases. Amorphous sulfur forms when the short chains are cooled so rapidly that the chains and rings cannot reform. In time, the bonds slowly form so that S_8 molecules are again present.

Discuss the gases SO_2 and SO_3, pointing out the oxidation states of +4 and +6 for sulfur in these molecules.

Show that elemental sulfur is used in some substances. Ask students to name chemical compounds that contain sulfur.

Demonstration 1: Plastic Sulfur
PURPOSE: To show some of the allotropic forms of sulfur
MATERIALS: Powdered sulfur, large test tube, beaker, water, burner, tongs
PROCEDURE: Wear goggles. Work in a hood because sulfur dioxide gas may be produced. Place about 1 inch of sulfur in the test tube. Observe its properties. Heat the sulfur until it melts into an orange liquid (*lambda*-sulfur). Continue to strongly heat

the test tube until a thick, dark red liquid forms (*mu*-sulfur). Quickly pour the molten sulfur into a beaker of cold water. Observe the formation of the brown, rubbery, plastic sulfur (amorphous sulfur). Allow the sulfur to cool for several minutes, remove with tongs, and note its elasticity. On standing, the plastic sulfur will revert back to its original form.

Demonstration 2: A Chemical Sunset
Purpose: To show the formation of colloidal sulfur
Materials: Glass container of approximately 2 L (an empty acid bottle works well), $Na_2S_2O_3 \cdot 5H_2O$, 6 M hydrochloric acid, slide projector, water
Procedure: Wear goggles. Dissolve approximately 15 g of $Na_2S_2O_3 \cdot 5H_2O$ in 2 L of water in the container. In a darkened room, shine the beam from a slide projector through the container onto a wall or screen. Pour about 10 mL of 6 M HCl into the container, and quickly swirl it to mix. Observe the color changes in the container and on the wall or screen. The reaction slowly produces colloidal sulfur that scatters light as it is being formed and produces a sequence of colors from yellow to red on the screen, just as the colors of a natural sunset are produced when light is scattered by gas molecules in the atmosphere.

$$2H^+(aq) + S_2O_3^{2-}(aq) \rightarrow H_2S_2O_3(aq)$$

$$H_2S_2O_3 \rightarrow H_2SO_3 + S(\text{colloidal})$$

27.2 Important Compounds of Sulfur

Discuss sulfur dioxide as both a chemical substance with great utility and as a byproduct of combustion that is an environmental hazard. Explain the equations that lead to the formation of SO_2. Discuss the oxidation states of sulfur in SO_2 and in the sulfite ion (both +4). Show that sulfur dioxide is the "anhydride" of sulfurous acid, H_2SO_3, because in both compounds sulfur has the same oxidation state.

Ask students to look at labels of substances such as dried fruits, cereals with fruit, and wines for evidence of treatment with sulfur dioxide. Discuss treatment of fresh fruits and vegetables with SO_2 to preserve them. Discuss papermaking, in which sulfites are used to dissolve the lignin, the sticky substance that holds the wood fibers together.

Acid precipitation is an important ecological issue. Assign students reports based on current articles. Discuss the strengths of the acids formed when carbon dioxide, sulfur dioxide, sulfur trioxide, and nitrogen dioxide dissolve in rainwater, and how each of these affect the pH of the precipitation.

Sulfuric acid is sometimes called the king of chemicals because of the enormous quantities of it that are consumed. Show students a reagent bottle of H_2SO_4 and have them read the label and note properties of the acid. Swirl the acid so they can observe the oily viscosity. Write the equations showing the stepwise ionization of the acid. Discuss the fact that the first ionization goes essentially to completion, and the equilibrium constant for the second ionization shows that it is about 10% ionized. Write formulas for salts of sulfuric acid such as Na_2SO_4 and $NaHSO_4$, explaining that it is possible to form both sulfate and hydrogen sulfate compounds.

Discuss the reactivity of sulfuric acid in three ways: as a hydrogen ion donor, as an oxidizing agent, and as a dehydrating agent. This variety of reactions makes it a very useful chemical substance. Write equations showing each of these types of behavior.

The majority of sulfuric acid in the United States is used to make fertilizers. Discuss the need for the enormous quantity of fertilizer that is produced each year. Show a lead storage battery that uses sulfuric acid as an electrolyte and discuss the fact that the density of the solution is used as a measure of the state of the battery.

Chemistry and the Environment: The Smell of Sulfur Compounds

Students might be surprised to think that the sense of smell is a chemical sense, and that the combination of atoms in the molecule can cause odors to form. Sulfur compounds are notorious for having bad smells. Discuss the benefit of adding compounds with strong odors to natural gas to allow gas leaks to be detected.

Demonstration 3: Dehydration of Sucrose
Purpose: To demonstrate the dehydrating ability of sulfuric acid
Materials: Sucrose (table sugar), 100-mL beaker, concentrated H_2SO_4
Procedure: Wear goggles. Work in a hood. Fill a small beaker about one-third full of sucrose. Set the beaker in a tray. Carefully pour 5-10 mL of concentrated sulfuric acid over the sugar. Do NOT stir. Stand back and watch. The sugar will start to char as the acid dehydrates the sugar; the carbon that remains forms a large "snake," along with steam and smoke. When cool, remove the carbon snake with tongs because it will contain some sulfuric acid. Rinse it thoroughly with water before discarding.

References and Resources

Books and Periodicals

Likens, Gene E. "Acid Rain," *Scientific American*, Vol. 241, 1979, p. 43.

Martin, J.A., P. Baudot, J.L. Manal, and M.F. Lejaille. "Synthesis of Sulfuric Acid by the Contact Process," *Journal of Chemical Education*, Vol. 52, 1975, p. 188.

Mohnen, Volker A. "The Challenge of Acid Rain," *Scientific American*, Vol. 259, 1988, p. 259.

Shakhashiri, Bassam Z. "Heat of Dilution of Sulfuric Acid." *Chemical Demonstrations*, Vol. 1, University of Wisconsin Press, 1983, p. 17.

Sheldon, Richard P. "Phosphate Rock," *Scientific American*, Vol. 246, 1982, p. 45.

Summerlin, Lee R. and James L. Ealy, Jr. "Exothermic Reaction: Sodium Sulfite and Bleach," *Chemical Demonstrations, A Sourcebook for Teachers*, Vol. 1, American Chemical Society, 1985.

Summerlin, Lee R., Christie L. Borgford, and Julie B. Ealy. "Acid Rain," *Chemical Demonstrations, A Sourcebook for Teachers*, Vol. 2, American Chemical Society, 1987, p. 165.

"Sulfuric Acid": film, University of Akron, Audio Visual Services.
"Sulfur and Its Compounds (2nd ed)": film and video, #1436, Coronet Films & Video.

Audiovisual Resources

"Sulfur": slides and cassette. Educational Images Ltd. Allotropes of sulfur, sulfur dioxide, and sulfuric acid discussed.
"Sulfur Dioxide and Sulfurous Acid": film, University of Akron, Audio Visual Services.

Computer Software

"The Periodic Table Videodisc: Reactions of the Elements." CAV type videodisc, Journal of Chemical Education Software, 1989. Shows action sequences and still shots of each element in its most common forms; reacting with air, water, acids, bases; and some common applications

Answers and Solutions

Reviewing Concepts

1. a. S_8
   b. It would complicate an equation by making it necessary to multiply all the coefficients by 8.
2. At the melting point, sulfur is a pale-yellow liquid that flows easily. As it is heated further, it becomes more viscous; above about 160°C, it hardly pours. The color at this time has changed from light yellow to reddish-brown to almost black. Between 190°C and the boiling point the fluidity increases again. These changes are the result of changing from S_8 rings of sulfur to long, entangled chains of sulfur atoms and then back to smaller groups of atoms. The color changes are due to the increasing number of unshared electrons.
3. From pyrite, FeS_2, and from the purification of coke oven gas, smelter gases, petroleum, or natural gas
4. a. Odorless b. A suffocating, choking odor
5. This was the result of a change in the sulfur from amorphous sulfur back to rhombic sulfur.
6. From volcanic gases and mineral waters, from the burning of fuels with sulfur as an impurity, and from the roasting of sulfide ores
7. $S(s) + O_2(g) \xrightarrow{\Delta} SO_2(g)$
   $2ZnS(s) + 3O_2(g) \xrightarrow{\Delta} 2ZnO(s) + 2SO_2$
8. a. $SO_2(aq) + NaOH(aq) \rightarrow NaHSO_3(aq)$
   b. $SO_2(aq) + 2NaOH(aq) \rightarrow Na_2SO_3(aq) + H_2O(\ell)$
9. a. The contact process
   b. Because the mixture of sulfur dioxide and oxygen comes in contact with the catalyst.
10. It ranks first among the substances produced by the U.S. chemical industry with over 37 metric tons produced annually.
11. To remove oxides from the surface of iron or steel before the metal is plated or coated with enamel
12. Sulfuric acid acts as a dehydrating agent to absorb the water formed in the reaction between the nitric acid and the glycerol, thus causing the reaction to proceed more readily.

Problems

1. $49.4 \text{ m}^3 \times \dfrac{10^6 \text{ cm}^3}{\text{m}^3} \times \dfrac{2.07 \text{ g}}{\text{cm}^3} \times \dfrac{1 \text{ kg}}{10^3 \text{ g}} \times \dfrac{1 \text{ t}}{10^3 \text{ kg}}$
   $= 102 \text{ t}$

2. $Na_2S: \dfrac{2.5 - 0.9}{2.5} \times 100\% = 64\%$

   $MgS: \dfrac{2.5 - 1.2}{2.5} \times 100\% = 52\%$

   $Al_2S_3: \dfrac{2.5 - 1.5}{2.5} \times 100\% = 40\%$

   $SiS_2: \dfrac{2.5 - 1.8}{2.5} \times 100\% = 28\%$

   $P_4S_{10}: \dfrac{2.5 - 2.1}{2.5} \times 100\% = 16\%$

   $SCl_2: \dfrac{3.0 - 2.5}{3.0} \times 100\% = 17\%$

3. Molar mass of $H_2SO_4 = 98.1$ g

   $\dfrac{2.02 \text{ g H}}{98.1 \text{ g H}_2SO_4} \times 100\% \text{ H}_2SO_4 = 2.06\% \text{ H}$

   $\dfrac{32.1 \text{ g S}}{98.1 \text{ g H}_2SO_4} \times 100\% \text{ H}_2SO_4 = 32.7\% \text{ S}$

   $\dfrac{64.0 \text{ g O}}{98.1 \text{ g H}_2SO_4} \times 100\% \text{ H}_2SO_4 = 65.2\% \text{ O}$

4. 100.0% cpd $- 86.6\%$ Pb $= 13.4\%$ S

   $86.6 \text{ g Pb} \times \dfrac{1 \text{ mol Pb}}{207 \text{ g Pb}} = 0.418 \text{ mol Pb}$

   $13.4 \text{ g S} \times \dfrac{1 \text{ mol S}}{32.1 \text{ g S}} = 0.417 \text{ mol S}$

   Pb:S $= 0.418:0.417$
   $= 1:1$
   Simplest formula, PbS

5. $S + O_2 \rightarrow SO_2$
   $2.5 \text{ kg S} \times \dfrac{10^3 \text{ g}}{\text{kg}} \times \dfrac{1 \text{ mol S}}{32 \text{ g S}} \times \dfrac{1 \text{ mol SO}_2}{1 \text{ mol S}}$
   $\times \dfrac{64 \text{ g SO}_2}{\text{mol SO}_2} \times \dfrac{1 \text{ kg}}{10^3 \text{ g}} = 5.0 \text{ kg SO}_2$

6. For each mole of ZnS roasted, two moles of H_2SO_4 may be theoretically produced.

   $450. \text{ t ZnS} \times \dfrac{10^6 \text{ g}}{\text{t}} \times \dfrac{1 \text{ mol ZnS}}{97.4 \text{ g ZnS}} \times \dfrac{2 \text{ mol H}_2SO_4}{1 \text{ mol ZnS}}$

$$\times \frac{98.1 \text{ g H}_2\text{SO}_4}{\text{mol H}_2\text{SO}_4} \times \frac{1 \text{ kg}}{10^3 \text{ g}} = 9.06 \times 10^4 \text{ kg H}_2\text{SO}_4$$

7. $\text{Na}_2\text{SO}_3 + \text{H}_2\text{SO}_4 \rightarrow \text{H}_2\text{O} + \text{SO}_2(g) + \text{Na}_2\text{SO}_4$

$$5.00 \text{ L SO}_2 \times \frac{1 \text{ mol}}{22.4 \text{ L}} \times \frac{1 \text{ mol Na}_2\text{SO}_3}{1 \text{ mol SO}_2}$$

$$\times \frac{126 \text{ g Na}_2\text{SO}_3}{\text{mol Na}_2\text{SO}_3} = 28.1 \text{ g Na}_2\text{SO}_3$$

8. **a.** $\overset{-2}{\text{H}_2\text{S}} + \overset{+4}{\text{SO}_2} \rightarrow \overset{0}{\text{S}} + \text{H}_2\text{O}$

   $2(\overset{-2}{\text{S}} \rightarrow \overset{0}{\text{S}} + 2e^-)$

   $\overset{+4}{\text{S}} + 4e^- \rightarrow \overset{0}{\text{S}}$

   $2\text{H}_2\text{S}(g) + \text{SO}_2(g) \rightarrow 3\text{S}(s) + 2\text{H}_2\text{O}(\ell)$

   b. $\overset{0}{\text{I}_2} + \overset{+4}{\text{Na}_2\text{SO}_3} + \text{H}_2\text{O} \rightarrow \overset{-1}{\text{NaI}} + \overset{+6}{\text{H}_2\text{SO}_4}$

   $2(\overset{0}{\text{I}} + e^- \rightarrow \overset{-1}{\text{I}^-})$

   $\overset{+4}{\text{S}} \rightarrow \overset{+6}{\text{S}} + 2e^-$

   $\text{I}_2(aq) + \text{Na}_2\text{SO}_3(aq) + \text{H}_2\text{O}(\ell) \rightarrow 2\text{NaI}(aq) + \text{H}_2\text{SO}_4(aq)$

   c. $\overset{0}{\text{Hg}} + \overset{+6}{\text{H}_2\text{SO}_4} \rightarrow \overset{+2}{\text{HgSO}_4} + \overset{+4}{\text{SO}_2} + \text{H}_2\text{O}$

   $\overset{0}{\text{Hg}} \rightarrow \overset{+2}{\text{Hg}^{2+}} + 2e^-$

   $\overset{+6}{\text{S}} + 2e^- \rightarrow \overset{+4}{\text{S}}$

   $\text{Hg}(\ell) + 2\text{H}_2\text{SO}_4(aq) \rightarrow \text{HgSO}_4(aq) + \text{SO}_2(g) + 2\text{H}_2\text{O}(\ell)$

   d. $\overset{-2}{\text{Cu}_2\text{S}} + \overset{0}{\text{O}_2} \rightarrow \overset{-2}{\text{Cu}_2\text{O}} + \overset{+4}{\text{SO}_2}$

   $\overset{-2}{\text{S}^{2-}} \rightarrow \overset{+4}{\text{S}} + 6e^-$

   $3(\overset{0}{\text{O}} + 2e^- \rightarrow \overset{-2}{\text{O}^{2-}})$

   $\text{Cu}_2\text{S} + \tfrac{3}{2}\text{O}_2 \rightarrow \text{Cu}_2\text{O} + \text{SO}_2$

   $2\text{Cu}_2\text{S}(s) + 3\text{O}_2(g) \rightarrow 2\text{Cu}_2\text{O}(s) + 2\text{SO}_2(g)$

   e. $\overset{+4}{\text{SO}_2} + \text{H}_2\text{O} + \overset{+7}{\text{KMnO}_4} \rightarrow \overset{+2}{\text{MnSO}_4} + \overset{+6}{\text{K}_2\text{SO}_4} + \overset{+6}{\text{H}_2\text{SO}_4}$

   $5(\overset{+4}{\text{S}} \rightarrow \overset{+6}{\text{S}} + 2e^-)$

   $2(\overset{+7}{\text{Mn}} + 5e^- \rightarrow \overset{+2}{\text{Mn}^{2+}})$

   $5\text{SO}_2(g) + 2\text{H}_2\text{O}(\ell) + 2\text{KMnO}_4(aq) \rightarrow 2\text{MnSO}_4(aq) + \text{K}_2\text{SO}_4(aq) + 2\text{H}_2\text{SO}_4(aq)$

9. $31.972\ 07\ u \times 0.950\ 18 = 30.379\ u$
   $32.971\ 46\ u \times 0.007\ 50 = 0.247\ u$
   $33.967\ 86\ u \times 0.042\ 15 = 1.432\ u$
   $35.967\ 09\ u \times 0.000\ 17 = \underline{0.006\ u}$
   $32.064\ u$, atomic mass

   Atomic mass = 32.064

10. From one mole of sulfur, two moles of sulfuric acid may be theoretically produced.

$$0.995 \times 50.0 \text{ kg S} \times \frac{10^3 \text{ g}}{1 \text{ kg}} \times \frac{1 \text{ mol S}}{32.1 \text{ S}}$$

$$\times \frac{2 \text{ mol H}_2\text{SO}_4}{1 \text{ mol S}} \times \frac{98.1 \text{ g H}_2\text{SO}_4}{\text{mol H}_2\text{SO}_4} \times \frac{1 \text{ kg}}{10^3 \text{ g}}$$

$$= 304 \text{ kg H}_2\text{SO}_4$$

11. **a.** $\text{Fe} + \text{H}_2\text{SO}_4 \rightarrow \text{FeSO}_4 + \text{H}_2$

$$70.0 \text{ kg Fe} \times \frac{10^3 \text{ g}}{1 \text{ kg}} \times \frac{1 \text{ mol Fe}}{55.8 \text{ kg Fe}} \times \frac{1 \text{ mol FeSO}_4}{1 \text{ mol Fe}}$$

$$\times \frac{152 \text{ g FeSO}_4}{\text{mol FeSO}_4} \times \frac{1 \text{ kg}}{10^3 \text{ g}} = 191 \text{ kg FeSO}_4$$

b. $70.0 \text{ kg Fe} \times \frac{10^3 \text{ g}}{1 \text{ kg}} \times \frac{1 \text{ mol Fe}}{55.8 \text{ Fe}} \times \frac{1 \text{ mol H}_2\text{SO}_4}{1 \text{ mol Fe}}$

$$\times \frac{98.1 \text{ g H}_2\text{SO}_4}{\text{mol H}_2\text{SO}_4} \times \frac{1 \text{ kg}}{10^3 \text{ g}} \times \frac{100\% \text{ H}_2\text{SO}_4}{95\% \text{ H}_2\text{SO}_4}$$

$$= 130. \text{ kg H}_2\text{SO}_4$$

12. $\text{Cu} + 2\text{H}_2\text{SO}_4 \rightarrow \text{CuSO}_4 + 2\text{H}_2\text{O} + \text{SO}_2$

$$100. \text{ g H}_2\text{SO}_4 \times \frac{1 \text{ mol H}_2\text{SO}_4}{98.1 \text{ g H}_2\text{SO}_4} = 1.02 \text{ mol H}_2\text{SO}_4$$

$$100. \text{ g Cu} \times \frac{1 \text{ mol Cu}}{63.5 \text{ g Cu}} = 1.57 \text{ mol Cu}$$

The copper is in excess.

$$100. \text{ g H}_2\text{SO}_4 \times \frac{1 \text{ mol H}_2\text{SO}_4}{98.1 \text{ g H}_2\text{SO}_4} \times \frac{1 \text{ mol SO}_2}{2 \text{ mol H}_2\text{SO}_4}$$

$$\times \frac{22.4 \text{ L}}{1 \text{ mol}} = 11.4 \text{ L SO}_2$$

13. $4\text{FeS}_2 + 11\text{O}_2 \rightarrow 2\text{Fe}_2\text{O}_3 + 8\text{SO}_2(g)$

$$1500. \text{ kg FeS}_2 \times \frac{10^3 \text{ g}}{\text{kg}} \times \frac{1 \text{ mol FeS}_2}{120.0 \text{ g FeS}_2}$$

$$\times \frac{8 \text{ mol SO}_2}{4 \text{ mol FeS}_2} \times \frac{22.4 \text{ L}}{\text{mol}} \times \frac{298\text{K}}{273\text{K}} \times \frac{760. \text{ mm}}{740. \text{ mm}}$$

$$= 6.28 \times 10^5 \text{ L SO}_2$$

14. $[\text{H}_3\text{O}^+] = \text{antilog}(-\text{pH}) = \text{antilog}(-2.10)$
    $= 7.9 \times 10^{-3}$

   $[\text{OH}^-] = \dfrac{1.0 \times 10^{-14}}{7.9 \times 10^{-3}}$

   $= 1.3 \times 10^{-12}$

15. $[\text{Ba}^{2+}][\text{SO}_4^{2-}] = 1.1 \times 10^{-10}$
    $0.01[\text{SO}_4^{2-}] = 1.1 \times 10^{-10}$
    $[\text{SO}_4^{2-}] = 1.1 \times 10^{-8}$

Critical Thinking

1. **a.** NO_2 is slightly soluble; CO_2 is moderately soluble; SO_2 is very soluble.
   b. The extent of solubility and degree of reactivity are similar.

2. It has a high temperature and is also a powerful oxidizing and dehydrating agent.

Videodisc Selections
from *The Periodic Table Videodisc* (JCE Software)

step forward

step reverse

pg 767 Reactions of sulfur

13145
sulfur + oxygen (m)

CHAPTER 27

pg 765 Sulfur

13133
sulfur (s)

pg 768 Uses of sulfur

36508
uses of sulfur (s)

Teacher's Notes

Chapter 27
Sulfur and Its Compounds

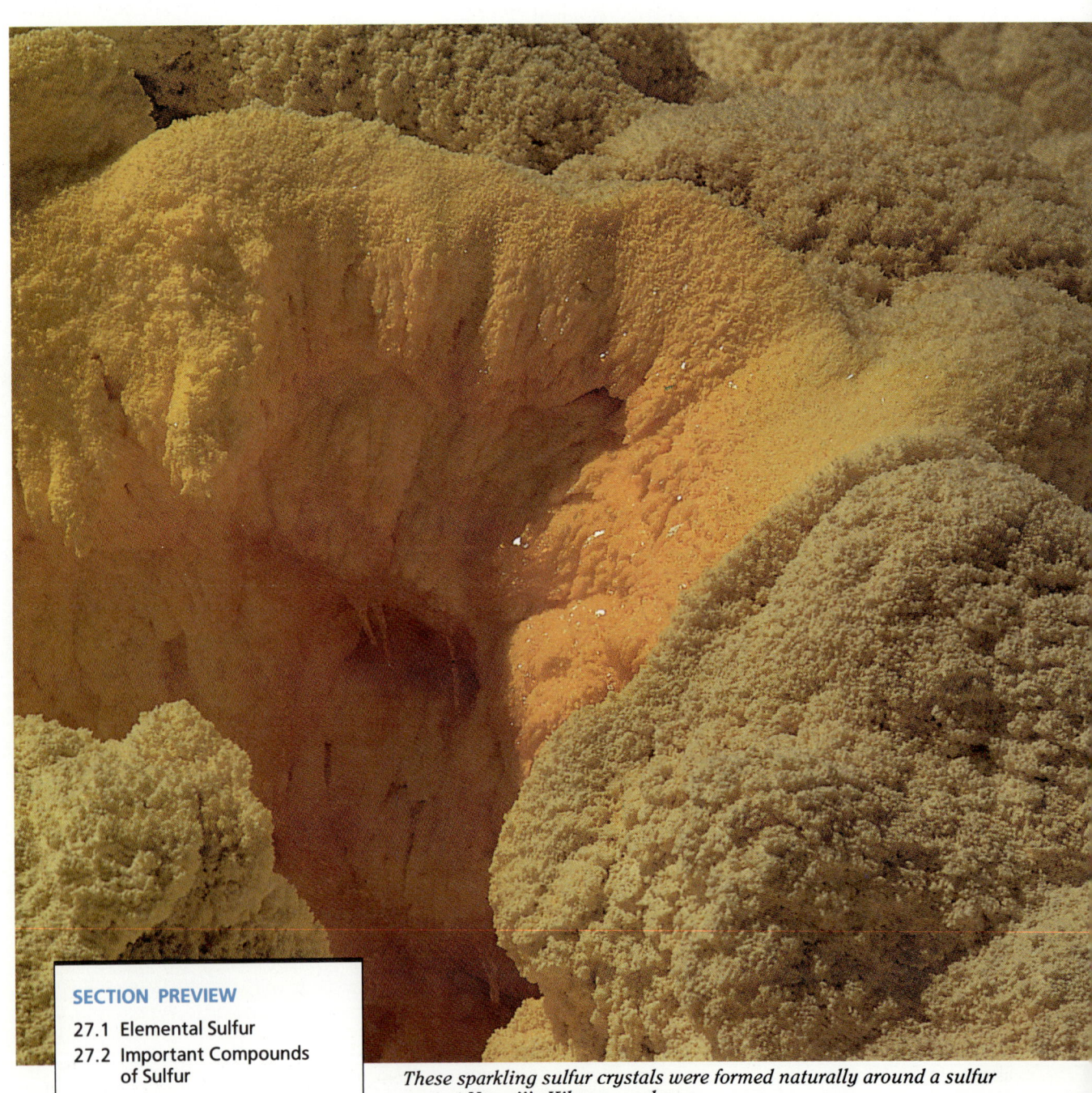

SECTION PREVIEW

27.1 Elemental Sulfur
27.2 Important Compounds of Sulfur

These sparkling sulfur crystals were formed naturally around a sulfur vent at Hawaii's Kilauea volcano.

27.1 Elemental Sulfur

Sulfur is one of the elements known since ancient times. It occurs in nature as the free element or combined with other elements in sulfides and sulfates. It is commonly found in the form of pale, yellow crystals. In the United States, huge deposits of nearly pure sulfur occur between 150 and 600 meters underground in Texas and Louisiana, near the Gulf of Mexico.

The Production of Sulfur

The sulfur beds in Texas and Louisiana are as much as 60 meters thick. A layer of quicksand between the surface of the ground and the sulfur makes it difficult to sink a shaft and mine the sulfur by ordinary methods.

The American chemist Herman Frasch (1851–1914) developed a method of obtaining the sulfur without sinking a shaft. *The* **Frasch process** *is a sulfur-mining process that uses a complex system of pipes to pump superheated water (170°C) under pressure into the sulfur deposit.* **Superheated water** *is water that has been heated under pressure to above the boiling point without conversion to steam.* The hot water then melts the sulfur, which is forced to the surface by compressed air.

SECTION OBJECTIVES

- Identify the principal deposits of sulfur.
- Describe the physical and chemical properties of sulfur.
- List several uses of sulfur.

An ancient name for sulfur is "brimstone," which means burning stone. Sulfur burns with a bright blue flame, forming the choking compound sulfur dioxide. It is often found in burning volcanos. This led people to think that the fires of hell must be like burning sulfur, brimstone.

Figure 27-1 Sulfur is found in many forms in nature.

Most people don't realize that sulfur itself is odorless. It is the compounds of sulfur that have the bad odors.

Some of the elemental sulfur produced in the United States is now obtained from *a compound of iron and sulfur called* **pyrite** *(FeS_2)* and from the purification of coke oven gas, smelter gases, petroleum, and natural gas.

Physical Properties of Elemental Sulfur

Common sulfur is a yellow, odorless solid. It is practically insoluble in water and is twice as dense as water. It dissolves readily in carbon disulfide and less readily in carbon tetrachloride.

Allotropes of sulfur Sulfur exists in several different solid and liquid allotropic forms. These forms are produced by different arrangements of groups of sulfur atoms.

Figure 27-2 The Frasch sulfur-mining process uses a system of nested pipes.

Sulfur and Its Compounds

Figure 27-3 Molecules of S_8 sulfur consist of eight-membered sulfur rings.

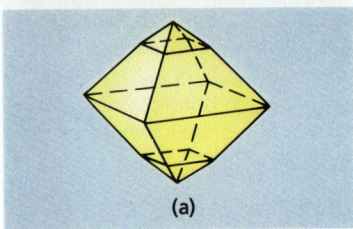

Figure 27-4 Rhombic crystals of sulfur are stable at room temperature.

Figure 27-5 Notice the long, needlelike shape of monoclinic sulfur crystals.

1. Rhombic sulfur (Alpha-sulfur, S_α). **Rhombic sulfur** *is the form of solid sulfur that is stable at ordinary temperatures.* Electron diffraction measurements and X-ray analysis give evidence that a crystal cell consists of sixteen S_8 molecules. These molecules consist of eight sulfur atoms bonded to each other by single covalent bonds in a puckered-ring structure, as shown in Figure 27-3. Rhombic crystals are shown in Figure 27-4.

2. Monoclinic sulfur (Beta-sulfur, S_β). **Monoclinic sulfur** *is a second form of solid sulfur that is less stable than rhombic sulfur. In this form, the eight-membered rings of sulfur are arranged in a monoclinic crystal pattern* as shown in Figure 27-5. This allotropic form is prepared by first melting some sulfur in a crucible at as low a temperature as possible. Next the sulfur is allowed to cool slowly, and crystals form. When such crystals cool below 95°C, they gradually change back into the rhombic form. The density of monoclinic sulfur is 1.96 g/cm³.

3. λ-sulfur (Lambda-sulfur). **Lambda-sulfur** *is the liquid allotropic form of sulfur produced at temperatures just above the melting point of 112.8 °C.* It flows easily and has a straw-yellow color. At this temperature, the eight-membered rings of sulfur atoms slip and roll over one another easily, giving this form of sulfur its fluidity.

4. μ-sulfur (Mu-sulfur). **Mu-sulfur** *is an allotropic form of sulfur that forms at temperatures above 160°C.* When sulfur is heated to approximately this temperature, it darkens to a reddish liquid and then turns almost black. The melted sulfur becomes so viscous that it does not flow. The heat gives enough energy to the sulfur atoms to break some of the eight-membered rings, forming chains of sulfur atoms as shown in Figure 27-6. When a ring of sulfur atoms breaks open, the sulfur atoms on either side of the break are each left with an unshared electron. These sulfur atoms form bonds with similar sulfur atoms from other open rings. In this way, long chains of μ-sulfur are formed. The dark color of μ-sulfur is due to the greater absorption of light by electrons from the broken ring structure.

The high viscosity of μ-sulfur is explained by the tangling of the sulfur-atom chains. As the temperature is raised above about 190°C, however, these chains break up into smaller groups of atoms. The mass then flows more easily. The color becomes still darker because more free electrons, which absorb more light, are produced by the breaking of chains.

Sulfur vapor, produced when sulfur boils at 444.6°C, also consists of S_8 molecules. If sulfur vapor is heated to a higher temperature, these molecules gradually dissociate into S_6 and then S_2 molecules. Monatomic molecules of sulfur are produced at very high temperatures.

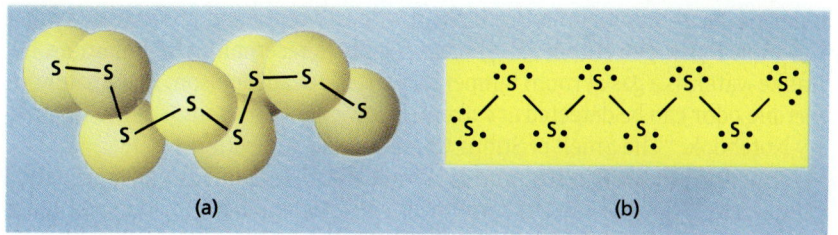

Figure 27-6 When sulfur is heated above its melting point, the eight-membered rings open, forming chains.

5. *Amorphous sulfur.* **Amorphous sulfur** *is a rubbery, dark brown or black mass made by pouring boiling sulfur into cold water.* When sulfur is boiling, the long tangled chains of μ-sulfur have largely broken down, and the molten sulfur is fluid again. Eight-membered rings of sulfur atoms and chains of sulfur atoms are in equilibrium. When this boiling mixture is suddenly cooled, the chains of μ-sulfur have no time to reform into rings. Instead, amorphous sulfur is produced.

A mass of amorphous sulfur soon loses its elasticity and becomes hard and brittle. For amorphous sulfur at room temperature, the successive changes into allotropic forms occur in reverse order. Eventually the amorphous sulfur again acquires the S_8 arrangement of stable rhombic sulfur.

Remind students that amorphous means without crystalline form.

Chemical Properties of Elemental Sulfur

At room temperature, sulfur is not very active chemically. When heated, sulfur combines with oxygen to produce sulfur dioxide.

$$S(s) + O_2(g) \xrightarrow{\Delta} SO_2(g)$$

Traces of sulfur trioxide (SO_3) also form when sulfur burns in the air. Sulfur can be made to combine with nonmetals such as hydrogen, carbon, and

Figure 27-7 Molten sulfur (bottom) becomes viscous and dark, but reverts to rhombic form (top) upon cooling.

TABLE 27-1	SOME PROPERTIES OF ELEMENTAL SULFUR
Property	Sulfur
Symbol	S
Atomic number	16
Atomic mass	32.06
Electron configuration	[Ne]$3s^2 3p^4$
Atomic radius (pm)	102
Ionic radius of S^{2-} (pm)	184
Oxidation numbers	−2, +4, +6
Electronegativity	2.5
Ionization energy (kJ/mol)	1000
Melting point (°C)	112.8
Boiling point (°C)	444.6
Density (g/cm^3)	2.07[a]
	1.96[b]
Color of element	yellow

[a] density for rhombic sulfur
[b] density for monoclinic sulfur

chlorine. Such compounds are formed with difficulty and are not very stable however. Hydrogen sulfide (H₂S) has a structure resembling that of water, but unlike water, is a gas at room temperature. It is a deadly poison. Fortunately, its terrible odor can be detected at extremely low concentrations (see the Chemistry Notebook, "The Smell of Sulfur Compounds," in Section 27.2).

The formulas SO₃, SO₂, and H₂S indicate that sulfur can have oxidation numbers of +6 or +4 when combined with oxygen and −2 when combined with hydrogen. Electron-dot formulas for these compounds are shown below. The actual molecules of sulfur trioxide and sulfur dioxide are resonance hybrids of these possible structures shown.

resonance structures of sulfur trioxide

resonance structures of sulfur dioxide

hydrogen sulfide

At least ten other oxides of sulfur are known, but only sulfur trioxide (SO₃) and sulfur dioxide (SO₂) are commercially important. The preparation and uses of sulfur dioxide are discussed in Section 27.2.

Sulfur combines directly with all metals except gold and platinum. Powdered zinc and sulfur combine vigorously, as shown in Figure 27-8. The heat produced when iron filings and sulfur unite causes the whole mass to glow red hot. Copper unites with hot sulfur vapor to form CuS and Cu₂S.

Uses of Elemental Sulfur

About 90% of the sulfur produced is used to make sulfuric acid. Sulfur is also used in making sulfur dioxide, carbon disulfide, and other sulfur compounds. Matches, fireworks, and black gunpowder all contain either sulfur or sulfur

Figure 27-8 Zinc dust and sulfur react dramatically when heated.

Chemistry and Health

Medicines Made from Sulfur

The medicinal properties of sulfur have been known for centuries. Sulfur can be used as an antiseptic and to kill parasites such as the mites that burrow in the skin of cattle, sheep, and humans to cause the skin disease called scabies.

Large deposits of sulfur in Sicily were mined by Roman slaves and prisoners. The workers seldom lived long because the mines were quite dangerous. The sulfur could be ignited easily, causing fires and the suffocating fumes of sulfur dioxide. The sulfur mines also contained hydrogen sulfide, a flammable, poisonous gas.

Three forms of pure sulfur are used in pharmacy and veterinary medicine. Sulfur that has been purified by sublimation is known as flowers of sulfur because the sulfur forms flowerlike pat-

terns. Washed sulfur is made by treating sublimed sulfur with ammonia to dissolve impurities and to remove traces of acid. Precipitated sulfur, also known as milk of sulfur, is made by boiling sulfur with lime and precipitating the sulfur.

Sulfur is also an important ingredient in a group of sulfonamide compounds known as sulfa drugs. Sulfa drugs, discovered in the late 1930s, were the first drugs to be proven safe and effective against many common bacterial infections.

---

compounds. Sulfur is used in the preparation of certain dyes, medicines, and **fungicides**, *compounds used to kill fungi.*

Sulfur is also used in making practical forms of rubber. Untreated rubber lacks strength, becomes brittle at cool temperatures, and tacky at warm temperatures. In a process called *vulcanization*, rubber is heated with sulfur causing the polymer chains of rubber to become cross linked. This makes the rubber stronger and flexible through a wider temperature range.

Section Review

1. In what form does sulfur occur in nature?
2. Where is sulfur found in the United States?
3. What are some of the physical and chemical properties of sulfur?
4. Name some of the uses of elemental sulfur.

1. In huge deposits underground. Occurs in combined form with other elements in sulfides and sulfates
2. In Texas and Louisiana, near the Gulf of Mexico
3. Accept all answers that are given in the subsections "Physical Properties of Elemental Sulfur" and "Chemical Properties of Elemental Sulfur."
4. In the manufacture of many sulfur compounds, including sulfuric acid, sulfur dioxide, and carbon disulfide. Also used in matches, fireworks, black gunpowder, medicines, and fungicides

27.2 Important Compounds of Sulfur

SECTION OBJECTIVES

- Cite the common occurrences and preparation of sulfur dioxide.
- Describe the physical properties, chemical properties, and uses of sulfur dioxide.
- Identify the contact process as the major method for the preparation of sulfuric acid.
- Describe the physical properties, chemical properties, and uses of sulfuric acid.

Sulfur dioxide is most harmful to individuals who have respiratory problems and to older people and children. High concentrations of sulfur dioxide in areas such as London, England, have been known to cause many deaths.

Because it is used in so many industrial processes, sulfur is essential to the chemical industry and our national economy. The variety and complexity of sulfur compounds is quite large. Several sulfur compounds are shown in Table 27-2. Two of the most important sulfur compounds are sulfur dioxide and sulfuric acid.

Sulfur Dioxide

Sulfur dioxide is a colorless gas sometimes found in large amounts in the atmosphere of densely populated urban areas. Although sulfur dioxide occurs in some volcanic gases and in some mineral waters, its primary source is the burning of coal and fuel oil that contains sulfur as an impurity. As these fuels are burned, the sulfur combines with oxygen to form sulfur dioxide. Coal and fuel oil of low sulfur content produce less sulfur dioxide air pollution than similar fuels of high sulfur content. In the United States and many other countries, environmental protection laws limit the amount of sulfur that can be emitted by coal-burning power plants.

The roasting of sulfide ores converts the sulfur in the ore to sulfur dioxide. In modern smelting plants, this sulfur dioxide is converted to sulfuric acid.

Preparation of sulfur dioxide *1. By burning sulfur.* The simplest way to prepare sulfur dioxide is to burn sulfur in air or in pure oxygen.

$$S(s) + O_2(g) \xrightarrow{\Delta} SO_2(g)$$

TABLE 27-2   COMMON COMPOUNDS OF SULFUR AND SOME OF THEIR PROPERTIES

Oxidation Number of Sulfur	Compound	Formula	Structure	Density	Melting Point (°C)	Boiling Point (°C)
−2	Hydrogen sulfide	H_2S	H–S–H	1.539 g/L	−85.5	−60.7
+4	Sulfur dioxide	SO_2	O=S=O	2.927 g/L	−72.7	−10
+6	Sulfur trioxide	SO_3	O=S(=O)–O	1.97 g/mL	16.83	44.8
+6	Sulfuric acid	H_2SO_4	H–O–S(=O)(=O)–O–H	1.841 g/mL	10.36	330

2. By roasting sulfides. Huge quantities of sulfur dioxide are produced by *heating sulfide ores in the presence of air to convert the sulfides to oxides*, a process known as **roasting.** The roasting of zinc sulfide ore is typical.

$$2ZnS(s) + 3O_2(g) \xrightarrow{\Delta} 2ZnO(s) + 2SO_2(g)$$

Sulfur dioxide is a by-product in this operation.

3. By the reduction of sulfuric acid. In one laboratory method of preparing this gas, copper is heated with concentrated sulfuric acid as shown in Figure 27-9.

$$Cu(s) + 2H_2SO_4(aq) \rightarrow CuSO_4(aq) + 2H_2O(\ell) + SO_2(g)$$

4. By the decomposition of sulfites. In this second laboratory method, sulfur dioxide is formed by the action of a strong acid on a sulfite.

$$Na_2SO_3(aq) + H_2SO_4(aq) \rightarrow Na_2SO_4(aq) + H_2O(\ell) + SO_2(g)$$

Physical properties of sulfur dioxide Pure sulfur dioxide is a colorless gas with a suffocating, choking odor. It is more than twice as dense as air and is very soluble in water. It is one of the easiest gases to liquefy, becoming liquid at room temperature under a pressure of about 3 atmospheres.

Figure 27-9 The reduction of concentrated sulfuric acid by copper produces sulfur dioxide. This reaction should be performed under a hood.

Laboratory methods of production of chemicals are not always the most economical, but are easiest to carry out.

Chemical properties of sulfur dioxide *1. It is an acid anhydride.* Sulfur dioxide dissolves in water and also reacts with the water.

$$SO_2(g) + 2H_2O(\ell) \rightleftarrows H_3O^+(aq) + HSO_3^-(aq)$$
$$HSO_3^-(aq) + H_2O(\ell) \rightleftarrows H_3O^+(aq) + SO_3^{2-}(aq)$$

These reactions partly account for the high solubility of sulfur dioxide in water. The solution is commonly called "sulfurous acid" although no significant amount of H_2SO_3 exists in the solution.

2. It is a stable gas. Sulfur dioxide does not burn. With a suitable catalyst and at a high temperature, it can be oxidized to sulfur trioxide.

$$2SO_2(g) + O_2(g) \rightleftarrows 2SO_3(g)$$

Uses for sulfur dioxide *1. For making sulfuric acid.* In the chemical industry, great quantities of sulfur dioxide are oxidized to sulfur trioxide. The sulfur trioxide is then combined with water, forming sulfuric acid.

2. As a preservative. Dried fruits such as apricots and prunes are treated with sulfur dioxide, which acts as a preservative. Sulfiting agents are also used to prolong the fresh appearance of salad bar fruits and vegetables. Some individuals are extremely allergic to these sulfur compounds and must avoid them.

3. For bleaching. The weakly acid water solution of sulfur dioxide does not harm the fibers of wool, silk, straw, or paper. It can thus be used to bleach these materials.

4. In preparing paper pulp. Sulfur dioxide in water reacts with limestone to form calcium hydrogen sulfite, $Ca(HSO_3)_2$. Wood chips are heated in calcium hydrogen sulfite solution as a first step in making paper (Figure 27-10). The hot solution dissolves the lignin, which binds the cellulose fibers of wood together. The cellulose fibers are left unchanged and are processed to form paper.

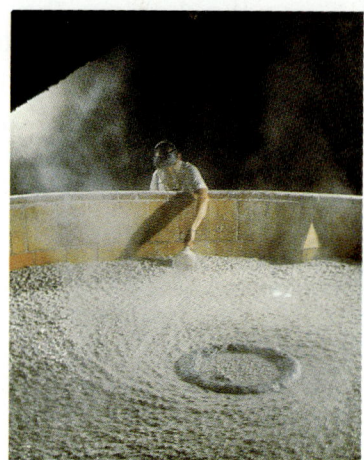

Figure 27-10 Sulfur dioxide is used to make paper pulp.

Chemistry and the Environment

The Smell of Sulfur Compounds

Nearly every compound of sulfur has a distinctive, unpleasant smell. Everyone is familiar with the pungent, nauseating odor of rotten or overcooked eggs. The culprit responsible for this memorable smell is hydrogen sulfide (H_2S).

Idling in traffic with the car windows open is another instance in which we are aware of hydrogen sulfide, the production of which is catalyzed by the platinum in the catalytic converter. Not only is H_2S repugnant, it is highly toxic. Luckily, our noses can detect it at low concentrations, so we can escape before the danger level is exceeded. Unfortunately, the phenomenon of smell fatigue is known to have caused the deaths of several researchers working on sulfur chemistry. They work "just another minute," then no longer smell the gas and mistakenly think that it is no longer present.

Two odorous groups of organic sulfur compounds are mercaptans, which contain an —SH group, and sulfides. Our sense of smell can detect these compounds in extremely low concentrations. Dimethyl sulfide, $(CH_3)_2S$, is often used as an additive to give a distinct odor to natural gas because of its chemical stability. Even smellier and more often used is *t*-butyl mercaptan (systematically named 2-methyl-2-propanethiol). It can be detected at levels of one molecule in a billion molecules of air or natural gas. Both of these organic sulfur compounds have an extremely unpleasant skunk-like odor. In fact, the skunk's defensive spray is a 4:3:3 mixture of the three sulfurous compounds shown above.

Sulfur-containing compounds are also the cause of the odor and taste in onions, garlic, and horseradish.

Acid precipitation Acid precipitation *forms when oxides of sulfur or nitrogen enter the air and dissolve in airborne water vapor producing acids and, as a result, lower the pH of precipitation below the average normal pH of 5.6.* Sulfur dioxide is produced when sulfur-containing coal is burned or when sulfide ores are smelted. Although the waste gases of newer industrial plants are treated to remove sulfur dioxide, millions of metric tons of it still enter the atmosphere annually. Sulfur dioxide reacts with oxygen and water vapor in the air to form sulfuric acid.

Oxides of nitrogen, principally nitrogen monoxide (NO) and nitrogen dioxide (NO_2) get into the air as combustion products of automobile engines and some generating processes for electricity. Nitrogen monoxide reacts with oxygen in the air to form nitrogen dioxide, which then reacts with water vapor in the air to form nitric acid (HNO_3).

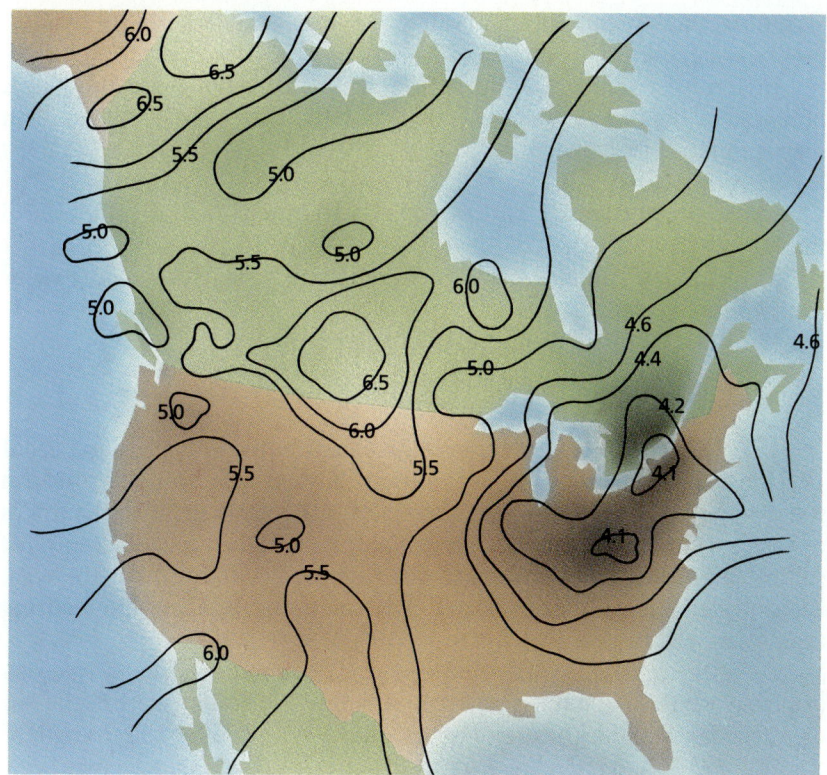

Figure 27-11 This map shows the variation in pH of the precipitation in North America. The areas of lowest precipitation pH correspond to areas within and downwind of heavy industrialization and urbanization.

Sulfur oxide and nitrogen oxides can be carried many miles through the atmosphere by the prevailing winds. A local acid-precipitation problem becomes widespread as pollutants travel across state and national boundaries. Figure 27-11 shows the pHs of rainfall in North America.

Sulfuric Acid

Fertilizer, steel, and petroleum industries consume huge amounts of sulfuric acid each year. Over 3×10^{10} kg of sulfuric acid, the so-called "king of chemicals," is produced each year in the United States.

Preparation of sulfuric acid Sulfuric acid is produced in the United States today by the **contact process**, *a method in which sulfur dioxide is prepared by burning sulfur or by roasting iron pyrite (FeS$_2$)*. Impurities that might combine with and ruin the catalyst used in the process are then removed from the sulfur dioxide gas. The purified sulfur dioxide is mixed with air and passed through heated iron pipes containing a catalyst, usually divanadium pentoxide (V$_2$O$_5$). This close "contact" of the sulfur dioxide and the catalyst gives the contact process its name. Sulfur dioxide and oxygen from the air are both absorbed on the surface of the catalyst where they react to form sulfur trioxide. The sulfur trioxide is then dissolved in approximately 98% sulfuric acid with which it combines readily to form pyrosulfuric acid (H$_2$S$_2$O$_7$).

$$SO_3(g) + H_2SO_4(\ell) \rightarrow H_2S_2O_7(\ell)$$

When diluted with water, pyrosulfuric acid yields sulfuric acid.

$$H_2S_2O_7(\ell) + H_2O(\ell) \rightarrow 2H_2SO_4(\ell)$$

Acid precipitation and its effects are also discussed in "Acid Precipitation and Lakes," Chemistry Notebook in Section 17.2. You can test for acid precipitation in the Desktop Investigation, "Testing for Acid Precipitation," in Section 17.3.

The SO$_3$ is dissolved in H$_2$SO$_4$, rather than in water because it dissolves much more quickly in H$_2$SO$_4$.

Figure 27-12 These trees in the high elevations of the Great Smoky Mountains have been damaged by acid precipitation.

Figure 27-13 This diagram summarizes the contact process for manufacturing sulfuric acid.

The acid converts the $Ca_3(PO_4)_2$ to the more soluble $Ca(H_2PO_4)_2$.

CAUTION:
Sulfuric acid burns the flesh severely. The burns are the result of its dehydrating action on the skin. Handling sulfuric acid requires constant care to prevent its contact with the skin.

Very pure, highly concentrated sulfuric acid is produced by the contact process, diagrammed in Figure 27-13.

In terms of tonnage, sulfuric acid ranks first among the substances produced by the United States chemical industry. Each year, over 37 million metric tons are prepared.

Chemical properties of sulfuric acid The properties of sulfuric acid make it useful in a variety of ways:

1. As an acid. Sulfuric acid is a diprotic acid. It ionizes in dilute water solution in two stages:

$$H_2SO_4 + H_2O \rightarrow H_3O^+ + HSO_4^-$$
$$HSO_4^- + H_2O \rightleftarrows H_3O^+ + SO_4^{2-}$$
$$K_a = 1.26 \times 10^{-2}$$

At 25°C, 0.1 M H_2SO_4 is completely ionized in the first stage and about 10% ionized in the second stage. Sulfuric acid reacts with hydroxides to form sulfates and hydrogen sulfates. It also reacts with metals and with the oxides of metals. Dilute sulfuric acid is more highly ionized than cold, concentrated sulfuric acid. Thus the dilute acid reacts more vigorously than the cold, concentrated acid with metals above hydrogen in the oxidizing and reducing agents series.

2. As an oxidizing agent. Hot, concentrated sulfuric acid is a vigorous oxidizing agent. The sulfur is reduced from the +6 oxidation state to the +4 or −2 oxidation state. The extent of the reduction depends on the strength of the acid and on the reducing agent used.

3. As a dehydrating agent. The strong attraction of sulfuric acid for water makes it a very good **dehydrating agent**. Gases that do not react with sulfuric acid can be dried by bubbling them through the concentrated acid. Sulfuric acid is such an active dehydrating agent that it takes hydrogen and oxygen directly from certain substances in the proportion in which they are found in water. For example, it dehydrates cellulose, $(C_6H_{10}O_5)_n$, and sucrose, $(C_{12}H_{22}O_{11})$ in this manner, leaving the carbon uncombined. The equation for this process with sucrose is

$$C_{12}H_{22}O_{11} + 11H_2SO_4 \rightarrow 12C + 11H_2SO_4 \cdot H_2O$$

When sulfuric acid is poured on sucrose, the sucrose rapidly decomposes and turns into a foamy, black carbon mass.

In some commercial chemical processes, water is formed as a byproduct, as in the production of nitroglycerin from glycerol and nitric acid.

$$C_3H_5(OH)_3 + 3HNO_3 \rightarrow C_3H_5(NO_3)_3 + 3H_2O$$

Sulfuric acid can be used as a dehydrating agent for the reaction above, absorbing the water as fast as it is formed and maintaining the reaction rate.

Uses of sulfuric acid About 60% of the sulfuric acid produced in the United States is used to make **superphosphate**, *a mixture of phosphate compounds that have been treated with sulfuric acid to increase their solubility and used as a fertilizer.*

Sulfuric acid is also used in making phosphoric and other acids, various sulfates, and many chemicals. The iron and steel industries use sulfuric acid to remove oxides from the surface of iron or steel before the metal is plated or coated with an enamel. In petroleum refining, sulfuric acid is used to remove certain organic impurities.

Sulfuric acid serves as a dehydrating agent in the production of smokeless powder and nitroglycerin. It is used in making photographic film, nitrocellulose plastics, and rayon, paints and pigments, cellophane, and thousands of other commercial articles. The electrolyte in lead storage batteries is dilute sulfuric acid.

Section Review

1. Cite two natural occurrences of sulfur dioxide.
2. List several physical and chemical properties of sulfur dioxide.
3. What are some of the uses of sulfur dioxide?
4. Write a balanced chemical equation for the preparation of sulfuric acid.
5. List some chemical properties of sulfuric acid.
6. Identify several uses of sulfuric acid.

Chapter Summary

- Sulfur occurs in nature as a free element or combined with other elements in sulfides and sulfates. The Frasch process is used to mine sulfur in Texas and Louisiana. It uses a system of pipes to pump superheated water under pressure into the sulfur deposit. Molten sulfur is then forced to the surface by compressed air. Elemental sulfur is also obtained from pyrite and from the purification of coke-oven gas, smelter gases, petroleum, and natural gas.
- Common sulfur is a yellow, odorless solid that is practically insoluble in water. It is soluble in carbon disulfide and in carbon tetrachloride. Many sulfur compounds have unpleasant odors.
  Sulfur exists in several allotropic forms. The solid allotropes are rhombic (Alpha-sulfur), monoclinic (Beta-sulfur), and amorphous sulfur. The liquid allotropes are λ-sulfur (Lambda-sulfur) and μ-sulfur (Mu-sulfur). Mu-sulfur is a viscous, dark brown form of molten sulfur that does not flow.
- Sulfur dioxide is a dense, suffocating gas that is easily liquefied and extremely soluble in water. Water solutions of sulfur dioxide are weakly acidic.
- Sulfur dioxide is used for making sulfuric acid and sulfites. It is also used for bleaching, as a preservative, and in preparing paper pulp.
- Oxides of sulfur and nitrogen react with water vapor in the air to form sulfuric acid and nitric acid. These airborne acids can cause acid precipitation.
- Sulfuric acid is made from sulfur dioxide, oxygen, and water by the contact process. In dilute form, it acts as an acid. When hot and concentrated, it is a vigorous oxidizing agent. It is also a good dehydrating agent. Due to its possible dehydrating action on skin, sulfuric acid must always be handled with care.
- Sulfuric acid is one of the most important chemicals used in industry. More tons of sulfuric acid are used each year than any other chemical. It is used in fertilizer manufacture, petroleum refining, explosives production, the making of lead storage batteries and a great many other products.

1. Volcanic gases, mineral waters
2. Colorless gas, choking odor, twice as dense as air, very soluble in water, easily liquefiable; reacts with water, oxidizes to sulfur trioxide at high temperature and in the presence of a suitable catalyst
3. Accept all answers given under the heading "Uses for sulfur dioxide."
4. $H_2S_2O_7(l) + H_2O(l) \rightarrow 2H_2SO_4(l)$
5. Accept answers given under the heading "Chemical properties of sulfuric acid."
6. Accept all uses given under the heading "Uses of sulfuric acid."

Chapter 27 Review

Sulfur and Its Compounds

VOCABULARY

acid precipitation (772)
amorphous sulfur (767)
contact process (773)
dehydrating agent (774)
Frasch process (765)
fungicide (769)
lambda-sulfur (766)
monoclinic sulfur (766)
mu-sulfur (766)
pyrite (765)
rhombic sulfur (766)
roasting (771)
superheated water (765)
superphosphate (774)

REVIEWING CONCEPTS

1. **a.** Give the molecular formula for rhombic sulfur.
   b. Why do you suppose this molecular formula is not usually used in equations? (27.1)
2. As sulfur is heated from the melting point to the boiling point, it undergoes changes in color and fluidity. Describe these changes and explain why they occur. (27.1)
3. In addition to Frasch process for mining sulfur, what other sources of sulfur are important in the United States? (27.1)
4. Describe the odor of
   a. sulfur
   b. sulfur dioxide (27.2)
5. A student prepared some nearly black amorphous sulfur in the laboratory. When it was examined the following week, it had become brittle and much lighter in color. Explain. (27.2)
6. Sulfur dioxide is an important air pollutant. Where does it come from? (27.2)
7. Write balanced equations for two commercial methods of preparing sulfur dioxide. (27.2)
8. Write balanced equations to show the formation from a water solution of sulfur dioxide and sodium hydroxide of
   a. sodium hydrogen sulfite
   b. sodium sulfite (27.2)
9. **a.** What process is used for the manufacture of sulfuric acid in the United States?
   b. How did the process get its name? (27.2)
10. Why is H_2SO_4 called the "king of chemicals"? (27.2)
11. For what purpose is sulfuric acid used in the iron and steel industries? (27.2)
12. Why are both nitric acid and sulfuric acid used in making nitroglycerin? (27.2)

PROBLEMS

1. A railroad gondola car has a capacity of 49.4 m³. How many metric tons of sulfur does it contain when loaded to capacity? The density of sulfur is 2.07 g/cm³.
2. Common sulfides of Period 3 elements are Na_2S, MgS, Al_2S_3, SiS_2, P_4S_{10}, and SCl_2. What is the percentage of ionic character of the bonds in each of these compounds?
3. Calculate the percentage composition (to three significant figures) of H_2SO_4.
4. A compound of lead and sulfur contains 86.6% lead. Determine the simplest formula of the compound.
5. How many kilograms of sulfur dioxide can be produced by burning 2.5 kg of pure sulfur?
6. A lead smelter processes 450. metric tons of zinc sulfide (ZnS) each day. If no sulfur dioxide is lost, how many kilograms of sulfuric acid could be made in the plant daily?
7. How many grams of sodium sulfite are needed to produce 5.00 L of sulfur dioxide at STP by reaction with sulfuric acid?
8. Balance the following oxidation-reduction equations:
   a. $H_2S(g) + SO_2(g) \rightarrow S(s) + H_2O(\ell)$
   b. $I_2(aq) + Na_2SO_3(aq) + H_2O(\ell) \rightarrow NaI(aq) + H_2SO_4(aq)$

c. $Hg(\ell) + H_2SO_4(aq) \rightarrow HgSO_4(aq) + SO_2(g) + H_2O(\ell)$
d. $Cu_2S(s) + O_2(g) \rightarrow Cu_2O(s) + SO_2(g)$
e. $SO_2(g) + H_2O(\ell) + KMnO_4(aq) \rightarrow MnSO_4(aq) + K_2SO_4(aq) + H_2SO_4(aq)$

9. Natural sulfur consists of 95.018% S-32 (31.972 07 u), 0.750% S-33 (32.971 46 u), 4.215% S-34 (33.967 86 u), and 0.017% S-36 (35.967 09 u). Calculate the atomic mass of sulfur.

10. How many kilograms of sulfuric acid can be prepared from 50.0 kg of sulfur that is 99.5% pure?

11. a. If 70.0 kg of scrap iron is added to a large vat of dilute sulfuric acid, how many kilograms of iron(II) sulfate can be produced?
    b. How many kilograms of 95% sulfuric acid are required?

12. A laboratory technician has on hand 100. g of copper but only 100. g of H_2SO_4. How many liters of sulfur dioxide at STP can the technician prepare?

13. How many liters of sulfur dioxide at 25°C and 740. mm pressure can be produced by roasting 1500. kg of iron pyrite, FeS_2?

14. A 0.01 N solution of sulfuric acid has a pH of 2.10. Calculate $[H_3O^+]$ and $[OH^-]$.

15. For barium sulfate, $K_{sp} = 1.1 \times 10^{-10}$. If a barium chloride solution is 0.01 M, what is the smallest sulfate ion concentration that can be detected by precipitation?

CRITICAL THINKING

1. **Inferring Conclusions**
   a. Using data from Appendix Table A-10, compare the solubility in water of CO_2, NO_2, and SO_2.
   b. What do these solubilities indicate about the extent to which these oxides react with water?

2. **Applying Ideas** Give two reasons why boiling concentrated sulfuric acid burns the flesh.

CROSS CONNECTIONS

1. **History** Consult references in the library and write a report about the history of human uses of sulfur.

2. **Medicine** Ask several veterinarians for the names and purposes of sulfur-containing medicines or drugs that they use to treat animals in their veterinary practice.

RESEARCH AND WRITING

1. Research London smog and Los Angeles smog in the library. Compare the histories of the air pollution problems of these two cities and report your findings to the class.

2. Describe recent research efforts by chemists and chemical engineers to reduce the amount of sulfur dioxide released during the combustion of coal, fuel oil, and gasoline.

PROJECT

1. Contact state agencies, universities, or library references in your area to learn if your area has acid precipitation. What is the approximate pH of precipitation in your community? Can you identify the nearest sources of sulfur oxides or nitrogen oxides in your area?

CHAPTER 28
The Halogens

PLANNING CHART
Approximate Time: 3/4 days

	SECTION/TEXT FEATURES		RESOURCES	

H O N O R S	**28.1 THE HALOGEN FAMILY** 779
	Periodic Table Videodisc 778E
	Section Review 781
	Chapter Review 792
	• Questions 1–2
			Teacher's Resource Binder
			  Lab Experiment 37 193
			  Section Review Worksheets 102
			Other Resources
			  Blackline Master 60	

H O N O R S	**28.2 FLUORINE** 781 *Periodic Table* Videodisc 778E Section Review 782 Chapter Review 792 • Questions 4–7		***Teacher's Resource Binder*** Section Review Worksheets 103	
H O N O R S	**28.3 CHLORINE** 782 *Periodic Table* Videodisc 778E Demo 1 778B Section Review 786 Chapter Review 792 • Questions 6–10, Problems 1, 3–5, 7, 9–11		***Teacher's Resource Binder*** Section Review Worksheets 104	

H O N O R S	**28.4 BROMINE** 787
	Desktop Investigation 789
	Periodic Table Videodisc 778E
	Section Review 789
	Chapter Review 792
	• Questions 11–12, Problems 2, 8
			Teacher's Resource Binder
			  Section Review Worksheets 105
			Other Resources
			  Shakhashiri Video Demo 32	

H O N O R S	**28.5 IODINE** 790 *Periodic Table* Videodisc 778E Section Review 790 Chapter Review 792 • Questions 13–15, Problem 6		***Teacher's Resource Binder*** Section Review Worksheets 106	
	CHAPTER END ACTIVITIES Critical Thinking, Cross Connections, Research and Writing, Projects		***Teacher's Resource Binder*** Test Booklet 137, 139 ***Computer Test Generator***	

28.1 The Halogen Family

An introduction to the halogen family might include discussing a number of important and familiar compounds that contain halogens. Ask students to name as many as they can. Show samples of, for example, salt, iodized salt, tincture of iodine, Teflon plastic, fluoride toothpaste, liquid bleach, and photographic film.

Have students develop a table of properties of the halogens based on their knowledge of periodic trends among the elements. Discuss the difference between the halogen atoms, halogen molecules, and halide ions. Use electron dot structures to show these differences and to show why the halogens occur as diatomic molecules.

28.2 Fluorine

Fluorine can react with more elements than any other element. Explain the reactivity of fluorine in terms of its high electronegativity. Emphasize the reasons for the high electronegativity and reactivity. Discuss the problem of finding containers to store fluorine gas.

Most students know that fluorides are added to toothpastes and drinking water to help prevent tooth decay.

Mention other fluorine compounds, such as Teflon and Freons. Students may want to read about Teflon in the Intra-Science feature that appears after this unit. Discuss the problems associated with the interaction of Freons with ozone and the implications of the depletion of the earth's ozone layer. An Intra-Science article about the ozone layer may be found after Unit 4.

28.3 Chlorine

Exhibit chlorides, table salt, iodized salt, salt substitutes, rock salt, halite crystals, and a model of NaCl. Show the equations for the laboratory production of chlorine, identifying the oxidation and reduction steps of each reaction.

Discuss the physical properties of chlorine. Most students will be familiar with the odor of chlorine from the familiar odor of chlorine bleach or a swimming pool. Point out the reactivity of chlorine, which is predicted by its high electron affinity and electronegativity.

Obtain bottles of swimming-pool disinfectants and commercial bleaches and ask students to examine their labels. Students may be interested in finding out the manner in which local water supplies are disinfected.

Explain the equations for the production of hydrogen chloride. Point out that hydrochloric acid is a water solution of the gas hydrogen chloride. Show students a bottle of reagent hydrochloric acid, and ask them to read the contents and the warnings on the label. Discuss the fact that the stomach contains 0.16 M hydrochloric acid to help digest foods. Remind students that chloride compounds are generally very soluble in water.

Chemistry and its Applications: PVC Pipes

The ingenious technology involved in this application of polyvinyl chloride pipe should appeal to the students. Discuss the term "polymer" and show that the plastic pipe contains the vinyl chloride grouping, ($-CH_2-CHCl-$), repeated many times. Other common forms of polyvinyl chloride are in credit cards, phonograph records, and floor tile.

One problem associated with polyvinyl chloride is its disposal. When this plastic is incinerated, hydrogen chloride gas can form, causing an environmental problem.

Demonstration 1: Electrolysis of Salt Solutions

PURPOSE: To show the formation of chlorine and hydrogen gases from the electrolysis of salt water
MATERIALS: U-tube, small 9-volt battery, carbon electrodes (can be pencil lead), wire and wire clips, sodium chloride, phenolphthalein
PROCEDURE: Perform this demonstration in a hood or with good ventilation for only a short time. Dissolve the salt in water, add a few drops of the phenolphthalein solution, and fill the U-tube. Use the wire and clips to attach the carbon electrodes to the battery, and put one in each side of the U-tube. Note the formation of chlorine gas at the anode and hydrogen gas at the cathode (negative terminal). Also note that hydroxide ions form at the cathode and cause the phenolphthalein solution to turn pink. Write the equations for the half-reactions that occur:

$$2Cl^-(aq) \rightarrow Cl_2(g) + 2e^-$$
$$2H_2O + 2e^- \rightarrow H_2(g) + 2OH^-(aq)$$

28.4 Bromine

Bromine is not as familiar to students as fluorine and chlorine. This is understandable since it is far less abundant. Discuss the physical properties of bromine, and show how they continue the trend shown by the halogens.

Desktop Investigation: Silver Halides

Photosensitive substances are interesting to students. You might suggest that students lay a wire gauze over their precipitates when they expose them to light. The grids of the wire gauze will show up as a "photo" on their precipitates.

To make 0.1 M AgNO$_3$ solution, dissolve 17 g of AgNO$_3$ in enough water to make a liter of solution. The AgNO$_3$ solution must be stored in a brown bottle to prevent its decomposition. This is a very sensitive reaction; if necessary you can use less silver nitrate to make your solution and still get good results. To make 0.2 M NaCl, make a liter of solution using 12 g of NaCl. To make the NaBr solution use 21 g of NaBr. To make the NaI solution, use 30 g of NaI.

Be sure to caution students to be careful with silver nitrate. If spilled on hands or clothing it causes long lasting stains that do not appear for several hours. Dispose of the precipitates in a hazardous waste landfill, or recycle for their silver content.

ANSWERS TO DISCUSSION QUESTIONS
1. The reactions in Procedure 1 all formed precipitates. However, they were different colors. AgCl is white, AgBr is yellow, and AgI is yellowish green. In Procedure 2, all the silver halides become dark when exposed to light. However, the rate at which the change takes place varies. AgI was the fastest, and AgCl was the slowest.
2. NaCl + AgNO$_3$ → AgCl(s) + NaNO$_3$
   NaBr + AgNO$_3$ → AgBr(s) + NaNO$_3$
   NaI + AgNO$_3$ → AgI(s) + NaNO$_3$

   $$2AgCl \xrightarrow{light} 2Ag + Cl_2(g)$$
   $$2AgBr \xrightarrow{light} 2Ag + Br_2$$
   $$2AgI \xrightarrow{light} 2Ag + I_2$$
3. AgI is used to make fast film, AgCl to make slow film.
4. Silver halides are made by reacting a sodium salt of a halide with silver nitrate. The ionic reaction goes to completion because of the formation of the insoluble silver halide. All silver halides are light-sensitive and turn dark when exposed to light. As silver iodide is the most light-sensitive, it decomposes first. As silver chloride is the least light-sensitive, it breaks down last.

28.5 Iodine

Show crystals of iodine, and comment on their metallic appearance. Discuss the fact that the dividing line between metals and nonmetals comes close to iodine. Predict the properties of iodine based on the periodic trends shown by the other metals.

Show that iodine does not dissolve well in water, but dissolves readily in water that contains iodide ions. Iodine dissolves readily in alcohol, forming a brown solution. Tincture of iodine is an alcohol solution of iodine that is used as a disinfectant. In nonpolar solvents, iodine forms a pink-purple color.

References and Resources

Books and Periodicals

Shakhashiri, Bassam A. "Photochemical Reaction of Hydrogen and Chlorine," p. 121; "Reaction of Sodium and Chlorine," p. 61; "Reaction of Antimony and Chlorine," p. 64; "Reaction of Iron and Chlorine," p. 66; "Reaction of Aluminum and Bromine," p. 68; "Reaction of Zinc and Iodine," p. 49; "Reaction of Metals and Hydrochloric Acid, p. 25. *Chemical Demonstrations*, Vol. 1, University of Wisconsin Press, 1983.

Summerlin, Lee R., and James Ealy. "Preparation of Chlorine Gas from Laundry Bleach," *Chemical Demonstrations, A Sourcebook for Teachers*, Vol. 1, American Chemical Society, 1985, p. 13.

Summerlin, Lee R., Christie L. Borgford, and Julie B. Ealy. "Electrolysis of Potassium Iodide," p. 196; "Colorful Effects of Hydrochloric Acid Dilution," p. 175; "Halogens Compete for Electrons," p. 60. *Chemical Demonstrations, A Sourcebook for Teachers*, Vol. 2, American Chemical Society, 1987.

Audiovisual Resources

"Bromine—Element from the Sea": *CHEM* Study 16 mm film, Ward's Modern Learning Aids. Shows production of elemental bromine

"Chemical Families": *CHEM* Study 16 mm film, Ward's Modern Learning Aids. Shows halogens reacting with alkali metals and with phosphorus, and the reaction of xenon with fluorine

"A Research Problem: Inert(?) Gas Compounds": *CHEM* Study 16 mm film, Ward's Modern Learning Aids. Shows the research to find the first compound of krypton and fluorine

"The Halogens (2nd ed.)": film or video, #1643, Coronet Films and Video.

Computer Software

"The Periodic Table Videodisc: Reactions of the Elements": CAV type videodisc, Journal of Chemical Education Software, 1989. Shows action sequences and still shots of each element in its most common forms; reactions with air, water, acids, bases; and some common applications

Answers and Solutions

Reviewing Concepts

1. **a.** Almost exclusively as ions
   b. They are active elements with high electronegativities.
2. **a.** Increase **b.** Increase **c.** Decrease
   d. Decrease **e.** Decrease **f.** Increase
   g. Increase **h.** Increase **i.** Remains constant
3. **a.** Carbon steel
   b. The iron becomes coated with iron fluoride, which resists the action of fluorine.
4. It is nonflammable, nontoxic, and has a moderately high critical temperature and critical pressure.
5. There are no elements more electronegative than fluorine; therefore, fluorine is assigned a negative oxidation state in all compounds.
6. **a.** Sodium chloride **b.** Sodium
7. When inhaled in small quantities, chlorine affects the nose and throat membranes. If inhaled in larger quantities, it may cause death.
8. **a.** Cleaning of metals to remove oxide coatings and other forms of tarnish
   b. It is essential in the process of digestion.
9. $4\overset{-1}{\text{HCl}}(aq) + \text{Ca}(\overset{+1}{\text{ClO}})_2(s) \rightarrow \text{CaCl}_2(aq) + 2\overset{0}{\text{Cl}_2}(g) + 2\text{H}_2\text{O}(\ell)$

Chloride ions in hydrochloric acid are oxidized. Chlorine in calcium hypochlorite is reduced.

10. **a.** Greenish-yellow
    b. The color of chlorine is imparted to it.
    c. Yes, the color disappears.
    d. The chlorine reacts with water to form hypochlorous acid and hydrochloric acid, which are colorless.
11. **a.** $\overset{+4}{\text{MnO}_2} + 4\overset{-1}{\text{HBr}} \xrightarrow{\Delta} \overset{+2}{\text{MnBr}_2} + 2\text{H}_2\text{O} + \overset{0}{\text{Br}_2}$
    b. $2\text{Na}\overset{-1}{\text{Br}} + 2\text{H}_2\text{SO}_4 + \overset{+4}{\text{MnO}_2} \xrightarrow{\Delta} \text{Na}_2\text{SO}_4 + \overset{+2}{\text{MnSO}_4} + 2\text{H}_2\text{O} + \overset{0}{\text{Br}_2}$
    c. See **a** and **b** above. In both reactions, bromine is oxidized and manganese is reduced.
    d. They are essentially the same.
12. CH_3Br, soil fumigant; $\text{C}_2\text{H}_4\text{Br}_2$, in antiknock gasoline. AgBr, in making photographic film; CaBr_2, in solution as a high-density fluid in oil well drilling.
13. At atmospheric pressure, iodine sublimes; that is, it passes directly from the solid to the vapor phase.
14. $:\ddot{\text{I}}: + e^- \rightarrow :\ddot{\text{I}}:^-$
15. **a.** Bluish black **b.** Dark brown **c.** Violet **d.** Violet

Problems

1. **a.** $\dfrac{3.21 \text{ g Cl}_2}{\text{L Cl}_2} \times 5.00 \text{ L Cl}_2 \times \dfrac{1 \text{ mol Cl}_2}{71.0 \text{ g Cl}_2} = 0.226 \text{ mol Cl}_2$

   b. $0.226 \text{ mol Cl}_2 \times \dfrac{6.02 \times 10^{23} \text{ molecules}}{\text{mol}} = 1.36 \times 10^{23} \text{ molecules Cl}_2$

2. Molar mass of $C_2H_4Br_2$ = 187.86 g/mol

$$\frac{159.81 \text{ g Br}}{187.86 \text{ g } C_2H_4Br_2} \times 100\% \text{ } C_2H_4Br_2 = 85.069\%$$

3. **a.** $Cl_2 + Ca(OH)_2 \rightarrow Ca(ClO)Cl + H_2O$

$$200. \text{ g Ca(ClO)Cl} \times \frac{1 \text{ mol Ca(ClO)Cl}}{127 \text{ g Ca(ClO)Cl}} \times \frac{1 \text{ mol Ca(OH)}_2}{1 \text{ mol Ca(ClO)Cl}} \times \frac{74.1 \text{ g Ca(OH)}_2}{\text{mol Ca(OH)}_2} = 117 \text{ g Ca(OH)}_2$$

b. $200. \text{ g Ca(ClO)Cl} \times \frac{1 \text{ mol Ca(ClO)Cl}}{127 \text{ g Ca(ClO)Cl}} \times \frac{1 \text{ mol Cl}_2}{1 \text{ mol Ca(ClO)Cl}} \times \frac{71.0 \text{ Cl}_2}{\text{mol Cl}_2} = 112 \text{ g Cl}_2$

4. $2NaCl \xrightarrow{elec.} 2Na + Cl_2(g)$

$$375 \text{ g NaCl} \times \frac{1 \text{ mol NaCl}}{58.4 \text{ g NaCl}} \times \frac{1 \text{ mol Cl}_2}{2 \text{ mol NaCl}} \times \frac{22.4 \text{ L}}{\text{mol}} = 71.9 \text{ L Cl}_2$$

5. $Zn + Cl_2(g) \rightarrow ZnCl_2$

$$0.500 \text{ L Cl}_2 \times \frac{1 \text{ mol}}{22.4 \text{ L}} \times \frac{1 \text{ mol ZnCl}_2}{1 \text{ mol Cl}_2} \times \frac{136 \text{ g ZnCl}_2}{\text{mol ZnCl}_2} = 3.04 \text{ g ZnCl}_2$$

6. **a.** $2Cl^- + F_2 \rightarrow Cl_2 + 2F^-$
   b. No reaction
   c. $2I^- + F_2 \rightarrow I_2 + 2F^-$
   d. $2Br^- + Cl_2 \rightarrow Br_2 + 2Cl^-$
   e. No reaction
   f. No reaction

7. Let x = fractional part due to Cl-35; $1 - x$ = fractional part due to Cl-37.
   $34.968\,85x + 36.965\,90(1 - x) = 35.453$
   $x = 0.7576$ or 75.76% Cl-35
   $1 - 0.7576 = 0.2424$ or 24.24% Cl-37

8. $Cl_2 + 2HBr \rightarrow 2HCl + Br_2$
   $\Delta H = \{[2 \text{ mol HCl} \times \Delta H_f(HCl)] + \Delta H_f(Br_2)\}$
   $- \{\Delta H_f(Cl_2) + [2 \text{ mol HBr} \times \Delta H_f(HBr)]\}$

$$= \left\{\left(2 \text{ mol HCl} \times \frac{-92.3 \text{ kJ}}{\text{mol HCl}}\right) + 0\right\}$$
$$- \left\{0 + \left(2 \text{ mol HBr} \times \frac{-36.40 \text{ kJ}}{\text{mol HBr}}\right)\right\} = -111.8 \text{ kJ}$$

9. $5 \times 250. \text{ mL Cl}_2 \times \frac{1 \text{ L}}{10^3 \text{ mL}} \times \frac{273 K}{295 K} \times \frac{730. \text{ mm}}{760 \text{ mm}} = 1.11 \text{ L Cl}_2$ at STP

   $4HCl + MnO_2 \rightarrow MnCl_2 + 2H_2O + Cl_2(g)$

   $1.11 \text{ L Cl}_2 \times \frac{1 \text{ mol}}{22.4 \text{ L}} \times \frac{4 \text{ mol HCl}}{\text{mol Cl}_2} \times \frac{36.5 \text{ g HCl}}{\text{mol HCl}} \times \frac{100\% \text{ acid}}{38\% \text{ HCl}} \times \frac{1 \text{ mL}}{1.20 \text{ g HCl}} = 15.9 \text{ mL conc. HCl}$

   $1.11 \text{ L Cl} \times \frac{1 \text{ mol}}{22.4 \text{ L}} \times \frac{1 \text{ mol MnO}_2}{\text{mol Cl}_2} \times \frac{86.9 \text{ g MnO}_2}{\text{mol MnO}_2} = 4.31 \text{ g MnO}_2$

10. $PV = nRT$
    $n = PV/RT$
    $n = (1 \text{ atm} \times 5.00 \text{ L})/(0.0821 \text{ L} \cdot \text{atm/mol} \cdot \text{K} \times 273K)$
    $n = 0.223$ mol
    At STP chlorine does not behave exactly as an ideal gas. This is probably because at STP, chlorine is so near its condensation temperature of $-34.6\,°C$.

11. 1 mol $CaCl_2$ = 111.0 g; 1 mol $CaCl_2$ yields 3 mol of ions.

    $$\frac{62.5 \text{ g CaCl}_2}{500. \text{ g H}_2O} \times \frac{10^3}{\text{kg}} \times \frac{1 \text{ mol CaCl}_2}{111.0 \text{ g CaCl}_2} \times \frac{3 \text{ mol ions}}{\text{mol CaCl}_2} \times \frac{-1.86 C° \cdot \text{kg H}_2O}{\text{mol ions}} = -6.28 C°$$

    $0°C - 6.28 C° = -6.28°C$

Critical Thinking

1. Because otherwise they would react with one another.
2. **a.** HF: The concentration of hydrogen-bonded hydronium ion—fluoride ion pairs is much higher than the concentrations of either separate hydronium ions and fluoride ions or hydrogen fluoride molecules. HCl, HBr, and HI: The concentrations of hydronium and halide ions are much higher than the concentration of hydrogen halide molecules.
   b. Hydrogen-bonded hydronium ion-fluoride ion pairs are more stable than either separated hydronium ions and fluoride ions or hydrogen fluoride molecules. Hydrogen chloride, hydrogen bromide, and hydrogen iodide molecules are less stable than hydronium ions and the corresponding halide ions.
3. Aluminum is not as highly electropositive a metal as sodium and magnesium.
4. Student answer

Videodisc Selections from *The Periodic Table Videodisc* (JCE Software)

step forward

step reverse

15750
chlorine + bases (m)

CHAPTER 28

pg 780 Halogens

14228
chlorine (s)

35171
bromine (s)

45067
iodine (s)

pg 781 Fluorine

36414
uses of fluorine (s)

pg 783 Reactions of chlorine

14234
chlorine + water (m)

14778
chlorine + acids (m)

pg 785

36514
uses of chlorine (s)

pg 788 Uses of bromine

36736
fire extinguishers (s)

36748
plastics, fire retardants (s)

35187
bromine + bases (m)

pg 790 Iodine

45079
iodine + bases (m)

36928
uses of iodine (s)

778E Chapter 28 Teacher's Notes

Teacher's Notes

Chapter 28
The Halogens

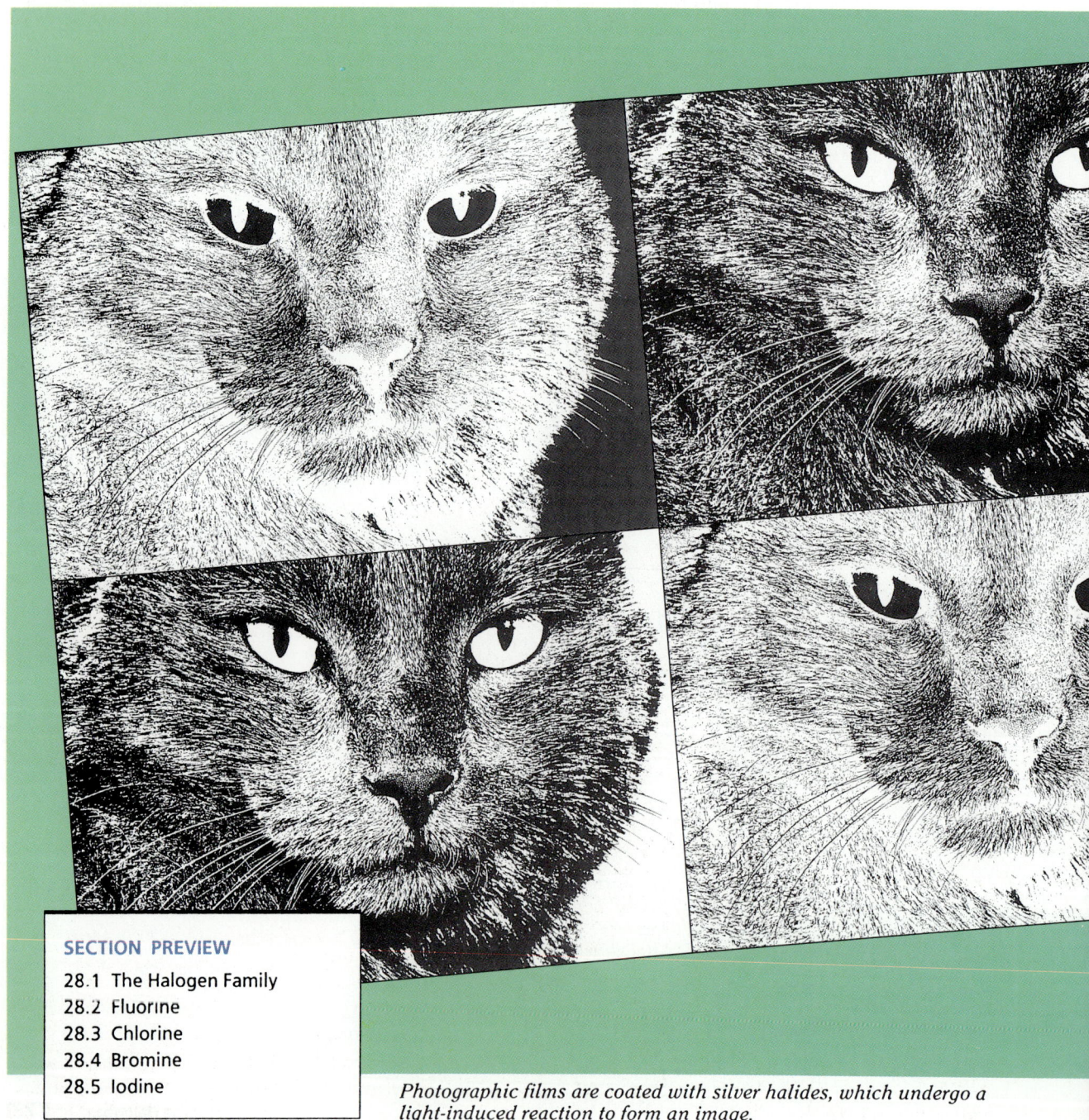

SECTION PREVIEW

28.1 The Halogen Family
28.2 Fluorine
28.3 Chlorine
28.4 Bromine
28.5 Iodine

Photographic films are coated with silver halides, which undergo a light-induced reaction to form an image.

28.1 The Halogen Family

The members of the Halogen family are the colorful, active, nonmetallic elements fluorine, chlorine, bromine, iodine, and astatine. *These elements constitute Group 17 of the Periodic Table and are referred to as* **halogens.** They are named *halogens* from the Greek words meaning *salt producer* because they form ionic salts with the alkali metals, salts that are among the most abundant soluble salts found in nature. Samples of the first four of these elements are shown in Figure 28-1.

SECTION OBJECTIVES

- Identify the members of the Halogen family.
- Describe the general characteristics of the halogen elements. Contrast the properties of fluorine to those of the other members of the Halogen family.

Figure 28-1 The halogens are colorful elements. *(Clockwise from lower left:)* Chlorine is a greenish-yellow gas, fluorine is a pale yellow gas, bromine is a reddish-brown liquid, and iodine is a dark violet crystal.

General Characteristics of the Halogens

Some of the properties of the halogens are shown in Table 28-1. The atoms of each halogen have seven electrons in the highest energy level. The addition of one electron to a halogen atom converts it to a halide ion having an outer octet of electrons.

$$:\!\ddot{\underset{..}{Cl}}\!: \;+\; e^- \;\longrightarrow\; :\!\ddot{\underset{..}{Cl}}\!:^-$$

Using a chlorine atom as an example, the halogens are all active elements that have high electronegativities. They are very rarely found free in nature. In the elemental state they exist as covalent, diatomic molecules.

Fluorine has the smallest atoms in the Halogen family. It has a strong attraction for electrons. Fluorine is the most highly electronegative element and the most active nonmetal. Because of these properties, fluorine cannot be prepared from its compounds by chemical oxidation. It must be prepared by electrolysis.

The other halogens, with increasingly larger atoms, are less electronegative than fluorine. As a result, the smaller, lighter halogens can replace and oxidize the larger, heavier halogens from compounds. Astatine is a synthetic radioactive halogen.

STUDY HINT

Reactions involving the replacement of halogens are described in Section 8.2.

The Halogens   **779**

Students might expect that hydrogen fluoride would be ionic because of the great difference in electronegativity between hydrogen and fluorine. However, in the gas phase the polar molecules become hydrogen-bonded to one another, forming a chain-like structure of HF units rather than an ionic crystal. In aqueous solution the negative fluorines bond readily to hydrogen atoms in water molecules, and the molecule does not dissociate.

Table 28-1 clearly shows regular change in properties that occurs in this family. This change in properties proceeds from the smallest and lightest atoms to the largest and heaviest.

Each of the halogens combines with hydrogen. The great electronegativity difference between hydrogen and fluorine explains why hydrogen fluoride molecules are so polar that they associate by hydrogen bonding. The remaining hydrogen halides have smaller electronegativity differences and do not show this property. All hydrogen halides are colorless gases that are ionized in water solution. Hydrofluoric acid is a weak acid because hydronium ions are strongly hydrogen bonded to fluoride ions. The other binary halogen acids are highly dissociated, strong acids.

TABLE 28-1 PROPERTIES OF THE HALOGEN ATOMS

	Fluorine	Chlorine	Bromine	Iodine	Astatine
Atomic number	9	17	35	53	85
Atomic mass	18.998 403	35.453	79.904	126.905	(210)*
Electron configuration	[He]$2s^2 2p^5$	[Ne]$3s^2 3p^5$	[Ar]$3d^{10} 4s^2 4p^5$	[Kr]$4d^{10} 5s^2 5p^5$	[Xe]$4f^{14} 5d^{10} 6s^2 6p^5$
Ionization energy (kJ/mol)	1681	1251	1140	1008	
Electron affinity (kJ/mol)	−322	−349	−325	−295	−270**
Bond energy X_2 (kJ/mol)	159	243	192	151	80**
Electronegativity	4.0	3.0	2.8	2.5	2.2
Atomic radius (pm)	71	99	114	133	140
Ionic radius of X^- (pm)	136	181	195	216	
Melting point (°C)	−219.62	−100.98	−7.2	113.5	302
Boiling point (°C)	−188.14	−34.6	58.78	184.35	337**
Density	1.70 g/L	3.21 g/L	3.12 g/mL	4.94 g/cm³	
Oxidation numbers	−1	−1,+1,+3,+5,+7	−1,+1,+3,+5,+7	−1,+1,+3,+5,+7	−1,+1,+3,+5,+7**
Reduction potential (V)	2.87	1.36	1.07	0.54	
Color of element	pale yellow	yellow-green	reddish-brown	dark violet	
Standard phase	gas	gas	liquid	solid	

* Atomic mass in parentheses is that of the most stable isotope.
**Estimated

STUDY HINT
Half-lives are covered in Section 29.2.

The Physical Phases of the Halogens

If you examine closely the data in Table 28-1, you will find that, under normal conditions, fluorine and chlorine are gases, bromine is a liquid, and iodine is a solid. Astatine is a radioactive element; its longest-lived isotope, astatine-210, has a half-life of 8.3 hours. Because a cyclotron must be used to synthesize astatine, it is very expensive to produce, and its chemistry has not been investigated extensively.

We can use our knowledge of periodicity to make certain predictions about astatine, however. Using trends of the periodic properties of the other halogens, we would expect astatine to have the smallest ionization energy among the halogens, and the highest melting point and boiling point. We would also predict astatine would be a dark-colored solid, that it would have a greater density than the other halogens, and that it would be less soluble in water than the other halogens.

Under ordinary conditions, all of the halogens exist as diatomic molecules. In these molecules, sharing a single pair of electrons gives each halogen atom a stable noble-gas electronic configuration. In the liquid and solid states, the molecules are held together by intermolecular forces (van der Waals forces). As the atomic number increases, the intermolecular forces tend to increase. We should thus expect an increase in the intermolecular forces between halogen molecules as we go from top to bottom within the Halogen family in the Periodic Table. The ionization energies of the halogen elements decrease, as would be expected, with increasing atomic numbers. Because iodine is the largest and most easily ionized of the naturally occurring halogens, the intermolecular forces between iodine molecules are the strongest. The strength of the intermolecular forces between halogen molecules explains the increase in the melting and boiling points from fluorine to iodine. Iodine molecules have 106 electrons, while bromine molecules have 70. The intermolecular force should thus be greater between iodine molecules than between bromine molecules, which explains the higher melting and boiling points for iodine than bromine.

Section Review

1. Which elements are known as the halogens, and what are their symbols?
2. List the general characteristics of the halogens.
3. Why does fluorine exhibit properties that are different than those of the other halogens?

1. Fluorine, F; chlorine, Cl; bromine, Br; iodine, I; astatine, At
2. Reactive, high electronegativities, high ionization energies, high electron affinities, low melting and boiling points, colorful, exist as covalent diatomic molecules in the elemental state
3. It has the smallest atoms in the halogen family and has a strong attraction for electrons. As a result, it cannot be prepared from its compounds by chemical reduction. Also, because of the great electronegative difference between hydrogen and fluorine, hydrogen-fluorine molecules are so polar that they associate by hydrogen bonding. The other hydrogen halides do not exhibit this property.

28.2 Fluorine

Because of its high electronegativity and this tendency to form compounds, fluorine is never found free in nature. In the earth's crust, fluorine exists mainly in the form of the minerals fluorspar (CaF_2), fluorapatite ($CaF_2 \cdot 3Ca_3(PO_4)_2$), and cryolite ($Na_3AlF_6$).

Preparation and Properties of Fluorine

Fluorine is prepared by electrolyzing a mixture of potassium fluoride and hydrogen fluoride. A steel and Monel metal electrolytic cell with a positive carbon electrode is used. **Monel metal** *is a corrosion-resistant alloy composed principally of nickel and copper.* The fluoride coating that forms on these metals protects them from further reaction.

Fluorine is the most active nonmetallic element. It unites with hydrogen explosively. It forms compounds with all elements except helium, neon, and argon. As the most electronegative element, fluorine is assigned an oxidation number of −1 in all compounds. There are no known positive oxidation states of fluorine. It forms salts known as fluorides. Fluorine reacts with gold and platinum slowly. Special carbon-steel containers are used to transport fluorine. These containers become coated with iron fluoride, which resists further action.

SECTION OBJECTIVES

- Cite the common occurrences of fluorine.
- Describe the properties and preparation of fluorine.
- Identify commercial uses of fluorine compounds.

Nothing forms compounds with helium, neon, or argon.

Since no other element pulls harder on electrons than fluorine, it cannot lose electrons or form positive oxidation states in ordinary chemical reactions.

The Halogens

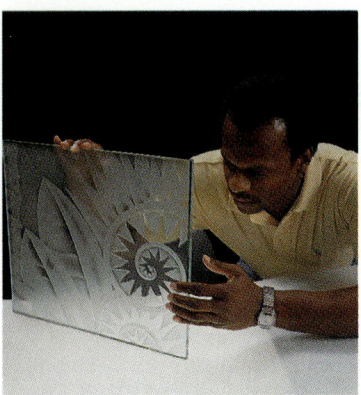

Figure 28-2 Hydrofluoric acid is used to etch glass.

1. Fluorspar, (CaF_2); fluorapatite, $Ca_5(PO_4)_3F$; cryolite, Na_3AlF_6
2. The position of fluorine in Table 20-3 indicates that it is the strongest common oxidizing agent known. This means that fluoride ions cannot be oxidized by other chemical oxidizing agents.
3. Fluorine is prepared by electrolyzing a mixture of potassium fluoride and hydrogen fluoride.
4. Accept all answers given in the subsection "Compounds of Fluorine and Their Uses."

SECTION OBJECTIVES

- Cite the principal occurrences of chlorine.
- Describe methods for the preparation of chlorine.
- Relate the properties of chlorine to its important commercial uses.
- Describe the preparation, properties, and uses of hydrogen chloride.

Compounds of Fluorine and Their Uses

The mineral fluorspar (CaF_2) is often used in preparing fluorine compounds. Sodium fluoride is used as a poison in pesticides for roaches, rats, and other vermin. A trace of sodium fluoride, or the less expensive sodium hexafluorosilicate (Na_2SiF_6), is added to drinking water in many areas to help prevent tooth decay. The fluoride ion concentration is carefully regulated to be 1 ppm. Fluorides have also been added to some toothpastes.

One of the freons, dichlorodifluoromethane (CCl_2F_2) is used as a refrigerant. It is odorless, nonflammable, and nontoxic. Its critical temperature is 111.7°C and critical pressure is 39.4 atm. In producing aluminum, melted cryolite Na_3AlF_6 is used as a solvent for aluminum oxide. Uranium is changed to uranium hexafluoride gas (UF_6) in the process for separating the uranium isotopes.

Fluorine combines with the noble gases krypton, xenon, and radon. Krypton difluoride (KrF_2) is prepared by passing electricity through a krypton-fluorine mixture. This process is carried out at the temperature of liquid nitrogen. Krypton difluoride is a volatile white solid that decomposes slowly at room temperature.

Three fluorides of xenon—XeF_2, XeF_4, and XeF_6—have been made. All are colorless crystalline solids at ordinary temperatures. XeF_6 is the most highly reactive. Each of the three compounds reacts with hydrogen to produce elemental xenon and hydrogen fluoride. With water, XeF_2 produces xenon, oxygen, and hydrogen fluoride. The other two fluorides react with water to yield xenon trioxide (XeO_3) a white, highly explosive solid.

Hydrofluoric acid (HF) is used as a catalyst in producing high-octane gasoline. It is also used in making synthetic cryolite for aluminum production. For many years hydrofluoric acid has been used for etching glass. Glassware is given a frosty appearance by exposing it to hydrogen fluoride fumes, as shown in Figure 28-2.

Section Review

1. What are the principal minerals in which fluorine is found?
2. What is the reason that fluorine does not exhibit positive oxidation states?
3. Describe the method used for the preparation of fluorine.
4. Name two fluorine compounds and identity their commercial uses.

28.3 Chlorine

Chlorine is rarely found uncombined in nature. Elemental chlorine is found in small amounts in some volcanic gases. Chlorides of sodium, potassium, and magnesium are fairly abundant. Common table salt, sodium chloride, is a widely distributed compound. It is found in sea water, in underground brines, and in rock salt deposits. Sodium chloride is the commercial source of chlorine.

You may be familiar with the use of chlorine for maintaining clean water in swimming pools. In fact chlorine is used worldwide for the purification of drinking water.

Preparation of Chlorine

The element chlorine was first isolated in 1774 by Carl Wilhelm Scheele, the codiscoverer of oxygen. The preparation of elemental chlorine involves the oxidation of chloride ions. Strong oxidizing agents are required. The following three methods can be used to isolate chlorine.

1. By the electrolysis of sodium chloride. Chlorine is most often prepared by the electrolysis of a saturated water solution of sodium chloride. At this concentration, hydrogen from the water is released at the negative electrode, and chlorine is set free at the positive electrode. The hydrogen and chlorine gases are kept separate from each other and from the solution by asbestos partitions. The sodium and hydroxide ions remaining in the solution are recovered as sodium hydroxide.

$$2NaCl(aq) + 2H_2O(\ell) \xrightarrow{\text{(electricity)}} 2NaOH(aq) + H_2(g) + Cl_2(g)$$

2. By the oxidation of hydrogen chloride. This method involves heating a mixture of manganese dioxide and concentrated hydrochloric acid. The manganese oxidizes half of the chloride ions in the reacting HCl to chlorine atoms. Manganese is reduced during the reaction from the +4 oxidation state to the +2 state.

$$MnO_2(s) + 4HCl(aq) \xrightarrow{\Delta} MnCl_2(aq) + 2H_2O(\ell) + Cl_2(g)$$

This was the method used by Scheele in first preparing chlorine. It is a useful laboratory preparation. An alternative process involves heating a mixture of manganese dioxide, sodium chloride, and sulfuric acid, as shown in Figure 28-3.

$$2NaCl(s) + 2H_2SO_4(aq) + MnO_2(s) \xrightarrow{\Delta}$$
$$Na_2SO_4(aq) + MnSO_4(aq) + 2H_2O(\ell) + Cl_2(g)$$

3. By the action of hydrochloric acid on calcium hypochlorite. Hydrochloric acid added drop by drop to calcium hypochlorite powder releases chlorine and forms calcium chloride and water.

$$4HCl(aq) + Ca(ClO)_2(s) \rightarrow CaCl_2(aq) + 2Cl_2(g) + 2H_2O(\ell)$$

Most chlorine compounds are soluble, so chlorine is found in sea water and deposits from sea water. Fluorine compounds are generally insoluble, so fluorine is found in mineral ores.

Figure 28-3 Chlorine may be prepared by heating a mixture of manganese dioxide, sodium chloride, and sulfuric acid. Chlorine is collected by upward displacement of air. The preparation should be performed under a hood.

The Halogens

Chemistry and its Applications

PVC Pipes

Lead was once used extensively for water pipes and tanks because it was inexpensive and slow to corrode. When the hazards of lead poisoning were recognized, clay was used instead. As our cities age, replacement of those aging clay pipes has become a major problem. How can the pipes be replaced without interrupting the water and sewage service? Engineers developed an interesting process using PVC.

PVC stands for "polyvinyl chloride," a polymeric form of vinyl chloride. This plastic is durable, cheap, and easy to manufacture into pipes that do not corrode. The PVC pipes are created from a continuous strip of PVC with special clamping edges. The PVC is fed as a spiral directly into the old pipes, forming a sealed tube that is flexible enough to expand up against the old pipe. This lightweight PVC tube is supported by the old pipes, so it doesn't have to be as thick as other kinds of pipe.

Physical and Chemical Properties of Chlorine

Chlorine is a toxic, highly reactive element. Its chemical properties, however, make it an extremely useful element.

Physical Properties At room temperature, chlorine is a greenish-yellow gas with a disagreeable, suffocating odor. It is about 2.5 times as dense as air. It is moderately soluble in water, forming a pale greenish-yellow solution. Chlorine is easily liquefied and is usually marketed in steel cylinders.

When inhaled in small quantities, chlorine affects the mucous membranes of the nose and throat. If inhaled in larger quantities, chlorine is so poisonous that it may cause death.

The odor of chlorine is familiar to everyone as the odor of chlorine bleaches and of swimming-pool disinfectant.

Chemical Properties The third energy level of a chlorine atom contains seven electrons. There are many reactions by which chlorine atoms can acquire an additional electron and complete the third-energy-level octet. Three reactions involving chlorine are given as examples.

1. Reaction with metals and metalloids. Most metals and metalloids react with chlorine directly. For example, when powdered antimony is sprinkled into a jar of moist chlorine, the two elements combine spontaneously, emitting sparks. Antimony trichloride is formed.

$$2Sb(s) + 3Cl_2(g) \rightarrow 2SbCl_3(s)$$

In a similar manner, hot metallic sodium burns in chlorine to form sodium chloride. Chlorine combines directly with such metals as copper, iron, zinc, and arsenic, if they are heated slightly.

2. Reaction with hydrogen. If hydrogen and chlorine are mixed in the dark, no reaction occurs. But such a mixture explodes violently if it is heated or exposed to sunlight. The heat or sunlight provides the activation energy. A jet of hydrogen that is burning in air will continue to burn if it is inserted into a bottle of chlorine.

$$H_2(g) + Cl_2(g) \xrightarrow{\Delta} 2HCl(g)$$

Chlorine does not support the combustion of wood or paper. A paraffin candle, however, continues to burn in chlorine with a smoky flame, as shown in Figure 28-4. In this reaction the hydrogen of the paraffin combines with the chlorine, forming hydrogen chloride. The carbon is left uncombined.

3. Reaction with water. A freshly prepared solution of chlorine in water has a greenish-yellow color. If it stands in sunlight for a few days, both the color and the strong chlorine odor will disappear. The chlorine combines with the water to form hypochlorous acid and hydrochloric acid. Hypochlorous acid is unstable and decomposes into hydrochloric acid and oxygen.

$$2H_2O(\ell) + 2Cl_2(g) \rightarrow 2HClO(aq) + 2HCl(aq)$$
$$2HClO(aq) \rightarrow 2HCl(aq) + O_2(g)$$

The instability of hypochlorous acid makes chlorine water a good oxidizing agent.

Figure 28-4 A paraffin candle will burn in chlorine. The hydrogen of the paraffin combines with the chlorine to form hydrogen chloride. The smoke is the carbon from the wax, which is left uncombined.

Uses of Chlorine

1. For bleaching. **Bleaching** is the operation by which color is partially or wholly removed from colored materials. Household bleaching solution is usually a solution of sodium hypochlorite. It is made by reacting liquid chlorine with sodium hydroxide solution. Chlorine is used as a bleaching agent by the pulp and paper industry.

CAUTION
Chlorine destroys silk or wool fibers. Never use commercial bleaches containing hypochlorites on silk or wool.

2. As a disinfectant. Since moist chlorine is a good oxidizing agent, it destroys bacteria. **Chlorinated lime** *is Ca(ClO)Cl.* Large quantities of Ca(ClO)Cl, are used as a **disinfectant** *which is a chemical agent that destroys microorganisms but not bacterial spores.* In city water systems, billions of gallons of water are treated with chlorine to kill disease-producing bacteria. The water in swimming pools is usually treated with chlorine. As part of the sewage purification process, chlorine is sometimes used to kill bacteria in sewage.

3. For making compounds. Because chlorine combines directly with many substances, it is used to produce a variety of compounds. Among these are aluminum chloride (Al_2Cl_6), used as a catalyst; carbon tetrachloride (CCl_4), a solvent; 1,1,2-trichloroethene (ClHC=CCl_2), a cleaning agent; and **polyvinyl chloride** which is *the polymer of vinyl chloride (CH_2CHCl).* Polyvinyl chloride is replacing steel and cast iron as the material of choice in plumbing and electrical pipe. Polyvinyl chloride has the advantage of being lightweight, strong, and free from corrosion.

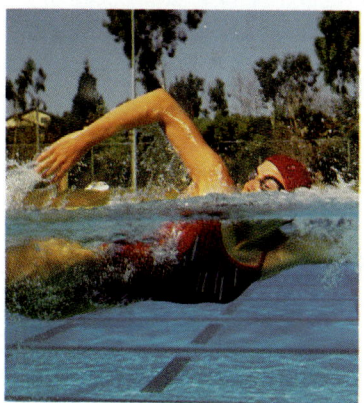

Figure 28-5 Chlorine is used to kill microorganisms in swimming pools.

Seventy percent of chlorine produced is used for the production of organic-chloride compounds such as polyvinyl chloride.

The Halogens 785

Figure 28-6 Large quantities of hydrochloric acid are used to clean metals.

STUDY HINT
Some of these processes are the chlorination of hydrocarbons described in Chapter 21.

1. Usually in the combined state as chlorides of sodium, potassium, and magnesium. Sodium chloride is found in salt brines underground, in rock-salt deposits, and in sea water as sodium chloride. Elemental chlorine is found in some volcanic gases.
2. Accept all answers given in the subsection "Preparation of Chlorine."
3. Accept all answers given in the subsection "Uses of Chlorine."
4. By treating sodium chloride with concentrated sulfuric acid
5. Colorless gas, sharp penetrating odor, denser than air, soluble in water. Hydrochloric acid reacts with metals above hydrogen in the activity series and with oxides of metals; it is used in preparing chlorides and in cleaning metals.

An Important Chlorine Compound: Hydrogen Chloride

Hydrogen chloride ranks in the top 30 among industrial chemicals in production. Hydrogen chloride dissolved in pure water is sold under the name hydrochloric acid. Its nontechnical name is muriatic acid. Both hydrogen chloride and hydrochloric acid are industrial chemicals of major importance.

Preparation of hydrogen chloride   In the laboratory, hydrogen chloride can be prepared by treating sodium chloride with concentrated sulfuric acid.

$$NaCl(s) + H_2SO_4(\ell) \rightarrow NaHSO_4(s) + HCl(g)$$

This same reaction, carried out at a higher temperature, is used commercially. If more NaCl is used at this higher temperature a second molecule of HCl can be produced per molecule of H_2SO_4.

$$2NaCl(s) + H_2SO_4(\ell) \xrightarrow{\Delta} Na_2SO_4(s) + 2HCl(g)$$

Another commercial preparation involves the direct union of hydrogen and chlorine. Both elements are obtained by the electrolysis of concentrated sodium chloride solution.

By far the most important commercial source of hydrogen chloride results from the formation of HCl as a by-product of chlorination processes.

Physical and chemical properties of hydrogen chloride   Hydrogen chloride is a colorless gas with a sharp, penetrating odor. It is denser than air and extremely soluble in water. One volume of water at 0°C dissolves more than 500 volumes of hydrogen chloride at standard pressure. Hydrogen chloride makes a fog with moist air. It is so soluble that it condenses water vapor from the air into tiny drops of hydrochloric acid.

Hydrogen chloride is a stable compound that does not burn. Some vigorous oxidizing agents react with it to form water and chlorine. The water solution of hydrogen chloride is hydrochloric acid. Concentrated hydrochloric acid contains about 38 percent hydrogen chloride by weight and is about 1.2 times as dense as water. Hydrochloric acid reacts with metals above hydrogen in the activity series and with the oxides of metals. It neutralizes hydroxides, forming salts and water.

Uses of hydrochloric acid   Hydrochloric acid is used in preparing certain chlorides and in cleaning metals. This cleaning involves removing oxides and other forms of tarnish.

Some hydrochloric acid is essential in the process of digestion. The concentration of hydrochloric acid in gastric juice is $0.16 M$.

Section Review

1. In what state is chlorine found in nature?
2. Describe a method for the preparation of chlorine, and write a balanced chemical equation that shows the reaction.
3. List one commercial use of chlorine and explain its use in terms of its properties.
4. How is hydrogen chloride prepared in the laboratory?
5. List some of the properties of hydrogen chloride and its uses as a water solution.

28.4 Bromine

Bromine is far less abundant in the earth's crust than the first two members of the Halogen family. It occurs mainly as potassium, magnesium, and sodium bromides in sea water, in underground brines, and in salt deposits. Water from brine wells in Michigan and Arkansas have relatively high concentrations of bromine and are a major source of the halogen. Seawater, which contains relatively low concentrations of bromine (about 85 parts per million), is not an economical source for this element. Bromine is toxic and reacts readily with many elements. It is the only nonmetal that is liquid under standard conditions.

SECTION OBJECTIVES
- Describe the preparation of elemental bromine.
- Describe some properties of elemental bromine.
- List some compounds of bromine and their commercial uses.

Preparation of Bromine

In the laboratory, bromine can be prepared by using manganese dioxide, sulfuric acid, and sodium bromide as shown in Figure 28-7.

$$2NaBr(s) + MnO_2(s) + 2H_2SO_4(aq) \xrightarrow{\Delta}$$
$$Na_2SO_4(aq) + MnSO_4(aq) + 2H_2O(\ell) + Br_2(g)$$

This is similar to the preparation of chlorine as discussed earlier in this section.

The commercial production of bromine from salt-well brines depends on the ability of chlorine to displace bromide ions from solution. Chlorine displaces bromide ions because it is more highly electronegative than bromine.

$$2Br^-(aq) + Cl_2(g) \rightarrow 2Cl^-(aq) + Br_2(\ell)$$

Figure 28-7 Bromine can be prepared by heating a mixture of sodium bromide, manganese dioxide, and sulfuric acid in a glass-stoppered retort. This preparation should be performed under a hood.

The Halogens 787

CAUTION
Use great care in handling bromine. It burns the flesh and forms wounds that heal slowly.

Physical and Chemical Properties of Bromine

Bromine is a dark, reddish-brown liquid that is about three times as dense as water. It evaporates readily, forming a vapor that burns the eyes and throat and has a very disagreeable odor. The name "bromine" is derived from the Greek word *bromos*, meaning stench. Bromine is moderately soluble in water. The reddish-brown water solution used in the laboratory is known as bromine water. Bromine dissolves readily in carbon tetrachloride and carbon disulfide and in water solutions of bromides.

Bromine unites directly with hydrogen to form hydrogen bromide, as shown in Figure 28-8. It combines with most metals to form bromides. When it is moist, bromine is a good bleaching agent. A water solution of bromine is a strong oxidizing agent that forms hydrobromic acid and oxygen in sunlight.

Figure 28-8 Hydrogen burns in an atmosphere of bromine vapor. Hydrogen bromide is formed.

Compounds of Bromine and Their Uses

Complex bromine-containing organic compounds are added to polyurethane and polystyrene plastics to make them flame retardant. Methyl bromide (CH_3Br) is a useful soil fumigant. Improved crop yields result from such treatment. Ethylene bromide ($C_2H_4Br_2$) increases the efficiency of tetraethyl lead in antiknock gasoline. The use of bromine in this application is becoming more limited since the United States Environmental Protection Agency has initiated a phaseout of the use of lead in gasoline. This phaseout started in the early 1980's. Ethylene bromide was formerly used to control insects in grain. This use has been stopped because ethylene bromide is believed to cause cancer.

Silver bromide (AgBr) is a yellowish solid. It is highly **photosensitive,** *meaning it is sensitive to light rays* and is widely used in making photographic film. A solution of calcium bromide ($CaBr_2$) is used as a

Desktop Investigation

Silver Halides

Question
How do you make silver halides, and what effect does light have on these halogen compounds?

Materials
sodium chloride, 0.2 M solution
sodium bromide, 0.2 M solution
sodium iodide, 0.2 M solution
silver nitrate, 0.1 M solution
three test tubes, 16 × 150 mm
medicine dropper
filter paper
ring stand with iron ring
beaker, 150 mL

Procedure
Remove all combustible materials from the work area. Remember to wear safety glasses and apron. **Caution:** Do not allow AgNO₃ to come in contact with your skin or clothing.
1. Add 5 mL of sodium chloride solution to the first test

tube. Add 5 mL of sodium bromide solution to the second test tube. Add 5 mL of sodium iodine solution to the third test tube. To each of the three test tubes, add 10 drops of silver nitrate solution. What is the result?
2. Immediately filter the contents of each test tube. (See diagram.) If necessary, flush the test tube with small amounts of water from a wash bottle to remove all of the precipitate. Unfold the filter papers and expose their contents to direct sunlight for several minutes. What is the result? Note the order in which the products show a change.
3. Your teacher will tell you how to dispose of your chemicals.

Discussion
1. In what ways were the reactions (a) similar and (b) different?
2. Write the equations for all reactions occurring.
3. Silver halides are used in black-and-white film. Which silver halide is probably used to make fast film (film that requires relatively little light)? Which is used to make slow film (film that requires a relatively large amount of light)?
4. Write a good conclusion for this experiment.

---

high-density fluid in oil-well drilling. The bromides of sodium and potassium are used in medicine as sedatives. Such medicines should not be used unless prescribed by a physician. Certain bromine compounds may be used as disinfectants. Many organic bromine compounds are useful intermediates in reactions forming amine, phenols, or ethers.

Section Review
1. Write a balanced chemical equation for the preparation of bromine in the laboratory.
2. What are some of the properties of bromine?
3. Name two compounds of bromine and identify their commercial uses.
4. Describe some ways in which bromine compounds are known to be dangerous to human health.

1. $2NaBr(s) + MnO_2(s) + 2H_2SO_4(aq) \xrightarrow{\Delta} Na_2SO_4(aq) + MnSO_4(aq) + 2H_2O(l) + Br_2(g)$
2. Accept all answers given in the subsection "Physical and Chemical Properties of Bromine," and in Table 28-1.
3. Accept all answers given in the subsection "Compounds of Bromine and Their Uses."
4. See "Compounds of Bromine and Their Uses."

The Halogens

Some seaweeds can concentrate the iodine found in the oceans. When the seaweed is dried, this "kelp" is used as a commercial source of iodine.

SECTION OBJECTIVES

- Describe a method for the preparation of iodine.
- List physical and chemical properties of iodine.
- Cite uses of iodine and some of its compounds.

Figure 28-9 Iodine may be prepared by heating sodium iodide and manganese dioxide with sulfuric acid. It should be done under a hood.

A lack of sufficient iodine in the human body can cause an enlarged thyroid gland, which is called a goiter. The potassium iodide added to salt helps prevent this condition.

1. $2NaI(s) + MnO_2(s) + 2H_2SO_4(aq) \xrightarrow{\Delta} Na_2SO_4(aq) + MnSO_4(aq) + 2H_2O(\ell) + I_2(g)$
2. Accept answers given in the subsection "Physical and Chemical Properties of Iodine," and in Table 28-1.
3. Accept all answers given in the subsection "Uses of Iodine and Its Compounds."

28.5 Iodine

Iodine, like bromine, is far less abundant in the earth's crust than are fluorine and chlorine. Iodine occurs principally as sodium iodate and periodate impurities in sodium nitrate deposits found in Chile. The largest source of iodine in the United States is underground brine from oil wells in California.

Preparation of Iodine

The laboratory preparation of iodine is similar to that of chlorine and bromine. An iodide is heated with manganese dioxide and sulfuric acid, as shown in Figure 28-9.

$$2NaI(s) + MnO_2(s) + 2H_2SO_4(aq) \xrightarrow{\Delta} Na_2SO_4(aq) + MnSO_4(aq) + 2H_2O(\ell) + I_2(g)$$

The iodine is driven off as a vapor and can be condensed as a solid on the walls of a cold dish or beaker.

Physical and Chemical Properties of Iodine

Iodine is a bluish-black crystalline solid. When heated, it sublimes (vaporizes without melting) and produces a violet-colored vapor. The odor of this vapor is irritating, resembling that of chlorine. Iodine is very slightly soluble in water. It is much more soluble in water solutions of sodium or potassium iodide. With these solutions it forms complex I_3^- ions. It dissolves readily in alcohol, forming a dark-brown solution. It is very soluble in carbon disulfide and carbon tetrachloride, giving a rich violet color.

Iodine is active chemically, though less so than either bromine or chlorine. It combines with metals to form iodides. It can also form compounds with all nonmetals except sulfur, selenium, and the noble gases. Iodine reacts with organic compounds just as chlorine and bromine do. But since the carbon-iodine bond is not strong, many of the compounds formed decompose readily.

Uses of Iodine and Its Compounds

Iodine is most useful as a disinfectant. Iodine that is "complexed" (loosely bonded) with certain organic compounds is used as an antiseptic on the skin. An **antiseptic** *is a substance that checks the growth of microorganisms on or in the body.* Detergents with which iodine is complexed make good combination cleaning and sanitizing agents. Iodine and some iodine compounds are used as catalysts in certain organic reactions.

Silver iodide is a light-sensitive compound used in photographic film. Potassium iodide is added to table salt to provide the iodine necessary for proper nutrition.

Section Review

1. Write a balanced chemical equation for the preparation of iodine.
2. What are some of the properties of elemental iodine?
3. Describe some of the uses of iodine and its compounds.

Chemistry in Detail

Discovering How to Discover Elements

The search for new elements is as old as chemistry, but in the early days, it was not very systematic. In 1807, Sir Humphry Davy built a powerful battery (see illustration), ran current through molten "caustic potash" (potassium hydroxide), and produced a new element, potassium. Because "caustic soda" (sodium hydroxide) was chemically similar, he treated it similarly and discovered sodium. Within a year, he had discovered four other elements!

Later in the same century, Dmitri Mendeleev discovered the Periodic Law: properties of elements occur in cycles. He believed that some elements had not been discovered because of gaps in some cycles. Mendeleev predicted several properties of these missing elements, and eventually he was shown to be correct.

Modern atomic theory correlates the properties of elements and their position on the Periodic Table with their electronic structure. This allows chemists to predict the properties of unknown elements with high precision. Such predictions help us to confirm that a new element has been discovered. For example, the elements technetium and promethium have been identified in the light of certain stars.

Chapter Summary

- The Halogen family consists of the highly electronegative elements fluorine, chlorine, bromine, iodine, and astatine.
- Fluorine compounds are used *(1)* as refrigerants *(2)* to help prevent tooth decay *(3)* in aluminum production *(4)* for separating uranium isotopes *(5)* in gasoline production and *(6)* for etching glass. Chlorine is used for bleaching, disinfecting, and making chlorine-containing compounds.
- Bromine compounds are used *(1)* to make plastics flame retardant *(2)* in agriculture *(3)* in antiknock gasoline *(4)* in oil-well drilling *(5)* as disinfectants *(6)* in medicine and *(7)* in photography.
- Iodine is used as an antiseptic and in organic synthesis. Iodides are used in photography and as nutrients.

Chapter 28 Review

The Halogens

VOCABULARY

antiseptic (790)
chlorinated lime (785)
disinfectant (785)
halogen (779)
Monel metal (781)
photosensitive (788)
polyvinyl chloride (785)

REVIEWING CONCEPTS

1. **a.** In what form are the halogens found in nature, as atoms, molecules, or ions?
   b. Why? (28.1)
2. As one does down the Halogen family on the Periodic Table, how do the members vary in
   a. atomic radius
   b. ionic radius
   c. first ionization energy
   d. electron affinity
   e. electronegativity
   f. melting point
   g. boiling point
   h. density
   i. principal oxidation number? (28.1)
3. **a.** What kind of container is used for fluorine?
   b. What chemical property makes this use possible? (28.2)
4. What properties make dichlorodifluoromethane useful as a refrigerant? (28.2)
5. All the halogens except fluorine exhibit positive oxidation states. Why is fluorine different? (28.2)
6. **a.** What compound is the commercial source of chlorine?
   b. For what other element is this compound the commercial source? (28.3)
7. What effects does chlorine have on the body? (28.3)
8. **a.** Describe an industrial use for hydrochloric acid.
   b. What is the function of hydrochloric acid in the human body? (28.3)
9. Write the balanced formula equation for the lab preparation of chlorine by the action of hydrochloric acid on calcium hypochlorite. Assign oxidation numbers. Tell what is oxidized and what is reduced. (28.3)
10. **a.** What is the color of freshly prepared chlorine water?
    b. Why?
    c. Does the color change as chlorine water is exposed to sunlight?
    d. Explain. (28.3)
11. Write the equation for the laboratory preparation of bromine from
    a. manganese dioxide and hydrobromic acid.
    b. manganese dioxide, sodium bromide, and sulfuric acid.
    c. Assign oxidation numbers in both equations and tell which element is oxidized and which element is reduced.
    d. How do these reactions compare? (28.4)
12. Write formulas for two organic and two inorganic bromine compounds and give a use for each. (28.4)
13. What phase change does iodine undergo that the other halogens do not? (28.5)
14. Write an equation using electron-dot symbols for the formation of an iodide ion from an iodine atom. (28.5)
15. Compare the colors of
    a. solid iodine
    b. iodine in alcohol
    c. iodine in carbon tetrachloride
    d. iodine vapor (28.5)

PROBLEMS

1. **a.** Given that chlorine has a density of 3.21 g/L how many moles of Cl_2 are there in 5.00 L of the gas at STP?
   b. How many Cl_2 molecules is this?

2. What is the mass percentage of bromine in ethylene bromide ($C_2H_4Br_2$)?

3. Chlorine reacts with calcium hydroxide to produce bleaching powder, Ca(ClO)Cl, and water.
   a. What mass of calcium hydroxide is required for making 200. g of bleaching powder?
   b. What mass of chlorine is also required?

4. How many liters of chlorine at STP can be obtained from 375 g of sodium chloride by electrolysis?

5. How many grams of zinc chloride can be produced from 0.500 L of chlorine at STP?

6. If a reaction occurs, complete and balance the following ionic equations:
   a. $Cl^- + F_2 \rightarrow$ _____.
   b. $F^- + Br_2 \rightarrow$ _____.
   c. $I^- + F_2 \rightarrow$ _____.
   d. $Br^- + Cl_2 \rightarrow$ _____.
   e. $Cl^- + I_2 \rightarrow$ _____.
   f. $Br^- + I_2 \rightarrow$ _____.

7. Natural chlorine consists of only the two isotopes Cl-35 (34.968 85 u) and Cl-37 (36.965 90 u). If the atomic mass is 35.453, what are the percentages of these two isotopes in naturally occurring chlorine?

8. Calculate for heat of reaction for:
$$\tfrac{1}{2}Cl_2 + HBr \rightarrow HCl + \tfrac{1}{2}Br_2$$
from the heat of formation data in Appendix Table A-13.

9. A laboratory experiment requires five 250.-mL bottles of chlorine measured at 22°C and 730. mm pressure. What volume of 38% hydrochloric acid (density 1.20 g/mL) and what mass of manganese dioxide will be required?

10. Using the ideal gas equation (Section 12.2), calculate the number of moles of Cl_2 in 5.00 L of the gas at STP. Why does your answer differ from that obtained in Problem 9?

11. A solution contains 62.5 g of $CaCl_2$ in 500. g of water. What is the theoretical freezing point of the solution?

CRITICAL THINKING

1. **Inferring Conclusions** In the commercial electrolysis of sodium chloride solution, the products chlorine, hydrogen, and sodium hydroxide are kept separated from each other. Why?

2. **Analyzing Information** The chemical reactions between water molecules and molecules of the hydrogen halides are reversible.
   a. Qualitatively, at equilibrium, what are the relative concentrations of the particles involved?
   b. What does this indicate about the relative stability of the hydrogen halide molecules compared with the stability of the ions that can be formed from them?

3. **Relating Ideas** Sodium chloride and magnesium chloride are ionic compounds, but aluminum chloride is molecular. Why?

4. **Applying Ideas** The improved health that results from using chlorine to kill disease-producing bacteria in municipal water supply systems has been recognized for many years. But chlorine can also react with natural substances in water to produce chloroform ($CHCl_3$), which causes cancer in laboratory animals. In what ways can scientists deal constructively with this dilemma?

RESEARCH AND WRITING

1. Investigate such factors as the hardness of water, temperature, alkalinity, and amount of sunlight a swimming pool receives and explain the chemistry involved. Report your findings to your class.

Chemistry and Careers

Metallurgist and Materials Scientist

Metallurgy is the science of metals, alloys, and their properties. Materials science is a combination of metallurgy and the study of modern ceramic, plastic, and composite materials. Both of these fields offer exciting opportunities.

If cars are your passion, then designing custom alloys for automobile parts is a career possibility. If bicycles are more your speed, why not improve the linkages and frame while keeping the materials as light as possible? Space technology is another rewarding area for future work as a metallurgist or materials scientist. The challenge is to keep spacecraft light to reduce fuel costs—yet strong enough to survive meteor impact—and sufficiently heat-resistant to withstand the heat of a fiery reentry through the atmosphere.

To work with metals and alloys, a metallurgist must understand the chemical and physical nature of metals. This requires a strong background in chemistry and engineering. A similar background is needed for understanding the complex nature of plastics and composites. In the current job market, a bachelor's degree in chemistry or metallurgy (with a minor in chemistry) is the minimum requirement. A doctorate enables you to conduct research on the new materials that will be needed for tomorrow's technology.

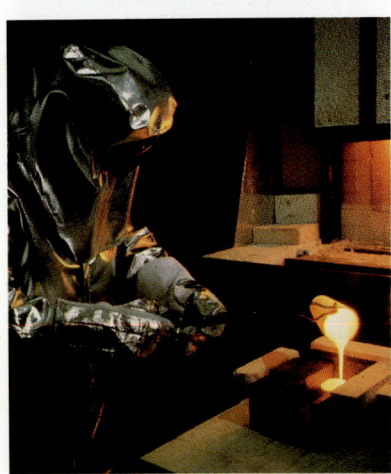

Chemical Supplier

Chemical suppliers must be salespersons, regulation and transportation experts, and applied chemists all at the same time. Marketing a chemical product requires good sales techniques, communications skills, and knowledge of the products. A technical background is helpful in answering questions about which products to use for a specific purpose and about possible side reactions and impurities.

Packaging and transporting chemicals also requires chemical knowledge. The shelf life of each chemical

being handled, stored, or shipped must be considered when arranging production and shipping dates. Chemical reactions with air, light, or other chemicals must be considered. Transportation of chemicals is subject to strict government regulation. Packaging must be clearly labeled with the name and hazardous nature of the chemical. Procedures in case of accidents, such as spills, should be clearly listed on the label.

An associate's degree, together with the ability to deal with people and to sell products, can get you an entry-level job with a chemical supplier. A bachelor's degree in business with a minor in chemistry can lead to a job as a salesperson or as a technical representative. The people most in demand have a master's degree in business administration with an undergraduate degree in chemistry.

UNIT 8 • FOCUS ON REASONING

College Entrance Exam Preparation

The following experimental data were collected by testing various reagents with solutions containing halide ions. Each test involved using five drops of the reagent solution to react with five drops of the halide ion solution to be tested. The last two tests involved unknown solutions. The first unknown contained a single halide ion. The second unknown contained two different halide ions.

1. Which silver halide is least soluble in other reagents?
   a. AgF
   b. AgCl
   c. AgBr
   d. AgI

2. What generalizations can be made about silver halides?
   a. they are all soluble
   b. they are all insoluble
   c. they are all soluble except AgF
   d. they are all insoluble except AgF

3. Determining the identity of the single unknown is based on the reaction of the unknown with
   a. $Ca(NO_3)_2$.
   b. starch.
   c. $AgNO_3 + NH_3$.
   d. $AgNO_3 + Na_2S_2O_3$.

4. The double unknown is most likely
   a. NaF and NaCl.
   b. NaBr and NaI.
   c. NaCl and NaBr.
   d. NaF and NaI.

			Reagents		
Halide Ions	$Ca(NO_3)_2$	$AgNO_3$	Precipitate from $AgNO_3$ reaction + NH_3(aq)	Precipitate from $AgNO_3$ reaction + $Na_2S_2O_3$	NaOCl + Starch
NaF	white ppt	clear	N/A	N/A	no change
NaCl	clear	white ppt	dissolves completely	dissolves completely	no change
NaBr	clear	pale-yellow ppt	dissolves slightly	dissolves completely	no change
NaI	clear	yellow-tart ppt	does not dissolve	does not dissolve	turns blue
Single unknown	white ppt	clear	N/A	N/A	no change
Double unknown	clear	yellowish ppt	dissolves slightly	dissolves slightly	burns blue

INTRA-SCIENCE

The Miracle Fluorocarbon Resin

The *Guinness Book of World Records* says that the most slippery material in the world is a fluorocarbon resin that was invented more than 50 years ago. Two smooth, flat surfaces of this resin can slide past each other more easily than can two equivalent surfaces of any other material.

The name of the substance is Teflon.* It is best known for keeping eggs from sticking to frying pans, but it does far more. It has been to the moon. It helps prevent corrosion on the Statue of Liberty. It insulates wires. It coats lightbulbs so that the glass doesn't shatter when they break. It is used for gaskets, seals, and bearings in cars.

Teflon was discovered by accident during research into refrigeration gases in 1938. Its molecular structure is based on a chain of carbon atoms, the same as all polymers. Unlike some other fluoropolymers, Teflon has a chain completely surrounded by fluorine atoms. The bond between carbon and fluorine is very strong, and the fluorine atoms shield the vulnerable carbon chain. This unusual structure gives Teflon its unique properties. In addition to being extremely slippery, it is inert to almost every known chemical. It is useful over a remarkably wide temperature range, from −250 to 300°C.

The Connection to Physics

It took scientists years to understand Teflon's chemical properties and work out production methods before exploring the potential physical applications. It was not commercially available until 1948 and did not become widely used until the late 1950s. As technology advances, engineers are finding more and more uses for Teflon.

The biggest single use of Teflon today is not in kitchen pans, although this is a widespread commercial application, but in electrical

wire and cable. In modern office buildings, miles of telephone and computer cables are insulated with Teflon. It also coats the logging cables that carry instruments to measure geological characteristics at various depths in oil wells.

Teflon keeps Detroit Lions football fans comfortable as they sit under the 10-acre fiberglass-fabric roof of the Pontiac Silverdome, which is coated with translucent and flame-resistant Teflon. It also helps preserve the Statue of Liberty. When the statue was restored in the early 1980s, Teflon replaced shellac-soaked asbestos as a galvanic corrosion insulator and lubricant between the copper skin and the inner stainless-steel framework.

*Teflon is a registered trademark of DuPont Company.

When the U.S. put a person on the moon, Teflon was part of the space suits, fuel tanks, linings for liquid oxygen systems, heat shields, and reentry nose cone. It is useful to the space program because it doesn't get brittle and remains flexible at low temperatures.

The Connection to Health Sciences

Teflon is almost completely inert, and therefore serves as an ideal material for surgical implants. More than a million people are functioning with artificial arteries or veins made of a material containing Teflon. It replaces knee ligaments, and heart patches, and improves suture materials because its inertness makes it compatible with living tissue.

Teflon powder has been injected directly into the vocal cords of people who have lost their voices as a result of extreme vocal cord relaxation. The fluoropolymer acts as an inert substance that supports, or fills out, the cords.

The Connection to Other Sciences

Teflon's versatility has led to its use in hundreds of everyday objects. Sandwiched as a membrane between two layers of nylon, Teflon helps make outdoor gear both waterproof and breathable. It is used in rainwear, ski clothing, running shoes and suits, and military uniforms for wet, cold weather. The membrane has nine billion pores per square inch. Each pore is 20 000 times larger than a molecule of water vapor. Because people sweat vapor, not drops,

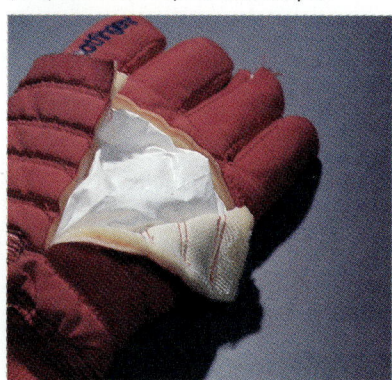

perspiration escapes through the membrane, but wind, snow, and rain cannot penetrate from the outside.

Other applications include soil- and stain-repellent products, laboratory equipment, pumps, films, molded parts, food-packaging machinery, pipes to measure groundwater contamination, solar collectors, straps, and fasteners. Sometimes, Teflon replaces other materials, simply because it can do the job better. As the material in pipes for systems monitoring groundwater, for instance, it

proved superior to glass, PVC (polyvinyl chloride), and stainless steel. It is the best available substance for protecting water samples against contamination, and it can withstand many years of exposure to chemicals, bacteria, and moisture.

The chemistry of fluorocarbon resins makes them difficult and expensive to put together. Because it is expensive, Teflon is used only where nothing else works, or where it is clearly superior to less expensive materials.

Unit 9

Chapter 29
Nuclear Chemistry 800

NUCLEAR REACTIONS

As a young man growing up in New York City, I was always fascinated by science fiction stories about space and traveling to other worlds. This made me wonder about the forces that would have to be understood and conquered to reach other planets and stars. My early curiosity convinced me to choose a career in physics so that I could learn about the fundamental nature of matter.

This unit introduces the concepts of nuclear reactions. Nuclear chemistry is the study of the interactions between atomic nuclei. The forces holding the nucleus together are much greater than those that hold electrons around atoms. Consequently, nuclear reactions involve a great deal more energy than chemical reactions.

For centuries, the way in which energy is produced by the sun and stars remained a mystery to scientists. It has only been during this century that we have discovered that this energy is released through the fusion of lightweight atoms. To study and understand the nuclear interactions involved in a fusion reaction, elaborate measuring instruments are needed. As a physicist working on the development of a fusion energy reactor, I am responsible for producing these "eyes and ears."

Donald H Priester
Donald H. Priester
Plasma Physicist
United States Department of Energy

Nuclear fusion is the source of the sun's energy.

the crystal. This method, called fission-track dating, can establish the age of materials from a few decades old to as old as the solar system.

Nuclear chain reaction The fissions of $^{235}_{92}U$ and $^{239}_{94}Pu$ can produce chain reactions. One neutron causes the fission of one $^{235}_{92}U$ nucleus. Two or three neutrons given off when this fission occurs can cause the fission of other $^{235}_{92}U$ nuclei. Again neutrons are emitted. These can cause the fission of still other $^{235}_{92}U$ nuclei. This is a chain reaction. *A* **chain reaction** *is one in which the material that starts the reaction is also one of the products.* It continues until all the $^{235}_{92}U$ atoms have split or until the neutrons fail to strike $^{235}_{92}U$ nuclei. See Figure 29-13.

Nuclear reactor *A* **nuclear reactor** *is a device in which the controlled fission of radioactive material produces new radioactive substances and energy.* One of the earliest nuclear reactors was built at Oak Ridge, Tennessee, in 1943. This reactor uses natural uranium. It has a lattice-type structure with blocks of graphite forming the framework. Spaced between the blocks of graphite are rods of uranium, encased in aluminum cans for protection. *Control rods* are inserted into the lattice. **Control rods** *are composed of neutron-absorbing steel and are used to limit the number of free neutrons.* The reactor is air-cooled.

The rods of uranium or uranium oxide are the nuclear fuel for the reactor. The energy released in the reactor comes from changes in the uranium nuclei. Graphite is the **moderator** *which slows down the fast neutrons produced by fission.* By doing so, it makes them more readily captured by a nucleus and thus more effective for producing additional nuclear changes. The

Figure 29-12 The yield of various nuclides of uranium-235. The products vary in mass number from 72 to 161. Observe the low-probability of fission products of nearly equal mass numbers.

Figure 29-13 An illustration of a nuclear chain reaction.

Figure 29-14 A nuclear power plant reactor core

mass of uranium in such a reactor is important. *The amount of uranium needed to provide the number of neutrons needed to sustain a chain reaction is called the* **critical mass.**

Two types of reactions occur in the fuel in such a reactor. Neutrons cause $^{235}_{92}U$ nuclei to undergo fission. The fast neutrons from this fission are slowed down as they pass through the graphite. Some strike other $^{235}_{92}U$ nuclei and continue the chain reaction. Other neutrons strike $^{238}_{92}U$ nuclei, starting the changes that finally produce plutonium. Great quantities of heat energy are released. For this reason the reactor must be cooled continuously by blowing air through tubes in the lattice. The rate of the reaction is controlled by the insertion or removal it the neutron-absorbing control rods. This type of reactor is now used to produce radioactive isotopes.

In nuclear power plants, the reactor is the source of heat energy. Pressurized water is both the moderator and the coolant. The heat from the reactor absorbed by the pressurized water is used to produce steam. This steam turns the turbines, which drive the electric generators. Present problems with nuclear power plant development include location and environmental requirements, safety of operation, and plant construction costs. The procurement and enrichment of uranium, as well as methods for storing or reprocessing used nuclear reactor fuel, are also being studied.

Figure 29-15 A cutaway of the Oak Ridge reactor

Nuclear Fusion

Nuclear stability can be increased by combining lightweight nuclei into heavier nuclei in the process of nuclear fusion.

Fusion reactions are the source of the sun's energy. There are two series of such reactions going on in the sun it is believed. One series occurs at the very hot center of the sun, and the other takes place in the cooler outer portion of the sun. These two reactions proceed by different pathways. Their net effect, however, is the combination of four hydrogen nuclei into a helium nucleus. A loss of mass occurs and a tremendous amount of energy is released. More energy is released per gram of fuel in a fusion reaction than in a fission reaction.

Current research indicates that fusion reactions may be controlled. Scientists are attempting to find ways to confine the nuclear fuel, usually ionized deuterium and tritium, and at the same time, make it hot enough and dense enough long enough for fusion to occur. Magnetic fields are usually used to confine the ionized fuel because, at the initial temperature required (about 10^8K), no known material container will do. Experiments using high-powered laser light or beams of particles, such as protons or lithium ions, as ways to start the fusion reaction are being conducted. If fusion reactions are to be a practical source of energy, the energy given off must be greater than the energy required to heat and compress the ionized fuel.

Section Review

1. Distinguish between *nuclear fission* and *nuclear fusion*.
2. Define *chain reaction*.
3. What is the function of the following in a nuclear reactor: *(a)* nuclear fuel *(b)* control rods *(c)* moderator *(d)* cooling fluid.

Figure 29-16 Glass pellet containing deuterium and tritium: The target contains deuterium-tritium fuel at over 100 atmospheres ready to be fused by the heat from a powerful laser.

Chapter Summary

- Protons and neutrons, called nucleons, are held together strongly by a short-range force of attraction known as the nuclear strong force.
- Radioactivity is the spontaneous breakdown of an unstable atomic nucleus with the release of particles and rays.
- Radioactive nuclides and their compounds have several unusual properties: *(1)* they affect the light-sensitive emulsion in a photographic film *(2)* they produce an electric charge in the surrounding air *(3)* they produce fluorescence with certain other compounds *(4)* their radiations have special physiological effects and *(5)* they undergo radioactive decay.
- The half-life of a radioactive nuclide is the length of time that it takes for half of a given number of atoms of the nuclide to decay.
- Three kinds of radiation given off by radioactive nuclides are: *(1)* alpha particles, which are helium nuclei, *(2)* beta particles, which are electrons, *(3)* gamma rays, which is high-energy electromagnetic radiation.
- The age of certain minerals and of carbon-containing materials can be estimated by the amounts of radioactive nuclides they contain.
- The difference between the sum of the masses of the nucleons in a nucleus and the actual mass of a nucleus is the nuclear mass defect.
- The four types of reactions that produce more stable nuclei are: *(1)* radioactive decay, *(2)* nuclear disintegration, *(3)* fission, and *(4)* fusion.
- A nuclear reactor is a device in which the controlled fission of radioactive material produces new radioactive substances and heat energy.
- A fusion reaction is one in which lightweight nuclei are combined into heavier nuclei. Fusion reactions produce the sun's heat and light.

1. Nuclear fission is a reaction in which a heavy nucleus splits into medium-weight fragments. Nuclear fusion is a process in which light-weight nuclei combine into heavier nuclei.
2. The material or energy that starts the reaction is also one of the products.
3. *(a)* Provides the energy in a nuclear reactor *(b)* used to absorb neutrons *(c)* slows down fast-moving neutrons produced by fission *(d)* fluid used to cool a reactor in which great amounts of heat energy are released

Chapter 29 Review

Nuclear Chemistry

VOCABULARY

alpha particle (809)
beta particle (809)
chain reaction (819)
control rod (819)
cosmic ray (816)
critical mass (820)
daughter nuclide (810)
fission (804)
fluoresce (804)
fusion (804)
gamma ray (809)
half-life (807)
moderator (819)
nuclear binding energy (802)
nuclear disintegration (804)
nuclear equation (810)
nuclear mass defect (802)
nuclear reactor (819)
nucleons (801)
parent nuclide (810)
radioactive decay (804)
radioactive tracer (817)
radioactivity (804)
strong nuclear force (801)
transmutation (810)
transuranium element (814)

REVIEWING CONCEPTS

1. Where on the Periodic Table are most of the radioactive nuclides located? (29.1)

2. a. How does binding energy per nuclear particle vary with mass number?
   b. How does binding energy per nuclear particle affect the stability of a nucleus? (29.1)

3. Describe two ways in which the number of protons and the number of neutrons in a nucleus affect its stability. (29.1)

4. Why can the radiation from a radioactive material affect photographic film even though the film is well wrapped in black paper? (29.2)

5. Make a chart that compares the following properties of alpha particles and beta particles: identity, mass number, charge, speed, and penetrating ability. (29.2)

6. a. What are gamma rays?
   b. How do scientists believe they are produced?
   c. How does their penetrating ability compare with that of alpha particles and beta particles? (29.2)

7. Explain the nuclear changes that occur when $^{239}_{94}Pu$ is produced from $^{238}_{92}U$. (29.2)

8. Name and briefly describe the four types of nuclear reactions that produce more stable nuclei. (29.2)

9. Write the nuclear equation for the reaction by which Chadwick discovered neutrons. (29.2)

10. Neutrons are more effective than protons or alpha particles for bombarding atomic nuclei. Why? (29.2)

11. Name the naturally occurring isotopes of uranium and indicate their relative abundance. (29.2)

12. How is the fission of a $^{235}_{92}U$ nucleus produced? (29.2)

13. Compare the radioactivity of uranium, radium, and polonium. (29.2)

14. What is the source of the alpha or beta particles emitted by a radioactive nuclide? (29.2)

15. What changes in atomic number, identity, and mass number occur when a radioactive nuclide gives off an alpha particle? (29.2)

16. What changes in atomic number, identity, and mass number occur when a radioactive nuclide gives off a beta particle? (29.2)

17. a. Radon-222 has a half-life of less than four days, yet it is continually found in very low concentrations in the earth's crust, surface waters, and atmosphere. Why?
    b. In what ways is radon useful?
    c. In what way may radon be harmful? (29.3)

18. a. Describe how the age of a radioactive mineral is estimated.
    b. What assumptions are necessary? (29.3)

19. What factors are currently affecting the construction of nuclear power plants? (29.4)
20. Describe the reaction that produces the sun's energy. (29.4)
21. How does the fission of $^{235}_{92}U$ produce a chain reaction? (29.4)
22. How is a uranium-graphite reactor constructed? (29.4)
23. What problems must be overcome before energy-producing controlled fusion reactions are a reality? (29.4)

PROBLEMS

1. The half-life of thorium-27 is 18.72 days. How many days are required for three-fourths of a given amount to decay?
2. The half-life of $^{234}_{91}Pa$ is 6.75 hours. How much of a given amount remains after 27.0 hours?
3. After 4806 years, how much of an original 0.250 g of $^{226}_{88}Ra$ remains?
4. The mass of a $^{7}_{3}Li$ nucleus is 7.014 36 u. Calculate the nuclear mass defect. The atomic masses of nuclear particles are given in Section 3.2.
5. Calculate the nuclear mass defect of $^{20}_{10}Ne$ if its nucleus has a mass of 19.986 95 u.
6. Write the nuclear equation for the release of an alpha particle by $^{210}_{84}Po$.
7. Write the nuclear equation for the release of a beta particle by $^{210}_{82}Pb$.
8. The parent nuclide of the thorium decay series is $^{232}_{90}Th$. The first four decays are: alpha particle emission, beta particle emission, beta particle emission, alpha particle emission. Write the nuclear equations for this series of emissions.
9. The half-life of $^{222}_{86}Rn$ is 3.823 days. What was the original mass of $^{222}_{86}Rn$ if 0.0500 g remains after 7.646 days?
10. The half-life of $^{239}_{94}Pu$ is 24 100 years. Of an original mass of 100 g, approximately how much remains after 10^5 years?
11. Calculate the nuclear mass defect per nuclear particle for $^{7}_{3}Li$ using the answer to Problem 4. Convert the mass in u to binding energy in megaelectron volts by using the relationship 1 u = 931 MeV.
12. Calculate the binding energy per nuclear particle of $^{238}_{92}U$ in MeV if the atomic mass of its nucleus is 238.0003 u.

CROSS CONNECTIONS

1. **Medicine** Research *positron-emission tomography,* a medical tool for diagnosis of many diseases. Investigate the chemistry involved in positron-emission transaxial tomography.
2. **Environmental Science** What progress is being made in devising environmentally safe methods for storing radioactive wastes?

RESEARCH AND WRITING

1. Investigate the history of the Manhattan Project.
2. Write a report on the findings in May 1972, at the nuclear fuel processing plant at Pierrelatte, France, regarding "Nature's Fission Reactor" at the Oklo mine in the Gabon Republic of West Africa.
3. Find out about the various fusion-energy research projects that are being conducted in the United States and other parts of the world. How close are the researchers to finding a method that will economically produce energy? What obstacles must still be overcome?

Nuclear Chemistry

Microscale Investigations

1: Conservation of Mass 830
2: Mixture Separation 832
3: Measuring the Specific Heats of Metals 834
4: Conductivity 837
5: Colored Precipitates 840
6: Pin-the-Formulas-on-the-Compounds 843
7: Generating and Collecting O_2 845
8: Generating and Collecting H_2 848
9: Testing Reaction Combinations of H_2 and O_2 851
10: "Wet" Dry Ice 854
11: Measuring the Triple Point Pressure of CO_2 855
12: Water Purification 858
13: Testing Water 860
14: Is It an Acid or a Base? 864
15: How Much Calcium Carbonate Is in an Eggshell? 866
16: Clock Reactions 869
17: Measuring K_a for Acetic Acid 871
18: Acid-Catalyzed Iodination of Acetone 873

MICROSCALE INVESTIGATIONS TEACHER'S GUIDE

The investigations that are part of this text are exclusively microscale investigations. If you prefer to use traditional techniques or seek more experiments, the *Modern Chemistry Laboratory Experiments* book provides 44 experiments, including 17 that can be performed in microscale.

Advantages of Microscale

◆ **reduced cost:** smaller amounts of chemicals are used (as little as 0.1% of a full-scale lab's needs), and polystyrene is often cheaper than glass

◆ **less breakage:** polystyrene doesn't break like glass

◆ **less danger:** with smaller amounts of chemicals reacting, smaller amounts of products are formed, smaller amounts of heat are evolved, and there are fewer chances for uncontrolled reactions

◆ **decreased chance of accidents:** with smaller amounts of chemicals, there is less to spill

◆ **decreased chance of misuse:** with smaller amounts of chemicals, there is less to be misused

◆ **less waste:** with smaller amounts, there is less excess

◆ **environmentally sensitive:** with smaller amounts of chemicals there is less to dispose, and waste is often already diluted to safer concentrations

◆ **less demand on your lab's ventilation system:** with smaller amounts of chemicals, smaller amounts of fumes are produced

◆ **requires less lab bench space:** the equipment takes up less room, so students don't need to move around as much, cutting down on accidents

◆ **requires less storage space:** both for equipment and chemicals

◆ **more experiments available:** experiments that were too expensive or dangerous often may now be done effectively in microscale

◆ **more techniques available:** take advantage of properties like surface tension that work most effectively in microscale

◆ **quicker prep-time:** small amounts of solution can serve a whole lab section of students, slightly larger quantities of some chemicals can be made that will serve classes for several years

◆ **quicker clean-up:** both for students and teacher

◆ **quicker for students to perform:** with less time needed for obtaining materials and cleaning up, they can re-do the experiment under different conditions or else have time for discussions, extensions, and inquiry-based learning

◆ **requires more care** and precision on part of students

INVESTIGATION TITLE	TEXT SECTION (Best match in boldface)
1: Conservation of Mass	1.2
2: Mixture Separation	**1.3**, 14.1
3: Measuring the Specific Heats of Metals	1.3, **2.2**, 6.4, 18.1, 25.1
4: Conductivity	1.3, **6.4**, 14.1, 15.3
5: Colored Precipitates	**7.1**, 8.2, 15.1, 16.2, 19.4, 25.1
6: Pin-the-Formulas-on-the-Compounds	**7.4**, 8.1, 9.2, 14.3, 15.1, 16.2, 17.3, 19.4, 25.1
7: Generating and Collecting O_2	1.2, 2.1, 8.2, **10.1**, 12.4
8: Generating and Collecting H_2	1.2, 2.1, 8.2, **10.2**, 12.4
9: Testing Reaction Combinations of H_2 and O_2	8.2, 9.2, 10.1, **12.3**, 18.3
10: "Wet" Dry Ice	10.4, 11.2, **13.3**
11: Measuring the Triple Point Pressure of CO_2	10.4, **11.3**, 12.2, 13.3, 19.1
12: Water Purification	1.2, 1.3, **14.1**
13: Testing Water	8.2, 14.2, **15.1**, 19.4
14: Is It an Acid or a Base?	14.3, **16.2**, 17.1
15: How Much Calcium Carbonate is in an Eggshell?	8.2, 9.2, 10.4, 16.5, **17.3**
16: Clock Reactions	8.2, 15.1, **18.4**, 20.1
17: Measuring K_a for Acetic Acid	14.1, 15.3, 16.1, **19.3**, 22.5
18: The Acid-Catalyzed Iodination of Acetone	8.2, 18.3, **22.4**

The Format of Microscale Investigations

The format of the teacher's guide to each investigation is closely matched to the pupil's investigation.

Pupil's Investigation	Teacher's Guide
Title	Title
Introduction	Introduction
Objectives keyed to process skills	
Materials	Detailed Preparation Instructions
Procedure	Safety Notes
	Disposal
Analysis and Interpretation Questions	Answers to Questions
Conclusion Questions	Answers to Conclusion Questions
Extension Questions and Activities	Answers to Extensions, where appropriate

In addition to providing safety cautions in the Teacher's Guide, specific safety hints are provided in the pupil's investigations **at the point of use.**

Safety in Your Laboratory

As a teacher, you must be the safety "expert" for your school in many ways. Along with keeping your students safe and avoiding costly accidents, the administration may look to you to provide assurances that liabilities are being minimized. While these investigations have been designed with safety in mind, as a teacher you must play an active role in fostering "safety-consciousness" in your students, your fellow faculty members, and the administration.

The pupil's versions of the investigations are preceded by a safety section describing important overall safety tips, the safety symbols used in the investigations, and detailed explanations of each of the hazards represented by the symbol. Safety equipment students should be familiar with (fire extinguishers, etc.) is listed in the following section and in the pupil's investigations.

There is some additional information included here to help you maintain a safe environment for laboratory learning. **However, this information is not all-inclusive.** Every school's laboratory situation is different, and no publication could possibly list safe practices for all situations that could possibly arise. **This information probably does not include all the legal requirements you will need to follow.** Although federal, state, and local laws can vary from place to place and from time to time, the information given should provide a safety base on which you can build. Other sources of more information for you may include your state science supervisors' office, state science safety handbooks, state and national science teachers' associations, and chemical supply catalogs. A *Total Science Safety System,* including software for safety reports, forms, and an expandable database of 400 chemicals is available from JaKel, Inc. (see p. T35 for address).

Be sure you are aware of any federal, state, or local laws that may cover your lab. Many, but not all, chemical safety regulations are made and enforced by the federal Occupational Safety and Health Agency (OSHA) or by an OSHA-approved state safety agency. An additional source of information on safety could be your district safety officer or the ten regional or 93 area offices of OSHA. Teachers should keep in mind that OSHA's regulations were primarily designed for private industry, but some states have extended the regulations to cover public schools as well.

In addition, one person from your school may need to be designated as the chemical contact person for your local fire department to contact in case of emergency. As a teacher, you should be certain to maintain a complete set of Material Safety Data Sheets (MSDS) which describe all the chemicals in your lab and storeroom, and the precautions to take in the event of a spill or accident. These may be required by law in your area, and are provided by some chemical suppliers with chemicals. A small amount of time and effort spent planning ahead like this can have a big payoff if an emergency happens.

Safety Equipment Your Lab Should Have (also listed in pupil's Investigations):

Review the use and location of this equipment with your students before they begin their first investigation.

◆ **Eye and Face Wash Fountains:** should be checked daily. If chemicals come in contact with the eyes, flush them for 15 minutes with water.

◆ **Safety Shower(s):** should be checked monthly. They should dilute material to a safe level in 15 seconds or less.

- **Fire Blankets: are the best way to smother flaming hair or clothes.** If the clothes are polyester, the safety shower is the best way to put out a fire.
- **Fire Extinguishers:** should be checked monthly. Most multi-purpose fire extinguishers put out ONLY class A (wood and paper, etc.), B (grease and oil), and C (electrical) fires, NOT class D (combustible) fires such as burning magnesium. Dry sand applied with a scoop may put out small class D fires.
- **First Aid Kit:** should also include one copy of the Red Cross book, *Standard First Aid*. It is recommended that teachers receive certification in First Aid from the American National Red Cross.
- **Spill Control Packages (sand buckets, etc.):** a variety of packages are available commercially. Along with sand, there should be easy access to substances which can be used to quickly neutralize acidic and basic spills.

Other Lab Safety Guidelines

- **Know the location and procedures for utilities shut-offs.** Label or color-code master shut-off switches. Be sure you will know what to do so well that you will be able to remember when under the pressures of an emergency.
- **Always perform an investigation PRIOR to allowing students to do it.** Look for potential hazards and alert students to possible dangers, and review all safety symbols and cautions specified in the investigation as well.
- **Be careful with volatile liquids.** NEVER allow open flames near volatile liquids such as ether, gasoline, alcohol, and carbon disulfide, and use them only in small quantities in well-ventilated rooms.
- Post the following numbers near the telephone for use in emergency.
  Rescue Squad:
  Fire:
  Police:
  Hospital:
  Duke Poison Control Center: 1-800-672-1697
  (The Duke Poison Control Center is open 24 hours daily to advise individuals and physicians on treatments relating to any type of poisoning.)
- **Make accident reports promptly, accurately and completely.** Your school or district should have a procedure and a standard form that ensures all necessary information is recorded.

- **Dispose of chemicals appropriately.** Some disposal hints are provided with the teacher's guide to each investigation, but be sure you are aware of the standards and laws of your community.
- **Have a thorough understanding of the potential hazards of materials, processes, and equipment in your lab.** The Material Safety Data Sheets can be an important first step in understanding chemical dangers.
- **Excuse students having medical problems that could be aggravated by conditions in the lab.** Contact lenses should not be allowed in the lab unless certified written permission is obtained from a student's doctor and parents.
- **Perform ongoing safety audits.** Keep track of what problems you have found, what you recommended, and how long it took for the problem to be solved.
- **Maintain a current inventory of chemicals.** Keep track of purchase dates and amounts, and keep your records current. Cut down on storage problems by buying only what you need for the school year.
- **Label all bottles prominently.** Use labeling materials that won't be affected by the reagent.
- **Lock all labs and storage facilities when not under direct supervision of a responsible person.** No studying, working, or experimenting should be done in the lab unless there is competent supervision. Eating, drinking and smoking are always prohibited in the lab.
- **Be sure to follow fire safety guidelines approved by your local fire department.**
- **Keep chemicals stored indoors in a designated chemical storage room.** An adequate room should be fire-safe, prevent accidental chemical exposure, and should protect against criminal and careless acts.
- **Be sure chemicals are stored in a safe arrangement.** Chemicals must be stored by type, NOT ALPHABETICALLY. Many dangerous reactions can occur if chemicals of different types are not isolated. There are several recommended shelf storage patterns for different classes of chemicals outlined in safety handbooks, chemical supply catalogs, and *Modern Chemistry Laboratory Experiments/ Teacher's Edition*.

Make sure your students have a thorough understanding of lab safety. After a discussion of your lab's safety features, you may want to give them a quiz similar to the one below, and make a perfect score a prerequisite for working in the lab. Later in the year, be sure they remember their safety procedures by having an unannounced drill before beginning an investigation. You can test their readiness for such accidents as chemical fires, acid spills, severe bleeding from cut glass, and other situations. Be sure they are ready to safely evacuate the building in case of fire or fire drills in the middle of an investigation.

Sample Safety Quiz

1) How do you operate this lab's fire extinguisher(s)?

    (Check your extinguisher(s) for answer.)

2) What should you do if your lab partner spilled acid on himself or herself?

    (Use the safety shower. Students should describe how it works.)

3) Describe how you would put out a fire in your lab partner's hair.

    (Use a fire blanket. Students should indicate where they are kept.)

4) Where do you put broken glassware?

    (Answer depends upon your decision.)

5) How would you clean up a spill of acid on the lab desk?

    (Answer may vary. Use neutralizing agent, and/or dilute, then wipe clean.)

6) Where is the number of the fire department kept?

    (This should be near one of your telephones.)

7) Where is the first aid kit?

    (Answer depends upon your lab's setup.)

8) What would you do if you got corrosive material in your eyes?

    (Students should describe working of eyewash station.)

9) Where should you put used chemicals after the lab?

    (Answer depends upon your lab's setup.)

10) When is it okay to work alone in the lab?

    (NEVER!)

INVESTIGATION 1: Conservation of Mass

Pages 830–832

Part I of this investigation should take about 10 minutes. Part II might take an additional 10–20 minutes, depending on how quickly the students catch on to the problem inherent in Part I: That the system is open, so the CO_2 product formed is able to escape.

Although the equipment list is given for students working in pairs, students may work in larger groups, or individually, if this is preferred.

MATERIALS

(for a class of thirty students working in pairs, includes amounts for Extensions)

- 4000 mL vinegar (5% acetic acid solution—includes extra for Extensions)
- 250 g baking soda (sodium bicarbonate—includes extra for Extensions)
- 30 clear plastic cups
- 15 triple-beam or electronic balances (fewer may be used if students take turns measuring mass)
- 15 2-L plastic soda bottles
- 15 hook-insert caps for bottle (use hot-melt glue to attach a wire hook (hand-made or bought from hardware store, etc.) to inside center of the cap, see Figure T1-A)
- 15 micro plungers (cut from graduated pipets with thread loops attached, see Figure T1-A)
- 15 graduated cylinders (100 mL or larger)
- 15 resealable (zippered) plastic bags *for Extension 2*

SAFETY NOTES

 Students are cautioned to wear goggles to protect their eyes from the acid solution, and this caution is even more appropriate for the instructor preparing for the lab, and pouring large quantities of acidic vinegar.

 Keep an acid neutralization package handy such as resealable (zippered) plastic bags filled with $NaHCO_3$, in case of spills.

DISPOSAL

All of the solutions and chemicals used in this investigation may be washed down the sink with an excess of water.

ANALYSIS & INTERPRETATION— PART I—ANSWERS

1. Reactants bubbled and fizzed, with the formation of a new gaseous substance. The baking soda seemed to disappear eventually.

2. Students' answers will vary, but typical results show a mass loss of 1.0–1.5 g for Part I.

3. The students' answers should indicate that although the mass loss appears to violate the law of conservation of mass, the law was not violated. The law probably still holds, because not all of the products were present for the second

Figure T1–A: The hook-insert cap and microplunger used in Investigation 1 are easy to make.

measurement of mass, since the bubbles that formed and popped were releasing a gas product.

4. Students' answers will vary. Possible Procedure for Part II:
   - Measure out 100 mL of vinegar into the bottle.
   - Tap the micro plunger into the baking soda, packing the bulb end with a plug of powder.
   - Hang the plunger on the hook of the cap, and carefully and tightly screw the cap back onto the bottle, with the plunger dangling over the vinegar inside the bottle.
   - Carefully place the bottle on the balance to determine the total initial mass of the system without disturbing the baking soda.
   - Shake the bottle to jar the baking soda loose, so that it falls and reacts with the vinegar.
   - When the reaction is finished, place the bottle back on the balance to determine the total final mass of the system.

ANALYSIS & INTERPRETATION — PART II — ANSWERS

1. Again, there was vigorous bubbling for the first few seconds. However, the fizzing noise was not very audible. The sides of the bottle become tight — hard to push in, and when the cap is removed, there is a hissing noise of gas escaping under high pressure.

2. The reactants were a liquid (vinegar) and a solid (baking soda). Students should recognize that one of the products of the reaction was a gas (the carbon dioxide seen forming the bubbles). The other product was a liquid (a solution of excess vinegar and sodium acetate).

3. Students' answers will vary, depending on how comfortable they are with the notations for chemical equations. Answers should include at least the following:

   baking soda + vinegar solution →
   bubbles of carbon dioxide gas + clear liquid
   (sodium acetate solution + water)

   Students familiar with chemical symbols may give the following as an answer:

   $$NaHCO_3(s) + HC_2H_3O_2(aq) \rightarrow$$
   $$CO_2(g) + NaC_2H_3O_2(aq) + H_2O(\ell)$$

CONCLUSIONS — ANSWERS

1. In Part I the system was left open, but it was closed in Part II, so that not even gas could escape.

2. Part II worked better because one of the products was a gas, and the bottle-and-cap system kept that product trapped, unlike the open cup, so that its mass could be measured.

EXTENSIONS — ANSWERS

1. Students' answers will vary. The following is a possible response:
   - Place the baking soda in one corner of the bag, using a twist-tie to pinch the corner, isolating it from the rest of the bag.
   - Add the vinegar to the rest of the bag.
   - Seal the bag.
   - Measure the mass of the bag.
   - Undo the twist-tie and mix the reactants.
   - Remeasure the mass at the end of the reaction. (Don't forget to include the twist-tie.)

2. Students' answers will vary. Some will think the bag will be just as successful, since it is a closed system. Other students might argue that the bag is more likely to leak than the bottle, so a mass loss is possible anyway.
   If the procedure is tested, a mass loss of 0.5–0.7 g will be found, **even if the bag is perfectly sealed and does not leak.** Students may be able to recognize that, unlike the bottle, the bag **inflates** during the reaction.
   Students may have trouble resolving the discrepancy, even if they notice the inflation of the bag. They may suggest that the CO_2 was able to escape (effuse) through the sides of the bag, but actually this loss is negligible.
   The real reason the mass appears to change is that a balance actually measures weight and not mass. Since the bag occupies more space after the reaction, it will displace more air, and according to Archimedes' Principle, it will appear to weigh slightly less, even though it has the same mass. (In other words, the bag will have a greater "buoyant force.") Thus, to verify the law of conservation of mass using a balance to measure mass, the system must be closed, and be rigid enough to have a fixed volume (like the soda bottle).

3. Just as in the reaction studied in this investigation, the burning of a log represents the formation of many products, not just ashes. The smoke, CO_2, and water vapor all escape as the log burns. Presumably if one had a way to measure the mass of all of the products, the mass would remain constant.

4. Students' answers will vary, but they should include a way to burn the log in a closed, rigid container that can withstand heat and flames without reacting. Students may also recognize that the amount of oxygen needed to burn with the log would require a very large container, unless the oxygen were purified from the air, or kept under high pressure. You may want to point out to them that this is very difficult and potentially dangerous because of the pressures that would build up after the reaction is completed as all of the carbon in the solid, dense log is converted to carbon dioxide gas. All things considered, it would be very difficult to perform this test.

INVESTIGATION TEACHER'S GUIDE

INVESTIGATION 2: Mixture Separation

Pages 832–833

This investigation is intended to be an open-ended, self-designed activity in which the students are presented only with the problem and a list of equipment available to them. They are required to develop and test their own procedures. Because of this, the lab generally takes a full 40–50 minutes on one day, and another 15-20 minutes on a successive day. To shorten the time necessary, either devote 15–20 minutes of class or lab time to a planning session for the investigation, or else require a written plan for the investigation as a homework assignment, prior to beginning the investigation. Even so, some of the separation procedures (especially those involving evaporation of a solvent) may require having equipment (petri dishes) setting overnight.

Since some students will take longer than others, teachers may want to have an optional activity available to occupy students who finish this investigation quickly.

Although the equipment list is given for students working in pairs, students may work in larger groups, or individually, if this is preferred. The dynamics of brain-storming among group members may make this a great cooperative learning activity.

There are hundreds of different procedures that students could use to separate and recover their mixtures with varying degrees of success, and for any procedural step, there are countless refinements and improvements possible.

It is best if the teacher avoids answering directly any of the students' questions. The best response is always one that directs the inquiry back toward the student, such as "I don't know; but maybe you can think of a way of answering that for yourself."

Teachers may wish to modify the list of additional equipment, but it should remain extensive, for it is meant to provide students with a wide range of choices and an opportunity for hands-on, minds-on creative problem solving. Remind students that they are not obligated to use all, or even most, of the items. What appears extraneous to one student may be a critical component in another student's procedure. To expand on this list, pay attention to the answers the students give for Conclusion #3. Some of their suggestions can be incorporated the next time you use this investigation.

If desired, the investigation can be made more quantitative by emphasizing concepts such as percent composition of a mixture and percent recovery. Students can measure the mass of the original sample (**after** the representative sample has been set aside) and the mass of each of the recovered components (**before** the representative sample has been placed in the appropriate well). This allows for comparative discussion on the effectiveness of techniques in providing high yields of recovered substances.

MATERIALS

(for a class of thirty students working in pairs, includes amounts for Extensions)

For the teacher to dispense:
60 g of mixture
   The mixture can be made by mixing about 15 g of each of the 4 components in a large resealable (zippered) plastic bag. It may be dispensed to the students in stoppered test tubes, each containing 1–2 g of the mixture.

At the lab stations:
15 petri dishes
15 8-well strips
15 micro funnels (made by cutting the middle section out of a jumbo pipet. See Figure T2-A)
15 tape dispensers (fewer will do if students share)

Additional equipment at central lab table:
roll of cellophane
50 test tubes (13 × 100)
15 rubber stoppers
roll of aluminum foil
10 wooden splints
10 plastic spoons
10 plastic forks
30 pipets
15 filter funnels
50 filter papers
10 forceps
roll of tissue paper
roll of paper towels
10 straws

Figure T2-A: Micro funnels can be made by cutting apart jumbo pipets.

10 paper clips
500 mL distilled water
15 magnets
15 cotton balls
4 wash basins (recovery bins for purified materials — see Disposal section below.)

SAFETY NOTES

 Students are cautioned to wear goggles to protect their eyes from unexpected accidents, but none of the materials used are particularly hazardous.

DISPOSAL

Keep the four basins in a central location for students to return their recovered materials, after they have set aside a portion in the appropriate well. Of course, the materials may be re-used next year, even if the students have not done a particularly good job in their separations, since you will be mixing the substances up again the next time the investigation is performed.

ANALYSIS & INTERPRETATION — ANSWERS

1. Students' answers will vary. They may mention as justifications for their estimations of success the presence (or lack of) impurities in the separated components and the amount recovered.

CONCLUSIONS — ANSWERS

1. Students' answers will vary.

2. Students' answers will vary, but a possible procedure is as follows:
- Spread mixture on a paper towel.
- Wrap a magnet in a piece of cellophane.
- Pass the magnet over and through the mixture, shaking loose non-iron granules. Repeat this technique to remove as much iron as possible.
- Unwrap the magnet, keeping the iron inside the cellophane. Iron is now separated and recovered.
- Fold a filter paper in quarters and place it in a filter funnel.
- Transfer the remaining mixture from the petri dish to a test tube. Add 4–5 mL of distilled water, stopper and shake for 30 seconds to dissolve the salt.
- Unstopper, swirl, and decant only the liquid from the test tube into the funnel, allowing the filtrate to drain back into the test tube. Add more water to the test tube and repeat.
- Set the petri dish aside overnight to allow the water to evaporate.
- Remove the filter paper from the funnel. It should contain the poppy seeds that were floating in the water. Allow them to dry overnight. Poppy seeds are now separated and recovered.
- Place a second piece of filter paper in the funnel, and hold the test tube upside down over the funnel. Use a plastic pipet to squirt a few milliliters of distilled water up into the test tube to flush out the sand. Allow the funnel to drain into the sink.
- Remove the filter paper from the funnel, and let it dry overnight. The sand should now be separated and recovered.
- After the water has evaporated from the petri dish, use the plastic spoon to scrape as much salt as possible from the inside of the dish. The salt has now been separated and recovered.

3. Students' answers will vary. Pay close attention to them, in case you want to add to your list of additional available equipment

4. Students' answers will vary, but may include: iron is attracted by a magnet, salt dissolves in water, poppy seeds float on water, sand sinks in water.

EXTENSIONS — ANSWERS

1. Students' answers will vary, but may include: comparing the densities of their samples to the densities of standard samples of the pure components, picking through small samples from within their samples and checking for visible impurities, and more.

2. Students' answers will vary, but possible answers are:
a) magnet
b) screen or sifter
c) flotation in water
d) try dissolving in a solvent that dissolves one but not the other
e) distillation
f) distillation

INVESTIGATION 3: Measuring the Specific Heat of Metals

Pages 834–837

This investigation can be completed in 60 minutes. If a 30-minute investigation is desired, consider omitting the calibration of the calorimeter (Procedure Steps 1 and 3–5; Analysis & Interpretation Questions 3–5, and omit the heat capacity of the calorimeter in other calculations). Or have students calibrate the calorimeters in one lab period, save them, and determine specific heats in another lab period. (Some teachers report acceptable results without calibrating the calorimeters, or even using lids!). Additional time will be necessary for the extensions.

If time allows, the calibration process should be repeated, to find an average value for the specific heat of the calorimeter, C'. The heat absorbed by the calorimeter, $q_{calorimeter}$, is given by $C' \times \Delta T$ and changes with each reaction as the change in temperature varies depending on the experiment.

Although the equipment list is given for students work-

INVESTIGATION TEACHER'S GUIDE

ing in pairs, students may work in larger groups, or individually, if this is preferred.

The use of calorimeters may also be an appropriate topic to accompany Chapter 18.

MATERIALS

(for a class of thirty students working in pairs, includes amounts for Extensions)

30 plastic foam cups (depending on the size used, you may need to adjust the amounts of water used in the calibration step)
15 pairs of scissors or tools to trim cups (fewer will do if students share)
15 thermometers (alcohol-filled, **not mercury**)
15 100 mL graduated cylinders (fewer will do if students share)
30 400 mL beakers
15 bunsen burners or hotplates
15 large test tubes (20 × 150)
15 balances (fewer will do if students share)
15 pairs of tongs
15 sets of metal samples—1.15 kg total for a class of 15 lab groups for each metal: Smaller amounts of metals can be used with this lab, if less precision is acceptable. Remember these samples are re-usable from lab section to lab section and year to year. Rather than reagent-grade quality, try to locate inexpensive sources, such as iron nails, discarded copper wire, and aluminum pellets. Each student or lab group will need a little less than 75 g. For unknown metals, choose a metal from the ones in Table 2–6: aluminum, iron, copper, silver, or lead.
15 ringstands
15 ring clamps
15 wire gauze squares (if using bunsen burners)
15 test tube clamps (if using bunsen burners)
boiling chips for 15 hot water baths
materials for improved calorimeter *for Extension #3*
glass beads—1.15 kg total for a class of 30 *for Extension #4*

SAFETY NOTES

Although students are cautioned to be careful with thermometers, they often break. The use of alcohol-filled thermometers instead of mercury is encouraged whenever possible, to avoid the hazards associated with mercury spills.

DISPOSAL

All of the metals used can be disposed of in a landfill which is approved for the disposal of chemical and hazardous wastes. But since they undergo no reactions in this lab, they can be saved and used year after year, if students are instructed to dry the metals well after each use.

ANALYSIS & INTERPRETATION—ANSWERS

1. The assumption that energy released by one part of the system is absorbed by another part of the system is based on the law of conservation of energy which states that energy can be converted from one form to another, but that energy cannot be created or destroyed.

2. Students should recognize that since the density of water is given as 1.00 g/mL, the mass of water used in grams will be equal to the volume of water used in milliliters. If directions were followed as written, that would be 50.0 g of cool water and 50.0 g of hot water.

3. Students' answers will vary. The following is a sample calculation:

Hot water:
50.0 g H_2O × 25.5 C° × 4.184 J/g H_2O·C° = 5330 J = 5.33 kJ lost

Cool water:
50.0 g H_2O × 24.5 C° × 4.184 J/g H_2O·C° = 5130 J = 5.13 kJ gained

4. Students' answers will vary. The following is a sample calculation:

5.33 kJ lost-hot H_2O − 5.13 kJ gained-cool H_2O
= 0.20 kJ gained by calorimeter

5. Students' answers will vary. The following is a sample calculation:

$$q_{calorimeter} = C' \times \Delta T; \quad C' = \frac{q_{calorimeter}}{\Delta T}$$

$$C' = \frac{0.20 \text{ kJ}}{24.5 \text{ C°}} = 8.2 \text{ J/C°}$$

6. The appropriate equations are in bold below:

$$\mathbf{q_{metal} = q_{water} + q_{calorimeter};}$$
$$\mathbf{q = m \times c_p \times \Delta T}$$

$m_{metal} \times c_{p,metal} \times \Delta T_{metal} = (m_{water} \times c_{p,water} \times \Delta T_{water})$
$\qquad + (C' \times \Delta T_{calorimeter})$

Because $\Delta T_{calorimeter} = \Delta T_{water}$

$$c_{metal} = \frac{\Delta T_{water}[(m_{water} \times c_{p,water}) + C']}{m_{metal} \times \Delta T_{metal}}$$

CONCLUSIONS—ANSWERS

1. Students' answers will vary. The following are sample data and sample calculations:

Calibration			
	Cool H_2O	Hot H_2O	Calorimeter
Initial Temp.	25.0°C	75.0°C	25.0°C
Final Temp.	49.5°C	49.5°C	49.5°C
Change in Temp.	24.5 C°	−25.5 C°	24.5 C°
Volume	50.0 mL	50.0 mL	(not needed)
Mass	50.0 g	50.0 g	(not needed)

$C' = 8.2$ J/C° (see calculations in answer to Analysis & Interpretation #5)

Aluminum Metal Specific Heat Test		
	H_2O	Metal
Initial Temp.	21.0°C	99.0°C
Final Temp.	28.2°C	28.2°C
Change in Temp.	7.2 C°	−70.8 C°
Mass	75.0 g	35.2 g

$$c_{p,\text{metal}} = \frac{\Delta T_{\text{water}}[(m_{\text{water}} \times c_{p,\text{water}}) + C']}{m_{\text{metal}} \times \Delta T_{\text{metal}}}$$

$$c_{p,\text{Al}} = \frac{7.2\,\text{C}°[(75.0\,\text{g} \times 4.184\,\text{J/g}\cdot\text{C}°) + 8.2\,\text{J/C}°]}{(35.2\,\text{g})(70.8\,\text{C}°)}$$

$$= 0.93\,\text{J/g}\cdot\text{C}°$$

Iron Metal Specific Heat Test

	H₂O	Metal
Initial Temp.	21.0°C	99.0°C
Final Temp.	28.1°C	28.1°C
Change in Temp.	7.1 C°	−70.9 C°
Mass	75.0 g	69.2 g

$$c_{p,\text{metal}} = \frac{\Delta T_{\text{water}}[(m_{\text{water}} \times c_{p,\text{water}}) + C']}{m_{\text{metal}} \times \Delta T_{\text{metal}}}$$

$$c_{p,\text{Fe}} = \frac{7.1\,\text{C}°[(75.0\,\text{g} \times 4.184\,\text{J/g}\cdot\text{C}°) + 8.2\,\text{J/C}°]}{(69.2\,\text{g})(70.8\,\text{C}°)}$$

$$= 0.47\,\text{J/g}\cdot\text{C}°$$

2. Students' answers will vary, depending upon the unknown metal chosen for them.

3. Students' answers will vary. The following sample calculations were made using values found in *CRC Handbook of Chemistry and Physics*:

Observed value, Al: 0.93 J/g·C°
Accepted value, Al: 0.90 J/g·C°

$$\text{Percent Error} = \frac{0.93 - 0.90}{0.90} = 3.3\%$$

Observed value, Fe: 0.47 J/g·C°;
Accepted value, Fe: 0.44 J/g·C°

$$\text{Percent Error} = \frac{0.47 - 0.44}{0.44} = 6.8\%$$

4. Metals have low specific heats in comparison to water. As a result, they are better conductors of heat than water. The same amount of heat will raise the temperature of a certain mass of metal higher than it would raise the temperature of the same mass of water.

5. The lower the heat capacity of the calorimeter, the more accurate the measurement of any change in heat content will be, since little heat will be lost to the environment. (Some students will note that even though calorimeter heat capacity can be used to adjust results, the change in heat of the calorimeter itself in different situations will vary more than the change of heat of water in different situations.)

EXTENSIONS — ANSWERS

1. Students' answers will vary depending upon the accuracy of the class measurements.

2. Students' answers will vary but may include: inaccuracy in mass or volume measurements, the escape of heat through the hole in the calorimeter to allow insertion of the thermometer, cooling of the metal while it is transferred to the calorimeter, and error introduced by the assumption that liquid water has a density of 1.0 g/mL (true only at 4°C).

3. Students' answers will vary, but suggestions for improvement may include: adjusting the water content of the calorimeter so that it is nearly all water with little air remaining, insulating or eliminating the thermometer hole in some way, and insulating the entire calorimeter more.

4. Students' answers will vary, but students should suggest that glass will have a specific heat between that of metals and water. Materials which might have similar specific heats include quartz, some ceramics, and sand.

5. There are many possible answers. Tiles on the space shuttle must have high specific heats to be able to withstand the heat of reentry to the earth's atmosphere. Pots, pans, and baking sheets must be made of material that have low specific heats and will conduct heat well. Energy-efficient homes should be made of substances with high specific heats that will help insulate against changes in temperature. Automobile coolants must also have high specific heats to absorb the maximum amount of heat from an engine with a given amount of coolant.

6. The oceans will cool much more slowly in winter than the land, so air moving onto land after passing over the oceans is not as cold as air travelling over the colder land masses. In the summer, the air moving over the oceans is not as hot as the air over the land because the oceans warm more slowly than the land. A small-scale version of this effect can also be observed when night and day are compared.

INVESTIGATION 4: Conductivity

Pages 837–840

This investigation should take about 30 minutes. If you wish to emphasize the role of planning and hypotheses in scientific work, have students try to narrow the unknowns based on the "Useful Information" given in the Analysis & Interpretation section before students begin the investigation. With this method, students will make hypotheses about what each unknown substance is, before they perform the conductivity tests. Such an approach will take more time. More time should also be allowed for the Extensions.

Although the equipment list is given for students working in pairs, students may work in larger groups, or individually, if this is preferred.

INVESTIGATION TEACHER'S GUIDE

MATERIALS

(for a class of thirty students working in pairs, includes amounts for Extensions)

15 sets of labeled containers (fewer sets will do if students share), each with a different unknown substance, labeled A–J. Keep track of which letter corresponds to which substance to check students' work. Once these containers are filled with the substances in question, they can be re-used from lab to lab and year to year.
- sodium sulfate
- sugar
- sulfur powder (50 g)
- tap water (50 mL)
- deionized (or distilled) water (50 mL)
- tin (foil — 50 g)
- tin(IV) oxide (100 g)
- graphite dust (50 g)
- yellow chalk dust (grind up 2 sticks of chalk)
- charcoal (grind up 2 or 3 chunks, or use dust often found in bottom of charcoal bag)

15 spot plates
15 dropper bottles of distilled water
15 spatulas
15 thin-stemmed pipets
paper towels
15 small glass stirring rods or toothpicks
15 LED conductivity testers (may be made by the teacher, or bought from Lab Aids. Lab Aids also sells a conductivity tester which emits a tone which students may find easier to use in distinguishing small differences. Have students test the apparatus before each investigation. You may want to have them test the metal faucets in the lab or other similar conductors.)

15 1-kΩ resistors ($\frac{1}{4}$ W)
15 LED's (light-emitting diodes)
15 9-V battery with battery clips
15 film canisters with lids
4.5 m of hook-up wire (22 GA, solid): cut into 15 10-cm lengths and 15 20-cm lengths
15 3.5-cm lengths of straw or tubing to fit over LED's
nail or ice-pick
electrician's tape
glue gun (optional)
15 iron nails *for Extension 1*
75 g wax *for Extension 1*
50 mL 1.0 M HCl solution (4.3 mL conc HCl in 50 mL of solution) *for Extension 1*
50 g NaCl *for Extension 1*

CONDUCTIVITY TESTER PREPARATION

1. Strip 1.5 cm of insulation from the ends of the hook-up wires and from the battery clip leads. Cut the resistor leads down to half of their length.

2. Poke two holes near the edge of the canister cap about as far apart as the two LED leads are. Insert the leads through the holes, and spread them apart slightly.

3. Connect the circuit as shown in Figure T4-A. **Be sure the *red* battery clip lead is connected to the *longer* LED lead.** The black battery clip lead should connect to the resistor, and the other end of the resistor should connect to the 10-cm length of wire. The shorter LED lead should be connected to the 20-cm wire. Bend the leads up and tape them against the inside of the lid, so that they are not touching each other and there are no exposed wires.

4. Wrap tape around the resistor and its exposed wires.

Figure T4-A: A simple conductivity tester can be made from common materials.

5. Poke two holes in the side of the film canister about 1 cm apart, near the bottom of the film canister. Insert the hook-up wires through these holes.

6. Connect the battery and resistor, and place them in the canister, snapping the lid shut. Adjust the protruding hook-up wires so they extend out about 5 cm. (Optional: place drops of hot-melt glue around the LED and the hook-up wires so that the canister is water-proof).

7. Wrap electrician's tape around the straw and place it over the LED to provide a "dark tunnel" through which the LED may be viewed for testing weak electrolytes.

SAFETY NOTES

Avoid open flames: sulfur powder and graphite powder can be fire hazards.

Sulfur powder can also irritate skin and other tissues, so wear gloves and goggles at all times.

DISPOSAL

All solutions can be washed down the sink with plenty of excess water, provided your school drains are connected to a sanitary sewer system with a treatment plant. The solid precipitates should be dried and buried in a landfill that is approved for chemical disposal. You may want to consider drying and re-using the materials that are not soluble in water. In addition, the tin has value as scrap metal, so you may want to recycle it instead of discarding it. The microplates or spot plates can be wiped with a clean paper towel, rinsed with tap water, rinsed with distilled water, and wiped once more with a clean paper towel.

ANALYSIS & INTERPRETATION — ANSWERS

1. Metallic bonding or other bonds involving free electrons (such as those between graphite sheets) are the types of bonding that should conduct in pure form. The free-moving electrons are the charged particles that carry the current.

2. Ionic substances that dissolve in water, and polar-covalent substances that ionize when dissolved in water are likely to conduct electricity when mixed with water. The ions themselves serve as free-moving charged particles to carry the current.

3. The nonpolar-covalent substances, and those ionic substances that do not dissolve in water are likely to be insulators in their pure form. The covalent bonds in nonpolar molecules hold the electrons too firmly for them to carry the current. In ionic substances that do not dissolve in water, the ions are firmly bound in a lattice, and are not available to move freely and carry the current.

4. Students' answers will vary. The following are some possible responses:

- Chalk is an ionic compound that does not dissolve in water, so it will not conduct.
- Sulfur is a molecule with nonpolar-covalent bonds, that will not conduct.
- Sodium sulfate is an ionic compound that will conduct after it is dissolved in water.
- Tin has metallic bonding, and will conduct electricity without being dissolved in water.
- Charcoal is a molecule with nonpolar-covalent bonds, that will not conduct.
- Graphite is a molecule mostly with nonpolar-covalent bonds, but it also has some free electrons between its sheets, so it will conduct slightly.
- Sugar is a molecule with nonpolar-covalent bonds, that will not conduct electricity even when dissolved in water since it does not form ions.
- Distilled or deionized water has only polar-covalent bonds that do not ionize appreciably when dissolved, so it will not conduct.
- Tap water will contain enough dissolved impurities that it should conduct electricity (or at least conduct better than distilled or deionized water).

CONCLUSIONS — ANSWERS

1. Students' answers will vary, depending upon the arrangement of unknowns. Along with the reasoning shown in Analysis & Interpretation #4, students may also make use of other physical properties (state of matter, color, etc.) to help them narrow down which unknown is which.

EXTENSIONS — ANSWERS

1. An iron nail should conduct, since it contains metallic bonds with free electrons. Liquid bromine is a nonpolar-covalently bonded molecule, so it should not conduct. Wax is another nonpolar-covalently bonded molecule which should not conduct. The hydrochloric acid molecule is polar-covalent, and in solution it will ionize, so it will conduct. When more water is added, it will conduct a little less. Table salt will not conduct in the solid form, but since it is composed of ionic bonds, it will conduct when dissolved in water.

2. Students' answers will vary, but they will probably suggest using tin or graphite, since both conduct electricity without being dissolved in a solution that could evaporate. Some students will note that other properties of metals discussed in Chapter 6, such as ductility and malleability, make tin the best choice.

3. Although pure water does not conduct electricity appreciably as shown from the test of distilled (or deionized) water, water with impurities conducts electricity very well, as was shown in the test of tap water. Lakes, pools, and other bodies of water are very likely to have impurities dissolved in them. As a result, there is a danger of electrocution if you are in the water and lightning strikes it, even if it strikes the water not very near to where you are.

INVESTIGATION 5: Colored Precipitates

Pages 840–843

Approximate time: less than 30 minutes. Allow more time for the additional activities suggested in the Extensions section. This investigation may also be of interest in discussing soluble and insoluble compounds, transition metals, or stoichiometric concepts.

Although the equipment list is given for students working in pairs, students may work in larger groups, or individually, if this is preferred.

MATERIALS

(for a class of thirty students working in pairs, includes amounts for Extensions)

Given the dangers of cobalt and nickel compounds in their pure forms, teachers may want to consider ordering the solutions pre-made, a fairly inexpensive way to minimize preparation hazards.

Make solutions in dropper bottles or containers that can be used for both storing and dispensing the solutions.

- 25 mL 0.1 M cobalt(II) chloride solution (0.60 g $CoCl_2 \cdot 6H_2O$ in 25 mL of solution)
- 25 mL 0.1 M copper(II) chloride solution (0.43 g $CuCl_2 \cdot 2H_2O$ in 25 mL of solution)
- 25 mL 0.1 M iron(III) chloride solution (0.68 g $FeCl_3 \cdot 6H_2O$ in 25 mL of solution)
- 25 mL 0.1 M nickel(II) chloride solution (0.60 g $NiCl_2 \cdot 6H_2O$ in 25 mL of solution)
- 125 mL 0.1 M sodium hydroxide solution (0.50 g NaOH in 125 mL of solution)
- 25 mL mystery solution I: 0.1 M iron(III) sulfate solution (1.40 g $Fe_2(SO_4)_3 \cdot 9H_2O$ in 25 mL of solution)
- 25 mL mystery solution II: 0.2 M nickel(II) sulfate solution (1.32 g $NiSO_4 \cdot 6H_2O$ in 25 mL of solution)
- 25 mL mystery solution III: 0.2 M copper(II) sulfate solution (1.25 g $CuSO_4 \cdot 5H_2O$ in 25 mL of solution)

Of course, teachers may opt to use only one or two of the mystery solutions, as resources or time constraints dictate.

- 120 thin-stem pipets (as few as 8 can be used for a class of 15 lab groups if students take turns acquiring reagents)
- 15 8-well flat-bottom strips *or* 105 small test tubes
- 105 pieces of filter paper *may be requested by students for Extension 3*
- 15 funnels *may be requested by students for Extension 3*
- 15 balances *may be requested by students for Extension 3* (fewer balances will do if students share)
- 1 roll of pH paper sensitive to 12–14 range *for Extension 4*

SAFETY NOTES

During preparation, be sure to wear both goggles and gloves, since hazardous substances may be handled in pure forms. If inhaled, the dust of nickel compounds is suspected to be carcinogenic, so be careful when preparing the solutions.

Keep a basic neutralization package handy, in case of spills.

DISPOSAL

Cobalt and Nickel Compounds and Solutions: are extremely toxic and accumulate in body tissues. By converting them into sulfides, which are insoluble, they may be buried in a landfill approved for the disposal of hazardous waste. Dissolve them in a minimum amount of water (for some compounds, this will require the addition of 6 M HCl.) Add a 3-fold molar excess of sodium sulfide to be sure the cobalt and nickel ions react completely with the sulfide, and stir occasionally for one hour. Test the pH of the solution with pH paper to be sure it is neutral or basic. (If not, add some 3.0 M NaOH solution to complete precipitation of the sulfide.) Filter and dry the sulfide. Place the metal sulfide in a plastic container, to be sealed and buried in a landfill approved for the disposal of hazardous waste. The solution left over after filtering probably contains excess sulfide ion, and should be disposed of by adding it to an excess of $Fe(NO_3)_3$ solution or $FeCl_3$ solution. A precipitate should be formed. Use pH paper to check the solution and neutralize it, if necessary, with Na_2CO_3. There will be some carbon dioxide escaping, and finally, some orange iron(III) hydroxide may also precipitate. Filter and dry the precipitated Fe_2S_3. Then place it in the plastic container with the other sulfides and bury it in a landfill approved for the disposal of hazardous waste. The remaining neutral solution may be washed down the drain with excess water.

Copper and Iron Compounds and Solutions and Sodium Hydroxide Solutions: can be neutralized with a small amount of weak acid (or even lemon juice or vinegar) and washed down the sink with plenty of excess water, provided your school drains are connected to a sanitary sewer system with a treatment plant. Precipitates may be gathered, placed in a container, and disposed of in a landfill approved for chemical waste.

ANALYSIS & INTERPRETATION—ANSWERS

1. The balanced chemical equations for the four reactions are:

$CoCl_2(aq) + 2\ NaOH(aq) \rightarrow Co(OH)_2(s) + 2\ NaCl(aq)$
$CuCl_2(aq) + 2\ NaOH(aq) \rightarrow Cu(OH)_2(s) + 2\ NaCl(aq)$
$FeCl_3(aq) + 3\ NaOH(aq) \rightarrow Fe(OH)_3(s)* + 3\ NaCl(aq)$
$NiCl_2(aq) + 2\ NaOH(aq) \rightarrow Ni(OH)_2(s) + 2\ NaCl(aq)$

* Actually, to be most accurate, the substance formed is FeOOH, or $FeOOH \cdot H_2O$ (FeO_3H_3).

2. The net-ionic equations for the four reactions are:
$$Co^{2+}(aq) + 2\ OH^-(aq) \rightarrow Co(OH)_2(s)$$
$$Cu^{2+}(aq) + 2\ OH^-(aq) \rightarrow Cu(OH)_2(s)$$
$$Fe^{3+}(aq) + 3\ OH^-(aq) \rightarrow Fe(OH)_3(s)$$
$$Ni^{2+}(aq) + 2\ OH^-(aq) \rightarrow Ni(OH)_2(s)$$

3. The reactions would not be any different if KOH had been used instead of NaOH, since each supplies the $OH^-(aq)$ needed for the reaction. The net-ionic equations for the KOH reactions would be the same as those of the NaOH reactions, since Na^+ (or K^+) plays no part in the reaction.

CONCLUSIONS — ANSWERS

1. Students' answers should be:
Mystery Solution I: $Fe^{3+}(aq) + OH^-(aq) \rightarrow Fe(OH)_3(s)$
Mystery Solution II: $Ni^{2+}(aq) + 2\ OH^-(aq) \rightarrow Ni(OH)_2(s)$
Mystery Solution III:
$Cu^{2+}(aq) + 2OH^-(aq) \rightarrow Cu(OH)_2(s)$

2. Students' answers will vary, but students should note that the color of the precipitate is caused by the positive ion forming the precipitate (cobalt, copper, iron, or nickel). The color can't be caused by the negative hydroxide ions since all the compounds formed were hydroxides, yet they had different colors.

EXTENSIONS — ANSWERS

1. If the drops are exactly the same size and concentrations remain the same, inspection of the balanced equation reveals that the cobalt, copper, and nickel solutions should take twice as much sodium hydroxide solution (10 drops) and the iron solution should take three times as much to react (15 drops). Since 20 drops of sodium hydroxide were used, there should be five drops of sodium hydroxide left unreacted after all of the iron chloride has reacted, and 10 drops left over after the other metal chlorides have reacted.

2. Students' answers may vary, but if they were able to tell that the mystery solutions had a higher concentration of the metallic cations, the evidence cited should include the larger amounts of precipitates formed when the mystery solutions reacted.

3. Students' answers will vary, but students may suggest filtering the precipitates, drying them, and measuring their mass. The mass of hydroxide formed by one of the standard "known" solutions will be proportional to its concentration, and the same proportion will hold for the mystery solutions.

4. The iron chloride solution should have the lowest pH and the smallest concentration of OH^- ions left unreacted at the end of the reaction, because each iron ion required three units of hydroxide to react, more than the other metals, as seen in the net-ionic equations. The others each required two units of hydroxide, and so their pH should be similar. (Note: if this lab were perfectly executed, the pH difference could still be hard to detect, since the solution from the iron reaction should have a pH near 12.3 and the others should have a pH of 12.8.)

5. Students may suggest that by using a pH meter to accurately measure the pH of the solution remaining after the reaction. By comparing the measurements of the reacted mystery solutions with those of the "known" standards, and calculating the concentration of the OH^- ions remaining, they could estimate the number of OH^- ions that reacted, and the number of positive ions reacting with those OH^- ions.

INVESTIGATION 6: Pin-the-Formulas-on-the-Compounds

Pages 843–845

Approximate time: 30 minutes. Allow more time for the additional activities suggested in the **Extensions** section. You might want to have the students perform the experiment more than once to check the precision of their microscale techniques.

Although the equipment list is given for students working in pairs, students may work in larger groups, or individually, if this is preferred.

This investigation may also be useful in familiarizing students with the basics of stoichiometry and molar concentrations of solutions.

The "mystery" chlorides are copper(II) chloride, $CuCl_2$, and iron(III) chloride, $FeCl_3$.

MATERIALS

(for a class of thirty students working in pairs, includes amounts for Extensions)

30 mL 0.1 M copper(II) chloride
30 mL 0.5 M copper(II) chloride *for Extension 3*
30 mL mystery concentration solution of copper(II) chloride *for Extension 4*
  To prepare 0.1 M solution alone: 0.43 g of $CuCl_2 \cdot 2H_2O$ in 30 mL solution
  To prepare 0.1 M solution from 0.5 M solution: 4.27 g of $CuCl_2 \cdot 2H_2O$ in 60 mL of solution to make 0.5 M. Take 6 mL of 0.5 M and dilute with water to make 30 mL of 0.1 M solution. After reserving at least 30 mL of the 0.5 M solution, dilute part of what remains to make the mystery concentration solution. (Remember to keep track of the dilution, so you can tell the students how close their concentration estimates are.)

30 mL 0.1 M iron(III) chloride
30 mL 0.5 M iron(III) chloride *for Extension 3*
30 mL mystery concentration of iron(III) chloride *for Extension 4*
  To prepare 0.1 M solution alone: 0.82 g of $FeCl_3 \cdot 6H_2O$ in 30 mL solution
  To prepare 0.1 M solution from 0.5 M solution: 8.17 g of

FeCl$_3 \cdot$6H$_2$O in 60 mL of solution to make 0.5 M. Take 6 mL of 0.5 M and dilute with water to make 30 mL of 0.1 M solution. After reserving at least 30 mL of the 0.5 M solution, dilute part of what remains to make the mystery concentration solution. (Remember to keep track of the dilution, so you can tell the students how close their concentration estimates are.)

150 mL 0.1 M sodium hydroxide (0.6 g NaOH in 150 mL of solution)

13 mL phenolphthalein solution (dissolve 0.13 g phenolphthalein in 6 mL denatured alcohol and add 7 mL water.)

60 fine-tipped dropper bulbs (May be re-used from class to class and year to year. Available commercially, or may be made out of thin-stem pipets. Stretch the pipets by hand and then cut them off. (See diagram after the master lists of chemicals and equipment for all investigations.) For more accurate drop sizes, place the dropper bulb in a tubing screw-clamp. Each turn of the screw will force out a reproducible drop.)

30 8-well strips

SAFETY NOTES

During preparation, be sure to wear both goggles and gloves, since hazardous substances may be handled in pure forms. Keep a base neutralization package handy, in case of spills.

DISPOSAL

All solutions can be neutralized with a small amount of very dilute HCl and washed down the sink with plenty of excess water, provided your school drains are connected to a sanitary sewer system with a treatment plant. The solid precipitates should be dried and buried in an approved landfill.

ANALYSIS & INTERPRETATION — ANSWERS

1. a) CuCl(aq) + NaOH(aq) → CuOH(s) + NaCl(aq)
b) CuCl$_2$(aq) + 2 NaOH(aq) → Cu(OH)$_2$(s) + 2 NaCl(aq)
c) FeCl$_2$(aq) + 2 NaOH(aq) → Fe(OH)$_2$(s) + 2 NaCl(aq)
d) FeCl$_3$(aq) + 3 NaOH(aq) → Fe(OH)$_3$(s)* + 3NaCl(aq)
* Actually, to be most accurate, the substance formed is FeOOH, or FeOOH$\cdot$H$_2$O (FeO$_3$H$_3$).

2. Equal sized drops will have equal numbers of formula units of the compounds in them.

3. Students' answers will vary, but if the drop sizes were kept equal, the copper chloride should require about ten drops on average, and the iron chloride should require about 15 drops on average.

CONCLUSIONS — ANSWERS

1. For every formula unit of copper chloride, two formula units of sodium hydroxide were needed. For every formula unit of iron chloride, three formula units of sodium hydroxide were needed.

2. Students should choose copper(II) chloride, CuCl$_2$, and iron(III) chloride, FeCl$_3$.

EXTENSIONS — ANSWERS

1. Students' answers will vary.

2. Students' answers may vary, but may include the following: Difficulty in making drops the same size, unfamiliarity with the end-point of the reaction, and large margin of error with such small samples. Students may suggest minimizing these errors by repeated trials, by using the same dropper for all solutions and washing it out very well in between uses, or by calibrating their drop sizes with a graduated cylinder. Comparison to a standard to identify the end point may be useful. Students familiar with pH paper may suggest using it instead of phenolphthalein. Another few trials may be run using test tubes and larger quantities (25 drops instead of 5 drops, etc.)

3. Students' answers will vary, but the 0.5 M copper chloride should take about 50 drops of 0.1 M NaOH, and the 0.5 M iron chloride should take about 75 drops of 0.1 M NaOH.

4. Students' answers will vary, but they should suggest that they could compare the amount of NaOH needed to react with a similar amount of "mystery concentration" solution to the amount needed to react with the solution of known concentration in the main procedure. The concentrations will be in the same proportion as the amounts of NaOH needed.

INVESTIGATION 7: Generating and Collecting O$_2$

Pages 845–848

Approximate time: 20–25 minutes. Although the equipment list is given for students working in pairs, students may work in larger groups, or individually, if this is preferred.

MATERIALS

(for a class of thirty students working in pairs, includes amounts for Extensions)

15 petri dish halves

15 bunsen burners or candles (Fewer can be used if students alternate their tests.)
15 micro O$_2$ generators
  15 vials with caps
  15 gradiated pipets
250 mL 3% H$_2$O$_2$ (store-bought hydrogen peroxide is fine, provided it is "fresh")
4–8 dropper bottles containing 1.0 M KI (0.83 g in 50 mL of solution or 0.42 g in 25 mL of solution will

make far more than your class will use, so keep it for next time. If dropper bottles are not available, the KI solution may be placed in cut-off thin-stem pipets. One bulb can hold about 80–120 drops, possibly enough for the entire class.

- 30 beakers (15 for water, other 15 in case students need to dispose of spent solutions)
- 105 test tubes (13 × 100)
- 105 1-holed rubber stoppers
- 105 pegs to fit holes in stoppers
- 30 wood splints
- 15 small tufts of steel wool
- 15 pairs of tongs or forceps
- tap water
- paper towels

MICRO GENERATOR PREPARATION

Cut the tapered tips off the graduated pipets, and they may be used as nozzles for the gas generators. Another way to make nozzles is to take a thin-stemmed pipet, and, with a steady pull, stretch it to a much narrower diameter. With this method, each stem can provide two nozzles. Punch or drill holes (or melt them with a hot nail) in the center of the vial cap and slip the tapered nozzles into the holes. If the nozzle segment is inserted until the flared-end section wedges into the hole, a relatively air-tight junction can be formed.

The concentrations listed above for the H_2O_2 solution (3%) and the KI solution (1.0 M) are meant only as guidelines. You may wish to adjust them as needed to moderate the reaction rate. Having the solutions at (or slightly above) room temperature will also help speed the reaction.

The petri dishes will catch the water displaced when the test tubes are filled with water.

SAFETY NOTES

Always be sure that the nozzles are clean, so that when gas is generated it can escape without building up dangerous amounts of pressure. Be sure to wear goggles to protect your eyes from the solutions used in this investigation.

DISPOSAL

After the lab, solutions may be disposed of by washing them down the sink with a 20-fold excess of water provided your school drains are connected to a sanitary sewer system with a treatment plant.

ANALYSIS & INTERPRETATION — ANSWERS

Test Tube #	Typical Observations
1	flame burns normally, then dies out . . . smoke fills the tube
2	flame burns much more intensely, then dies out
3	ember glows for a while, then dies out
4	glowing ember re-ignites, then dies out
5	ember glows for a while, then dies out
6	ember glows for a while, then dies out
7	steel wool ember brightens, glows white-hot, sparks fly

INVESTIGATION TEACHER'S GUIDE

1. The evidence for a chemical reaction is the appearance of bubbles.

2. $2\ H_2O_2(aq) \xrightarrow{KI} 2\ H_2O(\ell) + O_2(g)$.

3. Wood + oxygen → ash + carbon dioxide + water vapor

Students may also include heat or enthalpy evolved as a part of the equation.

CONCLUSIONS — ANSWERS

1. The wood splint was extinguished because the reaction between the wood and the oxygen in the air stopped when the oxygen ran out.

2. The wood splint burned less intensely in test tube #1 because it had only the amount of oxygen present in air (20%), so the oxygen ran out more quickly than in test tube #2, which had 100% oxygen. Even in the 100% oxygen test tube, oxygen was the limiting reactant, since the reaction ended before the wood was completely consumed.

3. The splint re-ignited in pure oxygen. Presumably the glowing ember was not hot enough to re-start the reaction and rekindle the flame in air, but it was hot enough to rekindle the flame in an environment of 100% oxygen.

4. Since both the mouth-upward and the mouth-downward test tubes behaved like the control test tube full of air, they apparently filled with air. (Note: there is no visible evidence that this occurred, since both gases are clear, colorless, and odorless.) Since there was not a noticeable difference between the two, apparently oxygen has a density near that of air. Some students will notice that this finding also implies that oxygen can diffuse either upward or downward.

5. As was the case in wood, steel wool burns much more readily in pure oxygen than in air. As discussed above, this is what would be expected, given that pure oxygen has five times as many oxygen molecules in a given volume available to react with the steel wool.

6. Students' answers may vary. Four chemical properties of oxygen that were observable in this investigation are:
- oxygen can be formed from hydrogen peroxide by a decomposition reaction
- oxygen reacts with wood, and is exhausted in the reaction (gets used up)
- oxygen supports combustion of an ember
- oxygen reacts with iron

7. Students' answers may vary. Four physical properties of oxygen that were observable in this investigation are:
- oxygen is colorless
- oxygen is odorless
- oxygen does not dissolve in water
- oxygen is about as dense as air

(Some students may also note that oxygen diffuses upward and downward.)

EXTENSIONS — ANSWERS

1. No, the wood splint would not burst into flame or form an ember by itself, because along with the reactant chemicals, a small amount of energy or heat is needed to start the reaction. Once the reaction has begun, the energy released during the exothermic reaction would serve as the activation energy for other molecules of wood and oxygen nearby, and so on.

2. Air appears to be less than half oxygen, since it supports combustion so much less than pure oxygen. Students' answers will vary.

3. Students' answers will vary, but students should make the connection between embers bursting into flame and flammable materials burning much more quickly in the test tubes of pure oxygen and the similarity with conditions in an environment that has more oxygen than air. Smoking is forbidden around oxygen equipment because of the fire hazard. Stray embers from the end of a cigarette that would ordinarily extinguish themselves and become ash will remain lit longer, possibly igniting other materials. Materials that would ordinarily not ignite easily will catch fire and burn rapidly. The cigarette will also burn much faster.

4. Students' answers will vary, depending on the topics chosen.

INVESTIGATION 8: Generating and Collecting H_2

Pages 848–850

This investigation should take 25–30 minutes for students to complete. Although the equipment list is given for students working in pairs, students may work in larger groups, or individually, if this is preferred.

MATERIALS

(for a class of thirty students working in pairs, includes amounts for Extensions)

15 petri dish halves
15 bunsen burners or candles (Fewer can be used if students alternate their tests.)
15 micro H_2 generators
   15 vials with caps
   15 graduated pipets or thin-stemmed pipets
30 g mossy zinc
250 mL of 1.0 M HCl (20.6 mL conc. HCl in 250 mL of solution)
30 beakers (15 for water, 15 for spent solution)
75 test tubes (13 × 100)
75 1-holed rubber stoppers (size 0 should fit the 13 × 100 test tubes)
75 pegs to fit holes in stoppers
30 wood splints
tap water
paper towels
15 pairs of tongs or forceps (optional)

MICRO GENERATOR PREPARATION:

Cut the tapered tips off the graduated pipets, and they may be used as nozzles for the gas generators. Another way to make nozzles is to take a thin-stemmed pipet, and, with a steady pull, stretch it to a much narrower diameter. With this method, each stem can provide two nozzles. Punch or drill holes (or melt them with a hot nail) in the center of the vial cap and slip the tapered nozzles into the holes. If this nozzle segment is inserted until the flared-end section wedges into the hole, a relatively air-tight junction can be formed.

Break up the zinc into pieces. Put about 2 g zinc into 15 vials. (The vials should be about $\frac{1}{6}$ full.)

Prepare the metal shortly before the lab. Read all safety notes that come with your supply of zinc. It should not be pulverized or powdered, since the metal dust can be flammable. Remember that if zinc dust gets damp in a closed container, an explosion may occur. Keep your zinc dry until lab time.

A few drops of $CuSO_4$ solution will help catalyze the Zn/HCl reaction. In addition, the HCl solution concentrations may be altered to moderate the reaction rate.

The petri dishes will catch the water displaced when the test tubes are filled with water.

SAFETY NOTES

Keep an acid neutralization package handy such as resealable (zippered) plastic bags filled with $NaHCO_3$, in case of spills.

As mentioned in the pupil's investigations, goggles must be worn to protect the eyes from the solutions used. The pop-test explosions are harmless.

DISPOSAL

After the lab, solutions may be disposed of by washing them down the sink with a 20-fold excess of water provided your school drains are connected to a sanitary sewer system with a treatment plant. The small amounts of zinc that remain may be collected, dried, and re-used or else they should be placed in a solution of 1.0 M HCl until completely reacted. The resulting solution may also be washed down the drain with a 20-fold excess of water.

ANALYSIS & INTERPRETATION—ANSWERS

Test Tube #	Typical Observations
1	small "pop" (can be repeated two or three times)
2	louder, sharper "POP" (only one pop possible)
3	glowing ember is extinguished, no "pop"
4	no "pop" at all
5	loud "pop" comparable to #2 (probably only one pop possible)

1. The evidence for a chemical reaction include the appearance of bubbles, and the slow disappearance of the zinc metal.

2. $HCl(aq) + Zn(s) \rightarrow ZnCl_2(aq) + H_2(g)$.

3. $O_2(g) + 2H_2(g) \rightarrow 2 H_2O(g)$. Students may also include heat or enthalpy evolved as a part of the equation.

CONCLUSIONS—ANSWERS

1. The pop-test for the second tube (with half as much hydrogen) was much louder than that of the first test tube (with pure hydrogen). As seen in the balanced reaction, the pop-test involves both hydrogen and oxygen. In the first tube, the pure hydrogen is reacting with what little air there is near the opening of the tube. The second test tube is half hydrogen and half air. The air provides a supply of oxygen throughout the tube, so that the hydrogen can react more quickly and thoroughly.

2. In test tube #3, the hydrogen did not ignite, but rather, it extinguished the ember. As mentioned in Section 10.2, one of the chemical properties of hydrogen is that it will combine with oxygen if it is heated above 800°C or ignited by a spark. Apparently, the ember did not provide a high enough temperature or enough energy to start the pop-test reaction.

3. Hydrogen gas is much less dense than air. When the tube was held mouth-up, as in #4, much of the hydrogen gas floated away. When the tube was held mouth-down, as in #5, much of the hydrogen gas remained. This would suggest that air is denser than the hydrogen.

Students may interpret the absence of a pop in #4 as indicating that there was no hydrogen left in the tube. In fact, it really means that not **enough** hydrogen was left to constitute a combustible mixture.

Students may also note that in #5, the pop was fairly loud, suggesting that some of the hydrogen had mixed with air, as in #2, rather than remaining perfectly pure, as in #1. Other students may think that hydrogen will not diffuse downward. A good demonstration would be to clamp a hydrogen-filled test tube mouth-downward on a ring stand and ask the students what would happen if they came back the following day and tried the pop-test. Some might believe that the hydrogen will still be there. Leave the tube set up overnight and try pop-testing it the next day.

4. Students' answers may vary. Four chemical properties of hydrogen that were observable in this investigation are:
- hydrogen is replaced in acids by zinc
- hydrogen reacts with oxygen in the air if ignited by a flame
- hydrogen will not ignite and react with oxygen in the air if a red-hot ember is used
- hydrogen alone cannot support combustion of an ember

5. Students' answers may vary. Four physical properties of hydrogen that were observable in this investigation are:
- hydrogen is colorless
- hydrogen is odorless
- hydrogen does not dissolve in water
- hydrogen is less dense than air

(Some students may also note that hydrogen diffuses very readily.)

6. Hydrogen's low density is its main advantage for using it to inflate objects. Its main disadvantage is its reactivity with air if there is a spark present. (Some students may also note that its ease of diffusion makes it difficult to contain in a lightweight container.)

EXTENSIONS—ANSWERS

1. Yes, the same spark could ignite an entire room full of H_2 and air. Once the reaction has begun, the energy released during the exothermic reaction would serve to start the reaction of other molecules of hydrogen and oxygen nearby, and so on. (This chain reaction takes place so quickly that it appears that the entire mixture explodes simultaneously. This is why quantities of hydrogen gas greater than microscale are so dangerous.)

2. Students' answers may vary, but they should suggest that a proportion of about 5 parts air to 2 parts hydrogen should be the most effective, since, in this proportion, the oxygen present in the air should react completely with the hydrogen, maximizing the production of the products: water vapor, heat, and noise.

3. Students' answers will vary. Some typical responses are shown in the table at the beginning of the Analysis & Interpretation answers. The reason why tube #2 and tube #5 will not pop again is that nearly all of the hydrogen will react on the first attempt. Tube #1 will pop repeatedly. First, the hydrogen near the opening of the tube will react with the air nearby. After the first pop, the hydrogen further down the tube will then be able to react with the air near the opening, and so on.

4. Students' answers may vary. The *World Book Encyclopedia* (in the "Airships" article) states that the *Hindenburg* had a volume of 199,980 m³. If the test tubes used were the 13 × 100 size, they had a volume of about 10 cm³. This would mean that there was about 2.0×10^{11} times as much hydrogen in the *Hindenburg* (200 billion times as much).

INVESTIGATION 9: Testing Reaction Combinations of H$_2$ and O$_2$

Pages 851–853

Approximate time: 30 minutes for the steps leading up to the rocket launch. Allow more time for students to experiment with the different variables involved in launching the rockets, and/or the additional activities suggested in the **Extensions** section.

Although the equipment list is given for students working in pairs, students may work in larger groups, or individually, if this is preferred.

This investigation can also be useful in discussing activation energy and thermodynamics, stoichiometry, or even rocketry and propulsion.

MATERIALS

(for a class of thirty students working in pairs, includes amounts for Extensions)

15 petri dish halves
15 micro O$_2$ generators and 15 micro H$_2$ generators
- 15 micro O$_2$ generators
  15 vials with caps (reusable)
  15 graduated pipets
  500 mL 3% H$_2$O$_2$ (store-bought hydrogen peroxide is fine, provided it is "fresh")
  4–8 dropper bottles containing 1.0 M KI (0.83 g in 50 mL of solution or 0.415 g in 25 mL of solution will make far more than your class will use, so keep it for next time. If dropper bottles are not available, the KI solution may be placed in cut-off thin-stem pipets. One bulb can hold about 80–120 drops, possibly enough for the entire class.)
- 15 micro H$_2$ generators
  15 vials with caps
  15 graduated pipets or thin-stemmed pipets
  30 g mossy zinc
  250 mL of 1.0 M HCl (20.6 mL conc. HCl in 250 mL of solution)

30 incremented collected bulbs (this includes extras for those that get lost!)
permanent fine-tipped marker or pen
4–5 piezoelectric sparking elements (3–4 students can easily share the same sparker)
  4–5 piezoelectric/butane charcoal/fireplace lighters (such as Aim 'n Flame, etc.)
  50 cm of 24 AWG speaker hook-up wire (solid not stranded: Radio Shack #278-1509)
  4–5 film canister lids
  electrician's tape
  hot glue gun and glue
  scissors
  wire strippers
  $\frac{1}{8}$″ drill or 6-D common nail
(If piezoelectric sparking elements are not available, the bulbs can be pop-tested with a bunsen burner. The danger of unexpected explosions and burning bulbs is greatly increased when a burner is used. If a burner is used, the rocket portion of the lab is best left omitted. Follow this procedure: Light the burner and adjust it to a cool flame. Once the bulb is filled, remove it from the nozzle, and hold it securely and horizontally, with its mouth 1 cm from the midsection of the flame. Gently squeeze a very small portion of the contents of the bulb into the flame.

 Avoid putting the bulb directly in the flame, it will melt and possibly burn. If the bulb does burn, quench the tip in a beaker of water.

30 beakers (15 for water, other 15 in case students need to dispose of spent solutions)
15 pairs of tongs or forceps (optional)
15 10-mL or 25-mL graduated cylinders
tap water
paper towels
refrigerator or ice bath to chill gases *for Extension #1*

MICRO GENERATOR PREPARATION

Cut the tapered tips off the graduated pipets, and they may be used as nozzles for the gas generators. Another way to make nozzles is to take a thin-stemmed pipet, and, with a steady pull, stretch it to a much narrower diameter. With this method, each stem can provide two nozzles. (See Figure T9-A.) Punch or drill holes (or melt them with a hot nail) in the center of the vial cap and slip the tapered nozzles into

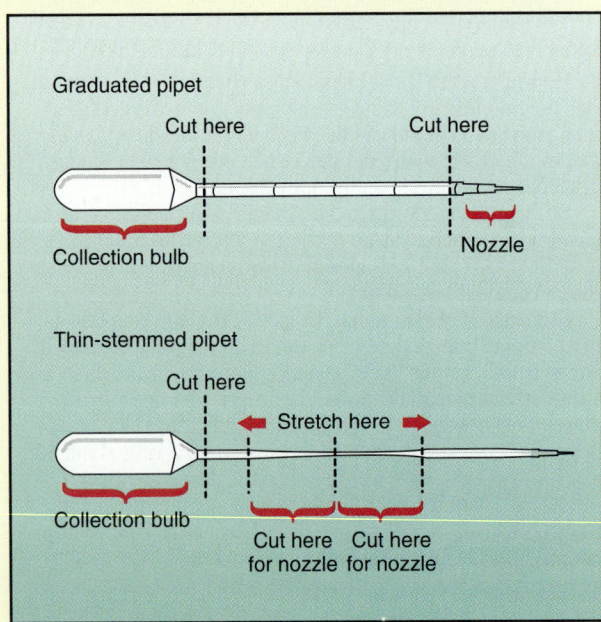

Figure T9–A: Collection bulbs and nozzles can easily be made from either graduated pipets or thin-stemmed pipets.

the holes. If this nozzle is inserted until the flared-end section wedges in the hole, a relatively airtight junction can be formed.

Break the zinc into pieces. Put about 2 g zinc into 15 vials. (The vials should be about $\frac{1}{6}$ full.)

Prepare the metal shortly before the lab. Read all safety notes that come with your supply of zinc. It should not be pulverized or powdered, since the metal dust can be flammable. Remember that if zinc dust gets damp in a closed container, an explosion may occur. Keep your zinc dry until lab time.

A few drops of $CuSO_4$ solution will help catalyze the Zn/HCl reaction. The concentrations listed above for the HCl solution (1.0 M), the H_2O_2 solution (3%) and the KI solution (1.0 M) are meant only as guidelines. You may wish to adjust them as needed to moderate the reaction rate. Having the solutions at (or slightly above) room temperature will also help speed the reaction.

Once built, the generator vials can be re-used from class to class and year to year.

COLLECTION BULB PREPARATION

Cut the bulbs from the pipets used to make nozzles. (See Figure T9-A.) To calibrate the bulbs, fill one completely with water. (If you have trouble filling the bulb completely, or the students do later, you may want to consider using thin-stemmed pipets to deliver the water into the bulbs instead of having the water squeezed up from the petri dish.)

Squeeze all the water out into a 10-mL graduated cylinder and record the volume. Empty the graduated cylinder and refill the bulb with water. Into the graduated cylinder, squeeze out of the bulb $\frac{1}{6}$ of the volume of water you recorded earlier. Mark the water level on the surface of the bulb when held mouth-downward. (You may have to dry the side of the bulb first.) Squeeze out a second sixth of the volume, and mark the level again. Repeat for the remainder of the water.

Once one bulb has been calibrated, rest a wood splint across it, with the splint's end even with the end of the bulb. Mark off the splint at the same places that the bulb is marked. Use the splint as a template for the rest of the bulbs. Once calibrated, the bulbs can be re-used from class to class and year to year.

PIEZOELECTRIC SPARKER ELEMENT PREPARATION:

Construct the sparker from a piezoelectric charcoal/fireplace lighter found in many stores for just three or four dollars. Once wired as described in the following instructions, it can be used in place of a Tesla coil for a wide variety of chemical demonstrations. If the wire and tape are removed, the sparker can be used as a bunsen burner lighter or for any other demonstration requiring a quick and safe light (Mg ribbons, exploding cans, etc.).

Using the scissors, split the end of a 10-cm length of speaker wire down the center for a distance of 3.5 cm. Strip the last 2.3 cm of insulation from both sides of the fork, as shown in Figure T9-B.

INVESTIGATION TEACHER'S GUIDE

Drill a $\frac{1}{8}''$ hole (or alternately, melt a hole with a hot 6D common nail) through the center of the film canister lid. Slide the wire through, and secure the lid in place about 4 cm from the unforked end of the wire with hot glue or other water-proof adhesive. The film canister lid serves both as launch support pad and as a splash guard.

Slide one of the two stripped wire ends down into the small butane nozzle of the charcoal lighter, all the way down, so that the insulation is against the nozzle. (Rotate the wire as you insert it.) Lay the other stripped end along the metal barrel of the lighter and secure it in place with 3–4 wrappings of electrician's tape.

To use, put the bulb over the wire, aim and pull the trigger. Rather than arcing at the butane nozzle, the device will arc at the tip of the wire (the path of least resistance).

SAFETY NOTES

During preparation, be sure to wear both goggles and gloves, since hazardous substances may be handled in concentrated forms.

If you divide the reagents up into several bottles of each and set up "reagent stations" in different parts of the lab, students won't have as far to walk after obtaining chemicals, and the likelihood of spills and accidents will be decreased.

Keep an acid neutralization package handy such as resealable (zippered) plastic bags filled with $NaHCO_3$, in case of spills.

The explosions in this investigation are harmless, and the rockets are soft enough to be safe. Students should avoid aiming them at each other, so choose a target for them such as a bull's eye drawn on the chalkboard or an open box at the end of the room.

DISPOSAL

After the lab, solutions may be disposed of by washing them down the sink with a 20-fold excess of water provided your school drains are connected to a sanitary sewer system with a treatment plant. The small amounts of zinc that remain may be collected, dried, and re-used, or else it should be placed in a solution of 1.0 M HCl until completely reacted. The resulting solution may also be washed down the drain with a 20-fold excess of water.

ANALYSIS & INTERPRETATION—ANSWERS

1. a) $2 H_2O_2(aq) \xrightarrow{KI} 2 H_2O(\ell) + O_2(g)$.
   b) $HCl(aq) + Zn(s) \rightarrow ZnCl_2(aq) + H_2(g)$.

2. The Zn will need to be replaced first, since it is actually being consumed as a reactant. The KI is acting as a catalyst for the first reaction, and theoretically would never need replacing.

Investigation Teacher's Guide

INVESTIGATION TEACHER'S GUIDE

Figure T9–B: When finished, this apparatus can be used instead of a Tesla coil in any chemical demonstration requiring a spark.

3. Students' answers may vary, but some of the primary techniques used in this investigation are as follows: The generators are filled so full that there is little room for air. Then the gas that comes out of the nozzle once the reaction starts is mostly pure oxygen or hydrogen, without air in it. The collection bulbs are filled with water, instead of air, so that the collected gas remains uncontaminated by air, once it is in the bulb. After the bulb is removed, it is promptly covered (with a finger) until testing, to prevent air from diffusing in and gas from diffusing out.

4. $O_2(g) + 2 H_2(g) \rightarrow 2 H_2O(g)$. Students may also include heat or enthalpy evolved as a part of the equation.

5. Students' answers may vary, depending upon their perceptions of loudness. However, all students should pick the $2:4$ O_2/H_2 ratio as being the loudest, with mixtures having greater and lesser concentrations of H_2 as being incrementally less loud. A sample bar graph is shown below:

CONCLUSIONS — ANSWERS

1. a) Oxygen does not pop or explode when tested. Oxygen supports combustion of other materials, but it is not itself combustible.
   b) Hydrogen is combustible, meaning that it reacts readily with oxygen. Pure hydrogen does not explode, since there is no oxygen present for it to react with. The slight pop that may have been heard results from hydrogen diffusing out of the collection bulb and mixing with oxygen from the air. (If the test of pure hydrogen is repeated over and over, after each test, more hydrogen has diffused out and more air (and oxygen) has diffused in. Eventually a combustible mixture is attained, and a flame can backfire into the bulb.)

2. Some mixtures contain so little of one of the reactants that the likelihood of having an O_2 particle collide with a H_2 particle and react to form H_2O is slight. This decreases the chances of the reaction propagating throughout the mixture, and the explosive chain reaction will not occur.

3. The most explosive mixture should be the $2:4$ mixture (which is equivalent to one part oxygen to two parts hydrogen). This is the same as the $1:2$ mole ratio in the balanced equation for the pop-test. It is most explosive because it allows for the maximum yield of the product (water) and thus of the energy released during the reaction that forms this product. The reactants are completely consumed when they are present in this ratio. In other words, all other mixtures represent cases where there are limiting reactants, so that although the O_2 and H_2 always react in a $1:2$ ratio, one of the two runs out first.

4. The O_2 and H_2 may collide as they mix in the bulb, but they do not react because a small amount of extra energy needs to be present for the reaction to occur. The spark supplies extra energy in the form of electricity or heat for the activation energy of the reaction to break the bonds and start the reaction.

EXTENSIONS — ANSWERS

1. According to the assumptions of the kinetic theory of matter, if the gas was chilled, the molecules would have less kinetic energy. With less kinetic energy, they are less likely to collide favorably and produce a reaction, so the activation energy which must be added is larger. In actual testing, this difference may not be noticeable. The spark or flame used will be enough to fill the increased activation energy needs of enough molecules to get the chain reaction going and cause an explosion, as before.

2. Students' answers will vary, but the class average for relative loudness should reach a maximum at the 2 parts O_2 : 4 parts H_2 mixture. The average should reach minimums for pure O_2 and pure H_2.

3. Students' answers will vary. Along with the ratio of the gases, the angle of the launch, the mass of the rocket, and the shape of the rocket can also play a part in the distance the rocket will travel. Although a 45° angle is best under ideal, frictionless conditions, with air resistance, an angle less than 45° will prove more effective. If the rocket has a larger mass, it will be less subject to air resistance, but will have a greater inertia to overcome. Streamlining the rocket (with tail-fins and a nose-cone) can also increase the flight. If some water is left in the bulb, it will greatly increase the flight, for it gives the expanding gases a propellant to push against. This principle is used in toy water rockets, which rely on pumping air into a plastic bulb partially filled with water.

4. The most efficient way to load the space shuttle's fuel is so that it has the same 1:2 (2:4) ratio of oxygen to hydrogen. This ratio is a **mole ratio**. If hydrogen and oxygen's molar masses are figured into the ratio to make a mass ratio, it will be about an 8:1 ratio. When applied to the 680 000. kg of fuel on the space shuttle, this means that 603 900 kg of it should be oxygen, and 76 100 kg of it should be hydrogen, to attain the same 1:2 mole ratio. Calculations are shown below:

$$\frac{1\ mol\ O_2}{2\ mol\ H_2} \times \frac{31.999\ g/mol\ O_2}{2.016\ g/mol\ H_2} = \frac{7.936\ g\ O_2}{1.000\ g\ H_2},$$
equivalent to 11.19% H_2, 88.81% O_2

$$680\ 000\ kg\ total \times \frac{11.19\%\ H_2}{100\%\ total} = 76\ 100\ kg\ H_2$$

$$680\ 000\ kg\ total \times \frac{88.81\%\ O_2}{100\%\ total} = 603\ 900\ kg\ O_2$$

5. Students' answers will vary. A sample calculation is shown below:

$$3.5\ mL\ bulb\ capacity \times \frac{1\ L}{1000\ mL} \times \frac{2\ parts\ O_2}{6\ parts\ total}$$
$$\times \frac{1.429\ g\ O_2}{1\ L\ O_2} = 1.7 \times 10^{-3}\ g\ O_2$$

603 900 kg O_2 in space shuttle = 6.039×10^8 g O_2;

$$\frac{6.039 \times 10^8\ g\ O_{2\,shuttle}}{1.7 \times 10^{-3}\ g\ O_{2\,bulb}} = 3.6 \times 10^{11}\ bulbs\ per\ shuttle$$

$$3.5\ mL\ bulb\ capacity \times \frac{1\ L}{1000\ mL} \times \frac{4\ parts\ H_2}{6\ parts\ total}$$
$$\times \frac{0.0899\ g\ H_2}{1\ L\ H_2} = 2.1 \times 10^{-4}\ g\ H_2$$

76 100 kg H_2 in space shuttle = 7.61×10^7 g H_2;

$$\frac{7.61 \times 10^7\ g\ H_{2\,shuttle}}{2.1 \times 10^{-4}\ g\ H_{2\,bulb}} = 3.6 \times 10^{11}\ bulbs\ per\ shuttle$$

In other words, the fuel in the shuttle is roughly equivalent to 360 billion of the little rockets the students were using. (Alert students will notice that the densities are given for 0°C. If an additional challenge is needed, introduce students to the volume-temperature relationship, and have them recalculate the density at room temperature, and recalculate their answers.)

6. Students' answers will vary, depending upon the resources available to them. Piezoelectricity is a special property shared by certain crystals. When they are under mechanical stress, their opposite surfaces become charged. If they are connected to a circuit, this separation of charges can cause current to flow through the surface.

When the trigger is pushed on the piezoelectric sparker, the crystal is stressed, and current flows along the outside of the "barrel" of the sparker. Ordinarily, it leaps across to the metal butane nozzle, which ignites the butane, causing a small flame. However, in this investigation, the ends of the two strands of speaker wire are much closer together than the nozzle and the barrel, so the spark will leap across at this point (the path of least resistance). If a bulb of hydrogen and oxygen is placed over this end of the speaker wire, the spark will ignite the hydrogen, causing a small explosion.

The piezoelectric effect is used in phonograph needles, and "crystal" headphones and microphones.

INVESTIGATION 10: "Wet" Dry Ice

Pages 854–855

Approximate time: 30 minutes. Allow more time for the additional activities suggested in the **Extensions** section.

Although the equipment list is given for students working in pairs, students may work in larger groups, or individually, if this is preferred.

This investigation may also be useful for familiarizing students with carbon dioxide and the behavior of gases.

MATERIALS

(for a class of thirty students working in pairs, includes amounts for Extensions)

65 g of dry ice, broken into small rice-grain-sized pieces
  Dry ice can often be ordered from ice cream stores or the businesses that make and sell regular ice cubes. Keep the dry ice wrapped in several layers of newspaper or other wrappings to help insulate it.

Dry ice can cause frostbite, so be sure to wear gloves if you handle it.

When chipping it into pieces with an ice pick, be sure to wear some sort of face shield, to protect your eyes.

Do not stick your head in a cooler or other container full of dry ice. There will not be enough oxygen for you to breathe.

15 transparent plastic cups
15 graduated pipets
tap water
15 pairs of scissors (fewer can be used if students share them)
15 sets of pliers (or C-clamps: if these aren't available in your school's science department, you may want to encourage some of your students to bring pliers or clamps from home. Although some students may want to use their fingers to hold the pipet shut, it may be difficult to achieve enough of a seal to build up the required pressure.)
15 sets of forceps

SAFETY NOTES

If students do not loosen the pliers as instructed, the pipet will burst, and the water will splash, possibly startling others.

DISPOSAL

Any remaining dry ice that does not need to be saved for later can be left to sublime. Be sure the room is well-ventilated, so that the CO_2 gas does not build up to intolerable levels. Water may be poured down the drains.

ANALYSIS & INTERPRETATION — ANSWERS

1. Student answers may vary. At first, the dry ice sublimes, but then it should gradually begin melting.

2. The sample was used up because every time the pliers were loosened, some CO_2 escaped as gas.

3. Students' answers will vary, but may include one or more of the following: The water in the cup provided the heat that the dry ice needed to absorb to sublime and melt. The water also re-absorbed the heat when the dry ice re-froze, keeping the system's temperature from fluctuating too much. This was necessary because plastic materials like the pipet bulb tend to get brittle when cold. If the experiment were performed outside of the water, condensation would form on the outside of the pipet bulb, obscuring the view. Since the glass is curved, the water slightly magnified the view of the dry ice in the pipet bulb. Finally, if the bulb had burst due to the pressure, the water would have caught the dry ice and the pieces of the bulb, muffling the little explosion.

4. The melting of dry ice occurs only at higher pressures, as one would expect from its phase diagram. As discussed in Section 6.5, carbon dioxide molecules are not polar, and do not form hydrogen bonds like water can. The intermolecular forces between CO_2 molecules are much weaker than those between water, which is polar, and can form hydrogen bonds. As a result, at any given pressure, the boiling point of CO_2 will be much lower than that of water.

CONCLUSIONS — ANSWERS

1. Students' answers will vary. All answers should show the dry ice starting at 1 atmosphere and $-78°C$, the sublimation point for dry ice. As the dry ice sublimed in the sealed bulb, the pressure and temperature increased. Students' charts should show a diagonal line across the phase diagram, both rightward and upward, that reaches the liquid sector of the diagram. As the bulb was unsealed, the pressure and temperature dropped back diagonally in the opposite direction. The overall pattern should be one of repeated peaks and valleys, with the peaks getting lower (due to less pressure), as time goes by.

EXTENSIONS — ANSWERS

1. If fewer pieces of dry ice were used, it would take longer to reach a pressure high enough for the dry ice to melt, if at all.

2. If more dry ice were used, the conditions for melting would be achieved very quickly, but the pressure would also build up so quickly that the bulb might explode or stretch. A large amount could absorb enough heat from the water as it sublimed and melted to turn the water into ice.

3. The pressure will continue to build as the dry ice melts and vaporizes until the bulb stretches out or blows apart, releasing the CO_2 gas.

INVESTIGATION TEACHER'S GUIDE

INVESTIGATION 11: Measuring the Triple Point Pressure of CO_2

Pages 855–858

Approximate time: 45 minutes. This time may be shortened by having the micro pressure gauges prepared beforehand. Allow more time for the additional activities suggested in the **Extensions** section. You may want to have the gauges prepared beforehand anyway, if students may not have the dexterity and coordination needed to prepare the gauges. Once prepared, the micro pressure gauges should be re-usable for other lab sections, and from year to year. However, a new drop of colored water will probably be necessary each time. This investigation may also be useful for helping students understand the relationship between pressure and temperature of gases.

Although the equipment list is given for students working in pairs, students may work in larger groups, or individually, if this is preferred.

MATERIALS

(for a class of thirty students working in pairs, includes amounts for Extensions)

65 g of dry ice, broken into small pieces
Dry ice can often be ordered from ice cream stores or businesses that make and sell regular ice cubes. Keep the dry ice wrapped in several layers of newspaper or other wrappings to help insulate it.

 Dry ice can cause frostbite, so be sure to wear heat and cold resistant gloves if you handle it.

 When chipping it into pieces with an ice pick, be sure to also wear a face shield, to prevent small pieces from harming your eyes.

 Do not stick your head in a cooler or other container full of dry ice. There will not be enough oxygen for you to breathe.

- 15 transparent plastic cups
- 15 graduated pipets
- 15 7-cm tubes (cut from a thin-stemmed pipet)
- 15 pieces of thread, each 15–20 cm long
- several fine-tipped dropper bulbs (Available commercially, or may be made out of thin-stemmed pipets. Stretch the pipets by hand and then cut them off. One of these will do if students share. These are re-usable)
- tap water
- 15 pairs of scissors (fewer will do if students share)
- 15 sets of pliers
- 15 sets of forceps
- 15 metric rulers (fewer will do if students share)
- several hot glue guns (one will do if students share)

 The tip of the glue gun is hot, so be careful!

- several fine-tipped markers (one will do if students share)
- dark colored water (one beaker with about 15 mL of water and several drops of food coloring will do if students share)

SAFETY NOTES

If students do not loosen the pliers as instructed, the pipet will burst, and the water will splash, possibly startling others.

DISPOSAL

Any remaining dry ice that does not need to be saved for later can be allowed to sublime. Be sure the room is well-ventilated, so that the CO_2 gas does not build up to intolerable levels. The water may be poured down the drains.

ANALYSIS & INTERPRETATION — ANSWERS

1. Before the dry ice melted, the pressure inside the bulb steadily increased as the dry ice sublimed. During the melting, pressure remained the same. After melting, when the pliers were loosened, the pressure dropped.

2. Students' graphs may vary, but should show a pattern of pressure rising, levelling off, dropping, and then repeating.

3. Students' answers will vary. If their work was fairly accurate, the measurement of the length of the air column when the dry ice began melting should be about $\frac{1}{5}$ the initial measurement. The initial pressure inside the tube was about equal to 1 atm, the pressure inside the room.

CONCLUSIONS — ANSWERS

1. Students' answers will vary. The accepted value for the triple-point pressure of CO_2 is 5.11 atm. Sample calculations follow:

$$P_1 \times V_1 = P_2 \times V_2; \quad P_2 = \frac{P_1 \times V_1}{V_2}$$

for micro pressure gauges, length is proportional to volume:

$$P_2 = \frac{P_1 \times \text{length}_1}{\text{length}_2}$$

$P_1 = 1$ atm; $\text{length}_1 = 6.2$ cm; $\text{length}_2 = 1.2$ cm

$$P_2 = \frac{1 \text{ atm} \times 6.2 \text{ cm}}{1.2 \text{ cm}} = 5.2 \text{ atm}$$

2. The pressure found experimentally should be the triple point pressure of CO_2. As the phase diagram shows, as the temperature increases, there is a minimum pressure at which CO_2 can form a liquid, and this minimum corresponds to the triple point. If the observation were made exactly at the point where melting began, it should be at the triple point.

3. Students' answers will vary.

EXTENSIONS — ANSWERS

1. The assumption is that the width and shape of the trapped air column are constant, so that volume is directly proportional to length.

2. Students' answers will vary but may include: Difficulty in precisely measuring the length of the air column, and calculation errors added by assuming pressure to be equal to exactly 1 atm. Suggested changes may include adding smaller calibration marks to the micro pressure gauge and reading the barometric pressure in the lab.

3. Students' answers will vary, but they should be based on the ideal gas law relationship:

$$n = \frac{P \cdot V}{R \cdot T}, \quad m = n \cdot \text{(molar mass)}, \text{ and } d = \frac{m}{V}$$

Sample Calculation:

$V = 4.1$ mL (measured by filling the spent pipet with water, then emptying the water into a graduated cylinder)

$$V = 4.1 \text{ mL} \times \frac{1 \text{ L}}{10^3 \text{ mL}} = 4.1 \times 10^{-3} \text{L}$$

$P = 5.11$ atm
$T = -56°C = 217$K
$R = 0.0821$ L · atm/mol · K

$$n = \frac{5.11 \text{ atm} \cdot 4.1 \times 10^{-3} \text{ L}}{217\text{K} \cdot 0.0821 \text{ L} \cdot \text{atm/mol} \cdot \text{K}} = 1.2 \times 10^{-3} \text{ mol}$$

$$m = 1.2 \times 10^{-3} \text{ mol} \cdot 44 \text{ g/mol} = 5.3 \times 10^{-2} \text{ g}$$

$$d = \frac{5.3 \times 10^{-2} \text{ g}}{4.1 \times 10^{-3} \text{ L}} = 13 \text{ g/L at } 5.11 \text{ atm and } -56°C$$

INVESTIGATION 12: Purifying Water

Pages 858–860

This investigation can easily be completed in a 30-35 minute period. If students also do clean-up and disposal as described in the Disposal section that follows, it may take an additional 15 minutes. You may want students to repeat the purification steps several times to practice the techniques, if time allows.

Although the equipment list is given for students working in pairs, students may work in larger groups, or individually, if this is preferred.

This investigation can also be used to explore the classification of matter as either a mixture or a pure substance.

MATERIALS

(for a class of 30 students working in pairs, includes amounts for Extensions)

75 mL foul water
  Use the following recipe: 60 mL tap water, 0.25 tsp sifted coffee grounds, 0.25 tsp sifted garlic powder, and 15 mL mineral oil. Mix the first three ingredients together in a 125 mL Erlenmeyer flask. Add the mineral oil, stopper, and shake. To make a convenient dispenser, slide a graduated beral pipet through a 1-holed rubber stopper that fits the flask mouth. Stopper the bulb, and shake vigorously. Squeeze the bulb and release, to draw up a sample the instant you stop shaking, and then dispense the sample. This can help provide a consistent mixture of ingredients and deliver a reproducible 3.0 mL of foul water as shown in Figure T12-A.

15 incremented wood splints
  To increment the wood splints, simply add 1.0 mL of water to one of the test tubes, place it upright on the table, and place the wooden splint alongside it. Draw a line level with the bottom of the meniscus. Add another milliliter, and draw another increment. Repeat for the entire length of the test tube. Use this wood splint as a template for the rest of the splints.

60 test tubes (recommended size: 13 × 100 or 10 × 80)

15 thin-stemmed pipets

15 separatory bulbs
  The separatory bulbs may be made by cutting off all but 4-5 mm of the stem from a graduated or micro-tip pipet, as shown in Figure T12-B.

15 long-stemmed micro funnels
  The micro-funnels may be made by cutting the top section off of the bulb of a jumbo-sized pipet, as shown in Figure T12-C.

15 test tube racks

15 lengths of wire (10-12 GA, 12 cm long — fewer will do if students share them)

15 small strainers (fine mesh — fewer will do if students share them)

15 plastic trays or petri dish halves (fewer will do if students share them)

15 plastic spoons (fewer will do if students share them)

3 small sheets of tissue paper (bathroom or facial tissue)

100 g sand (rinsed and dried — can be re-used. See Disposal section below.)

50 g activated charcoal (aquarium grade, crushed by mortar and pestle to a small granular consistency — can be re-used. See Disposal section below.)

flashlight or slide projector and screen (one will do for the class, if students take turns in testing their mixtures for the Tyndall effect to determine if it is a colloid or a solution.

2 wash basins (recovery bins for sand and charcoal — see Disposal section below.)

INVESTIGATION TEACHER'S GUIDE

Figure T12–A: Reproducible and consistent samples of foul water are delivered using a set up with a flask as shown (a). Either graduated or thin-stemmed pipets can be made into separatory bulbs (b). The micro funnels are made by trimming jumbo-sized pipets (c).

SAFETY NOTES

 Students are cautioned to wear goggles to protect their eyes from unexpected accidents, but none of the materials used are particularly hazardous.

DISPOSAL

The sand and charcoal can be recovered in wash basins. Fill the basins half way with water. To recover the sand, have students hold the sand filter funnel upside down over the basin, and squeeze the tip and roll it between their fingers to loosen the sand. A micro-tip pipet can be used to squirt water down through the opening to flush the sand into the basin. The sand may be washed, dried, and used again.

Recover the charcoal in a separate basin. Again, pinch and roll the tip of the pipet to loosen the charcoal, and then use the wire to push the tissue wad and the charcoal into the basin. Discard the tissue. Wash the charcoal, and dry it for several hours in the oven (this also helps re-activate it).

The test tubes can be washed and scrubbed well with soap and water. The oily residue in the first test tube should be absorbed onto paper towels. The towels may be placed in the trash can. Remember that oily paper towels can create a fire hazard, so be sure they are disposed of promptly.

ANALYSIS & INTERPRETATION — ANSWERS

1. The assumption being made is that they have the same thickness in the bottom and the same cross-sectional area, so that the length along the splint is directly proportional to the volume of liquid.

2. As discussed in Section 14.2, oil and water are immiscible because water is a polar substance and oil is nonpolar.

3. The coarse sand at the bottom of the funnel holds the fine sand in the funnel. The fine sand in the middle filters out small particles. The coarse sand in the top filters out the larger particles.

4. Students' answers will vary. Charcoal filtration was the slowest step, and with a variety of very large particles, it could take even longer. If the mixture was very impure, the charcoal may become so full of impurities that it is not able to adsorb all of them.

5. If new test tubes weren't used, the purified water would be put back into test tubes which could contain residues from the impure water, and could contaminate the sample.

CONCLUSIONS — ANSWERS

1. Students should report the correct arrangement of densities as: oil, water, sediment particles. In the separation step, the denser substances formed layers near the bottom of the bulb. In this case, the sediment particles were on the very bottom, with a layer of water (and impurities) on top. The oily material formed a layer on the top, indicating that it is the least dense of the substances.

2. The particles making the water cloudy should be larger than the particles making the water colored. Particles that color the water are likely to be in solution, and by definition are very small, but particles that cause cloudiness are likely to be parts of colloids or suspensions, and will be larger than those in solution. The cloudiness was eliminated in a step of the purification before than the color was eliminated.

3. Student' answers for the percentage will vary, but should be far less than 80% if the foul-water recipe was followed. The following is a sample calculation.

$$\frac{2.1 \text{ mL H}_2\text{O recovered}}{2.7 \text{ mL in foul water sample)}} \times 100 = 78\%$$

4. Students' answers will vary, but students should note that there is a trade-off between the purity of the water and the percentage recovered. The more pure the water, the less will likely be recovered, and vice versa.

Investigation Teacher's Guide

EXTENSIONS — ANSWERS

1. Students' answers will vary, but may include suggestions to perform the filtration steps twice in a row, or else to try to get additional pure water out of the discarded components, by repeating the experiment with the discarded materials.

2. Students' answers will vary.

3. Students' answers will vary, but students should note that if one of the liquids is boiled and becomes a gas, it will be considerably less dense than the other liquid. Then, if the gas and liquid components are separated, the purity of each should improve. Some students may also suggest that by cooling the separated gas, it could be condensed and returned to its original liquid state.

INVESTIGATION 13: Testing Water

Pages 860–864

Approximate time: 45 minutes. More time will be necessary for the extensions.

This investigation may also be useful in discussing solubilities, double-replacement reactions, and equilibria.

Although the equipment list is given for students working in pairs, students may work in larger groups, or individually, if this is preferred.

MATERIALS

(for a class of thirty students working in pairs, includes amounts for Extensions)

- 15 24-well microplate lids (usually available with microplates. A suitable substitute is a petri dish with 24 circles marked on its underside with marker. Other flat substitutes may be used, but students must be able to see the drops of water from the side to properly gauge their test results.)
- 150 labeled fine-tipped dropper bulbs, in 15 sets of 10 each with solutions as follows: (fewer bulbs will do if students share):
  - 15 mL solution 1 (reference): 0.1 g $CaSO_4$ (plaster of paris) and 2.0 g of $FeCl_3 \cdot 6H_2O$ in 15 mL of solution. **Note: this will make a saturated calcium sulfate solution. You may have some undissolved $CaSO_4$ in the bottom of the solution. Filter this out before giving it to the students.**
  - 15 mL solution 2: distilled water
  - 15 mL solution 3: tap water
  - 15 mL solution 4: bottled spring water
  - 15 mL solution 5: local river or lake water
  - 15 mL solution 6: solution X: Which ions and what amounts to use are left up to the teacher's discretion. Remember that calcium together with sulfate is only slightly soluble, so if you include both in your "solution X," you will need to filter it before giving it to the students. Be sure to keep track of which ions are added, so you can test the students' lab techniques.
  - 25 mL solution A: 0.5 M NaSCN solution (1.01 g NaSCN in 25 mL of solution)
  - 25 mL solution B: 0.5 M $Na_2C_2O_4$ solution (1.68 g $Na_2C_2O_4$ in 25 mL of solution)
  - 25 mL solution C: 0.1 M $AgNO_3$ solution (0.43 g $AgNO_3$ in 25 mL of solution)
  - 25 mL solution D: 0.5 M $Ba(NO_3)_2$ solution (3.27 g $Ba(NO_3)_2$ in 25 mL of solution)
- 15 sheets of white paper
- overhead projector (optional)
- paper towels
- 100 fine-tipped dropper bulbs for acquiring additional samples *for Extension 1*
- freezer *may be requested for Extension 5*
- 15 alcohol-filled thermometers *may be requested for Extension 5*
- 15 conductivity testers (see instructions for building probes in Equipment Notes immediately following equipment and chemical lists) *may be requested for Extension 5*
- 60 pieces of filter paper *may be requested for Extension 5*
- 15 funnels *may be requested for Extension 5*

Note: Students may also ask for other items not included here for Extension 5.

SAFETY NOTES

Since you may be handling chemicals in concentrated forms during preparation, it is important for you to follow the precautions listed in the student's instructions, including wearing safety goggles, gloves, and a lab apron during preparation.

DISPOSAL

All solutions can be washed down the sink with plenty of excess water, provided your school drains are connected to a sanitary sewer system with a treatment plant. The solid precipitates should be dried and buried in a landfill that is approved for chemical disposal. The microplate lids can be wiped with a clean paper towel, rinsed with tap water, rinsed with distilled water, and wiped once more with a clean paper towel.

ANALYSIS & INTERPRETATION — ANSWERS

1. Test A: $Fe^{3+}(aq) + NaSCN(aq) \rightarrow Na^+(aq) + Fe(SCN)^{2+}(aq)$, red color

INVESTIGATION TEACHER'S GUIDE

Test B: $Ca^{2+}(aq) + Na_2C_2O_2O_4(aq) \rightarrow 2Na^+(aq) + CaC_2O_4(s)$, precipitate
Test C: $AgNO_3(aq) + Cl^-(aq) \rightarrow AgCl(s) + NO_3^-(aq)$, precipitate
Test D: $SO_4^{2-}(aq) + Ba(NO_3)_2(aq) \rightarrow BaSO_4(s) + 2NO_3^-(aq)$, precipitate

2. Test A: $Fe^{3+}(aq) + SCN^-(aq) \rightarrow Fe(SCN)^{2+}(aq)$
Test B: $Ca^{2+}(aq) + C_2O_4^{2-}(aq) \rightarrow CaC_2O_4(s)$
Test C: $Ag^+(aq) + Cl^-(aq) \rightarrow AgCl(s)$
Test D: $SO_4^{2-}(aq) + Ba^{2+}(aq) \rightarrow BaSO_4(s)$

3. If any of the sample solutions were accidentally mixed, either by dipping the dropper in one and then another, or by being spilled, you could not be sure whether positive tests were due to one solution, the other, or both.

4. A control was necessary to show what the testing solutions look like when there is nothing for them to react with. The distilled water was chosen as a control, because it should have had none of the ions tested present in it.

CONCLUSIONS — ANSWERS

1. Students' answers will vary, depending upon the sources of the water samples. All students should indicate that the reference solution had all the ions present, and the distilled water had none of the ions present.

2. Students answers may vary, depending upon the sources of the water samples. The solution that showed the strongest test for calcium ions is likely to be "harder" than the other samples, since the presence of calcium ions is one cause of hardness in water. Some students may suggest that unless tests are also performed for magnesium ions, it will be difficult to be certain which is the hardest. (It may be interesting to test the students' predictions by taking the water samples, adding precise amounts of soap, and shaking them up inside a stoppered test tube.)

EXTENSIONS — ANSWERS

1. Students' answers will vary, although students may guess that salt water from an ocean and well water are likely to contain many ions. Rainwater and running water will probably contain fewer ions.

2. Most sodium compounds and nitrate compounds are soluble. The two test solutions for the cations (Fe^{3+} and Ca^{2+}) contained different sodium salts, but each of the salts was soluble. However, the calcium salt of oxalate precipitated, and was not soluble. Similarly, the two test solutions for the anions (Cl^- and SO_4^{2-}) contained nitrate salts, but each of these salts was soluble. However, other compounds of silver and barium, such as silver chloride and barium sulfate, precipitated out, and were not soluble.

3. Students' answers will vary, but students should recognize that by definition, the reference solution contained chloride ions, sulfate ions, iron(III) ions, and calcium(II) ions. Any precipitate left after evaporation would be a combination of all of these substances (and possibly others as well).

4. Students' answers will vary, but students should recognize that the sulfate ions discussed cannot exist alone. There must be an equal amount of positive charge provided by some unidentified cations in this solution. Thus, the white powder left after evaporation is probably an ionic compound consisting of the cation(s) and the sulfate ions.

5. Students' answers will vary. Some students will suggest getting larger samples, and precipitating out each of the ions in question, one at a time, filtering the precipitates, measuring their mass, and calculating the amounts of each ion present.

Others may suggest trying to use colligative properties such as conductivity, or freezing-point depression. Conductivity is probably best tested by comparing the solution in question to several standard dilutions of known concentrations of the substance in question. The best way to test freezing point depression is probably to freeze the solution in question, and then measure its **melting point.**

Students may note that the difficulty inherent with such attempts to solve this problem is that colligative properties only measure the concentration of dissolved particles in a solution (not necessarily only the concentration of specific particles such as chloride ions in a solution).

INVESTIGATION 14: Is It an Acid or a Base?

Pages 864–866

Approximate time: 45 minutes, assuming the students arrive in the lab with a plan that will work. This lab is an open-ended one, designed to encourage students to think actively in the laboratory, so the time required will vary from student to student. More time may be useful to some students, and more time will be necessary for the extensions.

You may even dedicate one lab period to planning and another to carrying out the plans. Point out that professional scientists often spend more of their time planning ex-periments than actually performing them. If students are able to create their own plan of attack, they will feel a real sense of accomplishment when the problem has been solved. Solid planning will also help prevent waste of time and chemicals.

If time constraints are a problem be sure that students write out a plan the night before the lab. You may want to consider giving more and more specific clues about the techniques that can be used every 15 minutes or so to help those

Investigation Teacher's Guide **826-28**

INVESTIGATION TEACHER'S GUIDE

who are having trouble. Encourage brainstorming and collective work, even among different lab groups.

The level of difficulty of this investigation can be controlled by whether or not you allow students to use the optional materials. Although the equipment list is given for students working in pairs, students may work in larger groups, or individually, if this is preferred. You need not have all of the materials listed below available for the students, and you may want to be ready for students to ask for other materials. It may be best to keep the additional materials ready, but out of sight, so that students have to think about what they want, rather than simply guessing by looking at what's available.

MATERIALS
(for a class of thirty students working in pairs, includes amounts for Extensions)

15 sets of eight unknown solutions stored in thin-stemmed pipets (total pipets: 120):
a convenient container for storing each set is an empty tape cassette box, which should hold a set of eight bulbs kept upside down, if parts of the long stems are cut off. (See Figure 14-A in the student's edition of the investigation.) Label the bulbs with a fine-tipped permanent marker. Keep track of which concentrations correspond to which numbered solutions. You may want to mix up solutions, so that solution #1 is not the same for all students, etc. Remember to add about 1 mL of phenolphthalein solution to the acid solutions before preparing individual sets of unknown solutions.
- 150 mL of 0.8 M HCl (10 mL of concentrated HCl in 150 mL of solution: 75 mL for unknowns, 75 used for next solution)
- 150 mL of 0.4 M HCl (75 mL of 0.8 M HCl (see previous step) in 150 mL of solution: 75 mL for unknowns, 75 used for next solution)
- 150 mL of 0.2 M HCl (75 mL of 0.4 M HCl (see previous step) in 150 mL of solution: 75 mL for unknowns, 75 used for next solution)
- 150 mL of 0.1 M HCl (75 mL of 0.2 M HCl (see previous step) in 150 mL of solution: 75 mL for unknowns, 75 kept aside as possible standard solution for students.
- 150 mL of 0.8 M NaOH (16 g NaOH in 150 mL of solution: 75 mL for unknowns, 75 used for next solution)
- 150 mL of 0.4 M NaOH (75 mL of 0.8 M NaOH (see previous step) in 150 mL of solution: 75 mL for unknowns, 75 used for next solution)
- 150 mL of 0.2 M NaOH (75 mL of 0.4 M NaOH (see previous step) in 150 mL of solution: 75 mL for unknowns, 75 used for next solution)
- 150 mL of 0.1 M HaOH (75 mL of 0.2 M NaOH (see previous step) in 150 mL of solution: 75 mL for unknowns, 75 kept aside as possible standard solution for students.

15 mL phenolphthalein solution (dissolve 0.15 g phenolphthalein in 7 mL denatured alcohol and add 8 mL water)
15 24-well microplates (or small test tubes—at least 210)
150 toothpicks to mix solutions

OPTIONAL MATERIALS
(Provide only as students ask for them and at your discretion)

Note: Students may also ask for other items not included here.

pH or litmus paper
phenolphthalein solution (left over from preparing acid solutions)
methyl orange indicator solution (dissolve 0.1 g in 50 mL denatured alcohol and add 50 mL of water)
30 spare thin-stemmed pipets
100 g $CaCO_3$
100 g $FeCl_3 \cdot 6H_2O$
100 g of iron filings
75 mL standard 0.1 M NaOH (made during unknown preparation)
75 mL standard 0.1 M HCl (made during unknown preparation)
distilled water
5 conductivity testers (see instructions for building testers in Equipment Notes immediately following equipment and chemical lists)

SAFETY NOTES

Since you may be handling acids and bases in concentrated forms during preparation, it is important for you to follow the precautions listed in the student's instructions, including wearing safety goggles, gloves, and a lab apron during preparation.

DISPOSAL

Na_2CO_3 solution (1.0 M) can be added to neutralize the acid solution, and vinegar or lemon juice can be added to the basic solution to neutralize it. Once neutralized, any solutions may be washed down the drain with plenty of excess water. Any precipitates formed can be dried, packed in appropriate containers, and disposed of in an approved landfill.

ANALYSIS & INTERPRETATION—ANSWERS

1. Students' answers will vary. Some properties of aqueous acids are that they react with active metals to liberate H_2 gas, that they change blue litmus red, change red (pink) phenolphthalein to colorless, and change methyl orange to red. They react with carbonates to liberate carbon dioxide. They can be neutralized by reacting them with bases. Some properties of aqueous bases are that they are neutralized by acids, that they change red litmus blue, change phenolphthalein red (pink), and change methyl orange yellow. Hydroxide-forming bases form insoluble hydroxides with most transition metals. Aqueous acids and aqueous bases conduct electricity.

2. Students' answers may vary, both in terms of content and detail. Some possibilities are listed below:
HCl(aq) + NaOH(aq) → NaCl(aq) + $H_2O(\ell)$
HCl(aq) + blue litmus → HCl(aq) + red litmus

INVESTIGATION TEACHER'S GUIDE

$2HCl(aq) + CaCO_3(aq) \rightarrow CaCl_2(aq) + H_2O(\ell) + CO_2(g)$

$6HCl(aq) + 2Fe(s) \rightarrow 2FeCl_3(aq) + 3H_2(g)$

$NaOH(aq)$ + clear phenolphthalein $\rightarrow NaOH(aq)$ + red (pink) phenolphthalein

$NaOH(aq)$ + red litmus $\rightarrow NaOH(aq)$ + blue litmus

$3NaOH(aq) + FeCl_3(aq) \rightarrow 3NaCl(aq) + Fe(OH)_3(s)$

CONCLUSIONS — ANSWERS

1. Students' answers will vary, depending upon the teacher's assignment of the numbers to the different unknown solutions.

2. Students' answers may vary. Students may have tried to measure the volumes of the standards that were needed to neutralize the solution in order to determine concentration, or they may have gauged concentration by comparing reaction rates.

EXTENSIONS — ANSWERS

1. Students' answers will vary, but should demonstrate an understanding of what the different properties of acids and bases are, and how they were manifested in this investigation.

2. One possible method of figuring out the solutions' identities with the smallest amount of materials would be to take several 8-drop samples of each of the solutions, and count how many (if any) drops of each of the other solutions are required to turn the phenolphthalein in the sample red, and how many (if any) are required to neutralize it again. The strongest base will be the solution that is able to turn the greatest number of solutions red. The strongest acid will be the solution that requires the most drops of the strongest base to neutralize. The other solutions should be easy to identify after the strongest of each type are determined. (Note: this could be a bit tricky on a 24-well microplate. The results of each test would have to be carefully recorded, since the plate would have to be cleaned off several times to make room for all tests (probably 56 would be necessary).

3. Students' answers will vary, but should involve some characteristic property, such as the use of acid-base indicators or paper, or a reaction, like that of acids and $CaCO_3$ or that of bases and $FeCl_3$. To clean up such a spill, students should recommend neutralizing the spill with a weak base if it is an acid spill or with a weak acid if it is a base spill. Students should note that the product will then be a solution of neutral salts, which can easily be mopped up and disposed of.

INVESTIGATION 15: How Much Calcium Carbonate is in an Eggshell?

Pages 866–869

Approximate time: 60 minutes, assuming the students are comfortable with the process and the calculations of a titration. If time is a problem, it may help to have the eggshells prepared earlier.

Although the equipment list is given for students working in pairs, students may work in larger groups, or individually, if this is preferred.

MATERIALS

(for a class of thirty students working in pairs, includes amounts for Extensions)

15 eggshells
15 balances (fewer will do if students share)
15 150-mL beakers
drying oven set at 110°C
15 dessicators (optional — fewer will do if students share)
15–20 pieces of weighing paper
15 mortars and pestles (fewer will do if students take turns grinding up eggshells)
15 50-mL micro solution bottles or small Erlenmeyer flasks
30 thin-stemmed pipets (once labeled as "acid" or "base", these can be re-used from class to class. For more reproducible drop sizes, place the pipet bulb in a tubing screw-clamp. Each turn of the screw will force out a reproducible drop.)
15 10-mL graduated cylinders (fewer will do if students take turns calibrating pipets)
13 mL phenolphthalein solution (dissolve 0.13 g phenolphthalein in 6 mL denatured alcohol and add 7 mL water)—
3–5 dropper bulbs for students to use to add phenolphthalein to their eggshell mixture
75 mL 2.00 M HCl (12.9 mL conc. HCl in 75 mL of solution)
150 mL 1.00 M NaOH (6.0 g NaOH in 150 mL of solution)
distilled water
15 shells (clam, oyster, snail, etc.) *for Extension #4*
15 pieces of classroom chalk *for Extension #6*

SAFETY NOTES

 Since you may be handling acids and bases in concentrated forms during preparation, it is important for you to follow the precautions listed in the student's instructions, including wearing safety goggles, gloves, and a lab apron during preparation.

DISPOSAL

Na_2CO_3 solution (1.0 M) can be added to neutralize the acid solution, and vinegar or lemon juice can be added to the basic solution to neutralize it. Once neutralized, the solutions may be washed down the drain with plenty of excess water.

INVESTIGATION TEACHER'S GUIDE

ANALYSIS & INTERPRETATION — ANSWERS

Sample data — graduated cylinder readings (pipet calibration: steps 3–5)

Students' answers will vary

Trial	initial— acid pipet	final— acid pipet	initial— base pipet	final— base pipet
1	5.00 mL	5.79 mL	5.00 mL	5.81 mL
2	5.79 mL	6.41 mL	5.81 mL	6.65 mL
3	6.41 mL	7.40 mL	6.65 mL	7.46 mL

Total vol. of drops — acid pipet: 2.40 mL
Average volume of each drop: 0.040 mL
Total vol. of drops — base pipet: 2.46 mL
Average volume of each drop: 0.041 mL

Sample data — titration (steps 7–12)

Students' answers will vary

Mass of entire eggshell	5.27 g
Mass of ground eggshell sample	0.11 g
Number of drops of 2.00 M HCl added	100 drops
Vol. of 2.00 M HCl added	4.00 mL
Number of drops of 1.00 M NaOH added	151 drops
Vol. of 1.00 M NaOH added	6.19 mL
Vol. of 2.00 M HCl reacting with NaOH	3.10 mL
Vol. of 2.00 M HCl reacting with eggshell	0.90 mL
# of moles HCl reacted with eggshell	1.8×10^{-3} mol
# of moles of $CaCO_3$ reacting with HCl	9.0×10^{-4} mol
Mass of $CaCO_3$ in eggshell sample	0.090 g
% of $CaCO_3$ in eggshell sample	82%

1. $CaCO_3 + 2HCl(aq) \rightarrow CO_2(g) + H_2O(\ell) + CaCl_2(aq)$

2. $HCl(aq) + NaOH(aq) \rightarrow NaCl(aq) + H_2O(\ell)$

3. Students' answers will vary. The following are sample calculations:

acid pipet calibration: 60 drops is equivalent to 2.40 mL;
$$\frac{2.40 \text{ mL}}{60 \text{ drops}} = 0.040 \text{ mL/drop}$$

base pipet calibration: 60 drops is equivalent to 2.46 mL;
$$\frac{2.46 \text{ mL}}{60 \text{ drops}} = 0.041 \text{ mL/drop}$$

100 drops of acid used $\times$ 0.040 mL/drop = 4.00 mL of acid used
151 drops of base used $\times$ 0.041 mL/drop = 6.19 mL of base used

4. In this neutralization reaction a mole of acid and a mole of base are equivalent. Therefore, the 2.00 M HCl is 2.00 N and the 1.00 M NaOH is 1.00 N.

5. Students' answers will vary. The following is a sample calculation: From Analysis & Interpretation #2, there is a one-to-one mole ratio between acid and base in the neutralization reaction.

$$M_{acid} \times V_{acid} = M_{base} \times V_{base}$$
$$(2.00 \text{ M})(X) = (1.00 \text{ M})(6.19 \text{ mL})$$
$$X = 3.10 \text{ mL HCl}$$

6. Students' answers will vary. The following is a sample calculation:

4.00 mL − 3.10 mL = 0.90 mL HCl reacted with calcium carbonate.

$$V_{acid} \times M_{acid} = \text{moles acid}$$

$$0.90 \text{ mL} \times \frac{1 \text{ L}}{1000 \text{ mL}} \times \frac{2.00 \text{ mol}}{1 \text{ L}} = 1.8 \times 10^{-3} \text{ mol HCl}$$

CONCLUSIONS — ANSWERS

1. Students' answers will vary, but should include the 2 mol HCl/1 mol $CaCO_3$ mole ratio. The following is a sample calculation:

$$1.8 \times 10^{-3} \text{ mol HCl} \times \frac{1 \text{ mol } CaCO_3}{2 \text{ mol HCl}} = 9.0 \times 10^{-4} \text{ mol } CaCO_3$$

2. The molar mass of $CaCO_3$ is 100.1 g/mol. Students' answers for the mass of calcium carbonate present will vary. The following is a sample calculation:

$$9.0 \times 10^{-4} \text{ mol } CaCO_3 \times \frac{100.1 \text{ g}}{\text{mol } CaCO_3} = 0.09 \text{ g } CaCO_3$$

3. Students' answers for the percentage will vary, but should be somewhat near the 95–97% range. The following is a sample calculation.

$$\frac{0.09 \text{ g } CaCO_3}{0.11 \text{ g eggshell}} = 82\%$$

4. Students' answers will vary. Depending upon lab technique, there may be wide variations in percent error in any given class. The following is a sample calculation.

$$\frac{97\% \text{ accepted value} - 82\% \text{ experimental value}}{97\% \text{ accepted value}} =$$

15% error

EXTENSIONS — ANSWERS

1. Students' answers will vary, but given that the average mass of a chicken eggshell is about 5 g, their estimates for mass should be somewhat near 4.75 g. The following is a sample calculation:

$$5.27 \text{ g total eggshell} \times \frac{82\% \ CaCO_3}{100\% \text{ total eggshell}}$$
$$= 4.32 \text{ g } CaCO_3$$

2. Students may find a slightly different percentage, but it should still be close to the 95–99% range (with allowance for some error) no matter where in the shell a sample is taken.

3. Students' answers will vary, but may include: Accidentally bypassing the end-point, inadequate cleaning of the eggshell, etc. Suggestions for improvement may include per-

forming the titration using a microtip pipet for greater accuracy, taking a larger sample of eggshell to test, and drying the eggshell until its mass is constant for several trials.

4. Students' answers will vary, depending on the type of shell used.

5. Students' answers will vary, but they should recognize that since PCB's can cause decreased calcium carbonate content in birds' eggs (as mentioned in the Introduction), the eggs will be weaker and more likely to break, so that fewer will survive to hatch into birds. This investigation measures the amount of calcium carbonate content. If the warden tests some eggshells from the bird sanctuary and compares them to normal healthy eggshells, and finds that the ones from the sanctuary have less calcium carbonate, this would be evidence that links PCB's to the declining bird population.

6. Students' answers will vary, depending on the type of chalk used. Most classroom chalks are mostly calcium carbonate.

INVESTIGATION 16: Clock Reactions

Pages 869–871

Approximate time: 30 minutes. Allow more time for the additional activities suggested in the **Extensions** section. The coordination of starting the reaction and checking times elapsed may require students to work in groups, even though the materials list is written for students working in pairs. Alternatively, you may want to challenge students by having them work individually. You might want to have the students perform the experiment more than once to check the precision of their microscale techniques.

This investigation may also be useful in discussing characteristics of solutions and oxidation-reduction reactions.

A simplified net equation is given for the reaction. The reaction actually proceeds through the following steps:

$$Na_2S_2O_5(aq) + H_2O(\ell) \rightarrow 2\ NaHSO_3(aq)$$
$$IO_3^-(aq) + 3\ HSO_3^-(aq) + 3\ H_2O(\ell) \rightarrow$$
$$I^-(aq) + 3\ SO_4^{2-}(aq) + 3\ H_3O^+(aq)$$
$$5\ I^-(aq) + 6\ H_3O^+(aq) + IO_3^-(aq) \rightarrow 3\ I_2(aq) + 9\ H_2O(\ell)$$
$$I_2(aq) + HSO_3^-(aq) + 4\ H_2O(\ell) \rightarrow$$
$$2\ I^-(aq) + SO_4^{2-}(aq) + 3\ H_3O^+(aq)$$

When the HSO_3^- is completely used up, the I_2 in the last equation will react with the starch indicator to produce the blue-black color.

MATERIALS

(for a class of thirty students working in pairs, includes amounts for Extensions)

250 mL solution A (Should be prepared less than two months prior to use:
  1 gram of water-soluble starch, 0.05 grams of $Na_2S_2O_5$ (sodium metabisulfate), 1.3 mL of $1M\ H_2SO_4$)
  (Make a paste of the starch with about 10 mL of water, add 225 mL of boiling water, boil for a few minutes, cool, and add the $Na_2S_2O_5$ and H_2SO_4 and enough water to make 250 mL of solution.)
250 mL solution B (1.07 g KIO_3 in 250 mL solution)
  (Note: Other clock reactions may be used, but they should have distinctive end-points. Be sure to test the reaction you use beforehand. If the reaction suggested here is too slow, add a few mg of $Na_2S_2O_5$ to solution A. If it is too fast, dilute solution B, KIO_3.)

distilled or deionized water
45 fine-tipped dropper bulbs (Available commercially, or may be made out of thin-stem pipets. Stretch the pipets by hand and then cut them off. (See diagram after the master lists of chemicals and equipment for all investigations.) For more accurate drop sizes, place the dropper bulb in a tubing screw-clamp. Each turn of the screw will force out a reproducible drop.)
30 8-well strips
timing device (wall clock with a second hand, or stopwatches)

SAFETY NOTES

 During preparation, be sure to wear both goggles and gloves, since hazardous substances may be handled in pure forms.

DISPOSAL

All solutions can be neutralized with a small amount of Na_2CO_3 and washed down the sink with plenty of excess water, provided your school drains are connected to a sanitary sewer system with a treatment plant.

ANALYSIS & INTERPRETATION—ANSWERS

1. The most concentrated solution (5 drops of B) was the fastest reaction. The least concentrated solution (1 drop of B, 4 drops of water) was the slowest reaction.

2. Student answers will vary, but should show a linear relationship for rate.

3. The reactions began at the same time because the shake-start mixes all of the solutions at once. This is important, because the reactions can be compared, since they all began at the same time, and are all at the same conditions, such as temperature.

CONCLUSIONS—ANSWERS

1. The effect of reactant concentration (for solution B) on reaction rate is being measured here.

2. The rate of the reaction is in direct proportion to the concentration of solution B. $R = k[B]$ or $R \propto [B]$.

EXTENSIONS — ANSWERS

1. Students' answers may vary.

2. Students' answers may vary, but may include the following: Difficulty in making drops the same size, imprecision of time measurements, and difficulty in shaking the solutions together. Students may suggest minimizing these errors by repeated trials, by using the same dropper for solutions A, B, and the water, and washing it out very well in between uses. The use of a standard to identify the end point may be useful.

3. Students' answers will vary, depending on the conditions in your lab.

4. According to collision theory, lowering the temperature decreases the number of collisions possible by decreasing the average kinetic energy of the particles of a substance. Students may suggest testing this by chilling the solutions before and/or after mixing.

5. Students may suggest using a similar dilution scheme with drops of water and solution A and keeping the amount of solution B constant. They will probably find another linear relationship is involved. $R = k[A][B]$.

INVESTIGATION 17: Measuring K_a for Acetic Acid

Pages 871–873

The lab should be quite easy for students to complete in 30 minutes. This should leave time for some of the additional activities suggested in the **Extensions** section. Another possibility could involve having the students perform the experiment more than once to check the precision of their microscale techniques.

Although the equipment list is given for students working in pairs, students may work in larger groups, or individually, if preferred.

MATERIALS

(for a class of thirty students working in pairs, includes amounts for Extensions)

50 mL 1.0 M hydrochloric acid (4.3 mL conc. HCl in 50 mL of solution)
50 mL 1.0 M acetic acid (2.3 mL glacial acetic acid in 50 mL of solution)
distilled or de-ionized water (be sure there is plenty accessible in case of spills)
paper towels
30 thin-stemmed pipets
15 24-well plates
15 LED conductivity testers (may be made by the teacher, or bought from Lab Aids. Lab Aids also sells a conductivity tester which emits a tone which students may find easier to use in distinguishing small differences of conductivity. Have students test the apparatus before each investigation. You may want to have them test the metal faucets in the lab or other similar conductors.)
15 1-kΩ resistors ($\frac{1}{4}$ W)
15 LED's (light-emitting diodes)
15 9-V batteries with battery clips
15 film canisters with lids
4.5 m of hook-up wire (22 GA, solid): cut into 15 10-cm lengths and 15 20-cm lengths
15 3.5-cm lengths of straw or tubing to fit over LED's
nail or ice-pick
electrician's tape
glue gun (optional)
50 mL 0.10 M acetic acid (5 mL of 1.0 M acetic acid solution in 50 mL of solution) *for Extension 3*
50 mL 0.50 M lactic acid (2.2 mL of 85% lab grade lactic acid in 50 mL of solution) *for Extension 4*

CONDUCTIVITY TESTER PREPARATION:

1. Strip 1.5 cm of insulation from the ends of the hook-up wires and from the battery clip leads. Cut the resistor leads down to half of their length.

2. Poke two holes near the edge of the canister cap about as far apart as the two LED leads are. Insert the leads through the holes, and spread them apart slightly.

3. Connect the circuit as shown in Figure T17-A. **Be sure the *red* battery clip lead is connected to the *longer* LED lead.** The black battery clip lead should connect to the resistor, and the other end of the resistor should connect to the 10-cm length of wire. The shorter LED lead should be connected to the 20-cm wire. Bend the leads up and tape them against the inside of the lid, so that they are not touching each other and there are no exposed wires.

4. Wrap tape around the resistor and its exposed wires.

5. Poke two holes in the side of the film canister about 1 cm apart, near the bottom of the film canister. Insert the hook-up wires through these holes.

6. Connect the battery and resistor, and place them in the canister, snapping the lid shut. Adjust the protruding hook-up wires so they extend out about 5 cm. (Optional: place drops of hot-melt glue around the LED and the hook-up wires so that the canister is water-proof).

INVESTIGATION TEACHER'S GUIDE

Figure T17–A: A simple conductivity tester can be made from common materials.

7. Wrap electrician's tape around the straw and place it over the LED to provide a "dark tunnel" through which the LED may be viewed for testing weak electrolytes.

SAFETY NOTES

Since the teacher will be dealing with very caustic substances in their pure form during preparation, goggles and gloves should be worn at all times. Keep an acid neutralization package handy, such as resealable (zippered) plastic bags filled with $NaHCO_3$, in case of spills.

DISPOSAL

All solutions can be neutralized with Na_2CO_3 and washed down the sink with plenty of excess water.

ANALYSIS & INTERPRETATION — ANSWERS

1. The conductivity of the acetic acid solution should be much less than that of the hydrochloric acid solution. Hydrochloric acid is a strong acid and completely ionizes, but acetic acid is a weak acid and does not. The greater number of ions in the HCl solution are responsible for the greater conductivity.

2. Students should find that either the 10^{-2} M solution or the 10^{-3} M solution will be the best match. Careful students will estimate that it falls between these values.

3. The estimates for $[H_3O^+]$ concentration should range between 10^{-2} M and 10^{-3} M, since these values match the values of the solutions with the same conductivity, and thus, the same number of ions and the same $[H_3O^+]$ concentration.

CONCLUSIONS — ANSWERS

1. Students values for K_a will vary depending on their estimates of $[H_3O^+]$ concentration. Those that estimated it as closer to 10^{-2} should calculate a value of 10^{-4}. Those who estimate it as closer to 10^{-3} should calculate a value of 10^{-6}. Sample Calculations (based on $[H_3O^+] = 10^{-3}$)

$$K_a = \frac{[H_3O^+][CH_3COO^-]}{[CH_3COOH]} = \frac{[10^{-3}][10^{-3}]}{[1.0 - 10^{-3}]}$$
$$\approx \frac{[10^{-3}][10^{-3}]}{[1.0]}$$
$$= 1 \times 10^{-6}$$

2. The conductivity of solutions is a function of the number of ions in solution. The ions come from the dissociation of the strong acid HCl, which we assume to be completely dissociated, and the weak acid CH_3COOH, which only partly dissociates. By matching the conductivity of 1.0 M acetic acid to a known concentration of HCl, we establish the concentration of ions of the two to be identical. Because both acids produce one proton and one anion, the concentration of ions is identical in each case.

EXTENSIONS — ANSWERS

1. Students' answers will vary.

2. Students' answers may vary, but may include the following: Difficulty in distinguishing differences in LED brightness from one solution to another, apparent conductivity of acetic acid lying between that of two of the standard solutions, imprecision in micropipet techniques, etc.

Suggestions for improving the lab may include making more intermediate solutions (such as 2.0×10^{-3},

Investigation Teacher's Guide   **826-34**

INVESTIGATION TEACHER'S GUIDE

4.0×10^{-3}, 6.0×10^{-3}, and 8.0×10^{-3}), or using a different or more precise method of measuring the concentration of H_3O^+ ions in solution (such as indicator paper or pH meters).

3. If an acetic acid solution of different concentration is used, the conductivity will vary. However, the K_a value should remain the same (for the same temperature) no matter what concentration is used.

4. Lactic acid has a larger K_a, so it will glow brighter in a conductivity tester. However, it should glow only about as bright as the tester does in the 0.01 M HCl solution, because $\sqrt{1.4 \times 10^{-4}} \approx 0.01$. Since the conductivity of the acetic acid solution was intermediate between that of the 0.01 M HCl and the 0.001 M HCl, the difference will not be very great.

5. The procedure will be very similar. Students will use the conductivity apparatus to find $[H_3O^+]$. However, the calculations will be different, since we are seeking concentration instead of K_a.

$$K_a = \frac{[H_3O^+][CH_3CHOHCOO^-]}{[CH_3CHOHCOOH]}$$

$$[CH_3CHOHCOOH] = \frac{[H_3O^+][CH_3CHOHCOO^-]}{K_a}$$

$$= \frac{[H_3O^+]^2}{K_a}$$

Solution concentration =
$[CH_3CHOHCOOH] + [CH_3CHOHCOO^-]$

Answers will vary for estimated solution concentration and percent error.

INVESTIGATION 18: Acid-Catalyzed Iodination of Acetone

Pages 873–875

The lab should be quite easy for students to complete in 30 minutes. More time should be allowed for the additional activities suggested in the **Extensions** section. The coordination required to mix the solutions and keep track of time elapsed may require students working in groups instead of in pairs. Alternatively, you may want to challenge students by having them work individually.

Along with providing an example of a organic replacement reaction, producing a product with two different functional groups, this investigation can also be useful in discussing kinetics and reaction mechanisms.

MATERIALS

(for a class of thirty students working in pairs, includes amounts for Extensions)

125 mL 4.0 M acetone (36.9 mL acetone in enough water to make 125 mL of solution)
125 mL 1.0 M HCl (10.3 mL conc. HCl in enough water to make 125 mL of solution)
125 mL 0.0012 M iodine (0.10 g in enough water to make 125 mL of a saturated solution)
 Iodine dissolves slowly, so make the solution one week in advance, or first dissolve the iodine in 5 mL of methanol, then add the water.
125 mL ~1% starch solution (Reagent-grade starch is not necessary for this reaction. Simply bring 125 mL of distilled water to a boil, spray in laundry starch for about 6 seconds, until a faint translucence is visible. Do not let the solution get cloudy. Some brands of laundry starch work better than others.)
15 24-well plates
75 thin-stemmed pipets (Far fewer pipets may be used if students share them to acquire reagents—**be certain to never use same pipet for two different reagents**)
white paper (needed only as background to observe color-change in reaction)
timing device with second hand (clock or stopwatches)

SAFETY NOTES

 During preparation, be sure to wear both goggles and gloves, since hazardous substances may be handled in pure forms.

DISPOSAL

The solutions should be dilute enough that they may be washed down the sink with plenty of excess water, provided your school drains are connected to a sanitary sewer system with a treatment plant.

ANALYSIS & INTERPRETATION—ANSWERS

1, 2, and 3

Reaction #	[HCl]	[I₂]	[Acetone]	Rate
1	0.2	0.00024	0.8	(will vary)*
2	0.2	0.00024	1.6	(will vary)*
3	0.4	0.00024	0.8	(will vary)*
4	0.2	0.00048	0.8	(will vary)*

*See Analysis & Interpretation #4

4. If students are accurate, rates for reactions 2 and 3 will be roughly the same and double the rates for reactions 1 and 4, which will also be roughly the same.

CONCLUSIONS — ANSWERS

1. Students answers may vary slightly, but should be close to:
$$Rate = k[\text{acetone}]^1[\text{HCl}]^1 \text{ or}$$
$$Rate = k[\text{acetone}]^1[\text{HCl}]^1[\text{I}_2]^0.$$
$$x = 1, y = 1, z = 0$$

2. Values for k will vary, depending upon lab conditions, but should probably be on the order of 10^{-5}.

3. Yes, some reactants have no effect on the rate, like I_2 in the reaction studied. This means only that this reactant does not play a part in the slowest (rate-determining) step of the reaction mechanism.

EXTENSIONS — ANSWERS

1. The reaction mechanism's slowest step must include only the acetone and the HCl catalyst, and I_2 does not play a role in this step.

2. Students' answers will vary, but the predicted times elapsed should be between one-half and one-third of the times elapsed for mixtures 1 and 4.

3. Students' answers will vary, but may include the following: Difficulty in making drops the same size, difficulty in determining precise end point, and imprecision of time measurements. Students may suggest minimizing these errors by repeated trials, by using the same dropper (washed out between uses) for the different solutions, or by using indicator standards for comparison.

4. Students' answers will vary.

5. Students' suggestions for reaction mechanisms will vary, but should include the absence of I_2 in the rate-determining step. A possible mechanism is:

Step 1 (fast) $CH_3-\overset{O}{\underset{\|}{C}}-CH_3 + H^+ \underset{}{\overset{H^+}{\rightleftarrows}} CH_3-\overset{O\cdots H^+}{\underset{\|}{C}}-CH_3$

Step 2 (slow) $CH_3-\overset{O\cdots H^+}{\underset{\|}{C}}-CH_3 \rightleftarrows CH_3-\overset{O-H^+}{\underset{|}{C}}=CH_2 + H^+$

Step 3 (fast) $CH_3-\overset{O-H^+}{\underset{|}{C}}=CH_2 + I_2 \rightarrow CH_3-\overset{O}{\underset{\|}{C}}-CH_2I + HI$

Modern Chemistry Microscale Investigations

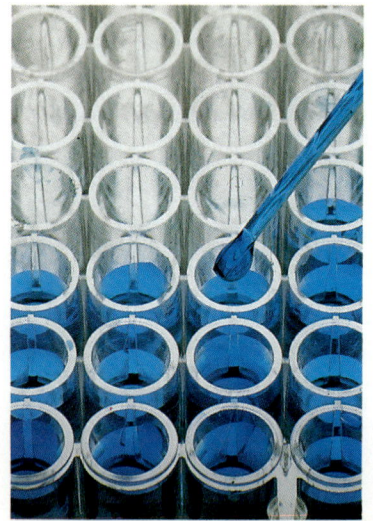

These investigations can give you some hands-on experience with some of the basic principles discussed in the text. Your teacher also may choose to supplement them with other lab procedures from the *Modern Chemistry Labortory Experiments* book.

Since these investigations are performed using microscale techniques with very small amounts of chemicals, they are much safer and easier for you to perform than a full-scale lab would be. However, this means that to get reliable data, you must be very careful and precise with your techniques.

Microscale investigations are also frequently faster than full-scale labs. If you make a mistake, you may be able to do the investigation over in the same lab period. In fact, some teachers will have everyone perform the investigation several times, so that they get practice and more reliable data.

The investigations are written with simple step-by-step procedures, but be sure to **read the whole investigation at least twice before you perform it in the laboratory. Pay very close attention to the safety notes in the investigation.** If you read through the questions at the end of the investigation before you perform it, you will know what aspects of the investigation you should focus on.

Most of the investigations relate to a specific topic in the text. You will be more successful if you make sure you understand the concepts being demonstrated before you perform the investigation.

Please Read This Safety Information

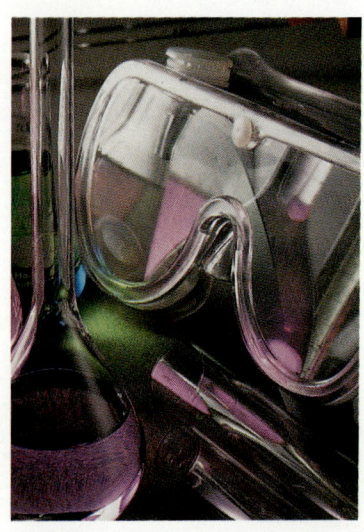

These investigations are designed to be safe for you, your classmates, and your teacher, but any chemicals can be dangerous if misused. You can help be certain that no one is harmed by these investigations by paying close attention to the safety notes in the investigation, and by performing the procedures exactly as written. Do not do anything different unless told to do so by your teacher.

Chemicals can also be hazardous for our environment. You can help protect the environment by following any instructions for chemical disposal. Don't assume that anything can be washed down the sink or tossed in the trashcan unless you've checked with your teacher.

There are some safety regulations that are important for all lab work, no matter what chemicals are being used. Keep in mind these rules can keep you from being hurt.

1. **Always wear safety goggles in the lab.** Even if you aren't working on an investigation at the time, lab rooms always contain many chemicals that can cause eye damage.

2. **No contact lenses in the lab.** Chemical vapors can get between the lenses and the eyes and cause irreparable eye damage.

3. **Read the investigation before entering the lab, and follow all instructions.** Review all safety precautions with your teacher before you begin. Only materials and equipment authorized by your teacher should be used.

4. **Be careful and stay at your lab station.** If you don't carry chemicals around, the likelihood of bumping into other students and spilling dangerous chemicals will be greatly decreased. Unattended experiments can result in accidents.

5. **Wear the right clothes for the job.** Long hanging necklaces, heavy jewelry, and bulky clothing can restrict your movement, cause you to knock things over, or get caught in flames. Long hair should be put up or pulled back. Cotton clothing is better than nylon, polyester or wool.

6. **Only books and notebooks for the investigation should be in the lab.** Purses, bookbags, backpacks and other textbooks should be kept in your desk or storage area.

7. **Food, beverages, and smoking are <u>NEVER</u> allowed in the lab.**

8. **<u>NEVER</u> taste chemicals. <u>NEVER</u> touch chemicals with your hands.**

9. **Be careful with Bunsen burners, hot plates, and flames.** Keep your head and clothing away from the flame. Light the burner only with a sparker. Be sure all gas valves and hot plates are turned off when you leave the lab.

10. **Know fire drill procedures and locations of exits.**

11. **Know the location and operation of all safety equipment such as safety showers, eye-wash stations, etc.**

12. **Keep your work area clean and tidy.** After an investigation, clean and wipe dry your lab area and all equipment.

13. **Wash your hands with soap after each investigation.**

14. **Heed and follow all safety cautions in the investigations.**

15. **All accidents should be reported immediately to your teacher, no matter how minor.**

16. **<u>NEVER</u> WORK ALONE IN THE LABORATORY.** You should always be under the supervision of your teacher.

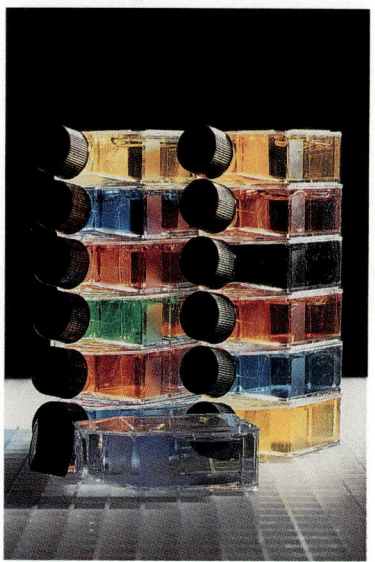

Safety Symbols in the Microscale Investigations

You should not be afraid to work with chemicals and equipment. Although they can be dangerous if misused, if they are used according to instructions, there is very little danger of harm to yourself or others. To remind you of this, and to let you know about added cautions, each investigation contains safety notes set off from the text in red type and a safety symbol to alert you to the type of hazards involved. The symbols shown and explained below are designed to help you and your teacher make quick references to the precautions that should be followed for the investigation.

SAFETY SYMBOLS

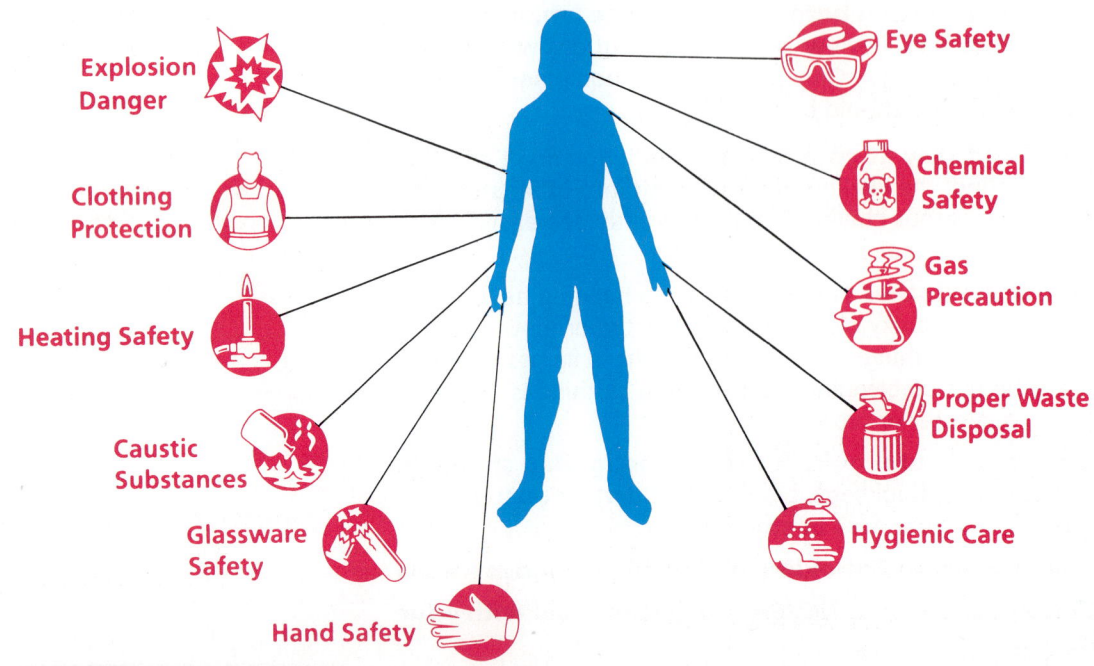

Safety Symbols With Precautions

 ### Caustic Substances

- Alert your teacher to any chemical spills.
- Do not let acids and bases touch your skin or clothing. If a substance gets on your skin, rinse immediately with cool water and alert your teacher.
- Wear your laboratory apron to protect your clothing.
- Never add water to acids; always add acids to water.
- When shaking or heating a container with chemicals, always point the container's opening away from yourself and other students.

 ### Chemical Safety

- Never taste any substance in the laboratory. Do not eat or drink from laboratory containers.
- Do not eat or drink in the laboratory.
- Properly label all containers containing chemicals.
- Never transfer substances with a mouth pipette or straw; use a suction bulb.
- Never return unused chemicals to the original container.

 ### Gas Precaution

- Do not inhale fumes directly. When instructed to smell a substance, wave fumes toward your nose and inhale gently.
- Use flammable liquids only in small amounts and in a well-ventilated room or under a fume hood.
- Always use a fume hood when working with toxic or flammable fumes.
- Do not breathe pure gases such as hydrogen, argon, helium, nitrogen, or high concentrations of carbon dioxide.

 ### Explosion Danger

- Use safety shields or screens if there is a potential danger of an explosion or implosion of apparatus.
- Never use an open flame when working with flammable liquids such as ether or alcohol.
- Follow a water bath procedure to heat solids. Never risk an explosion by heating materials directly.

Eye Safety

- Wear approved safety goggles in the laboratory.
- Make sure an emergency eye wash station is available in the laboratory.
- Never look directly at the sun, even for short periods of time. Laboratory goggles will not protect your eyes from the sun's rays.
- Do not wear contact lenses.

 ### Heating Safety

- Make sure that fire extinguishers and fire blankets are available in the laboratory.
- Tie back long hair and confine loose clothing.
- Wear safety goggles when working with flames.
- Never reach across an open flame.
- Use proper procedures when lighting Bunsen burners.
- Turn off hot plates, Bunsen burners, and other open flames when not in use.
- Heat flasks or beakers on a ringstand with a wire gauze between the glass and the flame.
- Store hot liquids only in heat-resistant glassware.
- Turn off gas valves when not in use.

 ### Proper Waste Disposal

- Clean up the laboratory after you are finished; dispose of paper toweling, etc.
- Follow your teacher's directions regarding proper procedures for waste disposal, especially for hazardous waste disposal.
- Place broken glass in a specially designated container.
- Dispose of hazardous wastes only in proper containers, never in an ordinary trash can.

 ### Glassware Safety

- Check the condition of glassware before and after using it. Inform your teacher about any broken, chipped, or cracked glassware; it should not be used.
- Air-dry glassware; do not dry by toweling. Do not use glassware that is not completely dry.
- Do not pick up broken glass with your bare hands.
- Never force glass tubing into rubber stoppers.
- Never place glassware near edges of your work surface.

 ### Hand Safety

- Use tongs when heating containers. Never hold or touch containers in your hand while heating them.
- Always allow heated materials to cool before handling them.

 ### Hygienic Care

- Always wash your hands after the laboratory.
- Keep your hands away from your face and mouth.

 ### Clothing Protection

- Wear laboratory aprons in the classroom.
- Confine loose clothing.
- Do not wear dangling jewelry.

Safety Equipment

Make sure you know where this equipment is in your lab and how to operate it **before** performing any investigations. You don't want to figure this out during an accident or an emergency.

Safety Goggles: Safety goggles will protect your eyes from chemical and particle injuries. You need to have your own pair. Goggles should be washed frequently and stored in a protected place.

Eye and Face Wash Fountains: If chemicals splash on or near the eyes, you should flush them with water for at least 15 minutes. Fountains should be checked for proper operation daily.

Safety Showers: Safety showers can be used when acids or caustic chemicals burn the skin or in case of clothing fires. They should dilute material to a safe level in 15 seconds or less. Check them monthly for proper operation.

Fire Blankets: Fire blankets are the best way to smother flaming hair or clothes. If the clothing is polyester, the best way to put out the fire is the safety shower.

Fire Extinguishers: Fire extinguishers are labeled to indicate which fires they can extinguish. Not all types of fires can be extinguished by the same fire extinguisher. You should be familiar with the operating instructions for your lab's fire extinguishers.

First Aid Kits: Emergency treatment for cuts, burns, or other minor injuries should be administered by your teacher only.

Sand Buckets and/or Spill Control Packages: Sand can be used to smother small fires, such as those contained in beakers. In addition, sand may be mixed with other substances that can neutralize acidic or basic spills. Ask your teacher which spill control packages you have, and how they should be used.

INVESTIGATION 1: Conservation of Mass

Introduction

The law of conservation of mass states that matter is neither created nor destroyed during a chemical reaction and that the mass of a system should therefore remain constant during any chemical process. To test how true this law is, we will examine a simple chemical reaction and compare the mass of the system before and after the reaction.

Objectives

- **Observe** signs of a chemical reaction
- **Infer** that a reaction has occurred
- **Compare** masses of reactants and products
- **Resolve** chemical discrepancies
- **Design** experiments
- **Relate** observations to law of conservation of mass

Materials

5% acetic acid solution (vinegar)
sodium bicarbonate (baking soda)
two clear plastic cups
balance
2-L plastic soda bottle
hook-insert cap for bottle
micro plunger
graduated cylinder

Procedure

 The vinegar solution used in the investigation could be hazardous to the eyes. Safety goggles must be worn at all times during this lab.

 Avoid getting any of the solution on your skin or clothing. Any acid spills should be immediately washed with excess water. Make sure you know where the neutralization packages for acid spills are kept.

Part I:

1 Make a data table like this one in your lab notebook:

	Initial Mass (g)	Final Mass (g)	Change in Mass (g)
Part I			
Observations of Reaction, Part I			
Part II			
Observations of Reaction, Part II			

Figure 1-A: The law of conservation of mass can be tested using some materials found in most homes.

2 Tap the micro plunger down into the baking soda, enough so that the bulb end is packed with a plug of the powder. (4–5 mL of baking soda should be enough to pack the bulb with a plug.)

3 Deliver the powder into one of the plastic cups. (Squeezing in on the sides of the micro plunger can help loosen the plug.)

4 Use the graduated cylinder to measure out 100 mL of vinegar into the second plastic cup.

5 Place both cups side by side on the balance to determine the total initial mass of the system. Record the mass in the data table.

6 Carefully pour the vinegar into the cup containing the baking soda, allowing vinegar to slowly run down the side of the cup's interior. Add just a little vinegar at a time to avoid having the reaction get out of control. Observe and record your observations about the reaction.

7 When the reaction has finished, place both cups back on the balance to determine the total final mass of the system to 0.01 g, calculate any change in mass, and record both in the data table.

Analysis & Interpretation — Part I

Write the answers to these questions in your lab notebook. The questions will help you organize your thinking to prepare for Part II.

1 Inferring Conclusions: What evidence was there that a chemical reaction occurred?

2 Organizing Data: How did the final mass of the system compare with the initial mass of the system?

3 Resolving Discrepancies: Considering your answer to Analysis & Interpretation #2, was the law of conservation of mass violated? (Hint: Another way to express the law of conservation of mass is to say that the mass of **all** of the products equals the mass of **all** of the reactants.)

4 Designing Experiments: Examine the plastic bottle and the hook-insert cap. Try to develop a modified procedure that will test the law of conservation of mass more accurately than the procedure in Part I.

Procedure — Part II

 Safety goggles must still be worn throughout Part II.

1 After you get your teacher to approve the procedure you designed in the Analysis & Interpretation section, implement it, using the same materials and quantities as in Part I, and using the bottle and hook-insert cap in place of the two cups. Record your data in the data table.

2 If you were successful in Step 1 above, and your results reflect the conservation of mass, proceed to finish the investigation. If not, find a lab group that was successful, and discuss with them what they did and why they did it. Test their procedure yourself to determine whether the results of the other group are reproducible.

3 Check with your teacher for instructions on how to dispose of your materials when finished. Remember to wash your hands thoroughly before leaving the laboratory.

Analysis & Interpretation — Part II

1 Inferring Conclusions: Was there any new evidence in Part II that a chemical reaction had occurred?

2 Organizing Ideas: Identify the state of matter for each reactant in Part II. Identify the state of matter for each product.

3 Organizing Ideas: Write a chemical reaction using words and/or chemical formulas describing the reaction you observed in this investigation. (Hint: There are several examples of chemical reactions written in Section 1.2. The chemical reaction used in this investigation is also discussed briefly in this section.)

Conclusions

1 Relating Ideas: What was different about the "systems" in Part I and Part II? Which difference led to the improved results between the tests?

2 Evaluating Methods: Why did the procedure for Part II work better than the procedure for Part I?

Extensions

1 Designing Experiments: How would you verify the law of conservation of mass if you had been given a resealable (zippered) plastic bag and a twist-tie instead of the bottle and hook-cap? Write your proposed procedure, step by step.

2 Predicting Outcomes: Would you have been as successful with the resealable plastic bag and twist tie? Why or why not? If time allows, try the procedure you wrote out and test your prediction. Report your results and try to explain any discrepancies you found. (Hint: What are the differences between what happened with the bag and what happened with the bottle?)

3 Applying Models: When a log burns, the resulting ash obviously has less mass than the initial log did. Explain whether this loss of mass violates the law of conservation of mass.

4 Designing Experiments: Design a procedure that would test the law of conservation of mass in the case of a burning log as in Extension #3 above.

INVESTIGATION 2: Mixture Separation

Introduction

Separation and recovery of pure substances are extremely important operations for scientific research and industry, but separating the different components in a mixture can often prove quite challenging. In this investigation, you will have an opportunity to design, develop, and implement your own procedure in an attempt to separate and recover the components of a mixture. The mixture contains the following four substances in dry granular form: salt, sand, iron filings, and poppy seeds.

Objectives

- **Observe** chemical and physical properties of a mixture
- **Relate** knowledge of properties to the task of purifying the mixture
- **Analyze** success of methods of purifying the mixture

Materials

At each lab station:
petri dish	clear tape
8-well strip	sample of mixture
micro funnel	

Additional Equipment (at central location):
cellophane	aluminum foil
test tubes	wood splints
rubber stoppers	plastic forks
plastic spoons	paper towels
pipets	plastic straws
filter funnels	paper clips
filter papers	distilled water
forceps	magnet
tissue paper	cotton balls

Procedure

 Safety goggles must be worn at all times during this investigation to protect your eyes.

1 Get a sample of the mixture from your teacher. Using the micro funnel, place a representative sample of it in the first well of the 8-well strip. Place a small piece of tape over the opening of the well.

2 Using any or all of the materials and additional equipment listed above, develop and implement your own procedure to separate out the mixture and recover all four components, each in as pure a state as possible. Before you begin, decide what you will do to separate the components. **Record this procedure (a step-by-step plan) in your lab notebook before starting.** In your lab notebook keep track of your observations, what you do, and in what order you do it.

 (Hint: It may help you to get done on time if you come up with a plan before you come to the lab. If any part of your procedure involves a long period of waiting, during which time the set-up can remain unattended, try to reach that point by the end of the class period. Write your name on the side of your petri dish and use it as a holding bin for your set-up, and ask your teacher where you can set it. On a subsequent day, you will have an opportunity to complete your procedure.)

3 As each component is separated off, use the micro funnel to place a representative portion of it into the appropriate well of the 8-well strip as shown in Figure 2-A and cover the well with a small piece of tape.

4 Place the remainder of each of the recovered components (after the well has been filled) in the appropriate recovery bin at the front of the lab room. Remember to wash your hands thoroughly before leaving the laboratory.

Analysis & Interpretation

1 Evaluating Methods: In your estimation, how successful were you (on a scale of 1–10) in separating and recovering each of the four components: sand, salt, iron filings and poppy seeds? Justify your estimations of your success, based on your observations.

Conclusions

1 Evaluating Methods: What made you decide to do your procedural steps in the order in which you did them? Would any order have worked?

2 Designing Experiments: If you were able to do the lab over again, what specific things would you do differently?

3 Designing Experiments: Name two materials or tools that weren't available that might have made your separation easier.

4 Applying Ideas: For each of the four components, describe a specific physical property that enabled you to separate it from the rest of the mixture.

Extensions

1 Evaluating Methods: What ways can you think of to determine the purity of each of your recovered components?

2 Designing Experiments: How would you separate out each of the following two-part mixtures?
a) lead filings and iron filings
b) sand and gravel
c) sand and finely ground styrofoam
d) salt and sugar
e) alcohol and water
f) nitrogen and oxygen

Figure 2-A: Once you have a sample of the unseparated mixture or a separated component, pour some into the appropriate well and seal it with tape.

First well: unseparated mixture
Second well: empty
Third well: sand
Fourth well: salt
Fifth well: iron filings
Sixth well: poppy seeds
Seventh well: empty
Eighth well: empty

INVESTIGATION 3: Measuring the Specific Heats of Metals

Introduction

Heat is a form of energy that cannot be measured directly. However, changes in heat can be determined by measuring changes in temperature. When a substance is heated, the heat gained, q, depends upon three important factors; the mass, m, of the substance in grams, the specific heat of the substance, c_p, and the substance's change in temperature, Δt. As discussed in Section 2.2, to calculate the amount of heat absorbed or lost by a substance, the following equation is used:

$$q = m \times c_p \times \Delta t$$

The specific heat of a substance is an intensive physical property, reflecting the type of bonding and intermolecular forces in the substance. In this investigation, you will measure the specific heats of several metals. First, you will need to make and calibrate a calorimeter, an instrument for measuring heat changes. You will start with a predetermined mass of water in your calorimeter. When the heated metal is added to the water, the change in temperature for the water will be measured. Using the specific heat of water (4.184 J/g·C°), you will be able to calculate the amount of heat gained by the water. This value will be equal to the amount of heat lost by the metal. If the temperature change and mass of the metal are known, its specific heat can be determined.

$$q_{metal} = q_{calorimeter}$$
$$mass_{metal} \times \Delta T_{metal} \times c_{p,metal} = mass_{water} \times \Delta T_{water} \times c_{p,water}$$

Calibrating the Calorimeter

When carrying out calorimeter experiments, it is impossible to have perfect insulation and prevent heat from escaping. The problem of heat loss is usually solved by calibrating the calorimeter before making measurements on a system.

When you add a known amount of hot water to a known amount of cool water in the calorimeter, the heat from the hot water will be transferred to the cool water and some heat will be lost to the surroundings. This relationship can be expressed mathematically as follows:

$$q_{hot\ water} = (q_{cold\ water} + q_{surroundings})$$

Substituting with the equation $q = m \times c_p \times \Delta T$ for each factor above:

$$m_{hot}(4.184\ J/g \cdot C°)\Delta T_{hot} = [m_{cool}(4.184\ J/g \cdot C°)\Delta T_{cold} + m_{calorimeter}(c_{p,calorimeter})\Delta T]$$

Since the mass of the calorimeter and the specific heat of the calorimeter do not change, their product, $m \times c_{p,calorimeter}$, is a constant, which is referred to as the "heat capacity of the calorimeter," or C'. This value will be used to correct the measurements made with the metals.

$$m_{hot}(4.184\ J/g \cdot C°)\Delta T_{hot} = [m_{cool}(4.184\ J/g \cdot C°)\Delta T_{cool} + C'\Delta T]$$

Objectives

- **Calibrate** a simple calorimeter
- **Relate** measurements of temperature to changes in heat content
- **Calculate** the specific heats of several metals
- **Determine** the identity of an unknown metal
- **Infer** information about the strength of intermolecular forces

Materials

2 plastic foam cups for calorimeter
scissors or tools to trim cups
thermometer
100 mL graduated cylinder
two 400 mL beakers
bunsen burner or hot-plate
large test tube
balance
tongs to handle hot beakers and metal
metal samples
   (for example: iron nails, copper wires, aluminum pellets)
unknown metal sample
ringstand
ring clamp (if using bunsen burner)
wire gauze square (if using bunsen burner)
test tube clamp (attaching to ringstand)
boiling chips

Procedure

 Be careful while transferring hot water. Wear goggles for safety.

 Use caution in working with glass thermometers, which can easily shatter.

1 Make a data table like the one below in your lab notebook:

Calibration			
	Cool H$_2$O	Hot H$_2$O	Calorimeter
Initial Temp.			
Final Temp.			
Change in Temp.			
Volume			(not needed)
Mass			(not needed)
C'=	(not needed)	(not needed)	

2 Construct a calorimeter from 2 plastic foam cups. Trim the lip of one cup and use that cup for the top of the calorimeter. Make a small hole in the top so a thermometer can be inserted, as shown in Figure 3-A. Place the calorimeter in one of the beakers to help prevent it from tipping over.

3 Put 50.0 mL of cold water in your calorimeter and allow the water to reach room temperature. Measure the temperature every minute until the temperature stays the same for three minutes. Record the volume and temperature in your data table as the initial temperature for the cool water and the calorimeter.

4 (Clamp the ring clamp to the ringstand if using a bunsen burner. Place the other beaker onto the wire gauze on top of the ring clamp if using a bunsen burner). Heat about 75 mL of water in the beaker to between 70–80°C, using the bunsen burner or hot-plate. Remove the water from the heat source and pour 50.0 mL into a graduated cylinder. Let the water sit a minute and stir to obtain a uniform temperature. Record the volume and temperature in your data table.

5 Immediately pour the hot water into the calorimeter with the cool water. Put on the lid and stir for 30 seconds.

 Be careful not to break a thermometer!

Record the highest temperature obtained by the mixture of hot and cool water as the final tem-

Figure 3-A: A simple calorimeter can be made from two plastic foam cups with water inside.

perature for the hot water, the cool water, and the calorimeter.

6 Empty the calorimeter and make one data table like the one below in your lab notebook for each of the metals you will be testing, including the unknown metal.

Metal Specific Heat Tests		
	H$_2$O	Metal
Initial Temp.		
Final Temp.		
Change in Temp.		
Mass		

7 Put 75.0 mL of water into the calorimeter and allow it to come to room temperature. Record the mass and temperature of the water in the second data table.

8 Fill a 400 mL beaker with water and add some boiling chips. Use enough metal to fit into the large test tube but remain below the surface of the water when the test tube is placed into the beaker. You need to use as much metal as possible. **Measure the mass of the metal as precisely**

Investigation 3: Measuring the Specific Heats of Metals   **835** as possible before placing it in the test tube. Record the mass in the data table.

9 Clamp the test tube to the ring stand, (over the ring clamp, if using a bunsen burner) as shown in Figure 3-B. Set the beaker upon the wire gauze on the ring clamp. Lower the test tube clamp so that the test tube is lowered into the beaker of water. Heat it until the water begins to boil. Allow the boiling to continue for 10 minutes. You can now assume the metal is at the boiling temperature of water. Record this temperature in the second data table.

10 Transfer the metal to the calorimeter, holding the test tube with the clamp.

 Be careful not to burn your skin with the hot water or the metal!

Put the top on the calorimeter. Mix or stir for 30 seconds. Now, record the highest temperature of the water.

11 Repeat Steps 6–10 as needed for other metals, including the unknown metal.

12 Check with your teacher for instructions on how to dispose of your chemicals when finished. Remember to wash your hands thoroughly before leaving the laboratory.

Analysis and Interpretation

1 Organizing Ideas: State the scientific law that is the basis for the assumption that the heat energy lost by the metal as it cools will be equal to the heat energy gained by the water and the calorimeter.

2 Organizing Data: Using the density value given for water in Table 2-4 of the textbook, calculate the masses of water used in the calibration step, and record your results in the first data table.

3 Organizing Data: Finish filling out the first data table by calculating the changes in temperature involved. Using the value of 4.184 J/g·C°, determine how much heat was lost by the hot water in the calibration portion of the experiment. How much heat was gained by the cool water?

4 Inferring Conclusions: Based on your answer to the previous question, how much heat was gained by the calorimeter?

5 Organizing Data: Calculate the heat capacity, C', of the calorimeter using the relationship discussed in the Introduction. Record your answer under the first data table.

6 Inferring Relationships: For the second part of the investigation, write an equation for the relationship between the heat lost by a metal, the heat gained by the water and the heat gained by the calorimeter. Solve for the specific heat of the metal by substituting the equation for specific heat ($q = m \times c_p \times \Delta T$) into the first equation. Show all work for each metal. (Hint: Be sure to include the heat capacity of the calorimeter, C', in the substituted equation.)

Conclusions

1 Inferring Conclusions: Finish filling in the second set of data tables and use the relationship from Analysis & Interpretation #6 to calculate the specific heat of each of the metals tested.

2 Evaluating Conclusions: Compare your value for the unknown metal to the values given in Table 2-6 in the text. What do you think your unknown is made of?

3 Evaluating Conclusions: Compare your values for specific heats to the table values that can be found in a handbook such as the *CRC Handbook of Chemistry and Physics* or *Lange's Handbook of Chemistry*. (You may need to convert the values given from calories to joules to use the handbook values. The conversion factor is the same as the specific heat of water: 1 cal = 4.184 J.) Calculate your percent error.

4 Analyzing Conclusions: How do the specific heats of the metals compare to the specific heat of water? What does that imply about the amount of heat needed to heat a certain mass of metal or water to the same temperature? Which one would absorb heat better?

5 Organizing Ideas: Are higher values or lower values better for the specific heat of a calorimeter? Why?

Extensions

1 Evaluating Methods: Share your data with other lab groups and calculate a class average for the different specific heats of the metals. Compare the averages to the figures in a chemical handbook and calculate the percent error for the class averages.

2 Designing Experiments: What are some likely sources of imprecision in this experiment? If you can think of ways to eliminate them, ask your teacher to approve your suggestion, and run more trials.

3 Designing Experiments: Design a different calorimeter. If your teacher approves of the design, construct it, and compare its performance to the plastic foam cups.

4 Predicting Outcomes: Predict the specific heat of glass. Name other substances that would have similar specific heats. Design an experiment to calculate the specific heat of glass beads. If your teacher approves your suggestion, perform the experiment and check your prediction.

5 Applying Ideas: How many practical applications can you think of that would relate to knowledge of a substance's specific heat?

6 Applying Ideas: Explain how the large specific heat of water is responsible for the moderating effects the oceans have on weather.

Figure 3-B: The metal samples are each heated by a hot-water bath using either a bunsen burner or a hot-plate.

INVESTIGATION 4: Conductivity

Introduction

For a substance to conduct an electrical current, the substance must possess free-moving charged particles. These charged particles may be delocalized electrons, such as those found in substances that form metallic bonds. They may be mobile ions, such as those found in dissolved (or molten) salts or in certain molecular substances that have polar-covalent bonds that can be easily broken to form ions when they are dissolved in water.

On the other hand, ionic substances in their solid form that do **not** dissolve in water are usually considered to be **insulators,** substances that do not conduct electricity. Similarly, substances with nonpolar-covalent bonds (and those whose polar-covalent bonds are **not** easily broken to form ions when they are dissolved in water) are also insulators.

The fact that a substance is a conductor in pure form, or conducts only in solution, or is an insulator, is a physical property of the substance. In fact, some scientists use quantitative measurements of conductivity of substances in their work. Tables of conductivities for various solutions and substances can be found in chemical reference works.

Electrical conductivity can provide clues about the inner structure and type of bonding found in a substance in terms of the types of chemical bonds discussed in Chapter 6. In this investigation, you will use qualitative measures of conductivity to determine the identities of unknown substances.

You will be given ten substances, identified by the letters A, B, C, D, E, F, G, H, I, and J. They correspond to these substances: sodium sulfate, sugar, sulfur, tap water, deionized (or distilled) water, tin, tin(IV) oxide, graphite, yellow chalk dust, and charcoal (not necessarily in that order). In this investigation, you will use conductivity results and other information to determine which unknown corresponds to which substance.

Objectives

- **Observe and Compare** conductivities of various substances
- **Relate** the type of bonding of a substance to its conductivity
- **Infer** the identities of unknown substances

Materials

10 labeled containers, each with a different unknown substance
hand-held conductivity tester
spot plate
dropper bottle of distilled water
spatula
thin-stemmed pipet
paper towels
small glass stirring rod or toothpicks

Procedure

 Safety goggles and gloves must be worn at all times during this investigation to protect your eyes and hands.

1 Examine each of the ten unknowns and record your observations regarding their state (solid or liquid), color, texture, etc in your lab notebook.

2 Test your conductivity apparatus to make sure it lights up when the two wires or leads touch different parts of a conducting substance. Your teacher will tell you what conducting substance to use.

3 Using a clean spatula for each solid and a clean pipet for each liquid, deliver small samples of each unknown into separate wells of a spot plate. **Keep track of which unknown is in which well! The spatula should be wiped with a clean paper towel between each use to prevent contamination. Similarly, the pipet should be wiped with a clean paper towel and rinsed with distilled water between each use.** Put any water used for rinsing into the sink, not back into the beaker.

4 Place the wire leads in good contact with the first unknown sample and note whether or not it conducts, as indicated by the light on the conductivity tester. (It may help to move the tester around a bit while watching the light on it.) If the two leads on the tester touch each other, however, the light will automatically shine. Do not mistake this for a positive test. Most pure substances are non-conductors, so don't expect a majority of your substances to conduct. For those that do conduct, compare on a scale from 0-4 which ones conduct the best. Some results may be faint, requiring careful observations. Record all observations in your lab notebook.

5 Repeat this technique for the other unknown samples. **Be sure to thoroughly wipe off the leads of the conductivity probe with a clean paper towel between each test.**

6 Draw up some fresh distilled water into the pipet and add 10 drops of it onto each of the ten unknowns, **holding the pipet's tip 1-2 cm above the unknowns.** Use a small glass stirring rod or toothpicks to stir each mixture. **Be sure to thoroughly wipe off the stirring rod or toothpicks with a clean paper towel after each use.**

Figure 4-A: When testing the conductivities of your unknown substances, be sure not to let the two leads from the probe touch each other.

7 Observe each of the wells and record any observations regarding the dissolving or mixing of the unknowns in your lab notebook.

8 Now, test the conductivity of each mixture, again wiping the leads off thoroughly between each test.

9 Check with your teacher for instructions on how to dispose of your unknowns when finished. Remember to wash your hands thoroughly before leaving the laboratory.

Analysis & Interpretation

1 Organizing Ideas: Which type(s) of bonding are likely to be involved in the unknowns that you would expect to be conductive in their pure form? Explain what free-moving charged particles are available in this (these) type(s) of bonding.

2 Organizing Ideas: Which type(s) of bonding are likely to be involved in the unknowns that were not conductive in their pure form, but conducted electricity when they were mixed with water? Explain what free-moving charged particles are available in this (these) type(s) of bonding.

3 Organizing Ideas: Which type(s) of bonding are likely to be involved in the unknowns that were not conductive in their pure form, or when they were mixed with water? Explain why there were no free-moving charged particles available in this (these) type(s) of bonding.

4 Organizing Data and Relating Ideas: Copy the chart at the end of this question into your lab notebook. Fill in the chart by using the following information to determine which of the types of bonding discussed in Analysis & Interpretation Questions #1-3 is likely to be present in each one of these substances, and whether or not the compounds are likely to conduct electricity. Also try to judge which will be the best conductors in pure form and dissolved in water, which will conduct only marginally, and which will not conduct at all.

Useful Information

- Chalk is mostly calcium carbonate ($CaCO_3$) and only dissolves slightly in water.
- Sulfur is a yellow molecular substance, arranged in S_8 ring molecules.
- Sodium sulfate is white and dissolves in water.
- Tin is a metal.
- Tin oxide is white and does not dissolve in water.
- Charcoal is a black, powdered, amorphous form of carbon with the atoms bonded to each other in a randomly organized way.
- Graphite is a black, powdered form of carbon, but the atoms are bonded to each other in systematic sheets with some free electrons bonding the sheets together.
- Sugar is a molecular substance that dissolves in water.
- Distilled or deionized water can be considered to be pure H_2O.
- Tap water contains many dissolved impurities, some molecular, some ionic.

Substance	Type of Bonding	Conductivity	Conductivity in H_2O	Matches Unknown #
Chalk				
Charcoal				
Graphite				
Sugar				
Sulfur				
Sodium Sulfate				
Tin				
Tin Oxide				
Pure H_2O				
Tap Water				

Investigation 4: Conductivity

Conclusions

1 Organizing Conclusions: List the identities of the unknowns A-J. Explain your reasoning for deciding which substance matched to which unknown.

Extensions

1 Designing Experiments and Predicting Outcomes: What sort of results do you think you would get and why would you get them if you tested the conductivities of the following materials:
 a) an iron nail
 b) liquid bromine, Br_2
 c) wax (a random arrangement of molecules with long, nonpolar chains of carbon and hydrogen atoms)
 d) a dilute hydrochloric acid solution HCl(aq)
 e) table salt, NaCl

How will the conductivity of each be affected by adding water to it? If time allows, test your predictions for a), c), d), and e). (Liquid bromine is too dangerous to work with in the laboratory.)

2 Applying Conclusions: If you were choosing a substance to be used to make wires to carry electricity from a generator to an appliance, and you could only choose from the substances used in the investigation, which one would you choose? What property(ies) makes it a good choice?

3 Applying Conclusions and Resolving Discrepancies: Whenever a thunderstorm starts to form, people swimming or boating always head for shore immediately. Explain why there is a danger from lightning in water, comparing your results using tap water and distilled (or deionized) water. Which one is most like lake or pool water?

INVESTIGATION 5: Colored Precipitates

Introduction

The binary ionic compounds studied in Section 7-1 have a variety of properties. Some of them can dissolve easily in water, and others do not. Some of them are brightly colored compounds. You can compare the characteristic properties of known substances with those of an unknown to determine the identity of an unknown.

In this investigation, you will use what you find out about characteristic colors and the formation of precipitates in replacement reactions to identify unknown substances used in a "double-replacement" reaction, as described in Section 8-2.

Objectives

- **Observe** chemical processes and interactions
- **Compare** different chemical and physical properties of chemicals
- **Relate** observations to identifications of unknown solutions
- **Infer** a conclusion from experimental data

Materials

0.1 mol/L $CoCl_2$ solution
0.1 mol/L $CuCl_2$ solution
0.1 mol/L $FeCl_3$ solution
0.1 mol/L $NiCl_2$ solution
0.1 mol/L NaOH solution
mystery solutions
8 thin-stem pipets
one 8-well flat-bottom strip *or* small test tubes

Procedure

The solutions used in the investigation are very hazardous to the eyes. Safety goggles must be worn at all times during this investigation.

Avoid getting any of the solutions on your skin or clothing. Any basic spills should be immediately washed with excess water. Make sure you know where the neutralization packages for basic spills are kept.

1 With one thin-stem pipet, place 20 drops of sodium hydroxide solution in the first seven of the eight wells in the strip (or if using test tubes, 40 drops). **For best results, try to keep all the drops in this investigation the same size.** You may want to prepare a piece of paper to place under the 8-well strip as shown in Figure 5-A to label the wells as follows:

Well #:	1	2	3	4	5	6	7
Label:	$CoCl_2$	$CuCl_2$	$FeCl_3$	$NiCl_2$	Mystery #1	Mystery #2	Mystery #3

2 To the first well, add 5 drops of cobalt(II) chloride solution. (If using test tubes, add 10 drops of each of the metal chloride solutions to each tube.) To the second, add 5 drops of copper(II) chloride solution. To the third, add 5 drops of iron(III) chloride. To the fourth, add 5 drops of nickel(II) chloride solution. Observe what happens. Record observations of the reactions and the physical properties of the substances formed in your lab notebook.

3 In the fifth well, add 5 drops of mystery solution #1. (If using test tubes, add 10 drops of each of the mystery solutions.) In the sixth well add 5 drops of mystery solution #2. In the seventh well, add 5 drops of mystery solution #3. Observe what happens. Record observations in your lab notebook about what kind of reaction you see and the physical properties of the substances formed.

4 Check with your teacher for instructions on how to dispose of your chemicals when finished. Remember to wash your hands thoroughly before leaving the laboratory.

Analysis & Interpretation

1 Organizing Ideas: The reactions in this lab are all double-replacement reactions. Write balanced

Figure 5-A: Use a blank sheet of paper under the 8-well strip to keep track of which solutions are in the wells.

Investigation 5: Colored Precipitates

chemical equations for the reactions occurring in the first four mixtures.

2 Organizing Ideas: As will be discussed in Chapter 15, a net-ionic equation is one which includes only those dissolved ions that undergo a reaction. For example, in writing a net-ionic equation for this chemical reaction, the following steps are followed:

- Write the original balanced equation

$$HCl(aq) + NaOH(aq) \rightarrow NaCl(aq) + H_2O(\ell)$$

- Rewrite the equation with all ions separated or *dissociated*. Formulas for precipitates, gases, or pure liquids formed should be written as usual:

$$H^+(aq) + Cl^-(aq) + Na^+(aq) + OH^-(aq) \rightarrow Na^+(aq) + Cl^-(aq) + H_2O(\ell)$$

- Cross out any ions that appear in the same form on both sides of the equation (in this case, Na^+ and Cl^-).

$$H^+(aq) + \cancel{Cl^-(aq)} + \cancel{Na^+(aq)} + OH^-(aq) \rightarrow \cancel{Na^+(aq)} + \cancel{Cl^-(aq)} + H_2O(\ell)$$

- Re-write the equation, omitting any ions that were in the same form on both sides of the equation and crossed out in the previous step.

$$H^+(aq) + OH^-(aq) \rightarrow H_2O(\ell)$$

This is a net ionic equation, which includes only those ions undergoing a chemical change in a reaction. (The Na^+ and Cl^- ions do not react at all, and remain unchanged.) Write a net-ionic equation for the reactions occurring in the first four mixtures.

3 Analyzing Results: Would the reactions have occurred any differently if potassium hydroxide (KOH) had been used instead of sodium hydroxide? Explain your answer. Hint: Write the net-ionic equations using KOH(aq) instead of NaOH(aq).

Conclusions

1 Inferring Conclusions: Write the three net-ionic equations from the Interpretation & Analysis section that you think match the reactions occurring with the three mystery solutions, and explain the reasons why you chose the reactions you did.

2 Relating Ideas: All the reactants in these reactions are solutions of ionic compounds. Do you think the characteristic colors of these reactions are caused by the cations (the positive ions) in the substances, or the anions (the negative ions)? Explain your reasoning.

Extensions

1 Predicting Outcomes: Using the balanced chemical equations you wrote in Analysis & Interpretation #1, and given that the concentrations of the reactants are the same (they all have the same number of moles in the same volume), how many drops of NaOH are needed to react completely with five drops of each of the metal chloride solutions, if all the drops are the same size? If you followed the directions, which substance would be left over in each of the wells after the reaction: excess unreacted NaOH solution or excess unreacted metal chloride solution?

2 Formulating Hypotheses: Can you tell if the mystery solutions had concentrations of positive ions that were greater than, less than, or the same concentration as those used as standards in the first part of the lab? Explain your reasoning.

3 Designing Experiments: Can you think of a way to use the products of these reactions to make an estimate of the concentration of the mystery solutions? Write a procedure for making this determination. If time permits and your teacher approves of your suggestions, test your procedure.

4 Predicting Outcomes: In the storeroom, your teacher probably has pH paper, special paper which turns different colors when it is dipped in a solution, depending upon the concentration of OH^- (and H_3O^+) ions dissolved in a solution. A solution with a higher pH has a greater concentration of OH^- ions, as will be discussed in detail later in Chapter 17. Which of the metal chloride solutions would have the smallest concentration of OH^- ions left over after the reaction (lowest pH)? Why? (Hint: Examine the net-ionic equations.) How would the other metal chloride solutions compare to each other? If time permits and your teacher approves, obtain some pH paper and test your predictions.

5 Designing Experiments: The NaOH solution used as a reactant in this reaction (0.1 mol/L) has a pH of 13.0. The pH is a mathematical measurement related to OH⁻ concentration in moles per liter as follows:

$$pH = [14.00 + \log(OH^- \text{ conc.})]$$

Describe how you could use a pH meter to determine an estimate of the concentration of the positive ions in the mystery solutions.

INVESTIGATION 6: Pin-the-Formulas-on-the-Compounds

Introduction

Chemists compare the chemical properties of known substances to help them identify unknown substances. A common method involves reacting the unknown substance with measured amounts of other reactants, in order to complete a replacement reaction.

In this investigation, you will determine the formulas of two metallic salts. One salt is copper chloride, but you will need to find out if it is copper(I) chloride or copper(II) chloride. Another is iron chloride, but is it iron(II) chloride or iron(III) chloride?

The reaction to be used is a double replacement reaction with sodium hydroxide, NaOH. Aqueous solutions of both copper and iron ions form precipitates when they are combined with a solution containing aqueous hydroxide ions.

Phenolphthalein is an indicator which turns bright pink or red in the presence of unreacted aqueous OH⁻ ions. It will be used to show the reaction is complete. When there is no more metal left to react with the OH⁻ ions being added, the OH⁻ ions will trigger this color-change.

Objectives

- **Observe** chemical replacement reactions
- **Use** reagent measurements to determine the formula of a chemical substance
- **Infer** a conclusion from experimental data
- **Apply** reaction stoichiometry concepts

Materials

copper chloride (0.1 mol/1 liter of solution)
iron chloride (0.1 mol/1 liter of solution)
NaOH (sodium hydroxide, 0.1 mol/1 liter of solution)
phenolphthalein indicator solution
four fine-tipped dropper bulbs
two 8-well flat-bottom strips
10 toothpicks

Procedure

The solutions used in the investigation are very hazardous to the eyes. Safety goggles must be worn at all times during this investigation.

Any basic spills should be immediately washed with excess water. Make sure you know where the neutralization packages for basic spills are kept. Avoid getting any of the solutions on your skin or clothing.

1 Copy the following data tables into your lab notebook.

Copper Chloride Solution (five drops/well)					
Well #	1	2	3	4	5
# Drops of NaOH					
Iron Chloride Solution (five drops/well)					
Well #	1	2	3	4	5
# Drops of NaOH					

2 Obtain four dropper bulbs. Label them: "Cu," "Fe," "NaOH," and "In."

3 Fill the bulb labeled "Cu" with copper chloride solution. Fill the bulb labeled "NaOH" with sodium hydroxide solution.

4 Using one 8-well strip, place five drops of copper chloride solution into each of the first five wells. **For best results, try to make all the drops in this investigation about the same size.**

5 Using the bulb labeled "In," put one drop of phenolphthalein indicator in each of the five wells.

6 In the first data table, record the number of drops of NaOH solution that need to be added to react completely with the copper chloride so that the pink or red color of the phenolphthalein just begins to show clearly as shown in Figure 6-A. **Add the NaOH one drop at a time, and then mix**

the well with a toothpick before adding another drop. Use a different toothpick for each well.

7 Fill the bulb labeled "Fe" with the iron chloride solution, and repeat Steps 3-6 for the iron chloride in the second 8-well strip. In the second data table, record the number of drops of NaOH solution that need to be added to react completely with the iron chloride so that the pink or red color of the phenolphthalein just begins to show clearly, as shown in Figure 6-A.

8 Check with your teacher for instructions on how to dispose of your chemicals when finished. Remember to wash your hands thoroughly before leaving the laboratory.

Analysis & Interpretation

1 Organizing Ideas: Write out the balanced chemical equations for a double replacement reaction between sodium hydroxide and
a) copper(I) chloride;
b) copper(II) chloride;
c) iron(II) chloride;
d) iron(III) chloride.

Figure 6-A: Drop by drop, add enough sodium hydroxide solution so that the solution of copper chloride turns red or pink. Copper hydroxide will also precipitate to the bottom of the well.

2 Organizing Ideas: You may have already noticed that each of the solutions used had 0.1 mole of formula units for every liter of solution. Assuming that all of the drops were the same size, how do the numbers of formula units in a drop of each solution compare?

3 Organizing Data: What was the average number of drops of NaOH needed to react with copper chloride? What was the average number of drops of NaOH needed to react with iron chloride?

Conclusions

1 Relating Ideas: Using your answer from Analysis & Interpretation #2 and #3, determine how many NaOH formula units were needed to react with each copper chloride formula unit in your investigation. How many NaOH formula units were needed to react with each iron chloride formula unit?

2 Inferring Conclusions: Compare your answers for Conclusion Question 1 to the balanced chemical equations from the Analysis and Interpretation section. Which chlorides of copper and iron were in the solutions you used?

Extensions

1 Evaluating Methods: Share your data with other lab groups and calculate a class average for the ratio of NaOH formula units to copper chloride formula units, and then calculate a class average for the ratio of NaOH formula units to iron chloride formula units. Compare it to your results. Using the class average ratio as the accepted value, calculate your percent error.

2 Designing Experiments: What are some likely sources of imprecision in this experiment? If you can think of ways to eliminate them, ask your teacher to approve your suggestion, and run more trials.

3 Predicting Outcomes: How many drops of NaOH would it take to react with a solution of the same type of copper chloride, but containing 0.5 mol/liter of solution? How many drops of NaOH would it take for a solution of the same type of iron chloride that contained 0.5 mol/liter of solution? If your teacher has made the solutions for you, test your predictions.

4 Designing Experiments: What if your teacher made a solution of the same kind of copper chloride (or iron chloride) but forgot (or won't tell you) what the mol/liter concentration was? Can you think of a way to use samples of it with your 0.1 mol/liter solution of NaOH to find out what the concentration of the mystery solution is? If your teacher has made mystery solutions for you, ask your teacher to approve your suggestion, and see if you can find out what the concentration is.

INVESTIGATION 7: Generating and Collecting O_2

Introduction

Oxygen is a very important gas for humans and other animal life because it is required for survival. It is also one of the most reactive gases found in large quantities in nature. Oxygen makes up about 20% of the atmosphere, and it can react with just about anything, provided that there is energy to help the reaction start. When oxygen reacts with another substance and large quantities of heat and light are generated, it is called burning or combustion. Oxygen is referred to as "supporting combustion," although it is not usually considered a combustible gas itself. But what is the nature of combustion? How does combustion in an environment of 100% oxygen compare to that of air (20% oxygen)? To examine these questions, you will generate and collect oxygen gas in this investigation, so you can test firsthand some of its chemical and physical properties.

Objectives

- **Observe** chemical processes and interactions in generating O_2
- **Collect** O_2 gas by water displacement
- **Relate** chemical and physical properties to observations
- **Apply** conclusions to different settings
- **Formulate** hypotheses
- **Compare** reactions of O_2 gas in different situations

Materials

petri dish bottom
Bunsen burner *or* candle
micro O_2 generator
7 test tubes
5 one-holed rubber stoppers
5 small pegs to fit stoppers
steel wool
3% H_2O_2 solution (hydrogen peroxide)
1.0 *M* KI
wood splints
tap water
two beakers (at least 100 mL)
forceps

Procedure

The solutions used in the investigation are hazardous to the eyes. Safety goggles must be worn at all times during this investigation.

A beaker of water should be kept nearby at all times, in case you need to quickly extinguish a burning wood splint.

1 Copy the following table into your lab notebook, and record all of your observations in it.

Test Tube #	Prediction	Observations
1 flame, air		
2 flame, O_2		
3 ember, air		
4 ember, O_2		
5 mouth-up, flame		
6 mouth-down, flame		
7 lit steel wool, O_2		

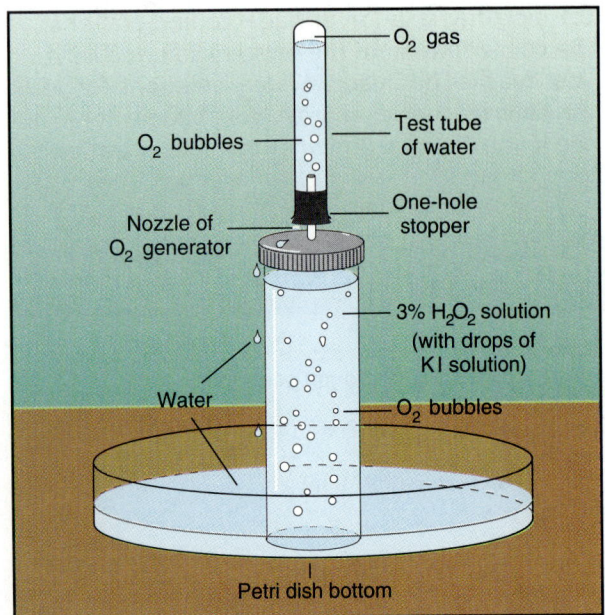

Figure 7-A: Collect oxygen gas using this O₂-generator.

test tube is completely filled with oxygen, remove it from the nozzle and plug the stopper hole with a small peg to prevent the collected gas from leaking out. Repeat this procedure for the other four test tubes (#4, #5, #6, and #7 — remember to leave #1 and #3 set aside and empty).

6 If the reaction slows down so that it takes more than a minute to collect the tube of oxygen gas, uncap the vial, decant the liquid into a large beaker, and replace it with fresh H_2O_2 solution. Then replace the cap and nozzle, and resume collecting gas.

For Steps 7-13, read the directions carefully, predict what you think might happen, and write your prediction down in the data table. Then follow the directions and record your observations alongside your predictions in the data table.

7 *Flaming wood splint and tube #1 (control):* Light the bunsen burner (or candle) and use it to light one end of a wood splint.

 Keep the gas-generating vial away from flame and wooden splint!

Take test tube #1 (control test tube — filled with air), hold it horizontally, and carefully insert the flaming wood splint. Record your observations in the data table.

8 *Flaming wood splint and tube #2 (oxygen):* Relight the wood splint, remove the stopper from test tube #2, hold the test tube horizontally, and carefully insert the burning wood splint. Record your observations in the data table.

9 *Glowing wood splint and tube #3 (control):* Light the end of the wood splint again, let it burn for 5-10 seconds, then **blow out the flame, but make sure a glowing ember persists on the tip of the wood splint.** Take test tube #3 (control test tube — filled with air), hold it horizontally, and carefully insert the glowing wood splint. Record your observations in the data table.

10 *Glowing wood splint and tube #4 (oxygen):* Repeat Step #9 above, using test tube #4, and holding the test tube horizontally. (Carefully unstopper the test tube before inserting the glowing wood splint.) Record your observations in the data table.

2 Fill one beaker full of water. Be sure both beakers are nearby, and clearly labeled, so you can tell quickly which one is full of water, and which will be used to dispose of spent solution.

3 Fill test tubes #2, #4, #5, #6 and #7 completely with water. Insert the one holed stoppers without creating any air bubbles in the test tubes. (Leave test tubes #1 and #3 empty and open in the test tube rack — they will serve as controls.)

4 The micro O₂ generator is a plastic vial capped with a lid having a nozzle.

 Make sure your O₂ generator cap has a nozzle and that the nozzle is not plugged in any way before continuing the investigating.

Remove the cap, and carefully add enough 3% H_2O_2 to fill the vial to within 1 cm of the top. Then add 3 or 4 drops of KI solution. Replace the cap and set the O₂ generator in the petri dish. Record your observations about the reaction in your lab notebook.

5 To collect the gases, place test tube #2 from Step 3 mouth downward over the nozzle of the generator, as shown in Figure 7-A. As soon as the

11 Unstopper test tube #5 and hold it **mouth upward** for two minutes. Then, try the glowing wood splint test (as described in Step #9). Record your observations in the data table.

12 Unstopper test tube #6 and hold it **mouth downward** for two minutes. Then try the glowing wood splint test (as described in Step #9). Record your observations in the data table.

13 Twist a few strands of steel wool into a single yarn-like string. Fold it in half, and hold the two loose ends together with a pair of forceps. Hold the folded bend in the bunsen burner (or candle) flame for a few seconds, then remove it and observe. Repeat this procedure, but this time quickly and carefully insert the glowing steel wool into test tube #7. Record your observations in the data table.

14 If time permits, (check with your teacher) refill the test tubes, first with water, and then with oxygen gas, where appropriate, and recheck any of the above observations. Record all observations in your lab notebook.

15 Check with your teacher for instructions on how to dispose of your chemicals when finished. Remember to wash your hands thoroughly before leaving the laboratory.

Analysis & Interpretation

1 Analyzing Information: What evidence was there that a reaction was occurring inside the generator?

2 Relating Ideas: Write a balanced chemical equation for the O_2 generator reaction. (Hint: include KI in the equation, but only as a catalyst. Refer to Table 8-2 for the correct way to symbolize a catalyst in a reaction.)

3 Relating Ideas: Write a word equation for the wood splint tests, using the following reactants and products: ash, wood, carbon dioxide, water vapor, and oxygen.

Conclusions

1 Inferring Conclusions and Relating Ideas: What caused the burning wood splint to eventually be extinguished in test tube #1? (Hint: refer to your word equation for the wood splint test from Analysis & Interpretation #3.)

2 Inferring Conclusions and Resolving Discrepancies: Compare your results for test tubes #1 and #2. Explain the differences between the two by explaining what the limiting reactant is for the wood-oxygen reaction.

3 Inferring Conclusions and Resolving Discrepancies: Compare your results for test tubes #3 and #4. Why didn't the ember in #3 behave more like the ember in #4? (Hint: which reaction was quicker and more thorough? Why?)

4 Inferring Conclusions and Resolving Discrepancies: Is oxygen gas more dense, less dense, or about equal in density to air? (Hint: compare your results for test tubes #5 and #6. What was different about how they were tested and the results of the tests?)

5 Inferring Conclusions and Resolving Discrepancies: How does the "burning" of steel wool in air compare to the burning of steel wool in pure O_2 (test tube #7)? Explain.

6 Organizing Conclusions: Using your observations of the gas itself, the generation reaction, and the wooden splint tests, list as many chemical properties of oxygen as you can. Explain what evidence you have for each one. (There are at least four properties that were observable in this investigation.)

7 Organizing Conclusions: Using your observations from this investigation, list as many of oxygen's physical properties as you can. Explain what evidence you have for each one. (There are at least four properties that were observable in this investigation.)

Extensions

1 Predicting Outcomes: If you were to put a plain wood splint (with no ember or flame) into a tube full of oxygen, would it burst into flame? Would an ember form? Explain your answer. (Hint: what is needed for a chemical reaction along with the reactant(s)?)

2 Applying Conclusions: Is air more or less than half oxygen? What evidence from your investigation supports your answer? How would you determine the percentage of oxygen in air more precisely?

3 Applying Conclusions: Why is there a strict

"no smoking" ban around oxygen equipment in hospitals, and why do flight attendants tell air travelers that the first thing to do in an emergency requiring the use of oxygen masks is to extinguish all smoking materials? Explain your answer using evidence from this investigation.

4 Applying Models and Designing Experiments: Two hundred years ago, scientists used a theory about the burning of materials such as wood that is very different from the one described by the word equation from Analysis & Interpretation #3. At the time, the nature of different gases was not fully understood, and the theory of atoms and elements was still in its infancy. At that time, scientists believed that combustible materials contained *phlogiston,* a substance that caused burning when it was absorbed into air. A English chemist named Joseph Priestley, who settled in Pennsylvania in 1794, is credited with isolating oxygen in 1771. He considered it to be "de-phlogisticated air." Research and write a report about the phlogiston theory, the evidence that eventually refuted it, or the life of Priestley. An alternative might be to design an experiment that would test the phlogiston theory, or describe the chemical reactions used by Priestley to isolate oxygen.

INVESTIGATION 8: Generating and Collecting H₂

Introduction

Hydrogen gas has the lowest density of all the substances found on earth. Once, hydrogen was used to buoy up blimps and dirigibles. Today, however, such aircraft use only helium, even though helium is less buoyant and considerably more difficult to obtain. Why was the switch made? What was wrong with using hydrogen? To answer these questions, you will generate and collect small quantities of hydrogen gas in this investigation, so you can test some of its properties first-hand.

Objectives

- **Observe** chemical processes and interactions in generating H₂
- **Collect** H₂ gas by water displacement
- **Relate** chemical and physical properties to observations
- **Apply** conclusions to larger settings
- **Formulate** hypotheses
- **Compare** reactions of H₂ gas in different situations

Materials

petri dish half
micro H₂ generator
5 small test tubes
5 one-holed rubber stoppers
5 small pegs (fitting holes in stoppers)
1.0 M HCl
tap water
Bunsen burner *or* candle
wood splints
two beakers (at least 100 mL)
forceps or tongs (optional)

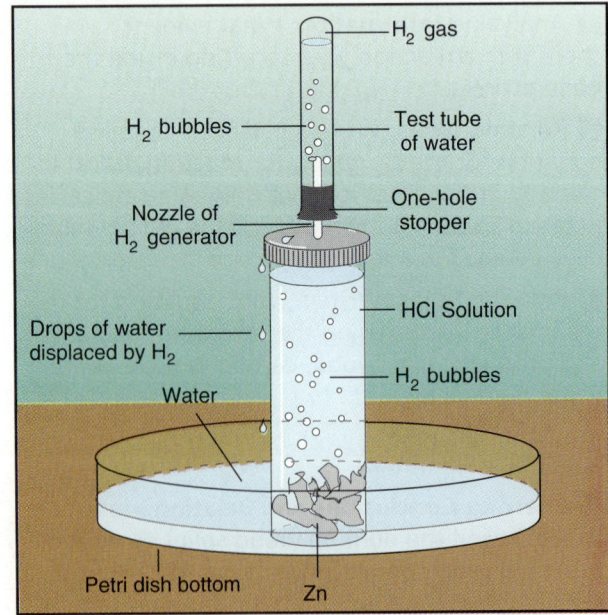

Figure 8-A: Collect hydrogen gas in the test tubes by water displacement. Invert the test tubes of water over the nozzle of the hydrogen generator.

Procedure

The acid solution used in the investigation is hazardous to the eyes. Safety goggles must be worn at all times during this investigation.

Any acid spills should be immediately washed with excess water. Make sure you know where the neutralization packages for acid spills are kept. A beaker of water should be kept nearby at all times, in case you need to quickly extinguish a burning wood splint or dilute an acid spill.

1 Copy the following table into your lab notebook, and record all of your observations in it.

Test Tube #	Prediction	Observations
1 flame, pure H_2		
2 flame, 50% H_2		
3 ember, pure H_2		
4 mouth-up, flame		
5 mouth-down, flame		

2 Fill one beaker with water. Be sure both beakers are nearby, and they are labeled clearly so you can tell quickly which is full of water, and which will be used to dispose of spent solution.

3 Fill test tubes #1, #3, #4, and #5 completely full of water. Fill test tube #2 only half way. Insert the one-holed stoppers into all five test tubes. Avoid creating any air bubbles in test tubes #1, #3, #4, or #5.

4 The micro H_2 generator is a plastic vial containing several pieces of zinc (Zn) metal, capped with a lid having a nozzle.

Make sure your H_2 generator cap has a nozzle and that the nozzle is not plugged in any way before continuing with the investigation.

Remove the cap and carefully add enough 1.0 *M* HCl to fill the vial to within 1 cm of the top. Replace the cap and set the H_2 generator in the petri dish. Observe the reaction, and record your observations in your lab notebook.

5 To collect the gas, place the first water-filled test tube from Step 3 upside down over the nozzle of the generator, as shown in the Figure 8-A. As soon as the test tube is completely filled with the hydrogen, remove it from the nozzle and plug the stopper hole with a small peg to prevent the collected gas from leaking out. Repeat this procedure for the other four test tubes.

6 If the reaction slows down so that it takes more than a minute to collect the tube of hydrogen gas, lift off the tube, uncap the vial, decant the remaining liquid into a large beaker, replace it with fresh solution, replace the cap and nozzle, and resume collecting gas.

For Steps 7–11, read the directions carefully, predict what you think might happen, and write your prediction down in the data table. Then follow the directions and record your observations alongside your predictions in the data table.

Because this investigation is carried out on a microscale level, the pop-tests used in Steps 7–11, though potentially loud, are completely safe. The test tubes may be held with forceps or tongs or else held in your hand. Check with your teacher to see how you should hold the test tubes for the pop-tests. Be sure to keep the gas-generating vial away from the flame and the wooden splint!

7 *Flaming wood splint and tube #1 (hydrogen):* Light the bunsen burner (or candle) and use it to light one end of a wood splint. Remove the stopper from test tube #1 and carefully insert the burning splint into the mouth of the test tube. Record your observations in the data table.

8 *Flaming wood splint and tube #2 (half hydrogen):* Repeat the pop-test described in Step #7 with test tube #2, which contains only half as much hydrogen as #1 did. Record your observations in the data table.

Investigation 8: Generating and Collecting H_2 849

9 *Glowing wood splint and tube #3 (hydrogen):* Use test tube #3 to repeat the pop-test, but this time, **blow out the flame but make sure a glowing ember persists on the tip of the wood splint as you insert it into the test tube.** Record your observations in the data table.

10 Unstopper test tube #4 and hold it **mouth upward** for 30 seconds; then try the pop-test. Record your observations in the data table.

11 Unstopper test tube #5 and hold it **mouth downward** for 30 seconds; then try the pop-test. Record your observations in the data table.

12 If time permits, (check with your teacher) refill the test tubes, first with water, and then with hydrogen gas, and recheck any of the above observations. Record all observations in your lab notebook.

13 Check with your teacher for instructions on how to dispose of your chemicals when finished. Remember to wash your hands thoroughly before leaving the laboratory.

Analysis & Interpretation

1 Analyzing Information: What evidence was there that a reaction was occurring inside the generator?

2 Relating Ideas: Write a balanced chemical equation for the H_2 generator reaction.

3 Relating Ideas: Write a balanced chemical equation for the reaction taking place during the pop-test. (Hint: water vapor is the product of the reaction).

Conclusions

1 Inferring Conclusions and Resolving Discrepancies: Compare your results for test tubes #1 and #2. The second test tube had only half as much hydrogen; was the pop-test only half as loud? Explain your answer, referring to the balanced chemical equation for the pop-test.

2 Inferring Conclusions and Resolving Discrepancies: Consider your results for test tube #3. Did the hydrogen ignite? Can you explain why you observed what you did? (Hint: refer to the balanced chemical equation for the pop test and the chemical properties of hydrogen described in Section 10.2.)

3 Inferring Conclusions and Resolving Discrepancies: Is hydrogen gas more dense or less dense than air? (Hint: compare your results for test tubes #4 and #5. What was different about how they were treated, and how did this affect their pop-tests?)

4 Organizing Conclusions: Using your observations of the gas itself in each trial, the generation reaction, and the pop-tests, list as many chemical properties of hydrogen as you can. Explain what evidence you have for each one. (There are at least four that were observable in this investigation.)

5 Organizing Conclusions: Using your observations from this investigation, list as many of hydrogen's physical properties as you can. Explain what evidence you have for each one. (There are at least four that were observable in this investigation.)

6 Applying Conclusions: List the advantages and disadvantages to using large quantities of hydrogen to inflate balloons, blimps, and dirigibles.

Extensions

1 Predicting Outcomes: Would the same flame that you used to test the tube of H_2 and air also be adequate to cause a room full of H_2 and air to explode, or would it require a very large flame? Explain your answer.

2 Predicting Outcomes and Designing Experiments: Air is about 20% oxygen. Use this information to predict what ratio of air to hydrogen will provide the loudest pop-test. Design an experiment which will test a range of proportions, including the one you predicted would be loudest. If your teacher approves of your suggestions, test your predictions.

3 Predicting Outcomes: For the test tubes having positive pop-tests, try repeating the pop-test on the same tube over and over. How many times will each tube pop? Can you explain why?

4 Applying Conclusions: Research the *Hindenburg* airship disaster. Write a report, and include in it a comparison of the amount of hydrogen in the *Hindenburg* and the amount of hydrogen present in one test-tube in this investigation. (Use water to determine the volume of the test tubes you used in the lab.)

INVESTIGATION 9: Testing Reaction Combinations of H_2 and O_2

Introduction

Hydrogen, H_2, and oxygen, O_2, are two gases that react with each other in a very quick, exothermic (heat-producing) manner. The explosiveness of this reaction is greatest when the hydrogen and oxygen are reacted in just the proper proportion. This reaction is used to power three of the rocket engines that carry the space shuttle into orbit. In this investigation, you will generate hydrogen and oxygen to test their explosive nature, first separately, and then in reactions of different proportions, collecting them using the "water displacement" method. After finding the most "powerful" reaction combination, you will use this reaction to launch a rocket across the room.

Objectives

- **Observe** chemical processes and interactions in generating H_2 and O_2
- **Collect** H_2 and O_2 gas by water displacement
- **Relate** chemical concepts such as stoichiometry, limiting reagents, activation energy, and thermodynamics to observations of chemical reactions
- **Infer** a conclusion from experimental data
- **Evaluate** methods

Materials

petri dish bottom
piezoelectric sparking element
micro O_2 generator
micro H_2 generator
1.0 M HCl
3% H_2O_2
1.0 M KI
calibrated collection bulb
two beakers (at least 100 mL)
tap water
tongs or forceps (optional)

Procedure

 Because this investigation is carried out on a microscale level, the explosions, though potentially loud, are completely safe.

 The solutions used in the investigation are hazardous to the eyes. Safety goggles must be worn at all times during this investigation.

Any acid spills should be immediately washed with excess water. Make sure you know where the neutralization packages for acid spills are kept. A beaker of water should be kept nearby at all times, in case you need to quickly extinguish a burning bulb of gas or dilute an acid spill.

1 Copy this data into your lab notebook, and record your observations in it.

Parts H_2	6	5	4	3	2	1	0
Parts O_2	0	1	2	3	4	5	6
Relative Loudness (rated from 0–10)							

2 Fill the petri dish bottom $\frac{3}{4}$ full of water. Fill one beaker with water. Be sure both beakers are nearby, and you can tell quickly which is full of water, and which will be used to dispose of spent solution.

3 The micro H_2 generator is a plastic vial containing several pieces of zinc (Zn) metal, capped with a lid having a nozzle.

 Make sure both generator caps have a nozzle and that the nozzle is not plugged in any way before continuing with the investigation.

Remove the cap and carefully add enough 1.0 M HCl to fill the vial to within 1 cm of the top. Replace the cap and set the H_2 generator in the petri dish of water. Observe the reaction, and record your observations in your lab notebook.

4 The micro O_2 generator is a plastic vial capped with a lid having a nozzle. Remove the cap, and carefully add enough 3% H_2O_2 to fill the vial to within 1 cm of the top. Then add 3 or 4 drops of KI solution to serve as a catalyst. Replace the cap and set the O_2 generator in the petri dish beside the H_2 generator. Record your observations about the reaction in your lab notebook.

5 Fill the collection bulb completely with water from the petri dish. The best method is to squeeze the bulb tightly, invert it into the petri dish, and try to draw up as much water as possible. Then hold the bulb mouth upward and squeeze out the remaining air. Without letting go of the bulb, invert it again into the petri dish and draw up the remainder of the water needed to fill up the bulb.

6 To collect hydrogen gas, place the water-filled collection bulb from Step 5 mouth downward over the nozzle of the H_2 generator, as shown in the Figure 9-A. As soon as it is completely filled with hydrogen, remove it from the nozzle and place a finger over the mouth of the bulb to prevent the collected gas from leaking out.

7 Test the bulb of hydrogen gas by moving your finger aside and inserting the wire tip of the piezoelectric sparker into the bulb.

Keep the piezoelectric sparker away from the gas-generating vials. Hold the bulb securely. (Your teacher will instruct you on whether you should hold the bulb with your hand, or tongs, or forceps.)

Pull the trigger of the piezoelectric sparker and observe. Record the loudness of the reaction (on a scale of 1–10) in the **first** square of the data table.

8 To collect oxygen gas, refill the bulb with water, but place the bulb mouth downward over the nozzle of the O_2 generator, and repeat Steps 6 and 7, pop-testing it with the piezoelectric sparker. Record the loudness of the reaction (on a scale of 1–10) in the **last** square of the data table.

9 Refill the bulb with water, and collect another bulb of O_2, but when the bulb is about $\frac{1}{6}$ full of O_2 (with $\frac{5}{6}$ water), move the bulb from the O_2 generator to the H_2 generator, and continue collecting. When the bulb is filled with gas, you will have a combination that is 1 part oxygen and 5 parts hydrogen. Remove the bulb from the nozzle, test it as in Step 7, and record the relative loudness of the reaction in the appropriate box in the data table (second from left).

10 Repeat Step 9 above, making the switch from the O_2 to the H_2 generator at different times, so you can test various proportions (2:4, 3:3, 4:2, and 5:1), to complete the data table. Try to determine the optimum (most explosive) combination.

11 If either generator reaction slows down so that it takes more than a minute to collect the bulb, lift off the bulb, uncap the vial, decant the remaining liquid into a large beaker, replace it with fresh solution(s) from the appropriate bottle(s), replace the cap, and resume collecting gas.

12 *Rocket Launch (ask your teacher before performing this step):* Collect the optimum (loudest) mixture one more time, and instead of holding onto the bulb, aim it outward at a target chosen by your teacher.

 Make sure nobody is in the line of fire.

Pull the trigger and launch your micro rocket. Can you think of ways to make your rocket go farther? Try them!

13 Check with your teacher for instructions on how to dispose of your chemicals when finished. Remember to wash your hands thoroughly before leaving the laboratory.

Analysis & Interpretation

1 **Relating Ideas:** Write a balanced equation for the reaction:

Figure 9-A: Hydrogen gas and oxygen gas are collected by water displacement.

852 Microscale Investigations

a) taking place inside the O_2 generator
b) taking place inside the H_2 generator.

2 Interpreting Ideas: Which do you think will be used up first: the KI solution in the O_2 generator or the Zn in the H_2 generator? Why?

3 Analyzing Methods: What laboratory techniques did you use to help ensure that the gas-collecting bulb contains only the desired gas (not air, etc.)?

4 Relating Ideas: Write a balanced equation for the reaction of O_2 and H_2.

5 Analyzing Results: Make a bar graph of the relative loudness produced by your pop-tests of the mixtures, with the loudness from 0–10 on the y-axis and trials 1–7 on the x-axis.

Conclusions

1 Inferring Conclusions and Relating Ideas: From your observations, state the relative combustibility of:
a) pure O_2.
b) pure H_2.

2 Resolving Discrepancies and Inferring Relationships: Were there any reaction combinations that produced no reaction at all? Explain what happened.

3 Inferring Conclusions and Inferring Relationships: What proportion of oxygen to hydrogen produced the most explosive reaction? Refer to the balanced chemical equation for the reaction, and the concepts of limiting reagents and maximum yields, to explain why that combination was most explosive.

4 Resolving Discrepancies and Relating Ideas: Why don't the O_2 and H_2 react immediately as soon as they mix in the collection bulb? (The individual O_2 and H_2 molecules are certainly colliding with one another.) What role does the spark (or flame) play?

Extensions

1 Predicting Outcomes: How would your results change if you performed the same experiment with the gases chilled? If time permits, go back and test your prediction. Explain your results.

2 Evaluating Information: Share your data with the rest of the class. Calculate the average relative loudness values for each mixture for the entire class, and make a bar graph with the results.

3 Analyzing Methods and Inferring Conclusions: What methods did you attempt for making your rocket fly farther? Which ones worked best? Explain why they worked.

4 Applying Models: The space shuttle carries 680 000. kg of fuel for its main engines. These engines use liquid hydrogen and liquid oxygen. Based on your results in this investigation, what is the most effective way to divide this mass: how much of it should be liquid oxygen, and how much liquid hydrogen?

5 Applying Conclusions: Measure how much volume is in your collection bulb by filling it with water, as if you were going to collect more gas, and then emptying it into a graduated cylinder. Calculate the volumes of hydrogen and oxygen that were present in the most effective combination. Using the density values for hydrogen and oxygen gas given in Chapter 10, calculate the masses of hydrogen and oxygen present in your collection bulb. Comparing these values to those from Extension #3 above, calculate how many of these bulbs of mixed hydrogen and oxygen would be equivalent to the fuel supply of the space shuttle.

6 Evaluating Methods: Research the phenomenon of piezoelectricity. Can you explain how the piezoelectric sparker works?

INVESTIGATION 10: "Wet" Dry Ice

Introduction

The partial phase diagram for carbon dioxide, Figure 10-A, shows that at normal room conditions, CO_2 is most stable as a gas. The phase diagram also explains why solid CO_2 earned the nickname "dry ice," for, if a sample is left out in a room, it will not melt to a liquid, but will instead immediately sublime to a gas, as shown by the arrow in Figure 10-A.

Although you may never have seen liquid CO_2 before, that doesn't mean it doesn't exist. It is simply not stable at 1 atm of pressure. To observe it, one must be at or above the triple point pressure of CO_2, which is labeled "X" in the diagram.

Objectives

- **Observe** a change of state while varying pressure
- **Relate** a phase diagram to observations of CO_2

Materials

4–5 g of dry ice, broken into rice-grain-sized pieces
transparent plastic cup
graduated pipet
tap water
scissors
pliers
forceps

Figure 10-A: A partial phase diagram for carbon dioxide shows the triple point.

Procedure

 Safety goggles must be worn at all times during this investigation.
 Dry ice is cold enough to cause frostbite, so heat and cold resistant gloves should be used if you handle it.

1 Place 2–3 very small pieces of dry ice on the table and observe them until they have completely sublimed.

2 Fill the plastic cup with tap water to a depth of 4–5 cm.

3 Cut the tapered end (tip) off the graduated pipet.

4 Carefully slide 8–10 pieces of dry ice down the stem and into the bulb, using forceps.

5 Using a pair of pliers, clamp the opening of the stem securely shut so that no gas escapes. Holding the tube by the pliers, lower the pipet into the cup just enough so that the bulb is submerged, as shown in Figure 10-B. From the side of the cup, observe the behavior of the dry ice.

6 As soon as the dry ice has begun to melt, quickly loosen the grip on the pliers, still holding the bulb in the water, and observe the CO_2.

7 Tighten the grip on the pliers again and observe.

8 Repeat steps 6 and 7 as many times as possible.

9 Check with your teacher for instructions on cleaning up your lab area when finished.

Analysis & Interpretation

1 Analyzing Results: What differences did you observe between the subliming and the melting of CO_2?

2 Analyzing Methods: As you melted the CO_2 sample over and over, why did it eventually get used up? Can you think of ways that might have made the sample last longer?

3 Analyzing Methods: What purpose(s) do you suppose the water in the cup served?

4 Resolving Discrepancies and Relating Ideas: What is the difference between the conditions under which dry ice and regular ice will melt? Explain how these differences are a result of the types of bonding and intermolecular forces present.

Conclusions

1 Interpreting Graphics and Inferring Conclusions: Sketch the phase diagram for CO_2 on another sheet of paper. Using it as a "map," diagram the "path" taken by your sample of CO_2 as it melted over and over.

Extensions

1 Predicting Outcomes: What would have happened if fewer pieces of dry ice (only 1 or 2) were placed inside the pipet bulb? If time permits, go back and test your prediction.

2 Predicting Outcomes: What might have happened if more pieces of dry ice (20 or 30, for example) had been placed inside the pipet bulb? How quickly would the process occur? If time permits, go back and test your prediction.

3 Predicting Outcomes: What would have happened if the grip on the pliers had not been released once the dry ice melted? If time permits, go back and test your prediction.

Figure 10-B: Clamp the end of the pipet shut with the pliers. Submerge the bulb with the dry ice in a clear cup of water.

INVESTIGATION 11: Measuring the Triple Point Pressure of CO_2

Introduction

Using some of the techniques from Investigation 10, "Wet" Dry Ice, and by making micro pressure gauges, the triple-point pressure for CO_2 can be determined. Examine the phase diagram with Investigation 10. In the diagram, the triple-point pressure is referred to as "X." The three phases of CO_2 can exist together at equilibrium only at this pressure and temperature of $-56°C$. In addition, this is the highest pressure where CO_2 will sublime, and the lowest pressure where liquid CO_2 can exist.

Objectives

- **Interpret** a phase diagram
- **Observe** changes in pressure during a phase change
- **Relate** pressure values to observations of air-column length
- **Determine** the triple-point pressure of CO_2
- **Infer** a conclusion from experimental data

Materials

micro pressure gauge:
7-cm tube (cut from thin-stemmed pipet)
15–20 cm of thread
one fine-tipped dropper bulb
scissors
metric ruler
hot glue gun
permanent fine-tipped marker
dark colored water
other materials:
4–5 g of dry ice, broken into rice-grain-sized pieces
transparent plastic cup
graduated pipet
pliers
forceps

Procedure

Safety goggles must be worn at all times during this investigation.

Dry ice is cold enough to cause frostbite, so heat and cold resistant gloves should be used if you handle it.

The tip of the hot glue gun is very hot, as is the glue. Be very careful!

1 Copy this data table into your lab notebook, and record your measurements in it.

	Trial 1	Trial 2	Trial 3	Trial 4	Trial 5
Initial reading (cm)					
CO_2 melt pt. reading (cm)					

2 Fill the plastic cup with about 4–5 cm of water. Cut the tapered tip off the graduated pipet. Your teacher may have prepared micro pressure gauges for you already. If so, go to Step 6. If not, Steps 3–5 will explain how to make a micro pressure gauge.

3 Take the 7-cm tube (cut from a thin-stemmed pipet), and place a small drop of hot glue on one end, sealing it. Once it has cooled, tie the thread around the tube below the seal, as shown in Figure 11-A. Trim away any excess glue so that this tube will easily pass through the end of the graduated pipet from Step 2.

4 Measuring from the inside edge of the glue, mark off every centimeter along the length of the tube with the fine-tipped permanent marker, numbering them for reference.

5 Using the fine-tipped dropper bulb, place a small drop of dark colored water inside the open end of the tube. Record its position on the scale from the inside edge of the drop to the inside end of the glue plug. This should be a measurement of the length of the trapped air column inside the tube, as shown in Figure 11-A. Note this measurement in the data table.

6 Carefully slide 8–10 pieces of dry ice down the stem and into the graduated pipet bulb from Step #2, as in Investigation 10. However, instead of clamping the opening of the graduated pipet shut, insert the micro pressure gauge, open-end downward, so that it hangs in the stem (not the bulb) of the graduated pipet, as shown in Figure 11-B. The thread of the micro pressure gauge should be hanging out of the end of the pipet. Use the pliers to clamp the pipet opening shut around the thread, so no gas can escape, and the micro pressure gauge remains suspended in the stem of the pipet.

Figure 11-A: Calibrate the micro pressure gauge by marking the tube at 1 cm intervals from the inside edge of the glue plug. When measuring the length of the air column, measure from the inside of the colored water drop.

Figure 11-B: This is the proper way to use your micro pressure gauge to measure the triple point of CO_2.

Analysis & Interpretation

1 Analyzing Information: What happened to the pressure inside the bulb before, during, and after the dry ice melted? Explain what evidence you have as a basis for your answer from this investigation.

2 Interpreting Graphics: Sketch a rough graph of pressure vs. time for the CO_2 in the bulb during the entire interval you observed. (Include the repetitive pressure changes.)

3 Organizing Data: What was the initial length of the air column inside the tube? What was the length of the air column at the instant the dry ice started to melt? What was the initial pressure (in atmospheres) inside the tube when you placed the drop of liquid in it?

Conclusions

1 Inferring Relationships: Using your answers from Question 3 in Analysis & Interpretation, calculate the pressure (in atmospheres) inside the pipet when the liquid CO_2 first appeared. Provide your data and formula, and show your work. (Hint: consider how pressure and volume of a gas sample are related.)

2 Relating Ideas: How does your answer to Question 1 above relate to the triple point pressure "X"? Justify your explanation using evidence from the phase diagram for CO_2.

3 Evaluating Methods: The accepted value for the triple point of CO_2 is 5.11 atm. Compare this value to your experimental value. Calculate your percent error.

Extensions

1 Evaluating Methods: When you used the micro pressure gauge, you measured the **length** of a trapped air column, with a glue plug on one end and a drop of water on the other. However, the calculations for gases are for a relationship between pressure and **volume,** not length. What assumptions are being made in using length of the trapped air column in place of volume?

2 Evaluating Methods and Designing Experiments: What possible sources of error can you identify in this procedure? If you can think of ways to eliminate them, ask your teacher to ap-

7 Holding the tube with the pliers, lower the bulb (not the stem) of the pipet until it is submerged in the cup of water, as shown in the Figure 11-B.

8 From the side of the cup, observe the movement of the drop of colored water in the micro pressure gauge before and while the CO_2 melts. Specifically, note the position in centimeters (as marked on the micro pressure gauge) of the top of the colored drop the instant the dry ice begins to melt. Then, quickly loosen the grip on the pliers.

9 Record what the drop's position was (when the dry ice began melting in your data table.) Now that the pliers are loosened, observe the CO_2 and the drop of colored water in the micro pressure gauge.

10 Tighten the grip on the pliers again and observe as before.

11 Repeat steps 8–10 as many times as possible.

12 Check with your teacher for instructions on how to dispose of your chemicals when finished.

prove your suggestion, run the procedure again, and calculate percent error again.

3 Inferring Relationships: Determine the volume of the pipet bulb. Assuming that the pressure measured on the pressure gauge is correct and the temperature is the triple point of CO_2, $-56°C$, calculate how many moles of CO_2 gas were present. Calculate how many grams this would be. What was the density of CO_2 gas under these conditions?

INVESTIGATION 12: Water Purification

Introduction

Knowledge of the chemical and physical properties of different substances is useful in many ways. One way to put this knowledge to work is in separating mixtures of substances. Very few substances are found naturally in a pure state, so separating mixtures, such as metal ore, petroleum, and impure water, into pure substances is frequently necessary. Often, a series of steps will be required to remove the different impurities in a mixture.

In this investigation, a sample of foul water will be purified using the following steps: oil-water-sediment separation, sand filtration, and charcoal adsorption/filtration. These three approaches are very similar to some of the approaches often used in practical settings. For example, some people have purchased charcoal filters that they attach to the water supply in their homes or apartments to remove impurities from the water that can cause bad odors or taste. Your objective is to recover as much pure water as possible.

Objectives

- **Observe** chemical and physical properties of a mixture
- **Categorize** mixtures as solutions, colloids, or suspensions
- **Relate** knowledge of properties to the task of purifying the mixture
- **Compare** means of purifying the mixture

Materials

foul water sample
incremented wood splint
four test tubes
one thin-stemmed pipet
separatory bulb
2 long-stemmed micro funnels
test tube rack
small piece of wire strainer
tray
spoon
tissue paper
sand
charcoal
flashlight or slide projector and screen

Procedure

 The impure water samples used in the investigation could be hazardous to the eyes. Safety goggles must be worn at all times during this investigation. Avoid getting any of the solutions on your skin or clothing.

1 Make a data table like this in your lab notebook:

	Before Treatment	After oil/H_2O separation	After sand	After charcoal
Color				
Clarity				
Odor				
Oil present?				
Solids present?				
Solution, Colloid, or Suspension?				
Volume				

2 Use a test tube to obtain a sample of foul water from your teacher. Carefully examine the properties of the foul water sample and record your observations in the data table. To measure the volume of the sample, place the test tube in the rack with the wood splint alongside it, and read to a tenth of a mL, as shown in Figure 12-A. (Hint: if you don't remember how to distinguish between solutions, colloids, and suspensions, there is a summary of their properties in Table 14-3 on page 410.)

3 Using the thin-stemmed pipet, draw up the entire foul water sample and deliver it into the separatory bulb. Turn the separatory bulb mouth-side down and set it back on top of the first test tube as shown in Figure 12-A. Surface tension should keep the liquid from spilling out. Let it stand undisturbed until two liquid layers have formed, and some sediment particles have settled into a layer on the bottom.

4 Keeping the separatory bulb mouth-side down, carefully squeeze out the first drop or two of sediments back into the first test tube. Then, deliver the water layer to another clean test tube, being careful not to allow any of the top oily layer to drip out. Finally squeeze out the remaining oil layer back into the first test tube.

5 Observe the properties and measure the volume of the water layer in the second test tube. Record this information in the data table. Set aside the water sample in the second test tube.

6 Place a spoonful of sand in a strainer and sift it for about 15 seconds over a plastic tray.

7 Place some coarse sand (left inside the strainer) and fine sand (from the tray) in layers in the bottom half of a long-stemmed micro funnel as shown in Figure 12-A.

8 Place the funnel in the third test tube, and carefully pour the water sample from Step 5 into it. When the filtration appears to be complete, remove the sand filter funnel (be sure to catch the last drops so you are retrieving as much water as possible).

9 Observe the properties and measure the volume of the filtered water sample. Record this information in the data table. Set aside the filtered water sample in the third test tube.

10 Tear off a piece of tissue paper about the size of a quarter, wad it up, and place it in the second long-stemmed micro funnel. Use a piece of wire to pack it down gently but securely into the tapered tip. Then add enough charcoal to fill the lower half of the stem as shown in Figure 12-A. Place the funnel in the fourth test tube.

11 Carefully pour the water from the third test tube into the funnel and allow it to filter through. (Hint: to help speed this process up, increase the pressure on the water by placing a finger or thumb securely over the top rim of the funnel and gently squeezing inward on the sides of the cut-off bulb as shown in Figure 12-A. This should help push the liquid down through the charcoal and tissue paper.)

12 When the charcoal filtration is complete, observe the properties and measure the final volume of your sample, and record this information

Figure 12-A: The measurement and purification procedures for this investigation are shown here.

Investigation 12: Water Purification

in your data table. Show your purified water sample to your teacher.

13 Check with your teacher for instructions on how to dispose of your materials when finished. Remember to wash your hands thoroughly before leaving the laboratory.

Analysis & Interpretation

1 Analyzing Methods: What assumption is being made about the test tubes in using the calibrated wood splint as a measuring stick for the volume of liquid in all of the test tubes?

2 Relating Ideas: Why don't oil and water mix?

3 Analyzing Methods: What is the purpose of the coarse sand at the bottom of the funnel? What is the purpose of the fine sand in the middle of the funnel? What is the purpose of the coarse sand at the top of the funnel?

4 Analyzing Methods: Since charcoal does the best job of removing the smaller impurities from the water, why not just use charcoal from the start and skip all of the other steps?

5 Analyzing Methods: Why were new test tubes used to hold the sample after each purification step?

Conclusions

1 Inferring Relationships: Rank the following in order from lowest density to highest density: water, sediment particles, oil. How do you know from this investigation what the order should be?

2 Inferring Conclusions: Which are larger: the particles that make the water appear cloudy (instead of transparent) or the particles that make the water colored (but transparent)? Explain your reasoning.

3 Evaluating Methods: Calculate the percentage of water recovered using the following equation:

$$\text{percentage recovered} = \frac{\text{volume of purified } H_2O}{\text{volume of initial sample}} \times 100$$

4 Evaluating Methods: Compare your results with those of other lab groups. How should each group's success with the techniques be judged?

Extensions

1 Designing Experiments: What ways can you think of to improve the percentage of water recovered or the purity of the water recovered? If your teacher approves of your suggestions, test your ideas on another sample of foul water.

2 Applying Ideas: Call your local treatment plant and find out how water is purified for the water supply where you live. What are the most common impurities present, and what steps are taken to separate them from the water? Prepare a report describing the system.

3 Analyzing Ideas: In the sedimentation step, a solid and a liquid were separated since one was more dense than the other. In the process called distillation, a mixture is gradually heated until one of the substances in the mixture reaches its boiling point and becomes a gas. Explain how you would use distillation to separate a mixture of two miscible liquids with different boiling points.

INVESTIGATION 13: Testing Water

Introduction

The presence of ions in an aqueous solution changes the physical and chemical properties of that solution so that it is no longer like pure water, even when the ions are present in small amounts.

For example, if a sample of water contains enough Mg^{2+} or Ca^{2+} ions, something unusual will happen when you use soap to wash something in it. Instead of forming lots of bubbles and lather, the soap will form an insoluble white substance that floats on top of the water. A scummy coating will be left behind after the water is drained. This problem is called hard water, and it is common where the water supply contains many minerals.

Aqueous solutions of other ions, such as Pb^{2+} and Co^{2+} can be poisonous, since they accumulate in body tissues. Some older houses still have lead pipes for plumbing. In many cases, the plumbing systems for such houses must be completely redone and converted to another type of pipe such as copper or polyvinyl chloride, because the lead metal in the pipes slowly ionizes and dissolves in the water carried to the house for drinking and cooking.

Because many sources of water are polluted, it is important to be able to test water for such impurities, to be sure it is clean enough for the use intended.

In this investigation, you will check a variety of water samples for the presence of four specific ions: Fe^{3+}, Ca^{2+}, Cl^-, and SO_4^{2-}, using the chemical properties of these ions in aqueous solutions. Because some of the water samples may contain these ions at very low concentrations, the results of these tests may be very faint and difficult to detect. Be sure to make very careful observations.

Objectives

- **Observe** chemical reactions involving aqueous solutions of ions
- **Relate** observations of chemical properties to the presence of ions
- **Infer** whether or not an ion is present in a water sample
- **Apply** concepts concerning aqueous solutions of ions

Materials

24-well microplate lid
10 labeled fine-tipped dropper bulbs, with solutions:
solution 1: reference (all ions)
solution 2: distilled water (no ions)
solution 3: tap water (may have ions)
solution 4: bottled spring water (may have ions)
solution 5: local river or lake water (may have ions)
solution 6: solution X, prepared by teacher (may have ions)
solution A: NaSCN solution (Fe^{3+} test)
solution B: $Na_2C_2O_4$ solution (Ca^{2+} test)
solution C: $AgNO_3$ solution (Cl^- test)
solution D: $Ba(NO_3)_2$ solution (SO_4^{2-} test)
white paper
paper towels
overhead projector (optional)

Procedure

Some of the solutions used in the investigation are very hazardous to the eyes and skin. Safety goggles, gloves, and aprons must be worn at all times during this investigation. Be sure you know the location and operation of the eyewash station.

Avoid getting any of the solutions on your skin or clothing. Rinse any spills immediately with large amounts of water.

1 Copy the following data table into your lab notebook, and record all of your observations in it:

Test for: Reacting with:	Fe^{3+} SCN^-	Ca^{2+} $C_2O_4^{2-}$	Cl^- Ag^+	SO_4^{2-} Ba^{2+}
Reference (all 4 ions)				
Distilled H_2O (control-no ions)				
Tap Water				
Bottled Spring Water				
River Water				
Solution X				

2 Place the 24-well plate lid in front of you on

Figure 13-A: Use labels on a sheet of paper underneath the 24-well plate lid to keep track of which solutions belong where.

Investigation 13: Testing Water

top of a white piece of paper or other white background aligned as shown in Figure 13-A. The coordinates shown will be used to designate the individual circles. For example, the one in the top right corner would be designated 1-D, and the one in the lower left corner would be designated 6-A. (You may want to label the columns and rows as shown in Figure 13-A.)

3 Place a drop of the solution from bulb 1 in circles 1-A, 1-B, 1-C, and 1-D (the top row). Solution 1 contains all four of the dissolved ions, so these drops will show what a **positive** test for each of the ions looks like. **Be careful to keep the solutions in the appropriate circles. Any spills will cause poor results.**

4 Place a drop of the solution from bulb 2 in circles 2-A, 2-B, 2-C, and 2-D (the second row). Solution 2 is distilled water, and should contain none of the ions, so it will be used to show what a **negative** test for each of the ions looks like.

5 Place a drop from bulb 3 in each of the circles in row 3, a drop from bulb 4 in each of the circles in row 4, and so on, with bulbs 5 and 6. These solutions may or may not contain the ions. Bulb 3 contains ordinary tap water. Bulb 4 contains bottled spring water. Bulb 5 contains local river or lake water. Bulb 6 contains solution X, which was prepared by your teacher.

6 Now that each circle contains a drop of solution to be analyzed, use the solutions in bulbs A, B, C, and D to test for the presence or absence of the ions.

7 Bulb A contains NaSCN (sodium thiocyanate), which will react with an Fe^{3+} present to form $Fe(SCN)^{2+}$, a complex which forms a brightly colored solution. **Holding the tip of the bulb 1–2 cm above the drop of water to be tested,** add one drop of this solution to the drop of reference solution in circle 1-A and one drop to the distilled water in circle 2-A below it. Circle 1-A should show a positive test, and circle 2-A should show a negative test. Record observations about what the positive and negative tests looked like in the data table in your lab notebook.

8 Use the NaSCN solution in bulb A to test the rest of the water drops in column A (leftmost column) of the plate, to determine whether any of them contain the Fe^{3+} ion. Record your observations in the data table in your lab notebook. For each of the tests where the ion was found to be present, specify whether it seemed to be present at a high, moderate, or low concentration. (If the evidence was quite visible, assume the ion was at a high concentration. If it was somewhat visible, assume the ion was at a moderate concentration. If the evidence was barely visible, assume a low concentration of the ion.)

9 Bulb B contains $Na_2C_2O_4$ (sodium oxalate), which will undergo a replacement reaction with the Ca^{2+} ion to form an insoluble precipitate. **Holding the tip of the bulb 1–2 cm above the drop of water to be tested,** add one drop of this solution to the drop of reference solution in the circles in column B (second column).

10 Record observations about what the positive and negative $Na_2C_2O_4$ tests looked like, and whether any of the solutions contained the Ca^{2+} ion in the data table in your lab notebook. For each of the positive tests, specify whether it

Figure 13-B: Be sure to look at the drops from the side of the well dish when you determine whether or not they test positive for an ion.

seemed to be present at a high, moderate, or low concentration. A black background may be useful for this test and the following tests.

11 Bulb C contains AgNO$_3$ (silver nitrate), which will undergo a replacement reaction with the Cl$^-$ ion to form an insoluble precipitate. **Holding the tip of the bulb 1–2 cm above the drop of water to be tested,** add one drop of this solution to the drop of reference solution in the circles in column C (third column).

12 Record observations about what the positive and negative AgNO$_3$ tests looked like, and whether any of the solutions contained the Cl$^-$ ion in the data table in your lab notebook. For each of the tests where the ion was found to be present, specify whether it seemed to be present at a high, moderate, or low concentration.

13 Bulb D contains Ba(NO$_3$)$_2$ (barium nitrate), which will undergo a replacement reaction with the SO$_4^{2-}$ ion to form an insoluble precipitate. **Holding the tip of the bulb 1–2 cm above the drop of water to be tested,** add one drop of this solution to the drop of reference solution in the circles in column D (rightmost column).

14 Record observations about what the positive and negative Ba(NO$_3$)$_2$ tests looked like, and whether any of the solutions contained the SO$_4^{2-}$ ion in the data table in your lab notebook. For each of the tests where the ion was found to be present, specify whether it seemed to be present at a high, moderate, or low concentration.

15 If there are some tests that are difficult to discern, place your plate on an overhead projector, if one is available. Examine the drops carefully for any signs of cloudiness. Be sure your line of vision is 10–15° above the plane of the lid as shown in Figure 13-B, so you are looking at the sides of the drops. Compare each drop tested to the control drops (distilled water) in row 2. If any signs of cloudiness are detected in a tested sample, it is due to the Tyndall effect, and should be considered a positive test result. Record any further results in your lab notebook.

16 Check with your teacher for instructions on how to dispose of your chemicals when finished. Remember to wash your hands thoroughly before leaving the laboratory.

Analysis & Interpretation

1 Organizing Ideas: Write the balanced chemical equations for each of the positive tests. Describe what each test looked like.

2 Organizing Ideas: Write the net-ionic equations for each of the positive tests. Hint: Net-ionic equations are discussed in Section 15.1.

3 Analyzing Methods: Why was it important to hold the dropper of testing solution 1–2 cm above each of the drops of water samples? Why was it important to be sure that the water samples were not spilled and accidentally mixed?

4 Analyzing Methods: Why was it important to use a control in the experiment? Why was distilled water used as the control?

Conclusions

1 Organizing Conclusions: List the solutions tested and the ions you found in each. Include notes on whether the concentration of each ion was high, moderate, or low.

2 Applying Conclusions and Predicting Outcomes: Based on your test results, predict which water sample would be the "hardest." Explain your reasoning.

Extensions

1 Applying Conclusions and Predicting Outcomes: Try to collect water samples from a variety of sources. If you do this investigation near a holiday time (such as spring break) and will be travelling, get a sample from each place you visit. Also acquire rainwater or melted snow, swimming pool water, still water from ponds or lakes, running water from creeks and streams, well water, and, if you're near the coast, some salt water from an ocean. Before testing the water, draw up a chart, and try to predict which ions you will find in each one.

2 Relating Ideas: Considering the substances used in the tests for this investigation, and your results, what can you say about the solubilities of most sodium compounds and most nitrate compounds? Explain your reasoning **without** referring to a chart of solubility rules.

3 Predicting Outcomes and Resolving Discrepancies: Suppose you decided to determine the amount of chloride ion present in the reference solution, by evaporating the water and measuring the mass of what was left behind. If you were left with 0.37 gram of white powder, could you safely assume that it was all chloride ions? Explain your reasoning.

4 Predicting Outcomes and Resolving Discrepancies: If you were given a sample of a solution that tested positive only for sulfate ions, and not for the other three ions considered in this investigation, and you evaporated away the water and were left with 0.21 grams of white powder, could you safely assume that it was all sulfate ions? Explain your reasoning.

5 Designing Experiments: What if you wanted to determine the actual concentrations of each of the ions in the solutions that tested positive? How would you do it? If there is time, and your teacher approves your suggestion, test your method out and see if it works. Hint: Consider the properties of ions in aqueous solutions that are discussed in Chapter 15.

INVESTIGATION 14: Is It an Acid or a Base?

Introduction

Professional scientists rarely have the luxury of doing work in the lab where they can follow an instruction manual that tells them everything they need to know and all that they need to do. Usually, scientists uncover a problem, think about ways to solve it, and test their plans. Although they do not always know if they are on the right track, the advantage to this approach is that there are often several ways to solve any single problem.

In this open-ended investigation, you will be given a set of eight colorless solutions. Four of them are acidic solutions (dilute hydrochloric acid) and four of the solutions are basic (dilute sodium hydroxide). In both the acidic and the basic set, the solutions have concentrations of 0.1 *M*, 0.2 *M*, 0.4 *M*, and 0.8 *M*. The acids have phenolphthalein added to them. (Phenolphthalein is colorless in acidic solutions, but red (pink) in basic solutions. For more about phenolphthalein, see Section 17.3.)

Your assignment is to write a plan that will determine which solutions are acids and which are bases, and then carry out your plan. You also will need to perform further experiments to rank the concentrations from weakest to strongest. As you plan your procedures, carefully consider the properties of acids and bases that were discussed in Chapter 16. It may help to predict what would happen to a solution of each type and concentration if you did the tests. Then you can compare your predictions to what happened for the unknown solutions.

You will not get refills of your unknowns if you use them all up during the tests. Therefore, be very careful in running your tests. Check with your teacher about what additional supplies (if any) will be available for you to use.

Objectives

- **Design** an experiment to solve a chemical problem
- **Relate** observations of chemical properties to identify unknowns
- **Infer** a conclusion from experimental data
- **Apply** acid-base concepts

Materials

set of eight labeled pipets containing solutions numbered 1–8
24-well microplate (or small test tubes)
toothpicks for mixing solutions
For other supplies, ask your teacher

Procedure

 The solutions used in the investigation are very hazardous to the eyes. Safety goggles and aprons must be worn at all times during this investigation. Be sure you know the location and operation of the eyewash station.

Avoid getting any of the solutions on your skin or clothing. Rinse any spills immediately with large amounts of water.

 As always, the solutions should not be tasted or touched.

1 Write the steps you will use to determine which solutions are acids and which solutions are bases in your lab notebook.

2 Make data tables like these in your lab notebook:

Unknown Solution #'s, in no particular order	Acids	Bases

Concentration	HCl	NaOH
0.1 M		
0.2 M		
0.4 M		
0.8 M		

3 Ask your teacher to approve your plan, and give you any additional supplies you will need.

4 Carry out your plan for determining which solutions are acids and which are bases. **As you perform your tests, be sure to avoid letting the tips of the storage pipets come into contact with other chemicals. Instead, squeeze drops out of the pipets onto the 24-well plate and then use these drops for your tests.** Record all observations in your lab notebook, and then record your results in the first table.

Figure 14-A: How would you figure out which of these pipets contain HCl solution and which contain NaOH?

5 Write your plan for determining the concentrations of the solutions in your lab notebook. Ask your teacher to approve your plan, and request any additional supplies you will need.

6 Carry out your plan for determining the concentrations of the solutions. Record all observations in your lab notebook, and record your results in the second data table.

7 Check with your teacher for instructions on how to dispose of your chemicals when finished. Remember to wash your hands thoroughly before leaving the laboratory.

Analysis & Interpretation

1 **Organizing Ideas:** Name some properties of aqueous acids and bases. Which ones did you use to determine which of the solutions were acids and which were bases?

2 **Organizing Ideas:** Write word equations that describe the tests you performed on the acids and the bases. If possible, also write the balanced chemical equation for the tests.
An example of a word equation: acid + blue litmus → ???

Conclusions

1 **Analyzing Conclusions:** List the solutions and their concentrations.

2 **Analyzing Conclusions:** Describe what test results led you to assign the classes (acid or base) and concentrations to the unknown solutions. Refer to your answer for Analysis & Interpretation #2 if possible.

Extensions

1 **Evaluating Methods:** Compare your results to another lab group's. Do you think that your teacher gave both groups the same set of solutions? (Is your solution #1 the same as their solution #1, and so on?) Explain your reasoning.

2 **Designing Experiments:** Although your teacher may have allowed you to use a variety of materials and methods to figure out the classes (acid or base) and concentrations of the solutions, there is a way you could do this using just the pipets of unknown solutions and the 24-well

Investigation 14: Is It an Acid or a Base?

plate. Write out a procedure that would identify the solutions using only these materials. If time allows, test your experiment, and compare the results you get this time with your first results.

3 Applying Conclusions: Imagine you are helping clean out the school's chemical storeroom. You find a spill coming from a large reagent bottle filled with a clear liquid. What tests would you do to quickly determine if the substance was acidic or basic? How would you clean up the spill if it was acidic or basic?

INVESTIGATION 15: How Much Calcium Carbonate is in an Eggshell?

Introduction

Birds have evolved a chemical process which allows them to rapidly produce the calcium carbonate ($CaCO_3$), that is required for eggshell formation. The shell provides a strong protective covering for the developing embryo. Research has shown that some chemicals like DDT and PCBs can decrease the amount of calcium carbonate in the eggshell, resulting in shells that are too thin and fragile to survive until hatching.

The calcium carbonate content of eggshells can easily be determined by means of an acid/base back-titration, using some of the same techniques and calculations described in Chapter 17. In this back-titration, a carefully measured excess of a strong acid will react with the calcium carbonate. The resulting solution will be titrated with a strong base to determine how much acid remained unreacted. From this measurement, the amount of acid that reacted with the eggshell and the amount of calcium carbonate it reacted with can be determined. Phenolphthalein will be used as an indicator to signal the end-point of the titration.

Objectives

- **Measure** the amount of calcium carbonate present in an eggshell
- **Relate** experimental titration measurements to balanced chemical equation
- **Infer** a conclusion from experimental data
- **Apply** reaction-stoichiometry concepts

Materials

eggshell
balance
100 mL beaker
drying oven
dessicator (optional)
weighing paper
mortar and pestle
50 mL micro solution bottle or small Erlenmeyer flask
two thin-stemmed pipets
10-mL graduated cylinder
phenolphthalein solution
fine-tipped dropper bulbs
2.00 M HCl
1.00 M NaOH
distilled water

Procedure

The solutions used in the investigation are very hazardous to the eyes. Safety goggles must be worn at all times during this investigation. Avoid getting any of the solutions on your skin or clothing.

The oven used in the investigation is very hot, so be careful, and use tongs to remove the beakers left in the oven.

1 Prepare the eggshell by removing the white and the yolk from the egg. Wash the inside of the shell with distilled water and carefully peel all the membranes from the inside of the shell. Place ALL the shell in a premassed beaker and dry the shell in the drying oven at 110°C for about 15 minutes. Continue with steps 2–5 while the eggshell is drying.

2 Make data and calculations tables like these in your lab notebook:

Graduated Cylinder Readings (Pipet Calibration: Steps 3–5)				
Trial	initial— acid pipet	final— acid pipet	initial— base pipet	final— base pipet
1				
2				
3				

Total vol. of drops—acid pipet: _____
Average volume of each drop: _____
Total vol. of drops—base pipet: _____
Average volume of each drop: _____

866 Microscale Investigations

Titration: Steps 7-12	
Mass of entire eggshell	
Mass of ground eggshell sample	
Number of drops of 2.00 M HCl added	100 drops
Vol. of 2.00 M HCl added	
Number of drops of 1.00 M NaOH added	
Vol. of 1.00 M NaOH added	
Vol. of 2.00 M HCl reacting with NaOH	
Vol. of 2.00 M HCl reacting with eggshell	
# of moles HCl reacted with eggshell	
# of moles of $CaCO_3$ reacting with HCl	
Mass of $CaCO_3$ in eggshell sample	
% of $CaCO_3$ in eggshell sample	

3 Put exactly 5.0 mL of water in the 10.0 mL graduated cylinder. Record this volume in the chart in your lab notebook. Fill the first thin-stemmed pipet with water. This pipet should be designated (perhaps even labeled) the acid pipet. **Do not use this pipet for the base solution.** Holding the pipet vertical, add 20 drops from it into the cylinder. **For the best results, keep the sizes of the drops as even as possible throughout this investigation.** Record the new volume of the graduated cylinder in the chart as Trial 1.

4 Without emptying the graduated cylinder, add an additional 20 drops from the pipet as before, and record the new volume as the final volume for Trial 2. Repeat this procedure once more for Trial 3.

5 Repeat Steps 3 and 4 for the second thin-stemmed pipet. This will now be the base pipet. **Do not use this pipet for the acid solution.**

6 Make sure that the three trials produce data that are similar to each other. If one is greatly at variance with the others, perform steps 3–5 over again. If you're still waiting for the 15 minutes that the eggshell needs to be in the drying oven, make the calculations needed, and fill in the total volume and volume per drop blanks below the table.

7 Place half of the shell into a clean mortar and grind to a very fine powder. This will save you time dissolving the eggshell. (If time permits, dry again and cool in desiccator.)

8 **Measure the mass of a piece of weighing paper.** Transfer about 0.1 g of ground eggshell to a piece of weighing paper, and measure the eggshell's mass as accurately as possible. Record the mass in the second data table. Place this eggshell sample into a clean 50 mL micro solution bottle (or Erlenmeyer flask).

9 Rinse the acid pipet with the 2.00 M HCl acid solution, and then empty the pipet into the sink.

Figure 15-A: Prepare the eggshell by cleaning it, drying it, and grinding it in a mortar and pestle. After checking its mass, react it with exactly 100 drops of 2.00 M HCl, and titrate the remaining solution with 1.00 M NaOH.

Investigation 15: How Much Calcium Carbonate is in an Eggshell?

Rinse the base pipet with the 1.00 M NaOH base solution, and then empty the pipet into the sink.

10 Fill the acid pipet once more with 2.00 M HCl. Holding the acid pipet vertical, add exactly 100 drops of 2.00 M HCl to the bottle (or flask) with the eggshell. Swirl gently for 3 to 4 minutes. Observe the reaction taking place. Wash down the sides of the flask with about 10 mL of distilled water. Add 4 drops of phenolphthalein solution with a fine-tipped dropper bulb. Record the number of drops of HCl used in the second data table in your lab notebook.

11 Titrate with 1.00 M NaOH: Fill the base pipet with the 1.00 M NaOH. Slowly add NaOH from the vertical base pipet into the bottle with the eggshell reaction mixture until the first instant a faint pink color persists in the bottle, even after the bottle is swirled gently. **Be sure to add the base drop by drop, and be certain the drops end up in the reaction mixture and not on the walls of the bottle or flask. Keep careful count of the number of drops used.** Record the number of drops of base used in the second data table.

12 Check with your teacher for instructions on how to dispose of your chemicals when finished. Remember to wash your hands thoroughly before leaving the laboratory.

Analysis & Interpretation

1 Organizing Ideas: The calcium carbonate in the eggshell sample undergoes a double-replacement reaction with the hydrochloric acid in Step 10. Write a balanced chemical equation for this reaction. (Hint: the gas observed was carbon dioxide.)

2 Organizing Ideas: Write the balanced chemical equation for the acid/base neutralization of the excess unreacted HCl with the NaOH.

3 Organizing Data: Make the necessary calculations from the first data table to find the number of mL's in each drop. Using this mL/drop ratio, convert the number of drops of each solution in the second data table to volume of each solution used.

4 Organizing Ideas: For the neutralization reaction, what is the normality of the HCl solution? What is the normality of the NaOH solution?

5 Organizing Data: Using the relationship between the molarity and volume of an amount of acid and the molarity and volume of the amount of base needed to neutralize it, calculate what volume of the HCl solution was neutralized by the NaOH, and record it in your table. (Hint: this relationship was discussed in Section 17.3.)

6 Analyzing Results: If the volume of HCl originally added is known and the volume of excess acid is determined by the titration, then the difference will be the amount of HCl that reacted with the $CaCO_3$. Calculate volume and the number of moles HCl that reacted with the $CaCO_3$, and record both in your table.

Conclusions

1 Organizing Data: Use the stoichiometry of the reaction in Analysis & Interpretation #1 to calculate the number of moles of calcium carbonate that reacted with the HCl, and record it in your table.

2 Organizing Data: Calculate the molar mass of calcium carbonate from a periodic table. In your table, record the mass of calcium carbonate present in your eggshell sample.

3 Organizing Data: Using your answer to Conclusions #2, calculate the percentage of calcium carbonate in your eggshell sample, and record it in your table.

4 Evaluating Methods: The percentage of calcium carbonate in a normal eggshell is in the range of 95–99%. Calculate the percent error for your measurement of the $CaCO_3$ content, assuming 97% $CaCO_3$ is the "accepted" value.

Extensions

1 Inferring Conclusions: Calculate an estimate of the mass of $CaCO_3$ present in the entire eggshell, based on your results for the sample of eggshell. (Hint: apply the percent composition of your sample to the mass of the entire eggshell.)

2 Evaluating Methods: Repeat the experiment, testing another section of the eggshell. Compare your results from this test with the results from your first trial.

3 Designing Experiments: What possible sources of error can you identify with this procedure? If you can think of ways to eliminate them, ask your teacher to approve your suggestion and run the procedure again.

4 Applying Ideas: Find out how much calcium carbonate is in another type of shell (clam, oyster, snail, etc.).

5 Applying Conclusions: If you were the warden at a bird sanctuary, and you suspected that the declines in bird sightings and population were linked to PCB contamination from a factory in a nearby town, explain how you could use this investigation to provide evidence for your conclusion.

6 Designing Experiments: Calcium carbonate is present in many substances, from limestone, marble, and the mineral calcite to chalk. In fact, chalk is composed of tiny shells from animals. How would you determine the amount of $CaCO_3$ present in a piece of classroom chalk? If your teacher approves your suggestion, perform the experiment.

INVESTIGATION 16: Clock Reactions

Introduction

A simple procedure, using micro-well strips and the phenomenon of surface tension, enables one to perform multiple simultaneous mixtures. Using microscale techniques, changes that affect the rates of a reaction can be explored.

The reaction investigated here is a special kind of electron-transfer reaction called a "reduction-oxidation" or "redox" reaction. Reactions of this type are studied extensively in Chapter 20 of the text. The net equation for this reaction is:

$$3 Na_2S_2O_5(aq) + 2 KIO_3(aq) + 3 H_2O(\ell) \xrightarrow{H^+} 2 KI(aq) + 6 NaHSO_4(aq)$$

The progress of the reaction will be measured using a starch indicator solution. When the $Na_2S_2O_5$ solution is used up, the build-up of I_2, an intermediate in the reaction, will cause the indicator to change to a blue-black color. In the investigation, the concentrations of reactants are discussed in terms of drops of "Solution A" and "Solution B." "Solution A" actually contains $Na_2S_2O_5$, the starch indicator solution, and dilute sulfuric acid to supply hydrogen ions needed to catalyze the reaction. "Solution B" contains KIO_3.

Objectives

- **Observe** chemical processes and interactions.
- **Compare** chemical reactions by measuring reaction rates
- **Relate** reaction rate concepts to observations
- **Infer** a conclusion from experimental data

Materials

Clock solution A
Clock solution B
distilled or de-ionized water
three fine-tipped dropper bulbs
two 8-well flat bottom strips
clock with second hand or stopwatch

Procedure

 Some of the solutions used in the investigation are very hazardous to the eyes. Safety goggles must be worn at all times during this investigation.

 Avoid getting any of the solutions on your skin or clothing.

1 Obtain three dropper bulbs. Label them: "A," "B," and "W."

2 Fill the bulb labeled "A" with solution A. Fill the bulb labeled "B" with solution B. Fill the bulb labeled "W" with distilled water.

3 Using one 8-well strip, place five drops of solution A into each of the first five wells. (Disregard the remaining three wells.) **For best results, try to make all of the drops about the same size.**

First Strip:	Well #1	Well #2	Well #3	Well #4	Well #5
# drops A	5	5	5	5	5

4 In the second 8-well strip, place drops of solution B and distilled water according to the following table. **For best results, try to make all the drops about the same size.**

Figure 16-A: To mix your solutions, hold the two 8-well strips together as shown, and shake vigorously once.

Second Strip:	Well #1	Well #2	Well #3	Well #4	Well #5
# drops B	5	4	3	2	1
# drops H_2O	0	1	2	3	4
Total # drops	5	5	5	5	5

5 Copy this data table into your lab notebook, and use it to record how long it takes each well to react completely. (If there are stopwatches, you will only need the row labeled "time elapsed.")

	Well #1	Well #2	Well #3	Well #4	Well #5
time reaction began					
time reaction ended					
time elapsed(s)					

6 Carefully invert one of the strips. The surface tension should keep the solutions from falling out of the wells. Place it well-to-well on top of the other strip as shown in Figure 16-A.

7 Holding the strips together tightly, record the exact time or set the stopwatch as you shake them down once, with a vigorous motion. This should effectively transfer all of the upper solutions down to mix with each of the corresponding lower ones.

8 Observe the lower wells. Note the sequence in which the solutions react, and record the time for each to turn a deep blue color.

9 Perform any calculations necessary to find the "time elapsed" for each solution.

10 Check with your teacher for instructions on how to dispose of your chemicals when finished. Remember to wash your hands thoroughly before leaving the laboratory.

Analysis & Interpretation

1 Analyzing Results: Which solution showed the quickest or fastest reaction? Which solution showed the slowest reaction?

2 Organizing Data: Make a graph of your results with number of drops of solution B on the *x*-axis and the time elapsed on the *y*-axis. Make a similar graph for drops of solution B vs. rate (1/time).

3 Analyzing Methods: How can you be sure all chemical reactions began at about the same time? Why is this important?

Conclusions

1 Applying Models: Section 18.4 lists several factors that can affect the rate of a reaction. Which one is being tested here?

2 Inferring Conclusions: Use your data and graph to determine the relationship between solution B concentration and the rate of this reaction. Describe this relationship in terms of a rate law.

Extensions

1 Evaluating Methods: Share your data with other lab groups and calculate a class average for the rate for each concentration of B. Compare them to your results. Using the class average rate as the accepted value, calculate your percent error.

2 Designing Experiments: What are some likely sources of imprecision in this experiment? If you can think of ways to eliminate them, ask your teacher to approve your suggestion, and run more trials.

3 Predicting Outcomes and Designing Experiments: What combination of drops would you use if you wanted the reaction to last exactly 2 minutes and 30 seconds? Design an experiment to test your answer. If your teacher approves your suggestion, perform the experiment and record these results. Make another graph including both the old and the new data.

4 Predicting Outcomes and Designing Experiments: How would the results of this experiment be affected if the reaction took place in a cold place? Design an experiment that tests your answer with materials at hand. If your teacher approves your suggestion, perform the experiment and record these results. Make a graph of these results, and compare it to the old data.

5 Designing Experiments: How would you determine the effect of solution A on the rate law? If your teacher approves of your suggestion, perform the experiment, and determine the rate law for this reaction.

INVESTIGATION 17: Measuring K_a for Acetic Acid

Introduction

As discussed in Section 19.3, the acid-ionization constant K_a provides important information about the strength of an acid. Strong acids, which are completely dissociated, have larger values for K_a than weak acids, which are only partly ionized. Properties that depend upon the ability of a substance to dissociate or ionize, such as conductivity and colligative properties, can be used to easily measure K_a.

Objectives

- **Observing** chemical processes and interactions
- **Relating** chemical measurements
- **Applying** defined terms based on observations
- **Inferring** a conclusion from experimental data

Materials

1.0 M hydrochloric acid
1.0 M acetic acid
distilled or de-ionized water
paper towels
thin-stemmed pipets
24-well plate
LED conductivity testers

Procedure

Safety goggles must be worn throughout this activity, because of the use of strong acids.

Any acid spills should be immediately washed with excess water. Make sure you know where the neutralization packages for acid spills are kept.

1 Obtain samples of 1.0 M HCl (aq) and 1.0 M CH₃COOH (aq).

2 Place 20 drops of HCl in a well of a 24-well plate. Then place 20 drops of CH₃COOH in an adjacent well. Note the location of each sample.

3 Test the samples with the conductivity tester, noting the relative intensity of the tester bulb for each solution. After testing, rinse off the tester's probes with distilled water. Also remove any excess moisture with a paper towel.

4 Place 18 drops of distilled water in six of the wells in your 24-well plate. Add two drops of 1.0 M HCl to the first well. (You now have a total of 20 drops of solution.) Mix the contents of this well thoroughly by picking the contents up in the pipet and returning them to the well, as shown in Figure 17-A.

5 Repeat this procedure of taking two drops of the previous dilution and placing it in the next well of water and mixing with a new pipet. Return any unused solution to the well from which it was taken. (You now have 1.0 M HCl in the well from Step 1, 0.10 M HCl in the first diluted well, and 0.010 M HCl in the second dilution.)

Figure 17-A: Use this technique to mix the contents of the wells so that several dilutions of HCl can be made.

6 Continue diluting in this manner until you have six successive dilutions. (The [H_3O^+] should now run from 1.0 M through 1.0×10^{-6} M.

7 Test each of the cells containing HCl and note the brightness of the tester bulb. After each test, rinse off the tester's probes with distilled water. When the brightness is about the same as that of the acetic acid, you have about the same hydronium ion concentrations. (At this point, the pH of the HCl is equal to that of the acetic acid.) NOTE: You may retest the acetic acid cell any time for a comparison. If the bulb glow is too faint to see, turn off the lights and/or build a light shield around your conductivity tester bulb. Be careful around the acid solutions!

8 Record the results of your observations by noting which HCl concentration causes the intensity of the bulb to most closely match that of the acetic acid. (If you match no single concentration, then estimate the value between the two concentrations which best match.)

9 All of the solutions may be washed down the sink, after they have been neutralized with Na_2CO_3. An excess of at least 20 times as much water should be used. (NOTE: the same disposal directions apply to the solutions in the **Extensions** below.)

Analysis & Interpretation

1 Resolving Discrepancies: How did the conductivity of the 1.0 M HCl solution compare to that of the 1.0 M CH_3COOH? Why do you think this was so?

2 Organizing Data: What is the [H_3O^+] of the HCl solution that most closely matches the acetic acid in conductivity?

3 Inferring Conclusions: What do you think the [H_3O^+] of the 1.0 M CH_3COOH was? Why?

Conclusions

1 Applying Models: Since $K_a = \dfrac{[H_3O^+][CH_3COO^-]}{[CH_3COOH]}$, calculate K_a for the acetic acid solution.

2 Applying Models: How is it that the solutions of HCl and acetic acid that showed the same degree of conductivity had different concentrations?

Extensions

1 Evaluating Methods: Compare the K_a value you have calculated with the value found on page 577 of this text. Calculate the percent error for this experiment.

2 Designing Experiments: What are the greatest sources of inaccuracy or imprecision for this experiment? Design an experiment which will reduce your percent error. If your teacher approves your suggestions, perform the experiment and recalculate K_a.

3 Predicting Outcomes: How will your results be affected if you tested the conductivity of a 0.10 M acetic acid solution instead of a 1.0 M acetic acid solution? What effect should it have on the K_a you calculate? If told to do so by your teacher, try the experiment again with an acetic acid solution of another concentration.

4 Predicting Outcomes: Lactic acid (HOOC-CHOHCH$_3$) has a K_a of 1.4×10^{-4}. Predict whether a solution of lactic acid will cause the conductivity tester to glow brighter or dimmer, compared to a solution of acetic acid with the same concentration? How noticeable would the difference be?

5 Designing Experiments: What changes would you need to make to use this experiment to calculate the concentration for a lactic acid solution of unknown concentration? If your teacher approves your suggestions, perform the experiment and calculate the concentration of the lactic acid. If your teacher gives you the actual concentration after you're done, calculate percent error.

INVESTIGATION 18: Acid-Catalyzed Iodination of Acetone

Introduction

Hydrogen atoms in a molecule of acetone can be replaced by iodine, a halogen, under certain conditions. The resulting compound has two functional groups, and is both a ketone and a halocarbon. The iodination of acetone proceeds according to the equation:

$$CH_3-\underset{\underset{}{\|}}{\overset{\overset{O}{\|}}{C}}-CH_3(aq) + I_2(aq) \xrightarrow{H^+} CH_3-\overset{\overset{O}{\|}}{C}-CH_2I(aq) + HI(aq)$$

The rate of the reaction depends on the hydrogen ion concentration in the solution (acting as a catalyst) as well as the concentration of acetone. The rate equation for the reaction is:

$$R = k[\text{acetone}]^x[\text{HCl}]^y[I_2]^z.$$

In this experiment you will measure the rate by noting the disappearance of I_2 using a starch indicator solution. The reaction is complete when the blue-black color disappears. By measuring the rate experimentally with differing concentrations of reactants, the values for x, y, and z can be determined. Note that these values can be determined only through experimentation.

Objectives

- **Observe** chemical processes in the iodination of acetone
- **Measure and compare** rates of chemical reactions
- **Relate** reaction rate concepts to observations
- **Infer** a conclusion from experimental data
- **Evaluate** methods

Materials

4.0 M acetone
1.0 M HCl
0.0012 M iodine
1% starch solution
distilled water
24-well plate
5 thin-stemmed pipets (1 for each solution, and 1 for H_2O)
white paper (1 sheet)
clock with seconds display or stopwatch
toothpicks

Procedure

The solutions used in the investigation are very hazardous to the eyes. Safety goggles must be worn at all times during this investigation. Avoid getting any of the solutions on your skin or clothing.

1 Place a piece of white paper underneath the 24-well plate.

2 Using four wells of the 24-well plate, mix the water, HCl, iodine, and starch solutions in the proportions and order indicated in the following table. **Be sure to use different pipets for the different solutions.**

3 Copy this table into your lab notebook, including the blank columns.

Reaction #	starch (+H_2O) (drops)	0.0012 M I_2 (drops)	1.0 M HCl (drops)	4.0 M acetone (drops)	Time (s)
1	10 + 10 H_2O	10	10		
2	10	10	10		
3	10	10	20		
4	10	20	10		

4 Check the time (or set the stopwatch) as you add the appropriate amount of acetone solution from the following table to the first well. **The**

acetone must be added last and you must keep track of the time elapsed. Stir with a toothpick to thoroughly to mix the reagents. (Later, add these drop numbers to the table you made.)

Reaction #	4.0 M acetone (drops)
1	10
2	20
3	10
4	10

5 Continue stirring until the dark blue color disappears, as shown in Figure 18-A. Record the time it took for the color to disappear in the data table.

6 Repeat this procedure for the other mixtures.

7 Check with your teacher for instructions on how to dispose of your chemicals when finished. Remember to wash your hands thoroughly before leaving the laboratory.

Analysis & Interpretation

1 Organizing Data: Make another table for calculations as follows:

Reaction #	[HCl]	[I$_2$]	[Acetone]	Rate
1				
2				
3				
4				

2 Organizing Data: Fill in the concentrations. In the first mixture, 10 drops of 1.0 M HCl were diluted to a solution having a total of 50 drops, so the new HCl concentration is $\frac{10}{50}$ of the original, or 0.20 M. The other concentrations can be determined using similar reasoning.

3 Organizing Data: Fill in the rates for the mixtures. Since the rate of the reaction is equal to the amount of a reactant consumed divided by the time elapsed, and the iodine concentration can be assumed to be zero at the end of the reaction, the average rate for this reaction can be determined as follows:

$$\text{Average Rate} = \frac{[I_2]_{final} - [I_2]_{initial}}{time} \approx \frac{0 - [I_2]_{initial}}{time}$$

4 Analyzing Information: Look at your data table and notice that in mixtures 2, 3, and 4, one substance's concentration changed, while the others remained the same. Use this information to determine how the reaction rate was affected when the concentration of one of the reactants was doubled while the concentration of all the other reactants stayed the same. Repeat for the determination for each of the other reactants also.

Figure 18-A: As the acetone is iodinated, the starch indicator changes color from blue-black to clear.

Conclusions

1 Inferring Conclusions: Write the rate law for the reaction in this experiment. What are the values of x, y, and z as determined in your experiment? (Hint: Remember, a reaction's rate should be related to the concentration of the reactants as follows:

$$R = k[\text{acetone}]^x[\text{HCl}]^y[\text{I}_2]^z.$$

If trials were run where the concentrations of I_2 and HCl stayed the same, but the concentration of acetone was different in one trial, the following relationships are true:

$$R_1 = k[\text{acetone}_1]^x[\text{HCl}]^y[\text{I}_2]^z \text{ and}$$
$$R_2 = k[\text{acetone}_2]^x[\text{HCl}]^y[\text{I}_2]^z$$

Since k, [HCl], and [I_2] are the same for both reactions, one can solve for x as follows:

$$\frac{R_1}{R_2} = \frac{k[\text{acetone}_1]^x[\text{HCl}]^y[\text{I}_2]^z}{k[\text{acetone}_2]^x[\text{HCl}]^y[\text{I}_2]^z} \text{ and so,}$$

$$\frac{R_1}{R_2} = \frac{[\text{acetone}_1]^x}{[\text{acetone}_2]^x}$$

Taking the logarithm of both sides,

$$\log\left(\frac{R_1}{R_2}\right) = x \log\left(\frac{[\text{acetone}_1]}{[\text{acetone}_2]}\right) \text{ and so,}$$

$$x = \left\{\frac{\log\left(\frac{R_1}{R_2}\right)}{\log\left(\frac{[\text{acetone}_1]}{[\text{acetone}_2]}\right)}\right\}$$

A similar calculation can determine the values of the other variables.)

2 Inferring Conclusions: Using your data, calculate a value of k for this reaction.

3 Resolving Discrepancies and Relating Ideas: Is it possible for the concentration of one of the reactants to have no effect on the rate? Explain your answer.

Extensions

1 Inferring Conclusions and Relating Ideas: Given what you have discovered about the rate law, what can you determine about the mechanism for this reaction?

2 Predicting Outcomes: What would the rate for the reaction be if the following amounts of the reactants were used? How much time would elapse before the reaction was complete?

Reaction #	1.0 M HCl (drops)	Starch (drops)	0.0012 M I_2 (drops)	4.0 M acetone (drops)
5	15	5	15	15

If time allows, perform this experiment and check your prediction. **Remember to add the acetone last, only after other ingredients are mixed.** Then, calculate the percent error for your calculation, considering your experimental results to be the accepted values.

3 Evaluating Methods and Designing Experiments: You may not have come up with whole numbers for x, y, and z. Can you think of possible sources of error or imprecision in this experiment? If you can think of ways to eliminate them, ask your teacher to approve your suggestion, and run more trials.

4 Evaluating Viewpoints: Share your data with the rest of the class. Calculate your percent error in determining x, y, and z, considering the class average for each to be the accepted value.

5 Formulating Hypotheses: Suggest a possible mechanism for this reaction.

Investigation 18: Acid-Catalyzed Iodination of Acetone

Appendix A

TABLE A-1 PHYSICAL CONSTANTS

Quantity	Symbol	Value
Atomic mass unit	u	$1.660\,5042 \times 10^{-27}$ kg
		$5.485\,803 \times 10^{-4}\,u$
Avogadro number	N_A	$6.022\,137 \times 10^{23}$/mol
Electron rest mass	m_e	$9.109\,389 \times 10^{-31}$ kg
		$5.4858 \times 10^{-4}\,u$
Molar gas constant	R	$8.205\,68 \times 10^{-2}$ L·atm/(mol·K)
		$8.314\,51$ J/(mol·K)
Molar volume of ideal gas at STP	V_m	$22.414\,10$ L/mol
Neutron rest mass	m_n	$1.674\,928\,6 \times 10^{-27}$ kg
		$1.008\,665\,u$
Normal boiling point of water	T_b	$373.15\text{K} = 100.0°\text{C}$
Normal freezing point of water	T_b	$273.15\text{K} = 0.00°\text{C}$
Planck's constant	h	$6.626\,076 \times 10^{-34}$ J·s
Proton rest mass	m_p	$1.672\,6231 \times 10^{-27}$ kg
		$1.007\,276\,u$
Speed of light in a vacuum	c	$2.997\,924\,58 \times 10^8$ m/s
Temperature of triple point of water		$273.16\text{K} = 0.01°\text{C}$

TABLE A-2 HEAT OF COMBUSTION

Substance	Formula	Phase	ΔH_c
hydrogen	H_2	g	-285.8
graphite	C	s	-393.51
carbon monoxide	CO	g	-283.0
methane	CH_4	g	-890.31
ethane	C_2H_6	g	-1559.8
propane	C_3H_8	g	-2219.9
butane	C_4H_{10}	g	-2878.5
pentane	C_5H_{12}	g	-3536.1
hexane	C_6H_{14}	ℓ	-4163.1
heptane	C_7H_{16}	ℓ	-4811.2
octane	C_8H_{18}	ℓ	-5450.5
ethene (ethylene)	C_2H_4	g	-1411.0
propene (propylene)	C_3H_6	g	-2051.0
ethyne (acetylene)	C_2H_2	g	-1299.6
benzene	C_6H_6	ℓ	-3267.5
toluene	C_7H_8	ℓ	-3908.7
naphthalene	$C_{10}H_8$	s	-5153.9
anthracene	$C_{14}H_{10}$	s	-7163.0
methanol	CH_3OH	ℓ	-726.34
ethanol	C_2H_5OH	ℓ	-1366.8
ether	$(C_2H_5)_2O$	ℓ	-2751.1
formaldehyde	CH_2O	g	-570.78
glucose	$C_6H_{12}O_6$	s	-2803.0
sucrose	$C_{12}H_{22}O_{11}$	s	-5640.9

ΔH_c = heat of combustion of the given substance. All values of ΔH_c are expressed as kJ/mol of substance oxidized to $H_2O(\ell)$ and/or $CO_2(g)$ at constant pressure and 25 °C. s = solid, ℓ = liquid, g = gas.

TABLE A-3 THE ELEMENTS—THEIR SYMBOLS, ATOMIC NUMBERS, AND ATOMIC MASSES

Name of element	Symbol	Atomic number	Atomic mass	Name of element	Symbol	Atomic number	Atomic mass
actinium	Ac	89	227.028	neon	Ne	10	20.179
aluminum	Al	13	26.981 54	neptunium	Np	93	237.048
americium	Am	95	[243]	nickel	Ni	28	58.69
antimony	Sb	51	121.75	niobium	Nb	41	92.9064
argon	Ar	18	39.948	nitrogen	N	7	14.0067
arsenic	As	33	74.9216	nobelium	No	102	[259]
astatine	At	85	[210]	osmium	Os	76	190.2
barium	Ba	56	137.33	oxygen	O	8	15.9994
berkelium	Bk	97	[247]	palladium	Pd	46	106.42
beryllium	Be	4	9.012 18	phosphorus	P	15	30.973 76
bismuth	Bi	83	208.980	platinum	Pt	78	195.08
boron	B	5	10.81	plutonium	Pu	94	[244]
bromine	Br	35	79.904	polonium	Po	84	[209]
cadmium	Cd	48	112.41	potassium	K	19	39.0983
calcium	Ca	20	40.08	praseodymium	Pr	59	140.908
californium	Cf	98	[251]	promethium	Pm	61	[145]
carbon	C	6	12.0111	protactinium	Pa	91	231.036
cerium	Ce	58	140.12	radium	Ra	88	226.025
cesium	Cs	55	132.905	radon	Rn	86	[222]
chlorine	Cl	17	35.453	rhenium	Re	75	186.207
chromium	Cr	24	51.996	rhodium	Rh	45	102.906
cobalt	Co	27	58.9332	rubidium	Rb	37	85.4678
copper	Cu	29	63.546	ruthenium	Ru	44	101.07
curium	Cm	96	[247]	samarium	Sm	62	150.36
dysprosium	Dy	66	162.50	scandium	Sc	21	44.9559
einsteinium	Es	99	[252]	selenium	Se	34	78.96
erbium	Er	68	167.26	silicon	Si	14	28.0855
europium	Eu	63	151.96	silver	Ag	47	107.868
fermium	Fm	100	[257]	sodium	Na	11	22.989 77
fluorine	F	9	18.998 403	strontium	Sr	38	87.62
francium	Fr	87	[223]	sulfur	S	16	32.06
gadolinium	Gd	64	157.25	tantalum	Ta	73	180.948
gallium	Ga	31	69.72	technetium	Tc	43	[98]
germanium	Ge	32	72.59	tellurium	Te	52	127.60
gold	Au	79	196.967	terbium	Tb	65	158.925
hafnium	Hf	72	178.49	thallium	Tl	81	204.383
helium	He	2	4.002 60	thorium	Th	90	232.038
holmium	Ho	67	164.930	thulium	Tm	69	168.934
hydrogen	H	1	1.007 94	tin	Sn	50	118.71
indium	In	49	114.82	titanium	Ti	22	47.88
iodine	I	53	126.905	tungsten	W	74	183.85
iridium	Ir	77	192.22	unnilennium	Une	109	[266?]
iron	Fe	26	55.847	unnilhexium	Unh	106	[263]
krypton	Kr	36	83.80	unniloctium	Uno	108	[265]
lanthanum	La	57	138.906	unnilpentium	Unp	105	[262]
lawrencium	Lr	103	[260]	unnilquadium	Unq	104	[261]
lead	Pb	82	207.2	unnilseptium	Uns	107	[262]
lithium	Li	3	6.941	uranium	U	92	238.029
lutetium	Lu	71	174.967	vanadium	V	23	50.9415
magnesium	Mg	12	24.305	xenon	Xe	54	131.29
manganese	Mn	25	54.9380	ytterbium	Yb	70	173.04
mendelevium	Md	101	[258]	yttrium	Y	39	88.9059
mercury	Hg	80	200.59	zinc	Zn	30	65.39
molybdenum	Mo	42	95.94	zirconium	Zr	40	91.224
neodymium	Nd	60	144.24				

A value given in brackets denotes the mass number of the most stable or most common isotope. The atomic masses of most of these elements are believed to have an error no greater than ±1 in the last digit given.

TABLE A–4 ELECTRON ARRANGEMENT OF THE ELEMENTS

	Sublevels	1s	2s	2p	3s	3p	3d	4s	4p	4d	4f	5s	5p	5d	5f	6s	6p	6d	6f	7s
1	hydrogen	1																		
2	helium	2																		
3	lithium	2	1																	
4	beryllium	2	2																	
5	boron	2	2	1																
6	carbon	2	2	2																
7	nitrogen	2	2	3																
8	oxygen	2	2	4																
9	fluorine	2	2	5																
10	neon	2	2	6																
11	sodium	2	2	6	1															
12	magnesium	2	2	6	2															
13	aluminum	2	2	6	2	1														
14	silicon	2	2	6	2	2														
15	phosphorus	2	2	6	2	3														
16	sulfur	2	2	6	2	4														
17	chlorine	2	2	6	2	5														
18	argon	2	2	6	2	6														
19	potassium	2	2	6	2	6		1												
20	calcium	2	2	6	2	6		2												
21	scandium	2	2	6	2	6	1	2												
22	titanium	2	2	6	2	6	2	2												
23	vanadium	2	2	6	2	6	3	2												
24	chromium	2	2	6	2	6	5	1												
25	manganese	2	2	6	2	6	5	2												
26	iron	2	2	6	2	6	6	2												
27	cobalt	2	2	6	2	6	7	2												
28	nickel	2	2	6	2	6	8	2												
29	copper	2	2	6	2	6	10	1												
30	zinc	2	2	6	2	6	10	2												
31	gallium	2	2	6	2	6	10	2	1											
32	germanium	2	2	6	2	6	10	2	2											
33	arsenic	2	2	6	2	6	10	2	3											
34	selenium	2	2	6	2	6	10	2	4											
35	bromine	2	2	6	2	6	10	2	5											
36	krypton	2	2	6	2	6	10	2	6											
37	rubidium	2	2	6	2	6	10	2	6			1								
38	strontium	2	2	6	2	6	10	2	6			2								
39	yttrium	2	2	6	2	6	10	2	6	1		2								
40	zirconium	2	2	6	2	6	10	2	6	2		2								
41	niobium	2	2	6	2	6	10	2	6	4		1								
42	molybdenum	2	2	6	2	6	10	2	6	5		1								
43	technetium	2	2	6	2	6	10	2	6	6		1								
44	ruthenium	2	2	6	2	6	10	2	6	7		1								
45	rhodium	2	2	6	2	6	10	2	6	8		1								
46	palladium	2	2	6	2	6	10	2	6	10										
47	silver	2	2	6	2	6	10	2	6	10		1								
48	cadmium	2	2	6	2	6	10	2	6	10		2								
49	indium	2	2	6	2	6	10	2	6	10		2	1							
50	tin	2	2	6	2	6	10	2	6	10		2	2							
51	antimony	2	2	6	2	6	10	2	6	10		2	3							
52	tellurium	2	2	6	2	6	10	2	6	10		2	4							
53	iodine	2	2	6	2	6	10	2	6	10		2	5							
54	xenon	2	2	6	2	6	10	2	6	10		2	6							

TABLE A-4 ELECTRON ARRANGEMENT OF THE ELEMENTS (cont'd)

	Sublevels	1s	2s	2p	3s	3p	3d	4s	4p	4d	4f	5s	5p	5d	5f	6s	6p	6d	6f	7s
55	cesium	2	2	6	2	6	10	2	6	10		2	6			1				
56	barium	2	2	6	2	6	10	2	6	10		2	6			2				
57	lanthanum	2	2	6	2	6	10	2	6	10		2	6	1		2				
58	cerium	2	2	6	2	6	10	2	6	10	1	2	6	1		2				
59	praseodymium	2	2	6	2	6	10	2	6	10	3	2	6			2				
60	neodymium	2	2	6	2	6	10	2	6	10	4	2	6			2				
61	promethium	2	2	6	2	6	10	2	6	10	5	2	6			2				
62	samarium	2	2	6	2	6	10	2	6	10	6	2	6			2				
63	europium	2	2	6	2	6	10	2	6	10	7	2	6			2				
64	gadolinium	2	2	6	2	6	10	2	6	10	7	2	6	1		2				
65	terbium	2	2	6	2	6	10	2	6	10	9	2	6			2				
66	dysprosium	2	2	6	2	6	10	2	6	10	10	2	6			2				
67	holmium	2	2	6	2	6	10	2	6	10	11	2	6			2				
68	erbium	2	2	6	2	6	10	2	6	10	12	2	6			2				
69	thulium	2	2	6	2	6	10	2	6	10	13	2	6			2				
70	ytterbium	2	2	6	2	6	10	2	6	10	14	2	6			2				
71	lutetium	2	2	6	2	6	10	2	6	10	14	2	6	1		2				
72	hafnium	2	2	6	2	6	10	2	6	10	14	2	6	2		2				
73	tantalum	2	2	6	2	6	10	2	6	10	14	2	6	3		2				
74	tungsten	2	2	6	2	6	10	2	6	10	14	2	6	4		2				
75	rhenium	2	2	6	2	6	10	2	6	10	14	2	6	5		2				
76	osmium	2	2	6	2	6	10	2	6	10	14	2	6	6		2				
77	iridium	2	2	6	2	6	10	2	6	10	14	2	6	7		2				
78	platinum	2	2	6	2	6	10	2	6	10	14	2	6	9		1				
79	gold	2	2	6	2	6	10	2	6	10	14	2	6	10		1				
80	mercury	2	2	6	2	6	10	2	6	10	14	2	6	10		2				
81	thallium	2	2	6	2	6	10	2	6	10	14	2	6	10		2	1			
82	lead	2	2	6	2	6	10	2	6	10	14	2	6	10		2	2			
83	bismuth	2	2	6	2	6	10	2	6	10	14	2	6	10		2	3			
84	polonium	2	2	6	2	6	10	2	6	10	14	2	6	10		2	4			
85	astatine	2	2	6	2	6	10	2	6	10	14	2	6	10		2	5			
86	radon	2	2	6	2	6	10	2	6	10	14	2	6	10		2	6			
87	francium	2	2	6	2	6	10	2	6	10	14	2	6	10		2	6			1
88	radium	2	2	6	2	6	10	2	6	10	14	2	6	10		2	6			2
89	actinium	2	2	6	2	6	10	2	6	10	14	2	6	10		2	6	1		2
90	thorium	2	2	6	2	6	10	2	6	10	14	2	6	10		2	6	2		2
91	protactinium	2	2	6	2	6	10	2	6	10	14	2	6	10	2	2	6	1		2
92	uranium	2	2	6	2	6	10	2	6	10	14	2	6	10	3	2	6	1		2
93	neptunium	2	2	6	2	6	10	2	6	10	14	2	6	10	4	2	6	1		2
94	plutonium	2	2	6	2	6	10	2	6	10	14	2	6	10	6	2	6			2
95	americium	2	2	6	2	6	10	2	6	10	14	2	6	10	7	2	6			2
96	curium	2	2	6	2	6	10	2	6	10	14	2	6	10	7	2	6	1		2
97	berkelium	2	2	6	2	6	10	2	6	10	14	2	6	10	9	2	6			2
98	californium	2	2	6	2	6	10	2	6	10	14	2	6	10	10	2	6			2
99	einsteinium	2	2	6	2	6	10	2	6	10	14	2	6	10	11	2	6			2
100	fermium	2	2	6	2	6	10	2	6	10	14	2	6	10	12	2	6			2
101	mendelevium	2	2	6	2	6	10	2	6	10	14	2	6	10	13	2	6			2
102	nobelium	2	2	6	2	6	10	2	6	10	14	2	6	10	14	2	6			2
103	lawrencium	2	2	6	2	6	10	2	6	10	14	2	6	10	14	2	6	1		2
104	unnilquadium	2	2	6	2	6	10	2	6	10	14	2	6	10	14	2	6	2		2
105	unnilpentium	2	2	6	2	6	10	2	6	10	14	2	6	10	14	2	6	3		2
106	unnilhexium	2	2	6	2	6	10	2	6	10	14	2	6	10	14	2	6	4		2
107	unnilseptium	2	2	6	2	6	10	2	6	10	14	2	6	10	14	2	6	5		2
108	unniloctium	2	2	6	2	6	10	2	6	10	14	2	6	10	14	2	6	6		2
109	unnilennium	2	2	6	2	6	10	2	6	10	14	2	6	10	14	2	6	7		2

TABLE A–5 COMMON ELEMENTS

Name	Symbol	Approx. mass	Common ox. nos.	Name	Symbol	Approx. mass	Common ox. nos.
aluminum	Al	27.0	+3	magnesium	Mg	24.3	+2
antimony	Sb	121.8	−3, +3, +5	manganese	Mn	54.9	+2, +3, +4, +6, +7
arsenic	As	74.9	−3, +3, +5	mercury	Hg	200.6	+1, +2
barium	Ba	137.3	+2	nickel	Ni	58.7	+2, +3
bismuth	Bi	209.0	+3	nitrogen	N	14.0	−3, +3, +5
bromine	Br	79.9	−1, +1, +3, +5, +7	oxygen	O	16.0	−2
calcium	Ca	40.1	+2	phosphorus	P	31.0	+3, +5
carbon	C	12.0	+2, +4	platinum	Pt	195.1	+2, +4
chlorine	Cl	35.5	−1, +1, +3, +5, +7	potassium	K	39.1	+1
chromium	Cr	52.0	+2, +3, +6	silicon	Si	28.1	+2, +4
cobalt	Co	58.9	+2, +3	silver	Ag	107.9	+1
copper	Cu	63.5	+1, +2	sodium	Na	23.0	+1
fluorine	F	19.0	−1	strontium	Sr	87.6	+2
gold	Au	197.0	+1, +3	sulfur	S	32.1	−2, +4, +6
hydrogen	H	1.0	−1, +1	tin	Sn	118.7	+2, +4
iodine	I	126.9	−1, +1, +3, +5, +7	titanium	Ti	47.9	+2, +3, +4
iron	Fe	55.8	+2, +3	tungsten	W	183.8	+6
lead	Pb	207.2	+2, +4	zinc	Zn	65.4	+2

TABLE A–6 COMMON IONS

Name	Symbol	Name	Symbol
aluminum	Al^{3+}	lead(II)	Pb^{2+}
ammonium	NH_4^+	magnesium	Mg^{2+}
barium	Ba^{2+}	mercury(I)	Hg_2^{2+}
calcium	Ca^{2+}	mercury(II)	Hg^{2+}
chromium(III)	Cr^{3+}	nickel(II)	Ni^{2+}
cobalt(II)	Co^{2+}	potassium	K^+
copper(I)	Cu^+	silver	Ag^+
copper(II)	Cu^{2+}	sodium	Na^+
hydronium	H_3O^+	tin(II)	Sn^{2+}
iron(II)	Fe^{2+}	tin(IV)	Sn^{4+}
iron(III)	Fe^{3+}	zinc	Zn^{2+}
acetate	$C_2H_3O_2^-$	hydrogen sulfate	HSO_4^-
bromide	Br^-	hydroxide	OH^-
carbonate	CO_3^{2-}	hypochlorite	ClO^-
chlorate	ClO_3^-	iodide	I^-
chloride	Cl^-	nitrate	NO_3^-
chlorite	ClO_2^-	nitrite	NO_2^-
chromate	CrO_4^{2-}	oxide	O^{2-}
cyanide	CN^-	perchlorate	ClO_4^-
dichromate	$Cr_2O_7^{2-}$	permanganate	MnO_4^-
fluoride	F^-	peroxide	O_2^{2-}
hexacyanoferrate(II)	$Fe(CN)_6^{4-}$	phosphate	PO_4^{3-}
hexacyanoferrate(III)	$Fe(CN)_6^{3-}$	sulfate	SO_4^{2-}
hydride	H^-	sulfide	S^{2-}
hydrogen carbonate	HCO_3^-	sulfite	SO_3^{2-}

TABLE A-7 WATER-VAPOR PRESSURE

Temperature (°C)	Pressure (mm Hg)	Temperature (°C)	Pressure (mm Hg)	Temperature (°C)	Pressure (mm Hg)
0.0	4.6	19.5	17.0	27.0	26.7
5.0	6.5	20.0	17.5	28.0	28.3
10.0	9.2	20.5	18.1	29.0	30.0
12.5	10.9	21.0	18.6	30.0	31.8
15.0	12.8	21.5	19.2	35.0	42.2
15.5	13.2	22.0	19.8	40.0	55.3
16.0	13.6	22.5	20.4	50.0	92.5
16.5	14.1	23.0	21.1	60.0	149.4
17.0	14.5	23.5	21.7	70.0	233.7
17.5	15.0	24.0	22.4	80.0	355.1
18.0	15.5	24.5	23.1	90.0	525.8
18.5	16.0	25.0	23.8	95.0	633.9
19.0	16.5	26.0	25.2	100.0	760.0

TABLE A-8 DENSITY OF GASES AT STP

Gas	Density (g/L)	Gas	Density (g/L)
air, dry	1.2929	hydrogen	0.0899
ammonia	0.771	hydrogen chloride	1.639
carbon dioxide	1.977	hydrogen sulfide	1.539
carbon monoxide	1.250	methane	0.716
chlorine	3.214	nitrogen	1.251
dinitrogen monoxide	1.977	nitrogen monoxide	1.340
ethyne (acetylene)	1.171	oxygen	1.429
helium	0.1785	sulfur dioxide	2.927

TABLE A-9 DENSITY OF WATER

Temperature (°C)	Density (g/mL)	Temperature (°C)	Density (g/mL)
0	0.999 84	24	0.997 30
2	0.999 94	26	0.996 78
3.98 (maximum)	0.999 973	30	0.995 65
4	0.999 97	40	0.992 22
6	0.999 94	50	0.988 04
8	0.999 85	60	0.983 20
10	0.999 70	70	0.977 77
14	0.999 24	80	0.971 79
16	0.998 94	90	0.965 31
20	0.998 20	100	0.958 36

TABLE A–10 SOLUBILITY OF GASES IN WATER

Volume of gas (reduced to STP) that can be dissolved in 1 volume of water at the temperature (°C) indicated.

Gas	0°	10°	20°	60°
air	0.029 18	0.022 84	0.018 68	0.012 16
ammonia	1130	870	680	200
carbon dioxide	1.713	1.194	0.878	0.359
carbon monoxide	0.035 37	0.028 16	0.023 19	0.014 88
chlorine	4.54	3.148	2.299	1.023
hydrogen	0.021 48	0.019 55	0.018 19	0.016 00
hydrogen chloride	512	475	442	339
hydrogen sulfide	4.670	3.399	2.582	1.190
methane	0.055 63	0.041 77	0.033 08	0.019 54
nitrogen	0.023 54	0.018 61	0.015 45	0.010 23
nitrogen dioxide	0.073 81	0.057 09	0.047 06	0.029 54
oxygen	0.048 89	0.038 02	0.031 02	0.019 46
sulfur dioxide	79.789	56.647	39.374	—

TABLE A–11 SOLUBILITY CHART

S = soluble in water. A = soluble in acids, insoluble in water. P = partially soluble in water, soluble in dilute acids.
I = insoluble in dilute acids and in water. a = slightly soluble in acids, insoluble in water. d = decomposes in water.

	acetate	bromide	carbonate	chlorate	chloride	chromate	hydroxide	iodide	nitrate	oxide	phosphate	silicate	sulfate	sulfide
aluminum	S	S	—	S	S	—	A	S	S	a	A	I	S	d
ammonium	S	S	S	S	S	S	—	S	S	—	S	—	S	S
barium	S	S	P	S	S	A	S	S	S	S	A	S	a	d
calcium	S	S	P	S	S	S	P	S	S	P	P	P	P	P
copper(II)	S	S	—	S	S	—	A	—	S	A	A	A	S	A
hydrogen	S	S	—	S	S	—	—	S	S	—	S	I	S	S
iron(II)	S	S	P	S	S	—	A	S	S	A	A	—	S	A
iron(III)	S	S	—	S	S	A	A	S	S	A	P	—	P	d
lead(II)	S	S	A	S	S	A	P	P	S	P	A	A	P	A
magnesium	S	S	P	S	S	S	A	S	S	A	P	A	S	d
manganese(II)	S	S	P	S	S	—	A	S	S	A	P	I	S	A
mercury(I)	P	A	A	S	a	P	—	A	S	A	A	—	P	I
mercury(II)	S	S	—	S	S	P	A	P	S	P	A	—	d	I
potassium	S	S	S	S	S	S	S	S	S	S	S	S	S	S
silver	P	a	A	S	a	P	—	I	S	P	A	—	P	A
sodium	S	S	S	S	S	S	S	S	S	S	S	S	S	S
strontium	S	S	P	S	S	P	S	S	S	S	A	A	P	S
tin(II)	d	S	—	S	S	A	A	S	d	A	A	—	S	A
tin(IV)	S	S	—	—	S	S	P	d	—	A	—	—	S	A
zinc	S	S	P	S	S	P	A	S	S	P	A	A	S	A

TABLE A-12 SOLUBILITY OF COMPOUNDS

Solubilities are given in grams of anhydrous solute that can be dissolved in 100 g of water at the temperature (°C) indicated.

Compound	Formula	0°	20°	60°	100°
aluminum sulfate	$Al_2(SO_4)_3$	31.2	36.4	59.2	89.0
ammonium chloride	NH_4Cl	29.4	37.2	55.3	77.3
ammonium nitrate	NH_4NO_3	118	192	421	871
ammonium sulfate	$(NH_4)_2SO_4$	70.6	75.4	88	103
barium carbonate	$BaCO_3$	$0.0016^{8°}$	$0.0022^{18°}$	—	0.0065
barium chloride	$BaCl_2$	31.2	35.8	46.2	59.4
barium hydroxide	$Ba(OH)_2$	1.67	3.89	20.94	$101.40^{80°}$
barium nitrate	$Ba(NO_3)_2$	4.95	9.02	20.4	34.4
barium sulfate	$BaSO_4$	0.000 115	0.000 24	—	0.000 413
cadmium sulfate	$CdSO_4$	75.4	76.6	81.8	60.8
calcium acetate	$Ca(C_2H_3O_2)_2$	37.4	34.7	32.7	29.7
calcium carbonate	$CaCO_3$	—	0.0012	—	0.002
calcium fluoride	CaF_2	$0.0016^{18°}$	$0.0017^{25°}$	—	—
calcium hydrogen carbonate	$Ca(HCO_3)_2$	16.15	16.60	17.50	18.40
calcium hydroxide	$Ca(OH)_2$	0.189	0.173	0.121	0.076
calcium sulfate	$CaSO_4$	0.176	$0.209^{30°}$	0.205	0.162
cerium sulfate	$Ce_2(SO_4)_3$	21.4	9.84	3.87	—
cesium nitrate	$CsNO_3$	9.33	23.0	83.8	197
copper(II) chloride	$CuCl_2$	68.6	73.0	96.5	120
copper(II) sulfate	$CuSO_4$	23.1	32.0	61.8	114
lead(II) chloride	$PbCl_2$	0.67	1.00	1.94	3.20
lead(II) nitrate	$Pb(NO_3)_2$	37.5	54.3	91.6	133
lithium chloride	$LiCl$	69.2	83	98.4	128
lithium sulfate	Li_2SO_4	36.1	34.8	32.6	29.9
magnesium hydroxide	$Mg(OH)_2$	—	$0.0009^{18°}$	—	—
magnesium sulfate	$MgSO_4$	22.0	33.7	54.6	68.3
mercury(I) chloride	Hg_2Cl_2	0.000 14	0.0002	$0.0007^{40°}$	—
mercury(II) chloride	$HgCl_2$	3.63	6.57	16.3	61.3
potassium aluminum sulfate	$KAl(SO_4)_2$	3.00	5.90	24.8	109
potassium bromide	KBr	53.6	65.3	85.5	104
potassium chlorate	$KClO_3$	3.3	7.3	23.8	56.3
potassium chloride	KCl	28.0	34.2	45.8	56.3
potassium chromate	K_2CrO_4	56.3	63.7	70.1	75.6
potassium iodide	KI	128	144	176	206
potassium nitrate	KNO_3	13.9	31.6	106	245
potassium permanganate	$KMnO_4$	2.83	6.34	22.1	—
potassium sulfate	K_2SO_4	7.4	11.1	18.2	24.1
silver acetate	$AgC_2H_3O_2$	0.73	1.05	1.93	$2.59^{80°}$
silver chloride	$AgCl$	0.000 07	0.000 194	$0.0005^{50°}$	0.002
silver nitrate	$AgNO_3$	122	216	440	733
sodium acetate	$NaC_2H_3O_2$	36.2	46.4	139	170.15
sodium chlorate	$NaClO_3$	79.6	95.9	137	204
sodium chloride	$NaCl$	35.7	35.9	37.1	39.2
sodium nitrate	$NaNO_3$	73.0	87.6	122	180
sucrose	$C_{12}H_{22}O_{11}$	179.2	203.9	287.3	487.2
ytterbium sulfate	$Yb_2(SO_4)_3$	44.2	$22.2^{30°}$	10.4	4.7

TABLE A–13 HEAT OF FORMATION

ΔH_f is heat of formation of the given substance from its elements. All values of ΔH_f are expressed as kJ/mol at 25°C. Negative values of ΔH_f indicate exothermic reactions. s = solid, ℓ = liquid, g = gas.

Substance	Phase	ΔH_f	Substance	Phase	ΔH_f
aluminum oxide (α)	s	−1675.7	lead(II) nitrate	s	−451.97
ammonia	g	−46.11	lead(II) sulfate	s	−919.94
ammonium chloride	s	−314.43	lithium chloride	s	−408.78
ammonium sulfate	s	−1180.85	lithium nitrate	s	−481.58
barium chloride	s	−860.06	lithium sulfate	s	−1434.4
barium nitrate	s	−991.86	magnesium chloride	s	−641.32
barium sulfate	s	−1465.2	magnesium oxide	s	−601.83
benzene	g	+82.93	magnesium sulfate	s	−1278.2
benzene	ℓ	+48.50	manganese(IV) oxide	s	−520.03
calcium carbonate	s	−1206.9	manganese(II) sulfate	s	−1065.2
calcium chloride	s	−795.0	mercury(I) chloride	s	−265.22
calcium hydroxide	s	−986.6	mercury(II) chloride	s	−224.3
calcium nitrate	s	−937.22	mercury(II) fulminate	s	+270
calcium oxide	s	−635.5	mercury(II) nitrate	s	−392
calcium sulfate	s	−1432.69	mercury(II) oxide (red)	s	−90.84
carbon (diamond)	s	+1.90	methane	g	−74.81
carbon (graphite)	s	0.00	nitrogen dioxide	g	+33.18
carbon dioxide	g	−393.51	nitrogen monoxide	g	+90.25
carbon disulfide	g	+117.07	dinitrogen monoxide	g	+82.05
carbon disulfide	ℓ	+89.70	dinitrogen pentoxide	g	+11.3
carbon monoxide	g	−110.53	dinitrogen pentoxide	s	−43.1
carbon tetrachloride	ℓ	−95.81	dinitrogen tetroxide	g	+9.16
copper(II) nitrate	s	−302.9	oxygen (O_2)	g	0.00
copper(II) oxide	s	−157.3	ozone (O_3)	g	+142.7
copper(II) sulfate	s	−771.36	diphosphorus pentoxide	s	−2984.0
copper(I) sulfide	s	−79.5	potassium bromide	s	−392.17
copper(II) sulfide	s	−53.1	potassium chloride	s	−435.87
ethane	g	−84.68	potassium hydroxide	s	−425.85
ethyne (acetylene)	g	+226.73	potassium iodide	s	−327.65
hydrogen (H_2)	g	0.00	potassium nitrate	s	−492.71
hydrogen bromide	g	−36.40	potassium sulfate	s	−1433.69
hydrogen chloride	g	−92.31	silicon dioxide (quartz)	s	−910.94
hydrogen fluoride	g	−271.1	silicon tetrachloride	ℓ	−687.01
hydrogen iodide	g	+26.48	silver chloride	s	−127.07
hydrogen oxide (water)	g	−241.82	silver nitrate	s	−124.39
hydrogen oxide (water)	ℓ	−285.83	silver sulfide	s	−32.6
hydrogen peroxide	g	−136.11	sodium bromide	s	−361.1
hydrogen peroxide	ℓ	−187.78	sodium chloride	s	−411.15
hydrogen sulfide	g	−20.63	sodium hydroxide	s	−426.73
iodine (I_2)	s	0.00	sodium nitrate	s	−467.9
iodine (I_2)	g	+62.44	sodium sulfate	s	−1484.49
iron (III) chloride	s	−399.49	sulfur dioxide	g	−296.83
iron (III) oxide	s	−824.2	sulfur trioxide	g	−395.72
iron (II, III) oxide	s	−1118.4	tin(IV) chloride	ℓ	−511.3
iron (II) sulfate	s	−928.4	zinc nitrate	s	−483.7
iron (II) sulfide	s	−100.0	zinc oxide	s	−358.28
lead (II) oxide	s	−218.07	zinc sulfate	s	−982.82
lead (IV) oxide	s	−277.4	zinc sulfide (sphalerite)	s	−205.98

TABLE A–14 PROPERTIES OF COMMON ELEMENTS

Name	Form/color at room temperature	Density (g/cm³)	Melting point (°C)	Boiling point (°C)	Common oxidation numbers
aluminum	silv metal	2.70	660.4	2467	+3
antimony	silv metal	6.69	630.7	1750	−3, +3, +5
argon	colorless gas	1.784*	−189.2	−185.7	0
arsenic	gray metal	5.73	817 (28 atm)	613 (sublimes)	−3, +3, +5
barium	silv metal	3.51	725	1640	+2
beryllium	gray metal	1.85	1280	2970	+2
bismuth	silv metal	9.75	271.3	1560	+3
boron	blk solid	2.34	2079	2550 (sublimes)	+3
bromine	red-br liquid	3.12	−7.2	58.8	−1, +1, +3, +5, +7
calcium	silv metal	1.55	839	1484	+2
carbon	diamond	3.51	3700	4200	+2, +4
	graphite	2.26	3620 (sublimes)	4200	
chlorine	grn-yel gas	3.214*	−101.0	−34.6	−1, +1, +3, +5, +7
chromium	silv metal	7.19	1860	2672	+2, +3, +6
cobalt	silv metal	8.9	1495	2870	+2, +3
copper	red metal	8.96	1083.4	2567	+1, +2
fluorine	yel gas	1.696*	−219.6	−188.1	−1
germanium	gray metalloid	5.32	937.4	2830	+4
gold	yel metal	19.3	1064.4	2807	+1, +3
helium	colorless gas	0.1785*	−272.2 (26 atm)	−268.9	0
hydrogen	colorless gas	0.08988*	−259.1	−252.9	−1, +1
iodine	bl-blk solid	4.93	113.5	184.4	−1, +1, +3, +5, +7
iron	silv metal	7.87	1535	2750	+2, +3
lead	silv metal	11.4	327.5	1740	+2, +4
lithium	silv metal	0.534	180.5	1347	+1
magnesium	silv metal	1.74	649	1090	+2
manganese	silv metal	7.3	1244	1962	+2, +3, +4, +6, +7
mercury	silv liquid	13.5	−38.8	356.6	+1, +2
neon	colorless gas	0.8999*	−248.7	−246.0	0
nickel	silv metal	8.90	1453	2732	+2, +3
nitrogen	colorless gas	1.251*	−209.9	−195.8	−3, +3, +5
oxygen	colorless gas	1.429*	−218.4	−183.0	−2
phosphorus	yel solid	1.82	44.1	280	+3, +5
platinum	silv metal	21.4	1772	3800	+2, +4
plutonium	silv metal	19.8	641	3232	+3, +4, +5, +6
potassium	silv metal	0.862	63.65	774	+1
radium	silv metal	5(?)	700	1140	+2
radon	colorless gas	9.73*	−71	−61.8	0
silicon	gray solid	2.33	1410	2355	+2, +4
silver	silv metal	10.5	961.9	2212	+1
sodium	silv metal	0.971	97.8	882.9	+1
strontium	silv metal	2.54	769	1384	+2
sulfur	yel solid	2.07	112.8	444.7	−2, +4, +6
tin	silv metal	7.31	232.0	2270	+2, +4
titanium	silv metal	4.54	1660	3287	+2, +3, +4
tungsten	gray metal	19.3	3410	5660	+6
uranium	silv metal	19.0	1132	3818	+4, +6
xenon	colorless gas	5.89*	−111.9	−107	0
zinc	silv metal	7.13	419.6	907	+2

*Densities of gases are given in grams/liter at STP.

APPENDIX B

Logarithms

Logarithms are exponents. They are useful when working with calculations in which numbers vary over a large range. The logarithm of a number is the exponent or the power to which the number must be raised in order to obtain that number. For example, if $x = z^y$, then by raising z to the y power, x is obtained. This can be said mathematically as: y is the log of x using base z or; $y = \log_z x$. For example, if 10 is used as the base, then $100 = 10^2$ or 2 is the log of 100 using base 10.

Logarithms that use 10 as the base are called common logarithms. Only common logarithms are used in this book so it is not necessary to write 10 as the base when using common logarithms in this course. For example, in the expression, $2 = \log 100$, it is understood that base 10 is being used.

Logarithms are exponents and follow the laws of exponents in multiplication and division. The logarithm of a product is the sum of the logarithms of the factors.

$$\log(a \times b) = \log(a) + \log(b)$$

The logarithm of a quotient is the logarithm of the dividend minus the logarithm of the divisor.

$$\log(a/b) = \log(a) - \log(b)$$

A logarithm is composed of two parts: the characteristic, or whole-number part; and the mantissa, or decimal part. For example, the logarithm of 219 is 2.340, where 2 is the characteristic and .340 is the mantissa.

To find the logarithm of a number, it is helpful to use the fact that

$$\log(a \times 10^b) = \log(a) + \log(10^b) = \log(a) + b$$

To illustrate this we will find the logarithm of 219 using four steps.

Step 1. Write the number in scientific notation.
$$219 = 2.19 \times 10^2$$

Step 2. Then take the log of both sides of this equation.

$$\log(219) = \log(2.19) + \log(10^2) = \log(2.19) + 2$$

Step 3. Find the log of 2.19 by consulting a logarithm table (Table B-1). Find 2.1 in the left column. Then, move to the right until you are under the column labeled 9. The log of 2.19 is .3404.

Step 4. Since we know the log of 2.19, we can fill in the missing information from Step 2 as follows.

$$\log(219) = 2 + .3404 = 2.340$$

Note that in logarithms, only the mantissa represents significant figures. Therefore, the log of 219 is 2.340 to 3 significant figures. The characteristic 2 is simply a factor of magnitude.

In many cases you are given the log of a number and are asked to find the number. The number whose log is given is called the antilog. How can you find an antilog? The antilog of a number is the number obtained by raising ten to the power indicated by the number. For example, the antilog of 2 is 100; that, is 10 raised to the power 2 is 100, or;

$$\text{antilog}(2) = 100$$

To find the antilog of a mixed number or fraction such as 10.32 or 0.189, the 4 steps shown above must essentially be reversed. This is illustrated in the following problem of finding the antilog of 1.32.

Step 1. Write the number as the sum of a fraction and a whole number.

$$1.32 = 0.32 + 1$$

Step 2. Take the antilog of the fraction and the whole number.

$$\text{antilog}(1.32) = \text{antilog}(0.32) + \text{antilog}(1)$$

Step 3. The antilog of 1 is 10; remember that $10 = 10^1$. Find the antilog of 0.32 by looking up 0.32 in the body of Table B-1. It is found in the column labeled 9 and the row labeled 2.0. The antilog of 0.32 is 2.09.

Step 4. Since the antilog of the sum of two numbers is equal to the product of the two antilogs, obtain the final antilog by multiplying the antilogs of the two numbers.

$$10 \times 2.09 = 20.9$$

TABLE B–1 COMMON LOGARITHMS

N	0	1	2	3	4	5	6	7	8	9	N	0	1	2	3	4	5	6	7	8	9
1.0	.0000	.0043	.0086	.0128	.0170	.0212	.0253	.0294	.0334	.0374	5.5	.7404	.7412	.7419	.7427	.7435	.7443	.7451	.7459	.7466	.7474
1.1	.0414	.0453	.0492	.0531	.0569	.0607	.0645	.0682	.0719	.0755	5.6	.7482	.7490	.7497	.7505	.7513	.7520	.7528	.7536	.7543	.7551
1.2	.0792	.0828	.0864	.0899	.0934	.0969	.1004	.1038	.1072	.1106	5.7	.7559	.7566	.7574	.7582	.7589	.7597	.7604	.7612	.7619	.7627
1.3	.1139	.1173	.1206	.1239	.1271	.1303	.1335	.1367	.1399	.1430	5.8	.7634	.7642	.7649	.7657	.7664	.7672	.7679	.7686	.7694	.7701
1.4	.1461	.1492	.1523	.1553	.1584	.1614	.1644	.1673	.1703	.1732	5.9	.7709	.7716	.7723	.7731	.7738	.7745	.7752	.7760	.7767	.7774
1.5	.1761	.1790	.1818	.1847	.1875	.1903	.1931	.1959	.1987	.2014	6.0	.7782	.7789	.7796	.7803	.7810	.7818	.7825	.7832	.7839	.7846
1.6	.2041	.2068	.2095	.2122	.2148	.2175	.2201	.2227	.2253	.2279	6.1	.7853	.7860	.7868	.7875	.7882	.7889	.7896	.7903	.7910	.7917
1.7	.2304	.2330	.2355	.2380	.2405	.2430	.2455	.2480	.2504	.2529	6.2	.7924	.7931	.7938	.7945	.7952	.7959	.7966	.7973	.7980	.7987
1.8	.2553	.2577	.2601	.2625	.2648	.2672	.2695	.2718	.2742	.2765	6.3	.7993	.8000	.8007	.8014	.8021	.8028	.8035	.8041	.8048	.8055
1.9	.2788	.2810	.2833	.2856	.2878	.2900	.2923	.2945	.2967	.2989	6.4	.8062	.8069	.8075	.8082	.8089	.8096	.8102	.8109	.8116	.8122
2.0	.3010	.3032	.3054	.3075	.3096	.3118	.3139	.3160	.3181	.3201	6.5	.8129	.8136	.8142	.8149	.8156	.8162	.8169	.8176	.8182	.8189
2.1	.3222	.3243	.3263	.3284	.3304	.3324	.3345	.3365	.3385	.3404	6.6	.8195	.8202	.8209	.8215	.8222	.8228	.8235	.8241	.8248	.8254
2.2	.3424	.3444	.3464	.3483	.3502	.3522	.3541	.3560	.3579	.3598	6.7	.8261	.8267	.8274	.8280	.8287	.8293	.8299	.8306	.8312	.8319
2.3	.3617	.3636	.3655	.3674	.3692	.3711	.3729	.3747	.3766	.3784	6.8	.8325	.8331	.8338	.8344	.8351	.8357	.8363	.8370	.8376	.8382
2.4	.3802	.3820	.3838	.3856	.3874	.3892	.3909	.3927	.3945	.3962	6.9	.8388	.8395	.8401	.8407	.8414	.8420	.8426	.8432	.8439	.8445
2.5	.3979	.3997	.4014	.4031	.4048	.4065	.4082	.4099	.4116	.4133	7.0	.8451	.8457	.8463	.8470	.8476	.8482	.8488	.8494	.8500	.8506
2.6	.4150	.4166	.4183	.4200	.4216	.4232	.4249	.4265	.4281	.4298	7.1	.8513	.8519	.8525	.8531	.8537	.8543	.8549	.8555	.8561	.8567
2.7	.4314	.4330	.4346	.4362	.4378	.4393	.4409	.4425	.4440	.4456	7.2	.8573	.7579	.8585	.8591	.8597	.8603	.8609	.8615	.8621	.8627
2.8	.4472	.4487	.4502	.4518	.4533	.4548	.4564	.4579	.4594	.4609	7.3	.8633	.8639	.8645	.8651	.8657	.8663	.8669	.8675	.8681	.8686
2.9	.4624	.4639	.4654	.4669	.4683	.4698	.4713	.4728	.4742	.4757	7.4	.8692	.8698	.8704	.8710	.8716	.8722	.8727	.8733	.8739	.8745
3.0	.4771	.4786	.4800	.4814	.4829	.4843	.4857	.4871	.4886	.4900	7.5	.8751	.8756	.8762	.8768	.8774	.8779	.8785	.8791	.8797	.8802
3.1	.4914	.4928	.4942	.4955	.4969	.4983	.4997	.5011	.5024	.5038	7.6	.8808	.8814	.8820	.8825	.8831	.8837	.8842	.8848	.8854	.8859
3.2	.5051	.5065	.5079	.5092	.5105	.5119	.5132	.5145	.5159	.5172	7.7	.8865	.8871	.8876	.8882	.8887	.8893	.8899	.8904	.8910	.8915
3.3	.5185	.5198	.5211	.5224	.5237	.5250	.5263	.5276	.5289	.5307	7.8	.8921	.8927	.8932	.8938	.8943	.8949	.8954	.8960	.8965	.8971
3.4	.5315	.5328	.5340	.5353	.5366	.5378	.5391	.5403	.5416	.5428	7.9	.8976	.8982	.8987	.8993	.8998	.9004	.9009	.9015	.9020	.9025
3.5	.5441	.5453	.5465	.5478	.5490	.5502	.5514	.5527	.5539	.5551	8.0	.9031	.9036	.9042	.9047	.9053	.9058	.9063	.9069	.9074	.9079
3.6	.5563	.5575	.5587	.5599	.5611	.5623	.5635	.5647	.5658	.5670	8.1	.9085	.9090	.9096	.9101	.9106	.9112	.9117	.9122	.9128	.9133
3.7	.5682	.5694	.5705	.5717	.5729	.5740	.5752	.5763	.5775	.5786	8.2	.9138	.9143	.9149	.9154	.9159	.9165	.9170	.9175	.9180	.9186
3.8	.5798	.5809	.5821	.5832	.5843	.5855	.5866	.5877	.5888	.5899	8.3	.9191	.9196	.9201	.9206	.9212	.9217	.9222	.9227	.9232	.9238
3.9	.5911	.5922	.5933	.5944	.5955	.5966	.5977	.5988	.5999	.6010	8.4	.9243	.9248	.9253	.9258	.9263	.9269	.9274	.9279	.9284	.9289
4.0	.6021	.6031	.6042	.6053	.6064	.6075	.6085	.6096	.6107	.6117	8.5	.9294	.9299	.9304	.9309	.9315	.9320	.9325	.9330	.9335	.9340
4.1	.6128	.6138	.6149	.6160	.6170	.6180	.6191	.6201	.6212	.6222	8.6	.9345	.9350	.9555	.9360	.9365	.9370	.9375	.9380	.9385	.9390
4.2	.6232	.6243	.6253	.6263	.6274	.6284	.6294	.6304	.6314	.6325	8.7	.9395	.9400	.9405	.9410	.9415	.9420	.9425	.9430	.9435	.9440
4.3	.6335	.6345	.6355	.6365	.6375	.6385	.6395	.6405	.6415	.6425	8.8	.9445	.9450	.9455	.9460	.9465	.9469	.9474	.9479	.9484	.9489
4.4	.6435	.6444	.6454	.6464	.6474	.6484	.6493	.6503	.6513	.6522	8.9	.9494	.9499	.9504	.9509	.9513	.9518	.9523	.9528	.9533	.9538
4.5	.6532	.6542	.6551	.6561	.6571	.6580	.6590	.6599	.6609	.6618	9.0	.9542	.9547	.9552	.9557	.9562	.9566	.9571	.9576	.9581	.9586
4.6	.6628	.6637	.6646	.6656	.6665	.6675	.6684	.6693	.6702	.6712	9.1	.9590	.9595	.9600	.9605	.9609	.9614	.9619	.9624	.9628	.9633
4.7	.6721	.6730	.6739	.6749	.6758	.6767	.6776	.6785	.6794	.6803	9.2	.9638	.9643	.9647	.9652	.9657	.9661	.9666	.9671	.9675	.9680
4.8	.6812	.6821	.6830	.6839	.6848	.6857	.6866	.6875	.6884	.6893	9.3	.9685	.9689	.9694	.9699	.9703	.9708	.9713	.9717	.9722	.9727
4.9	.6902	.6911	.6920	.6928	.6937	.6946	.6955	.6964	.6972	.6981	9.4	.9731	.9736	.9741	.9745	.9750	.9754	.9759	.9763	.9768	.9773
5.0	.6990	.6998	.7007	.7016	.7024	.7033	.7042	.7050	.7059	.7067	9.5	.9777	.9782	.9786	.9791	.9795	.9800	.9805	.9809	.9914	.9818
5.1	.7076	.7084	.7093	.7101	.7110	.7118	.7126	.7135	.7143	.7152	9.6	.9823	.9827	.9832	.9836	.9841	.9845	.9850	.9854	.9859	.9863
5.2	.7160	.7168	.7177	.7185	.7193	.7202	.7210	.7218	.7226	.7235	9.7	.9868	.9872	.9877	.9881	.9886	.9890	.9894	.9899	.9903	.9908
5.3	.7243	.7251	.7259	.7267	.7275	.7284	.7292	.7300	.7308	.7316	9.8	.9912	.9917	.9921	.9926	.9930	.9934	.9939	.9943	.9948	.9952
5.4	.7324	.7332	.7340	.7348	.7356	.7364	.7372	.7380	.7388	.7396	9.9	.9956	.9961	.9965	.9969	.9974	.9978	.9983	.9987	.9991	.9996

Appendix C
Study Skills for Chemistry

Your success in this course will depend on your ability to apply a number of simple strategies to learning the material. Studying chemistry can be difficult, but you can make it easier using some simple strategies for dealing with the concepts and problems. Becoming skillful in using these strategies will be your keys to success in this course.

Reading the Text

- **Select a quiet setting,** away from distractions so you can concentrate while reading.
- **Have a pencil and paper nearby to jot down notes and questions** you may have from your reading. Be sure to get these questions answered in class.
- **Read the assigned material before class,** so the class lecture makes sense.
- While reading, one of your tasks is to discern what information is important. **Use the Objectives in the beginning of each section as a list of what you need to know from the section.** Teachers generally make their tests based on the text objectives or their own objectives. Using the objectives to focus your reading can make your learning more efficient.

Taking Notes in Class

- **Be prepared to take notes during class.** Have your materials organized in a notebook. Separate papers can be easily lost.
- **Don't write down everything your teacher says.** Organize your notes as you are putting them down, so that you can make sense of your notes when you review them.
- **Recopying notes later is a waste of time** and does not help you learn material for a test. Do it right the first time.

Reviewing Class Notes

- **Review your notes as soon as possible after class.** Write down any questions you may have about the material covered that day. Be sure to get these questions answered the next day in class.
- **Do not wait until the test to review.** By then you will have forgotten a good portion of the material.
- **Be selective about what you memorize.** You cannot memorize everything in a chapter. First of all, it is too time consuming. Second, memorizing doesn't guarantee that you will be able to answer questions correctly on a test because the ability to memorize is not a measure of understanding. You should only memorize material that you understand. Using a technique called concept mapping can help you decide what material in a chapter is important and the best way to learn that material.

Concept Maps

A concept map presents key ideas, meanings, and relationships for the major concepts being studied. A concept map that is developed for a chapter can be thought of as a visual road map for learning the material in the chapter. This learning happens efficiently because you work with only the key ideas and how they fit together and maps are easier to remember than lots of text. The concept map on page 889 is built from Section 1.2 of your text.

This map was made from most of the vocabulary terms in Section 1.2. Vocabulary terms are generally labels for concepts. Notice that concepts are generally nouns. Concepts are linked using linking words to form propositions. A proposition is a phrase that gives meaning to the concept. For example, in the map shown "matter can be changed by energy" is a proposition. Take time to read through the map, so that you are familiar with the information presented. Look at the map and Section 1.2. Which would be easier to study?

Why Make Concept Maps?

Studying from a concept map is a more efficient use of your time. Studies show that people are better able to remember materials presented visually than to remember page after page of text. The concept map is better than an outline because you can see relationships among many ideas. Outlines are linear. They progress from major head through subheads with no way of linking the ideas from various sections of the outline.

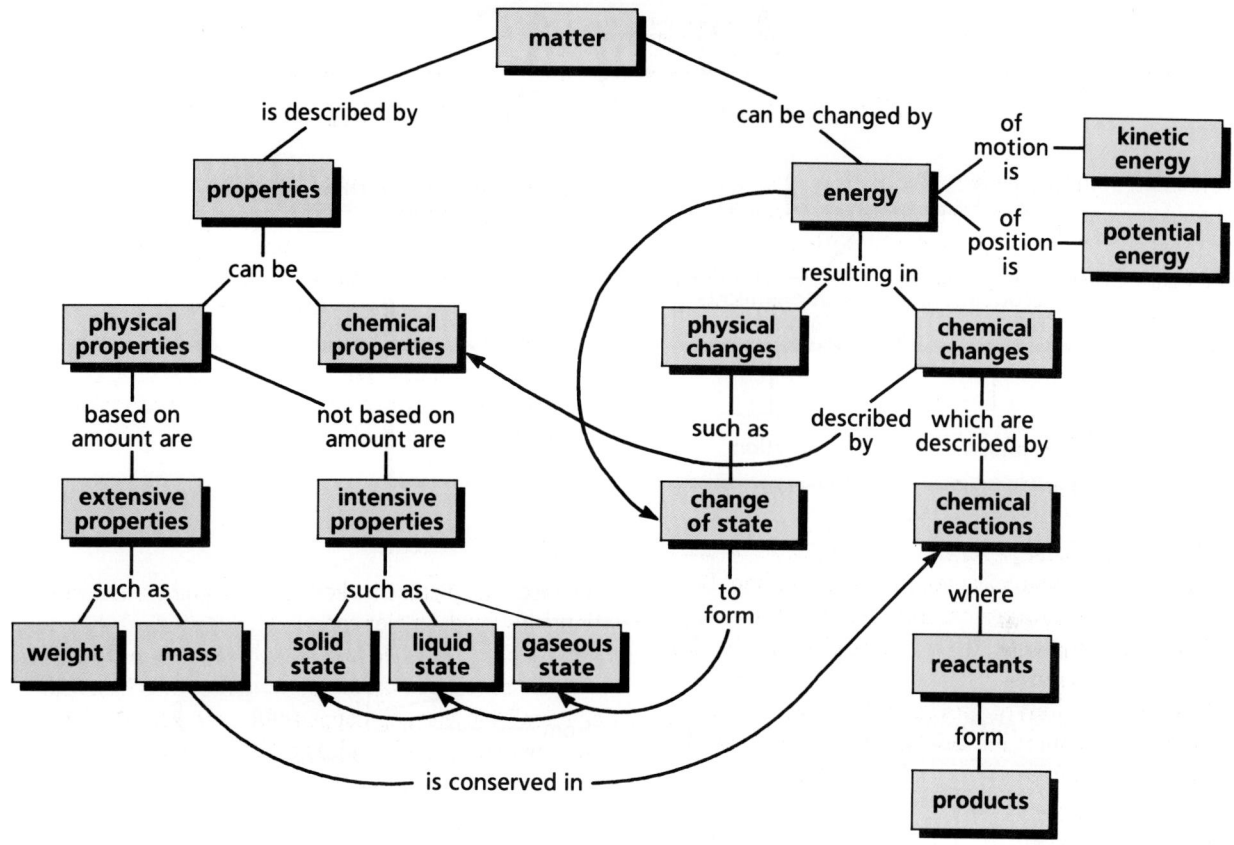

Making Concept Maps

- **Start by taking a text section and listing all the important concepts.** We'll use the boldface terms in Section 1.3.

  mixture
  heterogeneous mixture
  phase
  solution
  homogeneous mixture
  pure substance
  element
  chemical compound
  law of definite composition

- **From this list, group similar concepts together.** For example, one way to group these concepts would be into two groups—one that is related to mixtures, and the other that is related to pure substances.

  mixtures
  heterogeneous mixtures
  phase
  solution
  homogeneous mixture
  pure substance
  element
  chemical compound
  law of definite composition

- **Select a main concept for the map.** Notice that the title of Section 1.3 is *Classification of Matter*. We'll use "matter" as the main concept for this map. **Start building the map by placing the concepts according to their importance under the main concept, matter.** One way of arranging the concepts is shown at the top of page 890.

- **Now add the linking words that give meaning to the arrangement of concepts.** When adding the links, be sure that each proposition makes sense. To distinguish concepts from links, place your concepts in circles, ovals, or rectangles—like the ones in the maps shown.

The finished map on page 890 covers the main ideas of Section 1.3. The information on the map should be easier to remember than trying to outline or memorize all of Section 1.3. At first, making maps seems difficult. The process forces you to think about the meanings and relationships among the concepts. However, the time you devote to mapping will pay off when it comes time to review for an exam.

Modern Chemistry **889**

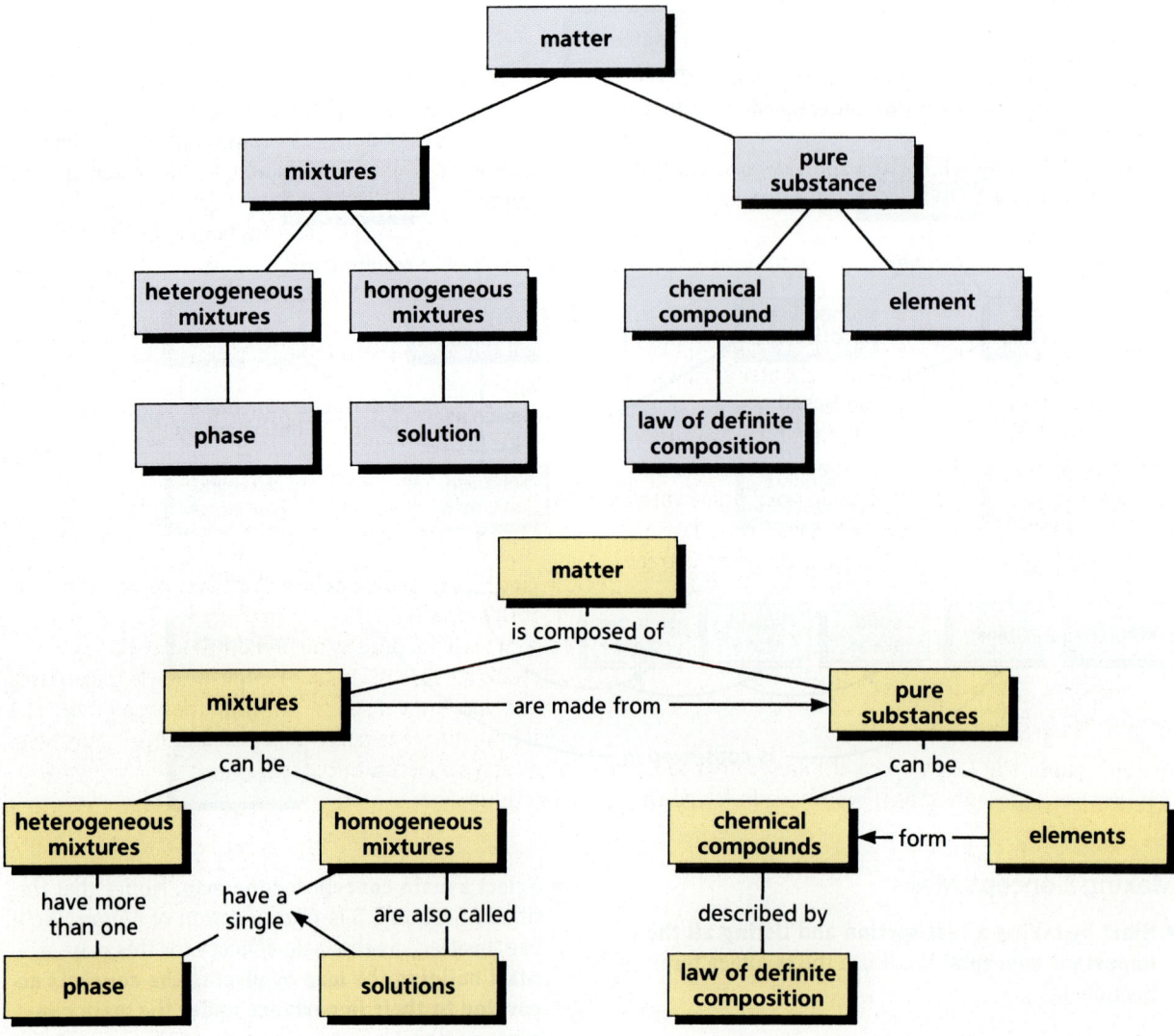

Working Problems

In addition to understanding concepts, the ability to solve problems in chemistry will be a large measure of your success in the course. You will generally spend a lot of time working problems in class and for homework. The ability to solve chemistry problems is a skill, and like any skill, it requires practice.

- The Sample Problems in the text provide road maps for solving certain types of problems. **Always review the Sample Problems in the chapter.** Cover the solution while trying to work the problem yourself.
- The four steps: ***Analyze, Plan, Compute,*** and ***Evaluate*** should be the steps you go through in working assigned problems.
- The problems in the chapter review are similar to the Sample Problems. If you can relate an assigned problem to one of the Sample Problems in the chapter, it shows that you understand the material.
- **Never spend more than 15 minutes trying to solve a problem.** If you have not been able to come up with a plan for the solution after 15 minutes, additional time spent will only cause you to be frustrated. What do you do? Get help! See your teacher or a classmate. Find out what it is that you don't understand.

- **Don't try to memorize the Sample Problems,** spend your time trying to understand how the solution develops.
- **Always look at your answer and ask yourself if it is reasonable and makes sense.** Check to be sure you have the correct units.

Homework

Your teacher will probably assign questions and problems from the section reviews and chapter reviews. The purpose of these assignments is to review what you have covered in class and to see if you can use the information in answering a question or solving a problem. As in reviewing class notes, do your homework as soon as possible after class. Do not wait until late at night when you are tired and likely to be very frustrated.

Exams

Reviewing for an Exam

- **Don't panic and don't cram.** It takes longer to learn if you are under pressure. If you have followed the strategies listed here and reviewed along the way, studying for the exam should be easy.
- **When looking over your notes and concept maps, recite ideas out loud.** There are two reasons for reciting.
  (1) You are hearing the information which is effective in helping you learn.
  (2) If you cannot recite the ideas, it should be a clue that you do not understand the materials and you should begin rereading or reviewing the material again.
- **Studying with a friend provides a good opportunity for recitation.** If you can explain ideas to your study partner, you know the material.

Taking an Exam

- **Get plenty of rest before the exam so you can think clearly.**
- **Start with the questions you know.** If you get stuck on a question, save it for later. As time passes, and you work through the exam, you may recall the information you need to answer a difficult question or solve a difficult problem.

Good luck!

GLOSSARY

A

absolute zero lowest possible temperature; −273.15°C or 0K (330)
accuracy the closeness of a measurement to the true or accepted value of the quantity measured (46)
acid anhydride an oxide that reacts with water to form an acid (480)
acid, Brønsted a molecule or ion that is a proton donor (462)
acid ionization constant the equilibrium constant for the ionization of an acid; K_a (577)
acid, Lewis an atom, ion, or molecule that is an electron-pair acceptor (463)
acid precipitation precipitation having a pH below about 5.6 (772)
acid, traditional a chemical compound that contains hydrogen and ionizes in aqueous solution to form hydrogen ions (462)
acid-base reaction, Brønsted the transfer of protons from one reactant (the acid) to another (the base) (473)
actinides the 14 elements with atomic numbers from 90 (thorium, Th) through 103 (lawrencium, Lr) (130)
activated carbon a form of amorphous carbon that is prepared in a way that gives it a very large internal surface area (638)
activated complex a transitional structure that results from an effective collision and that persists while bonds are breaking and new bonds are forming (545)
activation energy the energy needed to transform reactants into an activated complex (545)
activity series a list of elements organized according to the ease with which they undergo certain chemical reactions (251)
actual yield the measured amount of a product obtained from a reaction (276)
adenosine triphosphate the nucleotide that supplies the energy for many metabolic reactions; ATP (695)
alcohol a substituted hydrocarbon that contains one or more hydroxyl (—OH) units (665)
aldehyde a substituted hydrocarbon with a carbonyl group that has one bond attached to a hydrogen atom (674)
alkadienes straight- or branched-chain hydrocarbons in which two carbon atoms in each molecule are connected by two double covalent bonds (651)
alkali metals the elements of Group 1 of the Periodic Table (lithium, sodium, potassium, rubidium, cesium, and francium) (136)
alkaline a solution that contains the hydroxide anion (OH⁻) from a soluble base (471)
alkaline-earth metals the elements of Group 2 of the Periodic Table (beryllium, magnesium, calcium, strontium, barium, and radium) (136)
alkanes straight- or branched-chain hydrocarbons whose atoms are connected by single covalent bonds (643)
alkenes straight- or branched-chain hydrocarbons in which two carbon atoms in each molecule are connected by a double covalent bond (644)
alkyl group the neutral species formed when one hydrogen atom is removed from an alkane molecule (647)
alkyl halide an alkane in which one or more halogen atoms are substituted for a hydrogen atom; a halocarbon (670)
alkylation the process by which gaseous alkanes are combined with unsaturated hydrocarbons by heat in the presence of a catalyst (651)
alkynes straight- or branched-chain hydrocarbons in which two carbon atoms in each molecule are connected by a triple covalent bond (644)
allotrope one of the two or more forms of an element that have the same physical state (289)
allotropy the existence of two or more allotropes of an element (289)
alloy a solid solution in which the atoms of two or more metals are uniformly mixed (407)
alpha particle a helium nucleus emitted from the nucleus of a radioactive nuclide (809)
alumina anhydrous aluminum oxide, Al_2O_3 (757)
amine an ammonia base in which one or more of the three hydrogens is replaced by an alkyl or aryl group (694)
amino acid organic molecules containing nitrogen in the form of NH_2 and a carboxyl group (COOH) bonded to the same carbon atom (686)
amorphous carbon a carbon-containing residue, such as charcoal or activated carbon, that is black and has no crystalline shape (636)
amorphous solid a solid, such as plastic, in which the particles are arranged randomly (377)
amorphous sulfur a rubbery, dark brown or black mass made by pouring boiling sulfur into cold water (767)
amphoteric any species that can react as either an acid or a base (479)
anion a negative ion (151)
anode the electrode at which oxidation occurs in an electrochemical cell (612)
anthracite a coal believed to have been subjected to greater temperatures and pressures than other coals (637)
antiseptic a substance that checks the growth or action of microorganisms on or in the body (790)
aqueous solution a solution in which water is the solvent (439)
aromatic hydrocarbons hydrocarbons with six-membered carbon rings that exhibit resonance (644)
atmosphere of pressure (atm) average atmospheric pressure at sea level and 0°C; 1 atm = 760 mm Hg or 760 torr (326)
atom the smallest particle of an element that can exist either alone or in combination with other atoms (75)
atomic mass the mass of an atom expressed in atomic mass units (85)
atomic mass unit a unit of mass that is exactly $\frac{1}{12}$ the mass of the carbon-12 atom or $1.660\ 5402 \times 10^{-24}$ g (85)
atomic number the number of protons in the nucleus of an atom of an element (83)
atomic radius one-half of the distance between the nuclei of identical atoms joined in a molecule (143)
atomic structure the identity and arrangement of smaller particles within atoms (75)
Aufbau principle the rule that an electron occupies the lowest-energy orbital that can receive it (112)
auto-oxidation a process in which a substance acts as both an oxidizing and a reducing agent and is self-oxidizing and self-reducing (610)
average atomic mass the weighted average of the atomic masses of the naturally occurring isotopes of an element (86)
Avogadro's number $6.022\ 137 \times 10^{23}$; the number of particles in 1.0 mol of a pure substance (88)
Avogadro's principle equal volumes of gases at the same temperature and pressure contain equal numbers of molecules (344)

B

baking soda sodium hydrogen carbonate, $NaHCO_3$ (726)
barometer a device used to measure atmospheric pressure (324)
base, Brønsted a molecule or ion that is a proton acceptor (472)
base, Lewis an atom, ion, or molecule that is an electron-pair donor (473)
base, traditional a substance that contains hydroxide ions (OH⁻) and dissociates in water to give hydroxide ions (470)
base-pairing attractive pairing between nucleotides' bases (696)
basic anhydride oxide forming alkaline solutions with water (479)
bauxite a hydrated aluminum oxide ore (753)
beta particle an electron emitted from the nucleus of a radioactive nuclide (809)
binary acid an acid composed of hydrogen and a second element (465)
binary compounds compounds composed of two different elements (204)
biochemistry the study of chemical reactions that take place within living things (685)
bituminous coal a form of coal that appears to have been subjected to greater heat and pressure than subbituminous coal (637)

blast furnace a tall, cylindrical chamber in which iron oxide is reduced using coke, limestone, and a blast of hot air (739)
boiling the change of a liquid to a vapor, within the liquid and at its surface, when the equilibrium vapor pressure of the liquid equals the atmospheric pressure (387)
boiling point the temperature at which the equilibrium vapor pressure of a liquid is equal to the atmospheric pressure (387)
boiling-point elevation the difference between the boiling points of a pure solvent and a nonelectrolyte solution in that solvent (431)
bond energy the energy required to break a chemical bond and form neutral atoms (166)
bond length the average distance between two bonded atoms (166)
Boyle's law the volume of a fixed mass of gas varies inversely with pressure at constant temperature (327)
buffered solutions solutions that can resist changes in pH (578)

C

calorie the quantity of heat required to raise the temperature of 1 g of water through one Celsius degree, equivalent to 4.184 J (43)
calorimeter a device used to measure heats of reactions (527)
carbohydrate includes sugars, starches, and related compounds (689)
carbonyl group a group with the formula $C\!=\!O$ (674)
carboxyl group a group with the formula COOH (676)
carboxylic acid a hydrocarbon in which two of the hydrogen atoms on a terminal carbon atom have been replaced by a doubly bonded oxygen, and the third hydrogen atom has been replaced by a hydroxyl group, —OH (676)
catalysis activity by a catalyst (551)
catalyst a substance that speeds up a chemical reaction but is not permanently consumed in the reaction (551)
cathode the electrode at which reduction occurs in an electrochemical cell (612)
cation a positive ion (151)
caustic having destructive effects on skin, hair, and wool (724)
cellulose a polysaccharide of glucose monomers with hydroxyl links; the main building material for plants (690)
chain reaction a reaction in which the material or energy that starts the reaction is also one of the products (819)
changes of state changes between the gaseous, liquid, or solid states (14)
charcoal the residue that remains after the destructive distillation of wood (638)
Charles' law the volume of a fixed mass of gas varies directly with the Kelvin temperature at constant pressure (330)
chemical bond a link between atoms resulting from the mutual attraction of their nuclei for electrons (161)
chemical change or chemical reaction any change or reaction in which one or more substances are converted into different substances with different characteristic properties (15)
chemical compound a pure substance that can be decomposed into two or more simpler substances by an ordinary chemical change (22)
chemical equation a representation, with symbols and formulas, of the reactants and products in a chemical reaction (233)
chemical equilibrium a state of balance in which the rates of opposing chemical reactions are equal (564)
chemical equilibrium expression the equation for the equilibrium constant K (566)
chemical formula a shorthand representation of the composition of a substance using atomic symbols and numeric subscripts (164)
chemical kinetics the branch of chemistry that is concerned with reaction rates and reaction mechanisms (549)
chemical property the ability of a substance to undergo a change that alters its identity (14)
chemistry the study of the composition, properties, and structure of substances and the changes they undergo (5)
chitin a polysaccharide of repeating glucose units with a nitrogen-containing group that composes the "armor" of insects and other invertebrates (691)
chlorinated lime Ca(ClO)Cl; a disinfectant (785)
chlorohydrocarbons hydrocarbons in which chlorine atoms are substituted for one or more of the hydrogen atoms (640)
cholesterol the most common steroid, making up part of every animal cell membrane, brain, and nerve (693)

coal a natural dark-brown or black solid that consists mostly of amorphous carbon with various organic and some inorganic compounds (636)
coefficient a small whole number that appears in front of a formula in a chemical equation (234)
coenzyme a molecule that helps an enzyme bind to its substrate (688)
coke the solid residue left behind after volatile materials are removed from bituminous coal (637)
colligative property a property of a system that depends on the number of particles present but is independent of their nature (427)
collision theory the set of assumptions regarding collisions of particles and the reactions that result from those collisions (543)
colloid a mixture consisting of particles intermediate in size between those in solutions and those in suspensions (409)
combined gas law relationship between pressure, volume, and temperature of a gas when the amount of gas is fixed (333)
combustion reaction a reaction in which a large amount of light and heat energy is produced by combination with oxygen (250)
common-ion effect the phenomenon in which the addition of an ion common to two solutes brings about precipitation or reduced ionization (575)
complex carbohydrate a polymer formed by joining many monosaccharides; a polysaccharide (689)
complex ion an ionic species composed of a central metal ion combined with a specific number of polar molecules or ions (734)
composition stoichiometry calculations involving the mass relationships of elements in chemical compounds (261)
concentration refers to the number of particles per unit volume (384); a measure of the amount of solute in a given amount of solvent or solution (420)
condensation the process by which a gas changes to a liquid (383)
conjugate acid the species formed when a Brønsted base gains a proton (475)
conjugate base the species that remains after a Brønsted acid has given up a proton (475)
contact process a method in which sulfur dioxide is prepared by burning sulfur or by roasting iron pyrite, FeS_2, in contact with the catalyst divanadium pentoxide, V_2O_5 (773)
continuous spectrum a spectrum that includes all wavelengths within a given range (98)
control rod a rod of neutron-absorbing steel used to regulate the reaction in a nuclear reactor (819)
conversion factor a ratio of two different units that can be used to convert from one unit to the other (34)
coordination number the number of units covalently bonded to, or coordinated with, the central ion of a complex ion (734)
copper matte an intermediate product in copper refining, containing copper sulfides, iron sulfides, and other sulfides (746)
corundum a mineral containing a natural oxide of aluminum (756)
cosmic rays protons and other nuclei of very high energy that come to earth from outer space (816)
covalent bond a chemical bond resulting from the sharing of electrons between two atoms (161)
covalent hydride a compound consisting of hydrogen and one other more electronegative element (296)
cracking a process in which complex organic molecules are broken up into simpler molecules (649)
critical mass the mass of uranium that provides the number of neutrons needed to sustain a chain reaction (820)
critical point indicates the critical temperature and critical pressure of a substance (389)
critical pressure the lowest pressure required for a substance to exist as a liquid at the critical temperature; P_c (389)
critical temperature the temperature above which a substance cannot exist in the liquid state no matter how high the pressure exerted; t_c (389)
cryolite Na_3AlF_6, used in refining aluminum (753)
crystal a substance in which the particles are arranged in an orderly, geometric, repeating pattern (377)
crystal lattice the total three-dimensional array of points that describes the arrangement of the particles of a crystal (379)
crystalline solid a solid consisting of crystals (377)

D

Dalton's law of partial pressures the total pressure of a mixture of gases is equal to the sum of the partial pressures of the component gases (335)

daughter nuclide a nuclide that is the product of the sequential radioactive decay of a parent nuclide (810)

decomposition reaction a reaction in which a single compound produces two or more simpler substances (246)

degree Celsius the unit of temperature on the Celsius scale (41)

degree Kelvin the unit of temperature on the Kelvin scale; the fundamental SI unit for temperature (41)

dehydrating agent a substance that removes water from a material (774)

denature to destroy proteins by breaking up their three-dimensional structure with high temperatures (688)

denatured alcohol a mixture composed principally of ethanol and other poisonous and nauseating materials that have been added to make it unfit for consumption (668)

density a material's mass divided by its volume, or its mass per unit volume (37)

deoxyribonucleic acid nucleic acid of all organisms that contains the genetic information passed on to their offspring; DNA (695)

destructive distillation a process by which a complex carbon-containing material, such as coal, is heated in a closed container in the absence of air or oxygen (637)

diamagnetism the property of a substance that is weakly repelled by a magnetic field (111)

diamond the form of carbon that is a colorless crystalline solid with an extremely high density of 3.514 g/cm^3 (634)

diatomic molecule a molecule containing two atoms (164)

diffusion the gradual mixing of two substances because of their spontaneous, random motion (318)

dimer a molecule formed by the combination of two simpler molecules (653)

dipole equal but opposite charges separated by a short distance (187)

dipole–dipole forces forces of attraction between polar molecules (188)

diprotic acid an acid that can donate two protons per molecule (464)

directly proportional two variables that have a constant ratio (58)

disaccharide two monosaccharides joined together (689)

disinfectant a chemical agent that destroys microorganisms but not bacterial spores (785)

dissociation the separation of ions that occurs when an ionic compound dissolves (442)

dissolved describes a substance that, when mixed with another, is no longer of a different physical state (372)

domain small magnetized regions formed by groups of properly aligned atoms of ferromagnetic substances (738)

double bond a covalent bond produced by sharing two pairs of electrons between two atoms (170)

double helix structural shape of a DNA molecule, consisting of two nucleotide strands wound around each other (696)

double-replacement reaction (ionic reactions) a type of reaction in which the ions of two compounds exchange places in an aqueous solution to form two new compounds (248)

Downs cell apparatus that prepares elemental sodium by the electrolysis of fused sodium chloride (720)

ductility the state of being able to be drawn, pulled, or extruded through a small opening to produce a wire (181)

E

effervescence the rapid escape of a gas from a liquid in which it is dissolved (416)

effusion a process by which gas particles under pressure pass through a very small opening from one container to another (318)

elastic collision a collision between gas particles or between gas particles and container walls in which there is no net energy loss due to friction (315)

electrochemical cell a system of electrodes and electrolytes in which a spontaneous redox reaction is used as a source of electrical energy (611)

electrochemistry the branch of chemistry that deals with the electricity-related applications of redox reactions (611)

electrode a conductor used to establish electric contact with a nonmetallic part of a circuit, such as a solution of an electrolyte (611)

electrode potential the difference in potential between an electrode and the solution it rests in (618)

electrolysis the process in which an electric current is used to drive oxidation-reduction reactions (612); also the decomposition of a substance by an electric current (246)

electrolyte a substance that dissolves in water to give a solution that conducts electric current (406)

electrolytic cell a system of electrodes and electrolyte in which a current is used to drive a nonspontaneous redox reaction (612)

electromagnetic radiation a form of energy that exhibits wavelike behavior as it travels through space (97)

electromagnetic spectrum all electromagnetic radiation, arranged according to increasing wavelength (98)

electron affinity the energy change that occurs when an electron is acquired by a neutral atom (149)

electron configuration the arrangement of electrons in atoms (112)

electronegativity a measure of the ability of an atom in a chemical compound to attract electrons (153)

electrons negatively charged, subatomic particles (76)

electroplating an electrolytic process in which a metal is deposited on a surface (615)

element a pure substance that cannot be decomposed by a naturally occurring, ordinary chemical change (22)

emery a mineral containing a natural oxide of aluminum, used as an abrasive (756)

empirical formula the symbols for the elements of a compound with subscripts showing the smallest whole-number ratio of atoms (224)

end point the point in a titration where an indicator changes color (512)

endothermic describes processes that absorb heat (19)

energy the ability to cause change or to do work (12)

enthalpy the heat content of a system at constant pressure (528)

enthalpy change the amount of heat absorbed or lost by a system in a process carried out at constant pressure (528)

entropy the property that describes the disorder of a system; S (538)

enzyme a protein molecule that acts as a catalyst (688)

epinephrine an amine hormone that speeds up the heart, increases the blood supply to the muscles, and generally gets the body ready for "fight or flight" (694)

equilibrium a dynamic condition in which two opposing physical or chemical changes occur at equal rates in the same closed system (383)

equilibrium constant the ratio of the mathematical product of the concentrations of the products in a reversible reaction, at equilibrium, to the mathematical product of the concentrations of the reactants, each concentration being raised to the power that is the coefficient of the number of moles of that substance in the chemical equation; K (566)

equilibrium vapor pressure the pressure exerted by a vapor in equilibrium with its corresponding liquid at a given temperature (385)

equivalent mass of acid or base for an acid, the mass in grams that will provide one mole of protons in a reaction (492); for a base, the mass in grams that will react with one mole of protons (493)

equivalence point the point at which there are equivalent quantities of hydronium and hydroxide ions (512)

equivalents quantities of solutes that have equivalent combining capacity (491)

ester a compound with the formula RCOOR', in which R' is always an alkyl group and R is either a hydrogen atom or an alkyl group (679)

esterification the production of an ester by the reaction of an organic acid and an alcohol (679)

ether a compound in which two hydrocarbon groups are bonded to an atom of oxygen (673)

evaporation the process by which particles escape from the surface of a nonboiling liquid and enter the gaseous state (375)

excess reactant the substance that is not used up completely in a reaction (272)

excited state a state in which an atom has a higher potential energy than it has in its ground state (100)

exothermic describes processes that release heat (17)

extensive physical properties properties, including mass, length, and volume, that depend on the amount of matter present (14)

F

factor-label method a problem-solving method in which units of measurement are treated in calculations as if they were algebraic factors (34)

fat an ester, of plant or animal origin, of glycerol and long-carbon-chain acids (680)

fatty acids the simplest lipids; unbranched chains of carbon and hydrogen atoms, with a carboxyl group at one end (692)

ferromagnetism the property of certain metals whereby they are strongly attracted by a magnet (737)

fiber composite monocrystalline strands or fibers of one substance embedded in and held in place by another material (757)

fission the process in which a very heavy nucleus splits to form medium-weight nuclei (804)

fluid a substance that can flow and therefore conform to the outline of its container (372); also a liquid or a gas composed of particles that glide easily past one another (317)

fluoresce give off visible light after being exposed to sunlight (804)

formula equation a representation of the reactants and products of a chemical reaction by their symbols or formulas (234)

formula mass the sum of the atomic masses of all atoms represented in the formula of any compound or polyatomic ion (216)

formula unit the simplest collection of atoms from which a compound's formula can be established (175)

formyl group the functional group of an aldehyde (674)

fractional distillation distillation in which components of a mixture with different boiling points are collected separately (656)

Frasch process a method of obtaining sulfur from the ground by melting the deposit of sulfur with superheated water and then forcing it to the surface with compressed air (765)

free energy the combined enthalpy-entropy function of a system; G (538)

free-energy change the difference between the change in enthalpy, ΔH, and the product of the Kelvin temperature and the entropy change, $T\Delta S$, at a constant pressure and temperature; ΔG (538)

freezing the physical change of a liquid to a solid (388)

freezing point the temperature at which the solid and liquid are in equilibrium at 1 atm (101.3 kPa) pressure (388)

freezing-point depression the difference between the freezing points of a pure solvent and a nonelectrolyte solution in it (429)

frequency the number of waves that pass a given point in a specific amount of time, usually one second (98)

Friedel-Crafts reaction a reaction in which an alkyl group can be introduced into a benzene ring using an alkyl halide in the presence of anhydrous aluminum chloride (654)

fullerenes a spherically networked class of carbon allotropes (634)

functional group the atom or group of atoms bonded to a hydrocarbon responsible for the specific properties of the compound (639)

fundamental unit a unit that is defined by a physical standard of measurement (32)

fungicide compounds used to kill fungi (769)

fusion the process in which lightweight nuclei combine to form heavier, more stable nuclei (804)

G

gamma rays high-energy electromagnetic waves (809)

gas laws simple mathematical relationships between the volume, temperature, pressure, and quantity of a gas (322)

gaseous state the state of any matter that has neither a definite shape nor a definite volume (13)

Gay-Lussac's law the pressure of a fixed mass of gas varies directly with the Kelvin temperature at constant volume (332)

Gay-Lussac's law of combining volumes of gases at constant temperature and pressure, the volumes of gaseous reactants and products can be expressed as ratios of small whole numbers (344)

gene a small segment of a DNA chain where the genetic instructions for making a specific protein are encoded (697)

genetic engineering the deliberate manipulation of genetic material (700)

glycogen a glucose monomer that stores energy in animals (689)

glycolipids fatty acids combined with carbohydrates (692)

glycolysis one of a group of metabolic pathways by which animals turn their food into an energy supply (701)

graphite a soft, black crystalline form of carbon that is a fair conductor of electricity (634)

Graham's law of effusion (diffusion) the rates of effusion (or diffusion) of gases at the same temperature and pressure are inversely proportional to the square roots of their molar masses (362)

ground state the state of lowest energy of an atom (100)

groups or families the vertical columns of elements in the Periodic Table, numbered consecutively from 1 to 18 (23)

H

Haber process the catalytic synthesis of ammonia from nitrogen gas and hydrogen gas (302)

half-cell one of two portions of a voltaic cell, consisting of a metal electrode in contact with a solution of its ions (618)

half-life the length of time during which half of a given number of atoms of a radioactive nuclide decay (807)

half-reaction the reaction taking place at each electrode in an electrochemical cell (612)

halite rock salt; it is mined throughout the world (722)

Hall process a process in which aluminum is extracted by electrolyzing anhydrous aluminum oxide (refined bauxite) dissolved in molten cryolite (753)

halocarbon an alkane in which one or more halogen atoms are substituted for a hydrogen atom; an alkyl halide (670)

halogen any of the the colorful, active, nonmetallic elements fluorine, chlorine, bromine, iodine, and astatine that constitute Group 17 of the Periodic Table (779)

halogenation a chemical reaction by which a halogen is introduced into a compound by addition or substitution (654)

heat the sum of the kinetic energies of the particles in a sample of matter (40)

heat capacity the amount of heat energy needed to raise the temperature of a given sample of matter by 1C° (44)

heat of combustion heat released by the complete combustion of 1 mol of a substance; defined in terms of 1 mol of reactant (531)

heat of formation molar heat of formation (529)

heat of hydration the energy released when ions become surrounded by water molecules (441)

heat of reaction the quantity of heat released or absorbed during a chemical reaction (528)

heat of solution the amount of heat energy absorbed or released when a solute dissolves in a specific amount of solvent (417)

Henry's law the solubility of a gas in a liquid is directly proportional to the partial pressure of that gas above the liquid (415)

Hess's law the overall enthalpy change in a reaction is equal to the sum of the enthalpy changes for the individual steps in the process (531)

heterogeneous mixture a mixture whose composition and properties are not uniform but that differ from point to point in the mixture (19)

heterogeneous reaction a reaction involving reactants in two different phases (550)

heterogenous catalyst a catalyst whose phase is different from that of the reactants (551)

highest occupied energy level the main energy level with the highest principal quantum number that contains at least one electron (113)

homogeneous catalyst a catalyst in the same phase as all the reactants and products in a reaction system (551)

homogeneous mixture a mixture whose composition and properties are uniform throughout the mixture; also called a solution (20)

homogeneous reaction a reaction system that occurs in a single phase (541)

homologous series one in which adjacent members differ by a constant unit (645)

hormones chemical messengers, such as steroids, proteins, or amines, that elicit specific responses from target cells (693)

Hund's rule orbitals of equal energy are each occupied by one electron before any orbital is occupied by a second electron and all electrons in singly occupied orbitals must have the same spin (113)

hybrid orbitals   orbitals of equal energy produced by the combination of two or more orbitals on the same atom (186)
hybridization   the production of hybrid orbitals (185)
hydration   the solution process with water as the solvent (440)
hydrocarbons   the simplest organic compounds, composed only of carbon and hydrogen (639)
hydrogen bonding   the intermolecular attraction between a hydrogen atom and an unshared pair of electrons on a strongly electronegative atom (188)
hydrolysis   a reaction between water and the ions of a dissolved salt (580)
hydrolysis constant   the equilibrium constant for an ion hydrolysis reaction; K_h (582)
hydronium ion   the H_3O^+ ion (449)
hydroxide ion   a negative ion found in ionic compounds known as hydroxides (467)
hydroxyl group   the covalently bonded —OH group (467)
hypothesis   a testable statement (8)

I

ideal gas   an imaginary gas that conforms perfectly to all of the assumptions of the kinetic theory (315)
ideal gas constant   the constant R (351)
ideal gas law   a relationship between pressure, volume, temperature, and the number of moles of a gas (350)
immiscible   not mutually soluble (414)
indicators   weak acid or base dyes whose colors are sensitive to hydronium ion concentration (505)
inertia   resistance to change in motion (10)
inhibitor   a substance that slows down a process (552)
inner-shell electrons   electrons that are not in the highest-occupied energy level (114)
instability constant   the equilibrium constant for complex-ion equilibria (735)
intensive physical properties   properties that do not depend on the amount of matter present; they include melting point, boiling point, density, ductility, malleability, color, crystalline shape, and refractive index (14)
intermolecular forces   the forces of attraction between molecules (187)
inversely proportional   two variables that have a constant mathematical product (59)
ion   an atom or group of atoms that has a positive or negative charge (145)
ionic bond   the chemical bond resulting from electrostatic attraction between positive and negative ions (161)
ionic compound   a compound of positive and negative ions combined so that the positive and negative charges are equal (175)
ionic hydride   a compound consisting of hydrogen and one element that is less electronegative (296)
ionization   the formation of ions from solute molecules by the action of the solvent (448); any process that results in the formation of an ion (145)
ionization energy   the energy required to remove one electron from an atom of an element (145)
isomers   compounds that have the same molecular formula but different structures (639)
isotopes   atoms of the same element that have different masses (81)

J

joule   the derived SI unit for heat energy and all other forms of energy, such as potential energy, kinetic energy, and electrical energy; it is equivalent to $1 \text{ kg} \cdot m^2/s^2$ (43)

K

ketone   a substituted hydrocarbon with a carbonyl group in which the carbon atom of the carbonyl group is attached to two other carbon atoms (674)
kilogram   the SI standard unit for mass (33)
kinetic energy   the energy of an object in motion (12)
kinetic theory   the idea that particles of matter are always in motion and that this motion has consequences (315)

L

lambda-sulfur   the liquid allotropic form of sulfur produced at temperatures just above the melting point of sulfur, 112.8°C (766)
lanthanides or rare-earth elements   the 14 elements with atomic numbers from 58 (cerium, Ce) through 71 (lutetium, Lu) (130)
lattice energy   the energy released when one mole of an ionic crystalline compound is formed from gaseous ions (178)
law   a generalization that describes a wide variety of behaviors in nature (8)
law of conservation of energy   energy can be converted from one form to another, but it cannot be created or destroyed in ordinary chemical or physical changes (13)
law of conservation of mass   matter cannot be either created or destroyed in ordinary chemical or physical changes (12)
law of definite composition   a chemical compound contains the same elements in exactly the same proportions by mass regardless of the size of the sample or source of the compound (22)
law of mass action   the rate of a chemical reaction is directly proportional to the product of the concentrations of reacting substances, each raised to the appropriate power (556)
law of multiple proportions   if two or more different compounds are composed of the same two elements, the masses of the second element combined with a certain mass of the first element can be expressed as ratios of small whole numbers (73)
Le Chatelier's principle   when a physical or chemical system at equilibrium is disturbed by the application of a stress, it attains a new equilibrium position that minimizes the stress (384)
Lewis structure   a formula in which atomic symbols represent nuclei and inner-shell electrons, dots or dashes represent electron pairs in covalent bonds, and dots represent unshared electrons (169)
lignite   a low-grade form of coal sometimes called "brown coal" because of its brownish-black color (637)
limiting reactant   the reactant that limits the amounts of the other reactants that can combine, and the amount of product(s) formed (272)
line spectrum   the colored lines produced when an electron drops from a higher-energy orbit to a lower energy orbit (103)
lipids   various organic compounds not very soluble in water that have a high proportion of C—H bonds (691)
lipoprotein   a fatty acid combined with a protein (692)
liquid state   the state of any matter that has a definite volume but an indefinite shape (13)
London dispersion forces   intermolecular attractions resulting from the constant motion of electrons and the creation of instantaneous dipoles and induced dipoles (190)
lye   sodium hydroxide; converts some types of animal and vegetable matter into soluble materials (724)

M

macromolecules   large organic molecules (685)
magnetic quantum number   the quantum number that indicates the orientation of an orbital about the nucleus (110)
main-group elements   the elements of the s-and p-blocks (139)
malleability   the state of being able to be shaped or extended by beating with a hammer, rolling, or otherwise exerting physical pressure that results in a change in contour (181)
mass   a measure of the quantity of matter (11)
mass number   the total number of protons and neutrons in the nuclei of an isotope (83)
matter   anything that has mass and occupies space (11)
melting point   the temperature at which a solid becomes a liquid (378)
messenger RNA   the type of RNA that carries the coded information for a particular protein from DNA to a ribosome (698)
metabolic pathway   chemically linked steps in which a series of enzymes converts substrate molecules, step by step, into a final product (701)
metabolism   all of the biochemical reactions that occur within an organism (701)
metal   an element that is a good conductor of heat and electricity (24)
metallic bond   a chemical bond resulting from the attraction between positive ions and surrounding mobile electrons (180)

metalloid an element that has some properties characteristic of metals and others characteristic of nonmetals, and is a semiconductor (25)

meter the SI standard unit for length (33)

methyl group methane with one of its hydrogen atoms removed (647)

millimeters of mercury (mm Hg) a common unit of pressure (326)

mineral acids acids made from minerals; include sulfuric acid, nitric acid, hydrochloric acid, and phosphoric acid (465)

miscible mutually soluble (415)

mixture a combination of two or more kinds of matter, each of which retains its own composition and properties (19)

moderator a material that slows down neutrons (819)

molal boiling-point constant the boiling-point elevation of a solvent in a 1 molal solution of a nonvolatile, molecular solute (431)

molal freezing-point constant the freezing-point depression of the solvent in a 1 molal solution of a nonvolatile, molecular solute (429)

molality the concentration of a solution expressed in moles of solute per kilogram of solvent (423)

molar heat of formation heat of reaction released or absorbed when 1 mol of a compound is formed by combination of its elements; defined in terms of 1 mol of product (529)

molar heat of fusion the heat energy required to melt one mole of solid at its melting point (388)

molar heat of vaporization the amount of heat energy required to vaporize one mole of liquid at its boiling point (387)

molar mass the mass in grams of one mole of a pure substance (88)

molarity the number of moles of solute in one liter of solution (421)

mole the amount of a substance that contains the same number of particles as there are atoms in exactly 12 g of carbon-12 (88)

mole ratio a conversion factor that relates the number of moles of any two substances involved in a chemical reaction (263)

molecular compound a chemical compound whose simplest formula units are molecules (164)

molecular formula shows the types and numbers of atoms combined in a single molecule (164)

molecular genetics the branch of science that studies the biochemistry of genes (699)

molecule a group of two or more atoms held together by covalent bonds and able to exist independently (164)

monatomic ions ions formed from a single atom (203)

Monel metal an alloy of nickel and copper that is highly resistant to corrosion (781)

monoclinic sulfur a form of solid sulfur in which the eight-membered rings of sulfur are arranged in a monoclinic crystal pattern (766)

monomers the small molecules that join together to make a polymer (651)

monoprotic acid an acid that can donate one proton per molecule (463)

monosaccharides the monomers of carbohydrates, such as fructose and glucose; simple sugars (689)

mu-sulfur an allotropic form of sulfur that forms at temperatures above 160°C (766)

multiple bonds double and triple bonds (171)

N

neoprene a synthetic rubber that is resistant to heat and flames, as well as to oil and chemicals (659)

net ionic equation an equation that includes, for a reaction in aqueous solution, only the compounds and ions that change chemically (445)

neuron a nerve cell (695)

neurotransmitter a chemical messenger that carries a message across a synapse from one neuron to the next (695)

neutralization the reaction of hydronium ions and hydroxide ions to form water molecules (471)

neutrons electrically neutral, subatomic particles found in atomic nuclei (80)

Newton (N) the SI unit for force (323)

nitration a chemical reaction by which an —NO_2 group is introduced into an organic compound (654)

nitrogen fixation the conversion of atmospheric nitrogen gas (N_2) into nitrogen compounds that can be used by plants (300)

noble gas configuration an atomic configuration in which the outer main energy level is fully occupied by eight electrons (116)

noble gases the Group 18 elements, helium, neon, argon, krypton, xenon, and radon (116)

noble metals metals that show little chemical activity (736)

nomenclature methods of naming chemical compounds (206)

nonelectrolyte a substance that dissolves in water to give a solution that does not conduct an electric current (406)

nonmetal an element that is a poor conductor of heat and electricity (25)

nonpolar–covalent bond a covalent bond in which the bonding electrons are shared equally by the bonded atoms, with a resulting balanced distribution of electrical charge (162)

nonvolatile a substance that has little tendency to become a gas under the existing conditions (426)

normality the number of equivalents of solute per liter of solution; N (494)

nuclear binding energy the energy released when a nucleus is formed from its constituent particles (802)

nuclear disintegration the emission of a proton or neutron from a nucleus as a result of bombarding the nucleus with alpha particles, protons, deuterons, or neutrons (804)

nuclear equation an equation representing changes in the nuclei of atoms (810)

nuclear mass defect the difference between the mass of a nucleus and the sum of the masses of its constituent particles (802)

nuclear reactor a device in which controlled fission of radioactive material produces new radioactive substances and energy (819)

nucleic acids macromolecules that transmit genetic information (695)

nucleons the protons and neutrons in an atom (801)

nucleotide the monomer of a nucleic acid (695)

nucleus the positively charged, dense central portion of an atom that contains nearly all of its mass but takes up only an insignificant fraction of its volume (78)

nuclide the general term for any isotope of any element (84)

O

octet rule chemical compounds tend to form so that each atom, by gaining, losing, or sharing electrons, has 8 electrons (an octet) in its highest occupied energy level (168)

oil a liquid ester of glycerol and long carbon-chain acids having a plant or animal origin (680)

orbital a three-dimensional region about the nucleus in which a particular electron can be located (106)

orbital quantum number the quantum number that indicates the shape of an orbital (108)

organic acid an organic compound that contains an acid group, the carboxyl group (—COOH) (465)

organic chemistry the chemistry of carbon compounds (639)

oxidation a reaction in which the atoms or ions of an element lose one or more electrons and hence attain a more positive (or less negative) oxidation state (597)

oxidation numbers or oxidation states numbers assigned to the atoms in molecules that show the general distribution of electrons among bonded atoms (212)

oxide a compound containing the simple oxide ion (O^{2-}), which has an oxidation number of -2 (290)

oxidized a species that undergoes a change in oxidation state in a positive mathematical direction (597)

oxidizing agent a species that attains a more negative oxidation state during an oxidation-reduction reaction (608)

oxyacids acids containing hydrogen, oxygen, and a third element (211)

oxyanions polyatomic ions that contain oxygen (208)

P

pH the concentration of the hydronium ions of a solution, expressed as the negative of the common logarithm (499)

paraffin wax a mixture of hydrocarbons of the paraffin series (644)

parent nuclide the heaviest, most complex, naturally occurring nuclide in a decay series of radioactive nuclides (810)

partial pressure the pressure of each gas in a mixture (335)

pascal the pressure exerted by a force of one newton acting on an area of one square meter (326)

Pauli exclusion principle no two electrons in the same atom can have the same set of four quantum numbers (113)

peat a form of partially carbonized plant material that can be used as fuel (636)

peptide bond a bond joining amino acids in proteins (687)

percent by mass the number of grams of solute per 100 grams of solution (420)

percent composition the percent by mass of each element in a compound (222)

percent yield the ratio of the actual yield to the theoretical yield, multiplied by 100 percent (277)

periodic law the physical and chemical properties of the elements are periodic functions of their atomic numbers (129)

Periodic Table an arrangement of the elements in order of their atomic numbers so that elements with similar properties fall in the same column (130)

periods the horizontal rows of elements in the Periodic Table, numbered 1 to 7 from the top down (23)

peroxide a compound containing the O_2^{2-} anion, in which oxygen has the oxidation number -1 (290)

petroleum a complex mixture of hydrocarbons that varies greatly in composition (crude oil) (656)

petroleum coke a form of amorphous carbon produced by the destructive distillation of an oily residue that is a remnant of petroleum refining (638)

phase a portion of matter that has both the same chemical properties and the same physical properties (20)

phase diagram a graph of temperature versus pressure that indicates the conditions under which gaseous, liquid, and solid states of a substance exist (389)

phenyl group a —C_6H_5 group derived from benzene (654)

phospholipid a lipid in which one or two of the fatty acids are replaced by a phosphate group (693)

photochemical smog ozone and other irritating chemicals in the air that are the result of nitrogen oxides reacting either with unburned hydrocarbons from automobile exhausts or with atmospheric oxygen (299)

photoelectric effect the emission of electrons by certain metals when light shines on them (99)

photon a quantum of light, thought of as particles of radiation (100)

photosensitive sensitive to light rays (788)

photosynthesis the process by which plants use energy from the sun to combine carbon dioxide with water to produce glucose (303)

physical change any change in a property of matter that does not result in a change in identity (14)

physical property a property that can be observed or measured without altering the identity of a material (14)

pig iron iron recovered from a blast furnace (739)

plasma a gaseous system composed of positively-charged particles and negatively-charged electrons (13)

polar a bond that has an uneven distribution of charge (162)

polar-covalent bond a covalent bond with an unequal attraction for the shared electrons (162)

polyatomic ion a charged group of covalently bonded atoms (173)

polymer a large molecule made of repeating structural units (651)

polymerization the process that produces polymers (651)

polyprotic acid an acid that can donate more than one proton per molecule (464)

polysaccharide a polymer formed by joining many monosaccharides; a complex carbohydrate (689)

polyunsaturated fat fats and oils with double bonds in their fatty acid chains (692)

polyvinyl chloride a polymeric form of vinyl chloride (CH_2CHCl), also called PVC (785)

potential energy energy that an object has because of its position or composition (12)

precipitate a solid that separates from a solution (16)

precision the agreement among the numerical values of a set of measurements of the same quantity made in the same way (46)

pressure force per unit area on a surface (323)

primary standard a highly purified compound, when used in solution to check the concentration of the known solution in a titration (512)

principal quantum number the quantum number that indicates the main energy levels surrounding a nucleus; n (108)

products the new substance or substances produced by a chemical reaction (15) or by an enzyme reaction (688)

properties characteristics that enable us to distinguish one kind of matter from another (14)

proteins macromolecules formed by condensation reactions between monomers called amino acids (686)

protons subatomic particles that have a positive charge equal in magnitude to the negative charge of an electron and that are present in atomic nuclei (79)

pure substance a homogeneous sample of matter that has the same composition and characteristic properties throughout (20)

pyrite a compound of iron and sulfur (765)

Q

qualitative information nonnumerical information (8)

quantitative information numerical information (8)

quantum a finite quantity of energy that can be gained or lost by an atom (100)

quantum numbers numbers that specify the properties of atomic orbitals and their electrons (108)

quantum theory a mathematical description of the wave properties of electrons and other very small particles (105)

R

radioactive decay the release, by the nucleus, of an alpha or beta particle or gamma rays, forming a slightly lighter and more stable nucleus (804)

radioactive tracer radioactive atoms incorporated in substances so radiation detectors can follow movement of substances (817)

radioactivity the spontaneous disintegration in which small particles of matter (α or β particles) and/or electromagnetic radiation (γ rays) are emitted by unstable atomic nuclei (804)

rate law an equation that relates reaction rate and concentrations of reactants (553)

rate-determining step the lowest-rate step for a chemical reaction (551)

reactant(s) the substance(s) that undergo a chemical reaction (15)

reaction mechanism the step-by-step sequence of reactions by which the overall chemical change occurs (541)

reaction rate the change in concentration of reactants per unit time as a reaction proceeds (549)

reaction stoichiometry calculations involving the mass relationships among reactants and products in a chemical reaction (261)

real gas a gas that does not obey completely all the assumptions of the kinetic theory (319)

redox reaction oxidation–reduction reaction; any chemical process in which elements undergo a change in oxidation number (599)

reducing agent a species that attains a more positive oxidation state during an oxidation–reduction reaction (608)

reduction a reaction in which the atoms or ions of an element gain one or more electrons and hence attain a more negative (or less positive) oxidation state (597)

resonance the bonding in molecules that cannot be correctly represented by a single Lewis structure (174)

reversible reaction a chemical reaction in which the products can react to re-form the reactants (237)

rhombic sulfur the form of solid sulfur that is stable at ordinary temperatures (766)

ribonucleic acids nucleic acids that assist in protein synthesis by helping to coordinate the process of protein assembly; RNA (695)

ribosomes a structure within the cell, containing RNA and protein that serves as the site for protein assembly (698)

roasting heating sulfide ores in the presence of air (771)

ruby a gemstone of aluminum oxide contaminated with traces of impurities, including chromium (757)

S

salt an ionic compound composed of a metal cation of an aqueous base and the anion from an aqueous acid (472)

saponification the hydrolysis of fats or oils using a solution of a strong hydroxide (680)

sapphire a gemstone of aluminum oxide contaminated with traces of impurities, including titanium (757)

saturated fat a fat that contains only saturated fatty acids (692)

saturated fatty acid a fatty acid in which all carbon atoms of the hydrocarbon chain are filled with hydrogens (692)

saturated hydrocarbon a hydrocarbon in which each carbon atom in the molecule forms four single covalent bonds with other atoms (644)

saturated solution a solution that contains the maximum amount of dissolved solute possible at solution equilibrium under the existing conditions (412)

scientific method a logical approach to the solution of problems that lend themselves to investigation by observing, generalizing, theorizing, and testing (7)

scientific notation numbers written in the form $M \times 10^n$ where M is a number greater than or equal to 1 but less than 10 and n is an integer (52)

second the SI standard unit for time (34)

self-ionization the process that ionizes water; autoprotolysis (496)

semiconductor a substance with a slight electric conductivity that increases with temperature (758)

significant figures all digits in a measurement that are known with certainty, plus one final digit, which is uncertain or estimated (47)

simplest formula an expression of a compound that combines the symbols of the elements with subscripts showing the smallest whole-number ratio of the atoms (224)

single bond a covalent bond produced by sharing one pair of electrons between two atoms (169)

single-replacement reaction (displacement reaction) one element replaces a similar element in a compound (247)

slag the earthy impurities that remain when iron oxide is reduced in a blast furnace (739)

soda ash dried sodium carbonate, Na_2CO_3—an important industrial chemical (725)

solid state the state of any matter that has a definite shape and volume (13)

solubility the amount of a substance that is dissolved at solution equilibrium in a specific amount of solvent at a specified temperature (413)

solubility product constant the product of the molar concentrations of ions of a substance in a saturated solution, each raised to the power that is the coefficient of that ion in the chemical equation; K_{sp} (585)

soluble capable of being dissolved (405)

solute in a solution, the substance that is dissolved in solvent (406)

solution a homogeneous mixture of two or more substances in a single phase (20) (405)

solution equilibrium the physical state in which the opposing processes of dissolving and crystallizing of a solute occur at equal rates (412)

solvated a solute molecule surrounded by solvent molecules (418)

Solvay process the method for industrial production of sodium carbonate from brine and limestone (724)

solvent the dissolving medium in a solution (406)

specific heat the amount of heat energy required to raise the temperature of 1 g of a substance, 1C° (44)

spectator ions ions that do not take part in a chemical reaction and are found in solution both before and after the reaction (445)

spin quantum number the quantum number that indicates two possible states of an electron in an orbital, $+\frac{1}{2}$ or $-\frac{1}{2}$ (111)

standard electrode potentials a half-cell potential measured relative to a potential of zero for the standard hydrogen electrode (618)

standard molar volume of a gas the volume occupied by one mole of a gas at STP, 22.414 10 L (345)

standard solution a solution that contains a precisely known concentration of a solute (511)

standard temperature and pressure conditions (STP) agreed-upon standard conditions of exactly 1 atm pressure and 0°C (327)

standards of measurement objects or natural phenomena of constant value, easy to preserve and reproduce, and practical in size, that are used to define SI units (31)

starch a plant polysaccharide that is used as energy storage and as building material (689)

steroid a lipid with a "skeleton" structure consisting of four carbon rings (693)

stoichiometry the branch of chemistry dealing with mass relationships of elements in compounds and among reactants and products in chemical reactions (261)

strong acid an acid that is a strong electrolyte and ionizes 100% in aqueous solution (463)

strong electrolyte any solute that is present in dilute aqueous solution entirely as hydrated ions (449)

strong nuclear force the force that keeps the nucleons together in the nucleus (801)

structural formula a formula that indicates the exact number and types of atoms present in a molecule and how the atoms are bonded to each other (641)

structural isomers structural formulas for the various isomers with a specific molecular formula (641)

subbituminous coal a form of coal that has a higher percentage of carbon and a lower percentage of moisture than lignite (637)

sublimation the change of state from a solid directly to a gas (388)

substituted hydrocarbons hydrocarbons that are classified according to the functional groups they contain (639)

substrates the substance or substances on which an enzyme has its effect (688)

sulfation the addition of sulfur to alcohols (667)

supercooled liquid a substance that retains certain liquid properties even at temperatures at which it appears to be solid (378)

superheated water water heated under pressure to a temperature above its normal boiling point (765)

superoxide a binary compound containing the O_2^- anion, in which oxygen has the uncommon oxidation number $-\frac{1}{2}$ (291)

superphosphate a mixture of phosphate compounds that have been treated with sulfuric acid to increase their solubility and used as a fertilizer (774)

supersaturated solution a solution that contains more dissolved solute than a saturated solution under the same conditions (412)

surface tension a force, common to all liquids, that tends to pull adjacent parts of a liquid's surface together, thus making the surface less penetrable by solid bodies (374)

suspension a heterogeneous mixture of a solvent-like substance with particles that slowly settle out (408)

synapse the space between one neuron and the next (695)

synthesis reaction (composition reaction) a reaction in which two or more substances combine to form a new compound (244)

system the sample of matter being studied (383)

T

temperature a measure of the average kinetic energy of the particles in a sample of matter (40)

theoretical yield the maximum amount of a product that can be produced from a given amount of reactant (276)

theory a broad generalization that explains a body of facts or phenomena (9)

thermite a mixture of powdered aluminum and iron(III) oxide (755)

thermite reaction the reaction by which a metal is prepared from its oxide by reduction with aluminum (755)

thermochemical equation an equation that includes the quantity of heat released or absorbed during the reaction as written (527)

thermochemistry the study of the changes in heat energy that accompany chemical reactions and physical changes (527)

titration the controlled addition and measurement of the amount of a solution of known concentration that is required to react completely with a measured amount of a solution of unknown concentration (511)

torr a pressure unit named after Torricelli; 1 torr = 1 mm Hg (326)

transition element the ten subgroups of elements that intervene between Group 2 and Group 13, beginning with Period 4 (729)

transition interval the pH range over which an indicator changes color (505)

transmutation a change in the identity of a nucleus because of a change in the number of its protons (810)

transuranium element an element with more than 92 protons in its nucleus (814)

triple bond a covalent bond produced by the sharing of three pairs of electrons between two atoms (171)

triple point indicates the temperature and pressure conditions at which the solid, liquid, and vapor of a substance can coexist at equilibrium (389)

triprotic acid an acid able to donate three protons per molecule (464)

U

unit cell the smallest portion of a crystal lattice that reveals the three-dimensional pattern of the entire lattice (379)

unit of measurement a physical quantity of a defined size (31)

unsaturated describes an organic compound that has one or more double or triple covalent bonds in each molecule (650)

unsaturated fat a fat that contains at least one fatty acid group with double bonds (692)

unsaturated solution a solution that contains less solute than a saturated solution under the existing conditions (412)

unshared pair a lone pair of electrons that is not involved in bonding but instead belongs exclusively to one atom (169)

V

valence electrons the electrons available to be lost, gained, or shared in the formation of chemical compounds (152)

vaporization the process by which a liquid or solid changes to a gas (375)

variable a quantity that can change in value (58)

VSEPR (valence-shell electron-pair repulsion) theory electrostatic repulsion between the electron pairs surrounding an atom causes these pairs to be oriented as far apart as possible (182)

vinyl group the $-CH=CH_2$ group (653)

voltaic cell a device in which an electric current can be generated from two solutions separated by a porous partition that permits ions to pass (but that otherwise prevents mixing of the solutions) and which has two electrodes connected in a circuit (617)

volume the amount of space occupied by an object (36)

vulcanization the process by which rubber is heated with sulfur, which bonds with rubber molecules, making the rubber more elastic, less sticky and longer wearing (658)

W

wavelength the distance between corresponding points on adjacent waves (97)

wax a lipid made of esters of long fatty acids and long-chain monohydroxy alcohols (692)

weak acid an acid that is a weak electrolyte (464)

weak electrolyte a solute that yields a relatively low concentration of ions in aqueous solution (450)

weight a measure of the earth's gravitational attraction for matter (11)

word equation an equation in which the reactants and products in a chemical reaction are represented by words (234)

INDEX

Page numbers in boldface type refer to figures. Page numbers preceded by the letter *t* refer to tables.

A

absolute zero, 330
accuracy, 46–47, **46,** 59
acetic acid: anion of, 474; common-ion effect and, 576; conjugate base of, 475; ionization constant of, 577–578, t 578; preparation of, 677–678; properties of, 465, t 465, 469; structure of, 467; uses of, 469, 678, **678;** as weak acid, 477, **483**
acetone, **674,** 675–676
acetylene, 531
acid(s): binary acids, 467, t 467; Brønsted acids, 462, 486, 475–477, 508; in carbon dioxide preparation, 306; chemical reactions of, 483–485, **484, 485,** 486; conjugate, 475–478, t 478; decomposition of, 247; diprotic acids, 466; examples of common acids, 465–469; household, 476; in hydrogen preparation, 298; hydroxyl group, 465; indicators for, 505–510, t 505, **506–508, 510,** 516; ionization constant of, 576–579, t 578, 590; Lewis acids, 462–463, 486; mineral acids, 467; monoprotic acids, 463; organic acids, 467; oxyacids, 464–465, t 464, **465;** pH and, 499, 516; polyprotic acids, 466; properties of, 461, **461,** 486; relative strengths of, 477–478, t 478; strong acids, 463, t 466; traditional, 461–462, 486; triprotic acids, 466; types of, 211; weak acids, 466, t 466
acid anhydride(s), 480
acid-base reaction(s), 473, 486
acid precipitation, **291,** 299, 504, 509 772–773, 775
actinide(s), 130, 141
activated carbon, 638
activated complex, 545–548, **546,** 557
activation energy, 544–545, **545,** 557
activity series, 251–255, t 252, **253**
actual yield, 276
adenosine triphosphate (ATP), 695, 701–703, **701**
adiabatic demagnetization, 322
Agent Orange, 640
alcohol(s), 665–669, **667–667, 669,** 681
aldehyde(s), 674–675, **674, 675,** 681
alkadiene(s), 651
alkali metal(s), 136, 713–716, t 714, **716,** 726
alkaline(s), 471
alkaline-earth metal(s), 136–137, 717–719, **717,** t 718, 726
alkane(s), 643–649, **645,** t 645 **648,** 660
alkene(s), 644, 649–651, **650,** 660
alkyl group, 647, t 647
alkyl halide(s), 670
alkylation, 651
alkyne(s), 644, 651–653, **652,** 660
allotrope(s), 289
allotropy, 289
alloy(s), 407, **407**

alpha particle(s), 806, 809, **809,** t 809, 814
alum(s), 757–758, 762
alumina, 753, 757
aluminum: atomic radii of, 143; compounds of, 756–758, **757;** electron configuration of, t 117; properties of, 751, t 752, 753, 754–755, 762; thermite reaction, 755–756, **756,** 762
aluminum chloride, 449, 583
aluminum hydroxide, 481–482
aluminum nitrate, 248
aluminum oxide, 205, 263, 265, 480, 756–757
aluminum recycling, 754
aluminum sulfate, 201
alundum, 757
amine(s), 694, **695**
amino acid(s), 686, **687**
ammonia: aqueous solution of, 450; as base, 472–474; molality and, 423; occurrence of, 299, **299,** 300; preparation of, 301–302, **203;** reaction with hydrogen chloride, 364, **364;** structure and properties of, 184, **184,** 186, 299, 301, t 301; uses of, 301, 310
ammonium chloride, 301, 364, **364,** 472, 582–583, 611
ammonium ion(s), 173
ammonium nitrite, 299
amorphous carbon, 636–638
amorphous solid(s), 377, 381–382, 394
amorphous sulfur, 767, **767**
ampere, t 32
amphoteric, 479
amphoteric hydroxide(s), 480–483
amphoteric oxide(s), 480–483
analytical chemistry, 7
anhydride(s), 480
aniline, 474
anion(s), 151, 212, 464–465, t 464, 474
anion hydrolysis, 581–582, 590
anode, 612
antacid(s), 481
anthracite, 637
antimony, **25,** t 119, 760–761, 762
antiseptic, 790
aqueous salt solution(s), 614
argon, 116, t 117, 130
Aristotle, 71
aromatic hydrocarbon(s), 644, 653–655, **654,** 660
Arrhenius, Svante, 439, 461–462, 470
arsenic, **25,** 760, 762
arsenopyrite, **25**
astatine, 779, t 780
Aston, Francis William, 363
atm. *See* atmosphere of pressure (atm)
atmosphere of pressure (atm), 326, t 326
atom(s): excited state of, 100–101; Greek ideas on, 71; ground state of, 100–101; nuclear forces in, 80; quantum model of, 104–106; radii of, 143–144, **143–145;** sizes of, 81–82, 92; structure of, 75–82; subdivision of, 75. *See also* electron(s); neutron(s); proton(s)

atomic mass(es), 85–87, t 86, 92, 217, 801
atomic mass unit, 85
atomic nucleus: composition and structure of, 75, 78–80, **78,** 801–804, **802, 803;** discovery of, 77–78, 92; size of, 81–82, 92
atomic number(s), 83, 92, 803
atomic radii, 143–144, **143–145,** 155
atomic structure, 75
atomic symbol(s), 221
atomic theory: Dalton's model of, 71–72, 92, 221, 344; and definite composition of chemical compounds, **72,** 73; and law of conservation of mass, 72–73, **72,** 92; and law of multiple proportions, 73–74; quantum model of, 104–106
ATP. *See* adenosine triphosphate (ATP)
Aufbau principle, 112–113, **115,** 121
auto-oxidation, 610
autoprotolysis, 496
average atomic mass, 85–87, t 86, 92
Avogadro, Amadeo, 88, 217, 344–345
Avogadro's number, 88, 90, **90,** 92
Avogardo's principle, 344–345, **344,** 365
azeotrope(s), 451

B

baking soda, 726
barium hydride, 296
barium oxide, 291
barometer, 324, 325–326, 337
base(s): Brønsted bases, 472–473, **472,** t 473, 475–477, 486; chemical reactions of, 483–485, **484, 485,** 486; conjugate, 475–478, t 478; household, 476; indicators for, 505–510, t 505, **506–508, 510,** 516; Lewis bases, 473, t 473, 486; neutralization reactions, 471–472, **471, 472;** pH and, 499, 516; properties of, 470, **470,** 486; traditional, 486; traditional bases, 470–471, t 473; types of, 473–474, **474;** weak bases, 473–474
base-pairing, 696, **697**
basic anhydride(s), 479
bauxite, **218,** 753, 756
Becquerel, Henri, 804, **805,** 806
benzene(s), 415, 653–654
Bergius process, 658
beryl, 136
beryllium, 115, t 116, 136, 168
beryllium fluoride, 182
Berzelius, Jöns Jakob, 23, 221
beta particle(s), 809, **809,** t 809
binary acid(s), 211, t 211, 465, t 465
binary compound(s), 204–207, **205,** 209–210, **209,** t 210, 246, **246**
biochemistry, 7, 685
bituminous coal, 637
Black, Joseph, 260
black magnetite, 291
blast furnace, 739, **739**
bleaching, 785
block(s), in Periodic Table of Elements, 133–142, **133–136, 138, 139,** t **140, 141**

Modern Chemistry **901**

blue jeans, 610
blueprint(s), 743
Bohr, Niels, 102, 104, 105–106
boiling, 375, 387–388, **387, 388,** 394, 451
boiling point(s): and bond types, t 187, 189, 190; of crystals, t 380; of liquid, 375, 387–388, 394, 451; of solutions, 429, t 429
boiling-point elevation, 431, **431**
bond(s). *See* chemical bond(s)
bond energy: covalent bonds and, 166, t 166; and reaction heat, 535–536, **535,** 557; single and multiple covalent bonds and, 171, t 172
bond length(s), 166, t 166, 171, t 172
boron, t 116, 168
boron trifluoride, 463
Boyle, Robert, 296, 322, 327
Boyle's law, 327–329, **327, 328,** t 328, 337
bromine compounds of, 788–789; **788;** preparation of, 787, **787;** properties of, 25, **139,** 140, 375, **375, 779,** t 780, 788; single-replacement reactions with, 248; uses of, 788–789, 791
Brønsted, J. N., 462
Brønsted acid(s), 462, 486, 508
Brønsted acid-base reaction(s), 473, 475–477, 486
Brønsted base(s), 472–473, **472,** t 473, 475–477, 486
Brown, Robert, 409
Brownian motion, 409
buffered solution(s), 578, **579**
butadiene, 651
butane, 645

C

calcium, 137
calcium bromide, 788–789
calcium carbonate, 136, 247, 306, 485
calcium chloride, 442–443
calcium fluoride, 177–178, **178,** 585–586
calcium hydroxide, 245–246, 247, 301, 305, 473, 485
calcium hypochlorite, 570
calcium oxide, 245, 247
calcium sulfate, 211, 246
calculator(s), 501–502
calorie(s), 43, t 43, **527**
calorimeter, 527, **527**
cancer, 172, 806, 816, 818
candela, t 32
Cannizzaro, Stanislao, 127, 344
capillary action, 375
carbohydrate(s), 689–691, **689, 691,** 703
carbon: allotropic forms of, 634–636, **634,** 660; amorphous forms of, 636–638, 660; electron configuration of, t 116; structure and bonding of, 171, t 172, 633, **633,** 660; uses of, 140
carbon dioxide: chemical properties of, 304–305, 307; desktop investigation, 307; occurrence of, 303; physical properties of, 304, **304,** t 304, 307; preparation of, 234–235, **235,** 245, 247, 272, 305–306; structure of, 303–304, **203;** uses of, 303, 305, **305,** 310, 407, 416, **416**
carbon-14, 816
carbon monoxide: occurrence of, 306, 308; physical properties of, t 309; poisoning, 306, 308, 309; and preparation of hydrogen, 297; structure of, 308; uses of, 303, 309, 310
carbon tetrachloride, 414–415, **415,** 671
carbonate(s): in carbon dioxide preparation, 306; reactions of acids with, 485
carbonic acid, 247, 306, 574
carboxyl group, 676
carboxylic acid(s), 676–678, **677,** 681
catalysis, 551
catalyst(s), 236, 551–552, **553**
cathode, 612
cathode ray(s), 75–77, **76, 77**
cation(s), 151
cation hydrolysis, 582–583, 590
caustic, 724
Cavendish, Henry, 211, 296
cellulose, 690, **691**
Celsius, Anders, 41
Celsius temperature scale, 41–43, **42**
cesium, 25, 136, 154
Chadwick, Sir James, 813
chain reaction, 819, **819**
charcoal, 550–551, 638
Charles, Jacques, 329
Charles' law, 329–332, **329, 330,** t 330, t 331, 337
chemical(s), 6–7
chemical bond(s), 161–164, **161, 162,** 190. *See also* covalent bond(s); ionic bond(s); metallic bond(s)
chemical change(s). *See* chemical reaction(s)
chemical compound(s): chemical formulas for, 201–202, t 201, **202;** containing polyatomic ions, 208–209; definite composition of, 73; desktop investigation, 215; law of definite composition, 22–23, **22;** law of multiple proportions, 73–74; low and high heats of formation, 529; models of, 215; molar masses and, **216,** 217–220, **218;** nomenclature of, 206, **207;** numerical prefixes for, **209,** t 210; percent composition of, 222–223; systematic and common names of, t 201
chemical element(s). *See* element(s)
chemical equation(s): balancing, 240–243, **240, 241,** 249; coefficients in, 234; desktop investigation, 249; reading and writing, 233–238, 255; requirements for, 233–234; significance of, 238–240, **239;** symbols used in, t 236; word and formula equations, 234–238, **235**
chemical equilibrium, 564. *See also* equilibrium
chemical equilibrium expression, 566
chemical equivalent(s). *See* equivalent(s)
chemical formula(s): of binary ionic compounds, 204–207, **205,** 228; calculation of, 224–227; description of, 164; empirical, 224, 228; of familiar compounds, t 201; formula masses and, 216–217; molar masses and, **216,** 217–220, **218,** 228; molecular formulas, 226–227; molecular masses and, 217; oxidation numbers and, 214–216; percent composition and, 222–223; significance of, 201–202, **202,** 228; simplest, 224, 228
chemical kinetics, 549

chemical properties, 14–15, 27
chemical reaction(s): description of, 15; desktop investigation, 18; endothermic, 18–19; exothermic, 16–17, **17;** indications of, 16, **16, 17,** 27; spontaneous, 536–537; types of, 244–250
chemical solution(s). *See* solution(s)
chemical symbol(s), 23, t 23, 236–237, t 236
chemistry: branches of, 7, 27; description of, 5; descriptive, 25–26; in modern society, **5, 6, 6**
chitin, 691
chloramine, 254
chlorinated lime, 785
chlorine: bonding and, 162, **162,** 182; compounds of, 786, **786;** electron configurations of, t 117; preparation of, 782, **782;** properties of, **139,** 779, **779,** t 780, 784–785; single-replacement reactions, 248; uses of, 782, 785, **785,** 791
chlorofluorocarbon(s), 289, **289**
chlorohydrocarbon(s), 640
cholesteric crystal(s), 376
cholesterol, 693, **693**
chromium, 118, t 118
coal, 531, 550, 636–637, t 641, 658
cobalt, 118, t 118, 736–737, t 737
Cockcroft, J. D., 813–814
coefficient(s), 234
coenzyme(s), 688
coke, 637–638
colligative properties: of electrolyte solutions, 453–456, **455,** t 455, t 456; of solutions, 426–433, **426, 427, 431**
collision theory, 543–544, **544,** 557
colloid(s), 405, 408, 409, **409,** t 409, t 410
combined gas law, 333–334, 337
combustion reaction(s), 250, **250,** 530–531, **530,** t 530, 557
common-ion effect, 575–576, 590
complex carbohydrate(s), 689–690
complex ion(s), 733–736, **734,** t 735, t 736, 748
composition reaction(s). *See* synthesis reaction(s) composition stoichiometry, 261
compound(s). *See* binary compound(s); chemical compound(s); ionic compound(s); molecular compound(s)
computer(s), 760–762
computer chip(s), 761
concentration, 384, 420
concentration: description of, 384, 420; equilibrium and changes in, 569; and reaction rate, 550–551, **550**
condensation, 383–384
conductor(s), 611
configuration, *See* electron configuration(s)
conjugate acid(s), 475–478, t 478
conjugate base(s), 475–478, t 478
conservation of energy, law of, 13, 15, 27
conservation of mass, law of, 12, 15, 27, 72–73, **72,** 92, 233–234
contact process, 773
continuous spectrum, 98, **98,** 104, **105**
control rod(s), 819
conversion factor(s), 34–35, 56
Cooke, Josiah Parsons, Jr., 127
coordination number(s), 734

copper: activity series and, 252–253, **253**; electrode potentials and, 616–620; electrolytic refining of, 746, **746**; electron configuration of, t 118; oxidation-reduction reactions and, 608–609; properties of, 24, 26, **26**, 138, 744, 747; recovery of, 745–746, **745**; tests for copper ion, 747–748; uses of, 744, 747
copper chloride, **205**
copper hydroxide, 583
copper oxide, 484
copper sulfate, **406,** 484, 583, 747
corundum, 756
cosmic ray(s), 816
covalent bond(s): bond lengths and bond energies for, 166, t 166; description of, 161–162, 163; double, 170–171, 190; formation of, 165–167, **165, 167**; Lewis structures and, 169–170; multiple, 170–173, t 172, 190; octet rule, 167–168; single, 169, t 172; triple, 171, 190
covalent hydride(s), 296
covalent molecular crystal(s), 381
covalent network crystal(s), 380
cracking, 649
critical mass, 820
critical point(s), 389
critical pressure, 389
critical temperature, 389
crude oil, 656
cryolite, 753
crystal(s): binding forces in, 379–381, **381**; classification of, 379, **379,** 394; ionic crystals, **176–178**; liquid crystals, 376; melting points and boiling points of, t 380; as solids, 377
crystal lattice, 379, **379**
crystalline solid(s), 377, **378,** 379–381, **379,** t 380, **381,** 394
Curie, Marie, 804, **805,** 806
cycloalkane(s), 648–649

D
Dalton, John, 71–72, 73, 92, 221, 334, 344
Dalton's law of partial pressures, 334–336, **335**
Da Vinci, Leonardo, 290
daughter nuclide(s), 810
Davy, Sir Humphry, 247, 720, 791
DDT, 640
De Broglie, Louis, 104–105
Debye, Peter, 456
decay, in carbon dioxide preparation, 306
decomposition reaction(s), 246–247, **246,** 255
definite composition, law of, 22–23
dehydrating agent(s), 774
Democritus, 71
denaturation, 688
denatured alcohol, 668
density: of common materials, **37,** 39, t 39; derived SI units and, t 36, 37, 39; desktop investigation, 38
deoxyribonucleic acid (DNA), 695–698, **696-698,** 700, 703
destructive distillation, 637
deuterium, 80, 83, t 84, 821, **821**
diamagnetism, 111
diamond, 380, **381,** 634–635, **634,** 660

diatomic molecule(s), 164, 233, t 233
diborane, 224–227
dicarboxylic acid(s), 678
dichlorodifluoromethane, 671, 782
diet, 715
diffusion, 318, **318,** 360–362, **360, 362,** 364–365, 373, **373**
dimer(s), 653
dimethyl sulfide, 772
dipole(s), 187–188
dipole-dipole forces, 187–188, **188**
diprotic acid(s), 464
direct proportion, 58
disaccharide(s), 689, **689**
disinfectant(s), 785, 790
displacement reaction(s). *See* single-replacement reaction(s)
dissociation, 442–443
dissolving agent(s), 372–373
distillation, 428
DNA. *See* deoxyribonucleic acid (DNA)
Dorn, Freidrich Ernst, 130
double bond(s), 170–171, 190
double helix, 696
double-replacement reaction(s), 248, 250, **250,** 255
Downs cell, 720–721, **720**
dry cells(s), 611, **612**
dry ice, 305
ductility, 181

E
effervescence, 416
effusion, 318, 360–365, **360**
Einstein, Albert, 15, 101
Einstein's equation, 15, 813–814
elastic collision(s), 315
electrochemical cell(s), 611–612, **610, 611,** 622
electrochemistry: electrochemical cells, 611–612, 622; electrode potentials, 616–622, **617–169,** t 620; electrolytic cells, 612–615, **613–615**; rechargeable cells, 615–616
electrode(s), 611
electrode potential(s), 616–622, **617–619,** t 620
electrolysis: of aqueous salt solutions, 614–615; as decomposition reaction, 246–247; description of, 612, 622; in hydrogen preparation, 298; in oxygen preparation, 293; of water, 613–614
electrolyte(s): acids as, 461; apparent degree of ionization of solutions, 454–456, **455**; bases as, 470; coining of word, 439; colligative properties of solutions, 453–456, **455,** t 455, t 456; conductivity of solutions, 450–452, **452**; description of 406; freezing point depression of solutions, 453–454, t 455, t 456; in human bodies, 447; molecular electrolyte(s), 447–450, **448**; strong, 449; weak, 450
electrolyte solution(s), 447–448, **448,** 450–456, t 455, t 456
electrolytic cell(s), 612–615, **613–615**
electromagnetic radiation, 97–98, 100
electromagnetic spectrum, 98, **98, 99**
electron(s): Bohr's theory of, 102–106, **103**; charge and mass of, 76–77, 79; discovery of, 75–77, 92; highest occupied energy level and, 113–114; inner-shell electrons, 114; in main energy levels, 111; orbitals of, 106, 110, **110,** 121; orbits of, 103, 104; properties of, t 80; uncertainty principle and, 106–107; unshared pair of, 169; valence electrons, 152, t 152; wave properties of, 105–106, **106**; wave-particle model of, 97, 104–105
electron affinity, 149–151, **149, 150**
electron configuration(s): fifth-period elements, 119, t 119; first-period elements, 114, 115; fourth-period elements, 117–119, t 118; notations for, 113–114, 121; and Periodic Table, 132–142, **134–136,** 138, **139,** t 132, t 140, 141; rules governing, 112–113, **112,** 121; second-period elements, 115, t 116; seventh-period elements, 120; sixth-period elements, 120; third-period elements, 115–116, t 117
electron-configuration notation, 113
electron-dot notation, 113–114
electron-dot symbol(s). *See* Lewis structure(s)
electronegativity, 152–154, **153,** 156
electroplating, 615
element(s): activity series of, 251–255, t 252; atomic masses and abundances of, t 86, 87; classes of, 24–25; discovery of, 791; electron configurations of, 115–120, t 116, t 117, t 118, t 119; fifth-period, 119, t 119; first-period, 114, 115; fourth-period, 117–119, t 118; number of, 21–22; reactions wtih oxygen, 244–245; second-period, 115, t 116; seventh-period, 120; sixth-period, 120; symbols for, 23, t 23; third-period, 115–116, t 117. *See also* Periodic Table of Elements
emery, 756
empirical formula(s), 224, 228
end point of titration, 512
endothermic reaction, 18–19, **545**
energy: conservation of, 13, 27; interconversion of matter and energy, 15; kinetic, 12; potential, 12; in starting a car, 12, **13**
enthalpy, 528, 536–537, **537**
enthalpy change, 528
entropy, 537–538, **538,** 557
enzyme(s), 688, 703
epinephrine, 694, **694**
equation(s). *See* chemical equation(s)
equilibrium: change in concentration and, 569; change in pressure and, 569–572, **571, 572**; change in temperature and, 572–573; and changes of state, 383–384, **384**; common-ion effect and, 575–576, 590; description of, 383, 394, 564–565, **565,** 590; equilibrium constant, 565–568, **567,** t 567, 573, t 573, 590; equilibrium vapor pressure, 385–386, **386,** 394; Le Châtelier's principle and, 384–385, 394; reactions that run to completion, 573–575, **575**
equilibrium constant, 565–568, **567,** t 567, 573, t 573, 590
equilibrium vapor pressure, 385–386, **386,** 394
equivalence point, 512

equivalent(s), 491–493, **491**, 516
equivalent mass, 492–493, 516
ester(s), 679, **679, 680,** 681
esterification, 677, 679–680
ethane, 171, 645
ethanol, 414–415, **415,** 665, **666,** 667–669, **667**
ethene, 649
ether(s), 415, 673, 681
ethyl alcohol, 407, 665
ethylene bromide, 788
ethylene dichloride, 672
ethylene glycol, 407, 669, **669**
ethyne, 652–653
evaporation, 375
excess reactant(s), 272
excited state, 100–101
exothermic reaction, 16–17, **17, 545**
experiment(s), 8
extensive physical properties, 14

F

factor-label method, 34–36, 59, 56
Fahrenheit, Gabriel Daniel, 42
Fahrenheit temperature scale, 42–43, **42**
families of elements, 23
Faraday, Michael, 439
fat(s), 296, 414, 680–681, 692–693
fatty acid(s), 692, **692**
fermentation, 306
ferromagnetism, 111, 737
fertilizer(s), 301
fiber composite(s), 757
fire extinguisher(s), 305, **305**
Fischer-Tropsch process, 658
fission, 75, 804, 817, 819, **819**
flame test(s), 716
fluid(s), 372. *See also* liquid(s); water
fluoresce, 804, **805,** 807
fluorine: compounds of, 782; oxidation number, 213; preparation of, 781; properties of, 115, t 115, **139,** 154, 251, 779, **779,** t 780; reactions involving, 245, 245; uses of, 782, 791
fluorite, 177, **178**
fluorspar, 782
food(s): calories in, 43, t 43; carbon dioxide in, 305, 407, 416, **416;** recipes for, 273, **431**
formaldehyde, 172, 675, **675**
formic acid, 678
formula(s). *See* chemical formula(s); empirical formula(s); molecular formula(s); structural formula(s)
formula equation(s), 234–235, **235**
formula mass(es), 216–217
formula unit(s), 175, 201
formyl group, 674
fractional distillation, 656–657, t 656
francium, 154
Frasch, Herman, 765
Frasch process, 765, **765,** 775
free energy, 538–539, 557
free-energy change, 538–540, 557
freezing, 375–376, 388, 394
freezing point(s), 388, 429, t 429
freezing-point depression, 428–430, 453–454, t 455, t 456
Freon(s), 671, **672**

frequency, 98
Friedel-Crafts reaction, 654, 671
fuel(s), 531
functional group(s), 639–640
fundamental unit(s) of measurement, 32–34, t 32, t 33, 59
fungicide(s), 769
fusion, 804, 820–821, **821**

G

gallium, 25, 143
gamma ray(s), 809, **809,** t 809
gas(es): Avogadro's principle, 344–345, 365; Boyle's law, 327–329, **327, 328,** t 328, 337; changes of state, t 383; Charles' law, 329–332, **329, 330,** t 330, t 331, 337; collection by water displacement, 335–336; combined gas law, 333–334, 337; compressibility of, 317–318; Dalton's law of partial pressures, 334–336, **335;** density of, 347–348, **347;** deviations of real from ideal gases, 319, **319;** diffusibility of, 318, **318;** discoveries of, 130; expansiveness of, 317; fluidity of, 317; formation of, 250, 574–575; Gay-Lussac's law, 332–333, 337, 343–344, 366; ideal gas law, 350–355, **351,** t 352, 365; ideal gases, 315, 319, 337; kinetic theory of, 315–319, **315, 317–319,** 337; low density of, 317; measuring and comparing volumes of, 343–344, 366; molar mass of, 348–350; molar volume of, 345–347, **346,** 365; nature of, 317–318, 337; noble gases, 130, 140; pressure and, 322–327, **323–325,** t 326; pressure and temperature, 321, **321;** pressure and volume at constant temperature, 320; qualitative description of, 320–321; quantitative description of, 322–336, t 326, t 328, t 330, t 331, 337; real gases, 319, **319;** relationship between pressure and moles, 321; relationship between volume and moles, 321, **321;** solubility of, 415–416; solutions of, 407, t 407; stoichiometry of, 356–359; temperature and volume at constant pressure, 320–321, **320;** tests for, t 310; volume-mass and mass-volume calculations, 357–359; volume-volume calculations, 356–357; water-gas reaction, 535–536, **535.** *See also* names of specific gases
gas law(s): Avogadro's principle, 344–345, 365; Boyle's law, 327–329, t 328, 337; Charles' law, 329–332, t 330, t 331, 337; Dalton's law of partial pressures, 334–336; description of, 322; Gay-Lussac's law, 332–333, 337, 343–344, 366; ideal gas law, 350–355, t 352, 365
gaseous state, 13–14
gasoline, 414, 415, 646, 788
Gay-Lussac, Joseph, 332, 343–344, **343**
Gay-Lussac's law, 332–333, **332,** 337, 343–344, 366
Geiger, Hans, 77
Geiger counter(s), 807, **807**
gene(s), 697
genetic engineering, 699–700
germanium, 25, 752, t 752
Gibbs, Josiah Willard, 538

Gibbs Free Energy, 538
glycerol, 669, **669**
glycogen, 689
glycolipid(s), 692
glycolysis, 701, **701, 702,** 703
gold, 24, 138–139, 252, 744, t 744
Graham, Thomas, 409
Graham's law of effusion or diffusion, 360–362, **362,** 364–365, **364**
granite, 20, **20**
graphite, 380, **381,** 634, 635–636, **635, 636,** 660, 819
ground state, 100–101
group(s) of elements, 23
Guldberg, Cato Maximilian, 556

H

Haber, Fritz, 301
Haber process, 302, **302,** 570–571, **572**
Hahn, Otto, 75, 817
half-cell(s), 618
half-life, 807–808, 821
half-reaction(s), 612, 622, 620–621, t 620
halite, 722
Hall, Charles, 753
Hall process, 753
halocarbon(s), 670–672, **670, 672,** 681
halogen(s), 140, 248, 779–781, **779,** t 780, 791
halogenation, 654
heat: description of, 40, **41,** 59; heat capacity, 44; specific heat, 44–45, t 44; symbol for, 236; units of, 43, t 43; water-gas reaction, 535–536, **535**
heat capacity, 44
heat of combustion, 530–531, **530,** t 530, 557
heat of formation, 528–529, **529,** 556–557
heat of hydration, 441
heat of reaction, 527–528, **527, 528,** 531–536, **532,** 556, 557
heat of solution, 417–419, t, 419, 441–442, **442**
Heisenberg, Werner, 106–107
helium, t 84, 113–114, 130, 137, 345
hematite, 738
Henry, William, 416
Henry's law, 415–416
Heroult, Paul, 753
hertz, 98
Hess's law, 531–532
heterogeneous catalyst(s), 552
heterogeneous mixture(s), 19–20
heterogeneous reaction(s), 550
homogeneous catalyst(s), 551–552
homogeneous mixture(s), 20, **20, 21,** 27
homogeneous reaction(s), 541–542
homologous series, 645
hormone(s), 693–695, **694–695**
Hückel, E., 456
Hund's rule, 113
hybrid orbital(s), 186, t 187
hybridization, 185–186, **186,** t 187, 190
hydration, 440
hydrazine, 254
hydrobromic acid, 449

904 Index

hydrocarbon(s): alkanes, 643–649, t 645, 660; alkenes, 644, 649–651, 660; alkynes, 644, 651–653, 660; aromatic hydrocarbons, 644, 653–655, 660; description of, 250, 643, 660; groups of, 643–644, 660; in hydrogen preparation, 298; industrial and energy uses of, 655–660
hydrochloric acid: formation of, 449; as nonoxidizing acid, 483–484; and preparation of carbon dioxide, 306; and preparation of hydrogen, 297; properties of, 465, t 465, 467, 469; reactions including, 248, **248**, 549, 574, 603–604; as strong acid, 463, 476; uses of, 786, **786**
hydrofluoric acid, 782, **782**
hydrogen: chemical properties of, 296; covalent bonding, 162, **162**, 164–167, **165, 167**; diffusion of, 362; discovery of, 211, **211**; electron configurations of, 113–114, 137; history of, 296; isotopes of, 80–81, 83, t 84; mass of, 216, 345; molecules of, 182; occurrence of, 294–295; physical properties of, 295, t 295; preparation of, 248, **295**, 297, **297**, 648; reaction with chlorine, 785, **785**; reaction with metals, 296, 310; reaction with nonmetals, 296, 310; replacement by metal, 248, **248**; uses of, 294, **295**, 296, 310
hydrogen atom(s): Bohr model of, 102–103, **103**; bonding of, 162, **162**, 164–167, **165, 167**, 189, **189**; line emission spectra of, 103; spectrum, 99, 100–103, **103**
hydrogen bonding, 162, **162**, 164–167, **165, 167**, 188, **189**
hydrogen bromide, 449, 788, **788**
hydrogen chloride: as acid, 463, 472; ammonia and, 364, **364**; bonding of, 162, 164, 182; chemical equation for, 238–239, **239**; common-ion effect and, 576; description of, 786; double-replacement reaction and, 250; formation of, 296, 343–345; molecular electrolytes of, 448–449, **448**; oxidation and reduction and, 597–598
hydrogen fluoride, 212, 449–450, 475
hydrogen iodide, 449
hydrogen peroxide, **202**, 292, **292**, 610
hydrogen sulfate, 478
hydrogen sulfide, 250, 575, 772
hydroiodic acid, 449
hydrolysis, 580–583, **581–583**, 590
hydrolysis constant, 582
hydronium ion(s), 448–449, t 504
hydroxide(s), 467, 480–482, 485, t 504
hydroxyl group, 467
hyberbola, 59
hypochlorous acid, 570
hypothesis, 8

I

ice crystal(s), 391, **391**
ideal gas(es), 315, 319, 337
ideal gas constant, 351–352, t 352
ideal gas law, 350–355, **351**, t 352, 365
immiscible, 414
indicator(s), 505–510, t 505, **506–508, 510**, 516
inertia, 10–11
inhibitor(s), 552

inner-shell electron(s), 114
inorganic chemistry, 7
inorganic compound(s), 642–643, t 643
insecticide(s), 670, **670**
instability constant, 735
insulin, 687
integrated circuit(s), 760–761
intensive physical properties, 14
intermediate(s), 542–543
intermolecular forces, 187–190, **188–190**
inverse proportion, **58**, 59
iodine, **139**, 248, **779**, t 780, 790, **790**, 791
ion(s): complex, 733–736, **734**, t 735, t 736, 748; description of, 145; formation of negative, 150–151; formation of positive, 147–148, 151; in metals, 156; monatomic, 203–204, t 203, t 204; origin of word, 439; spectator ions, 445
ionic bond(s), 161–163, **161, 162**, 176–178, **176, 177**
ionic compound(s): binary, 204–207, **205**; compared with molecular compounds, 178–179, **178**, 190; description of, 175, **175**; dissociation and, 442–443; dissolving of, 440–443, **440–442**, 456; heat of solution for, 441–442, **442**; net ionic equations, 444–445; precipitation reactions, 444; solubility equilibria, 443–446, **444, 445**; solution process for, 440–441, **441**; theory of ionization, 439–440; very slightly soluble, 443–444
ionic crystal(s), 380
ionic hydride(s), 296
ionic radii, 151–152, **151**, 156
ionization: apparent degree of, 454–456, **455**; description of, 145, 448; theory of, 439–440; of water, 452
ionization constant: of water, 579–580; of weak acids, 576–579, t 578, 590
ionization energy: first ionization energies, 147–148, t 148; formation of positive ions, 147–148; group trends, 146–147; of metals, 155; period trends, 145–146, **146, 147**
iron: blast furnace and, 739, **739**; characteristics of, 252, 736–738, t 737; electron configuration of, 118, t 118; occurrence of, 738–741, **738–740**; oxides of, 741–742; pure iron, 741–743, **741, 742**; steel production and, 739–741, **740**; tests for iron ions, 742–743
iron chloride, 250, 575
iron nitrate, 248
iron oxide, 222, 245, 291, 309, 741
iron sulfate, 741–742
iron sulfide, 250, 574
isomer(s), 639
isotope(s): artificial, 85; atomic masses and abundances of, t 86, 87; description of, 72, 81, 92; designation of, 83; of helium, t 84; of hydrogen, 80–81, 83, t 84; relative atomic masses of, 85; separation of, 363

J

Janssen, Pierre, 130
Joliot-Curie, Frédéric, 85, 812
Joliot-Curie, Irène, 85, 812
joule, t 36, 43, t 43
journals. *See* scientific journals

K

Kelvin, William Thomson Lord, 41
Kelvin temperature scale, t 32, 41–42, **42**
ketone(s), 674–676, 681
kilogram(s), t 32, 33
kinetic energy, 12
kinetic theory: of gases, 315–319, **315, 317–319**, 337; of liquids, 371; of solids, 377
krypton, t 118, 130
krypton difluoride, 782

L

lambda-sulfur, 764
lanthanide(s), 130, 141
latex, 658
lattice energy, 178
lauryl alcohol, 667
Lavoisier, Antoine-Laurent, 245, 290, 296
law(s), 8–9. *See also* gas law(s); rate law(s); and names of specific laws
law of conservation of energy, 13, 15, 27
law of conservation of mass, 12, 15, 27, 72–73, **72**, 92, 233–234
law of definite composition, 22–23
law of mass action, 556
law of multiple proportions, 73–74
LCDs. *See* liquid-crystal display(s) (LCDs)
Le Châtelier, Henri Louis, 384
Le Châtelier's principle, 384–386, 394, 508, 569–570, 572, 575, 578, 590
lead nitrate, 250, **250**
lead oxide, **207**
Lewis, Gilbert Newton, 114, 168, 462, 463
Lewis acid(s), 462–463, 486
Lewis acid-base reaction(s), 473
Lewis base(s), 473, t 473, 486
Lewis structure(s), 169–170
light: electromagnetic spectrum of, 98, **98, 99**; interference and, 100, 102; as particles, 99–101, **100, 101**; speed of, 97; as waves, 97–98, 102
lignite, 637
lime, 245, 306
limestone, 136
limiting reactant(s), 272–276
line emission spectrum, 101, 103, **105**
lipid(s), 691–695, **692–695**, t 692, 703
lipoprotein(s), 692
liquid(s): boiling of, 387–388, 394, **387, 388**; capillary action of, 375; changes of state, t 383; density of, 372, **372**; diffusion ability, 373, **373**; dissolving ability of, 372–373; equilibrium vapor pressure of, 385–386, **386**, 394; evaporation and boiling, 375, **375**; fluidity of, 372; incompressibility of, 372 **373**; kinetic-theory description of, 371; particle model of, 371–376, **371**; properties of, 371–376, **372 373, 375**, 394; solutions of, 407, t 407; surface tension of, 374–375, **375**; tendency to solidify, 375; volume of, 372. *See also* fluid(s); water
liquid air, 293
liquid crystal(s), 376
liquid state, 13–14
liquid-crystal display(s) (LCDs), 376
liter(s), 37, **37**
lithium, 115, t 116

Modern Chemistry   **905**

lithium carbonate, 715
litmus, 461, 470, 505, 506
Lockyer, Joseph Norman, 296
London, Fritz, 189
London dispersion forces, 189–190, **190**
lone pair. *See* unshared pair
lye, 724

M

macromolecule(s), 685–686
magnesium, t 117, 136, 244–245, 299, 304–305, 441
magnesium bromide, 204
magnesium chloride, 175, 248, **248**
magnesium nitride, 299
magnesium oxide, 245, **245**
magnetic quantum number(s), 110–111, **110**
magnetism. *See* ferromagnetism
main-group element(s): description of, 139; valence electrons in, 152, t 152
malachite, **26,** 744
malleability, 181
manganese, 118, t 118, 600–601, 612
manganese dioxide, 292, **292,** 601, 603–604
manometer, **325,** 337
marble, 136, 306
Marsden, Ernest, 77
mass: conservation of, 12, 15, 27, 72–73, **72,** 92, 233–234; description of, 11, 27; percent by mass of solutions, 420–421; volume-mass and mass-volume calculations for gases, 357–359
mass action, law of, 556
mass number, 83, 92
mass-mass calculation(s), 262, 270–271
mass-mole calculation(s), 262, 269–270, **269**
matter: changes of state, 14, **14;** chemical and physical changes in, 16–19; classification of, 22; conservation of, *See* mass: conservation of; gaseous state of, 13–14; interconversion of matter and energy, 15; in law of conservation of mass, 12, 72; liquid state of, 13–14; mass and weight of, 11; mixtures of, 19–20; properties of, 10–11, 14–15, 27; pure substances of, 20–23, 27; solid state of, 13–14
measurement: accuracy and precision, 46–47, **46,** 59; calculations involving, 34–36, **35;** derived SI units, 36–39, t 36, **37;** of heat, 43–45, t 43, t 44; SI measurement system, 31–34, **31,** t 32, t 33, 59; significant figures and, 47–52, **47–49,** t 48, t 50; standards of, 31, **31–34;** of temperature, 41–43, **42;** units of measurement, 31–32
medication, 715
Meitner, Lise, 75, 817
melting, 378, 388, 394
melting point(s): of crystals, t 380; description of, 388; of solids, 378
Mendeleev, Dmitri, 127–129, **127, 128,** t 129, 791
mercaptan(s), 772
mercury, 24, 246, **246**
mercury oxide, 563–564
messenger RNA, 698
metabolic pathway(s), 701–703, **701, 702**

metabolism, 701–703
metal(s): atomic radii of, 155; description of, 141; ductility of, 181; electronegativity of, 156; heats of vaporization of, t 181; ionic formation in, 156; ionic radii of, 156; ionization energy of, 155; malleability of, 181; properties of, 24, **24,** 179–181, 190; reaction with acids, 483–484; reaction with chlorine, 784–785; reaction with hydrogen, 296, 310; reaction with nonmetals, 245; reaction of two metals, 245; replacement in compound by more reactive metal, 248; replacement of hydrogen by, 248. *See also* specific types of metals
metal carbonate(s), 247
metal hydroxide(s), 473, t 474
metal oxide(s), 484–485
metallic bond(s): description of, 163, 180; electron sea model of, 180, **180,** 190; formation of, 179–180, **180**
metallic chlorate(s), 247
metallic crystal(s), 380–381
metalloid(s): antimony, 760–761, 762; arsenic, 760, 762; description of, 25, **25,** 141; properties of, 751–753, **751,** t 752, 758, 762; reaction with chlorine, 784–785; silicon, 758–762, **759, 760**
meter(s), t 32, 33
methane, 185–186, **186,** 234–235, **235,** 297, 306, 645, 647, 656
methanol, 309, 667
methyl bromide, 788
methyl group, 647
methylpropane, 645
Millikan, Robert A., 76–77, 79
millimeter(s) of mercury (mm Hg), 326, t 326
mineral(s), 688, t 692
mineral acid(s), 465
miscible substance(s), 415
mixture(s): compared with pure substances, 21; description of, 405, **405;** heterogeneous, 19–20, 27; homogeneous, 20, **20, 21,** 27. *See also* colloid(s); solution(s); suspension(s)
mm Hg. *See* millimeter(s) of mercury (mm Hg)
model(s), construction of, 74
moderator, 820
molal boiling-point constant, t 429, 431
molal freezing-point constant, 429, t 429
molality, 423–425
molar heat of formation, 529
molar heat of fusion, 388
molar heat of vaporization, 387, 394
molar mass(es): of chemical compounds, 216, 217–218, **218;** as conversion factor, **87,** 88–89, **88, 90,** 218–220, **219,** 228; of gases, 348–350, 354–355; of solutes, 431–433
molar volume of a gas, 345–347, **346,** 365
molarity: description of, 421, **422;** formula for, 421–423; and normality, 495–496; and titration, 512–514, 516
mole(s), t 32, 88
mole ratio, 263–264, 278
mole-mass calculation(s), 262, 267–269
mole-mole calculation(s), 262, 266–267, **266**

molecular compound(s): binary, 209–210, **209,** t 210; compared with ionic compounds, 178–179, **178,** 190; description of, 164; hybridization and, 185–186, **186,** t 187, 190; intermolecular forces and, 187–190, **188–190;** naming of, 209–210, **209,** t 210; numerical prefixes for, **209,** t 210; VSEPR theory and, 182–185, **182,** t 183, **184,** 190
molecular electrolyte(s): hydronium ion, 448–449; solution process for, 447–448, **448;** strong and weak electrolytes, 449–450
molecular formula(s): atomic symbols and, 221; calculation of, 226–227; description of, 164
molecular genetics, 699
molecular mass(es), 217
molecular motion, 316
molecular weight, 217
molecule(s), diatomic, 164, 233, t 233
monatomic ion(s), 203–204, t 203, t 204
Monel metal, 781
monoclinic sulfur, 764, **764**
monomer(s), 651
monoprotic acid(s), 463
monosaccharide(s), 689
Moseley, Henry Gwyn-Jeffreys, 129–130
multiple bond(s), 170–173, t 172, 190
multiple proportions, law of, 73–74
mu-sulfur, 764, **765**
mutagen(s), 698
mutation, 698

N

natural gas, 306, 531, 655–656
nematic crystal(s), 376
neon, 115–116, t 116
neoprene, 659
neptunium, 814
net ionic equation(s), 444–445
neuron(s), **694,** 695
neurotransmitter(s), 695, **695**
neutral solution(s), 497
neutralization, 471–472, **471, 472,** 492–493, 516
neutron(s), 75, 80, t 80, 801–804, **803,** 814, 821
Newton, Sir Isaac, 326
Nicholson, William, 240
nickel, 118, t 118, 138, 252, 730, 736–737, t 737
nitration, 654
nitric acid, 463, 465, t 465, 467, 469, 484
nitrogen: binary compounds of, t 210; electron configuration of, t 116; occurrence of, 298; physical properties of, 298, t 298; preparation of, 299; structure and chemical properties of, 298–299; triple bond and, 171; uses of, 140, 298, **298,** 299, 310
nitrogen fixation, 300
nitrogen monoxide, 168
noble gas(es), 116, 130, **130,** 140
noble gas configuration, 116
nomenclature, 206, **207**
nonelectrolyte(s), 406, 439
nonmetal(s): description of, 25–27, **27;** reaction with hydrogen, 296, 310; reaction with metals, 245

906 Index

nonmetal oxide(s), 485
nonpolar-covalent bond(s), 162
nonvolatile substance(s), 426
normality, 494–496, **494,** t 494, 514–516
nuclear binding energy, 801–802, **802**
nuclear disintegration, 804, 812–813, **813**
nuclear equation, 810–811
nuclear fission, 658, 817, 819, **819**
nuclear forces, 80
nuclear fusion, 820–821, **821**
nuclear mass defect, 802
nuclear reaction(s), 804
nuclear reactor(s), 726, 819–820, **820,** 821
nucleic acid(s), 695–700, **696–698,** 703
nucleon(s), 801, 821
nucleotide(s), 695–696, **696, 697**
nucleus. *See* atomic nucleus
nuclide(s), 84
nylon, **677**

O

octane, 201
octane rating(s), 646
octet rule, 168, 190
oil(s), 296, 414, 680–681, 692–693
oil shale, 658
orbital(s), 106, 110, **110,** 121
orbital notation, 113
orbital quantum number(s), 108–109, **109,** t 109
organic acid(s), 465
organic chemistry, 7, 639
organic compound(s): compared with inorganic compounds, 642–643, t 643; description of, 639–640, 660; structural formulas and bonding, 640–642, 660
oxalic acid, 678
oxidation, 597, **597,** t 598, t 608, 622
oxidation number(s): assigning of, 212–214, t 598; formulas and names, 214–216; method for balancing redox equations, 602–605; for nonmetals, t 214; uses of, 213, 228
oxidation states, 213
oxidation-reduction reaction, 598–601, **599, 601,** t 608, **609, 610,** 622
oxide(s): amphoteric, 480–482; basic and acidic, 479–480, 486, t 482; chemical reactions of, 483–485, **484, 485,** 486; description of, 290, 310; synthesis reactions of, 245–246
oxidizing agent(s), 608–610, t 608, t 609, 622
oxyacid(s), 211, t 211, 464–465, t 464, **465**
oxyanion(s), 208, 211
oxygen: binary compounds of, t 210; chemical properties of, 290–291, **291;** electron configuration of, 115, t 116; history of, 290; mass of, 216, 345; molecular structure of, 289–290; occurrence of, 289; oxidation number of, 213; physical properties of, 290, t 290; preparation of, 246, **246,** 292–293, **292, 293;** reactions of elements with, 244–245, **245;** uses of, 140, 289, 291–292, **292,** 310
ozone: formation of, 293–294; importance of, 289, 310; nitrogen oxides and, 299; structure and properties of, 293, **294**

P

palladium, 138
paraffin wax, 644
paramagnetism, 111
parent nuclide(s), 810, **810**
partial pressure, 335
particle model: of liquids, 371–376, **371;** of solids, 377–379, **377**
pascal, 326, t 326
Pascal, Blaise, 326
Pauli exclusion principle, 113
Pauling, Linus, 153
PCBs, 640
peat, 636–637
peptide bond(s), 687
percent by mass, 420–421
percent composition, 222–223
percent yield, 276–278, **276**
perchloric acid, 463, 477
Percy, Marguerite, 137
period(s), 23, 132–133
periodic law, 129–130, 791
Periodic Table of Atomic Radii, 143, **144**
Periodic Table of Electron Affinities, **149**
Periodic Table of Elements: atomic numbers and, 83; chart of, **24, 134–135;** description of 23–24, 27; design of, 131; desktop investigation, 131; electron configurations and, 132–142, **132–136,** t 132, **138, 139,** t 140, **141;** history of, 127–130, **128,** t 129
Periodic Table of Ionic Radii, **151**
Periodic Table of Ionization Energy, 146
Periodic Table of the Electronegativities, **153**
periodicity, 130–131, **130**
permanganate, 609
peroxide(s), 290, 609
pesticide(s), 640
petroleum, 656–658, t 656, **657**
petroleum coke, 638
petroleum substitute(s), 658
pH: calculations involving, 500–504; of common materials, t 500; equation for, 499, 516; indicators and, 505–506, t 505, **506–508,** 508, 510, **510,** 516; relationship between hydronium ion and hydroxide ion concentrations and, 504, t 504; scale, 497–499 **499, 500;** transition interval and, 505–506, t 505
phase(s), 20, 383
phase diagram(s), 389–390, **389**
phenyl group, 654
phospholipid(s), 693
phosphoric acid, 465, t 465, 466–467, 469
phosphorous acid, 465
phosphorus, 26–27, **27,** t 117, 140, 245
photochemical smog, 299
photoelectric cell(s), 136
photoelectric effect, 99–100, 136
photon(s), 100, 121
photosensitive, 788
photosynthesis, 268, 303
physical change(s), 14, 18
physical chemistry, 7
physical properties, 14, 27
pig iron, 739
Planck, Max, 100
Planck's constant, 100

platinum, 138–139
plutonium, 814
polar bond(s), 162
polar-covalent bond(s), 162–163
pollution. *See* acid rain
polonium, 804–805
polyatomic ion(s), 173, 208–209, t 208
polyethylene(s), 651
polymer(s), 651
polymerization, 651
polyprotic acid(s), 464
polysaccharide(s), 689–690
polyunsaturated fat(s), 692
polyvinyl chloride, 784, 785
potassium, 136, 252, 621, 791
potassium bromide, 248
potassium chlorate, 217–218, 222, 247, 293
potassium chloride, 247, 444
potassium fluoride, 248
potassium hydroxide, 247, 471, 474
potassium iodide, 250, **250,** 790
potassium nitrate, 298, 444
potential energy, 12
precipitate(s), 16, **17,** 250, **250,** 575
precipitation, 444, 589
precision, 46–47, **46,** 59
pressure: Boyle's law, 327–329, **327, 328,** t 328, 337; combined gas law, 333–334; critical pressure, 389; Dalton's law of partial pressures, 334–336, **335;** equilibrium and changes in, 569–572, **571, 572;** and force, 323–324; Gay-Lussac's law, 332–333, **332,** 337; ideal gas law and, 352–354; measurement of, 324–325, **235;** partial pressure, 335; phase diagrams and, 389–390; solubility and, 415–416; standard temperature and pressure, 327; units of pressure, 325–326, t 326
Priestley, Joseph, 246, 290
primary standard(s), 512
principal quantum number(s), 108, t 109
product(s), in chemical equations, 15, 233, 688
propane, **357,** 645
proportional relationship(s), 58–59, **58**
protein(s), 686–688, t 687, **687, 688,** 698, 703
protium, 80, 83, t 84
proton(s), 75, 79, t 80, 801–804, **803,** 814, 821
Proust, Joseph Louis, 23, 222
pure substance(s), 20, 21–23, 27
PVC pipe(s), 784
pyrite, 765

Q

qualitative information, 8
quantitative information, 8
quantitative problem(s), 55–56, 59
quantum, 100
quantum number(s), 108–111, **109,** t 109, 121
quantum theory, 104–106, **106,** 121
quartz, 380
quicklime, 245, 306

R

radiation. *See* electromagnetic radiation
radiation exposure, 818

Modern Chemistry **907**

radioactive dating, 816
radioactive decay, 804, 807, 812
radioactive nuclide(s): artificially induced, 812–814, **813,** t 815; naturally occurring, 806–812, **807,** t 807, **809–811,** t 809; properties of, t 805, 806–808, 821; radiation given off by, 809, **809,** t 809, 821; series of, 809–812, **810;** stable nuclei from radioactive decay, 812
radioactive tracer(s), 817
radioactivity, 804, 821
radioisotope(s), 816–817
radium, 136, 804–806
radon, 130, 811, **811**
Ramsay, William, 130
Rankine Scale, 42
rare-earth element(s), 130
rate law(s), 552–557, **553**
rate-determining step, 551
reactant(s), in chemical equations, 15, 233
reaction(s). *See* chemical reaction(s)
reaction mechanism(s), 541–543, **542, 543,** 557
reaction rate: desktop investigation, 555; factors influencing, 549–552, **550, 552, 553,** 555, 557; rate laws, 552–557, **553**
reaction stoichiometry: mass-mass calculations, 262, 270–271; mass-mole calculations, 262, 269–270, **269;** mole-mass calculations, 262, 267–269; mole-mole calculations, 262, 266–267, **266;** problem-solving techniques, 261–265, **261–263**
real gas(es), 319, **319**
rechargeable cell(s), 615–616, **615, 616**
recipe(s), 273, **431**
redox equation(s): ion-electron method of balancing, 605–607; oxidation-number method of balancing, 602–605, **603**
redox reaction(s), 600–601
reducing agent(s), 608–610, t 608, t 609, 622
reduction, 597–598, t 608, 622
relative atomic mass, 85, 92
replacement reaction(s), 247–248, **248,** 250, **250**
resonance, 174
resonance hybrid(s), 174
respiration, in carbon dioxide preparation, 306
reversible reaction(s), 237, 563–564, **563**
rhombic sulfur, 764, **764**
ribonucleic acid (RNA), 695, 698, 703
ribosome(s), 698
RNA. *See* ribonucleic acid (RNA)
roasting, 770–771
rock salt, 722
rocket fuel, 254
roentgen, 818
rotation, 316
rounding off, in calculations, 49–50, t 50
rubber, 658–659, **658, 659**
ruby, 757, **757**
Rutherford, Ernest, 77–78, 92, 97, 812–813

S

salt. *See* sodium chloride
salt(s): description of, 211, 472, **472;** hydrolysis of, 580–583, **581–583,** 590
saltwater distillation, 428
saponification, 680
sapphire, 757, **757**
saturated fat(s), 692
saturated hydrocarbon(s), 644
saturated solution(s), 412
SBR, 659
scandium, 117, t 118
Scheele, Carl Wilhelm, 290, 783
Schrödinger, Erwin, 105
Schrödinger wave equation, 105
scientific journals, 264
scientific measurement. *See* measurement
scientific method, 7–10, **7–9,** 27
scientific notation: form of, 52–53; mathematics involving, 54; purpose of, 52, 59
sea water, 136, 428
second(s), t 32, 34
selenium, 752, t 752
self-ionization, of water, 496–497, **496,** 516
semiconductor(s), 25
SI measurement system: derived SI units, 36–39, t 36; description of, 31–32, 59; fundamental SI units, 32–34, t 32, **33,** t 33, **34**
significant figure(s), 47–52, **47–49,** t 48, t 50, 59
silicon, 25, t 117, 758–762, **759, 760**
silicon carbide, 380
silicone(s), 759
silver, t 119, 138–139, 744, t 744
silver bromide, 788
silver chloride, **17,** 444, **445,** 449, 575, 584–586, 599
silver halide(s), 789
silver iodide, 790
silver nitrate, **17,** 208, 252–253, 444, **445,** 575, 599
silver nitrate, 208
single bond(s), 169, t 172
single-replacement reaction(s), 247–248, **248,** 255
slag, 739
smectic crystal(s), 376
smog. *See* photochemical smog
soap(s), 680–681
soda ash, 725
sodium: compounds of, 722–726, **722,** t 723; **725;** electron configuration of, 115–116, t 117; formation of metallic bonds, 180; preparation of, 720–721, **720;** properties and uses of, 136, 721–722, **722**
sodium acetate, 413, 576
sodium carbonate, 485, 582, 724–726, **725**
sodium chloride: common-ion effect and, 575–576; description of, 722–723, **722;** dissolving of, 440–441, **440;** as electrolyte, 406; equilibrium and, 564; hydration of, 442; hydrochloric acid and, 211; as ionic compound, 175–177, **177, 205;** oxidation and, 597; oxidation-reduction reaction and, 599; solubility of, **409,** 413, 415; solutions of, **445**
sodium fluoride, 212, 245, 248, 782
sodium hydride, 296
sodium hydroxide, **205,** 247, 248, 420, 421, 471, 473, 485, 723–724
sodium hypochlorite, 254
sodium iodide, 248
sodium nitrate, 298
sodium pentothal, 715
sodium peroxide, 290, 293
sodium thiosulfate, 413
solid(s): amorphous, 377, 381–382, 394; changes of state, t 383; crystalline, 377; density of, 379; diffusion rate of, 379; incompressibility of, 379; kinetic theory description of, 377; melting point of, 378–379; nonfluidity of, 378; particle model of, 377–379, **377;** properties of, 377–379, 394; shape of, 377; solutions of, 407, t 407; volume of, 378
solid state, 13–14
solubility: calculations of, 587–589; description of, 411–413, 433; factors affecting, 413–417, **414–417,** t 417, 433; equilibrium of, 443–446, **444, 445,** 584–590, **585;** solubility-product constant, 585–587, t 588, 590
solubility equilibria, 443–446, **444, 445**
solubility-product constant, 585–587, t 588, 590
soluble, 405
solute(s), 406, 413–414, 431–433
solution(s): aqueous, 496–497; boiling-point elevation, 431, **431;** buffered solutions, 579; colligative properties of, 426–433, **426, 427, 431;** concentration of, 420–425; description of, 20, **20, 21,** 405–406; desktop investigation, 408; electrolyte, 447–448, 450–456, t 455, t 456; equivalents, 491–493, **491,** 516; factors affecting rate of dissolving, 410–411; freezing-point depression, 428–430; formation of, **418;** heats of, 417–419, t 419, 441–442, **442;** molality of, 423–425; molar mass of solute, 431–433; molarity of, 421–423, **422,** 495–496; molecular electrolyte(s), 447–448, **448;** nature of, 406, **406;** neutral, 497; normality, 494–496, **494,** t 494, 516; percent by mass, 420–421; and pH, 496–504; properties of, t 410; saturated, 412; solubility and, 411–417, **411–417,** t 417, 433; solution equilibrium, 412; standard solutions, 511, 516; supersaturated, 412–413; types of, 407, t 407; unsaturated, 412; vapor-pressure lowering, 426–427, **427**
solution equilibrium, 412
solvated, 418
Solvay, Ernest, 724
Solvay process, 724–726, **725**
solvent(s), 406, 413–414
specific heat, 44–45, t 44
spectator ion(s), 445
spectroscope, 103–104
spectroscopy, 103–104, **105**
spectrum, 104. *See also* electromagnetic spectrum; line emission spectrum
speed: of electromagnetic radiation, 97; formula for, 59, **59**
spin quantum number(s), 111
standard(s), of measurement, 31
standard electrode potential, 618–620, t 620, 622
standard molar volume of a gas, 345–347, **346,** 365

standard solution, 511, 516
standard temperature and pressure (STP), 327, 337
starch, 689–690
steel production, 291, **292,** 739–741, **740**
steroid(s), 693
Stock system of nomenclature, 206, 214–216
stoichiometry: applications of, 261; of gases, 356–359; ideal conditions and, 265–266; limiting reactants and, 272–276; mass-mass calculations, 262, 270–271; mass-mole calculations, 262, 269–270, **269**; mole-mass calculations, 262, 267–269; mole-mole calculations, 262, 266–267, **266**; percent yield, 276–278, **276**; reaction-stoichiometry problems, 261–265, **261–263**
STP. *See* standard temperature and pressure (STP)
strong acid(s), 463, t 466
strong electrolyte(s), 449
strontium, 136–137, 719
structural formula(s), 169, 640–642, **641, 642**
structural isomer(s), 641
Strutt, John William, 130
subatomic particle(s). *See* electron(s); neutron(s); proton(s)
subbituminous coal, 637
sublimation, 388–389
substituted hydrocarbon(s): alcohols, 665–669, 681; aldehydes, 674–675, 681; carboxylic acids, 676–678, 681; description of, 639, 681; esters, 679, 681; ethers, 673, 681; halocarbons, 670–672, 681; ketones, 674–676, 681
substrate(s), 688
sugar(s): disaccharides and polysaccharides, 689–690, **689**; heating of, 22, **22**; simple sugars, 689; solution with water, 20, **21,** 405, **406,** 410–412, **411,** 414, 564
sulfur: allotropes of, 765–767, **764, 765,** 775; chemical properties of, 767–768, t 767, **768**; compounds of, 770–775, t 770, **771, 773, 774**; electron configuration of, t 117; mixed with zinc, **17**; physical properties of, **764,** 765–767, **765,** 775, production of, 765, **765,** 775; synthesis reactions with, 245; uses of, 140, 769
sulfur dioxide, 245, 291, 770–771, **771,** 775
sulfur trioxide, 174, 247
sulfuric acid, 246, 247, 297, 464–468, t 465, **468,** 483–484, 773–775
superconductor(s), 382
supercooled liquid(s), 378–379
superoxide(s), 291
supersaturated solution(s), 412–413
surface tension, 374–375, **375**
suspension(s), 405, 408–409, t 410
swimming pool chemistry, 570
synapse, **694,** 695
synthesis reaction(s), 244–246, **245,** 255
system(s), 383

T

taconite, 738
Teflon, 672
tellurium, 752–753, t 752
temperature: absolute zero, 330; Charles' law, 329–332, **329, 330,** t 330, t 331, 337; combined gas law, 333–334; critical pressure, 389; description of, 40, **41,** 59; equilibrium and changes in, 572–573; Gay-Lussac's law, 332–333, **332,** 337; ideal gas law and, 352–354; low temperatures, 322; measurement of, 41–43, **47**; and molecular motion, 316; phase diagrams and, 389–390, **389**; solubility and, 416–417, t 417. *See also* boiling; freezing; melting
tetrachloromethane, 671
tetrafluoroethene, 671–672
theoretical yield, 276
theory, 9
thermite, 755
thermite reaction, 755–756, **756,** 762
thermochemical equation(s), 527
thermochemistry: description of, 527; heat of combustion, 530–531, **530,** t 530, 557; heat of formation, 528–529, **529,** 556–557; heat of reaction, 527–528, **527, 528,** 531–536, **532,** 556, 557
thermometer(s), 41, **47,** 322. *See also* temperature
Thomson, Sir John, 76, 77
tin, t 119
titanium, 117, t 118
titration, 510–516, **511**
torr, 326, t 326
Torricelli, Evangelista, 324–325, 326
transition element(s), 138, 729–731
transition interval, 505–506, t 505
transition metal(s): formation of colored compounds, 733, **733,** 748; formation of complex ions, 733–736, **734,** t 735, t 736, 748; oxidation states of, 731–732, t 732, 748; properties of, 729–731, **729,** 748
translation, 316
transmutation, 810
transmutation reaction, 810
transuranium element(s), 814, t 815
triple bond(s), 171, 190
triple point(s), 389
triprotic acid(s), 464
tritium, 80, 83, t 84, 821, **821**
tungsten, 359
Tyndall effect, 409

U

uncertainty principle, 106–107
unit(s) of measurement: calculations involving, 34–36; derived SI units, 36–39, t 36; SI measurement system, 31–34, t 32, t 33, 59; symbols for, 31–32
unit cell(s), 379
unsaturated, 650
unsaturated fat(s), 692
unsaturated solution(s), 412
unshared pair, 169
uranium, 245, 817, 819, **819**
uranium hexafluoride, 363
uranium-235, 363

V

valence electron(s), 152
van der Waals, Johannes, 319
vanadium, 117, t 118
vaporization, 375, 388, 394
variable(s), 58
vibration, 316
vinyl group, 653
vitamin(s), 688
voltaic cell(s), 616–617
volume: derived SI units and, 36–37, t 36, **37**; ideal gas law and, 352–354; volume-mass and mass-volume calculations for gases, 357–359; volume-volume calculations of gases, 356–357
VSEPR theory, 182–185, **182,** t 183, **184,** 190
vulcanization, 658

W

Waage, Peter, 556
Walton, E. T. S., 813–814
water: boiling point of, 387, **388**; distillation of, 428; electrolysis of, 21, **21,** 613–614; equilibrium vapor pressure of, 385–386, **386**; formation of, 250; formula for, 202, **202**; fresh water from the sea, 428; in hydrogen preparation, 298; ionization constant of, 579–580; ionization of, 452; mass of, 216–217; oxidation number of, 212; in oxygen preparation, 293, **293**; phase diagrams of, 389–390, **389**; physical properties of, **52,** 392–393, **393**; reaction with chlorine, 785; self-ionization of, 496–497, **496,** 516; structure of, 184–185, **184, 186,** 390–391, **391,** 394; uses of, 685, **685**; water-gas reaction, 535–536, **535**
wave properties: of electromagnetic radiation, 97; of electrons, 105–106, **106**; of light, 97–98, 102
wavelength, 97, 98
wax, 692, **693**
weak acid(s), 464–465, t 464
weak base(s), 473–474
weak electrolyte(s), 449
weight, 11, 27
Williamson synthesis, 673
word equation(s), 234

X

xenon, t 119, 130, 782

Z

zero(s): absolute zero, 330; significant figures involving, t 48
zinc: activity series and, 252; electrode potentials of, 616–620; electron configuration of, 118–119, t 118; as oxidizing agent, 608–609; preparation of hydrogen 297; reaction with copper, 252, **253**; reaction with hydrochloric acid, 549–550
zinc chloride, 611
zinc sulfate, 252

Modern Chemistry **909**

Photo Credits

Abbreviations used: (t)top; (c)center; (b)bottom; (l)left; (r)right.

Title Page: Background, Manfred Kage/Peter Arnold, Inc.; foreground, Paul Silverman/Fundamental Photographs.

Copyright Page: Page T2, HRW Photo by Dennis Fagan.

Table of Contents: T4(b) Runk/Schoenberger/Grant Heilman; T4(m), HRW Photo by Dennis Fagan; T4(t), FPG International; T4(tl), Barry L. Runk/Grant Heilman; T4(tr), Robert Brenner/Photo Edit; T5(b), FPG International; T5(t), CNRI/Science Photo Library/Photo Researchers, Inc.; T6(b), HRW Photo by Dennis Fagan; T6(l), HRW Photo by Dennis Fagan; T6(r), Luettge/Zefa/H. Armstrong Roberts; T6(t), Brian Parker/Tom Stack & Associates; T7(b), Richard Megna/Fundamental Photographs; T7(r), N.A.S.A.; T8(bl), HRW Photo/Gemological Institute of America; T8(br), HRW Photo by Dennis Fagan; T8(tl), HRW Photo by Russell Dian; T8(tm), Dr. Harold E. Edgerton/M.I.T.; T8(tr), Jeff Rotman; T9(b), from the CHEM Study Film: Chemical Bonding; T9(br), HRW Photo by Russell Dian; T9(m), SuperStock; T9(tr), Lawrence Livermore National Laboratory; T10(bm), SuperStock; T10(r), John Shaw; T10(tl), HRW Photo by Dennis Fagan; T11(b), Ken Cooper/The Image Bank; T11(r), HRW Photo by Dennis Fagan; T11(tl), HRW Photo by Russell Dian; T12(bl), L.V. Bergman & Associates; T12(br), David R. Frazier Photolibrary; T12(c), HRW Photo/Gemological Institute of America; T12(l), Ralph P. Earlandson; T12(tr), Bill Longcore/Photo Researchers, Inc.; T13(b), HRW Photo by Ken Karp; T13(l), HRW Photo by Dennis Fagan; T13(r), N.A.S.A.

Front Matter: T14, HRW Photo by Dennis Fagan; T14-T59(border), Peter Arnold, Inc.; T15(br), B.M. Shaub; T15(l), N.A.S.A./Goddard Space Center; T15(tr), FPG International; T16(l), Paul Silverman/Fundamental Photographs; T17, HRW Photo by Russell Dian; T18, Lawrence Livermore Laboratory/Science Photo Library/Custom Medical Stock Photo; T21, Lisa Davis Photojournalist; T22(br), Paul Shambroom/Photo Researchers, Inc.; T22(l), Courtesy, Mettler Instruments Corp.; T22(m), T22(tr), Barry L. Runk/Grant Heilman; T23, HRW Photo/Bethlehem/Grant Heilman; T25, Phillip Habib/Tony Stone Worldwide; T25(l), HRW Photo by Dennis Fagan; T27, Peter Berndt, M.D., P.A./Custom Medical Stock Photo; T28, Ted Horowitz/The Stock Market; T30, HRW Photo by Dennis Fagan; T37, HRW Photo by Russell Dian; T39, HRW Photo by Russell Dian; T45, Joe DeMaggio/The Stock Market.

Unit 1: Page 2, Courtesy of Bonnie F. McBride; 3, N.A.S.A.

Chapter 1: Page 4, Photri, Inc.; 5, Visuals Unlimited; 6(tl), Mark Godfrey/Pfizer, Inc.; 6(cl), Steve Elmore/Tom Stack & Associates; 6(cr), Yoav Levy/Phototake; 7, Hans Namuth/Photo Researchers, Inc.; 8(l), Lawrence Migdale/Photo Researchers, Inc.; 8(r), Stanley Fiegler/Visuals Unlimited; 11(l), HRW Photo by Steve McCarroll; 11(t), 11(c), HRW Photo by Dennis Fagan; 15, Jan Halaska/Photo Researchers, Inc.; 16(tl), Yoav Levy/Phototake; 16(bl), L.S. Stepanowicz/Panographics; 16(tcl), Yoav Levy/Phototake; 16(tc), Dr. E.R. Degginger; 16(tr), Barry L. Runk/Grant Heilman; 17(l), HRW Photo by Richard Haynes; 17(c), HRW Photo by Yoav Levy; 17(r), L.S. Stepanowicz/Panographics; 20(l), Doug Sokell/Tom Stack & Associates; 20(b), HRW Photo by Dennis Fagan; 21(t), HRW Photo by Dennis Fagan; 21(b), Runk/Schoenberger/Grant Heilman; 22, Dr. E.R. Degginger; 24, Leonard Lessen/Peter Arnold, Inc.; 25(l), Dr. E.R. Degginger; 25(r), Runk/Schoenberger/Grant Heilman; 26(l), Brian Parker/Tom Stack & Associates; 26(r), Brian Parker/Tom Stack & Associates; 27, Rich Treptow/Visuals Unlimited.

Chapter 2: Page 30, Philip Habib/Tony Stone Worldwide; 31, U.S. Department of Commerce; 34, HRW Photo by Dennis Fagan; 35, HRW Photo by John Langford; 37, 38, 46, HRW Photos by Dennis Fagan; 48, Photo courtesy of Mettler Instrument Corporation; 49, HRW Photo by Dennis Fagan; 64(t), Tom Pantages; 64(b), Yoav Levy/Phototake; 66(l), Barry L. Runk/Grant Heilman Photography; 66(r), Terry Wild Studio; 67(l), Stephen J. Krasemann/DRK Photo; 67(c), Robert Holland/DRK Photo; 67(b), Biophoto Associates/Science Source/Photo Researchers.

Unit 2: Page 68, HRW Photo by Robert Severi/Woodfin Camp; 69, Lawrence Livermore National Laboratories/Peter Arnold, Inc.

Chapter 3: Page 70, I.B.M. Research; 71, Tony Freeman/PhotoEdit; 81(l), M. Ohtsuki/Enrico Fermi Institute/University of Chicago; 81(r), IBM Watson/Peter Arnold, Inc.; 88, HRW Photo by Dennis Fagan.

Chapter 4: Page 96, Bill Longcore/Photo Researchers, Inc.; 98, L.V. Bergman & Associates; 100, from the CHEM Study Film: Chemical Bonding; 105, Sargent-Welch Scientific Company; 106, Margot Bergmann/Polytechnic Institute of Brooklyn.

Chapter 5: Page 126, © 1952 M.C. Escher/Cordon Art, Baarn, Holland; 127, Patrick Watson, Medichrome; 136, 138, 139(l), 139(c), Dr. E.R. Degginger; 139(r), L.S. Stepanowicz/Panographics; 141, L.V. Bergman & Associates.

Chapter 6: Page 160, Irving Geis/Photo Researchers, Inc.; 175, HRW Photo by Dennis Fagan; 178, L.V. Bergman & Associates; 181, HRW Photo by Russell Dian; 189(l), John Shaw; 189(r), A. Pasieka/Taurus Photos; 194(t), Lawrence Migdale; 194(b), SuperStock; 196(t), Runk/Schoenberger/Grant Heilman; 196(b), David Wolberg; 197(l), Jim Pickerell/FPG; 197(r), SIU/Photo Researchers, Inc.

Unit 3: Page 198, Courtesy Robert A. Cantini; 199, David R. Frazier Photolibrary.

Chapter 7: Page 200, Chemical Design, Ltd./Science Photo Library/Photo Researchers, Inc.; 202, HRW Photo by John King; 207, L.S. Stepanowicz/Panographics; 211, Dr. E.R. Degginger; 216, HRW Photo by Dennis Fagan; 218, HRW Photo by John King; 221, Lawrence Livermore National Laboratory.

Chapter 8: Page 232, Paul Silverman/Fundamental Photographs; 240, HRW Photo by Dennis Fagan; 241, 245, L.S. Stepanowicz/Panographics; 246, Dr. E.R. Degginger; 248, L.S. Stepanowicz/Panographics; 250(t), HRW Photo by Dennis Fagan; 250(b), HRW Photo by Rodney Jones Studios; 253, HRW Photo by Dennis Fagan; 254, N.A.S.A.

Chapter 9: Page 260, SuperStock; 261, HRW Photo by Dennis Fagan; 263, FPG; 264, HRW Photo by Dennis Fagan; 266, N.A.S.A.; 269, HRW Photo by Russell Dian; 270, Michal Heron/Woodfin Camp & Associates, Inc.; 272, Marvin E. Collins/Fran Heyl Associates; 276, Paul Shambroom/Photo Researchers; 282, David R. Frazier; 284(t), HRW Photo by Dennis Fagan; 284(b), Tom Bean/DRK Photo; 285(t), HRW Photo by Eric Beggs; 285(b), HRW Photo by John Langford.

Unit 4: Page 286, Courtesy of Susan Solomon; 287, David J. Hofman/University of Wyoming.

Chapter 10: Page 288, N.A.S.A. Goddard Space Center; 289, 291(t), HRW Photo by Dennis Fagan; 291(b), Hans Reinhard/Bruce Coleman, Inc.; 292(l), Mark Sherman/Bruce Coleman, Inc.; 292(cl), HRW Photo by John Langford; 292(rc), William E. Ferguson; 292(r), Glenn Short/Bruce Coleman, Inc.; 295(t), N.A.S.A.; 295(b), HRW Photo by Dennis Fagan; 297, B.M. Shaub; 298(t), Doug Wilson; 298(b), Yoav Levy/Phototake; 299, N.A.S.A.; 302, 304, HRW Photo by Dennis Fagan; 305, Carl Ogust/The Image Works; 307, HRW Photo by Dennis Fagan; 308, HRW Photo by Ken Lax; 309, Mark Miller/Southern Light.

Chapter 11: Page 314, Telegraph Colour Library/FPG International; 318, Charles D. Winters; 322, Oxford Instruments; 323, HRW Photo by Dennis Fagan; 329, Susan Van Etten/The Stock Shop.

Chapter 12: Page 342, Paul Silverman/Fundamental Photographs; 343, Daniel MacDonald/The Stock Shop; 347, HRW Photo by Dennis Fagan; 357, Hanson Carroll/Peter Arnold, Inc.; 359(l), Rich Treptow/Photo Researchers; 359(r), Strategic Minerals Corp.; 363, U.S. Department of Energy/Oak Ridge, Tennessee; 364, HRW Photo by Dennis Fagan.

Chapter 13: Page 370, Greg Pease/Tony Stone Worldwide; 372, HRW Photo by Russell Dian; 373, HRW Photo by Dennis Fagan; 375(t), Dr. Harold E. Edgerton/M.I.T.; 375(b), Dr. E.R. Degginger; 376, Caroline Brown/Fran Heyl Associates; 378(l), Runk/Schoenberger/Grant Heilman; 378(tr), HRW Photo by Dennis Fagan; 378(br), Arthur Singer/American Museum of Natural History; 381(l), Photri; 381(r), Breck P. Kent; 382, Schoning/Journalism Services; 398(l), HRW Photo by Dennis Fagan; 398(b), HRW Photo by Eric Beggs; 400(t), Yoav Levy/Phototake; 400(b), David R. Frazier; 401(t), Kelly Huffey/DRK Photo; 401(b), National Science Foundation.

Unit 5: Page 402, HRW Photo by Joe Traver; 403, Jeff Rotman.

Chapter 14: Page 404, SuperStock; 405(l), David R. Frazier; 405(lc), Dr. C.W. Dill/Texas A&M University; 405(rc), Dr. E.R. Degginger; 405(r), L.S. Stepanowicz/Panographics; 406, HRW Photo by Dennis Fagan; 407, Dr. E.R. Degginger; 409, HRW Photo by Richard Haynes; 411, HRW Photo by Dennis Fagan; 412, HRW Photo by Richard Haynes; 413, HRW Photo by John King; 416, HRW Photo by Dennis Fagan; 418(l), 418(r), HRW Photo by Richard Haynes; 422, HRW Photo by Dennis Fagan; 426, Tom Myers; 427, HRW Photo by Dennis Fagan; 428, HRW Photo by Paul Markow/Markow Southwest, Inc.; 431, HRW Photo by Dennis Fagan.

Chapter 15: Page 438, Paul Silverman/Fundamental Photographs; 441, 445, HRW Photo by Dennis Fagan; 451, HRW Photo by John Langford; 452, HRW Photos by Dennis Fagan.

Chapter 16: Page 460, SuperStock; 461, HRW Photo by Dennis Fagan; 462, HRW Photo by John Langford; 463, Yoav Levy/Phototake; 468, HRW Photo by Dennis Fagan; 470(tl), RED DEVIL® is a registered trademark of Boyle-Midway Household Products Inc./HRW Photo by Dennis Fagan; 470(b), 471, HRW Photo by Yoav Levy; 472(t), Dr. E.R. Degginger; 472(b), HRW Photo by Russell Dian; 474, HRW Photo by Yoav Levy; 478, HRW Photo by Dennis Fagan; 481(l), HRW Photo by Dennis Fagan; 481(r), HRW Photo by John Langford; 484(t), HRW Photo by Dennis Fagan; 484(b), HRW Photo by Dennis Fagan; 485(l), Gary Milburn/Tom Stack & Associates; 485(b), HRW Photo by Dennis Fagan.

Chapter 17: Page 490, Richard Megna/Fundamental Photographs; 494, 499, HRW Photo by Dennis Fagan; 504, Photo Researchers, Inc.; 506, HRW Photo by John King; 507, 508, 510(t), HRW Photo by Dennis Fagan; 510(b), Corning Glass Works; 512, Chromatics Ltd.; 514, HRW Photo by Dennis Fagan; 520(l), David R. Frazier; 520(r), HRW Photo by Russell Dian; 522(tl), Whitney Lane/The Image Bank; 522(tr), HRW Photo by Eric Beggs; 522(b), 522(br), Tom Myers; 523(t), David R. Frazier; 523(b), The Image Works.

Unit 6: Page 524, Courtesy Yuan T. Lee; 525, Yoav Levy/Phototake.

Chapter 18: Page 526, Erwin & Peggy Bauer/Bruce Coleman, Inc; 529, L.S. Stepanowicz/Panographics; 530, Dr. E.R. Degginger; 531, Billy Hustace/The Stock Shop; 537, K. Burke/Texaco Development Corp.; 550, Yoav Levy/Phototake; 552(l), Dr. E.R. Degginger; 552(r), HRW Photo by Eric Beggs.

Chapter 19: Page 562, David Madison; 567, Dr. E.R. Degginger; 570, HRW Photo by John Langford; 572, HRW Photo by Dennis Fagan; 574, Dr. E.R. Degginger; 575, HRW Photo by Dennis Fagan; 579, 581(t), HRW Photo by Dennis Fagan; 581(b), Sidney Moulds (SPL)/Science Source/Photo Researchers; 582, Michael Siegel/Phototake; 583, Dr. E.R. Degginger; 589, HRW Photo by John King.

Chapter 20: Page 596, Runk Schoenberger/Grant Heilman Photography; 597, Yoav Levy/Phototake; 599(t), HRW Photo by Yoav Levy; 599(b), HRW Photo by Dr. E.R. Degginger; 601(t), Richard L. Carlton/Photo Researchers; 601(b), HRW Photo by Dr. E.R. Degginger; 603, HRW Photo by Dennis Fagan; 609, Dr. E.R. Degginger; 610(t), HRW Photo by Dennis Fagan; 610(b), L.V. Bergman & Associates; 611, Robert J. Witkowski/Bruce Coleman, Inc.; 612, HRW Photo by Dennis Fagan; 614, PPG Industries, Inc.; 615, Allan Kaye/DRK Photo; 616, Gary C. Bublitz/Bublitz Photography; 621, HRW Photo by John Langford; 626(t), Mitch Kezar/Phototake; 626(b), David A. Wagner; 628(t), Yoav Levy/Phototake; 628(b), HRW Photo by Dennis Fagan; 629(t), Ken Lucas/Biological Photo Service; 629(b), Al Grotell.

Unit 7: Page 630, HRW Photo by Tom Collins; 631, G.I. Bernard/Oxford Scientific Films/Animals, Animals.

Chapter 21: Page 632, J. Bernholc, Q.-M. Zhang, J.-Y. Yi, C. Brabec, and Palmer, N.C. State University/N.C. Superconducting Center; 634, Smithsonian Institution; 636, Daniel MacDonald/The Stock Shop; 637(t), Painted by Rudolph F. Zallinger, Courtesy of the Peabody Museum of Natural History, Yale University; 637(b), Chris Bryan/Robert Harding Associates; 640, O. Franken/Sygma; 646, David R. Frazier; 652, HRW Photo by Dennis Fagan; 657, Erich Hartmann/Magnum Photos; 658, Marcus Brooke/FPG; 659(t), Larry Brownstein/The Stock Shop; 659(b), David R. Frazier Photolibrary.

Chapter 22: Page 664, Gary Gladstone/The Image Bank; 666, Mitch Kezar/Phototake; 667(l), Archer Daniels Midland; 667(r), Dennis Brack/Black Star; 668, HRW Photo by Dennis Fagan; 669(t), David R. Frazier; 669(b), HRW Photo by Dennis Fagan; 670, G.J. James/Biological Photo Service; 674, HRW Photo by Dennis Fagan; 675(t), Barry L. Runk/Grant Heilman Photography; 675(c), Grant Heilman Photography; 675(r), HRW Photo by Dennis Fagan; 677, Charles D. Winters; 678, Hoechst Celanese Corporation; 679, Nino Mascardi/The Image Bank; 680, E.I. DuPont de Nemours & Co., Inc.

Chapter 23: Page 684, Lawrence Livermore Laboratory/Science Photo Library/Custom Medical Stock Photos; 685, Ed Reschke; 691, Grant Heilman Photography; 693(t), Dr. Jeremy Burgess/Science Photo Library/Photo Researchers; 693(b), The National Heart, Lung and Blood Institute; 698, David M. Phillips/Visuals Unlimited; 699, R.L. Brinster and R.E. Hammer/School of Veterinary Medicine/University of Pennsylvania; 706(c), Alvis Upitis/The Image Bank; 706(c), Sepp Seitz/Woodfin Camp & Associates, Inc.; 706(b), Lawrence Migdale; 706(b), Brookhaven National Laboratory; 708(c), The Image Works; 708(b), Runk/Schoenberger/Grant Heilman Photography; 709(t), E.I. DuPont de Nemours & Co., Inc.; 709(r), Steven Lindow/University of California at Berkeley; 709(c), David R. Frazier Photolibrary; 709(r), Runk/Schoenberger/Grant Heilman Photography.

Unit 8: Page 710, Courtesy William J. Janhunen, Jr.; 711, David Muench.

Chapter 24: Page 712, Richard Megna/Fundamental Photographs; 715, David York/Medichrome; 717, Christopher Springman/The Stock Market; 719(l), Orin Cassill/The Stock Shop; 719(r), Phototake; 722(t), L.V. Bergman & Associates; 722(b), Bob and Clara Calhoun/Bruce Coleman, Inc.

Chapter 25: Page 728, SuperStock; 731(t), Robert Hessler/Scripps Institution of Oceanography; 731(b), Tom Myers; 732(l), Dr. E.R. Degginger; 732(r), Ross Lewis; 733, From the CHEM Study Film: Vanadium, Transition Element; 734(t), L.S. Stepanowicz/Panographics; 734(b), HRW Photo by Dennis Fagan; 740(t), Inland Steel Company; 740(t), Bethlehem Steel; 740(b), Inland Steel Company; 740(b), Bethlehem Steel; 741, Michael Melford/Peter Arnold; 742, HRW Photo by Dennis Fagan; 740(b), Sunshine Mining Co., Kellogg, Idaho; 745(r), Henkel Corporation; 746, ASARCO Incorporated.

Chapter 26: Page 750, John Garrett/Tony Stone Worldwide; 754, ALCOA; 756, Donald L. Miller/International Stock Photo; 757, HRW Photo/Gemological Institute of America; 759(tl), HRW Photo by Eric Beggs; 759(tr), J. Freis/Image Bank; 759(b), Chuck O'Rear/Woodfin Camp & Associates; 760(t), L.S. Stepanowicz/Panographics; 760(b), Energy Technology Visuals Collection; 761, Ken Cooper/The Image Bank, 762, Gazelle Microcircuits, Inc./C & I Photography.

Chapter 27: Page 764, David Cavagnaro; 765, Jacques Jangoux/Peter Arnold, 766(r), 766(b), Dr. E.R. Degginger; 767, Charles D. Winters; 768, Charles D. Winters; 769, HRW Photo by John Langford; 771(t), HRW Photo by John King; 771(b), William Rivelli/The Image Bank; 773, Lois Moulton/F-Stop Pictures.

Chapter 28: Page 778, E.R. Degginger/Bruce Coleman, Inc.; 778 (shells), David Muench; 779, Dr. E.R. Degginger; 782, HRW Photo by Eric Beggs; 784, CertainTeed Corporation; 785(t), HRW Photo by Yoav Levy; 785(b), C.J. Zimmerman/FPG; 786, Inland Steel Company; 788, From the CHEM Study Film: "Bromine—Element from The Sea"; 790, HRW Photo by Yoav Levy, 791, Historical Picture Service, Chicago; 794(t), HRW Photo by John Langford; 794(b), David R. Frazier; 796(t), HRW Photo by Dennis Fagan; 796(c), Douglas G. Ashley; 796(b), Ulrike Welsch/Photo Researchers; 797(t), N.A.S.A./Photo by Neil A. Armstrong; 797(b), HRW Photo by Dennis Fagan.

Unit 9: Page 798, Courtesy Donald H. Priester; 799, SuperStock.

Chapter 29: Page 800, Peter Berndt/Custom Medical Stock Photo; 805(tl), David R. Frazier Photolibrary; 805(tr), Matt Grimaldi; 805(b), Mark Schupack; 806(l), HRW Photo by Eric Beggs; 806(r), Mary Evans/Photo Researchers, Inc.; 811, Dr. Mark Shapiro, W.K. Kellogg Radiation Laboratory, California Institute of Technology; 816, Lester V. Bergman & Associates, Inc.; 818, David R. Frazier/Folio, Inc.; 820, Pierre Kopp/West Light; 821, Dan McCoy/Rainbow; 824, Tom Raymond/Medichrome/The Stock Shop; 826(t), Matt Meadows/Peter Arnold, Inc.; 824(b), Richard Megna/Fundamental Photographs; 827, Harry J. Przekop, Jr./Medichrome/The Stock Shop.

Art Credits

Accurate Art, Inc.—9, 17, 18, 22, 38, 41, 42, 47, 52, 59, 72, 76, 78, 79, 87, 101, 104, 106, 109, 110, 112, 115, 130, 131, 143, 145, 154, 161, 162, 165, 167, 176, 178, 182, 184, 186, 188, 189, 202, 205, 209, 215, 221, 235, 239, 240, 292, 293, 302, 316, 317, 318, 319, 324, 325, 328, 330, 332, 335, 346, 362, 371, 379, 384, 386, 391, 412, 416, 417, 418, 427, 440, 442, 448, 451, 452, 455, 467, 483, 491, 508, 511, 528, 532, 534, 543, 545, 546, 553, 554, 563, 565, 571, 585, 612, 613, 615, 617, 618, 619, 633, 634, 635, 638, 641, 645, 648, 650, 654, 657, 665, 687, 688, 689, 690, 691, 692, 693, 695, 696, 697, 698, 701, 702, 716, 720, 725, 738, 745, 756, 760, 766, 767, 774, 789, 802, 803, 807, 809, 810, 819.

Cooper, Holly—32, 307, 361, 476.

Erickson, Barry—13, 14, 47, 74, 82, 90, 99, 101, 102, 103, 107, 177, 187, 215, 219, 262, 294, 300, 303, 320, 321, 373, 374, 377, 393, 496, 508, 511, 527, 542, 544, 546, 572, 694, 730, 739, 740, 765, 772, 820.

Green & Williams—468, 783, 787, 830, 833, 839, 846, 848, 852, 859, 861, 862, 865, 872.

McElhaney, Gary—694.

Norby, Tom—177, 180, 351, 713, 773.

Powell, Terrell—vii, 75, 77, 315, 324, 344, 360, 387, 414, 415, 813, 819.

Signature Design Associates—249, 408, 428, 555.

Tracor, Inc.—547, 548.

Vantage Art, Inc.—835, 837, 841, 844, 854, 855, 856, 857, 867, 870, 874.

Focus on Reasoning Answers

Unit 1	Unit 2	Unit 3	Unit 4	Unit 5	Unit 6	Unit 7	Unit 8
1. C	1. A	1. B	1. A	1. C	1. D	1. D	1. D
2. C	2. A	2. A	2. C	2. C	2. B	2. D	2. D
3. C	3. A	3. B	3. B	3. D	3. C	3. D	3. A
4. C	4. C	4. D	4. C	4. B		4. C	4. B
5. C		5. C	5. B	5. C		5. B	
				6. A			

Periodic Table of the Elements

Legend (Carbon example):
- 6 — Atomic number
- C — Symbol
- Carbon — Name
- 12.0111 — Average Atomic Mass
- [He] 2s²2p² — Electron Configuration

Period	Group 1	Group 2	Group 3	Group 4	Group 5	Group 6	Group 7	Group 8	Group 9
1	1 **H** Hydrogen 1.007 94 1s¹								
2	3 **Li** Lithium 6.941 [He]2s¹	4 **Be** Beryllium 9.012 18 [He]2s²							
3	11 **Na** Sodium 22.989 77 [Ne]3s¹	12 **Mg** Magnesium 24.305 [Ne]3s²							
4	19 **K** Potassium 39.0983 [Ar]4s¹	20 **Ca** Calcium 40.08 [Ar]4s²	21 **Sc** Scandium 44.9559 [Ar]3d¹4s²	22 **Ti** Titanium 47.88 [Ar]3d²4s²	23 **V** Vanadium 50.9415 [Ar]3d³4s²	24 **Cr** Chromium 51.996 [Ar]3d⁵4s¹	25 **Mn** Manganese 54.9380 [Ar]3d⁵4s²	26 **Fe** Iron 55.847 [Ar]3d⁶4s²	27 **Co** Cobalt 58.9332 [Ar]3d⁷4s²
5	37 **Rb** Rubidium 85.4678 [Kr]5s¹	38 **Sr** Strontium 87.62 [Kr]5s²	39 **Y** Yttrium 88.9059 [Kr]4d¹5s²	40 **Zr** Zirconium 91.224 [Kr]4d²5s²	41 **Nb** Niobium 92.9064 [Kr]4d⁴5s¹	42 **Mo** Molybdenum 95.94 [Kr]4d⁵5s¹	43 **Tc** Technetium (98) [Kr]4d⁵5s²	44 **Ru** Ruthenium 101.07 [Kr]4d⁷5s¹	45 **Rh** Rhodium 102.906 [Kr]4d⁸5s¹
6	55 **Cs** Cesium 132.905 [Xe]6s¹	56 **Ba** Barium 137.33 [Xe]6s²	57 **La** Lanthanum 138.906 [Xe]5d¹6s²	72 **Hf** Hafnium 178.49 [Xe]4f¹⁴5d²6s²	73 **Ta** Tantalum 180.948 [Xe]4f¹⁴5d³6s²	74 **W** Tungsten 183.85 [Xe]4f¹⁴5d⁴6s²	75 **Re** Rhenium 186.207 [Xe]4f¹⁴5d⁵6s²	76 **Os** Osmium 190.2 [Xe]4f¹⁴5d⁶6s²	77 **Ir** Iridium 192.22 [Xe]4f¹⁴5d⁷6s²
7	87 **Fr** Francium (223) [Rn]7s¹	88 **Ra** Radium 226.025 [Rn]7s²	89 **Ac** Actinium 227.028 [Rn]6d¹7s²	104 **Unq*** (261) [Rn]5f¹⁴6d²7s²	105 **Unp*** (262) [Rn]5f¹⁴6d³7s²	106 **Unh*** (263) [Rn]5f¹⁴6d⁴7s²	107 **Uns*** (262) [Rn]5f¹⁴6d⁵7s²	108 **Uno*** (265) [Rn]5f¹⁴6d⁶7s²	109 **Une*** (266?) [Rn]5f¹⁴6d⁷7s²

* The systematic names and symbols for elements of atomic number greater than 103 will be used until the approval of trivial names by IUPAC.

Lanthanide Series

58 **Ce** Cerium 140.12 [Xe]4f¹5d¹6s²	59 **Pr** Praseodymium 140.908 [Xe]4f³6s²	60 **Nd** Neodymium 144.24 [Xe]4f⁴6s²	61 **Pm** Promethium (145) [Xe]4f⁵6s²	62 **Sm** Samarium 150.36 [Xe]4f⁶6s²

Actinide Series

90 **Th** Thorium 232.038 [Rn]6d²7s²	91 **Pa** Protactinium 231.036 [Rn]5f²6d¹7s²	92 **U** Uranium 238.029 [Rn]5f³6d¹7s²	93 **Np** Neptunium 237.048 [Rn]5f⁴6d¹7s²	94 **Pu** Plutonium (244) [Rn]5f⁶7s²